Contents in Brief

Glencoe

The McGraw-Hill Companies

The Princeton Review

The Standardized Test Practice features in this book were aligned and verified by The Princeton Review, the nation's leader in test preparation. Through its association with McGraw-Hill, The Princeton Review offers the best way to help students excel on standardized assessments.

The Princeton Review is not affiliated with Princeton University or Educational Testing Service.

Send all inquiries to:
Glencoe/McGraw-Hill
8787 Orion Place
Columbus, OH 43240-4027

ISBN: 0-07-860173-8 *(Florida Student Edition)*
ISBN: 0-07-860176-2 *(Florida Teacher Wraparound Edition)*

Printed in the United States of America.

2 3 4 5 6 7 8 9 10 055/071 12 11 10 09 08 07 06 05 04

GLENCOE MATHEMATICS

Florida Edition

Geometry

Teacher Wraparound Edition

Boyd Cummins Malloy

Carter Flores

McGraw Hill Glencoe

New York, New York
Columbus, Ohio
Chicago, Illinois
Peoria, Illinois
Woodland Hills, California

Teacher Handbook

Countdown to FCAT

Week 2

Monday

Extended Response On a Florida map with a scale of $\frac{3}{8}$ inch = 10 miles, lines are drawn to connect Taylor Creek, South Bay, and Lakeport. The distance between Lakeport and Taylor Creek is $1\frac{1}{8}$ inches, and the distance between Lakeport and South Bay is $1\frac{1}{2}$ inches. What is the distance across the lake, in miles, from Taylor Creek to South Bay? Show all your work. *(Lesson 1-3)* See margin.

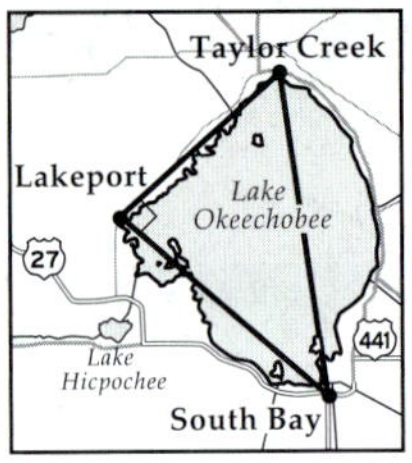

Tuesday

Multiple Choice Two angles of a triangle measure 40° and 30°. Which statement best describes the remaining angle? *(Lesson 1-4)* D

(A) Its measure is less than the measure of the sum of the first two angles.
(B) It is acute.
(C) It is a right angle.
(D) It is obtuse.

Gridded Response $\angle AOB \cong \angle COD$ and $m\angle COD = (2x - 10)°$. What is the value of x? *(Lesson 1-5)* 75

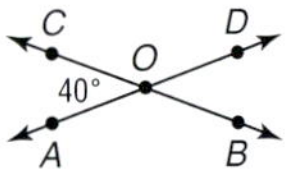

Wednesday

Multiple Choice Angles ROQ and SOT can best be described as what kind of angles? *(Lesson 1-5)* F

(F) vertical
(G) adjacent
(H) supplementary
(I) complementary

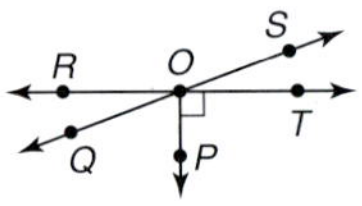

Gridded Response The length of a rectangle is two less than three times its width. The perimeter is 12 centimeters. Find the length of the rectangle in centimeters. *(Lesson 1-6)* 4

Thursday

Short Response Find the perimeter of triangle JKL if $J(-2, -1)$, $K(-2, -3)$, and $L(5, -1)$. Sketch and label the triangle. Show all your work. *(Lesson 1-6)*

$\overline{JK}$ has a length of 2 and $\overline{JL}$ has a length of 7. I can figure that out by counting squares. I must use the Pythagorean Theorem to find KL.

$2^2 + 7^2 = (KL)^2$
$4 + 49 = (KL)^2$
$(KL)^2 = 53$
$KL = \sqrt{53}$

The perimeter is the sum of the measures of the sides. That makes $2 + 7 + \sqrt{53}$ or $9 + \sqrt{53}$. See margin for sketch.

Friday

Multiple Choice When the graphs of $y = 2x$ and $y = -2x$ intersect, their intersection can be which of the following? *(Lesson 1-1)* A

I. a point
II. a line
III. a plane

(A) I only
(B) I and II
(C) II only
(D) I, II, or III

Countdown to FCAT

Monday
Extended Response Sample Answer

Using the Pythagorean Theorem, I can find the length of the line between Taylor Creek and South Bay on the map. Then I can change it to miles using the map scale. I would get $(\text{hypotenuse})^2 = \left(\frac{9}{8}\right)^2 + \left(\frac{3}{2}\right)^2 = \frac{81}{64} = \frac{9}{4} = \frac{225}{64}$. So the hypotenuse $= \sqrt{\frac{225}{64}}$ or $\frac{15}{8}$. Using that in a proportion with the map scale, I find that the distance between Taylor Creek and South Bay is 50 miles.

Thursday
Short Response Sample Answer

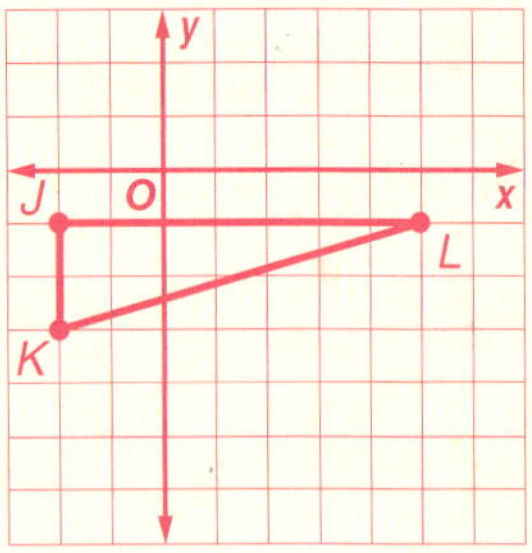

Benchmarks Addressed

Monday	MA.B.3.4.1
Tuesday	MA.B.1.4.2
Wednesday	MA.B.1.4.1, MA.C.2.4.1
Thursday	MA.B.3.4.1, MA.C.3.4.2
Friday	MA.C.3.4.2

Countdown to FCAT

Monday

Extended Response Sample Answer

In the figure, $\angle RSV$ and $\angle TSV$ form a linear pair. Angles that form a linear pair are supplementary. So, $m\angle RSV + m\angle TSV = 180°$. But I am given that $m\angle RSV = 120°$. By substituting, the equation becomes $120° + m\angle TSV = 180°$. Subtract 120° from each side to get $m\angle TSV = 60°$.

Thursday

Short Response Sample Answer

- Florida and California produce more than 100,000 tons of oranges.
- Florida and California produce more than 100,000 tons of oranges and grapefruit.
- Arizona does not produce more than 100,000 tons of oranges nor does it produce more than 100,000 tons of grapefruit. Texas does not produce more than 100,000 tons of oranges.

Countdown to FCAT

Week 3

Monday

Extended Response Write an informal proof or explain each step to find $m\angle VST$ if $m\angle RSV = 120°$. *(Lesson 2-6)* See margin.

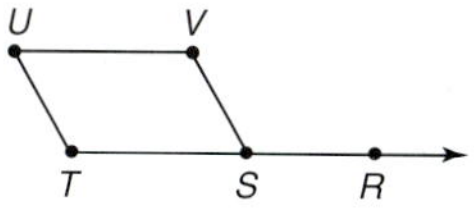

Tuesday

Multiple Choice Solve for x in the equation $\frac{2}{3}x + \frac{5}{9} = 3$. *(Prerequisite)* C

(A) $-\frac{5}{2}$ (B) $\frac{3}{11}$ (C) $\frac{11}{3}$ (D) $-\frac{2}{5}$

Gridded Response A beekeeper adds hives every year. The pattern for the first three years is shown. How many hives would there be in Year 4? *(Lesson 2-1)* 16

Wednesday

Multiple Choice Looking at the Venn diagram, how many students in the sophomore class at Howard High School have a dog, a cat, or both? *(Lesson 2-2)* F

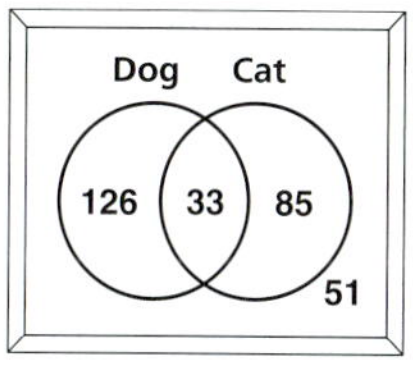

(F) 244 (G) 211 (H) 51 (I) 38

Gridded Response If $3x + 2 = 12$, then $x - \frac{1}{3} = \underline{\ ?\ }$. *(Prerequisite)* 3

Thursday

Short Response The Venn diagram shows those states producing more than 100,000 tons of the specified fruit. Answer the following questions. *(Lesson 2-2)* See margin.

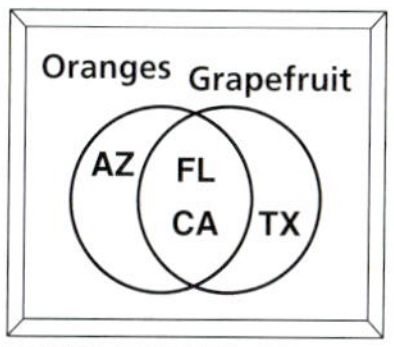

Source: www.nass.usda.gov

- Which states produce more than 100,000 tons of oranges?
- Which states produce more than 100,000 tons of oranges and grapefruit?
- Which states do not produce more than 100,000 tons of oranges or grapefruit? (FL = Florida, CA = California, TX = Texas, AZ = Arizona)

Friday

Multiple Choice A sampling of 133 high school students was asked questions about movies they had seen. How many students said they had seen at least two of the three movies shown in the Venn diagram? *(Lesson 2-2)* B

(A) 28 (B) 30 (C) 70 (D) 100

Benchmarks Addressed

Monday	MA.C.1.4.1
Tuesday	MA.A.3.4.2, MA.A.5.4.1
Wednesday	MA.E.1.4.1
Thursday	MA.E.1.4.1
Friday	MA.E.1.4.1

Countdown to FCAT

Week 4

Monday

Extended Response Given the statements in the box at the right for the proof below, write a reason for each statement. *(Lesson 2-7)*

Given: $\overline{AB} \cong \overline{RS}$, $\overline{BC} \cong \overline{ST}$

Prove: $\overline{AC} \cong \overline{RT}$

See margin.

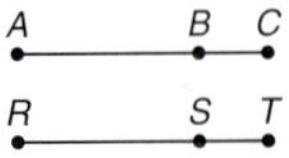

1. $\overline{AB} \cong \overline{RS}$	6. $AC = RS + ST$
2. $AB = RS$	7. $RS + ST = RT$
3. $\overline{BC} \cong \overline{ST}$	8. $AC = RT$
4. $BC = ST$	9. $\overline{AC} \cong \overline{RT}$
5. $AC = AB + BC$	

Tuesday

Multiple Choice Which statement is NOT true about the figure shown? *(Lesson 2-8)* **D**

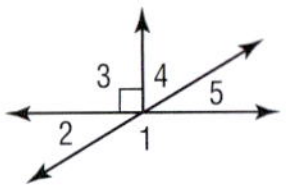

(A) $m\angle 3 = 90°$

(B) $\angle 4$ is complementary to $\angle 5$.

(C) $m\angle 2 = m\angle 5$

(D) $\angle 1$ is complementary to $\angle 5$.

Gridded Response If $m\angle 1 = 30°$, what is $m\angle 2 + m\angle 4$ in degrees? *(Lesson 2-8)* **300**

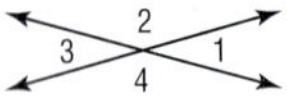

Wednesday

Multiple Choice If $m\angle 7 = (4x + 20)°$ and $m\angle 8 = (3x - 15)°$, what is the actual measure of $\angle 7$? *(Lesson 2-8)* **H**

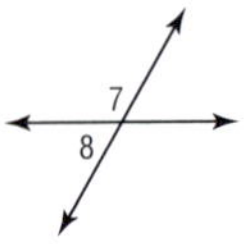

(F) 90° (G) 60°

(H) 120° (I) 75°

Gridded Response Find $m\angle 4$ in degrees if $m\angle 6 = 40°$ and $\angle 6$ is congruent to $\angle 3$. *(Lesson 2-8)* **50**

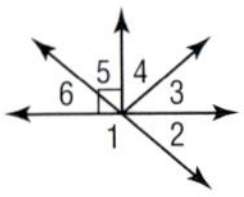

Thursday

Short Response Construct a Venn diagram depicting the following information. *(Lesson 2-2)* **See margin.**

50 students were surveyed about math courses they had taken. The results are shown.

- 17 had taken Algebra I.
- 17 had taken Algebra II.
- 23 had taken Geometry.
- 6 had taken Algebra I and Algebra II.
- 8 had taken Algebra I and Geometry.
- 10 had taken Algebra II and Geometry.
- 2 had taken all three math courses.

Friday

Multiple Choice Between Titusville and Melbourne, U.S. 1 and Interstate 95 run parallel to each other. Both are intersected by Highway 192 near Melbourne. If Highway 192 and Interstate 95 are perpendicular, why does $m\angle 1 = m\angle 2$? *(Lesson 2-8)* **A**

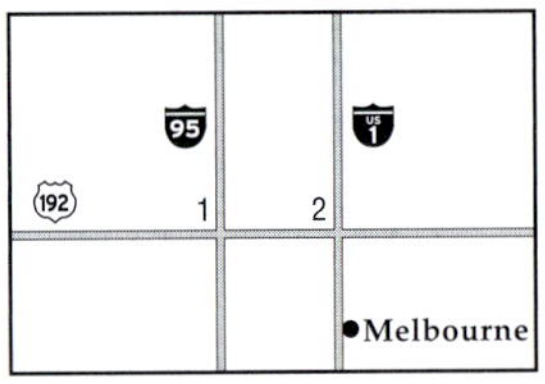

(A) Because the lines are perpendicular, the angles are right, and all right angles are congruent.

(B) Reflexive Property

(C) Because the lines are perpendicular, the angles are right, and all right angles are complementary.

(D) Perpendicular lines are always congruent.

Countdown to FCAT

Monday

Extended Response Sample Answer

Statements (Reasons)

1. $\overline{AB} \cong \overline{RS}$ (Given)
2. $AB = RS$ (Definition of congruent segments)
3. $\overline{BC} \cong \overline{ST}$ (Given)
4. $BC = ST$ (Definition of congruent segments)
5. $AC = AB + BC$ (Segment Addition Property)
6. $AC = RS + ST$ (Substitution)
7. $RS + ST = RT$ (Segment Addition Property)
8. $AC = RT$ (Transitive Property)
9. $\overline{AC} \cong \overline{RT}$ (Definition of Congruent Segments)

Thursday

Short Response Sample Answer

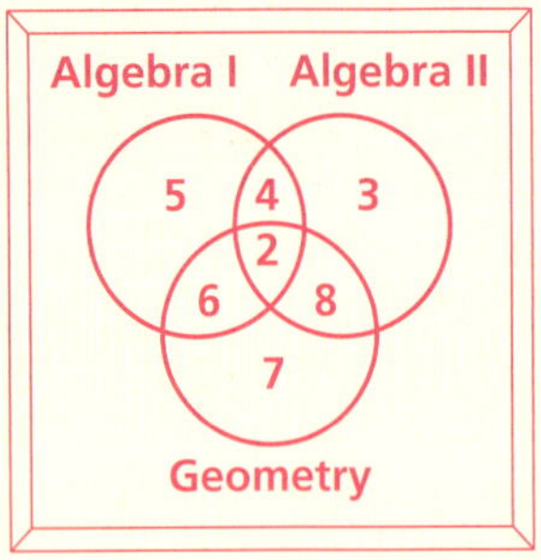

Benchmarks Addressed

Monday	MA.C.1.4.1
Tuesday	MA.C.2.4.1
Wednesday	MA.C.2.4.1
Thursday	MA.E.1.4.1
Friday	MA.C.2.4.1

Countdown to FCAT

Monday

Extended Response Sample Answer

I must first find the slope from the two points given. If $(x_1, y_1) = (1980, 10{,}470{,}000)$ and $(x_2, y_2) = (2000, 13{,}305{,}000)$, I get $m = \frac{13{,}305{,}000 - 10{,}470{,}000}{2000 - 1980}$ or 141,750. This means growth of 141,750 tons per year. The amount for 2020 would be 13,305,000 + 20(141,750) or 16,140,000.

Thursday

Short Response Sample Answer

$\angle 1 \cong \angle 14$ because they are alternate exterior angles of parallel lines cut by a transversal. So, $m\angle 1 = m\angle 14$, or $6x + 45 = 8x + 5$. Solving this for x gives me $x = 20$, making $m\angle 1 = 6(20) + 45$ or 165°. I also know that $\angle 1$ and $\angle 5$ are supplementary. This means $m\angle 5 = 15°$. I can substitute to write the equation $15° = 3(y - 10)°$. Solving this equation yields $y = 15$. So the answer to the question is that $x = 20$ and $y = 15$.

Countdown to FCAT

Week 5

Monday

THINK SOLVE EXPLAIN

Extended Response Citrus production in Florida increased from 10,470,000 tons of fruit produced in 1980 to 13,305,000 tons produced in 2000. If we can expect the same rate of growth for the next 20 years, how many tons of fruit will Florida produce in the year 2020? Show all your work or explain your answer. *(Lesson 3-3)* **See margin.**

Tuesday

Multiple Choice Find $m\angle 3$ if $m \parallel n$. *(Lesson 3-2)* **C**

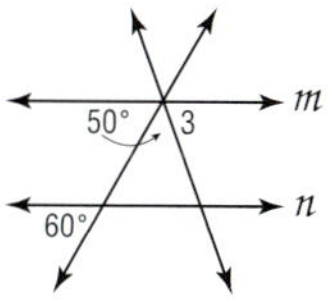

Ⓐ 130°
Ⓑ 110°
Ⓒ 70°
Ⓓ 40°

Gridded Response Find $m\angle SRT$ in degrees. *(Lesson 2-8)* **35**

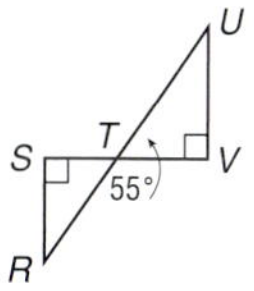

Wednesday

Multiple Choice Find all angles congruent to $\angle 7$ if $n \parallel m$. *(Lesson 3-2)* **H**

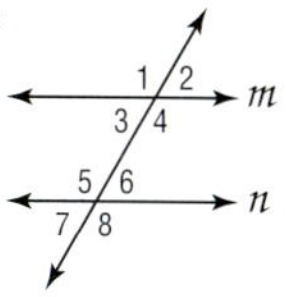

Ⓕ $\angle 8, \angle 5, \angle 4, \angle 1$
Ⓖ $\angle 8, \angle 4$
Ⓗ $\angle 6, \angle 3, \angle 2$
Ⓘ $\angle 5, \angle 3, \angle 1$

Gridded Response Find the slope of the line containing $A(-1, -4)$ and $B(3, -12)$. *(Lesson 3-3)* **−2**

Thursday

THINK SOLVE EXPLAIN

Short Response If $m\angle 1 = (6x + 45)°$, $m\angle 5 = 3(y - 10)°$, and $m\angle 14 = (8x + 5)°$, find x and y. Show and explain all your work. Assume $a \parallel b$. *(Lesson 3-2)* **See margin.**

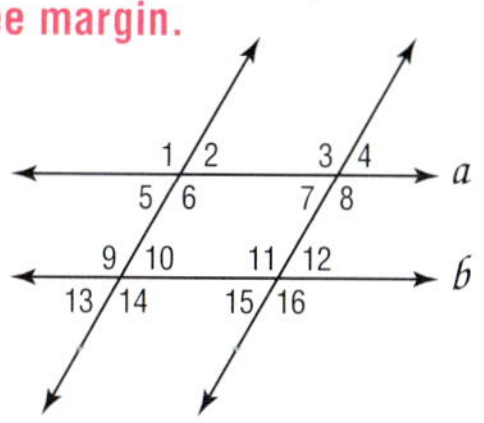

Friday

Multiple Choice What is the approximate distance in units from the northwest corner of the deck to the southwest corner of the garage? *(Lesson 1-3)* **B**

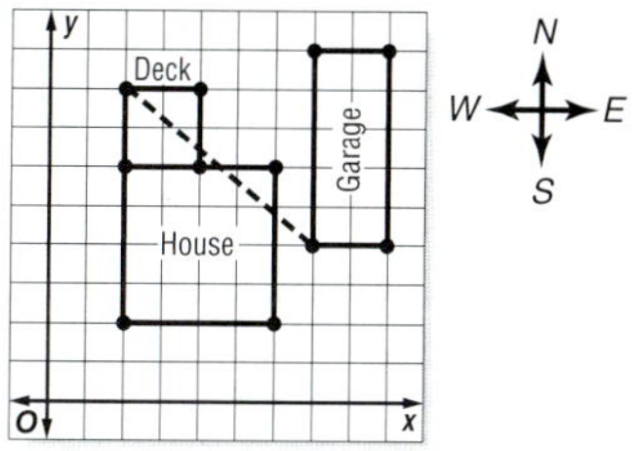

Ⓐ 5.4
Ⓑ 6.4
Ⓒ 9.1
Ⓓ 41

Benchmarks Addressed

Monday	MA.C.3.4.2, MA.D.1.4.1
Tuesday	MA.B.1.4.2, MA.C.2.4.1
Wednesday	MA.C.2.4.1, MA.C.3.4.2
Thursday	MA.C.2.4.1
Friday	MA.B.1.4.2

Week 6

Monday

THINK SOLVE EXPLAIN

Extended Response In 1920, the average life expectancy for a citizen of the U.S. was only 54 years. *(Lesson 3-3)*

Part A What was the rate of change between 1920 and 1960?

Part B Was it the same between 1960 and 1980?

Part C What statement can you make about the life expectancy in 2000, if the graph were to show it? Support your opinion using the graph.

Parts A–C: See margin.

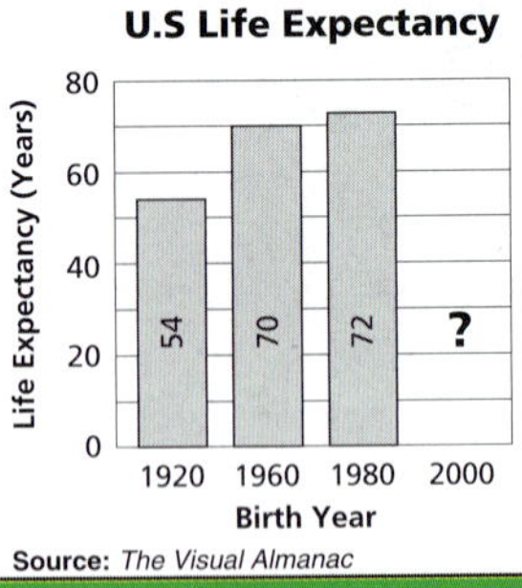

Source: *The Visual Almanac*

Tuesday

Multiple Choice Which is an equation for the line shown? *(Lesson 3-4)* **D**

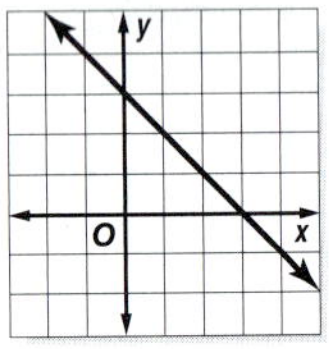

Ⓐ $y = 4x$

Ⓑ $y = -4x$

Ⓒ $y = -x - 4$

Ⓓ $y = -x + 4$

Gridded Response How many angles are supplementary to $\angle 1$ if $a \parallel b$ and $c \parallel d$? *(Lesson 3-2)* **8**

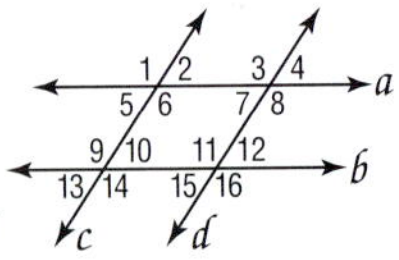

Wednesday

Multiple Choice What conclusion can be made about the figure? *(Lesson 3-5)* **G**

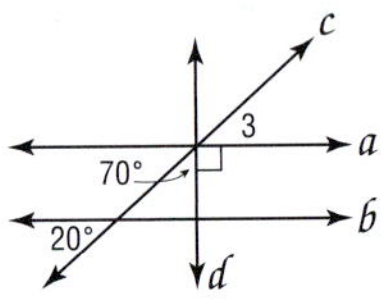

Ⓕ Lines a and b are not $\parallel$.

Ⓖ Lines a and b are $\parallel$.

Ⓗ Line a is $\perp$ to line b.

Ⓘ Line d is $\perp$ to line c.

Gridded Response Using the figure above, what is the measure in degrees of angle 3? *(Lesson 3-2)* **20**

Thursday

THINK SOLVE EXPLAIN

Short Response Determine whether $\overleftrightarrow{RS}$ and $\overleftrightarrow{TU}$ are parallel, perpendicular, or neither. Show all your work and explain your answer. *(Lesson 3-3)* **See margin.**

$R(-1, 4)$
$S(4, -6)$
$T(2, -7)$
$U(-5, 7)$

Friday

Multiple Choice Between which years is the greatest rate of change shown? The change can be positive or negative. *(Lesson 3-3)* **C**

Ⓐ 1980 and 1985 Ⓑ 1985 and 1990

Ⓒ 1990 and 1995 Ⓓ 1995 and 2000

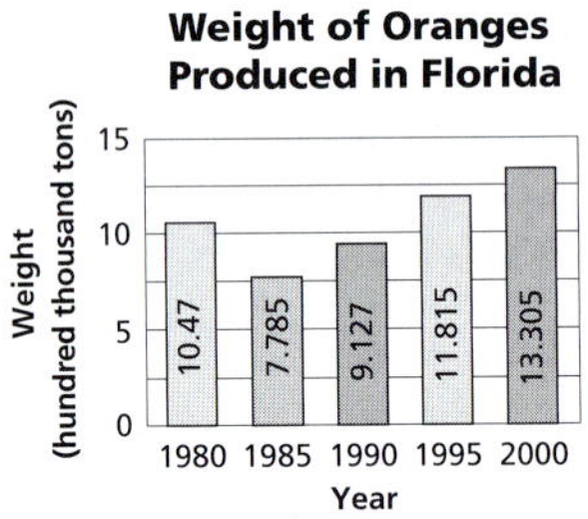

Source: www.nass.usda.gov

Benchmarks Addressed

Monday	MA.E.1.4.1
Tuesday	MA.D.1.4.1
Wednesday	MA.C.2.4.1
Thursday	MA.C.3.4.2
Friday	MA.E.1.4.1

Countdown to FCAT

Monday

Extended Response Sample Answer

Part A Using the Slope Formula, I can find the rate of change between the years shown. Between 1920 and 1960 it is $\frac{70-54}{40}$ or 0.4.

Part B Between 1960 and 1980 it is $\frac{72-70}{20}$ or 0.1. So the rate of change is not constant between these years.

Part C Due to medical advances, increasing population, and other variables, the life expectancy will probably continue to increase through 2000 and on into the 21st century. However, judging by this graph, I do not think it will continue at the rate of change first shown between 1920 and 1960. It will probably more closely match the rate for between 1960 and 1980.

Thursday

Short Response Sample Answer

$\overleftrightarrow{RS}$ and $\overleftrightarrow{TU}$ are parallel lines. I know this because I figured their slope, and they have the same slope.

$$m_1 = \frac{-6-4}{4-(-1)} = \frac{-10}{5} \text{ or } -2$$

$$m_2 = \frac{7-(-7)}{-5-2} = \frac{14}{-7} \text{ or } -2$$

Countdown to FCAT

Monday
Extended Response Sample Answer

The midpoint of $\overline{RS}$ $= \left(\frac{0-4}{2}, \frac{8+0}{2}\right) = (-2, 4)$ and the midpoint of $\overline{RT}$ $= \left(\frac{0+4}{2}, \frac{8-2}{2}\right) = (2, 3)$. Now I just need to find the distance between these two points. The distance $= \sqrt{(2+2)^2 + (3-4)^2}$ or $\sqrt{17}$. So the distance between the two midpoints is $\sqrt{17}$, which is approximately equal to 4.123.

Thursday
Short Response Sample Answer

Since both triangles are right, that is one angle they have the same. Also, $\angle AOC$ is congruent to $\angle ROS$ because they are vertical angles. If two of the angles of the two triangles are congruent, then the third one must be congruent also, because the sums of the measures of the angles in each triangle must be 180°. However, AAA is not a way to prove triangle congruence, so from this drawing, with no side lengths marked, you cannot tell if the triangles are congruent.

Countdown to FCAT

Week 7

Monday

Extended Response Find the distance between the midpoints of $\overline{RS}$ and $\overline{RT}$. Show all your work. *(Lesson 4-1)* **See margin.**

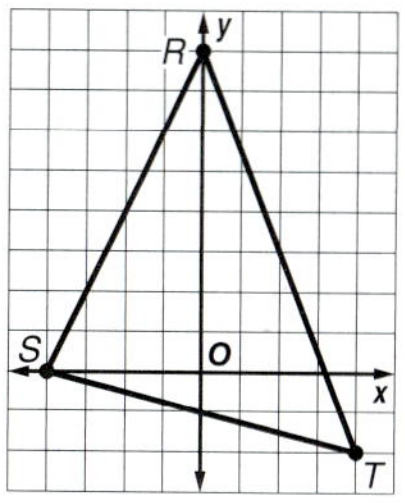

Tuesday

Multiple Choice Choose the word that does NOT describe the triangle. *(Lesson 4-1)* **C**

Ⓐ acute
Ⓑ isosceles
Ⓒ obtuse
Ⓓ equilateral

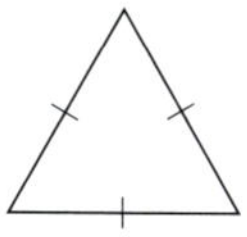

Gridded Response Solve for m in the equation $\frac{3}{4}m - 8 = 2$. *(Prerequisite)*

$\frac{40}{3}$ or 13.33

Wednesday

Multiple Choice If the legs of the triangle shown are doubled in length, what is the length of the resulting hypotenuse? *(Lesson 4-1)* **G**

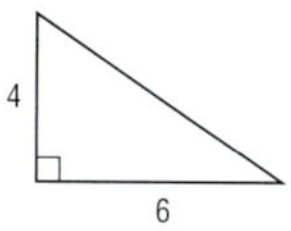

Ⓕ $10\sqrt{2}$ Ⓖ $4\sqrt{13}$
Ⓗ 20 Ⓘ $20\sqrt{2}$

Gridded Response In triangle ABC, $\overline{AB} \cong \overline{AC}$. If $AB = 3x + 4$ and $AC = 8x - 6$, how many times as large as x is AB? *(Lesson 4-1)* **5**

Thursday

Short Response Use the figure to determine whether $\triangle AOC \cong \triangle ROS$. Explain your reasoning. *(Lesson 4-5)* **See margin.**

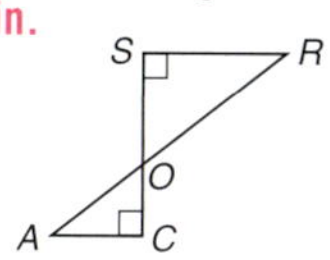

Friday

Multiple Choice Alltell Stadium, home of the Jacksonville Jaguars, seats a total of 73,800. Of those, 15% are club seats. How many club seats are in Alltel Stadium? *(Prerequisite)* **B**

Ⓐ 10,500 Ⓑ 11,070
Ⓒ 12,040 Ⓓ 13,020

Benchmarks Addressed

Monday	MA.C.3.4.2
Tuesday	MA.A.3.4.2, MA.C.2.4.1
Wednesday	MA.C.3.4.2, MA.D.1.4.2
Thursday	MA.C.1.4.1, MA.C.2.4.1
Friday	MA.A.3.4.3

Week 8

Monday

Extended Response Determine whether the two triangles are related by a *flip*, *slide*, or *turn*. Then verify that the triangles are congruent. Show all your work. *(Lesson 4-3)* $\triangle ABC$ is flipped over the y-axis to form $\triangle DEF$. To show congruence, I must show that the corresponding sides and the included angles of the two triangles are congruent. The distances can be found just by counting squares. $AB = DF = 3$ and $AC = DE = 6$. So, $\overline{AB} \cong \overline{DF}$ and $\overline{AC} \cong \overline{DE}$. $\angle A$ and $\angle D$ are formed by horizontal and vertical grid lines, which are perpendicular. That makes $\angle A \cong \angle D$ because all right angles are congruent. $\triangle ABC \cong \triangle DEF$ by SAS.

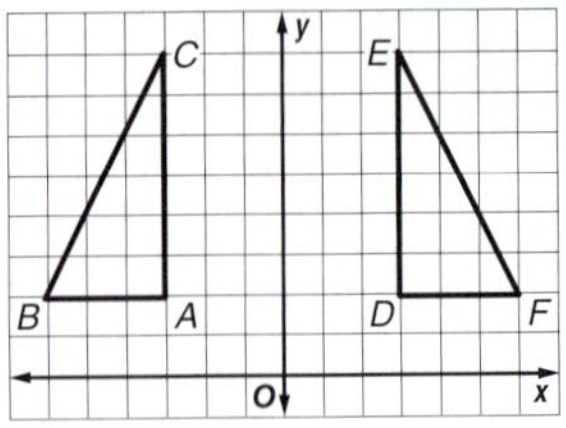

Tuesday

Multiple Choice Find $m\angle BCD$. *(Lesson 4-2)* C

Ⓐ 125° Ⓑ 75° Ⓒ 70° Ⓓ 55°

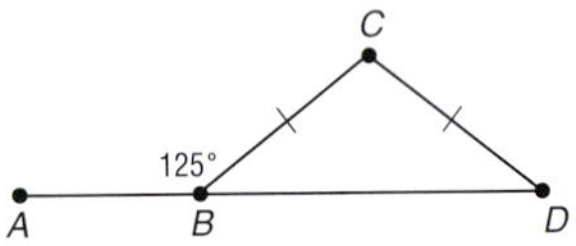

Gridded Response Find $m\angle 1$ in degrees. *(Lesson 4-2)* 102

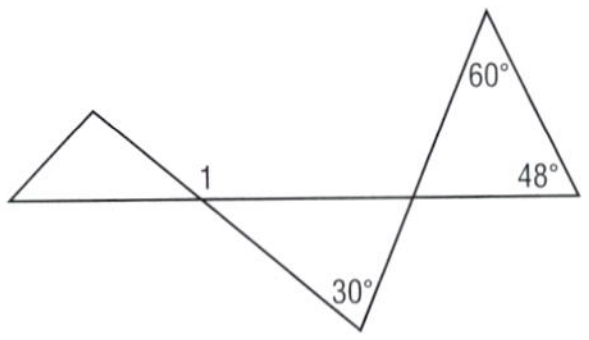

Wednesday

Multiple Choice The measures of the three angles of a triangle are as follows: $m\angle 2$ is six less than twice $m\angle 1$, and $m\angle 3$ is six less than three times $m\angle 1$. Find the three angle measures. *(Lesson 4-2)* G

Ⓕ 42°, 48°, 90° Ⓖ 32°, 58°, 90°
Ⓗ 30°, 60°, 90° Ⓘ 28°, 62°, 90°

Gridded Response Find the value of y. *(Lesson 4-2)* 70

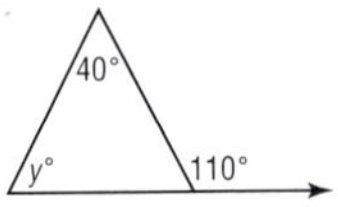

Thursday

Short Response Study the figures below.

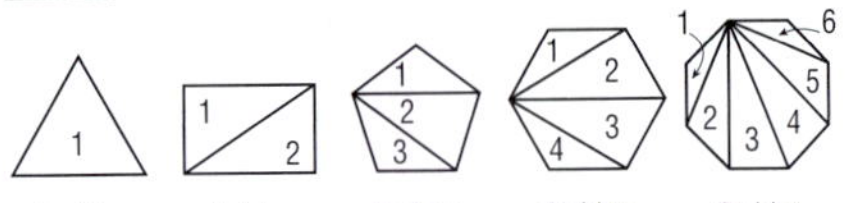

Explain why this statement is true.
For any n-sided polygon, the sum of the measures of its interior angles will always be $(n - 2)180°$. *(Lesson 4-2)*
See margin.

Friday

Multiple Choice What is the approximate grid distance between Tampa and Orlando if Tampa is located on the grid at (3, 0) and Orlando is located at (5, 1)? *(Lesson 1-4)* A

Ⓐ 2.24
Ⓑ 3.17
Ⓒ 5.21
Ⓓ 8.06

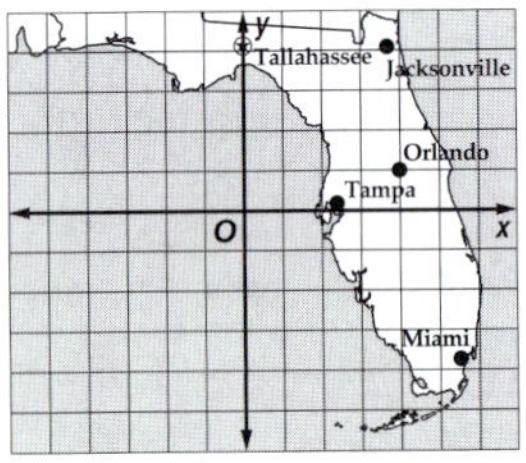

Benchmarks Addressed

Monday	MA.C.2.4.1
Tuesday	MA.C.1.4.1
Wednesday	MA.C.1.4.1
Thursday	MA.B.1.4.2
Friday	MA.C.3.4.2

Countdown to FCAT

Thursday

Short Response Sample Answer

The pattern in the figures is that if you draw segments from one vertex to each of the others to form triangles, you can then add up the number of triangles to get the total number of degrees in the polygon. Since there are always two segments connecting any vertex to the two adjacent vertices, the number of triangles in the interior will always be 2 less than the number of sides. So you multiply $(n - 2)$ times 180° to get that total.

Countdown to FCAT

Monday

Extended Response Sample Answer

Part A Since $m\angle DBC = 165 - 2x$, then $m\angle ABC = 180 - (165 - 2x)$ or $15 + 2x$. The measure of angle *ECB* also equals $165 - 2x$. Therefore, $m\angle ACB$ also equals $15 + 2x$. The measures of the angles of a triangle total 180°, so the equation can be written as follows.

$$2x + 2x + 15 + 2x + 15 = 180$$
$$6x + 30 = 180$$
$$6x = 150$$
$$x = 25$$

Part B Therefore, $m\angle BAC = 50°$, $m\angle ACB = 65°$, and $m\angle CBA = 65°$.

Thursday

Short Response Sample Answer

What I know is this.

$m\angle 6 > m\angle 4$ The Exterior Angle Inequality Theorem

$m\angle 6 > m\angle 7$ The Exterior Angle Inequality Theorem

$m\angle 6 > m\angle 2$ The Exterior Angle Inequality Theorem

$m\angle 6 > m\angle 7 + m\angle 8$ The Exterior Angle Inequality Theorem

Therefore, $m\angle 6 > m\angle 4$, $m\angle 7$, $m\angle 2$, and $m\angle 8$.

Week 9

Monday

Extended Response *(Lesson 5-2)*

Part A Find the value of x. **See margin.**

Part B Then find the measures of $\angle BAC$, $\angle ACB$, and $\angle CBA$. Show all your work. **See margin.**

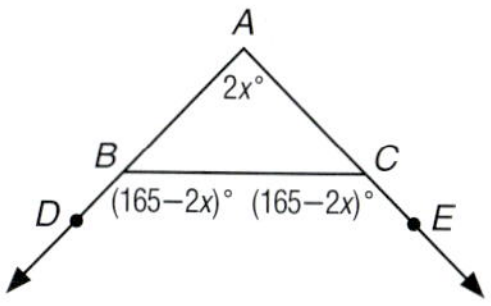

Tuesday

Multiple Choice Find the midpoint of the line with endpoints at (2, 2) and (−4, −1). *(Lesson 1-3)* **D**

Ⓐ $\left(-3, -\frac{1}{2}\right)$ Ⓑ $\left(-3, \frac{3}{2}\right)$

Ⓒ $\left(-1, -\frac{1}{2}\right)$ Ⓓ $\left(-1, \frac{1}{2}\right)$

Gridded Response What is the measure in degrees of the angle at Jacksonville? *(Lesson 5-3)* **95**

Wednesday

Multiple Choice Which lists the angles in the order of the angle measures from least to greatest? *(Lesson 5-2)* **F**

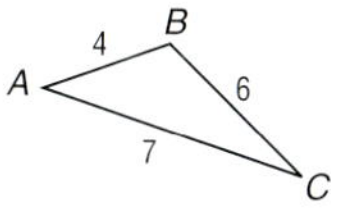

Ⓕ *C, A, B* Ⓖ *B, A, C*

Ⓗ *A, B, C* Ⓘ *B, C, A*

Gridded Response In $\triangle RST$, $\overline{RS} \cong \overline{RT}$, $m\angle SRT = (5x + 5)°$, and $m\angle RTS = (4x - 10)°$. Find the value of x. *(Lesson 4-1)* **15**

Thursday

Short Response Determine all angles with measures that are less than $m\angle 6$.

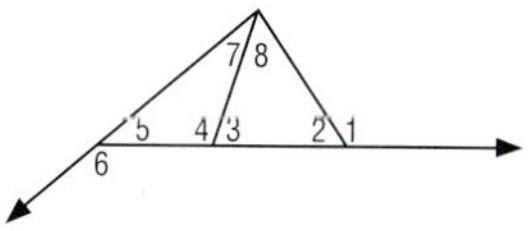

Show your work and explain your reasoning. *(Lesson 5-2)* **See margin.**

Friday

Multiple Choice Which statement is true regarding the figure? *(Lesson 5-4)* **B**

Ⓐ $CD > CE$

Ⓑ $CD < CE$

Ⓒ $AF > CE$

Ⓓ $AF > CD$

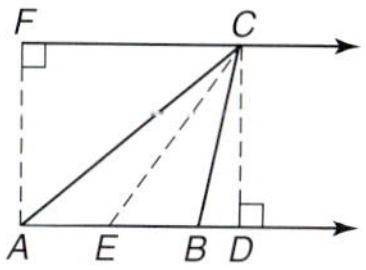

Benchmarks Addressed

Monday	MA.B.1.4.2, MA.C.1.4.1
Tuesday	MA.B.1.4.2, MA.C.3.4.2
Wednesday	MA.B.1.4.2, MA.C.1.4.1
Thursday	MA.B.1.4.2, MA.C.1.4.1
Friday	MA.C.1.4.1

Week 10

Monday

Extended Response The vertices of two triangles are $P(3, 3)$, $Q(3, 7)$, $R(10, 3)$, $S(-10, -3)$, $T(-3, -3)$, and $U(-3, -7)$. *(Lesson 4-4)*

Part A Name the congruence transformation relating $\triangle PQR$ and $\triangle TUS$. **See margin.**

Part B Verify that the two triangles are congruent. Show all your work. **See margin.**

Tuesday

Multiple Choice Which sentence is true regarding angles BCE and DCE? *(Lesson 5-5)* **C**

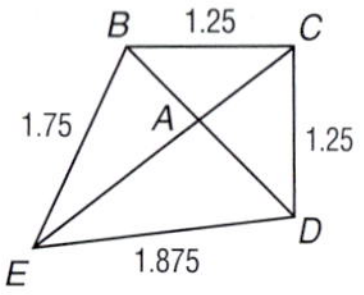

(A) $m\angle BCE = m\angle DCE$

(B) $m\angle BCE > m\angle DCE$

(C) $m\angle BCE < m\angle DCE$

(D) $m\angle BCE + m\angle DCE = 180°$

Gridded Response If 8 people enter a room and each shake hands once with every person in the room, how many different handshakes will occur? Use noncollinear points to find your answer. *(Lesson 2-3)* **28**

Wednesday

Multiple Choice If E is the midpoint of $\overline{AC}$, which statement below can be made with certainty? *(Lesson 5-4)* **G**

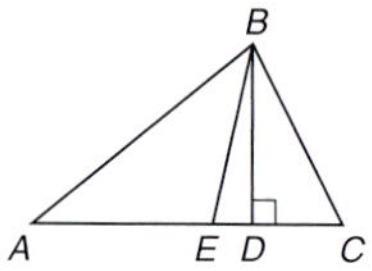

(F) $AE > CE$

(G) $BD < BE$

(H) $AD < DC$

(I) $BD > BE$

Gridded Response What is the slope of the line containing the points at $\left(-1, \frac{1}{2}\right)$ and $(1, -1)$? *(Lesson 3-3)*

$-\frac{3}{4}$

Thursday

Short Response Write an equation of the line that is perpendicular to $3x + 2y = 4$ and contains the point at $(3, -2)$. *(Lesson 3-4)* **The original equation is $3x + 2y = 4$. Writing it in slope-intercept form gives me $y = -\frac{3}{2}x + 2$. So the slope is $-\frac{3}{2}$. Since perpendicular lines have opposite reciprocal slopes, the slope of the new line will be $\frac{2}{3}$. I can use the slope-intercept form to find b.**

$$y = mx + b$$

$$-2 = \frac{2}{3}(3) + b$$

$$-4 = b$$

So the equation of the line perpendicular to $y = -\frac{3}{2}x + 2$ is $y = \frac{2}{3}x - 4$.

Friday

Multiple Choice What statement can be made with certainty about the angles formed when lines are drawn joining Tampa (T), Orlando (O), and Miami (M)? *(Lesson 5-4)* **A**

(A) $m\angle MTO > m\angle TOM$

(B) $m\angle TOM > m\angle MTO$

(C) $m\angle MTO = m\angle TOM$

(D) $m\angle OMT > m\angle TOM$

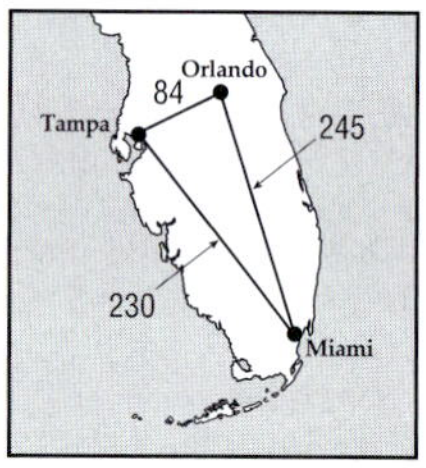

Countdown to FCAT

Monday

Extended Response Sample Answer

Part A The transformation is a turn.

Part B To verify congruence I can show that the lengths of corresponding sides and the included angles of the two triangles are congruent. The distances can be figured just by counting squares. $QP = TU = 4$ and $PR = ST = 7$. So, $\overline{QP} \cong \overline{TU}$ and $\overline{PR} \cong \overline{ST}$. $\angle T$ and $\angle P$ are formed by perpendicular grid lines so they are congruent because all right angles are congruent. Thus, I have $\triangle PQR \cong \triangle TUS$ by SAS.

Benchmarks Addressed

Monday	MA.C.2.4.1
Tuesday	MA.C.1.4.1
Wednesday	MA.C.1.4.1, MA.C.3.4.2
Thursday	MA.C.3.4.2
Friday	MA.B.1.4.2, MA.C.1.4.1

Monday

Extended Response Sample Answer

If the two pentagons are similar, then their corresponding sides must be proportional. I will look at corresponding sides that are whole numbers. One ratio is 4:2 or 2:1. The second set of sides has a ratio of 2:2 or 1:1. So, without having to finish each pair of corresponding sides, I know that the two pentagons are NOT similar.

Thursday

Short Response Sample Answer

Part A The ratio of the side measures is 6:8:10. It can be thought of as $6x:8x:10x$. So it might look like this.

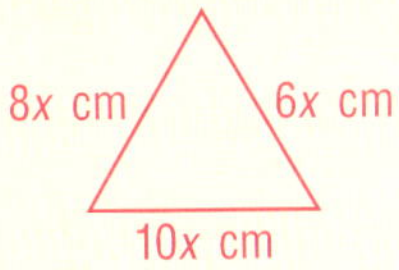

Part B The perimeter of the triangle is 96 cm, so $6x + 8x + 10x = 96$ or $24x = 96$, so $x = 4$. The sides are 24, 32, and 40. This means I have drawn the triangle incorrectly. It should be a right triangle because $24^2 + 32^2 = 40^2$.

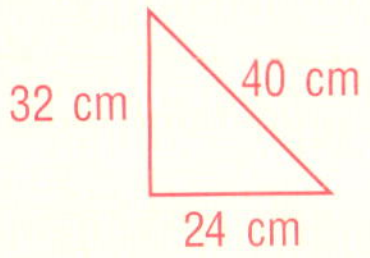

Countdown to FCAT

Week 11

Monday

Extended Response Explain why the two pentagons *are* or *are not* similar. *(Lesson 6-2)* **See margin.**

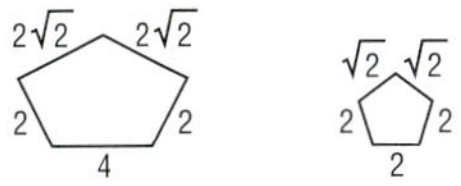

Tuesday

Multiple Choice What is the scale factor for the two similar rectangles? *(Lesson 6-2)* **C**

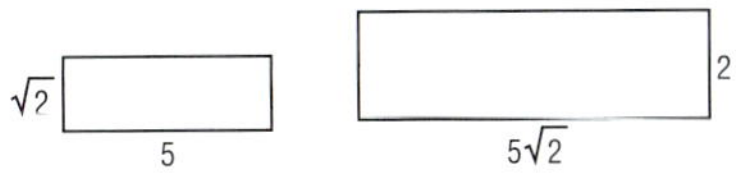

Ⓐ 5 Ⓑ 4 Ⓒ $\sqrt{2}$ Ⓓ 2

Gridded Response Solve for x: $\frac{x+3}{4} - 2 = \frac{2}{3}$. *(Prerequisite)* $\frac{23}{3}$ **or 7.667**

Wednesday

Multiple Choice Name the angles that are supplementary to $\angle 6$ given $\ell \parallel m$. *(Lesson 3-5)* **I**

Ⓕ 7, 3, 2
Ⓖ 5, 7, 8
Ⓗ 1, 2, 4, 8
Ⓘ 1, 4, 5, 8

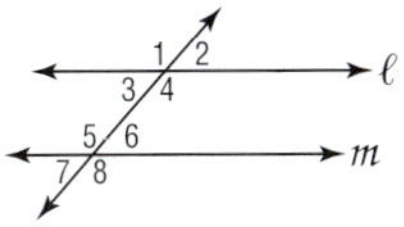

Gridded Response The scale on a Florida map is 0.75 inches = 100 miles. The distance on the map from Miami to Orlando is 1.725 inches. What is the distance in miles from Miami to Orlando? *(Lesson 6-2)* **230**

Thursday

Short Response A triangle has a perimeter of 96 centimeters. The ratio of the measures of its three sides is 6:8:10. *(Lesson 6-1)* **Part A: See margin.**

Part A Draw and label the triangle.

Part B Find the length of each of the three sides. **See margin.**

Friday

Multiple Choice Celeste wanted to rearrange her bedroom, so she drew a scale drawing to help plan the changes. If the scale she used was 3 centimeters = 1.5 feet, how many centimeters should she draw in her 5-foot long bookcases? *(Lesson 6-1)* **B**

Ⓐ 2.5 Ⓑ 10
Ⓒ 12 Ⓓ 15

Benchmarks Addressed

Monday	MA.A.3.4.3, MA.B.1.4.3
Tuesday	MA.A.3.4.3
Wednesday	MA.A.3.4.3, MA.B.1.4.3, MA.C.3.4.1
Thursday	MA.A.3.4.3, MA.B.1.4.3
Friday	MA.A.3.4.3, MA.B.1.4.3

Countdown to FCAT

Week 12

Monday

Extended Response The picture shown is enlarged, and then the enlargement is enlarged. The ratio of the original to the first enlargement is 3:4. The ratio of the first enlargement to the second is 2:3. Find the dimensions of both of the new pictures. Show all your work. *(Lesson 6-2)* **See margin.**

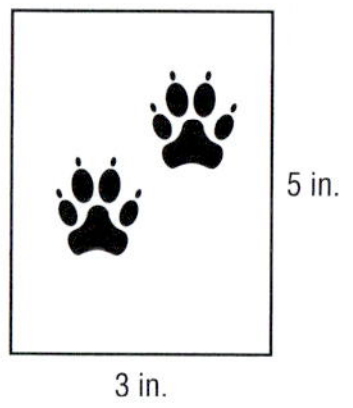

Tuesday

Multiple Choice Which theorem or postulate proves similarity of $\triangle RST$ and $\triangle RUV$? *(Lesson 6-3)* **A**

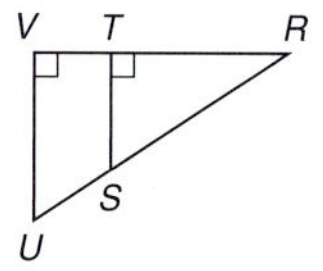

- (A) Angle-Angle Similarity
- (B) Side-Side-Side Similarity
- (C) Side-Angle-Side Similarity
- (D) Angle-Side-Angle Similarity

Gridded Response In the fall of 2001, the ratio of females to males at Florida State University (FSU) was approximately 4:3. If there were 35,500 students enrolled, to the nearest hundred, how many females were enrolled at FSU in Fall 2001? *(Lesson 6-2)* **20,300**

Wednesday

Multiple Choice What do you know for certain about $\overline{MN}$? *(Lesson 6-4)* **H**

- (F) $MN = \frac{1}{2}JL$
- (G) $MN = JL$
- (H) $\overline{MN} \parallel \overline{JL}$
- (I) $MN = 10$

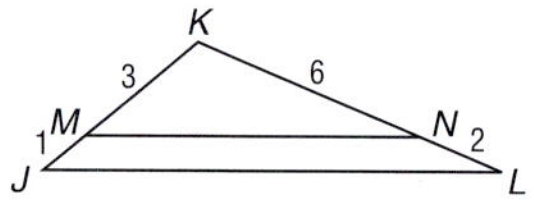

Gridded Response Find the length of $\overline{US}$ if $\overline{UV} \parallel \overline{RT}$. *(Lesson 6-4)* **6**

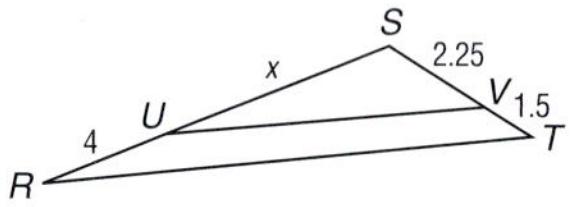

Thursday

Short Response If $\triangle ABC \sim \triangle ADE$, find the value of x and the scale factor. Show all your work. *(Lesson 6-3)* **See margin.**

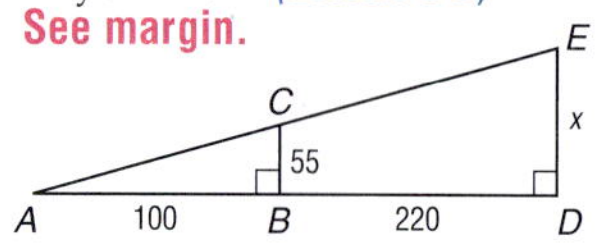

Friday

Multiple Choice Find x if $\overline{AB} \parallel \overline{EC}$. *(Lesson 6-4)* **B**

- (A) 12
- (B) 7
- (C) 5
- (D) 4

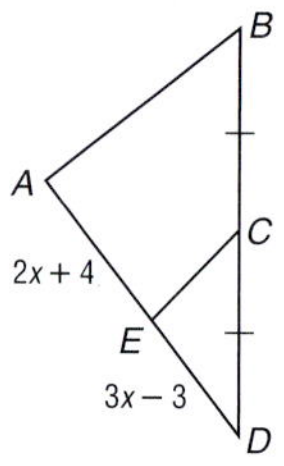

Countdown to FCAT

Monday

Extended Response Sample Answer

The original dimensions of the picture are 3" by 5". To find the new width for the 4:3 enlargement, I can write a proportion as $\frac{3}{x} = \frac{3}{4}$. Therefore, the first new width is 4". I will do the same for the length: $\frac{5}{x} = \frac{3}{4}$, resulting in $x = \frac{20}{3}$ or $6\frac{2}{3}$. Continuing with the same process and the next enlargement, I get the proportion for width as $\frac{4}{x} = \frac{2}{3}$ and $x = 6$ and for length, $\frac{\frac{20}{3}}{x} = \frac{2}{3}$ and $x = 10$.

I also could have done the whole thing in one step by combining the two enlargements of 4:3 and 3:2 and treating it as 12:6 or 2:1. The answers verify that this would work.

Thursday

Short Response Sample Answer

I can set up a proportion to find x, $\frac{100}{320} = \frac{55}{x}$, and cross multiply to find x. In this case, $x = 176$. The scale factor is $\frac{100}{320}$ or $\frac{55}{176}$, both of which equal 5:16.

Benchmarks Addressed

Monday	MA.A.3.4.3, MA.B.1.4.3
Tuesday	MA.B.1.4.3, MA.C.1.4.1, MA.C.2.4.1, MA.C.3.4.1, MA.D.2.4.2
Wednesday	MA.B.1.4.3, MA.C.1.4.1, MA.C.2.4.1, MA.C.3.4.1
Thursday	MA.A.3.4.3, MA.B.1.4.3
Friday	MA.A.3.4.3, MA.B.1.4.3, MA.C.1.4.1, MA.C.2.4.1, MA.C.3.4.1

Florida Geometry FL13

Countdown to FCAT

Monday

Extended Response Sample Answer

For the first rectangle, the width and length are 2 and 6. This can be found by plotting the points and counting squares. For the second rectangle, the width and length are 1 and 2. The lengths of the diagonals can be found using the Pythagorean Theorem.

$(d_1)^2 = 2^2 + 6^2$

$(d_1)^2 = 4 + 36$

$(d_1)^2 = 40$

$d_1 = \sqrt{40}$ or $2\sqrt{10}$

$(d_2)^2 = 1^2 + 2^2$

$(d_2)^2 = 1 + 4$

$(d_2)^2 = 5$

$d_2 = \sqrt{5}$

So the ratio of the two diagonals is $\frac{2\sqrt{10}}{\sqrt{5}}$ or $2\sqrt{2}:1$.

Thursday

Short Response Sample Answer

First, I need the lengths of the sides of the two triangles. I can plot the points and count squares for two of the sides and use the Pythagorean Theorem to figure the third. In $\triangle ABC$, the sides are 3, 4, and 5, so the perimeter is 12. In $\triangle DEF$, the sides are 6, 8, and 10, so the perimeter is 24. Therefore, the ratio of the two perimeters is 12 : 24 or 1 : 2.

Countdown to FCAT

Week 13

Monday

THINK SOLVE EXPLAIN

Extended Response Find the ratio of the lengths of the diagonals of rectangles *ABCD* and *RSTU* with the given vertex coordinates. Show all your work. *(Lesson 7-2)* **See margin.**

$A(-2, -2)$ $B(4, -2)$ $C(4, -4)$ $D(-2, -4)$ $R(0, 0)$ $S(0, 2)$ $T(1, 2)$ $U(1, 0)$

Tuesday

Multiple Choice Find the length of the hypotenuse of the given triangle. *(Lesson 1-3)* **B**

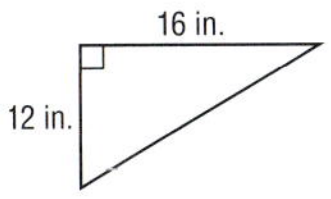

(A) 28 inches
(B) 20 inches
(C) 18 inches
(D) 15 inches

Gridded Response What is the measure of $\overline{RV}$? *(Lesson 7-1)* **2.4**

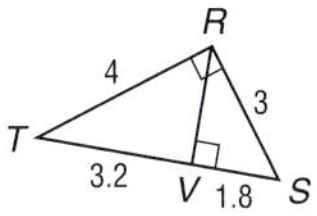

Wednesday

Multiple Choice Name two triangles that are similar to $\triangle AOD$. *(Lesson 7-1)* **F**

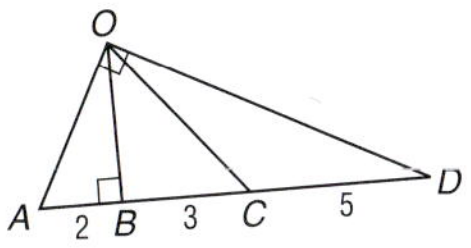

(F) $\triangle ABO$ and $\triangle OBD$
(G) $\triangle AOB$ and $\triangle COD$
(H) $\triangle BOD$ and $\triangle COD$
(J) $\triangle AOC$ and $\triangle COD$

Gridded Response Given $\triangle JKL \sim \triangle MNO$, find *x*. *(Lesson 6-5)* **15**

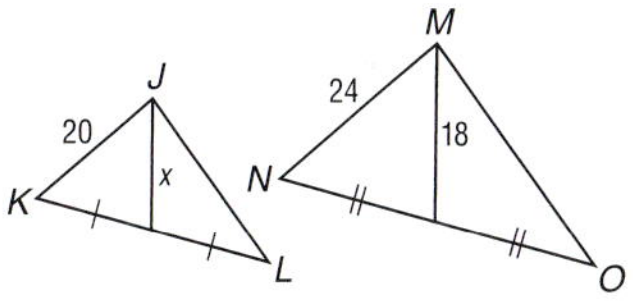

Thursday

THINK SOLVE EXPLAIN

Short Response Find the ratio of the perimeters of $\triangle ABC$ and $\triangle DEF$ if the coordinates for each are as follows. *(Lesson 6-3)* **See margin.**

$A(3, 3)$ $D(5, 0)$
$B(3, 7)$ $E(5, 8)$
$C(6, 3)$ $F(11, 0)$

Friday

Multiple Choice The ratio of undergraduate to graduate students at Florida State University was approximately 4.3 : 1 in 2001. If there were 35,500 students enrolled, how many were graduate students? *(Lesson 6-2)* **D**

(A) 28,801
(B) 20,097
(C) 13,398
(D) 6,699

Benchmarks Addressed

Monday	MA.A.3.4.3, MA.C.3.4.2
Tuesday	MA.B.1.4.3, MA.C.1.4.1, MA.C.2.4.1, MA.C.3.4.1, MA.C.3.4.2
Wednesday	MA.A.3.4.3, MA.B.1.4.3, MA.C.1.4.1, MA.C.2.4.1, MA.C.3.4.1
Thursday	MA.A.3.4.3, MA.B.1.4.1, MA.C.3.4.2
Friday	MA.A.3.4.3, MA.D.2.4.2

Week 14

Monday

Extended Response A ladder leans against the side of a house while gutters are being cleaned. The base of the ladder is 12 feet from the house. The top of the ladder rests 16 feet up the side of the house. Calculate the length of the ladder by first drawing a picture. Show all your work. *(Lesson 7-2)* See margin.

Tuesday

Multiple Choice The rectangle is reduced so that the ratio of the smaller length to the greater length is 1.5 : 2.25. What is the length of the longest side of the reduction? *(Lesson 6-2)* A

- (A) 4 centimeters
- (B) 5 centimeters
- (C) 6.5 centimeters
- (D) 8 centimeters

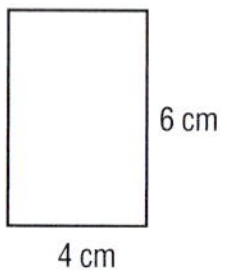

Gridded Response A rectangle has a width of 8 inches and a diagonal of 17 inches. What is the length in inches of this rectangle? *(Lesson 7-2)* 15

Wednesday

Multiple Choice Jackie and Marie left school at the same time and walked in two different directions. Jackie walked east for 12 blocks to the library. Marie walked south for 5 blocks to buy a CD. If the girls met halfway between the library and the CD store, how far would each girl have to walk from their destination until they meet? *(Lesson 7-2)* H

- (F) 13 blocks
- (G) 10 blocks
- (H) 6.5 blocks
- (I) 5 blocks

Gridded Response A sailboat is anchored 24 yards from a point on the shore. The top of the mast of the sailboat is 25 yards from the same point on the shore. How tall **in feet** is the mast? *(Lesson 7-2)* 21

Thursday

Short Response Which of the following is NOT a Pythagorean triple? Explain your answer. *(Lesson 7-2)* See margin.

5, 12, 13
21, 20, 29
16, 30, 32
15, 8, 17

Friday

Gridded Response The scale on a map of Orlando is 0.75 inch = 5 miles. The distance on the map from Epcot® to the planetarium is 2.75 inches. What is the distance to the nearest tenth of a mile from Epcot® to the planetarium? *(Lesson 6-2)* 18.3

Countdown to FCAT

Monday

Extended Response Sample Answer

Since the house and the ground make a right angle, this can be solved by using the Pythagorean Theorem. So the length of the ladder is $\sqrt{12^2 + 16^2} = \sqrt{144 + 256} = \sqrt{400}$ or 20. Therefore, the ladder is 20 feet long.

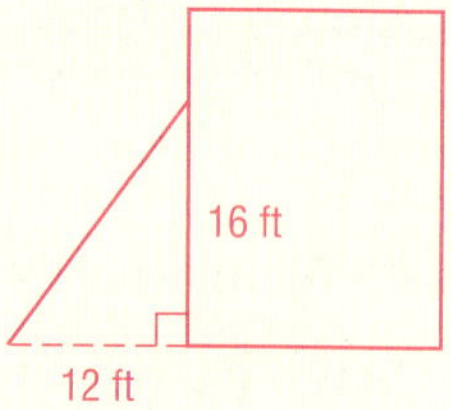

Thursday

Short Response Sample Answer

To check to see if each set is a Pythagorean triple, I have to determine whether the sum of squares of the two numbers equals the square of the third number.

$5^2 + 12^2 = 13^2$

$21^2 + 20^2 = 29^2$

$16^2 + 30^2 \neq 32^2$

So the third set is not a Pythagorean triple.

Benchmarks Addressed

Monday	MA.A.3.4.3, MA.C.3.4.1
Tuesday	MA.A.3.4.3, MA.B.1.4.3, MA.C.3.4.1
Wednesday	MA.B.2.4.1, MA.C.3.4.1
Thursday	MA.A.3.4.3, MA.A.5.4.1, MA.C.3.4.1
Friday	MA.A.3.4.3, MA.C.3.4.1

Florida Geometry FL15

Countdown to
FCAT

Monday
Extended Response Sample Answer

Parts A, B Using the formula to determine the measure of the interior angles of a regular polygon, I can find the number of sides in the new figure. But first I have to know the measure of an interior angle. The measure of an interior angle of a hexagon can be found by using the formula $s = \frac{180(6-2)}{6} = \frac{720}{6} = 120°$. I already know that a square has interior angles of 90°. So knowing that the sum of the three angles at C is 360°, I find $m\angle ACB = 360 - 120$ or 150°. Then I use the formula for the measure of an interior angle again.

$$s = \frac{180(n-2)}{n}$$

$$150 = \frac{180(n-2)}{n}$$

$$150n = 180n - 360$$

$$n = 12$$

So the new figure would be 12-sided.

Countdown to FCAT

Week 15

Monday

THINK SOLVE EXPLAIN

Extended Response If a regular hexagon and a square share a side, calculate, *(Lesson 8-1)*

Part A the measure of the exterior angle formed at their intersection ($\angle ACB$) and **See margin.**

Part B the number of sides of a regular polygon that has $\overline{AC}$ and $\overline{CB}$ as two of its sides. **See margin.**

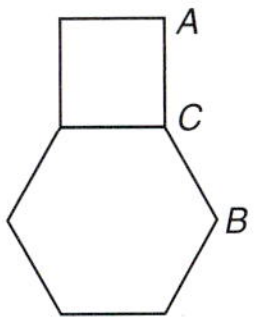

Tuesday

Multiple Choice If the measure of an interior angle of a regular polygon is 135°, how many sides does the polygon have? *(Lesson 8-1)* **B**

(A) 10 (B) 8
(C) 7 (D) 6

Gridded Response What is the value of x in the regular pentagon? *(Lesson 8-1)* **36**

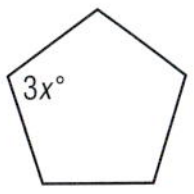

Wednesday

Multiple Choice Find x and y so that the quadrilateral is a parallelogram. *(Lesson 8-3)* **H**

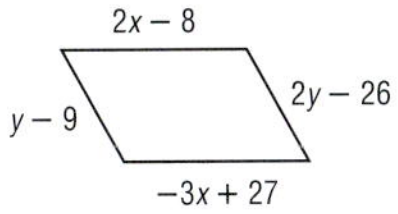

(F) $x = 8, y = 6$ (G) $x = 6, y = 8$
(H) $x = 7, y = 17$ (I) $x = 17, y = 7$

Gridded Response Find $m\angle 7$ in degrees. *(Lesson 8-1)* **60**

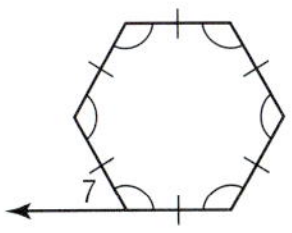

Thursday

THINK SOLVE EXPLAIN

Short Response Determine the coordinates of the point representing the intersection of the two diagonals of the parallelogram with vertices at $A(-2, -2)$, $B(-4, -4)$, $C(2, -4)$, and $D(4, -2)$. Show all yourwork. *(Lesson 8-2)*

Since I know that the diagonals of a parallelogram bisect each other, I can find the midpoint of one of the diagonals.

$$m = \left(\frac{4-4}{2}, \frac{-2-4}{2}\right) \text{ or } (0, -3)$$

Friday

Gridded Response Given $\square ABCD$, find the measure in degrees of $\angle BDC$ near Tampa. *(Lesson 8-3)* **75**

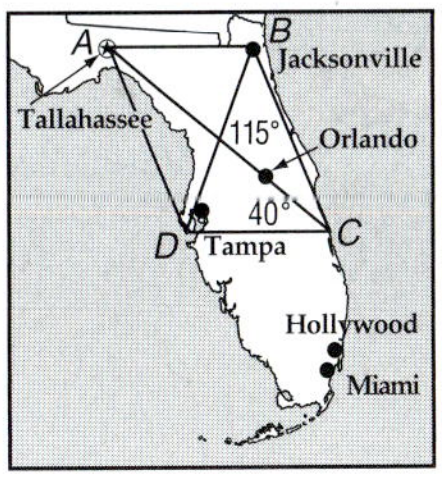

Benchmarks Addressed

Monday	MA.A.3.4.3, MA.B.1.4.2
Tuesday	MA.A.3.4.3, MA.B.1.4.2
Wednesday	MA.B.1.4.2, MA.C.1.4.1, MA.C.2.4.1, MA.C.3.4.1
Thursday	MA.C.3.4.2
Friday	MA.B.1.4.2, MA.C.2.4.1

Countdown to FCAT

Week 16

Monday

Extended Response A quadrilateral has vertices $R(-3, -3)$, $S(0, 2)$, $T(1, -2)$, and $U(-3, -6)$. *(Lesson 8-3)*

Part A Graph the quadrilateral. **See margin.**

Part B Then determine whether the quadrilateral is a parallelogram. Explain your answer. **See margin.**

Tuesday

Multiple Choice A parallelogram has vertices at (2, 3), (4, 5), and (6, 5). Which is an ordered pair for the fourth vertex? *(Lesson 3-8)* **A**

Ⓐ (4, 3) Ⓑ (5, 3) Ⓒ (6, 3) Ⓓ (3, 4)

Gridded Response Find the value of m in rectangle $JKLM$. *(Lesson 8-4)* **20**

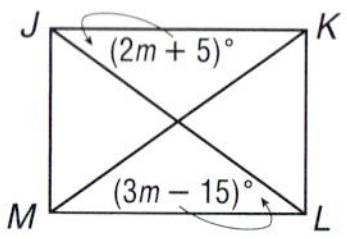

Wednesday

Multiple Choice If the perimeter of a square is 36 centimeters, what is its area? *(Lesson 8-5)* **H**

Ⓕ 72 square centimeters
Ⓖ 80 square centimeters
Ⓗ 81 square centimeters
Ⓘ 100 square centimeters

Gridded Response If the area of a square is 49 square centimeters, what is its perimeter in centimeters? *(Lesson 8-5)* **28**

Thursday

Short Response Show that the parallelogram with vertices at $A(-6, 3)$, $B(-6, 4)$, $C(-1, 4)$, and $D(-1, 3)$ is a rectangle. Explain your answer. *(Lesson 8-4)*

See margin for graph. To show that this parallelogram is a rectangle, I must show that one angle is a right angle. $\overline{AB}$ is vertical and $\overline{AD}$ is horizontal, so they are perpendicular. This means that $\angle BAD$ is a right angle. So the parallelogram is a rectangle.

Friday

Multiple Choice How many times as steep is the slope of the line representing U.S. citrus production as is the slope of the line representing Florida citrus production for the years 1998–1999? *(Lesson 3-6)* **D**

Citrus Production

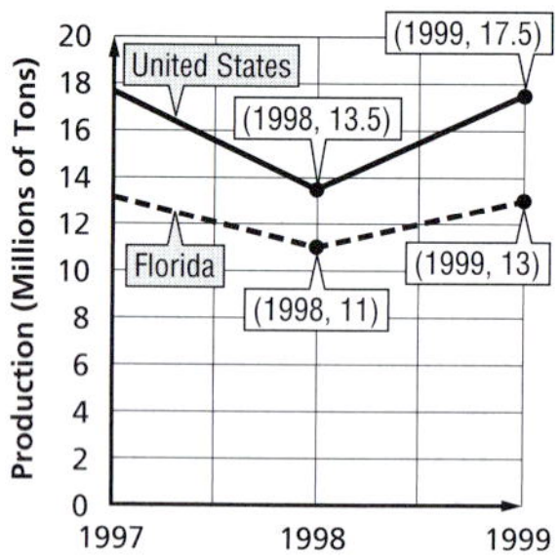

Source: www.nass.usda.gov

Ⓐ 4 times Ⓑ 3.5 times
Ⓒ 3 times Ⓓ 2 times

Countdown to FCAT

Monday

Extended Response Sample Answer

Part A

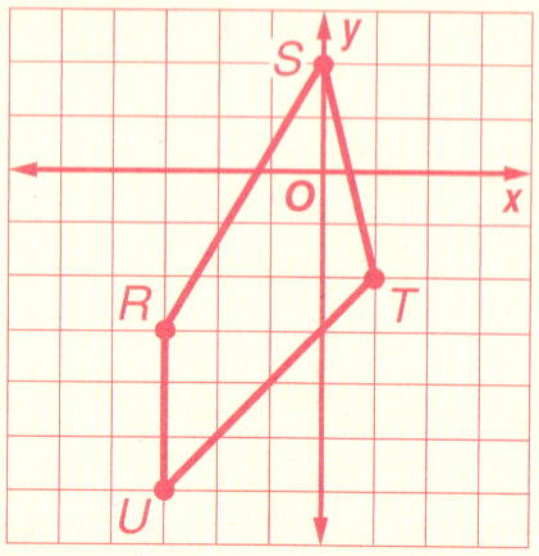

Part B To show whether the quadrilateral is a parallelogram or not, I must show that opposite sides are parallel—meaning having the same slope. $\overline{RU}$ is a vertical segment. $\overline{ST}$ is opposite $\overline{RU}$ and is not vertical, so they are not parallel. The quadrilateral is not a parallelogram.

Thursday

Short Response Sample Answer

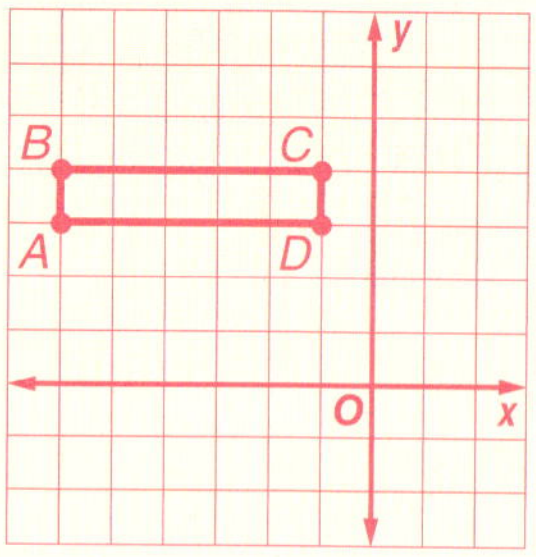

Benchmarks Addressed

Monday	MA.C.1.4.1, MA.C.2.4.1, MA.C.3.4.2
Tuesday	MA.C.2.4.1, MA.C.3.4.2
Wednesday	MA.B.1.4.1, MA.C.2.4.1
Thursday	MA.C.1.4.1, MA.C.2.4.1, MA.C.3.4.2
Friday	MA.C.3.4.2, MA.E.1.4.1

Florida Geometry FL17

Countdown to FCAT

Monday
Extended Response Sample Answer

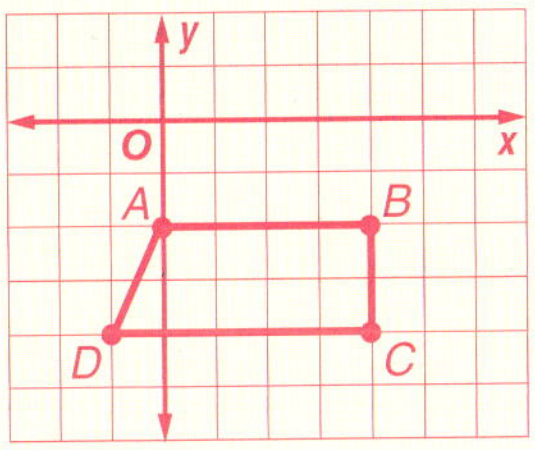

By plotting the points, it appears the figure is a trapezoid. But to verify for certain, I must check to see that the figure has one pair of parallel sides. The slope of $\overline{AB}$ is 0 and the slope of $\overline{DC}$ is also 0. Therefore, those two sides are parallel, but the other two are not. A four-sided figure with exactly one pair of parallel sides is a trapezoid.

Thursday
Short Response Sample Answer

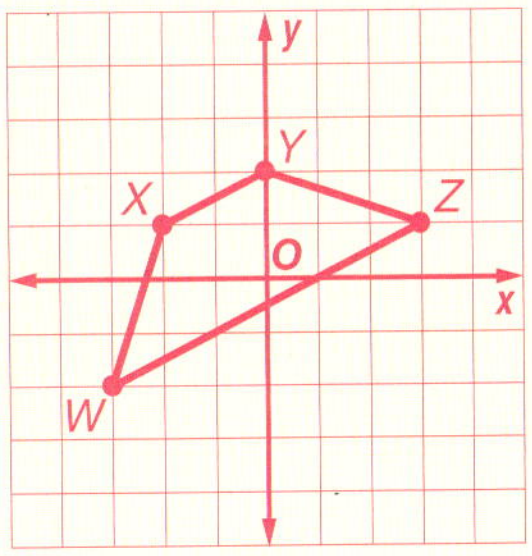

Countdown to FCAT

Week 17

Monday

Extended Response Plot $A(0, -2)$, $B(4, -2)$, $C(4, -4)$, and $D(-1, -4)$ and connect the points to form a quadrilateral. Label the quadrilateral as a *square, rectangle, rhombus,* or *trapezoid.* Be as specific as possible. Justify your answer. *(Lesson 8-6)* **See margin.**

Tuesday

Multiple Choice Determine the slope of the segment with endpoints at $(3, a)$ and $(2, 2a)$. *(Lesson 8-6)* **C**

Ⓐ $\frac{1}{a}$ Ⓑ a
Ⓒ $-a$ Ⓓ $-\frac{1}{a}$

Gridded Response If a right triangle is reflected in the y-axis, what is the y-coordinate of the point corresponding to $(4, 1)$? *(Lesson 9-1)* **1**

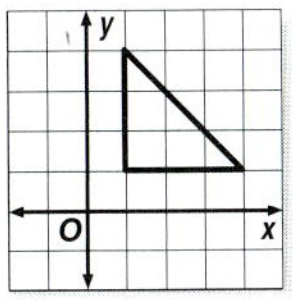

Wednesday

Multiple Choice What are the missing coordinates for the rhombus? *(Lesson 8-7)* **G**

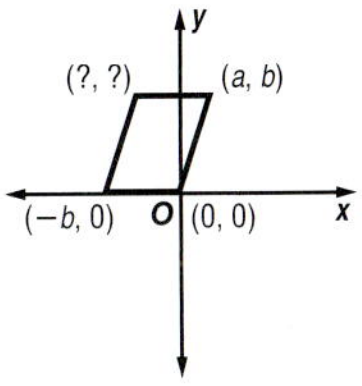

Ⓕ (a, b)
Ⓖ $(-a, b)$
Ⓗ $(-a, -b)$
Ⓘ $(a, -b)$

Gridded Response If the Florida flag shown were drawn using a scale of 1 inch $= 2\frac{2}{3}$ feet, what is the length of the flag in inches in the scale drawing? *(Lesson 6-2)* **$\frac{9}{8}$ or 1.125**

Thursday

Short Response Verify that quadrilateral $WXYZ$ is a trapezoid. Explain your reasoning. *(Lesson 8-6)*

$W(-3, -2)$
$X(-2, 1)$
$Y(0, 2)$
$Z(3, 1)$

I plotted the points. (See margin for graph.) The figure has to have at least one pair of parallel sides. The slope of $\overline{XY}$ is $\frac{2-1}{2}$ or $\frac{1}{2}$. The slope of $\overline{WZ}$ is $\frac{3}{6}$ or $\frac{1}{2}$ also. Therefore, the figure is a trapezoid. (You can see from plotting the figure that the other two sides are definitely not parallel.)

Friday

Multiple Choice The transformation shown represents all but a ___?___. *(Lesson 9-5)* **C**

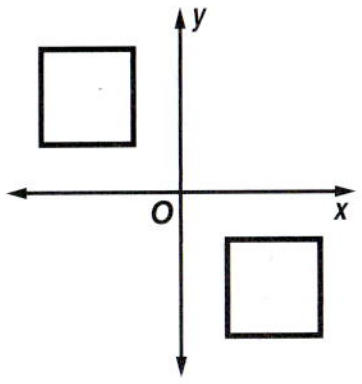

Ⓐ reflection
Ⓑ translation
Ⓒ dilation
Ⓓ rotation

Benchmarks Addressed

Monday	MA.C.1.4.1, MA.C.2.4.1, MA.C.3.4.2
Tuesday	MA.C.3.4.2
Wednesday	MA.B.1.4.2, MA.C.2.4.1
Thursday	MA.C.1.4.1, MA.C.2.4.1, MA.C.3.4.2
Friday	MA.C.2.4.1

FL18 Florida Geometry

Countdown to FCAT

Week 18

Monday

Extended Response A triangle has vertices at $(-6, -2)$, $(-3, -2)$, and $(-6, -5)$. Suppose the triangle is translated 3 units right and 4 units up. Draw the new figure and label each vertex. *(Lesson 9-2)* **See margin.**

Tuesday

Multiple Choice Which transformations best describe the figure below? *(Lesson 9-5)* **B**

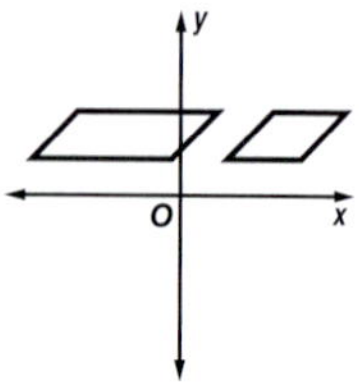

Ⓐ translation and reflection
Ⓑ translation and dilation
Ⓒ reflection and dilation
Ⓓ rotation and dilation

Gridded Response Ignoring the type and image on the emblem, how many lines of symmetry are there in the Florida state flag? *(Lesson 9-1)* **2**

Wednesday

Multiple Choice If the point at $(m, -n)$ is reflected in the y-axis, what are the coordinates of the new point? *(Lesson 9-1)* **G**

Ⓕ (m, n) Ⓖ $(-m, -n)$
Ⓗ $(-m, n)$ Ⓘ $(m, -n)$

Gridded Response The coordinates of the vertices of $\triangle ABC$ are $A(-4, -2)$, $B(-2, 0)$, and $C(-1, -2)$. If $\triangle ABC$ is rotated 180° in point O, what is the x-coordinate of point C'? *(Lesson 9-3)* **2**

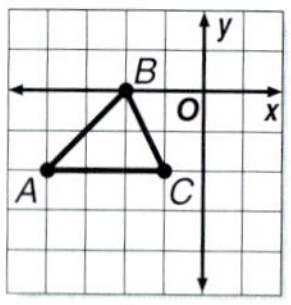

Thursday

Short Response Draw and label the points for the reflection of the right triangle in the x-axis. *(Lesson 9-1)* **See margin.**

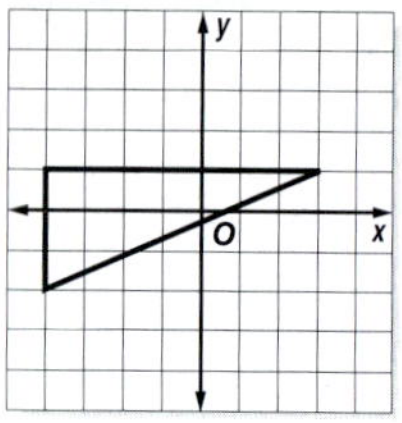

Friday

Multiple Choice Choose the statement that is an accurate description of the transformations. *(Lesson 9-5)* **C**

Ⓐ A dilation can be thought of as two reflections.
Ⓑ Two translations across parallel lines is the same as one reflection.
Ⓒ Two reflections in the same direction across parallel lines is the same as a translation.
Ⓓ A rotation is equivalent to two translations.

Benchmarks Addressed

Monday	MA.C.2.4.1, MA.C.3.4.2
Tuesday	MA.C.2.4.1, MA.C.3.4.2
Wednesday	MA.C.2.4.1, MA.C.3.4.2
Thursday	MA.C.2.4.1, MA.C.3.4.2
Friday	MA.C.2.4.1, MA.C.3.4.2

Countdown to FCAT

Monday

Extended Response Sample Answer

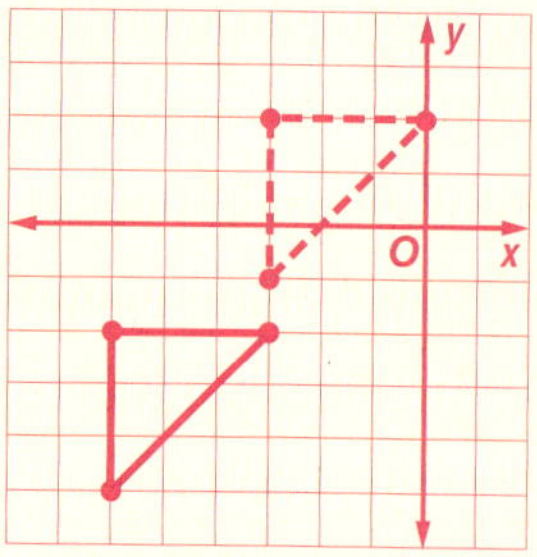

If the triangle shown is translated three units right and four units up, the resulting vertices are $(-3, -1)$, $(-3, 2)$, and $(0, 2)$. Those can be found by counting on the coordinate grid or by adding 3 to each of the x-coordinates and adding 4 to each of the y-coordinates.

Thursday

Short Response Sample Answer

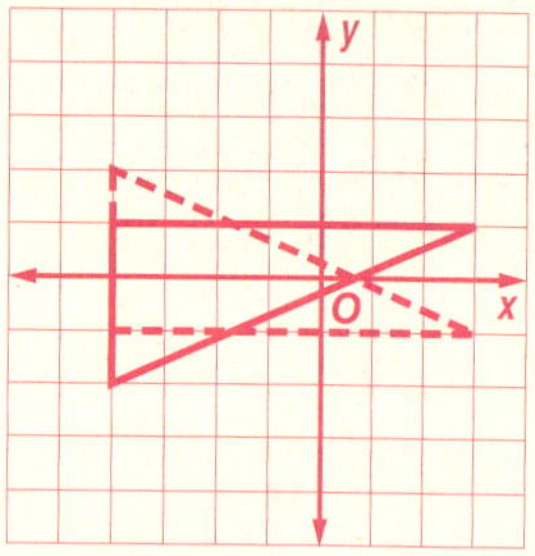

Since the coordinates of the original triangle were $(-4, -2)$, $(-4, 1)$, and $(3, 1)$, the coordinates of the new triangle reflected in the x-axis are $(-4, 2)$, $(-4, -1)$, and $(3, -1)$. All of the x-coordinates stayed the same, but the y-coordinates were all multiplied by -1.

Countdown to FCAT

Monday
Extended Response Sample Answer

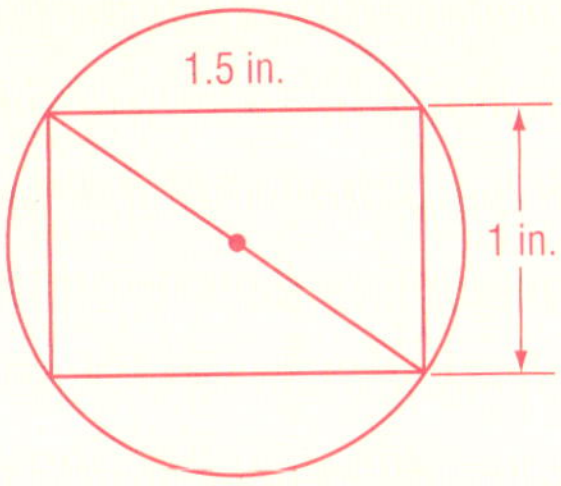

If the dimensions of the rectangle are 1 inch by 1.5 inches, I need to find the diagonal of the rectangle, which will be the same measure as the diameter of the circle. Using the Pythagorean Theorem, I find that the diagonal is 1.8 inches. Then using the formula for the circumference of a circle, I get $C = 1.8\pi$, which is approximately equal to 5.66 inches.

Thursday
Short Response Sample Answer

The circumference of the circle uses the diameter. In this case, the diameter is the same as the diagonal of the rectangle or the hypotenuse of the right triangle whose sides are 6 and 8. Since 6 and 8 are part of a Pythagorean triple, I know that the length of the hypotenuse is 10 and the circumference is 10 times pi or 10π.

Countdown to FCAT

Week 19

Monday

Extended Response A rectangle with dimensions 1.5 inches by 1 inch is inscribed in a circle. Calculate the circumference of the circle to the nearest hundredth. Show all your work. *(Lesson 10-1)* **See margin.**

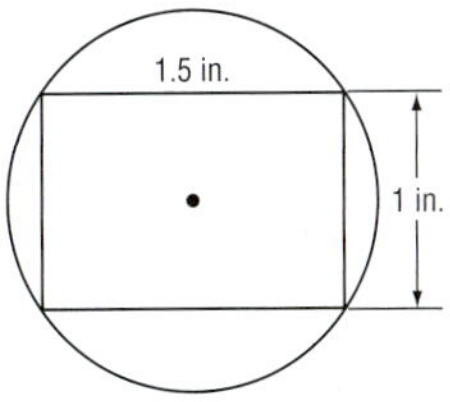

Tuesday

Multiple Choice Solve $F = \frac{9}{5}C + 32$ for C. *(Prerequisite)* **B**

A $\frac{9}{5}(F - 32)$ B $\frac{5}{9}(F - 32)$

C $\frac{9}{5}(F + 32)$ D $\frac{5}{9}(F + 32)$

Gridded Response The diameter of $\odot A$ is 8 centimeters, and the diameter of $\odot C$ is 5 centimeters. What is the length in centimeters of $\overline{AB}$? *(Lesson 10-1)* **1.5**

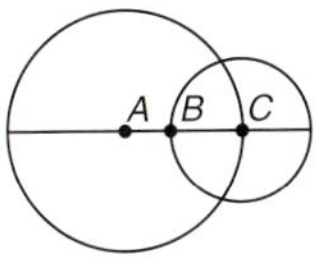

Wednesday

Multiple Choice What is the ratio of the circumference of circle R to the circumference of circle S? *(Lesson 10-1)* **H**

F $2:5$
G $4:5\pi$
H $4:5$
I $4:10$

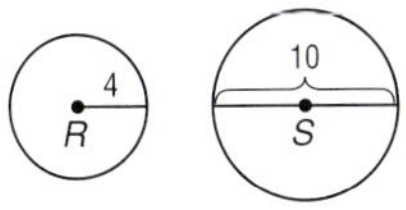

Gridded Response Two hundred Key West residents were surveyed as to their ice cream preference. How many people prefer vanilla? *(Lesson 10-2)* **80**

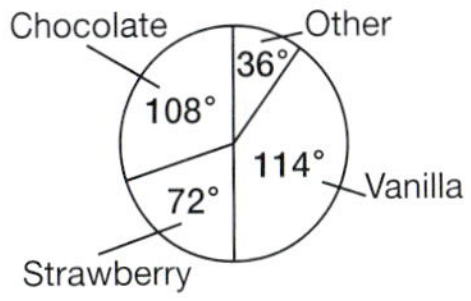

Thursday

Short Response Find the exact circumference of circle A. *(Lesson 10-1)* **See margin.**

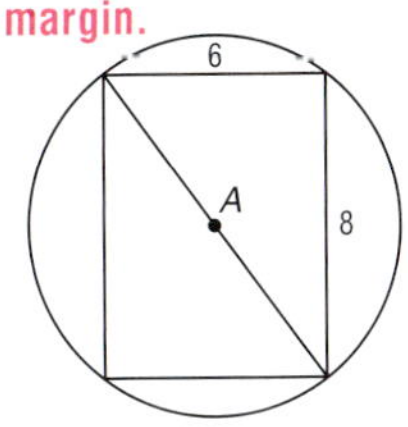

Friday

Gridded Response Find $m\angle AOD$ in degrees. *(Lesson 10-2)* **135**

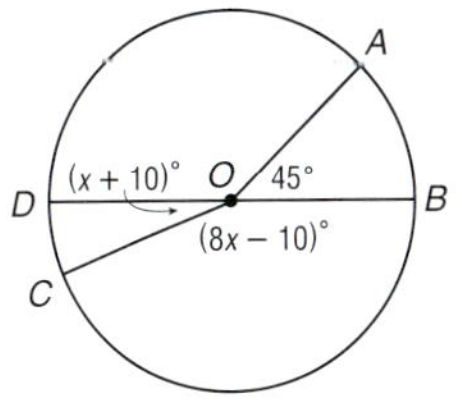

FL20 Florida Geometry

Benchmarks Addressed

Monday	MA.B.1.4.1
Tuesday	MA.B.1.4.1, MA.B.1.4.2, MA.D.1.4.1
Wednesday	MA.B.1.4.2, MA.E.1.4.1
Thursday	MA.B.1.4.1
Friday	MA.B.1.4.2

Countdown to FCAT

Week 20

Monday

THINK SOLVE EXPLAIN **Extended Response** Explain why the two triangles are not similar. *(Lesson 6-2)* **See margin.**

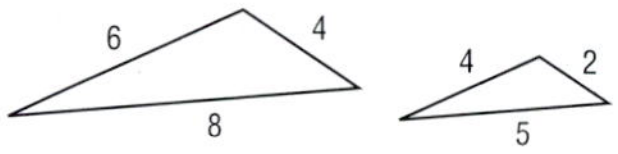

Tuesday

Multiple Choice What is the approximate length of the line segment with endpoints at (−3, 4) and (5, 2)? *(Lesson 1-3)* **A**

Ⓐ 8.25 Ⓑ 6.32
Ⓒ 5.27 Ⓓ 5.01

Gridded Response How many dots are in the 5th stage of the sequence? *(Lesson 2-1)* **15**

Wednesday

Multiple Choice Given $a \parallel b$, find all angles supplementary to $\angle 3$. *(Lesson 3-2)* **I**

Ⓕ 1, 2, 4
Ⓖ 2, 6, 7
Ⓗ 1, 2, 5, 8
Ⓘ 1, 4, 5, 8

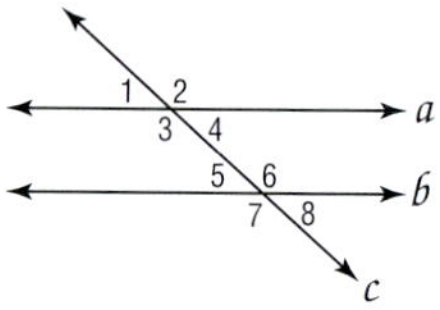

Gridded Response If the legs of the triangle shown are cut in half, what is the length to the nearest hundredth unit of the resulting hypotenuse? *(Lesson 4-1)* **5.83**

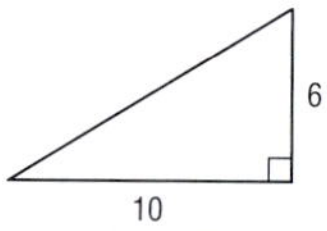

Thursday

THINK SOLVE EXPLAIN **Short Response** List all angles with measures less than $m\angle 2$. Show your work and explain your reasoning. *(Lesson 5-2)* **See margin.**

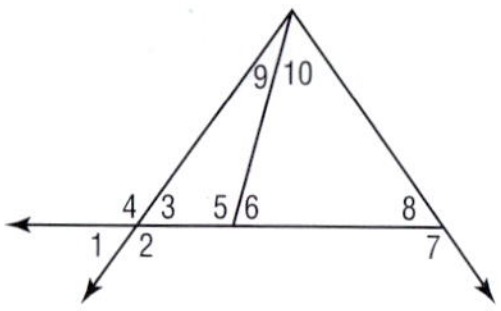

Friday

Gridded Response What is the grid distance to the nearest hundredth between Tampa and Tallahassee if Tampa is located on the grid at (3, 0) and Tallahassee is located at (0, 5)? *(Lesson 1-3)* **5.83**

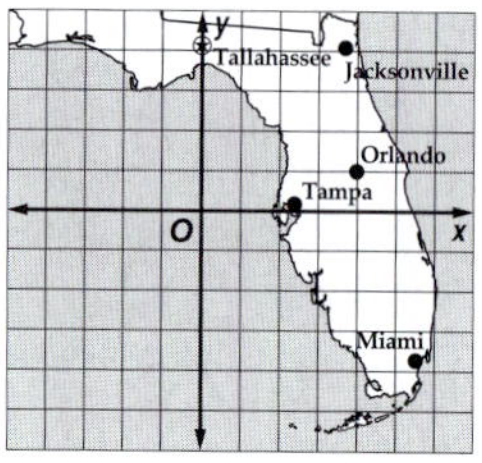

Benchmarks Addressed

Monday	MA.A.3.4.3, MA.B.1.4.3
Tuesday	MA.B.3.4.1, MA.C.3.4.2
Wednesday	MA.A.5.4.1, MA.C.2.4.1, MA.D.1.4.2
Thursday	MA.B.1.4.2, MA.C.1.4.1
Friday	MA.C.3.4.2

Countdown to FCAT

Monday

Extended Response Sample Answer

If the two triangles are similar, then their corresponding sides must be proportional. The first pair of corresponding sides I will look at are the short sides. The ratio is 4:2 or 2:1. So if the triangles are similar, all the other corresponding sides should be in a ratio of 2:1. The ratio of the longest sides is 8:5. This does not equal 2:1. Therefore, without looking at the third sides, I know that the two triangles are not similar.

Thursday

Short Response Sample Answer

What I know is this.

$m\angle 2 > m\angle 5$ Exterior Angle Inequality Theorem

$m\angle 2 > m\angle 9$ Exterior Angle Inequality Theorem

$m\angle 2 > m\angle 8$ Exterior Angle Inequality Theorem

$m\angle 2 > m\angle 9 + m\angle 10$ Exterior Angle Inequality Theorem

Therefore, $m\angle 2 > m\angle 5$, $m\angle 8$, $m\angle 9$, and $m\angle 10$.

Countdown to FCAT

Monday

Extended Response Sample Answer

Use the formula for the area of a parallelogram.

$A = bh$ Area of a parallelogram

$30 = (3x + 1)x$ Substitution

$30 = 3x^2 + x$ Distributive Property

$0 = 3x^2 + x - 30$ Subtraction Property

$0 = (3x + 10)(x - 3)$ Factor.

$x = -\frac{10}{3}$ or $x = 3$

The only solution that makes sense is $x = 3$. Therefore, the height of the parallelogram is 3 and the length is 10.

Thursday

Short Response Sample Answer

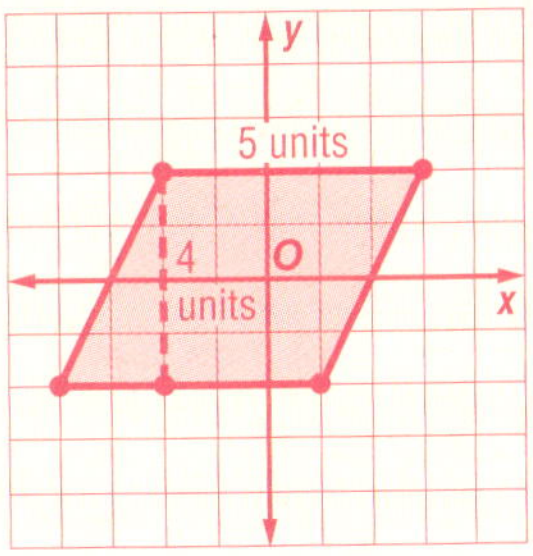

By sketching the figure, I can see that the height of the parallelogram is 4 and the base is 5. Therefore, the area is 4 times 5 or 20 square units.

Countdown to FCAT

Week 21

Monday

Extended Response Find the height and base of the parallelogram with an area of 30 square units. Show all your work. *(Lesson 11-1)* See margin.

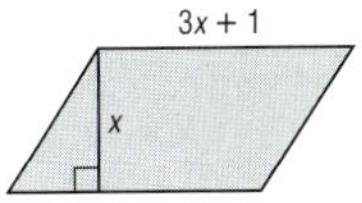

Tuesday

Multiple Choice What is the ratio of the area of △ABE to the area of ▱ABCD? *(Lesson 11-2)* D

Ⓐ 1:2 Ⓑ 2:5
Ⓒ 1:4 Ⓓ 1:10

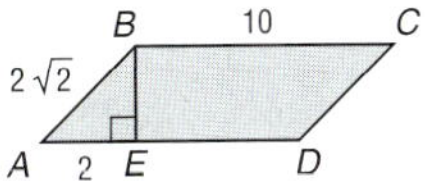

Gridded Response What is the area of △ABC in square units? *(Lesson 11-2)* 7.5

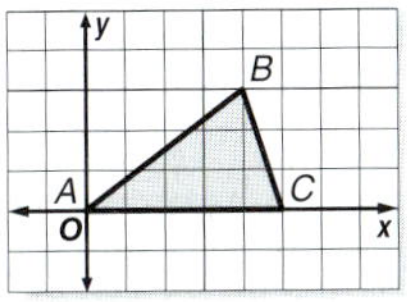

Wednesday

Multiple Choice Find the area of the shaded region to the nearest thousandth square unit. *(Lesson 11-3)* F

Ⓕ 0.658
Ⓖ 1.525
Ⓗ 2.405
Ⓘ 3.063

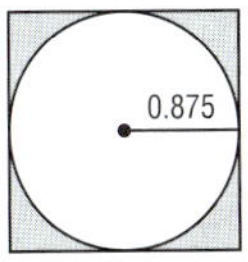

Gridded Response What is the maximum area in square units of a rectangle that has a perimeter of 30 units and side measures that are whole numbers? *(Lesson 11-1)* 56

Thursday

Short Response Find the area of the parallelogram with vertices at (−2, 2), (3, 2), (−4, −2), and (1, −2). Sketch the figure. Show all your work. *(Lesson 11-1)* See margin.

Friday

Multiple Choice If the diameter of the seal in the center of the Florida state flag is $\frac{1}{3}$ the total length of the flag, what is the area of the seal on a flag measuring 6 feet long by 4 feet wide? *(Lesson 11-3)* A

Ⓐ 3.14 feet Ⓑ 6.28 feet
Ⓒ 12.56 feet Ⓓ 24 feet

Benchmarks Addressed

Monday	MA.B.1.4.1, MA.C.3.4.1, MA.D.1.4.1
Tuesday	MA.A.3.4.3, MA.B.1.4.1
Wednesday	MA.A.3.4.3, MA.B.1.4.1, MA.C.3.4.1
Thursday	MA.B.1.4.1, MA.C.3.4.1, MA.C.3.4.2
Friday	MA.A.3.4.3, MA.B.1.4.1

Week 22

Monday

Extended Response Calculate the area of the octagon. Show all your work. *(Lesson 11-1)* **See margin.**

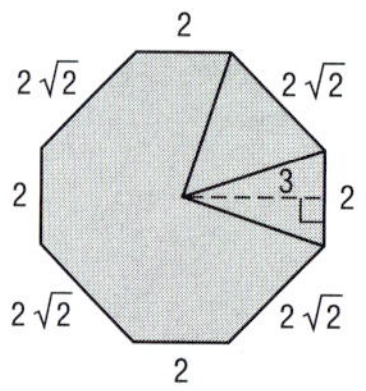

Tuesday

Multiple Choice
Calculate the area of the hexagon in square units. *(Lesson 11-1)* **D**

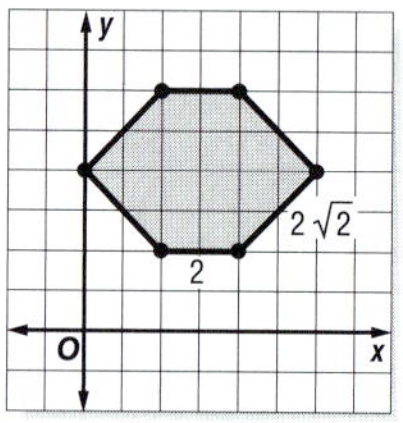

Ⓐ $4\sqrt{2}$
Ⓑ 6
Ⓒ 9
Ⓓ 16

Gridded Response
How many square feet of carpet are needed to carpet the entire room shown? *(Lesson 11-1)* **325**

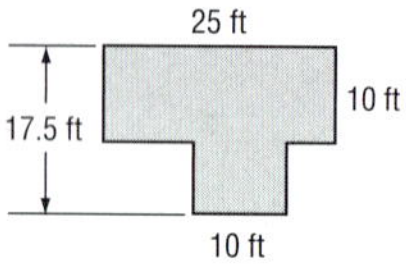

Wednesday

Multiple Choice What is the ratio of the area of the trapezoid to the area of the right triangle? *(Lesson 11-2)* **F**

Ⓕ 3:4
Ⓖ 2:3
Ⓗ 1:2
Ⓘ 4:3

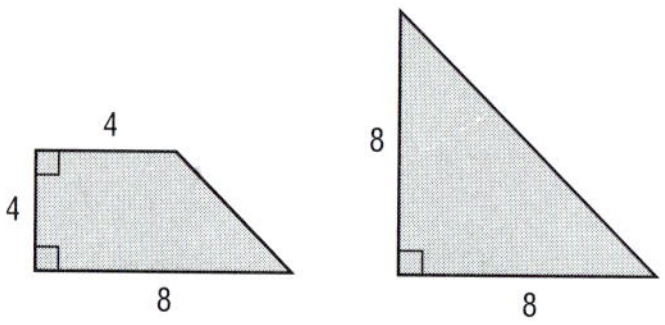

Gridded Response If trapezoid *RSTU* has an area of 8 square units, find the length of the shorter base in units if the longer base is 6 units and the height is 2 units. *(Lesson 11-2)* **2**

Thursday

Short Response A trapezoidal placemat drawn on a coordinate grid has vertices at (−1, 3), (2, 3), (3, 0), and (−3, 0). A rectangular placemat drawn on the same grid has vertices at (−1, 3), (2, 3), (2, −2), and (−1, −2). *(Lesson 11-2)*

Part A Which placemat will require more material to make?

Part B How much more material will be required?

See margin for graph.

Part A *A* of trapezoid $= \frac{1}{2}(3)(6 + 3)$ or 13.5. *A* of rectangle $= 3(5)$ or 15. So the rectangle will require more material to make.

Part B It will take 1.5 more units of material to make the rectangular placemats.

Friday

Gridded Response Of the 200 Key West residents surveyed, what percent prefers strawberry ice cream? *(Lesson 10-2)* **20**

Ice Cream Preferences in Key West

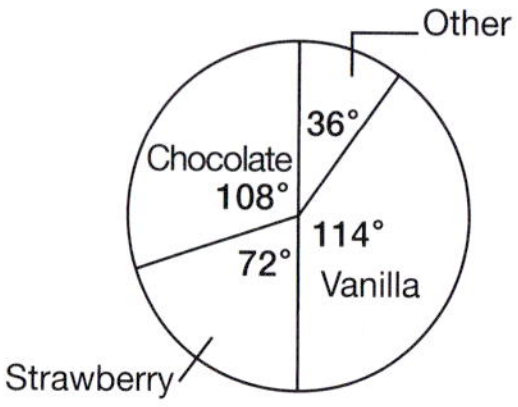

Benchmarks Addressed

Monday	MA.A.3.4.3, MA.B.1.4.1
Tuesday	MA.A.3.4.3, MA.B.1.4.1, MA.C.3.4.2
Wednesday	MA.A.3.4.3, MA.B.1.4.1
Thursday	MA.A.3.4.3, MA.B.1.4.1
Friday	MA.A.3.4.3, MA.B.1.4.2, MA.E.1.4.1

Countdown to FCAT

Monday

Extended Response Sample Answer

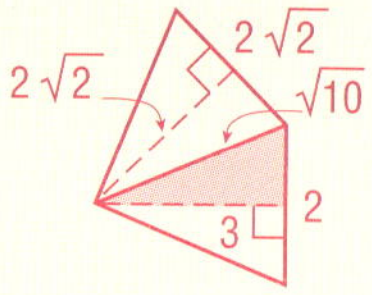

There would be 8 triangles in the octagon. Four have bases of 2 units and 4 have bases of $2\sqrt{2}$ units. The area of a triangle with base of 2 units is $A = \frac{1}{2}(2)(3)$ or 3. I have to find the height of the other triangle. The hypotenuse of the shaded right triangle is $\sqrt{3^2 + 1^2}$ or $\sqrt{10}$ units long. The height of this triangle can be found using the Pythagorean Theorem: $(\sqrt{2})^2 + x^2 = (\sqrt{10})^2$, so $x = 2\sqrt{2}$.

Now I can find the area: $A = \frac{1}{2}(2\sqrt{2})(2\sqrt{2})$ or 4. There are 4 of each triangle, so the total area is $4(3) + 4(4)$ or 28 square units.

Thursday

Short Response Sample Answer

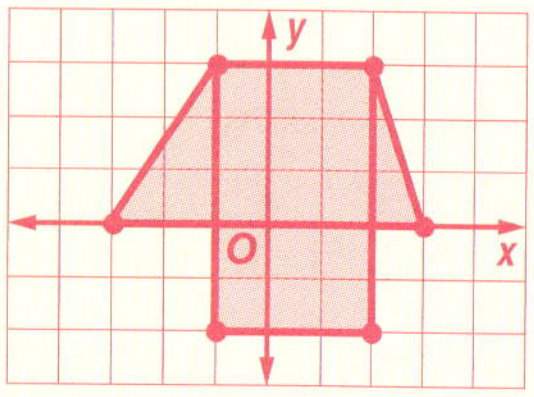

Countdown to FCAT

Thursday

Short Response Sample Answer

Part A

Use the formula for the surface area of a cube if $\ell = 5$, $w = 5$, and $h = 5$.

$$SA = 2(\ell w) + 2(\ell h) + 2(hw)$$
$$= 2(25) + 2(25) + 2(25)$$
$$= 150 \text{ square inches}$$

Use the formula for the surface area of a cylinder if $r = 3$ and $h = 6$.

$$SA = 2\pi rh + 2\pi r^2$$
$$= 2\pi(3)(6) + 2\pi(3)^2$$
$$\approx 113.10 + 56.55$$
$$\approx 169.65 \text{ square inches}$$

The cylinder requires more paper.

Part B It will take 169.65 − 150 or 19.65 square inches more adhesive paper to cover the cylinder than to cover the cube.

Countdown to FCAT

Week 23

Monday

THINK SOLVE EXPLAIN **Extended Response** If one gallon of paint will coat 1750 square feet of surface, how many coats of paint can be applied to the figure shown using only 1 gallon of paint? *(Lesson 12-3)* The surface area of the solid shown is $2(5 \times 8) + 2(8 \times 15) + 2(5 \times 15) = 80 + 240 + 150$ or 470. Since one gallon of paint will coat 1750 square feet of surface, $\frac{1750}{470}$ or about 3.72 coats. This means that 3 full coats and part of a 4th coat of paint can be applied to the figure from one gallon of paint.

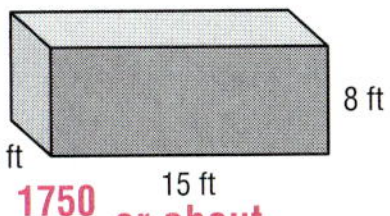

Tuesday

Multiple Choice If the flat figure shown were folded on each dark line to form a solid, which solid would result? *(Lesson 11-2)* A

(A) square pyramid
(B) triangular pyramid
(C) rectangular prism
(D) triangular prism

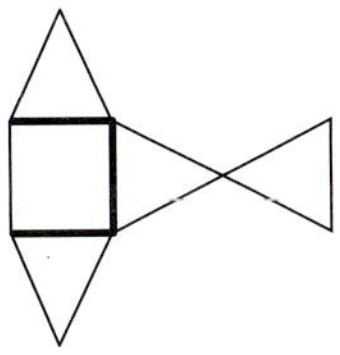

Gridded Response What is the surface area in square centimeters of a square pyramid with a base of 4 centimeters by 4 centimeters and a slant height of 8 centimeters? *(Lesson 12-5)* 80

Wednesday

Multiple Choice What is the surface area of the right circular cylinder to the nearest hundredth? *(Lesson 12-4)* G

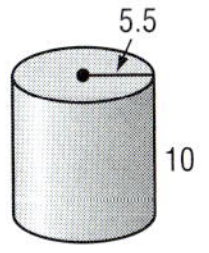

(F) 267.83 (G) 535.65
(H) 580.65 (I) 1071.30

Gridded Response The surface area of a cube is 96 square inches. What is the length in inches of each edge? *(Lesson 12-3)* 4

Thursday

THINK SOLVE EXPLAIN **Short Response** Jaime is covering two small containers with adhesive paper. The first is a cube measuring 5 inches on each edge. The second is a right circular cylinder with a diameter of 6 inches and a height of 6 inches. *(Lesson 12-5)*

Part A Which container will require the most adhesive paper?

Part B How much more? Show all your work. Parts A & B: See margin.

Friday

Multiple Choice What is the approximate ratio of the surface area of the ball used by the Miami Heat (basketball) to that of the ball used by the Tampa Bay Devil Rays (baseball)? Round to the nearest whole number. *(Lesson 12-7)* C

(A) 9:1 (B) 10:1
(C) 11:1 (D) 12:1

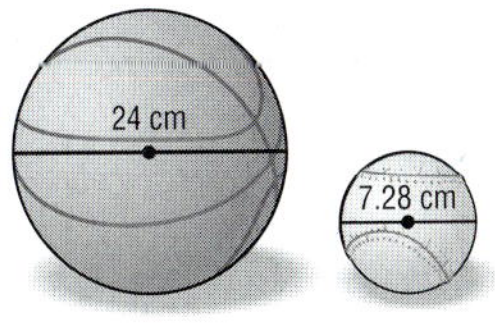

Benchmarks Addressed

Monday	MA.A.3.4.3, MA.B.1.4.1
Tuesday	MA.B.1.4.1
Wednesday	MA.A.3.4.3, MA.B.1.4.1
Thursday	MA.B.1.4.1
Friday	MA.B.1.4.1

Week 24

Monday

Extended Response A sphere sits in a cubic container, just touching each side. If the diameter of the sphere is 10 inches and the cube is 10 inches to an edge, how much air space exists between the box and the ball? Give your answer to the nearest tenth. *(Lesson 13-3)* **See margin.**

Tuesday

Multiple Choice What is the volume of the right circular cone shown? *(Lesson 13-2)* **C**

Ⓐ $\frac{3\pi}{a}$ inches

Ⓑ $\frac{3a}{\pi}$ inches

Ⓒ $\frac{\pi}{3a}$ inches

Ⓓ $\frac{\pi a}{3}$ inches

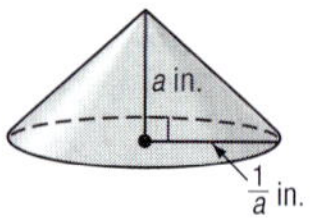

Gridded Response A cube has a volume of 216 cubic centimeters. What is the length of an edge of the cube in centimeters? *(Lesson 13-1)* **6**

Wednesday

Multiple Choice What is the ratio of surface area to volume for a cube if the length of an edge is 5? *(Lesson 13-1)* **I**

Ⓕ 3:4 Ⓖ 4:3

Ⓗ 5:4 Ⓘ 6:5

Gridded Response What is the area in square units of the shaded region of the flag flown in Florida during the first Spanish occupation in the 1500s? *(Lesson 11-2)* **6**

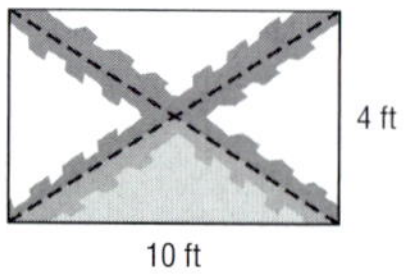

Thursday

Short Response What is the difference in the volumes of the two right circular cylinders shown? Keep your answer in exact form. Show all your work. *(Lesson 13-1)* **See margin.**

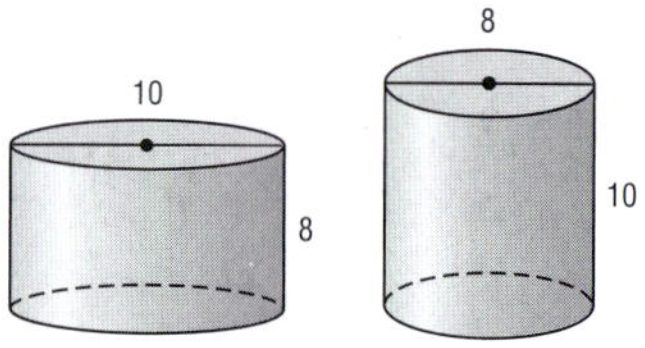

Friday

Multiple Choice What is the ratio of the volume of a right circular cylinder to the volume of a right circular cone with the same radius and height? *(Lesson 13-2)* **B**

Ⓐ 4:1 Ⓑ 3:1

Ⓒ 1:3 Ⓓ 1:4

Benchmarks Addressed

Monday	MA.A.3.4.3, MA.B.1.4.1
Tuesday	MA.A.3.4.3, MA.B.1.4.1
Wednesday	MA.A.3.4.3, MA.B.1.4.1
Thursday	MA.A.3.4.3, MA.B.1.4.1
Friday	MA.A.3.4.3, MA.B.1.4.1

Countdown to FCAT

Monday

Extended Response Sample Answer

The volume of the cubic container can be found with the formula $V = \ell wh$ if $\ell = 10$, $w = 10$, and $h = 10$.

$V = \ell wh$

$= 10 \cdot 10 \cdot 10$

$= 1000$

The volume of the sphere sitting inside the cube can be found with the formula $V = \frac{4}{3}\pi r^3$ if $r = 5$.

$V = \frac{4}{3}\pi r^3$

$= \frac{4}{3}\pi(5)^3$

≈ 523.3

The difference in the two volumes would represent the air space left over, which is $1000 - 523.3$ or 476.7 cubic inches of air space.

Thursday

Short Response Sample Answer

The volume of the first cylinder can be found with the formula $V = \pi r^2 h$, which in this case is $V = \pi(5^2)(8)$ or 200π. The volume of the second cylinder is $V = \pi(4^2)(10)$ or 160π. So the difference between the two volumes is 40π.

Countdown to FCAT

Monday

Extended Response Sample Answer

From the diagram, two of the walls have dimensions of 10 by 24, so the area of those two walls would be 2(10)(24) or 480 square feet. The area of the third wall is (30)(10) or 300 square feet. So the total surface area of the plastering region is 780 square feet. The plaster is priced in square yards, so 780 square feet must be converted to square yards by dividing by 9. That gives $86\frac{2}{3}$ square yards at \$0.95 per square yard, or a total cost of \$82.33.

Thursday

Short Response Sample Answer

To find the volume of the cylinder, I need to know the radius, which would be half the diameter or 2.5 inches. Now I can use the formula to find the volume of the cylinder: $V = \pi r^2 h = \pi(2.5)^2(12)$ or about 235.62 in^3. The volume of the rectangular carton is $V = \ell wh$ or 10(5)(5), which is 250 in^3. The rectangular carton holds more than the cylinder.

Countdown to FCAT

Week 25

Monday

Extended Response How much would it cost to plaster the three walls shown of a newly built double-car garage that is 24 feet long, 30 feet wide, and 10 feet high, if the plaster costs \$0.95 per square yard? *(Lesson 12-3)* **See margin.**

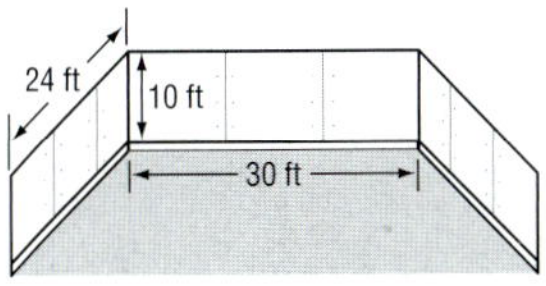

Tuesday

Multiple Choice A tennis court measuring 36 feet by 78 feet is being recoated. If it takes 2 gallons of topcoat material to cover the shaded region below, approximately how many gallons are needed for the entire job? *(Lesson 11-2)* **B**

(A) 15 (B) 12
(C) 10 (D) 9

Gridded Response The shaded area in the figure below is a planar cross-section of a rectangular prism. What is the area of this cross-section to the nearest square foot? *(Lesson 12-2)* **894**

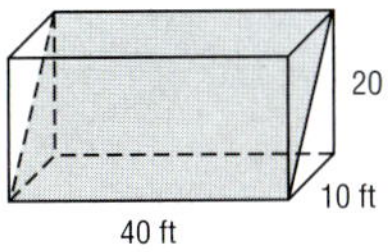

Wednesday

Multiple Choice Based only on the information given in the diagram, which of the following proves that lines a and b are parallel? *(Lesson 3-5)* **F**

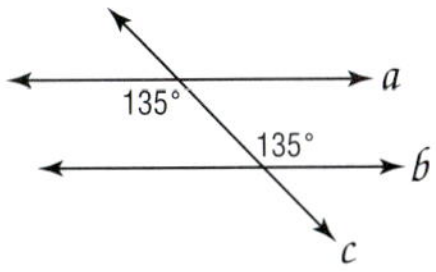

(F) Alternate interior angles are congruent.
(G) Corresponding angles are congruent.
(H) Same side interior angles are supplementary.
(I) Alternate exterior angles are congruent.

Gridded Response Manatee Springs in Levy County, Florida, produces a water flow volume of 50–150 million gallons of water each day. What is the maximum water flow in million gallons from the springs in one week? *(Prerequisite)* **1050**

Thursday

Short Response A company manufacturing small packing pellets can ship in either cylinders or rectangular cartons, both of which cost the same. The right circular cylinder has a height of 12 inches and a diameter of 5 inches, while the rectangular carton has a height of 5 inches, a width of 5 inches, and a length of 10 inches. Which container will hold the most packing pellets? Show your work. *(Lesson 13-1)* **See margin.**

Friday

Multiple Choice Approximately how many times as great is the volume of a basketball than that of a baseball? Round to the nearest whole number. *(Lesson 13-3)* **A**

(A) 36 (B) 30
(C) 28 (D) 24

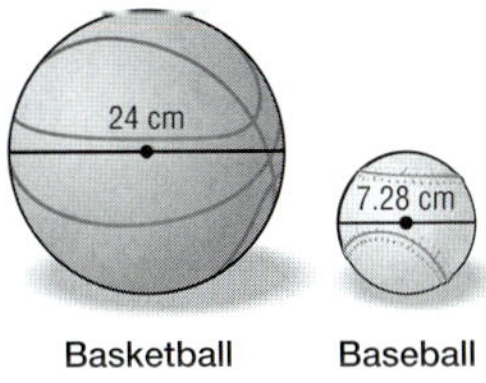

Benchmarks Addressed

Monday	MA.B.1.4.1, MA.B.2.4.1
Tuesday	MA.A.3.4.3, MA.A.4.4.1, MA.C.2.4.2
Wednesday	MA.C.1.4.1, MA.C.2.4.1, MA.C.3.4.2
Thursday	MA.A.3.4.3, MA.B.1.4.1
Friday	MA.A.3.4.3, MA.B.1.4.1

AF861098

FEATURES AND BENEFITS
Geometry ©2004

Curriculum and Instructional Strategies	. . . presents a coherent curriculum that effectively organizes and integrates important mathematical ideas.	
	• Chapters are grouped by units to bring depth to geometry concepts.	iii
	• Lessons are divided into two related objectives to allow teachers flexibility in presenting the lesson.	133
Student Support	. . . is provided throughout the text to help all students succeed in geometry.	
	• Foldables™ Study Organizers help students actively organize key concepts and create their own review materials.	125
	• Key Concept and Concept Summary boxes help students identify main concepts.	128, 134
	• Study Tips in the margins help students understand new material.	126–127
	• Homework Help in the margin of the exercise sets links homework exercises to corresponding examples within the lesson.	129
Reading / Writing In Mathematics	. . . strategies and activities are essential for student success in mathematics.	
	• Reading Math Study Tips help clarify mathematical terms.	126
	• Reading Mathematics pages help students learn to read effectively in mathematics and make connections to everyday meanings of terms.	81
	• Practice in vocabulary usage in each lesson and at the end of each chapter builds reading and writing skills.	167
	• Writing in Math exercises require students to summarize what they have learned in the lesson.	130, 149
Daily Intervention	. . . opportunities are provided throughout the program.	
	• Prerequisite Skills at the beginning of each chapter and in each lesson assess student readiness.	125
	• The Student Handbook contains review and practice of prerequisite skills.	727
	• Daily Intervention features provide suggestions for addressing various learning styles and helping students who are having difficulty.	128
	• A variety of Online Study Tools are readily accessible to students.	1
Test Preparation and Assessment	. . . provides targeted practice for local, state, and national tests.	
	• Standardized Test Practice questions appear in each lesson.	131, 138
	• Standardized Test Practice Examples help students learn how to approach test questions.	135
	• Two pages of Standardized Test Practice at the end of each chapter include multiple-choice, short-response/grid-in, and open-ended questions.	172–173
	• Interactive Standardized Test Practice is available in the Online Study Tools.	173
Staff Development	. . . features are available to assist new teachers and those teaching outside of their primary subject area.	
"Sticky Notes" in Chapter 1 provide a "walk-through" of key features. pp. 4–65	• Mathematical Connections and Background provides an overview of the mathematics in the chapter and links to prior knowledge and future topics.	124C, 124D
	• Building on Prior Knowledge links what students have previously learned to the content of the current lesson.	126
	• Tips for New Teachers provide helpful suggestions for classroom management, teaching techniques, and assessment.	128
Partnerships	. . . strengthen the relevance of applications, projects, and assessment.	
	• USA TODAY Snapshots®, provide current topics and data in graphs, charts, and tables and enhance the unit WebQuest projects.	143
	• The Princeton Review aligns and verifies our testing materials to ensure that they meet the tough standards of state and national testing agencies.	135

Authors

Cindy J. Boyd
Mathematics Teacher
Albilene High School
Abilene, Texas

Jerry Cummins
Past President
National Council of Supervisors of Mathematics (NCSM)
Western Springs, Illinois

Carol Malloy
Associate Professor, Math Education
The University of North Carolina at Chapel Hill
Chapel Hill, North Carolina

John Carter
Director of Mathematics
Adlai E. Stevenson High School
Lincolnshire, Illinois

Alfinio Flores, Ph.D.
Professor
Arizona State University
Tempe, Arizona

Contributing Authors

USA TODAY
The USA TODAY Snapshots®, created by USA TODAY®, help students make the connection between real life and mathematics.

Dinah Zike
Educational Consultant
Dinah-Might Activities, Inc.
San Antonio, TX

Content Consultants

Each of the Content Consultants reviewed every chapter and gave suggestions for improving the effectiveness of the mathematics instruction.

Content Consultants

Ruth M. Casey
Mathematics Teacher/
Department Chair
Anderson County High School
Lawrenceburg, KY

Gilbert Cuevas
Professor of Mathematics Education
University of Miami
Coral Gables, FL

Alan G. Foster
Former Mathematics
Department Chair
Addison Trail High School
Addison, IL

Linda M. Hayek
Curriculum Facilitator
Ralston Public Schools
Omaha, NE

Berchie Holliday
Educational Consultant
Silver Spring, MD

Joseph Kavanaugh
Academic Head for Mathematics
Scotia-Glenville Senior High School
Scotia, NY

Yvonne Medina Mojica
Mathematics Coach
Verdugo Hills High School, Los Angeles
Unified School District
Tujunga, CA

Teacher Reviewers

Each Teacher Reviewer reviewed at least two chapters of the Student Edition, giving feedback and suggestions for improving the effectiveness of the mathematics instruction.

Liza Allen
Math Teacher
Conway High School West
Conway, AR

Molly M. Andaya
Mathematics Teacher
Ellensburg H.S.
Ellensburg, WA

David J. Armstrong
Mathematics Facilitator
Huntington Beach UHSD
Huntington Beach, CA

Jerry C. Bencivenga
Consultant
Connecticut VTSS
Connecticut SDE
Middletown, CT

Patrick M. Blake
Math Department Chairperson
Ritenour H.S.
St. Louis, MO

Donna L. Burns
Math Department Chairperson
Los Angeles H.S.
Los Angeles, CA

Nita Carpenter
Mathematics Teacher
Okemos H.S.
Okemos, MI

Vincent Ciraulo
Supervisor of Mathematics
J.P. Stevens H.S.
Edison, NJ

Keitha Cleveland
Mathematics Teacher
E.S. Aiken Optional School
Alexandria, LA

Janice Garner Coffer
Math Department Chairperson
Page H.S.
Greensboro, NC

Karyn S. Cummins
Mathematics Department Head
Franklin Central H.S.
Indianapolis, IN

Christine B. Denardo
Math Teacher and Department Chair
Blacksburg H.S.
Blacksburg, VA

Joseph N. Di Cioccio
Mathematics Teacher/Department
Chair
Mahopac H.S.
Mahopac, NY

Robert A. Di Dio, M.S.Ed., P.D.
Assistant Principal
IS 192X
Bronx, NY

James S. Emery
Geometry Teacher
Beaverton H.S.
Beaverton, MI

Teacher Reviewers

Nancy S. Falls
Math Department Chair/
Geometry Teacher
Northern York County H.S.
Dillsburg, PA

Jane Fasullo
Retired High School Teacher
Ward Melville H.S.
E. Setauket, NY

Susan Fischbein
Math Department Chair
Desert Mountain H.S.
Scottsdale, AZ

Dolores Fischer
Mathematics Teacher
Adlai E. Stevenson H.S.
Lincolnshire, IL

Joyce E. Fisher
Math Teacher
La Cueva H.S.
Albuquerque, NM

Candace Frewin
Teacher on Special Assignment
Pinellas County Schools
Largo, FL

Karen L. George
Math Teacher
Taunton H.S.
Taunton, MA

Douglas E. Hall
Geometry Teacher
Chaparral H.S.
Las Vegas, NV

Cynthia D. Hodges
Teacher
Shoemaker H.S.
Killeen, TX

Brian J. Johnson
K–12 Curriculum Specialist
Bay City Public Schools
Bay City, MI

Melissa L. Jones
Mathematics Teacher
Bexley H.S.
Bexley, OH

Nancy Lee Keen
Geometry Teacher
Martinsville H.S.
Martinsville, IN

John R. Kennedy
Mathematics Department Chairman
Derby H.S.
Derby, KS

Sharon Kenner
Algebra/Geometry Teacher
Coconut Creek H.S.
Coconut Creek, FL

Julie Kolb
Mathematics Teacher
Leesville Road H.S.
Raleigh, NC

Gary Kubina
Math Department Head
Citronelle H.S.
Citronelle, AL

Jenita Lyons
Mathematics Teacher
William M. Raines H.S.
Jacksonville, FL

Debra D. McCoy
Math Teacher
Hixson H.S.
Chattanooga, TN

Delia Dee Miller
Geometry/Algebra II Teacher
Caddo Parish Magnet H.S.
Shreveport, LA

Kenneth E. Montgomery
Mathematics Teacher
Tri-Cities H.S.
East Point, GA

Cynthia Orkin
Teacher/Department Chairperson
Brookville H.S.
Lynchburg, VA

Susan M. Parece
Mathematics Teacher
Plymouth South H.S.
Plymouth, MA

Mike Patterson
High School Mathematics Teacher
Advanced Technologies Academy
Las Vegas, NV

Cynthia W. Poché
Math Teacher
Salmen H.S.
Slidell, LA

David E. Rader
K–12 Math/Science Supervisor
Wissahickon School District
Ambler, PA

Monique Siedschlag
Math Teacher
Thoreau H.S.
Thoreau, NM

Frank Louis Sparks
Curriculum Design and Support
Specialist–Mathematics
New Orleans Public Schools
New Orleans, LA

Alice Brady Sprinkle
Math Teacher
Broughton H.S.
Raleigh, NC

Joy F. Stanford
Mathematics Chair
Booker T. Washington Magnet H.S.
Montgomery, AL

Dora Swart
Math Teacher & Department Chair
W.F. West H.S.
Chehalis, WA

David Tate
Mathematics Teacher
Prairie Grove H.S.
Prairie Grove, AR

Kathy Thirkell
Math Curriculum Coordinator
Lewis-Palmer H.S.
Monument, CO

Marylu Tyndell
Teacher of Mathematics
Colts Neck H.S.
Colts Neck, NJ

Sarah L. Waldrop
Math Department Chairperson
Forestview H.S.
Gastonia, NC

Christine A. Watts
Mathematics Department Chairman
Monmouth H.S.
Monmouth, IL

René Wilkins
Mathematics Teacher
Fairhope H.S.
Fairhope, AL

Rosalyn Zeid
Mathematics Supervisor
Union Township Public Schools
Union, NJ

Teacher Handbook

Table of Contents

Designed to be in more

The consistent instructional design of each lesson strengthens student learning—from the very beginning to the very end.

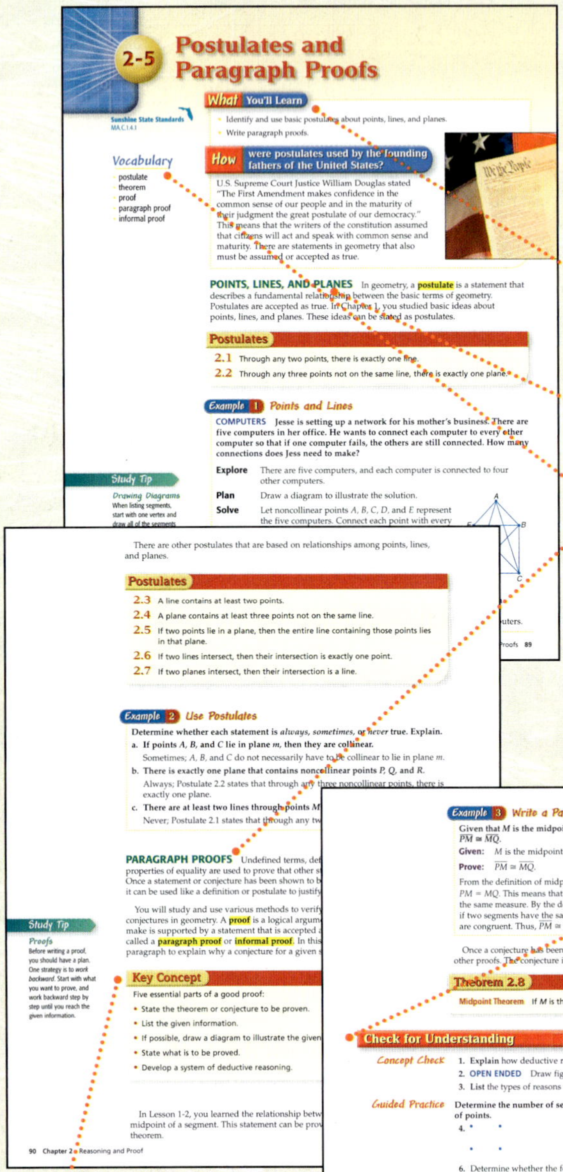

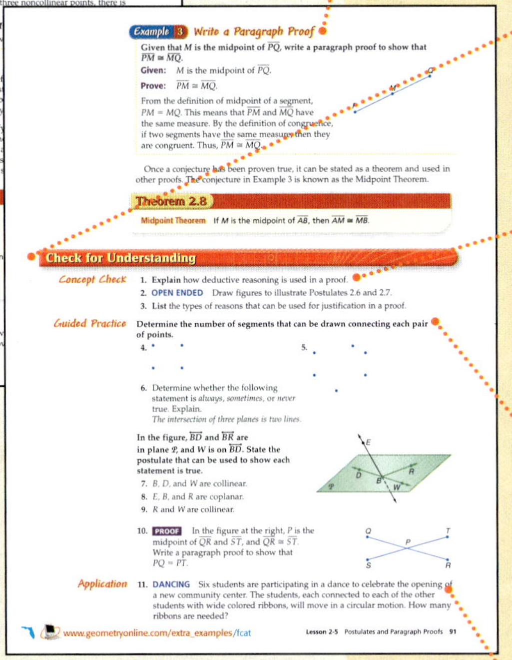

- **What You'll Learn** previews the topics to come, and engaging Why or How questions incorporate **problem-based learning** in your classroom.
- **Two-part lesson structure** gives you the flexibility to teach the two related lesson objectives together or separately.
- *Vocabulary* terms are listed at the beginning of each lesson for easy reference, and they're also highlighted in yellow within the lesson.
- **Completely worked-out examples** with clear explanations are paralleled by the Guided Practice and Practice and Apply exercises that follow.

Check for Understanding

You can use this portion of exercises in class to ensure that all students understand the concepts.

- *Concept Check* exercises give students opportunities to define, describe, and explain the mathematical concepts they've just learned.
- *Guided Practice* presents a representative sample of the exercises in the Practice and Apply section. A key in the **Teacher Wraparound Edition** correlates the exercises with examples.
- *Application* problems provide problem-based learning opportunities.

Key Concept boxes use words, symbols, models, and examples to illustrate new rules, properties, and definitions, so students can build their reading skills as they build their math skills. **Concept Summary** boxes provide a concise overview of key topics.

effective,

ways than one.

Practice and Apply

- **Skill Exercises** correspond to the Guided Practice exercises and are structured so that students practice the same concepts whether they are assigned odd- or even-numbered problems. Homework Help is provided so students can refer to examples in the lesson as they complete the exercises.

- **Applications** give students frequent opportunities to apply concepts to both real-life and mathematical situations.

- **CRITICAL THINKING** exercises in each lesson require students to explain, make conjectures, and prove mathematical relationships.

- *Standardized Test Practice* (A) (B) (C) (D) questions provide students with ongoing opportunities to sharpen their test-taking skills.

Maintain Your Skills

- *Mixed Review* includes spiraled, cumulative exercises from the two previous lessons as well as earlier lessons.

- *Getting Ready for the Next Lesson* exercises give students the chance to preview prerequisite skills for the coming lesson. A reference is provided should students need additional help.

Practice and Apply

Homework Help

For Exercises	See Examples
12–15	1
16–21	2
22–28	3

Extra Practice See page 757.

Determine the number of segments that can be drawn connecting each pair of points.

12. 

13. 

14. 

15. 

Determine whether the following statements are *always*, *sometimes*, or *never* true. Explain.

16. Three points determine a plane.
17. Points G and H are in plane $\mathcal{X}$. Any point collinear with G and H is in plane $\mathcal{X}$.
18. The intersection of two planes can be a point.
19. Points S, T, and U determine three lines.
20. Points A and B lie in at least one plane.
21. If line ℓ lies in plane $\mathcal{P}$ and line m lies in plane Q, then lines ℓ and m lie in plane $\mathcal{R}$.

Career Choices

Detective A police detective gathers facts and collects evidence for use in criminal cases. The facts and evidence are used together to prove a suspect's guilt in court.

Online Research For information about a career as a detective, visit: www.geometryonline.com/careers

In the figure at the right, $\overleftrightarrow{AC}$ and $\overleftrightarrow{BD}$ lie in plane $\mathcal{J}$, and $\overleftrightarrow{BY}$ and $\overleftrightarrow{CX}$ lie in plane $\mathcal{K}$. State the postulate that can be used to show each statement is true.

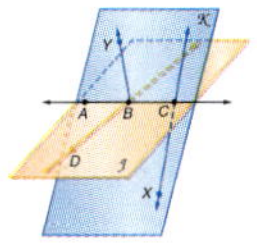

22. C and D are collinear.
23. $\overleftrightarrow{XB}$ lies in plane $\mathcal{K}$.
24. Points A, C, and X are coplanar.
25. $\overline{AD}$ lies in plane $\mathcal{J}$.
26. X and Y are collinear.
27. Points Y, D, and C are coplanar.
28. **PROOF** Point C is the midpoint of $\overline{AB}$ and B is the midpoint of $\overline{CD}$. Prove that $\overline{AC} \cong \overline{BD}$.
29. **MODELS** Faith's teacher asked her to make a figure showing the number of lines and planes formed from four points that are noncollinear and noncoplanar. Faith decided to make a mobile of straws, pipe cleaners, and colored sheets of tissue paper. She plans to glue the paper to the straws and connect the straws together to form a group of connected planes. How many planes and lines will she have?
30. **CAREERS** Many professions use deductive reasoning and paragraph proofs. For example, a police officer uses deductive reasoning investigating a traffic accident and then writes the findings in a report. List a profession, and describe how it can use paragraph proofs.
31. **CRITICAL THINKING** You know that three noncollinear points lie in a single plane. In Exercise 29, you found the number of planes defined by four noncollinear points. What are the least and greatest number of planes defined by five noncollinear points?
32. **WRITING IN MATH** Answer the question that was posed at the beginning of the lesson.

    How are postulates used in literature?

    Include the following in your answer:
    - an example of a postulate in historic United States' documents, and
    - an example of a postulate in mathematics.

FCAT Practice *Standardized Test Practice*

33. Which statement cannot be true?
    (A) A plane can be determined using three noncollinear points.
    (B) Two lines intersect at exactly one point.
    (C) At least two lines can contain the same two points.
    (D) A midpoint divides a segment into two congruent segments.
34. **ALGEBRA** For all values of x, $(8x^4 - 2x^2 + 3x - 5) - (2x^4 + x^3 + 3x + 5) =$
    (A) $6x^4 - x^3 - 2x^2 - 10$.
    (B) $6x^4 - 3x^2 + 6x - 10$.
    (C) $6x^4 + x^3 - 2x^2 + 6x$.
    (D) $6x^4 - 3x^2$.

Maintain Your Skills

Mixed Review

35. Determine whether statement (3) follows from statements (1) and (2) by the Law of Detachment or the Law of Syllogism. If it does, state which law was used. If it does not, write *invalid*. *(Lesson 2-4)*
    (1) Part-time jobs require 20 hours of work per week.
    (2) Jamie has a part-time job.
    (3) Jamie works 20 hours per week.

Write the converse, inverse, and contrapositive of each conditional statement. Determine whether each related conditional is *true* or *false*. If a statement is false, find a counterexample. *(Lesson 2-3)*

36. If you have access to the Internet at your house, then you have a computer.
37. If $\triangle ABC$ is a right triangle, one of its angle measures is greater than 90.
38. **BIOLOGY** Use a Venn diagram to illustrate the following statement. *If an animal is a butterfly, then it is an arthropod.* *(Lesson 2-2)*

Use the Distance Formula to find the distance between each pair of points. *(Lesson 1-3)*

39. $D(3, 3)$, $F(4, -1)$
40. $M(0, 2)$, $N(-5, 5)$
41. $P(-8, 2)$, $Q(1, -3)$
42. $R(-5, 12)$, $S(2, 1)$

Getting Ready for the Next Lesson

PREREQUISITE SKILL **Solve each equation.** *(To review **solving equations**, see pages 737 and 738.)*

43. $m - 17 = 8$
44. $3y = 57$
45. $\frac{y}{6} + 12 = 14$
46. $-t + 3 = 27$
47. $8n - 39 = 41$
48. $-6x + 33 = 0$

www.geometryonline.com/self_check_quiz/fcat

Lesson 2-5 Postulates and Paragraph Proofs 93

Accomplish more

Glencoe Geometry provides so many resources for lesson planning and teaching that you can create a complete, customized course in geometry quickly...and easily.

This is where you start.

The **Teacher Wraparound Edition** is your key to all of the teaching resources in *Glencoe Geometry*. In addition to teaching suggestions, additional examples, and answers, the Teacher Wraparound Edition provides a guide for all of the print and software materials available for each lesson.

FAST FILE Chapter Resource Masters contain all of the core supplements you'll need to begin teaching a chapter of *Glencoe Geometry*. Each chapter booklet features convenient tabs for easy filing.

- **Vocabulary Builder** helps students locate and define key vocabulary words from the chapter.
- **Proof Builder** helps students learn and understand theorems and postulates from the chapter.
- **Study Guide and Intervention** for each objective summarizes key concepts and provides practice.
- **Skills Practice** provides ample exercises to help students develop basic computational skills, lesson by lesson.
- **Practice** mimics the computational and verbal problems in each lesson at an average level.
- **Reading to Learn Mathematics** provides students with various reading strategies to master the mathematics presented in each lesson.
- **Enrichment** activities extend students' knowledge and widen their appreciation of how mathematics relates to the world around them.
- **Assessment** options for each chapter include six forms of chapter tests, assessment tasks, quizzes, mid-chapter test, cumulative review, and standardized test practice.

Reading and Writing

WebQuest and Project Resources include teacher notes and answers for the Internet WebQuest projects as well as other long-term projects that can be used with *Glencoe Geometry*.

Reading and Writing in the Mathematics Classroom features suggestions and activities for including reading as an integral part of the mathematics curriculum as well as differentiated approaches to teaching mathematics that promote English learning and inclusion.

Teaching Mathematics with Foldables™ offers guidelines for using Foldables interactive study organizers in your class. The booklet was written by Foldables creator Dinah Zike.

More information on options for reading and writing in Glencoe Geometry is available on pages T6–T7.

Applications

School-to-Career Masters feature activities that show how mathematics relates to various careers.

Graphing Calculator and Computer Masters include activities to incorporate the TI-83 Plus calculator and computer applications such as spreadsheets and The Geometer's Sketchpad into your geometry course.

Real-World Transparencies and Masters feature colorful transparencies with accompanying student worksheets to show how mathematics relates to real-world topics.

than you'd ever imagine

in less time than you'd ever believe

Assessment and Intervention

5-Minute Check Transparencies with Standardized Test Practice include a transparency for each lesson that evaluates what students have learned in the previous lesson. Each transparency also includes a standardized test practice question.

Closing the Gap for Absent Students provides an easy-to-use summary of all the materials you have covered in the chapter in a format that can be posted or distributed to students who have missed class.

Guide to Daily Intervention offers suggestions for daily assessment and tips on how to help students succeed.

Prerequisite Skills Workbook: Remediation and Intervention includes worksheets to review the arithmetic skills needed in geometry.

Staff Development

Answer Key Transparencies provide answers to Student Edition exercises.

Lesson Planning Guide features a daily resource guide for planning your curriculum, as well as pacing for block scheduling.

Solutions Manual includes completely worked-out solutions for all exercises in the Student Edition.

Using the Internet in the Mathematics Classroom provides guidelines for using the Internet, as well as a guide to additional mathematics resources available on the Internet.

Teaching Geometry with Manipulatives features activities and teaching suggestions to help you present geometric concepts with manipulatives and hands-on materials.

Technology Support for Teachers

Glencoe offers many timesaving software products to help you develop creative classroom presentations...fast.

TeacherWorks All-in-One Lesson Planner and Resource Center CD-ROM includes a lesson planner and interactive Teacher Edition, so you can customize lesson plans and reproduce classroom resources quickly and easily, from just about anywhere.

Answer Key Maker software allows you to customize answer keys for your assignments from the Student Edition exercises.

Interactive Chalkboard CD-ROM includes fully worked-out examples, the 5-Minute Check Transparencies, and Your Turn problems in a customizable Microsoft® PowerPoint® format.

And more... *Additional technology products and Internet resources for students, teachers, and parents are discussed on pages T6–T13 and T17.*

HELP your students become fluent

Glencoe Geometry makes it easy for you to incorporate constructive reading and writing strategies into every class you teach.

Reading Mathematics

Prefixes

Many of the words used in mathematics use the same prefixes as other everyday words. Understanding the meaning of the prefixes can help you understand the terminology better.

Prefix	Meaning	Everyday Words	Meaning
bi-	2	bicycle	a 2-wheeled vehicle
		bipartisan	involving members of 2 political parties
tri-	3	triangle	closed figure with 3 sides
		tricycle	a 3-wheeled vehicle
		triplet	one of 3 children born at the same time
quad-	4	quadrilateral	closed figure with 4 sides
		quadriceps	muscles with 4 parts
		quadruple	four times as many
penta-	5	pentagon	closed figure with 5 sides
		pentathlon	athletic contest with 5 events
hexa-	6	hexagon	closed figure with 6 sides
hept-	7	heptagon	closed figure with 7 sides
oct-	8	octagon	closed figure with 8 sides
		octopus	animal with 8 legs
dec-	10	decagon	closed figure with 10 sides
		decade	a period of 10 years
		decathlon	athletic contest with 10 events

Several pairs of words in the chart have different prefixes, but the same root word. *Pentathlon* and *decathlon* are both athletic contests. *Heptagon* and *octagon* are both closed figures. Knowing the meaning of the root of the term as well as the prefix can help you learn vocabulary.

Reading to Learn

Use a dictionary to find the meanings of the prefix and root for each term. Then write a definition of the term.

1. bisector **2.** polygon **3.** equilateral
4. concentric **5.** circumscribe **6.** collinear

7. **RESEARCH** Use a dictionary to find the meanings of the prefix and root of *circumference*.

8. **RESEARCH** Use a dictionary or the Internet to find as many words as you can with the prefix *poly-* and the definition of each.

594 Chapter 11 Areas of Polygons and Circles

Reading Mathematics activities help students master new mathematics vocabulary words and develop technical reading skills so they can understand and apply the language of math in their daily lives.

In the Student Edition

Foldables™ Study Organizers at the beginning of each chapter provide students with tools for organizing what they are reading and studying.

Reading Math Study Tips appear throughout each chapter to help students learn and use the language of geometry.

Writing in Math questions in every lesson require students to use critical thinking skills to develop their answers.

Vocabulary terms are listed at the beginning of each lesson and highlighted when defined.

The **Vocabulary and Concept Check** in each Study Guide and Review checks students' understanding of the key concepts of the chapter.

Key Concepts are illustrated using Words, Symbols, Models, and Examples, as appropriate. This approach improves reading comprehension by using multiple representations.

WebQuest Internet Projects give students the chance to work through long-term projects to develop their research and creative writing skills.

in the Language of MATHEMATICS

In the Teacher Wraparound Edition

Study Notebook suggestions provide motivational ideas to help students create study notebooks that are thorough and effective.

Concept Check questions require students to describe, write, and explain the mathematical concepts they have learned in each lesson.

Modeling, Speaking, and **Writing** in every lesson require students to summarize what they have learned by responding to open-ended prompts.

ELL Resources highlight features and activities that help English-Language Learners grasp content.

Differentiated Instruction features help students at all points on the learning spectrum develop their reading, writing, and comprehension skills.

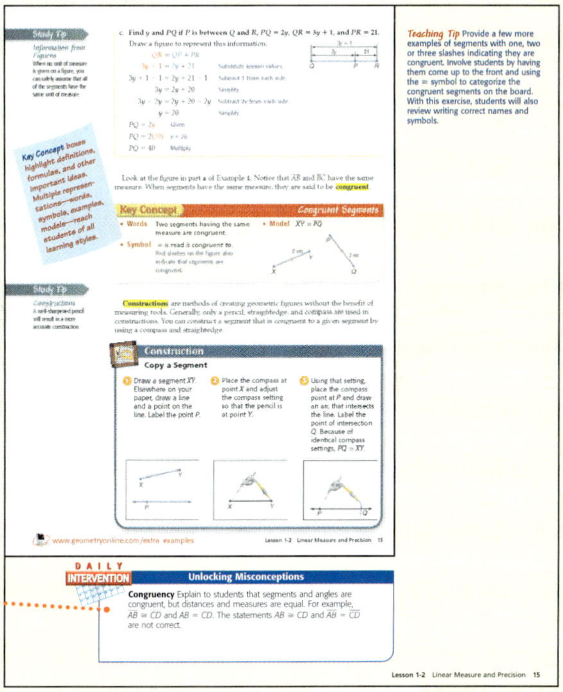

Study Tip

Information from Figures

When no unit of measure is given on a figure, you can safely assume that all of the segments have the same unit of measure.

c. Find y and PQ if P is between Q and R, $PQ = 2y$, $QR = 3y + 1$, and $PR = 21$.

Draw a figure to represent this information.

$QR = QP + PR$

$3y + 1 = 2y + 21$ Substitute known values.

$3y + 1 - 1 = 2y + 21 - 1$ Subtract 1 from each side.

$3y = 2y + 20$ Simplify.

$3y - 2y = 2y + 20 - 2y$ Subtract 2y from each side.

$y = 20$ Simplify.

$PQ = 2y$ Given

$PQ = 2(20)$ $y = 20$

$PQ = 40$ Multiply.

Key Concept boxes highlight definitions, formulas, and other important ideas. Multiple representations—words, symbols, examples, models—reach students of all learning styles.

Look at the figure in part a of Example 4. Notice that $\overline{AB}$ and $\overline{BC}$ have the same measure. When segments have the same measure, they are said to be **congruent**.

Key Concept *Congruent Segments*

- **Words** Two segments having the same measure are congruent.
- **Symbol** ≅ is read *is congruent to*. Red slashes on the figure also indicate that segments are congruent.
- **Model** $\overline{XY} \cong \overline{PQ}$

Study Tip

Constructions

A well-sharpened pencil will result in a more accurate construction.

Constructions are methods of creating geometric figures without the benefit of measuring tools. Generally, only a pencil, straightedge, and compass are used in constructions. You can construct a segment that is congruent to a given segment by using a compass and straightedge.

Construction

Copy a Segment

1. Draw a segment $\overline{XY}$. Elsewhere on your paper, draw a line and a point on the line. Label the point P.
2. Place the compass at point X and adjust the compass setting so that the pencil is at point Y.
3. Using that setting, place the compass point at P and draw an arc that intersects the line. Label the point of intersection Q. Because of identical compass settings, $\overline{PQ} \cong \overline{XY}$.

www.geometryonline.com/extra_examples

Lesson 1-2 Linear Measure and Precision 15

Teaching Tip Provide a few more examples of segments with one, two or three slashes indicating they are congruent. Involve students by having them come up to the front and using the ≅ symbol to categorize the congruent segments on the board. With this exercise, students will also review writing correct names and symbols.

DAILY INTERVENTION — **Unlocking Misconceptions**

Congruency Explain to students that segments and angles are congruent, but distances and measures are equal. For example, $\overline{AB} \cong \overline{CD}$ and $AB = CD$. The statements $AB \cong CD$ and $\overline{AB} = \overline{CD}$ are not correct.

Lesson 1-2 Linear Measure and Precision 15

Technology Support

Interactive Student Edition (ISE) includes the entire text, formatted like the hardbound book, so students can study from just about anywhere—no book required. With the ISE, students can also access interactive assessments, links to the Internet, and many more exciting tools.

www.geometryonline.com/vocabulary_review is a Glencoe site that provides online study tools for reviewing the vocabulary of each chapter.

Vocabulary PuzzleMaker software creates crossword, jumble, and word search puzzles using vocabulary lists that you can customize.

Multimedia Applications: Virtual Activities CD-ROM provides in-depth interactive activities that help students explore the main concepts of each chapter in a real-world setting.

Additional Resources

Chapter Resource Masters

- Vocabulary Builder
- Proof Builder
- Reading to Learn Mathematics

Teaching Mathematics with Foldables™

Reading and Writing in the Mathematics Classroom

WebQuest and Project Resources

For more information on these products, see pp. T4–T5.

With these TOOLS, you'll always know

Whether you need daily intervention resources integrated right into the program, or supplemental materials for after school and summer school programs, *Glencoe Geometry* puts it all right at your fingertips!

Prerequisite Skills

Students often struggle in geometry because they have not mastered the prerequisite skills needed to be successful. *Glencoe Geometry* provides several opportunities to check student skills and determine which students need additional review and practice.

- The **Prerequisite Skills** at the beginning of every chapter help students identify and practice the skills they'll need for each new concept.
- Additional **prerequisite skills practice** is provided at the end of each lesson and includes page references to help students get extra review whenever they need it. More prerequisite skill practice appears in the Student Handbook section at the back of the Student Edition.
- The **Prerequisite Skills Workbook** provides extra practice on the basic skills needed for success in geometry.

Daily Intervention Opportunities

Guide to Daily Intervention offers suggestions for using Glencoe materials to intercept students who are having difficulties and prescribe a system of reinforcement to promote student success.

The **Chapter Resource Masters** include several types of worksheets that can be used for daily intervention in each lesson. For a description of each worksheet, see page T4.

- **Study Guide and Intervention***
- **Skills Practice***
- **Practice***
- **Reading to Learn Mathematics**

* *Each of these types of worksheets is also available as a **consumable workbook**.*

The **Student Edition** contains additional problems to help students master each lesson before completing the chapter assessment.

- **Extra Practice**, located in the back of the Student Edition, provides additional, immediate practice with the concepts from each lesson.
- **Mixed Problem Solving and Proof**, also in the back of the Student Edition, includes numerous proofs and verbal problems to help students reinforce their skills in problem solving and proof.

who needs EXTRA HELP.

And you'll be able to *DELIVER* it.

Technology Resources for Intervention

In addition to print resources, Glencoe offers a variety of timesaving technology tools to help students build their math skills more effectively.

GeomPASS: Tutorial Plus CD-ROM provides an interactive, self-paced tutorial for a complete geometry curriculum. The 25 lessons are correlated directly to *Glencoe Geometry*. Each lesson, or concept, includes a pretest, tutorial, guided practice, and posttest. Students' answers to the pretests automatically determine whether they need the tutorial for each concept, so students can take responsibility for their own learning—without taking teacher time for grading.

Online Study Tools include comprehensive review and intervention tools that are available anytime, anyplace simply by logging on to

www.geometryonline.com.

Self-check quizzes are available for every lesson, and immediate feedback helps students check their progress and find specific pages and examples in the Student Edition whenever they need extra review. These Online Study Tools also include extra examples, chapter tests, standardized test practice, and vocabulary review.

Additional Teacher Resources

The following materials are available to help you determine which students need intervention and allow you to develop strategies for giving students the help they need. For a description of each feature, see page T5.

- **5-Minute Check Transparencies with Standardized Test Practice**
- **Daily Intervention** features in the Teacher Wraparound Edition
- **Closing the Gap for Absent Students**

Give ASSESSMENT

Glencoe Geometry gives you all the tools you need to prepare students for success—including Standardized Test Practice prepared exclusively for Glencoe by The Princeton Review and the powerful TestCheck and Worksheet Builder CD-ROM.

Student Edition

Glencoe has partnered with The Princeton Review to ensure that all of our materials meet state and national standards. Every lesson contains two Standardized Test Practice questions, and every chapter contains a completely worked-out standardized test example as well as two full pages of Standardized Test Practice with Test-Taking Tips.

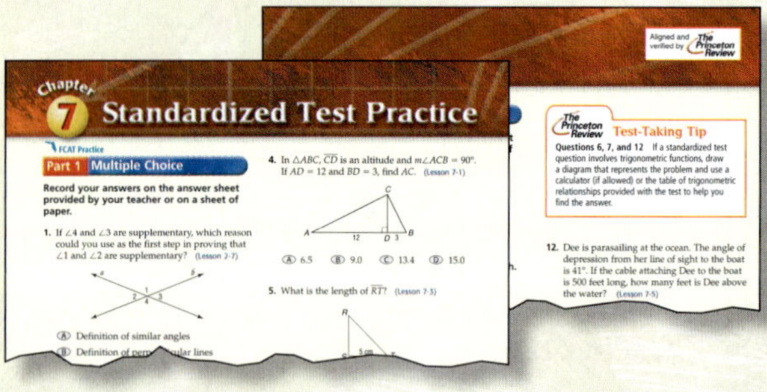

Teacher Wraparound Edition

An **Open-Ended Assessment** activity is provided in each lesson in the margin of the Teacher Wraparound Edition.

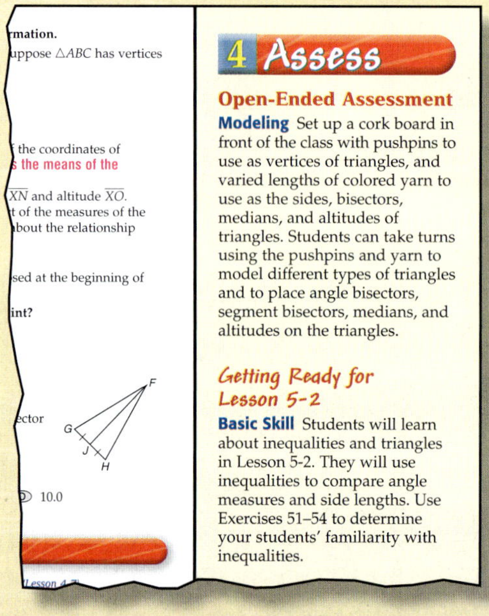

Chapter Study Guide and Review provides Vocabulary and Concept review—a Glencoe exclusive—and Lesson-by-Lesson Review, all at the point of use for students.

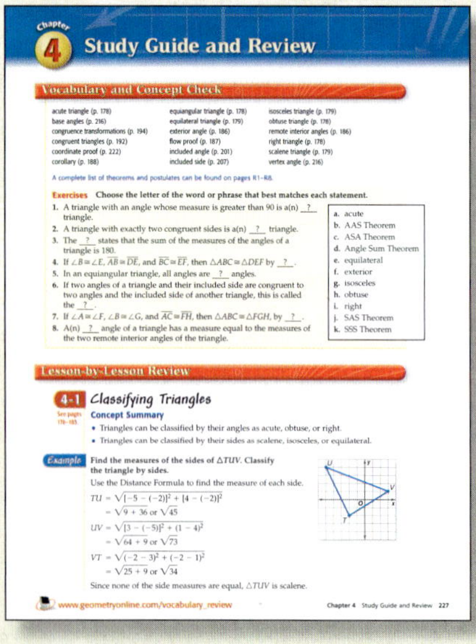

Practice Quizzes (2 per chapter) and a **Practice Test** for each chapter provide the variety of practice questions students need to succeed on tests.

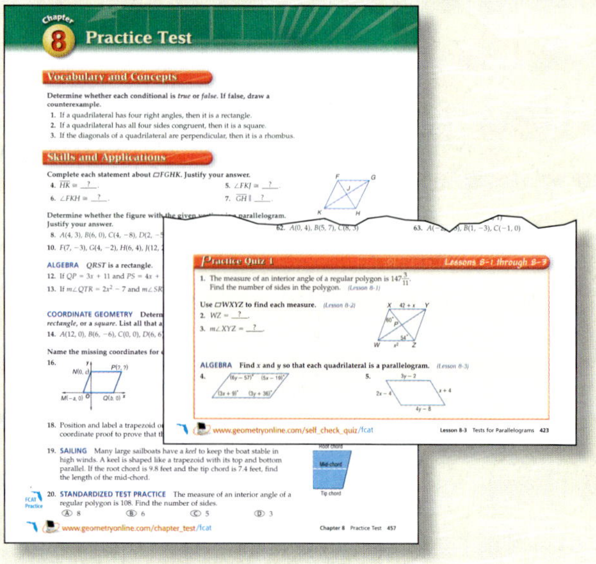

the extra attention it deserves, without the extra prep time.

Teacher Classroom Resources

5-Minute Check Transparencies with Standardized Test Practice provide full-size transparencies with questions covering the previous lesson or chapter. Standardized Test Practice Questions are also included.

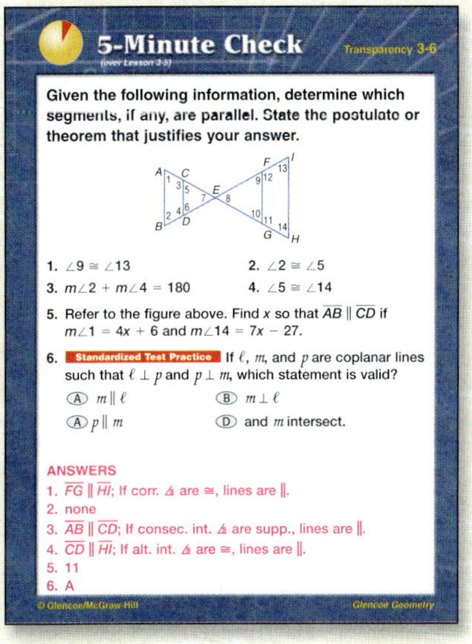
5-Minute Check (over Lesson 2-5) Transparency 3-6

Given the following information, determine which segments, if any, are parallel. State the postulate or theorem that justifies your answer.

1. $\angle 9 \cong \angle 13$
2. $\angle 2 \cong \angle 5$
3. $m\angle 2 + m\angle 4 = 180$
4. $\angle 5 \cong \angle 14$
5. Refer to the figure above. Find x so that $\overline{AB} \parallel \overline{CD}$ if $m\angle 1 = 4x + 6$ and $m\angle 14 = 7x - 27$.
6. Standardized Test Practice If ℓ, m, and p are coplanar lines such that $\ell \perp p$ and $p \perp m$, which statement is valid?
   (A) $m \parallel \ell$ (B) $m \perp \ell$ (A) $p \parallel m$ (D) and m intersect.

ANSWERS
1. $\overline{FG} \parallel \overline{HI}$; If corr. ∠s are ≅, lines are ∥.
2. none
3. $\overline{AB} \parallel \overline{CD}$; If consec. int. ∠s are supp., lines are ∥.
4. $\overline{CD} \parallel \overline{HI}$; If alt. int. ∠s are ≅, lines are ∥.
5. 11
6. A

© Glencoe/McGraw-Hill Glencoe Geometry

Technology Support

TestCheck and Worksheet Builder (Win/Mac) is a networkable CD-ROM with three modules:

- **Worksheet Builder** to create customized tests and worksheets.
- **Student Module** to take tests on screen.
- **Management System** to keep student records.

It also contains built-in state and national correlations.

MindJogger Videoquizzes present chapter-by-chapter review sessions in a game show format to make review more interesting and active to students…especially great for reluctant readers.

Online Study Tools

- Self-Check Quizzes
- Vocabulary Review
- Chapter Test Practice
- Standardized Test Practice

Assessment Options in the Chapter Resource Masters

These assessment resources are available for each chapter in *Glencoe Geometry*.

- 6 Chapter Tests
- Open-Ended Assessment with Scoring Rubric
- Vocabulary Test and Review **Glencoe Exclusive!**
- 4 Quizzes
- Mid-Chapter Test
- Cumulative Review
- 2-page Standardized Test Practice

Unit Tests, Semester Tests, and a **Final Test** are also available at point of use in the Chapter Resource Masters.

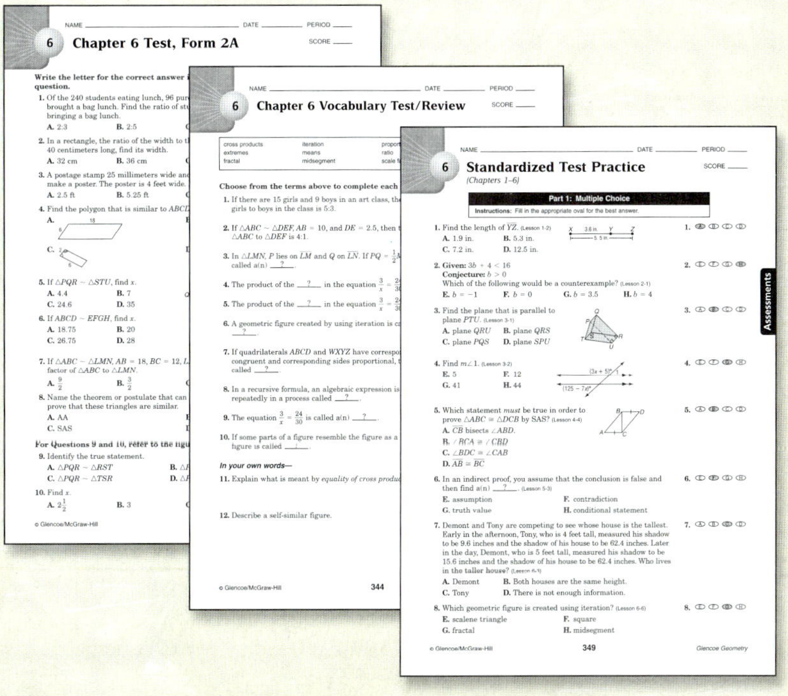

Introducing our new partner

USA TODAY® Education

USA TODAY Snapshots®

This is the same up-to-date data you know so well. But now, in an exclusive partnership with Glencoe/McGraw-Hill, USA TODAY® Education has brought its powerful, one-of-a-kind perspective and dynamic content to the pages of *Glencoe Geometry.* USA TODAY Snapshots® explode off the page to make geometry come alive with current, relevant data.

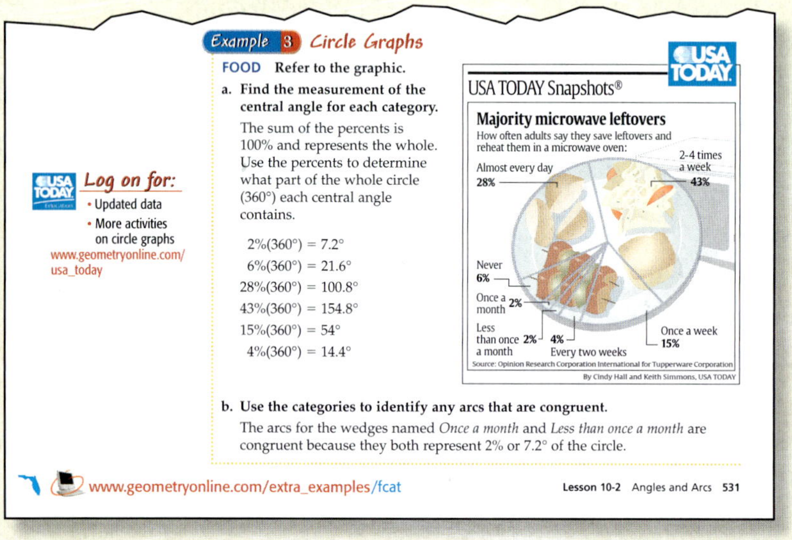

Example 3 Circle Graphs

FOOD Refer to the graphic.

a. **Find the measurement of the central angle for each category.**

The sum of the percents is 100% and represents the whole. Use the percents to determine what part of the whole circle (360°) each central angle contains.

2%(360°) = 7.2°
6%(360°) = 21.6°
28%(360°) = 100.8°
43%(360°) = 154.8°
15%(360°) = 54°
4%(360°) = 14.4°

b. **Use the categories to identify any arcs that are congruent.**

The arcs for the wedges named *Once a month* and *Less than once a month* are congruent because they both represent 2% or 7.2° of the circle.

Log on for:
- Updated data
- More activities on circle graphs

www.geometryonline.com/usa_today

www.geometryonline.com/extra_examples/fcat

Lesson 10-2 Angles and Arcs 531

- www.geometryonline.com/usa_today provides additional activities related to the topics presented in the USA TODAY Snapshots®.
- www.education.usatoday.com, USA TODAY® K-12 Education's Website offers resources and interactive features connected to each day's newspaper. *Experience Today*, USA TODAY®'s daily lesson plan, is available on the site and delivered daily to subscribers. This plan provides instruction for integrating USA TODAY® graphics and key editorial features into your mathematics classroom.

Stay current with additional charts and graphs with USA TODAY®. Log on to www.education.usatoday.com, or call USA TODAY® at (800) 757-TEACH.

WebQuest: Online Projects

www.geometryonline.com/webquest gives students the chance to work through a long-term pro-ject to enable them to develop their research, creative writing, and presentation skills.

- WebQuests often utilize USA TODAY Snapshots® or USA TODAY® articles.
- Special features in the Student Edition prompt students to complete each stage of their WebQuest.
- Parents can use the guided instruction to help students become familiar with the Internet in a safe, productive manner.

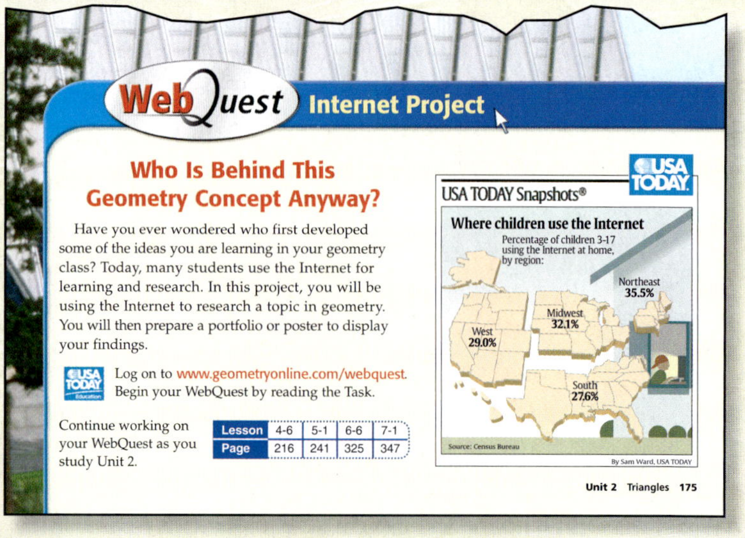

WebQuest Internet Project

Who Is Behind This Geometry Concept Anyway?

Have you ever wondered who first developed some of the ideas you are learning in your geometry class? Today, many students use the Internet for learning and research. In this project, you will be using the Internet to research a topic in geometry. You will then prepare a portfolio or poster to display your findings.

Log on to www.geometryonline.com/webquest. Begin your WebQuest by reading the Task.

Continue working on your WebQuest as you study Unit 2.

Lesson	4-6	5-1	6-6	7-1
Page	216	241	325	347

Unit 2 Triangles 175

The INTERNET: One TOOL. Endless possibilities.

Many of your students may already be familiar with the Internet, but may not have discovered the full potential of this powerful research tool. With *Glencoe Geometry*, your students can use the Internet to build their geometry skills. And you can access a wide variety of resources to help you plan classes, extend lessons, even meet professional development requirements.

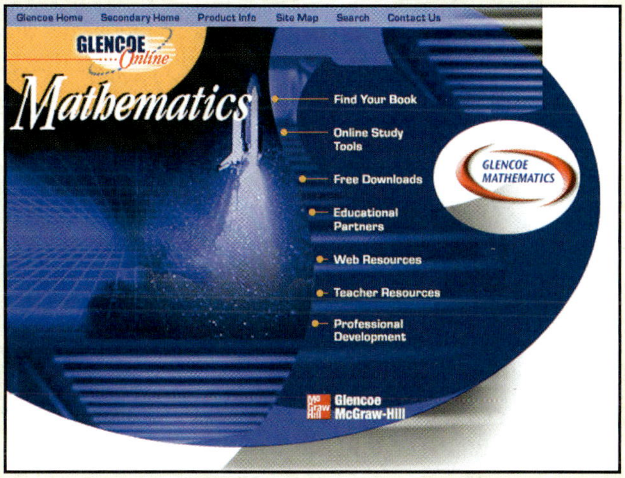

For Students

Online Study Tools, referenced on the Student Edition pages, are keyed specifically to *Glencoe Geometry*.

- www.geometryonline.com/extra_examples features additional fully worked-out examples.
- www.geometryonline.com/self_check_quiz allows students to check their progress in each lesson.
- www.geometryonline.com/vocabulary_review lets students check their vocabulary comprehension.
- www.geometryonline.com/chapter_test provides additional practice in test taking.
- www.geometryonline.com/standardized_test simulates questions that appear on standardized and proficiency tests.

Other Online Resources

- www.geometryonline.com/webquest offers an online research project.
- www.geometryonline.com/usa_today provides additional activities related to the topics presented in the USA TODAY Snapshots®.
- www.geometryonline.com/data_update features links to updated statistical data presented in exercises.
- www.geomctryonline.com/careers offers information about career opportunities.
- www.geometryonline.com/other_calculator_keystrokes provides keystroke instructions for various calculators to accompany graphing calculator activities and exercises in the Student Edition.

For Teachers

Powerful tools to make your job easier

- Vocabulary Review Games
- Problem of the Week Activities
- USA TODAY® K–12 Education daily lesson plans
- Sharing Ideas with Other Teachers
- Cool Math Links
- State and National Resources
- The Princeton Review

Staff Development Sites

- NCTM links
- Teaching Today link
- McGraw-Hill Learning Network link
- Cooperative learning suggestions
- Using the Internet in the Mathematics Classroom

For Parents

Help parents get involved with their child's learning

- Involving Parents and Community in the Mathematics Classroom

DISCOVER how a simple sheet of paper can CHANGE the way your students THINK about math... forever.

Students love Foldables™ because they're fun. Teachers love them because they're effective.

Foldables are easy-to-make, three-dimensional interactive graphic organizers that students create out of simple sheets of paper. These unique hands-on tools for learning and reviewing were created exclusively for *Glencoe Geometry* by teaching specialist Dinah Zike.

Building Prereading Skills

At the beginning of each chapter, students construct one of a variety of Foldables. Each Foldable helps students create an interactive strategy for organizing what they read and observe. As they work through each chapter, students add more detail to their Foldable until they have created a comprehensive, interactive snapshot of the key concepts and vocabulary of the chapter.

Reading and Writing

Each Foldable helps students practice basic reading and writing skills, find and report main ideas, organize information, review key vocabulary terms, and more.

Review and Reinforcement

The completed Foldable is a comprehensive overview of the chapter concepts—perfect for preparing for chapter, unit, and even end-of-course tests.

Assessment

Foldables present an ideal opportunity to probe the depth of your students' understanding of chapter concepts. You'll get detailed feedback on what your students know and what misconceptions they may have.

Staff Development

Teaching Mathematics with Foldables™ equips teachers to extend the use of Foldables in their classrooms by exploring the different Foldable formats and providing suggestions for using them throughout the mathematics curriculum.

Now, you can reach every student as if he or she were your *only* student.

As an educator, you know that every student has his or her own unique style of learning. The challenge is to develop a curriculum that meets the needs of every student and ensures that everyone in your class has the same chance at success. *Glencoe Geometry* provides you with a wide range of tools to help you reach every student you teach.

Student Edition

Prerequisite Skills at the beginning of each chapter and end of each lesson assess student readiness. In the back of the Student Edition, you can also find a list of topics covered in previous courses.

Hands-on study organizers called **Foldables** and **Geometry Activities** accommodate the visual/spatial learner.

Each lesson is **organized by objective**, so you can break the lesson into parts and vary the pacing according to your students' needs.

Key Concepts and **Concept Summaries** focus students on key mathematical concepts in each lesson.

Mixed Review in each lesson allows students to use skills taught in previous chapters.

Teacher Classroom Resources

Study Guide and Intervention Masters provide worked-out examples for each objective as well as practice problems. Available in Spanish.

Two types of **Practice Masters** are provided for each lesson. **Skills Practice** allows students who are progressing at a slower pace to practice the concept using easier problems, while the **Practice** provides more challenging problems for students who are moving at a regular or faster pace.

The **Prerequisite Skills Workbook** contains skills necessary for success in geometry.

Teacher Wraparound Edition

Differentiated Instruction strategies focus on eight learning styles: verbal/linguistic, logical, visual/spatial, auditory/musical, kinesthetic, interpersonal, intrapersonal, and naturalist.

Assignment Guides are organized by objective as well as by level, making it easy to assign homework.

ELL This ELL symbol shows you which activities are appropriate for English Language Learners.

Technology Support

GeomPASS: Tutorial Plus is a CD-ROM that provides step-by-step, self-paced instruction correlated to *Glencoe Geometry*.

www.geometryonline.com

Online Study Tools

- Extra Examples
- Self-Check Quizzes
- Vocabulary Review

It's Staff Development,

As professional development continues to take on greater importance for educators across the country, teachers are constantly looking for easy-to-use tools to help them stay abreast of current trends and issues. At Glencoe, we know how valuable your time is, so we've developed a variety of staff development tools to help you meet your district's requirements.

Teacher Wraparound Edition

Mathematical Connections and Background found at the beginning of each chapter gives you an overview of the mathematics skills required in each lesson. Information about prior knowledge as well as future connections lets you see the continuity of instruction.

Building on Prior Knowledge provides you with information that links what students have previously learned to the content of the lesson.

Tips for New Teachers offers helpful suggestions for such things as classroom management, assessment, teaching techniques, and more.

Teaching Tips can be found not only in the margins but also on the reduced student pages at point of use.

Teacher Classroom Resources

Glencoe Mathematics Staff Development Series is a series of publications that allows you to stay current with issues that affect your teaching effectiveness. The series is intended to help you implement new mathematics strategies and enhance your classroom performance.

Available in print

- *Using the Internet in the Mathematics Classroom*
- *Reading and Writing in the Mathematics Classroom*
- *Teaching Mathematics with Foldables™*
- *Teaching Geometry with Manipulatives*

Available online at www.math.glencoe.com

- *Graphing Calculators in the Mathematics Classroom*
- *Cooperative Learning in the Mathematics Classroom*
- *Alternative Assessment in the Mathematics Classroom*
- *Involving Parents and the Community in the Mathematics Classroom*

made convenient.

Technology Support

At www.math.glencoe.com, you'll find:

- a Staff Development site that addresses current issues in education.
- a Teacher Forum that allows teachers to discuss issues and ideas with colleagues.
- a State and National Resources site that links to math and math education resources, nationally and by state.

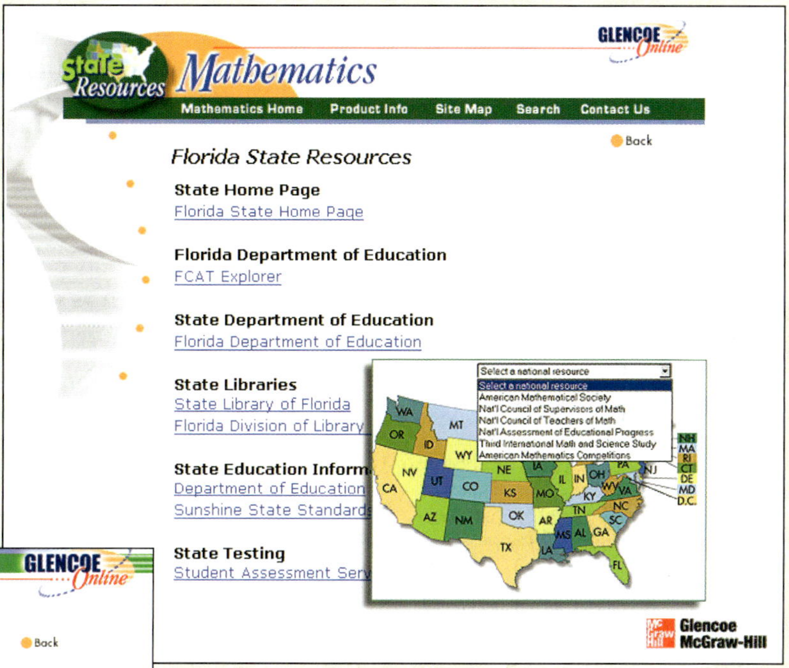

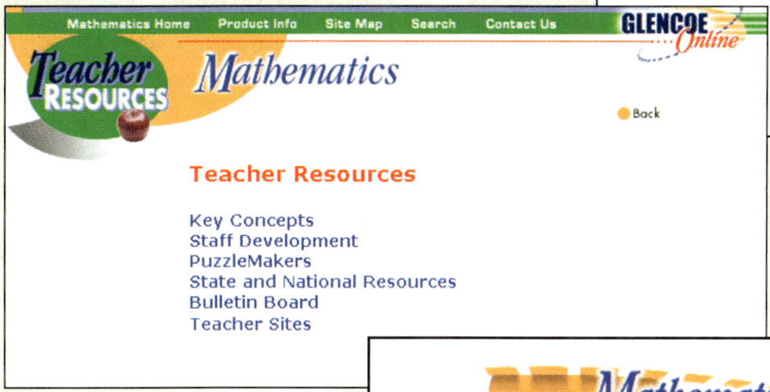

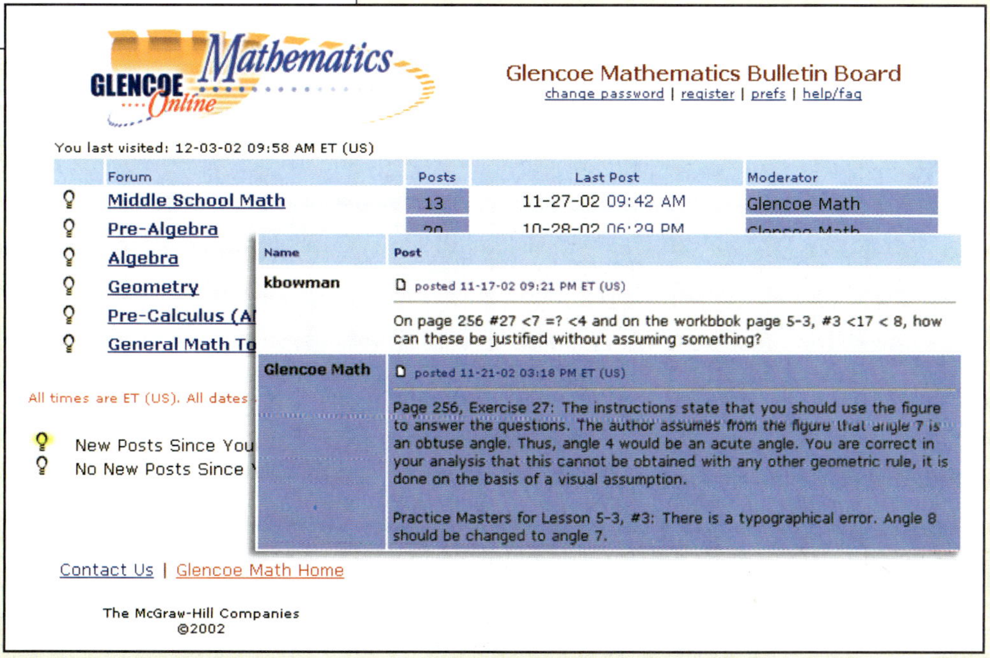

Glencoe Mathematics Programs—Research-Based and

Glencoe's mathematics programs are the product of ongoing classroom and educational research activities involving students, teachers, curriculum supervisors, administrators, parents, and college-level mathematics educators, mathematicians, and researchers.

SOUND

Prior to the publication of any Glencoe mathematics program, the following initial research is completed.

- Monitoring of national and state changes and trends such as graduation requirements, standardized test exams, and the latest NCTM and NAEP reports.
- Incorporating the most current and applicable educational research in which reported results show significant improvement on student learning and achievement.
- Analyzing returns from independently contracted mailing and telephone surveys.

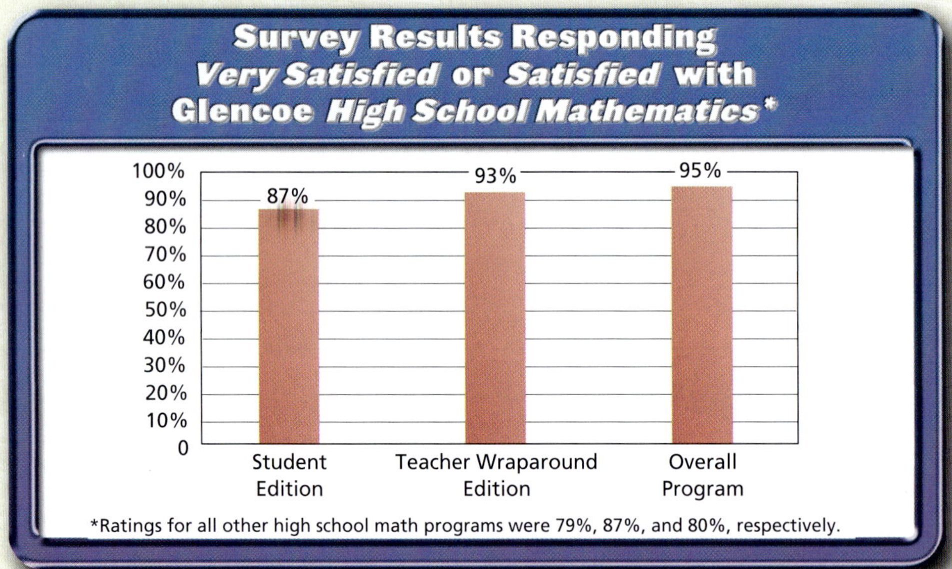

Source: High School Mathematics Longitudinal Survey, 1997, 1999, 2001

- Reviewing all comments and correspondence on appropriate prior editions in terms of specific lessons. This helps Glencoe to build in staff development support, which makes the programs easy to implement from the first day of use.

PROVEN

Prior to the publication of *Glencoe Geometry*, extensive research was conducted using manuscript and pre-publication versions of the program.

- Nationwide discussion groups were conducted, which involved mathematics teachers, department chairpersons, supervisors, and educational learning specialists.

tested to ensure success.

- Face-to-face interviews was carried out with mathematics teachers.
- Reviewers and consultants reviewed *Glencoe Geometry* manuscripts for accuracy, content development, and thoroughness. The Princeton Review, a leader in test expertise, reviewed the assessment strand to ensure test validity.
- An independent research company organized and conducted blind focus groups with high school geometry teachers in various cities. The teachers' reactions and comments were recorded and used for improvements in the Student Edition.
- Follow-up interviews, observations, and surveys of users of Glencoe mathematics programs are continuously conducted and monitored.

EXPERIMENTAL STUDY RESULTS

At the beginning of the 2003-2004 school year, Glencoe conducted an experimental study in which each participating school had two classes of geometry, a "control group" and an "experimental group." Students in the experimental groups used the *Glencoe Geometry* program, and students in the control groups used other geometry programs.

Top-Line Results

- At every school, the experimental group (students using the *Glencoe Geometry*) had post-test scores that were *significantly* higher than pre-test scores.
- The experimental groups had significantly greater growth from pre-test to post-test, compared with students in the control groups.
- For students in the experimental group, *Glencoe Geometry* was equally effective with boys and girls and with minority and non- minority students.

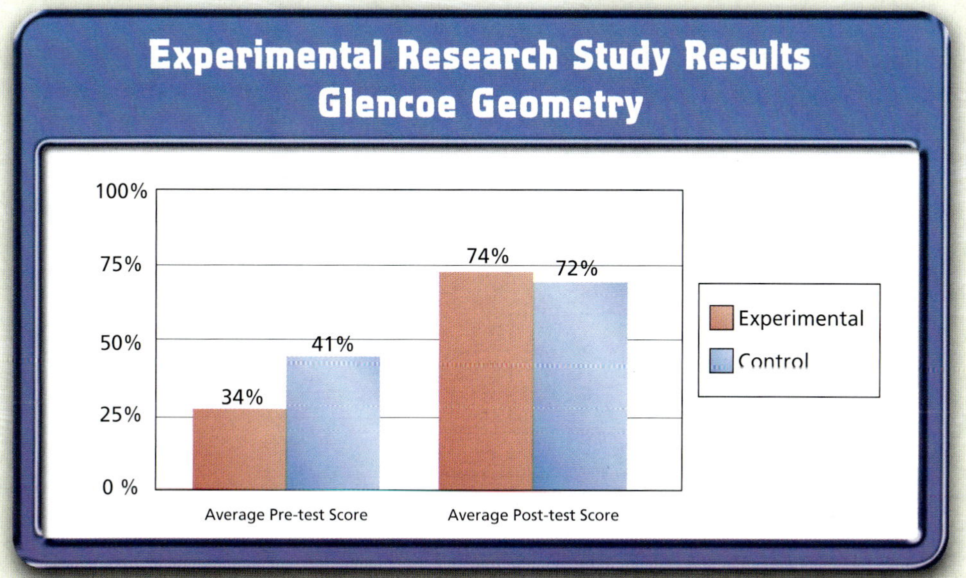

For more details of Glencoe's research, please contact us at www.math.glencoe.com.

Planning Your

Glencoe Geometry and the accompanying support materials allow you to create a geometry course that meets the needs of each class of students. The charts shown on these two pages offer general suggestions for pacing your students through the book for average and advanced levels. Pacing for both standard class periods and block schedule class periods is given. A more detailed pacing chart appears on interleaf page A preceding each chapter in the ***Teacher Wraparound Edition.***

The total number of days in each level of pacing is less than the typical 180-day school year and 90-day semester to allow for flexibility in planning due to testing, school cancellation, or shortened class periods.

AVERAGE PACING

Average Pacing is for those students who have a fairly good mathematical preparation for geometry. You may want to use one of the six chapter tests provided in the Chapter Resource Masters as a pretest to determine how well your students are prepared for each chapter. If you find that they are well prepared, consider using the Study Guide and Review at the end of the chapter as a one-day lesson and proceed to the next chapter.

If your students are better prepared for geometry, you may want to spend less time in the earlier chapters in order to have more time to explore later chapters.

Year-Long Schedule
45–50 minute periods

Grading Period	Chapter	Days
1	1	13
	2	15
	3	13
2	4	15
	5	13
	6	12
3	7	10
	8	14
	9	11
4	10	16
	11	10
	12	10
	13	9
	Total	161

Block Schedule
90 minute periods

Chapter	Days
1	6.5
2	8
3	8
4	8
5	7
6	6
7	5
8	7
9	6
10	8
11	5
12	5
13	4.5
Total	84

Modifying Average Pacing for Basic Students

For those students who are less prepared for geometry, spend less time on Units 2 and 3 (Chapters 4–10) by deemphasizing proof.

Geometry Course

ADVANCED PACING

Advanced Pacing is for those students who have a strong preparation for geometry. In advanced pacing, not as much time is required for the earlier chapters. This allows for more time to be spent on lessons and activities in Chapters 7–13 that are considered optional for average pacing.

To assess students' knowledge prior to each chapter, you may use one of the six chapter tests in the Chapter Resource Masters as a pretest.

Year-Long Schedule
45–50 minute periods

Grading Period	Chapter	Days
1	1	10
	2	12
	3	11
2	4	14
	5	11
	6	14
3	7	14
	8	14
	9	14
4	10	16
	11	10
	12	10
	13	11
	Total	161

Block Schedule
90 minute periods

Chapter	Days
1	5.5
2	6
3	6
4	7
5	6
6	7
7	7
8	7
9	7
10	8
11	5
12	5
13	5.5
Total	82

Daily Planning

- A more detailed Suggested Pacing chart appears in the interleaf preceding each chapter in the ***Teacher Wraparound Edition.***
- The ***Lesson Planning Guide*** offers further suggestions for the materials to be covered each day and how to adapt these for Block Scheduling.
- ***TeacherWorks: All in One Lesson Planner and Resource Center*** CD-ROM enables you to customize an entire course of study to meet your specific needs.

Planning Your Geometry Course ...continued

TWO-YEAR PACING

Two-year Pacing is for those students who want to take geometry, but find the abstract concepts difficult to grasp. This pacing allows students to cover the same material, work the same problems, and complete the same proofs as students using the average or advanced pacing. Students will be able to spend more time on each concept and will have more time to complete hands-on labs and activities that help develop and internalize the abstract concepts presented in this course. More time will also give students a better opportunity to learn the fundamentals of mathematical proof, which are crucial to a more rigorous and abstract understanding of mathematics.

This pacing plan allows at least two days for each lesson and allows four days for review and assessment for each chapter. The plan allows one day for each practice quiz. The plan also allows one or two days for each activity or investigation.

In year one, students cover the first eight chapters of the book. Four days are allowed at the end of each semester for review.

In year two, students begin by reviewing the first eight chapters. One week is allowed for the review of each chapter. If less time is needed for review of these chapters, more time will be available for covering the more difficult concepts and activities in the last five chapters of the book. Five days are allowed at the end of each semester for review.

Year One Schedule

Grading Period	Chapter	Days
1	1	22
	2	20
2	3	19
	4	21
	Semester Review	4
3	5	18
	6	20
4	7	20
	8	21
	Semester Review	4
	Total	**169**

Year Two Schedule

Grading Period	Chapter	Days
1	1 (Review)	5
	2 (Review)	5
	3 (Review)	5
	4 (Review)	5
	5 (Review)	5
	6 (Review)	5
	7 (Review)	5
	8 (Review)	5
2	9	24
	10-1 through 10-4	9
	Semester Review	5
3	10-5 through 10-8	15
	11	20
4	12	22
	13	20
	Semester Review	5
	Total	**160**

Implementing the NCTM *Principles and Standards*

In 1989, the National Council of Teachers of Mathematics (NCTM) published their *Curriculum and Evaluation Standards for School Mathematics,* which gave mathematics teachers their first set of goals toward a national mathematics curriculum. Teachers and supervisors have embraced these Standards and developed state standards based on this framework. In 2000, the National Council of Teachers of Mathematics published a revision of these guidelines entitled *NCTM Principles and Standards for School Mathematics.*

NCTM Principles for School Mathematics	Glencoe Geometry
Equity *Excellence in mathematics education requires equity—high expectations and strong support for all students.*	Glencoe's product line encourages high achievement at every level. Numerous teacher support materials provide activities for **differentiated instruction,** promotion of **reading and writing, pacing** for individual levels of achievement, and **daily intervention.**
Curriculum *A curriculum is more than a collection of activities: it must be coherent, focused on important mathematics, and well articulated across the grades.*	Glencoe authors developed a philosophy and scope and sequence to ensure a continuum of mathematical learning that builds on **prior knowledge** and extends concepts toward more **advanced mathematical thinking.**
Teaching *Effective mathematics teaching requires understanding what students know and need to learn and then challenging and supporting them to learn it well.*	Glencoe offers a plethora of teacher support materials. A comprehensive *Teacher Wraparound Edition* provides **mathematical background,** teaching tips, resource management guidelines, and **tips for new teachers.**
Learning *Students must learn mathematics with understanding, actively building new knowledge from experience and prior knowledge.*	The *Teacher Wraparound Edition* includes instruction on building from prior knowledge with materials in each interleaf and in **Building On Prior Knowledge** features. **Find the Error** and **Unlocking Misconception** teaching tips help to evaluate how students are thinking and learning.
Assessment *Assessment should support the learning of important mathematics and furnish useful information to both teachers and students.*	The **Practice Quizzes** and the **Chapter Practice Test** provide ways for students to check their own progress. Online Study Tools, such as **Self-Check Quizzes,** offer a unique way for students with Internet access to monitor their progress. The assessment tools in the *Chapter Resource Masters* contain different levels and formats for tests, as well as intermediate opportunities for assessment.
Technology *Technology is essential in teaching and learning mathematics; it influences the mathematics that is taught and enhances students' learning.*	The *Student Edition* includes opportunities to utilize graphing calculators, spreadsheets, and geometry software in the exploration of geometry concepts. The *Teacher Wraparound Edition* offers teaching tips on using technology. *Graphing Calculator and Computer Masters* has additional activities. Glencoe's Web site is constantly updated to meet the needs of students and teachers in excelling in mathematics education.

NCTM Standards for School Mathematics

The Standards portion of the *NCTM Principles and Standards for School Mathematics* center upon ten areas of mathematics curriculum development. The number assigned to each standard is for easy reference and is not part of each standard's official title.

Instructional programs from prekindergarten through grade 12 should enable all students to:

1 Numbers and Operations

- Understand numbers, ways of representing numbers, relationships among numbers, and number systems
- Understand the meaning of operations and how they relate to each other
- Compute fluently and make reasonable estimates

Pages: 20–36, 62–66, 282–323, 325–331, 342–348

2 Algebra

- Understand patterns, relations, and functions
- Represent and analyze mathematical situations and structures using algebraic symbols
- Use mathematical models to represent and understand quantitative relationships
- Analyze change in various contexts

Pages: 13–19, 21–43, 45–50, 94–100, 133–157, 159–164, 238–273, 282–287, 289–331, 404–437, 439–451, 463–488, 490–511, 688–706

3 Geometry

- Analyze characteristics and properties of two- and three-dimensional geometric shapes and develop mathematical arguments about geometric relationships
- Specify locations and describe spatial relationships using coordinate geometry and other representational systems
- Apply transformations and use symmetry to analyze mathematical situations
- Use visualization, spatial reasoning, and geometric modeling to solve problems

Pages: 6–52, 75–87, 89–114, 126–166, 178–226, 236–273, 289–331, 342–391, 404–451, 462–511, 522–580, 595–627, 636–677, 688–719

4 Measurement

- Understand measurable attributes of objects and the units, systems, and processes of measurement
- Apply appropriate techniques, tools, and formulas to detemine measurements

Pages: 13–28, 51–52, 522–528, 536–543, 569–580

5 Data Analysis and Probability

- Formulate questions that can be addressed with data and collect, organize, and display relevant data to answer them
- Select and use appropriate statistical methods to analyze data
- Develop and evaluate inferences and predictions that are based on data
- Understand and apply basic concepts of probability

Pages: 20, 622–627

Instructional programs from prekindergarten through grade 12 should enable all students to:

6 Problem Solving

- Build new mathematical knowledge through problem solving
- Solve problems that arise in mathematics and in other contexts
- Apply and adapt a variety of appropriate strategies to solve problems
- Monitor and reflect on the process of mathematical problem solving

Pages: 6–43, 45–52, 62–114, 126–131, 133–164, 178–213, 216–226, 238–273, 282–331, 342–391, 404–451, 463–488, 490–511, 522–580, 595–627, 636–676, 688–719

7 Reasoning and Proof

- Recognize reasoning and proof as fundamental aspects of mathematics
- Make and investigate mathematical conjectures
- Develop and evaluate mathematical arguments and proofs
- Select and use various types of reasoning and methods of proof

Pages: 51–52, 62–114, 132–138, 151–157, 165–166, 200–226, 255–273, 417–423, 431–437, 439–451

8 Communication

- Organize and consolidate their mathematical thinking through communication
- Communicate their mathematical thinking coherently and clearly to peers, teachers, and others
- Analyze and evaluate the mathematical thinking and strategies of others
- Use the language of mathematics to express mathematical ideas precisely

Pages: 6–43, 45–52, 62–87, 89–114, 126–157, 159–166, 178–183, 185–213, 216–226, 238–273, 282–323, 325–331, 342–348, 350–383, 385–390, 404–409, 411–437, 439–451, 463–488, 490–511, 522–558, 561–580, 595–627, 636–677, 688–694, 696–719

9 Connections

- Recognize and use connections among mathematical ideas
- Understand how mathematical ideas build on one another to produce a coherent whole
- Recognize and apply mathematics in contexts outside of mathematics

Pages: 6–19, 21–27, 29–43, 45–50, 62–87, 89–114, 126–131, 133–157, 159–164, 178–183, 185–213, 216–226, 238–254, 261–273, 282–287, 289–323, 325–331, 342–348, 350–383, 385–390, 404–409, 411–437, 439–451, 463–488, 490–511, 522–558, 561–580, 595–627, 636–677, 688–694, 696–719

10 Representation

- Create and use representations to organize, record, and communicate mathematical ideas
- Select, apply, and translate among mathematical representations to solve problems
- Use representations to model and interpret physical, social, and mathematical phenomena

Pages: 6–19, 21–43, 45–50, 62–87, 89–114, 126–131, 133–157, 159–164, 178–226, 236 273, 282–287, 289–323, 325–331, 342–348, 350–383, 385–390, 404–409, 411–437, 439–451, 463–488, 490–511, 522–558, 561–580, 595–627, 636–677, 688–694, 696–719

Lines and Angles 2

Chapter 1 Points, Lines, Planes, and Angles 4

WebQuest Internet Project

Lesson 1-5, p. 37

Chapter 2 Reasoning and Proof 60

All

Lesson 2-3, p. 79

Chapter 3 Parallel and Perpendicular Lines 124

All

Lesson 3-6, p. 163

UNIT 2

Triangles 174

Chapter 4 Congruent Triangles 176

WebQuest Internet Project

Lesson 4-4, p. 204

Prerequisite Skills

FOLDABLES™ Study Organizer 177

Reading and Writing Mathematics

Standardized Test Practice

USA TODAY Snapshots 206

Constructions 200, 202, 207, 214

Chapter 5 Relationships in Triangles 234

Prerequisite Skills

FOLDABLES™

Reading and Writing Mathematics

Standardized Test Practice

USA TODAY

Lesson 5-5, p. 267

Unit 2

Chapter 6 Proportions and Similarity 280

Prerequisite Skills

FOLDABLES™ Study Organizer 281

Reading and Writing Mathematics

Standardized Test Practice

USA TODAY Snapshots 296

Constructions 311, 314

Lesson 6-3, p. 305

Chapter 7 Right Triangles and Trigonometry 340

Lesson 7-2, p. 350

UNIT 3

Quadrilaterals and Circles

Chapter 8 Quadrilaterals

WebQuest Internet Project

- Introduction 401
- Follow-Ups 444, 469, 527
- Culmination 580

Lesson 8-4, p. 429

Prerequisite Skills

- Getting Started 403
- Getting Ready for the Next Lesson 409, 416, 423, 430, 437, 445

FOLDABLES™ Study Organizer 403

Reading and Writing Mathematics

- Hierarchy of Polygons 446
- Reading Math Tips 411, 432
- Writing in Math 409, 416, 422, 430, 436, 444, 451

Standardized Test Practice

- Multiple Choice 409, 413, 414, 416, 423, 430, 437, 445, 451, 457, 458
- Short Response/Grid In 409, 416, 444, 459
- Open Ended 459

USA TODAY Snapshots 411

Constructions 425, 433, 435, 438, 441, 444

Chapter 9 Transformations 460

Lesson 9-5, p. 495

Chapter 10 Circles 520

Lesson 10-3, p. 541

UNIT 4

Area and Volume 590

Lesson 11-4, p. 617

Chapter 12 Surface Area 634

Prerequisite Skills

FOLDABLES™

Reading and Writing Mathematics

Standardized Test Practice

Lesson 12-5, p. 660

Chapter 13 Volume 686

Lesson 13-1, p. 693

Student Handbook

Skills

Reference

Student & Parent One-Stop Internet Resources

Need extra help or information? Log on to math.glencoe.com or any of the web addresses below to learn more.

www.fl.geometryonline.com

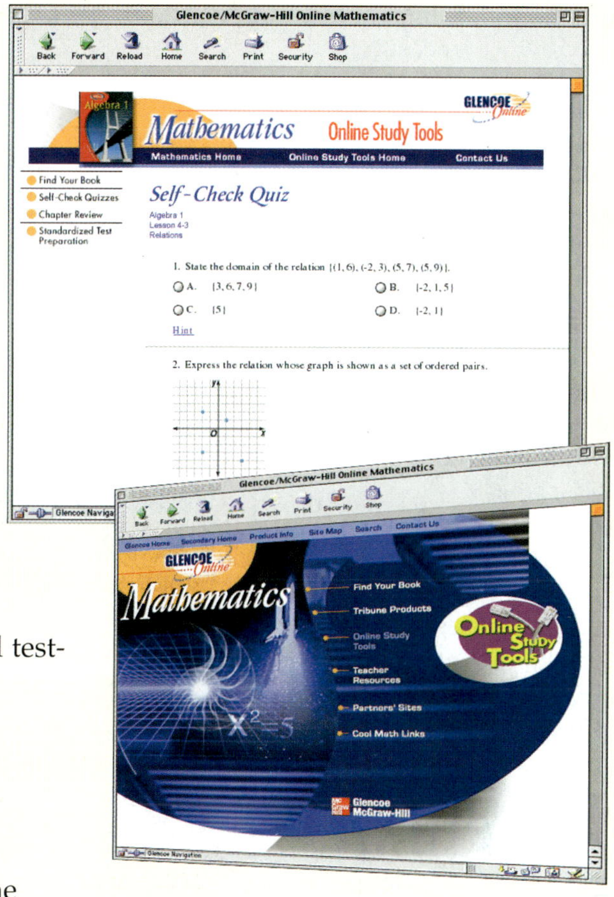

Online Study Tools

- www.geometryonline.com/extra_examples shows you additional worked-out examples that mimic the ones in your book.
- www.geometryonline.com/self_check_quiz provides you with a practice quiz for each lesson that grades itself.
- www.geometryonline.com/vocabulary_review lets you check your understanding of the terms and definitions used in each chapter.
- www.geometryonline.com/chapter_test allows you to take a self-checking test before the actual test.
- www.geometryonline.com/standardized_test is another way to brush up on your standardized test-taking skills.

Research Options

- www.geometryonline.com/webquest walks you step-by-step through a long-term project using the web. One WebQuest for each unit is explored using the mathematics from that unit.
- www.geometryonline.com/usa_today provides activities related to the concept of the lesson as well as up-to-date Snapshot data.
- www.geometryonline.com/careers links you to additional information about interesting careers.
- www.geometryonline.com/data_update links you to the most current data available for subjects such as basketball and family.

Calculator Help

- www.geometryonline.com/other_calculator_keystrokes provides you with instructions for using Cabri Jr. on the TI-83 Plus and keystrokes other than the TI-83 Plus used in your textbook.

UNIT 1 Notes

Introduction

In this unit, students will be introduced to points, lines, and angles. Accuracy of measurement will be explored, and the concept of congruency will be introduced. Students will also learn to apply inductive and deductive reasoning to situations in preparation for writing proofs in later chapters.

Students will participate in an in-depth exploration of parallel and perpendicular lines. They will review slope, the equations for a line, and ways to calculate distance between points.

Assessment Options

Unit 1 Test Pages 181–182 of the *Chapter 3 Resource Masters* may be used as a test or review for Unit 1. This assessment contains both multiple-choice and short answer items.

TestCheck and Worksheet Builder

This CD-ROM can be used to create additional unit tests and review worksheets.

UNIT 1

Lines and Angles

Lines and angles are all around us and can be used to model and describe real-world situations. In this unit, you will learn about lines, planes, and angles and how they can be used to prove theorems.

WebQuest Internet Project

When Is Weather Normal?

Source: *USA TODAY,* October 8, 2000

"Climate normals are a useful way to describe the average weather of a location. Several statistical measures are computed as part of the normals, including measures of central tendency, such as mean or median, of dispersion or how spread out the values are, such as the standard deviation or inter-quartile range, and of frequency or probability of occurrence." In this project, you will explore how latitude, longitude, and *degree distance* relate to differences in temperature for pairs of U.S. cities.

USA TODAY Snapshots®

Coldest cities in the USA

City	Mean temperature
International Falls, Minn.	36.4
Duluth, Minn.	38.2
Caribou, Maine	38.9
Marquette, Mich.	39.2
Sault Ste. Marie, Mich.	39.7
Williston, N.D.	40.1
Fargo, N.D.	40.5
Alamosa, Colo.	41.2
Bismarck, N.D.	41.3
St. Cloud, Minn.	41.4

Source: Planet101.com By Lori Joseph and Keith Simmons, USA TODAY

Log on to www.geometryonline.com/webquest. Begin your WebQuest by reading the Task.

Continue working on your WebQuest as you study Unit 1.

Lesson	1-3	2-1	3-5
Page	23	65	155

WebQuest Internet Project

A WebQuest is an online project in which students do research on the Internet, gather data, and make presentations using word processing, graphing, page-making, or presentation software. In each chapter, students advance to the next step in their WebQuest. At the end of Chapter 3, the project culminates with a presentation of their findings.

Teaching notes and sample answers are available in the *WebQuest and Project Resources.*

USA TODAY Education Teaching Suggestions

Have students study the USA TODAY Snapshot.

- Ask them how they would calculate the difference between the mean temperature in Duluth and the mean temperature in Fargo. **Subtract Duluth's temperature from Fargo's temperature.**
- What conclusion can be drawn about the locations of the coldest cities? **They tend to be in the northern states.**
- Point out to students that in their WebQuest they will be making conjectures about the relationships between latitude, longitude, degree distance, and the mean temperature.

Additional USA TODAY Snapshots appearing in Unit 1:

Chapter 1 Taking care of business (p. 16)

Chapter 2 Latest CD rates (p. 63)

Chapter 3 Median age continues to rise (p. 143)

Unit 1 Lines and Angles **3**

Chapter 1

Points, Lines, Planes, and Angles

Chapter Overview and Pacing

LESSON OBJECTIVES	PACING (days) Regular Basic/Average	Regular Advanced	Block Basic/Average	Block Advanced
1-1 Points, Lines, and Planes *(pp. 6–12)* • Identify and model points, lines, and planes. • Identify collinear and coplanar points and intersecting lines and planes in space.	1	1	0.5	0.5
1-2 Linear Measure and Precision *(pp. 13–20)* • Measure segments and determine accuracy of measurement. • Compute with measures. ***Follow-Up:*** Relate probability to segment measure.	2 (with 1-2 Follow-Up)	2 (with 1-2 Follow-Up)	1 (with 1-2 Follow-Up)	1 (with 1-2 Follow-Up)
1-3 Distance and Midpoints *(pp. 21–28)* • Find the distance between two points. • Find the midpoint of a segment. ***Follow-Up:*** Use a model to demonstrate the Pythagorean Theorem.	2 (with 1-3 Follow-Up)	1	1 (with 1-3 Follow-Up)	1
1-4 Angle Measure *(pp. 29–36)* • Measure and classify angles. • Identify and use congruent angles and the bisector of an angle.	2	1	1	0.5
1-5 Angle Relationships *(pp. 37–44)* • Identify and use special pairs of angles. • Identify perpendicular lines. ***Follow-Up:*** Use a compass and a straightedge to construct perpendicular lines.	2 (with 1-5 Follow-Up)	1 (with 1-5 Follow-Up)	1 (with 1-5 Follow-Up)	0.5 (with 1-5 Follow-Up)
1-6 Polygons *(pp. 45–52)* • Identify and name polygons. • Find perimeters of polygons. ***Follow-Up:*** Use the Geometer's Sketchpad® to draw and investigate polygons.	2 (with 1-6 Follow-Up)	2 (with 1-6 Follow-Up)	1 (with 1-6 Follow-Up)	1 (with 1-6 Follow-Up)
Study Guide and Practice Test *(pp. 53–57)* **Standardized Test Practice** *(p. 58–59)*	1	1	0.5	0.5
Chapter Assessment	1	1	0.5	0.5
TOTAL	**13**	**10**	**6.5**	**5.5**

Pacing suggestions for the entire year can be found on pages T20–T21.

Chapter Resource Manager

Timesaving Tools
TeacherWorks™
All-In-One Planner and Resource Center
See pages T5 and T21.

Chapter 1 Resource Masters: Study Guide and Intervention	Practice (Skills and Average)	Reading to Learn Mathematics	Enrichment	Assessment	Prerequisite Skills Workbook	Applications*	5-Minute Check Transparencies	Interactive Chalkboard	GeomPASS: Tutorial Plus (lessons)	Materials
1–2	3–4	5	6			SC 1 GCC 17	1-1	1-1		grid paper, straightedge, index cards, scissors, tape, tracing paper
7–8	9–10	11	12	51	3–4, 15–18, 73–78, 83–84		1-2	1-2	3	customary and metric rulers, compass
13–14	15–16	17	18	51, 53	7–8, 33–34, 77–80, 83–86	GCC 18	1-3	1-3	4	grid paper, compass, customary and metric rulers (*Follow-Up:* grid paper, scissors)
19–20	21–22	23	24		81–82	SC 2	1-4	1-4		protractor, compass, straightedge, patty paper or tracing paper, customary and metric rulers
25–26	27–28	29	30	52	85–86		1-5	1-5	5	patty paper, protractor (*Follow-Up:* straightedge, compass)
31–32	33–34	35	36	52			1-6	1-6		grid paper
				37–50, 54–56						

**Key to Abbreviations:* GCC = Graphing Calculator and Computer Masters
SC = School-to-Career Masters

Chapter 1

Mathematical Connections and Background

Continuity of Instruction

Prior Knowledge

In previous courses, students learned how to graph and label points on a coordinate plane. They also learned how to add and subtract fractions and integers. Students have used the Pythagorean Theorem. They may be able to find the perimeter of a three- or four-sided figure.

This Chapter

Students learn about points, lines, and planes, the building blocks of geometry. Line segments, rays, angles, and polygons are also introduced in this chapter. Students explore congruent segments and angles and learn to construct them with a compass and straightedge. Students expand on their knowledge of the Pythagorean Theorem to master the Distance Formula and use the Midpoint Formula to find the midpoint of a segment. Students also compute the perimeter of a given polygon.

Future Connections

Students will apply their knowledge of angle measure in Chapter 3 and their knowledge of congruence of triangles in Chapter 4. They will explore the sum of the measures of the interior angles of a polygon in Chapter 9.

1-1 Points, Lines, and Planes

In geometry, a point is a location without shape or size. It is named by a capital letter, such as *A*. It is drawn as a dot.

A line contains points and has no thickness or width. Points on the same line are collinear, and there is exactly one line through any two points. The intersection of two lines is a point. A line can be named either by a lowercase script letter or by the letters of two points on the line. For example, a line could have the name *n* or $\overleftrightarrow{AB}$ or $\overleftrightarrow{BA}$.

A plane is a flat surface made of points. A plane has no depth and extends infinitely in all directions. Points on the same plane are coplanar, and the intersection of two planes is a line. A plane can be named by a capital script letter or by the letters naming three noncollinear points. For example, a plane with points *A*, *B*, and *C* could have the name $\mathcal{N}$, *ABC*, *ACB*, *BAC*, *BCA*, *CAB*, or *CBA*. A plane is usually drawn as a shaded parallelogram.

1-2 Linear Measure and Precision

A line cannot be measured because it extends infinitely in each direction. A line segment, however, has two endpoints and can be measured. Measures are expressed as real numbers. This means that you can perform calculations with measures.

Two segments with the same measure are said to be congruent. The symbol for congruence is ≅. Red slashes on the figure also indicate that segments are congruent. You can construct a segment congruent to a given segment using a compass and a straightedge.

The precision of any measurement depends on the smallest unit available on the measuring tool. If a ruler is divided into centimeters, you can only measure to the nearest centimeter accurately. The precision for this measurement is 0.5 centimeters. So a measurement of 85 centimeters with this ruler means the true measurement falls between 84.5 and 85.5 centimeters. Relative error is the ratio of the half-unit difference in precision to the entire measure. The relative error is expressed as a percent. The smaller the relative error of a measurement, the more accurate the measure is.

1-3 Distance and Midpoints

The coordinates of the endpoints of a segment can be used to find the length of the segment. On a number line, the distance between the endpoints is the absolute value of their difference. On a coordinate plane, you can use the Distance Formula or the Pythagorean Theorem to calculate the distance between two points. The Distance Formula for two points (x_1, y_1) and (x_2, y_2) is $d = \sqrt{(x_2 - x_1)^2 + (y_2 - y_1)^2}$.

The Pythagorean Theorem states that in a right triangle with sides a and b, and hypotenuse c, $a^2 + b^2 = c^2$.

The midpoint of a segment is the point halfway between its endpoints. On a number line, the coordinate of a midpoint of a segment whose endpoints have coordinates a and b is the sum of a and b divided by 2. To find the midpoint of a segment on the coordinate plane, use the Midpoint Formula, $\left(\frac{x_1 + x_2}{2}, \frac{y_1 + y_2}{2}\right)$.

Any segment, line, or plane that intersects a segment at its midpoint is called a segment bisector. You can construct a line that bisects a segment using a compass and a straightedge.

1-4 Angle Measure

An angle is the intersection of two noncollinear rays at a common endpoint. The common endpoint is called the vertex, and the rays are the sides of the angle. An angle can be named by a single letter or by three letters: a point on one side, the vertex, and a point on the other side. For example, an angle could have the name $\angle A$ or $\angle BAC$ if point B is on one side and point C is on the other. However, if there are other angles in the figure that have A as their vertex, the angle cannot be named $\angle A$.

An angle is measured in degrees. Ninety degrees is an important marker in angle measure. An angle measuring exactly 90° is a right angle. Angles less than 90° are acute, and those greater than 90° are obtuse. Angles that have the same measure are congruent. You can construct an angle congruent to a given angle using a compass and a straightedge.

In this textbook, the degree measure of an angle is represented by m so the degree measure of $\angle A$ that measures 75° is written as $m\angle A = 75$.

A ray that divides an angle into two congruent angles is called an angle bisector. A compass and straightedge can be used to construct an angle bisector even if you do not know the measure of the angle.

1-5 Angle Relationships

Certain pairs of angles have special relationships. Adjacent angles are two angles that lie in the same plane, have a common vertex and a common side, but have no common interior points. Vertical angles are two nonadjacent angles formed by two intersecting lines. All vertical angles are congruent. A linear pair is a pair of adjacent angles whose noncommon sides are opposite rays.

Complementary angles are a pair of angles whose angle measures have a sum of 90. Supplementary angles are two angles whose measures have a sum of 180. The two angles in a complementary or supplementary pair do not need to have any points in common. Note that all linear pairs are supplementary as well.

If two lines intersect to form four right angles, the lines are said to be perpendicular. Segments and rays can be perpendicular to lines or to other segments and rays. The symbol for perpendicular is ⊥, so if line a is perpendicular to line b, we write $a \perp b$.

Not all lines that appear to be perpendicular are perpendicular, however. A red ⌝ symbol is used in a figure to denote perpendicular lines. Figures are not always drawn to reflect total accuracy of the situation.

1-6 Polygons

Polygons appear everywhere in our world. In geometry, a polygon is defined as a closed figure formed by a finite number of coplanar segments. The sides of the figure that have a common endpoint are noncollinear, and each side intersects exactly two other sides, but only at their endpoints. A polygon is named by the letters of its vertices, written in clockwise or counterclockwise order.

Polygons can be concave or convex. If no points of the lines are in the interior of the figure, it is convex. A convex polygon in which all sides and angles are congruent is called a regular polygon.

The perimeter of a polygon is the sum of the lengths of its sides. You may need to use the Distance Formula to calculate the lengths of the sides of some polygons that are graphed on a coordinate grid.

Chapter 1

DAILY INTERVENTION and Assessment

	Type	Student Edition	Teacher Resources	Technology/Internet
INTERVENTION	Ongoing	Prerequisite Skills, pp. 5, 19, 27, 36, 43 Practice Quiz 1, p. 19 Practice Quiz 2, p. 36	5-Minute Check Transparencies *Prerequisite Skills Workbook,* pp. 3–4, 7–8, 15–18, 33–34, 73–86 Quizzes, *CRM* pp. 51–52 Mid-Chapter Test, *CRM* p. 53 Study Guide and Intervention, *CRM* pp. 1–2, 7–8, 13–14, 19–20, 25–26, 31–32	GeomPASS: Tutorial Plus www.geometryonline.com/self_check_quiz www.geometryonline.com/extra_examples
	Mixed Review	pp. 19, 27, 36, 43, 50	Cumulative Review, *CRM* p. 54	
	Error Analysis	Find the Error, pp. 9, 48 Common Misconceptions, p. 22	Find the Error, *TWE* pp. 9, 48 Unlocking Misconceptions, *TWE* pp. 15, 23, 47 Tips for New Teachers, *TWE* pp. 8, 31, 39	
ASSESSMENT	Standardized Test Practice	pp. 11, 19, 23, 25, 27, 35, 43, 50, 57, 58	*TWE* pp. 58–59 Standardized Test Practice, *CRM* pp. 55–56	Standardized Test Practice CD-ROM www.geometryonline.com/standardized_test
	Open-Ended Assessment	Writing in Math, pp. 11, 19, 27, 35, 43, 50 Open Ended, pp. 9, 17, 25, 33, 41, 48 Standardized Test, p. 59	Modeling: *TWE* pp. 11, 36 Speaking: *TWE* pp. 27, 43 Writing: *TWE* pp. 19, 50 Open-Ended Assessment, *CRM* p. 49	
	Chapter Assessment	Study Guide, pp. 53–56 Practice Test, p. 57	Multiple-Choice Tests (Forms 1, 2A, 2B), *CRM* pp. 37–42 Free-Response Tests (Forms 2C, 2D, 3), *CRM* pp. 43–48 Vocabulary Test/Review, *CRM* p. 50	TestCheck and Worksheet Builder (see below) MindJogger Videoquizzes www.geometryonline.com/vocabulary_review www.geometryonline.com/chapter_test

Key to Abbreviations: TWE = Teacher Wraparound Edition; CRM = Chapter Resource Masters

Additional Intervention Resources

The Princeton Review's *Cracking the SAT & PSAT*

The Princeton Review's *Cracking the ACT*

TestCheck and Worksheet Builder

This **networkable** software has three modules for intervention and assessment flexibility:

- **Worksheet Builder** to make worksheet and tests
- **Student Module** to take tests on screen (optional)
- **Management System** to keep student records (optional)

Special banks are included for SAT, ACT, TIMSS, NAEP, and End-of-Course tests.

Intervention Technology

GeomPASS: Tutorial Plus CD-ROM offers a complete, self-paced geometry curriculum.

Geometry Lesson	GeomPASS Lesson
1-2	3 *Understanding Basic Geometric Terms I*
1-3	4 *Segment Relationships*
1-5	5 *Understanding Basic Geometric Terms II*

Intervention at Home

Log on for student study help.

- For each lesson in the Student Edition, there are Extra Examples and Self-Check Quizzes.
  www.geometryonline.com/extra_examples
  www.geometryonline.com/self_check_quiz
- For chapter review, there is vocabulary review, test practice, and standardized test practice.
  www.geometryonline.com/vocabulary_review
  www.geometryonline.com/chapter_test
  www.geometryonline.com/standardized_test

For more information on Intervention and Assessment, see pp. T8–T11.

Reading and Writing in Mathematics

Glencoe Geometry provides numerous opportunities to incorporate reading and writing into the mathematics classroom.

Student Edition

- Foldables Study Organizer, p. 5
- Concept Check questions require students to verbalize and write about what they have learned in the lesson. (pp. 9, 16, 25, 33, 41, 48)
- Reading Mathematics, p. 12
- Writing in Math questions in every lesson, pp. 11, 19, 27, 35, 43, 50
- Reading Study Tip, pp. 6, 29, 45, 46
- WebQuest, p. 23

Teacher Wraparound Edition

- Foldables Study Organizer, pp. 5, 53
- Study Notebook suggestions, pp. 9, 12, 17, 20, 25, 28, 33, 41, 44, 48
- Modeling activities, pp. 11, 36
- Speaking activities, pp. 27, 43
- Writing activities, pp. 19, 50
- ELL Resources, pp. 4, 10, 12, 18, 26, 35, 42, 49, 53

Additional Resources

- Vocabulary Builder worksheets require students to define and give examples for key vocabulary terms as they progress through the chapter. (*Chapter 1 Resource Masters*, pp. vii-viii)
- Reading to Learn Mathematics master for each lesson (*Chapter 1 Resource Masters*, pp. 5, 11, 17, 23, 29, 35)
- *Vocabulary PuzzleMaker* software creates crossword, jumble, and word search puzzles using vocabulary lists that you can customize.
- *Teaching Mathematics with Foldables* provides suggestions for promoting cognition and language.
- *Reading Strategies for the Mathematics Classroom*
- *WebQuest and Project Resources*

For more information on Reading and Writing in Mathematics, see pp. T6–T7.

Chapter 1 Notes

What You'll Learn

Have students read over the list of objectives and make a list of any words with which they are not familiar.

Why It's Important

Point out to students that this is only one of many reasons why each objective is important. Others are provided in the introduction to each lesson.

The chart below correlates the objectives for each lesson to the NCTM Standards 2000. There is also space for you to reference your state and/or local objectives.

Lesson	NCTM Standards	Local Objectives
1-1	3, 6, 8, 9, 10	
1-2	2, 3, 4, 6, 8, 9, 10	
1-2 Follow-Up	1, 3, 4, 5, 6, 8	
1-3	1, 2, 3, 4, 6, 8, 9, 10	
1-3 Follow-Up	1, 2, 3, 4, 6, 8, 10	
1-4	1, 2, 3, 6, 8, 9, 10	
1-5	2, 3, 6, 8, 9, 10	
1-5 Follow-Up	3	
1-6	2, 3, 6, 8, 9, 10	
1-6 Follow-Up	3, 4, 6, 7, 8	

Key to NCTM Standards:

1=Number & Operations, 2=Algebra, 3=Geometry, 4=Measurement, 5=Data Analysis & Probability, 6=Problem Solving, 7=Reasoning & Proof, 8=Communication, 9=Connections, 10=Representation

Chapter 1 Points, Lines, Planes, and Angles

What You'll Learn

- **Lesson 1-1** Identify and model points, lines, and planes.
- **Lesson 1-2** Measure segments and determine accuracy of measurements.
- **Lesson 1-3** Calculate the distance between points and find the midpoint of a segment.
- **Lessons 1-4 and 1-5** Measure and classify angles and identify angle relationships.
- **Lesson 1-6** Identify polygons and find their perimeters.

Key Vocabulary

- line segment (p. 13)
- congruent (p. 15)
- segment bisector (p. 24)
- angle bisector (p. 32)
- perpendicular (p. 40)

Why It's Important

Points, lines, and planes are the basic building blocks used in geometry. They can be used to describe real-world objects. For example, a kite can model lines, angles, and planes in two and three dimensions. *You will explore the angles formed by the structure of a kite in Lesson 1-2.*

Vocabulary Builder

ELL

The Key Vocabulary list introduces students to some of the main vocabulary terms included in this chapter. For a more thorough vocabulary list with pronunciations of new words, give students the Vocabulary Builder worksheets found on pages vii and viii of the *Chapter 1 Resource Masters*. Encourage them to complete the definition of each term as they progress through the chapter. You may suggest that they add these sheets to their study notebooks for future reference when studying for the Chapter 1 test.

Getting Started

Prerequisite Skills To be successful in this chapter, you'll need to master these skills and be able to apply them in problem-solving situations. Review these skills before beginning Chapter 1.

For Lesson 1-1 1–4. See margin. — Graph Points

Graph and label each point in the coordinate plane. *(For review, see pages 728 and 729.)*

1. $A(3, -2)$ **2.** $B(4, 0)$ **3.** $C(-4, -4)$ **4.** $D(-1, 2)$

For Lesson 1-2 — Add and Subtract Fractions

Find each sum or difference.

5. $\frac{3}{4} + \frac{3}{8}$ $1\frac{1}{8}$ **6.** $2\frac{5}{16} + 5\frac{1}{8}$ $7\frac{7}{16}$ **7.** $\frac{7}{8} - \frac{9}{16}$ $\frac{5}{16}$ **8.** $11\frac{1}{2} - 9\frac{7}{16}$ $2\frac{1}{16}$

For Lessons 1-3 through 1-5 — Operations With Integers

Evaluate each expression. *(For review, see pages 734 and 735.)*

9. $2 - 17$ -15 **10.** $23 - (-14)$ 37 **11.** $[-7 - (-2)]^2$ 25 **12.** $9^2 + 13^2$ 250

For Lesson 1-6 — Find Perimeter

Find the perimeter of each figure. *(For review, see pages 732 and 733.)*

13. 20 in.

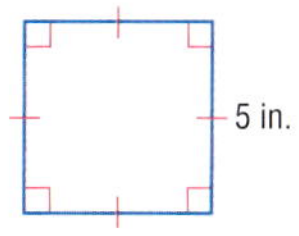

14. 17 ft

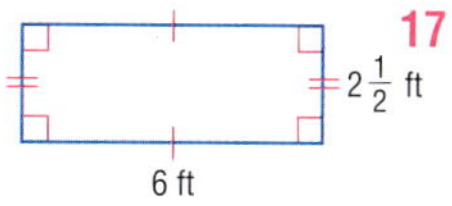

15. 24.6 m

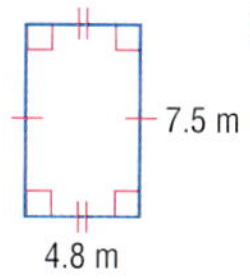

FOLDABLES™ Study Organizer

Lines and Angles Make this Foldable to collect examples of and notes about lines and angles. Begin with a sheet of paper.

Step 1 Fold

Step 2 Fold Again

Step 3 Cut

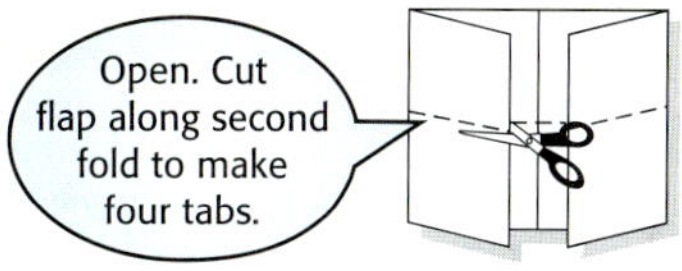

Step 4 Label

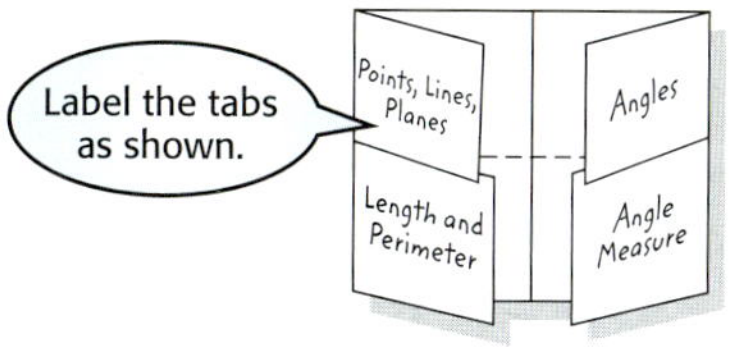

Reading and Writing As you read and study the chapter, record examples and notes from each lesson under the appropriate tab.

Getting Started

This section provides a review of the basic concepts needed before beginning Chapter 1. Page references are included for additional student help.

Additional review is provided in the *Prerequisite Skills Workbook,* pages 3–4, 7–8, 15–18, 33–34, and 73–86.

Prerequisite Skills in the Getting Ready for the Next Lesson section at the end of each exercise set review a skill needed in the next lesson.

For Lesson	Prerequisite Skill
1-3	Evaluating Expressions, p. 19
1-4	Solving Equations, p. 27
1-5	Solving Equations, p. 36
1-6	Evaluating Expressions, p. 43

Answers

1–4.

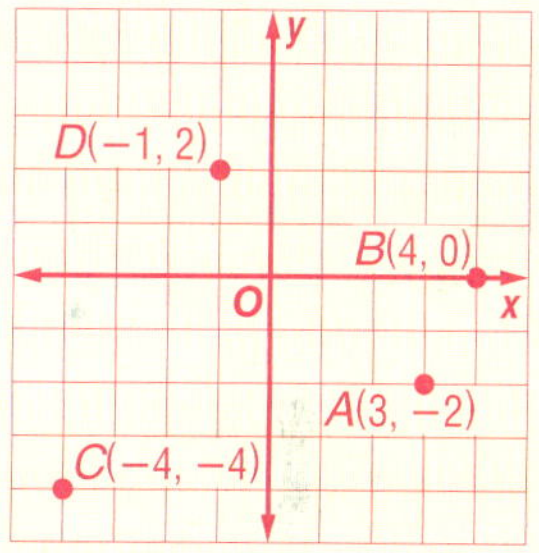

Each chapter opens with Prerequisite Skills practice for lessons in the chapter. More Prerequisite Skill practice can be found at the end of each lesson.

FOLDABLES™ Study Organizer

For more information about Foldables, see *Teaching Mathematics with Foldables.*

Main Ideas and Note-Taking Use this Foldable for student writing about points, lines, planes, and angles. Note-taking is a skill that is based upon listening or reading for main ideas and then recording those ideas for future reference. Under the tabs of their Foldables, have students take notes about what they need to know to identify and measure line segments and angles. Encourage students to apply these concepts by drawing and measuring angles and line segments, and writing about the process.

Foldables™ are a unique way to enhance students' study skills. Encourage students to add to their Foldable as they work through the chapter, and use it to review for their chapter test.

1-1 Lesson Notes

1 Focus

5-Minute Check Transparency 1-1 Use as a quiz or review of previous course materials.

Mathematical Background notes are available for this lesson on p. 4C.

Why do chairs sometimes wobble?

Ask students:

- What do the feet of the three-legged stool represent? **points that lie in the same plane**
- Why does a four-legged chair sometimes wobble on a flat surface? **At least one of the feet represents a point that does not lie on the same plane as the other three feet, or points.**
- If you were to connect the points represented by the feet of the three-legged stool, what geometric shape would you have? **a triangle**

Lessons open with a question that is designed to engage students in the mathematics of the lesson. These opening problems should also help to answer the question "When am I ever going to use this?"

A Four-step Teaching Plan shows you how to Focus, Teach, Practice/Apply, and Assess each lesson.

1-1 Points, Lines, and Planes

What You'll Learn

- Identify and model points, lines, and planes.
- Identify collinear and coplanar points and intersecting lines and planes in space.

Why do chairs sometimes wobble?

Have you ever noticed that a four-legged chair sometimes wobbles, but a three-legged stool never wobbles? This is an example of points and how they lie in a plane. All geometric shapes are made of points. In this book, you will learn about those shapes and their characteristics.

Vocabulary

- point
- line
- collinear
- plane
- coplanar
- undefined term
- space
- locus

NAME POINTS, LINES, AND PLANES

You are familiar with the terms *plane*, *line*, and *point* from algebra. You graph on a coordinate *plane*, and ordered pairs represent *points* on *lines*. In geometry, these terms have similar meanings.

Unlike objects in the real world that model these shapes, points, lines, and planes do not have any actual size.

- A **point** is simply a location.
- A **line** is made up of points and has no thickness or width. Points on the same line are said to be **collinear**.
- A **plane** is a flat surface made up of points. Points that lie on the same plane are said to be **coplanar**. A plane has no depth and extends infinitely in all directions.

Points are often used to name lines and planes.

Study Tip

Reading Math

The word *noncollinear* means not collinear or not lying on the same line. Likewise, *noncoplanar* means not lying in the same plane.

Notice that the letters can be in any order when naming lines or planes.

Key Concept *Points, Lines, and Planes*

	Point	Line	Plane
Model:			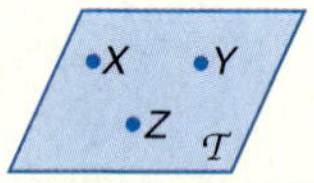
Drawn:	as a dot	with an arrowhead at each end	as a shaded, slanted 4-sided figure
Named by:	a capital letter	the letters representing two points on the line or a lowercase script letter	a capital script letter or by the letters naming three noncollinear points
Facts	A point has neither shape nor size.	There is exactly one line through any two points.	There is exactly one plane through any three noncollinear points.
Words/ Symbols	point P	line n, line AB or $\overleftrightarrow{AB}$, line BA or $\overleftrightarrow{BA}$	plane $\mathcal{T}$, plane XYZ, plane XZY, plane YXZ, plane YZX, plane ZXY, plane ZYX

Resource Manager

Workbook and Reproducible Masters

Chapter 1 Resource Masters

- Study Guide and Intervention, pp. 1–2
- Skills Practice, p. 3
- Practice, p. 4
- Reading to Learn Mathematics, p. 5
- Enrichment, p. 6

Graphing Calculator and Computer Masters, p. 17

School-to-Career Masters, p. 1

Teaching Geometry With Manipulatives Masters, pp. 1, 17, 26, 27

Transparencies

5-Minute Check Transparency 1-1
Real-World Transparency 1
Answer Key Transparencies

Technology

Interactive Chalkboard

Example 1 Name Lines and Planes

Use the figure to name each of the following.

a. a line containing point *A*

The line can be named as line ℓ.

There are four points on the line. Any two of the points can be used to name the line.

$\overleftrightarrow{AB}$ $\overleftrightarrow{BA}$ $\overleftrightarrow{AC}$ $\overleftrightarrow{CA}$ $\overleftrightarrow{AD}$ $\overleftrightarrow{DA}$
$\overleftrightarrow{BC}$ $\overleftrightarrow{CB}$ $\overleftrightarrow{BD}$ $\overleftrightarrow{DB}$ $\overleftrightarrow{CD}$ $\overleftrightarrow{DC}$

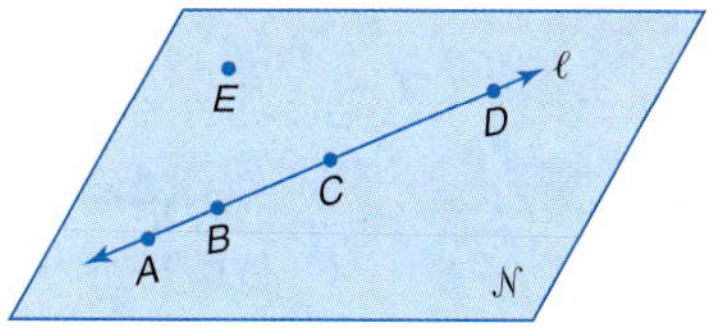

b. a plane containing point *C*

The plane can be named as plane $\mathcal{N}$.

You can also use the letters of any three *noncollinear* points to name the plane.
plane *ABE* plane *ACE* plane *ADE* plane *BCE* plane *BDE* plane *CDE*

The letters of each of these names can be reordered to create other acceptable names for this plane. For example, *ABE* can also be written as *AEB*, *BEA*, *BAE*, *EBA*, and *EAB*. In all, there are 36 different three-letter names for this plane.

Example 2 Model Points, Lines, and Planes

VISUALIZATION **Name the geometric shapes modeled by the picture.**

The pencil point models point *A*.

The blue rule on the paper models line *BC*.

The edge of the paper models line *BD*.

The sheet of paper models plane *ADC*.

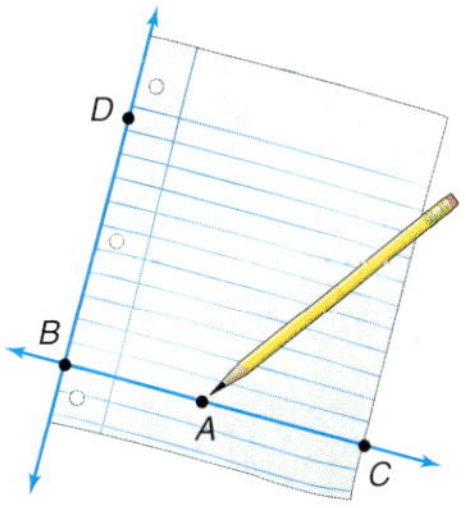

In geometry, the terms *point*, *line*, and *plane* are considered **undefined terms** because they have only been explained using examples and descriptions. Even though they are undefined, these terms can still be used to define other geometric terms and properties. For example, two lines intersect in a point. In the figure at the right, point *P* represents the intersection of $\overleftrightarrow{AB}$ and $\overleftrightarrow{CD}$. Lines can intersect planes, and planes can intersect each other.

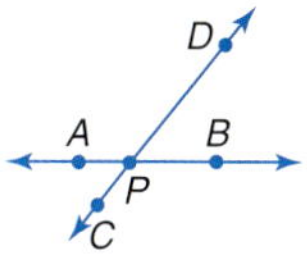

Study Tip

Naming Points
Points on the coordinate plane are named using *ordered pairs*. Point *G* can be named as $G(-1, -3)$.

Example 3 Draw Geometric Figures

Draw and label a figure for each relationship.

a. ALGEBRA Lines *GH* and *JK* intersect at *L* for $G(-1, -3)$, $H(2, 3)$, $J(-3, 2)$, and $K(2, -3)$ on a coordinate plane. Point *M* is coplanar with these points, but not collinear with $\overleftrightarrow{GH}$ or $\overleftrightarrow{JK}$.

Graph each point and draw $\overleftrightarrow{GH}$ and $\overleftrightarrow{JK}$.

Label the intersection point as *L*.

There are an infinite number of points that are coplanar with *G*, *H*, *J*, *K*, and *L*, but are not collinear with $\overleftrightarrow{GH}$ or $\overleftrightarrow{JK}$. In the graph, one such point is $M(-4, 0)$.

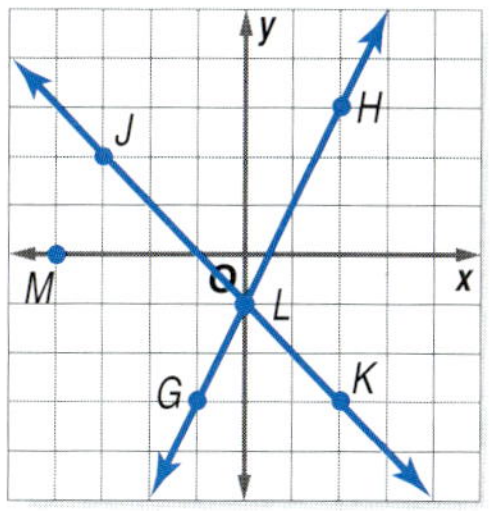

www.geometryonline.com/extra_examples/fcat

2 Teach

NAME POINTS, LINES, AND PLANES

In-Class Examples

Power Point®

1 Use the figure to name each of the following.

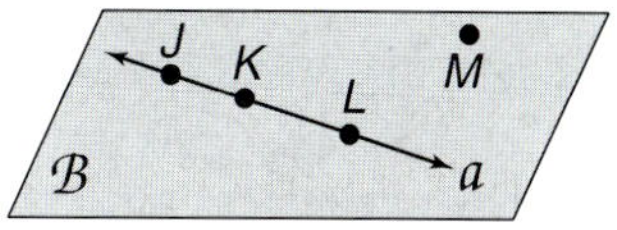

a. a line containing point *K* **line *a*, $\overleftrightarrow{JK}$, $\overleftrightarrow{JL}$, $\overleftrightarrow{KJ}$, $\overleftrightarrow{KL}$, $\overleftrightarrow{LJ}$, $\overleftrightarrow{LK}$**

b. a plane containing point *L* **plane $\mathcal{B}$, plane *JKM*, plane *KLM*, plane *JLM*. Reorder the letters in these names to create 15 other acceptable names.**

2 **VISUALIZATION** Name the geometric shape modeled by each object.

a. the long hand on a clock **line segment**

b. a 10 × 12 patio **plane**

c. the location where the corner of a driveway meets the road **point**

3 Draw and label a figure for each relationship.

a. plane $\mathcal{R}$ containing lines $\overleftrightarrow{AB}$ and $\overleftrightarrow{DE}$ intersect at point *P*. Add point *C* on plane $\mathcal{R}$ so that it is not collinear with $\overleftrightarrow{AB}$ or $\overleftrightarrow{DE}$.

Sample answer:

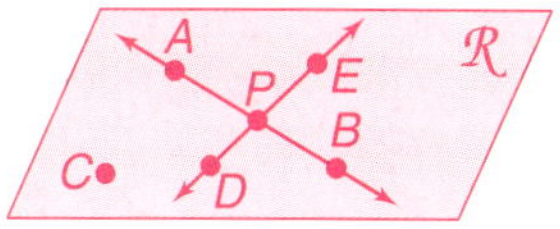

b. $\overleftrightarrow{QR}$ on a coordinate plane contains $Q(-2, 4)$ and $R(4, -4)$. Add point *T* so that *T* is collinear with these points.

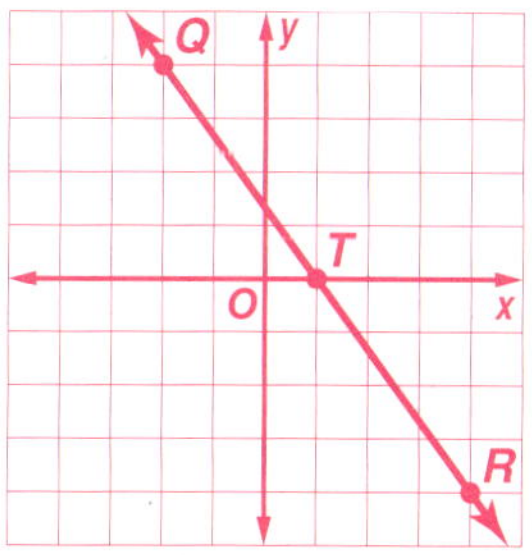

DAILY INTERVENTION

Differentiated Instruction

Naturalist Explain how points, lines, and planes exist in nature. For example, planes can model leaves, lily pads, and the surface of a pond; lines can model spider webs, sunbeams, tree trunks, and the edge of a riverbed.

POINTS, LINES, AND PLANES IN SPACE

In-Class Example Power Point®

Teaching Tip Explain to students that creating and interpreting three-dimensional drawings is vital to such fields as architecture, engineering, and computer gaming.

4

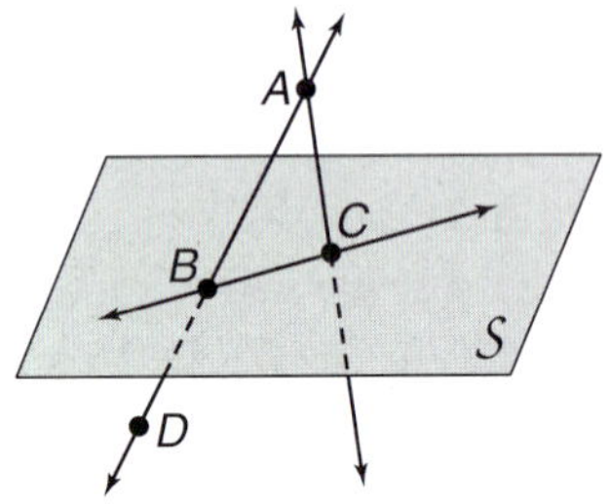

a. How many planes appear in this figure? **two**

b. Name three points that are collinear. ***A, B,* and *D***

c. Are points A, B, C, and D coplanar? Explain. **Points *A, B, C,* and *D* all lie in plane *ABC*, so they are coplanar.**

d. At what point do $\overleftrightarrow{DB}$ and $\overleftrightarrow{CA}$ intersect? ***A***

Tips for New Teachers

Intervention Remind students that lines and planes extend infinitely and discuss examples that demonstrate this concept. For example, draw two lines so that they would intersect if extended, and ask students if the lines intersect.

New teachers, or teachers new to teaching mathematics, may especially appreciate the Tips for New Teachers.

---

b. $\overleftrightarrow{TU}$ **lies in plane** $\mathcal{Q}$ **and contains point** R.

Draw a surface to represent plane $\mathcal{Q}$ and label it.
Draw a line anywhere on the plane.
Draw dots on the line for points T and U.
Since $\overleftrightarrow{TU}$ contains R, point R lies on $\overleftrightarrow{TU}$.
Draw a dot on $\overleftrightarrow{TU}$ and label it R.

The locations of points T, R, and U are totally arbitrary.

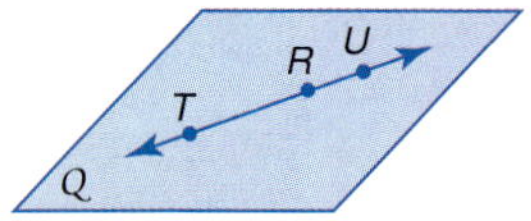

POINTS, LINES, AND PLANES IN SPACE **Space** is a boundless, three-dimensional set of all points. Space can contain lines and planes.

Study Tip

Three-Dimensional Drawings
Because it is impossible to show space or an entire plane in a figure, edged shapes with different shades of color are used to represent planes. If the lines are hidden from view, the lines or segments are shown as dashed lines or segments.

Geometry Activities use manipulatives and models to help students learn key concepts. There are teacher notes for every Geometry Activity in the Student Edition.

Example 4 *Interpret Drawings*

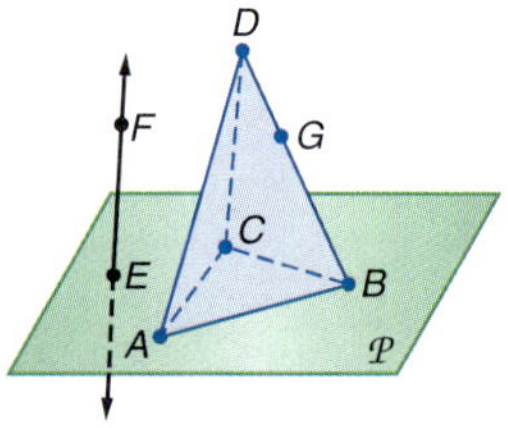

a. How many planes appear in this figure?

There are four planes: plane $\mathcal{P}$, plane ADB, plane BCD, plane ACD

b. Name three points that are collinear.

Points D, B, and G are collinear.

c. Are points G, A, B, and E coplanar? Explain.

Points A, B, and E lie in plane $\mathcal{P}$, but point G does not lie in plane $\mathcal{P}$. Thus, they are not coplanar. Points A, G, and B lie in a plane, but point E does not lie in plane AGB.

d. At what point do $\overleftrightarrow{EF}$ and $\overleftrightarrow{AB}$ intersect?

$\overleftrightarrow{EF}$ and $\overleftrightarrow{AB}$ do not intersect. $\overleftrightarrow{AB}$ lies in plane $\mathcal{P}$, but only point E of $\overleftrightarrow{EP}$ lies in $\mathcal{P}$.

Sometimes it is difficult to identify collinear or coplanar points in space unless you understand what a drawing represents. In geometry, a model is often helpful in understanding what a drawing is portraying.

Geometry Activity

Modeling Intersecting Planes

- Label one index card as $\mathcal{Q}$ and another as $\mathcal{R}$.
- Hold the two index cards together and cut a slit halfway through both cards.

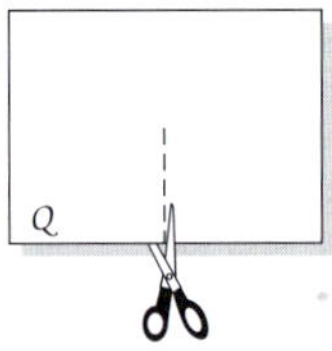

- Hold the cards so that the slits meet and insert one card into the slit of the other. Use tape to hold the cards together.

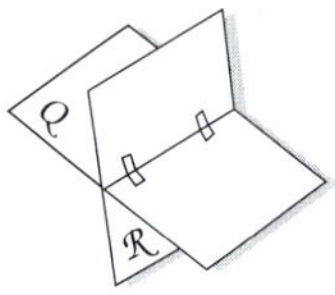

- Where the two cards meet models a line. Draw the line and label two points, C and D, on the line.

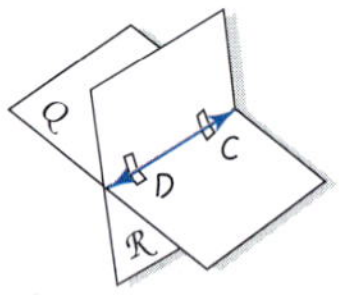

3. On $\overleftrightarrow{CD}$; see students' work.

Analyze **4. See students' work.**

1. Draw a point F on your model so that it lies in $\mathcal{Q}$ but not in $\mathcal{R}$. Can F lie on $\overleftrightarrow{DC}$? **no**
2. Draw point G so that it lies in $\mathcal{R}$, but not in $\mathcal{Q}$. Can G lie on $\overleftrightarrow{DC}$? **no**
3. If point H lies in both $\mathcal{Q}$ and $\mathcal{R}$, where would it lie? Draw point H on your model.
4. Draw a sketch of your model on paper. Label all points, lines, and planes appropriately.

Geometry Activity

Materials: 2 index cards, straightedge, scissors, tape

- Suggest that students draw and label any line AB in plane $\mathcal{Q}$ so that neither A nor B lies in $\overleftrightarrow{DC}$ before assembling the model.
- Have students determine whether $\overleftrightarrow{AB}$ intersects $\overleftrightarrow{CD}$.
- Students should visualize that A, B, C, and D are contained in plane $\mathcal{Q}$, but only C and D are contained in plane $\mathcal{R}$.

Check for Understanding

Concept Check

2. See students' work; sample answer: Two lines intersect at a point.

3. Micha; the points must be noncollinear to determine a plane.

4. Sample answers: line p; plane $\mathcal{R}$

1. **Name** three undefined terms from this lesson. point, line, plane
2. **OPEN ENDED** Fold a sheet of paper. Open the paper and fold it again in a different way. Open the paper and label the geometric figures you observe. Describe the figures.
3. **FIND THE ERROR** Raymond and Micha were looking for patterns to determine how many ways there are to name a plane given a certain number of points.

Raymond	Micha
If there are 4 points, then there are 4 · 3 · 2 ways to name the plane.	If there are 5 noncollinear points, then there are 5 · 4 · 3 ways to name the plane.

Who is correct? Explain your reasoning.

Guided Practice

GUIDED PRACTICE KEY	
Exercises	Examples
4	1
5, 6	3
7–9	4
10–12	2

4. Use the figure at the right to name a line containing point B and a plane containing points D and C.

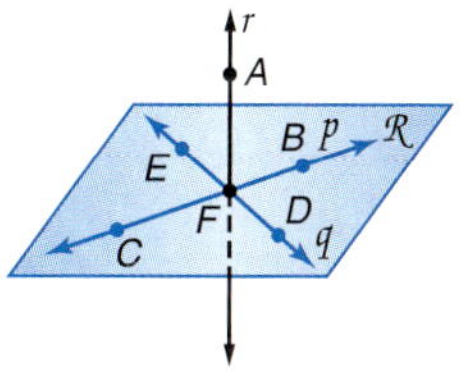

Draw and label a figure for each relationship.

5–6. See p. 59A.

5. A line in a coordinate plane contains $X(3, -1)$, $Y(-3, -4)$, and $Z(-1, -3)$ and a point W that does not lie on $\overleftrightarrow{XY}$.
6. Plane Q contains lines r and s that intersect in P.

Refer to the figure.

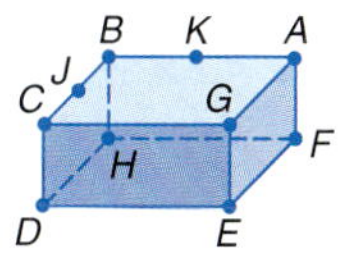

7. How many planes are shown in the figure? 6
8. Name three points that are collinear. A, K, B or B, J, C
9. Are points A, C, D, and J coplanar? Explain.

9. No; A, C, and J lie in plane ABC, but D does not.

Application

VISUALIZATION **Name the geometric term modeled by each object.**

10. 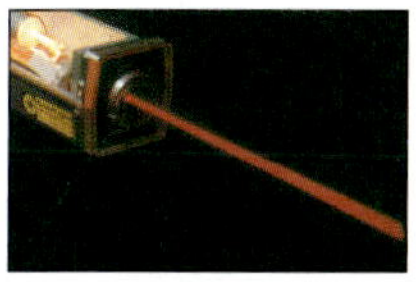 line
11. a pixel on a computer screen point
12. a ceiling plane

★ indicates increased difficulty

Practice and Apply

Homework Help	
For Exercises	See Examples
13–18	1
21–28	3
30–37	4
38–46	2

Extra Practice
See page 754.

Refer to the figure.

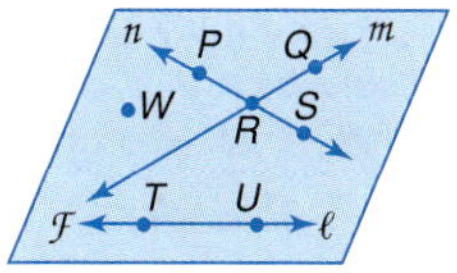

13. Name a line that contains point P. n
14. Name the plane containing lines n and m. $\mathcal{F}$
15. Name the intersection of lines n and m. R
16. Name a point not contained in lines ℓ, m, or n. W
17. What is another name for line n? Sample answer: $\overleftrightarrow{PR}$
18. Does line ℓ intersect line m or line n? Explain.

18. Yes, it intersects both m and n when all three lines are extended.

MAPS **For Exercises 19 and 20, refer to the map, and use the following information.** A map represents a plane. Points on this plane are named using a letter/number combination. 19. (D, 9)

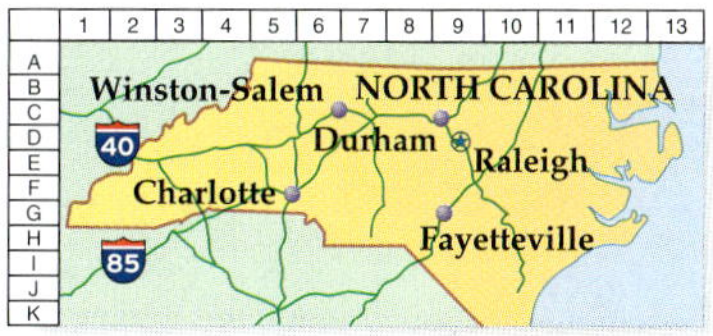

19. Name the point where Raleigh is located.
20. What city is located at (F, 5)? Charlotte

3 Practice/Apply

Study Notebook

Have students—

- add the definitions/examples of the vocabulary terms to their Vocabulary Builder worksheets for Chapter 1.
- write the steps of the four-step problem-solving plan in their study notebooks. In addition to these steps, students should include examples of what each step means.
- include any other item(s) that they find helpful in mastering the skills in this lesson.

DAILY INTERVENTION

FIND THE ERROR Note that the only difference in the two explanations is the word *non-collinear*. Stress the importance of proper terminology when trying to communicate geometric ideas.

About the Exercises...

Organization by Objective

- **Name Points, Lines, and Planes:** 13–18, 21–28, 38–46
- **Points, Lines, and Planes in Space:** 30–37

Odd/Even Assignments

Exercises 13–18, 21–28, and 30–46 are structured so that students practice the same concepts whether they are assigned odd or even problems.

Alert! Exercise 50 requires the Internet or other research materials.

Assignment Guide

Basic: 13–25 odd, 29–35 odd, 39–47 odd, 51, 53–57, 60–65

Average: 13–53 odd, 54–57, 60–65 (optional: 58, 59)

Advanced: 14–54 even, 55–59 (optional: 60–65)

Interactive Chalkboard
PowerPoint® Presentations

This CD-ROM is a customizable Microsoft® PowerPoint® presentation that includes:

- Step-by-step, dynamic solutions of each In-Class Example from the Teacher Wraparound Edition
- Additional, Try These exercises for each example
- The 5-Minute Check Transparencies
- Hot links to Glencoe Online Study Tools

Study Guide and Intervention, p. 1 (shown) and p. 2

Name Points, Lines, and Planes In geometry, a **point** is a location, a **line** contains points, and a **plane** is a flat surface that contains points and lines. If points are on the same line, they are **collinear**. If points on are the same plane, they are **coplanar**.

Example **Use the figure to name each of the following.**

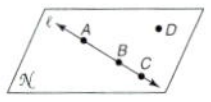

a. a line containing point *A*

The line can be named as ℓ. Also, any two of the three points on the line can be used to name it.

$\overleftrightarrow{AB}$, $\overleftrightarrow{AC}$, or $\overleftrightarrow{BC}$

b. a plane containing point *D*

The plane can be named as plane 𝒩 or can be named using three noncollinear points in the plane, such as plane *ABD*, plane *ACD*, and so on.

Exercises

Refer to the figure.

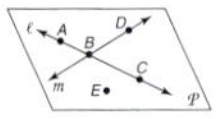

1. Name a line that contains point *A*. $\overleftrightarrow{AB}$, $\overleftrightarrow{AC}$, $\overleftrightarrow{BC}$, or ℓ
2. What is another name for line *m*? $\overleftrightarrow{BD}$
3. Name a point not on $\overleftrightarrow{AC}$. D or E
4. Name the intersection of $\overleftrightarrow{AC}$ and $\overleftrightarrow{DB}$. B
5. Name a point not on line ℓ or line *m*. E

Draw and label a plane *Q* for each relationship.

6. $\overleftrightarrow{AB}$ is in plane *Q*.
7. $\overleftrightarrow{ST}$ intersects $\overleftrightarrow{AB}$ at *P*.
8. Point *X* is collinear with points *A* and *P*.
9. Point *Y* is not collinear with points *T* and *P*.
10. Line ℓ contains points *X* and *Y*.

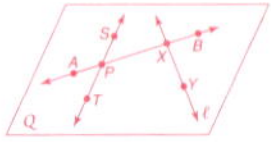

Answers for Exercises 6–10

Skills Practice, p. 3 and Practice, p. 4 (shown)

Refer to the figure.

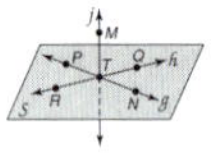

1. Name a line that contains points *T* and *P*.
   g, $\overleftrightarrow{TP}$, $\overleftrightarrow{TN}$, $\overleftrightarrow{NP}$
2. Name a line that intersects the plane containing points *Q*, *N*, and *P*.
   j or $\overleftrightarrow{MT}$
3. Name the plane that contains $\overleftrightarrow{TN}$ and $\overleftrightarrow{QR}$. Sample answer: plane 𝒮

Draw and label a figure for each relationship. Sample answers are given.

4. $\overleftrightarrow{AK}$ and $\overleftrightarrow{CG}$ intersect at point *M* in plane 𝒯.

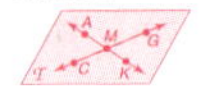

5. A line contains *L*(−4, −4) and *M*(2, 3). Line *q* is in the same coordinate plane but does not intersect $\overleftrightarrow{LM}$. Line *q* contains point *N*.

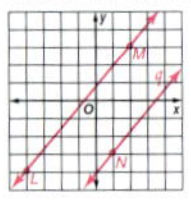

Refer to the figure.

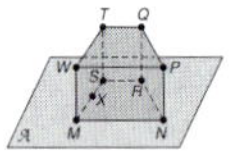

6. How many planes are shown in the figure? 6
7. Name three collinear points. S, X, M
8. Are points *N*, *R*, *S*, and *W* coplanar? Explain.
   No; sample answer: points N, R, and S lie in plane 𝒜, but point W does not.

VISUALIZATION Name the geometric term(s) modeled by each object.

9. 

plane and line

10. 

point

11. 

lines

12. a car antenna
    line and point
13. a library card
    plane

Reading to Learn Mathematics, p. 5

ELL

Pre-Activity **Why do chairs sometimes wobble?**

Read the introduction to Lesson 1-1 at the top of page 6 in your textbook.

- Find three pencils of different lengths and hold them upright on your desk so that the three pencil points do not lie along a single line. Can you place a flat sheet of paper or cardboard so that it touches all three pencil points? yes
- How many ways can you do this if you keep the pencil points in the same position? one
- How will your answer change if there are four pencil points? Sample answer: It may not be possible to place the paper to touch all four points.

Reading the Lesson

1. Complete each sentence.
   a. Points that lie on the same lie are called __collinear__ points.
   b. Points that do not lie in the same plane are called __noncoplanar__ points.
   c. There is exactly one __line__ through any two points.
   d. There is exactly one __plane__ through any three noncollinear points.
2. Refer to the figure at the right. Indicate whether each statement is true or false.

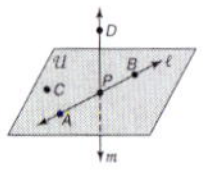

   a. Points *A*, *B*, and *C* are collinear. false
   b. The intersection of plane *ABC* and line *m* is point *P*. true
   c. Line ℓ and line *m* do not intersect. false
   d. Points *A*, *P*, and *B* can be used to name plane 𝒰. false
   e. Line ℓ lies in plane *ACB*. true
3. Complete the figure at the right to show the following relationship: Lines ℓ, *m*, and *n* are coplanar and lie in plane *Q*. Lines ℓ and *m* intersect at point *P*. Line *n* intersects line *m* at *R*, but does not intersect line ℓ.

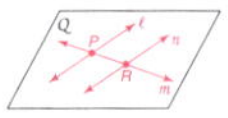

Helping You Remember

4. Recall or look in a dictionary to find the meaning of the prefix *co-*. What does this prefix mean? How can it help you remember the meaning of *collinear*?
   Sample answer: The prefix *co-* means *together*. The word *collinear* contains the word *line*, so *collinear* means *together on a line*.

Draw and label a figure for each relationship. 21–28. See p. 59A.

21. Line *AB* intersects plane *Q* at *W*.
22. Point *T* lies on $\overleftrightarrow{WR}$.
23. Points *Z*(4, 2), *R*(−4, 2), and *S* are collinear, but points *Q*, *Z*, *R*, and *S* are not.
24. The coordinates for points *C* and *R* are (−1, 4) and (6, 4), respectively. $\overleftrightarrow{RS}$ and $\overleftrightarrow{CD}$ intersect at *P*(3, 2).
25. Lines *a*, *b*, and *c* are coplanar, but do not intersect.
26. Lines *a*, *b*, and *c* are coplanar and meet at point *F*.
★ 27. Point *C* and line *r* lie in ℳ. Line *r* intersects line *s* at *D*. Point *C*, line *r*, and line *s* are not coplanar.
★ 28. Planes 𝒜 and ℬ intersect in line *s*. Plane 𝒞 intersects 𝒜 and ℬ, but does not contain *s*.

29. **ALGEBRA** Name at least four ordered pairs for which the sum of coordinates is −2. Graph them and describe the graph. See p. 59A for graph; points that seem collinear; Sample answer: (0, −2), (1, −3), (2, −4), (3, −5).

Refer to the figure.

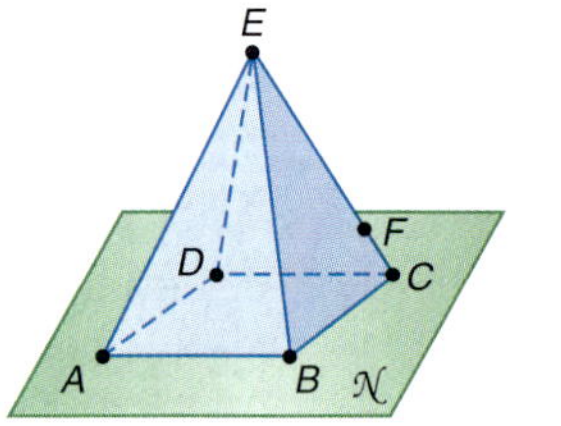

30. How many planes are shown in the figure? 5
31. How many planes contain points *B*, *C*, and *E*? 1
32. Name three collinear points. E, F, C
33. Where could you add point *G* on plane 𝒩 so that *A*, *B*, and *G* would be collinear?
34. Name a point that is not coplanar with *A*, *B*, and *C*. E, F
35. Name four points that are coplanar.
★ 36. Name an example that shows that three points are always coplanar, but four points are not always coplanar.
★ 37. Name the intersection of plane 𝒩 and the plane that contains points *A*, *E*, and *C*. $\overleftrightarrow{AC}$

33. anywhere on $\overleftrightarrow{AB}$
35. A, B, C, D or E, F, C, B
36. Sample answer: points E, A, and B are coplanar, but points E, A, B, and C are not.
42. two planes intersecting in a line
44. intersecting lines

VISUALIZATION **Name the geometric term(s) modeled by each object.**

38. 
point

39. 
lines

40. 
plane

41. a table cloth plane
42. a partially-opened newspaper
43. a star in the sky point
44. woven threads in a piece of cloth
45. a knot in a string point
46. satellite dish signal line

ONE-POINT PERSPECTIVE **One-point perspective drawings use lines to convey depth in a picture. Lines representing horizontal lines in the real object can be extended to meet at a single point called the *vanishing point*.**

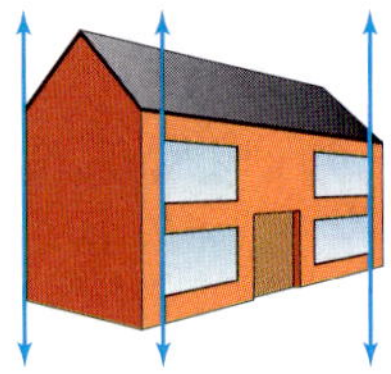

47. Trace the figure at the right. Draw all of the vertical lines. Several are already drawn for you.
48. Draw and extend the horizontal lines to locate the vanishing point and label it. 47–48. See p. 59A.
49. Draw a one-point perspective of your classroom or a room in your house. See students' work.

Career Choices

Engineering Technician

Engineering technicians or drafters use perspective to create drawings used in construction, and manufacturing. Technicians must have knowledge of math, science, and engineering.

Online Research
For information about a career as an engineering technician, visit: www.geometryonline.com/careers

Enrichment, p. 6

Points and Lines on a Matrix

A **matrix** is a rectangular array of rows and columns. Points and lines on a matrix are not defined in the same way as in Euclidean geometry. A **point** on a matrix is a dot, which can be small or large. A **line** on a matrix is a path of dots that "line up." Between two points on a line there may or may not be other points. Three examples of lines are shown at the upper right. The broad line can be thought of as a single line or as two narrow lines side by side.

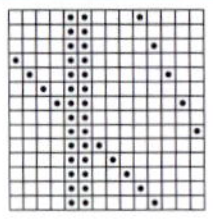

Dot-matrix printers for computers used dots to form characters. The dots are often called **pixels**. The matrix at the right shows how a dot-matrix printer might print the letter P.

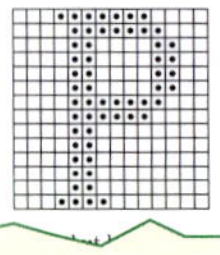

Answers may vary. Sample answers are shown.

Draw points on each matrix to create the given figures.

There is a Study Guide and Intervention, Skills Practice, Practice, Reading to Learn Mathematics, and Enrichment Master for every lesson in the Student Edition. These masters can be found in the Chapter Resource Masters.

50. **RESEARCH** Use the Internet or other research resources to investigate one-point perspective drawings in which the vanishing point is in the center of the picture. How do they differ from the drawing for Exercises 47–49? Sample answer: The image is rotated so that the front or back plane is not angled.

TWO-POINT PERSPECTIVE **Two-point perspective drawings also use lines to convey depth, but two sets of lines can be drawn to meet at two vanishing points.**

51. Trace the outline of the house. Draw all of the vertical lines. Sample vertical lines are shown.

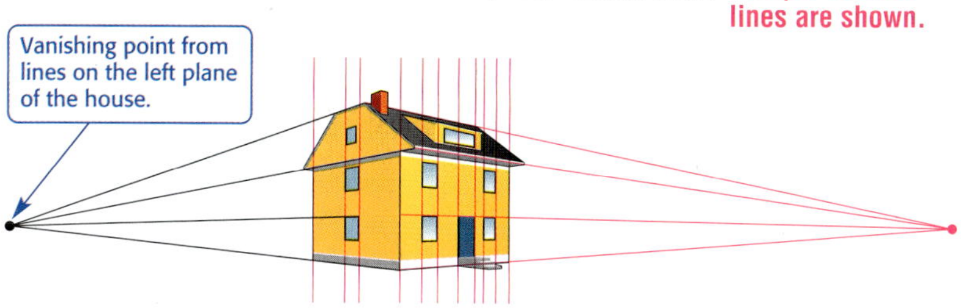

52. See picture.

52. Draw and extend the lines on your sketch representing horizontal lines in the real house to identify the vanishing point on the right plane in this figure.

53. Which types of lines seem unaffected by any type of perspective drawing? vertical

Standardized Test Practice exercises were created to closely parallel those on actual state proficiency tests and college entrance exams.

54. **CRITICAL THINKING** Describe a real-life example of three lines in space that do not intersect each other and no two lines lie in the same plane. Sample answer: the paths flown by airplanes flying in formation

55. **WRITING IN MATH** Answer the question that was posed at the beginning of the lesson. See margin.

Why do chairs sometimes wobble?

Include the following in your answer:

- an explanation of how the chair legs relate to points in a plane, and
- how many legs would create a chair that does not wobble.

FCAT Practice
Standardized Test Practice
Ⓐ Ⓑ Ⓒ Ⓓ

56. Four lines are coplanar. What is the greatest number of intersection points that can exist? C

Ⓐ 4 Ⓑ 5 Ⓒ 6 Ⓓ 7

57. **ALGEBRA** If $2 + x = 2 - x$, then $x = ?$ B

Ⓐ -1 Ⓑ 0 Ⓒ 1 Ⓓ 2

Extending the Lesson

58–59. See margin for graphs.

Another way to describe a group of points is called a locus. A locus is a set of points that satisfy a particular condition.

58. Find five points that satisfy the equation $4 - x = y$. Graph them on a coordinate plane and describe the geometric figure they suggest. a line

59. Find ten points that satisfy the inequality $y > -2x + 1$. Graph them on a coordinate plane and describe the geometric figure they suggest. part of the coordinate plane above the line $y = -2x + 1$

Getting Ready for the Next Lesson

BASIC SKILL **Replace each ● with >, <, or = to make a true statement.**

60. $\frac{1}{2}$ in. ● $\frac{3}{8}$ in. >

61. $\frac{4}{16}$ in. ● $\frac{1}{4}$ in. =

62. $\frac{4}{5}$ in. ● $\frac{6}{10}$ in. >

63. 10 mm ● 1 cm =

64. 2.5 cm ● 28 mm <

65. 0.025 cm ● 25 mm <

www.geometryonline.com/self_check_quiz/fcat

4 Assess

Open-Ended Assessment

Modeling Discuss how points, lines, and planes are modeled by the objects students see and use every day. Examples could be pinpoints, pencils, and bulletin boards. Have students come up with examples to demonstrate for the class.

Getting Ready for Lesson 1-2

Basic Skill Students will learn about linear measure and precision in Lesson 1-2. They will use fractions, decimals, and units of measure to accurately evaluate the lengths of objects. Use Exercises 60–65 to determine your students' familiarity with comparing measurements.

Answers

55. Sample answer: Chairs wobble because all four legs do not touch the floor at the same time. Answers should include the following.

- The ends of the legs represent points. If all points lie in the same plane, the chair will not wobble.
- Because it only takes three points to determine a plane, a chair with three legs will never wobble.

58.

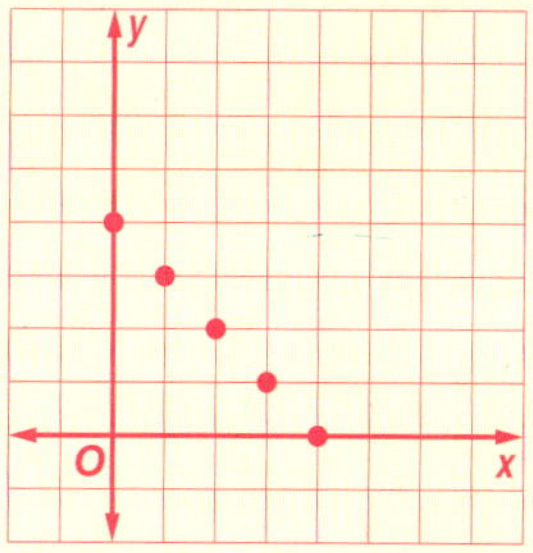

59.

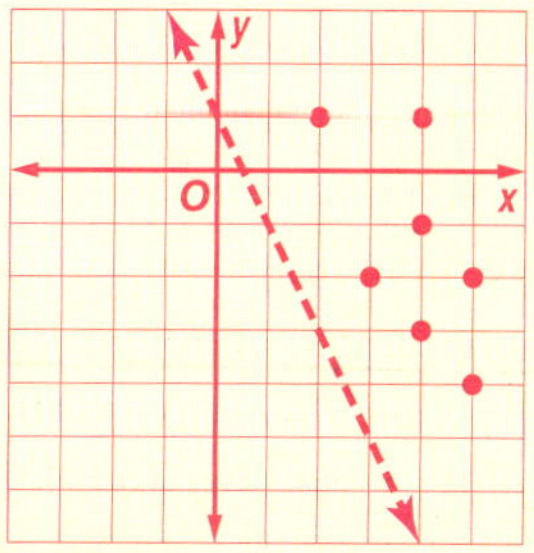

Reading Mathematics

Getting Started

Have students practice their speaking and communicating skills by reading the descriptions of the examples aloud. Then pair students and allow them to discuss how they would describe the figures in Exercises 1–3.

Teach

- Remind students that a single uppercase letter can name a plane or a point, and a single lowercase letter can name a line. Also, two uppercase letters name lines, and three uppercase letters name planes. Explain that learning these rules for proper naming is a fundamental step in communicating geometric ideas.
- Encourage students to use geometric verbs, such as *intersects* and *contains* freely and to try to avoid using other words to describe the figures.
- For extra practice, students can make a game of describing other figures for their partners to draw.

Assess

Study Notebook

Ask students to summarize what they have learned about describing geometric figures effectively by using proper naming techniques and geometric terms.

ELL English Language Learners may benefit from writing key concepts from this activity in their Study Notebooks in their native language and then in English.

Reading Mathematics

Describing What You See

Figures play an important role in understanding geometric concepts. It is helpful to know what words and phrases can be used to describe figures. Likewise, it is important to know how to read a geometric description and be able to draw the figure it describes.

The figures and descriptions below help you visualize and write about points, lines, and planes.

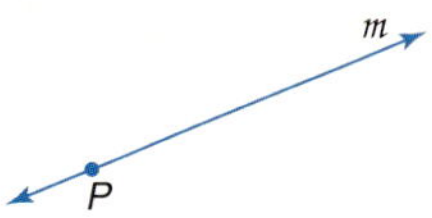

Point P is on m.
Line m contains P.
Line m passes through P.

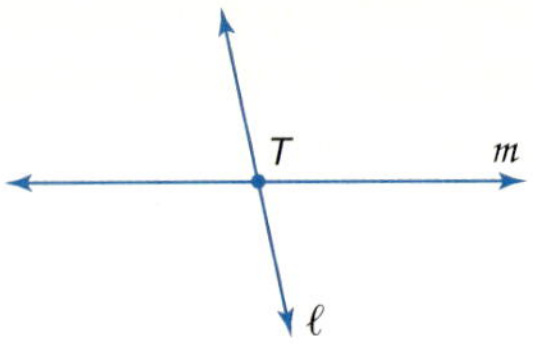

Lines ℓ and m intersect in T.
Point T is the intersection of ℓ and m.
Point T is on m. Point T is on ℓ.

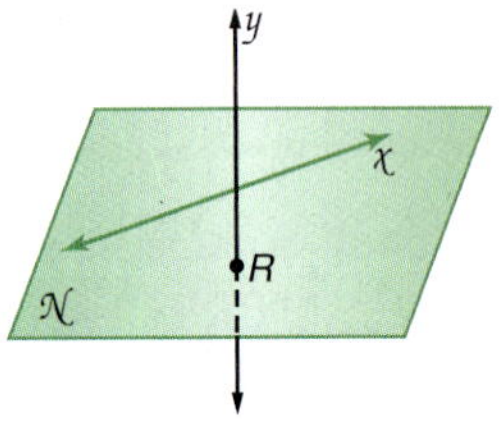

Line x and point R are in $\mathcal{N}$.
Point R lies in $\mathcal{N}$.
Plane $\mathcal{N}$ contains R and x.
Line y intersects $\mathcal{N}$ at R.
Point R is the intersection of y with $\mathcal{N}$.
Lines y and x do not intersect.

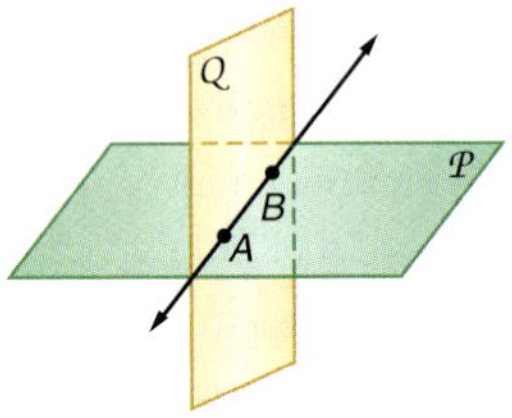

$\overleftrightarrow{AB}$ is in $\mathcal{P}$ and $\mathcal{Q}$.
Points A and B lie in both $\mathcal{P}$ and $\mathcal{Q}$.
Planes $\mathcal{P}$ and $\mathcal{Q}$ both contain $\overleftrightarrow{AB}$.
Planes $\mathcal{P}$ and $\mathcal{Q}$ intersect in $\overleftrightarrow{AB}$.
$\overleftrightarrow{AB}$ is the intersection of $\mathcal{P}$ and $\mathcal{Q}$.

Reading Mathematics features help students learn and use the language of mathematics.

Reading to Learn

Write a description for each figure. **1–4. See margin.**

1.

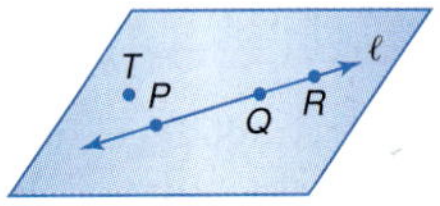

2.

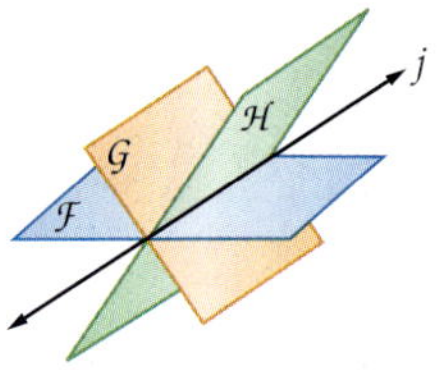

3.

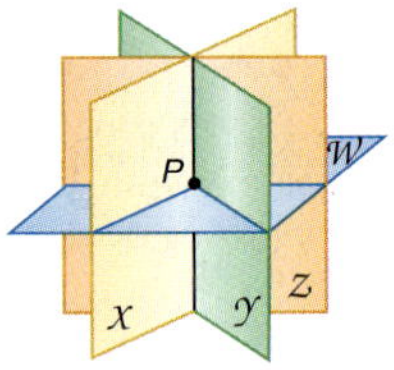

4. Draw and label a figure for the statement *Planes $\mathcal{A}$, $\mathcal{B}$, and $\mathcal{C}$ do not intersect.*

Answers

1. Points P, Q, and R lie on ℓ. Point T is not collinear with P, Q, and R.
2. Planes $\mathcal{F}$, $\mathcal{G}$, and $\mathcal{H}$ intersect in Line j.
3. The intersection of planes $\mathcal{W}$, $\mathcal{X}$, $\mathcal{Y}$, and $\mathcal{Z}$ is point P.
4. 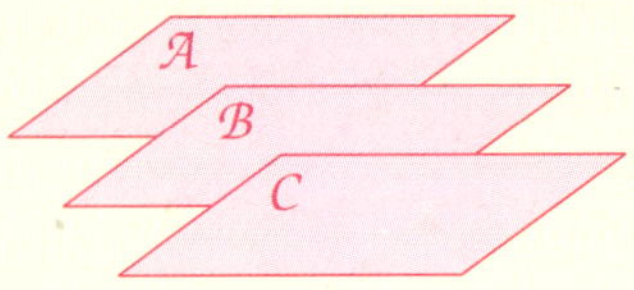

1-2 Linear Measure and Precision

Sunshine State Standards
MA.B.3.4.1

Vocabulary
- line segment
- precision
- betweenness of points
- between
- congruent
- construction
- relative error

What You'll Learn
- Measure segments and determine accuracy of measurement.
- Compute with measures.

Why are units of measure important?

When you look at the sign, you probably assume that the unit of measure is miles. However, if you were in France, this would be 17 kilometers, which is a shorter distance than 17 miles. Units of measure give us points of reference when evaluating the sizes of objects.

MEASURE LINE SEGMENTS Unlike a line, a **line segment**, or *segment*, can be measured because it has two endpoints. A segment with endpoints A and B can be named as $\overline{AB}$ or $\overline{BA}$. The length or measure of $\overline{AB}$ is written as AB. The length of a segment is only as precise as the smallest unit on the measuring device.

Study Tip

Using a Ruler
On a ruler, the smallest unit is frequently labeled as cm, mm, or 16th of an inch. The zero point on a ruler may not be clearly marked. For some rulers, zero is the left edge of the ruler. On others, it may be a line farther in on the scale. If it is not clear where zero is, align the endpoint on 1 and subtract 1 from the measurement at the other endpoint.

Example 1 Length in Metric Units

Find the length of $\overline{CD}$ using each ruler.

a.

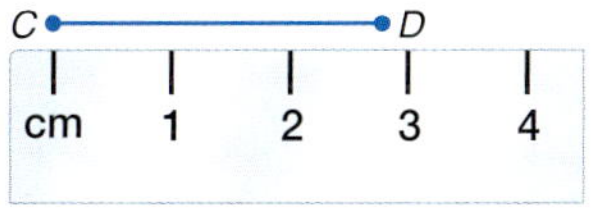

The ruler is marked in centimeters. Point D is closer to the 3-centimeter mark than to 2 centimeters. Thus, $\overline{CD}$ is about 3 centimeters long.

b.

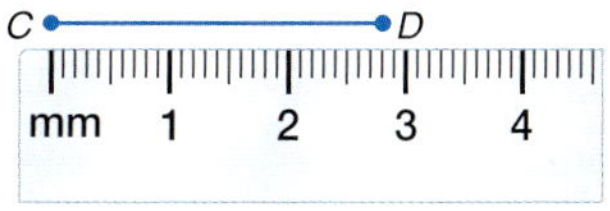

The long marks are centimeters, and the shorter marks are millimeters. There are 10 millimeters for each centimeter. Thus, $\overline{CD}$ is about 28 millimeters long.

Example 2 Length in Customary Units

Find the length of $\overline{AB}$ using each ruler.

a.

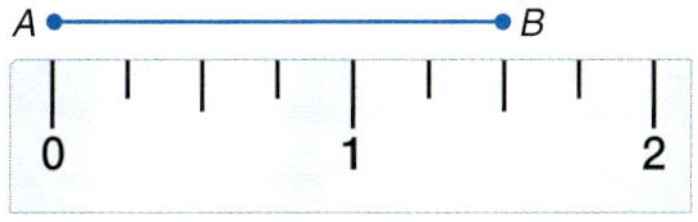

Each inch is divided into fourths. The long marks are half-inch increments. Point B is closer to the $1\frac{2}{4}$-inch mark. Thus, $\overline{AB}$ is about $1\frac{2}{4}$ or $1\frac{1}{2}$ inches long.

b.

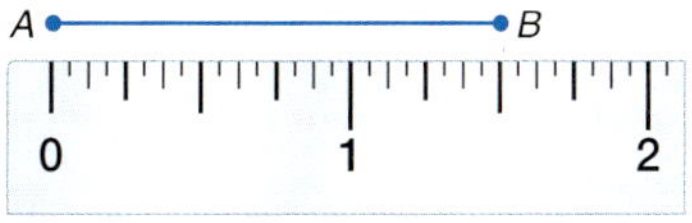

Each inch is divided into sixteenths. Point B is closer to the $1\frac{8}{16}$-inch mark. Thus, $\overline{AB}$ is about $1\frac{8}{16}$ or $1\frac{1}{2}$ inches long.

...amples illustrate all of the concepts taught in the lesson and closely mirror the exercises in the Guided Practice and Practice and Apply sections.

1-2 Lesson Notes

1 Focus

5-Minute Check Transparency 1-2 Use as a quiz or review of Lesson 1-1.

Mathematical Background notes are available for this lesson on p. 4C.

Why are units of measure important?

Ask students:
- What type of unit is displayed on the sign, metric or customary? **No units are displayed on the sign so it could be metric or customary.**
- Which is longer, 17 miles or 17 kilometers? **17 mi**

2 Teach

MEASURE LINE SEGMENTS

In-Class Example Power Point®

1 Use a metric ruler to draw each segment.

a. Draw $\overline{LM}$ that is 42 millimeters long. **See students' work.**

b. Draw $\overline{QR}$ that is 5 centimeters long. **See students' work.**

Resource Manager

Workbook and Reproducible Masters

Chapter 1 Resource Masters
- Study Guide and Intervention, pp. 7–8
- Skills Practice, p. 9
- Practice, p. 10
- Reading to Learn Mathematics, p. 11
- Enrichment, p. 12
- Assessment, p. 51

Prerequisite Skills Workbook, pp. 3–4, 15–18, 73–78, 83–84

Teaching Geometry With Manipulatives Masters, p. 17

Transparencies
5-Minute Check Transparency 1-2
Answer Key Transparencies

Technology
GeomPASS: Tutorial Plus, Lesson 2
Interactive Chalkboard

In-Class Examples

Power Point®

2 Use a customary ruler to draw each segment.

a. Draw $\overline{DE}$ that is 3 inches long. **See students' work.**

b. Draw $\overline{FG}$ that is $2\frac{3}{4}$ inches long. **See students' work.**

Teaching Tip You may want to invite a science teacher to talk with students about the roles of precision, accuracy, and significant digits in scientific measurements.

3 **PRECISION** Find the precision for each measurement. Explain its meaning.

a. $32\frac{3}{4}$ inches **$\frac{1}{8}$ in.; $32\frac{5}{8}$ in. to $32\frac{7}{8}$ in.**

b. 15 millimeters **0.5 mm; 14.5 mm to 15.5 mm**

CALCULATE MEASURES

In-Class Example

Power Point®

4 **a.** Find LM. **1.4 cm**

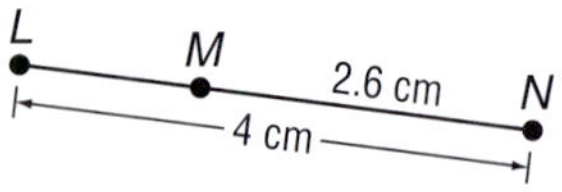

b. Find XZ. **$7\frac{1}{8}$ in.**

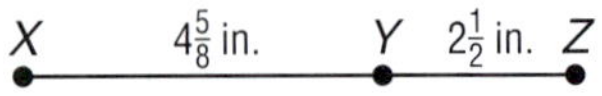

c. Find x and ST if T is between S and U, $ST = 7x$, $SU = 45$, and $TU = 5x - 3$.

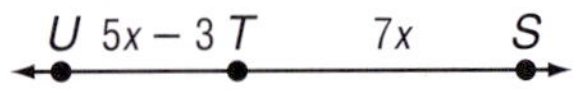

$x = 4$, $ST = 28$

In-Class Examples, which are included for every example in the Student Edition, exactly parallel the examples in the text. Teaching Tips about the Student Edition examples in the Student Edition are included where appropriate.

Study Tip

Units of Measure

A measurement of 38.0 centimeters on a ruler with millimeter marks means a measurement of 380 millimeters. So the actual measurement is between 379.5 millimeters and 380.5 millimeters, not 37.5 centimeters and 38.5 centimeters. For a measurement of $2\frac{1}{2}$ inches, you can only assume that the ruler has half-inch increments. A measurement of $2\frac{2}{4}$ inches indicates that the ruler has one-fourth inch increments.

Study Tip

Comparing Measures

Because measures are real numbers, you can compare measures. If X, Y, and Z are collinear in that order, then one of these statements is true. $XY = YZ$, $XY > YZ$, or $XY < YZ$.

The **precision** of any measurement depends on the smallest unit available on the measuring tool. The measurement should be precise to within 0.5 unit of measure. For example, in part **a** of Example 1, 3 centimeters means that the actual length is no less than 2.5 centimeters, but no more than 3.5 centimeters.

Measurements of 28 centimeters and 28.0 centimeters indicate different precision in measurement. A measurement of 28 centimeters means that the ruler is divided into centimeters. However, a measurement of 28.0 centimeters indicates that the ruler is divided into millimeters.

Example 3 Precision

Find the precision for each measurement. Explain its meaning.

a. 5 millimeters

The measurement is precise to within 0.5 millimeter. So, a measurement of 5 millimeters could be 4.5 to 5.5 millimeters.

b. $8\frac{1}{2}$ inches

The measuring tool is divided into $\frac{1}{2}$-inch increments. Thus, the measurement is precise to within $\frac{1}{2}\left(\frac{1}{2}\right)$ or $\frac{1}{4}$ inch. Therefore, the measurement could be between $8\frac{1}{4}$ inches and $8\frac{3}{4}$ inches.

CALCULATE MEASURES

Measures are real numbers, so all arithmetic operations can be used with them. You know that the whole usually equals the sum of its parts. That is also true of line segments in geometry.

Recall that for any two real numbers a and b, there is a real number n between a and b such that $a < n < b$. This relationship also applies to points on a line and is called **betweenness of points**. Point M is **between** points P and Q if and only if P, Q, and M are collinear and $PM + MQ = PQ$.

Example 4 Find Measurements

a. Find AC.

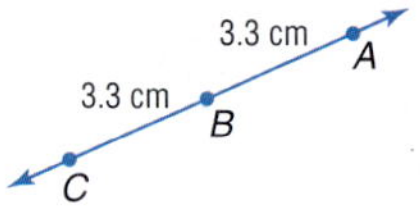

AC is the measure of $\overline{AC}$.

Point B is between A and C. AC can be found by adding AB and BC.

$AB + BC = AC$ Sum of parts = whole

$3.3 + 3.3 = AC$ Substitution

$6.6 = AC$ Add.

So, $\overline{AC}$ is 6.6 centimeters long.

b. Find DE.

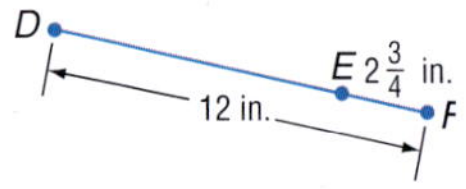

DE is the measure of $\overline{DE}$.

$DE + EF = DF$ Sum of parts = whole

$DE + 2\frac{3}{4} = 12$ Substitution

$DE + 2\frac{3}{4} - 2\frac{3}{4} = 12 - 2\frac{3}{4}$ Subtract $2\frac{3}{4}$ from each side.

$DE = 9\frac{1}{4}$ Simplify.

So, $\overline{DE}$ is $9\frac{1}{4}$ inches long.

DAILY INTERVENTION

Differentiated Instruction

Kinesthetic Students can physically participate in techniques of measuring, accuracy, and the betweenness of points by grouping in threes, standing in designated spots, and using a yardstick or meterstick to measure distances between them, add distances together, and find unknown distances. They can model examples in the book or create new scenarios. If the floors or walls are tiled, they can also measure distances with one tile representing one unit increment.

Study Tip

Information from Figures
When no unit of measure is given on a figure, you can safely assume that all of the segments have the same unit of measure.

c. **Find y and PQ if P is between Q and R, $PQ = 2y$, $QR = 3y + 1$, and $PR = 21$.**

Draw a figure to represent this information.

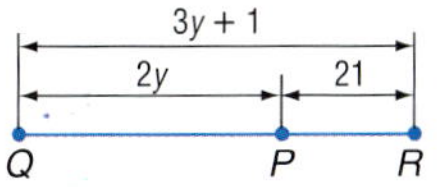

$QR = QP + PR$	
$3y + 1 = 2y + 21$	Substitute known values.
$3y + 1 - 1 = 2y + 21 - 1$	Subtract 1 from each side.
$3y = 2y + 20$	Simplify.
$3y - 2y = 2y + 20 - 2y$	Subtract $2y$ from each side.
$y = 20$	Simplify.
$PQ = 2y$	Given
$PQ = 2(20)$	$y = 20$
$PQ = 40$	Multiply.

Key Concept boxes highlight definitions, formulas, and other important ideas. Multiple representations—words, symbols, examples, models—reach students of all learning styles.

Look at the figure in part **a** of Example 4. Notice that $\overline{AB}$ and $\overline{BC}$ have the same measure. When segments have the same measure, they are said to be **congruent**.

Key Concept — *Congruent Segments*

Words Two segments having the same measure are congruent.

Symbol $\cong$ is read *is congruent to*. Red slashes on the figure also indicate that segments are congruent.

Model $\overline{XY} \cong \overline{PQ}$

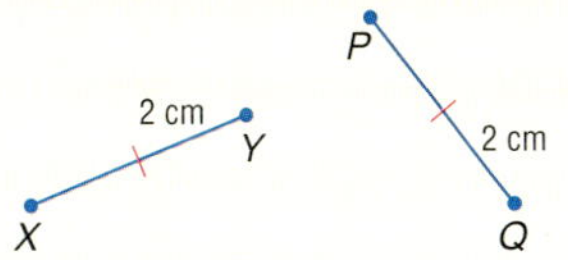

Study Tip

Constructions
A well-sharpened pencil will result in a more accurate construction.

Constructions are methods of creating geometric figures without the benefit of measuring tools. Generally, only a pencil, straightedge, and compass are used in constructions. You can construct a segment that is congruent to a given segment by using a compass and straightedge.

Construction

Copy a Segment

1. Draw a segment $\overline{XY}$. Elsewhere on your paper, draw a line and a point on the line. Label the point P.
2. Place the compass at point X and adjust the compass setting so that the pencil is at point Y.
3. Using that setting, place the compass point at P and draw an arc that intersects the line. Label the point of intersection Q. Because of identical compass settings, $\overline{PQ} \cong \overline{XY}$.

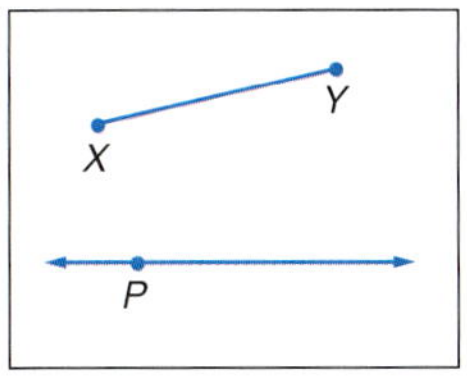

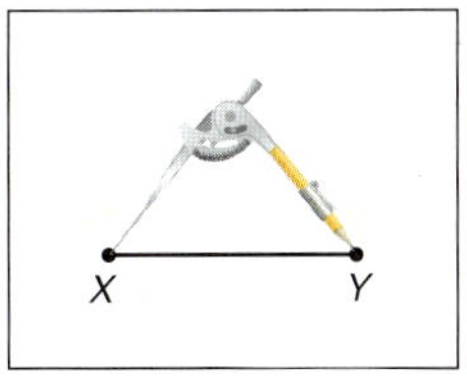

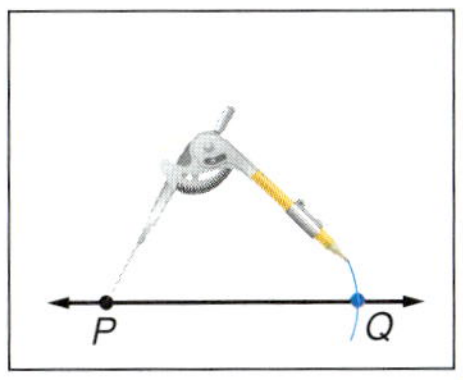

Teaching Tip Provide a few more examples of segments with one, two or three slashes indicating they are congruent. Involve students by having them come up to the front and using the $\cong$ symbol to categorize the congruent segments on the board. With this exercise, students will also review writing correct names and symbols.

DAILY INTERVENTION — **Unlocking Misconceptions**

Congruency Explain to students that segments and angles are congruent, but distances and measures are equal. For example, $\overline{AB} \cong \overline{CD}$ and $AB = CD$. The statements $AB \cong CD$ and $\overline{AB} = \overline{CD}$ are not correct.

Lesson 1-2 Linear Measure and Precision 15

In-Class Example

Power Point®

5 **FONTS** The Arial font is often used because it is easy to read. Study the word *time* shown in Arial type. Each letter can be broken into individual segments. The letter T would have two segments, a short horizontal segment, and a long vertical segment. Assume that all segments overlap when they meet. Which segments are congruent?

TIME

The five vertical segments in the letters T, I, M, and E are congruent. The four horizontal segments in T and E are congruent. The two diagonal segments in the letter M are congruent.

Answer

1. Align the 0 point on the ruler with the leftmost endpoint of the segment. Align the edge of the ruler along the segment. Note where the rightmost endpoint falls on the scale and read the closest eighth of an inch measurement.

Example 5 Congruent Segments

TIME MANAGEMENT **In the graph at the right, suppose a segment was drawn along the top of each bar. Which categories would have segments that are congruent? Explain.**

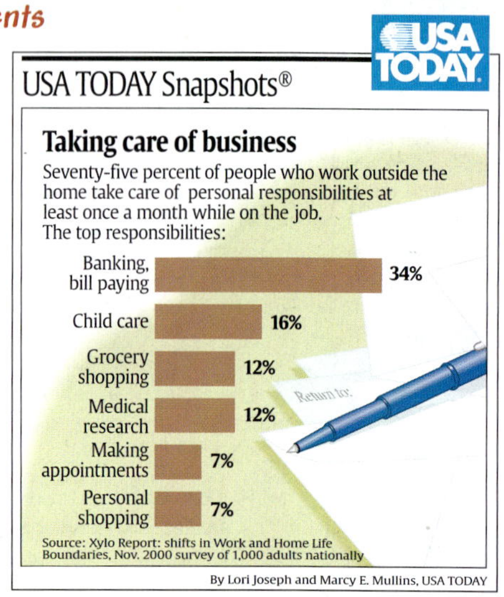

The segments on the bars for grocery shopping and medical research would be congruent because they both have the same length, representing 12%.

The segments on bars for making appointments and personal shopping would be congruent because they have the same length, representing 7%.

USA TODAY Education

Log on for:
- Updated data
- More activities on comparison using percent

www.geometryonline.com/usa_today

Glencoe's exclusive partnership with USA TODAY provides actual USA TODAY Snapshots® USA TODAY that illustrate mathematical concepts.

Check for Understanding

Concept Check

1. **Describe** how to measure a segment with a ruler that is divided into eighths of an inch. See margin.
2. **OPEN ENDED** Name or draw some geometric figures that have congruent segments. Sample answers: rectangle, square, equilateral triangle

Guided Practice

GUIDED PRACTICE KEY	
Exercises	**Examples**
3, 4	1, 2
5, 6	3
7–10	4
11	5

5. 0.5 m; 14 m could be 13.5 to 14.5 m

6. $\frac{1}{8}$ in.; $3\frac{1}{4}$ in. could be $3\frac{1}{8}$ to $3\frac{3}{8}$ in.

Find the length of each line segment or object.

3. 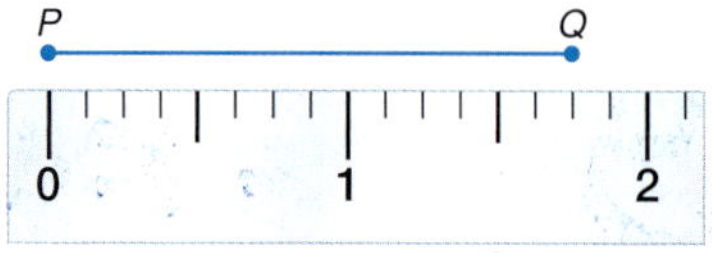

$1\frac{3}{4}$ in.

4. 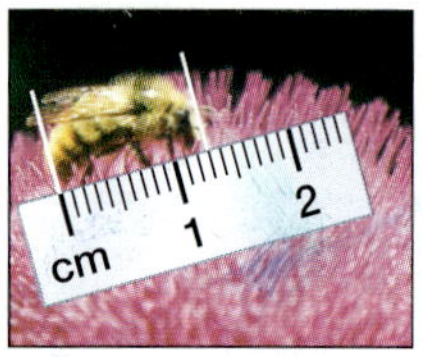

1.3 cm

5. Find the precision for a measurement of 14 meters. Explain its meaning.
6. Find the precision for a measurement of $3\frac{1}{4}$ inches. Explain its meaning.

Find the measurement of each segment. Assume that each figure is not drawn to scale.

7. $\overline{EG}$ 3.7 cm

8. $\overline{XY}$ $1\frac{3}{8}$ in.
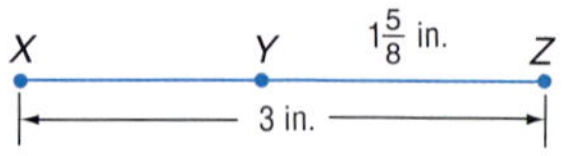

Find the value of the variable and *LM* if *L* is between *N* and *M*.

9. $NL = 5x$, $LM = 3x$, and $NL = 15$ $x = 3$; $LM = 9$
10. $NL = 6x - 5$, $LM = 2x + 3$, and $NM = 30$ $x = 4$; $LM = 11$

Application

11. **KITES** Kite making has become an art form using numerous shapes and designs for flight. The figure at the right is known as a *diamond kite*. The measures are in inches. Name all of the congruent segments in the figure. $\overline{BC} \cong \overline{CD}$, $\overline{BE} \cong \overline{ED}$, $\overline{BA} \cong \overline{DA}$

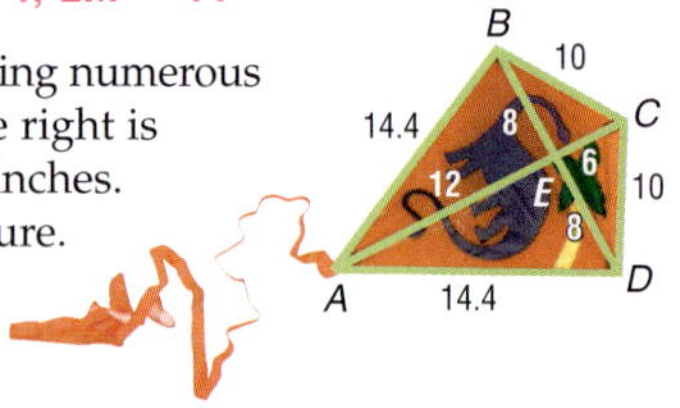

16 Chapter 1 Points, Lines, Planes, and Angles

★ indicates increased difficulty

Practice and Apply

Homework Help

For Exercises	See Examples
12–15	1, 2
16–21	3
22–33	4
34–39	5

Extra Practice
See page 754.

Find the length of each line segment or object. **13. 4.5 cm or 45 mm**

12. **$1\frac{5}{16}$ in.**

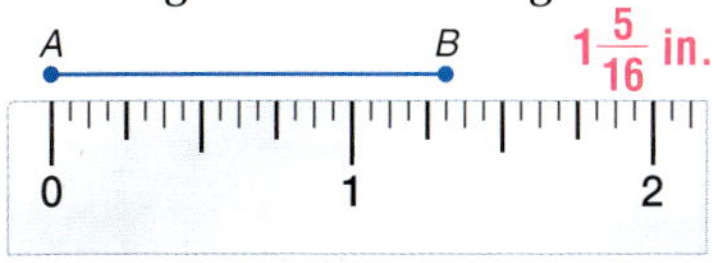

13.

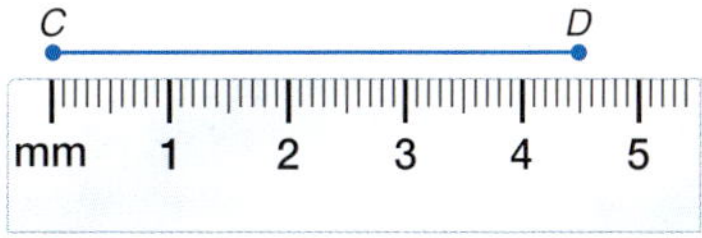

14.

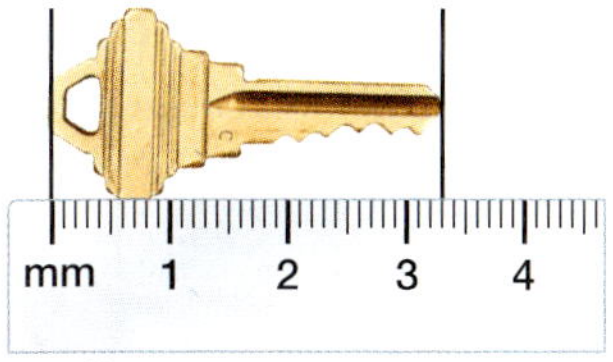

3.3 cm or 33 mm

15.

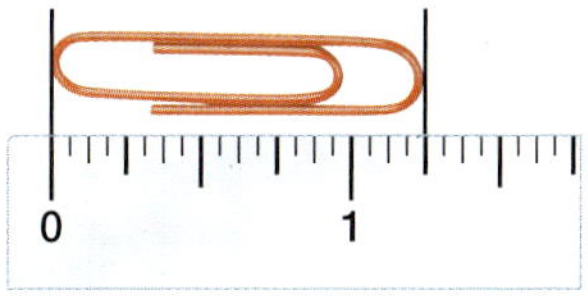

$1\frac{1}{4}$ in.

Find the precision for each measurement. Explain its meaning.

16. 80 in.
17. 22 mm
18. $16\frac{1}{2}$ in.
19. 308 cm **0.5 cm; 307.5 to 308.5 cm**
★20. 3.75 meters **5 mm; 3745 to 3755 mm**
★21. $3\frac{1}{4}$ ft **$\frac{1}{8}$ ft; $3\frac{1}{8}$ to $3\frac{3}{8}$ ft**

16. $\frac{1}{2}$ in.; $79\frac{1}{2}$ to $80\frac{1}{2}$ in.
17. 0.5 mm; 21.5 to 22.5 mm
18. $\frac{1}{4}$ in.; $16\frac{1}{4}$ to $16\frac{3}{4}$ in.

Find the measurement of each segment.

22. AC **29.5 mm**

23. XZ **$1\frac{1}{4}$ in.**

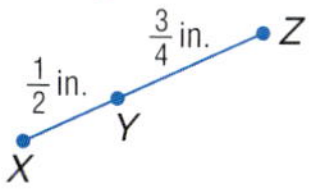

24. QR **$1\frac{15}{16}$ in.**

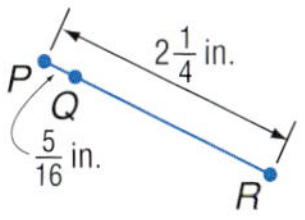

25. ST **2.8 cm**

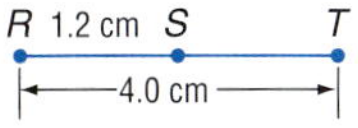

★26. WX **2.4 cm**

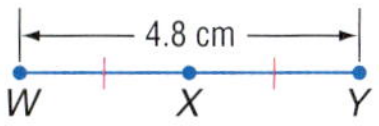

★27. BC **$1\frac{1}{4}$ in.**

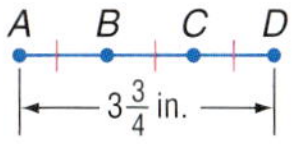

Find the value of the variable and ST if S is between R and T.

28. $RS = 7a$, $ST = 12a$, $RS = 28$
29. $RS = 12$, $ST = 2x$, $RT = 34$
30. $RS = 2x$, $ST = 3x$, $RT = 25$
31. $RS = 16$, $ST = 2x$, $RT = 5x + 10$
32. $RS = 3y + 1$, $ST = 2y$, $RT = 21$ **$y = 4$; $ST = 8$**
★33. $RS = 4y - 1$, $ST = 2y - 1$, $RT = 5y$ **$y = 2$; $ST = 3$**

28. $a = 4$; $ST = 48$
29. $x = 11$; $ST = 22$
30. $x = 5$; $ST = 15$
31. $x = 2$; $ST = 4$

Use the figures to determine whether each pair of segments is congruent.

34. $\overline{AB}$, $\overline{CD}$ **yes**

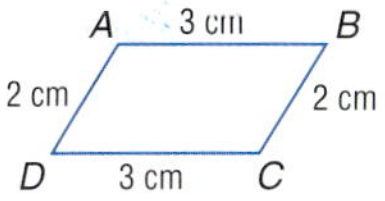

35. $\overline{EF}$, $\overline{FG}$ **no**

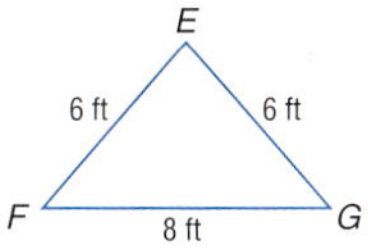

36. $\overline{NP}$, $\overline{LM}$ **no**

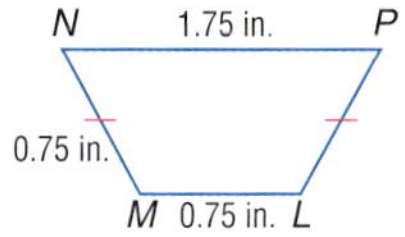

37. $\overline{WX}$, $\overline{XY}$ **yes**

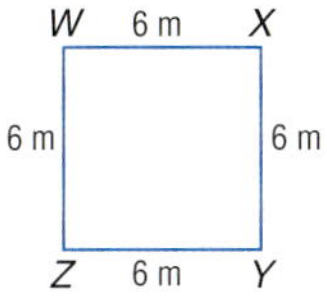

★38. $\overline{CH}$, $\overline{CM}$ **not from the information given**

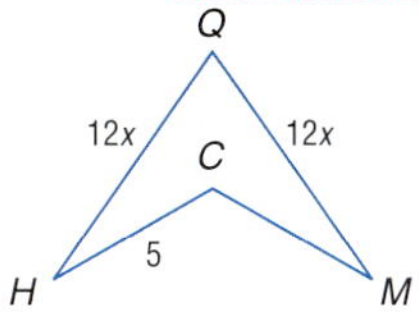

★39. $\overline{TR}$, $\overline{SU}$ **yes**

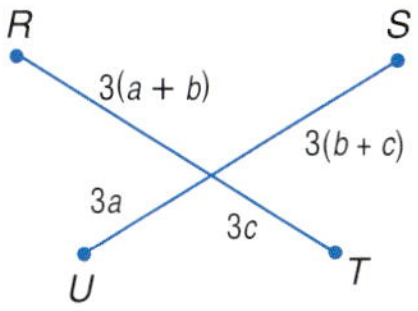

3 Practice/Apply

Study Notebook

Have students—

- add the definitions/examples of the vocabulary terms to their Vocabulary Builder worksheets for Lesson 1-2.
- include any other item(s) that they find helpful in mastering the skills in this lesson.

About the Exercises...

Organization by Objective

- **Measure Line Segments:** 12–21, 42–47
- **Calculate Measures:** 22–33, 34–41, 48–49

Odd/Even Assignments

Exercises 12–39 are structured so that students practice the same concepts whether they are assigned odd or even problems.

Assignment Guide

Basic: 13–19 odd, 23, 25, 29, 31, 35, 37, 41, 43–45, 50, 51, 54–65

Average: 13–49 odd, 50–51, 56–65 (optional: 52–55)

Advanced: 12–48 even, 50–61 (optional: 62–65)

All: Practice Quiz 1 (1–5)

The Assignment Guides provide suggestions for exercises that are appropriate for basic, average, or advanced students. Many of the homework exercises are paired, so that students can do the odds one day and the evens the next day.

More About. . .

Recreation

There are more than 3300 state parks, historic sites, and natural areas in the United States. Most of the parks are open year round to visitors.

Source: *Parks Directory of the United States*

44. 98.45 million to 98.55 million visitors

40. **MUSIC** A CD has a single spiral track of data, circling from the inside of the disc to the outside. Use a metric ruler to determine the full width of a music CD. 12 cm

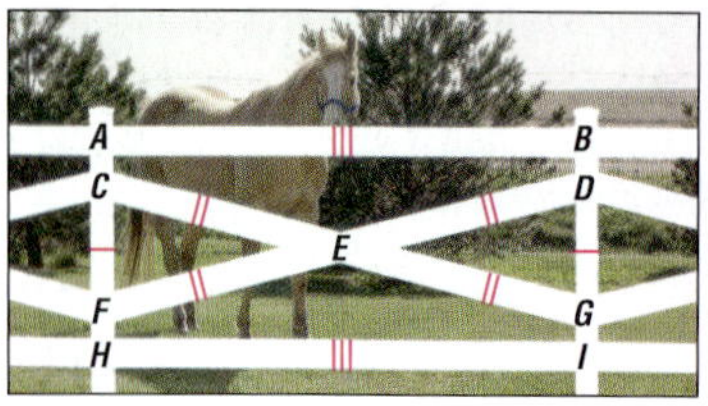

41. **DOORS** Name all segments in the crossbuck pattern in the picture that appear to be congruent.
$\overline{CF} \cong \overline{DG}$, $\overline{AB} \cong \overline{HI}$, $\overline{CE} \cong \overline{ED} \cong \overline{EF} \cong \overline{EG}$

42. **CRAFTS** Martin makes pewter figurines and wants to know how much molten pewter he needs for each mold. He knows that when a solid object with a volume of 1 cubic centimeter is submerged in water, the water level rises 1 milliliter. Martin pours 200 mL of water in a measuring cup, completely submerges a figurine in it, and watches it rise to 343 mL. What is the maximum amount of molten pewter, in cubic centimeters, Martin would need to make a figurine? Explain. 144 cm^3; 343 mL could be actually as much as 343.5 mL and 200 mL as little as 199.5 mL; $343.5 - 199.5 = 144$.

RECREATION For Exercises 43–45, refer to the graph that shows the states with the greatest number of visitors to state parks in a recent year.

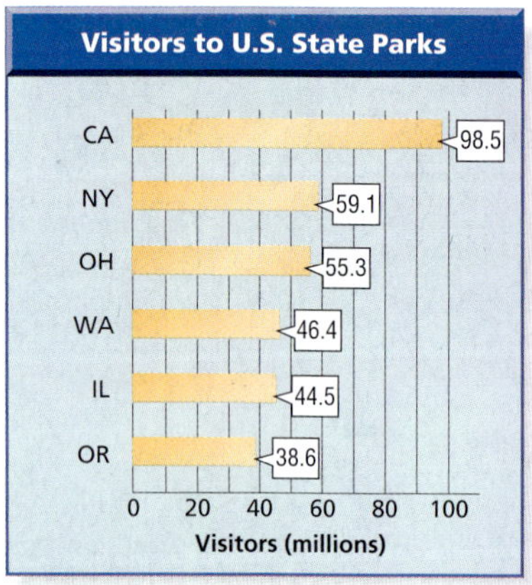

Source: National Association of Park Directors

43. To what number can the precision of the data be measured? 50,000 visitors

44. Find the precision for the California data.

45. Can you be sure that 1.9 million more people visited Washington state parks than Illinois state parks? Explain. See margin.

Online Research **Data Update** Find the current park data for your state and determine the precision of its measure. Visit www.geometryonline.com/data_update to learn more.

PERIMETER For Exercises 46 and 47, use the following information.
The **perimeter** of a geometric figure is the sum of the lengths of its sides. Pablo used a ruler divided into centimeters and measured the sides of a triangle as 3 centimeters, 5 centimeters, and 6 centimeters. Use what you know about the accuracy of any measurement to answer each question.

★ 46. What is the least possible perimeter of the triangle? Explain.

★ 47. What is the greatest possible perimeter of the triangle? Explain.

46. 12.5 cm; Each measurement is accurate within 0.5 cm, so the least perimeter is 2.5 cm + 4.5 cm + 5.5 cm.

47. 15.5 cm; Each measurement is accurate within 0.5 cm, so the greatest perimeter is 3.5 cm + 5.5 cm + 6.5 cm.

48–49. See margin.

CONSTRUCTION For Exercises 48 and 49, refer to the figure.

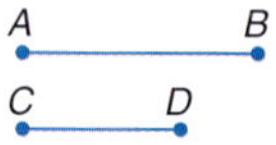

48. Construct a segment whose measure is $4(CD)$.

★ 49. Construct a segment that has length $3(AB) - 2(CD)$.

50. **CRITICAL THINKING** **Significant digits** represent the accuracy of a measurement.
- Nonzero digits are always significant.
- In whole numbers, zeros are significant if they fall between nonzero digits.
- In decimal numbers greater than or equal to 1, every digit is significant.
- In decimal numbers less than 1, the first nonzero digit and every digit to its right are significant.

For example, 600.070 has six significant digits, but 0.0210 has only three. How many significant digits are there in each measurement below?

a. 83,000 miles 2 b. 33,002 miles 5 c. 450.0200 liters 7

Study Guide and Intervention, p. 7 (shown) and p. 8

Measure Line Segments A part of a line between two endpoints is called a **line segment**. The lengths of $\overline{MN}$ and $\overline{RS}$ are written as MN and RS. When you measure a segment, the precision of the measurement is half of the smallest unit on the ruler.

Example 1 **Find the length of $\overline{MN}$.**

The long marks are centimeters, and the shorter marks are millimeters. The length of $\overline{MN}$ is 3.4 centimeters. The measurement is accurate to within 0.5 millimeter, so $\overline{MN}$ is between 3.35 centimeters and 3.45 centimeters long.

Example 2 **Find the length of $\overline{RS}$.**

The long marks are inches and the short marks are quarter inches. The length of $\overline{RS}$ is about $1\frac{3}{4}$ inches. The measurement is accurate to within one half of a quarter inch, or $\frac{1}{8}$ inch, so $\overline{RS}$ is between $1\frac{5}{8}$ inches and $1\frac{7}{8}$ inches long.

Exercises

Find the length of each line segment or object.

1. 2.5 cm 2. $1\frac{1}{4}$ in. 3. $2\frac{1}{4}$ in. 4. 1.7 cm

Find the precision for each measurement.

5. 10 in. $\frac{1}{2}$ in. 6. 32 mm 0.5 mm 7. 44 cm 0.5 cm

8. 2 ft $\frac{1}{2}$ ft or 6 in. 9. 3.5 mm 0.05 mm 10. $2\frac{1}{2}$ yd $\frac{1}{4}$ yd or 9 in.

Skills Practice, p. 9 and Practice, p. 10 (shown)

Find the length of each line segment or object.

1. $1\frac{11}{16}$ in. 2. 42 mm

Find the precision for each measurement.

3. 120 meters 0.5 m 4. $7\frac{1}{4}$ inches $\frac{1}{8}$ in. 5. 30.0 millimeters 0.5 mm

Find the measurement of each segment.

6. $\overline{PS}$ (PQ = 18.4 cm, QS = 4.7 cm) 23.1 cm 7. $\overline{AD}$ (AC = $2\frac{3}{8}$ in., CD = $1\frac{1}{4}$ in.) $3\frac{5}{8}$ in. 8. $\overline{WX}$ (XY = 89.6 cm, WY = 100 cm) 10.4cm

Find the value of the variable and KL if K is between J and L.

9. $JK = 6r$, $KL = 3r$, and $JL = 27$ 3; 9

10. $JK = 2s$, $KL = s + 2$, and $JL = 5s - 10$ 6; 8

Use the figures to determine whether each pair of segments is congruent.

11. $\overline{TU}$, $\overline{SW}$ no 12. $\overline{AD}$, $\overline{BC}$ yes 13. $\overline{GF}$, $\overline{FE}$ no

14. **CARPENTRY** Jorge used the figure at the right to make a pattern for a mosaic he plans to inlay on a tabletop. Name all of the congruent segments in the figure.
$\overline{BC} \cong \overline{FE}$, $\overline{AB} \cong \overline{CD} \cong \overline{DE} \cong \overline{FA}$

Reading to Learn Mathematics, p. 11 ELL

Pre-Activity **Why are units of measure important?**

Read the introduction to Lesson 1-2 at the top of page 13 in your textbook.

- The basic unit of length in the metric system is the meter. How many meters are there in one kilometer? 1000
- Do you think it would be easier to learn the relationships between the different units of length in the customary system (used in the United States) or in the metric system? Explain your answer. Sample answer: The metric system is easier because you can change between the different units by just moving the decimal point.

Reading the Lesson

1. Explain the difference between a *line* and a *line segment* and why one of these can be measured, while the other cannot.
Sample answer: A line is infinite. Since it has no endpoints, a line does not have a definite length and cannot be measured. A line segment has two endpoints, so it has a definite length and can be measured.

2. What is the smallest length marked on a 12-inch ruler? Sample answer: $\frac{1}{16}$ in.
What is the smallest length marked on a centimeter ruler? 1 mm

3. Find the precision of each measurement.
a. 15 cm 0.5 cm
b. 15.0 cm 0.05 cm

4. Refer to the figure at the right. Which one of the following statements is true? Explain your answer.
$\overline{AB} = \overline{CD}$ $\overline{AB} \cong \overline{CD}$
$\overline{AB} \cong \overline{CD}$; Sample answer: The two segments are congruent because they have the same measure or length. They are not equal because they are not the same segment.

5. Suppose that S is a point on $\overline{VW}$ and S is not the same point as V or W. Tell whether each of the following statements is *always*, *sometimes*, or *never* true.
a. $VS = SW$ sometimes
b. S is between V and W. always
c. $VS + VW = SW$ never

Helping You Remember

6. A good way to remember terms used in mathematics is to relate them to everyday words you know. Give three words that are used outside of mathematics that can help you remember that there are 100 centimeters in a meter. Sample answer: cent, century, centennial

Enrichment, p. 12

Points Equidistant from Segments

The distance from a point to a segment is zero if the point is on the segment. Otherwise, it is the length of the shortest segment from the point to the segment.

A figure is a **locus** if it is the set of all points that satisfy a set of conditions. The locus of all points that are $\frac{1}{4}$ inch from the segment AB is shown by two dashed segments with semicircles at both ends.

1. Suppose A, B, C, and D are four different points, and consider the locus of all points x units from $\overline{AB}$ *and* x units from $\overline{CD}$. Use any unit you find convenient. The locus can take different forms. Sketch at least three possibilities. List some of the things that seem to affect the form of the locus. Sample answers are shown.

Answer

45. No; the number of visitors to Washington state parks could be as low as 46.35 million or as high as 46.45 million. The visitors to Illinois state parks could be as low as 44.45 million or as high as 44.55 million visitors. The difference in visitors could be as high as 2.0 million.

51. **WRITING IN MATH** Answer the question that was posed at the beginning of the lesson. See margin.

Why are units of measure important?

Include the following in your answer.

- an example of how measurements might be misinterpreted, and
- what measurements you can assume from a figure.

Extending the Lesson

RELATIVE ERROR Accuracy is an indication of **relative error**. The relative error is the ratio of the half-unit difference in precision to the entire measure. The relative error is expressed as a percent. The smaller the relative error of a measurement, the more accurate the measure is. For 11 inches, the relative error can be found as follows.

$$\frac{\text{allowable error}}{\text{measure}} = \frac{0.5 \text{ in.}}{11 \text{ in.}} \approx 0.045 \text{ or } 4.5\%$$ Divide and convert to percent.

Determine the relative error for each measurement.

52. 27 ft 1.9%
53. $14\frac{1}{2}$ in. 1.7%
★54. 42.3 cm 0.1%
★55. 63.7 km 0.08%

FCAT Practice
Standardized Test Practice

56. The pipe shown is divided into five equal sections. How many feet long is the pipe? B

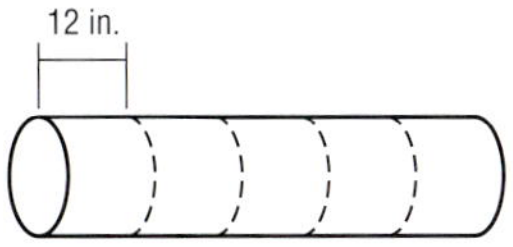

Ⓐ 2.4 ft Ⓑ 5 ft
Ⓒ 28.8 ft Ⓓ 60 ft

57. **ALGEBRA** Forty percent of a collection of 80 tapes are jazz tapes, and the rest are blues tapes. How many blues tapes are in the collection? D

Ⓐ 32 Ⓑ 40 Ⓒ 42 Ⓓ 48

Maintain Your Skills

Mixed Review

Refer to the figure at the right. *(Lesson 1-1)*

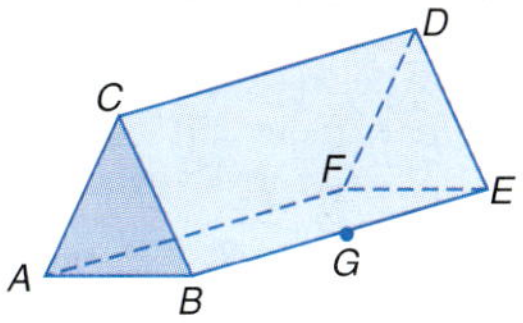

58. Name three collinear points. *B, G, E*
59. Name two planes that contain points B and C.
60. Name another point in plane DFA. C
61. How many planes are shown? 5

59. Sample answer: planes *ABC* and *BCD*

Getting Ready for the Next Lesson

PREREQUISITE SKILL **Evaluate each expression if $a = 3$, $b = 8$, and $c = 2$.**
*(To review **evaluating expressions**, see page 736.)*

62. $2a + 2b$ 22
63. $ac + bc$ 22
64. $\frac{a-c}{2}$ $\frac{1}{2}$
65. $\sqrt{(c-a)^2}$ 1

Practice Quiz 1

Lessons 1-1 and 1-2

For Exercises 1–3, refer to the figure. *(Lesson 1-1)*

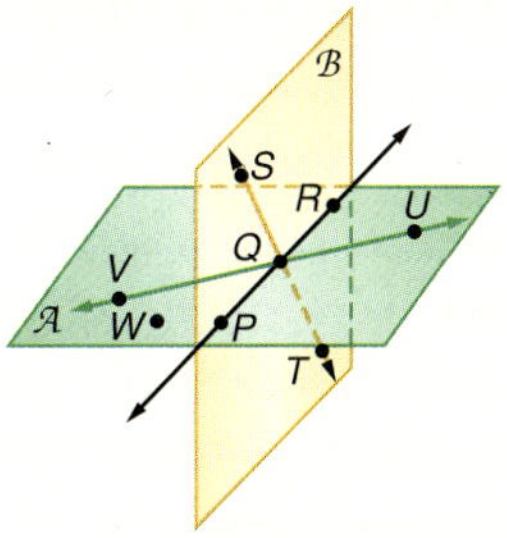

1. Name the intersection of planes $\mathcal{A}$ and $\mathcal{B}$. $\overleftrightarrow{PR}$
2. Name another point that is collinear with points S and Q. T
3. Name a line that is coplanar with $\overrightarrow{VU}$ and point W. $\overleftrightarrow{PR}$

Given that R is between S and T, find each measure. *(Lesson 1-2)*

4. $RS = 6$, $TR = 4.5$, $TS =$ __?__. 10.5
5. $TS = 11.75$, $TR = 3.4$, $RS =$ __?__. 8.35

www.geometryonline.com/self_check_quiz/fcat

Answers

48.

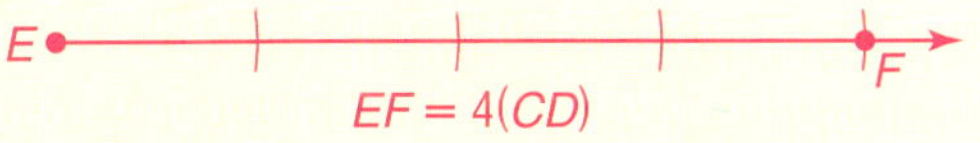

49.

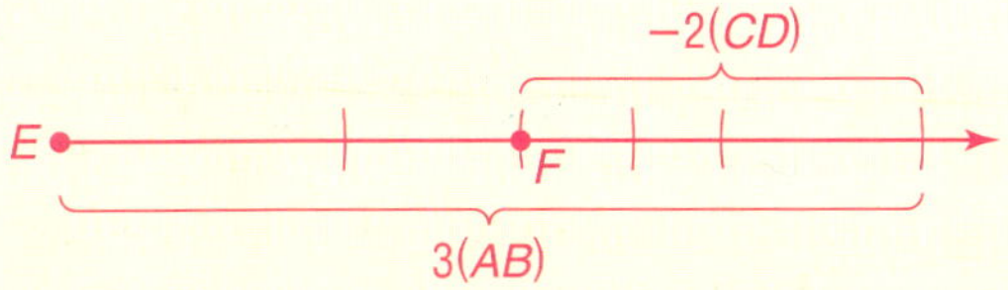

4 Assess

Open-Ended Assessment

Writing Students can practice using rulers and compasses to draw and label lines, planes, simple geometric shapes, and line segments. Using measuring tools, they can explore the concepts of precision, accuracy, and congruency and write simple descriptions of their own work or the work of a partner.

Getting Ready for Lesson 1-3

Prerequisite Skill Students will learn about distance and midpoints in Lesson 1-3. They will use algebraic formulas to find the distance between two points and the midpoint of a segment. Use Exercises 62–65 to determine your students' familiarity with evaluating algebraic expressions.

Assessment Options

Practice Quiz 1 The quiz provides students with a brief review of the concepts and skills in Lessons 1-1 and 1-2. Lesson numbers are given to the right of the exercises or instruction lines so students can review concepts not yet mastered.

Quiz (Lessons 1-1 and 1-2) is available on p. 51 of the *Chapter 1 Resource Masters*.

Answer

51. Sample answer: Units of measure are used to differentiate between size and distance, as well as for precision. Answers should include the following.

- When a measurement is stated, you do not know the precision of the instrument used to make the measure. Therefore, the actual measure could be greater or less than that stated.
- You can assume equal measures when segments are shown to be congruent.

Lesson 1-2 Linear Measure and Precision 19

Geometry Activity

A Follow-Up of Lesson 1-2

Getting Started

Objective Find the probability that a point lies on a segment.

Teach

- Explain to students that this activity is about finding the possibility that an arbitrarily chosen point lies on a particular part of a given segment. This is why they do not see point Q on $\overline{RT}$ in the example or any of the arbitrary points in the exercises.
- Remind students that probability is at most 1. If students calculate a probability that is greater than 1, they have most likely switched the numerator and denominator of their fraction during the calculation. Point out that the total number of possible outcomes, the larger value of the two, is always in the denominator.

Assess

Exercises 1–3 extend the example to include three divisions of the segment instead of two. Students should remember that the top number always has to be less than or equal to the bottom number in the probability fraction.

Exercises 4–5 are extensions of the activity, requiring more logical thought and discussion.

Study Notebook

Ask students to summarize what they have learned about using the measures of line segments to find probabilities.

Geometry Activity

A Follow-Up of Lesson 1-2

Probability and Segment Measure

You may remember that probability is often expressed as a fraction.

$$\text{Probability } (P) \text{ of an event} = \frac{\text{number of favorable outcomes}}{\text{total number of possible outcomes}}$$

To find the probability that a point lies on a segment, you need to calculate the length of the segment.

Activity

Assume that point Q is contained in $\overline{RT}$. Find the probability that Q is contained in $\overline{RS}$.

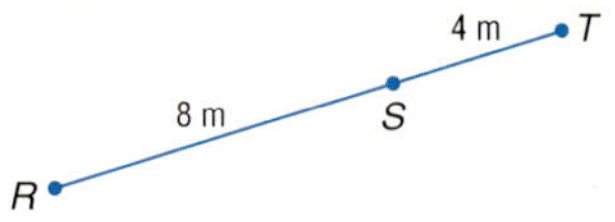

Collect Data

- Find the measures of all segments in the figure.
- $RS = 8$ and $ST = 4$, so $RT = RS + ST$ or 12.
- While a point has no dimension, the segment that contains it does have one dimension, length. To calculate the probability that a point, randomly selected, is in a segment contained by another segment, you must compare their lengths.

$$P(Q \text{ lies in } \overline{RS}) = \frac{RS}{RT}$$

$$= \frac{8}{12} \quad RS = 8 \text{ and } RT = 12$$

$$= \frac{2}{3} \quad \text{Simplify.}$$

The probability that Q is contained in $\overline{RS}$ is $\frac{2}{3}$.

Analyze

For Exercises 1–3, refer to the figure at the right.

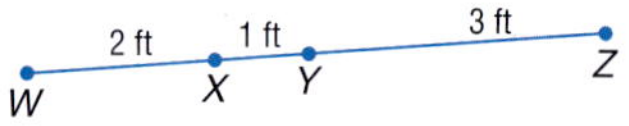

1. Point J is contained in $\overline{WZ}$. What is the probability that J is contained in $\overline{XY}$? $\frac{1}{6}$
2. Point R is contained in $\overline{WZ}$. What is the probability that R is contained in $\overline{YZ}$? $\frac{1}{2}$
3. Point S is contained in $\overline{WY}$. What is the probability that S is contained in $\overline{XY}$? $\frac{1}{3}$

Make a Conjecture

For Exercises 4–5, refer to the figure for Exercises 1–3.

4. Point T is contained in both $\overline{WY}$ and $\overline{XZ}$. What do you think is the probability that T is contained in $\overline{XY}$? Explain. **1; $\overline{XY}$ contains all points that lie on both $\overline{WY}$ and $\overline{XZ}$.**
5. Point U is contained in $\overline{WX}$. What do you think is the probability that U is contained in $\overline{YZ}$? Explain. **0; If point U lies on $\overline{WX}$, it cannot lie on $\overline{YZ}$.**

20 Chapter 1 Points, Lines, Planes, and Angles

Resource Manager

Teaching Geometry with Manipulatives

- p. 28 (student recording sheet)

1-3 Distance and Midpoints

What You'll Learn

- Find the distance between two points.
- Find the midpoint of a segment.

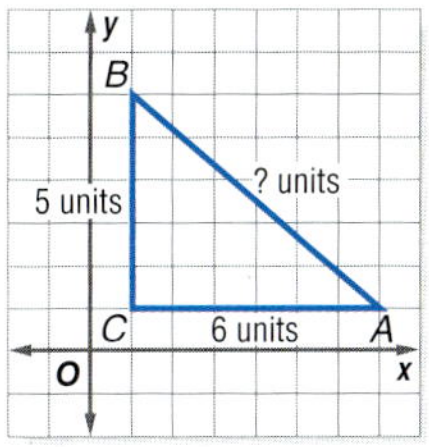

How can you find the distance between two points without a ruler?

Whenever you connect two points on a number line or on a plane, you have graphed a line segment. Distance on a number line is determined by counting the units between the two points. On a coordinate plane, you can use the Pythagorean Theorem to find the distance between two points. In the figure, to find the distance from A to B, use $(AC)^2 + (CB)^2 = (AB)^2$.

Sunshine State Standards
MA.C.3.4.2

Vocabulary
- midpoint
- segment bisector

TEACHING TIP
This book does not present simplifying radicals until it has some significance in Chapter 7. If you wish students to simplify radicals, refer to pages 744–745 for review.

Vocabulary words are listed at the beginning of the lesson and are highlighted in yellow at point of use.

DISTANCE BETWEEN TWO POINTS The coordinates of the endpoints of a segment can be used to find the length of the segment. Because the distance from A to B is the same as the distance from B to A, the order in which you name the endpoints makes no difference.

Key Concept — Distance Formulas

- **Number line**

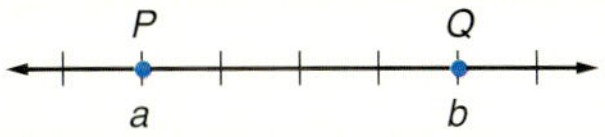

$PQ = |b - a|$ or $|a - b|$

- **Coordinate Plane**

The distance d between two points with coordinates (x_1, y_1) and (x_2, y_2) is given by $d = \sqrt{(x_2 - x_1)^2 + (y_2 - y_1)^2}$.

Example 1 Find Distance on a Number Line

Use the number line to find CD.

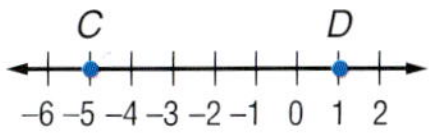

The coordinates of C and D are -5 and 1.

$CD = |-5 - 1|$ Distance Formula

$= |-6|$ or 6 Simplify.

Study Tip

Pythagorean Theorem
Recall that the Pythagorean Theorem is often expressed as $a^2 + b^2 = c^2$, where a and b are the measures of the shorter sides (legs) of a right triangle, and c is the measure of the longest side (hypotenuse) of a right triangle.

You can use the Pythagorean Theorem to find the distance between two points on a coordinate plane.

Example 2 Find Distance on a Coordinate Plane

Find the distance between $R(5, 1)$ and $S(-3, -3)$.

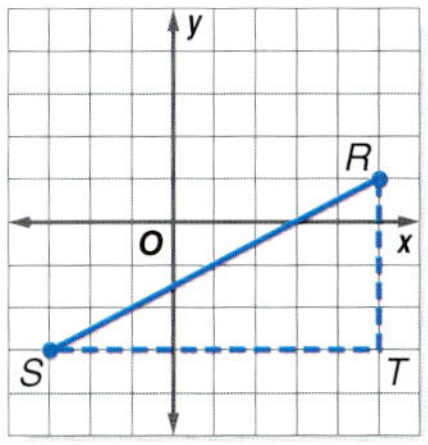

Method 1 Pythagorean Theorem

Use the gridlines to form a triangle so you can use the Pythagorean Theorem.

$(RS)^2 = (RT)^2 + (ST)^2$ Pythagorean Theorem

$(RS)^2 = 4^2 + 8^2$ $RT = 4$ units, $ST = 8$ units

$(RS)^2 = 80$ Simplify.

$RS = \sqrt{80}$ Take the square root of each side.

1-3 Lesson Notes

1 Focus

5-Minute Check Transparency 1-3 Use as a quiz or review of Lesson 1-2.

Mathematical Background notes are available for this lesson on p. 4D.

How can you find the distance between two points without a ruler?

Ask students:

- On a coordinate plane, what can you do to find the length of a horizontal or vertical segment? **Count the units between two points.**
- If you move the triangle in the figure in any direction, flip it, or rotate it, do the measures of its segments change? **no**

2 Teach

DISTANCE BETWEEN TWO POINTS

In-Class Example Power Point®

1 Use the number line to find QR. **3**

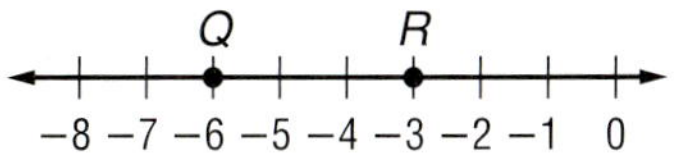

Resource Manager

Workbook and Reproducible Masters

Chapter 1 Resource Masters
- Study Guide and Intervention, pp. 13–14
- Skills Practice, p. 15
- Practice, p. 16
- Reading to Learn Mathematics, p. 17
- Enrichment, p. 18
- Assessment, pp. 51, 53

Graphing Calculator and Computer Masters, p. 18
Prerequisite Skills Workbook, pp. 7–8, 33–34, 79–80, 83–86
Teaching Geometry With Manipulatives Masters, pp. 1, 17, 29, 30, 31

Transparencies
5-Minute Check Transparency 1-3
Answer Key Transparencies

Technology
GeomPASS: Tutorial Plus, Lesson 3
Interactive Chalkboard
Multimedia Applications: Virtual Activities

In-Class Example

Power Point®

2 Find the distance between $E(-4, 1)$ and $F(3, -1)$.

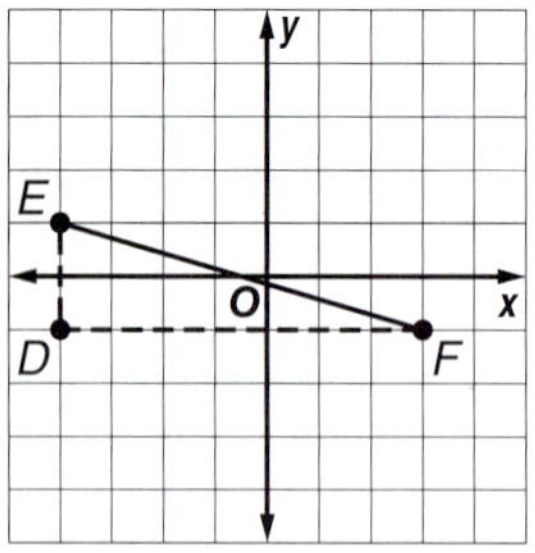

$\sqrt{53} \approx 7.28$

Study Tip

Distance Formula
The Pythagorean Theorem is used to develop the Distance Formula. You will learn more about the Pythagorean Theorem in Lesson 7-2.

Method 2 Distance Formula

$d = \sqrt{(x_2 - x_1)^2 + (y_2 - y_1)^2}$ Distance Formula

$RS = \sqrt{(-3 - 5)^2 + (-3 - 1)^2}$ $(x_1, y_1) = (5, 1)$ and $(x_2, y_2) = (-3, -3)$

$RS = \sqrt{(-8)^2 + (-4)^2}$ Simplify.

$RS = \sqrt{80}$ Simplify.

The distance from R to S is $\sqrt{80}$ units. You can use a calculator to find that $\sqrt{80}$ is approximately 8.94.

MIDPOINT OF A SEGMENT The **midpoint** of a segment is the point halfway between the endpoints of the segment. If X is the midpoint of $\overline{AB}$, then $AX = XB$.

Geometry Activity

Midpoint of a Segment

Model

- Graph points $A(5, 5)$ and $B(-1, 5)$ on grid paper. Draw $\overline{AB}$.
- Hold the paper up to the light and fold the paper so that points A and B match exactly. Crease the paper slightly.
- Open the paper and put a point where the crease intersects $\overline{AB}$. Label this midpoint as C.
- Repeat the first three steps using endpoints $X(-4, 3)$ and $Y(2, 7)$. Label the midpoint Z.

Make a Conjecture

1. What are the coordinates of point C? **(2, 5)**
2. What are the lengths of $\overline{AC}$ and $\overline{CB}$? **Both are 3 units long.**
3. What are the coordinates of point Z? **(−1, 5)**
4. What are the lengths of $\overline{XZ}$ and $\overline{ZY}$?
5. Study the coordinates of points A, B, and C. Write a rule that relates these coordinates. Then use points X, Y, and Z to verify your conjecture.

4. Both are $\sqrt{13}$ units or about 3.6 units long.

5. Sample answer: The x-coordinate of the midpoint is one half the sum of the x-coordinates of the endpoints. The y-coordinate of the midpoint is one half the sum of the y-coordinates of the endpoints.

The points found in the activity are both midpoints of their respective segments.

Study Tip

Common Misconception
The Distance Formula and the Midpoint Formula do not use the same relationship among the coordinates.

Key Concept — Midpoint

Words	The midpoint M of $\overline{PQ}$ is the point between P and Q such that $PM = MQ$.	
	Number Line	**Coordinate Plane**
Symbols	The coordinate of the midpoint of a segment whose endpoints have coordinates a and b is $\frac{a+b}{2}$.	The coordinates of the midpoint of a segment whose endpoints have coordinates (x_1, y_1) and (x_2, y_2) are $\left(\frac{x_1+x_2}{2}, \frac{y_1+y_2}{2}\right)$.
Models	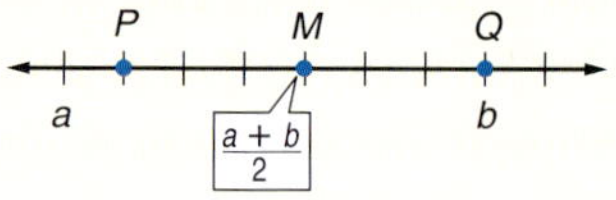	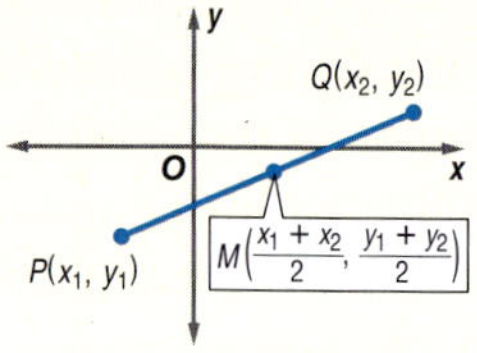

Geometry Activity

Materials: grid paper

- When students first fold the paper to match the points, ask them if A and B are the same distance from the crease. Students should recognize that they are dividing their original segment into two equal parts.
- For Exercise 5, ask students to add the x-coordinates of A and B and write this sum above the x-coordinate of C. Repeat for the y-coordinates. Have students apply this entire technique for the x- and y-coordinates of X, Y, and Z.

Example 3 Find Coordinates of Midpoint

a. TEMPERATURE Find the coordinate of the midpoint of $\overline{PQ}$.

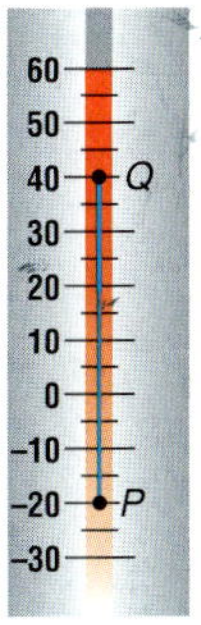

The coordinates of P and Q are -20 and 40.

Let M be the midpoint of $\overline{PQ}$.

$$M = \frac{-20 + 40}{2} \quad a = -20,\ b = 40$$
$$= \frac{20}{2} \text{ or } 10 \quad \text{Simplify.}$$

b. Find the coordinates of M, the midpoint of $\overline{PQ}$, for $P(-1, 2)$ and $Q(6, 1)$.

Let P be (x_1, y_1) and Q be (x_2, y_2).

$$M\left(\frac{x_1 + x_2}{2}, \frac{y_1 + y_2}{2}\right) = M\left(\frac{-1 + 6}{2}, \frac{2 + 1}{2}\right) \quad (x_1, y_1) = (-1, 2), (x_2, y_2) = (6, 1)$$
$$= M\left(\frac{5}{2}, \frac{3}{2}\right) \text{ or } M\left(2\frac{1}{2}, 1\frac{1}{2}\right) \quad \text{Simplify.}$$

You can also find the coordinates of the endpoint of a segment if you know the coordinates of its other endpoint and its midpoint.

Example 4 Find Coordinates of Endpoint

Find the coordinates of X if $Y(-2, 2)$ is the midpoint of $\overline{XZ}$ and Z has coordinates (2, 8).

Let Z be (x_2, y_2) in the Midpoint Formula.

$$Y(-2, 2) = Y\left(\frac{x_1 + 2}{2}, \frac{y_1 + 8}{2}\right) \quad (x_2, y_2) = (2, 8)$$

Write two equations to find the coordinates of X.

$$-2 = \frac{x_1 + 2}{2} \qquad\qquad 2 = \frac{y_1 + 8}{2}$$
$$-4 = x_1 + 2 \quad \text{Multiply each side by 2.} \qquad 4 = y_1 + 8 \quad \text{Multiply each side by 2.}$$
$$-6 = x_1 \quad \text{Subtract 2 from each side.} \qquad -4 = y_1 \quad \text{Subtract 8 from each side.}$$

he coordinates of X are $(-6, -4)$.

Example 5 Use Algebra to Find Measures

Multiple-Choice Test Item

What is the measure of $\overline{BC}$ if B is the midpoint of $\overline{AC}$?

Ⓐ −5 Ⓑ 8

Ⓒ 17 Ⓓ 27

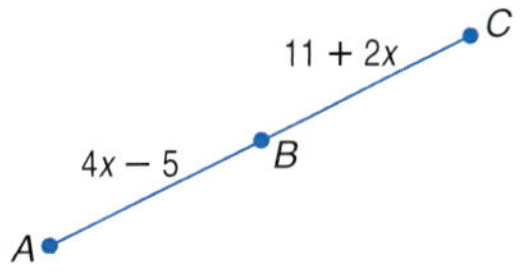

Read the Test Item

You know that B is the midpoint of $\overline{AC}$, and the figure gives algebraic measures for $\overline{AB}$ and $\overline{BC}$. You are asked to find the measure of $\overline{BC}$.

(continued on the next page)

www.geometryonline.com/extra_examples/fcat

WebQuest

Latitude and longitude form another coordinate system. The latitude, longitude, degree distance, and monthly high temperature can be used to create several different scatter plots. Visit www.geometryonline.com/webquest to continue work on your WebQuest project.

ch chapter contains example that gives students practice in solving problems on standardized tests. Standardized Test Practice suggestions give students additional methods for achieving success on standardized tests.

... ractice

Standardized Test Practice

Ⓐ Ⓑ Ⓒ Ⓓ

The Princeton Review

Test-Taking Tip

Eliminate Possibilities You can sometimes eliminate choices by looking at the answer. In this test item, you can eliminate choice A because measures cannot be negative.

MIDPOINT OF A SEGMENT

In-Class Examples

Power Point®

Teaching Tip Explain to students that they can use the given points in any order when they find the midpoint because addition is commutative.

3 a. The coordinates on a number line of J and K are -12 and 16, respectively. Find the coordinate of the midpoint of $\overline{JK}$. **2**

b. Find the coordinates of the midpoint of $\overline{GH}$ for $G(8, -6)$ and $H(-14, 12)$. **(−3, 3)**

4 Find the coordinates of D if $E(-6, 4)$ is the midpoint of $\overline{DF}$ and F has coordinates $(-5, -3)$. **(−7, 11)**

5 What is the measure of $\overline{PR}$ if Q is the midpoint of $\overline{PR}$? **D**

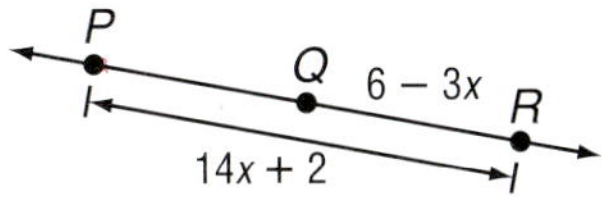

A $\frac{1}{2}$ **B** 4

C $4\frac{1}{2}$ **D** 9

DAILY INTERVENTION

Unlocking Misconceptions

Midpoints Students often assume that a segment has a midpoint if the point is drawn close to the middle of the segment, which sometimes results in their solving a problem incorrectly. Advise students that a point near the center of a segment must not be assumed to be the midpoint.

Daily Intervention notes help you help students when they need it most. Unlocking Misconceptions suggestions help you analyze where students make common errors so you can point these trouble spots out to them.

Teaching Tip Explain and demonstrate that a segment with points, A, B, and C, and slashes marked on $\overline{AB}$ and $\overline{BC}$ indicates that B is the *midpoint* and *bisector* of $\overline{AC}$.

Tips for New Teachers

Constructions Constructions may also be done with a safety compass, which resembles a rotating ruler. The steps in the construction are the same.

Solve the Test Item

Because B is the midpoint, you know that $AB = BC$. Use this equation and the algebraic measures to find a value for x.

$AB = BC$	Definition of midpoint
$4x - 5 = 11 + 2x$	$AB = 4x - 5$, $BC = 11 + 2x$
$4x = 16 + 2x$	Add 5 to each side.
$2x = 16$	Subtract $2x$ from each side.
$x = 8$	Divide each side by 2.

Now substitute 8 for x in the expression for BC.

$BC = 11 + 2x$	Original measure
$BC = 11 + 2(8)$	$x = 8$
$BC = 11 + 16$ or 27	Simplify.

The answer is D.

Study Tip

Segment Bisectors There can be an infinite number of bisectors, and each must contain the midpoint of the segment.

Any segment, line, or plane that intersects a segment at its midpoint is called a **segment bisector**. In the figure at the right, M is the midpoint of $\overline{AB}$. Plane $\mathcal{N}$, $\overline{MD}$, $\overleftrightarrow{RM}$, and point M are all bisectors of $\overline{AB}$. We say that they *bisect* $\overline{AB}$.

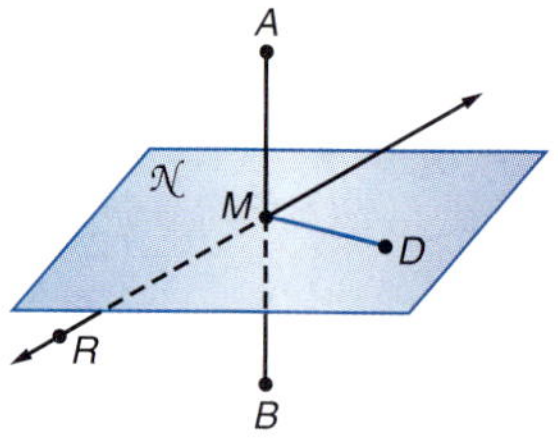

You can construct a line that bisects a segment without measuring to find the midpoint of the given segment.

Construction

Bisect a Segment

1. Draw a segment and name it $\overline{XY}$. Place the compass at point X. Adjust the compass so that its width is greater than $\frac{1}{2}XY$. Draw arcs above and below $\overline{XY}$.

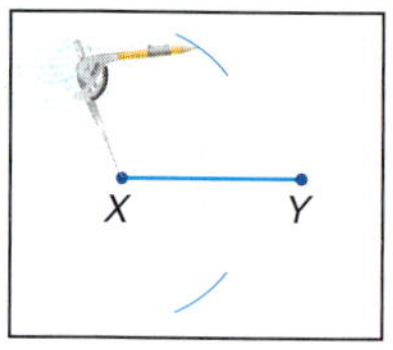

2. Using the same compass setting, place the compass at point Y and draw arcs above and below $\overline{XY}$ intersect the two arcs previously drawn. Label the points of the intersection of the arcs as P and Q.

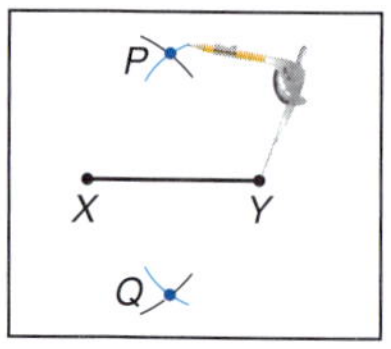

3. Use a straightedge to draw $\overline{PQ}$. Label the point where it intersects $\overline{XY}$ as M. Point M is the midpoint of $\overline{XY}$, and $\overline{PQ}$ is a bisector of $\overline{XY}$. Also $XM = MY = \frac{1}{2}XY$.

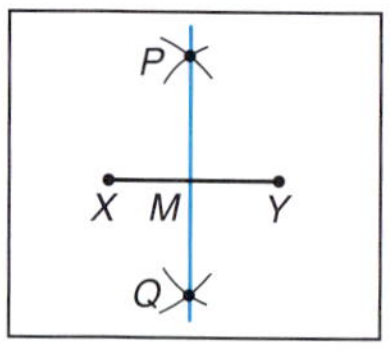

DAILY INTERVENTION **Differentiated Instruction**

Visual/Spatial Hold a meterstick up for students to see so that the marked side is facing away from them. Ask a volunteer to mark on the back of the stick about where they visualize the middle of the meterstick to be. Have a second volunteer verify the first student's mark or add another mark. Place a pen upright on the 50-cm mark so that it shows exactly where the midpoint of the meterstick is and compare to the students' marks. Explain how people can use spatial skills to very closely identify the exact middle of many objects.

Daily Intervention notes help you help students when they need it most. Differentiated Instruction suggestions are keyed to eight commonly-accepted learning styles.

Check for Understanding

Concept Check

1. **Explain** three ways to find the midpoint of a segment. **See margin.**
2. **OPEN ENDED** Draw a segment. Construct the bisector of the segment and use a millimeter ruler to check the accuracy of your construction. **See margin.**

Guided Practice

GUIDED PRACTICE KEY	
Exercises	Examples
3, 4	1
5, 6	2
7–10	3
11	4
12	5

Use the number line to find each measure.

3. AB **8**
4. CD **7**

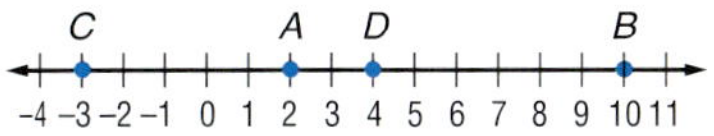

5. Use the Pythagorean Theorem to find the distance between $X(7, 11)$ and $Y(-1, 5)$. **5. 10**
6. Use the Distance Formula to find the distance between $D(2, 0)$ and $E(8, 6)$. $\sqrt{72}$

Use the number line to find the coordinate of the midpoint of each segment.

7. $\overline{RS}$ **−6**
8. $\overline{UV}$ **1.5**

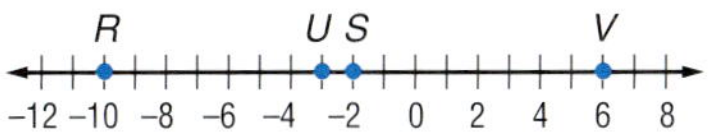

Find the coordinates of the midpoint of a segment having the given endpoints.

9. $X(-4, 3)$, $Y(-1, 5)$ **(−2.5, 4)**
10. $A(2, 8)$, $B(-2, 2)$ **(0, 5)**

11. Find the coordinates of A if $B(0, 5.5)$ is the midpoint of $\overline{AC}$ and C has coordinates $(-3, 6)$. **(3, 5)**

FCAT Practice
Standardized Test Practice
A B C D

12. Point M is the midpoint of $\overline{AB}$. What is the value of x in the figure? **B**
   (A) 1.5 (B) 5
   (C) 5.5 (D) 11

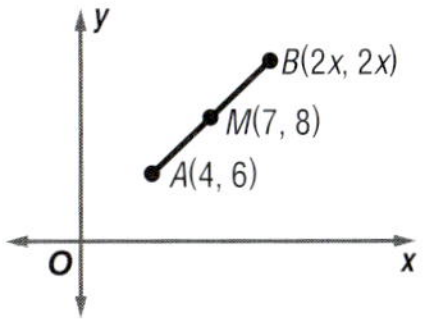

★ indicates increased difficulty

Practice and Apply

Homework Help	
For Exercises	See Examples
13–18	1
19–28	2
29, 30	5
31–42	3
43–45	4

Extra Practice
See page 754.

Use the number line to find each measure.

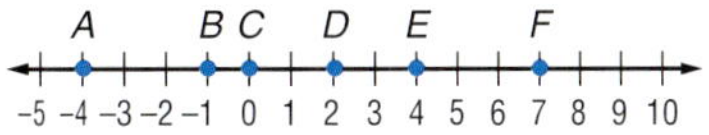

13. DE **2**
14. CF **7**
15. AB **3**
16. AC **4**
17. AF **11**
18. BE **5**

Use the Pythagorean Theorem to find the distance between each pair of points.

19. $A(0, 0)$, $B(8, 6)$ **10**
20. $C(-10, 2)$, $D(-7, 6)$ **5**
21. $E(-2, -1)$, $F(3, 11)$ **13**
22. $G(-2, -6)$, $H(6, 9)$ **17**

Use the Distance Formula to find the distance between each pair of points.

23. $J(0, 0)$, $K(12, 9)$ **15**
24. $L(3, 5)$, $M(7, 9)$ $\sqrt{32} \approx 5.7$
25. $S(-3, 2)$, $T(6, 5)$ $\sqrt{90} \approx 9.5$
26. $U(2, 3)$, $V(5, 7)$ **5**

27. $\sqrt{61} \approx 7.8$

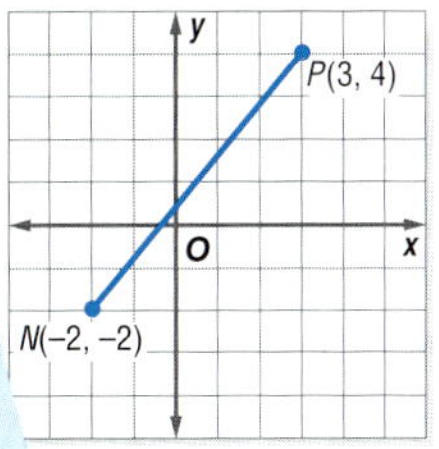

28. $\sqrt{40} \approx 6.3$

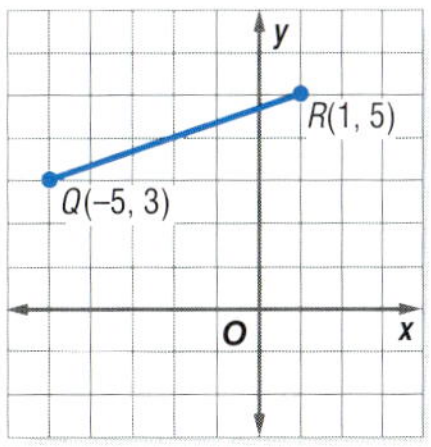

Homework Help charts show students which examples to refer to if they need additional practice. Extra Practice for every lesson is provided on pages 754–781.

Answer

1. Sample answers: (1) Use one of the Midpoint Formulas if you know the coordinates of the endpoints. (2) Draw a segment and fold the paper so that the endpoints match to locate the middle of the segment. (3) Use a compass and straightedge to construct the bisector of the segment.

2. Sample answer:

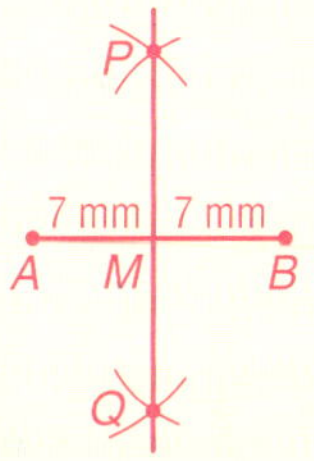

3 Practice/Apply

Study Notebook

Have students—

- add the definitions/examples of the vocabulary terms to their Vocabulary Builder worksheets for Lesson 1-3.
- include a chart listing each method for finding the length of a segment on a number line or in a coordinate plane and each method for finding the midpoint of a segment on a number line, in a coordinate plane and for a segment with algebraic measures.
- include any other item(s) that they find helpful in mastering the skills in this lesson.

About the Exercises…

Organization by Objective

- **Distance Between Two Points:** 13–28
- **Midpoint of a Segment:** 29–45

Odd/Even Assignments

Exercises 13–28 and 31–42 are structured so that students practice the same concepts whether they are assigned odd or even problems.

Alert! Exercises 48–49 require spreadsheet software.

Assignment Guide

Basic: 13–39 odd, 43, 47, 51, 53, 54–68

Average: 13–53 odd, 54–68

Advanced: 14–52 even, 54–62 (optional: 63–68)

Study Guide and Intervention, p. 13 (shown) and p. 14

Distance Between Two Points

Distance on a Number Line	Distance in the Coordinate Plane
$AB = \lvert b - a \rvert$ or $\lvert a - b \rvert$	Pythagorean Theorem: $a^2 + b^2 = c^2$ Distance Formula: $d = \sqrt{(x_2 - x_1)^2 + (y_2 - y_1)^2}$

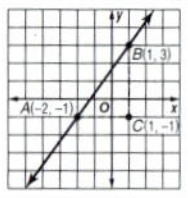

Example 1 Find AB.

$AB = \lvert(-4) - 2\rvert$
$= \lvert -6 \rvert$
$= 6$

Example 2 Find the distance between $A(-2, -1)$ and $B(1, 3)$.

Pythagorean Theorem

$(AB)^2 = (AC)^2 + (BC)^2$
$(AB)^2 = (3)^2 + (4)^2$
$(AB)^2 = 25$
$AB = \sqrt{25}$
$= 5$

Distance Formula

$d = \sqrt{(x_2 - x_1)^2 + (y_2 - y_1)^2}$
$AB = \sqrt{(1 - (-2))^2 + (3 - (-1))^2}$
$AB = \sqrt{(3)^2 + (4)^2}$
$= \sqrt{25}$
$= 5$

Exercises

Use the number line to find each measure.

1. BD 6
2. DG 9
3. AF 12
4. EF 3
5. BG 15
6. AG 17
7. BE 7
8. DE 1

Use the Pythagorean Theorem to find the distance between each pair of points.

9. $A(0, 0)$, $B(6, 8)$ 10
10. $R(-2, 3)$, $S(3, 15)$ 13
11. $M(1, -2)$, $N(9, 13)$ 17
12. $E(-12, 2)$, $F(-9, 6)$ 5

Use the Distance Formula to find the distance between each pair of points.

13. $A(0, 0)$, $B(15, 20)$ 25
14. $O(-12, 0)$, $P(-8, 3)$ 5
15. $C(11, -12)$, $D(6, 2)$ $\sqrt{221} \approx 14.9$
16. $E(-2, 10)$, $F(-4, 3)$ $\sqrt{53} \approx 7.3$

Skills Practice, p. 15 and Practice, p. 16 (shown)

Use the number line to find each measure.

1. VW 4
2. TV 5
3. ST 3
4. SV 8

Use the Pythagorean Theorem to find the distance between each pair of points.

5.

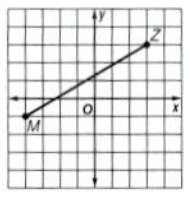

$\sqrt{65} \approx 8.1$

6.

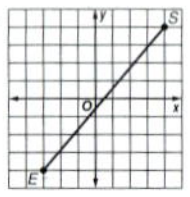

$\sqrt{113} \approx 10.6$

Use the Distance Formula to find the distance between each pair of points.

7. $L(-7, 0)$, $Y(5, 9)$ 15
8. $U(1, 3)$, $B(4, 6)$ $\sqrt{18} \approx 4.2$

Use the number line to find the coordinate of the midpoint of each segment.

9. $\overline{RT}$ 1
10. $\overline{QR}$ −4
11. $\overline{ST}$ $2\frac{1}{2}$
12. $\overline{PR}$ $-5\frac{1}{2}$

Find the coordinates of the midpoint of a segment having the given endpoints.

13. $K(-9, 3)$, $H(5, 7)$ $(-2, 5)$
14. $W(-12, -7)$, $T(-8, -4)$ $(-10, -5.5)$

Find the coordinates of the missing endpoint given that E is the midpoint of $\overline{DF}$.

15. $F(5, 8)$, $E(4, 3)$ $D(3, -2)$
16. $F(2, 9)$, $E(-1, 6)$ $D(-4, 3)$
17. $D(-3, -8)$, $E(1, -2)$ $F(5, 4)$

18. **PERIMETER** The coordinates of the vertices of a quadrilateral are $R(-1, 3)$, $S(3, 3)$, $T(5, -1)$, and $U(-2, -1)$. Find the perimeter of the quadrilateral. Round to the nearest tenth. 19.6 units

Reading to Learn Mathematics, p. 17 ELL

Pre-Activity **How can you find the distance between two points without a ruler?**

Read the introduction to Lesson 1-3 at the top of page 21 in your textbook.

- Look at the triangle in the introduction to this lesson. What is the special name for $\overline{AB}$ in this triangle? hypotenuse
- Find AB in this figure. Write your answer both as a radical and as a decimal number rounded to the nearest tenth. $\sqrt{61}$ units; 7.8 units

Reading the Lesson

1. Match each formula or expression in the first column with one of the names in the second column.

a. $d = \sqrt{(x_2 - x_1)^2 + (y_2 - y_1)^2}$ ii — i. Pythagorean Theorem
b. $\frac{a + b}{2}$ v — ii. Distance Formula in the Coordinate Plane
c. $XY = \lvert a - b \rvert$ iv — iii. Midpoint of a Segment in the Coordinate Plane
d. $c^2 = a^2 + b^2$ i — iv. Distance Formula on a Number Line
e. $\left(\frac{x_1 + x_2}{2}, \frac{y_1 + y_2}{2}\right)$ iii — v. Midpoint of a Segment on a Number Line

2. Fill in the steps to calculate the distance between the points $M(4, -3)$ and $N(-2, 7)$.

Let $(x_1, y_1) = (4, -3)$. Then $(x_2, y_2) = (\underline{-2}, \underline{7})$.

$d = \sqrt{(\underline{x_2} - \underline{x_1})^2 + (\underline{y_2} - \underline{y_1})^2}$
$MN = \sqrt{(\underline{-2} - \underline{4})^2 + (\underline{7} - \underline{-3})^2}$
$MN = \sqrt{(\underline{-6})^2 + (\underline{10})^2}$
$MN = \sqrt{\underline{36} + \underline{100}}$
$MN = \sqrt{\underline{136}}$

Find a decimal approximation for MN to the nearest hundredth. 11.66

Helping You Remember

3. A good way to remember a new formula in mathematics is to relate it to one you already know. If you forget the Distance Formula, how can you use the Pythagorean Theorem to find the distance d between two points on a coordinate plane? Sample answer: If the segment determined by the points is neither horizontal nor vertical, draw a right triangle that has the segment as its hypotenuse. The horizontal side will have length $\lvert x_2 - x_1 \rvert$ and the vertical side will have length $\lvert y_2 - y_1 \rvert$. By the Pythagorean Theorem, $d^2 = \lvert x_2 - x_1 \rvert^2 + \lvert y_2 - y_1 \rvert^2 = (x_2 - x_1)^2 + (y_2 - y_1)^2$.

PERIMETER For Exercises 29 and 30, use the following information.

The perimeter of a figure is the sum of the lengths of its sides.

29. The vertices of a triangle are located at $X(-2, -1)$, $Y(2, 5)$, and $Z(4, 3)$. What is the perimeter of this triangle? Round to the nearest tenth. 17.3 units

30. What is the perimeter of a square whose vertices are $A(-4, -3)$, $B(-5, 1)$, $C(-1, 2)$, and $D(0, -2)$? 16.5 units

Use the number line to find the coordinate of the midpoint of each segment.

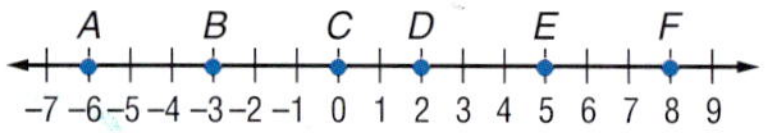

31. $\overline{AC}$ −3
32. $\overline{DF}$ 5
33. $\overline{CE}$ 2.5
34. $\overline{BD}$ −0.5
35. $\overline{AF}$ 1
36. $\overline{BE}$ 1

Find the coordinates of the midpoint of a segment having the given endpoints.

37. $A(8, 4)$, $B(12, 2)$ (10, 3)
38. $C(9, 5)$, $D(17, 4)$ (13, 4.5)
39. $E(-11, -4)$, $F(-9, -2)$ (−10, −3)
40. $G(4, 2)$, $H(8, -6)$ (6, −2)
★ 41. $J(3.4, 2.1)$, $K(7.8, 3.6)$ (5.6, 2.85)
★ 42. $L(-1.4, 3.2)$, $M(2.6, -5.4)$ (0.6, −1.1)

Find the coordinates of the missing endpoint given that S is the midpoint of $\overline{RT}$.

43. $T(-4, 3)$, $S(-1, 5)$
44. $T(2, 8)$, $S(-2, 2)$
★ 45. $R\left(\frac{2}{3}, -5\right)$, $S\left(\frac{5}{3}, 3\right)$

GEOGRAPHY For Exercises 46 and 47, use the following information.

The geographic center of Texas is located northeast of Brady at (31.1°, 99.3°), which represent north latitude and west longitude. El Paso is located near the western border of Texas at (31.8°, 106.4°).

46. If El Paso is one endpoint of a segment and the geographic center is its midpoint, find the latitude and longitude of the other endpoint. (30.4°, 92.2°)

47. Use an atlas or the Internet to find a city near this location. LaFayette, LA

SPREADSHEETS For Exercises 48 and 49, refer to the information at the left and use the following information.

Spreadsheets can be used to perform calculations quickly. Values are used in formulas by using a specific cell name. For example, the value of x_1 below is used in a formula using its cell name, A2. Special commands are used to perform some operations. For example, $\sqrt{x_1 - x_2}$ would be written as =SQRT(A2 − C2). The spreadsheet below can be used to calculate the distance between two points.

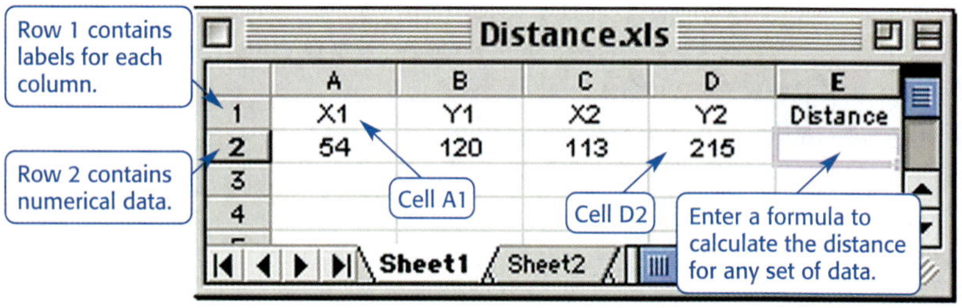

★ 48. Write a formula for cell E2 that could be used to calculate the distance between (x_1, y_1) and (x_2, y_2). Sample answer: =SQRT((A2−C2)^2+(B2−D2)^2)

★ 49. Find the distance between each pair of points to the nearest tenth.

a. (54, 120), (113, 215) 111.8
b. (68, 153), (175, 336) 212.0
c. (421, 454), (502, 798) 353.4
d. (837, 980), (612, 625) 420.3
e. (1967, 3), (1998, 24) 37.4
f. (4173.5, 34.9), (2080.6, 22.4) 2092.9

43. $R(2, 7)$
44. $R(-6, -4)$
45. $T\left(\frac{8}{3}, 11\right)$

Study Tip

Latitude and Longitude

Actual distances on Earth are calculated along the curve of Earth's surface. However, when the two points are located relatively close together, you can apply the concepts of plane geometry to approximate coordinates and distances.

Study Tip

Spreadsheets

Spreadsheets often use special commands to perform operations. For example, to find the square root of a number, use the SQRT command. To raise a number to a power, x^2 for example, write it as x^2.

Enrichment, p. 18

Lengths on a Grid

Evenly-spaced horizontal and vertical lines form a grid.

You can easily find segment lengths on a grid if the endpoints are grid-line intersections. For horizontal or vertical segments, simply count squares. For diagonal segments, use the Pythagorean Theorem (proven in Chapter 7). This theorem states that in any right triangle, if the length of the longest side (the side opposite the right angle) is c and the two shorter sides have lengths a and b, then $c^2 = a^2 + b^2$.

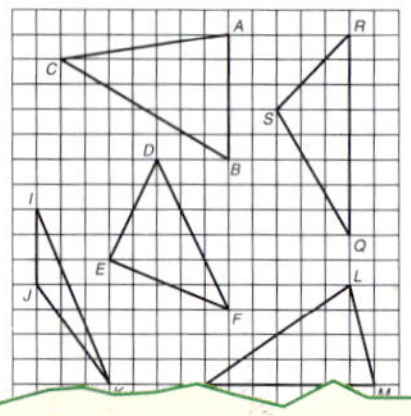

Example Find the measure of $\overline{EF}$ on the grid at the right. Locate a right triangle with $\overline{EF}$ as its longest side

ELL notations throughout the chapter indicate items that can assist English-Language Learners.

ENLARGEMENT **For Exercises 50–53, use the following information.**
The coordinates of the vertices of a triangle are $A(1, 3)$, $B(6, 10)$, and $C(11, 18)$.

50. Find the perimeter of $\triangle ABC$. **≈ 36.1**
51. Suppose each coordinate is multiplied by 2. What is the perimeter of this triangle? **≈ 72.1**
52. Find the perimeter of the triangle when the coordinates are multiplied by 3. **≈ 108.2**
53. Make a conjecture about the perimeter of a triangle when the coordinates of its vertices are multiplied by the same positive factor. **Sample answer: The perimeter increases by the same factor.**

54. **CRITICAL THINKING** In the figure, $\overline{GE}$ bisects $\overline{BC}$, and $\overline{GF}$ bisects $\overline{AB}$. $\overline{GE}$ is a horizontal segment, and $\overline{GF}$ is a vertical segment.

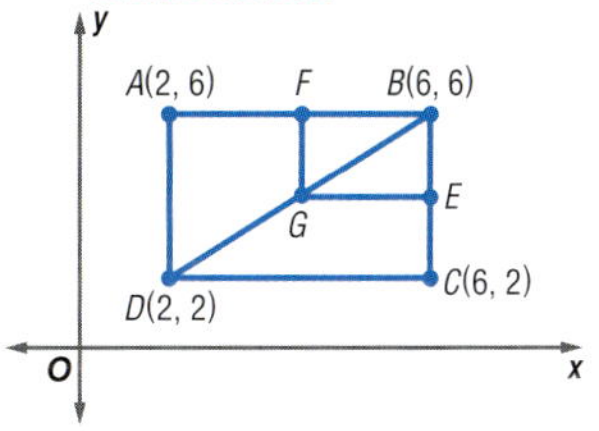

a. Find the coordinates of points F and E.
b. Name the coordinates of G and explain how you calculated them.
c. Describe what relationship, if any, exists between $\overline{DG}$ and $\overline{GB}$. Explain.

54a. $F(4, 6)$, $E(6, 4)$
54b. $G(4, 4)$; it has the same x-coordinate as F and the same y-coordinate as E.
54c. $\overline{DG} \cong \overline{GB}$; you can use the Distance Formula to find that $DG = GB$.

55. **CRITICAL THINKING** $\overline{WZ}$ has endpoints $W(-3, -8)$ and $Z(5, 12)$. Point X lies between W and Z, such that $WX = \frac{1}{4}WZ$. Find the coordinates of X. **(−1, −3)**

56. **WRITING IN MATH** Answer the question that was posed at the beginning of the lesson. **See margin.**

How can you find the distance between two points without a ruler?

Include the following in your answer:
- how to use the Pythagorean Theorem and the Distance Formula to find the distance between two points, and
- the length of $\overline{AB}$ from the figure on page 21.

FCAT Practice
Standardized Test Practice

57. Find the distance between points at (6, 11) and (−2, −4). **B**
    (A) 16 units (B) 17 units (C) 18 units (D) 19 units

58. **ALGEBRA** Which equation represents the following problem? **A**
    Fifteen minus three times a number equals negative twenty-two. Find the number.
    (A) $15 - 3n = -22$ (B) $3n - 15 = -22$
    (C) $3(15 - n) = -22$ (D) $3(n - 15) = -22$

Maintain Your Skills

Mixed Review **Find the measurement of each segment.** *(Lesson 1-2)*

59. $\overline{WY}$ **$4\frac{1}{4}$ in.**

60. $\overline{BC}$ **5.5 cm**

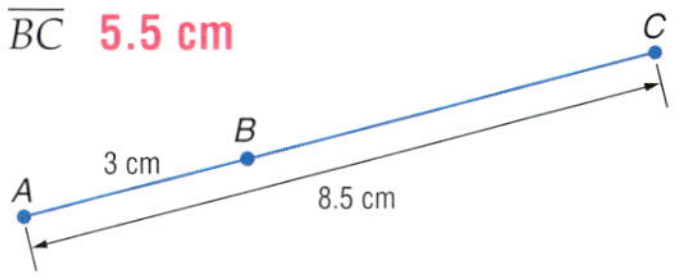

Draw and label a figure for each relationship. *(Lesson 1-1)* **61–62. See margin.**

61. four noncollinear points A, B, C, and D that are coplanar
62. line m that intersects plane $\mathcal{A}$ and line n in plane $\mathcal{A}$

Getting Ready for the Next Lesson **PREREQUISITE SKILL** **Solve each equation.** *(To review* ***solving equations****, see page 737.)*

63. $2k = 5k - 30$ **10**
64. $14x - 31 = 12x + 8$ **19.5**
65. $180 - 8t = 90 + 2t$ **9**
66. $12m + 7 = 3m + 52$ **5**
67. $8x + 7 = 5x + 20$ **$\frac{13}{3}$**
68. $13n - 18 = 5n + 32$ **6.25**

64. 19.5

4 Assess

Open-Ended Assessment

Speaking Separate students into three groups and assign each group one of the three important concepts in this lesson: the Distance Formula, the Pythagorean Theorem, and the Midpoint Formula. Allow each group to discuss the concepts behind its particular formula, including how to use it and why it works, and then summarize their findings for the rest of the class.

Getting Ready for Lesson 1-4

Prerequisite Skill Students will learn about measuring angles in Lesson 1-4. They will use algebra to find angle measures. Use Exercises 63–68 to determine your students' familiarity with solving algebraic equations.

Assessment Options

Quiz (Lesson 1-3) is available on p. 51 of the *Chapter 1 Resource Masters.*

Mid-Chapter Test (Lessons 1-1 through 1-3) is available on p. 53 of the *Chapter 1 Resource Masters.*

By having your students complete the Getting Ready exercises, you can target specific skills they will need for the next lesson.

61. Sample answer:

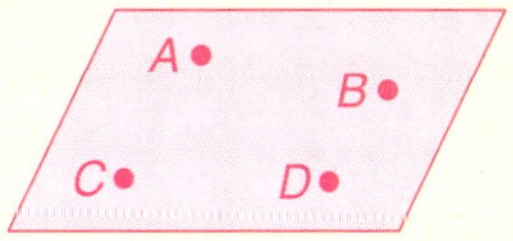

62.

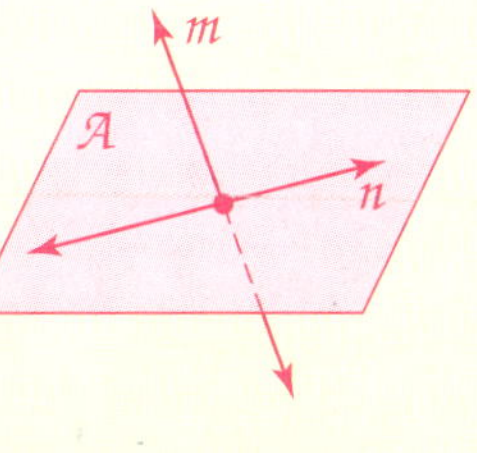

Answers

56. Sample answer: You can copy the segment onto a coordinate plane and then use either the Pythagorean Theorem or the Distance Formula to find its length. Answers should include the following.
- To use the Pythagorean Theorem, draw a vertical segment from one endpoint and a horizontal segment from the other endpoint to form a triangle. Use the measures of these segments as a and b in the formula $a^2 + b^2 = c^2$. Then solve for c. To use the Distance Formula, assign the coordinates of the endpoints of the segment as (x_1, y_1) and (x_2, y_2). Then use them in $d = \sqrt{(x_2 - x_1)^2 + (y_2 - y_1)^2}$ to find the length of the segment.
- $\sqrt{61} \approx 7.8$ units

Geometry Activity

A Follow-Up of Lesson 1-3

Getting Started

Objective Use grid paper to model the Pythagorean Theorem.

Materials
grid paper
scissors
straightedge

Teach

- Suggest that students plot the lower left vertex of the triangle as the first point, place the second point 12 units to the right of the first, and the third point 5 units up from the second. Then they can use a straightedge to connect the points and easily replicate the right triangle in the activity.
- Point out that the longest side of the right triangle has to be the one opposite the right angle, as they will see when they relate the sides of the triangle in the exercises.

Assess

After **Exercises 1–4,** students should have a visual understanding of how the Pythagorean Theorem works.

Exercises 5 allows students to reinforce the activity and further illustrate the Pythagorean Theorem.

Exercise 6 is an extension of the activity and alerts students to consider how the Pythagorean Theorem works with isosceles triangles.

Study Notebook

Ask students to summarize what they have learned from this activity about the Pythagorean Theorem.

Geometry Activity

A Follow-Up of Lesson 1-3

Modeling the Pythagorean Theorem

In Chapter 7, you will formally write a verification of the Pythagorean Theorem, but this activity will suggest that the Pythagorean Theorem holds for any right triangle. Remember that a right triangle is a triangle with a right angle, and that a right angle measures 90°.

Make a Model

- Draw right triangle *ABC* in the center of a piece of grid paper.

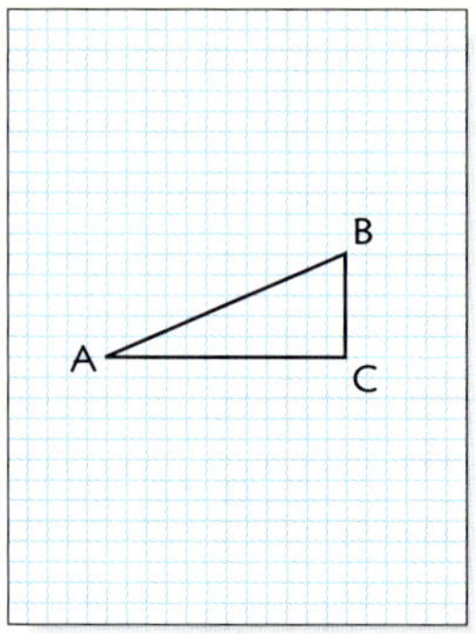

- Use another piece of grid paper to draw a square that is 5 units on each side, a square that is 12 units on each side, and a square that is 13 units on each side. Use colored pencils to shade each of these squares. Cut out the squares. Label them as 5 × 5, 12 × 12, and 13 × 13 respectively.
- Place the squares so that a side of the square matches up with a side of the right triangle.

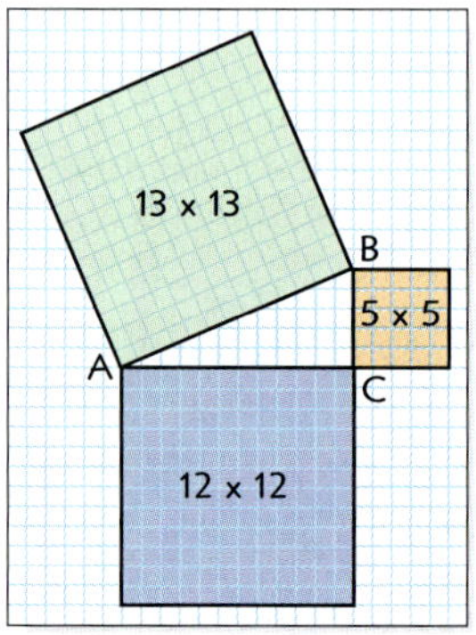

Analyze

1. Determine the number of grid squares in each square you drew. **25, 144, 169**
2. How do the numbers of grid squares relate? **$25 + 144 = 169$**
3. If $AB = c$, $BC = a$, and $AC = b$, write an expression to describe each of the squares. **a^2, b^2, c^2**
4. How does this expression compare with what you know about the Pythagorean Theorem? **The formula for the Pythagorean Theorem can be expressed as $a^2 + b^2 = c^2$.**

Make a Conjecture

5. Repeat the activity for triangles with each of the side measures listed below. What do you find is true of the relationship of the squares on the sides of the triangle? **All of these fit the $a^2 + b^2 = c^2$ pattern.**

   a. 3, 4, 5 **b.** 8, 15, 17 **c.** 6, 8, 10

6. Repeat the activity with a right triangle whose shorter sides are both 5 units long. How could you determine the number of grid squares in the larger square? **The number of grid squares is $5^2 + 5^2$, which is 50 grid squares.**

Resource Manager

Teaching Geometry with Manipulatives
- p. 34 (student recording sheet)
- p. 1 (master for grid paper)

Glencoe Mathematics Classroom Manipulative Kit
- scissors
- straightedge

28 Chapter 1 Points, Lines, Planes, and Angles

1-4 Angle Measure

What You'll Learn

- Measure and classify angles.
- Identify and use congruent angles and the bisector of an angle.

How big is a degree?

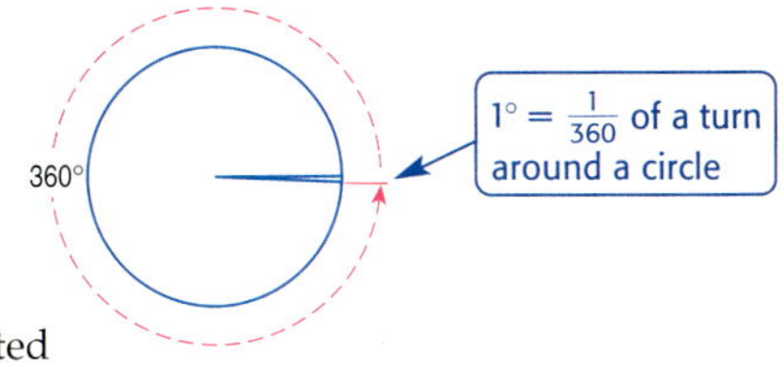

One of the first references to the measure now known as a degree came from astronomer Claudius Ptolemy. He based his observations of the solar system on a unit that resulted from dividing the circumference, or the distance around, a circle into 360 parts. This later became known as a **degree**. In this lesson, you will learn to measure angles in degrees.

Vocabulary

- degree
- ray
- opposite rays
- angle
- sides
- vertex
- interior
- exterior
- right angle
- acute angle
- obtuse angle
- angle bisector

Study Tip

Reading Math

Opposite rays are also known as a *straight angle*. Its measure is 180°. Unless otherwise specified, the term *angle* in this book means a nonstraight angle.

MEASURE ANGLES A **ray** is part of a line. It has one endpoint and extends indefinitely in one direction. Rays are named stating the endpoint first and then any other point on the ray. The figure at the right shows ray *EF*, which can be symbolized as $\overrightarrow{EF}$. This ray could also be named as $\overrightarrow{EG}$, but not as $\overrightarrow{FE}$ because *F* is not the endpoint of the ray.

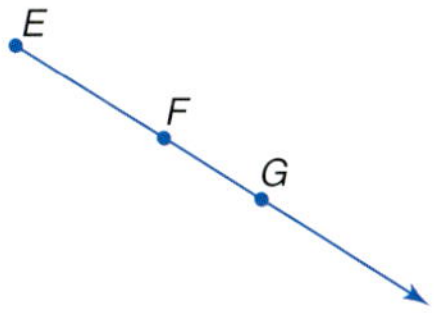

If you choose a point on a line, that point determines exactly two rays called **opposite rays**. Line *m*, shown below, is separated into two opposite rays, $\overrightarrow{PQ}$ and $\overrightarrow{PR}$. Point *P* is the common endpoint of those rays. $\overrightarrow{PQ}$ and $\overrightarrow{PR}$ are collinear rays.

An **angle** is formed by two *noncollinear* rays that have a common endpoint. The rays are called **sides** of the angle. The common endpoint is the **vertex**.

Key Concept — Angle

- **Words** An angle is formed by two noncollinear rays that have a common endpoint.
- **Symbols** $\angle A$, $\angle BAC$, $\angle CAB$, $\angle 4$
- **Model**

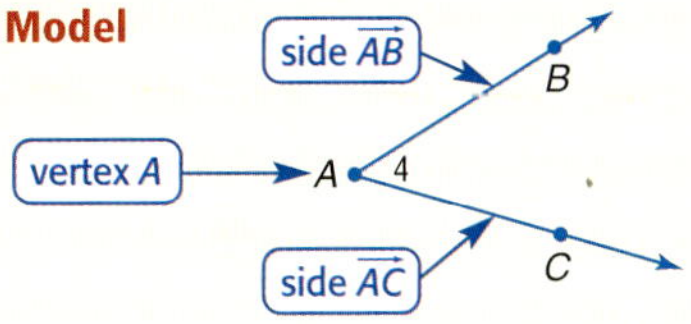

An angle divides a plane into three distinct parts.

- Points *A*, *D*, and *E* lie on the angle.
- Points *C* and *B* lie in the **interior** of the angle.
- Points *F* and *G* lie in the **exterior** of the angle.

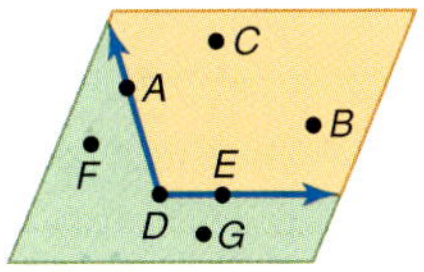

1-4 Lesson Notes

1 Focus

5-Minute Check Transparency 1-4 Use as a quiz or review of Lesson 1-3.

Mathematical Background notes are available for this lesson on p. 4D.

How big is a degree?

Ask students:

- Does the size of a degree depend on the size of a circle? Explain. **No. Every circle can either contain or be contained in another circle. A degree is always $\frac{1}{360}$ of a turn around any circle.**
- How many multiples of 60 degrees are there in a circle? **6**

The Resource Manager lists all of the resources available for the lesson, including workbooks, blackline masters, transparencies, and technology.

Resource Manager

Workbook and Reproducible Masters

Chapter 1 Resource Masters

- Study Guide and Intervention, pp. 19–20
- Skills Practice, p. 21
- Practice, p. 22
- Reading to Learn Mathematics, p. 23
- Enrichment, p. 24

School-to-Career Masters, p. 2

Prerequisite Skills Workbook, pp. 81–82

Teaching Geometry With Manipulatives Masters, pp. 16, 17, 35, 36, 37, 38

Transparencies

5-Minute Check Transparency 1-4

Answer Key Transparencies

Technology

Interactive Chalkboard

2 Teach

MEASURE ANGLES

In-Class Examples

Power Point®

1

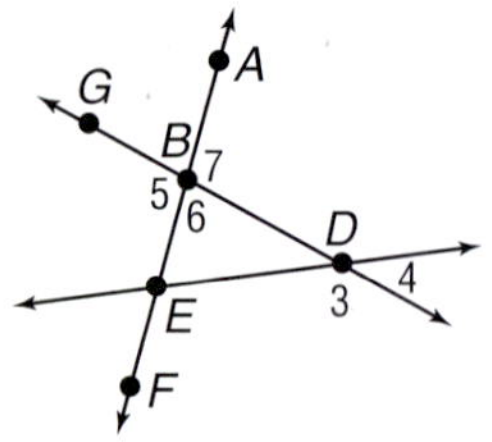

a. Name all angles that have B as a vertex. **∠5, ∠6, ∠7, ∠ABG**

b. Name the sides of ∠5.
$\overrightarrow{BG}$ and $\overrightarrow{BE}$ or $\overrightarrow{BF}$

c. Write another name for ∠6.
∠EBD, ∠FBD, ∠DBF, or ∠DBE

2 Measure each angle named and classify it as *right*, *acute*, or *obtuse*.

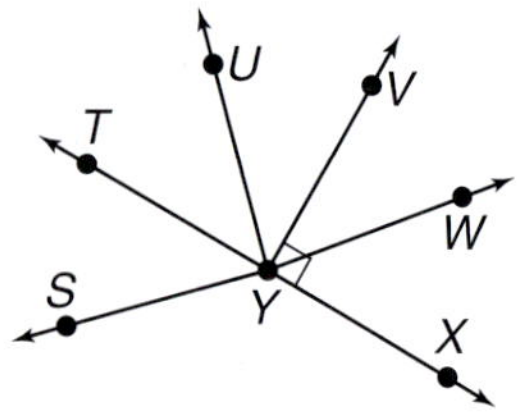

a. ∠TYV **90, right**

b. ∠WYT **130, obtuse**

c. ∠TYU **45, acute**

Teaching Tip Ask students to recall how they used a construction to copy a segment in Lesson 1-2. Explain that to copy a segment, they needed only to copy its length; but in order to copy an angle, they must determine how far apart the rays of the angle are at any given point, which is why the construction method is so different. Point out that the first time they opened the compass to copy a line segment, they used a specified length; however, the first time they open the compass to copy an angle, they select an arbitrary length. They will adjust the compass to a specified length to make the second arc in the angle construction.

Study Tips offer students helpful information about the topics they are studying.

Study Tip

Naming Angles
You can name an angle by a single letter *only* when there is one angle shown at that vertex.

Example 1 Angles and Their Parts

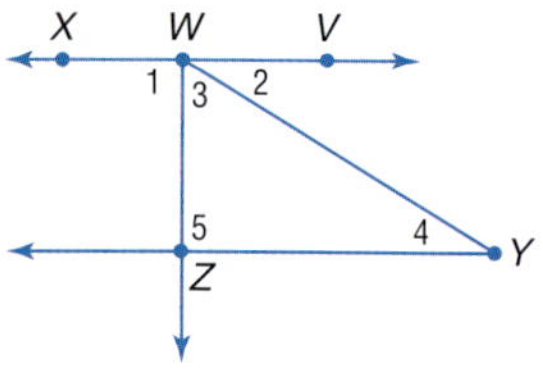

a. Name all angles that have W as a vertex.
∠1, ∠2, ∠3, ∠XWY, ∠ZWV

b. Name the sides of ∠1.
$\overrightarrow{WZ}$ and $\overrightarrow{WX}$ are the sides of ∠1.

c. Write another name for ∠WYZ.
∠4, ∠Y, and ∠ZYW are other names for ∠WYZ.

To measure an angle, you can use a *protractor*. Angle PQR is a 65 degree (65°) angle. We say that the *degree measure* of ∠PQR is 65, or simply $m\angle PQR = 65$.

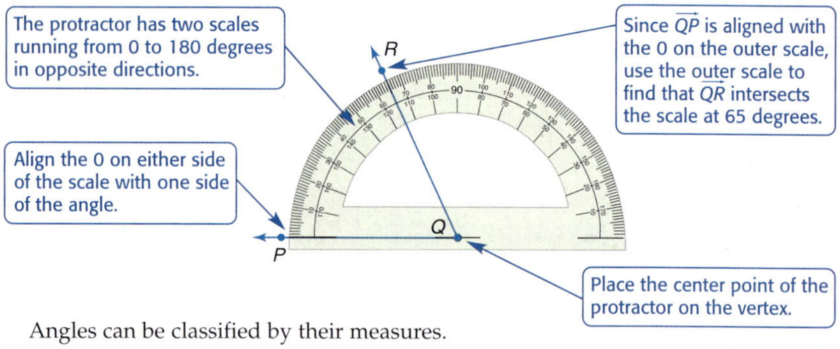

Angles can be classified by their measures.

Study Tip

Classifying Angles
The corner of a piece of paper is a right angle. Use the corner to determine if an angle's measure is greater than 90 or less than 90.

Key Concept — *Classify Angles*

Name	right angle	acute angle	obtuse angle
Measure	$m\angle A = 90$	$m\angle B < 90$	$180 > m\angle C > 90$
Model	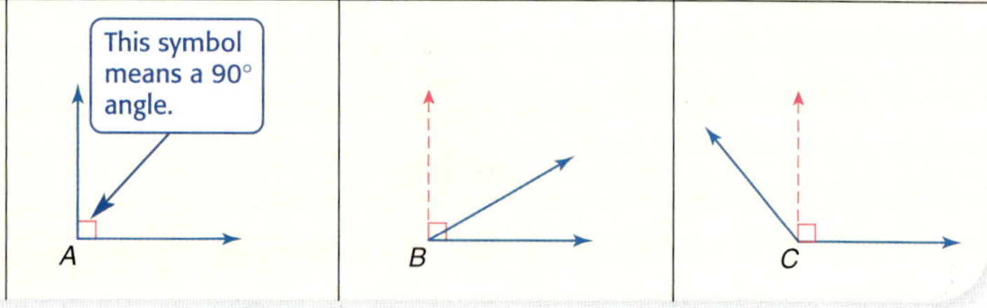		

Example 2 Measure and Classify Angles

Measure each angle named and classify it as *right*, *acute*, or *obtuse*.

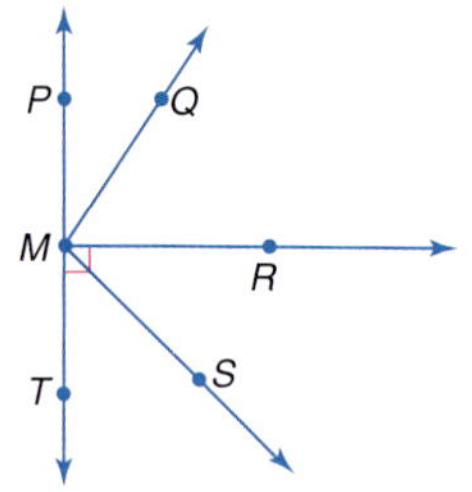

a. ∠PMQ
Use a protractor to find that $m\angle PMQ = 30$.
$30 < 90$, so ∠PMQ is an acute angle.

b. ∠PMR
∠PMR is marked with a right angle symbol, so measuring is not necessary; $m\angle PMR = 90$.

c. ∠QMS
Use a protractor to find that $m\angle QMS = 110$. ∠QMS is an obtuse angle.

DAILY INTERVENTION — **Differentiated Instruction**

Auditory/Musical A metronome is a tool used to keep a constant tempo in music. It is composed of a pendulum that swings back and forth at varying speeds. The fulcrum of the pendulum acts as a vertex of the angle through which the pendulum swings. Demonstrate this by holding two pens at an angle in one hand and tapping another pen between the first two, creating a series of "ticks."

CONGRUENT ANGLES Just as segments that have the same measure are congruent, angles that have the same measure are congruent.

Key Concept — *Congruent Angles*

- **Words** Angles that have the same measure are congruent angles. Arcs on the figure also indicate which angles are congruent.
- **Symbols** $\angle NMP \cong \angle QMR$
- **Model**

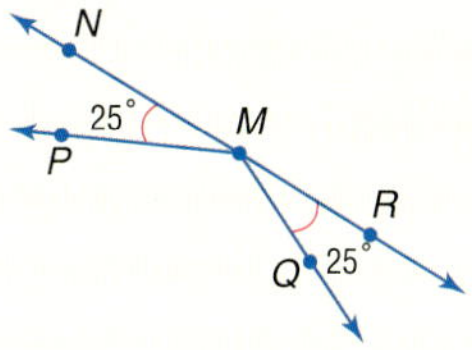

You can construct an angle congruent to a given angle without knowing the measure of the angle.

Construction

Copy an Angle

1. Draw an angle like $\angle P$ on your paper. Use a straightedge to draw a ray on your paper. Label its endpoint T.

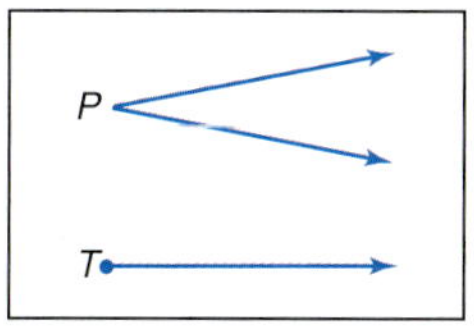

2. Place the tip of the compass at point P and draw a large arc that intersects both sides of $\angle P$. Label the points of intersection Q and R.

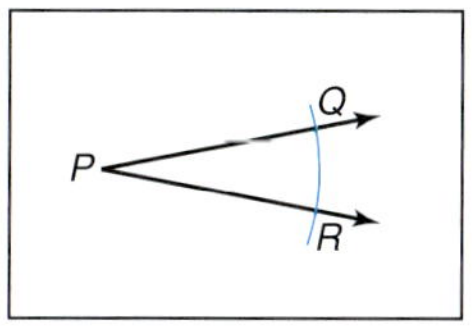

3. Using the same compass setting, put the compass at T and draw a large arc that intersects the ray. Label the point of intersection S.

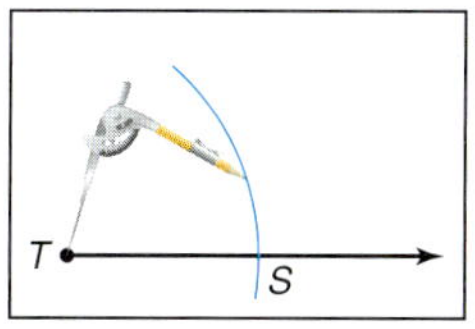

4. Place the point of your compass on R and adjust so that the pencil tip is on Q.

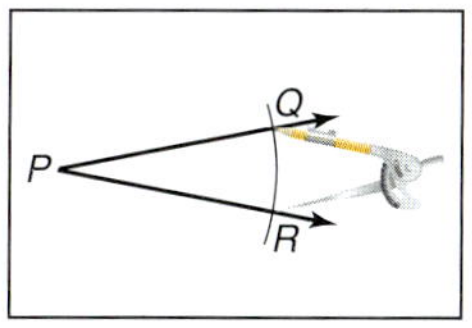

5. Without changing the setting, place the compass at S and draw an arc to intersect the larger arc you drew in Step 3. Label the point of intersection U.

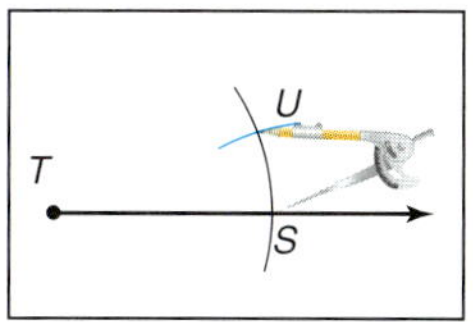

6. Use a straightedge to draw $\overrightarrow{TU}$.

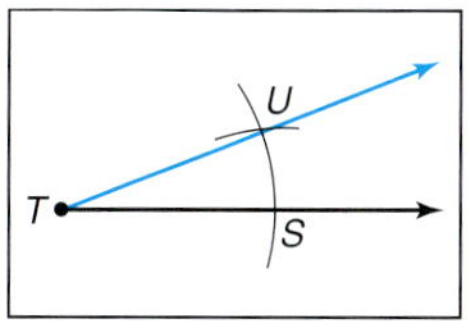

www.geometryonline.com/extra_examples/fcat

CONGRUENT ANGLES

Tips for New Teachers

Building on Prior Knowledge Always try to relate new material to what students have previously learned. Relate angle vocabulary to what they used with segments, such as segments of same length are congruent, angles with same measure are congruent, and that the symbol to write this relation for both segments and angles is $\cong$. Likewise, the bisector of a segment cuts it in half, and the bisector of an angle cuts it in half.

Teaching Tip Some students may feel they need to draw figures as small as they see them printed in the book. Encourage students to use a whole piece of paper for the construction. Draw the angle so it takes up much of the top half of the paper. Then they can construct the copy on the bottom half. Working with larger figures also makes working with a compass easier until students get more proficient with constructions.

In-Class Example

Power Point®

3 INTERIOR DESIGN Wall stickers of standard shapes are often used to provide a stimulating environment for a young child's room. A five-pointed star sticker is shown with vertices labeled. Find $m\angle GBH$ and $m\angle HCI$ if $\angle GBH \cong \angle HCI$, $m\angle GBH = 2x + 5$, and $m\angle HCI = 3x - 10$.

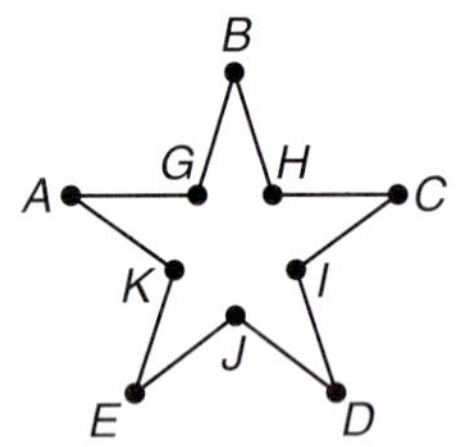

$m\angle GBH = m\angle HCI = 35$

Example 3 Use Algebra to Find Angle Measures

GARDENING A trellis is often used to provide a frame for vining plants. Some of the angles formed by the slats of the trellis are congruent angles. In the figure, $\angle ABC \cong \angle DBF$. If $m\angle ABC = 6x + 2$ and $m\angle DBF = 8x - 14$, find the actual measurements of $\angle ABC$ and $\angle DBF$.

$\angle ABC \cong \angle DBF$	Given
$m\angle ABC = m\angle DBF$	Definition of congruent angles
$6x + 2 = 8x - 14$	Substitution
$6x + 16 = 8x$	Add 14 to each side.
$16 = 2x$	Subtract 6x from each side.
$8 = x$	Divide each side by 2.

Use the value of x to find the measure of one angle.

$m\angle ABC = 6x + 2$	Given
$= 6(8) + 2$	$x = 8$
$= 48 + 2$ or 50	Simplify.

Since $m\angle ABC = m\angle DBF$, $m\angle DBF = 50$.

Both $\angle ABC$ and $\angle DBF$ measure 50°.

Study Tip

Checking Solutions

Check that you have computed the value of x correctly by substituting the value into the expression for $\angle DBF$. If you don't get the same measure as $\angle ABC$, you have made an error.

Geometry Activity

Bisect an Angle

Make a Model

- Draw any $\angle XYZ$ on patty paper or tracing paper.
- Fold the paper through point Y so that $\overrightarrow{YX}$ and $\overrightarrow{YZ}$ are aligned together.
- Open the paper and label a point on the crease in the interior of $\angle XYZ$ as point W.

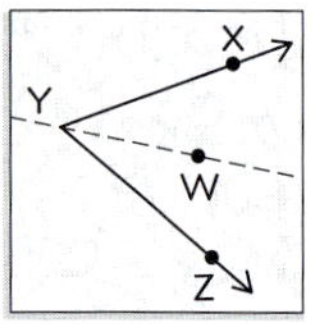

Analyze the Model

1. What seems to be true about $\angle XYW$ and $\angle WYZ$? **They are congruent.**
2. Measure $\angle XYZ$, $\angle XYW$, and $\angle WYZ$. **See students' work.**
3. You learned about a segment bisector in Lesson 1-3. Write a sentence to explain the term *angle bisector*.

3. A segment bisector separates a segment into two congruent segments; an angle bisector separates an angle into two congruent angles.

Study Tip

Adding Angle Measures

Just as with segments, when a line divides an angle into smaller angles, the sum of the measures of the smaller angles equals the measure of the largest angle. So in the figure, $m\angle RPS = m\angle RPQ + m\angle QPS$.

A ray that divides an angle into two congruent angles is called an **angle bisector**. If $\overrightarrow{PQ}$ is the angle bisector of $\angle RPS$, then point Q lies in the interior of $\angle RPS$ and $\angle RPQ \cong \angle QPS$.

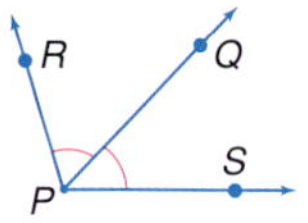

You can construct the angle bisector of any angle without knowing the measure of the angle.

Geometry Activity

Materials: patty paper or tracing paper, straightedge, protractor

- Suggest that students extend the rays for their angles. This will help later when students measure their angles with protractors. Lines should also be dark enough to see when the students are folding the paper.
- Have students repeat the activity using a different kind of angle. For example, a student who drew an acute angle would now draw an obtuse angle and vice versa.

32 Chapter 1 Points, Lines, Planes, and Angles

Construction

Bisect an Angle

1. Draw an angle on your paper. Label the vertex as A. Put your compass at point A and draw a large arc that intersects both sides of $\angle A$. Label the points of intersection B and C.
2. With the compass at point B, draw an arc in the interior of the angle.
3. Keeping the same compass setting, place the compass at point C and draw an arc that intersects the arc drawn in Step 2.
4. Label the point of intersection D. Draw $\overrightarrow{AD}$. $\overrightarrow{AD}$ is the bisector of $\angle A$. Thus, $m\angle BAD = m\angle DAC$ and $\angle BAD \cong \angle DAC$.

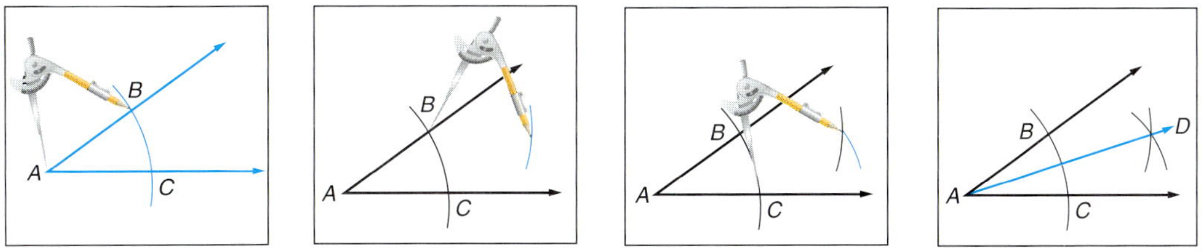

Check for Understanding

Concept Check

1. **Determine** whether all right angles are congruent. **Yes; they all have the same measure.**
2. **OPEN ENDED** Draw and label a figure to show $\overline{PR}$ that bisects $\angle SPQ$ and $\overrightarrow{PT}$ that bisects $\angle SPR$. Use a protractor to measure each angle. **See margin.**
3. **Write** a statement about the measures of congruent angles A and Z. $m\angle A = m\angle Z$

Guided Practice

GUIDED PRACTICE KEY	
Exercises	Examples
4–6	1
7, 8, 11	2
9–10	3

For Exercises 4 and 5, use the figure at the right.

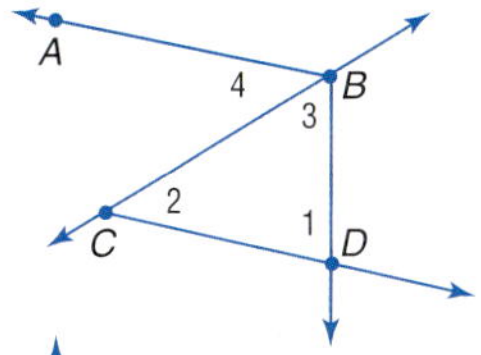

4. Name the vertex of $\angle 2$. **C**
5. Name the sides of $\angle 4$. $\overrightarrow{BA}$, $\overrightarrow{BC}$
6. Write another name for $\angle BDC$. $\angle CDB$, $\angle 1$

Measure each angle and classify as *right*, *acute*, or *obtuse*.

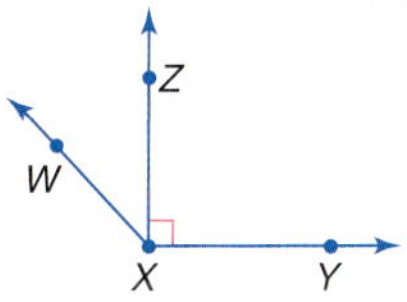

7. $\angle WXY$ **135°, obtuse**
8. $\angle WXZ$ **45°, acute**

ALGEBRA In the figure, $\overrightarrow{QP}$ and $\overrightarrow{QR}$ are opposite rays, and $\overrightarrow{QT}$ bisects $\angle RQS$.

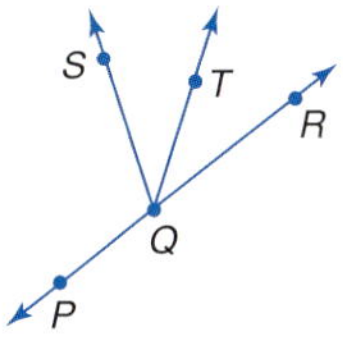

9. If $m\angle RQT = 6x + 5$ and $m\angle SQT = 7x - 2$, find $m\angle RQT$. **47**
10. Find $m\angle TQS$ if $m\angle RQS = 22a - 11$ and $m\angle RQT = 12a - 8$. **22**

Application

11. **ORIGAMI** The art of origami involves folding paper at different angles to create designs and three-dimensional figures. One of the folds in origami involves folding a strip of paper so that the lower edge of the strip forms a right angle with itself. Identify each numbered angle as *right*, *acute*, or *obtuse*. $\angle 1$, **right;** $\angle 2$, **acute;** $\angle 3$, **obtuse**

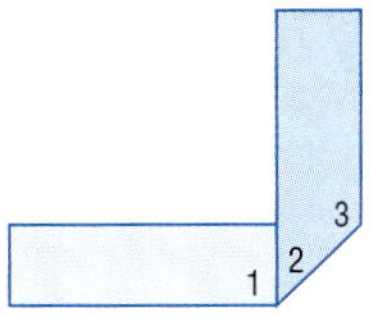

Check for Understanding exercises are intended to be completed in class. Concept Check exercises ensure that students understand the concepts in the lesson. The other exercises are representative of the exercises used for homework.

Teaching Tip Advise students that in Step 3 they can draw longer arcs than pictured in the figure to ensure that the arcs intersect.

3 Practice/Apply

Study Notebook

Have students—
- add the definitions/examples of the vocabulary terms to their Vocabulary Builder worksheets for Lesson 1-4.
- include any other item(s) that they find helpful in mastering the skills in this lesson.

Study Notebook tips offer suggestions for helping your students keep notes they can use to study this chapter.

Answer

2. Sample answer:

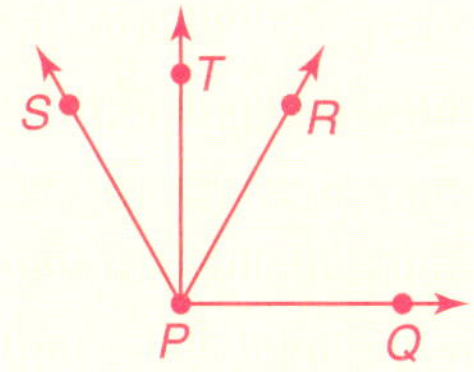

$m\angle QPR = 60$;
$m\angle QPT = 90$;
$m\angle QPS = 120$

About the Exercises...

Organization by Objective
- **Measure Angles:** 12–33
- **Congruent Angles:** 34–39

Odd/Even Assignments

Exercises 12–39 are structured so that students practice the same concepts whether they are assigned odd or even problems.

Assignment Guide

Basic: 13–37 odd, 41–43, 44–66

Average: 13–43 odd, 44–66

Advanced: 12–42 even, 44–60 (optional: 61–66)

All: Practice Quiz 2 (1–5)

Answers

40. The angle at which the dogs must turn to get the scent of the article they wish to find is an acute angle.

41. Sample answer: *Acute* can mean something that is sharp or having a very fine tip like a pen, a knife, or a needle. *Obtuse* means not pointed or blunt, so something that is obtuse would be wide.

46. 3 rays: $(3 \cdot 2) \div 2 = 3$ angles; 4 rays: $(4 \cdot 3) \div 2 = 6$ angles; 5 rays: $(5 \cdot 4) \div 2 = 10$ angles; 6 rays: $(6 \cdot 5) \div 2 = 15$ angles

49. Sample answer: A degree is $\frac{1}{360}$ of a circle. Answers should include the following.
   - Place one side of the angle to coincide with 0 on the protractor and the vertex of the angle at the center point of the protractor. Observe the point at which the other side of the angle intersects the scale of the protractor.
   - See students' work.

Every effort is made to show the Answers to exercises (1) on the reduced Student Edition page, or (2) in the margin of the Teacher's Wraparound Edition. However, answers that do not fit in either of these places can be found in pages at the end of each chapter.

★ indicates increased difficulty

Practice and Apply

Homework Help

For Exercises	See Examples
12–27	1
28–33	2
34–39	3

Extra Practice
See page 755.

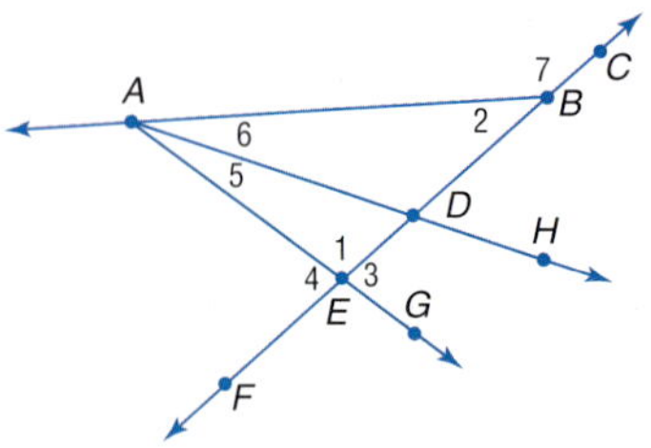

Name the vertex of each angle.

12. $\angle 1$ **E**
13. $\angle 2$ **B**
14. $\angle 6$ **A**
15. $\angle 5$ **A**

Name the sides of each angle.

16. $\angle ADB$ $\overrightarrow{DA}$, $\overrightarrow{DB}$
17. $\angle 6$ $\overrightarrow{AB}$, $\overrightarrow{AD}$
18. $\angle 3$ $\overrightarrow{ED}$, $\overrightarrow{EG}$
19. $\angle 5$ $\overrightarrow{AD}$, $\overrightarrow{AE}$

Write another name for each angle.

20. $\angle 7$ $\angle ABC$, $\angle CBA$
21. $\angle AEF$ $\angle FEA$, $\angle 4$
22. $\angle ABD$ $\angle 2$, $\angle DBA$, $\angle EBA$, $\angle ABE$, $\angle FBA$, $\angle ABF$
23. $\angle 1$ $\angle AED$, $\angle DEA$, $\angle AEB$, $\angle BEA$, $\angle AEC$, $\angle CEA$
24. Name a point in the interior of $\angle GAB$. **D, H**
25. Name an angle with vertex B that appears to be acute. $\angle 2$
26. Name a pair of angles that share exactly one point. **Sample answer:** $\angle 4$, $\angle 3$
27. If $\overrightarrow{AD}$ bisects $\angle EAB$ and $m\angle EAB = 60$, find $m\angle 5$ and $m\angle 6$. **30, 30**

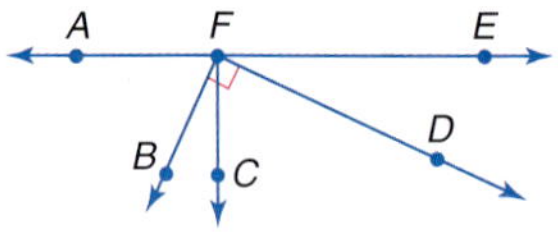

Measure each angle and classify it as *right*, *acute*, or *obtuse*.

28. $\angle BFD$ **90°, right**
29. $\angle AFB$ **60°, acute**
30. $\angle DFE$ **30°, acute**
31. $\angle EFC$ **90°, right**
32. $\angle AFD$ **150°, obtuse**
33. $\angle EFB$ **120°, obtuse**

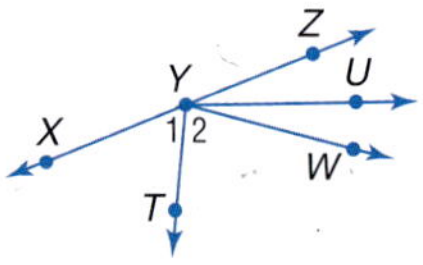

ALGEBRA **In the figure, $\overrightarrow{YX}$ and $\overrightarrow{YZ}$ are opposite rays. $\overrightarrow{YU}$ bisects $\angle ZYW$, and $\overrightarrow{YT}$ bisects $\angle XYW$.**

34. If $m\angle ZYU = 8p - 10$ and $m\angle UYW = 10p - 20$, find $m\angle ZYU$. **30**
35. If $m\angle 1 = 5x + 10$ and $m\angle 2 = 8x - 23$, find $m\angle 2$. **65**
36. If $m\angle 1 = y$ and $m\angle XYW = 6y - 24$, find y. **6**
37. If $m\angle WYZ = 82$ and $m\angle ZYU = 4r + 25$, find r. **4**

★ 38. If $m\angle WYX = 2(12b + 7)$ and $m\angle ZYU = 9b - 1$, find $m\angle UYW$. **35**

★ 39. If $\angle ZYW$ is a right angle and $m\angle ZYU = 13a - 7$, find a. **4**

40. **DOG TRACKING** A dog is *tracking* when it is following the scent trail left by a human being or other animal that has passed along a certain route. One of the training exercises for these dogs is a tracking trail. The one shown is called an acute tracking trail. Explain why it might be called this. **See margin.**

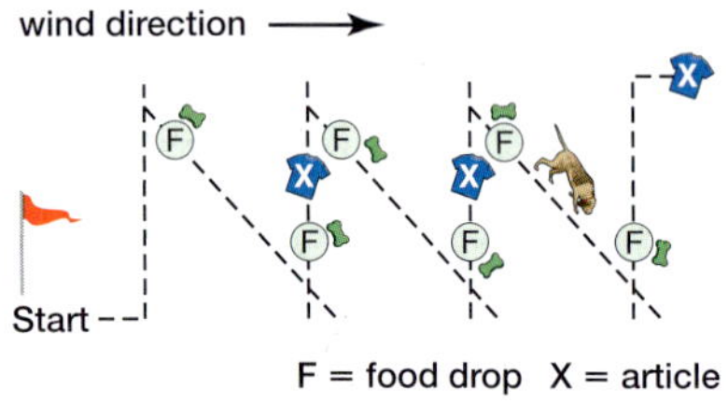

41. **LANGUAGE** The words *obtuse* and *acute* have other meanings in the English language. Look these words up and write how the everyday meaning relates to the mathematical meaning. **See margin.**

42. **PATTERN BLOCKS** Pattern blocks can be arranged to fit in a circular pattern without leaving spaces. Remember that the measurement around a full circle is 360°. Determine the angle measure of the numbered angles shown below.

More About. . .

Physics

A ripple tank is a large glass-bottomed tank of water. A light is placed above the water, and a white sheet of paper is placed below the tank. Because rays of light undergo bending as they pass through the troughs and crests of the water, there is a pattern of light and dark spots on the white sheet of paper. These model the wave.

43. **PHYSICS** A ripple tank can be used to study the behavior of waves in two dimensions. As a wave strikes a barrier, it is reflected. The angle of incidence and the angle of reflection are congruent. In the diagram at the right, if $m\angle IBR = 62$, find the angle of reflection and $m\angle IBA$. **31; 59**

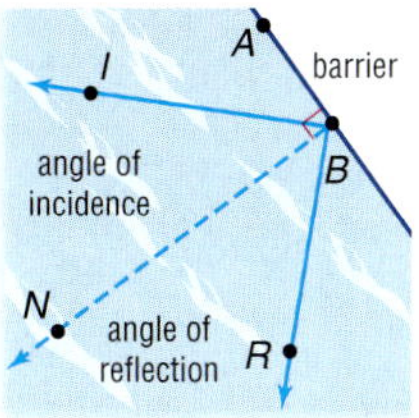

44. **CRITICAL THINKING** How would you compare the size of $\angle P$ and $\angle Q$? Explain. **You can only compare the measures of the angles. The arcs indicate both measures are the same regardless of the length of the rays.**

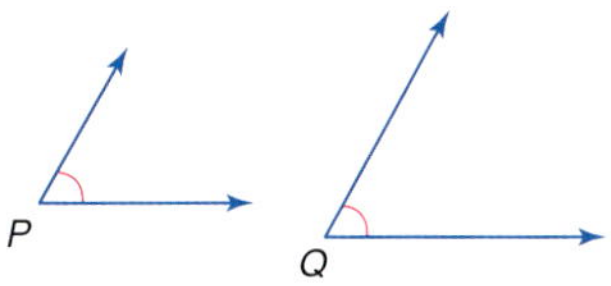

CRITICAL THINKING **For Exercises 45–48, use the following information.**
Each figure below shows noncollinear rays with a common endpoint.

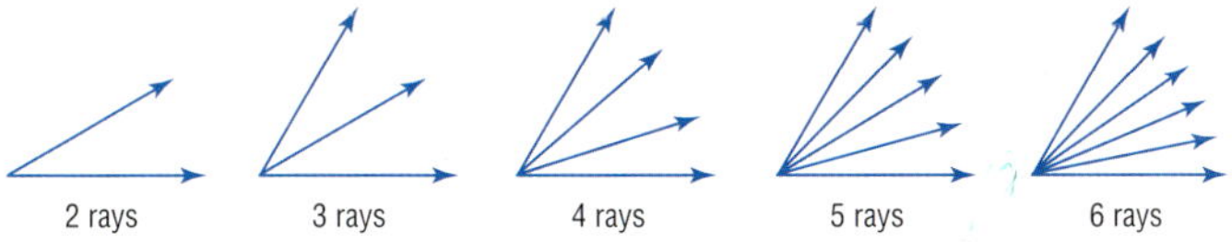

45. Count the number of angles in each figure. **1, 3, 6, 10, 15**

46. Describe the pattern between the number of rays and the number of angles.

46. See margin.

47. **Make a conjecture** of the number of angles that are formed by 7 noncollinear rays and by 10 noncollinear rays. **21, 45**

48. Write a formula for the number of angles formed by n noncollinear rays with a common endpoint. $a = \frac{n(n-1)}{2}$, **for a = number of angles and n = number of rays**

49. **WRITING IN MATH** Answer the question that was posed at the beginning of the lesson. **See margin.**

How big is a degree?

Include the following in your answer:

- how to find degree measure with a protractor, and
- drawings of several angles and their degree measures.

FCAT Practice
Standardized Test Practice
Ⓐ Ⓑ Ⓒ Ⓓ

50. If $\overrightarrow{BX}$ bisects $\angle ABC$, which of the following are true? **D**

Ⓐ $m\angle ABX = m\angle XBC$
Ⓑ $m\angle ABX = \frac{1}{2}m\angle ABC$
Ⓒ $\frac{1}{2}m\angle ABC = m\angle XBC$
Ⓓ all of these

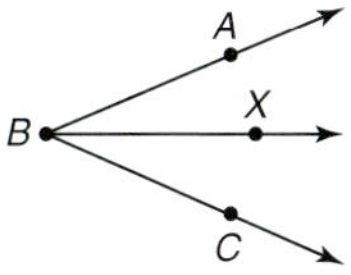

51. **ALGEBRA** Solve $5n + 4 = 7(n + 1) - 2n$. **C**

Ⓐ 0 Ⓑ −1 Ⓒ no solution Ⓓ all numbers

www.geometryonline.com/self_check_quiz/fcat

Enrichment, p. 24

Angle Relationships

Angles are measured in degrees (°). Each degree of an angle is divided into 60 minutes (′), and each minute of an angle is divided into 60 seconds (″).

60′ = 1°
60″ = 1′
$67\frac{1}{2}° = 67°30'$
70.4° = 70°24′
90° = 89°60′

Two angles are complementary if the sum of their measures is 90°. Find the complement of each of the following angles.

1. 35°15′ 2. 27°16′ 3. 15°54′

Study Guide and Intervention, p. 19 (shown) and p. 20

Measure Angles If two noncollinear **rays** have a common endpoint, they form an **angle**. The rays are the **sides** of the angle. The common endpoint is the **vertex**. The angle at the right can be named as $\angle A$, $\angle BAC$, $\angle CAB$, or $\angle 1$.

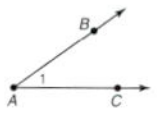

A **right angle** is an angle whose measure is 90. An **acute angle** has measure less than 90. An **obtuse angle** has measure greater than 90 but less than 180.

Example 1

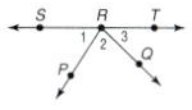

a. Name all angles that have *R* as a vertex.
Three angles are $\angle 1$, $\angle 2$, and $\angle 3$. For other angles, use three letters to name them: $\angle SRQ$, $\angle PRT$, and $\angle SRT$.

b. Name the sides of $\angle 1$.
$\overrightarrow{RS}$, $\overrightarrow{RP}$

Example 2 **Measure each angle and classify it as *right*, *acute*, or *obtuse*.**

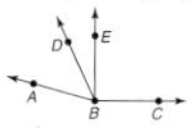

a. $\angle ABD$
Using a protractor, $m\angle ABD = 50$.
$50 < 90$, so $\angle ABD$ is an acute angle.

b. $\angle DBC$
Using a protractor, $m\angle DBC = 115$.
$180 > 115 > 90$, so $\angle DBC$ is an obtuse angle.

c. $\angle EBC$
Using a protractor, $m\angle EBC = 90$.
$\angle EBC$ is a right angle.

Exercises

Refer to the figure.

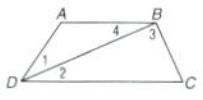

1. Name the vertex of $\angle 4$. **B**
2. Name the sides of $\angle BDC$. $\overrightarrow{DB}$, $\overrightarrow{DC}$
3. Write another name for $\angle DBC$. **$\angle 3$ or $\angle CBD$**

Measure each angle in the figure and classify it as *right*, *acute*, or *obtuse*.

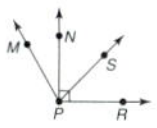

4. $\angle MPR$ **120; obtuse**
5. $\angle RPN$ **90; right**
6. $\angle NPS$ **45; acute**

Skills Practice, p. 21 and Practice, p. 22 (shown)

For Exercises 1–10, use the figure at the right.

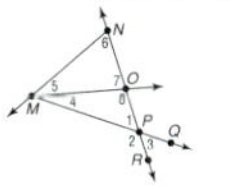

Name the vertex of each angle.

1. $\angle 5$ **M**
2. $\angle 3$ **P**
3. $\angle 8$ **O**
4. $\angle NMP$ **M**

Name the sides of each angle.

5. $\angle 6$
$\overrightarrow{NM}$, $\overrightarrow{NO}$ or $\overrightarrow{NP}$ or $\overrightarrow{NR}$
6. $\angle 2$
$\overrightarrow{PR}$, $\overrightarrow{PM}$
7. $\angle MOP$
$\overrightarrow{OM}$, $\overrightarrow{OP}$ or $\overrightarrow{OR}$
8. $\angle OMN$
$\overrightarrow{MO}$, $\overrightarrow{MN}$

Write another name for each angle.

9. $\angle QPR$
$\angle 3$, $\angle RPQ$
10. $\angle 1$
$\angle MPO$, $\angle OPM$, $\angle MPN$, $\angle NPM$

Measure each angle and classify it as *right*, *acute*, or *obtuse*.

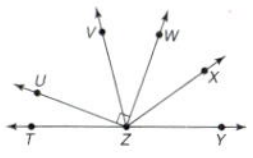

11. $\angle UZW$
90°, right
12. $\angle YZW$
70°, acute
13. $\angle TZW$
110°, obtuse
14. $\angle UZT$
20°, acute

ALGEBRA In the figure, $\overrightarrow{CB}$ and $\overrightarrow{CD}$ are opposite rays, $\overrightarrow{CE}$ bisects $\angle DCF$, and $\overrightarrow{CG}$ bisects $\angle FCB$.

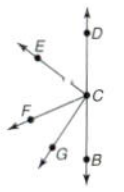

15. If $m\angle DCE = 4x + 15$ and $m\angle ECF = 6x - 5$, find $m\angle DCE$. **55**
16. If $m\angle FCG = 9x + 3$ and $m\angle GCB = 13x - 9$, find $m\angle GCB$. **30**
17. **TRAFFIC SIGNS** The diagram shows a sign used to warn drivers of a school zone or crossing. Measure and classify each numbered angle.
$m\angle 1 = 90$, right angle; $m\angle 2 = 130$, obtuse

Reading to Learn Mathematics, p. 23 ELL

Pre-Activity **How big is a degree?**

Read the introduction to Lesson 1-4 at the top of page 29 in your textbook.

- A semicircle is half a circle. How many degrees are there in a semicircle? **180**
- How many degrees are there in a quarter circle? **90**

Reading the Lesson

1. Match each description in the first column with one of the terms in the second column. Some terms in the second column may be used more than once or not at all.

a. a figure made up of two noncollinear rays with a common endpoint **4**	1. vertex
b. angles whose degree measures are less than 90 **8**	2. angle bisector
c. angles that have the same measure **6**	3. opposite rays
d. angles whose degree measures are between 90 and 180 **5**	4. angle
e. a tool used to measure angles **10**	5. obtuse angles
f. the common endpoint of the rays that form an angle **1**	6. congruent angles
g. a ray that divides an angle into two congruent angles **2**	7. right angles
	8. acute angles
	9. compass
	10. protractor

2. Use the figure to name each of the following.

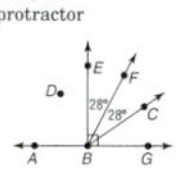

a. a right angle **$\angle ABE$ or $\angle EBG$**
b. an obtuse angle **$\angle ABF$ or $\angle ABC$**
c. an acute angle **$\angle EBF$, $\angle FBC$, $\angle CBG$, $\angle EBC$, or $\angle FBG$**
d. a point in the interior of $\angle EBC$ **F**
e. a point in the exterior of $\angle EBA$ **F, C, or G**
f. the angle bisector of $\angle EBC$ **BF**
g. a point on $\angle CBE$ **C, B, or E**
h. the sides of $\angle ABF$ $\overrightarrow{BA}$ **and** $\overrightarrow{BF}$
i. a pair of opposite rays $\overrightarrow{BA}$ **and** $\overrightarrow{BG}$
j. the common vertex of all angles shown in the figure **B**
k. a pair of congruent angles **$\angle EBF$ and $\angle FBC$, or $\angle ABE$ and $\angle EBG$**
l. the angle with the greatest measure **$\angle ABG$**

Helping You Remember

3. A good way to remember related geometric ideas is to compare them and see how they are alike and how they are different. Give some similarities and differences between *congruent segments* and *congruent angles*.
Sample answer: Congruent segments and congruent angles are alike because they both involve a pair of figures with the same measure. They are different because congruent segments have the same length, which can be measured in units such as inches or centimeters, while congruent angles have the same degree measure.

4 Assess

Open-Ended Assessment

Modeling Have students connect two strips of cardboard or two craftsticks with a brad. Then call out types of angles and have students move the sides to form that angle. Extend the assessment to have students try to estimate angle measurements such as 30°, 45°, and 60°.

Each lesson ends with Open-Ended Assessment strategies for closing the lesson. These include writing, modeling, and speaking.

Getting Ready for Lesson 1-5

Prerequisite Skill Students will learn about adjacent, vertical, complementary, and supplementary angles in Lesson 1-5. They will apply solving algebraic equations to finding measures of unknown angles. Use Exercises 61–66 to determine your students' familiarity with solving more complex algebraic equations.

Assessment Options

Practice Quiz 2 The quiz provides students with a brief review of the concepts and skills in Lessons 1-3 and 1-4. Lesson numbers are given to the right of the exercises or instruction lines so students can review concepts not yet mastered.

Maintain Your Skills

Mixed Review

Find the distance between each pair of points. Then find the coordinates of the midpoint of the line segment between the points. *(Lesson 1-3)*

52. $A(2, 3)$, $B(5, 7)$ **5; (3.5, 5)**

53. $C(-2, 0)$, $D(6, 4)$ **$\sqrt{80} \approx 8.9$; (2, 2)**

54. $E(-3, -2)$, $F(5, 8)$ **$\sqrt{164} \approx 12.8$; (1, 3)**

Find the measurement of each segment. *(Lesson 1-2)*

55. $\overline{WX}$ **$9\frac{2}{3}$ ft**

56. $\overline{YZ}$ **11.4 mm**

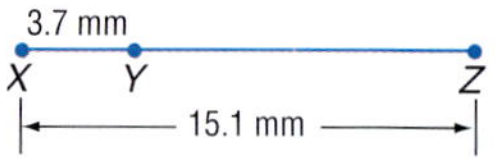

57. Find PQ if Q lies between P and R, $PQ = 6x - 5$, $QR = 2x + 7$, and $PQ = QR$. *(Lesson 1-2)* **13**

Refer to the figure at the right. *(Lesson 1-1)*

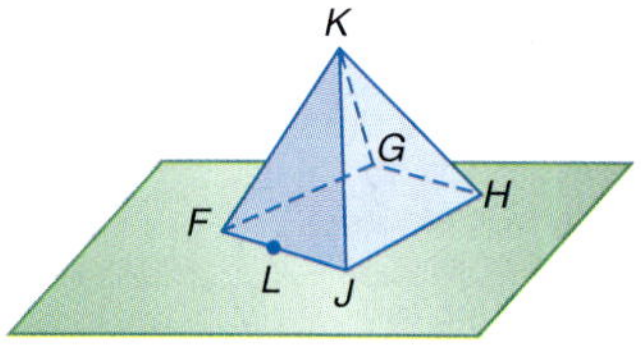

58. How many planes are shown? **5**

59. Name three collinear points. **F, L, J**

60. Name a point coplanar with J, H, and F. **G or L**

Getting Ready for the Next Lesson

PREREQUISITE SKILL **Solve each equation.**
*(To review **solving equations**, see pages 737 and 738.)*

61. $14x + (6x - 10) = 90$ **5**

62. $2k + 30 = 180$ **75**

63. $180 - 5y = 90 - 7y$ **−45**

64. $90 - 4t = \frac{1}{4}(180 - t)$ **12**

65. $(6m + 8) + (3m + 10) = 90$ **8**

66. $(7n - 9) + (5n + 45) = 180$ **12**

Practice Quiz 2 — *Lessons 1-3 and 1-4*

Find the coordinates of the midpoint of each segment. Then find the distance between the endpoints. *(Lesson 1-3)*

1.

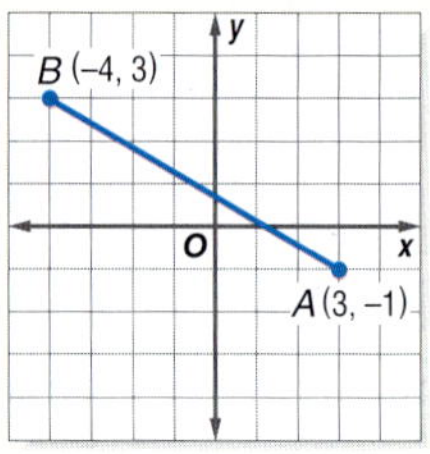

$\left(-\frac{1}{2}, 1\right)$; $\sqrt{65} \approx 8.1$

2.

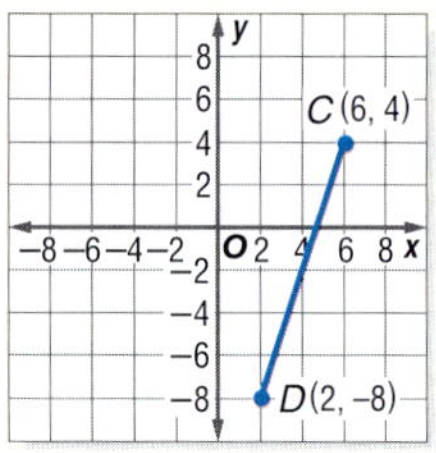

(4, −2); $\sqrt{160} \approx 12.6$

3.

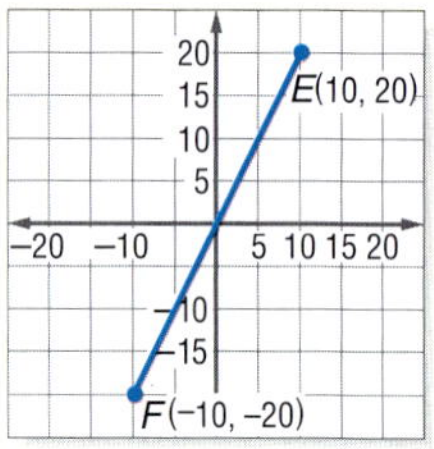

(0, 0); $\sqrt{2000} \approx 44.7$

In the figure, $\overrightarrow{XP}$ and $\overrightarrow{XT}$ are opposite rays. Given the following conditions, find the value of a and the measure of the indicated angle. *(Lesson 1-4)*

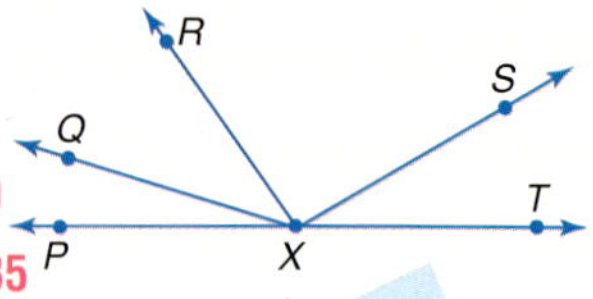

4. $m\angle SXT = 3a - 4$, $m\angle RXS = 2a + 5$, $m\angle RXT = 111$; $m\angle RXS$ **22; 49**

5. $m\angle QXR = a + 10$, $m\angle QXS = 4a - 1$, $m\angle RXS = 91$; $m\angle QXS$ **34; 135**

Two Quizzes in each chapter review skills and concepts presented in previous lessons.

1-5 Angle Relationships

What You'll Learn

- Identify and use special pairs of angles.
- Identify perpendicular lines.

What kinds of angles are formed when streets intersect?

When two lines intersect, four angles are formed. In some cities, more than two streets might intersect to form even more angles. All of these angles are related in special ways.

Vocabulary

- adjacent angles
- vertical angles
- linear pair
- complementary angles
- supplementary angles
- perpendicular

PAIRS OF ANGLES

Certain pairs of angles have special names.

Key Concept — Angle Pairs

- **Words** **Adjacent angles** are two angles that lie in the same plane, have a common vertex, and a common side, but no common interior points.
- **Examples** $\angle ABC$ and $\angle CBD$

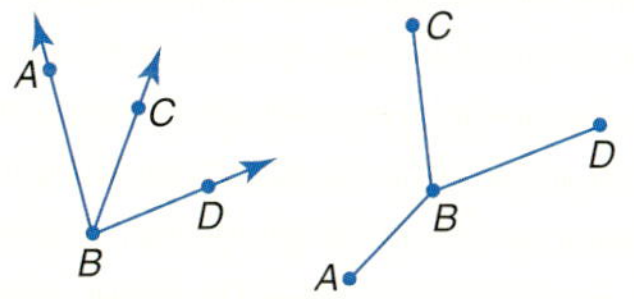

- **Nonexamples** $\angle ABC$ and $\angle ABD$ $\angle ABC$ and $\angle BCD$

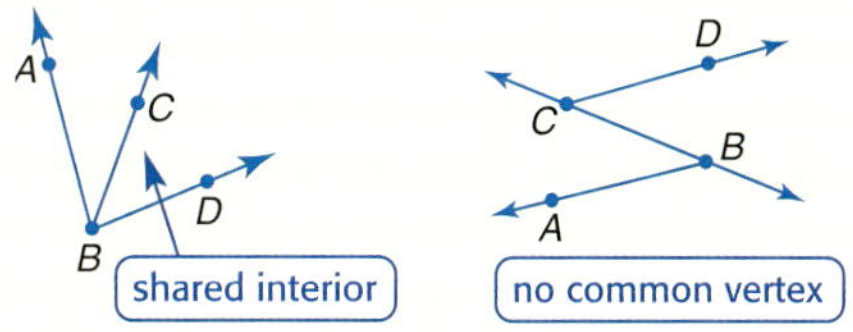

- **Words** **Vertical angles** are two nonadjacent angles formed by two intersecting lines.
- **Examples** $\angle AEB$ and $\angle CED$
  $\angle AED$ and $\angle BEC$

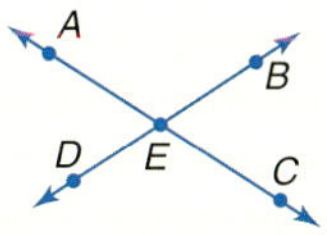

- **Nonexample** $\angle AED$ and $\angle BEC$

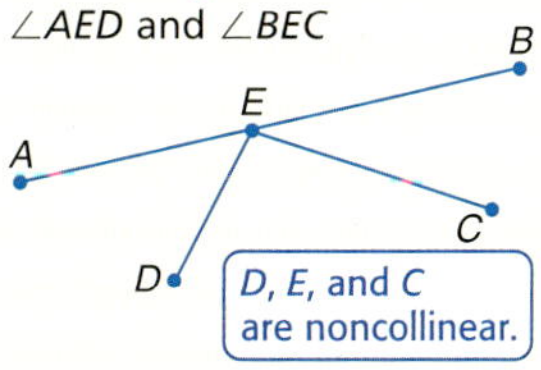

- **Words** A **linear pair** is a pair of adjacent angles whose noncommon sides are opposite rays.
- **Example** $\angle BED$ and $\angle BEC$

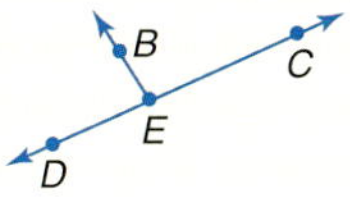

- **Nonexample**

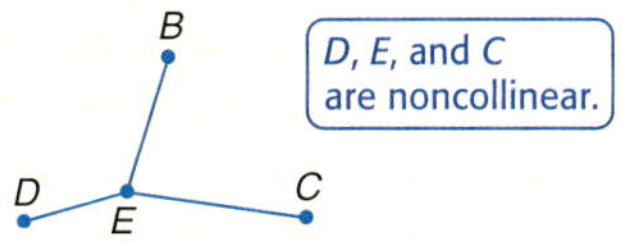

1-5 Lesson Notes

1 Focus

5-Minute Check Transparency 1-5 Use as a quiz or review of Lesson 1-4.

Mathematical Background notes are available for this lesson on p. 4D.

What kinds of angles are formed when streets intersect?

Ask students:

- What do the streets in the overhead picture model? **rays forming the sides of angles**
- What does an intersection model? **the vertex of one or more angles**
- Would it be easier to maneuver a car to make a turn through an intersection with an obtuse angle or an acute angle? Explain. **Obtuse angle; you would only have to turn the wheel slightly to make a turn through an obtuse angle, but you would have to slow down and turn the wheel more to make a turn through an acute angle.**

Resource Manager

Workbook and Reproducible Masters

Chapter 1 Resource Masters

- Study Guide and Intervention, pp. 25–26
- Skills Practice, p. 27
- Practice, p. 28
- Reading to Learn Mathematics, p. 29
- Enrichment, p. 30
- Assessment, p. 52

Prerequisite Skills Workbook, pp. 85–86

Teaching Geometry With Manipulatives Masters, pp. 16, 39

Transparencies

5-Minute Check Transparency 1-5
Answer Key Transparencies

Technology

GeomPASS: Tutorial Plus, Lesson 4
Interactive Chalkboard

2 Teach

PAIRS OF ANGLES

In-Class Example

Power Point®

Teaching Tip Encourage students to look for angles that are composed of other angles, such as angles ∠*VZT*, ∠*TZX*, and ∠*YZW* in Example 1.

1 Refer to the figure in Example 1. Name an angle pair that satisfies each condition.

a. two angles that form a linear pair ∠*VZY* and ∠*VZX* or ∠*VZY* and ∠*YZW* or ∠*YZT* and ∠*TZX* or ∠*WZX* and ∠*XZV* or ∠*YZW* and ∠*XZW*

b. two acute vertical angles ∠*VZY* and ∠*XZW*

Answers

4. ∠*ACD* and ∠*ECB*, ∠*DCB* and ∠*ACE*; measures for each pair of vertical angles should be the same.
5. ∠*ACD* and ∠*DCB*, ∠*DCB* and ∠*BCE*, ∠*BCE* and ∠*ECA*, ∠*ECA* and ∠*ACD*; measures for each linear pair should add to 180.
6. Sample answers: The measures of vertical angles are equal or vertical angles are congruent. The sum of the measures of a linear pair is 180 or angles that form a linear pair are supplementary

Example 1 Identify Angle Pairs

Name an angle pair that satisfies each condition.

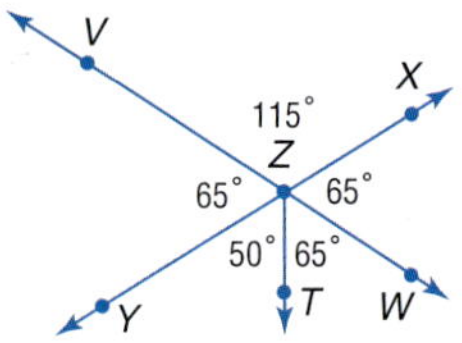

a. two obtuse vertical angles

∠*VZX* and ∠*YZW* are vertical angles.

They each have measures greater than 90°, so they are obtuse.

b. two acute adjacent angles

There are four acute angles shown.

Adjacent acute angles are ∠*VZY* and ∠*YZT*, ∠*YZT* and ∠*TZW*, and ∠*TZW* and ∠*WZX*.

The measures of angles formed by intersecting lines also have a special relationship.

Study Tip

Patty Paper

Patty paper is the squares of paper used to separate hamburger patties. You can also use waxed paper or tracing paper.

Geometry Activity

Angle Relationships

Make a Model

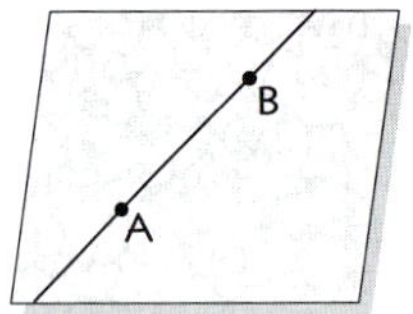

Step 1

Fold a piece of patty paper so that it makes a crease across the paper. Open the paper, trace the crease with a pencil, and name two points on the crease *A* and *B*.

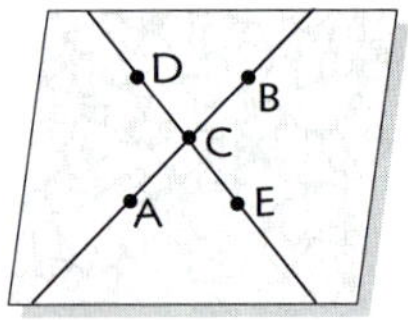

Step 2

Fold the paper again so that the crease intersects $\overline{AB}$ between the two labeled points. Open the paper, trace this crease, and label the intersection *C*. Label two other points, *D* and *E*, on the second crease so that *C* is between *D* and *E*.

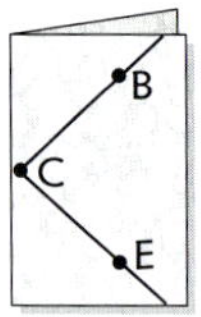

Step 3

Fold the paper again through point *C* so that $\overrightarrow{CB}$ aligns with $\overrightarrow{CD}$.

Analyze the Model 1. ∠*BCE* ≅ ∠*DCA* 3. See students' work.

1. What do you notice about ∠*BCE* and ∠*DCA* when you made the last fold?
2. Fold the paper again through *C* so that $\overrightarrow{CB}$ aligns with $\overrightarrow{CE}$. What do you notice? ∠*DCB* ≅ ∠*ACE*
3. Use a protractor to measure each angle. Label the measure on your model.
4. Name pairs of vertical angles and their measures. 4–6. See margin.
5. Name linear pairs of angles and their measures.
6. Compare your results with those of your classmates. Write a "rule" about the measures of vertical angles and another about the measures of linear pairs.

The Geometry Activity suggests that all vertical angles are congruent. It also supports the concept that the sum of the measures of a linear pair is 180.

There are other angle relationships that you may remember from previous math courses. These are complementary angles and supplementary angles.

Geometry Activity

Materials: patty paper, straightedge, protractor

- Ask students to trace the crease made by the third fold. Students should recognize that this crease provides angle bisectors for angles *DCB* and *ACE*.
- This activity allows students to become more familiar with the terms *linear pair, adjacent angles*, and *vertical angles*. Encourage students to discuss these terms freely while engaged in the activity.

38 Chapter 1 Points, Lines, Planes, and Angles

Key Concept — *Angle Relationships*

- **Words** **Complementary angles** are two angles whose measures have a sum of 90.
- **Examples**
  $\angle 1$ and $\angle 2$ are complementary.
  $\angle PQR$ and $\angle XYZ$ are complementary.

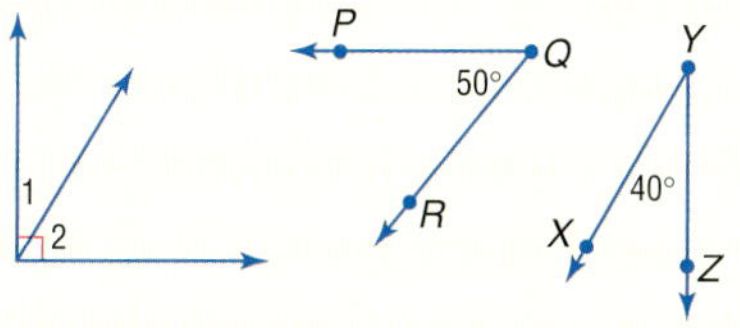

- **Words** **Supplementary angles** are two angles whose measures have a sum of 180.
- **Examples**
  $\angle EFH$ and $\angle HFG$ are supplementary.
  $\angle M$ and $\angle N$ are supplementary.

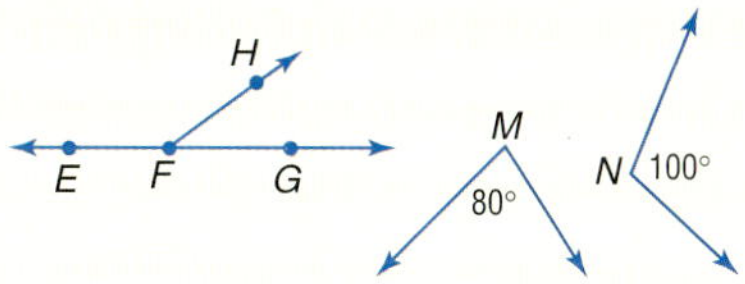

Remember that angle measures are real numbers. So, the operations for real numbers and algebra can be used with angle measures.

Example 2 Angle Measure

ALGEBRA Find the measures of two complementary angles if the difference in the measures of the two angles is 12.

Explore The problem relates the measures of two complementary angles. You know that the sum of the measures of complementary angles is 90.

Plan Draw two figures to represent the angles.

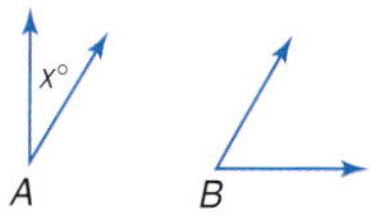

Let the measure of one angle be x.
If $m\angle A = x$, then because $\angle A$ and $\angle B$ are complementary, $m\angle B + x = 90$ or $m\angle B = 90 - x$.

The problem states that the difference of the two angle measures is 12, or $m\angle B - m\angle A = 12$.

Solve

$m\angle B - m\angle A = 12$	Given
$(90 - x) - x = 12$	$m\angle A = x$, $m\angle B = 90 - x$
$90 - 2x = 12$	Simplify.
$-2x = -78$	Subtract 90 from each side.
$x = 39$	Divide each side by -2.

Use the value of x to find each angle measure.

$m\angle A = x$	$m\angle B = 90 - x$
$m\angle A = 39$	$m\angle B = 90 - 39$ or 51

Examine Add the angle measures to verify that the angles are complementary.

$$m\angle A + m\angle B = 90$$
$$39 + 51 = 90$$
$$90 = 90$$

www.geometryonline.com/extra_examples/fcat

Study Tip

Complementary and Supplementary Angles
While the other angle pairs in this lesson share at least one point, complementary and supplementary angles need not share any points.

In-Class Example

Power Point®

2 **ALGEBRA** Find the measures of two supplementary angles if the measure of one angle is 6 less than five times the measure of the other angle. **31, 149**

Tips for New Teachers

Problem Solving The 4-step plan can be used to solve any problem. Encourage students to mentally attempt the steps even if they do not write them down. The Explore stage is also called the *Read* stage. Stress that good planning makes the solving stage easier. Remind students to always examine their solutions for reasonableness.

DAILY INTERVENTION — Differentiated Instruction

Logical/Mathematical Have students list each angle relationship presented in the Key Concept sections of this lesson on pp. 37 and 39. Then, they can write one or two sentences in their own words to describe each relationship and provide an example. For extra practice, have students analyze the figures found throughout the lesson and determine which angle relationships are or are not present in them.

PERPENDICULAR LINES

In-Class Example

Power Point®

3 **ALGEBRA** Find x so that $\overleftrightarrow{KO} \perp \overleftrightarrow{HM}$. 7

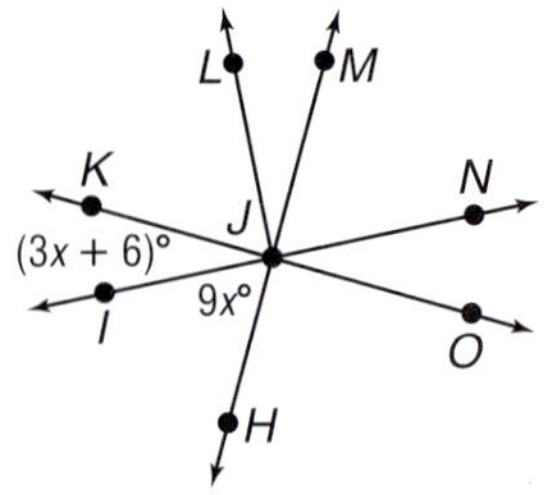

Teaching Tip Stress that students should systematically write down all the information they know about each problem and logically use each given fact to progress toward the solution. Caution students that some information they need to solve a problem may be contained in the figure and not described in the problem statement.

Concept Check

Have pairs of students sketch simple figures demonstrating each angle relationship in this lesson. For example, a student can suggest that a partner draw and label a simple figure with $\angle LMP$ adjacent to $\angle PMQ$. To make things more challenging, suggest that the students include one, two, and then three angle relationships in their suggestions to one another.

Study Tip

Interpreting Figures

Never assume that two lines are perpendicular because they appear to be so in the figure. The only sure way to know if they are perpendicular is if the right angle symbol is present or if the problem states angle measures that allow you to make that conclusion.

PERPENDICULAR LINES Lines that form right angles are **perpendicular**. The following statements are also true when two lines are perpendicular.

Key Concept — Perpendicular Lines

- Perpendicular lines intersect to form four right angles.
- Perpendicular lines intersect to form congruent adjacent angles.
- Segments and rays can be perpendicular to lines or to other line segments and rays.
- The right angle symbol in the figure indicates that the lines are perpendicular.
- **Symbol** $\perp$ is read *is perpendicular to.*

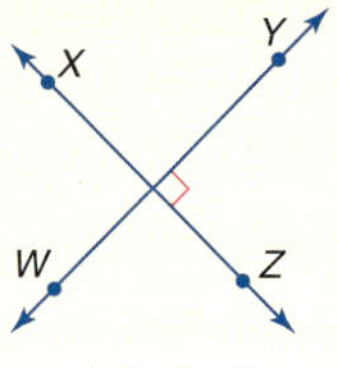

Example 3 Perpendicular Lines

ALGEBRA **Find x and y so that $\overleftrightarrow{BE}$ and $\overleftrightarrow{AD}$ are perpendicular.**

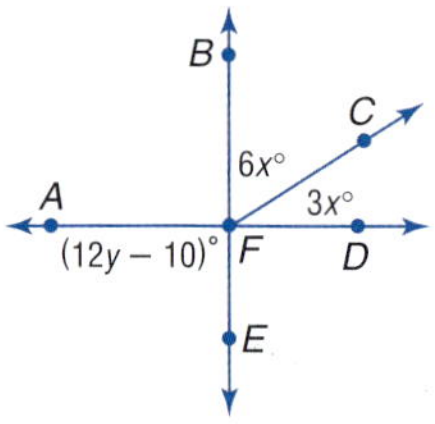

If $\overleftrightarrow{BE} \perp \overleftrightarrow{AD}$, then $m\angle BFD = 90$ and $m\angle AFE = 90$.

To find x, use $\angle BFC$ and $\angle CFD$.

$m\angle BFD = m\angle BFC + m\angle CFD$	Sum of parts = whole
$90 = 6x + 3x$	Substitution
$90 = 9x$	Add.
$10 = x$	Divide each side by 9.

To find y, use $\angle AFE$.

$m\angle AFE = 12y - 10$	Given
$90 = 12y - 10$	Substitution
$100 = 12y$	Add 10 to each side.
$\frac{25}{3} = y$	Divide each side by 12, and simplify.

Study Tip

Naming Figures

The list of statements that can be assumed is not a complete list. There are more special pairs of angles than those listed. Also remember that all figures except points usually have more than one way to name them.

While two lines may appear to be perpendicular in a figure, you cannot assume this is true unless other information is given. In geometry, figures are used to depict a situation. They are not drawn to reflect total accuracy of the situation. There are certain relationships you can assume to be true, but others that you cannot.

Study the figure at the right and then compare the lists below.

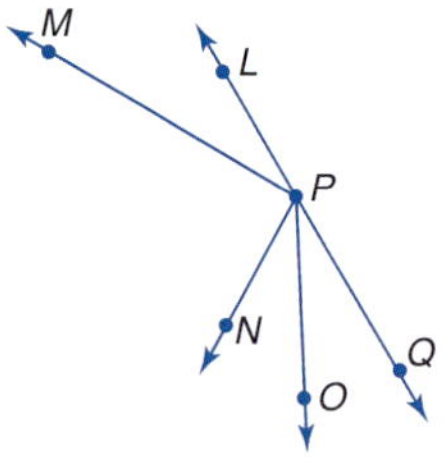

Can Be Assumed	Cannot Be Assumed
All points shown are coplanar.	Perpendicular lines: $\overleftrightarrow{PN} \perp \overleftrightarrow{PM}$
L, P, and Q are collinear.	Congruent angles: $\angle QPO \cong \angle LPM$
$\overrightarrow{PM}$, $\overrightarrow{PN}$, $\overrightarrow{PO}$, and $\overleftrightarrow{LQ}$ intersect at P.	$\angle QPO \cong \angle OPN$
P is between L and Q.	$\angle OPN \cong \angle LPM$
N is in the interior of MPO.	Congruent segments: $\overline{LP} \cong \overline{PQ}$
$\angle LPM$ and $\angle MPN$ are adjacent angles.	$\overline{PQ} \cong \overline{PO}$
$\angle LPN$ and $\angle NPQ$ are a linear pair.	$\overline{PO} \cong \overline{PN}$
$\angle QPO$ and $\angle OPL$ are supplementary.	$\overline{PN} \cong \overline{PL}$

Answers

1.

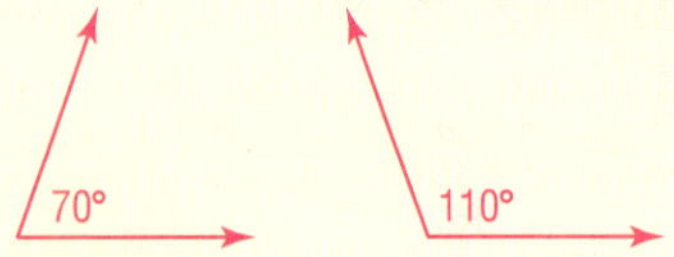

2. Sample answer: When two angles form a linear pair, then their noncommon sides form a straight angle, which measures 180. When the sum of the measures of two angles is 180, then the angles are supplementary.

40 Chapter 1 Points, Lines, Planes, and Angles

Example 4 Interpret Figures

Determine whether each statement can be assumed from the figure below.

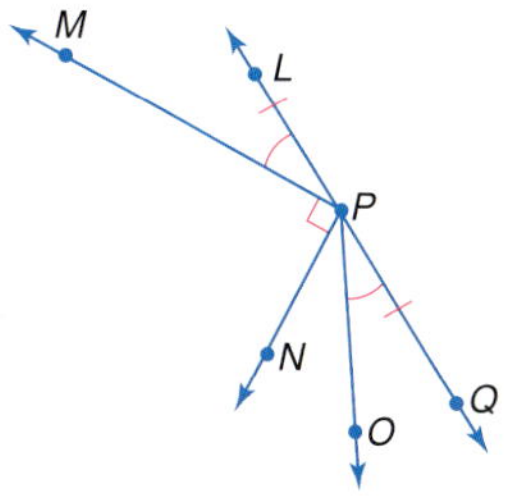

a. ∠LPM and ∠MPO are adjacent angles.

Yes; they share a common side and vertex and have no interior points in common.

b. ∠OPQ and ∠LPM are complementary.

No; they are congruent, but we do not know anything about their exact measures.

c. ∠LPO and ∠QPO are a linear pair.

Yes; they are adjacent angles whose noncommon sides are opposite rays.

Check for Understanding

Concept Check

1–2. See margin.

1. **OPEN ENDED** Draw two angles that are supplementary, but not adjacent.
2. **Explain** the statement *If two adjacent angles form a linear pair, they must be supplementary.*
3. **Write** a sentence to explain why a linear pair of angles is called *linear*. Sample answer: The noncommon sides of a linear pair of angles form a straight line.

Guided Practice

GUIDED PRACTICE KEY	
Exercises	Examples
4, 5	1
6, 10	2
7	3
8, 9	4

For Exercises 4 and 5, use the figure at the right and a protractor.

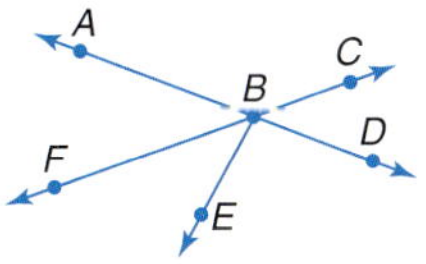

Sample answer:

4. Name two acute vertical angles. ∠ABF, ∠CBD
5. Name two obtuse adjacent angles. Sample answer: ∠ABC, ∠CBE
6. The measure of the supplement of an angle is 60 less than three times the measure of the complement of the angle. Find the measure of the angle. 15
7. Lines p and q intersect to form adjacent angles 1 and 2. If $m\angle 1 = 3x + 18$ and $m\angle 2 = -8y - 70$, find the values of x and y so that p is perpendicular to q. $x = 24$, $y = -20$

Determine whether each statement can be assumed from the figure. Explain.

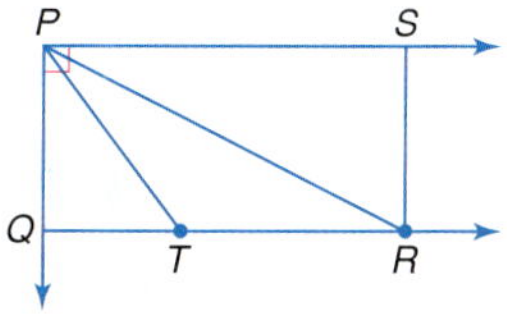

8. ∠SRP and ∠PRT are complementary.
9. ∠QPT and ∠TPR are adjacent, but neither complementary or supplementary. See margin.

8. No; while ∠SRT appears to be a right angle, no information verifies this.

Application

10. **SKIING** Alisa Camplin won the gold medal in the 2002 Winter Olympics with a triple-twisting, double backflip jump in the women's freestyle skiing event. While she is in the air, her skis are positioned like intersecting lines. If ∠4 measures 60°, find the measures of the other angles. $m\angle 1 = 120$, $m\angle 2 = 60$, $m\angle 3 = 120$

Answers

9. Yes; they share a common side and vertex, so they are adjacent. Since $\overline{PR}$ falls between $\overline{PQ}$ and $\overline{PS}$, $m\angle QPR < 90$, so the two angles cannot be complementary or supplementary.

In-Class Example

Power Point®

4 Determine whether each statement can be assumed from the figure below. Explain.

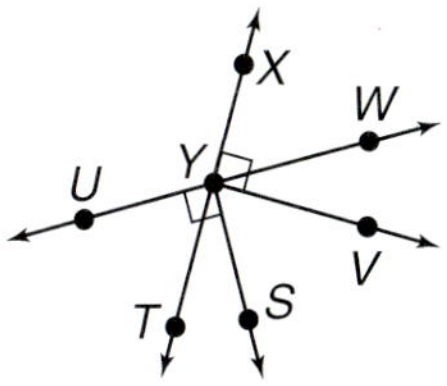

a. $m\angle VYT = 90$ Yes; lines VY and TX are perpendicular.

b. ∠TYW and ∠TYU are supplementary. Yes; they form a linear pair of angles.

c. ∠VYW and ∠TYS are adjacent angles. No; they do not share a common side.

3 Practice/Apply

Study Notebook

Have students—

- add the definitions/examples of the vocabulary terms to their Vocabulary Builder worksheets for Lesson 1-5.
- include the Can be Assumed/Cannot be Assumed chart on p. 40.
- include any other item(s) that they find helpful in mastering the skills in this lesson.

About the Exercises...

Organization by Objective

- **Pairs of Angles:** 11–26
- **Perpendicular Lines:** 27–35

Odd/Even Assignments

Exercises 11–30 are structured so that students practice the same concepts whether they are assigned odd or even problems.

Assignment Guide

Basic: 11–27 odd, 31–37 odd, 39–42, 44–62

Average: 11–37 odd, 39–42, 44–62 (optional: 43)

Advanced: 12–38 even, 39–57 (optional: 58–62)

Study Guide and Intervention, p. 25 (shown) and p. 26

Pairs of Angles **Adjacent angles** are angles in the same plane that have a common vertex and a common side, but no common interior points. **Vertical angles** are two nonadjacent angles formed by two intersecting lines. A pair of adjacent angles whose noncommon sides are opposite rays is called a **linear pair**.

Example **Identify each pair of angles as *adjacent angles, vertical angles,* and/or as a *linear pair.***

a. 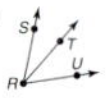

$\angle SRT$ and $\angle TRU$ have a common vertex and a common side, but no common interior points. They are adjacent angles.

b. 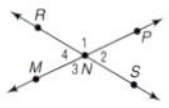

$\angle 1$ and $\angle 3$ are nonadjacent angles formed by two intersecting lines. They are vertical angles. $\angle 2$ and $\angle 4$ are also vertical angles.

c. 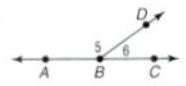

$\angle 6$ and $\angle 5$ are adjacent angles whose noncommon sides are opposite rays. The angles form a linear pair.

d. 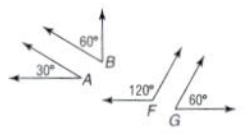

$\angle A$ and $\angle B$ are two angles whose measures have a sum of 90. They are complementary. $\angle F$ and $\angle G$ are two angles whose measures have a sum of 180. They are supplementary.

Exercises

Identify each pair of angles as *adjacent*, *vertical*, and/or as a *linear pair*.

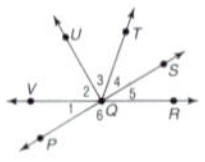

1. $\angle 1$ and $\angle 2$ adjacent
2. $\angle 1$ and $\angle 6$ linear pair; adjacent
3. $\angle 1$ and $\angle 5$ vertical
4. $\angle 3$ and $\angle 2$ adjacent

For Exercises 5–7, refer to the figure at the right.

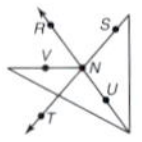

5. Identify two obtuse vertical angles. $\angle RNT$ and $\angle SNU$
6. Identify two acute adjacent angles. $\angle RNV$ and $\angle VNT$ or $\angle VNT$ and $\angle TNU$
7. Identify an angle supplementary to $\angle TNU$. $\angle UNS$ or $\angle TNR$
8. Find the measures of two complementary angles if the difference in their measures is 18. 36 and 54

Skills Practice, p. 27 and Practice, p. 28 (shown)

For Exercises 1–4, use the figure at the right and a protractor.

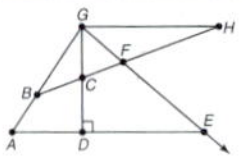

1. Name two obtuse vertical angles. Sample answer: $\angle GFH$, $\angle CFE$
2. Name a linear pair whose vertex is B. $\angle GBC$, $\angle CBA$
3. Name an angle not adjacent to but complementary to $\angle FGC$. $\angle FED$
4. Name an angle adjacent and supplementary to $\angle DCB$. $\angle BCG$ or $\angle DCH$
5. Two angles are complementary. The measure of one angle is 21 more than twice the measure of the other angle. Find the measures of the angles. 23, 67
6. If a supplement of an angle has a measure 78 less than the measure of the angle, what are the measures of the angles? 129, 51

ALGEBRA For Exercises 7–8, use the figure at the right.

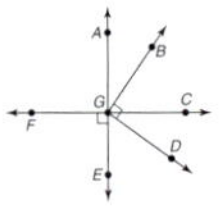

7. If $m\angle FGE = 5x + 10$, find x so that $\overleftrightarrow{FC} \perp \overleftrightarrow{AE}$. 16
8. If $m\angle BGC = 16x - 4$ and $m\angle CGD = 2x + 13$, find x so that $\angle BGD$ is a right angle. 4.5

Determine whether each statement can be assumed from the figure. Explain.

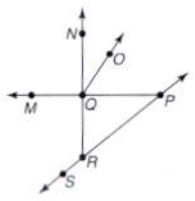

9. $\angle NQO$ and $\angle OQP$ are complementary. No; $m\angle NQP$ is not known to be 90.
10. $\angle SRQ$ and $\angle QRP$ is a linear pair. Yes; they are adjacent angles whose noncommon sides are opposite rays.
11. $\angle MQN$ and $\angle MQR$ are vertical angles. No; the angles are adjacent.
12. STREET MAPS Darren sketched a map of the cross streets nearest to his home for his friend Miguel. Describe two different angle relationships between the streets. Sample answer: Beacon ⊥ Main; Olive divides two of the angles formed by Bacon and Main into pairs of complementary angles.

Reading to Learn Mathematics, p. 29 ELL

Pre-Activity **What kinds of angles are formed when streets intersect?**

Read the introduction to Lesson 1-5 at the top of page 37 in your textbook.

- How many separate angles are formed if three lines intersect at a common point? (Do not use an angle whose interior includes part of another angle.) 6
- How many separate angles are formed if n lines intersect at a common point? (Do not count an angle whose interior includes part of another angle.) $2n$

Reading the Lesson

1. Name each of the following in the figure at the right.

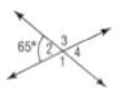

   a. two pairs of congruent angles $\angle 1$ and $\angle 3$, $\angle 2$ and $\angle 4$
   b. a pair of acute vertical angles $\angle 2$ and $\angle 4$
   c. a pair of obtuse vertical angles $\angle 1$ and $\angle 3$
   d. four pairs of adjacent angles $\angle 1$ and $\angle 2$, $\angle 2$ and $\angle 3$, $\angle 3$ and $\angle 4$, $\angle 4$ and $\angle 1$
   e. two pairs of vertical angles $\angle 1$ and $\angle 3$, $\angle 2$ and $\angle 4$
   f. four linear pairs $\angle 1$ and $\angle 2$, $\angle 2$ and $\angle 3$, $\angle 3$ and $\angle 4$, $\angle 4$ and $\angle 1$
   g. four pairs of supplementary angles $\angle 1$ and $\angle 2$, $\angle 2$ and $\angle 3$, $\angle 3$ and $\angle 4$, $\angle 4$ and $\angle 1$
2. Tell whether each statement is *always*, *sometimes*, or *never* true.
   a. If two angles are adjacent angles, they form a linear pair. sometimes
   b. If two angles form a linear pair, they are complementary. never
   c. If two angles are supplementary, they are congruent. sometimes
   d. If two angles are complementary, they are adjacent. sometimes
   e. When two perpendicular lines intersect, four congruent angles are formed. always
   f. Vertical angles are supplementary. sometimes
   g. Vertical angles are complementary. sometimes
   h. The two angles in a linear pair are both acute. never
   i. If two angles form a linear pair, one is acute and the other is obtuse. sometimes
3. Complete each sentence.
   a. If two angles are supplementary and x is the measure of one of the angles, then the measure of the other angle is $180 - x$.
   b. If two angles are complementary and x is the measure of one of the angles, then the measure of the other angle is $90 - x$.

Helping You Remember

4. Look up the nonmathematical meaning of *supplementary* in your dictionary. How can this definition help you to remember the meaning of supplementary angles? Sample answer: *Supplementary* means something added to complete a thing. An angle and its supplement can be joined to obtain a linear pair.

★ indicates increased difficulty

Practice and Apply

Homework Help

For Exercises	See Examples
11–16	1
17–22	2
27–30	3
31–35	4

Extra Practice See page 755.

13. $\angle UWT$, $\angle TWY$
14. $\angle VXU$, $\angle WYT$

For Exercises 11–16, use the figure at the right and a protractor.

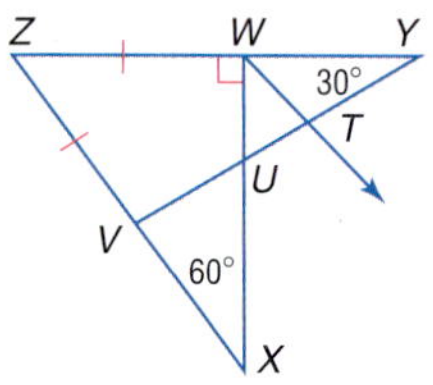

11. Name two acute vertical angles. $\angle WUT$, $\angle VUX$
12. Name two obtuse vertical angles. $\angle WUV$, $\angle XUT$
13. Name a pair of complementary adjacent angles.
14. Name a pair of complementary nonadjacent angles.
15. Name a linear pair whose vertex is T. $\angle WTY$, $\angle WTU$
16. Name an angle supplementary to $\angle UVZ$. $\angle UVX$
17. Rays PQ and QR are perpendicular. Point S lies in the interior of $\angle PQR$. If $m\angle PQS = 4 + 7a$ and $m\angle SQR = 9 + 4a$, find $m\angle PQS$ and $m\angle SQR$. 53, 37
18. The measures of two complementary angles are $16z - 9$ and $4z + 3$. Find the measures of the angles. 67.8, 22.2
19. Find $m\angle T$ if $m\angle T$ is 20 more than four times its supplement. 148
20. The measure of an angle's supplement is 44 less than the measure of the angle. Find the measure of the angle and its supplement. 112, 68
21. Two angles are supplementary. One angle measures 12° more than the other. Find the measures of the angles. 84, 96
22. The measure of $\angle 1$ is five less than four times the measure of $\angle 2$. If $\angle 1$ and $\angle 2$ form a linear pair, what are their measures? 37, 143

Determine whether each statement is *sometimes*, *always*, or *never* true.

23. If two angles are supplementary and one is acute, the other is obtuse. always
24. If two angles are complementary, they are both acute angles. always
25. If $\angle A$ is supplementary to $\angle B$ and $\angle B$ is supplementary to $\angle C$, then $\angle A$ is supplementary to $\angle C$. sometimes
26. If $\overline{PN} \perp \overline{PQ}$, then $\angle NPQ$ is acute. never

ALGEBRA **For Exercises 27–29, use the figure at the right.**

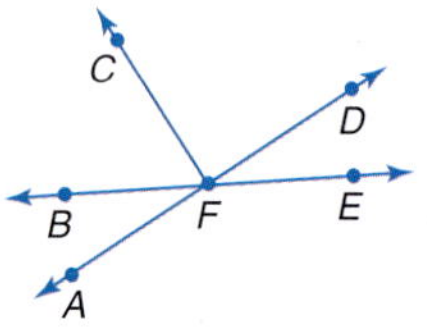

27. If $m\angle CFD = 12a + 45$, find a so that $\overrightarrow{FC} \perp \overrightarrow{FD}$. 3.75
28. If $m\angle AFB = 8x - 6$ and $m\angle BFC = 14x + 8$, find the value of x so that $\angle AFC$ is a right angle. 4
★ 29. If $\angle BFA = 3r + 12$ and $m\angle DFE = -8r + 210$, find $m\angle AFE$. 114

★ 30. $\angle L$ and $\angle M$ are complementary angles. $\angle N$ and $\angle P$ are complementary angles. If $m\angle L = y - 2$, $m\angle M = 2x + 3$, $m\angle N = 2x - y$, and $m\angle P = x - 1$, find the values of x, y, $m\angle L$, $m\angle M$, $m\angle N$, and $m\angle P$. 36, 17, 15, 75, 55, 35

Determine whether each statement can be assumed from the figure. Explain.

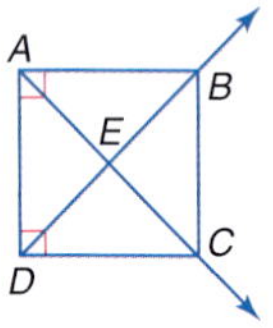

31. $\angle DAB$ is a right angle.
32. $\angle AEB \cong \angle DEC$ Yes; they are vertical angles.
33. $\angle ADB$ and $\angle BDC$ are complementary.
34. $\angle DAE \cong \angle ADE$
35. $\overline{AB} \perp \overline{BC}$ No; we do not know $m\angle ABC$.

31. Yes; the symbol denotes that $\angle DAB$ is a right angle.

33. Yes; the sum of their measures is $m\angle ADC$, which is 90.

34. No; there is no indication of the measures of these angles.

36. **LANGUAGE** Look up the words *complementary* and *complimentary*. Discuss the differences and which has a mathematical meaning. See margin.
37. **CRITICAL THINKING** A counterexample is used to show that a statement is not necessarily true. Find a counterexample for the statement *Supplementary angles form linear pairs.* See margin.

Enrichment, p. 30

Curve Stitching

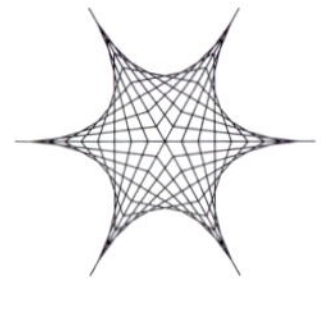

The star design at the right was created by a method known as **curve stitching**. Although the design appears to contain curves, it is made up entirely of line segments.

To begin the star design, draw a 60° angle. Mark eight equally-spaced points on each ray, and number the points as shown below. Then connect pairs of points that have the same number.

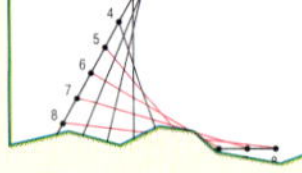

Answer

36. Sample answer: *Complementary* means serving to fill out or complete, while *complimentary* means given as a courtesy or favor. *Complementary* has the mathematical meaning of an angle completing the measure to make 90.

More About . . .

Stained Glass

In the 13th century, artists began using geometric patterns and shapes to create complex, colorful windows.

Source: www.infoplease.com

38. **STAINED GLASS** Stained glass patterns often have several pieces of glass held together by lead at a central point. In the pattern at the right, determine which segments are perpendicular. $\overline{AK} \perp \overline{KD}$, $\overline{KD} \perp \overline{KF}$, $\overline{KE} \perp \overline{KG}$, $\overline{AF} \perp \overline{KD}$

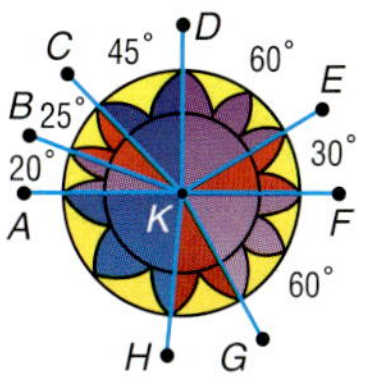

39. **CRITICAL THINKING** In the figure, $\angle WUT$ and $\angle XUV$ are vertical angles, $\overline{YU}$ is the bisector of $\angle WUT$, and $\overline{UZ}$ is the bisector of $\angle TUV$. Write a convincing argument that $\overline{YU} \perp \overline{UZ}$. See margin.

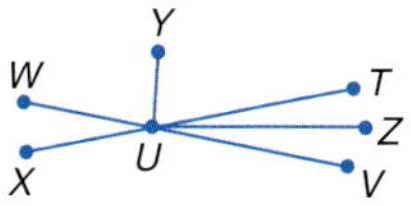

40. **WRITING IN MATH** Answer the question that was posed at the beginning of the lesson. See margin.

What kinds of angles are formed when streets intersect?

Include the following in your answer.
- the types of angles that might be formed by two intersecting lines, and
- a sketch of intersecting streets with angle measures marked and all angle pairs identified.

FCAT Practice
Standardized Test Practice
Ⓐ Ⓑ Ⓒ Ⓓ

41. Which statement is true of the figure? A

Ⓐ $x > y$ Ⓑ $x < y$

Ⓒ $x = y$ Ⓓ cannot be determined

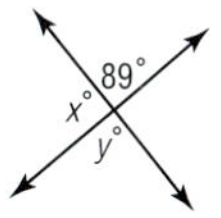

42. **SHORT RESPONSE** The product of 4, 5, and 6 is equal to twice the sum of 10 and what number? 50

Extending the Lesson

43. The concept of perpendicularity can be extended to include planes. If a line, line segment, or ray is perpendicular to a plane, it is perpendicular to every line, line segment, or ray in that plane that intersects it. In the figure at the right, $\overleftrightarrow{AB} \perp \mathcal{E}$. Name all pairs of perpendicular lines.
$\ell \perp \overleftrightarrow{AB}$, $m \perp \overleftrightarrow{AB}$, $n \perp \overleftrightarrow{AB}$

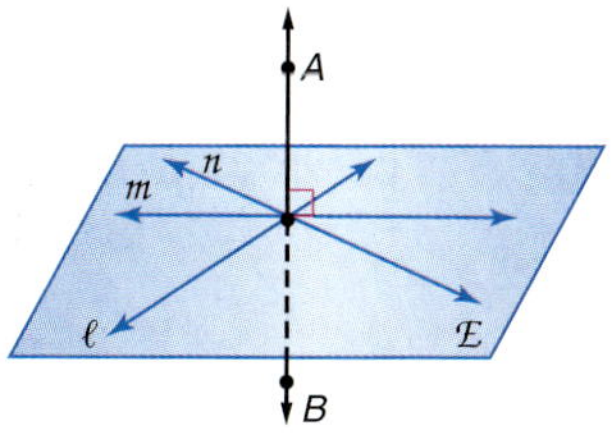

Maintain Your Skills

Mixed Review

Measure each angle and classify it as *right*, *acute*, or *obtuse*. *(Lesson 1-4)*

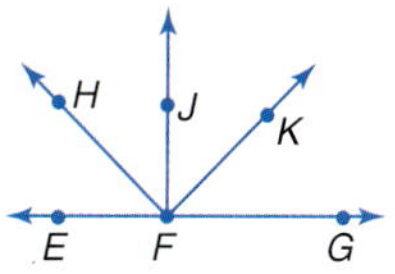

44. $\angle KFG$ acute
45. $\angle HFG$ obtuse
46. $\angle HFK$ right
47. $\angle JFE$ right
48. $\angle HFJ$ acute
49. $\angle EFK$ obtuse

Find the distance between each pair of points. *(Lesson 1-3)*

50. $A(3, 5), B(0, 1)$ 5
51. $C(5, 1), D(5, 9)$ 8
52. $E(-2, -10), F(-4, 10)$
53. $G(7, 2), H(-6, 0)$
54. $J(-8, 9), K(4, 7)$
55. $L(1, 3), M(3, -1)$

52. $\sqrt{404} \approx 20.1$
53. $\sqrt{173} \approx 13.2$
54. $\sqrt{148} \approx 12.2$
55. $\sqrt{20} \approx 4.5$

Find the value of the variable and *QR* if *Q* is between *P* and *R*. *(Lesson 1-2)*

56. $PQ = 1 - x$, $QR = 4x + 17$, $PR = -3x$ $x = -3$, $QR = 5$
57. $PR = 7n + 8$, $PQ = 4n - 3$, $QR = 6n + 2$ $n = 3$, $QR = 20$

Getting Ready for the Next Lesson

PREREQUISITE SKILL **Evaluate each expression if $\ell = 3$, $w = 8$, and $s = 2$.**
*(To review **evaluating expressions**, see page 736.)*

58. $2\ell + 2w$ 22
59. ℓw 24
60. $4s$ 8
61. $\ell w + ws$ 40
62. $s(\ell + w)$ 22

www.geometryonline.com/self_check_quiz/fcat

4 Assess

Open-Ended Assessment

Speaking Describing angle relationships in figures provides an opportunity for students to practice making correct assumptions and properly assessing the information given in the figures. Call on students to describe some of the figures presented in problems and examples in this lesson.

Getting Ready for Lesson 1-6

Prerequisite Skill Students will find perimeters of regular and irregular polygons in Lesson 1-6. They will have to find two or more unknown lengths. Use Exercises 58–62 to determine your students' familiarity with evaluating expressions with multiple variables.

Assessment Options

Quiz (Lessons 1-4 and 1-5) is available on p. 52 of the *Chapter 1 Resource Masters*.

Answers

37. Sample answer:

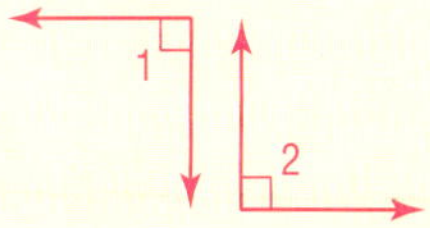

39. Because $\angle WUT$ and $\angle TUV$ are supplementary, let $m\angle WUT = x$ and $m\angle TUV = 180 - x$. A bisector creates measures that are half of the original angle, so $m\angle YUT = \frac{1}{2}m\angle WUT$ or $\frac{x}{2}$ and $m\angle TUZ = \frac{1}{2}m\angle TUV$ or $\frac{180 - x}{2}$. Then $m\angle YUZ = m\angle YUT + m\angle TUZ$ or $\frac{x}{2} + \frac{180 - x}{2}$. This sum simplifies to $\frac{180}{2}$ or 90. Because $m\angle YUZ = 90$, $\overline{YU} \perp \overline{UZ}$.

40. Sample answer: The types of angles formed depends on how the streets intersect. There may be as few as two angles or many more if there are more than two lines intersecting. Answers should include the following.
- linear pairs, vertical angles, adjacent angles
- See students' work.

Geometry Activity

A Follow-Up of Lesson 1-5

Getting Started

Objective Construct perpendiculars.

Materials
compass   straightedge

Teach

- Point out to students that it is helpful if they place point C in Activity 1 and point Z in Activity 2 in an appropriate position so that they have enough room to use their compasses comfortably.
- Stress the importance of keeping the compass setting stable when performing the construction, or the result may not be correct.

Assess

Exercise 1 gives students the opportunity to practice the two methods of constructing perpendiculars.

After successfully completing **Exercise 2,** students will have a better understanding as to how and why these construction methods work.

Study Notebook

Ask students to summarize what they have learned about constructing perpendiculars to lines. Have students select which method they like better and explain why.

Answer

2. The first step of the construction locates two points on the line. Then the process is very similar to the construction through a point on a line.

Geometry Activity

A Follow-Up of Lesson 1-5

Constructing Perpendiculars

You can use a compass and a straightedge to construct a line perpendicular to a given line through a point on the line, or through a point *not* on the line.

Activity 1 Perpendicular Through a Point on the Line

Construct a line perpendicular to line *n* and passing through point *C* on *n*.

1. Place the compass at point C. Using the same compass setting, draw arcs to the right and left of C, intersecting line *n*. Label the points of intersection *A* and *B*.

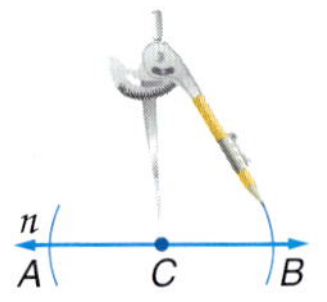

2. Open the compass to a setting greater than *AC*. Put the compass at point *A* and draw an arc above line *n*.

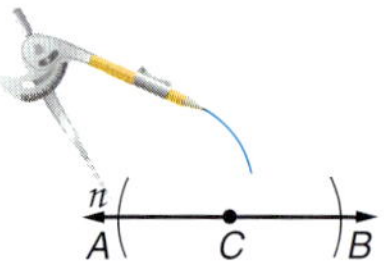

3. Using the same compass setting as in Step 2, place the compass at point *B* and draw an arc intersecting the arc drawn in Step 2. Label the point of intersection *D*.

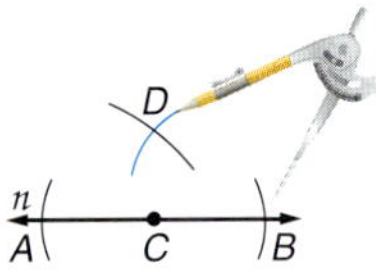

4. Use a straightedge to draw $\overleftrightarrow{CD}$.

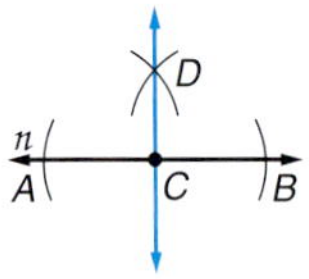

Activity 2 Perpendicular Through a Point not on the Line

Construct a line perpendicular to line *m* and passing through point *Z* not on *m*.

1. Place the compass at point Z. Draw an arc that intersects line *m* in two different places. Label the points of intersection X and Y.

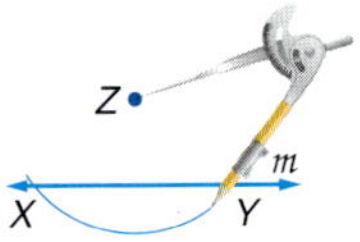

2. Open the compass to a setting greater than $\frac{1}{2}$ *XY*. Put the compass at point *X* and draw an arc below line *m*.

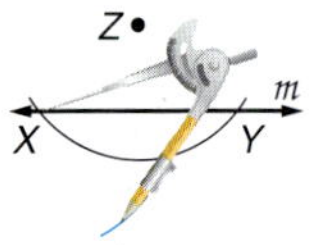

3. Using the same compass setting, place the compass at point *Y* and draw an arc intersecting the arc drawn in Step 2. Label the point of intersection *A*.

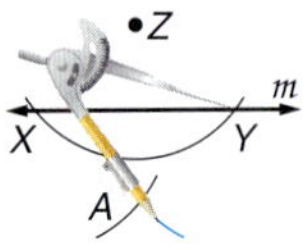

4. Use a straightedge to draw $\overleftrightarrow{ZA}$.

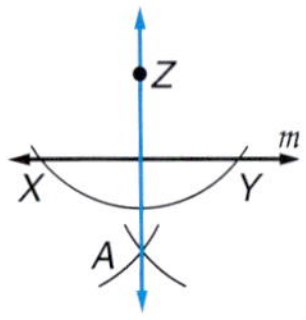

Model and Analyze

1. Draw a line and construct a line perpendicular to it through a point on the line. Repeat with a point not on the line. See students' work.
2. How is the second construction similar to the first one? See margin.

Resource Manager

Teaching Geometry with Manipulatives
- p. 40 (student recording sheet)

Glencoe Mathematics Classroom Manipulative Kit
- safety compass
- straightedge

1-6 Polygons

Sunshine State Standards
MA.B.3.4.1

What You'll Learn

- Identify and name polygons.
- Find perimeters of polygons.

How are polygons related to toys?

There are numerous types of building sets that connect sticks to form various shapes. Whether they are made of plastic, wood, or metal, the sticks represent segments. When the segments are connected, they form angles. The sticks are connected to form closed figures that in turn are connected to make a model of a real-world object.

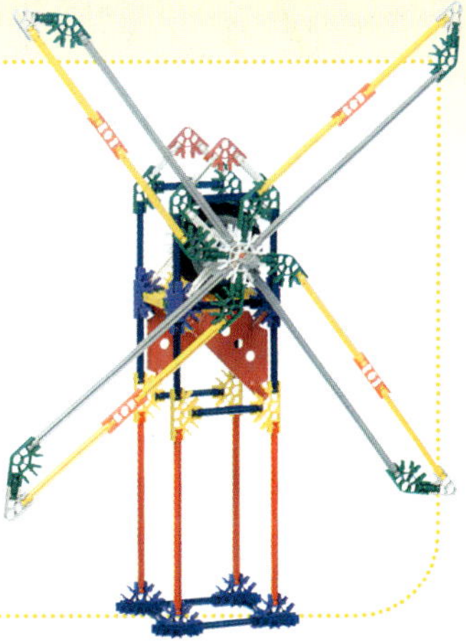

Vocabulary

- polygon
- concave
- convex
- *n*-gon
- regular polygon
- perimeter

POLYGONS Each closed figure shown in the toy is a **polygon**. A polygon is a closed figure whose sides are all segments. The sides of each angle in a polygon are called *sides* of the polygon, and the vertex of each angle is a *vertex* of the polygon.

Study Tip

Reading Math
The plural of vertex is *vertices*.

Key Concept — Polygon

- **Words** A polygon is a closed figure formed by a finite number of coplanar segments such that
  (1) the sides that have a common endpoint are noncollinear, and
  (2) each side intersects exactly two other sides, but only at their endpoints.
- **Symbol** A polygon is named by the letters of its vertices, written in consecutive order.
- **Examples**

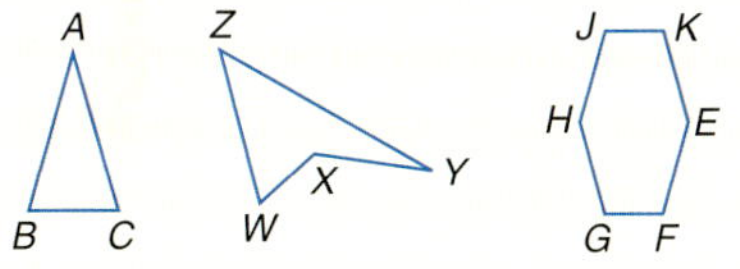

polygons *ABC*, *WXYZ*, *EFGHJK*

- **Nonexamples**

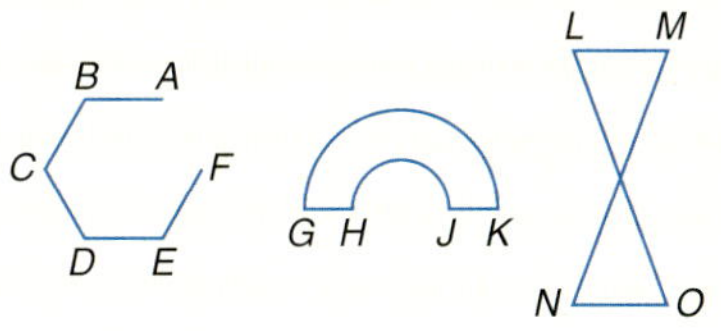

Polygons can be **concave** or **convex**. Suppose the line containing each side is drawn. If any of the lines contain any point in the interior of the polygon, then it is concave. Otherwise it is convex.

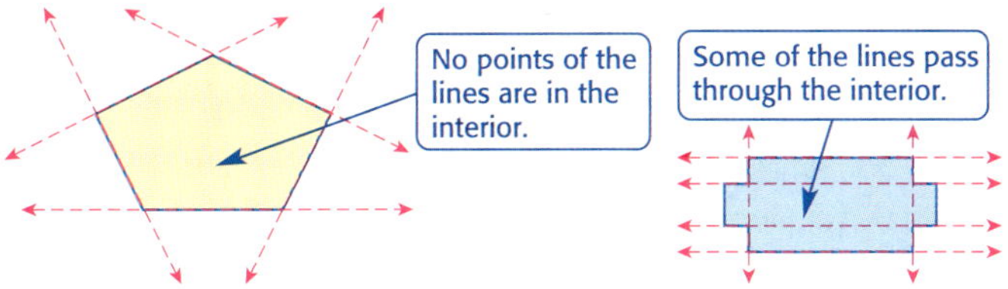

convex polygon concave polygon

1-6 Lesson Notes

1 Focus

5-Minute Check Transparency 1-6 Use as a quiz or review of Lesson 1-5.

Mathematical Background notes are available for this lesson on p. 4D.

Questions are provided at the beginning of each lesson to help you use the problem provided there to engage and inform students.

How are polygons related to toys?

Ask students:

- What shape would you make most often with these toys if you wanted to build large models? Why? **Sample answer: Some students may have noticed that triangular constructions are more stable than other shapes.**
- Have you learned anything about angles that would help you build better models with construction toys? **Accept all reasonable answers.**

Resource Manager

Workbook and Reproducible Masters

Chapter 1 Resource Masters

- Study Guide and Intervention, pp. 31–32
- Skills Practice, p. 33
- Practice, p. 34
- Reading to Learn Mathematics, p. 35
- Enrichment, p. 36
- Assessment, p. 52

Teaching Geometry With Manipulatives Masters, p. 1

Transparencies

5-Minute Check Transparency 1-6
Answer Key Transparencies

Technology

Interactive Chalkboard

2 Teach

POLYGONS

In-Class Example

Power Point®

1 Name each polygon by the number of sides. Then classify it as *convex* or *concave*, *regular* or *irregular*.

a.

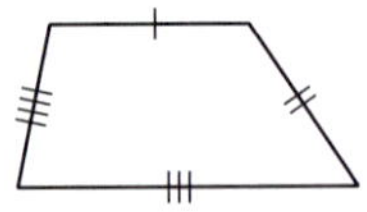

quadrilateral, convex, irregular

b.

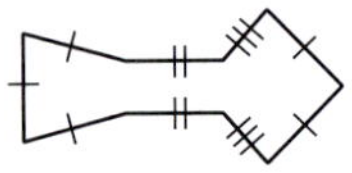

nonagon, concave, irregular

PERIMETER

Teaching Tip While there are formulas for the perimeters of a few special shapes, stress that the perimeter can *always* be found by adding the measures of all the sides.

Study Tip

Reading Math
The term *polygon* is derived from a Greek word meaning *many angles*. Since *hexa-* means 6, you would think *hexagon* means 6 angles, and you would be correct. Every polygon has the same number of angles as it does sides.

You are already familiar with many polygon names, such as triangle, square, and rectangle. In general, polygons can be classified by the number of sides they have. A polygon with n sides is an **n-gon**. The table lists some common names for various categories of polygon.

Number of Sides	Polygon
3	triangle
4	quadrilateral
5	pentagon
6	hexagon
7	heptagon
8	octagon
9	nonagon
10	decagon
12	dodecagon
n	n-gon

A convex polygon in which all the sides are congruent and all the angles are congruent is called a **regular polygon**. Octagon $PQRSTUVW$ below is a regular octagon.

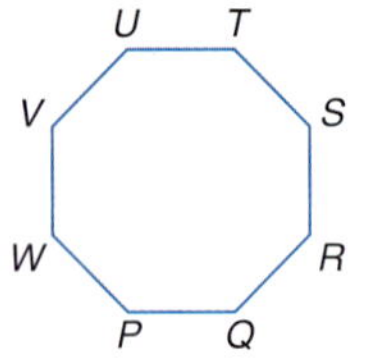

Example 1 Identify Polygons

Name each polygon by its number of sides. Then classify it as *convex* or *concave* and *regular* or *irregular*.

a.

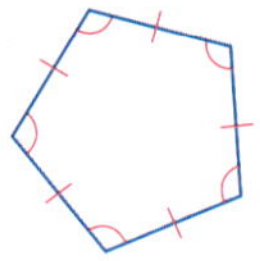

There are 5 sides, so this is a pentagon. No line containing any of the sides will pass through the interior of the pentagon, so it is convex.

The sides are congruent, and the angles are congruent. It is regular.

b.

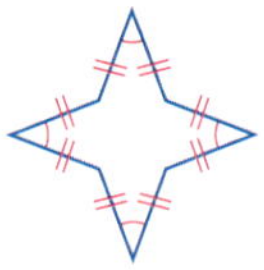

There are 8 sides, so this is an octagon. A line containing any of the sides will pass through the interior of the octagon, so it is concave.

The sides are congruent. However, since it is concave, it cannot be regular.

PERIMETER The **perimeter** of a polygon is the sum of the lengths of its sides, which are segments. Some shapes have special formulas, but they are all derived from the basic definition of perimeter.

Key Concept — Perimeter

- **Words** The perimeter P of a polygon is the sum of the lengths of the sides of a polygon.
- **Examples**

triangle	square	rectangle
$P = a + b + c$	$P = s + s + s + s$	$P = \ell + w + \ell + w$
	$P = 4s$	$P = 2\ell + 2w$

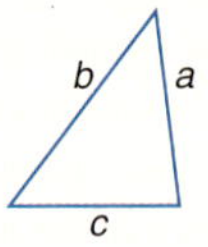

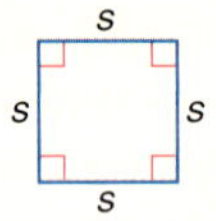

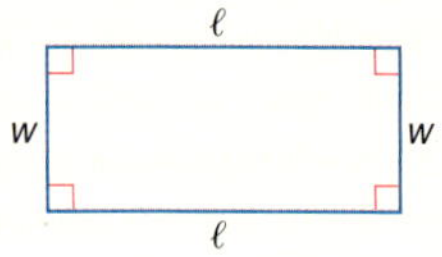

Teacher to Teacher

Joy F. Stanford, Booker T. Washington Magnet High School   Montgomery, AL

When discussing concave and convex polygons, I illustrate the difference by placing a rubber band around a concave polygon and then a convex polygon. It will stretch to touch every side if the figure is convex, and it will not touch all of the sides of the concave polygon.

Teacher to Teacher features contain teaching suggestions from teachers who are creatively teaching Geometry in their classrooms.

Example 2 Find Perimeter

GARDENING **A landscape designer is putting black plastic edging around a rectangular flower garden that has length 5.7 meters and width 3.8 meters. The edging is sold in 5-meter lengths.**

a. Find the perimeter of the garden and determine how much edging the designer should buy.

$P = 2\ell + 2w$

$= 2(5.7) + 2(3.8)$ $\ell = 5.7, w = 3.8$

$= 11.4 + 7.6$ or 19

The perimeter of the garden is 19 meters.

The designer needs to buy 20 meters of edging.

b. Suppose the length and width of the garden are tripled. What is the effect on the perimeter and how much edging should the designer buy?

The new length would be 3(5.7) or 17.1 meters.

The new width would be 3(3.8) or 11.4 meters.

$P = 2\ell + 2w$

$= 2(17.1) + 2(11.4)$ or 57

Compare the original perimeter to this measurement.

$57 = 3(19)$ meters

So, when the lengths of the sides of the rectangle are tripled, the perimeter also triples. The designer needs to buy 60 meters of edging.

You can use the Distance Formula to find the perimeter of a polygon graphed on a coordinate plane.

Example 3 Perimeter on the Coordinate Plane

COORDINATE GEOMETRY **Find the perimeter of triangle PQR if $P(-5, 1)$, $Q(-1, 4)$, and $R(-6, -8)$.**

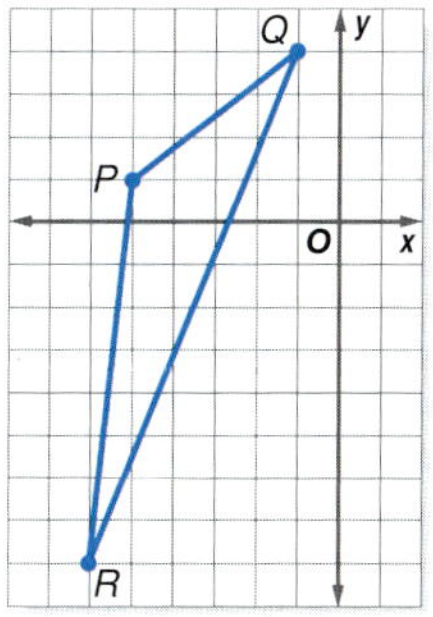

Use the Distance Formula,

$d = \sqrt{(x_2 - x_1)^2 + (y_2 - y_1)^2}$,

to find PQ, QR, and PR.

$PQ = \sqrt{[-1 - (-5)]^2 + (4 - 1)^2}$

$= \sqrt{4^2 + 3^2}$

$= \sqrt{25}$ or 5

$QR = \sqrt{[-6 - (-1)]^2 + (-8 - 4)^2}$

$= \sqrt{(-5)^2 + (-12)^2}$

$= \sqrt{169}$ or 13

$PR = \sqrt{[-6 - (-5)]^2 + (-8 - 1)^2}$

$= \sqrt{(-1)^2 + (-9)^2}$

$= \sqrt{82} \approx 9.1$

The perimeter of triangle PQR is $5 + 13 + \sqrt{82}$ or about 27.1 units.

Study Tip

Look Back
To review the **Distance Formula**, see Lesson 1-3.

www.geometryonline.com/extra_examples/fcat

In-Class Examples

Power Point®

2 **CONSTRUCTION** A masonry company is contracted to lay three layers of decorative brick along the foundation for a new house given the dimensions below.

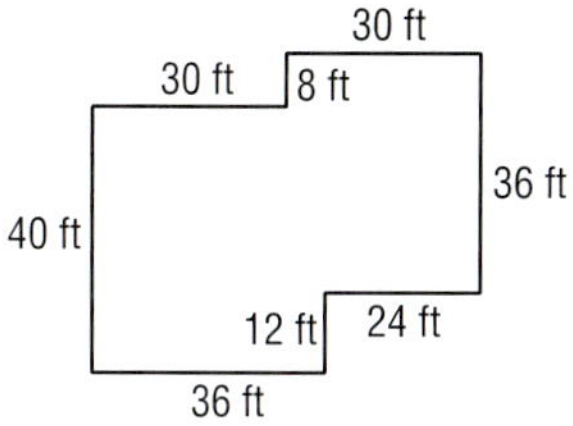

a. Find the perimeter of the foundation and determine how many bricks the company will need to complete the job. Assume that one brick is 8 inches long. **216 ft; 972 bricks**

b. The builder realizes he accidentally halved the size of the foundation in part *a*, so he reworks the drawing with the correct dimensions. How will this affect the perimeter of the house and the number of bricks the masonry company needs? **The perimeter and the number of bricks needed are doubled.**

3 Find the perimeter of pentagon $ABCDE$ with $A(0, 4)$, $B(4, 0)$, $C(3, -4)$, $D(-3, -4)$, and $E(-3, 1)$.

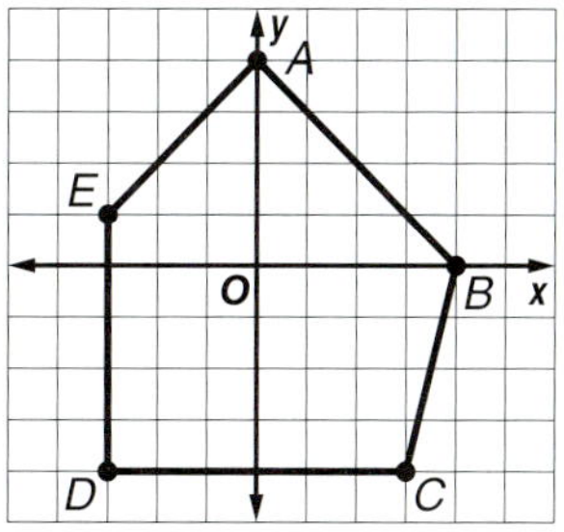

about 25 units

DAILY INTERVENTION

Unlocking Misconceptions

Finding the Correct Answer Students sometimes think they have solved the problem when they find the value of the variable. In Example 2a, a common mistake is to say that the answer is 19, but the problem also asks for the amount of edging the designer should buy.

In-Class Example

Power Point®

4 The width of a rectangle is 5 less than twice its length. The perimeter is 80 centimeters. Find the length of each side. **$\ell = 15$ cm, $w = 25$ cm**

DAILY INTERVENTION

FIND THE ERROR Remind students that the sides of a polygon can only intersect other sides at their endpoints.

3 Practice/Apply

Study Notebook

Have students—
- add the definitions/examples of the vocabulary terms to their Vocabulary Builder worksheets for Lesson 1-6.
- include any other item(s) that they find helpful in mastering the skills in this lesson.

Find the Error exercises help students identify and address common errors before they occur.

About the Exercises...

Organization by Objective
- **Polygons:** 12–18
- **Perimeter:** 19–34

Odd/Even Assignments

Exercises 12–34 are structured so that students practice the same concepts whether they are assigned odd or even problems.

Assignment Guide

Basic: 13–23 odd, 27–35 odd, 36–44

Average: 13–35 odd, 36–44

Advanced: 12–34 even, 36–44

You can also use algebra to find the lengths of the sides if the perimeter is known.

Study Tip

Equivalent Measures
In Example 5, the dimensions $\frac{1}{4}$ foot by $\frac{3}{4}$ foot can also be expressed as 3 inches by 9 inches.

Example 4 Use Perimeter to Find Sides

ALGEBRA **The length of a rectangle is three times the width. The perimeter is 2 feet. Find the length of each side.**

Let w represent the width. Then the length is $3w$.

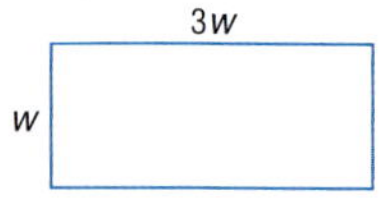

$P = 2\ell + 2w$ Perimeter formula for rectangle

$2 = 2(3w) + 2w$ $\ell = 3w$

$2 = 8w$ Simplify.

$\frac{1}{4} = w$ Divide each side by 8.

The width is $\frac{1}{4}$ foot. By substituting $\frac{1}{4}$ for w, the length $3w$ becomes $3\left(\frac{1}{4}\right)$ or $\frac{3}{4}$ foot.

Check for Understanding

Concept Check

1. **OPEN ENDED** Explain how you would find the length of a side of a regular decagon if the perimeter is 120 centimeters. **Divide the perimeter by 10.**

2. **FIND THE ERROR** Saul and Tiki were asked to draw quadrilateral $WXYZ$ with $m\angle Z = 30$.

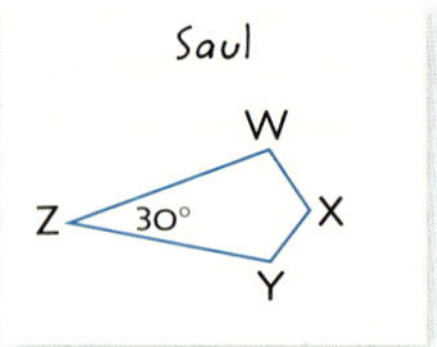

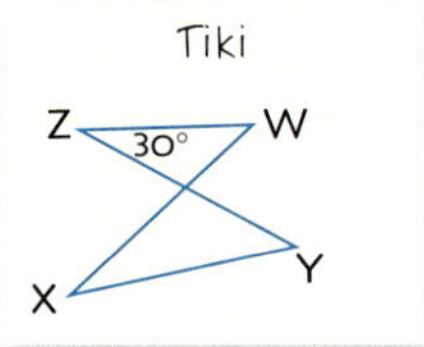

Who is correct? Explain your reasoning. **Saul; Tiki's figure is not a polygon.**

3. **Write** a formula for the perimeter of a triangle with congruent sides of length s. **3. $P = 3s$**

4. **Draw** a concave pentagon and explain why it is concave. **See margin.**

Guided Practice

GUIDED PRACTICE KEY	
Exercises	Examples
5, 6	1
7, 8, 11	2
9	3
10	4

Name each polygon by its number of sides. Then classify it as *convex* or *concave* and *regular* or *irregular*.

5. 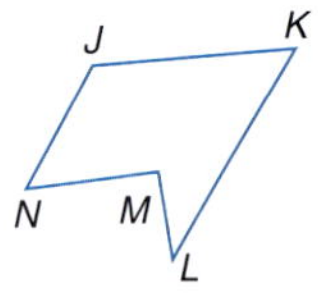

pentagon; concave; irregular

6. 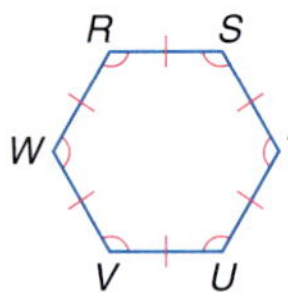

hexagon; convex; regular

For Exercises 7 and 8, use pentagon *LMNOP*.

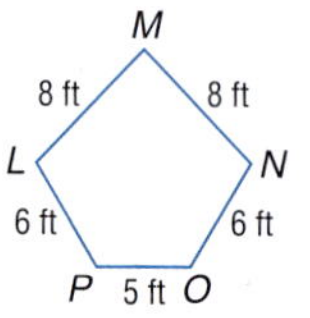

7. Find the perimeter of pentagon $LMNOP$. **33 ft**

8. Suppose the length of each side of pentagon $LMNOP$ is doubled. What effect does this have on the perimeter? **8. It doubles.**

9. **COORDINATE GEOMETRY** A polygon has vertices $P(-3, 4)$, $Q(0, 8)$, $R(3, 8)$, and $S(0, 4)$. Find the perimeter of $PQRS$. **9. 16 units**

10. **ALGEBRA** Quadrilateral $ABCD$ has a perimeter of 95 centimeters. Find the length of each side if $AB = 3a + 2$, $BC = 2(a - 1)$, $CD = 6a + 4$, and $AD = 5a - 5$. **10. 20, 10, 40, 25**

Application

11. **HISTORIC LANDMARKS** The Pentagon building in Arlington, Virginia, is so named because of its five congruent sides. Find the perimeter of the outside of the Pentagon if one side is 921 feet long. **4605 ft**

DAILY INTERVENTION **Differentiated Instruction**

Interpersonal Separate students into groups to complete Exercises 15–18. Allow groups to choose a familiar traffic sign that is not included in these exercises. The groups can name and classify their signs with one person stating whether the polygon is convex or concave, another deciding if it is regular or not, and so on. Have each group determine a reasonable perimeter for each sign. As an extension, groups can make models of their signs from poster boards to hang in the classroom.

★ indicates increased difficulty

Practice and Apply

Homework Help

For Exercises	See Examples
12–18	1
19–25	2
26–28	3
29–34	4

Extra Practice
See page 755.

Name each polygon by its number of sides. Then classify it as *convex* or *concave* and *regular* or *irregular*.

12. 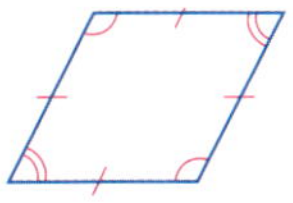
quadrilateral; convex; irregular

13. 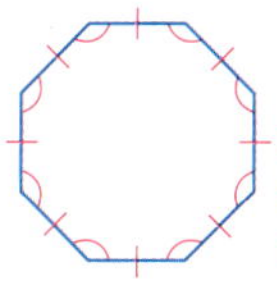
octagon; convex; regular

14. 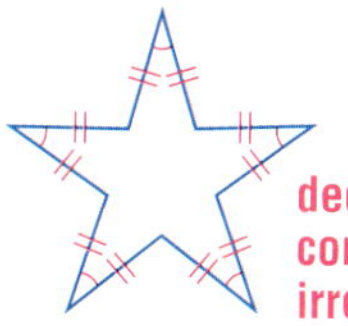
decagon; concave; irregular

TRAFFIC SIGNS **Identify the shape of each traffic sign.**

15. school zone

pentagon

16. caution or warning

quadrilateral

17. yield

triangle

18. railroad

dodecagon

Find the perimeter of each figure.

19. 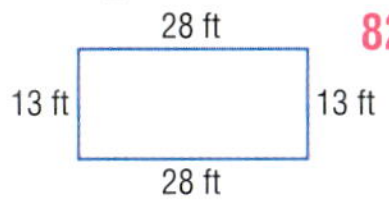

82 ft

20. 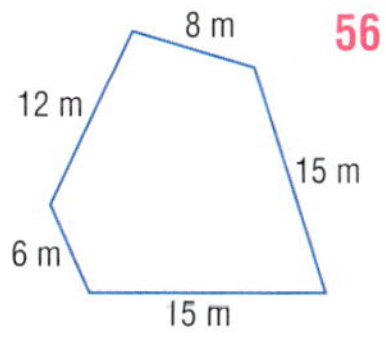

56 m

21. 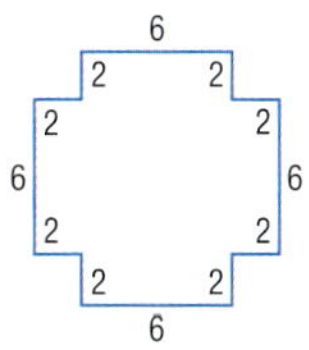

40 units

22. What is the effect on the perimeter of the figure in Exercise 19 if each measure is multiplied by 4? The perimeter is multiplied by 4.

23. What is the effect on the perimeter of the figure in Exercise 20 if each measure is tripled? The perimeter is tripled.

24. What is the effect on the perimeter of the figure in Exercise 21 if each measure is divided by 2? The perimeter is divided by 2.

★ 25. The perimeter of an n-gon is 12.5 meters. Find the perimeter of the n-gon if the length of each of its n sides is multiplied by 10. 125 m

COORDINATE GEOMETRY **Find the perimeter of each polygon.**

26. rectangle with vertices $A(-1, 1)$, $B(3, 4)$, $C(6, 0)$, and $D(2, -3)$ 20 units

27. hexagon with vertices $P(-2, 3)$, $Q(3, 3)$, $R(7, 0)$, $S(3, -3)$, $T(-2, -3)$, and $U(-6, 0)$

28. pentagon with vertices $V(3, 0)$, $W(-2, 12)$, $X(-10, -3)$, $Y(-8, -12)$, and $Z(-2, -12)$

27. 30 units 28. 58.2 units

ALGEBRA **Find the length of each side of the polygon for the given perimeter.**

29. $P = 90$ centimeters
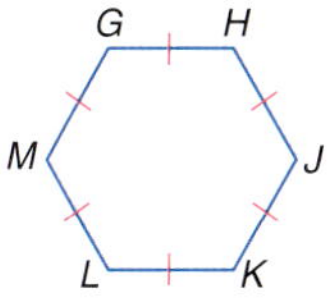

30. $P = 14$ miles
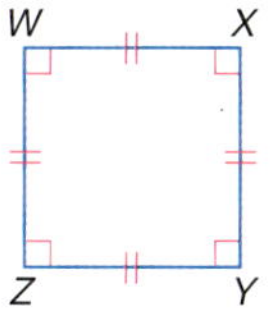

31. $P = 31$ units
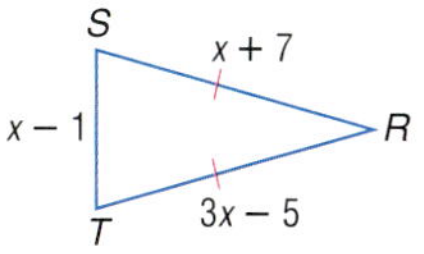

29. All are 15 cm.
30. All are 3.5 mi.
31. 13 units, 13 units, 5 units

More About. . .

Traffic Signs

The shape of a traffic sign is determined by the use of the sign. For example, tall rectangle shaped signs are used for posting speed limits, parking restrictions, and other regulations. Wide rectangle shaped signs are used to provide drivers with information.

Source: U.S. Federal Highway Administration

www.geometryonline.com/self_check_quiz/fcat

Answer

4. 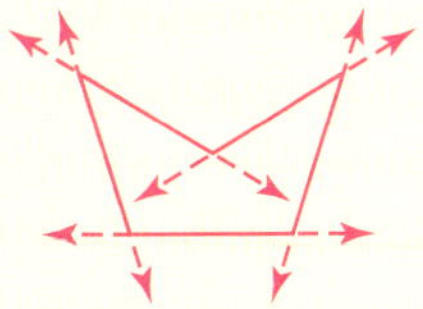

Sample answer: Some of the lines containing the sides pass through the interior of the pentagon.

Enrichment, p. 36

Perimeter and Area of Irregular Shapes

Two formulas that are used frequently in mathematics are perimeter and area of a rectangle.

Perimeter: $P = 2\ell + 2w$
Area: $A = \ell w$, where ℓ is the length and w is the width

However, many figures are combinations of two or more rectangles creating **irregular shapes**. To find the area of an irregular shape, it helps to separate the shape into rectangles, calculate the formula for each rectangle, then find the sum of the areas.

Example **Find the area of the figure at the right.**

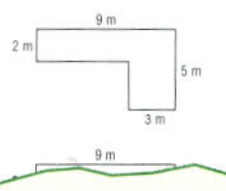

Separate the figure into two rectangles.

$A = \ell w$
$A_1 = 9 \cdot 2$ $A_2 = 3 \cdot 3$
$= 18$ $= 9$

Study Guide and Intervention, p. 31 (shown) and p. 32

Polygons A **polygon** is a closed figure formed by a finite number of coplanar line segments. The sides that have a common endpoint must be noncollinear and each side intersects exactly two other sides at their endpoints. A polygon is named according to its number of sides. A **regular polygon** has congruent sides and congruent angles. A polygon can be **concave** or **convex**.

Example **Name each polygon by its number of sides. Then classify it as *concave* or *convex* and *regular* or *irregular*.**

a. 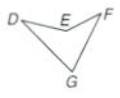

The polygon has 4 sides, so it is a quadrilateral. It is concave because part of $\overline{DE}$ or $\overline{EF}$ lies in the interior of the figure. Because it is concave, it cannot have all its angles congruent and so it is irregular.

b. 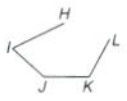

The figure is not closed, so it is not a polygon.

c. 

The polygon has 5 sides, so it is a pentagon. It is convex. All sides are congruent and all angles are congruent, so it is a regular pentagon.

d. 

The figure has 8 congruent sides and 8 congruent angles. It is convex and is a regular octagon.

Exercises

Name each polygon by its number of sides. Then classify it as *concave* or *convex* and *regular* or *irregular*.

1. 
hexagon; convex; regular

2. 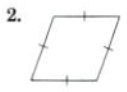
quadrilateral; convex; irregular

3. 
pentagon; concave; irregular

4. 
triangle; convex; irregular

5. 
pentagon; concave; irregular

6. 
octagon; concave; irregular

Skills Practice, p. 33 and Practice, p. 34 (shown)

Name each polygon by its number of sides and then classify it as *convex* or *concave* and *regular* or *irregular*.

1. 
hexagon; concave; irregular

2. 
nonagon; convex; regular

3. 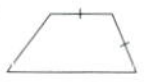
quadrilateral; convex; irregular

Find the perimeter of each figure.

4. 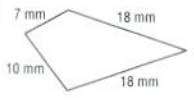

53 mm

5. 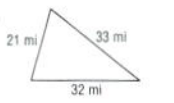

86 mi

6. 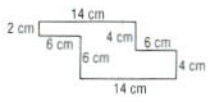

56 cm

COORDINATE GEOMETRY **Find the perimeter of each polygon.**

7. quadrilateral $OPQR$ with vertices $O(-3, 2)$, $P(1, 5)$, $Q(6, 4)$, and $R(5, -2)$
≈25.1 units

8. pentagon $STUVW$ with vertices $S(0, 0)$, $T(3, -2)$, $U(2, -5)$, $V(-2, -5)$, and $W(-3, -2)$
≈17.5 units

ALGEBRA **Find the length of each side of the polygon for the given perimeter.**

9. $P = 26$ inches
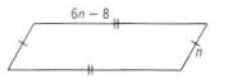

3 in., 3 in., 10 in., 10 in.

10. $P = 39$ centimeters
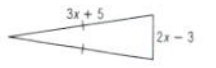

17 cm, 17 cm, 5 cm

11. $P = 89$ feet
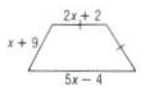

18 ft, 18 ft, 36 ft, 17 ft

SEWING **For Exercises 12–13, use the following information.**
Jasmine plans to sew fringe around the scarf shown in the diagram.

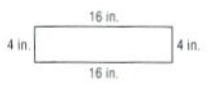

12. How many inches of fringe does she need to purchase?
40 in.

13. If Jasmine doubles the width of the scarf, how many inches of fringe will she need?
48 in.

Reading to Learn Mathematics, p. 35

ELL

Pre-Activity **How are polygons related to toys?**

Read the introduction to Lesson 1-6 at the top of page 45 in your textbook.

Name four different shapes that can each be formed by four sticks connected to form a closed figure. Assume you have sticks with a good variety of lengths.
Sample answer: square, rectangle, parallelogram, trapezoid

Reading the Lesson

1. Tell why each figure is *not* a polygon.

a. not closed
b. curved (not all made up of segments)
c. Sides intersect at a point that is not an endpoint.

2. Name each polygon by its number of sides. Then classify it as *convex* or *concave* and *regular* or *not regular*.

a. pentagon, convex, regular
b. quadrilateral, concave, not regular
c. quadrilateral, convex, not regular

3. What is another name for a regular quadrilateral? a square

4. Match each polygon in the first column with the formula in the second column that can be used to find its perimeter. (s represents the length of each side of a regular polygon.)

a. regular dodecagon iv — i. $P = 8s$
b. square vi — ii. $P = 6s$
c. regular hexagon ii — iii. $P = a + b + c$
d. rectangle v — iv. $P = 12s$
e. regular octagon i — v. $P = 2\ell + 2w$
f. triangle iii — vi. $P = 4s$

Helping You Remember

5. One way to remember the meaning of a term is to explain it to another person. How would you explain to a friend what a regular polygon is?
Sample answer: A regular polygon looks the same no matter what part you look at. The sides are the same length, and the angles are the same size.

Lesson 1-6 Polygons 49

4 Assess

Open-Ended Assessment

Writing Draw several polygons on the board. Have volunteers label each figure, classify it, use a ruler to measure the sides, and calculate the perimeter.

Assessment Options

Quiz (Lessons 1-6) is available on p. 52 of the *Chapter 1 Resource Masters*.

Assessment Options lists the quizzes and tests that are available in the Chapter Resource Masters.

Answer

37. Sample answer: Some toys use pieces to form polygons. Others have polygon-shaped pieces that connect together. Answers should include the following.

- triangles, quadrilaterals, pentagons
- 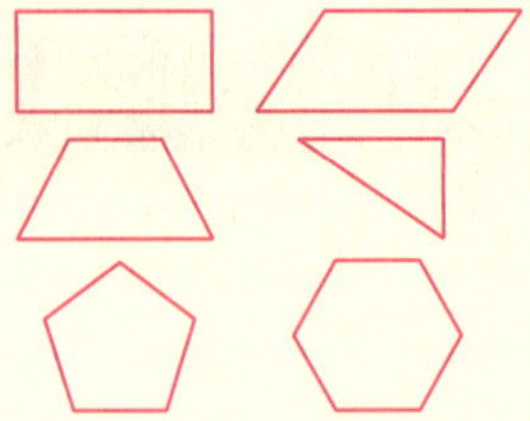

ALGEBRA Find the length of each side of the polygon for the given perimeter.

32. $P = 84$ meters

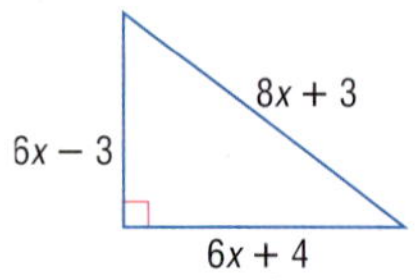

21 m, 28 m, 35 m

33. $P = 42$ inches

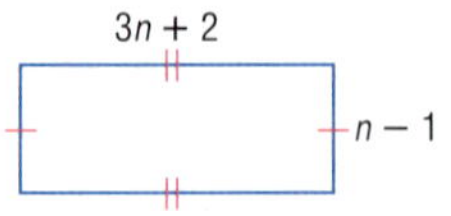

4 in., 4 in., 17 in., 17 in.

34. $P = 41$ yards

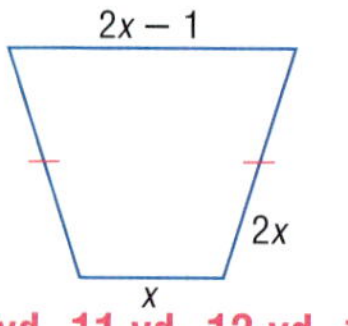

6 yd, 11 yd, 12 yd, 12 yd

35. **NETS** *Nets* are patterns that form a three-dimensional figure when cut out and folded. The net at the right makes a rectangular box. What is the perimeter of the net? 52 units

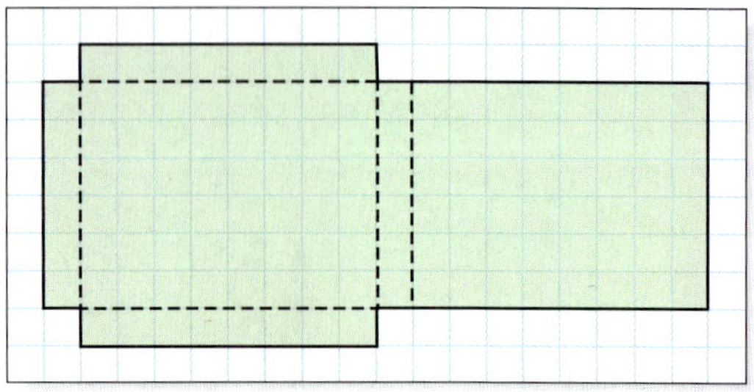

36. **CRITICAL THINKING** Use grid paper to draw all possible rectangles with length and width that are whole numbers and with a perimeter of 12. Record the number of grid squares contained in each rectangle.
   a. What do you notice about the rectangle with the greatest number of squares?
   b. The perimeter of another rectangle is 36. What would be the dimensions of the rectangle with the greatest number of squares? 9×9

36a. It is a square with side length of 3 units.

37. **WRITING IN MATH** Answer the question that was posed at the beginning of the lesson. See margin.

How are polygons related to toys?

Include the following in your answer:
- names of the polygons shown in the picture of the toy structure, and
- sketches of other polygons that could be formed with construction toys with which you are familiar.

FCAT Practice
Standardized Test Practice Ⓐ Ⓑ Ⓒ Ⓓ

38. **SHORT RESPONSE** A farmer fenced all but one side of a square field. If he has already used $3x$ meters of fence, how many meters will he need for the last side? x m

39. **ALGEBRA** If $5n + 5 = 10$, what is the value of $11 - n$? D

Ⓐ −10 Ⓑ 0 Ⓒ 5 Ⓓ 10

Maintain Your Skills

Mixed Review

Determine whether each statement is *always, sometimes,* or *never true*. *(Lesson 1-5)*

40. Two angles that form a linear pair are supplementary. always

41. If two angles are supplementary, then one of the angles is obtuse. sometimes

In the figure, $\overrightarrow{AM}$ bisects $\angle LAR$, and $\overrightarrow{AS}$ bisects $\angle MAR$. *(Lesson 1-4)*

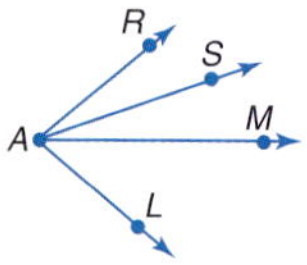

42. If $m\angle MAR = 2x + 13$ and $m\angle MAL = 4x - 3$, find $m\angle RAL$. 58

43. If $m\angle RAL = x + 32$ and $m\angle MAR = x - 31$, find $m\angle LAM$. 63

44. Find $m\angle LAR$ if $m\angle RAS = 25 - 2x$ and $m\angle SAM = 3x + 5$. 68

Geometry Software Investigation

A Follow-Up of Lesson 1-6

Measuring Polygons

You can use The Geometer's Sketchpad® to draw and investigate polygons. It can be used to find the measures of the sides and the perimeter of a polygon. You can also find the measures of the angles in a polygon.

Step 1 *Draw $\triangle ABC$.*

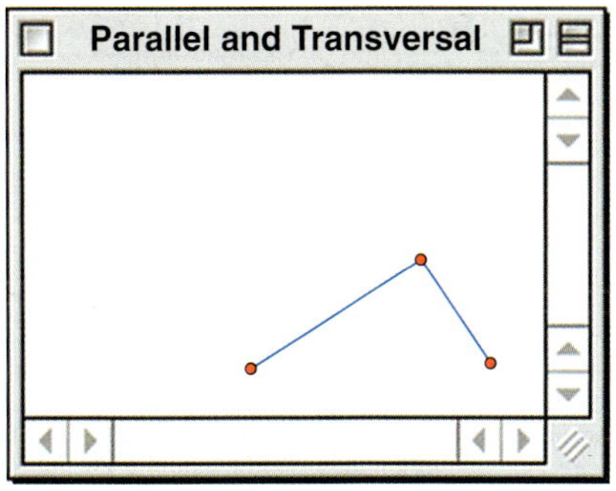

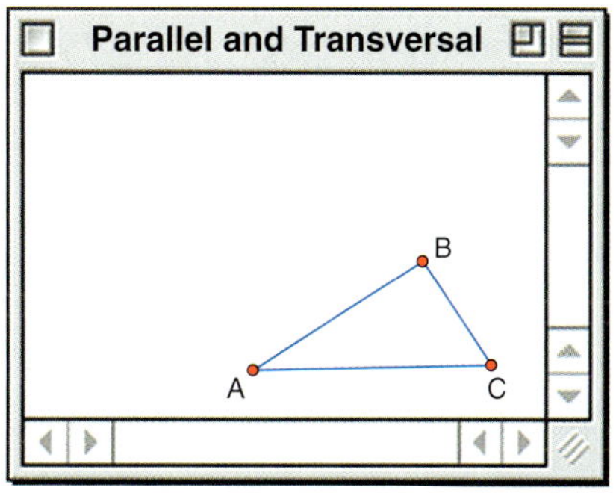

- Select the segment tool from the toolbar, and click to set the first endpoint A of side $\overline{AB}$. Then drag the cursor and click again to set the other endpoint B.
- Click on point B to set the endpoint of $\overline{BC}$. Drag the cursor and click to set point C.
- Click on point C to set the endpoint of $\overline{CA}$. Then move the cursor to highlight point A. Click on A to draw $\overline{CA}$.
- Use the pointer tool to click on points A, B, and C. Under the **Display** menu, select **Show Labels** to label the vertices of your triangle.

Step 2 *Find AB, BC, and CA.*

- Use the pointer tool to select $\overline{AB}$, $\overline{BC}$, and $\overline{CA}$.
- Select the **Length** command under the **Measure** menu to display the lengths of $\overline{AB}$, $\overline{BC}$, and $\overline{CA}$.

$AB = 5.30$
$BC = 3.80$
$CA = 6.54$

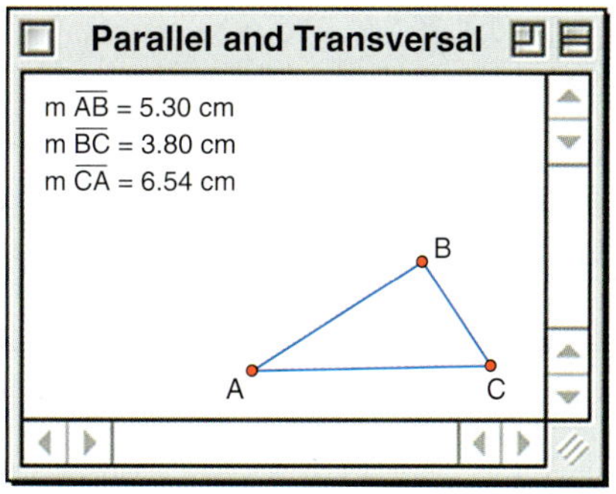

Geometry Software Investigation

A Follow-Up of Lesson 1-6

Getting Started

Software This activity can be done with any dynamic software that allows construction of figures and measurement. Some of these include Cabri, Cabri Junior on the TI-83 Plus, and Cabri or Sketchpad on the TI-92 Plus.

Measuring Polygons Explain to students that their measurements will not be the same as the screen shown in the activity.

Teach

- Remind students to examine the measures provided by the software. An unreasonable measurement may mean the student has not selected the item correctly.
- Before students complete Question 4, ask them to predict what will happen to the perimeter of the quadrilateral based on what they have gathered from similar examples they have seen thus far in Lesson 1-6.

Geometry Software Investigations empower students to use dynamic software to discover geometric relationships.

Step 3 *Find the perimeter of $\triangle ABC$.*

- Use the pointer tool to select points A, B, and C.
- Under the **Construct** menu, select **Triangle Interior**. The triangle will now be shaded.
- Select the triangle interior using the pointer.
- Choose the **Perimeter** command under the **Measure** menu to find the perimeter of $\triangle ABC$. The perimeter of $\triangle ABC$ is 15.64 centimeters.

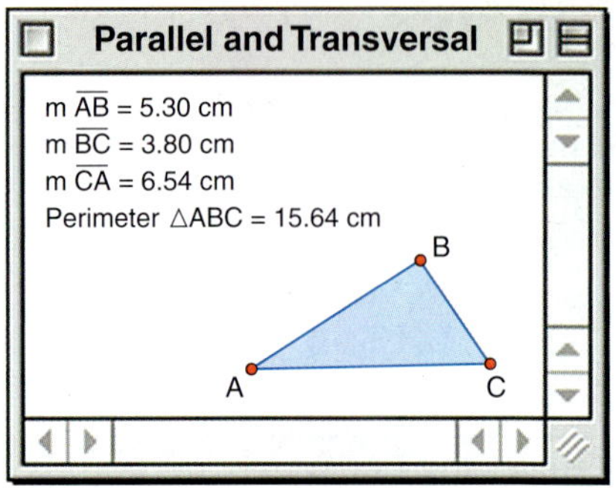

Step 4 *Find $m\angle A$, $m\angle B$, and $m\angle C$.*

- Recall that $\angle A$ can also be named $\angle BAC$ or $\angle CAB$. Use the pointer to select points B, A, and C in order.
- Select the **Angle** command from the **Measure** menu to find $m\angle A$.
- Select points A, B, and C. Find $m\angle B$.
- Select points A, C, and B. Find $m\angle C$.

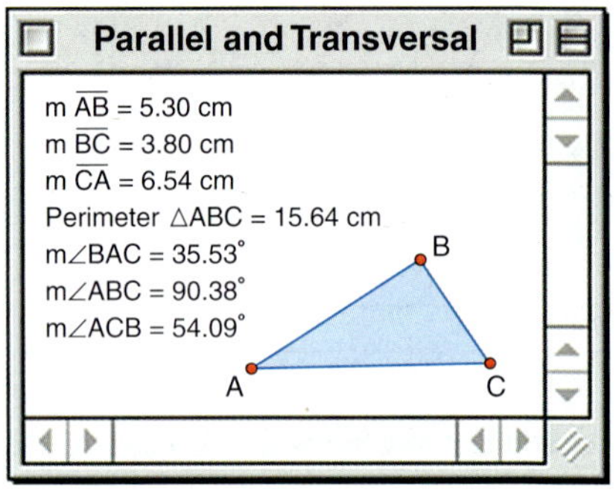

Analyze

1. The sum of the side measures equals the perimeter measure.

1. Add the side measures you found in Step 2. Compare this sum to the result of Step 3. How do these compare?
2. What is the sum of the angle measures of $\triangle ABC$? **180**
3. Repeat the activities for each convex polygon. **See students' work.**
   a. irregular quadrilateral **b.** square **c.** pentagon **d.** hexagon
4. Draw another quadrilateral and find its perimeter. Then enlarge your figure using the **Dilate** command under the **Transform** menu. How does changing the sides affect the perimeter? **See margin.**
5. Compare your results with those of your classmates. **See students' work.**

Make a Conjecture

6–10. See margin.

6. Make a conjecture about the sum of the measures of the angles in any triangle.
7. What is the sum of the measures of the angles of a quadrilateral? pentagon? hexagon?
8. Make a conjecture about how the sums of the measures of the angles of polygons are related to the number of sides.
9. Test your conjecture on other polygons. Does your conjecture hold for these polygons? Explain.
10. When the sides of a polygon are changed by a common factor, does the perimeter of the polygon change by the same factor as the sides? Explain.

Assess

Exercises 1–5 prepare students for making conjectures about and finding relationships among the information they gathered in the four activities.

In **Exercises 6–10,** students derive the formula that calculates the sum of the interior angles of a polygon with n sides. They also recognize that changing the sides of a polygon by a common factor causes the perimeter to be changed by the same factor.

Answers

4. Sample answer: When the lengths of the sides are doubled, the perimeter is doubled.
6. Sample answer: The sum of the measures of the angles of a triangle is 180.
7. Sample answer: The sum of the measures of the angles of a quadrilateral is 360; pentagon = 540; hexagon = 720
8. Sample answer: The sum of the measures of the angles of polygons increases by 180 for each additional side.
9. yes; sample answer: triangle: 3 sides, angle measure sum: 180; quadrilateral: 4 sides, angle measure sum: 180 + 180 = 360; pentagon: 5 sides, angle measure sum: 360 + 180 = 540; hexagon: 6 sides, angle measure sum: 540 + 180 = 720
10. Yes; sample answer: If the sides of a polygon are a, b, c, and d, then its perimeter is $a + b + c + d$. If each of the sides are increased by a factor of n then the sides measure na, nb, nc, and nd, and the perimeter is $na + nb + nc + nd$. By factoring, the perimeter is $n(a + b + c + d)$, which is the original perimeter increased by the same factor as the sides.

52 Chapter 1 Points, Lines, Planes, and Angles

Study Guide and Review

Vocabulary and Concept Check

- acute angle (p. 30)
- adjacent angles (p. 37)
- angle (p. 29)
- angle bisector (p. 32)
- between (p. 14)
- betweenness of points (p. 14)
- collinear (p. 6)
- complementary angles (p. 39)
- concave (p. 45)
- congruent (p. 15)
- construction (p. 15)
- convex (p. 45)
- coplanar (p. 6)
- degree (p. 29)
- exterior (p. 29)
- interior (p. 29)
- line (p. 6)
- line segment (p. 13)
- linear pair (p. 37)
- locus (p. 11)
- midpoint (p. 22)
- *n*-gon (p. 46)
- obtuse angle (p. 30)
- opposite rays (p. 29)
- perimeter (p. 46)
- perpendicular (p. 40)
- plane (p. 6)
- point (p. 6)
- polygon (p. 45)
- precision (p. 14)
- ray (p. 29)
- regular polygon (p. 46)
- relative error (p. 19)
- right angle (p. 30)
- segment bisector (p. 24)
- sides (p. 29)
- space (p. 8)
- supplementary angles (p. 39)
- undefined terms (p. 7)
- vertex (p. 29)
- vertical angles (p. 37)

Exercises Choose the letter of the term that best matches each figure.

1. 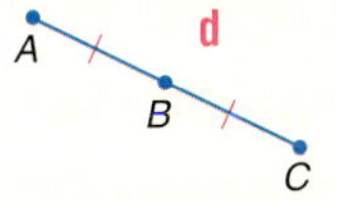

d

2. 

h

3. 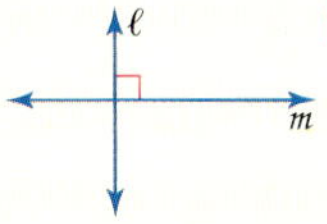

f

4. 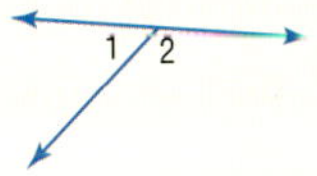

e

5. 

b

6. •Q g

a. line
b. ray
c. complementary angles
d. midpoint
e. supplementary angles
f. perpendicular
g. point
h. line segment

Lesson-by-Lesson Review

1-1 Points, Lines, and Planes

See pages 6–11.

Concept Summary

- A line is determined by two points.
- A plane is determined by three noncollinear points.

Example **Use the figure to name a plane containing point *N*.**

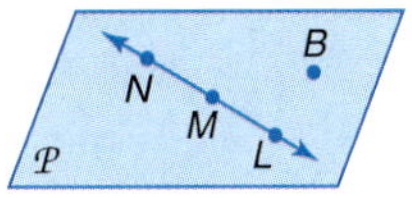

The plane can be named as plane 𝒫.

You can also use any three noncollinear points to name the plane as plane *BNM*, plane *MBL*, or plane *NBL*.

Exercises **Refer to the figure.** *See Example 1 on page 7.*

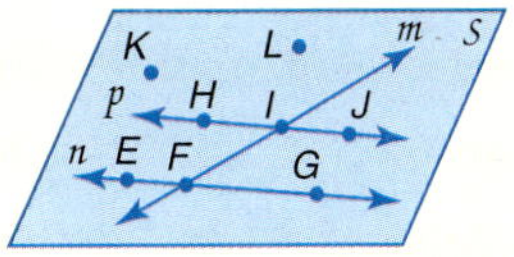

7. Name a line that contains point *I*. *p* or *m*
8. Name a point that is not in lines *n* or *p*. *K* or *L*
9. Name the intersection of lines *n* and *m*. *F*
10. Name the plane containing points *E*, *J*, and *L*. *S*

www.geometryonline.com/vocabulary_review

Chapter 1 Study Guide and Review

Vocabulary and Concept Check

- This alphabetical list of vocabulary terms in Chapter 1 includes a page reference where each term was introduced.
- **Assessment** A vocabulary test/review for Chapter 1 is available on p. 50 of the *Chapter 1 Resource Masters*.

Lesson-by-Lesson Review

For each lesson,

- the main ideas are summarized,
- additional examples review concepts, and
- practice exercises are provided.

Vocabulary PuzzleMaker

ELL The Vocabulary PuzzleMaker software improves students' mathematics vocabulary using four puzzle formats—crossword, scramble, word search using a word list, and word search using clues. Students can work on a computer screen or from a printed handout.

MindJogger Videoquizzes

ELL MindJogger Videoquizzes provide an alternative review of concepts presented in this chapter. Students work in teams in a game show format to gain points for correct answers. The questions are presented in three rounds.

Round 1 Concepts (5 questions)
Round 2 Skills (4 questions)
Round 3 Problem Solving (4 questions)

FOLDABLES™ Study Organizer

For more information about Foldables, see *Teaching Mathematics with Foldables.*

Have students look through the chapter to make sure they have included notes and examples in their Foldables for each lesson of Chapter 1.

Encourage students to refer to their Foldables while completing the Study Guide and Review and to use them in preparing for the Chapter Test.

Answers

11.

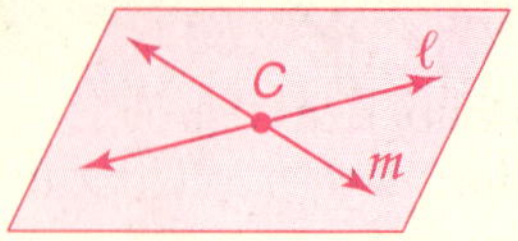

12.

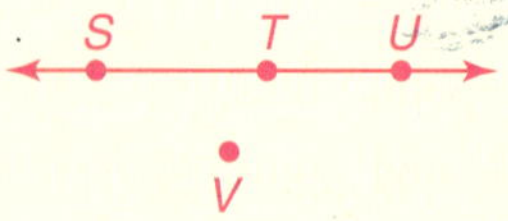

Key concepts from the lesson, one or two examples, and several practice problems are included in the Lesson-by-Lesson Review.

Draw and label a figure for each relationship. *See Example 3 on pages 7–8.*

11. Lines ℓ and m are coplanar and meet at point C. **11–12. See margin.**

12. Points S, T, and U are collinear, but points S, T, U, and V are not.

1-2 Linear Measure and Precision

See pages 13–19.

Concept Summary

- The precision of any measurement depends on the smallest unit available on the measuring device.
- The measure of a line segment is the sum of the measures of its parts.

Example **Use the figure to find JK.**

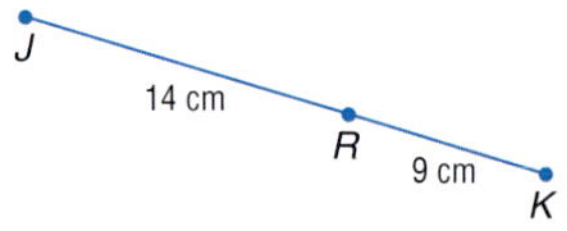

$JK = JR + RK$ — Sum of parts = whole

$= 14 + 9$ or 23 — Substitution

So, $\overline{JK}$ is 23 centimeters long.

Exercises **Find the value of the variable and PB, if P is between A and B.** *See Example 4 on pages 14 and 15.* **13. $x = 6$, $PB = 18$ 14. $c = 1.5$, $PB = 3$**

13. $AP = 7$, $PB = 3x$, $AB = 25$

14. $AP = 4c$, $PB = 2c$, $AB = 9$

15. $AP = s + 2$, $PB = 4s$, $AB = 8s - 7$ **$s = 3$, $PB = 12$**

16. $AP = -2k$, $PB = k + 6$, $AB = 11$ **$k = -5$, $PB = 1$**

Determine whether each pair of segments is congruent. *See Example 5 on page 16.*

17. $\overline{HI}$, $\overline{KJ}$ **yes**

18. $\overline{AB}$, $\overline{AC}$ **no**

19. $\overline{VW}$, $\overline{WX}$ **not enough information**

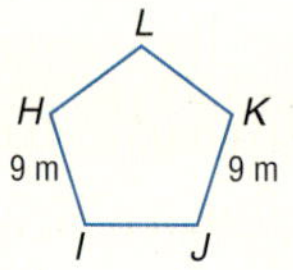

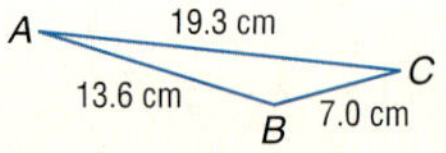

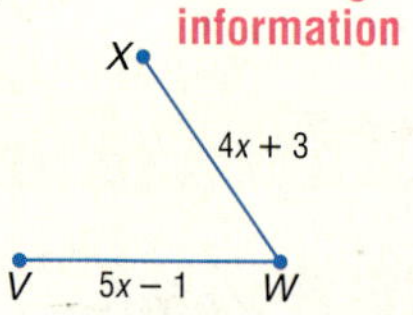

1-3 Distance and Midpoints

See pages 21–27.

Concept Summary

- Distances can be determined on a number line or the coordinate plane by using the Distance Formulas.
- The midpoint of a segment is the point halfway between the segment's endpoints.

Example **Find the distance between $A(3, -4)$ and $B(-2, -10)$.**

$d = \sqrt{(x_2 - x_1)^2 + (y_2 - y_1)^2}$ — Distance Formula

$AB = \sqrt{(-2 - 3)^2 + [-10 - (-4)]^2}$ — $(x_1, y_1) = (3, -4)$ and $(x_2, y_2) = (-2, -10)$

$= \sqrt{(-5)^2 + (-6)^2}$ — Simplify.

$= \sqrt{61}$ or about 7.8 — Simplify.

54 Chapter 1 Points, Lines, Planes, and Angles

Study Guide and Review

Exercises **Find the distance between each pair of points.**
See Example 2 on pages 21–22.

20. $A(1, 0), B(-3, 2)$ $\sqrt{20} \approx 4.5$

21. $G(-7, 4), L(3, 3)$ $\sqrt{101} \approx 10.0$

22. $J(0, 0), K(4, -1)$ $\sqrt{17} \approx 4.1$

23. $M(-4, 16), P(-6, 19)$ $\sqrt{13} \approx 3.6$

Find the coordinates of the midpoint of a segment having the given endpoints.
See Example 3 on page 23.

24. $D(0, 0), E(22, -18)$ **(11, −9)**

25. $U(-6, -3), V(12, -7)$ **(3, −5)**

26. $P(2, 5), Q(-1, -1)$ **(0.5, 2)**

27. $R(3.4, -7.3), S(-2.2, -5.4)$ **(0.6, −6.35)**

1-4 Angle Measure

See pages 29–36.

Concept Summary

- Angles are classified as acute, right, or obtuse according to their measure.
- An angle bisector is a ray that divides an angle into two congruent angles.

Examples

a. Name all angles that have *B* as a vertex.

$\angle 6, \angle 4, \angle 7, \angle ABD, \angle EBC$

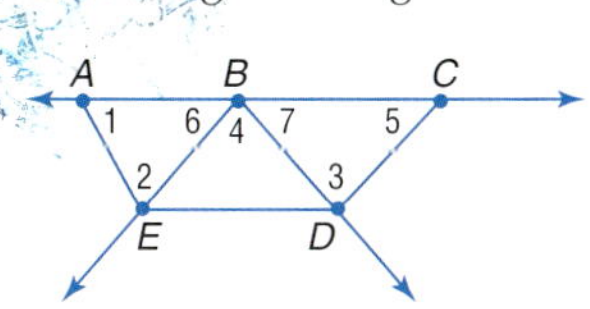

b. Name the sides of ∠2.

$\overline{EA}$ and $\overline{EB}$

Exercises **For Exercises 28–30, refer to the figure at the right.** *See Example 1 on page 30.*

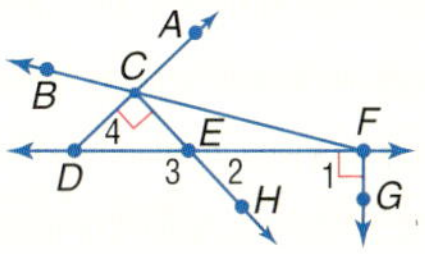

28. Name the vertex of ∠4. ***D***

29. Name the sides of ∠1. $\overrightarrow{FE}, \overrightarrow{FG}$

30. Write another name for ∠3. **∠*DEH***

Measure each angle and classify it as *right, acute,* or *obtuse*. *See Example 2 on page 30.*

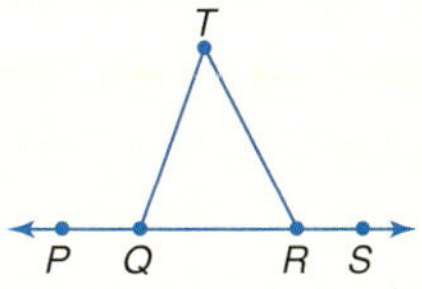

31. $\angle SQT$ **70°, acute**

32. $\angle PQT$ **110°, obtuse**

33. $\angle T$ **50°, acute**

34. $\angle PRT$ **70°, acute**

In the figure, $\overrightarrow{XW}$ bisects $\angle YXZ$ and $\overrightarrow{XV}$ bisects $\angle YXW$.
See Example 3 on page 32.

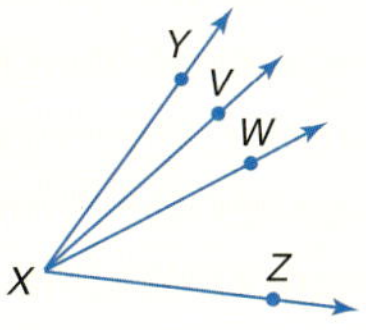

35. If $m\angle YXV = 3x$ and $m\angle VXW = 2x + 6$, find $m\angle YXW$. **36**

36. If $m\angle YXW = 12x - 10$ and $m\angle WXZ = 8(x + 1)$, find $m\angle YXZ$. **88**

37. If $m\angle YXZ = 9x + 17$ and $m\angle WXZ = 7x - 9$, find $m\angle YXW$. **40**

Chapter 1 Study Guide and Review 55

Chapter 1 **For More ...**

- Extra Practice, see pages 754–755.
- Mixed Problem Solving, see page 782.

1-5 Angle Relationships

See pages 37–43.

Concept Summary

- There are many special pairs of angles, such as adjacent angles, vertical angles, complementary angles, and linear pairs.

Example **Find the value of x so that $\overleftrightarrow{AC}$ and $\overleftrightarrow{BD}$ are perpendicular.**

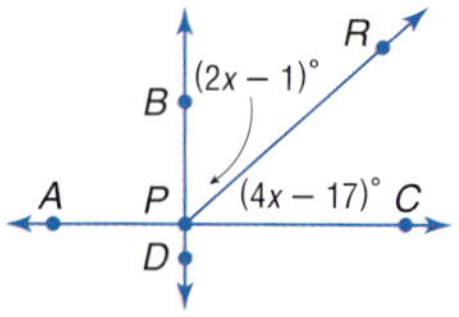

$m\angle BPC = m\angle BPR + m\angle RPC$	Sum of parts = whole
$90 = 2x - 1 + 4x - 17$	Substitution
$108 = 6x$	Simplify.
$18 = x$	Divide each side by 6.

Exercises **For Exercises 38–41, use the figure at the right.**
See Examples 1 and 3 on pages 38 and 40.

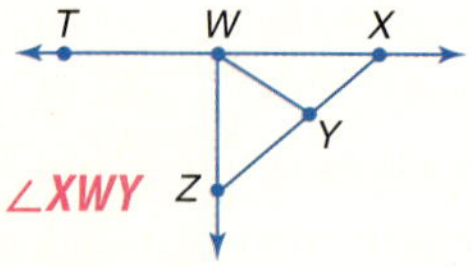

38. Name two obtuse angles. **∠TWY, ∠WYX**
39. Name a linear pair whose angles have vertex W. **∠TWY, ∠XWY**
40. If $m\angle TWZ = 2c + 36$, find c so that $\overline{TW} \perp \overline{WZ}$. **27**
41. If $m\angle ZWY = 4k - 2$, and $m\angle YWX = 5k + 11$, find k so that $\angle ZWX$ is a right angle. **9**

1-6 Polygons

See pages 45–50.

Concept Summary

- A polygon is a closed figure made of line segments.
- The perimeter of a polygon is the sum of the lengths of its sides.

Example **Find the perimeter of the hexagon.**

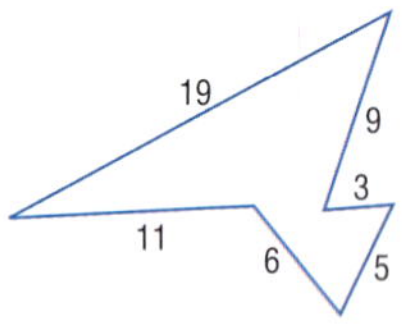

$P = s_1 + s_2 + s_3 + s_4 + s_5 + s_6$	Definition of perimeter
$= 19 + 9 + 3 + 5 + 6 + 11$ or 53	Substitution

Exercises **Name each polygon by its number of sides. Then classify it as *convex* or *concave* and *regular* or *irregular*.** *See Example 1 on page 46.*

42. 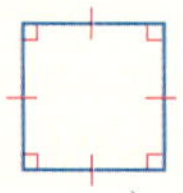
quadrilateral; convex; regular

43. 
not a polygon

44. 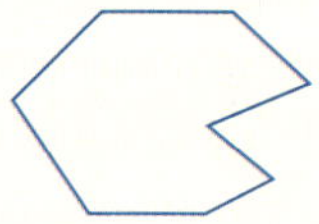
octagon; concave; irregular

Find the perimeter of each polygon. *See Example 3 on page 47.*

45. hexagon $ABCDEF$ with vertices $A(1, 2)$, $B(5, 1)$, $C(9, 2)$, $D(9, 5)$, $E(5, 6)$, $F(1, 5)$ **≈ 22.5 units**
46. rectangle $WXYZ$ with vertices $W(-3, 5)$, $X(7, 1)$, $Y(5, -4)$, $Z(-5, 0)$ **≈ 32.3 units**

56 Chapter 1 Points, Lines, Planes, and Angles

Chapter 1 Practice Test

Vocabulary and Concepts

Determine whether each statement is *true* or *false*.

1. A plane contains an infinite number of lines. **true**
2. If two angles are congruent, then their measures are equal. **true**
3. The sum of two complementary angles is 180. **false**
4. Two angles that form a linear pair are supplementary. **true**

Skills and Applications

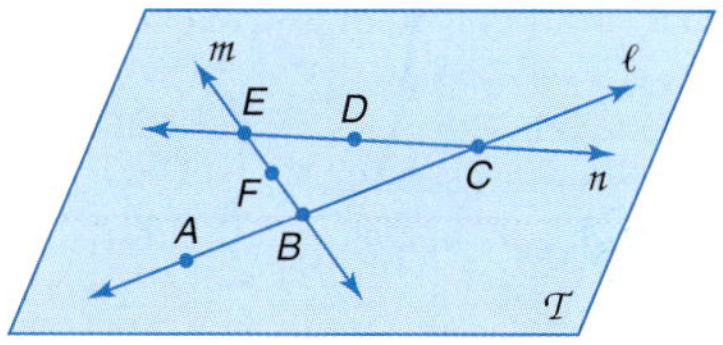

For Exercises 5–7, refer to the figure at the right.

5. Name the line that contains points B and F. m
6. Name a point not contained in lines ℓ or m. **D**
7. Name the intersection of lines ℓ and n. **C**

Find the value of the variable and VW if V is between U and W.

8. $UV = 2$, $VW = 3x$, $UW = 29$ $x = 9$; $VW = 27$
9. $UV = r$, $VW = 6r$, $UW = 42$ $r = 6$; $VW = 36$
10. $UV = 4p - 3$, $VW = 5p$, $UW = 15$ $p = 2$; $VW = 10$
11. $UV = 3c + 29$, $VW = -2c - 4$, $UW = -4c$ $c = -5$; $VW = 6$

Find the distance between each pair of points.

12. $G(0, 0)$, $H(-3, 4)$ **5**
13. $N(5, 2)$, $K(-2, 8)$ $\sqrt{85} \approx 9.2$
14. $A(-4, -4)$, $W(-2, 2)$ $\sqrt{40} \approx 6.3$

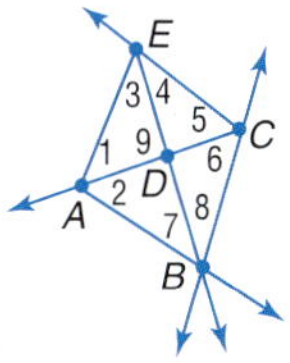

For Exercises 15–18, refer to the figure at the right. 17. $\angle ABD$ or $\angle ABE$

15. Name the vertex of $\angle 6$. **C**
16. Name the sides of $\angle 4$. $\overrightarrow{EC}$, $\overrightarrow{ED}$
17. Write another name for $\angle 7$.
18. Write another name for $\angle ADE$. $\angle 9$
19. **ALGEBRA** The measures of two supplementary angles are $4r + 7$ and $r - 2$. Find the measures of the angles. **147, 33**
20. Two angles are complementary. One angle measures 26 degrees more than the other. Find the measures of the angles. **32, 58**

Find the perimeter of each polygon.

21. triangle PQR with vertices $P(-6, -3)$, $Q(1, -1)$, and $R(1, -5)$ ≈ 18.6 units
22. pentagon $ABCDE$ with vertices $A(-6, 2)$, $B(-4, 7)$, $C(0, 4)$, $D(0, 0)$, and $E(-4, -3)$ ≈ 24.8 units

DRIVING **For Exercises 23 and 24, use the following information and the diagram.**

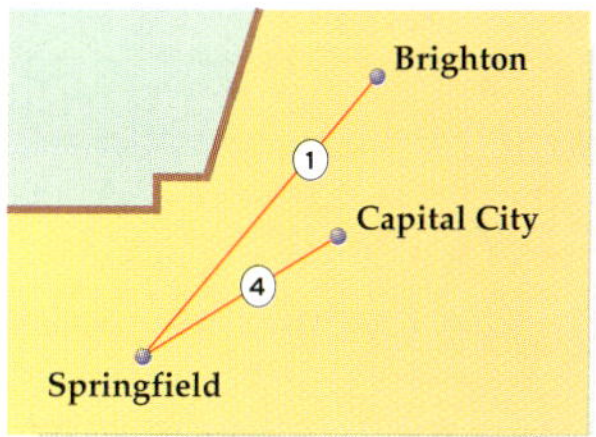

The city of Springfield is 5 miles west and 3 miles south of Capital City, while Brighton is 1 mile east and 4 miles north of Capital City. Highway 1 runs straight between Brighton and Springfield; Highway 4 connects Springfield and Capital City.

23. Find the length of Highway 1. ≈ 9.2 mi
24. How long is Highway 4? ≈ 5.8 mi
25. **STANDARDIZED TEST PRACTICE** Which of the following figures is *not* a polygon? **C**

Ⓐ 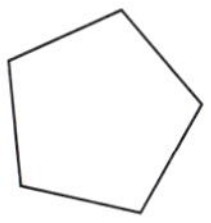

Ⓑ 

Ⓒ 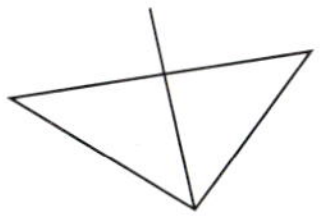

Ⓓ 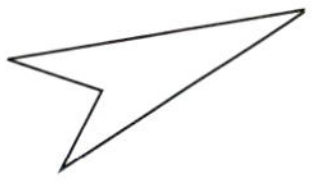

www.geometryonline.com/chapter_test/fcat

Chapter 1 Practice Test

Assessment Options

Vocabulary Test A vocabulary test/review for Chapter 1 can be found on p. 50 of the *Chapter 1 Resource Masters*.

Chapter Tests There are six Chapter 1 Tests and an Open-Ended Assessment task available in the *Chapter 1 Resource Masters*.

Chapter 1 Tests			
Form	**Type**	**Level**	**Pages**
1	MC	basic	37–38
2A	MC	average	39–40
2B	MC	average	41–42
2C	FR	average	43–44
2D	FR	average	45–46
3	FR	advanced	47–48

MC = multiple-choice questions
FR = free-response questions

Open-Ended Assessment
Performance tasks for Chapter 1 can be found on p. 49 of the *Chapter 1 Resource Masters*. A sample scoring rubric for these tasks appears on p. A25.

TestCheck and Worksheet Builder

This **networkable software** has three modules for assessment.

- **Worksheet Builder** to make worksheets and tests.
- **Student Module** to take tests on-screen.
- **Management System** to keep student records.

Portfolio Suggestion

Introduction Sometimes it helps to briefly summarize key concepts and methods presented throughout a chapter before proceeding to a new chapter.

Ask Students On a sheet of paper, create up to three geometric figures that demonstrate some of the concepts and terms learned in Chapter 1. Be sure to apply proper labels and include all classifying and identifying information about the figure(s). List all definitions and construction techniques separately or as they apply to the figures you create. Place this sheet of paper in your portfolio.

Chapter 1 Standardized Test Practice

These two pages contain practice questions in the various formats that can be found on the most frequently given standardized tests.

A practice answer sheet for these two pages can be found on p. A1 of the *Chapter 1 Resource Masters.*

Standardized Test Practice Student Recording Sheet, p. A1

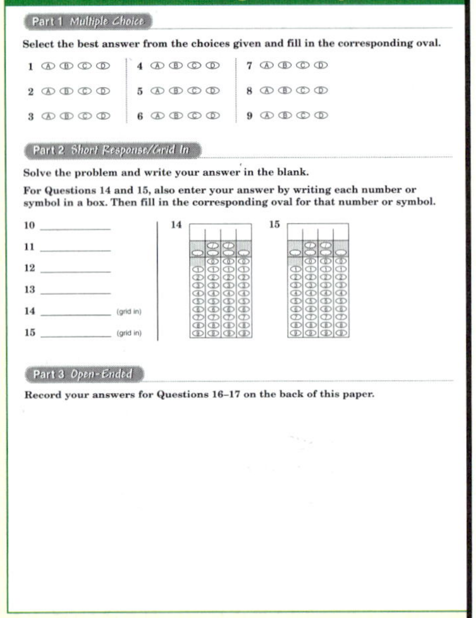
Part 1 Multiple Choice

Select the best answer from the choices given and fill in the corresponding oval.

1 Ⓐ Ⓑ Ⓒ Ⓓ 4 Ⓐ Ⓑ Ⓒ Ⓓ 7 Ⓐ Ⓑ Ⓒ Ⓓ
2 Ⓐ Ⓑ Ⓒ Ⓓ 5 Ⓐ Ⓑ Ⓒ Ⓓ 8 Ⓐ Ⓑ Ⓒ Ⓓ
3 Ⓐ Ⓑ Ⓒ Ⓓ 6 Ⓐ Ⓑ Ⓒ Ⓓ 9 Ⓐ Ⓑ Ⓒ Ⓓ

Part 2 Short Response/Grid In

Solve the problem and write your answer in the blank.

For Questions 14 and 15, also enter your answer by writing each number or symbol in a box. Then fill in the corresponding oval for that number or symbol.

10 ______
11 ______
12 ______
13 ______
14 ______ (grid in)
15 ______ (grid in)

Part 3 Open-Ended

Record your answers for Questions 16–17 on the back of this paper.

Additional Practice

See pp. 55–56 in the *Chapter 1 Resource Masters* for additional standardized test practice.

The items on the Standardized Test Practice pages were created to closely parallel those on actual state proficiency tests and college entrance exams, like PSAT, ACT and SAT.

Chapter 1 Standardized Test Practice

FCAT Practice

Part 1 Multiple Choice

Record your answers on the answer sheet provided by your teacher or on a sheet of paper.

1. During a science experiment, Juanita recorded that she blinked 11 times in one minute. If this is a normal count and Juanita wakes up at 7 A.M. and goes to bed at 10 P.M., how many times will she blink during the time she is awake? (Prerequisite Skill) **C**

Ⓐ 165 Ⓑ 660
Ⓒ 9900 Ⓓ 15,840

2. Find $-\sqrt{0.0225}$. (Prerequisite Skill) **A**

Ⓐ -0.15 Ⓑ -0.015
Ⓒ 0.015 Ⓓ 0.15

3. Simplify $\frac{2x^2 + 12x + 16}{2x + 4}$. (Prerequisite Skill) **B**

Ⓐ 24 Ⓑ $x + 4$
Ⓒ $4x + 12$ Ⓓ $4x^3 + x^2 + 20$

4. If two planes intersect, their intersection can be **A**

I. a line.
II. three noncollinear points.
III. two intersecting lines. (Lesson 1-1)

Ⓐ I only Ⓑ II only
Ⓒ III only Ⓓ I and II only

5. Before sonar technology, sailors determined the depth of water using a device called a sounding line. A rope with a lead weight at the end was marked in intervals called fathoms. Each fathom was equal to 6 feet. Suppose a specific ocean location has a depth of 55 fathoms. What would this distance be in yards? (Lesson 1-2) **B**

Ⓐ $9\frac{1}{6}$ yd Ⓑ 110 yd
Ⓒ 165 yd Ⓓ 330 yd

6. An 18-foot ladder leans against the side of a house so that the bottom of the ladder is 6 feet from the house. To the nearest foot, how far up the side of the house does the top of the ladder reach? (Lesson 1-3) **C**

Ⓐ 12 ft
Ⓑ 14 ft
Ⓒ 17 ft
Ⓓ 19 ft

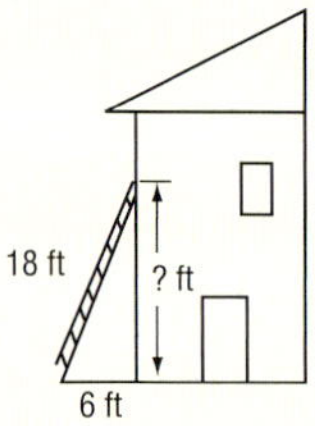

7. Ray BD is the bisector of $\angle ABC$. If $m\angle ABD = 2x + 14$ and $m\angle CBD = 5x - 10$, what is the measure of $\angle ABD$? (Lesson 1-5) **C**

Ⓐ 8 Ⓑ 16
Ⓒ 30 Ⓓ 40

8. If $m\angle DEG$ is $6\frac{1}{2}$ times $m\angle FEG$, what is $m\angle DEG$? (Lesson 1-6) **D**

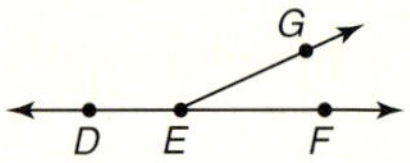

Ⓐ 24 Ⓑ 78
Ⓒ 130 Ⓓ 156

9. Kaitlin and Henry are participating in a treasure hunt. They are on the same straight path, walking toward each other. When Kaitlin reaches the Big Oak, she will turn 115° onto another path that leads to the treasure. At what angle will Henry turn when he reaches the Big Oak to continue on to the treasure? (Lesson 1-6) **D**

Ⓐ 25° Ⓑ 35°
Ⓒ 55° Ⓓ 65°

The Princeton Review

Log On for Test Practice

The Princeton Review offers additional test-taking tips and practice problems at their web site. Visit www.princetonreview.com or www.review.com

TestCheck and Worksheet Builder

Special banks of standardized test questions similar to those on the SAT, ACT, TIMSS 8, NAEP 8, and state proficiency tests can be found on this CD-ROM.

Aligned and verified by The Princeton Review

FCAT Practice

Part 2 Short Response/Grid In

Record your answers on the answer sheet provided by your teacher or on a sheet of paper.

10. Simplify $-2x + 6 + 4x^2 + x + x^2 - 5$. (Prerequisite Skill) $5x^2 - x + 1$

11. Solve the system of equations. (Prerequisite Skill)
$2y = 3x + 8$
$y = 2x + 3$ (2, 7)

12. In rectangle $ABCD$, vertices A, B, and C have the coordinates $(-4, -1)$, $(-4, 4)$, and $(3, 4)$, respectively. Plot A, B, and C and find the coordinates of vertex D. (Lesson 1-1) See margin for graph; $D(3, -1)$.

13. The endpoints of a line segment are $(2, -1)$ and $(-4, 3)$. What are the coordinates of its midpoint? (Lesson 1-3) $(-1, 1)$

14. The 200-meter race starts at point A, loops around the track, and finishes at point B. The track coach starts his stopwatch when the runners begin at point A and crosses the interior of the track so he can be at point B to time the runners as they cross the finish line. To the nearest meter, how long is $\overline{AB}$? (Lesson 1-3) 125

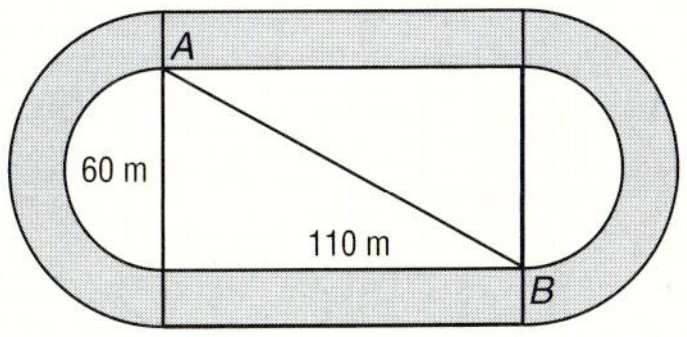

15. Mr. Lopez wants to cover the walls of his unfinished basement with pieces of plasterboard that are 8 feet high, 4 feet wide, and $\frac{1}{4}$ inch thick. If the basement measures 24 feet wide, 16 feet long, and 8 feet tall, how many pieces of plasterboard will he need to cover all four walls? (Lesson 1-4) 20

The Princeton Review Test-Taking Tip

Question 8 Most standardized tests allow you to write in the test booklet or on scrap paper. To avoid careless errors, work out your answers on paper rather than in your head. For example, in Question 8, you can write an equation, such as $x + 6\frac{1}{2}x = 180$ and solve for x. All of your solution can be examined for accuracy when written down.

FCAT Practice

Part 3 Open Ended

Record your answers on a sheet of paper. Show your work.

16. Tami is creating a sun catcher to hang in her bedroom window. She makes her design on grid paper so that she can etch the glass appropriately before painting it.

a. Graph the vertices of the glass if they are located at $(4, 0)$, $(-4, 0)$, $(0, -4)$, and $(0, 4)$. (Prerequisite Skill) See margin.

b. Tami is putting a circle on the glass so that it touches the edge at the midpoint of each side. Find the coordinates of these midpoints. (Lesson 1-3) $(2, 2)$, $(2, -2)$, $(-2, -2)$, $(-2, 2)$

17. William Sparrow and his father are rebuilding the roof of their barn. They first build a system of rafters to support the roof. The angles formed by each beam are shown in the diagram.

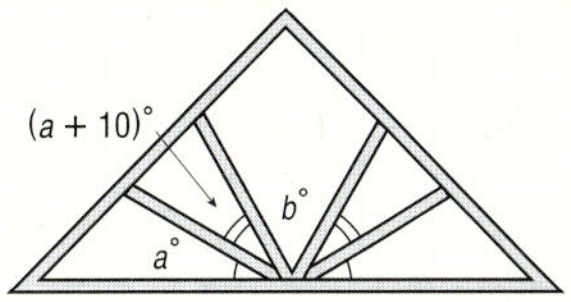

a. If $a = 25$, what is the measure of the five angles formed by the beams? Justify your answer. (Lesson 1-6) See margin.

b. Classify each of the angles formed by the beams. (Lesson 1-5) All are acute.

www.geometryonline.com/standardized_test/fcat

Answers

12.

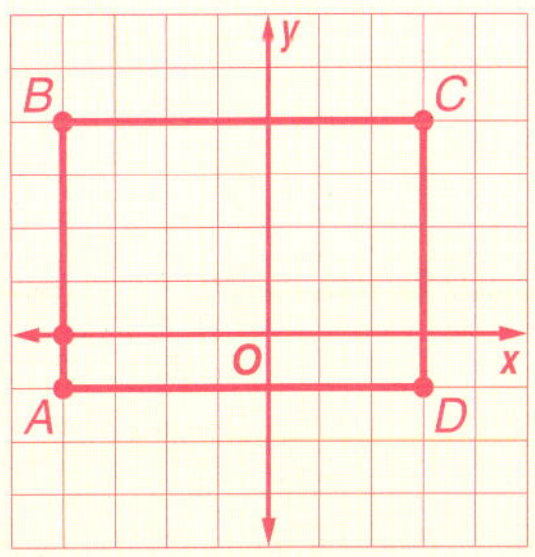

16a.

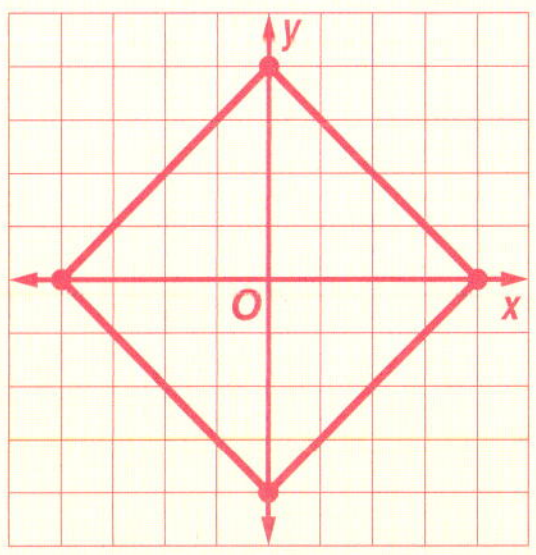

Evaluating Open-Ended Assessment Questions

Open-Ended Assessment questions are graded by using a multilevel rubric that guides you in assessing a student's knowledge of a particular concept.

Goal for Question 16: Graph a figure on a coordinate plane and locate the midpoints of its sides.

Goal for Question 17: Find angle measures and classify angles.

Sample Scoring Rubric: The following rubric is a sample scoring device. You may wish to add more detail to this sample to meet your individual scoring needs.

Score	Criteria
4	A correct solution that is supported by well-developed, accurate explanations
3	A generally correct solution, but may contain minor flaws in reasoning or computation
2	A partially correct interpretation and/or solution to the problem
1	A correct solution with no supporting evidence or explanation
0	An incorrect solution indicating no mathematical understanding of the concept or task, or no solution is given

17a. You are given that $a = 25$. The measure of the other angle marked with a single arc is also 25 because the arcs tell us that the angles are congruent. Both of the angles marked with double arcs have a measure of $a + 10$, which is $25 + 10$ or 35. In the remaining angle, $b = 180 - (25 + 35 + 25 + 35)$ or 60 because the angles can be combined to form linear pairs, which are supplementary.

Pages 9–11, Lesson 1-1

5. Sample answer:

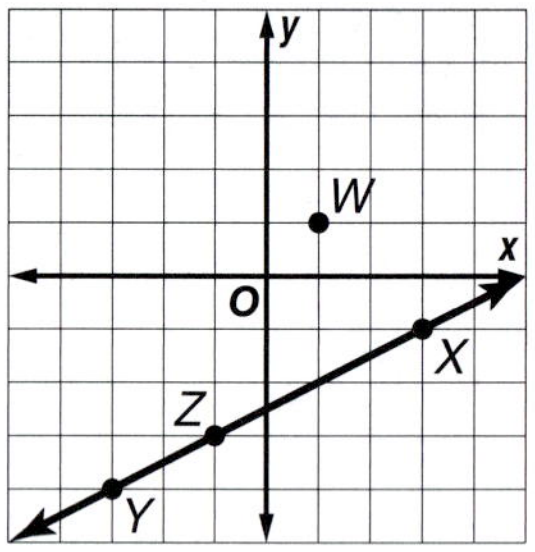

6.

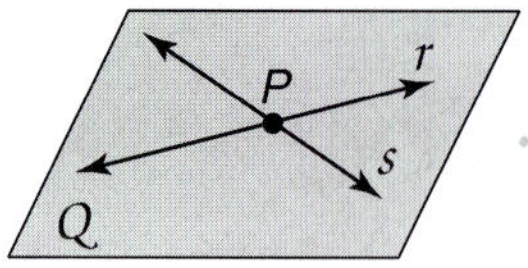

21.

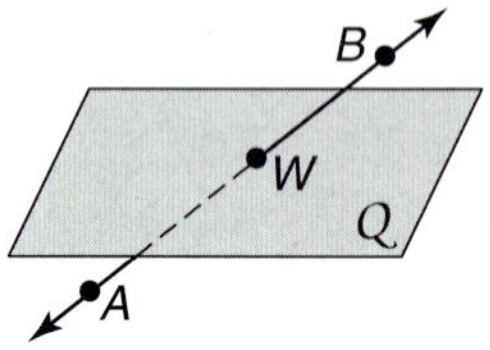

22.

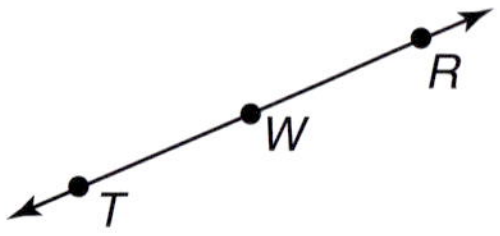

23. Sample answer:

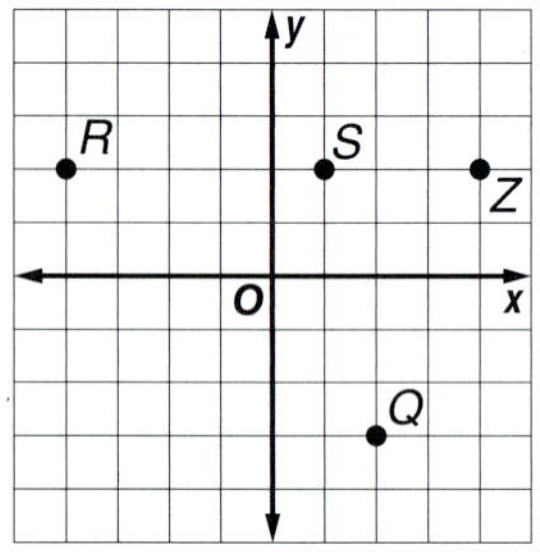

24. Sample answer:

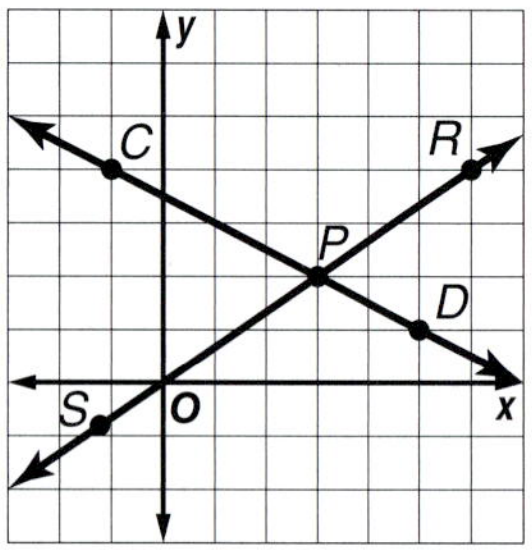

25.

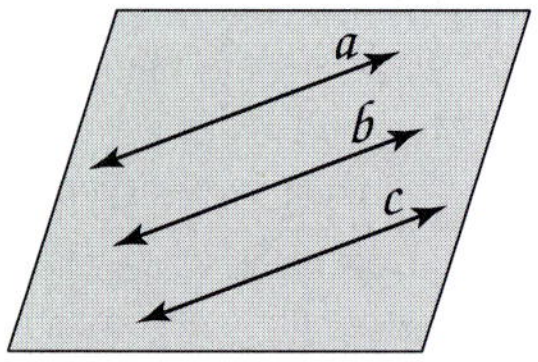

26.

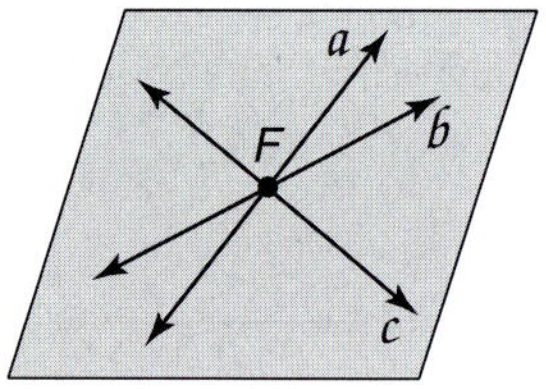

27.

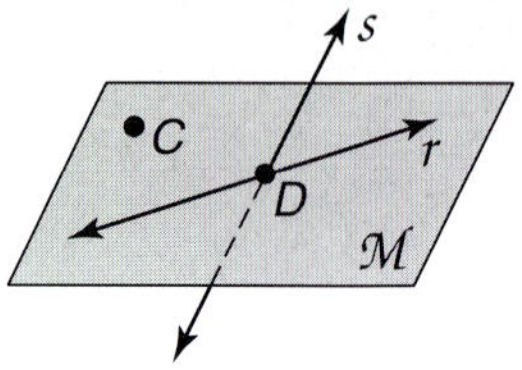

28.

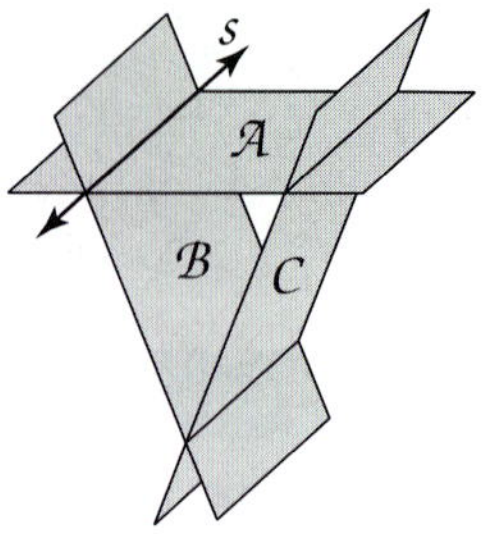

29. Sample answer:

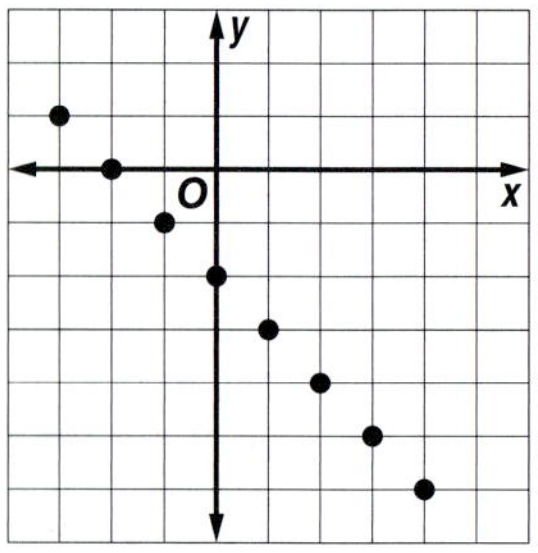

47.

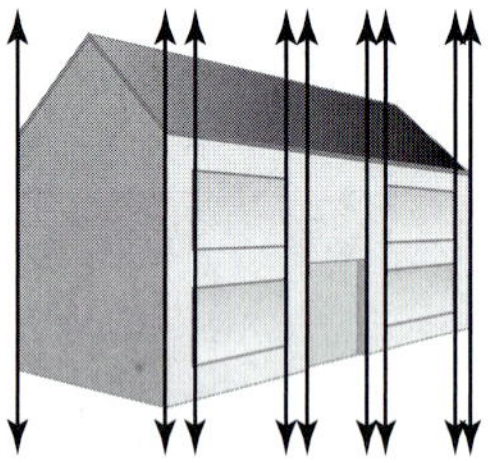

48.

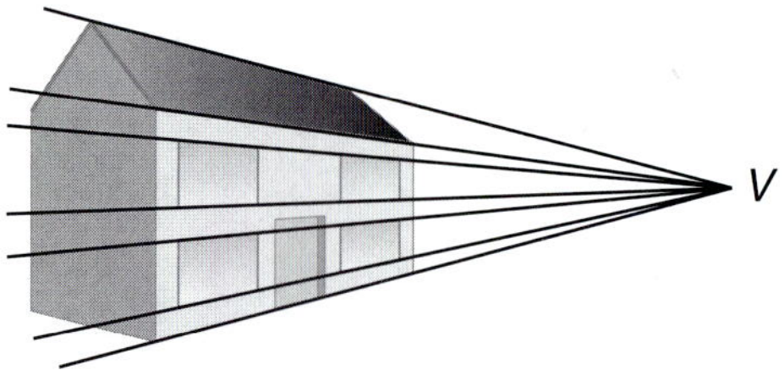

Notes

Chapter 2

Reasoning and Proof

Chapter Overview and Pacing

LESSON OBJECTIVES

LESSON OBJECTIVES	PACING (days) Regular Basic/Average	Regular Advanced	Block Basic/Average	Block Advanced
2-1 Inductive Reasoning and Conjecture *(pp. 62–66)* • Make conjectures based on inductive reasoning. • Find counterexamples.	1	1	0.5	0.5
2-2 Logic *(pp. 67–74)* • Determine truth values of conjunctions and disjunctions. • Construct truth tables.	2	1	1	0.5
2-3 Conditional Statements *(pp. 75–80)* • Analyze statements in if-then form. • Write the converse, inverse, and contrapositive of if-then statements.	2 (with 2-3 Follow-Up)	1 (with 2-3 Follow-Up)	1 (with 2-3 Follow-Up)	0.5 (with 2-3 Follow-Up)
2-4 Deductive Reasoning *(pp. 82–88)* • Use the Law of Detachment. • Use the Law of Syllogism. ***Follow-Up:*** Use a table and deductive reasoning to solve a logic problem.	1 (with 2-4 Follow-Up)	1 (with 2-4 Follow-Up)	0.5 (with 2-3 Follow-Up)	0.5 (with 2-3 Follow-Up)
2-5 Postulates and Paragraph Proofs *(pp. 89–93)* • Identify and use basic postulates about points, lines, and planes. • Write paragraph proofs.	1.5	1	1	0.5
2-6 Algebraic Proof *(pp. 94–100)* • Use algebra to write two-column proofs. • Use properties of equality in geometry proofs.	1.5	1	1	0.5
2-7 Proving Segment Relationships *(pp. 101–106)* • Write proofs involving segment addition. • Write proofs involving segment congruence.	2	2	1	1
2-8 Proving Angle Relationships *(pp. 107–114)* • Write proofs involving supplementary and complementary angles. • Write proofs involving congruent and right angles.	2	2	1	1
Study Guide and Practice Test *(pp. 115–121)* **Standardized Test Practice** *(pp. 122–123)*	1	1	0.5	0.5
Chapter Assessment	1	1	0.5	0.5
TOTAL	**15**	**12**	**8**	**6**

Pacing suggestions for the entire year can be found on pages T20–T21.

Timesaving Tools

TeacherWorks™

All-In-One Planner and Resource Center

See pages T5 and T21.

Chapter Resource Manager

Chapter 2 Resource Masters

Study Guide and Intervention	Practice (Skills and Average)	Reading to Learn Mathematics	Enrichment	Assessment	Prerequisite Skills Workbook	Applications*	5-Minute Check Transparencies	Interactive Chalkboard	GeomPASS: Tutorial Plus (lessons)	Materials
57–58	59–60	61	62			GCC 19	2-1	2-1		
63–64	65–66	67	68	119		GCC 20	2-2	2-2		protractor
69–70	71–72	73	74				2-3	2-3	6	
75–76	77–78	79	80	119, 121		SC 3	2-4	2-4		
81–82	83–84	85	86				2-5	2-5		
87–88	89–90	91	92	120	41–44, 83–86, 93–94		2-6	2-6		
93–94	95–96	97	98		89–90, 101–104	SC 4	2-7	2-7		
99–100	101–102	103	104	120	81–82, 85–86		2-8	2-8		patty paper, protractor, paper
				105–118, 122–124						

**Key to Abbreviations:* GCC = Graphing Calculator and Computer Masters
SC = School-to-Career Masters

Chapter 2 Mathematical Connections and Background

Continuity of Instruction

Prior Knowledge

In algebra, students learned to solve for a variable. In Chapter 1, students learned about points, lines and planes. They were introduced to adjacent and vertical angles, complementary and supplementary angles, and right angles. They can identify congruent segments and angles as well as perpendicular lines.

This Chapter

In this chapter, students explore methods of reasoning and learn to apply those methods to geometry. They make conjectures, determine the truth values of compound statements, and construct truth tables. They also analyze conditional statements and write related conditionals. The terms postulate and theorem are introduced. Algebraic properties of equality are applied to geometry, enabling students to write formal and informal proofs proving segment and angle relationships.

Future Connections

In Chapter 4, students will build on their knowledge of proofs when they investigate triangles. Students will also write proofs in Chapter 7. Thinking logically is a crucial skill for daily living. Logic is used to make informed choices and to examine a statement for truth.

2-1 Inductive Reasoning and Conjecture

A conjecture is an educated guess based on known information. Examining several specific situations to arrive at a conjecture is called inductive reasoning. Meteorologists use inductive reasoning to predict weather conditions.

Just because a conjecture is true in most circumstances does not make it a true conjecture. If just one example contradicts the conjecture, the conjecture is not true. The false example is called a counterexample.

2-2 Logic

A statement is any sentence that is either true or false, but not both. The truth or falsity of a statement is called its truth value. The negation of a statement has the opposite meaning as well as an opposite truth value. This means that if a statement is represented by p, then *not* p is the negation of the statement. You could also write *not* p as $\sim p$.

Two or more statements can be joined to form a compound statement. A conjunction is a compound statement formed by joining two or more statements with the word *and*. The symbol $\wedge$ can be used instead of *and*. Two or more statements can also be joined to form a disjunction. A disjunction is a compound statement formed by joining two or more statements with the word *or*. You might see the symbol $\vee$ instead of the word *or* in a disjunction.

Conjunctions and disjunctions can be illustrated with Venn diagrams. Truth tables can also be helpful in evaluating the truth values of statements. A truth table will show that a conjunction is true only when both statements are true. A disjunction, on the other hand, is true unless both statements are false.

2-3 Conditional Statements

A conditional statement is a statement that can be written in if-then form: *if* p, *then* q. The phrase immediately following the word *if* is called the hypothesis. The phrase immediately following the word *then* is called the conclusion. An arrow pointing to the right is written between p and q to symbolize an if-then situation. A conditional statement is true in all cases except where the hypothesis is true and the conclusion is false.

Related conditionals are statements constructed from an if-then statement. A converse statement is formed by exchanging the hypothesis and the conclusion: *if* q, *then* p. An inverse statement is formed by negating both the hypothesis and the conclusion of the original statement: if $\sim p$, *then* $\sim q$. A contrapositive is formed by negating both the hypothesis and the conclusion of the converse statement: *if* $\sim q$, *then* $\sim p$.

2-4 Deductive Reasoning

Deductive reasoning uses facts, rules, definitions, or properties to reach logical conclusions. A form of deductive reasoning that is used to draw conclusions from true conditional statements is called the Law of Detachment. This law states that if $p \rightarrow q$ is true and p is true, then q is also true.

The Law of Syllogism is another law of logic. It states that if $p \rightarrow q$ and $q \rightarrow r$ are true, then $p \rightarrow r$ is also true. You may see the similarity between this law and the Transitive Property of Equality from algebra.

2-5 Postulates and Paragraph Proofs

In geometry, a postulate is a statement that describes a fundamental relationship between the basic terms of geometry. Postulates are accepted as true without proof. Several postulates based on the relationship among points, lines, and planes were introduced in Chapter 1, but were not labeled as postulates.

In this course, you will learn to use various methods to justify the truth of a statement or conjecture. Once a statement or conjecture has been shown to be true, it is called a theorem. A theorem can be used like a definition or postulate to justify that other statements are true.

A proof is a logical argument in which each statement you make is supported by a statement that is accepted as true. One type of proof is called a paragraph or informal proof. It is a written explanation of why a conjecture for a given situation is true. A good proof states the theorem or conjecture to be proven. It lists the given information and, if possible, supplies a diagram to illustrate the given information. The proof states what is to be proved and develops a system of deductive reasoning.

2-6 Algebraic Proof

In algebra, you learned to use properties of equality to solve algebraic equations and to verify relationships. These properties can be used to justify each step when solving an equation. A group of algebraic steps used to solve problems form a deductive argument. This argument can be demonstrated by writing the solution to the equation in the first column and listing the property justifying each step in the second column.

In geometry, a similar format is used to prove conjectures and theorems. A two-column, or formal, proof contains statements and reasons organized in two columns. Each step is called a statement, and the properties that justify each step are called reasons.

2-7 Proving Segment Relationships

As you learned in Chapter 1, a segment can be measured, and measures can be used in calculations because they are real numbers. One postulate about segments is called the Ruler Postulate. It states that the points on any line or line segment can be paired with real numbers so that, given any two points A and B on a line, A corresponds to 0, and B corresponds to a positive real number. That number is the length of the segment. Another postulate states that if point B lies between points A and C on the same line, $AB + BC = AC$. The converse statement holds true as well.

The Reflexive, Symmetric, and Transitive Properties of Equality can be used to write proofs about segment congruence. The theorem resulting from the proofs states that congruence of segments is reflexive, symmetric, and transitive.

2-8 Proving Angle Relationships

This lesson introduces postulates and theorems about angle relationships. The Protractor Postulate states, "Given $\overrightarrow{AB}$ and a number r between 0 and 180, there is exactly one ray with endpoint A, extending on either side of $\overrightarrow{AB}$, such that the measure of the angle formed is r." The Angle Addition Postulate states that if R is in the interior of $\angle PQS$, then $m\angle PQR + m\angle RQS = m\angle PQS$. If $m\angle PQR + m\angle RQS = m\angle PQS$, then R is in the interior of $\angle PQS$. This postulate can be used with other angle relationships to prove other theorems relating to angles.

Some of these theorems relate to supplementary and complementary angles. Another theorem extends the Reflexive, Transitive, and Symmetric properties to angle congruence. There is also a series of theorems about perpendicular lines and right angles.

Chapter 2

DAILY INTERVENTION and Assessment

INTERVENTION | ASSESSMENT

Type	Student Edition	Teacher Resources	Technology/Internet
Ongoing	Prerequisite Skills, pp. 61, 80, 87, 93, 100, 106 Practice Quiz 1, p. 80 Practice Quiz 2, p. 100	5-Minute Check Transparencies *Prerequisite Skills Workbook*, pp. 41–44, 81–86, 89–90, 93–94, 101–104 Quizzes, *CRM* pp. 119–120 Mid-Chapter Test, *CRM* p. 121 Study Guide and Intervention, *CRM* pp. 57–58, 63–64, 69–70, 75–76, 81–82, 87–88, 93–94, 99–100	GeomPASS: Tutorial Plus www.geometryonline.com/self_check_quiz www.geometryonline.com/extra_examples
Mixed Review	pp. 66, 74, 80, 93, 100, 106, 114	Cumulative Review, *CRM* p. 122	
Error Analysis	Find the Error, pp. 84, 111 Common Misconceptions, p. 76	Find the Error, *TWE* pp. 84, 111 Unlocking Misconceptions, *TWE* p. 91 Tips for New Teachers, *TWE* p. 70	
Standardized Test Practice	pp. 66, 74, 80, 86, 87, 93, 96, 97, 99, 106, 114, 121, 122	*TWE* pp. 122–123 Standardized Test Practice, *CRM* pp. 123–124	Standardized Test Practice CD-ROM www.geometryonline.com/standardized_test
Open-Ended Assessment	Writing in Math, pp. 66, 74, 79, 86, 93, 99, 106, 114, 123 Open Ended, pp. 63, 71, 78, 84, 91, 97, 103, 111 Standardized Test, p. 123	Modeling: *TWE* pp. 74, 87, 106 Speaking: *TWE* pp. 80, 93 Writing: *TWE* pp. 66, 100, 114 Open-Ended Assessment, *CRM* p. 117	
Chapter Assessment	Study Guide, pp. 115–120 Practice Test, p. 121	Multiple-Choice Tests (Forms 1, 2A, 2B), *CRM* pp. 105–110 Free-Response Tests (Forms 2C, 2D, 3), *CRM* pp. 111–116 Vocabulary Test/Review, *CRM* p. 118	TestCheck and Worksheet Builder (see below) MindJogger Videoquizzes www.geometryonline.com/vocabulary_review www.geometryonline.com/chapter_test

Key to Abbreviations: TWE = Teacher Wraparound Edition; CRM = Chapter Resource Masters

Additional Intervention Resources

The Princeton Review's *Cracking the SAT & PSAT*

The Princeton Review's *Cracking the ACT*

TestCheck and Worksheet Builder

This **networkable** software has three modules for intervention and assessment flexibility:

- **Worksheet Builder** to make worksheet and tests
- **Student Module** to take tests on screen (optional)
- **Management System** to keep student records (optional)

Special banks are included for SAT, ACT, TIMSS, NAEP, and End-of-Course tests.

Intervention Technology

GeomPASS: Tutorial Plus CD-ROM offers a complete, self-paced geometry curriculum.

Geometry Lesson	GeomPASS Lesson
2-3	6 *If-Then Statements and Converses*

Intervention at Home

Log on for student study help.

- For each lesson in the Student Edition, there are Extra Examples and Self-Check Quizzes.
  www.geometryonline.com/extra_examples
  www.geometryonline.com/self_check_quiz
- For chapter review, there is vocabulary review, test practice, and standardized test practice.
  www.geometryonline.com/vocabulary_review
  www.geometryonline.com/chapter_test
  www.geometryonline.com/standardized_test

For more information on Intervention and Assessment, see pp. T8–T11.

Reading and Writing in Mathematics

Glencoe Geometry provides numerous opportunities to incorporate reading and writing into the mathematics classroom.

Student Edition

- Foldables Study Organizer, p. 61
- Concept Check questions require students to verbalize and write about what they have learned in the lesson. (pp. 63, 71, 78, 84, 91, 97, 103, 111)
- Reading Mathematics, p. 81
- Writing in Math questions in every lesson, pp. 66, 74, 79, 86, 93, 99, 106, 114
- Reading Study Tip, p. 75
- WebQuest, p. 65

Teacher Wraparound Edition

- Foldables Study Organizer, pp. 61, 115
- Study Notebook suggestions, pp. 64, 72, 78, 81, 84, 88, 91, 97, 104, 111
- Modeling activities, pp. 74, 87, 106
- Speaking activities, pp. 80, 93
- Writing activities, pp. 66, 100, 114
- Differentiated Instruction (Verbal/Linguistic), p. 83
- ELL Resources, pp. 60, 65, 73, 79, 81, 83, 86, 92, 99, 105, 113, 115

Additional Resources

- Vocabulary Builder worksheets require students to define and give examples for key vocabulary terms as they progress through the chapter. (*Chapter 2 Resource Masters,* pp. vii-viii)
- Proof Builder helps students learn and understand theorems and postulates from the chapter. (*Chapter 2 Resource Masters,* pp. ix–x)
- Reading to Learn Mathematics master for each lesson (*Chapter 2 Resource Masters*, pp. 61, 67, 73, 79, 85, 91, 97, 103)
- *Vocabulary PuzzleMaker* software creates crossword, jumble, and word search puzzles using vocabulary lists that you can customize.
- *Teaching Mathematics with Foldables* provides suggestions for promoting cognition and language.
- *Reading Strategies for the Mathematics Classroom*
- *WebQuest and Project Resources*

For more information on Reading and Writing in Mathematics, see pp. T6–T7.

Chapter 2 Notes

What You'll Learn

Have students read over the list of objectives and make a list of any words with which they are not familiar.

Why It's Important

Point out to students that this is only one of many reasons why each objective is important. Others are provided in the introduction to each lesson.

Lesson	NCTM Standards	Local Objectives
2-1	1, 6, 7, 8, 9, 10	
2-2	6, 7, 8, 9, 10	
2-3	3, 6, 7, 8, 9, 10	
2-4	3, 6, 7, 8, 9, 10	
2-4 Follow-Up	6, 7	
2-5	3, 6, 7, 8, 9, 10	
2-6	2, 3, 6, 7, 8, 9, 10	
2-7	3, 6, 7, 8, 9, 10	
2-8	3, 6, 7, 8, 9, 10	

Key to NCTM Standards:

1=Number & Operations, 2=Algebra, 3=Geometry, 4=Measurement, 5=Data Analysis & Probability, 6=Problem Solving, 7=Reasoning & Proof, 8=Communication, 9=Connections, 10=Representation

Chapter 2 Reasoning and Proof

What You'll Learn

- **Lessons 2-1 through 2-3** Make conjectures, determine whether a statement is true or false, and find counterexamples for statements.
- **Lesson 2-4** Use deductive reasoning to reach valid conclusions.
- **Lessons 2-5 and 2-6** Verify algebraic and geometric conjectures using informal and formal proof.
- **Lessons 2-7 and 2-8** Write proofs involving segment and angle theorems.

Key Vocabulary

- inductive reasoning (p. 62)
- deductive reasoning (p. 82)
- postulate (p. 89)
- theorem (p. 90)
- proof (p. 90)

Why It's Important

Logic and reasoning are used throughout geometry to solve problems and reach conclusions. There are many professions that rely on reasoning in a variety of situations. Doctors, for example, use reasoning to diagnose and treat patients. *You will investigate how doctors use reasoning in Lesson 2-4.*

Vocabulary Builder ELL

The Key Vocabulary list introduces students to some of the main vocabulary terms included in this chapter. For a more thorough vocabulary list with pronunciations of new words, give students the Vocabulary Builder worksheets found on pages vii and viii of the *Chapter 2 Resource Masters*. Encourage them to complete the definition of each term as they progress through the chapter. You may suggest that they add these sheets to their study notebooks for future reference when studying for the Chapter 2 test.

Getting Started

Prerequisite Skills To be successful in this chapter, you'll need to master these skills and be able to apply them in problem-solving situations. Review these skills before beginning Chapter 2.

For Lesson 2-1 — Evaluate Expressions

Evaluate each expression for the given value of n. *(For review, see page 736.)*

1. $3n - 2; n = 4$ **10**
2. $(n + 1) + n; n = 6$ **13**
3. $n^2 - 3n; n = 3$ **0**
4. $180(n - 2); n = 5$ **540**
5. $n\left(\frac{n}{2}\right); n = 10$ **50**
6. $\frac{n(n-3)}{2}; n = 8$ **20**

For Lessons 2-6 through 2-8 — Solve Equations

Solve each equation. *(For review, see pages 737 and 738.)*

7. $6x - 42 = 4x$ **21**
8. $8 - 3n = -2 + 2n$ **2**
9. $3(y + 2) = -12 + y$ **−9**
10. $12 + 7x = x - 18$ **−5**
11. $3x + 4 = \frac{1}{2}x - 5$ $-\frac{18}{5}$
12. $2 - 2x = \frac{2}{3}x - 2$ $\frac{3}{2}$

For Lesson 2-8 — Adjacent and Vertical Angles

For Exercises 13–14, refer to the figure at the right. *(For review, see Lesson 1-5.)*

13. If $m\angle AGB = 4x + 7$ and $m\angle EGD = 71$, find x. **16**

14. If $m\angle BGC = 45$, $m\angle CGD = 8x + 4$, and $m\angle DGE = 15x - 7$, find x. **6**

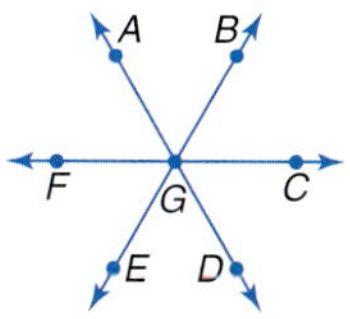

FOLDABLES™ Study Organizer

Reasoning and Proof Make this Foldable to help you organize information from this chapter. Begin with four sheets of $8\frac{1}{2}$-inch by 11-inch grid paper.

Step 1 Fold and Cut

Fold each sheet of grid paper in half along the width. Then cut along the crease.

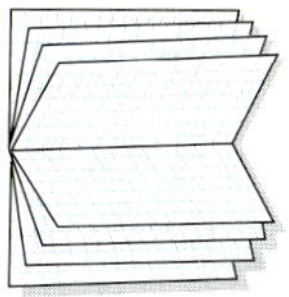

Step 2 Staple

Staple the eight half-sheets together to form a booklet.

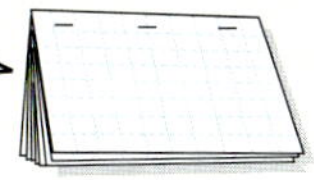

Step 3 Cut Tabs

Cut seven lines from the bottom of the top sheet, six lines from the second sheet, and so on.

Step 4 Label

Label each of the tabs with a lesson number. Add the chapter title to the first tab.

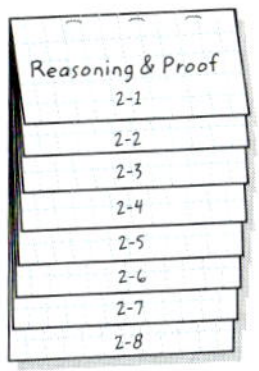

Reading and Writing As you read and study each lesson, use the corresponding page to write proofs and record examples of when you used logical reasoning in your daily life.

Getting Started

This section provides a review of the basic concepts needed before beginning Chapter 2. Page references are included for additional student help.

Additional review is provided in the *Prerequisite Skills Workbook*, pages 41–44, 81–86, 89–90, 93–94, 101–104.

Prerequisite Skills in the Getting Ready for the Next Lesson section at the end of each exercise set review a skill needed in the next lesson.

For Lesson	Prerequisite Skill
2-3	Evaluating Algebraic Expressions, p. 74
2-4	Solving Equations, p. 80
2-5	Information from Figures, p. 87
2-6	Solving Equations, p. 93
2-7	Segment Measures, p. 100
2-8	Complementary and Supplementary Angles, p. 106

FOLDABLES™ Study Organizer

For more information about Foldables, see *Teaching Mathematics with Foldables.*

Organization of Data Use this Foldable for student writing about reasoning and proofs. After students make their Foldable, have them label the tabs to correspond to the eight lessons in this chapter. Students use their Foldable to take notes, define terms, record concepts, write statements in if-then form, and write paragraph proofs. On the back of the Foldable, have students record examples of ways in which they use reasoning and proofs in their daily lives. Note how columnists and authors present their reasoning and ways in which they try to prove or disprove their points of view.

2-1 Lesson Notes

1 Focus

5-Minute Check Transparency 2-1 Use as a quiz or review of Chapter 1.

Mathematical Background notes are available for this lesson on p. 60C.

How can inductive reasoning help predict weather conditions?

Ask students:

- What are normal temperatures for the month of January? **Sample answer: The temperatures in January are usually in the 30s or 40s. (Answers will vary in different parts of the country.)**
- How do people benefit from the inductive reasoning techniques of meteorologists? **Sample answers: People can plan for outdoor events a few days in advance; they can dress appropriately for daily weather conditions and carry weather-related items, such as umbrellas, sunglasses, and so on.**

2-1 Inductive Reasoning and Conjecture

What You'll Learn

- Make conjectures based on inductive reasoning.
- Find counterexamples.

Vocabulary

- conjecture
- inductive reasoning
- counterexample

How can inductive reasoning help predict weather conditions?

Meteorologists use science and weather patterns to make predictions about future weather conditions. They are able to make accurate educated guesses based on past weather patterns.

MAKE CONJECTURES A **conjecture** is an educated guess based on known information. Examining several specific situations to arrive at a conjecture is called inductive reasoning. **Inductive reasoning** is reasoning that uses a number of specific examples to arrive at a plausible generalization or prediction.

Example 1 Patterns and Conjecture

The numbers represented below are called *triangular numbers*. Make a conjecture about the next triangular number based on the pattern.

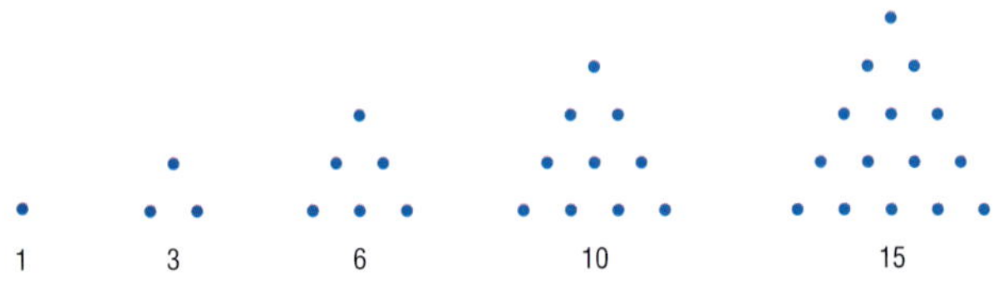

Observe: Each triangle is formed by adding another row of dots.

Find a Pattern: 1 → 3 → 6 → 10 → 15 (+2, +3, +4, +5)

The numbers increase by 2, 3, 4, and 5.

Conjecture: The next number will increase by 6.
So, it will be 15 + 6 or 21.

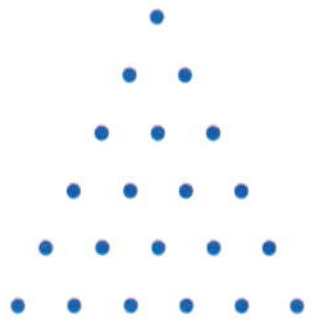

Study Tip

Conjectures
List your observations and identify patterns before you make a conjecture.

In Chapter 1, you learned some basic geometric concepts. These concepts can be used to make conjectures in geometry.

Resource Manager

Workbook and Reproducible Masters

Chapter 2 Resource Masters
- Study Guide and Intervention, pp. 57–58
- Skills Practice, p. 59
- Practice, p. 60
- Reading to Learn Mathematics, p. 61
- Enrichment, p. 62

Graphing Calculator and Computer Masters, p. 19

Transparencies

5-Minute Check Transparency 2-1
Answer Key Transparencies

Technology

Interactive Chalkboard

Example 2 Geometric Conjecture

For points P, Q, and R, $PQ = 9$, $QR = 15$, and $PR = 12$. Make a conjecture and draw a figure to illustrate your conjecture.

Given: points P, Q, and R; $PQ = 9$, $QR = 15$, and $PR = 12$

Examine the measures of the segments. Since $PQ + PR \neq QR$, the points cannot be collinear.

Conjecture: P, Q, and R are noncollinear.

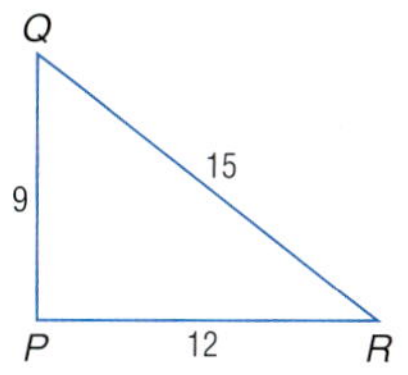

FIND COUNTEREXAMPLES A conjecture based on several observations may be true in most circumstances, but false in others. It takes only one false example to show that a conjecture is not true. The false example is called a **counterexample**.

Example 3 Find a Counterexample

FINANCE **Find a counterexample for the following statement based on the graph.**

The rates for CDs are at least 1.5% less than the rates a year ago.

Examine the graph. The statement is true for 6-month, 1-year, and $2\frac{1}{2}$-year CDs. However, the difference in the rate for a 5-year CD is 0.74% less, which is less than 1.5%. The statement is false for a 5-year certificate of deposit. Thus, the change in the 5-year rate is a counterexample to the original statement.

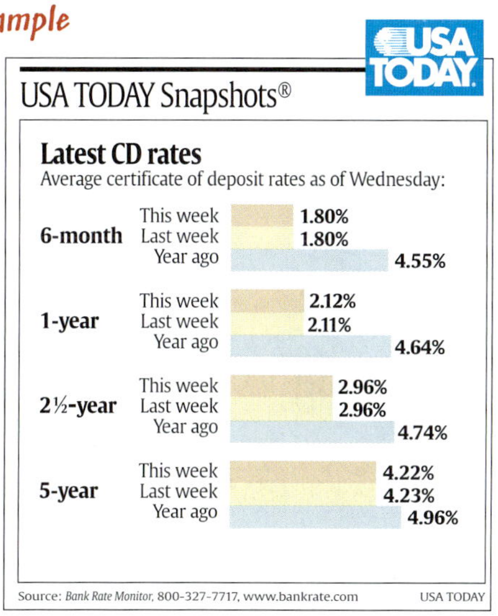

Log on for:
- Updated data
- More on finding counterexamples

www.geometryonline.com/usa_today

Check for Understanding

Concept Check

1–3. See p. 123A.

1. **Write** an example of a conjecture you have made outside of school.
2. **Determine** whether the following conjecture is *always*, *sometimes*, or *never* true based on the given information.
   Given: collinear points D, E, and F
   Conjecture: $DE + EF = DF$
3. **OPEN ENDED** Write a statement. Then find a counterexample for the statement.

www.geometryonline.com/extra_examples/fcat

2 Teach

MAKE CONJECTURES

In-Class Examples Power Point®

Teaching Tip Tell students to test all fundamental operations, including powers and roots, when they are looking for patterns in a series of numbers. Advise students that sometimes two operations can be used.

1 Make a conjecture about the next number based on the pattern. 2, 4, 12, 48, 240 **1440**

2 For points L, M, and N, $LM = 20$, $MN = 6$, and $LN = 14$. Make a conjecture and draw a figure to illustrate your conjecture.

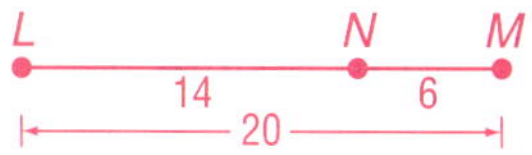

Conjecture: L, M, and N are collinear.

FIND COUNTEREXAMPLES

In-Class Example Power Point®

3 **UNEMPLOYMENT** Based on the table showing unemployment rates for various cities in Kansas, find a counterexample for the following statement.

The unemployment rate is highest in the cities with the most people.

County	Civilian Labor Force	Rate
Shawnee	90,254	3.1%
Jefferson	9,937	3.0%
Jackson	8,915	2.8%
Douglas	55,730	3.2%
Osage	10,182	4.0%
Wabaunsee	3,575	3.0%
Pottawatomie	11,025	2.1%

Source: Labor Market Information Services–Kansas Department of Human Resources

Osage has only 10,182 people on its civilian labor force, and it has a higher rate of unemployment than Shawnee, which has 90,254 people on its civilian labor force.

Interactive Chalkboard
PowerPoint® Presentations

This CD-ROM is a customizable Microsoft® PowerPoint® presentation that includes:
- Step-by-step, dynamic solutions of each In-Class Example from the Teacher Wraparound Edition
- Additional, Try These exercises for each example
- The 5-Minute Check Transparencies
- Hot links to Glencoe Online Study Tools

3 Practice/Apply

Study Notebook

Have students—
- add the definitions/examples of the vocabulary terms to their Vocabulary Builder worksheets for Chapter 2.
- include any other item(s) that they find helpful in mastering the skills in this lesson.

About the Exercises...

Organization by Objective
- **Make Conjectures:** 11–28
- **Find Counterexamples:** 29–36

Odd/Even Assignments
Exercises 11–36 are structured so that students practice the same concepts whether they are assigned odd or even problems.

Assignment Guide

Basic: 11–41 odd, 43–67
Average: 11–41 odd, 43–67
Advanced: 12–40 even, 41–64 (optional: 65–67)

Answers

6. $PQ = TU$

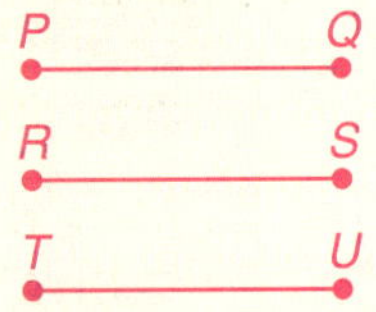

7. A, B, C, and D are noncollinear.

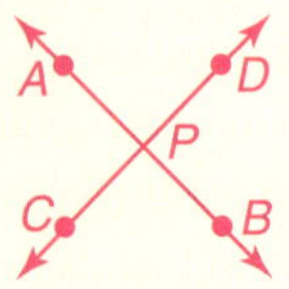

Guided Practice

GUIDED PRACTICE KEY

Exercises	Examples
4–5	1
6–7	2
8–9	3

Make a conjecture about the next item in each sequence.

4. 

5. $-8, -5, -2, 1, 4$ 7 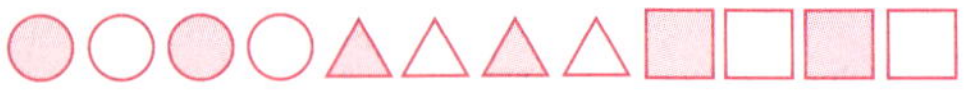

Make a conjecture based on the given information. Draw a figure to illustrate your conjecture. 6–7. See margin.

6. $PQ = RS$ and $RS = TU$

7. $\overleftrightarrow{AB}$ and $\overleftrightarrow{CD}$ intersect at P.

Determine whether each conjecture is *true* or *false*. Give a counterexample for any false conjecture.

8. **Given:** x is an integer.
   Conjecture: $-x$ is negative. False; if $x = -2$, then $-x = -(-2)$ or 2.

9. **Given:** $WXYZ$ is a rectangle.
   Conjecture: $WX = YZ$ and $WZ = XY$ true

Application

10. **HOUSES** Most homes in the northern United States have roofs made with steep angles. In the warmer areas of the southern states, homes often have flat roofs. Make a conjecture about why the roofs are different. Sample answer: Snow will not stick on a roof with a steep angle.

★ indicates increased difficulty

Practice and Apply

Homework Help

For Exercises	See Examples
11–20	1
21–28	2
29–36	3

Extra Practice
See page 756.

Make a conjecture about the next item in each sequence.

11. 

12. 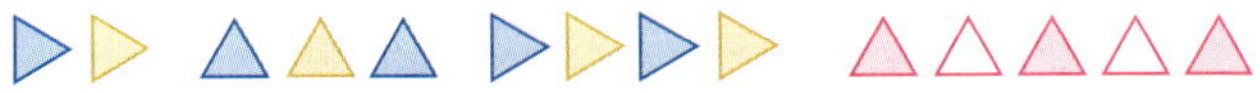

13. 1, 2, 4, 8, 16 32

14. 4, 6, 9, 13, 18 24

15. $\frac{1}{3}, 1, \frac{5}{3}, \frac{7}{3}, 3$ $\frac{11}{3}$

16. $1, \frac{1}{2}, \frac{1}{4}, \frac{1}{8}, \frac{1}{16}$ $\frac{1}{32}$

17. $2, -6, 18, -54$ 162

18. $-5, 25, -125, 625$ -3125

Make a conjecture about the number of blocks in the next item of each sequence.

19. 

20. 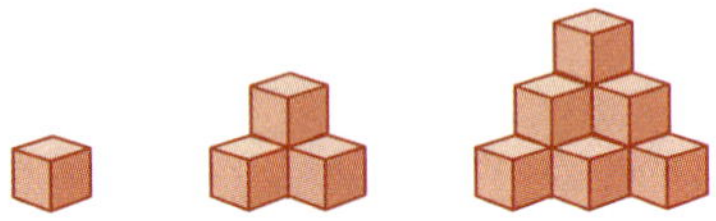

19–20. See p. 123A.

Make a conjecture based on the given information. Draw a figure to illustrate your conjecture. 21–28. See p. 123A for figures.

21. Lines ℓ and m are perpendicular.

22. $A(-2, -11)$, $B(2, 1)$, $C(5, 10)$

23. $\angle 3$ and $\angle 4$ are a linear pair.

24. $\overrightarrow{BD}$ is an angle bisector of $\angle ABC$.

25. $P(-1, 7)$, $Q(6, -2)$, $R(6, 5)$

26. $HIJK$ is a square.

27. $PQRS$ is a rectangle. $PQ = SR$, $QR = PS$

28. $\angle B$ is a right angle in $\triangle ABC$. $(AB)^2 + (BC)^2 = (AC)^2$

21. Lines ℓ and m form four right angles.
22. A, B, and C are collinear.
23. $\angle 3$ and $\angle 4$ are supplementary.
24. $\angle ABD \cong \angle DBC$.
25. $\triangle PQR$ is a scalene triangle.
26. $HI = IJ = JK = KH$

DAILY INTERVENTION

Differentiated Instruction

Naturalist Students can practice brainstorming conjectures and finding counterexamples in nature. For example have students consider the statement, "If plants don't receive water daily, they will not survive." A counterexample would be a cactus, which can go weeks without water. Nature topics could include plants, animals, predator/food supply relationships, insects, weather, and so on.

WebQuest

You can use scatter plots to make conjectures about the relationships between latitude, longitude, degree distance, and the monthly high temperature.
Visit www.geometryonline.com/WebQuest to continue work on your WebQuest project.

Determine whether each conjecture is *true* or *false*. Give a counterexample for any false conjecture.

29. **Given:** $\angle 1$ and $\angle 2$ are complementary angles.
Conjecture: $\angle 1$ and $\angle 2$ form a right angle. **False; see margin for counterexample.**

30. **Given:** $m + y \geq 10$, $y \geq 4$
Conjecture: $m \leq 6$ **False; if $y = 7$ and $m = 5$, then $7 + 5 \geq 10$ and $5 \geq 4$, but $7 \not\leq 6$.**

31. **Given:** points W, X, Y, and Z
Conjecture: W, X, Y, and Z are noncollinear. **False; see margin for counterexample.**

32. **Given:** $A(-4, 8)$, $B(3, 8)$, $C(3, 5)$
Conjecture: $\triangle ABC$ is a right triangle. **true**

33. **Given:** n is a real number.
Conjecture: n^2 is a nonnegative number. **true**

34. **Given:** $DE = EF$
Conjecture: E is the midpoint of $\overline{DF}$. **False; D, E, and F do not have to be collinear.**

35. **Given:** $JK = KL = LM = MJ$
Conjecture: $JKLM$ forms a square. **False; $JKLM$ may not have a right angle.**

36. **Given:** noncollinear points R, S, and T
Conjecture: $\overline{RS}$, $\overline{ST}$, and $\overline{RT}$ form a triangle. **true**

More About. . .

Music

The average medium-sized piano has about 230 strings. Each string has about 165 pounds of tension. That's a combined tension of about 18 tons.

Source: www.pianoworld.com

37. **MUSIC** Many people learn to play the piano by ear. This means that they first learned how to play without reading music. What process did they use? **trial and error, a process of inductive reasoning**

CHEMISTRY **For Exercises 38–40, use the following information.**

Hydrocarbons are molecules composed of only carbon (C) and hydrogen (H) atoms. The simplest hydrocarbons are called alkanes. The first three alkanes are shown below.

Alkanes			
Compound Name	Methane	Ethane	Propane
Chemical Formula	CH_4	C_2H_6	C_3H_8
Structural Formula	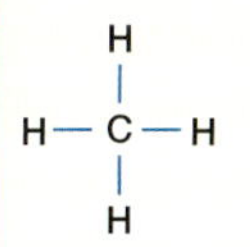	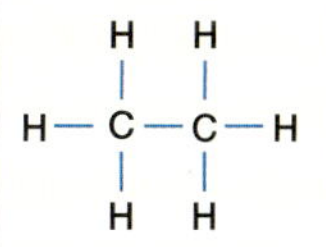	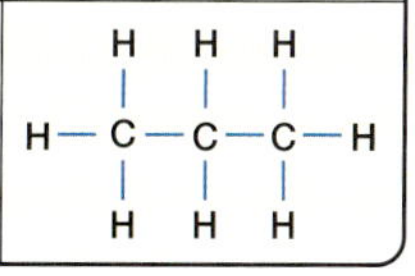

38. Make a conjecture about butane, which is the next compound in the group. Write its structural formula.

39. Write the chemical formula for the 7th compound in the group. C_7H_{16}

★**40.** Develop a rule you could use to find the chemical formula of the nth substance in the alkane group. C_nH_{2n+2}

38. Butane will have 4 carbon atoms and 10 hydrogen atoms.

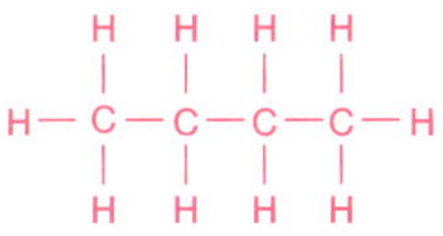

41. **CRITICAL THINKING** The expression $n^2 - n + 41$ has a prime value for $n = 1$, $n = 2$, and $n = 3$. Based on this pattern, you might conjecture that this expression always generates a prime number for any positive integral value of n. Try different values of n to test the conjecture. Answer *true* if you think the conjecture is always true. Answer *false* and give a counterexample if you think the conjecture is false. **false; $n = 41$**

www.geometryonline.com/self_check_quiz/fcat

Answers

29.

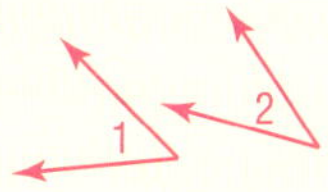

31.

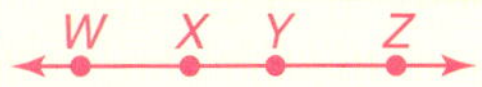

Study Guide and Intervention, p. 57 (shown) and p. 58

Make Conjectures A **conjecture** is a guess based on analyzing information or observing a pattern. Making a conjecture after looking at several situations is called **inductive reasoning**.

Example 1 **Make a conjecture about the next number in the sequence 1, 3, 9, 27, 81.**

Analyze the numbers:
Notice that each number is a power of 3.

1	3	9	27	81
3^0	3^1	3^2	3^3	3^4

Conjecture: The next number will be 3^5 or 243.

Example 2 **Make a conjecture about the number of small squares in the next figure.**

Observe a pattern: The sides of the squares have measures 1, 2, and 3 units.
Conjecture: For the next figure, the side of the square will be 4 units, so the figure will have 16 small squares.

Exercises

Describe the pattern. Then make a conjecture about the next number in the sequence.

1. −5, 10, −20, 40 Pattern: Each number is −2 times the previous number. Conjecture: The next number is −80.
2. 1, 10, 100, 1000 Pattern: Each number is 10 times the previous number. Conjecture: The next number is 10,000.
3. $1, \frac{6}{5}, \frac{7}{5}, \frac{8}{5}$ Pattern: Each number is $\frac{1}{5}$ more than the previous number. Conjecture: The next number is $\frac{9}{5}$.

Make a conjecture based on the given information. Draw a figure to illustrate your conjecture. 4–7. Sample answers are given.

4. $A(-1, -1)$, $B(2, 2)$, $C(4, 4)$
Points A, B, and C are collinear.

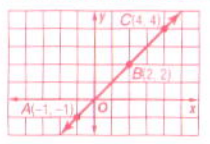

5. $\angle 1$ and $\angle 2$ form a right angle.
$\angle 1$ and $\angle 2$ are complementary.

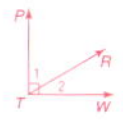

6. $\angle ABC$ and $\angle DBE$ are vertical angles.
$\angle ABC$ and $\angle DBE$ are congruent.

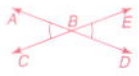

7. $\angle E$ and $\angle F$ are right angles.
$\angle E$ and $\angle F$ are congruent.

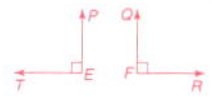

Skills Practice, p. 59 and Practice, p. 60 (shown)

Make a conjecture about the next item in each sequence.

1.

2. 5, −10, 15, −20 25
3. $-2, 1, -\frac{1}{2}, \frac{1}{4}, -\frac{1}{8}$ $\frac{1}{16}$
4. 12, 6, 3, 1.5, 0.75 0.375

Make a conjecture based on the given information. Draw a figure to illustrate your conjecture. 5–8. Sample answers are given.

5. $\angle ABC$ is a right angle.
$\overrightarrow{BA} \perp \overrightarrow{BC}$

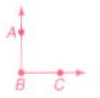

6. Point S is between R and T.
$RS + ST = RT$

7. P, Q, R, and S are noncollinear and $\overline{PQ} \cong \overline{QR} \cong \overline{RS} \cong \overline{SP}$.
The segments form a square.

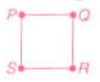

8. $ABCD$ is a parallelogram.
$AB = CD$ and $BC = AD$.

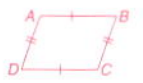

Determine whether each conjecture is true or false. Give a counterexample for any false conjecture.

9. Given: S, T, and U are collinear and $ST = TU$.
Conjecture: T is the midpoint of $\overline{SU}$.
true

10. Given: $\angle 1$ and $\angle 2$ are adjacent angles.
Conjecture: $\angle 1$ and $\angle 2$ form a linear pair.
False; $\angle 1$ and $\angle 2$ could each measure 60°.

11. Given: $\overline{GH}$ and $\overline{JK}$ form a right angle and intersect at P.
Conjecture: $\overline{GH} \perp \overline{JK}$
true

12. **ALLERGIES** Each spring, Rachel starts sneezing when the pear trees on her street blossom. She reasons that she is allergic to pear trees. Find a counterexample to Rachel's conjecture.
Sample answer: Rachel could be allergic to other types of plants that blossom when the pear trees blossom.

Reading to Learn Mathematics, p. 61 ELL

Pre-Activity **How can inductive reasoning help predict weather conditions?**

Read the introduction to Lesson 2-1 at the top of page 62 in your textbook.

- What kind of weather patterns do you think meteorologists look at to help predict the weather? Sample answer: patterns of high and low temperatures, including heat spells and cold spells; patterns of precipitation, including wet spells and dry spells
- What is a factor that might contribute to long-term changes in the weather? Sample answer: global warming due to high usage of fossil fuels

Reading the Lesson

1. Explain in your own words the relationship between a conjecture, a counterexample, and inductive reasoning.
Sample answer: A conjecture is an educated guess based on specific examples or information. A counterexample is an example that shows that a conjecture is false. Inductive reasoning is the process of making a conjecture based on specific examples or information.

2. Make a conjecture about the next item in each sequence.
a. 5, 9, 13, 17 21
b. $1, \frac{1}{3}, \frac{1}{9}, \frac{1}{27}$ $\frac{1}{81}$
c. 0, 1, 3, 6, 10 15
d. 8, 3, −2, −7 −12
e. 1, 8, 27, 64 125
f. 1, −2, 4, −8 16
g.

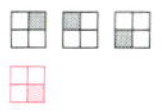

h.

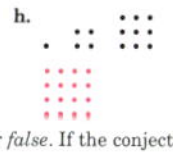

3. State whether each conjecture is *true* or *false*. If the conjecture is false, give a counterexample.
a. The sum of two odd integers is even.
true
b. The product of an odd integer and an even integer is odd.
False; sample answer: $5 \cdot 8 = 40$, which is even.
c. The opposite of an integer is a negative integer. False; sample answer: The opposite of the integer −5 is 5, which is a positive integer.
d. The perfect squares (squares of whole numbers) alternate between odd and even.
true

Helping You Remember

4. Write a short sentence that can help you remember why it only takes one counterexample to prove that a conjecture is false.
Sample answer: True means *always* true.

Enrichment, p. 62

Counterexamples

When you make a conclusion after examining several specific cases, you have used **inductive reasoning**. However, you must be cautious when using this form of reasoning. By finding only one **counterexample**, you disprove the conclusion.

Example **Is the statement $\frac{1}{x} \leq 1$ true when you replace x with 1, 2, and 3? Is the statement true for all reals? If possible, find a counterexample.**

$\frac{1}{1} = 1$, $\frac{1}{2} < 1$, and $\frac{1}{3} < 1$. But when $x = \frac{1}{2}$, then $\frac{1}{x} = 2$. This counterexample shows that the statement is not always true.

Answer each question.

1. The coldest day of the year in Chicago occurred in January for five straight years. Is it safe to conclude that the …

2. Suppose John misses the school bus four Tuesdays in a row. Can you safely conclude that John misses the …

4 Assess

Open-Ended Assessment

Writing Ask students to write five conjectures about school rules or activities. Then have students swap papers with a partner and try to come up with a counterexample for each conjecture. An example statement could be: *Students must attend school Monday through Friday.* A counterexample for this would be a holiday or a snow day.

Getting Ready for Lesson 2-2

Basic Skill Students will learn about logic statements in Lesson 2-2. They will determine the truth value of various situations. Use Exercises 65–67 to determine your students' familiarity with determining which elements make a statement true.

Answer

42. Sample answer: By past experience, when dark clouds appear, there is a chance of rain. Answers should include the following.
    - When there is precipitation in the summer, it is usually rain because the temperature is above freezing. When the temperature is below freezing, as in the winter, ice or snow forms.
    - See students' work.

42. **WRITING IN MATH** Answer the question that was posed at the beginning of the lesson. See margin.

How can inductive reasoning help predict weather conditions?

Include the following in your answer:
- an explanation as to how a conjecture about a weather pattern in the summer might be different from a similar weather pattern in the winter, and
- a conjecture about tomorrow's weather based on your local weather over the past several days.

FCAT Practice
Standardized Test Practice

43. What is the next term in the sequence 1, 1, 2, 3, 5, 8? **C**

(A) 11 (B) 12 (C) 13 (D) 14

44. **ALGEBRA** If the average of six numbers is 18 and the average of three of the numbers is 15, then what is the sum of the remaining three numbers? **D**

(A) 21 (B) 45 (C) 53 (D) 63

Maintain Your Skills

Mixed Review Name each polygon by its number of sides and then classify it as *convex* or *concave* and *regular* or *irregular*. *(Lesson 1-6)*

45. 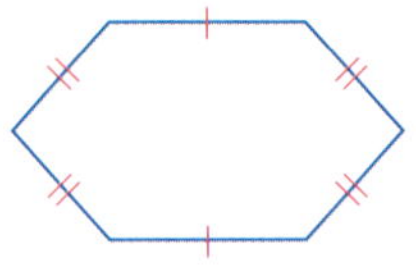
hexagon, convex, irregular

46. 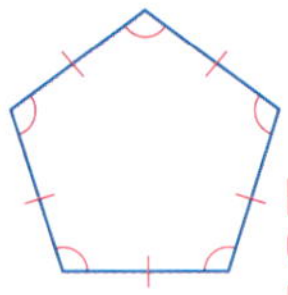
pentagon, convex, regular

47. 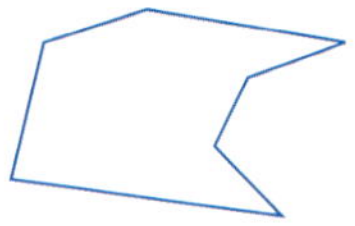
heptagon, concave, irregular

Determine whether each statement can be assumed from the figure. Explain. *(Lesson 1-5)*

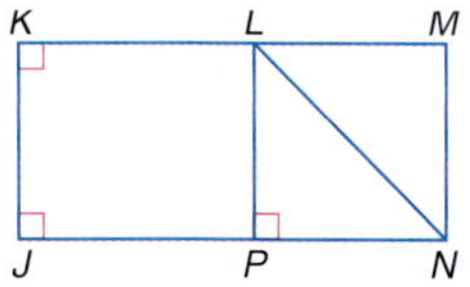

48. $\angle KJN$ is a right angle.
49. $\angle PLN \cong \angle NLM$
50. $\angle PNL$ and $\angle MNL$ are complementary.
51. $\angle KLN$ and $\angle MLN$ are supplementary.
52. $\angle KLP$ is a right angle.

48. Yes; the symbol denotes that $\angle KJN$ is a right angle.
49. No; we do not know anything about the angle measures.
50. No; we do not know whether $\angle MNP$ is a right angle.
51. Yes; they form a linear pair.
52. Yes; since the other three angles in rectangle $KLPJ$ are right angles, $\angle KLP$ must also be a right angle.

Find the coordinates of the midpoint of a segment having the given endpoints. *(Lesson 1-3)*

53. $\overline{AB}$ for $A(-1, 3)$, $B(5, -5)$ **(2, −1)**
54. $\overline{CD}$ for $C(4, 1)$, $D(-3, 7)$ **(0.5, 4)**
55. $\overline{FG}$ for $F(4, -9)$, $G(-2, -15)$ **(1, −12)**
56. $\overline{HJ}$ for $H(-5, -2)$, $J(7, 4)$ **(1, 1)**
57. $\overline{KL}$ for $K(8, -1.8)$, $L(3, 6.2)$ **(5.5, 2.2)**
58. $\overline{MN}$ for $M(-1.5, -6)$, $N(-4, 3)$ **(−2.75, −1.5)**

Find the value of the variable and MP, if P is between M and N. *(Lesson 1-2)*

59. $MP = 7x$, $PN = 3x$, $PN = 24$ **8; 56**
60. $MP = 2c$, $PN = 9c$, $PN = 63$ **7; 14**
61. $MP = 4x$, $PN = 5x$, $MN = 36$ **4; 16**
62. $MP = 6q$, $PN = 6q$, $MN = 60$ **5; 30**
63. $MP = 4y + 3$, $PN = 2y$, $MN = 63$ **10; 43**
64. $MP = 2b - 7$, $PN = 8b$, $MN = 43$ **5; 3**

Getting Ready for the Next Lesson

BASIC SKILL Determine which values in the given replacement set make the inequality true.

65. $x + 2 > 5$ **4, 5**
$\{2, 3, 4, 5\}$
66. $12 - x < 0$ **13, 14**
$\{11, 12, 13, 14\}$
67. $5x + 1 > 25$ **5, 6, 7**
$\{4, 5, 6, 7\}$

2-2 Logic

What You'll Learn

- Determine truth values of conjunctions and disjunctions.
- Construct truth tables.

How does logic apply to school?

When you answer true-false questions on a test, you are using a basic principle of logic. For example, refer to the map, and answer *true* or *false*.

Raleigh is a city in North Carolina.

You know that there is only one correct answer, either true or false.

Vocabulary

- statement
- truth value
- negation
- compound statement
- conjunction
- disjunction
- truth table

DETERMINE TRUTH VALUES A **statement**, like the true-false example above, is any sentence that is either true or false, but not both. Unlike a conjecture, we know that a statement is either true or false. The truth or falsity of a statement is called its **truth value**.

Statements are often represented using a letter such as p or q. The statement above can be represented by p.

p: Raleigh is a city in North Carolina. This statement is true.

The **negation** of a statement has the opposite meaning as well as an opposite truth value. For example, the negation of the statement above is *not p*.

not p: Raleigh is not a city in North Carolina. In this case, the statement is false.

Study Tip

Statements

A mathematical statement with one or more variables is called an *open sentence*. The truth value of an open sentence cannot be determined until values are assigned to the variables. A statement with only numeric values is a *closed sentence*.

Key Concept — *Negation*

- **Words** If a statement is represented by p, then *not p* is the negation of the statement.
- **Symbols** $\sim p$, read *not p*

Two or more statements can be joined to form a **compound statement**. Consider the following two statements.

p: Raleigh is a city in North Carolina.
q: Raleigh is the capital of North Carolina.

The two statements can be joined by the word *and*.

p and q: Raleigh is a city in North Carolina, *and* Raleigh is the capital of North Carolina.

2-2 Lesson Notes

1 Focus

5-Minute Check Transparency 2-2 Use as a quiz or review of Lesson 2-1.

Mathematical Background notes are available for this lesson on p. 60C.

How does logic apply to school?

Ask students:

- Determine whether the following statement is *true* or *false*: "South Carolina borders North Carolina, Georgia, and Tennessee." **false**
- Locate Wilmington on the map of North Carolina. Is this a coastal or inland city? Make a conjecture about whether you might find ocean or lakes in Wilmington given its location on the map. **Coastal; sample answer: You would find ocean in Wilmington, North Carolina.**

Resource Manager

Workbook and Reproducible Masters

Chapter 2 Resource Masters
- Study Guide and Intervention, pp. 63–64
- Skills Practice, p. 65
- Practice, p. 66
- Reading to Learn Mathematics, p. 67
- Enrichment, p. 68
- Assessment, p. 119

Graphing Calculator and Computer Masters, p. 20
Teaching Geometry With Manipulatives Masters, p. 16

Transparencies

5-Minute Check Transparency 2-2
Answer Key Transparencies

Technology

Interactive Chalkboard

2 Teach

DETERMINE TRUTH VALUES

In-Class Example Power Point®

1 Use the following statements to write a compound statement for each conjunction. Then find its truth value.

p: One foot is 14 inches.
q: September has 30 days.
r: A plane is defined by three noncollinear points.

a. *p* and *q*
One foot is 14 inches, and September has 30 days; false.

b. $r \wedge p$
A plane is defined by three noncollinear points, and one foot is 14 inches; false.

c. $\sim q \wedge r$
September does not have 30 days, and a plane is defined by three noncollinear points; false.

d. $\sim p \wedge r$
A foot is not 14 inches, and a plane is defined by three noncollinear points; true.

The statement formed by joining *p* and *q* is an example of a conjunction.

Key Concept — *Conjunction*

- **Words** A **conjunction** is a compound statement formed by joining two or more statements with the word *and*.
- **Symbols** $p \wedge q$, read *p and q*

A conjunction is true only when both statements in it are true. Since it is true that Raleigh is in North Carolina and it is the capital, the conjunction is also true.

Example 1 *Truth Values of Conjunctions*

Use the following statements to write a compound statement for each conjunction. Then find its truth value.

p: **January 1 is the first day of the year.**
q: $-5 + 11 = -6$
r: **A triangle has three sides.**

a. ***p*** **and** ***q***

January is the first day of the year, and $-5 + 11 = -6$.
p and *q* is false, because *p* is true and *q* is false.

b. $r \wedge p$

A triangle has three sides, and January 1 is the first day of the year.
$r \wedge p$ is true, because *r* is true and *p* is true.

c. ***p*** **and not** ***r***

January 1 is the first day of the year, and a triangle does not have three sides.
p and not *r* is false, because *p* is true and not *r* is false.

d. $\sim q \wedge r$

$-5 + 11 \neq -6$, and a triangle has three sides
$\sim q \wedge r$ is true because $\sim q$ is true and *r* is true.

Study Tip

Negations
The negation of a statement is not necessarily false. It has the opposite truth value of the original statement.

Statements can also be joined by the word *or*. This type of statement is a disjunction. Consider the following statements.

p: Ahmed studies chemistry.

q: Ahmed studies literature.

p or *q*: Ahmed studies chemistry, *or* Ahmed studies literature.

Key Concept — *Disjunction*

- **Words** A **disjunction** is a compound statement formed by joining two or more statements with the word *or*.
- **Symbols** $p \vee q$, read *p or q*

A disjunction is true if at least one of the statements is true. In the case of p or q above, the disjunction is true if Ahmed either studies chemistry or literature or both. The disjunction is false only if Ahmed studies neither chemistry nor literature.

Example 2 Truth Values of Disjunctions

Use the following statements to write a compound statement for each disjunction. Then find its truth value.

p: **100 ÷ 5 = 20**

q: **The length of a radius of a circle is twice the length of its diameter.**

r: **The sum of the measures of the legs of a right triangle equals the measure of the hypotenuse.**

a. p **or** q

100 ÷ 5 = 20, or the length of a radius of a circle is twice the length of its diameter.

p or q is true because p is true. It does not matter that q is false.

b. $q \vee r$

The length of a radius of a circle is twice the length of its diameter, or the sum of the measures of the legs of a right triangle equals the measure of the hypotenuse.

$q \vee r$ is false since neither statement is true.

Study Tip

Venn Diagrams
The size of the overlapping region in a Venn Diagram does not indicate how many items fall into that category.

Conjunctions can be illustrated with Venn diagrams. Refer to the statement at the beginning of the lesson. The Venn diagram at the right shows that Raleigh (R) is represented by the *intersection* of the set of cities in North Carolina and the set of state capitals. In other words, Raleigh must be in the set containing cities in North Carolina and in the set of state capitals.

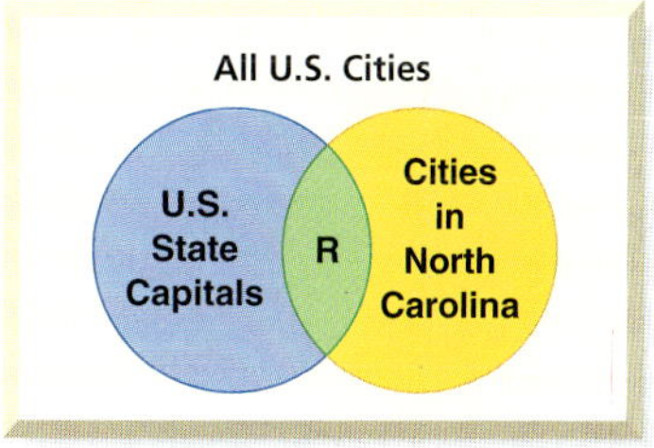

A disjunction can also be illustrated with a Venn diagram. Consider the following statements.

p: Jerrica lives in a U.S. state capital.

q: Jerrica lives in a North Carolina city.

$p \vee q$: Jerrica lives in a U.S. state capital, or Jerrica lives in a North Carolina city.

In the Venn diagrams, the disjunction is represented by the *union* of the two sets. The union includes all U.S. capitals and all cities in North Carolina. The city in which Jerrica lives could be located in any of the three regions of the union.

The three regions represent

A U.S. state capitals excluding the capital of North Carolina,

B cities in North Carolina excluding the state capital, and

C the capital of North Carolina, which is Raleigh.

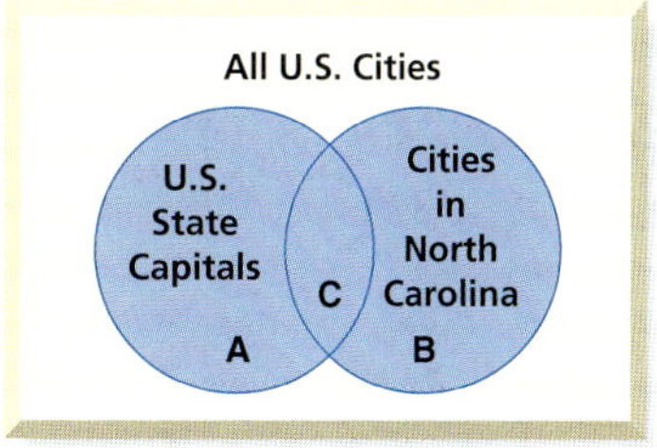

www.geometryonline.com/extra_examples/fcat

In-Class Example

Power Point®

2 Use the following statements to write a compound statement for each disjunction. Then find its truth value.

p: $\overline{AB}$ is proper notation for "line AB."

q: Centimeters are metric units.

r: 9 is a prime number.

a. p or q
$\overline{AB}$ is proper notation for "line AB," or centimeters are metric units; true.

b. $q \vee r$
Centimeters are metric units, or 9 is a prime number; true.

Lesson 2-2 Logic 69

In-Class Example

Power Point®

3 **DANCING** The Venn diagram shows the number of students enrolled in Monique's Dance School for tap, jazz, and ballet classes.

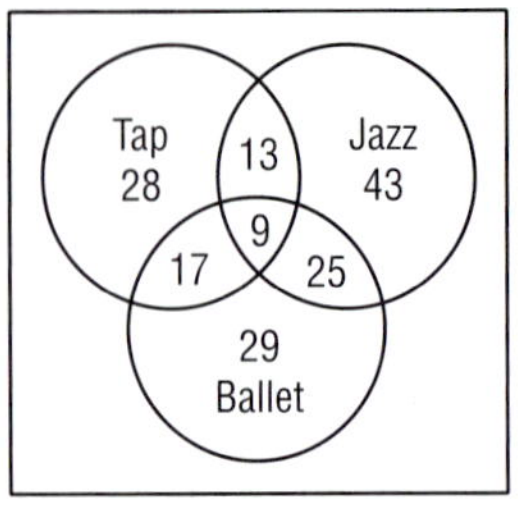

a. How many students are enrolled in all three classes? **9**

b. How many students are enrolled in tap or ballet? **121**

c. How many students are enrolled in jazz and ballet and not tap? **25**

TRUTH TABLES

Tips for New Teachers

Intervention Tell students that truth tables must display all combinations of Ts and Fs to exhaust all possible outcomes, so for each statement, p, q, and r, they will need to mix the occurrences of *true* and *false*. First, they should determine the number of rows they need. Then fill the top half of the p column with Ts and the other half with Fs. For the q column, they can alternate T and F the whole way down. If they need an r column, they can alternate sets of 2 Ts and 2 Fs all the way down, and so on. Assure students that as long as they initially set up the basic structure correctly, they should be able to produce the rest of the table quite easily.

Venn diagrams can be used to solve real-world problems involving conjunctions and disjunctions.

Example 3 Use Venn Diagrams

RECYCLING **The Venn diagram shows the number of neighborhoods that have a curbside recycling program for paper or aluminum.**

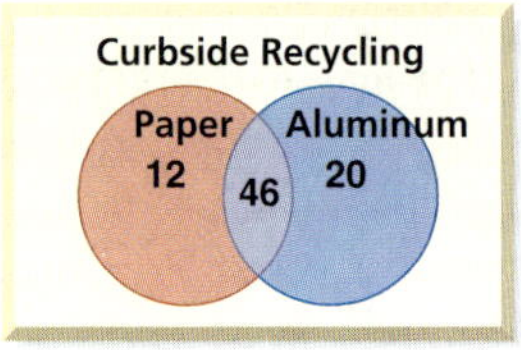

a. How many neighborhoods recycle both paper and aluminum?

The neighborhoods that have paper and aluminum recycling are represented by the intersection of the sets. There are 46 neighborhoods that have paper and aluminum recycling.

b. How many neighborhoods recycle paper or aluminum?

The neighborhoods that have paper or aluminum recycling are represented by the union of the sets. There are 12 + 46 + 20 or 78 neighborhoods that have paper or aluminum recycling.

c. How many neighborhoods recycle paper and not aluminum?

The neighborhoods that have paper and not aluminum recycling are represented by the nonintersecting portion of the paper region. There are 12 neighborhoods that have paper and not aluminum recycling.

TRUTH TABLES A convenient method for organizing the truth values of statements is to use a **truth table**.

Negation	
p	$\sim p$
T	F
F	T

If p is a true statement, then $\sim p$ is a false statement.
If p is a false statement, then $\sim p$ is a true statement.

Truth tables can also be used to determine truth values of compound statements.

Study Tip

Tautology
A compound sentence is a *tautology* if its truth value is always true. For example, "It is snowing or it is not snowing" is a tautology.

Conjunction		
p	q	$p \wedge q$
T	T	T
T	F	F
F	T	F
F	F	F

A conjunction is true only when both statements are true.

Disjunction		
p	q	$p \vee q$
T	T	T
T	F	T
F	T	T
F	F	F

A disjunction is false only when both statements are false.

You can use the truth values for negation, conjunction, and disjunction to construct truth tables for more complex compound statements.

Answers

1. The conjunction (p and q) is represented by the intersection of the two circles.

2a. Sample answer: October has 31 days or $-5 + 3 = -8$.

2b. Sample answer: A square has five right angles and the Postal Service does not deliver mail on Sundays.

2c. Sample answer: July 5th is not a national holiday.

3. A conjunction is a compound statement using the word *and*, while a disjunction is a compound statement using the word *or*.

70 Chapter 2 Reasoning and Proof

Example 4 Construct Truth Tables

Construct a truth table for each compound statement.

a. $p \wedge \sim q$

Step 1 Make columns with the headings p, q, $\sim q$, and $p \wedge \sim q$.

Step 2 List the possible combinations of truth values for p and q.

Step 3 Use the truth values of q to determine the truth values of $\sim q$.

Step 4 Use the truth values for p and $\sim q$ to write the truth values for $p \wedge \sim q$.

Step 1 →

p	q	$\sim q$	$p \wedge \sim q$
T	T	F	F
T	F	T	T
F	T	F	F
F	F	T	F

Step 2 (columns p and q) Step 3 (column $\sim q$) Step 4 (column $p \wedge \sim q$)

b. $\sim p \vee \sim q$

p	q	$\sim p$	$\sim q$	$\sim p \vee \sim q$
T	T	F	F	F
T	F	F	T	T
F	T	T	F	T
F	F	T	T	T

Study Tip

Truth Tables Use the Fundamental Counting Principle to determine the number of rows necessary.

c. $(p \wedge q) \vee r$

Make columns for p, q, $p \wedge q$, r, and $(p \wedge q) \vee r$.

p	q	$p \wedge q$	r	$(p \wedge q) \vee r$
T	T	T	T	T
T	F	F	T	T
T	T	T	F	T
T	F	F	F	F
F	T	F	T	T
F	F	F	T	T
F	T	F	F	F
F	F	F	F	F

Check for Understanding

Concept Check

1–3. See margin.

1. **Describe** how to interpret the Venn diagram for $p \wedge q$.

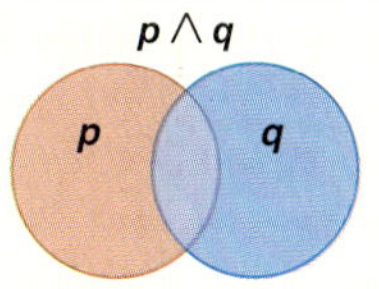

2. **OPEN ENDED** Write a compound statement for each condition.
   a. a true disjunction
   b. a false conjunction
   c. a true statement that includes a negation
3. **Explain** the difference between a conjunction and a disjunction.

In-Class Example

PowerPoint®

4 **Teaching Tip** Tell students that they could potentially interchange the columns for p, q, and r in 4c, and as long as they correctly fill in the columns for $p \wedge q$ and $(p \wedge q) \vee r$, the overall outcome would be the same. They would still end up with 5 Ts and 3 Fs, but the Ts and Fs would be in a different order.

Construct a truth table for each compound statement.

a. $\sim p \vee q$

p	q	$\sim p$	$\sim p \vee q$
T	T	F	T
T	F	F	F
F	T	T	T
F	F	T	T

b. $p \vee (\sim q \wedge r)$

p	q	r	$\sim q$	$\sim q \wedge r$	$p \vee (\sim q \wedge r)$
T	T	T	F	F	T
T	F	T	T	T	T
T	T	F	F	F	T
T	F	F	T	F	T
F	T	T	F	F	F
F	F	T	T	T	T
F	T	F	F	F	F
F	F	F	T	F	F

c. $(p \vee q) \wedge \sim r$

p	q	r	$\sim r$	$p \vee q$	$(p \vee q) \wedge \sim r$
T	T	T	F	T	F
T	F	T	F	T	F
T	T	F	T	T	T
T	F	F	T	T	T
F	T	T	F	T	F
F	F	T	F	F	F
F	T	F	T	T	T
F	F	F	T	F	F

DAILY INTERVENTION

Differentiated Instruction

Logical/Mathematical Have students examine the relationship between the number of simple statements (p, q, and r) and the number of rows necessary to exhaust all possible combinations in a truth table. Point out that for Example 4a, there are 2 statements and 4 rows; for Example 4c, there are 3 statements and 8 rows. Ask students to form a conjecture about how many rows would be needed for 4, 5, and n statements. Similarly, students can examine the relationship between the number of circles and the number of intersecting areas of a Venn diagram.

3 Practice/Apply

Study Notebook

Have students—

- add the definitions/examples of the vocabulary terms to their Vocabulary Builder worksheets for Chapter 2.
- include an example of a Venn diagram and an example of a truth table.
- include any other item(s) that they find helpful in mastering the skills in this lesson.

About the Exercises...

Organization by Objective

- **Determine Truth Values:** 18–29, 42–48
- **Truth Tables:** 30–41

Odd/Even Assignments

Exercises 18–40 are structured so that students practice the same concepts whether they are assigned odd or even problems.

Alert! Exercises 48–50 require the Internet or other research materials.

Assignment Guide

Basic: 19–37 odd, 41–51 odd, 52–73

Average: 19–51 odd, 52–73

Advanced: 18–50 even, 51–52, 54–69 (optional: 70–73)

Guided Practice

GUIDED PRACTICE KEY	
Exercises	Examples
4–6	1
7–9	2
10–14	3
15–17	4

Use the following statements to write a compound statement for each conjunction and disjunction. Then find its truth value. 4–9. See margin for statements.

p: $9 + 5 = 14$
q: February has 30 days.
r: A square has four sides.

4. p and q false
5. p and r true
6. $q \wedge r$ false
7. p or $\sim q$ true
8. $q \vee r$ true
9. $\sim p \vee \sim r$ false

10. Copy and complete the truth table.

p	q	$\sim q$	$p \wedge \sim q$
T	T	F	F
T	F	T	T
F	T	F	F
F	F	T	F

Construct a truth table for each compound statement. 11–14. See p. 123A.

11. $p \wedge q$
12. $q \vee r$
13. $\sim p \wedge r$
14. $(p \vee q) \vee r$

Application

AGRICULTURE For Exercises 15–17, refer to the Venn diagram that represents the states producing more than 100 million bushels of corn or wheat per year.

Grain Production

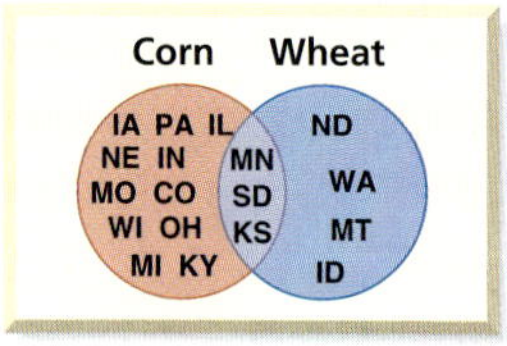

Source: U.S. Department of Agriculture

15. How many states produce more than 100 million bushels of corn? 14
16. How many states produce more than 100 million bushels of wheat? 7
17. How many states produce more than 100 million bushels of corn and wheat? 3

★ indicates increased difficulty

Practice and Apply

Homework Help	
For Exercises	See Examples
18–29	1, 2
30–41	4
42–48	3

Extra Practice See page 756.

Use the following statements to write a compound statement for each conjunction and disjunction. Then find its truth value. 18–29. See p. 123A for statements.

p: $\sqrt{-64} = 8$
q: An equilateral triangle has three congruent sides.
r: $0 < 0$
s: An obtuse angle measures greater than 90° and less than 180°.

18. p and q false
19. p or q true
20. p and r false
21. r and s false
22. q or r true
23. q and s true
24. $p \wedge s$ false
25. $q \wedge r$ false
26. $r \vee p$ false
27. $s \vee q$ true
28. $(p \wedge q) \vee s$ true
29. $s \vee (q$ and $r)$ true

Copy and complete each truth table.

30.

p	q	$\sim p$	$\sim p \vee q$
T	T	F	T
T	F	F	F
F	T	T	T
F	F	T	T

31.

p	q	$\sim p$	$\sim q$	$\sim p \wedge \sim q$
T	T	F	F	F
T	F	F	T	F
F	T	T	F	F
F	F	T	T	T

Answers

4. 9 + 5 = 14 and February has 30 days.
5. 9 + 5 = 14 and a square has four sides.
6. February has 30 days and a square has four sides.
7. 9 + 5 = 14 or February does not have 30 days.
8. February has 30 days or a square has four sides.
9. $9 + 5 \neq 14$ or a square does not have four sides.

45. Level of Participation Among 310 Students

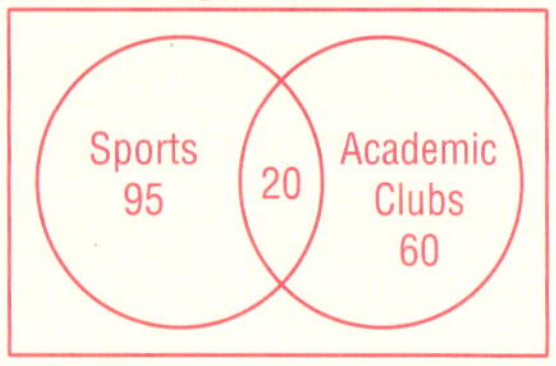

★ 32. Copy and complete the truth table.

p	q	r	$p \vee q$	$(p \vee q) \wedge r$
T	T	T	T	T
T	T	F	T	F
T	F	T	T	T
T	F	F	T	F
F	T	T	T	T
F	T	F	T	F
F	F	T	F	F
F	F	F	F	F

Construct a truth table for each compound statement. 33–40. See pp. 123A–123B.

33. q and r
34. p or q
35. p or r
36. p and q
37. $q \wedge \sim r$
38. $\sim p \wedge \sim q$
★ 39. $\sim p \vee (q \wedge \sim r)$
★ 40. $p \wedge (\sim q \vee \sim r)$

MUSIC For Exercises 41–44, use the following information.
A group of 400 teens were asked what type of music they listened to. They could choose among pop, rap, and country. The results are shown in the Venn diagram.

Music Preference

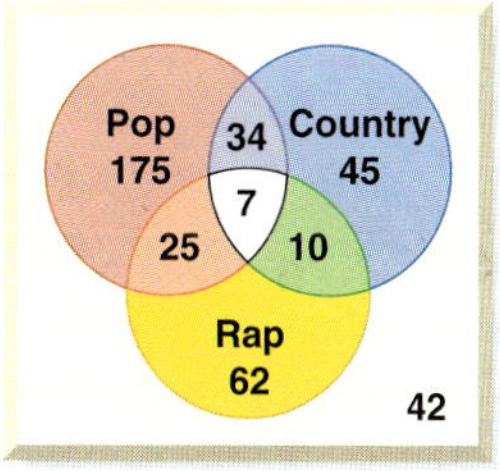

41. How many teens said that they listened to none of these types of music? 42
42. How many said that they listened to all three types of music? 7
43. How many said that they listened to only pop and rap music? 25
44. How many teens said that they listened to pop, rap, or country music? 358

SCHOOL For Exercises 45–47, use the following information.
In a school of 310 students, 80 participate in academic clubs, 115 participate in sports, and 20 students participate in both.

45. Make a Venn diagram of the data. See margin.
46. How many students participate in either clubs or sports? 175
47. How many students do not participate in either clubs or sports? 135

RESEARCH For Exercises 48–50, use the Internet or another resource to determine whether each statement about cities in New York is *true* or *false*.

48. Albany is not located on the Hudson river. false
49. Either Rochester or Syracuse is located on Lake Ontario. true
50. It is false that Buffalo is located on Lake Erie. false

CRITICAL THINKING For Exercises 51 and 52, use the following information.
All members of Team A also belong to Team B, but only some members of Team B also belong to Team C. Teams A and C have no members in common.

51. Draw a Venn diagram to illustrate the situation. See margin.
52. Which of the following statements is true? b
   a. If a person is a member of Team C, then the person is not a member of Team A.
   b. If a person is not a member of Team B, then the person is not a member of Team A.
   c. No person that is a member of Team A can be a member of Team C.

More About. . .

School

Nationwide, approximately 80% of high school seniors participate in extracurricular activities. Athletics, performing arts, and clubs are the most popular.

Source: *National Center for Education Statistics*

51.

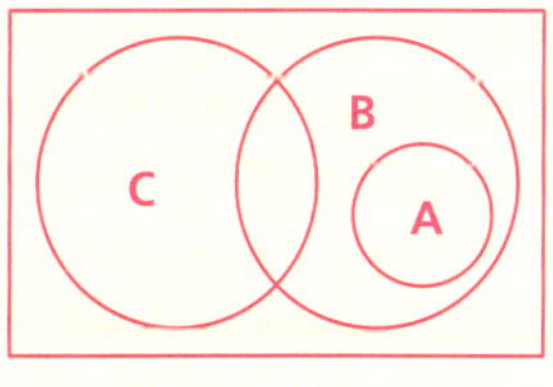

Enrichment, p. 68

Letter Puzzles

An **alphametic** is a computation puzzle using letters instead of digits. Each letter represents one of the digits 0–9, and two different letters cannot represent the same digit. Some alphametic puzzles have more than one answer.

Example **Solve the alphametic puzzle at the right.**

```
  FOUR
+  ONE
  FIVE
```

Since R + E = E, the value of R must be 0. Notice that the thousands digit must be the same in the first addend and the sum. Since the value of I is 9 or less, O must be 4 or less. Use trial and error to find values that work.

F = 8, O = 3, U = 1, R = 0

N = 4, E = 7, I = 6, and V = 5.

```
  8310
+  347
  8657
```

Can you find other solutions to this puzzle?

Find a value for each letter in each alphametic. Sample answers are shown

9703

Study Guide and Intervention, p. 63 (shown) and p. 64

Determine Truth Values A **statement** is any sentence that is either true or false. The truth or falsity of a statement is its **truth value**. A statement can be represented by using a letter. For example,

Statement p: Chicago is a city in Illinois. The truth value of statement p is true.

Several statements can be joined in a **compound statement**.

Statement p and statement q joined by the word *and* is a **conjunction**.	Statement p and statement q joined by the word *or* is a **disjunction**.	**Negation:** *not p* is the negation of the statement p.
Symbols: $p \wedge q$ (Read: *p and q*)	Symbols: $p \vee q$ (Read: *p or q*)	Symbols: $\sim p$ (Read: *not p*)
The conjunction $p \wedge q$ is true only when both p and q are true.	The disjunction $p \vee q$ is true if p is true, if q is true, or if both are true.	The statements p and $\sim p$ have opposite truth values.

Example 1 **Write a compound statement for each conjunction. Then find its truth value.**

p: An elephant is a mammal.
q: A square has four right angles.

a. $p \wedge q$
Join the statements with *and*: An elephant is a mammal and a square has four right angles. Both parts of the statement are true so the compound statement is true.

b. $\sim p \wedge q$
$\sim p$ is the statement "An elephant is not a mammal." Join $\sim p$ and q with the word *and*: An elephant is not a mammal and a square has four right angles. The first part of the compound statement, $\sim p$, is false. Therefore the compound statement is false.

Example 2 **Write a compound statement for each disjunction. Then find its truth value.**

p: A diameter of a circle is twice the radius.
q: A rectangle has four equal sides.

a. $p \vee q$
Join the statements p and q with the word *or*: A diameter of a circle is twice the radius or a rectangle has four equal sides. The first part of the compound statement, p, is true, so the compound statement is true.

b. $\sim p \vee q$
Join $\sim p$ and q with the word *or*: A diameter of a circle is not twice the radius or a rectangle has four equal sides. Neither part of the disjunction is true, so the compound statement is false.

Exercises

Write a compound statement for each conjunction and disjunction. Then find its truth value.

p: $10 + 8 = 18$ q: September has 30 days. r: A rectangle has four sides.

1. p and q 10 + 8 = 18 and September has 30 days; true.
2. p or r 10 + 8 = 18 or a rectangle has four sides; true.
3. q or r September has 30 days or a rectangle has four sides; true.
4. q and $\sim r$ September has 30 days and a rectangle does not have four sides; false.

Skills Practice, p. 65 and Practice, p. 66 (shown)

Use the following statements to write a compound statement for each conjunction and disjunction. Then find its truth value.

p: 60 seconds = 1 minute
q: Congruent supplementary angles each have a measure of 90.
r: $-12 + 11 < -1$

1. $p \wedge q$ 60 seconds = 1 minute and congruent supplementary angles each have a measure of 90; true.
2. $q \vee r$ Congruent supplementary angles each have a measure of 90 or $-12 + 11 < -1$; true.
3. $\sim p \vee q$ 60 seconds ≠ 1 minute or congruent supplementary angles each have a measure of 90; true.
4. $\sim p \wedge \sim r$ 60 seconds ≠ 1 minute and $-12 + 11 \geq -1$; false.

Copy and complete each truth table.

5.

p	q	$\sim p$	$\sim q$	$\sim p \vee \sim q$
T	T	F	F	F
T	F	F	T	T
F	T	T	F	T
F	F	T	T	T

6.

p	q	$\sim p$	$\sim p \vee q$	$p \wedge (\sim p \vee q)$
T	T	F	T	T
T	F	F	F	F
F	T	T	T	F
F	F	T	T	F

Construct a truth table for each compound statement.

7. $q \vee (p \wedge \sim q)$

p	q	$\sim q$	$p \wedge \sim q$	$q \vee (p \wedge \sim q)$
T	T	F	F	T
T	F	T	T	T
F	T	F	F	T
F	F	T	F	F

8. $\sim q \wedge (\sim p \vee q)$

p	q	$\sim p$	$\sim q$	$\sim p \vee q$	$\sim q \wedge (\sim p \vee q)$
T	T	F	F	T	F
T	F	F	T	F	F
F	T	T	F	T	F
F	F	T	T	T	T

SCHOOL For Exercises 9 and 10, use the following information.
The Venn diagram shows the number of students in the band who work after school or on the weekends.

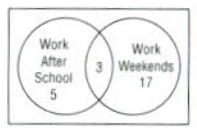

9. How many students work after school and on weekends? 3
10. How many students work after school or on weekends? 25

Reading to Learn Mathematics, p. 67 ELL

Pre-Activity **How does logic apply to school?**

Read the introduction to Lesson 2-2 at the top of page 67 in your textbook.

How can you use logic to help you answer a multiple-choice question on a standardized test if you are not sure of the correct answer? Sample answer: Eliminate the choices that you know are wrong. Then choose the one you think is most likely correct from the ones that are left.

Reading the Lesson

1. Supply one or two words to complete each sentence.
   a. Two or more statements can be joined to form a __compound__ statement.
   b. A statement that is formed by joining two statements with the word *or* is called a __disjunction__.
   c. The truth or falsity of a statement is called its __truth value__.
   d. A statement that is formed by joining two statements with the word *and* is called a __conjunction__.
   e. A statement that has the opposite truth value and the opposite meaning from a given statement is called the __negation__ of the statement.
2. Use *true* or *false* to complete each sentence.
   a. If a statement is true, then its negation is __false__.
   b. If a statement is false, then its negation is __true__.
   c. If two statements are both true, then their conjunction is __true__ and their disjunction is __true__.
   d. If two statements are both false, then their conjunction is __false__ and their disjunction is __false__.
   e. If one statement is true and another is false, then their conjunction is __false__ and their disjunction is __true__.
3. Consider the following statements:
   p: Chicago is the capital of Illinois. q: Sacramento is the capital of California.
   Write each statement symbolically and then find its truth value.
   a. Sacramento is not the capital of California. $\sim q$; false
   b. Sacramento is the capital of California and Chicago is not the capital of Illinois. $q \wedge \sim p$; true

Helping You Remember

4. Prefixes can often help you to remember the meaning of words or to distinguish between similar words. Use your dictionary to find the meanings of the prefixes *con* and *dis* and explain how these meanings can help you remember the difference between a *conjunction* and a *disjunction*. Sample answer: *Con* means *together* and *dis* means *apart*, so a *conjunction* is an *and* (or *both together*) statement and a *disjunction* is an *or* statement.

4 Assess

Open-Ended Assessment

Modeling Have students model a Venn diagram and a truth table with buttons or chips. For the Venn diagram, students can draw two large overlapping circles on a piece of paper and label them Science and English. Then they can place buttons on the diagram to represent the number of students in the class who like one, the other, or both subjects. Students can also draw a grid and use white buttons for *true* and black buttons for *false* to model one of the truth tables in the lesson.

Getting Ready for Lesson 2-3

Prerequisite Skill Students will learn about conditional statements in Lesson 2-3. They will substitute the hypothesis and conclusion for the *if* and *then* parts of statements. Use Exercises 70–73 to determine your students' familiarity with substituting numbers for variables in algebraic expressions.

Assessment Options

Quiz (Lessons 2-1 and 2-2) is available on p. 119 of the *Chapter 2 Resource Masters*.

Answer

53. Sample answer: Logic can be used to eliminate false choices on a multiple choice test. Answers should include the following.
   - Math is my favorite subject and drama club is my favorite activity.
   - See students' work.

53. **WRITING IN MATH** Answer the question that was posed at the beginning of the lesson. **See margin.**

How does logic apply to school?

Include the following in your answer:

- an example of a conjunction using statements about your favorite subject and your favorite extracurricular activity, and
- a Venn diagram showing various characteristics of the members of your geometry class (for example, male/female, grade in school, and so on).

FCAT Practice

Standardized Test Practice Ⓐ Ⓑ Ⓒ Ⓓ

54. Which statement about $\triangle ABC$ has the same truth value as $AB = BC$? **A**

Ⓐ $m\angle A = m\angle C$ Ⓑ $m\angle A = m\angle B$

Ⓒ $AC = BC$ Ⓓ $AB = AC$

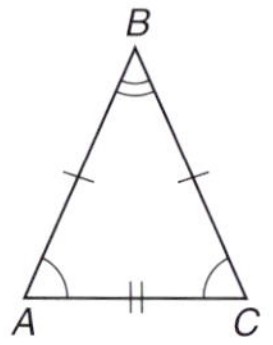

55. **ALGEBRA** If the sum of two consecutive even integers is 78, which number is the greater of the two integers? **C**

Ⓐ 36 Ⓑ 38

Ⓒ 40 Ⓓ 42

Maintain Your Skills

Mixed Review Make a conjecture about the next item in each sequence. *(Lesson 2-1)*

56. 3, 5, 7, 9 **11**
57. 1, 3, 9, 27 **81**
58. $6, 3, \frac{3}{2}, \frac{3}{4}$ $\mathbf{\frac{3}{8}}$
59. 17, 13, 9, 5 **1**
60. 64, 16, 4, 1 $\mathbf{\frac{1}{4}}$
61. 5, 15, 45, 135 **405**

COORDINATE GEOMETRY Find the perimeter of each polygon. Round answers to the nearest tenth. *(Lesson 1-6)*

63. 34.4
64. 30.4

62. triangle ABC with vertices $A(-6, 7)$, $B(1, 3)$, and $C(-2, -7)$ **33.1**
63. square $DEFG$ with vertices $D(-10, -9)$, $E(-5, -2)$, $F(2, -7)$, and $G(-3, -14)$
64. quadrilateral $HIJK$ with vertices $H(5, -10)$, $I(-8, -9)$, $J(-5, -5)$, and $K(-2, -4)$
65. hexagon $LMNPQR$ with vertices $L(2, 1)$, $M(4, 5)$, $N(6, 4)$, $P(7, -4)$, $Q(5, -8)$, and $R(3, -7)$ **29.5**

Measure each angle and classify it as *right*, *acute*, or *obtuse*. *(Lesson 1-4)*

66. $\angle ABC$ **145°, obtuse**
67. $\angle DBC$ **55°, acute**
68. $\angle ABD$ **90°, right**

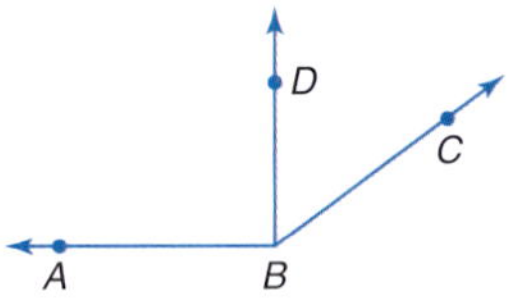

69. **FENCING** Michelle wanted to put a fence around her rectangular garden. The front and back measured 35 feet each, and the sides measured 75 feet each. If she wanted to make sure that she had enough feet of fencing, how much should she buy? *(Lesson 1-2)* **222 ft**

Getting Ready for the Next Lesson **PREREQUISITE SKILL** Evaluate each expression for the given values.
*(To review **evaluating algebraic expressions**, see page 736.)*

70. $5a - 2b$ if $a = 4$ and $b = 3$ **14**
71. $4cd + 2d$ if $c = 5$ and $d = 2$ **44**
72. $4e + 3f$ if $e = -1$ and $f = -2$ **−10**
73. $3g^2 + h$ if $g = 8$ and $h = -8$ **184**

Conditional Statements

What You'll Learn

- Analyze statements in if-then form.
- Write the converse, inverse, and contrapositive of if-then statements.

How are conditional statements used in advertisements?

Advertisers often lure consumers into purchasing expensive items by convincing them that they are getting something for free in addition to their purchase.

Vocabulary

- conditional statement
- if-then statement
- hypothesis
- conclusion
- related conditionals
- converse
- inverse
- contrapositive
- logically equivalent

IF-THEN STATEMENTS The statements above are examples of conditional statements. A **conditional statement** is a statement that can be written in *if-then form*. The first example above can be rewritten to illustrate this.

If you buy a car, *then* you get $1500 cash back.

Key Concept — *If-Then Statement*

- **Words** An **if-then statement** is written in the form *if p, then q*. The phrase immediately following the word *if* is called the **hypothesis**, and the phrase immediately following the word *then* is called the **conclusion**.
- **Symbols** $p \to q$, read *if p then q*, or *p implies q*.

Example 1 Identify Hypothesis and Conclusion

Identify the hypothesis and conclusion of each statement.

a. If points A, B, and C lie on line ℓ, then they are collinear.

If points A, B, and C lie on line ℓ, then they are collinear.
(hypothesis: "points A, B, and C lie on line ℓ"; conclusion: "they are collinear")

Hypothesis: points A, B, and C lie on line ℓ
Conclusion: they are collinear

b. The Tigers will play in the tournament if they win their next game.

Hypothesis: the Tigers win their next game
Conclusion: they will play in the tournament

Study Tip

Reading Math
The word *if* is not part of the hypothesis. The word *then* is not part of the conclusion.

Identifying the hypothesis and conclusion of a statement is helpful when writing statements in if-then form.

2-3 Lesson Notes

1 Focus

5-Minute Check Transparency 2-3 Use as a quiz or review of Lesson 2-2.

Mathematical Background notes are available for this lesson on p. 60C.

How are conditional statements used in advertisements?

Ask students:

- Use the advertisements to answer the following questions: What happens if you buy a new car? How can you get a free phone? **You get $1500 cash back; enroll in phone service for one year.**
- How effective are these types of advertisements? Why? **Very effective; people like to get something for free, even if they have to pay for something else.**

Resource Manager

Workbook and Reproducible Masters

Chapter 2 Resource Masters

- Study Guide and Intervention, pp. 69–70
- Skills Practice, p. 71
- Practice, p. 72
- Reading to Learn Mathematics, p. 73
- Enrichment, p. 74

Teaching Geometry With Manipulatives Masters, p. 43

Transparencies

5-Minute Check Transparency 2-3
Real-World Transparency 2
Answer Key Transparencies

Technology

GeomPASS: Tutorial Plus, Lesson 6
Interactive Chalkboard

IF-THEN STATEMENTS

In-Class Examples

Power Point®

1 Identify the hypothesis and conclusion of each statement.

a. If a polygon has 6 sides, then it is a hexagon. **Hypothesis: a polygon has 6 sides; Conclusion: it is a hexagon**

b. Tamika will advance to the next level of play if she completes the maze in her computer game. **Hypothesis: Tamika completes the maze in her computer game; Conclusion: she will advance to the next level of play.**

2 Identify the hypothesis and conclusion of each statement. Then write each statement in the if-then form.

a. Distance is positive. **Hypothesis: a distance is determined; Conclusion: it is positive; If a distance is determined, then it is positive.**

b. A five-sided polygon is a pentagon. **Hypothesis: a polygon has five sides; Conclusion: it is a pentagon; If a polygon has five sides, then it is a pentagon.**

Teaching Tip Tell students to use parentheses to identify the hypothesis and conclusion in each situation. Explain that if the hypothesis in the situation matches the hypothesis in the original statement, students can mark a T over the parentheses; if not, they can mark an F. They can do the same for the conclusions.

3 Determine the truth value of the following statement for each set of conditions. *If Yukon rests for 10 days, his ankle will heal.*

a. Yukon rests for 10 days, and he still has a hurt ankle. **false**

b. Yukon rests for 3 days, and he still has a hurt ankle. **true**

c. Yukon rests for 10 days, and he does not have a hurt ankle anymore. **true**

d. Yukon rests for 7 days, and he does not have a hurt ankle anymore. **true**

Study Tip

If-Then Statements

When you write a statement in if-then form, identify the condition that causes the result as the hypothesis. The result is the conclusion.

Example 2 Write a Conditional in If-Then Form

Identify the hypothesis and conclusion of each statement. Then write each statement in if-then form.

a. An angle with a measure greater than 90 is an obtuse angle.

Hypothesis: an angle has a measure greater than 90

Conclusion: it is an obtuse angle

If an angle has a measure greater than 90, then it is an obtuse angle.

b. Perpendicular lines intersect.

Sometimes you must add information to a statement. In this case, it is necessary to know that perpendicular lines come in pairs.

Hypothesis: two lines are perpendicular

Conclusion: they intersect

If two lines are perpendicular, then they intersect.

Recall that the truth value of a statement is either true or false. The hypothesis and conclusion of a conditional statement, as well as the conditional statement itself, can also be true or false.

Example 3 Truth Values of Conditionals

SCHOOL **Determine the truth value of the following statement for each set of conditions.**

If you get 100% on your test, then your teacher will give you an A.

a. You get 100%; your teacher gives you an A.

The hypothesis is true since you got 100%, and the conclusion is true because the teacher gave you an A. Since what the teacher promised is true, the conditional statement is true.

b. You get 100%; your teacher gives you a B.

The hypothesis is true, but the conclusion is false. Because the result is not what was promised, the conditional statement is false.

c. You get 98%; your teacher gives you an A.

The hypothesis is false, and the conclusion is true. The statement does not say what happens if you do not get 100% on the test. You could still get an A. It is also possible that you get a B. In this case, we cannot say that the statement is false. Thus, the statement is true.

d. You get 85%; your teacher gives you a B.

As in part c, we cannot say that the statement is false. Therefore, the conditional statement is true.

Study Tip

Common Misconception

A true hypothesis does not necessarily mean that a conditional is true. Likewise, a false conclusion does not guarantee that a conditional is false.

The resulting truth values in Example 3 can be used to create a truth table for conditional statements. Notice that a conditional statement is true in all cases except where the hypothesis is true and the conclusion is false.

p	q	$p \to q$
T	T	T
T	F	F
F	T	T
F	F	T

Teacher to Teacher

Nancy Lee Keen   Martinsville High School, Martinsville, IN

To develop the concept of conditional statements, I made posters of each of the four related conditionals. I wrote the hypotheses on yellow poster board, the conclusions on blue poster board, and NOT on red poster board. As we introduced each type of conditional, we placed the posters in the correct order.

CONVERSE, INVERSE, AND CONTRAPOSITIVE Other statements based on a given conditional statement are known as **related conditionals**.

Key Concept — Related Conditionals

Statement	Formed by	Symbols	Examples
Conditional	given hypothesis and conclusion	$p \to q$	If two angles have the same measure, then they are congruent.
Converse	exchanging the hypothesis and conclusion of the conditional	$q \to p$	If two angles are congruent, then they have the same measure.
Inverse	negating both the hypothesis and conclusion of the conditional	$\sim p \to \sim q$	If two angles do not have the same measure, then they are not congruent.
Contrapositive	negating both the hypothesis and conclusion of the converse statement	$\sim q \to \sim p$	If two angles are not congruent, then they do not have the same measure.

If a given conditional is true, the converse and inverse are not necessarily true. However, the contrapositive of a true conditional is always true, and the contrapositive of a false conditional is always false. Likewise, the converse and inverse of a conditional are either both true or both false.

Statements with the same truth values are said to be **logically equivalent**. So, a conditional and its contrapositive are logically equivalent as are the converse and inverse of a conditional. These relationships are summarized below.

p	q	Conditional $p \to q$	Converse $q \to p$	Inverse $\sim p \to \sim q$	Contrapositive $\sim q \to \sim p$
T	T	T	T	T	T
T	F	F	T	T	F
F	T	T	F	F	T
F	F	T	T	T	T

Study Tip

Contrapositive
The relationship of the truth values of a conditional and its contrapositive is known as the Law of Contrapositive.

Example 4 Related Conditionals

Write the converse, inverse, and contrapositive of the statement *Linear pairs of angles are supplementary*. Determine whether each statement is *true* or *false*. If a statement is false, give a counterexample.

First, write the conditional in if-then form.

Conditional: If two angles form a linear pair, then they are supplementary. The conditional statement is true.

Write the converse by switching the hypothesis and conclusion of the conditional.

Converse: If two angles are supplementary, then they form a linear pair. The converse is false. $\angle ABC$ and $\angle PQR$ are supplementary, but are not a linear pair.

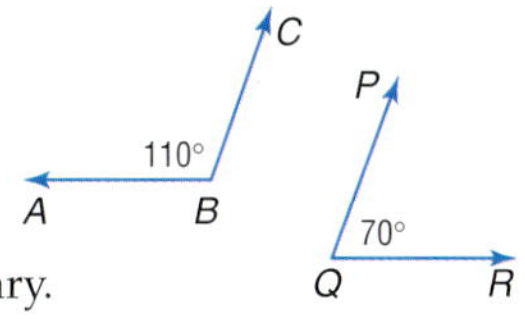

Inverse: If two angles do not form a linear pair, then they are not supplementary. The inverse is false. $\angle ABC$ and $\angle PQR$ do not form a linear pair, but they are supplementary.

The contrapositive is the negation of the hypothesis and conclusion of the converse.

Contrapositive: If two angles are not supplementary, then they do not form a linear pair. The contrapositive is true.

Study Tip

Contrapositive
To write the contrapositive of a conditional, it is helpful to first write the converse of the conditional.

CONVERSE, INVERSE, AND CONTRAPOSITIVE

In-Class Examples

Power Point®

4 Write the converse, inverse, and contrapositive of the statement *All squares are rectangles*. Determine whether each statement is *true* or *false*. If a statement is false, give a counterexample.
Conditional: If a shape is a square, then it is a rectangle. Converse: If a shape is a rectangle, then it is a square. False; a rectangle with $\ell = 2$ and $w = 4$ is not a square. Inverse: If a shape is not a square, then it is not a rectangle. False; a 4-sided polygon with side lengths 2, 2, 4, and 4 is not a square. Contrapositive: If a shape is not a rectangle, then it is not a square. true

Concept Check

In Lesson 2-2, p and q represented simple statements, not necessarily related to one another. In this lesson, they become the hypothesis and conclusion of a conditional statement. Make sure students know that separately, p and q are still simple statements, but they now have an interdependent relationship. Before moving on, students should feel very comfortable identifying the hypothesis and conclusion, determining the truth value of each one separately, and determining their combined truth value in various forms of conditional statements.

DAILY INTERVENTION

Differentiated Instruction

Kinesthetic Provide index cards for each student labeled "Hypothesis," "Conclusion," and "Implies" (or an arrow pointing to the right). Give each student two cards labeled "Not" in red ink. Ask students to use the cards to form a conditional, a converse, an inverse, and a contrapositive. Students should respond by placing the cards in the correct position and order to reflect the requests. Students can also use the cards to work some examples or exercises in this lesson by writing the parts of conditional statements on corresponding cards.

Lesson 2-3 Conditional Statements 77

3 Practice/Apply

Study Notebook

Have students—
- add the definitions/examples of the vocabulary terms to their Vocabulary Builder worksheets for Chapter 2.
- include a simplified version of the Related Conditionals chart and the truth table on page 77.
- include any other item(s) that they find helpful in mastering the skills in this lesson.

About the Exercises...

Organization by Objective
- **If-Then Statements:** 16–39
- **Converse, Inverse, and Contrapositive:** 40–45

Odd/Even Assignments

Exercises 16–45 are structured so that students practice the same concepts whether they are assigned odd or even problems.

Assignment Guide

Basic: 17–47 odd, 48–68
Average: 17–47 odd, 48–68
Advanced: 16–48 even, 50–65 (optional: 66–68)
All: Quiz 1 (1–5)

Check for Understanding

Concept Check

1–3. See margin.

1. **Explain** why writing a conditional statement in if-then form is helpful.
2. **OPEN ENDED** Write an example of a conditional statement.
3. **Compare and contrast** the inverse and contrapositive of a conditional.

Guided Practice

GUIDED PRACTICE KEY	
Exercises	Examples
4–6	1
7–9, 15	2
10–12	3
13, 14	4

Identify the hypothesis and conclusion of each statement. 4–6. See margin.

4. If it rains on Monday, then I will stay home.
5. If $x - 3 = 7$, then $x = 10$.
6. If a polygon has six sides, then it is a hexagon.

Write each statement in if-then form.

7. A 32-ounce pitcher holds a quart of liquid. *7. If a pitcher is a 32-ounce pitcher, then it holds a quart of liquid.*
8. The sum of the measures of supplementary angles is 180. *8. If two angles are supplementary, then the sum of the measures of the angles is 180.*
9. An angle formed by perpendicular lines is a right angle. *9. If an angle is formed by perpendicular lines, then it is a right angle.*

Determine the truth value of the following statement for each set of conditions.

If you drive faster than 65 miles per hour on the interstate, then you will receive a speeding ticket.

10. You drive 70 miles per hour, and you receive a speeding ticket. true
11. You drive 62 miles per hour, and you do not receive a speeding ticket. true
12. You drive 68 miles per hour, and you do not receive a speeding ticket. false

Write the converse, inverse, and contrapositive of each conditional statement. Determine whether each related conditional is *true* or *false*. If a statement is false, find a counterexample. 13–14. See margin.

13. If plants have water, then they will grow. True
14. Flying in an airplane is safer than riding in a car.

Application

15. **FORESTRY** In different regions of the country, different variations of trees dominate the landscape. In Colorado, aspen trees cover high areas of the mountains. In Florida, cypress trees rise from swamps. In Vermont, maple trees are prevalent. Write these conditionals in if-then form. See p. 123B.

Practice and Apply

Homework Help	
For Exercises	See Examples
16–21	1
22–27	2
28–39	3
40–45	4

Extra Practice
See page 756.

Identify the hypothesis and conclusion of each statement. 16–21. See p. 123B.

16. If $2x + 6 = 10$, then $x = 2$.
17. If you are a teenager, then you are at least 13 years old.
18. If you have a driver's license, then you are at least 16 years old.
19. If three points lie on a line, then they are collinear.
20. "If a man hasn't discovered something that he will die for, he isn't fit to live." *(Martin Luther King, Jr., 1963)*
21. If the measure of an angle is between 0 and 90, then the angle is acute.

Write each statement in if-then form. 22–27. See p. 123B.

22. Get a free visit with a one-year fitness plan.
23. Math teachers love to solve problems.
24. "I think, therefore I am." (*Descartes*)
25. Adjacent angles have a common side.
26. Vertical angles are congruent.
27. Equiangular triangles are equilateral.

Answers

1. Writing a conditional in if-then form is helpful so that the hypothesis and conclusion are easily recognizable.
2. Sample answer: If you eat your peas, then you will have dessert.
3. In the inverse, you negate both the hypothesis and the conclusion of the conditional. In the contrapositive, you negate the hypothesis and the conclusion of the converse.
4. H: it rains on Monday; C: I will stay home
5. H: $x - 3 = 7$; C: $x = 10$
6. H: a polygon has six sides; C: it is a hexagon
13. Converse: If plants grow, then they have water; true. Inverse: If plants do not have water, then they will not grow; true. Contrapositive: If plants do not grow, then they do not have water. False; they may have been killed by overwatering.
14. Converse: If you are safer than riding in a car, then you are flying in an airplane. False; there are other places that are safer than riding in a car. Inverse: If you are not flying in an airplane, then you are not safer than riding in a car. False; there are other places that are safer than riding in a car. Contrapositive: If you are not safer than riding in a car, then you are not flying in an airplane; true.

Determine the truth value of the following statement for each set of conditions.

If you are over 18 years old, then you vote in all elections.

28. You are 19 years old and you vote. true
29. You are 16 years old and you vote. true
30. You are 21 years old and do not vote. false
31. You are 17 years old and do not vote. true
32. Your sister is 21 years old and votes. true
33. Your dad is 45 years old and does not vote. false

In the figure, P, Q, and R are collinear, P and A lie in plane $\mathcal{M}$, and Q and B lie in plane $\mathcal{N}$. Determine the truth value of each statement.

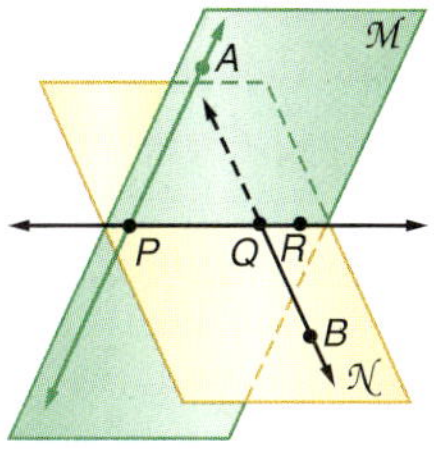

34. P, Q, and R lie in plane $\mathcal{M}$. true
35. $\overleftrightarrow{QB}$ lies in plane $\mathcal{N}$. true
36. Q lies in plane $\mathcal{M}$. true
37. P, Q, A, and B are coplanar. false
38. $\overleftrightarrow{AP}$ contains Q. false
39. Planes $\mathcal{M}$ and $\mathcal{N}$ intersect at $\overleftrightarrow{RQ}$. true

Write the converse, inverse, and contrapositive of each conditional statement. Determine whether each related conditional is *true* or *false*. If a statement is false, find a counterexample. 40–45. See p. 123B.

40. If you live in Dallas, then you live in Texas.
41. If you exercise regularly, then you are in good shape.
42. The sum of two complementary angles is 90.
43. All rectangles are quadrilaterals.
44. All right angles measure 90.
45. Acute angles have measures less than 90.

46. Sample answer: In Alaska, if it is summer, then there are more hours of daylight than darkness. In Alaska, if it is winter, then there are more hours of darkness than daylight.

More About. . .

Seasons

At the poles, sunlight may shine continuously for six months during spring and summer, but never rises more than 23.5° above the horizon. During the other six months of the year, the poles are in darkness.

Source: *U.S. Geological Survey*

SEASONS **For Exercises 46 and 47, use the following information.**
Due to the movement of Earth around the sun, summer days in Alaska have more hours of daylight than darkness, and winter days have more hours of darkness than daylight.

46. Write two true conditional statements in if-then form for summer days and winter days in Alaska.
47. Write the converse of the two true conditional statements. State whether each is *true* or *false*. If a statement is false, find a counterexample. See p. 123B.
48. **CRITICAL THINKING** Write a false conditional statement. Is it possible to insert the word *not* into your conditional to make it true? If so, write the true conditional.
49. **WRITING IN MATH** Answer the question that was posed at the beginning of the lesson. See margin.

How are conditional statements used in advertisements?

Include the following in your answer:
- an example of a conditional statement in if-then form, and
- an example of a conditional statement that is not in if-then form.

www.geometryonline.com/self_check_quiz

Answer

49. Conditional statements can be used to describe how to get a discount, rebate, or refund. Sample answers should include the following.
- If you are not 100% satisfied, then return the product for a full refund.
- Wearing a seatbelt reduces the risk of injuries.

Enrichment, p. 74

Venn Diagrams

A type of drawing called a **Venn diagram** can be useful in explaining conditional statements. A Venn diagram uses circles to represent sets of objects.

Consider the statement "All rabbits have long ears." To make a Venn diagram for this statement, a large circle is drawn to represent all animals with long ears. Then a smaller circle is drawn inside the first to represent all rabbits. The Venn diagram shows that every rabbit is included in the group of long-eared animals.

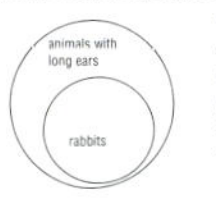

The set of rabbits is called a **subset** of the set of long-eared animals.

The Venn diagram can also explain how to write the statement, "All rabbits have long ears," in if-then form. Every rabbit is in the group of long-eared animals, so if an animal is a rabbit, then it has long ears.

For each statement, draw a Venn diagram. Then write the sentence in if-then form.

Study Guide and Intervention, p. 69 (shown) and p. 70

If-then Statements An if-then statement is a statement such as "If you are reading this page, then you are studying math." A statement that can be written in if-then form is called a **conditional statement**. The phrase immediately following the word *if* is the **hypothesis**. The phrase immediately following the word *then* is the **conclusion**.

A conditional statement can be represented in symbols as $p \rightarrow q$, which is read "p implies q" or "if p, then q."

Example 1 **Identify the hypothesis and conclusion of the statement.**

If $\angle X \cong \angle R$ and $\angle R \cong \angle S$ (hypothesis), then $\angle X \cong \angle S$ (conclusion).

Example 2 **Identify the hypothesis and conclusion. Write the statement in if-then form.**

You receive a free pizza with 12 coupons.

If you have 12 coupons (hypothesis), then you receive a free pizza (conclusion).

Exercises

Identify the hypothesis and conclusion of each statement.

1. If it is Saturday, then there is no school. H: it is Saturday; C: there is no school
2. If $x - 8 = 32$, then $x = 40$. H: $x - 8 = 32$; C: $x = 40$
3. If a polygon has four right angles, then the polygon is a rectangle. H: a polygon has four right angles; C: the polygon is a rectangle

Write each statement in if-then form.

4. All apes love bananas. If an animal is an ape, then it loves bananas.
5. The sum of the measures of complementary angles is 90. If two angles are complementary, then the sum of their measures is 90.
6. Collinear points lie on the same line. If points are collinear, then they lie on the same line.

Determine the truth value of the following statement for each set of conditions.
If it does not rain this Saturday, we will have a picnic.

7. It rains this Saturday, and we have a picnic. true
8. It rains this Saturday, and we don't have a picnic. true
9. It doesn't rain this Saturday, and we have a picnic. true
10. It doesn't rain this Saturday, and we don't have a picnic. false

Skills Practice, p. 71 and Practice, p. 72 (shown)

Identify the hypothesis and conclusion of each statement.

1. If $3x + 4 = -5$, then $x = -3$. H: $3x + 4 = -5$; C: $x = -3$
2. If you take a class in television broadcasting, then you will film a sporting event. H: you take a class in television broadcasting; C: you will film a sporting event

Write each statement in if-then form.

3. "Those who do not remember the past are condemned to repeat it." (*George Santayana*) If you do not remember the past, then you are condemned to repeat it.
4. Adjacent angles share a common vertex and a common side. If two angles are adjacent, then they share a common vertex and a common side.

Determine the truth value of the following statement for each set of conditions.
If DVD players are on sale for less than $100, then you buy one.

5. DVD players are on sale for $95 and you buy one. true
6. DVD players are on sale for $100 and you do not buy one. true
7. DVD players are not on sale for under $100 and you do not buy one. true
8. Write the converse, inverse, and contrapositive of the conditional statement. Determine whether each statement is true or false. If a statement is false, find a counterexample.
If $(-8)^2 > 0$, then $-8 > 0$.
Converse: If $-8 > 0$, then $(-8)^2 > 0$; true.
Inverse: If $(-8)^2 \leq 0$, then $-8 \leq 0$; true.
Contrapositive: If $-8 \leq 0$, then $(-8)^2 \leq 0$; false.

SUMMER CAMP **For Exercises 9 and 10, use the following information.**
Older campers who attend Woodland Falls Camp are expected to work. Campers who are juniors wait on tables.

9. Write a conditional statement in if-then form. Sample answer: If you are a junior, then you wait on tables.
10. Write the converse of your conditional statement. If you wait on tables, then you are a junior.

Reading to Learn Mathematics, p. 73 ELL

Pre-Activity **How are conditional statements used in advertisements?**

Read the introduction to Lesson 2-3 at the top of page 75 in your textbook.

Does the second advertising statement in the introduction mean that you will not get a free phone if you sign a contract for only six months of service? Explain your answer. No; it only tells you what happens if you sign up for one year.

Reading the Lesson

1. Identify the hypothesis and conclusion of each statement.
   a. If you are a registered voter, then you are at least 18 years old. Hypothesis: you are a registered voter; Conclusion: you are at least 18 years old
   b. If two integers are even, their product is even. Hypothesis: two integers are even; Conclusion: their product is even
2. Complete each sentence.
   a. The statement that is formed by replacing both the hypothesis and the conclusion of a conditional with their negations is the ___inverse___.
   b. The statement that is formed by exchanging the hypothesis and conclusion of a conditional is the ___converse___.
3. Consider the following statement:
   You live in North America if you live in the United States.
   a. Write this conditional statement in if-then form and give its truth value. If the statement is false, give a counterexample. If you live in the United States, then you live in North America; false: You live in Hawaii.
   b. Write the inverse of the given conditional statement in if-then form and give its truth value. If the statement is false, give a counterexample. If you do not live in the United States, then you do not live in North America; false; sample answer: You live in Mexico.
   c. Write the contrapositive of the given conditional statement in if-then form and give its truth value. If the statement is false, give a counterexample. If you do not live in North America, then you do not live in the United States; false: You live in Hawaii.
   d. Write the converse of the given conditional statement in if-then form and give its truth value. If the statement is false, give a counterexample. If you live in North America, then you live in the United States; false; sample answer: You live in Canada.

Helping You Remember

4. When working with a conditional statement and its three related conditionals, what is an easy way to remember which statements are logically equivalent to each other? Sample answer: The two statements whose names contain *verse* (the converse and the inverse) are a logically equivalent pair. The other two (the original conditional and the contrapositive) are the other logically equivalent pair.

4 Assess

Open-Ended Assessment

Speaking Students can practice their speaking skills by identifying parts of statements and translating statements into the converse, inverse, and contrapositive aloud.

Getting Ready for Lesson 2-4

Prerequisite Skill Students will learn about deductive reasoning in Lesson 2-4. They will apply concepts of solving equations to deductive-reasoning techniques. Use Exercises 66–68 to determine your students' familiarity with solving equations.

Assessment Options

Practice Quiz 1 The quiz provides students with a brief review of the concepts and skills in Lessons 2-1 through 2-3. Lesson numbers are given to the right of the exercises or instruction lines so students can review concepts not yet mastered.

Answers

52. George Washington was the first president of the United States and a hexagon has 5 sides.
53. A hexagon has five sides or $60 \times 3 = 18$.
54. George Washington was the first president of the United States or a hexagon has five sides.
55. A hexagon doesn't have five sides or $60 \times 3 = 18$.
56. George Washington was the first president of the United States and a hexagon doesn't have five sides.
57. George Washington was not the first president of the United States and $60 \times 3 \neq 18$.

58.

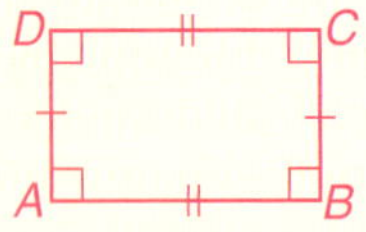

FCAT Practice
Standardized Test Practice

50. Which statement has the same truth value as the following statement? **C**
If Ava and Willow are classmates, then they go to the same school.
   - (A) If Ava and Willow go to the same school, then they are classmates.
   - (B) If Ava and Willow are not classmates, then they do not go to the same school.
   - (C) If Ava and Willow do not go to the same school, then they are not classmates.
   - (D) If Ava and Willow go to the same school, then they are not classmates.

51. **ALGEBRA** In a history class with 32 students, the ratio of girls to boys is 5 to 3. How many more girls are there than boys? **B**
   - (A) 2 (B) 8 (C) 12 (D) 20

Maintain Your Skills

Mixed Review

Use the following statements to write a compound statement for each conjunction and disjunction. Then find its truth value. *(Lesson 2-2)* **52–57. See margin.**

p: George Washington was the first president of the United States.
q: A hexagon has five sides.
r: $60 \times 3 = 18$

52. $p \wedge q$ **false**
53. $q \vee r$ **false**
54. $p \vee q$ **true**
55. $\sim q \vee r$ **true**
56. $p \wedge \sim q$ **true**
57. $\sim p \wedge \sim r$ **false**

Make a conjecture based on the given information. Draw a figure to illustrate your conjecture. *(Lesson 2-1)* **58–61. See margin for sample figures.**

58. $ABCD$ is a rectangle. **58. $AB = CD$; $AD = BC$**
59. In $\triangle FGH$, $m\angle F = 45$, $m\angle G = 67$, $m\angle H = 68$. **59. The sum of the measures of the angles in a triangle is 180.**
60. $J(-3, 2)$, $K(1, 8)$, $L(5, 2)$ **60. $\triangle JKL$ has two sides congruent.**
61. In $\triangle PQR$, $m\angle PQR = 90$ **$\angle PQR$ is a right angle.**

Use the Distance Formula to find the distance between each pair of points. *(Lesson 1-3)*

62. $C(-2, -1)$, $D(0, 3)$ $\sqrt{20} \approx 4.5$
63. $J(-3, 5)$, $K(1, 0)$ $\sqrt{41} \approx 6.4$
64. $P(-3, -1)$, $Q(2, -3)$ $\sqrt{29} \approx 5.4$
65. $R(1, -7)$, $S(-4, 3)$ $\sqrt{125} \approx 11.2$

Getting Ready for the Next Lesson

PREREQUISITE SKILL **Identify the operation used to change Equation (1) to Equation (2).** *(To review **solving equations**, see pages 737 and 738.)* **66–68. See margin.**

66. (1) $3x + 4 = 5x - 8$
    (2) $3x = 5x - 12$
67. (1) $\frac{1}{2}(a - 5) = 12$
    (2) $a - 5 = 24$
68. (1) $8p = 24$
    (2) $p = 3$

Practice Quiz 1 — *Lessons 2-1 through 2-3*

Determine whether each conjecture is *true* or *false*. Give a counterexample for any false conjecture. *(Lesson 2-1)*

1. **Given:** $WX = XY$
   Conjecture: W, X, and Y are collinear.
   False; see p. 123B for counterexample.
2. **Given:** $\angle 1$ and $\angle 2$ are complementary. $\angle 2$ and $\angle 3$ are complementary.
   Conjecture: $m\angle 1 = m\angle 3$ **true**

Construct a truth table for each compound statement. *(Lesson 2-2)* **3–4. See p. 123B.**

3. $\sim p \wedge q$
4. $p \vee (q \wedge r)$
5. Write the converse, inverse, and contrapositive of the following conditional statement. Determine whether each related conditional is *true* or *false*. If a statement is false, find a counterexample. *(Lesson 2-3)*
If two angles are adjacent, then the angles have a common vertex. **See p. 123C.**

59.

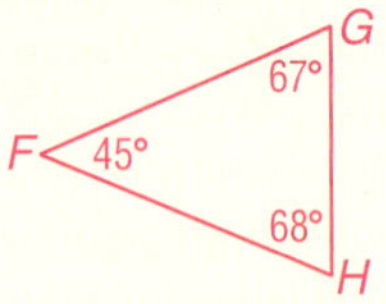

60.

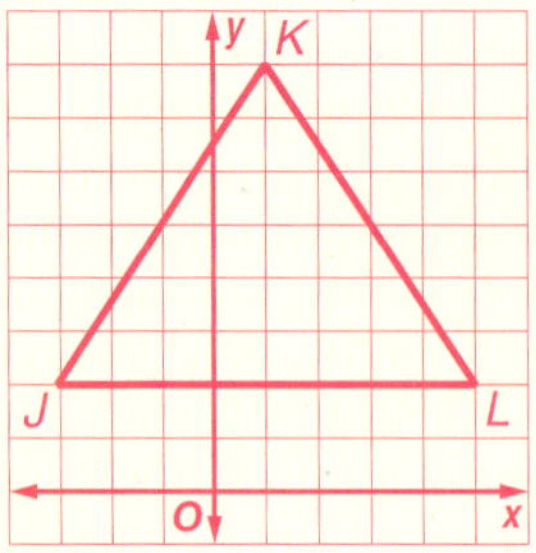

61.

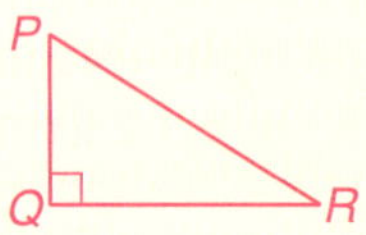

66. Subtract 4 from each side.
67. Multiply each side by 2.
68. Divide each side by 8.

Biconditional Statements

Ashley began a new summer job, earning $10 an hour. If she works over 40 hours a week, she earns time and a half, or $15 an hour. If she earns $15 an hour, she has worked over 40 hours a week.

p: Ashley earns $15 an hour
q: Ashley works over 40 hours a week

$p \rightarrow q$: If Ashley earns $15 an hour, she has worked over 40 hours a week.
$q \rightarrow p$: If Ashley works over 40 hours a week, she earns $15 an hour.

In this case, both the conditional and its converse are true. The conjunction of the two statements is called a **biconditional**.

Key Concept — *Biconditional Statement*

- **Words** A biconditional statement is the conjunction of a conditional and its converse.
- **Symbols** $(p \rightarrow q) \wedge (q \rightarrow p)$ is written $(p \leftrightarrow q)$ and read p *if and only if* q

If and only if can be abbreviated *iff*.

So, the biconditional statement is as follows.

$p \leftrightarrow q$: Ashley earns $15 an hour *if and only if* she works over 40 hours a week.

Examples

Write each biconditional as a conditional and its converse. Then determine whether the biconditional is *true* or *false*. If false, give a counterexample.

a. Two angle measures are complements if and only if their sum is 90.

Conditional: If two angle measures are complements, then their sum is 90.
Converse: If the sum of two angle measures is 90, then they are complements.
Both the conditional and the converse are true, so the biconditional is true.

b. $x > 9$ **iff** $x > 0$

Conditional: If $x > 9$, then $x > 0$.
Converse: If $x > 0$, then $x > 9$.
The conditional is true, but the converse is not. Let $x = 2$. Then $2 > 0$ but $2 \not> 9$. So, the biconditional is false.

Reading to Learn 1–5. See margin.

Write each biconditional as a conditional and its converse. Then determine whether the biconditional is *true* or *false*. If false, give a counterexample.

1. A calculator will run if and only if it has batteries.
2. Two lines intersect if and only if they are not vertical.
3. Two angles are congruent if and only if they have the same measure.
4. $3x - 4 = 20$ iff $x = 7$.
5. A line is a segment bisector if and only if it intersects the segment at its midpoint.

Reading Mathematics

Getting Started

Explain that true biconditional statements are extremely helpful for writing proofs because they can be used forwards or backwards. Tell students, however, that they are not as common as regular conditionals, and students will need to be wary and thorough when they are determining whether a biconditional is true or false.

Teach

Biconditional Statements
Students can also get some extra practice writing the inverse and contrapositive of each biconditional statement.

Assess

Study Notebook

Ask students to summarize what they have learned about biconditional statements.

ELL English Language Learners may benefit from writing key concepts from this activity in their Study Notebooks in their native language and then in English.

Answers

1. Conditional: If a calculator runs, then it has batteries. Converse: If a calculator has batteries, then it will run. False; a calculator may be solar powered.
2. Conditional: If two lines intersect, then they are not vertical. Converse: If two lines are not vertical, then they intersect. False; two parallel horizontal lines will not intersect.
3. Conditional: If two angles are congruent, then they have the same measure. Converse: If two angles have the same measure, then they are congruent. true
4. Conditional: If $3x - 4 = 20$, then $x = 7$. Converse: If $x = 7$, then $3x - 4 = 20$. False; $3x - 4 = 17$ when $x = 7$.
5. Conditional: If a line is a segment bisector, then it intersects the segment at its midpoint. Converse: If a line intersects a segment at its midpoint, then it is a segment bisector. true

2-4 Lesson Notes

1 Focus

5-Minute Check Transparency 2-4 Use as a quiz or review of Lesson 2-3.

Mathematical Background notes are available for this lesson on p. 60D.

How does deductive reasoning apply to health?

Ask students:

- If you have a mass of 57.8 kg, what dose will a doctor give you based on the chart? **350 mg**
- What might happen if a patient used *inductive* reasoning to form a conjecture about the dose of an antidepressant based on the chart above? Is this a safe method for health situations? **Sample answer: The dose of the antidepressant might be much less than that of the antibiotic for the same weight, so the patient could potentially overdose on the antidepressant; no.**

2-4 Deductive Reasoning

What You'll Learn

- Use the Law of Detachment.
- Use the Law of Syllogism.

How does deductive reasoning apply to health?

When you are ill, your doctor may prescribe an antibiotic to help you get better. Doctors may use a dose chart like the one shown to determine the correct amount of medicine you should take.

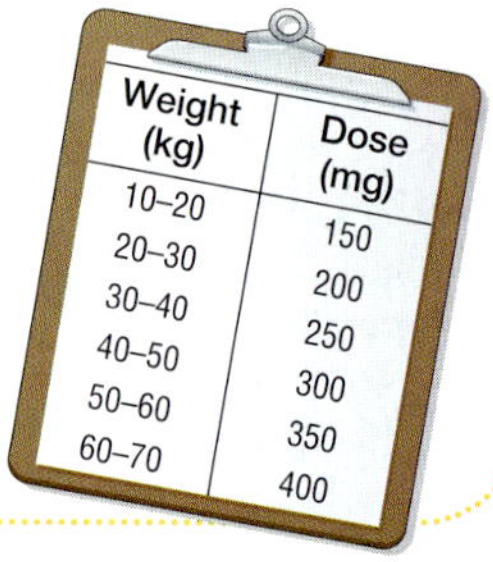

Vocabulary

- deductive reasoning
- Law of Detachment
- Law of Syllogism

LAW OF DETACHMENT The process that doctors use to determine the amount of medicine a patient should take is called **deductive reasoning**. Unlike inductive reasoning, which uses examples to make a conjecture, deductive reasoning uses facts, rules, definitions, or properties to reach logical conclusions.

A form of deductive reasoning that is used to draw conclusions from true conditional statements is called the **Law of Detachment**.

Key Concept — *Law of Detachment*

- **Words** If $p \rightarrow q$ is true and p is true, then q is also true.
- **Symbols** $[(p \rightarrow q) \wedge p] \rightarrow q$

Study Tip

Validity
When you apply the Law of Detachment, make sure that the conditional is true before you test the validity of the conclusion.

Example 1 *Determine Valid Conclusions*

The following is a true conditional. Determine whether each conclusion is valid based on the given information. Explain your reasoning.

If a ray is an angle bisector, then it divides the angle into two congruent angles.

a. Given: $\overrightarrow{BD}$ **bisects** $\angle ABC$.

Conclusion: $\angle ABD \cong \angle CBD$

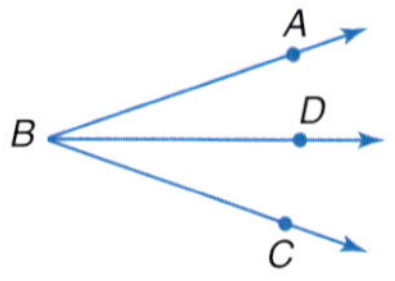

The hypothesis states that $\overrightarrow{BD}$ is the bisector of $\angle ABC$. Since the conditional is true and the hypothesis is true, the conclusion is valid.

b. Given: $\angle PQT \cong \angle RQS$

Conclusion: $\overrightarrow{QS}$ **and** $\overrightarrow{QT}$ **are angle bisectors.**

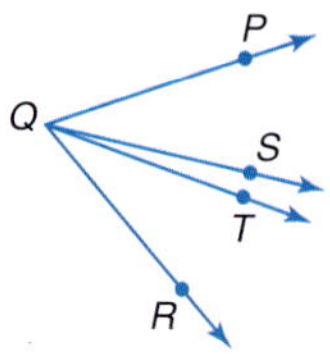

Knowing that a conditional statement and its conclusion are true does not make the hypothesis true. An angle bisector divides an angle into two separate congruent angles. In this case, the given angles are not separated by one ray. Instead, they overlap. The conclusion is not valid.

Resource Manager

Workbook and Reproducible Masters

Chapter 2 Resource Masters
- Study Guide and Intervention, pp. 75–76
- Skills Practice, p. 77
- Practice, p. 78
- Reading to Learn Mathematics, p. 79
- Enrichment, p. 80
- Assessment, pp. 119, 121

School-to-Career Masters, p. 3
Teaching Geometry With Manipulatives Masters, p. 47

Transparencies

5-Minute Check Transparency 2-4
Answer Key Transparencies

Technology

Interactive Chalkboard

LAW OF SYLLOGISM Another law of logic is the **Law of Syllogism**. It is similar to the Transitive Property of Equality.

Key Concept *Law of Syllogism*

- **Words** If $p \rightarrow q$ and $q \rightarrow r$ are true, then $p \rightarrow r$ is also true.
- **Symbols** $[(p \rightarrow q) \wedge (q \rightarrow r)] \rightarrow (p \rightarrow r)$

Study Tip

Conditional Statements

Label the hypotheses and conclusions of a series of statements before applying the Law of Syllogism.

Example 2 Determine Valid Conclusions From Two Conditionals

CHEMISTRY Use the Law of Syllogism to determine whether a valid conclusion can be reached from each set of statements.

a. (1) If the symbol of a substance is Pb, then it is lead.

(2) The atomic number of lead is 82.

Let p, q, and r represent the parts of the statement.

p: the symbol of a substance is Pb

q: it is lead

r: the atomic number is 82

Statement (1): $p \rightarrow q$

Statement (2): $q \rightarrow r$

Since the given statements are true, use the Law of Syllogism to conclude $p \rightarrow r$. That is, *If the symbol of a substance is Pb, then its atomic number is 82.*

b. (1) Water can be represented by H_2O.

(2) Hydrogen (H) and oxygen (O) are in the atmosphere.

There is no valid conclusion. While both statements are true, the conclusion of each statement is not used as the hypothesis of the other.

Example 3 Analyze Conclusions

Determine whether statement (3) follows from statements (1) and (2) by the Law of Detachment or the Law of Syllogism. If it does, state which law was used. If it does not, write *invalid*.

a. (1) Vertical angles are congruent.

(2) If two angles are congruent, then their measures are equal.

(3) If two angles are vertical, then their measures are equal.

p: two angles are vertical

q: they are congruent

r: their measures are equal

Statement (3) is a valid conclusion by the Law of Syllogism.

b. (1) If a figure is a square, then it is a polygon.

(2) Figure A is a polygon.

(3) Figure A is a square.

Statement (1) is true, but statement (3) does not follow from statement (2). Not all polygons are squares.

Statement (3) is invalid.

www.geometryonline.com/extra_examples/fcat

2 Teach

LAW OF DETACHMENT

In-Class Example Power Point®

1 The following is a true conditional. Determine whether each conclusion is valid based on the given information. Explain your reasoning.

If two segments are congruent and the second segment is congruent to a third segment, then the first segment is also congruent to the third segment.

a. Given: $\overline{WX} \cong \overline{UV}$; $\overline{UV} \cong \overline{RT}$
Conclusion: $\overline{WX} \cong \overline{RT}$ **true**

b. Given: $\overline{UV}$; $\overline{WX} \cong \overline{RT}$
Conclusion: $\overline{WX} \cong \overline{UV}$ and $\overline{UV} \cong \overline{RT}$ **false**

LAW OF SYLLOGISM

In-Class Example Power Point®

2 **PROM** Use the Law of Syllogism to determine whether a valid conclusion can be reached from each set of statements.

a. (1) If Salline attends the prom, she will go with Mark.
(2) Mark is a 17-year-old student. **not valid**

b. (1) If Mel and his date eat at the Peddler Steakhouse before going to the prom, they will miss the senior march.
(2) The Peddler Steakhouse stays open until 10 P.M.
not valid

In-Class Example 3 is on p. 84.

DAILY INTERVENTION **Differentiated Instruction** **ELL**

Verbal/Linguistic Have students write a paragraph to explain and provide an example for the Law of Detachment. Repeat for the Law of Syllogism. Then students can write another paragraph to point out similarities and differences between the two laws. They can place their written explanations in their study notebooks.

In-Class Example

3 Determine whether statement (3) follows from statements (1) and (2) by the Law of Detachment or the Law of Syllogism. If it does, state which law was used. If it does not, write *invalid*.

a. (1) If the sum of the squares of two sides of a triangle is equal to the square of the third side, then the triangle is a right triangle. (2) For $\triangle XYZ$, $(XY)^2 + (YZ)^2 = (ZX)^2$. (3) $\triangle XYZ$ is a right triangle. **Law of Detachment**

b. (1) If Ling wants to participate in the wrestling competition, he will have to meet an extra three times a week to practice. (2) If Ling adds anything extra to his weekly schedule, he cannot take karate lessons. (3) If Ling wants to participate in the wrestling competition, he cannot take karate lessons. **Law of Syllogism**

3 Practice/Apply

Study Notebook

Have students—
- add the definitions/examples of the vocabulary terms to their Vocabulary Builder worksheets for Chapter 2.
- include any other item(s) that they find helpful in mastering the skills in this lesson.

DAILY INTERVENTION

FIND THE ERROR Explain that if you have $p \rightarrow q$ and $p \rightarrow r$, you cannot assume $q \rightarrow r$, as Suzanne tries to do. Tell students that the same hypothesis can imply two different conclusions, but the conclusions are not related to one another.

Check for Understanding

Concept Check

1. **OPEN ENDED** Write an example to illustrate the correct use of the Law of Detachment. **1–3. See margin.**

2. **Explain** how the Transitive Property of Equality is similar to the Law of Syllogism.

3. **FIND THE ERROR** An article in a magazine states that if you get seasick, then you will get dizzy. It also says that if you get seasick, you will get an upset stomach. Suzanne says that this means that if you get dizzy, then you will get an upset stomach. Lakeisha says that she is wrong. Who is correct? Explain.

Guided Practice

GUIDED PRACTICE KEY	
Exercises	Examples
4, 5	1
6, 7	2
8, 9	3

Determine whether the stated conclusion is valid based on the given information. If not, write *invalid*. Explain your reasoning.

If two angles are vertical angles, then they are congruent.

4. **Given:** $\angle A$ and $\angle B$ are vertical angles.
   Conclusion: $\angle A \cong \angle B$ **valid**

5. **Given:** $\angle C \cong \angle D$
   Conclusion: $\angle C$ and $\angle D$ are vertical angles. **Invalid; congruent angles do not have to be vertical.**

Use the Law of Syllogism to determine whether a valid conclusion can be reached from each set of statements. If a valid conclusion is possible, write it. If not, write *no conclusion*.

6. If you are 18 years old, you are in college.
   You are in college. **no conclusion**

7. The midpoint divides a segment into two congruent segments.
   If two segments are congruent, then their measures are equal.

7. The midpoint of a segment divides it into two segments with equal measures.

Determine whether statement (3) follows from statements (1) and (2) by the Law of Detachment or the Law of Syllogism. If it does, state which law was used. If it does not, write *invalid*. **8. valid; Law of Syllogism**

8. (1) If Molly arrives at school at 7:30 A.M., she will get help in math.
   (2) If Molly gets help in math, then she will pass her math test.
   (3) If Molly arrives at school at 7:30 A.M., then she will pass her math test.

9. (1) Right angles are congruent.
   (2) $\angle X \cong \angle Y$
   (3) $\angle X$ and $\angle Y$ are right angles. **invalid**

Application

INSURANCE **For Exercises 10 and 11, use the following information.**
An insurance company advertised the following monthly rates for life insurance.

If you are a:	Premium for $30,000 Coverage	Premium for $50,000 Coverage
Female, age 35	$14.35	$19.00
Male, age 35	$16.50	$21.63
Female, age 45	$21.63	$25.85
Male, age 45	$23.75	$28.90

10. If Ann is 35 years old and she wants to purchase $30,000 of insurance from this company, then what is her premium? **$14.35**

11. If Terry paid $21.63 for life insurance, can you conclude that Terry is 35? Explain. **No; Terry could be a man or a woman. She could be 45 and have purchased $30,000 of life insurance.**

Answers

1. Sample answer: a: If it is rainy, the game will be cancelled. b: It is rainy. c: The game will be cancelled.
2. Transitive Property of Equality: $a = b$ and $b = c$ implies $a = c$. Law of Syllogism: a implies b and b implies c implies a implies c. Each statement establishes a relationship between a and c through their relationships to b.
3. Lakeisha; if you are dizzy, that does not necessarily mean that you are seasick and thus have an upset stomach.

Practice and Apply

Homework Help

For Exercises	See Examples
12–19	1
20–23	2
24–29	3

Extra Practice
See page 757.

13. Valid; since 5 and 7 are odd, the Law of Detachment indicates that their sum is even.

14. Valid; since 11 and 23 are odd, the Law of Detachment indicates that their sum is even.

16. Valid; *A*, *B*, and *C* are noncollinear, and by definition three noncollinear points determine a plane.

17. Invalid; *E*, *F*, and *G* are not necessarily noncollinear.

18. Invalid; the hypothesis is false as there are only two points.

19. Valid; the vertices of a triangle are noncollinear, and therefore determine a plane.

21. If the measure of an angle is less than 90, then it is not obtuse.

22. If *X* is the midpoint of $\overline{YZ}$, then $\overline{YX} \cong \overline{XZ}$.

For Exercises 12–19, determine whether the stated conclusion is valid based on the given information. If not, write *invalid*. Explain your reasoning.

If two numbers are odd, then their sum is even.

12. Given: The sum of two numbers is 22.
Conclusion: The two numbers are odd. invalid; $10 + 12 = 22$

13. Given: The numbers are 5 and 7.
Conclusion: The sum is even.

14. Given: 11 and 23 are added together.
Conclusion: The sum of 11 and 23 is even.

15. Given: The numbers are 2 and 6.
Conclusion: The sum is odd. Invalid; the sum is even.

If three points are noncollinear, then they determine a plane.

16. Given: *A*, *B*, and *C* are noncollinear.
Conclusion: *A*, *B*, and *C* determine a plane.

17. Given: *E*, *F*, and *G* lie in plane *M*.
Conclusion: *E*, *F*, and *G* are noncollinear.

18. Given: *P* and *Q* lie on a line.
Conclusion: *P* and *Q* determine a plane.

19. Given: $\triangle XYZ$
Conclusion: *X*, *Y*, and *Z* determine a plane.

Use the Law of Syllogism to determine whether a valid conclusion can be reached from each set of statements. If a valid conclusion is possible, write it. If not, write *no conclusion*.

20. If you spend money on it, then it is a business.
If you spend money on it, then it is fun. no conclusion

21. If the measure of an angle is less than 90, then it is acute.
If an angle is acute, then it is not obtuse.

22. If *X* is the midpoint of segment *YZ*, then $YX = XZ$.
If the measures of two segments are equal, then they are congruent.

23. If two lines intersect to form a right angle, then they are perpendicular.
Lines ℓ and *m* are perpendicular. no conclusion

Determine whether statement (3) follows from statements (1) and (2) by the Law of Detachment or the Law of Syllogism. If it does, state which law was used. If it does not, write *invalid*.

24. (1) In-line skaters live dangerously.
(2) If you live dangerously, then you like to dance.
(3) If you are an in-line skater, then you like to dance. yes; Law of Syllogism

25. (1) If the measure of an angle is greater than 90, then it is obtuse.
(2) $m\angle ABC > 90$
(3) $\angle ABC$ is obtuse. yes; Law of Detachment

26. (1) Vertical angles are congruent.
(2) $\angle 3 \cong \angle 4$
(3) $\angle 3$ and $\angle 4$ are vertical angles. invalid

27. (1) If an angle is obtuse, then it cannot be acute.
(2) $\angle A$ is obtuse.
(3) $\angle A$ cannot be acute. yes; Law of Detachment

About the Exercises...

Organization by Objective
- **Law of Detachment:** 12–19
- **Law of Syllogism:** 20–29

Odd/Even Assignments
Exercises 12–29 are structured so that students practice the same concepts whether they are assigned odd or even problems.

Alert! Exercise 31 requires the Internet or other research materials.

Assignment Guide

Basic: 13–31 odd, 32–58

Average: 13–31 odd, 32–58

Advanced: 12–30 even, 32, 34–55 (optional: 56–58)

Determine whether statement (3) follows from statements (1) and (2) by the Law of Detachment or the Law of Syllogism. If it does, state which law was used. If it does not, write *invalid.*

28. (1) If you drive safely, then you can avoid accidents.
(2) Tika drives safely.
(3) Tika can avoid accidents. **yes; Law of Detachment**

29. (1) If you are a customer, then you are always right.
(2) If you are a teenager, then you are always right.
(3) If you are a teenager, then you are a customer. **invalid**

30. then he could hear the grating noise of the fish canneries

More About. . .

Literature

The Pulitzer Prize is awarded annually for outstanding contributions in the fields of journalism, literature, drama, and music.

Source: www.pulitzer.org

30. **LITERATURE** John Steinbeck, a Pulitzer Prize winning author, lived in Monterey, California, for part of his life. In 1945, he published the book, *Cannery Row*, about many of his local working-class heroes from Monterey. If you visited Cannery Row in Monterey during the 1940s, then you could hear the grating noise of the fish canneries. Write a valid conclusion to the following hypothesis.

If John Steinbeck lived in Monterey in 1941, . . .

31. **SPORTS** In the 2002 Winter Olympics, Canadian speed skater Catriona Le May Doan won her second Olympic title in 500-meter speed skating. Ms. Doan was in the last heat for the second round of that race. Use the two true conditional statements to reach a valid conclusion about Ms. Doan's 2002 competition.

(1) If Catriona Le May Doan skated her second 500 meters in 37.45 seconds, then she would beat the time of Germany's Monique Garbrecht-Enfeldt.

(2) If Ms. Doan beat the time of Monique Garbrecht-Enfeldt, then she would win the race. **If Catriona Le May Doan skated her second 500 meters in 37.45 seconds, then she would win the race.**

Online Research **Data Update** Use the Internet or another resource to find the winning times for other Olympic events. Write statements using these times that can lead to a valid conclusion. Visit www.geometryonline.com/data_update to learn more.

32. Sample answer: Stacey assumed that the conditional statement was true.

32. **CRITICAL THINKING** An advertisement states that "If you like to ski, then you'll love Snow Mountain Resort." Stacey likes to ski, but when she went to Snow Mountain Resort, she did not like it very much. If you know that Stacey saw the ad, explain how her reasoning was flawed.

33. **WRITING IN MATH** Answer the question that was posed at the beginning of the lesson. **See p. 123C.**

How does deductive reasoning apply to health?

Include the following in your answer:

- an explanation of how doctors may use deductive reasoning to prescribe medicine, and
- an example of a doctor's uses of deductive reasoning to diagnose an illness, such as strep throat or chickenpox.

FCAT Practice

Standardized Test Practice

Ⓐ Ⓑ Ⓒ Ⓓ

34. Based on the following statements, which statement must be true? **C**

I If Yasahiro is an athlete and he gets paid, then he is a professional athlete.
II Yasahiro is not a professional athlete.
III Yasahiro is an athlete.

Ⓐ Yasahiro is an athlete and he gets paid.
Ⓑ Yasahiro is a professional athlete or he gets paid.
Ⓒ Yasahiro does not get paid.
Ⓓ Yasahiro is not an athlete.

Study Guide and Intervention, p. 75 (shown) and p. 76

Law of Detachment **Deductive reasoning** is the process of using facts, rules, definitions, or properties to reach conclusions. One form of deductive reasoning that draws conclusions from a true conditional $p \to q$ and a true statement p is called the **Law of Detachment**.

Law of Detachment	If $p \to q$ is true and p is true, then q is true.
Symbols	$[(p \to q)] \wedge p] \to q$

Example **The statement *If two angles are supplementary to the same angle, then they are congruent* is a true conditional. Determine whether each conclusion is valid based on the given information. Explain your reasoning.**

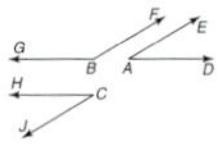

a. **Given: ∠A and ∠C are supplementary to ∠B.**
Conclusion: ∠A is congruent to ∠C.
The statement *∠A and ∠C are supplementary to ∠B* is the hypothesis of the conditional. Therefore, by the Law of Detachment, the conclusion is true.

b. **Given: ∠A is congruent to ∠C.**
Conclusion: ∠A and ∠C are supplementary to ∠B.
The statement *∠A is congruent to ∠C* is not the hypothesis of the conditional, so the Law of Detachment cannot be used. The conclusion is not valid.

Exercises

Determine whether each conclusion is valid based on the true conditional given. If not, write *invalid*. Explain your reasoning.

If two angles are complementary to the same angle, then the angles are congruent.

1. **Given:** ∠A and ∠C are complementary to ∠B.
Conclusion: ∠A is congruent to ∠C.
The given statement is the hypothesis of the conditional statement. Since the conditional is true, the conclusion ∠A ≅ ∠C is true.

2. **Given:** ∠A ≅ ∠C
Conclusion: ∠A and ∠C are complements of ∠B.
The given statement is not the hypothesis of the conditional. Therefore, the conclusion is invalid.

3. **Given:** ∠E and ∠F are complementary to ∠G.
Conclusion: ∠E and ∠F are vertical angles.
While the given statement is the hypothesis of the conditional statement, the statement that ∠E and ∠F are vertical angles is not the conclusion of the conditional. The conclusion is invalid.

Skills Practice, p. 77 and Practice, p. 78 (shown)

Determine whether the stated conclusion is valid based on the given information. If not, write *invalid*. Explain your reasoning.

If a point is the midpoint of a segment, then it divides the segment into two congruent segments.

1. **Given:** R is the midpoint of $\overline{QS}$.
Conclusion: $\overline{QR} \cong \overline{RS}$
Valid; since R is the midpoint of $\overline{QS}$, the Law of Detachment indicates that it divides $\overline{QS}$ into two congruent segments.

2. **Given:** $\overline{AB} \cong \overline{BC}$
Conclusion: B divides $\overline{AC}$ into two congruent segments.
Invalid; the points A, B, and C may not be collinear, and if they are not, then B will not be the midpoint of $\overline{AC}$.

Use the Law of Syllogism to determine whether a valid conclusion can be reached from each set of statements. If a valid conclusion is possible, write it.

3. If two angles form a linear pair, then the two angles are supplementary.
If two angles are supplementary, then the sum of their measures is 180.
If two angles form a linear pair, then the sum of their measures is 180.

4. If a hurricane is Category 5, then winds are greater than 155 miles per hour.
If winds are greater than 155 miles per hour, then trees, shrubs, and signs are blown down.
If a hurricane is Category 5, then trees, shrubs, and signs are blown down.

Determine whether statement (3) follows from statements (1) and (2) by the Law of Detachment or the Law of Syllogism. If it does, state which law was used. If it does not, write *invalid*.

5. (1) If a whole number is even, then its square is divisible by 4.
(2) The number I am thinking of is an even whole number.
(3) The square of the number I am thinking of is divisible by 4.
yes; Law of Detachment

6. (1) If the football team wins its homecoming game, then Conrad will attend the school dance the following Friday.
(2) Conrad attends the school dance on Friday.
(3) The football team won the homecoming game.
invalid

7. **BIOLOGY** If an organism is a parasite, then it survives by living on or in a host organism. If a parasite lives in or on a host organism, then it harms its host. What conclusion can you draw if a virus is a parasite?
If a virus is a parasite, then it harms its host.

Reading to Learn Mathematics, p. 79 ELL

Pre-Activity **How does deductive reasoning apply to health?**

Read the introduction to Lesson 2-4 at the top of page 82 in your textbook.

Suppose a doctor wants to use the dose chart in your textbook to prescribe an antibiotic, but the only scale in her office gives weights in pounds. How can she use the fact that 1 kilogram is about 2.2 pounds to determine the correct dose for a patient? Sample answer: The doctor can divide the patient's weight in pounds by 2.2 to find the equivalent mass in kilograms. She can then use the dose chart.

Reading the Lesson

If *s*, *t*, and *u* are three statements, match each description from the list on the left with a symbolic statement from the list on the right.

1. negation of t e	a. $s \vee u$
2. conjunction of s and u g	b. $[(s \to t) \wedge s] \to t$
3. converse of $s \to t$ h	c. $\sim s \to \sim u$
4. disjunction of s and u a	d. $\sim u \to \sim s$
5. Law of Detachment b	e. $\sim t$
6. contrapositive of $s \to t$ j	f. $[(u \to t) \wedge (t \to s)] \to (u \to s)$
7. inverse of $s \to u$ c	g. $s \wedge u$
8. contrapositive of $s \to u$ d	h. $t \to s$
9. Law of Syllogism f	i. t
10. negation of $\sim t$ i	j. $\sim t \to \sim s$

11. Determine whether statement (3) follows from statements (1) and (2) by the Law of Detachment or the Law of Syllogism. If it does, state which law was used. If it does not, write *invalid*.
a. (1) Every square is a parallelogram.
(2) Every parallelogram is a polygon.
(3) Every square is a polygon. yes; Law of Syllogism
b. (1)If two lines that lie in the same plane do not intersect, they are parallel.
(2) Lines ℓ and m lie in plane $\mathcal{U}$ and do not intersect.
(3) Lines ℓ and m are parallel. yes; Law of Detachment
c. (1) Perpendicular lines intersect to form four right angles.
(2) ∠A, ∠B, ∠C, and ∠D are four right angles.
(3) ∠A, ∠B, ∠C, and ∠D are formed by intersecting perpendicular lines. invalid

Helping You Remember

12. A good way to remember something is to explain it to someone else. Suppose that a classmate is having trouble remembering what the Law of Detachment means?
Sample answer: The word *detach* means to take something off of another thing. The Law of Detachment says that when a conditional and its hypothesis are both true, you can detach the conclusion and feel confident that it too is a true statement.

Enrichment, p. 80

Valid and Faulty Arguments

Consider the statements at the right. What conclusions can you make?

(1) Boots is a cat.
(2) Boots is purring.
(3) A cat purrs if it is happy.

From statements 1 and 3, it is correct to conclude that Boots purrs if it is happy. However, it is faulty to conclude from only statements 2 and 3 that Boots is happy. The if-then form of statement 3 is *If a cat is happy, then it purrs.*

Advertisers often use faulty logic in subtle ways to help sell their products. By studying the arguments, you can decide whether the argument is valid or faulty.

Decide if each argument is valid or faulty.

1. (1) If you buy Tuff Cote luggage, it will survive airline travel.
Tuff Cote lu

2. (1) If you buy Tuff Cote luggage, it will survive airline travel.

35. **ALGEBRA** At a restaurant, a diner uses a coupon for 15% off the cost of one meal. If the diner orders a meal regularly priced at \$16 and leaves a tip of 20% of the discounted meal, how much does she pay in total? **B**

Ⓐ \$15.64 Ⓑ \$16.32 Ⓒ \$16.80 Ⓓ \$18.72

Maintain Your Skills

Mixed Review

36. If you try Casa Fiesta, then you're looking for a fast, easy way to add some fun to your family's menu.

37. They are a fast, easy way to add fun to your family's menu.

38. No; the conclusion is implied.

ADVERTISING **For Exercises 36–38, use the following information.** *(Lesson 2-3)*

Advertising writers frequently use if-then statements to relay a message and promote their product. An ad for a type of Mexican food reads, *If you're looking for a fast, easy way to add some fun to your family's menu, try Casa Fiesta.*

36. Write the converse of the conditional.

37. What do you think the advertiser wants people to conclude about Casa Fiesta products?

38. Does the advertisement say that Casa Fiesta adds fun to your family's menu?

Construct a truth table for each compound statement. *(Lesson 2-2)* **39–42. See p. 123C.**

39. $q \wedge r$ 40. $\sim p \vee r$ 41. $p \wedge (q \vee r)$ 42. $p \vee (\sim q \wedge r)$

For Exercises 43–47, refer to the figure at the right. *(Lesson 1-5)*

44. Sample answer: ∠KHJ and ∠DHG

45. Sample answer: ∠JHK and ∠DHK

46. congruent, adjacent, supplementary, linear pair

43. Which angle is complementary to ∠FDG? **∠HDG**

44. Name a pair of vertical angles.

45. Name a pair of angles that are noncongruent and supplementary.

46. Identify ∠FDH and ∠CDH as *congruent, adjacent, vertical, complementary, supplementary,* and/or a *linear pair.*

47. Can you assume that $\overline{DC} \cong \overline{CK}$? Explain. **Yes, slashes on the segments indicate that they are congruent.**

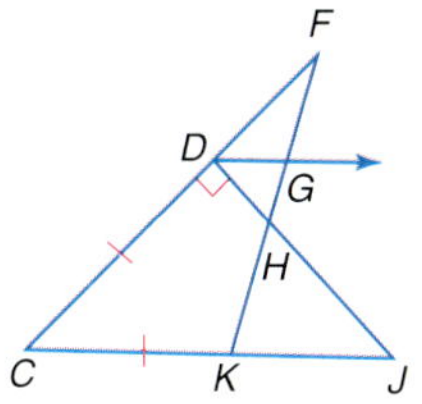

Use the Pythagorean Theorem to find the distance between each pair of points. *(Lesson 1-3)*

48. $A(1, 5), B(-2, 9)$ **5** 49. $C(-4, -2), D(2, 6)$ **10**

50. $F(7, 4), G(1, 0)$ **$\sqrt{52} \approx 7.2$** 51. $M(-5, 0), N(4, 7)$ **$\sqrt{130} \approx 11.4$**

For Exercises 52–55, draw and label a figure for each relationship. *(Lesson 1-1)*

52. $\overleftrightarrow{FG}$ lies in plane $\mathcal{M}$ and contains point H. **52–55. See margin.**

53. Lines r and s intersect at point W.

54. Line ℓ contains P and Q, but does not contain R.

55. Planes $\mathcal{A}$ and $\mathcal{B}$ intersect in line n.

Getting Ready for the Next Lesson

56–58. See margin.

PREREQUISITE SKILL **Write what you can assume about the segments or angles listed for each figure.** *(To review **information from figures**, see Lesson 1-5.)*

56. $\overline{AM}, \overline{CM}, \overline{CN}, \overline{BN}$ 57. ∠1, ∠2 58. ∠4, ∠5, ∠6

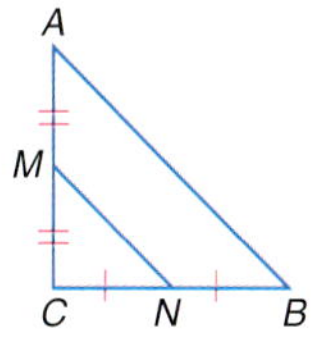

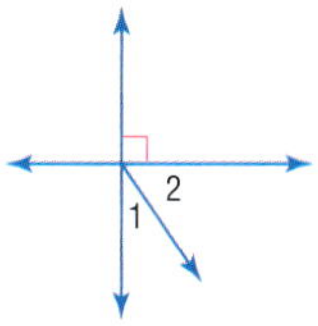

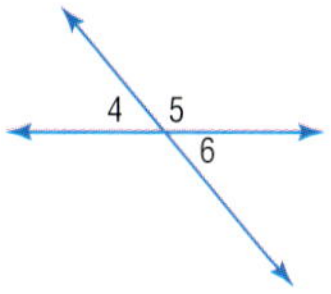

www.geometryonline.com/self_check_quiz/fcat

4 Assess

Open-Ended Assessment

Modeling Students can use shapes of paper to model the Law of Detachment and the Law of Syllogism. For example, you could provide students with two yellow squares of laminated construction paper marked with *p*, two blue triangles marked with *q*, and two red circles marked with *r*. Students can physically arrange the shapes to represent the symbolic relationships of the two laws. They could also secure these shapes in their study notebooks for a colorful reminder of these laws.

Getting Ready for Lesson 2-5

Prerequisite Skill Students will learn about postulates and paragraph proofs in Lesson 2-5. They will extract information from figures to test postulates and write proofs. Use Exercises 56–58 to determine your students' familiarity with information from figures.

Assessment Options

Quiz (Lessons 2-3 and 2-4) is available on p. 119 of the *Chapter 2 Resource Masters.*

Mid-Chapter Test (Lessons 2-1 through 2-4) is available on p. 121 of the *Chapter 2 Resource Masters.*

Answers

52.

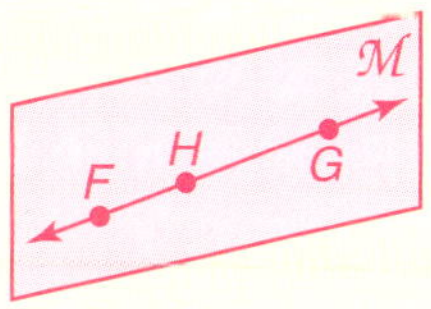

53.

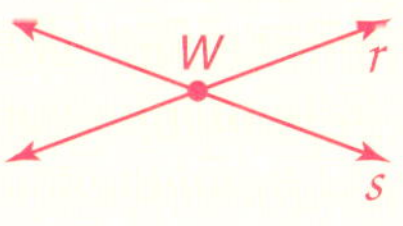

54.

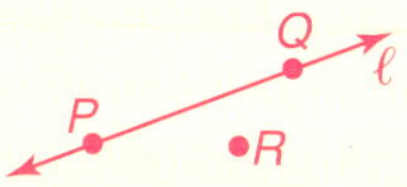

55.

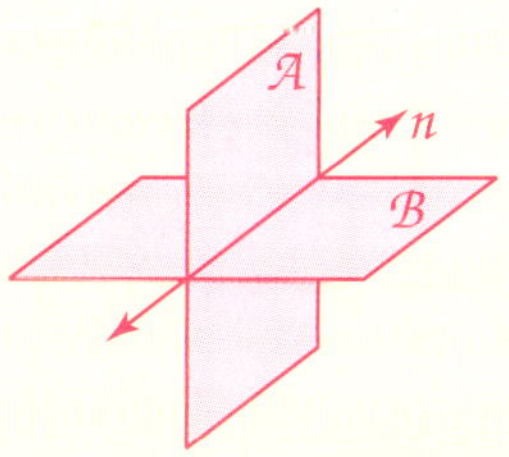

56. Sample answer: $\overline{CM} \cong \overline{AM}$, $\overline{CN} \cong \overline{BN}$, $AM = CM$, $CN = BN$, M is midpoint of $\overline{AC}$, N is midpoint of $\overline{BC}$.

57. Sample answer: ∠1 and ∠2 are complementary, $m\angle 1 + m\angle 2 = 90$.

58. Sample answer: ∠4 and ∠5 are supplementary, $m\angle 4 + m\angle 5 = 180$, ∠5 and ∠6 are supplementary, $m\angle 5 + m\angle 6 = 180$, $\angle 4 \cong \angle 6$, $m\angle 4 = m\angle 6$.

Geometry Activity

A Follow-Up of Lesson 2-4

Getting Started

You could provide students with a sheet of three blank logic tables in which they would fill in the titles of the rows and columns, or ask students to use a ruler to draw the tables as they go along.

Objective Apply deductive reasoning by using matrix logic tables to solve problems.

Teach

- When students are considering the second observation, tell them they can think in terms of boxes to mark out or boxes to leave open. The only two minerals that are harder than glass are feldspar and jasper, so these boxes are left open in the columns B and E. Then they can place an X in the rest of the boxes in these columns.
- For Exercise 2, advise students to draw a model with six adjacent boxes representing the apartments. After marking all obvious information, students can use the model to place each person in the correct apartment.

Assess

In **Exercises 1 and 2** students practice their thinking and deductive reasoning skills by working more logic tables.

Geometry Activity

A Follow-Up of Lesson 2-4

Matrix Logic

Deductive reasoning can be used in problem-solving situations. One method of solving problems uses a table. This method is called **matrix logic**.

Example

GEOLOGY On a recent test, Rashaun was given five different mineral samples to identify, along with the chart at right. Rashaun observed the following.

- **Sample C is brown.**
- **Samples B and E are harder than glass.**
- **Samples D and E are red.**

Identify each of the samples.

Mineral	Color	Hardness
Biotite	brown or black	softer than glass
Halite	white	softer than glass
Hematite	red	softer than glass
Feldspar	white, pink, or green	harder than glass
Jaspar	red	harder than glass

Make a table to organize the information. Mark each false condition with an X and each true condition with a √. The first observation is that Sample C is brown. Only one of the minerals, biotite, is brown, so place a check in the box that corresponds to biotite and Sample C. Then place an X in each of the other boxes in the same column and row.

Sample	A	B	C	D	E
Biotite	X	X	√	X	X
Halite		X	X	X	X
Hematite		X	X		X
Feldspar			X	X	X
Jaspar	X	X	X	X	√

The second observation is that Samples B and E are harder than glass. Place an X in each box for minerals that are softer than glass. The third observation is that Samples D and E are red. Mark the boxes accordingly. Notice that Sample E has an X in all but one box. Place a check mark in the remaining box, and an X in all other boxes in that row.

Sample	A	B	C	D	E
Biotite	X	X	√	X	X
Halite	√	X	X	X	X
Hematite	X	X	X	√	X
Feldspar	X	√	X	X	X
Jaspar	X	X	X	X	√

Then complete the table. Sample A is Halite, Sample B is Feldspar, Sample C is Biotite, Sample D is Hematite, and Sample E is Jaspar.

Exercises

1. Nate, John, and Nick just began after-school jobs. One works at a veterinarian's office, one at a computer store, and one at a restaurant. Nate buys computer games on the way to work. Nick is allergic to cat hair. John receives free meals at his job. Who works at which job? **Nate, veterinarian's office; John, restaurant; Nick, computer store**

2. Six friends live in consecutive apartments on the same side of their apartment building. Anita lives in apartment C. Kelli's apartment is just past Scott's. Anita's closest neighbors are Eric and Ava. Scott's apartment is not A through D. Eric's apartment is before Ava's. If Roberto lives in one of the apartments, who lives in which apartment? **A, Roberto; B, Eric; C, Anita; D, Ava; E, Scott; F, Kelli**

Resource Manager

Teaching Geometry with Manipulatives

- p. 46 (student recording sheet)

Study Notebook

Ask students to summarize what they have learned about using deductive reasoning in problem-solving situations.

2-5 Postulates and Paragraph Proofs

What You'll Learn

- Identify and use basic postulates about points, lines, and planes.
- Write paragraph proofs.

Sunshine State Standards
MA.C.1.4.1

Vocabulary

- postulate
- theorem
- proof
- paragraph proof
- informal proof

How were postulates used by the founding fathers of the United States?

U.S. Supreme Court Justice William Douglas stated "The First Amendment makes confidence in the common sense of our people and in the maturity of their judgment the great postulate of our democracy." This means that the writers of the constitution assumed that citizens will act and speak with common sense and maturity. There are statements in geometry that also must be assumed or accepted as true.

POINTS, LINES, AND PLANES In geometry, a **postulate** is a statement that describes a fundamental relationship between the basic terms of geometry. Postulates are accepted as true. In Chapter 1, you studied basic ideas about points, lines, and planes. These ideas can be stated as postulates.

Postulates

2.1 Through any two points, there is exactly one line.

2.2 Through any three points not on the same line, there is exactly one plane.

Example 1 *Points and Lines*

COMPUTERS **Jesse is setting up a network for his mother's business. There are five computers in her office. He wants to connect each computer to every other computer so that if one computer fails, the others are still connected. How many connections does Jesse need to make?**

Explore There are five computers, and each computer is connected to four other computers.

Plan Draw a diagram to illustrate the solution.

Solve Let noncollinear points A, B, C, D, and E represent the five computers. Connect each point with every other point. Then, count the number of segments.

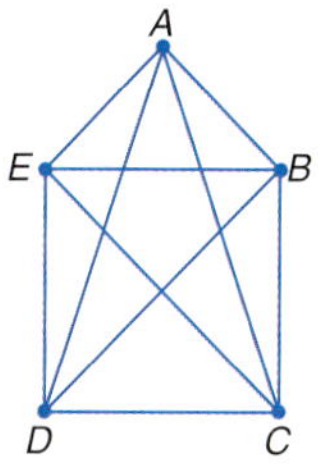

Between every two points there is exactly one segment. So, the connection between computer A and computer B is the same as the connection between computer B and computer A. For the five points, ten segments can be drawn.

Examine In the figure, $\overline{AB}$, $\overline{AC}$, $\overline{AD}$, $\overline{AE}$, $\overline{BC}$, $\overline{BD}$, $\overline{BE}$, $\overline{CD}$, $\overline{CE}$, and $\overline{DE}$ each represent a connection between two computers. There are ten segments, so there will be ten connections among the five computers.

Study Tip

Drawing Diagrams
When listing segments, start with one vertex and draw all of the segments from that vertex. Then move on to the other vertices until all possible segments have been drawn.

2-5 Lesson Notes

1 Focus

5-Minute Check Transparency 2-5 Use as a quiz or review of Lesson 2-4.

Mathematical Background notes are available for this lesson on p. 60D.

How were postulates used by the founding fathers of the United States?

Ask students:

- How would you interpret the words of William Douglas? **Sample answer: The constitution assumes that people have the maturity to handle the responsibilities of democracy.**
- Do you think the founding fathers set up a test to find out whether or not people do have common sense and maturity? **Sample answer: No; they assumed that people did.**

2 Teach

POINTS, LINES, AND PLANES

In-Class Example Power Point®

1 **SNOW CRYSTALS** Some snow crystals are shaped like regular hexagons. How many lines must be drawn to interconnect all vertices of a hexagonal snow crystal? **15**

Resource Manager

Workbook and Reproducible Masters

Chapter 2 Resource Masters

- Study Guide and Intervention, pp. 81–82
- Skills Practice, p. 83
- Practice, p. 84
- Reading to Learn Mathematics, p. 85
- Enrichment, p. 86

Teaching Geometry With Manipulatives Masters, p. 8

Transparencies

5-Minute Check Transparency 2-5
Answer Key Transparencies

Technology

Interactive Chalkboard

There are other postulates that are based on relationships among points, lines, and planes.

Postulates

2.3 A line contains at least two points.

2.4 A plane contains at least three points not on the same line.

2.5 If two points lie in a plane, then the entire line containing those points lies in that plane.

2.6 If two lines intersect, then their intersection is exactly one point.

2.7 If two planes intersect, then their intersection is a line.

Example 2 Use Postulates

Determine whether each statement is *always, sometimes,* or *never* true. Explain.

a. If points A, B, and C lie in plane $\mathcal{M}$, then they are collinear.

Sometimes; A, B, and C do not necessarily have to be collinear to lie in plane $\mathcal{M}$.

b. There is exactly one plane that contains noncollinear points P, Q, and R.

Always; Postulate 2.2 states that through any three noncollinear points, there is exactly one plane.

c. There are at least two lines through points M and N.

Never; Postulate 2.1 states that through any two points, there is exactly one line.

PARAGRAPH PROOFS Undefined terms, definitions, postulates, and algebraic properties of equality are used to prove that other statements or conjectures are true. Once a statement or conjecture has been shown to be true, it is called a **theorem**, and it can be used like a definition or postulate to justify that other statements are true.

You will study and use various methods to verify or prove statements and conjectures in geometry. A **proof** is a logical argument in which each statement you make is supported by a statement that is accepted as true. One type of proof is called a **paragraph proof** or **informal proof**. In this type of proof, you write a paragraph to explain why a conjecture for a given situation is true.

Study Tip

Proofs
Before writing a proof, you should have a plan. One strategy is to *work backward*. Start with what you want to prove, and work backward step by step until you reach the given information.

Key Concept — *Proofs*

Five essential parts of a good proof:

- State the theorem or conjecture to be proven.
- List the given information.
- If possible, draw a diagram to illustrate the given information.
- State what is to be proved.
- Develop a system of deductive reasoning.

In Lesson 1-2, you learned the relationship between segments formed by the midpoint of a segment. This statement can be proven, and the result stated as a theorem.

In-Class Example (Power Point®)

Teaching Tip Tell students that most postulates are very obvious and make very good sense, but they do not have a formal proof behind them. Nonetheless, students are to accept them as true and use them to prove other statements and theorems.

2 Determine whether each statement is *always*, *sometimes*, or *never* true. Explain.

a. If plane $\mathcal{T}$ contains $\overleftrightarrow{EF}$ and $\overleftrightarrow{EF}$ contains point G, then plane $\mathcal{T}$ contains point G. Always; Postulate 2.5 states that if two points lie in a plane, then the entire line containing those points lies in the plane.

b. For $\overleftrightarrow{XY}$, if X lies in plane $\mathcal{Q}$ and Y lies in plane $\mathcal{R}$, then plane $\mathcal{Q}$ intersects plane $\mathcal{R}$. Sometimes; planes $\mathcal{Q}$ and $\mathcal{R}$ can be parallel, and $\overleftrightarrow{XY}$ can intersect both planes.

c. $\overleftrightarrow{GH}$ contains three noncollinear points. Never; noncollinear points do not lie on the same line by definition.

Building on Prior Knowledge

Students learned basic principles about points, lines, and planes in Chapter 1. In this lesson, they will revisit those concepts in the form of postulates that they can use to write informal proofs and paragraph proofs.

Answers

1. Deductive reasoning is used to support claims that are made in a proof.
2. 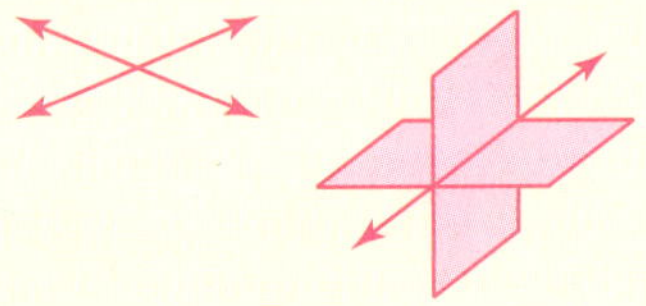
3. postulates, theorems, algebraic properties, definitions

DAILY INTERVENTION — **Differentiated Instruction**

Intrapersonal Tell students to read quietly over the postulates and examples in this lesson and note the differences in the postulate statements and the statements they are to write proofs for. Advise students to go through the text and their study notebooks to compile a list of useful information they could use to write the proofs in this lesson.

90 Chapter 2 Reasoning and Proof

Example 3 Write a Paragraph Proof

Given that M is the midpoint of $\overline{PQ}$, write a paragraph proof to show that $\overline{PM} \cong \overline{MQ}$.

Given: M is the midpoint of $\overline{PQ}$.

Prove: $\overline{PM} \cong \overline{MQ}$.

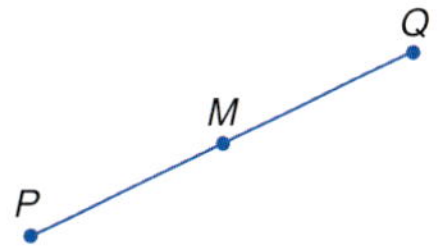

From the definition of midpoint of a segment, $PM = MQ$. This means that $\overline{PM}$ and $\overline{MQ}$ have the same measure. By the definition of congruence, if two segments have the same measure, then they are congruent. Thus, $\overline{PM} \cong \overline{MQ}$.

Once a conjecture has been proven true, it can be stated as a theorem and used in other proofs. The conjecture in Example 3 is known as the Midpoint Theorem.

Thm 2-1

~~Theorem 2.8~~

Midpoint Theorem If M is the midpoint of $\overline{AB}$, then $\overline{AM} \cong \overline{MB}$.

Check for Understanding

Concept Check

1. **Explain** how deductive reasoning is used in a proof. **1–3. See margin.**
2. **OPEN ENDED** Draw figures to illustrate Postulates 2.6 and 2.7.
3. **List** the types of reasons that can be used for justification in a proof.

Guided Practice

GUIDED PRACTICE KEY	
Exercises	Examples
4–5, 11	1
6	2
7–10	3

Determine the number of segments that can be drawn connecting each pair of points.

4. **6**

5. **15**

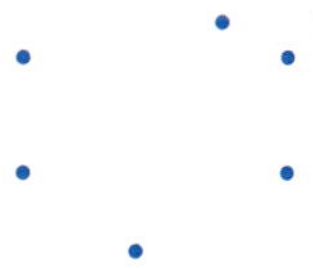

6. Determine whether the following statement is *always*, *sometimes*, or *never* true. Explain. **See p. 123C.**
   The intersection of three planes is two lines.

In the figure, $\overleftrightarrow{BD}$ and $\overleftrightarrow{BR}$ are in plane $\mathcal{P}$, and W is on $\overleftrightarrow{BD}$. State the postulate or definition that can be used to show each statement is true.

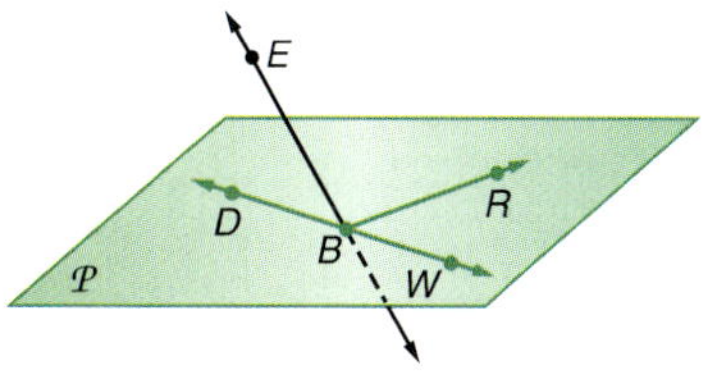

7. B, D, and W are collinear.
8. E, B, and R are coplanar.
9. R and W are collinear.

7. definition of collinear.
8. Through any three points not on the same line, there is exactly one plane.
9. Through any two points, there is exactly one line.

10. **PROOF** In the figure at the right, P is the midpoint of $\overline{QR}$ and $\overline{ST}$, and $\overline{QR} \cong \overline{ST}$. Write a paragraph proof to show that $PQ = PT$. **See p. 123C.**

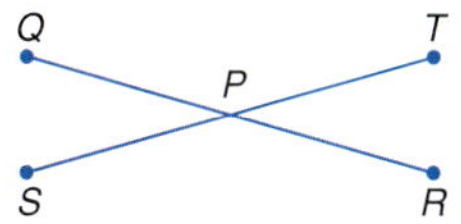

Application

11. **DANCING** Six students are participating in a dance to celebrate the opening of a new community center. The students, each connected to each of the other students with wide colored ribbons, will move in a circular motion. How many ribbons are needed? **15 ribbons**

www.geometryonline.com/extra_examples/fcat

PARAGRAPH PROOFS

In-Class Example

PowerPoint®

3 Given $\overleftrightarrow{AC}$ intersecting $\overleftrightarrow{CD}$, write a paragraph proof to show that A, C, and D determine a plane.

$\overleftrightarrow{AC}$ and $\overleftrightarrow{CD}$ must intersect at C because if two lines intersect, then their intersection is exactly one point. Point A is on $\overleftrightarrow{AC}$ and point D is on $\overleftrightarrow{CD}$. Therefore points A and D are not collinear. Therefore ACD is a plane as it contains three points not on the same line.

3 Practice/Apply

Study Notebook

Have students—

- add the definitions/examples of the vocabulary terms to their Vocabulary Builder worksheets for Chapter 2.
- include any other item(s) that they find helpful in mastering the skills in this lesson.

About the Exercises...

Organization by Objective

- **Points, Lines, and Planes:** 12–21
- **Paragraph Proofs:** 22–28

Odd/Even Assignments

Exercises 12–27 are structured so that students practice the same concepts whether they are assigned odd or even problems.

Assignment Guide

Basic: 13–19 odd, 23–31 odd, 33–48

Average: 13–31 odd, 33–48

Advanced: 12–30 even, 31–42 (optional: 43–48)

DAILY INTERVENTION

Unlocking Misconceptions

Writing Proofs Explain to students that a common mistake in writing proofs is skipping a step or assuming a step that should be included in the proof. Sometimes, the missed step can be quite obvious, but it still has to be included. Tell students to make a habit of listing each piece of information with a separate explanation for each and to avoid using two reasons or postulates for the same statement when they are writing proofs.

Study Guide and Intervention, p. 81 (shown) and p. 82

Points, Lines, and Planes In geometry, a **postulate** is a statement that is accepted as true. Postulates describe fundamental relationships in geometry.

Postulate: Through any two points, there is exactly one line.
Postulate: Through any three points not on the same line, there is exactly one plane.
Postulate: A line contains at least two points.
Postulate: A plane contains at least three points not on the same line.
Postulate: If two points lie in a plane, then the line containing those points lies in the plane.
Postulate: If two lines intersect, then their intersection is exactly one point.
Postulate: If two planes intersect, then their intersection is a line.

Example **Determine whether each statement is *always*, *sometimes*, or *never* true.**

a. There is exactly one plane that contains points *A*, *B*, and *C*.
Sometimes; if *A*, *B*, and *C* are collinear, they are contained in many planes. If they are noncollinear, then they are contained in exactly one plane.

b. Points *E* and *F* are contained in exactly one line.
Always; the first postulate states that there is exactly one line through any two points.

c. Two lines intersect in two distinct points *M* and *N*.
Never; the intersection of two lines is one point.

Exercises

Use postulates to determine whether each statement is *always*, *sometimes*, or *never* true.

1. A line contains exactly one point. never
2. Noncollinear points *R*, *S*, and *T* are contained in exactly one plane. always
3. Any two lines ℓ and m intersect. sometimes
4. If points *G* and *H* are contained in plane $\mathcal{M}$, then $\overline{GH}$ is perpendicular to plane $\mathcal{M}$. never
5. Planes $\mathcal{R}$ and $\mathcal{S}$ intersect in point *T*. never
6. If points *A*, *B*, and *C* are noncollinear, then segments $\overline{AB}$, $\overline{BC}$, and $\overline{CA}$ are contained in exactly one plane. always

In the figure, $\overline{AC}$ and $\overline{DE}$ are in plane Q and $\overline{AC} \parallel \overline{DE}$. State the postulate that can be used to show each statement is true.

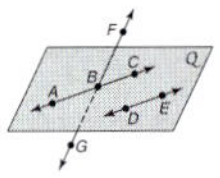

7. Exactly one plane contains points *F*, *B*, and *E*. Through any three points not on the same line, there is exactly one plane.
8. $\overleftrightarrow{BE}$ lies in plane Q. If two points lie in a plane, then the line containing those points lies in the plane.

Skills Practice, p. 83 and Practice, p. 84 (shown)

Determine the number of line segments that can be drawn connecting each pair of points.

1.  21

2.  28

Determine whether the following statements are *always*, *sometimes*, or *never* true. Explain.

3. The intersection of two planes contains at least two points.
Always; the intersection of two planes is a line, and a line contains at least two points.

4. If three planes have a point in common, then they have a whole line in common.
Sometimes; they might have only that single point in common.

In the figure, line m and $\overrightarrow{TQ}$ lie in plane $\mathcal{A}$. State the postulate that can be used to show that each statement is true.

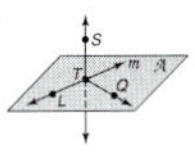

5. *L*, *T*, and line m lie in the same plane.
Postulate 2.5: If two points lie in a plane, then the entire line containing those points lies in that plane.

6. Line m and $\overline{ST}$ intersect at *T*.
Postulate 2.6: If two lines intersect, then their intersection is exactly one point.

7. In the figure, *E* is the midpoint of $\overline{AB}$ and $\overline{CD}$, and $AB = CD$. Write a paragraph proof to prove that $\overline{AE} \cong \overline{ED}$.

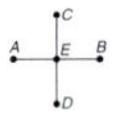

Given: *E* is the midpoint of $\overline{AB}$ and $\overline{CD}$
$AB = CD$
Prove: $\overline{AE} \cong \overline{ED}$
Proof: Since *E* is the midpoint of $\overline{AB}$ and $\overline{CD}$, we know by the Midpoint Theorem, that $\overline{AE} \cong \overline{EB}$ and $\overline{CE} \cong \overline{ED}$. By the definition of congruent segments, $AE = EB = \frac{1}{2}AB$ and $CE = ED = \frac{1}{2}CD$. Since $AB = CD$, $\frac{1}{2}AB = \frac{1}{2}CD$ by the Multiplication Property. So $AE = ED$, and by the definition of congruent segments, $\overline{AE} \cong \overline{ED}$.

8. **LOGIC** Points *A*, *B*, and *C* are not collinear. Points *B*, *C*, and *D* are not collinear. Points *A*, *B*, *C*, and *D* are not coplanar. Describe two planes that intersect in line *BC*.
the plane that contains *A*, *B*, and *C* and the plane that contains *B*, *C*, and *D*

Reading to Learn Mathematics, p. 85 ELL

Pre-Activity **How are postulates used by the founding fathers of the United States?**

Read the introduction to Lesson 2-5 at the top of page 89 in your textbook.

Postulates are often described as statements that are so basic and so clearly correct that people will be willing to accept them as true without asking for evidence or proof. Give a statement about numbers that you think most people would accept as true without evidence. Sample answer: Every number is equal to itself.

Reading the Lesson

1. Determine whether each of the following is a *correct* or *incorrect* statement of a geometric postulate. If the statement is incorrect, replace the underlined words to make the statement correct.
   a. A plane contains at least <u>two points</u> that do not lie on the same line. incorrect; three points
   b. If <u>two planes</u> intersect, then the intersection is a line. correct
   c. Through any <u>four points</u> not on the same line, there is exactly one plane. incorrect; three points
   d. A line contains at least <u>one point</u>. incorrect; two points
   e. If two lines <u>are parallel</u>, then their intersection is exactly one point. incorrect; intersect
   f. Through any two points, there is <u>at most</u> one line. incorrect; exactly

2. Determine whether each statement is *always*, *sometimes*, or *never* true. If the statement is not always true, explain why.
   a. If two planes intersect, their intersection is a line. always
   b. The midpoint of a segment divides the segment into two congruent segments. always
   c. There is exactly one plane that contains three collinear points. never; Sample answer: There are infinitely many planes if the three points are collinear, but only one plane if the points are noncollinear.
   d. If two lines intersect, their intersection is one point. always

3. Use the walls, floor, and ceiling of your classroom to describe a model for each of the following geometric situations.
   a. two planes that intersect in a line Sample answer: two adjacent walls that intersect at an edge of both walls in the corner of the room
   b. two planes that do not intersect Sample answer: the ceiling and the floor (or two opposite walls)
   c. three planes that intersect in a point Sample answer: the floor (or ceiling) and two adjacent walls that intersect at a corner of the floor (or ceiling)

Helping You Remember

4. A good way to remember a new mathematical term is to relate it to a word you already know. Explain how the idea of a mathematical *theorem* is related to the idea of a scientific *theory*. Sample answer: Scientists do experiments to prove theories; mathematicians use deductive reasoning to prove theorems. Both processes involve using evidence to show that certain statements are true.

★ indicates increased difficulty

Practice and Apply

Homework Help

For Exercises	See Examples
12–15	1
16–21	2
22–28	3

Extra Practice
See page 757.

Determine the number of segments that can be drawn connecting each pair of points.

12.  6

13.  10

14. 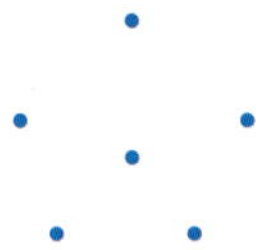 15

15. 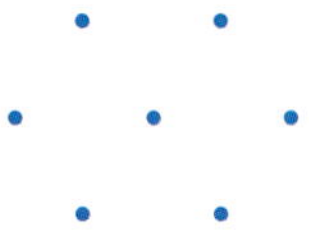 21

Determine whether the following statements are *always*, *sometimes*, or *never* true. Explain. 16–21. See p. 123C.

16. Three points determine a plane.
17. Points *G* and *H* are in plane $\mathcal{X}$. Any point collinear with *G* and *H* is in plane $\mathcal{X}$.
18. The intersection of two planes can be a point.
19. Points *S*, *T*, and *U* determine three lines.
★ 20. Points *A* and *B* lie in at least one plane.
★ 21. If line ℓ lies in plane $\mathcal{P}$ and line m lies in plane Q, then lines ℓ and m lie in plane $\mathcal{R}$.

In the figure at the right, $\overleftrightarrow{AC}$ and $\overleftrightarrow{BD}$ lie in plane $\mathcal{J}$, and $\overleftrightarrow{BY}$ and $\overleftrightarrow{CX}$ lie in plane $\mathcal{K}$. State the postulate that can be used to show each statement is true. 22–27. See p. 123C.

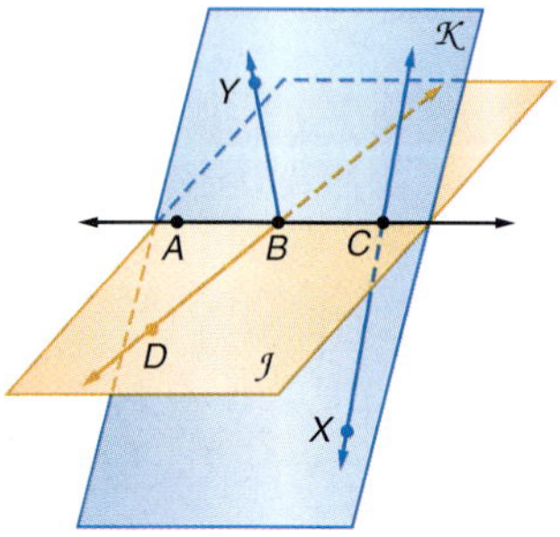

22. *C* and *D* are collinear.
23. $\overleftrightarrow{XB}$ lies in plane $\mathcal{K}$.
24. Points *A*, *C*, and *X* are coplanar.
25. $\overleftrightarrow{AD}$ lies in plane $\mathcal{J}$.
26. *X* and *Y* are collinear.
27. Points *Y*, *D*, and *C* are coplanar.
28. **PROOF** Point *C* is the midpoint of $\overline{AB}$ and *B* is the midpoint of $\overline{CD}$. Prove that $\overline{AC} \cong \overline{BD}$. See margin.

29. **MODELS** Faith's teacher asked her to make a figure showing the number of lines and planes formed from four points that are noncollinear and noncoplanar. Faith decided to make a mobile of straws, pipe cleaners, and colored sheets of tissue paper. She plans to glue the paper to the straws and connect the straws together to form a group of connected planes. How many planes and lines will she have? She will have 4 different planes and 6 lines.

30. **CAREERS** Many professions use deductive reasoning and paragraph proofs. For example, a police officer uses deductive reasoning investigating a traffic accident and then writes the findings in a report. List a profession, and describe how it can use paragraph proofs. Sample answer: Lawyers make final arguments, which is a speech that uses deductive reasoning, in court cases.

Career Choices

Detective

A police detective gathers facts and collects evidence for use in criminal cases. The facts and evidence are used together to prove a suspect's guilt in court.

Online Research
For information about a career as a detective, visit: www.geometryonline.com/careers

Enrichment, p. 86

Logic Problems

The following problems can be solved by eliminating possibilities. It may be helpful to use charts such as the one shown in the first problem. Mark an X in the chart to eliminate a possible answer.

Solve each problem.

1. Nancy, Olivia, Mario, and Kenji each have one piece of fruit in their school lunch. They have a peach, an orange, a banana, and an apple. Mario does not have a peach or a banana. Olivia and Mario just came from class with the student who has an apple. Kenji and Nancy are sitting next to the student who has a banana. Nancy does not have a peach. Which student has each piece of fruit?

	Nancy	Olivia	Mario	Kenji
		X	X	

2. Victor, Leon, Kasha, and Sheri each play one instrument. They play the viola, clarinet, trumpet, and flute. Sheri does not play the flute. Kasha lives near the student who plays flute and the one who plays trumpet. Leon does not play a brass or wind instrument. Which student plays each instrument?
Victor—flute,
Leon—viola,
Kasha—clarinet,

31. **CRITICAL THINKING** You know that three noncollinear points lie in a single plane. In Exercise 29, you found the number of planes defined by four noncollinear points. What are the least and greatest number of planes defined by five noncollinear points? **one, ten**

32. **WRITING IN MATH** Answer the question that was posed at the beginning of the lesson. **See margin.**

How are postulates used in literature?

Include the following in your answer:
- an example of a postulate in historic United States' documents, and
- an example of a postulate in mathematics.

FCAT Practice

Standardized Test Practice

33. Which statement cannot be true? **C**

(A) A plane can be determined using three noncollinear points.
(B) Two lines intersect at exactly one point.
(C) At least two lines can contain the same two points.
(D) A midpoint divides a segment into two congruent segments.

34. **ALGEBRA** For all values of x, $(8x^4 - 2x^2 + 3x - 5) - (2x^4 + x^3 + 3x + 5) =$ **A**

(A) $6x^4 - x^3 - 2x^2 - 10$.
(B) $6x^4 - 3x^2 + 6x - 10$.
(C) $6x^4 + x^3 - 2x^2 + 6x$.
(D) $6x^4 - 3x^2$.

Maintain Your Skills

Mixed Review

35. Determine whether statement (3) follows from statements (1) and (2) by the Law of Detachment or the Law of Syllogism. If it does, state which law was used. If it does not, write *invalid*. *(Lesson 2-4)* **yes; Law of Detachment**

(1) Part-time jobs require 20 hours of work per week.
(2) Jamie has a part-time job.
(3) Jamie works 20 hours per week.

Write the converse, inverse, and contrapositive of each conditional statement. Determine whether each related conditional is *true* or *false*. If a statement is false, find a counterexample. *(Lesson 2-3)* **36–37. See margin.**

36. If you have access to the Internet at your house, then you have a computer.

37. If $\triangle ABC$ is a right triangle, one of its angle measures is greater than 90.

38. **BIOLOGY** Use a Venn diagram to illustrate the following statement. *If an animal is a butterfly, then it is an arthropod.* *(Lesson 2-2)* **See p. 123C.**

Use the Distance Formula to find the distance between each pair of points. *(Lesson 1-3)*

39. $D(3, 3), F(4, -1)$ **$\sqrt{17} \approx 4.1$**

40. $M(0, 2), N(-5, 5)$ **$\sqrt{34} \approx 5.8$**

41. $P(-8, 2), Q(1, -3)$ **$\sqrt{106} \approx 10.3$**

42. $R(-5, 12), S(2, 1)$ **$\sqrt{170} \approx 13.0$**

Getting Ready for the Next Lesson

PREREQUISITE SKILL **Solve each equation.**
*(To review **solving equations**, see pages 737 and 738.)*

43. $m - 17 = 8$ **25**

44. $3y = 57$ **19**

45. $\frac{y}{6} + 12 = 14$ **12**

46. $-t + 3 = 27$ **−24**

47. $8n - 39 = 41$ **10**

48. $-6x + 33 = 0$ **$\frac{11}{2}$**

www.geometryonline.com/self_check_quiz/fcat

28. Given: C is the midpoint of $\overline{AB}$. B is the midpoint of $\overline{CD}$.
Prove: $\overline{AC} \cong \overline{BD}$
Proof: We are given that C is the midpoint of $\overline{AB}$, and B is the midpoint of $\overline{CD}$. By the definition of midpoint $\overline{AC} \cong \overline{CB}$ and $\overline{CB} \cong \overline{BD}$. Using the definition of congruent segments, $AC = CB$, and $CB = BD$. $AC = BD$ by the Transitive Property of Equality. Thus, $\overline{AC} \cong \overline{BD}$ by the definition of congruent segments.

4 Assess

Open-Ended Assessment

Speaking Have students choose a paragraph proof they wrote and explain each step aloud.

Getting Ready for Lesson 2-6

Prerequisite Skill Students will learn about algebraic proof in Lesson 2-6. Use Exercises 43–48 to determine your students' familiarity with solving equations.

Answers

32. Sample answer: The forms and structures of different types of writing are accepted as valid, such as the structure of a poem. Answers should include the following.
- The Declaration of Independence, "We hold these truths to be self-evident, ..."
- Through any two points, there is exactly one line.

36. Converse: If you have a computer, then you have access to the Internet at your house. False; you can have a computer and not have access to the Internet. Inverse: If you do not have access to the Internet at your house, then you do not have a computer. False; it is possible to not have access to the Internet and still have a computer. Contrapositive: If you do not have a computer, then you do not have access to the Internet at your house. False; you could have Internet access through your television or wireless phone.

37. Converse: If $\triangle ABC$ has an angle with measure greater than 90, then $\triangle ABC$ is a right triangle. False; the triangle would be obtuse. Inverse: If $\triangle ABC$ is not a right triangle, none of its angle measures are greater than 90. False; it could be an obtuse triangle. Contrapositive: If $\triangle ABC$ does not have an angle measure greater than 90, $\triangle ABC$ is not a right triangle. False; $m\angle ABC$ could still be 90 and $\triangle ABC$ be a right triangle.

2-6 Lesson Notes

1 Focus

5-Minute Check Transparency 2-6 Use as a quiz or review of Lesson 2-5.

Mathematical Background notes are available for this lesson on p. 60D.

How is mathematical evidence similar to evidence in law?

Ask students:

- In math, what one thing do you need to prove a statement is false? What evidence can a lawyer use to prove that someone is innocent (or has been *falsely* accused)? **a counterexample; sample answer: an alibi**
- How does the use of evidence in law differ from its use in math? **Sample answer: Lawyers present evidence to sway opinions, sometimes without knowing the truth. A mathematician presents evidence to prove factual statements.**

2-6 Algebraic Proof

Sunshine State Standards
MA.C.1.4.1

Vocabulary

- deductive argument
- two-column proof
- formal proof

What You'll Learn

- Use algebra to write two-column proofs.
- Use properties of equality in geometry proofs.

How is mathematical evidence similar to evidence in law?

Lawyers develop their cases using logical arguments based on evidence to lead a jury to a conclusion favorable to their case. At the end of a trial, a lawyer will make closing remarks summarizing the evidence and testimony that they feel proves their case. These closing arguments are similar to a proof in mathematics.

ALGEBRAIC PROOF In algebra, you learned to use properties of equality to solve algebraic equations and to verify relationships. Algebra is a system of mathematics with sets of numbers, operations, and properties that allow you to perform algebraic operations.

Study Tip

Commutative and Associative Properties
Throughout this text, we shall assume the Commutative and Associative Properties for addition and multiplication.

Concept Summary *Properties of Equality for Real Numbers*

Reflexive Property	For every number a, $a = a$.
Symmetric Property	For all numbers a and b, if $a = b$, then $b = a$.
Transitive Property	For all numbers a, b, and c, if $a = b$ and $b = c$, then $a = c$.
Addition and Subtraction Properties	For all numbers a, b, and c, if $a = b$, then $a + c = b + c$ and $a - c = b - c$.
Multiplication and Division Properties	For all numbers a, b, and c, if $a = b$, then $a \cdot c = b \cdot c$ and if $c \neq 0$, $\frac{a}{c} = \frac{b}{c}$.
Substitution Property	For all numbers a and b, if $a = b$, then a may be replaced by b in any equation or expression
Distributive Property	For all numbers a, b, and c, $a(b + c) = ab + ac$.

The properties of equality can be used to justify each step when solving an equation. A group of algebraic steps used to solve problems form a **deductive argument**.

Example 1 *Verify Algebraic Relationships*

Solve $3(x - 2) = 42$.

Algebraic Steps	Properties
$3(x - 2) = 42$	Original equation
$3x - 6 = 42$	Distributive Property
$3x - 6 + 6 = 42 + 6$	Addition Property
$3x = 48$	Substitution Property
$\frac{3x}{3} = \frac{48}{3}$	Division Property
$x = 16$	Substitution Property

Resource Manager

Workbook and Reproducible Masters

Chapter 2 Resource Masters
- Study Guide and Intervention, pp. 87–88
- Skills Practice, p. 89
- Practice, p. 90
- Reading to Learn Mathematics, p. 91
- Enrichment, p. 92
- Assessment, p. 120

Prerequisite Skills Workbook, pp. 41–44, 83–86, 93–94
Teaching Geometry With Manipulatives Masters, p. 8

Transparencies

5-Minute Check Transparency 2-6
Answer Key Transparencies

Technology

Interactive Chalkboard
Multimedia Applications: Virtual Activities

Example 1 is a proof of the conditional statement *If $5x + 3(x - 2) = 42$, then $x = 6$.* Notice that the column on the left is a step-by-step process that leads to a solution. The column on the right contains the reason for each statement.

In geometry, a similar format is used to prove conjectures and theorems. A **two-column proof**, or **formal proof**, contains statements and reasons organized in two columns. In a two-column proof, each step is called a *statement*, and the properties that justify each step are called *reasons*.

Example 2 Write a Two-Column Proof

Write a two-column proof.

a. If $3\left(x - \frac{5}{3}\right) = 1$, then $x = 2$

Statements	Reasons
1. $3\left(x - \frac{5}{3}\right) = 1$	1. Given
2. $3x - 3\left(\frac{5}{3}\right) = 1$	2. Distributive Property
3. $3x - 5 = 1$	3. Substitution
4. $3x - 5 + 5 = 1 + 5$	4. Addition Property
5. $3x = 6$	5. Substitution
6. $\frac{3x}{3} = \frac{6}{3}$	6. Division Property
7. $x = 2$	7. Substitution

Study Tip

Mental Math If your teacher permits you to do so, some steps may be eliminated by performing mental calculations. For example, in part **a** of Example 2, statements 4 and 6 could be omitted. Then the reason for statements 5 would be Addition Property and Division Property for statement 7.

b. Given: $\frac{7}{2} - n = 4 - \frac{1}{2}n$

Prove: $n = -1$

Proof:

Statements	Reasons
1. $\frac{7}{2} - n = 4 - \frac{1}{2}n$	1. Given
2. $2\left(\frac{7}{2} - n\right) = 2\left(4 - \frac{1}{2}n\right)$	2. Multiplication Property
3. $7 - 2n = 8 - n$	3. Distributive Property
4. $7 - 2n + n = 8 - n + n$	4. Addition Property
5. $7 - n = 8$	5. Substitution
6. $7 - n - 7 = 8 - 7$	6. Subtraction Property
7. $-n = 1$	7. Substitution
8. $\frac{-n}{-1} = \frac{1}{-1}$	8. Division Property
9. $n = -1$	9. Substitution

GEOMETRIC PROOF Since geometry also uses variables, numbers, and operations, many of the properties of equality used in algebra are also true in geometry. For example, segment measures and angle measures are real numbers, so properties from algebra can be used to discuss their relationships. Some examples of these applications are shown below.

Property	Segments	Angles
Reflexive	$AB = AB$	$m\angle 1 = m\angle 1$
Symmetric	If $AB = CD$, then $CD = AB$.	If $m\angle 1 = m\angle 2$, then $m\angle 2 = m\angle 1$.
Transitive	If $AB = CD$ and $CD = EF$, then $AB = EF$.	If $m\angle 1 = m\angle 2$ and $m\angle 2 = m\angle 3$, then $m\angle 1 = m\angle 3$.

www.geometryonline.com/extra_examples/fcat

2 Teach

ALGEBRAIC PROOFS

In-Class Examples Power Point®

1 Solve $2(5 - 3a) - 4(a + 7) = 92$.

$2(5 - 3a) - 4(a + 7) = 92$ (Original eqn.)
$10 - 6a - 4a - 28 = 92$ (Distr. Prop.)
$-18 - 10a = 92$ (Subst. Prop.)
$-18 - 10a + 18 = 92 + 18$ (Add. Prop.)
$-10a = 110$ (Subst. Prop.)
$\frac{-10a}{-10} = \frac{110}{-10}$ (Div. Prop.)
$a = -11$ (Subst. Prop.)

Teaching Tip Explain that since students may have different preferences when solving algebraic equations, their proofs might vary slightly from the examples. For example, one student may distribute a variable first, while another uses addition or subtraction. Assure students that as long as they use properties of equality appropriately, their proofs will be correct.

2 Write a two-column proof for each of the following.

a. If $\frac{7d + 3}{4} = 6$, then $d = 3$.

Statements (Reasons)

1. $\frac{7d + 3}{4} = 6$ (Given)
2. $4\left(\frac{7d + 3}{4}\right) = 4(6)$ (Mult. Prop.)
3. $7d + 3 = 24$ (Substitution Prop.)
4. $7d + 3 - 3 = 24 - 3$ (Subtr. Prop.)
5. $7d = 21$ (Substitution Prop.)
6. $\frac{7d}{7} = \frac{21}{7}$ (Div. Prop.)
7. $d = 3$ (Substitution Prop.)

(continued on the next page)

FCAT Practice

Standardized Test Practice

Ⓐ Ⓑ Ⓒ Ⓓ

Example 3 Justify Geometric Relationships

Multiple-Choice Test Item

If $\overline{AB} \cong \overline{CD}$, and $\overline{CD} \cong \overline{EF}$, then which of the following is a valid conclusion?

I $AB = CD$ and $CD = EF$

II $\overline{AB} \cong \overline{EF}$

III $AB = EF$

Ⓐ I only Ⓑ I and II

Ⓒ I and III Ⓓ I, II, and III

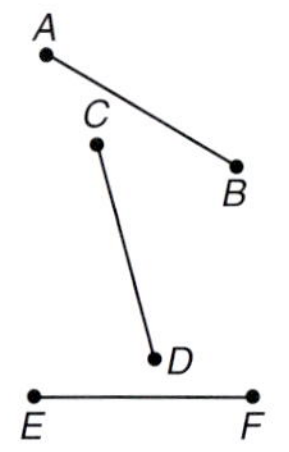

The Princeton Review

Test-Taking Tip

More than one statement may be correct. Work through each problem completely before indicating your answer.

Read the Test Item

Determine whether the statements are true based on the given information.

Solve the Test Item

Statement I:

Examine the given information, $\overline{AB} \cong \overline{CD}$ and $\overline{CD} \cong \overline{EF}$. From the definition of congruent segments, if $\overline{AB} \cong \overline{CD}$ and $\overline{CD} \cong \overline{EF}$, then $AB = CD$ and $CD = EF$. Thus, Statement I is true.

Statement II:

By the definition of congruent segments, if $AB = EF$, then $\overline{AB} \cong \overline{EF}$. Statement II is true also.

Statement III:

If $AB = CD$ and $CD = EF$, then $AB = EF$ by the Transitive Property. Thus, Statement III is true.

Because Statements I, II, and III are true, choice D is correct.

In Example 3, each conclusion was justified using a definition or property. This process is used in geometry to verify and prove statements.

Example 4 Geometric Proof

TIME **On a clock, the angle formed by the hands at 2:00 is a 60° angle. If the angle formed at 2:00 is congruent to the angle formed at 10:00, prove that the angle at 10:00 is a 60° angle.**

Given: $m\angle 2 = 60$
$\angle 2 \cong \angle 10$

Prove: $m\angle 10 = 60$

Proof:

Statements	Reasons
1. $m\angle 2 = 60$ $\angle 2 \cong \angle 10$	1. Given
2. $m\angle 2 = m\angle 10$	2. Definition of congruent angles
3. $60 = m\angle 10$	3. Substitution
4. $m\angle 10 = 60$	4. Symmetric Property

b. If $3p - \frac{9}{5} = \frac{11}{10} + \frac{p}{10}$, then $p = 1$.

Statements (Reasons)

1. $3p - \frac{9}{5} = \frac{11}{10} + \frac{p}{10}$ (Given)
2. $10\left(3p - \frac{9}{5}\right) = 10\left(\frac{11}{10} + \frac{p}{10}\right)$ (Mult. Prop.)
3. $30p - 18 = 11 + p$ (Distr. Prop.)
4. $30p - p - 18 = 11 + p - p$ (Subtr. Prop.)
5. $29p - 18 = 11$ (Substitution)
6. $29p - 18 + 18 = 11 + 18$ (Add. Prop.)
7. $29p = 29$ (Substitution)
8. $\frac{29p}{29} = \frac{29}{29}$ (Div. Prop.)
9. $p = 1$ (Substitution)

GEOMETRIC PROOFS

In-Class Examples

Power Point®

3 If $GH + JK = ST$ and $\overline{ST} \cong \overline{RP}$, then which of the following is a valid conclusion? **B**

I. $GH + JK = RP$
II. $PR = TS$
III. $GH + JK = ST + RP$

A I only **B** I and II
C I and III **D** I, II, and III

4 **SEA LIFE** A starfish has five arms. If the length of arm 1 is 22 cm, and arm 1 is congruent to arm 2, and arm 2 is congruent to arm 3, prove that arm 3 has length 22 cm. **We are given arm 1 ≅ arm 2 and arm 2 ≅ arm 3, so by the definition of congruence, the measure of arm 1 = the measure of arm 2 and the measure of arm 2 = the measure of arm 3. By the Transitive Property of Equality, we know that the measure of arm 1 = the measure of arm 3. We can then substitute 22 cm for the measure of arm 1 to prove that the measure of arm 3 is 22 cm.**

DAILY INTERVENTION

Differentiated Instruction

Interpersonal Let groups of students work one or two selected problems from Exercises 24–29 on p. 98. Stipulate that each group member should contribute at least one step of the proof. Encourage groups to brainstorm beforehand to determine the properties they will use and the order they will use them in. Allow the groups to check and compare their proofs when they are done to see if any two groups found different ways to prove the same statement.

Check for Understanding

Concept Check

1. **OPEN ENDED** Write a statement that illustrates the Substitution Property of Equality. **1–2. See margin.**
2. **Describe** the parts of a two-column proof.
3. **State** the part of a conditional that is related to the *Given* statement of a proof. What part is related to the *Prove* statement? **hypothesis; conclusion**

Guided Practice

GUIDED PRACTICE KEY	
Exercises	Examples
4–7	1
8	3
9, 10, 12	2
11	4

State the property that justifies each statement.

4. If $2x = 5$, then $x = \frac{5}{2}$ **Division Property**
5. If $\frac{x}{2} = 7$, then $x = 14$. **Multiplication Property**
6. If $x = 5$ and $b = 5$, then $x = b$. **Substitution Property**
7. If $XY - AB = WZ - AB$, then $XY = WZ$. **Addition Property**
8. Solve $\frac{x}{2} + 4x - 7 = 11$. List the property that justifies each step. **See margin.**
9. Complete the following proof.

   Given: $5 - \frac{2}{3}x = 1$

   Prove: $x = 6$

   Proof:

Statements	Reasons
a. ? $5 - \frac{2}{3}x = 1$	a. Given
b. $3\left(5 - \frac{2}{3}x\right) = 3(1)$	b. ? **Mult. Prop.**
c. $15 - 2x = 3$	c. ? **Dist. Prop.**
d. ? $-2x = -12$	d. Subtraction Prop.
e. $x = 6$	e. ? **Div. Prop.**

PROOF **Write a two-column proof.** **10–12. See pp. 123C–123D.**

10. Prove that if $25 = -7(y - 3) + 5y$, then $-2 = y$.
11. If rectangle $ABCD$ has side lengths $AD = 3$ and $AB = 10$, then $AC = BD$.
12. The Pythagorean Theorem states that in a right triangle ABC, $c^2 = a^2 + b^2$. Prove that $a = \sqrt{c^2 - b^2}$.

FCAT Practice

Standardized Test Practice

13. **ALGEBRA** If $8 + x = 12$, then $4 - x =$? . **C**

    (A) 28 (B) 24 (C) 0 (D) 4

Practice and Apply

Homework Help	
For Exercises	See Examples
15, 16, 20	1
14, 17–19, 21	2
22–27	3
28, 29	4

Extra Practice
See page 757.

State the property that justifies each statement.

14. If $m\angle A = m\angle B$ and $m\angle B = m\angle C$, $m\angle A = m\angle C$. **Trans. Prop.**
15. If $HJ + 5 = 20$, then $HJ = 15$. **Subt. Prop.**
16. If $XY + 20 = YW$ and $XY + 20 = DT$, then $YW = DT$. **Substitution**
17. If $m\angle 1 + m\angle 2 = 90$ and $m\angle 2 = m\angle 3$, then $m\angle 1 + m\angle 3 = 90$. **Substitution**
18. If $\frac{1}{2}AB = \frac{1}{2}EF$, then $AB = EF$. **Div. or Mult. Prop.**
19. $AB = AB$ **Reflexive Property**

3 Practice/Apply

Study Notebook

Have students—

- add the definitions/examples of the vocabulary terms to their Vocabulary Builder worksheets for Chapter 2.
- include a sample algebraic proof and a sample geometric proof.
- include any other item(s) that they find helpful in mastering the skills in this lesson.

About the Exercises...

Organization by Objective

- **Algebraic Proofs:** 14–21
- **Geometric Proofs:** 22–29

Odd/Even Assignments

Exercises 14–29 are structured so that students practice the same concepts whether they are assigned odd or even problems.

Assignment Guide

Basic: 15–31 odd, 32–33, 35–51

Average: 15–31 odd, 32–33, 35–51

Advanced: 14–30 even, 32–48 (optional: 49–51)

All: Quiz 2 (1–5)

8. **Given:** $\frac{x}{2} + 4x - 7 = 11$

   Prove: $x = 4$

   Proof:

   Statements (Reasons)

   1. $\frac{x}{2} + 4x - 7 = 11$ (Given)
   2. $2\left(\frac{x}{2} + 4x - 7\right) = 2(11)$ (Mult. Prop.)
   3. $x + 8x - 14 = 22$ (Dist.Prop.)
   4. $9x - 14 = 22$ (Substitution)
   5. $9x = 36$ (Add. Prop.)
   6. $x = 4$ (Div. Prop.)

Answers

1. Sample answer: If $x = 2$ and $x + y = 6$, then $2 + y = 6$.
2. given and prove statements and two columns, one of statements and one of reasons

Answers

32. Given: $E_k = hf + W$

Prove: $f = \dfrac{E_k - W}{h}$

Proof:

Statements (Reasons)
1. $E_k = hf + W$ (Given)
2. $E_k - W = hf$ (Subt. Prop.)
3. $\dfrac{E_k - W}{h} = f$ (Div. Prop.)
4. $f = \dfrac{E_k - W}{h}$ (Sym. Prop.)

36. Sample answer: Lawyers use evidence and testimony as reasons for justifying statements and actions. All of the evidence and testimony are linked together to prove a lawyer's case, much as in a proof in mathematics. Answers should include the following.

- Evidence is used to verify facts from witnesses or materials.
- Postulates, theorems, definitions, and properties can be used to justify statements made in mathematics.

20. If $2\left(x - \frac{3}{2}\right) = 5$, which property can be used to support the statement $2x - 3 = 5$? **Dist. Prop.**

21. Which property allows you to state $m\angle 4 = m\angle 5$, if $m\angle 4 = 35$ and $m\angle 5 = 35$? **Substitution**

22. If $\frac{1}{2}AB = \frac{1}{2}CD$, which property can be used to justify the statement $AB = CD$? **Div. or Mult. Prop.**

23. Which property could be used to support the statement $EF = JK$, given that $EF = GH$ and $GH = JK$? **Transitive Prop.**

Complete each proof.

24. Given: $\dfrac{3x + 5}{2} = 7$

Prove: $x = 3$

Proof:

Statements	Reasons
a. $\dfrac{3x + 5}{2} = 7$	a. ___?___ **Given**
b. ___?___ $2\left(\dfrac{3x + 5}{2}\right) = 2(7)$	b. Mult. Prop.
c. $3x + 5 = 14$	c. ___?___ **Substitution**
d. $3x = 9$	d. ___?___ **Subt. Prop.**
e. ___?___ $x = 3$	e. Div. Prop.

25. Given: $2x - 7 = \frac{1}{3}x - 2$

Prove: $x = 3$

Proof:

Statements	Reasons
a. ___?___ $2x - 7 = \frac{1}{3}x - 2$	a. Given
b. ___?___ $3(2x - 7) = 3\left(\frac{1}{3}x - 2\right)$	b. Mult. Prop.
c. $6x - 21 = x - 6$	c. ___?___ **Dist. Prop.**
d. ___?___ $5x - 21 = -6$	d. Subt. Prop.
e. $5x = 15$	e. ___?___ **Add. Prop.**
f. ___?___ $x = 3$	f. Div. Prop.

More About. . .

Physics

A gymnast exhibits kinetic energy when performing on the balance beam. The movements and flips show the energy that is being displayed while the gymnast is moving.

Source: www.infoplease.com

PROOF Write a two-column proof. **26–31. See p. 123D.**

26. If $4 - \frac{1}{2}a = \frac{7}{2} - a$, then $a = -1$.

27. If $-2y + \frac{3}{2} = 8$, then $y = -\frac{13}{4}$.

28. If $-\frac{1}{2}m = 9$, then $m = -18$.

29. If $5 - \frac{2}{3}z = 1$, then $z = 6$.

30. If $XZ = ZY$, $XZ = 4x + 1$, and $ZY = 6x - 13$, then $x = 7$.

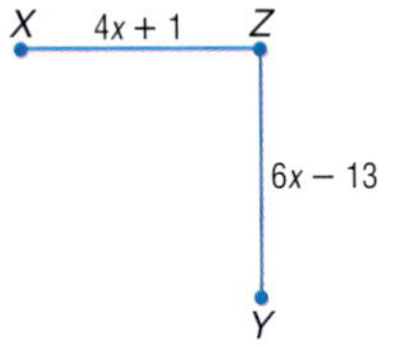

31. If $m\angle ACB = m\angle ABC$, then $m\angle XCA = m\angle YBA$.

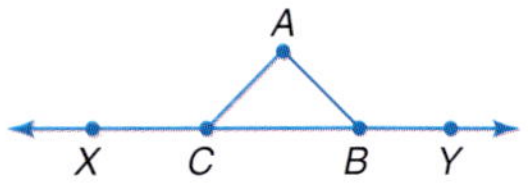

32. PHYSICS Kinetic energy is the energy of motion. The formula for kinetic energy is $E_k = h \cdot f + W$, where h represents Planck's Constant, f represents the frequency of its photon, and W represents the work function of the material being used. Solve this formula for f and justify each step. **See margin.**

33. **GARDENING** Areas in the southwest and southeast have cool but mild winters. In these areas, many people plant pansies in October so that they have flowers outside year-round. In the arrangement of pansies shown, the walkway divides the two sections of pansies into four beds that are the same size. If $m\angle ACB = m\angle DCE$, what could you conclude about the relationship among $\angle ACB$, $\angle DCE$, $\angle ECF$, and $\angle ACG$? **All of the angle measures would be equal.**

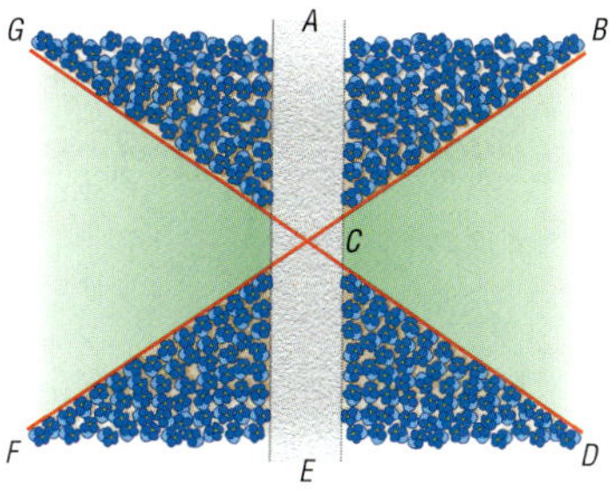

CRITICAL THINKING **For Exercises 34 and 35, use the following information.** Below is a family tree of the Gibbs family. Clara, Carol, Cynthia, and Cheryl are all daughters of Lucy. Because they are sisters, they have a transitive and symmetric relationship. That is, Clara is a sister of Carol, Carol is a sister of Cynthia, so Clara is a sister of Cynthia.

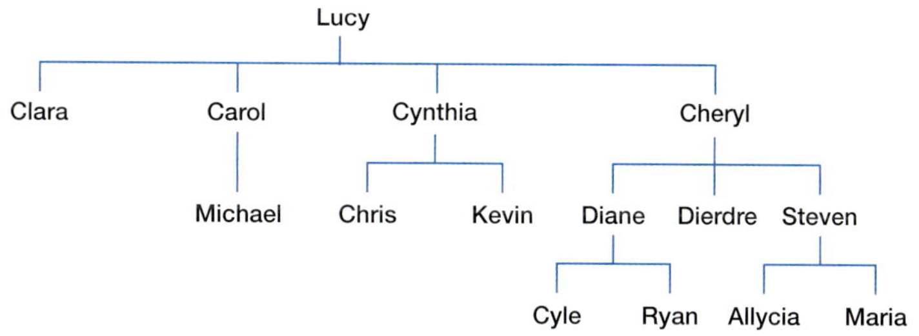

34. Sample answers are: Michael has a symmetric relationship of first cousin with Chris, Kevin, Diane, Dierdre, and Steven. Diane, Dierdre, and Steve have a symmetric and transitive relationship of sibling. Any direct line from bottom to top has a transitive descendent relationship.

34. What other relationships in a family have reflexive, symmetric, or transitive relationships? Explain why. Remember that the child or children of each person are listed beneath that person's name. Consider relationships such as first cousin, ancestor or descendent, aunt or uncle, sibling, or any other relationship.

35. Construct your family tree on one or both sides of your family and identify the reflexive, symmetric, or transitive relationships. **See students' work.**

36. **WRITING IN MATH** Answer the question that was posed at the beginning of the lesson. **See margin.**

How is mathematical evidence similar to evidence in law?

Include the following in your answer:
- a description of how evidence is used to influence jurors' conclusions in court, and
- a description of the evidence used to make conclusions in mathematics.

Standardized Test Practice Ⓐ Ⓑ Ⓒ Ⓓ

37. In $\triangle PQR$, $m\angle P = m\angle Q$ and $m\angle R = 2(m\angle Q)$. Find $m\angle P$ if $m\angle P + m\angle Q + m\angle R = 180$. **B**

Ⓐ 30 Ⓑ 45
Ⓒ 60 Ⓓ 90

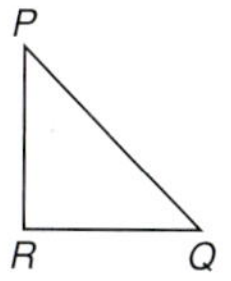

38. **ALGEBRA** If 4 more than x is 5 less than y, what is x in terms of y? **B**

Ⓐ $y - 1$ Ⓑ $y - 9$ Ⓒ $y + 9$ Ⓓ $y - 5$

www.geometryonline.com/self_check_quiz

Enrichment, p. 92

Symmetric, Reflexive, and Transitive Properties

Equality has three important properties.

Reflexive	$a = a$
Symmetric	If $a = b$, then $b = a$.
Transitive	If $a = b$ and $b = c$, then $a = c$.

Other relations have some of the same properties. Consider the relation "is next to" for objects labeled X, Y, and Z. Which of the properties listed above are true for this relation?

X is next to X. *False*
If X is next to Y, then Y is next to X. *True*
If X is next to Y and Y is next to Z, then X is next to Z. *False*

Only the symmetric property is true for the relation "is next to."

For each relation, state which properties (*symmetric, reflexive,* ...ue.

Study Guide and Intervention, p. 87 (shown) and p. 88

Algebraic Proof The following properties of algebra can be used to justify the steps when solving an algebraic equation.

Property	Statement
Reflexive	For every number a, $a = a$.
Symmetric	For all numbers a and b, if $a = b$ then $b = a$.
Transitive	For all numbers a, b, and c, if $a = b$ and $b = c$ then $a = c$.
Addition and Subtraction	For all numbers a, b, and c, if $a = b$ then $a + c = b + c$ and $a - c = b - c$.
Multiplication and Division	For all numbers a, b, and c, if $a = b$ then $a \cdot c = b \cdot c$, and if $c \neq 0$ then $\frac{a}{c} = \frac{b}{c}$.
Substitution	For all numbers a and b, if $a = b$ then a may be replaced by b in any equation or expression.
Distributive	For all numbers a, b, and c, $a(b + c) = ab + ac$.

Example **Solve $6x + 2(x - 1) = 30$.**

Algebraic Steps	Properties
$6x + 2(x - 1) = 30$	Given
$6x + 2x - 2 = 30$	Distributive Property
$8x - 2 = 30$	Substitution
$8x - 2 + 2 = 30 + 2$	Addition Property
$8x = 32$	Substitution
$\frac{8x}{8} = \frac{32}{8}$	Division Property
$x = 4$	Substitution

Exercises

Complete each proof.

1. **Given:** $\frac{4x + 6}{2} = 9$
Prove: $x = 3$

Statements	Reasons
a. $\frac{4x + 6}{2} = 9$	a. Given
b. $2\left(\frac{4x + 6}{2}\right) = 2(9)$	b. Mult. Prop.
c. $4x + 6 = 18$	c. Subst.
d. $4x + 6 - 6 = 18 - 6$	d. Subtr. Prop.
e. $4x = 12$	e. Substitution
f. $\frac{4x}{4} = \frac{12}{4}$	f. Div. Prop.
g. $x = 3$	g. Substitution

2. **Given:** $4x + 8 = x + 2$
Prove: $x = -2$

Statements	Reasons
a. $4x + 8 = x + 2$	a. Given
b. $4x + 8 - x = x + 2 - x$	b. Subtr. Prop.
c. $3x + 8 = 2$	c. Substitution
d. $3x + 8 - 8 = 2 - 8$	d. Subtr. Prop.
e. $3x = -6$	e. Substitution
f. $\frac{3x}{3} = \frac{-6}{3}$	f. Div. Prop.
g. $x = -2$	g. Substitution

Skills Practice, p. 89 and Practice, p. 90 (shown)

PROOF Write a two-column proof.

1. If $m\angle ABC + m\angle CBD = 90$, $m\angle ABC = 3x - 5$, and $m\angle CBD = \frac{x + 1}{2}$, then $x = 27$.

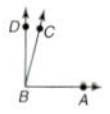

Given: $m\angle ADC + m\angle CBD = 90$
$m\angle ABC = 3x - 5$
$m\angle CBD = \frac{x + 1}{2}$
Prove: $x = 27$
Proof:

Statements	Reasons
1. $m\angle ABC + m\angle CBD = 90$, $m\angle ABC = 3x - 5$, $m\angle CBD = \frac{x + 1}{2}$	1. Given
2. $3x - 5 + \frac{x + 1}{2} = 90$	2. Substitution Property
3. $(2)(3x - 5) + (2)\left(\frac{x + 1}{2}\right) = (2)90$	3. Multiplication Property
4. $6x - 10 + x + 1 = 180$	4. Substitution Property
5. $7x - 9 = 180$	5. Substitution Property
6. $7x - 9 + 9 = 180 + 9$	6. Addition Property
7. $7x = 189$	7. Substitution Property
8. $\frac{7x}{7} = \frac{189}{7}$	8. Division Property
9. $x = 27$	9. Substitution Property

2. **FINANCE** The formula for simple interest is $I = prt$, where I is interest, p is principal, r is rate, and t is time. Solve the formula for r and justify each step.

Given: $I = prt$
Prove: $r = \frac{I}{pt}$
Proof:

Statements	Reasons
1. $I = prt$	1. Given
2. $\frac{I}{pt} = \frac{prt}{pt}$	2. Division Property
3. $\frac{I}{pt} = r$	3. Substitution Property
4. $r = \frac{I}{pt}$	4. Symmetric Property

Reading to Learn Mathematics, p. 91 ELL

Pre-Activity **How is mathematical evidence similar to evidence in law?**

Read the introduction to Lesson 2-6 at the top of page 94 in your textbook.

What are some of the things that lawyers might use in presenting their closing arguments to a trial jury in addition to evidence gathered prior to the trial and testimony heard during the trial? **Sample answer: They might tell the jury about laws related to the case, court rulings, and precedents set by earlier trials.**

Reading the Lesson

1. Name the property illustrated by each statement.
   a. If $a = 4.75$ and $4.75 = b$, then $a = b$. **Transitive Property of Equality**
   b. If $x = y$, then $x + 8 = y + 8$. **Addition Property of Equality**
   c. $5(12 + 19) = 5 \cdot 12 + 5 \cdot 19$ **Distributive Property**
   d. If $x = 5$, then x may be replaced with 5 in any equation or expression. **Substitution Property of Equality**
   e. If $x = y$, then $8x = 8y$. **Multiplication Property of Equality**
   f. If $x = 23.45$, then $23.45 = x$. **Symmetric Property of Equality**
   g. If $5x = 7$, then $x = \frac{7}{5}$. **Division Property of Equality**
   h. If $x = 12$, then $x - 3 = 9$. **Subtraction Property of Equality**

2. Give the reason for each statement in the following two-column proof.
Given: $5(n - 3) = 4(2n - 7) - 14$
Prove: $n = 9$

Statements	Reasons
1. $5(n - 3) = 4(2n - 7) - 14$	1. Given
2. $5n - 15 = 8n - 28 - 14$	2. Distributive Property
3. $5n - 15 = 8n - 42$	3. Substitution Property
4. $5n - 15 + 15 = 8n - 42 + 15$	4. Addition Property
5. $5n = 8n - 27$	5. Substitution Property
6. $5n - 8n = 8n - 27 - 8n$	6. Subtraction Property
7. $-3n = -27$	7. Substitution Property
8. $\frac{-3n}{-3} = \frac{-27}{-3}$	8. Division Property
9. $n = 9$	9. Substitution Property

Helping You Remember

3. A good way to remember mathematical terms is to relate them to words you already know. Give an everyday word that is related in meaning to the mathematical term *reflexive* and explain how this word can help you to remember the Reflexive Property and to distinguish it from the Symmetric and Transitive Properties. **Sample answer: Reflection: If you look at your reflection, you see *yourself*. The Reflexive Property says that every number is equal to *itself*. The Reflexive Property involves only one number, while the Symmetric and Transitive Properties each involve two or three numbers.**

4 Assess

Open-Ended Assessment

Writing Select some statements to prove and write them on the board. Have different volunteers come up to the board and write one statement and reason to advance the proof until the students have proven the original statement.

Getting Ready for Lesson 2-7

Prerequisite Skill Students will learn about proving segment relationships in Lesson 2-7. They will learn about segment addition and will use segment measures to prove segment congruence. Use Exercises 51–53 to determine your students' familiarity with segment measures.

Assessment Options

Practice Quiz 2 The quiz provides students with a brief review of the concepts and skills in Lessons 2-4 through 2-6. Lesson numbers are given to the right of the exercises or instruction lines so students can review concepts not yet mastered.

Quiz (Lessons 2-5 and 2-6) is available on p. 120 of the *Chapter 2 Resource Masters*.

Answers

43. If people are happy, then they rarely correct their faults.
44. If you don't know where you are going, then you will probably end up somewhere else.
45. If a person is a champion, then the person is afraid of losing.
46. If we would have new knowledge, then we must get a whole new world of questions.

Maintain Your Skills

Mixed Review

39. **CONSTRUCTION** There are four buildings on the Medfield High School Campus, no three of which stand in a straight line. How many sidewalks need to be built so that each building is directly connected to every other building? *(Lesson 2-5)* **6**

Determine whether the stated conclusion is valid based on the given information. If not, write *invalid*. Explain your reasoning. *A number is divisible by 3 if it is divisible by 6.* *(Lesson 2-4)*

40. **Given:** 24 is divisible by 6. **Conclusion:** 24 is divisible by 3.
41. **Given:** 27 is divisible by 3. **Conclusion:** 27 is divisible by 6.
42. **Given:** 85 is not divisible by 3. **Conclusion:** 85 is not divisible by 6.

40. Valid; since 24 is divisible by 6, the Law of Detachment says it is divisible by 3.
41. Invalid; 27 ÷ 6 = 4.5, which is not an integer.
42. Valid; since 85 is not divisible by 3, the contrapositive of the statement and the Law of Detachment say that 85 is not divisible by 6.

Write each statement in if-then form. *(Lesson 2-3)* **43–46. See margin.**

43. "Happy people rarely correct their faults." (*La Rochefoucauld*)
44. "If you don't know where you are going, you will probably end up somewhere else." (*Laurence Peters*)
45. "A champion is afraid of losing." (*Billie Jean King*)
46. "If we would have new knowledge, we must get a whole new world of questions." (*Susanne K. Langer*)

Find the precision for each measurement. *(Lesson 1-2)*

47. 13 feet $\frac{1}{2}$ **ft**
48. 5.9 meters **0.05 m**
49. 74 inches **0.5 in.**
50. 3.1 kilometers **0.05 km**

Getting Ready for the Next Lesson

PREREQUISITE SKILL **Find the measure of each segment.**
*(To review **segment measures**, see Lesson 1-2.)*

51. $\overline{KL}$ **11**

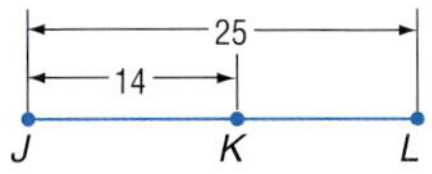

52. $\overline{QS}$ **28**

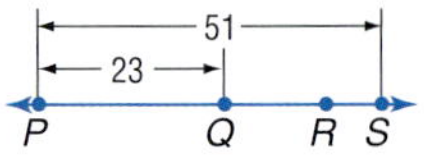

53. $\overline{WZ}$ **47**

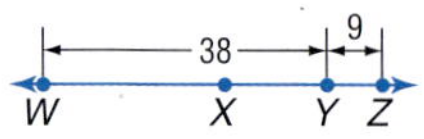

Practice Quiz 2 — Lessons 2-4 through 2-6

1. Determine whether statement (3) follows from statements (1) and (2) by the Law of Detachment or the Law of Syllogism. If it does, state which law was used. If it does not, write *invalid*. *(Lesson 2-4)*
   (1) If n is an integer, then n is a real number.
   (2) n is a real number.
   (3) n is an integer. **invalid**

In the figure at the right, *A*, *B*, and *C* are collinear. Points *A*, *B*, *C*, and *D* lie in plane *N*. State the postulate or theorem that can be used to show each statement is true. *(Lesson 2-5)* **2–4. See margin.**

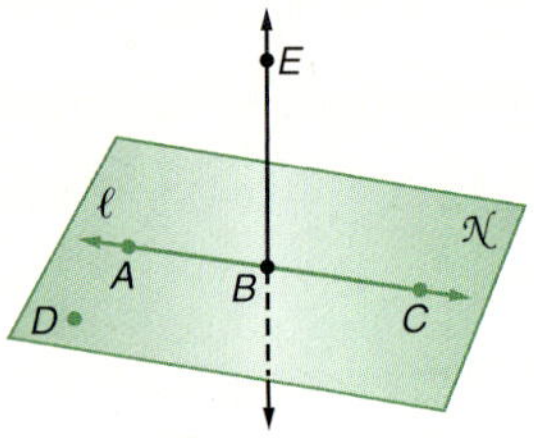

2. *A*, *B*, and *D* determine plane *N*.
3. $\overleftrightarrow{BE}$ intersects $\overleftrightarrow{AC}$ at *B*.
4. ℓ lies in plane *N*.
5. **PROOF** If $2(n - 3) + 5 = 3(n - 1)$, prove that $n = 2$. *(Lesson 2-6)* **See p. 123E.**

100 Chapter 2 Reasoning and Proof

Answers

Practice Quiz 2

2. Through any three points not on the same line, there is exactly one plane.
3. If two lines intersect, then their intersection is exactly one point.
4. If two points lie in a plane, then the entire line containing those points lies in that plane.

2-7 Proving Segment Relationships

Sunshine State Standards
MA.C.2.4.1, MA.C.1.4.1

What You'll Learn

- Write proofs involving segment addition.
- Write proofs involving segment congruence.

How can segment relationships be used for travel?

When leaving San Diego, the pilot said that the flight would be about 360 miles to Phoenix before continuing on to Dallas. When the plane left Phoenix, the pilot said that the flight would be flying about 1070 miles to Dallas.

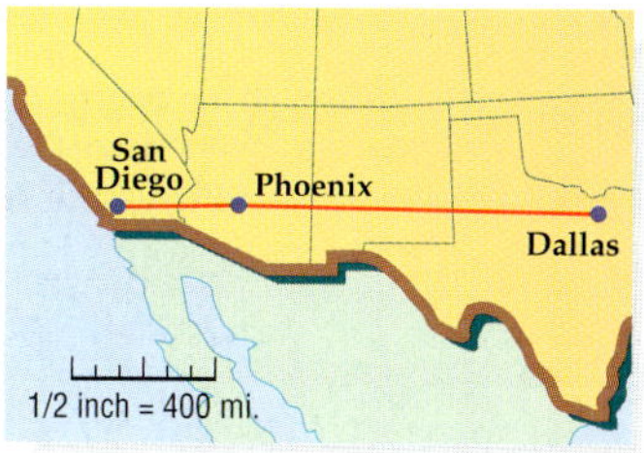

SEGMENT ADDITION In Lesson 1-2, you measured segments with a ruler by placing the mark for zero on one endpoint, then finding the distance to the other endpoint. This illustrates the **Ruler Postulate.**

Postulate 2.8

Ruler Postulate The points on any line or line segment can be paired with real numbers so that, given any two points A and B on a line, A corresponds to zero, and B corresponds to a positive real number.

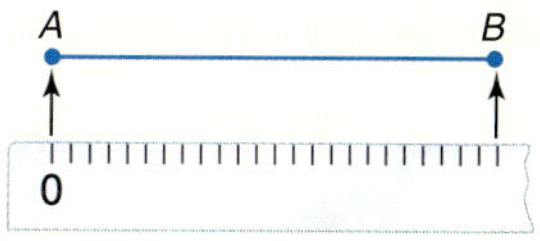

The Ruler Postulate can be used to further investigate line segments.

Geometry Software Investigation

Adding Segment Measures

Construct a Figure

- Use The Geometer's Sketchpad to construct $\overline{AC}$.
- Place point B on $\overline{AC}$.
- Find AB, BC, and AC.

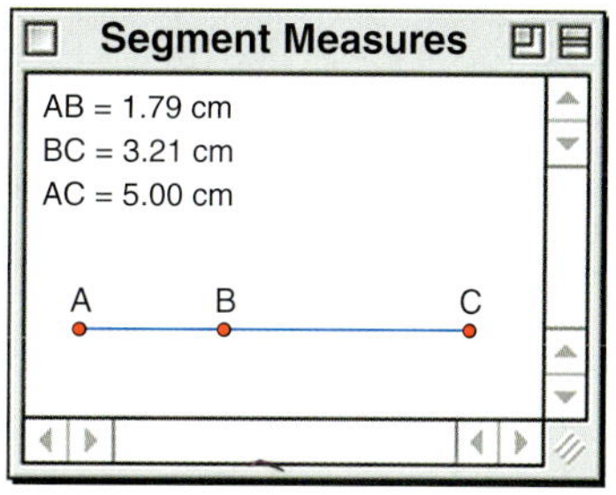

Analyze the Model

1. What is the sum $AB + BC$?
2. Move B. Find AB, BC and AC. What is the sum of $AB + BC$?
3. Repeat moving B, measuring the segments, and finding the sum $AB + BC$ three times. Record your results. 1–3. See students' work. The sum $AB + BC$ should always equal AC.

Make a Conjecture

4. What is true about the relationship of AB, BC, and AC? $AB + BC = AC$
5. Is it possible to place B on $\overline{AC}$ so that this relationship is not true? no

2-7 Lesson Notes

1 Focus

5-Minute Check Transparency 2-7 Use as a quiz or review of Lesson 2-6.

Mathematical Background notes are available for this lesson on p. 60D.

How can segment relationships be used for travel?

Ask students:

- Use a ruler to measure the length in millimeters from San Diego to Phoenix and from Phoenix to Dallas. about 11 mm; about 32 mm
- Given the pilot's information, how many miles will Janelle be flying from San Diego to Dallas? 1430
- How are segment lengths helpful for air travel? Sample answer: Pilots can use segment lengths to calculate distance, flight time and necessary fuel requirements.

Resource Manager

Workbook and Reproducible Masters

Chapter 2 Resource Masters
- Study Guide and Intervention, pp. 93–94
- Skills Practice, p. 95
- Practice, p. 96
- Reading to Learn Mathematics, p. 97
- Enrichment, p. 98

School-to-Career Masters, p. 4
Prerequisite Skills Workbook, pp. 89–90, 101–104
Teaching Geometry With Manipulatives Masters, p. 8

Transparencies
5-Minute Check Transparency 2-7
Answer Key Transparencies

Technology
Interactive Chalkboard

2 Teach

SEGMENT ADDITION

In-Class Example

Power Point®

1 Prove the following. Use the figure from Example 1 in the Student Edition.

Given: $PR = QS$

Prove: $PQ = RS$

Statements (Reasons)

1. $PR = QS$ (Given)
2. $PR - QR = QS - QR$ (Subtr. Prop.)
3. $PR - QR = PQ$; $QS - QR = RS$ (Seg. Add. Post.)
4. $PQ = RS$ (Substitution)

Teaching Tip Tell students that with each new lesson, they are accumulating more postulates and theorems that they can use for writing proofs. Encourage students to practice using these concepts as much as possible before moving on to the next lesson to strengthen their ability to recall important facts for proof-writing skills.

Study Tip

Betweenness

In general, the definition of *between* is that *B* is between *A* and *C* if *A*, *B*, and *C* are collinear and $AB + BC = AC$.

Examine the measures *AB*, *BC*, and *AC* in the Geometry Activity. Notice that wherever *B* is placed between *A* and *C*, $AB + BC = AC$. This suggests the following postulate.

Postulate 2.9

Segment Addition Postulate If *B* is between *A* and *C*, then $AB + BC = AC$.

If $AB + BC = AC$, then *B* is between *A* and *C*.

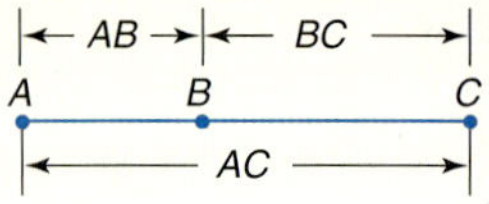

Example 1 Proof With Segment Addition

Prove the following.

Given: $PQ = RS$

Prove: $PR = QS$

Proof:

Statements	Reasons
1. $PQ = RS$	**1.** Given
2. $PQ + QR = QR + RS$	**2.** Addition Property
3. $PQ + QR = PR$ $QR + RS = QS$	**3.** Segment Addition Postulate
4. $PR = QS$	**4.** Substitution

SEGMENT CONGRUENCE In Lesson 2-5, you learned that once a theorem is proved, it can be used in proofs of other theorems. One theorem we can prove is similar to properties of equality from algebra.

Theorem 2.2 *Segment Congruence*

Congruence of segments is reflexive, symmetric, and transitive.

Reflexive Property $\overline{AB} \cong \overline{AB}$

Symmetric Property If $\overline{AB} \cong \overline{CD}$, then $\overline{CD} \cong \overline{AB}$.

Transitive Property If $\overline{AB} \cong \overline{CD}$, and $\overline{CD} \cong \overline{EF}$, then $\overline{AB} \cong \overline{EF}$.

You will prove the first two properties in Exercises 10 and 24.

Proof *Transitive Property of Congruence*

Given: $\overline{MN} \cong \overline{PQ}$

$\overline{PQ} \cong \overline{RS}$

Prove: $\overline{MN} \cong \overline{RS}$

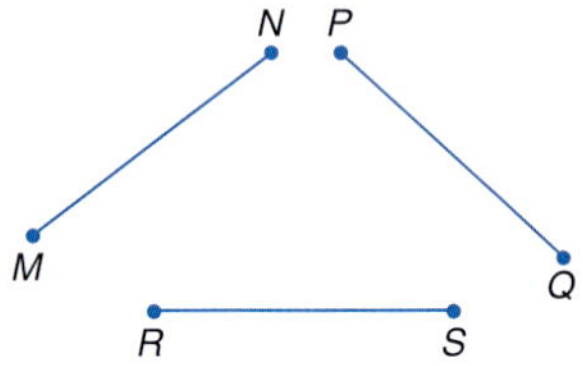

Proof:

Method 1 Paragraph Proof

Since $\overline{MN} \cong \overline{PQ}$ and $\overline{PQ} \cong \overline{RS}$, $MN = PQ$ and $PQ = RS$ by the definition of congruent segments. By the Transitive Property of Equality, $MN = RS$. Thus, $\overline{MN} \cong \overline{RS}$ by the definition of congruent segments.

Geometry Software Investigation

Adding Segment Measures Have students repeat the activity for different lengths of $\overline{AC}$. Students can also construct $\overline{AC}$ vertically and at various diagonals. Tell students that this activity provides several examples to substantiate the Segment Addition Postulate.

Method 2 Two-Column Proof

Statements	Reasons
1. $\overline{MN} \cong \overline{PQ}, \overline{PQ} \cong \overline{RS}$	1. Given
2. $MN = PQ, PQ = RS$	2. Definition of congruent segments
3. $MN = RS$	3. Transitive Property
4. $\overline{MN} \cong \overline{RS}$	4. Definition of congruent segments

The theorems about segment congruence can be used to prove segment relationships.

Example 2 Proof With Segment Congruence

Prove the following.

Given: $\overline{JK} \cong \overline{KL}, \overline{HJ} \cong \overline{GH}, \overline{KL} \cong \overline{HJ}$

Prove: $\overline{GH} \cong \overline{JK}$

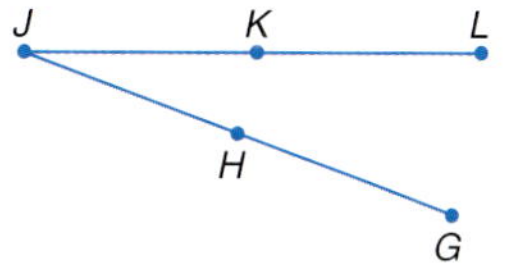

Proof:

Statements	Reasons
1. $\overline{JK} \cong \overline{KL}, \overline{KL} \cong \overline{HJ}$	1. Given
2. $\overline{JK} \cong \overline{HJ}$	2. Transitive Property
3. $\overline{HJ} \cong \overline{GH}$	3. Given
4. $\overline{JK} \cong \overline{GH}$	4. Transitive Property
5. $\overline{GH} \cong \overline{JK}$	5. Symmetric Property

Check for Understanding

Concept Check

1. Choose two cities from a United States road map. Describe the distance between the cities using the Reflexive Property. See margin.
2. OPEN ENDED Draw three congruent segments, and illustrate the Transitive Property using these segments. See margin.
3. **Describe** how to determine whether a point B is between points A and C. If A, B, and C are collinear and $AB + BC = AC$, then B is between A and C.

Guided Practice

Justify each statement with a property of equality or a property of congruence.

4. $\overline{XY} \cong \overline{XY}$ Reflexive
5. If $\overline{GH} \cong \overline{MN}$, then $\overline{MN} \cong \overline{GH}$. Symmetric
6. If $AB = AC + CB$, then $AB - AC = CB$. Subtraction

GUIDED PRACTICE KEY	
Exercises	Examples
6	1
4, 5, 7–10	2

7. Copy and complete the proof.

Given: $\overline{PQ} \cong \overline{RS}, \overline{QS} \cong \overline{ST}$

Prove: $\overline{PS} \cong \overline{RT}$

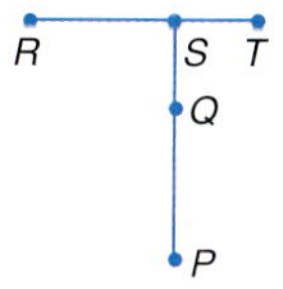

Proof:

Statements	Reasons
a. ___?___ , ___?___	a. Given
b. $PQ = RS, QS = ST$	b. ___?___ Def. of ≅ segs.
c. $PS = PQ + QS, RT = RS + ST$	c. ___?___ Segment Addition Post.
d. ___?___ $PQ + QS = RS + ST$	d. Addition Property
e. ___?___ $PS = RT$	e. Substitution
f. $\overline{PS} \cong \overline{RT}$	f. ___?___ Def. of ≅ segs.

7a. $\overline{PQ} \cong \overline{RS}$, $\overline{QS} \cong \overline{ST}$

www.geometryonline.com/extra_examples/fcat

SEGMENT CONGRUENCE

In-Class Example

Power Point®

2 Prove the following.

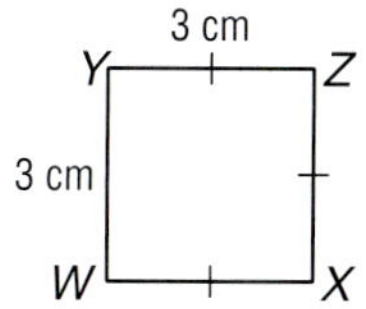

Given: $WY = YZ$
$\overline{YZ} \cong \overline{XZ}$
$\overline{XZ} \cong \overline{WX}$

Prove: $\overline{WX} \cong \overline{WY}$

Proof:

Statements (Reasons)

1. $WY = YZ$ (Given)
2. $\overline{WY} \cong \overline{YZ}$ (Def. of ≅ Segs.)
3. $\overline{YZ} \cong \overline{XZ}$; $\overline{XZ} \cong \overline{WX}$ (Given)
4. $\overline{WY} \cong \overline{WX}$ (Trans. Prop.)
5. $\overline{WX} \cong \overline{WY}$ (Symmetric)

Answers

1. Sample answer: The distance from Cleveland to Chicago is the same as the distance from Cleveland to Chicago.
2. Sample answer: If $\overline{AB} \cong \overline{XY}$ and $\overline{XY} \cong \overline{PQ}$, then $\overline{AB} \cong \overline{PQ}$.

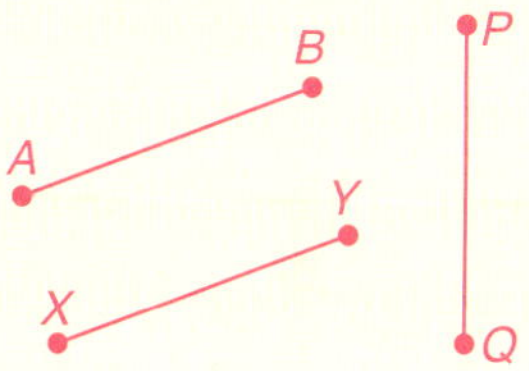

DAILY INTERVENTION

Differentiated Instruction

Visual/Spatial When students are first examining figures to determine the steps necessary for writing proofs, encourage them to use their spatial skills to locate obvious and hidden congruent segments and parts of segments that qualify for the Segment Addition Postulate. Advise students to use the given information to mark the figures so they can easily refer to the relationships in the figures while they are writing their proofs.

3 Practice/Apply

Study Notebook

Have students—
- add the definitions/examples of the vocabulary terms to their Vocabulary Builder worksheets for Chapter 2.
- include a sample proof using segment addition and one using segment congruence.
- include any other item(s) that they find helpful in mastering the skills in this lesson.

About the Exercises...

Organization by Objective
- **Segment Addition:** 14, 16, 17
- **Segment Congruence:** 12, 13, 15, 18–24

Odd/Even Assignments

Exercises 12–23 are structured so that students practice the same concepts whether they are assigned odd or even problems.

Assignment Guide

Basic: 13–27 odd, 29–45

Average: 13–27 odd, 29–45

Advanced: 12–26 even, 27–39 (optional: 40–45)

Answer

10. Given: $\overline{AB} \cong \overline{CD}$
Prove: $\overline{CD} \cong \overline{AB}$

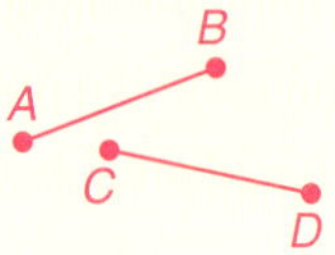

Proof:
Statements (Reasons)
1. $\overline{AB} \cong \overline{CD}$ (Given)
2. $AB = CD$ (Def. of $\cong$ segs.)
3. $CD = AB$ (Symmetric Prop.)
4. $\overline{CD} \cong \overline{AB}$ (Def. of $\cong$ segs.)

PROOF For Exercises 8–10, write a two-column proof. 8–9. See p. 123E.

8. Given: $\overline{AP} \cong \overline{CP}$
$\overline{BP} \cong \overline{DP}$
Prove: $\overline{AB} \cong \overline{CD}$

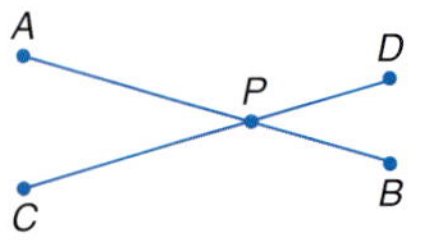

9. Given: $\overline{HI} \cong \overline{TU}$
$\overline{HJ} \cong \overline{TV}$
Prove: $\overline{IJ} \cong \overline{UV}$

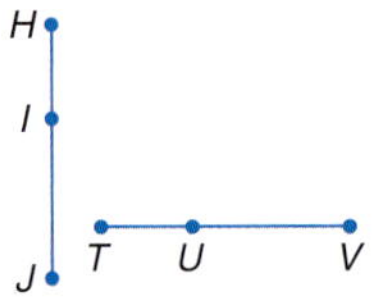

10. Symmetric Property of Congruence (Theorem 2.2) See margin.

Application

11. **GEOGRAPHY** Aberdeen in South Dakota and Helena, Miles City, and Missoula, all in Montana, are connected in a straight line by interstate highways. Missoula is 499 miles from Miles City and 972 miles from Aberdeen. Aberdeen is 473 miles from Miles City and 860 miles from Helena. Between which cities does Helena lie? **Helena is between Missoula and Miles City.**

Practice and Apply

Homework Help

For Exercises	See Examples
14, 16, 17	1
12, 13, 15, 18–24	2

Extra Practice
See page 758.

Justify each statement with a property of equality or a property of congruence.

12. If $\overline{JK} \cong \overline{LM}$, then $\overline{LM} \cong \overline{JK}$. **Symmetric**
13. If $AB = 14$ and $CD = 14$, then $AB = CD$. **Substitution**
14. If W, X, and Y are collinear, in that order, then $WY = WX + XY$. **Segment Addition**
15. If $\overline{MN} \cong \overline{PQ}$ and $\overline{PQ} \cong \overline{RS}$, then $\overline{MN} \cong \overline{RS}$. **Transitive**
16. If $EF = TU$ and $GH = VW$, then $EF + GH = TU + VW$. **Addition**
17. If $JK + MN = JK + QR$, then $MN = QR$. **Subtraction**

18. Copy and complete the proof.
Given: $\overline{AD} \cong \overline{CE}$, $\overline{DB} \cong \overline{EB}$
Prove: $\overline{AB} \cong \overline{CB}$

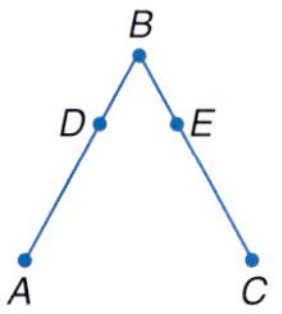

Proof:

Statements	Reasons
a. ? $\overline{AD} \cong \overline{CE}$, $\overline{DB} \cong \overline{EB}$	a. Given
b. $AD = CE$, $DB = EB$	b. ? Def. of $\cong$ segs.
c. $AD + DB = CE + EB$	c. ? Add. Prop.
d. ?	d. Segment Addition Postulate
e. $AB = CB$	e. ? Substitution
f. $\overline{AB} \cong \overline{CB}$	f. ? Def. of $\cong$ segs.

18d. $AB = AD + DB$, $CB = CE + EB$

PROOF Write a two-column proof. 19–20. See p. 123E.

19. If $\overline{XY} \cong \overline{WZ}$ and $\overline{WZ} \cong \overline{AB}$, then $\overline{XY} \cong \overline{AB}$.

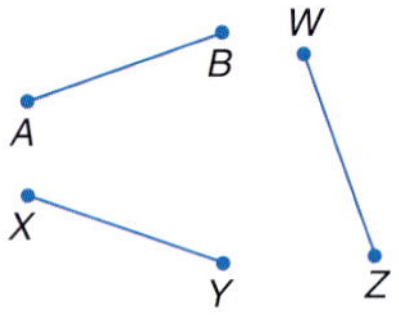

20. If $\overline{AB} \cong \overline{AC}$ and $\overline{PC} \cong \overline{QB}$, then $\overline{AP} \cong \overline{AQ}$.

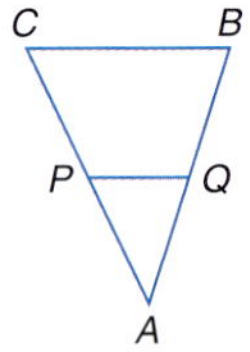

104 Chapter 2 Reasoning and Proof

21. Copy and complete the proof.

Given: $\overline{WY} \cong \overline{ZX}$
A is the midpoint of $\overline{WY}$.
A is the midpoint of $\overline{ZX}$.

Prove: $\overline{WA} \cong \overline{ZA}$

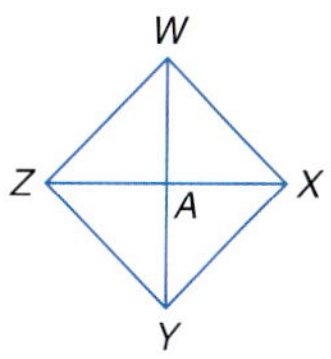

Proof:

Statements	Reasons
a. $\overline{WY} \cong \overline{ZX}$; A is the midpoint of $\overline{WY}$. A is the midpoint of $\overline{ZX}$.	a. ? Given
b. $WY = ZX$	b. ? Def. of ≅ segs.
c. ? $WA = AY$, $ZA = AX$	c. Definition of midpoint
d. $WY = WA + AY$, $ZX = ZA + AX$	d. ? Segment Addition Post.
e. $WA + AY = ZA + AX$	e. ? Substitution
f. $WA + WA = ZA + ZA$	f. ? Substitution
g. $2WA = 2ZA$	g. ? Substitution
h. ? $WA = ZA$	h. Division Property
i. $\overline{WA} \cong \overline{ZA}$	i. ? Def. of ≅ segs.

PROOF For Exercises 22–24, write a two-column proof. 22–24. See p. 123E.

22. If $\overline{LM} \cong \overline{PN}$ and $\overline{XM} \cong \overline{XN}$, then $\overline{LX} \cong \overline{PX}$.

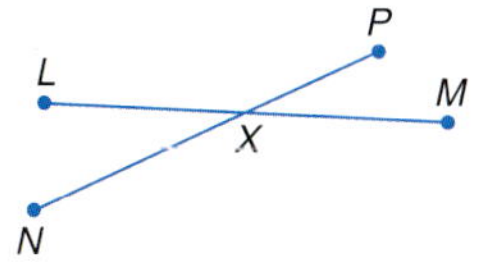

23. If $AB = BC$, then $AC = 2BC$.

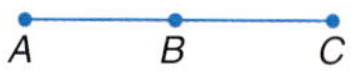

24. Reflexive Property of Congruence (Theorem 2.2)

25. **DESIGN** The front of a building has a triangular window. If $\overline{AB} \cong \overline{DE}$ and C is the midpoint of $\overline{BD}$, prove that $\overline{AC} \cong \overline{CE}$. See p. 123E.

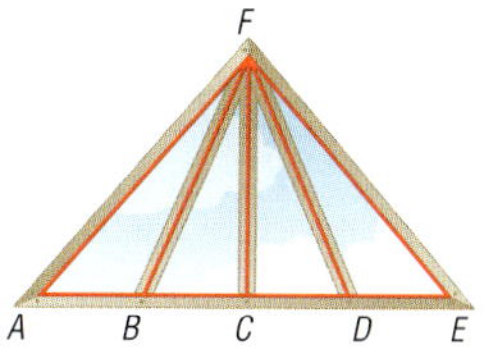

26. **LIGHTING** The light fixture in Gerrard Hall of the University of North Carolina is shown at the right. If $\overline{AB} \cong \overline{EF}$ and $\overline{BC} \cong \overline{DE}$, prove that $\overline{AC} \cong \overline{DF}$. See p. 123F.

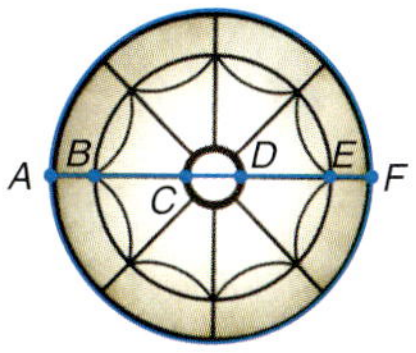

27. **CRITICAL THINKING** Given that $\overline{LN} \cong \overline{RT}$, $\overline{RT} \cong \overline{QO}$, $\overline{LQ} \cong \overline{NO}$, $\overline{MP} \cong \overline{NO}$, S is the midpoint of $\overline{RT}$, M is the midpoint of $\overline{LN}$, and P is the midpoint of $\overline{QO}$, list three statements that you could prove using the postulates, theorems, and definitions that you have learned. See p. 123F.

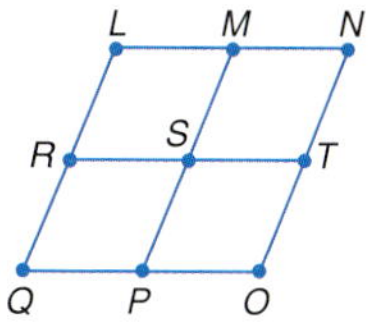

More About. . .

Design

Windows come in many different shapes and sizes. Some commonly used shapes for windows are circles, rectangles, squares, triangles, pentagons, and octagons.

Source: www.pella.com

www.geometryonline.com/self_check_quiz/fcat

Study Guide and Intervention, p. 93 (shown) and p. 94

Segment Addition Two basic postulates for working with segments and lengths are the Ruler Postulate, which establishes number lines, and the Segment Addition Postulate, which describes what it means for one point to be between two other points.

Ruler Postulate	The points on any line or line segment can be paired with real numbers so that, given any two points A and B on a line, A corresponds to zero and B corresponds to a positive real number.
Segment Addition Postulate	B is between A and C if and only if $AB + BC = AC$.

Example **Write a two-column proof.**

Given: Q is the midpoint of $\overline{PR}$.
R is the midpoint of $\overline{QS}$.

Prove: $PR = QS$

Statements	Reasons
1. Q is the midpoint of $\overline{PR}$.	1. Given
2. $PQ = QR$	2. Definition of midpoint
3. R is the midpoint of $\overline{QS}$.	3. Given
4. $QR = RS$	4. Definition of midpoint
5. $PQ + QR = QR + RS$	5. Addition Property
6. $PQ + QR = PR$, $QR + RS = QS$	6. Segment Addition Postulate
7. $PR = QS$	7. Substitution

Exercises

Complete each proof.

1. **Given:** $BC = DE$
Prove: $AB + DE = AC$

Statements	Reasons
a. $BC = DE$	a. Given
b. $AB + BC = AC$	b. Seg. Add. Post.
c. $AB + DE = AC$	c. Substitution

2. **Given:** Q is between P and R, R is between Q and S, $PR = QS$.
Prove: $PQ = RS$

Statements	Reasons
a. Q is between P and R.	a. Given
b. $PQ + QR = PR$	b. Seg. Add. Post.
c. R is between Q and S.	c. Given
d. $QR + RS = QS$	d. Seg. Add. Post.
e. $PR = QS$	e. Given
f. $PQ + QR = QR + RS$	f. Substitution
g. $PQ + QR - QR = QR + RS - QR$	g. Subtraction Prop.
h. $PQ = RS$	h. Substitution

Skills Practice, p. 95 and Practice, p. 96 (shown)

Complete the following proof.

1. **Given:** $\overline{AB} \cong \overline{DE}$
B is the midpoint of $\overline{AC}$.
E is the midpoint of $\overline{DF}$.
Prove: $\overline{BC} \cong \overline{EF}$
Proof:

Statements	Reasons
a. $\overline{AB} \cong \overline{DE}$; B is the midpoint of $\overline{AC}$. E is the midpoint of $\overline{DF}$.	a. Given
b. $AB = DE$	b. Definition of ≅ segments
c. $AB = BC$; $DE = EF$	c. Definition of Midpoint
d. $AC = AB + BC$; $DF = DE + EF$	d. Segment Addition Postulate
e. $AB + BC = DE + EF$	e. Substitution Property
f. $AB + BC = AB + EF$	f. Substitution Property
g. $AB + BC - AB = AB + EF - AB$	g. Subtraction Property
h. $BC = EF$	h. Substitution Property
i. $\overline{BC} \cong \overline{EF}$	i. Definition of ≅ segments

2. **TRAVEL** Refer to the figure. DeAnne knows that the distance from Grayson to Apex is the same as the distance from Redding to Pine Bluff. Prove that the distance from Grayson to Redding is equal to the distance from Apex to Pine Bluff.

Given: $\overline{GA} \cong \overline{RP}$
Prove: $\overline{GR} \cong \overline{AP}$
Proof:

Statements	Reasons
1. $\overline{GA} \cong \overline{RP}$	1. Given
2. $GA = RP$	2. Definition of ≅ segments
3. $GA + AR = AR + RP$	3. Addition Property
4. $GR = GA + AR$, $AP = AR + RP$	4. Segment Addition Postulate
5. $GR = AP$	5. Substitution Property
6. $\overline{GR} \cong \overline{AP}$	6. Definition of ≅ segments

Reading to Learn Mathematics, p. 97 ELL

Pre-Activity **How can segment relationships be used for travel?**

Read the introduction to Lesson 2-7 at the top of page 101 in your textbook.

- What is the total distance that the plane will fly to get from San Diego to Dallas? 1430 mi
- Before leaving home, a passenger used a road atlas to determine that the distance between San Diego and Dallas is about 1350 miles. Why is the flying distance greater than that? Sample answer: Phoenix is not on a straight line between San Diego and Dallas, so the stop added to the distance traveled. A nonstop flight would have been shorter.

Reading the Lesson

1. If E is between Y and S, which of the following statements are *always* true? B, E
A. $YS + ES = YE$ B. $YS - ES = YE$
C. $YE > ES$ D. $YE \cdot ES = YS$
E. $SE + EY = SY$ F. E is the midpoint of $\overline{YS}$.

2. Give the reason for each statement in the following two-column proof.
Given: C is the midpoint of $\overline{BD}$.
D is the midpoint of $\overline{CE}$.
Prove: $\overline{BD} \cong \overline{CE}$

Statements	Reasons
1. C is the midpoint of $\overline{BD}$.	1. Given
2. $BC = CD$	2. Definition of midpoint
3. D is the midpoint of $\overline{CE}$.	3. Given
4. $CD = DE$	4. Definition of midpoint
5. $BC = DE$	5. Transitive Property of Equality
6. $BC + CD = CD + DE$	6. Addition Property of Equality
7. $BC + CD = BD$; $CD + DE = CE$	7. Segment Addition Postulate
8. $BD = CE$	8. Substitution Property
9. $\overline{BD} \cong \overline{CE}$	9. Def. of ≅ segments

Helping You Remember

3. One way to keep the names of related postulates straight in your mind is to associate something in the name of the postulate with the content of the postulate. How can you use this idea to distinguish between the Ruler Postulate and the Segment Addition Postulate? Sample answer: There are two words in "Ruler Postulate" and three words in "Segment Addition Postulate." The statement of the Ruler Postulate mentions two points, and the statement of the Segment Addition Postulate mentions three points.

Enrichment, p. 98

Geometry Crossword Puzzle

4 Assess

Open-Ended Assessment

Modeling Make a "Reasons/Statements" board with Velcro in positions where you could place given information, statements and reasons. Create three or four proofs using segment addition and segment congruence, and write the given information and each statement and reason on a separate rectangular piece of poster board (large enough to read from the back of the classroom). Affix Velcro to the back of the boards so they can be easily placed on the R/S board. Place the given information at the top of the R/S board. Have students select each statement, match it with its corresponding reason and place it in the correct order on the R/S board.

Getting Ready for Lesson 2-8

Prerequisite Skill In Lesson 2-8, students will apply properties of supplementary and complementary angles to prove angle relationships. Use Exercises 40–45 to determine your students' familiarity with complementary and supplementary angles.

Answers

28. Sample answer: You can use segment addition to find the total distance between two destinations by adding the distances of various points in between. Answers should include the following.
    - A passenger can add the distance from San Diego to Phoenix and the distance from Phoenix to Dallas to find the distance from San Diego to Dallas.
    - The Segment Addition Postulate can be useful if you are traveling in a straight line.

28. **WRITING IN MATH** Answer the question that was posed at the beginning of the lesson. See margin.

How can segment relationships be used for travel?

Include the following in your answer:
- an explanation of how a passenger can use the distances the pilot announced to find the total distance from San Diego to Dallas, and
- an explanation of why the Segment Addition Postulate may or may not be useful when traveling.

FCAT Practice
Standardized Test Practice

29. If P is the midpoint of $\overline{BC}$ and Q is the midpoint of $\overline{AD}$, what is PQ? B

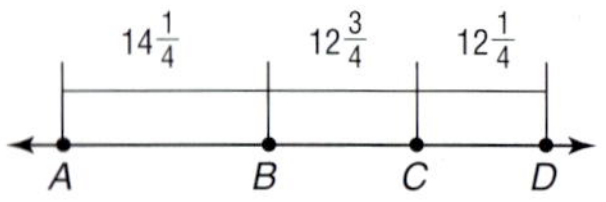

Ⓐ $\frac{1}{2}$ Ⓑ 1
Ⓒ 2 Ⓓ $2\frac{1}{2}$

30. **GRID IN** A refreshment stand sells a large tub of popcorn for twice the price of a box of popcorn. If 60 tubs were sold for a total of \$150 and the total popcorn sales were \$275, how many boxes of popcorn were sold? 100

Maintain Your Skills

Mixed Review **State the property that justifies each statement.** *(Lesson 2-6)*

31. If $m\angle P + m\angle Q = 110$ and $m\angle R = 110$, then $m\angle P + m\angle Q = m\angle R$. Substitution
32. If $x(y + z) = a$, then $xy + xz = a$. Dist. Prop.
33. If $n - 17 = 39$, then $n = 56$. Add. Prop.
34. If $cv = md$ and $md = 15$, then $cv = 15$. Trans. Prop.

Determine whether the following statements are *always, sometimes,* or *never* true. Explain. *(Lesson 2-5)* 35–38. See margin for explanations.

35. A midpoint divides a segment into two noncongruent segments. never
36. Three lines intersect at a single point. sometimes
37. The intersection of two planes forms a line. always
38. Three single points determine three lines. sometimes

39. If the perimeter of rectangle $ABCD$ is 44 centimeters, find x and the dimensions of the rectangle. *(Lesson 1-6)* 3; 9 cm by 13 cm

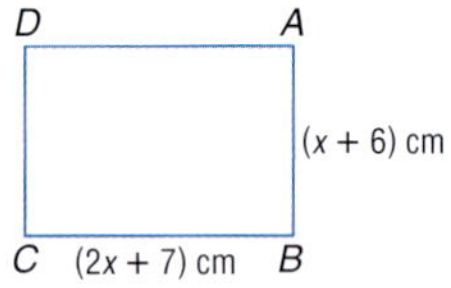

Getting Ready for the Next Lesson **PREREQUISITE SKILL** **Find x.**
*(To review **complementary and supplementary angles**, see Lesson 1-5.)*

40. 30
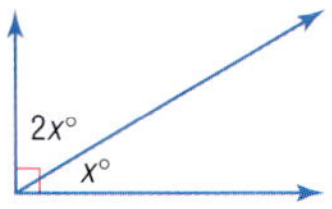

41. 15
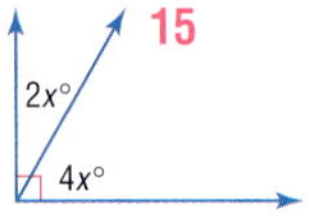

42. 22
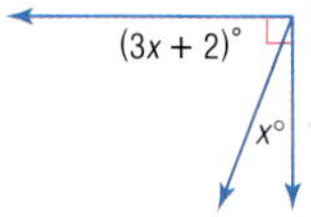

43. 45
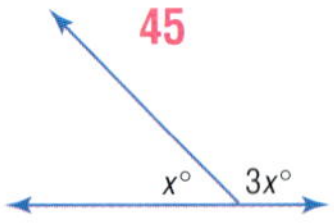

44. 5
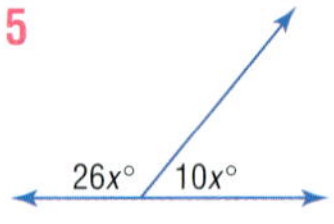

45. 25
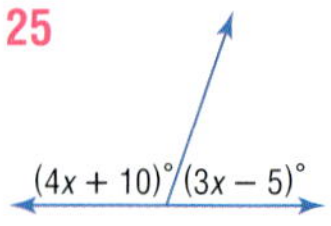

35. The midpoint of a segment divides it into two congruent segments.
36. If the lines have a common intersection point, then it is a single point.
37. If two planes intersect, they intersect in a line.
38. If the points are noncollinear, then they lie on three distinct lines.

2-8 Proving Angle Relationships

Sunshine State Standards
MA.C.2.4.1, MA.C.1.4.1

What You'll Learn

- Write proofs involving supplementary and complementary angles.
- Write proofs involving congruent and right angles.

How do scissors illustrate supplementary angles?

Notice that when a pair of scissors is opened, the angle formed by the two blades, ∠1, and the angle formed by a blade and a handle, ∠2, are a linear pair. Likewise, the angle formed by a blade and a handle, ∠2, and the angle formed by the two handles, ∠3, also forms a linear pair.

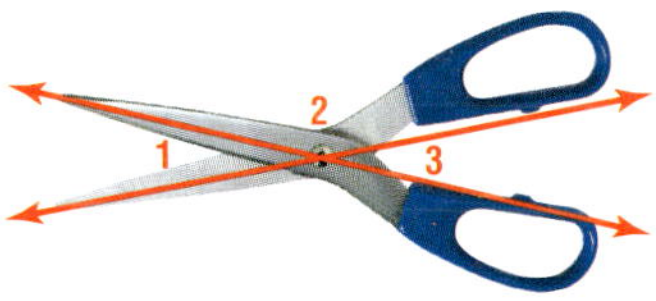

SUPPLEMENTARY AND COMPLEMENTARY ANGLES Recall that when you measure angles with a protractor, you position the protractor so that one of the rays aligns with zero degrees and then determine the position of the second ray. This illustrates the Protractor Postulate.

Postulate 2.10

Protractor Postulate Given $\overrightarrow{AB}$ and a number r between 0 and 180, there is exactly one ray with endpoint A, extending on either side of $\overrightarrow{AB}$, such that the measure of the angle formed is r.

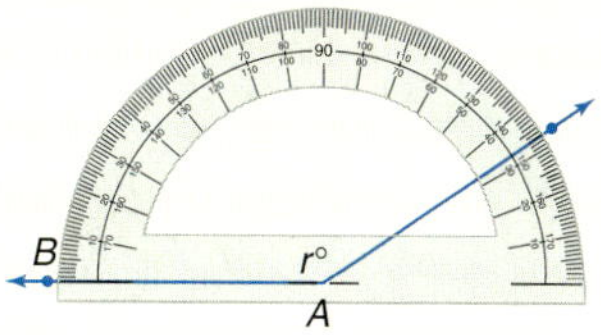

More About. . .

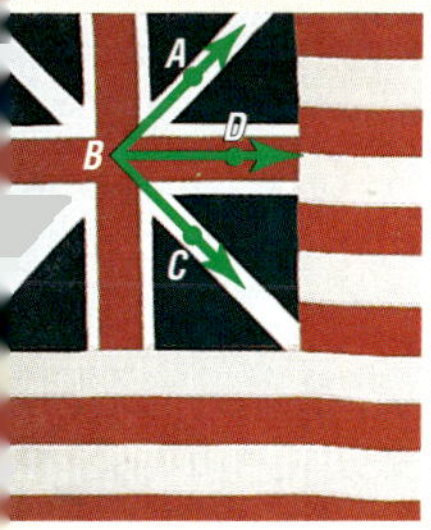

History The Grand Union flag was the first flag used by the colonial United States that resembles the current flag. It was made up of thirteen stripes with the flag of Great Britain in the corner.

Source: www.usflag.org

In Lesson 2-7, you learned about the Segment Addition Postulate. A similar relationship exists between the measures of angles.

Postulate 2.11

Angle Addition Postulate If R is in the interior of $\angle PQS$, then $m\angle PQR + m\angle RQS = m\angle PQS$.

If $m\angle PQR + m\angle RQS = m\angle PQS$, then R is in the interior of $\angle PQS$.

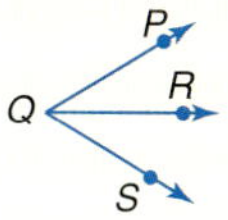

Example 1 Angle Addition

HISTORY **The Grand Union Flag at the left contains several angles. If $m\angle ABD = 44$ and $m\angle ABC = 88$, find $m\angle DBC$.**

$m\angle ABD + m\angle DBC = m\angle ABC$	Angle Addition Postulate
$44 + m\angle DBC = 88$	$m\angle ABD = 44$, $m\angle ABC = 88$
$m\angle DBC = 44$	Subtraction Property

2-8 Lesson Notes

1 Focus

5-Minute Check Transparency 2-8 Use as a quiz or review of Lesson 2-7.

Mathematical Background notes are available for this lesson on p. 60D.

How do scissors illustrate supplementary angles?

Ask students:

- In the figure, label ∠4 vertical to ∠2 and name all pairs of supplementary angles. **∠1 and ∠2, ∠2 and ∠3, ∠3 and ∠4, ∠4 and ∠1**
- Use a protractor to measure angles 1 and 2. What is the sum of these two measures? **about 40°; about 140°; 180°.**
- Will the same angles still form linear pairs if the scissors were opened wider? narrower? **yes; yes**

Resource Manager

Workbook and Reproducible Masters

Chapter 2 Resource Masters
- Study Guide and Intervention, pp. 99–100
- Skills Practice, p. 101
- Practice, p. 102
- Reading to Learn Mathematics, p. 103
- Enrichment, p. 104
- Assessment, p. 120

Prerequisite Skills Workbook, pp. 81–82, 85–86

Teaching Geometry With Manipulatives Masters, pp. 8, 16, 48

Transparencies

5-Minute Check Transparency 2-8
Answer Key Transparencies

Technology

Interactive Chalkboard

2 Teach

SUPPLEMENTARY AND COMPLEMENTARY ANGLES

In-Class Examples

Power Point®

1 **TIME** At 4 o' clock, the angle between the hour and minute hands of a clock is 120°. If the second hand stops where it bisects the angle between the hour and minute hands, what are the measures of the angles between the minute and second hands and between the second and hour hands? **They are both 60° by the definition of angle bisector and the Angle Addition Postulate.**

2 If $\angle 1$ and $\angle 2$ form a linear pair and $m\angle 2 = 166$, find $m\angle 1$. **14**

Study Tip

Look Back
To review **supplementary** and **complementary angles**, see Lesson 1-5.

The Angle Addition Postulate can be used with other angle relationships to provide additional theorems relating to angles.

Theorems

2.3 Supplement Theorem If two angles form a linear pair, then they are supplementary angles.

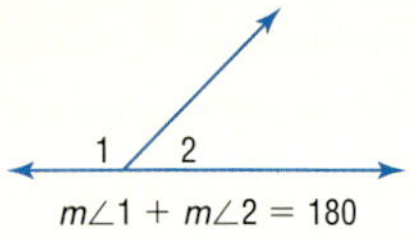

2.4 Complement Theorem If the noncommon sides of two adjacent angles form a right angle, then the angles are complementary angles.

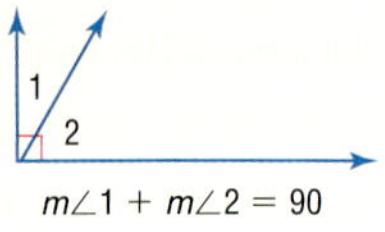

You will prove Theorems 2.3 and 2.4 in Exercises 10 and 11.

Example 2 Supplementary Angles

If $\angle 1$ and $\angle 2$ form a linear pair and $m\angle 2 = 67$, find $m\angle 1$.

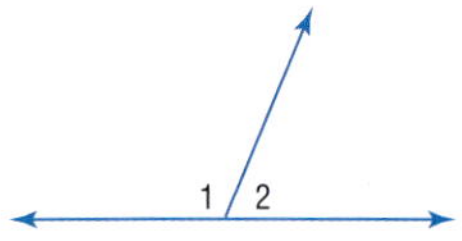

$m\angle 1 + m\angle 2 = 180$ Supplement Theorem

$m\angle 1 + 67 = 180$ $m\angle 2 = 67$

$m\angle 1 = 113$ Subtraction Property

CONGRUENT AND RIGHT ANGLES The properties of algebra that applied to the congruence of segments and the equality of their measures also hold true for the congruence of angles and the equality of their measures.

Theorem 2.5

Congruence of angles is reflexive, symmetric, and transitive.

Reflexive Property $\angle 1 \cong \angle 1$

Symmetric Property If $\angle 1 \cong \angle 2$, then $\angle 2 \cong \angle 1$.

Transitive Property If $\angle 1 \cong \angle 2$, and $\angle 2 \cong \angle 3$, then $\angle 1 \cong \angle 3$.

You will prove the Reflexive and Transitive Properties of Angle Congruence in Exercises 26 and 27.

TEACHING TIP
The Symmetric Property is often assumed in proofs to condense the number of steps in a proof. The rigor of proof is left up to the teacher, but we will assume symmetric property statements in future chapters.

Proof Symmetric Property of Congruence

Given: $\angle A \cong \angle B$

Prove: $\angle B \cong \angle A$

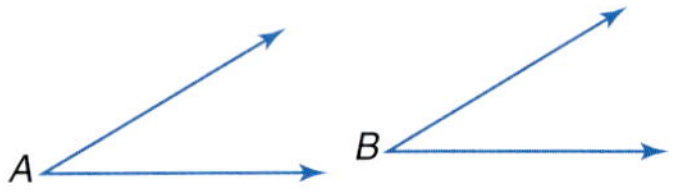

Paragraph Proof:

We are given $\angle A \cong \angle B$. By the definition of congruent angles, $m\angle A = m\angle B$. Using the Symmetric Property, $m\angle B = m\angle A$. Thus, $\angle B \cong \angle A$ by the definition of congruent angles.

Algebraic properties can be applied to prove theorems for congruence relationships involving supplementary and complementary angles.

DAILY INTERVENTION

Differentiated Instruction

Auditory/Musical Ask students to close their books. Read Theorems 2.3–2.13 aloud for students one by one. After each one, ask students to discuss how they know the theorem is true and how they might use the theorem in a proof.

Theorems

2.6 Angles supplementary to the same angle or to congruent angles are congruent.

Abbreviation: $\measuredangle$ suppl. to same $\angle$ or $\cong$ $\measuredangle$ are $\cong$.

Example: If $m\angle 1 + m\angle 2 = 180$ and $m\angle 2 + m\angle 3 = 180$, then $\angle 1 \cong \angle 3$.

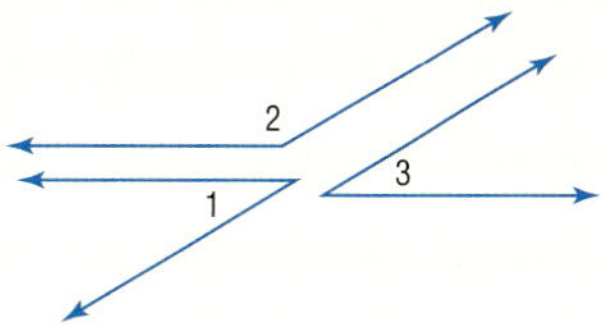

2.7 Angles complementary to the same angle or to congruent angles are congruent.

Abbreviation: $\measuredangle$ compl. to same $\angle$ or $\cong$ $\measuredangle$ are $\cong$.

Example: If $m\angle 1 + m\angle 2 = 90$ and $m\angle 2 + m\angle 3 = 90$, then $\angle 1 \cong \angle 3$.

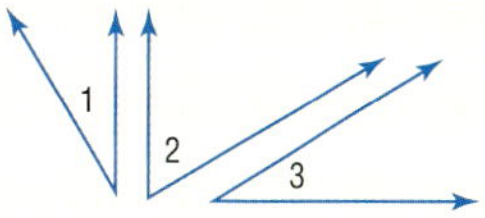

You will prove Theorem 2.6 in Exercise 6.

Proof *Theorem 2.7*

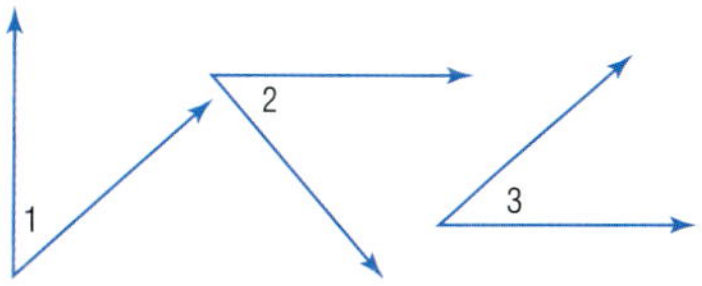

Given: $\angle 1$ and $\angle 3$ are complementary.
$\angle 2$ and $\angle 3$ are complementary.

Prove: $\angle 1 \cong \angle 2$

Proof:

Statements	Reasons
1. $\angle 1$ and $\angle 3$ are complementary. $\angle 2$ and $\angle 3$ are complementary.	1. Given
2. $m\angle 1 + m\angle 3 = 90$; $m\angle 2 + m\angle 3 = 90$	2. Definition of complementary angles
3. $m\angle 1 + m\angle 3 = m\angle 2 + m\angle 3$	3. Substitution
4. $m\angle 3 = m\angle 3$	4. Reflective Property
5. $m\angle 1 = m\angle 2$	5. Subtraction Property
6. $\angle 1 \cong \angle 2$	6. Definition of congruent angles

comp to same ∠ are ≅

Example 3 *Use Supplementary Angles*

In the figure, $\angle 1$ and $\angle 2$ form a linear pair and $\angle 2$ and $\angle 3$ form a linear pair. Prove that $\angle 1$ and $\angle 3$ are congruent.

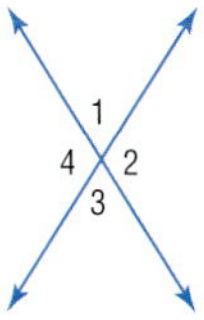

vertical angles ≅

Given: $\angle 1$ and $\angle 2$ form a linear pair.
$\angle 2$ and $\angle 3$ form a linear pair.

Prove: $\angle 1 \cong \angle 3$

Proof:

Statements	Reasons
1. $\angle 1$ and $\angle 2$ form a linear pair. $\angle 2$ and $\angle 3$ form a linear pair.	1. Given
2. $\angle 1$ and $\angle 2$ are supplementary. $\angle 2$ and $\angle 3$ are supplementary.	2. Supplement Theorem
3. $\angle 1 \cong \angle 3$	3. $\measuredangle$ suppl. to same $\angle$ or $\cong$ $\measuredangle$ are $\cong$.

www.geometryonline.com/extra_examples/fcat

CONGRUENT AND RIGHT ANGLES

In-Class Example

Power Point®

3 In the figure, $\angle 1$ and $\angle 4$ form a linear pair, and $m\angle 3 + m\angle 1 = 180$. Prove that $\angle 3$ and $\angle 4$ are congruent.

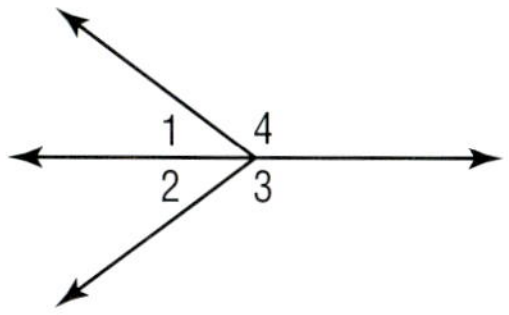

Statements (Reasons)

1. $m\angle 3 + m\angle 1 = 180$; $\angle 1$ and $\angle 4$ form a linear pair. (Given)
2. $\angle 1$ and $\angle 4$ are supplementary. (Linear pairs are suppl.)
3. $\angle 3$ and $\angle 1$ are supplementary. (Def. of suppl. $\measuredangle$)
4. $\angle 3 \cong \angle 4$ ($\measuredangle$ supplementary to same $\angle$ are $\cong$.)

In-Class Example

Power Point®

Teaching Tip Tell students to read problems carefully so they can be sure to provide the information requested. For this example, point out that students are to find angle measures, not just the value of the variable; however, they have to use the value of the variable to find the answer.

4 If ∠1 and ∠2 are vertical angles and $m\angle 1 = d - 32$ and $m\angle 2 = 175 - 2d$, find $m\angle 1$ and $m\angle 2$. 37; 37

Study Tip

Look Back
To review **vertical angles**, see Lesson 1-5.

Note that in Example 3, ∠1 and ∠3 are vertical angles. The conclusion in the example is a proof for the following theorem.

Theorem 2.8

Vertical Angles Theorem If two angles are vertical angles, then they are congruent.
Abbreviation: Vert. ∠s are ≅.

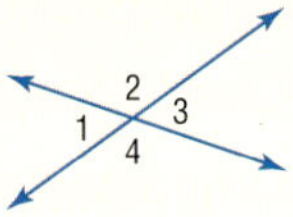

$\angle 1 \cong \angle 3$ and $\angle 2 \cong \angle 4$

Example 4 Vertical Angles

If ∠1 and ∠2 are vertical angles and $m\angle 1 = x$ and $m\angle 2 = 228 - 3x$, find $m\angle 1$ and $m\angle 2$.

$\angle 1 \cong \angle 2$	Vertical Angles Theorem
$m\angle 1 = m\angle 2$	Definition of congruent angles
$x = 228 - 3x$	Substitution
$4x = 228$	Add $3x$ to each side.
$x = 57$	Divide each side by 4.

$m\angle 1 = x = 57$ $\quad$ $m\angle 2 = m\angle 1 = 57$

The theorems you have learned can be applied to right angles. You can create right angles and investigate congruent angles by paper folding.

Geometry Activity

Right Angles

Make a Model

- Fold the paper so that one corner is folded downward.
- Fold along the crease so that the top edge meets the side edge.
- Unfold the paper and measure each of the angles formed.
- Repeat the activity three more times.

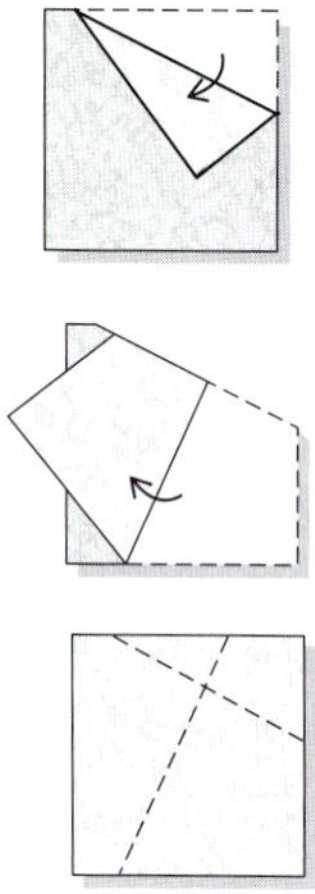

Analyze the Model 1. The lines are perpendicular.

1. What do you notice about the lines formed?
2. What do you notice about each pair of adjacent angles? They are congruent and they form linear pairs.
3. What are the measures of the angles formed? 90

Make a Conjecture 4. They form right angles.

4. What is true about perpendicular lines?
5. What is true about all right angles? They all measure 90 and are congruent.

The following theorems support the conjectures you made in the Geometry Activity.

Geometry Activity

Materials: paper, protractor

- When students are repeating the activity, tell them to use different folds from the right and the left sides of the paper each time.
- Ask students what they notice about each pair of vertical angles (they are congruent and form right angles).

Theorems — Right Angles

2.9 Perpendicular lines intersect to form four right angles.

2.10 All right angles are congruent.

2.11 Perpendicular lines form congruent adjacent angles.

2.12 If two angles are congruent and supplementary, then each angle is a right angle.

2.13 If two congruent angles form a linear pair, then they are right angles.

Check for Understanding

Concept Check

1. **FIND THE ERROR** Tomas and Jacob wrote equations involving the angle measures shown.

Tomas
$m\angle ABE + m\angle EBC = m\angle ABC$

Jacob
$m\angle ABE + m\angle FBC = m\angle ABC$

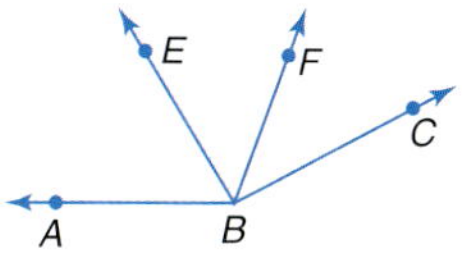

Who is correct? Explain your reasoning. **1–2. See margin.**

2. **OPEN ENDED** Draw three congruent angles. Use these angles to illustrate the Transitive Property for angle congruence.

Guided Practice

GUIDED PRACTICE KEY	
Exercises	Examples
4	1
5	2
6	3
7–11	4

Find the measure of each numbered angle.

3. $m\angle 1 = 65$

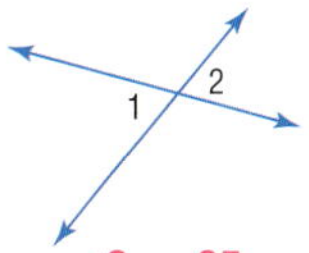

$m\angle 2 = 65$

4. $\angle 6$ and $\angle 8$ are complementary. $m\angle 8 = 47$

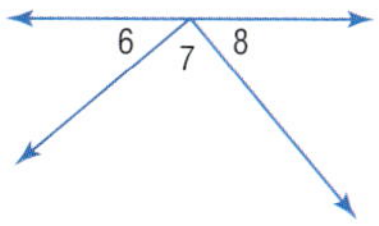

$m\angle 6 = 43$, $m\angle 7 = 90$

5. $m\angle 11 = x - 4$, $m\angle 12 = 2x - 5$

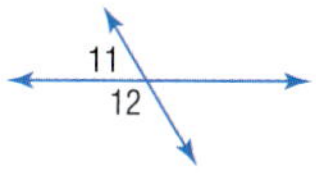

$m\angle 11 = 59$, $m\angle 12 = 121$

6. **PROOF** Copy and complete the proof of Theorem 2.6.

Given: $\angle 1$ and $\angle 2$ are supplementary.
$\angle 3$ and $\angle 4$ are supplementary.
$\angle 1 \cong \angle 4$

Prove: $\angle 2 \cong \angle 3$

Proof:

Statements	Reasons
a. $\angle 1$ and $\angle 2$ are supplementary. $\angle 3$ and $\angle 4$ are supplementary. $\angle 1 \cong \angle 4$	a. ? **Given**
b. $m\angle 1 + m\angle 2 = 180$ $m\angle 3 + m\angle 4 = 180$	b. ? **Def. of suppl. ∠s**
c. $m\angle 1 + m\angle 2 = m\angle 3 + m\angle 4$	c. ? **Substitution**
d. $m\angle 1 = m\angle 4$	d. ? **Def. of ≅ ∠s**
e. $m\angle 2 = m\angle 3$	e. ? **Subtr. Prop.**
f. $\angle 2 \cong \angle 3$	f. ? **Def. of ≅ ∠s**

Answers

1. Tomas; Jacob's answer left out the part of $\angle ABC$ represented by $\angle EBF$.

2. Sample answer: If $\angle 1 \cong \angle 2$ and $\angle 2 \cong \angle 3$, then $\angle 1 \cong \angle 3$.

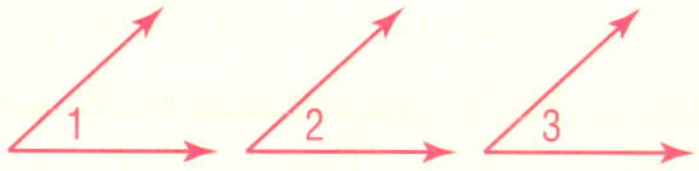

3 Practice/Apply

Study Notebook

Have students—
- add the definitions/examples of the vocabulary terms to their Vocabulary Builder worksheets for Chapter 2.
- include an example each of a proof involving supplementary, complementary, congruent, and right angles.
- include any other item(s) that they find helpful in mastering the skills in this lesson.

DAILY INTERVENTION

FIND THE ERROR Explain that when two angle measures are added using the Angle Addition Postulate, they must share a common ray. Students can note that the common ray in Tomas's answer is $\overrightarrow{BE}$, and a combination of these letters appears in both angles that are being added ($A\underline{BE}$ and $\underline{EB}C$).

About the Exercises...

Organization by Objective
- **Supplementary and Complementary Angles:** 16–18
- **Congruent and Right Angles:** 19–39

Odd/Even Assignments
Exercises 20–36 and 42–43 are structured so that students practice the same concepts whether they are assigned odd or even problems.

Assignment Guide

Basic: 17–41 odd, 42–55
Average: 17–41 odd, 42–55
Advanced: 16–42 even, 44–55

Answers

7. Given: $\overrightarrow{VX}$ bisects $\angle WVY$, $\overrightarrow{VY}$ bisects $\angle XVZ$.

Prove: $\angle WVX \cong \angle YVZ$

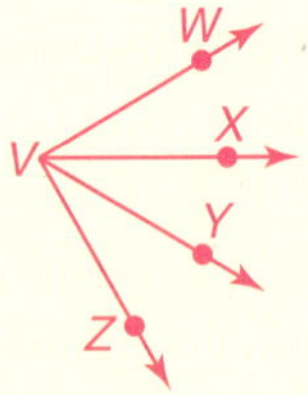

Proof:

Statements (Reasons)

1. $\overrightarrow{VX}$ bisects $\angle WVY$; $\overrightarrow{VY}$ bisects $\angle XVZ$. (Given)
2. $\angle WVX \cong \angle XVY$ (Def. of $\angle$ bisector)
3. $\angle XVY \cong \angle YVZ$ (Def. of $\angle$ bisector)
4. $\angle WVX \cong \angle YVZ$ (Tran. Prop.)

10. Given: Two angles form a linear pair.

Prove: The angles are supplementary

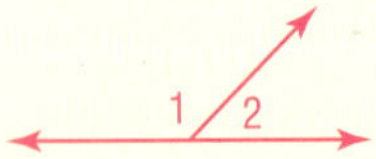

Paragraph Proof: When two angles form a linear pair, the resulting angle is a straight angle whose measure is 180. By definition, two angles are supplementary if the sum of their measures is 180. By the Angle Addition Postulate, $m\angle 1 + m\angle 2 = 180$. Thus, if two angles form a linear pair, then the angles are supplementary.

11. Given: $\angle ABC$ is a right angle.

Prove: $\angle 1$ and $\angle 2$ are complementary angles.

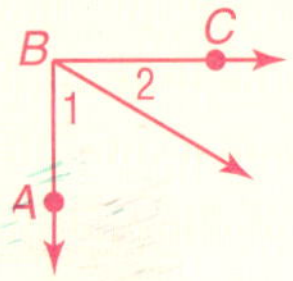

Proof:

Statements (Reasons)

1. $\angle ABC$ is a right angle. (Given)
2. $m\angle ABC = 90$ (Def. of rt. $\angle$)
3. $m\angle ABC = m\angle 1 + m\angle 2$ ($\angle$ Add. Post.)
4. $90 = m\angle 1 + m\angle 2$ (Subst.)
5. $\angle 1$ and $\angle 2$ are complementary angles. (Def. of comp. $\angle$s)

7. **PROOF** Write a two-column proof. See margin.

Given: $\overrightarrow{VX}$ bisects $\angle WVY$. $\overrightarrow{VY}$ bisects $\angle XVZ$.

Prove: $\angle WVX \cong \angle YVZ$

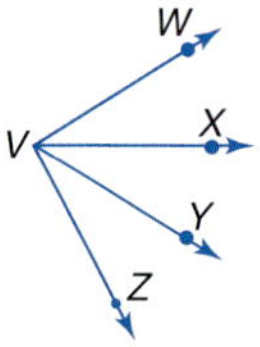

Determine whether the following statements are *always*, *sometimes*, or *never* true.

8. Two angles that are nonadjacent are __?__ vertical. sometimes
9. Two angles that are congruent are __?__ complementary to the same angle. sometimes

PROOF **Write a proof for each theorem.** 10–11. See margin.

10. Supplement Theorem
11. Complement Theorem

Application

ALGEBRA **For Exercises 12–15, use the following information.**

$\angle 1$ and $\angle X$ are complementary,
$\angle 2$ and $\angle X$ are complementary,
$m\angle 1 = 2n + 2$, and $m\angle 2 = n + 32$.

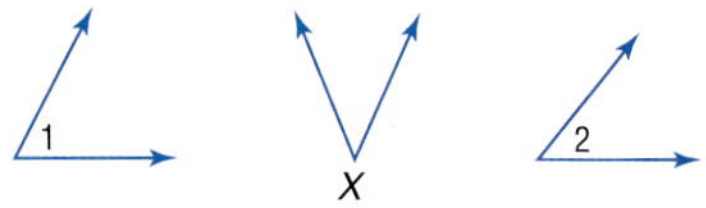

12. Find n. 30
13. Find $m\angle 1$. 62
14. What is $m\angle 2$? 62
15. Find $m\angle X$. 28

Practice and Apply

Homework Help	
For Exercises	See Examples
16–18	1, 2
19–24	4
25–39	3

Extra Practice See page 758.

Find the measure of each numbered angle.

16. $m\angle 2 = 67$ $m\angle 1 = 113$

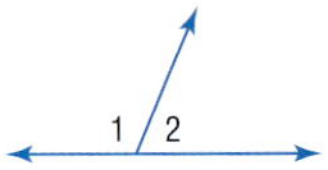

17. $m\angle 3 = 38$ $m\angle 4 = 52$

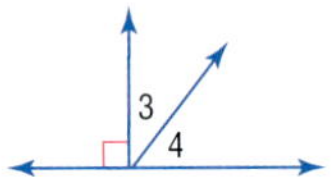

18. $\angle 7$ and $\angle 8$ are complementary. $\angle 5 \cong \angle 8$ and $m\angle 6 = 29$.

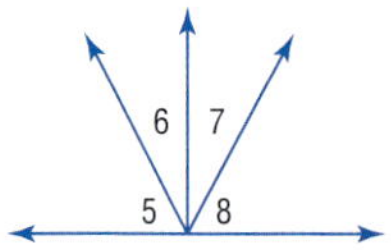

18. $m\angle 5 = 61$, $m\angle 7 = 29$, $m\angle 8 = 61$

19. $m\angle 9 = 2x - 4$, $m\angle 10 = 2x + 4$

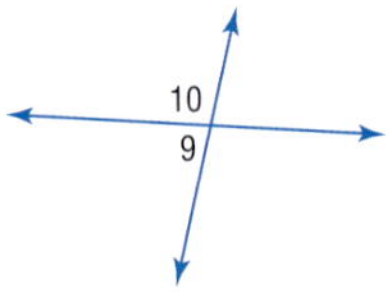

$m\angle 9 = 86$, $m\angle 10 = 94$

20. $m\angle 11 = 4x$, $m\angle 12 = 2x - 6$

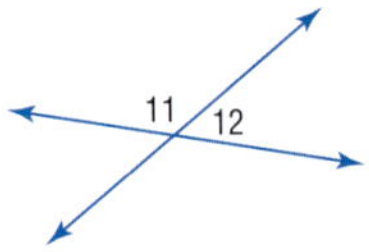

$m\angle 11 = 124$, $m\angle 12 = 56$

21. $m\angle 13 = 2x + 94$, $m\angle 14 = 7x + 49$

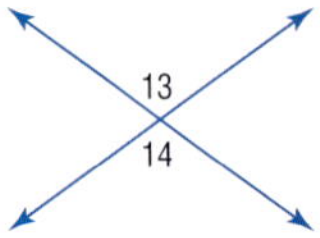

$m\angle 13 = 112$, $m\angle 14 = 112$

22. $m\angle 15 = x$, $m\angle 16 = 6x - 290$

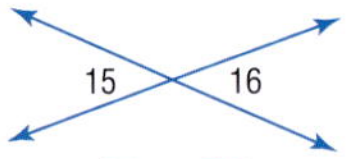

$m\angle 15 = 58$, $m\angle 16 = 58$

23. $m\angle 17 = 2x + 7$, $m\angle 18 = x + 30$

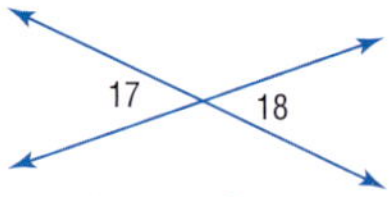

$m\angle 17 = 53$, $m\angle 18 = 53$

24. $m\angle 19 = 100 + 20x$, $m\angle 20 = 20x$

$m\angle 19 = 140$, $m\angle 20 = 40$

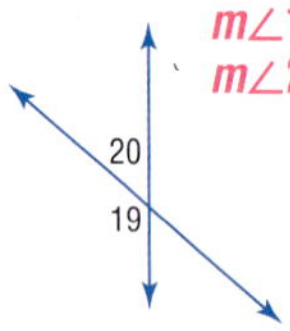

25. Prove that congruence of angles is reflexive. See p. 123F.

26. Write a proof of the Transitive Property of Angle Congruence. See p. 123F.

Determine whether the following statements are *always*, *sometimes*, or *never* true.

27. Two angles that are complementary __?__ form a right angle. sometimes

28. Two angles that are vertical are __?__ nonadjacent. always

29. Two angles that form a right angle are __?__ complementary. always

30. Two angles that form a linear pair are __?__ congruent. sometimes

31. Two angles that are supplementary are __?__ congruent. sometimes

32. Two angles that form a linear pair are __?__ supplementary. always

PROOF **Use the figure to write a proof of each theorem.** 33–37. See p. 123F.

33. Theorem 2.9
34. Theorem 2.10
35. Theorem 2.11
36. Theorem 2.12
37. Theorem 2.13

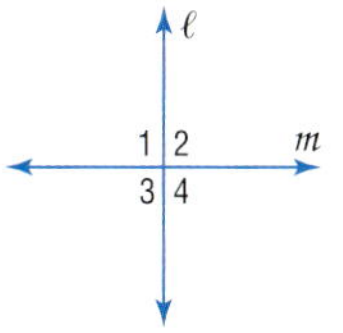

PROOF **Write a two-column proof.** 38–39. See p. 123G.

38. **Given:** $\angle ABD \cong \angle YXZ$
Prove: $\angle CBD \cong \angle WXZ$

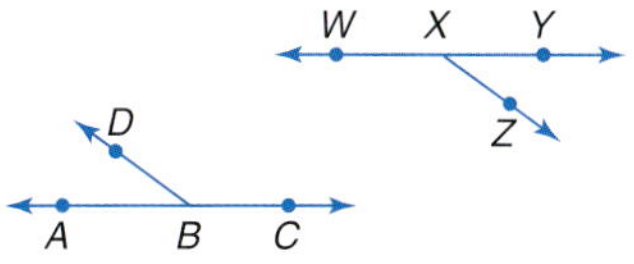

39. **Given:** $m\angle RSW = m\angle TSU$
Prove: $m\angle RST = m\angle WSU$

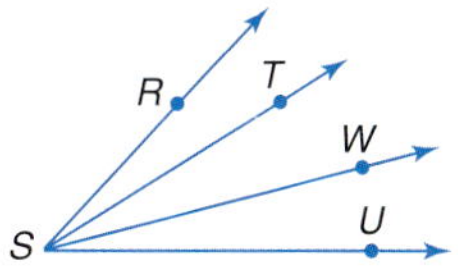

40. **RIVERS** Tributaries of rivers sometimes form a linear pair of angles when they meet the main river. The Yellowstone River forms the linear pair $\angle 1$ and $\angle 2$ with the Missouri River. If $m\angle 1$ is 28, find $m\angle 2$. 152

41. **HIGHWAYS** Near the city of Hopewell, Virginia, Route 10 runs perpendicular to Interstate 95 and Interstate 295. Show that the angles at the intersections of Route 10 with Interstate 95 and Interstate 295 are congruent. Because the lines are perpendicular, the angles formed are right angles. All right angles are congruent. Therefore, $\angle 1$ is congruent to $\angle 2$.

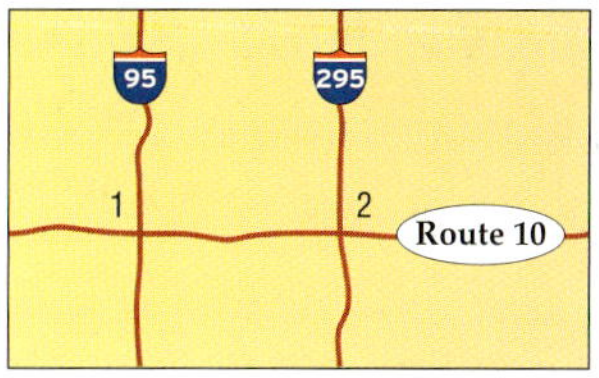

42. **CRITICAL THINKING** What conclusion can you make about the sum of $m\angle 1$ and $m\angle 4$ if $m\angle 1 = m\angle 2$ and $m\angle 3 = m\angle 4$? Explain. See margin.

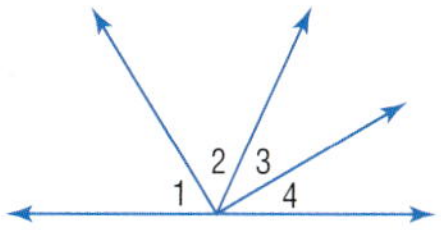

More About...

Highways

Interstate highways that run from north to south are odd-numbered with the lowest numbers in the west. East-west interstates are even-numbered, and begin in the south.

Source: www.infoplease.com

www.geometryonline.com/self_check_quiz/fcat

Answer

42. $m\angle 1 + m\angle 4 = 90;$

$$m\angle 1 + m\angle 2 + m\angle 3 + m\angle 4 = 180$$
$$m\angle 1 + m\angle 1 + m\angle 4 + m\angle 4 = 180$$
$$2(m\angle 1) + 2(m\angle 4) = 180$$
$$2(m\angle 1 + m\angle 4) = 180$$
$$m\angle 1 + m\angle 4 = 90$$

Enrichment, p. 104

Bisecting a Hidden Angle

The vertex of $\angle BAD$ at the right is hidden in a region. Within the region, you are not allowed to use a compass. Can you bisect the angle?

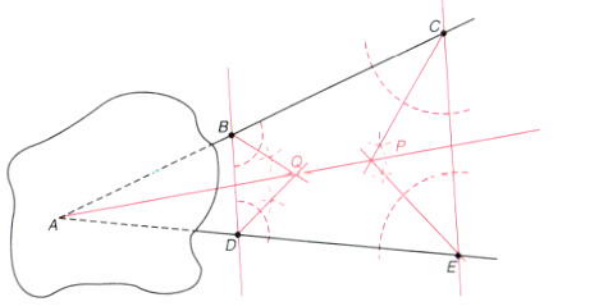

Follow these instructions to bisect $\angle BAD$.

1. Use a straightedge to draw lines ... and BD.

Study Guide and Intervention, p. 99 (shown) and p. 100

Supplementary and Complementary Angles There are two basic postulates for working with angles. The Protractor Postulate assigns numbers to angle measures, and the Angle Addition Postulate relates parts of an angle to the whole angle.

Protractor Postulate	Given $\overrightarrow{AB}$ and a number r between 0 and 180, there is exactly one ray with endpoint A, extending on either side of $\overrightarrow{AB}$, such that the measure of the angle formed is r.
Angle Addition Postulate	R is in the interior of $\angle PQS$ if and only if $m\angle PQR + m\angle RQS = m\angle PQS$.

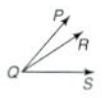

The two postulates can be used to prove the following two theorems.

Supplement Theorem	If two angles form a linear pair, then they are supplementary angles. If $\angle 1$ and $\angle 2$ form a linear pair, then $m\angle 1 + m\angle 2 = 180$.
Complement Theorem	If the noncommon sides of two adjacent angles form a right angle, then the angles are complementary angles. If $\overrightarrow{GF} \perp \overrightarrow{GH}$, then $m\angle 3 + m\angle 4 = 90$.

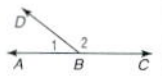

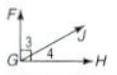

Example 1 **If $\angle 1$ and $\angle 2$ form a linear pair and $m\angle 2 = 115$, find $m\angle 1$.**

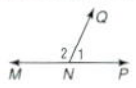

$m\angle 1 + m\angle 2 = 180$ Suppl. Theorem
$m\angle 1 + 115 = 180$ Substitution
$m\angle 1 = 65$ Subtraction Prop.

Example 2 **If $\angle 1$ and $\angle 2$ form a right angle and $m\angle 2 = 20$, find $m\angle 1$.**

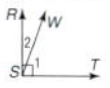

$m\angle 1 + m\angle 2 = 90$ Compl. Theorem
$m\angle 1 + 20 = 90$ Substitution
$m\angle 1 = 70$ Subtraction Prop.

Exercises

Find the measure of each numbered angle.

1. 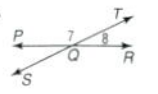

$m\angle 7 = 5x + 5$,
$m\angle 8 = x - 5$
$m\angle 7 = 155$,
$m\angle 8 = 25$

2. 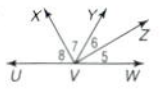

$m\angle 5 = 5x$, $m\angle 6 = 4x + 6$,
$m\angle 7 = 10x$,
$m\angle 8 = 12x - 12$
$m\angle 5 = 30$, $m\angle 6 = 30$,
$m\angle 7 = 60$, $m\angle 8 = 60$

3. 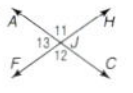

$m\angle 11 = 11x$,
$m\angle 12 = 10x + 10$
$m\angle 11 = 110$,
$m\angle 12 = 110$,
$m\angle 13 = 70$

Skills Practice, p. 101 and Practice, p. 102 (shown)

Find the measure of each numbered angle.

1. $m\angle 1 = x + 10$
$m\angle 2 = 3x + 18$

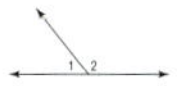

$m\angle 1 = 48$,
$m\angle 2 = 132$

2. $m\angle 4 = 2x - 5$
$m\angle 5 = 4x - 13$

$m\angle 3 = 90$, $m\angle 4 = 31$,
$m\angle 5 = 59$

3. $m\angle 6 = 7x - 24$
$m\angle 7 = 5x + 14$

$m\angle 6 = 109$,
$m\angle 7 = 109$

Determine whether the following statements are *always*, *sometimes*, or *never* true.

4. Two angles that are supplementary are complementary.
never

5. Complementary angles are congruent.
sometimes

6. Write a two-column proof.
Given: $\angle 1$ and $\angle 2$ form a linear pair.
$\angle 2$ and $\angle 3$ are supplementary.
Prove: $\angle 1 \cong \angle 3$

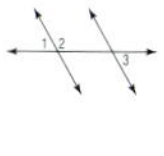

Proof:

Statements	Reasons
1. $\angle 1$ and $\angle 2$ form a linear pair. $\angle 2$ and $\angle 3$ are supplementary.	1. Given
2. $\angle 1$ and $\angle 2$ are supplementary.	2. Supplement Theorem
3. $\angle 1 \cong \angle 3$	3. △ suppl. to the same ∠ or ≅ △ are ≅.

7. **STREETS** Refer to the figure. Barton Road and Olive Tree Lane form a right angle at their intersection. Tryon Street forms a 57° angle with Olive Tree Lane. What is the measure of the acute angle Tryon Street forms with Barton Road? 33

Reading to Learn Mathematics, p. 103 ELL

Pre-Activity **How do scissors illustrate supplementary angles?**

Read the introduction to Lesson 2-8 at the top of page 107 in your textbook.

Is it possible to open a pair of scissors so that the angles formed by the two blades, a blade and a handle, and the two handles, are all congruent? If so, explain how this could happen. Sample answer: Yes; open the scissors so that the two blades are perpendicular. Then all the angles will be right angles and will be congruent.

Reading the Lesson

1. Complete each sentence to form a statement that is always true.
   a. If two angles form a linear pair, then they are adjacent and __supplementary__.
   b. If two angles are complementary to the same angle, then they are __congruent__.
   c. If D is a point in the interior of $\angle ABC$, then $m\angle ABC = m\angle ABD +$ __$m\angle DBC$__.
   d. Given $\overrightarrow{RS}$ and a number x between __0__ and __180__, there is exactly one ray with endpoint R, extended on either side of RS, such that the measure of the angle formed is x.
   e. If two angles are congruent and supplementary, then each angle is a(n) __right__ angle.
   f. __Perpendicular__ lines form congruent adjacent angles.
   g. "Every angle is congruent to itself" is a statement of the __Reflexive__ Property of angle congruence.
   h. If two congruent angles form a linear pair, then the measure of each angle is __90__.
   i. If the noncommon sides of two adjacent angles form a right angle, then the angles are __complementary__.

2. Determine whether each statement is *always*, *sometimes*, or *never* true.
   a. Supplementary angles are congruent. sometimes
   b. If two angles form a linear pair, they are complementary. never
   c. Two vertical angles are supplementary. sometimes
   d. Two adjacent angles form a linear pair. sometimes
   e. Two vertical angles form a linear pair. never
   f. Complementary angles are congruent. sometimes
   g. Two angles that are congruent to the same angle are congruent to each other. always
   h. Complementary angles are adjacent angles. sometimes

Helping You Remember

3. A good way to remember something is to explain it to someone else. Suppose that a classmate thinks that two angles can only be *vertical* angles if one angle lies above the other. How can you explain to him the meaning of vertical angles, using the word *vertex* in your explanation? Sample answer: Two angles are *vertical* angles if they share the same *vertex* and their sides are opposite rays. It doesn't matter how the angles are positioned.

4 Assess

Open-Ended Assessment

Writing Give students a list of theorems from this chapter. Have students choose a theorem and write a proof of it with their books closed.

Assessment Options

Quiz (Lessons 2-7 and 2-8) is available on p. 120 of the *Chapter 2 Resource Masters.*

Answers

43. Two angles that are supplementary to the same angle are congruent. Answers should include the following.
    - $\angle 1$ and $\angle 2$ are supplementary; $\angle 2$ and $\angle 3$ are supplementary.
    - $\angle 1$ and $\angle 3$ are vertical angles, and are therefore congruent.
    - If two angles are complementary to the same angle, then the angles are congruent.

46. Given: G is between F and H. H is between F and J.
Prove: $FG + GJ = FH + HJ$

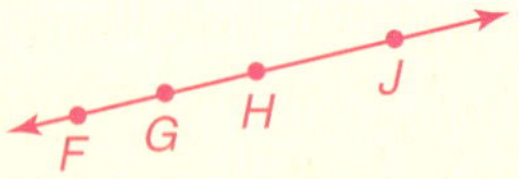

Proof:
Statements (Reasons)
1. G is between F and H; H is between F and J. (Given)
2. $FG + GJ = FJ$, $FH + HJ = FJ$ (Seg. Add. Post.)
3. $FJ = FH + HJ$ (Sym. Prop.)
4. $FG + GJ = FH + HJ$ (Transitive Prop.)

47. Given: X is the midpoint of $\overline{WY}$.
Prove: $WX + YZ = XZ$

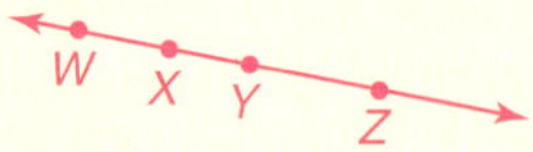

Proof:
Statements (Reasons)
1. X is the midpoint of $\overline{WY}$. (Given)
2. $WX = XY$ (Def. of midpoint)
3. $XY + YZ = XZ$ (Segment Add. Post.)
4. $WX + YZ = XZ$ (Substitution)

43. WRITING IN MATH Answer the question that was posed at the beginning of the lesson. **See margin.**

How do scissors illustrate supplementary angles?

Include the following in your answer:
- a description of the relationship among $\angle 1$, $\angle 2$, and $\angle 3$,
- an example of another way that you can tell the relationship between $\angle 1$ and $\angle 3$, and
- an explanation of whether this relationship is the same for two angles complementary to the same angle.

FCAT Practice
Standardized Test Practice

44. The measures of two complementary angles are in the ratio 4:1. What is the measure of the smaller angle? **B**

Ⓐ 15 Ⓑ 18 Ⓒ 24 Ⓓ 36

45. **ALGEBRA** T is the set of all positive numbers n such that $n < 50$ and $\sqrt{n}$ is an integer. What is the median of the members of set T? **B**

Ⓐ 4 Ⓑ 16 Ⓒ 20 Ⓓ 25

Maintain Your Skills

Mixed Review

PROOF **Write a two-column proof.** *(Lesson 2-7)* **46–47. See margin.**

46. **Given:** G is between F and H.
H is between G and J.
Prove: $FG + GJ = FH + HJ$

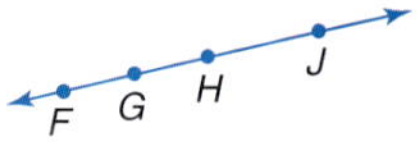

47. **Given:** X is the midpoint of $\overline{WY}$.
Prove: $WX + YZ = XZ$

48. **PHOTOGRAPHY** Film is fed through a camera by gears that catch the perforation in the film. The distance from the left edge of the film, A, to the right edge of the image, C, is the same as the distance from the left edge of the image, B, to the right edge of the film, D. Show that the two perforated strips are the same width. *(Lesson 2-6)* **See p. 123G.**

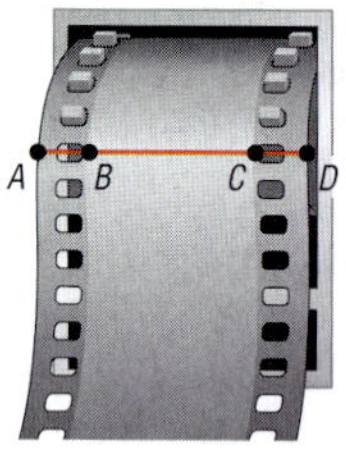

For Exercises 49–55, refer to the figure at the right.
(Lesson 1-4)

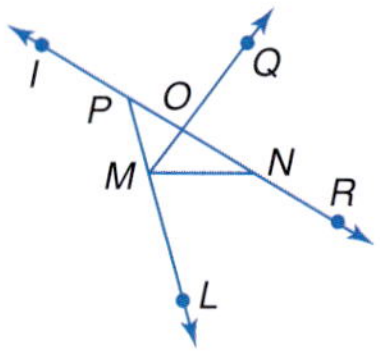

49. Name two angles that have N as a vertex. **$\angle ONM$, $\angle MNR$**
50. If $\overrightarrow{MQ}$ bisects $\angle PMN$, name two congruent angles. **50. $\angle PMQ \cong \angle QMN$**
51. Name a point in the interior of $\angle LMQ$. **N or R**
52. List all the angles that have O as the vertex. **52. $\angle POQ$, $\angle QON$, $\angle NOM$, $\angle MOP$**
53. Does $\angle QML$ appear to be acute, obtuse, right, or straight? **obtuse**
54. Name a pair of opposite rays. **Sample answer: $\overrightarrow{NR}$ and $\overrightarrow{NP}$**
55. List all the angles that have $\overrightarrow{MN}$ as a side. **$\angle NML$, $\angle NMP$, $\angle NMO$, $\angle RNM$, $\angle ON$**

Answers (page 115)

9.

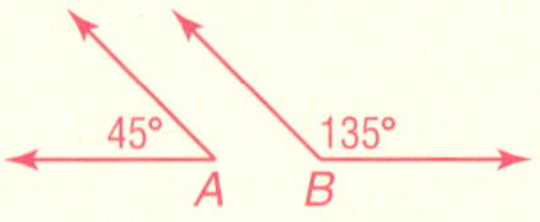

10.

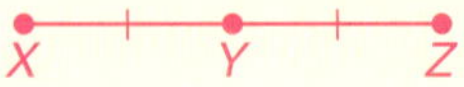

11.

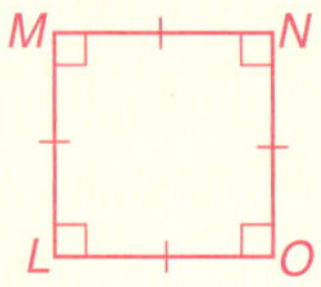

Chapter 2 Study Guide and Review

Vocabulary and Concept Check

- biconditional (p. 81)
- compound statement (p. 67)
- conclusion (p. 75)
- conditional statement (p. 75)
- conjecture (p. 62)
- conjunction (p. 68)
- contrapositive (p. 77)
- converse (p. 77)
- counterexample (p. 63)
- deductive argument (p. 94)
- deductive reasoning (p. 82)
- disjunction (p. 68)
- formal proof (p. 95)
- hypothesis (p. 75)
- if-then statement (p. 75)
- inductive reasoning (p. 62)
- informal proof (p. 90)
- inverse (p. 77)
- Law of Detachment (p. 82)
- Law of Syllogism (p. 83)
- logically equivalent (p. 77)
- matrix logic (p. 88)
- negation (p. 67)
- paragraph proof (p. 90)
- postulate (p. 89)
- proof (p. 90)
- related conditionals (p. 77)
- statement (p. 67)
- theorem (p. 90)
- truth table (p. 70)
- truth value (p. 67)
- two-column proof (p. 95)

A complete list of postulates and theorems can be found on pages R1–R8.

Exercises **Choose the correct term to complete each sentence.**

1. A (*counterexample*, *conjecture*) is an educated guess based on known information.
2. The truth or falsity of a statement is called its (*conclusion*, *truth value*).
3. Two or more statements can be joined to form a (*conditional*, *compound*) statement.
4. A conjunction is a compound statement formed by joining two or more statements using (*or*, *and*).
5. The phrase immediately following the word *if* in a conditional statement is called the (*hypothesis*, *conclusion*).
6. The (*inverse*, *converse*) is formed by exchanging the hypothesis and the conclusion.
7. (*Theorems*, *Postulates*) are accepted as true without proof.
8. A paragraph proof is a (an) (*informal proof*, *formal proof*).

Lesson-by-Lesson Review

2-1 Inductive Reasoning and Conjecture

See pages 62–66.

Concept Summary

- Conjectures are based on observations and patterns.
- Counterexamples can be used to show that a conjecture is false.

Example **Given that points *P*, *Q*, and *R* are collinear, determine whether the conjecture that *Q* is between *P* and *R* is *true* or *false*. If the conjecture is false, give a counterexample.**

The figure at the right can be used to disprove the conjecture. In this case, *R* is between *P* and *Q*. Since we can find a counterexample, the conjecture is false.

Exercises **Make a conjecture based on the given information. Draw a figure to illustrate your conjecture.** *See Example 2 on page 63.* **9–11. See margin for figures.**

9. $\angle A$ and $\angle B$ are supplementary. **$m\angle A + m\angle B = 180$**
10. *X*, *Y*, and *Z* are collinear and $XY = YZ$. ***Y* is the midpoint of $\overline{XZ}$.**
11. In quadrilateral *LMNO*, $LM = LO = MN = NO$, and $m\angle L = 90$. ***LMNO* is a square.**

www.geometryonline.com/vocabulary_review

FOLDABLES™ Study Organizer

For more information about Foldables, see *Teaching Mathematics with Foldables.*

Have students look through the chapter to make sure they have included notes and examples in their Foldables for each lesson of Chapter 2.

Encourage students to refer to their Foldables while completing the Study Guide and Review and to use them in preparing for the Chapter Test.

Chapter 2 Study Guide and Review

Vocabulary and Concept Check

- This alphabetical list of vocabulary terms in Chapter 2 includes a page reference where each term was introduced.
- **Assessment** A vocabulary test/review for Chapter 2 is available on p. 118 of the *Chapter 2 Resource Masters*.

Lesson-by-Lesson Review

For each lesson,

- the main ideas are summarized,
- additional examples review concepts, and
- practice exercises are provided.

Vocabulary PuzzleMaker

ELL The Vocabulary PuzzleMaker software improves students' mathematics vocabulary using four puzzle formats—crossword, scramble, word search using a word list, and word search using clues. Students can work on a computer screen or from a printed handout.

MindJogger Videoquizzes

ELL MindJogger Videoquizzes provide an alternative review of concepts presented in this chapter. Students work in teams in a game show format to gain points for correct answers. The questions are presented in three rounds.

Round 1 Concepts (5 questions)
Round 2 Skills (4 questions)
Round 3 Problem Solving (4 questions)

Answers

12. $-1 > 0$ and in a right triangle with right angle C, $a^2 + b^2 = c^2$.
13. In a right triangle with right angle C, $a^2 + b^2 = c^2$ or the sum of the measures of two supplementary angles is 180.
14. The sum of the measures of two supplementary angles is 180 and $-1 > 0$.
15. $-1 > 0$, and in a right triangle with right angle C, $a^2 + b^2 = c^2$, or the sum of the measures of two supplementary angles is 180.
16. In a right triangle with right angle C, $a^2 + b^2 = c^2$, or $-1 > 0$ or the sum of the measures of two supplementary angles is 180.
17. In a right triangle with right angle C, $a^2 + b^2 = c^2$ and the sum of the measures of two supplementary angles is 180, and $-1 > 0$.
18. Converse: If an angle is obtuse, then it measures 120. False; the measure could be any value between 90 and 180. Inverse: If an angle measure does not equal 120, then it is not obtuse. False; the measure could be any value other than 120 between 90 and 180. Contrapositive: If an angle is not obtuse, then its measure does not equal 120; true.
19. Converse: If a month has 31 days, then it is March. False; July has 31 days. Inverse: If a month is not March, then it does not have 31 days. False; July has 31 days. Contrapositive: If a month does not have 31 days, then it is not March; true.
20. Converse: If a point lies on the y-axis, then its ordered pair has 0 for its x-coordinate; true. Inverse: If an ordered pair does not have 0 for its x-coordinate, then the point does not lie on the y-axis; true. Contrapositive: If a point does not lie on the y-axis, then its ordered pair does not have 0 for its x-coordinate; true.

2-2 Logic

See pages 67–74.

Concept Summary

- The negation of a statement has the opposite truth value of the original statement.
- Venn diagrams and truth tables can be used to determine the truth values of statements.

Example **Use the following statements to write a compound statement for each conjunction. Then find its truth value.**

p: $\sqrt{15} = 5$ q: **The measure of a right angle equals 90.**

a. p and q

$\sqrt{15} = 5$, and the measure of a right angle equals 90.

p and q is false because p is false and q is true.

b. $p \vee q$

$\sqrt{15} = 5$, or the measure of a right angle equals 90.

$p \vee q$ is true because q is true. It does not matter that p is false.

Exercises **Use the following statements to write a compound statement for each conjunction. Then find its truth value.** *See Examples 1 and 2 on pages 68 and 69.*

p: $-1 > 0$ q: **In a right triangle with right angle C, $a^2 + b^2 = c^2$.**

r: **The sum of the measures of two supplementary angles is 180.**

12. p and q false
13. q or r true
14. $r \wedge p$ false
15. $p \wedge (q \vee r)$ false
16. $q \vee (p \vee r)$ true
17. $(q \wedge r) \wedge p$ false

12–17. See margin for statements.

2-3 Conditional Statements

See pages 75–80.

Concept Summary

- Conditional statements are written in if-then form.
- Form the converse, inverse, and contrapositive of an if-then statement by using negations and by exchanging the hypothesis and conclusion.

Example **Identify the hypothesis and conclusion of the statement *The intersection of two planes is a line.* Then write the statement in if-then form.**

Hypothesis: two planes intersect

Conclusion: their intersection is a line

If two planes intersect, then their intersection is a line.

Exercises **Write the converse, inverse, and contrapositive of each conditional statement. Determine whether each related conditional is *true* or *false*. If a statement is false, find a counterexample.** *See Example 4 on page 77.* 18–20. See margin.

18. If an angle measure equals 120, then the angle is obtuse.
19. If the month is March, then it has 31 days.
20. If an ordered pair for a point has 0 for its x-coordinate, then the point lies on the y-axis.

Study Guide and Review

Determine the truth value of the following statement for each set of conditions.

If the temperature is at most 0°C, then water freezes. *See Example 3 on page 76.*

21. The temperature is −10°C, and water freezes. **true**

22. The temperature is 15°C, and water freezes. **true**

23. The temperature is −2°C, and water does not freeze. **false**

24. The temperature is 30°C, and water does not freeze. **true**

2-4 Deductive Reasoning

See pages 82–87.

Concept Summary

- The Law of Detachment and the Law of Syllogism can be used to determine the truth value of a compound statement.

Example

Use the Law of Syllogism to determine whether a valid conclusion can be reached from the following statements.

(1) If a body in our solar system is the Sun, then it is a star.
(2) Stars are in constant motion.

p: **a body in our solar system is the sun**
q: **it is a star**
r: **stars are in constant motion**

Statement (1): $p \to q$ Statement (2): $q \to r$

Since the given statements are true, use the Law of Syllogism to conclude $p \to r$. That is, *If a body in our solar system is the Sun, then it is in constant motion.*

Exercises **Determine whether the stated conclusion is valid based on the given information. If not, write *invalid*. Explain your reasoning.** *See Example 1 on page 82.*

If two angles are adjacent, then they have a common vertex.

25. Given: ∠1 and ∠2 are adjacent angles. **Valid; by definition, adjacent angles**
Conclusion: ∠1 and ∠2 have a common vertex. **have a common vertex.**

26. Given: ∠3 and ∠4 have a common vertex. **Invalid; vertical angles also have**
Conclusion: ∠3 and ∠4 are adjacent angles. **a common vertex.**

Determine whether statement (3) follows from statements (1) and (2) by the Law of Detachment or the Law of Syllogism. If it does, state which law was used. If it does not follow, write *invalid*. *See Example 3 on page 83.*

27. (1) If a student attends North High School, then the student has an ID number.
(2) Josh Michael attends North High School.
(3) Josh Michael has an ID number. **yes; Law of Detachment**

28. (1) If a rectangle has four congruent sides, then it is a square.
(2) A square has diagonals that are perpendicular.
(3) A rectangle has diagonals that are perpendicular. **invalid**

29. (1) If you like pizza with everything, then you'll like Cardo's Pizza. **yes; Law**
(2) If you like Cardo's Pizza, then you are a pizza connoisseur. **of Syllogism**
(3) If you like pizza with everything, then you are a pizza connoisseur.

2-5 Postulates and Paragraph Proofs

See pages 89–93.

Concept Summary

- Use undefined terms, definitions, postulates, and theorems to prove that statements and conjectures are true.

Example **Determine whether the following statement is *always, sometimes,* or *never* true. Explain.** ***Two points determine a line.***

According to a postulate relating to points and lines, two points determine a line. Thus, the statement is always true.

Exercises **Determine whether the following statements are *always, sometimes,* or *never* true. Explain.** *See Example 2 on page 90.* **30–37. See margin.**

30. The intersection of two lines can be a line.
31. If P is the midpoint of $\overline{XY}$, then $XP = PY$.
32. If $MX = MY$, then M is the midpoint of XY.
33. Three points determine a line.
34. Points Q and R lie in at least one plane.
35. If two angles are right angles, they are adjacent.
36. An angle is congruent to itself.
37. Vertical angles are adjacent.

38. **PROOF** Write a paragraph proof to prove that if M is the midpoint of $\overline{AB}$ and Q is the midpoint of $\overline{AM}$, then $AQ = \frac{1}{4}AB$. **See margin.**

2-6 Algebraic Proof

See pages 94–100.

Concept Summary

- The properties of equality used in algebra can be applied to the measures of segments and angles to verify and prove statements.

Example

Given: $2x + 6 = 3 + \frac{5}{3}x$

Prove: $x = -9$

Proof:

Statements	Reasons
1. $2x + 6 = 3 + \frac{5}{3}x$	1. Given
2. $3(2x + 6) = 3\left(3 + \frac{5}{3}x\right)$	2. Multiplication Property
3. $6x + 18 = 9 + 5x$	3. Distributive Property
4. $6x + 18 - 5x = 9 + 5x - 5x$	4. Subtraction Property
5. $x + 18 = 9$	5. Substitution
6. $x + 18 - 18 = 9 - 18$	6. Subtraction Property
7. $x = -9$	7. Substitution

Answers

30. Never; the intersection of two lines is a point.
31. Always; if P is the midpoint of $\overline{XY}$, then $\overline{XP} \cong \overline{PY}$. By definition of congruent segments, $XP = PY$.
32. Sometimes; if M, X, and Y are collinear.
33. Sometimes; if the points are collinear.
34. Always; there is exactly one line through Q and R. The line lies in at least one plane.
35. Sometimes; if the right angles form a linear pair.
36. Always; the Reflexive Property states that $\angle 1 \cong \angle 1$.
37. Never; adjacent angles must share a common side, and vertical angles do not.
38. If M is the midpoint of $\overline{AB}$, then $AM = \frac{1}{2}(AB)$. Since Q is the midpoint of $\overline{AM}$, $AQ = \frac{1}{2}AM$ or $\frac{1}{2}\left(\frac{1}{2}(AB)\right) = \frac{1}{4}AB$.

Exercises State the property that justifies each statement. *See Example 1 on page 94.*

39. If $3(x + 2) = 6$, then $3x + 6 = 6$. Dist. Prop.
40. If $10x = 20$, then $x = 2$. Div. Prop.
41. If $AB + 20 = 45$, then $AB = 25$. Subt. Prop.
42. If $3 = CD$ and $CD = XY$, then $3 = XY$. Transitive Prop.

PROOF Write a two-column proof. *See Examples 2 and 4 on pages 95 and 96.*

43. If $5 = 2 - \frac{1}{2}x$, then $x = -6$.
44. If $x - 1 = \frac{x - 10}{-2}$, then $x = 4$.
45. If $AC = AB$, $AC = 4x + 1$, and $AB = 6x - 13$, then $x = 7$.
46. If $MN = PQ$ and $PQ = RS$, then $MN = RS$.

43–46. See margin.

2-7 Proving Segment Relationships

See pages 101–106.

Concept Summary

- Use properties of equality and congruence to write proofs involving segments.

Example Write a two-column proof.

Given: $QT = RT$, $TS = TP$

Prove: $QS = RP$

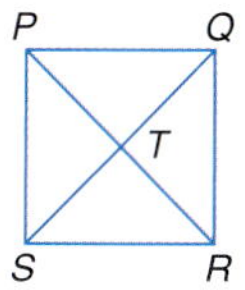

Proof:

Statements	Reasons
1. $QT = RT$, $TS = TP$	1. Given
2. $QT + TS = RT + TS$	2. Addition Property
3. $QT + TS = RT + TP$	3. Substitution
4. $QT + TS = QS$, $RT + TP = RP$	4. Segment Addition Postulate
5. $QS = RP$	5. Substitution

Exercises Justify each statement with a property of equality or a property of congruence. *See Example 1 on page 102.*

47. $PS = PS$ Reflexive Prop.
48. If $XY = OP$, then $OP = XY$. Symmetric Prop.
49. If $AB - 8 = CD - 8$, then $AB = CD$. Add. Prop.
50. If $EF = GH$ and $GH = LM$, then $EF = LM$. Transitive Prop.
51. If $2(XY) = AB$, then $XY = \frac{1}{2}(AB)$. Div. or Mult. Prop.
52. If $AB = CD$, then $AB + BC = CD + BC$. Add. Prop.

46. Given: $MN = PQ$, $PQ = RS$
Prove: $MN = RS$

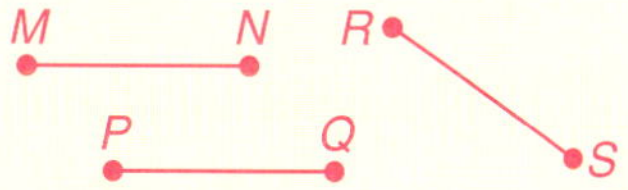

Proof:
Statements (Reasons)
1. $MN = PQ$, $PQ = RS$ (Given)
2. $MN = RS$ (Transitive Prop.)

Answers

43. Given: $5 = 2 - \frac{1}{2}x$
Prove: $x = -6$
Proof:
Statements (Reasons)
1. $5 = 2 - \frac{1}{2}x$ (Given)
2. $5 - 2 = 2 - \frac{1}{2}x - 2$ (Subt. Prop.)
3. $3 = -\frac{1}{2}x$ (Substitution)
4. $-2(3) = -2\left(-\frac{1}{2}x\right)$ (Mult. Prop)
5. $-6 = x$ (Substitution)
6. $x = -6$ (Sym. Prop.)

44. Given: $x - 1 = \frac{x - 10}{-2}$
Prove: $x = 4$
Proof:
Statements (Reasons)
1. $x - 1 = \frac{x - 10}{-2}$ (Given)
2. $-2(x - 1) = -2\left(\frac{x - 10}{-2}\right)$ (Mult. Prop.)
3. $-2x + 2 = x - 10$ (Dist. Prop.)
4. $-2x + 2 - 2 = x - 10 - 2$ (Subt. Prop.)
5. $-2x = x - 12$ (Substitution)
6. $-2x - x = x - 12 - x$ (Subt. Prop.)
7. $-3x = -12$ (Substitution)
8. $\frac{-3x}{-3} = \frac{-12}{-3}$ (Div. Prop.)
9. $x = 4$ (Substitution)

45. Given: $AC = AB$, $AC = 4x + 1$, $AB = 6x - 13$
Prove: $x = 7$

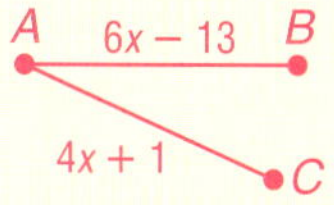

Proof:
Statements (Reasons)
1. $AC = AB$, $AC = 4x + 1$, $AB = 6x - 13$ (Given)
2. $4x + 1 = 6x - 13$ (Subst.)
3. $4x + 1 - 1 = 6x - 13 - 1$ (Subt. Prop.)
4. $4x = 6x - 14$ (Subst.)
5. $4x - 6x = 6x - 14 - 6x$ (Subt. Prop.)
6. $-2x = -14$ (Subst.)
7. $\frac{-2x}{-2} = \frac{-14}{-2}$ (Div. Prop.)
8. $x = 7$ (Subst.)

PROOF **Write a two-column proof.** *See Examples 1 and 2 on pages 102 and 103.*

53. Given: $BC = EC, CA = CD$
Prove: $BA = DE$

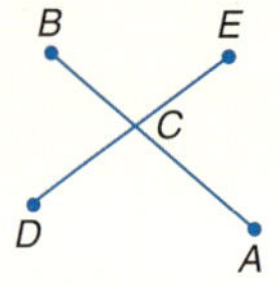

54. Given: $AB = CD$
Prove: $AC = BD$

53–54. See p. 123G.

2-8 Proving Angle Relationships

See pages 107–114.

Concept Summary

- The properties of equality and congruence can be applied to angle relationships.

Example **Find the measure of each numbered angle.**

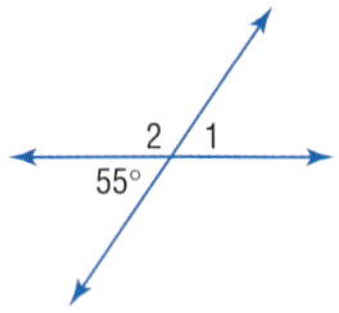

$m\angle 1 = 55$, since $\angle 1$ is a vertical angle to the 55° angle. $\angle 2$ and the 55° angle form a linear pair.

$55 + m\angle 2 = 180$ Def. of supplementary ∠s
$m\angle 2 = 125$ Subtract 55 from each side.

Exercises **Find the measure of each numbered angle.** *See Example 2 on page 108.*

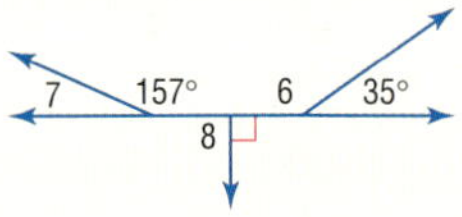

55. $m\angle 6$ 145
56. $m\angle 7$ 23
57. $m\angle 8$ 90

58. **PROOF** **Copy and complete the proof.** *See Example 3 on page 109.*

Given: $\angle 1$ and $\angle 2$ form a linear pair.
$m\angle 2 = 2(m\angle 1)$
Prove: $m\angle 1 = 60$
Proof:

Statements	Reasons
a. $\angle 1$ and $\angle 2$ form a linear pair.	a. ? Given
b. $\angle 1$ and $\angle 2$ are supplementary.	b. ? Supplement Theorem
c. ? $m\angle 1 + m\angle 2 = 180$	c. Definition of supplementary angles
d. $m\angle 2 = 2(m\angle 1)$	d. ? Given
e. ? $m\angle 1 + 2(m\angle 1) = 180$	e. Substitution
f. ? $3(m\angle 1) = 180$	f. Substitution
g. $\frac{3(m\angle 1)}{3} = \frac{180}{3}$	g. ? Division Property
h. ? $m\angle 1 = 60$	h. Substitution

Answers (page 121)

1. Sample answer: Formal is the two-column proof, informal can be paragraph proofs.
2. Sample answer: You can use a counterexample.
3. Sample answer: statements and reasons to justify statements
7. $-3 > 2$ and $3x = 12$ when $x = 4$.
8. $-3 > 2$ or $3x = 12$ when $x = 4$.
9. $-3 > 2$, or $3x = 12$ when $x = 4$ and an equilateral triangle is also equiangular.
10. H: you eat an apple a day; C: the doctor will stay away; If you eat an apple a day, then the doctor will stay away. Converse: If the doctor stays away, then you eat an apple a day. Inverse: If you do not eat an apple a day, then the doctor will not stay away. Contrapositive: If the doctor does not stay away, then you do not eat an apple a day.
11. H: a stone is rolling; C: it gathers no moss; If a stone is rolling, then it gathers no moss. Converse: If a stone gathers no moss, then it is rolling. Inverse: If a stone is not rolling, then it gathers moss. Contrapositive: If a stone gathers moss, then it is not rolling.
16. Given: $y = 4x + 9$; $x = 2$
Prove: $y = 17$
Proof:
Statements (Reasons)
1. $y = 4x + 9$; $x = 2$ (Given)
2. $y = 4(2) + 9$ (Substitution)
3. $y = 8 + 9$ (Substitution)
4. $y = 17$ (Substitution)
17. Given: $AM = CN$, $MB = ND$

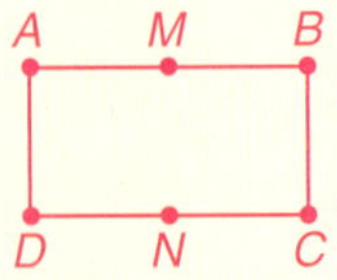

Prove: $AB = CD$

Paragraph Proof:
We are given that $AM = CN$, $MB = ND$. By the Addition Property, $AM + MB = CN + MB$. By Substitution, $AM + MB = CN + ND$. Using the Segment Addition Postulate, $AB = AM + MB$, and $CD = CN + ND$. Then, by Substitution $AB = CD$.

18. H: you are a hard-working person; C: you deserve a great vacation; If you are a hard-working person, then you deserve a great vacation.

Chapter 2 Practice Test

Vocabulary and Concepts

1. **Explain** the difference between formal and informal proofs. **1–3. See margin.**
2. **Explain** how you can prove that a conjecture is false.
3. **Describe** the parts of a two-column proof.

Skills and Applications

Determine whether each conjecture is *true* or *false*. Explain your answer and give a counterexample for any false conjecture.

4. **Given:** $\angle A \cong \angle B$
   Conjecture: $\angle B \cong \angle A$
   true; Symmetric Prop.
5. **Given:** y is a real number
   Conjecture: $-y > 0$
   false; $y = 2$
6. **Given:** $3a^2 = 48$
   Conjecture: $a = 4$
   false; $a = -4$

Use the following statements to write a compound statement for each conjunction or disjunction. Then find its truth value. **7–9. See margin for statements.**

p: $-3 > 2$ q: $3x = 12$ when $x = 4$. r: An equilateral triangle is also equiangular.

7. p and q false
8. p or q true
9. $p \vee (q \wedge r)$ true

Identify the hypothesis and conclusion of each statement and write each statement in if-then form. Then write the converse, inverse, and contrapositive of each conditional. **10–11. See margin.**

10. An apple a day keeps the doctor away.
11. A rolling stone gathers no moss.
12. Determine whether statement (3) follows from statements (1) and (2) by the Law of Detachment or the Law of Syllogism. If it does, state which law was used. If it does not, write *invalid.*
    (1) Perpendicular lines intersect.
    (2) Lines m and n are perpendicular.
    (3) Lines m and n intersect. **valid; Law of Detachment**

Find the measure of each numbered angle.

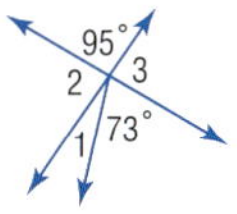

13. $\angle 1$ 22
14. $\angle 2$ 85
15. $\angle 3$ 85

16–17. See margin.

16. Write a two-column proof.
    If $y = 4x + 9$ and $x = 2$, then $y = 17$.
17. Write a paragraph proof.
    Given: $AM = CN$, $MB = ND$
    Prove: $AB = CD$

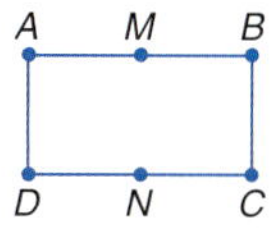

18. **ADVERTISING** Identify the hypothesis and conclusion of the following statement, then write it in if-then form. *Hard working people deserve a great vacation.* **See margin.**
19. **STANDARDIZED TEST PRACTICE** If two planes intersect, their intersection can be **A**
    I a line. II three noncollinear points. III two intersecting lines.
    (A) I only (B) II only (C) III only (D) I and II only

www.geometryonline.com/chapter_test/fcat

Chapter 2 Practice Test

Assessment Options

Vocabulary Test A vocabulary test/review for Chapter 2 can be found on p. 118 of the *Chapter 2 Resource Masters*.

Chapter Tests There are six Chapter 2 Tests and an Open-Ended Assessment task available in the *Chapter 2 Resource Masters*.

Chapter 2 Tests			
Form	**Type**	**Level**	**Pages**
1	MC	basic	105–106
2A	MC	average	107–108
2B	MC	average	109–110
2C	FR	average	111–112
2D	FR	average	113–114
3	FR	advanced	115–116

MC = multiple-choice questions
FR = free-response questions

Open-Ended Assessment
Performance tasks for Chapter 2 can be found on p. 117 of the *Chapter 2 Resource Masters*. A sample scoring rubric for these tasks appears on p. A31.

TestCheck and Worksheet Builder

This **networkable software** has three modules for assessment.

- **Worksheet Builder** to make worksheets and tests.
- **Student Module** to take tests on-screen.
- **Management System** to keep student records.

Portfolio Suggestion

Introduction In a chapter of diverse material, highlight concepts that are important but may not be used often or regularly, so that they will remain fresh in students' minds.

Ask Students Search the chapter for items you found the most difficult. Record these items in your portfolio and write about how you were able to master the concepts. If you are still having difficulty with the concepts, write about the steps you could take to better your understanding of them.

Chapter 2 Standardized Test Practice

These two pages contain practice questions in the various formats that can be found on the most frequently given standardized tests.

A practice answer sheet for these two pages can be found on p. A1 of the *Chapter 2 Resource Masters*.

Standardized Test Practice Student Recording Sheet, p. A1

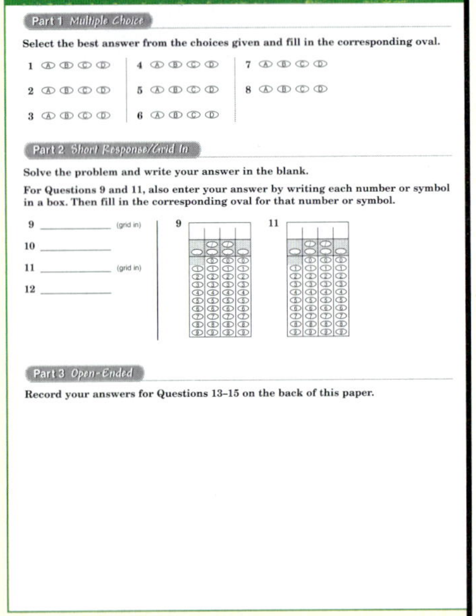
Part 1 Multiple Choice

Select the best answer from the choices given and fill in the corresponding oval.

1 Ⓐ Ⓑ Ⓒ Ⓓ 4 Ⓐ Ⓑ Ⓒ Ⓓ 7 Ⓐ Ⓑ Ⓒ Ⓓ

2 Ⓐ Ⓑ Ⓒ Ⓓ 5 Ⓐ Ⓑ Ⓒ Ⓓ 8 Ⓐ Ⓑ Ⓒ Ⓓ

3 Ⓐ Ⓑ Ⓒ Ⓓ 6 Ⓐ Ⓑ Ⓒ Ⓓ

Part 2 Short Response/Grid In

Solve the problem and write your answer in the blank.

For Questions 9 and 11, also enter your answer by writing each number or symbol in a box. Then fill in the corresponding oval for that number or symbol.

9 ________ (grid in)

10 ________

11 ________ (grid in)

12 ________

Part 3 Open-Ended

Record your answers for Questions 13–15 on the back of this paper.

Additional Practice

See pp. 123–124 in the *Chapter 2 Resource Masters* for additional standardized test practice.

Chapter 2 Standardized Test Practice

FCAT Practice

Part 1 Multiple Choice

Record your answers on the answer sheet provided by your teacher or on a sheet of paper.

1. Arrange the numbers $|-7|, \frac{1}{7}, \sqrt{7}, -7^2$ in order from least to greatest. (Prerequisite Skill) **D**

Ⓐ $|-7|, \sqrt{7}, \frac{1}{7}, -7^2$

Ⓑ $-7^2, |-7|, \frac{1}{7}, \sqrt{7}$

Ⓒ $|-7|, \frac{1}{7}, \sqrt{7}, -7^2$

Ⓓ $-7^2, \frac{1}{7}, \sqrt{7}, |-7|$

2. Points A and B lie on the line $y = 2x - 3$. Which of the following are coordinates of a point noncollinear with A and B? (Lesson 1-1) **C**

Ⓐ (7, 11) Ⓑ (4, 5)

Ⓒ (–2, –10) Ⓓ (–5, –13)

3. Dana is measuring distance on a map. Which of the following tools should Dana use to make the most accurate measurement? (Lesson 1-2) **A**

Ⓐ centimeter ruler Ⓑ protractor

Ⓒ yardstick Ⓓ calculator

4. Point E is the midpoint of $\overline{DF}$. If $DE = 8x - 3$ and $EF = 3x + 7$, what is x? (Lesson 1-3) **B**

Ⓐ 1 Ⓑ 2 Ⓒ 4 Ⓓ 13

5. What is the relationship between $\angle ACF$ and $\angle DCF$? (Lesson 1-6) **A**

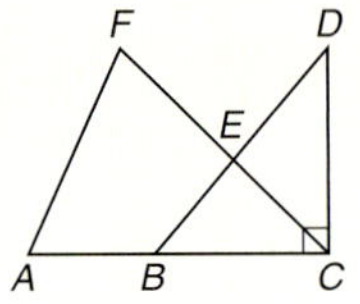

Ⓐ complementary angles

Ⓑ congruent angles

Ⓒ supplementary angles

Ⓓ vertical angles

6. Which of the following is an example of inductive reasoning? (Lesson 2-1) **C**

Ⓐ Carlos learns that the measures of all acute angles are less than 90. He conjectures that if he sees an acute angle, its measure will be less than 90.

Ⓑ Carlos reads in his textbook that the measure of all right angles is 90. He conjectures that the measure of each right angle in a square equals 90.

Ⓒ Carlos measures the angles of several triangles and finds that their measures all add up to 180. He conjectures that the sum of the measures of the angles in any triangle is always 180.

Ⓓ Carlos knows that the sum of the measures of the angles in a square is always 360. He conjectures that if he draws a square, the sum of the measures of the angles will be 360.

7. Which of the following is the contrapositive of the statement *If Rick buys hamburgers for lunch, then Denzel buys French fries and a large soda*? (Lesson 2-2) **A**

Ⓐ If Denzel does not buy French fries and a large soda, then Rick does not buy hamburgers for lunch.

Ⓑ If Rick does not buy hamburgers for lunch, then Denzel does not buy French fries and a large soda.

Ⓒ If Denzel buys French fries and a large soda, then Rick buys hamburgers for lunch.

Ⓓ If Rick buys hamburgers for lunch, then Denzel does not buy French fries and a large soda.

8. Which property could justify the first step in solving $3 \times \frac{14x + 6}{8} = 18$? (Lesson 2-5) **A**

Ⓐ Division Property of Equality

Ⓑ Substitution Property of Equality

Ⓒ Addition Property of Equality

Ⓓ Transitive Property of Equality

The Princeton Review

***Log On* for Test Practice**

The Princeton Review offers additional test-taking tips and practice problems at their web site. Visit www.princetonreview.com or www.review.com

TestCheck and Worksheet Builder

Special banks of standardized test questions similar to those on the SAT, ACT, TIMSS 8, NAEP 8, and state proficiency tests can be found on this CD-ROM.

Aligned and verified by The Princeton Review

FCAT Practice

Part 2 Short Response/Grid In

Record your answers on the answer sheet provided by your teacher or on a sheet of paper.

9. Two cheerleaders stand at opposite corners of a football field. What is the shortest distance between them, to the nearest yard? (Lesson 1-3) 131 yd

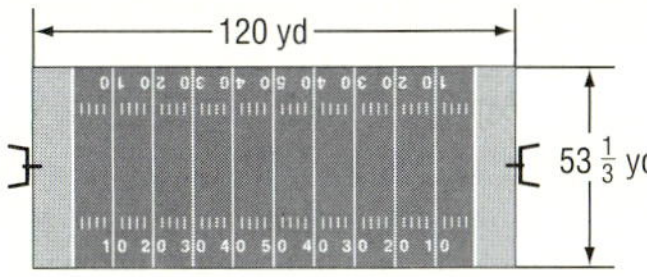

10. Consider the conditional *If I call in sick, then I will not get paid for the day.* Based on the original conditional, what is the name of the conditional *If I do not call in sick, then I will get paid for the day?* (Lesson 2-2) inverse

11. Examine the following statements.

p: Martina drank a cup of soy milk.
q: A cup is 8 ounces.
r: Eight ounces of soy milk contains 300 milligrams of calcium.

Using the Law of Syllogism, how many milligrams of calcium did Martina get from drinking a cup of soy milk? (Lesson 2-4) 300

12. In the following proof, what property justifies statement c? (Lesson 2-7) Segment Addition Postulate

Given: $\overline{AC} \cong \overline{MN}$

Prove: $AB + BC = MN$

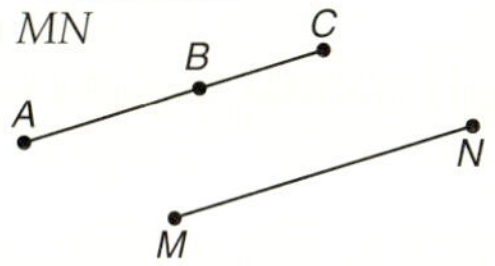

Proof:

Statements	Reasons
a. $\overline{AC} \cong \overline{MN}$	a. Given
b. $AC = MN$	b. Definition of congruent segments
c. $AC = AB + BC$	c. ___?___
d. $AC + BC = MN$	d. Substitution

www.geometryonline.com/standardized_test/fcat

The Princeton Review **Test-Taking Tip**

Question 6 When answering a multiple-choice question, always read every answer choice and eliminate those you decide are definitely wrong. This way, you may deduce the correct answer. Remember that *deductive* reasoning involves drawing a conclusion from a general rule, and that *inductive* reasoning involves making a general rule after observing several examples.

FCAT Practice

Part 3 Open Ended

THINK SOLVE EXPLAIN

Record your answers on a sheet of paper. Show your work.

13. In any right triangle, the sum of the squares of the lengths of the legs equals the square of the length of the hypotenuse. From a single point in her yard, Marti measures and marks distances of 18 feet and 24 feet for two sides of her garden. Explain how Marti can ensure that the two sides of her garden form a right angle. (Lesson 1-3) See margin.

14. A farmer needs to make a 100-square-foot rectangular enclosure for her chickens. She wants to save money by purchasing the least amount of fencing possible to enclose the area. (Lesson 1-4) a–c. See margin.

a. What whole-number dimensions, to the nearest yard, will require the least amount of fencing?

b. Explain your procedure for finding the dimensions that will require the least amount of fencing.

c. Explain how the amount of fencing required to enclose the area changes as the dimensions change.

15. Given: $\angle 1$ and $\angle 3$ are vertical angles. $m\angle 1 = 3x + 5$, $m\angle 3 = 2x + 8$

Prove: $m\angle 1 = 14$ (Lesson 2-8) See p. 123G.

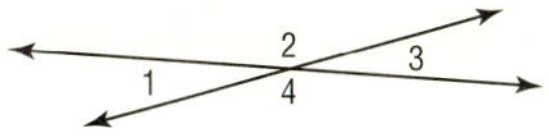

Evaluating Open-Ended Assessment Questions

Open-Ended Assessment questions are graded by using a multilevel rubric that guides you in assessing a student's knowledge of a particular concept.

Goal: Find measures and prove an angle measure.

Sample Scoring Rubric: The following rubric is a sample scoring device. You may wish to add more detail to this sample to meet your individual scoring needs.

Score	Criteria
4	A correct solution that is supported by well-developed, accurate explanations
3	A generally correct solution, but may contain minor flaws in reasoning or computation
2	A partially correct interpretation and/or solution to the problem
1	A correct solution with no supporting evidence or explanation
0	An incorrect solution indicating no mathematical understanding of the concept or task, or no solution is given

Answers

13. Sample answer: Marti can measure a third distance *c*, the distance between the ends of the two sides, and make sure it satisfies the equation $a^2 + b^2 = c^2$.

14a. 10 yd by 10 yd

14b. Sample answer: Make a list of all possible whole-number lengths and widths that will form a 100-square-foot area. Then find the perimeter of each rectangle. Choose the length and width combination that has the smallest perimeter.

14c. As the length and width get closer to having the same measure as one another, the amount of fencing required decreases.

Pages 63–66, Lesson 2-1

1. Sample answer: After the news is over, it's time for dinner.
2. Sometimes; the conjecture is true when *E* is between *D* and *F*; otherwise it is false.
3. Sample answer: When it is cloudy, it rains. Counterexample: It is often cloudy and it does not rain.

19. 30

20. 20

21.

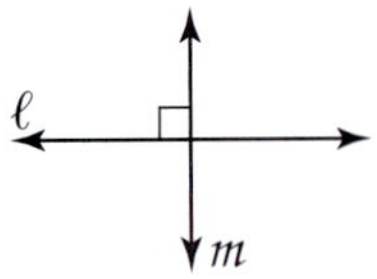

22.

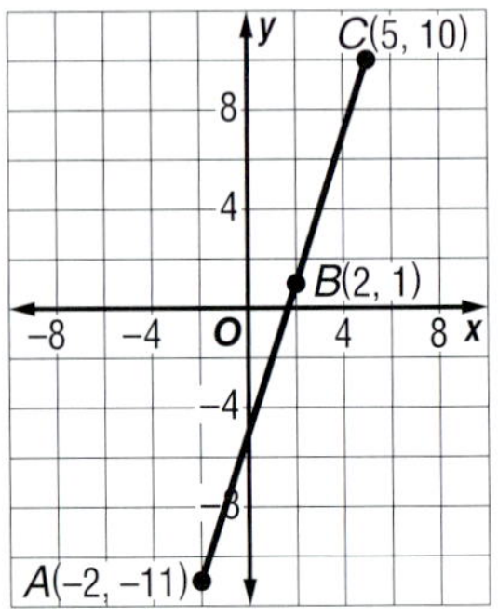

23.

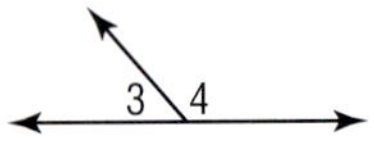

24.

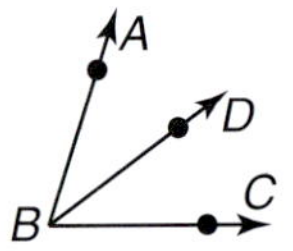

25.

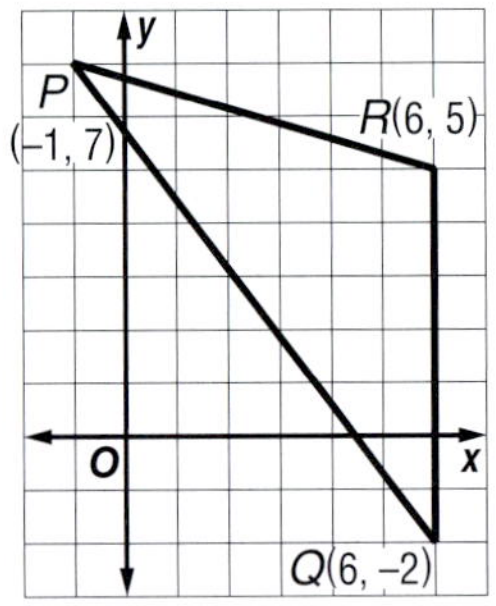

26.

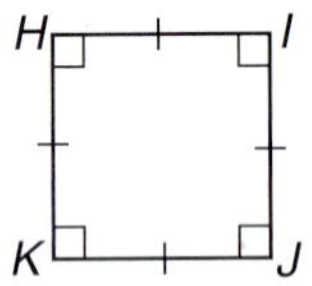

27.

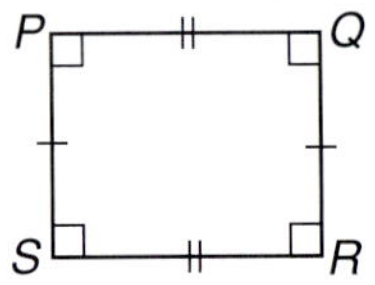

28.

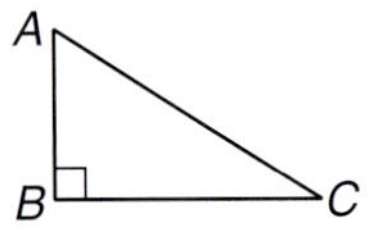

Pages 71–74, Lesson 2-2

11.

p	q	$p \wedge q$
T	T	T
T	F	F
F	T	F
F	F	F

12.

q	r	$q \vee r$
T	T	T
T	F	T
F	T	T
F	F	F

13.

p	r	$\sim p$	$\sim p \wedge r$
T	T	F	F
T	F	F	F
F	T	T	T
F	F	T	F

14.

p	q	r	$p \vee q$	$(p \vee q) \vee r$
T	T	T	T	T
T	T	F	T	T
T	F	T	T	T
T	F	F	T	T
F	T	T	T	T
F	T	F	T	T
F	F	T	F	T
F	F	F	F	F

18. $\sqrt{-64} = 8$ and an equilateral triangle has three congruent sides.
19. $\sqrt{-64} = 8$ or an equilateral triangle has three congruent sides.
20. $\sqrt{-64} = 8$ and $0 < 0$.
21. $0 < 0$ and an obtuse angle measures greater than 90° and less than 180°.
22. An equilateral triangle has three congruent sides or $0 < 0$.
23. An equilateral triangle has three congruent sides and an obtuse angle measures greater than 90° and less than 180°.
24. $\sqrt{-64} = 8$ and an obtuse angle measures greater than 90° and less than 180°.
25. An equilateral triangle has three congruent sides and $0 < 0$.
26. $0 < 0$ or $\sqrt{-64} = 8$
27. An obtuse angle measures greater than 90° and less than 180° or an equilateral triangle has three congruent sides.
28. $\sqrt{-64} = 8$ and an equilateral triangle has three congruent sides, or an obtuse angle measures greater than 90° and less than 180°.
29. An obtuse angle measures greater than 90° and less than 180°, or an equilateral triangle has three congruent sides and $0 < 0$.

33.

q	r	q and r
T	T	T
T	F	F
F	T	F
F	F	F

34.

p	q	p or q
T	T	T
T	F	T
F	T	T
F	F	F

35.

p	r	p or r
T	T	T
T	F	T
F	T	T
F	F	F

36.

p	q	p and q
T	T	T
T	F	F
F	T	F
F	F	F

37.

q	r	$\sim r$	$q \wedge \sim r$
T	T	F	F
T	F	T	T
F	T	F	F
F	F	T	F

38.

p	q	$\sim p$	$\sim q$	$\sim p \wedge \sim q$
T	T	F	F	F
T	F	F	T	F
F	T	T	F	F
F	F	T	T	T

39.

p	q	r	$\sim p$	$\sim r$	$q \wedge \sim r$	$\sim p \vee (q \wedge \sim r)$
T	T	T	F	F	F	F
T	T	F	F	T	T	T
T	F	T	F	F	F	F
T	F	F	F	T	F	F
F	T	T	T	F	F	T
F	T	F	T	T	T	T
F	F	T	T	F	F	T
F	F	F	T	T	F	T

40.

p	q	r	$\sim q$	$\sim r$	$\sim q \vee \sim r$	$p \wedge (\sim q \vee \sim r)$
T	T	T	F	F	F	F
T	T	F	F	T	T	T
T	F	T	T	F	T	T
T	F	F	T	T	T	T
F	T	T	F	F	F	F
F	T	F	F	T	T	F
F	F	T	T	F	T	F
F	F	F	T	T	T	F

Pages 78–80, Lesson 2-3

15. If you are in Colorado, then aspen trees cover high areas of the mountains. If you are in Florida, then cypress trees rise from the swamps. If you are in Vermont, then maple trees are prevalent.

16. H: $2x + 6 = 10$, C: $x = 2$

17. H: you are a teenager; C: you are at least 13 years old

18. H: you have a driver's license; C: you are at least 16 years old

19. H: three points lie on a line; C: the points are collinear

20. H: a man hasn't discovered something he will die for; C: he isn't fit to live

21. H: an angle measures between 0 and 90; C: the angle is acute

22. If you buy a 1-year fitness plan, then you get a free visit.

23. If you are a math teacher, then you love to solve problems.

24. If I think, then I am.

25. If two angles are adjacent, then they have a common side.

26. If two angles are vertical, then they are congruent.

27. If two triangles are equiangular, then they are equilateral.

40. Converse: If you live in Texas, then you live in Dallas. False; you could live in Austin. Inverse: If you don't live in Dallas, then you don't live in Texas. False; you could live in Austin. Contrapositive: If you don't live in Texas, then you don't live in Dallas; true.

41. Converse: If you are in good shape, then you exercise regularly; true. Inverse: If you do not exercise regularly, then you are not in good shape; true. Contrapositive: If you are not in good shape, then you do not exercise regularly. False; an ill person may exercise a lot, but still not be in good shape.

42. Converse: If the sum of two angles is 90, then they are complementary; true. Inverse: If two angles are not complementary, then their sum is not 90; true. Contrapositive: If the sum of two angles is not 90, then they are not complementary; true.

43. Converse: If a figure is a quadrilateral, then it is a rectangle; false, rhombus. Inverse: If a figure is not a rectangle, then it is not a quadrilateral; false, rhombus. Contrapositive: If a figure is not a quadrilateral, then it is not a rectangle; true.

44. Converse: If an angle has a measure of 90, then it is a right angle; true. Inverse: If an angle is not a right angle, then its measure is not 90; true. Contrapositive: If an angle does not have a measure of 90, then it is not a right angle; true.

45. Converse: If an angle has measure less than 90, then it is acute; true. Inverse: If an angle is not acute, then its measure is not less than 90; true. Contrapositive: If an angle's measure is not less than 90, then it is not acute; true.

47. Sample answer: In Alaska, if there are more hours of daylight than darkness, then it is summer; true. In Alaska, if there are more hours of darkness than daylight, then it is winter; true.

Page 80, Practice Quiz 1

1.

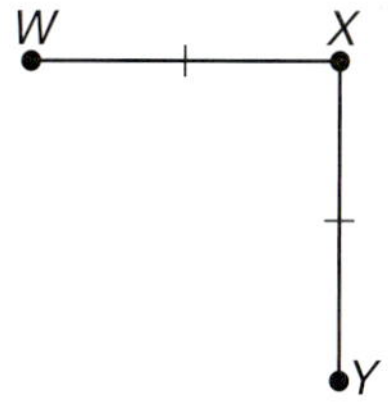

3.

p	q	$\sim p$	$\sim p \wedge q$
T	T	F	F
T	F	F	F
F	T	T	T
F	F	T	F

4.

p	q	r	$q \wedge r$	$p \vee (q \wedge r)$
T	T	T	T	T
T	T	F	F	T
T	F	T	F	T
T	F	F	F	T
F	T	T	T	T
F	T	F	F	F
F	F	T	F	F
F	F	F	F	F

5. Converse: If two angles have a common vertex, then the angles are adjacent. False; $\angle ABD$ is not adjacent to $\angle ABC$.

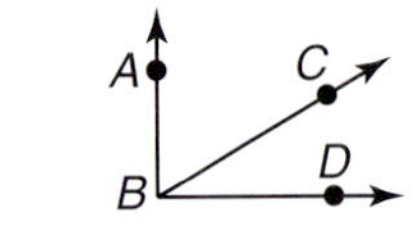

Inverse: If two angles are not adjacent, then they do not have a common vertex. False, $\angle ABC$ and $\angle DBE$ have a common vertex and are not adjacent.

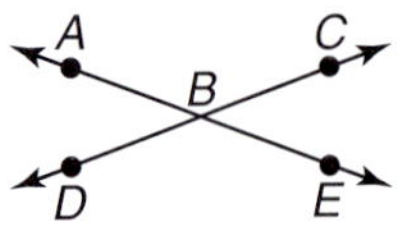

Contrapositive: If two angles do not have a common vertex, then they are not adjacent; true.

Pages 84–87, Lesson 2-4

33. Sample answer: Doctors and nurses use charts to assist in determining medications and their doses for patients. Answers should include the following.

- Doctors need to note a patient's symptoms to determine which medication to prescribe, then determine how much to prescribe based on weight, age, severity of the illness, and so on.
- Doctors use what is known to be true about diseases and when symptoms appear, then deduce that the patient has a particular illness.

39.

q	r	$q \wedge r$
T	T	T
T	F	F
F	T	F
F	F	F

40.

p	r	$\sim p$	$\sim p \vee r$
T	T	F	T
T	F	F	F
F	T	T	T
F	F	T	T

41.

p	q	r	$q \vee r$	$p \wedge (q \vee r)$
T	T	T	T	T
T	T	F	T	T
T	F	T	T	T
T	F	F	F	F
F	T	T	T	F
F	T	F	T	F
F	F	T	T	F
F	F	F	F	F

42.

p	q	r	$\sim q$	$\sim q \wedge r$	$p \vee (\sim q \wedge r)$
T	T	T	F	F	T
T	T	F	F	F	T
T	F	T	T	T	T
T	F	F	T	F	T
F	T	T	F	F	F
F	T	F	F	F	F
F	F	T	T	T	T
F	F	F	T	F	F

Pages 91–93, Lesson 2-5

6. Sometimes; if the planes have a common intersection, then their intersection is one line.

10. Since P is the midpoint of $\overline{QR}$ and $\overline{ST}$, $PQ = PR = \frac{1}{2}QR$ and $PS = PT = \frac{1}{2}ST$ by the definition of midpoint. We are given $\overline{QR} \cong \overline{ST}$ so $QR = ST$ by the definition of congruent segments. By the Multiplication Property, $\frac{1}{2}QR = \frac{1}{2}ST$. So, by substitution, $PQ = PT$.

16. Sometimes; the three points cannot be on the same line.

17. Always; if two points lie in a plane, then the entire line containing those points lies in that plane.

18. Never; the intersection of a line and a plane can be a point, but the intersection of two planes is a line.

19. Sometimes; the three points cannot be on the same line.

20. Always; one plane contains at least three points, so it must contain two.

21. Sometimes; ℓ and m could be skew, so they would not lie in the same plane.

22. Postulate 2.1; through any two points, there is exactly one line.

23. Postulate 2.5; if two points lie in a plane, then the entire line containing those points lies in that plane.

24. Postulate 2.2; through any three points not on the same line, there is exactly one plane.

25. Postulate 2.5; if two points lie in a plane, then the entire line containing those points lies in the plane.

26. Postulate 2.1; through any two points, there is exactly one line.

27. Postulate 2.2; through any three points not on the same line, there is exactly one plane.

38.

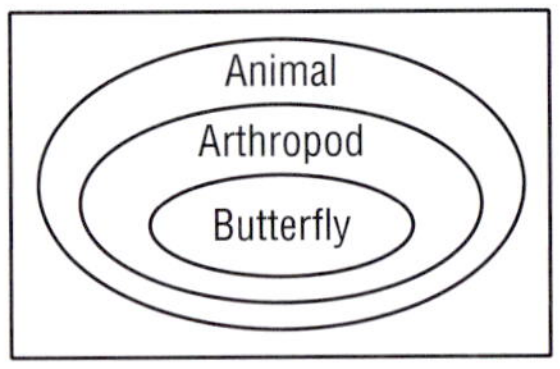

Pages 97–100, Lesson 2-6

10. Given: $25 = -7(y - 3) + 5y$

Prove: $-2 = y$

Proof:

Statements (Reasons)

1. $25 = -7(y - 3) + 5y$ (Given)
2. $25 = -7y + 21 + 5y$ (Dist. Prop.)
3. $25 = -2y + 21$ (Substitution)
4. $4 = -2y$ (Subt. Prop.)
5. $-2 = y$ (Div. Prop.)

11. Given: Rectangle $ABCD$, $AD = 3$, $AB = 10$

Prove: $AC = BD$

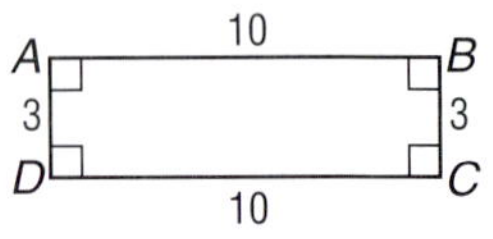

Proof:

Statements (Reasons)

1. Rectangle $ABCD$, $AD = 3$, $AB = 10$ (Given)
2. Draw segments AC and DB. (Two points determine a line.)
3. $\triangle ABC$ and $\triangle BCD$ are right triangles. (Def. of rt. $\triangle$)
4. $AC = \sqrt{3^2 + 10^2}$, $DB = \sqrt{3^2 + 10^2}$ (Pythag. Th.)
5. $AC = BD$ (Substitution)

12. Given: $c^2 = a^2 + b^2$

Prove: $a = \sqrt{c^2 - b^2}$

Proof:

Statements (Reasons)

1. $c^2 = a^2 + b^2$ (Given)
2. $c^2 - b^2 = a^2$ (Subt. Prop.)
3. $a^2 = c^2 - b^2$ (Reflexive Prop.)
4. $\sqrt{a^2} = \sqrt{c^2 - b^2}$ (Square Root Prop.)
5. $a = \sqrt{c^2 - b^2}$ (Square Root Prop.)

26. Given: $4 - \frac{1}{2}a = \frac{7}{2} - a$

Prove: $a = -1$

Proof:

Statements (Reasons)

1. $4 - \frac{1}{2}a = \frac{7}{2} - a$ (Given)
2. $2\left(4 - \frac{1}{2}a\right) = 2\left(\frac{7}{2} - a\right)$ (Mult. Prop.)
3. $8 - a = 7 - 2a$ (Dist. Prop.)
4. $1 - a = -2a$ (Subt. Prop.)
5. $1 = -1a$ (Add. Prop.)
6. $-1 = a$ (Div. Prop.)
7. $a = -1$ (Symmetric Prop.)

27. Given: $-2y + \frac{3}{2} = 8$

Prove: $y = -\frac{13}{4}$

Proof:

Statements (Reasons)

1. $-2y + \frac{3}{2} = 8$ (Given)
2. $2\left(-2y + \frac{3}{2}\right) = 2(8)$ (Mult. Prop.)
3. $-4y + 3 = 16$ (Dist. Prop.)
4. $-4y = 13$ (Subt. Prop.)
5. $y = -\frac{13}{4}$ (Div. Prop.)

28. Given: $-\frac{1}{2}m = 9$

Prove: $m = -18$

Proof:

Statements (Reasons)

1. $-\frac{1}{2}m = 9$ (Given)
2. $-2\left(-\frac{1}{2}m\right) = -2(9)$ (Mult. Prop.)
3. $m = -18$ (Substitution)

29. Given: $5 - \frac{2}{3}z = 1$

Prove: $z = 6$

Proof:

Statements (Reasons)

1. $5 - \frac{2}{3}z = 1$ (Given)
2. $3\left(5 - \frac{2}{3}z\right) = 3(1)$ (Mult. Prop.)
3. $15 - 2x = 3$ (Dist. Prop.)
4. $15 - 2x - 15 = 3 - 15$ (Subt. Prop.)
5. $-2x = -12$ (Substitution)
6. $\frac{-2x}{-2} = \frac{-12}{-2}$ (Div. Prop.)
7. $x = 6$ (Substitution)

30. Given: $XZ = ZY$, $XZ = 4x + 1$, and $ZY = 6x - 13$

Prove: $x = 7$

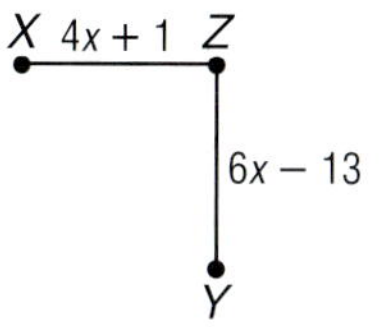

Proof:

Statements (Reasons)

1. $XZ = ZY$, $XZ = 4x + 1$, and $ZY = 6x - 13$ (Given)
2. $4x + 1 = 6x - 13$ (Substitution)
3. $4x + 1 - 4x = 6x - 13 - 4x$ (Subt. Prop.)
4. $1 = 2x - 13$ (Substitution)
5. $1 + 13 = 2x - 13 + 13$ (Add. Prop.)
6. $14 = 2x$ (Substitution)
7. $\frac{14}{2} = \frac{2x}{2}$ (Div. Prop.)
8. $7 = x$ (Substitution)
9. $x = 7$ (Symmetric Prop.)

31. Given: $m\angle ACB = m\angle ABC$

Prove: $m\angle XCA = m\angle YBA$

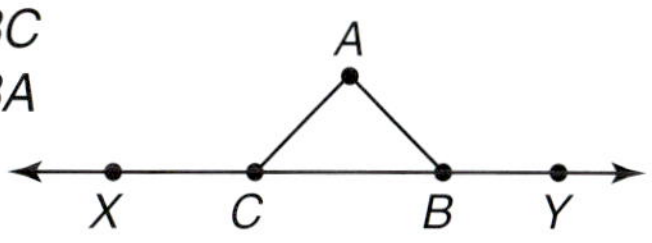

Proof:

Statements (Reasons)

1. $m\angle ACB = m\angle ABC$ (Given)
2. $m\angle XCA + m\angle ACB = 180$, $m\angle YBA + m\angle ABC = 180$ (Def. of supp. $\angle$s)
3. $m\angle XCA + m\angle ACB = m\angle YBA + m\angle ABC$ (Substitution)
4. $m\angle XCA + m\angle ACB = m\angle YBA + m\angle ACB$ (Substitution)
5. $m\angle XCA = m\angle YBA$ (Subt. Prop.)

Page 100, Practice Quiz 2

5. Given: $2(n - 3) + 5 = 3(n - 1)$
Prove: $n = 2$

Proof:
Statements (Reasons)

1. $2(n - 3) + 5 = 3(n - 1)$ (Given)
2. $2n - 6 + 5 = 3n - 3$ (Dist. Prop.)
3. $2n - 1 = 3n - 3$ (Substitution)
4. $2n - 1 - 2n = 3n - 3 - 2n$ (Subt. Prop.)
5. $-1 = n - 3$ (Substitution)
6. $-1 + 3 = n - 3 + 3$ (Add. Prop.)
7. $2 = n$ (Substitution)
8. $n = 2$ (Symmetric Prop.)

Pages 103–106, Lesson 2-7

8. Given: $\overline{AP} \cong \overline{CP}$
$\overline{BP} \cong \overline{DP}$
Prove: $\overline{AB} \cong \overline{CD}$

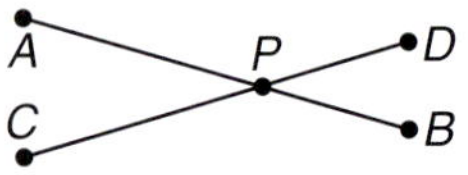

Proof:
Statements (Reasons)

1. $\overline{AP} \cong \overline{CP}$ and $\overline{BP} \cong \overline{DP}$ (Given)
2. $AP = CP$ and $BP = DP$ (Def. of $\cong$ segs.)
3. $AP + PB = AB$ (Seg. Add. Post.)
4. $CP + DP = AB$ (Substitution)
5. $CP + PD = CD$ (Seg. Add. Post.)
6. $AB = CD$ (Transitive Prop.)
7. $\overline{AB} \cong \overline{CD}$ (Def. of $\cong$ segs.)

9. Given: $\overline{HI} \cong \overline{TU}$ and
$\overline{HJ} \cong \overline{TV}$
Prove: $\overline{IJ} \cong \overline{UV}$

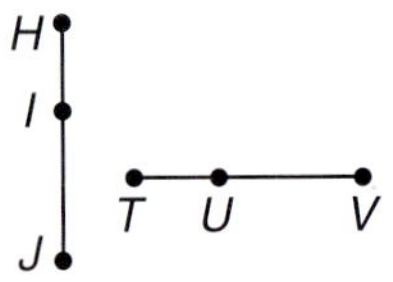

Proof:
Statements (Reasons)

1. $\overline{HI} \cong \overline{TU}$ and $\overline{HJ} \cong \overline{TV}$ (Given)
2. $HI = TU$ and $HJ = TV$ (Def. of $\cong$ segs.)
3. $HI + IJ = HJ$ (Seg. Add. Post.)
4. $TU + IJ = TV$ (Substitution)
5. $TU + UV = TV$ (Seg. Add. Post.)
6. $TU + IJ = TU + UV$ (Substitution)
7. $TU = TU$ (Reflexive Prop.)
8. $IJ = UV$ (Subt. Prop.)
9. $\overline{IJ} \cong \overline{UV}$ (Def. of $\cong$ segs.)

19. Given: $\overline{XY} \cong \overline{WZ}$ and $\overline{WZ} \cong \overline{AB}$
Prove: $\overline{XY} \cong \overline{AB}$

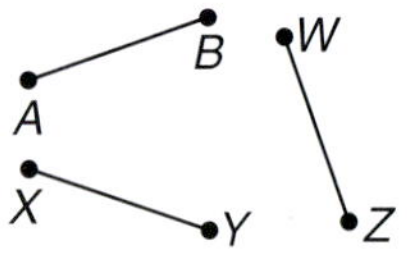

Proof:
Statements (Reasons)

1. $\overline{XY} \cong \overline{WZ}$ and $\overline{WZ} \cong \overline{AB}$ (Given)
2. $XY = WZ$ and $WZ = AB$ (Def. of $\cong$ segs.)
3. $XY = AB$ (Transitive Prop.)
4. $\overline{XY} \cong \overline{AB}$ (Def. of $\cong$ segs.)

20. Given: $\overline{AB} \cong \overline{AC}$ and $\overline{PC} \cong \overline{QB}$
Prove: $\overline{AP} \cong \overline{AQ}$

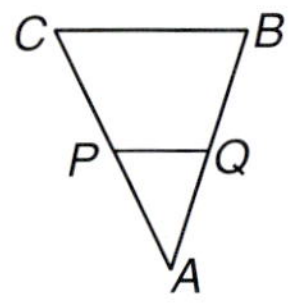

Proof:
Statements (Reasons)

1. $\overline{AB} \cong \overline{AC}$ and $\overline{PC} \cong \overline{QB}$ (Given)
2. $AB = AC$, $PC = QB$ (Def. of $\cong$ segs.)
3. $AB = AQ + QB$, $AC = AP + PC$ (Seg. Add. Post.)
4. $AQ + QB = AP + PC$ (Substitution)
5. $AQ + QB = AP + QB$ (Substitution)
6. $QB = QB$ (Reflexive Prop.)
7. $AP = AQ$ (Subt. Prop.)
8. $\overline{AP} \cong \overline{AQ}$ (Def. of $\cong$ segs.)

22. Given: $\overline{LM} \cong \overline{PN}$ and
$\overline{XM} \cong \overline{XN}$
Prove: $\overline{LX} \cong \overline{PX}$

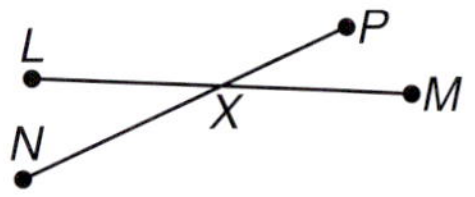

Proof:
Statements (Reasons)

1. $\overline{LM} \cong \overline{PN}$ and $\overline{XM} \cong \overline{XN}$ (Given)
2. $LM = PN$ and $XM = XN$ (Def. of $\cong$ segs.)
3. $LM = LX + XM$, $PN = PX + XN$ (Seg. Add. Post.)
4. $LX + XM = PX + XN$ (Substitution)
5. $LX + XN = PX + XN$ (Substitution)
6. $XN = XN$ (Reflexive Prop.)
7. $LX = PX$ (Subt. Prop.)
8. $\overline{LX} \cong \overline{PX}$ (Def. of $\cong$ segs.)

23. Given: $AB = BC$
Prove: $AC = 2BC$

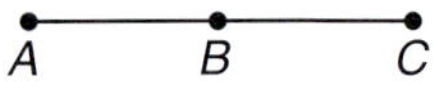

Proof:
Statements (Reasons)

1. $AB = BC$ (Given)
2. $AC = AB + BC$ (Seg. Add. Post.)
3. $AC = BC + BC$ (Substitution)
4. $AC = 2BC$ (Substitution)

24. Given: $\overline{AB}$
Prove: $\overline{AB} \cong \overline{AB}$

Proof:
Statements (Reasons)

1. $\overline{AB}$ (Given)
2. $AB = AB$ (Reflexive Prop.)
3. $\overline{AB} \cong \overline{AB}$ (Def. of $\cong$ segs.)

25. Given: $\overline{AB} \cong \overline{DE}$, C is the midpoint of $\overline{BD}$.
Prove: $\overline{AC} \cong \overline{CE}$

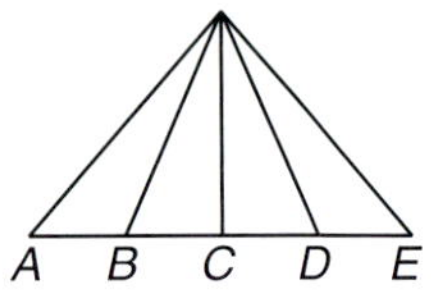

Proof:
Statements (Reasons)

1. $\overline{AB} \cong \overline{DE}$, C is the midpoint of $\overline{BD}$. (Given)
2. $BC = CD$ (Def. of midpoint)
3. $AB = DE$ (Def. of $\cong$ segs.)
4. $AB + BC = CD + DE$ (Add. Prop.)
5. $AB + BC = AC$, $CD + DE = CE$ (Seg. Add. Post.)
6. $AC = CE$ (Substitution)
7. $\overline{AC} \cong \overline{CE}$ (Def. of $\cong$ segs.)

26. Given: $\overline{AB} \cong \overline{EF}$ and $\overline{BC} \cong \overline{DE}$
Prove: $\overline{AC} \cong \overline{DF}$

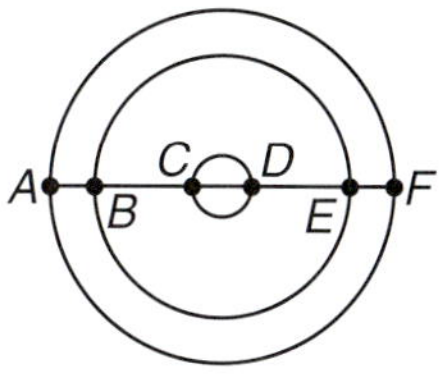

Proof:
Statements (Reasons)

1. $\overline{AB} \cong \overline{EF}$ and $\overline{BC} \cong \overline{DE}$ (Given)
2. $AB = EF$ and $BC = DE$ (Def. of $\cong$ segs.)
3. $AB + BC = DE + EF$ (Add. Prop.)
4. $AC = AB + BC$, $DF = DE + EF$ (Seg. Add. Post.)
5. $AC = DF$ (Substitution)
6. $\overline{AC} \cong \overline{DF}$ (Def. of $\cong$ segs.)

27. Sample answers: $\overline{LN} \cong \overline{QO}$ and $\overline{LM} \cong \overline{MN} \cong \overline{RS} \cong \overline{ST} \cong \overline{QP} \cong \overline{PO}$

Pages 111–114, Lesson 2-8

25. Given: $\angle A$
Prove: $\angle A \cong \angle A$

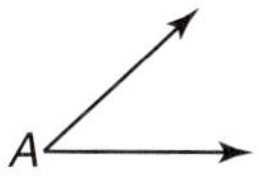

Proof:
Statements (Reasons)

1. $\angle A$ is an angle. (Given)
2. $m\angle A = m\angle A$ (Reflexive Prop)
3. $\angle A \cong \angle A$ (Def. of $\cong$ angles)

26. Given: $\angle 1 \cong \angle 2$, $\angle 2 \cong \angle 3$
Prove: $\angle 1 \cong \angle 3$

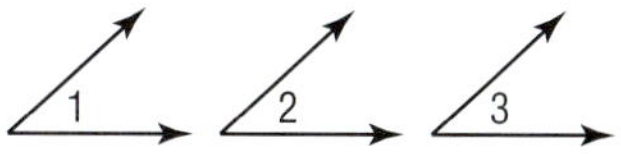

Proof:
Statements (Reasons)

1. $\angle 1 \cong \angle 2$, $\angle 2 \cong \angle 3$ (Given)
2. $m\angle 1 = m\angle 2$, $m\angle 2 = m\angle 3$ (Def. of $\cong$ angles)
3. $m\angle 1 = m\angle 3$ (Trans. Prop.)
4. $\angle 1 \cong \angle 3$ (Def. of $\cong$ angles)

33. Given: $\ell \perp m$
Prove: $\angle 2$, $\angle 3$, $\angle 4$ are rt. $\angle$s

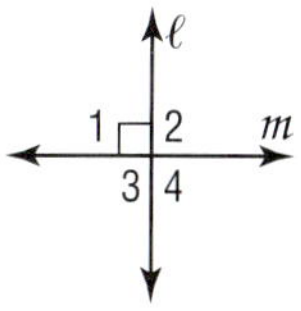

Proof:
Statements (Reasons)

1. $\ell \perp m$ (Given)
2. $\angle 1$ is a right angle. (Def. of $\perp$)
3. $m\angle 1 = 90$ (Def. of rt. $\angle$s)
4. $\angle 1 \cong \angle 4$ (Vert. $\angle$s are $\cong$)
5. $m\angle 1 = m\angle 4$ (Def. of $\cong$ $\angle$s)
6. $m\angle 4 = 90$ (Substitution)
7. $\angle 1$ and $\angle 2$ form a linear pair; $\angle 3$ and $\angle 4$ form a linear pair. (Def. of linear pair)
8. $m\angle 1 + m\angle 2 = 180$, $m\angle 4 + m\angle 3 = 180$ (Linear pairs are supplementary.)
9. $90 + m\angle 2 = 180$, $90 + m\angle 3 = 180$ (Substitution)
10. $m\angle 2 = 90$, $m\angle 3 = 90$ (Subt. Prop.)
11. $\angle 2$, $\angle 3$, $\angle 4$ are rt. $\angle$s. (Def. of rt. $\angle$s (steps 6, 10))

34. Given: $\angle 1$ and $\angle 2$ are rt. $\angle$s.
Prove: $\angle 1 \cong \angle 2$

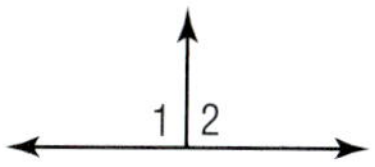

Proof:
Statements (Reasons)

1. $\angle 1$ and $\angle 2$ are rt. $\angle$s. (Given)
2. $m\angle 1 = 90$, $m\angle 2 = 90$ (Def. of rt. $\angle$s)
3. $m\angle 1 = m\angle 2$ (Substitution)
4. $\angle 1 \cong \angle 2$ (Def. of $\cong$ angles)

35. Given: $\ell \perp m$
Prove: $\angle 1 \cong \angle 2$

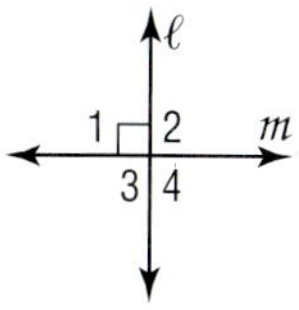

Proof:
Statements (Reasons)

1. $\ell \perp m$ (Given)
2. $\angle 1$ and $\angle 2$ are rt. $\angle$s. ($\perp$ lines intersect to form 4 rt. $\angle$s.)
3. $\angle 1 \cong \angle 2$ (All rt. $\angle$s are $\cong$.)

36. Given: $\angle 1 \cong \angle 2$, $\angle 1$ and $\angle 2$ are supplementary.
Prove: $\angle 1$ and $\angle 2$ are rt. $\angle$s.

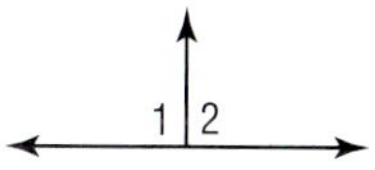

Proof:
Statements (Reasons)

1. $\angle 1 \cong \angle 2$, $\angle 1$ and $\angle 2$ are supplementary. (Given)
2. $m\angle 1 + m\angle 2 = 180$ (Def. of supplementary $\angle$s)
3. $m\angle 1 = m\angle 2$ (Def. of $\cong$ angle)
4. $m\angle 1 + m\angle 1 = 180$ (Substitution)
5. $2(m\angle 1) = 180$ (Add. Prop.)
6. $m\angle 1 = 90$ (Div. Prop.)
7. $m\angle 2 = 90$ (Substitution (steps 3, 6))
8. $\angle 1$ and $\angle 2$ are rt. $\angle$s. (Def. of rt. $\angle$s)

37. Given: $\angle ABD \cong \angle CBD$, $\angle ABD$ and $\angle CBD$ form a linear pair.
Prove: $\angle ABD$ and $\angle CBD$ are rt. $\angle$s.

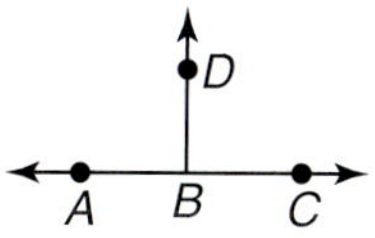

Proof:
Statements (Reasons)

1. $\angle ABD \cong \angle CBD$, $\angle ABD$ and $\angle CBD$ form a linear pair. (Given)
2. $\angle ABD$ and $\angle CBD$ are supplementary. (Linear pairs are supplementary.)
3. $\angle ABD$ and $\angle CBD$ are rt. $\angle$s. (If $\angle$s are $\cong$ and suppl., they are rt. $\angle$s.)

38. Given: $\angle ABD \cong \angle YXZ$
Prove: $\angle CBD \cong \angle WXZ$

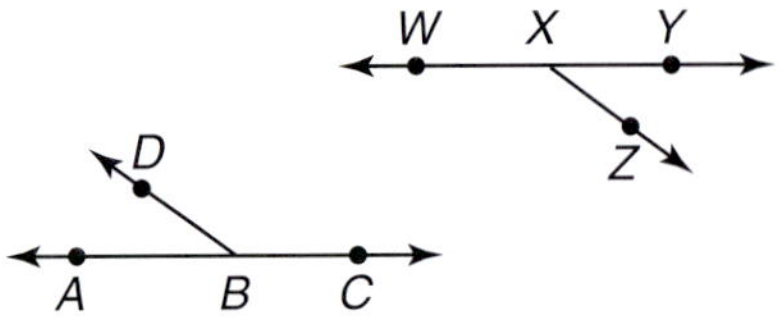

Proof:
Statements (Reasons)

1. $\angle ABD \cong \angle YXZ$; $\angle ABD$ and $\angle CBD$ form a linear pair; $\angle YXZ$ and $\angle WXZ$ form a linear pair. (Given; from the figure)
2. $m\angle ABD + m\angle CBD = 180$, $m\angle YXZ + m\angle WXZ = 180$ (Linear pairs are supplementary.)
3. $m\angle ABD + m\angle CBD = m\angle YXZ + m\angle WXZ$ (Subst.)
4. $m\angle ABD = m\angle YXZ$ (Def. of $\cong$ ∡s)
5. $m\angle YXZ + m\angle CBD = m\angle YXZ + m\angle WXZ$ (Subst.)
6. $m\angle YXZ = m\angle YXZ$ (Reflexive Prop.)
7. $m\angle CBD = m\angle WXZ$ (Subt. Prop.)
8. $\angle CBD \cong \angle WXZ$ (Def. of $\cong$ ∡s)

39. Given: $m\angle RSW = m\angle TSU$
Prove: $m\angle RST = m\angle WSU$

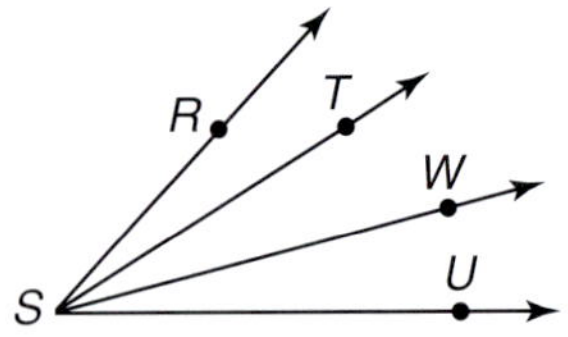

Proof:
Statements (Reasons)

1. $m\angle RSW = m\angle TSU$ (Given)
2. $m\angle RSW = m\angle RST + m\angle TSW$, $m\angle TSU = m\angle TSW + m\angle WSU$ (Angle Addition Postulate)
3. $m\angle RST + m\angle TSW = m\angle TSW + m\angle WSU$ (Substitution)
4. $m\angle TSW = m\angle TSW$ (Reflexive Prop.)
5. $m\angle RST = m\angle WSU$ (Subt. Prop.)

48. Given: $AC = BD$
Prove: $AB = CD$

Proof:
Statements (Reasons)

1. $AC = BD$ (Given)
2. $AB + BC = AC$, $BC + CD = BD$ (Segment Addition Postulate)
3. $BC = BC$ (Reflexive Prop.)
4. $AB + BC = BC + CD$ (Substitution (2 and 3))
5. $AB = CD$ (Subt. Prop.)

Page 115-120, Chapter 2 Study Guide and Review

53. Given: $BC = EC$, $CA = CD$
Prove: $BA = DE$

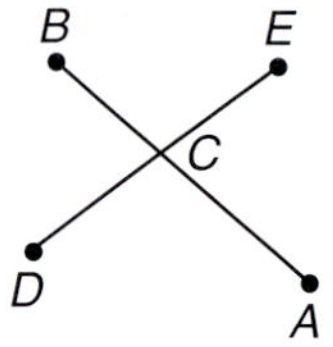

Proof:
Statements (Reasons)

1. $BC = EC$, $CA = CD$ (Given)
2. $BC + CA = EC + CA$ (Add. Prop.)
3. $BC + CA = EC + CD$ (Substitution)
4. $BC + CA = BA$, $EC + CD = DE$ (Seg. Add. Post.)
5. $BA = DE$ (Substitution)

54. Given: $AB = CD$
Prove: $AC = BD$

Proof:
Statements (Reasons)

1. $AB = CD$ (Given)
2. $BC = BC$ (Reflexive Prop.)
3. $AB + BC = CD + BC$ (Add. Prop.)
4. $AB + BC = AC$, $CD + BC = BD$ (Seg. Add. Post.)
5. $AC = BD$ (Substitution)

Page 123, Chapter 2 Standardized Test Practice

15. Given: $\angle 1$ and $\angle 3$ are vertical angles.
$m\angle 1 = 3x + 5$, $m\angle 3 = 2x + 8$
Prove: $m\angle 1 = 14$

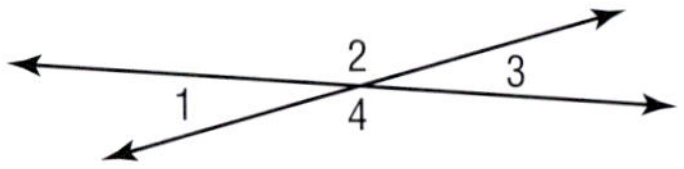

Proof:
Statements (Reasons)

a. $\angle 1$ and $\angle 3$ are vertical angles; $m\angle 1 = 3x + 5$, $m\angle 3 = 2x + 8$ (Given)
b. $\angle 1 \cong \angle 3$ (Vert. ∡s are $\cong$.)
c. $m\angle 1 = m\angle 3$ (Def. of $\cong$ ∡s)
d. $3x + 5 = 2x + 8$ (Substitution)
e. $x + 5 = 8$ (Subt. Prop.)
f. $x = 3$ (Subt. Prop.)
g. $m\angle 1 = 3(3) + 5$ (Substitution)
h. $m\angle 1 = 14$ (Substitution)

Notes

Chapter 3

Parallel and Perpendicular Lines

Chapter Overview and Pacing

LESSON OBJECTIVES

	PACING (days)			
	Regular		Block	
	Basic/ Average	Advanced	Basic/ Average	Advanced
3-1 Parallel Lines and Transversals *(pp. 126–131)* • Identify the relationships between two lines or two planes. • Name angles formed by a pair of lines and a transversal.	2	1	1	0.5
3-2 Angles and Parallel Lines *(pp. 133–138)* ***Preview:*** Use The Geometer's Sketchpad to investigate the measures of angles formed by two parallel lines and a transversal. • Use the properties of parallel lines to determine congruent angles. • Use algebra to find angle measures.	2 (with 3-2 Preview)	2 (with 3-2 Preview)	1.5 (with 3-2 Preview)	1 (with 3-2 Preview)
3-3 Slopes of Lines *(pp. 139–144)* • Find slopes of lines. • Use slope to identify parallel and perpendicular lines.	1.5	1	1	0.5
3-4 Equations of Lines *(pp. 145–150)* • Write an equation of a line given information. • Solve problems by writing equations.	1.5	1	1	0.5
3-5 Proving Lines Parallel *(pp. 151–157)* • Recognize angle conditions that occur with parallel lines. • Prove that two lines are parallel based on given angle relationships.	2	2	1	1
3-6 Perpendiculars and Distance *(pp. 158–166)* ***Preview:*** Use a graphing calculator to determine the points of intersection of a transversal and two parallel lines. • Find the distance between a point and a line. • Find the distance between parallel lines. ***Follow-Up:*** Compare plane Euclidean geometry and spherical geometry.	2	2 (with 3-6 Preview and 3-6 Follow-Up)	1	1.5 (with 3-6 Preview and 3-6 Follow-Up)
Study Guide and Practice Test *(pp. 167–171)* **Standardized Test Practice** *(p. 172–173)*	1	1	1	0.5
Chapter Assessment	1	1	0.5	0.5
TOTAL	13	11	8	6

Pacing suggestions for the entire year can be found on pages T20–T21.

Timesaving Tools
TeacherWorks™
All-In-One Planner and Resource Center
See pages T5 and T21.

Chapter Resource Manager

Chapter 3 Resource Masters: Study Guide and Intervention	Chapter 3 Resource Masters: Practice (Skills and Average)	Chapter 3 Resource Masters: Reading to Learn Mathematics	Chapter 3 Resource Masters: Enrichment	Chapter 3 Resource Masters: Assessment	Prerequisite Skills Workbook	Applications*	5-Minute Check Transparencies	Interactive Chalkboard	GeomPASS: Tutorial Plus (lessons)	Materials
125–126	127–128	129	130				3-1	3-1		ruler, compass
131–132	133–134	135	136	175	83–84	SC 5	3-2	3-2		
137–138	139–140	141	142	175, 177	7–8, 33–34, 77–78	SC 6 GCC 21	3-3	3-3	7	grid paper, straightedge
143–144	145–146	147	148		3–4		3-4	3-4	8	grid paper, straightedge
149–150	151–152	153	154	176	3–4		3-5	3-5		compass, straightedge
155–156	157–158	159	160	176	1–4	GCC 22	3-6	3-6		(*Preview:* TI-83 Plus graphing calculator) grid paper, compass, straightedge
				161–172, 178–182						

**Key to Abbreviations:* GCC = Graphing Calculator and Computer Masters
SC = School-to-Career Masters

Chapter 3

Mathematical Connections and Background

Continuity of Instruction

Prior Knowledge

In previous courses, students wrote and solved equations with one or more variables. In Chapter 1, they identified points, lines, and planes. Congruent angles were introduced. In Chapter 2, students wrote paragraph and two-column proofs.

This Chapter

In this chapter students identify the special angle relationships that result when a transversal intersects parallel lines. Slope and forms for the equation of a line are reviewed. Students solve problems by writing linear equations and use slope to determine whether two lines are parallel, perpendicular, or neither. Students expand their understanding of parallel and perpendicular lines to find the distance between a point and a line and between two parallel lines.

Future Connections

In Chapter 6, students find relationships among segments of the transversal cut off by parallel lines. Chapter 8 shows how parallel lines are used to identify the various quadrilaterals. Writing and solving linear equations are crucial mathematical skills that students will draw on in their future studies.

3-1 Parallel Lines and Transversals

Coplanar lines that do not intersect are called parallel lines. Planes that do not intersect are called parallel planes. The notation $\|$ is used to show parallelism. Noncoplanar lines are called skew lines.

A line that intersects two or more lines in a plane at different points is called a transversal. The intersection of these lines creates a variety of angle relationships. Angles on the exterior of the figure are called exterior angles. Interior angles are inside the two lines that the transversal intersects. Consecutive interior angles are interior angles on the same side of the transversal. Alternate exterior angles are exterior angles on opposite sides of the transversal. Alternate interior angles are on the interior, on opposite sides of the transversal. To identify corresponding angles, look at each intersection individually rather than at the figure as a whole. Each intersection creates four angles. Each angle has a corresponding angle in the other intersection.

3-2 Angles and Parallel Lines

When a transversal intersects a pair of parallel lines, the corresponding angles are congruent. This postulate is called the Corresponding Angles Postulate. In this same situation, alternate interior angles and alternate exterior angles are also congruent. Furthermore, each pair of consecutive interior angles is supplementary.

The Perpendicular Transversal Theorem states that, in a plane, if a transversal is perpendicular to one of two parallel lines, it is also perpendicular to the other. Students use their knowledge of how transversals create congruent and supplementary angles to calculate angle measures.

3-3 Slopes of Lines

The slope of a line is the ratio of its vertical rise to its horizontal run. The slope of a vertical line is undefined, and the slope of a horizontal line is zero. Two nonvertical lines have the same slope if and only if they are parallel. Two nonvertical lines are perpendicular if and only if the product of their slopes is -1. This means that you can use slope to identify parallel and perpendicular lines. You can also use slope to graph parallel and perpendicular lines.

3-4 Equations of Lines

As you learned in algebra, the equation of a nonvertical and nonhorizontal line includes variables, one for values on the x-axis and one for values on the y-axis. This lesson presents two basic forms for the equations of lines. One is called the slope-intercept form. It is written as $y = mx + b$, where m is slope and b is the y-intercept. The point-slope form is the second form. It is written as $y - y_1 = m(x - x_1)$, where (x_1, y_1) are the coordinates of any point contained in the line.

You can write linear equations to solve real-world problems. Slope often represents a rate of change. This rate can be used to determine cost or other information.

3-5 Proving Lines Parallel

Lines can be proved parallel if certain angle conditions are met. If two lines in a plane are cut by a transversal so that corresponding angles are congruent, then the lines are parallel. This postulate justifies the construction of parallel lines. A tranversal is drawn through a given point to intersect a given line. The given point becomes the vertex for constructing an angle congruent to the one formed by the line and the transversal. Using a compass and straightedge, copy the given angle. The result is a pair of parallel lines cut by a transversal. This construction leads to the Parallel Postulate: If given a line and a point not on the line, then there exists exactly one line through the point that is parallel to the given line.

Since parallel lines create pairs of angles with special relationships, those pairs of angles can be used to prove that lines are parallel. Some of the conditions that verify parallel lines are:

- congruent corresponding angles,
- congruent alternate exterior angles,
- congruent alternate interior angles,
- consecutive interior angles that are supplementary, and
- lines that are perpendicular to the same line.

3-6 Perpendiculars and Distance

The distance from a line to a point not on the line is the length of the segment perpendicular to the line from the point. This is the shortest distance from the point to the line. You can construct a perpendicular segment using a compass and straightedge.

Distance can also be used to determine parallel lines. Two lines in a plane are parallel if they are everywhere equidistant. Equidistant means that the distance between two lines is always the same. To find the distance between two parallel lines, measure the length of a perpendicular segment whose endpoints lie on each of the two lines. You only need to measure in one place because the distance remains consistent. This also means that if two lines are equidistant from a third line, then the two lines are parallel to each other.

Chapter 3

DAILY INTERVENTION and Assessment

	Type	Student Edition	Teacher Resources	Technology/Internet
INTERVENTION	Ongoing	Prerequisite Skills, pp. 125, 131, 138, 144, 150, 157 Practice Quiz 1, p. 138 Practice Quiz 2, p. 150	5-Minute Check Transparencies *Prerequisite Skills Workbook*, pp. 1–4, 7–8, 33–34, 77–78, 83–84 Quizzes, *CRM* pp. 175–176 Mid-Chapter Test, *CRM* p. 177 Study Guide and Intervention, *CRM* pp. 125–126, 131–132, 137–138, 143–144, 149–150, 155–156	GeomPASS: Tutorial Plus www.geometryonline.com/self_check_quiz www.geometryonline.com/extra_examples
	Mixed Review	pp. 131, 138, 144, 150, 157, 164	Cumulative Review, *CRM* p. 178	
	Error Analysis	Find the Error, pp. 128, 142 Common Misconceptions, p. 140	Find the Error, *TWE* pp. 129, 142 Unlocking Misconceptions, *TWE* p. 135 Tips for New Teachers, *TWE* pp. 128, 153	
ASSESSMENT	Standardized Test Practice	pp. 131, 135, 136, 138, 144, 149, 157, 164, 171, 172, 173	*TWE* pp. 172–173 Standardized Test Practice, *CRM* pp. 179–180	Standardized Test Practice CD-ROM www.geometryonline.com/standardized_test
	Open-Ended Assessment	Writing in Math, pp. 130, 138, 144, 149, 157, 164 Open Ended, pp. 128, 136, 142, 147, 154, 162 Standardized Test, p. 173	Modeling: *TWE* pp. 131, 164 Speaking: *TWE* pp. 138, 150 Writing: *TWE* pp. 144, 157 Open-Ended Assessment, *CRM* p. 173	
	Chapter Assessment	Study Guide, pp. 167–170 Practice Test, p. 171	Multiple-Choice Tests (Forms 1, 2A, 2B), *CRM* pp. 161–166 Free-Response Tests (Forms 2C, 2D, 3), *CRM* pp. 167–172 Vocabulary Test/Review, *CRM* p. 174	TestCheck and Worksheet Builder (see below) MindJogger Videoquizzes www.geometryonline.com/vocabulary_review www.geometryonline.com/chapter_test

Key to Abbreviations: TWE = Teacher Wraparound Edition; CRM = Chapter Resource Masters

Additional Intervention Resources

The Princeton Review's *Cracking the SAT & PSAT*

The Princeton Review's *Cracking the ACT*

TestCheck and Worksheet Builder

This **networkable** software has three modules for intervention and assessment flexibility:

- **Worksheet Builder** to make worksheet and tests
- **Student Module** to take tests on screen (optional)
- **Management System** to keep student records (optional)

Special banks are included for SAT, ACT, TIMSS, NAEP, and End-of-Course tests.

Intervention Technology

GeomPASS: Tutorial Plus CD-ROM offers a complete, self-paced geometry curriculum.

Geometry Lesson	GeomPASS Lesson
3-3	7 *Slope*
3-4	8 *Writing Equations of Lines*

Intervention at Home

Log on for student study help.

- For each lesson in the Student Edition, there are Extra Examples and Self-Check Quizzes.
  www.geometryonline.com/extra_examples
  www.geometryonline.com/self_check_quiz
- For chapter review, there is vocabulary review, test practice, and standardized test practice.
  www.geometryonline.com/vocabulary_review
  www.geometryonline.com/chapter_test
  www.geometryonline.com/standardized_test

For more information on Intervention and Assessment, see pp. T8–T11.

Reading and Writing in Mathematics

Glencoe Geometry provides numerous opportunities to incorporate reading and writing into the mathematics classroom.

Student Edition

- Foldables Study Organizer, p. 125
- Concept Check questions require students to verbalize and write about what they have learned in the lesson. (pp. 128, 136, 142, 147, 154, 162)
- Writing in Math questions in every lesson, pp. 130, 138, 144, 149, 157, 164
- Reading Study Tip, p. 126
- WebQuest, pp. 155, 164

Teacher Wraparound Edition

- Foldables Study Organizer, pp. 125, 167
- Study Notebook suggestions, pp. 129, 136, 142, 147, 154, 162, 166
- Modeling activities, pp. 131, 164
- Speaking activities, pp. 138, 150
- Writing activities, pp. 144, 157
- ELL Resources, pp. 124, 130, 137, 143, 148, 155, 163, 167

Additional Resources

- Vocabulary Builder worksheets require students to define and give examples for key vocabulary terms as they progress through the chapter. (*Chapter 3 Resource Masters*, pp. vii-viii)
- Proof Builder helps students learn and understand theorems and postulates from the chapter. (*Chapter 3 Resource Masters*, pp. ix–x)
- Reading to Learn Mathematics master for each lesson (*Chapter 3 Resource Masters*, pp. 129, 135, 141, 147, 153, 159)
- *Vocabulary PuzzleMaker* software creates crossword, jumble, and word search puzzles using vocabulary lists that you can customize.
- *Teaching Mathematics with Foldables* provides suggestions for promoting cognition and language.
- *Reading Strategies for the Mathematics Classroom*
- *WebQuest and Project Resources*

For more information on Reading and Writing in Mathematics, see pp. T6–T7.

Chapter 3 Notes

What You'll Learn

Have students read over the list of objectives and make a list of any words with which they are not familiar.

Why It's Important

Point out to students that this is only one of many reasons why each objective is important. Others are provided in the introduction to each lesson.

Lesson	NCTM Standards	Local Objectives
3-1	3, 6, 8, 9, 10	
3-2 Preview	3, 7, 8	
3-2	2, 3, 6, 7, 8, 9, 10	
3-3	2, 3, 6, 8, 9, 10	
3-4	2, 3, 6, 8, 9, 10	
3-5	2, 3, 6, 7, 8, 9, 10	
3-6 Preview	3, 6	
3-6	2, 3, 6, 8, 9, 10	
3-6 Follow-Up	3, 7, 8	

Key to NCTM Standards:

1=Number & Operations, 2=Algebra, 3=Geometry, 4=Measurement, 5=Data Analysis & Probability, 6=Problem Solving, 7=Reasoning & Proof, 8=Communication, 9=Connections, 10=Representation

Chapter 3 Parallel and Perpendicular Lines

What You'll Learn

- **Lessons 3-1, 3-2, and 3-5** Identify angle relationships that occur with parallel lines and a transversal, and identify and prove lines parallel from given angle relationships.
- **Lessons 3-3 and 3-4** Use slope to analyze a line and to write its equation.
- **Lesson 3-6** Find the distance between a point and a line and between two parallel lines.

Key Vocabulary

- parallel lines (p. 126)
- transversal (p. 127)
- slope (p. 139)
- equidistant (p. 160)

Why It's Important

The framework of a wooden roller coaster is composed of millions of feet of intersecting lumber that often form parallel lines and transversals. Roller coaster designers, construction managers, and carpenters must know the relationships of angles created by parallel lines and their transversals to create a safe and stable ride.
You will find how measures of angles are used in carpentry and construction in Lesson 3-2.

Vocabulary Builder

ELL

The Key Vocabulary list introduces students to some of the main vocabulary terms included in this chapter. For a more thorough vocabulary list with pronunciations of new words, give students the Vocabulary Builder worksheets found on pages vii and viii of the *Chapter 3 Resource Masters*. Encourage them to complete the definition of each term as they progress through the chapter. You may suggest that they add these sheets to their study notebooks for future reference when studying for the Chapter 3 test.

Getting Started

Prerequisite Skills To be successful in this chapter, you'll need to master these skills and be able to apply them in problem-solving situations. Review these skills before beginning Chapter 3.

For Lesson 3-1 — Naming Segments

Name all of the lines that contain the given point.
(For review, see Lesson 1-1.)

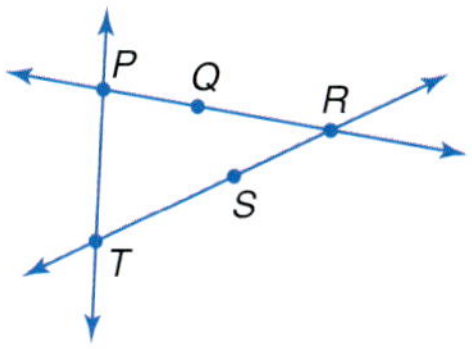

1. Q $\overleftrightarrow{PQ}$
2. R $\overleftrightarrow{PR}$ or $\overleftrightarrow{RS}$
3. S $\overleftrightarrow{ST}$
4. T $\overleftrightarrow{TR}$ or $\overleftrightarrow{TP}$

1–4. Sample answers are given.

For Lessons 3-2 and 3-5 — Congruent Angles

Name all angles congruent to the given angle.
(For review, see Lesson 1-4.)

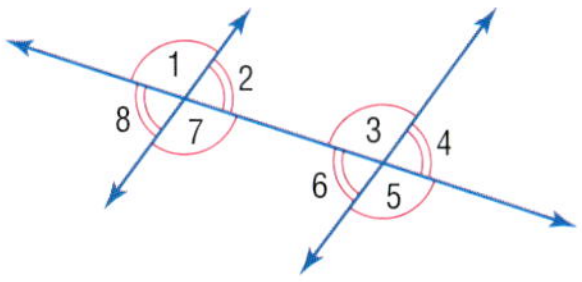

5. $\angle 2$ $\angle 4, \angle 6, \angle 8$
6. $\angle 5$ $\angle 1, \angle 3, \angle 7$
7. $\angle 3$ $\angle 1, \angle 5, \angle 7$
8. $\angle 8$ $\angle 2, \angle 4, \angle 6$

For Lessons 3-3 and 3-4 — Equations of Lines

For each equation, find the value of y for the given value of x. *(For review, see pages 736 and 738.)*

9. $y = 7x - 12$, for $x = 3$ 9
10. $y = -\frac{2}{3}x + 4$, for $x = 8$ $-\frac{4}{3}$
11. $2x - 4y = 18$, for $x = 6$ $-\frac{3}{2}$

FOLDABLES™ Study Organizer

Parallel and Perpendicular Lines Make this Foldable to help you organize information from this chapter. Begin with one sheet of $8\frac{1}{2}$-inch by 11-inch paper.

Step 1 Fold

Fold in half matching the short sides.

Step 2 Fold Again

Unfold and fold the long side up 2 inches to form a pocket.

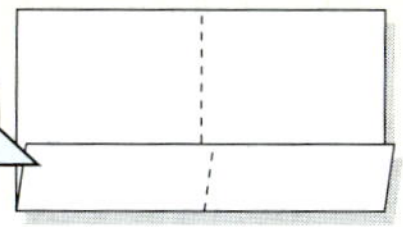

Step 3 Staple or Glue

Staple or glue the outer edges to complete the pocket.

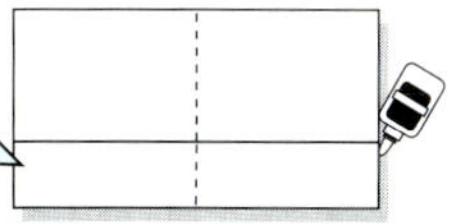

Step 4 Label

Label each side as shown. Use index cards to record examples.

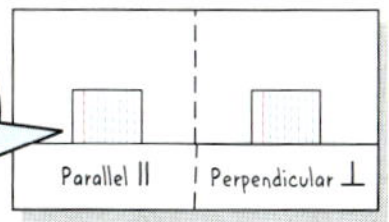

Reading and Writing As you read and study the chapter, write examples and notes about parallel and perpendicular lines on index cards. Place the cards in the appropriate pocket.

Getting Started

This section provides a review of the basic concepts needed before beginning Chapter 3. Page references are included for additional student help.

Additional review is provided in the *Prerequisite Skills Workbook,* pages 1–4, 7–8, 33–34, 77–78, 83–84.

Prerequisite Skills in the Getting Ready for the Next Lesson section at the end of each exercise set review a skill needed in the next lesson.

For Lesson	Prerequisite Skill
3-2	Finding measures of linear pairs, p. 131
3-3	Simplifying expressions, p. 138
3-4	Solving equations, p. 144
3-5	Finding measures of angles formed by two lines and a transversal, p. 150
3-6	Using the Distance Formula, p. 157

FOLDABLES™ Study Organizer

For more information about Foldables, see *Teaching Mathematics with Foldables.*

Organization of Data Use this Foldable for student writing about parallel and perpendicular lines. Students will need study cards, either 3″ × 5″ index cards, or sheets of notebook paper cut into quarter sections. As students learn about parallel lines and transversals in Lesson 3-1, have them draw angles formed by a pair of lines and a transversal on one side of their card and describe in writing what they have drawn on the other side. Store this card in the Parallel Lines pocket of the Foldable. Continue through the chapter using the study cards to take notes, draw examples, and record and define the vocabulary words and concepts presented in each lesson.

3-1 Lesson Notes

1 Focus

5-Minute Check Transparency 3-1 Use as a quiz or review of Chapter 2.

Mathematical Background notes are available for this lesson on p. 124C.

How are parallel lines and planes used in architecture?

Ask students:

- What would happen if the top of a door were not parallel to the top of the doorway? **The door would not fit into the opening.**
- What are some of the parallel planes in a stairway? **The tops of the stairs (the treads) are parallel planes, and the vertical part of the stairs (the risers) are parallel planes.**

Building on Prior Knowledge

In Chapter 1, students identified and labeled points, lines, and planes, and measured and classified angles. In this lesson, they identify intersecting lines in space, and classify pairs of angles formed when one line intersects two other lines.

3-1 Parallel Lines and Transversals

Sunshine State Standards
MA.C.2.4.1

What You'll Learn

- Identify the relationships between two lines or two planes.
- Name angles formed by a pair of lines and a transversal.

Vocabulary

- parallel lines
- parallel planes
- skew lines
- transversal
- consecutive interior angles
- alternate exterior angles
- alternate interior angles
- corresponding angles

How are parallel lines and planes used in architecture?

Architect Frank Lloyd Wright designed many buildings using basic shapes, lines, and planes. His building at the right has several examples of parallel lines, parallel planes, and skew lines.

RELATIONSHIPS BETWEEN LINES AND PLANES Lines ℓ and m are coplanar because they lie in the same plane. If the lines were extended indefinitely, they would not intersect. Coplanar lines that do not intersect are called **parallel lines**. Segments and rays contained within parallel lines are also parallel.

Study Tip

Reading Math
The term *parallel* and the notation $\parallel$ are used for lines, segments, rays, and planes. The symbol $\nparallel$ means *is not parallel to.*

The symbol $\parallel$ means *is parallel to*. Arrows are used in diagrams to indicate that lines are parallel. In the figure, the arrows indicate that $\overleftrightarrow{PQ}$ is parallel to $\overleftrightarrow{RS}$.

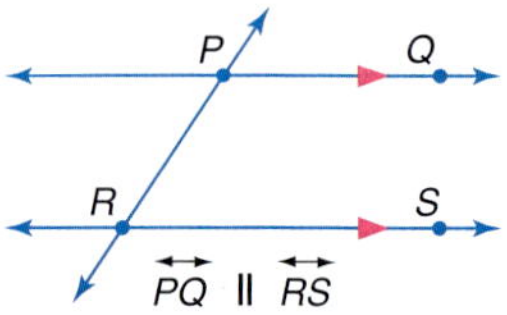

Similarly, two planes can intersect or be parallel. In the photograph above, the roofs of each level are contained in **parallel planes**. The walls and the floor of each level lie in intersecting planes.

Geometry Activity

Draw a Rectangular Prism

A rectangular prism can be drawn using parallel lines and parallel planes.

Step 1 Draw two parallel planes to represent the top and bottom of the prism.

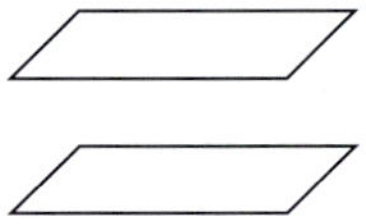

Step 2 Draw the edges. Make any hidden edges of the prism dashed.

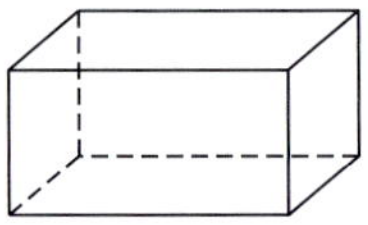

Step 3 Label the vertices.

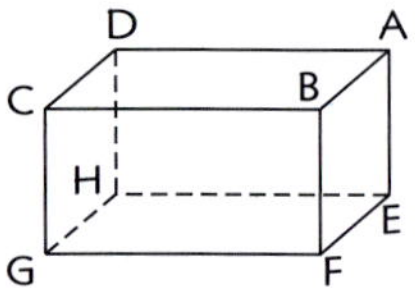

Analyze

1. Identify the parallel planes in the figure. **ABC and EFG, BCG and ADH, and ABF and DCG**
2. Name the planes that intersect plane *ABC* and name their intersections. **plane ABF, $\overleftrightarrow{AB}$; plane DCG, $\overleftrightarrow{DC}$; plane ADH, $\overleftrightarrow{AD}$; plane BCG, $\overleftrightarrow{BC}$**
3. Identify all segments parallel to $\overline{BF}$. **$\overline{AE}$, $\overline{CG}$, and $\overline{DH}$**

Resource Manager

Workbook and Reproducible Masters

Chapter 3 Resource Masters
- Study Guide and Intervention, pp. 125–126
- Skills Practice, p. 127
- Practice, p. 128
- Reading to Learn Mathematics, p. 129
- Enrichment, p. 130

Teaching Geometry With Manipulatives Masters, p. 52

Transparencies
5-Minute Check Transparency 3-1
Answer Key Transparencies

Technology
Interactive Chalkboard

Notice that in the Geometry Activity, $\overline{AE}$ and $\overline{GF}$ do not intersect. These segments are not parallel since they do not lie in the same plane. Lines that do not intersect and are not coplanar are called **skew lines**. Segments and rays contained in skew lines are also skew.

TEACHING TIP

Remind students that the planes can be named in more ways than those shown in the example.

Example 1 Identify Relationships

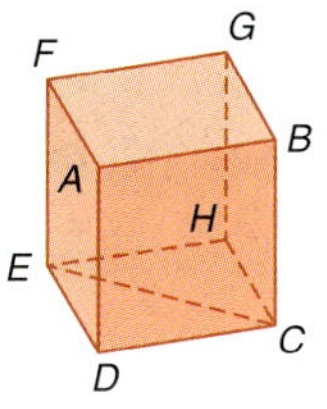

a. Name all planes that are parallel to plane *ABG*.

plane *CDE*

b. Name all segments that intersect $\overline{CH}$.

$\overline{BC}$, $\overline{CD}$, $\overline{CE}$, $\overline{EH}$, and $\overline{GH}$

c. Name all segments that are parallel to $\overline{EF}$.

$\overline{AD}$, $\overline{BC}$, and $\overline{GH}$

d. Name all segments that are skew to $\overline{BG}$.

$\overline{AD}$, $\overline{CD}$, $\overline{CE}$, $\overline{EF}$, and $\overline{EH}$

Study Tip

Identifying Segments

Use the segments drawn in the figure even though other segments exist.

ANGLE RELATIONSHIPS

In the drawing of the railroad crossing, notice that the tracks, represented by line *t*, intersect the sides of the road, represented by lines *m* and *n*. A line that intersects two or more lines in a plane at different points is called a **transversal**.

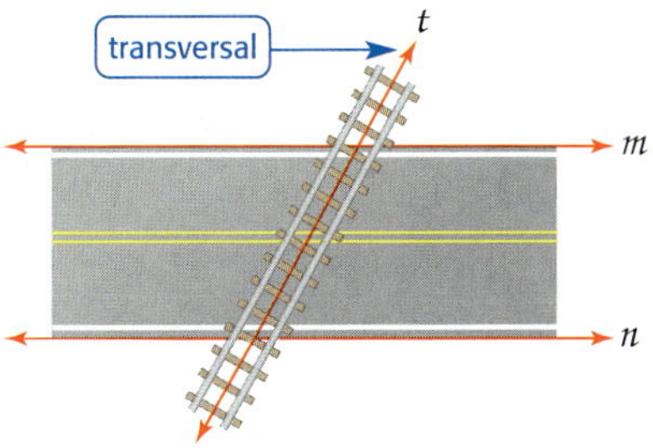

Study Tip

Transversals

The lines that the transversal intersects need not be parallel.

Example 2 Identify Transversals

AIRPORTS **Some of the runways at O'Hare International Airport are shown below. Identify the sets of lines to which each given line is a transversal.**

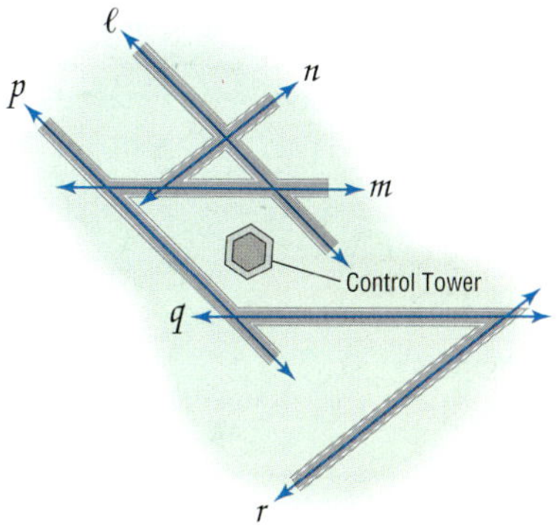

a. line *q*

If the lines are extended, line *q* intersects lines *ℓ*, *n*, *p*, and *r*.

b. line *m*

lines *ℓ*, *n*, *p*, and *r*

c. line *n*

lines *ℓ*, *m*, *p*, and *q*

d. line *r*

lines *ℓ*, *m*, *p*, and *q*

In the drawing of the railroad crossing above, notice that line *t* forms eight angles with lines *m* and *n*. These angles are given special names, as are specific pairings of these angles.

www.geometryonline.com/extra_examples/fcat

2 Teach

RELATIONSHIPS BETWEEN LINES AND PLANES

In-Class Example Power Point®

1 Refer to the figure in Example 1.

a. Name all planes that are parallel to plane *AEF*. **plane *BHG***

b. Name all segments that intersect $\overline{AF}$. **$\overline{EF}$, $\overline{GF}$, $\overline{DA}$, and $\overline{BA}$**

c. Name all segments that are parallel to $\overline{DC}$. **$\overline{AB}$, $\overline{FG}$, and $\overline{EH}$**

d. Name all segments that are skew to $\overline{AD}$. **$\overline{FG}$, $\overline{GB}$, $\overline{EH}$, $\overline{EC}$, and $\overline{CH}$**

Teaching Tip You may ask students to identify other segments that exist for the points given, but are not drawn in the figure.

ANGLE RELATIONSHIPS

In-Class Example Power Point®

2 **BUS STATION** Some of a bus station's driveways are shown. Identify the sets of lines to which each given line is a transversal.

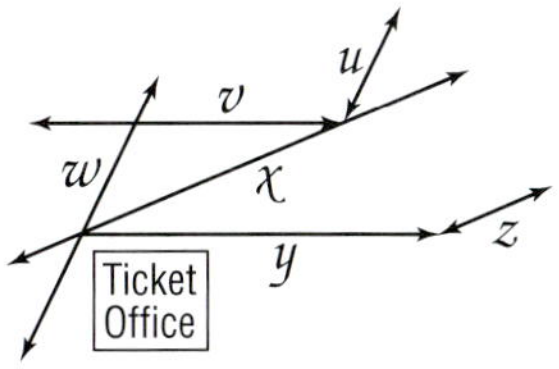

a. line *v* **If the lines are extended, line *v* intersects lines *u*, *w*, *x*, and *z*.**

b. line *y* **lines *u*, *w*, *x*, *z***

c. line *u* **lines *v*, *x*, *y*, *z***

d. line *w* **lines *v*, *x*, *y*, *z***

Geometry Activity

Materials: ruler

- To help students visualize the prism, ask them to name the left and right faces, the front and back, and the top and bottom. Also, ask them to name the segments determined by the corners of each face.
- It may help some students visualize the prism if you use different colors to shade some of the faces of the prism.

In-Class Example

Power Point®

3 Refer to the figure in Example 3. Identify each pair of angles as *alternate interior*, *alternate exterior*, *corresponding*, or *consecutive interior* angles.

a. ∠7 and ∠3 **corresponding**

b. ∠8 and ∠2 **alternate exterior**

c. ∠4 and ∠11 **corresponding**

d. ∠7 and ∠1 **alternate exterior**

e. ∠3 and ∠9 **alternate interior**

f. ∠7 and ∠10 **consecutive interior**

Teaching Tip In Example 3a, suggest that students use a finger or pencil to block out line *c* while they examine ∠1 and ∠7; in 3b, they can block out line *b* while they examine ∠2 and ∠10.

Tips for New Teachers

Intervention To help students distinguish between a transversal and the other two lines, draw a figure formed by three intersecting lines like the one for Exercise 2. Label the three lines as well as the 12 angles. Have students select one line as the transversal and then identify pairs of angles that are alternate interior, alternate exterior, corresponding, and consecutive interior. Then they should select a different line as the transversal and identify appropriate pairs of angles.

Answer

1. Sample answer: The bottom and top of a cylinder are contained in parallel planes.

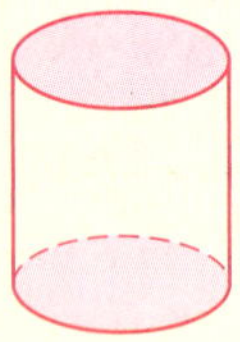

Study Tip

Same Side Interior Angles Consecutive interior angles are also called *same side interior angles*.

Key Concept — Transversals and Angles

Name	Angles
exterior angles	∠1, ∠2, ∠7, ∠8
interior angles	∠3, ∠4, ∠5, ∠6
consecutive interior angles	∠3 and ∠6, ∠4 and ∠5
alternate exterior angles	∠1 and ∠7, ∠2 and ∠8
alternate interior angles	∠3 and ∠5, ∠4 and ∠6
corresponding angles	∠1 and ∠5, ∠2 and ∠6, ∠3 and ∠7, ∠4 and ∠8

Transversal p intersects lines q and r.

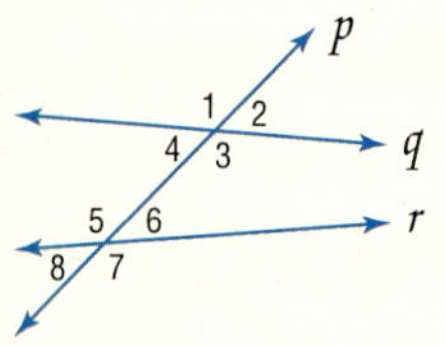

Example 3 Identify Angle Relationships

Refer to the figure below. Identify each pair of angles as *alternate interior*, *alternate exterior*, *corresponding*, or *consecutive interior* angles.

a. ∠1 and ∠7
alternate exterior

b. ∠2 and ∠10
corresponding

c. ∠8 and ∠9
consecutive interior

d. ∠3 and ∠12
corresponding

e. ∠4 and ∠10
alternate interior

f. ∠6 and ∠11
alternate exterior

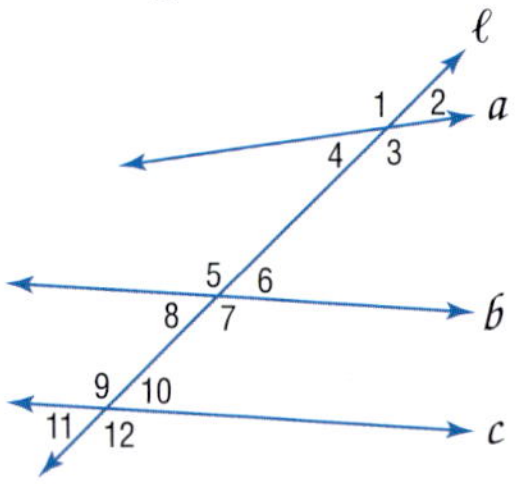

Check for Understanding

Concept Check

1. **OPEN ENDED** Draw a solid figure with parallel planes. Describe which parts of the figure are parallel. **See margin.**

2. **FIND THE ERROR** Juanita and Eric are naming alternate interior angles in the figure at the right. One of the angles must be ∠4.

Juanita	Eric
∠4 and ∠9	∠4 and ∠10
∠4 and ∠6	∠4 and ∠5

Who is correct? Explain your reasoning.

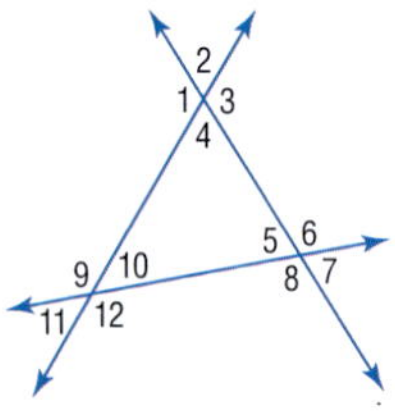

2. Juanita; Eric has listed interior angles, but they are not alternate interior angles.

3. **Describe** a real-life situation in which parallel lines seem to intersect.
Sample answer: looking down railroad tracks

GUIDED PRACTICE KEY

Exercises	Examples
4–6, 18–20	1
7–10, 21	2
11–17	3

Guided Practice

For Exercises 4–6, refer to the figure at the right.

4. Name all planes that intersect plane *ADM*.
5. Name all segments that are parallel to $\overline{CD}$.
6. Name all segments that intersect $\overline{KL}$.
$\overline{BK}$, $\overline{CL}$, $\overline{JK}$, $\overline{LM}$, $\overline{BL}$, $\overline{KM}$

4. *ABC*, *JKL*, *ABK*, *CDM*

5. $\overline{AB}$, $\overline{JK}$, $\overline{LM}$

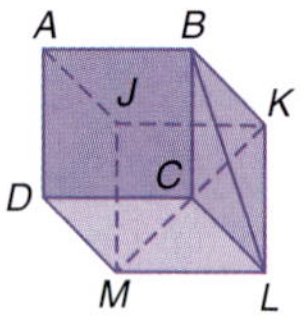

DAILY INTERVENTION — Differentiated Instruction

Visual/Spatial In the first part of this lesson students have to visualize three dimensional figures drawn on a flat page. Encourage students with strong visual/spatial skills to help interpret these figures to other students.

128 Chapter 3 Parallel and Perpendicular Lines

Identify the pairs of lines to which each given line is a transversal.

7. p — *q* and *r*, *q* and *t*, *r* and *t*
8. r — *p* and *q*, *p* and *t*, *q* and *t*
9. q — *p* and *r*, *p* and *t*, *r* and *t*
10. t — *p* and *q*, *p* and *r*, *q* and *r*

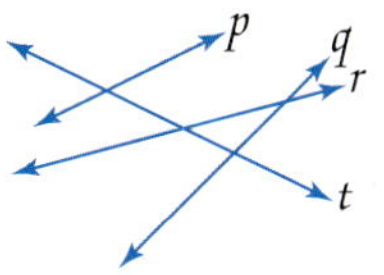

Identify each pair of angles as *alternate interior*, *alternate exterior*, *corresponding*, or *consecutive interior* angles.

11. ∠7 and ∠10 alt. int.
12. ∠1 and ∠5 corr.
13. ∠4 and ∠6 cons. int.
14. ∠8 and ∠1 alt. ext.

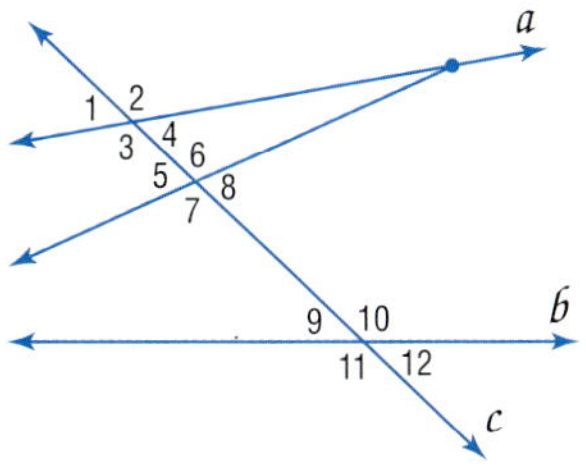

Name the transversal that forms each pair of angles. Then identify the special name for the angle pair.

15. ∠3 and ∠10 *p*; cons. int.
16. ∠2 and ∠12 *p*; alt. ext.
17. ∠ 8 and ∠ 14 *q*; alt. int.

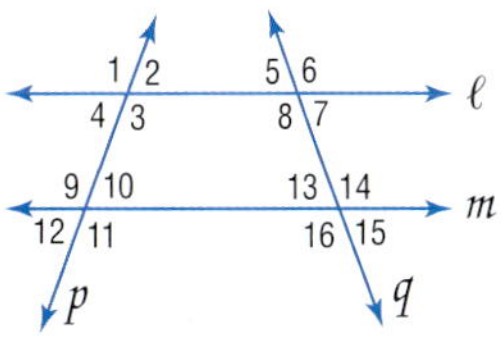

Application

18–21. See margin.

MONUMENTS For Exercises 18–21, refer to the photograph of the Lincoln Memorial.

18. Describe a pair of parallel lines found on the Lincoln Memorial.
19. Find an example of parallel planes.
20. Locate a pair of skew lines.
21. Identify a transversal passing through a pair of lines.

Practice and Apply

Homework Help

For Exercises	See Examples
22–27	1
28–31	2
32–47	3

Extra Practice
See page 758.

For Exercises 22–27, refer to the figure at the right. 22–27. See margin.

22. Name all segments parallel to $\overline{AB}$.
23. Name all planes intersecting plane *BCR*.
24. Name all segments parallel to $\overline{TU}$.
25. Name all segments skew to $\overline{DE}$.
26. Name all planes intersecting plane *EDS*.
27. Name all segments skew to $\overline{AP}$.

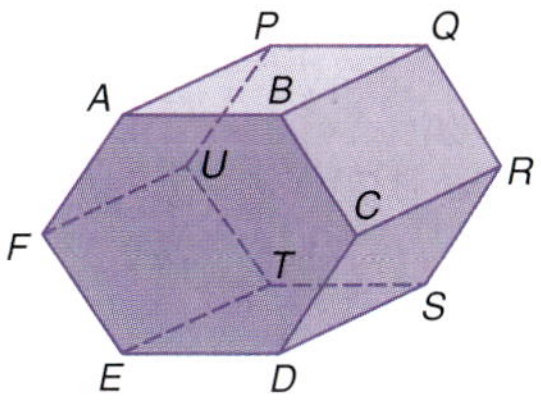

Identify the pairs of lines to which each given line is a transversal.

28. a — *b* and *c*, *b* and *r*, *r* and *c*
29. b — *a* and *c*, *a* and *r*, *r* and *c*
30. c — *a* and *b*, *a* and *r*, *b* and *r*
31. r — *a* and *b*, *a* and *c*, *b* and *c*

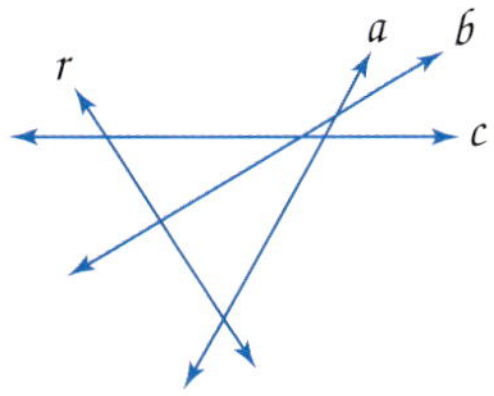

Answers

18. The pillars form parallel lines.
19. The roof and the floor are parallel planes.
20. One of the west pillars and the base on the east side form skew lines.
21. The top of the memorial "cuts" the pillars.
22. $\overline{DE}$, $\overline{PQ}$, $\overline{ST}$
23. *ABC*, *ABQ*, *PQR*, *CDS*, *APU*, *DET*
24. $\overline{BC}$, $\overline{EF}$, $\overline{QR}$
25. $\overline{AP}$, $\overline{BQ}$, $\overline{CR}$, $\overline{FU}$, $\overline{PU}$, $\overline{QR}$, $\overline{RS}$, $\overline{TU}$
26. *ABC*, *AFU*, *BCR*, *CDS*, *EFU*, *PQR*
27. $\overline{BC}$, $\overline{CD}$, $\overline{DE}$, $\overline{EF}$, $\overline{QR}$, $\overline{RS}$, $\overline{ST}$, $\overline{TU}$

3 Practice/Apply

Study Notebook

Have students—

- add the definitions/examples of the vocabulary terms to their Vocabulary Builder worksheets for Chapter 3.
- include the Key Concepts from p. 128.
- include any other item(s) that they find helpful in mastering the skills in this lesson.

DAILY INTERVENTION

FIND THE ERROR Ask students to label the three lines *a*, *b*, and *c*. Then for each pair of alternate interior angles, students can tell which line is the transversal and whether the two named angles are on opposite sides of that transversal.

About the Exercises...

Organization by Objective

- **Relationships Between Lines and Planes:** 22–27, 48–52
- **Angle Relationships:** 28–47

Odd/Even Assignments

Exercises 22–47 are structured so that students practice the same concepts whether they are assigned odd or even problems.

Assignment Guide

Basic: 23–47 odd, 49–51, 53–75

Average: 23–47 odd, 49–51, 53–75

Advanced: 22–48 even, 49–69 (optional: 70–75)

Identify each pair of angles as *alternate interior, alternate exterior, corresponding,* or *consecutive interior* angles.

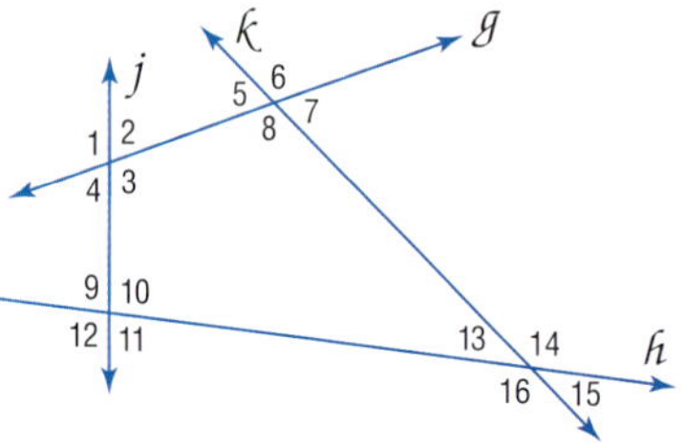

32. ∠2 and ∠10 **32. corr.**
33. ∠1 and ∠11 **alt. ext.**
34. ∠5 and ∠3 **34. alt. int.**
35. ∠6 and ∠14 **corr.**
36. ∠5 and ∠15 **36. alt. ext.**
37. ∠11 and ∠13 **37. alt. int.**
38. ∠8 and ∠3 **cons. int.**
39. ∠9 and ∠4 **cons. int.**

Study Tip

Make a Sketch

Use patty paper or tracing paper to copy the figure. Use highlighters or colored pencils to identify the lines that compose each pair of angles.

Name the transversal that forms each pair of angles. Then identify the special name for the angle pair.

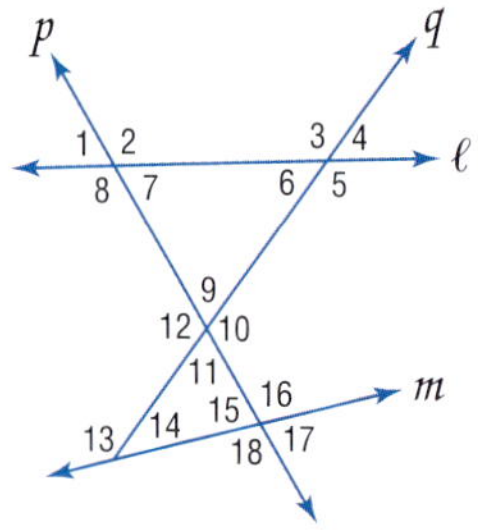

40. ∠2 and ∠9
41. ∠7 and ∠15 **p; alt. int.**
42. ∠13 and ∠17
43. ∠8 and ∠4 **ℓ; alt. ext.**
44. ∠14 and ∠16
45. ∠6 and ∠14 **q; alt. int.**
46. ∠8 and ∠6 **ℓ; corr.**
47. ∠14 and ∠15 **m; cons. int.**

40. p; corr. 42. m; alt. ext. 44. m; corr.

48. **AVIATION** Airplanes heading eastbound are assigned an altitude level that is an odd number of thousands of feet. Airplanes heading westbound are assigned an altitude level that is an even number of thousands of feet. If one airplane is flying northwest at 34,000 feet and another airplane is flying east at 25,000 feet, describe the type of lines formed by the paths of the airplanes. Explain your reasoning. **Skew lines; the planes are flying in different directions and at different altitudes.**

STRUCTURES **For Exercises 49–51, refer to the drawing of the gazebo at the right.**

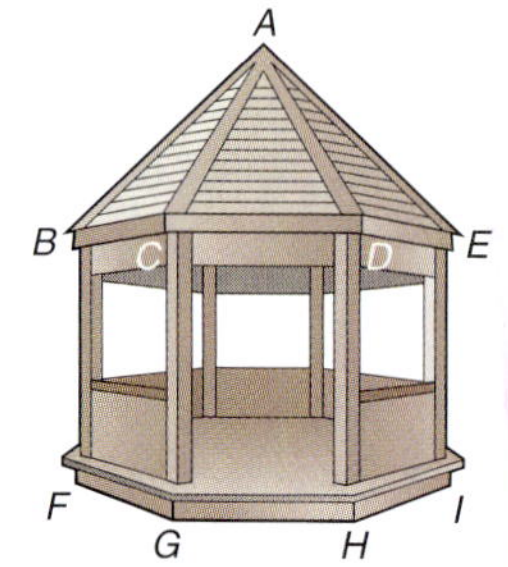

49. Name all labeled segments parallel to $\overline{BF}$. **$\overline{CG}$, $\overline{DH}$, $\overline{EI}$**
50. Name all labeled segments skew to $\overline{AC}$. **50. $\overline{DE}$, $\overline{FG}$, $\overline{HI}$, $\overline{GH}$, $\overline{BF}$, $\overline{DH}$, $\overline{EI}$**
51. Are any of the planes on the gazebo parallel to plane *ADE*? Explain. **No; plane *ADE* will intersect all the planes if they are extended.**
52. **COMPUTERS** The word *parallel* when used with computers describes processes that occur simultaneously, or devices, such as printers, that receive more than one bit of data at a time. Find two other examples for uses of the word *parallel* in other subject areas such as history, music, or sports. **52. Sample answers: parallel bars in gymnastics, parallel port on a computer, parallel events, parallel voices in a choir, latitude parallels on a map**

CRITICAL THINKING **Suppose there is a line ℓ and a point *P* not on the line.**

53. In space, how many lines can be drawn through *P* that do not intersect ℓ? **53. infinite number**
54. In space, how many lines can be drawn through *P* that are parallel to ℓ? **1**

55. **WRITING IN MATH** Answer the question that was posed at the beginning of the lesson. **See margin.**

How are parallel lines and planes used in architecture?

Include the following in your answer:

- a description of where you might expect to find examples of parallel lines and parallel planes, and
- an example of skew lines and nonparallel planes.

Study Guide and Intervention, p. 125 (shown) and p. 126

Relationships Between Lines and Planes When two lines lie in the same plane and do not intersect, they are **parallel**. Lines that do not intersect and are not coplanar are **skew lines**. In the figure, ℓ is parallel to m, or $\ell \parallel m$. You can also write $\overleftrightarrow{PQ} \parallel \overleftrightarrow{RS}$. Similarly, if two planes do not intersect, they are **parallel planes**.

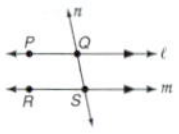

Example

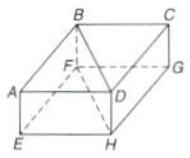

a. **Name all planes that are parallel to plane *ABD*.**
plane *EFH*

b. **Name all segments that are parallel to $\overline{CG}$.**
$\overline{BF}$, $\overline{DH}$, and $\overline{AE}$

c. **Name all segments that are skew to $\overline{EH}$.**
$\overline{BF}$, $\overline{CG}$, $\overline{BD}$, $\overline{CD}$, and $\overline{AB}$

Exercises

For Exercises 1–3, refer to the figure at the right.

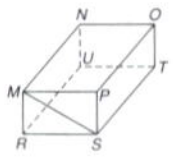

1. Name all planes that intersect plane *OPT*.
MNO, MPS, NOT, RST
2. Name all segments that are parallel to $\overline{NU}$.
$\overline{OT}$, $\overline{PS}$, $\overline{MR}$
3. Name all segments that intersect $\overline{MP}$.
$\overline{MR}$, $\overline{MN}$, $\overline{MS}$, $\overline{PS}$, $\overline{PO}$

For Exercises 4–7, refer to the figure at the right.

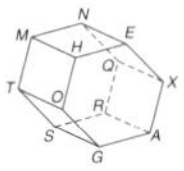

4. Name all segments parallel to $\overline{QX}$.
$\overline{RA}$, $\overline{SG}$, $\overline{TO}$, $\overline{MH}$, $\overline{NE}$
5. Name all planes that intersect plane *MHE*.
MHO, NEX, HEX, MNQ, SGO, RAX
6. Name all segments parallel to $\overline{QR}$.
$\overline{AX}$, $\overline{HO}$, $\overline{MT}$
7. Name all segments skew to $\overline{AG}$.
$\overline{ST}$, $\overline{TM}$, $\overline{NQ}$, $\overline{QR}$, $\overline{TO}$, $\overline{MH}$, $\overline{NE}$, $\overline{QX}$

Skills Practice, p. 127 and Practice, p. 128 (shown)

For Exercises 1–4, refer to the figure at the right.

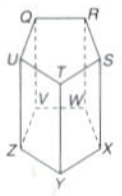

1. Name all planes that intersect plane *STX*.
TUY, RSW, STU, VWX, QUV, QVW
2. Name all segments that intersect $\overline{QU}$. **$\overline{QR}$, $\overline{QV}$, $\overline{TU}$, $\overline{UZ}$**
3. Name all segments that are parallel to $\overline{XY}$. **$\overline{ST}$**
4. Name all segments that are skew to $\overline{VW}$. **$\overline{QU}$, $\overline{RS}$, $\overline{ST}$, $\overline{SX}$, $\overline{TU}$, $\overline{TY}$, $\overline{UZ}$**

Identify the sets of lines to which each given line is a transversal.

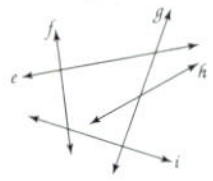

5. e
f and g, f and h, f and i, g and h, g and i, h and i
6. h
e and f, e and g, e and i, f and g, f and i, g and i

Identify each pair of angles as *alternate interior, alternate exterior, corresponding,* or *consecutive interior* angles.

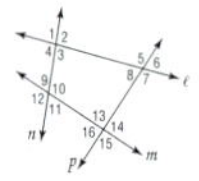

7. ∠9 and ∠13
corresponding
8. ∠6 and ∠16
alternate exterior
9. ∠3 and ∠10
consecutive interior
10. ∠8 and ∠14
alternate interior

Name the transversal that forms each pair of angles. Then identify the special name for the angle pair.

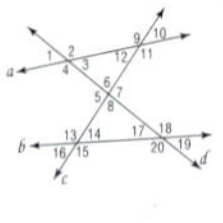

11. ∠2 and ∠12
a; alternate interior
12. ∠6 and ∠18
d; corresponding
13. ∠13 and ∠19
b; alternate exterior
14. ∠11 and ∠7
c; consecutive interior

FURNITURE **For Exercises 15–16, refer to the drawing of the end table.**

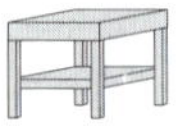

15. Find an example of parallel planes. **Sample answer: the top of the table and the bottom shelf**
16. Find an example of parallel lines. **Sample answer: the table legs**

Reading to Learn Mathematics, p. 129 ELL

Pre-Activity **How are parallel lines and planes used in architecture?**

Read the introduction to Lesson 3-1 at the top of page 126 in your textbook.

- Give an example of parallel lines that can be found in your classroom.
Sample answers: edges of floor along opposite walls; vertical edges of a door
- Give an example of parallel planes that can be found in your classroom.
Sample answers: ceiling and floor; opposite walls

Reading the Lesson

1. Write a geometrical term that matches each definition.
   a. two planes that do not intersect **parallel planes**
   b. lines that are not coplanar and do not intersect **skew lines**
   c. two coplanar lines that do not intersect **parallel lines**
   d. a line that intersects two or more lines in a plane at different points **transversal**
   e. a pair of angles determined by two lines and a transversal consisting of an interior angle and an exterior angle that have different vertices and that lie on the same side of the transversal **corresponding angles**

2. Refer to the figure at the right. Give the special name for each angle pair.

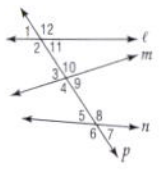

   a. ∠3 and ∠5 **corresponding angles**
   b. ∠6 and ∠12 **alternate exterior angles**
   c. ∠4 and ∠8 **alternate interior angles**
   d. ∠2 and ∠3 **consecutive interior angles**
   e. ∠8 and ∠12 **corresponding angles**
   f. ∠5 and ∠9 **alternate interior angles**
   g. ∠4 and ∠10 **vertical angles**
   h. ∠6 and ∠7 **linear pair**

Helping You Remember

3. A good way to remember new mathematical terms is to relate them to words that you use in everyday life. Many words start with the prefix *trans-*, which is a Latin root meaning *across*. List four English words that start with *trans-*. How can the meaning of this prefix help you remember the meaning of *transversal*?
Sample answer: Translate, transfer, transport, transcontinental; a transversal is a line that goes across two or more other lines.

Enrichment, p. 130

Perspective Drawings

To draw three-dimensional objects, artists make **perspective drawings** such as the ones shown. To indicate depth in a perspective drawing, some parallel lines are drawn as converging lines. The dotted lines in the figures below each extend to a **vanishing point**, or spot where parallel lines appear to meet.

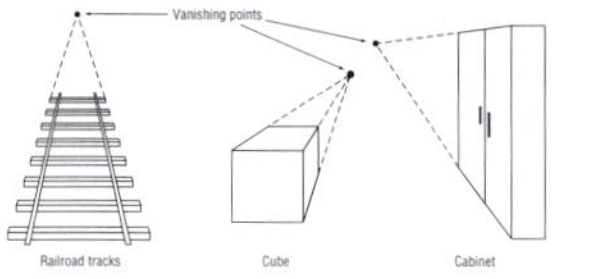

Draw lines to locate the vanishing point in each drawing of a box.

1. 2. 3.

Answer

55. Sample answer: Parallel lines and planes are used in architecture to make structures that will be stable. Answers should include the following.

- **Opposite walls should form parallel planes; the floor may be parallel to the ceiling.**
- **The plane that forms a stairway will not be parallel to some of the walls.**

FCAT Practice

Standardized Test Practice

56. ∠3 and ∠5 are __?__ angles. **A**

Ⓐ alternate interior
Ⓑ alternate exterior
Ⓒ consecutive interior
Ⓓ corresponding

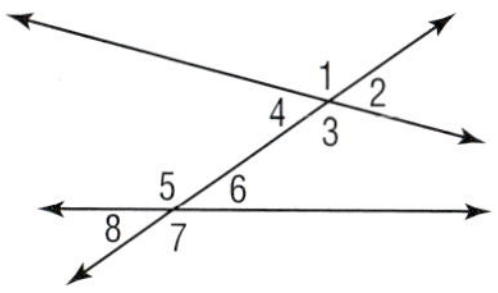

57. **GRID IN** Set M consists of all multiples of 3 between 13 and 31. Set P consists of all multiples of 4 between 13 and 31. What is one possible number in P but NOT in M? **16, 20, or 28**

Maintain Your Skills

Mixed Review

58. **PROOF** Write a two-column proof. *(Lesson 2-8)*
Given: $m\angle ABC = m\angle DFE$, $m\angle 1 = m\angle 4$
Prove: $m\angle 2 = m\angle 3$ **See p. 173A.**

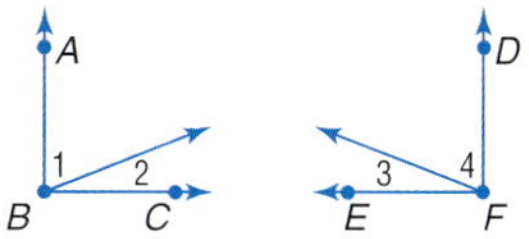

59. **PROOF** Write a paragraph proof. *(Lesson 2-7)*
Given: $\overline{PQ} \cong \overline{ZY}$, $\overline{QR} \cong \overline{XY}$
Prove: $\overline{PR} \cong \overline{XZ}$ **See margin.**

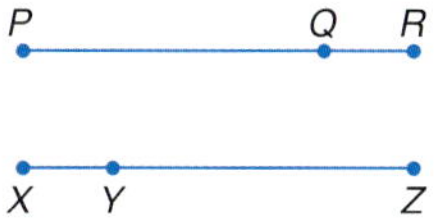

Determine whether a valid conclusion can be reached from the two true statements using the Law of Detachment or the Law of Syllogism. If a valid conclusion is possible, state it and the law that is used. If a valid conclusion does not follow, write *no conclusion*. *(Lesson 2-4)*

60. (1) If two angles are vertical, then they do not form a linear pair.
(2) If two angles form a linear pair, then they are not congruent. **no conclusion**

61. (1) If an angle is acute, then its measure is less than 90.
(2) ∠EFG is acute. **m∠EFG is less than 90; Detachment.**

Find the distance between each pair of points. *(Lesson 1-3)*

62. $A(-1, -8), B(3, 4)$
63. $C(0, 1), D(-2, 9)$
64. $E(-3, -12), F(5, 4)$
65. $G(4, -10), H(9, -25)$ **$\sqrt{250} \approx 15.81$**
66. $J\left(1, \frac{1}{4}\right), K\left(-3, -\frac{7}{4}\right)$ **$\sqrt{20} \approx 4.47$**
67. $L\left(-5, \frac{8}{5}\right), M\left(5, -\frac{2}{5}\right)$ **$\sqrt{104} \approx 10.20$**

62. $\sqrt{160} \approx 12.65$
63. $\sqrt{68} \approx 8.25$
64. $\sqrt{320} \approx 17.89$

Draw and label a figure for each relationship. *(Lesson 1-1)* **68–69. See margin.**

68. $\overleftrightarrow{AB}$ perpendicular to $\overleftrightarrow{MN}$ at point P
69. line ℓ contains R and S but not T

Getting Ready for the Next Lesson

PREREQUISITE SKILL **State the measures of linear pairs of angles in each figure.**
*(To review **linear pairs**, see Lesson 2-6.)*

70. **50, 130**

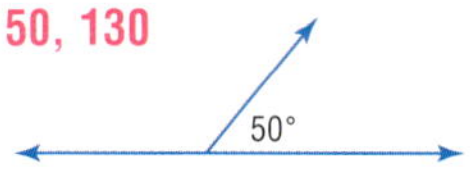

71. **90, 90**

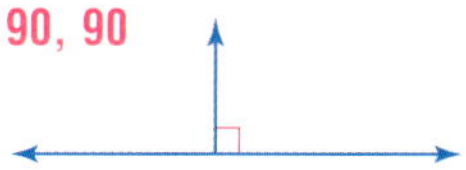

72. **60, 120**

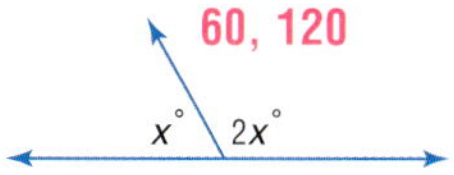

73. **72, 108**

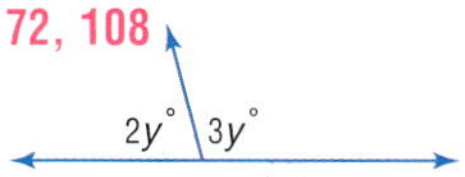

74.

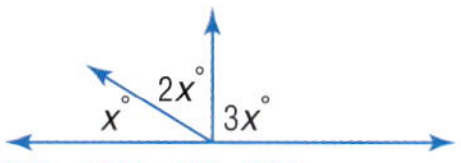

30, 150; 90, 90

75. **76, 104**

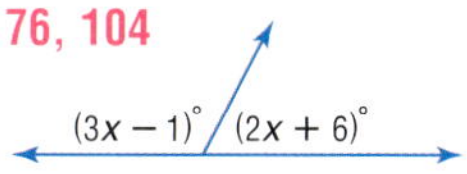

www.geometryonline.com/self_check_quiz/fcat

4 Assess

Open-Ended Assessment

Modeling Have students model two lines and a transversal with uncooked spaghetti. They can then indicate pairs of angles such as alternate interior angles, alternate exterior angles, corresponding angles, and consecutive interior angles using counters or pieces of colored candy.

Getting Ready for Lesson 3-2

Prerequisite Skill Students will use parallel lines to find congruent angles in Lesson 3-2. They will use linear pairs to find measures of supplementary angles. Use Exercises 70–75 to determine your students' familiarity with linear pairs.

Answers

59. Given: $\overline{PQ} \cong \overline{ZY}$, $\overline{QR} \cong \overline{XY}$
Prove: $\overline{PR} \cong \overline{XZ}$

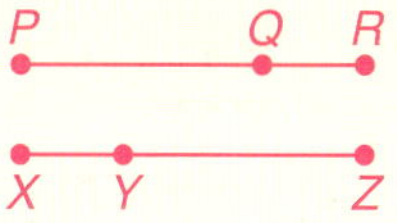

Proof:
Since $\overline{PQ} \cong \overline{ZY}$ and $\overline{QR} \cong \overline{XY}$, $PQ = ZY$ and $QR = XY$ by the definition of congruent segments. By the Addition Property, $PQ + QR = ZY + XY$. Using the Segment Addition Postulate, $PR = PQ + QR$ and $XZ = XY + YZ$. By substitution, $PR = XZ$. Because the measures are equal, $\overline{PR} \cong \overline{XZ}$ by the definition of congruent segments.

68.

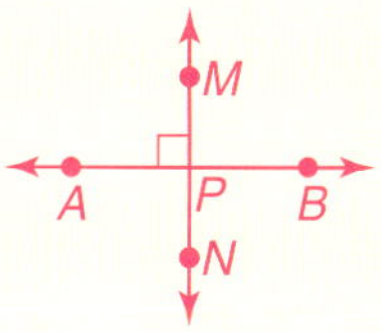

69.

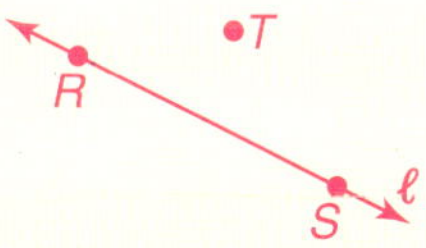

This CD-ROM is a customizable Microsoft® PowerPoint® presentation that includes:

- Step-by-step, dynamic solutions of each In-Class Example from the Teacher Wraparound Edition
- Additional, Try These exercises for each example
- The 5-Minute Check Transparencies
- Hot links to Glencoe Online Study Tools

Geometry Software Investigation

A Preview of Lesson 3-2

Angles and Parallel Lines

You can use The Geometer's Sketchpad to investigate the measures of angles formed by two parallel lines and a transversal.

Step 1 *Draw parallel lines.*

- Place two points A and B on the screen.
- Construct a line through the points.
- Place point C so that it does not lie on $\overleftrightarrow{AB}$.
- Construct a line through C parallel to $\overleftrightarrow{AB}$.
- Place point D on this line.

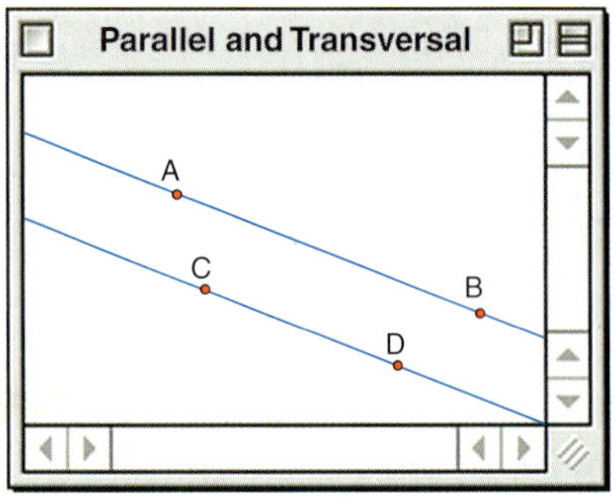

Step 2 *Construct a transversal.*

- Place point E on $\overleftrightarrow{AB}$ and point F on $\overleftrightarrow{CD}$.
- Construct $\overleftrightarrow{EF}$ as a transversal through $\overleftrightarrow{AB}$ and $\overleftrightarrow{CD}$.
- Place points G and H on $\overleftrightarrow{EF}$, as shown.

Step 3 *Measure angles.*

- Measure each angle.

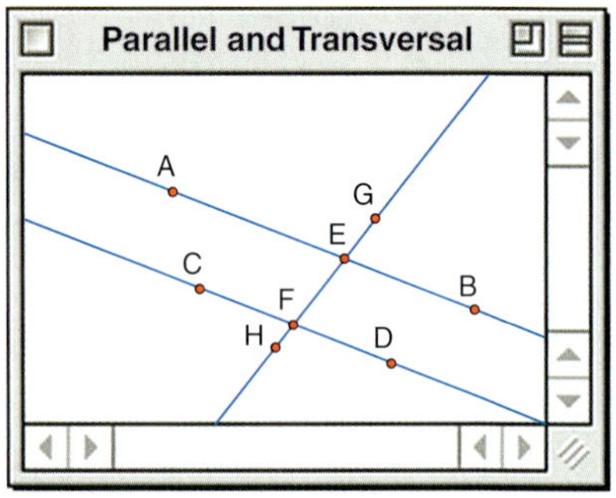

Analyze 1–2. See margin.

1. List pairs of angles by the special names you learned in Lesson 3-1.
2. Which pairs of angles listed in Exercise 1 have the same measure?
3. What is the relationship between consecutive interior angles? **They are supplementary.**

Make a Conjecture

4. Make a conjecture about the following pairs of angles formed by two parallel lines and a transversal. Write your conjecture in if-then form. **See margin.**
   a. corresponding angles
   b. alternate interior angles
   c. alternate exterior angles
   d. consecutive interior angles
5. Rotate the transversal. Are the angles with equal measures in the same relative location as the angles with equal measures in your original drawing? **Yes; the angle pairs show the same relationship.**
6. Test your conjectures by rotating the transversal and analyzing the angles. **See students' work.**
7. Rotate the transversal so that the measure of any of the angles is 90. **See margin.**
   a. What do you notice about the measures of the other angles?
   b. Make a conjecture about a transversal that is perpendicular to one of two parallel lines.

Geometry Software Investigation

A Preview of Lesson 3-2

Getting Started

Creating Parallel Lines This activity uses software to create a line parallel to the given line. Then, after students add a transversal, they can use the figure to identify pairs of congruent angles and pairs of supplementary angles.

Teach

- Be sure students know how to drag and move lines so they can have the transversal intersect the parallel lines at various angles.
- Be sure students use 3-letter names for the angles so their references to particular angles are clear and understood.

Assess

Exercise 4 Encourage students to write their conjectures in full sentences. This helps them communicate their conjectures to other students and understand their own notes later.

Answers

1. corr.: $\angle AEG$ and $\angle CFE$, $\angle AEF$ and $\angle CFH$, $\angle BEG$ and $\angle DFE$, $\angle BEF$ and $\angle DFH$; cons. int.: $\angle AEF$ and $\angle CFE$, $\angle BEF$ and $\angle DFE$; alt. int.: $\angle AEF$ and $\angle DFE$, $\angle BEF$ and $\angle CFE$; alt. ext.: $\angle AEG$ and $\angle DFH$, $\angle BEG$ and $\angle CFH$
2. corr.: $\angle AEG$ and $\angle CFE$, $\angle AEF$ and $\angle CFH$, $\angle BEG$ and $\angle DFE$, $\angle BEF$ and $\angle DFH$; alt. int.: $\angle AEF$ and $\angle DFE$, $\angle BEF$ and $\angle CFE$; alt. ext.: $\angle AEG$ and $\angle DFH$, $\angle BEG$ and $\angle CFH$

4a. If two parallel lines are cut by a transversal, then corresponding angles are congruent.
4b. If two parallel lines are cut by a transversal, then alternate interior angles are congruent.
4c. If two parallel lines are cut by a transversal, then alternate exterior angles are congruent.
4d. If two parallel lines are cut by a transversal, then consecutive interior angles are supplementary.
7a. Sample answer: All of the angles measure 90°.
7b. Sample answer: If two parallel lines are cut by a transversal so that it is perpendicular to one of the lines, then the transversal is perpendicular to the other line.

132 Chapter 3 Parallel and Perpendicular Lines

3-2 Angles and Parallel Lines

Sunshine State Standards
MA.C.2.4.1

What You'll Learn

- Use the properties of parallel lines to determine congruent angles.
- Use algebra to find angle measures.

How can angles and lines be used in art?

In the painting, the artist uses lines and transversals to create patterns. The figure on the painting shows two parallel lines with a transversal passing through them. There is a special relationship between the angle pairs formed by these lines.

The Order of Tradition II by T.C. Stuart

PARALLEL LINES AND ANGLE PAIRS In the figure above, ∠1 and ∠2 are corresponding angles. When the two lines are parallel, there is a special relationship between these pairs of angles.

Postulate 3.1

Corresponding Angles Postulate If two parallel lines are cut by a transversal, then each pair of corresponding angles is congruent.

Examples: ∠1 ≅ ∠5, ∠2 ≅ ∠6, ∠3 ≅ ∠7, ∠4 ≅ ∠8

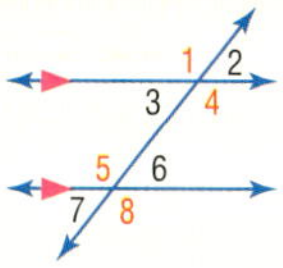

Study Tip
Look Back
To review **vertical angles**, see Lesson 1-6.

Example 1 Determine Angle Measures

In the figure, $m\angle 3 = 133$. Find $m\angle 5$.

∠3 ≅ ∠7	Corresponding Angles Postulate
∠7 ≅ ∠5	Vertical Angles Theorem
∠3 ≅ ∠5	Transitive Property
$m\angle 3 = m\angle 5$	Definition of congruent angles
$133 = m\angle 5$	Substitution

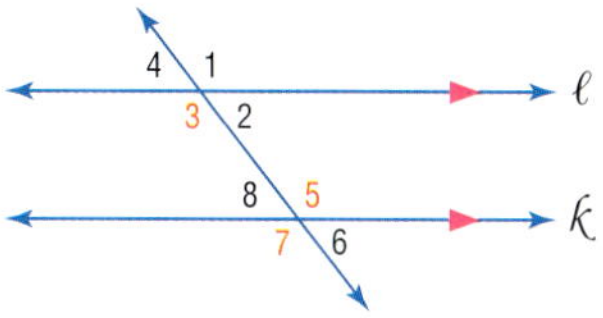

In Example 1, alternate interior angles 3 and 5 are congruent. This suggests another special relationship between angles formed by two parallel lines and a transversal. Other relationships are summarized in Theorems 3.1, 3.2, and 3.3.

3-2 Lesson Notes

1 Focus

5-Minute Check Transparency 3-2 Use as a quiz or review of Lesson 3-1.

Mathematical Background notes are available for this lesson on p. 124C.

How can angles and lines be used in art?

Ask students:

- In the painting, how does the artist use lines? **Sample answer: to separate regions of color**
- What other lines appear parallel? **Accept all reasonable answers.**

2 Teach

PARALLEL LINES AND ANGLE PAIRS

In-Class Example PowerPoint®

1 In the figure, $x \parallel y$ and $m\angle 11 = 51$. Find $m\angle 16$. **51**

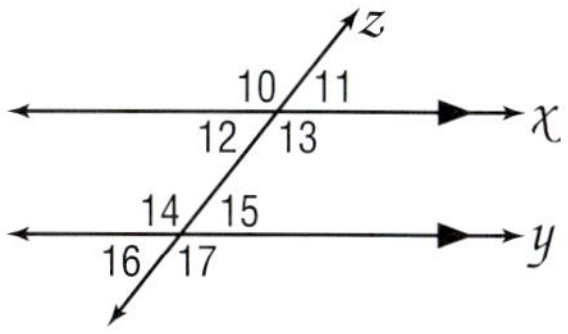

Resource Manager

Workbook and Reproducible Masters

Chapter 3 Resource Masters
- Study Guide and Intervention, pp. 131–132
- Skills Practice, p. 133
- Practice, p. 134
- Reading to Learn Mathematics, p. 135
- Enrichment, p. 136
- Assessment, p. 175

School-to-Career Masters, p. 5
Prerequisite Skills Workbook, pp. 83–84
Teaching Geometry With Manipulatives Masters, p. 53

Transparencies
5-Minute Check Transparency 3-2
Answer Key Transparencies

Technology
Interactive Chalkboard

Key Concept — Parallel Lines and Angle Pairs

Theorem	Examples
3.1 Alternate Interior Angles Theorem If two parallel lines are cut by a transversal, then each pair of alternate interior angles is congruent.	$\angle 4 \cong \angle 5$ $\angle 3 \cong \angle 6$
3.2 Consecutive Interior Angles Theorem If two parallel lines are cut by a transversal, then each pair of consecutive interior angles is supplementary.	$m\angle 4 + m\angle 6 = 180$ $m\angle 3 + m\angle 5 = 180$
3.3 Alternate Exterior Angles Theorem If two parallel lines are cut by a transversal, then each pair of alternate exterior angles is congruent.	$\angle 1 \cong \angle 8$ $\angle 2 \cong \angle 7$

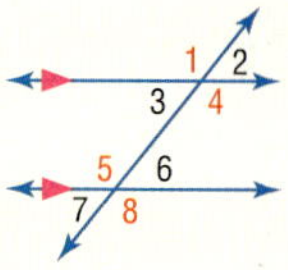

You will prove Theorems 3.2 and 3.3 in Exercises 40 and 39, respectively.

Proof — *Theorem 3.1*

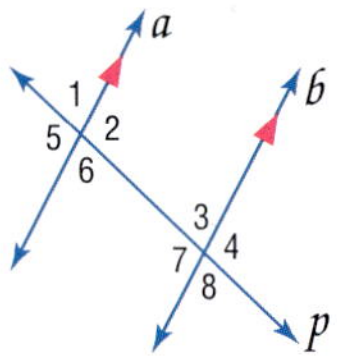

Given: $a \parallel b$; p is a transversal of a and b.

Prove: $\angle 2 \cong \angle 7$, $\angle 3 \cong \angle 6$

Paragraph Proof: We are given that $a \parallel b$ with a transversal p. By the Corresponding Angles Postulate, $\angle 2 \cong \angle 4$ and $\angle 8 \cong \angle 6$. Also, $\angle 4 \cong \angle 7$ and $\angle 3 \cong \angle 8$ because vertical angles are congruent. Therefore, $\angle 2 \cong \angle 7$ and $\angle 3 \cong \angle 6$ since congruence of angles is transitive.

A special relationship occurs when the transversal is a perpendicular line.

Theorem 3.4

Perpendicular Transversal Theorem In a plane, if a line is perpendicular to one of two parallel lines, then it is perpendicular to the other.

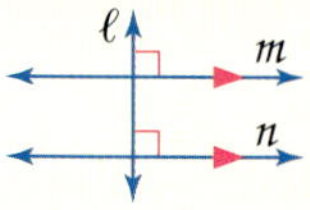

Proof — *Theorem 3.4*

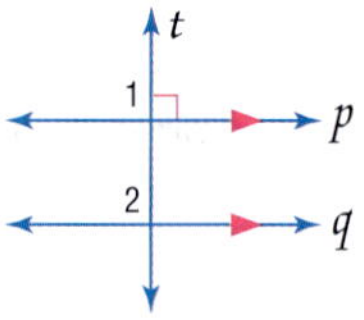

Given: $p \parallel q$, $t \perp p$

Prove: $t \perp q$

Proof:

Statements	Reasons
1. $p \parallel q$, $t \perp p$	1. Given
2. $\angle 1$ is a right angle.	2. Definition of $\perp$ lines
3. $m\angle 1 = 90$	3. Definition of right angle
4. $\angle 1 \cong \angle 2$	4. Corresponding Angles Postulate
5. $m\angle 1 = m\angle 2$	5. Definition of congruent angles
6. $m\angle 2 = 90$	6. Substitution Property
7. $\angle 2$ is a right angle.	7. Definition of right angles
8. $t \perp q$	8. Definition of $\perp$ lines

DAILY INTERVENTION — Differentiated Instruction

Kinesthetic Mark two parallel lines and a transversal on the floor. Have pairs of students stand in angles that are congruent or supplementary, and have them explain whether their angles are alternate interior, alternate exterior, and so on.

134 Chapter 3 Parallel and Perpendicular Lines

FCAT Practice
Standardized Test Practice

Example 2 Use an Auxiliary Line

Grid-In Test Item

What is the measure of $\angle GHI$?

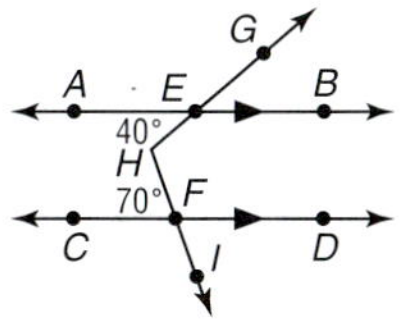

Read the Test Item

You need to find $m\angle GHI$. Be sure to identify it correctly on the figure.

Solve the Test Item

Draw $\overleftrightarrow{JK}$ through H parallel to $\overleftrightarrow{AB}$ and $\overleftrightarrow{CD}$.

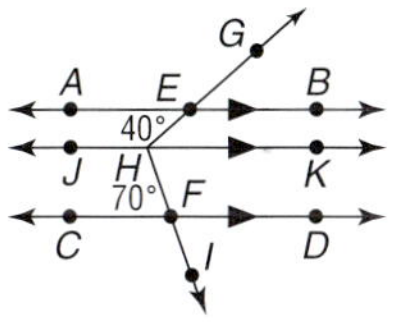

$\angle EHK \cong \angle AEH$	Alternate Interior Angles Theorem
$m\angle EHK = m\angle AEH$	Definition of congruent angles
$m\angle EHK = 40$	Substitution
$\angle FHK \cong \angle CFH$	Alternate Interior Angles Theorem
$m\angle FHK = m\angle CFH$	Definition of congruent angles
$m\angle FHK = 70$	Substitution
$m\angle GHI = m\angle EHK + m\angle FHK$	Angle Addition Postulate
$= 40 + 70$ or 110	$m\angle EHK = 40$, $m\angle FHK = 70$

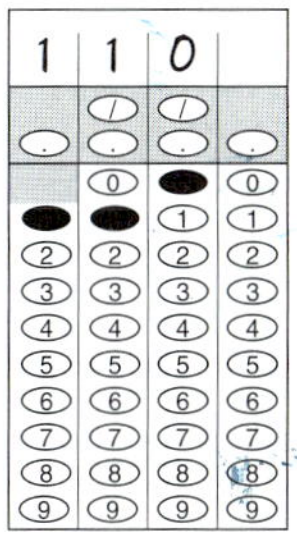

Write each digit of 110 in a column of the grid. Then shade in the corresponding bubble in each column.

Test-Taking Tip

Make a Drawing If you are allowed to write in your test booklet, sketch your drawings near the question to keep your work organized. Do not make any marks on the answer sheet except your answers.

ALGEBRA AND ANGLE MEASURES

Angles formed by two parallel lines and a transversal can be used to find unknown values.

Example 3 Find Values of Variables

ALGEBRA **If $m\angle 1 = 3x + 40$, $m\angle 2 = 2(y - 10)$, and $m\angle 3 = 2x + 70$, find x and y.**

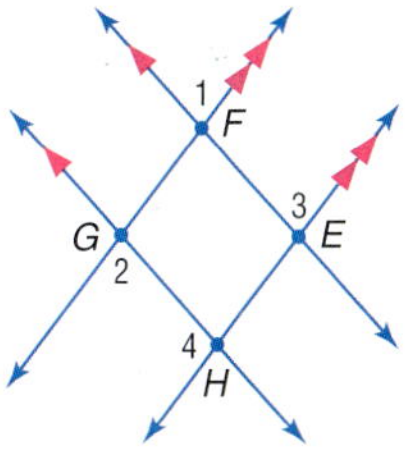

- Find x.

Since $\overleftrightarrow{FG} \parallel \overleftrightarrow{EH}$, $\angle 1 \cong \angle 3$ by the Corresponding Angles Postulate.

$m\angle 1 = m\angle 3$	Definition of congruent angles
$3x + 40 = 2x + 70$	Substitution
$x = 30$	Subtract $2x$ and 40 from each side.

- Find y.

Since $\overleftrightarrow{FE} \parallel \overleftrightarrow{GH}$, $\angle 1 \cong \angle 2$ by the Alternate Exterior Angles Theorem.

$m\angle 1 = m\angle 2$	Definition of congruent angles
$3x + 40 = 2(y - 10)$	Substitution
$3(30) + 40 = 2(y - 10)$	$x = 30$
$130 = 2y - 20$	Simplify.
$150 = 2y$	Add 20 to each side.
$75 = y$	Divide each side by 2.

www.geometryonline.com/extra_examples/fcat

ALGEBRA AND ANGLE MEASURES

In-Class Examples

Power Point®

2 What is the measure of $\angle RTV$? **125**

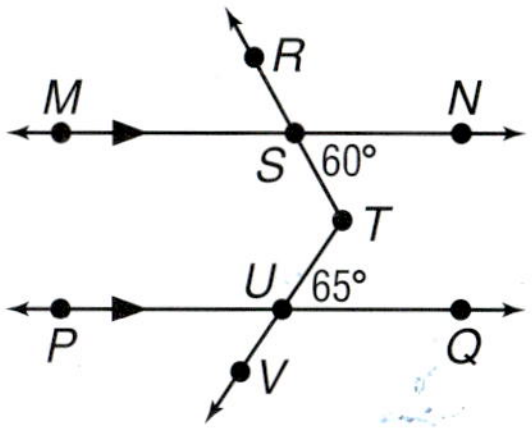

Teaching Tip In Example 3, draw the figure and label the angles with their algebraic expressions. Ask students which pair of angles can be used to write an equation with just one variable. They should realize that $m\angle 1$ and $m\angle 3$ can both be represented by expressions involving x, so they can write an equation with just one variable.

3 **ALGEBRA** If $m\angle 5 = 2x - 10$, $m\angle 6 = 4(y - 25)$, and $m\angle 7 = x + 15$, find x and y.

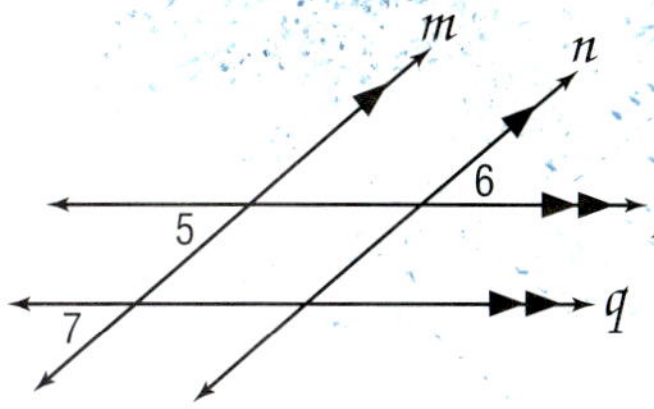

$x = 25$, $y = 35$

DAILY INTERVENTION

Unlocking Misconceptions

After students complete this lesson, they may think that whenever they see two lines cut by a transversal that the pairs of angles are congruent or supplementary. They should realize that they cannot assume, just from a figure, that lines are parallel and thus, that angles are congruent or supplementary.

3 Practice/Apply

Study Notebook

Have students—
- add the definitions/examples of the vocabulary terms to their Vocabulary Builder worksheets for Chapter 3.
- include a figure and statement of the Perpendicular Transversal Theorem and an example of adding an auxiliary line to a figure.
- include any other item(s) that they find helpful in mastering the skills in this lesson.

About the Exercises...

Organization by Objective
- **Parallel Lines and Angle Pairs:** 14–31
- **Algebra and Angle Measures:** 32–38

Odd/Even Assignments
Exercises 14–37 are structured so that students practice the same concepts whether they are assigned odd or even problems.

Assignment Guide

Basic: 15–27 odd, 35, 38, 39, 43–59

Average: 15–37 odd, 39–59

Advanced: 14–38 even, 40–54 (optional: 55–59)

All: Practice Quiz 1 (1–5)

Check for Understanding

Concept Check

1. **Determine** whether $\angle 1$ is *always*, *sometimes*, or *never* congruent to $\angle 2$. Explain. **1–2. See margin.**
2. **OPEN ENDED** Use a straightedge and protractor to draw a pair of parallel lines cut by a transversal so that one pair of corresponding angles measures 35°.
3. **Determine** the minimum number of angle measures you would have to know to find the measures of all of the angles in the figure for Exercise 1. **1**
4. **State** the postulate or theorem that allows you to conclude $\angle 3 \cong \angle 5$ in the figure at the right. **Alternate Interior Angles Theorem**

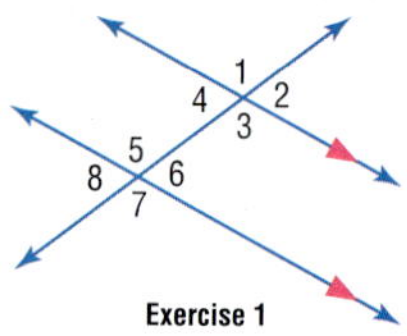

Exercise 1

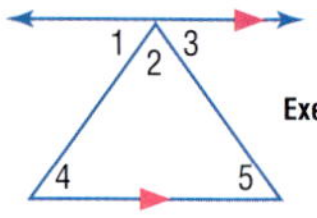

Exercise 4

Guided Practice

GUIDED PRACTICE KEY	
Exercises	**Examples**
5–10	1
13	2
11, 12	3

In the figure, $m\angle 3 = 110$ and $m\angle 12 = 55$. Find the measure of each angle.

5. $\angle 1$ **110**
6. $\angle 6$ **110**
7. $\angle 2$ **70**
8. $\angle 10$ **55**
9. $\angle 13$ **55**
10. $\angle 15$ **55**

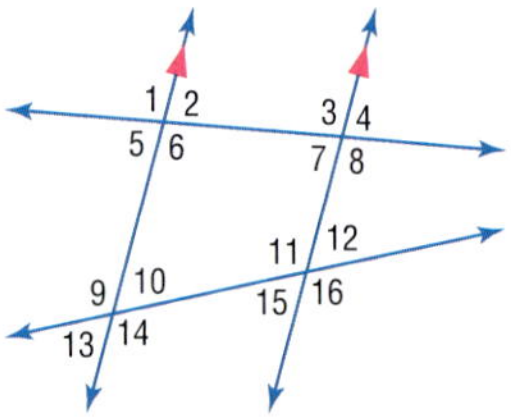

Find x and y in each figure.

11. 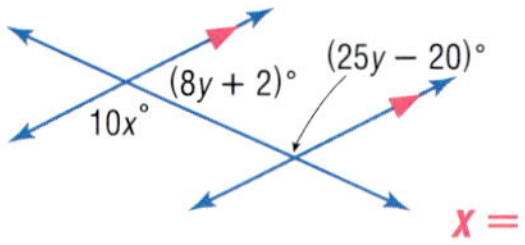

x = 13, y = 6

12. 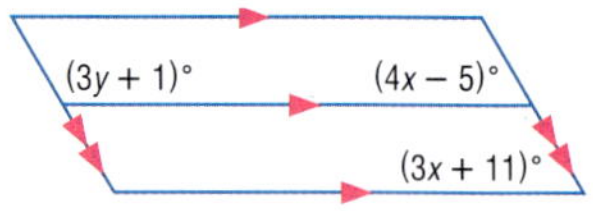

x = 16, y = 40

Standardized Test Practice

13. **SHORT RESPONSE** Find $m\angle 1$. **67**

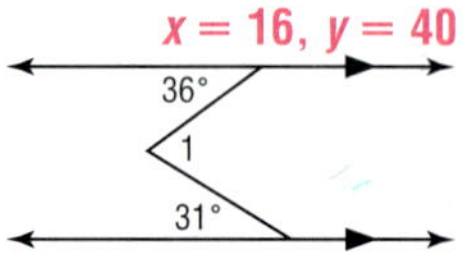

★ indicates increased difficulty

Practice and Apply

Homework Help

For Exercises	See Examples
14–31	1, 2
32–37	3

Extra Practice See page 759.

In the figure, $m\angle 9 = 75$. Find the measure of each angle.

14. $\angle 3$ **75**
15. $\angle 5$ **75**
16. $\angle 6$ **105**
17. $\angle 8$ **105**
18. $\angle 11$ **75**
19. $\angle 12$ **105**

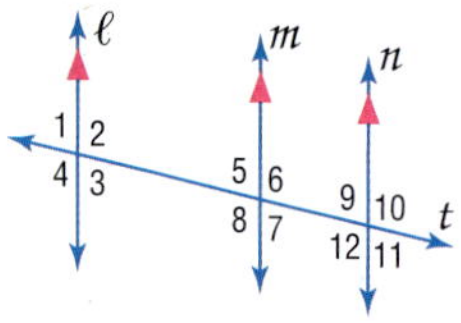

In the figure, $m\angle 3 = 43$. Find the measure of each angle.

20. $\angle 2$ **137**
21. $\angle 7$ **43**
22. $\angle 10$ **137**
23. $\angle 11$ **43**
24. $\angle 13$ **43**
25. $\angle 16$ **137**

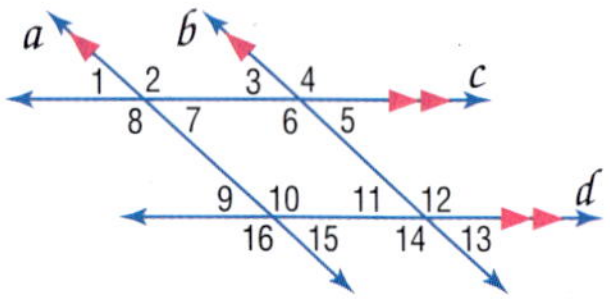

triangle sum p. 185

In the figure, $m\angle 1 = 50$ and $m\angle 3 = 60$. Find the measure of each angle.

26. $\angle 4$ **50**
27. $\angle 5$ **60**
★28. $\angle 2$ **110**
★29. $\angle 6$ **70**
★30. $\angle 7$ **110**
★31. $\angle 8$ **120**

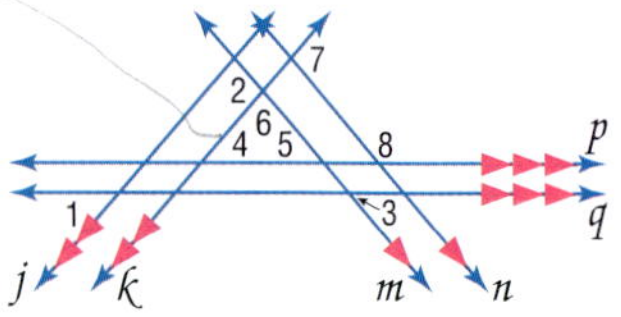

136 Chapter 3 Parallel and Perpendicular Lines

Answers

1. Sometimes; if the transversal is perpendicular to the parallel lines, then $\angle 1$ and $\angle 2$ are right angles and are congruent.

2. 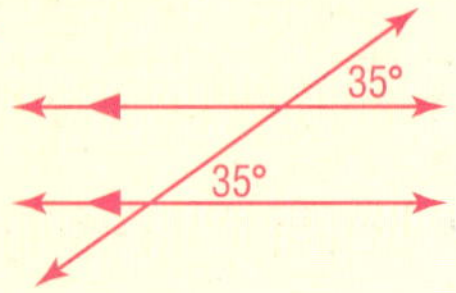

41. Given: $\ell \perp m$, $m \parallel n$
Prove: $\ell \perp n$
Proof: Since $\ell \perp m$, we know that $\angle 1 \cong \angle 2$, because perpendicular lines form congruent right angles. Then by the Corresponding Angles Postulate, $\angle 1 \cong \angle 3$ and $\angle 2 \cong \angle 4$. By the definition of congruent angles, $m\angle 1 = m\angle 2$, $m\angle 1 = m\angle 3$ and $m\angle 2 = m\angle 4$. By substitution, $m\angle 3 = m\angle 4$. Because $\angle 3$ and $\angle 4$ form a congruent linear pair, they are right angles. By definition, $\ell \perp n$.

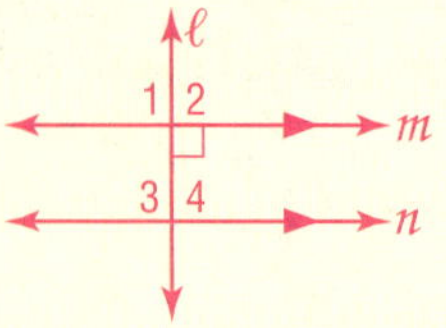

Find x and y in each figure.

32. 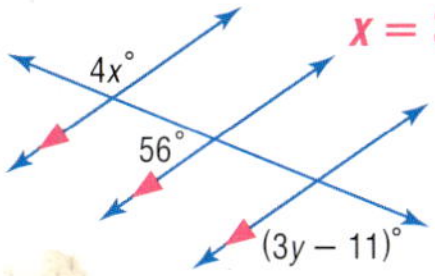

$x = 31$, $y = 45$

★ 33. 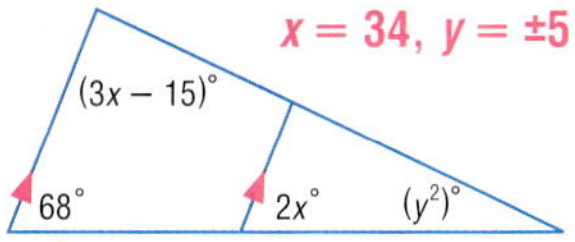

$x = 34$, $y = \pm 5$

Find $m\angle 1$ in each figure.

34. $m\angle 1 = 107$ 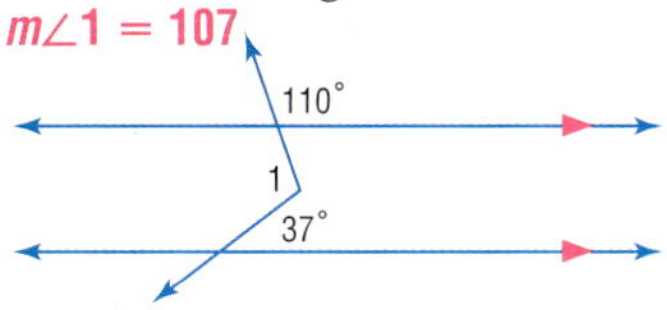

35. 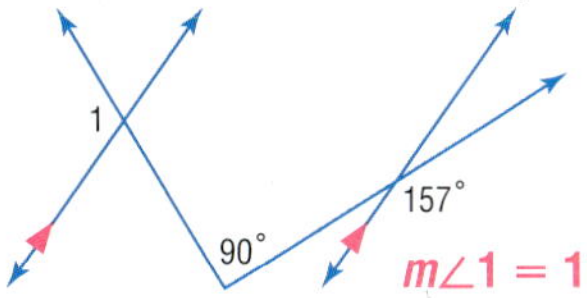

$m\angle 1 = 113$

Find x, y, and z in each figure.

★ 36. 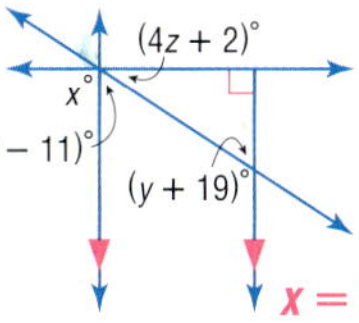

$x = 90$, $y = 15$, $z = 13.5$

★ 37. 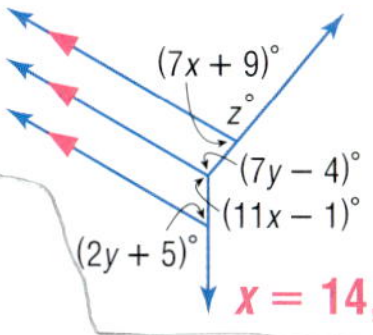

$x = 14$, $y = 11$, $z = 73$

38. **CARPENTRY** Anthony is building a picnic table for his patio. He cut one of the legs at an angle of 40°. At what angle should he cut the other end to ensure that the top of the table is parallel to the ground? **140°**

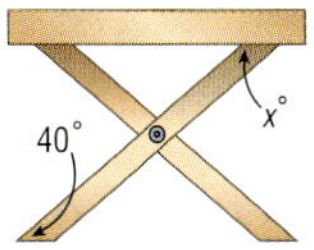

More About. . .

Construction

In 2001, the United States spent about $30 billion for federal highway projects.

Source: U.S. Dept. of Transportation

39. **PROOF** Copy and complete the proof of Theorem 3.3.

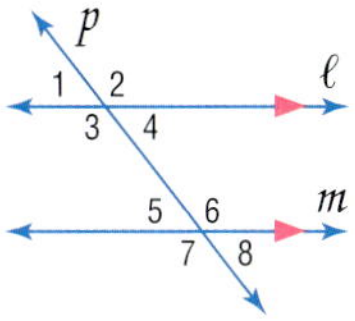

Given: $\ell \parallel m$

Prove: $\angle 1 \cong \angle 8$

$\angle 2 \cong \angle 7$

Proof:

Statements	Reasons
1. $\ell \parallel m$	1. ? Given
2. $\angle 1 \cong \angle 5$, $\angle 2 \cong \angle 6$	2. ? Corresponding Angles Postulate
3. $\angle 5 \cong \angle 8$, $\angle 6 \cong \angle 7$	3. ? Vertical Angles Theorem
4. $\angle 1 \cong \angle 8$, $\angle 2 \cong \angle 7$	4. ? Transitive Property

40. **PROOF** Write a two-column proof of Theorem 3.2. See p. 173A.

41. **PROOF** Write a paragraph proof of Theorem 3.4. See margin.

42. **CONSTRUCTION** Parallel drainage pipes are laid on each side of Polaris Street. A pipe under the street connects the two pipes. The connector pipe makes a 65° angle as shown. What is the measure of the angle it makes with the pipe on the other side of the road? **115**

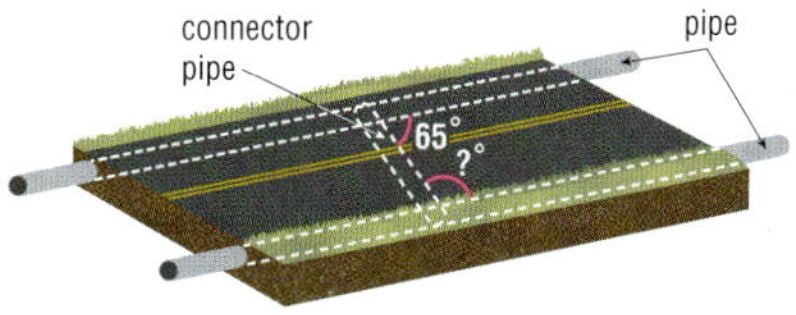

www.geometryonline.com/self_check_quiz/fcat

Study Guide and Intervention, p. 131 (shown) and p. 132

Parallel Lines and Angle Pairs When two parallel lines are cut by a transversal, the following pairs of angles are congruent.

- corresponding angles
- alternate interior angles
- alternate exterior angles

Also, consecutive interior angles are supplementary.

Example **In the figure, $m\angle 2 = 75$. Find the measures of the remaining angles.**

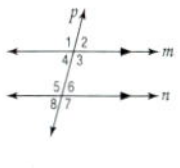

$m\angle 1 = 105$ $\angle 1$ and $\angle 2$ form a linear pair.
$m\angle 3 = 105$ $\angle 3$ and $\angle 2$ form a linear pair.
$m\angle 4 = 75$ $\angle 4$ and $\angle 2$ are vertical angles.
$m\angle 5 = 105$ $\angle 5$ and $\angle 3$ are alternate interior angles.
$m\angle 6 = 75$ $\angle 6$ and $\angle 2$ are corresponding angles.
$m\angle 7 = 105$ $\angle 7$ and $\angle 3$ are corresponding angles.
$m\angle 8 = 75$ $\angle 8$ and $\angle 6$ are vertical angles.

Exercises

In the figure, $m\angle 3 = 102$. Find the measure of each angle.

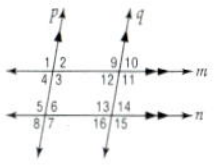

1. $\angle 5$ 102
2. $\angle 6$ 78
3. $\angle 11$ 102
4. $\angle 7$ 102
5. $\angle 15$ 102
6. $\angle 14$ 78

In the figure, $m\angle 9 = 80$ and $m\angle 5 = 68$. Find the measure of each angle.

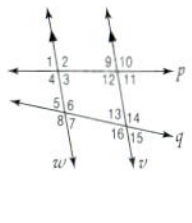

7. $\angle 12$ 100
8. $\angle 1$ 80
9. $\angle 4$ 100
10. $\angle 3$ 80
11. $\angle 7$ 68
12. $\angle 16$ 112

Skills Practice, p. 133 and Practice, p. 134 (shown)

In the figure, $m\angle 2 = 92$ and $m\angle 12 = 74$. Find the measure of each angle.

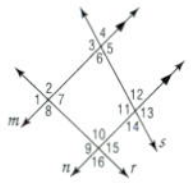

1. $\angle 10$ 92
2. $\angle 8$ 92
3. $\angle 9$ 88
4. $\angle 5$ 106
5. $\angle 11$ 106
6. $\angle 13$ 106

Find x and y in each figure.

7. 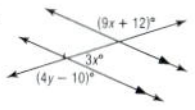

$x = 14$, $y = 37$

8. 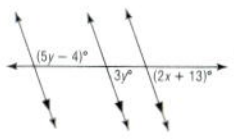

$x = 28$, $y = 23$

Find $m\angle 1$ in each figure.

9. 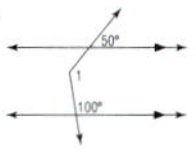

130

10. 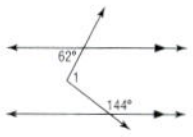

98

11. **PROOF** Write a paragraph proof of Theorem 3.3.
Given: $\ell \parallel m$, $m \parallel n$
Prove: $\angle 1 \cong \angle 12$

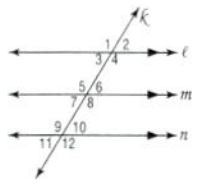

Sample proof:
It is given that $\ell \parallel m$, so $\angle 1 \cong \angle 8$ by the Alternate Exterior Angles Theorem. Since it is given that $m \parallel n$, $\angle 8 \cong \angle 12$ by the Corresponding Angles Postulate. Therefore, $\angle 1 \cong \angle 12$, since congruence of angles is transitive.

12. **FENCING** A diagonal brace strengthens the wire fence and prevents it from sagging. The brace makes a 50° angle with the wire as shown. Find y. 130

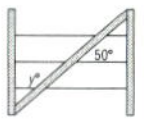

Reading to Learn Mathematics, p. 135 ELL

Pre-Activity **How can angles and lines be used in art?**

Read the introduction to Lesson 3-2 at the top of page 133 in your textbook.

- Your textbook shows a painting that contains two parallel lines and a transversal. What is the name for $\angle 1$ and $\angle 2$? corresponding angles
- What is the relationship between these two angles? They are congruent.

Reading the Lesson

1. Choose the correct word to complete each sentence.
   a. If two parallel lines are cut by a transversal, then alternate exterior angles are congruent (congruent/complementary/supplementary).
   b. If two parallel lines are cut by a transversal, then corresponding angles are congruent (congruent/complementary/supplementary).
   c. If parallel lines are cut by a transversal, then consecutive interior angles are supplementary (congruent/complementary/supplementary).
   d. In a plane, if a line is perpendicular to one of two parallel lines, then it is perpendicular (parallel/perpendicular/skew) to the other.

Use the figure for Exercises 2 and 3.

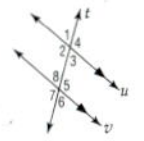

2. a. Name four pairs of vertical angles.
   $\angle 1$ and $\angle 3$, $\angle 2$ and $\angle 4$, $\angle 5$ and $\angle 7$, $\angle 6$ and $\angle 8$
   b. Name all angles that form a linear pair with $\angle 7$. $\angle 6$, $\angle 8$
   c. Name all angles that are congruent to $\angle 1$. $\angle 3$, $\angle 6$, $\angle 8$
   d. Name all angles that are congruent to $\angle 4$. $\angle 2$, $\angle 5$, $\angle 7$
   e. Name all angles that are supplementary to $\angle 3$. $\angle 2$, $\angle 4$, $\angle 5$, $\angle 7$
   f. Name all angles that are supplementary to $\angle 2$. $\angle 1$, $\angle 3$, $\angle 6$, $\angle 8$
3. Which conclusion(s) could you make about lines u and v if $m\angle 4 = m\angle 1$? B, D
   A. $t \parallel u$ B. $t \perp u$ C. $v \perp u$ D. $v \perp t$ E. $v \parallel t$

Helping You Remember

4. How can you use an everyday meaning of the adjective *alternate* to help you remember the types of angle pairs for two lines and a transversal?
Sample answer: One meaning of *alternate* is "obtained by switching back and forth from one thing to another." The angle pairs in this lesson all use angles with different vertices, and those whose names contain the adjective *alternate* can be located in a figure by switching from one side of the transversal to the other. The pairs whose names do not include the word *alternate* are found on the same side of the transversal.

Enrichment, p. 136

More Optical Illusions

In drawings, diagonal lines may create the illusion of depth. For example, the figure at the right can be thought of as picturing a flat figure or a cube. The optical illusions on this page involve depth perception.

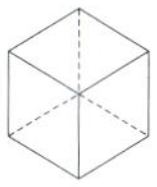

Answer each question.

1. How many cubes do you see in the drawing? 5 or 6

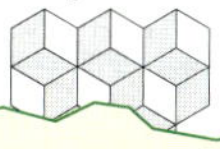

2. Can this figure show an actual object? no

4 Assess

Open-Ended Assessment

Speaking Working in small groups, have students take turns describing to their group a pair of angles that are congruent when two parallel lines are cut by a transversal.

Getting Ready for Lesson 3-3

Prerequisite Skill Students will learn about slopes of lines in Lesson 3-3. To find slope, they will simplify a fraction whose numerator and denominator contain differences. Use Exercises 55–59 to familiarize your students with simplifying a fraction whose numerator and denominator contain differences.

Assessment Options

Practice Quiz 1 The quiz provides students with a brief review of the concepts and skills in Lessons 3-1 and 3-2. Lesson numbers are given to the right of the exercises or instruction lines so students can review concepts not yet mastered.

Quiz (Lessons 3-1 and 3-2) is available on p. 175 of the *Chapter 3 Resource Masters*.

Answers

43. $\angle 2$ and $\angle 6$ are consecutive interior angles for the same transversal, which makes them supplementary because $\overline{WX} \parallel \overline{YZ}$. $\angle 4$ and $\angle 6$ are not necessarily supplementary because $\overline{WZ}$ may not be parallel to $\overline{XY}$.

44. Sample answer: Angles and lines are used in art to show depth, and to create realistic objects. Answers should include the following.
   - Rectangular shapes are made by drawing parallel lines and perpendiculars.
   - M.C. Escher and Pablo Picasso use lines and angles in their art.

43. **CRITICAL THINKING** Explain why you can conclude that $\angle 2$ and $\angle 6$ are supplementary, but you cannot state that $\angle 4$ and $\angle 6$ are necessarily supplementary. **See margin.**

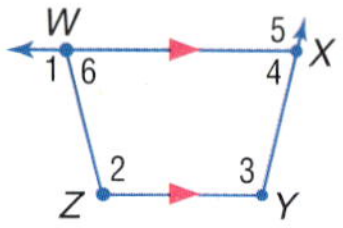

44. **WRITING IN MATH** Answer the question that was posed at the beginning of the lesson. **See margin.**

How can angles and lines be used in art?

Include the following in your answer:
- a description of how angles and lines are used to create patterns, and
- examples from two different artists that use lines and angles.

FCAT Practice
Standardized Test Practice

45. Line ℓ is parallel to line m. What is the value of x? **C**

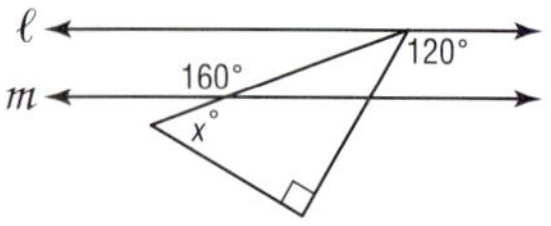

Ⓐ 30 Ⓑ 40 Ⓒ 50 Ⓓ 60

46. **ALGEBRA** If $ax = bx + c$, then what is the value of x in terms of a, b, and c? **C**

Ⓐ $\frac{c}{a+b}$ Ⓑ $\frac{b}{a+c}$ Ⓒ $\frac{c}{a-b}$ Ⓓ $\frac{b+c}{a}$

Maintain Your Skills

Mixed Review

For Exercises 47–50, refer to the figure at the right. *(Lesson 3-1)*

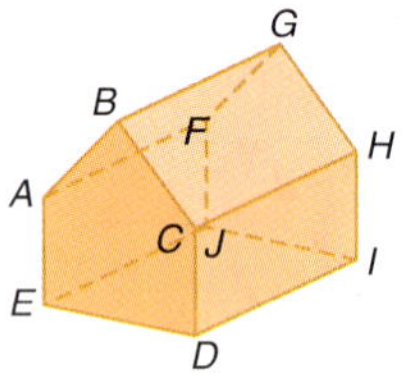

47. Name all segments parallel to $\overline{AB}$. $\overline{FG}$
48. Name all segments skew to $\overline{CH}$. **48.** $\overline{AB}$, $\overline{DE}$, $\overline{FG}$, $\overline{IJ}$, $\overline{AE}$, $\overline{FJ}$
49. Name all planes parallel to *AEF*. **CDH**
50. Name all segments intersecting $\overline{GH}$. $\overline{BG}$, $\overline{CH}$, $\overline{FG}$, $\overline{HI}$

Find the measure of each numbered angle. *(Lesson 2-8)*

51. **56**

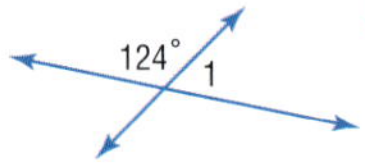

52. **53**

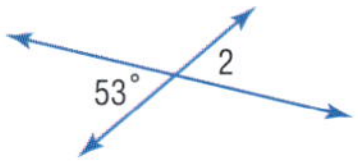

Identify the hypothesis and conclusion of each statement. *(Lesson 2-3)*

53. If it rains this evening, then I will mow the lawn tomorrow. **53. H: it rains this evening; C: I will mow the lawn tomorrow**
54. A balanced diet will keep you healthy. **H: you eat a balanced diet; C: it will keep you healthy**

Getting Ready for the Next Lesson

PREREQUISITE SKILL Simplify each expression.
*(To review **simplifying expressions**, see pages 735 and 736.)*

55. $\frac{7-9}{8-5}$ $-\frac{2}{3}$ 56. $\frac{-3-6}{2-8}$ $\frac{3}{2}$ 57. $\frac{14-11}{23-15}$ $\frac{3}{8}$ 58. $\frac{15-23}{14-11}$ $-\frac{8}{3}$ 59. $\frac{2}{9}\cdot\left(-\frac{18}{5}\right)$ $-\frac{4}{5}$

Practice Quiz 1 — Lessons 3-1 and 3-2

State the transversal that forms each pair of angles. Then identify the special name for the angle pair. *(Lesson 3-1)*

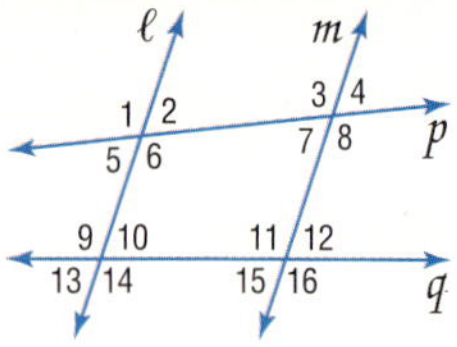

1. $\angle 1$ and $\angle 8$ p; **alt. ext.** 2. $\angle 6$ and $\angle 10$ ℓ; **cons. int.** 3. $\angle 11$ and $\angle 14$ q; **alt. int.**

Find the measure of each angle if $\ell \parallel m$ and $m\angle 1 = 105$. *(Lesson 3-2)*

4. $\angle 6$ **105** 5. $\angle 4$ **75**

138 Chapter 3 Parallel and Perpendicular Lines

3-3 Slopes of Lines

Sunshine State Standards
MA.C.3.4.2

Vocabulary

- slope
- rate of change

What You'll Learn

- Find slopes of lines.
- Use slope to identify parallel and perpendicular lines.

How is slope used in transportation?

Traffic signs are often used to alert drivers to road conditions. The sign at the right indicates a hill with a 6% *grade.* This means that the road will rise or fall 6 feet vertically for every 100 horizontal feet traveled.

SLOPE OF A LINE The **slope** of a line is the ratio of its vertical rise to its horizontal run.

$$\text{slope} = \frac{\text{vertical rise}}{\text{horizontal run}}$$

In a coordinate plane, the slope of a line is the ratio of the change along the y-axis to the change along the x-axis.

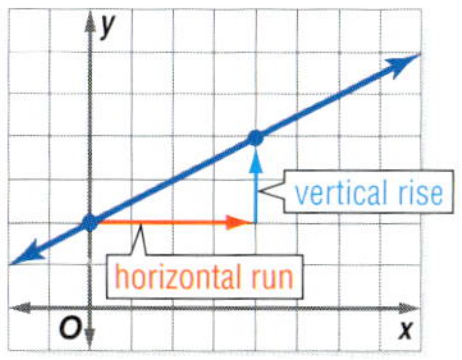

TEACHING TIP

Slope is sometimes expressed as $\frac{\Delta y}{\Delta x}$, read *delta y over delta x*, which means the change in *y* values over the change in *x* values.

Key Concept — *Slope*

The slope m of a line containing two points with coordinates (x_1, y_1) and (x_2, y_2) is given by the formula

$$m = \frac{y_2 - y_1}{x_2 - x_1}, \text{ where } x_1 \neq x_2.$$

The slope of a line indicates whether the line rises to the right, falls to the right, or is horizontal. The slope of a vertical line, where $x_1 = x_2$, is undefined.

Example 1 *Find the Slope of a Line*

Find the slope of each line.

a.

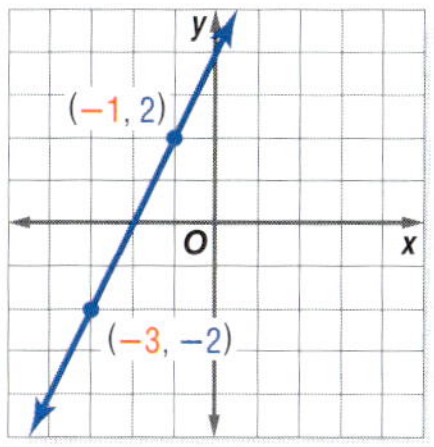

Use the $\frac{\text{rise}}{\text{run}}$ method.

From (−3, −2) to (−1, 2), go up 4 units and right 2 units.

$$\frac{\text{rise}}{\text{run}} = \frac{4}{2} \text{ or } 2$$

b.

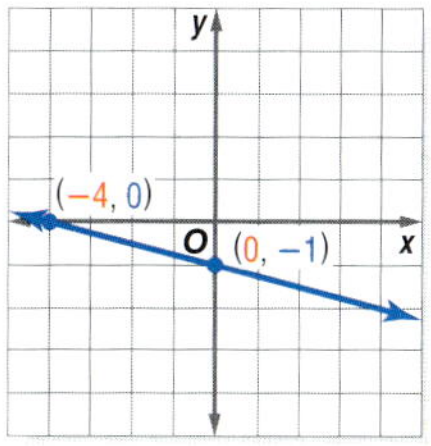

Use the slope formula.

Let (−4, 0) be (x_1, y_1) and (0, −1) be (x_2, y_2).

$$m = \frac{y_2 - y_1}{x_2 - x_1}$$

$$= \frac{-1 - 0}{0 - (-4)} \text{ or } -\frac{1}{4}$$

Study Tip

Slope

Lines with positive slope *rise* as you move from left to right, while lines with negative slope *fall* as you move from left to right.

3-3 Lesson Notes

1 *Focus*

5-Minute Check Transparency 3-3 Use as a quiz or review of Lesson 3-2.

Mathematical Background notes are available for this lesson on p. 124C.

How is slope used in transportation?

Ask students:

- Why would a road or train track wind its way up a mountain instead of going directly toward the top? **A path going directly toward the top might be too steep for a car or train.**
- To reach the same height, is it easier to push a wheelchair up a long ramp or a short ramp? **A long ramp is easier because the climb is less steep, even though you travel farther.**

Resource Manager

Workbook and Reproducible Masters

Chapter 3 Resource Masters
- Study Guide and Intervention, pp. 137–138
- Skills Practice, p. 139
- Practice, p. 140
- Reading to Learn Mathematics, p. 141
- Enrichment, p. 142
- Assessment, pp. 175, 177

Graphing Calculator and Computer Masters, p. 21
School-to-Career Masters, p. 6
Prerequisite Skills Workbook, pp. 3–4, 7–8, 33–34, 77–78
Teaching Geometry With Manipulatives Masters, pp. 1, 17

Transparencies

5-Minute Check Transparency 3-3
Real-World Transparency 3
Answer Key Transparencies

Technology

GeomPASS: Tutorial Plus, Lesson 7
Interactive Chalkboard
Multimedia Applications: Virtual Activities

2 Teach

SLOPE OF A LINE

In-Class Example Power Point®

1 Find the slope of each line.

a.

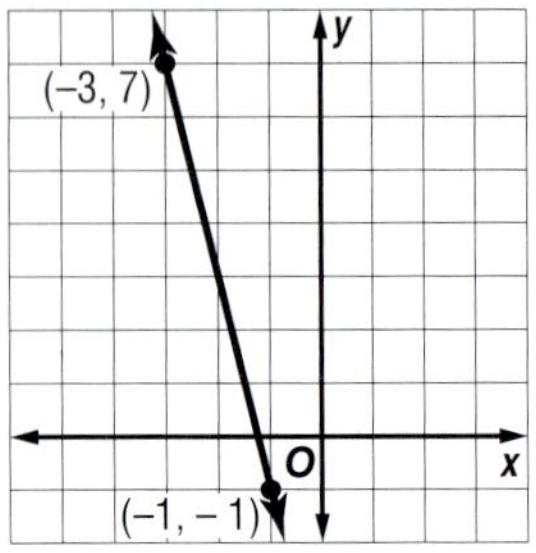

$\frac{-8}{2}$ or -4

b.

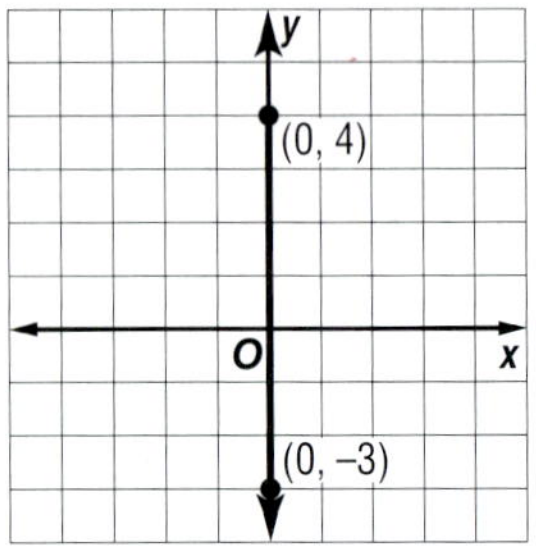

$\frac{-7}{0}$ or undefined

c.

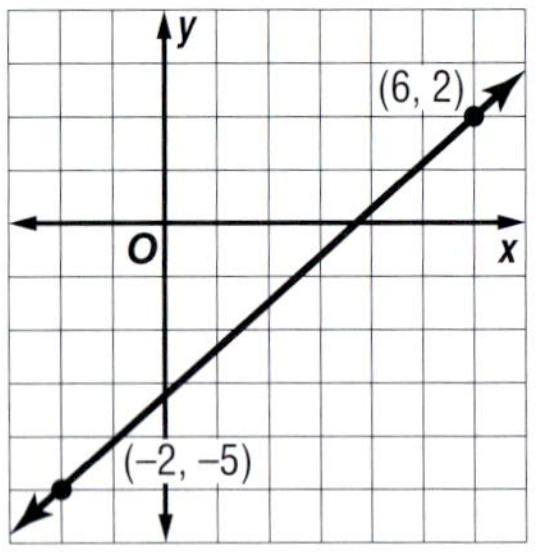

$\frac{7}{8}$

d.

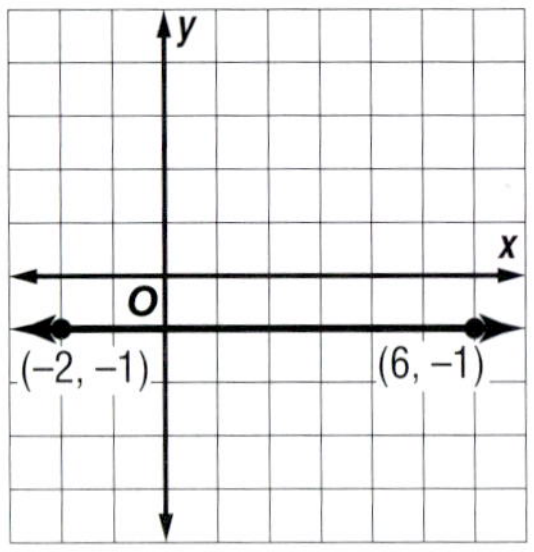

$\frac{0}{8}$ or zero

Teaching Tip After explaining Example 1, let students verify that they could switch (x_1, y_1) and (x_2, y_2) and get the same result. In part b, they would get $\frac{0-(-1)}{-4-0} = -\frac{1}{4}$.

Study Tip

Common Misconception

A line with a slope of 0 is a horizontal line. The slope of a vertical line is undefined.

c.

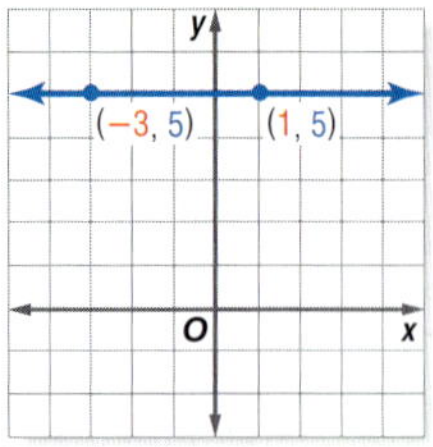

$$m = \frac{y_2 - y_1}{x_2 - x_1}$$
$$= \frac{5-5}{-3-1}$$
$$= \frac{0}{-4} \text{ or } 0$$

d.

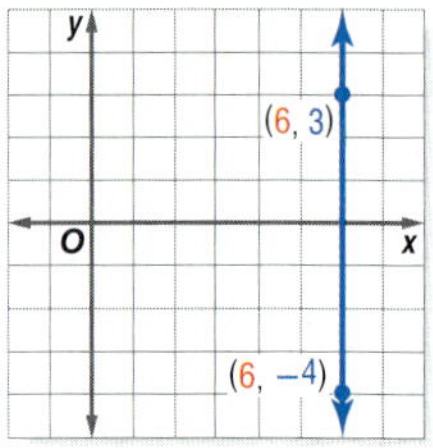

$$m = \frac{y_2 - y_1}{x_2 - x_1}$$
$$= \frac{3-(-4)}{6-6}$$
$$= \frac{7}{0}, \text{ which is undefined}$$

The slope of a line can be used to identify the coordinates of any point on the line. It can also be used to describe a rate of change. The **rate of change** describes how a quantity is changing over time.

Example 2 Use Rate of Change to Solve a Problem

RECREATION **Between 1990 and 2000, annual sales of inline skating equipment increased by an average rate of \$92.4 million per year. In 2000, the total sales were \$1074.4 million. If sales increase at the same rate, what will the total sales be in 2008?**

Let $(x_1, y_1) = (2000, 1074.4)$ and $m = 92.4$.

$m = \frac{y_2 - y_1}{x_2 - x_1}$	Slope formula
$92.4 = \frac{y_2 - 1074.4}{2008 - 2000}$	$m = 92.4$, $y_1 = 1074.4$, $x_1 = 2000$, and $x_2 = 2008$
$92.4 = \frac{y_2 - 1074.4}{8}$	Simplify.
$739.2 = y_2 - 1074.4$	Multiply each side by 8.
$1813.6 = y_2$	Add 1074.4 to each side.

The coordinates of the point representing the sales for 2008 are (2008, 1813.6). Thus, the total sales in 2008 will be about \$1813.6 million.

PARALLEL AND PERPENDICULAR LINES

Examine the graphs of lines ℓ, m, and n. Lines ℓ and m are parallel, and n is perpendicular to ℓ and m. Let's investigate the slopes of these lines.

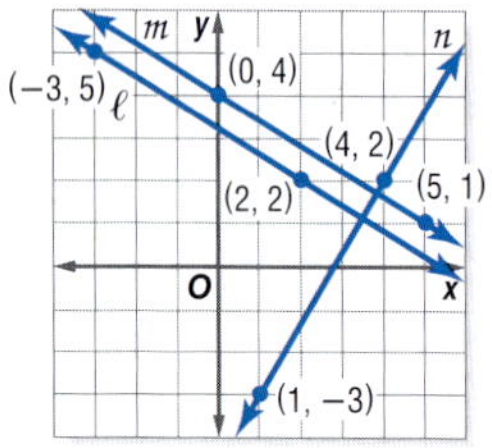

slope of ℓ	slope of m	slope of n
$m = \frac{2-5}{2-(-3)}$	$m = \frac{1-4}{5-0}$	$m = \frac{2-(-3)}{4-1}$
$= -\frac{3}{5}$	$= -\frac{3}{5}$	$= \frac{5}{3}$

Because lines ℓ and m are parallel, their slopes are the same. Line n is perpendicular to lines ℓ and m, and its slope is the opposite reciprocal of the slopes of ℓ and m; that is, $-\frac{3}{5} \cdot \frac{5}{3} = -1$. These results suggest two important algebraic properties of parallel and perpendicular lines.

140 Chapter 3 Parallel and Perpendicular Lines

Study Tip

Look Back
To review **if and only if statements**, see Reading Mathematics, page 81.

Postulates — Slopes of Parallel and Perpendicular Lines

3.2 Two nonvertical lines have the same slope if and only if they are parallel.

3.3 Two nonvertical lines are perpendicular if and only if the product of their slopes is −1.

Example 3 Determine Line Relationships

Determine whether $\overleftrightarrow{AB}$ and $\overleftrightarrow{CD}$ are *parallel*, *perpendicular*, or *neither*.

a. $A(-2, -5), B(4, 7), C(0, 2), D(8, -2)$

Find the slopes of $\overleftrightarrow{AB}$ and $\overleftrightarrow{CD}$.

slope of $\overleftrightarrow{AB} = \frac{7 - (-5)}{4 - (-2)} = \frac{12}{6}$ or 2

slope of $\overleftrightarrow{CD} = \frac{-2 - 2}{8 - 0} = -\frac{4}{8}$ or $-\frac{1}{2}$

The product of the slopes is $2\left(-\frac{1}{2}\right)$ or −1. So, $\overleftrightarrow{AB}$ is perpendicular to $\overleftrightarrow{CD}$.

Study Tip

Graphing Calculator
One way to make perpendicular lines appear perpendicular on a graphing calculator is to use the Zoom Square command.

b. $A(-8, -7), B(4, -4), C(-2, -5), D(1, 7)$

slope of $\overleftrightarrow{AB} = \frac{-4 - (-7)}{4 - (-8)} = \frac{3}{12}$ or $\frac{1}{4}$

slope of $\overleftrightarrow{CD} = \frac{7 - (-5)}{1 - (-2)} = \frac{12}{3}$ or 4

The slopes are not the same, so $\overleftrightarrow{AB}$ and $\overleftrightarrow{CD}$ are not parallel. The product of the slopes is $4\left(\frac{1}{4}\right)$ or 1. So, $\overleftrightarrow{AB}$ and $\overleftrightarrow{CD}$ are neither parallel nor perpendicular.

The relationships of the slopes of lines can be used to graph a line parallel or perpendicular to a given line.

Example 4 Use Slope to Graph a Line

Graph the line that contains $P(-2, 1)$ and is perpendicular to $\overleftrightarrow{JK}$ with $J(-5, -4)$ and $K(0, -2)$.

First, find the slope of $\overleftrightarrow{JK}$.

$m = \frac{y_2 - y_1}{x_2 - x_1}$ Slope formula

$= \frac{-2 - (-4)}{0 - (-5)}$ Substitution

$= \frac{2}{5}$ Simplify.

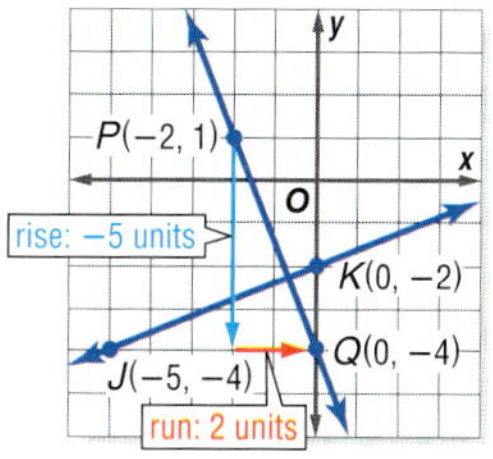

The product of the slopes of two perpendicular lines is −1.

Since $\frac{2}{5}\left(-\frac{5}{2}\right) = -1$, the slope of the line perpendicular to $\overleftrightarrow{JK}$ through $P(-2, 1)$ is $-\frac{5}{2}$.

Graph the line. Start at (−2, 1). Move down 5 units and then move right 2 units. Label the point Q. Draw $\overleftrightarrow{PQ}$.

Study Tip

Negative Slopes
To help determine direction with negative slopes, remember that $-\frac{5}{2} = \frac{-5}{2} = \frac{5}{-2}$.

www.geometryonline.com/extra_examples/fcat

PARALLEL AND PERPENDICULAR LINES

In-Class Examples

Power Point®

2 **RECREATION** For one manufacturer of camping equipment, between 1990 and 2000 annual sales increased by $7.4 million per year. In 2000, the total sales were $85.9 million. If sales increase at the same rate, what will be the total sales in 2010? **about $159.9 million**

3 Determine whether $\overleftrightarrow{FG}$ and $\overleftrightarrow{HJ}$ are *parallel*, *perpendicular*, or *neither*.

a. $F(1, -3), G(-2, -1), H(5, 0), J(6, 3)$ **neither**

b. $F(4, 2), G(6, -3), H(-1, 5), J(-3, 10)$ **parallel**

4 Graph the line that contains $Q(5, 1)$ and is parallel to $\overleftrightarrow{MN}$ with $M(-2, 4)$ and $N(2, 1)$.

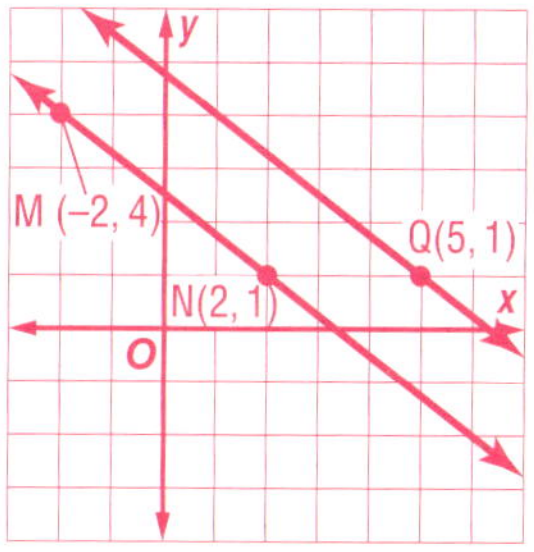

Concept Check

Ask students to describe lines that have a positive slope, a negative slope, and a slope of zero. Students should be able to explain that when a line rises to the right the slope is positive, when it falls to the right the slope is negative, and when a line is horizontal the slope is 0.

DAILY INTERVENTION — Differentiated Instruction

Interpersonal Have each student write a number on a card to represent the slope of a line. Students should briefly pair off, and each student write the slope of a line that is parallel to or perpendicular to the other student's line. Each student decides whether the two numbers represent slopes of parallel or perpendicular lines. Then students form different pairs.

3 Practice/Apply

Study Notebook

Have students—
- add the definitions/examples of the vocabulary terms to their Vocabulary Builder worksheets for Chapter 3.
- include slope, rate of change, the slopes of horizontal and vertical lines, and the relationship between the slopes of two lines that are parallel or perpendicular.
- include any other item(s) that they find helpful in mastering the skills in this lesson.

DAILY INTERVENTION **FIND THE ERROR** Point out that Lori should start by writing the general formula. Then when she replaces the variables, she will more likely include the subtraction signs.

About the Exercises...

Organization by Objective
- **Slope of a Line:** 15–18, 25–32, 42, 43
- **Parallel and Perpendicular Lines:** 19–24, 33–41, 44–46

Odd/Even Assignments
Exercises 15–38 are structured so that students practice the same concepts whether they are assigned odd or even problems.

Assignment Guide
Basic: 15–37 odd, 39–41, 47–72

Average: 15–37 odd, 39–41, 43, 47–72

Advanced: 16–38 even, 39–42, 44–69 (optional: 70–72)

Check for Understanding

Concept Check

1. **Describe** what type of line is perpendicular to a vertical line. What type of line is parallel to a vertical line? **horizontal; vertical**
2. **FIND THE ERROR** Curtis and Lori calculated the slope of the line containing $A(15, 4)$ and $B(-6, -13)$.

2. Curtis; Lori added the coordinates instead of finding the difference.

Curtis
$$m = \frac{4-(-13)}{15-(-6)} = \frac{17}{21}$$

Lori
$$m = \frac{4-13}{15-6} = -\frac{9}{11}$$

Who is correct? Explain your reasoning.

3. **OPEN ENDED** Give an example of a line whose slope is 0 and an example of a line whose slope is undefined. **horizontal line, vertical line**

Guided Practice

4. Determine the slope of the line that contains $A(-4, 3)$ and $B(-2, -1)$. **−2**

GUIDED PRACTICE KEY	
Exercises	Examples
4–7	1
12–14	2
8, 9	3
10, 11	4

Find the slope of each line.

5. ℓ $-\frac{1}{2}$
6. m $\frac{2}{3}$
7. any line perpendicular to ℓ **2**

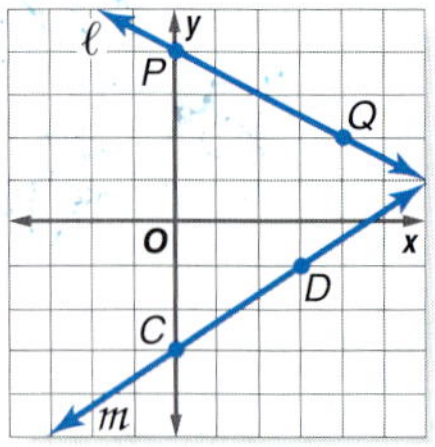

Exercises 5–7

Determine whether $\overleftrightarrow{GH}$ and $\overleftrightarrow{RS}$ are *parallel*, *perpendicular*, or *neither*.

8. $G(14, 13)$, $H(-11, 0)$, $R(-3, 7)$, $S(-4, -5)$ **neither**
9. $G(15, -9)$, $H(9, -9)$, $R(-4, -1)$, $S(3, -1)$ **parallel**

Graph the line that satisfies each condition. **10–11. See margin.**

10. slope = 2, contains $P(1, 2)$
11. contains $A(6, 4)$, perpendicular to $\overleftrightarrow{MN}$ with $M(5, 0)$ and $N(1, 2)$

Application

MOUNTAIN BIKING **For Exercises 12–14, use the following information.**
A certain mountain bike trail has a section of trail with a grade of 8%.

12. What is the slope of the hill? $-\frac{2}{25}$ **or** $\frac{2}{25}$
13. After riding on the trail, a biker is 120 meters below her original starting position. If her starting position is represented by the origin on a coordinate plane, what are the possible coordinates of her current position?
14. How far has she traveled down the hill? Round to the nearest meter. **1505 m**

13. (1500, −120) or (−1500, −120)

★ indicates increased difficulty

Practice and Apply

Homework Help

For Exercises	See Examples
15–18, 25–32	1
19–24	3
33–38	4
42, 43	2

Extra Practice
See page 759.

Determine the slope of the line that contains the given points.

15. $A(0, 2)$, $B(7, 3)$ $\frac{1}{7}$
16. $C(-2, -3)$, $D(-6, -5)$ $\frac{1}{2}$
17. $W(3, 2)$, $X(4, -3)$ **−5**
18. $Y(1, 7)$, $Z(4, 3)$ $-\frac{4}{3}$

19. perpendicular 20. parallel 21. neither 22. perpendicular

Determine whether $\overleftrightarrow{PQ}$ and $\overleftrightarrow{UV}$ are *parallel*, *perpendicular*, or *neither*.

19. $P(-3, -2)$, $Q(9, 1)$, $U(3, 6)$, $V(5, -2)$
20. $P(-4, 0)$, $Q(0, 3)$, $U(-4, -3)$, $V(8, 6)$
21. $P(-10, 7)$, $Q(2, 1)$, $U(4, 0)$, $V(6, 1)$
22. $P(-9, 2)$, $Q(0, 1)$, $U(-1, 8)$, $V(-2, -1)$
23. $P(1, 1)$, $Q(9, 8)$, $U(-6, 1)$, $V(2, 8)$ **parallel**
24. $P(5, -4)$, $Q(10, 0)$, $U(9, -8)$, $V(5, -13)$ **neither**

Answers

10.

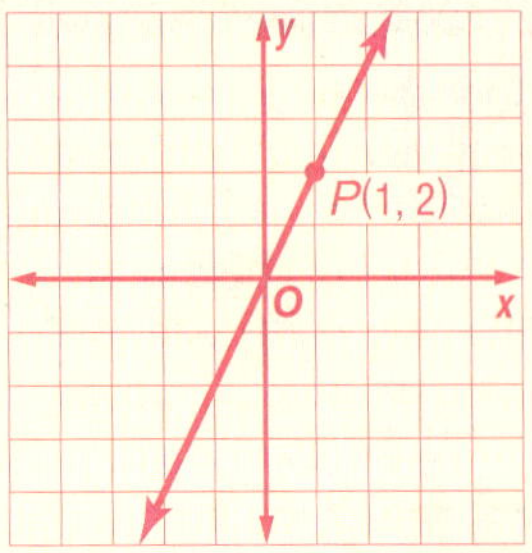

11.

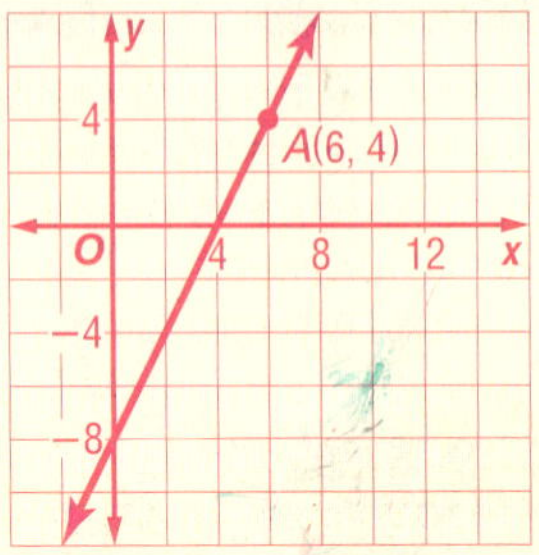

42.

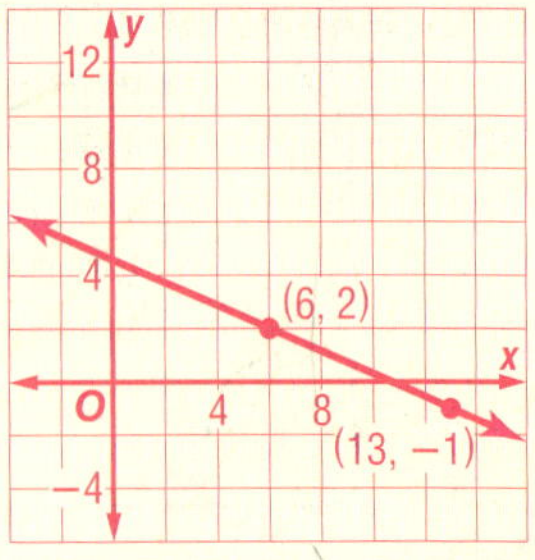

Find the slope of each line.

25. $\overleftrightarrow{AB}$ −3
26. $\overleftrightarrow{PQ}$ $\frac{9}{5}$
27. $\overleftrightarrow{LM}$ 6
28. $\overleftrightarrow{EF}$ 0
29. a line parallel to $\overleftrightarrow{LM}$ 6
30. a line perpendicular to $\overleftrightarrow{PQ}$ $-\frac{5}{9}$
31. a line perpendicular to $\overleftrightarrow{EF}$ undefined
32. a line parallel to $\overleftrightarrow{AB}$ −3

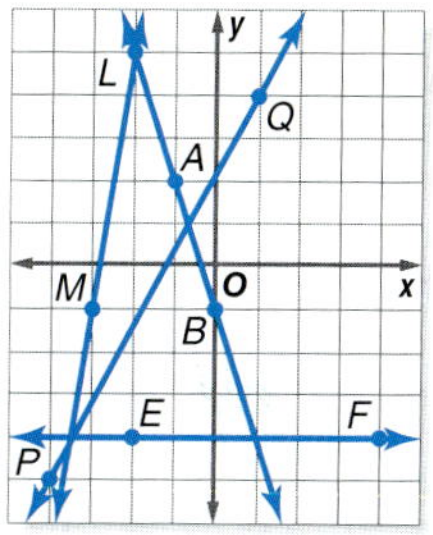

Graph the line that satisfies each condition.

33–38. See p. 173A.

33. slope = −4, passes through $P(-2, 1)$
34. contains $A(-1, -3)$, parallel to $\overleftrightarrow{CD}$ with $C(-1, 7)$ and $D(5, 1)$
35. contains $M(4, 1)$, perpendicular to $\overleftrightarrow{GH}$ with $G(0, 3)$ and $H(-3, 0)$
36. slope $= \frac{2}{5}$, contains $J(-7, -1)$
37. contains $Q(-2, -4)$, parallel to $\overleftrightarrow{KL}$ with $K(2, 7)$ and $L(2, -12)$
38. contains $W(6, 4)$, perpendicular to $\overleftrightarrow{DE}$ with $D(0, 2)$ and $E(5, 0)$.

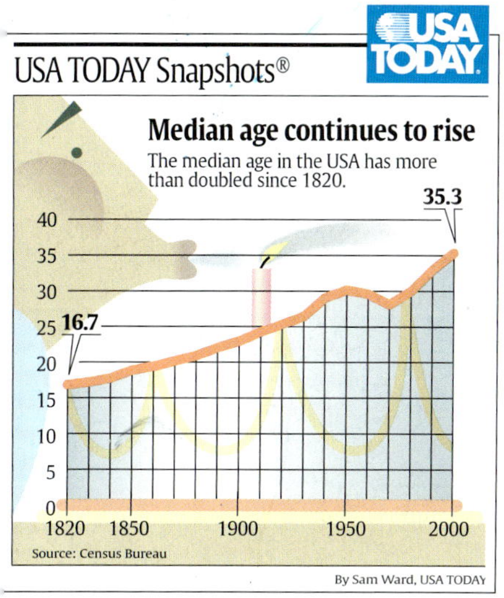

POPULATION For Exercises 39–41, refer to the graph.

39. Estimate the annual rate of change of the median age from 1970 to 2000. Sample answer: 0.24
40. If the median age continues to increase at the same rate, what will be the median age in 2010? Sample answer: 37.7
41. Suppose that after 2000, the median age increases by $\frac{1}{3}$ of a year anually. In what year will the median age be 40.6? 2016

Online Research **Data Update** Use the Internet or other resource to find the median age in the United States for years after 2000. Does the median age increase at the same rate as it did in years leading up to 2000? Visit www.geometryonline.com/data_update to learn more.

★ 42. Determine the value of x so that a line containing $(6, 2)$ and $(x, -1)$ has a slope of $-\frac{3}{7}$. Then graph the line. 13; See margin for graph.

★ 43. Find the value of x so that the line containing $(4, 8)$ and $(2, -1)$ is perpendicular to the line containing $(x, 2)$ and $(-4, 5)$. Graph the lines. $\frac{19}{2}$; See margin for graph.

COMPUTERS For Exercises 44–46, refer to the graph at the right.

44. What is the rate of change between 1998 and 2000? 13% per year
45. If the percent of classrooms with Internet access increases at the same rate as it did between 1999 and 2000, in what year will 90% of classrooms have Internet access? 2001
46. Will the graph continue to rise indefinitely? Explain. No; the graph can only rise until it reaches 100%.

Instructional Classrooms with Internet Access

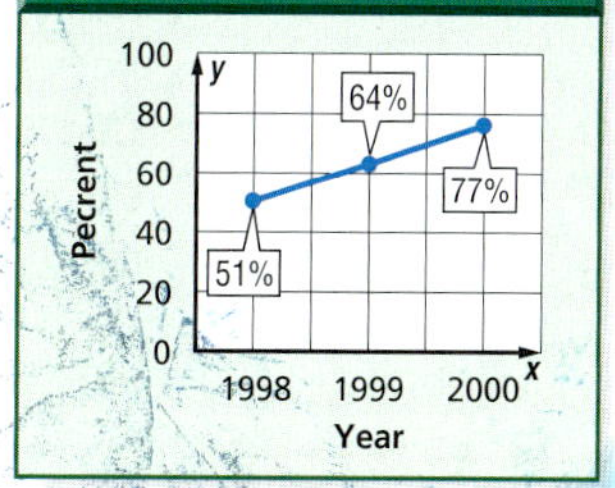

Source: U.S. Census Bureau

www.geometryonline.com/self_check_quiz/fcat

Answer

43.

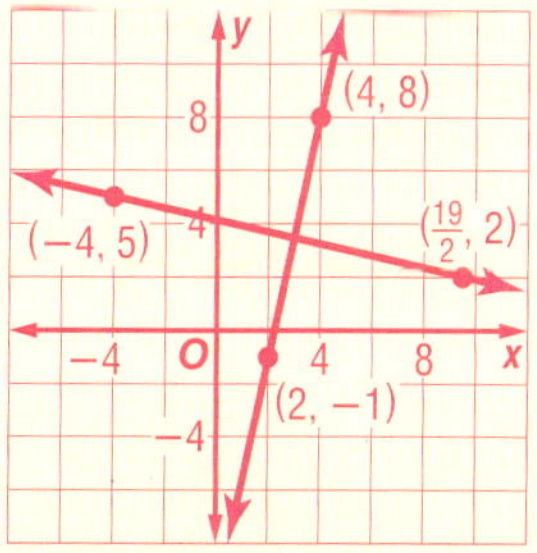

Study Guide and Intervention, p. 137 (shown) and p. 138

Slope of a Line The slope m of a line containing two points with coordinates (x_1, y_1) and (x_2, y_2) is given by the formula $m = \frac{y_2 - y_1}{x_2 - x_1}$, where $x_1 \neq x_2$.

Example **Find the slope of each line.**

For line p, let (x_1, y_1) be $(1, 2)$ and (x_2, y_2) be $(-2, -2)$.

$$m = \frac{y_2 - y_1}{x_2 - x_1} = \frac{-2 - 2}{-2 - 1} \text{ or } \frac{4}{3}$$

For line q, let (x_1, y_1) be $(2, 0)$ and (x_2, y_2) be $(-3, 2)$.

$$m = \frac{y_2 - y_1}{x_2 - x_1} = \frac{2 - 0}{-3 - 2} \text{ or } -\frac{2}{5}$$

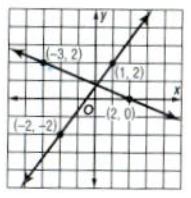

Exercises

Determine the slope of the line that contains the given points.

1. $J(0, 0)$, $K(-2, 8)$ −4
2. $R(-2, -3)$, $S(3, -5)$ $-\frac{2}{5}$
3. $L(1, -2)$, $N(-6, 3)$ $-\frac{5}{7}$
4. $P(-1, 2)$, $Q(-9, 6)$ $-\frac{1}{2}$
5. $T(1, -2)$, $U(6, -2)$ 0
6. $V(-2, 10)$, $W(-4, -3)$ $\frac{13}{2}$

Find the slope of each line.

7. $\overleftrightarrow{AB}$ 3
8. $\overleftrightarrow{CD}$ −2
9. $\overleftrightarrow{EM}$ undefined
10. $\overleftrightarrow{AE}$ 0
11. $\overleftrightarrow{EH}$ $\frac{2}{5}$
12. $\overleftrightarrow{BM}$ $-\frac{1}{2}$

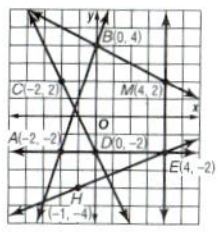

Skills Practice, p. 139 and Practice, p. 140 (shown)

Determine the slope of the line that contains the given points.

1. $B(-4, 4)$, $R(0, 2)$ $-\frac{1}{2}$
2. $I(-2, -9)$, $P(2, 4)$ $\frac{13}{4}$

Find the slope of each line.

3. $\overleftrightarrow{LM}$ $\frac{2}{3}$
4. $\overleftrightarrow{GR}$ $-\frac{2}{5}$
5. a line parallel to $\overleftrightarrow{GR}$ $-\frac{2}{5}$
6. a line perpendicular to $\overleftrightarrow{PS}$ $-\frac{1}{2}$

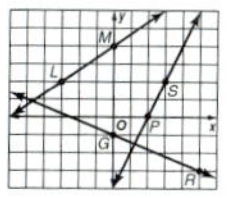

Determine whether $\overleftrightarrow{KM}$ and $\overleftrightarrow{ST}$ are *parallel*, *perpendicular*, or *neither*.

7. $K(-1, -8)$, $M(1, 6)$, $S(-2, -6)$, $T(2, 10)$ neither
8. $K(-5, -2)$, $M(5, 4)$, $S(-3, 6)$, $T(3, -4)$ perpendicular
9. $K(-4, 10)$, $M(2, -8)$, $S(1, 2)$, $T(4, -7)$ parallel
10. $K(-3, -7)$, $M(3, -3)$, $S(0, 4)$, $T(6, -5)$ perpendicular

Graph the line that satisfies each condition.

11. slope $= -\frac{1}{2}$, contains $U(2, -2)$

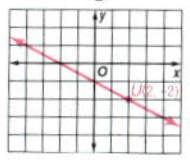

12. slope $= \frac{4}{3}$, contains $P(-3, -3)$

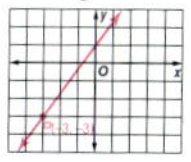

13. contains $B(-4, 2)$, parallel to $\overleftrightarrow{FG}$ with $F(0, -3)$ and $G(4, -2)$

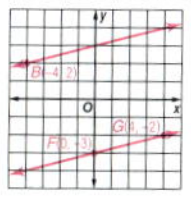

14. contains $Z(-3, 0)$, perpendicular to $\overleftrightarrow{EK}$ with $E(-2, 4)$ and $K(2, -2)$

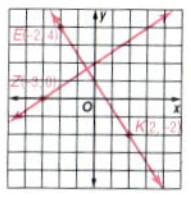

15. **PROFITS** After Take Two began renting DVDs at their video store, business soared. Between 2000 and 2003, profits increased at an average rate of $12,000 per year. Total profits in 2003 were $46,000. If profits continue to increase at the same rate, what will the total profit be in 2009? $118,000

Reading to Learn Mathematics, p. 141 ELL

Pre-Activity **How is slope used in transportation?**

Read the introduction to Lesson 3-3 at the top of page 139 in your textbook.

- If you are driving uphill on a road with a 4% grade, how many feet will the road rise for every 1000 horizontal feet traveled? 40 ft
- If you are driving downhill on a road with a 7% grade, how many meters will the road fall for every 500 meters traveled? 35 m

Reading the Lesson

1. Which expressions can be used to represent the slope of the line containing points (x_1, y_1) and (x_2, y_2)? Assume that no denominator is zero. A, C, F

A. $\frac{\Delta y}{\Delta x}$ B. $\frac{\text{horizontal run}}{\text{vertical rise}}$ C. $\frac{y_2 - y_1}{x_2 - x_1}$ D. $\frac{\text{change in } x}{\text{change in } y}$

E. $\frac{y_2 - y_1}{x_1 - x_2}$ F. $\frac{y_1 - y_2}{x_1 - x_2}$ G. $\frac{x_2 - x_1}{y_2 - y_1}$ H. $\frac{y_2 - x_2}{y_1 - x_1}$

2. Match the description of a line from the first column with the description of its slope from the second column.

Type of Line	Slope
a. a horizontal line ii	i. a negative number
b. a line that rises from left to right iv	ii. 0
c. a vertical line iii	iii. undefined
d. a line that falls from left to right i	iv. a positive number

3. Find the slope of each line.
   a. a line parallel to a line with slope $\frac{3}{4}$ $\frac{3}{4}$
   b. a line perpendicular to the x-axis undefined slope
   c. a line perpendicular to a line with slope 5 $-\frac{1}{5}$
   d. a line parallel to the x-axis 0
   e. y-axis undefined slope

Helping You Remember

4. A good way to remember something is to explain it to someone else. Suppose your friend thinks that perpendicular lines (if neither line is vertical) have slopes that are reciprocals of each other. How could you explain to your friend that this is incorrect and give her a good way to remember the correct relationship?

Sample answer: In order for two lines (neither one vertical) to meet at right angles, one must go upward from left to right and the other must go downward, so their slopes must have opposite signs. Reciprocals have the same sign. The product of the slopes must be −1, not 1. Remember: The slopes are *opposite* reciprocals.

Enrichment, p. 142

Slopes and Polygons

In coordinate geometry, the slopes of two lines determine if the lines are parallel or perpendicular. This knowledge can be useful when working with polygons.

1. The coordinates of the vertices of a triangle are $A(-6, 4)$, $B(8, 6)$, and $C(4, -4)$. Graph $\triangle ABC$.
2. J, K, and L are midpoints of $\overline{AB}$, $\overline{BC}$, and $\overline{AC}$, respectively. Find the coordinates of J, K, and L. Draw $\triangle JKL$.
   $J(1, 5)$, $K(6, 1)$, $L(-1, 0)$
3. Which segments appear to be parallel?
   $\overline{AB}$ and $\overline{LK}$, $\overline{BC}$ and $\overline{JL}$, $\overline{AC}$ and $\overline{JK}$
4. Show that the segments named in Exercise 3 are parallel by finding the slopes of all six segments.
   $\overline{AB}$: $\frac{1}{7}$; $\overline{LK}$: $\frac{1}{7}$; $\overline{BC}$: $\frac{5}{2}$; $\overline{JL}$: $\frac{5}{2}$; $\overline{AC}$: $-\frac{4}{5}$; $\overline{JK}$: $-\frac{4}{5}$

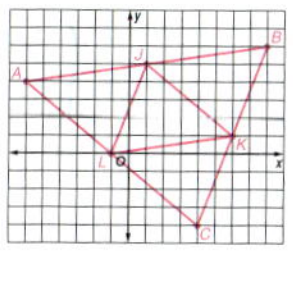

4 Assess

Open-Ended Assessment

Writing Have students write a paragraph explaining how to use the slopes of two lines to determine whether they are perpendicular.

Getting Ready for Lesson 3-4

Prerequisite Skill Students will work with equations of lines in Lesson 3-4. They will solve an equation in two variables for one of the variables. Use Exercises 70–72 to determine your students' familiarity with solving an equation for a particular variable.

Assessment Options

Quiz (Lesson 3-3) is available on p. 175 of the *Chapter 3 Resource Masters*.

Mid-Chapter Test (Lessons 3-1 through 303) is available on p. 177 of the *Chapter 3 Resource Masters*.

Answers

48. Sample answer: Slope is used when driving through hills to determine how fast to go. Answers should include the following.
   - Drivers should be notified of the grade so that they can adjust their speed accordingly. A positive slope indicates that the driver must speed up, while a negative slope indicates that the driver should slow down.
   - An escalator must be at a steep enough slope to be efficient, but also must be gradual enough to ensure comfort.

63. *H*, *I*, and *J* are noncollinear.

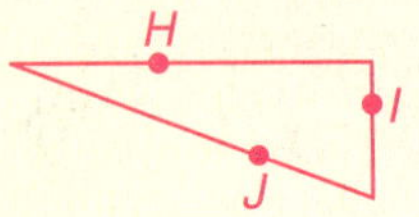

64. $XZ + ZY = XY$

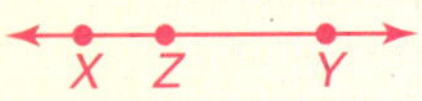

47. **CRITICAL THINKING** The line containing the point $(5 + 2t, -3 + t)$ can be described by the equations $x = 5 + 2t$ and $y = -3 + t$. Write the slope-intercept form of the equation of this line. $y = \frac{1}{2}y - \frac{11}{2}$

48. **WRITING IN MATH** Answer the question that was posed at the beginning of the lesson. See margin.

How is slope used in transportation?

Include the following in your answer:
- an explanation of why it is important to display the grade of a road, and
- an example of slope used in transportation other than roads.

FCAT Practice
Standardized Test Practice

49. Find the slope of a line perpendicular to the line containing (−5, 1) and (−3, −2). **C**

Ⓐ $-\frac{2}{3}$ Ⓑ $-\frac{3}{2}$ Ⓒ $\frac{2}{3}$ Ⓓ $\frac{3}{2}$

50. **ALGEBRA** The winning sailboat completed a 24-mile race at an average speed of 9 miles per hour. The second-place boat finished with an average speed of 8 miles per hour. How many minutes longer than the winner did the second-place boat take to finish the race? **A**

Ⓐ 20 min Ⓑ 33 min Ⓒ 60 min Ⓓ 120 min

Maintain Your Skills

Mixed Review

In the figure, $\overline{QR} \parallel \overline{TS}$, $\overleftrightarrow{QT} \parallel \overleftrightarrow{RS}$, and $m\angle 1 = 131$. Find the measure of each angle. *(Lesson 3-2)*

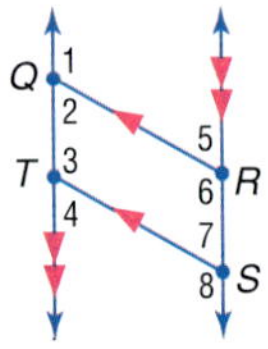

51. $\angle 6$ 131
52. $\angle 7$ 49
53. $\angle 4$ 49
54. $\angle 2$ 49
55. $\angle 5$ 49
56. $\angle 8$ 131

State the transversal that forms each pair of angles. Then identify the special name for each angle pair. *(Lesson 3-1)*

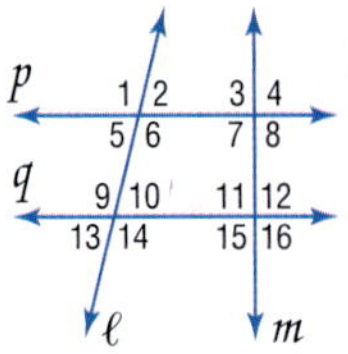

57. $\angle 1$ and $\angle 14$ ℓ; alt. ext.
58. $\angle 2$ and $\angle 10$ ℓ; corr.
59. $\angle 3$ and $\angle 6$ p; alt. int.
60. $\angle 14$ and $\angle 15$ q; cons. int.
61. $\angle 7$ and $\angle 12$ m; alt. int.
62. $\angle 9$ and $\angle 11$ q; corr.

Make a conjecture based on the given information. Draw a figure to illustrate your conjecture. *(Lesson 2-1)*

63–65. See margin.

63. Points *H*, *I*, and *J* are each located on different sides of a triangle.
64. Collinear points *X*, *Y*, and *Z*; *Z* is between *X* and *Y*.
65. $R(3, -4)$, $S(-2, -4)$, and $T(0, -4)$

Classify each angle as *right*, *acute*, or *obtuse*. *(Lesson 1-4)*

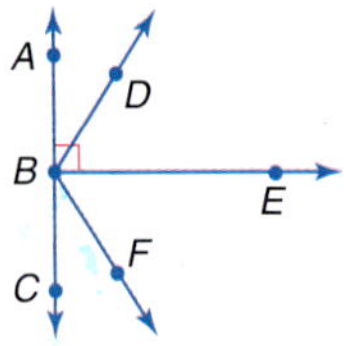

66. $\angle ABD$ acute
67. $\angle DBF$ obtuse
68. $\angle CBE$ right
69. $\angle ABF$ obtuse

Getting Ready for the Next Lesson

PREREQUISITE SKILL Solve each equation for *y*.
*(To review **solving equations**, see pages 737 and 738.)*

70. $2x + y = 7$ $y = -2x + 7$
71. $2x + 4y = -5$ $y = -\frac{1}{2}x - \frac{5}{4}$
72. $5x - 2y + 4 = 0$ $y = \frac{5}{2}x + 2$

65. *R*, *S*, and *T* are collinear.

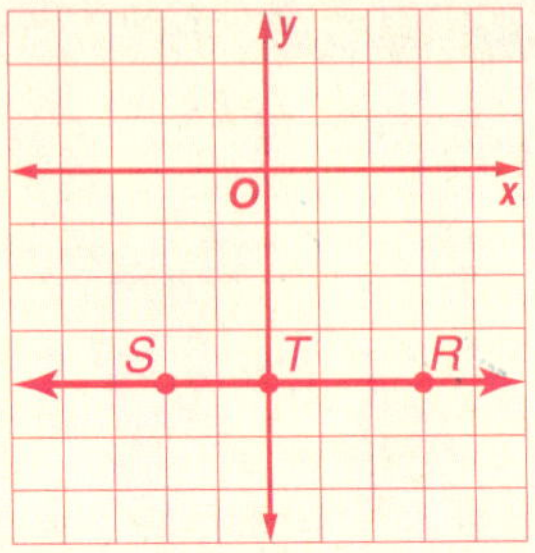

144 Chapter 3 Parallel and Perpendicular Lines

3-4 Equations of Lines

What You'll Learn

- Write an equation of a line given information about its graph.
- Solve problems by writing equations.

Vocabulary
- slope-intercept form
- point-slope form

How can the equation of a line describe the cost of cellular telephone service?

A certain cellular phone company charges a flat rate of \$19.95 per month for service. All calls are charged \$0.07 per minute of air time t. The total charge C for a month can be represented by the equation $C = 0.07t + 19.95$.

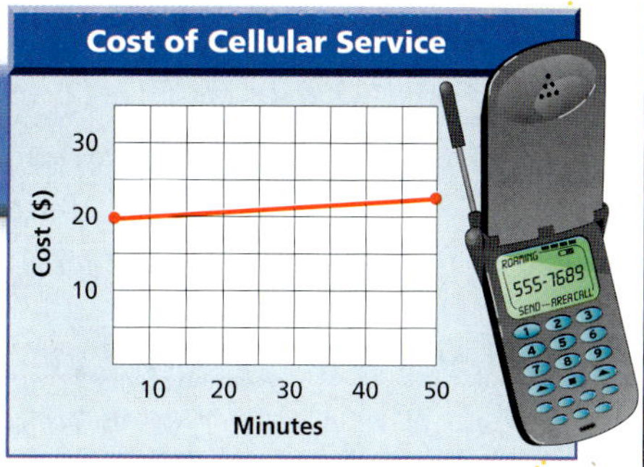

WRITE EQUATIONS OF LINES You may remember from algebra that an equation of a line can be written given any of the following:

- the slope and the y-intercept,
- the slope and the coordinates of a point on the line, or
- the coordinates of two points on the line.

The graph of $C = 0.07t + 19.95$ has a slope of 0.07, and it intersects the y-axis at 19.95. These two values can be used to write an equation of the line. The **slope-intercept form** of a linear equation is $y = mx + b$, where m is the slope of the line and b is the y-intercept.

slope

$$y = mx + b \qquad C = 0.07t + 19.95$$

y-intercept

Example 1 Slope and y-Intercept

Write an equation in slope-intercept form of the line with slope of −4 and y-intercept of 1.

$y = mx + b$ Slope-intercept form

$y = -4x + 1$ $m = -4, b = 1$

The slope-intercept form of the equation of the line is $y = -4x + 1$.

Another method used to write an equation of a line is the point-slope form of a linear equation. The **point-slope form** is $y - y_1 = m(x - x_1)$, where (x_1, y_1) are the coordinates of any point on the line and m is the slope of the line.

given point (x_1, y_1)

$$y - y_1 = m(x - x_1)$$

slope

3-4 Lesson Notes

1 Focus

5-Minute Check Transparency 3-4 Use as a quiz or review of Lesson 3-3.

Mathematical Background notes are available for this lesson on p. 124D.

How can the equation of a line describe the cost of cellular telephone service?

Ask students:

- If you use your cellular phone heavily each month, are you better off with a large monthly fee and a small per-minute fee, or the other way around? **A large monthly fee and a small per-minute fee is better for heavy cell-phone use.**
- What does it mean if the graph of the equation for a cellular service fee goes through the origin? **It means that the fixed fee is \$0; the bill is \$0 if no calls are made.**

Resource Manager

Workbook and Reproducible Masters

Chapter 3 Resource Masters
- Study Guide and Intervention, pp. 143–144
- Skills Practice, p. 145
- Practice, p. 146
- Reading to Learn Mathematics, p. 147
- Enrichment, p. 148

Teaching Geometry With Manipulatives Masters, pp. 1, 17, 54, 55

Transparencies

5-Minute Check Transparency 3-4
Answer Key Transparencies

Technology

GeomPASS: Tutorial Plus, Lesson 8
Interactive Chalkboard

2 Teach

WRITE EQUATIONS OF LINES

In-Class Examples

Power Point®

1 Write an equation in slope-intercept form of the line with slope of 6 and y-intercept of -3. $y = 6x - 3$

2 Write an equation in point-slope form of the line whose slope is $-\frac{3}{5}$ that contains $(-10, 8)$. $y - 8 = -\frac{3}{5}(x + 10)$

3 Write an equation in slope-intercept form for a line containing $(4, 9)$ and $(-2, 0)$. $y = \frac{3}{2}x + 3$

4 Write an equation in slope-intercept form for a line containing $(1, 7)$ that is perpendicular to the line $y = -\frac{1}{2}x + 1$. $y = 2x + 5$

WRITE EQUATIONS TO SOLVE PROBLEMS

In-Class Example

Power Point®

5 **RENTAL COSTS** An apartment complex charges \$525 per month plus a \$750 security deposit.

a. Write an equation to represent the total annual cost A for r months of rent. $A = 525r + 750$

b. Compare this rental cost to a complex which charges a \$200 security deposit but \$600 per month for rent. If a person expects to stay in an apartment for one year, which complex offers the better rate? The first complex offers the better rate: one year costs \$7050 instead of \$7400.

Study Tip

Choosing Forms of Linear Equations

If you are given a point on a line and the slope of the line, use point-slope form. Otherwise, use slope-intercept form.

Example 2 Slope and a Point

Write an equation in point-slope form of the line whose slope is $-\frac{1}{2}$ that contains $(3, -7)$.

$y - y_1 = m(x - x_1)$	Point-slope form
$y - (-7) = -\frac{1}{2}(x - 3)$	$m = -\frac{1}{2}$, $(x_1, y_1) = (3, -7)$
$y + 7 = -\frac{1}{2}(x - 3)$	Simplify.

The point-slope form of the equation of the line is $y + 7 = -\frac{1}{2}(x - 3)$.

Both the slope-intercept form and the point-slope form require the slope of a line in order to write an equation. There are occasions when the slope of a line is not given. In cases such as these, use two points on the line to calculate the slope. Then use the point-slope form to write an equation.

Study Tip

Writing Equations

Note that the point-slope form of an equation is different for each point used. However, the slope-intercept form of an equation is unique.

Example 3 Two Points

Write an equation in slope-intercept form for line ℓ.

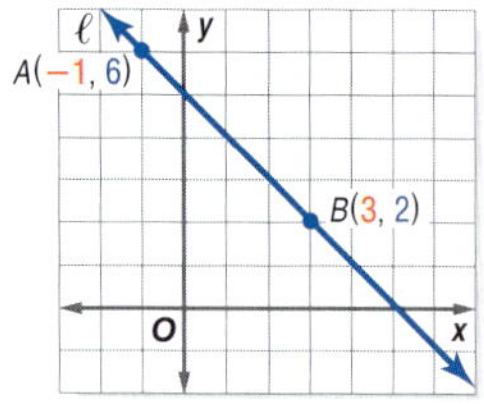

Find the slope of ℓ by using $A(-1, 6)$ and $B(3, 2)$.

$m = \frac{y_2 - y_1}{x_2 - x_1}$	Slope formula
$= \frac{2 - 6}{3 - (-1)}$	$x_1 = -1, x_2 = 3, y_1 = 6, y_2 = 2$
$= -\frac{4}{4}$ or -1	Simplify.

Now use the point-slope form and either point to write an equation.

Using Point A:

$y - y_1 = m(x - x_1)$	Point-slope form
$y - 6 = -1[x - (-1)]$	$m = -1$, $(x_1, y_1) = (-1, 6)$
$y - 6 = -1(x + 1)$	Simplify.
$y - 6 = -x - 1$	Distributive Property
$y = -x + 5$	Add 6 to each side.

Using Point B:

$y - y_1 = m(x - x_1)$	Point-slope form
$y - 2 = -1(x - 3)$	$m = -1$, $(x_1, y_1) = (3, 2)$
$y - 2 = -x + 3$	Distributive Property
$y = -x + 5$	Add 2 to each side.

Example 4 One Point and an Equation

Write an equation in slope-intercept form for a line containing (2, 0) that is perpendicular to the line $y = -x + 5$.

Since the slope of the line $y = -x + 5$ is -1, the slope of a line perpendicular to it is 1.

$y - y_1 = m(x - x_1)$	Point-slope form
$y - 0 = 1(x - 2)$	$m = 1$, $(x_1, y_1) = (2, 0)$
$y = x - 2$	Distributive Property

DAILY INTERVENTION

Differentiated Instruction

Auditory/Musical Have students write lyrics that describe how to set up the slope-intercept form and point-slope form of an equation. They can write the lyrics to be sung to a melody or spoken in a hip-hop cadence.

146 Chapter 3 Parallel and Perpendicular Lines

WRITE EQUATIONS TO SOLVE PROBLEMS Many real-world situations can be modeled using linear equations. In many business applications, the slope represents a rate.

Example 5 Write Linear Equations

CELL PHONE COSTS **Martina's current cellular phone plan charges \$14.95 per month and \$0.10 per minute of air time.**

a. Write an equation to represent the total monthly cost C for t minutes of air time.

For each minute of air time, the cost increases \$0.10. So, the rate of change, or slope, is 0.10. The y-intercept is located where 0 minutes of air time are used, or \$14.95.

$C = mt + b$ Slope-intercept form

$C = 0.10t + 14.95$ $m = 0.10$, $b = 14.95$

The total monthly cost can be represented by the equation $C = 0.10t + 14.95$.

b. Compare her current plan to the plan presented at the beginning of the lesson. If she uses an average of 40 minutes of air time each month, which plan offers the better rate?

Evaluate each equation for $t = 40$.

Current plan: $C = 0.10t + 14.95$

$= 0.10(40) + 14.95$ $t = 40$

$= 18.95$ Simplify.

Alternate plan: $C = 0.07t + 19.95$

$= 0.07(40) + 19.95$ $t = 40$

$= 22.75$ Simplify.

Given her average usage, Martina's current plan offers the better rate.

Check for Understanding

Concept Check 1–3. See margin.

1. **Explain** how you would write an equation of a line whose slope is $-\frac{2}{5}$ that contains $(-2, 8)$.
2. **Write** equations in slope-intercept form for two lines that contain $(-1, -5)$.
3. **OPEN ENDED** Graph a line that is not horizontal or vertical on the coordinate plane. Write the equation of the line.

Guided Practice

GUIDED PRACTICE KEY	
Exercises	Examples
4–6	1
7–9	2
10, 11	3
12	4
13, 14	5

Write an equation in slope-intercept form of the line having the given slope and y-intercept.

4. $m = \frac{1}{2}$, y-intercept: 4 $y = \frac{1}{2}x + 4$
5. $m = -\frac{3}{5}$, intercept at $(0, -2)$ $y = -\frac{3}{5}x - 2$
6. $m = 3$, y-intercept: -4 $y = 3x - 4$

Write an equation in point-slope form of the line having the given slope that contains the given point.

7. $m = \frac{3}{2}$, $(4, -1)$ $y + 1 = \frac{3}{2}(x - 4)$
8. $m = 3$, $(7, 5)$ $y - 5 = 3(x - 7)$
9. $m = 1.25$, $(20, 137.5)$ $y - 137.5 = 1.25(x - 20)$

www.geometryonline.com/extra_examples/fcat

3 Practice/Apply

Study Notebook

Have students—
- add the definitions/examples of the vocabulary terms to their Vocabulary Builder worksheets for Chapter 3.
- include the slope-intercept and point-slope forms of an equation and the steps for finding either form of an equation given two points or given a point and the slope.
- include any other item(s) that they find helpful in mastering the skills in this lesson.

About the Exercises...

Organization by Objective
- **Write Equations of Lines:** 15–44
- **Write Equations to Solve Problems:** 45–51

Odd/Even Assignments

Exercises 15–44 are structured so that students practice the same concepts whether they are assigned odd or even problems.

Assignment Guide

Basic: 15–41 odd, 45, 46–49, 52–71

Average: 15–45 odd, 46–49, 52–71

Advanced: 16–44 even, 50–67 (optional: 68–71)

All: Practice Quiz 2 (1–10)

Answers

1. Sample answer: Use the point-slope form where $(x_1, y_1) = (-2, 8)$ and $m = -\frac{2}{5}$.
2. Sample answer: $y = 2x - 3$, $y = -x - 6$
3. Sample answer:

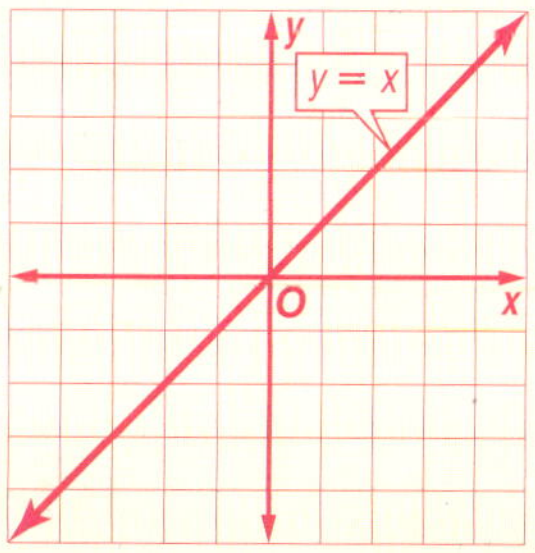

Study Guide and Intervention, p. 143 (shown) and p. 144

Write Equations of Lines You can write an equation of a line if you are given any of the following:

- the slope and the y-intercept,
- the slope and the coordinates of a point on the line, or
- the coordinates of two points on the line.

If m is the slope of a line, b is its y-intercept, and (x_1, y_1) is a point on the line, then:

- the **slope-intercept form** of the equation is $y = mx + b$,
- the **point-slope form** of the equation is $y - y_1 = m(x - x_1)$.

Example 1 **Write an equation in slope-intercept form of the line with slope −2 and y-intercept 4.**

$y = mx + b$ Slope-intercept form

$y = -2x + 4$ $m = -2, b = 4$

The slope-intercept form of the equation of the line is $y = -2x + 4$.

Example 2 **Write an equation in point-slope form of the line with slope $-\frac{3}{4}$ that contains (8, 1).**

$y - y_1 = m(x - x_1)$ Point-slope form

$y - 1 = -\frac{3}{4}(x - 8)$ $m = -\frac{3}{4}, (x_1, y_1) = (8, 1)$

The point-slope form of the equation of the line is $y - 1 = -\frac{3}{4}(x - 8)$.

Exercises

Write an equation in slope-intercept form of the line having the given slope and y-intercept.

1. m: 2, y-intercept: −3 $y = 2x - 3$
2. m: $-\frac{1}{2}$, y-intercept: 4 $y = -\frac{1}{2}x + 4$
3. m: $\frac{1}{4}$, y-intercept: 5 $y = \frac{1}{4}x + 5$
4. m: 0, y-intercept: −2 $y = -2$
5. m: $-\frac{5}{3}$, y-intercept: $\frac{1}{3}$ $y = -\frac{5}{3}x + \frac{1}{3}$
6. m: −3, y-intercept: −8 $y = -3x - 8$

Write an equation in point-slope form of the line having the given slope that contains the given point.

7. $m = \frac{1}{2}$, (3, −1) $y + 1 = \frac{1}{2}(x - 3)$
8. $m = -2$, (4, −2) $y + 2 = -2(x - 4)$
9. $m = -1$, (−1, 3) $y - 3 = -(x + 1)$
10. $m = \frac{1}{4}$, (−3, −2) $y + 2 = \frac{1}{4}(x + 3)$
11. $m = -\frac{5}{2}$, (0, −3) $y + 3 = -\frac{5}{2}x$
12. $m = 0$, (−2, 5) $y - 5 = 0$

Skills Practice, p. 145 and Practice, p. 146 (shown)

Write an equation in slope-intercept form of the line having the given slope and y-intercept.

1. m: $\frac{2}{3}$, y-intercept: −10 $y = \frac{2}{3}x - 10$
2. m: $-\frac{7}{9}$, $\left(0, -\frac{1}{2}\right)$ $y = -\frac{7}{9}x - \frac{1}{2}$
3. m: 4.5, (0, 0.25) $y = 4.5x + 0.25$

Write equations in point-slope form and slope-intercept form of the line having the given slope and containing the given point.

4. m: $\frac{3}{2}$, (4, 6) $y - 6 = \frac{3}{2}(x - 4)$, $y = \frac{3}{2}x$
5. m: $-\frac{6}{5}$, (−5, −2) $y + 2 = -\frac{6}{5}(x + 5)$, $y = -\frac{6}{5}x - 8$
6. m: 0.5, (7, −3) $y + 3 = 0.5(x - 7)$, $y = 0.5x - 6.5$
7. m: −1.3, (−4, 4) $y - 4 = -1.3(x + 4)$, $y = -1.3x - 1.2$

Write an equation in slope-intercept form for each line.

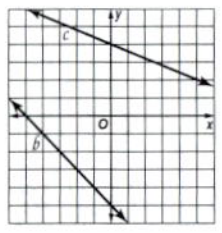

8. b $y = -x - 5$
9. c $y = -\frac{2}{5}x + 4$
10. parallel to line b, contains (3, −2) $y = -x + 1$
11. perpendicular to line c, contains (−2, −4) $y = \frac{5}{2}x + 1$

Write an equation in slope-intercept form for the line that satisfies the given conditions.

12. $m = -\frac{4}{9}$, y-intercept = 2 $y = -\frac{4}{9}x + 2$
13. $m = 3$, contains (2, −3) $y = 3x - 9$
14. x-intercept is −6, y-intercept is 2 $y = \frac{1}{3}x + 2$
15. x-intercept is 2, y-intercept is −5 $y = \frac{5}{2}x - 5$
16. passes through (2, −4) and (5, 8) $y = 4x - 12$
17. contains (−4, 2) and (8, −1) $y = -\frac{1}{4}x + 1$
18. COMMUNITY EDUCATION A local community center offers self-defense classes for teens. A $25 enrollment fee covers supplies and materials and open classes cost $10 each. Write an equation to represent the total cost of x self-defense classes at the community center. $C = 10x + 25$

Reading to Learn Mathematics, p. 147 ELL

Pre-Activity **How can the equation of a line describe the cost of cellular telephone service?**

Read the introduction to Lesson 3-4 at the top of page 145 in your textbook. If the rates for your cellular phone plan are described by the equation in your textbook, what will be the total charge (excluding taxes and fees) for a month in which you use 50 minutes of air time? $23.45

Reading the Lesson

1. Identify what each formula represents.
   a. $y - y_1 = m(x - x_1)$ point-slope form of an equation
   b. $m = \frac{y_2 - y_1}{x_2 - x_1}$ slope of a line
   c. $y = mx + b$ slope-intercept form of an equation
2. Write the point-slope form of the equation for each line.
   a. line with slope $-\frac{1}{2}$ containing (−2, 5) $y - 5 = -\frac{1}{2}(x + 2)$
   b. line containing (−4.5, −6.5) and parallel to a line with slope 0.5 $y + 6.5 = 0.5(x + 4.5)$
3. Which one of the following correctly describes the y-intercept of a line? C
   A. the y-coordinate of the point where the line intersects the x-axis
   B. the x-coordinate of the point where the line intersects the y-axis
   C. the y-coordinate of the point where the line crosses the y-axis
   D. the x-coordinate of the point where the line crosses the x-axis
   E. the ratio of the change in y-coordinates to the change in x-coordinates
4. Find the slope and y-intercept of each line.
   a. $y = 2x - 7$ slope = 2; y-intercept = −7
   b. $x + y = 8.5$ slope = −1; y-intercept = 8.5
   c. $2.4x - y = 4.8$ slope = 2.4; y-intercept = −4.8
   d. $y - 7 = x + 12$ slope = 1; y-intercept = 19
   e. $y + 5 = -2(x + 6)$ slope = −2; y-intercept = −17

Helping You Remember

5. A good way to remember something new is to relate it to something you already know. How can the slope formula help you to remember the equation for the point-slope form of a line? Sample answer: The slope of a line through (x, y) and (x_1, y_1) is given by $\frac{y - y_1}{x - x_1} = m$. Multiply each side of the slope formula by $x_2 - x_1$. The result will be the point-slope form.

Refer to the figure at the right. Write an equation in slope-intercept form for each line.

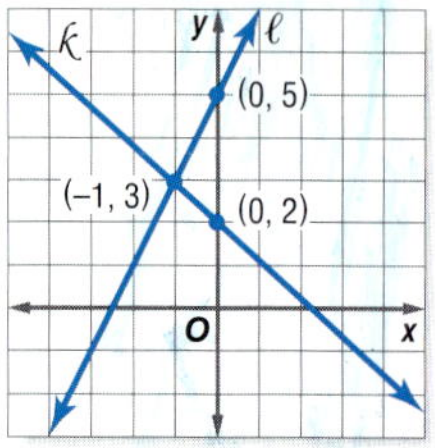

10. ℓ $y = 2x + 5$
11. k $y = -x + 2$
12. the line parallel to ℓ that contains (4, 4) $y = 2x - 4$

Application

INTERNET For Exercises 13–14, use the following information. 13. $y = 39.95$, $y = 0.95x + 4.95$

Justin's current Internet service provider charges a flat rate of $39.95 per month for unlimited access. Another provider charges $4.95 per month for access and $0.95 for each hour of connection.

13. Write an equation to represent the total monthly cost for each plan.
14. If Justin is online an average of 60 hours per month, should he keep his current plan, or change to the other plan? Explain. He should keep his current plan, based on his average usage.

★ indicates increased difficulty

Practice and Apply

Homework Help	
For Exercises	See Examples
15–20, 35, 36	1
21–26	2
27–30, 37–42	3
31–34, 43, 44	4
45–51	5

Extra Practice See page 759.

Write an equation in slope-intercept form of the line having the given slope and y-intercept. 15. $y = \frac{1}{6}x - 4$ 16. $y = \frac{2}{3}x + 8$ 17. $y = \frac{5}{8}x - 6$

15. m: $\frac{1}{6}$, y-intercept: −4
16. m: $\frac{2}{3}$, (0, 8)
17. m: $\frac{5}{8}$, (0, −6)
18. m: $\frac{2}{9}$, y-intercept: $\frac{1}{3}$ $y = \frac{2}{9}x + \frac{1}{3}$
19. m: −1, b: −3 $y = -x - 3$
20. m: $-\frac{1}{12}$, b: 1 $y = -\frac{1}{12}x + 1$

Write an equation in point-slope form of the line having the given slope that contains the given point. 21–26. See margin.

21. $m = 2$, (3, 1)
22. $m = -5$, (4, 7)
23. $m = -\frac{4}{5}$, (−12, −5)
24. $m = \frac{1}{16}$, (3, 11)
25. $m = 0.48$, (5, 17.12)
26. $m = -1.3$, (10, 87.5)

Write an equation in slope-intercept form for each line. 31. $y = -x + 5$

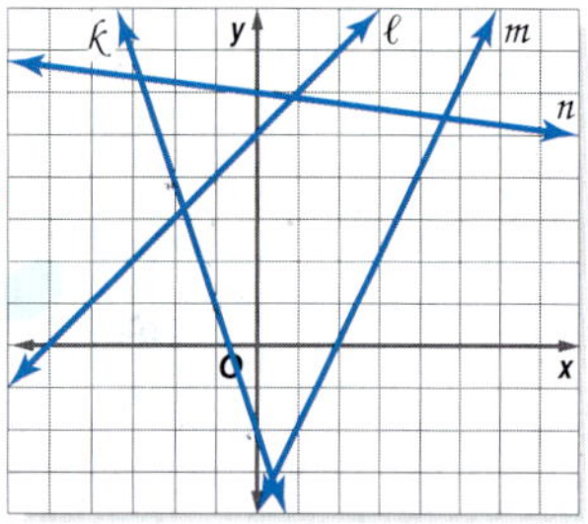

27. k $y = -3x - 2$
28. ℓ $y = x + 5$
29. m $y = 2x - 4$
30. n $y = -\frac{1}{8}x + 6$
31. perpendicular to line ℓ, contains (−1, 6)
32. parallel to line k, contains (7, 0) 32. $y = -3x + 21$
33. parallel to line n, contains (0, 0) $y = -\frac{1}{8}x$
34. perpendicular to line m, contains (−3, −3) $y = -\frac{1}{2}x - \frac{9}{2}$

Write an equation in slope-intercept form for the line that satisfies the given conditions.

35. $m = -3$, y-intercept = 5 $y = -3x + 5$
36. $m = 0$, y-intercept = 6 $y = 6$
37. x-intercept = 5, y-intercept = 3 37. $y = -\frac{3}{5}x + 3$
38. contains (4, −1) and (−2, −1) 38. $y = -1$
39. contains (−5, −3) and (10, −6) 39. $y = -\frac{1}{5}x - 4$
40. x-intercept = 5, y-intercept = −1 40. $y = \frac{1}{5}x - 1$
41. contains (−6, 8) and (−6, −4) no slope-intercept form, $x = -6$
42. contains (−4, −1) and (−8, −5) $y = x + 3$

★ 43. Write an equation of the line that contains (7, −2) and is parallel to $2x - 5y = 8$. 43. $y = \frac{2}{5}x - \frac{24}{5}$

★ 44. What is an equation of the line that is perpendicular to $2y + 2 = -\frac{7}{4}(x - 7)$ and contains (−2, −3)? $y = \frac{8}{7}x - \frac{5}{7}$

Enrichment, p. 148

Absolute Zero

All matter is made up of atoms and molecules that are in constant motion. Temperature is one measure of this motion. Absolute zero is the theoretical temperature limit at which the motion of the molecules and atoms of a substance is the least possible.

Experiments with gaseous substances yield data that allow you to estimate just how cold absolute zero is. For any gas of a constant volume, the pressure, expressed in a unit called atmospheres, varies linearly as the temperature. That is, the pressure P and the temperature t are related by an equation of the form $P = mt + b$, where m and b are real numbers.

1. Sketch a graph for the data in the table.

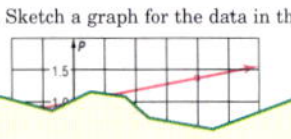

t (in °C)	P (in atmospheres)
−25	0.91

Answers

21. $y - 1 = 2(x - 3)$
22. $y - 7 = -5(x - 4)$
23. $y + 5 = -\frac{4}{5}(x + 12)$
24. $y - 11 = \frac{1}{16}(x - 3)$

45. **JOBS** Ann MacDonald is a salesperson at a discount appliance store. She earns \$50 for each appliance that she sells plus a 5% commission on the price of the appliance. Write an equation that represents what she earned in a week in which she sold 15 appliances. **$y = 0.05x + 750$, where x = total price of appliances sold**

BUSINESS For Exercises 46–49, use the following information.
The Rainbow Paint Company sells an average of 750 gallons of paint each day.

46. How many gallons of paint will they sell in x days? **$750x$**
47. The store has 10,800 gallons of paint in stock. Write an equation in slope-intercept form that describes how many gallons of paint will be on hand after x days if no new stock is added. **$y = -750x + 10{,}800$**
48. Draw a graph that represents the number of gallons of paint on hand at any given time. **See margin.**
49. If it takes 4 days to receive a shipment of paint from the manufacturer after it is ordered, when should the store manager order more paint so that the store does not run out? **in 10 days**

More About. . .

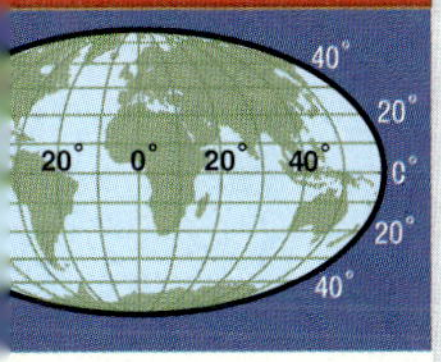

Maps

Global coordinates are usually stated latitude, the angular distance north or south of the equator, and longitude, the angular distance east or west of the prime meridian.

Source: www.worldatlas.com

MAPS For Exercises 50 and 51, use the following information.
Many states have county maps that are grid-like in nature. If a map of Texas is placed on a coordinate plane with the western tip at the origin, the county lines can be represented by equations. Jeff Davis, Pecos, and Brewster Counties meet at (130, −70), and Jeff Davis, Reeves, and Pecos Counties meet at (120, −60).

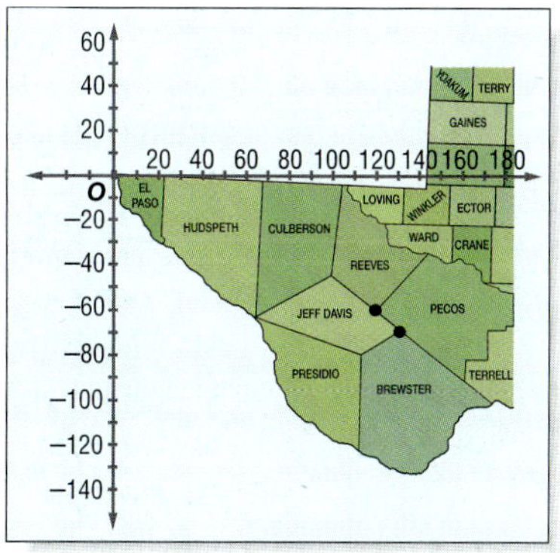

50. Write an equation in slope-intercept form that models the county line between Jeff Davis and Reeves Counties. **$y = -x + 60$**
51. The line separating Reeves and Pecos Counties runs perpendicular to the Jeff Davis/Reeves County line. Write an equation in slope-intercept form of the line that contains the Reeves/Pecos County line. **$y = x - 180$**

52. **CRITICAL THINKING** The point-slope form of an equation of a line can be rewritten as $y = m(x - x_1) + y_1$. Describe how the graph of $y = m(x - x_1) + y_1$ is related to the graph of $y = mx$. **See margin.**

53. **WRITING IN MATH** Answer the question that was posed at the beginning of the lesson. **See margin.**

How can the equation of a line describe cellular telephone service?

Include the following in your answer:
- an explanation of how the fee for air time affects the equation, and
- a description of how you can use equations to compare various plans.

FCAT Practice
Standardized Test Practice Ⓐ Ⓑ Ⓒ Ⓓ

54. What is the slope of a line perpendicular to the line represented by $2x - 8y = 16$? **A**

Ⓐ -4 Ⓑ -2 Ⓒ $-\frac{1}{4}$ Ⓓ $\frac{1}{4}$

55. **ALGEBRA** What are all of the values of y for which $y^2 < 1$? **B**

Ⓐ $y < -1$ Ⓑ $-1 < y < 1$ Ⓒ $y > -1$ Ⓓ $y < 1$

www.geometryonline.com/self_check_quiz/fcat

Answers

48.

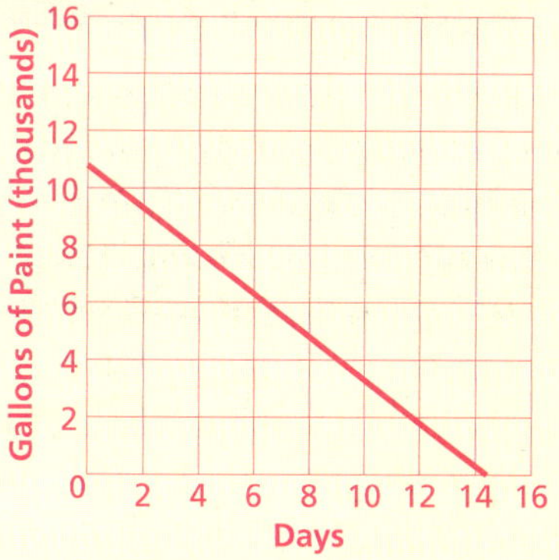

52. The equation $y = mx$ is the special case of $y = m(x - x_1) + y_1$ when $x_1 = y_1 = 0$.

53. Sample answer: In the equation of a line, the b value indicates the fixed rate, while the mx value indicates charges based on usage. Answers should include the following.
- The fee for air time can be considered the slope of the equation.
- We can find where the equations intersect to see where the plans would be equal.

Answers

25. $y - 17.12 = 0.48(x - 5)$
26. $y - 87.5 = -1.3(x - 10)$

Lesson 3-4 Equations of Lines 149

4 Assess

Open-Ended Assessment

Speaking Have students discuss the two forms of equations presented in this lesson. Ask them which form they would use for a given set of information and why they would use that form.

Getting Ready for Lesson 3-5

Prerequisite Skill Students will work with pairs of angles formed by a transversal in Lesson 3-5. They will identify pairs of angles such as alternate interior angles and corresponding angles. Use Exercises 68–71 to determine your students' familiarity with pairs of angles formed by a transversal.

Assessment Options

Practice Quiz 2 The quiz provides students with a brief review of the concepts and skills in Lessons 3-3 and 3-4. Lesson numbers are given to the right of the exercises or instruction lines so students can review concepts not yet mastered.

Answer

65. Given: $AC = DF$, $AB = DE$
Prove: $BC = EF$

Proof:
Statements (Reasons)

1. $AC = DF$, $AB = DE$ (Given)
2. $AC = AB + BC$; $DF = DE + EF$ (Segment Addition Postulate)
3. $AB + BC = DE + EF$ (Substitution Property)
4. $BC = EF$ (Subtraction Property)

Maintain Your Skills

Mixed Review

Determine the slope of the line that contains the given points. (Lesson 3-3)

56. $A(0, 6)$, $B(4, 0)$ $-\frac{3}{2}$
57. $G(8, 1)$, $H(8, -6)$ undefined
58. $E(6, 3)$, $F(-6, 3)$ 0

In the figure, $m\angle 1 = 58$, $m\angle 2 = 47$, and $m\angle 3 = 26$. Find the measure of each angle. (Lesson 3-2)

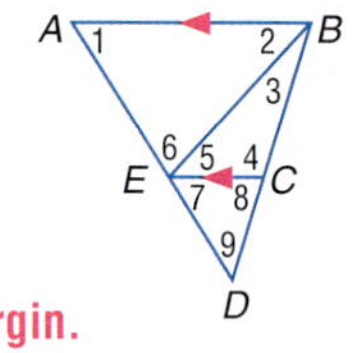

59. $\angle 7$ 58
60. $\angle 5$ 47
61. $\angle 6$ 75
62. $\angle 4$ 107
63. $\angle 8$ 73
64. $\angle 9$ 49

65. **PROOF** Write a two-column proof. (Lesson 2-6) See margin.

Given: $AC = DF$
$AB = DE$

Prove: $BC = EF$

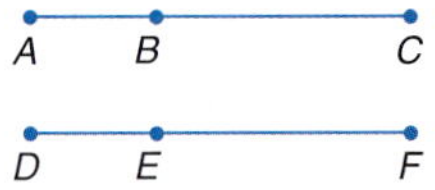

Find the perimeter of $\triangle ABC$ to the nearest hundredth, given the coordinates of its vertices. (Lesson 1-6)

66. $A(10, -6)$, $B(-2, -8)$, $C(-5, -7)$ 30.36
67. $A(-3, 2)$, $B(2, -9)$, $C(0, -10)$ 26.69

Getting Ready for the Next Lesson

PREREQUISITE SKILL **In the figure at the right, lines s and t are intersected by the transversal m. Name the pairs of angles that meet each description.**
*(To review **angles formed by two lines and a transversal**, see Lesson 3-1.)*

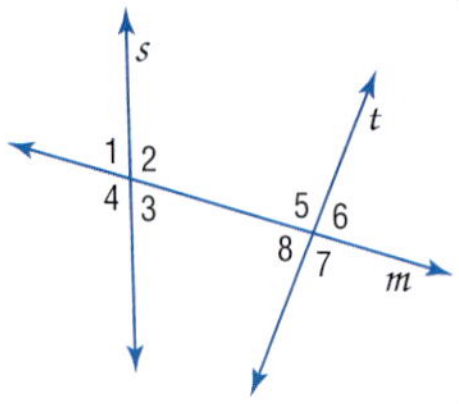

68. consecutive interior angles $\angle 2$ and $\angle 5$, $\angle 3$ and $\angle 8$
69. corresponding angles
70. alternate exterior angles $\angle 1$ and $\angle 7$, $\angle 4$ and $\angle 6$
71. alternate interior angles $\angle 2$ and $\angle 8$, $\angle 3$ and $\angle 5$

69. $\angle 1$ and $\angle 5$, $\angle 2$ and $\angle 6$, $\angle 4$ and $\angle 8$, $\angle 3$ and $\angle 7$

Practice Quiz 2 — Lessons 3-3 and 3-4

Determine whether $\overleftrightarrow{AB}$ and $\overleftrightarrow{CD}$ are *parallel*, *perpendicular*, or *neither*. (Lesson 3-3)

1. $A(3, -1)$, $B(6, 1)$, $C(-2, -2)$, $D(2, 4)$ neither
2. $A(-3, -11)$, $B(3, 13)$, $C(0, -6)$, $D(8, -8)$ perpendicular

For Exercises 3–8, refer to the graph at the right. Find the slope of each line. (Lesson 3-3)

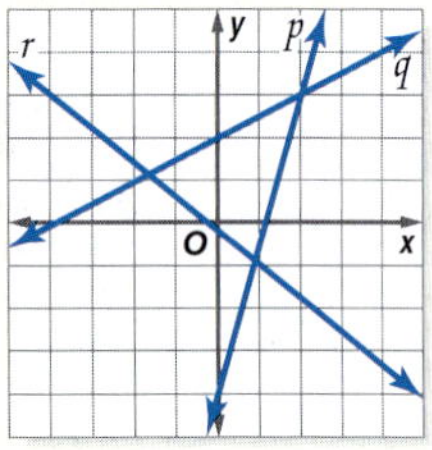

3. p $\frac{7}{2}$
4. a line parallel to q $\frac{1}{2}$
5. a line perpendicular to r $\frac{5}{4}$

Write an equation in slope-intercept form for each line. (Lesson 3-4)

6. q $y = \frac{1}{2}x + 2$
7. parallel to r, contains $(-1, 4)$ $y = -\frac{4}{5}x + \frac{16}{5}$
8. perpendicular to p, contains $(0, 0)$ $y = -\frac{2}{7}x$

Write an equation in point-slope form for the line that satisfies the given condition. (Lesson 3-4)

9. parallel to $y = -\frac{1}{4}x + 2$, contains $(5, -8)$ $y + 8 = -\frac{1}{4}(x - 5)$
10. perpendicular to $y = -3$, contains $(-4, -4)$ $0 = x + 4$

150 Chapter 3 Parallel and Perpendicular Lines

3-5 Proving Lines Parallel

Sunshine State Standards
MA.C.3.4.2

What You'll Learn

- Recognize angle conditions that occur with parallel lines.
- Prove that two lines are parallel based on given angle relationships.

How do you know that the sides of a parking space are parallel?

Have you ever been in a tall building and looked down at a parking lot? The parking lot is full of line segments that appear to be parallel. The workers who paint these lines must be certain that they are parallel.

IDENTIFY PARALLEL LINES When each stripe of a parking space intersects the center line, the angles formed are corresponding angles. If the lines are parallel, we know that the corresponding angles are congruent. Conversely, if the corresponding angles are congruent, then the lines must be parallel.

Postulate 3.4

If two lines in a plane are cut by a transversal so that corresponding angles are congruent, then the lines are parallel.

Abbreviation: *If corr. ∡s are ≅, then lines are ∥.*

Examples: If $\angle 1 \cong \angle 5$, $\angle 2 \cong \angle 6$, $\angle 3 \cong \angle 7$, or $\angle 4 \cong \angle 8$, then $m \parallel n$.

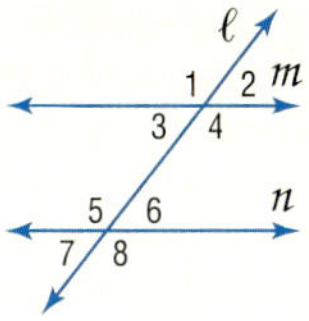

Postulate 3.4 justifies the construction of parallel lines.

Construction

Parallel Line Through a Point Not on Line

1. Use a straightedge to draw a line. Label two points on the line as *M* and *N*. Draw a point *P* that is not on $\overleftrightarrow{MN}$. Draw $\overleftrightarrow{PM}$.

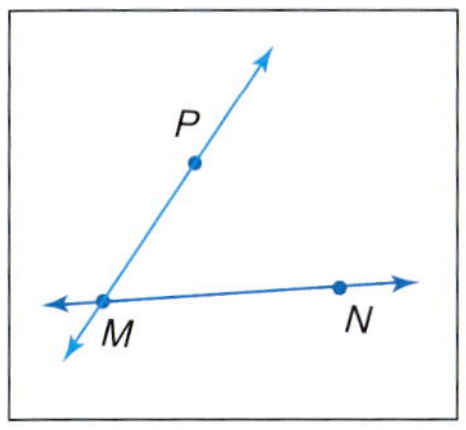

2. Copy $\angle PMN$ so that *P* is the vertex of the new angle. Label the intersection points *Q* and *R*.

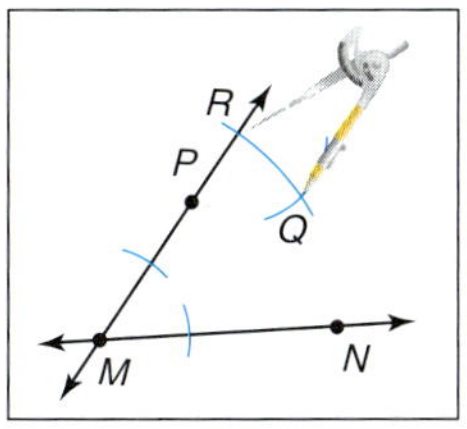

3. Draw $\overleftrightarrow{PQ}$. Because $\angle RPQ \cong \angle PMN$ by construction and they are corresponding angles, $\overleftrightarrow{PQ} \parallel \overleftrightarrow{MN}$.

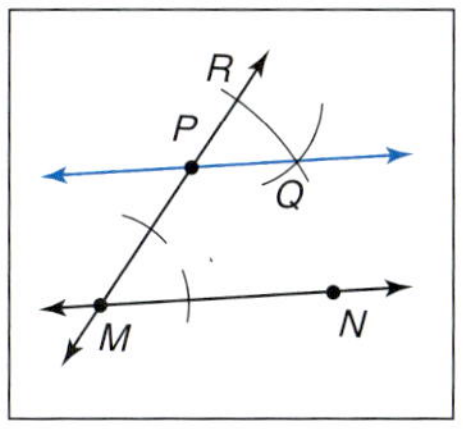

Study Tip

Look Back
To review **copying angles**, see Lesson 1-4.

3-5 Lesson Notes

1 Focus

5-Minute Check Transparency 3-5 Use as a quiz or review of Lesson 3-4.

Mathematical Background notes are available for this lesson on p. 124D.

How do you know that the sides of a parking space are parallel?

Ask students:

- When parallel parking, how do you know that the two lines marking the front and back of the parking space are parallel? **They are both perpendicular to the curb, so they are parallel to each other.**
- Why are there no-parking areas on each side of the lines of a parking space in a handicapped parking area? **The cars or vans need room for a wheelchair to enter or exit the side of the vehicle.**
- Are the lines for the sides of a parking space always perpendicular to the line for the front of the space? **No, when parking places are angled to the curb, the front is not perpendicular to the sides.**

Resource Manager

Workbook and Reproducible Masters

Chapter 3 Resource Masters
- Study Guide and Intervention, pp. 149–150
- Skills Practice, p. 151
- Practice, p. 152
- Reading to Learn Mathematics, p. 153
- Enrichment, p. 154
- Assessment, p. 176

Prerequisite Skills Workbook, pp. 3–4, 7–8

Teaching Geometry With Manipulatives Masters, pp. 8, 16, 57

Transparencies

5-Minute Check Transparency 3-5
Answer Key Transparencies

Technology

Interactive Chalkboard

2 Teach

IDENTIFY PARALLEL LINES

In-Class Examples

Power Point®

1 Determine which lines, if any, are parallel.

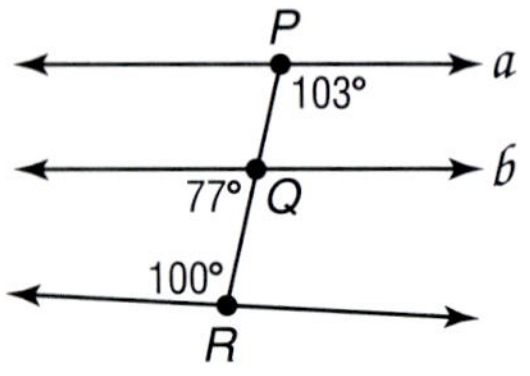

$a \parallel b$

2 **ALGEBRA** Find x and $m\angle ZYN$ so that $\overleftrightarrow{PQ} \parallel \overleftrightarrow{MN}$.

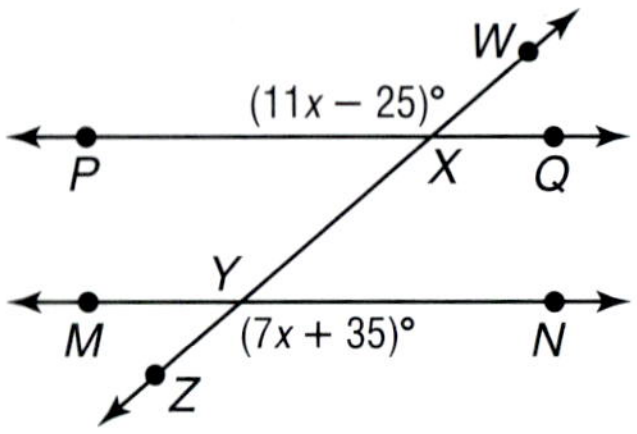

$x = 15$, $m\angle ZYN = 140$

The construction establishes that there is *at least* one line through P that is parallel to $\overleftrightarrow{MN}$. In 1795, Scottish physicist and mathematician John Playfair provided the modern version of Euclid's Parallel Postulate, which states there is *exactly* one line parallel to a line through a given point not on the line.

Postulate 3.5

Parallel Postulate If given a line and a point not on the line, then there exists exactly one line through the point that is parallel to the given line.

Parallel lines with a transversal create many pairs of congruent angles. Conversely, those pairs of congruent angles can determine whether a pair of lines is parallel.

Key Concept *Proving Lines Parallel*

	Theorems	Examples
3.5	If two lines in a plane are cut by a transversal so that a pair of alternate exterior angles is congruent, then the two lines are parallel. **Abbreviation:** *If alt. ext. ∡ are ≅, then lines are ∥.*	If $\angle 1 \cong \angle 8$ or if $\angle 2 \cong \angle 7$, then $m \parallel n$.
3.6	If two lines in a plane are cut by a transversal so that a pair of consecutive interior angles is supplementary, then the lines are parallel. **Abbreviation:** *If cons. int. ∡ are suppl., then lines are ∥.*	If $m\angle 3 + m\angle 5 = 180$ or if $m\angle 4 + m\angle 6 = 180$, then $m \parallel n$.
3.7	If two lines in a plane are cut by a transversal so that a pair of alternate interior angles is congruent, then the lines are parallel. **Abbreviation:** *If alt. int. ∡ are ≅, then lines are ∥.*	If $\angle 3 \cong \angle 6$ or if $\angle 4 \cong \angle 5$, then $m \parallel n$.
3.8	In a plane, if two lines are perpendicular to the same line, then they are parallel. **Abbreviation:** *If 2 lines are ⊥ to the same line, then lines are ∥.*	If $\ell \perp m$ and $\ell \perp n$, then $m \parallel n$.

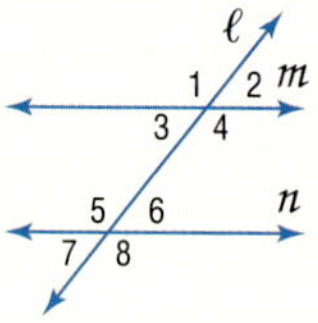

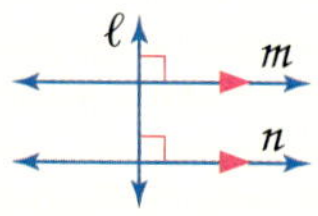

Example 1 *Identify Parallel Lines*

In the figure, $\overline{BG}$ bisects $\angle ABH$. Determine which lines, if any, are parallel.

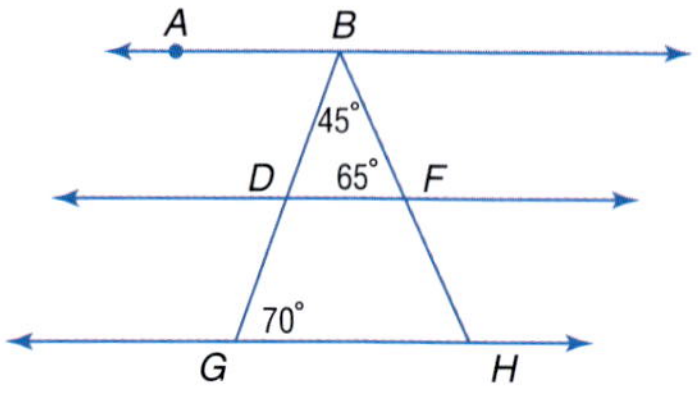

- The sum of the angle measures in a triangle must be 180, so $m\angle BDF = 180 - (45 + 65)$ or 70.
- Since $\angle BDF$ and $\angle BGH$ have the same measure, they are congruent.
- Congruent corresponding angles indicate parallel lines. So, $\overleftrightarrow{DF} \parallel \overleftrightarrow{GH}$.
- $\angle ABD \cong \angle DBF$, because $\overline{BG}$ bisects $\angle ABH$. So, $m\angle ABD = 45$.
- $\angle ABD$ and $\angle BDF$ are alternate interior angles, but they have different measures so they are not congruent.
- Thus, $\overleftrightarrow{AB}$ is not parallel to $\overleftrightarrow{DF}$ or $\overleftrightarrow{GH}$.

Teacher to Teacher

Cynthia W. Poché, Salmen High School Slidell, LA

I require "sticky notes." We bookmark important pages for easy reference. For example, we mark the page for proving lines parallel. We also label the top of the sticky note.

152 Chapter 3 Parallel and Perpendicular Lines

Angle relationships can be used to solve problems involving unknown values.

Example 2 Solve Problems with Parallel Lines

ALGEBRA **Find x and $m\angle RSU$ so that $m \parallel n$.**

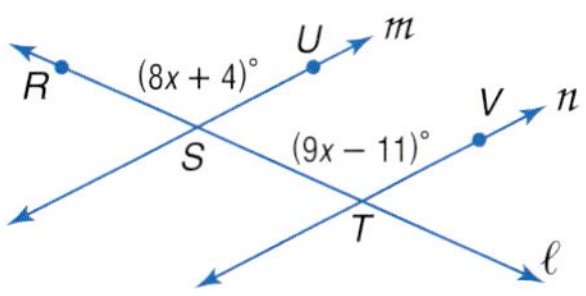

Explore From the figure, you know that $m\angle RSU = 8x + 4$ and $m\angle STV = 9x - 11$. You also know that $\angle RSU$ and $\angle STV$ are corresponding angles.

Plan For line m to be parallel to line n, the corresponding angles must be congruent. So, $m\angle RSU = m\angle STV$. Substitute the given angle measures into this equation and solve for x. Once you know the value of x, use substitution to find $m\angle RSU$.

Solve

$m\angle RSU = m\angle STV$	Corresponding angles
$8x + 4 = 9x - 11$	Substitution
$4 = x - 11$	Subtract 8x from each side.
$15 = x$	Add 11 to each side.

Now use the value of x to find $m\angle RSU$.

$m\angle RSU = 8x + 4$	Original equation
$= 8(15) + 4$	$x = 15$
$= 124$	Simplify.

Examine Verify the angle measure by using the value of x to find $m\angle STV$. That is, $9x - 11 = 9(15) - 11$ or 124. Since $m\angle RSU = m\angle STV$, $\angle RSU \cong \angle STV$ and $m \parallel n$.

PROVE LINES PARALLEL The angle pair relationships formed by a transversal can be used to prove that two lines are parallel.

Study Tip

Proving Lines Parallel

When proving lines parallel, be sure to check for congruent corresponding angles, alternate interior angles, alternate exterior angles, or supplementary consecutive interior angles.

Example 3 Prove Lines Parallel

Given: $r \parallel s$
$\angle 5 \cong \angle 6$

Prove: $\ell \parallel m$

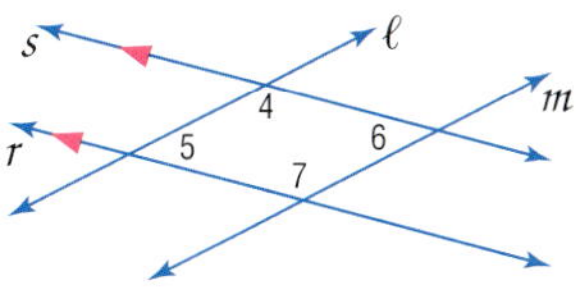

Proof:

Statements	Reasons
1. $r \parallel s$, $\angle 5 \cong \angle 6$	1. Given
2. $\angle 4$ and $\angle 5$ are supplementary.	2. Consecutive Interior Angle Theorem
3. $m\angle 4 + m\angle 5 = 180$	3. Definition of supplementary angles
4. $m\angle 5 = m\angle 6$	4. Definition of congruent angles
5. $m\angle 4 + m\angle 6 = 180$	5. Substitution Property (=)
6. $\angle 4$ and $\angle 6$ are supplementary.	6. Definition of supplementary angles
7. $\ell \parallel m$	7. If cons. int. ∡ are suppl., then lines are ∥.

www.geometryonline.com/extra_examples/fcat

PROVE LINES PARALLEL

In-Class Examples

Power Point®

3 Use the figure in Example 3 in the Student Edition.
Given: $\ell \parallel m$ and $\angle 4 \cong \angle 7$,
Prove: $r \parallel s$

Statements (Reasons)

1. $\ell \parallel m$, $\angle 4 \cong \angle 7$ (Given)
2. $\angle 4$ and $\angle 6$ are suppl. (Consec. Int. Angle Th.)
3. $m\angle 4 + m\angle 6 = 180$ (Def. of suppl. ∡)
4. $m\angle 4 = m\angle 7$ (Def. of ≅ ∡)
5. $m\angle 7 + m\angle 6 = 180$ (Subst.)
6. $\angle 7$ and $\angle 6$ are suppl. (Def. of suppl. ∡)
7. $r \parallel s$ (If cons. int. ∡ are suppl., then lines are ∥.)

Teaching Tip Have students confirm that the slopes of g and f would not change if they use the ordered pairs in reverse order.

slope of f: $m = \frac{0-4}{3-6}$ or $\frac{4}{3}$

slope of g: $m = \frac{0-4}{-3-0}$ or $\frac{4}{3}$

4 Determine whether $p \parallel q$.

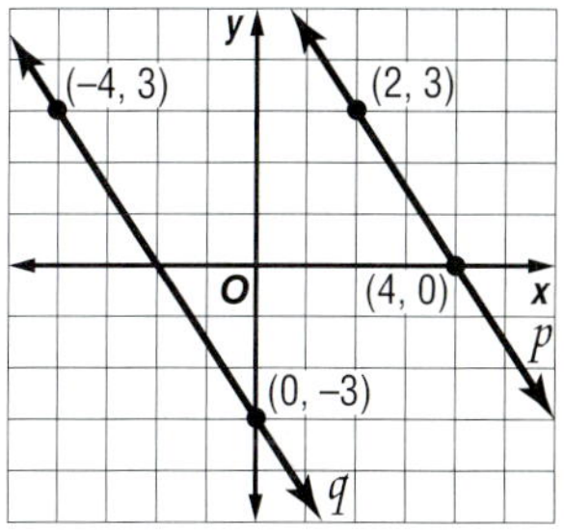

slope of p: $m = \frac{3-0}{2-4}$ or $-\frac{3}{2}$

slope of q: $m = \frac{3-(-3)}{-4-0}$ or $-\frac{3}{2}$

Since the slopes are equal, $p \parallel q$.

Tips for New Teachers

Intervention Many students think at first that the postulate and theorems in this lesson are the same as those in Lesson 3-2. Help them focus on the difference that in this lesson they are *concluding* that lines are parallel (the *then* clause) while in Lesson 3-2 they were *starting* with parallel lines (the *if* clause).

DAILY INTERVENTION

Differentiated Instruction

Logical Ask students to compare the theorems and postulates in this lesson to those in Lesson 3-2. Then ask them to explain any connections in logic that they find. Have them report their findings to the class.

3 Practice/Apply

Study Notebook

Have students—
- add the definitions/examples of the vocabulary terms to their Vocabulary Builder worksheets for Chapter 3.
- include the steps for constructing a line parallel to a given line through a point not on the line, as well as the conditions under which you can conclude that two lines are parallel.
- include any other item(s) that they find helpful in mastering the skills in this lesson.

About the Exercises...

Organization by Objective
- **Identify Parallel Lines:** 13–24, 26–31
- **Prove Lines Parallel:** 25, 32–39

Odd/Even Assignments
Exercises 13–24 and 26–31 are structured so that students practice the same concepts whether they are assigned odd or even problems.

Alert! Exercise 43 requires the Internet or other research materials.

Assignment Guide
Basic: 13–31 odd, 39, 41, 44–64

Average: 13–41 odd, 42, 44–64

Advanced: 14–30 even, 32–42 even, 43–61 (optional: 62–64)

In Lesson 3-3, you learned that parallel lines have the same slope. You can use the slopes of lines to prove that lines are parallel.

Example 4 Slope and Parallel Lines

Determine whether $g \parallel f$.

slope of f: $m = \frac{4-0}{6-3}$ or $\frac{4}{3}$

slope of g: $m = \frac{4-0}{0-(-3)}$ or $\frac{4}{3}$

Since the slopes are the same, $g \parallel f$.

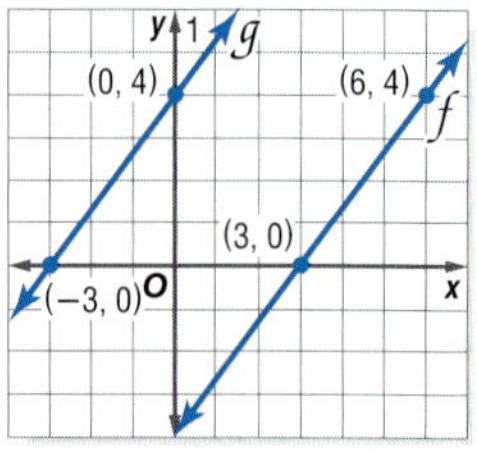

Check for Understanding

Concept Check

1–3. See margin.

1. **Summarize** five different methods to prove that two lines are parallel.
2. **Find a counterexample** for the following statement.
If lines ℓ and m are cut by transversal t so that consecutive interior angles are congruent, then lines ℓ and m are parallel and t is perpendicular to both lines.
3. **OPEN ENDED** Describe two situations in your own life in which you encounter parallel lines. How could you verify that the lines are parallel?

Guided Practice

GUIDED PRACTICE KEY	
Exercises	Examples
4–7	1
8, 9	2
10, 12	3
11	4

5. $\ell \parallel m$; alt. int. ∡

Given the following information, determine which lines, if any, are parallel. State the postulate or theorem that justifies your answer.

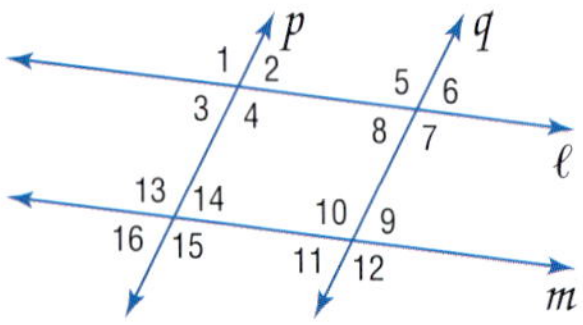

4. $\angle 16 \cong \angle 3$ $\ell \parallel m$; corr. ∡
5. $\angle 4 \cong \angle 13$
6. $m\angle 14 + m\angle 10 = 180$ $p \parallel q$; cons. int. ∡
7. $\angle 1 \cong \angle 7$ $p \parallel q$; alt. ext. ∡

Find x so that $\ell \parallel m$.

8. 9

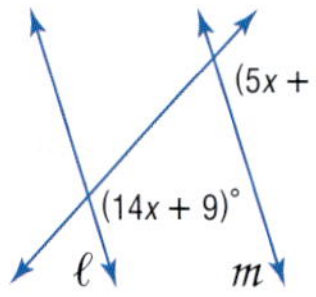

9. 11.375

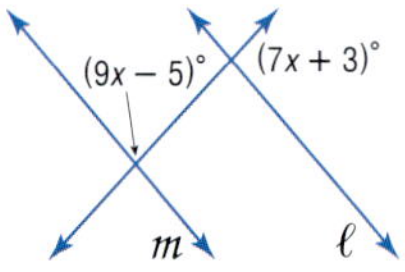

10. **PROOF** Write a two-column proof of Theorem 3.5. See p. 173A.

11. Determine whether $p \parallel q$.
The slope of $\overleftrightarrow{CD}$ is $\frac{1}{8}$, and the slope of $\overleftrightarrow{AB}$ is $\frac{1}{7}$. The slopes are not equal, so the lines are not parallel.

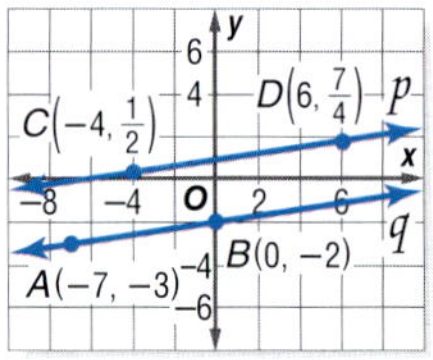

Application

12. **PHYSICS** The Hubble Telescope gathers parallel light rays and directs them to a central focal point. Use a protractor to measure several of the angles shown in the diagram. Are the lines parallel? Explain how you know.
Yes; sample answer: Pairs of alternate interior angles are congruent.

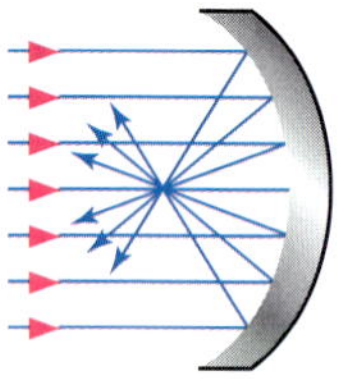

Answers

1. Sample answer: Use a pair of alt. ext. ∡ that are congruent and cut by transversal; show that a pair of consecutive interior angles are suppl.; show that alt. int. ∡ are ≅; show two lines are ⊥ to same line; show corresponding ∡ are ≅.

2. Sample answer:

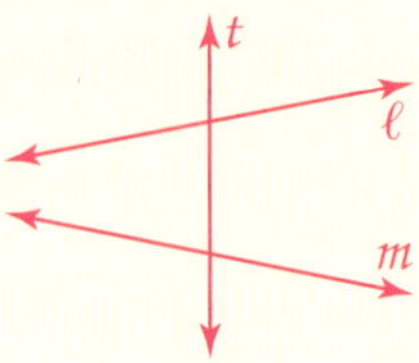

3. Sample answer: A basketball court has parallel lines, as does a newspaper. The edges should be equidistant along the entire line.

Practice and Apply

Homework Help

For Exercises	See Examples
13–24	1
26–31	2
25, 32–37	3
38–39	4

Extra Practice
See page 760.

Justifications:
18. ≅ corr. ∡
19. ≅ alt. int. ∡
22. ≅ alt. int. ∡
23. suppl. consec. int. ∡

Given the following information, determine which lines, if any, are parallel. State the postulate or theorem that justifies your answer.

13. $\angle 2 \cong \angle 8$ $a \parallel b$; ≅ alt. int. ∡
14. $\angle 9 \cong \angle 16$ none
15. $\angle 2 \cong \angle 10$ $\ell \parallel m$; ≅ corr. ∡
16. $\angle 6 \cong \angle 15$ none

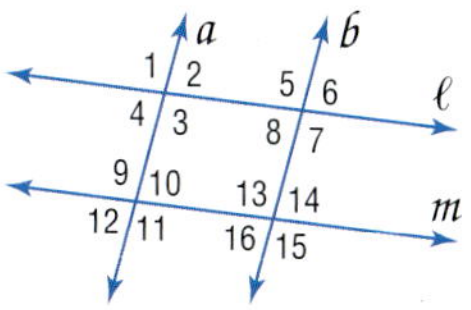

17. $\angle AEF \cong \angle BFG$ $\overleftrightarrow{AE} \parallel \overleftrightarrow{BF}$; ≅ corr. ∡
18. $\angle EAB \cong \angle DBC$ $\overleftrightarrow{AE} \parallel \overleftrightarrow{BF}$
19. $\angle EFB \cong \angle CBF$ $\overleftrightarrow{AC} \parallel \overleftrightarrow{EG}$
20. $m\angle GFD + m\angle CBD = 180$ $\overleftrightarrow{AC} \parallel \overleftrightarrow{EG}$; suppl. consec. int. ∡

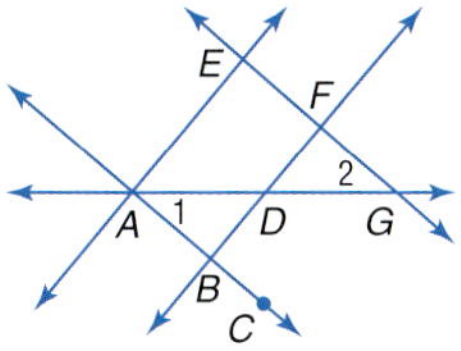

21. $\angle HLK \cong \angle JML$ $\overleftrightarrow{HS} \parallel \overleftrightarrow{JT}$; ≅ corr. ∡
22. $\angle PLQ \cong \angle MQL$ $\overleftrightarrow{HS} \parallel \overleftrightarrow{JT}$
23. $m\angle MLP + \angle RPL = 180$ $\overleftrightarrow{KN} \parallel \overleftrightarrow{PR}$
24. $\overleftrightarrow{HS} \perp \overleftrightarrow{PR}, \overleftrightarrow{JT} \perp \overleftrightarrow{PR}$ $\overleftrightarrow{HS} \parallel \overleftrightarrow{JT}$; 2 lines ⊥ the same line

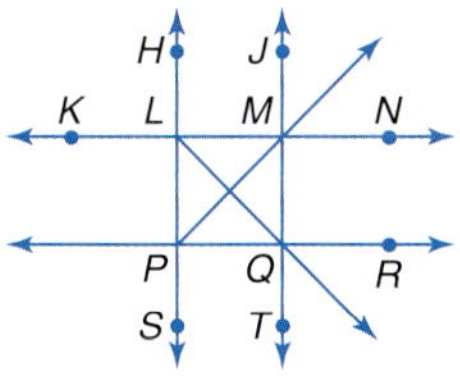

WebQuest

Latitude lines are parallel, and longitude lines appear parallel in certain locations on Earth. Visit www.geometryonline.com/webquest to continue work on your WebQuest project.

25. **PROOF** Copy and complete the proof of Theorem 3.8.

Given: $\ell \perp t$
$m \perp t$

Prove: $\ell \parallel m$

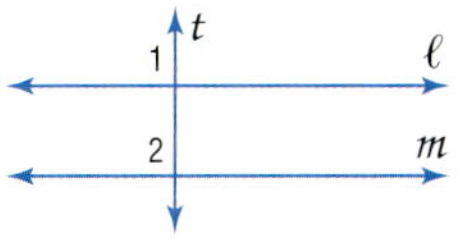

Proof:

Statements	Reasons
1. $\ell \perp t$, $m \perp t$	1. ___?___ Given
2. $\angle 1$ and $\angle 2$ are right angles.	2. ___?___ Definition of perpendicular
3. $\angle 1 \cong \angle 2$	3. ___?___ All rt. ∡ are ≅.
4. $\ell \parallel m$	4. ___?___ If corr. ∡ are ≅, then lines are ∥.

Find x so that $\ell \parallel m$.

26. 16

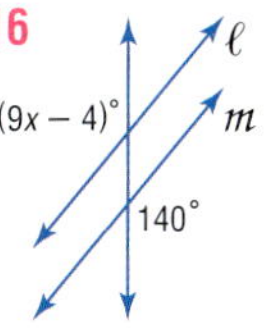

27. 15

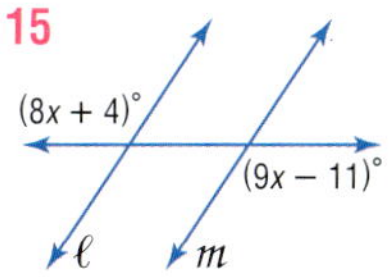

28. 13

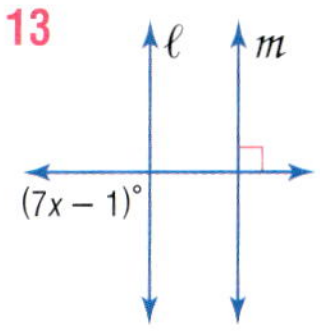

29. –8

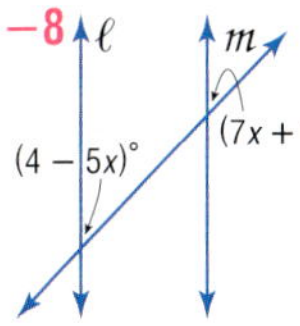

30. 9

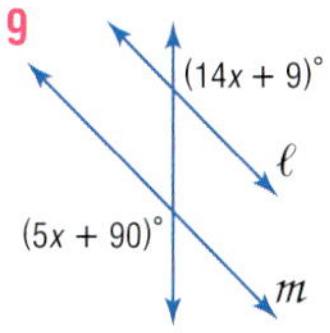

31. 21.6

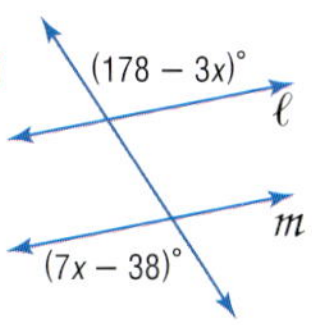

32. **PROOF** Write a two-column proof of Theorem 3.6. See p. 173A.
33. **PROOF** Write a paragraph proof of Theorem 3.7. See p. 173A.

Enrichment, p. 154

Scrambled-Up Proof

The reasons necessary to complete the following proof are scrambled up below. To complete the proof, number the reasons to match the corresponding statements.

Given: $\overline{CD} \perp \overline{BE}$
$\overline{AB} \perp \overline{BE}$
$\overline{AD} \cong \overline{CE}$
$\overline{BD} \cong \overline{DE}$

Prove: $\overline{AD} \parallel \overline{CE}$

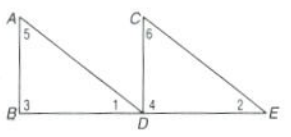

Proof:

Statements	Reasons
1. $\overline{CD} \perp \overline{BE}$	Definition of Right Triangle 4

Study Guide and Intervention, p. 149 (shown) and p. 150

Identify Parallel Lines If two lines in a plane are cut by a transversal and certain conditions are met, then the lines must be parallel.

If	then
• corresponding angles are congruent, • alternate exterior angles are congruent, • consecutive interior angles are supplementary, • alternate interior angles are congruent, or • two lines are perpendicular to the same line,	the lines are parallel.

Example 1 **If $m\angle 1 = m\angle 2$, determine which lines, if any, are parallel.**

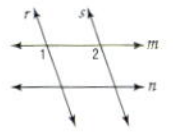

Since $m\angle 1 = m\angle 2$, then $\angle 1 \cong \angle 2$. $\angle 1$ and $\angle 2$ are congruent corresponding angles, so $r \parallel s$.

Example 2 **Find x and $m\angle ABC$ so that $m \parallel n$.**

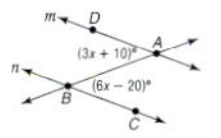

We can conclude that $m \parallel n$ if alternate interior angles are congruent.

$$m\angle DAB = m\angle CDA$$
$$3x + 10 = 6x - 20$$
$$10 = 3x - 20$$
$$30 = 3x$$
$$10 = x$$
$$m\angle ABC = 6x - 20$$
$$= 6(10) - 20 \text{ or } 40$$

Exercises

Find x so that $\ell \parallel m$.

1. 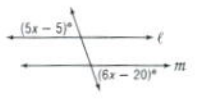

15

2. 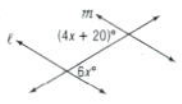

10

3. 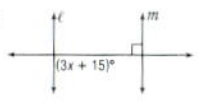

25

4. 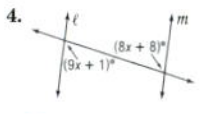

7

5. 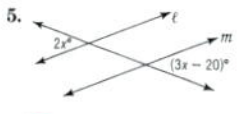

20

6. 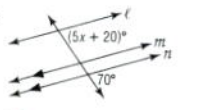

10

Skills Practice, p. 151 and Practice, p. 152 (shown)

Given the following information, determine which lines, if any, are parallel. State the postulate or theorem that justifies your answer.

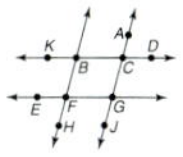

1. $m\angle BCG + m\angle FGC = 180$ $\overleftrightarrow{BD} \parallel \overleftrightarrow{EG}$; cons. int. ∡
2. $\angle CBF \cong \angle GFH$ $\overleftrightarrow{BD} \parallel \overleftrightarrow{EG}$; corr. ∡
3. $\angle EFB \cong \angle FBC$ $\overleftrightarrow{BD} \parallel \overleftrightarrow{EG}$; alt. int. ∡
4. $\angle ACD \cong \angle KBF$ $\overleftrightarrow{AJ} \parallel \overleftrightarrow{BH}$; alt. ext. ∡

Find x so that $\ell \parallel m$.

5. 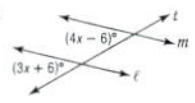

12

6. 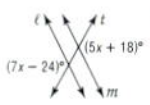

21

7. 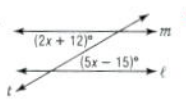

9

8. **PROOF** Write a two-column proof.
Given: $\angle 2$ and $\angle 3$ are supplementary.
Prove: $\overline{AB} \parallel \overline{CD}$

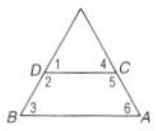

Proof:

Statements	Reasons
1. $\angle 2$ and $\angle 3$ are supplementary.	1. Given
2. $\overleftrightarrow{AB} \parallel \overleftrightarrow{CD}$	2. If consec. int ∡ are suppl., then lines are ∥.
3. $\overline{AB} \parallel \overline{CD}$	3. Segments contained in ∥ lines are ∥.

9. **LANDSCAPING** The head gardener at a botanical garden wants to plant rosebushes in parallel rows on either side of an existing footpath. How can the gardener ensure that the rows are parallel?
Sample answer: If the gardener digs each row at a 90° angle to the footpath, each row will be perpendicular to the footpath. If each of the rows is perpendicular to the footpath, then the rows are parallel.

Reading to Learn Mathematics, p. 153 ELL

Pre-Activity **How do you know that the sides of a parking space are parallel?**

Read the introduction to Lesson 3-5 at the top of page 151 in your textbook.

How can the workers who are striping the parking spaces in a parking lot check to see if the sides of the spaces are parallel? Sample answer: Use a T-square or other device for forming right angles to lay out perpendicular segments cut from string or rope connecting the two sides of a parking space at both ends and in the middle. Measure the three perpendicular segments. If they are all the same length, the sides of the parking space are parallel.

Reading the Lesson

1. Choose the word or phrase that best completes each sentence.
   a. If two coplanar lines are cut by a transversal so that corresponding angles are congruent, then the lines are ___parallel___ (parallel/perpendicular/skew).
   b. In a plane, if two lines are perpendicular to the same line, then they are ___parallel___ (perpendicular/parallel/skew).
   c. For a line and a point not on the line, there exists ___exactly one___ (at least one/exactly one/at most one) line through the point that is parallel to the given line.
   d. If two coplanar lines are cut by a transversal so that consecutive interior angles are ___supplementary___ (complementary/supplementary/congruent), then the lines are parallel.
   e. If two coplanar lines are cut by a transversal so that alternate interior angles are congruent, then the lines are ___parallel___ (perpendicular/parallel/skew).
2. Which of the following conditions verify that $p \parallel q$? A, C, F, G

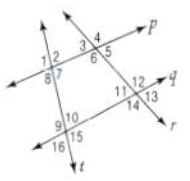

   A. $\angle 6 \cong \angle 12$ B. $\angle 2 \cong \angle 4$
   C. $\angle 8 \cong \angle 16$ D. $\angle 11 \cong \angle 13$
   E. $\angle 6$ and $\angle 7$ are supplementary. F. $\angle 1 \cong \angle 15$
   G. $\angle 7$ and $\angle 10$ are supplementary. H. $\angle 4 \cong \angle 16$

Helping You Remember

3. A good way to remember something new is to draw a picture. How can a sketch help you to remember the Parallel Postulate?
Sample answer: Draw a line with a ruler or straightedge and choose a point not on the line. Try to draw lines through the point that are parallel to the line you originally drew. You will see that there is exactly one way to do this.

Answers

34. Given: $\angle 2 \cong \angle 1$, $\angle 1 \cong \angle 3$
Prove: $\overline{ST} \parallel \overline{UV}$
Proof:
Statements (Reasons)
1. $\angle 2 \cong \angle 1$, $\angle 1 \cong \angle 3$ (Given)
2. $\angle 2 \cong \angle 3$ (Trans. Prop.)
3. $\overline{ST} \parallel \overline{UV}$ (If alt. int. ∠s are $\cong$, lines are $\parallel$.)

35. Given: $\overline{AD} \perp \overline{CD}$, $\angle 1 \cong \angle 2$
Prove: $\overline{BC} \perp \overline{CD}$
Proof:
Statements (Reasons)
1. $\overline{AD} \perp \overline{CD}$, $\angle 1 \cong \angle 2$ (Given)
2. $\overline{AD} \parallel \overline{BC}$ (If alt. int. ∠s are $\cong$, lines are $\parallel$.)
3. $\overline{BC} \perp \overline{CD}$ (Perpendicular Transversal Theorem)

36. Given: $\overline{JM} \parallel \overline{KN}$, $\angle 1 \cong \angle 2$, $\angle 3 \cong \angle 4$
Prove: $\overline{KM} \parallel \overline{LN}$
Proof:
Statements (Reasons)
1. $\overline{JM} \parallel \overline{KN}$, $\angle 1 \cong \angle 2$, $\angle 3 \cong \angle 4$ (Given)
2. $\angle 1 \cong \angle 3$ (If lines are $\parallel$, corr. ∠s are $\cong$.)
3. $\angle 2 \cong \angle 4$ (Substitution)
4. $\overline{KM} \parallel \overline{LN}$ (If corr. ∠s are $\cong$, lines are $\parallel$.)

37. Given: $\angle RSP \cong \angle PQR$, $\angle QRS$ and $\angle PQR$ are supplementary.
Prove: $\overline{PS} \parallel \overline{QR}$
Proof:
Statements (Reasons)
1. $\angle RSP \cong \angle PQR$, $\angle QRS$ and $\angle PQR$ are suppl. (Given)
2. $m\angle RSP = m\angle PQR$ (Def. of $\cong$ ∠s)
3. $m\angle QRS + m\angle PQR = 180$ (Definition of suppl. ∠s)
4. $m\angle QRS + m\angle RSP = 180$ (Substitution)
5. $\angle QRS$ and $\angle RSP$ are suppl. (Def. of suppl. ∠s)
6. $\overline{PS} \parallel \overline{QR}$ (If cons. int. ∠s are suppl., lines ∠s are $\parallel$.)

40. When he measures the angle that each picket makes with the 2 by 4, he is measuring corresponding angles. When all of the corresponding angles are congruent, the pickets must be parallel.

41. The 10-yard lines will be parallel because they are all perpendicular to the sideline and two or more lines perpendicular to the same line are parallel.

More About. . .

John Playfair
In 1795, John Playfair published his version of Euclid's *Elements*. In his edition, Playfair standardized the notation used for points and figures and introduced algebraic notation for use in proofs.
Source: mathworld.wolfram.com

PROOF Write a two-column proof for each of the following. 34–37. See margin.

34. Given: $\angle 2 \cong \angle 1$
$\angle 1 \cong \angle 3$
Prove: $\overline{ST} \parallel \overline{UV}$

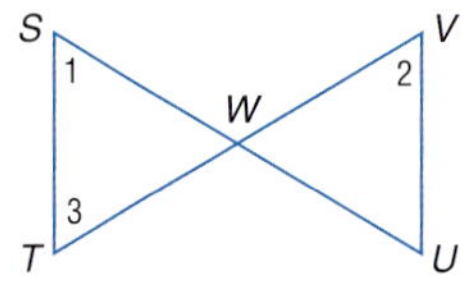

35. Given: $\overline{AD} \perp \overline{CD}$
$\angle 1 \cong \angle 2$
Prove: $\overline{BC} \perp \overline{CD}$

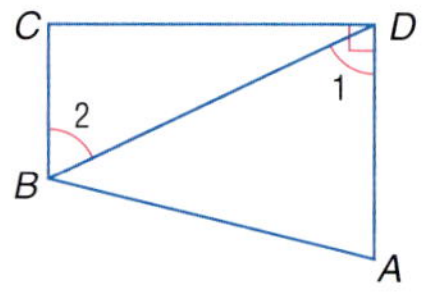

36. Given: $\overline{JM} \parallel \overline{KN}$
$\angle 1 \cong \angle 2$
$\angle 3 \cong \angle 4$
Prove: $\overline{KM} \parallel \overline{LN}$

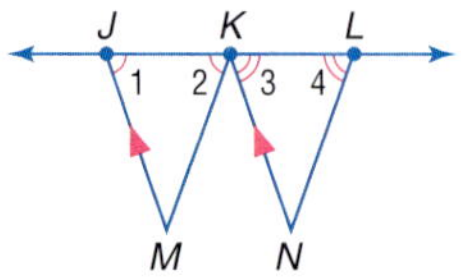

37. Given: $\angle RSP \cong \angle PQR$
$\angle QRS$ and $\angle PQR$ are supplementary.
Prove: $\overline{PS} \parallel \overline{QR}$

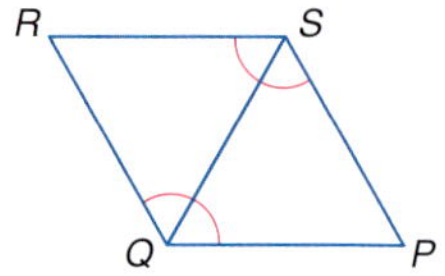

Determine whether each pair of lines is parallel. Explain why or why not.

38.

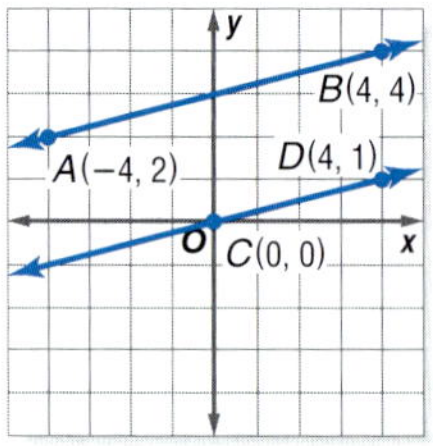

Yes; the slopes are the same.

39.

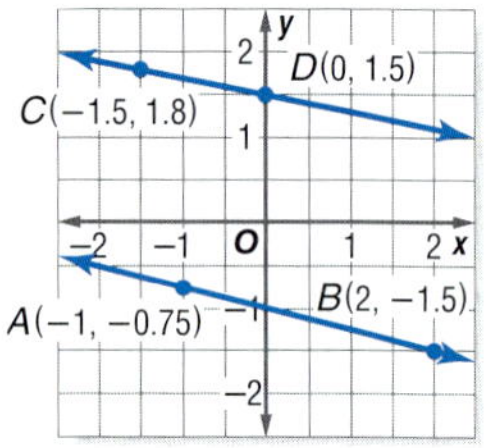

No; the slopes are not the same.

40. **HOME IMPROVEMENT** To build a fence, Jim positioned the fence posts and then placed a 2×4 board at an angle between the fence posts. As he placed each picket, he measured the angle that the picket made with the 2×4. Why does this ensure that the pickets will be parallel?

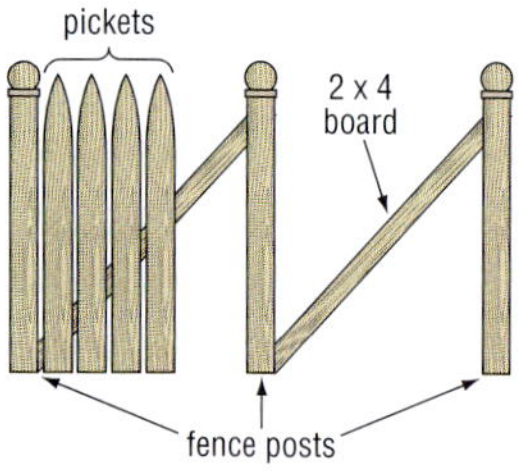

41. **FOOTBALL** When striping the practice football field, Mr. Hawkinson first painted the sidelines. Next he marked off 10-yard increments on one sideline. He then constructed lines perpendicular to the sidelines at each 10-yard mark. Why does this guarantee that the 10-yard lines will be parallel?

42. **CRITICAL THINKING** When Adeel was working on an art project, he drew a four-sided figure with two pairs of opposite parallel sides. He noticed some patterns relating to the angles in the figure. List as many patterns as you can about a 4-sided figure with two pairs of opposite parallel sides. See margin.

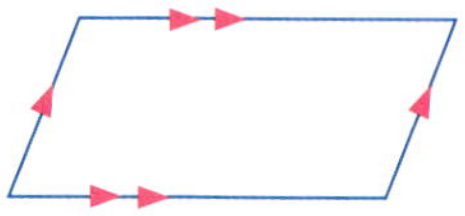

43. **RESEARCH** Use the Internet or other resource to find mathematicians like John Playfair who discovered new concepts and proved new theorems related to parallel lines. Briefly describe their discoveries. See students' work.

44. WRITING IN MATH Answer the question that was posed at the beginning of the lesson. **See margin.**

How do you know that the sides of a parking space are parallel?

Include the following in your answer:

- a comparison of the angles at which the lines forming the edges of a parking space strike the centerline, and
- a description of the type of parking spaces that form congruent consecutive interior angles.

FCAT Practice

Standardized Test Practice Ⓐ Ⓑ Ⓒ Ⓓ

45. In the figure, line ℓ is parallel to line m. Line n intersects both ℓ and m. Which of the following lists includes all of the angles that are supplementary to $\angle 1$? **B**

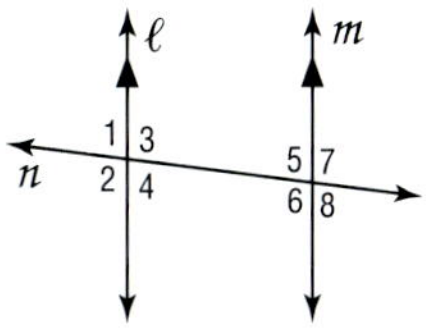

Ⓐ angles 2, 3, and 4 Ⓑ angles 2, 3, 6, and 7

Ⓒ angles 4, 5, and 8 Ⓓ angles 3, 4, 7, and 8

46. **ALGEBRA** Kendra has at least one quarter, one dime, one nickel, and one penny. If she has three times as many pennies as nickels, the same number of nickels as dimes, and twice as many dimes as quarters, then what is the least amount of money she could have? **D**

Ⓐ \$0.41 Ⓑ \$0.48 Ⓒ \$0.58 Ⓓ \$0.61

Maintain Your Skills

Mixed Review **Write an equation in slope-intercept form for the line that satisfies the given conditions.** *(Lesson 3-4)*

47. $m = 0.3$, y-intercept is -6 $y = 0.3x - 6$

48. $m = \frac{1}{3}$, contains $(-3, -15)$ $y = \frac{1}{3}x - 14$

49. contains $(5, 7)$ and $(-3, 11)$ $y = -\frac{1}{2}x + \frac{19}{2}$

50. perpendicular to $y = \frac{1}{2}x - 4$, contains $(4, 1)$ $y = -2x + 9$

Find the slope of each line. *(Lesson 3-3)*

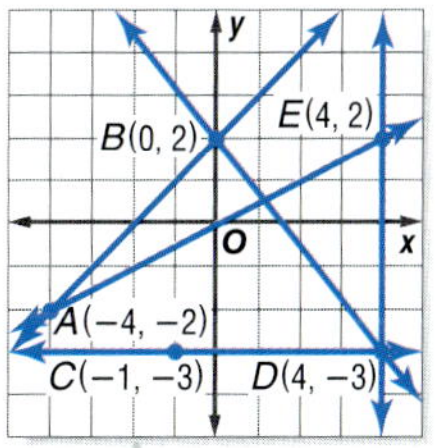

51. $\overleftrightarrow{BD}$ $-\frac{5}{4}$

52. $\overleftrightarrow{CD}$ 0

53. $\overleftrightarrow{AB}$ 1

54. $\overleftrightarrow{EO}$ $\frac{1}{2}$

55. any line parallel to $\overleftrightarrow{DE}$ **undefined**

56. any line perpendicular to $\overleftrightarrow{BD}$ $\frac{4}{5}$

57–60. See margin.

Construct a truth table for each compound statement. *(Lesson 2-2)*

57. p and q

58. p or $\sim q$

59. $\sim p \wedge q$

60. $\sim p \wedge \sim q$

61. **CARPENTRY** A carpenter must cut two pieces of wood at angles so that they fit together to form the corner of a picture frame. What type of angles must he use to make sure that a corner results? *(Lesson 1-5)* **complementary angles**

Getting Ready for the Next Lesson **PREREQUISITE SKILL** **Use the Distance Formula to find the distance between each pair of points.** *(To review* ***the Distance Formula****, see Lesson 1-4.)*

62. $(2, 7), (7, 19)$ **13**

63. $(8, 0), (-1, 2)$ $\sqrt{85} \approx 9.22$

64. $(-6, -4), (-8, -2)$ $\sqrt{8} \approx 2.83$

www.geometryonline.com/self_check_quiz/fcat

4 Assess

Open-Ended Assessment

Writing Have students make a poster to show the step-by-step procedure for constructing a line parallel to a given line through an external point, or illustrating the techniques learned so far for proving lines parallel.

Getting Ready for Lesson 3-6

Prerequisite Skill Students will learn about the distance from a point to a line using the Distance Formula in Lesson 3-6. Use Exercises 62–64 to determine your students' familiarity with the Distance Formula.

Assessment Options

Quiz (Lessons 3-4 and 3-5) is available on p. 176 of the *Chapter 3 Resource Masters.*

Answers

57.

p	q	p and q
T	T	T
T	F	F
F	T	F
F	F	F

58.

p	q	$\sim q$	p or $\sim q$
T	T	F	T
T	F	T	T
F	T	F	F
F	F	T	T

59.

p	q	$\sim p$	$\sim p \wedge q$
T	T	F	F
T	F	F	F
F	T	T	T
F	F	T	F

60.

p	q	$\sim p$	$\sim q$	$\sim p \wedge \sim q$
T	T	F	F	F
T	F	F	T	F
F	T	T	F	F
F	F	T	T	T

Answers

42. Consecutive angles are supplementary; opposite angles are congruent; the sum of the measures of the angles is 360.

44. Sample answer: They should appear to have the same slope. Answers should include the following.

- The corresponding angles must be equal in order for the lines to be parallel.
- The parking lot spaces have right angles.

Graphing Calculator Investigation

A Preview of Lesson 3-6

Getting Started

To find the points where a transversal t intersects lines a and b, the steps are to solve two linear systems, one comprised of the equations for lines a and t and the other system comprised of the equations for lines b and t.

Teach

- In order to check their answers, students must check the first ordered pair in the equations for lines a and t and the second ordered pair in the equations for lines b and t.
- Emphasize that unless the solutions are rational numbers, the intersection points the calculator finds are approximations and not exact.

Assess

Encourage students to use the TRACE feature as an additional check for their ordered pairs.

Graphing Calculator Investigation

A Preview of Lesson 3-6

Points of Intersection

You can use a TI-83 Plus graphing calculator to determine the points of intersection of a transversal and two parallel lines.

Example

Parallel lines ℓ and m are cut by a transversal t. The equations of ℓ, m, and t are $y = \frac{1}{2}x - 4$, $y = \frac{1}{2}x + 6$, and $y = -2x + 1$, respectively. Use a graphing calculator to determine the points of intersection of t with ℓ and m.

Step 1 Enter the equations in the Y= list and graph in the standard viewing window.

KEYSTROKES: [Y=] 1 [÷] 2 [X,T,θ,n] [–] 4 [ENTER] 1 [÷] 2 [X,T,θ,n] [+] 6 [ENTER] [(–)] 2 [X,T,θ,n] [+] 1 [ZOOM] 6

Step 2 Use the CALC menu to find the points of intersection.

- Find the intersection of ℓ and t.

  KEYSTROKES: [2nd] [CALC] 5 [ENTER] [▼] [ENTER] [ENTER]

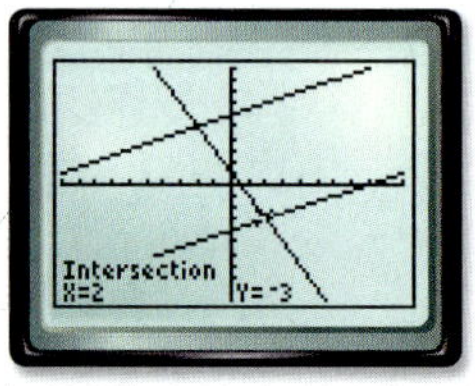

[–10, 10] scl: 1 by [–10, 10] scl: 1

Lines ℓ and t intersect at (2, –3).

- Find the intersection of m and t.

  KEYSTROKES: [2nd] [CALC] 5 [▼] [ENTER] [ENTER] [ENTER]

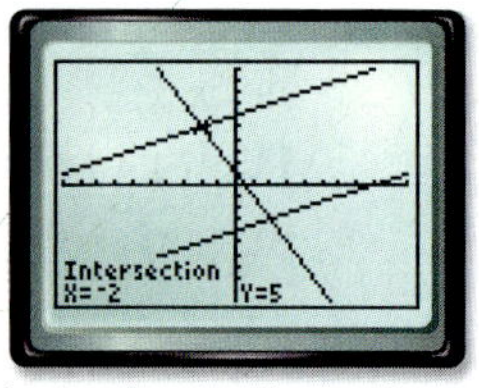

[–10, 10] scl: 1 by [–10, 10] scl: 1

Lines m and t intersect at (–2, 5).

Exercises

Parallel lines a and b are cut by a transversal t. Use a graphing calculator to determine the points of intersection of t with a and b. Round to the nearest tenth. 5. (1.0, –1.2), (3.4, –4.3)

1. a: $y = 2x - 10$
b: $y = 2x - 2$
t: $y = -\frac{1}{2}x + 4$
(5.6, 1.2), (2.4, 2.8)

2. a: $y = -x - 3$
b: $y = -x + 5$
t: $y = x - 6$
(1.5, –4.5), (5.5, –0.5)

3. a: $y = 6$
b: $y = 0$
t: $x = -2$ (–2, 6), (–2, 0)

4. a: $y = -3x + 1$
b: $y = -3x - 3$
t: $y = \frac{1}{3}x + 8$
(–2.1, 7.3), (–3.3, 6.9)

5. a: $y = \frac{4}{5}x - 2$
b: $y = \frac{4}{5}x - 7$
t: $y = -\frac{5}{4}x$

6. a: $y = -\frac{1}{6}x + \frac{2}{3}$
b: $y = -\frac{1}{6}x + \frac{5}{12}$
t: $y = 6x + 2$
(–0.2, 0.7), (–0.3, 0.5)

www.geometryonline.com/other_calculator_keystrokes

3-6 Perpendiculars and Distance

What You'll Learn

- Find the distance between a point and a line.
- Find the distance between parallel lines.

Sunshine State Standards
MA.C.2.4.1

Vocabulary
- equidistant

How does the distance between parallel lines relate to hanging new shelves?

When preparing to install new shelf brackets, it is important that the vertical bracing be parallel in order for the shelves to line up appropriately. One technique for ensuring that the braces are parallel is to install the first brace and then use a carpenter's square to measure and mark two or more points the same distance from the first brace. You can then align the second brace with those marks.

DISTANCE FROM A POINT TO A LINE In Lesson 3-5, you learned that if two lines are perpendicular to the same line, then they are parallel. The carpenter's square is used to construct a line perpendicular to each pair of shelves. The space between each pair of shelves is measured along the perpendicular segment. This is to ensure that the shelves are parallel. This relationship is an example of using lines and perpendicular segments to determine distance. The shortest segment from a point to a line is the perpendicular segment from the point to the line.

Key Concept — *Distance Between a Point and a Line*

- **Words** The distance from a line to a point not on the line is the length of the segment perpendicular to the line from the point.
- **Model**

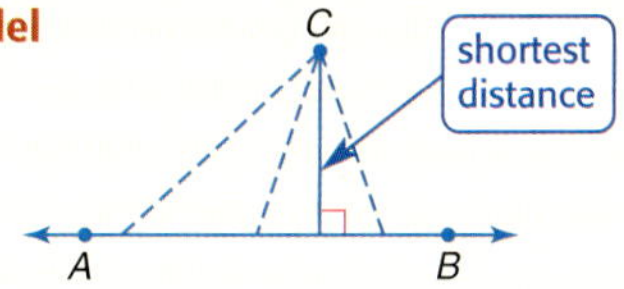

Study Tip

Measuring the Shortest Distance
You can use tools like the corner of a piece of paper or your book to help draw a right angle.

Example 1 *Distance from a Point to a Line*

Draw the segment that represents the distance from P to $\overleftrightarrow{AB}$.

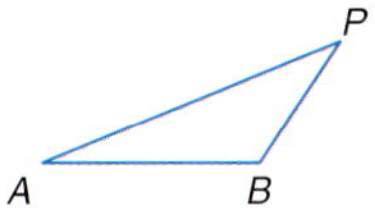

Since the distance from a line to a point not on the line is the length of the segment perpendicular to the line from the point, extend $\overline{AB}$ and draw $\overline{PQ}$ so that $\overline{PQ} \perp \overleftrightarrow{AB}$.

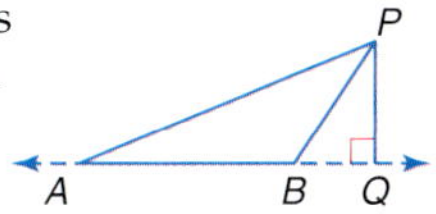

When you draw a perpendicular segment from a point to a line, you can guarantee that it is perpendicular by using the construction of a line perpendicular to a line through a point not on that line.

3-6 Lesson Notes

1 Focus

5-Minute Check Transparency 3-6 Use as a quiz or review of Lesson 3-5.

Mathematical Background notes are available for this lesson on p. 124D.

How does the distance between parallel lines relate to hanging new shelves?

Ask students:

- If the bracing is not lined up, what will be wrong with the installed shelf? **The shelf will not be level and items placed on the shelf might slide off.**
- What is the relationship between the distance between two parallel lines and the length of the shortest segment that connects the lines? **The distance between the two parallel lines is the same as the length of the shortest segment that connects the lines.**

Resource Manager

Workbook and Reproducible Masters

Chapter 3 Resource Masters
- Study Guide and Intervention, pp. 155–156
- Skills Practice, p. 157
- Practice, p. 158
- Reading to Learn Mathematics, p. 159
- Enrichment, p. 160
- Assessment, p. 176

Graphing Calculator and Computer Masters, p. 22
Prerequisite Skills Workbook, pp. 1–4
Teaching Geometry With Manipulatives Masters, pp. 1, 8, 17, 58, 60

Transparencies

5-Minute Check Transparency 3-6
Answer Key Transparencies

Technology

GeomPASS: Tutorial Plus, Lesson 9
Interactive Chalkboard

2 Teach

DISTANCE FROM A POINT TO A LINE

In-Class Examples Power Point®

1 Copy the figure from Example 1 in the Student Edition. Draw the segment that represents the distance from A to $\overleftrightarrow{BP}$.

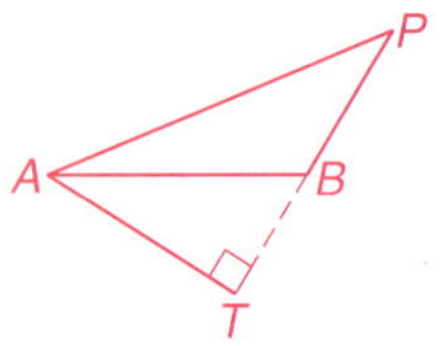

Teaching Tip Let students practice the construction on plain paper before using grid paper. That way the grid lines will not get in the way of the construction lines.

2 Construct a line perpendicular to line s through $V(1, 5)$ not on s. Then find the distance from V to s.

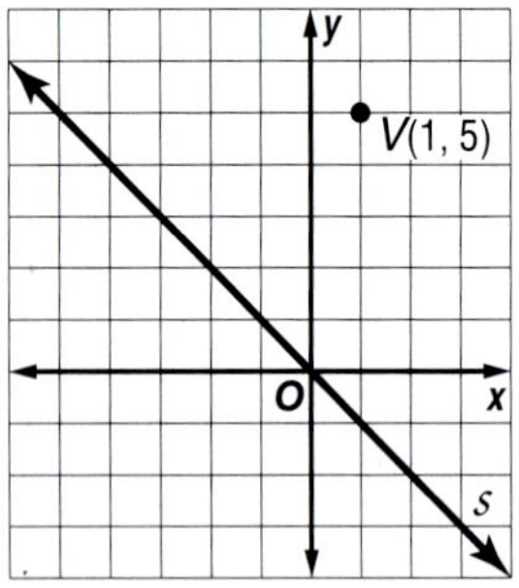

Sample answer:

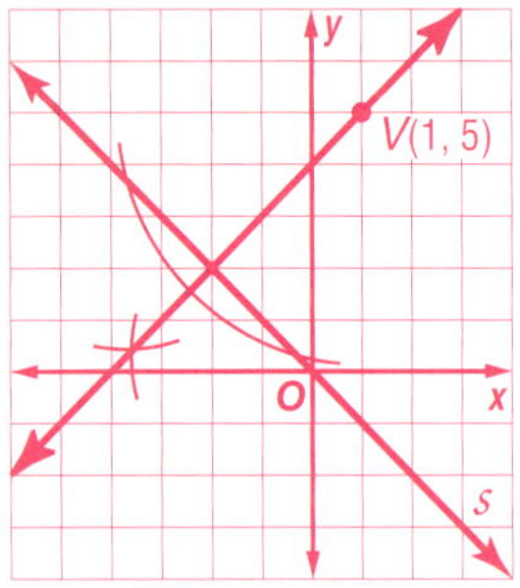

$d = \sqrt{18}$ or about 4.24 units

Example 2 Construct a Perpendicular Segment

COORDINATE GEOMETRY **Line ℓ contains points $(-6, -9)$ and $(0, -1)$. Construct a line perpendicular to line ℓ through $P(-7, -2)$ not on ℓ. Then find the distance from P to ℓ.**

1. Graph line ℓ and point P. Place the compass point at point P. Make the setting wide enough so that when an arc is drawn, it intersects ℓ in two places. Label these points of intersection A and B.

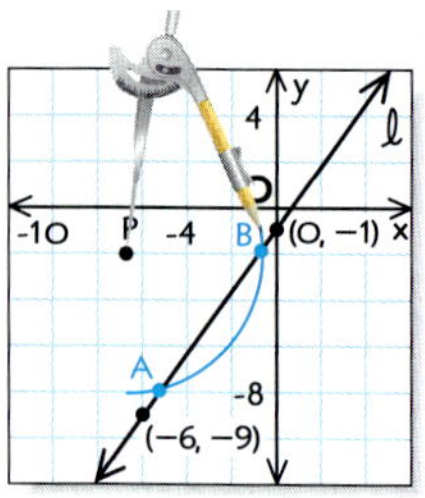

2. Put the compass at point A and draw an arc below line ℓ. (*Hint:* Any compass setting greater than $\frac{1}{2}AB$ will work.)

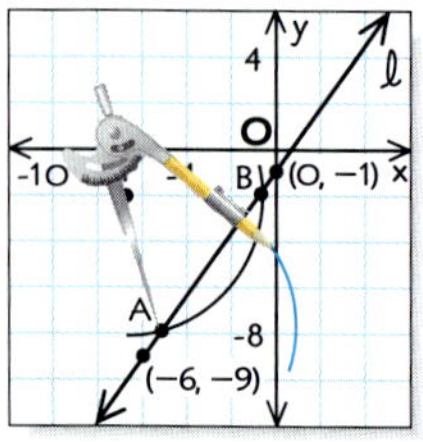

3. Using the same compass setting, put the compass at point B and draw an arc to intersect the one drawn in step 2. Label the point of intersection Q.

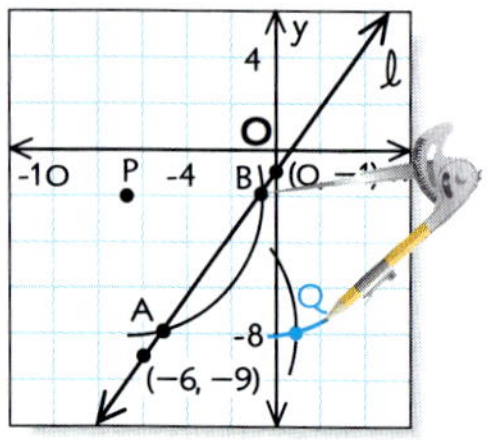

4. Draw $\overleftrightarrow{PQ}$. $\overleftrightarrow{PQ} \perp \ell$. Label point R at the intersection of $\overleftrightarrow{PQ}$ and ℓ. *Use the slopes of $\overleftrightarrow{PQ}$ and ℓ to verify that the lines are perpendicular.*

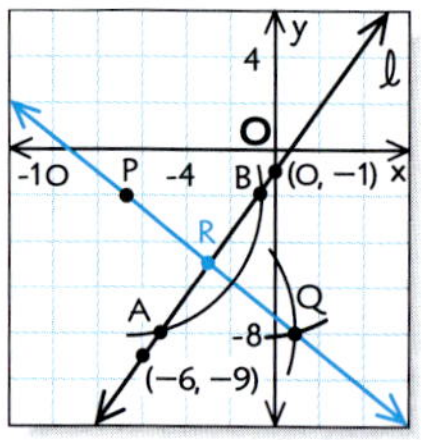

The segment constructed from point $P(-7, -2)$ perpendicular to the line ℓ, appears to intersect line ℓ at $R(-3, -5)$. Use the Distance Formula to find the distance between point P and line ℓ.

$$d = \sqrt{(x_2 - x_1)^2 + (y_2 - y_1)^2}$$
$$= \sqrt{(-7 - (-3))^2 + (-2 - (-5))^2}$$
$$= \sqrt{25} \text{ or } 5$$

The distance between P and ℓ is 5 units.

Study Tip

Distance
Note that the distance from a point to the x-axis can be determined by looking at the y-coordinate and the distance from a point to the y-axis can be determined by looking at the x-coordinate.

DISTANCE BETWEEN PARALLEL LINES

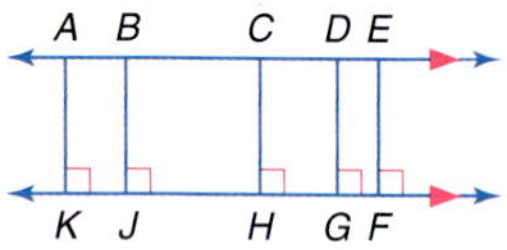

$AK = BJ = CH = DG = EF$

By definition, two parallel lines do not intersect. An alternate definition states that two lines in a plane are parallel if they are everywhere **equidistant**. Equidistant means that the distance between two lines measured along a perpendicular line to the lines is always the same. To measure the distance between two parallel lines, measure the length of the perpendicular segment whose endpoints lie on each of the two lines.

Key Concept — *Distance Between Parallel Lines*

The distance between two parallel lines is the distance between one of the lines and any point on the other line.

Recall that a *locus* is the set of all points that satisfy a given condition. Parallel lines can be described as the locus of points in a plane equidistant from a given line.

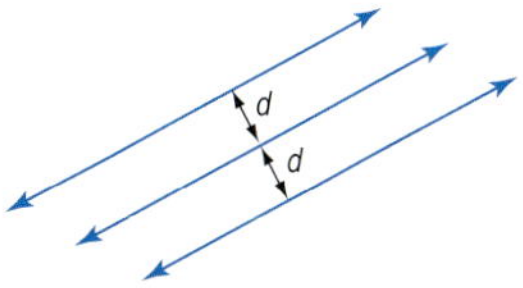

Study Tip

Look Back
To review **locus**, see Lesson 1-1.

Theorem 3.9

In a plane, if two lines are equidistant from a third line, then the two lines are parallel to each other.

Example 3 *Distance Between Lines*

Find the distance between the parallel lines ℓ and m whose equations are $y = -\frac{1}{3}x - 3$ and $y = -\frac{1}{3}x + \frac{1}{3}$, respectively.

You will need to solve a system of equations to find the endpoints of a segment that is perpendicular to both ℓ and m. The slope of lines ℓ and m is $-\frac{1}{3}$.

- First, write an equation of a line p perpendicular to ℓ and m. The slope of p is the opposite reciprocal of $-\frac{1}{3}$, or 3. Use the y-intercept of line ℓ, (0, −3), as one of the endpoints of the perpendicular segment.

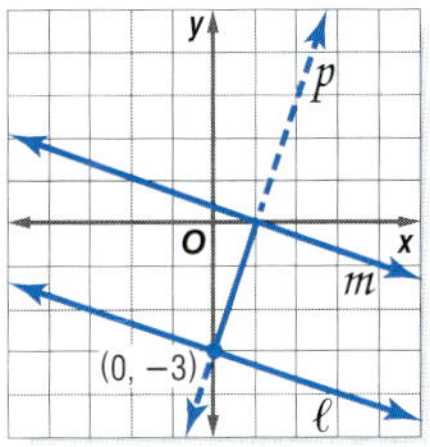

$y - y_1 = m(x - x_1)$ — Point-slope form

$y - (-3) = 3(x - 0)$ — $x_1 = 0$, $y_1 = -3$, $m = 3$

$y + 3 = 3x$ — Simplify.

$y = 3x - 3$ — Subtract 3 from each side.

- Next, use a system of equations to determine the point of intersection of line m and p.

m: $y = -\frac{1}{3}x + \frac{1}{3}$

p: $y = 3x - 3$

$-\frac{1}{3}x + \frac{1}{3} = 3x - 3$ — Substitute $-\frac{1}{3}x + \frac{1}{3}$ for y in the second equation.

$-\frac{1}{3}x - 3x = -3 - \frac{1}{3}$ — Group like terms on each side.

$-\frac{10}{3}x = -\frac{10}{3}$ — Simplify on each side.

$x = 1$ — Divide each side by $-\frac{10}{3}$.

$y = 3(1) - 3$ — Substitute 1 for x in the equation for p.

$y = 0$ — Simplify.

The point of intersection is (1, 0).

- Then, use the Distance Formula to determine the distance between (0, −3) and (1, 0).

$d = \sqrt{(x_2 - x_1)^2 + (y_2 - y_1)^2}$ — Distance Formula

$= \sqrt{(0 - 1)^2 + (-3 - 0)^2}$ — $x_2 = 0$, $x_1 = 1$, $y_2 = -3$, $y_1 = 0$

$= \sqrt{10}$ — Simplify.

The distance between the lines is $\sqrt{10}$ or about 3.16 units.

www.geometryonline.com/extra_examples

DISTANCE BETWEEN PARALLEL LINES

In-Class Example

Power Point®

3 Find the distance between the parallel lines a and b whose equations are $y = 2x + 3$ and $y = 2x - 3$, respectively. **The distance between the lines is $\sqrt{7.2}$ or about 2.7 units.**

Building on Prior Knowledge

Example 3 builds on concepts from algebra and from Chapter 1 as well as what they have learned in Chapter 3. You may want to review solving systems of equations by using pp. 742–743.

DAILY INTERVENTION

Differentiated Instruction

Naturalist Ask students to identify a straight path in a park, playground, or field. They should select any point in the park that is not on the path, and go from that point directly to the path. Then they should verify that their movement to the path was perpendicular to the path.

3 Practice/Apply

Study Notebook

Have students—
- add the definitions/examples of the vocabulary terms to their Vocabulary Builder worksheets for Chapter 3.
- include the distance from a point to a line, the distance between parallel lines, equidistant, and the steps of constructing a line perpendicular to a given point from a point not on the line.
- include any other item(s) that they find helpful in mastering the skills in this lesson.

About the Exercises...

Organization by Objective
- **Distance From a Point to a Line:** 11–18, 25–27
- **Distance Between Parallel Lines:** 19–24

Odd/Even Assignments
Exercises 11–24 are structured so that students practice the same concepts whether they are assigned odd or even problems.

Assignment Guide

Basic: 11–21 odd, 25–29 odd, 32–44

Average: 11–31 odd, 32–44

Advanced: 12–30 even, 31–44

Check for Understanding

Concept Check

1–3. See margin.

1. **Explain** how to construct a segment between two parallel lines to represent the distance between them.
2. **OPEN ENDED** Make up a problem involving an everyday situation in which you need to find the distance between a point and a line or the distance between two lines. For example, find the shortest path from the patio of a house to a garden to minimize the length of a walkway and material used in its construction.
3. **Compare and contrast** three different methods that you can use to show that two lines in a plane are parallel.

Guided Practice

GUIDED PRACTICE KEY	
Exercises	Examples
4, 5, 10	1
6	2
7, 8	3
9	1–3

Copy each figure. Draw the segment that represents the distance indicated.

4. L to $\overleftrightarrow{KN}$

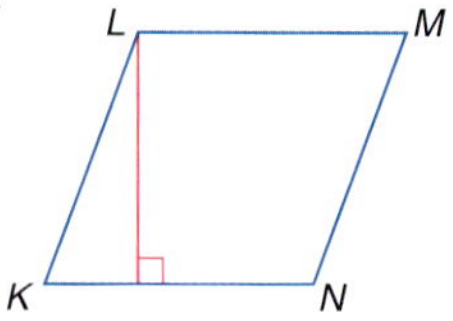

5. D to $\overleftrightarrow{AE}$

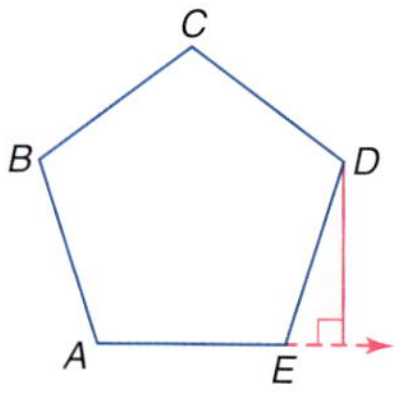

6. **COORDINATE GEOMETRY** Line ℓ contains points (0, 0) and (2, 4). Draw line ℓ. Construct a line perpendicular to ℓ through $A(2, -6)$. Then find the distance from A to ℓ. **See margin for graph; $d = \sqrt{20} \approx 4.47$**

Find the distance between each pair of parallel lines.

7. $y = \frac{3}{4}x - 1$
   $y = \frac{3}{4}x + \frac{1}{8}$ **0.9**

8. $x + 3y = 6$
   $x + 3y = -14$ **$\sqrt{40} \approx 6.32$**

9. Graph the line whose equation is $y = -\frac{3}{4}x + \frac{1}{4}$. Construct a perpendicular segment through $P(2, 5)$. Then find the distance from P to the line. **See margin.**

Application

10. **UTILITIES** Housing developers often locate the shortest distance from a house to the water main so that a minimum of pipe is required to connect the house to the water supply. Copy the diagram, and draw a possible location for the pipe. **See margin.**

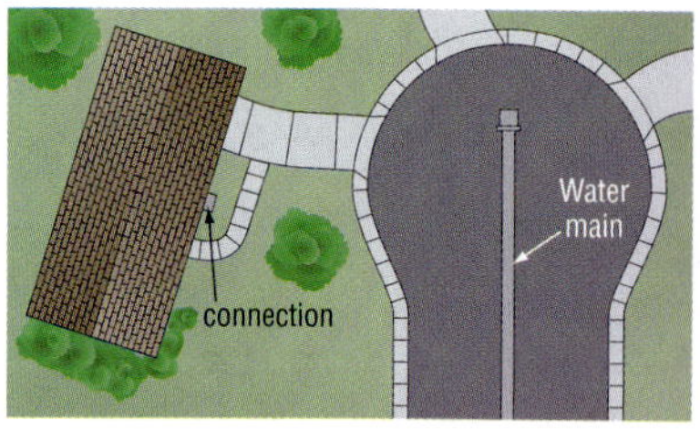

★ indicates increased difficulty

Practice and Apply

Homework Help	
For Exercises	See Examples
11–16	1
17, 18	2
19–24	3
25–27	1–2

Extra Practice
See page 760.

Copy each figure. Draw the segment that represents the distance indicated.

11. C to $\overleftrightarrow{AD}$

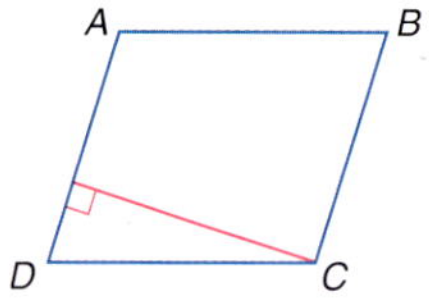

12. K to $\overleftrightarrow{JL}$

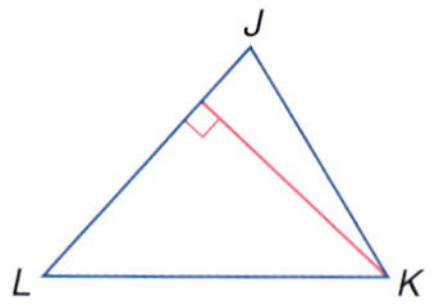

13. Q to $\overleftrightarrow{RS}$

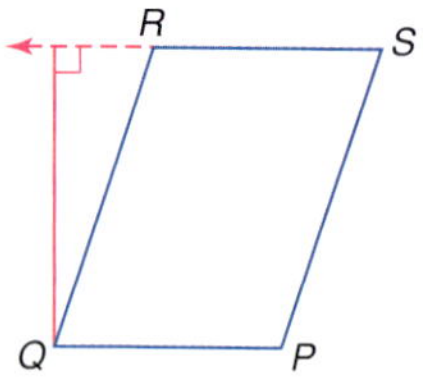

Answers

1. Construct a perpendicular line between them.
2. Sample answer: You are hiking and need to find the shortest path to a shelter.
3. Sample answer: Measure distances at different parts; compare slopes; measure angles. Finding slopes is the most readily available method.

6.

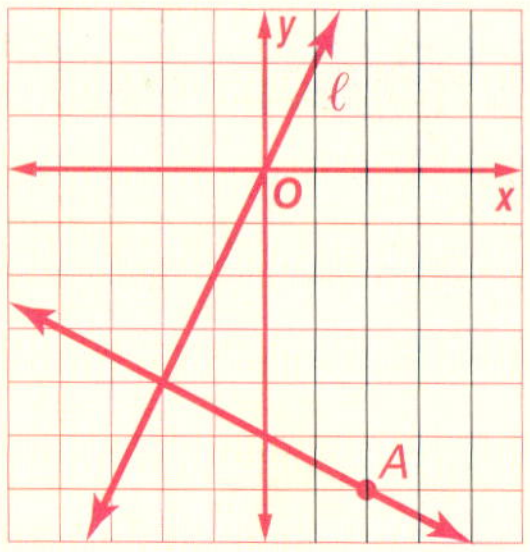

9. 5 units;

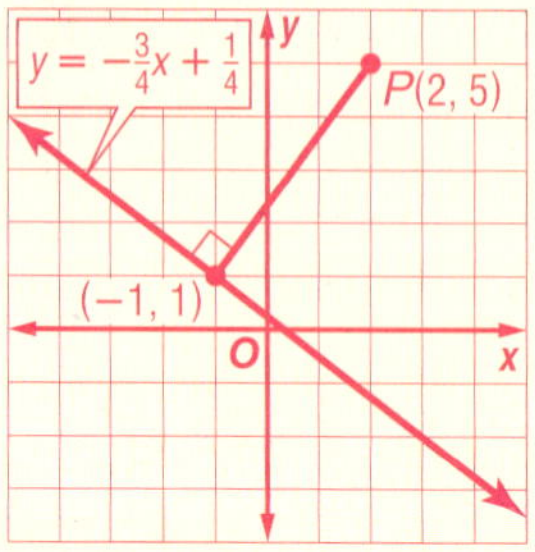

10.

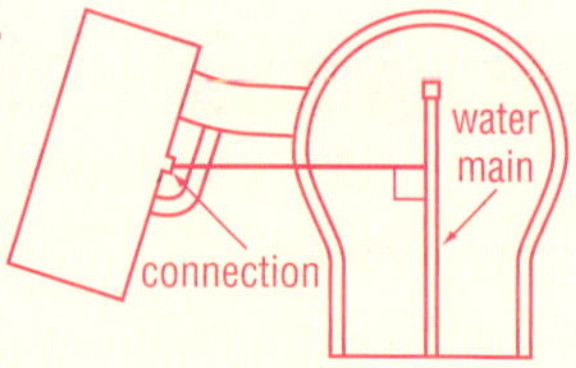

Copy each figure. Draw the segment that represents the distance indicated.

14. Y to $\overleftrightarrow{WX}$ **15.** G to $\overleftrightarrow{HJ}$ **16.** W to $\overleftrightarrow{UV}$

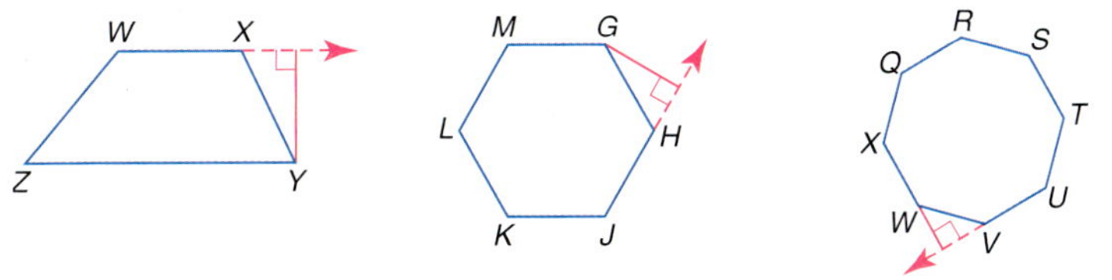

COORDINATE GEOMETRY Construct a line perpendicular to ℓ through P. Then find the distance from P to ℓ. 17–18. See p. 173B.

17. Line ℓ contains points $(-3, 0)$ and $(3, 0)$. Point P has coordinates $(4, 3)$.

18. Line ℓ contains points $(0, -2)$ and $(1, 3)$. Point P has coordinates $(-4, 4)$.

Find the distance between each pair of parallel lines.

19. $y = -3$, $y = 1$ **4**

20. $x = 4$, $x = -2$ **6**

21. $y = 2x + 2$, $y = 2x - 3$ $\sqrt{5}$

22. $y = 4x$, $y = 4x - 17$ $\sqrt{17}$

★ **23.** $y = 2x - 3$, $2x - y = -4$ $\sqrt{9.8}$

★ **24.** $y = -\frac{3}{4}x - 1$, $3x + 4y = 20$ $\frac{24}{5}$

Graph each line. Construct a perpendicular segment through the given point. Then find the distance from the point to the line. 25–27. See p. 173B for graphs.

25. $y = 5, (-2, 4)$ **1**

26. $y = 2x + 2, (-1, -5)$ $\sqrt{5}$

27. $2x - 3y = -9, (2, 0)$ $\sqrt{13}$

28. **PROOF** Write a paragraph proof of Theorem 3.9. See p. 173B.

29. **INTERIOR DESIGN** Theresa is installing a curtain rod on the wall above the window. In order to ensure that the rod is parallel to the ceiling, she measures and marks 9 inches below the ceiling in several places. If she installs the rod at these markings centered over the window, how does she know the curtain rod will be parallel to the ceiling? **It is everywhere equidistant from the ceiling.**

Career Choices

Interior Designer

Interior designers compute areas and volumes, work with scale models, and create budgets, in addition to being artistic. Usually two years of college and two years of practical experience are necessary before a designer can take a licensing exam.

Online Research For information about a career as an interior designer, visit: www.geometryonline.com/careers

30. **CONSTRUCTION** When framing a wall during a construction project, carpenters often use a plumb line. A *plumb line* is a string with a weight called a *plumb bob* attached on one end. The plumb line is suspended from a point and then used to ensure that wall studs are vertical. How does the plumb line help to find the distance from a point to the floor? See margin.

★ **31.** **ALGEBRA** In the coordinate plane, if a line has equation $ax + by = c$, then the distance from a point (x_1, y_1) is given by $\frac{|ax_1 + by_1 - c|}{\sqrt{a^2 + b^2}}$. Determine the distance from $(4, 6)$ to the line whose equation is $3x + 4y = 6$. **6**

32. **CRITICAL THINKING** Draw a diagram that represents each description.

a. Point P is equidistant from two parallel lines.

b. Point P is equidistant from two intersecting lines.

c. Point P is equidistant from two parallel planes.

d. Point P is equidistant from two intersecting planes.

e. A line is equidistant from two parallel planes.

f. A plane is equidistant from two other planes that are parallel. See p. 173B.

Answer

30. The plumb line will be vertical and will be perpendicular to the floor. The shortest distance from a point to the floor will be along the plumb line.

Enrichment, p. 160

Parallelism in Space

In space geometry, the concept of parallelism must be extended to include two planes and a line and a plane.

Definition: Two planes are parallel if and only if they do not intersect.

Definition: A line and a plane are parallel if and only if they do not intersect.

Thus, in space, two lines can be intersecting, parallel, or skew while two planes or a line and a plane can only be intersecting or parallel. In the figure at the right, $t \perp \mathcal{M}$, $t \perp \mathcal{P}$, $\mathcal{P} \parallel \mathcal{H}$, and ℓ and n are skew.

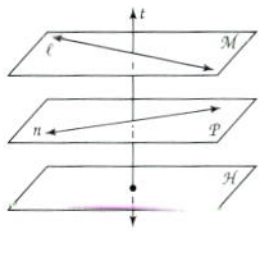

The following five statements are theorems about parallel planes.

Theorem: Two planes perpendicular to the same line are parallel.

Theorem: Two planes parallel to the same plane are parallel.

Theorem: A line perpendicular to one of two parallel planes is perpendicular to the other.

Study Guide and Intervention, p. 155 (shown) and p. 156

Distance From a Point to a Line When a point is not on a line, the distance from the point to the line is the length of the segment that contains the point and is perpendicular to the line.

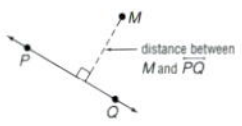

Example **Draw the segment that represents the distance from E to $\overleftrightarrow{AF}$.**

Extend $\overleftrightarrow{AF}$. Draw $\overline{EG} \perp \overleftrightarrow{AF}$.

$\overline{EG}$ represents the distance from E to $\overleftrightarrow{AF}$.

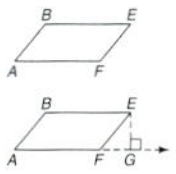

Exercises

Draw the segment that represents the distance indicated.

1. C to $\overleftrightarrow{AB}$

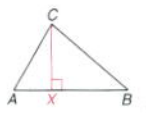

2. D to $\overleftrightarrow{AB}$

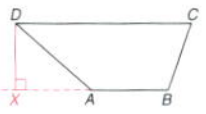

3. T to $\overleftrightarrow{RS}$

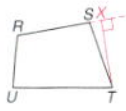

4. S to $\overleftrightarrow{PQ}$

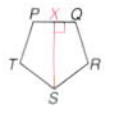

5. S to $\overleftrightarrow{QR}$

6. S to $\overleftrightarrow{RT}$

Skills Practice, p. 157 and Practice, p. 158 (shown)

Draw the segment that represents the distance indicated.

1. O to $\overleftrightarrow{MN}$

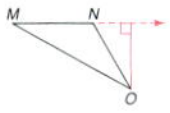

2. A to $\overleftrightarrow{DC}$

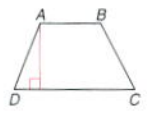

3. T to $\overleftrightarrow{VU}$

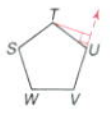

Construct a line perpendicular to ℓ through B. Then find the distance from B to ℓ.

4.

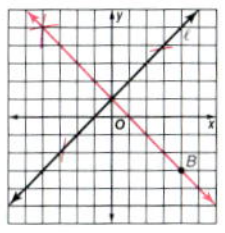

$4\sqrt{2}$

5.

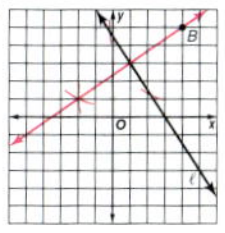

$\sqrt{13}$

Find the distance between each pair of parallel lines.

6. $y = -x$, $y = -x - 4$ $2\sqrt{2}$

7. $y = 2x + 7$, $y = 2x - 3$ $2\sqrt{5}$

8. $y = 3x + 12$, $y = 3x - 18$ $3\sqrt{10}$

9. Graph the line $y = -x + 1$. Construct a perpendicular segment through the point at $(-2, -3)$. Then find the distance from the point to the line. $3\sqrt{2}$

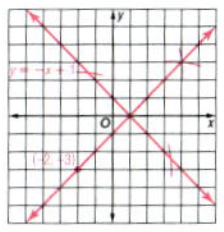

10. **CANOEING** Bronson and a friend are going to carry a canoe across a flat field to the bank of a straight canal. Describe the shortest path they can use.

Sample answer: The shortest path would be a perpendicular segment from where they are to the bank of the canal.

Reading to Learn Mathematics, p. 159 ELL

Pre-Activity **How does the distance between parallel lines relate to hanging new shelves?**

Read the introduction to Lesson 3-6 at the top of page 159 in your textbook.

Name three examples of situations in home construction where it would be important to construct parallel lines.

Sample answer: opposite walls of a room, planks of hardwood flooring, tops and bottoms of cabinets

Reading the Lesson

1. Fill in the blank with a word or phrase to complete each sentence.

a. The distance from a line to a point not on the line is the length of the segment __perpendicular__ to the line from the point.

b. Two coplanar lines are parallel if they are everywhere __equidistant__.

c. In a plane, if two lines are both equidistant from a third line, then the two lines are __parallel__ to each other.

d. The distance between two parallel lines measured along a perpendicular to the two lines is always __the same__.

e. To measure the distance between two parallel lines, measure the distance between one of the lines and any point on the __other line__.

2. On each figure, draw the segment that represents the distance indicated.

a. P to ℓ

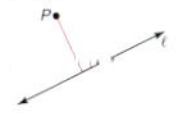

b. D to $\overleftrightarrow{AB}$

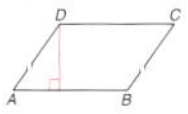

c. E to $\overleftrightarrow{FG}$

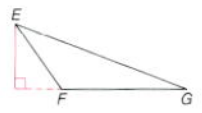

d. U to $\overleftrightarrow{RV}$

Helping You Remember

3. A good way to remember a new word is to relate it to words that use the same root. Use your dictionary to find the meaning of the Latin root *aequus*. List three words other than equal and equidistant that are derived from this root and give the meaning of each.

Sample answer: *Aequus* means *even*, *fair*, or *equal*. *Equinox* means one of the two times of year when day and night are of equal length. *Equity* means being just or fair. *Equivalent* means being equal in value or meaning.

4 Assess

Open-Ended Assessment

Modeling Have students work in groups. They should place a long, thin object such as a yardstick (or broomstick) on the ground, and mark a point on the ground. Then they discuss how to find the distance from the point to the line represented by the yardstick. Each student in the group should measure the distance, and the members should compare their answers.

Assessment Options

Quiz (Lesson 3-6) is available on p. 176 of the *Chapter 3 Resource Masters.*

Answers

33. Sample answer: We want new shelves to be parallel so they will line up. Answers should include the following.
    - After making several points, a slope can be calculated, which should be the same slope as the original brace.
    - Building walls requires parallel lines.

44. Given: $NL = NM$, $AL = BM$
    Prove: $NA = NB$
    Proof:
    Statements (Reasons)
    1. $NL = NM$, $AL = BM$ (Given)
    2. $NL = NA + AL$, $NM = NB + BM$ (Segment Addition Post.)
    3. $NA + AL = NB + BM$ (Substitution)
    4. $NA + BM = NB + BM$ (Substitution)
    5. $NA = NB$ (Subtraction Property)

33. **WRITING IN MATH** Answer the question that was posed at the beginning of the lesson. See margin.

How does the distance between parallel lines relate to hanging new shelves?

Include the following in your answer:
- an explanation of why marking several points equidistant from the first brace will ensure that the braces are parallel, and
- a description of other types of home improvement projects that require that two or more elements are parallel.

FCAT Practice
Standardized Test Practice

34. **GRID IN** Segment AB is perpendicular to segment BD. Segment AB and segment CD bisect each other at point X. If $AB = 16$ and $CD = 20$, what is the length of $\overline{BD}$? 6

35. **ALGEBRA** A coin was flipped 24 times and came up heads 14 times and tails 10 times. If the first and the last flips were both heads, what is the greatest number of consecutive heads that could have occurred? D

    (A) 7 (B) 9 (C) 10 (D) 13

Maintain Your Skills

Mixed Review **Given the following information, determine which lines, if any, are parallel. State the postulate or theorem that justifies your answer.** *(Lesson 3-5)*

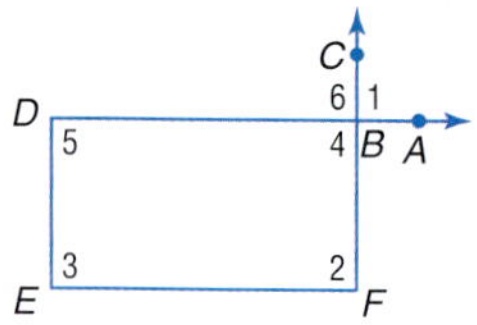

36. $\angle 5 \cong \angle 6$ $\overleftrightarrow{DE} \parallel \overleftrightarrow{CF}$; alt. int. ∠s
37. $\angle 6 \cong \angle 2$ $\overleftrightarrow{DA} \parallel \overleftrightarrow{EF}$; corr. ∠s
38. $\angle 1$ and $\angle 2$ are supplementary. $\overleftrightarrow{DA} \parallel \overleftrightarrow{EF}$; $\angle 1 \cong \angle 4$ and cons. int ∠s are suppl.

Write an equation in slope-intercept form for each line. *(Lesson 3-4)*

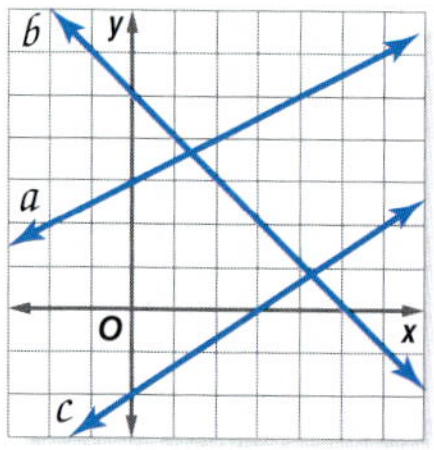

39. a $y = \frac{1}{2}x + 3$
40. b $y = -x + 5$
41. c $y = \frac{2}{3}x - 2$
42. perpendicular to line a, contains $(-1, -4)$ $y = -2x - 6$
43. parallel to line c, contains $(2, 5)$ $y = \frac{2}{3}x + \frac{11}{3}$

44. **PROOF** Write a two-column proof. *(Lesson 2-7)*

    Given: $NL = NM$
    $AL = BM$
    Prove: $NA = NB$ See margin.

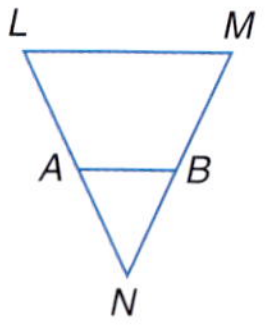

WebQuest Internet Project *When Is Weather Normal?*

It's time to complete your project. Use the information and data you have gathered about climate and locations on Earth to prepare a portfolio or Web page. Be sure to include graphs and/or tables in the presentation.

www.geometryonline.com/webquest

164 Chapter 3 Parallel and Perpendicular Lines

Geometry Activity

A Follow-Up of Lesson 3-6

Non-Euclidean Geometry

So far in this text, we have studied **plane Euclidean geometry**, which is based on a system of points, lines, and planes. In **spherical geometry**, we study a system of points, great circles (lines), and spheres (planes). Spherical geometry is one type of **non-Euclidean geometry**.

Plane Euclidean Geometry

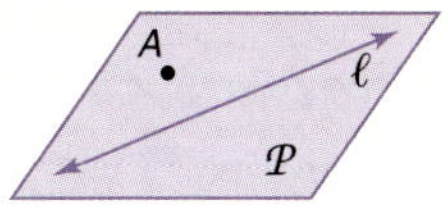

Plane $\mathcal{P}$ contains line ℓ and point A not on ℓ.

Spherical Geometry

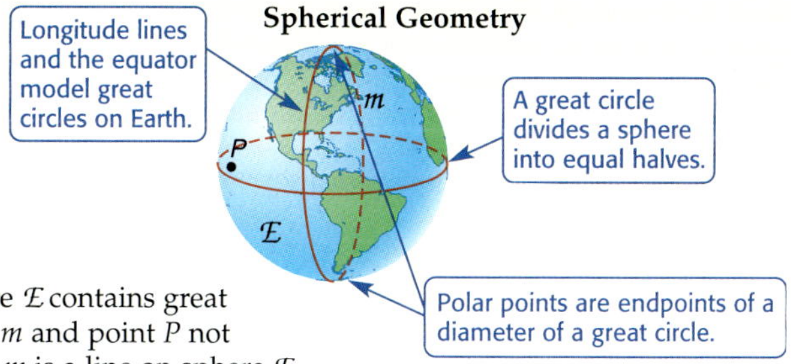

Sphere $\mathcal{E}$ contains great circle m and point P not on m. m is a line on sphere $\mathcal{E}$.

The table below compares and contrasts *lines* in the system of plane Euclidean geometry and *lines* (great circles) in spherical geometry.

Plane Euclidean Geometry Lines on the Plane	Spherical Geometry Great Circles (Lines) on the Sphere
1. A line segment is the shortest path between two points.	**1.** An arc of a great circle is the shortest path between two points.
2. There is a unique line passing through any two points.	**2.** There is a unique great circle passing through any pair of nonpolar points.
3. A line goes on infinitely in two directions.	**3.** A great circle is finite and returns to its original starting point.
4. If three points are collinear, exactly one is between the other two.	**4.** If three points are collinear, any one of the three points is between the other two.
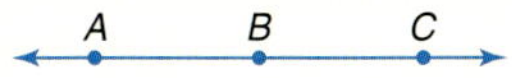 B is between A and C.	A is between B and C. B is between A and C. C is between A and B. 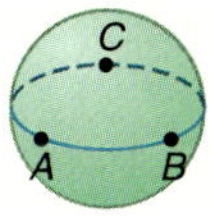

In spherical geometry, Euclid's first four postulates and their related theorems hold true. However, theorems that depend on the parallel postulate (Postulate 5) may not be true.

In Euclidean geometry parallel lines lie in the same plane and never intersect. In spherical geometry, the sphere is the plane, and a great circle represents a line. Every great circle containing A intersects ℓ. Thus, there exists no line through point A that is parallel to ℓ.

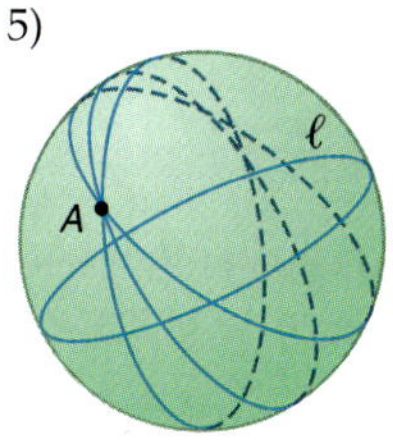

(continued on the next page)

Geometry Activity

A Follow-Up of Lesson 3-6

Getting Started

Spherical geometry is interesting as an alternative interpretation for points, lines, and planes. But it also is a practical and vital subject for people involved with global transportation and satellite tracking.

Teach

- You can use a sphere and a piece of string to show that, for any two points on the sphere, a string stretched between the points ("the distance between the points") will be part of a great circle of the sphere.
- Euclid's five postulates are:
  1. Between any two points there exists exactly one line.
  2. A straight line segment can be continued indefinitely in either direction.
  3. It is possible to construct a circle with any point as its center with radius of any length.
  4. All right angles are congruent to each other.
  5. For every line ℓ and every point P not on ℓ, there is exactly one line m through P that is parallel to ℓ.

Resource Manager

Teaching Geometry with Manipulatives

- p. 63 (student recording sheet)
- GeomPASS: Tutorial Plus, Lesson 9

Every great circle of a sphere intersects all other great circles on that sphere in exactly two points. In the figure at the right, one possible line through point A intersects line ℓ at P and Q.

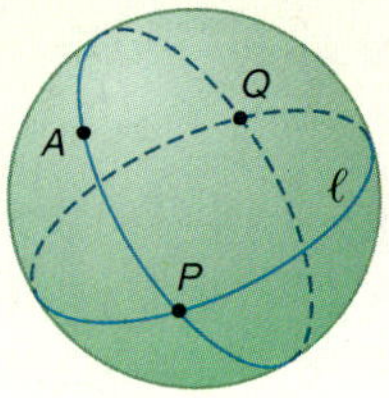

If two great circles divide a sphere into four congruent regions, the lines are perpendicular to each other at their intersection points. Each longitude circle on Earth intersects the equator at right angles.

Example *Compare Plane and Spherical Geometries*

For each property listed from plane Euclidean geometry, write a corresponding statement for spherical geometry.

a. Perpendicular lines intersect at one point.

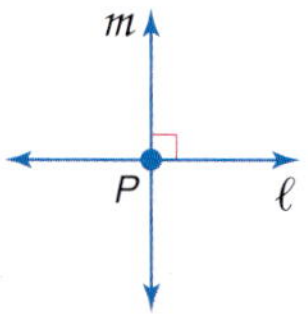

Perpendicular great circles intersect at two points.

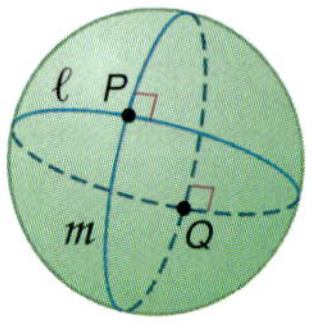

b. Perpendicular lines form four right angles.

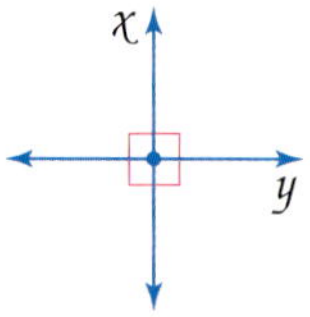

Perpendicular great circles form eight right angles.

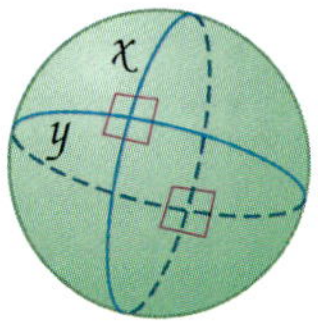

Exercises

For each property from plane Euclidean geometry, write a corresponding statement for spherical geometry. **1–7. See margin.**

1. A line goes on infinitely in two directions.
2. A line segment is the shortest path between two points.
3. Two distinct lines with no point of intersection are parallel.
4. Two distinct intersecting lines intersect in exactly one point.
5. A pair of perpendicular straight lines divides the plane into four infinite regions.
6. Parallel lines have infinitely many common perpendicular lines.
7. There is only one distance that can be measured between two points.

If spherical points are restricted to be nonpolar points, determine if each statement from plane Euclidean geometry is also *true* in spherical geometry. If *false*, explain your reasoning. **8–10. See margin.**

8. Any two distinct points determine exactly one line.
9. If three points are collinear, exactly one point is between the other two.
10. Given a line ℓ and point P not on ℓ, there exists exactly one line parallel to ℓ passing through P.

Assess

Exercises 1–7 Be sure students understand that while a term such as *segment* has different meanings in the two geometries, other terms such as *intersect*, *parallel*, and *perpendicular* will have the same meaning in each geometry.

Exercise 2 Students should understand that a line segment in spherical geometry must be a part of a line, so it would be a part of a great circle.

Study Notebook

Ask students to summarize what they have learned about how points, lines, and planes in spherical geometry are similar to and different from those terms in Euclidean geometry.

Answers

1. The great circle is finite.
2. A curved path on the great circle passing through two points is the shortest distance between the two points.
3. There exist no parallel lines.
4. Two distinct great circles intersect in exactly two points.
5. A pair of perpendicular great circles divides the sphere into four finite congruent regions.
6. There exist no parallel lines.
7. There are two distances between two points.
8. true
9. False; in spherical geometry, if three points are collinear, any point can be between the other two.
10. False; in spherical geometry, there are no parallel lines.

Study Guide and Review

Vocabulary and Concept Check

alternate exterior angles (p. 128)
alternate interior angles (p. 128)
consecutive interior angles (p. 128)
corresponding angles (p. 128)
equidistant (p. 160)
non-Euclidean geometry (p. 165)
parallel lines (p. 126)
parallel planes (p. 126)
plane Euclidean geometry (p. 165)
point-slope form (p. 145)
rate of change (p. 140)
skew lines (p. 127)
slope (p. 139)
slope-intercept form (p. 145)
spherical geometry (p. 165)
transversal (p. 127)

A complete list of postulates and theorems can be found on pages R1–R8.

Exercises **Refer to the figure and choose the term that best completes each sentence.**

1. Angles 4 and 5 are *(consecutive, alternate)* interior angles.
2. The distance from point A to line n is the length of the segment *(perpendicular, parallel)* to line n through A.
3. If $\angle 4$ and $\angle 6$ are supplementary, lines m and n are said to be *(parallel, intersecting)* lines.
4. Line ℓ is a *(slope-intercept, transversal)* for lines n and m.
5. $\angle 1$ and $\angle 8$ are *(alternate interior, alternate exterior)* angles.
6. If $n \parallel m$, $\angle 6$ and $\angle 3$ are *(supplementary, congruent)*.
7. Angles 5 and 3 are *(consecutive, alternate)* interior angles.

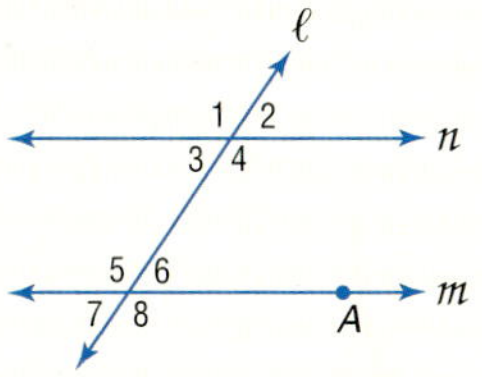

Lesson-by-Lesson Review

3-1 Parallel Lines and Transversals

See pages 126–131.

Concept Summary

- Coplanar lines that do not intersect are called *parallel*.
- When two lines are cut by a transversal, there are many angle relationships.

Example **Identify each pair of angles as *alternate interior, alternate exterior, corresponding,* or *consecutive interior* angles.**

a. $\angle 7$ and $\angle 3$
corresponding

b. $\angle 4$ and $\angle 6$
consecutive interior

c. $\angle 7$ and $\angle 2$
alternate exterior

d. $\angle 3$ and $\angle 6$
alternate interior

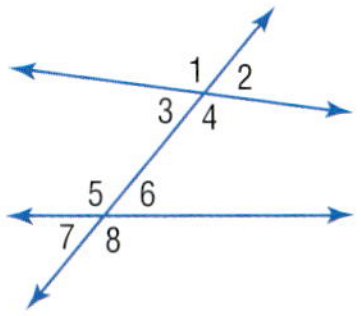

Exercises **Identify each pair of angles as *alternate interior, alternate exterior, corresponding,* or *consecutive interior* angles.** *See Example 3 on page 128.*

8. $\angle 10$ and $\angle 6$ corr.
9. $\angle 5$ and $\angle 12$ alt. ext.
10. $\angle 8$ and $\angle 10$ cons. int.
11. $\angle 1$ and $\angle 9$ corr.
12. $\angle 3$ and $\angle 6$ alt. int.
13. $\angle 5$ and $\angle 3$ cons. int.
14. $\angle 2$ and $\angle 7$ alt. ext.
15. $\angle 8$ and $\angle 9$ alt. int.

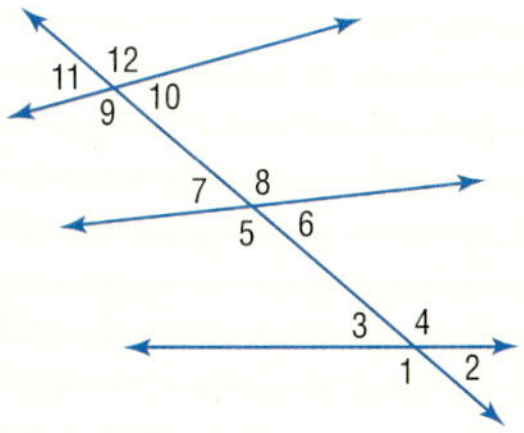

www.geometryonline.com/vocabulary_review

FOLDABLES™ Study Organizer

For more information about Foldables, see *Teaching Mathematics with Foldables.*

Have students look through the index cards they added to their Foldables while studying Chapter 3.

Have them edit and/or combine information on the cards as necessary. Remind students to include algebraic examples as well as geometry examples in their notes.

Encourage students to refer to their Foldables while completing the Study Guide and Review and to use them in preparing for the Chapter Test.

Chapter 3 Study Guide and Review

Vocabulary and Concept Check

- This alphabetical list of vocabulary terms in Chapter 3 includes a page reference where each term was introduced.
- **Assessment** A vocabulary test/review for Chapter 3 is available on p. 174 of the *Chapter 3 Resource Masters.*

Lesson-by-Lesson Review

For each lesson,

- the main ideas are summarized,
- additional examples review concepts, and
- practice exercises are provided.

Vocabulary PuzzleMaker

ELL The Vocabulary PuzzleMaker software improves students' mathematics vocabulary using four puzzle formats—crossword, scramble, word search using a word list, and word search using clues. Students can work on a computer screen or from a printed handout.

MindJogger Videoquizzes

ELL MindJogger Videoquizzes provide an alternative review of concepts presented in this chapter. Students work in teams in a game show format to gain points for correct answers. The questions are presented in three rounds.

Round 1 Concepts (5 questions)
Round 2 Skills (4 questions)
Round 3 Problem Solving (4 questions)

3-2 Angles and Parallel Lines

See pages 133–138.

Concept Summary

- Pairs of congruent angles formed by parallel lines and a transversal are corresponding angles, alternate interior angles, and alternate exterior angles.
- Pairs of consecutive interior angles are supplementary.

Example **In the figure, $m\angle 1 = 4p + 15$, $m\angle 3 = 3p - 10$, and $m\angle 4 = 6r + 5$. Find the values of p and r.**

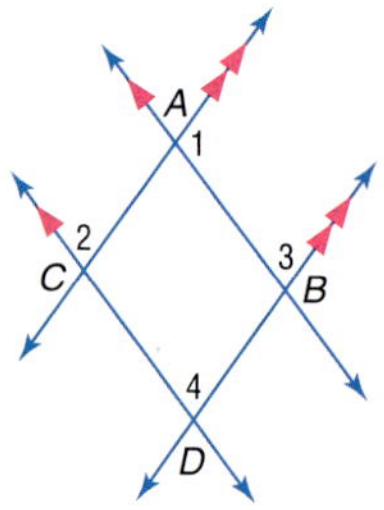

- Find p.

Since $\overleftrightarrow{AC} \parallel \overleftrightarrow{BD}$, $\angle 1$ and $\angle 3$ are supplementary by the Consecutive Interior Angles Theorem.

$m\angle 1 + m\angle 3 = 180$	Definition of supplementary angles
$(4p + 15) + (3p - 10) = 180$	Substitution
$7p + 5 = 180$	Simplify.
$p = 25$	Solve for p.

- Find r.

Since $\overleftrightarrow{AB} \parallel \overleftrightarrow{CD}$, $\angle 4 \cong \angle 3$ by the Corresponding Angles Postulate.

$m\angle 4 = m\angle 3$	Definition of congruent angles
$6r + 5 = 3(25) - 10$	Substitution, $p = 25$
$6r + 5 = 65$	Simplify.
$r = 10$	Solve for x.

Exercises **In the figure, $m\angle 1 = 53$. Find the measure of each angle.** *See Example 1 on page 133.*

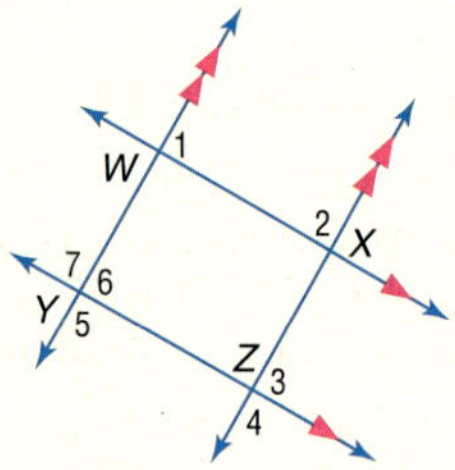

16. $\angle 2$ 127 **17.** $\angle 3$ 53

18. $\angle 4$ 127 **19.** $\angle 5$ 127

20. $\angle 6$ 53 **21.** $\angle 7$ 127

22. In the figure, $m\angle 1 = 3a + 40$, $m\angle 2 = 2a + 25$, and $m\angle 3 = 5b - 26$. Find a and b.
See Example 3 on page 135. $a = 23$, $b = 27$

3-3 Slopes of Lines

See pages 139–144.

Concept Summary

- The slope of a line is the ratio of its vertical rise to its horizontal run.
- Parallel lines have the same slope, while perpendicular lines have slopes whose product is -1.

Example **Determine whether $\overleftrightarrow{KM}$ and $\overleftrightarrow{LN}$ are *parallel, perpendicular,* or *neither* for $K(-3, 3)$, $M(-1, -3)$, $L(2, 5)$, and $N(5, -4)$.**

slope of $\overleftrightarrow{KM}$: $m = \frac{-3 - 3}{-1 - (-3)}$ or -3 slope of $\overleftrightarrow{LN}$: $m = \frac{-4 - 5}{5 - 2}$ or -3

The slopes are the same. So $\overleftrightarrow{KM}$ and $\overleftrightarrow{LN}$ are parallel.

168 Chapter 3 Parallel and Perpendicular Lines

Exercises **Determine whether $\overleftrightarrow{AB}$ and $\overleftrightarrow{CD}$ are *parallel*, *perpendicular*, or *neither*.** *See Example 3 on page 141.* **23–26. See margin.**

23. $A(-4, 1), B(3, -1), C(2, 2), D(0, 9)$
24. $A(6, 2), B(2, -2), C(-1, -4), D(5, 2)$
25. $A(1, -3), B(4, 5), C(1, -1), D(-7, 2)$
26. $A(2, 0), B(6, 3), C(-1, -4), D(3, -1)$

Graph the line that satisfies each condition. *See Example 4 on page 141.*

27. contains (2, 3) and is parallel to $\overleftrightarrow{AB}$ with $A(-1, 2)$ and $B(1, 6)$
28. contains (−2, −2) and is perpendicular to $\overleftrightarrow{PQ}$ with $P(5, 2)$ and $Q(3, -4)$

27–28. See margin.

3-4 Equations of Lines

See pages 145–150.

Concept Summary

In general, an equation of a line can be written if you are given:

- slope and the y-intercept
- the slope and the coordinates of a point on the line, or
- the coordinates of two points on the line.

Example **Write an equation in slope-intercept form of the line that passes through (2, −4) and (−3, 1).**

Find the slope of the line.

$m = \dfrac{y_2 - y_1}{x_2 - x_1}$ Slope Formula

$= \dfrac{1-(-4)}{-3-2}$ $(x_1, y_1) = (2, -4)$, $(x_2, y_2) = (-3, 1)$

$= \dfrac{5}{-5}$ or -1 Simplify.

Now use the point-slope form and either point to write an equation.

$y - y_1 = m(x - x_1)$ Point-slope form

$y - (-4) = -1(x - 2)$ $m = -1$, $(x_1, y_1) = (2, -4)$

$y + 4 = -x + 2$ Simplify.

$y = -x - 2$ Subtract 4 from each side.

Exercises **Write an equation in slope-intercept form of the line that satisfies the given conditions.** *See Examples 1–3 on pages 145 and 146.* **29–34. See margin.**

29. $m = 2$, contains (1, −5)
30. contains (2, 5) and (−2, −1)
31. $m = -\frac{2}{7}$, y-intercept $= 4$
32. $m = -\frac{3}{2}$, contains (2, −4)
33. $m = 5$, y-intercept $= -3$
34. contains (3, −1) and (−4, 6)

3-5 Proving Lines Parallel

See pages 151–157.

Concept Summary

When lines are cut by a transversal, certain angle relationships produce parallel lines.

- congruent corresponding angles
- congruent alternate exterior angles
- congruent alternate interior angles
- supplementary consecutive interior angles

Answers

23. neither
24. parallel
25. perpendicular
26. parallel
27.

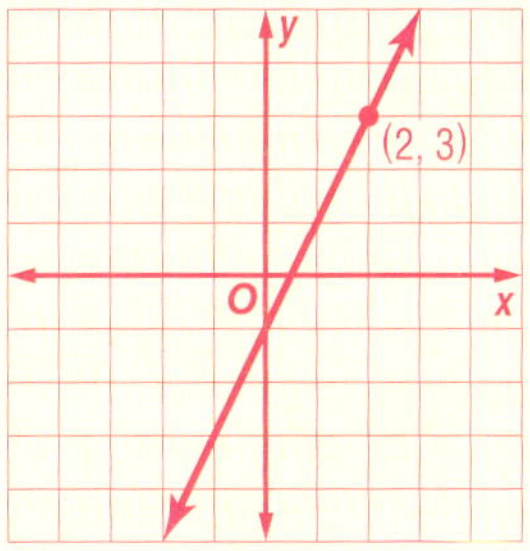

28.

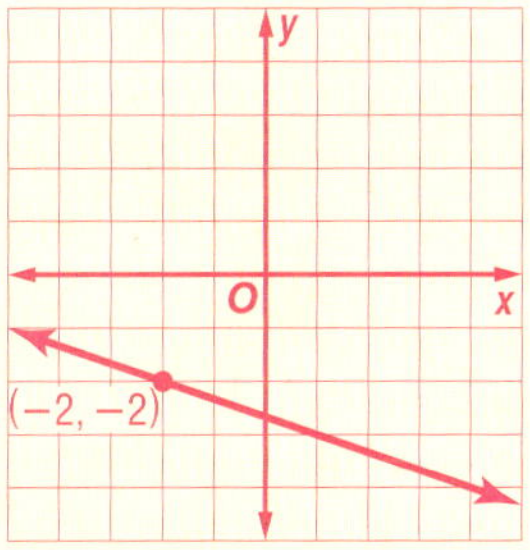

29. $y = 2x - 7$
30. $y = \frac{3}{2}x + 2$
31. $y = -\frac{2}{7}x + 4$
32. $y = -\frac{3}{2}x - 1$
33. $y = 5x - 3$
34. $y = -x + 2$

Chapter 3 Study Guide and Review 169

• Extra Practice, see pages 758–760.
• Mixed Problem Solving, see page 784.

Answers (p. 171)

12.

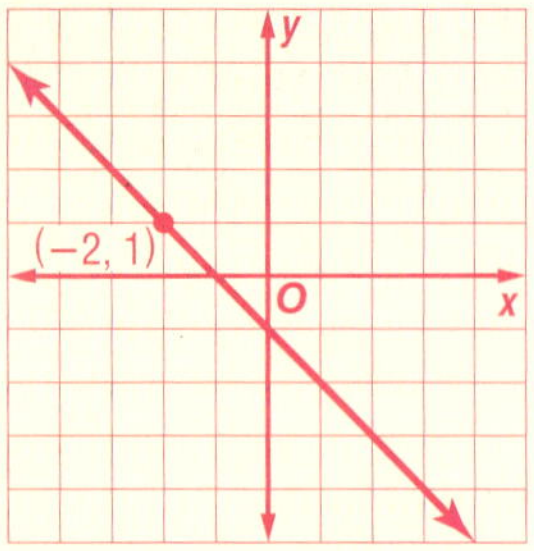

13.

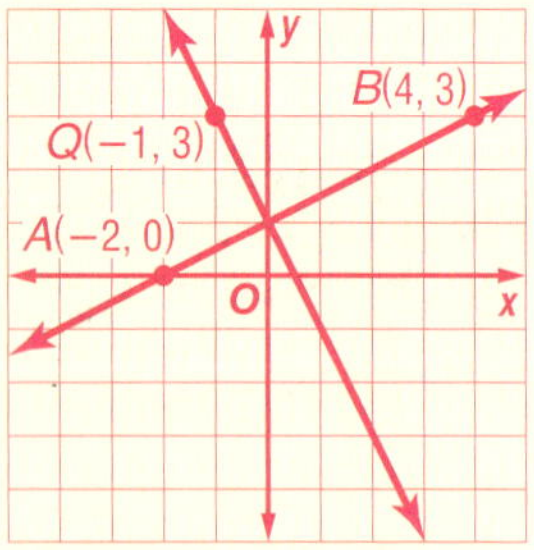

14.

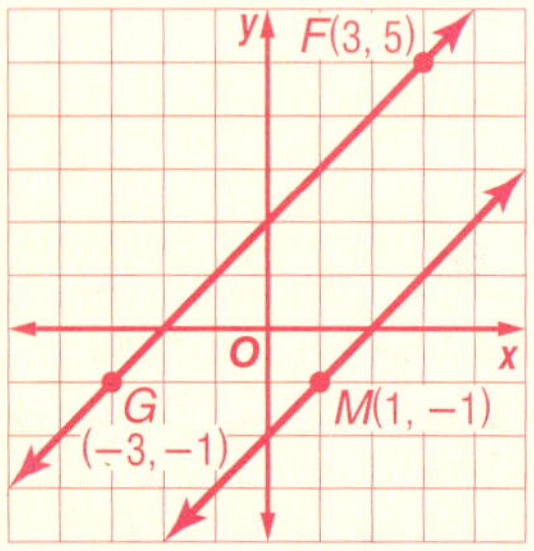

15.

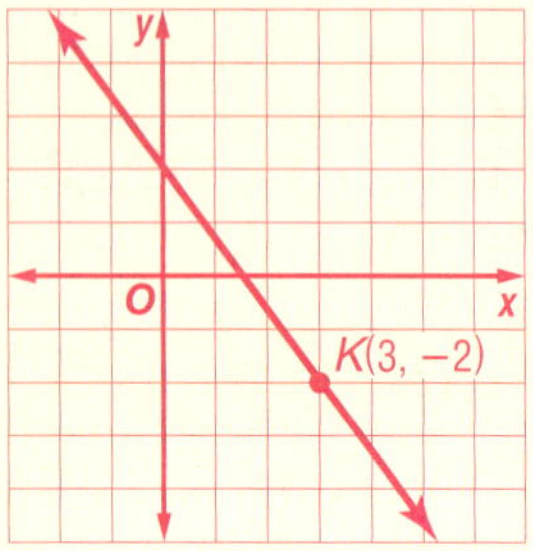

Example If $\angle 1 \cong \angle 8$, which lines if any are parallel?

$\angle 1$ and $\angle 8$ are alternate exterior angles for lines r and s. These lines are cut by the transversal p. Since the angles are congruent, lines r and s are parallel by Theorem 3.5.

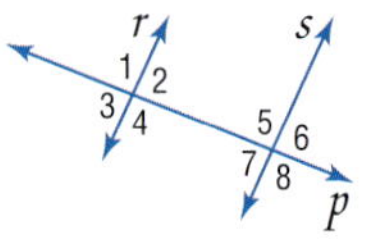

post or
Theorem
not needed

Exercises **Given the following information, determine which lines, if any, are parallel. State the postulate or theorem that justifies your answer.**

See Example 1 on page 152.

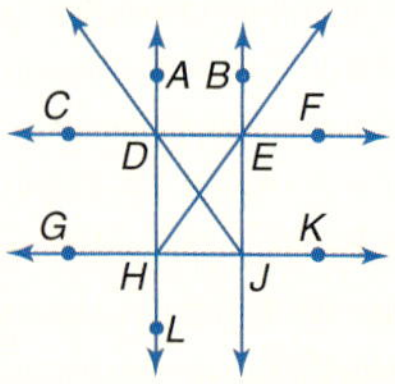

35. $\angle GHL \cong \angle EJK$ **$\overleftrightarrow{AL} \parallel \overleftrightarrow{BJ}$, alt. ext. ∠s ≅**
36. $m\angle ADJ + m\angle DJE = 180$ **$\overleftrightarrow{AL} \parallel \overleftrightarrow{BJ}$, cons. int. ∠s suppl.**
37. $\overleftrightarrow{CF} \perp \overleftrightarrow{AL}, \overleftrightarrow{GK} \perp \overleftrightarrow{AL}$ **$\overleftrightarrow{CF} \parallel \overleftrightarrow{GK}$, 2 lines ⊥ same line**
38. $\angle DJE \cong \angle HDJ$ **$\overleftrightarrow{AL} \parallel \overleftrightarrow{BJ}$, alt. int. ∠s ≅**
39. $m\angle EJK + m\angle JEF = 180$ **$\overleftrightarrow{CF} \parallel \overleftrightarrow{GK}$, cons. int. ∠s suppl.**
40. $\angle GHL \cong \angle CDH$ **$\overleftrightarrow{CF} \parallel \overleftrightarrow{GK}$, corr. ∠s ≅**

3-6 Perpendiculars and Distance

See pages 159–164.

OMIT

Concept Summary

- The distance between a point and a line is measured by the perpendicular segment from the point to the line.

Example **Find the distance between the parallel lines q and r whose equations are $y = x - 2$ and $y = x + 2$, respectively.**

- The slope of q is 1. Choose a point on line q such as $P(2, 0)$. Let line k be perpendicular to q through P. The slope of line k is -1. Write an equation for line k.

$y = mx + b$	Slope-intercept form
$0 = (-1)(2) + b$	$y = 0, m = -1, x = 2$
$2 = b$	Solve for b. An equation for k is $y = -x + 2$.

- Use a system of equations to determine the point of intersection of k and r.

$y = x + 2$		Substitute 2 for y in the original equation.
$y = -x + 2$		$2 = -x + 2$
$2y = 4$	Add the equations.	$x = 0$ Solve for x.
$y = 2$	Divide each side by 2.	The point of intersection is (0, 2).

- Now use the Distance Formula to determine the distance between (2, 0) and (0, 2).

$$d = \sqrt{(x_2 - x_1)^2 + (y_2 - y_1)^2} = \sqrt{(2 - 0)^2 + (0 - 2)^2} = \sqrt{8}$$

The distance between the lines is $\sqrt{8}$ or about 2.83 units.

Exercises **Find the distance between each pair of parallel lines.**

See Example 3 on page 161.

41. $y = 2x - 4, y = 2x + 1$ **$\sqrt{5}$**
42. $y = \frac{1}{2}x, y = \frac{1}{2}x + 5$ **$\sqrt{20}$**

170 Chapter 3 Parallel and Perpendicular Lines

Chapter 3 Practice Test

Vocabulary and Concepts

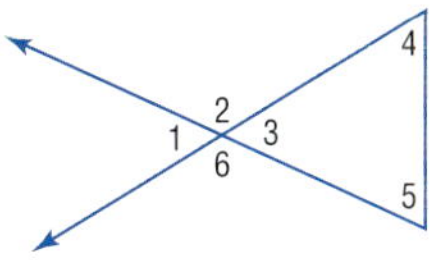

1. Write an equation of a line that is perpendicular to $y = 3x - \frac{2}{7}$. **Sample answer: $y = -\frac{1}{3}x + 1$**
2. Name a theorem that can be used to prove that two lines are parallel.
3. Find all the angles that are supplementary to ∠1. **∠2, ∠6**

2. Sample answer: If alt. int. ∠s are ≅, then lines are ∥.

Skills and Applications

In the figure, $m\angle 12 = 64$. Find the measure of each angle.

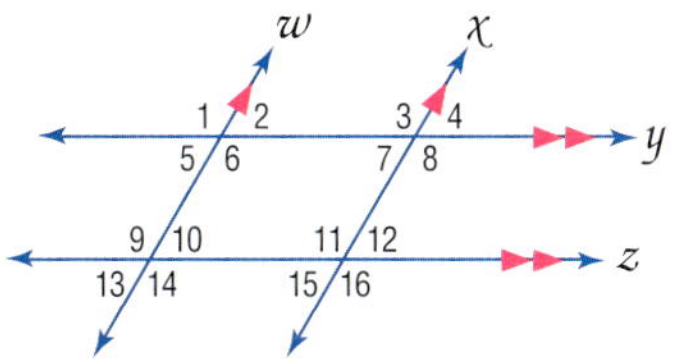

4. ∠8 **116**
5. ∠13 **64**
6. ∠7 **64**
7. ∠11 **116**
8. ∠3 **116**
9. ∠4 **64**
10. ∠9 **116**
11. ∠5 **64**

Graph the line that satisfies each condition. **12–15. See margin.**

12. slope = −1, contains $P(-2, 1)$
13. contains $Q(-1, 3)$ and is perpendicular to $\overleftrightarrow{AB}$ with $A(-2, 0)$ and $B(4, 3)$
14. contains $M(1, -1)$ and is parallel to $\overleftrightarrow{FG}$ with $F(3, 5)$ and $G(-3, -1)$
15. slope $= -\frac{4}{3}$, contains $K(3, -2)$

For Exercises 16–21, refer to the figure at the right. Find each value if $p \parallel q$.

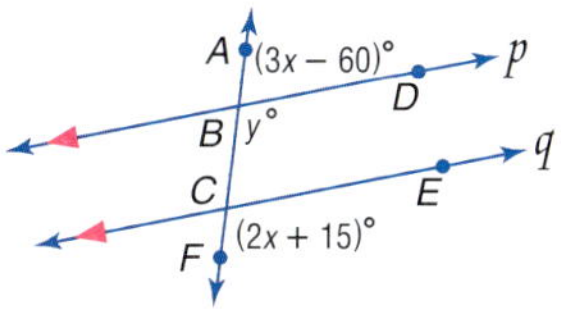

16. x **45**
17. y **105**
18. $m\angle FCE$ **105**
19. $m\angle ABD$ **75**
20. $m\angle BCE$ **75**
21. $m\angle CBD$ **105**

Find the distance between each pair of parallel lines.

22. $y = 2x - 1$, $y = 2x + 9$ **$\sqrt{20} \approx 4.47$**
23. $y = -x + 4$, $y = -x - 2$ **$\sqrt{18} \approx 4.24$**

24. **COORDINATE GEOMETRY** Detroit Road starts in the center of the city, and Lorain Road starts 4 miles west of the center of the city. Both roads run southeast. If these roads are put on a coordinate plane with the center of the city at (0, 0), Lorain Road is represented by the equation $y = -x - 4$ and Detroit Road is represented by the equation $y = -x$. How far away is Lorain Road from Detroit Road? **about 2.83 mi**

25. **STANDARDIZED TEST PRACTICE** In the figure at the right, which cannot be true if $m \parallel \ell$ and $m\angle 1 = 73$? **B**

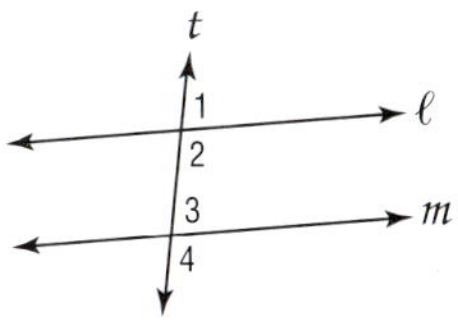

Ⓐ $m\angle 4 > 73$ Ⓑ $\angle 1 \cong \angle 4$
Ⓒ $m\angle 2 + m\angle 3 = 180$ Ⓓ $\angle 3 \cong \angle 1$

www.geometryonline.com/chapter_test/fcat

Portfolio Suggestion

Introduction Two important terms in this chapter are *parallel* and *perpendicular*. Students used those terms when they explored angles formed when a transversal intersected two parallel lines.

Ask Students to make an art design that includes parallel lines and a transversal. Have students label angles in their design with letters or color codes and write a key describing the kinds of angle relationships shown. Have students add their art designs to their portfolios.

Chapter 3 Practice Test

Assessment Options

Vocabulary Test A vocabulary test/review for Chapter 3 can be found on p. 174 of the *Chapter 3 Resource Masters*.

Chapter Tests There are six Chapter 3 Tests and an Open-Ended Assessment task available in the *Chapter 3 Resource Masters*.

Chapter 3 Tests

Form	Type	Level	Pages
1	MC	basic	161–162
2A	MC	average	163–164
2B	MC	average	165–166
2C	FR	average	167–168
2D	FR	average	169–170
3	FR	advanced	171–172

MC = multiple-choice questions
FR = free-response questions

Open-Ended Assessment Performance tasks for Chapter 3 can be found on p. 173 of the *Chapter 3 Resource Masters*. A sample scoring rubric for these tasks appears on p. A25.

Unit 1 Test A unit test/review can be found on pp. 181–182 of the *Chapter 3 Resource Masters*.

TestCheck and Worksheet Builder

This **networkable software** has three modules for assessment.

- **Worksheet Builder** to make worksheets and tests.
- **Student Module** to take tests on-screen.
- **Management System** to keep student records.

Chapter 3 Standardized Test Practice

These two pages contain practice questions in the various formats that can be found on the most frequently given standardized tests.

A practice answer sheet for these two pages can be found on p. A1 of the *Chapter 3 Resource Masters*.

Standardized Test Practice Student Recording Sheet, p. A1

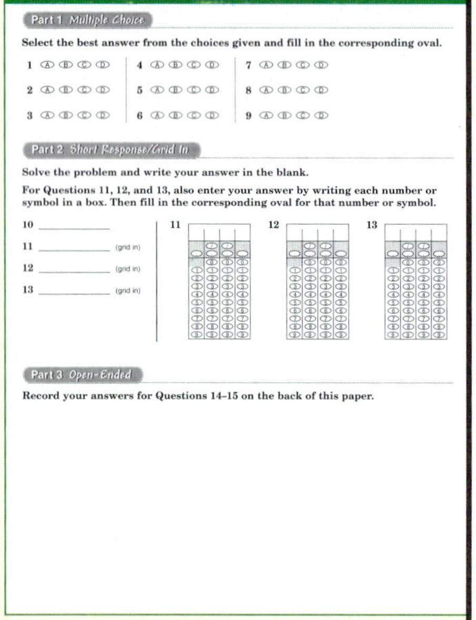
Part 1 Multiple Choice

Select the best answer from the choices given and fill in the corresponding oval.

1 Ⓐ Ⓑ Ⓒ Ⓓ 4 Ⓐ Ⓑ Ⓒ Ⓓ 7 Ⓐ Ⓑ Ⓒ Ⓓ

2 Ⓐ Ⓑ Ⓒ Ⓓ 5 Ⓐ Ⓑ Ⓒ Ⓓ 8 Ⓐ Ⓑ Ⓒ Ⓓ

3 Ⓐ Ⓑ Ⓒ Ⓓ 6 Ⓐ Ⓑ Ⓒ Ⓓ 9 Ⓐ Ⓑ Ⓒ Ⓓ

Part 2 Short Response/Grid In

Solve the problem and write your answer in the blank.

For Questions 11, 12, and 13, also enter your answer by writing each number or symbol in a box. Then fill in the corresponding oval for that number or symbol.

10 ______

11 ______ (grid in)

12 ______ (grid in)

13 ______ (grid in)

Part 3 Open-Ended

Record your answers for Questions 14–15 on the back of this paper.

Additional Practice

See pp. 179–180 in the *Chapter 3 Resource Masters* for additional standardized test practice.

Chapter 3 Standardized Test Practice

FCAT Practice

Part 1 Multiple Choice

Record your answers on the answer sheet provided by your teacher or on a sheet of paper.

1. Jahaira needed 2 meters of fabric to reupholster a chair in her bedroom. If Jahaira can only find a centimeter ruler, how much fabric should she cut? (Prerequisite Skill) **B**

Ⓐ 20 cm Ⓑ 200 cm

Ⓒ 2000 cm Ⓓ 20,000 cm

2. A fisherman uses a coordinate grid marked in miles to locate the nets cast at sea. How far apart are nets *A* and *B*? (Lesson 1-3) **C**

Ⓐ 3 mi

Ⓑ $\sqrt{28}$ mi

Ⓒ $\sqrt{65}$ mi

Ⓓ 11 mi

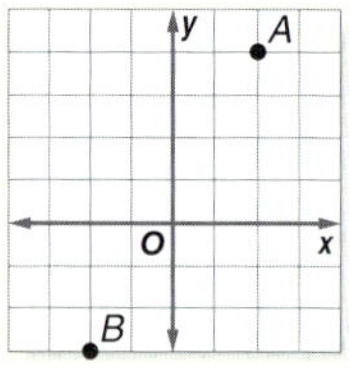

3. If $\angle ABC \cong \angle CBD$, which statement must be true? (Lesson 1-5) **A**

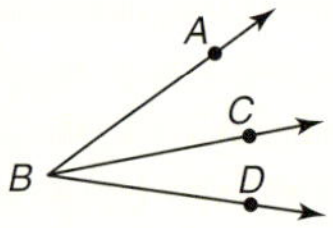

Ⓐ Segment *BC* bisects $\angle ABD$.

Ⓑ $\angle ABD$ is a right angle.

Ⓒ $\angle ABC$ and $\angle CBD$ are supplementary.

Ⓓ Segments *AB* and *BD* are perpendicular.

4. Valerie cut a piece of wood at a 72° angle for her project. What is the degree measure of the supplementary angle on the leftover piece of wood? (Lesson 1-6) **D**

Ⓐ 18 Ⓑ 78 Ⓒ 98 Ⓓ 108

5. A pan balance scale is often used in science classes. What is the value of *x* to balance the scale if one side weighs $4x + 4$ units and the other weighs $6x - 8$ units? (Lesson 2-3) **D**

Ⓐ 1 Ⓑ 2 Ⓒ 3 Ⓓ 6

Use the diagram below of a tandem bicycle frame for Questions 6 and 7.

6. The diagram shows the two posts on which seats are placed and several crossbars. Which term describes $\angle 6$ and $\angle 5$? (Lesson 3-1) **C**

Ⓐ alternate exterior angles

Ⓑ alternate interior angles

Ⓒ consecutive interior angles

Ⓓ corresponding angles

7. The quality control manager for the bicycle manufacturer wants to make sure that the two seat posts are parallel. Which angles can she measure to determine this? (Lesson 3-5) **B**

Ⓐ $\angle 2$ and $\angle 3$ Ⓑ $\angle 1$ and $\angle 3$

Ⓒ $\angle 4$ and $\angle 8$ Ⓓ $\angle 5$ and $\angle 7$

8. Which is the equation of a line that is perpendicular to the line $4y - x = 8$? (Lesson 3-3) **C**

Ⓐ $y = -\frac{1}{4}x - 2$ Ⓑ $y = \frac{1}{4}x + 2$

Ⓒ $y = -4x - 15$ Ⓓ $y = 4x + 15$

9. The graph of $y = 2x - 5$ is shown at the right. How would the graph be different if the number 2 in the equation was replaced with a 4? (Lesson 3-4) **C**

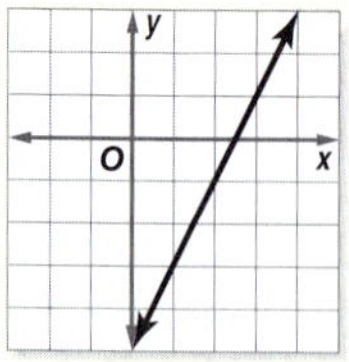

Ⓐ parallel to the line shown above, but shifted two units higher

Ⓑ parallel to the line shown above, but shifted two units lower

Ⓒ have a steeper slope, but intercept the *y*-axis at the same point

Ⓓ have a less steep slope, but intercept the *y*-axis at the same point

TestCheck and Worksheet Builder

Special banks of standardized test questions similar to those on the SAT, ACT, TIMSS 8, NAEP 8, and state proficiency tests can be found on this CD-ROM.

Aligned and verified by The Princeton Review

FCAT Practice

Part 2 Short Response/Grid In

Record your answers on the answer sheet provided by your teacher or on a sheet of paper.

10. What statement is needed in step 2 to complete this proof? (Lesson 2-4)

Given: $\frac{4x-6}{3} = 10$ $3\left(\frac{4x-6}{3}\right) = 3(10)$

Prove: $x = 9$

Statements	Reasons
1. $\frac{4x-6}{3} = 10$	1. Given
2. ?	2. Multiplication Property
3. $4x - 6 = 30$	3. Simplify.
4. $4x = 36$	4. Addition Property
5. $x = 9$	5. Division Property

The director of a high school marching band draws a diagram of a new formation as shown below. In the figure, $\overleftrightarrow{AB} \parallel \overleftrightarrow{CD}$. Use the figure for Questions 11 and 12.

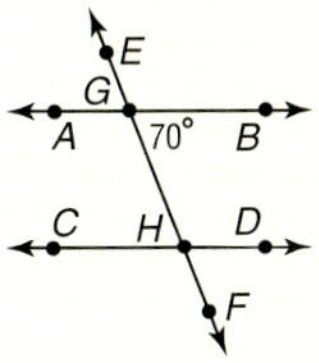

11. During the performance, a flag holder stands at point H, facing point F, and rotates right until she faces point C. What angle measure describes the flag holder's rotation? (Lesson 3-2) 110

12. Band members march along segment CH, turn left at point H, and continue to march along $\overline{HG}$. What is $m\angle CHG$ created by these marchers? (Lesson 3-2) 70

13. What is the slope of a line containing points (3, 4) and (9, 6)? (Lesson 3-3) $\frac{1}{3}$

www.geometryonline.com/standardized_test/fcat

The Princeton Review **Test-Taking Tip**

Question 13 Many standardized tests provide a reference sheet that includes formulas you may use. Quickly review the sheet before you begin so that you know what formulas are available.

FCAT Practice

Part 3 Open Ended

Record your answers on a sheet of paper. Show your work. 14. See margin.

14. To get a player out who was running from third base to home, Kahlil threw the ball a distance of 120 feet, from second base toward home plate. Did the ball reach home plate? Explain and show your calculations to justify your answer. (Lesson 1-3)

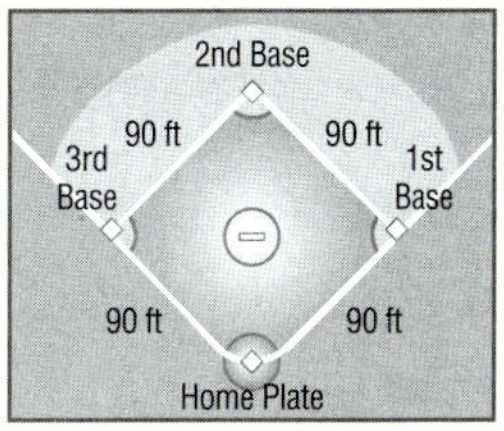

15. Brad's family has subscribed to cable television for 4 years, as shown below.

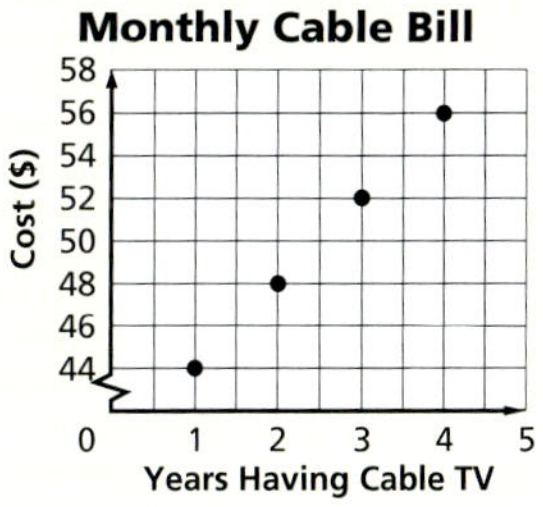

a. Find the slope of a line connecting the points on the graph. (Lesson 3-4) 4

b. Describe what the slope of the line represents. (Lesson 3-4) See margin.

c. If the trend continues, how much will the cable bill be in the tenth year? (Lesson 3-4) \$80

Evaluating Open-Ended Assessment Questions

Open-Ended Assessment questions are graded by using a multilevel rubric that guides you in assessing a student's knowledge of a particular concept.

Goal for Question 14: Determine the distance from 2nd base to home plate, and whether Kahlil's throw was long enough to reach home.

Goal for Question 15: Write a linear equation to represent data points on a graph, interpret the slope of that equation in terms of the data, and use the equation to make a prediction.

Sample Scoring Rubric: The following rubric is a sample scoring device. You may wish to add more detail to this sample to meet your individual scoring needs.

Score	Criteria
4	A correct solution that is supported by well-developed, accurate explanations
3	A generally correct solution, but may contain minor flaws in reasoning or computation
2	A partially correct interpretation and/or solution to the problem
1	A correct solution with no supporting evidence or explanation
0	An incorrect solution indicating no mathematical understanding of the concept or task, or no solution is given

Answers

14. The ball did not reach home plate. The distance between second base and home plate forms the hypotenuse of a right triangle, with second base to third base as one leg, and third base to home plate as the other leg. The Pythagorean Theorem is used to find the distance between second base and home plate. Since $a^2 + b^2 = c^2$, $90^2 + 90^2 = c^2$, or $8{,}100 + 8{,}100 = c^2$. So $c^2 = 16{,}200$, or $c = \sqrt{16{,}200}$. Then $c = 127.3$ ft. Since the ball traveled 120 ft and the distance from second base to home plate is 127.3 ft, the ball did not make it to home plate.

15b. The slope represents the increase in the average monthly cable bill each year.

Pages 128–131, Lesson 3-1

58. Given: $m\angle ABC = m\angle DFE$, $m\angle 1 = m\angle 4$

Prove: $m\angle 2 = m\angle 3$

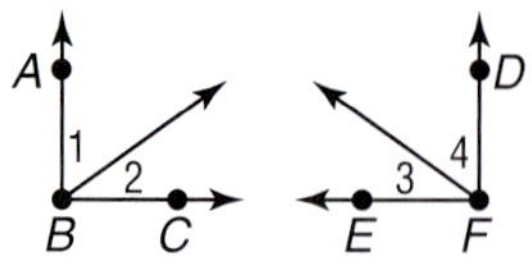

Proof:

Statements	Reasons
1. $m\angle ABC = m\angle DFE$ $m\angle 1 = m\angle 4$	1. Given
2. $m\angle ABC = m\angle 1 + m\angle 2$ $m\angle DFE = m\angle 3 + m\angle 4$	2. Angle Addition Postulate
3. $m\angle 1 + m\angle 2 = m\angle 3 + m\angle 4$	3. Substitution Property
4. $m\angle 4 + m\angle 2 = m\angle 3 + m\angle 4$	4. Substitution Property
5. $m\angle 2 = m\angle 3$	5. Subtraction Property

Pages 136–138, Lesson 3-2

40. Given: $m \parallel n$, ℓ is a transversal.

Prove: $\angle 1$ and $\angle 2$ are supplementary; $\angle 3$ and $\angle 4$ are supplementary.

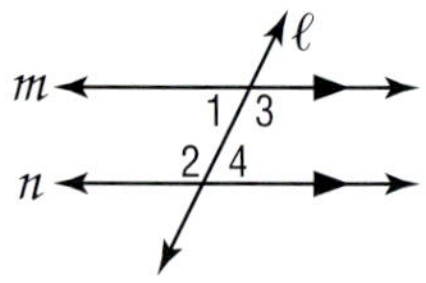

Proof:

Statements	Reasons
1. $m \parallel n$, ℓ is a transversal.	1. Given
2. $\angle 1$ and $\angle 3$ form a linear pair; $\angle 2$ and $\angle 4$ form a linear pair	2. Definition of linear pair
3. $\angle 1$ and $\angle 3$ are supplementary; $\angle 2$ and $\angle 4$ are supplementary	3. If 2 angles form a linear pair, then they are supplementary.
4. $\angle 1 \cong \angle 4$, $\angle 2 \cong \angle 3$	4. Alt. int. $\angle$s $\cong$
5. $\angle 1$ and $\angle 2$ are supplementary; $\angle 3$ and $\angle 4$ are supplementary.	5. Substitution

Pages 142–144, Lesson 3-3

33.

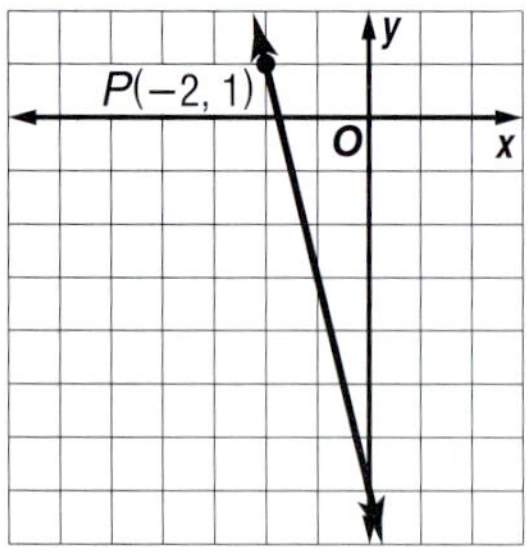

34.

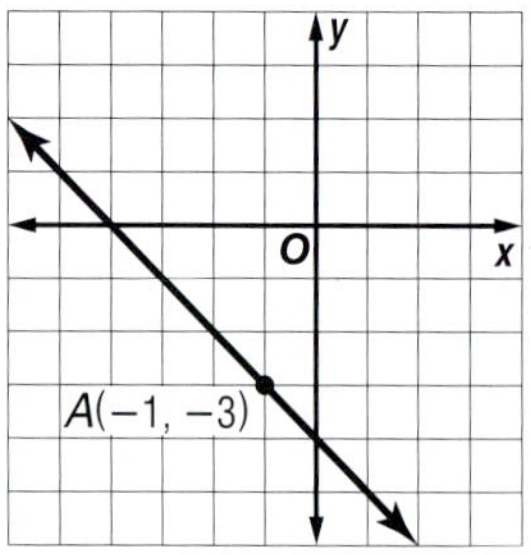

35.

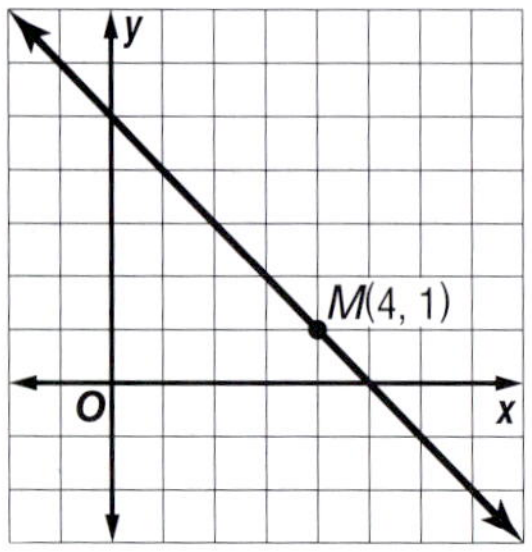

36.

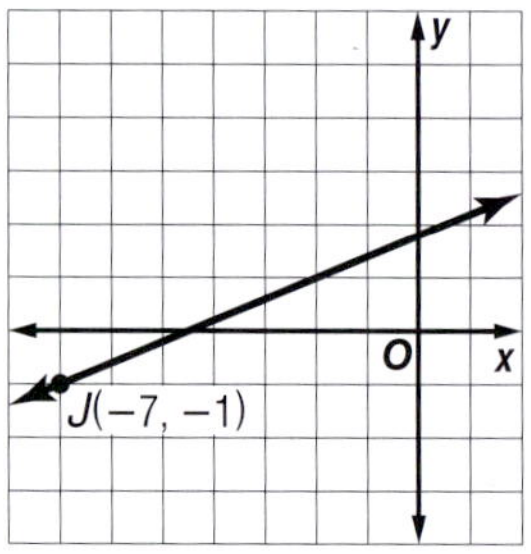

37.

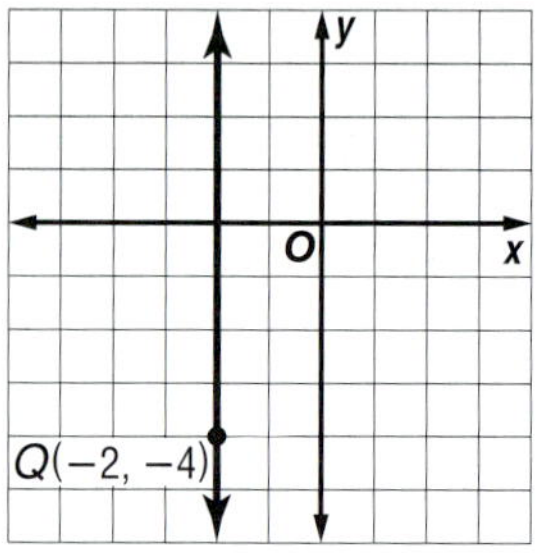

38.

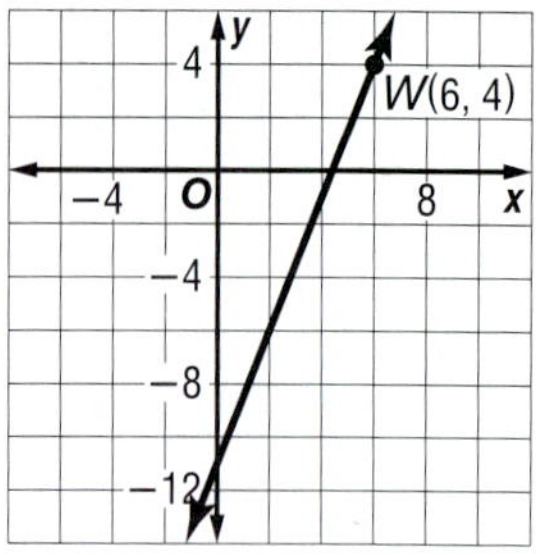

Pages 154–157, Lesson 3-5

10. Given: $\angle 1 \cong \angle 2$

Prove: $\ell \parallel m$

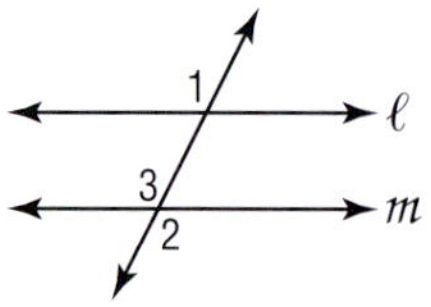

Proof:

Statements	Reasons
1. $\angle 1 \cong \angle 2$	1. Given
2. $\angle 2 \cong \angle 3$	2. Vertical angles are congruent.
3. $\angle 1 \cong \angle 3$	3. Trans. Prop. of $\cong$
4. $\ell \parallel m$	4. If corr. $\angle$s are $\cong$, then lines are $\parallel$.

32. Given: $\angle 1$ and $\angle 2$ are supplementary.

Prove: $\ell \parallel m$

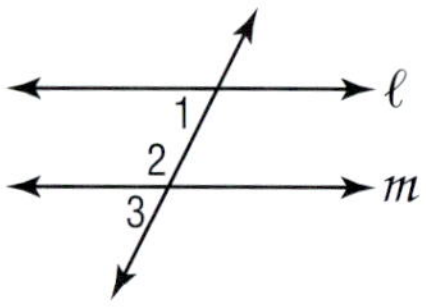

Proof:

Statements	Reasons
1. $\angle 1$ and $\angle 2$ are supplementary.	1. Given
2. $\angle 2$ and $\angle 3$ form a linear pair.	2. Definition of linear pair
3. $\angle 2$ and $\angle 3$ are supplementary.	3. Supplement Th.
4. $\angle 1 \cong \angle 3$	4. $\angle$s suppl. to same $\angle$ are $\cong$.
5. $\ell \parallel m$	5. If corr. $\angle$s are $\cong$, then lines are $\parallel$.

33. Given: $\angle 4 \cong \angle 6$

Prove: $\ell \parallel m$

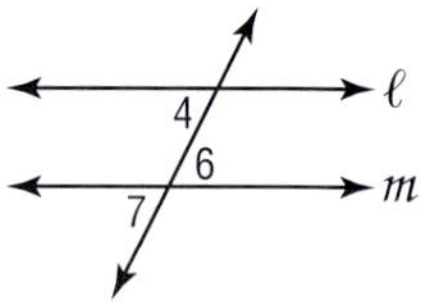

Proof: We know that $\angle 4 \cong \angle 6$. Because $\angle 6$ and $\angle 7$ are vertical angles, they are congruent. By the Transitive Property of Congruence, $\angle 4 \cong \angle 7$. Since $\angle 4$ and $\angle 7$ are corresponding angles, and they are congruent, $\ell \parallel m$.

Pages 162–164, Lesson 3-6

17. $d = 3$;

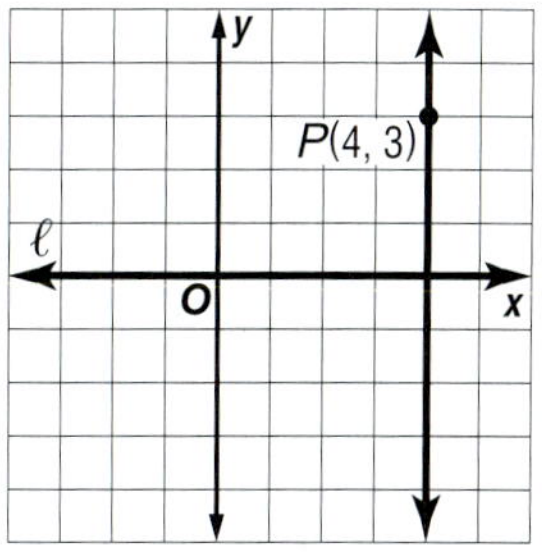

18. $d = \sqrt{26}$;

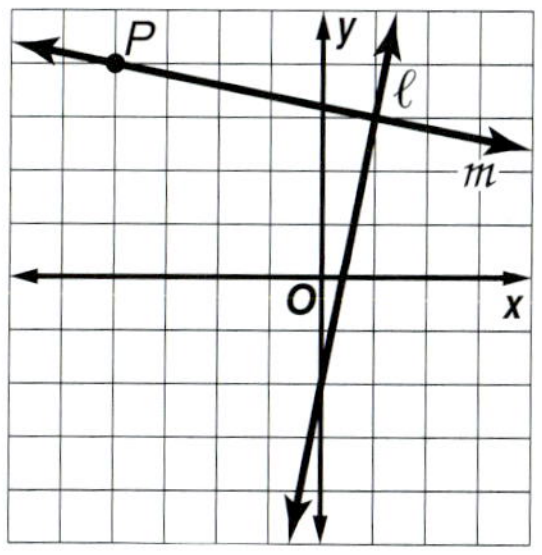

25.

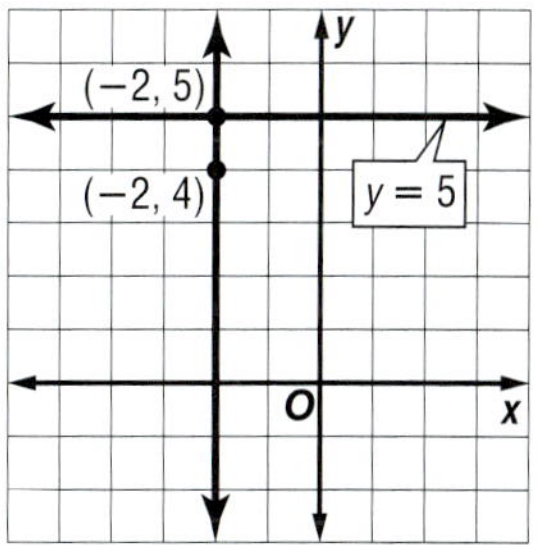

26.

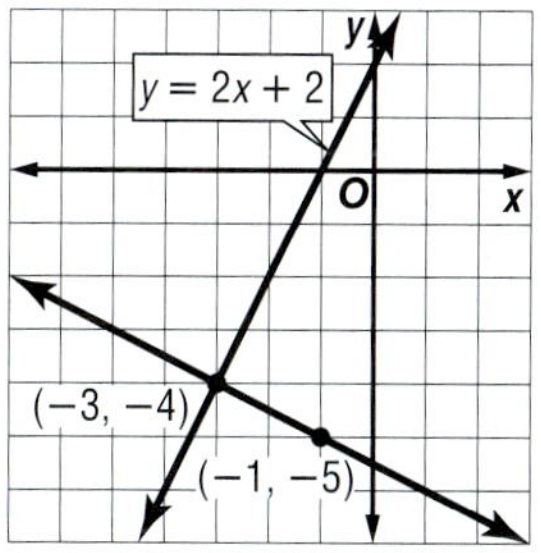

27.

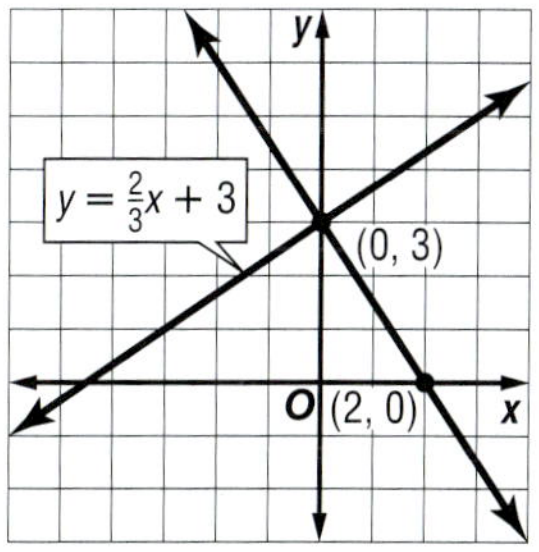

28. Given: ℓ is equidistant to m.
n is equidistant to m.
Prove: $\ell \parallel n$

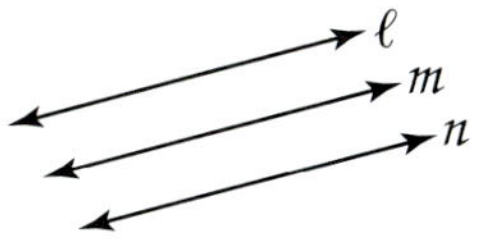

Paragraph proof: We are given that ℓ is equidistant to m, and n is equidistant to m. By definition of equidistant, ℓ is parallel to m, and n is parallel to m. By definition of parallel lines, slope of ℓ = slope of m, and slope of n = slope of m. By substitution, slope of ℓ = slope of n. Then, by definition of parallel lines, $\ell \parallel n$.

32a.

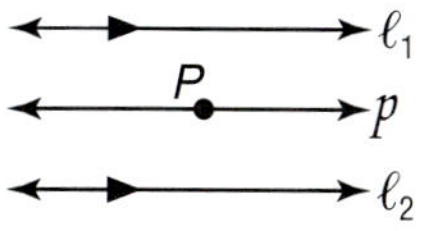

32b.

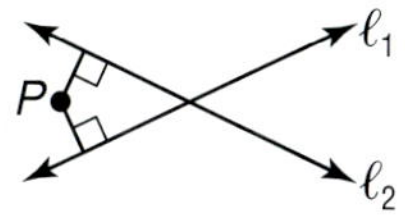

32c.

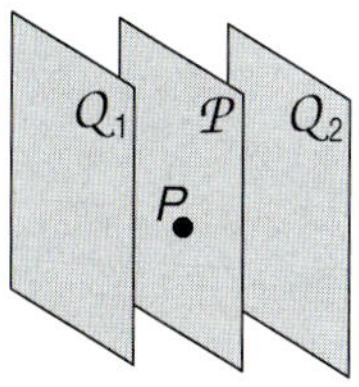

32d.

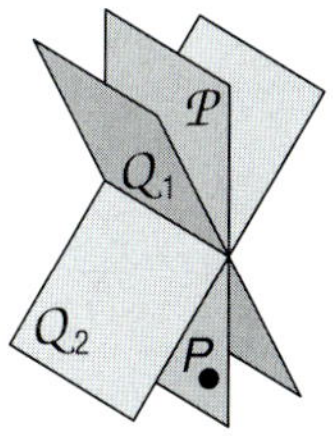

32e.

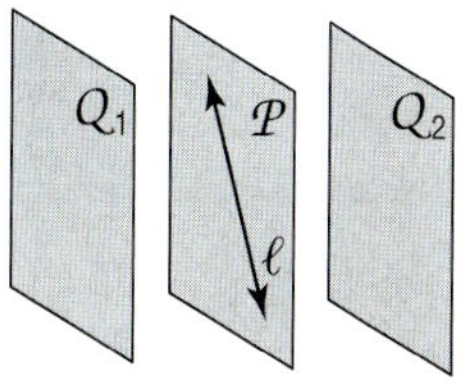

32f.

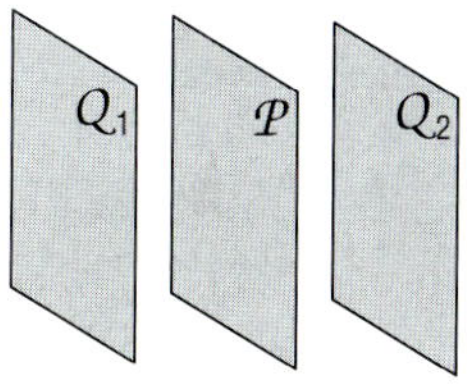

UNIT 2 Notes

Introduction

In this unit students identify and classify triangles by various methods. They learn how to test for and prove triangle congruence and how to write coordinate proofs. Bisectors, medians, and altitudes of triangles are introduced. Students explore inequalities and triangles and learn to use indirect proof.

Students apply their knowledge of ratios and proportions to similar figures and scale factors. They explore proportional parts of triangles and proportional relationships between similar triangles. Finally, students focus on right triangles. They use the Pythagorean Theorem and are introduced to trigonometric ratios.

About the Photograph The large photograph is the Air Force Academy Chapel in Colorado Springs, Colorado. Its 17-spire structure is designed to resemble aircraft in synchronized flight. Have students use the Internet to find out more about this building.

Assessment Options

Unit 2 Test Pages 413–414 of the *Chapter 7 Resource Masters* may be used as a test or review for Unit 2. This assessment contains both multiple-choice and short answer items.

TestCheck and Worksheet Builder

This CD-ROM can be used to create additional unit tests and review worksheets.

UNIT 2 Triangles

You can use triangles and their properties to model and analyze many real-world situations. In this unit, you will learn about relationships in and among triangles, including congruence and similarity.

174 Unit 2 Triangles

WebQuest Internet Project

Who Is Behind This Geometry Concept Anyway?

Have you ever wondered who first developed some of the ideas you are learning in your geometry class? Today, many students use the Internet for learning and research. In this project, you will be using the Internet to research a topic in geometry. You will then prepare a portfolio or poster to display your findings.

Log on to www.geometryonline.com/webquest. Begin your WebQuest by reading the Task.

Continue working on your WebQuest as you study Unit 2.

Lesson	4-6	5-1	6-6	7-1
Page	216	241	325	347

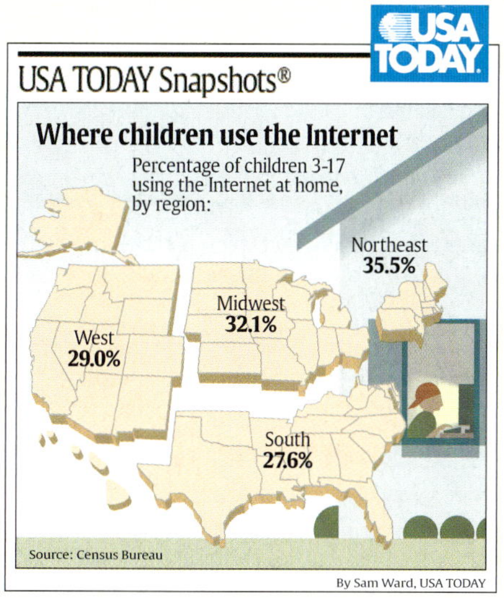

USA TODAY Education Teaching Suggestions

Have students study the USA TODAY Snapshot.

- Ask them which region of the United States has the lowest percentage of children using the Internet. **South**
- Have students speculate about why the Northeast has the greatest percentage of children using the Internet.
- What percentage of children in the Midwest use the Internet? **32.1%**

Additional USA TODAY Snapshots appearing in Unit 2:

Chapter 4 Gross Domestic Product slides in 2001 (p. 206)

Chapter 5 Sources of college information (p. 259)

Chapter 6 Workplace manners declining (p. 296)

Chapter 7 Bruins bring skills to Major League Soccer (p. 347)

WebQuest Internet Project

A WebQuest is an online project in which students do research on the Internet, gather data, and make presentations using word processing, graphing, page-making, or presentation software. In each chapter, students advance to the next step in their WebQuest. At the end of Chapter 7, the project culminates with a presentation of their findings.

Teaching notes and sample answers are available in the *WebQuest and Project Resources.*

Chapter 4

Congruent Triangles

Chapter Overview and Pacing

LESSON OBJECTIVES	PACING (days) Regular		Block	
	Basic/ Average	Advanced	Basic/ Average	Advanced
4-1 Classifying Triangles *(pp. 178–183)* • Identify and classify triangles by angles. • Identify and classify triangles by sides.	1	1	0.5	0.5
4-2 Angles of Triangles *(pp. 184–191)* ***Preview:*** Use a model to find the relationships among the measures of the interior angles of a triangle. • Apply the Angle Sum Theorem. • Apply the Exterior Angle Theorem.	2 (with 4-2 Preview)	1	1 (with 4-2 Preview)	0.5
4-3 Congruent Triangles *(pp. 192–198)* • Name and label corresponding parts of congruent triangles. • Identify congruence transformations.	2	2	1	1
4-4 Proving Congruence—SSS, SAS *(pp. 200–206)* • Use the SSS Postulate to test for triangle congruence. • Use the SAS Postulate to test for triangle congruence.	2	2	1	1
4-5 Proving Congruence—ASA, AAS *(pp. 207–215)* • Use the ASA Postulate to test for triangle congruence. • Use the AAS Theorem to test for triangle congruence. ***Follow-Up:*** Use models to explore congruence in right triangles.	2 (with 4-5 Follow-Up)	2 (with 4-5 Follow-Up)	1 (with 4-5 Follow-Up)	1 (with 4-5 Follow-Up)
4-6 Isosceles Triangles *(pp. 216–221)* • Use properties of isosceles triangles. • Use properties of equilateral triangles.	2	2	1	1
4-7 Triangles and Coordinate Proof *(pp. 222–226)* • Position and label triangles for use in coordinate proofs. • Write coordinate proofs.	2	2	1	1
Study Guide and Practice Test *(pp. 227–231)* **Standardized Test Practice** *(pp. 232–233)*	1	1	1	0.5
Chapter Assessment	1	1	0.5	0.5
TOTAL	**15**	**14**	**8**	**7**

Pacing suggestions for the entire year can be found on pages T20–T21.

Timesaving Tools
TeacherWorks™
All-In-One Planner and Resource Center
See pages T5 and T21.

Chapter Resource Manager

Study Guide and Intervention (Chapter 4 Resource Masters)	Practice (Skills and Average) (Chapter 4 Resource Masters)	Reading to Learn Mathematics (Chapter 4 Resource Masters)	Enrichment (Chapter 4 Resource Masters)	Assessment (Chapter 4 Resource Masters)	Prerequisite Skills Workbook	Applications*	5-Minute Check Transparencies	Interactive Chalkboard	GeomPASS: Tutorial Plus (lessons)	Materials
183–184	185–186	187	188		1–2		4-1	4-1		patty paper, protractor, grid paper
189–190	191–192	193	194	239	81–84	GCC 23	4-2	4-2		(*Preview:* protractor, scissors) straightedge
195–196	197–198	199	200			SC 7	4-3	4-3		straightedge
201–202	203–204	205	206	239, 241	1–2		4-4	4-4		straightedge, compass, scissors
207–208	209–210	211	212				4-5	4-5	10	straightedge, compass, scissors, patty paper, protractor, ruler (*Follow-Up:* ruler, protractor)
213–214	215–216	217	218	240		SC 8	4-6	4-6		ruler, scissors, protractor, tracing paper
219–220	221–222	223	224	240			4-7	4-7		grid paper, straightedge
				225–238, 242–244						

**Key to Abbreviations:* GCC = Graphing Calculator and Computer Masters
SC = School-to-Career Masters

Chapter 4

Mathematical Connections and Background

Continuity of Instruction

Prior Knowledge

In Chapter 1, students calculated distance between points on the coordinate plane using the Distance Formula. In Chapter 3, they learned that an intersection between parallel lines and a transversal creates a variety of congruent and supplementary angles.

This Chapter

In this chapter, students prove triangles congruent using SSS, SAS, ASA, and AAS. They also learn how to write two new types of proofs, the flow proof and the coordinate proof. Students classify triangles according to their angles or sides and apply the Angle Sum Theorem and the Exterior Angle Theorem. The special properties of isosceles and equilateral triangles are introduced, and students are expected to use those properties in proofs. Students also position and label triangles for use in coordinate proofs.

Future Connections

Students will need the knowledge gained from this chapter to master the skills taught later in this course. They must understand triangle congruence to be successful scholars of trigonometry, a precursor to calculus and all courses in higher math.

4-1 Classifying Triangles

Triangles can be classified based on their angle measures. In an acute triangle, all of the angles are acute. In an obtuse triangle, one of the angles is obtuse. In a right triangle, one angle measures 90. When all of the angles of an acute triangle are congruent, it is called an equiangular triangle.

Triangles can also be classified according to their number of congruent sides. No two sides of a scalene triangle are congruent. At least two sides of an isosceles triangle are congruent. All of the sides of an equilateral triangle are congruent. Equilateral triangles are a special kind of isosceles triangle.

4-2 Angles of Triangles

The Angle Sum Theorem states that the sum of the measures of the interior angles of a triangle is always 180. This theorem can be applied to any triangle. It also leads to the Third Angle Theorem: If two angles of one triangle are congruent to two angles of a second triangle, then the third angles of the triangles are congruent.

Each angle of a triangle has an exterior angle, which is formed by one side of the triangle and the extension of another side. The interior angles of the triangle not adjacent to a given exterior angle are called remote interior angles. The measure of an exterior angle of a triangle is equal to the sum of the measures of the two remote interior angles. This is the Exterior Angle Theorem.

This lesson also introduces flow proofs. A flow proof organizes a series of statements in logical order. Arrows are used to indicate the order of the statements. Flow proofs can be written horizontally or vertically.

4-3 Congruent Triangles

Two triangles are congruent if and only if their corresponding parts are congruent. Certain transformations, including a slide, flip, and turn, do not affect congruence. These transformations are called congruence transformations.

Congruence of triangles, like that of angles and segments, is reflexive, symmetric, and transitive.

4-4 Proving Congruence—SSS, SAS

In this lesson you will construct a triangle in which three sides are congruent to the three sides of a given triangle. This activity demonstrates the Side-Side-Side Postulate. Also written as SSS, it states that if the sides of one triangle are congruent to the sides of a second triangle, then the triangles are congruent.

You will also construct a triangle in which two sides are congruent to two sides of a given triangle and the included angle is congruent to the included angle in the given triangle. This activity demonstrates the Side-Angle-Side Postulate, also written SAS. It states that if two sides and the included angle of one triangle are congruent to two sides and the included angle of another triangle, then the triangles are congruent.

4-5 Proving Congruence—ASA, AAS

The Angle-Side-Angle Postulate, written as ASA, works because the measures of two angles of a triangle and the side between them form a unique triangle. The postulate states that if two angles and the included side of one triangle are congruent to two angles and the included side of another triangle, then the triangles are congruent.

The Angle-Angle-Side, or AAS, Theorem follows from the ASA Postulate: If two angles and a nonincluded side of one triangle are congruent to the corresponding two angles and side of a second triangle, then the two triangles are congruent.

Right triangles have their own theorems to prove congruence. One of those is the LL Congruence Theorem, which is the SAS Postulate applied to right triangles. It states that if the legs of one right triangle are congruent to the corresponding legs of another right triangle, then the triangles are congruent. The HA Theorem is based on the AAS Theorem: If the hypotenuse and acute angle of one right triangle are congruent to the hypotenuse and corresponding acute angle of another right triangle, then the two triangles are congruent. The LA Theorem states that if one leg and an acute angle of one right triangle are congruent to the corresponding leg and acute angle of another right triangle, then the triangles are congruent. The HL Postulate is based on SSA, a test that only works for right triangles. It states that if the hypotenuse and leg of one right triangle are congruent to the hypotenuse and corresponding leg of another right triangle, then the triangles are congruent.

4-6 Isosceles Triangles

Isosceles triangles have special terminology for their parts. The angle formed by the congruent sides is called the vertex angle. The two angles formed by the base and one of the congruent sides are called base angles. The congruent sides are called legs. Isosceles triangles also have special properties recognized in Isosceles Triangle Theorem and its converse: If two sides of a triangle are congruent, then the angles opposite those sides are congruent.

This theorem leads to corollaries about the angles of an equilateral triangle. The first states that a triangle is equilateral if and only if it is equiangular. The second states that each angle of an equilateral triangle measures 60°.

4-7 Triangles and Coordinate Proof

The coordinate plane can be used in combination with algebra in a new method of proof called coordinate proof. Before beginning a coordinate proof, you will need to place the figure in the coordinate plane. It is important that you use coordinates that make computation as simple as possible. Using the origin as a vertex or center will help, and you should place at least one side of a polygon on an axis. If possible, keep the figure within the first quadrant.

Once the triangle is placed, you can proceed with the proof. The Distance Formula, Slope Formula, and Midpoint Formula are often used in coordinate proof.

Chapter 4

DAILY INTERVENTION and Assessment

	Type	Student Edition	Teacher Resources	Technology/Internet
INTERVENTION	Ongoing	Prerequisite Skills, pp. 177, 183, 191, 198, 206, 213, 221 Practice Quiz 1, p. 198 Practice Quiz 2, p. 221	5-Minute Check Transparencies *Prerequisite Skills Workbook*, pp. 1–2, 81–84 Quizzes, *CRM* pp. 239–240 Mid-Chapter Test, *CRM* p. 241 Study Guide and Intervention, *CRM* pp. 183–184, 189–190, 195–196, 201–202, 207–208, 213–214, 219–220	GeomPASS: Tutorial Plus www.geometryonline.com/self_check_quiz www.geometryonline.com/extra_examples
	Mixed Review	pp. 183, 191, 198, 206, 213, 221, 226	Cumulative Review, *CRM* p. 242	
	Error Analysis	Find the Error, pp. 188, 203 Common Misconceptions, p. 178	Find the Error, *TWE* pp. 189, 203 Unlocking Misconceptions, *TWE* p. 202 Tips for New Teachers, *TWE* p. 210	
	Standardized Test Practice	pp. 183, 191, 198, 206, 213, 217, 219, 221, 226, 231, 232, 233	*TWE* pp. 232–233 Standardized Test Practice, *CRM* pp. 243–244	Standardized Test Practice CD-ROM www.geometryonline.com/standardized_test
ASSESSMENT	Open-Ended Assessment	Writing in Math, pp. 183, 191, 198, 205, 213, 221, 226 Open Ended, pp. 180, 188, 195, 203, 210, 219, 224 Standardized Test, p. 233	Modeling: *TWE* p. 198 Speaking: *TWE* pp. 183, 206, 221, 226 Writing: *TWE* pp. 191, 213 Open-Ended Assessment, *CRM* p. 237	
	Chapter Assessment	Study Guide, pp. 227–230 Practice Test, p. 231	Multiple-Choice Tests (Forms 1, 2A, 2B), *CRM* pp. 225–230 Free-Response Tests (Forms 2C, 2D, 3), *CRM* pp. 231–236 Vocabulary Test/Review, *CRM* p. 238	TestCheck and Worksheet Builder (see below) MindJogger Videoquizzes www.geometryonline.com/vocabulary_review www.geometryonline.com/chapter_test

Key to Abbreviations: TWE = Teacher Wraparound Edition; CRM = Chapter Resource Masters

Additional Intervention Resources

The Princeton Review's *Cracking the SAT & PSAT*

The Princeton Review's *Cracking the ACT*

TestCheck and Worksheet Builder

This **networkable** software has three modules for intervention and assessment flexibility:

- **Worksheet Builder** to make worksheet and tests
- **Student Module** to take tests on screen (optional)
- **Management System** to keep student records (optional)

Special banks are included for SAT, ACT, TIMSS, NAEP, and End-of-Course tests.

Intervention Technology

GeomPASS: Tutorial Plus CD-ROM offers a complete, self-paced geometry curriculum.

Geometry Lesson	GeomPASS Lesson
4-5	**10** *Solving Problems with Congruent Triangles*

Intervention at Home

Log on for student study help.

- For each lesson in the Student Edition, there are Extra Examples and Self-Check Quizzes.
  www.geometryonline.com/extra_examples
  www.geometryonline.com/self_check_quiz
- For chapter review, there is vocabulary review, test practice, and standardized test practice.
  www.geometryonline.com/vocabulary_review
  www.geometryonline.com/chapter_test
  www.geometryonline.com/standardized_test

For more information on Intervention and Assessment, see pp. T8–T11.

Reading and Writing in Mathematics

Glencoe Geometry provides numerous opportunities to incorporate reading and writing into the mathematics classroom.

Student Edition

- Foldables Study Organizer, p. 177
- Concept Check questions require students to verbalize and write about what they have learned in the lesson. (pp. 180, 188, 195, 203, 210, 219, 224)
- Reading Mathematics, p. 199
- Writing in Math questions in every lesson, pp. 183, 191, 198, 205, 213, 221, 226
- Reading Study Tip, pp. 186, 207
- WebQuest, p. 216

Teacher Wraparound Edition

- Foldables Study Organizer, pp. 177, 227
- Study Notebook suggestions, pp. 181, 184, 188, 194, 199, 203, 210, 214, 219, 224
- Modeling activities, p. 198
- Speaking activities, pp. 183, 206, 221, 226
- Writing activities, pp. 191, 213
- ELL Resources, pp. 176, 182, 190, 197, 199, 205, 212, 220, 225, 227

Additional Resources

- Vocabulary Builder worksheets require students to define and give examples for key vocabulary terms as they progress through the chapter. (*Chapter 4 Resource Masters,* pp. vii-viii)
- Proof Builder helps students learn and understand theorems and postulates from the chapter. (*Chapter 4 Resource Masters,* pp. ix–x)
- Reading to Learn Mathematics master for each lesson (*Chapter 4 Resource Masters*, pp. 187, 193, 199, 205, 211, 217, 223)
- *Vocabulary PuzzleMaker* software creates crossword, jumble, and word search puzzles using vocabulary lists that you can customize.
- *Teaching Mathematics with Foldables* provides suggestions for promoting cognition and language.
- *Reading Strategies for the Mathematics Classroom*
- *WebQuest and Project Resources*

For more information on Reading and Writing in Mathematics, see pp. T6–T7.

Chapter 4 Notes

What You'll Learn

Have students read over the list of objectives and make a list of any words with which they are not familiar.

Why It's Important

Point out to students that this is only one of many reasons why each objective is important. Others are provided in the introduction to each lesson.

Lesson	NCTM Standards	Local Objectives
4-1	3, 6, 8, 9, 10	
4-2 Preview	3, 6, 10	
4-2	3, 6, 8, 9, 10	
4-3	3, 6, 8, 9, 10	
4-4	3, 6, 7, 8, 9, 10	
4-5	3, 6, 7, 8, 9, 10	
4-5 Follow-Up	3, 7, 10	
4-6	3, 6, 7, 8, 9, 10	
4-7	3, 6, 7, 8, 9, 10	

Key to NCTM Standards:

1=Number & Operations, 2=Algebra, 3=Geometry, 4=Measurement, 5=Data Analysis & Probability, 6=Problem Solving, 7=Reasoning & Proof, 8=Communication, 9=Connections, 10=Representation

Chapter 4 Congruent Triangles

What You'll Learn

- **Lesson 4-1** Classify triangles.
- **Lesson 4-2** Apply the Angle Sum Theorem and the Exterior Angle Theorem.
- **Lesson 4-3** Identify corresponding parts of congruent triangles.
- **Lessons 4-4 and 4-5** Test for triangle congruence using SSS, SAS, ASA, and AAS.
- **Lesson 4-6** Use properties of isosceles and equilateral triangles.
- **Lesson 4-7** Write coordinate proofs.

Key Vocabulary

- exterior angle (p. 186)
- flow proof (p. 187)
- corollary (p. 188)
- congruent triangles (p. 192)
- coordinate proof (p. 222)

Why It's Important

Triangles are found everywhere you look. Triangles with the same size and shape can even be found on the tail of a whale. *You will learn more about orca whales in Lesson 4-4.*

Vocabulary Builder

ELL

The Key Vocabulary list introduces students to some of the main vocabulary terms included in this chapter. For a more thorough vocabulary list with pronunciations of new words, give students the Vocabulary Builder worksheets found on pages vii and viii of the *Chapter 4 Resource Masters*. Encourage them to complete the definition of each term as they progress through the chapter. You may suggest that they add these sheets to their study notebooks for future reference when studying for the Chapter 4 test.

Getting Started

Prerequisite Skills To be successful in this chapter, you'll need to master these skills and be able to apply them in problem-solving situations. Review these skills before beginning Chapter 4.

For Lesson 4-1 — Solve Equations

Solve each equation. *(For review, see pages 737 and 738.)*

1. $2x + 18 = 5$ $-6\frac{1}{2}$ **2.** $3m - 16 = 12$ $9\frac{1}{3}$ **3.** $4y + 12 = 16$ 1

4. $10 = 8 - 3z$ $-\frac{2}{3}$ **5.** $6 = 2a + \frac{1}{2}$ $2\frac{3}{4}$ **6.** $\frac{2}{3}b + 9 = -15$ -36

For Lessons 4-2, 4-4, and 4-5 — Congruent Angles

Name the indicated angles or pairs of angles if $p \parallel q$ and $m \parallel \ell$. *(For review, see Lesson 3-1.)*

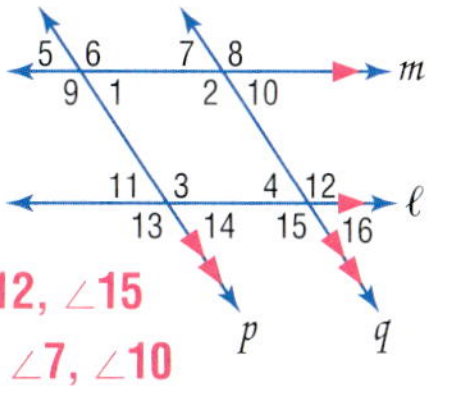

7. angles congruent to ∠8 ∠2, ∠12, ∠15, ∠6, ∠9, ∠3, ∠13

8. angles congruent to ∠13 ∠2, ∠12, ∠15, ∠6, ∠9, ∠3, ∠8

9. angles supplementary to ∠1 ∠6, ∠9, ∠3, ∠13, ∠2, ∠8, ∠12, ∠15

10. angles supplementary to ∠12 ∠4, ∠16, ∠11, ∠14, ∠5, ∠1, ∠7, ∠10

For Lessons 4-3 and 4-7 — Distance Formula

Find the distance between each pair of points. Round to the nearest tenth. *(For review, see Lesson 1-3.)*

11. $(6, 8), (-4, 3)$ 11.2 **12.** $(-15, 12), (6, 18)$ 21.8 **13.** $(11, -8), (-3, -4)$ 14.6

FOLDABLES™ Study Organizer

Triangles Make this Foldable to help you organize your notes about triangles. Begin with two sheets of grid paper.

Step 1 Fold

Fold each sheet of paper along the diagonal.

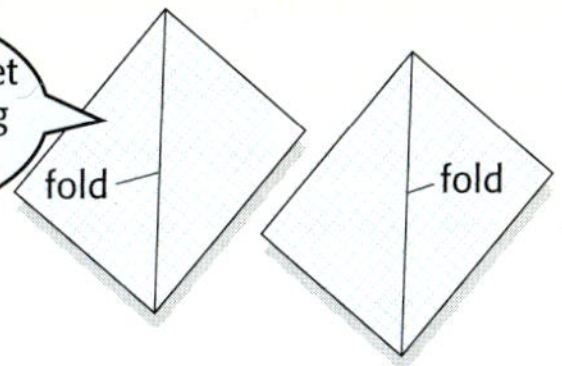

Step 2 Cut

Cut 1-inch slits into the folds of one sheet. For the other sheet, cut the fold in the middle, leaving one inch on each side.

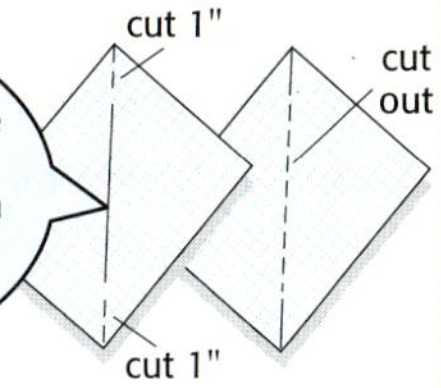

Step 3 Roll up and Slip

Roll up the slitted sheet and slip it through the hole of the other sheet.

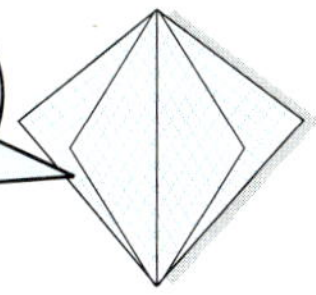

Step 4 Label

Label each page with a different heading dealing with triangles.

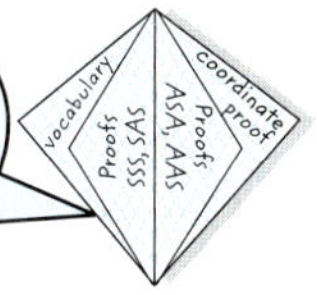

Reading and Writing As you read and study the chapter, use your journal for sketches and examples of terms associated with triangles and sample proofs.

Getting Started

This section provides a review of the basic concepts needed before beginning Chapter 4. Page references are included for additional student help.

Additional review is provided in the *Prerequisite Skills Workbook,* pages 1–2, 81–84.

Prerequisite Skills in the Getting Ready for the Next Lesson section at the end of each exercise set review a skill needed in the next lesson.

For Lesson	Prerequisite Skill
4-2	Angles Formed by Parallel Lines and a Transversal, p. 183
4-3	Properties of Congruence, p. 191
4-4	Distance Formula, p. 198
4-5	Bisectors of Segments and Angles, p. 206
4-6	Classification of triangles by sides, p. 213
4-7	Finding Midpoints, p. 221

FOLDABLES™ Study Organizer

For more information about Foldables, see *Teaching Mathematics with Foldables.*

Writer's Journal Use this Foldable for student writing about triangles. After students make their Foldable journal, have them label the pages to correspond to the seven lessons in this chapter. Students use their Foldable to take notes, define terms, record concepts, and write examples. Writer's journals can also be used by students to record the direction and progress of learning, describe positive and negative experiences during learning, write about personal associations and experiences called to mind during learning, and to list examples of ways in which new knowledge has or will be used in their daily lives.

4-1 Lesson Notes

1 Focus

5-Minute Check Transparency 4-1 Use as a quiz or review of Chapter 3.

Mathematical Background notes are available for this lesson on p. 176C.

Why are triangles important in construction?

Ask students:

- Why does the shape of a triangle offer good support for a roof? **As gravity and the weight of roof material act on the two top legs of the triangle, the third leg prevents them from collapsing or spreading apart. Triangular supports within the truss offer even more support.**
- How would you classify the angles of a triangular truss with angle measures 96°, 42°, and 42°? **obtuse, acute, acute**

4-1 Classifying Triangles

Sunshine State Standards
MA.A.3.4.3, MA.C.3.4.2

Vocabulary

- acute triangle
- obtuse triangle
- right triangle
- equiangular triangle
- scalene triangle
- isosceles triangle
- equilateral triangle

What You'll Learn

- Identify and classify triangles by angles.
- Identify and classify triangles by sides.

Why are triangles important in construction?

Many structures use triangular shapes as braces for construction. The roof sections of houses are made of triangular trusses that support the roof and the house.

CLASSIFY TRIANGLES BY ANGLES Recall that a triangle is a three-sided polygon. Triangle *ABC*, written $\triangle ABC$, has parts that are named using the letters *A*, *B*, and *C*.

- The sides of $\triangle ABC$ are $\overline{AB}$, $\overline{BC}$, and $\overline{CA}$.
- The vertices are *A*, *B*, and *C*.
- The angles are $\angle ABC$ or $\angle B$, $\angle BCA$ or $\angle C$, and $\angle BAC$ or $\angle A$.

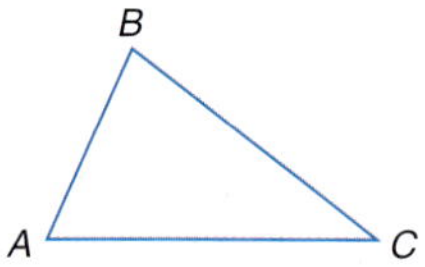

There are two ways to classify triangles. One way is by their angles. All triangles have at least two acute angles, but the third angle is used to classify the triangle.

Study Tip

Common Misconceptions
These classifications are distinct groups. For example, a triangle cannot be right and acute.

Key Concept — Classifying Triangles by Angles

In an **acute triangle**, all of the angles are acute.

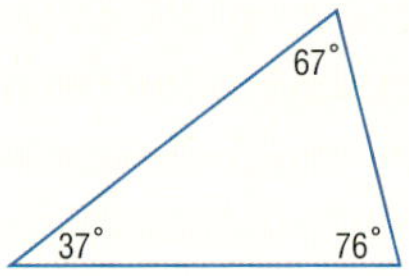

all angle measures < 90

In an **obtuse triangle**, one angle is obtuse.

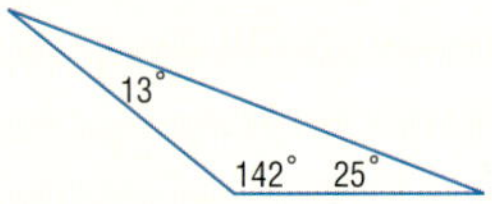

one angle measure > 90

In a **right triangle**, one angle is right.

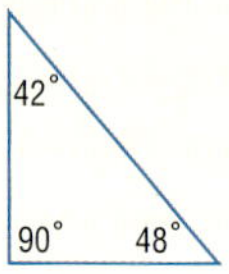

one angle measure = 90

An acute triangle with all angles congruent is an **equiangular triangle**.

Example 1 Classify Triangles by Angles

ARCHITECTURE The roof of this house is made up of three different triangles. Use a protractor to classify $\triangle DFH$, $\triangle DFG$, and $\triangle HFG$ as *acute*, *equiangular*, *obtuse*, or *right*.

$\triangle DFH$ has all angles with measures less than 90, so it is an acute triangle. $\triangle DFG$ and $\triangle HFG$ both have one angle with measure equal to 90. Both of these are right triangles.

Resource Manager

Workbook and Reproducible Masters

Chapter 4 Resource Masters
- Study Guide and Intervention, pp. 183–184
- Skills Practice, p. 185
- Practice, p. 186
- Reading to Learn Mathematics, p. 187
- Enrichment, p. 188

Prerequisite Skills Workbook, pp. 1–2
Teaching Geometry With Manipulatives Masters, pp. 1, 16, 69

Transparencies

5-Minute Check Transparency 4-1
Answer Key Transparencies

Technology

Interactive Chalkboard

CLASSIFY TRIANGLES BY SIDES Triangles can also be classified according to the number of congruent sides they have. To indicate that sides of a triangle are congruent, an equal number of hash marks are drawn on the corresponding sides.

Study Tip

Double Classification
Triangles can be classified by both angles and sides. For example, a scalene triangle can be acute, right, or obtuse.

Key Concept — *Classifying Triangles by Sides*

No two sides of a **scalene triangle** are congruent.	At least two sides of an **isosceles triangle** are congruent.	All of the sides of an **equilateral triangle** are congruent.

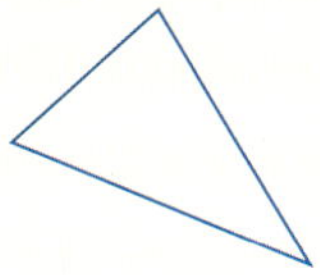
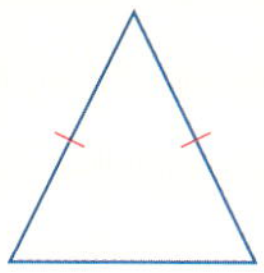
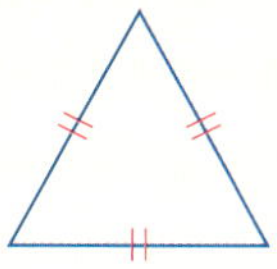

An equilateral triangle is a special kind of isosceles triangle.

Geometry Activity

Equilateral Triangles

Model

- Align three pieces of patty paper as indicated. Draw a dot at *X*.
- Fold the patty paper through *X* and *Y* and through *X* and *Z*.

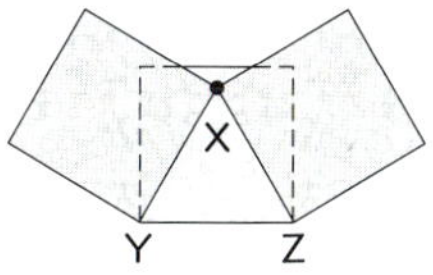

1. Yes, all edges of the paper are an equal length.

Analyze 2–3. See students' work.

1. Is $\triangle XYZ$ equilateral? Explain.
2. Use three pieces of patty paper to make a triangle that is isosceles, but not equilateral.
3. Use three pieces of patty paper to make a scalene triangle.

Example 2 Classify Triangles by Sides

Identify the indicated type of triangle in the figure.

a. isosceles triangles

Isosceles triangles have at least two sides congruent. So, $\triangle ABD$ and $\triangle EBD$ are isosceles.

b. scalene triangles

Scalene triangles have no congruent sides. $\triangle AEB$, $\triangle AED$, $\triangle ACB$, $\triangle ACD$, $\triangle BCE$, and $\triangle DCE$ are scalene.

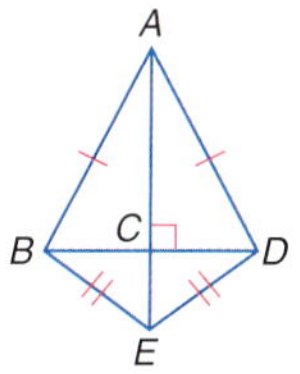

Example 3 Find Missing Values

ALGEBRA **Find x and the measure of each side of equilateral triangle RST if $RS = x + 9$, $ST = 2x$, and $RT = 3x - 9$.**

Since $\triangle RST$ is equilateral, each side has the same length. So $RS = ST$.

$x + 9 = 2x$ Substitution

$9 = x$ Subtract x from each side.

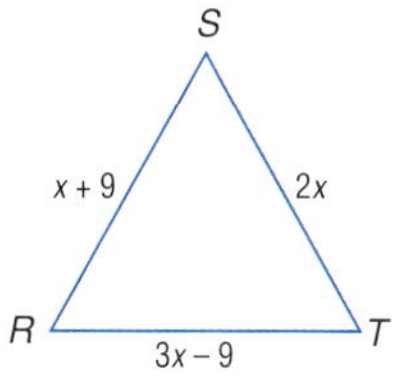

Next, substitute to find the length of each side.

$RS = x + 9$	$ST = 2x$	$RT = 3x - 9$
$= 9 + 9$	$= 2(9)$	$= 3(9) - 9$
$= 18$	$= 18$	$= 27 - 9$ or 18

For $\triangle RST$, $x = 9$ and the measure of each side is 18.

www.geometryonline.com/extra_examples/fcat

Geometry Activity

Materials: patty paper, pencil

- Have students use a protractor to measure the angles of $\triangle XYZ$, and ask them to consider whether an equilateral triangle could be obtuse or right.
- Students can fold a corner of the patty paper to form a right triangle. They can also use a ruler and protractor to draw and label different combinations of triangles, such as an acute scalene triangle, an obtuse isosceles triangle, and so on.

2 Teach

CLASSIFY TRIANGLES BY ANGLES

In-Class Example Power Point®

1 **ARCHITECTURE** The triangular truss below is modeled for steel construction. Use a protractor to classify $\triangle JMN$, $\triangle JKO$, and $\triangle OLN$ as *acute*, *equiangular*, *obtuse* or *right*.

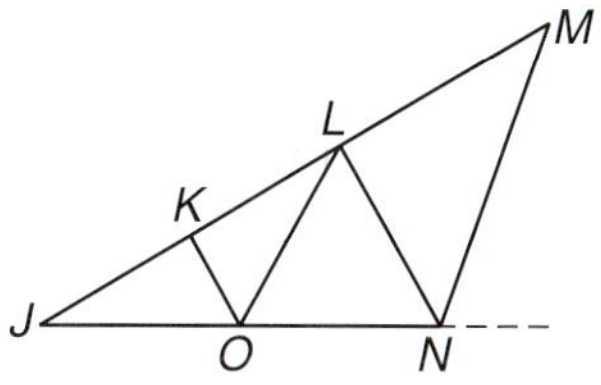

$\triangle JMN$ is obtuse. $\triangle JKO$ is right. $\triangle OLN$ is equiangular.

CLASSIFY TRIANGLES BY SIDES

In-Class Examples Power Point®

Teaching Tip Explain to students that all equilateral triangles are isosceles, but not all isosceles triangles are equilateral.

2 Identify the indicated triangles in the figure if $\overline{UV} \cong \overline{VX} \cong \overline{UX}$.

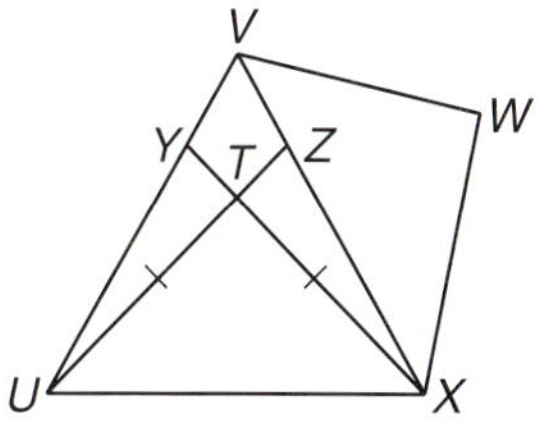

a. isosceles triangles $\triangle UTX$, $\triangle UVX$

b. scalene triangles $\triangle VYX$, $\triangle ZTX$, $\triangle VZU$, $\triangle YTU$, $\triangle VWX$, $\triangle ZUX$, $\triangle YXU$

3 **ALGEBRA** Find d and the measure of each side of equilateral triangle KLM if $KL = d + 2$, $LM = 12 - d$, and $KM = 4d - 13$. $d = 5$ and the measure of each side is 7.

In-Class Example

Power Point®

4 COORDINATE GEOMETRY Find the measures of the sides of $\triangle RST$. Classify the triangle by sides.

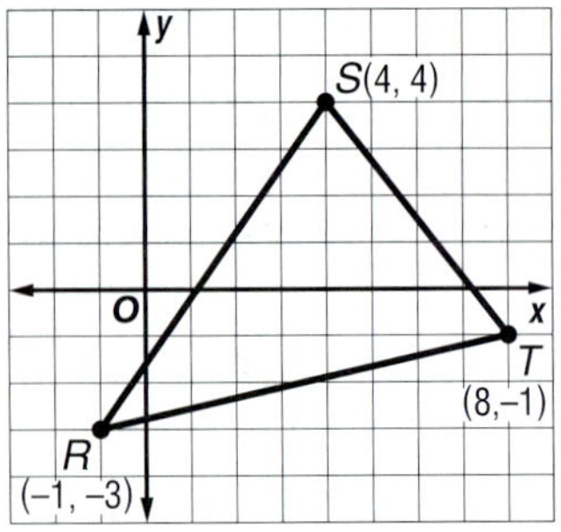

$RS = \sqrt{74}$; $ST = \sqrt{41}$; $RT = \sqrt{85}$; $\triangle RST$ is scalene.

Answers

1. Triangles are classified by sides and angles. For example, a triangle can have a right angle and have no two sides congruent.
2. Sample answer:

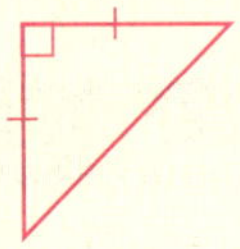

Study Tip

Look Back To review the **Distance Formula**, see Lesson 1-3.

Example 4 Use the Distance Formula

COORDINATE GEOMETRY **Find the measures of the sides of $\triangle DEC$. Classify the triangle by sides.**

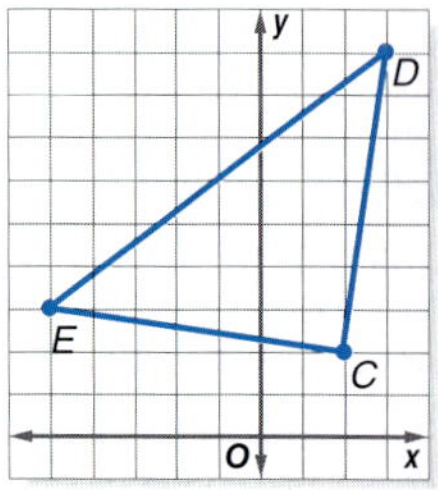

Use the Distance Formula to find the lengths of each side.

$$EC = \sqrt{(-5-2)^2 + (3-2)^2} = \sqrt{49+1} = \sqrt{50}$$

$$ED = \sqrt{(-5-3)^2 + (3-9)^2} = \sqrt{64+36} = \sqrt{100}$$

$$DC = \sqrt{(3-2)^2 + (9-2)^2} = \sqrt{1+49} = \sqrt{50}$$

Since $\overline{EC}$ and $\overline{DC}$ have the same length, $\triangle DEC$ is isosceles.

Check for Understanding

Concept Check

1. **Explain** how a triangle can be classified in two ways. 1–2. See margin.
2. **OPEN ENDED** Draw a triangle that is isosceles and right.

Determine whether each of the following statements is *always*, *sometimes*, or *never* true. Explain. 3. Always; equiangular triangles have three acute angles.

3. Equiangular triangles are also acute.
4. Right triangles are acute. Never; right triangles have one right angle and acute triangles have all acute angles.

Guided Practice

GUIDED PRACTICE KEY	
Exercises	Examples
5–8	1
9, 10	3
11	4
12	2

Use a protractor to classify each triangle as *acute*, *equiangular*, *obtuse*, or *right*.

5. obtuse

6. equiangular

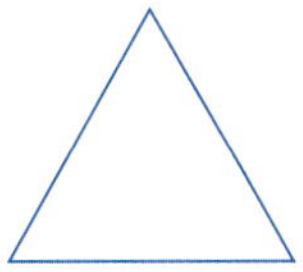

7. Identify the obtuse triangles if $\angle MJK \cong \angle KLM$, $m\angle MJK = 126$, and $m\angle JNM = 52$.

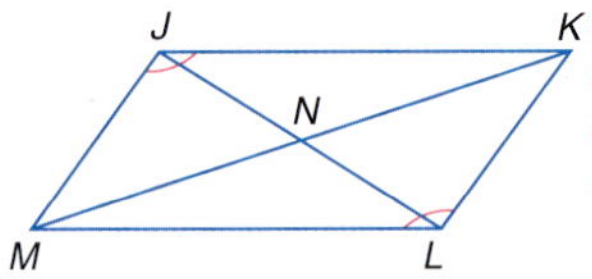

$\triangle MJK$, $\triangle KLM$, $\triangle JKN$, $\triangle LMN$

8. Identify the right triangles if $\overline{IJ} \parallel \overline{GH}$, $\overline{GH} \perp \overline{DF}$, and $\overline{GI} \perp \overline{EF}$.

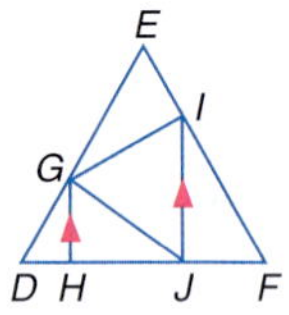

$\triangle GHD$, $\triangle GHJ$, $\triangle IJF$, $\triangle EIG$

9. **ALGEBRA** Find x, JM, MN, and JN if $\triangle JMN$ is an isosceles triangle with $\overline{JM} \cong \overline{MN}$.

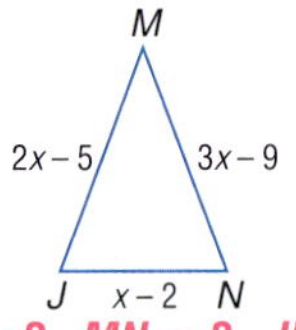

$x = 4$, $JM = 3$, $MN = 3$, $JN = 2$

10. **ALGEBRA** Find x, QR, RS, and QS if $\triangle QRS$ is an equilateral triangle.

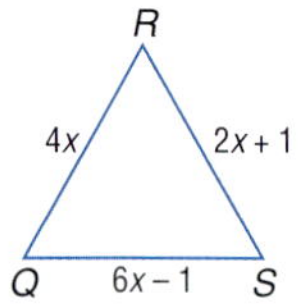

$x = \frac{1}{2}$, $QR = 2$, $RS = 2$, $QS = 2$

DAILY INTERVENTION

Differentiated Instruction

Naturalist Certain forms of algae are triangular in structure. Three-sided leaves are said to have a triangular shape. Some wings of birds and insects are triangular. Blue spruce trees grow in a triangular shape. Cats have triangular ears. Students can use these examples, find more throughout the chapter, or come up with their own ideas, and classify triangles found in nature.

11. Find the measures of the sides of $\triangle TWZ$ with vertices at $T(2, 6)$, $W(4, -5)$, and $Z(-3, 0)$. Classify the triangle. $TW = \sqrt{125}$, $WZ = \sqrt{74}$, $TZ = \sqrt{61}$; scalene

Application

12. **QUILTING** The star-shaped composite quilting square is made up of four different triangles. Use a ruler to classify the four triangles by sides. 8 scalene triangles (green), 8 isosceles triangles in the middle (blue), 4 isosceles triangles around the middle (yellow) and 4 isosceles at the corners of the square (purple)

★ indicates increased difficulty

Practice and Apply

Homework Help

For Exercises	See Examples
13–18	1
19, 21–25	1, 2
26–29	3
30, 31	2
32–37, 40, 41	4

Extra Practice
See page 760.

Use a protractor to classify each triangle as *acute*, *equiangular*, *obtuse*, or *right*.

13. 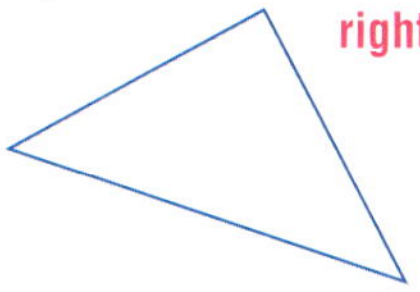 right

14. 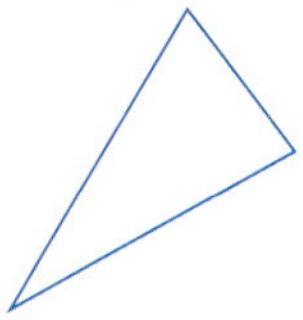 acute

15. 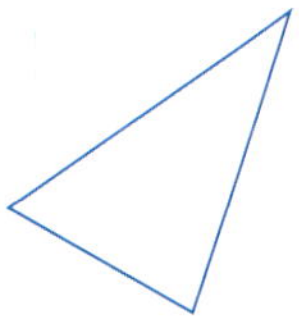 acute

16. 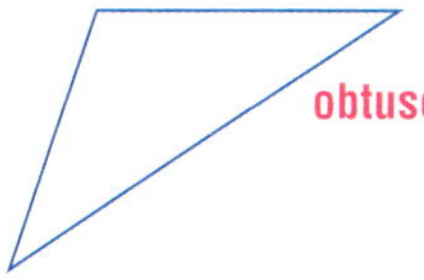 obtuse

17. 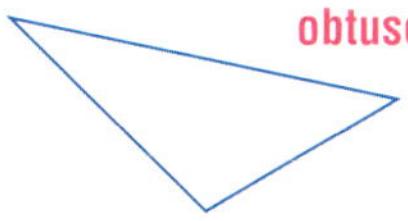 obtuse

18. 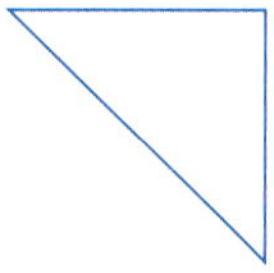 right

19. **ASTRONOMY** On May 5, 2002, Venus, Saturn, and Mars were aligned in a triangular formation. Use a protractor or ruler to classify the triangle formed by sides and angles. equilateral, equiangular

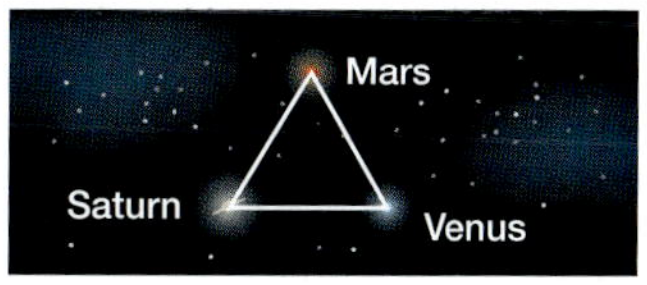

20. **RESEARCH** Use the Internet or other resource to find out how astronomers can predict planetary alignment. See students' work.

21. **ARCHITECTURE** The restored and decorated Victorian houses in San Francisco are called the "Painted Ladies." Use a protractor to classify the triangles indicated in the photo by sides and angles. isosceles, acute

More About. . .

Architecture

The Painted Ladies are located in Alamo Square. The area is one of 11 designated historic districts in San Francisco.

Source: www.sfvisitor.org

22. $\triangle AGB$, $\triangle AGC$, $\triangle DGB$, $\triangle DGC$

Identify the indicated type of triangles in the figure if $\overline{AB} \cong \overline{BD} \cong \overline{DC} \cong \overline{CA}$ and $\overline{BC} \perp \overline{AD}$.

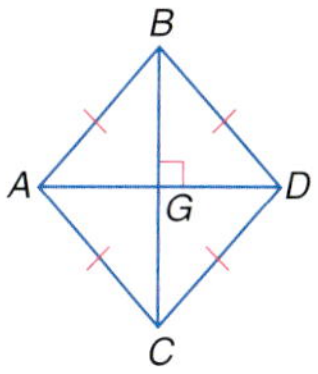

22. right
23. obtuse $\triangle BAC$, $\triangle CDB$
24. scalene $\triangle AGB$, $\triangle AGC$, $\triangle DGB$, $\triangle DGC$
25. isosceles $\triangle ABD$, $\triangle ACD$, $\triangle BAC$, $\triangle CDB$

ALGEBRA **Find x and the measure of each side of the triangle.**

26. $\triangle GHJ$ is isosceles, with $\overline{HG} \cong \overline{JG}$, $GH = x + 7$, $GJ = 3x - 5$, and $HJ = x - 1$. $x = 6$, $GH = 13$, $GJ = 13$, $HJ = 5$

27. $\triangle MPN$ is equilateral with $MN = 3x - 6$, $MP = x + 4$, and $NP = 2x - 1$.

28. $\triangle QRS$ is equilateral. QR is two less than two times a number, RS is six more than the number, and QS is ten less than three times the number.

29. $\triangle JKL$ is isosceles with $\overline{KJ} \cong \overline{LJ}$. JL is five less than two times a number. JK is three more than the number. KL is one less than the number. Find the measure of each side. $x = 8$, $JL = 11$, $JK = 11$, $KL = 7$

27. $x = 5$, $MN = 9$, $MP = 9$, $NP = 9$

28. $x = 8$, $QR = 14$, $RS = 14$, $QS = 14$

3 Practice/Apply

Study Notebook

Have students—
- add the definitions/examples of the vocabulary terms to their Vocabulary Builder worksheets for Chapter 4.
- include any other item(s) that they find helpful in mastering the skills in this lesson.

About the Exercises...

Organization by Objective
- **Classify Triangles by Angles:** 13–18, 19, 21
- **Classify Triangles by Sides:** 19, 21, 22–39

Odd/Even Assignments
Exercises 13–18, 22–29, and 32–41 are structured so that students practice the same concepts whether they are assigned odd or even problems.

Alert! Exercise 20 requires the Internet or other research materials.

Assignment Guide

Basic: 13–37 odd, 42–57
Average: 13–41 odd, 42–57
Advanced: 14–40 even, 42–52 (optional: 53–57)

Interactive Chalkboard
PowerPoint® Presentations

This CD-ROM is a customizable Microsoft® PowerPoint® presentation that includes:
- Step-by-step, dynamic solutions of each In-Class Example from the Teacher Wraparound Edition
- Additional, Try These exercises for each example
- The 5-Minute Check Transparencies
- Hot links to Glencoe Online Study Tools

Study Guide and Intervention, p. 183 (shown) and p. 184

Classify Triangles by Angles One way to classify a triangle is by the measures of its angles.

• If *one* of the angles of a triangle is an obtuse angle, then the triangle is an **obtuse triangle**.
• If *one* of the angles of a triangle is a right angle, then the triangle is a **right triangle**.
• If *all three* of the angles of a triangle are acute angles, then the triangle is an **acute triangle**.
• If all three angles of an acute triangle are congruent, then the triangle is an **equiangular triangle**.

Example **Classify each triangle.**

a.
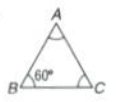

All three angles are congruent, so all three angles have measure 60°. The triangle is an equiangular triangle.

b.
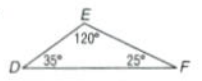

The triangle has one angle that is obtuse. It is an obtuse triangle.

c.
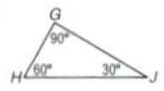

The triangle has one right angle. It is a right triangle.

Exercises

Classify each triangle as *acute*, *equiangular*, *obtuse*, or *right*.

1.
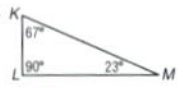

right

2.
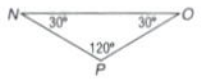

obtuse

3.
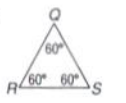

equiangular

4.
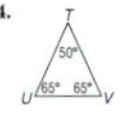

acute

5.
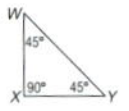

right

6.
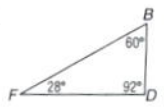

obtuse

Skills Practice, p. 185 and Practice, p. 186 (shown)

Use a protractor to classify each triangle as *acute*, *equiangular*, *obtuse*, or *right*.

1.

obtuse

2.

acute

3.

right

Identify the indicated type of triangles if $\overline{AB} \cong \overline{AD} \cong \overline{BD} \cong \overline{DC}$, $\overline{BE} \cong \overline{ED}$, $\overline{AB} \perp \overline{BC}$, and $\overline{ED} \perp \overline{DC}$.

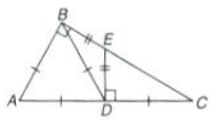

4. right
$\triangle ABC$, $\triangle CDE$

5. obtuse
$\triangle BED$, $\triangle BDC$

6. scalene
$\triangle ABC$, $\triangle CDE$

7. isosceles
$\triangle ABD$, $\triangle BED$, $\triangle BDC$

ALGEBRA Find x and the measure of each side of the triangle.

8. $\triangle FGH$ is equilateral with $FG = x + 5$, $GH = 3x - 9$, and $FH = 2x - 2$.
$x = 7$, $FG = 12$, $GH = 12$, $FH = 12$

9. $\triangle LMN$ is isosceles, $\angle L$ is the vertex angle, $LM = 3x - 2$, $LN = 2x + 1$, and $MN = 5x - 2$.
$x = 3$, $LM = 7$, $LN = 7$, $MN = 13$

Find the measures of the sides of $\triangle KPL$ and classify each triangle by its sides.

10. $K(-3, 2)$ $P(2, 1)$, $L(-2, -3)$
$KP = \sqrt{26}$, $PL = 4\sqrt{2}$, $LK = \sqrt{26}$; isosceles

11. $K(5, -3)$, $P(3, 4)$, $L(-1, 1)$
$KP = \sqrt{53}$, $PL = 5$, $LK = 2\sqrt{13}$; scalene

12. $K(-2, -6)$, $P(-4, 0)$, $L(3, -1)$
$KP = 2\sqrt{10}$, $PL = 5\sqrt{2}$, $LK = 5\sqrt{2}$; isosceles

13. **DESIGN** Diana entered the design at the right in a logo contest sponsored by a wildlife environmental group. Use a protractor. How many right angles are there? 5

Reading to Learn Mathematics, p. 187

ELL

Pre-Activity **Why are triangles important in construction?**

Read the introduction to Lesson 4-1 at the top of page 178 in your textbook.

- Why are triangles used for braces in construction rather than other shapes? Sample answer: Triangles lie in a plane and are rigid shapes.
- Why do you think that isosceles triangles are used more often than scalene triangles in construction? Sample answer: Isosceles triangles are symmetrical.

Reading the Lesson

1. Supply the correct numbers to complete each sentence.
   a. In an obtuse triangle, there are __2__ acute angle(s), __0__ right angle(s), and __1__ obtuse angle(s).
   b. In an acute triangle, there are __3__ acute angle(s), __0__ right angle(s), and __0__ obtuse angle(s).
   c. In a right triangle, there are __2__ acute angle(s), __1__ right angle(s), and __0__ obtuse angle(s).
2. Determine whether each statement is *always*, *sometimes*, or *never* true.
   a. A right triangle is scalene. sometimes
   b. An obtuse triangle is isosceles. sometimes
   c. An equilateral triangle is a right triangle. never
   d. An equilateral triangle is isosceles. always
   e. An acute triangle is isosceles. sometimes
   f. A scalene triangle is obtuse. sometimes
3. Describe each triangle by as many of the following words as apply: *acute*, *obtuse*, *right*, *scalene*, *isosceles*, or *equilateral*.

a.
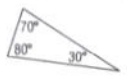

acute, scalene

b.

obtuse, isosceles

c.
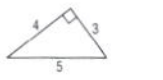

right, scalene

Helping You Remember

4. A good way to remember a new mathematical term is to relate it to a nonmathematical definition of the same word. How is the use of the word *acute*, when used to describe *acute pain*, related to the use of the word *acute* when used to describe an *acute angle* or an *acute triangle*? Sample answer: Both are related to the meaning of *acute* as *sharp*. An *acute pain* is a sharp pain, and an *acute angle* can be thought of as an angle with a sharp point. In an *acute triangle* all of the angles are acute.

★ 30. **CRYSTAL** The top of the crystal bowl shown is circular. The diameter at the top of the bowl is $\overline{MN}$. P is the midpoint of $\overline{MN}$, and $\overline{OP} \perp \overline{MN}$. If $MN = 24$ and $OP = 12$, determine whether $\triangle MPO$ and $\triangle NPO$ are equilateral.
No, $MO = NO = \sqrt{288}$

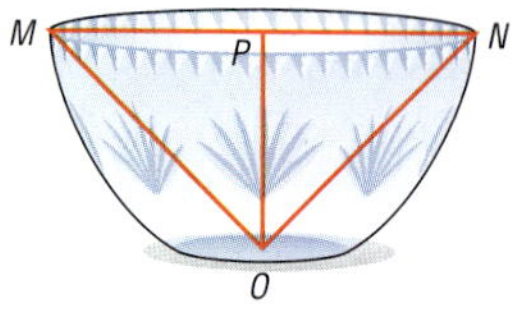

31. **MAPS** The total distance from Nashville, Tennessee, to Cairo, Illinois, to Lexington, Kentucky, and back to Nashville, Tennessee, is 593 miles. The distance from Cairo to Lexington is 81 more miles than the distance from Lexington to Nashville. The distance from Cairo to Nashville is 40 miles less than the distance from Nashville to Lexington. Classify the triangle formed by its sides. Scalene; it is 184 miles from Lexington to Nashville, 265 miles from Cairo to Lexington, and 144 miles from Cairo to Nashville.

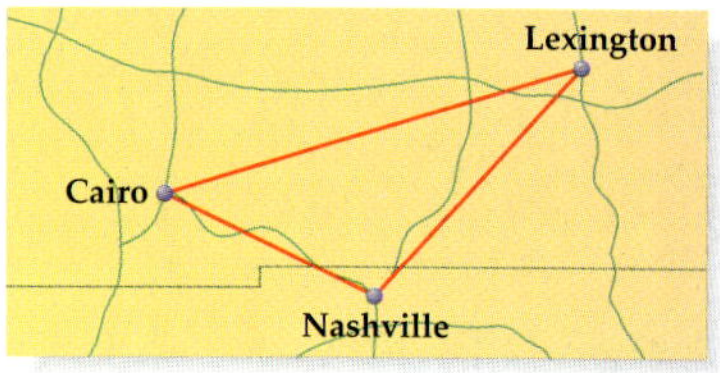

Find the measures of the sides of $\triangle ABC$ and classify each triangle by its sides.

32. $A(5, 4)$, $B(3, -1)$, $C(7, -1)$

33. $A(-4, 1)$, $B(5, 6)$, $C(-3, -7)$

34. $A(-7, 9)$, $B(-7, -1)$, $C(4, -1)$

35. $A(-3, -1)$, $B(2, 1)$, $C(2, -3)$

36. $A(0, 5)$, $B(5\sqrt{3}, 2)$, $C(0, -1)$

37. $A(-9, 0)$, $B\left(-5, 6\sqrt{3}\right)$, $C(-1, 0)$

32. $AB = \sqrt{29}$, $BC = 4$, $AC = \sqrt{29}$; isosceles

33. $AB = \sqrt{106}$, $BC = \sqrt{233}$, $AC = \sqrt{65}$; scalene

34. $AB = 10$, $BC = 11$, $AC = \sqrt{221}$; scalene

35. $AB = \sqrt{29}$, $BC = 4$, $AC = \sqrt{29}$; isosceles

36. $AB = \sqrt{84}$, $BC = \sqrt{84}$, $AC = 6$; isosceles

37. $AB = \sqrt{124}$, $BC = \sqrt{124}$, $AC = 8$; isosceles

★ 38. **PROOF** Write a two-column proof to prove that $\triangle EQL$ is equiangular. See p. 233A.

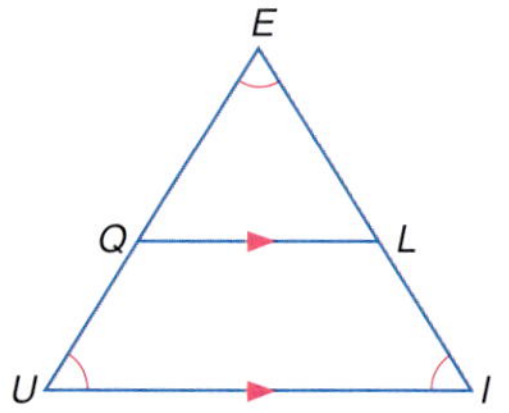

★ 39. **PROOF** Write a paragraph proof to prove that $\triangle RPM$ is an obtuse triangle if $m\angle NPM = 33$. See p. 233A.

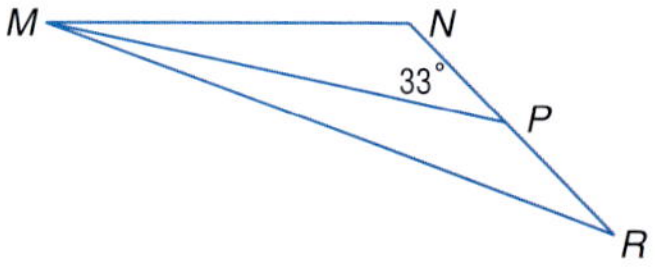

★ 40. **COORDINATE GEOMETRY** Show that S is the midpoint of $\overline{RT}$ and U is the midpoint of $\overline{TV}$. See p. 233A.

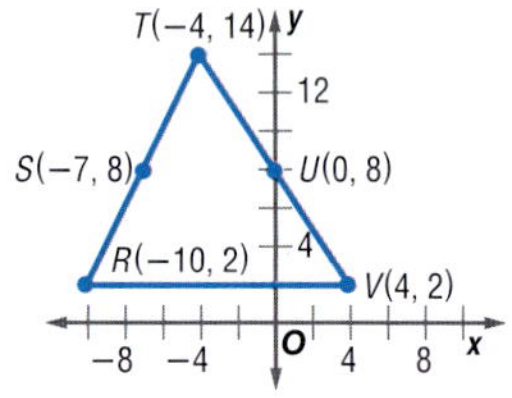

★ 41. **COORDINATE GEOMETRY** Show that $\triangle ADC$ is isosceles. See p. 233A.

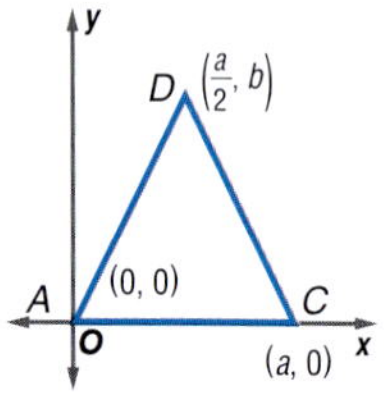

42. **CRITICAL THINKING** $\overline{KL}$ is a segment representing one side of isosceles right triangle KLM, with $K(2, 6)$, and $L(4, 2)$. $\angle KLM$ is a right angle, and $\overline{KL} \cong \overline{LM}$. Describe how to find the coordinates of vertex M and name these coordinates. Use the Distance Formula and Slope Formula; (0, 0) or (8, 4).

Enrichment, p. 188

Reading Mathematics

When you read geometry, you may need to draw a diagram to make the text easier to understand.

Example **Consider three points, A, B, and C on a coordinate grid. The y-coordinates of A and B are the same. The x-coordinate of B is greater than the x-coordinate of A. Both coordinates of C are greater than the corresponding coordinates of B. Is triangle ABC acute, right, or obtuse?**

To answer this question, first draw a sample triangle that fits the description.

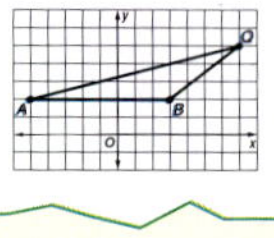

Side AB must be a horizontal segment because the y-coordinates are the same. Point C must be located to the right and up from point B.

From the diagram you can see that triangle ABC must be obtuse.

43. WRITING IN MATH Answer the question that was posed at the beginning of the lesson. **See margin.**

Why are triangles important in construction?

Include the following in your answer:

- describe how to classify triangles, and
- if one type of triangle is used more often in architecture than other types.

FCAT Practice
Standardized Test Practice

44. Classify $\triangle ABC$ with vertices $A(-1, 1)$, $B(1, 3)$, and $C(3, -1)$. **C**

Ⓐ scalene acute Ⓑ equilateral Ⓒ isosceles acute Ⓓ isosceles right

45. **ALGEBRA** Find the value of y if the mean of x, y, 15, and 35 is 25 and the mean of x, 15, and 35 is 27. **B**

Ⓐ 18 Ⓑ 19 Ⓒ 31 Ⓓ 36

Maintain Your Skills

Mixed Review

Graph each line. Construct a perpendicular segment through the given point. Then find the distance from the point to the line. *(Lesson 3-6)*

46. $y = x + 2$, $(2, -2)$ $\sqrt{18}$ 47. $x + y = 2$, $(3, 3)$ $\sqrt{8}$ 48. $y = 7$, $(6, -2)$ **9**

46–48. See margin for graphs.

Find x so that $p \parallel q$. *(Lesson 3-5)*

49. **15**

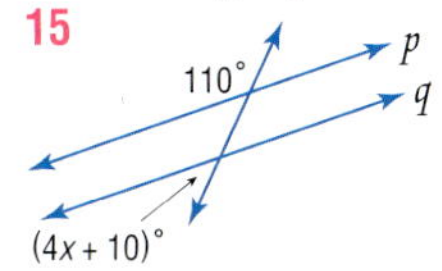

50. **45**

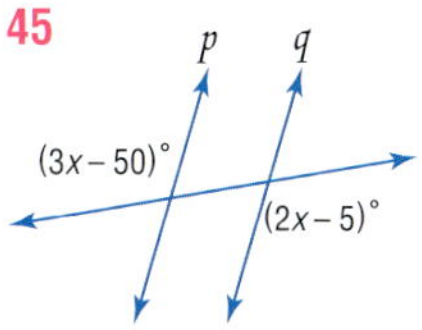

51. **44**

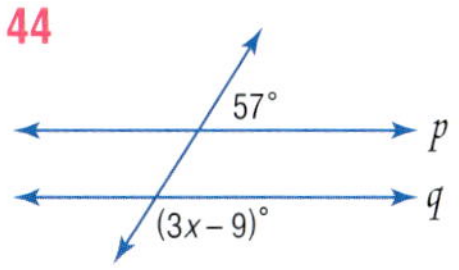

For this proof, the reasons in the right column are not in the proper order. Reorder the reasons to properly match the statements in the left column. *(Lesson 2-6)*

52. **Given:** $3x - 4 = x - 10$ **Order should be: 1. Given 2. Subtraction Property 3. Addition Property 4. Division Property**

Prove: $x = -3$

Proof:

Statements	Reasons
a. $3x - 4 = x - 10$	1. Subtraction Property
b. $2x - 4 = -10$	2. Division Property
c. $2x = -6$	3. Given
d. $x = -3$	4. Addition Property

Getting Ready for the Next Lesson

PREREQUISITE SKILL **In the figure, $\overline{AB} \parallel \overline{RQ}$, $\overline{BC} \parallel \overline{PR}$, and $\overline{AC} \parallel \overline{PQ}$. Name the indicated angles or pairs of angles.**

*(To review **angles formed by parallel lines and a transversal**, see Lessons 3-1 and 3-2.)*

53. three pairs of alternate interior angles
54. six pairs of corresponding angles
55. all angles congruent to $\angle 3$ **$\angle 6$, $\angle 9$, and $\angle 12$**
56. all angles congruent to $\angle 7$ **$\angle 1$, $\angle 4$, and $\angle 10$**
57. all angles congruent to $\angle 11$ **$\angle 2$, $\angle 5$, and $\angle 8$**

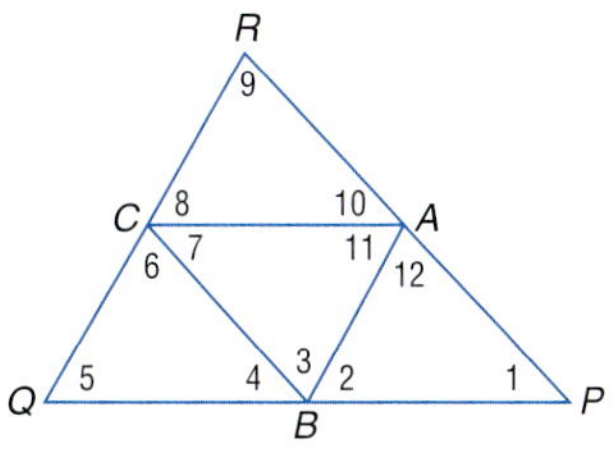

53. any three: $\angle 2$ and $\angle 11$, $\angle 3$ and $\angle 6$, $\angle 4$ and $\angle 7$, $\angle 3$ and $\angle 12$, $\angle 7$ and $\angle 10$, $\angle 8$ and $\angle 11$

54. $\angle 1$ and $\angle 4$, $\angle 1$ and $\angle 10$, $\angle 5$ and $\angle 2$, $\angle 5$ and $\angle 8$, $\angle 9$ and $\angle 12$, $\angle 6$, $\angle 9$ and $\angle 12$

www.geometryonline.com/self_check_quiz/fcat

Answers

43. Sample answer: Triangles are used in construction as structural support. Answers should include the following.

- **Triangles can be classified by sides and angles. If the measure of each angle is less than 90, the triangle is acute. If the measure of one angle is greater than 90, the triangle is obtuse. If one angle equals 90°, the triangle is right. If each angle has the same measure, the triangle is equiangular. If no two sides are congruent, the triangle is scalene. If at least two sides are congruent, it is isosceles. If all of the sides are congruent, the triangle is equilateral.**
- **Isosceles triangles seem to be used more often in architecture and construction.**

4 Assess

Open-Ended Assessment

Speaking Ask students to call out the classifications of selected triangles from the book or on the board by both angles and sides. Provide angle measures, side measures, congruency tick-marks, or right angle symbols for figures on the board. Label the triangles, and encourage students to use proper terminology to refer to the triangle, its angles, and its segments.

Getting Ready for Lesson 4-2

Prerequisite Skill Students will learn about angles of triangles in Lesson 4-2. They will use angle relationships with the Angle Sum Theorem and the Exterior Angle Theorem to find angle measures. Use Exercises 53–57 to determine your students' familiarity with angles formed by parallel lines and a transversal.

46.

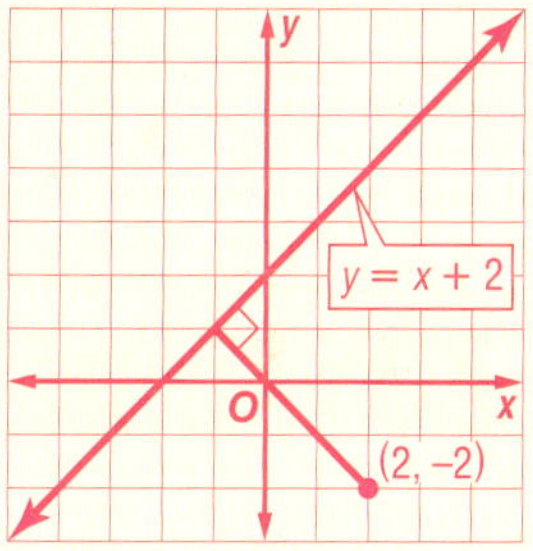

47.

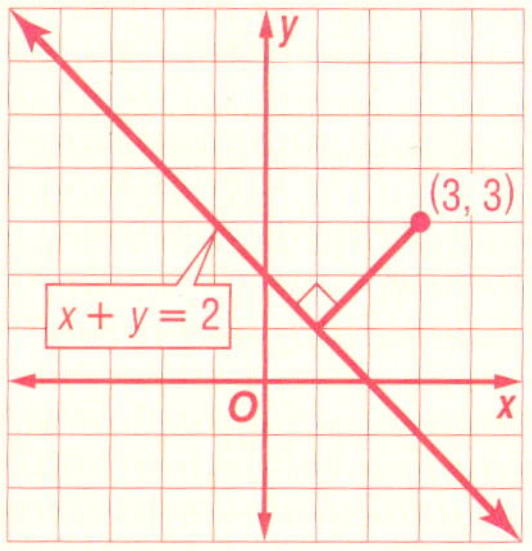

48.

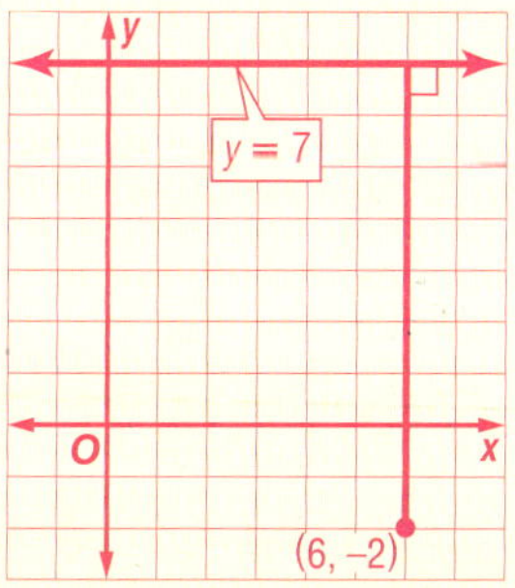

Geometry Activity

A Preview of Lesson 4-2

Getting Started

Objective Find the relationships among the measures of the interior and exterior angles of a triangle.

Materials
protractor scissors

Teach

- Advise students to label the obtuse angle B when they are first working through Activity 1. They can also repeat Activity 1 using one of the acute angles as angle B to further verify concepts.
- As an extension, students can repeat Activity 1 using an acute triangle or a right triangle. They can also cut along the segments DF, FE, and DE and arrange angles A, B, and C over their congruent counterparts in $\triangle DEF$ to see that the angles have equal measures.

Assess

In **Exercises 1–12** students determine angle measures of the triangles used in this activity, find relationships, and make conjectures that will lead them to the Angle Sum Theorem and the Exterior Angle Theorem.

Study Notebook

Ask students to summarize what they have learned about the relationships among the measures of the interior and exterior angles of triangles.

Geometry Activity

A Preview of Lesson 4-2

Angles of Triangles

There are special relationships among the angles of a triangle.

Activity 1 **Find the relationship among the measures of the interior angles of a triangle.**

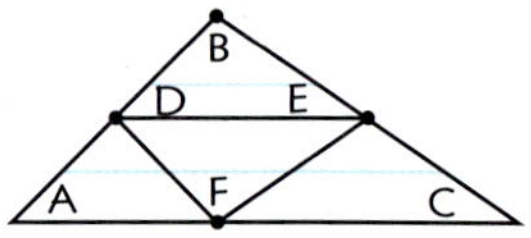

Step 1 Draw an obtuse triangle and cut it out. Label the vertices A, B, and C.

Step 2 Find the midpoint of $\overline{AB}$ by matching A to B. Label this point D.

Step 3 Find the midpoint of $\overline{BC}$ by matching B to C. Label this point E.

Step 4 Draw $\overline{DE}$.

Step 5 Fold $\triangle ABC$ along $\overline{DE}$. Label the point where B touches $\overline{AC}$ as F.

Step 6 Draw $\overline{DF}$ and $\overline{FE}$. Measure each angle.

Analyze the Model

Describe the relationship between each pair.

1. $\angle A$ and $\angle DFA$ **congruent** 2. $\angle B$ and $\angle DFE$ **congruent** 3. $\angle C$ and $\angle EFC$ **congruent**
4. What is the sum of the measures of $\angle DFA$, $\angle DFE$, and $\angle EFC$? **180**
5. What is the sum of the measures of $\angle A$, $\angle B$, and $\angle C$? **180**
6. **Make a conjecture** about the sum of the measures of the angles of any triangle. **The sum of the measures of the angles of any triangle is 180.**

In the figure at the right, $\angle 4$ is called an *exterior angle* of the triangle. $\angle 1$ and $\angle 2$ are the *remote interior angles* of $\angle 4$.

Activity 2 **Find the relationship among the interior and exterior angles of a triangle.**

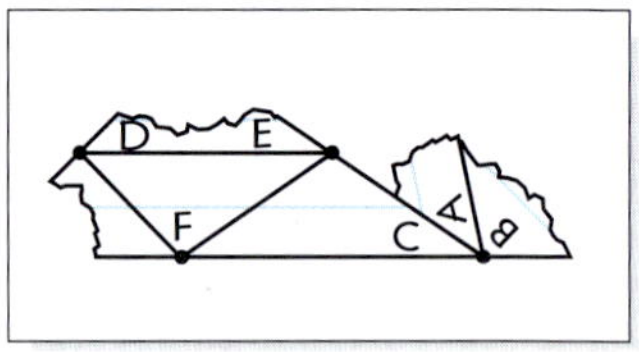

Step 1 Trace $\triangle ABC$ from Activity 1 onto a piece of paper. Label the vertices.

Step 2 Extend $\overline{AC}$ to draw an exterior angle at C.

Step 3 Tear $\angle A$ and $\angle B$ off the triangle from Activity 1.

Step 4 Place $\angle A$ and $\angle B$ over the exterior angle.

Analyze the Model

7. $m\angle A + m\angle B$ is the measure of the exterior angle at C.

7. **Make a conjecture** about the relationship of $\angle A$, $\angle B$, and the exterior angle at C.
8. Repeat the steps for the exterior angles of $\angle A$ and $\angle B$. **See students' work.**
9. Is your conjecture true for all exterior angles of a triangle? **yes**
10. Repeat Activity 2 with an acute triangle. **10–11. See students' work.**
11. Repeat Activity 2 with a right triangle.
12. **Make a conjecture** about the measure of an exterior angle and the sum of the measures of its remote interior angles. **The measure of an exterior angle is equal to the sum of measures of the two remote interior angles.**

Resource Manager

Teaching Geometry with Manipulatives
- p. 70 (student recording sheet)
- p. 16 (protractor)

Glencoe Mathematics Classroom Manipulative Kit
- protractor
- scissors

4-2 Angles of Triangles

Sunshine State Standards
MA.C.1.4.1

What You'll Learn

- Apply the Angle Sum Theorem.
- Apply the Exterior Angle Theorem.

Vocabulary

- exterior angle
- remote interior angles
- flow proof
- corollary

How are the angles of triangles used to make kites?

The Drachen Foundation coordinates the annual Miniature Kite Contest. This kite won second place in the Most Beautiful Kite category in 2001. The overall dimensions are 10.5 centimeters by 9.5 centimeters. The wings of the beetle are triangular.

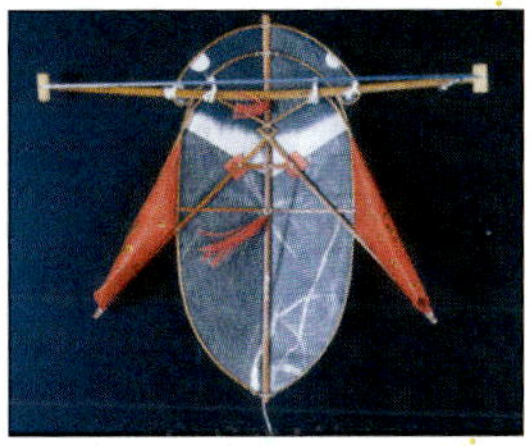

ANGLE SUM THEOREM If the measures of two of the angles of a triangle are known, how can the measure of the third angle be determined? The Angle Sum Theorem explains that the sum of the measures of the angles of any triangle is always 180.

Theorem 4.1

Angle Sum Theorem The sum of the measures of the angles of a triangle is 180.

Example: $m\angle W + m\angle X + m\angle Y = 180$

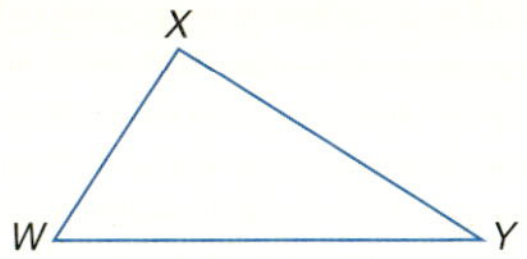

Study Tip

Look Back
Recall that sometimes extra lines have to be drawn to complete a proof. These are called *auxiliary lines.*

Proof — Angle Sum Theorem

Given: $\triangle ABC$

Prove: $m\angle C + m\angle 2 + m\angle B = 180$

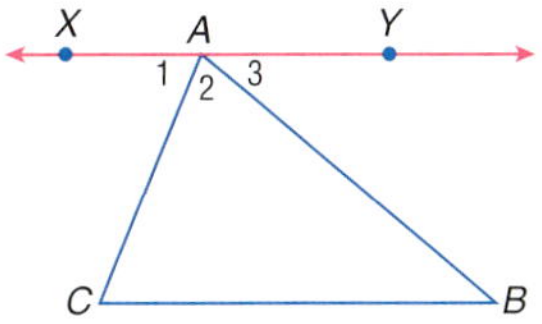

Proof:

Statements	Reasons
1. $\triangle ABC$	1. Given
2. Draw $\overleftrightarrow{XY}$ through A parallel to $\overline{CB}$.	2. Parallel Postulate P. 3.5
3. $\angle 1$ and $\angle CAY$ form a linear pair.	3. Def. of a linear pair
4. $\angle 1$ and $\angle CAY$ are supplementary.	4. If 2 $\angle$s form a linear pair, they are supplementary. Thm 2.3
5. $m\angle 1 + m\angle CAY = 180$	5. Def. of suppl. $\angle$s
6. $m\angle CAY = m\angle 2 + m\angle 3$	6. Angle Addition Postulate P 2.11
7. $m\angle 1 + m\angle 2 + m\angle 3 = 180$	7. Substitution
8. $\angle 1 \cong \angle C$, $\angle 3 \cong \angle B$	8. Alt. Int. $\angle$s Theorem Thm 3.1
9. $m\angle 1 = m\angle C$, $m\angle 3 = m\angle B$	9. Def. of $\cong$ $\angle$s
10. $m\angle C + m\angle 2 + m\angle B = 180$	10. Substitution

4-2 Lesson Notes

1 Focus

5-Minute Check Transparency 4-2 Use as a quiz or review of Lesson 4-1.

Mathematical Background notes are available for this lesson on p. 176C.

How are the angles of triangles used to make kites?

Ask students:

- Assuming that the wings are equal in size and the angle between the two wings is 90°, what type of triangle is formed if you draw a line to connect one wing-tip to the other wing-tip? **right isosceles triangle**
- Are the wings of a real beetle perfectly triangular in shape? **no**

Resource Manager

Workbook and Reproducible Masters

Chapter 4 Resource Masters
- Study Guide and Intervention, pp. 189–190
- Skills Practice, p. 191
- Practice, p. 192
- Reading to Learn Mathematics, p. 193
- Enrichment, p. 194
- Assessment, p. 239

Graphing Calculator and Computer Masters, p. 23
Prerequisite Skills Workbook, pp. 81–84
Teaching Geometry With Manipulatives Masters, p. 71

Transparencies

5-Minute Check Transparency 4-2
Answer Key Transparencies

Technology

Interactive Chalkboard

2 Teach

ANGLE SUM THEOREM

In-Class Example

Power Point®

1 Find the missing angle measures.

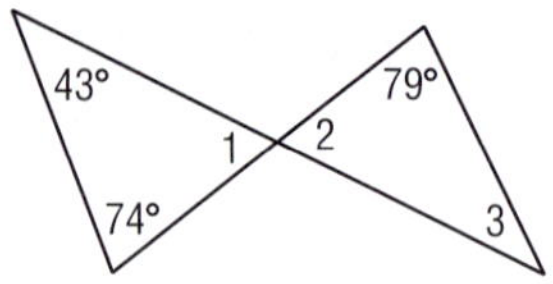

$m\angle 1 = 63$; $m\angle 2 = 63$; $m\angle 3 = 38$

If we know the measures of two angles of a triangle, we can find the measure of the third.

Example 1 Interior Angles

Find the missing angle measures.

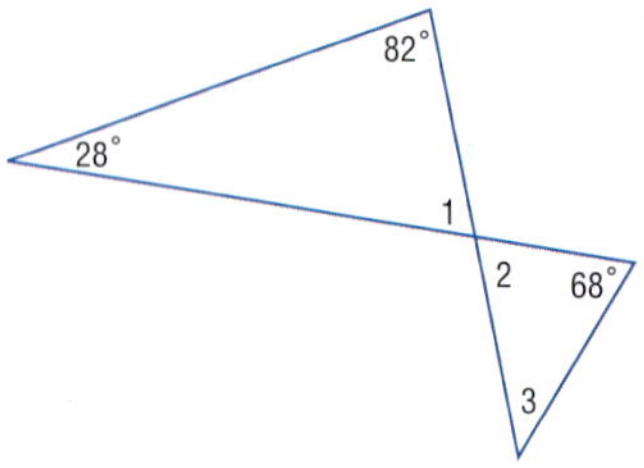

Find $m\angle 1$ first because the measures of two angles of the triangle are known.

$m\angle 1 + 28 + 82 = 180$ Angle Sum Theorem

$m\angle 1 + 110 = 180$ Simplify.

$m\angle 1 = 70$ Subtract 110 from each side.

$\angle 1$ and $\angle 2$ are congruent vertical angles. So $m\angle 2 = 70$.

$m\angle 3 + 68 + 70 = 180$ Angle Sum Theorem

$m\angle 3 + 138 = 180$ Simplify.

$m\angle 3 = 42$ Subtract 138 from each side.

Therefore, $m\angle 1 = 70$, $m\angle 2 = 70$, and $m\angle 3 = 42$.

The Angle Sum Theorem leads to a useful theorem about the angles in two triangles.

Theorem 4.2

Third Angle Theorem If two angles of one triangle are congruent to two angles of a second triangle, then the third angles of the triangles are congruent.

Example: If $\angle A \cong \angle F$ and $\angle C \cong \angle D$, then $\angle B \cong \angle E$.

You will prove this theorem in Exercise 44.

EXTERIOR ANGLE THEOREM

Each angle of a triangle has an exterior angle. An **exterior angle** is formed by one side of a triangle and the extension of another side. The interior angles of the triangle not adjacent to a given exterior angle are called **remote interior angles** of the exterior angle.

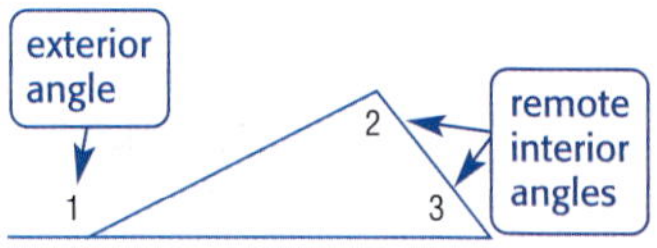

Study Tip

Reading Math

Remote means far away and *interior* means inside. The remote interior angles are the interior angles farthest from the exterior angle.

Theorem 4.3

Exterior Angle Theorem The measure of an exterior angle of a triangle is equal to the sum of the measures of the two remote interior angles.

Example: $m\angle YZP = m\angle X + m\angle Y$

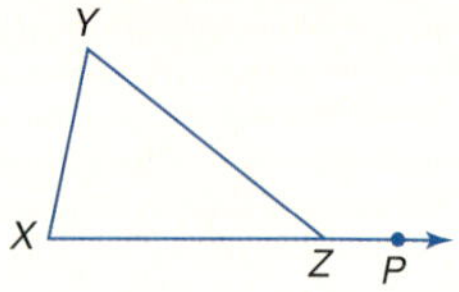

DAILY INTERVENTION

Differentiated Instruction

Visual/Spatial Tell students that the Angle Sum Theorem and Exterior Angle Theorem are both based on the idea that a straight angle measures 180°. Show that if they cut the angles of any triangle and place them right next to one another, they form a straight line. This visually demonstrates why the sum of the interior angles of a triangle measures 180°.

186 Chapter 4 Congruent Triangles

We will use a flow proof to prove this theorem. A **flow proof** organizes a series of statements in logical order, starting with the given statements. Each statement is written in a box with the reason verifying the statement written below the box. Arrows are used to indicate how the statements relate to each other.

Proof — Exterior Angle Theorem

Write a flow proof of the Exterior Angle Theorem.

Given: $\triangle ABC$

Prove: $m\angle CBD = m\angle A + m\angle C$

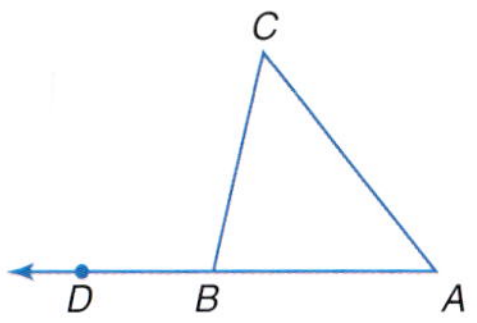

Flow Proof:

$\triangle ABC$ — Given → $m\angle A + m\angle ABC + \angle C = 180$ — Angle Sum Theorem

$\angle CBD$ and $\angle ABC$ form a linear pair. — Definition of linear pair → $\angle CBD$ and $\angle ABC$ are supplementary. — If 2 ∠s form a linear pair, they are supplementary. → $m\angle CBD + m\angle ABC = 180$ — Definition of supplementary

→ $m\angle A + m\angle ABC + m\angle C = m\angle CBD + m\angle ABC$ — Substitution Property

→ $m\angle A + m\angle C = m\angle CBD$ — Subtraction Property

Example 2 — Exterior Angles

Find the measure of each numbered angle in the figure.

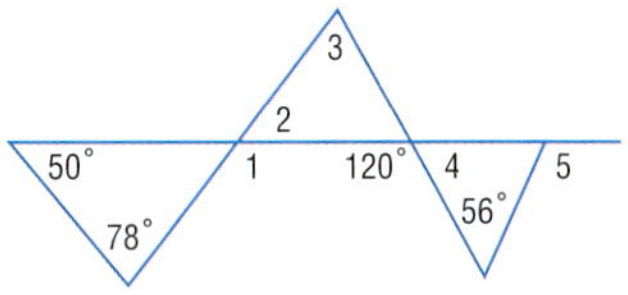

$m\angle 1 = 50 + 78$ — Exterior Angle Theorem

$= 128$ — Simplify.

$m\angle 1 + m\angle 2 = 180$ — If 2 ∠s form a linear pair, they are suppl.

$128 + m\angle 2 = 180$ — Substitution

$m\angle 2 = 52$ — Subtract 128 from each side.

$m\angle 2 + m\angle 3 = 120$ — Exterior Angle Theorem

$52 + m\angle 3 = 120$ — Substitution

$m\angle 3 = 68$ — Subtract 52 from each side.

$120 + m\angle 4 = 180$ — If 2 ∠s form a linear pair, they are suppl.

$m\angle 4 = 60$ — Subtract 120 from each side.

$m\angle 5 = m\angle 4 + 56$ — Exterior Angle Theorem

$= 60 + 56$ — Substitution

$= 116$ — Simplify.

Therefore, $m\angle 1 = 128$, $m\angle 2 = 52$, $m\angle 3 = 68$, $m\angle 4 = 60$, and $m\angle 5 = 116$.

www.geometryonline.com/extra_examples/fcat

EXTERIOR ANGLE THEOREM

In-Class Example — Power Point®

2 Find the measure of each numbered angle in the figure.

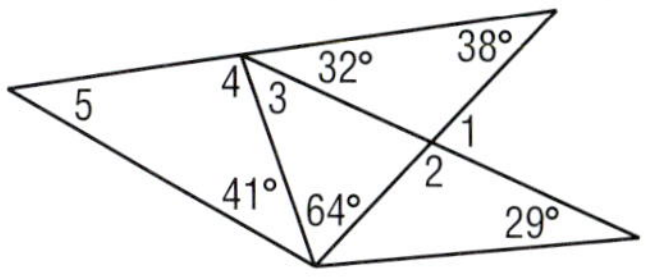

$m\angle 1 = 70$, $m\angle 2 = 110$, $m\angle 3 = 46$, $m\angle 4 = 102$, and $m\angle 5 = 37$.

Building on Prior Knowledge

In Chapter 3, students used angle relationships to find angle measures. In this lesson, students will apply their knowledge of vertical angles, supplementary angles, and complementary angles along with the Angle Sum Theorem and the Exterior Angle Theorem to find angle measures in figures.

In-Class Example

Power Point®

3 **GARDENING** The flower bed shown is in the shape of a right triangle. Find $m\angle A$ if $m\angle C$ is 20.

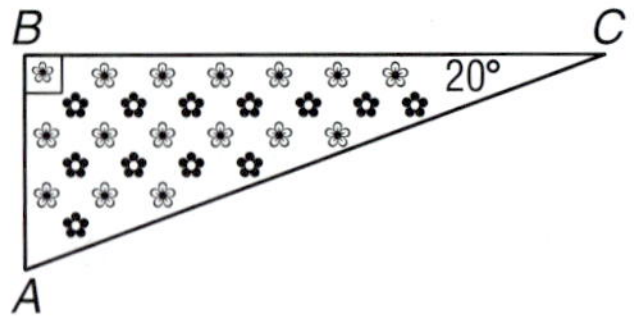

$m\angle A = 70$

3 Practice/Apply

Study Notebook

Have students—

- add the definitions/examples of the vocabulary terms to their Vocabulary Builder worksheets for Chapter 4.
- include any other item(s) that they find helpful in mastering the skills in this lesson.

Answer

1. Sample answer: $\angle 2$ and $\angle 3$ are the remote interior angles of exterior $\angle 1$.

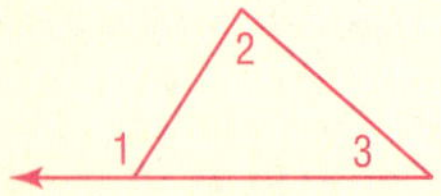

A statement that can be easily proved using a theorem is often called a **corollary** of that theorem. A corollary, just like a theorem, can be used as a reason in a proof.

Corollaries

4.1 The acute angles of a right triangle are complementary.

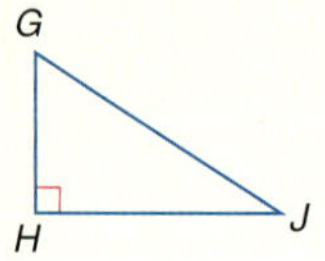

Example: $m\angle G + m\angle J = 90$

4.2 There can be at most one right or obtuse angle in a triangle.

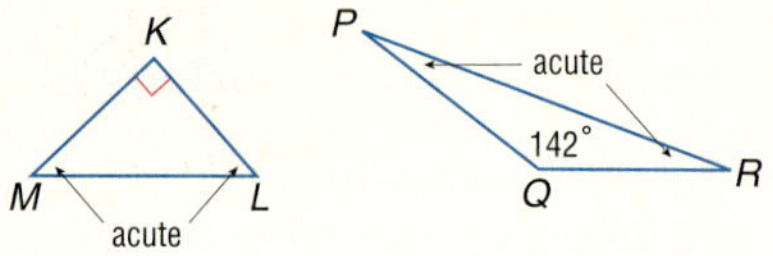

You will prove Corollaries 4.1 and 4.2 in Exercises 42 and 43.

Example 3 Right Angles

SKI JUMPING **Ski jumper Simon Ammann of Switzerland forms a right triangle with his skis and his line of sight. Find $m\angle 2$ if $m\angle 1$ is 27.**

Use Corollary 4.1 to write an equation.

$m\angle 1 + m\angle 2 = 90$

$27 + m\angle 2 = 90$ Substitution

$m\angle 2 = 63$ Subtract 27 from each side.

Check for Understanding

Concept Check

1. **OPEN ENDED** Draw a triangle. Label one exterior angle and its remote interior angles. **See margin.**

2. **FIND THE ERROR** Najee and Kara are discussing the Exterior Angle Theorem.

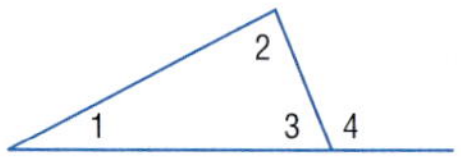

Najee	Kara
$m\angle 1 + m\angle 2 = m\angle 4$	$m\angle 1 + m\angle 2 + m\angle 4 = 180$

Who is correct? Explain your reasoning.

2. Najee; the sum of the measures of the remote interior angles is equal to the measure of the corresponding exterior angle.

Guided Practice

GUIDED PRACTICE KEY	
Exercises	Examples
3–4	1
5–7	2
8–10	3

Find the missing angle measure.

3. **43**

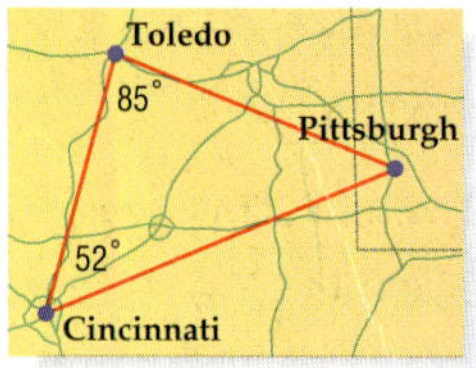

4. **99**

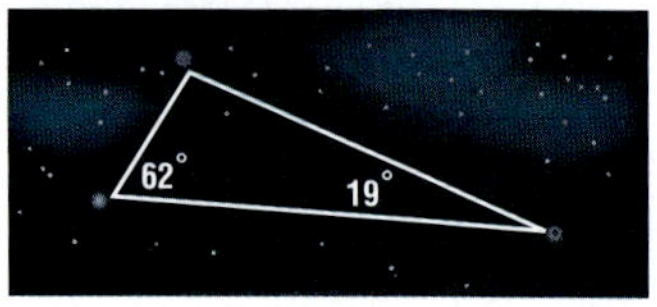

Find each measure.

5. $m\angle 1$ **55**
6. $m\angle 2$ **33**
7. $m\angle 3$ **147**

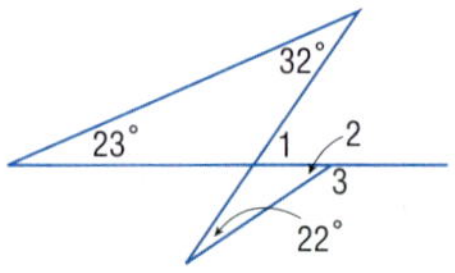

188 Chapter 4 Congruent Triangles

Find each measure.

8. $m\angle 1$ **65**

9. $m\angle 2$ **25**

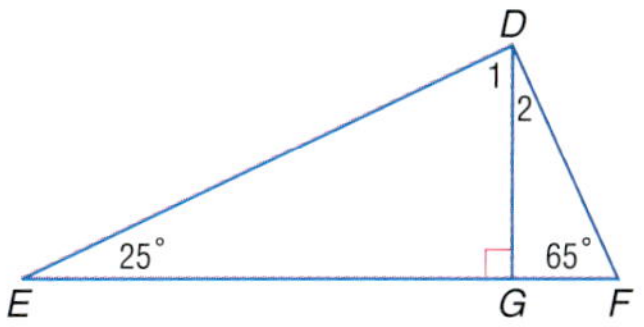

Application

10. **SKI JUMPING** American ski jumper Eric Bergoust forms a right angle with his skis. If $m\angle 2 = 70$, find $m\angle 1$. **20**

★ indicates increased difficulty

Practice and Apply

Homework Help

For Exercises	See Examples
11–17	1
18–31	2
32–35	3
36–38	2

Extra Practice

See page 761.

Find the missing angle measures.

11. **93**

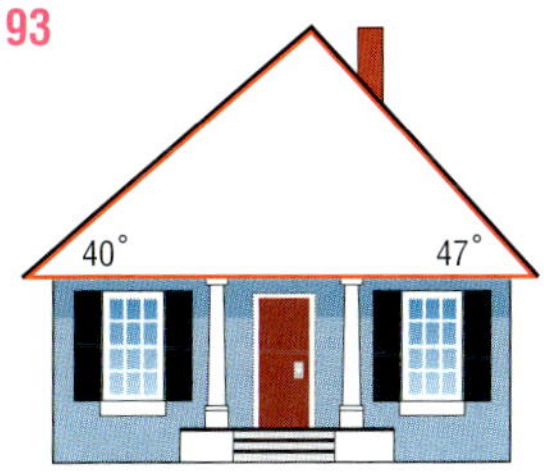

12. **70.5, 70.5**

13. **65, 65**

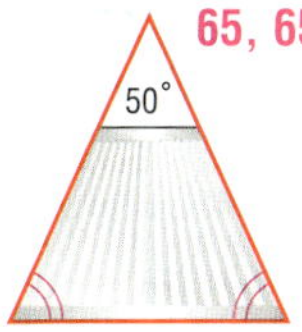

14. **63**

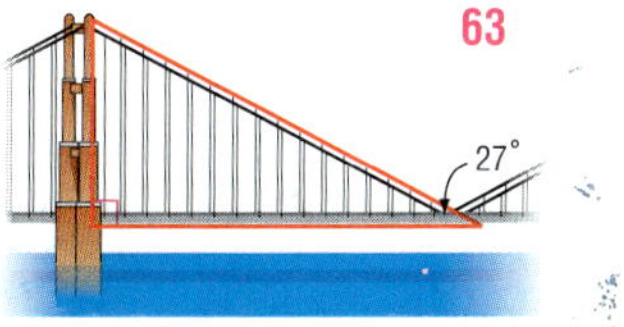

Find each measure.

15. $m\angle 1$ **76**

16. $m\angle 2$ **76**

17. $m\angle 3$ **49**

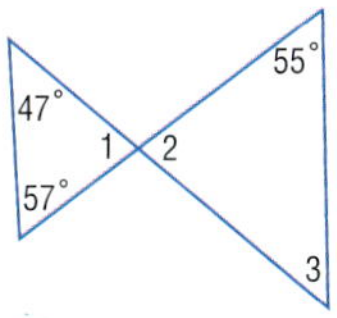

Find each measure if $m\angle 4 = m\angle 5$.

18. $m\angle 1$ **64**

19. $m\angle 2$ **53**

20. $m\angle 3$ **116**

21. $m\angle 4$ **32**

22. $m\angle 5$ **32**

23. $m\angle 6$ **44**

24. $m\angle 7$ **89**

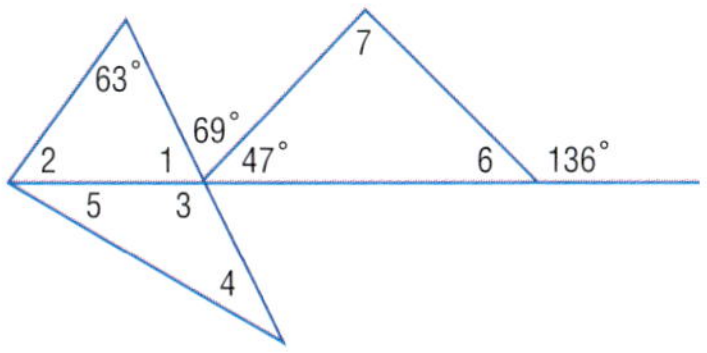

Find each measure.

25. $m\angle 1$ **123**

26. $m\angle 2$ **28**

27. $m\angle 3$ **14**

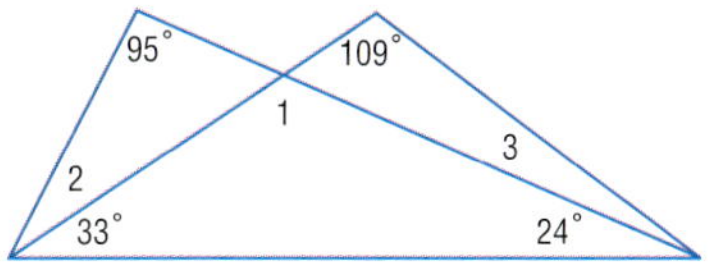

DAILY INTERVENTION **FIND THE ERROR**

Explain to students that Kara's equation could only be true in the special case when $m\angle 3 = m\angle 4 = 90$.

About the Exercises...

Organization by Objective

- **Angle Sum Theorem:** 11–17
- **Exterior Angle Theorem:** 18–38

Odd/Even Assignments

Exercises 11–38 are structured so that students practice the same concepts whether they are assigned odd or even problems.

Assignment Guide

Basic: 11–35 odd, 41, 43, 45, 46–64

Average: 11–45 odd, 46–64

Advanced: 12–44 even, 45–58 (optional: 59–64)

Lesson 4-2 Angles of Triangles 189

Study Guide and Intervention, p. 189 (shown) and p. 190

Angle Sum Theorem If the measures of two angles of a triangle are known, the measure of the third angle can always be found.

Angle Sum Theorem	The sum of the measures of the angles of a triangle is 180. In the figure at the right, $m\angle A + m\angle B + m\angle C = 180$.

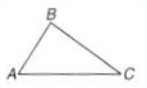

Example 1 **Find $m\angle T$.**

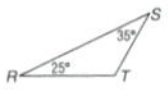

$m\angle R + m\angle S + m\angle T = 180$ Angle Sum Theorem
$25 + 35 + m\angle T = 180$ Substitution
$60 + m\angle T = 180$ Add.
$m\angle T = 120$ Subtract 60 from each side.

Example 2 **Find the missing angle measures.**

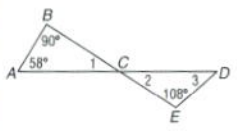

$m\angle 1 + m\angle A + m\angle B = 180$ Angle Sum Theorem
$m\angle 1 + 58 + 90 = 180$ Substitution
$m\angle 1 + 148 = 180$ Add.
$m\angle 1 = 32$ Subtract 148 from each side.
$m\angle 2 = 32$ Vertical angles are congruent.
$m\angle 3 + m\angle 2 + m\angle E = 180$ Angle Sum Theorem
$m\angle 3 + 32 + 108 = 180$ Substitution
$m\angle 3 + 140 = 180$ Add.
$m\angle 3 = 40$ Subtract 140 from each side.

Exercises

Find the measure of each numbered angle.

1. 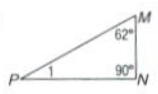 $m\angle 1 = 28$
2. 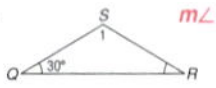 $m\angle 1 = 120$
3. 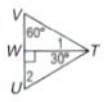 $m\angle 1 = 30$, $m\angle 2 = 60$
4. 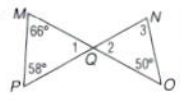 $m\angle 1 = 56$, $m\angle 2 = 56$, $m\angle 3 = 74$
5. 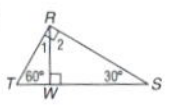 $m\angle 1 = 30$, $m\angle 2 = 60$
6. 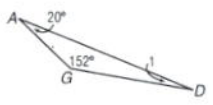 $m\angle 1 = 8$

Skills Practice, p. 191 and Practice, p. 192 (shown)

Find the missing angle measures.

1. 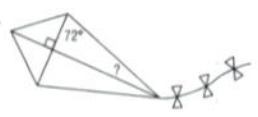 18
2.  85

Find the measure of each angle.

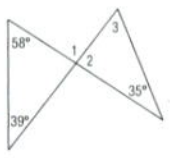

3. $m\angle 1$ 97
4. $m\angle 2$ 83
5. $m\angle 3$ 62

Find the measure of each angle.

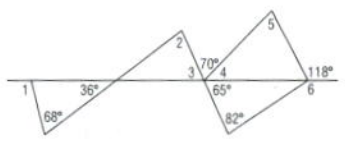

6. $m\angle 1$ 104
7. $m\angle 4$ 45
8. $m\angle 3$ 65
9. $m\angle 2$ 79
10. $m\angle 5$ 73
11. $m\angle 6$ 147

Find the measure of each angle if $\angle BAD$ and $\angle BDC$ are right angles and $m\angle ABC = 84$.

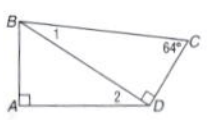

12. $m\angle 1$ 26
13. $m\angle 2$ 32

14. **CONSTRUCTION** The diagram shows an example of the Pratt Truss used in bridge construction. Use the diagram to find $m\angle 1$. 55

Reading to Learn Mathematics, p. 193 ELL

Pre-Activity **How are the angles of triangles used to make kites?**

Read the introduction to Lesson 4-2 at the top of page 185 in your textbook.

The frame of the simplest kind of kite divides the kite into four triangles. Describe these four triangles and how they are related to each other. Sample answer: There are two pairs of right triangles that have the same size and shape.

Reading the Lesson

1. Refer to the figure.

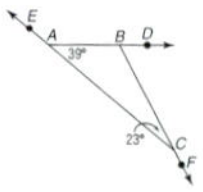

a. Name the three interior angles of the triangle. (Use three letters to name each angle.) $\angle BAC$, $\angle ABC$, $\angle BCA$
b. Name three exterior angles of the triangle. (Use three letters to name each angle.) $\angle EAB$, $\angle DBC$, $\angle FCA$
c. Name the remote interior angles of $\angle EAB$. $\angle ABC$, $\angle BCA$
d. Find the measure of each angle without using a protractor.
i. $\angle DBC$ 62 ii. $\angle ABC$ 118 iii. $\angle ACF$ 157 iv. $\angle EAB$ 141

2. Indicate whether each statement is *true* or *false*. If the statement is false, replace the underlined word or number with a word or number that will make the statement true.
a. The acute angles of a right triangle are supplementary. false; complementary
b. The sum of the measures of the angles of any triangle is 100. false; 180
c. A triangle can have at most one right angle or acute angle. false; obtuse
d. If two angles of one triangle are congruent to two angles of another triangle, then the third angles of the triangles are congruent. true
e. The measure of an exterior angle of a triangle is equal to the difference of the measures of the two remote interior angles. false; sum
f. If the measures of two angles of a triangle are 62 and 93, then the measure of the third angle is 35. false; 25
g. An exterior angle of a triangle forms a linear pair with an interior angle of the triangle. true

Helping You Remember

3. Many students remember mathematical ideas and facts more easily if they see them demonstrated visually rather than having them stated in words. Describe a visual way to demonstrate the Angle Sum Theorem.
Sample answer: Cut off the angles of a triangle and place them side-by-side on one side of a line so that their vertices meet at a common point. The result will show three angles whose measures add up to 180.

More About. . .

Speed Skating

Catriona Lemay Doan is the first Canadian to win a Gold medal in the same event in two consecutive Olympic games.

Source: www.catrionalemaydoan.com

SPEED SKATING For Exercises 28–31, use the following information.

Speed skater Catriona Lemay Doan of Canada forms at least two sets of triangles and exterior angles as she skates. Use the measures of given angles to find each measure.

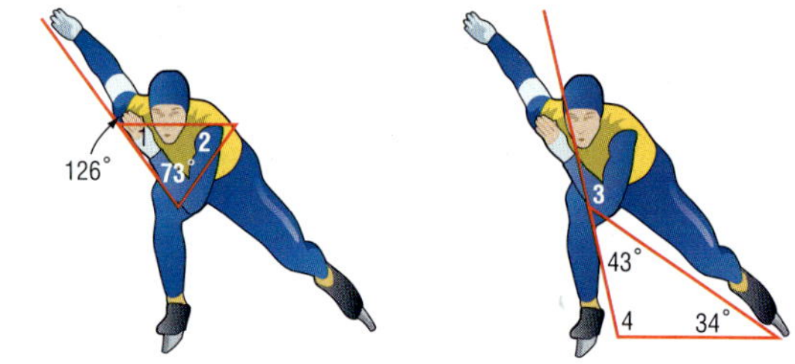

28. $m\angle 1$ 54
29. $m\angle 2$ 53
30. $m\angle 3$ 137
31. $m\angle 4$ 103

Online Research **Data Update** Use the Internet or other resource to find the world record in speed skating. Visit www.geometryonline.com/data_update to learn more.

Find each measure if $m\angle DGF = 53$ and $m\angle AGC = 40$.

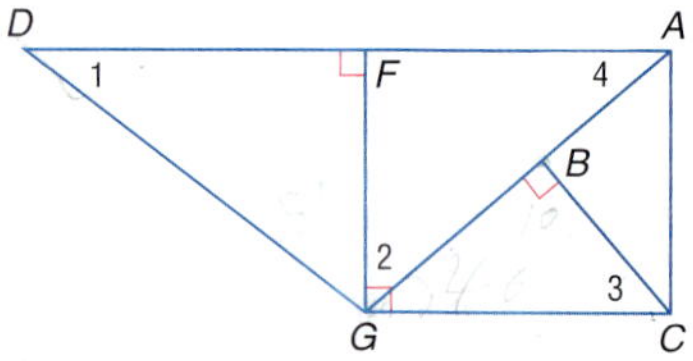

32. $m\angle 1$ 37
33. $m\angle 2$ 50
34. $m\angle 3$ 50
35. $m\angle 4$ 40

HOUSING For Exercises 36–38, use the following information.

The two braces for the roof of a house form triangles. Find each measure.

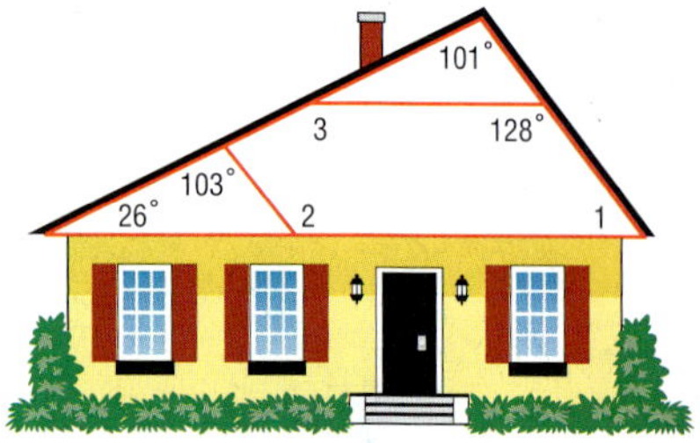

★ 36. $m\angle 1$ 53
★ 37. $m\angle 2$ 129
★ 38. $m\angle 3$ 153

★ **PROOF** For Exercises 39–44, write the specified type of proof. 39–44. See pp. 233A–233B.

39. flow proof
Given: $\angle FGI \cong \angle IGH$
$\overline{GI} \perp \overline{FH}$
Prove: $\angle F \cong \angle H$

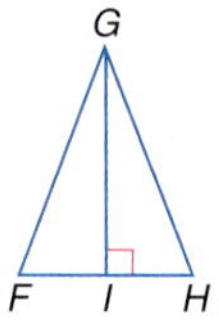

★ 40. two-column
Given: $ABCD$ is a quadrilateral.
Prove: $m\angle DAB + m\angle B + m\angle BCD + m\angle D = 360$

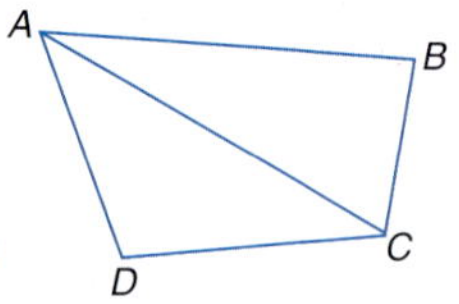

41. two-column proof of Theorem 4.3
42. flow proof of Corollary 4.1
43. paragraph proof of Corollary 4.2
44. two-column proof of Theorem 4.2

45. **CRITICAL THINKING** $\overrightarrow{BA}$ and $\overrightarrow{BC}$ are opposite rays. The measures of $\angle 1$, $\angle 2$, and $\angle 3$ are in a 4:5:6 ratio. Find the measure of each angle. $m\angle 1 = 48$, $m\angle 2 = 60$, $m\angle 3 = 72$

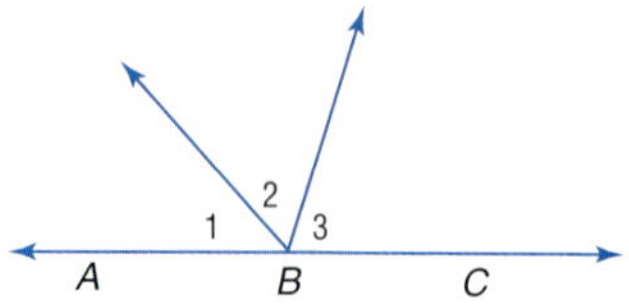

Enrichment, p. 194

Finding Angle Measures in Triangles

You can use algebra to solve problems involving triangles.

Example **In triangle ABC, $m\angle A$, is twice $m\angle B$, and $m\angle C$ is 8 more than $m\angle B$. What is the measure of each angle?**

Write and solve an equation. Let $x = m\angle B$.

$m\angle A + m\angle B + m\angle C = 180$
$2x + x + (x + 8) = 180$
$4x + 8 = 180$
$4x = 172$
$x = 43$

So, $m\angle A = 2(43)$ or 86, $m\angle B = 43$, and $m\angle C = 43 + 8$ or 51.

Solve each problem.

1. In triangle DEF, $m\angle E$ is three times $m\angle D$, and $m\angle F$ is 9 less than $m\angle E$. ...sure of each...
2. In triangle RST, $m\angle T$ is 5 more than $m\angle R$, and $m\angle S$ is 10 less than $m\angle T$. ...

46. **WRITING IN MATH** Answer the question that was posed at the beginning of the lesson. **See margin.**

How are the angles of triangles used to make kites?

Include the following in your answer:

- if two angles of two triangles are congruent, how you can find the measure of the third angle, and
- if one angle measures 90, describe the other two angles.

FCAT Practice

Standardized Test Practice

47. In the triangle, what is the measure of $\angle Z$? **A**

(A) 18 (B) 24

(C) 72 (D) 90

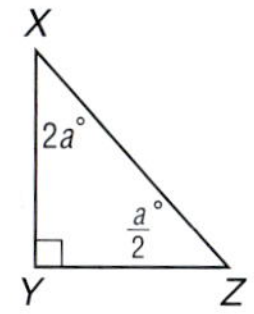

48. **ALGEBRA** The measure of the second angle of a triangle is three times the measure of the first, and the measure of the third angle is 25 more than the measure of the first. Find the measure of each angle. **B**

(A) 25, 85, 70 (B) 31, 93, 56 (C) 39, 87, 54 (D) 42, 54, 84

Maintain Your Skills

Mixed Review

Identify the indicated type of triangle if $\overline{BC} \cong \overline{AD}$, $\overline{EB} \cong \overline{EC}$, $\overline{AC}$ bisects $\overline{BD}$, and $m\angle AED = 125$. *(Lesson 4-1)*

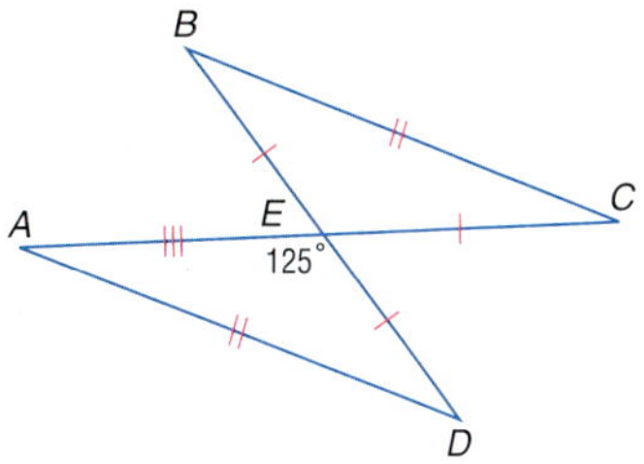

49. scalene triangles **$\triangle AED$**
50. obtuse triangles **$\triangle BEC$, $\triangle AED$**
51. isosceles triangles **$\triangle BEC$**

Find the distance between each pair of parallel lines. *(Lesson 3-6)*

53. $\sqrt{20}$ units

52. $y = x + 6, y = x - 10$ **$\sqrt{128}$ units**
53. $y = -2x + 3, y = -2x - 7$
54. $4x - y = 20, 4x - y = 3$ **$\sqrt{17}$ units**
55. $2x - 3y = -9, 2x - 3y = -6$ **$\frac{\sqrt{117}}{13}$ units**

Find x, y, and z in each figure. *(Lesson 3-2)*

56.

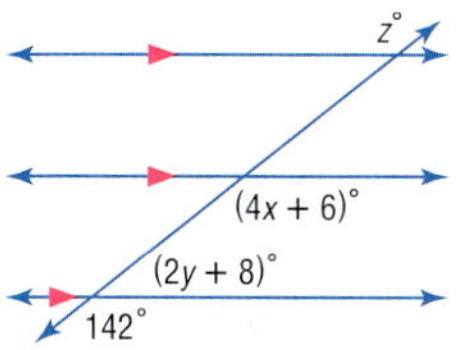

$x = 34$, $y = 15$, $z = 142$

57.

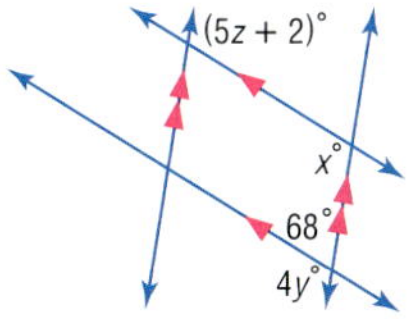

$x = 112$, $y = 28$, $z = 22$

58.

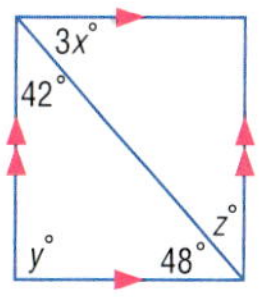

$x = 16$, $y = 90$, $z = 42$

Getting Ready for the Next Lesson

PREREQUISITE SKILL **Name the property of congruence that justifies each statement.** *(To review **properties of congruence**, see Lessons 2-5 and 2-6.)*

59. $\angle 1 \cong \angle 1$ and $\overline{AB} \cong \overline{AB}$. **reflexive**
60. If $\overline{AB} \cong \overline{XY}$, then $\overline{XY} \cong \overline{AB}$. **symmetric**
61. If $\angle 1 \cong \angle 2$, then $\angle 2 \cong \angle 1$. **symmetric**
62. If $\angle 2 \cong \angle 3$ and $\angle 3 \cong \angle 4$, then $\angle 2 \cong \angle 4$. **transitive**
63. If $\overline{PQ} \cong \overline{XY}$ and $\overline{XY} \cong \overline{HK}$, then $\overline{PQ} \cong \overline{HK}$. **transitive**
64. If $\overline{AB} \cong \overline{CD}$, $\overline{CD} \cong \overline{PQ}$, and $\overline{PQ} \cong \overline{XY}$, then $\overline{AB} \cong \overline{XY}$. **transitive**

www.geometryonline.com/self_check_quiz/fcat

4 Assess

Open-Ended Assessment

Writing Draw an acute triangle with two angles that measure 44° and 56°, an obtuse triangle with angles 110° and 40°, and an isosceles triangle sitting on a line with two angles measuring 75°. Ask students to use the theorems in this lesson to find the missing angle measures in each triangle and then write a paragraph summarizing how they found the measures.

Getting Ready for Lesson 4-3

Prerequisite Skill Students will learn about congruent triangles in Lesson 4-3. They will use properties of congruent segments and angles to identify corresponding parts of congruent triangles and to prove congruency between a triangle and its transformed image. Use Exercises 59–64 to determine your students' familiarity with properties of congruence for segments and angles.

Assessment Options

Quiz (Lessons 4-1 and 4-2) is available on p. 239 of the *Chapter 4 Resource Masters.*

Answer

46. Sample answer: The shape of a kite is symmetric. If triangles are used on one side of the kite, congruent triangles are used on the opposite side. The wings of this kite are made from congruent right triangles. Answers should include the following.

- **By the Third Angle Theorem, if two angles of two congruent triangles are congruent, then the third angles of each triangle are congruent.**
- **If one angle measures 90, the other two angles are both acute.**

4-3 Lesson Notes

1 Focus

5-Minute Check Transparency 4-3 Use as a quiz or review of Lesson 4-2.

Mathematical Background notes are available for this lesson on p. 176C.

Why are triangles used in bridges?

Ask students:

- What types of triangles do you notice in the construction of the bridge? **acute triangles**
- What do you notice about the size and shape of the triangles? **The triangles appear to be the same size and shape.**

4-3 Congruent Triangles

Sunshine State Standards
MA.C.1.4.1, MA.C.2.4.1

What You'll Learn

- Name and label corresponding parts of congruent triangles.
- Identify congruence transformations.

Vocabulary

- congruent triangles
- congruence transformations

Why are triangles used in bridges?

In 1930, construction started on the West End Bridge in Pittsburgh, Pennsylvania. The arch of the bridge is trussed, not solid. Steel rods are arranged in a triangular web that lends structure and stability to the bridge.

Study Tip

Congruent Triangles
Always list the names of congruent triangles in the order of corresponding parts.

CORRESPONDING PARTS OF CONGRUENT TRIANGLES Triangles that are the same size and shape are **congruent triangles**. Each triangle has three angles and three sides. If all six of the corresponding parts of two triangles are congruent, then the triangles are congruent.

If $\triangle ABC$ is congruent to $\triangle EFG$, the vertices of the two triangles correspond in the same order as the letters naming the triangles.

$$\triangle ABC \cong \triangle EFG$$

This correspondence of vertices can be used to name the corresponding congruent sides and angles of the two triangles.

$\angle A \cong \angle E$ $\quad$ $\angle B \cong \angle F$ $\quad$ $\angle C \cong \angle G$

$\overline{AB} \cong \overline{EF}$ $\quad$ $\overline{BC} \cong \overline{FG}$ $\quad$ $\overline{AC} \cong \overline{EG}$

The corresponding sides and angles can be determined from any congruence statement by following the order of the letters.

Study Tip

Congruent Parts
In congruent triangles, congruent sides are opposite congruent angles.

Key Concept — *Definition of Congruent Triangles (CPCTC)*

Two triangles are congruent if and only if their corresponding parts are congruent.

CPCTC stands for *corresponding parts of congruent triangles are congruent*. "If and only if" is used to show that both the conditional and its converse are true.

Resource Manager

Workbook and Reproducible Masters

Chapter 4 Resource Masters

- Study Guide and Intervention, pp. 195–196
- Skills Practice, p. 197
- Practice, p. 198
- Reading to Learn Mathematics, p. 199
- Enrichment, p. 200

School-to-Career Masters, p. 7
Teaching Geometry With Manipulatives Masters, pp. 72, 73

Transparencies

5-Minute Check Transparency 4-3
Answer Key Transparencies

Technology

Interactive Chalkboard

Example 1 Corresponding Congruent Parts

FURNITURE DESIGN The seat and legs of this stool form two triangles. Suppose the measures in inches are $QR = 12$, $RS = 23$, $QS = 24$, $RT = 12$, $TV = 24$, and $RV = 23$.

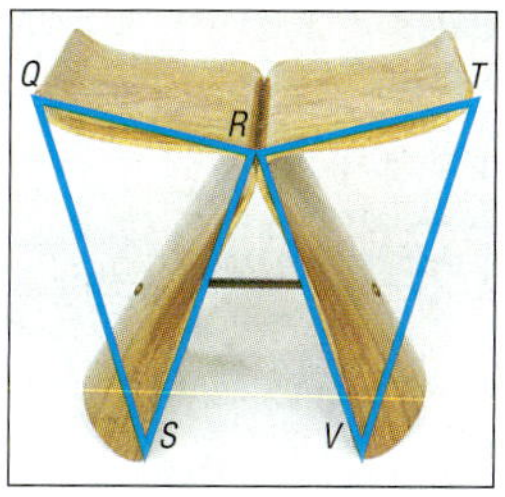

a. Name the corresponding congruent angles and sides.

$\angle Q \cong \angle T$ $\quad \angle QRS \cong \angle TRV$ $\quad \angle S \cong \angle V$

$\overline{QR} \cong \overline{TR}$ $\quad \overline{RS} \cong \overline{RV}$ $\quad \overline{QS} \cong \overline{TV}$

b. Name the congruent triangles.

$\triangle QRS \cong \triangle TRV$

Like congruence of segments and angles, congruence of triangles is reflexive, symmetric, and transitive.

Theorem 4.4 Properties of Triangle Congruence

Congruence of triangles is reflexive, symmetric, and transitive.

Reflexive	Symmetric	Transitive
$\triangle JKL \cong \triangle JKL$	If $\triangle JKL \cong \triangle PQR$, then $\triangle PQR \cong \triangle JKL$.	If $\triangle JKL \cong \triangle PQR$, and $\triangle PQR \cong \triangle XYZ$, then $\triangle JKL \cong \triangle XYZ$

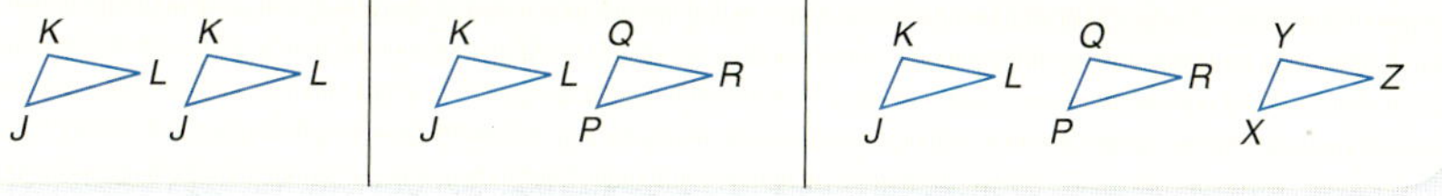

You will prove the symmetric and reflexive parts of Theorem 4.4 in Exercises 33 and 35, respectively.

Proof Theorem 4.4 (Transitive)

Given: $\triangle ABC \cong \triangle DEF$
$\triangle DEF \cong \triangle GHI$

Prove: $\triangle ABC \cong \triangle GHI$

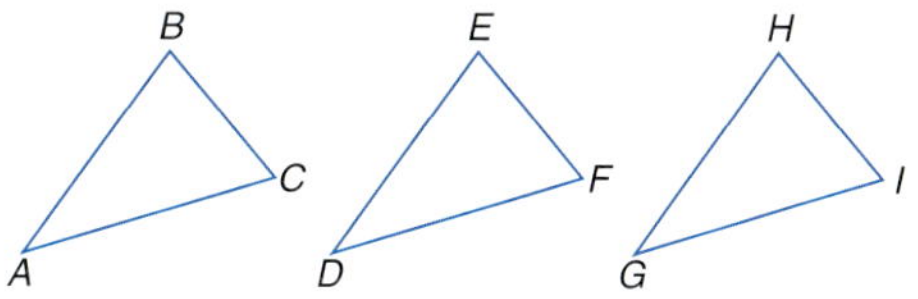

Proof:

Statements	Reasons
1. $\triangle ABC \cong \triangle DEF$	1. Given
2. $\angle A \cong \angle D$, $\angle B \cong \angle E$, $\angle C \cong \angle F$ $\overline{AB} \cong \overline{DE}$, $\overline{BC} \cong \overline{EF}$, $\overline{AC} \cong \overline{DF}$	2. CPCTC
3. $\triangle DEF \cong \triangle GHI$	3. Given
4. $\angle D \cong \angle G$, $\angle E \cong \angle H$, $\angle F \cong \angle I$ $\overline{DE} \cong \overline{GH}$, $\overline{EF} \cong \overline{HI}$, $\overline{DF} \cong \overline{GI}$	4. CPCTC
5. $\angle A \cong \angle G$, $\angle B \cong \angle H$, $\angle C \cong \angle I$	5. Congruence of angles is transitive.
6. $\overline{AB} \cong \overline{GH}$, $\overline{BC} \cong \overline{HI}$, $\overline{AC} \cong \overline{GI}$	6. Congruence of segments is transitive.
7. $\triangle ABC \cong \triangle GHI$	7. Def. of $\cong \triangle$s

2 Teach

CORRESPONDING PARTS OF CONGRUENT TRIANGLES

In-Class Example

Power Point®

Teaching Tip Students can use tick marks on sides and angles to help visually organize the corresponding parts of congruent triangles.

1 ARCHITECTURE A tower roof is composed of congruent triangles all converging toward a point at the top.

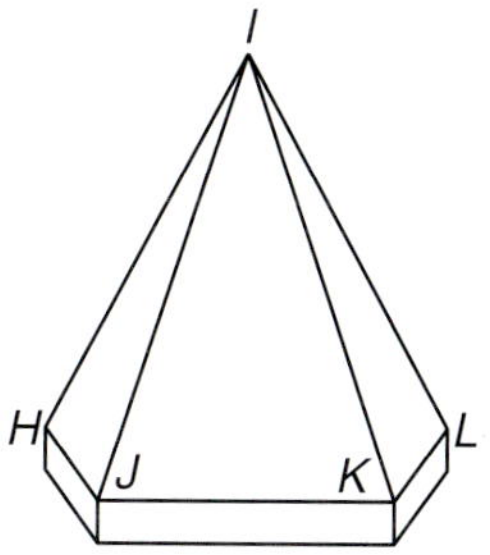

a. Name the corresponding congruent angles and sides of $\triangle HIJ$ and $\triangle LIK$.

$\angle HJI \cong \angle LKI$; $\angle ILK \cong \angle IHJ$; $\angle HIJ \cong \angle LIK$; $\overline{HI} \cong \overline{LI}$; $\overline{HJ} \cong \overline{LK}$; $\overline{JI} \cong \overline{KI}$

b. Name the congruent triangles. $\triangle HIJ \cong \triangle LIK$

DAILY INTERVENTION

Differentiated Instruction

Auditory/Musical Explain to students that congruency can be appealing to both the eyes and the ears. Point out that if students use beats to model two congruent equilateral triangles, they could use three equally-spaced drum beats for the first and then repeat the exact same rhythm for the second. An isosceles beat could consist of two quick beats and one slow beat or vice versa. Tell students that often in music, a "congruent" rhythm is used throughout a song. A popular example is the song, "Louie, Louie."

IDENTIFY CONGRUENCE TRANSFORMATIONS

In-Class Example Power Point®

Teaching Tip In Chapter 9, students will be introduced to the formal names for the slide, flip, and turn transformations.

2 **COORDINATE GEOMETRY**

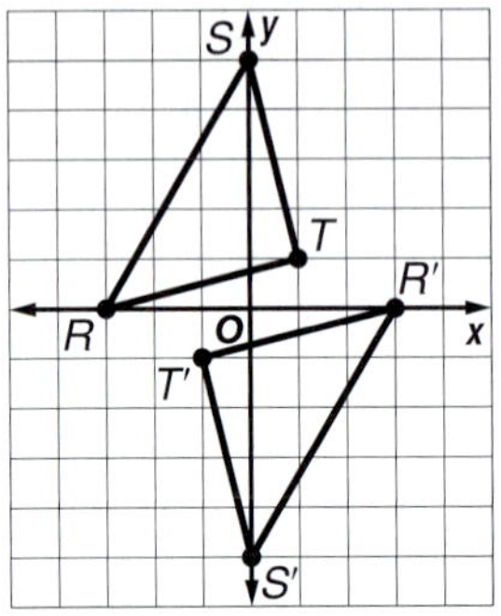

a. Verify that $\triangle RST \cong \triangle R'S'T'$.

$RS = R'S' = \sqrt{34}$

$ST = S'T' = \sqrt{17}$

$TR = T'R' = \sqrt{17}$

b. Name the congruence transformation for $\triangle RST$ and $\triangle R'S'T'$. turn

3 Practice/Apply

Study Notebook

Have students—

- add the definitions/examples of the vocabulary terms to their Vocabulary Builder worksheets for Chapter 4.
- include an example from each theorem introduced in this lesson.
- include any other item(s) that they find helpful in mastering the skills in this lesson.

Study Tip

Naming Congruent Triangles
There are six ways to name each pair of congruent triangles.

Study Tip

Transformations
Not all of the transformations preserve congruence. Only transformations that do not change the size or shape of the triangle are congruence transformations.

IDENTIFY CONGRUENCE TRANSFORMATIONS In the figures below, $\triangle ABC$ is congruent to $\triangle DEF$. If you *slide* $\triangle DEF$ up and to the right, $\triangle DEF$ is still congruent to $\triangle ABC$.

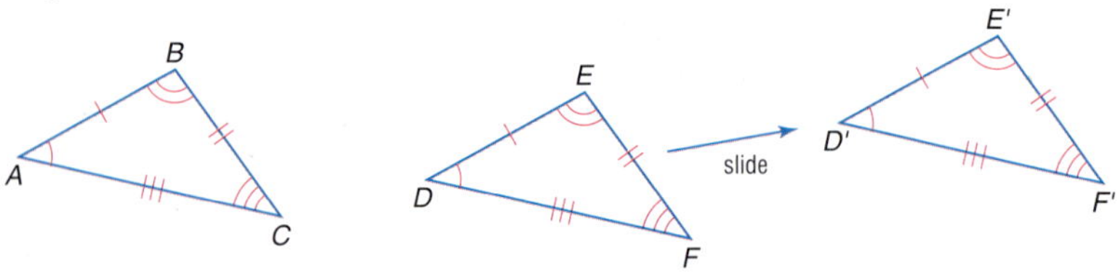

The congruency does not change whether you *turn* $\triangle DEF$ or *flip* $\triangle DEF$. $\triangle ABC$ is still congruent to $\triangle DEF$.

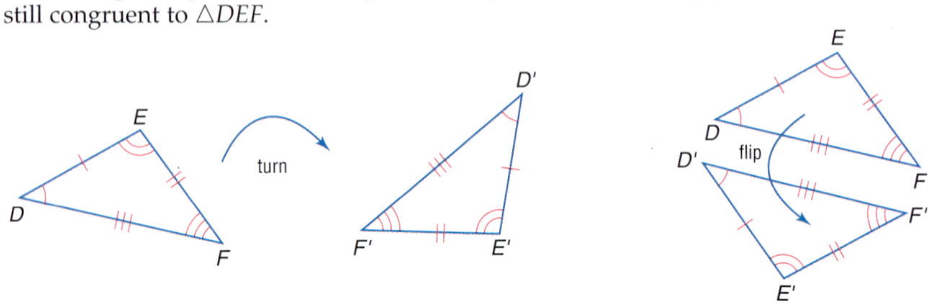

If you slide, flip, or turn a triangle, the size and shape do not change. These three transformations are called **congruence transformations**.

Example 2 Transformations in the Coordinate Plane

COORDINATE GEOMETRY **The vertices of $\triangle CDE$ are $C(-5, 7)$, $D(-8, 6)$, and $E(-3, 3)$. The vertices of $\triangle C'D'E'$ are $C'(5, 7)$, $D'(8, 6)$, and $E'(3, 3)$.**

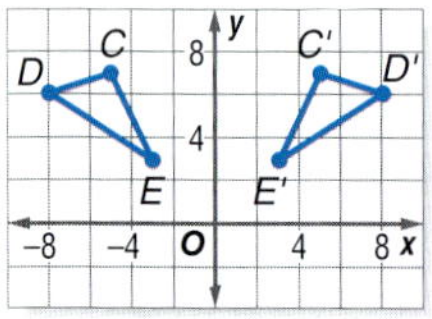

a. Verify that $\triangle CDE \cong \triangle C'D'E'$.

Use the Distance Formula to find the length of each side in the triangles.

$DC = \sqrt{[-8 - (-5)]^2 + (6 - 7)^2}$
$= \sqrt{9 + 1}$ or $\sqrt{10}$

$D'C' = \sqrt{(8 - 5)^2 + (6 - 7)^2}$
$= \sqrt{9 + 1}$ or $\sqrt{10}$

$DE = \sqrt{[-8 - (-3)]^2 + (6 - 3)^2}$
$= \sqrt{25 + 9}$ or $\sqrt{34}$

$D'E' = \sqrt{(8 - 3)^2 + (6 - 3)^2}$
$= \sqrt{25 + 9}$ or $\sqrt{34}$

$CE = \sqrt{[-5 - (-3)]^2 + (7 - 3)^2}$
$= \sqrt{4 + 16}$ or $\sqrt{20}$

$C'E' = \sqrt{(5 - 3)^2 + (7 - 3)^2}$
$= \sqrt{4 + 16}$ or $\sqrt{20}$

The lengths of the corresponding sides of two triangles are equal. Therefore, by the definition of congruence, $\overline{DC} \cong \overline{D'C'}$, $\overline{DE} \cong \overline{D'E'}$, and $\overline{CE} \cong \overline{C'E'}$.

Use a protractor to measure the angles of the triangles. You will find that the measures are the same.

In conclusion, because $\overline{DC} \cong \overline{D'C'}$, $\overline{DE} \cong \overline{D'E'}$, and $\overline{CE} \cong \overline{C'E'}$, $\angle D \cong \angle D'$, $\angle C \cong \angle C'$, and $\angle E \cong \angle E'$, $\triangle DCE \cong \triangle D'C'E'$.

b. Name the congruence transformation for $\triangle CDE$ and $\triangle C'D'E'$.

$\triangle C'D'E'$ is a flip of $\triangle CDE$.

Check for Understanding

Concept Check

1. **Explain** how slides, flips, and turns preserve congruence. **See margin.**
2. **OPEN ENDED** Draw a pair of congruent triangles and label the congruent sides and angles. **See margin.**

Guided Practice

GUIDED PRACTICE KEY	
Exercises	Examples
3–6, 8	1
7	2

Identify the congruent triangles in each figure.

3. $\triangle AFC \cong \triangle DFB$

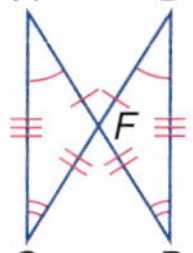

4. $\triangle HJT \cong \triangle TKH$

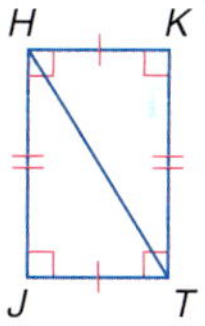

5. If $\triangle WXZ \cong \triangle STJ$, name the congruent angles and congruent sides.

5. $\angle W \cong \angle S$, $\angle X \cong \angle T$, $\angle Z \cong \angle J$, $\overline{WX} \cong \overline{ST}$, $\overline{XZ} \cong \overline{TJ}$, $\overline{WZ} \cong \overline{SJ}$

6. **QUILTING** In the quilt design, assume that angles and segments that appear to be congruent are congruent. Indicate which triangles are congruent. **See margin.**

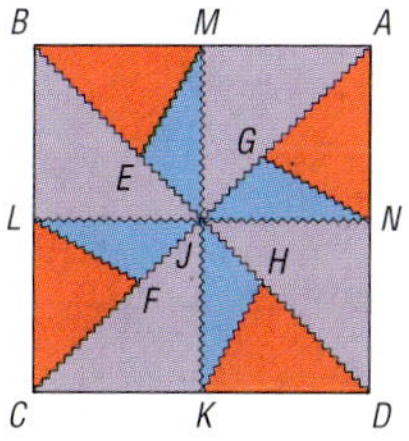

7. The coordinates of the vertices of $\triangle QRT$ and $\triangle Q'R'T'$ are $Q(-4, 3)$, $Q'(4, 3)$, $R(-4, -2)$, $R'(4, -2)$, $T(-1, -2)$, and $T'(1, -2)$. Verify that $\triangle QRT \cong \triangle Q'R'T'$. Then name the congruence transformation. **See margin.**

Application

8. **GARDENING** This garden lattice will be covered with morning glories in the summer. Wesley wants to save two triangular areas for artwork. If $\triangle GHJ \cong \triangle KLP$, name the corresponding congruent angles and sides.
$\angle G \cong \angle K$, $\angle H \cong \angle L$, $\angle J \cong \angle P$, $\overline{GH} \cong \overline{KL}$, $\overline{HJ} \cong \overline{LP}$, $\overline{GJ} \cong \overline{KP}$

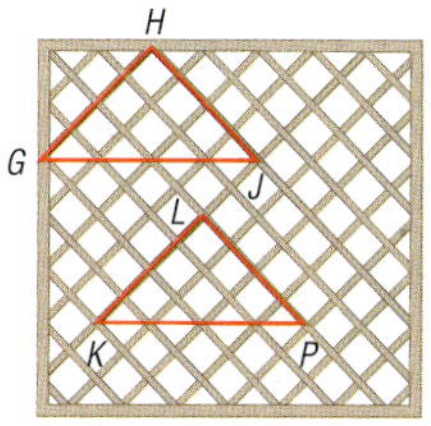

★ indicates increased difficulty

Practice and Apply

Homework Help	
For Exercises	See Examples
9–22, 27–35	1
23–26	2

Extra Practice
See page 761.

Identify the congruent triangles in each figure.

9. $\triangle CFH \cong \triangle JKL$

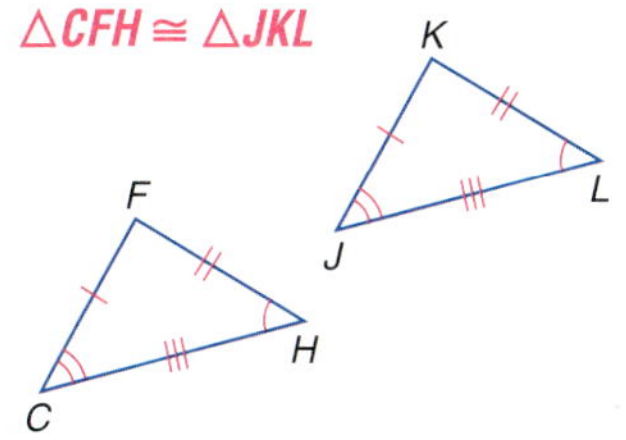

10. $\triangle RSV \cong \triangle TSV$

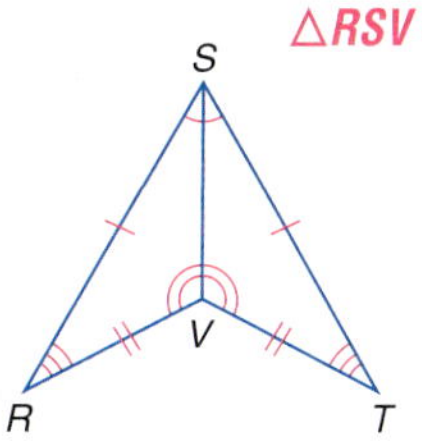

11. $\triangle WPZ \cong \triangle QVS$

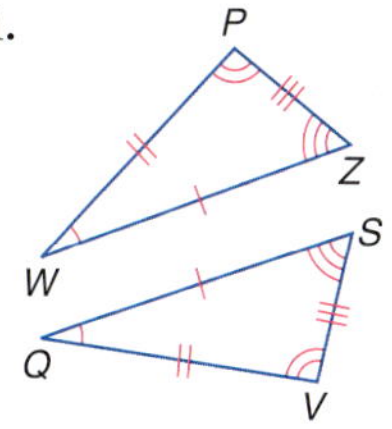

12. $\triangle EFH \cong \triangle GHF$

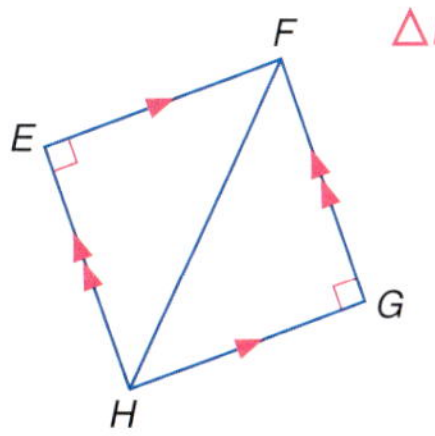

Name the congruent angles and sides for each pair of congruent triangles.

13–16. See margin.

13. $\triangle TUV \cong \triangle XYZ$
14. $\triangle CDG \cong \triangle RSW$
15. $\triangle BCF \cong \triangle DGH$
16. $\triangle ADG \cong \triangle HKL$

About the Exercises...

Organization by Objective

- **Corresponding Parts of Congruent Triangles:** 9–22, 27–35
- **Identify Congruence Transformations:** 23–26

Odd/Even Assignments

Exercises 9–35 are structured so that students practice the same concepts whether they are assigned odd or even problems.

Assignment Guide

Basic: 9–19 odd, 23, 27–33, 36–51

Average: 9–35 odd, 36–51

Advanced: 10–36 even, 37–48 (optional: 49–51)

All: Quiz 1 (1–5)

Answers

1. The sides and the angles of the triangle are not affected by a congruence transformation, so congruence is preserved.
2. Sample answer:

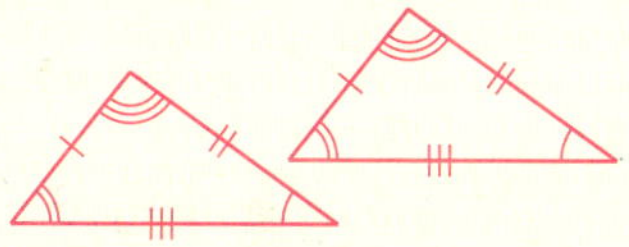

6. $\triangle BME$, $\triangle ANG$, $\triangle DKH$, $\triangle CLF$; $\triangle EMJ$, $\triangle GNJ$, $\triangle HKJ$, $\triangle FLJ$; $\triangle BAJ$, $\triangle ADJ$, $\triangle DCJ$, $\triangle CBJ$; $\triangle BCD$, $\triangle ADC$, $\triangle CBA$, $\triangle DAB$; $\triangle BLJ$, $\triangle AMJ$, $\triangle JND$, $\triangle JKC$, $\triangle BMJ$, $\triangle ANJ$, $\triangle JKD$, $\triangle JLC$
7. $QR = 5$, $Q'R' = 5$, $RT = 3$, $R'T' = 3$, $QT = \sqrt{34}$, and $Q'T' = \sqrt{34}$. Use a protractor to confirm that the corresponding angles are congruent; flip.
13. $\angle T \cong \angle X$, $\angle U \cong \angle Y$, $\angle V \cong \angle Z$, $\overline{TU} \cong \overline{XY}$, $\overline{UV} \cong \overline{YZ}$, $\overline{TV} \cong \overline{XZ}$
14. $\angle C \cong \angle R$, $\angle D \cong \angle S$, $\angle G \cong \angle W$, $\overline{CD} \cong \overline{RS}$, $\overline{DG} \cong \overline{SW}$, $\overline{CG} \cong \overline{RW}$
15. $\angle B \cong \angle D$, $\angle C \cong \angle G$, $\angle F \cong \angle H$, $\overline{BC} \cong \overline{DG}$, $\overline{CF} \cong \overline{GH}$, $\overline{BF} \cong \overline{DH}$
16. $\angle A \cong \angle H$, $\angle D \cong \angle K$, $\angle G \cong \angle L$, $\overline{AD} \cong \overline{HK}$, $\overline{DG} \cong \overline{KL}$, $\overline{AG} \cong \overline{HL}$

Answers

22. Flip; $PQ = 2$, $P'Q' = 2$, $QV = 4$, $Q'V' = 4$, $PV = \sqrt{20}$, and $P'V' = \sqrt{20}$. Use a protractor to confirm that the corresponding angles are congruent.

23. Flip; $MN = 8$, $M'N' = 8$, $NP = 2$, $N'P' = 2$, $MP = \sqrt{68}$, and $M'P' = \sqrt{68}$. Use a protractor to confirm that the corresponding angles are congruent.

24. Slide; $GH = \sqrt{10}$, $G'H' = \sqrt{10}$, $HF = \sqrt{8}$, $H'F' = \sqrt{8}$, $GF = \sqrt{10}$, and $G'F' = \sqrt{10}$. Use a protractor to confirm that the corresponding angles are congruent.

25. Turn; $JK = \sqrt{40}$, $J'K' = \sqrt{40}$, $KL = \sqrt{29}$, $K'L' = \sqrt{29}$, $JL = \sqrt{17}$, and $J'L' = \sqrt{17}$. Use a protractor to confirm that the corresponding angles are congruent.

26. False; $\angle A \cong \angle X$, $\angle B \cong \angle Y$, and $\angle C \cong \angle Z$ but the corresponding sides are not congruent.

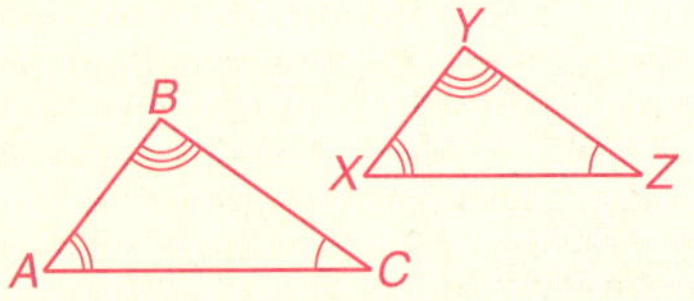

27.

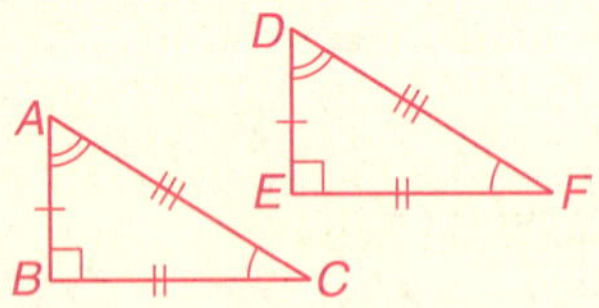

29.

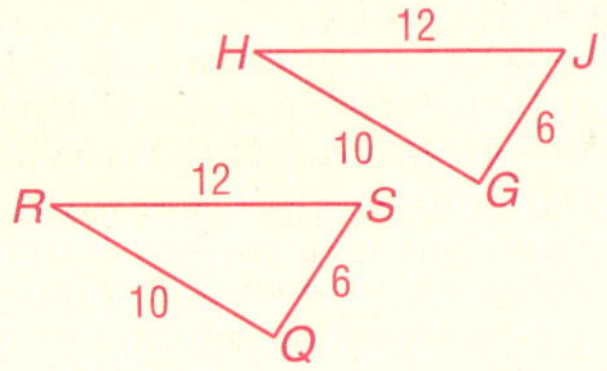

31.

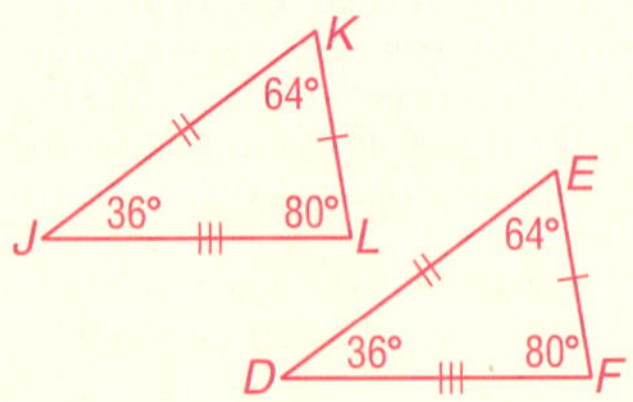

18. △s 1–4, △s 5–12, △s 13–20

19. △s 1, 5, 6, and 11, △s 3, 8, 10, and 12, △s 2, 4, 7, and 9

Assume that segments and angles that appear to be congruent in the numbered triangles are congruent. Indicate which triangles are congruent.

17.

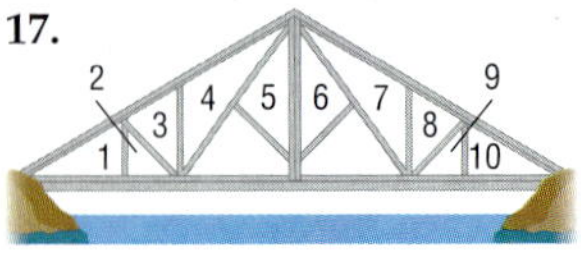

18.

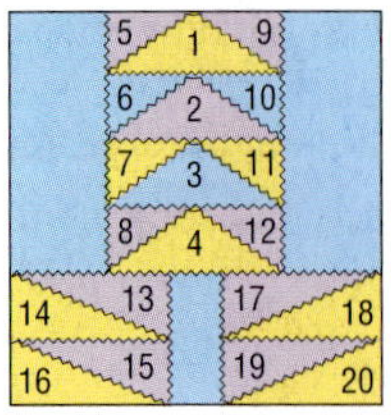

19.

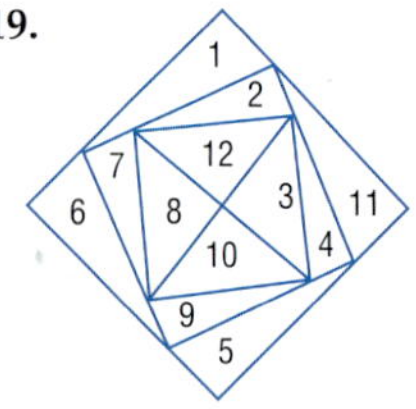

21. We need to know that all of the corresponding angles are congruent and that the other corresponding sides are congruent.

★ 20. All of the small triangles in the figure at the right are congruent. Name three larger congruent triangles. $\triangle UFS$, $\triangle TDV$, $\triangle ACB$

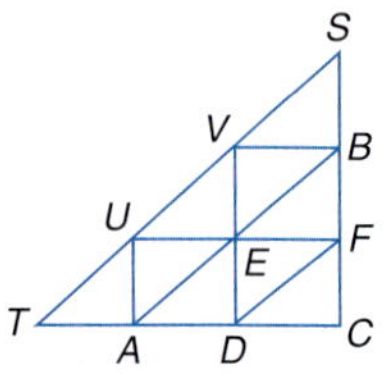

More About...

Mosaics

A mosaic is composed of glass, marble, or ceramic pieces often arranged in a pattern. The pieces, or *tesserae*, are set in cement. Mosaics are used to decorate walls, floors, and gardens.

Source: www.dimosaico.com

21. **MOSAICS** ★ The picture at the left is the center of a Roman mosaic. Because the four triangles connect to a square, they have at least one side congruent to a side in another triangle. What else do you need to know to conclude that the four triangles are congruent?

Verify that each of the following preserves congruence and name the congruence transformation. 22–25. See margin for verification.

22. $\triangle PQV \cong \triangle P'Q'V'$ flip

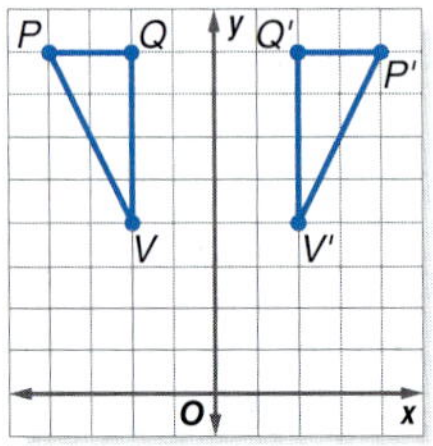

23. $\triangle MNP \cong \triangle M'N'P'$ flip

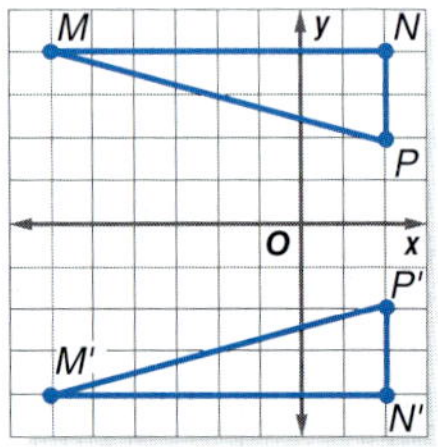

★ 24. $\triangle GHF \cong \triangle G'H'F'$ slide

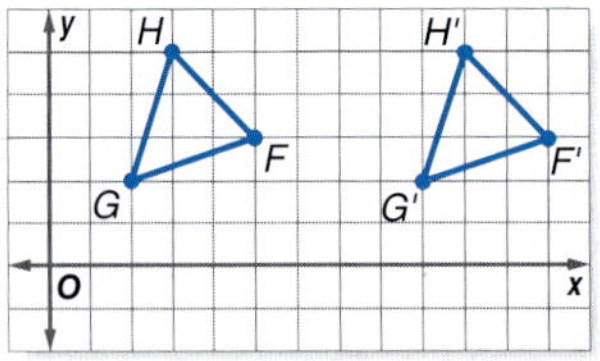

★ 25. $\triangle JKL \cong \triangle J'K'L'$ turn

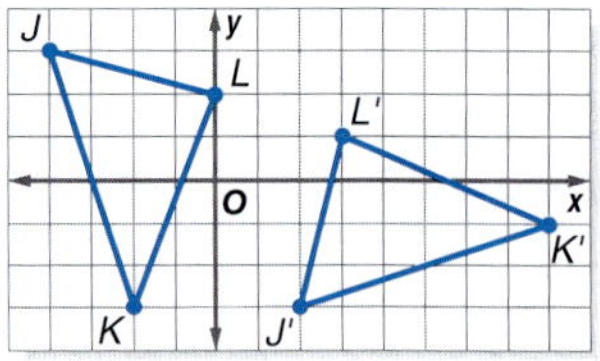

26–27. See margin for drawings.

Determine whether each statement is *true* or *false*. Draw an example or counterexample for each.

26. Two triangles with corresponding congruent angles are congruent. false

27. Two triangles with angles and sides congruent are congruent. true

★ 28. **UMBRELLAS** Umbrellas usually have eight congruent triangular sections with ribs of equal length. Are the statements $\triangle JAD \cong \triangle IAE$ and $\triangle JAD \cong \triangle EAI$ both correct? Explain.
Both statements are correct because the spokes are the same length, $\overline{EA} \cong \overline{IA}$, and $\overline{AE} \cong \overline{AI}$.

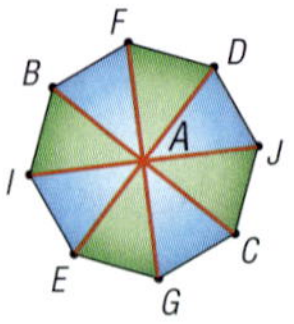

35. Given: $\triangle DEF$
Prove: $\triangle DEF \cong \triangle DEF$

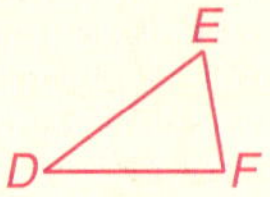

Proof:

$\triangle DEF$ — Given

$\overline{DE} \cong \overline{DE}$, $\overline{EF} \cong \overline{EF}$, $\overline{DF} \cong \overline{DF}$ — Congruence of segments is reflexive.

$\angle D \cong \angle D$, $\angle E \cong \angle E$, $\angle F \cong \angle F$ — Congruence of $\angle$s is reflexive.

$\triangle DEF \cong \triangle DEF$ — Def. of $\cong$ △s

ALGEBRA For Exercises 29 and 30, use the following information.
$\triangle QRS \cong \triangle GHJ$, $RS = 12$, $QR = 10$, $QS = 6$, and $HJ = 2x - 4$.

29. Draw and label a figure to show the congruent triangles. **See margin.**

30. Find x. **8**

ALGEBRA For Exercises 31 and 32, use the following information.
$\triangle JKL \cong \triangle DEF$, $m\angle J = 36$, $m\angle E = 64$, and $m\angle F = 3x + 52$.

31. Draw and label a figure to show the congruent triangles. **See margin.**

32. Find x. $\frac{28}{3}$

33. **PROOF** The statements below can be used to prove that *congruence of triangles is symmetric*. Use the statements to construct a correct flow proof. Provide the reasons for each statement. **See p. 233B.**

Given: $\triangle RST \cong \triangle XYZ$

Prove: $\triangle XYZ \cong \triangle RST$

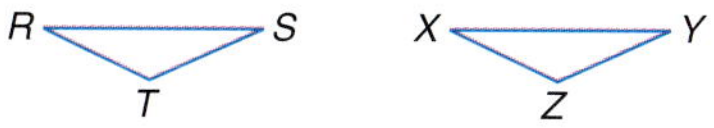

Flow Proof:

- $\angle X \cong \angle R$, $\angle Y \cong \angle S$, $\angle Z \cong \angle T$, $\overline{XY} \cong \overline{RS}$, $\overline{YZ} \cong \overline{ST}$, $\overline{XZ} \cong \overline{RT}$ — ___?___
- $\angle R \cong \angle X$, $\angle S \cong \angle Y$, $\angle T \cong \angle Z$, $\overline{RS} \cong \overline{XY}$, $\overline{ST} \cong \overline{YZ}$, $\overline{RT} \cong \overline{XZ}$ — ___?___
- $\triangle RST \cong \triangle XYZ$ — ___?___
- $\triangle XYZ \cong \triangle RST$ — ___?___

34. **PROOF** Copy the flow proof and provide the reasons for each statement.

Given: $\overline{AB} \cong \overline{CD}$, $\overline{AD} \cong \overline{CB}$, $\overline{AD} \perp \overline{DC}$, $\overline{AB} \perp \overline{BC}$, $\overline{AD} \parallel \overline{BC}$, $\overline{AB} \parallel \overline{CD}$

Prove: $\triangle ACD \cong \triangle CAB$

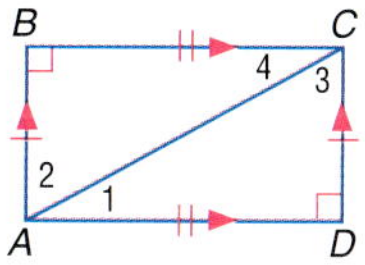

Flow Proof:

- $\overline{AB} \cong \overline{CD}$ — a. ___?___
- $\overline{AD} \cong \overline{CB}$ — b. ___?___
- $\overline{AC} \cong \overline{CA}$ — c. ___?___
- $\overline{AD} \perp \overline{DC}$ — d. ___?___ → $\angle D$ is a rt. $\angle$. — e. ___?___
- $\overline{AB} \perp \overline{BC}$ — f. ___?___ → $\angle B$ is a rt. $\angle$. — g. ___?___
- $\angle D \cong \angle B$ — h. ___?___
- $\overline{AD} \parallel \overline{BC}$ — i. ___?___ → $\angle 1 \cong \angle 4$ — j. ___?___
- $\overline{AB} \parallel \overline{CD}$ — k. ___?___ → $\angle 2 \cong \angle 3$ — l. ___?___
- $\triangle ACD \cong \triangle CAB$ — m. ___?___

34.
a. Given
b. Given
c. Congruence of segments is symmetric.
d. Given
e. Def. of ⊥ lines
f. Given
g. Def. of ⊥ lines
h. All right ∠s are ≅.
i. Given
j. Alt. int. ∠s are ≅.
k. Given
l. Alt. int. ∠s are ≅.
m. Def. of ≅ △s

★ 35. **PROOF** Write a flow proof to prove *Congruence of triangles is reflexive*. (Theorem 4.4) **See margin.**

36. **CRITICAL THINKING** $\triangle RST$ is isosceles with $RS = RT$, M, N, and P are midpoints of their sides, $\angle S \cong \angle MPS$, and $\overline{NP} \cong \overline{MP}$. What else do you need to know to prove that $\triangle SMP \cong \triangle TNP$? **$\angle SMP \cong \angle TNP$, $\angle MPS \cong \angle NPT$**

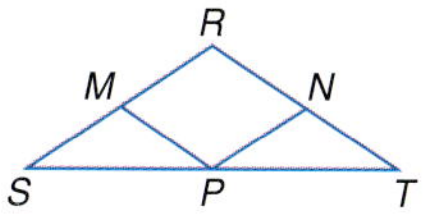

www.geometryonline.com/self_check_quiz/fcat

Study Guide and Intervention, p. 195 (shown) and p. 196

Corresponding Parts of Congruent Triangles
Triangles that have the same size and same shape are **congruent triangles**. Two triangles are congruent if and only if all three pairs of corresponding angles are congruent and all three pairs of corresponding sides are congruent. In the figure, $\triangle ABC \cong \triangle RST$.

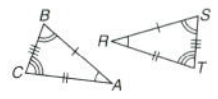

Example **If $\triangle XYZ \cong \triangle RST$, name the pairs of congruent angles and congruent sides.**
$\angle X \cong \angle R$, $\angle Y \cong \angle S$, $\angle Z \cong \angle T$
$\overline{XY} \cong \overline{RS}$, $\overline{XZ} \cong \overline{RT}$, $\overline{YZ} \cong \overline{ST}$

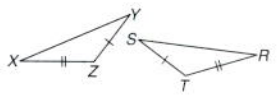

Exercises

Identify the congruent triangles in each figure.

1. 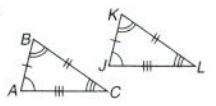
$\triangle ABC \cong \triangle JKL$

2. 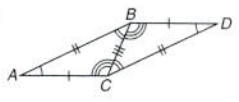
$\triangle ABC \cong \triangle DCB$

3. 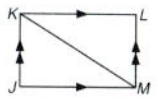
$\triangle JKM \cong \triangle LMK$

Name the corresponding congruent angles and sides for the congruent triangles.

4. 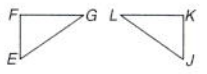
$\angle E \cong \angle J$; $\angle F \cong \angle K$; $\angle G \cong \angle L$; $\overline{EF} \cong \overline{JK}$; $\overline{EG} \cong \overline{JL}$; $\overline{FG} \cong \overline{KL}$

5. 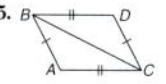
$\angle A \cong \angle D$; $\angle ABC \cong \angle DCB$; $\angle ACB \cong \angle DBC$; $\overline{AB} \cong \overline{DC}$; $\overline{AC} \cong \overline{DB}$; $\overline{BC} \cong \overline{CB}$

6. 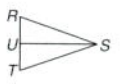
$\angle R \cong \angle T$; $\angle RSU \cong \angle TSU$; $\angle RUS \cong \angle TUS$; $\overline{RU} \cong \overline{TU}$; $\overline{RS} \cong \overline{TS}$; $\overline{SU} \cong \overline{SU}$

Skills Practice, p. 197 and Practice, p. 198 (shown)

Identify the congruent triangles in each figure.

1. 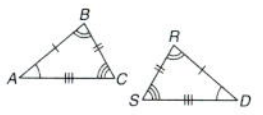
$\triangle ABC \cong \triangle DRS$

2. 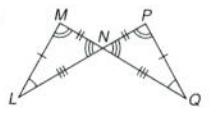
$\triangle LMN \cong \triangle QPN$

Name the congruent angles and sides for each pair of congruent triangles.

3. $\triangle GKP \cong \triangle LMN$
$\angle G \cong \angle L$, $\angle K \cong \angle M$, $\angle P \cong \angle N$; $\overline{GK} \cong \overline{LM}$, $\overline{KP} \cong \overline{MN}$, $\overline{GP} \cong \overline{LN}$

4. $\triangle ANC \cong \triangle RBV$
$\angle A \cong \angle R$, $\angle N \cong \angle B$, $\angle C \cong \angle V$; $\overline{AN} \cong \overline{RB}$, $\overline{NC} \cong \overline{BV}$, $\overline{AC} \cong \overline{RV}$

Verify that each of the following transformations preserves congruence, and name the congruence transformation.

5. $\triangle PST \cong \triangle P'S'T'$

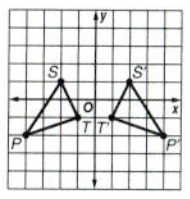

$PS = \sqrt{13}$, $P'S' = \sqrt{13}$, $ST = \sqrt{5}$, $S'T' = \sqrt{5}$, $PT = \sqrt{10}$, $P'T' = \sqrt{10}$, $\angle P \cong \angle P'$, $\angle S \cong \angle S'$, $\angle T \cong \triangle T'$; flip

6. $\triangle LMN \cong \triangle L'M'N'$

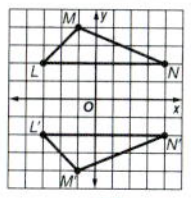

$LM = 2\sqrt{2}$, $L'M' = 2\sqrt{2}$, $MN = \sqrt{29}$, $M'N' = \sqrt{29}$, $LN = 7$, $L'N' = 7$, $\angle L \cong \angle L'$, $\angle M \cong \angle M'$, $\angle N \cong \angle N'$; flip

QUILTING For Exercises 7 and 8, refer to the quilt design.

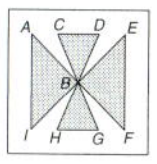

7. Indicate the triangles that appear to be congruent.
$\triangle ABI \cong \triangle EBF$, $\triangle CBD \cong \triangle HBG$

8. Name the congruent angles and congruent sides of a pair of congruent triangles.
Sample answer: $\angle A \cong \angle E$, $\angle ABI \cong \angle EBF$, $\angle I \cong \angle F$; $\overline{AB} \cong \overline{EB}$, $\overline{BI} \cong \overline{BF}$, $\overline{AI} \cong \overline{EF}$

Reading to Learn Mathematics, p. 199 ELL

Pre-Activity **Why are triangles used in bridges?**

Read the introduction to Lesson 4-3 at the top of page 192 in your textbook.

In the bridge shown in the photograph in your textbook, diagonal braces were used to divide squares into two isosceles right triangles. Why do you think these braces are used on the bridge? Sample answer: The diagonal braces make the structure stronger and prevent it from being deformed when it has to withstand a heavy load.

Reading the Lesson

1. If $\triangle RST \cong \triangle UWV$, complete each pair of congruent parts.
$\angle R \cong$ $\angle U$ $\qquad$ $\angle S$ $\cong \angle W$ $\qquad$ $\angle T \cong$ $\angle V$
$\overline{RT} \cong$ $\overline{UV}$ $\qquad$ $\overline{RS}$ $\cong \overline{UW}$ $\qquad$ $\overline{ST}$ $\cong \overline{WV}$

2. Identify the congruent triangles in each diagram.

a. 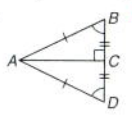
$\triangle ABC \cong \triangle ADC$

b. 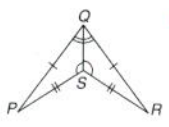
$\triangle PQS \cong \triangle RQS$

c. 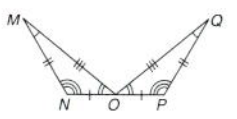
$\triangle MNO \cong \triangle QPO$

d. 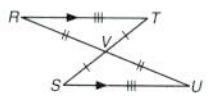
$\triangle RTV \cong \triangle USV$

3. Determine whether each statement says that congruence of triangles is *reflexive*, *symmetric*, or *transitive*.
a. If the first of two triangles is congruent to the second triangle, then the second triangle is congruent to the first. symmetric
b. If there are three triangles for which the first is congruent to the second and the second is congruent to the third, then the first triangle is congruent to the third. transitive
c. Every triangle is congruent to itself. reflexive

Helping You Remember

4. A good way to remember something is to explain it to someone else. Your classmate Ben is having trouble writing congruence statements for triangles because he thinks he has to match up three pairs of sides and three pairs of angles. How can you help him understand how to write correct congruence statements more easily? Sample answer: Write the three vertices of one triangle in any order. Then write the corresponding vertices of the second triangle in the same order. If the angles are written in the correct correspondence, the sides will automatically be in the correct correspondence also.

Enrichment, p. 200

Transformations in The Coordinate Plane

The following statement tells one way to map preimage points to image points in the coordinate plane.

$(x, y) \rightarrow (x + 6, y - 3)$

This can be read, "The point with coordinates (x, y) is mapped to the point with coordinates $(x + 6, y - 3)$." With this transformation, for example, $(3, 5)$ is mapped to $(3 + 6, 5 - 3)$ or $(9, 2)$. The figure shows how the triangle ABC is mapped to triangle XYZ.

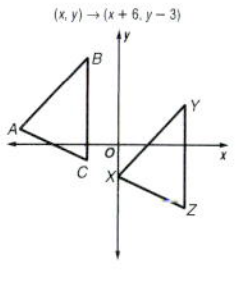

1. Does the transformation above appear to be a congruence transformation? Explain your answer. Yes; the transformation slides the figure to the lower right without changing its size or shape.

4 Assess

Open-Ended Assessment

Modeling Ask students to name examples of how congruent triangles are modeled in objects or in nature. Then they can name the corresponding congruent angles and sides and determine the congruence transformations applied to the triangles. The umbrella for Exercise 28 on p. 196 is an example of rotated triangles, and students can name corresponding parts.

Getting Ready for Lesson 4-4

Prerequisite Skill Students will learn about proving congruence using SSS and SAS in Lesson 4-4. They will use the Distance Formula to find side lengths of triangles in a coordinate plane. Use Exercises 49–51 to determine your students' familiarity with the Distance Formula.

Assessment Options

Practice Quiz 1 The quiz provides students with a brief review of the concepts and skills in Lessons 4-1 through 4-3. Lesson numbers are given to the right of the exercises or instruction lines so students can review concepts not yet mastered.

Answers

37. Sample answer: Triangles are used in bridge design for structure and support. Answers should include the following.
- The shape of the triangle does not matter.
- Some of the triangles used in the bridge supports seem to be congruent.

37. **WRITING IN MATH** Answer the question that was posed at the beginning of the lesson. See margin.

Why are triangles used in bridges?

Include the following in your answer:
- whether the shape of the triangle matters, and
- whether the triangles appear congruent.

FCAT Practice
Standardized Test Practice

38. Determine which statement is true given $\triangle ABC \cong \triangle XYZ$. B

Ⓐ $\overline{BC} \cong \overline{ZX}$ Ⓑ $\overline{AC} \cong \overline{XZ}$ Ⓒ $\overline{AB} \cong \overline{YZ}$ Ⓓ cannot be determined

39. **ALGEBRA** Find the length of $\overline{DF}$ if $D(-5, 4)$ and $F(3, -7)$. D

Ⓐ $\sqrt{5}$ Ⓑ $\sqrt{13}$ Ⓒ $\sqrt{57}$ Ⓓ $\sqrt{185}$

Maintain Your Skills

Mixed Review Find x. (Lesson 4-2)

40. 75

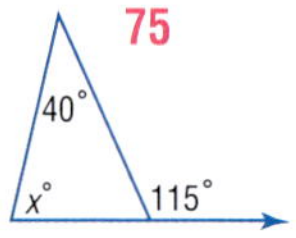

41. 58

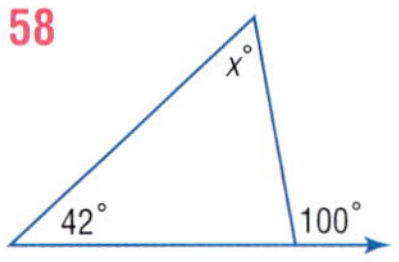

42. 75

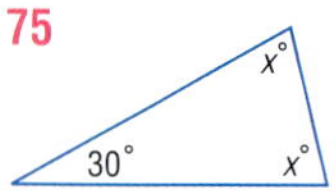

Find x and the measure of each side of the triangle. (Lesson 4-1)

43. $\triangle BCD$ is isosceles with $\overline{BC} \cong \overline{CD}$, $BC = 2x + 4$, $BD = x + 2$, and $CD = 10$.
43. $x = 3$, $BC = 10$, $CD = 10$, $BD = 5$

44. Triangle HKT is equilateral with $HK = x + 7$ and $HT = 4x - 8$.
$x = 5$; $HK = 12$, $HT = 12$, $KT = 12$

Write an equation in slope-intercept form for the line that satisfies the given conditions. (Lesson 3-4)

45. contains (0, 3) and (4, −3) 45. $y = -\frac{3}{2}x + 3$

46. $m = \frac{3}{4}$, y-intercept $= 8$ $y = \frac{3}{4}x + 8$

47. parallel to $y = -4x + 1$; contains $(-3, 1)$ $y = -4x - 11$

48. $m = -4$, contains $(-3, 2)$ $y = -4x - 10$

Getting Ready for the Next Lesson PREREQUISITE SKILL **Find the distance between each pair of points.**
(To review the **Distance Formula**, see Lesson 1-4.)

49. $(-1, 7)$, $(1, 6)$ $\sqrt{5}$

50. $(8, 2)$, $(4, -2)$ $\sqrt{32}$

51. $(3, 5)$, $(5, 2)$ $\sqrt{13}$

Practice Quiz 1 — Lessons 4-1 through 4-3

1. Identify the isosceles triangles in the figure, if $\overline{FH}$ and $\overline{DG}$ are congruent perpendicular bisectors. (Lesson 4-1) $\triangle DFJ$, $\triangle GJF$, $\triangle HJG$, $\triangle DJH$,

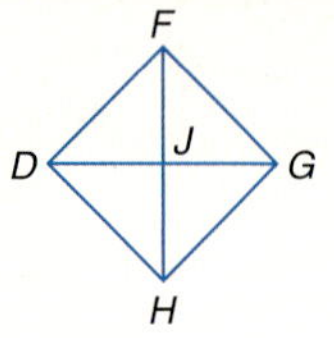

ALGEBRA $\triangle ABC$ **is equilateral with** $AB = 2x$, $BC = 4x - 7$, **and** $AC = x + 3.5$. (Lesson 4-1)

2. Find x. $x = 3.5$

3. Find the measure of each side. $AB = BC = AC = 7$

4. Find the measure of each numbered angle. (Lesson 4-2)
$m\angle 1 = 60$, $m\angle 2 = 110$, $m\angle 3 = 49$

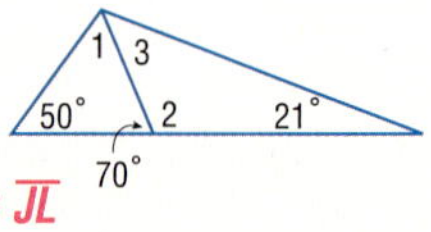

5. $\angle M \cong \angle J$, $\angle N \cong \angle K$, $\angle P \cong \angle L$; $\overline{MN} \cong \overline{JK}$, $\overline{NP} \cong \overline{KL}$, $\overline{MP} \cong \overline{JL}$

5. If $\triangle MNP \cong \triangle JKL$, name the corresponding congruent angles and sides. (Lesson 4-3)

198 Chapter 4 Congruent Triangles

Making Concept Maps

When studying a chapter, it is wise to record the main topics and vocabulary you encounter. In this chapter, some of the new vocabulary words were *triangle, acute triangle, obtuse triangle, right triangle, equiangular triangle, scalene triangle, isosceles triangle,* and *equilateral triangle.* The triangles are all related by the size of the angles or the number of congruent sides.

A graphic organizer called a *concept map* is a convenient way to show these relationships. A concept map is shown below for the different types of triangles. The main ideas are in boxes. Any information that describes how to move from one box to the next is placed along the arrows.

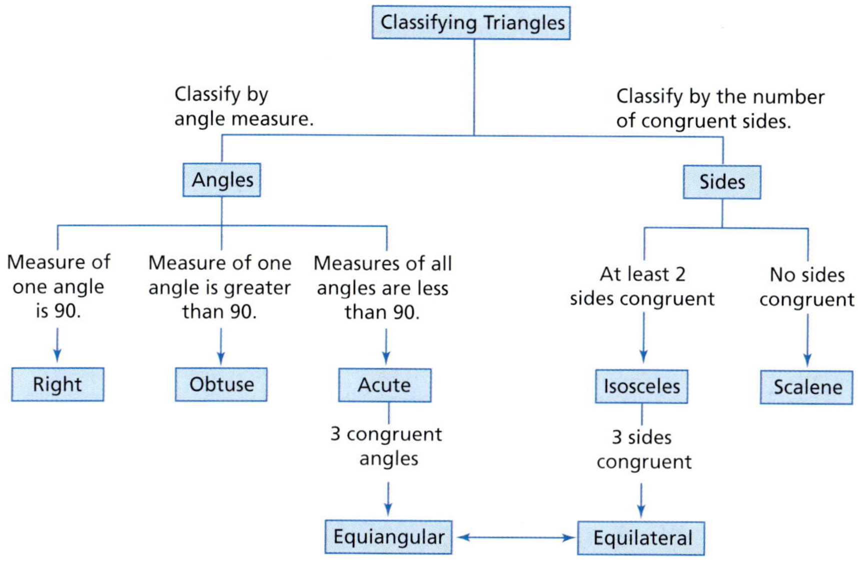

Reading to Learn

1. Describe how to use the concept map to classify triangles by their side lengths. **See margin.**
2. In $\triangle ABC$, $m\angle A = 48$, $m\angle B = 41$, and $m\angle C = 91$. Use the concept map to classify $\triangle ABC$. **obtuse**
3. Identify the type of triangle that is linked to both classifications. **equiangular or equilateral**

Reading Mathematics

Getting Started

Advise students to create their own version of this concept map and place it in their study notebooks. Encourage them to use colored pencils or highlighters to group related items.

Teach

Making Concept Maps
Students should discern from the concept map that if all the angle measures of a triangle are congruent, then it must also be acute, equilateral, and isosceles. Similarly, if all sides of a triangle are congruent, then it must also be acute, equiangular, and isosceles.

Assess

Study Notebook

Ask students to summarize what they have learned about using concept maps to review chapter material and enhance their knowledge of chapter concepts.

ELL English Language Learners may benefit from writing key concepts from this activity in their Study Notebooks in their native language and then in English.

Answer

1. Sample answer: If side lengths are given, determine the number of congruent sides and name the triangle. Some isosceles triangles are equilateral triangles.

4-4 Lesson Notes

1 Focus

5-Minute Check Transparency 4-4 Use as a quiz or review of Lesson 4-3.

Mathematical Background notes are available for this lesson on p. 176D.

How do land surveyors use congruent triangles?

Ask students:

- What does it mean for two triangles to be congruent? **All three corresponding sides and all three corresponding angles are congruent.**
- Would two congruent triangles have the same perimeter? Explain. **Yes, the three corresponding sides are congruent so the sum of the measures of the sides of the triangles would be equal.**

4-4 Proving Congruence—SSS, SAS

Sunshine State Standards
MA.A.3.4.3, MA.B.3.4.1, MA.C.1.4.1, MA.C.2.4.1, MA.C.3.4.2

Vocabulary
- included angle

What You'll Learn
- Use the SSS Postulate to test for triangle congruence.
- Use the SAS Postulate to test for triangle congruence.

How do land surveyors use congruent triangles?

Land surveyors mark and establish property boundaries. To check a measurement, they mark out a right triangle and then mark a second triangle that is congruent to the first.

SSS POSTULATE Is it always necessary to show that all of the corresponding parts of two triangles are congruent to prove that the triangles are congruent? In this lesson, we will explore two other methods to prove that triangles are congruent.

Construction

Congruent Triangles Using Sides

1. Draw a triangle and label the vertices X, Y, and Z.

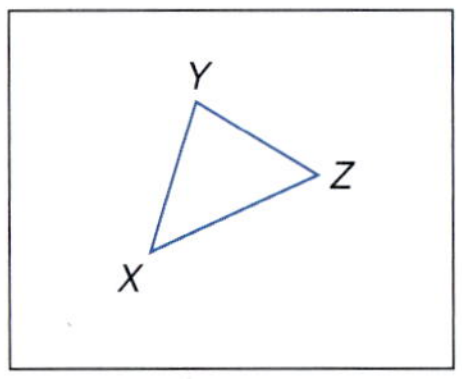

2. Use a straightedge to draw any line ℓ and select a point R. Use a compass to construct $\overline{RS}$ on ℓ such that $\overline{RS} \cong \overline{XZ}$.

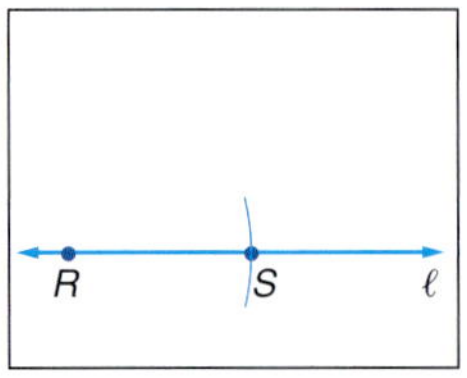

3. Using R as the center, draw an arc with radius equal to XY.

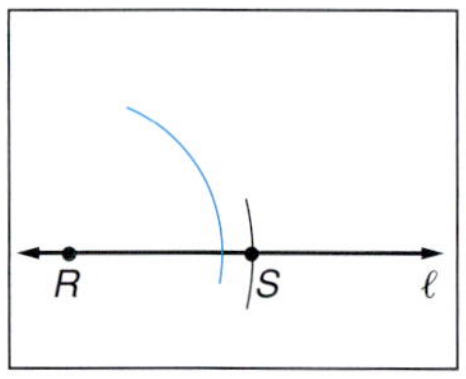

4. Using S as the center, draw an arc with radius equal to YZ.

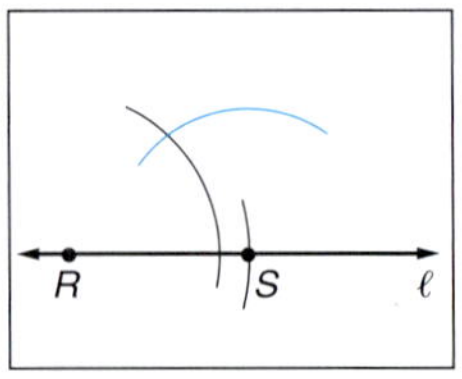

5. Let T be the point of intersection of the two arcs. Draw $\overline{RT}$ and $\overline{ST}$ to form $\triangle RST$.

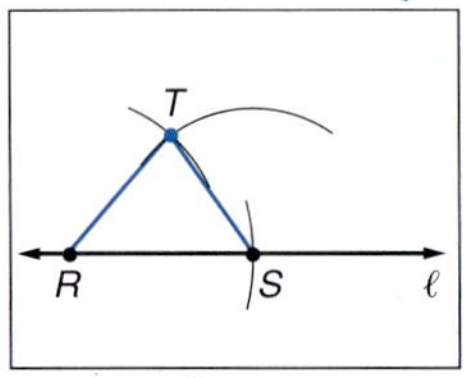

6. Cut out $\triangle RST$ and place it over $\triangle XYZ$. How does $\triangle RST$ compare to $\triangle XYZ$? **$\triangle RST \cong \triangle XYZ$**

In the construction, you see that if the corresponding sides of two triangles are congruent, then the triangles are congruent. This is the Side-Side-Side Postulate, and is written as SSS.

Resource Manager

Workbook and Reproducible Masters

Chapter 4 Resource Masters
- Study Guide and Intervention, pp. 201–202
- Skills Practice, p. 203
- Practice, p. 204
- Reading to Learn Mathematics, p. 205
- Enrichment, p. 206
- Assessment, pp. 239, 241

Prerequisite Skills Workbook, pp. 1–2
Teaching Geometry With Manipulatives Masters, pp. 8, 75

Transparencies

5-Minute Check Transparency 4-4
Answer Key Transparencies

Technology

Interactive Chalkboard

Postulate 4.1

Side-Side-Side Congruence If the sides of one triangle are congruent to the sides of a second triangle, then the triangles are congruent.

Abbreviation: *SSS*

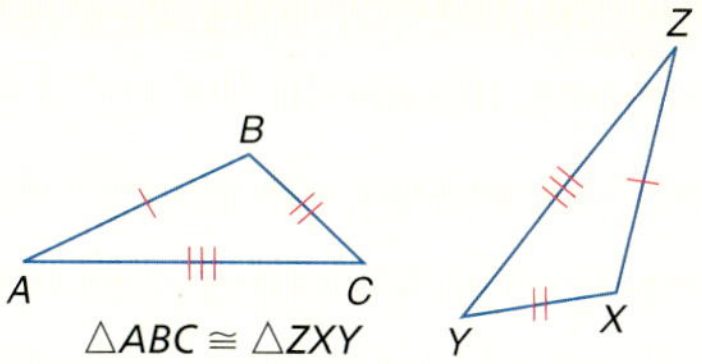

$\triangle ABC \cong \triangle ZXY$

Example 1 Use SSS in Proofs

MARINE BIOLOGY **The tail of an orca whale can be viewed as two triangles that share a common side. Write a two-column proof to prove that $\triangle BYA \cong \triangle CYA$ if $\overline{AB} \cong \overline{AC}$ and $\overline{BY} \cong \overline{CY}$.**

Given: $\overline{AB} \cong \overline{AC}$; $\overline{BY} \cong \overline{CY}$

Prove: $\triangle BYA \cong \triangle CYA$

Proof:

Statements	Reasons
1. $\overline{AB} \cong \overline{AC}$; $\overline{BY} \cong \overline{CY}$	1. Given
2. $\overline{AY} \cong \overline{AY}$	2. Reflexive Property
3. $\triangle BYA \cong \triangle CYA$	3. SSS

Example 2 SSS on the Coordinate Plane

COORDINATE GEOMETRY **Determine whether $\triangle RTZ \cong \triangle JKL$ for $R(2, 5)$, $Z(1, 1)$, $T(5, 2)$, $L(-3, 0)$, $K(-7, 1)$, and $J(-4, 4)$. Explain.**

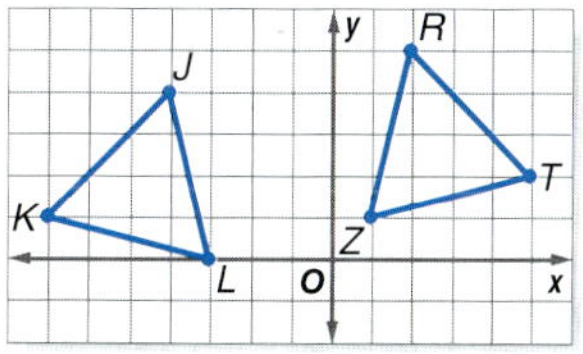

Use the Distance Formula to show that the corresponding sides are congruent.

$RT = \sqrt{(2-5)^2 + (5-2)^2} = \sqrt{9+9}$ or $\sqrt{18}$

$JK = \sqrt{[-4-(-7)]^2 + (4-1)^2} = \sqrt{9+9}$ or $\sqrt{18}$

$TZ = \sqrt{(5-1)^2 + (2-1)^2} = \sqrt{16+1}$ or $\sqrt{17}$

$KL = \sqrt{[-7-(-3)]^2 + (1-0)^2} = \sqrt{16+1}$ or $\sqrt{17}$

$RZ = \sqrt{(2-1)^2 + (5-1)^2} = \sqrt{1+16}$ or $\sqrt{17}$

$JL = \sqrt{[-4-(-3)]^2 + (4-0)^2} = \sqrt{1+16}$ or $\sqrt{17}$

$RT = JK$, $TZ = KL$, and $RZ = JL$. By definition of congruent segments, all corresponding segments are congruent. Therefore, $\triangle RTZ \cong \triangle JKL$ by SSS.

SAS POSTULATE Suppose you are given the measures of two sides and the angle they form, called the **included angle**. These conditions describe a unique triangle. Two triangles in which corresponding sides and the included pairs of angles are congruent provide another way to show that triangles are congruent.

www.geometryonline.com/extra_examples/fcat

2 Teach

SSS POSTULATE

In-Class Examples Power Point®

Teaching Tip Explain to students that when naming congruent triangles, it is customary to use triangle names in the same order as their congruent parts. $\triangle BYA \cong \triangle CYA$ uses appropriate order to signify the corresponding sides and angles that are congruent in the two triangles. It would be incorrect to write $\triangle BAY \cong \triangle CYA$.

1 **ENTOMOLOGY** The wings of one type of moth form two triangles. Write a two-column proof to prove that $\triangle FEG \cong \triangle HIG$ if $\overline{EI} \cong \overline{FH}$, $\overline{FE} \cong \overline{HI}$, and G is the midpoint of both $\overline{EI}$ and $\overline{FH}$.

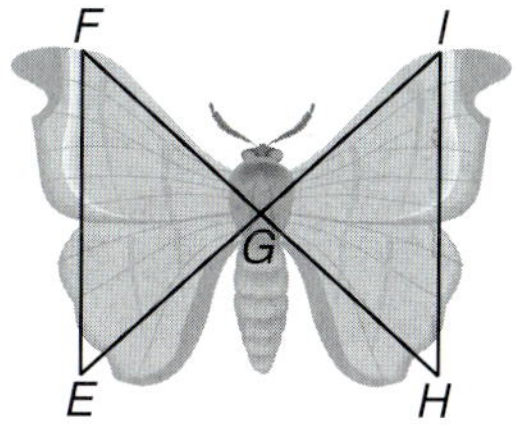

Statements (Reasons)

1. $\overline{FE} \cong \overline{HI}$; G is the midpoint of $\overline{EI}$; G is the midpoint of $\overline{FH}$. (Given)
2. $\overline{FG} \cong \overline{HG}$; $\overline{EG} \cong \overline{IG}$. (Midpoint Theorem)
3. $\triangle FEG \cong \triangle HIG$. (SSS)

2 **COORDINATE GEOMETRY** Determine whether $\triangle WDV \cong \triangle MLP$. Explain.

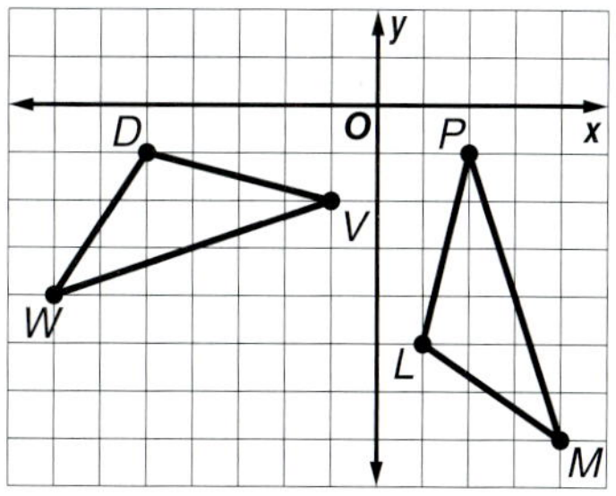

$WD = ML$, $DV = LP$, and $VW = PM$. By definition of congruent segments, all corresponding segments are congruent. Therefore, $\triangle WDV \cong \triangle MLP$ by SSS.

DAILY INTERVENTION — Differentiated Instruction

Logical/Mathematical Students can use a systematic approach to write the proofs for problems and examples in this lesson. Have students start by looking for possible methods of proof using SSS or SAS. Then they should examine the problem to determine how much necessary information is given and how they can find any other information that they need for the proof. Finally, they can draw on prior knowledge of midpoints, distances, angle relationships, and so on, to extract any other necessary information and compile the facts for the final proof.

SAS POSTULATE

In-Class Examples

Power Point®

3 Write a proof for the following.

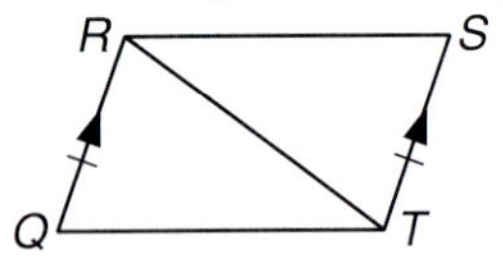

Given: $\overline{RQ} \parallel \overline{TS}$
$\overline{RQ} \cong \overline{TS}$

Prove: $\triangle QRT \cong \triangle STR$

Proof:
Statements (Reasons)

1. $\overline{RQ} \parallel \overline{TS}$, $\overline{RQ} \cong \overline{TS}$ (Given)
2. $\angle QRT \cong \angle STR$ (Alt. int. ∠s are ≅.)
3. $\overline{RT} \cong \overline{TR}$ (Reflexive Property)
4. $\triangle QRT \cong \triangle STR$ (SAS)

4 Determine which postulate can be used to prove that the triangles are congruent. If it is not possible to prove that they are congruent, write *not possible*.

a.

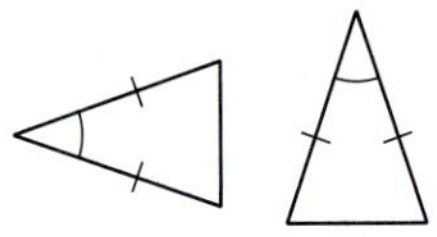

SAS

b.

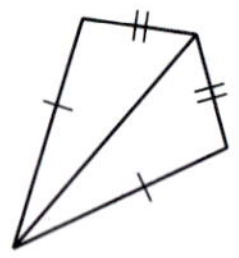

SSS

Postulate 4.2

Side-Angle-Side Congruence If two sides and the included angle of one triangle are congruent to two sides and the included angle of another triangle, then the triangles are congruent.

Abbreviation: *SAS*

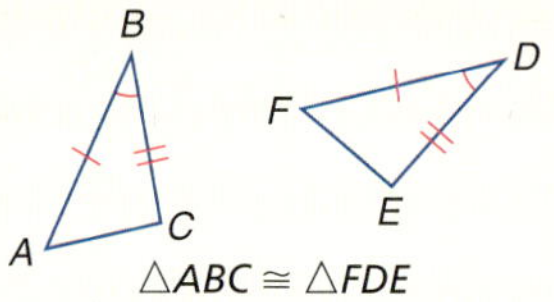

$\triangle ABC \cong \triangle FDE$

You can also construct congruent triangles given two sides and the included angle.

Construction

Congruent Triangles using Two Sides and the Included Angle

1. Draw a triangle and label its vertices *A*, *B*, and *C*.
2. Select a point *K* on line *m*. Use a compass to construct $\overline{KL}$ on *m* such that $\overline{KL} \cong \overline{BC}$.
3. Construct an angle congruent to ∠*B* using $\overrightarrow{KL}$ as a side of the angle and point *K* as the vertex.
4. Construct $\overline{JK}$ such that $\overline{JK} \cong \overline{AB}$. Draw $\overline{JL}$ to complete △*JKL*.

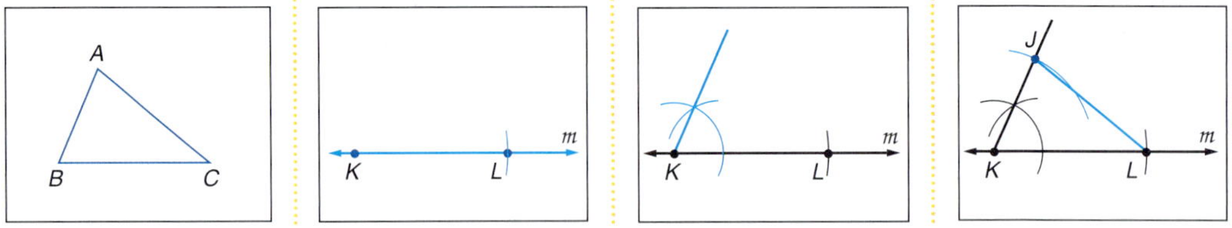

5. Cut out △*JKL* and place it over △*ABC*. How does △*JKL* compare to △*ABC*? **△*JKL* ≅ △*ABC***

Study Tip

Flow Proofs
Flow proofs can be written vertically or horizontally.

Example 3 *Use SAS in Proofs*

Write a flow proof.

Given: *X* is the midpoint of $\overline{BD}$.
X is the midpoint of $\overline{AC}$.

Prove: $\triangle DXC \cong \triangle BXA$

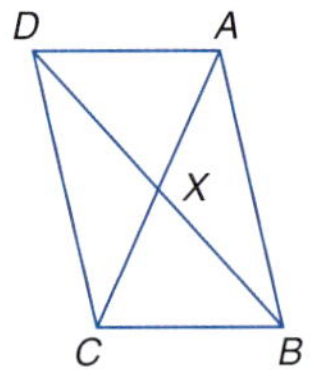

Flow Proof:

- *X* is the midpoint of $\overline{DB}$. (**Given**) → $\overline{DX} \cong \overline{BX}$ (Midpoint Theorem) → $\triangle DXC \cong \triangle BXA$
- *X* is the midpoint of $\overline{AC}$. (**Given**) → $\overline{CX} \cong \overline{AX}$ (Midpoint Theorem) → $\triangle DXC \cong \triangle BXA$ (SAS)
- $\angle DXC \cong \angle BXA$ (Vertical ∠s are ≅.) → $\triangle DXC \cong \triangle BXA$

DAILY INTERVENTION — Unlocking Misconceptions

Figures Point out that figures will not always be marked and that it is up to students to draw on their knowledge of geometric concepts to prove congruence. Stress the importance of using only information that is given and not forming any assumptions about two figures just because they appear to be congruent.

Example 4 Identify Congruent Triangles

Determine which postulate can be used to prove that the triangles are congruent. If it is not possible to prove that they are congruent, write *not possible*.

a.

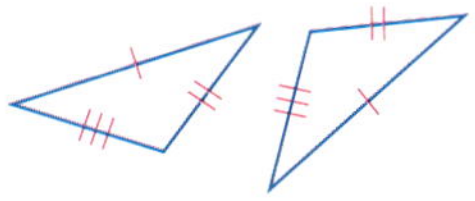

Each pair of corresponding sides are congruent. The triangles are congruent by the SSS Postulate.

b.

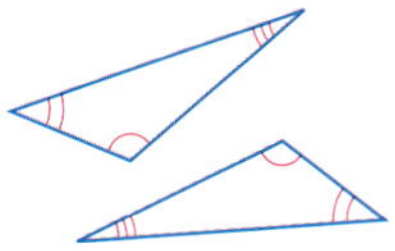

The triangles have three pairs of corresponding angles congruent. This does not match the SSS Postulate or the SAS Postulate. It is *not possible* to prove the triangles congruent.

Check for Understanding

Concept Check

1. **OPEN ENDED** Draw a triangle and label the vertices. Name two sides and the included angle. **See margin.**

2. **FIND THE ERROR** Carmelita and Jonathan are trying to determine whether $\triangle ABC$ is congruent to $\triangle DEF$.

Carmelita

$\triangle ABC \cong \triangle DEF$ by SAS

Jonathan

Congruence cannot be determined.

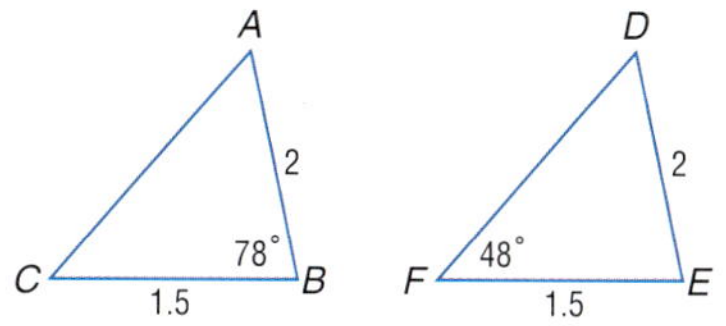

Who is correct and why? **Jonathan; the measure of $\angle DEF$ is needed to use SAS.**

Guided Practice

GUIDED PRACTICE KEY	
Exercises	Examples
3–4	2
5–6	3
7–8	4
9	1

Determine whether $\triangle EFG \cong \triangle MNP$ given the coordinates of the vertices. Explain.

3. $E(-4, -3), F(-2, 1), G(-2, -3), M(4, -3), N(2, 1), P(2, -3)$ **3–4. See margin.**

4. $E(-2, -2), F(-4, 6), G(-3, 1), M(2, 2), N(4, 6), P(3, 1)$

5. Write a flow proof.

Given: $\overline{DE}$ and $\overline{BC}$ bisect each other.

Prove: $\triangle DGB \cong \triangle EGC$

6. Write a two-column proof.

Given: $\overline{KM} \parallel \overline{JL}$, $\overline{KM} \cong \overline{JL}$

Prove: $\triangle JKM \cong \triangle MLJ$

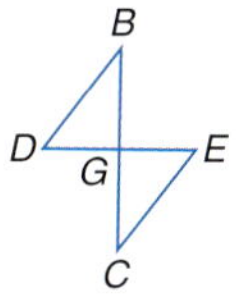

Exercise 5

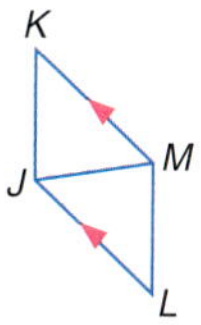

Exercise 6

5–6. See p. 233B.

Determine which postulate can be used to prove that the triangles are congruent. If it is not possible to prove that they are congruent, write *not possible*.

7. **SAS**

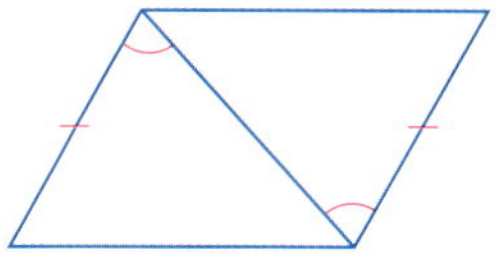

8. **SSS**

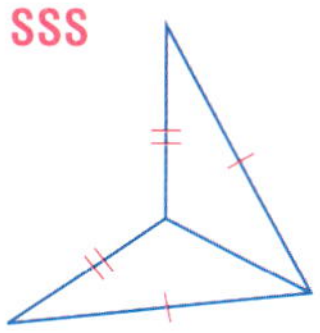

3 Practice/Apply

Study Notebook

Have students—
- add the definitions/examples of the vocabulary terms to their Vocabulary Builder worksheets for Chapter 4.
- include a simple example of a proof using SSS and a proof using SAS.
- include any other item(s) that they find helpful in mastering the skills in this lesson.

DAILY INTERVENTION

FIND THE ERROR Even though the figures look congruent and $\angle E$ appears to have the same measure as $\angle B$, this cannot be assumed based on the information given in the figures. Carmelita's answer would have been correct if it could be shown that $m\angle E = 78$.

About the Exercises...

Organization by Objective
- **SSS Postulate:** 10–13, 20–21, 28–29
- **SAS Postulate:** 14–19

Odd/Even Assignments

Exercises 10–27 are structured so that students practice the same concepts whether they are assigned odd or even problems.

Assignment Guide

Basic: 11–25 odd, 28–47

Average: 11–27 odd, 28–47

Advanced: 10–26 even, 28–43 (optional: 44–47)

Answer

1. Sample answer: In $\triangle QRS$, $\angle R$ is the included angle of the sides $\overline{QR}$ and $\overline{RS}$.

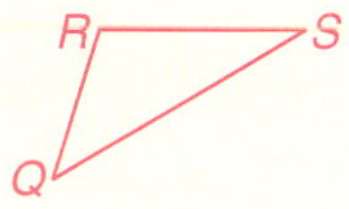

3. $EG = 2$, $MP = 2$, $FG = 4$, $NP = 4$, $EF = \sqrt{20}$, and $MN = \sqrt{20}$. The corresponding sides have the same measure and are congruent. $\triangle EFG \cong \triangle MNP$ by SSS.

4. $EG = \sqrt{10}$, $FG = \sqrt{26}$, $EF = \sqrt{68}$, $MP = \sqrt{2}$, $NP = \sqrt{26}$, and $MN = \sqrt{20}$. The corresponding sides are not congruent, so the triangles are not congruent.

Application

9. **PRECISION FLIGHT** The United States Navy Flight Demonstration Squadron, the Blue Angels, fly in a formation that can be viewed as two triangles with a common side. Write a two-column proof to prove that $\triangle SRT \cong \triangle QRT$ if T is the midpoint of $\overline{SQ}$ and $\overline{SR} \cong \overline{QR}$. **See margin.**

★ indicates increased difficulty

Practice and Apply

Homework Help

For Exercises	See Examples
10–13	2
14–19	3
20–21, 28–29	1
22–27	4

Extra Practice
See page 761.

Determine whether $\triangle JKL \cong \triangle FGH$ given the coordinates of the vertices. Explain.

10. $J(-3, 2)$, $K(-7, 4)$, $L(-1, 9)$, $F(2, 3)$, $G(4, 7)$, $H(9, 1)$
11. $J(-1, 1)$, $K(-2, -2)$, $L(-5, -1)$, $F(2, -1)$, $G(3, -2)$, $H(2, 5)$
12. $J(-1, -1)$, $K(0, 6)$, $L(2, 3)$, $F(3, 1)$, $G(5, 3)$, $H(8, 1)$
13. $J(3, 9)$, $K(4, 6)$, $L(1, 5)$, $F(1, 7)$, $G(2, 4)$, $H(-1, 3)$

10–13. See margin.

Write a flow proof. **14–19. See p. 233C.**

14. **Given:** $\overline{AE} \cong \overline{FC}$, $\overline{AB} \cong \overline{BC}$, $\overline{BE} \cong \overline{BF}$
Prove: $\triangle AFB \cong \triangle CEB$

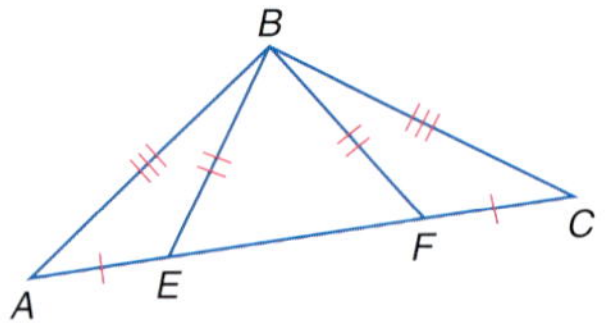

15. **Given:** $\overline{RQ} \cong \overline{TQ} \cong \overline{YQ} \cong \overline{WQ}$, $\angle RQY \cong \angle WQT$
Prove: $\triangle QWT \cong \triangle QYR$

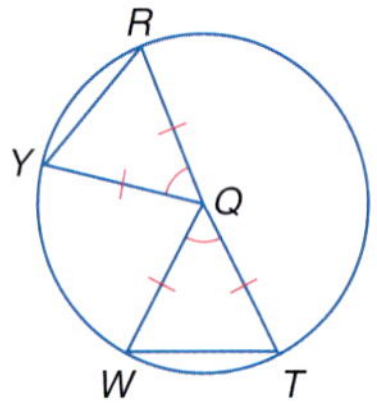

Write a two-column proof.

16. **Given:** $\triangle CDE$ is isosceles. G is the midpoint of $\overline{CE}$.
Prove: $\triangle CDG \cong \triangle EDG$

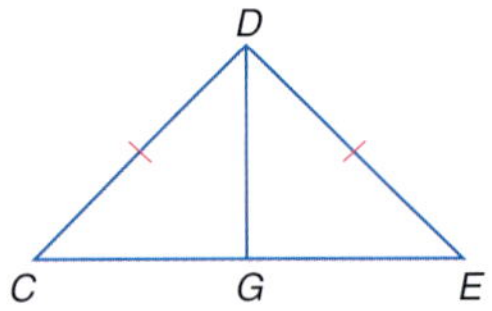

17. **Given:** $\triangle MRN \cong \triangle QRP$, $\angle MNP \cong \angle QPN$
Prove: $\triangle MNP \cong \triangle QPN$

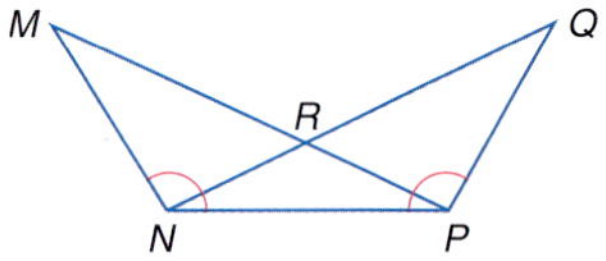

18. **Given:** $\overline{AC} \cong \overline{GC}$, $\overline{EC}$ bisects $\overline{AG}$.
Prove: $\triangle GEC \cong \triangle AEC$

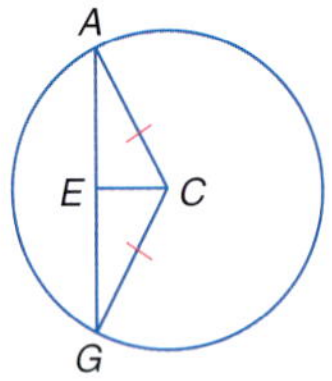

★ 19. **Given:** $\triangle GHJ \cong \triangle LKJ$
Prove: $\triangle GHL \cong \triangle LKG$

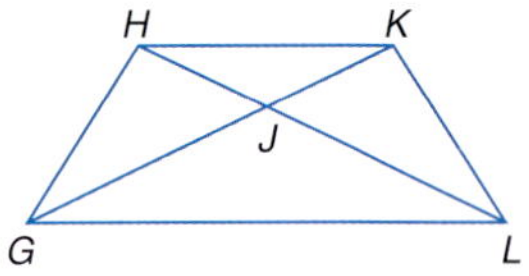

Answers

9. Given: T is the midpoint of $\overline{SQ}$. $\overline{SR} \cong \overline{QR}$
Prove: $\triangle SRT \cong \triangle QRT$

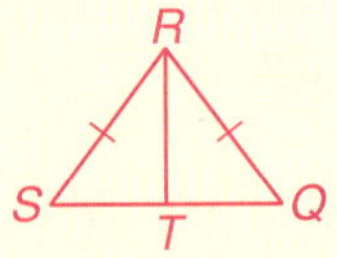

Proof:

Statements (Reasons)

1. T is the midpoint of $\overline{SQ}$. (Given)
2. $\overline{ST} \cong \overline{TQ}$ (Def. of midpoint)
3. $\overline{SR} \cong \overline{QR}$ (Given)
4. $\overline{RT} \cong \overline{RT}$ (Reflexive Prop.)
5. $\triangle SRT \cong \triangle QRT$ (SSS)

10. $JK = \sqrt{20}$, $KL = \sqrt{61}$, $JL = \sqrt{53}$, $FG = \sqrt{20}$, $GH = \sqrt{61}$, and $FH = \sqrt{53}$. Each pair of corresponding sides have the same measure so they are congruent. $\triangle JKL \cong \triangle FGH$ by SSS.

11. $JK = \sqrt{10}$, $KL = \sqrt{10}$, $JL = \sqrt{20}$, $FG = \sqrt{2}$, $GH = \sqrt{50}$, and $FH = 6$. The corresponding sides are not congruent so $\triangle JKL$ is not congruent to $\triangle FGH$.

12. $JK = \sqrt{50}$, $KL = \sqrt{13}$, $JL = 5$, $FG = \sqrt{8}$, $GH = \sqrt{13}$, and $FH = 5$. The corresponding sides are not congruent so $\triangle JKL$ is not congruent to $\triangle FGH$.

13. $JK = \sqrt{10}$, $KL = \sqrt{10}$, $JL = \sqrt{20}$, $FG = \sqrt{10}$, $GH = \sqrt{10}$, and $FH = \sqrt{20}$. Each pair of corresponding sides have the same measure so they are congruent. $\triangle JKL \cong \triangle FGH$ by SSS.

29. Sample answer: The properties of congruent triangles help land surveyors double check measurements. Answers should include the following.
- If each pair of corresponding angles and sides are congruent, the triangles are congruent by definition. If two pairs of corresponding sides and the included angle are congruent, the triangles are congruent by SAS. If each pair of corresponding sides are congruent, the triangles are congruent by SSS.
- Sample answer: Architects also use congruent triangles when designing buildings.

204 Chapter 4 Congruent Triangles

20–21. See p. 233C.

20. **CATS** A cat's ear is triangular in shape. Write a two-column proof to prove $\triangle RST \cong \triangle PNM$ if $\overline{RS} \cong \overline{PN}$, $\overline{RT} \cong \overline{MP}$, $\angle S \cong \angle N$, and $\angle T \cong \angle M$.

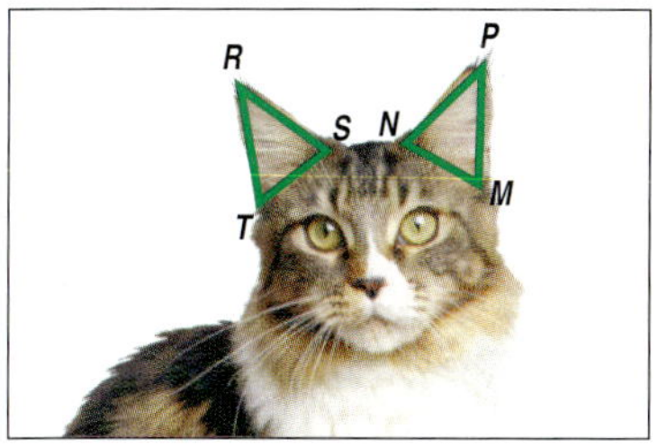

21. **GEESE** This photograph shows a flock of geese flying in formation. Write a two-column proof to prove that $\triangle EFG \cong \triangle HFG$, if $\overline{EF} \cong \overline{HF}$ and G is the midpoint of $\overline{EH}$.

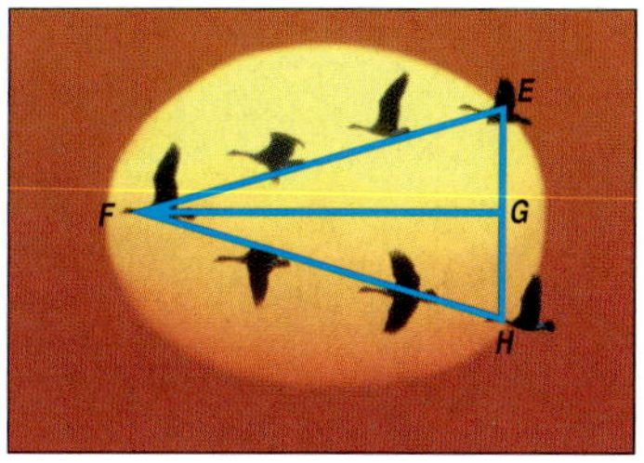

Determine which postulate can be used to prove that the triangles are congruent. If it is not possible to prove that they are congruent, write *not possible*.

22. 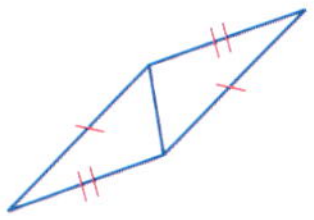 SSS

23. 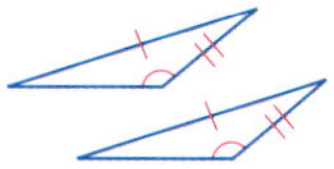 not possible

24. 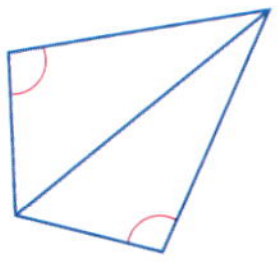 not possible

25. 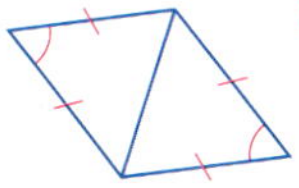 SSS or SAS

More About. . .

Baseball

The infield is a square 90 feet on each side.

Source: www.mlb.com

BASEBALL **For Exercises 26 and 27, use the following information.**
A baseball diamond is a square with four right angles and all sides congruent.

★ 26. Write a two-column proof to prove that the distance from first base to third base is the same as the distance from home plate to second base.

★ 27. Write a two-column proof to prove that the angle formed by second base, home plate, and third base is the same as the angle formed by second base, home plate, and first base.

26–27. See pp. 233C–233D.

28. **CRITICAL THINKING** Devise a plan and write a two-column proof for the following.
Given: $\overline{DE} \cong \overline{FB}$, $\overline{AE} \cong \overline{FC}$, See p. 233D.
$\overline{AE} \perp \overline{DB}$, $\overline{CF} \perp \overline{DB}$
Prove: $\triangle ABD \cong \triangle CDB$

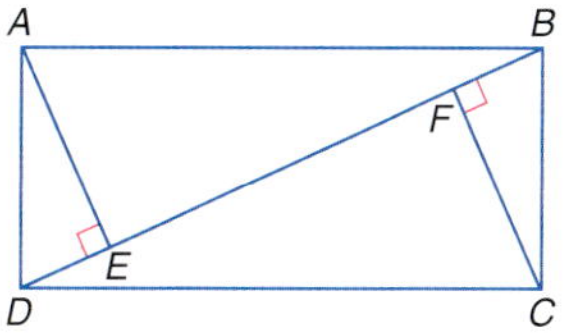

29. **WRITING IN MATH** Answer the question that was posed at the beginning of the lesson. See margin.

How do land surveyors use congruent triangles?

Include the following in your answer:
- description of three methods to prove triangles congruent, and
- another example of a career that uses properties of congruent triangles.

Study Guide and Intervention, p. 201 (shown) and p. 202

SSS Postulate You know that two triangles are congruent if corresponding sides are congruent and corresponding angles are congruent. The Side-Side-Side (SSS) Postulate lets you show that two triangles are congruent if you know only that the sides of one triangle are congruent to the sides of the second triangle.

SSS Postulate	If the sides of one triangle are congruent to the sides of a second triangle, then the triangles are congruent.

Example **Write a two-column proof.**
Given: $\overline{AB} \cong \overline{DB}$ and C is the midpoint of $\overline{AD}$.
Prove: $\triangle ABC \cong \triangle DBC$

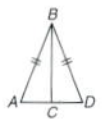

Statements	Reasons
1. $\overline{AB} \cong \overline{DB}$	1. Given
2. C is the midpoint of $\overline{AD}$.	2. Given
3. $\overline{AC} \cong \overline{DC}$	3. Definition of midpoint
4. $\overline{BC} \cong \overline{BC}$	4. Reflexive Property of ≅
5. $\triangle ABC \cong \triangle DBC$	5. SSS Postulate

Exercises

Write a two-column proof.

1. 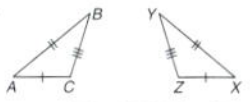

Given: $\overline{AB} \cong \overline{XY}$, $\overline{AC} \cong \overline{XZ}$, $\overline{BC} \cong \overline{YZ}$
Prove: $\triangle ABC \cong \triangle XYZ$

Statements	Reasons
1. $\overline{AB} \cong \overline{XY}$, $\overline{AC} \cong \overline{XZ}$, $\overline{BC} \cong \overline{YZ}$	1. Given
2. $\triangle ABC \cong \triangle XYZ$	2. SSS Post.

2. 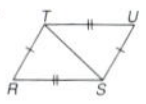

Given: $\overline{RS} \cong \overline{UT}$, $\overline{RT} \cong \overline{US}$
Prove: $\triangle RST \cong \triangle UTS$

Statements	Reasons
1. $\overline{RS} \cong \overline{UT}$, $\overline{RT} \cong \overline{US}$	1. Given
2. $\overline{ST} \cong \overline{TS}$	2. Refl. Prop.
3. $\triangle RST \cong \triangle UTS$	3. SSS Post.

Skills Practice, p. 203 and Practice, p. 204 (shown)

Determine whether $\triangle DEF \cong \triangle PQR$ given the coordinates of the vertices. Explain.

1. $D(-6, 1)$, $E(1, 2)$, $F(-1, -4)$, $P(0, 5)$, $Q(7, 6)$, $R(5, 0)$
$DE = 5\sqrt{2}$, $PQ = 5\sqrt{2}$, $EF = 2\sqrt{10}$, $QR = 2\sqrt{10}$, $DF = 5\sqrt{2}$, $PR = 5\sqrt{2}$. $\triangle DEF \cong \triangle PQR$ by SSS since corresponding sides have the same measure and are congruent.

2. $D(-7, -3)$, $E(-4, -1)$, $F(-2, -5)$, $P(2, -2)$, $Q(5, -4)$, $R(0, -5)$
$DE = \sqrt{13}$, $PQ = \sqrt{13}$, $EF = 2\sqrt{5}$, $QR = \sqrt{26}$, $DF = \sqrt{29}$, $PR = \sqrt{13}$. Corresponding sides are not congruent, so $\triangle DEF$ is not congruent to $\triangle PQR$.

3. Write a flow proof.
Given: $\overline{RS} \cong \overline{TS}$
V is the midpoint of $\overline{RT}$.
Prove: $\triangle RSV \cong \triangle TSV$

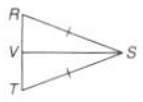

Proof:

$\overline{RS} \cong \overline{TS}$ (Given) → $\triangle RSV \cong \triangle TSV$ (SSS)
$\overline{SV} \cong \overline{SV}$ (Reflexive Property) → $\triangle RSV \cong \triangle TSV$
V is the midpoint of $\overline{RT}$. (Given) → $\overline{RV} \cong \overline{VT}$ (Definition of midpoint) → $\triangle RSV \cong \triangle TSV$

Determine which postulate can be used to prove that the triangles are congruent. If it is not possible to prove that they are congruent, write *not possible*.

4.  not possible

5. 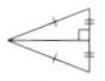 SAS or SSS

6. 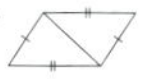 SSS

7. **INDIRECT MEASUREMENT** To measure the width of a sinkhole on his property, Harmon marked off congruent triangles as shown in the diagram. How does he know that the lengths $A'B'$ and AB are equal?

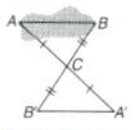

Since $\angle ACB$ and $\angle A'CB'$ are vertical angles, they are congruent. In the figure, $\overline{AC} \cong \overline{A'C}$ and $\overline{BC} \cong \overline{B'C}$. So $\triangle ABC \cong \triangle A'B'C$ by SAS. By CPCTC, the lengths $A'B'$ and AB are equal.

Reading to Learn Mathematics, p. 205 ELL

Pre-Activity **How do land surveyors use congruent triangles?**

Read the introduction to Lesson 4-4 at the top of page 200 in your textbook.

Why do you think that land surveyors would use congruent right triangles rather than other congruent triangles to establish property boundaries?
Sample answer: Land is usually divided into rectangular lots, so their boundaries meet at right angles.

Reading the Lesson

1. Refer to the figure.

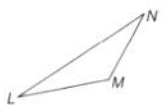

a. Name the sides of $\triangle LMN$ for which $\angle L$ is the included angle.
$\overline{LM}$, $\overline{LN}$

b. Name the sides of $\triangle LMN$ for which $\angle N$ is the included angle.
$\overline{NL}$, $\overline{NM}$

c. Name the sides of $\triangle LMN$ for which $\angle M$ is the included angle.
$\overline{ML}$, $\overline{MN}$

2. Determine whether you have enough information to prove that the two triangles in each figure are congruent. If so, write a congruence statement and name the congruence postulate that you would use. If not, write *not possible*.

a. 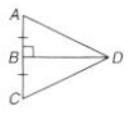

$\triangle ABD \cong \triangle CBD$; SAS

b. 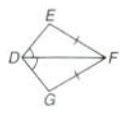

not possible

c. $\overline{EH}$ and $\overline{DG}$ bisect each other.
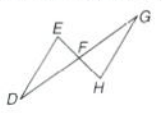

$\triangle DEF \cong \triangle GHF$; SAS

d. 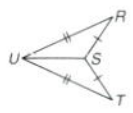

$\triangle RSU \cong \triangle TSU$; SSS

Helping You Remember

3. Find three words that explain what it means to say that two triangles are congruent and that can help you recall the meaning of the SSS Postulate.
Sample answer: Congruent triangles are triangles that are the *same size* and *shape*, and the SSS Postulate ensures that two triangles with three corresponding sides congruent will be the same size and shape.

Enrichment, p. 206

Congruent Parts of Regular Polygonal Regions

Congruent figures are figures that have exactly the same size and shape. There are many ways to divide regular polygonal regions into congruent parts. Three ways to divide an equilateral triangular region are shown. You can verify that the parts are congruent by tracing one part, then rotating, sliding, or reflecting that part on top of the other parts.

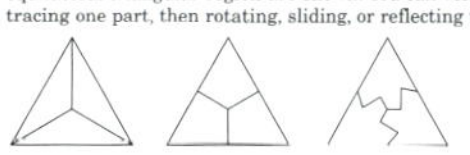

1. Divide each square into four congruent parts. Use three different ways. Sample answers are shown.

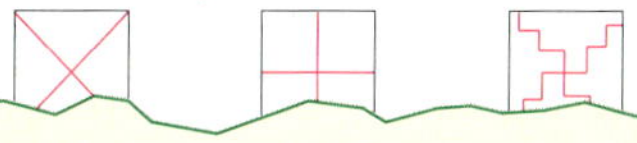

4 Assess

Open-Ended Assessment

Speaking Ask students to explain in their own words how they can use SSS and SAS to prove triangle congruence. Then students can explain how they would approach proving triangle congruence for some of the exercises in Lesson 4-4.

Getting Ready for Lesson 4-5

Prerequisite Skill Students will learn about proving congruence with ASA and AAS in Lesson 4-5. They will apply the Angle Sum Theorem and concepts of angle and segment bisection toward proving triangle congruence. Use Exercises 44–47 to determine your students' familiarity with bisectors of segments and angles.

Assessment Options

Quiz (Lessons 4-3 and 4-4) is available on p. 239 of the *Chapter 4 Resource Masters*.

Mid-Chapter Test (Lessons 4-1 through 4-4) is available on p. 241 of the *Chapter 4 Resource Masters*.

FCAT Practice
Standardized Test Practice

30. Which of the following statements about the figure is true? **C**
    Ⓐ $90 > a + b$
    Ⓑ $a + b > 90$
    Ⓒ $a + b = 90$
    Ⓓ $a > b$

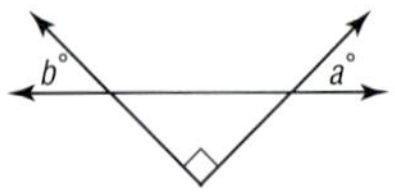

31. Classify the triangle with the measures of the angles in the ratio 3:6:7. **B**
    Ⓐ isosceles Ⓑ acute Ⓒ obtuse Ⓓ right

Maintain Your Skills

Mixed Review Identify the congruent triangles in each figure. *(Lesson 4-3)*

32.
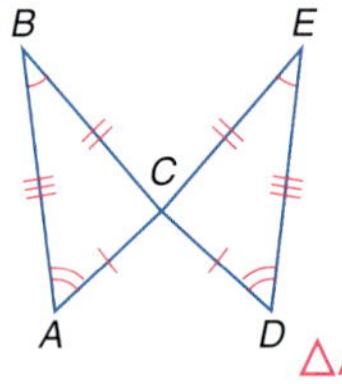

△ACB ≅ △DCE

33.
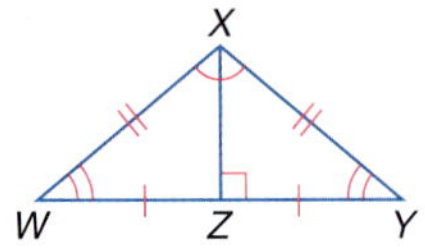

△WXZ ≅ △YXZ

34.
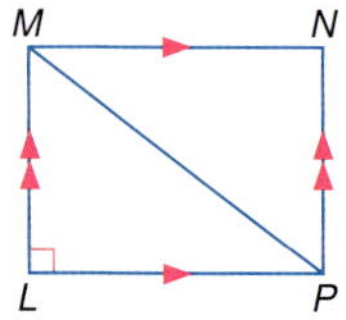

△LMP ≅ △NPM

Find each measure if $\overline{PQ} \perp \overline{QR}$. *(Lesson 4-2)*

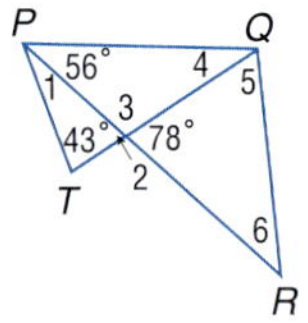

35. $m\angle 2$ **78**
36. $m\angle 3$ **102**
37. $m\angle 5$ **68**
38. $m\angle 4$ **22**
39. $m\angle 1$ **59**
40. $m\angle 6$ **34**

For Exercises 41–43, use the graphic at the right. *(Lesson 3-3)*

41. Find the rate of change from first quarter to the second quarter. **−1**
42. Find the rate of change from the second quarter to the third quarter. **−1.4**
43. Compare the rate of change from the first quarter to the second, and the second quarter to the third. Which had the greater rate of change? **There is a steeper rate of decline from the second quarter to the third.**

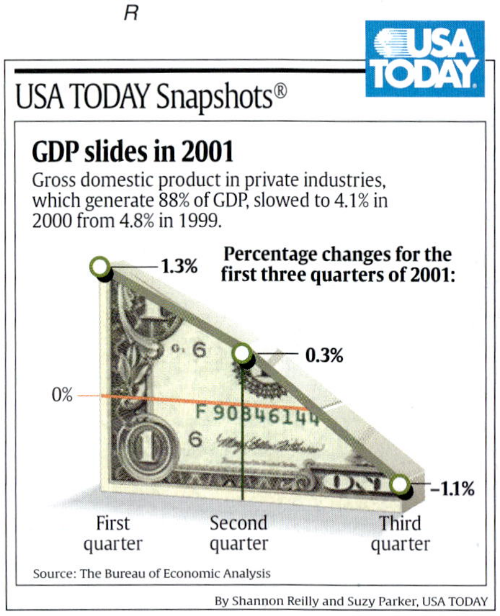

Getting Ready for the Next Lesson

PREREQUISITE SKILL $\overrightarrow{BD}$ and $\overrightarrow{AE}$ are angle bisectors and segment bisectors. Name the indicated segments and angles.

(To review ***bisectors of segments and angles****, see Lessons 1-5 and 1-6.)*

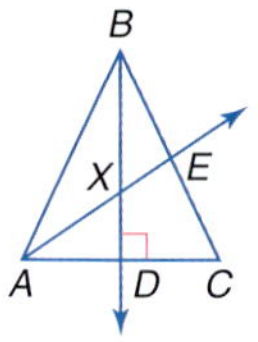

44. a segment congruent to $\overline{EC}$ **$\overline{BE}$**
45. an angle congruent to $\angle ABD$ **$\angle CBD$**
46. an angle congruent to $\angle BDC$ **$\angle BDA$**
47. a segment congruent to $\overline{AD}$ **$\overline{CD}$**

4-5 Proving Congruence—ASA, AAS

Sunshine State Standards
MA.C.1.4.1, MA.C.2.4.1

Vocabulary
- included side

What You'll Learn
- Use the ASA Postulate to test for triangle congruence.
- Use the AAS Theorem to test for triangle congruence.

How are congruent triangles used in construction?

The Bank of China Tower in Hong Kong has triangular trusses for structural support. These trusses form congruent triangles. In this lesson, we will explore two additional methods of proving triangles congruent.

ASA POSTULATE Suppose you were given the measures of two angles of a triangle and the side between them, the **included side**. Do these measures form a unique triangle?

Construction

Congruent Triangles Using Two Angles and Included Side

1. Draw a triangle and label its vertices *A*, *B*, and *C*.
2. Draw any line *m* and select a point *L*. Construct $\overline{LK}$ such that $\overline{LK} \cong \overline{CB}$.
3. Construct an angle congruent to $\angle C$ at *L* using $\overrightarrow{LK}$ as a side of the angle.
4. Construct an angle congruent to $\angle B$ at *K* using $\overrightarrow{LK}$ as a side of the angle. Label the point where the new sides of the angles meet *J*.

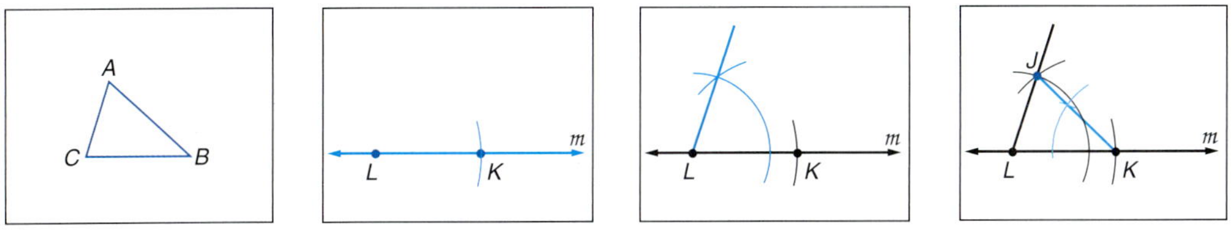

5. Cut out $\triangle JKL$ and place it over $\triangle ABC$. How does $\triangle JKL$ compare to $\triangle ABC$? $\triangle JKL \cong \triangle ABC$

This construction leads to the Angle-Side-Angle Postulate, written as ASA.

Study Tip
Reading Math
The included side refers to the side that each of the angles share.

Postulate 4.3

Angle-Side-Angle Congruence If two angles and the included side of one triangle are congruent to two angles and the included side of another triangle, then the triangles are congruent.

Abbreviation: *ASA*

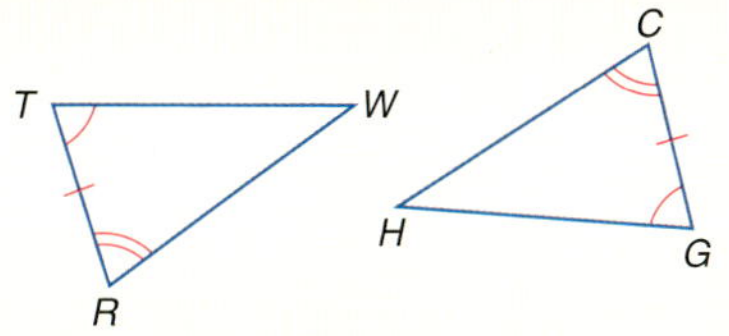

$\triangle RTW \cong \triangle CGH$

4-5 Lesson Notes

1 Focus

5-Minute Check Transparency 4-5 Use as a quiz or review of Lesson 4-4.

Mathematical Background notes are available for this lesson on p. 176D.

How are congruent triangles used in construction?

Ask students:
- How do the congruent triangles in the trusses contribute to the appearance of the structure? **Sample answer: They make the structure visually appealing.**
- How would using congruent triangles make the structure easier to assemble? **The triangles could be manufactured in bulk and construction workers could place any triangle in any location.**

Resource Manager

Workbook and Reproducible Masters

Chapter 4 Resource Masters
- Study Guide and Intervention, pp. 207–208
- Skills Practice, p. 209
- Practice, p. 210
- Reading to Learn Mathematics, p. 211
- Enrichment, p. 212

Teaching Geometry With Manipulatives Masters, pp. 8, 16, 17, 77

Transparencies
5-Minute Check Transparency 4-5
Real-World Transparency 4
Answer Key Transparencies

Technology
GeomPASS: Tutorial Plus, Lesson 10
Interactive Chalkboard
Multimedia Applications: Virtual Activities

2 Teach

ASA POSTULATE

In-Class Examples

Power Point®

1 Write a paragraph proof.

Given: L is the midpoint of $\overline{WE}$.
$\overline{WR} \parallel \overline{ED}$

Prove: $\triangle WRL \cong \triangle EDL$

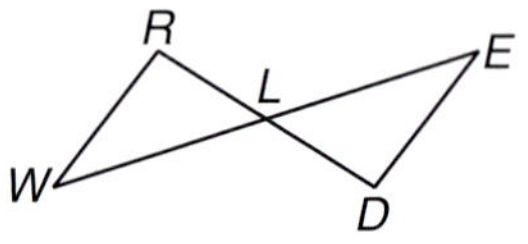

Proof: $\angle W \cong \angle E$ because alternate interior angles are congruent. By the Midpoint Theorem, $\overline{WL} \cong \overline{EL}$. Since vertical angles are congruent, $\angle WLR \cong \angle ELD$. $\triangle WRL \cong \triangle EDL$ by ASA.

Example 1 Use ASA in Proofs

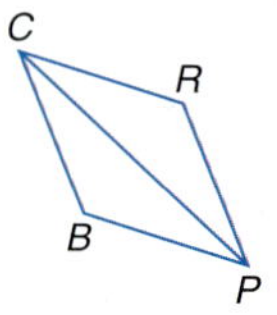

Write a paragraph proof.

Given: $\overline{CP}$ bisects $\angle BCR$ and $\angle BPR$.

Prove: $\triangle BCP \cong \triangle RCP$

Proof:

Since $\overline{CP}$ bisects $\angle BCR$ and $\angle BPR$, $\angle BCP \cong \angle RCP$ and $\angle BPC \cong \angle RPC$. $\overline{CP} \cong \overline{CP}$ by the Reflexive Property. By ASA, $\triangle BCP \cong \triangle RCP$.

AAS THEOREM Suppose you are given the measures of two angles and a nonincluded side. Is this information sufficient to prove two triangles congruent?

Geometry Activity

Angle-Angle-Side Congruence

Model

1. Draw a triangle on a piece of patty paper. Label the vertices *A*, *B*, and *C*.
2. Copy $\overline{AB}$, $\angle B$, and $\angle C$ on another piece of patty paper and cut them out.
3. Assemble them to form a triangle in which the side is not the included side of the angles.

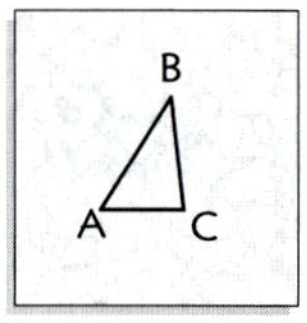

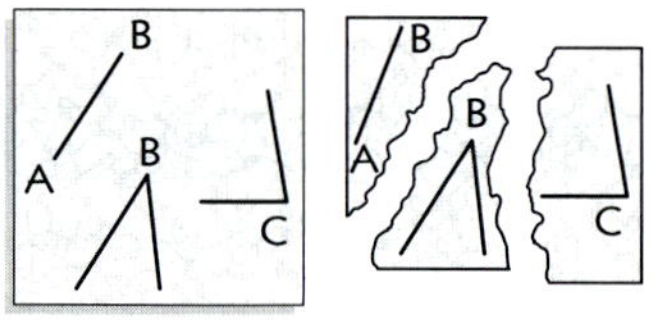

Analyze 1. They are congruent.

1. Place the original $\triangle ABC$ over the assembled figure. How do the two triangles compare?
2. **Make a conjecture** about two triangles with two angles and the nonincluded side of one triangle congruent to two angles and the nonincluded side of the other triangle. The triangles are congruent.

In the activity, you see that two triangles that have two pairs of corresponding angles congruent and congruent nonincluded sides are congruent.

Theorem 4.5

Angle-Angle-Side Congruence If two angles and a nonincluded side of one triangle are congruent to the corresponding two angles and side of a second triangle, then the two triangles are congruent.

Abbreviation: AAS

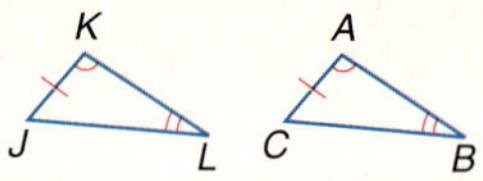

Example: $\triangle JKL \cong \triangle CAB$

Study Tip

ASA and AAS
While ASA and AAS are related, they are not interchangeable.

Proof Theorem 4.5

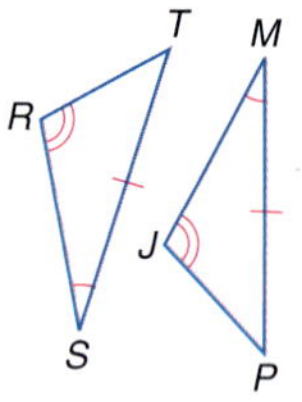

Given: $\angle M \cong \angle S$, $\angle J \cong \angle R$, $\overline{MP} \cong \overline{ST}$

Prove: $\triangle JMP \cong \triangle RST$

Proof:

Statements	Reasons
1. $\angle M \cong \angle S$, $\angle J \cong \angle R$, $\overline{MP} \cong \overline{ST}$	1. Given
2. $\angle P \cong \angle T$	2. Third Angle Theorem
3. $\triangle JMP \cong \triangle RST$	3. ASA

Geometry Activity

Materials: patty paper, straightedge, scissors

- When students are copying $\angle B$ and $\angle C$, tell them to extend the sides so that they are longer than the sides in the original triangle. Explain that these sides represent unknown side lengths.
- In Step 2, make sure students copy $\angle C$ and not $\angle A$.

Example 2 Use AAS in Proofs

Write a flow proof.

Given: $\angle EAD \cong \angle EBC$

$\overline{AD} \cong \overline{BC}$

Prove: $\overline{AE} \cong \overline{BE}$

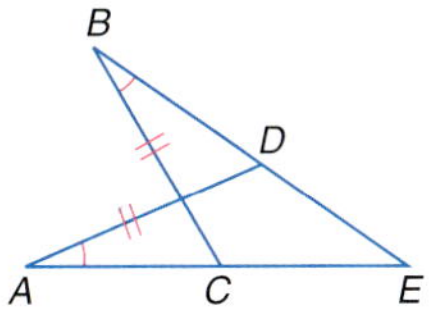

Flow Proof:

$\angle EAD \cong \angle EBC$ (Given) →
$\overline{AD} \cong \overline{BC}$ (Given) → $\triangle ADE \cong \triangle BCE$ (AAS) → $\overline{AE} \cong \overline{BE}$ (CPCTC)
$\angle E \cong \angle E$ (Reflexive Property) →

Study Tip

Overlapping Triangles
When triangles overlap, it is a good idea to draw each triangle separately and label the congruent parts.

You have learned several methods for proving triangle congruence. The Concept Summary lists ways to help you determine which method to use.

Concept Summary — Methods to Prove Triangle Congruence

Definition of Congruent Triangles	All corresponding parts of one triangle are congruent to the corresponding parts of the other triangle.
SSS	The three sides of one triangle must be congruent to the three sides of the other triangle.
SAS	Two sides and the included angle of one triangle must be congruent to two sides and the included angle of the other triangle.
ASA	Two angles and the included side of one triangle must be congruent to two angles and the included side of the other triangle.
AAS	Two angles and a nonincluded side of one triangle must be congruent to two angles and side of the other triangle.

Career Choices

Architect

About 28% of architects are self-employed. Architects design a variety of buildings including offices, retail spaces, and schools.

Online Research For information about a career as an architect, visit: www.geometryonline.com/careers

Example 3 Determine if Triangles Are Congruent

ARCHITECTURE **This glass chapel was designed by Frank Lloyd Wright's son, Lloyd Wright. Suppose the redwood supports, $\overline{TU}$ and $\overline{TV}$, measure 3 feet, $TY = 1.6$ feet, and $m\angle U$ and $m\angle V$ are 31. Determine whether $\triangle TYU \cong \triangle TYV$. Justify your answer.**

Explore We are given three measurements of each triangle. We need to determine whether the two triangles are congruent.

Plan Since $m\angle U = m\angle V$, $\angle U \cong \angle V$.
Likewise, $TU = TV$ so $\overline{TU} \cong \overline{TV}$, and $TY = TY$ so $\overline{TY} \cong \overline{TY}$. Check each possibility using the five methods you know.

Solve We are given information about side-side-angle (SSA). This is not a method to prove two triangles congruent.

(continued on the next page)

AAS THEOREM

In-Class Examples

Power Point®

2 Write a proof.

Given: $\angle NKL \cong \angle NJM$

$\overline{KL} \cong \overline{JM}$

Prove: $\overline{LN} \cong \overline{MN}$

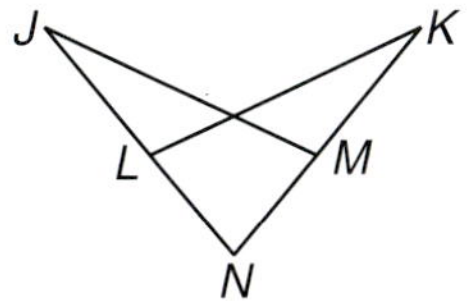

Proof:

Statements (Reasons)

1. $\angle N \cong \angle N$ (Reflex. Prop. of $\angle$s)
2. $\angle NKL \cong \angle NJM$ (Given)
3. $\overline{KL} \cong \overline{JM}$ (Given)
4. $\triangle JNM \cong \triangle KNL$ (AAS)
5. $\overline{LN} \cong \overline{MN}$ (CPCTC)

3 **STANCES** When Ms. Gomez puts her hands on her hips, she forms two triangles with her upper body and arms. Suppose her arm lengths AB and DE measure 9 inches, and AC and EF measure 11 inches. Also suppose that you are given that $\overline{BC} \cong \overline{DF}$. Determine whether $\triangle ABC \cong \triangle EDF$. Justify your answer.

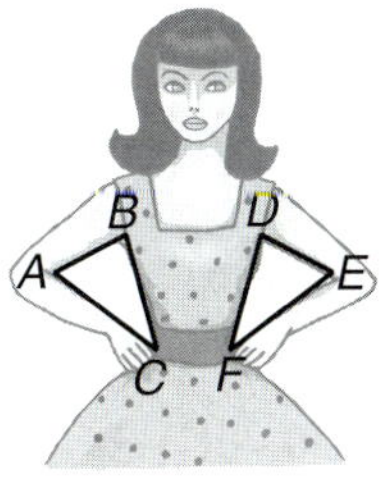

With $\overline{BC} \cong \overline{DF}$, you could use SSS to prove $\triangle ABC \cong \triangle EDF$.

DAILY INTERVENTION — **Differentiated Instruction**

Intrapersonal Ask students to study the proofs for the examples in this lesson and note the properties that recur, such as the reflexive properties of angles and segments, bisectors, midpoints, and so on. Students can start a list of helpful tools and things to watch for when they are working proofs and include recurring properties, theorems, formulas and methods that they can refer to in later lessons. They can also look at the order of the steps in paragraph proofs, flow proofs, and two-column proofs for similarities and differences.

Tips for New Teachers

Intervention A student may ask about proving congruence with AAA. Explain that while congruent triangles do share three congruent angles, AAA is not a possible tool for proving congruence of triangles because two triangles with three corresponding congruent angles can be *similar* but not *congruent*. Provide students with an example of two different-sized similar triangles.

3 Practice/Apply

Study Notebook

Have students—
- add the definitions/examples of the vocabulary terms to their Vocabulary Builder worksheets for Chapter 4.
- include any other item(s) that they find helpful in mastering the skills in this lesson.

Answers

1. Two triangles can have corresponding congruent angles without corresponding congruent sides. $\angle A \cong \angle D$, $\angle B \cong \angle E$, and $\angle C \cong \angle F$. However, $\overline{AB} \ncong \overline{DE}$, so $\triangle ABC \ncong \triangle DEF$.

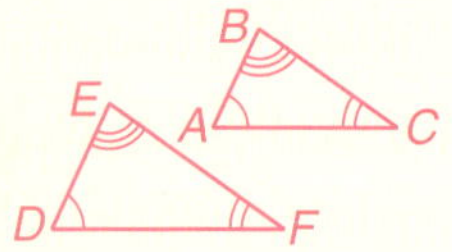

2. Sample answer: In $\triangle ABC$, $\overline{AB}$ is the included side of $\angle A$ and $\angle B$.

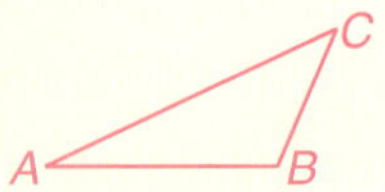

Examine Use a compass, protractor, and ruler to draw a triangle with the given measurements. For simplicity of measurement, we will use centimeters instead of feet, so the measurements of the construction and those of the support beams will be proportional.

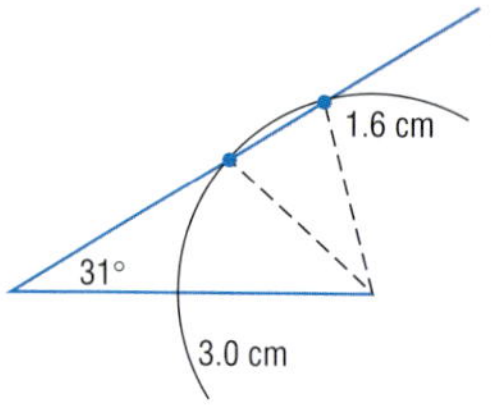

- Draw a segment 3.0 centimeters long.
- At one end, draw an angle of 31°. Extend the line longer than 3.0 centimeters.
- At the other end of the segment, draw an arc with a radius of 1.6 centimeters such that it intersects the line.

Notice that there are two possible segments that could determine the triangle. Since the given measurements do not lead to a unique triangle, we cannot show that the triangles are congruent.

Check for Understanding

Concept Check

1–2. See margin.

1. **Find a counterexample** to show why AAA (Angle-Angle-Angle) cannot be used to prove triangle congruence.
2. **OPEN ENDED** Draw a triangle and label the vertices. Name two angles and the included side.
3. **Explain** why AAS is a theorem, not a postulate. AAS can be proven using the Third Angle Theorem. Postulates are accepted as true without proof.

Guided Practice

GUIDED PRACTICE KEY	
Exercises	Examples
4, 6	1
5, 7	2
8	3

Write a flow proof. 4–5. See p. 233D.

4. **Given:** $\overline{GH} \parallel \overline{KJ}$, $\overline{GK} \parallel \overline{HJ}$
Prove: $\triangle GJK \cong \triangle JGH$

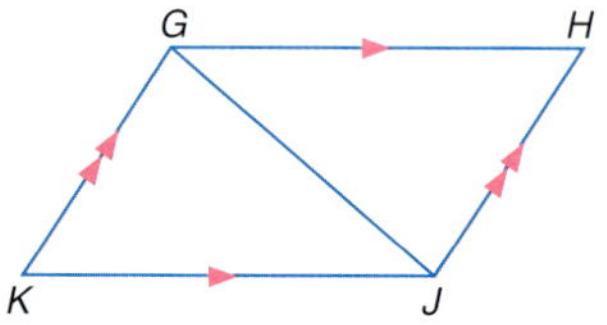

5. **Given:** $\overline{XW} \parallel \overline{YZ}$, $\angle X \cong \angle Z$
Prove: $\triangle WXY \cong \triangle YZW$

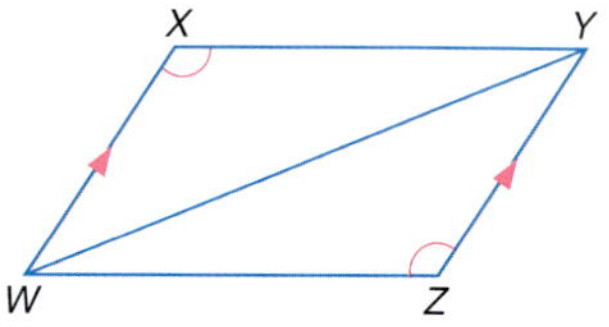

Write a paragraph proof. 6–7. See p. 233D.

6. **Given:** $\overline{QS}$ bisects $\angle RST$; $\angle R \cong \angle T$.
Prove: $\triangle QRS \cong \triangle QTS$

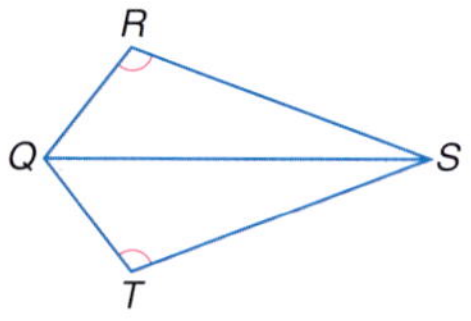

7. **Given:** $\angle E \cong \angle K$, $\angle DGH \cong \angle DHG$
$\overline{EG} \cong \overline{KH}$
Prove: $\triangle EGD \cong \triangle KHD$

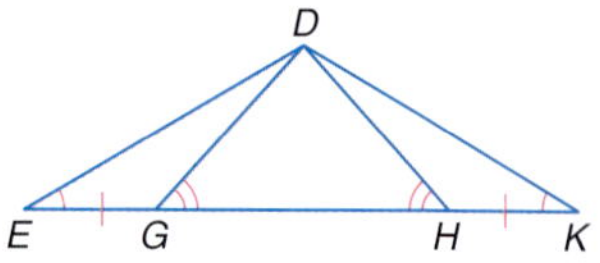

Application

8. **PARACHUTES** Suppose $\overline{ST}$ and $\overline{ML}$ each measure 7 feet, $\overline{SR}$ and $\overline{MK}$ each measure 5.5 feet, and $m\angle T = m\angle L = 49$. Determine whether $\triangle SRT \cong \triangle MKL$. Justify your answer. See margin.

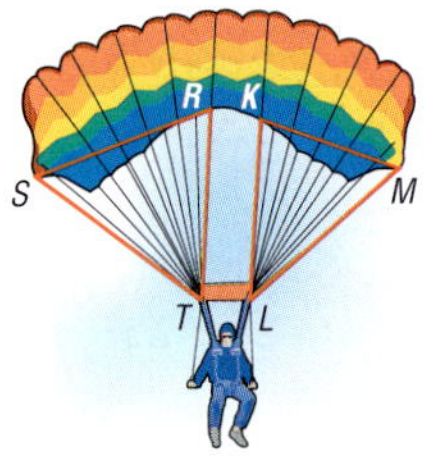

8. This cannot be determined. The information given cannot be used with any of the triangle congruence postulates, theorems or the definition of congruent triangles. By construction, two different triangles can be shown with the given information. Therefore, it cannot be determined if $\triangle SRT \cong \triangle MKL$.

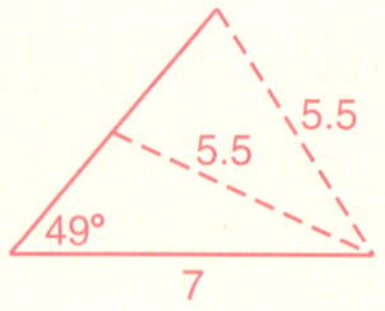

210 Chapter 4 Congruent Triangles

★ indicates increased difficulty

Practice and Apply

Homework Help

For Exercises	See Examples
9, 11, 14, 15–18	2
10, 12, 13, 19, 20	1
21–28	3

Extra Practice
See page 762.

Write a flow proof. 9–14. See pp. 233D–233E.

9. Given: $\overline{EF} \parallel \overline{GH}$, $\overline{EF} \cong \overline{GH}$
Prove: $\overline{EK} \cong \overline{KH}$

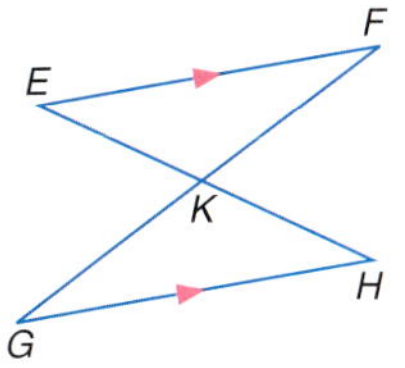

10. Given: $\overline{DE} \parallel \overline{JK}$, $\overline{DK}$ bisects $\overline{JE}$.
Prove: $\triangle EGD \cong \triangle JGK$

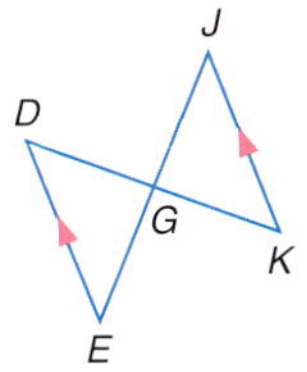

11. Given: $\angle V \cong \angle S$, $\overline{TV} \cong \overline{QS}$
Prove: $\overline{VR} \cong \overline{SR}$

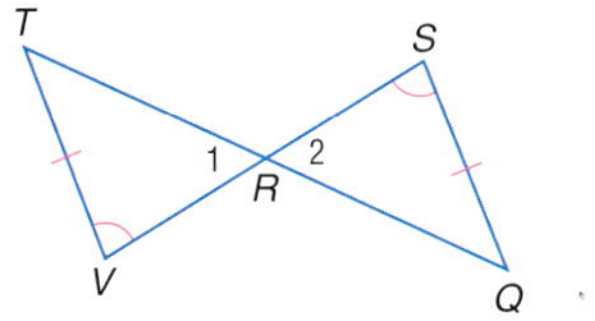

12. Given: $\overline{EJ} \parallel \overline{FK}$, $\overline{JG} \parallel \overline{KH}$, $\overline{EF} \cong \overline{GH}$
Prove: $\triangle EJG \cong \triangle FKH$

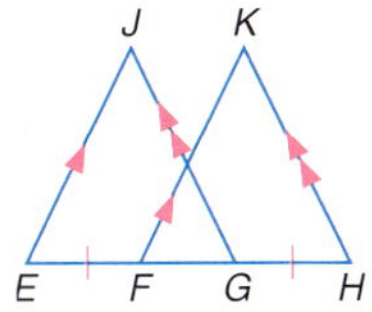

13. Given: $\overline{MN} \cong \overline{PQ}$, $\angle M \cong \angle Q$
$\angle 2 \cong \angle 3$
Prove: $\triangle MLP \cong \triangle QLN$

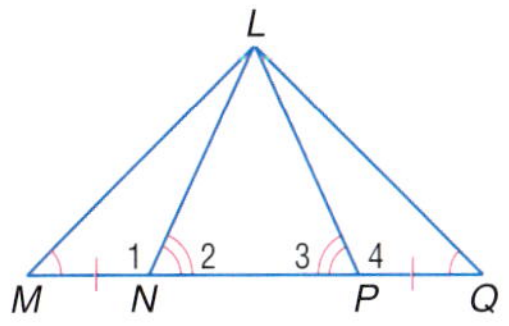

14. Given: Z is the midpoint of $\overline{CT}$.
$\overline{CY} \parallel \overline{TE}$
Prove: $\overline{YZ} \cong \overline{EZ}$

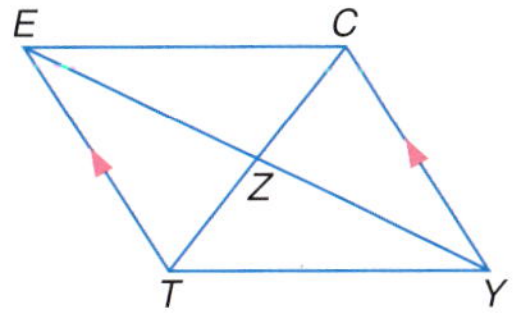

Write a paragraph proof. 15–18. See margin.

15. Given: $\angle NOM \cong \angle POR$,
$\overline{NM} \perp \overline{MR}$
$\overline{PR} \perp \overline{MR}$, $\overline{NM} \cong \overline{PR}$
Prove: $\overline{MO} \cong \overline{OR}$

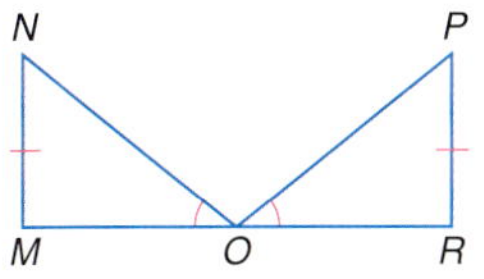

16. Given: $\overline{DL}$ bisects $\overline{BN}$,
$\angle XLN \cong \angle XDB$
Prove: $\overline{LN} \cong \overline{DB}$

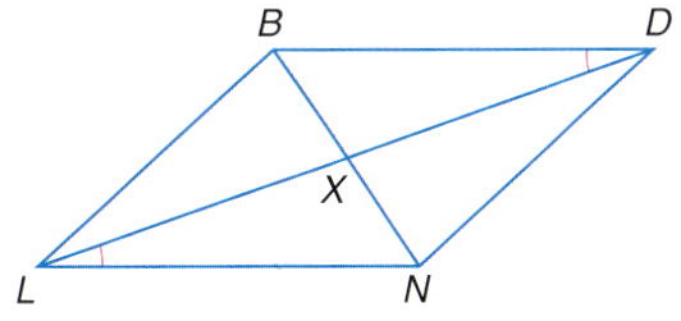

17. Given: $\angle F \cong \angle J$, $\angle E \cong \angle H$
$\overline{EC} \cong \overline{GH}$
Prove: $\overline{EF} \cong \overline{HJ}$

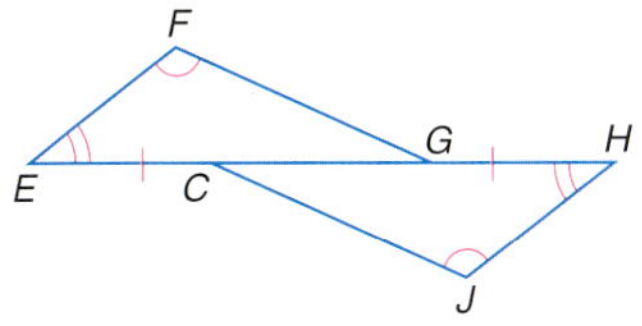

18. Given: $\overline{TX} \parallel \overline{SY}$
$\angle TXY \cong \angle TSY$
Prove: $\triangle TSY \cong \triangle YXT$

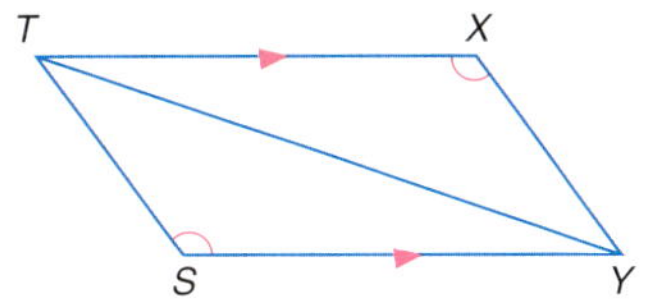

About the Exercises...

Organization by Objective

- **ASA Postulate:** 10, 12, 13, 19, 20
- **AAS Theorem:** 9, 11, 14–18, 21–28

Odd/Even Assignments
Exercises 9–28 are structured so that students practice the same concepts whether they are assigned odd or even problems.

Assignment Guide

Basic: 9–17 odd, 21–29 odd, 30–41
Average: 9–29 odd, 30–41
Advanced: 10–28 even, 29–38 (optional: 39–41)

Answers

15. Given: $\angle NOM \cong \angle POR$, $\overline{NM} \perp \overline{MR}$, $\overline{PR} \perp \overline{MR}$, $\overline{NM} \cong \overline{PR}$
Prove: $\overline{MO} \cong \overline{OR}$

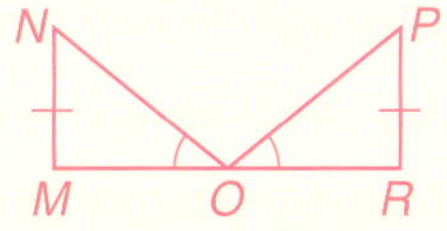

Proof: Since $\overline{NM} \perp \overline{MR}$ and $\overline{PR} \perp \overline{MR}$, $\angle M$ and $\angle R$ are right angles. $\angle M \cong \angle R$ because all right angles are congruent. We know that $\angle NOM \cong \angle POR$ and $\overline{NM} \cong \overline{PR}$. By AAS, $\triangle NMO \cong \triangle PRO$. $\overline{MO} \cong \overline{OR}$ by CPCTC.

16. Given: $\overline{DL}$ bisects $\overline{BN}$.
$\angle XLN \cong \angle XDB$
Prove: $\overline{LN} \cong \overline{DB}$

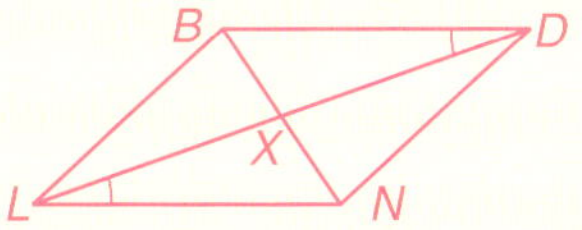

Proof: Since $\overline{DL}$ bisects $\overline{BN}$, $\overline{BX} \cong \overline{XN}$. $\angle XLN \cong \angle XDB$. $\angle LXN \cong \angle DXB$ because vertical angles are congruent. $\triangle LXN \cong \triangle DXB$ by AAS. $\overline{LN} \cong \overline{DB}$ by CPCTC.

17. Given: $\angle F \cong \angle J$, $\angle E \cong \angle H$
$\overline{EC} \cong \overline{GH}$
Prove: $\overline{EF} \cong \overline{HJ}$

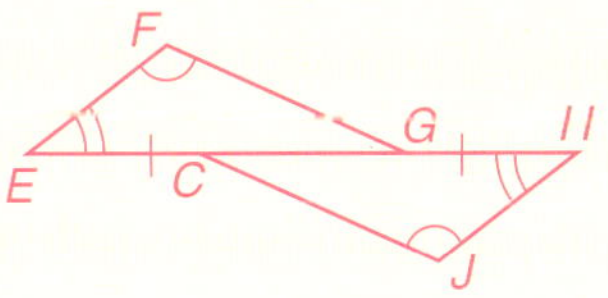

Proof: We are given that $\angle F \cong \angle J$, $\angle E \cong \angle H$, and $\overline{EC} \cong \overline{GH}$. By the Reflexive Property, $\overline{CG} \cong \overline{CG}$. Segment addition results in $EG = EC + CG$ and $CH = CG + GH$. By the definition of congruence, $EC = GH$ and $CG = CG$. Substitute to find $EG = CH$. By AAS, $\triangle EFG \cong \triangle HJC$. By CPCTC, $\overline{EF} \cong \overline{HJ}$.

18. Given: $\overline{TX} \parallel \overline{SY}$
$\angle TXY \cong \angle TSY$
Prove: $\triangle TSY \cong \triangle YXT$

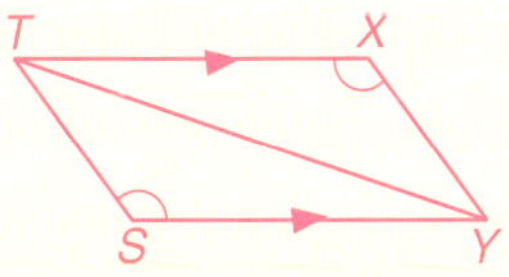

Proof: Since $\overline{TX} \parallel \overline{SY}$, $\angle YTX \cong \angle TYS$ by Alternate Interior Angles Theorem. $\overline{TY} \cong \overline{TY}$ by the Reflexive Property. Given $\angle TXY \cong \angle TSY$, $\triangle TSY \cong \triangle YXT$ by AAS.

Study Guide and Intervention, p. 207 (shown) and p. 208

ASA Postulate The Angle-Side-Angle (ASA) Postulate lets you show that two triangles are congruent.

ASA Postulate	If two angles and the included side of one triangle are congruent to two angles and the included side of another triangle, then the triangles are congruent.

Example **Find the missing congruent parts so that the triangles can be proved congruent by the ASA Postulate. Then write the triangle congruence.**

a. Two pairs of corresponding angles are congruent, $\angle A \cong \angle D$ and $\angle C \cong \angle F$. If the included sides $\overline{AC}$ and $\overline{DF}$ are congruent, then $\triangle ABC \cong \triangle DEF$ by the ASA Postulate.

b. $\angle R \cong \angle Y$ and $\overline{SR} \cong \overline{XY}$. If $\angle S \cong \angle X$, then $\triangle RST \cong \triangle YXW$ by the ASA Postulate.

Exercises

What corresponding parts must be congruent in order to prove that the triangles are congruent by the ASA Postulate? Write the triangle congruence statement.

1. $\overline{DC} \cong \overline{BC}$; $\triangle CDE \cong \triangle CBA$
2. $\overline{WY} \cong \overline{WY}$; $\angle XYW \cong \angle ZYW$; $\triangle WXY \cong \triangle WZY$
3. $\angle ABE \cong \angle CBD$; $\triangle ABE \cong \triangle CBD$
4. $\overline{BD} \cong \overline{DB}$; $\angle ADB \cong \angle CBD$; $\triangle ABD \cong \triangle CDB$
5. $\overline{ST} \cong \overline{VT}$; $\triangle RST \cong \triangle UVT$
6. $\angle ACB \cong \angle E$; $\triangle ABC \cong \triangle CDE$

Skills Practice, p. 209 and Practice, p. 210 (shown)

1. Write a flow proof.
   Given: S is the midpoint of $\overline{QT}$. $\overline{QR} \parallel \overline{TU}$
   Prove: $\triangle QSR \cong \triangle TSU$
   Sample proof:
   S is the midpoint of $\overline{QT}$ (Given) → $\overline{QS} \cong \overline{TS}$ (Def of midpoint) → $\triangle QSR \cong \triangle TSU$ (ASA)
   $\overline{QR} \parallel \overline{TU}$ (Given) → $\angle Q \cong \angle T$ (Alt. Int. ∠s are ≅) → $\triangle QSR \cong \triangle TSU$
   $\angle QSR \cong \angle TSU$ (Vertical ∠s are ≅) → $\triangle QSR \cong \triangle TSU$

2. Write a paragraph proof.
   Given: $\angle D \cong \angle F$; $\overline{GE}$ bisects $\angle DEF$.
   Prove: $\overline{DG} \cong \overline{FG}$
   Proof: Since it is given that $\overline{GE}$ bisects $\angle DEF$, $\angle DEG \cong \angle FEG$ by the definition of an angle bisector. It is given that $\angle D \cong \angle F$. By the Reflexive Property, $\overline{GE} \cong \overline{GE}$. So $\triangle DEG \cong \triangle FEG$ by AAS. Therefore $\overline{DG} \cong \overline{FG}$ by CPCTC.

ARCHITECTURE **For Exercises 3 and 4, use the following information.**
An architect used the window design in the diagram when remodeling an art studio. $\overline{AB}$ and $\overline{CB}$ each measure 3 feet.

3. Suppose D is the midpoint of $\overline{AC}$. Determine whether $\triangle ABD \cong \triangle CBD$. Justify your answer.
   Since D is the midpoint of $\overline{AC}$, $\overline{AD} \cong \overline{CD}$ by the definition of midpoint. $\overline{AB} \cong \overline{CB}$ by the definition of congruent segments. By the Reflexive Property, $\overline{BD} \cong \overline{BD}$. So $\triangle ABD \cong \triangle CBD$ by SSS.
4. Suppose $\angle A \cong \angle C$. Determine whether $\triangle ABD \cong \triangle CBD$. Justify your answer.
   We are given $\overline{AB} \cong \overline{CB}$ and $\angle A \cong \angle C$. $\overline{BD} \cong \overline{BD}$ by the Reflexive Property. Since SSA cannot be used to prove that triangles are congruent, we cannot say whether $\triangle ABD \cong \triangle CBD$.

Reading to Learn Mathematics, p. 211 ELL

Pre-Activity **How are congruent triangles used in construction?**
Read the introduction to Lesson 4-5 at the top of page 207 in your textbook.
Which of the triangles in the photograph in your textbook appear to be congruent? Sample answer: The four right triangles are congruent to each other. The two obtuse isosceles triangles are congruent to each other.

Reading the Lesson

1. Explain in your own words the difference between how the ASA Postulate and the AAS Theorem are used to prove that two triangles are congruent.
   Sample answer: In ASA, you use two pairs of congruent angles and the *included* congruent sides. In AAS, you use two pairs of congruent angles and a pair of *nonincluded* congruent sides.
2. Which of the following conditions are sufficient to prove that two triangles are congruent? B, D, E, G, H
   A. Two sides of one triangle are congruent to two sides of the other triangle.
   B. The three sides of one triangles are congruent to the three sides of the other triangle.
   C. The three angles of one triangle are congruent to the three angles of the other triangle.
   D. All six corresponding parts of two triangles are congruent.
   E. Two angles and the included side of one triangle are congruent to two sides and the included angle of the other triangle.
   F. Two sides and a nonincluded angle of one triangle are congruent to two sides and a nonincluded angle of the other triangle.
   G. Two angles and a nonincluded side of one triangle are congruent to two angles and the corresponding nonincluded side of the other triangle.
   H. Two sides and the included angle of one triangle are congruent to two sides and the included angle of the other triangle.
   I. Two angles and a nonincluded side of one triangle are congruent to two angles and a nonincluded side of the other triangle.
3. Determine whether you have enough information to prove that the two triangles in each figure are congruent. If so, write a congruence statement and name the congruence postulate or theorem that you would use. If not, write *not possible*.
   a. $\triangle AEB \cong \triangle DEC$; AAS
   b. T is the midpoint of $\overline{RU}$. $\triangle RST \cong \triangle UVT$; ASA

Helping You Remember

4. A good way to remember mathematical ideas is to summarize them in a general statement. If you want to prove triangles congruent by using three pairs of corresponding parts, what is a good way to remember which combinations of parts will work?
   Sample answer: At least one pair of corresponding parts must be sides. If you use two pairs of sides and one pair of angles, the angles must be the included angles. If you use two pairs of angles and one pair of sides, then the sides must both be included by the angles or must both be corresponding nonincluded sides.

Write a two-column proof. 19–20. See pp. 233E–233F.

★ 19. **Given:** $\angle MYT \cong \angle NYT$
$\angle MTY \cong \angle NTY$
Prove: $\triangle RYM \cong \triangle RYN$

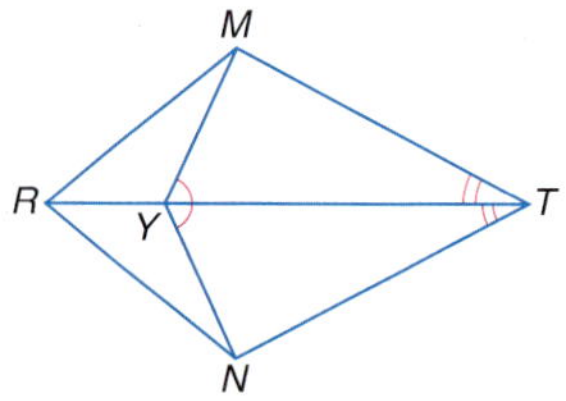

★ 20. **Given:** $\triangle BMI \cong \triangle KMT$
$\overline{IP} \cong \overline{PT}$
Prove: $\triangle IPK \cong \triangle TPB$

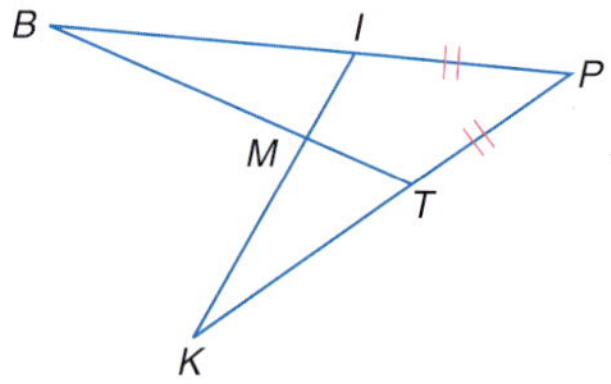

GARDENING **For Exercises 21 and 22, use the following information.**
Beth is planning a garden. She wants the triangular sections, $\triangle CFD$ and $\triangle HFG$, to be congruent. F is the midpoint of $\overline{DG}$, and $DG = 16$ feet. 21–22. See p. 233E.

21. Suppose $\overline{CD}$ and $\overline{GH}$ each measure 4 feet and the measure of $\angle CFD$ is 29. Determine whether $\triangle CFD \cong \triangle HFG$. Justify your answer.
22. Suppose F is the midpoint of $\overline{CH}$, and $\overline{CH} \cong \overline{DG}$. Determine whether $\triangle CFD \cong \triangle HFG$. Justify your answer.

More About. . .

Kites

The largest kite ever flown was 210 feet long and 72 feet wide.

Source: *Guinness Book of World Records*

KITES **For Exercises 23 and 24, use the following information.**
Austin is building a kite. Suppose JL is 2 feet, JM is 2.7 feet, and the measure of $\angle NJM$ is 68. 23–24. See p. 233F.

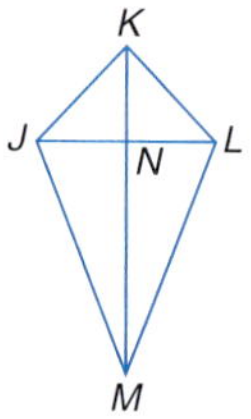

23. If N is the midpoint of $\overline{JL}$ and $\overline{KM} \perp \overline{JL}$, determine whether $\triangle JKN \cong \triangle LKN$. Justify your answer.
24. If $\overline{JM} \cong \overline{LM}$ and $\angle NJM \cong \angle NLM$, determine whether $\triangle JNM \cong \triangle LNM$. Justify your answer.

Complete each congruence statement and the postulate or theorem that applies.

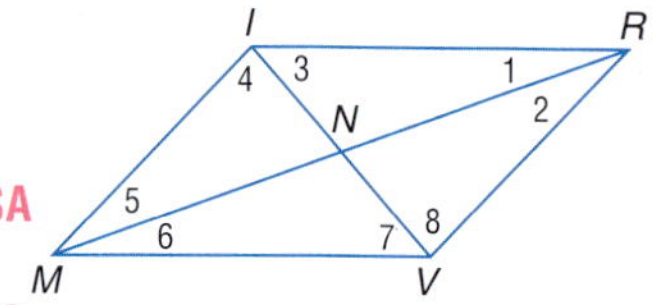

25. If $\overline{IM} \cong \overline{RV}$ and $\angle 2 \cong \angle 5$, then $\triangle INM \cong \triangle$ _?_ by _?_. $\triangle VNR$, AAS or ASA
26. If $\overline{IR} \parallel \overline{MV}$ and $\overline{IR} \cong \overline{MV}$, then $\triangle IRN \cong \triangle$ _?_ by _?_. $\triangle VMN$, ASA or AAS
27. If $\overline{IV}$ and $\overline{RM}$ bisect each other, then $\triangle RVN \cong \triangle$ _?_ by _?_. $\triangle MIN$, SAS
28. If $\angle MIR \cong \angle RVM$ and $\angle 1 \cong \angle 6$, then $\triangle MRV \cong \triangle$ _?_ by _?_. $\triangle RMI$, AAS or ASA

29. **CRITICAL THINKING** Aiko wants to estimate the distance between herself and a duck. She adjusts the visor of her cap so that it is in line with her line of sight to the duck. She keeps her neck stiff and turns her body to establish a line of sight to a point on the ground. Then she paces out the distance to the new point. Is the distance from the duck the same as the distance she just paced out? Explain your reasoning. See margin.

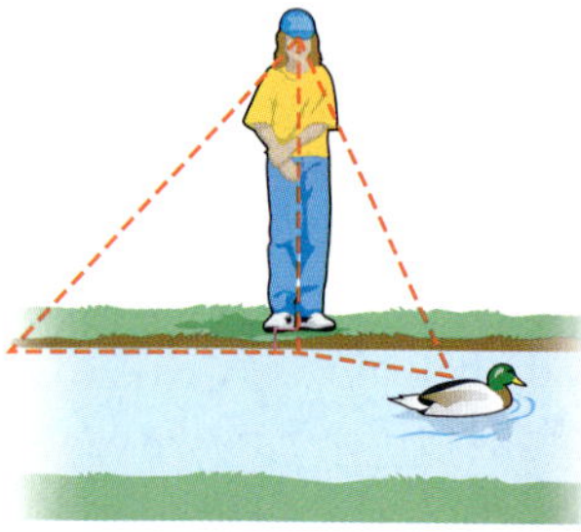

Enrichment, p. 212

Congruent Triangles in the Coordinate Plane

If you know the coordinates of the vertices of two triangles in the coordinate plane, you can often decide whether the two triangles are congruent. There may be more than one way to do this.

1. Consider $\triangle ABD$ and $\triangle CDB$ whose vertices have coordinates $A(0, 0)$, $B(2, 5)$, $C(9, 5)$, and $D(7, 0)$. Briefly describe how you can use what you know about congruent triangles and the coordinate plane to show that $\triangle ABD \cong \triangle CDB$. You may wish to make a sketch to help get you started.
   Sample answer: Show that the slopes of $\overline{AB}$ and $\overline{CD}$ are equal and that the slopes of $\overline{AD}$ and $\overline{BC}$ are equal. Conclude that $\overline{AB} \parallel \overline{CD}$ and $\overline{BC} \parallel \overline{AD}$. Use the angle relationships for parallel lines and a transversal and the fact that $\overline{BD}$ is a common side for the triangles to conclude that $\triangle ABD \cong \triangle CDB$ by ASA.

Answer

29. Since Aiko is perpendicular to the ground, two right angles are formed and right angles are congruent. The angles of sight are the same and her height is the same for each triangle. The triangles are congruent by ASA. By CPCTC, the distances are the same. The method is valid.

30. **WRITING IN MATH** Answer the question that was posed at the beginning of the lesson. **See margin.**

How are congruent triangles used in construction?

Include the following in your answer:
- explain how to determine whether the triangles are congruent, and
- why it is important that triangles used for structural support are congruent.

FCAT Practice
Standardized Test Practice

31. In $\triangle ABC$, $\overline{AD}$ and $\overline{DC}$ are angle bisectors and $m\angle B = 76$. What is the measure of $\angle ADC$? **D**

Ⓐ 26 Ⓑ 52
Ⓒ 76 Ⓓ 128

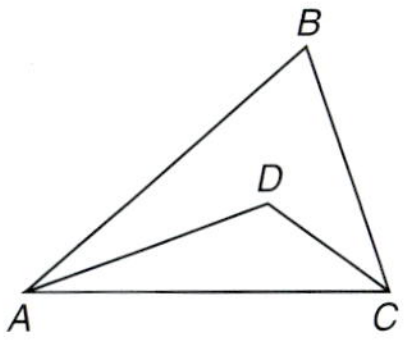

32. **ALGEBRA** For a positive integer x, 1 percent of x percent of 10,000 equals **A**

Ⓐ x. Ⓑ $10x$. Ⓒ $100x$. Ⓓ $1000x$.

Maintain Your Skills

Mixed Review **Write a flow proof.** *(Lesson 4-4)* **33–34. See p. 233F.**

33. **Given:** $\overline{BA} \cong \overline{DE}$, $\overline{DA} \cong \overline{BE}$
Prove: $\triangle BEA \cong \triangle DAE$

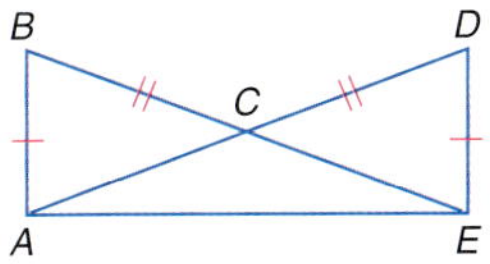

34. **Given:** $\overline{XZ} \perp \overline{WY}$
$\overline{XZ}$ bisects $\overline{WY}$.
Prove: $\triangle WZX \cong \triangle YZX$

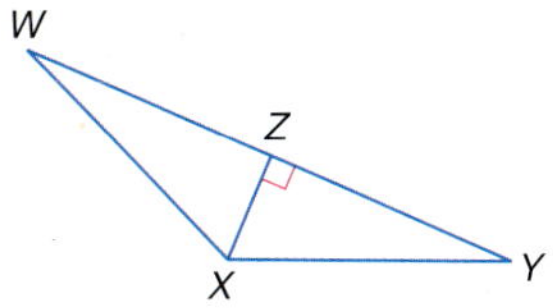

35–36. See margin.

Verify that each of the following preserves congruence and name the congruence transformation. *(Lesson 4-3)*

35.

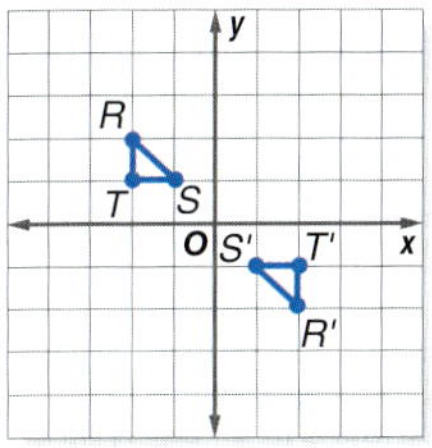

36.

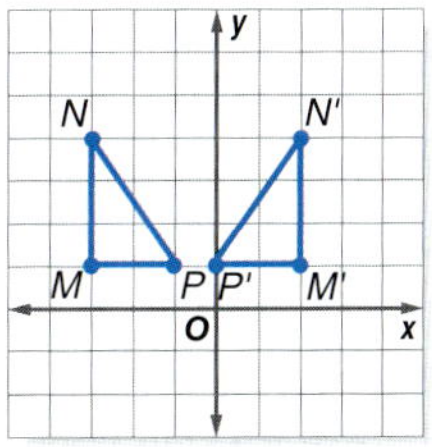

Write each statement in if-then form. *(Lesson 2-3)*

37. Happy people rarely correct their faults.

37. If people are happy, then they rarely correct their faults.

38. A champion is afraid of losing. **If a person is a champion, then he or she is afraid of losing.**

Getting Ready for the Next Lesson

PREREQUISITE SKILL **Classify each triangle according to its sides.**
*(To review **classification by sides**, see Lesson 4-1.)*

39.

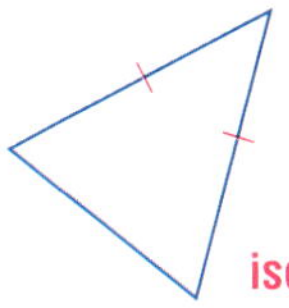

isosceles

40.

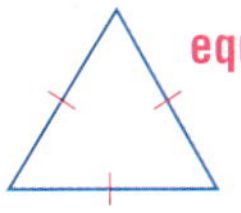

equilateral

41. **isosceles**

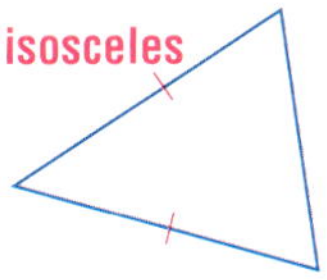

www.geometryonline.com/self_check_quiz/fcat

4 Assess

Open-Ended Assessment

Writing Have students practice writing different versions of proofs for each example. For Example 1, students can write a flow proof and a two-column proof.

Getting Ready for Lesson 4-6

Prerequisite Skill Students will learn about isosceles triangles in Lesson 4-6. They will use congruence postulates and theorems when writing proofs. Use Exercises 39–41 to determine your students' familiarity with the classification of triangles by sides.

Answers

30. Sample answer: The triangular trusses support the structure. Answers should include the following.
- **To determine whether two triangles are congruent, information is needed about consecutive side-angle-side, side-side-side, angle-side-angle, angle-angle-side, or about each angle and each side.**
- **Triangles that are congruent will support weight better because the pressure will be evenly divided.**

35. Turn; $RS = \sqrt{2}$, $R'S' = \sqrt{2}$, $ST = 1$, $S'T' = 1$, $RT = 1$, $R'T' = 1$. Use a protractor to confirm that the corresponding angles are congruent.

36. Flip; $MP = 2$, $M'P' = 2$, $MN = 3$, $M'N' = 3$, $NP = \sqrt{13}$, $N'P' = \sqrt{13}$. Use a protractor to confirm that the corresponding angles are congruent.

Teacher to Teacher

Karyn S. Cummins, Franklin Central High School Indianapolis, IN

After proofs are introduced I write up several simple proofs on card stock. I then cut the statements and reasons apart, give them to the students and have them reconstruct them.

Geometry Activity

A Follow-Up of Lesson 4-5

Getting Started

Objective Explore congruence in right triangles.

Materials

ruler   protractor

Teach

- Explain that right triangles are unique and typically have special relationships. Students will want to check for these relationships when they are working on proofs.
- Remind students that the right triangle theorems do not work for acute or obtuse triangles, but only for right triangles.

Assess

Exercises 1–3 guide students through SAS, ASA, and AAS, and introduce LL, HA, and LA. **Exercises 4–6** demonstrate that SSA works with right triangles and forms the HL Postulate. **Exercises 7–11** use the right triangle congruence theorems and postulate in proofs.

Study Notebook

Ask students to summarize what they have learned about congruence in right triangles. Tell students to list each method with a brief description.

Geometry Activity

A Follow-Up of Lesson 4-5

Sunshine State Standards
MA.C.1.4.1, MA.C.2.4.1

Congruence in Right Triangles

In Lessons 4-4 and 4-5, you learned theorems and postulates to prove triangles congruent. Do these theorems and postulates apply to right triangles?

Activity 1 Triangle Congruence

Model

Study each pair of right triangles.

a. **b.** **c.**

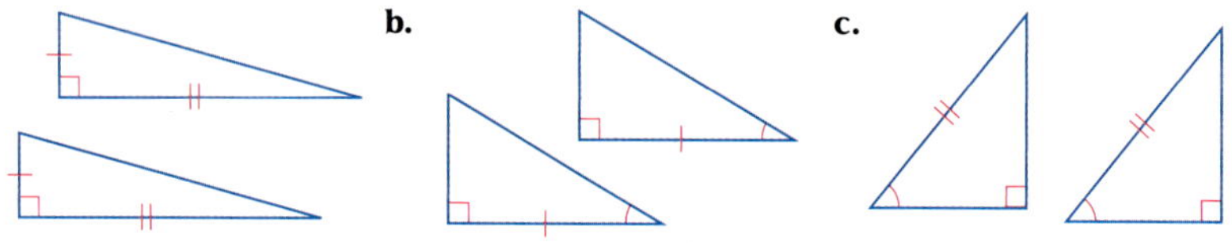

Analyze

1. Is each pair of triangles congruent? If so, which congruence theorem or postulate applies? **yes; a. SAS, b. ASA, c. AAS**
2. Rewrite the congruence rules from Exercise 1 using *leg*, (L), or *hypotenuse*, (H), to replace *side*. Omit the *A* for any right angle since we know that all right triangles contain a right angle and all right angles are congruent. **a. LL, b. LA, c. HA**
3. **Make a conjecture** If you know that the corresponding legs of two right triangles are congruent, what other information do you need to declare the triangles congruent? Explain. **None; two pairs of legs congruent is sufficient for proving right triangles congruent.**

In Lesson 4-5, you learned that SSA is not a valid test for determining triangle congruence. Can SSA be used to prove right triangles congruent?

Activity 2 SSA and Right Triangles

Make a Model

How many right triangles exist that have a hypotenuse of 10 centimeters and a leg of 7 centimeters?

Step 1 Draw $\overline{XY}$ so that $XY = 7$ centimeters.

Step 2 Use a protractor to draw a ray from Y that is perpendicular to $\overline{XY}$.

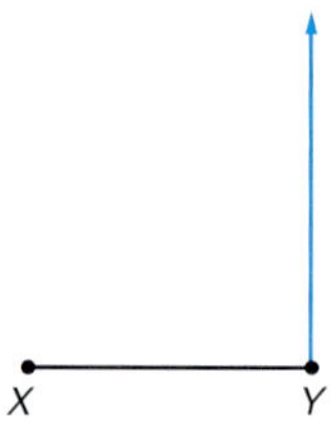

Step 3 Open your compass to a width of 10 centimeters. Place the point at X and draw a long arc to intersect the ray.

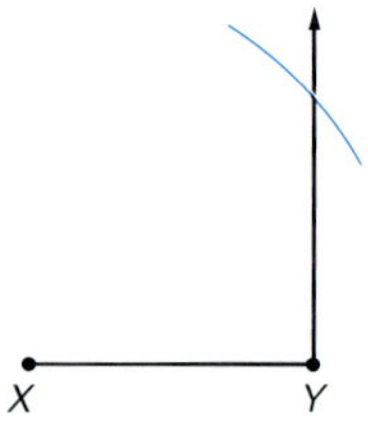

Step 4 Label the intersection Z and draw $\overline{XZ}$ to complete $\triangle XYZ$.

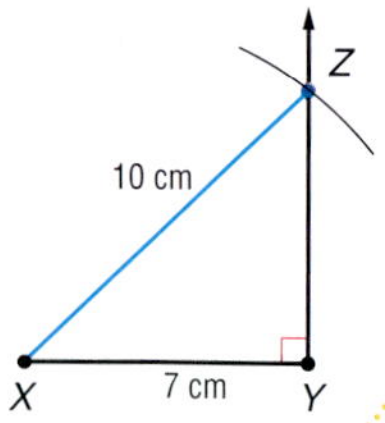

Resource Manager

Teaching Geometry with Manipulatives
- p. 78 (student recording sheet)
- p. 16 (protractor)
- p. 17 (ruler)

Glencoe Mathematics Classroom Manipulative Kit
- protractor
- ruler

Analyze

4. Does the model yield a unique triangle? **yes**
5. Can you use the lengths of the hypotenuse and a leg to show right triangles are congruent? **yes**
6. **Make a conjecture** about the case of SSA that exists for right triangles. **SSA is a valid test of congruence for right triangles.**

The two activities provide evidence for four ways to prove right triangles congruent.

Key Concept — *Right Triangle Congruence*

Theorem	Abbreviation	Example
4.6 Leg-Leg Congruence If the legs of one right triangle are congruent to the corresponding legs of another right triangle, then the triangles are congruent.	LL	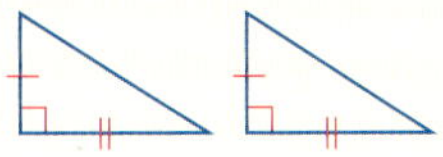
4.7 Hypotenuse-Angle Congruence If the hypotenuse and acute angle of one right triangle are congruent to the hypotenuse and corresponding acute angle of another right triangle, then the two triangles are congruent.	HA	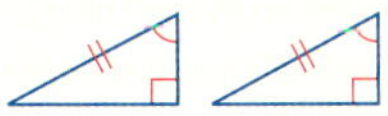
4.8 Leg-Angle Congruence If one leg and an acute angle of one right triangle are congruent to the corresponding leg and acute angle of another right triangle, then the triangles are congruent.	LA	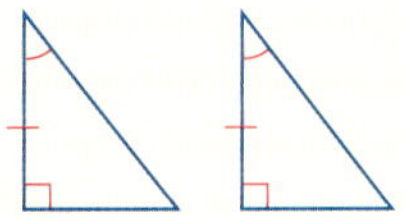
Postulate		
4.4 Hypotenuse-Leg Congruence If the hypotenuse and a leg of one right triangle are congruent to the hypotenuse and corresponding leg of another right triangle, then the triangles are congruent.	HL	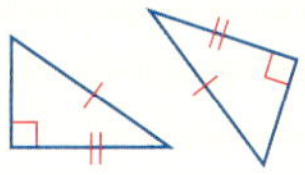

PROOF **Write a paragraph proof of each theorem.** **7–9. See margin.**

7. Theorem 4.6
8. Theorem 4.7
9. Theorem 4.8 (*Hint*: There are two possible cases.)

Use the figure to write a two-column proof. **10–11. See p. 233F.**

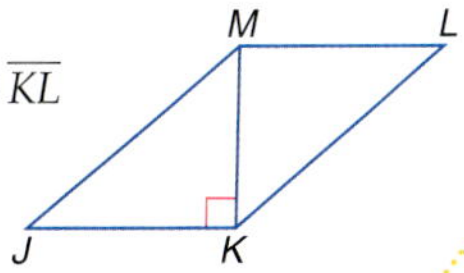

10. **Given:** $\overline{ML} \perp \overline{MK}$, $\overline{JK} \perp \overline{KM}$, $\angle J \cong \angle L$
    Prove: $\overline{JM} \cong \overline{KL}$
11. **Given:** $\overline{JK} \perp \overline{KM}$, $\overline{JM} \cong \overline{KL}$, $\overline{ML} \parallel \overline{JK}$
    Prove: $\overline{ML} \cong \overline{JK}$

Answers

7. Given: $\triangle DEF$ and $\triangle RST$ are right triangles.
$\angle E$ and $\angle S$ are right angles.
$\overline{EF} \cong \overline{ST}$
$\overline{ED} \cong \overline{SR}$
Prove: $\triangle DEF \cong \triangle RST$

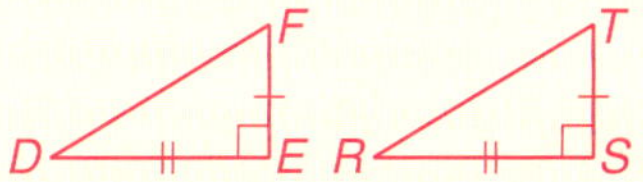

Proof: We are given that $\overline{EF} \cong \overline{ST}$, $\overline{ED} \cong \overline{SR}$, and $\angle E$ and $\angle S$ are right angles. Since all right angles are congruent, $\angle E \cong \angle S$. Therefore, by SAS, $\triangle DEF \cong \triangle RST$.

8. Given: $\triangle ABC$ and $\triangle XYZ$ are right triangles.
$\angle A$ and $\angle X$ are right angles.
$\overline{BC} \cong \overline{YZ}$
$\angle B \cong \angle Y$
Prove: $\triangle ABC \cong \triangle XYZ$

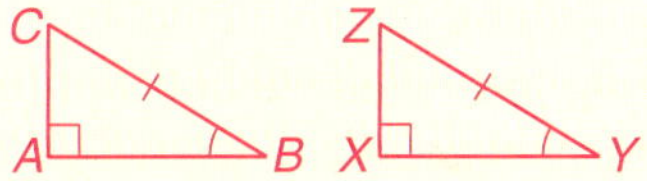

Proof: We are given that $\triangle ABC$ and $\triangle XYZ$ are right triangles with right angles $\angle A$ and $\angle X$, $\overline{BC} \cong \overline{YZ}$, and $\angle B \cong \angle Y$. Since all right angles are congruent, $\angle A \cong \angle X$. Therefore, $\triangle ABC \cong \triangle XYZ$ by AAS.

9. Case 1:
Given: $\triangle ABC$ and $\triangle DEF$ are right triangles.
$\overline{AC} \cong \overline{DF}$, $\angle C \cong \angle F$
Prove: $\triangle ABC \cong \triangle DEF$

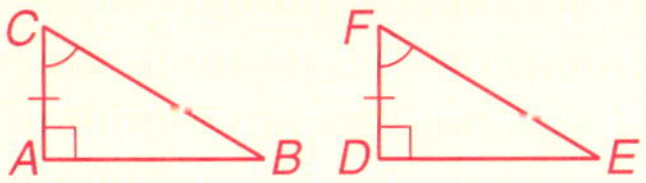

Proof: It is given that $\triangle ABC$ and $\triangle DEF$ are right triangles, $\overline{AC} \cong \overline{DF}$, $\angle C \cong \angle F$. By the definition of right triangles, $\angle A$ and $\angle D$ are right angles. Thus, $\angle A \cong \angle D$ since all right angles are congruent. $\triangle ABC \cong \triangle DEF$ by ASA.

Case 2:
Given: $\triangle ABC$ and $\triangle DEF$ are right triangles.
$\overline{AC} \cong \overline{DF}$, $\angle B \cong \angle E$
Prove: $\triangle ABC \cong \triangle DEF$

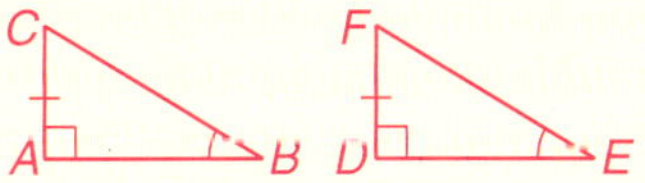

Proof: It is given that $\triangle ABC$ and $\triangle DEF$ are right triangles, $\overline{AC} \cong \overline{DF}$, and $\angle B \cong \angle E$. By the definition of right triangle, $\angle A$ and $\angle D$ are right angles. Thus, $\angle A \cong \angle D$ since all right angles are congruent. $\triangle ABC \cong \triangle DEF$ by AAS.

4-6 Lesson Notes

1 Focus

5-Minute Check Transparency 4-6 Use as a quiz or review of Lesson 4-5.

Mathematical Background notes are available for this lesson on p. 176D.

How are triangles used in art?

Ask students:

- How would the painting's overall appearance change if you removed or covered the triangles? **The painting would appear more blended and based on curves, circles, and ovals without the stark triangles.**
- Describe where isosceles triangles appear in the art. **Accept all reasonable answers.**

4-6 Isosceles Triangles

Sunshine State Standards
MA.C.1.4.1, MA.C.2.4.1

Vocabulary

- vertex angle
- base angles

WebQuest

You can use properties of triangles to prove Thales of Miletus' important geometric ideas. Visit www.geometryonline.com/webquest to continue work on your WebQuest project.

What You'll Learn

- Use properties of isosceles triangles.
- Use properties of equilateral triangles.

How are triangles used in art?

The art of Lois Mailou Jones, a twentieth-century artist, includes paintings and textile design, as well as book illustration. Notice the isosceles triangles in this painting, *Damballah*.

PROPERTIES OF ISOSCELES TRIANGLES In Lesson 4-1, you learned that isosceles triangles have two congruent sides. Like the right triangle, the parts of an isosceles triangle have special names.

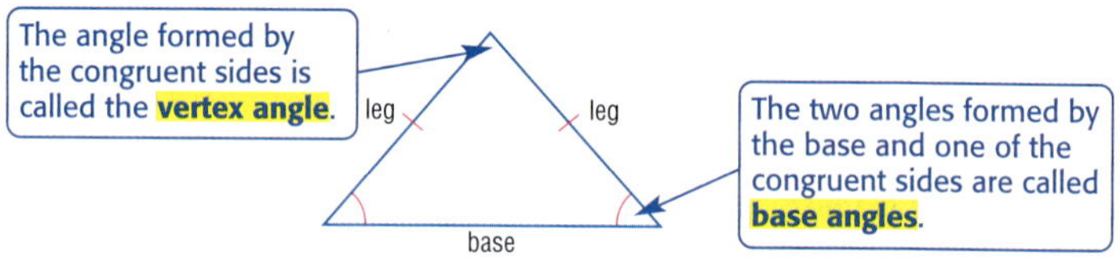

In this activity, you will investigate the relationship of the base angles and legs of an isosceles triangle.

Geometry Activity

Isosceles Triangles

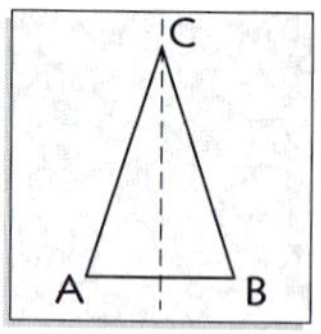

Model

- Draw an acute triangle on patty paper with $\overline{AC} \cong \overline{BC}$.
- Fold the triangle through C so that A and B coincide.

Analyze **2, 3. They are congruent.**

1. What do you observe about $\angle A$ and $\angle B$? **$\angle A \cong \angle B$**
2. Draw an obtuse isosceles triangle. Compare the base angles.
3. Draw a right isosceles triangle. Compare the base angles.

The results of the Geometry Activity suggest Theorem 4.9.

Theorem 4.9

Isosceles Triangle Theorem If two sides of a triangle are congruent, then the angles opposite those sides are congruent.

Example: If $\overline{AB} \cong \overline{CB}$, then $\angle A \cong \angle C$.

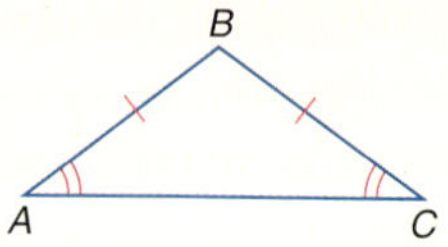

Resource Manager

Workbook and Reproducible Masters

Chapter 4 Resource Masters

- Study Guide and Intervention, pp. 213–214
- Skills Practice, p. 215
- Practice, p. 216
- Reading to Learn Mathematics, p. 217
- Enrichment, p. 218
- Assessment, p. 240

School-to-Career Masters, p. 8

Teaching Geometry With Manipulatives Masters, pp. 8, 80, 81

Transparencies

5-Minute Check Transparency 4-6
Answer Key Transparencies

Technology

Interactive Chalkboard

Example 1 Proof of Theorem

Write a two-column proof of the Isosceles Triangle Theorem.

Given: $\triangle PQR$, $\overline{PQ} \cong \overline{RQ}$

Prove: $\angle P \cong \angle R$

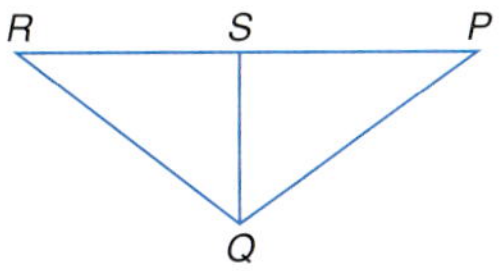

Proof:

Statements	Reasons
1. Let S be the midpoint of $\overline{PR}$.	1. Every segment has exactly one midpoint.
2. Draw an auxiliary segment $\overline{QS}$.	2. Two points determine a line.
3. $\overline{PS} \cong \overline{RS}$	3. Midpoint Theorem
4. $\overline{QS} \cong \overline{QS}$	4. Congruence of segments is reflexive.
5. $\overline{PQ} \cong \overline{RQ}$	5. Given
6. $\triangle PQS \cong \triangle RQS$	6. SSS
7. $\angle P \cong \angle R$	7. CPCTC

Example 2 Find the Measure of a Missing Angle

Multiple-Choice Test Item

If $\overline{GH} \cong \overline{HK}$, $\overline{HJ} \cong \overline{JK}$, and $m\angle GJK = 100$, what is the measure of $\angle HGK$?

(A) 10 (B) 15 (C) 20 (D) 25

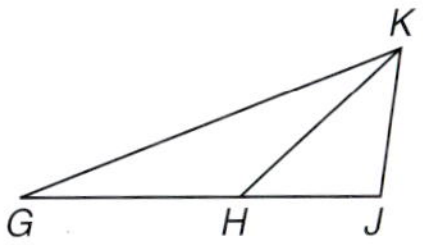

Read the Test Item

$\triangle GHK$ is isosceles with base $\overline{GK}$. Likewise, $\triangle HJK$ is isosceles with base $\overline{HK}$.

Solve the Test Item

Step 1 The base angles of $\triangle HJK$ are congruent. Let $x = m\angle KHJ = m\angle HKJ$.

$m\angle KHJ + m\angle HKJ + m\angle HJK = 180$ Angle Sum Theorem

$x + x + 100 = 180$ Substitution

$2x + 100 = 180$ Add.

$2x = 80$ Subtract 100 from each side.

$x = 40$ So, $m\angle KHJ = m\angle HKJ = 40$.

Step 2 $\angle GHK$ and $\angle KHJ$ form a linear pair. Solve for $m\angle GHK$.

$m\angle KHJ + m\angle GHK = 180$ Linear pairs are supplementary.

$40 + m\angle GHK = 180$ Substitution

$m\angle GHK = 140$ Subtract 40 from each side.

Step 3 The base angles of $\triangle GHK$ are congruent. Let y represent $m\angle HGK$ and $m\angle GKH$.

$m\angle GHK + m\angle HGK + m\angle GKH = 180$ Angle Sum Theorem

$140 + y + y = 180$ Substitution

$140 + 2y = 180$ Add.

$2y = 40$ Subtract 140 from each side.

$y = 20$ Divide each side by 2.

The measure of $\angle HGK$ is 20. Choice C is correct.

www.geometryonline.com/extra_examples/fcat

FCAT Practice
Standardized Test Practice
(A) (B) (C) (D)

The Princeton Review

Test-Taking Tip

Diagrams Label the diagram with the given information. Use your drawing to plan the next step in solving the problem.

2 Teach

PROPERTIES OF ISOSCELES TRIANGLES

In-Class Examples

Power Point®

1 Write a two-column proof.

Given: $AB = CB = BD$, $\angle ACB \cong \angle BCD$

Prove: $\angle A \cong \angle D$

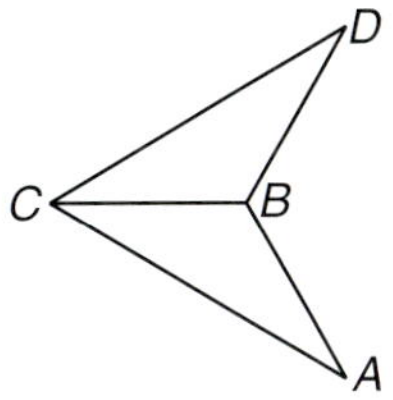

Proof:
Statements (Reasons)

1. $AB = CB = BD$ (Given)
2. $\overline{AB} \cong \overline{CB} \cong \overline{BD}$ (Def. of $\cong$ seg.)
3. $\triangle ABC$ and $\triangle BCD$ are isosceles. (Def. of isos. $\triangle$)
4. $\angle A \cong \angle ACB$, $\angle BCD \cong \angle D$ (Isos. $\triangle$ Th.)
5. $\angle ACB \cong \angle BCD$ (Given)
6. $\angle A \cong \angle D$ (Transitive Prop.)

2 If $\overline{DE} \cong \overline{CD}$, $\overline{BC} \cong \overline{AC}$, and $m\angle CDE = 120$, what is the measure of $\angle BAC$? D

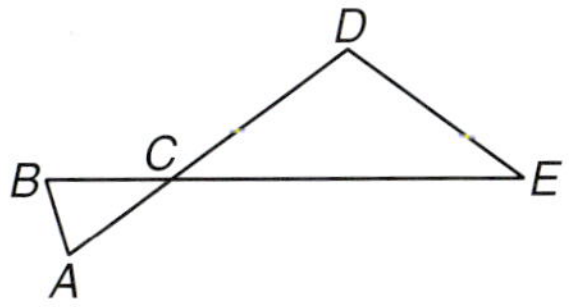

A 45.5 **B** 57.5
C 68.5 **D** 75

Geometry Activity

Materials: paper, scissors, ruler

- You may wish to provide students with rectangular dot paper to help them draw accurate isosceles triangles.
- Ask students to name the legs, base, vertex angle, and base angles of their triangles.
- Have students repeat the activity starting with a line segment and two congruent angles drawn at each end of the segment.

In-Class Example

3

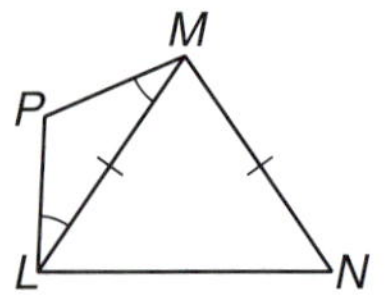

a. Name two congruent angles. $\angle MLN$ and $\angle MNL$

b. Name two congruent segments. $\overline{PL}$ and $\overline{PM}$

PROPERTIES OF EQUILATERAL TRIANGLES

In-Class Example

4 Copy the figure in Example 4 and then draw $\overline{EJ}$ so that $\overline{EJ}$ bisects $\angle 2$, and J lies on $\overline{FG}$.

a. Find $m\angle HEJ$ and $m\angle EJH$. 15; 75

b. Find $m\angle EJG$. 105

The converse of the Isosceles Triangle Theorem is also true.

Theorem 4.10

If two angles of a triangle are congruent, then the sides opposite those angles are congruent.

Abbreviation: *Conv. of Isos. △ Th.*

Example: If $\angle D \cong \angle F$, then $\overline{DE} \cong \overline{FE}$.

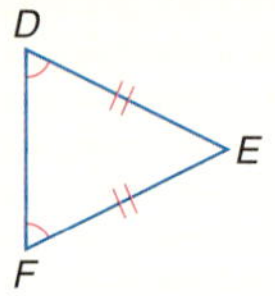

You will prove Theorem 4.10 in Exercise 33.

Example 3 Congruent Segments and Angles

a. Name two congruent angles.

$\angle AFC$ is opposite $\overline{AC}$ and $\angle ACF$ is opposite $\overline{AF}$, so $\angle AFC \cong \angle ACF$.

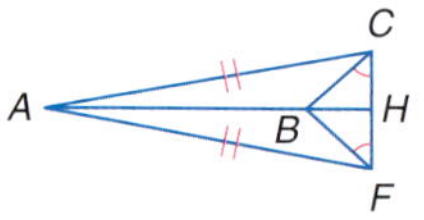

b. Name two congruent segments.

By the converse of the Isosceles Triangle Theorem, the sides opposite congruent angles are congruent. So, $\overline{BC} \cong \overline{BF}$.

PROPERTIES OF EQUILATERAL TRIANGLES Recall that an equilateral triangle has three congruent sides. The Isosceles Triangle Theorem also applies to equilateral triangles. This leads to two corollaries about the angles of an equilateral triangle.

Study Tip

Classifying Triangles
Equilateral triangles are also isosceles and have all of the properties of isosceles triangles.

Corollaries

4.3 A triangle is equilateral if and only if it is equiangular.

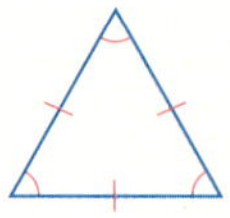

4.4 Each angle of an equilateral triangle measures 60°.

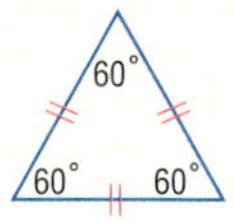

You will prove Corollaries 4.3 and 4.4 in Exercises 31 and 32.

Example 4 Use Properties of Equilateral Triangles

$\triangle EFG$ is equilateral, and $\overline{EH}$ bisects $\angle E$.

a. Find $m\angle 1$ and $m\angle 2$.

Each angle of an equilateral triangle measures 60°. So, $m\angle 1 + m\angle 2 = 60$. Since the angle was bisected, $m\angle 1 = m\angle 2$. Thus, $m\angle 1 = m\angle 2 = 30$.

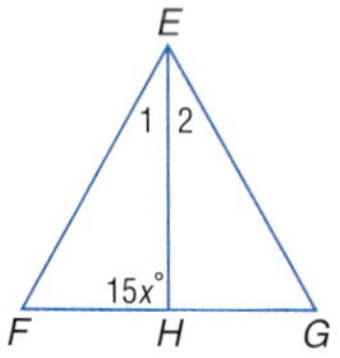

b. ALGEBRA Find x.

$m\angle EFH + m\angle 1 + m\angle EHF = 180$	Angle Sum Theorem
$60 + 30 + 15x = 180$	$m\angle EFH = 60$, $m\angle 1 = 30$, $m\angle EFH = 15x$
$90 + 15x = 180$	Add.
$15x = 90$	Subtract 90 from each side.
$x = 6$	Divide each side by 15.

DAILY INTERVENTION — Differentiated Instruction

Interpersonal Have groups of students work Exercise 7 on p. 219. Some group members can provide the statements, and other group members can provide corresponding reasons. Encourage groups to discuss the properties of isosceles and equilateral triangles while they are figuring out the proofs.

Check for Understanding

Concept Check

1. **Explain** how many angles in an isosceles triangle must be given to find the measures of the other angles. **1–3. See margin.**
2. **Name** the congruent sides and angles of isosceles $\triangle WXZ$ with base $\overline{WZ}$.
3. **OPEN ENDED** Describe a method to construct an equilateral triangle.

Guided Practice

GUIDED PRACTICE KEY

Exercises	Examples
4–5	3
6	4
7	1
8	2

Refer to the figure.

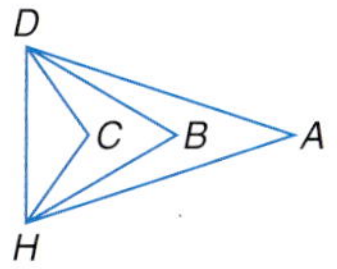

4. If $\overline{AD} \cong \overline{AH}$, name two congruent angles. $\angle ADH \cong \angle AHD$
5. If $\angle BDH \cong \angle BHD$, name two congruent segments. $\overline{BH} \cong \overline{BD}$
6. **ALGEBRA** Triangle GHF is equilateral with $m\angle F = 3x + 4$, $m\angle G = 6y$, and $m\angle H = 19z + 3$. Find x, y, and z. $\frac{56}{3}$, 10, 3

Write a two-column proof.

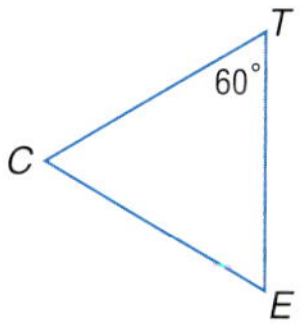

7. **Given:** $\triangle CTE$ is isosceles with vertex $\angle C$. $m\angle T = 60$
   Prove: $\triangle CTE$ is equilateral.
   See p. 233F.

FCAT Practice
Standardized Test Practice

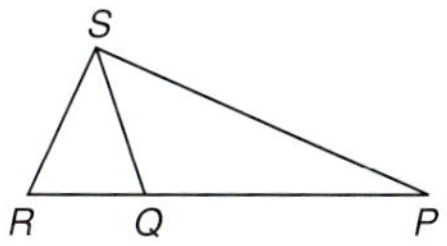

8. If $\overline{PQ} \cong \overline{QS}$, $\overline{QR} \cong \overline{RS}$, and $m\angle PRS = 72$, what is the measure of $\angle QPS$? **A**

   (A) 27 (B) 54 (C) 63 (D) 72

★ indicates increased difficulty

Practice and Apply

Homework Help

For Exercises	See Examples
9–14	3
15–22, 27–28, 34–37	4
23–26, 38–39	2
29–33	1

Extra Practice
See page 762.

Refer to the figure.

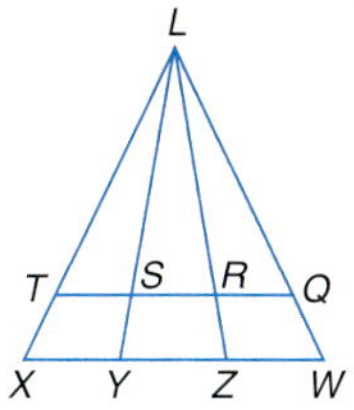

9. If $\overline{LT} \cong \overline{LR}$, name two congruent angles. $\angle LTR \cong \angle LRT$
10. If $\overline{LX} \cong \overline{LW}$, name two congruent angles. $\angle LXW \cong \angle LWX$
11. If $\overline{SL} \cong \overline{QL}$, name two congruent angles. $\angle LSQ \cong \angle LQS$
12. If $\angle LXY \cong \angle LYX$, name two congruent segments. $\overline{LX} \cong \overline{LY}$
13. If $\angle LSR \cong \angle LRS$, name two congruent segments. $\overline{LS} \cong \overline{LR}$
14. If $\angle LYW \cong \angle LWY$, name two congruent segments. $\overline{LY} \cong \overline{LW}$

$\triangle KLN$ and $\triangle LMN$ are isosceles and $m\angle JKN = 130$. Find each measure.

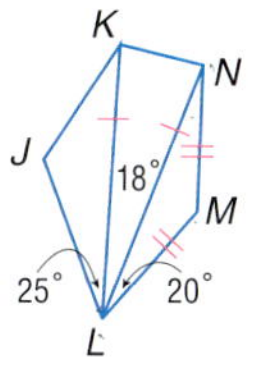

15. $m\angle LNM$ 20
16. $m\angle M$ 140
17. $m\angle LKN$ 81
18. $m\angle J$ 106

$\triangle DFG$ and $\triangle FGH$ are isosceles, $m\angle FDH = 28$ and $\overline{DG} \cong \overline{FG} \cong \overline{FH}$. Find each measure.

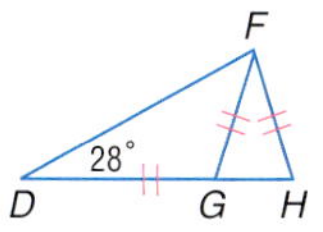

19. $m\angle DFG$ 28
20. $m\angle DGF$ 124
21. $m\angle FGH$ 56
22. $m\angle GFH$ 68

3 Practice/Apply

Study Notebook

Have students—
- add the definitions/examples of the vocabulary terms to their Vocabulary Builder worksheets for Chapter 4.
- include a concept map with all methods of proving triangle congruence accompanied by helpful theorems, postulates, properties, and formulas.
- include any other item(s) that they find helpful in mastering the skills in this lesson.

About the Exercises...

Organization by Objective
- **Properties of Isosceles Triangles:** 9–14, 23–26, 29–33, 38–39
- **Properties of Equilateral Triangles:** 15–22, 27–28, 34–37

Odd/Even Assignments
Exercises 9–39 are structured so that students practice the same concepts whether they are assigned odd or even problems.

Assignment Guide

Basic: 9–39 odd, 40–54
Average: 9–39 odd, 40–54
Advanced: 10–40 even, 41–51 (optional: 52–54)
All: Quiz 2 (1–5)

Answers

1. The measure of only one angle must be given in an isosceles triangle to determine the measures of the other two angles.
2. $\overline{WX} \cong \overline{ZX}$, $\angle W \cong \angle Z$
3. Sample answer: Draw a line segment. Set your compass to the length of the line segment and draw an arc from each endpoint. Draw segments from the intersection of the arcs to each endpoint.

Study Guide and Intervention, p. 213 (shown) and p. 214

Properties of Isosceles Triangles An **isosceles triangle** has two congruent sides. The angle formed by these sides is called the **vertex angle**. The other two angles are called **base angles**. You can prove a theorem and its converse about isosceles triangles.

- If two sides of a triangle are congruent, then the angles opposite those sides are congruent. **(Isosceles Triangle Theorem)**
- If two angles of a triangle are congruent, then the sides opposite those angles are congruent.

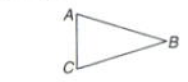

If $\overline{AB} \cong \overline{CB}$, then $\angle A \cong \angle C$.
If $\angle A \cong \angle C$, then $\overline{AB} \cong \overline{CB}$.

Example 1 **Find x.**

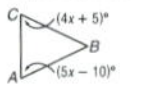

$BC = BA$, so

$m\angle A = m\angle C$.	Isos. Triangle Theorem
$5x - 10 = 4x + 5$	Substitution
$x - 10 = 5$	Subtract 4x from each side.
$x = 15$	Add 10 to each side.

Example 2 **Find x.**

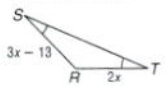

$m\angle S = m\angle T$, so

$SR = TR$.	Converse of Isos. △ Thm.
$3x - 13 = 2x$	Substitution
$3x = 2x + 13$	Add 13 to each side.
$x = 13$	Subtract 2x from each side.

Exercises

Find x.

1. 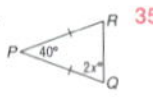 35

2. 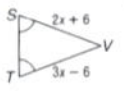 12

3. 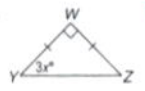 15

4. 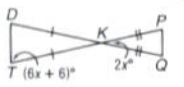 12

5. 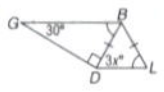 20

6. 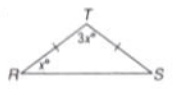 36

7. Write a two-column proof.
Given: $\angle 1 \cong \angle 2$
Prove: $\overline{AB} \cong \overline{CB}$

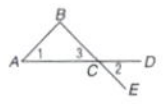

Statements	Reasons
1. $\angle 1 \cong \angle 2$	1. Given
2. $\angle 2 \cong \angle 3$	2. Vertical angles are congruent.
3. $\angle 1 \cong \angle 3$	3. Transitive Property of ≅
4. $\overline{AB} \cong \overline{CB}$	4. If two angles of a triangle are ≅, then the sides opposite the angles are ≅.

Skills Practice, p. 215 and Practice, p. 216 (shown)

Refer to the figure.

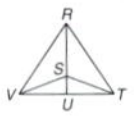

1. If $\overline{RV} \cong \overline{RT}$, name two congruent angles. $\angle RTV \cong \angle RVT$
2. If $\overline{RS} \cong \overline{SV}$, name two congruent angles. $\angle SVR \cong \angle SRV$
3. If $\angle SRT \cong \angle STR$, name two congruent segments. $\overline{ST} \cong \overline{SR}$
4. If $\angle STV \cong \angle SVT$, name two congruent segments. $\overline{ST} \cong \overline{SV}$

Triangles GHM and HJM are isosceles, with $\overline{GH} \cong \overline{MH}$ and $\overline{HJ} \cong \overline{MJ}$. Triangle KLM is equilateral, and $m\angle HMK = 50$. Find each measure.

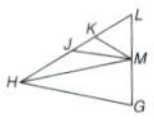

5. $m\angle KML$ 60
6. $m\angle HMG$ 70
7. $m\angle GHM$ 40
8. If $m\angle HJM = 145$, find $m\angle MHJ$. 17.5
9. If $m\angle G = 67$, find $m\angle GHM$. 46

10. Write a two-column proof.
Given: $\overline{DE} \parallel \overline{BC}$
$\angle 1 \cong \angle 2$
Prove: $\overline{AB} \cong \overline{AC}$

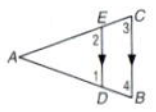

Proof:

Statements	Reasons
1. $\overline{DE} \parallel \overline{BC}$	1. Given
2. $\angle 1 \cong \angle 4$ $\angle 2 \cong \angle 3$	2. Corr. ∠s are ≅.
3. $\angle 1 \cong \angle 2$	3. Given
4. $\angle 3 \cong \angle 4$	4. Congruence of ∠s is transitive.
5. $\overline{AB} \cong \overline{AC}$	5. If 2 ∠s of a △ are ≅, then the sides opposite those ∠s are ≅.

11. **SPORTS** A pennant for the sports teams at Lincoln High School is in the shape of an isosceles triangle. If the measure of the vertex angle is 18, find the measure of each base angle. 81, 81

Reading to Learn Mathematics, p. 217 ELL

Pre-Activity **How are triangles used in art?**

Read the introduction to Lesson 4-6 at the top of page 216 in your textbook.

- Why do you think that isosceles and equilateral triangles are used more often than scalene triangles in art? Sample answer: Their symmetry is pleasing to the eye.
- Why might isosceles right triangles be used in art? Sample answer: Two congruent isosceles right triangles can be placed together to form a square.

Reading the Lesson

1. Refer to the figure.

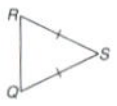

   a. What kind of triangle is $\triangle QRS$? isosceles
   b. Name the legs of $\triangle QRS$. $\overline{QS}$, $\overline{RS}$
   c. Name the base of $\triangle QRS$. $\overline{QR}$
   d. Name the vertex angle of $\triangle QRS$. $\angle S$
   e. Name the base angles of $\triangle QRS$. $\angle Q$, $\angle R$

2. Determine whether each statement is *always*, *sometimes*, or *never* true.
   a. If a triangle has three congruent sides, then it has three congruent angles. always
   b. If a triangle is isosceles, then it is equilateral. sometimes
   c. If a right triangle is isosceles, then it is equilateral. never
   d. The largest angle of an isosceles triangle is obtuse. sometimes
   e. If a right triangle has a 45° angle, then it is isosceles. always
   f. If an isosceles triangle has three acute angles, then it is equilateral. sometimes
   g. The vertex angle of an isosceles triangle is the largest angle of the triangle. sometimes

3. Give the measures of the three angles of each triangle.
   a. an equilateral triangle 60, 60, 60
   b. an isosceles right triangle 45, 45, 90
   c. an isosceles triangle in which the measure of the vertex angle is 70 70, 55, 55
   d. an isosceles triangle in which the measure of a base angle is 70 70, 70, 40
   e. an isosceles triangle in which the measure of the vertex angle is twice the measure of one of the base angles 90, 45, 45

Helping You Remember

4. If a theorem and its converse are both true, you can often remember them most easily by combining them into an "if-and-only-if" statement. Write such a statement for the Isosceles Triangle Theorem and its converse. Sample answer: Two sides of a triangle are congruent if and only if the angles opposite those sides are congruent.

In the figure, $\overline{JM} \cong \overline{PM}$ and $\overline{ML} \cong \overline{PL}$.

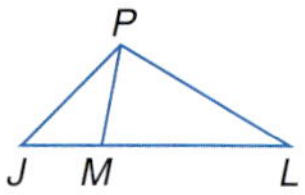

23. If $m\angle PLJ = 34$, find $m\angle JPM$. **36.5**
24. If $m\angle PLJ = 58$, find $m\angle PJL$. **30.5**

In the figure, $\overline{GK} \cong \overline{GH}$ and $\overline{HK} \cong \overline{KJ}$.

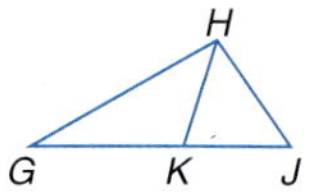

25. If $m\angle HGK = 28$, find $m\angle HJK$. **38**
26. If $m\angle HGK = 42$, find $m\angle HJK$. **34.5**

Triangle LMN is equilateral, and $\overline{MP}$ bisects $\overline{LN}$.

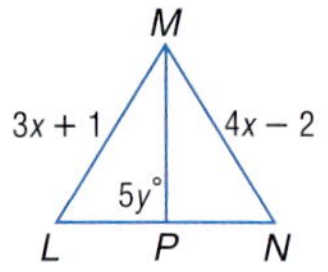

27. Find x and y. **$x = 3$; $y = 18$**
28. Find the measure of each side of $\triangle LMN$. **10**

PROOF **Write a two-column proof.** **29–33. See pp. 233F–233G.**

29. **Given:** $\triangle XKF$ is equilateral.
$\overline{XJ}$ bisects $\angle X$.
Prove: J is the midpoint of $\overline{KF}$.

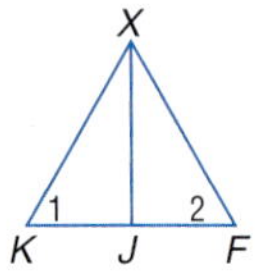

★ 30. **Given:** $\triangle MLP$ is isosceles.
N is the midpoint of $\overline{MP}$.
Prove: $\overline{LN} \perp \overline{MP}$

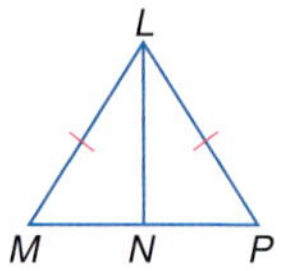

31. Corollary 4.3
32. Corollary 4.4
33. Theorem 4.10

More About. . .

Design

Spaceship Earth is a completely spherical *geodesic dome* that is covered with 11,324 triangular aluminum and plastic alloy panels.

Source: disneyworld.disney.go.com

34. **DESIGN** The basic structure covering Spaceship Earth at the Epcot Center in Orlando, Florida, is a triangle. Describe the minimum requirement to show that these triangles are equilateral. **The minimum requirement is that two angles measure 60°.**

ALGEBRA **Find x.**

35. 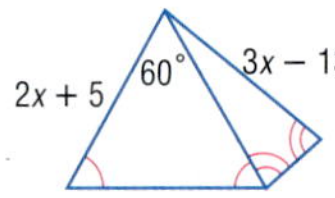 **18**

36. 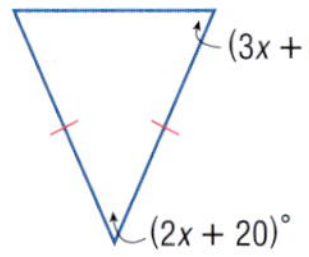 **18**

37. 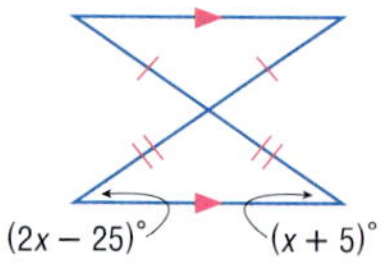 **30**

ARTISANS **For Exercises 38 and 39, use the following information.**
This geometric sign from the Grassfields area in Western Cameroon (Western Africa) uses approximations of isosceles triangles within and around two circles. **38–39. See p. 233G.**

38. Trace the figure. Identify and draw one isosceles triangle from each set in the sign.
39. Describe the similarities between the different triangles.

40. **CRITICAL THINKING** In the figure, $\triangle ABC$ is isosceles, $\triangle DCE$ is equilateral, and $\triangle FCG$ is isosceles. Find the measures of the five numbered angles at vertex C. **$m\angle 1 = 18$, $m\angle 2 = 17$, $m\angle 3 = 26$, $m\angle 4 = 17$, $m\angle 5 = 18$**

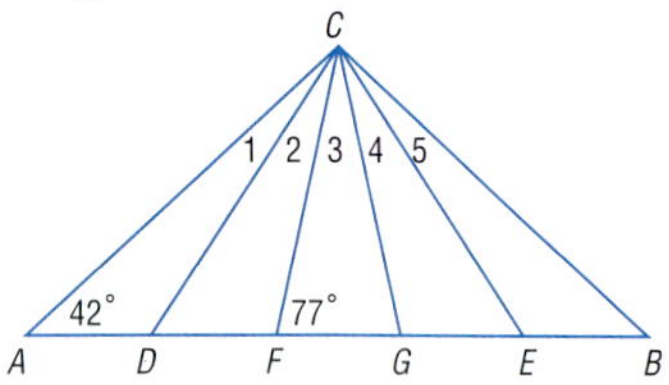

Enrichment, p. 218

Triangle Challenges

Some problems include diagrams. If you are not sure how to solve the problem, begin by using the given information. Find the measures of as many angles as you can, writing each measure on the diagram. This may give you more clues to the solution.

1. Given: $BE = BF$, $\angle BFG \cong \angle BEF \cong \angle BED$, $m\angle BFE = 82$ and $ABFG$ and $BCDE$ each have opposite sides parallel and congruent.
Find $m\angle ABC$. 148

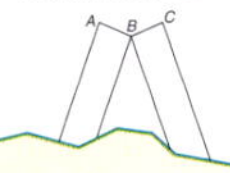

2. Given: $AC = AD$, and $\overline{AB} \perp \overline{BD}$, $m\angle DAC = 44$ and $\overline{CE}$ bisects $\angle ACD$.
Find $m\angle DEC$. 78

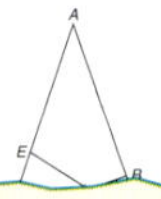

41. **WRITING IN MATH** Answer the question that was posed at the beginning of the lesson. See margin.

How are triangles used in art?

Include the following in your answer:

- at least three other geometric shapes frequently used in art, and
- a description of how isosceles triangles are used in the painting.

FCAT Practice

Standardized Test Practice

42. Given right triangle XYZ with hypotenuse $\overline{XY}$, YP is equal to YZ. If $m\angle PYZ = 26$, find $m\angle XZP$. A

Ⓐ 13   Ⓑ 26   Ⓒ 32   Ⓓ 64

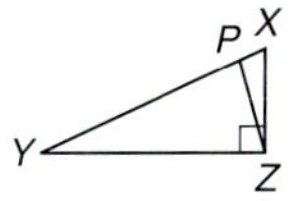

43. **ALGEBRA** A segment is drawn from (3, 5) to (9, 13). What are the coordinates of the midpoint of this segment? D

Ⓐ (3, 4)   Ⓑ (12, 18)   Ⓒ (6, 8)   Ⓓ (6, 9)

Maintain Your Skills

Mixed Review **Write a paragraph proof.** *(Lesson 4-5)* 44–45. See p. 233G.

44. **Given:** $\angle N \cong \angle D$, $\angle G \cong \angle I$, $\overline{AN} \cong \overline{SD}$
Prove: $\triangle ANG \cong \triangle SDI$

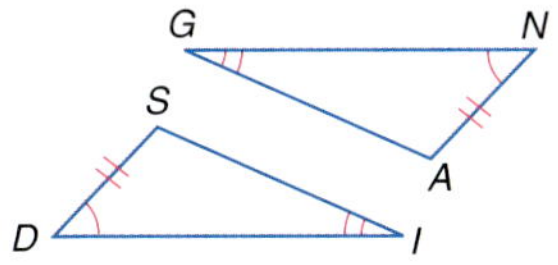

45. **Given:** $\overline{VR} \perp \overline{RS}$, $\overline{UT} \perp \overline{SU}$, $\overline{RS} \cong \overline{US}$
Prove: $\triangle VRS \cong \triangle TUS$

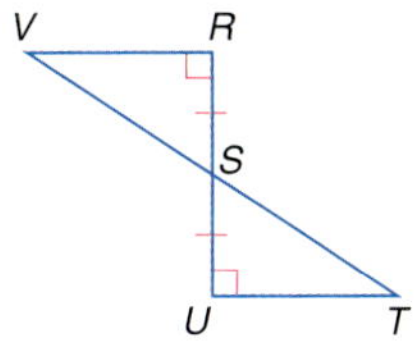

Determine whether $\triangle QRS \cong \triangle EGH$ given the coordinates of the vertices. Explain. *(Lesson 4-4)* 46–47. See margin.

46. $Q(-3, 1)$, $R(1, 2)$, $S(-1, -2)$, $E(6, -2)$, $G(2, -3)$, $H(4, 1)$

47. $Q(1, -5)$, $R(5, 1)$, $S(4, 0)$, $E(-4, -3)$, $G(-1, 2)$, $H(2, 1)$

Construct a truth table for each compound statement. *(Lesson 2-2)*

48–51. See p. 233G.

48. a and b 49. $\sim p$ or $\sim q$ 50. k and $\sim m$ 51. $\sim y$ or z

Getting Ready for the Next Lesson

PREREQUISITE SKILL **Find the coordinates of the midpoint of the segment with the given endpoints.** *(To review **finding midpoints**, see Lesson 1-5.)*

52. $A(2, 15)$, $B(7, 9)$ (4.5, 12)   53. $C(-4, 6)$, $D(2, -12)$ (−1, −3)   54. $E(3, 2.5)$, $F(7.5, 4)$ (5.25, 3.25)

Practice Quiz 2

Lessons 4-4 through 4-6

1. Determine whether $\triangle JML \cong \triangle BDG$ given that $J(-4, 5)$, $M(-2, 6)$, $L(-1, 1)$, $B(-3, -4)$, $D(-4, -2)$, and $G(1, -1)$. *(Lesson 4-4)* See p. 233G.

2. Write a two-column proof to prove that $\overline{AJ} \cong \overline{EH}$, given $\angle A \cong \angle H$, $\angle AEJ \cong \angle HJE$. *(Lesson 4-5)* See p. 233G.

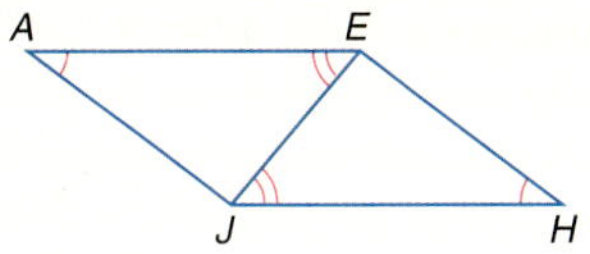

$\triangle WXY$ and $\triangle XYZ$ are isosceles and $m\angle XYZ = 128$. Find each measure. *(Lesson 4-6)*

3. $m\angle XWY$ 52   4. $m\angle WXY$ 76   5. $m\angle YZX$ 26

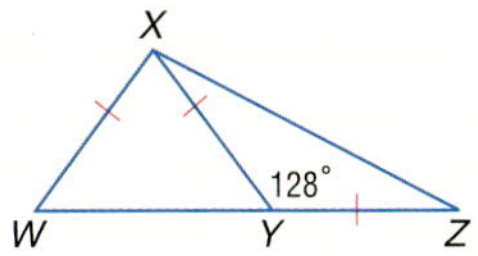

4 Assess

Open-Ended Assessment

Speaking Have students come up with examples of how isosceles and equilateral triangles are used in paintings, ceramics, and decorative architecture. Students can talk about where the base and legs of isosceles triangles are typically situated in architecture, and they can discuss how geometry can help builders determine how much material they will need. They can also discuss the visual effect that isosceles and equilateral triangles have in different art forms.

Getting Ready for Lesson 4-7

Prerequisite Skill Students will learn about triangles and coordinate proof in Lesson 4-7. They will prove congruency using distances and midpoints in coordinate planes. Use Exercises 52–54 to determine your students' familiarity with finding midpoints.

Assessment Options

Practice Quiz 2 The quiz provides students with a brief review of the concepts and skills in Lessons 4-4 through 4-6. Lesson numbers are given to the right of the exercises or instruction lines so students can review concepts not yet mastered.

Quiz (Lessons 4-5 and 4-6) is available on p. 240 of the *Chapter 4 Resource Masters*.

Answer

41. Sample answer: Artists use angles, lines and shapes to create visual images. Answers should include the following.

- Rectangle, squares, rhombi, and other polygons are used in many works of art.
- There are two rows of isosceles triangles in the painting. One row has three congruent isosceles triangles. The other row has six congruent isosceles triangles.

46. $QR = \sqrt{17}$, $RS = \sqrt{20}$, $QS = \sqrt{13}$, $EG = \sqrt{17}$, $GH = \sqrt{20}$, and $EH = \sqrt{13}$. Each pair of corresponding sides have the same measure so they are congruent. $\triangle QRS \cong \triangle EGH$ by SSS.

47. $QR = \sqrt{52}$, $RS = \sqrt{2}$, $QS = \sqrt{34}$, $EG = \sqrt{34}$, $GH = \sqrt{10}$, and $EH = \sqrt{52}$. The corresponding sides are not congruent so $\triangle QRS$ is not congruent to $\triangle EGH$.

4-7 Lesson Notes

1 Focus

5-Minute Check Transparency 4-7 Use as a quiz or review of Lesson 4-6.

Mathematical Background notes are available for this lesson on p. 176D.

How can the coordinate plane be useful in proofs?

Ask students:

- If $d = 3$ and $BC = 5$, what are the coordinates of B and C? Classify $\triangle ABC$. **(6, 0); (3, 4); isosceles**
- If you draw $\overline{CD}$ where D is the midpoint of $\overline{AB}$, what method(s) could you use to prove $\triangle ACD \cong \triangle BCD$? **the Distance Formula and SSS**

2 Teach

POSITION AND LABEL TRIANGLES

In-Class Example PowerPoint®

1 Position and label right triangle XYZ with leg $\overline{XZ}$ d units long on the coordinate plane.

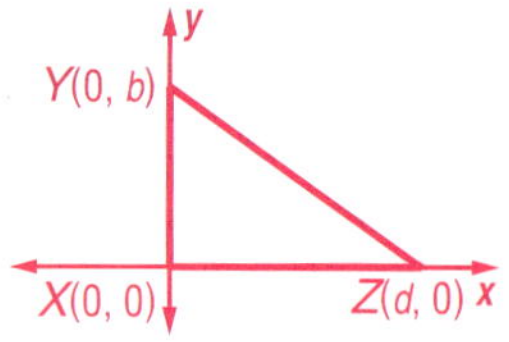

4-7 Triangles and Coordinate Proof

Sunshine State Standards
MA.B.3.4.1, MA.C.1.4.1, MA.C.3.4.2

Vocabulary
- coordinate proof

What You'll Learn

- Position and label triangles for use in coordinate proofs.
- Write coordinate proofs.

How can the coordinate plane be useful in proofs?

In this chapter, we have used several methods of proof. You have also used the coordinate plane to identify characteristics of a triangle. We can combine what we know about triangles in the coordinate plane with algebra in a new method of proof called *coordinate proof.*

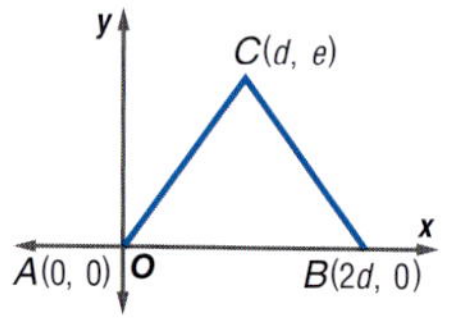

POSITION AND LABEL TRIANGLES

Coordinate proof uses figures in the coordinate plane and algebra to prove geometric concepts. The first step in writing a coordinate proof is the placement of the figure on the coordinate plane.

Study Tip

Placement of Figures
The guidelines apply to any polygon placed on the coordinate plane.

Key Concept *Placing Figures on the Coordinate Plane*

1. Use the origin as a vertex or center of the figure.
2. Place at least one side of a polygon on an axis.
3. Keep the figure within the first quadrant if possible.
4. Use coordinates that make computations as simple as possible.

Example 1 Position and Label a Triangle

Position and label isosceles triangle JKL on a coordinate plane so that base $\overline{JK}$ is a units long.

- Use the origin as vertex J of the triangle.
- Place the base of the triangle along the positive x-axis.
- Position the triangle in the first quadrant.
- Since K is on the x-axis, its y-coordinate is 0. Its x-coordinate is a because the base of the triangle is a units long.
- Since $\triangle JKL$ is isosceles, the x-coordinate of L is halfway between 0 and a or $\frac{a}{2}$. We cannot determine the y-coordinate in terms of a, so call it b.

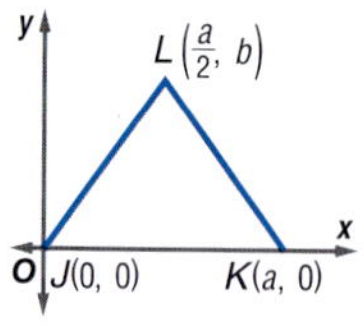

Example 2 Find the Missing Coordinates

Name the missing coordinates of isosceles right $\triangle EFG$.

Vertex F is positioned at the origin; its coordinates are (0, 0). Vertex E is on the y-axis, and vertex G is on the x-axis. So $\angle EFG$ is a right angle. Since $\triangle EFG$ is isosceles, $\overline{EF} \cong \overline{GF}$. The distance from E to F is a units. The distance from F to G must be the same. So, the coordinates of G are $(a, 0)$.

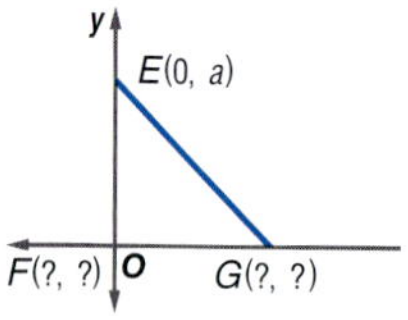

Resource Manager

Workbook and Reproducible Masters

Chapter 4 Resource Masters
- Study Guide and Intervention, pp. 219–220
- Skills Practice, p. 221
- Practice, p. 222
- Reading to Learn Mathematics, p. 223
- Enrichment, p. 224
- Assessment, p. 240

Teaching Geometry With Manipulatives Masters, pp. 1, 2, 8

Transparencies
5-Minute Check Transparency 4-7
Answer Key Transparencies

Technology
Interactive Chalkboard

WRITE COORDINATE PROOFS After the figure has been placed on the coordinate plane and labeled, we can use coordinate proof to verify properties and to prove theorems. The Distance Formula, Slope Formula, and Midpoint Formula are often used in coordinate proof.

Example 3 Coordinate Proof

Write a coordinate proof to prove that the measure of the segment that joins the vertex of the right angle in a right triangle to the midpoint of the hypotenuse is one-half the measure of the hypotenuse.

The first step is to position and label a right triangle on the coordinate plane. Place the right angle at the origin and label it A. Use coordinates that are multiples of 2 because the Midpoint Formula takes half the sum of the coordinates.

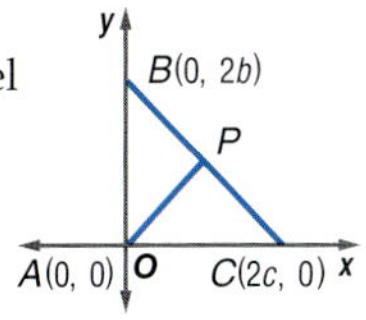

Given: right $\triangle ABC$ with right $\angle BAC$
P is the midpoint of $\overline{BC}$.

Prove: $AP = \frac{1}{2}BC$

Proof:

By the Midpoint Formula, the coordinates of P are $\left(\frac{0 + 2c}{2}, \frac{2b + 0}{2}\right)$ or (c, b).

Use the Distance Formula to find AP and BC.

$$AP = \sqrt{(c - 0)^2 + (b - 0)^2} = \sqrt{c^2 + b^2}$$

$$BC = \sqrt{(2c - 0)^2 + (0 - 2b)^2}$$

$$BC = \sqrt{4c^2 + 4b^2} \text{ or } 2\sqrt{c^2 + b^2}$$

$$\frac{1}{2}BC = \sqrt{c^2 + b^2}$$

Therefore, $AP = \frac{1}{2}BC$.

Example 4 Classify Triangles

ARROWHEADS **Write a coordinate proof to prove that this arrowhead is shaped like an isosceles triangle. The arrowhead is 3 inches long and 1.5 inches wide.**

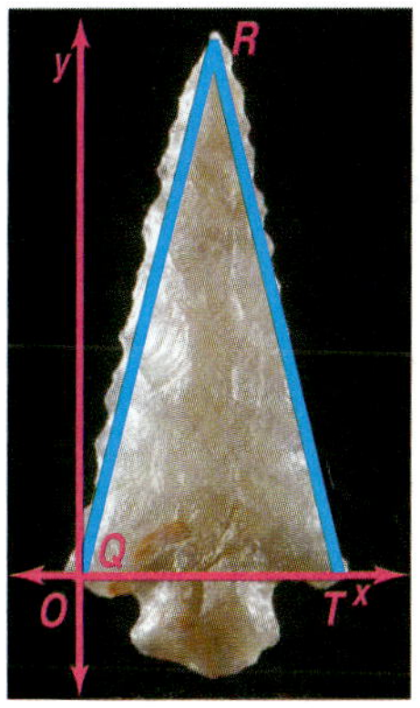

The first step is to label the coordinates of each vertex. Q is at the origin, and T is at (1.5, 0). The y-coordinate of R is 3. The x-coordinate is halfway between 0 and 1.5 or 0.75. So, the coordinates of R are (0.75, 3).

If the legs of the triangle are the same length, the triangle is isosceles. Use the Distance Formula to determine the lengths of QR and RT.

$$QR = \sqrt{(0.75 - 0)^2 + (3 - 0)^2} = \sqrt{0.5625 + 9} \text{ or } \sqrt{9.5625}$$

$$RT = \sqrt{(1.5 - 0.75)^2 + (0 - 3)^2} = \sqrt{0.5625 + 9} \text{ or } \sqrt{9.5625}$$

Since each leg is the same length, $\triangle QRT$ is isosceles. The arrowhead is shaped like an isosceles triangle.

Study Tip

Vertex Angle
Remember from the Geometry Activity on page 216 that an isosceles triangle can be folded in half. Thus, the x-coordinate of the vertex angle is the same as the x-coordinate of the midpoint of the base.

www.geometryonline.com/extra_examples/fcat

DAILY INTERVENTION **Differentiated Instruction**

Kinesthetic You can mark a coordinate plane on a corkboard. You can demonstrate for students, or volunteers can practice placing different figures on the coordinate plane using pushpins for vertices and string for sides.

In-Class Example Power Point®

2 Name the missing coordinates of isosceles right $\triangle QRS$.

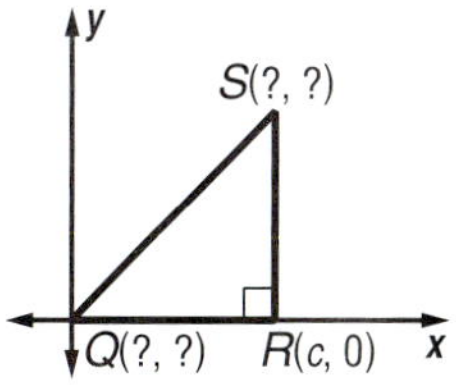

$Q(0, 0)$; $S(c, c)$

WRITE COORDINATE PROOFS

In-Class Examples Power Point®

3 Write a coordinate proof to prove that the segment that joins the vertex angle of an isosceles triangle to the midpoint of its base is perpendicular to the base.

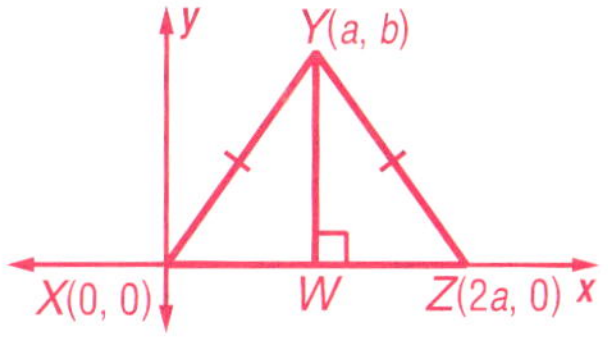

Midpoint of $\overline{XZ}$ is $(a, 0)$. Slope of $\overline{YW}$ is undefined. Slope of $\overline{XZ}$ is 0. So, $\overline{YW} \perp \overline{XZ}$.

Teaching Tip Advise students that they may want to place a figure using numeric coordinates first and then translate to variable coordinates to write their proofs.

4 **DRAFTING** Write a coordinate proof to prove that this drafter's tool is shaped like a right triangle. The length of one side is 10 inches and the length of another side is 5.75 inches.

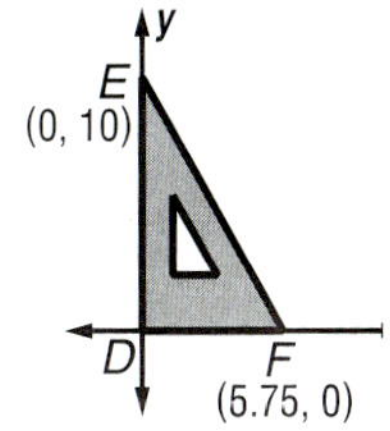

Slope of $\overline{ED}$ is undefined. Slope of $\overline{DF}$ is 0. $\overline{ED} \perp \overline{DF}$. $\triangle DEF$ is a right triangle. The drafter's tool is shaped like a right triangle.

Concept Check

Groups of students can work together writing coordinate proofs with one student placing the triangle, another student labeling the coordinates, and so on. Tell groups to discuss labeling variables with multiples of two to make midpoint computation easier and using points on the coordinate axes as much as possible to simplify formulas. Groups can also discuss limitations of coordinate proofs such as the difficulty of placing and labeling an equilateral triangle, and so on.

3 Practice/Apply

Study Notebook

Have students—

- *add the definitions/examples of the vocabulary terms to their Vocabulary Builder worksheets for Chapter 4.*
- *include any other item(s) that they find helpful in mastering the skills in this lesson.*

About the Exercises...

Organization by Objective

- **Position and Label Triangles:** 10–24
- **Write Coordinate Proofs:** 25–33

Odd/Even Assignments

Exercises 10–33 are structured so that students practice the same concepts whether they are assigned odd or even problems.

Assignment Guide

Basic: 11–37 odd, 38–47

Average: 11–37 odd, 38–47

Advanced: 10–36 even, 37–47

Check for Understanding

Concept Check

1. **Explain** how to position a triangle on the coordinate plane to simplify a proof.
2. **OPEN ENDED** Draw a scalene right triangle on the coordinate plane for use in a coordinate proof. Label the coordinates of each vertex. **1–2. See margin.**

Guided Practice

GUIDED PRACTICE KEY	
Exercises	Examples
3–4	1
5–7	2
8	3
9	4

Position and label each triangle on the coordinate plane. **3–4. See margin.**

3. isosceles $\triangle FGH$ with base $\overline{FH}$ that is $2b$ units long
4. equilateral $\triangle CDE$ with sides a units long

Find the missing coordinates of each triangle.

5. **P(0, b)**

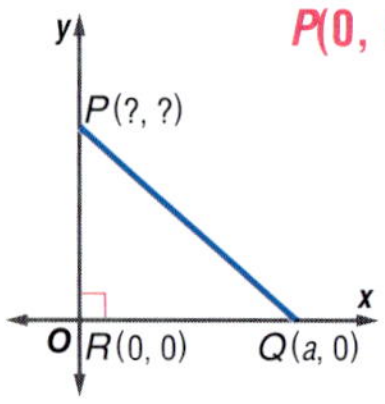

6. **Q(−2a, 0)**

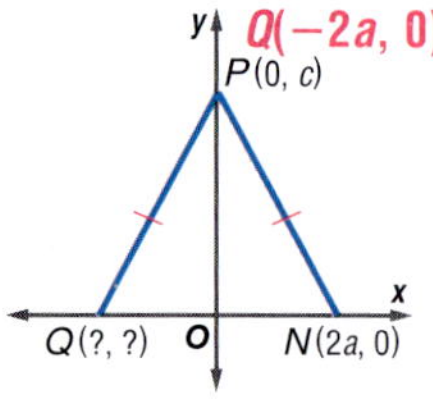

7. **N(0, b), Q(a, 0)**

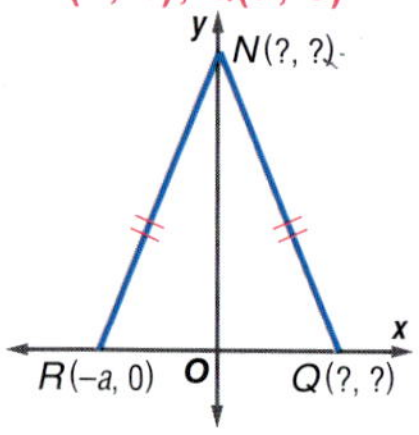

8. Write a coordinate proof for the following statement.
The midpoint of the hypotenuse of a right triangle is equidistant from each of the vertices. **See p. 233H.**

Application

9. **TEPEES** Write a coordinate proof to prove that the tepee is shaped like an isosceles triangle. Suppose the tepee is 8 feet tall and 4 feet wide. **See p. 233H.**

Practice and Apply

Homework Help	
For Exercises	See Examples
10–15	1
16–24	2
25–29	3
30–33	4

Extra Practice
See page 762.

Position and label each triangle on the coordinate plane.

10. isosceles $\triangle QRT$ with base $\overline{QR}$ that is b units long **10–15. See p. 233H.**
11. equilateral $\triangle MNP$ with sides $2a$ units long
12. isosceles right $\triangle JML$ with hypotenuse $\overline{JM}$ and legs c units long
13. equilateral $\triangle WXZ$ with sides $\frac{1}{2}b$ units long
14. isosceles $\triangle PWY$ with a base $\overline{PW}$ that is $(a + b)$ units long
15. right $\triangle XYZ$ with hypotenuse $\overline{XZ}$, $ZY = 2(XY)$, and $\overline{XY}$ b units long

Find the missing coordinates of each triangle.

16.

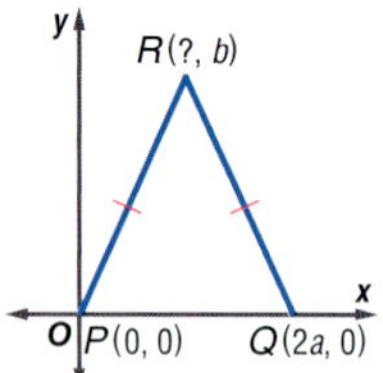

R(a, b)

17.

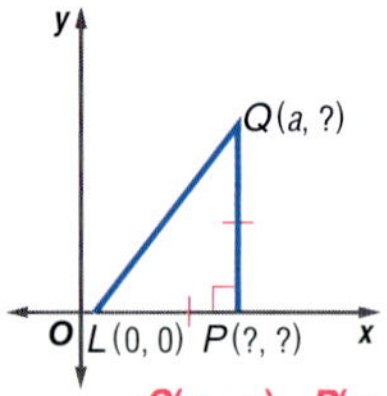

Q(a, a), P(a, 0)

18.

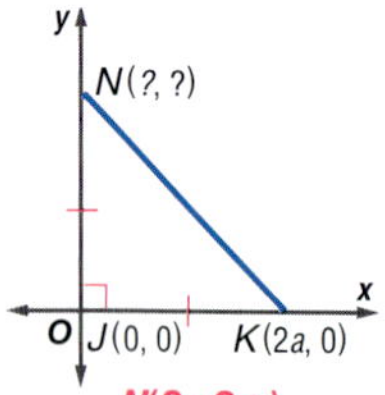

N(0, 2a)

Answers

1. Place one vertex at the origin, place one side of the triangle on the positive x-axis. Label the coordinates with expressions that will simplify the computations.

2. Sample answer:

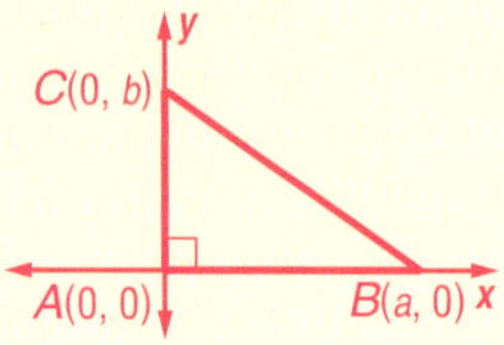

19. 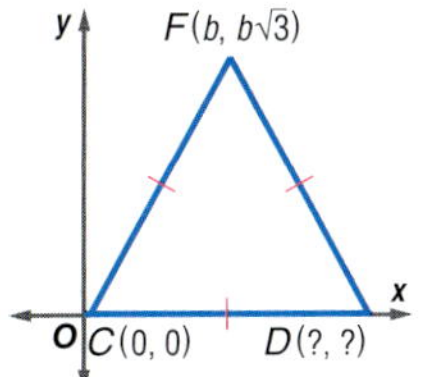

$D(2b, 0)$

20. 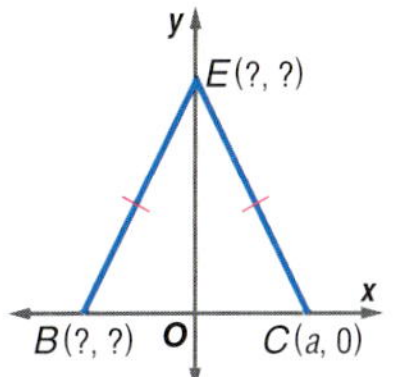

$B(-a, 0)$, $E(0, b)$

21. 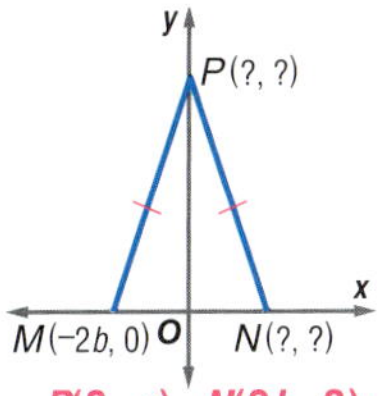

$P(0, c)$, $N(2b, 0)$

22. 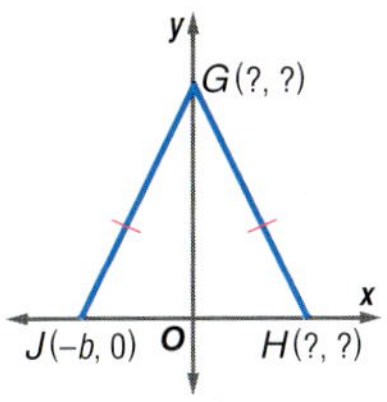

$G(0, c)$, $H(b, 0)$

23. 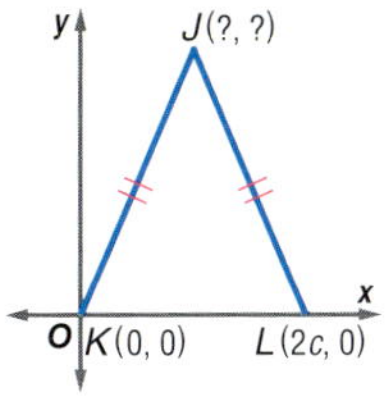

$J(c, b)$

24. 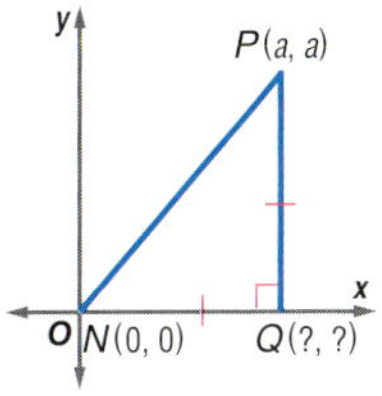

$Q(a, 0)$

More About. . .

Steeplechase

The Steeplechase is a horse race two to four miles long that focuses on jumping hurdles. The rails of the fences vary in height.

Source: www.steeplechasetimes.com

Write a coordinate proof for each statement. 25–28. See pp. 233H–233I.

25. The segments joining the vertices to the midpoints of the legs of an isosceles triangle are congruent.

26. The three segments joining the midpoints of the sides of an isosceles triangle form another isosceles triangle.

27. If a line segment joins the midpoints of two sides of a triangle, then it is parallel to the third side.

28. If a line segment joins the midpoints of two sides of a triangle, then its length is equal to one-half the length of the third side.

29. **STEEPLECHASE** Write a coordinate proof to prove that triangles ABD and FBD are congruent. $\overline{BD}$ is perpendicular to $\overline{AF}$, and B is the midpoint of the upper bar of the hurdle. See p. 233I.

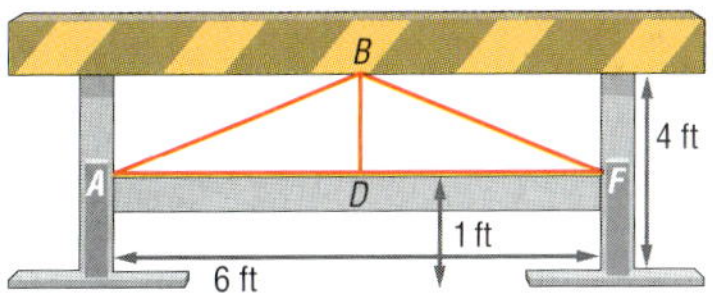

NAVIGATION **For Exercises 30 and 31, use the following information.**
A motor boat is located 800 yards east of the port. There is a ship 800 yards to the east, and another ship 800 yards to the north of the motor boat.

30. Write a coordinate proof to prove that the port, motor boat, and the ship to the north form an isosceles right triangle.

31. Write a coordinate proof to prove that the distance between the two ships is the same as the distance from the port to the northern ship. 30–31. See p. 233I.

HIKING **For Exercises 32 and 33, use the following information.**
Tami and Juan are hiking. Tami hikes 300 feet east of the camp and then hikes 500 feet north. Juan hikes 500 feet west of the camp and then 300 feet north.

32. Write a coordinate proof to prove that Juan, Tami, and the camp form a right triangle. See p. 233I.

33. Find the distance between Tami and Juan. $\sqrt{680{,}000}$ or about 824.6 ft

www.geometryonline.com/self_check_quiz/fcat

Answers

3. Sample answer:

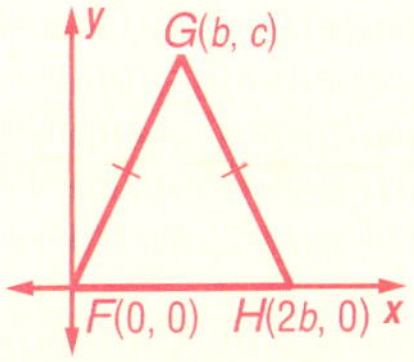

4. Sample answer:

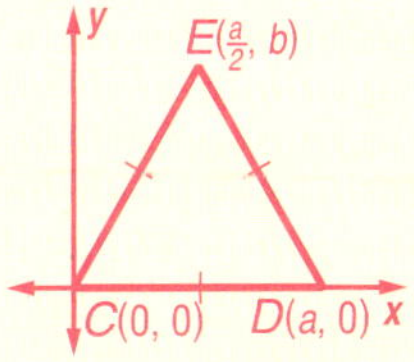

Enrichment, p. 224

How Many Triangles?

Each puzzle below contains many triangles. Count them carefully. Some triangles overlap other triangles.

How many triangles are there in each figure?

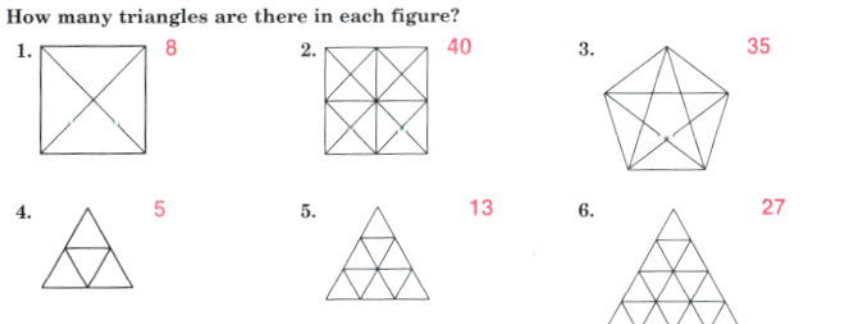

1. 8
2. 40
3. 35
4. 5
5. 13
6. 27

Study Guide and Intervention, p. 219 (shown) and p. 220

Position and Label Triangles A coordinate proof uses points, distances, and slopes to prove geometric properties. The first step in writing a coordinate proof is to place a figure on the coordinate plane and label the vertices. Use the following guidelines.

1. Use the origin as a vertex or center of the figure.
2. Place at least one side of the polygon on an axis.
3. Keep the figure in the first quadrant if possible.
4. Use coordinates that make the computations as simple as possible.

Example **Position an equilateral triangle on the coordinate plane so that its sides are a units long and one side is on the positive x-axis.**
Start with $R(0, 0)$. If RT is a, then another vertex is $T(a, 0)$.
For vertex S, the x-coordinate is $\frac{a}{2}$. Use b for the y-coordinate, so the vertex is $S\left(\frac{a}{2}, b\right)$.

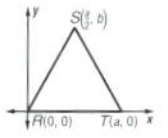

Exercises

Find the missing coordinates of each triangle.

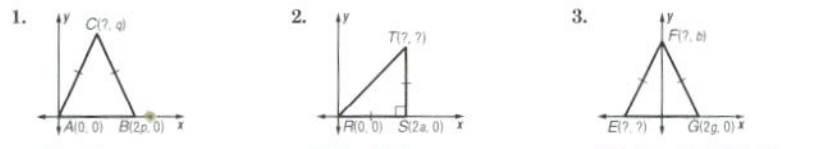

1. $C(p, q)$
2. $T(2a, 2a)$
3. $E(-2g, 0)$; $F(0, b)$

Position and label each triangle on the coordinate plane. Sample answers are given.

4. isosceles triangle $\triangle RST$ with base $\overline{RS}$ $4a$ units long
5. isosceles right $\triangle DEF$ with legs e units long
6. equilateral triangle $\triangle EQI$ with vertex $Q(0, a)$ and sides $2b$ units long

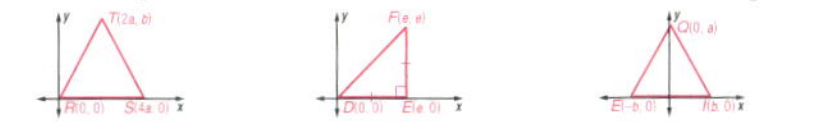

Skills Practice, p. 221 and Practice, p. 222 (shown)

Position and label each triangle on the coordinate plane. Sample answers are given.

1. equilateral $\triangle SWY$ with sides $\frac{1}{4}a$ long
2. isosceles $\triangle BLP$ with base $\overline{BL}$ $3b$ units long
3. isosceles right $\triangle DGJ$ with hypotenuse $\overline{DJ}$ and legs $2a$ units long

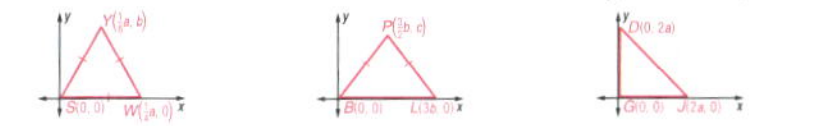

Find the missing coordinates of each triangle.

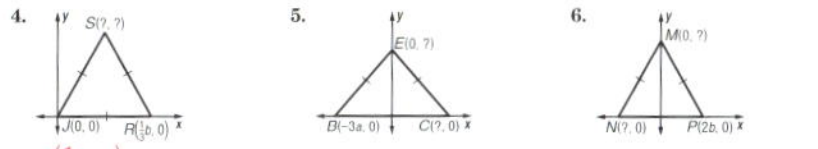

4. $S\left(\frac{1}{6}b, c\right)$
5. $C(3a, 0)$, $E(0, c)$
6. $M(0, c)$, $N(-2b, 0)$

NEIGHBORHOODS For Exercises 7 and 8, use the following information.
Karina lives 6 miles east and 4 miles north of her high school. After school she works part time at the mall in a music store. The mall is 2 miles west and 3 miles north of the school.

7. Write a coordinate proof to prove that Karina's high school, her home, and the mall are at the vertices of a right triangle.

Given: $\triangle SKM$
Prove: $\triangle SKM$ is a right triangle.

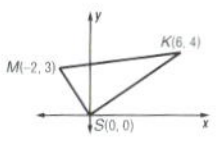

Proof:

Slope of $SK = \frac{4-0}{6-0}$ or $\frac{2}{3}$

Slope of $SM = \frac{3-0}{-2-0}$ or $-\frac{3}{2}$

Since the slope of $\overline{SM}$ is the negative reciprocal of the slope of $\overline{SK}$, $\overline{SM} \perp \overline{SK}$. Therefore, $\triangle SKM$ is right triangle.

8. Find the distance between the mall and Karina's home.

$KM = \sqrt{(-2-6)^2 + (3-4)^2} = \sqrt{64+1} = \sqrt{65}$ or ≈ 8.1 miles

Reading to Learn Mathematics, p. 223 ELL

Pre-Activity **How can the coordinate plane be useful in proofs?**

Read the introduction to Lesson 4-7 at the top of page 222 in your textbook.

From the coordinates of A, B, and C in the drawing in your textbook, what do you know about $\triangle ABC$? Sample answer: $\triangle ABC$ is isosceles with $\angle C$ as the vertex angle.

Reading the Lesson

1. Find the missing coordinates of each triangle.

a. $R(0, b)$, $S(0, 0)$, $T\left(a, \frac{b}{2}\right)$

b. $D(0, 0)$, $E(0, a)$, $F(a, a)$

2. Refer to the figure.

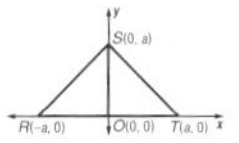

a. Find the slope of $\overline{SR}$ and the slope of $\overline{ST}$. 1; −1
b. Find the product of the slopes of $\overline{SR}$ and $\overline{ST}$. What does this tell you about $\overline{SR}$ and $\overline{ST}$? −1; $\overline{SR} \perp \overline{ST}$
c. What does your answer from part b tell you about $\triangle RST$? Sample answer: $\triangle RST$ is a right triangle with $\angle S$ as the right angle.
d. Find SR and ST. What does this tell you about $\overline{SR}$ and $\overline{ST}$? $SR = \sqrt{2a^2}$ or $a\sqrt{2}$; $ST = \sqrt{2a^2}$ or $a\sqrt{2}$; $\overline{SR} \cong \overline{ST}$
e. What does your answer from part d tell you about $\triangle RST$? Sample answer: $\triangle RST$ is isosceles with $\angle RST$ as the vertex angle.
f. Combine your answers from parts c and e to describe $\triangle RST$ as completely as possible. Sample answer: $\triangle RST$ is an isosceles right triangle. $\angle RST$ is the right angle and is also the vertex angle.
g. Find $m\angle SRT$ and $m\angle STR$. 45; 45
h. Find $m\angle OSR$ and $m\angle OST$. 45; 45

Helping You Remember

3. Many students find it easier to remember mathematical formulas if they can put them into words in a compact way. How can you use this approach to remember the slope and midpoint formulas easily?
Sample answer: Slope Formula: change in y over change in x; Midpoint Formula: average of x-coordinates, average of y-coordinates

4 Assess

Open-Ended Assessment

Speaking Have students speak about how they would place certain figures in a coordinate plane and how they would label the vertices. Students can discuss different ideas about placement and how they can simplify coordinate proofs by using the origin and simple labeling techniques.

Assessment Options

Quiz (Lesson 4-7) is available on p. 240 of the *Chapter 4 Resource Masters*.

Answers

37. $AB = 4a$

$AC = \sqrt{(0 - (-2a))^2 + (2a - 0)^2}$

$= \sqrt{4a^2 + 4a^2}$ or $\sqrt{8a^2}$

$CB = \sqrt{(0 - 2a)^2 + (2a - 0)^2}$

$= \sqrt{4a^2 + 4a^2}$ or $\sqrt{8a^2}$

Slope of $\overline{AC} = \dfrac{2a - 0}{0 - (-2a)}$ or 1;

slope of $\overline{CB} = \dfrac{2a - 0}{0 - 2a}$ or −1.

$\overline{AB} \perp \overline{CB}$ and $\overline{AC} \cong \overline{CB}$, so $\triangle ABC$ is a right isosceles triangle.

38. Sample answer: Placing the figures on the coordinate plane is useful in proofs. We can use coordinate geometry to prove theorems and verify properties. Answers should include the following.
- flow proof, two-column proofs, paragraph proofs, informal proofs, and coordinate proofs
- Sample answer: The Isosceles Triangle Theorem can be proved using coordinate proof.

34. $(a, 0)$ or $(0, b)$
35. $(2a, 0)$

Find the coordinates of point Z so $\triangle XYZ$ is the indicated type of triangle. Point X has coordinates (0, 0) and Y has coordinates (a, b).

34. right triangle with right angle Z

35. isosceles triangle with base $\overline{XZ}$

36. scalene triangle Sample answer: $(c, 0)$

37. CRITICAL THINKING Classify $\triangle ABC$ by its angles and its sides. Explain. See margin.

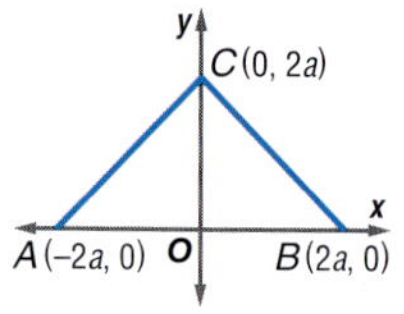

38. WRITING IN MATH Answer the question that was posed at the beginning of the lesson. See margin.

How can the coordinate plane be useful in proofs?

Include the following in your answer:
- types of proof, and
- a theorem from this chapter that could be proved using a coordinate proof.

FCAT Practice
Standardized Test Practice

39. What is the length of the segment whose endpoints are at (1, −2) and (−3, 1)? C

(A) 3 (B) 4 (C) 5 (D) 6

40. ALGEBRA What are the coordinates of the midpoint of the line segment whose endpoints are (−5, 4) and (−2, −1)? B

(A) (3, 3) (B) (−3.5, 1.5) (C) (−1.5, 2.5) (D) (3.5, −2.5)

Maintain Your Skills

Mixed Review **Write a two-column proof.** *(Lessons 4-5 and 4-6)* 41–44. See pp. 233I–233J.

41. Given: $\angle 3 \cong \angle 4$
Prove: $\overline{QR} \cong \overline{QS}$

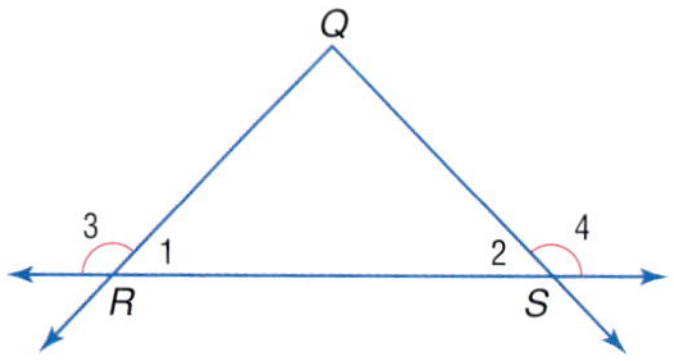

42. Given: isosceles triangle JKN with vertex $\angle N$, $\overline{JK} \parallel \overline{LM}$
Prove: $\triangle NML$ is isosceles.

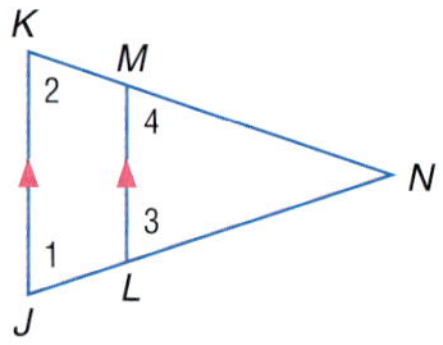

43. Given: $\overline{AD} \cong \overline{CE}$, $\overline{AD} \parallel \overline{CE}$
Prove: $\triangle ABD \cong \triangle EBC$

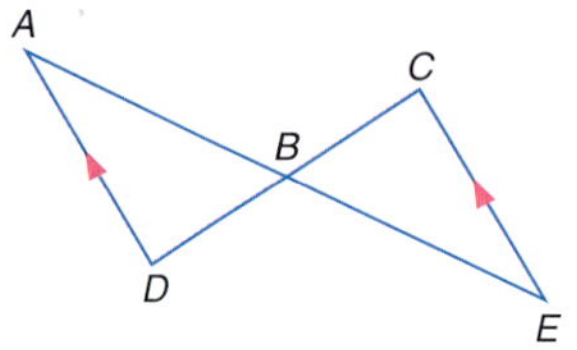

44. Given: $\overline{WX} \cong \overline{XY}$, $\angle V \cong \angle Z$
Prove: $\overline{WV} \cong \overline{YZ}$

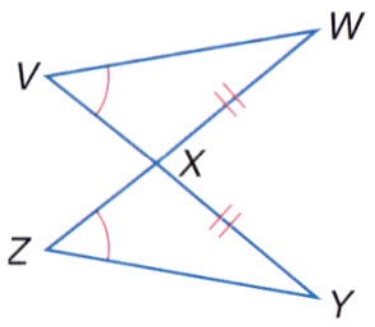

State which lines, if any, are parallel. State the postulate or theorem that justifies your answer. *(Lesson 3-5)*

45. $\overline{BC} \parallel \overline{AD}$; if alt. int. ∡ are ≅, lines are ∥
46. $h \parallel j$; Sample answer: cons. int. ∡ suppl.
47. $\ell \parallel m$; 2 lines ⊥ to same line are ∥.

45.

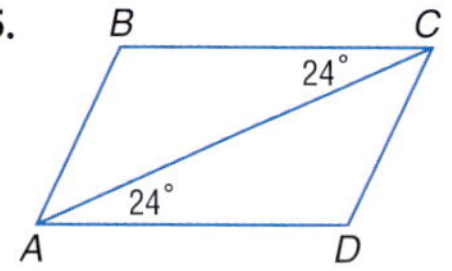

46.

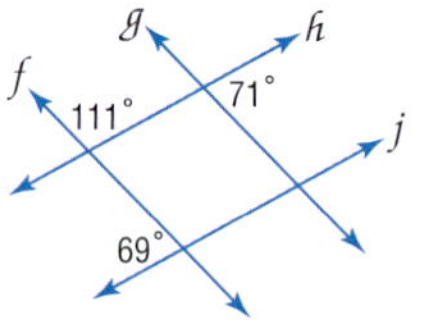

47.

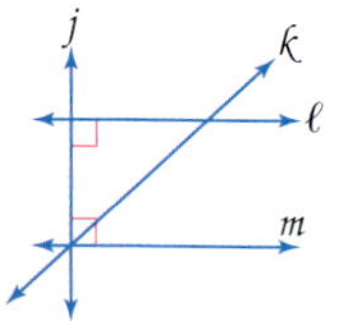

apter 4

Study Guide and Review

ocabulary and Concept Check

acute triangle (p. 178)	equiangular triangle (p. 178)	isosceles triangle (p. 179)
base angles (p. 216)	equilateral triangle (p. 179)	obtuse triangle (p. 178)
congruence transformations (p. 194)	exterior angle (p. 186)	remote interior angles (p. 186)
congruent triangles (p. 192)	flow proof (p. 187)	right triangle (p. 178)
coordinate proof (p. 222)	included angle (p. 201)	scalene triangle (p. 179)
corollary (p. 188)	included side (p. 207)	vertex angle (p. 216)

A complete list of theorems and postulates can be found on pages R1–R8.

Exercises **Choose the letter of the word or phrase that best matches each statement.**

1. A triangle with an angle whose measure is greater than 90 is a(n) __?__ triangle. h
2. A triangle with exactly two congruent sides is a(n) __?__ triangle. g
3. The __?__ states that the sum of the measures of the angles of a triangle is 180. d
4. If $\angle B \cong \angle E$, $\overline{AB} \cong \overline{DE}$, and $\overline{BC} \cong \overline{EF}$, then $\triangle ABC \cong \triangle DEF$ by __?__. j
5. In an equiangular triangle, all angles are __?__ angles. a
6. If two angles of a triangle and their included side are congruent to two angles and the included side of another triangle, this is called the __?__. c
7. If $\angle A \cong \angle F$, $\angle B \cong \angle G$, and $\overline{AC} \cong \overline{FH}$, then $\triangle ABC \cong \triangle FGH$, by __?__. b
8. A(n) __?__ angle of a triangle has a measure equal to the measures of the two remote interior angles of the triangle. f

a. acute
b. AAS Theorem
c. ASA Theorem
d. Angle Sum Theorem
e. equilateral
f. exterior
g. isosceles
h. obtuse
i. right
j. SAS Theorem
k. SSS Theorem

esson-by-Lesson Review

4-1 Classifying Triangles

See pages 178–183.

Concept Summary

- Triangles can be classified by their angles as acute, obtuse, or right.
- Triangles can be classified by their sides as scalene, isosceles, or equilateral.

Example **Find the measures of the sides of $\triangle TUV$. Classify the triangle by sides.**

Use the Distance Formula to find the measure of each side.

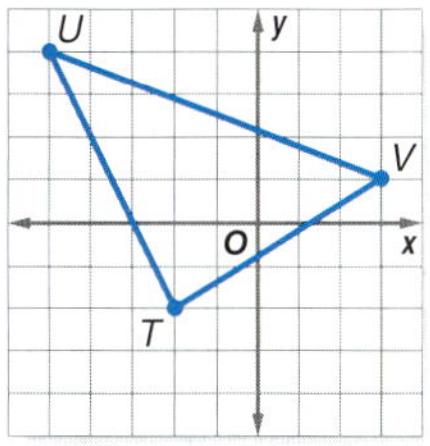

$$TU = \sqrt{[-5-(-2)]^2 + [4-(-2)]^2} = \sqrt{9+36} \text{ or } \sqrt{45}$$

$$UV = \sqrt{[3-(-5)]^2 + (1-4)^2} = \sqrt{64+9} \text{ or } \sqrt{73}$$

$$VT = \sqrt{(-2-3)^2 + (-2-1)^2} = \sqrt{25+9} \text{ or } \sqrt{34}$$

Since none of the side measures are equal, $\triangle TUV$ is scalene.

www.geometryonline.com/vocabulary_review

Chapter 4

Study Guide and Review

Vocabulary and Concept Check

- This alphabetical list of vocabulary terms in Chapter 4 includes a page reference where each term was introduced.
- **Assessment** A vocabulary test/review for Chapter 4 is available on p. 238 of the *Chapter 4 Resource Masters*.

Lesson-by-Lesson Review

For each lesson,

- the main ideas are summarized,
- additional examples review concepts, and
- practice exercises are provided.

Vocabulary PuzzleMaker

ELL The Vocabulary PuzzleMaker software improves students' mathematics vocabulary using four puzzle formats—crossword, scramble, word search using a word list, and word search using clues. Students can work on a computer screen or from a printed handout.

MindJogger Videoquizzes

ELL MindJogger Videoquizzes provide an alternative review of concepts presented in this chapter. Students work in teams in a game show format to gain points for correct answers. The questions are presented in three rounds.

Round 1 Concepts (5 questions)
Round 2 Skills (4 questions)
Round 3 Problem Solving (4 questions)

FOLDABLES™ Study Organizer

For more information about Foldables, see *Teaching Mathematics with Foldables.*

Have students look through the chapter to make sure they have included notes and examples in their Foldables for each lesson of Chapter 4.

Encourage students to refer to their Foldables while completing the Study Guide and Review and to use them in preparing for the Chapter Test.

Answers

15. $\angle E \cong \angle D$, $\angle F \cong \angle C$, $\angle G \cong \angle B$, $\overline{EF} \cong \overline{DC}$, $\overline{FG} \cong \overline{CB}$, $\overline{GE} \cong \overline{BD}$
16. $\angle FGC \cong \angle DLC$, $\angle GCF \cong \angle LCD$, $\angle GFC \cong \angle LDC$, $\overline{GC} \cong \overline{LC}$, $\overline{CF} \cong \overline{CD}$, $\overline{FG} \cong \overline{DL}$
17. $\angle KNC \cong \angle RKE$, $\angle NCK \cong \angle KER$, $\angle CKN \cong \angle ERK$, $\overline{NC} \cong \overline{KE}$, $\overline{CK} \cong \overline{ER}$, $\overline{KN} \cong \overline{RK}$

Exercises Classify each triangle by its angles and by its sides if $m\angle ABC = 100$. *See Examples 1 and 2 on pages 178 and 179.*

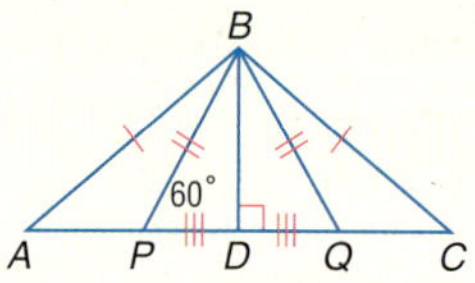

9. $\triangle ABC$ **10.** $\triangle BDP$ **11.** $\triangle BPQ$

9. obtuse, isosceles 10. right, scalene 11. equiangular, equilateral

4-2 Angles of Triangles

See pages 185–191.

Concept Summary

- The sum of the measures of the angles of a triangle is 180.
- The measure of an exterior angle is equal to the sum of the measures of the two remote interior angles.

Example If $\overline{TU} \perp \overline{UV}$ and $\overline{UV} \perp \overline{VW}$, find $m\angle 1$.

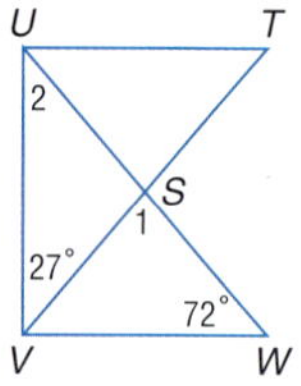

$m\angle 1 + 72 + m\angle TVW = 180$ Angle Sum Theorem

$m\angle 1 + 72 + (90 - 27) = 180$ $m\angle TVW = 90 - 27$

$m\angle 1 + 135 = 180$ Simplify.

$m\angle 1 = 45$ Subtract 135 from each side.

Exercises Find each measure. *See Example 1 on page 186.*

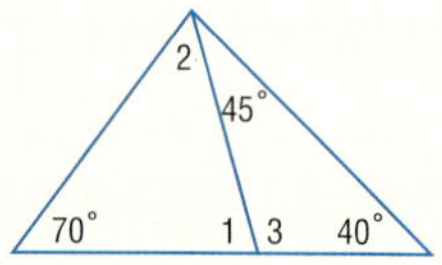

12. $m\angle 1$ 85 **13.** $m\angle 2$ 25 **14.** $m\angle 3$ 95

4-3 Congruent Triangles

See pages 192–198.

Concept Summary

- Two triangles are congruent when all of their corresponding parts are congruent.

Example If $\triangle EFG \cong \triangle JKL$, name the corresponding congruent angles and sides.

$\angle E \cong \angle J$, $\angle F \cong \angle K$, $\angle G \cong \angle L$, $\overline{EF} \cong \overline{JK}$, $\overline{FG} \cong \overline{KL}$, and $\overline{EG} \cong \overline{JL}$.

Exercises Name the corresponding angles and sides for each pair of congruent triangles. *See Example 1 on page 193.* 15–17. See margin.

15. $\triangle EFG \cong \triangle DCB$ **16.** $\triangle LCD \cong \triangle GCF$ **17.** $\triangle NCK \cong \triangle KER$

4-4 Proving Congruence—SSS, SAS

See pages 200–206.

Concept Summary

- If all of the corresponding sides of two triangles are congruent, then the triangles are congruent (SSS).
- If two corresponding sides of two triangles and the included angle are congruent, then the triangles are congruent (SAS).

228 Chapter 4 Congruent Triangles

Example Determine whether $\triangle ABC \cong \triangle TUV$. Explain.

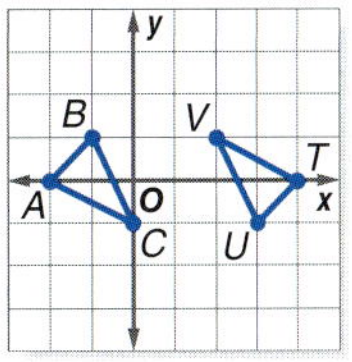

$AB = \sqrt{[-1-(-2)]^2 + (1-0)^2}$
$= \sqrt{1+1}$ or $\sqrt{2}$

$TU = \sqrt{(3-4)^2 + (-1-0)^2}$
$= \sqrt{1+1}$ or $\sqrt{2}$

$BC = \sqrt{[0-(-1)]^2 + (-1-1)^2}$
$= \sqrt{1+4}$ or $\sqrt{5}$

$UV = \sqrt{(2-3)^2 + [1-(-1)]^2}$
$= \sqrt{1+4}$ or $\sqrt{5}$

$CA = \sqrt{(-2-0)^2 + [0-(-1)]^2}$
$= \sqrt{4+1}$ or $\sqrt{5}$

$VT = \sqrt{(4-2)^2 + (0-1)^2}$
$= \sqrt{4+1}$ or $\sqrt{5}$

By the definition of congruent segments, all corresponding sides are congruent. Therefore, $\triangle ABC \cong \triangle TUV$ by SSS.

Exercises **Determine whether $\triangle MNP \cong \triangle QRS$ given the coordinates of the vertices. Explain.** *See Example 2 on page 201.* **18–19. See margin.**

18. $M(0, 3), N(-4, 3), P(-4, 6), Q(5, 6), R(2, 6), S(2, 2)$
19. $M(3, 2), N(7, 4), P(6, 6), Q(-2, 3), R(-4, 7), S(-6, 6)$

4-5 Proving Congruence—ASA, AAS

See pages 207–213.

Concept Summary

- If two pairs of corresponding angles and the included sides of two triangles are congruent, then the triangles are congruent (ASA).
- If two pairs of corresponding angles and a pair of corresponding nonincluded sides of two triangles are congruent, then the triangles are congruent (AAS).

Example **Write a proof.**

Given: $\overline{JK} \parallel \overline{MN}$; L is the midpoint of $\overline{KM}$.

Prove: $\triangle JLK \cong \triangle NLM$

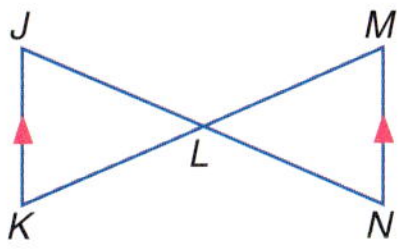

Flow proof:

- $\overline{JK} \parallel \overline{MN}$ (Given) → $\angle JKL \cong \angle LMN$ (Alt. int. ∠s are ≅.) → $\triangle JLK \cong \triangle NLM$
- L is the midpoint of $\overline{KM}$. (Given) → $\overline{KL} \cong \overline{ML}$ (Midpoint Theorem) → $\triangle JLK \cong \triangle NLM$
- $\angle JLK \cong \angle NLM$ (Vertical ∠s are ≅.) → $\triangle JLK \cong \triangle NLM$ (ASA)

Exercises **For Exercises 20 and 21, use the figure. Write a two-column proof for each of the following.** *See Example 2 on page 209.* **20–21. See margin.**

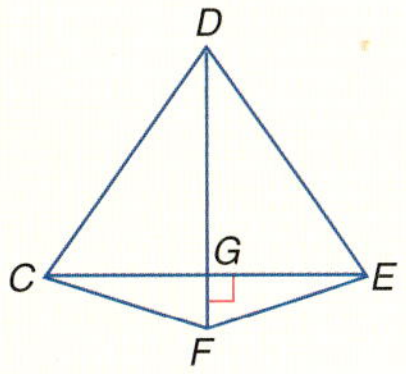

20. **Given:** $\overline{DF}$ bisects $\angle CDE$. $\overline{CE} \perp \overline{DF}$
Prove: $\triangle DGC \cong \triangle DGE$

21. **Given:** $\triangle DGC \cong \triangle DGE$, $\triangle GCF \cong \triangle GEF$
Prove: $\triangle DFC \cong \triangle DFE$

Answers

18. $MN = 4$, $NP = 3$, $MP = 5$, $QR = 3$, $RS = 4$, and $QS = 5$. Each pair of corresponding sides does not have the same measure. Therefore, $\triangle MNP$ is not congruent to $\triangle QRS$. $\triangle MNP$ is congruent to $\triangle SRQ$.

19. $MN = \sqrt{20}$, $NP = \sqrt{5}$, $MP = 5$, $QR = \sqrt{20}$, $RS = \sqrt{5}$, and $QS = 5$. Each pair of corresponding sides has the same measure. Therefore, $\triangle MNP \cong \triangle QRS$ by SSS.

20. Given: $\overline{DF}$ bisects $\angle CDE$, $\overline{CE} \perp \overline{DF}$.
Prove: $\triangle DGC \cong \triangle DGE$

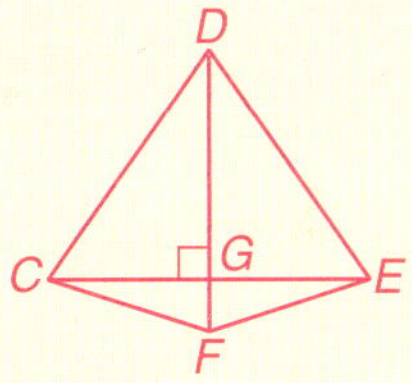

Proof:

Statements (Reasons)

1. $\overline{DF}$ bisects $\angle CDE$, $\overline{CE} \perp \overline{DF}$. (Given)
2. $\overline{DG} \cong \overline{DG}$ (Reflexive Prop.)
3. $\angle CDF \cong \angle EDF$ (Def. of ∠ bisector)
4. $\angle DGC$ is a rt. ∠; $\angle DGE$ is a rt. ∠ (Def. of ⊥ segments)
5. $\angle DGC \cong \angle DGE$ (All rt. ∠s are ≅.)
6. $\triangle DGC \cong \triangle DGE$ (ASA)

21. Given: $\triangle DGC \cong \triangle DGE$, $\triangle GCF \cong \triangle GEF$
Prove: $\triangle DFC \cong \triangle DFE$

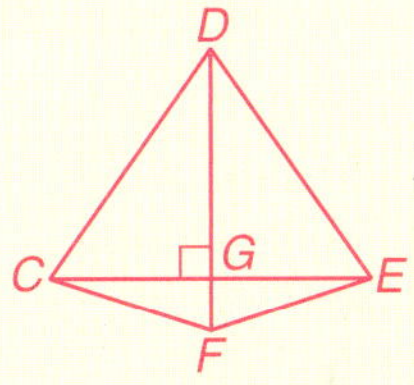

Proof:

Statements (Reasons)

1. $\triangle DGC \cong \triangle DGE$, $\triangle GCF \cong \triangle GEF$ (Given)
2. $\angle CDG \cong \angle EDG$, $\overline{CD} \cong \overline{ED}$, and $\angle CFD \cong \angle EFD$ (CPCTC)
3. $\triangle DFC \cong \triangle DFE$ (AAS)

4-6 Isosceles Triangles

See pages 216–221.

Concept Summary

- Two sides of a triangle are congruent if and only if the angles opposite those sides are congruent.
- A triangle is equilateral if and only if it is equiangular.

Example **If $\overline{FG} \cong \overline{GJ}$, $\overline{GJ} \cong \overline{JH}$, $\overline{FJ} \cong \overline{FH}$, and $m\angle GJH = 40$, find $m\angle H$.**

$\triangle GHJ$ is isosceles with base $\overline{GH}$, so $\angle JGH \cong \angle H$ by the Isosceles Triangle Theorem. Thus, $m\angle JGH = m\angle H$.

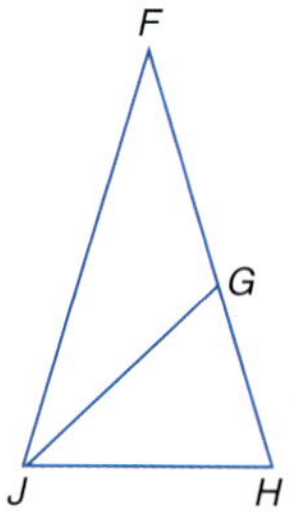

$m\angle GJH + m\angle JGH + m\angle H = 180$	Angle Sum Theorem
$40 + 2(m\angle H) = 180$	Substitution
$2(m\angle H) = 140$	Subtract 40 from each side.
$m\angle H = 70$	Divide each side by 2.

Exercises **For Exercises 22–25, refer to the figure at the right.**
See Example 2 on page 217.

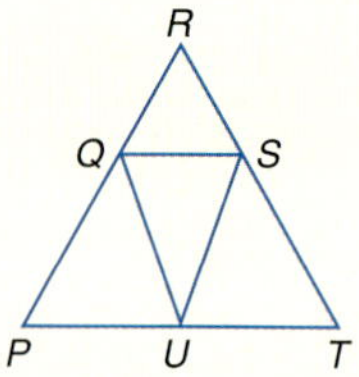

22. If $\overline{PQ} \cong \overline{UQ}$ and $m\angle P = 32$, find $m\angle PUQ$. **32**
23. If $\overline{PQ} \cong \overline{UQ}$, $\overline{PR} \cong \overline{RT}$, and $m\angle PQU = 40$, find $m\angle R$. **40**
24. If $\overline{RQ} \cong \overline{RS}$ and $m\angle RQS = 75$, find $m\angle R$. **30**
25. If $\overline{RQ} \cong \overline{RS}$, $\overline{RP} \cong \overline{RT}$, and $m\angle RQS = 80$, find $m\angle P$. **80**

4-7 Triangles and Coordinate Proof

See pages 222–226.

Concept Summary

- Coordinate proofs use algebra to prove geometric concepts.
- The Distance Formula, Slope Formula, and Midpoint Formula are often used in coordinate proof.

Example **Position and label isosceles right triangle ABC with legs of length a units on the coordinate plane.**

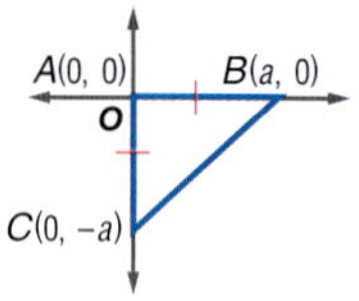

- Use the origin as the vertex of $\triangle ABC$ that has the right angle.
- Place each base along an axis.
- Since B is on the x-axis, its y-coordinate is 0. Its x-coordinate is a because the leg $\overline{AB}$ of the triangle is a units long.
- Since $\triangle ABC$ is isosceles, C should also be a distance of a units from the origin. Its coordinates are $(0, -a)$.

Exercises **Position and label each triangle on the coordinate plane.**
See Example 1 on page 222. **26–28. See margin.**

26. isosceles $\triangle TRI$ with base $\overline{TI}$ $4a$ units long
27. equilateral $\triangle BCD$ with side length $6m$ units long
28. right $\triangle JKL$ with leg lengths of a units and b units

Answers

26. Sample answer:

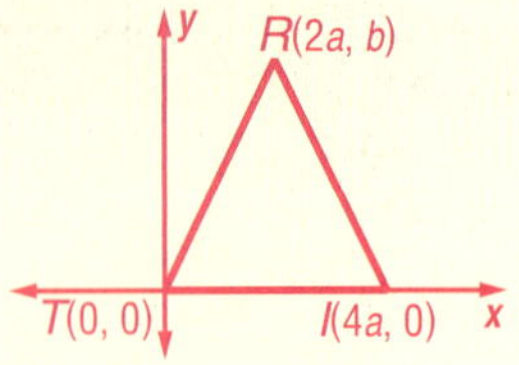

27. Sample answer:

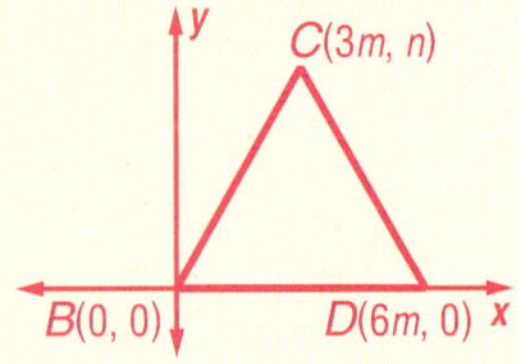

28. Sample answer:

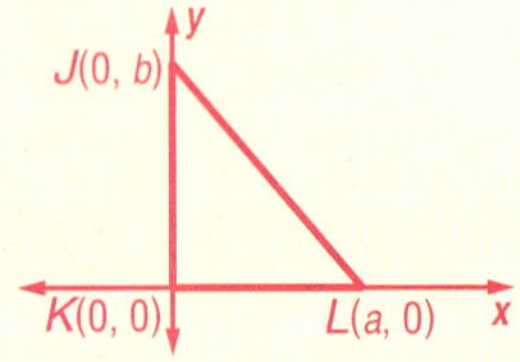

Answers (p. 231)

10. $\angle D \cong \angle P$, $\angle E \cong \angle Q$, $\angle F \cong \angle R$, $\overline{DE} \cong \overline{PQ}$, $\overline{EF} \cong \overline{QR}$, $\overline{DF} \cong \overline{PR}$
11. $\angle F \cong \angle H$, $\angle M \cong \angle N$, $\angle G \cong \angle J$, $\overline{FM} \cong \overline{HN}$, $\overline{MG} \cong \overline{NJ}$, $\overline{FG} \cong \overline{HJ}$
12. $\angle X \cong \angle Z$, $\angle Y \cong \angle Y$, $\angle Z \cong \angle X$, $\overline{XY} \cong \overline{ZY}$, $\overline{YZ} \cong \overline{YX}$, $\overline{XZ} \cong \overline{ZX}$
13. $JK = \sqrt{10}$, $KL = \sqrt{17}$, $JL = 5$, $MN = \sqrt{80}$, $NP = \sqrt{53}$, $MP = \sqrt{221}$. Corresponding sides are not congruent, so $\triangle JKL$ is not congruent to $\triangle MNP$.
14. Given: $\triangle JKM \cong \triangle JNM$
    Prove: $\triangle JKL \cong \triangle JNL$

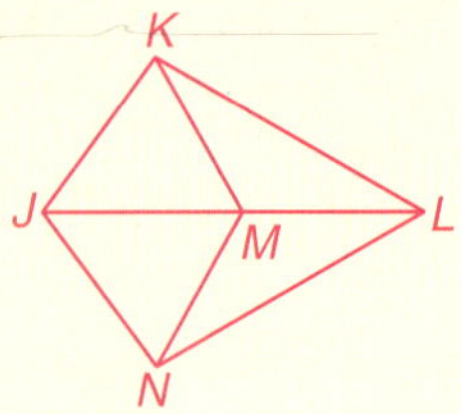

Proof:

$\angle JKM \cong \angle JNM$ (Given)
↓
$\overline{JK} \cong \overline{JN}$, $\angle KJL \cong \angle NJL$ (CPCTC)
↓
$\triangle JKL \cong \triangle JNL$ (SAS) ← $\overline{JL} \cong \overline{JL}$ (Reflexive Prop.)

Chapter 4

Practice Test

Vocabulary and Concepts

Choose the letter of the type of triangle that best matches each phrase.

1. triangle with no sides congruent **b**
2. triangle with at least two sides congruent **a**
3. triangle with all sides congruent **c**

a. isosceles
b. scalene
c. equilateral

Skills and Applications

Identify the indicated triangles in the figure if $\overline{PB} \perp \overline{AD}$ and $\overline{PA} \cong \overline{PC}$. **6. $\triangle PBA$, $\triangle PBC$, $\triangle PBD$**

4. obtuse **$\triangle PCD$** 5. isosceles **$\triangle PAC$** 6. right

Find the measure of each angle in the figure.

7. $m\angle 1$ **80** 8. $m\angle 2$ **105** 9. $m\angle 3$ **25**

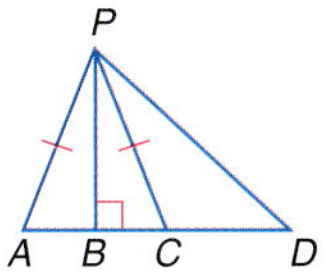

Questions 4–6

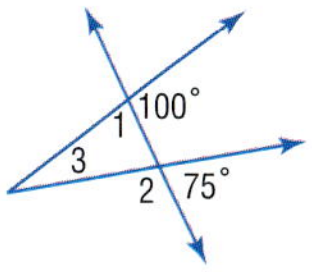

Questions 7–9

Name the corresponding angles and sides for each pair of congruent triangles. **10–12. See margin.**

10. $\triangle DEF \cong \triangle PQR$ 11. $\triangle FMG \cong \triangle HNJ$ 12. $\triangle XYZ \cong \triangle ZYX$

13. Determine whether $\triangle JKL \cong \triangle MNP$ given $J(-1, -2)$, $K(2, -3)$, $L(3, 1)$, $M(-6, -7)$, $N(-2, 1)$, and $P(5, 3)$. Explain. **See margin.**

14. Write a flow proof. **See margin.**
 Given: $\triangle JKM \cong \triangle JNM$
 Prove: $\triangle JKL \cong \triangle JNL$

In the figure, $\overline{FJ} \cong \overline{FH}$ and $\overline{GF} \cong \overline{GH}$.

15. If $m\angle JFH = 34$, find $m\angle J$. **73**
16. If $m\angle GHJ = 152$ and $m\angle G = 32$, find $m\angle JFH$. **24**

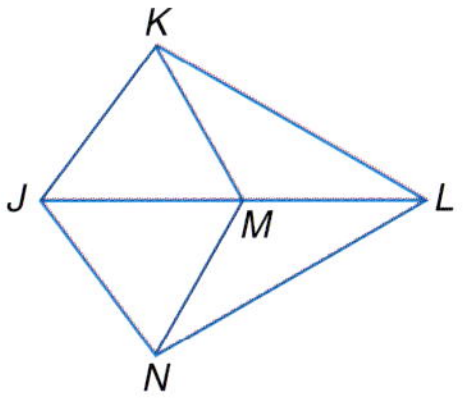

Question 14

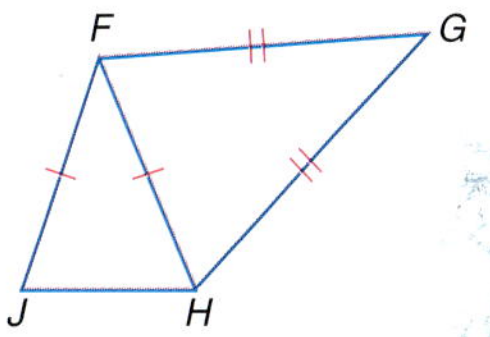

Questions 15–16

17. **LANDSCAPING** A landscaper designed a garden shaped as shown in the figure. The landscaper has decided to place point B 22 feet east of point A, point C 44 feet east of point A, point E 36 feet south of point A, and point D 36 feet south of point C. The angles at points A and C are right angles. Prove that $\triangle ABE \cong \triangle CBD$. **See p. 233J.**

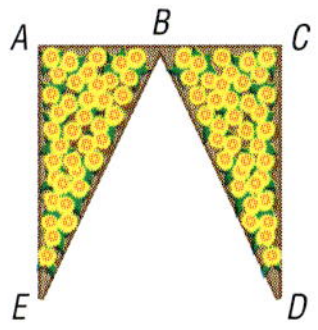

18. **STANDARDIZED TEST PRACTICE** In the figure, $\triangle FGH$ is a right triangle with hypotenuse $\overline{FH}$ and $GJ = GH$. What is the measure of $\angle JGH$? **C**
 Ⓐ 104 Ⓑ 62 Ⓒ 56 Ⓓ 28

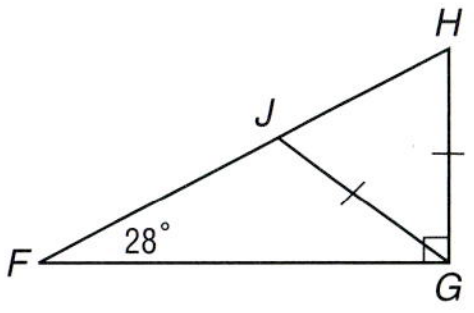

www.geometryonline.com/chapter_test/fcat

Chapter 4 Practice Test

Assessment Options

Vocabulary Test A vocabulary test/review for Chapter 4 can be found on p. 238 of the *Chapter 4 Resource Masters.*

Chapter Tests There are six Chapter 4 Tests and an Open-Ended Assessment task available in the *Chapter 4 Resource Masters.*

Chapter 4 Tests			
Form	**Type**	**Level**	**Pages**
1	MC	basic	225–226
2A	MC	average	227–228
2B	MC	average	229–230
2C	FR	average	231–232
2D	FR	average	233–234
3	FR	advanced	235–236

MC = multiple-choice questions
FR = free-response questions

Open-Ended Assessment
Performance tasks for Chapter 4 can be found on p. 237 of the *Chapter 4 Resource Masters.* A sample scoring rubric for these tasks appears on p. A28.

TestCheck and Worksheet Builder

This **networkable software** has three modules for assessment.

- **Worksheet Builder** to make worksheets and tests.
- **Student Module** to take tests on-screen.
- **Management System** to keep student records.

Portfolio Suggestion

Introduction As students progress with writing geometric proofs, it can help for them to have an example of their own work to refer to for self-encouragement or to review for method or style.

Ask Students Go back through your notes and problems that you worked and find a well-organized, clear and concise proof that you composed. Place an example of this proof in your portfolio, and write a paragraph explaining why you selected this particular example. Highlight important concepts and methods that you used to form your proof.

These two pages contain practice questions in the various formats that can be found on the most frequently given standardized tests.

A practice answer sheet for these two pages can be found on p. A1 of the *Chapter 4 Resource Masters.*

Standardized Test Practice Student Recording Sheet, p. A1

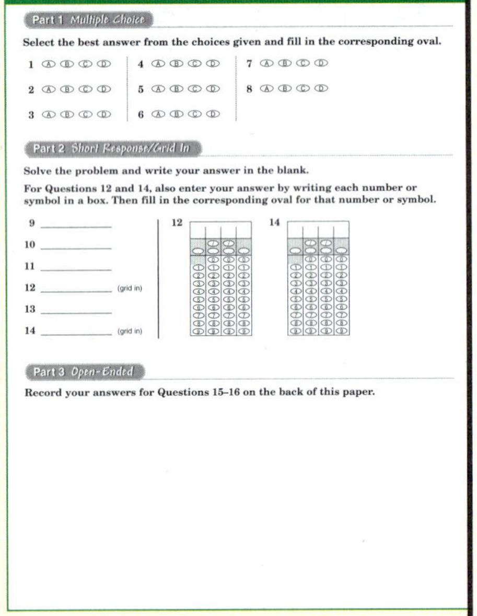
Part 1 Multiple Choice

Select the best answer from the choices given and fill in the corresponding oval.

1 Ⓐ Ⓑ Ⓒ Ⓓ 4 Ⓐ Ⓑ Ⓒ Ⓓ 7 Ⓐ Ⓑ Ⓒ Ⓓ

2 Ⓐ Ⓑ Ⓒ Ⓓ 5 Ⓐ Ⓑ Ⓒ Ⓓ 8 Ⓐ Ⓑ Ⓒ Ⓓ

3 Ⓐ Ⓑ Ⓒ Ⓓ 6 Ⓐ Ⓑ Ⓒ Ⓓ

Part 2 Short Response/Grid In

Solve the problem and write your answer in the blank.

For Questions 12 and 14, also enter your answer by writing each number or symbol in a box. Then fill in the corresponding oval for that number or symbol.

9 ______

10 ______

11 ______

12 ______ (grid in)

13 ______

14 ______ (grid in)

Part 3 Open-Ended

Record your answers for Questions 15–16 on the back of this paper.

Additional Practice

See pp. 243–244 in the *Chapter 4 Resource Masters* for additional standardized test practice.

Chapter 4 Standardized Test Practice

FCAT Practice

Part 1 Multiple Choice

Record your answers on the answer sheet provided by your teacher or on a sheet of paper.

1. In 2002, Capitol City had a population of 2010, and Shelbyville had a population of 1040. If Capitol City grows at a rate of 150 people a year and Shelbyville grows at a rate of 340 people a year, when will the population of Shelbyville be greater than that of Capitol City? (Prerequisite Skill) **B**

Ⓐ 2004 Ⓑ 2008

Ⓒ 2009 Ⓓ 2012

2. Which unit is most appropriate for measuring liquid in a bottle? (Lesson 1-2) **C**

Ⓐ grams Ⓑ feet

Ⓒ liters Ⓓ meters

3. A 9-foot tree casts a shadow on the ground. The distance from the top of the tree to the end of the shadow is 12 feet. To the nearest foot, how long is the shadow? (Lesson 1-3) **B**

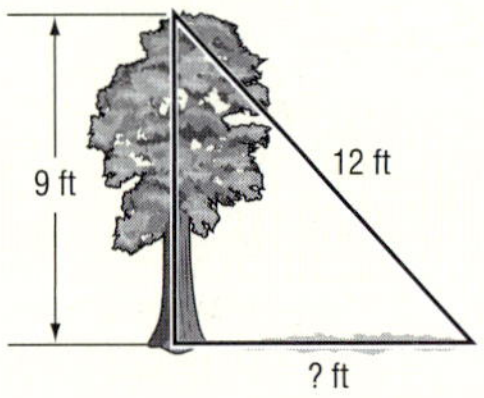

Ⓐ 7 ft

Ⓑ 8 ft

Ⓒ 10 ft

Ⓓ 13 ft

4. Which of the following is the inverse of the statement *If it is raining, then Kamika carries an umbrella?* (Lesson 2-2) **D**

Ⓐ If Kamika carries an umbrella, then it is raining.

Ⓑ If Kamika does not carry an umbrella, then it is not raining.

Ⓒ If it is not raining, then Kamika carries an umbrella.

Ⓓ If it is not raining, then Kamika does not carry an umbrella.

5. Students in a math classroom simulated stock trading. Kris drew the graph below to model the value of his shares at closing. The graph that modeled the value of Mitzi's shares was parallel to the one Kris drew. Which equation might represent the line for Mitzi's graph? (Lesson 3-3) **D**

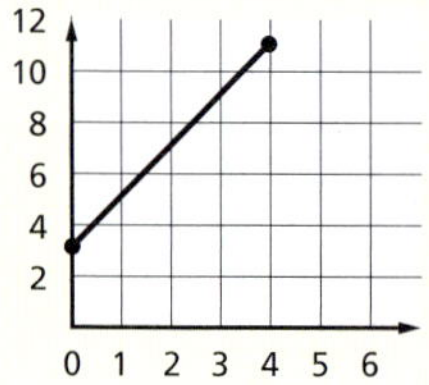

Ⓐ $-2x - y = 1$

Ⓑ $x - 2y = 1$

Ⓒ $x + 2y = 1$

Ⓓ $2x - y = 1$

6. What is $m\angle EFG$? (Lesson 4-2) **B**

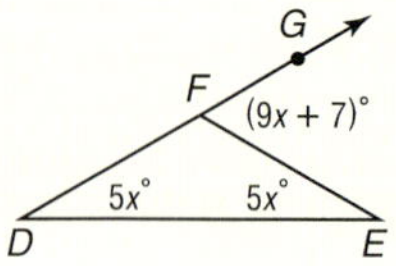

Ⓐ 35

Ⓑ 70

Ⓒ 90

Ⓓ 110

7. In the figure, $\triangle ABD \cong \triangle CBD$. If A has the coordinates $(-2, 4)$, what are the coordinates of C? (Lesson 4-3) **C**

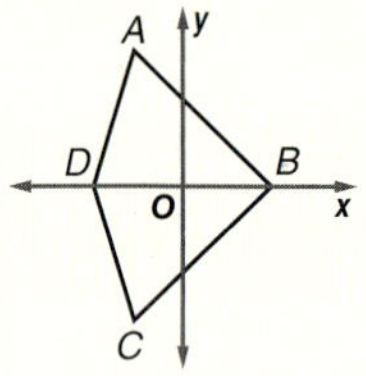

Ⓐ $(-4, -2)$

Ⓑ $(-4, 2)$

Ⓒ $(-2, -4)$

Ⓓ $(2, -4)$

8. The wings of some butterflies can be modeled by triangles as shown. If $\overline{AC} \cong \overline{DC}$ and $\angle ACB \cong \angle ECD$, which additional statements are needed to prove that $\triangle ACB \cong \triangle ECD$? (Lesson 4-4) **A**

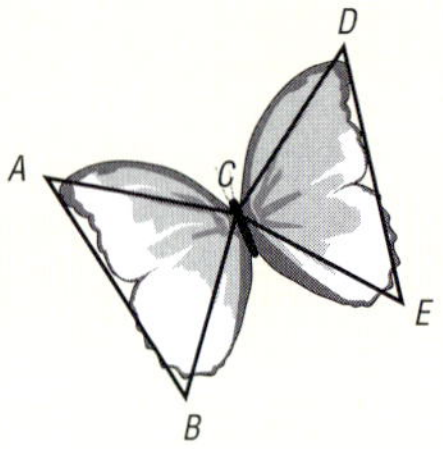

Ⓐ $\overline{BC} \cong \overline{CE}$

Ⓑ $\overline{AB} \cong \overline{ED}$

Ⓒ $\angle BAC \cong \angle CED$

Ⓓ $\angle ABC \cong \angle CDE$

The Princeton Review

Log On for Test Practice

The Princeton Review offers additional test-taking tips and practice problems at their web site. Visit www.princetonreview.com or www.review.com

TestCheck and Worksheet Builder

Special banks of standardized test questions similar to those on the SAT, ACT, TIMSS 8, NAEP 8, and state proficiency tests can be found on this CD-ROM.

Aligned and verified by The Princeton Review

FCAT Practice

Part 2 Short Response/Grid In

Record your answers on the answer sheet provided by your teacher or on a sheet of paper.

9. Find the product $3s^2(2s^3 - 7)$. (Prerequisite Skill) $6s^5 - 21s^2$

10. After a long workout, Brian noted, "If I do not drink enough water, then I will become dehydrated." He then made another statement, "If I become dehydrated, then I did not drink enough water." How is the second statement related to the original statement? (Lesson 2-2) **converse**

11. On a coordinate map, the towns of Creston and Milford are located at $(-1, -1)$ and $(1, 3)$, respectively. A third town, Dixville, is located at $(x, -1)$ so that Creston and Dixville are endpoints of the base of the isosceles triangle formed by the three locations. What is the value of x? (Lesson 4-1) **3**

12. A watchtower, built to help prevent forest fires, was designed as an isosceles triangle. If the side of the tower meets the ground at a 105° angle, what is the measure of the angle at the top of the tower? (Lesson 4-2) **30**

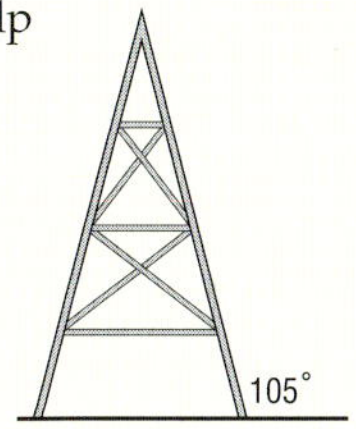

13. During a synchronized flying show, airplane A and airplane D are equidistant from the ground. They descend at the same angle to land at points B and E, respectively. Which postulate would prove that $\triangle ABC \cong \triangle DEF$? (Lesson 4-4) **ASA**

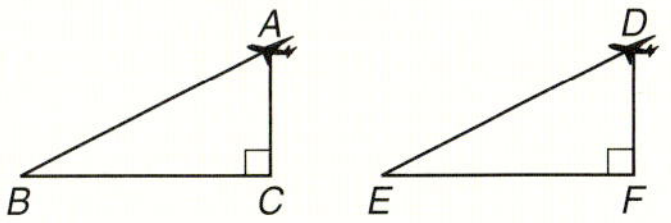

14. $\triangle ABC$ is an isosceles triangle with $\overline{AB} \cong \overline{BC}$, and the measure of vertex angle B is three times $m\angle A$. What is $m\angle C$? (Lesson 4-6) **36**

www.geometryonline.com/standardized_test/fcat

The Princeton Review Test-Taking Tip

Question 8

- If you are not permitted to write in your test booklet, make a sketch of the figure on scrap paper.
- Mark the figure with all of the information you know so that you can determine the congruent triangles more easily.
- Make a list of postulates or theorems that you might use for this case.

FCAT Practice

Part 3 Open Ended

Record your answers on a sheet of paper. Show your work.

15. Train tracks a and b are parallel lines although they appear to come together to give the illusion of distance in a drawing. All of the railroad ties are parallel to each other.

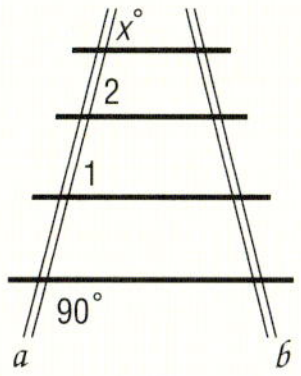

a. What is the value of x? (Lesson 3-1) **90**

b. What is the relationship between the tracks and the ties that run across the tracks? (Lesson 1-5) **perpendicular lines**

c. What is the relationship between $\angle 1$ and $\angle 2$? Explain. (Lesson 3-2) **See margin.**

16. The measures of the angles of $\triangle ABC$ are $5x$, $4x - 1$, and $3x + 13$.

a. Draw a figure to illustrate $\triangle ABC$. (Lesson 4-1) **See margin.**

b. Find the measure of each angle of $\triangle ABC$. Explain. (Lesson 4-2) **70, 55, and 55 See margin for explanation.**

c. Prove that $\triangle ABC$ is an isosceles triangle. (Lesson 4-6) **See margin.**

Evaluating Open-Ended Assessment Questions

Open-Ended Assessment questions are graded by using a multilevel rubric that guides you in assessing a student's knowledge of a particular concept.

Goal: Students find angle measures, describe angle relationships, and prove that a triangle is isosceles.

Sample Scoring Rubric: The following rubric is a sample scoring device. You may wish to add more detail to this sample to meet your individual scoring needs.

Score	Criteria
4	A correct solution that is supported by well-developed, accurate explanations
3	A generally correct solution, but may contain minor flaws in reasoning or computation
2	A partially correct interpretation and/or solution to the problem
1	A correct solution with no supporting evidence or explanation
0	An incorrect solution indicating no mathematical understanding of the concept or task, or no solution is given

Answers

15c. They are congruent. Sample answers: Both are right angles; they are supplementary angles; they are corresponding angles.

16a. Sample answer:

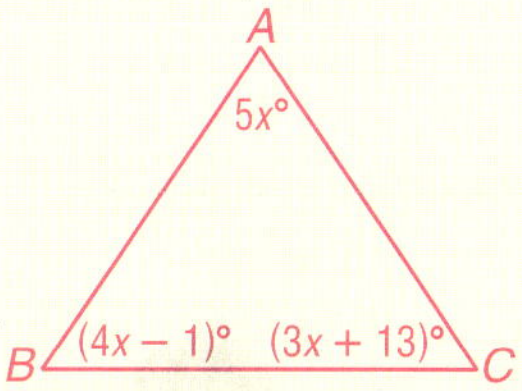

16b. From the Angle Sum Theorem, we know that $m\angle A + m\angle B + m\angle C = 180$. Substituting the given measures, $5x + 4x - 1 + 3x + 13 = 180$. Solve for x to find that $x = 14$. Substitute 14 for x to find the measures: $5x = 5(14)$ or 70, $4x - 1 = 4(14) - 1$ or 55, and $3x + 13 = 3(14) + 13$ or 55.

16c. If two angles of a $\triangle$ are $\cong$, then the sides opposite those angles are $\cong$ (Converse of the Isosceles $\triangle$ Theorem). Since two sides of this $\triangle$ are $\cong$, it is an isosceles $\triangle$ (Definition of Isosceles $\triangle$).

Page 180–184 Lesson 4-1

38. Given: $\triangle EUI$ is equiangular.
$\overline{QL} \parallel \overline{UI}$
Prove: $\triangle EQL$ is equiangular.

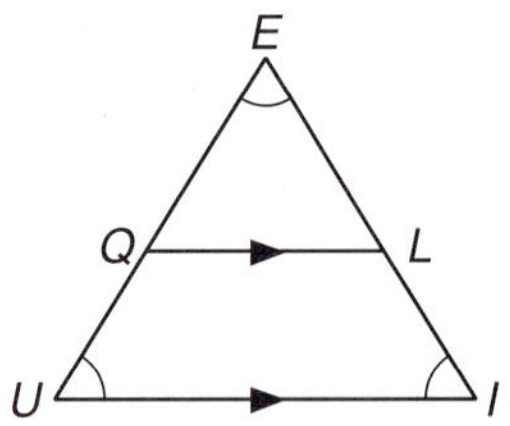

Proof:

Statements (Reasons)

1. $\triangle EUI$ is equiangular. $\overline{QL} \parallel \overline{UI}$ (Given)
2. $\angle E \cong \angle EUI \cong \angle EIU$ (Def. of equiangular $\triangle$)
3. $\angle EUI \cong \angle EQL$, $\angle EIU \cong \angle ELQ$ (Corr. ∠s Post.)
4. $\angle E \cong \angle E$ (Reflexive Prop.)
5. $\angle E \cong \angle EQL \cong \angle ELQ$ (Trans. Prop.)
6. $\triangle EQL$ is equiangular. (Def. of equiangular $\triangle$)

39. Given: $m\angle NPM = 33$
Prove: $\triangle RPM$ is obtuse.

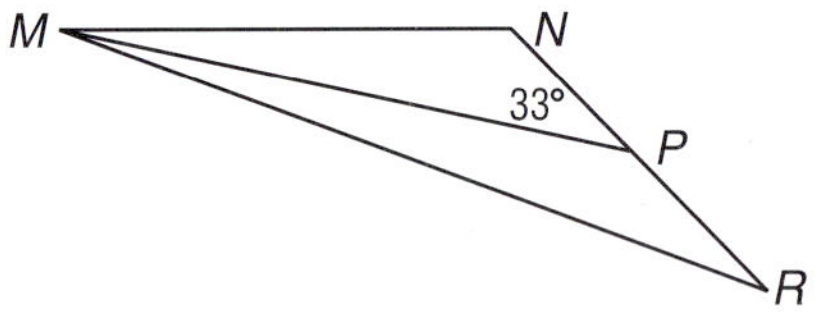

Proof: $\angle NPM$ and $\angle RPM$ form a linear pair. $\angle NPM$ and $\angle RPM$ are supplementary because if two angles form a linear pair, then they are supplementary. So, $m\angle NPM + m\angle RPM = 180$. It is given that $m\angle NPM = 33$. By substitution, $33 + m\angle RPM = 180$. Subtract to find that $m\angle RPM = 147$. $\angle RPM$ is obtuse by definition. $\triangle RPM$ is obtuse by definition.

40. $TS = \sqrt{(-7-(-4))^2 + (8-14)^2} = \sqrt{9+36}$ or $\sqrt{45}$

$SR = \sqrt{(-10-(-7))^2 + (2-8)^2} = \sqrt{9+36}$ or $\sqrt{45}$

S is the midpoint of $\overline{RT}$.

$UT = \sqrt{(0-(-4))^2 + (8-14)^2} = \sqrt{16+36}$ or $\sqrt{52}$

$VU = \sqrt{(4-0)^2 + (2-8)^2} = \sqrt{16+36}$ or $\sqrt{52}$

U is the midpoint of $\overline{TV}$.

41. $$AD = \sqrt{\left(0 - \frac{a}{2}\right)^2 + (0-b)^2} = \sqrt{\left(-\frac{a}{2}\right)^2 + (-b)^2} = \sqrt{\frac{a^2}{4} + b^2}$$

$$CD = \sqrt{\left(a - \frac{a}{2}\right)^2 + (0-b)^2} = \sqrt{\left(\frac{a}{2}\right)^2 + (-b)^2} = \sqrt{\frac{a^2}{4} + b^2}$$

$AD = CD$, so $\overline{AD} \cong \overline{CD}$. $\triangle ADC$ is isosceles by definition.

Pages 188–191, Lesson 4-2

39. Given: $\angle FGI \cong \angle IGH$
$\overline{GI} \perp \overline{FH}$
Prove: $\angle F \cong \angle H$

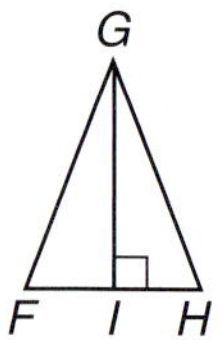

Proof:

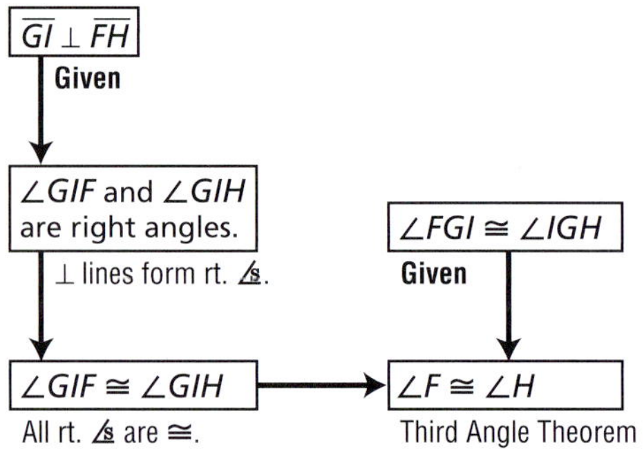

40. Given: $ABCD$ is a quadrilateral.
Prove: $m\angle DAB + m\angle B + m\angle BCD + m\angle D = 360$

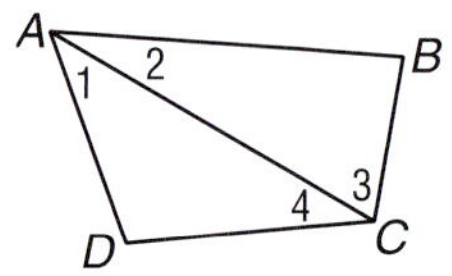

Proof:

Statements (Reasons)

1. $ABCD$ is a quadrilateral. (Given)
2. $m\angle 2 + m\angle 3 + m\angle B = 180$;
   $m\angle 1 + m\angle 4 + m\angle D = 180$ (∠ Sum Theorem)
3. $m\angle 2 + m\angle 3 + m\angle B + m\angle 1 + m\angle 4 + m\angle D = 360$ (Addition Prop.)
4. $m\angle DAB = m\angle 1 + m\angle 2$; $m\angle BCD = m\angle 3 + m\angle 4$ (∠ Addition)
5. $m\angle DAB + m\angle B + m\angle BCD + m\angle D = 360$ (Substitution)

41. Given: $\triangle ABC$
Prove: $m\angle CBD = m\angle A + m\angle C$

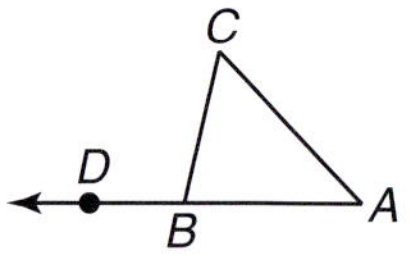

Proof:

Statements (Reasons)

1. $\triangle ABC$ (Given)
2. $\angle CBD$ and $\angle ABC$ form a linear pair. (Def. of linear pair)
3. $\angle CBD$ and $\angle ABC$ are supplementary. (If 2 ∠s form a linear pair, they are suppl.)
4. $m\angle CBD + m\angle ABC = 180$ (Def. of suppl.)
5. $m\angle A + m\angle ABC + m\angle C = 180$ (∠ Sum Theorem)
6. $m\angle A + m\angle ABC + m\angle C = m\angle CBD + m\angle ABC$ (Substitution)
7. $m\angle A + m\angle C = m\angle CBD$ (Subtraction Prop.)

42. Given: $\triangle RST$
$\angle R$ is a right angle.
Prove: $\angle S$ and $\angle T$ are complementary.

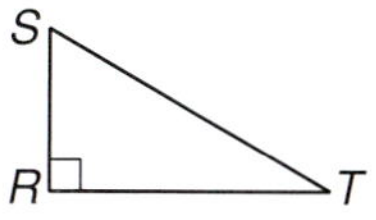

Proof:

$\angle R$ is a rt. $\angle$. (Given)
→ $m\angle R = 90$ (Def. of rt. $\angle$)
$m\angle R + m\angle S + m\angle T = 180$ (Angle Sum Theorem)
→ $90 + m\angle S + m\angle T = 180$ (Substitution)
→ $m\angle S + m\angle T = 90$ (Subtraction Prop.)
→ $\angle S$ and $\angle T$ are complementary (Def. of complementary $\angle$s)

43. Given: $\triangle MNO$
$\angle M$ is a right angle.
Prove: There can be at most one right angle in a triangle.

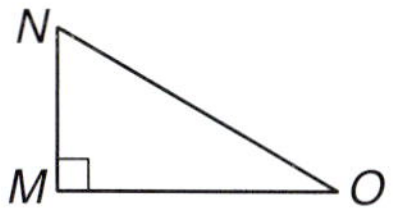

Proof: In $\triangle MNO$, $\angle M$ is a right angle. $m\angle M + m\angle N + m\angle O = 180$. $m\angle M = 90$, so $m\angle N + m\angle O = 90$. If $\angle N$ were a right angle, then $m\angle O = 0$. But that is impossible, so there cannot be two right angles in a triangle.

Given: $\triangle PQR$
$\angle P$ is obtuse.
Prove: There can be at most one obtuse angle in a triangle.

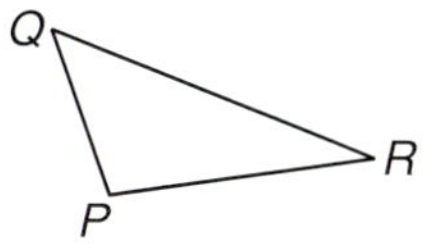

Proof: In $\triangle PQR$, $\angle P$ is obtuse. So $m\angle P > 90$. $m\angle P + m\angle Q + m\angle R = 180$. It must be that $m\angle Q + m\angle R > 90$. So, $\angle Q$ and $\angle R$ must be acute.

44. Given: $\angle A \cong \angle D$
$\angle B \cong \angle E$
Prove: $\angle C \cong \angle F$

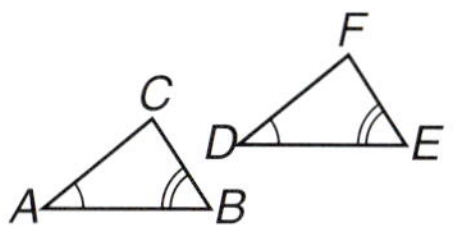

Proof:

Statements (Reasons)

1. $\angle A \cong \angle D$, $\angle B \cong \angle E$ (Given)
2. $m\angle A = m\angle D$, $m\angle B = m\angle E$ (Def. of $\cong$ $\angle$s)
3. $m\angle A + m\angle B + m\angle C = 180$, $m\angle D + m\angle E + m\angle F = 180$ (Angle Sum Theorem)
4. $m\angle A + m\angle B + m\angle C = m\angle D + m\angle E + m\angle F$ (Transitive Prop.)
5. $m\angle D + m\angle E + m\angle C = m\angle D + m\angle E + m\angle F$ (Substitution Prop.)
6. $m\angle C = m\angle F$ (Subtraction Prop.)
7. $\angle C \cong \angle F$ (Def. of $\cong$ $\angle$s)

Pages 195–198, Lesson 4-3

33. Given: $\triangle RST \cong \triangle XYZ$
Prove: $\triangle XYZ \cong \triangle RST$

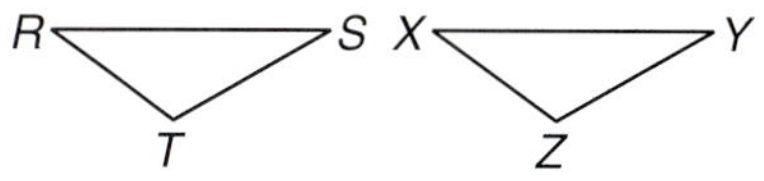

Proof:

$\triangle RST \cong \triangle XYZ$ (Given)
→ $\angle R \cong \angle X$, $\angle S \cong \angle Y$, $\angle T \cong \angle Z$, $\overline{RS} \cong \overline{XY}$, $\overline{ST} \cong \overline{YZ}$, $\overline{RT} \cong \overline{XZ}$ (CPCTC)
→ $\angle X \cong \angle R$, $\angle Y \cong \angle S$, $\angle Z \cong \angle T$, $\overline{XY} \cong \overline{RS}$, $\overline{YZ} \cong \overline{ST}$, $\overline{XZ} \cong \overline{RT}$ (Congruence of $\angle$s and segments is symmetric.)
→ $\triangle XYZ \cong \triangle RST$ (Def. of $\cong$ $\triangle$s)

Pages 203–206, Lesson 4-4

5. Given: $\overline{DE}$ and $\overline{BC}$ bisect each other.
Prove: $\triangle DGB \cong \triangle EGC$

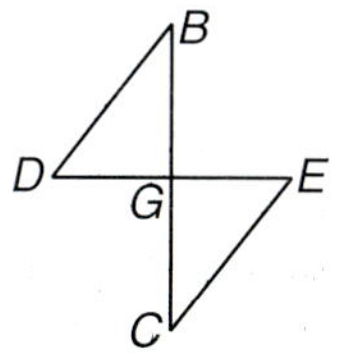

Proof:

$\overline{DG}$ and $\overline{GE}$ bisect each other. (Given)
→ $\overline{DE} \cong \overline{BC}$, $\overline{BG} \cong \overline{GC}$ (Def. of bisector of segments)
$\angle DGB \cong \angle EGC$ (Vertical $\angle$s are $\cong$.)
→ $\triangle DGB \cong \triangle EGC$ (SAS)

6. Given: $\overline{KM} \parallel \overline{JL}$, $\overline{KM} \cong \overline{JL}$
Prove: $\triangle JKM \cong \triangle MLJ$

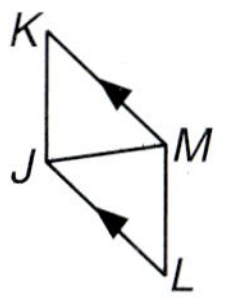

Proof:

Statements (Reasons)

1. $\overline{KM} \parallel \overline{JL}$, $\overline{KM} \cong \overline{JL}$ (Given)
2. $\angle KMJ \cong \angle LJM$ (Alt. Int. $\angle$s Theorem)
3. $\overline{JM} \cong \overline{JM}$ (Reflexive Prop.)
4. $\triangle JKM \cong \triangle MLJ$ (SAS)

14. Given: $\overline{AE} \cong \overline{FC}$, $\overline{AB} \cong \overline{BC}$, $\overline{BE} \cong \overline{BF}$
Prove: $\triangle AFB \cong \triangle CEB$

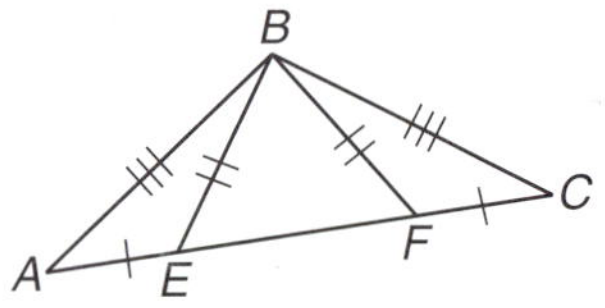

Proof:

- $\overline{AE} \cong \overline{FC}$ — Given
- ↓ $AE = FC$ — Def. of ≅ seg.
- ↓ $AE + EF = EF + FC$ — Addition Prop. (← $EF = EF$ — Reflexive Prop.)
- $AE = EC$ — Substitution (← $AE + EF = AF$, $EF + FC = EC$ — Seg. Addition Post.)
- ↓ $\overline{AF} \cong \overline{EC}$ — Def. of ≅ seg.
- ↓ $\triangle AFB \cong \triangle CEB$ — SAS (← $\overline{AB} \cong \overline{BC}$, $\overline{BE} \cong \overline{BF}$ — Given)

15. Given: $\overline{RQ} \cong \overline{TQ} \cong \overline{YQ} \cong \overline{WQ}$
$\angle RQY \cong \angle WQT$
Prove: $\triangle QWT \cong \triangle QYR$

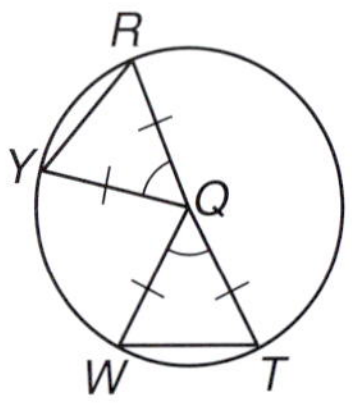

Proof:

- $\overline{RQ} \cong \overline{TQ} \cong \overline{YQ} \cong \overline{WQ}$ — Given
- $\angle RQY \cong \angle WQT$ — Given
- ↓ $\triangle QWT \cong \triangle QYR$ — SAS

16. Given: $\triangle CDE$ is an isosceles triangle.
G is the midpoint of $\overline{CE}$.
Prove: $\triangle CDG \cong \triangle EDG$

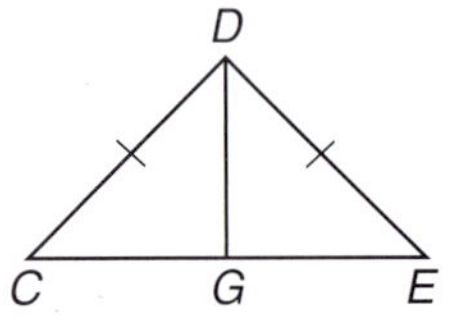

Proof:
Statements (Reasons)

1. $\triangle CDE$ is an isosceles triangle, G is the midpoint of $\overline{CE}$. (Given)
2. $\overline{CD} \cong \overline{DE}$ (Def. of isos. △)
3. $\overline{CG} \cong \overline{GE}$ (Midpoint Th.)
4. $\overline{DG} \cong \overline{DG}$ (Reflexive Prop.)
5. $\triangle CDG \cong \triangle EDG$ (SSS)

17. Given: $\triangle MRN \cong \triangle QRP$
$\angle MNP \cong \angle QPN$
Prove: $\triangle MNP \cong \triangle QPN$

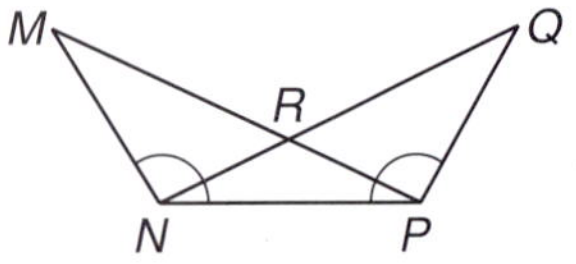

Proof:
Statements (Reasons)

1. $\triangle MRN \cong \triangle QRP$, $\angle MNP \cong \angle QPN$ (Given)
2. $\overline{MN} \cong \overline{QP}$ (CPCTC)
3. $\overline{NP} \cong \overline{NP}$ (Reflexive Prop.)
4. $\triangle MNP \cong \triangle QPN$ (SAS)

18. Given: $\overline{AC} \cong \overline{GC}$
$\overline{EC}$ bisects $\overline{AG}$.
Prove: $\triangle GEC \cong \triangle AEC$

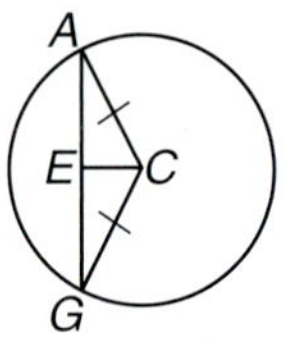

Proof:
Statements (Reasons)

1. $\overline{AC} \cong \overline{GC}$, $\overline{EC}$ bisects $\overline{AG}$. (Given)
2. $\overline{AE} \cong \overline{EG}$ (Def. of segment bisector)
3. $\overline{EC} \cong \overline{EC}$ (Reflexive Prop.)
4. $\triangle GEC \cong \triangle AEC$ (SSS)

19. Given: $\triangle GHJ \cong \triangle LKJ$
Prove: $\triangle GHL \cong \triangle LKG$

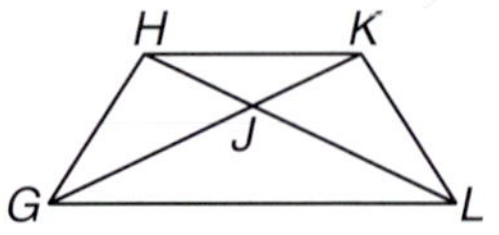

Proof:
Statements (Reasons)

1. $\triangle GHJ \cong \triangle LKJ$ (Given)
2. $\overline{HJ} \cong \overline{KJ}$, $\overline{GJ} \cong \overline{LJ}$, $\overline{GH} \cong \overline{LK}$ (CPCTC)
3. $HJ = KJ$, $GJ = LJ$ (Def. of ≅ segments)
4. $HJ + LJ = KJ + JG$ (Addition Prop.)
5. $KJ + GJ = KG$; $HJ + LJ = HL$ (Segment Addition)
6. $\overline{KG} = \overline{HL}$ (Substitution)
7. $\overline{KG} \cong \overline{HL}$ (Def. of ≅ segments)
8. $\overline{GL} \cong \overline{GL}$ (Reflexive Prop.)
9. $\triangle GHL \cong \triangle LKG$ (SSS)

20. Given: $\overline{RS} \cong \overline{PN}$, $\overline{RT} \cong \overline{MP}$
$\angle S \cong \angle N$, $\angle T \cong \angle M$
Prove: $\triangle RST \cong \triangle PNM$

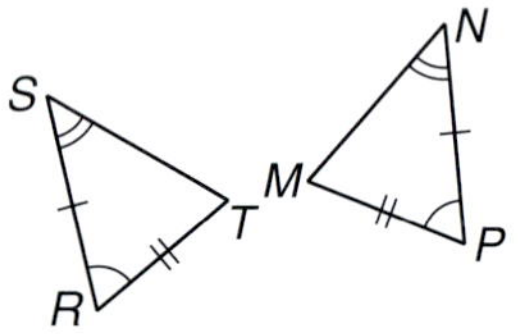

Proof:
Statements (Reasons)

1. $\overline{RS} \cong \overline{PN}$, $\overline{RT} \cong \overline{MP}$ (Given)
2. $\angle S \cong \angle N$, and $\angle T \cong \angle M$ (Given)
3. $\angle R \cong \angle P$ (Third Angle Theorem)
4. $\triangle RST \cong \triangle MNP$ (SAS)

21. Given: $\overline{EF} \cong \overline{HF}$
G is the midpoint of $\overline{EH}$.
Prove: $\triangle EFG \cong \triangle HFG$

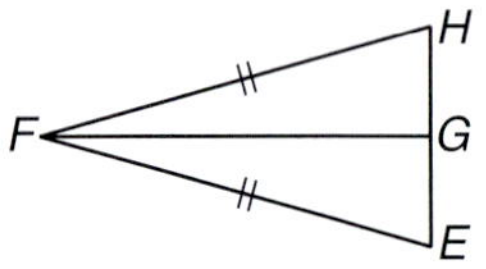

Proof:
Statements (Reasons)

1. $\overline{EF} \cong \overline{HF}$; G is the midpoint of $\overline{EH}$. (Given)
2. $\overline{EG} \cong \overline{GH}$ (Def. of midpoint)
3. $\overline{FG} \cong \overline{FG}$ (Reflexive Prop.)
4. $\triangle EFG \cong \triangle HFG$ (SSS)

26. Given: $\overline{TS} \cong \overline{SF} \cong \overline{FH} \cong \overline{HT}$
$\angle TSF$, $\angle SFH$, $\angle FHT$, and $\angle HTS$ are right angles.
Prove: $\overline{HS} \cong \overline{TF}$

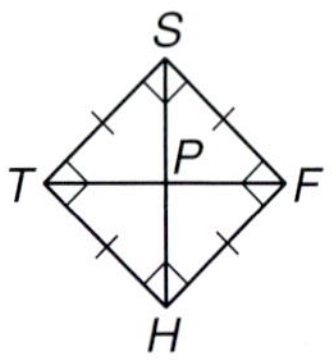

Proof:
Statements (Reasons)

1. $\overline{TS} \cong \overline{SF} \cong \overline{FH} \cong \overline{HT}$ (Given)
2. $\angle TSF$, $\angle SFH$, $\angle FHT$, and $\angle HTS$ are right angles. (Given)
3. $\angle STH \cong \angle THF$ (All right ∠s are ≅.)
4. $\triangle STH \cong \triangle THF$ (SAS)
5. $\overline{HS} \cong \overline{TF}$ (CPCTC)

27. Given: $\overline{TS} \cong \overline{SF} \cong \overline{FH} \cong \overline{HT}$
$\angle TSF$, $\angle SFH$, $\angle FHT$, and $\angle HTS$ are right angles.
Prove: $\angle SHT \cong \angle SHF$

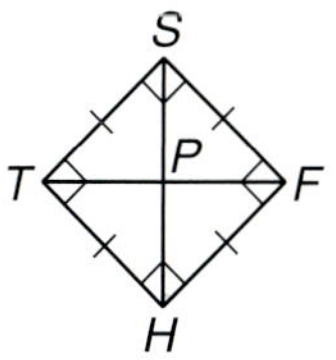

Proof:

Statements (Reasons)

1. $\overline{TS} \cong \overline{SF} \cong \overline{FH} \cong \overline{HT}$ (Given)
2. $\angle TSF$, $\angle SFH$, $\angle FHT$, and $\angle HTS$ are right angles. (Given)
3. $\angle STH \cong \angle SFH$ (All right ∠s are ≅.)
4. $\triangle STH \cong \triangle SFH$ (SAS)
5. $\angle SHT \cong \angle SHF$ (CPCTC)

28. Given: $\overline{DE} \cong \overline{FB}$, $\overline{AE} \cong \overline{FC}$, $\overline{AE} \perp \overline{DB}$, $\overline{CF} \perp \overline{DB}$
Prove: $\triangle ABD \cong \triangle CDB$

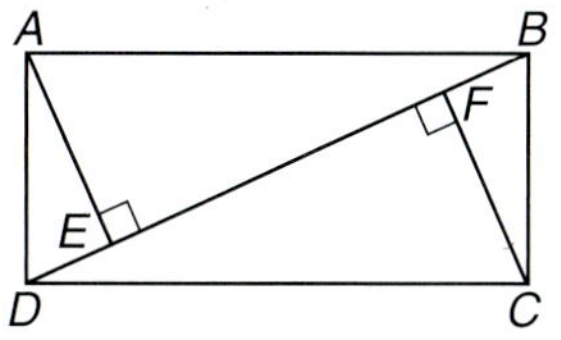

Plan: First use SAS to show that $\triangle ADE \cong \triangle CBF$. Next use CPCTC and Reflexive Property for segments to show $\triangle ABD \cong \triangle CDB$.

Proof:

Statements (Reasons)

1. $\overline{DE} \cong \overline{FB}$, $\overline{AE} \cong \overline{FC}$ (Given)
2. $\overline{AE} \perp \overline{DB}$, $\overline{CF} \perp \overline{DB}$ (Given)
3. $\angle AED$ is a right angle. $\angle CFB$ is a right angle. (⊥ lines form right ∠s.)
4. $\angle AED \cong \angle CFB$ (All right angles are ≅.)
5. $\triangle ADE \cong \triangle CBF$ (SAS)
6. $\overline{AD} \cong \overline{BC}$ (CPCTC)
7. $\overline{DB} \cong \overline{DB}$ (Reflexive Prop. for Segments)
8. $\angle CBD \cong \angle ADB$ (CPCTC)
9. $\triangle ABD \cong \triangle CDB$ (SAS)

Pages 210–213, Lesson 4-5

4. Given: $\overline{GH} \cong \overline{KJ}$, $\overline{GK} \cong \overline{HJ}$
Prove: $\triangle GJK \cong \triangle JGH$
Proof:

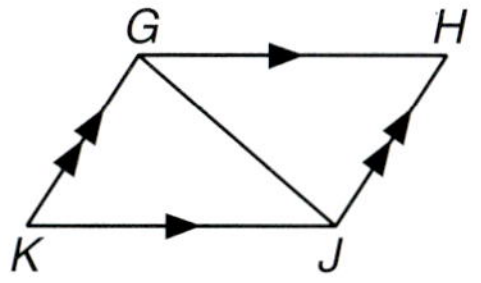

- $\overline{GH} \parallel \overline{KS}$ (Given) → $\angle HGJ \cong \angle KJG$ (Alt. int. ∠s are ≅.)
- $\overline{GK} \parallel \overline{HJ}$ (Given) → $\angle KGJ \cong \angle HJG$ (Alt. int. ∠s are ≅.)
- $\overline{GJ} \cong \overline{GJ}$ (Reflexive Property)
- → $\triangle WXY \cong \triangle YZW$ (ASA)

5. Given: $\overline{XW} \parallel \overline{YZ}$, $\angle X \cong \angle Z$
Prove: $\triangle WXY \cong \triangle YZW$
Proof:

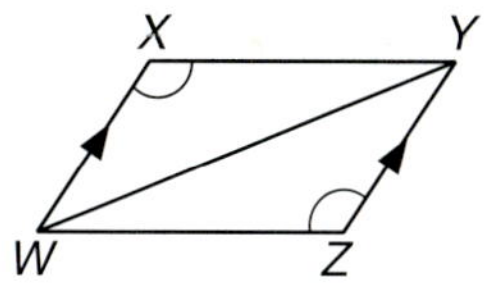

- $\overline{XW} \parallel \overline{YZ}$ (Given) → $\angle XWY \cong \angle ZYW$ (Alt. int. ∠s are ≅.)
- $\angle X \cong \angle Z$ (Given)
- $\overline{WY} \cong \overline{WY}$ (Reflexive Property)
- → $\triangle WXY \cong \triangle YZW$ (AAS)

6. Given: $\overline{QS}$ bisects $\angle RST$; $\angle R \cong \angle T$.
Prove: $\triangle QRS \cong \triangle QTS$

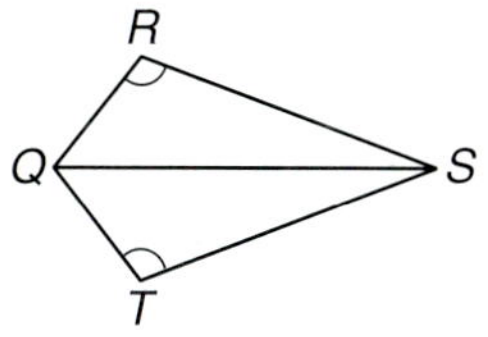

Proof: We are given that $\angle R \cong \angle T$ and $\overline{QS}$ bisects $\angle RST$, so by definition of angle bisector, $\angle RSQ \cong \angle TSQ$. By the Reflexive Property, $\overline{QS} \cong \overline{QS}$. $\triangle QRS \cong \triangle QTS$ by AAS.

7. Given: $\angle E \cong \angle K$, $\angle DGH \cong \angle DHG$, $\overline{EG} \cong \overline{KH}$
Prove: $\triangle EGD \cong \triangle KHD$

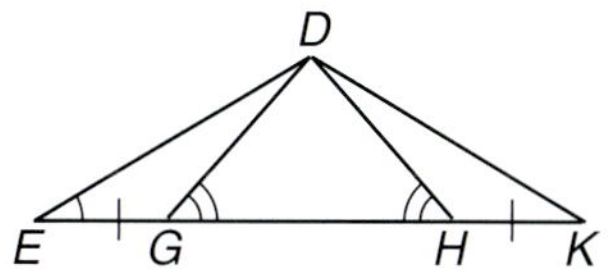

Proof: Since $\angle EGD$ and $\angle DGH$ are linear pairs, the angles are supplementary. Likewise, $\angle KHD$ and $\angle DHG$ are supplementary. We are given that $\angle DGH \cong \angle DHG$. Angles supplementary to congruent angles are congruent so $\angle EGD \cong \angle KHD$. Since we are given that $\angle E \cong \angle K$ and $\overline{EG} \cong \overline{KH}$, $\triangle EGD \cong \triangle KHD$ by ASA.

9. Given: $\overline{EF} \parallel \overline{GH}$, $\overline{EF} \cong \overline{GH}$
Prove: $\overline{EK} \cong \overline{KH}$
Proof:

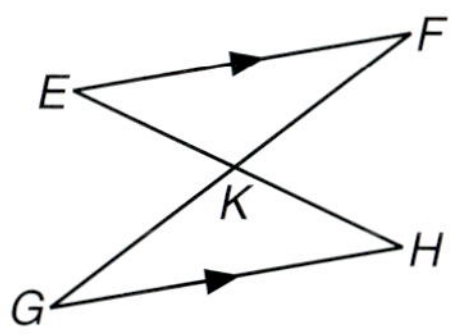

- $\overline{EF} \parallel \overline{GH}$ (Given) → $\angle E \cong \angle H$ (Alt. int. ∠s are ≅.)
- $\overline{EF} \cong \overline{GH}$ (Given)
- $\angle GKH \cong \angle EKF$ (Vert. ∠s are ≅.)
- → $\triangle EKF \cong \triangle HKG$ (AAS)
- → $\overline{EK} \cong \overline{KH}$ (CPCTC)

10. Given: $\overline{DE} \parallel \overline{JK}$, $\overline{DK}$ bisects $\overline{JE}$.
Prove: $\triangle EGD \cong \triangle JGK$
Proof:

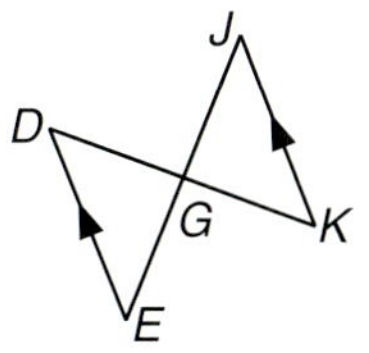

- $\overline{DK}$ bisects $\overline{JE}$. (Given) → $\overline{EG} \cong \overline{GJ}$ (Def. of seg. bisector)
- $\overline{DE} \parallel \overline{JK}$ (Given) → $\angle E \cong \angle J$ (Alt. int. ∠s are ≅.)
- $\angle DGE \cong \angle KGJ$ (Vert. ∠s are ≅.)
- → $\triangle EGD \cong \triangle JGK$ (ASA)

11. Given: $\angle V \cong \angle S$, $\overline{TV} \cong \overline{QS}$
Prove: $\overline{VR} \cong \overline{SR}$
Proof:

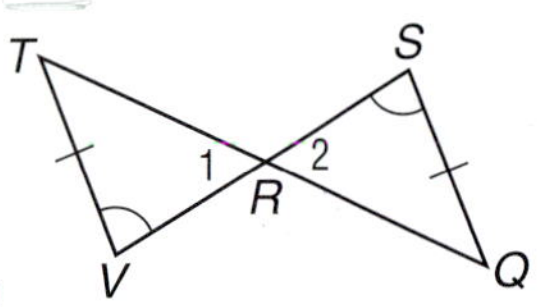

- $\angle V \cong \angle S$, $\overline{TV} \cong \overline{QS}$ (Given)
- $\angle 1 \cong \angle 2$ (Vert. ∠s are ≅.)
- → $\triangle TRV \cong \triangle QRS$ (AAS)
- → $\overline{VR} \cong \overline{SR}$ (CPCTC)

12. Given: $\overline{EJ} \parallel \overline{FK}$, $\overline{JG} \parallel \overline{KH}$, $\overline{EF} \cong \overline{GH}$
Prove: $\triangle EJG \cong \triangle FKH$
Proof:

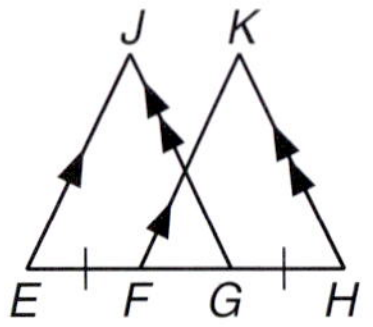

- $\overline{EF} \cong \overline{GH}$ (Given)
- → $EF = GH$ (Def. of ≅ seg.)
- → $EF + FG = FG + GH$ (Addition Prop.) ← $FG = FG$ (Reflexive Prop.)
- → $EG = FH$ (Substitution) ← $EF + FG = EG$, $FG + GH = FH$ (Seg. Addition Post.)
- → $\overline{EG} \cong \overline{FH}$ (Def. of ≅ seg.)
- $\overline{EJ} \parallel \overline{FK}$, $\overline{JG} \parallel \overline{KH}$ (Given) → $\angle JEG \cong \angle KFH$, $\angle JGE \cong \angle HFK$ (Corr. ∠s are ≅.)
- → $\triangle EJG \cong \triangle FKH$ (SAS)

13. Given: $\overline{MN} \cong \overline{PQ}$, $\angle M \cong \angle Q$
$\angle 2 \cong \angle 3$
Prove: $\triangle MLP \cong \triangle QLN$
Proof:

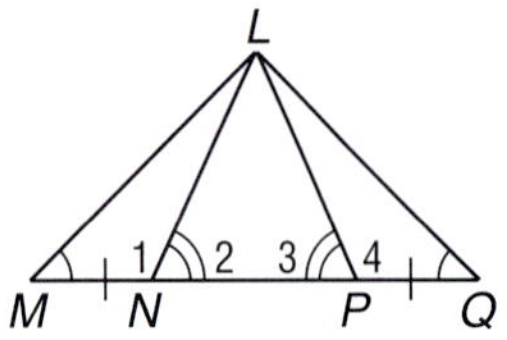

- $\overline{MN} \cong \overline{PQ}$ (Given)
- → $MN = PQ$ (Def. of ≅ seg.)
- → $MN + NP = NP + PQ$ (Addition Prop.) ← $NP = NP$ (Reflexive Prop.)
- → $MP = NQ$ (Substitution) ← $MN + NP = MP$, $NP + PQ = NQ$ (Seg. Addition Post.)
- → $\overline{MP} \cong \overline{NQ}$ (Def. of ≅ seg.)
- → $\triangle MLP \cong \triangle QLN$ (ASA) ← $\angle M \cong \angle Q$, $\angle 2 \cong \angle 3$ (Given)

14. Given: Z is the midpoint of $\overline{CT}$.
$\overline{CY} \parallel \overline{TE}$
Prove: $\overline{YZ} \cong \overline{EZ}$
Proof:

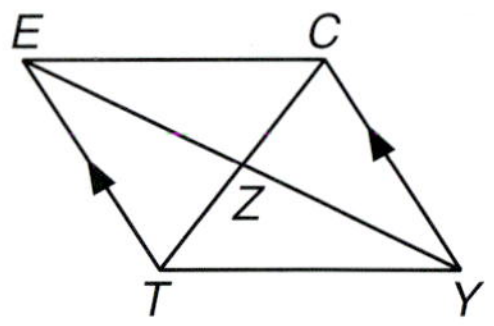

- Z is the midpoint of $\overline{CT}$. (Given) → $\overline{TZ} \cong \overline{CZ}$ (Midpt. Th.)
- $\overline{CY} \perp \overline{TE}$ (Given) → $\angle ETC \cong \angle YCT$, $\angle TEY \cong \angle CYE$ (Alt. int. ∠s are ≅.)
- → $\triangle EZT \cong \triangle YZC$ (AAS)
- → $\overline{YZ} \cong \overline{EZ}$ (CPCTC)

19. Given: $\angle MYT \cong \angle NYT$
$\angle MTY \cong \angle NTY$
Prove: $\triangle RYM \cong \triangle RYN$

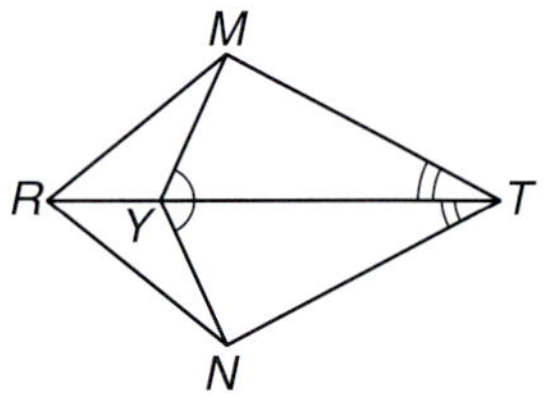

Proof:
Statements (Reasons)

1. $\angle MYT \cong \angle NYT$, $\angle MTY \cong \angle NTY$ (Given)
2. $\overline{YT} \cong \overline{YT}$, $\overline{RY} \cong \overline{RY}$ (Reflexive Property)
3. $\triangle MYT \cong \triangle NYT$ (ASA)
4. $\overline{MY} \cong \overline{NY}$ (CPCTC)
5. $\angle RYM$ and $\angle MYT$ are a linear pair; $\angle RYN$ and $\angle NYT$ are a linear pair (Def. of linear pair)
6. $\angle RYM$ and $\angle MYT$ are supplementary and $\angle RYN$ and $\angle NYT$ are supplementary. (Suppl.Th.)
7. $\angle RYM \cong \angle RYN$ (∠s suppl. to ≅ ∠s are ≅.)
8. $\triangle RYM \cong \triangle RYN$ (SAS)

20. Given: $\triangle BMI \cong \triangle KMT$
$\overline{IP} \cong \overline{PT}$
Prove: $\triangle IPK \cong \triangle TPB$

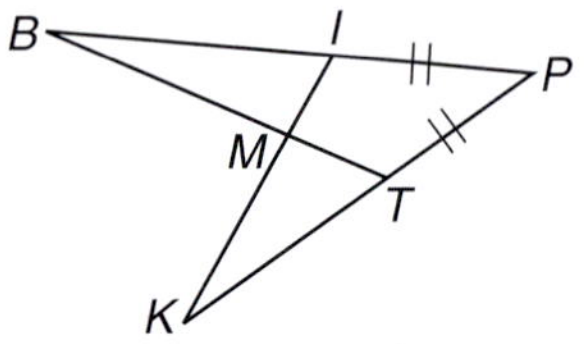

Proof:
Statements (Reasons)

1. $\triangle BMI \cong \triangle KMT$ (Given)
2. $\angle B \cong \angle K$ (CPCTC)
3. $\overline{IP} \cong \overline{PT}$ (Given)
4. $\angle P \cong \angle P$ (Reflexive Prop.)
5. $\triangle IPK \cong \triangle TPB$ (AAS)

21. $\overline{CD} \cong \overline{GH}$, because the segments have the same measure. $\angle CFD \cong \angle HFG$ because vertical angles are congruent. Since F is the midpoint of $\overline{DG}$, $\overline{DF} \cong \overline{FG}$. It cannot be determined whether $\triangle CFD \cong \triangle HFG$. The information given does not lead to a unique triangle.

22. Since F is the midpoint of $\overline{DG}$, $\overline{DF} \cong \overline{FG}$. F is also the midpoint of $\overline{CH}$, so $\overline{CF} \cong \overline{FH}$. Since $\overline{DG} \cong \overline{CH}$, $\overline{DF} \cong \overline{CF}$ and $\overline{FG} \cong \overline{FH}$. $\angle CFD \cong \angle HFG$ because vertical angles are congruent. $\triangle CFD \cong \triangle HFG$ by SAS.

23. Since N is the midpoint of $\overline{JL}$, $\overline{JN} \cong \overline{NL}$. $\angle JNK \cong \angle LNK$ because perpendicular lines form right angles and right angles are congruent. By the Reflexive Property, $\overline{KN} \cong \overline{KN}$. $\triangle JKN \cong \triangle LKN$ by SAS.

24. It is given that $\overline{JM} \cong \overline{LM}$ and $\angle NJM \cong \angle NLM$. By the Reflexive Property, $\overline{NM} \cong \overline{NM}$. It cannot be determined whether $\triangle JNM \cong \triangle LNM$. The information given does not lead to a unique triangle.

33. Given: $\overline{BA} \cong \overline{DE}$, $\overline{DA} \cong \overline{BE}$
Prove: $\triangle BEA \cong \triangle DAE$

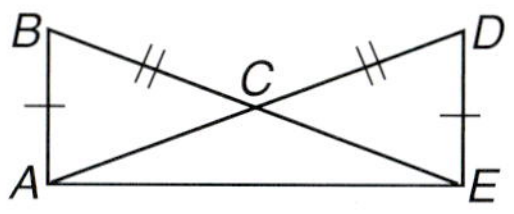

Proof:

$\overline{DA} \cong \overline{BE}$ (Given) → $\triangle BEA \cong \triangle DAE$
$\overline{BA} \cong \overline{DE}$ (Given) → $\triangle BEA \cong \triangle DAE$
$\overline{AE} \cong \overline{AE}$ (Reflexive Prop.) → $\triangle BEA \cong \triangle DAE$
$\triangle BEA \cong \triangle DAE$ (ASA)

34. Given: $\overline{XZ} \perp \overline{WY}$, $\overline{XZ}$ bisects $\overline{WY}$.
Prove: $\triangle WZX \cong \triangle YZX$

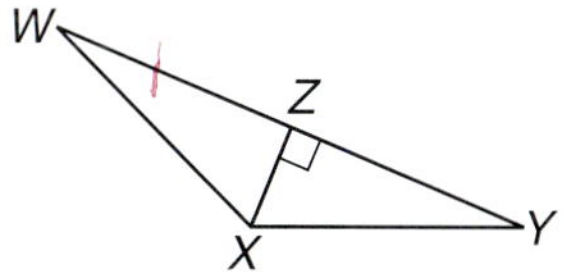

Proof:

$\overline{XZ} \perp \overline{WY}$ (Given) → $\angle WZX$ and $\angle YZX$ are rt $\angle$s. (Def. of $\perp$ lines.) → $\angle WZX \cong \angle YZX$ (All rt. $\angle$s are $\cong$.)
$\overline{XZ}$ bisects $\overline{WY}$. (Given) → $\overline{WZ} \cong \overline{ZY}$ (Def. of seg. bisector)
$\overline{XZ} \cong \overline{XZ}$ (Reflexive Prop.)
$\angle WZX \cong \angle YZX$, $\overline{WZ} \cong \overline{ZY}$, $\overline{XZ} \cong \overline{XZ}$ → $\triangle WZX \cong \triangle YZX$ (SAS)

Pages 214–215, Geometry Activity

10. Given: $\overline{ML} \perp \overline{MK}$, $\overline{JK} \perp \overline{KM}$, $\angle J \cong \angle L$
Prove: $\overline{JM} \cong \overline{KL}$

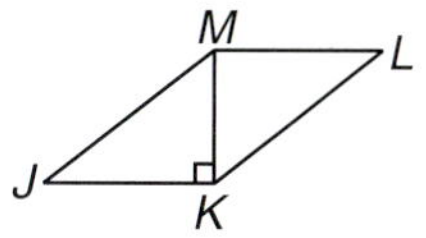

Proof:

Statements (Reasons)

1. $\overline{ML} \perp \overline{MK}$, $\overline{JK} \perp \overline{KM}$, $\angle J \cong \angle L$ (Given)
2. $\angle LMK$ and $\angle JKM$ are rt. $\angle$s ($\perp$ lines form 4 rt. $\angle$s.)
3. $\triangle LMK$ and $\triangle JKM$ are rt. $\triangle$s (Def. of rt. $\triangle$)
4. $\overline{MK} \cong \overline{MK}$ (Reflexive Property)
5. $\triangle LMK \cong \triangle JKM$ (LA)
6. $\overline{JM} \cong \overline{KL}$ (CPCTC)

11. Given: $\overline{JK} \perp \overline{KM}$, $\overline{JM} \cong \overline{KL}$, $\overline{ML} \parallel \overline{JK}$
Prove: $\overline{ML} \cong \overline{JK}$

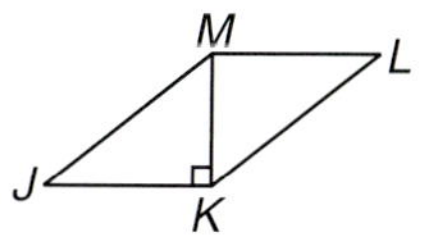

Proof:

Statements (Reasons)

1. $\overline{JK} \perp \overline{KM}$, $\overline{JM} \cong \overline{KL}$, $\overline{ML} \parallel \overline{JK}$ (Given)
2. $\angle JKM$ is a rt. $\angle$ ($\perp$ lines form rt. $\angle$s.)
3. $\overline{KM} \perp \overline{ML}$ (Perpendicular Transversal Th.)
4. $\angle LMK$ is a rt. $\angle$ ($\perp$ lines form rt. $\angle$s.)
5. $\overline{MK} \cong \overline{MK}$ (Reflexive Property)
6. $\triangle JMK \cong \triangle LMK$ (HL)
7. $\overline{ML} \cong \overline{JK}$ (CPCTC)

Pages 219–221, Lesson 4-6

7. Given: $\triangle CTE$ is isosceles with vertex $\angle C$. $m\angle T = 60$
Prove: $\triangle CTE$ is equilateral.

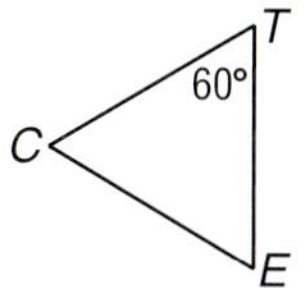

Proof:

Statements (Reasons)

1. $\triangle CTE$ is isosceles with vertex $\angle C$. (Given)
2. $\overline{CT} \cong \overline{CE}$ (Def. of isosceles $\triangle$)
3. $\angle E \cong \angle T$ (Isosceles $\triangle$ Th.)
4. $m\angle E = m\angle T$ (Def. of $\cong$ $\angle$s)
5. $m\angle T = 60$ (Given)
6. $m\angle E = 60$ (Substitution)
7. $m\angle C + m\angle E + m\angle T = 180$ (Angle Sum Theorem)
8. $m\angle C + 60 + 60 = 180$ (Substitution)
9. $m\angle C = 60$ (Subtraction)
10. $\triangle CTE$ is equiangular. (Def. of equiangular $\triangle$)
11. $\triangle CTE$ is equilateral. (Equiangular $\triangle$s are equilateral.)

29. Given: $\triangle XKF$ is equilateral. $\overline{XJ}$ bisects $\angle X$.
Prove: J is the midpoint of $\overline{KF}$.

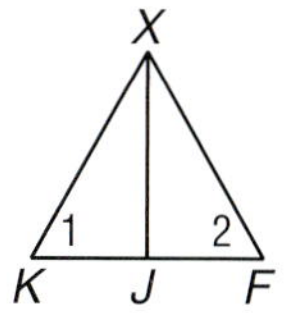

Proof:

Statements (Reasons)

1. $\triangle XKF$ is equilateral. (Given)
2. $\angle 1 \cong \angle 2$ (Equilateral $\triangle$s are equiangular.)
3. $\overline{KX} \cong \overline{FX}$ (Definition of equilateral $\triangle$)
4. $\overline{XJ}$ bisects $\angle X$ (Given)
5. $\angle KXJ \cong \angle FXJ$ (Def. of $\angle$ bisector)
6. $\triangle KXJ \cong \triangle FXJ$ (ASA)
7. $\overline{KJ} \cong \overline{JF}$ (CPCTC)
8. J is the midpoint of $\overline{KF}$. (Def. of midpoint)

30. Given: $\triangle MLP$ is isosceles.
N is the midpoint of $\overline{MP}$.
Prove: $\overline{LN} \perp \overline{MP}$

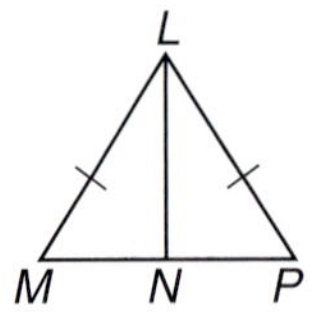

Proof:
Statements (Reasons)

1. $\triangle MLP$ is isosceles. (Given)
2. $\overline{ML} \cong \overline{LP}$ (Definition of isosceles $\triangle$)
3. $\angle M \cong \angle P$ (Isosceles $\triangle$ Theorem)
4. N is the midpoint of $\overline{MP}$. (Given)
5. $\overline{MN} \cong \overline{NP}$ (Midpoint Theorem)
6. $\triangle MNL \cong \triangle PNL$ (SAS)
7. $\angle LNM \cong \angle LNP$ (CPCTC)
8. $m\angle LNM = m\angle LNP$ ($\cong$ $\angle$s have equal measures.)
9. $\angle LNM$ and $\angle LNP$ are a linear pair (Def. of linear pair)
10. $m\angle LNM + m\angle LNP = 180$ (Sum of measures of linear pair of $\angle$s $= 180$)
11. $2m\angle LNM = 180$ (Substitution)
12. $m\angle LNM = 90$ (Division)
13. $\angle LNM$ is a right angle. (Definition of right $\angle$)
14. $\overline{LN} \perp \overline{MP}$ (Definition of $\perp$)

31. Case I:
Given: $\triangle ABC$ is an equilateral triangle.
Prove: $\triangle ABC$ is an equiangular triangle.

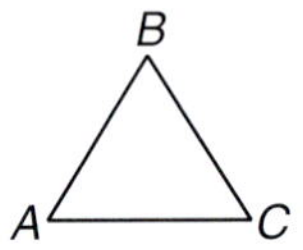

Proof:
Statements (Reasons)

1. $\triangle ABC$ is an equilateral triangle. (Given)
2. $\overline{AB} \cong \overline{AC} \cong \overline{BC}$ (Def. of equilateral $\triangle$)
3. $\angle A \cong \angle B \cong \angle C$ (Isosceles $\triangle$ Th.)
4. $\triangle ABC$ is an equiangular triangle. (Def. of equiangular $\triangle$)

Case II:
Given: $\triangle ABC$ is an equiangular triangle.
Prove: $\triangle ABC$ is an equilateral triangle.

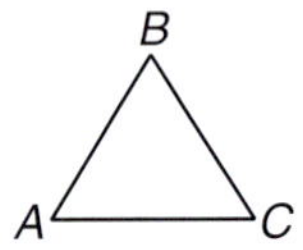

Proof:
Statements (Reasons)

1. $\triangle ABC$ is an equiangular triangle. (Given)
2. $\angle A \cong \angle B \cong \angle C$ (Def. of equiangular $\triangle$)
3. $\overline{AB} \cong \overline{AC} \cong \overline{BC}$ (If 2 $\angle$s of a $\triangle$ are $\cong$, then the sides opp. those $\angle$s are $\cong$.)
4. $\triangle ABC$ is an equilateral triangle. (Def. of equiangular $\triangle$)

32. Given: $\triangle MNO$ is an equilateral triangle.
Prove: $m\angle M = m\angle N = m\angle O = 60$

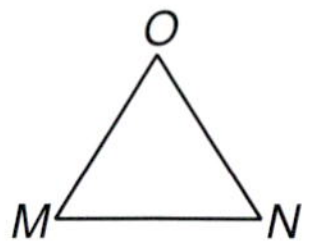

Proof:
Statements (Reasons)

1. $\triangle MNO$ is an equilateral triangle. (Given)
2. $\overline{MN} \cong \overline{MO} \cong \overline{NO}$ (Def. of equilateral $\triangle$)
3. $\angle M \cong \angle N \cong \angle O$ (Isosceles $\triangle$ Th.)
4. $m\angle M = m\angle N = m\angle O$ (Def. of $\cong$ $\angle$s)
5. $m\angle M + m\angle N + m\angle O = 180$ ($\angle$ Sum Theorem)
6. $3m\angle M = 180$ (Substitution)
7. $m\angle M = 60$ (Division prop.)
8. $m\angle M = m\angle N = m\angle O = 60$ (Substitution)

33. Given: $\triangle ABC$
$\angle A \cong \angle C$
Prove: $\overline{AB} \cong \overline{CB}$
Proof:

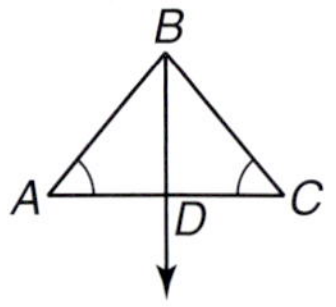

Statements (Reasons)

1. Let $\overrightarrow{BD}$ bisect $\angle ABC$. (Protractor Post.)
2. $\angle ABD \cong \angle CBD$ (Def. bisector)
3. $\angle A \cong \angle C$ (Given)
4. $\overline{BD} \cong \overline{BD}$ (Reflexive Prop.)
5. $\triangle ABD \cong \triangle CBD$ (AAS)
6. $\overline{AB} \cong \overline{CB}$ (CPCTC)

38. There are two sets of 12 isosceles triangles. One black set forms a circle with their bases on the outside of the circle. Another black set encircles a circle in the middle.

39. The triangles in each set appear to be acute.

44. Given: $\angle N \cong \angle D$, $\angle G \cong \angle I$, $\overline{AN} \cong \overline{SD}$
Prove: $\triangle ANG \cong \triangle SDI$

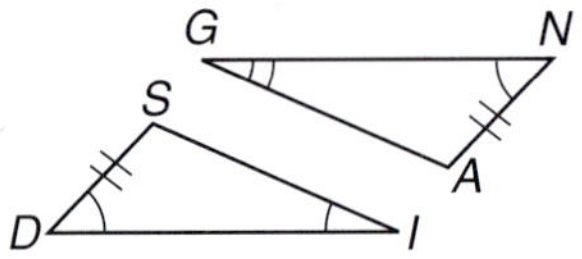

Proof: We are given $\angle N \cong \angle D$ and $\angle G \cong \angle I$. By the Third Angle Theorem, $\angle A \cong \angle S$. We are also given $\overline{AN} \cong \overline{SD}$. $\triangle ANG \cong \triangle SDI$ by ASA.

45. Given: $\overline{VR} \perp \overline{RS}$, $\overline{UT} \perp \overline{SU}$,
$\overline{RS} \cong \overline{US}$
Prove: $\triangle VRS \cong \triangle TUS$

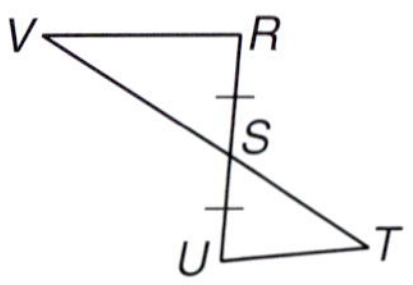

Proof: We are given that $\overline{VR} \perp \overline{RS}$, $\overline{UT} \perp \overline{SU}$, and $\overline{RS} \cong \overline{US}$. Perpendicular lines form four right angles, so $\angle R$ and $\angle U$ are right angles. $\angle R \cong \angle U$ because all right angles are congruent. $\angle RSV \cong \angle UST$ since vertical angles are congruent. Therefore, $\triangle VRS \cong \triangle TUS$ by ASA.

48.

a	b	a and b
T	T	T
T	F	F
F	T	F
F	F	F

49.

p	q	$\sim p$	$\sim q$	$\sim p$ or $\sim q$
T	T	F	F	F
T	F	F	T	T
F	T	T	F	T
F	F	T	T	T

50.

k	m	$\sim m$	k and $\sim m$
T	T	F	F
T	F	T	T
F	T	F	F
F	F	T	F

51.

y	z	$\sim y$	$\sim y$ or z
T	T	F	T
T	F	F	F
F	T	T	T
F	F	T	T

Page 221, Practice Quiz 2

1. $JM = \sqrt{5}$, $ML = \sqrt{26}$, $JL = 5$, $BD = \sqrt{5}$, $DG = \sqrt{26}$, and $BG = 5$. Each pair of corresponding sides have the same measure so they are congruent. $\triangle JML \cong \triangle BDG$ by SSS.
2. **Given:** $\angle A \cong \angle H$, $\angle AEJ \cong \angle HJE$
   Prove: $\overline{AJ} \cong \overline{EH}$

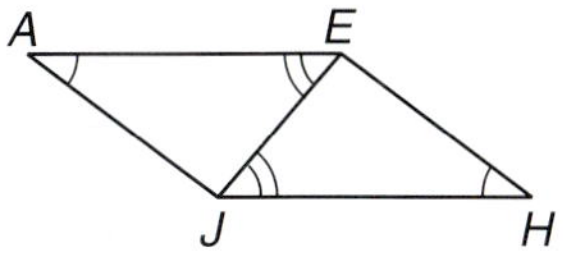

Proof:
Statements (Reasons)

1. $\angle A \cong \angle H$, $\angle AEJ \cong \angle HJE$ (Given)
2. $\overline{EJ} \cong \overline{EJ}$ (Reflexive Prop.)
3. $\triangle AEJ \cong \triangle HJE$ (AAS)
4. $\overline{AJ} \cong \overline{EH}$ (CPCTC)

Pages 224–226, Lesson 4-7

8. **Given:** $\triangle ABC$ is a right triangle with hypotenuse $\overline{BC}$. M is the midpoint of $\overline{BC}$.
   Prove: M is equidistant from the vertices.

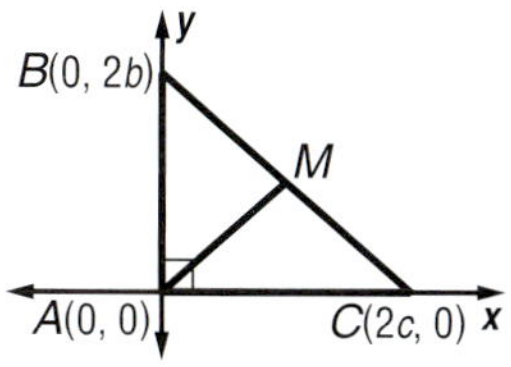

Proof: The coordinates of M, the midpoint of $\overline{BC}$, will be $\left(\frac{2c}{2}, \frac{2b}{2}\right) = (c, b)$. The distance from M to each of the vertices can be found using the Distance Formula.

$MB = \sqrt{(c-0)^2 + (b-2b)^2} = \sqrt{c^2 + b^2}$

$MC = \sqrt{(c-2c)^2 + (b-0)^2} = \sqrt{c^2 + b^2}$

$MA = \sqrt{(c-0)^2 + (b-0)^2} = \sqrt{c^2 + b^2}$

Thus, $MB = MC = MA$, and M is equidistant from the vertices.

9. **Given:** $\triangle ABC$
   Prove: $\triangle ABC$ is isosceles.
   Proof: Use the Distance Formula to find AB and BC.

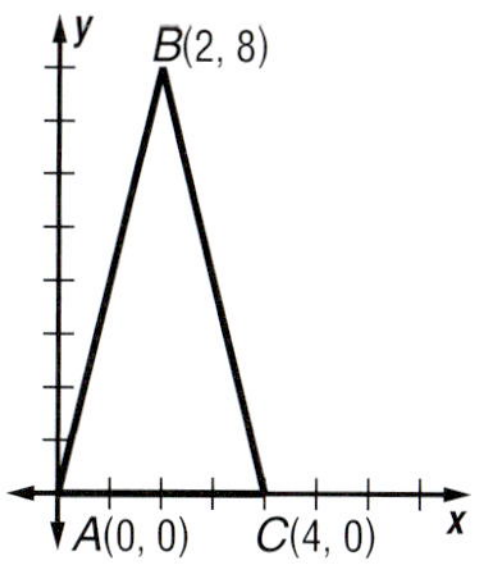

$AB = \sqrt{(2-0)^2 + (8-0)^2}$

$= \sqrt{4 + 64}$ or $\sqrt{68}$

$BC = \sqrt{(4-2)^2 + (0-8)^2}$

$= \sqrt{4 + 64}$ or $\sqrt{68}$

Since $AB = BC$, $\overline{AB} \cong \overline{BC}$. Since the legs are congruent, $\triangle ABC$ is isosceles.

10. Sample answer:

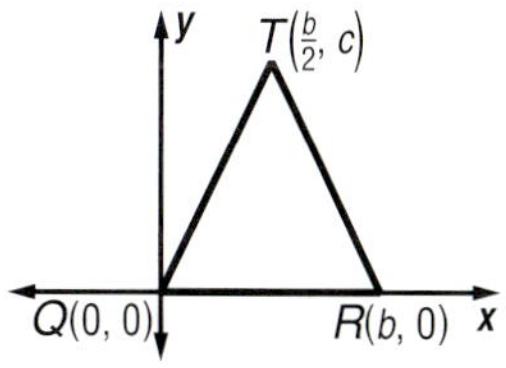

11. Sample answer:

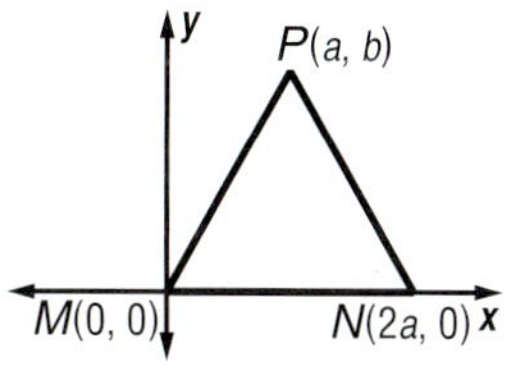

12. Sample answer:

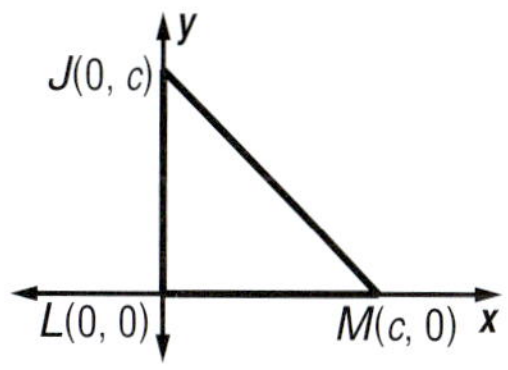

13. Sample answer:

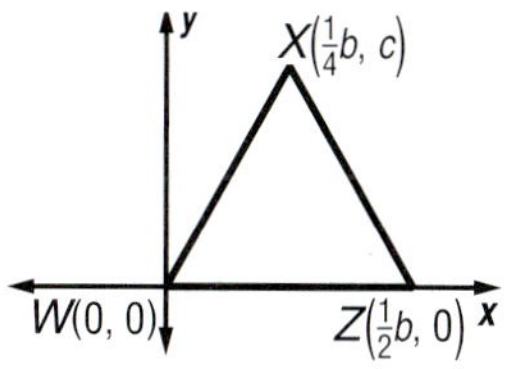

14. Sample answer:

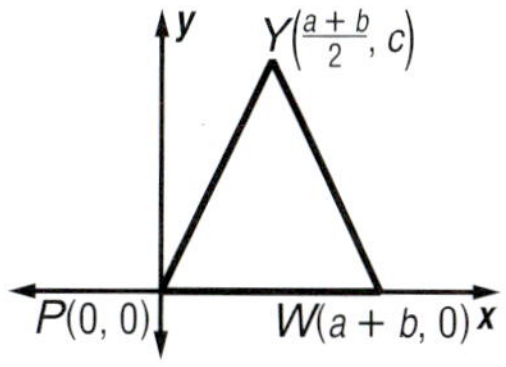

15. Sample answer:

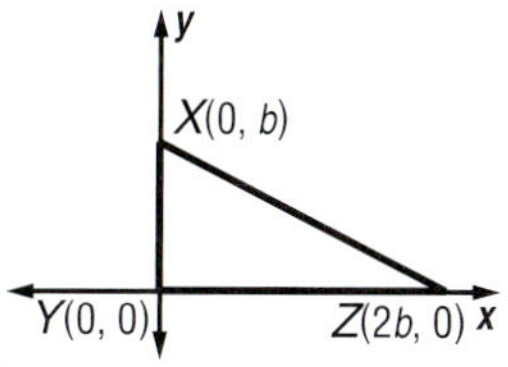

25. **Given:** isosceles $\triangle ABC$ with $\overline{AC} \cong \overline{BC}$
    R and S are midpoints of legs $\overline{AC}$ and $\overline{BC}$.
    Prove: $\overline{AS} \cong \overline{BR}$

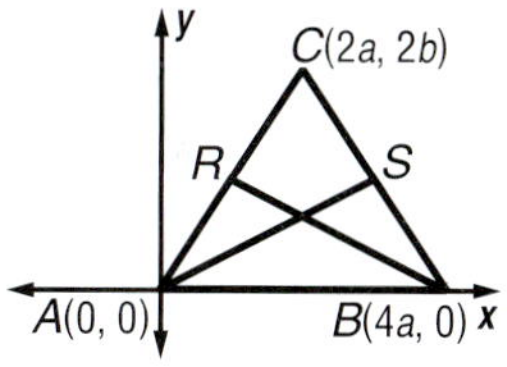

Proof: The coordinates of R are $\left(\frac{2a+0}{2}, \frac{2b+0}{2}\right)$ or (a, b).

The coordinates of S are $\left(\frac{2a+4a}{2}, \frac{2b+0}{2}\right)$ or $(3a, b)$.

$BR = \sqrt{(4a-a)^2 + (0-b)^2}$

$= \sqrt{(3a)^2 + (-b)^2}$ or $\sqrt{9a^2 + b^2}$

$AS = \sqrt{(3a-0)^2 + (b-0)^2}$

$= \sqrt{(3a)^2 + (b)^2}$ or $\sqrt{9a^2 + b^2}$

Since $BR = AS$, $\overline{AS} \cong \overline{BR}$.

26. Given: isosceles triangle ABC
$\overline{AB} \cong \overline{BC} \cong \overline{AC}$
R, S, and T are midpoints of their respective sides.

Prove: $\triangle RST$ is isosceles.

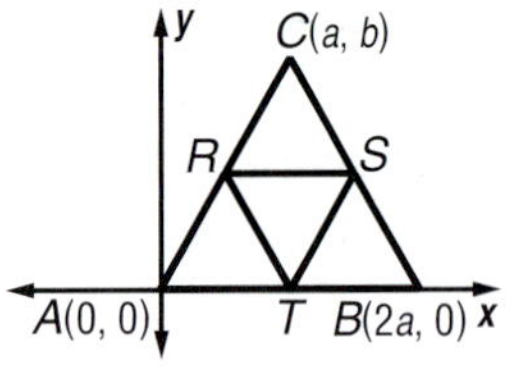

Proof:

Midpoint R is $\left(\frac{a+0}{2}, \frac{b+0}{2}\right)$ or $\left(\frac{a}{2}, \frac{b}{2}\right)$.

Midpoint S is $\left(\frac{a+2a}{2}, \frac{b+0}{2}\right)$ or $\left(\frac{3a}{2}, \frac{b}{2}\right)$.

Midpoint T is $\left(\frac{2a+0}{2}, \frac{0+0}{2}\right)$ or $(a, 0)$.

$$RT = \sqrt{\left(\frac{a}{2} - a\right)^2 + \left(\frac{b}{2} - 0\right)^2} = \sqrt{\left(-\frac{a}{2}\right)^2 + \left(\frac{b}{2}\right)^2}$$
$$= \sqrt{\left(\frac{a}{2}\right)^2 + \left(\frac{b}{2}\right)^2}$$
$$ST = \sqrt{\left(\frac{3a}{2} - a\right)^2 + \left(\frac{b}{2} - 0\right)^2}$$
$$= \sqrt{\left(\frac{a}{2}\right)^2 + \left(\frac{b}{2}\right)^2}$$

$RT = ST$ and $\overline{RT} \cong \overline{ST}$ and $\triangle RST$ is isosceles.

27. Given: $\triangle ABC$
S is the midpoint of $\overline{AC}$.
T is the midpoint of $\overline{BC}$.

Prove: $\overline{ST} \parallel \overline{AB}$

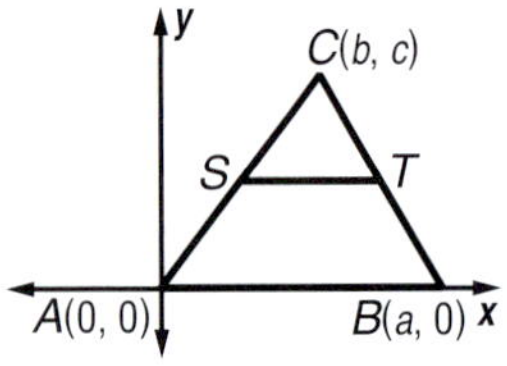

Proof:

Midpoint S is $\left(\frac{b+0}{2}, \frac{c+0}{2}\right)$ or $\left(\frac{b}{2}, \frac{c}{2}\right)$.

Midpoint T is $\left(\frac{a+b}{2}, \frac{0+c}{2}\right)$ or $\left(\frac{a+b}{2}, \frac{c}{2}\right)$.

$$\text{Slope of } \overline{ST} = \frac{\frac{c}{2} - \frac{c}{2}}{\frac{a+b}{2} - \frac{b}{2}} = \frac{0}{\frac{a}{2}} \text{ or } 0.$$
$$\text{Slope of } \overline{AB} = \frac{0-0}{a-0} = \frac{0}{a} \text{ or } 0.$$

$\overline{ST}$ and $\overline{AB}$ have the same slope so $\overline{ST} \parallel \overline{AB}$.

28. Given: $\triangle ABC$
S is the midpoint of $\overline{AC}$.
T is the midpoint of $\overline{BC}$.

Prove: $ST = \frac{1}{2}AB$

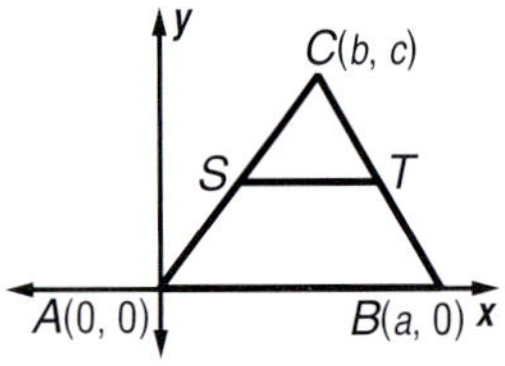

Proof:

$$ST = \sqrt{\left(\frac{a+b}{2} - \frac{b}{2}\right)^2 + \left(\frac{c}{2} - \frac{c}{2}\right)^2}$$
$$= \sqrt{\left(\frac{a}{2}\right)^2 + 0^2}$$
$$= \sqrt{\left(\frac{a}{2}\right)^2} \text{ or } \frac{a}{2}$$
$$AB = \sqrt{(a-0)^2 + (0-0)^2}$$
$$= \sqrt{a^2 + 0^2} \text{ or } a$$
$$ST = \frac{1}{2}AB$$

29. Given: $\triangle ABD$, $\triangle FBD$
$AF = 6$, $BD = 3$

Prove: $\triangle ABD \cong \triangle FBD$

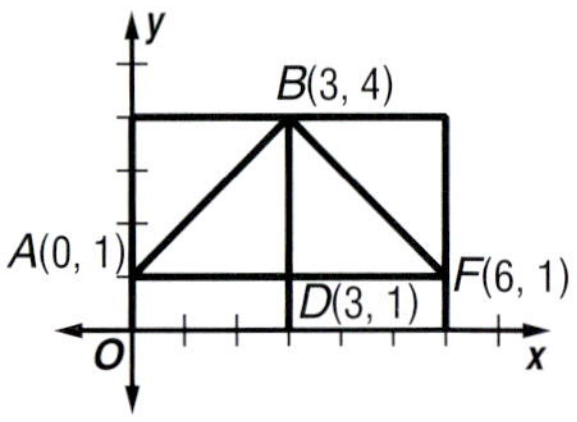

Proof: $\overline{BD} \cong \overline{BD}$ by the Reflexive Property.

$AD = \sqrt{(3-0)^2 + (1-1)^2} = \sqrt{9+0}$ or 3

$DF = \sqrt{(6-3)^2 + (4-1)^2} = \sqrt{9+0}$ or 3

Since $AD = DF$, $\overline{AD} \cong \overline{DF}$.

$AB = \sqrt{(3-0)^2 + (4-1)^2} = \sqrt{9+9}$ or $3\sqrt{2}$

$BF = \sqrt{(6-3)^2 + (1-4)^2} = \sqrt{9+9}$ or $3\sqrt{2}$

Since $AB = BF$, $\overline{AB} \cong \overline{BF}$.

$\triangle ABD \cong \triangle FBD$ by SSS.

30. Given: $\triangle BPR$
$PR = 800$, $BR = 800$

Prove: $\triangle BPR$ is an isosceles right triangle.

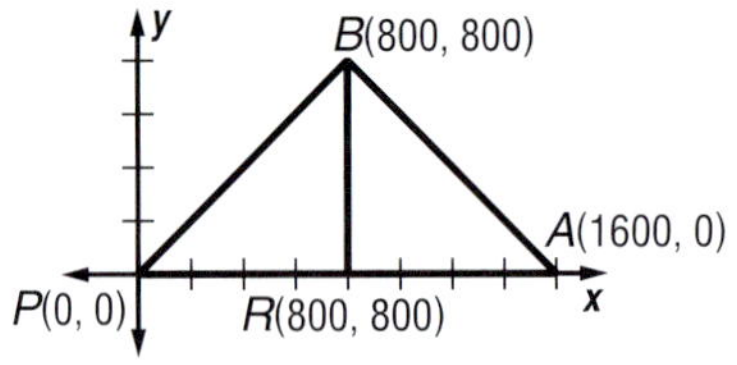

Proof:

Since PR and BR have the same measure, $\overline{PR} \cong \overline{BR}$.

The slope of $PR = \frac{0-0}{800-0}$ or 0.

The slope of $BR = \frac{800-0}{800-800}$, which is undefined.

$\overline{PR} \perp \overline{BR}$, so $\angle PRB$ is a right angle. $\triangle BPR$ is an isosceles right triangle.

31. Given: $\triangle BPR$, $\triangle BAR$
$PR = 800$, $BR = 800$, $RA = 800$

Prove: $\overline{PB} \cong \overline{BA}$

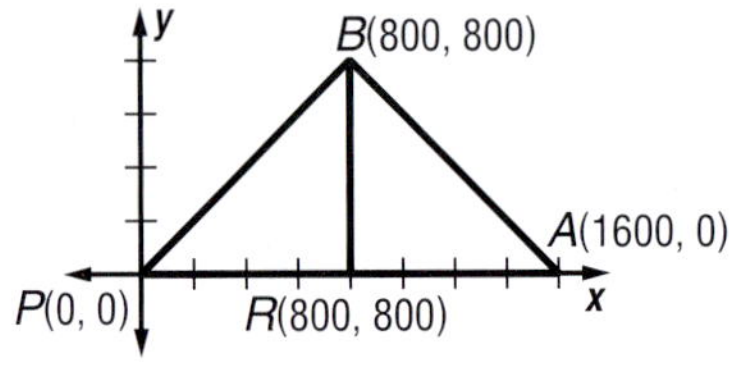

Proof:

$PB = \sqrt{(800-0)^2 + (800-0)^2}$ or $\sqrt{1,280,000}$

$BA = \sqrt{(800-1600)^2 + (800-0)^2}$ or $\sqrt{1,280,000}$

$PB = BA$, so $\overline{PB} \cong \overline{BA}$.

32. Given: $\triangle JCT$
Prove: $\triangle JCT$ is a right triangle.

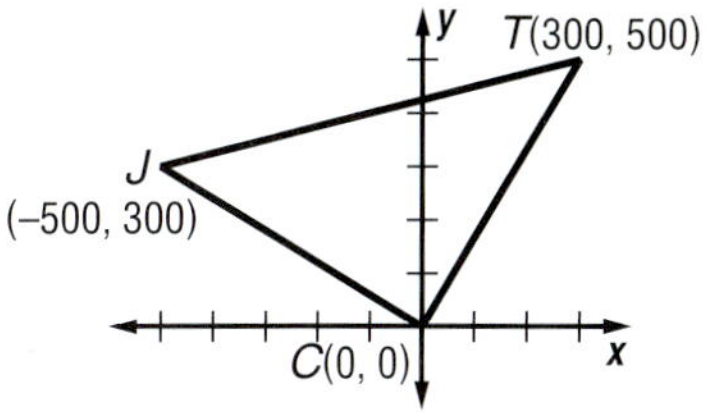

Proof:

The slope of $\overline{JC} = \dfrac{300 - 0}{-500 - 0}$ or $-\dfrac{3}{5}$.

The slope of $\overline{TC} = \dfrac{500 - 0}{300 - 0}$ or $\dfrac{5}{3}$.

The slope of $\overline{TC}$ is the opposite reciprocal of the slope of $\overline{JC}$. $\overline{JC} \perp \overline{TC}$, so $\angle TCJ$ is a right angle. So $\triangle JCT$ is a right triangle.

41. Given: $\angle 3 \cong \angle 4$
Prove: $\overline{QR} \cong \overline{QS}$

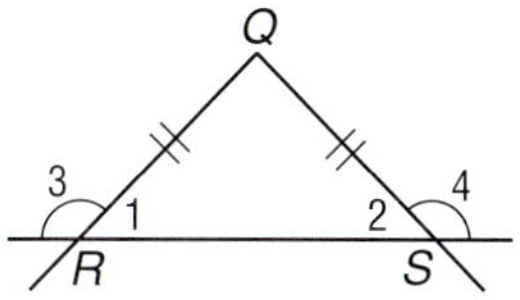

Proof:
Statements (Reasons)

1. $\angle 3 \cong \angle 4$ (Given)
2. $\angle 2$ and $\angle 4$ form a linear pair. $\angle 1$ and $\angle 3$ form a linear pair. (Def. of linear pair)
3. $\angle 2$ and $\angle 4$ are supplementary. $\angle 1$ and $\angle 3$ are supplementary. (If 2 $\angle$s form a linear pair, then they are suppl.)
4. $\angle 2 \cong \angle 1$ (Angles that are suppl. to $\cong$ $\angle$s are $\cong$.)
5. $\overline{QR} \cong \overline{QS}$ (If 2 $\angle$s of a $\triangle$ are $\cong$, then the sides opposite those $\angle$s are $\cong$.)

42. Given: isosceles $\triangle JKN$ with vertex $\angle N$, $\overline{JK} \parallel \overline{LM}$
Prove: $\triangle NML$ is isosceles

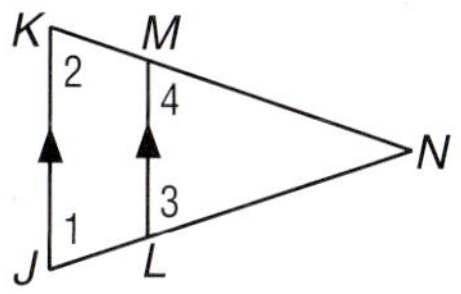

Proof:
Statements (Reasons)

1. isosceles triangle JKN with vertex $\angle N$ (Given)
2. $\overline{NJ} \cong \overline{NK}$ (Def. of isosceles triangle)
3. $\angle 2 \cong \angle 1$ (Isosceles Triangle Theorem)
4. $\overline{JK} \parallel \overline{LM}$ (Given)
5. $\angle 1 \cong \angle 3$, $\angle 4 \cong \angle 2$ (Corr. $\angle$s Post.)
6. $\angle 2 \cong \angle 3$, $\angle 4 \cong \angle 1$ (Congruence of $\angle$s is transitive. Statements 3 and 5)
7. $\angle 4 \cong \angle 3$ (Congruence of $\angle$s is transitive. Statements 3 and 6)
8. $\overline{LN} \cong \overline{MN}$ (If 2 $\angle$s of a $\triangle$ are $\cong$, then the sides opp. those $\angle$s are $\cong$.)
9. $\triangle NML$ is an isosceles triangle. (Def. of isosceles $\triangle$)

43. Given: $\overline{AD} \cong \overline{CE}$, $\overline{AD} \parallel \overline{CE}$
Prove: $\triangle ABD \cong \triangle EBC$

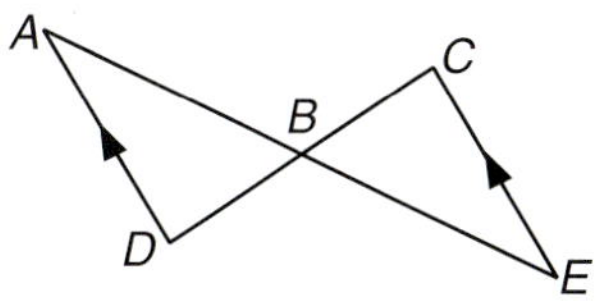

Proof:
Statements (Reasons)

1. $\overline{AD} \parallel \overline{CE}$ (Given)
2. $\angle A \cong \angle E$, $\angle D \cong \angle C$ (Alt. Int. $\angle$s Theorem)
3. $\overline{AD} \cong \overline{CE}$ (Given)
4. $\triangle ABD \cong \triangle EBC$ (ASA)

44. Given: $\overline{WX} \cong \overline{XY}$, $\angle V \cong \angle Z$
Prove: $\overline{WV} \cong \overline{YZ}$

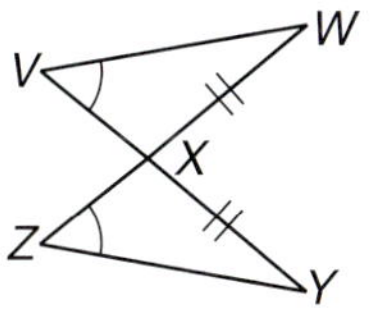

Proof:
Statements (Reasons)

1. $\overline{WX} \cong \overline{XY}$, $\angle V \cong \angle Z$ (Given)
2. $\angle WXV \cong \angle YXZ$ (Vertical $\angle$s $\cong$)
3. $\triangle WXV \cong \triangle YXZ$ (AAS)
4. $\overline{WV} \cong \overline{YZ}$ (CPCTC)

Page 231, Chapter 4 Practice Test

17. Given: $\triangle ABE$, $\triangle BCE$
$AB = 22$, $AC = 44$,
$AE = 36$, $CD = 36$
$\angle A$ and $\angle C$ are right angles.
Prove: $\triangle ABE \cong \triangle CBD$

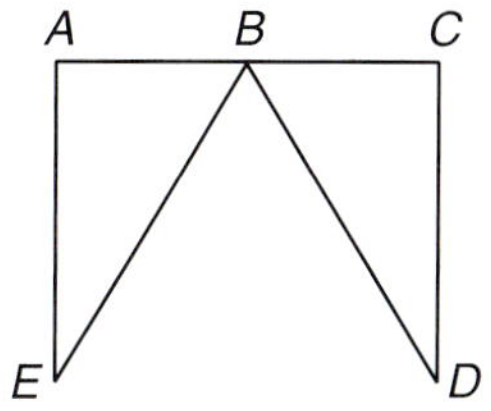

Proof: We are given that $AB = 22$ and $AC = 44$. By the Segment Addition Postulate,

$AB + BC = AC$
$22 + BC = 44$ Substitution
$BC = 22$ Subtract 22 from each side.

Since $AB = BC$, then by the definition of congruent segments, $\overline{AB} \cong \overline{BC}$.

We are given that $AE = 36$ and $CD = 36$. Then also by the definition of congruent segments, $\overline{AE} \cong \overline{CD}$.

We are additionally given that both $\angle A$ and $\angle C$ are right angles. Since all right angles are congruent, $\angle A \cong \angle C$.

Since $\overline{AB} \cong \overline{BC}$, $\angle A \cong \angle C$, and $\overline{AE} \cong \overline{CD}$, then by SAS, $\triangle ABE \cong \triangle CBD$.

Chapter 5

Relationships in Triangles

Chapter Overview and Pacing

LESSON OBJECTIVES	PACING (days) Regular Basic/Average	Regular Advanced	Block Basic/Average	Block Advanced
5-1 Bisectors, Medians, and Altitudes *(pp. 236–245)* ***Preview:*** Construct perpendicular bisectors, medians, altitudes, and angle bisectors for triangles. • Identify and use perpendicular bisectors and angle bisectors in triangles. • Identify and use medians and altitudes in triangles.	3 (with 5.1 Preview)	2 (with 5.1 Preview)	1.5 (with 5.1 Preview)	1.5 (with 5.1 Preview)
5-2 Inequalities and Triangles *(pp. 247–254)* • Recognize and apply properties of inequalities to the measures of angles of a triangle. • Recognize and apply properties of inequalities to the relationships between angles and sides of a triangle.	2	1	1	0.5
5-3 Indirect Proof *(pp. 255–260)* • Use indirect proof with algebra. • Use indirect proof with geometry.	2	2	1	1
5-4 The Triangle Inequality *(pp. 261–266)* • Apply the Triangle Inequality Theorem. • Determine the shortest distance between a point and a line.	2	2	1	1
5-5 Inequalities Involving Two Triangles *(pp. 267–273)* • Apply the SAS Inequality. • Apply the SSS Inequality.	2	2	1	1
Study Guide and Practice Test *(pp. 274–277)* **Standardized Test Practice** *(p. 278–279)*	1	1	1	0.5
Chapter Assessment	1	1	0.5	0.5
TOTAL	**13**	**11**	**7**	**6**

Pacing suggestions for the entire year can be found on pages T20–T21.

Timesaving Tools
TeacherWorks™
All-In-One Planner and Resource Center
See pages T5 and T21.

Chapter Resource Manager

Chapter 5 Resource Masters: Study Guide and Intervention	Chapter 5 Resource Masters: Practice (Skills and Average)	Chapter 5 Resource Masters: Reading to Learn Mathematics	Chapter 5 Resource Masters: Enrichment	Chapter 5 Resource Masters: Assessment	Prerequisite Skills Workbook	Applications*	5-Minute Check Transparencies	Interactive Chalkboard	GeomPASS: Tutorial Plus (lessons)	Materials
245–246	247–248	249	250		7–8, 41–42	SC 9	5-1	5-1		(*Preview:* straightedge, compass) straightedge, grid paper
251–252	253–254	255	256	289	9–10, 13–14, 87–88	GCC 25	5-2	5-2		ruler, protractor
257–258	259–260	261	262	289, 291	15–20, 25–26, 87–88		5-3	5-3		
263–264	265–266	267	268	290	15–16, 87–88, 107–108	SC 10	5-4	5-4		
269–270	271–272	273	274	290	87–88		5-5	5-5	11	
				275–288, 292–294						

**Key to Abbreviations:* GCC = Graphing Calculator and Computer Masters
SC = School-to-Career Masters

Chapter 5

Mathematical Connections and Background

Continuity of Instruction

Prior Knowledge

In previous courses, students solved problems involving inequalities and learned the properties of inequalities for real numbers. The Midpoint Formula and the Distance Formula, introduced in Chapter 1, are relied upon in this chapter. Students learned to apply deductive reasoning in Chapter 2. In Chapter 4 students learned about the properties of congruent triangles, including the Exterior Angle Theorem.

This Chapter

In this chapter students expand their knowledge of triangles and their properties. Bisectors, medians, and altitudes are identified and explored. Students apply properties of inequalities relating to the measures of angles and sides of a triangle and then extend those properties to two triangles. Students also use indirect proof with algebra and geometry.

Future Connections

Chapter 6 requires students to find medians and angle bisectors of triangles. In Chapter 7 students will use altitudes to solve problems about triangles.

5-1 Bisectors, Medians, and Altitudes

In the Preview to this lesson, students construct the perpendicular bisectors, medians, altitudes, and angle bisectors of a triangle using only a straightedge and a compass. A perpendicular bisector of a side of a triangle is a line, segment, or ray that passes through the midpoint of the side and is perpendicular to the side. Perpendicular bisectors have special properties. Any point on the perpendicular bisector of a segment is equidistant from the endpoints of the segment. The converse of this statement is also true. The point of concurrency of the perpendicular bisectors of a triangle is called the circumcenter. The circumcenter of a triangle is equidistant from the vertices of the triangle.

Angle bisectors also have special properties. Any point on the angle bisector is equidistant from the sides of the angle, and any point in the interior of an angle that is equidistant from the sides of the angle lies on the angle bisector. The intersection of the angle bisectors of a triangle is called the incenter. The incenter of a triangle is equidistant from the sides of the triangle.

A median is a line segment whose endpoints are a vertex of a triangle and the midpoint of the side opposite the vertex. The point of concurrency for the medians of a triangle is called a centroid. The centroid of a triangle is located two-thirds of the distance from a vertex to the midpoint of the side opposite the vertex on a median.

An altitude of a triangle is a segment perpendicular to a side of the triangle that has a vertex as one endpoint and a point on the line containing the side opposite the vertex as the other endpoint. The intersection of the altitudes of a triangle is called the orthocenter.

5-2 Inequalities and Triangles

In algebra, students learned the concept of inequality: For any real numbers a and b, $a > b$ if and only if there is a positive number c such that $a = b + c$. Students also studied several properties of inequalities for real numbers. In this lesson students apply these concepts to angles.

The Exterior Angle Inequality Theorem states that if an angle is an exterior angle of a triangle, then its measure is greater than the measure of either of its corresponding remote interior angles. Another inequality theorem in geometry is based on the relationship between a side and the vertex opposite that side. If one side of a triangle is longer than another side, then the angle opposite the longer side has a greater measure than the angle opposite the shorter side. The converse is also true: if one angle of a triangle has a greater measure than another angle, then the side opposite the greater angle is longer than the side opposite the lesser angle.

5-3 Indirect Proof

Indirect proof, or proof by contradiction, is a method of proving that a statement is true by first assuming that it is false. As the next step in the indirect proof, it is shown that this assumption leads to a contradiction of the hypothesis or some other fact, such as a definition, postulate, theorem, or corollary. Finally, the assumption that the original statement is rejected is false because it leads to a contradiction. So, the original statement is accepted as true. Indirect proof can be used in both algebra and geometry.

5-4 The Triangle Inequality

The Triangle Inequality Theorem states that the sum of the lengths of any two sides of a triangle is greater than the length of the third side. This theorem can be used to determine whether three segments can form a triangle.

The perpendicular segment from a point to a line is the shortest segment from the point to the line. This theorem can be proved using the Exterior Angle Inequality Theorem and leads to a corollary, that the perpendicular segment from a point to a plane is the shortest segment from the point to the plane.

5-5 Inequalities Involving Two Triangles

This lesson extends Theorem 5.10 to two triangles. That theorem states that if two sides of a triangle are congruent to two sides of another triangle and the included angle in one triangle has a greater measure than the included angle in the other, then the third side of the first triangle is longer than the third side of the second triangle. This is called the SAS Inequality, or Hinge, Theorem. The converse of this theorem is the SSS Inequality Theorem: If two sides of a triangle are congruent to two sides of another triangle and the third side in one triangle is longer than the third side in the other, then the angle between the pair of congruent sides in the first triangle is greater than the corresponding angle in the second triangle.

Chapter 5

DAILY INTERVENTION and Assessment

	Type	Student Edition	Teacher Resources	Technology/Internet
INTERVENTION	Ongoing	Prerequisite Skills, pp. 235, 260 Practice Quiz 1, p. 254 Practice Quiz 2, p. 266	5-Minute Check Transparencies *Prerequisite Skills Workbook,* pp. 7–10, 13–20, 25–26, 41–42, 87–88, 107–108 Quizzes, *CRM* pp. 289–290 Mid-Chapter Test, *CRM* p. 291 Study Guide and Intervention, *CRM* pp. 245–246, 251–252, 257–258, 263–264, 269–270	GeomPASS: Tutorial Plus www.geometryonline.com/self_check_quiz www.geometryonline.com/extra_examples
	Mixed Review	pp. 245, 254, 260, 266, 273	Cumulative Review, *CRM* p. 292	
	Error Analysis	Find the Error, pp. 251, 263 Common Misconceptions, p. 238	Find the Error, *TWE* pp. 251, 263 Unlocking Misconceptions, *TWE* p. 256 Tips for New Teachers, *TWE* p. 239	
	Standardized Test Practice	pp. 245, 253, 260, 262, 264, 265, 273, 277, 278, 279	*TWE* pp. 278–279 Standardized Test Practice, *CRM* pp. 293–294	Standardized Test Practice CD-ROM www.geometryonline.com/standardized_test
	Open-Ended Assessment	Writing in Math, pp. 245, 253, 260, 265, 273 Open Ended, pp. 242, 251, 257, 263, 270 Standardized Test, p. 279	Modeling: *TWE* pp. 245, 260 Speaking: *TWE* pp. 254, 273 Writing: *TWE* pp. 266 Open-Ended Assessment, *CRM* p. 287	
ASSESSMENT	Chapter Assessment	Study Guide, pp. 274–275 Practice Test, p. 277	Multiple-Choice Tests (Forms 1, 2A, 2B), *CRM* pp. 275–280 Free-Response Tests (Forms 2C, 2D, 3), *CRM* pp. 281–286 Vocabulary Test/Review, *CRM* p. 288	TestCheck and Worksheet Builder (see below) MindJogger Videoquizzes www.geometryonline.com/vocabulary_review www.geometryonline.con/chapter_test

Key to Abbreviations: TWE = Teacher Wraparound Edition; CRM = Chapter Resource Masters

Additional Intervention Resources

The Princeton Review's *Cracking the SAT & PSAT*

The Princeton Review's *Cracking the ACT*

TestCheck and Worksheet Builder

This **networkable** software has three modules for intervention and assessment flexibility:

- **Worksheet Builder** to make worksheet and tests
- **Student Module** to take tests on screen (optional)
- **Management System** to keep student records (optional)

Special banks are included for SAT, ACT, TIMSS, NAEP, and End-of-Course tests.

Intervention Technology

GeomPASS: Tutorial Plus CD-ROM offers a complete, self-paced geometry curriculum.

Geometry Lesson	GeomPASS Lesson
5-5	**11** *Inequalities in Triangles*

Intervention at Home

Log on for student study help.

- For each lesson in the Student Edition, there are Extra Examples and Self-Check Quizzes.
  www.geometryonline.com/extra_examples
  www.geometryonline.com/self_check_quiz
- For chapter review, there is vocabulary review, test practice, and standardized test practice.
  www.geometryonline.com/vocabulary_review
  www.geometryonline.com/chapter_test
  www.geometryonline.com/standardized_test

For more information on Intervention and Assessment, see pp. T8–T11.

Reading and Writing in Mathematics

Glencoe Geometry provides numerous opportunities to incorporate reading and writing into the mathematics classroom.

Student Edition

- Foldables Study Organizer, p. 235
- Concept Check questions require students to verbalize and write about what they have learned in the lesson. (pp. 242, 251, 257, 263, 270)
- Reading Mathematics, p. 246
- Writing in Math questions in every lesson, pp. 245, 253, 260, 265, 273
- Reading Study Tip, p. 238
- WebQuest, p. 241

Teacher Wraparound Edition

- Foldables Study Organizer, pp. 235, 274
- Study Notebook suggestions, pp. 237, 242, 246, 251, 258, 263, 271
- Modeling activities, pp. 245, 260
- Speaking activities, pp. 254, 273
- Writing activities, p. 266
- Differentiated Instruction (Verbal/Linguistic), p. 248
- ELL Resources, pp. 234, 244, 246, 248, 253, 259, 265, 272, 274

Additional Resources

- Vocabulary Builder worksheets require students to define and give examples for key vocabulary terms as they progress through the chapter. (*Chapter 5 Resource Masters,* pp. vii-viii)
- Proof Builder helps students learn and understand theorems and postulates from the chapter. (*Chapter 5 Resource Masters,* pp. ix–x)
- Reading to Learn Mathematics master for each lesson (*Chapter 5 Resource Masters*, pp. 249, 255, 261, 267, 273)
- *Vocabulary PuzzleMaker* software creates crossword, jumble, and word search puzzles using vocabulary lists that you can customize.
- *Teaching Mathematics with Foldables* provides suggestions for promoting cognition and language.
- *Reading Strategies for the Mathematics Classroom*
- *WebQuest and Project Resources*

For more information on Reading and Writing in Mathematics, see pp. T6–T7.

Chapter 5 Notes

What You'll Learn

Have students read over the list of objectives and make a list of any words with which they are not familiar.

Why It's Important

Point out to students that this is only one of many reasons why each objective is important. Others are provided in the introduction to each lesson.

Lesson	NCTM Standards	Local Objectives
5-1 Preview	3, 10	
5-1	2, 3, 6, 8, 9, 10	
5-2	2, 3, 6, 8, 9, 10	
5-3	2, 3, 6, 7, 8, 10	
5-4	2, 3, 6, 7, 8, 9, 10	
5-5	2, 3, 6, 7, 8, 9, 10	

Key to NCTM Standards:

1=Number & Operations, 2=Algebra, 3=Geometry, 4=Measurement, 5=Data Analysis & Probability, 6=Problem Solving, 7=Reasoning & Proof, 8=Communication, 9=Connections, 10=Representation

Chapter 5 Relationships in Triangles

What You'll Learn

- **Lesson 5-1** Identify and use perpendicular bisectors, angle bisectors, medians, and altitudes of triangles.
- **Lesson 5-2** Apply properties of inequalities relating to the measures of angles and sides of triangles.
- **Lesson 5-3** Use indirect proof with algebra and geometry.
- **Lessons 5-4 and 5-5** Apply the Triangle Inequality Theorem and SAS and SSS inequalities.

Key Vocabulary

- perpendicular bisector (p. 238)
- median (p. 240)
- altitude (p. 241)
- indirect proof (p. 255)

Why It's Important

There are several relationships among the sides and angles of triangles. These relationships can be used to compare the length of a person's stride and the rate at which that person is walking or running. *In Lesson 5-5, you will learn how to use the measure of the sides of a triangle to compare stride and rate.*

Vocabulary Builder

ELL

The Key Vocabulary list introduces students to some of the main vocabulary terms included in this chapter. For a more thorough vocabulary list with pronunciations of new words, give students the Vocabulary Builder worksheets found on pages vii and viii of the *Chapter 5 Resource Masters*. Encourage them to complete the definition of each term as they progress through the chapter. You may suggest that they add these sheets to their study notebooks for future reference when studying for the Chapter 5 test.

Getting Started

Prerequisite Skills To be successful in this chapter, you'll need to master these skills and be able to apply them in problem-solving situations. Review these skills before beginning Chapter 5.

For Lesson 5-1 — **Midpoint of a Segment**

Find the coordinates of the midpoint of a segment with the given endpoints.
(For review, see Lesson 1-3.)

1. $A(-12, -5)$, $B(4, 15)$ **(−4, 5)**
2. $C(-22, -25)$, $D(10, 10)$ **(−6, −7.5)**
3. $E(19, -7)$, $F(-20, -3)$ **(−0.5, −5)**

For Lesson 5-2 — **Exterior Angle Theorem**

Find the measure of each numbered angle if $\overline{AB} \perp \overline{BC}$. *(For review, see Lesson 4-2.)*

4. $\angle 1$ **76**
5. $\angle 2$ **68**
6. $\angle 3$ **76**
7. $\angle 4$ **40**
8. $\angle 5$ **64**
9. $\angle 6$ **26**
10. $\angle 7$ **140**
11. $\angle 8$ **14**

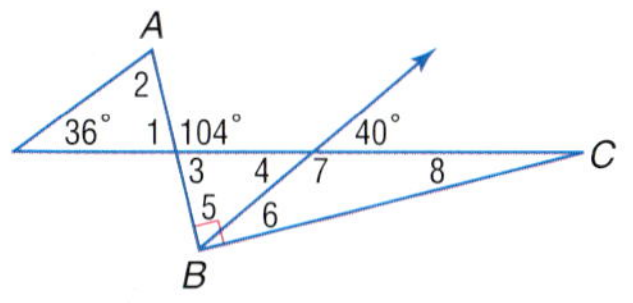

For Lesson 5-3 — **Deductive Reasoning**

Determine whether a valid conclusion can be reached from the two true statements using the Law of Detachment. If a valid conclusion is possible, state it. If a valid conclusion does not follow, write *no conclusion*. *(For review, see Lesson 2-4.)*

12. (1) If the three sides of one triangle are congruent to the three sides of a second triangle, then the triangles are congruent.
(2) $\triangle ABC$ and $\triangle PQR$ are congruent. **no conclusion**

13. (1) The sum of the measures of the angles of a triangle is 180.
(2) Polygon JKL is a triangle. **The sum of the measures of the angles of polygon *JKL* is 180.**

FOLDABLES™ Study Organizer

Relationships in Triangles Make this Foldable to help you organize information about relations in triangles. Begin with one sheet of notebook paper.

Step 1 Fold

Fold lengthwise to the holes.

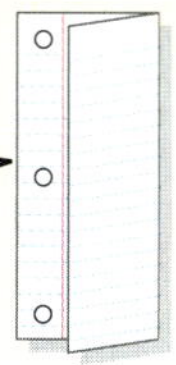

Step 2 Cut

Cut along the lines to create five tabs.

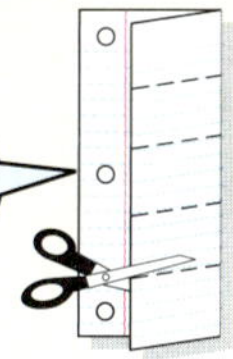

Step 3 Label

Label the edge. Then label the tabs using lesson numbers.

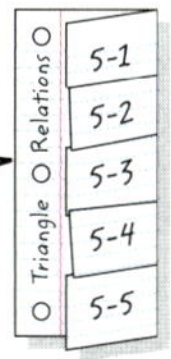

Reading and Writing As you read and study each lesson, write notes and examples under the appropriate tab.

Getting Started

This section provides a review of the basic concepts needed before beginning Chapter 5. Page references are included for additional student help.

Additional review is provided in the *Prerequisite Skills Workbook*, pages 7–10, 13–20, 25–26, 41–42, 87–88, 107–108.

Prerequisite Skills in the Getting Ready for the Next Lesson section at the end of each exercise set review a skill needed in the next lesson.

For Lesson	Prerequisite Skill
5-4	Meaning of Inequalities, p. 260
5-5	Solving Inequalities, p. 266

FOLDABLES™ Study Organizer

For more information about Foldables, see *Teaching Mathematics with Foldables.*

Descriptive Writing and Organizing Data Students use their Foldable to take notes, define terms, record concepts, and write proofs. After students make their Foldable, have them label the tabs to correspond to the five lessons in this chapter. At the end of each lesson, ask students to write a descriptive paragraph sharing their experiences with the concepts, vocabulary, reasoning, theorems, and graphics presented.

Geometry Activity

A Preview of Lesson 5-1

Getting Started

Objective Construct perpendicular bisectors, medians, altitudes, and angle bisectors for triangles.

Materials
compass straightedge

Special Segments The activity demonstrates four different constructions on an acute scalene triangle. Students could use patty paper to draw and trace an acute scalene triangle with the same side lengths, angle measures, and orientation in three different places on one sheet of paper. When students are finished with the constructions, they can see the differences among the bisectors, medians, and altitudes for the same triangle.

Teach

- As students are drawing the two congruent triangles to prove perpendicular bisection in Construction 1, tell them they can use point P or point Q because both sets of arcs are drawn with the same compass setting.
- When students are repeating the four constructions for an isosceles triangle in Exercise 9, have some students draw obtuse isosceles triangles and some draw acute isosceles triangles, and then let classmates compare.

Answer

1.

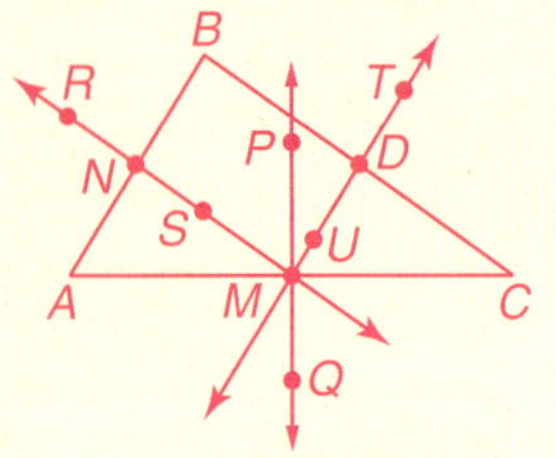

Geometry Activity

Bisectors, Medians, and Altitudes

You can use the constructions for midpoint, perpendiculars, and angle bisectors to construct special segments in triangles.

Construction 1 **Construct the bisector of a side of a triangle.**

1. Draw a triangle like $\triangle ABC$. Adjust the compass to an opening greater than $\frac{1}{2}AC$. Place the compass at vertex A, and draw an arc above and below $\overline{AC}$.

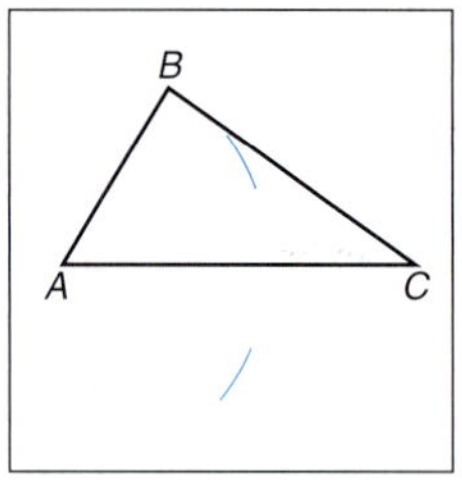

2. Using the same compass settings, place the compass at vertex C. Draw an arc above and below $\overline{AC}$. Label the points of intersection of the arcs P and Q.

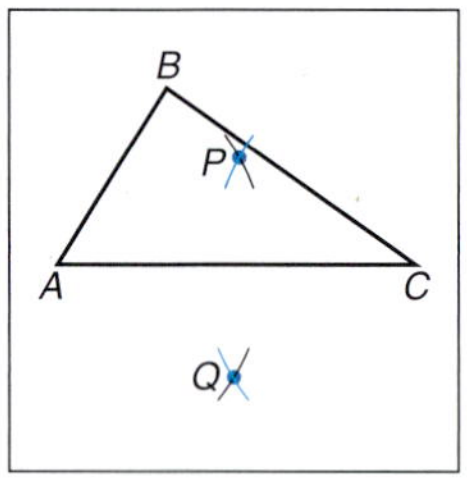

3. Use a straightedge to draw $\overleftrightarrow{PQ}$. Label the point where $\overleftrightarrow{PQ}$ bisects $\overline{AC}$ as M.

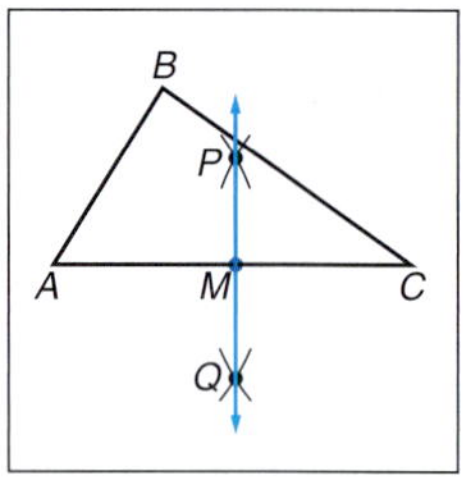

$\overline{AM} \cong \overline{MC}$ by construction and $\overline{PM} \cong \overline{PM}$ by the Reflexive Property. $\overline{AP} \cong \overline{CP}$ because the arcs were drawn with the same compass setting. Thus, $\triangle APM \cong \triangle CPM$ by SSS. By CPCTC, $\angle PMA \cong \angle PMC$. A linear pair of congruent angles are right angles. So $\overleftrightarrow{PQ}$ is not only a bisector of $\overline{AC}$, but a perpendicular bisector. **1. See margin.**

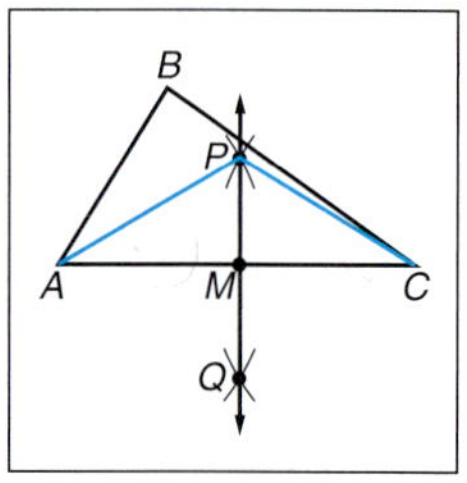

1. Construct the perpendicular bisectors for the other two sides.
2. What do you notice about the intersection of the perpendicular bisectors? **They intersect at the same point.**

Construction 2 **Construct a median of a triangle.**

1. Draw intersecting arcs above and below $\overline{BC}$. Label the points of intersection R and S.

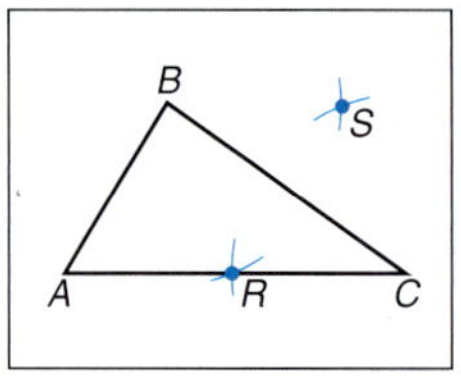

2. Use a straightedge to find the point where $\overline{RS}$ intersects $\overline{BC}$. Label the midpoint M.

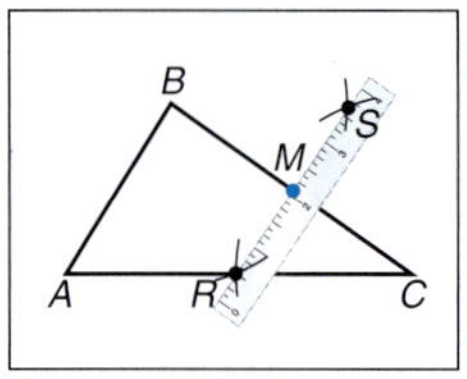

3. Draw a line through A and M. $\overline{AM}$ is a median of $\triangle ABC$.

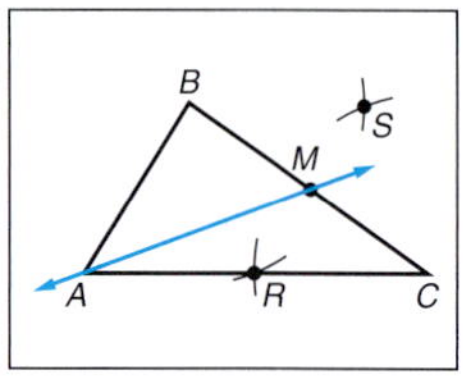

3. Construct the medians of the other two sides. **See margin.**
4. What do you notice about the medians of a triangle? **They intersect at the same point.**

Resource Manager

Teaching Geometry with Manipulatives
- p. 87 (student recording sheet)

Glencoe Mathematics Classroom Manipulative Kit
- compass

Construction 3 **Construct an altitude of a triangle.**

1. Place the compass at vertex *B* and draw two arcs intersecting $\overline{AC}$. Label the points where the arcs intersect the side *X* and *Y*.

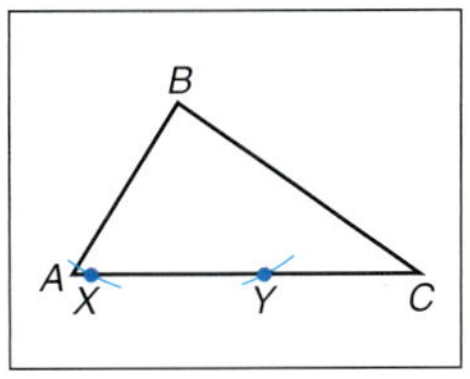

2. Adjust the compass to an opening greater than $\frac{1}{2}XY$. Place the compass on *X* and draw an arc above $\overline{AC}$. Using the same setting, place the compass on *Y* and draw an arc above $\overline{AC}$. Label the intersection of the arcs *H*.

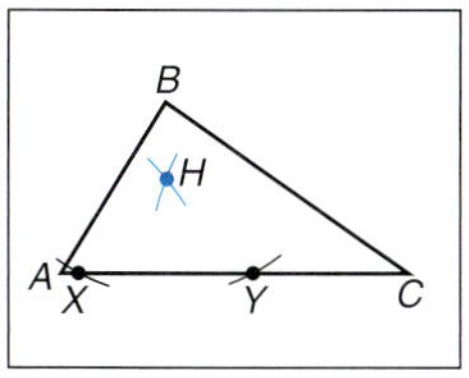

3. Use a straightedge to draw $\overleftrightarrow{BH}$. Label the point where $\overleftrightarrow{BH}$ intersects $\overline{AC}$ as *D*. $\overline{BD}$ is an altitude of $\triangle ABC$ and is perpendicular to $\overline{AC}$.

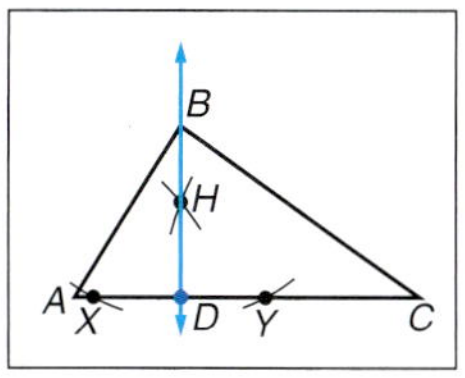

5. Construct the altitudes to the other two sides. (*Hint:* You may need to extend the lines containing the sides of your triangle.) **See margin.**
6. What observation can you make about the altitudes of your triangle? **They intersect at the same point.**

Construction 4 **Construct an angle bisector of a triangle.**

1. Place the compass on vertex *A*, and draw arcs through $\overline{AB}$ and $\overline{AC}$. Label the points where the arcs intersect the sides as *J* and *K*.

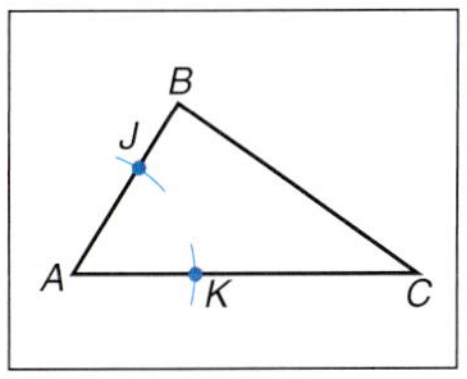

2. Place the compass on *J*, and draw an arc. Then place the compass on *K* and draw an arc intersecting the first arc. Label the intersection *L*.

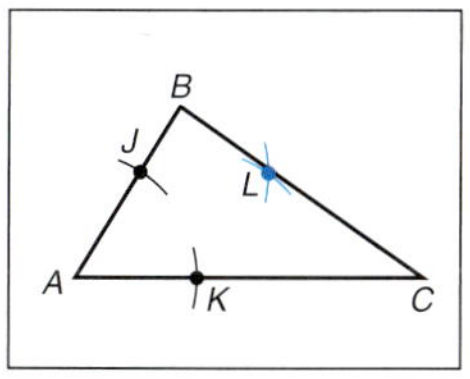

3. Use a straightedge to draw $\overrightarrow{AL}$. $\overrightarrow{AL}$ is an angle bisector of $\triangle ABC$.

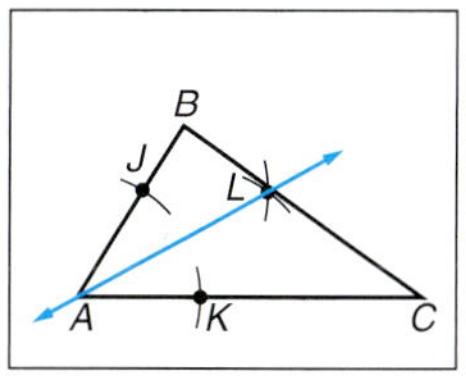

7. Construct the angle bisectors for the other two angles. **See margin.**
8. What do you notice about the angle bisectors? **They intersect at the same point.**

Analyze

9. Repeat the four constructions for each type of triangle. **See students' work.**
   a. obtuse scalene **b.** right scalene **c.** isosceles **d.** equilateral

Make a Conjecture **10–11. See margin.**

10. Where do the lines intersect for acute, obtuse, and right triangles?
11. Under what circumstances do the special lines of triangles coincide with each other?

Assess

Exercises 1–8 guide students through each construction set, having them note that bisectors, medians, and altitudes for triangles all intersect in one point. In **Exercise 9**, students repeat the activity for selected types of triangles. Students use **Exercises 10 and 11** to analyze and form conjectures about the special lines for all the types of triangles they used in the activity.

Study Notebook

Ask students to summarize what they have learned about bisectors, medians, and altitudes and their points of intersection.

10. Acute: all intersect inside the triangle; obtuse: perpendicular bisectors and altitudes intersect outside the triangle; medians and angle bisectors intersect inside the triangle; right: perpendicular bisectors intersect on the hypotenuse, medians intersect inside the triangle, altitudes intersect on the vertex of the right angle, and angle bisectors intersect inside the triangle.

11. For an isosceles triangle, the perpendicular bisector and median of the side opposite the vertex are the same as the altitude from the vertex angle and the angle bisector of the vertex angle. In an equilateral triangle, the perpendicular bisector and median of each side is the same as the altitude to each side and the angle bisector of the angle opposite each side.

Answers

3. Sample answer:

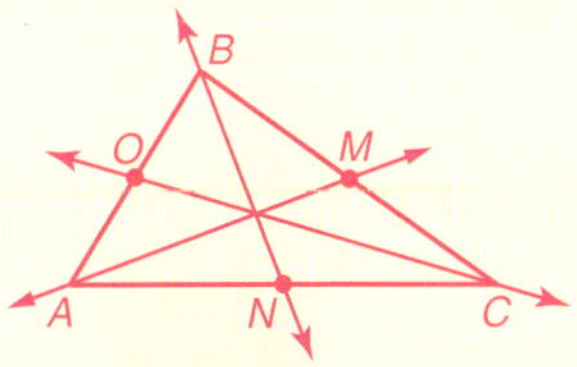

5.

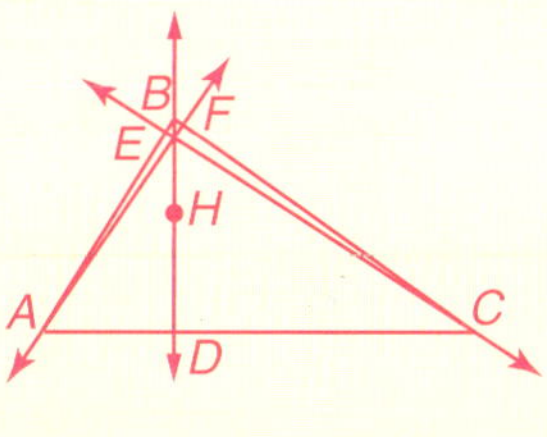

7.

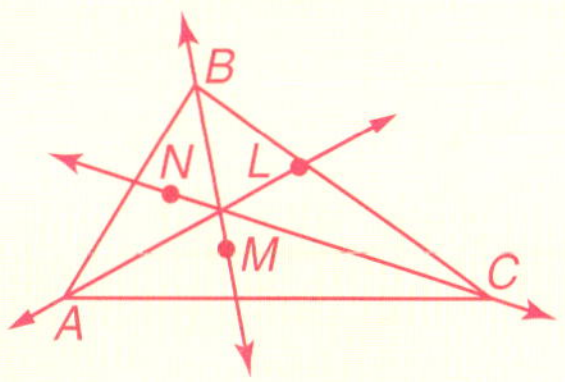

5-1 Lesson Notes

1 Focus

5-Minute Check Transparency 5-1 Use as a quiz or review of Chapter 4.

Mathematical Background notes are available for this lesson on p. 234C.

How can you balance a paper triangle on a pencil point?

Ask students:

- Without using math, how would you balance a paper triangle on a pencil point? What position are you locating on the triangle? **Place the triangle on the point and slightly move it in different directions until it does not fall; center of gravity.**
- Why does the intersection of a triangle's medians locate its center of gravity? **A segment can be balanced on its midpoint, and the three medians of a triangle intersect at a point that represents the midpoint of each side.**

5-1 Bisectors, Medians, and Altitudes

See Surfer and the spotter H. Jacobs Geometry

What You'll Learn

- Identify and use perpendicular bisectors and angle bisectors in triangles.
- Identify and use medians and altitudes in triangles.

Vocabulary

- perpendicular bisector
- concurrent lines
- point of concurrency
- circumcenter
- incenter
- median
- centroid
- altitude
- orthocenter

How can you balance a paper triangle on a pencil point?

Acrobats and jugglers often balance objects while performing their acts. These skilled artists need to find the center of gravity for each object or body position in order to keep balanced. The center of gravity for any triangle can be found by drawing the *medians* of a triangle and locating the point where they intersect.

PERPENDICULAR BISECTORS AND ANGLE BISECTORS The first construction you made in the Geometry Activity on pages 236 and 237 was the perpendicular bisector of a side of a triangle. A **perpendicular bisector** of a side of a triangle is a line, segment, or ray that passes through the midpoint of the side and is perpendicular to that side.

Perpendicular bisectors of segments have some special properties. These properties are described in the following theorems.

Study Tip

Reading Math
Equidistant means "the same distance."

Study Tip

Common Misconception
Note that Theorem 5.2 states the point is on the perpendicular bisector. It does not say that any line containing that point is a perpendicular bisector.

Theorems — *Points on Perpendicular Bisectors*

5.1 Any point on the perpendicular bisector of a segment is equidistant from the endpoints of the segment.

Example: If $\overline{AB} \perp \overline{CD}$ and $\overline{AB}$ bisects $\overline{CD}$, then $AC = AD$ and $BC = BD$.

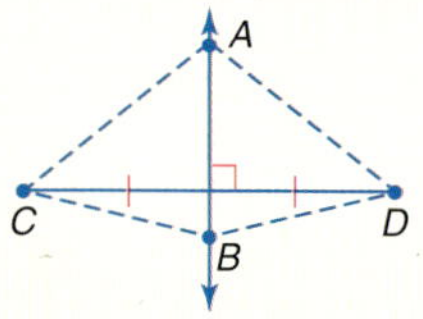

5.2 Any point equidistant from the endpoints of a segment lies on the perpendicular bisector of the segment.

Example: If $AC = AD$, then A lies on the perpendicular bisector of $\overline{CD}$.
If $BC = BD$, then B lies on the perpendicular bisector of $\overline{CD}$.

You will prove Theorems 5.1 and 5.2 in Exercises 10 and 31, respectively.

Recall that a locus is the set of all points that satisfy a given condition. A perpendicular bisector can be described as the locus of points in a plane equidistant from the endpoints of a given segment.

Since a triangle has three sides, there are three perpendicular bisectors in a triangle. The perpendicular bisectors of a triangle intersect at a common point. When three or more lines intersect at a common point, the lines are called **concurrent lines**, and their point of intersection is called the **point of concurrency**. The point of concurrency of the perpendicular bisectors of a triangle is called the **circumcenter**.

Resource Manager

Workbook and Reproducible Masters

Chapter 5 Resource Masters
- Study Guide and Intervention, pp. 245–246
- Skills Practice, p. 247
- Practice, p. 248
- Reading to Learn Mathematics, p. 249
- Enrichment, p. 250

School-to-Career Masters, p. 9
Prerequisite Skills Workbook, pp. 7–8, 41–42
Teaching Geometry With Manipulatives Masters, pp. 1, 89, 90, 91, 93

Transparencies

5-Minute Check Transparency 5-1
Answer Key Transparencies

Technology

Interactive Chalkboard

Theorem 5.3

Circumcenter Theorem The circumcenter of a triangle is equidistant from the vertices of the triangle.

Example: If J is the circumcenter of $\triangle ABC$, then $AJ = BJ = CJ$.

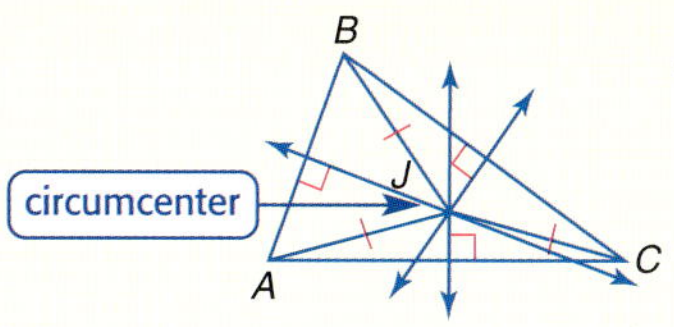

Proof *Theorem 5.3*

Given: ℓ, m, and n are perpendicular bisectors of $\overline{AB}$, $\overline{AC}$, and $\overline{BC}$, respectively.

Prove: $AJ = BJ = CJ$

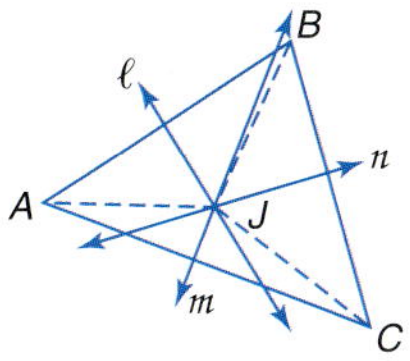

Paragraph Proof:

Since J lies on the perpendicular bisector of $\overline{AB}$, it is equidistant from A and B. By the definition of equidistant, $AJ = BJ$. The perpendicular bisector of $\overline{BC}$ also contains J. Thus, $BJ = CJ$. By the Transitive Property of Equality, $AJ = CJ$. Thus, $AJ = BJ = CJ$.

Study Tip

Bisectors
In most triangles, a perpendicular bisector will not pass through the vertex opposite a given side. Likewise, an angle bisector usually will not bisect the side opposite an angle.

Another special line, segment, or ray in triangles is an angle bisector.

Example 1 Use Angle Bisectors

Given: $\overline{PX}$ bisects $\angle QPR$, $\overline{XY} \perp \overline{PQ}$ and $\overline{XZ} \perp \overline{PR}$

Prove: $\overline{XY} \cong \overline{XZ}$

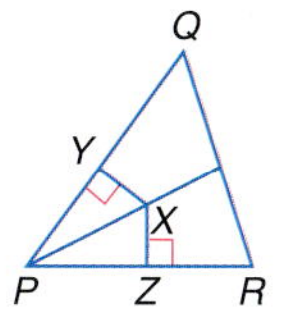

Proof:

Statements	**Reasons**
1. $\overline{PX}$ bisects $\angle QPR$, $\overline{XY} \perp \overline{PQ}$, and $\overline{XZ} \perp \overline{PR}$.	1. Given
2. $\angle YPX \cong \angle ZPX$	2. Definition of angle bisector
3. $\angle PYX$ and $\angle PZX$ are right angles.	3. Definition of perpendicular
4. $\angle PYX \cong \angle PZX$	4. Right angles are congruent.
5. $\overline{PX} \cong \overline{PX}$	5. Reflexive Property
6. $\triangle PYX \cong \triangle PZX$	6. AAS
7. $\overline{XY} \cong \overline{XZ}$	7. CPCTC

Study Tip

Locus
An angle bisector can be described as the locus of points in a plane equidistant from the sides of an angle.

In Example 1, XY and XZ are lengths representing the distance from X to each side of $\angle QPR$. This is a proof of Theorem 5.4.

Theorems *Points on Angle Bisectors*

5.4 Any point on the angle bisector is equidistant from the sides of the angle.

5.5 Any point equidistant from the sides of an angle lies on the angle bisector.

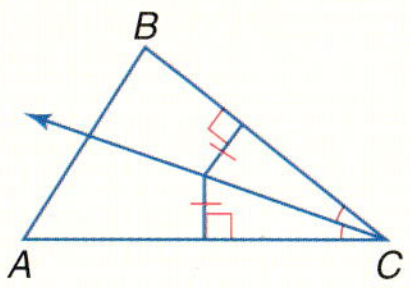

You will prove Theorem 5.5 in Exercise 32.

2 Teach

PERPENDICULAR BISECTORS AND ANGLE BISECTORS

Teaching Tip As you discuss the perpendicular bisector theorems, draw a line that intersects $\overline{CD}$ and contains point A to demonstrate that a line other than the perpendicular bisector can also contain a point equidistant from the endpoints of a segment. Tell students that there could also be a line through point A that is parallel to $\overline{CD}$.

Teaching Tip Explain that a circumcenter does not necessarily have to lie in the interior of a triangle and draw an obtuse isosceles triangle with angles 10°, 10°, and 160° to demonstrate this.

In-Class Example

Power Point®

1 **Given:** $m\angle F = 80$ and $m\angle E = 30$; $\overline{DG}$ bisects $\angle EDF$.

Prove: $m\angle DGE = 115$

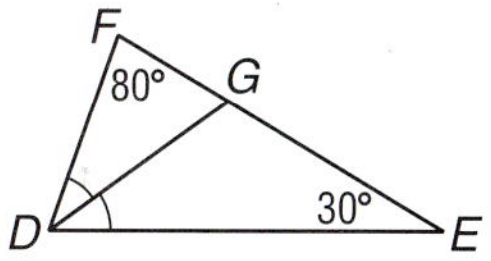

Statements (Reasons)

1. $m\angle F = 80$, $m\angle DEF = 30$, and $\overline{DG}$ bisects $\angle EDF$. (Given)
2. $m\angle EDF + m\angle E + m\angle F = 180$ ($\angle$ Sum Theorem)
3. $m\angle EDF + 30 + 80 = 180$ (Substitution)
4. $m\angle EDF = 180 - 110 = 70$ (Subtraction Prop.)
5. $m\angle GDE = 35$ (Def. of $\angle$ bisector)
6. $m\angle GDE + m\angle E + m\angle DGE = 180$ ($\angle$ Sum Theorem)
7. $35 + 30 + m\angle DGE = 180$ (Substitution)
8. $m\angle DGE = 180 - 65 = 115$ (Subtraction Prop.)

Tips for New Teachers

Intervention Expect some students to find the concepts and vocabulary in this lesson very confusing. Allow extra time for all the concepts in this lesson. After each concept, suggest that students add to a class poster that illustrates the different concepts and facts about them. Also review concepts as you complete them, discussing their similarities and differences with terms they could be most easily confused with.

MEDIANS AND ALTITUDES

In-Class Example Power Point®

2 **ALGEBRA** Points U, V, and W are the midpoints of $\overline{YZ}$, $\overline{ZX}$, and $\overline{XY}$, respectively. Find a, b, and c.

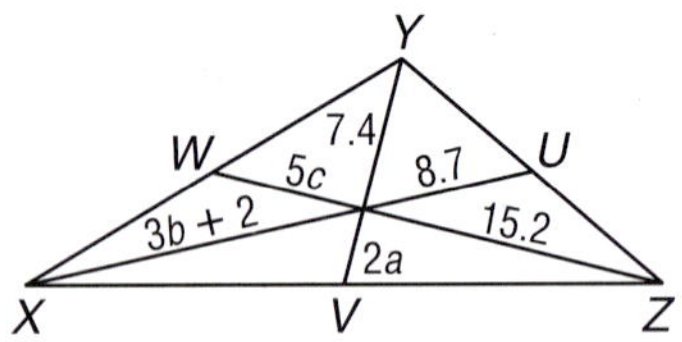

$a = 1.85$; $b = 5.13$; $c = 1.52$

As with perpendicular bisectors, there are three angle bisectors in any triangle. The angle bisectors of a triangle are concurrent, and their point of concurrency is called the **incenter** of a triangle.

Theorem 5.6

Incenter Theorem The incenter of a triangle is equidistant from each side of the triangle.

Example: If K is the incenter of $\triangle ABC$, then $KP = KQ = KR$.

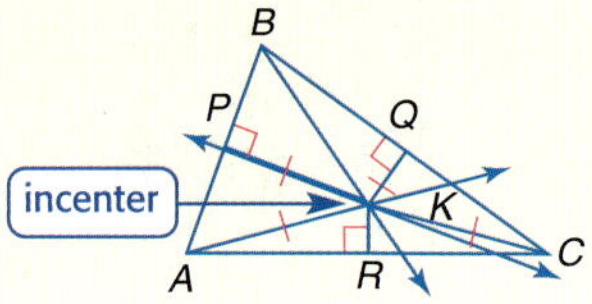

You will prove Theorem 5.6 in Exercise 33.

Study Tip

Medians as Bisectors

Because the median contains the midpoint, it is also a bisector of the side of the triangle.

MEDIANS AND ALTITUDES A **median** is a segment whose endpoints are a vertex of a triangle and the midpoint of the side opposite the vertex. Every triangle has three medians.

The medians of a triangle also intersect at a common point. The point of concurrency for the medians of a triangle is called a **centroid**. The centroid is the point of balance for any triangle.

Theorem 5.7

Centroid Theorem The centroid of a triangle is located two thirds of the distance from a vertex to the midpoint of the side opposite the vertex on a median.

Example: If L is the centroid of $\triangle ABC$, $AL = \frac{2}{3}AE$, $BL = \frac{2}{3}BF$, and $CL = \frac{2}{3}CD$.

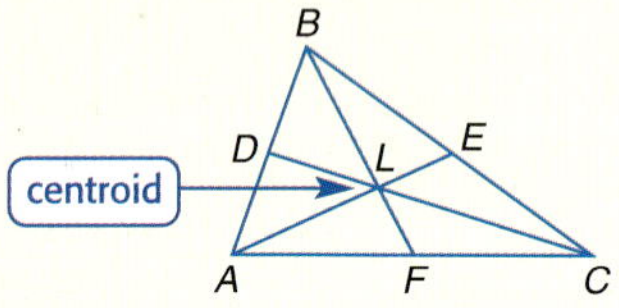

Example 2 ***Segment Measures***

ALGEBRA **Points S, T, and U are the midpoints of $\overline{DE}$, $\overline{EF}$, and $\overline{DF}$, respectively. Find x, y, and z.**

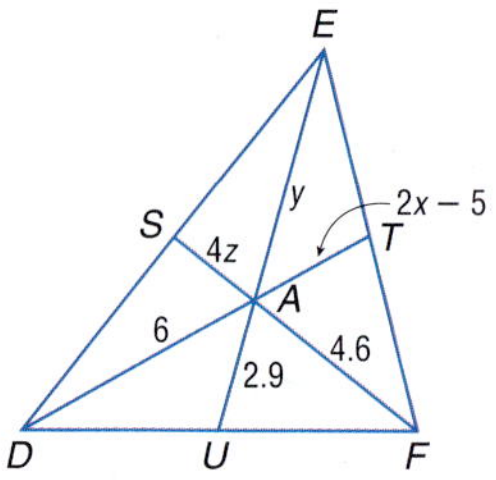

Study Tip

Eliminating Fractions

You could also multiply the equation $DA = \frac{2}{3}DT$ by 3 to eliminate the denominator.

• Find x.

$DT = DA + AT$	Segment Addition Postulate
$= 6 + (2x - 5)$	Substitution
$= 2x + 1$	Simplify.
$DA = \frac{2}{3}DT$	Centroid Theorem
$6 = \frac{2}{3}[2x + 1]$	$DA = 6$, $DT = 2x + 1$
$18 = 4x + 2$	Multiply each side by 3 and simplify.
$16 = 4x$	Subtract 2 from each side.
$4 = x$	Divide each side by 4.

DAILY INTERVENTION

Differentiated Instruction

Visual/Spatial Tell students to divide a sheet of paper in four sections labeled Circumcenter, Incenter, Centroid, and Orthocenter. Have students draw a copy of the same triangle in each section of the paper and use their spatial skills to determine the approximate position of the circumcenter, incenter, centroid, and orthocenter of the triangle. Then students can use metric rulers, compasses, and protractors to see how close their approximations are.

• Find y.

$EA = \frac{2}{3}EU$ Centroid Theorem

$y = \frac{2}{3}(y + 2.9)$ $EA = y, EU = y + 2.9$

$3y = 2y + 5.8$ Multiply each side by 3 and simplify.

$y = 5.8$ Subtract $2y$ from each side.

• Find z.

$FA = \frac{2}{3}FS$ Centroid Theorem

$4.6 = \frac{2}{3}(4.6 + 4z)$ $FA = 4.6, FS = 4.6 + 4z$

$13.8 = 9.2 + 8z$ Multiply each side by 3 and simplify.

$4.6 = 8z$ Subtract 9.2 from each side.

$0.575 = z$ Divide each side by 8.

An **altitude** of a triangle is a segment from a vertex to the line containing the opposite side and perpendicular to the line containing that side. Every triangle has three altitudes. The intersection point of the altitudes of a triangle is called the **orthocenter**.

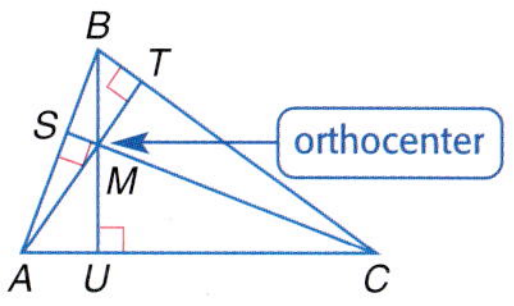

WebQuest

Finding the orthocenter can be used to help you construct your own nine-point circle. Visit www.geometryonline.com/webquest to continue work on your WebQuest project.

If the vertices of a triangle are located on a coordinate plane, you can use a system of equations to find the coordinates of the orthocenter.

Example 3 Use a System of Equations to Find a Point

COORDINATE GEOMETRY **The vertices of $\triangle JKL$ are $J(1, 3)$, $K(2, -1)$, and $L(-1, 0)$. Find the coordinates of the orthocenter of $\triangle JKL$.**

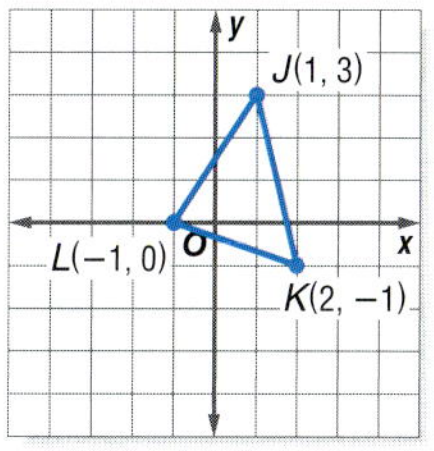

• Find an equation of the altitude from J to $\overline{KL}$.

The slope of $\overline{KL}$ is $-\frac{1}{3}$, so the slope of the altitude is 3.

$(y - y_1) = m(x - x_1)$ Point-slope form

$(y - 3) = 3(x - 1)$ $x_1 = 1, y_1 = 3, m = 3$

$y - 3 = 3x - 3$ Distributive Property

$y = 3x$ Add 3 to each side.

• Next, find an equation of the altitude from K to $\overline{JL}$. The slope of $\overline{JL}$ is $\frac{3}{2}$, so the slope of the altitude to $\overline{JL}$ is $-\frac{2}{3}$.

$(y - y_1) = m(x - x_1)$ Point-slope form

$(y + 1) = -\frac{2}{3}(x - 2)$ $x_1 = 2, y_1 = -1, m = -\frac{2}{3}$

$y + 1 = -\frac{2}{3}x + \frac{4}{3}$ Distributive Property

$y = -\frac{2}{3}x + \frac{1}{3}$ Subtract 1 from each side.

(continued on the next page)

In-Class Example

Power Point®

3 COORDINATE GEOMETRY The vertices of $\triangle HIJ$ are $H(1, 2)$, $I(-3, -3)$, and $J(-5, 1)$. Find the coordinates of the orthocenter of $\triangle HIJ$.

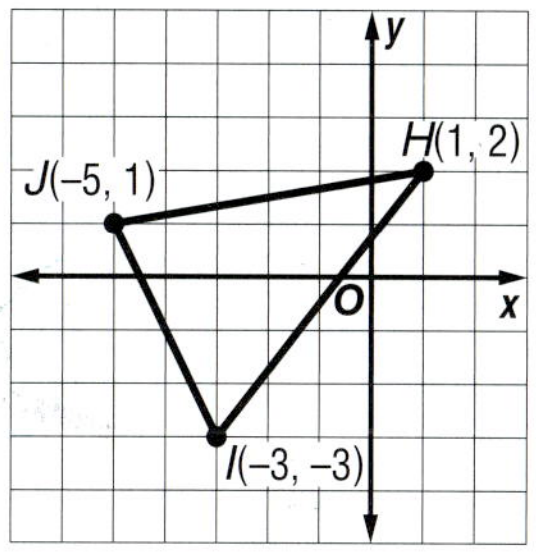

$\left(-3\frac{6}{13}, -\frac{3}{13}\right)$

Interactive Chalkboard

PowerPoint® Presentations

This CD-ROM is a customizable Microsoft® PowerPoint® presentation that includes:

- Step-by-step, dynamic solutions of each In-Class Example from the Teacher Wraparound Edition
- Additional, Try These exercises for each example
- The 5-Minute Check Transparencies
- Hot links to Glencoe Online Study Tools

3 Practice/Apply

Study Notebook

Have students—
- add the definitions/examples of the vocabulary terms to their Vocabulary Builder worksheets for Chapter 5.
- include constructions from the Geometry Activity previewing Lesson 5-1.
- include any other item(s) that they find helpful in mastering the skills in this lesson.

About the Exercises...

Organization by Objective
- **Perpendicular and Angle Bisectors:** 10–12, 31–33
- **Medians and Altitudes:** 7–9, 13–16, 21–30

Odd/Even Assignments

Exercises 11–30 are structured so that students practice the same concepts whether they are assigned odd or even problems.

Assignment Guide

Basic: 7–29 odd, 35–39 odd, 40–54

Average: 7–39 odd, 40–54

Advanced: 8–38 even, 39–50 (optional: 51–54)

Study Tip

Graphing Calculator
Once you have two equations, you can graph the two lines and use the Intersect option on the Calc menu to determine where the two lines meet.

- Then, solve a system of equations to find the point of intersection of the altitudes.

Find x.

$y = -\frac{2}{3}x + \frac{1}{3}$ Equation of altitude from K

$3x = -\frac{2}{3}x + \frac{1}{3}$ Substitution, $y = 3x$

$9x = -2x + 1$ Multiply each side by 3.

$11x = 1$ Add $2x$ to each side.

$x = \frac{1}{11}$ Divide each side by 11.

Replace x with $\frac{1}{11}$ in one of the equations to find the y-coordinate.

$y = 3\left(\frac{1}{11}\right)$ $x = \frac{1}{11}$

$y = \frac{3}{11}$ Multiply.

The coordinates of the orthocenter of $\triangle JKL$ are $\left(\frac{1}{11}, \frac{3}{11}\right)$.

You can also use systems of equations to find the coordinates of the circumcenter and the centroid of a triangle graphed on a coordinate plane.

Concept Summary — *Special Segments in Triangles*

Name	Type	Point of Concurrency
perpendicular bisector	line, segment, or ray	circumcenter
angle bisector	line, segment, or ray	incenter
median	segment	centroid
altitude	segment	orthocenter

Check for Understanding

Concept Check

1–3. See margin.

1. **Compare and contrast** a perpendicular bisector and a median of a triangle.

2. **OPEN ENDED** Draw a triangle in which the circumcenter lies outside the triangle.

3. **Find a counterexample** to the statement *An altitude and an angle bisector of a triangle are never the same segment.*

Guided Practice

GUIDED PRACTICE KEY	
Exercises	Examples
4	3
5	1
6	2

4. **COORDINATE GEOMETRY** The vertices of $\triangle ABC$ are $A(-3, 3)$, $B(3, 2)$, and $C(1, -4)$. Find the coordinates of the circumcenter. $\left(-\frac{17}{38}, -\frac{7}{38}\right)$

5. **PROOF** Write a two-column proof.
Given: $\overline{XY} \cong \overline{XZ}$
$\overline{YM}$ and $\overline{ZN}$ are medians.
Prove: $\overline{YM} \cong \overline{ZN}$ See p. 279A.

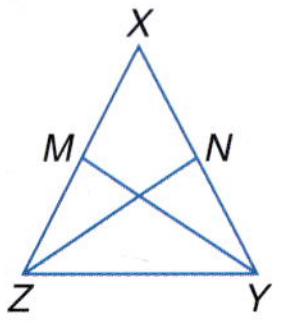

Application

6. **ALGEBRA** Lines ℓ, m, and n are perpendicular bisectors of $\triangle PQR$ and meet at T. If $TQ = 2x$, $PT = 3y - 1$, and $TR = 8$, find x, y, and z.
$x = 4$, $y = 3$, $z = 3$

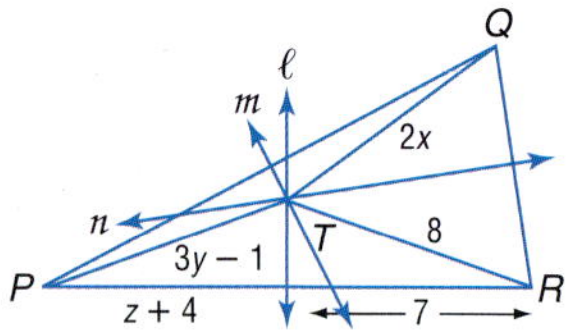

Answers

1. Sample answer: Both pass through the midpoint of a side. A perpendicular bisector is perpendicular to the side of a triangle, and does not necessarily pass through the vertex opposite the side, while a median does pass through the vertex and is not necessarily perpendicular to the side.

2. Sample answer:

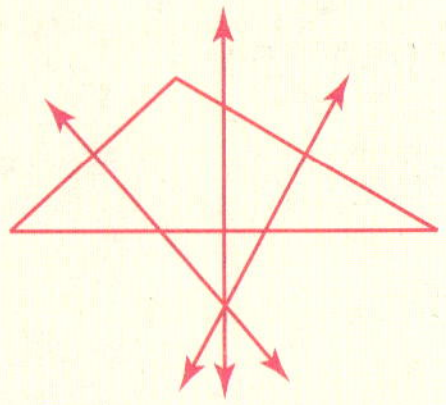

3. Sample answer: An altitude and angle bisector of a triangle are the same segment in an equilateral triangle.

242 Chapter 5 Relationships in Triangles

★ indicates increased difficulty

Practice and Apply

Homework Help

For Exercises	See Examples
10–12, 31–33	1
13–16, 21–26	2
7–9, 27–30	3

Extra Practice
See page 763.

COORDINATE GEOMETRY The vertices of $\triangle DEF$ are $D(4, 0)$, $E(-2, 4)$, and $F(0, 6)$. Find the coordinates of the points of concurrency of $\triangle DEF$.

7. centroid $\left(\frac{2}{3}, 3\frac{1}{3}\right)$
8. orthocenter $\left(-\frac{4}{5}, 4\frac{4}{5}\right)$
9. circumcenter $\left(1\frac{2}{5}, 2\frac{3}{5}\right)$

10. **PROOF** Write a paragraph proof of Theorem 5.1.
Given: $\overline{CD}$ is the perpendicular bisector of $\overline{AB}$. E is a point on $\overline{CD}$. See margin.
Prove: $EB = EA$

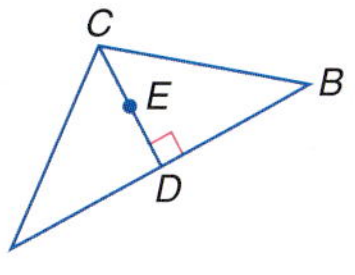

PROOF Write a two-column proof. 11–12. See margin.

11. **Given:** $\triangle UVW$ is isosceles with vertex angle UVW. $\overline{YV}$ is the bisector of $\angle UVW$.
Prove: $\overline{YV}$ is a median.

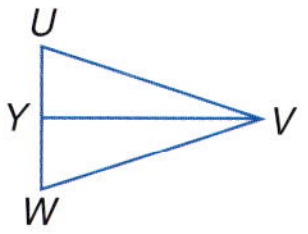

12. **Given:** $\overline{GL}$ is a median of $\triangle EGH$. $\overline{JM}$ is a median of $\triangle IJK$. $\triangle EGH \cong \triangle IJK$
Prove: $\overline{GL} \cong \overline{JM}$

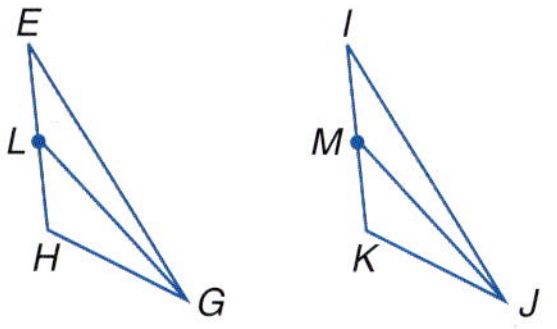

13. **ALGEBRA** Find x and $m\angle 2$ if $\overline{MS}$ is an altitude of $\triangle MNQ$, $m\angle 1 = 3x + 11$, and $m\angle 2 = 7x + 9$. $x = 7$, $m\angle 2 = 58$
14. **ALGEBRA** If $\overline{MS}$ is a median of $\triangle MNQ$, $QS = 3a - 14$, $SN = 2a + 1$, and $m\angle MSQ = 7a + 1$, find the value of a. Is $\overline{MS}$ also an altitude of $\triangle MNQ$? Explain. 15; no; because $m\angle MSQ = 106$

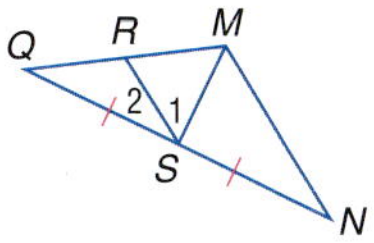

Exercises 13 and 14

15. **ALGEBRA** If $\overline{WP}$ is a median and an angle bisector, $AP = 3y + 11$, $PH = 7y - 5$, $m\angle HWP = x + 12$, $m\angle PAW = 3x - 2$, and $m\angle HWA = 4x - 16$, find x and y. Is $\overline{WP}$ also an altitude? Explain. $x = 20$, $y = 4$; yes; because $m\angle WPA = 90$
16. **ALGEBRA** If $\overline{WP}$ is a perpendicular bisector, $m\angle WHA = 8q + 17$, $m\angle HWP = 10 + q$, $AP = 6r + 4$, and $PH = 22 + 3r$, find r, q, and $m\angle HWP$. $r = 6$, $q = 7$, $m\angle HWP = 17$

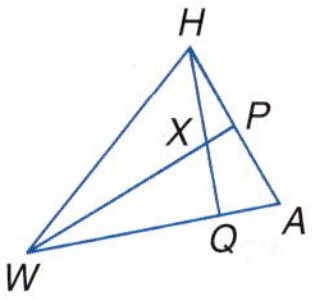

Exercises 15 and 16

State whether each sentence is *always, sometimes,* or *never* true. 17. always

17. The three medians of a triangle intersect at a point in the interior of the triangle.
18. The three altitudes of a triangle intersect at a vertex of the triangle. sometimes
19. The three angle bisectors of a triangle intersect at a point in the exterior of the triangle. never
20. The three perpendicular bisectors of a triangle intersect at a point in the exterior of the triangle. sometimes

Answers

10. **Given:** $\overline{CD}$ is the $\perp$ bisector of $\overline{AB}$. E is a point on $\overline{CD}$.
Prove: $EB = EA$

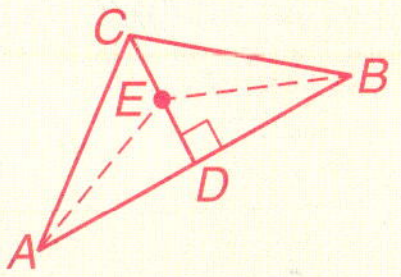

Proof: $\overline{CD}$ is the $\perp$ bisector of $\overline{AB}$. By definition of $\perp$ bisector, D is the midpoint of $\overline{AB}$. Thus, $\overline{AD} \cong \overline{BD}$ by the Midpoint Theorem. $\angle CDA$ and $\angle CDB$ are right angles by definition of perpendicular. Since all right angles are congruent, $\angle CDA \cong \angle CDB$. Since E is a point on $\overline{CD}$, $\angle EDA$ and $\angle EDB$ are right angles and are congruent. By the Reflexive Property, $\overline{ED} \cong \overline{ED}$. Thus, $\triangle EDA \cong \triangle EDB$ by SAS. $\overline{EB} \cong \overline{EA}$ because CPCTC, and by definition of congruence, $EB = EA$.

11. **Given:** $\triangle UVW$ is isosceles with vertex angle UVW. $\overline{YV}$ is the bisector of $\angle UVW$.
Prove: $\overline{YV}$ is a median.

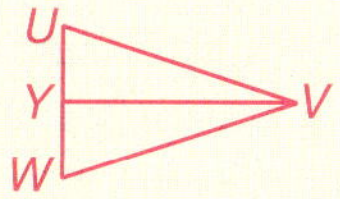

Proof:
Statements (Reasons)
1. $\triangle UVW$ is an isosceles triangle with vertex angle UVW, $\overline{YV}$ is the bisector of $\angle UVW$. (Given)
2. $\overline{UV} \cong \overline{WV}$ (Def. of isosceles $\triangle$)
3. $\angle UVY \cong \angle WVY$ (Def. of angle bisector)
4. $\overline{YV} \cong \overline{YV}$ (Reflexive Property)
5. $\triangle UVY \cong \triangle WVY$ (SAS)
6. $\overline{UY} \cong \overline{WY}$ (CPCTC)
7. Y is the midpoint of $\overline{UW}$. (Def. of midpoint)
8. $\overline{YV}$ is a median. (Def. of median)

12. **Given:** $\overline{GL}$ is a median of $\triangle EGH$. $\overline{JM}$ is a median of $\triangle IJK$. $\triangle EGH \cong \triangle IJK$
Prove: $\overline{GL} \cong \overline{JM}$

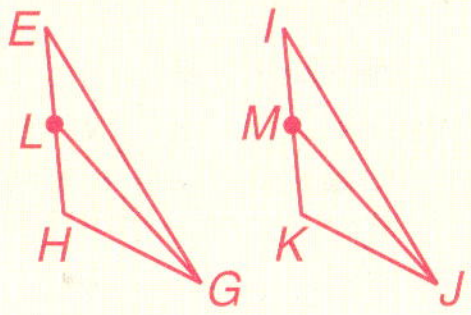

Proof:
Statements (Reasons)
1. $\overline{GL}$ is a median of $\triangle EGH$, $\overline{JM}$ is a median of $\triangle IJK$, and $\triangle EGH \cong \triangle IJK$. (Given)
2. $\overline{GH} \cong \overline{JK}$, $\angle GHL \cong \angle JKM$, $\overline{EH} \cong \overline{IK}$ (CPCTC)
3. $EH = IK$ (Def. of $\cong$)
4. $\overline{EL} \cong \overline{LH}$, $\overline{IM} \cong \overline{MK}$ (Def. of median)
5. $EL = LH$, $IM = MK$ (Def of $\cong$)
6. $EL + LH = EH$, $IM + MK = IK$ (Segment Addition Postulate)
7. $EL + LH = IM + MK$ (Substitution)
8. $LH + LH = MK + MK$ (Substitution)
9. $2LH = 2MK$ (Addition Prop.)
10. $LH = MK$ (Division Prop.)
11. $\overline{LH} \cong \overline{MK}$ (Def of $\cong$)
12. $\triangle GHL \cong \triangle JKM$ (SAS)
13. $\overline{GL} \cong \overline{JM}$ (CPCTC)

Lesson 5-1 Bisectors, Medians, and Altitudes 243

21. **ALGEBRA** Find x if $\overline{PS}$ is a median of $\triangle PQR$. **2**

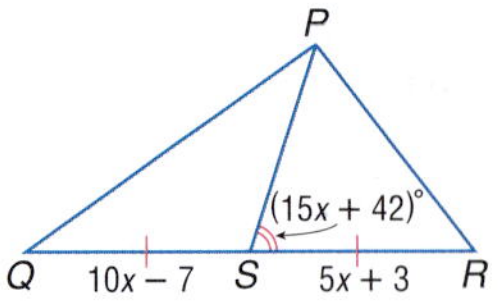

22. **ALGEBRA** Find x if $\overline{AD}$ is an altitude of $\triangle ABC$. **24**

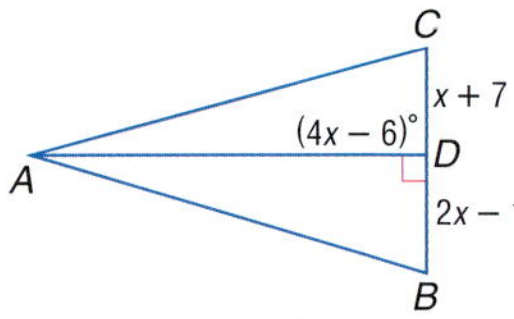

ALGEBRA **For Exercises 23–26, use the following information.**

In $\triangle PQR$, $ZQ = 3a - 11$, $ZP = a + 5$, $PY = 2c - 1$, $YR = 4c - 11$, $m\angle PRZ = 4b - 17$, $m\angle ZRQ = 3b - 4$, $m\angle QYR = 7b + 6$, and $m\angle PXR = 2a + 10$.

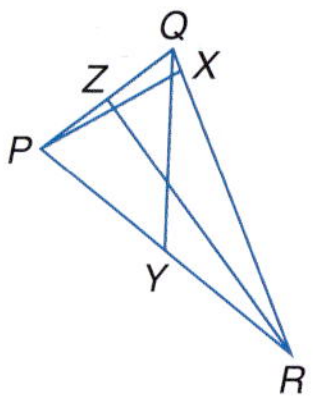

23. $\overline{PX}$ is an altitude of $\triangle PQR$. Find a. **40**
24. If $\overline{RZ}$ is an angle bisector, find $m\angle PRZ$. **$m\angle PRZ = 35$**
25. Find PR if $\overline{QY}$ is a median. **$PR = 18$**
26. If $\overline{QY}$ is a perpendicular bisector of $\overline{PR}$, find b. **12**

COORDINATE GEOMETRY **For Exercises 27–30, use the following information.**

$R(3, 3)$, $S(-1, 6)$, and $T(1, 8)$ are the vertices of $\triangle RST$, and $\overline{RX}$ is a median.

27. What are the coordinates of X? **(0, 7)**
28. Find RX. **5 units**
29. Determine the slope of $\overleftrightarrow{RX}$. $-\frac{4}{3}$
30. Is $\overline{RX}$ an altitude of $\triangle RST$? Explain. **No; the product of the slopes of $\overline{ST}$ and $\overline{RX}$ is not −1. Thus, the segments are not perpendicular.**

PROOF **Write a two-column proof for each theorem.** **31–33. See p. 279A.**

31. ★ Theorem 5.2

Given: $\overline{CA} \cong \overline{CB}$
$\overline{AD} \cong \overline{BD}$

Prove: C and D are on the perpendicular bisector of $\overline{AB}$.

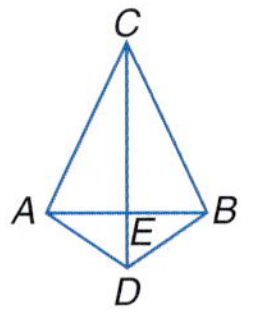

★ 32. Theorem 5.5

★ 33. Theorem 5.6

More About. . .

Orienteering

The International Orienteering Federation World Cup consists of a series of nine races held throughout the world, in which the runners compete for points based on their completion times.

Source: www.orienteering.org

34. **ORIENTEERING** Orienteering is a competitive sport, originating in Sweden, that tests the skills of map reading and cross-country running. Competitors race through an unknown area to find various checkpoints using only a compass and topographical map. On an amateur course, clues were given to locate the first flag.

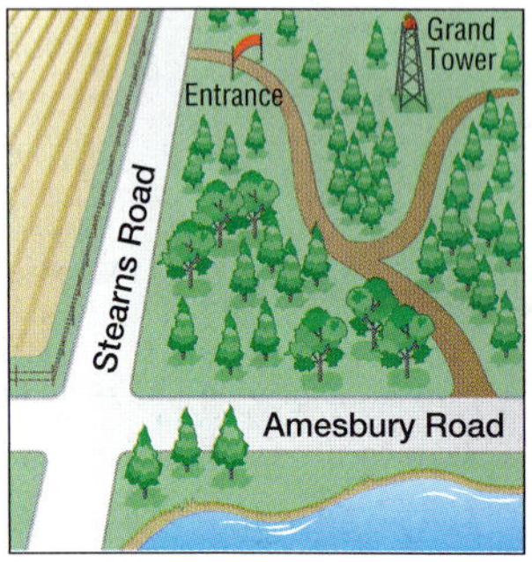

- The flag is as far from the Grand Tower as it is from the park entrance.
- If you run from Stearns Road to the flag or from Amesbury Road to the flag, you would run the same distance.

Describe how to find the first flag. **See margin.**

Study Guide and Intervention, p. 245 (shown) and p. 246

Perpendicular Bisectors and Angle Bisectors A **perpendicular bisector** of a side of a triangle is a line, segment, or ray that is perpendicular to the side and passes through its midpoint. Another special segment, ray, or line is an **angle bisector**, which divides an angle into two congruent angles.

Two properties of perpendicular bisectors are:

(1) a point is on the perpendicular bisector of a segment if and only if it is equidistant from the endpoints of the segment, and

(2) the three perpendicular bisectors of the sides of a triangle meet at a point, called the **circumcenter** of the triangle, that is equidistant from the three vertices of the triangle.

Two properties of angle bisectors are:

(1) a point is on the angle bisector of an angle if and only if it is equidistant from the sides of the angle, and

(2) the three angle bisectors of a triangle meet at a point, called the **incenter** of the triangle, that is equidistant from the three sides of the triangle.

Example 1 **$\overrightarrow{BD}$ is the perpendicular bisector of $\overline{AC}$. Find x.**

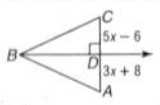

$\overrightarrow{BD}$ is the perpendicular bisector of $\overline{AC}$, so $AD = DC$.

$3x + 8 = 5x - 6$
$14 = 2x$
$7 = x$

Example 2 **$\overrightarrow{MR}$ is the angle bisector of $\angle NMP$. Find x if $m\angle 1 = 5x + 8$ and $m\angle 2 = 8x - 16$.**

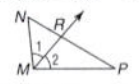

$\overrightarrow{MR}$ is the angle bisector of $\angle NMP$, so $m\angle 1 = m\angle 2$.

$5x + 8 = 8x - 16$
$24 = 3x$
$8 = x$

Exercises

Find the value of each variable.

1. 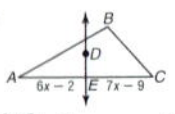

$\overleftrightarrow{DE}$ is the perpendicular bisector of $\overline{AC}$.
$x = 7$

2. 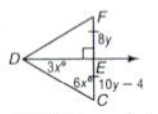

$\triangle CDF$ is equilateral.
$x = 10$; $y = 2$

3. 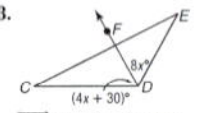

$\overrightarrow{DF}$ bisects $\angle CDE$.
$x = 7.5$

4. For what kinds of triangle(s) can the perpendicular bisector of a side also be an angle bisector of the angle opposite the side? **isosceles triangle, equilateral triangle**

5. For what kind of triangle do the perpendicular bisectors intersect in a point outside the triangle? **obtuse triangle**

Skills Practice, p. 247 and Practice, p. 248 (shown)

ALGEBRA In $\triangle ABC$, $\overline{BF}$ is the angle bisector of $\angle ABC$, $\overline{AE}$, $\overline{BF}$, and $\overline{CD}$ are medians, and P is the centroid.

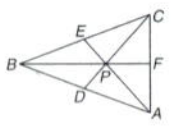

1. Find x if $DP = 4x - 3$ and $CP = 30$. **4.5**
2. Find y if $AP = y$ and $EP = 18$. **36**
3. Find z if $FP = 5z + 10$ and $BP = 42$. **2.2**
4. If $m\angle ABC = x$ and $m\angle BAC = m\angle BCA = 2x - 10$, is $\overline{BF}$ an altitude? Explain.
Yes; since $x = 40$ and $\overline{BF}$ is an angle bisector, it follows that $m\angle BAF = 70$ and $m\angle ABF = 20$. So $m\angle AFB = 90$, and $\overline{BF} \perp \overline{AC}$.

ALGEBRA In $\triangle PRS$, $\overline{PT}$ is an altitude and $\overline{PX}$ is a median.

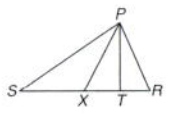

5. Find RS if $RX = x + 7$ and $SX = 3x - 11$.
32
6. Find RT if $RT = x - 6$ and $m\angle PTR = 8x - 6$.
6

ALGEBRA In $\triangle DEF$, $\overline{GI}$ is a perpendicular bisector.

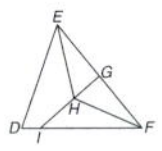

7. Find x if $EH = 16$ and $FH = 6x - 5$.
3.5
8. Find y if $EG = 3.2y - 1$ and $FG = 2y + 5$.
5
9. Find z if $m\angle EGH = 12z$.
7.5

COORDINATE GEOMETRY The vertices of $\triangle STU$ are $S(0, 1)$, $T(4, 7)$, and $U(8, -3)$. Find the coordinates of the points of concurrency of $\triangle STU$.

10. orthocenter $\left(\frac{5}{4}, \frac{3}{2}\right)$
11. centroid $\left(4, \frac{5}{3}\right)$
12. circumcenter $\left(\frac{43}{8}, \frac{7}{4}\right)$ or (5.375, 1.75)

13. **MOBILES** Nabuko wants to construct a mobile out of flat triangles so that the surfaces of the triangles hang parallel to the floor when the mobile is suspended. How can Nabuko be certain that she hangs the triangles to achieve this effect?
She needs to hang each triangle from its center of gravity or centroid, which is the point at which the three medians of the triangle intersect.

Reading to Learn Mathematics, p. 249 ELL

Pre-Activity **How can you balance a paper triangle on a pencil point?**

Read the introduction to Lesson 5-1 at the top of page 238 in your textbook.

Draw any triangle and connect each vertex to the midpoint of the opposite side to form the three medians of the triangle. Is the point where the three medians intersect the midpoint of each of the medians? **Sample answer: No; the intersection point appears to be more than halfway from each vertex to the midpoint of the opposite side.**

Reading the Lesson

1. Underline the correct word or phrase to complete each sentence.
   a. Three or more lines that intersect at a common point are called (parallel/perpendicular/concurrent) lines.
   b. Any point on the perpendicular bisector of a segment is (parallel to/congruent to/equidistant from) the endpoints of the segment.
   c. A(n) (altitude/angle bisector/median/perpendicular bisector) of a triangle is a segment drawn from a vertex of the triangle perpendicular to the line containing the opposite side.
   d. The point of concurrency of the three perpendicular bisectors of a triangle is called the (orthocenter/circumcenter/centroid/incenter).
   e. Any point in the interior of an angle that is equidistant from the sides of that angle lies on the (median/angle bisector/altitude).
   f. The point of concurrency of the three angle bisectors of a triangle is called the (orthocenter/circumcenter/centroid/incenter).

2. In the figure, E is the midpoint of $\overline{AB}$, F is the midpoint of $\overline{BC}$, and G is the midpoint of $\overline{AC}$.

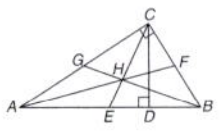

   a. Name the altitudes of $\triangle ABC$. **$\overline{AC}$, $\overline{BC}$, $\overline{CD}$**
   b. Name the medians of $\triangle ABC$. **$\overline{AF}$, $\overline{BG}$, $\overline{CE}$**
   c. Name the centroid of $\triangle ABC$. **H**
   d. Name the orthocenter of $\triangle ABC$. **C**
   e. If $AF = 12$ and $CE = 9$, find AH and HE. **$AH = 8$, $HE = 3$**

Helping You Remember

3. A good way to remember something is to explain it to someone else. Suppose that a classmate is having trouble remembering whether the center of gravity of a triangle is the orthocenter, the centroid, the incenter, or the circumcenter of the triangle. Suggest a way to remember which point it is. **Sample answer: The terms *centroid* and *center of gravity* mean the same thing and in both terms, the letters "cent" come at the beginning of the terms.**

Enrichment, p. 250

Inscribed and Circumscribed Circles

The three angle bisectors of a triangle intersect in a single point called the **incenter**. This point is the center of a circle that just touches the three sides of the triangle. Except for the three points where the circle touches the sides, the circle is inside the triangle. The circle is said to be inscribed in the triangle.

1. With a compass and a straightedge, construct the inscribed circle for $\triangle PQR$ by following the steps below.
   Step 1 Construct the bisectors of $\angle P$ and $\angle Q$. Label the point where the bisectors meet A.
   Step 2 Construct a perpendicular segment from A to $\overline{RQ}$. Use the letter B to label the point where the perpendicular segment intersects $\overline{RQ}$.
   Step 3 Use a compass to draw the circle with center at A and radius $\overline{AB}$.

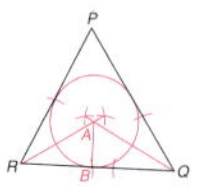

Construct the inscribed circle in each triangle.

2. 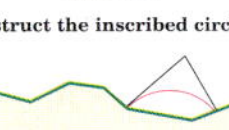

3. 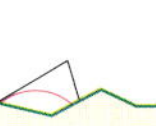

Answer

34. The flag is located at the intersection of the angle bisector between Amesbury and Stearns Roads and the perpendicular bisector of the segment joining Grand Tower and the park entrance.

244 Chapter 5 Relationships in Triangles

STATISTICS **For Exercises 35–38, use the following information.**

The *mean* of a set of data is an average value of the data. Suppose $\triangle ABC$ has vertices $A(16, 8)$, $B(2, 4)$, and $C(-6, 12)$.

35. Find the mean of the x-coordinates of the vertices. **4**
36. Find the mean of the y-coordinates of the vertices. **8**
37. Graph $\triangle ABC$ and its medians. **See margin.**
38. Make a conjecture about the centroid and the means of the coordinates of the vertices. **The centroid has the same coordinates as the means of the vertices' coordinates.**

39. The altitude will be the same for both triangles, and the bases will be congruent, so the areas will be equal.

39. **CRITICAL THINKING** Draw any $\triangle XYZ$ with median $\overline{XN}$ and altitude $\overline{XO}$. Recall that the area of a triangle is one-half the product of the measures of the base and the altitude. What conclusion can you make about the relationship between the areas of $\triangle XYN$ and $\triangle XZN$?

40. **WRITING IN MATH** Answer the question that was posed at the beginning of the lesson. **See margin.**

How can you balance a paper triangle on a pencil point?

Include the following in your answer:

- which special point is the center of gravity, and
- a construction showing how to find this point.

FCAT Practice
Standardized Test Practice

41. In $\triangle FGH$, which type of segment is $\overline{FJ}$? **C**

Ⓐ angle bisector Ⓑ perpendicular bisector
Ⓒ median Ⓓ altitude

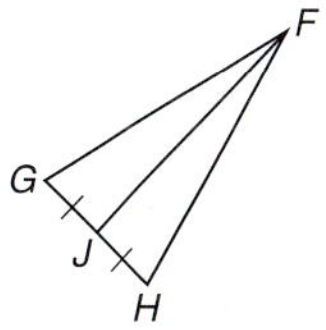

42. **ALGEBRA** If $xy \neq 0$ and $3x = 0.3y$, then $\frac{y}{x} = \underline{\ ?\ }$. **D**

Ⓐ 0.1 Ⓑ 1.0 Ⓒ 3.0 Ⓓ 10.0

Maintain Your Skills

Mixed Review **Position and label each triangle on the coordinate plane.** *(Lesson 4-7)*

43. equilateral $\triangle ABC$ with base $\overline{AB}$ n units long **43–45. See margin.**
44. isosceles $\triangle DEF$ with congruent sides $2a$ units long and base a units long
45. right $\triangle GHI$ with hypotenuse $\overline{GI}$, HI is three times GH, and GH is x units long

For Exercises 46–49, refer to the figure at the right. *(Lesson 4-6)*

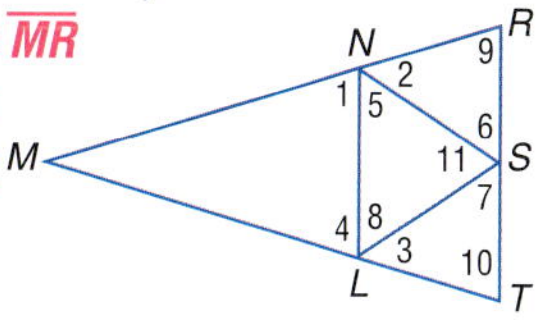

46. If $\angle 9 \cong \angle 10$, name two congruent segments. $\overline{MT} \cong \overline{MR}$
47. If $\overline{NL} \cong \overline{SL}$, name two congruent angles. $\angle 5 \cong \angle 11$
48. If $\overline{LT} \cong \overline{LS}$, name two congruent angles. $\angle 7 \cong \angle 10$
49. If $\angle 1 \cong \angle 4$, name two congruent segments. $\overline{ML} \cong \overline{MN}$
50. **INTERIOR DESIGN** Stacey is installing a curtain rod on the wall above the window. To ensure that the rod is parallel to the ceiling, she measures and marks 6 inches below the ceiling in several places. If she installs the rod at these markings centered over the window, how does she know the curtain rod will be parallel to the ceiling? *(Lesson 3-6)* **It is everywhere equidistant.**

Getting Ready for the Next Lesson **BASIC SKILL** **Replace each ● with < or > to make each sentence true.**

51. $\frac{3}{8}$ ● $\frac{5}{16}$ **>** 52. 2.7 ● $\frac{5}{3}$ **>** 53. -4.25 ● $-\frac{19}{4}$ **>** 54. $-\frac{18}{25}$ ● $-\frac{19}{27}$ **<**

www.geometryonline.com/self_check_quiz/fcat

4 Assess

Open-Ended Assessment

Modeling Set up a cork board in front of the class with pushpins to use as vertices of triangles, and varied lengths of colored yarn to use as the sides, bisectors, medians, and altitudes of triangles. Students can take turns using the pushpins and yarn to model different types of triangles and to place angle bisectors, segment bisectors, medians, and altitudes on the triangles.

Getting Ready for Lesson 5-2

Basic Skill Students will learn about inequalities and triangles in Lesson 5-2. They will use inequalities to compare angle measures and side lengths. Use Exercises 51–54 to determine your students' familiarity with inequalities.

43. Sample answer:

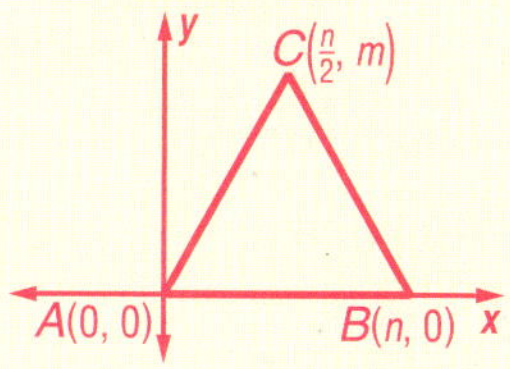

44. Sample answer:

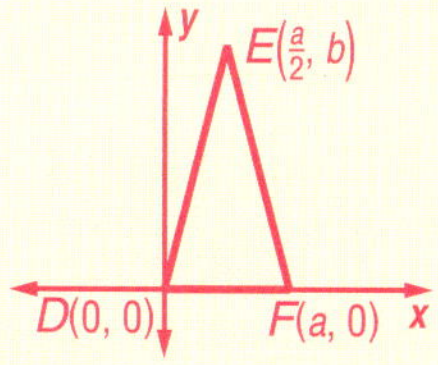

45. Sample answer:

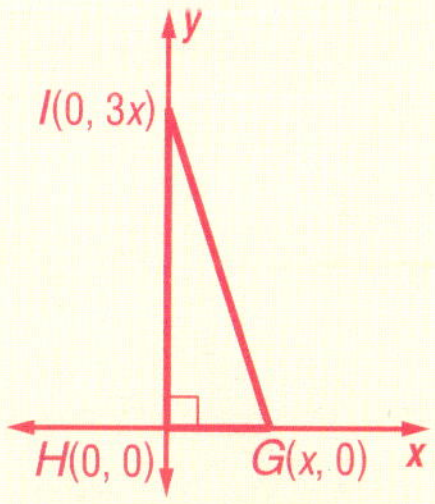

Answers

37.

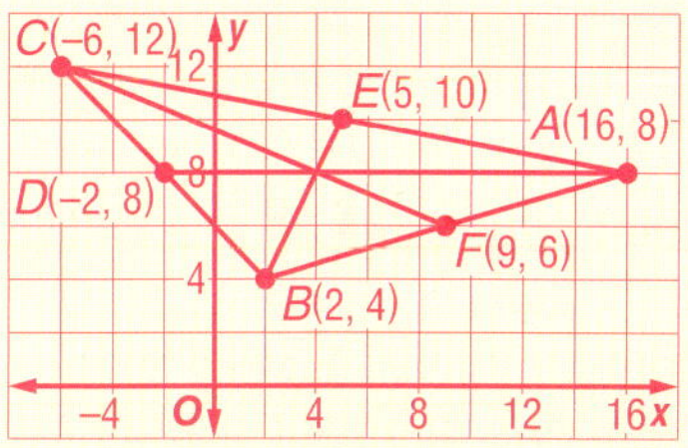

40. Sample answer: You can balance a triangle on a pencil point by locating the center of gravity of the triangle. Answers should include the following.

- **centroid**
-

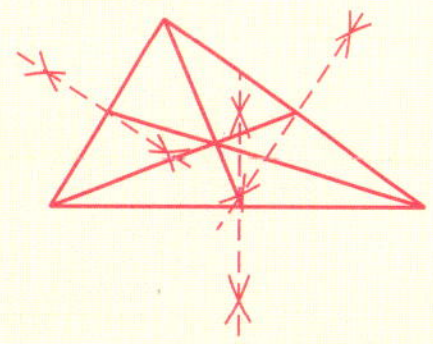

Reading Mathematics

Getting Started

To enhance their understanding and organize concepts, students will relate geometric terms to literal terms.

Teach

Math Words and Everyday Words The two geometric meanings students are most likely to confuse are median and perpendicular bisectors in triangles because both of these actually "bisect" the sides of a triangle. Point out that medians have to connect the vertices with their opposite sides, as paved medians connect two opposite lanes of traffic (even though the orientation is different). Explain that perpendicular bisectors for triangles will *perpendicularly* bisect the sides.

Assess

Study Notebook

Ask students to summarize what they have learned about the relationships between the mathematical meanings and everyday meanings of median, altitude, and bisector.

ELL English Language Learners may benefit from writing key concepts from this activity in their Study Notebooks in their native language and then in English.

Reading Mathematics

Math Words and Everyday Words

Several of the words and terms used in mathematics are also used in everyday language. The everyday meaning can help you to better understand the mathematical meaning and help you remember each meaning. This table shows some words used in this chapter with the everyday meanings and the mathematical meanings.

Word	Everyday Meaning	Geometric Meaning	
median	a paved or planted strip dividing a highway into lanes according to direction of travel	a segment of a triangle that connects the vertex to the midpoint of the opposite side	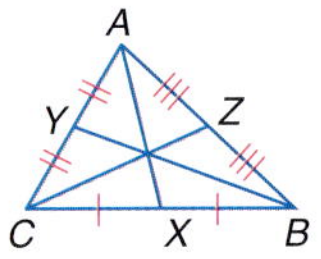
altitude	the vertical elevation of an object above a surface	a segment from a vertex of a triangle that is perpendicular to the line containing the opposite side	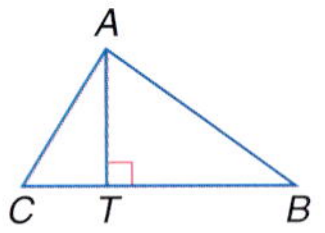
bisector	something that divides into two usually equal parts	a segment that divides an angle or a side into two parts of equal measure	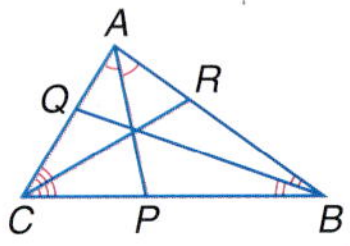

Source: *Merriam-Webster Collegiate Dictionary*

Notice that the geometric meaning is more specific, but related to the everyday meaning. For example, the everyday definition of *altitude* is elevation, or height. In geometry, an altitude is a segment of a triangle perpendicular to the base through the vertex. The length of an altitude is the height of the triangle.

Reading to Learn

1. How does the mathematical meaning of *median* relate to the everyday meaning? **See margin.**
2. **RESEARCH** Use a dictionary or other sources to find alternate definitions of *vertex*. **See margin.**
3. **RESEARCH** *Median* has other meanings in mathematics. Use the Internet or other sources to find alternate definitions of this term. **See margin.**
4. **RESEARCH** Use a dictionary or other sources to investigate definitions of *segment*. **See margin.**

Answers

1. Sample answer: A median of a triangle is a segment that has one endpoint at a vertex and the other at the midpoint of the opposite side; the everyday meaning says it is a paved or planted strip in the middle of a highway.
2. Sample answer: the intersection of two or more lines or curves, the top of the head, the highest point
3. Sample answer: in a trapezoid, the segment joining the midpoints of the legs; the middle value of a set of data that has been arranged into an ordered sequence
4. Sample answer: a separate piece of something; a portion cut off from a geometric figure by one or more points, lines, or planes.

5-2 Inequalities and Triangles

Sunshine State Standards
MA.C.1.4.1

What You'll Learn

- Recognize and apply properties of inequalities to the measures of angles of a triangle.
- Recognize and apply properties of inequalities to the relationships between angles and sides of a triangle.

How can you tell which corner is bigger?

Sam is delivering two potted trees to be used on a patio. The instructions say for the trees to be placed in the two largest corners of the patio. All Sam has is a diagram of the triangular patio that shows the measurements 45 feet, 48 feet, and 51 feet. Sam can find the largest corner because the measures of the angles of a triangle are related to the measures of the sides opposite them.

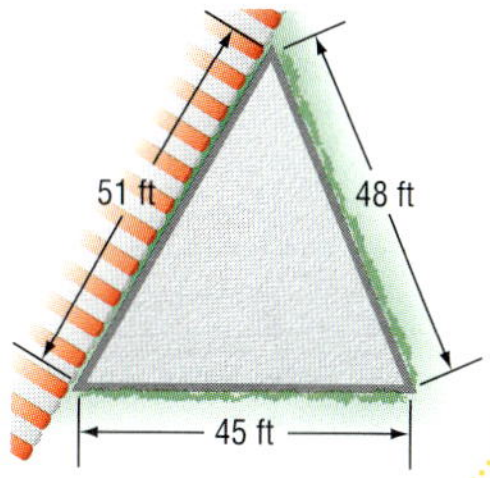

ANGLE INEQUALITIES In algebra, you learned about the inequality relationship between two real numbers. This relationship is often used in proofs.

Key Concept *Definition of Inequality*

For any real numbers a and b, $a > b$ if and only if there is a positive number c such that $a = b + c$.

Example: If $6 = 4 + 2$, $6 > 4$ and $6 > 2$.

The properties of inequalities you studied in algebra can be applied to the measures of angles and segments.

Properties of Inequalities for Real Numbers	
	For all numbers a, b, and c
Comparison Property	$a < b$, $a = b$, or $a > b$
Transitive Property	**1.** If $a < b$ and $b < c$, then $a < c$. **2.** If $a > b$ and $b > c$, then $a > c$.
Addition and Subtraction Properties	**1.** If $a > b$, then $a + c > b + c$ and $a - c > b - c$. **2.** If $a < b$, then $a + c < b + c$ and $a - c < b - c$.
Multiplication and Division Properties	**1.** If $c > 0$ and $a < b$, then $ac < bc$ and $\frac{a}{c} < \frac{b}{c}$. **2.** If $c > 0$ and $a > b$, then $ac > bc$ and $\frac{a}{c} > \frac{b}{c}$. **3.** If $c < 0$ and $a < b$, then $ac > bc$ and $\frac{a}{c} > \frac{b}{c}$. **4.** If $c < 0$ and $a > b$, then $ac < bc$ and $\frac{a}{c} < \frac{b}{c}$.

5-2 Lesson Notes

1 Focus

5-Minute Check Transparency 5-2 Use as a quiz or review of Lesson 5-1.

Mathematical Background notes are available for this lesson on p. 234C.

How can you tell which corner is bigger?

Ask students:

- Just by looking at the diagram and eyeballing the size of the corners, is it easy to tell which corners are bigger? **no**
- What else could Sam use to determine the size of the corners if the diagram is drawn to scale? **a protractor**

Resource Manager

Workbook and Reproducible Masters

Chapter 5 Resource Masters
- Study Guide and Intervention, pp. 251–252
- Skills Practice, p. 253
- Practice, p. 254
- Reading to Learn Mathematics, p. 255
- Enrichment, p. 256
- Assessment, p. 289

Graphing Calculator and Computer Masters, p. 25
Prerequisite Skills Workbook, pp. 9–10, 13–14, 87–88
Teaching Geometry With Manipulatives Masters, pp. 8, 16, 17, 94

Transparencies

5-Minute Check Transparency 5-2
Answer Key Transparencies

Technology

Interactive Chalkboard

Example 1 Compare Angle Measures

Determine which angle has the greatest measure.

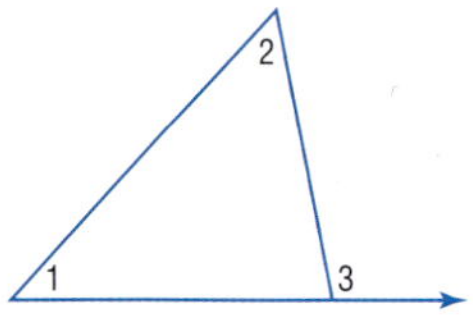

Explore Compare the measure of $\angle 3$ to the measures of $\angle 1$ and $\angle 2$.

Plan Use properties and theorems of real numbers to compare the angle measures.

Solve Compare $m\angle 1$ to $m\angle 3$.
By the Exterior Angle Theorem, $m\angle 3 = m\angle 1 + m\angle 2$. Since angle measures are positive numbers and from the definition of inequality, $m\angle 3 > m\angle 1$.
Compare $m\angle 2$ to $m\angle 3$.
Again, by the Exterior Angle Theorem, $m\angle 3 = m\angle 1 + m\angle 2$. The definition of inequality states that if $m\angle 3 = m\angle 1 + m\angle 2$, then $m\angle 3 > m\angle 2$.

Examine $m\angle 3$ is greater than $m\angle 1$ and $m\angle 2$. Therefore, $\angle 3$ has the greatest measure.

The results from Example 1 suggest that the measure of an exterior angle is always greater than either of the measures of the remote interior angles. This relationship can be stated as a theorem.

Study Tip

Symbols for Angles and Inequalities
The symbol for angle ($\angle$) looks similar to the symbol for less than ($<$), especially when handwritten. Be careful to write the symbols correctly in situations where both are used.

Theorem 5.8

Exterior Angle Inequality Theorem If an angle is an exterior angle of a triangle, then its measure is greater than the measure of either of its corresponding remote interior angles.

Example: $m\angle 4 > m\angle 1$

$m\angle 4 > m\angle 2$

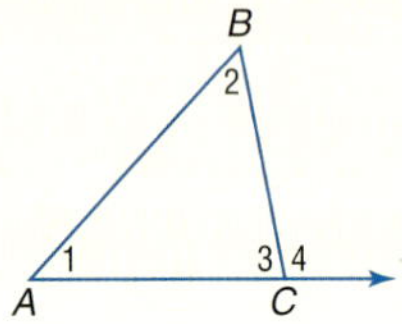

The proof of Theorem 5.8 is in Lesson 5-3.

Example 2 Exterior Angles

Use the Exterior Angle Inequality Theorem to list all of the angles that satisfy the stated condition.

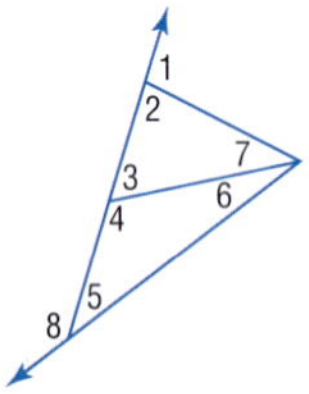

a. all angles whose measures are less than $m\angle 8$
By the Exterior Angle Inequality Theorem, $m\angle 8 > m\angle 4$, $m\angle 8 > m\angle 6$, $m\angle 8 > m\angle 2$, and $m\angle 8 > m\angle 6 + m\angle 7$. Thus, the measures of $\angle 4$, $\angle 6$, $\angle 2$, and $\angle 7$ are all less than $m\angle 8$.

b. all angles whose measures are greater than $m\angle 2$
By the Exterior Angle Inequality Theorem, $m\angle 8 > m\angle 2$ and $m\angle 4 > m\angle 2$. Thus, the measures of $\angle 4$ and $\angle 8$ are greater than $m\angle 2$.

ANGLE-SIDE RELATIONSHIPS Recall that if two sides of a triangle are congruent, then the angles opposite those sides are congruent. In the following Geometry Activity, you will investigate the relationship between sides and angles when they are not congruent.

2 Teach

ANGLE INEQUALITIES

In-Class Examples

Power Point®

1 Determine which angle has the greatest measure.

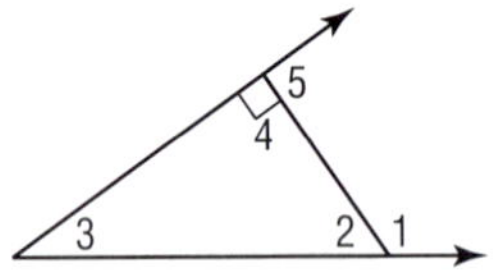

$\angle 1$ has the greatest measure.

Teaching Tip Tell students that exterior angles can be located within groups of triangles as well, such as angles 3 and 4 in the figure in Example 2.

2 Use the Exterior Angle Inequality to list all of the angles that satisfy the stated condition.

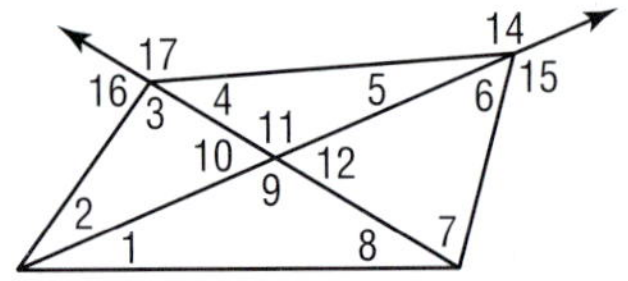

a. all angles whose measures are less than $m\angle 14$ $\angle 4$, $\angle 11$, $\angle 9$, $\angle 3$, $\angle 2$, $\angle 6$, $\angle 7$

b. all angles whose measures are greater than $m\angle 5$ $\angle 10$, $\angle 16$, $\angle 12$, $\angle 15$, $\angle 17$

DAILY INTERVENTION — Differentiated Instruction ELL

Verbal/Linguistic Ask students to summarize the proof of Theorem 5.9 using their own words in paragraph form. Tell them they do not have to use the exact order of the formal proof, but they should have a logical flow from the beginning to the end of the paragraph. Instead of using formal reasons, students could explain the concepts of the properties, definitions, postulates, and theorems used in the proof.

248 Chapter 5 Relationships in Triangles

Geometry Activity

Inequalities for Sides and Angles of Triangles

Model

- Draw an acute scalene triangle, and label the vertices *A*, *B*, and *C*.

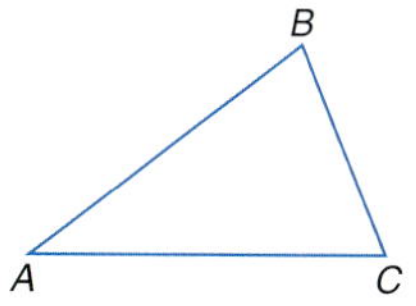

- Measure each side of the triangle. Record the measures in a table.

Side	Measure
$\overline{BC}$	
$\overline{AC}$	
$\overline{AB}$	

- Measure each angle of the triangle. Record each measure in a table.

Angle	Measure
$\angle A$	
$\angle B$	
$\angle C$	

Analyze

1. Describe the measure of the angle opposite the longest side in terms of the other angles. **Sample answer: It is the greatest measure.**
2. Describe the measure of the angle opposite the shortest side in terms of the other angles. **Sample answer: It is the least measure.**
3. Repeat the activity using other triangles. **See students' work.**

Make a Conjecture

4. What can you conclude about the relationship between the measures of sides and angles of a triangle?

4. Sample answer: The measures of the angles opposite the sides are in the same order as the lengths of the respective sides.

The Geometry Activity suggests the following theorem.

Theorem 5.9

If one side of a triangle is longer than another side, then the angle opposite the longer side has a greater measure than the angle opposite the shorter side.

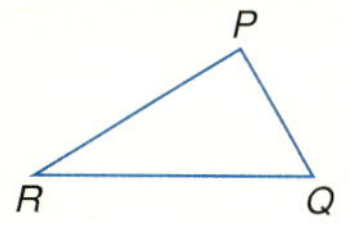

Study Tip

Theorem 5.9
The longest side in a triangle is opposite the largest angle in that triangle.

Proof *Theorem 5.9*

Given: $\triangle PQR$
$PQ < PR$
$\overline{PN} \cong \overline{PQ}$

Prove: $m\angle R < m\angle PQR$

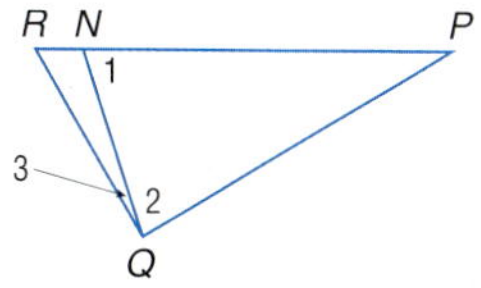

(continued on the next page)

www.geometryonline.com/extra_examples/fcat

Geometry Activity

Materials: ruler, protractor

- Have students set up inequalities for the side lengths and angle measures, such as $m\angle B > m\angle C > m\angle A$ or $AB > AC > BC$.
- Students can also cut a triangle from a piece of paper to measure and record its sides and angles.

ANGLE-SIDE RELATIONSHIPS

In-Class Examples

Power Point®

3 Determine the relationship between the measures of the given angles.

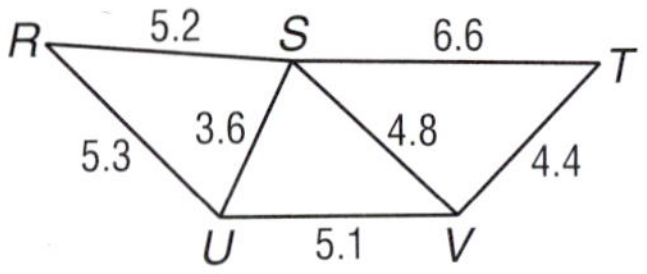

a. $\angle RSU$, $\angle SUR$
$m\angle RSU > m\angle SUR$

b. $\angle TSV$, $\angle STV$
$m\angle TSV < m\angle STV$

c. $\angle RSV$, $\angle RUV$
$m\angle RSV > m\angle RUV$

4 **HAIR ACCESSORIES** Ebony is following directions for folding a handkerchief to make a bandana for her hair. After she folds the handkerchief in half, the directions tell her to tie the two smaller angles of the triangle under her hair. If she folds the handkerchief with the dimensions shown, which two ends should she tie?

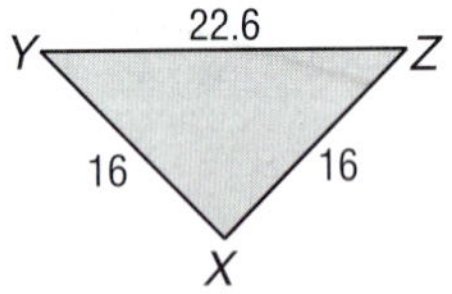

the ends marked *Y* and *Z*

Proof:

Statements	Reasons
1. $\triangle PQR$, $PQ < PR$, $\overline{PN} \cong \overline{PQ}$	1. Given
2. $\angle 1 \cong \angle 2$	2. Isosceles Triangle Theorem
3. $m\angle 1 = m\angle 2$	3. Definition of congruent angles
4. $m\angle R < m\angle 1$	4. Exterior Angle Inequality Theorem
5. $m\angle 2 + m\angle 3 = m\angle PQR$	5. Angle Addition Postulate
6. $m\angle 2 < m\angle PQR$	6. Definition of inequality
7. $m\angle 1 < m\angle PQR$	7. Substitution Property of Equality
8. $m\angle R < m\angle PQR$	8. Transitive Property of Inequality

Example 3 Side-Angle Relationships

Determine the relationship between the measures of the given angles.

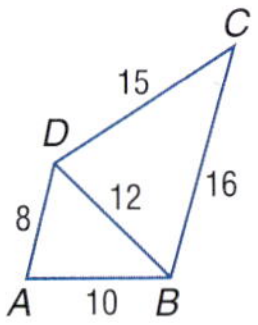

a. $\angle ADB$, $\angle DBA$

The side opposite $\angle ADB$ is longer than the side opposite $\angle DBA$, so $m\angle ADB > m\angle DBA$.

b. $\angle CDA$, $\angle CBA$

$$m\angle DBA < m\angle ADB$$
$$m\angle CBD < m\angle CDB$$
$$m\angle DBA + m\angle CBD < m\angle ADB + m\angle CDB$$
$$m\angle CBA < m\angle CDA$$

More About...

Treehouses

The strength of the tree is the most important concern when building a treehouse. It is important to look for a tree that has branches thick and strong.

Source: www.treehouses.com

The converse of Theorem 5.9 is also true.

Theorem 5.10

If one angle of a triangle has a greater measure than another angle, then the side opposite the greater angle is longer than the side opposite the lesser angle.

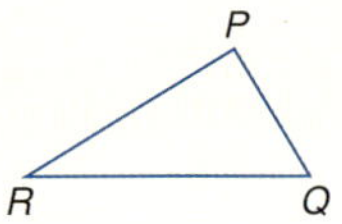

You will prove Theorem 5.10 in Lesson 5-3, Exercise 26.

Example 4 Angle-Side Relationships

TREEHOUSES **Mr. Jackson is constructing the framework for part of a treehouse for his daughter. He plans to install braces at the ends of a certain floor support as shown. Which supports should he attach to *A* and *B*?**

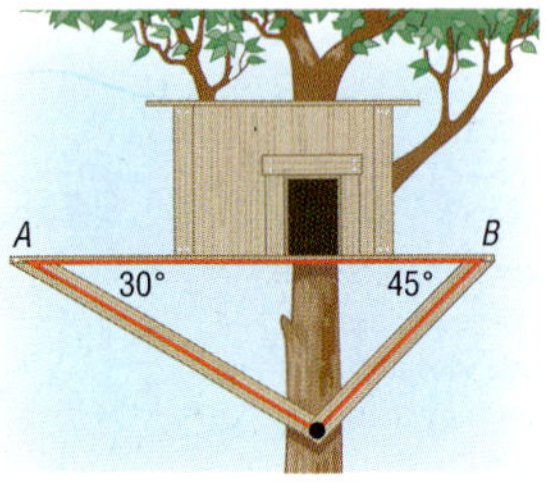

Theorem 5.9 states that if one angle of a triangle has a greater measure, then the side opposite that angle is longer than the side opposite the other angle. Therefore, Mr. Jackson should attach the longer brace at the end marked *A* and the shorter brace at the end marked *B*.

Teacher to Teacher

Douglas E. Hall, Chaparral High School — Las Vegas, NV

I have used the Study Guide and Practice worksheets as notes for my students. This helps the students to be able to work on example problems without waiting for them to copy a diagram, sketch, or geometric figure. It also helps to organize student notes and saves time.

Check for Understanding

Concept Check

1. **State** whether the following statement is *always, sometimes,* or *never* true. *In △JKL with right angle J, if m∠J is twice m∠K, then the side opposite ∠J is twice the length of the side opposite ∠K.* **never**

2. **OPEN ENDED** Draw △*ABC*. List the angle measures and side lengths of your triangle from greatest to least. **See margin.**

3. **FIND THE ERROR** Hector and Grace each labeled △*QRS*.

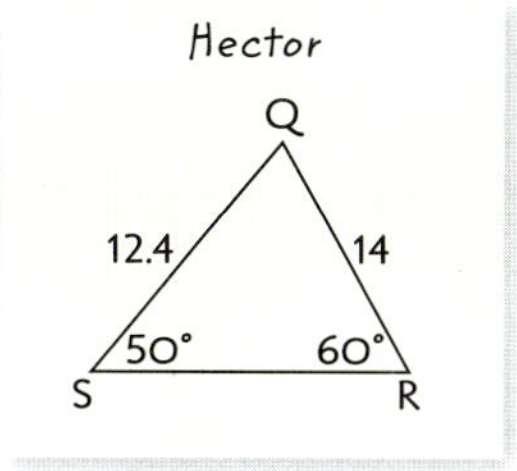

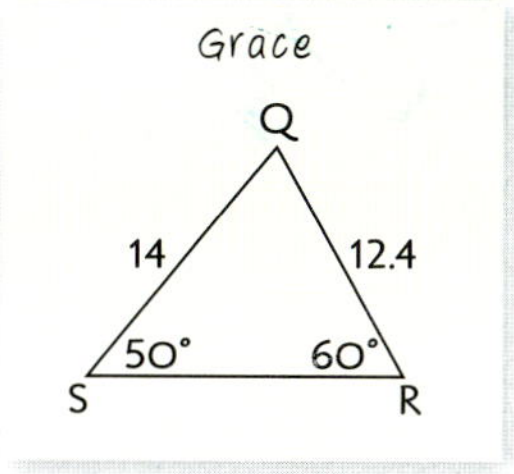

Who is correct? Explain. **Grace; she placed the shorter side with the smaller angle and the longer side with the larger angle.**

Guided Practice

GUIDED PRACTICE KEY	
Exercises	Examples
4–6	1
7–9	2
10–12	3
13–16	4

Determine which angle has the greatest measure.

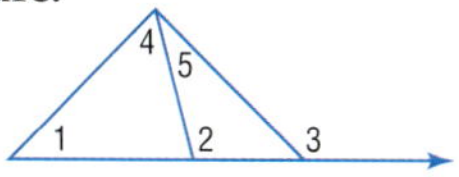

4. ∠1, ∠2, ∠4 **∠2**
5. ∠2, ∠3, ∠5 **∠3**
6. ∠1, ∠2, ∠3, ∠4, ∠5 **∠3**

Use the Exterior Angle Inequality Theorem to list all angles that satisfy the stated condition.

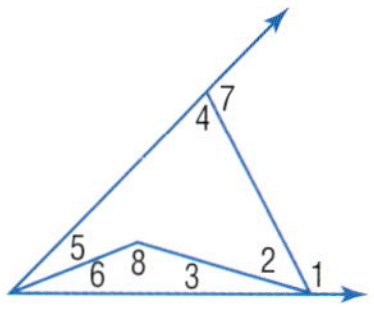

7. all angles whose measures are less than *m*∠1 **∠4, ∠5, ∠6**
8. all angles whose measures are greater than *m*∠6 **∠1, ∠7**
9. all angles whose measures are less than *m*∠7 **∠2, ∠3, ∠5, ∠6**

Determine the relationship between the measures of the given angles.

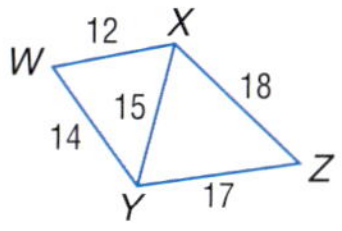

10. ∠*WXY*, ∠*XYW* **m∠WXY > m∠XYW**
11. ∠*XZY*, ∠*XYZ* **m∠XZY < m∠XYZ**
12. ∠*WYX*, ∠*XWY* **m∠WYX < m∠XWY**

Determine the relationship between the lengths of the given sides.

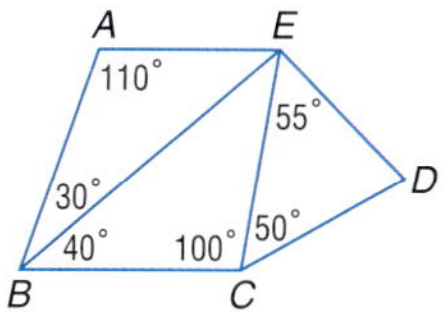

13. $\overline{AE}$, $\overline{EB}$ **AE < EB**
14. $\overline{CE}$, $\overline{CD}$ **CE > CD**
15. $\overline{BC}$, $\overline{EC}$ **BC = EC**

Application

16. **BASEBALL** During a baseball game, the batter hits the ball to the third baseman and begins to run toward first base. At the same time, the runner on first base runs toward second base. If the third baseman wants to throw the ball to the nearest base, to which base should he throw? Explain. **See margin.**

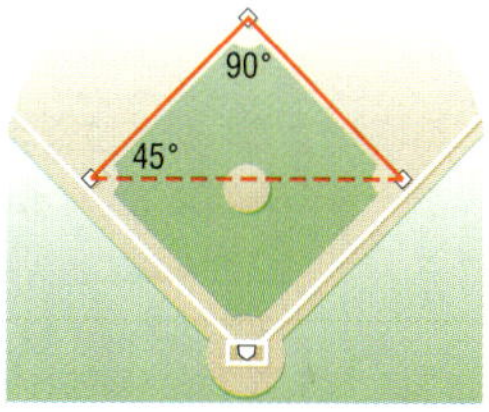

Answers

2. Sample answer: $m\angle CAB$, $m\angle ACB$, $m\angle ABC$; $\overline{BC}$, $\overline{AB}$, $\overline{AC}$

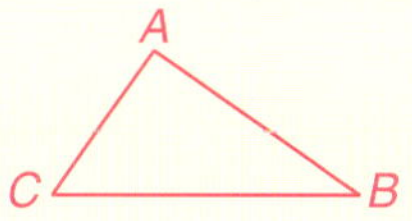

16. Second base; the angle opposite the side from third base to second base is smaller than the angle opposite the side from third to first. Therefore, the distance from third to second is shorter than the distance from third to first.

3 Practice/Apply

Study Notebook

Have students—
- add the definitions/examples of the vocabulary terms to their Vocabulary Builder worksheets for Chapter 5.
- include the definition of inequality and examples to demonstrate the theorems.
- include any other item(s) that they find helpful in mastering the skills in this lesson.

DAILY INTERVENTION

FIND THE ERROR Reinforce that the longer and shorter sides must be *opposite* the larger and smaller angles respectively, not adjacent to them.

About the Exercises...

Organization by Objective
- **Angle Inequalities:** 17–28
- **Angle-Side Relationships:** 29–42

Odd/Even Assignments

Exercises 17–42 are structured so that students practice the same concepts whether they are assigned odd or even problems.

Assignment Guide

Basic: 17–47 odd, 51, 53–66

Average: 17–51 odd, 53–66

Advanced: 18–50 even, 51–63 (optional: 64–66)

All: Quiz 1 (1–10)

Answers

35. Given: $\overline{JM} \cong \overline{JL}$, $\overline{JL} \cong \overline{KL}$
Prove: $m\angle 1 > m\angle 2$

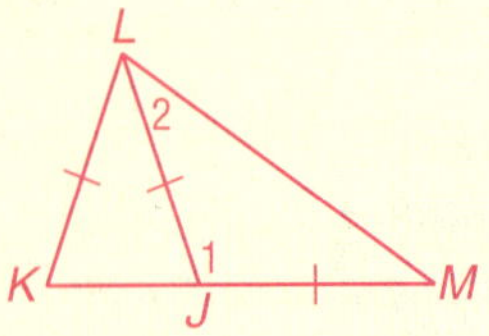

Statements (Reasons)

1. $\overline{JM} \cong \overline{JL}$, $\overline{JL} \cong \overline{KL}$ (Given)
2. $\angle LKJ \cong \angle LJK$ (Isosceles $\triangle$ Theorem)
3. $m\angle LKJ = m\angle LJK$ (Def. of $\cong$)
4. $m\angle 1 > m\angle LKJ$ (Ext. $\angle$ Inequality Theorem)
5. $m\angle 1 > m\angle LJK$ (Substitution)
6. $m\angle LJK > m\angle 2$ (Ext. $\angle$ Inequality Theorem)
7. $m\angle 1 > m\angle 2$ (Trans. Prop. of Inequality)

36. Given: $\overline{PR} \cong \overline{PQ}$; $QR > QP$
Prove: $m\angle P > m\angle Q$

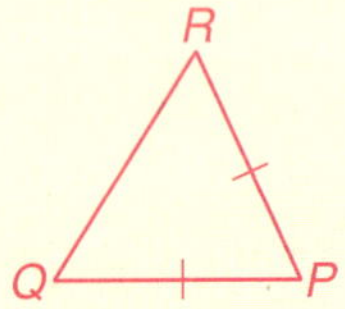

Statements (Reasons)

1. $QR > QP$ (Given)
2. $m\angle P > m\angle R$ (If one side of a $\triangle$ is longer than another, then the $\angle$ opp. the longer side is greater than the $\angle$ opposite the shorter side.)
3. $\overline{PR} \cong \overline{PQ}$ (Given)
4. $\angle Q \cong \angle R$ (Isosceles $\triangle$ Theorem)
5. $m\angle Q = m\angle R$ (Def. of $\cong$)
6. $m\angle P > m\angle Q$ (Substitution)

★ indicates increased difficulty

Practice and Apply

Homework Help

For Exercises	See Examples
17–22	1
23–28	2
30–35	3
38–43	4

Extra Practice
See page 763.

Determine which angle has the greatest measure.

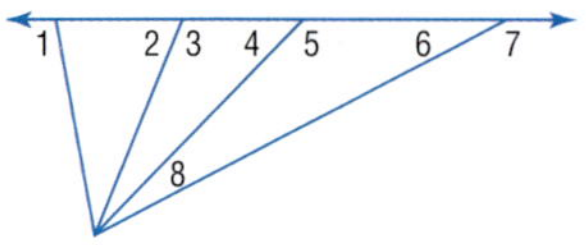

17. $\angle 1, \angle 2, \angle 4$ $\angle 1$
18. $\angle 2, \angle 4, \angle 6$ $\angle 2$
19. $\angle 3, \angle 5, \angle 7$ $\angle 7$
20. $\angle 1, \angle 2, \angle 6$ $\angle 1$
21. $\angle 5, \angle 7, \angle 8$ $\angle 7$
22. $\angle 2, \angle 6, \angle 8$ $\angle 2$

Use the Exterior Angle Inequality Theorem to list all angles that satisfy the stated condition.

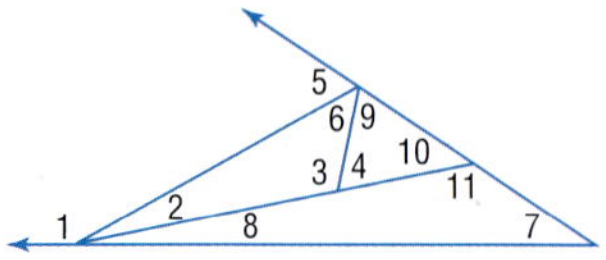

23. all angles whose measures are less than $m\angle 5$ (23. $\angle 2, \angle 7, \angle 8, \angle 10$)
24. all angles whose measures are greater than $m\angle 6$ $\angle 1, \angle 4, \angle 11$
25. all angles whose measures are greater than $m\angle 10$ $\angle 3, \angle 5$

Use the Exterior Angle Inequality Theorem to list all angles that satisfy the stated condition.

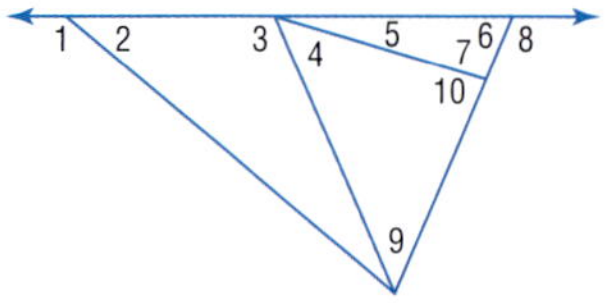

26. all angles whose measures are less than $m\angle 1$ (26. $\angle 3, \angle 6, \angle 9$)
27. all angles whose measures are greater than $m\angle 9$ (27. $\angle 8, \angle 7, \angle 3, \angle 1$)
28. all angles whose measures are less than $m\angle 8$ $\angle 2, \angle 4, \angle 5, \angle 9$

Determine the relationship between the measures of the given angles.

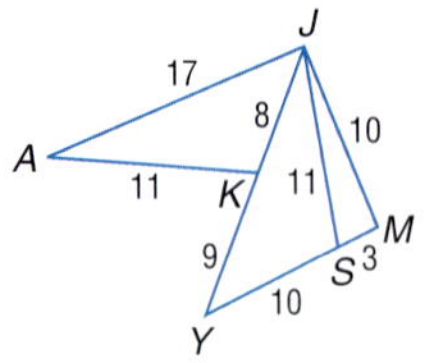

29. $\angle KAJ, \angle AJK$ (29. $m\angle KAJ < m\angle AJK$)
30. $\angle MJY, \angle JYM$ (30. $m\angle MJY > m\angle JYM$)
31. $\angle SMJ, \angle MJS$ (31. $m\angle SMJ > m\angle MJS$)
32. $\angle AKJ, \angle JAK$ (32. $m\angle AKJ > m\angle JAK$)
33. $\angle MYJ, \angle JMY$ $m\angle MYJ < m\angle JMY$
34. $\angle JSY, \angle JYS$ $m\angle JSY > m\angle JYS$

PROOF Write a two-column proof. 35–36. See margin.

35. Given: $\overline{JM} \cong \overline{JL}$
$\overline{JL} \cong \overline{KL}$
Prove: $m\angle 1 > m\angle 2$

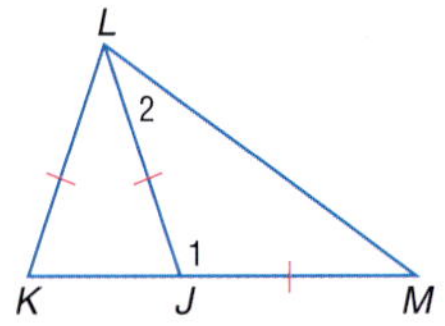

36. Given: $\overline{PR} \cong \overline{PQ}$
$QR > QP$
Prove: $m\angle P > m\angle Q$

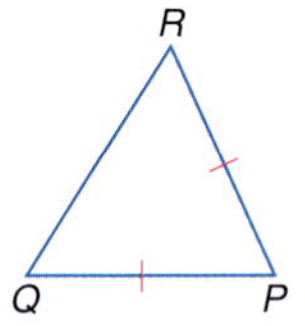

Determine the relationship between the lengths of the given sides.

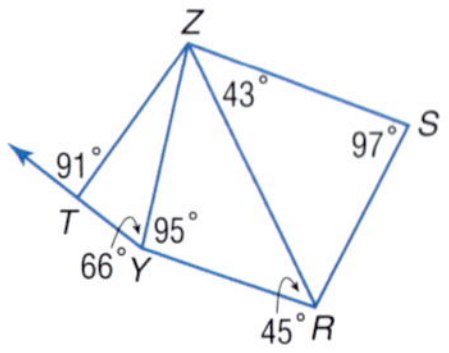

37. $\overline{ZY}, \overline{YR}$ $ZY > YR$
38. $\overline{SR}, \overline{ZS}$ $SR > ZS$
39. $\overline{RZ}, \overline{SR}$ $RZ > SR$
40. $\overline{ZY}, \overline{RZ}$ $ZY < RZ$
41. $\overline{TY}, \overline{ZY}$ $TY < ZY$
42. $\overline{TY}, \overline{ZT}$ $TY < ZT$

43. **COORDINATE GEOMETRY** Triangle KLM has vertices $K(3, 2)$, $L(-1, 5)$, and $M(-3, -7)$. List the angles in order from the least to the greatest measure. $\angle M, \angle L, \angle K$

★ 44. If $AB > AC > BC$ in $\triangle ABC$ and $\overline{AM}$, $\overline{BN}$, and $\overline{CO}$ are the medians of the triangle, list AM, BN, and CO in order from least to greatest. CO, BN, AM

252 Chapter 5 Relationships in Triangles

45. **TRAVEL** A plane travels from Des Moines to Phoenix, on to Atlanta, and then completes the trip directly back to Des Moines as shown in the diagram. Write the lengths of the legs of the trip in order from greatest to least. **Phoenix to Atlanta, Des Moines to Phoenix, Atlanta to Des Moines**

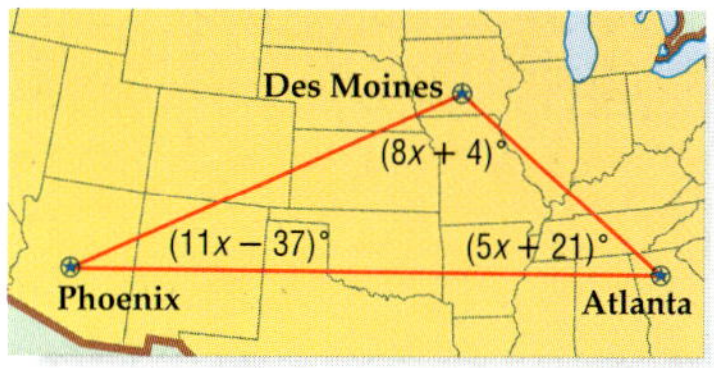

More About. . .

Travel

One sixth of adult Americans have never flown in a commercial aircraft.

Source: U.S. Bureau of Transportation Statistics

ALGEBRA **Find the value of n. List the sides of $\triangle PQR$ in order from shortest to longest for the given angle measures.**

46. $m\angle P = 9n + 29$, $m\angle Q = 93 - 5n$, $m\angle R = 10n + 2$ **4; $\overline{PQ}$, $\overline{QR}$, $\overline{PR}$**

47. $m\angle P = 12n - 9$, $m\angle Q = 62 - 3n$, $m\angle R = 16n + 2$ **5; $\overline{PR}$, $\overline{QR}$, $\overline{PQ}$**

48. $m\angle P = 9n - 4$, $m\angle Q = 4n - 16$, $m\angle R = 68 - 2n$ **12; $\overline{PR}$, $\overline{PQ}$, $\overline{QR}$**

49. $m\angle P = 3n + 20$, $m\angle Q = 2n + 37$, $\angle R = 4n + 15$ **12; $\overline{QR}$, $\overline{PR}$, $\overline{PQ}$**

50. $m\angle P = 4n + 61$, $m\angle Q = 67 - 3n$, $\angle R = n + 74$ **−11; $\overline{QR}$, $\overline{PQ}$, $\overline{PR}$**

★ 51. **DOORS** The wedge at the right is used as a door stopper. The values of x and y are in inches. Write an inequality relating x and y. Then solve the inequality for y in terms of x.

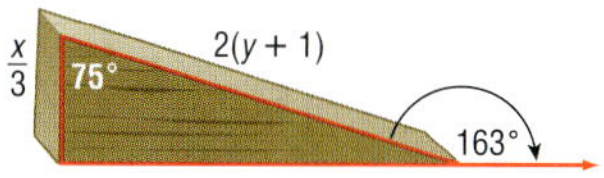

51. $2(y + 1) > \frac{x}{3}$, $y > \frac{x-6}{6}$

★ 52. **PROOF** Write a paragraph proof for the following statement.

If a triangle is not isosceles, then the measure of the median to any side of the triangle is greater than the measure of the altitude to that side. **See p. 279A.**

53. **CRITICAL THINKING** Write and solve an inequality for x.

$3x + 15 > 4x + 7 > 0$, $-\frac{7}{4} < x < 8$

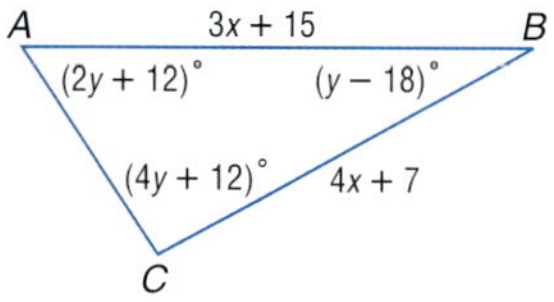

54. **WRITING IN MATH** Answer the question that was posed at the beginning of the lesson. **See margin.**

How can you tell which corner is bigger?

Include the following in your answer:

- the name of the theorem or postulate that lets you determine the comparison of the angle measures, and
- which angles in the diagram are the largest.

FCAT Practice

Standardized Test Practice

Ⓐ Ⓑ Ⓒ Ⓓ

55. In the figure at the right, what is the value of p in terms of m and n? **A**

Ⓐ $m + n - 180$

Ⓑ $m + n + 180$

Ⓒ $m - n + 360$

Ⓓ $360 - (m - n)$

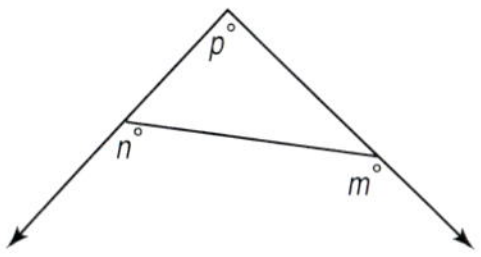

56. **ALGEBRA** If $\frac{1}{2}x - 3 = 2\left(\frac{x-1}{5}\right)$, then $x =$ __?__. **D**

Ⓐ 11 Ⓑ 13 Ⓒ 22 Ⓓ 26

www.geometryonline.com/self_check_quiz/fcat

Study Guide and Intervention, p. 251 (shown) and p. 252

Angle Inequalities Properties of inequalities, including the Transitive, Addition, Subtraction, Multiplication, and Division Properties of Inequality, can be used with measures of angles and segments. There is also a Comparison Property of Inequality.

For any real numbers a and b, either $a < b$, $a = b$, or $a > b$.

The Exterior Angle Theorem can be used to prove this inequality involving an exterior angle.

Exterior Angle Inequality Theorem	If an angle is an exterior angle of a triangle, then its measure is greater than the measure of either of its corresponding remote interior angles.	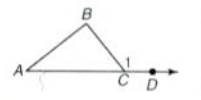 $m\angle 1 > m\angle A$, $m\angle 1 > m\angle B$

Example **List all angles of $\triangle EFG$ whose measures are less than $m\angle 1$.**

The measure of an exterior angle is greater than the measure of either remote interior angle. So $m\angle 3 < m\angle 1$ and $m\angle 4 < m\angle 1$.

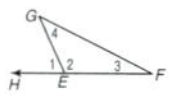

List all angles that satisfy the stated condition.

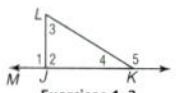

Exercises 1–2

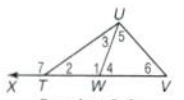

Exercises 3–8

1. all angles whose measures are less than $m\angle 1$ **∠3, ∠4**
2. all angles whose measures are greater than $m\angle 3$ **∠1, ∠5**
3. all angles whose measures are less than $m\angle 1$ **∠5, ∠6**
4. all angles whose measures are greater than $m\angle 1$ **∠7**
5. all angles whose measures are less than $m\angle 7$ **∠1, ∠3, ∠5, ∠6, ∠TUV**
6. all angles whose measures are greater than $m\angle 2$ **∠4**
7. all angles whose measures are greater than $m\angle 5$ **∠1, ∠7, ∠TUV**
8. all angles whose measures are less than $m\angle 4$ **∠2, ∠3**
9. all angles whose measures are less than $m\angle 1$ **∠4, ∠5, ∠7, ∠NPR**
10. all angles whose measures are greater than $m\angle 4$ **∠1, ∠8, ∠OPN, ∠ROQ**

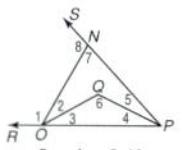

Exercises 9–10

Skills Practice, p. 253 and Practice, p. 254 (shown)

Determine which angle has the greatest measure.

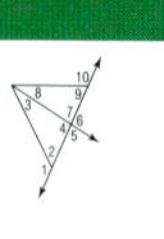

1. ∠1, ∠3, ∠4 **∠1**
2. ∠4, ∠8, ∠9 **∠4**
3. ∠2, ∠3, ∠7 **∠7**
4. ∠7, ∠8, ∠10 **∠10**

Use the Exterior Angle Inequality Theorem to list all angles that satisfy the stated condition.

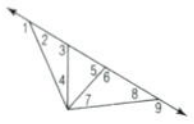

5. all angles whose measures are less than $m\angle 1$ **∠3, ∠4, ∠5, ∠7, ∠8**
6. all angles whose measures are less than $m\angle 3$ **∠5, ∠7, ∠8**
7. all angles whose measures are greater than $m\angle 7$ **∠1, ∠3, ∠5, ∠9**
8. all angles whose measures are greater than $m\angle 2$ **∠6, ∠9**

Determine the relationship between the measures of the given angles.

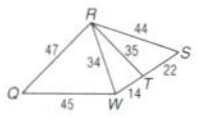

9. $m\angle QRW$, $m\angle RWQ$ **$m\angle QRW < \angle RWQ$**
10. $m\angle RTW$, $m\angle TWR$ **$m\angle RTW < \angle TWR$**
11. $m\angle RST$, $m\angle TRS$ **$m\angle RST > \angle TRS$**
12. $m\angle WQR$, $m\angle QRW$ **$m\angle WQR < \angle QRW$**

Determine the relationship between the lengths of the given sides.

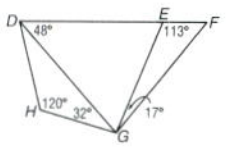

13. $\overline{DH}$, $\overline{GH}$ **$DH > GH$**
14. $\overline{DE}$, $\overline{DG}$ **$DE < DG$**
15. $\overline{EG}$, $\overline{FG}$ **$EG < FG$**
16. $\overline{DE}$, $\overline{EG}$ **$DE > EG$**

17. **SPORTS** The figure shows the position of three trees on one part of a Frisbee™ course. At which tree position is the angle between the trees the greatest? **2**

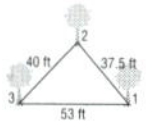

Reading to Learn Mathematics, p. 255 ELL

Pre-Activity **How can you tell which corner is bigger?**

Read the introduction to Lesson 5-2 at the top of page 247 in your textbook.

- Which side of the patio is opposite the largest corner? **the 51-foot side**
- Which side of the patio is opposite the smallest corner? **the 45-foot side**

Reading the Lesson

1. Name the property of inequality that is illustrated by each of the following.
   a. If $x > 8$ and $8 > y$, then $x > y$. **Transitive Property**
   b. If $x < y$, then $x - 7.5 < y - 7.5$. **Subtraction Property**
   c. If $x > y$, then $-3x < -3y$. **Multiplication Property**
   d. If x is any real number, $x > 0$, $x = 0$, or $x < 0$. **Comparison Property**
2. Use the definition of inequality to write an *equation* that shows that each inequality is true.
   a. $20 > 12$ **$20 = 12 + 8$** b. $101 > 99$ **$101 = 99 + 2$**
   c. $8 > -2$ **$8 = -2 + 10$** d. $7 > -7$ **$7 = -7 + 14$**
   e. $-11 > -12$ **$-11 = -12 + 1$** f. $-30 > -45$ **$-30 = -45 + 15$**
3. In the figure, $m\angle IJK = 45$ and $m\angle H > m\angle I$.

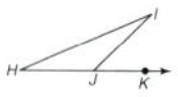

   a. Arrange the following angles in order from largest to smallest: ∠I, ∠IJK, ∠H, ∠IJH **∠IJH, ∠IJK, ∠H, ∠I**
   b. Arrange the sides of $\triangle HIJ$ in order from shortest to longest. **$\overline{HJ}$, $\overline{IJ}$, $\overline{HI}$**
   c. Is $\triangle HIJ$ an acute, right, or obtuse triangle? Explain your reasoning. **Obtuse; sample answer: ∠IJH is obtuse because $m\angle IJH = 180 - m\angle IJK = 135$. Therefore, $\triangle HIJ$ is obtuse because it has an obtuse angle.**
   d. Is $\triangle HIJ$ scalene, isosceles, or equilateral? Explain your reasoning. **Scalene; sample answer: the three angles of $\triangle HIJ$ all have different measures, so the sides opposite them must have different lengths.**

Helping You Remember

4. A good way to remember a new geometric theorem is to relate it to a theorem you learned earlier. Explain how the Exterior Angle Inequality Theorem is related to the Exterior Angle Theorem, and why the Exterior Angle Inequality Theorem must be true if the Exterior Angle Theorem is true. **Sample answer: The Exterior Angle Theorem says that the measure of an exterior angle of a triangle is equal to the sum of the measures of the two remote interior angles, while the Exterior Angle Inequality Theorem says that the measure of an exterior angle is greater than the measure of either remote interior angle. If a number is equal to the sum of two positive numbers, it must be greater than each of those two numbers.**

Answer

54. Sample answer: The largest corner is opposite the longest side. Answers should include the following.

- **the Exterior Angle Inequality Theorem**
- **the angle opposite the side that is 51 feet long**

Enrichment, p. 256

Construction Problem

The diagram below shows segment AB adjacent to a closed region. The problem requires that you construct another segment XY to the right of the closed region such that points A, B, X, and Y are collinear. You are not allowed to touch or cross the closed region with your compass or straightedge.

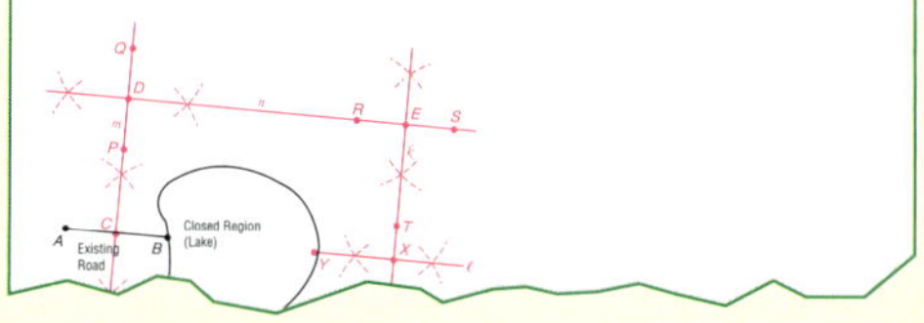

4 Assess

Open-Ended Assessment

Speaking Students can practice speaking and communicating with geometric terminology. For Exercise 23 on p. 252, have a volunteer explain that the measures of angles 10 and 2 are less than $m\angle 5$ because they are *corresponding remote interior angles* for $\angle 5$. Select examples from the lesson or practice exercises and call on different students to discuss angle inequalities and angle-side relationships using geometric terminology. Be sure students properly name angles and sides and use terms like *greater/lesser measure* for angles and *longer/shorter* or *greater/lesser measure* for sides.

Getting Ready for Lesson 5-3

Prerequisite Skill Students will learn about indirect proof in Lesson 5-3. They will show that assumptions are false in order to prove conclusions true. Use Exercises 64–66 to determine your students' familiarity with determining if equations are true or false.

Assessment Options

Practice Quiz 1 The quiz provides students with a brief review of the concepts and skills in Lessons 5-1 and 5-2. Lesson numbers are given to the right of the exercises or instruction lines so students can review concepts not yet mastered.

Quiz (Lessons 5-1 and 5-2) is available on p. 289 of the *Chapter 5 Resource Masters*.

Maintain Your Skills

Mixed Review

ALGEBRA For Exercises 57–59, use the following information. (Lesson 5-1)
Two vertices of $\triangle ABC$ are $A(3, 8)$ and $B(9, 12)$. $\overline{AD}$ is a median with D at $(12, 3)$.

57. What are the coordinates of C? **(15, −6)**
58. Is $\overline{AD}$ an altitude of $\triangle ABC$? Explain. **no; $-\frac{5}{9}(-3) \neq -1$**
59. The graph of point E is at $(6, 6)$. $\overline{EF}$ intersects $\overline{BD}$ at F. If F is at $\left(10\frac{1}{2}, 7\frac{1}{2}\right)$, is $\overline{EF}$ a perpendicular bisector of $\overline{BD}$? Explain. **Yes; $\frac{1}{3}(-3) = -1$, and F is the midpoint of $\overline{BD}$.**

STATISTICS For Exercises 60 and 61, refer to the graph at the right. (Lesson 4-7)

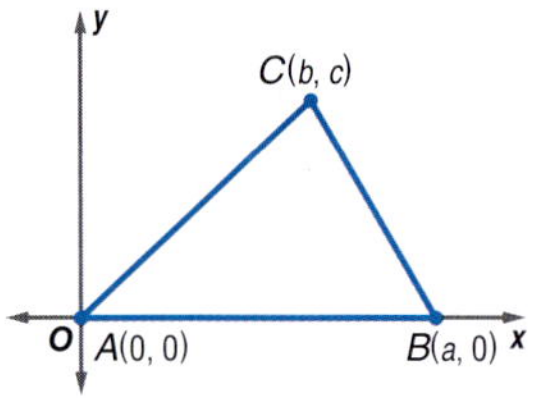

60. $D\left(\frac{a+b}{3}, \frac{c}{3}\right)$

60. Find the coordinates of D if the x-coordinate of D is the mean of the x-coordinates of the vertices of $\triangle ABC$ and the y-coordinate is the mean of the y-coordinates of the vertices of $\triangle ABC$.
61. Prove that D is the intersection of the medians of $\triangle ABC$. **See margin.**

Name the corresponding congruent angles and sides for each pair of congruent triangles. (Lesson 4-3) **62–64. See margin.**

62. $\triangle TUV \cong \triangle XYZ$
63. $\triangle CDG \cong \triangle RSW$
64. $\triangle BCF \cong \triangle DGH$

65. Find the value of x so that the line containing points at $(x, 2)$ and $(-4, 5)$ is perpendicular to the line containing points at $(4, 8)$ and $(2, -1)$. (Lesson 3-3) **9.5**

Getting Ready for the Next Lesson

BASIC SKILL Determine whether each equation is *true* or *false* if $a = 2$, $b = 5$, and $c = 6$. (To review **evaluating expressions**, see page 736.)

66. $2ab = 20$ **true**
67. $c(b - a) = 15$ **false**
68. $a + c > a + b$ **true**

Practice Quiz 1 — Lessons 5-1 and 5-2

ALGEBRA Use $\triangle ABC$. (Lesson 5-1)

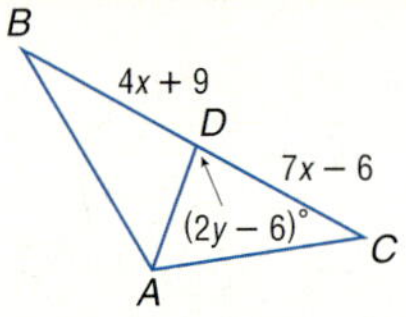

1. Find x if $\overline{AD}$ is a median of $\triangle ABC$. **5**
2. Find y if $\overline{AD}$ is an altitude of $\triangle ABC$. **48**

State whether each statement is *always, sometimes,* or *never* true. (Lesson 5-1)

3. The medians of a triangle intersect at one of the vertices of the triangle. **never**
4. The angle bisectors of a triangle intersect at a point in the interior of the triangle. **always**
5. The altitudes of a triangle intersect at a point in the exterior of the triangle. **sometimes**
6. The perpendicular bisectors of a triangle intersect at a point on the triangle. **sometimes**

7. Describe a triangle in which the angle bisectors all intersect in a point outside the triangle. If no triangle exists, write *no triangle*. (Lesson 5-1) **no triangle**

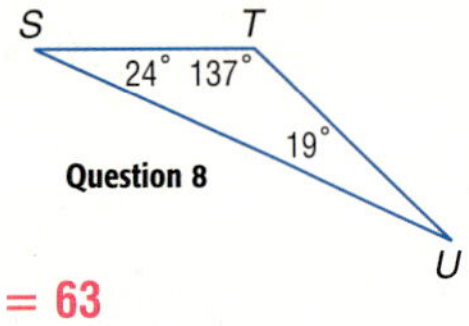

Question 8

8. List the sides of $\triangle STU$ in order from longest to shortest. (Lesson 5-2) **$\overline{SU}$, $\overline{TU}$, $\overline{ST}$**

ALGEBRA In $\triangle QRS$, $m\angle Q = 3x + 20$, $m\angle R = 2x + 37$, and $m\angle S = 4x + 15$. (Lesson 5-2)

9. Determine the measure of each angle. **$m\angle Q = 56$, $m\angle R = 61$, $m\angle S = 63$**
10. List the sides in order from shortest to longest. **$\overline{RS}$, $\overline{QS}$, $\overline{QR}$**

Answers

61. Label the midpoints of $\overline{AB}$, $\overline{BC}$, and $\overline{CA}$ as E, F, and G respectively. Then the coordinates of E, F, and G are $\left(\frac{a}{2}, 0\right)$, $\left(\frac{a+b}{2}, \frac{c}{2}\right)$, and $\left(\frac{b}{2}, \frac{c}{2}\right)$ respectively. The slope of $\overline{AF} = \frac{c}{a+b}$, and the slope of $\overline{AD} = \frac{c}{a+b}$, so D is on $\overline{AF}$. The slope of $\overline{BG} = \frac{c}{b-2a}$, and the slope of $\overline{BD} = \frac{c}{b-2a}$, so D is on $\overline{BG}$. The slope of $\overline{CE} = \frac{2c}{2b-a}$, and the slope of $\overline{CD} = \frac{2c}{2b-a}$, so D is on $\overline{CE}$. Since D is on $\overline{AF}$, $\overline{BG}$, and $\overline{CE}$, it is the intersection point of the three segments.

62. $\angle T \cong \angle X$, $\angle U \cong \angle Y$, $\angle V \cong \angle Z$, $\overline{TU} \cong \overline{XY}$, $\overline{UV} \cong \overline{YZ}$, $\overline{TV} \cong \overline{XZ}$

63. $\angle C \cong \angle R$, $\angle D \cong \angle S$, $\angle G \cong \angle W$, $\overline{CD} \cong \overline{RS}$, $\overline{DG} \cong \overline{SW}$, $\overline{CG} \cong \overline{RW}$

64. $\angle B \cong \angle D$, $\angle C \cong \angle G$, $\angle F \cong \angle H$, $\overline{BC} \cong \overline{DG}$, $\overline{CF} \cong \overline{GH}$, $\overline{BF} \cong \overline{DH}$

254 Chapter 5 Relationships in Triangles

5-3 Indirect Proof

Sunshine State Standards
MA.C.1.4.1

What You'll Learn

- Use indirect proof with algebra.
- Use indirect proof with geometry.

How is indirect proof used in literature?

In *The Adventure of the Blanched Soldier*, Sherlock Holmes describes his detective technique, stating, "That process starts upon the supposition that when you have eliminated all which is impossible, then whatever remains, . . . must be the truth." The method Sherlock Holmes uses is an example of *indirect reasoning*.

Vocabulary

- indirect reasoning
- indirect proof
- proof by contradiction

INDIRECT PROOF WITH ALGEBRA The proofs you have written so far use direct reasoning, in which you start with a true hypothesis and prove that the conclusion is true. When using **indirect reasoning**, you assume that the conclusion is false and then show that this assumption leads to a contradiction of the hypothesis, or some other accepted fact, such as a definition, postulate, theorem, or corollary. Since all other steps in the proof are logically correct, the assumption has been proven false, so the original conclusion must be true. A proof of this type is called an **indirect proof** or a **proof by contradiction**.

Study Tip

Truth Value of a Statement
Recall that a statement must be either true or false. To review **truth values**, see Lesson 2-2.

The following steps summarize the process of an indirect proof.

Key Concept — *Steps for Writing an Indirect Proof*

1. Assume that the conclusion is false.
2. Show that this assumption leads to a contradiction of the hypothesis, or some other fact, such as a definition, postulate, theorem, or corollary.
3. Point out that because the false conclusion leads to an incorrect statement, the original conclusion must be true.

Example 1 *Stating Conclusions*

State the assumption you would make to start an indirect proof of each statement.

a. $AB \neq MN$
$AB = MN$

b. $\triangle PQR$ **is an isosceles triangle.**
$\triangle PQR$ is not an isosceles triangle.

c. $x < 4$
If $x < 4$ is false, then $x = 4$ or $x > 4$. In other words, $x \geq 4$.

d. If 9 is a factor of n**, then 3 is a factor of** n**.**
The conclusion of the conditional statement is *3 is a factor of n*. The negation of the conclusion is *3 is not a factor of n*.

5-3 Lesson Notes

1 Focus

5-Minute Check Transparency 5-3 Use as a quiz or review of Lesson 5-2.

Mathematical Background notes are available for this lesson on p. 234D.

How is indirect proof used in literature?

Ask students:

- Why do you think Sherlock Holmes' method is called *indirect* reasoning? **Sample answer: Because he does not take direct steps to prove that something is true.**
- What does Sherlock Holmes need to know in order to use the method of indirect reasoning as described in the example above? **In order to eliminate all which is impossible, he will need to know all the possibilities.**

Resource Manager

Workbook and Reproducible Masters

Chapter 5 Resource Masters
- Study Guide and Intervention, pp. 257–258
- Skills Practice, p. 259
- Practice, p. 260
- Reading to Learn Mathematics, p. 261
- Enrichment, p. 262
- Assessment, pp. 289, 291

Prerequisite Skills Workbook, pp. 15–20, 25–26, 87–88

Transparencies

5-Minute Check Transparency 5-3
Answer Key Transparencies

Technology

Interactive Chalkboard
Multimedia Applications: Virtual Activities

2 Teach

INDIRECT PROOF WITH ALGEBRA

In-Class Examples

Power Point®

1 State the assumption you would make to start an indirect proof of each statement.

a. $\overline{EF}$ is not a perpendicular bisector. $\overline{EF}$ **is a perpendicular bisector.**

b. $3x = 4y + 1$ $3x \neq 4y + 1$

c. $m\angle 1$ is less than or equal to $m\angle 2$. $m\angle 1 > m\angle 2$

d. If B is the midpoint of $\overline{LH}$ and $LH = 26$, then $\overline{BH}$ is congruent to $\overline{LB}$. $\overline{BH}$ **is not congruent to** $\overline{LB}$**.**

2 **Given:** $\frac{1}{2y + 4} = 20$

Prove: $y \neq -2$

Assume that $y = -2$. Substitute -2 for y in the equation and find that $2(-2) + 4 = 0$. The denominator cannot be 0. So, $y \neq -2$.

Indirect proofs can be used to prove algebraic concepts.

Example 2 Algebraic Proof

Given: $2x - 3 > 7$

Prove: $x > 5$

Indirect Proof:

Step 1 Assume that $x \leq 5$. That is, assume that $x < 5$ or $x = 5$.

Step 2 Make a table with several possibilities for x given that $x < 5$ or $x = 5$.
This is a contradiction because when $x < 5$ or $x = 5$, $2x - 3 \leq 7$.

x	$2x - 3$
1	−1
2	1
3	4
4	5
5	7

Step 3 In both cases, the assumption leads to the contradiction of a known fact. Therefore, the assumption that $x \leq 5$ must be false, which means that $x > 5$ must be true.

Indirect reasoning and proof can be used in everyday situations.

Example 3 Use Indirect Proof

SHOPPING Lawanda bought two skirts for just over \$60, before tax. A few weeks later, her friend Tiffany asked her how much each skirt cost. Lawanda could not remember the individual prices. Use indirect reasoning to show that at least one of the skirts cost more than \$30.

Given: The two skirts cost more than \$60.

Prove: At least one of the skirts cost more than \$30.
That is, if $x + y > 60$, then either $x > 30$ or $y > 30$.

Indirect Proof:

Step 1 Assume that neither skirt costs more than \$30. That is, $x \leq 30$ and $y \leq 30$.

Step 2 If $x \leq 30$ and $y \leq 30$, then $x + y \leq 60$. This is a contradiction because we know that the two skirts cost more than \$60.

Step 3 The assumption leads to the contradiction of a known fact. Therefore, the assumption that $x \leq 30$ and $y \leq 30$ must be false. Thus, at least one of the skirts had to have cost more than \$30.

More About. . .

Shopping

The West Edmonton Mall in Edmonton, Alberta, Canada, is the world's largest entertainment and shopping center, with an area of 5.3 million square feet. The mall houses an amusement park, water park, ice rink, and aquarium, along with over 800 stores and services.

Source: www.westedmall.com

INDIRECT PROOF WITH GEOMETRY

Indirect reasoning can be used to prove statements in geometry.

Example 4 Geometry Proof

Given: $\ell \nparallel m$

Prove: $\angle 1 \not\cong \angle 3$

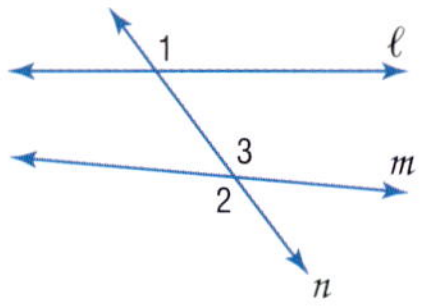

Indirect Proof:

Step 1 Assume that $\angle 1 \cong \angle 3$.

DAILY INTERVENTION — Unlocking Misconceptions

Algebraic Proofs Point out that students are very used to working forward to solve equations and inequalities, and they may be tempted to solve algebraic problems as a step in writing indirect proofs. Tell students that although this method works, it is not representative of an indirect proof, and they should avoid solving the algebraic problems in this lesson. Rather, they should use methods similar to the steps demonstrated for Example 2.

Step 2 $\angle 1$ and $\angle 3$ are corresponding angles. If two lines are cut by a transversal so that corresponding angles are congruent, the lines are parallel. This means that $\ell \parallel m$. However, this contradicts the given statement.

Step 3 Since the assumption leads to a contradiction, the assumption must be false. Therefore, $\angle 1 \not\cong \angle 3$.

Indirect proofs can also be used to prove theorems.

Proof *Exterior Angle Inequality Theorem*

Given: $\angle 1$ is an exterior angle of $\triangle ABC$.

Prove: $m\angle 1 > m\angle 3$ and $m\angle 1 > m\angle 4$

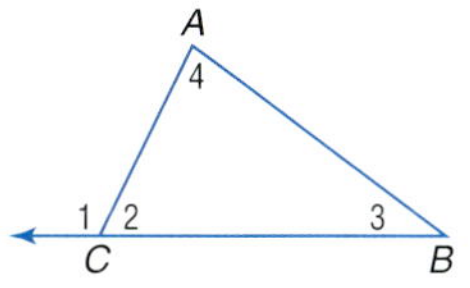

Indirect Proof:

Step 1 Make the assumption that $m\angle 1 \not> m\angle 3$ or $m\angle 1 \not> m\angle 4$. In other words, $m\angle 1 \leq m\angle 3$ or $m\angle 1 \leq m\angle 4$.

Step 2 You only need to show that the assumption $m\angle 1 \leq m\angle 3$ leads to a contradiction as the argument for $m\angle 1 \leq m\angle 4$ follows the same reasoning.

$m\angle 1 \leq m\angle 3$ means that either $m\angle 1 = m\angle 3$ or $m\angle 1 < m\angle 3$.

Case 1: $m\angle 1 = m\angle 3$

$m\angle 1 = m\angle 3 + m\angle 4$ — Exterior Angle Theorem

$m\angle 3 = m\angle 3 + m\angle 4$ — Substitution

$0 = m\angle 4$ — Subtract $m\angle 3$ from each side.

This contradicts the fact that the measure of an angle is greater than 0, so $m\angle 1 < m\angle 3$.

Case 2: $m\angle 1 < m\angle 3$

By the Exterior Angle Theorem, $m\angle 1 = m\angle 3 + m\angle 4$. Since angle measures are positive, the definition of inequality implies $m\angle 1 > m\angle 3$ and $m\angle 1 > m\angle 4$. This contradicts the assumption.

Step 3 In both cases, the assumption leads to the contradiction of a theorem or definition. Therefore, the assumption that $m\angle 1 > m\angle 3$ and $m\angle 1 > m\angle 4$ must be true.

Check for Understanding

Concept Check

1. **Explain** how contradicting a known fact means that an assumption is false.
2. **Compare and contrast** indirect proof and direct proof. See margin. See p. 279A.
3. **OPEN ENDED** State a conjecture. Then write an indirect proof to prove your conjecture. See p. 279B.

1. If a statement is shown to be false, then its opposite must be true.

www.geometryonline.com/extra_examples/fcat

DAILY INTERVENTION **Differentiated Instruction**

Interpersonal Tell groups of three students to work three exercises. Have them choose two exercises from the 19–22 set and one from the 23–26 set. Each group member should take turns providing Step 1, Step 2, or Step 3 for each exercise worked. Encourage the groups to discuss the more difficult exercises to gather ideas before writing their proofs.

In-Class Example — Power Point®

Teaching Tip Tell students to translate word problems into equations or inequalities before writing the indirect proof.

3 **CLASSES** Marta signed up for three classes at a community college for a little under \$156. There was an administration fee of \$15, but the class costs varied. How can you show that at least one class cost less than \$47?

Given: Marta spent less than \$156. Prove: At least one of the classes x cost less than \$47. That is, if $3x + 15 < 156$, then $x < 47$. Step 1: Assume $x \geq 47$. Step 2: $47 + 47 + 47 + 15 \geq 156$. Step 3: This contradicts the statement that the total cost was less than \$156, so the assumption that $x \geq 47$ must be false. Therefore, one class must cost less than 47.

INDIRECT PROOF WITH GEOMETRY

In-Class Example — Power Point®

4 **Given:** $\triangle JKL$ with side lengths 5, 7, and 8 as shown

Prove: $m\angle K < m\angle L$

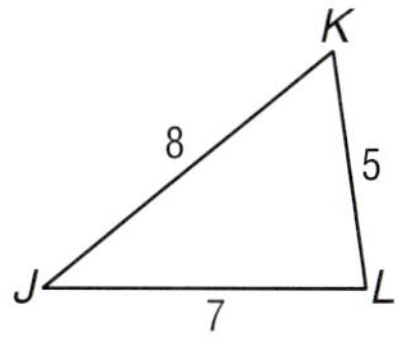

Step 1: Assume $m\angle K \geq m\angle L$. Step 2: By angle-side relationships, $JL \geq JK$. Step 3: This contradicts the given side lengths so the assumption $m\angle K \geq m\angle L$ must be false. Therefore $m\angle K < m\angle L$.

3 Practice/Apply

Study Notebook

Have students—

- add the definitions/examples of the vocabulary terms to their Vocabulary Builder worksheets for Chapter 5.
- include any other item(s) that they find helpful in mastering the skills in this lesson.

About the Exercises...

Organization by Objective

- **Indirect Proof with Algebra:** 14, 15, 19, 20, 23
- **Indirect Proof with Geometry:** 13, 16–18, 21, 22, 24–26

Odd/Even Assignments

Exercises 13–26 are structured so that students practice the same concepts whether they are assigned odd or even problems.

Alert! Exercise 32 requires the Internet or other research materials.

Assignment Guide

Basic: 13–33 odd, 35–48

Average: 13–33 odd, 35–48

Advanced: 14–32 even, 33–45 (optional: 46–48)

Guided Practice

GUIDED PRACTICE KEY	
Exercises	Examples
4–6	1
7, 8	2
12	3
9–11	4

Write the assumption you would make to start an indirect proof of each statement.

4. If $5x < 25$, then $x < 5$. $x \geq 5$
5. Two lines that are cut by a transversal so that alternate interior angles are congruent are parallel. **The lines are not parallel.**
6. If the alternate interior angles formed by two lines and a transversal are congruent, the lines are parallel. **The lines are not parallel.**

PROOF **Write an indirect proof.** **7–10. See p. 279B.**

7. **Given:** $a > 0$
   Prove: $\frac{1}{a} > 0$
8. **Given:** n is odd.
   Prove: n^2 is odd.
9. **Given:** $\triangle ABC$
   Prove: There can be no more than one obtuse angle in $\triangle ABC$.
10. **Given:** $m \nparallel n$
    Prove: Lines m and n intersect at exactly one point.
11. **PROOF** Use an indirect proof to show that the hypotenuse of a right triangle is the longest side. **See margin.**

Application

12. **BICYCLING** The Tour de France bicycle race takes place over several weeks in various stages throughout France. During two stages of the 2002 Tour de France, riders raced for just over 270 miles. Prove that at least one of the stages was longer than 135 miles. **See p. 279B.**

Practice and Apply

Homework Help	
For Exercises	See Examples
13–18	1
19, 20, 23	2, 3
21, 22, 24, 25	4

Extra Practice
See page 763.

Write the assumption you would make to start an indirect proof of each statement.

13. $\overline{PQ} \cong \overline{ST}$ $\overline{PQ} \ncong \overline{ST}$
14. If $3x > 12$, then $x > 4$. $x \not> 4$
15. If a rational number is any number that can be expressed as $\frac{a}{b}$, where a and b are integers, and $b \neq 0$, 6 is a rational number. **15. 6 cannot be expressed as $\frac{a}{b}$.**
16. A median of an isosceles triangle is also an altitude. **16. A median of an isosceles triangle is not an altitude.**
17. Points P, Q, and R are collinear. **Points P, Q, and R are noncollinear.**
18. The angle bisector of the vertex angle of an isosceles triangle is also an altitude of the triangle. **The angle bisector of the vertex angle of an isosceles triangle is not an altitude of the triangle.**

PROOF **Write an indirect proof.** **19–22. See p. 279B–279C.**

19. **Given:** $\frac{1}{a} < 0$
    Prove: a is negative.
20. **Given:** n^2 is even.
    Prove: n^2 is divisible by 4.
21. **Given:** $\overline{PQ} \cong \overline{PR}$
    $\angle 1 \ncong \angle 2$
    Prove: $\overline{PZ}$ is not a median of $\triangle PQR$.
22. **Given:** $m\angle 2 \neq m\angle 1$
    Prove: $\ell \nparallel m$

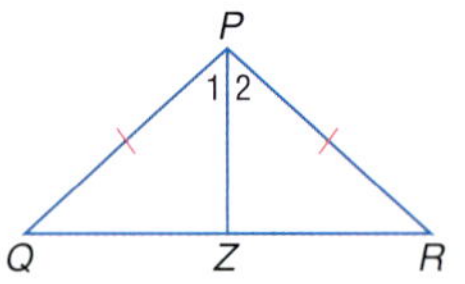

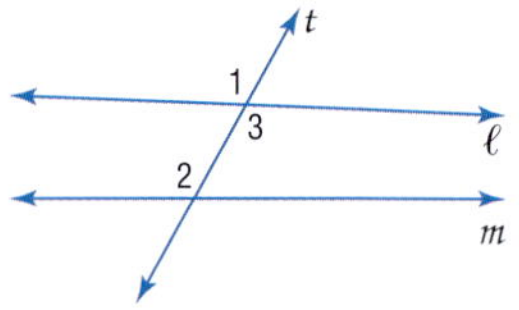

Answer

11. **Given:** $\triangle ABC$ is a right triangle; $\angle C$ is a right angle.
Prove: $AB > BC$ and $AB > AC$

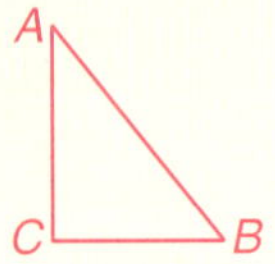

Proof:

Step 1: Assume that the hypotenuse of a right triangle is not the longest side. That is, $AB < BC$ or $AB < AC$.

Step 2: If $AB < BC$, then $m\angle C < m\angle A$. Since $m\angle C = 90$, $m\angle A > 90$. So, $m\angle C + m\angle A > 180$. By the same reasoning, $m\angle C + m\angle B > 180$.

Step 3: Both relationships contradict the fact that the sum of the measures of the angles of a triangle equals 180. Therefore, the hypotenuse must be the longest side of a right triangle.

258 Chapter 5 Relationships in Triangles

PROOF Write an indirect proof. 23–26. See p. 279C.

23. If $a > 0$, $b > 0$, and $a > b$, then $\frac{a}{b} > 1$.

24. If two sides of a triangle are not congruent, then the angles opposite those sides are not congruent.

25. **Given:** $\triangle ABC$ and $\triangle ABD$ are equilateral.
$\triangle ACD$ is not equilateral.
Prove: $\triangle BCD$ is not equilateral.

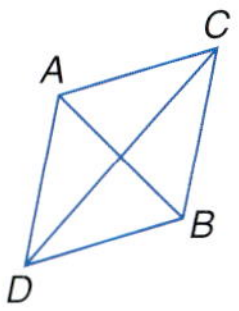

26. Theorem 5.10
Given: $m\angle A > m\angle ABC$
Prove: $BC > AC$

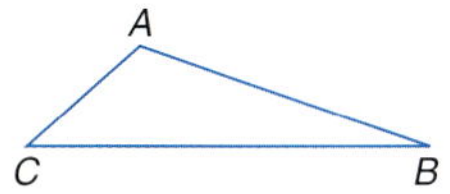

27. **TRAVEL** Ramon drove 175 miles from Seattle, Washington, to Portland, Oregon. It took him three hours to complete the trip. Prove that his average driving speed was less than 60 miles per hour. See p. 279C.

EDUCATION For Exercises 28–30, refer to the graphic at the right.

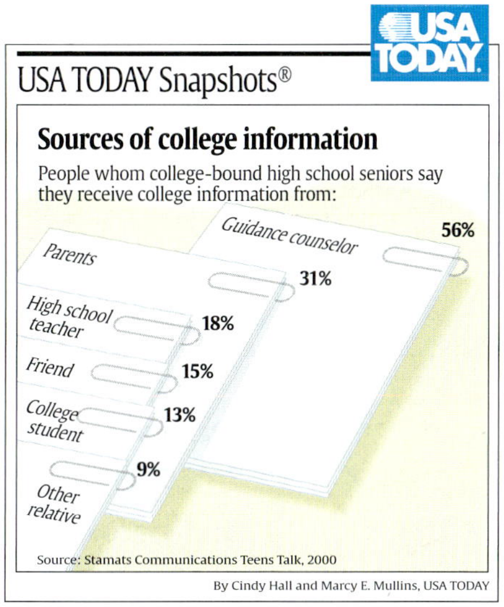

28. Prove the following statement.
The majority of college-bound seniors stated that they received college information from a guidance counselor. See p. 279C.

29. If 1500 seniors were polled for this survey, verify that 225 said they received college information from a friend.

29. $1500 \cdot 15\% \stackrel{?}{=} 225$
$1500 \cdot 0.15 \stackrel{?}{=} 225$
$225 = 225$

30. Did more seniors receive college information from their parents or from teachers and friends? Explain. teachers and friends; $15\% + 18\% = 33\%$, $33\% > 31\%$

31. **LAW** During the opening arguments of a trial, a defense attorney stated, "My client is innocent. The police report states that the crime was committed on November 6 at approximately 10:15 A.M. in San Diego. I can prove that my client was on vacation in Chicago with his family at this time. A verdict of not guilty is the only possible verdict." Explain whether this is an example of indirect reasoning.

31. Yes; if you assume the client was at the scene of the crime, it is contradicted by his presence in Chicago at that time. Thus, the assumption that he was present at the crime is false.

32. **RESEARCH** Use the Internet or other resource to write an indirect proof for the following statement.
In the Atlantic Ocean, the percent of tropical storms that developed into hurricanes over the past five years varies from year to year. See students' work.

www.geometryonline.com/self_check_quiz/fcat

Enrichment, p. 262

More Counterexamples

Some statements in mathematics can be proven false by **counterexamples.** Consider the following statement.

For any numbers a and b, $a - b = b - a$.

You can prove that this statement is false in general if you can find one example for which the statement is false.

Let $a = 7$ and $b = 3$. Substitute these values in the equation above.

$7 - 3 \stackrel{?}{=} 3 - 7$
$4 \neq -4$

In general, for any numbers a and b, the statement $a - b = b - a$ is false. You can make the equivalent verbal statement: subtraction is *not* a commutative operation.

In each of the following exercises a, b, and c are any numbers. Prove that the statement is false by counterexample. Sample answers are given.

Study Guide and Intervention, p. 257 (shown) and p. 258

Indirect Proof with Algebra One way to prove that a statement is true is to assume that its conclusion is false and then show that this assumption leads to a contradiction of the hypothesis, a definition, postulate, theorem, or other statement that is accepted as true. That contradiction means that the conclusion cannot be false, so the conclusion must be true. This is known as **indirect proof**.

Steps for Writing an Indirect Proof
1. Assume that the conclusion is false.
2. Show that this assumption leads to a contradiction.
3. Point out that the assumption must be false, and therefore, the conclusion must be true.

Example **Given:** $3x + 5 > 8$
Prove: $x > 1$

Step 1 Assume that x is not greater than 1. That is, $x = 1$ or $x < 1$.
Step 2 Make a table for several possibilities for $x = 1$ or $x < 1$. The contradiction is that when $x = 1$ or $x < 1$, then $3x + 5$ is not greater than 8.
Step 3 This contradicts the given information that $3x + 5 > 8$. The assumption that x is not greater than 1 must be false, which means that the statement "$x > 1$" must be true.

x	$3x + 5$
1	8
0	5
−1	2
−2	−1
−3	−4

Exercises

Write the assumption you would make to start an indirect proof of each statement.

1. If $2x > 14$, then $x > 7$. $x \leq 7$
2. For all real numbers, if $a + b > c$, then $a > c - b$. $a \leq c - b$

Complete the proof.
Given: n is an integer and n^2 is even.
Prove: n is even.

3. Assume that n is not even. That is, assume n is odd.
4. Then n can be expressed as $2a + 1$ by the meaning of odd number.
5. $n^2 = (2a + 1)^2$ Substitution
6. $= (2a + 1)(2a + 1)$ Multiply.
7. $= 4a^2 + 4a + 1$ Simplify.
8. $= 2(2a^2 + 2a) + 1$ Distributive Property
9. $2(2a^2 + 2a) + 1$ is an odd number. This contradicts the given that n^2 is even, so the assumption must be false.
10. Therefore, n is even.

Skills Practice, p. 259 and Practice, p. 260 (shown)

Write the assumption you would make to start an indirect proof of each statement.

1. $\overline{BD}$ bisects $\angle ABC$.
$\overline{BD}$ does not bisect $\angle ABC$.

2. $RT = TS$
$RT \neq TS$

PROOF Write an indirect proof.

3. **Given:** $-4x + 2 < -10$
Prove: $x > 3$
Proof:
Step 1: Assume $x \leq 3$.
Step 2: If $x \leq 3$, then $-4x \geq -12$. But $-4x \geq -12$ implies that $-4x + 2 \geq -10$, which contradicts the given inequality.
Step 3: Since the assumption that $x \leq 3$ leads to a contradiction, it must be true that $x > 3$.

4. **Given:** $m\angle 2 + m\angle 3 \neq 180$
Prove: $a \nparallel b$

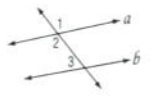

Proof:
Step 1: Assume $a \parallel b$.
Step 2: If $a \parallel b$, then the consecutive interior angles $\angle 2$ and $\angle 3$ are supplementary. Thus $m\angle 2 + m\angle 3 = 180$. This contradicts the given statement that $m\angle 2 + m\angle 3 \neq 180$.
Step 3: Since the assumption leads to a contradiction, the statement $a \parallel b$ must be false. Therefore, $a \nparallel b$ must be true.

5. **PHYSICS** Sound travels through air at about 344 meters per second when the temperature is 20°C. If Enrique lives 2 kilometers from the fire station and it takes 5 seconds for the sound of the fire station siren to reach him, how can you prove indirectly that it is not 20°C when Enrique hears the siren?
Assume that it is 20°C when Enrique hears the siren, then show that at this temperature it will take more than 5 seconds for the sound of the siren to reach him. Since the assumption is false, you will have proved that it is not 20°C when Enrique hears the siren.

Reading to Learn Mathematics, p. 261 ELL

Pre-Activity How is indirect proof used in literature?

Read the introduction to Lesson 5-3 at the top of page 255 in your textbook.

How could the author of a murder mystery use indirect reasoning to show that a particular suspect is not guilty? Sample answer: Assume that the person is guilty. Then show that this assumption contradicts evidence that has been gathered about the crime.

Reading the Lesson

1. Supply the missing words to complete the list of steps involved in writing an indirect proof.
Step 1 Assume that the conclusion is false.
Step 2 Show that this assumption leads to a contradiction of the hypothesis or some other fact, such as a definition, postulate, theorem, or corollary.
Step 3 Point out that the assumption must be false and, therefore, the conclusion must be true.

2. State the assumption that you would make to start an indirect proof of each statement.
a. If $-6x > 30$, then $x < -5$. $x \geq -5$
b. If n is a multiple of 6, then n is a multiple of 3. n is not a multiple of 3.
c. If a and b are both odd, then ab is odd. ab is even.
d. If a is positive and b is negative, then ab is negative. ab is greater than or equal to 0.
e. If F is between E and D, then $EF + FD = ED$. $EF + FD \neq ED$
f. In a plane, if two lines are perpendicular to the same line, then they are parallel. Two lines are not parallel.
g. Refer to the figure.

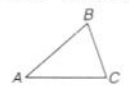

If $AB = AC$, then $m\angle B = m\angle C$.
$m\angle B \neq m\angle C$

h. Refer to the figure.

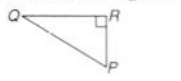

In $\triangle PQR$, $PR + QR > QP$.
$PR + QR \leq QP$

Helping You Remember

3. A good way to remember a new concept in mathematics is to relate it to something you have already learned. How is the process of indirect proof related to the relationship between a conditional statement and its contrapositive? Sample answer: The contrapositive of the conditional statement $p \rightarrow q$ is the statement $\sim q \rightarrow \sim p$. In an indirect proof of a conditional statement $p \rightarrow q$, you assume that q is false and show that this implies that p is false, that is, you show that $\sim q \rightarrow \sim p$ is true. Because a statement is logically equivalent to its contrapositive, proving the contrapositive is true is a way of proving the original conditional is true.

4 Assess

Open-Ended Assessment

Modeling Allow students to use string, masking tape, protractors, and rulers to model indirect proofs for geometric examples. For Exercise 22 in the Practice and Apply Section, students can start by assuming ℓ is parallel to m and tape two lengths of string parallel to each other on their desks (Step 1). Then ask students to try to place a third string as a transversal such that $m\angle 2 \neq m\angle 1$. When students realize that any string placed yields $m\angle 2 = m\angle 1$ (Step 2), tell them to adjust the first two strings so that their assumption is false and then place a third string so that $m\angle 2 \neq m\angle 1$ (Step 3).

Getting Ready for Lesson 5-4

Prerequisite Skill Students will learn about the Triangle Inequality in Lesson 5-4. They will determine the truth of inequalities to prove that three measures can or cannot be sides of triangles. Use Exercises 46–48 to determine your students' familiarity with determining whether an inequality is true or false.

Assessment Options

Quiz (Lesson 5-3) is available on p. 289 of the *Chapter 5 Resource Masters.*

Mid-Chapter Test (Lessons 5-1 through 5-3) is available on p. 291 of the *Chapter 5 Resource Masters.*

33. **CRITICAL THINKING** Recall that a rational number is any number that can be expressed in the form $\frac{a}{b}$, where a and b are integers with no common factors and $b \neq 0$, or as a terminating or repeating decimal. Use indirect reasoning to prove that $\sqrt{2}$ is not a rational number. **See p. 279C.**

34. **WRITING IN MATH** Answer the question that was posed at the beginning of the lesson. **See margin.**

How is indirect proof used in literature?

Include the following in your answer:
- an explanation of how Sherlock Holmes used indirect proof, and
- an example of indirect proof used every day.

FCAT Practice
Standardized Test Practice

35. Which statement about the value of x is not true? **D**

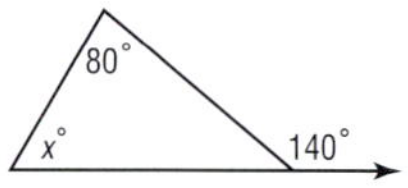

Ⓐ $x = 60$ Ⓑ $x < 140$
Ⓒ $x + 80 = 140$ Ⓓ $x < 60$

36. **PROBABILITY** A bag contains 6 blue marbles, 8 red marbles, and 2 white marbles. If three marbles are removed at random and no marble is returned to the bag after removal, what is the probability that all three marbles will be red? **A**

Ⓐ $\frac{1}{10}$ Ⓑ $\frac{1}{8}$ Ⓒ $\frac{3}{8}$ Ⓓ $\frac{1}{2}$

Maintain Your Skills

Mixed Review

For Exercises 37 and 38, refer to the figure at the right.
(Lesson 5-2)

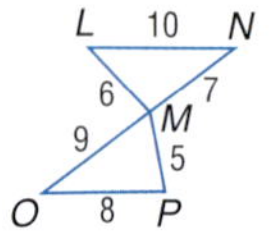

37. Which angle in $\triangle MOP$ has the greatest measure? **$\angle P$**
38. Name the angle with the least measure in $\triangle LMN$. **$\angle N$**

PROOF **Write a two-column proof.** *(Lesson 5-1)* **39–41. See p. 279D.**

39. If an angle bisector of a triangle is also an altitude of the triangle, then the triangle is isosceles.
40. The median to the base of an isosceles triangle bisects the vertex angle.
41. Corresponding angle bisectors of congruent triangles are congruent.

42. **ASTRONOMY** The Big Dipper is a part of the larger constellation Ursa Major. Three of the brighter stars in the constellation form $\triangle RSA$. If $m\angle R = 41$ and $m\angle S = 109$, find $m\angle A$. *(Lesson 4-2)* **30**

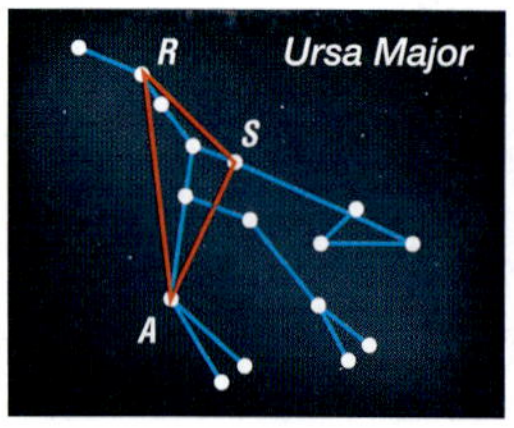

Write an equation in point-slope form of the line having the given slope that contains the given point.
(Lesson 3-4)

43. $m = 2, (4, 3)$ 44. $m = -3, (2, -2)$ 45. $m = 11, (-4, -9)$

43. $y - 3 = 2(x - 4)$
44. $y + 2 = -3(x - 2)$
45. $y + 9 = 11(x + 4)$

Getting Ready for the Next Lesson

PREREQUISITE SKILL **Determine whether each inequality is *true* or *false*.**
*(To review the **meaning of inequalities**, see pages 739 and 740.)*

46. $19 - 10 < 11$ **true** 47. $31 - 17 < 12$ **false** 48. $38 + 76 > 109$ **true**

Answer

34. Sample answer: Indirect proof is sometimes used in mystery novels. Answers should include the following.
- Sherlock Holmes would disprove all possibilities except the actual solution to a mystery.
- medical diagnosis, trials, scientific research

5-4 The Triangle Inequality

Sunshine State Standards
MA.C.1.4.1

What You'll Learn

- Apply the Triangle Inequality Theorem.
- Determine the shortest distance between a point and a line.

How can you use the Triangle Inequality Theorem when traveling?

Chuck Noland travels between Chicago, Indianapolis, and Columbus as part of his job. Mr. Noland lives in Chicago and needs to get to Columbus as quickly as possible. Should he take a flight that goes from Chicago to Columbus, or a flight that goes from Chicago to Indianapolis, then to Columbus?

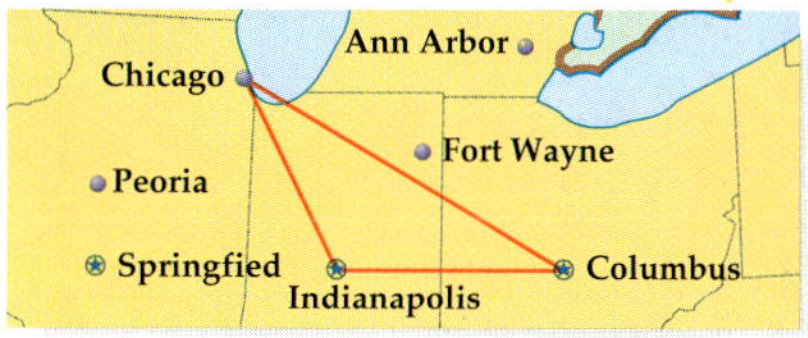

THE TRIANGLE INEQUALITY In the example above, if you chose to fly directly from Chicago to Columbus, you probably reasoned that a straight route is shorter. This is an example of the Triangle Inequality Theorem.

Theorem 5.11

Triangle Inequality Theorem The sum of the lengths of any two sides of a triangle is greater than the length of the third side.

Examples:
$AB + BC > AC$
$BC + AC > AB$
$AC + AB > BC$

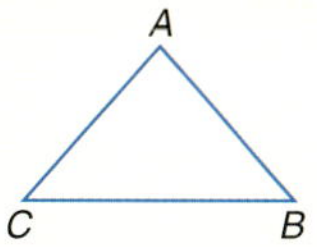

You will prove Theorem 5.11 in Exercise 40.

The Triangle Inequality Theorem can be used to determine whether three segments can form a triangle.

Study Tip

Inequality
If the sum of the smallest number and the middle number is greater than the largest number, then each combination of inequalities are true.

Example 1 *Identify Sides of a Triangle*

Determine whether the given measures can be the lengths of the sides of a triangle.

a. 2, 4, 5

Check each inequality.

$2 + 4 \stackrel{?}{>} 5$ $2 + 5 \stackrel{?}{>} 4$ $4 + 5 \stackrel{?}{>} 2$

$6 > 5 \checkmark$ $7 > 4 \checkmark$ $9 > 2 \checkmark$

All of the inequalities are true, so 2, 4, and 5 can be the lengths of the sides of a triangle.

b. 6, 8, 14

$6 + 8 \stackrel{?}{>} 14$

$14 \not> 14$

Because the sum of two measures equals the measure of the third side, the sides cannot form a triangle.

5-4 Lesson Notes

1 Focus

5-Minute Check Transparency 5-4 Use as a quiz or review of Lesson 5-3.

Mathematical Background notes are available for this lesson on p. 234D.

How **can you use the Triangle Inequality Theorem when traveling?**

Ask students:

- Which route is shorter? Chicago to Columbus direct
- How would you prove that this route is shorter? Accept all reasonable answers.

2 Teach

THE TRIANGLE INEQUALITY

In-Class Example Power Point®

1 Determine whether the given measures can be the lengths of the sides of a triangle.

a. $6\frac{1}{2}, 6\frac{1}{2}, 14\frac{1}{2}$ no

b. 6.8, 7.2, 5.1 yes

Resource Manager

Workbook and Reproducible Masters

Chapter 5 Resource Masters
- Study Guide and Intervention, pp. 263–264
- Skills Practice, p. 265
- Practice, p. 266
- Reading to Learn Mathematics, p. 267
- Enrichment, p. 268
- Assessment, p. 290

School-to-Career Masters, p. 10
Prerequisite Skills Workbook, pp. 15–16, 87–88, 107–108
Teaching Geometry With Manipulatives Masters, pp. 8, 95, 96

Transparencies

5-Minute Check Transparency 5-4
Answer Key Transparencies

Technology

Interactive Chalkboard

In-Class Example

Power Point®

Teaching Tip When students are checking triangles for valid side lengths, tell them they have to check the relationships for *all* three sides and not just two. For Example 2, 8 + 14 = 22, which is greater than 6, and 6 + 14 = 20, which is greater than 8, but the third measure does not work for $YZ = 6$.

2 In $\triangle PQR$, $PQ = 7.2$ and $QR = 5.2$. Which measure cannot be PR? **D**

A 7 **B** 9
C 11 **D** 13

Building on Prior Knowledge

In Chapter 3, students learned about properties of perpendicular lines. In Lesson 5-2, students learned about the Exterior Angle Theorem and angle-side relationships. Students combine these concepts in this lesson to prove that a perpendicular segment is the shortest distance from a point to a line.

When you know the lengths of two sides of a triangle, you can determine the range of possible lengths for the third side.

FCAT Practice
Standardized Test Practice

Example 2 Determine Possible Side Length

Multiple-Choice Test Item

In $\triangle XYZ$, $XY = 8$, and $XZ = 14$.
Which measure cannot be YZ?

(A) 6 (B) 10
(C) 14 (D) 18

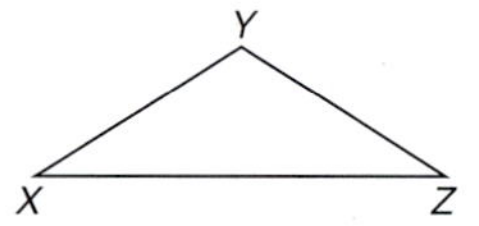

Read the Test Item

You need to determine which value is not valid.

Solve the Test Item

Solve each inequality to determine the range of values for YZ.

Let $YZ = n$.

$XY + XZ > YZ$	$XY + YZ > XZ$	$YZ + XZ > XY$
$8 + 14 > n$	$8 + n > 14$	$n + 14 > 8$
$22 > n$ or $n < 22$	$n > 6$	$n > -6$

Graph the inequalities on the same number line.

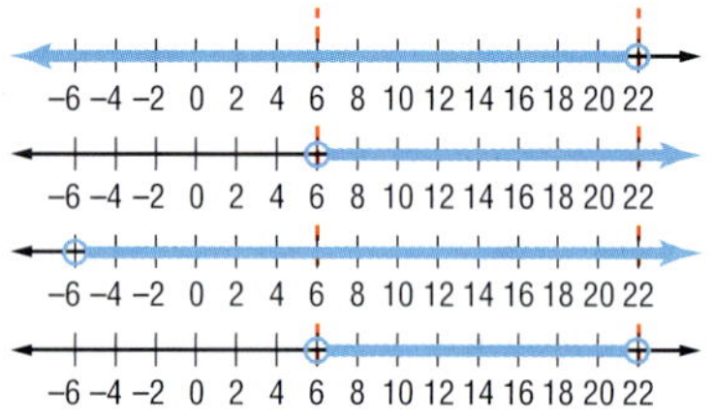

The range of values that fit all three inequalities is $6 < n < 22$.

Examine the answer choices. The only value that does not satisfy the compound inequality is 6 since $6 = 6$. Thus, the answer is choice A.

The Princeton Review

Test-Taking Tip

Testing Choices If you are short on time, you can test each choice to find the correct answer and eliminate any remaining choices.

DISTANCE BETWEEN A POINT AND A LINE

Recall that the distance between point P and line ℓ is measured along a perpendicular segment from the point to the line. It was accepted without proof that $\overline{PA}$ was the shortest segment from P to ℓ. The theorems involving the relationships between the angles and sides of a triangle can now be used to prove that a perpendicular segment is the shortest distance between a point and a line.

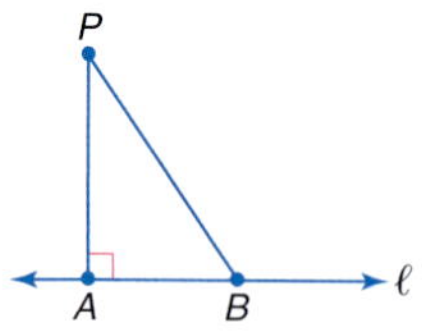

Theorem 5.12

The perpendicular segment from a point to a line is the shortest segment from the point to the line.

Example: $\overline{PQ}$ is the shortest segment from P to $\overleftrightarrow{AB}$.

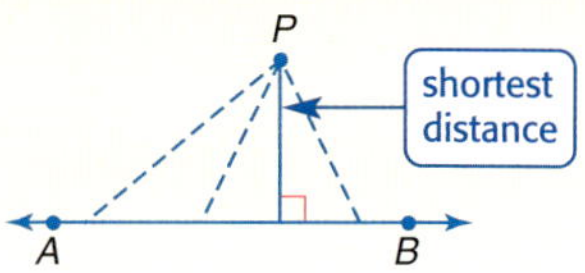

DAILY INTERVENTION

Differentiated Instruction

Naturalist Explain that even naturally occurring triangles must follow the principles presented in this lesson. Ask students to find examples to explore and test the Triangle Inequality Theorem, such as bird beaks, leaves, star constellations, animal tracks, and so on. The stars Vega, Deneb, and Altair form a right triangle, called the "Summer Triangle." Students can research to find estimated distances between the stars and verify that even in nature, the theorem is true.

Example 3 Prove Theorem 5.12

Given: $\overline{PA} \perp \ell$
$\overline{PB}$ is any nonperpendicular segment from P to ℓ.

Prove: $PB > PA$

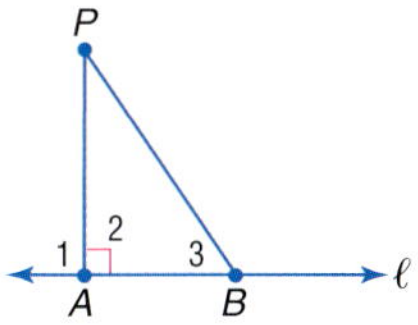

Proof:

Statements	Reasons
1. $\overline{PA} \perp \ell$	1. Given
2. $\angle 1$ and $\angle 2$ are right angles.	2. $\perp$ lines form right angles.
3. $\angle 1 \cong \angle 2$	3. All right angles are congruent.
4. $m\angle 1 = m\angle 2$	4. Definition of congruent angles
5. $m\angle 1 > m\angle 3$	5. Exterior Angle Inequality Theorem
6. $m\angle 2 > m\angle 3$	6. Substitution Property
7. $PB > PA$	7. If an angle of a triangle is greater than a second angle, then the side opposite the greater angle is longer than the side opposite the lesser angle.

Study Tip

Shortest Distance to a Line
If a line is horizontal, the shortest distance from a point to that line will be along a vertical line. Likewise, the shortest distance from a point to a vertical line lies along a horizontal line.

Corollary 5.1 follows directly from Theorem 5.12.

Corollary 5.1

The perpendicular segment from a point to a plane is the shortest segment from the point to the plane.

Example:

$\overline{QP}$ is the shortest segment from P to Plane $\mathcal{M}$.

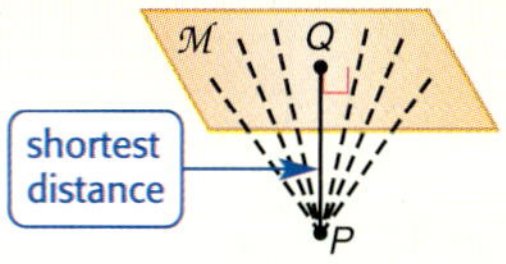

You will prove Corollary 5.1 in Exercise 12.

Check for Understanding

Concept Check

1. **Explain** why the distance between two nonhorizontal parallel lines on a coordinate plane cannot be found using the distance between their y-intercepts. **See margin.**

2. **FIND THE ERROR** Jameson and Anoki drew $\triangle EFG$ with $FG = 13$ and $EF = 5$. They each chose a possible measure for GE.

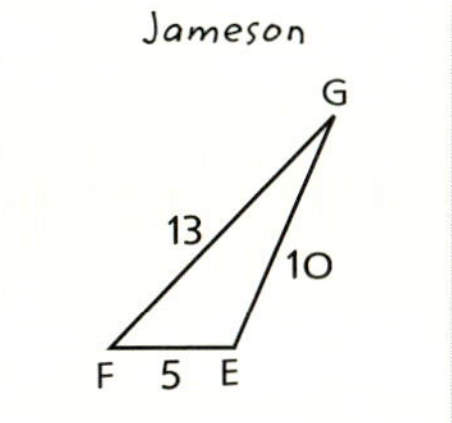

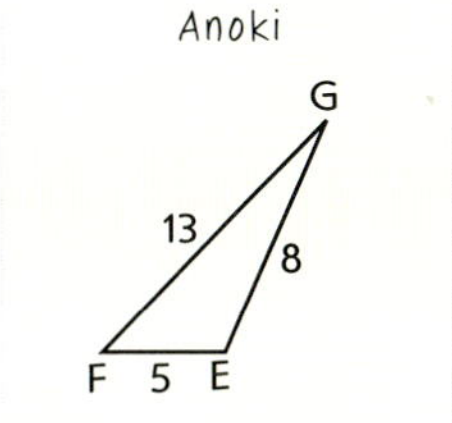

Who is correct? Explain. **Jameson; $5 + 10 > 13$ but $5 + 8 \not> 13$.**

3. **OPEN ENDED** Find three numbers that can be the lengths of the sides of a triangle and three numbers that cannot be the lengths of the sides of a triangle. Justify your reasoning with a drawing. **See margin.**

www.geometryonline.com/extra_examples/fcat

DISTANCE BETWEEN A POINT AND A LINE

In-Class Example

Power Point®

3 **Given:** Line $t \perp \overline{HJ}$ through point J
Point K lies on t.

Prove: $KJ < KH$

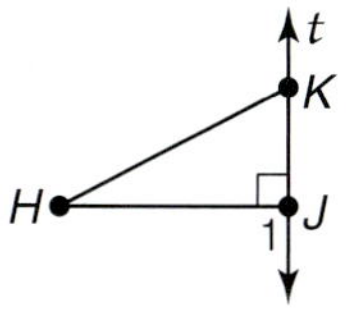

Statements (Reasons)

1. $t \perp \overline{HJ}$ (Given)
2. $\angle HJK$ and $\angle 1$ are right ∠s. ($\perp$ lines form rt. ∠s)
3. $\angle HJK \cong \angle 1$ (All rt. ∠s are $\cong$)
4. $m\angle HJK = m\angle 1$ (Def. of $\cong$ ∠s)
5. $m\angle KHJ < m\angle 1$ (Exterior ∠ Inequality Theorem)
6. $m\angle KHJ < m\angle HJK$ (Subst.)
7. $KJ < KH$ (If an angle of a triangle is greater than a second angle, then the side opposite the greater angle is longer than the side opposite the lesser angle.)

3 Practice/Apply

Study Notebook

Have students—
- *add the definitions/examples of the vocabulary terms to their Vocabulary Builder worksheets for Chapter 5.*
- *include any other item(s) that they find helpful in mastering the skills in this lesson.*

DAILY INTERVENTION

FIND THE ERROR Point out that in order to represent correctly the points and measurements for Anoki's figure, point E must lie on $\overline{GF}$ between points G and F, and three points contained on one line do not form a triangle.

Answers

1. Sample answer: If the lines are not horizontal, then the segment connecting their y-intercepts is not perpendicular to either line. Since distance is measured along a perpendicular segment, this segment cannot be used.

3. Sample answer: 2, 3, 4 and 1, 2, 3

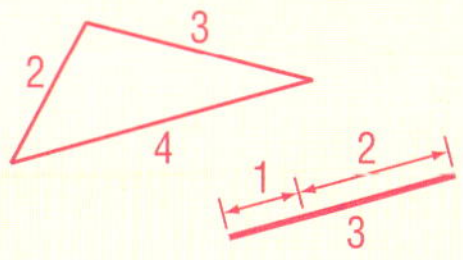

About the Exercises...

Organization by Objective

- **The Triangle Inequality:** 4–11, 13
- **Distance Between a Point and a Line:** 12

Odd/Even Assignments

Exercises 14–39 and 41–44 are structured so that students practice the same concepts whether they are assigned odd or even problems.

Assignment Guide

Basic: 15–45 odd, 52–62

Average: 15–49 odd, 50–62

Advanced: 14–50 even, 52–59 (optional: 60–62)

All: Quiz 2 (1–10)

Answers

12. Given: $\overline{PQ} \perp$ plane $\mathcal{M}$

Prove: $\overline{PQ}$ is the shortest segment from P to plane $\mathcal{M}$.

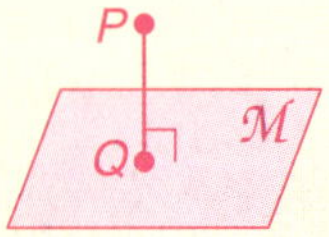

Proof:
By definition, $\overline{PQ}$ is perpendicular to plane $\mathcal{M}$ if it is perpendicular to every line in $\mathcal{M}$ that intersects it. But since the perpendicular segment from a point to a line is the shortest segment from the point to the line, that perpendicular segment is the shortest segment from the point to each of these lines. Therefore, $\overline{PQ}$ is the shortest segment from P to $\mathcal{M}$.

Guided Practice

GUIDED PRACTICE KEY	
Exercises	Examples
4–7	1
8–11, 13	2
12	3

Determine whether the given measures can be the lengths of the sides of a triangle. Write *yes* or *no*. Explain.

4. 5, 4, 3 yes; $3 + 4 > 5$
5. 5, 15, 10 no; $5 + 10 \ngtr 15$
6. 30.1, 0.8, 31 no; $30.1 + 0.8 \ngtr 31$
7. 5.6, 10.1, 5.2 yes; $5.2 + 5.6 > 10.1$

Find the range for the measure of the third side of a triangle given the measures of two sides.

8. 7 and 12 $5 < n < 19$
9. 14 and 23 $9 < n < 37$
10. 22 and 34 $12 < n < 56$
11. 15 and 18 $3 < n < 33$

12. **PROOF** Write a proof for Corollary 5.1. See margin.
Given: $\overline{PQ} \perp$ plane $\mathcal{M}$
Prove: $\overline{PQ}$ is the shortest segment from P to plane $\mathcal{M}$.

FCAT Practice
Standardized Test Practice

13. An isosceles triangle has a base 10 units long. If the congruent sides have whole number measures, what is the least possible length of the sides? B
Ⓐ 5 Ⓑ 6 Ⓒ 17 Ⓓ 21

Practice and Apply

Homework Help	
For Exercises	See Examples
14–25	1
26–37	2
38–40	3

Extra Practice
See page 764.

Determine whether the given measures can be the lengths of the sides of a triangle. Write *yes* or *no*. Explain. 16. yes; $8 + 8 > 15$

14. 1, 2, 3 no; $1 + 2 \ngtr 3$
15. 2, 6, 11 no; $2 + 6 \ngtr 11$
16. 8, 8, 15
17. 13, 16, 29
18. 18, 32, 21
19. 9, 21, 20
20. 5, 17, 9 no; $5 + 9 \ngtr 17$
21. 17, 30, 30
22. 8.4, 7.2, 3.5
23. 4, 0.9, 4.1 yes; $0.9 + 4 > 4.1$
24. 14.3, 12, 2.2 no; $2.2 + 12 \ngtr 14.3$
25. 0.18, 0.21, 0.52 no; $0.18 + 0.21 \ngtr 0.52$

17. no; $13 + 16 \ngtr 29$
18. yes; $18 + 21 > 32$
19. yes; $9 + 20 > 21$
21. yes; $17 + 30 > 30$
22. yes; $3.5 + 7.2 > 8.4$

Find the range for the measure of the third side of a triangle given the measures of two sides.

26. 5 and 11 $6 < n < 16$
27. 7 and 9 $2 < n < 16$
28. 10 and 15 $5 < n < 25$
29. 12 and 18 $6 < n < 30$
30. 21 and 47 $26 < n < 68$
31. 32 and 61 $29 < n < 93$
32. 30 and 30 $0 < n < 60$
33. 64 and 88 $24 < n < 152$
34. 57 and 55 $2 < n < 112$
35. 75 and 75 $0 < n < 150$
36. 78 and 5 $73 < n < 83$
37. 99 and 2 $97 < n < 101$

PROOF Write a two-column proof. 38–40. See p. 279D.

38. **Given:** $\angle B \cong \angle ACB$
Prove: $AD + AB > CD$

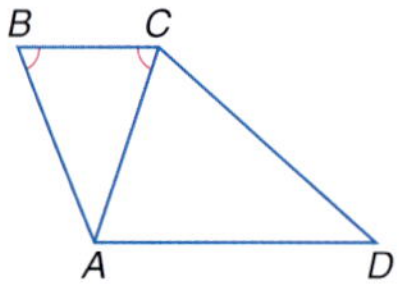

39. **Given:** $\overline{HE} \cong \overline{EG}$
Prove: $HE + FG > EF$

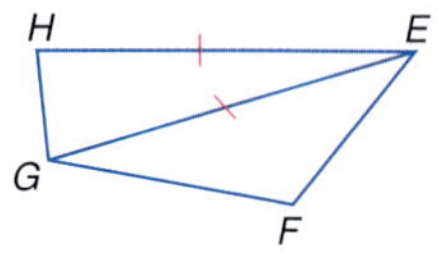

40. **Given:** $\triangle ABC$
Prove: $AC + BC > AB$ (Triangle Inequality Theorem)
(*Hint:* Draw auxiliary segment $\overline{CD}$, so that C is between B and D and $\overline{CD} \cong \overline{AC}$.)

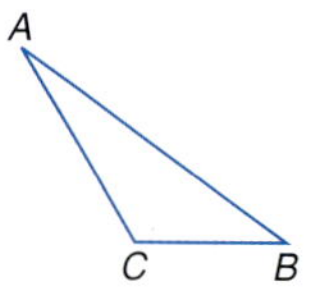

48. m is either 15 ft or 16 ft; n is 14 ft, 15 ft, or 16 ft. The possible triangles that can be made from sides with those measures are (2 ft, 15 ft, 14 ft), (2 ft, 15 ft, 15 ft), (2 ft, 15 ft, 16 ft), (2 ft, 16 ft, 16 ft).

50. Sample answer: The length of any side of a triangle is greater than the differences between the lengths of the other two sides.

Paragraph Proof: By the Triangle Inequality Theorem, for $\triangle ABC$ with side measures a, b, and c, $a + b > c$, $b + c > a$, and $c + a > b$. Using the Subtraction Property of Inequality, $a > c - b$, $b > a - c$, and $c > b - a$.

41. yes; $AB + BC > AC$, $AB + AC > BC$, $AC + BC > AB$

42. yes; $LM + MN > LN$, $LM + LN > MN$, $LN + MN > LM$

ALGEBRA **Determine whether the given coordinates are the vertices of a triangle. Explain.**

41. $A(5, 8)$, $B(2, -4)$, $C(-3, -1)$

42. $L(-24, -19)$, $M(-22, 20)$, $N(-5, -7)$

43. $X(0, -8)$, $Y(16, -12)$, $Z(28, -15)$ no; $XY + YZ = XZ$

44. $R(1, -4)$, $S(-3, -20)$, $T(5, 12)$ no; $RS + RT = ST$

CRAFTS **For Exercises 45 and 46, use the following information.**
Carlota has several strips of trim she wishes to use as a triangular border for a section of a decorative quilt she is going to make. The strips measure 3 centimeters, 4 centimeters, 5 centimeters, 6 centimeters, and 12 centimeters.

45. How many different triangles could Carlota make with the strips? 4

46. How many different triangles could Carlota make that have a perimeter that is divisible by 3? 2

More About. . .

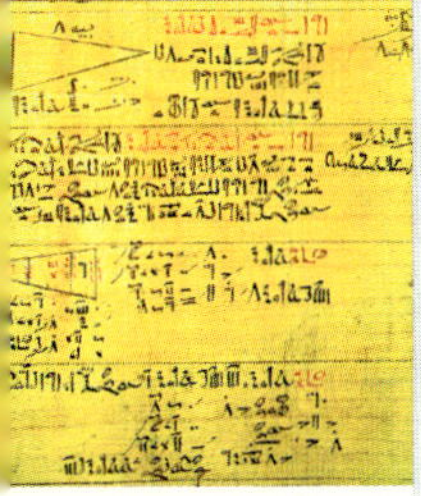

History

Ancient Egyptians used pieces of flattened, dried papyrus reed as paper. Surviving examples include the Rhind Papyrus and the Moscow Papyrus, from which we have attained most of our knowledge about Egyptian mathematics.

Source: www.aldokkan.com

47. **HISTORY** The early Egyptians used to make triangles by using a rope with knots tied at equal intervals. Each vertex of the triangle had to occur at a knot. How many different triangles can be formed using the rope below? 3

PROBABILITY **For Exercises 48 and 49, use the following information.**
One side of a triangle is 2 feet long. Let m represent the measure of the second side and n represent the measure of the third side. Suppose m and n are whole numbers and that $14 < m < 17$ and $13 < n < 17$.

48. List the measures of the sides of the triangles that are possible. See margin.

49. What is the probability that a randomly chosen triangle that satisfies the given conditions will be isosceles? $\frac{1}{2}$

50. **CRITICAL THINKING** State and prove a theorem that compares the measures of each side of a triangle with the differences of the measures of the other two sides. See margin.

51. **WRITING IN MATH** Answer the question that was posed at the beginning of the lesson. See margin.

How can you use the Triangle Inequality when traveling?

Include the following in your answer:
- an example of a situation in which you might want to use the greater measures, and
- an explanation as to why it is not always possible to apply the Triangle Inequality when traveling.

Standardized Test Practice

52. If two sides of a triangle measure 12 and 7, which of the following cannot be the perimeter of the triangle? D

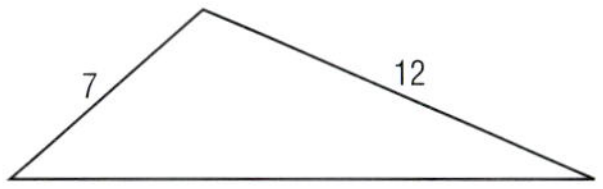

(A) 29 (B) 34
(C) 37 (D) 38

53. **ALGEBRA** How many points of intersection exist if the equations $(x - 5)^2 + (y - 5)^2 = 4$ and $y = -x$ are graphed on the same coordinate plane? A

(A) none (B) one (C) two (D) three

www.geometryonline.com/self_check_quiz

51. Sample answer: You can use the Triangle Inequality Theorem to verify the shortest route between two locations. Answers should include the following.
- A longer route might be better if you want to collect frequent flier miles.
- A straight route might not always be available.

Study Guide and Intervention, p. 263 (shown) and p. 264

The Triangle Inequality If you take three straws of lengths 8 inches, 5 inches, and 1 inch and try to make a triangle with them, you will find that it is not possible. This illustrates the Triangle Inequality Theorem.

Triangle Inequality Theorem	The sum of the lengths of any two sides of a triangle is greater than the length of the third side.

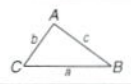

Example **The measures of two sides of a triangle are 5 and 8. Find a range for the length of the third side.**

By the Triangle Inequality, all three of the following inequalities must be true.

$5 + x > 8$ $\quad 8 + x > 5$ $\quad 5 + 8 > x$
$x > 3$ $\quad x > -3$ $\quad 13 > x$

Therefore x must be between 3 and 13.

Exercises

Determine whether the given measures can be the lengths of the sides of a triangle. Write *yes* or *no*.

1. 3, 4, 6 yes
2. 6, 9, 15 no
3. 8, 8, 8 yes
4. 2, 4, 5 yes
5. 4, 8, 16 no
6. 1.5, 2.5, 3 yes

Find the range for the measure of the third side given the measures of two sides.

7. 1 and 6 $5 < n < 7$
8. 12 and 18 $6 < n < 30$
9. 1.5 and 5.5 $4 < n < 7$
10. 82 and 8 $74 < n < 90$

11. Suppose you have three different positive numbers arranged in order from least to greatest. What single comparison will let you see if the numbers can be the lengths of the sides of a triangle?
Find the sum of the two smaller numbers. If that sum is greater than the largest number, then the three numbers can be the lengths of the sides of a triangle.

Skills Practice, p. 265 and Practice, p. 266 (shown)

Determine whether the given measures can be the lengths of the sides of a triangle. Write *yes* or *no*.

1. 9, 12, 18 yes
2. 8, 9, 17 no
3. 14, 14, 19 yes
4. 23, 26, 50 no
5. 32, 41, 63 yes
6. 2.7, 3.1, 4.3 yes
7. 0.7, 1.4, 2.1 no
8. 12.3, 13.9, 25.2 yes

Find the range for the measure of the third side of a triangle given the measures of two sides.

9. 6 and 19 $13 < n < 25$
10. 7 and 29 $22 < n < 36$
11. 13 and 27 $14 < n < 40$
12. 18 and 23 $5 < n < 41$
13. 25 and 38 $13 < n < 63$
14. 31 and 39 $8 < n < 70$
15. 42 and 6 $36 < n < 48$
16. 54 and 7 $47 < n < 61$

ALGEBRA **Determine whether the given coordinates are the vertices of a triangle. Explain.**

17. $R(1, 3)$, $S(4, 0)$, $T(10, -6)$
No; $RS = 3\sqrt{2}$, $ST = 6\sqrt{2}$, and $RT = 9\sqrt{2}$, so $RS + ST = RT$.

18. $W(2, 6)$, $X(1, 6)$, $Y(4, 2)$
Yes; $WX = 1$, $XY = 5$, and $WY = 2\sqrt{5}$, so $WX + XY > WY$, $WX + WY > XY$, and $WY + XY > WX$.

19. $P(-3, 2)$, $L(1, 1)$, $M(9, -1)$
No; $PL = \sqrt{17}$, $LM = 2\sqrt{17}$, and $PM = 3\sqrt{17}$, so $PL + LM = PM$.

20. $B(1, 1)$, $C(6, 5)$, $D(4, -1)$
Yes; $BC = \sqrt{41}$, $CD = 2\sqrt{10}$, and $BD = \sqrt{13}$, so $BC + CD > BD$, $BC + BD > CD$, and $BD + CD > BC$.

21. **GARDENING** Ha Poong has 4 lengths of wood from which he plans to make a border for a triangular-shaped herb garden. The lengths of the wood borders are 8 inches, 10 inches, 12 inches, and 18 inches. How many different triangular borders can Ha Poong make? 3

Reading to Learn Mathematics, p. 267 ELL

Pre-Activity **How can you use the Triangle Inequality Theorem when traveling?**

Read the introduction to Lesson 5-4 at the top of page 261 in your textbook.

In addition to the greater distance involved in flying from Chicago to Columbus through Indianapolis rather than flying nonstop, what are two other reasons that it would take longer to get to Columbus if you take two flights rather than one? Sample answer: time needed for an extra takeoff and landing; layover time in Indianapolis between the two flights

Reading the Lesson

1. Refer to the figure.

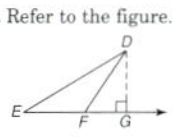

Which statements are true? C, D, F

A. $DE > EF + FD$
B. $DE = EF + FD$
C. $EG = EF + FG$
D. $ED + DG > EG$
E. The shortest distance from D to $\overleftrightarrow{EG}$ is DF.
F. The shortest distance from D to $\overleftrightarrow{EG}$ is DG.

2. Complete each sentence about $\triangle XYZ$.

a. If $XY = 8$ and $YZ = 11$, then the range of values for XZ is _3_ $< XZ <$ _19_.
b. If $XY = 13$ and $XZ = 25$, then YZ must be between _12_ and _38_.
c. If $\triangle XYZ$ is isosceles with $\angle Z$ as the vertex angle, and $XZ = 8.5$, then the range of values for XY is _0_ $< XY <$ _17_.
d. If $XZ = a$ and $YZ = b$, with $b < a$, then the range for XY is _$a - b$_ $< XY <$ _$a + b$_.

Helping You Remember

3. A good way to remember a new theorem is to state it informally in different words. How could you restate the Triangle Inequality Theorem?
Sample answer: The side that connects one vertex of a triangle to another is a shorter path between the two vertices than the path that goes through the third vertex.

Enrichment, p. 268

Constructing Triangles

The measurements of the sides of a triangle are given. If a triangle having sides with these measurements is not possible, then write *impossible*. If a triangle is possible, draw it and measure each angle with a protractor.

1. $AR = 5$ cm $\quad m\angle A = 30$
$RT = 3$ cm $\quad m\angle R = 90$
$AT = 6$ cm $\quad m\angle T = 60$

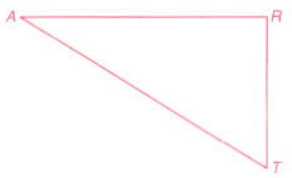

2. $PI = 8$ cm $\quad m\angle P =$
$IN = 3$ cm $\quad m\angle I =$
$PN = 2$ cm $\quad m\angle N =$
impossible

4 Assess

Open-Ended Assessment

Writing To enhance understanding of the lesson concepts, tell students to rewrite the theorems and corollary from this lesson in their own words and create and draw their own examples for each item. Have students place their writing in their study notebooks for future reference.

Getting Ready for Lesson 5-5

Prerequisite Skill Students will learn about inequalities involving two triangles in Lesson 5-5. They will solve inequalities as steps of proofs. Use Exercises 60–62 to determine your students' familiarity with solving inequalities.

Assessment Options

Practice Quiz 2 The quiz provides students with a brief review of the concepts and skills in Lessons 5-3 and 5-4. Lesson numbers are given to the right of the exercises or instruction lines so students can review concepts not yet mastered.

Quiz (Lesson 5-4) is available on p. 290 of the *Chapter 5 Resource Masters*.

Maintain Your Skills

Mixed Review

54. **PROOF** Write an indirect proof. *(Lesson 5-3)* **See margin.**

Given: P is a point not on line ℓ.
Prove: $\overleftrightarrow{PQ}$ is the only line through P perpendicular to ℓ.

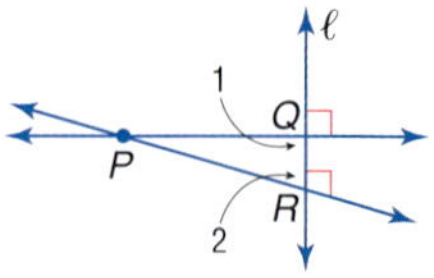

ALGEBRA **List the sides of $\triangle PQR$ in order from longest to shortest if the angles of $\triangle PQR$ have the given measures.** *(Lesson 5-2)*

55. $m\angle P = 7x + 8$, $m\angle Q = 8x - 10$, $m\angle R = 7x + 6$ **$\overline{QR}, \overline{PQ}, \overline{PR}$**

56. $m\angle P = 3x + 44$, $m\angle Q = 68 - 3x$, $m\angle R = x + 61$ **$\overline{PQ}, \overline{QR}, \overline{PR}$**

Determine whether $\triangle JKL \cong \triangle PQR$ given the coordinates of the vertices. Explain. *(Lesson 4-4)* **57–59. See margin.**

57. $J(0, 5)$, $K(0, 0)$, $L(-2, 0)$, $P(4, 8)$, $Q(4, 3)$, $R(6, 3)$

58. $J(6, 4)$, $K(1, -6)$, $L(-9, 5)$, $P(0, 7)$, $Q(5, -3)$, $R(15, 8)$

59. $J(-6, -3)$, $K(1, 5)$, $L(2, -2)$, $P(2, -11)$, $Q(5, -4)$, $R(10, -10)$

Getting Ready for the Next Lesson

PREREQUISITE SKILL **Solve each inequality.**
*(To review **solving inequalities**, see pages 739 and 740.)*

60. $3x + 54 < 90$ **$x < 12$**

61. $8x - 14 < 3x + 19$ **$x < 6.6$**

62. $4x + 7 < 180$ **$x < 43.25$**

Practice Quiz 2 — *Lessons 5-3 and 5-4*

Write the assumption you would make to start an indirect proof of each statement. *(Lesson 5-3)*

1. The number 117 is divisible by 13. **The number 117 is not divisible by 13.**
2. $m\angle C < m\angle D$ **$m\angle C \geq m\angle D$**

Write an indirect proof. *(Lesson 5-3)* **3–5. See pp. 279D–279E.**

3. If $7x > 56$, then $x > 8$.
4. **Given:** $\overline{MO} \cong \overline{ON}$, $\overline{MP} \not\cong \overline{NP}$
   Prove: $\angle MOP \not\cong \angle NOP$

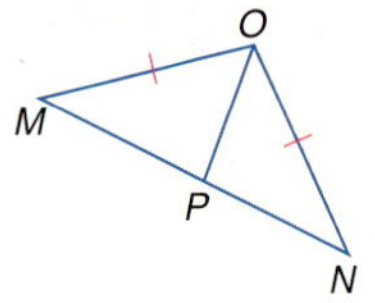

5. **Given:** $m\angle ADC \neq m\angle ADB$
   Prove: $\overline{AD}$ is not an altitude of $\triangle ABC$.

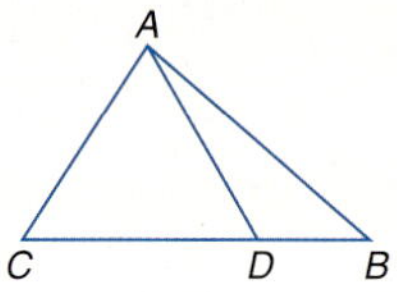

Determine whether the given measures can be the lengths of the sides of a triangle. Write *yes* or *no*. Explain. *(Lesson 5-4)*

6. 7, 24, 25 **yes; $7 + 24 > 25$**
7. 25, 35, 60 **no; $25 + 35 \not> 60$**
8. 20, 3, 18 **yes; $3 + 18 > 20$**
9. 5, 10, 6 **yes; $5 + 6 > 10$**
10. If the measures of two sides of a triangle are 57 and 32, what is the range of possible measures of the third side? *(Lesson 5-4)* **$25 < n < 89$**

Answers

54. Given: P is a point not on line ℓ.
Prove: $\overleftrightarrow{PQ}$ is the only line through P perpendicular to ℓ.

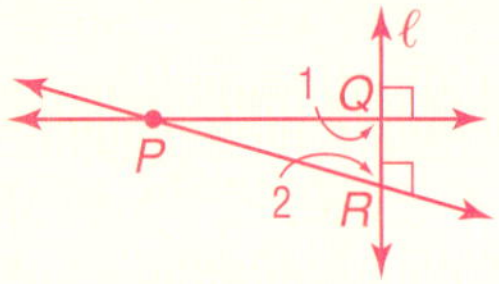

Statements (Reasons)

1. $\overleftrightarrow{PQ}$ is not the only line through P perpendicular to ℓ. (Assump.)
2. $\angle 1$ and $\angle 2$ are right angles. ($\perp$ lines form 4 rt. ∠s.)
3. $m\angle 1 = 90$, $m\angle 2 = 90$ (Def. of rt. $\angle$)
4. $m\angle 1 + m\angle 2 + m\angle QPR = 180$ (The sum of ∠s in a $\triangle$ is 180.)
5. $90 + 90 + m\angle QPR = 180$ (Substitution)
6. $m\angle QPR = 0$ (Subtraction Property)

This contradicts the fact that the measure of an angle is greater than 0. Thus, $\overleftrightarrow{PQ}$ is the only line through P perpendicular to ℓ.

57. $JK = 5$, $KL = 2$, $JL = \sqrt{29}$, $PQ = 5$, $QR = 2$, and $PR = \sqrt{29}$. The corresponding sides have the same measure and are congruent. $\triangle JKL \cong \triangle PQR$ by SSS.

58. $JK = \sqrt{125}$, $KL = \sqrt{221}$, $JL = \sqrt{226}$, $PQ = \sqrt{125}$, $QR = \sqrt{221}$, and $PR = \sqrt{226}$. The corresponding sides have the same measure and are congruent. $\triangle JKL \cong \triangle PQR$ by SSS.

59. $JK = \sqrt{113}$, $KL = \sqrt{50}$, $JL = \sqrt{65}$, $PQ = \sqrt{58}$, $QR = \sqrt{61}$, and $PR = \sqrt{65}$. The corresponding sides are not congruent, the triangles are not congruent.

5-5 Inequalities Involving Two Triangles

Sunshine State Standards
MA.C.1.4.1

What You'll Learn

- Apply the SAS Inequality.
- Apply the SSS Inequality.

How does a backhoe work?

Many objects, like a backhoe, have two fixed arms connected by a joint or hinge. This allows the angle between the arms to increase and decrease. As the angle changes, the distance between the endpoints of the arms changes as well.

SAS INEQUALITY The relationship of the arms and the angle between them illustrates the following theorem.

Study Tip

SAS Inequality
The SAS Inequality Theorem is also called the *Hinge Theorem*.

Theorem 5.13

SAS Inequality/Hinge Theorem If two sides of a triangle are congruent to two sides of another triangle and the included angle in one triangle has a greater measure than the included angle in the other, then the third side of the first triangle is longer than the third side of the second triangle.

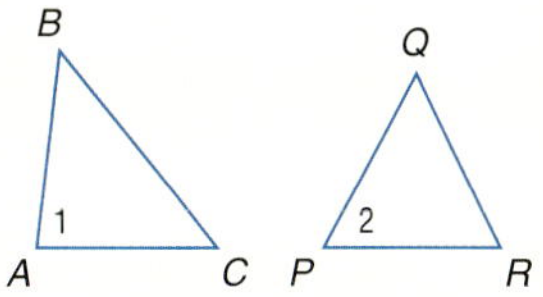

Example: Given $\overline{AB} \cong \overline{PQ}$, $\overline{AC} \cong \overline{PR}$, if $m\angle 1 > m\angle 2$, then $BC > QR$.

Proof *SAS Inequality Theorem*

Given: $\triangle ABC$ and $\triangle DEF$
$\overline{AC} \cong \overline{DF}$, $\overline{BC} \cong \overline{EF}$
$m\angle F > m\angle C$

Prove: $DE > AB$

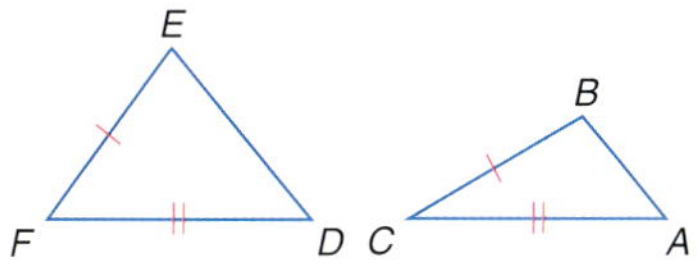

Proof:

We are given that $\overline{AC} \cong \overline{DF}$ and $\overline{BC} \cong \overline{EF}$. We also know that $m\angle F > m\angle C$. Draw auxiliary ray FZ such that $m\angle DFZ = m\angle C$ and that $\overline{ZF} \cong \overline{BC}$. This leads to two cases.

Case 1: If Z lies on $\overline{DE}$, then $\triangle FZD \cong \triangle CBA$ by SAS. Thus, $ZD = BA$ by CPCTC and the definition of congruent segments. By the Segment Addition Postulate, $DE = EZ + ZD$. Also, $DE > ZD$ by the definition of inequality. Therefore, $DE > AB$ by the Substitution Property.

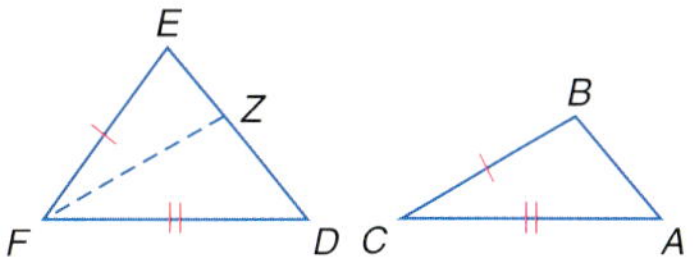

5-5 Lesson Notes

1 Focus

5-Minute Check Transparency 5-5 Use as a quiz or review of Lesson 5-4.

Mathematical Background notes are available for this lesson on p. 234D.

How does a backhoe work?

Ask students:

- If you decreased the angle between the hinged arms of the backhoe, what would happen to the distance between the endpoints of the arms? **The distance would decrease.**
- Name some other objects that are jointed or hinged. What do you notice about these objects? **Sample answers: Scissors, pliers, can openers, nail clippers, tripods, standing picture frames, the hands of a clock, and so on; each object has a specific purpose, and its joints or hinges are vital to that purpose.**

Resource Manager

Workbook and Reproducible Masters

Chapter 5 Resource Masters
- Study Guide and Intervention, pp. 269–270
- Skills Practice, p. 271
- Practice, p. 272
- Reading to Learn Mathematics, p. 273
- Enrichment, p. 274
- Assessment, p. 290

Prerequisite Skills Workbook, pp. 87–88
Teaching Geometry With Manipulatives Masters, p. 8

Transparencies

5-Minute Check Transparency 5-5
Real-World Transparency 5
Answer Key Transparencies

Technology

GeomPASS: Tutorial Plus, Lesson 11
Interactive Chalkboard

2 Teach

SAS INEQUALITY

In-Class Example Power Point®

1 **Given:** $\overline{KL} \parallel \overline{JH}$

$m\angle JKH + m\angle HKL <$
$m\angle JHK + m\angle KHL$
$JK = HL$

Prove: $JH < KL$

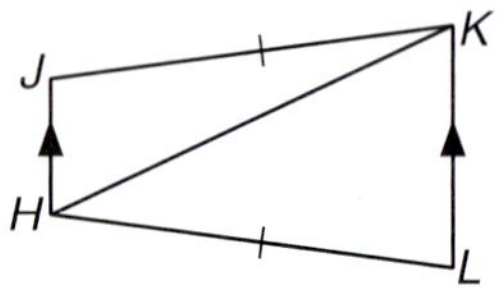

Statements (Reasons)

1. $m\angle JKH + m\angle HKL <$ $m\angle JHK + m\angle KHL$ (Given)
2. $m\angle HKL = m\angle JHK$ (Alt. Int. ∠s are ≅.)
3. $m\angle JKH + m\angle JHK <$ $m\angle JHK + m\angle KHL$ (Subst.)
4. $m\angle JKH < m\angle KHL$ (Subtr.)
5. $JK = HL$ (Given)
6. $HK = HK$ (Reflexive Prop.)
7. $JH < KL$ (SAS Inequality)

Case 2: If Z does not lie on $\overline{DE}$, then let the intersection of $\overline{FZ}$ and $\overline{ED}$ be point T. Now draw another auxiliary segment $\overline{FV}$ such that V is on $\overline{DE}$ and $\angle EFV \cong \angle VFZ$.

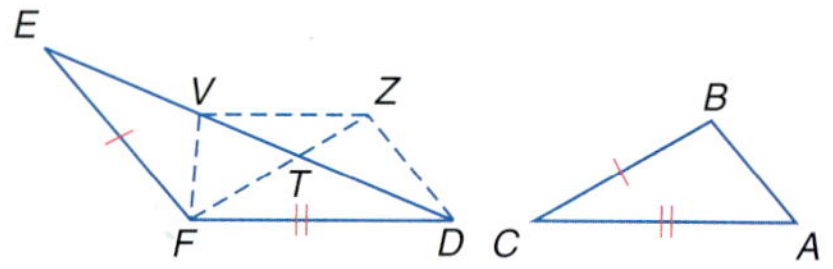

Since $\overline{FZ} \cong \overline{BC}$ and $\overline{BC} \cong \overline{EF}$, we have $\overline{FZ} \cong \overline{EF}$ by the Transitive Property. Also $\overline{VF}$ is congruent to itself by the Reflexive Property. Thus, $\triangle EFV \cong \triangle ZFV$ by SAS. By CPCTC, $\overline{EV} \cong \overline{ZV}$ or $EV = ZV$. Also, $\triangle FZD \cong \triangle CBA$ by SAS. So, $\overline{ZD} \cong \overline{BA}$ by CPCTC or $ZD = BA$. In $\triangle VZD$, $VD + ZV > ZD$ by the Triangle Inequality Theorem. By substitution, $VD + EV > ZD$. Since $ED = VD + EV$ by the Segment Addition Postulate, $ED > ZD$. Using substitution, $ED > BA$ or $DE > AB$.

Example 1 Use SAS Inequality in a Proof

Write a two-column proof.

Given: $\overline{YZ} \cong \overline{XZ}$
Z is the midpoint of $\overline{AC}$.
$m\angle CZY > m\angle AZX$
$\overline{BY} \cong \overline{BX}$

Prove: $BC > AB$

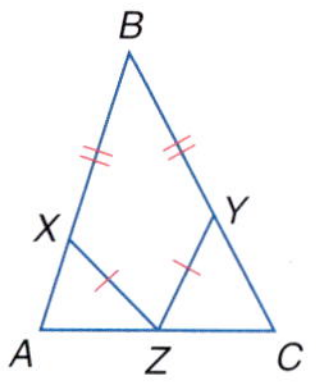

Proof:

Statements	Reasons
1. $\overline{YZ} \cong \overline{XZ}$ Z is the midpoint of $\overline{AC}$. $m\angle CZY > m\angle AZX$ $\overline{BY} \cong \overline{BX}$	1. Given
2. $CZ = AZ$	2. Definition of midpoint
3. $CY > AX$	3. SAS Inequality
4. $BY = BX$	4. Definition of congruent segments
5. $CY + BY > AX + BX$	5. Addition Property
6. $BC = CY + BY$ $AB = AX + BX$	6. Segment Addition Postulate
7. $BC > AB$	7. Substitution Property

SSS INEQUALITY The converse of the SAS Inequality Theorem is the SSS Inequality Theorem.

Theorem 5.14

SSS Inequality If two sides of a triangle are congruent to two sides of another triangle and the third side in one triangle is longer than the third side in the other, then the angle between the pair of congruent sides in the first triangle is greater than the corresponding angle in the second triangle.

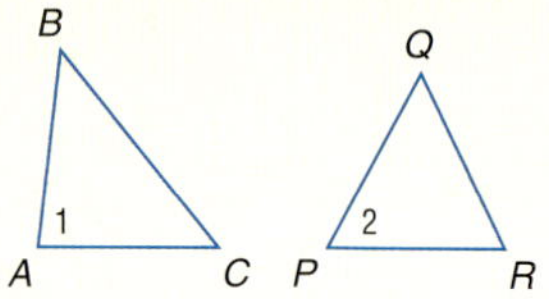

Example: Given $\overline{AB} \cong \overline{PQ}$, $\overline{AC} \cong \overline{PR}$, if $BC > QR$, then $m\angle 1 > m\angle 2$.

You will prove Theorem 5.14 in Exercise 24.

DAILY INTERVENTION **Differentiated Instruction**

Logical/Mathematical Tell students that the inequality theorems in this lesson are extremely logical, so students can rely on reasoning skills to remember them. Encourage students to examine the two theorems for similarities. Explain that students can simply remember that a longer side will always be opposite a larger angle, and a shorter side will always be opposite a smaller angle. Also, both theorems involve two triangles with an angle included between two congruent sides.

Example 2 Prove Triangle Relationships

Given: $\overline{AB} \cong \overline{CD}$
$\overline{AB} \parallel \overline{CD}$
$CD > AD$

Prove: $m\angle AOB > m\angle BOC$

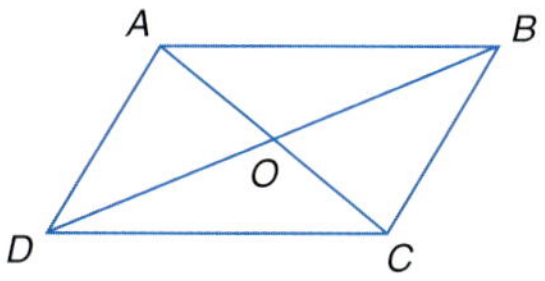

Flow Proof:

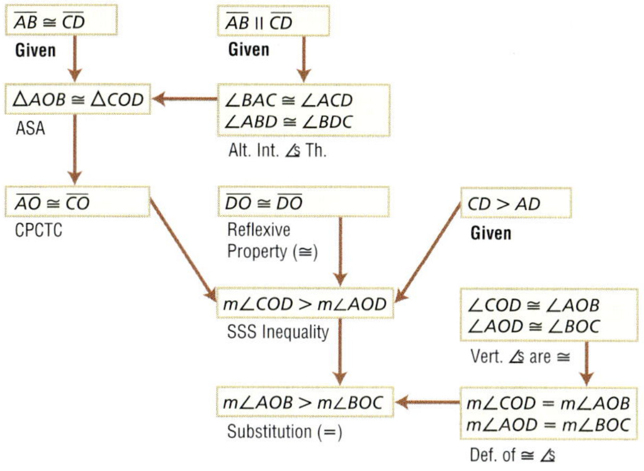

You can use algebra to relate the measures of the angles and sides of two triangles.

Example 3 Relationships Between Two Triangles

Write an inequality using the information in the figure.

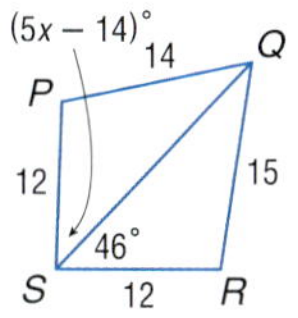

a. Compare $m\angle QSR$ and $m\angle QSP$.

In $\triangle PQS$ and $\triangle RQS$, $\overline{PS} \cong \overline{RS}$, $\overline{QS} \cong \overline{QS}$, and $QR > QP$. The SAS Inequality allows us to conclude that $m\angle QSR > m\angle QSP$.

b. Find the range of values containing x.

By the SSS Inequality, $m\angle QSR > m\angle QSP$, or $m\angle QSP < m\angle QSR$.

$m\angle QSP < m\angle QSR$	SSS Inequality
$5x - 14 < 46$	Substitution
$5x < 60$	Add 14 to each side.
$x < 12$	Divide each side by 5.

Also, recall that the measure of any angle is always greater than 0.

$5x - 14 > 0$	
$5x > 14$	Add 14 to each side.
$x > \frac{14}{5}$ or 2.8	Divide each side by 5.

The two inequalities can be written as the compound inequality $2.8 < x < 12$.

SSS INEQUALITY

In-Class Examples

Power Point®

Teaching Tip Tell students to place congruence marks and other helpful items on figures before writing a proof and during the proof-writing process as relationships are determined.

2 **Given:** $ST = PQ$
$SR = QR$
$ST = \frac{2}{3}SP$

Prove: $m\angle SRP > m\angle PRQ$

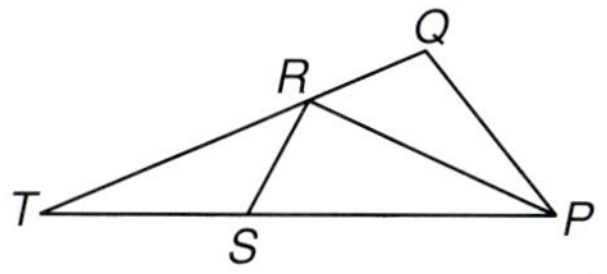

Statements (Reasons)

1. $SR = QR$ (Given)
2. $PR = PR$ (Reflexive)
3. $ST = PQ$ (Given)
4. $ST = \frac{2}{3}SP$; $SP > ST$ (Given)
5. $SP > PQ$ (Substitution)
6. $m\angle SRP > m\angle PRQ$ (SSS Inequality)

3

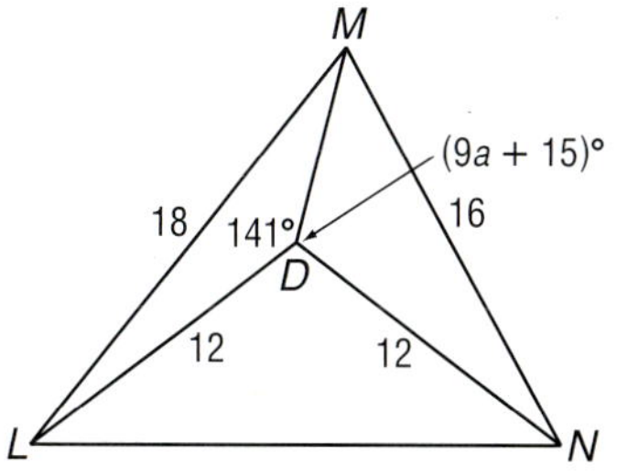

a. Write an inequality relating $m\angle LDM$ to $m\angle MDN$. $m\angle LDM > m\angle MDN$

b. Find the the range of values containing a. $-\frac{5}{3} < a < 14$

Teaching Tip In Example 3, make sure students understand that there is an exact value for x, not multiple values, that falls in the given range.

Concept Check

Have students draw three sets of two triangles with two congruent sides and measure and mark the sides and angles that satisfy the SAS Inequality Theorem and the SSS Inequality Theorem. Students can also check single triangles for the Triangle Inequality Theorem and extend the ends of the triangle and check the measurements for the Exterior Angle Inequality Theorem as well.

In-Class Example

Power Point®

4 **HEALTH** Doctors use a straight-leg-raising test to determine the amount of pain felt in a person's back. The patient lies flat on the examining table, and the doctor raises each leg until the patient experiences pain in the back area. Nitan can tolerate the doctor raising his right leg 35° and his left leg 65° from the table. Which foot can Nitan raise higher above the table? **his left foot**

Answers

1. Sample answer: A pair of scissors illustrates the SSS inequality. As the distance between the tips of the scissors decreases, the angle between the blades decreases, allowing the blades to cut.
2. The SSS Inequality Theorem compares the angle between two sides of a triangle for which the two sides are congruent and the third side is different. The SSS Postulate states that two triangles that have three sides congruent are congruent.

More About. . .

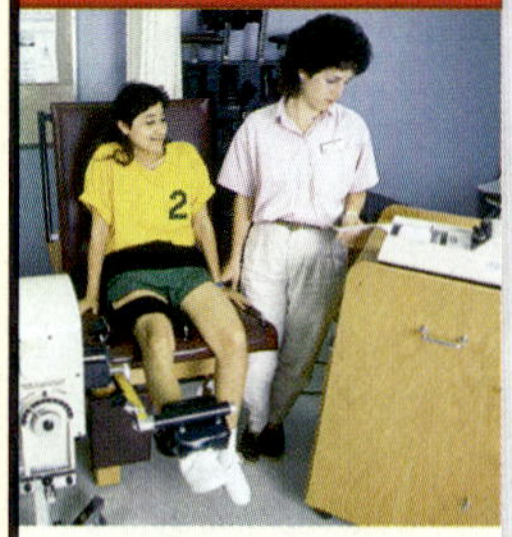

Health

Physical therapists help their patients regain range of motion after an illness or injury. They also teach patients exercises to help prevent injury.

Source: www.apta.org

Inequalities involving triangles can be used to describe real-world situations.

Example 4 Use Triangle Inequalities

HEALTH **Range of motion describes the amount that a limb can be moved from a straight position. To determine the range of motion of a person's forearm, determine the distance from his or her wrist to the shoulder when the elbow is bent as far as possible. Suppose Jessica can bend her left arm so her wrist is 5 inches from her shoulder and her right arm so her right wrist is 3 inches from her shoulder. Which of Jessica's arms has the greater range of motion? Explain.**

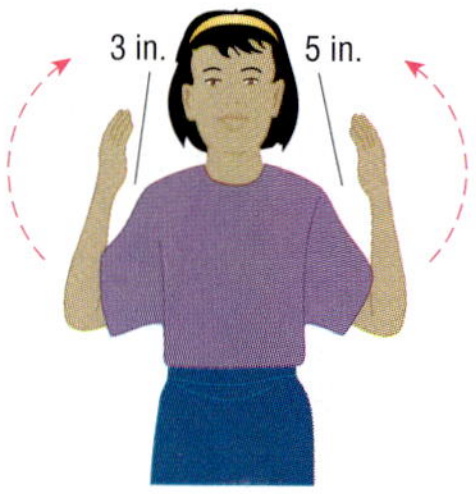

The distance between the wrist and shoulder is smaller on her right arm. Assuming that both her arms have the same measurements, the SSS inequality tells us that the angle formed at the elbow is smaller on the right arm. This means that the right arm has a greater range of motion.

Check for Understanding

Concept Check

1–2. See margin.

1. **OPEN ENDED** **Describe** a real-world object that illustrates either SAS or SSS inequality.
2. **Compare and contrast** the SSS Inequality Theorem to the SSS Postulate for triangle congruence.

Guided Practice

Write an inequality relating the given pair of angles or segment measures.

3. AB, CD **$AB < CD$**

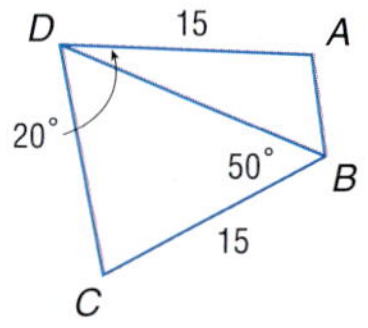

4. $m\angle PQS$, $m\angle RQS$ **$m\angle PQS > m\angle RQS$**

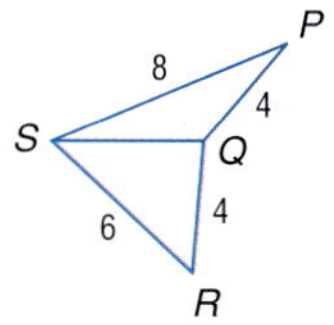

Write an inequality to describe the possible values of x.

5. **$\frac{7}{3} < x < 6$**

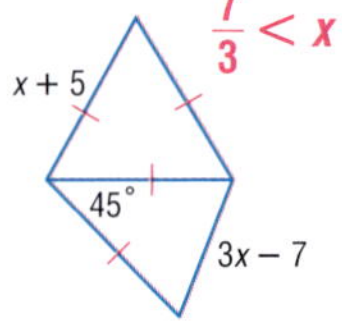

6. **$-\frac{4}{7} < x < \frac{136}{7}$**

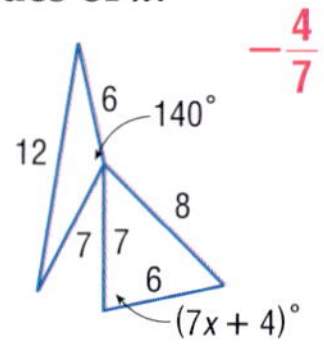

PROOF **Write a two-column proof.** **7–8. See p. 279E.**

GUIDED PRACTICE KEY	
Exercises	Examples
7, 8	1–2
3–6	3
9	4

7. **Given:** $\overline{PQ} \cong \overline{SQ}$
Prove: $PR > SR$

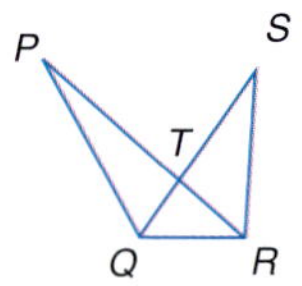

8. **Given:** $\overline{TU} \cong \overline{US}$
$\overline{US} \cong \overline{SV}$
Prove: $ST > UV$

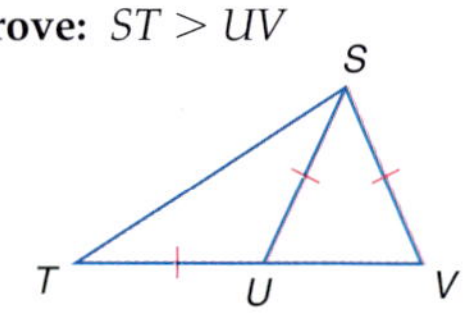

270 Chapter 5 Relationships in Triangles

Application

9. **TOOLS** A lever is used to multiply the force applied to an object. One example of a lever is a pair of pliers. Use the SAS or SSS Inequality to explain how to use a pair of pliers. **See margin.**

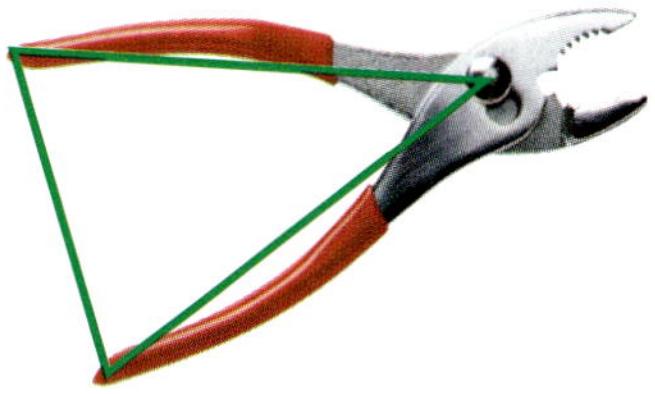

★ indicates increased difficulty

Practice and Apply

Homework Help	
For Exercises	See Examples
20–24	1–2
10–19	3
25, 26	4

Extra Practice
See page 764.

Write an inequality relating the given pair of angles or segment measures.

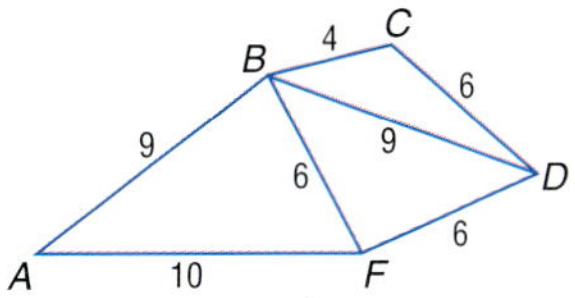

10. AB, FD **$AB > FD$**
11. $m\angle BDC$, $m\angle FDB$ **$m\angle BDC < m\angle FDB$**
12. $m\angle FBA$, $m\angle DBF$ **$m\angle FBA > m\angle DBF$**

Write an inequality relating the given pair of angles or segment measures.

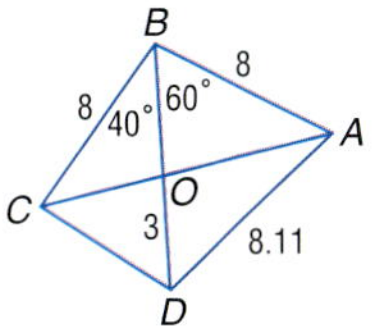

13. AD, DC **$AD > DC$**
14. OC, OA **$OC < OA$**
15. $m\angle AOD$, $m\angle AOB$ **$m\angle AOD > m\angle AOB$**

Write an inequality to describe the possible values of x.

16. **$x > 4$**

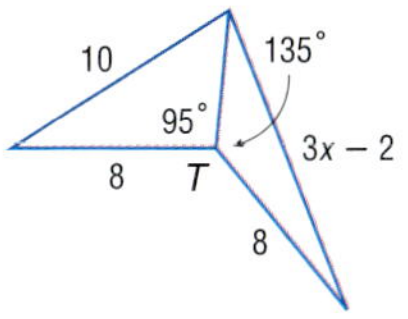

17. **$4 < x < 10$**

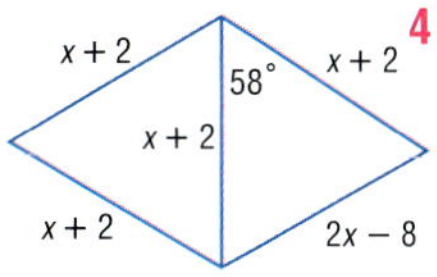

★ 18. In the figure, $\overline{AM} \cong \overline{MB}$, $AC > BC$, $m\angle 1 = 5x + 20$ and $m\angle 2 = 8x - 100$. Write an inequality to describe the possible values of x.
$12.5 < x < 40$

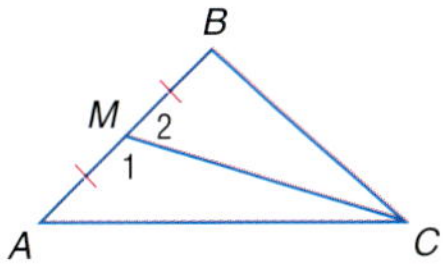

★ 19. In the figure, $m\angle RVS = 15 + 5x$, $m\angle SVT = 10x - 20$, $RS < ST$, and $\angle RTV \cong \angle TRV$. Write an inequality to describe the possible values of x. **$7 < x < 20$**

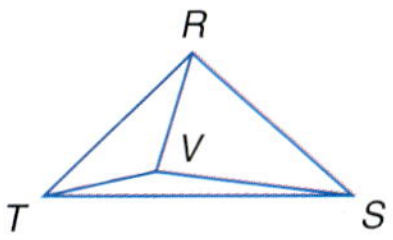

PROOF **Write a two-column proof.** **20–21. See p. 279E.**

20. **Given:** $\triangle ABC$
$\overline{AB} \cong \overline{CD}$
Prove: $BC > AD$

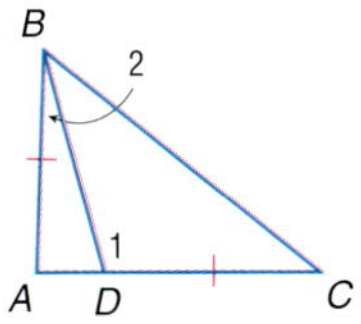

21. **Given:** $\overline{PQ} \cong \overline{RS}$
$QR < PS$
Prove: $m\angle 3 < m\angle 1$

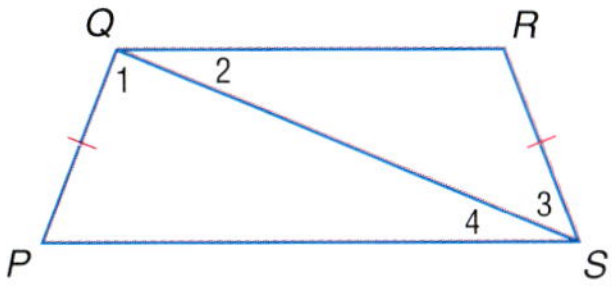

3 Practice/Apply

Study Notebook

Have students—
- add the definitions/examples of the vocabulary terms to their Vocabulary Builder worksheets for Chapter 5.
- include explanations and examples of the SAS Inequality Theorem and the SSS Inequality Theorem.
- include any other item(s) that they find helpful in mastering the skills in this lesson.

About the Exercises...

Organization by Objective
- **SAS Inequality:** 21–25
- **SSS Inequality:** 11–20, 26, 27

Odd/Even Assignments
Exercises 11–24 are structured so that students practice the same concepts whether they are assigned odd or even problems.

Assignment Guide

Basic: 11–17 odd, 21, 31–45
Average: 11–31 odd, 32–45
Advanced: 10–26, 27–44

Answer

9. Sample answer: The pliers are an example of the SAS inequality. As force is applied to the handles, the angle between them decreases causing the distance between them to decrease. As the distance between the ends of the pliers decreases, more force is applied to a smaller area.

Study Guide and Intervention, p. 269 (shown) and p. 270

SAS Inequality The following theorem involves the relationship between the sides of two triangles and an angle in each triangle.

SAS Inequality/Hinge Theorem	If two sides of a triangle are congruent to two sides of another triangle and the included angle in one triangle has a greater measure than the included angle in the other, then the third side of the first triangle is longer than the third side of the second triangle.

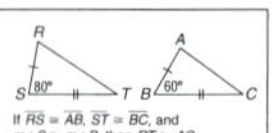

Example **Write an inequality relating the lengths of $\overline{CD}$ and $\overline{AD}$.**

Two sides of $\triangle BCD$ are congruent to two sides of $\triangle BAD$ and $m\angle CBD > m\angle ABD$. By the SAS Inequality/Hinge Theorem, $CD > AD$.

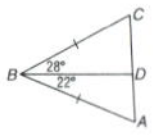

Exercises

Write an inequality relating the given pair of segment measures.

1. MR, RP
   $MR > RP$
2. AD, CD
   $AD > CD$
3. EG, HK
   $EG < HK$
4. MR, PR
   $MR > PR$

Write an inequality to describe the possible values of x.

5. $x > 12.5$
6. $x > 1.6$

Skills Practice, p. 271 and Practice, p. 272 (shown)

Write an inequality relating the given pair of angles or segment measures.

1. AB, BK
   $AB > BK$
2. ST, SR
   $ST > SR$
3. $m\angle CDF$, $m\angle EDF$
   $m\angle CDF < m\angle EDF$
4. $m\angle R$, $m\angle T$
   $m\angle R < m\angle T$
5. Write a two-column proof.
   Given: G is the midpoint of $\overline{DF}$.
   $m\angle 1 > m\angle 2$
   Prove: $ED > EF$
   Proof:

Statements	Reasons
1. G is the midpoint of $\overline{DF}$.	1. Given
2. $\overline{DG} \cong \overline{FG}$	2. Definition of midpoint
3. $\overline{EG} \cong \overline{EG}$	3. Reflexive Property
4. $m\angle 1 > m\angle 2$	4. Given
5. $ED > EF$	5. SAS Inequality

6. **TOOLS** Rebecca used a spring clamp to hold together a chair leg she repaired with wood glue. When she opened the clamp, she noticed that the angle between the handles of the clamp decreased as the distance between the handles of the clamp decreased. At the same time, the distance between the gripping ends of the clamp increased. When she released the handles, the distance between the gripping end of the clamp decreased and the distance between the handles increased. Is the clamp an example of the SAS or SSS Inequality?
   SAS Inequality

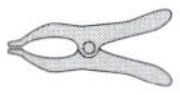

Reading to Learn Mathematics, p. 273 ELL

Pre-Activity **How does a backhoe work?**

Read the introduction to Lesson 5-5 at the top of page 267 in your textbook.

What is the main kind of task that backhoes are used to perform?
Backhoes are used mainly for digging.

Reading the Lesson

1. Refer to the figure. Write a conclusion that you can draw from the given information. Then name the theorem that justifies your conclusion.
   a. $\overline{LM} \cong \overline{OP}$, $\overline{MN} \cong \overline{PQ}$, and $LN > OQ$
   $m\angle M > m\angle P$; SSS Inequality Theorem
   b. $\overline{LM} \cong \overline{OP}$, $\overline{MN} \cong \overline{PQ}$, and $m\angle P < m\angle M$
   $OQ < LN$ (or $LN > OQ$); SAS Inequality Theorem (or Hinge Theorem)
   c. $LM = 8$, $LN = 15$, $OP = 8$, $OQ = 15$, $m\angle L = 22$, and $m\angle O = 21$
   $MN > PQ$; SAS Inequality Theorem (or Hinge Theorem)
2. In the figure, $\triangle EFG$ is isosceles with base $\overline{FG}$ and F is the midpoint of $\overline{DG}$. Determine whether each of the following is a valid conclusion that you can draw based on the given information. (Write *valid* or *invalid*.) If the conclusion is valid, identify the definition, property, postulate, or theorem that supports it.
   a. $\angle 3 \cong \angle 4$ valid; Isosceles Triangle Theorem
   b. $DF = GF$ valid; definition of midpoint
   c. $\triangle DEF$ is isosceles. invalid
   d. $m\angle 3 > m\angle 1$ valid; Exterior Angle Inequality Theorem
   e. $m\angle 2 > m\angle 4$ valid; Exterior Angle Inequality Theorem
   f. $m\angle 2 > m\angle 3$ valid; Substitution Property (using conclusions from parts a and e)
   g. $DE > EG$ valid; SAS Inequality Theorem (or Hinge Theorem)
   h. $DE > FG$ invalid

Helping You Remember

3. A good way to remember something is to think of it in concrete terms. How can you illustrate the Hinge Theorem with everyday objects? Sample answer: Put two pencils on a desktop so that the erasers touch. As you increase or decrease the measure of the angle formed by the pencils, the distance between the points of the pencils increases or decreases accordingly.

PROOF Write a two-column proof. 22–23. See pp. 279E–279F.

★22. **Given:** $\overline{PR} \cong \overline{PQ}$
$SQ > SR$
Prove: $m\angle 1 < m\angle 2$

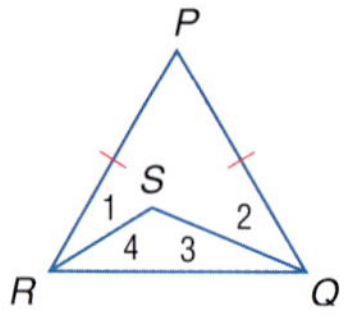

★23. **Given:** $\overline{ED} \cong \overline{DF}$
$m\angle 1 > m\angle 2$
D is the midpoint of $\overline{CB}$.
$\overline{AE} \cong \overline{AF}$
Prove: $AC > AB$

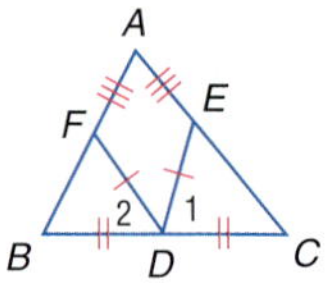

24. **PROOF** Use an indirect proof to prove the SSS Inequality Theorem (Theorem 5.14).
Given: $\overline{RS} \cong \overline{UW}$
$\overline{ST} \cong \overline{WV}$
$RT > UV$
Prove: $m\angle S > m\angle W$ See p. 279F.

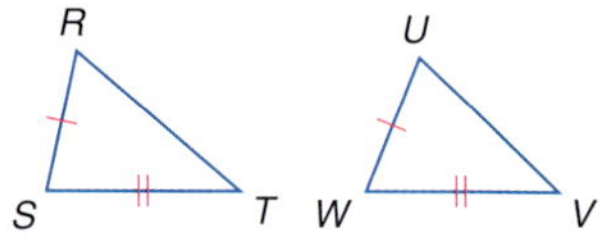

25. **DOORS** Open a door slightly. With the door open, measure the angle made by the door and the door frame. Measure the distance from the end of the door to the door frame. Open the door wider, and measure again. How do the measures compare?

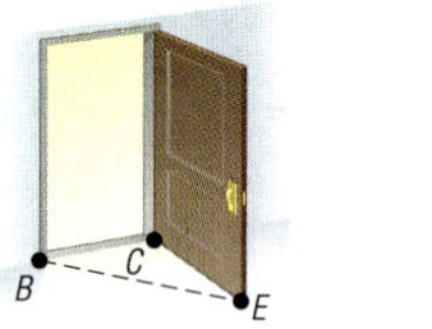

25. As the door is opened wider, the angle formed increases and the distance from the end of the door to the door frame increases.

26. **LANDSCAPING** When landscapers plant new trees, they usually brace the tree using a stake tied to the trunk of the tree. Use the SAS or SSS Inequality to explain why this is an effective method for supporting a newly planted tree. See margin.

Career Choices

Landscape Architect

Landscape architects design the settings of buildings and parklands by arranging both the location of the buildings and the placement of plant life so the site is functional, beautiful, and environmentally friendly.

Online Research
For information about a career as a landscape architect, visit:
www.geometryonline.com/careers

27. **CRITICAL THINKING** The SAS Inequality states that the base of an isosceles triangle gets longer as the measure of the vertex angle increases. Describe the effect of changing the measure of the vertex angle on the measure of the altitude. See margin.

BIOLOGY **For Exercises 28–30, use the following information.**

The velocity of a person walking or running can be estimated using the formula $v = \frac{0.78s^{1.67}}{h^{1.17}}$, where v is the velocity of the person in meters per second, s is the length of the stride in meters, and h is the height of the hip in meters.

28. Find the velocities of two people that each have a hip height of 0.85 meters and whose strides are 1.0 meter and 1.2 meters. 0.94 m/s, 1.28 m/s
29. Copy and complete the table at the right for a person whose hip height is 1.1 meters.
30. Discuss how the stride length is related to either the SAS Inequality of the SSS Inequality. As the length of the stride increases, the angle formed at the hip increases.

Stride (m)	Velocity (m/s)
0.25	0.07
0.50	0.22
0.75	0.43
1.00	0.70
1.25	1.01
1.50	1.37

Enrichment, p. 274

Drawing a Diagram

It is useful and often necessary to draw a diagram of the situation being described in a problem. The visualization of the problem is helpful in the process of problem solving.

Example **The roads connecting the towns of Kings, Chana, and Holcomb form a triangle. Davis Junction is located in the interior of this triangle. The distances from Davis Junction to Kings, Chana, and Holcomb are 3 km, 4 km, and 5 km, respectively. Jane begins at Holcomb and drives directly to Chana, then to Kings, and then back to Holcomb. At the end of her trip, she figures she has traveled 25 km altogether. Has she figured the distance correctly?**

To solve this problem, a diagram can be drawn. Based on this diagram and the Triangle Inequality Theorem, the distance from Holcomb to Chana is less than 9 km. Similarly, the distance from Chana to Kings is less than 7 km, and the distance fr[illegible]gs to Holcomb is l[illegible]than 8 km.

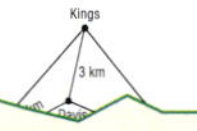

31. WRITING IN MATH Answer the question that was posed at the beginning of the lesson. See margin.

How does a backhoe work?

Include the following in your answer:

- a description of the angle between the arms as the backhoe operator digs, and
- an explanation of how the distance between the ends of the arms is related to the angle between them.

FCAT Practice
Standardized Test Practice
Ⓐ Ⓑ Ⓒ Ⓓ

32. If $\overline{DC}$ is a median of $\triangle ABC$ and $m\angle 1 > m\angle 2$, which of the following statements is *not* true? B

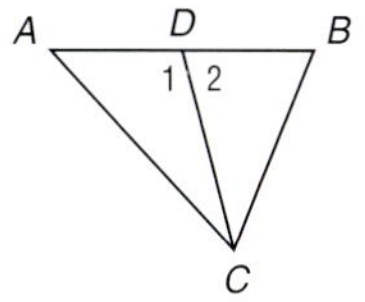

Ⓐ $AD = BD$ Ⓑ $m\angle ADC = m\angle BDC$

Ⓒ $AC > BC$ Ⓓ $m\angle 1 > m\angle B$

33. **ALGEBRA** A student bought four college textbooks that cost \$99.50, \$88.95, \$95.90, and \$102.45. She paid one half of the total amount herself and borrowed the rest from her mother. She repaid her mother in 4 equal monthly payments. How much was each of the monthly payments? B

Ⓐ \$24.18 Ⓑ \$48.35 Ⓒ \$96.70 Ⓓ \$193.40

Maintain Your Skills

Mixed Review

Determine whether the given measures can be the lengths of the sides of a triangle. Write *yes* or *no*. Explain. *(Lesson 5-4)*

34. 25, 1, 21 no; $1 + 21 \not> 25$

35. 16, 6, 19 yes; $16 + 6 > 19$

36. 8, 7, 15 no; $8 + 7 \not> 15$

Write the assumption you would make to start an indirect proof of each statement. *(Lesson 5-3)*

37. $\overline{AD}$ is a median of $\triangle ABC$. $\overline{AD}$ is not a median of $\triangle ABC$.

38. If two altitudes of a triangle are congruent, then the triangle is isosceles. The triangle is not isosceles.

Write a proof. *(Lesson 4-5)* 39–40. See p. 279F.

39. **Given:** $\overline{AD}$ bisects $\overline{BE}$.
$\overline{AB} \parallel \overline{DE}$
Prove: $\triangle ABC \cong \triangle DEC$

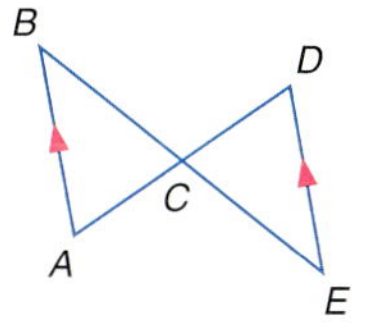

40. **Given:** $\overline{OM}$ bisects $\angle LMN$.
$\overline{LM} \cong \overline{MN}$
Prove: $\triangle MOL \cong \triangle MON$

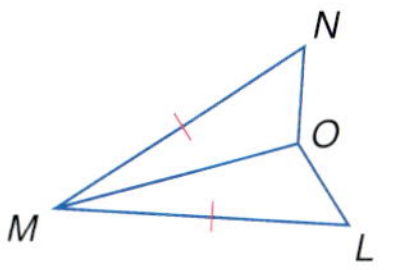

Find the measures of the sides of $\triangle EFG$ with the given vertices and classify each triangle by its sides. *(Lesson 4-1)* 41–44. See margin.

41. $E(4, 6)$, $F(4, 11)$, $G(9, 6)$

42. $E(-7, 10)$, $F(15, 0)$, $G(-2, -1)$

43. $E(16, 14)$, $F(7, 6)$, $G(-5, -14)$

44. $E(9, 9)$, $F(12, 14)$, $G(14, 6)$

45. **ADVERTISING** An ad for Wildflowers Gift Boutique says *When it has to be special, it has to be Wildflowers.* Catalina needs a special gift. Does it follow that she should go to Wildflowers? Explain. *(Lesson 2-4)* yes, by the Law of Detachment

4 Assess

Open-Ended Assessment

Speaking Select or create examples of proofs using the SAS Inequality Theorem and the SSS Inequality Theorem. For each example, allow students to volunteer and discuss different methods of using the given information to prove the conclusion. Then call on different students to give statements and reasons in the order necessary to advance the proof.

Assessment Options

Quiz (Lesson 5-5) is available on p. 290 of the *Chapter 5 Resource Masters.*

Answers

26. By the SAS Inequality Theorem, if the tree started to lean, one of the angles of the triangle formed by the tree, the ground, and the stake would change, and the side opposite that angle would change as well. However, with the stake in the ground and fixed to the tree, none of the sides of the triangle can change length. Thus, none of the angles can change. This ensures that the tree will stay straight.

27. As the vertex angle increases, the base angles decrease. Thus, as the base angles decrease, the altitude of the triangle decreases.

31. Sample answer: A backhoe digs when the angle between the two arms decreases and the shovel moves through the dirt. Answers should include the following.
- As the operator digs, the angle between the arms decreases.
- The distance between the ends of the arms increases as the angle between the arms increases, and decreases as the angle decreases.

41. $EF = 5$, $FG = \sqrt{50}$, $EG = 5$; isosceles

42. $EF = \sqrt{584}$, $FG = \sqrt{290}$, $EG = \sqrt{146}$; scalene

43. $EF = \sqrt{145}$, $FG = \sqrt{544}$, $EG = 35$; scalene

44. $EF = \sqrt{34}$, $FG = \sqrt{68}$, $EG = \sqrt{34}$; isosceles

Chapter 5 Study Guide and Review

Vocabulary and Concept Check

- This alphabetical list of vocabulary terms in Chapter 5 includes a page reference where each term was introduced.
- **Assessment** A vocabulary test/review for Chapter 5 is available on p. 288 of the *Chapter 5 Resource Masters*.

Lesson-by-Lesson Review

For each lesson,

- the main ideas are summarized,
- additional examples review concepts, and
- practice exercises are provided.

Vocabulary PuzzleMaker

ELL The Vocabulary PuzzleMaker software improves students' mathematics vocabulary using four puzzle formats—crossword, scramble, word search using a word list, and word search using clues. Students can work on a computer screen or from a printed handout.

MindJogger Videoquizzes

ELL MindJogger Videoquizzes provide an alternative review of concepts presented in this chapter. Students work in teams in a game show format to gain points for correct answers. The questions are presented in three rounds.

Round 1 Concepts (5 questions)
Round 2 Skills (4 questions)
Round 3 Problem Solving (4 questions)

Chapter 5 Study Guide and Review

Vocabulary and Concept Check

altitude (p. 241)	incenter (p. 240)	orthocenter (p. 241)
centroid (p. 240)	indirect proof (p. 255)	perpendicular bisector (p. 238)
circumcenter (p. 238)	indirect reasoning (p. 255)	point of concurrency (p. 238)
concurrent lines (p. 238)	median (p. 240)	proof by contradiction (p. 255)

For a complete list of postulates and theorems, see pages R1–R8.

Exercises **Choose the correct term to complete each sentence.**

1. All of the angle bisectors of a triangle meet at the (*incenter*, *circumcenter*).
2. In $\triangle RST$, if point P is the midpoint of $\overline{RS}$, then $\overline{PT}$ is a(n) (*angle bisector*, *median*).
3. The theorem that the sum of the lengths of two sides of a triangle is greater than the length of the third side is the (*Triangle Inequality Theorem*, *SSS Inequality*).
4. The three medians of a triangle intersect at the (*centroid*, *orthocenter*).
5. In $\triangle JKL$, if point H is equidistant from $\overrightarrow{KJ}$ and $\overrightarrow{KL}$, then $\overrightarrow{HK}$ is an (*angle bisector*, *altitude*).
6. The circumcenter of a triangle is the point where all three (*perpendicular bisectors*, *medians*) of the sides of the triangle intersect.
7. In $\triangle ABC$, if $\overleftrightarrow{AK} \perp \overleftrightarrow{BC}$, $\overleftrightarrow{BK} \perp \overleftrightarrow{AC}$, and $\overleftrightarrow{CK} \perp \overleftrightarrow{AB}$, then K is the (*orthocenter*, *incenter*) of $\triangle ABC$.

Lesson-by-Lesson Review

5-1 Bisectors, Medians, and Altitudes

See pages 238–245.

Concept Summary

- The perpendicular bisectors, angle bisectors, medians, and altitudes of a triangle are all special segments in triangles.

Example **Points P, Q, and R are the midpoints of $\overline{JK}$, $\overline{KL}$, and $\overline{JL}$, respectively. Find x.**

$KD = \frac{2}{3}(KR)$	Centroid Theorem
$6x + 23 = \frac{2}{3}(6x + 51)$	Substitution
$6x + 23 = 4x + 34$	Simplify.
$2x = 11$	Subtract $4x + 23$ from each side.
$x = \frac{11}{2}$	Divide each side by 2.

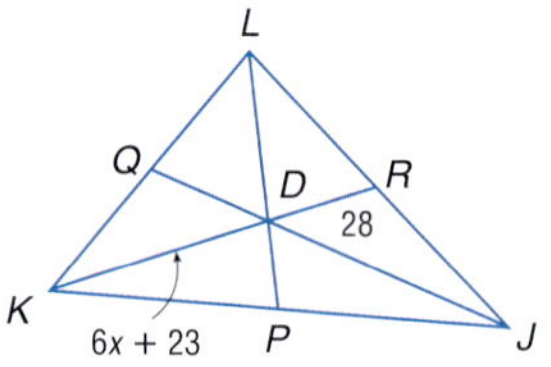

Exercises **In the figure, $\overline{CP}$ is an altitude, $\overline{CQ}$ is the angle bisector of $\angle ACB$, and R is the midpoint of $\overline{AB}$.**

See Example 2 on pages 240 and 241.

8. Find $m\angle ACQ$ if $m\angle ACB = 123 - x$ and $m\angle QCB = 42 + x$. **55**
9. Find AB if $AR = 3x + 6$ and $RB = 5x - 14$. **72**
10. Find x if $m\angle APC = 72 + x$. **18**

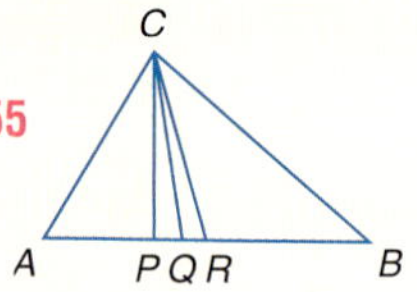

www.geometryonline.com/vocabulary_review

FOLDABLES™ Study Organizer

For more information about Foldables, see *Teaching Mathematics with Foldables.*

Have students look through the chapter to make sure they have included notes and examples in their Foldables for each lesson of Chapter 5.

Encourage students to refer to their Foldables while completing the Study Guide and Review and to use them in preparing for the Chapter Test.

274 Chapter 5 Relationships in Triangles

5-2 Inequalities and Triangles

See pages 247–254.

Concept Summary

- The largest angle in a triangle is opposite the longest side, and the smallest angle is opposite the shortest side.

Example **Use the Exterior Angle Theorem to list all angles with measures less than $m\angle 1$.**

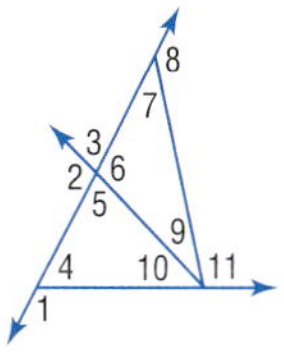

By the Exterior Angle Theorem, $m\angle 5 < m\angle 1$, $m\angle 10 < m\angle 1$, $m\angle 7 < m\angle 1$, and $m\angle 9 + m\angle 10 < m\angle 1$. Thus, the measures of $\angle 5$, $\angle 10$, $\angle 7$, and $\angle 9$ are all less than $m\angle 1$.

Exercises **Determine the relationship between the measures of the given angles.** *See Example 3 on page 250.*

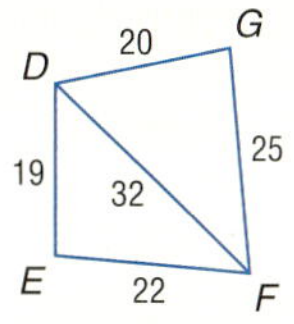

11. $\angle DEF$ and $\angle DFE$ $m\angle DEF > m\angle DFE$
12. $\angle GDF$ and $\angle DGF$ $m\angle GDF > m\angle DGF$
13. $\angle DEF$ and $\angle FDE$ $m\angle DEF > m\angle FDE$

Determine the relationship between the lengths of the given sides. *See Example 4 on page 250.*

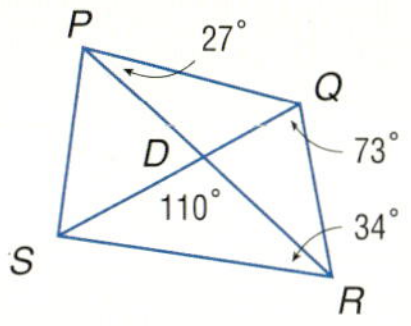

14. $\overline{SR}, \overline{SD}$ $SR > SD$
15. $\overline{DQ}, \overline{DR}$ $DQ < DR$
16. $\overline{PQ}, \overline{QR}$ $PQ > QR$
17. $\overline{SR}, \overline{SQ}$ $SR > SQ$

5-3 Indirect Proof

See pages 255–260.

Concept Summary

- In an indirect proof, the conclusion is assumed to be false, and a contradiction is reached.

Example **State the assumption you would make to start an indirect proof of the statement $AB < AC + BC$.**

If AB is not less than $AC + BC$, then either $AB > AC + BC$ or $AB = AC + BC$. In other words, $AB \geq AC + BC$.

Exercises **State the assumption you would make to start an indirect proof of each statement.** *See Example 1 on page 255.*

18. $\sqrt{2}$ is an irrational number. $\sqrt{2}$ is a rational number.
19. If two sides and the included angle are congruent in two triangles, then the triangles are congruent. The triangles are not congruent.
20. **FOOTBALL** Miguel plays quarterback for his high school team. This year, he completed 101 passes in the five games in which he played. Prove that, in at least one game, Miguel completed more than 20 passes. See margin.

Answers

20. Assume that Miguel completed at most 20 passes in each of the five games in which he played. If we let p_1, p_2, p_3, p_4, and p_5 be the number of passes Miguel completed in games 1, 2, 3, 4, and 5, respectively, then $p_1 + p_2 + p_3 + p_4 + p_5 =$ the total number of passes Miguel completed $= 101$.

Because we have assumed that he completed at most 20 passes in each of the five games, $p_1 \leq 20$ *and* $p_2 \leq 20$ *and* $p_3 \leq 20$ *and* $p_4 \leq 20$ *and* $p_5 \leq 20$.

Then, by a property of inequalities, $p_1 + p_2 + p_3 + p_4 + p_5 \leq 20 + 20 + 20 + 20 + 20$ or 100 passes.

But this says that Miguel completed at most 100 passes this season, which contradicts the information we were given, that he completed 101 passes. So our assumption must be false. Thus, Miguel completed more than 20 passes in at least one game this season.

Chapter 5 Study Guide and Review 275

• Extra Practice, see pages 763 and 764.
• Mixed Problem Solving, see page 786.

Answers (page 277)

12. Assume that Marcus spent less than one half hour on a teleconference every day. If we let t_1, t_2, and t_3 be the time spent on a teleconference on days 1, 2, and 3, respectively, then $t_1 + t_2 + t_3$ = the total amount of time over the three days spent on the teleconference.

Because he spent less than a half hour every day on a teleconference, $t_1 < 0.5$ *and* $t_2 < 0.5$ *and* $t_3 < 0.5$.

Then, by a property of inequalities, $t_1 + t_2 + t_3 < 0.5 + 0.5 + 0.5$ or 1.5 hours.

But this says that Marcus spent less than one and one-half hours on a teleconference over the three days, which contradicts the information we were given. So we must abandon our assumption. Thus, Marcus spent at least one half-hour on a teleconference, on at least one of the three days.

5-4 The Triangle Inequality

See pages 261–266.

Concept Summary

- The sum of the lengths of any two sides of a triangle is greater than the length of the third side.

Example **Determine whether 7, 6, and 14 can be the measures of the sides of a triangle.**

Check each inequality.

$7 + 6 \overset{?}{>} 14$ $\quad$ $7 + 14 \overset{?}{>} 6$ $\quad$ $6 + 14 \overset{?}{>} 7$

$13 \not> 14$ $\quad$ $21 > 6$ ✓ $\quad$ $20 > 7$ ✓

Because the inequalities are not true in all cases, the sides cannot form a triangle.

Exercises **Determine whether the given measures can be the lengths of the sides of a triangle. Write *yes* or *no*. Explain.** *See Example 1 on page 261.*

21. 7, 20, 5 **no; $7 + 5 \not> 20$**
22. 16, 20, 5 **yes; $5 + 16 > 20$**
23. 18, 20, 6 **yes; $6 + 18 > 20$**

5-5 Inequalities Involving Two Triangles

See pages 267–273.

Concept Summary

- SAS Inequality: In two triangles, if two sides are congruent, then the measure of the included angle determines which triangle has the longer third side.
- SSS Inequality: In two triangles, if two sides are congruent, then the length of the third side determines which triangle has the included angle with the greater measure.

Example **Write an inequality relating LM and MN.**

In $\triangle LMP$ and $\triangle NMP$, $\overline{LP} \cong \overline{NP}$, $\overline{PM} \cong \overline{PM}$, and $m\angle LPM > m\angle NPM$. The SAS Inequality allows us to conclude that $LM > MN$.

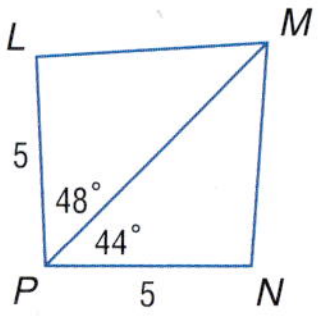

Exercises **Write an inequality relating the given pair of angles or segment measures.** *See Example 3 on page 269.*

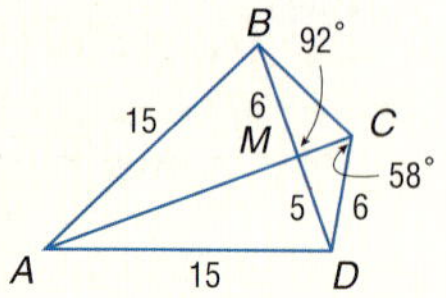

24. $m\angle BAC$ and $m\angle DAC$ **$m\angle BAC > m\angle DAC$**

25. BC and MD **$BC > MD$**

Write an inequality to describe the possible values of x.
See Example 3 on page 269.

26. **$-20 < x < 21$**

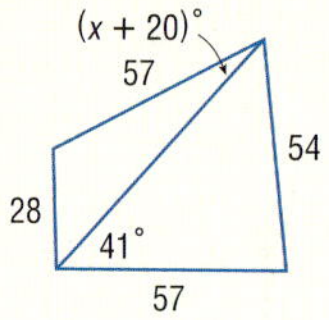

27. **$x > 7$**

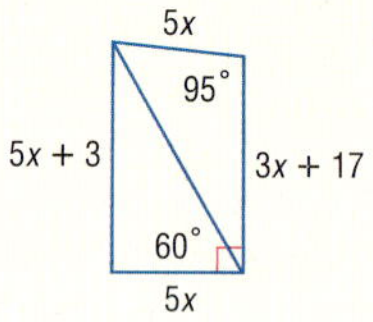

276 Chapter 5 Relationships in Triangles

Chapter 5 Practice Test

Vocabulary and Concepts

Choose the letter that best matches each description.

1. point of concurrency of the angle bisectors of a triangle **b**
2. point of concurrency of the altitudes of a triangle **c**
3. point of concurrency of the perpendicular bisectors of a triangle **a**

a. circumcenter
b. incenter
c. orthocenter
d. centroid

Skills and Applications

In $\triangle GHJ$, $HP = 5x - 16$, $PJ = 3x + 8$, $m\angle GJN = 6y - 3$, $m\angle NJH = 4y + 23$, and $m\angle HMG = 4z + 14$.

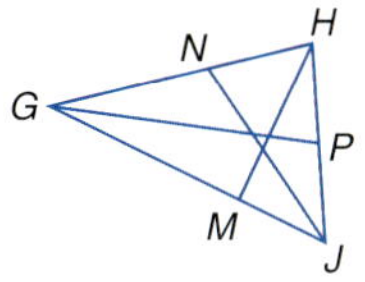

4. $\overline{GP}$ is a median of $\triangle GHJ$. Find HJ. **88**
5. Find $m\angle GJH$ if $\overline{JN}$ is an angle bisector. **150**
6. If $\overline{HM}$ is an altitude of $\angle GHJ$, find the value of z. **19**

Refer to the figure at the right. Determine which angle has the greatest measure.

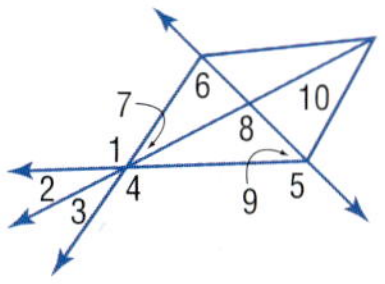

7. $\angle 8, \angle 5, \angle 7$ **$\angle 5$**
8. $\angle 6, \angle 7, \angle 8$ **$\angle 8$**
9. $\angle 1, \angle 6, \angle 9$ **$\angle 1$**

Write the assumption you would make to start an indirect proof of each statement.

10. If n is a natural number, then $2^n + 1$ is odd. **$2^n + 1$ is even.**
11. Alternate interior angles are congruent. **Alternate interior angles are not congruent.**

12. **BUSINESS** Over the course of three days, Marcus spent one and one-half hours on a teleconference for his marketing firm. Use indirect reasoning to show that, on at least one day, Marcus spent at least one half-hour on a teleconference. **See margin.**

Find the range for the measure of the third side of a triangle given the measures of two sides.

13. 1 and 14 **$13 < n < 15$**
14. 14 and 11 **$3 < n < 25$**
15. 13 and 19 **$6 < n < 32$**

Write an inequality for the possible values of x.

16. **$0 < x < 7$**

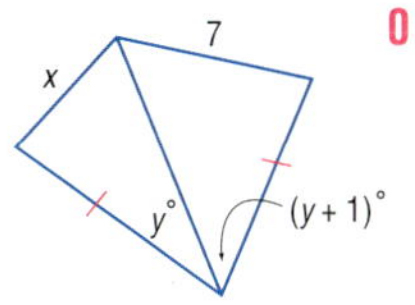

17. **$4 < x < 9$**

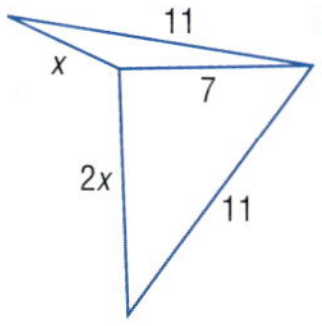

18.

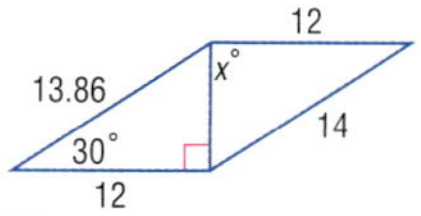

$90 < x < 180$

19. **GEOGRAPHY** The distance between Atlanta and Cleveland is about 554 miles. The distance between Cleveland and New York City is about 399 miles. Use the Triangle Inequality Theorem to find the possible values of the distance between New York and Atlanta. **155 mi $< d <$ 953 mi**

20. **STANDARDIZED TEST PRACTICE** A given triangle has two sides with measures 8 and 11. Which of the following is *not* a possible measure for the third side? **A**

Ⓐ 3 Ⓑ 7 Ⓒ 12 Ⓓ 18

www.geometryonline.com/chapter_test/fcat

Chapter 5 Practice Test

Assessment Options

Vocabulary Test A vocabulary test/review for Chapter 5 can be found on p. 288 of the *Chapter 5 Resource Masters.*

Chapter Tests There are six Chapter 5 Tests and an Open-Ended Assessment task available in the *Chapter 5 Resource Masters.*

Chapter 5 Tests

Form	Type	Level	Pages
1	MC	basic	275–276
2A	MC	average	277–278
2B	MC	average	279–280
2C	FR	average	281–282
2D	FR	average	283–284
3	FR	advanced	285–286

MC = multiple-choice questions
FR = free-response questions

Open-Ended Assessment
Performance tasks for Chapter 5 can be found on p. 287 of the *Chapter 5 Resource Masters.* A sample scoring rubric for these tasks appears on p. A22.

TestCheck and Worksheet Builder

This **networkable software** has three modules for assessment.

- **Worksheet Builder** to make worksheets and tests.
- **Student Module** to take tests on-screen.
- **Management System** to keep student records.

Portfolio Suggestion

Introduction When a chapter presents material that is very closely linked, it can help for students to organize each concept and understand it separately.

Ask Students Make a chart listing all the important concepts presented in this chapter with a short explanation of each item in your own words. Have students add the chart and explanations to their portfolios.

Chapter 5 Standardized Test Practice

These two pages contain practice questions in the various formats that can be found on the most frequently given standardized tests.

A practice answer sheet for these two pages can be found on p. A1 of the *Chapter 5 Resource Masters*.

Standardized Test Practice Student Recording Sheet, p. A1

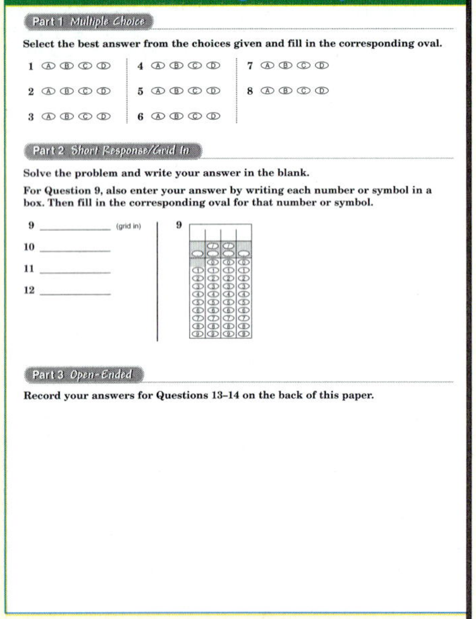
Part 1 Multiple Choice

Select the best answer from the choices given and fill in the corresponding oval.

1 Ⓐ Ⓑ Ⓒ Ⓓ 4 Ⓐ Ⓑ Ⓒ Ⓓ 7 Ⓐ Ⓑ Ⓒ Ⓓ

2 Ⓐ Ⓑ Ⓒ Ⓓ 5 Ⓐ Ⓑ Ⓒ Ⓓ 8 Ⓐ Ⓑ Ⓒ Ⓓ

3 Ⓐ Ⓑ Ⓒ Ⓓ 6 Ⓐ Ⓑ Ⓒ Ⓓ

Part 2 Short Response/Grid In

Solve the problem and write your answer in the blank.

For Question 9, also enter your answer by writing each number or symbol in a box. Then fill in the corresponding oval for that number or symbol.

9 ________ (grid in) 9

10 ________

11 ________

12 ________

Part 3 Open-Ended

Record your answers for Questions 13–14 on the back of this paper.

Additional Practice

See pp. 293–294 in the *Chapter 5 Resource Masters* for additional standardized test practice.

Chapter 5 Standardized Test Practice

FCAT Practice

Part 1 Multiple Choice

Record your answers on the answer sheet provided by your teacher or on a sheet of paper.

1. Tamara works at a rug and tile store after school. The ultra-plush carpet has 80 yarn fibers per square inch. How many fibers are in a square yard? (Prerequisite Skill) D

Ⓐ 2,880 Ⓑ 8,640

Ⓒ 34,560 Ⓓ 103,680

2. What is the perimeter of the figure? (Lesson 1-4) B

Ⓐ 20 units

Ⓑ 46 units

Ⓒ 90 units

Ⓓ 132 units

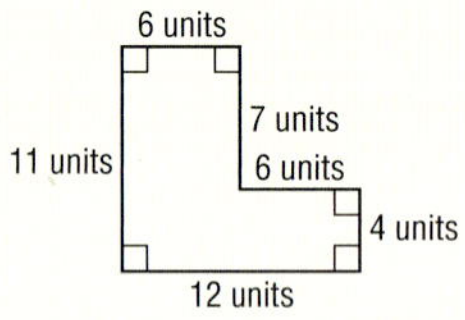

3. Which is a correct statement about the conditional and its converse below? (Lesson 2-2) B

Statement: If the measure of an angle is 50°, then the angle is an acute angle.

Converse: If an angle is an acute angle, then the measure of the angle is 50°.

Ⓐ The statement and its converse are both true.

Ⓑ The statement is true, but its converse is false.

Ⓒ The statement and its converse are both false.

Ⓓ The statement is false, but its converse is true.

4. Six people attend a meeting. When the meeting is over, each person exchanges business cards with each of the other people. Use noncollinear points to determine how many exchanges are made. (Lesson 2-3) B

Ⓐ 6 Ⓑ 15 Ⓒ 36 Ⓓ 60

For Questions 5 and 6, refer to the figure below.

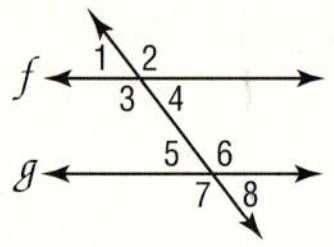

5. What is the term used to describe $\angle 4$ and $\angle 5$? (Lesson 3-1) B

Ⓐ alternate exterior angles

Ⓑ alternate interior angles

Ⓒ consecutive interior angles

Ⓓ corresponding angles

6. Given that lines f and g are not parallel, what assumption can be made to prove that $\angle 3$ is not congruent to $\angle 7$? (Lesson 5-2) B

Ⓐ $f \parallel g$ Ⓑ $\angle 3 \cong \angle 7$

Ⓒ $\angle 3 \cong \angle 2$ Ⓓ $m\angle 3 \cong m\angle 7$

7. $\overline{QT}$ is a median of $\triangle PST$, and $\overline{RT}$ is an altitude of $\triangle PST$. Which of the following line segments is shortest? (Lesson 5-4) C

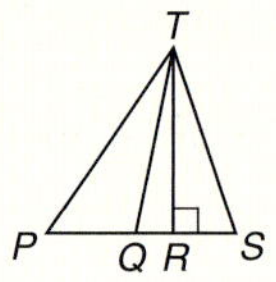

Ⓐ $\overline{PT}$ Ⓑ $\overline{QT}$ Ⓒ $\overline{RT}$ Ⓓ $\overline{ST}$

8. A paleontologist found the tracks of an animal that is now extinct. Which of the following lengths could be the measures of the three sides of the triangle formed by the tracks? (Lesson 5-4) A

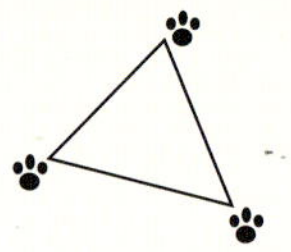

Ⓐ 2, 9, 10 Ⓑ 5, 8, 13

Ⓒ 7, 11, 20 Ⓓ 9, 13, 26

The Princeton Review ***Log On*** **for Test Practice** The Princeton Review offers additional test-taking tips and practice problems at their web site. Visit www.princetonreview.com or www.review.com

TestCheck and Worksheet Builder

Special banks of standardized test questions similar to those on the SAT, ACT, TIMSS 8, NAEP 8, and state proficiency tests can be found on this CD-ROM.

Aligned and verified by The Princeton Review

FCAT Practice

Part 2 Short Response/Grid In

Record your answers on the answer sheet provided by your teacher or on a sheet of paper.

9. The top of an access ramp to a building is 2 feet higher than the lower end of the ramp. If the lower end is 24 feet from the building, what is the slope of the ramp? (Lesson 3-3) $\frac{1}{12}$

10. The ramps at a local skate park vary in steepness. Find x. (Lesson 4-2) **35**

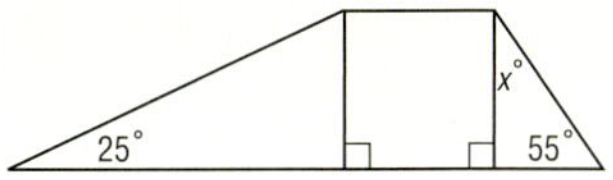

For Questions 11 and 12, refer to the graph below.

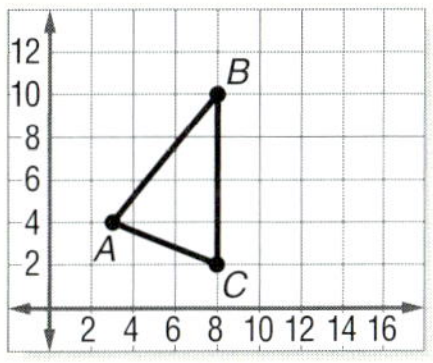

11. During a soccer game, a player stands near the goal at point A. The goalposts are located at points B and C. The goalkeeper stands at point P on the goal line $\overline{BC}$ so that $\overline{AP}$ forms a median. What is the location of the goalkeeper? (Lesson 5-1) **(8, 6)**

12. A defender positions herself on the goal line $\overline{BC}$ at point T to assist the goalkeeper. If $\overline{AT}$ forms an altitude of $\triangle ABC$, what is the location of defender T? (Lesson 5-1) **(8, 4)**

13. What postulate or theorem could you use to prove that the measure of $\angle QRT$ is greater than the measure of $\angle SRT$? (Lesson 5-5)
SSS Inequality

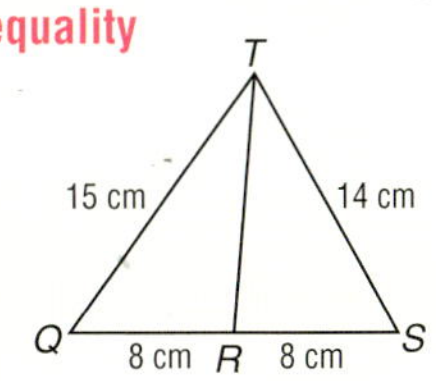

The Princeton Review Test-Taking Tip

Questions 7, 10, and 11 Review any terms that you have learned before you take a test. Remember that a median is a segment that connects a vertex of a triangle to the midpoint of the opposite side. An altitude is a perpendicular segment from a vertex to the opposite side.

FCAT Practice

Part 3 Open Ended

Record your answers on a sheet of paper. Show your work.

14. Kendell is purchasing a new car stereo for \$200. He agreed to pay the same amount each month until the \$200 is paid. Kendell made the graph below to help him figure out when the amount will be paid.
(Lesson 3-3) **14a–b. See margin.**

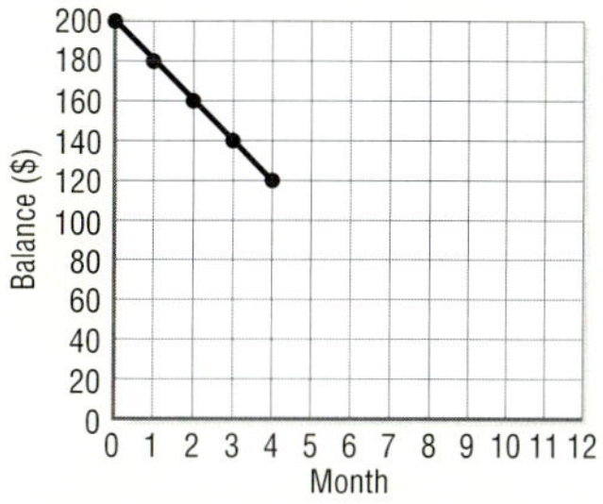

a. Use the slope of the line to write an argument that the line intersects the x-axis at (10, 0).

b. What does the point (10, 0) represent?

15. The vertices of $\triangle ABC$ are $A(-3, 1)$, $B(0, -2)$, and $C(3, 4)$. **15a–e. See margin.**

a. Graph $\triangle ABC$. (Prerequisite Skill)

b. Use the Distance Formula to find the length of each side to the nearest tenth. (Lesson 1-3)

c. What type of triangle is $\triangle ABC$? How do you know? (Lesson 4-1)

d. Prove $\angle A \cong \angle B$. (Lesson 4-6)

e. Prove $m\angle A > m\angle C$. (Lesson 5-3)

www.geometryonline.com/standardized_test/fcat

Evaluating Open-Ended Assessment Questions

Open-Ended Assessment questions are graded by using a multilevel rubric that guides you in assessing a student's knowledge of a particular concept.

Goal: Use the graph of a line and slope to determine when a product will be paid in full.

Sample Scoring Rubric: The following rubric is a sample scoring device. You may wish to add more detail to this sample to meet your individual scoring needs.

Score	Criteria
4	A correct solution that is supported by well-developed, accurate explanations
3	A generally correct solution, but may contain minor flaws in reasoning or computation
2	A partially correct interpretation and/or solution to the problem
1	A correct solution with no supporting evidence or explanation
0	An incorrect solution indicating no mathematical understanding of the concept or task, or no solution is given

Answers

14a. From the points (0, 200) and (4, 120), the slope of the line is $\frac{200-120}{0-4} = -20$. Find slope again using one of the given points and (10, 0); the slope is $\frac{120-0}{4-10} = -20$. Since the slope is the same, (10, 0) must be on the original line. Students may check by drawing an extension of the line and will see that it goes through (10, 0).

14b. The point (10, 0) shows that on the tenth payment Kendell's balance will be \$0, so the amount will be paid in full.

15a.

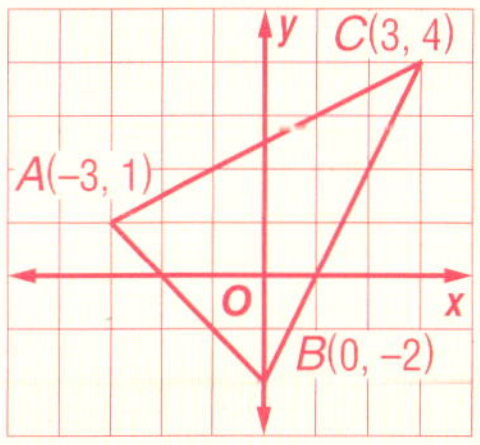

15b. $AB = 4.2$; $BC = 6.7$; $AC = 6.7$

15c. isosceles triangle because $\overline{BC}$ is congruent to $\overline{AC}$

15d. According to the Isosceles Triangle Theorem, if two sides of a triangle are congruent, then the angles opposite those sides are congruent.

15e. If one side of a triangle is longer than another side, the angle opposite the longer side has a greater measure than the angle opposite the shorter side. Since $\overline{BC}$ is longer than $\overline{AB}$, $m\angle A > m\angle C$.

Pages 242–245, Lesson 5-1

5. Given: $\overline{XY} \cong \overline{XZ}$
$\overline{YM}$ and $\overline{ZN}$ are medians.
Prove: $\overline{YM} \cong \overline{ZN}$

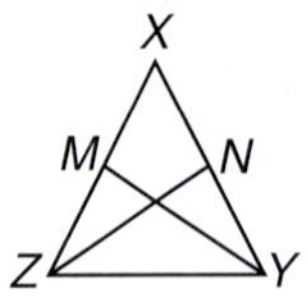

Proof:
Statements (Reasons)

1. $\overline{XY} \cong \overline{XZ}$, $\overline{YM}$ and $\overline{ZN}$ are medians. (Given)
2. M is the midpoint of $\overline{XZ}$. N is the midpoint of $\overline{XY}$. (Def. of median)
3. $XY = XZ$ (Def. of $\cong$)
4. $\overline{XM} \cong \overline{MZ}$, $\overline{XN} \cong \overline{NY}$ (Midpoint Theorem)
5. $XM = MZ$, $XN = NY$ (Def. of $\cong$)
6. $XM + MZ = XZ$, $XN + NY = XY$ (Segment Addition Postulate)
7. $XM + MZ = XN + NY$ (Substitution)
8. $MZ + MZ = NY + NY$ (Substitution)
9. $2MZ = 2NY$ (Addition Property)
10. $MZ = NY$ (Division Property)
11. $\overline{MZ} \cong \overline{NY}$ (Def. of $\cong$)
12. $\angle XZY \cong \angle XYZ$ (Isosceles Triangle Theorem)
13. $\overline{YZ} \cong \overline{YZ}$ (Reflexive Property)
14. $\triangle MYZ \cong \triangle NZY$ (SAS)
15. $\overline{YM} \cong \overline{ZN}$ (CPCTC)

31. Given: $\overline{CA} \cong \overline{CB}$, $\overline{AD} \cong \overline{BD}$
Prove: C and D are on the perpendicular bisector of $\overline{AB}$

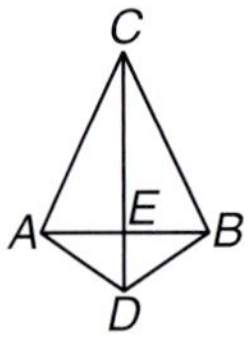

Proof:
Statements (Reasons)

1. $\overline{CA} \cong \overline{CB}$, $\overline{AD} \cong \overline{BD}$ (Given)
2. $\overline{CD} \cong \overline{CD}$ (Congruence of segments is reflexive.)
3. $\triangle ACD \cong \triangle BCD$ (SSS)
4. $\angle ACD \cong \angle BCD$ (CPCTC)
5. $\overline{CE} \cong \overline{CE}$ (Congruence of segments is reflexive.)
6. $\triangle CEA \cong \triangle CEA$ (SAS)
7. $\overline{AE} \cong \overline{BE}$ (CPCTC)
8. E is the midpoint of $\overline{AB}$ (Def. of midpoint)
9. $\angle CEA \cong \angle CEB$ (CPCTC)
10. $\angle CEA$ and $\angle CEB$ form a linear pair. (Def. of linear pair)
11. $\angle CEA$ and $\angle CEB$ are supplementary. (Supplement Theorem)
12. $m\angle CEA + m\angle CEB = 180$ (Def. of supplementary)
13. $m\angle CEA + m\angle CEA = 180$ (Substitution Prop.)
14. $2m\angle CEA = 180$ (Substitution Prop.)
15. $m\angle CEA = 90$ (Division Prop.)
16. $\angle CEA$ and $\angle CEB$ are rt. $\angle$s (Def. of rt. $\angle$)
17. $\overline{CD} \perp \overline{AB}$ (Def. of $\perp$)
18. $\overline{CD}$ is the perpendicular bisector of $\overline{AB}$ (Def. of $\perp$ bisector)
19. C and D are on the perpendicular bisector of $\overline{AB}$ (Def. of point on a line)

32. Given: $\angle BAC$, P is in the interior of $\angle BAC$, $PD = PE$
Prove: $\overrightarrow{AP}$ is the angle bisector of $\angle BAC$

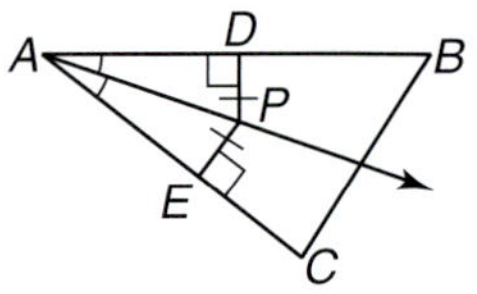

Proof:
Statements (Reasons)

1. $\angle BAC$, P is in the interior of $\angle BAC$, $PD = PE$ (Given)
2. $\overline{PD} \cong \overline{PE}$ (Def. of $\cong$)
3. $\overline{PD} \perp \overline{AB}$, $\overline{PE} \perp \overline{AC}$ (Distance from a point to a line is measured along $\perp$ segment from the point to the line.)
4. $\angle ADP$ and $\angle AEP$ are rt. $\angle$s (Def. of $\perp$)
5. $\triangle ADP$ and $\triangle AEP$ are rt. $\triangle$s (Def. of rt. $\triangle$)
6. $\overline{AP} \cong \overline{AP}$ (Reflexive Property)
7. $\triangle ADP \cong \triangle AEP$ (HL)
8. $\angle DAP \cong \angle EAP$ (CPCTC)
9. $\overrightarrow{AP}$ is the angle bisector of $\angle BAC$ (Def. of $\angle$ bisector)

33. Given: $\triangle ABC$, angle bisectors $\overrightarrow{AD}$, $\overrightarrow{BE}$, and $\overrightarrow{CF}$, $\overline{KP} \perp \overline{AB}$, $\overline{KQ} \perp \overline{BC}$, $\overline{KR} \perp \overline{AC}$
Prove: $KP = KQ = KR$

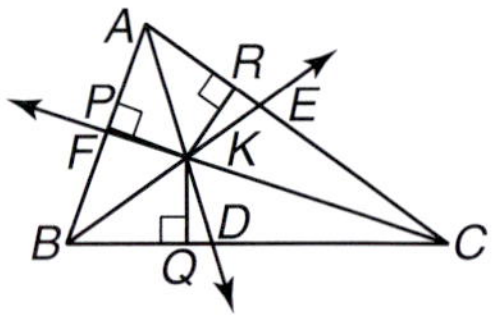

Proof:
Statements (Reasons)

1. $\triangle ABC$, angle bisectors $\overrightarrow{AD}$, $\overrightarrow{BE}$, and $\overrightarrow{CF}$, $\overline{KP} \perp \overline{AB}$, $\overline{KQ} \perp \overline{BC}$, $\overline{KR} \perp \overline{AC}$ (Given)
2. $KP = KQ$, $KQ = KR$, $KP = KR$ (Any point on the $\angle$ bisector is equidistant from the sides of the angle.)
3. $KP = KQ = KR$ (Transitive Property)

Pages 251–254, Lesson 5-2

52. Given: $\triangle ABC$ is scalene; $\overline{AM}$ is the median from A to $\overline{BC}$; $\overline{AT}$ is the altitude from A to $\overline{BC}$.
Prove: $AM > AT$

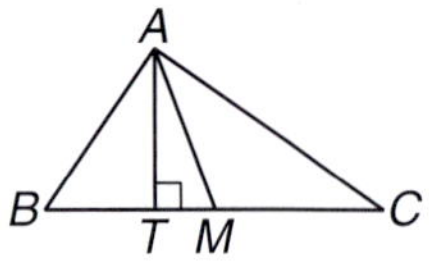

Proof:
$\angle ATB$ and $\angle ATM$ are right angles by the definition of altitude and $m\angle ATB = m\angle ATM$ because all right angles are congruent. By the Exterior Angle Inequality Theorem, $m\angle ATB > m\angle AMT$. So, $m\angle ATM > m\angle AMT$ by Substitution. If one angle of a triangle has a greater measure than another angle, then the side opposite the greater angle is longer than the side opposite the lesser angle. Thus, $AM > AT$.

Pages 257–260, Lesson 5-3

2. Sample answer: Indirect proofs are proved using the contrapositive, showing $\sim Q \to \sim P$. In a direct proof, it would be shown that $P \to Q$. For example, indirect reasoning can be used to prove that a person is not guilty of a crime by assuming the person is guilty, then contradicting evidence to show that the person could not have committed the crime.

3. Sample answer: $\triangle ABC$ is scalene.

Given: $\triangle ABC$; $AB \neq BC$; $BC \neq AC$; $AB \neq AC$

Prove: $\triangle ABC$ is scalene.

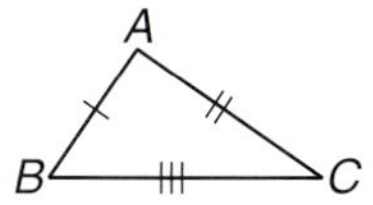

Proof:

Step 1: Assume $\triangle ABC$ is not scalene.

Case 1: $\triangle ABC$ is isosceles.
If $\triangle ABC$ is isosceles, then $AB = BC$, $BC = AC$, or $AB = AC$.
This contradicts the given information, so $\triangle ABC$ is not isosceles.

Case 2: $\triangle ABC$ is equilateral.
In order for a triangle to be equilateral, it must also be isosceles, and Case 1 proved that $\triangle ABC$ is not isosceles. Thus, $\triangle ABC$ is not equilateral.

Therefore, $\triangle ABC$ is scalene.

7. **Given:** $a > 0$

Prove: $\frac{1}{a} > 0$

Proof:

Step 1: Assume $\frac{1}{a} \ngtr 0$.

Step 2: $\frac{1}{a} \leq 0$; $a \cdot \frac{1}{a} \leq 0 \cdot a$, $1 \leq 0$

Step 3: The conclusion that $1 \ngtr 0$ is false, so the assumption that $\frac{1}{a} \ngtr 0$ must be false. Therefore, $a > 0$.

8. **Given:** n is odd.

Prove: n^2 is odd.

Proof:

Step 1: Assume n^2 is even.

Step 2: n is odd, so n can be expressed as $2a + 1$.

$n^2 = (2a + 1)^2$	Substitution
$= (2a + 1)(2a + 1)$	Multiply.
$= 4a^2 + 4a + 1$	Simplify.
$= 2(2a^2 + 2a) + 1$	Distributive Property

Step 3: $2(2a^2 + 2a) + 1$ is an odd number. This contradicts the assumption, so the assumption must be false. Thus n^2 is odd.

9. **Given:** $\triangle ABC$

Prove: There can be no more than one obtuse angle in $\triangle ABC$.

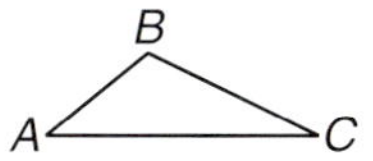

Proof:

Step 1: Assume that there can be more than one obtuse angle in $\triangle ABC$.

Step 2: An obtuse angle has a measure greater than 90. Suppose $\angle A$ and $\angle B$ are obtuse angles. Then $m\angle A + m\angle B > 180$ and $m\angle A + m\angle B + m\angle C > 180$.

Step 3: The conclusion contradicts the fact that the sum of the measures of the angles of a triangle equals 180. Thus, there can be at most one obtuse angle in $\triangle ABC$.

10. **Given:** $m \nparallel n$

Prove: Lines m and n intersect at exactly one point.

Proof:

Case 1: m and n intersect at more than one point.

Step 1: Assume that m and n intersect at more than one point.

Step 2: Lines m and n intersect at points P and Q. Both lines m and n contain P and Q.

Step 3: By postulate, there is exactly one line through any two points. Thus the assumption is false, and lines m and n intersect in no more than one point.

Case 2: m and n do not intersect.

Step 1: Assume that m and n do not intersect.

Step 2: If lines m and n do not intersect, then they are parallel.

Step 3: This conclusion contradicts the given information. Therefore the assumption is false, and lines m and n intersect in at least one point. Combining the two cases, lines m and n intersect in no more than one point and in no less than one point. So lines m and n intersect in exactly one point.

12. **Given:** $x + y > 270$,

Prove: $x > 135$ or $y > 135$.

Proof:

Step 1: Assume $x \leq 135$ and $y \leq 135$.

Step 2: $x + y \leq 270$

Step 3: This contradicts the fact that $x + y > 270$. Therefore, at least one of the stages was longer than 135 miles.

19. **Given:** $\frac{1}{a} < 0$

Prove: a is negative.

Proof:

Step 1: Assume $a > 0$. $a \neq 0$ since that would make $\frac{1}{a}$ undefined.

Step 2: $\frac{1}{a} < 0$

$a\left(\frac{1}{a}\right) < 0 \cdot a$

$1 < 0$

Step 3: $1 > 0$, so the assumption must be false. Thus, a must be negative.

20. **Given:** n^2 is even.

Prove: n^2 is divisible by 4.

Proof:

Step 1: Assume n^2 is not divisible by 4. In other words, 4 is not a factor of n^2.

Step 2: If the square of a number is even, then the number is also even. So, if n^2 is even, n must be even. Let $n = 2a$.

$n - 2a$

$n^2 = (2a)^2$ or $4a^2$

Step 3: 4 is a factor of n^2, which contradicts the assumption.

21. Given: $\overline{PQ} \cong \overline{PR}$
$\angle 1 \not\cong \angle 2$
Prove: $\overline{PZ}$ is not a median of $\triangle PQR$.

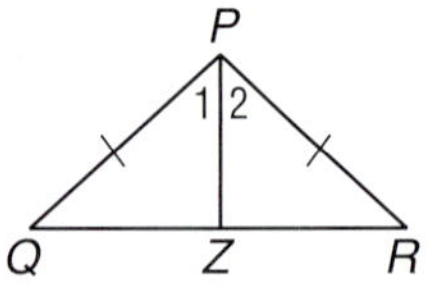

Proof:
Step 1: Assume $\overline{PZ}$ is a median of $\triangle PQR$.
Step 2: If $\overline{PZ}$ is a median of $\triangle PQR$, then Z is the midpoint of $\overline{QR}$, and $\overline{QZ} \cong \overline{RZ}$. $\overline{PZ} \cong \overline{PZ}$ by the Reflexive Property. $\triangle PZQ \cong \triangle PZR$ by SSS. $\angle 1 \cong \angle 2$ by CPCTC.
Step 3: This conclusion contradicts the given fact $\angle 1 \not\cong \angle 2$. Thus, $\overline{PZ}$ is not a median of $\triangle PQR$.

22. Given: $m\angle 2 \neq m\angle 1$
Prove: $\ell \nparallel m$

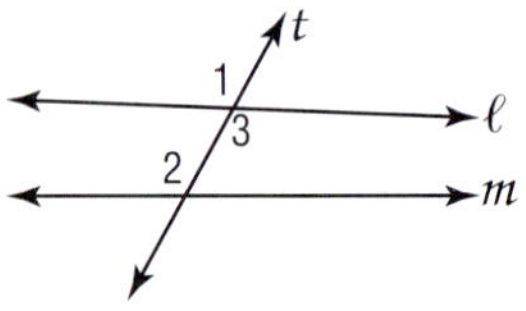

Proof:
Step 1: Assume that $\ell \parallel m$.
Step 2: If $\ell \parallel m$, then $\angle 1 \cong \angle 2$ because they are corresponding angles. Thus, $m\angle 1 = m\angle 2$.
Step 3: This contradicts the given fact that $m\angle 1 \neq m\angle 2$. Thus the assumption $\ell \parallel m$ is false. Therefore, $\ell \nparallel m$.

23. Given: $a > 0$, $b > 0$, and $a > b$
Prove: $\frac{a}{b} > 1$
Step 1: Assume that $\frac{a}{b} \leq 1$.
Step 2: Case 1 Case 2

Case 1	Case 2
$\frac{a}{b} < 1$	$\frac{a}{b} = 1$
$a < b$	$a = b$

Step 3: The conclusion of both cases contradicts the given fact $a > b$. Thus, $\frac{a}{b} > 1$.

24. Given: $\overline{AB} \not\cong \overline{AC}$
Prove: $\angle 1 \not\cong \angle 2$

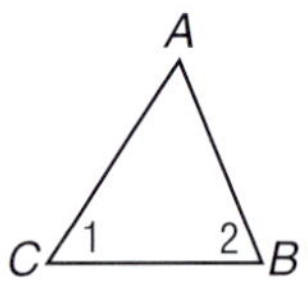

Proof:
Step 1: Assume that $\angle 1 \cong \angle 2$.
Step 2: If $\angle 1 \cong \angle 2$, then the sides opposite the angles are congruent. Thus $\overline{AB} \cong \overline{AC}$.
Step 3: The conclusion contradicts the given information. Thus $\angle 1 \cong \angle 2$ is false. Therefore, $\angle 1 \not\cong \angle 2$.

25. Given: $\triangle ABC$ and $\triangle ABD$ are equilateral.
$\triangle ACD$ is not equilateral.
Prove: $\triangle BCD$ is not equilateral.

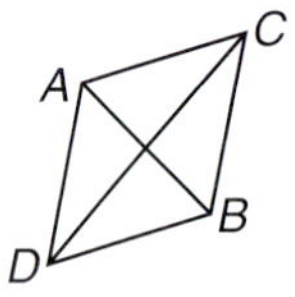

Proof:
Step 1: Assume that $\triangle BCD$ is an equilateral triangle.
Step 2: If $\triangle BCD$ is an equilateral triangle, then $\overline{BC} \cong \overline{CD} \cong \overline{DB}$. Since $\triangle ABC$ and $\triangle ABD$ are equilateral triangles, $\overline{AC} \cong \overline{AB} \cong \overline{BC}$ and $\overline{AD} \cong \overline{AB} \cong \overline{DB}$. By the Transitive Property, $\overline{AC} \cong \overline{AD} \cong \overline{CD}$. Therefore, $\triangle ACD$ is an equilateral triangle.
Step 3: This conclusion contradicts the given information. Thus, the assumption is false. Therefore, $\triangle BCD$ is not an equilateral triangle.

26. Given: $m\angle A > m\angle ABC$
Prove: $BC > AC$

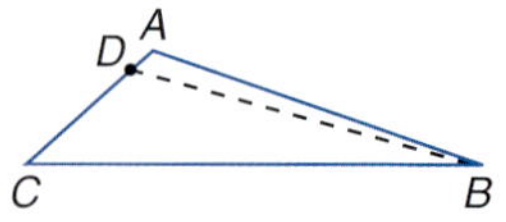

Proof:
Assume $BC \not> AC$. By the Comparison Property, $BC = AC$ or $BC < AC$.
Case 1: If $BC = AC$, then $\angle ABC \cong \angle A$ by the Isosceles Triangle Theorem. (If two sides of a triangle are congruent, then the angles opposite those sides are congruent.) But, $\angle ABC \cong \angle A$ contradicts the given statement that $m\angle A > m\angle ABC$. So, $BC \neq AC$.
Case 2: If $BC < AC$, then there must be a point D between A and C so that $\overline{DC} \cong \overline{BC}$. Draw the auxiliary segment $\overline{BD}$. Since $DC = BC$, by the Isosceles Triangle Theorem $\angle BDC \cong \angle DBC$. Now $\angle BDC$ is an exterior angle of $\triangle BAD$ and by the Exterior Angles Inequality Theorem (the measure of an exterior angle of a triangle is greater than the measure of either corresponding remote interior angle) $m\angle BDC > m\angle A$. By the Angle Addition Postulate, $m\angle ABC = m\angle ABD + m\angle DBC$. Then by the definition of inequality, $m\angle ABC > m\angle DBC$. By Substitution and the Transitive Property of Inequality, $m\angle ABC > m\angle A$. But this contradicts the given statement that $m\angle A > m\angle ABC$. In both cases, a contradiction was found, and hence our assumption must have been false. Therefore, $BC > AC$.

27. Use $r = \frac{d}{t}$, $t = 3$, and $d = 175$.
Proof:
Step 1: Assume that Ramon's average speed was greater than or equal to 60 miles per hour, $r \geq 60$.
Step 2:

Case 1	Case 2
$r = 60$	$r > 60$
$60 \stackrel{?}{=} \frac{175}{3}$	$\frac{175}{3} \stackrel{?}{>} 60$
$60 \neq 58.3$	$58.3 \not> 60$

Step 3: The conclusions are false, so the assumption must be false. Therefore, Ramon's average speed was less than 60 miles per hour.

28. A majority is greater than half or 50%.
Proof:
Step 1: Assume that the percent of college-bound seniors receiving information from guidance counselors is less than 50%.
Step 2: By examining the graph, you can see that 56% of college-bound seniors received information from guidance counselors.
Step 3: Since 56% > 50%, the assumption is false. Therefore, a majority of college-bound seniors received information from guidance counselors.

33. Proof:

Step 1: Assume that $\sqrt{2}$ is a rational number.

Step 2: If $\sqrt{2}$ is a rational number, it can be written as $\frac{a}{b}$, where a and b are relatively prime integers, and $b \neq 0$. If $\sqrt{2} = \frac{a}{b}$, then $2 = \frac{a^2}{b^2}$, and $2b^2 = a^2$. Thus a^2 is an even number, as is a. Because a is even it can be written as $2n$.

$2b^2 = a^2$

$2b^2 = (2n)^2$

$2b^2 = 4n^2$

$b^2 = 2n^2$

b^2 is an even number. So, b is also an even number.

Step 3: Because b and a are both even numbers, they have a common factor of 2. This contradicts the definition of rational numbers. Therefore, $\sqrt{2}$ is not rational.

39. Given: $\overline{CD}$ is an angle bisector. $\overline{CD}$ is an altitude.
Prove: $\triangle ABC$ is isosceles.

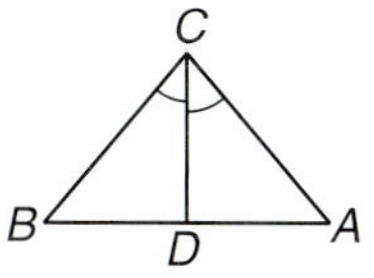

Statements (Reasons)

1. $\overline{CD}$ is an angle bisector. $\overline{CD}$ is an altitude. (Given)
2. $\angle ACD \cong \angle BCD$ (Def. of $\angle$ bisector)
3. $\overline{CD} \perp \overline{AB}$ (Def. of altitude)
4. $\angle CDA$ and $\angle CDB$ are rt. $\angle$s ($\perp$ lines form 4 rt. $\angle$s.)
5. $\angle CDA \cong \angle CDB$ (All rt. $\angle$s are $\cong$.)
6. $\overline{CD} \cong \overline{CD}$ (Reflexive Prop.)
7. $\triangle ACD \cong \triangle BCD$ (ASA)
8. $\overline{AC} \cong \overline{BC}$ (CPCTC)
9. $\triangle ACB$ is isosceles. (Def. of isosceles $\triangle$)

40. Given: $\overline{QT}$ is a median. $\triangle QRS$ is isosceles with base $\overline{RS}$.
Prove: $\overline{QT}$ bisects $\angle SQR$

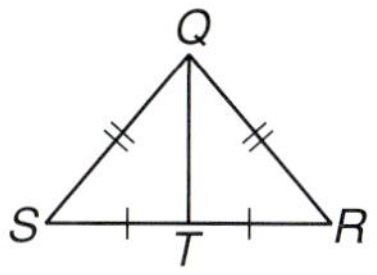

Statements (Reasons)

1. $\overline{QT}$ is a median. $\triangle QRS$ is isosceles with base $\overline{RS}$. (Given)
2. $\overline{RT} \cong \overline{ST}$ (Def. of median)
3. $\overline{QR} \cong \overline{QS}$ (Def. of isosceles $\triangle$)
4. $\overline{QT} \cong \overline{QT}$ (Reflexive Prop.)
5. $\triangle QRT \cong \triangle QST$ (SSS)
6. $\angle SQT \cong \angle RQT$ (CPCTC)
7. $\overline{QT}$ bisects $\angle SQR$ (Def. of $\angle$ bisector)

41. Given: $\triangle ABC \cong \triangle DEF$; $\overline{BG}$ is an angle bisector of $\angle ABC$. $\overline{EH}$ is an angle bisector of $\angle DEF$.
Prove: $\overline{BG} \cong \overline{EH}$

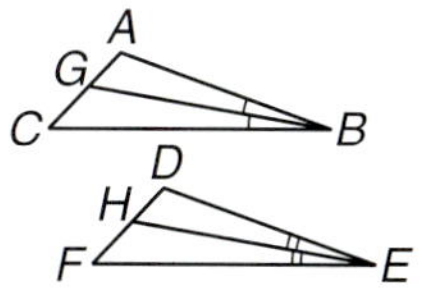

Statements (Reasons)

1. $\triangle ABC \cong \triangle DEF$ (Given)
2. $\angle A \cong \angle D$, $\overline{AB} \cong \overline{DE}$, $\angle ABC \cong \angle DEF$ (CPCTC)
3. $\overline{BG}$ is an angle bisector of $\angle ABC$. $\overline{EH}$ is an angle bisector of $\angle DEF$. (Given)
4. $\angle ABG \cong \angle GBC$, $\angle DEH \cong \angle HEF$ (Def. of $\angle$ bisector)
5. $m\angle ABC = m\angle DEF$ (Def. of $\cong$ $\angle$s)
6. $m\angle ABG = m\angle GBC$, $m\angle DEH = m\angle HEF$ (Def. of $\cong$ $\angle$s)
7. $m\angle ABC = m\angle ABG + m\angle GBC$, $m\angle DEF = m\angle DEH + m\angle HEF$ (Angle Addition Property)
8. $m\angle ABC = m\angle ABG + m\angle ABG$, $m\angle DEF = m\angle DEH + m\angle DEH$ (Substitution)
9. $m\angle ABG + m\angle ABG = m\angle DEH + m\angle DEH$ (Substitution)
10. $2m\angle ABG = 2m\angle DEH$ (Substitution)
11. $m\angle ABG = m\angle DEH$ (Division)
12. $\angle ABG \cong \angle DEH$ (Def. of $\cong$ $\angle$s)
13. $\triangle ABG \cong \triangle DEH$ (ASA)
14. $\overline{BG} \cong \overline{EH}$ (CPCTC)

Pages 263–266, Lesson 5-4

38. Given: $\angle B \cong \angle ACB$
Prove: $AD + AB > CD$

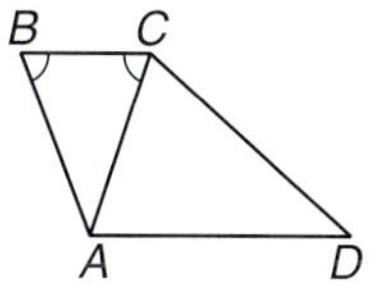

Statements (Reasons)

1. $\angle B \cong \angle ACB$ (Given)
2. $\overline{AB} \cong \overline{AC}$ (If two $\angle$s are $\cong$, the sides opposite the two $\angle$s are $\cong$.)
3. $AB = AC$ (Def. of $\cong$ segments)
4. $AD + AC > CD$ (Triangle Inequality)
5. $AD + AB > CD$ (Substitution)

39. Given: $\overline{HE} \cong \overline{EG}$
Prove: $HE + FG > EF$

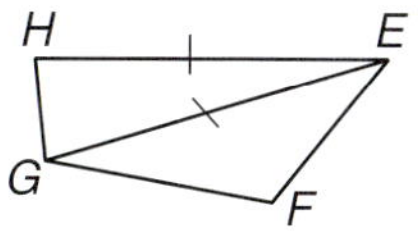

Statements (Reasons)

1. $\overline{HE} \cong \overline{EG}$ (Given)
2. $HE = EG$ (Def. of $\cong$ segments)
3. $EG + FG > EF$ (Triangle Inequality)
4. $HE + FG > EF$ (Substitution)

40. Given: $\triangle ABC$
Prove: $AC + BC > AB$

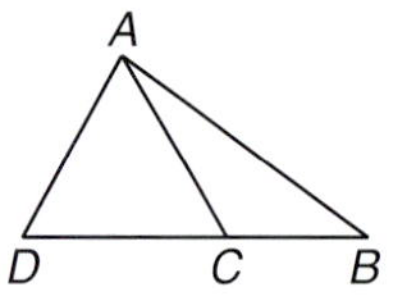

Statements (Reasons)

1. Construct $\overline{CD}$ so that C is between B and D and $\overline{CD} \cong \overline{AC}$. (Ruler Postulate)
2. $CD = AC$ (Definition of $\cong$)
3. $\angle CAD \cong \angle ADC$ (Isosceles Triangle Theorem)
4. $m\angle CAD = m\angle ADC$ (Definition of $\cong$ angles)
5. $m\angle BAC + m\angle CAD = m\angle BAD$ ($\angle$ Addition Post.)
6. $m\angle BAC + m\angle ADC = m\angle BAD$ (Substitution)
7. $m\angle ADC < m\angle BAD$ (Definition of inequality)
8. $AB < BD$ (If an angle of a triangle is greater than a second angle, then the side opposite the greater angle is longer than the side opposite the lesser angle.)
9. $BD = BC + CD$ (Segment Addition Postulate)
10. $AB < BC + CD$ (Substitution)
11. $AB < BC + AC$ (Substitution (Steps 2, 10))

Page 266, Practice Quiz

3. Step 1: Assume that $x \leq 8$.

Step 2: $7x > 56$

$x > 8$

Step 3: The solution of $7x > 56$ contradicts the assumption. Thus, $x \leq 8$ must be false. Therefore, $x > 8$.

4. Given: $\overline{MO} \cong \overline{ON}$, $\overline{MP} \not\cong \overline{NP}$
Prove: $\angle MOP \not\cong \angle NOP$

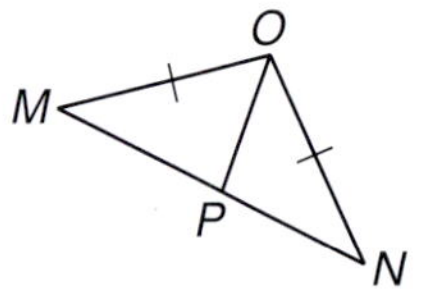

Step 1: Assume that $\angle MOP \cong \angle NOP$.

Step 2: We know that $\overline{MO} \cong \overline{ON}$, and $\overline{OP} \cong \overline{OP}$ by the Reflexive Property. If $\angle MOP \cong \angle NOP$, then $\triangle MOP \cong \triangle NOP$ by SAS. Then, $\overline{MP} \cong \overline{NP}$ by CPCTC.

Step 3: The conclusion that $\overline{MP} \cong \overline{NP}$ contradicts the given information. Thus, the assumption is false. Therefore, $\angle MOP \not\cong \angle NOP$.

5. Given: $m\angle ADC \neq m\angle ADB$
Prove: $\overline{AD}$ is not an altitude of $\triangle ABC$.

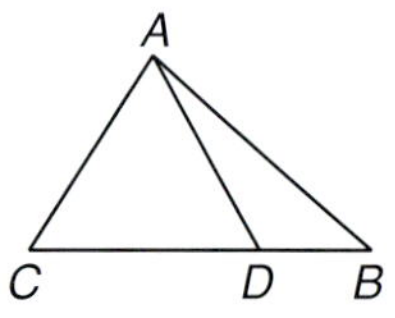

Statements (Reasons)

1. $\overline{AD}$ is an altitude of $\triangle ABC$. (Assumption)
2. $\angle ADC$ and $\angle ADB$ are right angles. (Def. of altitude)
3. $\angle ADC \cong \angle ADB$ (All rt. $\angle$s are $\cong$.)
4. $m\angle ADC = m\angle ADB$ (Def. of $\cong$ $\angle$s)

This contradicts the given information that $m\angle ADC \neq m\angle ADB$. Thus, $\overline{AD}$ is not an altitude of $\triangle ABC$.

Pages 270–273, Lesson 5-5

7. Given: $\overline{PQ} \cong \overline{SQ}$
Prove: $PR > SR$

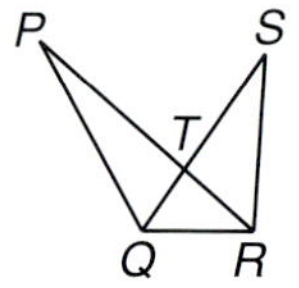

Statements (Reasons)

1. $\overline{PQ} \cong \overline{SQ}$ (Given)
2. $\overline{QR} \cong \overline{QR}$ (Reflexive Property)
3. $m\angle PQR = m\angle PQS + m\angle SQR$ ($\angle$ Addition Post.)
4. $m\angle PQR > m\angle SQR$ (Def. of inequality)
5. $PR > SR$ (SAS Inequality)

8. Given: $\overline{TU} \cong \overline{US}$; $\overline{US} \cong \overline{SV}$
Prove: $ST > UV$

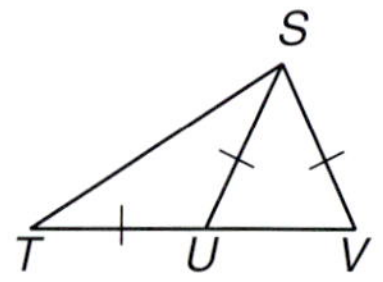

Statements (Reasons)

1. $\overline{TU} \cong \overline{US}$; $\overline{US} \cong \overline{SV}$ (Given)
2. $m\angle SUT > m\angle USV$ (Ext. $\angle$ Inequality Theorem)
3. $ST > UV$ (SAS Inequality)

20. Given: $\triangle ABC$, $\overline{AB} \cong \overline{CD}$
Prove: $BC > AD$

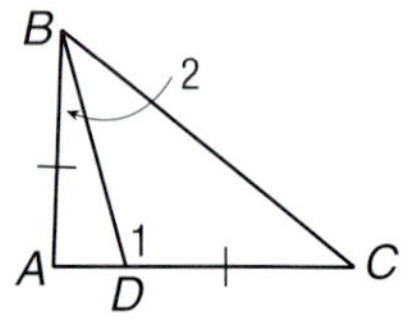

Statements (Reasons)

1. $\triangle ABC$, $\overline{AB} \cong \overline{CD}$ (Given)
2. $\overline{BD} \cong \overline{BD}$ (Reflexive Property)
3. $m\angle 1 = m\angle 2$ (If an $\angle$ is an ext. $\angle$ of a $\triangle$, then its measure is greater the measure of either remote int. $\angle$.)
4. $BC > AD$ (SAS Inequality)

21. Given: $\overline{PQ} \cong \overline{RS}$, $QR < PS$
Prove: $m\angle 3 < m\angle 1$

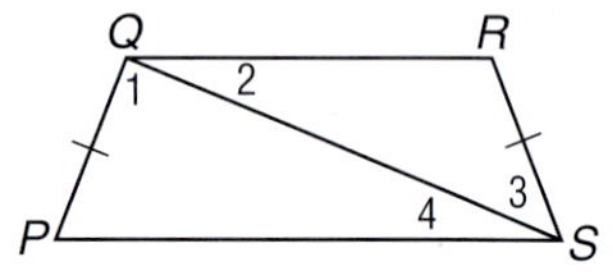

Statements (Reasons)

1. $\overline{PQ} \cong \overline{RS}$ (Given)
2. $\overline{QS} \cong \overline{QS}$ (Reflexive Property)
3. $QR < PS$ (Given)
4. $m\angle 3 < m\angle 1$ (SSS Inequality)

22. Given: $\overline{PR} \cong \overline{PQ}$, $SQ > SR$
Prove: $m\angle 1 < m\angle 2$

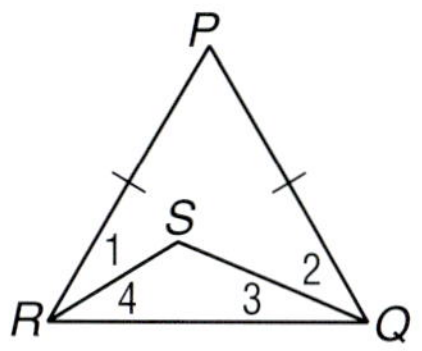

Statements (Reasons)

1. $\overline{PR} \cong \overline{PQ}$ (Given)
2. $\angle PRQ \cong \angle PQR$ (If two sides of $\triangle$ are $\cong$, the angles opposite the sides are $\cong$.)
3. $m\angle PRQ = m\angle 1 + m\angle 4$, $m\angle PQR = 2 + m\angle 3$ (Angle Add. Post.)
4. $m\angle PRQ = m\angle PQR$ (Def. of $\cong$ angles)
5. $m\angle 1 + m\angle 4 = m\angle 2 + m\angle 3$ (Substitution)
6. $SQ > SR$ (Given)
7. $m\angle 4 > m\angle 3$ (If one side of a $\triangle$ is longer than another side, then the $\angle$ opposite the longer side is greater than the $\angle$ opposite the shorter side.)
8. $m\angle 4 = m\angle 3 + x$ (Def. of inequality)
9. $m\angle 1 + m\angle 3 + x = m\angle 2 + m\angle 3$ (Substitution)
10. $m\angle 1 + x = m\angle 2$ (Subtraction Prop.)
11. $m\angle 1 < m\angle 2$ (Def. of inequality)

23. Given: $\overline{ED} \cong \overline{DF}$; $m\angle 1 > m\angle 2$;
D is the midpoint of $\overline{CB}$;
$\overline{AE} \cong \overline{AF}$.
Prove: $AC > AB$

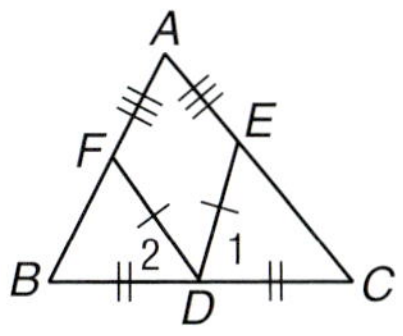

Statements (Reasons)

1. $\overline{ED} \cong \overline{DF}$; D is the midpoint of $\overline{CB}$. (Given)
2. $CD = BD$ (Def. of midpoint)
3. $\overline{CD} \cong \overline{BD}$ (Def. of $\cong$ segments)
4. $m\angle 1 > m\angle 2$ (Given)
5. $EC > FB$ (SAS Inequality)
6. $\overline{AE} \cong \overline{AF}$ (Given)
7. $AE = AF$ (Def. of $\cong$ segments)
8. $AE + EC > AE + FB$ (Add. Prop. of Inequality)
9. $AE + EC > AF + FB$ (Substitution Prop. of Inequality)
10. $AE + EC = AC$, $AF + FB = AB$ (Segment Add. Post.)
11. $AC > AB$ (Substitution)

24. Given: $\overline{RS} \cong \overline{UW}$, $\overline{ST} \cong \overline{WV}$, $RT > UV$
Prove: $m\angle S > m\angle W$

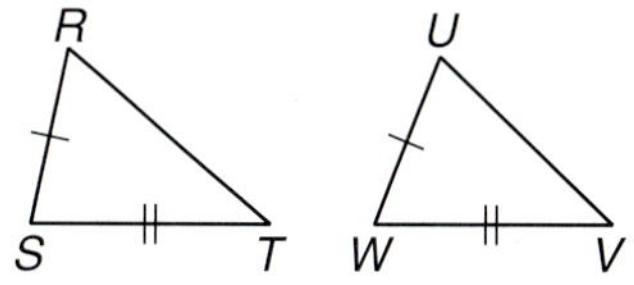

Indirect Proof

Step 1: Assume $m\angle S \leq m\angle W$.

Step 2: If $m\angle S \leq m\angle W$, then either $m\angle S < m\angle W$ or $m\angle S = m\angle W$.

Case 1: If $m\angle S < m\angle W$, then $RT < UV$ by the SAS Inequality.

Case 2: If $m\angle S = m\angle W$, then $\triangle RST \cong \triangle UVW$ by SAS, and $\overline{RT} \cong \overline{UV}$ by CPCTC. Thus $RT = UV$.

Step 3: Both cases contradict the given $RT > UV$. Therefore, the assumption must be false, and the conclusion, $m\angle S < m\angle W$, must be true.

39. Given: $\overline{AD}$ bisects $\overline{BE}$; $\overline{AB} \parallel \overline{DE}$.
Prove: $\triangle ABC \cong \triangle DEC$

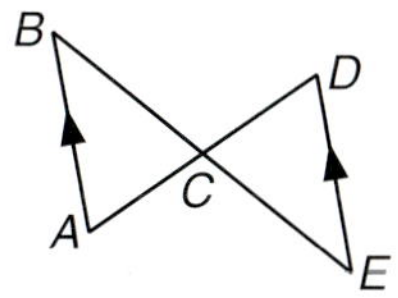

Statements (Reasons)

1. $\overline{AD}$ bisects $\overline{BE}$; $\overline{AB} \parallel \overline{DE}$. (Given)
2. $\overline{BC} \cong \overline{EC}$ (Def. of seg. bisector)
3. $\angle B \cong \angle E$ (Alt. int. $\angle$s Thm.)
4. $\angle BCA \cong \angle ECD$ (Vert. $\angle$s are $\cong$.)
5. $\triangle ABC \cong \triangle DEC$ (ASA)

40. Given: $\overline{OM}$ bisects $\angle LMN$;
$\overline{LM} \cong \overline{MN}$.
Prove: $\triangle MOL \cong \triangle MON$

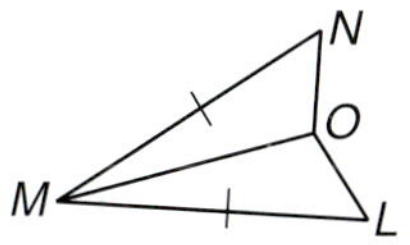

Statements (Reasons)

1. $\overline{OM}$ bisects $\angle LMN$; $\overline{LM} \cong \overline{MN}$. (Given)
2. $\angle LMO \cong \angle NMO$ (Def. of $\angle$ bisector)
3. $\overline{OM} \cong \overline{OM}$ (Reflexive Prop.)
4. $\triangle MOL \cong \triangle MON$ (SAS)

Chapter 6

Proportions and Similarity

Chapter Overview and Pacing

LESSON OBJECTIVES

	PACING (days)			
	Regular		Block	
	Basic/ Average	Advanced	Basic/ Average	Advanced
6-1 Proportions *(pp. 282–288)* • Write ratios. • Use properties of proportions. ***Follow-Up:*** Create the Fibonacci sequence using a spreadsheet.	1	2 (with 6-1 Follow-Up)	0.5	1 (with 6-1 Follow-Up)
6-2 Similar Polygons *(pp. 289–297)* • Identify similar figures. • Solve problems involving scale factors.	1	2	0.5	1
6-3 Similar Triangles *(pp. 298–306)* • Identify similar triangles. • Use similar triangles to solve problems.	2	2	1	1
6-4 Parallel Lines and Proportional Parts *(pp. 307–315)* • Use proportional parts of triangles. • Divide a segment into parts.	2	2	1	1
6-5 Parts of Similar Triangles *(pp. 316–323)* • Recognize and use proportional relationships of corresponding perimeters of similar triangles. • Recognize and use proportional relationships of corresponding angle bisectors, altitudes, and medians of similar triangles.	2	2	1	1
6-6 Fractals and Self-Similarity *(pp. 324–331)* ***Preview:*** Draw a Sierpinski Triangle. • Recognize and describe characteristics of fractals. • Identify nongeometric iteration.	2 (with 6-6 Preview)	2 (with 6-6 Preview)	1 (with 6-6 Preview)	1 (with 6-6 Preview)
Study Guide and Practice Test *(pp. 332–337)* **Standardized Test Practice** *(pp. 338–339)*	1	1	0.5	0.5
Chapter Assessment	1	1	0.5	0.5
TOTAL	**12**	**14**	**6**	**7**

Pacing suggestions for the entire year can be found on pages T20–T21.

Timesaving Tools
TeacherWorks™
All-In-One Planner and Resource Center
See pages T5 and T21.

Chapter Resource Manager

Chapter 6 Resource Masters										
Study Guide and Intervention	Practice (Skills and Average)	Reading to Learn Mathematics	Enrichment	Assessment	Prerequisite Skills Workbook	Applications*	5-Minute Check Transparencies	Interactive Chalkboard	GeomPASS: Tutorial Plus (lessons)	Materials
295–296	297–298	299	300		19–20, 23–24, 29–32, 37–38, 69–72, 91–92	GCC 27 SC 11	6-1	6-1		grid paper, straightedge
301–302	303–304	305	306	345	89–90, 95–96	SC 12	6-2	6-2		grid paper, straightedge
307–308	309–310	311	312	345, 347	25–26, 33–36, 53–54		6-3	6-3	12	protractor, centimeter ruler
313–314	315–316	317	318				6-4	6-4		straightedge, compass
319–320	321–322	323	324	346	31–32		6-5	6-5		
325–326	327–328	329	330	346		GCC 28	6-6	6-6		(*Preview:* isometric dot paper) straightedge, grid paper
				331–344, 348–350						

**Key to Abbreviations:* GCC = Graphing Calculator and Computer Masters
SC = School-to-Career Masters

Chapter 6

Mathematical Connections and Background

Continuity of Instruction

Prior Knowledge

In previous courses students solved rational equations and evaluated exponential expressions. In Chapter 3 students found the slope of a line and proved that lines are parallel. They also determined the measures of angles when a transversal intersects a pair of parallel lines.

This Chapter

In this chapter students extend their knowledge of ratios and proportions to similar figures. The term scale factor is introduced, and students solve problems using cross products of proportions. They also learn to recognize and use proportional parts and relationships to solve problems involving similar triangles and parallel lines. Finally, students are introduced to fractals and nongeometric iteration.

Future Connections

Chapter 7 builds on students' knowledge of ratios and proportions when trigonometric ratios are introduced. Students will need to recognize and use proportional parts and relationships in the real world to read maps accurately, work with blueprints, or adapt recipes.

6-1 Proportions

A ratio is a comparison of two quantities. The ratio of a to b can be expressed $\frac{a}{b}$, where b is not zero, or as $a:b$. The ratio of two corresponding quantities is called a scale factor. For example, a model might be $\frac{1}{8}$ the size of the actual object.

An equation stating that two ratios are equal is a proportion. Equivalent fractions are set equal to each other to form a proportion. Every proportion has two cross products. To find the cross products, multiply the numerator of the first fraction by the denominator of the second fraction, and the numerator of the second fraction by the denominator of the first fraction. One cross product forms the extremes, the other the means. The product of the means equals the product of the extremes. In other words, for any numbers a and c and any nonzero numbers b and d, $\frac{a}{b} = \frac{c}{d}$ if and only if $ad = bc$.

6-2 Similar Polygons

When figures have the same shape but may be different in size, they are similar figures. Two polygons are similar if and only if their corresponding angles are congruent and the measures of their corresponding sides are proportional. The ratio of the lengths of two corresponding sides of two similar polygons is called the scale factor. The use of scale factors will produce similar figures. This is called dilation.

6-3 Similar Triangles

For triangles to be similar, their corresponding angles must be congruent and the measures of their corresponding sides must be proportional. However, triangles can be proved similar without knowing the measures of every angle and side. The Angle-Angle Similarity Postulate states that if two angles of one triangle are congruent to two angles of another triangle, then the triangles are similar. If the measures of the corresponding sides of two triangles are proportional, then the triangles are similar (Side-Side-Side Similarity Theorem). If the measures of two sides of a triangle are proportional to the measures of two corresponding sides of another triangle and the included angles are congruent, then the triangles are similar (Side-Angle-Side Similarity Theorem).

Similarity of triangles is reflexive, symmetric, and transitive. Similar triangles can be used to find measurements. The most common example of this is the "shadow problem," in which you are asked to determine the height of a building based on the height of something whose height is known (a flagpole, for example) by measuring the shadows cast by the two objects at the same time of day.

6-4 Parallel Lines and Proportional Parts

The Triangle Proportionality Theorem states that if a line is parallel to one side of a triangle and intersects the other two sides in two distinct points, then it separates these sides into segments of proportional lengths. The converse of this theorem also holds true. If a line intersects two sides of a triangle and separates the sides into corresponding segments of proportional lengths, then the line is parallel to the third side. These theorems can be extended to three or more parallel lines. If three or more parallel lines intersect two transversals, then they cut off the transversals proportionally. If three or more parallel lines cut off congruent segments on one transversal, then they cut off congruent segments on every transversal.

One notable case of a line parallel to one side of a triangle is the midsegment. A midsegment is a segment whose endpoints are the midpoints of two sides of the triangle. A midsegment is parallel to one side of the triangle, and its length is one-half the length of that side.

6-5 Parts of Similar Triangles

If two triangles are similar, then the perimeters are proportional to the measures of corresponding sides. This proportionality relationship is also seen in the altitudes of similar triangles. If two triangles are similar, then the measures of the corresponding altitudes are proportional to the measures of the corresponding sides. In fact, this relationship holds true for angle bisectors and medians as well.

There is also a relationship between an angle bisector and the side of the triangle opposite the angle. An angle bisector in a triangle separates the opposite side into segments that have the same ratio as the other two sides.

6-6 Fractals and Self-Similarity

The word *fractal* was originally applied to things in nature that are irregular in shape. Similar patterns can be created on a computer using a process called *iteration*. This means that the same procedure is repeated over and over, potentially infinitely. The word *fractal* is now used to describe a geometric figure that was created using iteration. It is infinite in structure.

One characteristic of fractals is that they have self-similar shapes. In other words, the same geometrical characteristics are seen from a distance or at extreme magnification. A figure is strictly self-similar if any of its parts, no matter where they are located or what size is seen, contain the same figure as the whole. The Sierpinski Triangle is an example of a strictly self-similar pattern.

The process of iteration is not always applied to geometric shapes. Iterative processes can be translated into formulas or algebraic equations. These are called recursive formulas.

Chapter 6

DAILY INTERVENTION and Assessment

	Type	Student Edition	Teacher Resources	Technology/Internet
INTERVENTION	Ongoing	Prerequisite Skills, pp. 281, 287, 297, 306, 315, 323 Practice Quiz 1, p. 306 Practice Quiz 2, p. 323	5-Minute Check Transparencies *Prerequisite Skills Workbook*, pp. 19–20, 23–26, 29–38, 43–44, 53–54, 69–72, 89–92, 95–96 Quizzes, *CRM* pp. 345–346 Mid-Chapter Test, *CRM* p. 347 Study Guide and Intervention, *CRM* pp. 295–296, 301–302, 307–308, 313–314, 319–320, 325–326	GeomPASS: Tutorial Plus www.geometryonline.com/self_check_quiz www.geometryonline.com/extra_examples
	Mixed Review	pp. 287, 297, 306, 315, 323, 331	Cumulative Review, *CRM* p. 348	
	Error Analysis	Find the Error, pp. 284, 292, 301 Common Misconceptions, p. 284, 290, 326	Find the Error, *TWE* pp. 285, 293, 301 Unlocking Misconceptions, *TWE* p. 319 Tips for New Teachers, *TWE* pp. 283, 284, 289, 300	
	Standardized Test Practice	pp. 282, 285, 287, 297, 305, 314, 315, 322, 331, 337, 338, 339	*TWE* pp. 338–339 Standardized Test Practice, *CRM* pp. 349–350	Standardized Test Practice CD-ROM www.geometryonline.com/standardized_test
ASSESSMENT	Open-Ended Assessment	Writing in Math, pp. 286, 297, 305, 314, 322, 330 Open Ended, pp. 284, 292, 301, 311, 319, 328 Standardized Test, p. 339	Modeling: *TWE* pp. 306, 323, 331 Speaking: *TWE* p. 287 Writing: *TWE* pp. 297, 315 Open-Ended Assessment, *CRM* p. 343	
	Chapter Assessment	Study Guide, pp. 332–336 Practice Test, p. 337	Multiple-Choice Tests (Forms 1, 2A, 2B), *CRM* pp. 331–336 Free-Response Tests (Forms 2C, 2D, 3), *CRM* pp. 337–342 Vocabulary Test/Review, *CRM* p. 344	TestCheck and Worksheet Builder (see below) MindJogger Videoquizzes www.geometryonline.com/vocabulary_review www.geometryonline.com/chapter_test

Key to Abbreviations: TWE = Teacher Wraparound Edition; CRM = Chapter Resource Masters

Additional Intervention Resources

The Princeton Review's *Cracking the SAT & PSAT*

The Princeton Review's *Cracking the ACT*

TestCheck and Worksheet Builder

This **networkable** software has three modules for intervention and assessment flexibility:

- **Worksheet Builder** to make worksheet and tests
- **Student Module** to take tests on screen (optional)
- **Management System** to keep student records (optional)

Special banks are included for SAT, ACT, TIMSS, NAEP, and End-of-Course tests.

Intervention Technology

GeomPASS: Tutorial Plus CD-ROM offers a complete, self-paced geometry curriculum.

Geometry Lesson	GeomPASS Lesson
6-3	**12** *Solving Problems with Similar Triangles*

Intervention at Home

Log on for student study help.

- For each lesson in the Student Edition, there are Extra Examples and Self-Check Quizzes.
  www.geometryonline.com/extra_examples
  www.geometryonline.com/self_check_quiz
- For chapter review, there is vocabulary review, test practice, and standardized test practice.
  www.geometryonline.com/vocabulary_review
  www.geometryonline.com/chapter_test
  www.geometryonline.com/standardized_test

For more information on Intervention and Assessment, see pp. T8–T11.

Reading and Writing in Mathematics

Glencoe Geometry provides numerous opportunities to incorporate reading and writing into the mathematics classroom.

Student Edition

- Foldables Study Organizer, p. 281
- Concept Check questions require students to verbalize and write about what they have learned in the lesson. (pp. 284, 292, 301, 311, 319, 328)
- Writing in Math questions in every lesson, pp. 286, 297, 305, 314, 322, 330
- Reading Study Tip, p. 283
- WebQuest, p. 325

Teacher Wraparound Edition

- Foldables Study Organizer, pp. 281, 332
- Study Notebook suggestions, pp. 285, 293, 301, 311, 319, 324, 328
- Modeling activities, pp. 306, 323, 331
- Speaking activities, p. 287
- Writing activities, pp. 297, 315
- ELL Resources, pp. 280, 286, 294, 304, 312, 320, 330, 332

Additional Resources

- Vocabulary Builder worksheets require students to define and give examples for key vocabulary terms as they progress through the chapter. (*Chapter 6 Resource Masters,* pp. vii-viii)
- Proof Builder helps students learn and understand theorems and postulates from the chapter. (*Chapter 6 Resource Masters,* pp. ix–x)
- Reading to Learn Mathematics master for each lesson (*Chapter 6 Resource Masters*, pp. 299, 305, 311, 317, 323, 329)
- *Vocabulary PuzzleMaker* software creates crossword, jumble, and word search puzzles using vocabulary lists that you can customize.
- *Teaching Mathematics with Foldables* provides suggestions for promoting cognition and language.
- *Reading Strategies for the Mathematics Classroom*
- *WebQuest and Project Resources*

For more information on Reading and Writing in Mathematics, see pp. T6–T7.

Chapter 6 Notes

What You'll Learn

Have students read over the list of objectives and make a list of any words with which they are not familiar.

Why It's Important

Point out to students that this is only one of many reasons why each objective is important. Others are provided in the introduction to each lesson.

Lesson	NCTM Standards	Local Objectives
6-1	1, 2, 6, 8, 9, 10	
6-1 Follow-Up	1, 6, 8	
6-2	1, 2, 3, 6, 8, 9, 10	
6-3	1, 2, 3, 6, 8, 9, 10	
6-4	1, 2, 3, 6, 8, 9, 10	
6-5	1, 2, 3, 6, 8, 9, 10	
6-6 Preview	2, 3, 6	
6-6	1, 2, 3, 6, 8, 9, 10	

Key to NCTM Standards:

1=Number & Operations, 2=Algebra, 3=Geometry, 4=Measurement, 5=Data Analysis & Probability, 6=Problem Solving, 7=Reasoning & Proof, 8=Communication, 9=Connections, 10=Representation

Chapter 6 Proportions and Similarity

What You'll Learn

- **Lessons 6-1, 6-2, and 6-3** Identify similar polygons, and use ratios and proportions to solve problems.
- **Lessons 6-4 and 6-5** Recognize and use proportional parts, corresponding perimeters, altitudes, angle bisectors, and medians of similar triangles to solve problems.
- **Lesson 6-6** Identify the characteristics of fractals and nongeometric iteration.

Key Vocabulary

- proportion (p. 283)
- cross products (p. 283)
- similar polygons (p. 289)
- scale factor (p. 290)
- midsegment (p. 308)

Why It's Important

Similar figures are used to represent various real-world situations involving a scale factor for the corresponding parts. For example, photography uses similar triangles to calculate distances from the lens to the object and to the image size. *You will use similar triangles to solve problems about photography in Lesson 6-5.*

Vocabulary Builder

ELL

The Key Vocabulary list introduces students to some of the main vocabulary terms included in this chapter. For a more thorough vocabulary list with pronunciations of new words, give students the Vocabulary Builder worksheets found on pages vii and viii of the *Chapter 6 Resource Masters*. Encourage them to complete the definition of each term as they progress through the chapter. You may suggest that they add these sheets to their study notebooks for future reference when studying for the Chapter 6 test.

Getting Started

Prerequisite Skills To be successful in this chapter, you'll need to master these skills and be able to apply them in problem-solving situations. Review these skills before beginning Chapter 6.

For Lesson 6-1, 6-3, and 6-4 — Solve Rational Equations

Solve each equation. *(For review, see pages 737 and 738.)*

1. $\frac{2}{3}y - 4 = 6$ **15** **2.** $\frac{5}{6} = \frac{x-4}{12}$ **14** **3.** $\frac{4}{3} = \frac{y+2}{y-1}$ **10** **4.** $\frac{2y}{4} = \frac{32}{y}$ **±8**

For Lesson 6-2 — Slopes of Lines

Find the slope of the line given the coordinates of two points on the line.
(For review, see Lesson 3-3.)

5. (3, 5) and (0, −1) **2** **6.** (−6, −3) and (2, −3) **0** **7.** (−3, 4) and (2, −2) $-\frac{6}{5}$

For Lesson 6-5 — Show Lines Parallel

Given the following information, determine whether $a \parallel b$. State the postulate or theorem that justifies your answer.
(For review, see Lesson 3-5.)

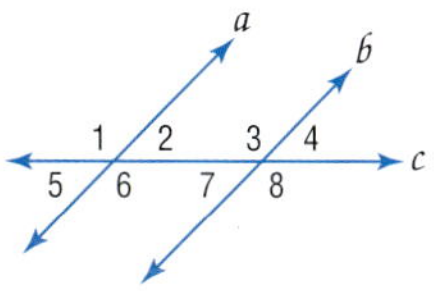

8. $\angle 1 \cong \angle 8$ **yes; ≅ alt. ext. ∠s**

9. $\angle 3 \cong \angle 6$ **yes; ≅ alt. int. ∠s**

10. $\angle 5 \cong \angle 3$ **no**

For Lesson 6-6 — Evaluate Expressions

Evaluate each expression for $n = 1, 2, 3,$ and 4. *(For review, see page 736.)*

11. 2^n **2, 4, 8, 16** **12.** $n^2 - 2$ **−1, 2, 7, 14** **13.** $3^n - 2$ **1, 7, 25, 79**

FOLDABLES™ Study Organizer

Proportions and Similarity Make this Foldable to record information about proportions and similarity in this chapter. Begin with one sheet of 11-inch × 17-inch paper.

Step 1 Punch and Fold

Fold lengthwise. Leave space to punch holes so it can be placed in your binder.

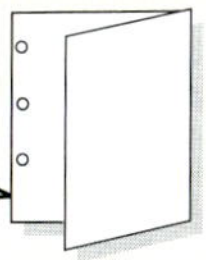

Step 2 Divide

Open the flap and draw lines to divide the inside into six equal parts.

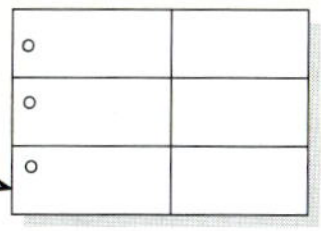

Step 3 Label

Label each part using the lesson numbers.

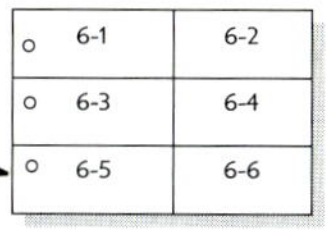

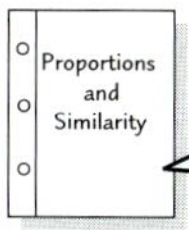

Put the name of the chapter on the front flap.

Reading and Writing As you read and study the chapter, use the Foldable to write down questions you have about the concepts in each lesson. Leave room to record the answers to your questions.

Getting Started

This section provides a review of the basic concepts needed before beginning Chapter 6. Page references are included for additional student help.

Additional review is provided in the *Prerequisite Skills Workbook,* pages 19–20, 23–26, 29–38, 43–44, 53–54, 69–72, 89–92, 95–96.

Prerequisite Skills in the Getting Ready for the Next Lesson section at the end of each exercise set review a skill needed in the next lesson.

For Lesson	Prerequisite Skill
6-2	Distance Formula, p. 287
6-3	Angles and Parallel Lines, p. 297
6-4	Coordinates of Midpoints, p. 306
6-5	Corresponding Congruent Parts, p. 315
6-6	Patterns, p. 323

FOLDABLES™ Study Organizer

For more information about Foldables, see *Teaching Mathematics with Foldables.*

Organization of Data Use this Foldable for student notes, problem solving, and descriptions. After students make their Foldable, have them label the sections to correspond to the six lessons in this chapter. As students read and work through each lesson of this chapter, have them record questions that arise on their Foldable. As students read and learn more about proportions and similarity, encourage students to answer their own questions. Self-questioning is a strategy that helps students to answer their own questions and stay focused during reading and skills practice.

6-1 Lesson Notes

1 Focus

5-Minute Check Transparency 6-1 Use as a quiz or review of Chapter 5.

Mathematical Background notes are available for this lesson on p. 280C.

How do artists use ratios?

Ask students:

- What geometric shape did artist Louis Comfort Tiffany use as the background for the flowers and vines? **rectangles**
- What is another example of the use of rectangles in a design? **Sample answer: Architects use rectangles in the designs of buildings.**
- What is an example of a scale model? **Sample answer: a model airplane**

6-1 Proportions

Sunshine State Standards
MA.A.3.4.3, MA.B.1.4.3, MA.C.3.4.1

What You'll Learn

- Write ratios.
- Use properties of proportions.

How do artists use ratios?

Stained-glass artist Louis Comfort Tiffany used geometric shapes in his designs. In a portion of *Clematis Skylight* shown at the right, rectangular shapes are used as the background for the flowers and vines. Tiffany also used ratio and proportion in the design of this piece.

Vocabulary

- ratio
- proportion
- cross products
- extremes
- means

WRITE RATIOS A **ratio** is a comparison of two quantities. The ratio of a to b can be expressed $\frac{a}{b}$, where b is not zero. This ratio can also be written as $a:b$.

Study Tip

Unit Ratios
Ratios should be written in simplest form. A ratio in which the denominator is 1 is called a *unit ratio*.

Example 1 Write a Ratio

SOCCER **The U.S. Census Bureau surveyed 8218 schools nationally about their girls' soccer programs. They found that 270,273 girls participated in a high school soccer program in the 1999–2000 school year. Find the ratio of girl soccer players per school to the nearest tenth.**

To find this ratio, divide the number of girl athletes by the number of schools.

$$\frac{\text{number of girl soccer players}}{\text{number of schools}} = \frac{270{,}273}{8{,}218} \text{ or about } 32.9$$

32.9 can be written as $\frac{32.9}{1}$. So, the ratio for this survey was 32.9 girl soccer players for each school in the survey.

Extended ratios can be used to compare three or more numbers. The expression $a:b:c$ means that the ratio of the first two numbers is $a:b$, the ratio of the last two numbers is $b:c$, and the ratio of the first and last numbers is $a:c$.

FCAT Practice
Standardized Test Practice
A B C D

Example 2 Extended Ratios in Triangles

Multiple-Choice Test Item

In a triangle, the ratio of the measures of three sides is 4:6:9, and its perimeter is 190 inches. Find the length of the longest side of the triangle. (A) 10 in. (B) 60 in. (C) 90 in. (D) 100 in.

Read the Test Item

You are asked to apply the ratio to the three sides of the triangle and the perimeter to find the longest side.

Resource Manager

Workbook and Reproducible Masters

Chapter 6 Resource Masters
- Study Guide and Intervention, pp. 295–296
- Skills Practice, p. 297
- Practice, p. 298
- Reading to Learn Mathematics, p. 299
- Enrichment, p. 300

Graphing Calculator and Computer Masters, p. 27
School-to-Career Masters, p. 11
Prerequisite Skills Workbook, pp. 19–20, 23–24, 29–32, 37–38, 69–72, 91–92
Teaching Geometry With Manipulatives Masters, pp. 1, 17

Transparencies

5-Minute Check Transparency 6-1
Real-World Transparency 6
Answer Key Transparencies

Technology

Interactive Chalkboard

Solve the Test Item

Recall that equivalent fractions can be found by multiplying the numerator and the denominator by the same number. So, $2:3 = \frac{2}{3} \cdot \frac{x}{x}$ or $\frac{2x}{3x}$. Thus, we can rewrite $4:6:9$ as $4x:6x:9x$ and use those measures for the sides of the triangle. Write an equation to represent the perimeter of the triangle as the sum of the measures of its sides.

$4x + 6x + 9x = 190$	Perimeter
$19x = 190$	Combine like terms.
$x = 10$	Divide each side by 19.

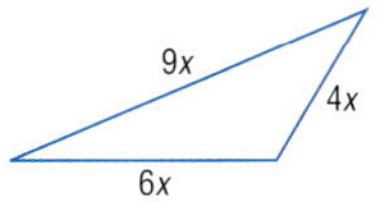

Use this value of x to find the measures of the sides of the triangle.

$4x = 4(10)$ or 40 inches

$6x = 6(10)$ or 60 inches

$9x = 9(10)$ or 90 inches

The longest side is 90 inches. The answer is C.

CHECK Add the lengths of the sides to make sure that the perimeter is 190.

$40 + 60 + 90 = 190$ ✓

The Princeton Review

Test-Taking Tip

Extended ratio problems require a variable to be the common factor among the terms of the ratio. This will enable you to write an equation to solve the problem.

USE PROPERTIES OF PROPORTIONS

An equation stating that two ratios are equal is called a **proportion**. Equivalent fractions set equal to each other form a proportion. Since $\frac{2}{3}$ and $\frac{6}{9}$ are equivalent fractions, $\frac{2}{3} = \frac{6}{9}$ is a proportion.

Every proportion has two **cross products**. The cross products in $\frac{2}{3} = \frac{6}{9}$ are 2 times 9 and 3 times 6. The **extremes** of the proportion are 2 and 9. The **means** are 3 and 6.

extremes → $\frac{2}{3} = \frac{6}{9}$ ← means

cross product of extremes → $2(9) = 3(6)$ ← cross product of means

$18 = 18$

The product of the means equals the product of the extremes, so the cross products are equal. Consider the general case.

$\frac{a}{b} = \frac{c}{d}$	$b \neq 0, d \neq 0$
$(bd)\frac{a}{b} = (bd)\frac{c}{d}$	Multiply each side by the common denominator, bd.
$da = bc$	Simplify.
$ad = bc$	Commutative Property

Study Tip

Reading Mathematics

When a proportion is written using colons, it is read using the word *to* for the colon. For example, 2:3 is read *2 to 3*. The means are the inside numbers, and the extremes are the outside numbers.

extremes
2:3 = 6:9
means

Key Concept — *Property of Proportions*

- **Words** For any numbers a and c and any nonzero numbers b and d, $\frac{a}{b} = \frac{c}{d}$ if and only if $ad = bc$.
- **Examples** $\frac{4}{5} = \frac{12}{15}$ if and only if $4 \cdot 15 = 5 \cdot 12$.

To *solve a proportion* means to find the value of the variable that makes the proportion true.

www.geometryonline.com/extra_examples/fcat

2 Teach

WRITE RATIOS

In-Class Examples Power Point®

1 The total number of students who participate in sports programs at Central High School is 520. The total number of students in the school is 1850. Find the athlete-to-student ratio to the nearest tenth. **0.3**

2 In a triangle, the ratio of the measures of three sides is 5:12:13, and the perimeter is 90 centimeters. Find the measure of the shortest side of the triangle. **A**

A 15 cm **B** 18 cm
C 36 cm **D** 39 cm

Tips for New Teachers

In Example 2, point out that while ratios often appear as fractions, a ratio is a comparison of two quantities. That means that if the quantities have units, the numerator and denominator should be in the same unit.

USE PROPERTIES OF PROPORTIONS

Teaching Tip Students may have previous experience with solving equations by cross-multiplying. Emphasize that a proportion is a statement of equality between ratios. Because the product of the means equals the product of extremes, cross multiplying can be used to solve a proportion.

DAILY INTERVENTION — **Differentiated Instruction**

Logical/Mathematical Stress that the proportion $\frac{a}{b} = \frac{c}{d}$ can also be written as $\frac{c}{d} = \frac{a}{b}$. Ask students to find the product of the means and the product of the extremes for each proportion. Then use the Commutative Property of Multiplication to verify that in each case, $ad = bc$.

Tips for New Teachers

Some students may have difficulty writing proportions correctly. For word problems, make sure they write each ratio comparing like units correctly before they state that two ratios are equal to each other.

In-Class Examples

Power Point®

3 Solve each proportion.

a. $\frac{6}{18.2} = \frac{9}{y}$ **27.3**

b. $\frac{4x-5}{3} = \frac{-26}{6}$ **−2**

4 A boxcar on a train has a length of 40 feet and a width of 9 feet. A scale model is made with a length of 16 inches. Find the width of the model. **3.6 in.**

Example 3 Solve Proportions by Using Cross Products

Solve each proportion.

a. $\frac{3}{5} = \frac{x}{75}$

$\frac{3}{5} = \frac{x}{75}$	Original proportion
$3(75) = 5x$	Cross products
$225 = 5x$	Multiply.
$45 = x$	Divide each side by 5.

b. $\frac{3x-5}{4} = \frac{-13}{2}$

$\frac{3x-5}{4} = \frac{-13}{2}$	Original proportion
$(3x-5)2 = 4(-13)$	Cross products
$6x - 10 = -52$	Simplify.
$6x = -42$	Add 10 to each side.
$x = -7$	Divide each side by 6.

Proportions can be used to solve problems involving two objects that are said to be *in proportion*. This means that if you write ratios comparing the measures of all parts of one object with the measures of comparable parts of the other object, a true proportion would always exist.

Study Tip

Common Misconception
The proportion shown in Example 4 is not the only correct proportion. There are four equivalent proportions:
$\frac{a}{b} = \frac{c}{d}$, $\frac{a}{d} = \frac{c}{b}$, $\frac{b}{a} = \frac{d}{c}$, and $\frac{b}{c} = \frac{d}{a}$.
All of these have identical cross products.

Example 4 Solve Problems Using Proportions

AVIATION **A twinjet airplane has a length of 78 meters and a wingspan of 90 meters. A toy model is made in proportion to the real airplane. If the wingspan of the toy is 36 centimeters, find the length of the toy.**

Because the toy airplane and the real plane are in proportion, you can write a proportion to show the relationship between their measures. Since both ratios compare meters to centimeters, you need not convert all the lengths to the same unit of measure.

$$\frac{\text{plane's length (m)}}{\text{model's length (cm)}} = \frac{\text{plane's wingspan (m)}}{\text{model's wingspan (cm)}}$$

$\frac{78}{x} = \frac{90}{36}$	Substitution
$(78)(36) = x \cdot 90$	Cross products
$2808 = 90x$	Multiply.
$31.2 = x$	Divide each side by 90.

The length of the model would be 31.2 centimeters.

Check for Understanding

Concept Check

1. **Explain** how you would solve $\frac{28}{48} = \frac{21}{x}$. **Cross multiply and divide by 28.**
2. **OPEN ENDED** Write two possible proportions having the extremes 5 and 8. **2. Sample answer:** $\frac{5}{4} = \frac{10}{8}$, $\frac{5}{10} = \frac{4}{8}$
3. **FIND THE ERROR** Madeline and Suki are solving $\frac{15}{x} = \frac{3}{4}$.

Madeline	Suki
$\frac{15}{x} = \frac{3}{4}$	$\frac{15}{x} = \frac{3}{4}$
$45 = 4x$	$60 = 3x$
$11.25 = x$	$20 = x$

Who is correct? Explain your reasoning. **Suki; Madeline did not find the cross products correctly.**

Interactive Chalkboard
PowerPoint® Presentations

This CD-ROM is a customizable Microsoft® PowerPoint® presentation that includes:

- Step-by-step, dynamic solutions of each In-Class Example from the Teacher Wraparound Edition
- Additional, Try These exercises for each example
- The 5-Minute Check Transparencies
- Hot links to Glencoe Online Study Tools

Guided Practice

GUIDED PRACTICE KEY	
Exercises	Examples
4, 5	1
6–8	3
9, 10	2
11	4

4. **HOCKEY** A hockey player scored 9 goals in 12 games. Find the ratio of goals to games. **3:4**

5. **SCULPTURE** A replica of *The Thinker* is 10 inches tall. A statue of *The Thinker,* located in front of Grawemeyer Hall on the Belnap Campus of the University of Louisville in Kentucky, is 10 feet tall. What is the ratio of the replica to the statue in Louisville? $\frac{1}{12}$

Solve each proportion.

6. $\frac{x}{5} = \frac{11}{35}$ $\frac{11}{7}$ 7. $\frac{2.3}{4} = \frac{x}{3.7}$ **2.1275** 8. $\frac{x-2}{2} = \frac{4}{5}$ **3.6**

24 parts

9. The ratio of the measures of three sides of a triangle is 9:8:7, and its perimeter is 144 units. Find the measure of each side of the triangle. **54, 48, 42**

10. The ratio of the measures of three angles of a triangle 5:7:8. Find the measure of each angle of the triangle. **45, 63, 72**

FCAT Practice
Standardized Test Practice

11. **GRID IN** The scale on a map indicates that 1.5 centimeters represent 200 miles. If the distance on the map between Norfolk, Virginia, and Atlanta, Georgia, measures 2.4 centimeters, how many miles apart are the cities? **320**

★ indicates increased difficulty

Practice and Apply

Homework Help	
For Exercises	See Examples
12–17, 23, 25	1
18–22	2
26, 27	4
28–35	3

Extra Practice
See page 764.

12. **BASEBALL** A designated hitter made 8 hits in 10 games. Find the ratio of hits to games. **4:5**

13. **SCHOOL** There are 76 boys in a sophomore class of 165 students. Find the ratio of boys to girls. **76:89**

14. **CURRENCY** In a recent month, 208 South African rands were equivalent to 18 United States dollars. Find the ratio of rands to dollars. **104:9**

15. **EDUCATION** In the 2000–2001 school year, Arizona State University had 44,125 students and 1747 full-time faculty members. What was the ratio of the students to each teacher rounded to the nearest tenth? **25.3:1**

16. Use the number line at the right to determine the ratio of *AC* to *BH*. **1:3**

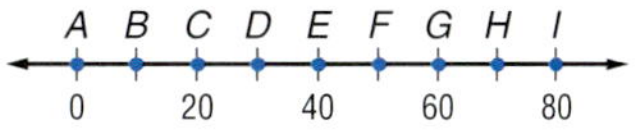

17. A cable that is 42 feet long is divided into lengths in the ratio of 3:4. What are the two lengths into which the cable is divided? **18 ft, 24 ft**

Find the measures of the angles of each triangle.

18. The ratio of the measures of the three angles is 2:5:3. **36, 90, 54**

19. The ratio of the measures of the three angles is 6:9:10. **43.2, 64.8, 72**

Find the measures of the sides of each triangle.

20. The ratio of the measures of three sides of a triangle is 8:7:5. Its perimeter is 240 feet. **96 ft, 84 ft, 60 ft**

21. The ratio of the measures of the sides of a triangle is 3:4:5. Its perimeter is 72 inches. **18 in., 24 in., 30 in.**

★ 22. The ratio of the measures of three sides of a triangle are $\frac{1}{2}:\frac{1}{3}:\frac{1}{5}$, and its perimeter is 6.2 centimeters. Find the measure of each side of the triangle. **3 cm, 2 cm, 1.2 cm**

$\frac{15}{30}$ $\frac{10}{30}$ $\frac{6}{30}$

$6.2 \div \frac{31}{30}$

$6.2 \cdot \frac{30}{31}$

6.

3 Practice/Apply

Study Notebook

Have students—

- add the definitions/examples of the vocabulary terms to their Vocabulary Builder worksheets for Chapter 6.
- include a model for the product of the means equals the product of the extremes, as shown on p. 283.
- include any other item(s) that they find helpful in mastering the skills in this lesson.

DAILY INTERVENTION

FIND THE ERROR Madeline made a common error when solving $\frac{15}{x} = \frac{3}{4}$. She multiplied the numerators together and the denominators together instead of using the cross products property. Stress that students use the cross products property correctly to solve proportions.

About the Exercises...

Organization by Objective

- **Write Ratios:** 12–23, 25
- **Use Properties of Proportions:** 26–35

Odd/Even Assignments

Exercises 12–35 are structured so that students practice the same concepts whether they are assigned odd or even problems.

Assignment Guide

Basic: 13–35 odd, 38–57

Average: 13–37 odd, 38–57

Advanced: 12–38 even, 40–53 (optional: 54–57)

Study Guide and Intervention, p. 295 (shown) and p. 296

Write Ratios A **ratio** is a comparison of two quantities. The ratio a to b, where b is not zero, can be written as $\frac{a}{b}$ or $a{:}b$. The ratio of two quantities is sometimes called a **scale factor**. For a scale factor, the units for each quantity are the same.

Example 1 **In 2002, the Chicago Cubs baseball team won 67 games out of 162. Write a ratio for the number of games won to the total number of games played.**
To find the ratio, divide the number of games won by the total number of games played. The result is $\frac{67}{162}$, which is about 0.41. The Chicago Cubs won about 41% of their games in 2002.

Example 2 **A doll house that is 15 inches tall is a scale model of a real house with a height of 20 feet. What is the ratio of the height of the doll house to the height of the real house?**
To start, convert the height of the real house to inches.
20 feet × 12 inches per foot = 240 inches
To find the ratio or scale factor of the heights, divide the height of the doll house by the height of the real house. The ratio is 15 inches:240 inches or 1:16. The height of the doll house is $\frac{1}{16}$ the height of the real house.

Exercises

1. In the 2002 Major League baseball season, Sammy Sosa hit 49 home runs and was at bat 556 times. Find the ratio of home runs to the number of times he was at bat. $\frac{49}{556}$
2. There are 182 girls in the sophomore class of 305 students. Find the ratio of girls to total students. $\frac{182}{305}$
3. The length of a rectangle is 8 inches and its width is 5 inches. Find the ratio of length to width. $\frac{8}{5}$
4. The sides of a triangle are 3 inches, 4 inches, and 5 inches. Find the scale factor between the longest and the shortest sides. $\frac{5}{3}$
5. The length of a model train is 18 inches. It is a scale model of a train that is 48 feet long. Find the scale factor. $\frac{1}{32}$

Skills Practice, p. 297 and Practice, p. 298 (shown)

1. **NUTRITION** One ounce of cheddar cheese contains 9 grams of fat. Six of the grams of fat are saturated fats. Find the ratio of saturated fats to total fat in an ounce of cheese. 2:3
2. **FARMING** The ratio of goats to sheep at a university research farm is 4:7. The number of sheep at the farm is 28. What is the number of goats? 16
3. **ART** Edward Hopper's oil on canvas painting *Nighthawks* has a length of 60 inches and a width of 30 inches. A print of the original has a length of 2.5 inches. What is the width of the print? 1.25 in.

Solve each proportion.

4. $\frac{5}{8} = \frac{x}{12}$ 7.5
5. $\frac{x}{1.12} = \frac{1}{5}$ 0.224
6. $\frac{6x}{27} = \frac{4}{3}$ 6
7. $\frac{x+2}{3} = \frac{8}{9}$ $\frac{2}{3}$
8. $\frac{3x-5}{4} = \frac{-5}{7}$ $\frac{5}{7}$
9. $\frac{x-2}{4} = \frac{x+4}{2}$ −10

Find the measures of the sides of each triangle.

10. The ratio of the measures of the sides of a triangle is 3:4:6, and its perimeter is 104 feet. 24 ft, 32 ft, 48 ft
11. The ratio of the measures of the sides of a triangle is 7:9:12, and its perimeter is 84 inches. 21 in., 27 in., 36 in.
12. The ratio of the measures of the sides of a triangle is 6:7:9, and its perimeter is 77 centimeters. 21 cm, 24.5 cm, 31.5 cm

Find the measures of the angles in each triangle.

13. The ratio of the measures of the angles is 4:5:6. 48, 60, 72
14. The ratio of the measures of the angles is 5:7:8. 45, 63, 72
15. **BRIDGES** The span of the Benjamin Franklin suspension bridge in Philadelphia, Pennsylvania, is 1750 feet. A model of the bridge has a span of 42 inches. What is the ratio of the span of the model to the span of the actual Benjamin Franklin Bridge? $\frac{1}{500}$

Reading to Learn Mathematics, p. 299 ELL

Pre-Activity **How do artists use ratios?**

Read the introduction to Lesson 6-1 at the top of page 282 in your textbook.

Estimate the ratio of length to width for the background rectangles in Tiffany's Clematis Skylight. Sample answer: about 5 to 2

Reading the Lesson

1. Match each description in the first column with a word or phrase from the second column.

a. The ratio of two corresponding quantities iv	i. proportion
b. r and u in the equation $\frac{r}{s} = \frac{t}{u}$ v	ii. cross products
c. a comparison of two quantities vi	iii. means
d. ru and st in the equation $\frac{r}{s} = \frac{t}{u}$ ii	iv. scale factor
e. an equation stating that two ratios are equal i	v. extremes
f. s and t in the equation $\frac{r}{s} = \frac{t}{u}$ iii	vi. ratio

2. If m, n, p, and q are nonzero numbers such that $\frac{m}{n} = \frac{p}{q}$, which of the following statements could be *false*? B, C, E

A. $np = mq$ B. $\frac{p}{n} = \frac{q}{m}$
C. $mp = nq$ D. $qm = pn$
E. $\frac{m}{n} = \frac{q}{p}$ F. $\frac{q}{p} = \frac{n}{m}$
G. $m{:}p = n{:}q$ H. $m{:}n = p{:}q$

3. Write two proportions that match each description.
   a. Means are 5 and 8; extremes are 4 and 10.
   any two of $\frac{4}{5} = \frac{8}{10}, \frac{4}{8} = \frac{5}{10}, \frac{10}{5} = \frac{8}{4}, \frac{10}{8} = \frac{5}{4}$
   b. Means are 5 and 4; extremes are positive integers that are different from means.
   any two of $\frac{2}{5} = \frac{4}{10}, \frac{2}{4} = \frac{5}{10}, \frac{1}{4} = \frac{5}{20}, \frac{1}{5} = \frac{4}{20}, \frac{10}{5} = \frac{4}{2}, \frac{10}{4} = \frac{5}{2}, \frac{20}{4} = \frac{5}{1}$, or $\frac{20}{5} = \frac{4}{1}$

Helping You Remember

4. Sometimes it is easier to remember a mathematical idea if you put it into words without using any mathematical symbols. How can you use this approach to remember the concept of equality of cross products? Sample answer: The product of the means equals the product of the extremes.

LITERATURE **For Exercises 23 and 24, use the following information.**
Throughout Lewis Carroll's book, *Alice's Adventures in Wonderland*, Alice's size changes. Her normal height is about 50 inches tall. She comes across a door, about 15 inches high, that leads to a garden. Alice's height changes to 10 inches so she can visit the garden.

23. Find the ratio of the height of the door to Alice's height in Wonderland. $\frac{3}{2}$
24. How tall would the door have been in Alice's normal world? **about 75 in.**

More About. . .

Entertainment

In the model, Lincoln is 8 inches tall. In the theater, Lincoln is 6 feet 4 inches tall (his actual adult height).

Source: Disney

25. **ENTERTAINMENT** Before actual construction of the Great Moments with Mr. Lincoln exhibit, Walt Disney and his design company built models that were in proportion to the displays they planned to build. What is the ratio of the height of the model of Mr. Lincoln compared to his actual height? **2:19**

ICE CREAM **For Exercises 26 and 27, use the following information.**
There were approximately 255,082,000 people in the United States in a recent year. According to figures from the United States Census, they consumed about 4,183,344,800 pounds of ice cream that year.

26. If there were 276,000 people in the city of Raleigh, North Carolina, about how much ice cream might they have been expected to consume? **4,526,400 lb**
27. Find the approximate consumption of ice cream per person. **16.4 lb**

Online Research **Data Update** Use the Internet or other resource to find the population of your community. Determine how much ice cream you could expect to be consumed each year in your community. Visit www.geometryonline.com/data_update to learn more.

ALGEBRA **Solve each proportion.**

28. $\frac{3}{8} = \frac{x}{5}$ $\frac{15}{8}$
29. $\frac{a}{5.18} = \frac{1}{4}$ 1.295
30. $\frac{3x}{23} = \frac{48}{92}$ 4
31. $\frac{13}{49} = \frac{26}{7x}$ 14
32. $\frac{2x-13}{28} = \frac{-4}{7}$ $-\frac{3}{2}$
33. $\frac{4x+3}{12} = \frac{5}{4}$ 3
34. $\frac{b+1}{b-1} = \frac{5}{6}$ −11
★ 35. $\frac{3x-1}{2} = \frac{-2}{x+2}$ $-1, -\frac{2}{3}$

PHOTOGRAPHY **For Exercises 36 and 37, use the following information.**
José reduced a photograph that is 21.3 centimeters by 27.5 centimeters so that it would fit in a 10-centimeter by 10-centimeter area.

★ 36. Find the maximum dimensions of the reduced photograph. **7.75 cm by 10 cm**
★ 37. What percent of the original length is the length of the reduced photograph? **about 36%**

38. **CRITICAL THINKING** The ratios of the lengths of the sides of three polygons are given below. Make a conjecture about identifying each type of polygon. **parallelogram**
    a. 2:2:3 **isosceles triangle**
    b. 3:3:3:3 **square or rhombus**
    c. 4:5:4:5 **rectangle or parallelogram**

39. **WRITING IN MATH** Answer the question that was posed at the beginning of the lesson. **See margin.**

How do artists use ratios?

Include the following in your answer:
- four rectangles from the photo that appear to be in proportion, and
- an estimate in inches of the ratio of the width of the skylight to the length of the skylight given that the dimensions of the rectangle in the bottom left corner are approximately 3.5 inches by 5.5 inches.

286 Chapter 6 Proportions and Similarity

Enrichment, p. 300

Golden Rectangles

Use a straightedge, compass, and the instructions below to construct a golden rectangle.

Check students' construction marks. The final figure should look like the one below.

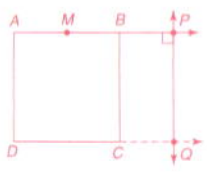

1. Construct square $ABCD$ with sides of 2 cm.
2. Construct the midpoint of $\overline{AB}$. Call the midpoint M.
3. Draw $\overleftrightarrow{AB}$. Set your compass at an opening equal to MC. Use M as the center to draw an arc that intersects $\overleftrightarrow{AB}$. Call the point of intersection P.
4. Construct a line through P that is perpendicular to $\overleftrightarrow{AB}$.
5. Draw $\overleftrightarrow{DC}$ so that it intersects the perpendicular line in step 4. Call the intersection point Q. $APQD$ is a **golden rectangle** because the ratio of its length to its width is 1.618. Check this conclusion by finding the value of $\frac{QP}{AP}$. Rectangles whose sides ...

FCAT Practice
Standardized Test Practice

40. **SHORT RESPONSE** In a golden rectangle, the ratio of the length of the rectangle to its width is approximately 1.618:1. Suppose a golden rectangle has a length of 12 centimeters. What is its width to the nearest tenth? **7.4 cm**

41. **ALGEBRA** A breakfast cereal contains wheat, rice, and oats in the ratio 3:1:2. If the manufacturer makes a mixture using 120 pounds of oats, how many pounds of wheat will be used? **D**

(A) 60 lb (B) 80 lb (C) 120 lb (D) 180 lb

Maintain Your Skills

Mixed Review

In the figure, $\overline{SO}$ is a median of $\triangle SLN$, $\overline{OS} \cong \overline{NP}$, $m\angle 1 = 3x - 50$, and $m\angle 2 = x + 30$. Determine whether each statement is *always*, *sometimes*, or *never* true. *(Lesson 5-5)*

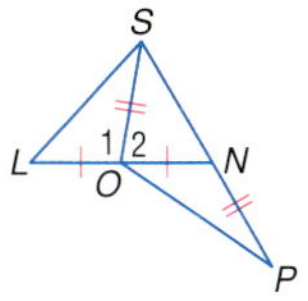

42. $LS > SN$ **always**
43. $SN < OP$ **always**
44. $x = 45$ **never**

Find the range for the measure of the third side of a triangle given the measures of two sides. *(Lesson 5-4)*

45. 16 and 31 **$15 < x < 47$** 46. 26 and 40 **$14 < x < 66$** 47. 11 and 23 **$12 < x < 34$**

48. **COORDINATE GEOMETRY** Given $\triangle STU$ with vertices $S(0, 5)$, $T(0, 0)$, and $U(-2, 0)$ and $\triangle XYZ$ with vertices $X(4, 8)$, $Y(4, 3)$, and $Z(6, 3)$, show that $\triangle STU \cong \triangle XYZ$. *(Lesson 4-4)* **$ST = 5$, $TU = 2$, $SU = \sqrt{29}$, $XY = 5$, $YZ = 2$, $XZ = \sqrt{29}$; $\triangle STU \cong \triangle XYZ$ by SSS.**

Graph the line that satisfies each condition. *(Lesson 3-3)* **49–52. See margin.**

49. $m = \frac{3}{5}$ and contains $P(-3, -4)$
50. contains $A(5, 3)$ and $B(-1, 8)$
51. parallel to $\overleftrightarrow{JK}$ with $J(-1, 5)$ and $K(4, 3)$ and contains $E(2, 2)$
52. contains $S(8, 1)$ and is perpendicular to $\overleftrightarrow{QR}$ with $Q(6, 2)$ and $R(-4, -6)$

53. **MAPS** On a U.S. map, there is a scale that lists kilometers on the top and miles on the bottom.

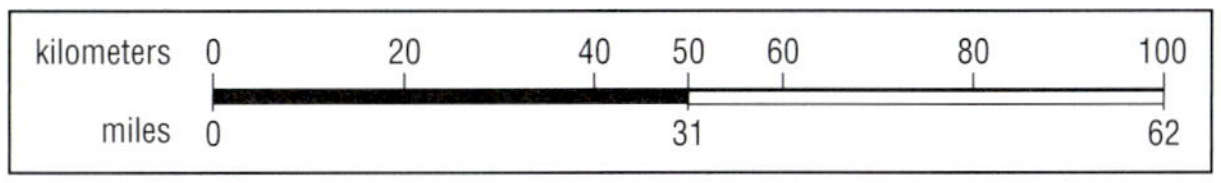

Suppose $\overline{AB}$ and $\overline{CD}$ are segments on this map. If $AB = 100$ kilometers and $CD = 62$ miles, is $\overline{AB} \cong \overline{CD}$? Explain. *(Lesson 2-7)* **Yes; 100 km and 62 mi are the same length, so $AB = CD$. By the definition of congruency, $\overline{AB} \cong \overline{CD}$.**

Getting Ready for the Next Lesson

PREREQUISITE SKILL **Find the distance between each pair of points to the nearest tenth.** *(To review the **Distance Formula**, see Lesson 1-3.)*

54. $A(12, 3)$, $B(-8, 3)$ **20.0**
55. $C(0, 0)$, $D(5, 12)$ **13.0**
56. $E\left(\frac{4}{5}, -1\right)$, $F\left(2, \frac{-1}{2}\right)$ **1.3**
57. $G\left(3, \frac{3}{7}\right)$, $H\left(4, -\frac{2}{7}\right)$ **1.2**

www.geometryonline.com/self_check_quiz/fcat

4 Assess

Open-Ended Assessment

Speaking Have students summarize how to solve a proportion. **Sample answer: Set the product of the means equal to the product of the extremes and solve.**

Getting Ready for Lesson 6-2

Prerequisite Skill Students will learn about similar polygons in Lesson 6-2. They will use the Distance Formula to prove that the lengths of the sides of two similar triangles are proportional. Use Exercises 54–57 to determine your students' familiarity with the Distance Formula.

49.

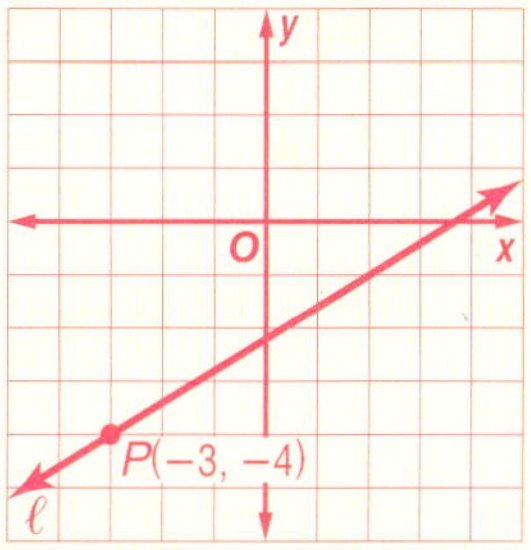

50.

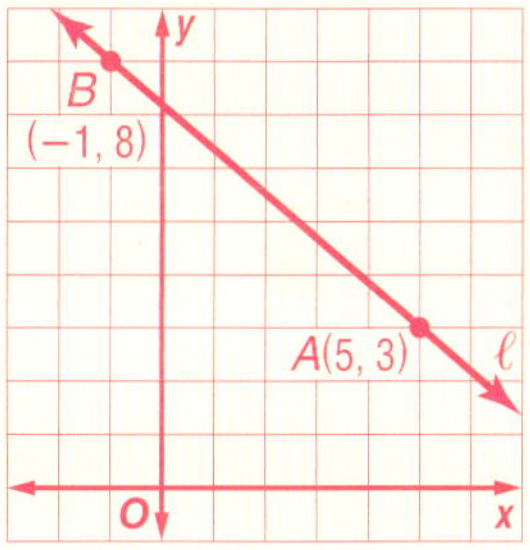

51.

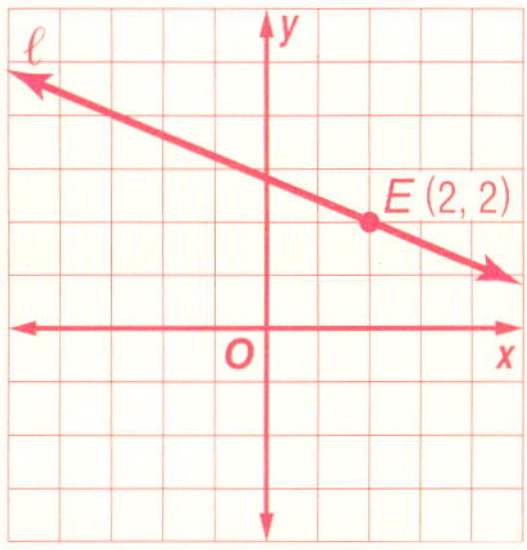

52.

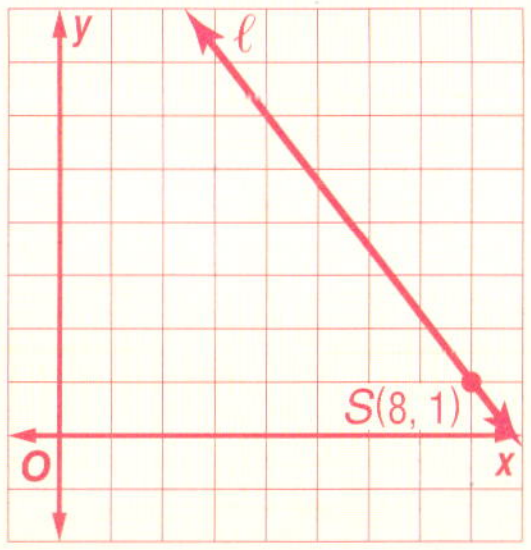

Answers

39. **Sample answer: It appears that Tiffany used rectangles with areas that were in proportion as a background for this artwork. Answers should include the following.**
- **The center column pieces are to the third column from the left pieces as the pieces from the third column are to the pieces in the outside column.**
- **The dimensions are approximately 24 inches by 34 inches.**

Spreadsheet Investigation

A Follow-Up of Lesson 6-1

Getting Started

Cells in Formulas Before students begin typing the formulas, make sure they know that instead of typing a cell name, such as A3, students can use the mouse to click on the cell they are referencing and it will be entered into the formula automatically.

Teach

- Students may not be familiar with sequences that are neither arithmetic nor geometric.
- Have students compare their spreadsheets to that shown on the student page before extending the sequence.
- If students work in pairs, make sure both have a turn entering data.
- Interested students may wish to research other sequences on the Internet. Or they may wish to research how the Fibonacci sequence can be found in sunflowers and pineapples.

Assess

Students may need to extend the sequence on their spreadsheets before they notice that the ratio is approaching 1.618.

Spreadsheet Investigation

A Follow-Up of Lesson 6-1

Fibonacci Sequence and Ratios

The Fibonacci sequence is a set of numbers that begins with 1 as its first and second terms. Each successive term is the sum of the two numbers before it. This sequence continues on indefinitely.

term	1	2	3	4	5	6	7
Fibonacci number	1	1	2	3	5	8	13
			↑ 1 + 1	↑ 1 + 2	↑ 2 + 3	↑ 3 + 5	↑ 5 + 8

Example

Use a spreadsheet to create twenty terms of the Fibonacci sequence. Then compare each term with its preceding term.

Step 1 Enter the column headings in rows 1 and 2.

Step 2 Enter 1 into cell A3. Then insert the formula =A3 + 1 in cell A4. Copy this formula down the column. This will automatically calculate the number of the term.

Step 3 In column B, we will record the Fibonacci numbers. Enter 1 in cells B3 and B4 since you do not have two previous terms to add. Then insert the formula =B3 + B4 in cell B5. Copy this formula down the column.

Step 4 In column C, we will find the ratio of each term to its preceding term. Enter 1 in cell C3 since there is no preceding term. Then enter =B4/B3 in cell C4. Copy this formula down the column.

Fibonacci Table

	A	B	C
1	term	Fibonacci number	ratio
2	n	F(n)	F(n+1)/F(n)
3	1	1	1
4	2	1	1
5	3	2	2
6	4	3	1.5
7	5	5	1.666666667
8	6	8	1.6
9	7	13	1.625

Sheet1 / Sheet2

Exercises

1. What happens to the Fibonacci number as the number of the term increases? **It increases also.**
2. What pattern of odd and even numbers do you notice in the Fibonacci sequence? **odd-odd-even**
3. As the number of terms gets greater, what pattern do you notice in the ratio column? **It approaches 1.618.**
4. Extend the spreadsheet to calculate fifty terms of the Fibonacci sequence. Describe any differences in the patterns you described in Exercises 1–3. **The increase in terms confirms the original observations.**

The rectangle that most humans perceive to be pleasing to the eye has a width to length ratio of about 1:1.618. This is called the *golden ratio,* and the rectangle is called the *golden rectangle.* This type of rectangle is visible in nature and architecture. The Fibonacci sequence occurs in nature in patterns that are also pleasing to the human eye, such as in sunflowers, pineapples, and tree branch structure.

5. **MAKE A CONJECTURE** How might the Fibonacci sequence relate to the golden ratio? **As the number of terms increases, the ratio of each term to its preceding term approaches the golden ratio.**

6-2 Similar Polygons

Sunshine State Standards
MA.A.3.4.3, MA.C.1.4.4, MA.B.1.4.3, MA.C.2.4.1, NC.C.1.4.1, NC.C.2.4.1,

Vocabulary

- similar polygons
- scale factor

What You'll Learn

- Identify similar figures.
- Solve problems involving scale factors.

How do artists use geometric patterns?

M.C. Escher (1898–1972) was a Dutch graphic artist known for drawing impossible structures, spatial illusions, and repeating interlocking geometric patterns. The image at the right is a print of Escher's *Circle Limit IV*, which is actually a woodcutting. It includes winged images that have the same shape, but are different in size. Also note that there are not only similar dark images but also similar light images.

Circle Limit IV, M.C. Escher (1960)

IDENTIFY SIMILAR FIGURES When polygons have the same shape but may be different in size, they are called **similar polygons**.

Study Tip

Similarity and Congruence

If two polygons are congruent, they are also similar. All of the corresponding angles are congruent, and the lengths of the corresponding sides have a ratio of 1:1.

Key Concept — *Similar Polygons*

- **Words** Two polygons are similar if and only if their corresponding angles are congruent and the measures of their corresponding sides are proportional.
- **Symbol** ~ is read *is similar to*
- **Example**

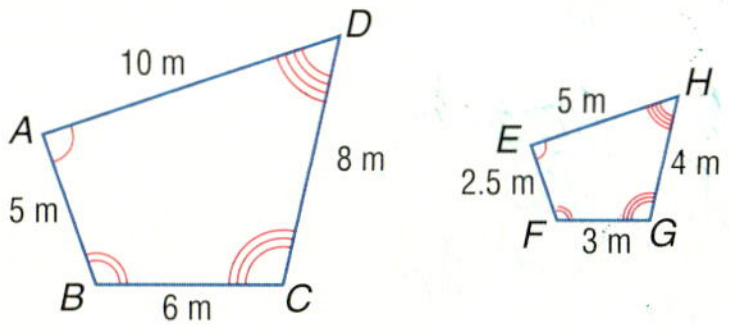

The order of the vertices in a similarity statement is important. It identifies the corresponding angles and the corresponding sides.

similarity statement	congruent angles	corresponding sides
$ABCD \sim EFGH$	$\angle A \cong \angle E$ $\angle B \cong \angle F$ $\angle C \cong \angle G$ $\angle D \cong \angle H$	$\frac{AB}{EF} = \frac{BC}{FG} = \frac{CD}{GH} = \frac{DA}{HE}$

Like congruent polygons, similar polygons may be repositioned so that corresponding parts are easy to identify.

6-2 Lesson Notes

1 Focus

5-Minute Check Transparency 6-2 Use as a quiz or review of Lesson 6-1.

Mathematical Background notes are available for this lesson on p. 280C.

How do artists use geometric patterns?

Ask students:

- How many shapes are shown in the pattern? **2**
- What is the measure of the angle between the figures at the center of the graphic? **120**
- Are all the black figures congruent? **No, they are the same shape, but different sizes.**

Tips for New Teachers

Students may have difficulty identifying the corresponding parts of similar triangles. Point out that the smallest size angles must correspond, as well as the middle size angles and the largest angles. Draw and label several pairs of triangles and ask students to practice writing similarity statements by naming corresponding vertices.

Resource Manager

Workbook and Reproducible Masters

Chapter 6 Resource Masters
- Study Guide and Intervention, pp. 301–302
- Skills Practice, p. 303
- Practice, p. 304
- Reading to Learn Mathematics, p. 305
- Enrichment, p. 306
- Assessment, p. 345

School-to-Career Masters, p. 12
Prerequisite Skills Workbook, pp. 89–90, 95–96
Teaching Geometry With Manipulatives Masters, pp. 1, 17

Transparencies

5-Minute Check Transparency 6-2
Answer Key Transparencies

Technology

Interactive Chalkboard
Multimedia Applications: Virtual Activities

2 Teach

IDENTIFY SIMILAR FIGURES

In-Class Example

Power Point®

1 Determine whether each pair of figures is similar. Justify your answer.

a.

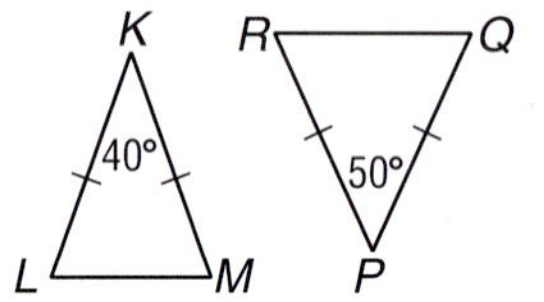

None of the corresponding angles are congruent, so the triangles are not similar.

b.

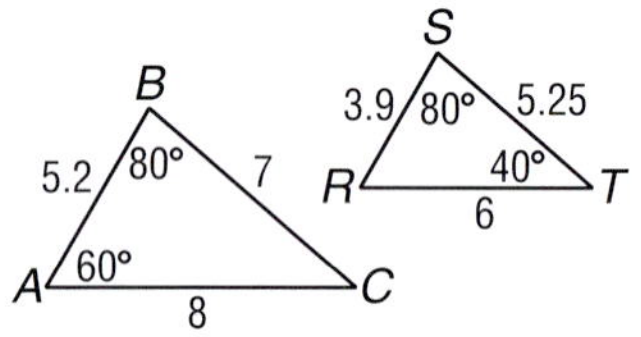

$\angle B \cong \angle S$; $m\angle C = 40$ by the Angle Sum Th.; $m\angle R = 60$ by the Angle Sum Th.; therefore $\angle C \cong \angle T$ and $\angle A \cong \angle R$. All the corresponding angles are congruent. $\frac{AC}{RT} = \frac{AB}{RS} = \frac{BC}{ST} = \frac{4}{3}$. The ratio of the measures of the corresponding sides are equal and the corresponding angles are congruent, so $\triangle ABC \sim \triangle RST$.

SCALE FACTORS

In-Class Example

Power Point®

2 An architect prepared a 12-inch model of a skyscraper to look like a real 1100-foot building. What is the scale factor of the model compared to the real building? $\frac{1}{1100}$

Study Tip

Identifying Corresponding Parts
Using different colors to circle letters of congruent angles may help you identify corresponding parts.

Study Tip

Common Misconception
When two figures have vertices that are in alphabetical order, this does not mean that the corresponding vertices in the similarity statement will follow alphabetical order.

Example 1 Similar Polygons

Determine whether each pair of figures is similar. Justify your answer.

a.

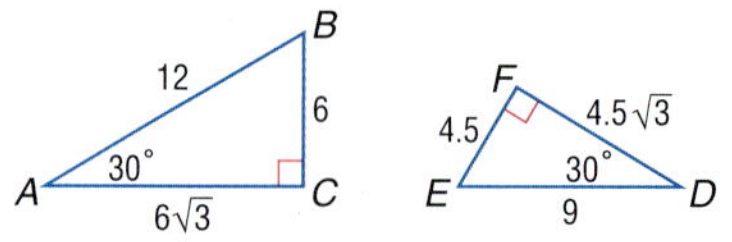

All right angles are congruent, so $\angle C \cong \angle F$.
Since $m\angle A = m\angle D$, $\angle A \cong \angle D$.
By the Third Angle Theorem, $\angle B \cong \angle E$.
Thus, all corresponding angles are congruent.

Now determine whether corresponding sides are proportional.

Sides opposite 90° angle	Sides opposite 30° angle	Sides opposite 60° angle
$\frac{AB}{DE} = \frac{12}{9}$ or $1.\overline{3}$	$\frac{BC}{EF} = \frac{6}{4.5}$ or $1.\overline{3}$	$\frac{AC}{DF} = \frac{6\sqrt{3}}{4.5\sqrt{3}}$ or $1.\overline{3}$

The ratios of the measures of the corresponding sides are equal, and the corresponding angles are congruent, so $\triangle ABC \sim \triangle DEF$.

b.

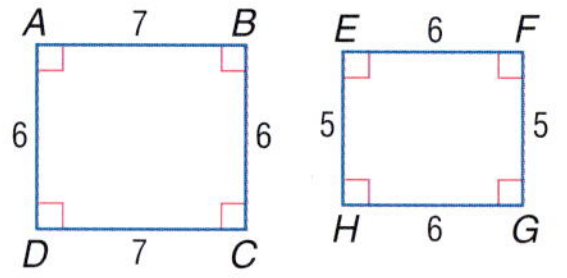

Both rectangles have all right angles and right angles are congruent.
$\frac{AB}{EF} = \frac{7}{6}$ and $\frac{BC}{FG} = \frac{6}{5}$, but $\frac{AB}{EF} \neq \frac{BC}{FG}$ because $\frac{7}{6} \neq \frac{6}{5}$. The rectangles are not similar.

SCALE FACTORS When you compare the lengths of corresponding sides of similar figures, you usually get a numerical ratio. This ratio is called the **scale factor** for the two figures. Scale factors are often given for models of real-life objects.

Example 2 Scale Factor

MOVIES **Some special effects in movies are created using miniature models. In a recent movie, a model sports-utility vehicle (SUV) 22 inches long was created to look like a real $14\frac{2}{3}$-foot SUV. What is the scale factor of the model compared to the real SUV?**

Before finding the scale factor you must make sure that both measurements use the same unit of measure.

$$14\tfrac{2}{3}(12) = 176 \text{ inches}$$

$$\frac{\text{length of model}}{\text{length of real SUV}} = \frac{22 \text{ inches}}{176 \text{ inches}} = \frac{1}{8}$$

The ratio comparing the two lengths is $\frac{1}{8}$ or 1:8. The scale factor is $\frac{1}{8}$, which means that the model is $\frac{1}{8}$ the length of the real SUV.

DAILY INTERVENTION

Differentiated Instruction

Visual/Spatial Show students how to be consistent when analyzing figures for similarity. For example, they may choose to always compare the figure on the left to the figure on the right. Show students ways to organize their work so that they reference corresponding vertices in the correct order.

290 Chapter 6 Proportions and Similarity

When finding the scale factor for two similar polygons, the scale factor will depend on the order of comparison.

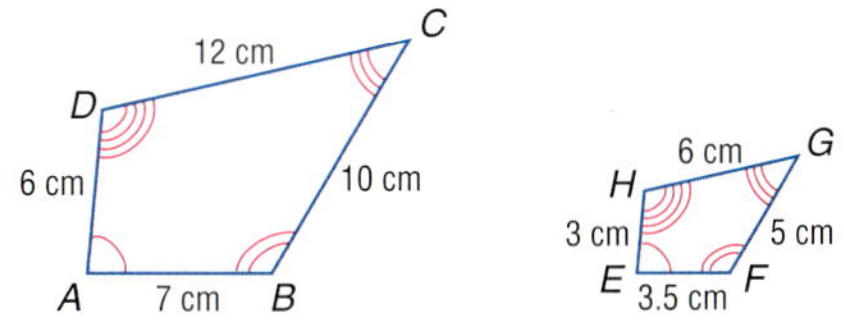

- The scale factor of quadrilateral *ABCD* to quadrilateral *EFGH* is 2.
- The scale factor of quadrilateral *EFGH* to quadrilateral *ABCD* is $\frac{1}{2}$.

Example 3 Proportional Parts and Scale Factor

The two polygons are similar.

a. Write a similarity statement. Then find *x*, *y*, and *UT*.

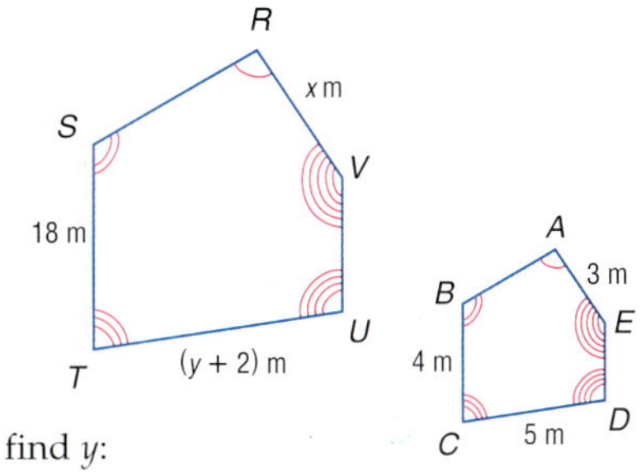

Use the congruent angles to write the corresponding vertices in order.

polygon $RSTUV \sim$ polygon $ABCDE$

Now write proportions to find x and y.

To find x:

$\frac{ST}{BC} = \frac{VR}{EA}$	Similarity proportion
$\frac{18}{4} = \frac{x}{3}$	$ST = 18$, $BC = 4$; $VR = x$, $EA = 3$
$18(3) = 4(x)$	Cross products
$54 = 4x$	Multiply.
$13.5 = x$	Divide each side by 4.

To find y:

$\frac{ST}{BC} = \frac{UT}{DC}$	Similarity proportion
$\frac{18}{4} = \frac{y + 2}{5}$	$ST = 18$, $BC = 4$; $UT = y + 2$, $EA = 3$
$18(5) = 4(y + 2)$	Cross products
$90 = 4y + 8$	Multiply.
$82 = 4y$	Subtract 8 from each side.
$20.5 = y$	Divide each side by 4.

$UT = y + 2$, so $UT = 20.5 + 2$ or 22.5.

b. Find the scale factor of polygon *RSTUV* to polygon *ABCDE*.

The scale factor is the ratio of the lengths of any two corresponding sides.

$\frac{ST}{BC} = \frac{18}{4}$ or $\frac{9}{2}$

You can use scale factors to produce similar figures.

Example 4 Enlargement of a Figure

Triangle *ABC* is similar to $\triangle XYZ$ with a scale factor of $\frac{2}{3}$. If the lengths of the sides of $\triangle ABC$ are 6, 8, and 10 inches, what are the lengths of the sides of $\triangle XYZ$?

Write proportions for finding side measures.

$\frac{\triangle ABC \to 6}{\triangle XYZ \to x} = \frac{2}{3}$	$\frac{\triangle ABC \to 8}{\triangle XYZ \to y} = \frac{2}{3}$	$\frac{\triangle ABC \to 10}{\triangle XYZ \to z} = \frac{2}{3}$
$18 = 2x$	$24 = 2y$	$30 = 2z$
$9 = x$	$12 = y$	$15 = z$

The lengths of the sides of $\triangle XYZ$ are 9, 12, and 15 inches.

Study Tip

Checking Solutions

To verify the scale factor, find the ratio of two other corresponding sides.

Building on Prior Knowledge

In Chapter 4, students learned that congruent triangles have corresponding angles and sides congruent. In this lesson, emphasize that if two figures are congruent, their scale factor is 1.

In-Class Examples

Power Point®

Teaching Tip Point out that students should not write a proportion containing both x and y since they will not be able to solve for both variables. Tell them to use only one variable in a proportion.

3 The two polygons are similar.

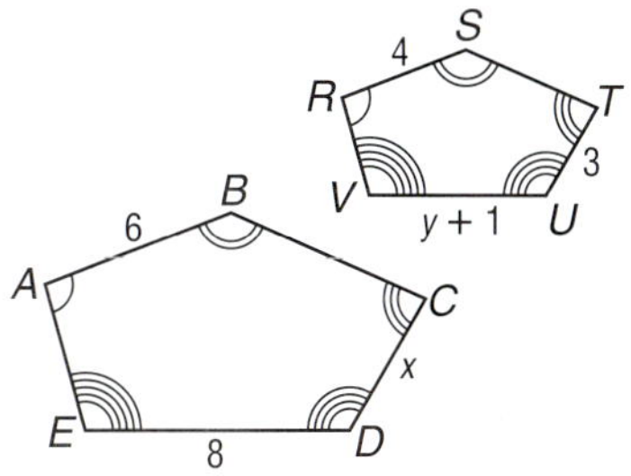

a. Write a similarity statement. Then find x, y, and UV. $ABCDE \sim RSTUV$; $x = \frac{9}{2}$; $y = \frac{13}{3}$; $UV = \frac{16}{3}$

b. Find the scale factor of polygon $ABCDE$ to polygon $RSTUV$. $\frac{3}{2}$

4 Rectangle $WXYZ$ is similar to rectangle $PQRS$ with a scale factor of 1.5. If the length and width of $PQRS$ are 10 meters and 4 meters, respectively, what are the length and width of rectangle $WXYZ$? $\ell = 15$ m, $w = 6$ m

Teacher to Teacher

Mike Patterson, Advanced Technologies Academy — Las Vegas, NV

I use currency exchange as a way to teach ratios. Every year I have students bring in currency from foreign countries and we determine its value compared to the U.S. dollar. I am always amazed at how well students handle ratios when it involves money. The students really enjoy this lesson because they are able to see many different forms of money and discuss where they have traveled or lived around the world. I use this every year. It's a big hit!

In-Class Example

Power Point®

5 The scale on the map of a city is $\frac{1}{4}$ inch equals 2 miles. On the map, the width of the city at its widest point is $3\frac{3}{4}$ inches. The city hosts a bicycle race across town at its widest point. Tashawna bikes at 10 miles per hour. How long will it take her to complete the race? **3 h**

Answers

1. Both students are correct. One student has inverted the ratio and reversed the order of the comparison.
2. A rectangle with consecutive sides of 4 in. and 12 in. would not have sides proportional to a rectangle with consecutive sides of 6 in. and 8 in. because $\frac{4}{6} \neq \frac{12}{8}$.
3. If two polygons are congruent, then they are similar. All of the corresponding angles are congruent, and the ratio of measures of the corresponding sides is 1. Two similar figures have congruent angles, and the sides are in proportion, but not always congruent. If the scale factor is 1, then the figures are congruent.

10. See students' drawings. The drawings will be similar since the measures of the corresponding sides will be proportional and the corresponding angles will be congruent.

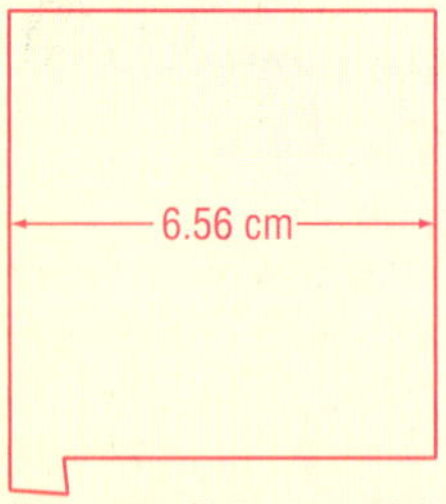

Figure not shown actual size.

11. *ABCF* is similar to *EDCF* since they are congruent.

Example 5 Scale Factors on Maps

MAPS **The scale on the map of New Mexico is 2 centimeters = 160 miles. The distance on the map across New Mexico from east to west through Albuquerque is 4.1 centimeters. How long would it take to drive across New Mexico if you drove at an average of 60 miles per hour?**

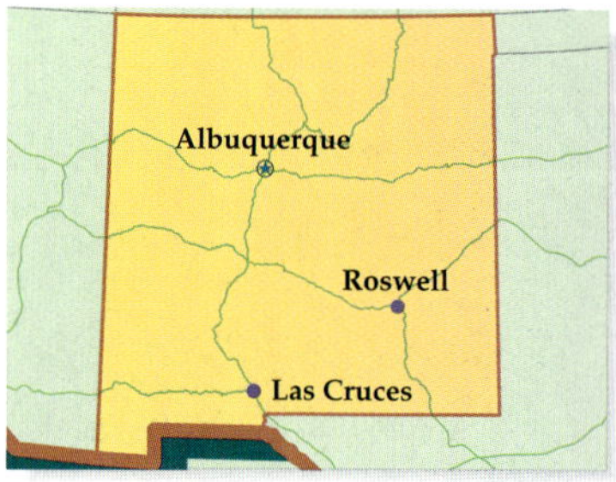

Explore Every 2 centimeters represents 160 miles. The distance across the map is 4.1 centimeters.

Plan Create a proportion relating the measurements to the scale to find the distance in miles. Then use the formula $d = rt$ to find the time.

Solve

centimeters → / miles → $\frac{2}{160} = \frac{4.1}{x}$ ← centimeters / ← miles

$2x = 656$ Cross products

$x = 328$ Divide each side by 2.

The distance across New Mexico is approximately 328 miles.

$d = rt$

$328 = 60t$ $d = 328$ and $r = 60$

$\frac{328}{60} = t$ Divide each side by 60.

$5\frac{7}{15} = t$ Simplify.

It would take $5\frac{7}{15}$ hours or about 5 hours and 28 minutes to drive across New Mexico at an average of 60 miles per hour.

Examine To determine whether the answer is reasonable, reexamine the scale. If 2 centimeters = 160 miles, then 4 centimeters = 320 miles. The map is about 4 centimeters wide, so the distance across New Mexico is about 320 miles. The answer is about 5.5 hours and at 60 miles per hour, the trip would be 330 miles. The two distances are close estimates, so the answer is reasonable.

Study Tip

Units of Time
Remember that there are 60 minutes in an hour. When rewriting $\frac{328}{60}$ as a mixed number, you could also write $5\frac{28}{60}$, which means 5 hours 28 minutes.

Check for Understanding

Concept Check

1. **FIND THE ERROR** Roberto and Garrett have calculated their scale factor for two similar triangles.

Roberto	Garrett
$\frac{AB}{PQ} = \frac{8}{10} = \frac{4}{5}$	$\frac{PQ}{AB} = \frac{10}{8} = \frac{5}{4}$

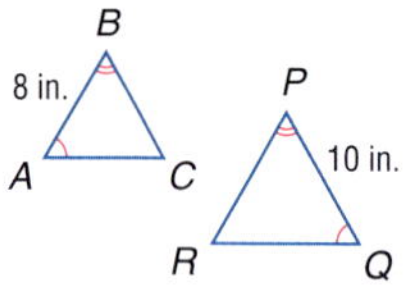

Who is correct? Explain your reasoning. **See margin.**

2. **Find a counterexample** for the statement *All rectangles are similar.*

3. **OPEN ENDED** Explain whether two polygons that are congruent are also similar. Then explain whether two polygons that are similar are also congruent. **2–3. See margin.**

Guided Practice

GUIDED PRACTICE KEY	
Exercises	Examples
4–5	1
6–7	2, 3
8–9	4
10	5

4. Yes; because $\angle P \cong \angle Q \cong \angle R \cong \angle G \cong \angle H \cong \angle I$ and $\frac{PQ}{GH} = \frac{QR}{HI} = \frac{RP}{IG} = \frac{3}{7}$.

5. Yes; $\angle A \cong \angle E$, $\angle B \cong \angle F$, $\angle C \cong \angle G$, $\angle D \cong \angle H$ and $\frac{AD}{EH} = \frac{DC}{HG} = \frac{CB}{GF} = \frac{BA}{FE} = \frac{2}{3}$.

Determine whether each pair of figures is similar. Justify your answer.

4.

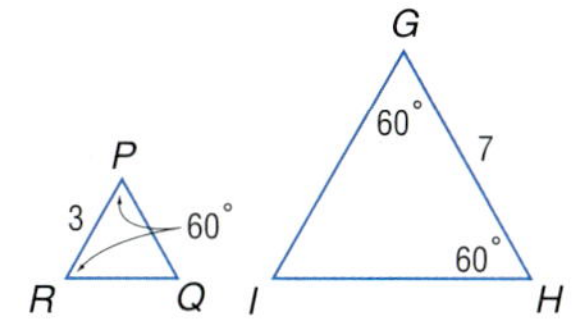

5.

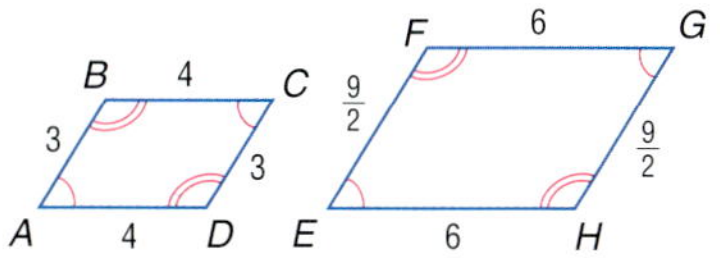

Each pair of polygons is similar. Write a similarity statement, and find *x*, the measure(s) of the indicated side(s), and the scale factor.

6. $\overline{DF}$ $\triangle ACB \sim \triangle DFE$; 14; 14; $\frac{3}{2}$

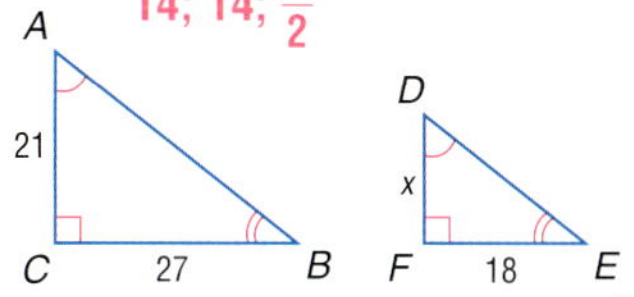

7. $\overline{FE}$, $\overline{EH}$, and $\overline{GF}$

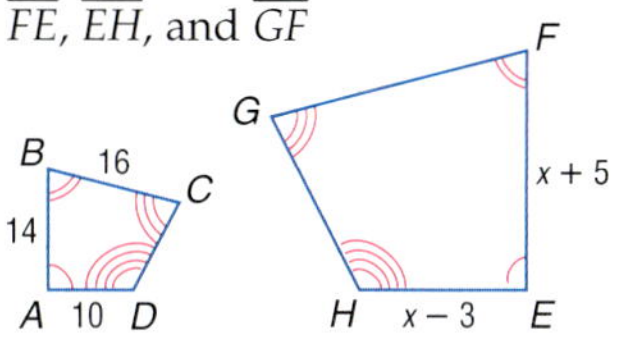

polygon $ABCD \sim$ polygon $EFGH$; 23; 28; 20; 32; $\frac{1}{2}$

8. A rectangle with length 60 centimeters and height 40 centimeters is reduced so that the new rectangle is similar to the original and the scale factor is $\frac{1}{4}$. Find the length and height of the new rectangle. **15 cm, 10 cm**

9. A triangle has side lengths of 3 meters, 5 meters, and 4 meters. The triangle is enlarged so that the larger triangle is similar to the original and the scale factor is 5. Find the perimeter of the larger triangle. **60 m**

Application

10. **MAPS** Refer to Example 5 on page 292. Draw the state of New Mexico using a scale of 2 centimeters = 100 miles. Is your drawing similar to the one in Example 4? Explain how you know. **See margin.**

★ indicates increased difficulty

Practice and Apply

Homework Help	
For Exercises	See Examples
11–14	1
15–20, 34–39	2, 3
21–23	4
24–26	5

Extra Practice See page 765.

11–14. See margin.

Determine whether each pair of figures is similar. Justify your answer.

11.

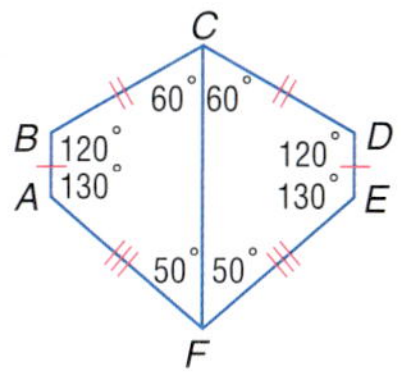

12.

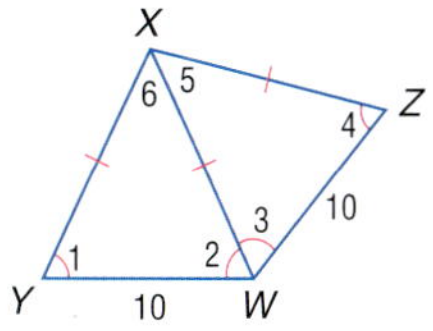

13.

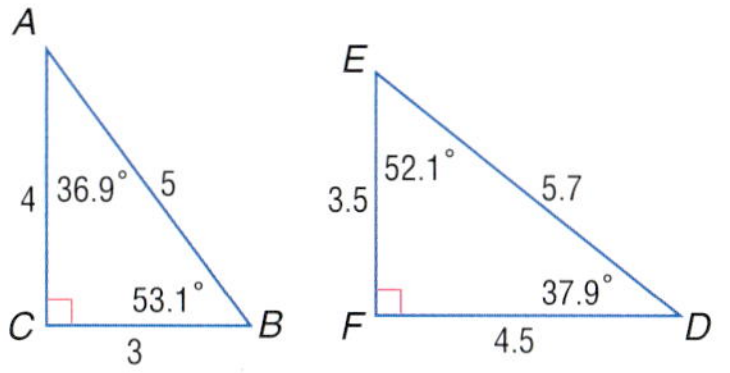

14.

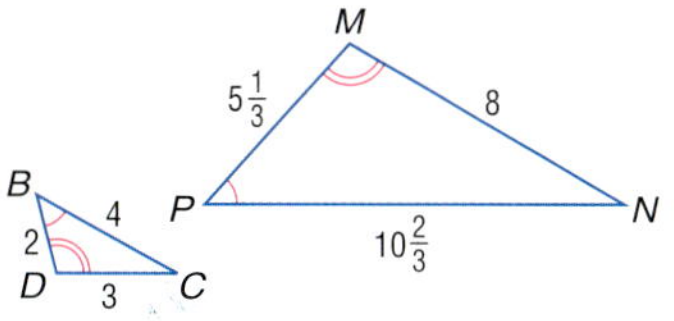

3 Practice/Apply

Study Notebook

Have students—

- add the definitions/examples of the vocabulary terms to their Vocabulary Builder worksheets for Chapter 6.
- include a review of the Slope and Distance Formulas.
- include any other item(s) that they find helpful in mastering the skills in this lesson.

DAILY INTERVENTION **FIND THE ERROR** Watch for students who have difficulty with Exercise 1 because they do not match corresponding sides correctly. Tell students that the similarity statement will tell them which triangle is compared to the other, so that the scale factor is more easily recognized.

About the Exercises...

Organization by Objective

- **Identify Similar Figures:** 11–14
- **Solve Problems with Scale Factors:** 15–26, 34–39

Odd/Even Assignments

Exercises 11–39 are structured so that students practice the same concepts whether they are assigned odd or even problems.

Alert! Exercises 24–26 require the use of a ruler with both customary and metric units.

Assignment Guide

Basic: 11–47 odd, 51–57 odd, 59–63, 70–87 (optional: 64–69)

Average: 11–57 odd, 59–63, 70–87 (optional: 64–69)

Advanced: 10–52 even, 53–56, 58–79 (optional: 80–87)

Answers

12. $\triangle XYW \sim \triangle XWZ$. $\angle 6 \cong \angle 5$ because if two angles of one triangle are congruent to two angles of a second triangle, then the third angles are congruent. The ratio of the corresponding sides is 1.

13. $\triangle ABC$ is not similar to $\triangle DEF$. $\angle A \not\cong \angle D$.

14. $\triangle BCD \sim \triangle PNM$. $\angle B \cong \angle P$, $\angle D \cong \angle M$, and $\angle C \cong \angle N$ because if two angles of one triangle are congruent to two angles of a second triangle, then the third angles are congruent. $\frac{DB}{MP} = \frac{BC}{PN} = \frac{CD}{NM} = \frac{3}{8}$.

Study Guide and Intervention, p. 301 (shown) and p. 302

Identify Similar Figures

Example 1 **Determine whether the triangles are similar.**

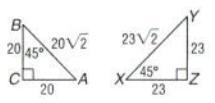

Two polygons are similar if and only if their corresponding angles are congruent and their corresponding sides are proportional.

$\angle C \cong \angle Z$ because they are right angles, and $\angle B \cong \angle X$. By the Third Angle Theorem, $\angle A \cong \angle Y$.

For the sides, $\frac{BC}{XZ} = \frac{20}{23}$, $\frac{BA}{XY} = \frac{20\sqrt{2}}{23\sqrt{2}} = \frac{20}{23}$, and $\frac{AC}{YZ} = \frac{20}{23}$.

The side lengths are proportional. So $\triangle BCA \sim \triangle XZY$.

Example 2 **Is polygon $WXYZ \sim$ polygon $PQRS$?**

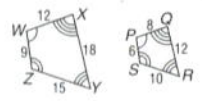

For the sides, $\frac{WX}{PQ} = \frac{12}{8} = \frac{3}{2}$, $\frac{XY}{QR} = \frac{18}{12} = \frac{3}{2}$, $\frac{YZ}{RS} = \frac{15}{10} = \frac{3}{2}$, and $\frac{ZW}{SP} = \frac{9}{6} = \frac{3}{2}$. So corresponding sides are proportional.

Also, $\angle W \cong \angle P$, $\angle X \cong \angle Q$, $\angle Y \cong \angle R$, and $\angle Z \cong \angle S$, so corresponding angles are congruent. We can conclude that polygon $WXYZ \sim$ polygon $PQRS$.

Exercises

Determine whether each pair of figures is similar. If they are similar, give the ratio of corresponding sides.

1. 
equilateral triangles
yes; $\frac{5}{8}$

2. 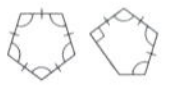
no

3. 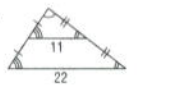
yes; $\frac{1}{2}$

4. 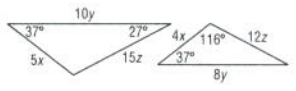
yes; $\frac{5}{4}$

Skills Practice, p. 303 and Practice, p. 304 (shown)

Determine whether each pair of figures is similar. Justify your answer.

1. 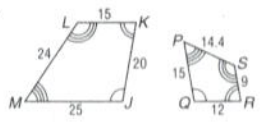
$JKLM \sim QRSP$; $\angle J \cong \angle Q$, $\angle K \cong \angle R$, $\angle L \cong \angle S$, $\angle M \cong \angle P$, and $\frac{JK}{QR} = \frac{KL}{RS} = \frac{LM}{SP} = \frac{MJ}{PQ} = \frac{5}{3}$ or 1.67

2. 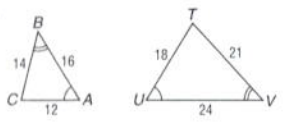
$\triangle ABC \sim \triangle UVT$; $\angle A \cong \angle U$, $\angle B \cong \angle V$, and $\angle C \cong \angle T$ by the Third Angle Theorem and $\frac{AB}{UV} = \frac{BC}{VT} = \frac{CA}{TU} = \frac{2}{3}$.

Each pair of polygons is similar. Write a similarity statement, and find x, the measure(s) of indicated side(s), and the scale factor.

3. $\overline{LM}$ and $\overline{MN}$

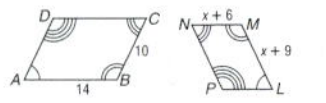

$ABCD \sim LMNP$; $\frac{3}{2}$; $10\frac{1}{2}$; $7\frac{1}{2}$; $\frac{4}{3}$

4. $\overline{DE}$ and $\overline{DF}$

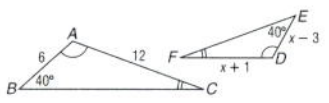

$\triangle ABC \sim \triangle DEF$; 7; 4; 8; $\frac{3}{2}$

5. **COORDINATE GEOMETRY** Triangle ABC has vertices $A(0, 0)$, $B(-4, 0)$, and $C(-2, 4)$. The coordinates of each vertex are multiplied by 3 to create $\triangle AEF$. Show that $\triangle AEF$ is similar to $\triangle ABC$.
$AB = 4$, $BC = CA = 2\sqrt{5}$ and $AE = 12$, $EF = FA = 6\sqrt{5}$. $\frac{AB}{AE} = \frac{BC}{EF} = \frac{CA}{FA} = \frac{1}{3}$; $\angle A \cong \angle A$ and since $\overline{BC} \parallel \overline{EF}$, $\angle B \cong \angle E$ and $\angle C \cong \angle F$. Since the corresponding angles are congruent and the corresponding sides are proportional, $\triangle ABC \sim \triangle AEF$.

6. **INTERIOR DESIGN** Graham used the scale drawing of his living room to decide where to place furniture. Find the dimensions of the living room if the scale in the drawing is 1 inch = 4.5 feet.
18 ft by 11 ft 3 in.

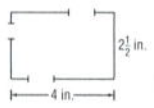

Reading to Learn Mathematics, p. 305 ELL

Pre-Activity **How do artists use geometric patterns?**

Read the introduction to Lesson 6-2 at the top of page 289 in your textbook.

- Describe the figures that have similar shapes. Sample answer: Light and dark figures are arranged in alternating patterns. Figures that have their feet pointing toward the center of the circle have the same shape but different sizes.
- What happens to the figures as your eyes move from the center to the outer edge? Sample answer: The figures become smaller.

Reading the Lesson

1. Complete each sentence.
   a. Two polygons that have exactly the same shape, but not necessarily the same size, are similar.
   b. Two polygons are congruent if they have exactly the same shape and the same size.
   c. Two polygons are similar if their corresponding angles are congruent and their corresponding sides are proportional.
   d. Two polygons are congruent if their corresponding angles are congruent and their corresponding sides are congruent.
   e. The ratio of the lengths of corresponding sides of two similar figures is called the scale factor.
   f. Multiplying the coordinates of all points of a figure in the coordinate plane by a scale factor to get a similar figure is called a dilation.
   g. If two polygons are similar with a scale factor of 1, then the polygons are congruent.
2. Determine whether each statement is *always*, *sometimes*, or *never* true.
   a. Two similar triangles are congruent. sometimes
   b. Two equilateral triangles are congruent. sometimes
   c. An equilateral triangle is similar to a scalene triangle. never
   d. Two rectangles are similar. sometimes
   e. Two isosceles right triangles are congruent. sometimes
   f. Two isosceles right triangles are similar. always
   g. A square is similar to an equilateral triangle. never
   h. Two acute triangles are similar. sometimes
   i. Two rectangles in which the length is twice the width are similar. always
   j. Two congruent polygons are similar. always

Helping You Remember

3. A good way to remember a new mathematical vocabulary term is to relate it to words used in everyday life. The word *scale* has many meanings in English. Give three phrases that include the word *scale* in a way that is related to proportions. Sample answer: scale drawing, scale model, scale of miles (on a map)

15. **ARCHITECTURE** The replica of the Eiffel Tower at an amusement park is $350\frac{2}{3}$ feet tall. The actual Eiffel Tower is 1052 feet tall. What is the scale factor comparing the amusement park tower to the actual tower? $\frac{1}{3}$

16. **PHOTOCOPYING** Mr. Richardson walked to a copier in his office, made a copy of his proposal, and sent the original to one of his customers. When Mr. Richardson looked at his copy before filing it, he saw that the copy had been made at an 80% reduction. He needs his filing copy to be the same size as the original. What enlargement scale factor must he use on the first copy to make a second copy the same size as the original? 1.25 or 125%

Each pair of polygons is similar. Write a similarity statement, and find x, the measures of the indicated sides, and the scale factor.

17. $\overline{AB}$ and $\overline{CD}$

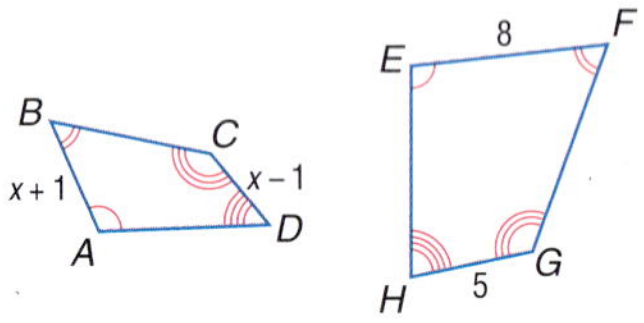

18. $\overline{AC}$ and $\overline{CE}$

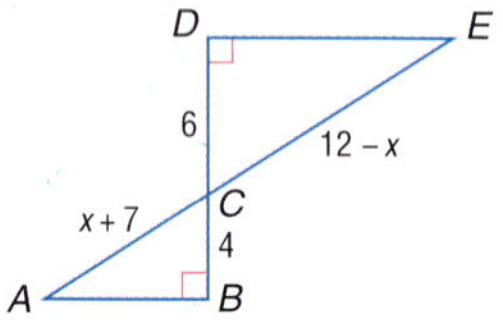

19. $\overline{BC}$ and $\overline{ED}$

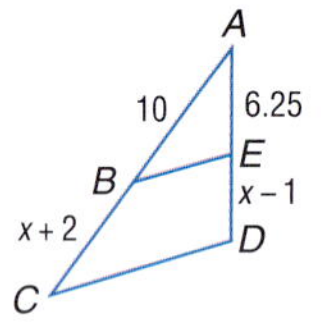

$\triangle ABE \sim \triangle ACD$; 6; $BC = 8$; $ED = 5$; $\frac{5}{9}$

20. $\overline{GF}$ and $\overline{EG}$

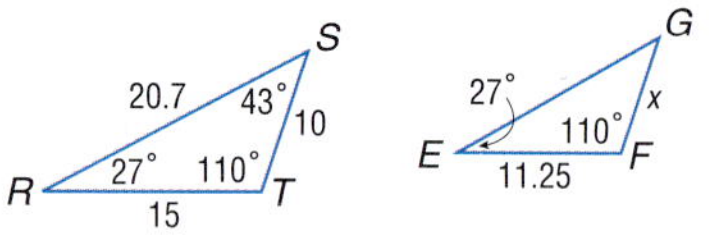

17. polygon $ABCD \sim$ polygon $EFGH$; $\frac{13}{3}$; $AB = \frac{16}{3}$; $CD = \frac{10}{3}$; $\frac{2}{3}$

18. $\triangle ABC \sim \triangle EDC$; $\frac{3}{5}$; $AC = 7\frac{3}{5}$; $CE = 11\frac{2}{5}$; $\frac{2}{3}$

20. $\triangle RST \sim \triangle EGF$; 7.5, $GF = 7.5$; $EG = 15.525$; $\frac{4}{3}$

PHOTOGRAPHY **For Exercises 21–23, use the following information.**

A picture is enlarged by a scale factor of $\frac{5}{4}$ and then enlarged again by the same factor.

21. If the original picture was 2.5 inches by 4 inches, what were its dimensions after both enlargements? about 3.9 in. by 6.25 in.

22. Write an equation describing the enlargement process. $E = \frac{5}{4}\left(\frac{5}{4}x\right)$

23. By what scale factor was the original picture enlarged? $\frac{25}{16}$

More About. . .

Sports

Crew Stadium in Columbus, Ohio, was specifically built for Major League Soccer. The dimensions of the field are about 69 meters by 105 meters.

Source: www.MLSnet.com

SPORTS **Make a scale drawing of each playing field using the given scale.**

24. Use the information about the soccer field in Crew Stadium. Use the scale 1 millimeter = 1 meter.

25. A basketball court is 84 feet by 50 feet. Use the scale $\frac{1}{4}$ inch = 4 feet.

26. A tennis court is 36 feet by 78 feet. Use the scale $\frac{1}{8}$ inch = 1 foot.
24–26. See margin.

Determine whether each statement is *always*, *sometimes*, or *never* true.

27. Two congruent triangles are similar. always
28. Two squares are similar. always
29. A triangle is similar to a quadrilateral. never
30. Two isosceles triangles are similar. sometimes
31. Two rectangles are similar. sometimes
32. Two obtuse triangles are similar. sometimes
33. Two equilateral triangles are similar. always

Enrichment, p. 306

Constructing Similar Polygons

Here are four steps for constructing a polygon that is similar to and with sides twice as long as those of an existing polygon.

Step 1 Choose any point either inside or outside the polygon and label it O.
Step 2 Draw rays from O through each vertex of the polygon.
Step 3 For vertex V, set the compass to length OV. Then locate a new point V' on ray OV such that $VV' = OV$. Thus, $OV' = 2(OV)$.
Step 4 Repeat Step 3 for each vertex. Connect points V', W', X' and Y' to form the new polygon.

Two constructions of polygons similar to and with sides twice those of $VWXY$ are shown below. Notice that the placement of point O does not affect the size or shape of $V'W'X'Y'$, only its location.

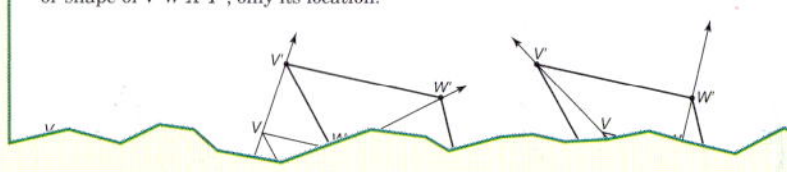

Each pair of polygons is similar. Find x and y. Round to the nearest hundredth if necessary.

34. **91; 30**

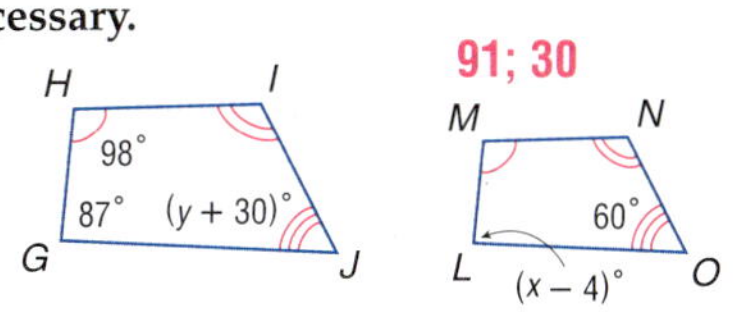

35. **30; 70**

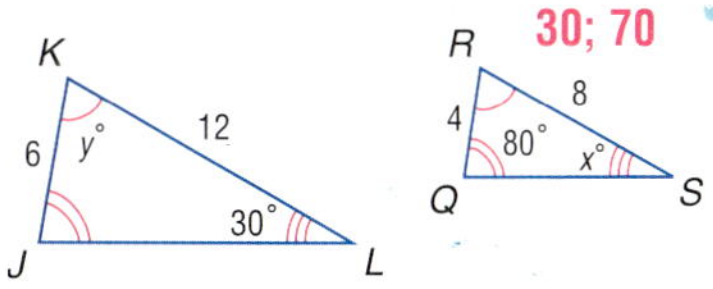

36. **10; 7**

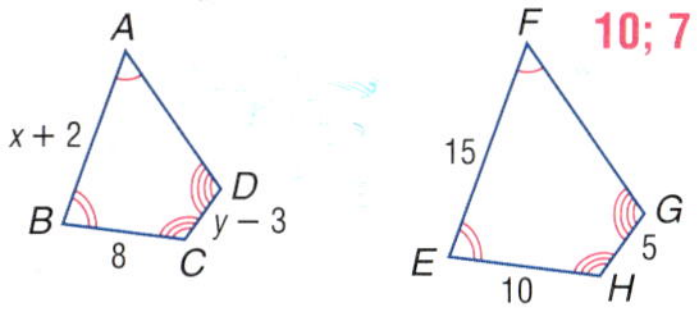

37. **27; 14**

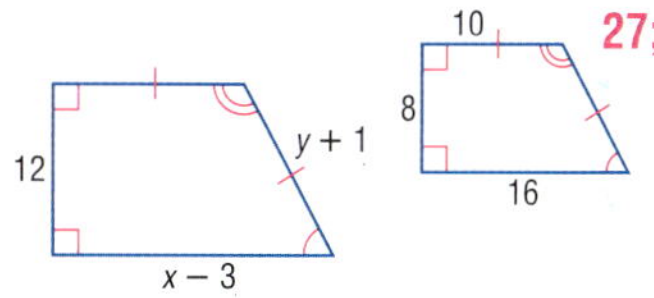

38. **8; 5**

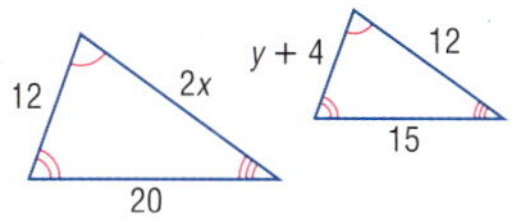

39. **71.05; 48.45**

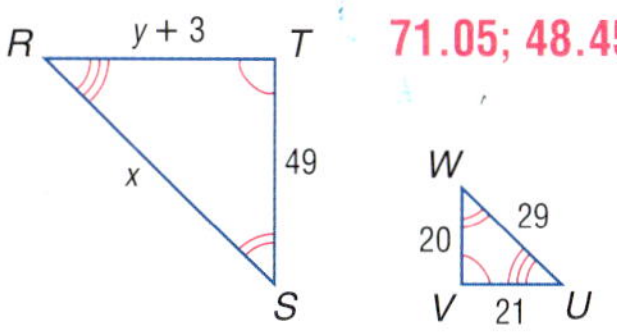

For Exercises 40–47, use the following information to find each measure.
Polygon $ABCD \sim$ polygon $AEFG$, $m\angle AGF = 108$, $GF = 14$, $AD = 12$, $DG = 4.5$, $EF = 8$, and $AB = 26$.

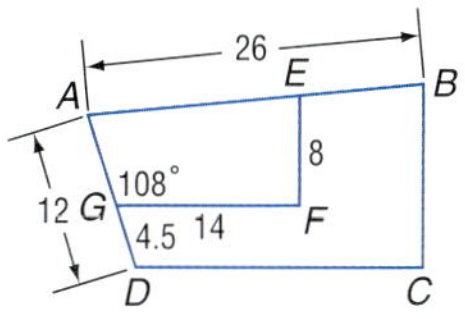

40. scale factor of trapezoid $ABCD$ to trapezoid $AEFG$ $\frac{8}{5}$
41. AG **7.5**
42. DC **22.4**
43. $m\angle ADC$ **108**
44. BC **12.8**
45. perimeter of trapezoid $ABCD$ **73.2**
46. perimeter of trapezoid $AEFG$ **45.75**
47. ratio of the perimeter of polygon $ABCD$ to the perimeter of polygon $AEFG$ $\frac{8}{5}$

48. Determine which of the following right triangles are similar. Justify your answer.

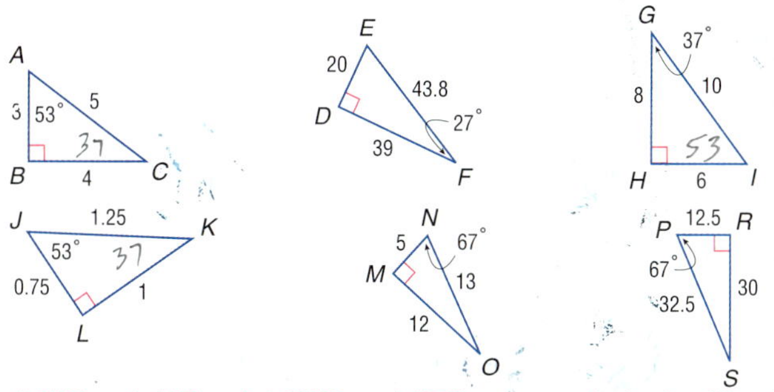

△ABC ∼ △IHG ∼ △JLK and △NMO ∼ △PRS; see margin for justification.

COORDINATE GEOMETRY Graph the given points. Draw polygon $ABCD$ and $\overline{MN}$. Find the coordinates for vertices L and P such that $ABCD \sim NLPM$.

★ 49. $A(2, 0)$, $B(4, 4)$, $C(0, 4)$, $D(-2, 0)$; $M(4, 0)$, $N(12, 0)$

★ 50. $A(-7, 1)$, $B(2, 5)$, $C(7, 0)$, $D(-2, -4)$; $M(-3, 1)$, $N\left(-\frac{11}{2}, \frac{7}{2}\right)$

49–50. See margin for graphs.

49. $L(16, 8)$ and $P(8, 8)$ or $L(16, -8)$ and $P(8, -8)$

50. $L\left(-1, \frac{11}{2}\right)$, $P\left(\frac{3}{2}, 3\right)$ or $L\left(-10, \frac{3}{2}\right)$, $P\left(-\frac{15}{2}, -1\right)$

Answers

24.

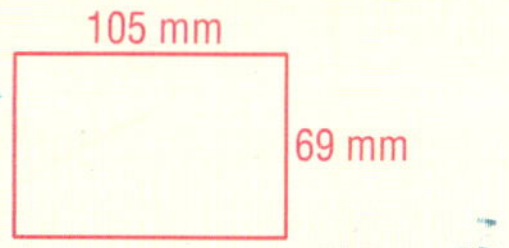

Figure not shown actual size.

25.

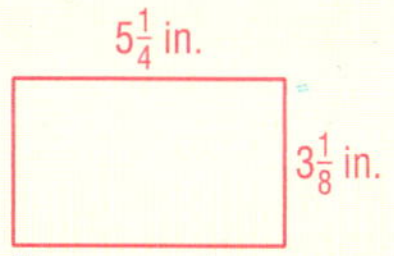

Figure not shown actual size.

26.

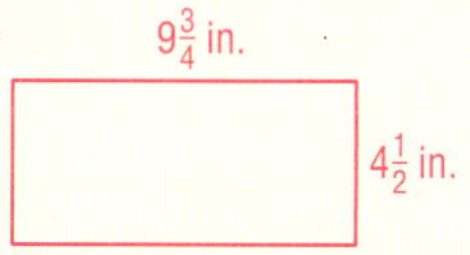

Figure not shown actual size.

48. $\triangle ABC \sim \triangle IHG \sim \triangle JLK$;
$\angle A \cong \angle I \cong \angle J$
$\angle B \cong \angle H \cong \angle L$
$\angle C \cong \angle G \cong \angle K$
$\frac{AC}{IG} = \frac{BC}{HG} = \frac{AB}{IH} = \frac{1}{2}$
$\frac{AC}{JK} = \frac{BC}{LK} = \frac{AB}{JL} = 4$
$\frac{IG}{JK} = \frac{HG}{LK} = \frac{IH}{JL} = 8$
The corresponding angles are congruent, and the measures of their corresponding sides are proportional. $\triangle NMO \sim \triangle PRS$; $\angle N \cong \angle P$, $\angle M \cong \angle R$, $\angle O \cong \angle S$
$\frac{PS}{NO} = \frac{RS}{MO} = \frac{PR}{NM} = 2.5$
The corresponding angles are congruent, and the measures of the corresponding sides are proportional.

49.

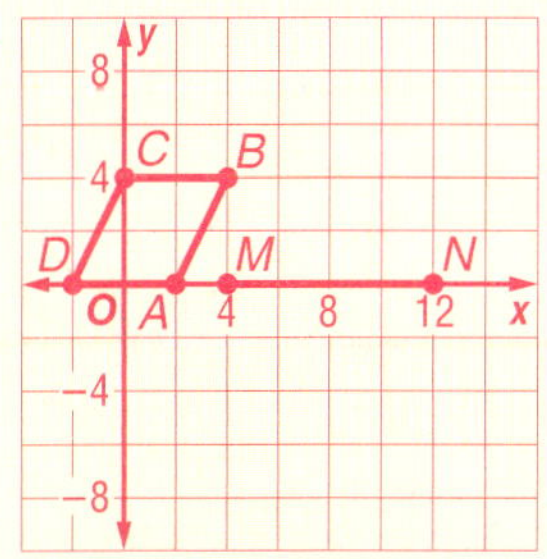

50.

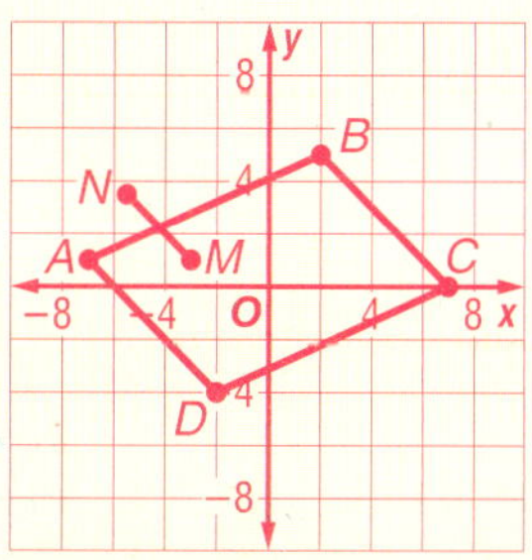

Answers

61. Sample answer: Artists use geometric shapes in patterns to create another scene or picture. The included objects have the same shape but are different sizes. Answers should include the following.

- The objects are enclosed within a circle. The objects seem to go on and on.
- Each "ring" of figures has images that are approximately the same width, but vary in number and design.

CONSTRUCTION **For Exercises 51 and 52, use the following information.**
A floor plan is given for the first floor of a new house. One inch represents 24 feet. Use the information in the plan to find the dimensions.

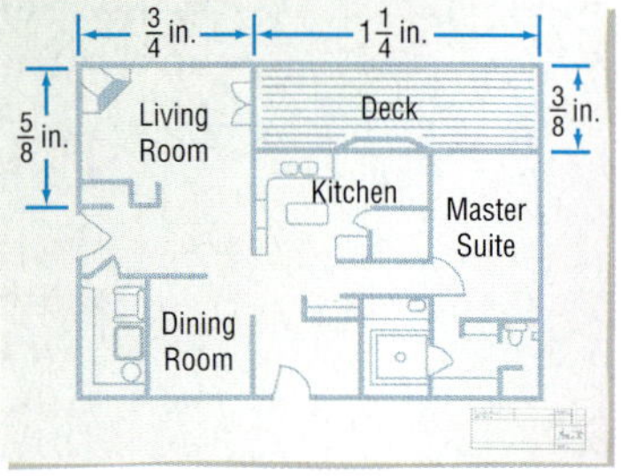

51. living room **18 ft by 15 ft**
52. deck **30 ft by 9 ft**

CRITICAL THINKING **For Exercises 53–55, use the following information.**
The area A of a rectangle is the product of its length ℓ and width w. Rectangle $ABCD$ is similar to rectangle $WXYZ$ with sides in a ratio of 4:1.

53. What is the ratio of the areas of the two rectangles? **16:1**
54. Suppose the dimension of each rectangle is tripled. What is the new ratio of the sides of the rectangles? **4:1**
55. What is the ratio of the areas of these larger rectangles? **16:1**

STATISTICS **For Exercises 56–58, refer to the graphic, which uses rectangles to represent percents.**

56. Are the rectangles representing 36% and 18% similar? Explain.
57. What is the ratio of the areas of the rectangles representing 36% and 18% if area = length × width? Compare the ratio of the areas to the ratio of the percents.
58. Use the graph to make a conjecture about the overall changes in the level of professional courtesy in the workplace in the past five years.

56. No; the corresponding sides are not in proportion. The ratio of the widths is 1 to 1 but the ratio of the heights is 2 to 1.
57. 2 : 1; the two ratios are the same.
58. Sample answer: The difference between increase and decrease is 8%, so the level of courtesy is only slightly decreased.

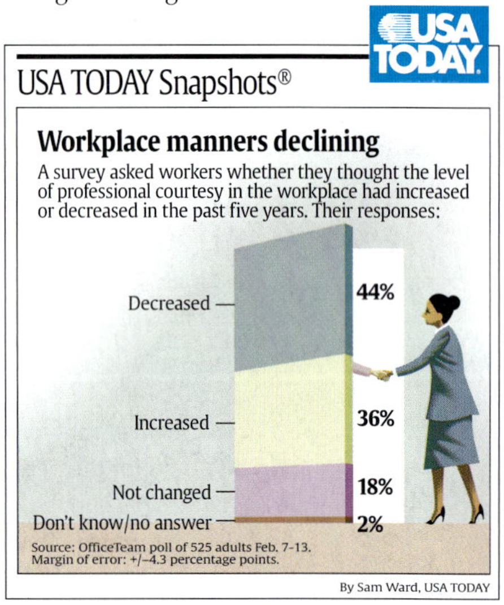

CRITICAL THINKING **For Exercises 59 and 60, $\triangle ABC \sim \triangle DEF$.**

59. Show that the perimeters of $\triangle ABC$ and $\triangle DEF$ have the same ratio as their corresponding sides.
60. If 6 units are added to the lengths of each side, are the new triangles similar? **No, the sides are no longer proportional.**

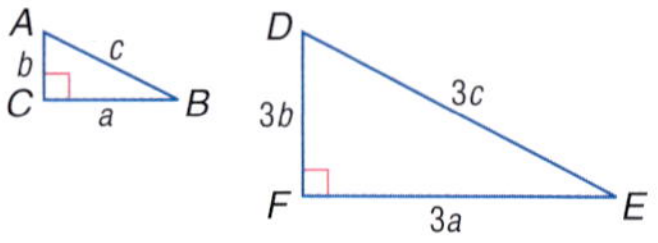

59. $\frac{a}{3a} = \frac{b}{3b} = \frac{c}{3c} = \frac{a + b + c}{3(a + b + c)} = \frac{1}{3}$

61. **WRITING IN MATH** Answer the question that was posed at the beginning of the lesson. **See margin.**

How do artists use geometric patterns?

Include the following in your answer:
- why Escher called the picture *Circle Limit IV*, and
- how one of the light objects and one of the dark objects compare in size.

FCAT Practice

Standardized Test Practice

62. In a history class with 32 students, the ratio of girls to boys is 5 to 3. How many more girls are there than boys? **B**

(A) 2 (B) 8 (C) 12 (D) 15

63. **ALGEBRA** Find x. **D**

(A) 4.2 (B) 4.65
(C) 5.6 (D) 8.4

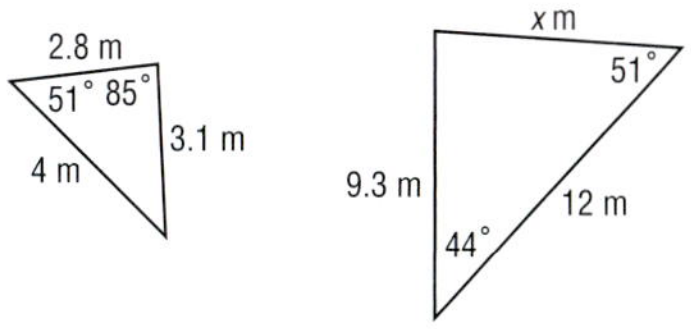

Extending the Lesson

Scale factors can be used to produce similar figures. The resulting figure is an enlargement or reduction of the original figure depending on the scale factor.

Triangle ABC has vertices $A(0, 0)$, $B(8, 0)$, and $C(2, 7)$. Suppose the coordinates of each vertex are multiplied by 2 to create the similar triangle $A'B'C'$.

64. Find the coordinates of the vertices of $\triangle A'B'C'$. **$A'(0, 0)$, $B'(16, 0)$, $C'(4, 14)$**
65. Graph $\triangle ABC$ and $\triangle A'B'C'$. **See margin.**
66. Use the Distance Formula to find the measures of the sides of each triangle.
67. Find the ratios of the sides that appear to correspond.
68. How could you use slope to determine if angles are congruent? **See margin.**
69. Is $\triangle ABC \sim \triangle A'B'C'$? Explain your reasoning. **The sides are proportional and the angles are congruent, so the triangles are similar.**

66. $AB = 8$, $BC = \sqrt{85}$, $AC = \sqrt{53}$; $A'B' = 16$, $B'C' = 2\sqrt{85}$, $A'C' = 2\sqrt{53}$

67. $\frac{AB}{A'B'} = \frac{AC}{A'C'} = \frac{BC}{B'C'} = \frac{1}{2}$

Maintain Your Skills

Mixed Review

Solve each proportion. *(Lesson 6-1)*

70. $\frac{b}{7.8} = \frac{2}{3}$ **5.2**
71. $\frac{c-2}{c+3} = \frac{5}{4}$ **−23**
72. $\frac{2}{4y+5} = \frac{-4}{y}$ **$-\frac{10}{9}$**

Use the figure to write an inequality relating each pair of angle or segment measures. *(Lesson 5-5)*

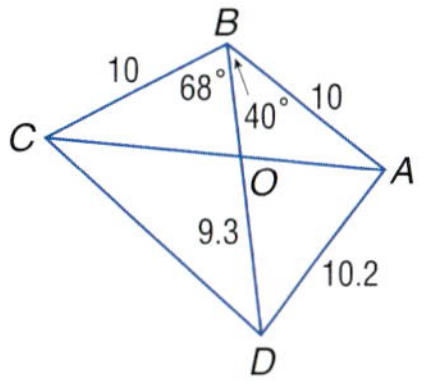

73. OC, AO **$OC > AO$**
74. $m\angle AOD$, $m\angle AOB$ **$m\angle AOD < m\angle AOB$**
75. $m\angle ABD$, $m\angle ADB$ **$m\angle ABD > m\angle ADB$**

Find x. *(Lesson 4-2)*

76. **93**

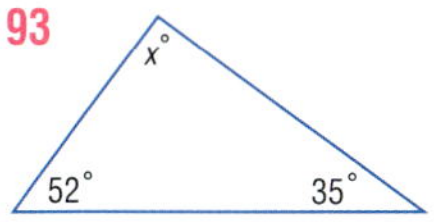

77. **91**

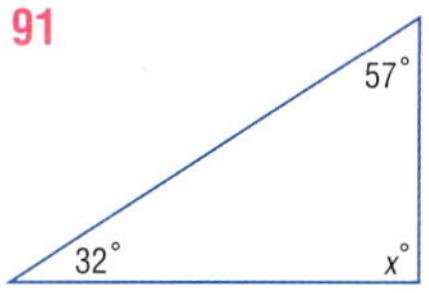

78. **115**

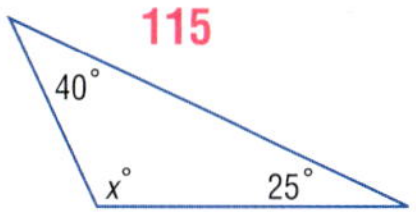

79. Suppose two parallel lines are cut by a transversal and $\angle 1$ and $\angle 2$ are alternate interior angles. Find $m\angle 1$ and $m\angle 2$ if $m\angle 1 = 10x - 9$ and $m\angle 2 = 9x + 3$. *(Lesson 3-2)* **$m\angle 1 = m\angle 2 = 111$**

Getting Ready for the Next Lesson

PREREQUISITE SKILL **In the figure, $\overline{AB} \parallel \overline{CD}$, $\overline{AC} \parallel \overline{BD}$, and $m\angle 4 = 118$. Find the measure of each angle.** *(To review **angles and parallel lines**, see Lesson 3-2.)*

80. $\angle 1$ **118**
81. $\angle 2$ **62**
82. $\angle 3$ **62**
83. $\angle 5$ **118**
84. $\angle ABD$ **118**
85. $\angle 6$ **62**
86. $\angle 7$ **62**
87. $\angle 8$ **118**

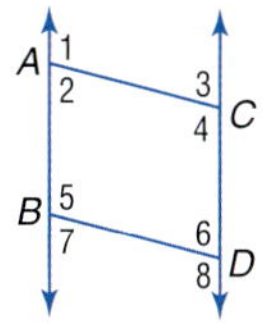

www.geometryonline.com/self_check_quiz/fcat

4 Assess

Open-Ended Assessment

Writing Ask students to state all facts they know about two triangles that are similar.

Getting Ready for Lesson 6-3

Prerequisite Skill Students will learn about similar triangles in Lesson 6-3. They will use parallel lines to find the congruent angles of similar triangles. Use Exercises 80–87 to determine your students' familiarity with parallel lines.

Assessment Options

Quiz (Lessons 6-1 and 6-2) is available on p. 345 of the *Chapter 6 Resource Masters.*

Answers

65.

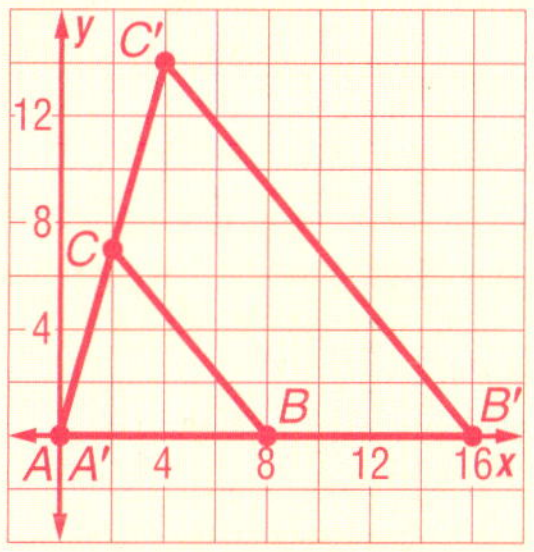

68. You could use the slope formula to find that $\overline{BC} \parallel \overline{B'C'}$. Thus, $\angle ABC \cong \angle A'B'C'$ and $\angle ACB \cong \angle A'C'B'$ because of corresponding angles. $\angle A \cong \angle A'$ because of the Third Angle Theorem.

6-3 Lesson Notes

1 Focus

5-Minute Check Transparency 6-3 Use as a quiz or review of Lesson 6-2.

Mathematical Background notes are available for this lesson on p. 280C.

How do engineers use geometry?

Ask students:

- Which geometric shape is used most often in the construction of the Eiffel tower? **triangle**
- Why did Gustave Eiffel choose this shape in his construction? **Triangular shapes result in rigid construction.**
- What is another structure that uses this shape in its construction? **Sample answer: high-voltage electrical tower**

6-3 Similar Triangles

Sunshine State Standards
MA.A.3.4.3, MA.B.1.4.3, MA.B.2.4.1, MA.C.1.4.1, MA.C.2.4.1, MA.C.3.4.1

What You'll Learn

- Identify similar triangles.
- Use similar triangles to solve problems.

How do engineers use geometry?

The Eiffel Tower was built in Paris for the 1889 world exhibition by Gustave Eiffel. Eiffel (1832–1923) was a French engineer who specialized in revolutionary steel constructions. He used thousands of triangles, some the same shape but different in size, to build the Eiffel Tower because triangular shapes result in rigid construction.

More About. . .

Eiffel Tower

The Eiffel Tower weighs 7000 tons, but the pressure per square inch it applies on the ground is only equivalent to that of a chair with a person seated in it.

Source: www.eiffel-tower.com

IDENTIFY SIMILAR TRIANGLES In Chapter 4, you learned several tests to determine whether two triangles are congruent. There are also tests to determine whether two triangles are similar.

Geometry Activity

Similar Triangles

Collect Data

- Draw $\triangle DEF$ with $m\angle D = 35$, $m\angle F = 80$, and $DF = 4$ centimeters.
- Draw $\triangle RST$ with $m\angle T = 35$, $m\angle S = 80$, and $ST = 7$ centimeters.
- Measure $\overline{EF}$, $\overline{ED}$, $\overline{RS}$, and $\overline{RT}$.
- Calculate the ratios $\frac{FD}{ST}$, $\frac{EF}{RS}$, and $\frac{ED}{RT}$.

Analyze the Data

1. What can you conclude about all of the ratios? **They equal about 0.57.**
2. Repeat the activity with two more triangles with the same angle measures, but different side measures. Then repeat the activity with a third pair of triangles. Are all of the triangles similar? Explain.
3. What are the minimum requirements for two triangles to be similar?

2. Yes, all sides are in the same ratio.
3. Sample answer: Either all sides proportional or two corresponding angles congruent.

The previous activity leads to the following postulate.

Postulate 6.1

Angle-Angle (AA) Similarity If the two angles of one triangle are congruent to two angles of another triangle, then the triangles are similar.

Example: $\angle P \cong \angle T$ and $\angle Q \cong \angle S$, so $\triangle PQR \sim \triangle TSU$.

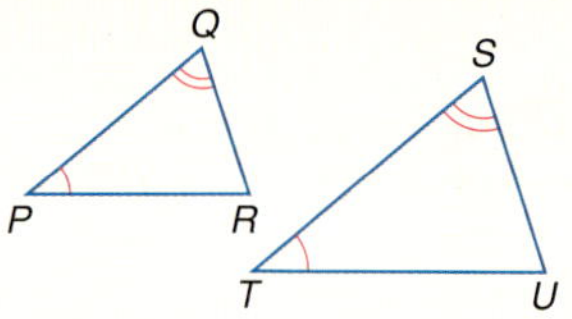

You can use the AA Similarity Postulate to prove two theorems that also verify triangle similarity.

Resource Manager

Workbook and Reproducible Masters

Chapter 6 Resource Masters
- Study Guide and Intervention, pp. 307–308
- Skills Practice, p. 309
- Practice, p. 310
- Reading to Learn Mathematics, p. 311
- Enrichment, p. 312
- Assessment, pp. 345, 347

Prerequisite Skills Workbook, pp. 25–26, 33–36, 43–44, 53–54

Teaching Geometry With Manipulatives Masters, pp. 16, 17, 102, 103, 104, 107

Transparencies

5-Minute Check Transparency 6-3
Answer Key Transparencies

Technology

GeomPASS: Tutorial Plus, Lesson 12
Interactive Chalkboard

Theorems

6.1 Side-Side-Side (SSS) Similarity If the measures of the corresponding sides of two triangles are proportional, then the triangles are similar.

Example: $\frac{PQ}{ST} = \frac{QR}{SU} = \frac{RP}{UT}$, so $\triangle PQR \sim \triangle TSU$.

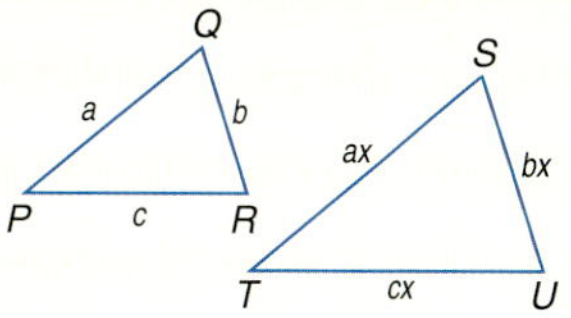

6.2 Side-Angle-Side (SAS) Similarity If the measures of two sides of a triangle are proportional to the measures of two corresponding sides of another triangle and the included angles are congruent, then the triangles are similar.

Example: $\frac{PQ}{ST} = \frac{QR}{SU}$ and $\angle Q \cong \angle S$, so $\triangle PQR \sim \triangle TSU$.

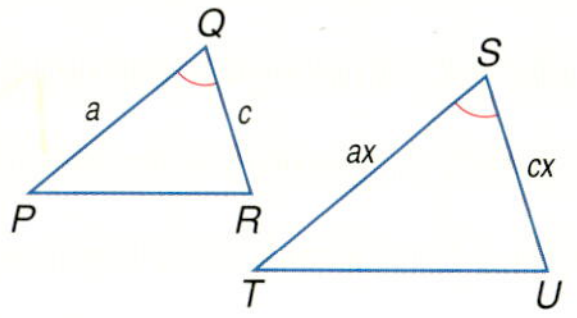

You will prove Theorem 6.2 in Exercise 34.

Proof Theorem 6.1

Given: $\frac{PQ}{AB} = \frac{QR}{BC} = \frac{RP}{CA}$

Prove: $\triangle BAC \sim \triangle QPR$

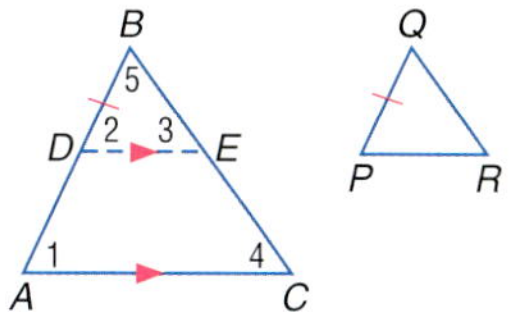

Locate D on $\overline{AB}$ so that $\overline{DB} \cong \overline{PQ}$ and draw $\overline{DE}$ so that $\overline{DE} \parallel \overline{AC}$.

Paragraph Proof:

Since $\overline{DB} \cong \overline{PQ}$, $DB = PQ$. $\frac{PQ}{AB} = \frac{QR}{BC} = \frac{RP}{CA}$ becomes $\frac{DB}{AB} = \frac{QR}{BC} = \frac{RP}{CA}$. Since $\overline{DE} \parallel \overline{AC}$, $\angle 2 \cong \angle 1$ and $\angle 3 \cong \angle 4$. By AA Similarity, $\triangle BDE \sim \triangle BAC$.

By the definition of similar polygons, $\frac{DB}{AB} = \frac{BE}{BC} = \frac{ED}{CA}$. By substitution, $\frac{QR}{BC} = \frac{BE}{BC}$ and $\frac{RP}{CA} = \frac{ED}{CA}$. This means that $QR = BE$ and $RP = ED$ or $\overline{QR} \cong \overline{BE}$ and $\overline{RP} \cong \overline{ED}$. With these congruences and $\overline{DB} \cong \overline{PQ}$, $\triangle BDE \cong \triangle QPR$ by SSS. By CPCTC, $\angle B \cong \angle Q$ and $\angle 2 \cong \angle P$. But $\angle 2 \cong \angle A$, so $\angle A \cong \angle P$. By AA Similarity, $\triangle BAC \sim \triangle QPR$.

Study Tip

Overlapping Triangles

When two triangles overlap, draw them separately so the corresponding parts are in the same position on the paper. Then write the corresponding angles and sides.

Example 1 Determine Whether Triangles Are Similar

In the figure, $\overline{FG} \cong \overline{EG}$, $BE = 15$, $CF = 20$, $AE = 9$, and $DF = 12$. Determine which triangles in the figure are similar.

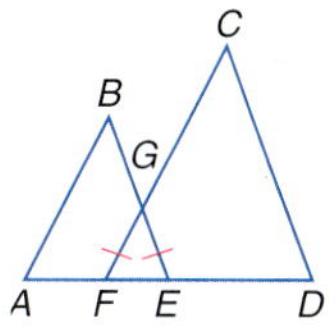

Triangle FGE is an isosceles triangle. So, $\angle GFE \cong \angle GEF$. If the measures of the corresponding sides that include the angles are proportional, then the triangles are similar.

$\frac{AE}{DF} = \frac{9}{12}$ or $\frac{3}{4}$ and $\frac{BE}{CF} = \frac{15}{20}$ or $\frac{3}{4}$

By substitution, $\frac{AE}{DF} = \frac{BE}{CF}$. So, by SAS Similarity, $\triangle ABE \sim \triangle DCF$.

www.geometryonline.com/extra_examples/fcat

2 Teach

IDENTIFY SIMILAR TRIANGLES

In-Class Example

Power Point®

Teaching Tip You may want to point out the similarities and differences between the triangle congruence postulates and theorems in Chapter 4 and the similarity postulates and theorems in this chapter. Emphasize that although only two pairs of corresponding angles need to be congruent for two triangles to be similar, all three pairs of corresponding sides must be proportional.

1 In the figure, $\overline{AB} \parallel \overline{DC}$, $BE = 27$, $DE = 45$, $AE = 21$, and $CE = 35$. Determine which triangles in the figure are similar. $\triangle ABE \sim \triangle CDE$

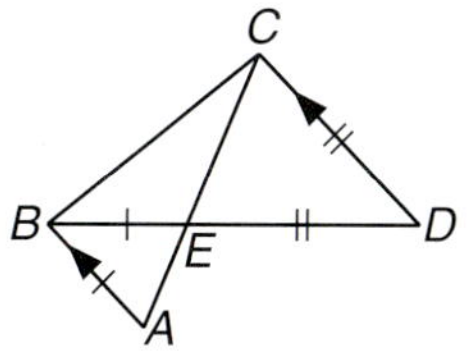

Geometry Activity

Materials: protractor; centimeter ruler

- You may want to do this activity in groups of three students. Ask one student to draw the triangles. Ask a second student to measure the segments, while a third student records the measurements.
- If you have time, have students rotate roles and repeat the activity with different triangles.

USE SIMILAR TRIANGLES

In-Class Examples

Power Point®

2 **ALGEBRA** Given $\overline{RS} \parallel \overline{UT}$, $RS = 4$, $RQ = x + 3$, $QT = 2x + 10$, $UT = 10$, find RQ and QT. **8; 20**

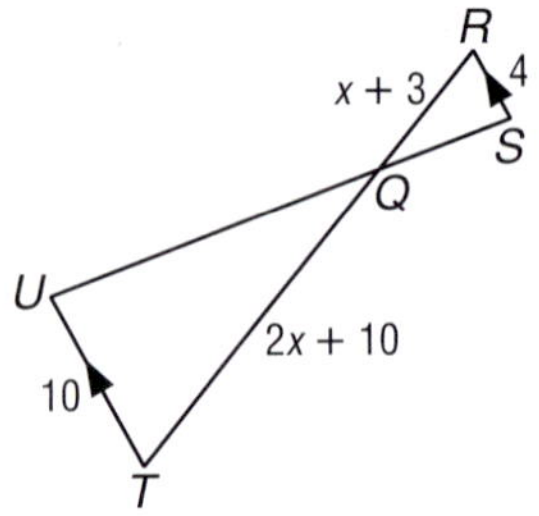

3 **INDIRECT MEASUREMENT** Josh wanted to measure the height of the Sears Tower in Chicago. He used a 12-foot light pole and measured its shadow at 1 P.M. The length of the shadow was 2 feet. Then he measured the length of Sears Tower's shadow and it was 242 feet at that time. What is the height of the Sears Tower?

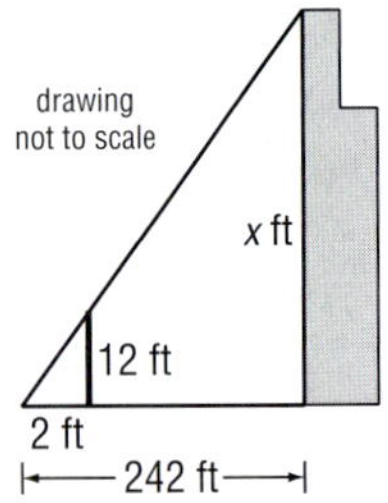

1452 ft (actual height: 1450 feet)

Tips for New Teachers

Point out that shadows and similar triangles are commonly used for indirectly measuring the heights of objects or structures that otherwise would be too tall to measure. This includes tall buildings and tall trees.

Like the congruence of triangles, similarity of triangles is reflexive, symmetric, and transitive.

Theorem 6.3

Similarity of triangles is reflexive, symmetric, and transitive.

Examples:

Reflexive: $\triangle ABC \sim \triangle ABC$

Symmetric: If $\triangle ABC \sim \triangle DEF$, then $\triangle DEF \sim \triangle ABC$.

Transitive: If $\triangle ABC \sim \triangle DEF$ and $\triangle DEF \sim \triangle GHI$, then $\triangle ABC \sim \triangle GHI$.

You will prove Theorem 6.3 in Exercise 38.

USE SIMILAR TRIANGLES Similar triangles can be used to solve problems.

Example 2 Parts of Similar Triangles

ALGEBRA **Find AE and DE.**

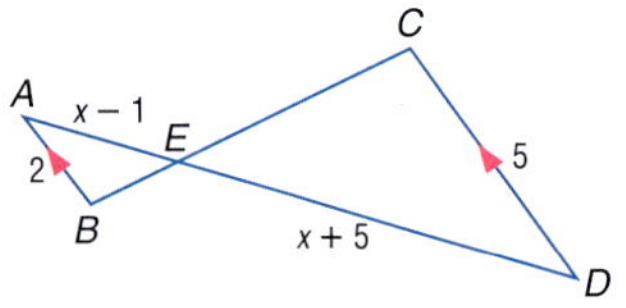

Since $\overline{AB} \parallel \overline{CD}$, $\angle BAE \cong \angle CDE$ and $\angle ABE \cong \angle DCE$ because they are the alternate interior angles. By AA Similarity, $\triangle ABE \sim \triangle DCE$. Using the definition of similar polygons, $\frac{AB}{DC} = \frac{AE}{DE}$.

$\frac{AB}{DC} = \frac{AE}{DE}$

$\frac{2}{5} = \frac{x-1}{x+5}$ Substitution

$2(x+5) = 5(x-1)$ Cross products

$2x + 10 = 5x - 5$ Distributive Property

$-3x = -15$ Subtract $5x$ and 10 from each side.

$x = 5$ Divide each side by -3.

Now find AE and ED. $AE = x - 1 = 5 - 1$ or 4 $\quad ED = x + 5 = 5 + 5$ or 10

Study Tip

Shadow Problems

In shadow problems, we assume that a right triangle is formed by the sun's ray from the top of the object to the end of the shadow.

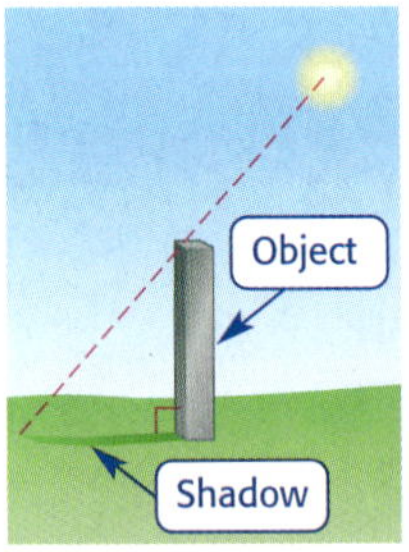

Similar triangles can be used to find measurements indirectly.

Example 3 Find a Measurement

INDIRECT MEASUREMENT **Nina was curious about the height of the Eiffel Tower. She used a 1.2 meter model of the tower and measured its shadow at 2 P.M. The length of the shadow was 0.9 meter. Then she measured the Eiffel Tower's shadow, and it was 240 meters. What is the height of the Eiffel Tower?**

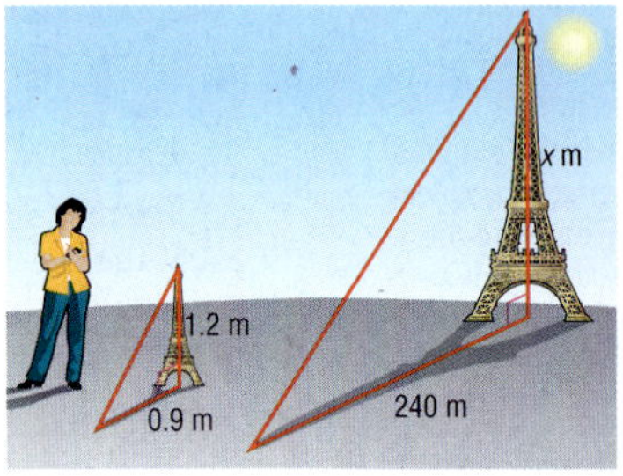

Assuming that the sun's rays form similar triangles, the following proportion can be written.

$$\frac{\text{height of the Eiffel Tower (m)}}{\text{height of the model tower (m)}} = \frac{\text{Eiffel Tower shadow length (m)}}{\text{model shadow length (m)}}$$

DAILY INTERVENTION **Differentiated Instruction**

Interpersonal Have students choose a partner. Ask each pair of students to measure the height of a school building by using their own shadows and similar triangles.

Now substitute the known values and let x be the height of the Eiffel Tower.

$\frac{x}{1.2} = \frac{240}{0.9}$ Substitution

$x \cdot 0.9 = 1.2(240)$ Cross products

$0.9x = 288$ Simplify.

$x = 320$ Divide each side by 0.9.

The Eiffel Tower is 320 meters tall.

Check for Understanding

Concept Check

1. **Compare and contrast** the tests to prove triangles similar with the tests to prove triangles congruent. **1–2. See margin.**

2. **OPEN ENDED** Is it possible that $\triangle ABC$ is not similar to $\triangle RST$ and that $\triangle RST$ is not similar to $\triangle EFG$, but that $\triangle ABC$ is similar to $\triangle EFG$? Explain.

3. **FIND THE ERROR** Alicia and Jason were writing proportions for the similar triangles shown at the right.

Alicia	Jason
$\frac{r}{k} = \frac{s}{m}$	$\frac{r}{k} = \frac{m}{s}$
$rm = ks$	$rs = km$

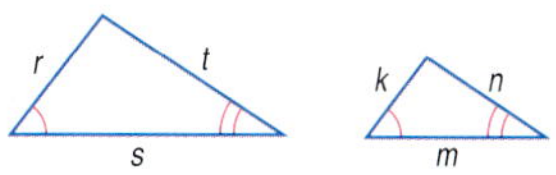

Who is correct? Explain your reasoning.

3. Alicia; while both have corresponding sides in a ratio, Alicia has them in proper order with the numerators from the same triangle.

Guided Practice

ALGEBRA **Identify the similar triangles. Find x and the measures of the indicated sides.**

GUIDED PRACTICE KEY	
Exercises	Examples
4–5	2
6–8	1
9	3

4. DE **$\triangle ABC \sim \triangle DEF$; $x = 9$; $DE = 9$**

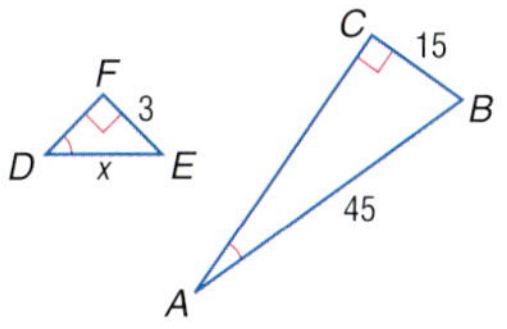

5. AB and DE **$\triangle ABC \sim \triangle DEF$; $x = 10$; $AB = 10$; $DE = 6$**

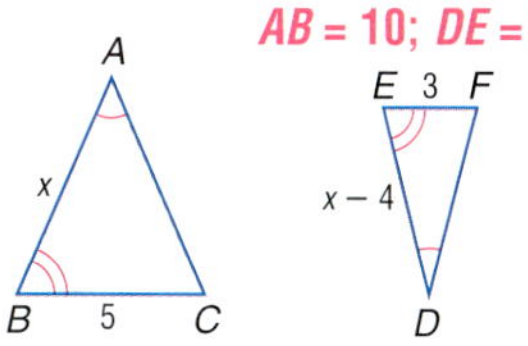

Determine whether each pair of triangles is similar. Justify your answer.

6.

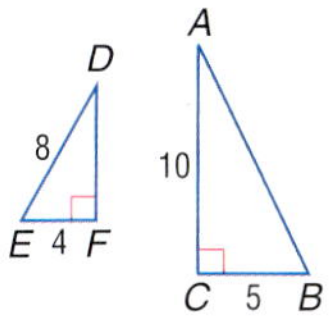

No; corresponding sides are not proportional.

7.

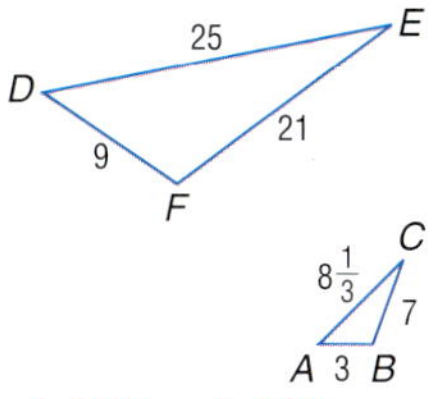

yes; $\triangle DEF \sim \triangle ACB$ by SSS Similarity

8.

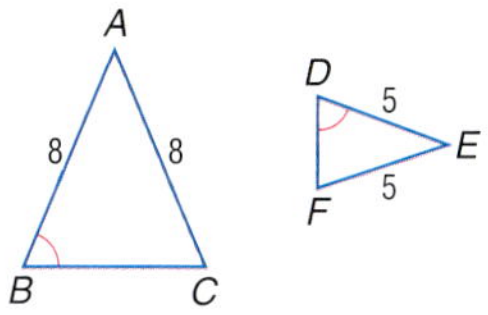

yes; $\triangle ABC \sim \triangle EDF$ by AA Similarity

Application

9. **INDIRECT MEASUREMENT** A cell phone tower in a field casts a shadow of 100 feet. At the same time, a 4 foot 6 inch post near the tower casts a shadow of 3 feet 4 inches. Find the height of the tower in feet and inches. (*Hint*: Make a drawing.) **135 ft**

3 Practice/Apply

Study Notebook

Have students—

- add the definitions/examples of the vocabulary terms to their Vocabulary Builder worksheets for Chapter 6.
- write the AA Similarity Postulate, the SSS Similarity Theorem, and the SAS Similarity Theorem in their notebooks. Ask them to include an example of each.
- include any other item(s) that they find helpful in mastering the skills in this lesson.

DAILY INTERVENTION FIND THE ERROR

In Exercise 3, caution students to write the ratios carefully for the proportions between the corresponding sides of two similar triangles.

Answers

1. Sample answer: Two triangles are congruent by the SSS, SAS, and ASA Postulates and the AAS Theorem. In these triangles, corresponding parts must be congruent. Two triangles are similar by AA Similarity, SSS Similarity, and SAS Similarity. In similar triangles, the sides are proportional and the angles are congruent. Congruent triangles are always similar triangles. Similar triangles are congruent only when the scale factor for the proportional sides is 1. SSS and SAS are common relationships for both congruence and similarity.

2. Yes; suppose $\triangle RST$ has angles that measure 46°, 54°, and 80°, $\triangle ABC$ has angles that measure 39°, 63°, and 78°, and $\triangle EFG$ has angles that measure 39°, 63°, and 78°. So $\triangle ABC$ is not similar to $\triangle RST$ and $\triangle RST$ is not similar to $\triangle EFG$, but $\triangle ABC$ is similar to $\triangle EFG$.

About the Exercises...

Organization by Objective

- **Identify Similar Triangles:** 10–17, 20–21
- **Use Similar Triangles to Solve Problems:** 19–20, 22–23, 26–32, 39, 41

Odd/Even Assignments

Exercises 10–41 are structured so that students practice the same concepts whether they are assigned odd or even problems.

Assignment Guide

Basic: 11–21 odd, 22, 23, 25, 29–43 odd, 47–61

Average: 11–21 odd, 22, 23, 25–43 odd, 47–61

Advanced: 10–20 even, 22, 23, 24–40 even, 41, 42, 44–58 (optional: 59–61)

All: Quiz 1 (1–5)

Answer

22. 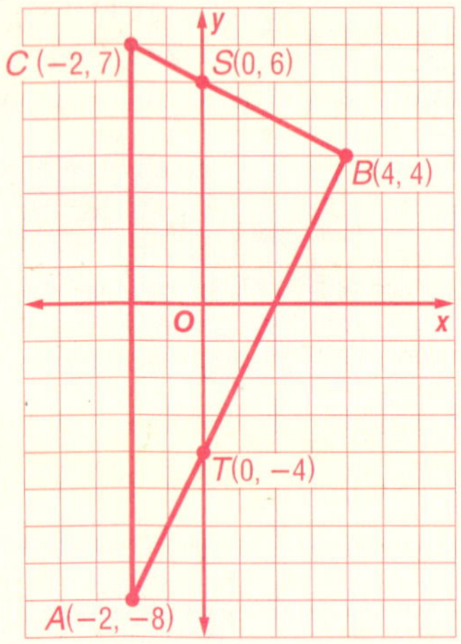

$AB = \sqrt{6^2 + 12^2} = \sqrt{180}$ or $6\sqrt{5}$; $BC = \sqrt{6^2 + 3^2} = \sqrt{45}$ or $3\sqrt{5}$; $CA = |7 - (-8)| = 15$; $ST = |6 - (-4)| = 10$; $TB = \sqrt{8^2 + 4^2} = \sqrt{80}$ or $4\sqrt{5}$; $BS = \sqrt{2^2 + 4^2} = \sqrt{20}$ or $2\sqrt{5}$. $\frac{CA}{ST} = \frac{15}{10}$ or $\frac{3}{2}$, $\frac{AB}{TB} = \frac{6\sqrt{5}}{4\sqrt{5}}$ or $\frac{3}{2}$, and $\frac{BC}{BS} = \frac{3\sqrt{5}}{2\sqrt{5}}$ or $\frac{3}{2}$. Since $\frac{CA}{ST} = \frac{AB}{TB} = \frac{BC}{BS}$, $\triangle ABC \sim \triangle TBS$ by SSS Similarity.

★ indicates increased difficulty

Practice and Apply

Homework Help	
For Exercises	**See Examples**
10–17, 26–27	1
18–21, 28–31	2
38–41	3

Extra Practice
See page 765.

Determine whether each pair of triangles is similar. Justify your answer.

10. 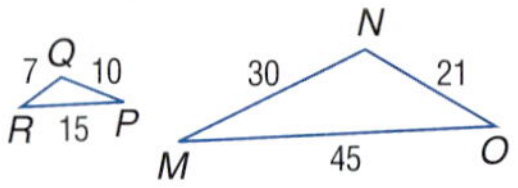

yes; $\triangle MNO \sim \triangle PQR$ by SSS Similarity

11. 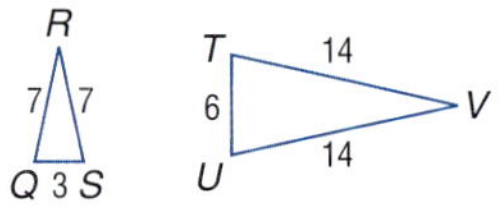

yes; $\triangle QRS \sim \triangle TVU$ by SSS Similarity

12. 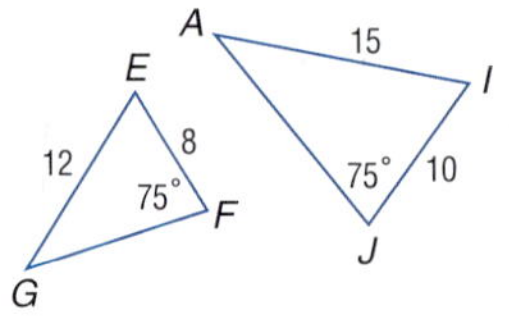

not enough information to determine

13. 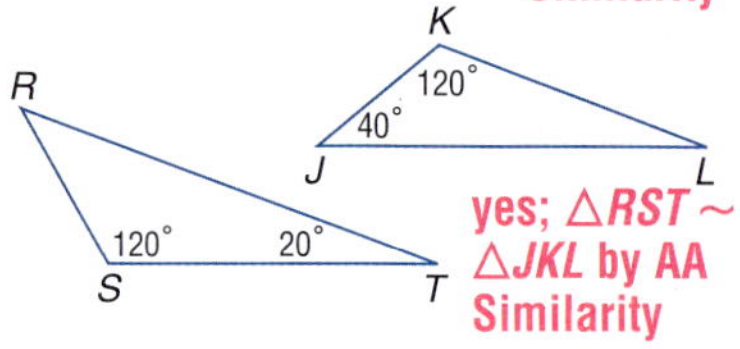

yes; $\triangle RST \sim \triangle JKL$ by AA Similarity

14. 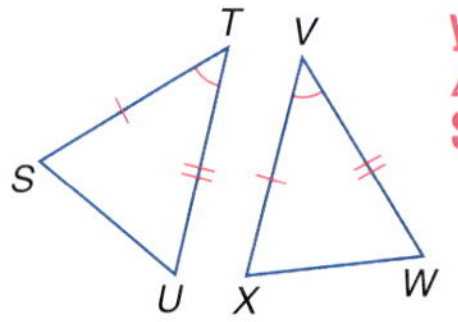

yes; $\triangle STU \sim \triangle XVW$ by SAS Similarity

15. 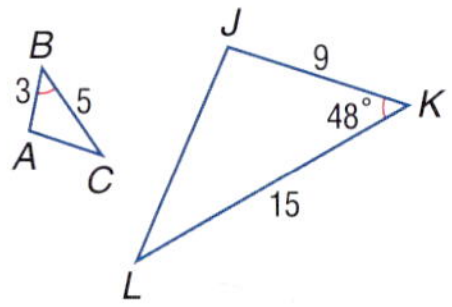

yes; $\triangle ABC \sim \triangle JKL$ by SAS Similarity

16. 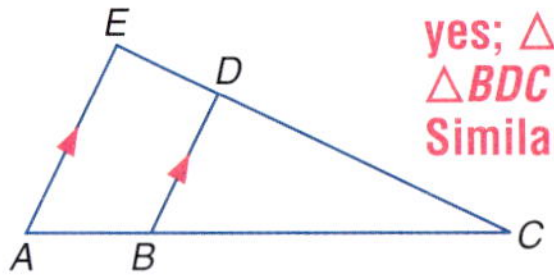

yes; $\triangle AEC \sim \triangle BDC$ by AA Similarity

17. 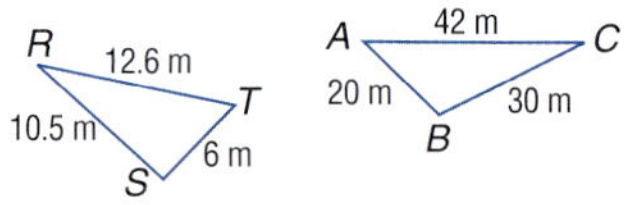

No; sides are not proportional.

ALGEBRA **Identify the similar triangles, and find x and the measures of the indicated sides.**

18. *AB* and *BC*

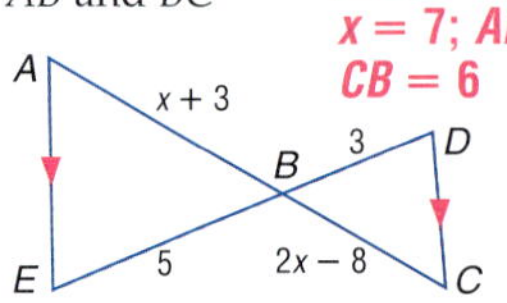

$\triangle ABE \sim \triangle CBD$; $x = 7$; $AB = 10$; $CB = 6$

19. *AB* and *AC*

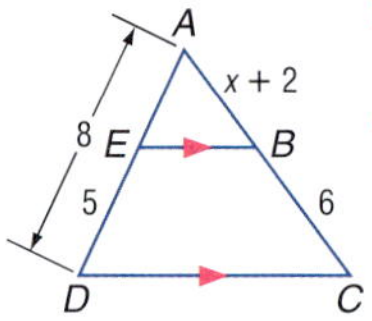

$\triangle ABE \sim \triangle ACD$; $x = \frac{8}{5}$; $AB = 3\frac{3}{5}$; $AC = 9\frac{3}{5}$

20. *BD* and *EC*

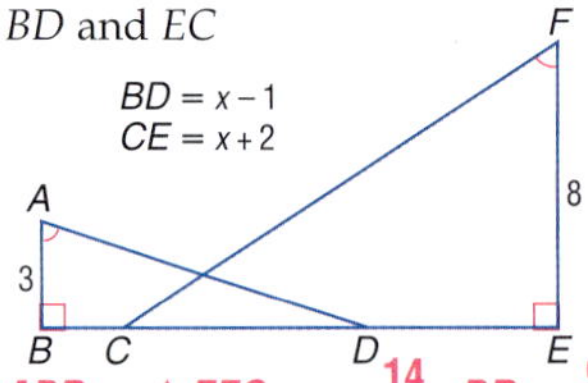

$\triangle ABD \sim \triangle FEC$; $x = \frac{14}{5}$; $BD = \frac{9}{5}$; $EC = \frac{24}{5}$

21. *AB* and *AS*

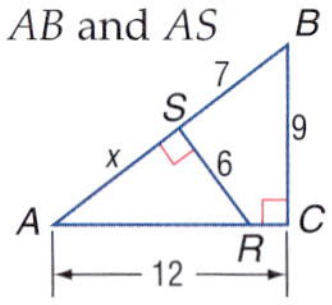

$\triangle ABC \sim \triangle ARS$; $x = 8$; 15; 8

COORDINATE GEOMETRY **Triangles *ABC* and *TBS* have vertices $A(-2, -8)$, $B(4, 4)$, $C(-2, 7)$, $T(0, -4)$, and $S(0, 6)$.**

22. Graph the triangles and prove that $\triangle ABC \sim \triangle TBS$. See margin.

23. Find the ratio of the perimeters of the two triangles. $\frac{3}{2}$

Identify each statement as *true* or *false*. If false, state why.

24. For every pair of similar triangles, there is only one correspondence of vertices that will give you correct angle correspondence and segment proportions.

24. False; this is not true for equilateral or isosceles triangles.

25. If $\triangle ABC \sim \triangle EFG$ and $\triangle ABC \sim \triangle RST$, then $\triangle EFG \sim \triangle RST$. true

Identify the similar triangles in each figure. Explain your answer.

★26. 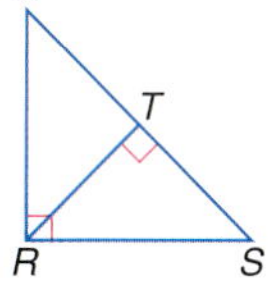

$\triangle QRS \sim \triangle QTR$ (AA Similarity), $\triangle QRS \sim \triangle RTS$ (AA Similarity), $\triangle QTR \sim \triangle RTS$ (Transitive Prop.)

★27. 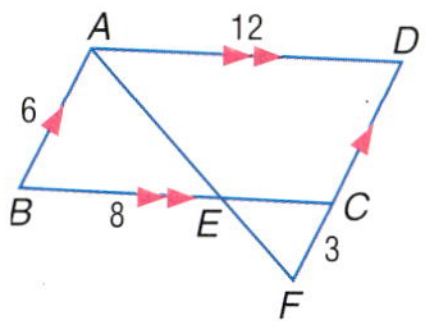

$\triangle EAB \sim \triangle EFC \sim \triangle AFD$: AA Similarity

Use the given information to find each measure.

29. $KP = 5$, $KM = 15$, $MR = 13\frac{1}{3}$, $ML = 20$, $MN = 12$, $PR = 16\frac{2}{3}$

28. If $\overline{PR} \parallel \overline{WX}$, $WX = 10$, $XY = 6$, $WY = 8$, $RY = 5$, and $PS = 3$, find PY, SY, and PQ. $PY = 5$, $SY = 4$, $PQ = 6$

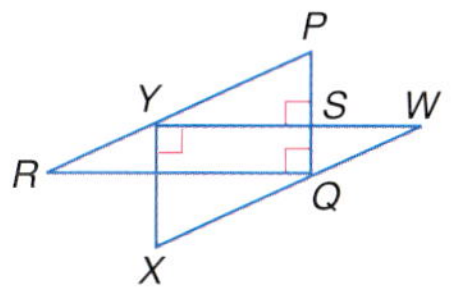

29. If $\overline{PR} \parallel \overline{KL}$, $KN = 9$, $LN = 16$, $PM = 2(KP)$, find KP, KM, MR, ML, MN, and PR.

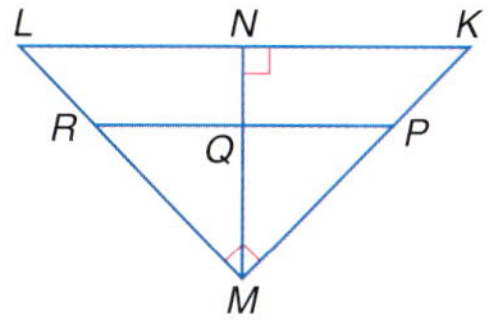

30. $m\angle YIZ = 50$, $m\angle JHI = 70$, $m\angle JIH = 50$, $m\angle J = 60$, $m\angle JHG = 110$

31. $m\angle TUV = 43$, $m\angle R = 43$, $m\angle RSU = 47$, $m\angle SUV = 47$

30. If $\frac{IJ}{XJ} = \frac{HJ}{YJ}$, $m\angle WXJ = 130$, and $m\angle WZG = 20$, find $m\angle YIZ$, $m\angle JHI$, $m\angle JIH$, $m\angle J$, and $m\angle JHG$.

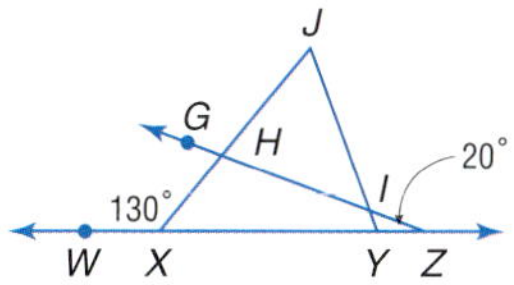

31. If $\angle RST$ is a right angle, $\overline{SU} \perp \overline{RT}$, $\overline{UV} \perp \overline{ST}$, and $m\angle RTS = 47$, find $m\angle TUV$, $m\angle R$, $m\angle RSU$, and $m\angle SUV$.

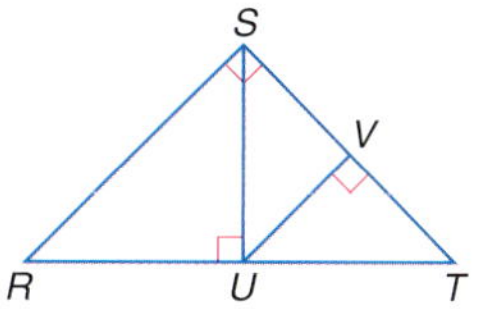

32. **HISTORY** The Greek mathematician Thales was the first to measure the height of a pyramid by using geometry. He showed that the ratio of a pyramid to a staff was equal to the ratio of one shadow to the other. If a pace is about 3 feet, approximately how tall was the pyramid at that time? 478 ft

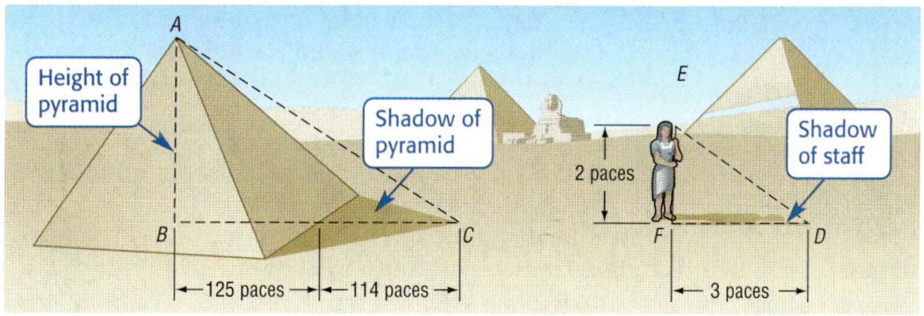

33. In the figure at the right, what relationship must be true of x and y for $\overline{BD}$ and $\overline{AE}$ to be parallel? Explain. $x = y$; if $\overline{BD} \parallel \overline{AE}$, then $\triangle BCD \sim \triangle ACE$ by AA Similarity and $\frac{BC}{AC} = \frac{DC}{EC}$. Thus, $\frac{2}{4} = \frac{x}{x+y}$. Cross multiply and solve for y, yielding $y = x$.

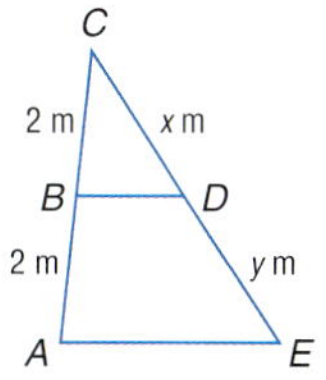

Enrichment, p. 312

Ratio Puzzles with Triangles

If you know the perimeter of a triangle and the ratios of the sides, you can find the lengths of the sides.

Example **The perimeter of a triangle is 84 units. The sides have lengths r, s, and t. The ratio of s to r is 5:3, and the ratio of t to r is 2:1. Find the length of each side.**

Since both ratios contain r, rewrite one or both ratios to make r the same. You can write the ratio of t to r as 6:3. Now you can write a three-part ratio.

$r:s:t = 3:5:6$

There is a number x such that $r = 3x$, $s = 5x$, and $t = 6x$. Since you know the perimeter, 84, you can use algebra to find the lengths of the sides.

$$r + s + t = 84$$
$$3x + 5x + 6x = 84$$
$$14x = 84$$
$$x = 6$$

$\ldots x = 36$

Study Guide and Intervention, p. 307 (shown) and p. 308

Identify Similar Triangles Here are three ways to show that two triangles are similar.

AA Similarity	Two angles of one triangle are congruent to two angles of another triangle.
SSS Similarity	The measures of the corresponding sides of two triangles are proportional.
SAS Similarity	The measures of two sides of one triangle are proportional to the measures of two corresponding sides of another triangle, and the included angles are congruent.

Example 1 **Determine whether the triangles are similar.**

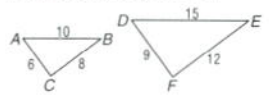

$\frac{AC}{DF} = \frac{6}{9} = \frac{2}{3}$

$\frac{BC}{EF} = \frac{8}{12} = \frac{2}{3}$

$\frac{AB}{DE} = \frac{10}{15} = \frac{2}{3}$

$\triangle ABC \sim \triangle DEF$ by SSS Similarity.

Example 2 **Determine whether the triangles are similar.**

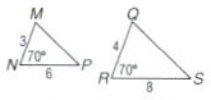

$\frac{3}{4} = \frac{6}{8}$, so $\frac{MN}{QR} = \frac{NP}{RS}$.

$m\angle N = m\angle R$, so $\angle N \cong \angle R$.

$\triangle NMP \sim \triangle RQS$ by SAS Similarity.

Exercises

Determine whether each pair of triangles is similar. Justify your answer.

1. 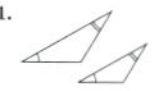

yes; AA Similarity

2. 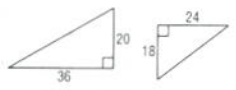

no; $\frac{36}{24} \neq \frac{20}{18}$

3. 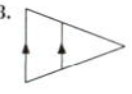

yes; AA Similarity

4. 

yes; SAS Similarity

5. 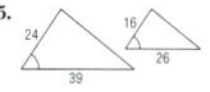

yes; SAS Similarity

6. 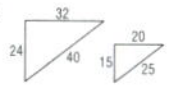

yes; SSS Similarity

Skills Practice, p. 309 and Practice, p. 310 (shown)

Determine whether each pair of triangles is similar. Justify your answer.

1. 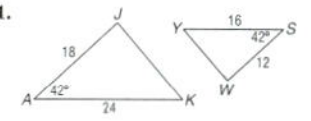

yes; $\triangle JAK \sim \triangle WSY$; SAS Similarity

2. 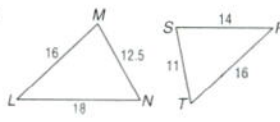

No; the sides are not proportional.

ALGEBRA Identify the similar triangles, and find x and the measures of the indicated sides.

3. $\overline{LM}$ and $\overline{QP}$

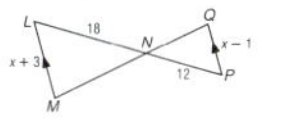

$\triangle LMN \sim \triangle PQN$; 9; 12; 8

4. $\overline{NL}$ and $\overline{ML}$

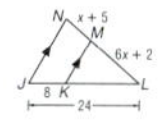

$\triangle JLN \sim \triangle KLM$; 2; 21; 14

Use the given information to find each measure.

5. If $\overline{TS} \parallel \overline{QR}$, $TS = 6$, $PS = x + 7$, $QR = 8$, and $SR = x - 1$, find PS and PR.

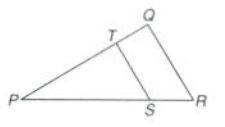

$PS = 12$; $PR = 16$

6. If $\overline{EF} \parallel \overline{HI}$, $EF = 3$, $EG = x + 1$, $HI = 4$, and $HG = x + 3$, find EG and HG.

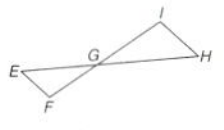

$EG = 6$; $HG = 8$

INDIRECT MEASUREMENT For Exercises 7 and 8, use the following information.

A lighthouse casts a 128-foot shadow. A nearby lamppost that measures 5 feet 3 inches casts an 8-foot shadow.

7. Write a proportion that can be used to determine the height of the lighthouse.
Sample answer: If x = height of lighthouse, $\frac{x}{5.25} = \frac{128}{8}$.

8. What is the height of the lighthouse? 84 ft

Reading to Learn Mathematics, p. 311 ELL

Pre-Activity How do engineers use geometry?

Read the introduction to Lesson 6-3 at the top of page 298 in your textbook.

- What does it mean to say that triangular shapes result in rigid construction? Sample answer: A triangular structure will not change its shape or size if you push on its sides without breaking or bending them.
- What would happen if the shapes used in the construction were quadrilaterals? Sample answer: The structure might collapse, because quadrilaterals are not rigid.

Reading the Lesson

1. State whether each condition guarantees that two triangles are *congruent* or *similar*. If the condition guarantees that the triangles are both similar and congruent, write *congruent*. If there is not enough information to guarantee that the triangles will be congruent or similar, write *neither*.
   a. Two sides and the included angle of one triangle are congruent to two sides and the included angle of the other triangle. congruent
   b. The measures of all three pairs of corresponding sides are proportional. similar
   c. Two angles of one triangle are congruent to two angles of the other triangle. similar
   d. Two angles and a nonincluded side of one triangle are congruent to two angles and the corresponding nonincluded side of the other triangle. congruent
   e. The measures of two sides of a triangle are proportional to the measures of two corresponding sides of another triangle, and the included angles are congruent. similar
   f. The three sides of one triangle are congruent to the three sides of the other triangle. congruent
   g. The three angles of one triangle are congruent to the three angles of the other triangle. similar
   h. One acute angle of a right triangle is congruent to one acute angle of another right triangle. similar
   i. The measures of two sides of a triangle are proportional to the measures of two sides of another triangle. neither

2. Identify each of the following as an example of a *reflexive*, *symmetric*, or *transitive* property.
   a. If $\triangle RST \sim \triangle UVW$, then $\triangle UVW \sim \triangle RST$. symmetric
   b. If $\triangle RST \sim \triangle UVW$ and $\triangle UVW \sim \triangle OPQ$, then $\triangle RST \sim \triangle OPQ$. transitive
   c. $\triangle RST \sim \triangle RST$ reflexive

Helping You Remember

3. A good way to remember something is to explain it to someone else. Suppose one of your classmates is having trouble understanding the difference between SAS for congruent triangles and SAS for similar triangles. How can you explain the difference to him? Sample answer: In SAS for congruence, corresponding sides are congruent, while in SAS for similarity, corresponding sides are proportional. (In both, the included angles are congruent.)

Lesson 6-3 Similar Triangles 303

Answers

34. Given: $\angle B \cong \angle E$; $\overline{QP} \parallel \overline{BC}$; $\overline{QP} \cong \overline{EF}$; $\frac{AB}{DE} = \frac{BC}{EF}$

Prove: $\triangle ABC \sim \triangle DEF$

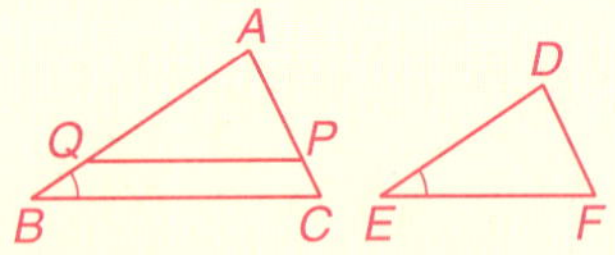

Proof:

Statements (Reasons)

1. $\angle B \cong \angle E$, $\overline{QP} \parallel \overline{BC}$; $\overline{QP} \cong \overline{EF}$; $\frac{AB}{DE} = \frac{BC}{EF}$ (Given)
2. $\angle APQ \cong \angle C$, $\angle AQP \cong \angle B$ (Corres. ≰s Post.)
3. $\angle AQP \cong \angle E$ (Transitive Prop.)
4. $\triangle ABC \sim \triangle AQP$ (AA Similarity)
5. $\frac{AB}{AQ} = \frac{BC}{QP}$ (Def. of $\sim \triangle$s)
6. $AB \cdot QP = AQ \cdot BC$, $AB \cdot EF = DE \cdot BC$ (Cross products)
7. $QP = EF$ (Def. of $\cong$ segments)
8. $AB \cdot EF = AQ \cdot BC$ (Substitution)
9. $AQ \cdot BC = DE \cdot BC$ (Substitution)
10. $AQ = DE$ (Div. Prop.)
11. $\overline{AQ} \cong \overline{DE}$ (Def. of $\cong$ segments)
12. $\triangle AQP \cong \triangle DEF$ (SAS)
13. $\angle APQ \cong \angle F$ (CPCTC)
14. $\angle C \cong \angle F$ (Transitive Prop.)
15. $\triangle ABC \sim \triangle DEF$ (AA Similarity)

35. Statements (Reasons)

1. $\overline{LP} \parallel \overline{MN}$ (Given)
2. $\angle PLN \cong \angle LNM$, $\angle LPM \cong \angle PMN$ (Alt. Int. $\angle$ Thm.)
3. $\triangle LPJ \sim \triangle NMJ$ (AA Similarity)
4. $\frac{LJ}{JN} = \frac{PJ}{JM}$ (Corr. sides of $\sim \triangle$s are proportional.)

36a. Prove: $\triangle ABH \sim \triangle DCB$

Proof: $\angle AHB$, $\angle AJC$, and $\angle EBC$ are right angles because perpendicular lines form right angles. Since all right angles are congruent, $\angle AHB \cong \angle AJC \cong \angle EBC$. Since $\angle A \cong \angle A$ by the Reflexive Property, $\triangle ABH \sim \triangle ACJ$, by AA Similarity. Likewise, since $\angle C \cong \angle C$, $\triangle ACJ \sim \triangle DCB$. By the Transitive Property, $\triangle ABH \sim \triangle DCB$.

34–37. See margin.

PROOF **For Exercises 34–38, write the type of proof specified.**

34. Write a two-column proof to show that if the measures of two sides of a triangle are proportional to the measures of two corresponding sides of another triangle and the included angles are congruent, then the triangles are similar. (Theorem 6.2)

35. a two-column proof

Given: $\overline{LP} \parallel \overline{MN}$

Prove: $\frac{LJ}{JN} = \frac{PJ}{JM}$

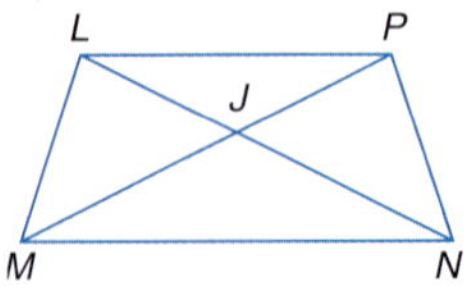

36. a paragraph proof

Given: $\overline{EB} \perp \overline{AC}$, $\overline{BH} \perp \overline{AE}$, $\overline{CJ} \perp \overline{AE}$

Prove: **a.** $\triangle ABH \sim \triangle DCB$

b. $\frac{BC}{BE} = \frac{BD}{BA}$

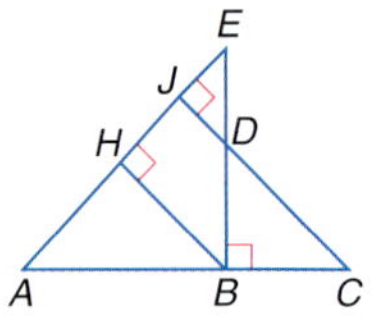

37. a two-column proof to show that if the measures of the legs of two right triangles are proportional, the triangles are similar

38. a two-column proof to prove that similarity of triangles is reflexive, symmetric, and transitive. (Theorem 6.3) See p. 339A.

39. SURVEYING Mr. Glover uses a carpenter's square, an instrument used to draw right angles, to find the distance across a stream. The carpenter's square models right angle *NOL*. He puts the square on top of a pole that is high enough to sight along $\overline{OL}$ to point *P* across the river. Then he sights along $\overline{ON}$ to point *M*. If *MK* is 1.5 feet and *OK* = 4.5 feet, find the distance *KP* across the stream. 13.5 ft

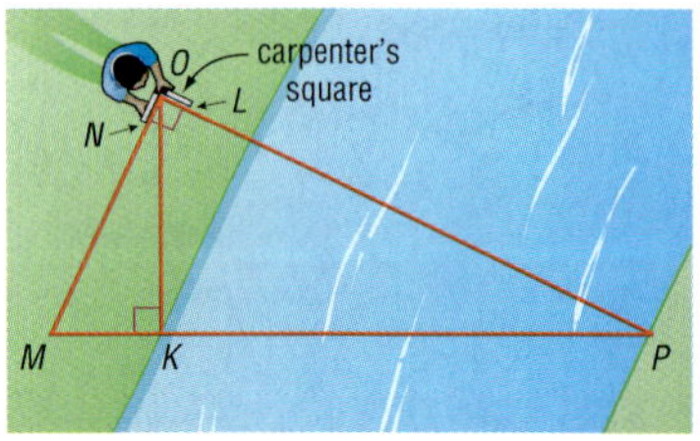

★**40.** The lengths of three sides of triangle *ABC* are 6 centimeters, 4 centimeters, and 9 centimeters. Triangle *DEF* is similar to triangle *ABC*. The length of one of the sides of triangle *DEF* is 36 centimeters. What is the greatest perimeter possible for triangle *DEF*? 171 cm

More About. . .

Towers

Completed in 1999, the Jin Mao Tower is 88 stories tall and sits on a 6-story podium, making it the tallest building in China. It is structurally engineered to tolerate typhoon winds and earthquakes.

Source: www.som.com

TOWERS **For Exercises 41 and 42, use the following information.**
To estimate the height of the Jin Mao Tower in Shanghai, a tourist sights the top of the tower in a mirror that is 87.6 meters from the tower. The mirror is on the ground and faces upward. The tourist is 0.4 meter from the mirror, and the distance from his eyes to the ground is about 1.92 meters.

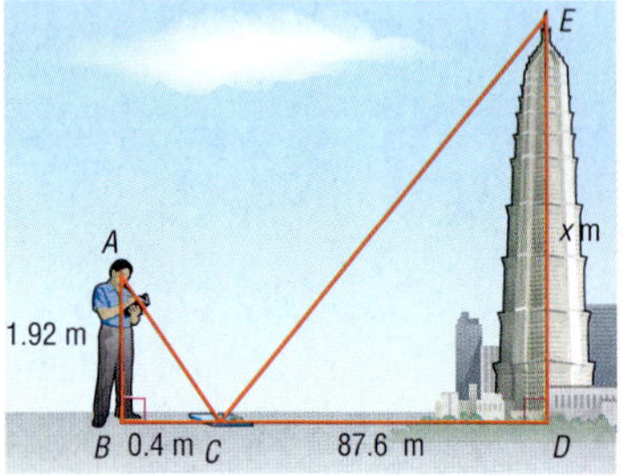

41. How tall is the tower? about 420.5 m

42. Why is the mirror reflection a better way to indirectly measure the tower than by using shadows? It is difficult to measure shadows within a city.

36b. From part a, $\angle A \cong \angle CDB$ by definition of similar triangles. $\angle ABE \cong \angle DBC$ because all right angles are congruent. Thus, $\triangle ABE \sim \triangle DBC$ by AA Similarity. $\frac{BC}{BE} = \frac{BD}{BA}$ from definition of similar triangles.

Career Choices

Forester Foresters plan and manage the growing, protection, and harvesting of trees. A 4-year degree in forestry stressing mathematics, science, computer science, communication skills, and technical forestry is required for a license in many states.

Online Research For information about a career as a forester, visit www.geometryonline.com/careers

43. **FORESTRY** A hypsometer as shown can be used to estimate the height of a tree. Bartolo looks through the straw to the top of the tree and obtains the readings given. Find the height of the tree. **10.75 m**

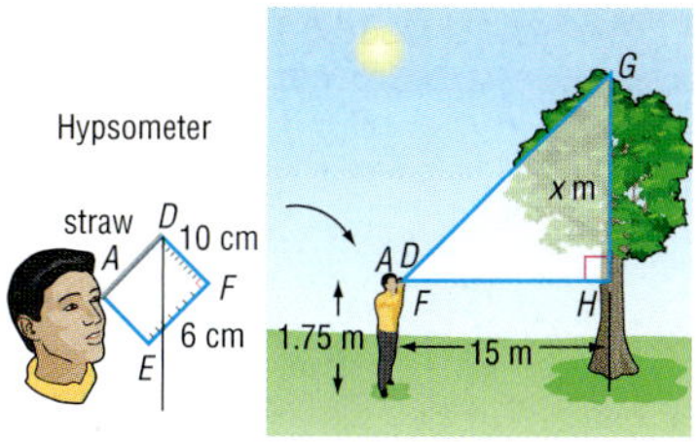

44. **CRITICAL THINKING** Suppose you know the height of a flagpole on the beach of the Chesapeake Bay and that it casts a shadow 4 feet long at 2:00 (EST). You also know the height of a flagpole on the shoreline of Lake Michigan whose shadow is hard to measure at 1:00 (CST). Since 2:00 (EST) = 1:00 (CST), you propose the following proportion of heights and lengths to find the length of the shadow of the Michigan flagpole. Explain whether this proportion will give an accurate measure.

$$\frac{\text{height of Chesapeake flagpole}}{\text{shadow of Chesapeake flagpole}} = \frac{\text{height of Michigan flagpole}}{\text{shadow of Michigan flagpole}}$$

No; the towns are on different latitudinal lines, so the sun is at a different angle to the two buildings.

COORDINATE GEOMETRY **For Exercises 45 and 46, use the following information.**

The coordinates of $\triangle ABC$ are $A(-10, 6)$, $B(-2, 4)$, and $C(-4, -2)$. Point $D(6, 2)$ lies on $\overleftrightarrow{AB}$.

★ 45. Graph $\triangle ABC$, point D, and draw $\overline{BD}$. **See margin.**

★ 46. Where should a point E be located so that $\triangle ABC \sim \triangle ADE$? **(2, −10)**

47. **CRITICAL THINKING** The altitude $\overline{CD}$ from the right angle C in triangle ABC forms two triangles. Triangle ABC is similar to the two triangles formed, and the two triangles formed are similar to each other. Write three similarity statements about these triangles. Why are the triangles similar to each other? **$\triangle ABC \sim \triangle ACD$; $\triangle ABC \sim \triangle CBD$; $\triangle ACD \sim \triangle CBD$; they are similar by AA Similarity.**

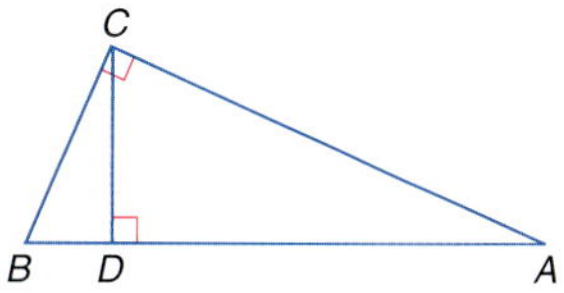

48. **WRITING IN MATH** Answer the question that was posed at the beginning of the lesson. **See margin.**

How do engineers use geometry?

Include the following in your answer:
- why engineers use triangles in construction, and
- why you think the pressure applied to the ground from the Eiffel Tower was so small.

FCAT Practice
Standardized Test Practice

49. If $\overline{EB} \parallel \overline{DC}$, find x. **A**

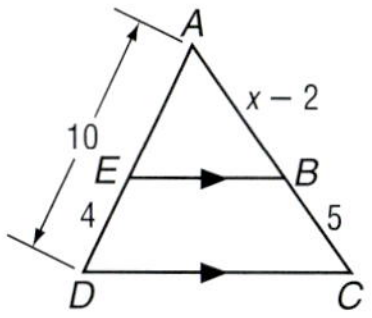

Ⓐ 9.5　Ⓑ 5
Ⓒ 4　Ⓓ 2

50. **ALGEBRA** Solve $\frac{x+3}{6} = \frac{x}{x-2}$. **B**

Ⓐ 6 or 1　Ⓑ 6 or −1
Ⓒ 3 or 2　Ⓓ −3 or 2

Answers

37. Given: $\triangle BAC$ and $\triangle EDF$ are right triangles. $\frac{AB}{DE} = \frac{AC}{DF}$

Prove: $\triangle ABC \sim \triangle DEF$

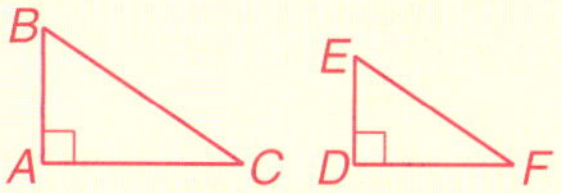

Proof:

Statements (Reasons)
1. $\triangle BAC$ and $\triangle EDF$ are right triangles. (Given)
2. $\angle BAC$ and $\angle EDF$ are right ∠s. (Def. of rt. △)
3. $\angle BAC \cong \angle EDF$ (All rt. ∠s are ≅.)
4. $\frac{AB}{DE} = \frac{AC}{DF}$ (Given)
5. $\triangle ABC \sim \triangle DEF$ (SAS Similarity)

45.

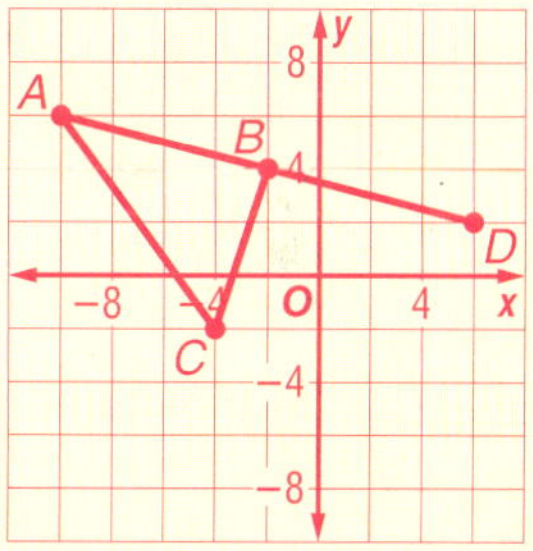

48. Sample answer: Engineers use triangles, some the same shape, but different in size, to complete a project. Answers should include the following.
- Engineers use triangles in construction because they are rigid shapes.
- With the small ground pressure, the tower does not sink, shift, lean, or fall over.

Lesson 6-3 Similar Triangles 305

4 Assess

Open-Ended Assessment

Modeling Ask students to demonstrate with models how similar triangles can be used to find the height of a tall tree. Sample answer: Line up the shadow of a stick and the shadow of the tree so that similar triangles can be used to model the height of the tree. Set up and solve a proportion between the quantities that can be measured.

Getting Ready for Lesson 6-4

Prerequisite Skill Students will learn about parallel lines and proportional parts in Lesson 6-4. They will find the midpoint of a segment and apply it to find the midsegment of a triangle. Use Exercises 59–61 to determine your students' familiarity with finding the midpoint of a segment.

Assessment Options

Practice Quiz 1 The quiz provides students with a brief review of the concepts and skills in Lessons 6-1 through 6-3. Lesson numbers are given to the right of the exercises or instruction lines so students can review concepts not yet mastered.

Quiz (Lesson 6-3) is available on p. 345 of the *Chapter 6 Resource Masters*.

Mid-Chapter Test (Lessons 6-1 through 6-3) is available on p. 347 of the *Chapter 6 Resource Masters*.

Maintain Your Skills

Mixed Review

51. $PQRS \sim ABCD$; 1.6; 1.4; 1.1; $\frac{1}{2}$
52. $\triangle EFG \sim \triangle XYZ$; 5; 30; 18.75; $\frac{4}{3}$

Each pair of polygons is similar. Write a similarity statement, find x, the measures of the indicated sides, and the scale factor. *(Lesson 6-2)*

51. $\overline{BC}$, $\overline{PS}$

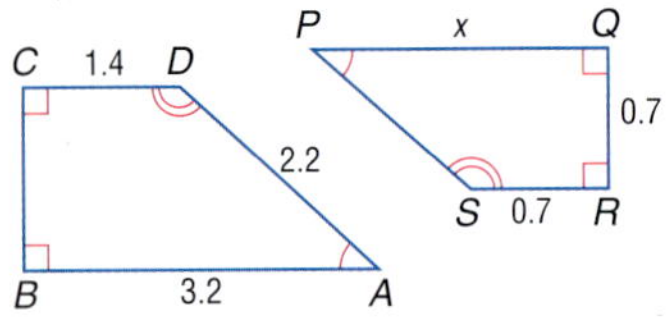

52. $\overline{EF}$, $\overline{XZ}$

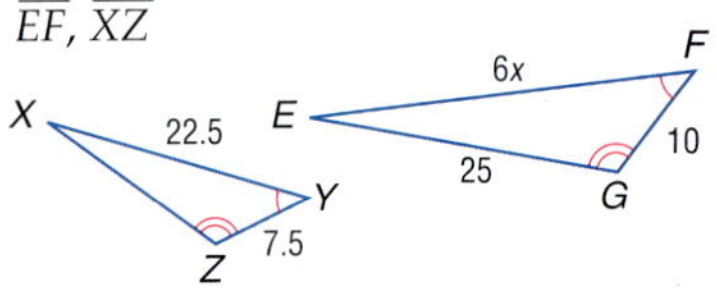

Solve each proportion. *(Lesson 6-1)*

53. $\frac{1}{y} = \frac{3}{15}$ **5**

54. $\frac{6}{8} = \frac{7}{b}$ $\frac{28}{3}$

55. $\frac{20}{28} = \frac{m}{21}$ **15**

56. $\frac{16}{7} = \frac{9}{s}$ $\frac{63}{16}$

57. **COORDINATE GEOMETRY** $\triangle ABC$ has vertices $A(-3, -9)$, $B(5, 11)$, and $C(9, -1)$. $\overline{AT}$ is a median from A to $\overline{BC}$. Determine whether $\overline{AT}$ is an altitude. *(Lesson 5-1)* **No; $\overline{AT}$ is not perpendicular to $\overline{BC}$.**

58. **ROLLER COASTERS** The sign in front of the Electric Storm roller coaster states *ALL riders must be at least 54 inches tall to ride.* If Adam is 5 feet 8 inches tall, can he ride the Electric Storm? Which law of logic leads you to this conclusion? *(Lesson 2-4)* **yes; Law of Detachment**

Getting Ready for the Next Lesson

PREREQUISITE SKILL **Find the coordinates of the midpoint of the segment whose endpoints are given.** *(To review **finding coordinates of midpoints**, see Lesson 1-3.)*

59. (2, 15), (9, 11) **(5.5, 13)**

60. (−4, 4), (2, −12) **(−1, −4)**

61. (0, 8), (7, −13) **(3.5, −2.5)**

Practice Quiz 1 — Lessons 6-1 through 6-3

Determine whether each pair of figures is similar. Justify your answer. *(Lesson 6-2)*

1.

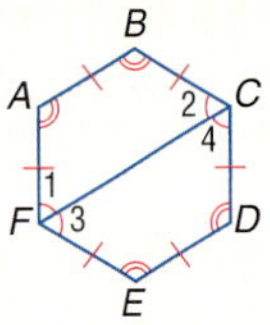

yes; $\angle A \cong \angle E$, $\angle B \cong \angle D$, $\angle 1 \cong \angle 3$, $\angle 2 \cong \angle 4$, and $\frac{AB}{ED} = \frac{BC}{DC} = \frac{AF}{EF} = \frac{FC}{FC} = 1$

2.

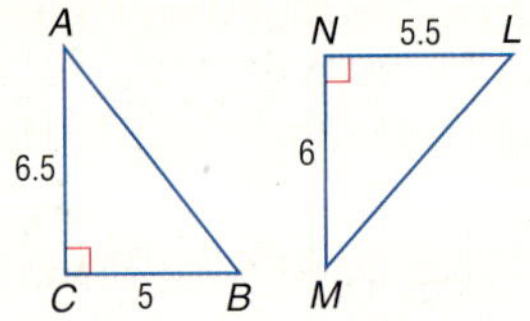

no; $\frac{6.5}{6} \neq \frac{5}{5.5}$ and $\frac{6.5}{5.5} \neq \frac{5}{6}$

Identify the similar triangles. Find x and the measures of the indicated sides. *(Lesson 6-3)*

3. $\overline{AE}$, $\overline{DE}$

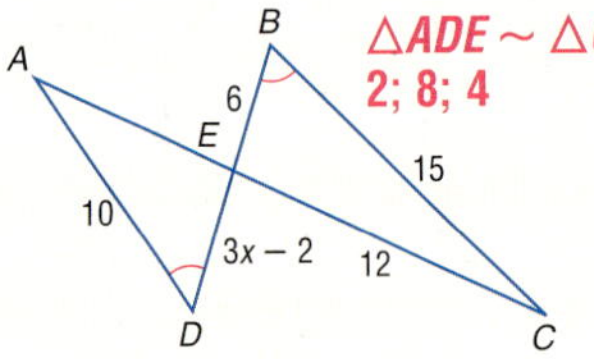

$\triangle ADE \sim \triangle CBE$; 2; 8; 4

4. $\overline{PT}$, $\overline{ST}$

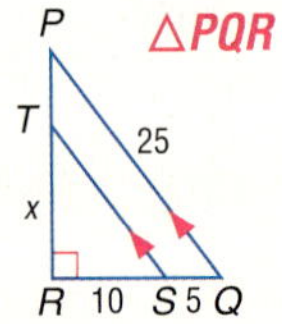

$\triangle PQR \sim \triangle TSR$; $\frac{40}{3}$; $\frac{20}{3}$; $\frac{50}{3}$

5. **MAPS** The scale on a map shows that 1.5 centimeters represents 100 miles. If the distance on the map from Atlanta, Georgia, to Los Angeles, California, is 29.2 centimeters, approximately how many miles apart are the two cities? *(Lesson 6-1)* **1947 mi**

6-4 Parallel Lines and Proportional Parts

What You'll Learn

- Use proportional parts of triangles.
- Divide a segment into parts.

Sunshine State Standards
MA.A.3.4.3, MA.B.1.4.3, MA.C.1.4.1, MA.C.2.4.1, MA.C.3.4.1

Vocabulary
- midsegment

How do city planners use geometry?

Street maps frequently have parallel and perpendicular lines. In Chicago, because of Lake Michigan, Lake Shore Drive runs at an angle between Oak Street and Ontario Street. City planners need to take this angle into account when determining dimensions of available land along Lake Shore Drive.

PROPORTIONAL PARTS OF TRIANGLES Nonparallel transversals that intersect parallel lines can be extended to form similar triangles. So the sides of the triangles are proportional.

Theorem 6.4

Triangle Proportionality Theorem If a line is parallel to one side of a triangle and intersects the other two sides in two distinct points, then it separates these sides into segments of proportional lengths.

Example: If $\overline{BD} \parallel \overline{AE}$, $\frac{BA}{CB} = \frac{DE}{CD}$.

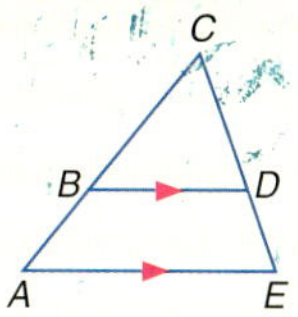

Study Tip

Overlapping Triangles
Trace two copies of $\triangle ACE$. Cut along $\overline{BD}$ to form $\triangle BCD$. Now $\triangle ACE$ and $\triangle BCD$ are no longer overlapping. Place the triangles side-by-side to compare corresponding angles and sides.

Proof *Theorem 6.4*

Given: $\overline{BD} \parallel \overline{AE}$

Prove: $\frac{BA}{CB} = \frac{DE}{CD}$

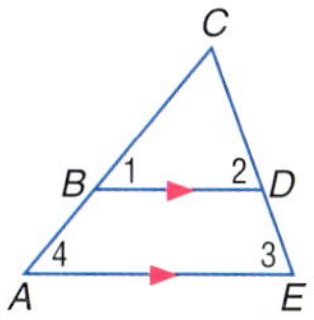

Paragraph Proof:

Since $\overline{BD} \parallel \overline{AE}$, $\angle 4 \cong \angle 1$ and $\angle 3 \cong \angle 2$ because they are corresponding angles. Then, by AA Similarity, $\triangle ACE \sim \triangle BCD$. From the definition of similar polygons, $\frac{CA}{CB} = \frac{CE}{CD}$. By the Segment Addition Postulate, $CA = BA + CB$ and $CE = DE + CD$. Substituting for CA and CE in the ratio, we get the following proportion.

$$\frac{BA + CB}{CB} = \frac{DE + CD}{CD}$$

$$\frac{BA}{CB} + \frac{CB}{CB} = \frac{DE}{CD} + \frac{CD}{CD} \quad \text{Rewrite as a sum.}$$

$$\frac{BA}{CB} + 1 = \frac{DE}{CD} + 1 \quad \frac{CB}{CB} = 1 \text{ and } \frac{CD}{CD} = 1$$

$$\frac{BA}{CB} = \frac{DE}{CD} \quad \text{Subtract 1 from each side.}$$

6-4 Lesson Notes

1 Focus

5-Minute Check Transparency 6-4 Use as a quiz or review of Lesson 6-3.

Mathematical Background notes are available for this lesson on p. 280D.

How do city planners use geometry?

Ask students:

- What kind of lines frequently are used to make up street maps? **parallel and perpendicular lines**
- Why does Lake Shore Drive run at an angle to the vertical lines? **Because it follows the lake shore.**
- Which street is perpendicular to most of the streets shown in the map? **Michigan Avenue**

Resource Manager

Workbook and Reproducible Masters

Chapter 6 Resource Masters
- Study Guide and Intervention, pp. 313–314
- Skills Practice, p. 315
- Practice, p. 316
- Reading to Learn Mathematics, p. 317
- Enrichment, p. 318

Teaching Geometry With Manipulatives Masters, pp. 17, 108

Transparencies
5-Minute Check Transparency 6-4
Answer Key Transparencies

Technology
Interactive Chalkboard

2 Teach

PROPORTIONAL PARTS OF TRIANGLES

In-Class Examples

Power Point®

Teaching Tip In Example 1, point out that without the Triangle Proportionality Theorem, the length of $\overline{LG}$ could be found by solving $\frac{30}{21} = \frac{x+6}{x}$.

1 In $\triangle RST$, $\overline{RT} \parallel \overline{VU}$, $SV = 3$, $VR = 8$, and $UT = 12$. Find SU.

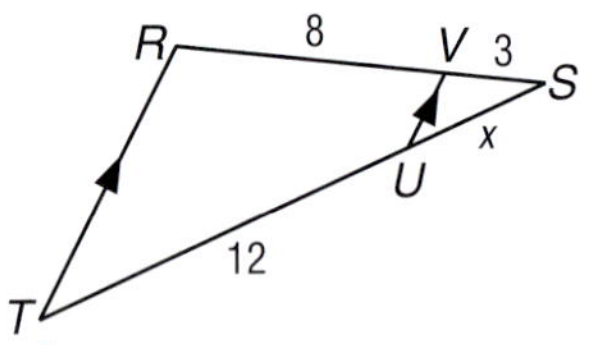

$4\frac{1}{2}$

2 In $\triangle DEF$, $DH = 18$, $HE = 36$, and $DG = \frac{1}{2}GF$. Determine whether $\overline{GH} \parallel \overline{FE}$. Explain.

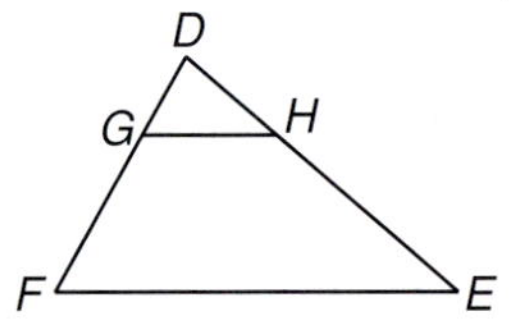

Yes; from the given information, $\frac{DG}{GF} = \frac{DH}{HE}$. Since the segments have proportional lengths, $\overline{GH} \parallel \overline{FE}$.

Study Tip

Using Fractions
You can also rewrite $\frac{9}{21}$ as $\frac{3}{7}$. Then use your knowledge of fractions to find the missing denominator.

$$\times 2$$
$$\frac{3}{7} = \frac{6}{?}$$
$$\times 2$$

The correct denominator is 14.

Example 1 Find the Length of a Side

In $\triangle EFG$, $\overline{HL} \parallel \overline{EF}$, $EH = 9$, $HG = 21$, and $FL = 6$. Find LG.

From the Triangle Proportionality Theorem, $\frac{EH}{HG} = \frac{FL}{LG}$.

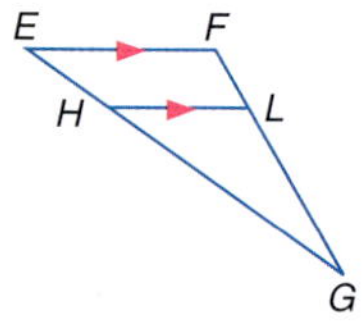

Substitute the known measures.

$\frac{9}{21} = \frac{6}{LG}$

$9(LG) = (21)6$ Cross products

$9(LG) = 126$ Multiply.

$LG = 14$ Divide each side by 9.

Proportional parts of a triangle can also be used to prove the converse of Theorem 6.4.

Theorem 6.5

Converse of the Triangle Proportionality Theorem

If a line intersects two sides of a triangle and separates the sides into corresponding segments of proportional lengths, then the line is parallel to the third side.

Example: If $\frac{BA}{CB} = \frac{DE}{CD}$, then $\overline{BD} \parallel \overline{AE}$.

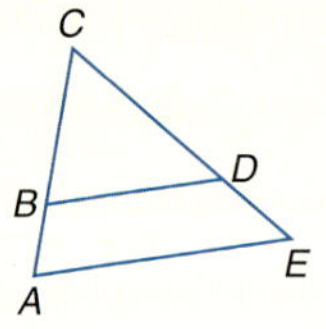

You will prove Theorem 6.5 in Exercise 38.

Example 2 Determine Parallel Lines

In $\triangle HKM$, $HM = 15$, $HN = 10$, and $\overline{HJ}$ is twice the length of $\overline{JK}$. Determine whether $\overline{NJ} \parallel \overline{MK}$. Explain.

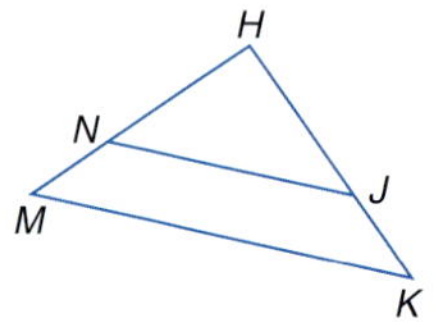

$HM = HN + NM$ Segment Addition Postulate

$15 = 10 + NM$ $HM = 15$, $HN = 10$

$5 = NM$ Subtract 10 from each side.

In order to show $\overline{NJ} \parallel \overline{MK}$, we must show that $\frac{HN}{NM} = \frac{HJ}{JK}$. $HN = 10$ and $NM = HM - HN$ or 5. So $\frac{HN}{NM} = \frac{10}{5}$ or 2. Let $JK = x$. Then $HJ = 2x$. So, $\frac{HJ}{JK} = \frac{2x}{x}$ or 2. Thus, $\frac{HN}{NM} = \frac{HJ}{JK} = 2$. Since the sides have proportional lengths, $\overline{NJ} \parallel \overline{MK}$.

A **midsegment** of a triangle is a segment whose endpoints are the midpoints of two sides of the triangle.

Theorem 6.6

Triangle Midsegment Theorem A midsegment of a triangle is parallel to one side of the triangle, and its length is one-half the length of that side.

Example: If B and D are midpoints of $\overline{AC}$ and $\overline{EC}$ respectively, $\overline{BD} \parallel \overline{AE}$ and $BD = \frac{1}{2}AE$.

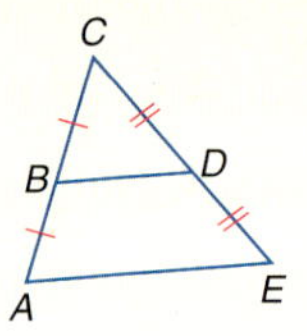

You will prove Theorem 6.6 in Exercise 39.

308 Chapter 6 Proportions and Similarity

Example 3 Midsegment of a Triangle

Triangle ABC has vertices $A(-4, 1)$, $B(8, -1)$, and $C(-2, 9)$. $\overline{DE}$ is a midsegment of $\triangle ABC$.

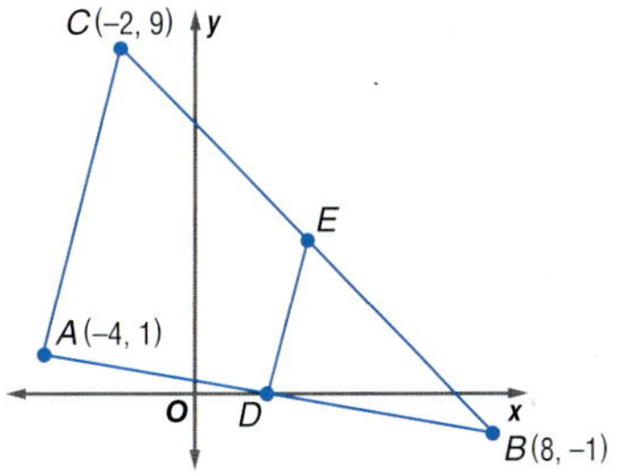

a. Find the coordinates of D and E.

Use the Midpoint Formula to find the midpoints of $\overline{AB}$ and $\overline{CB}$.

$$D\left(\frac{-4+8}{2}, \frac{1+(-1)}{2}\right) = D(2, 0)$$

$$E\left(\frac{-2+8}{2}, \frac{9+(-1)}{2}\right) = E(3, 4)$$

b. Verify that $\overline{AC}$ is parallel to $\overline{DE}$.

If the slopes of $\overline{AC}$ and $\overline{DE}$ are equal, $\overline{AC} \parallel \overline{DE}$.

$$\text{slope of } \overline{AC} = \frac{9-1}{-2-(-4)} \text{ or } 4$$

$$\text{slope of } \overline{DE} = \frac{4-0}{3-2} \text{ or } 4$$

Because the slopes of $\overline{AC}$ and $\overline{DE}$ are equal, $\overline{AC} \parallel \overline{DE}$.

c. Verify that $DE = \frac{1}{2}AC$.

First, use the Distance Formula to find AC and DE.

$$AC = \sqrt{[-2-(-4)]^2 + (9-1)^2} = \sqrt{4+64} = \sqrt{68}$$

$$DE = \sqrt{(3-2)^2 + (4-0)^2} = \sqrt{1+16} = \sqrt{17}$$

$$\frac{DE}{AC} = \frac{\sqrt{17}}{\sqrt{68}} = \sqrt{\frac{1}{4}} \text{ or } \frac{1}{2}$$

If $\frac{DE}{AC} = \frac{1}{2}$, then $DE = \frac{1}{2}AC$.

Study Tip

Look Back
To review the **Distance and Midpoint Formulas**, see Lesson 1-3.

DIVIDE SEGMENTS PROPORTIONALLY We have seen that parallel lines cut the sides of a triangle into proportional parts. Three or more parallel lines also separate transversals into proportional parts. If the ratio is 1, they separate the transversals into congruent parts.

Corollaries

6.1 If three or more parallel lines intersect two transversals, then they cut off the transversals proportionally.

Example: If $\overleftrightarrow{DA} \parallel \overleftrightarrow{EB} \parallel \overleftrightarrow{FC}$, then $\frac{AB}{BC} = \frac{DE}{EF}$, $\frac{AC}{DF} = \frac{BC}{EF}$, and $\frac{AC}{BC} = \frac{DF}{EF}$.

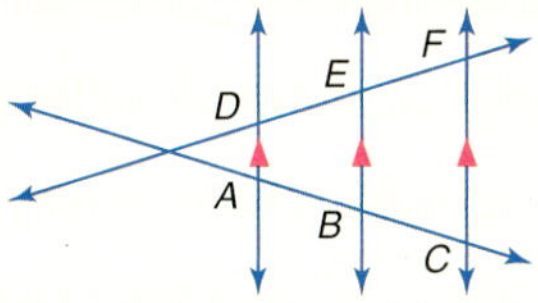

6.2 If three or more parallel lines cut off congruent segments on one transversal, then they cut off congruent segments on every transversal.

Example: If $\overline{AB} \cong \overline{BC}$, then $\overline{DE} \cong \overline{EF}$.

Study Tip

Three Parallel Lines
Corollary 6.1 is a special case of Theorem 6.4. In some drawings, the transversals are not shown to intersect. But, if extended, they will intersect and therefore, form triangles with each parallel line and the transversals.

www.geometryonline.com/extra_examples/fcat

In-Class Example

Power Point®

3 Triangle ABC has vertices $A(-2, 2)$, $B(2, 4)$, and $C(4, -4)$. $\overline{DE}$ is a midsegment of $\triangle ABC$.

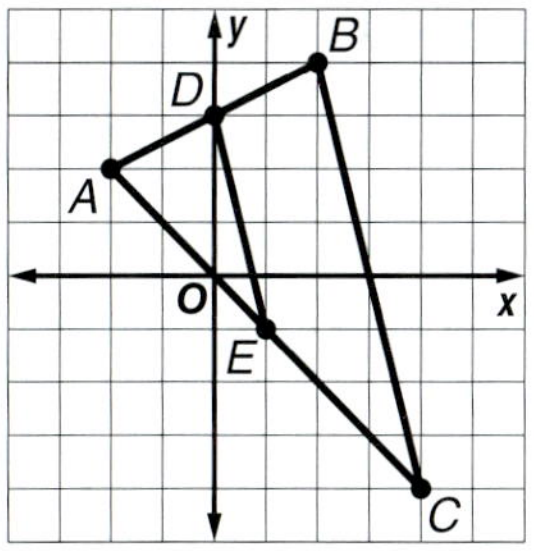

a. Find the coordinates of D and E. **$D(0, 3)$ $E(1, -1)$**

b. Verify that $\overline{BC} \parallel \overline{DE}$. **Since the slope of $\overline{BC} = -4$ and the slope of $\overline{DE} = -4$, $\overline{BC} \parallel \overline{DE}$.**

c. Verify that $DE = \frac{1}{2}BC$.
$BC = \sqrt{68} = 2\sqrt{17}$; $DE = \sqrt{17}$. Therefore $DE = \frac{1}{2}BC$.

Teaching Tip Remind students that corollaries are easily proved theorems that follow from a given theorem. You may wish to have students volunteer to present a verbal proof of each corollary.

DIVIDE SEGMENTS PROPORTIONALLY

In-Class Examples

Power Point®

Teaching Tip Watch for students who think parallel lines divide transversals into congruent parts instead of proportional parts.

4 In the figure, Larch, Maple, and Nuthatch Streets are all parallel. The figure shows the distances in city blocks that the streets are apart. Find x.

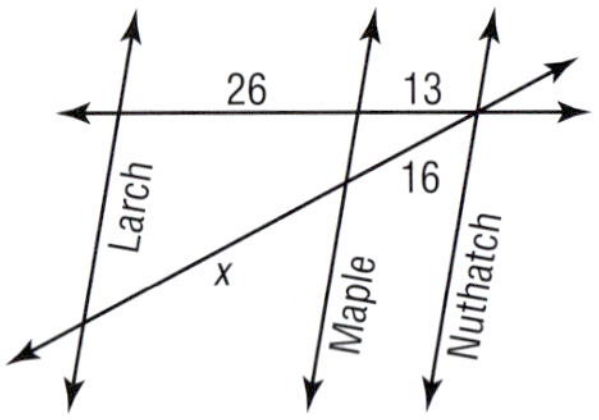

32

5 Find x and y.

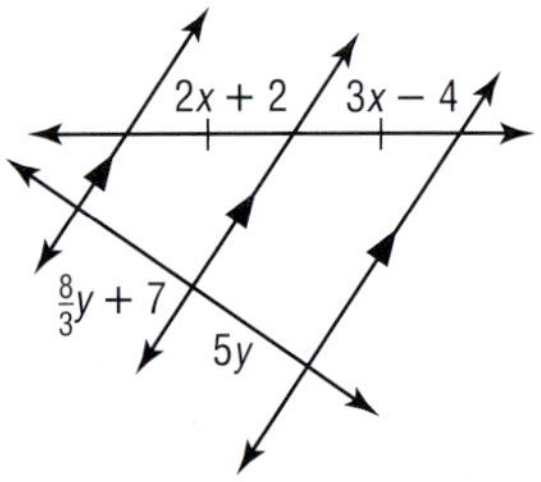

$x = 6$; $y = 3$

More About...

Maps Modern map-making techniques involve using images taken from space to produce accurate representations on paper. In February 2000, the crew of the space shuttle *Endeavor* performed a radar topography mission, collecting a trillion radar images of 46 million square miles of Earth.

Source: www.nima.mil

Example 4 Proportional Segments

MAPS **Refer to the map showing part of Lake Shore Drive in Chicago at the beginning of the lesson. The streets from Oak Street to Ontario Street are all parallel to each other. The distance from Oak Street to Ontario along Michigan Avenue is about 3800 feet. The distance between the same two streets along Lake Shore Drive is about 4430 feet. If the distance from Delaware Place to Walton Street along Michigan Avenue is about 411 feet, what is the distance between those streets along Lake Shore Drive?**

Make a sketch of the streets in the problem. Notice that the streets form the bottom portion of a triangle that is cut by parallel lines. So you can use the Triangle Proportionality Theorem.

$\frac{\text{Michigan Ave. Delaware to Walton}}{\text{Oak to Ontario}} = \frac{\text{Lake Shore Drive Delaware to Walton}}{\text{Oak to Ontario}}$	Triangle Proportionality Theorem
$\frac{411}{3800} = \frac{x}{4430}$	Substitution
$3800 \cdot x = 411(4430)$	Cross products
$3800x = 1{,}820{,}730$	Multiply.
$x = 479$	Divide each side by 3800.

The distance from Delaware Place to Oak Street along Lake Shore Drive is about 479 feet.

Example 5 Congruent Segments

Find x and y.

To find x:

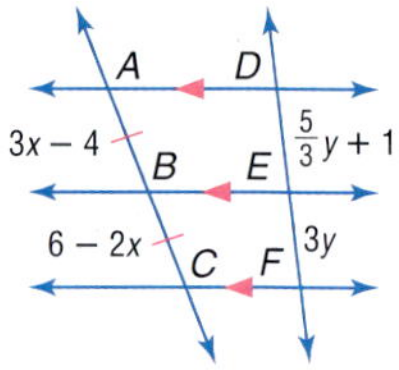

$AB = BC$	Given
$3x - 4 = 6 - 2x$	Substitution
$5x - 4 = 6$	Add $2x$ to each side.
$5x = 10$	Add 4 to each side.
$x = 2$	Divide each side by 5.

To find y:

$\overline{DE} \cong \overline{EF}$	Parallel lines that cut off congruent segments on one transversal cut off congruent segments on every transversal.
$DE \cong EF$	Definition of congruent segments
$3y = \frac{5}{3}y + 1$	Substitution
$9y = 5y + 3$	Multiply each side by 3 to eliminate the denominator.
$4y = 3$	Subtract $5y$ from each side.
$y = \frac{3}{4}$	Divide each side by 4.

DAILY INTERVENTION — Differentiated Instruction

Kinesthetic Have your students use string, masking tape, and a tiled floor to mark off congruent segments on parallel lines made with masking tape on the floor. Use the string to show that if three or more parallel lines cut off congruent segments on one transversal, they cut off congruent segments on another transversal.

310 Chapter 6 Proportions and Similarity

It is possible to separate a segment into two congruent parts by constructing the perpendicular bisector of a segment. However, a segment cannot be separated into three congruent parts by constructing perpendicular bisectors. To do this, you must use parallel lines and the similarity theorems from this lesson.

Construction

Trisect a Segment

1. Draw $\overline{AB}$ to be trisected. Then draw $\overrightarrow{AM}$.

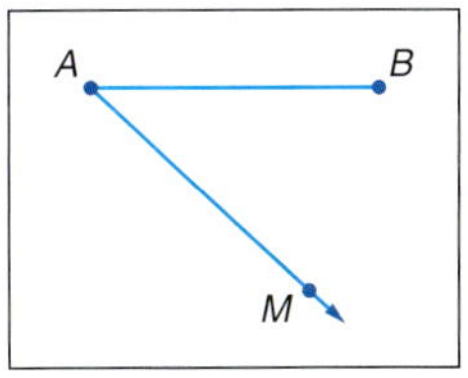

2. With the compass at A, mark off an arc that intersects $\overrightarrow{AM}$ at X. Use the same compass setting to construct $\overline{XY}$ and $\overline{YZ}$ congruent to $\overline{AX}$.

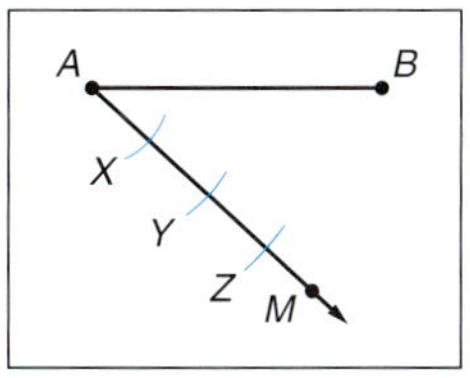

3. Draw $\overline{ZB}$. Then construct lines through Y and X that are parallel to $\overline{ZB}$. Label the intersection points on $\overline{AB}$ as P and Q.

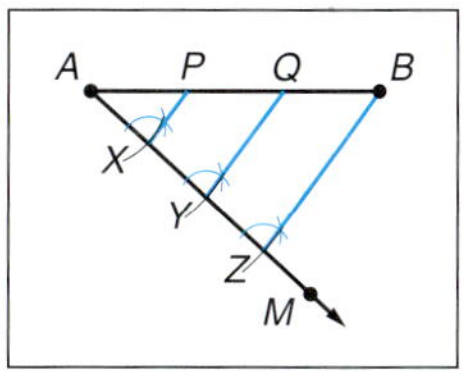

Conclusion: Because parallel lines cut off congruent segments on transversals, $\overline{AP} \cong \overline{PQ} \cong \overline{QB}$.

3 Practice/Apply

Study Notebook

Have students—
- add the definitions/examples of the vocabulary terms to their Vocabulary Builder worksheets for Chapter 6.
- include the construction of how to trisect a segment.
- include any other item(s) that they find helpful in mastering the skills in this lesson.

Check for Understanding

Concept Check

1. **Explain** how you would know if a line that intersects two sides of a triangle is parallel to the third side. **1–3. See margin.**
2. **OPEN ENDED** Draw two segments that are intersected by three lines so that the parts are proportional. Then draw a counterexample.
3. **Compare and contrast** Corollary 6.1 and Corollary 6.2.

Guided Practice

GUIDED PRACTICE KEY

Exercises	Examples
4–5	1
6–8	3
9–10	2
11–12	5
13	4

For Exercises 4 and 5, refer to $\triangle RST$.

4. If $RL = 5$, $RT = 9$, and $WS = 6$, find RW. **7.5**
5. If $TR = 8$, $LR = 3$, and $RW = 6$, find WS. **10**

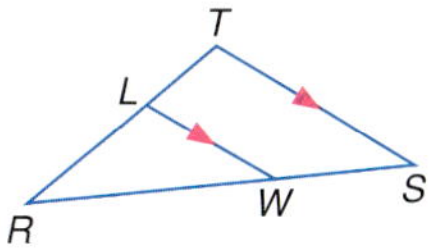

COORDINATE GEOMETRY **For Exercises 6–8, use the following information.** Triangle ABC has vertices $A(-2, 6)$, $B(-4, 0)$, and $C(10, 0)$. $\overline{DE}$ is a midsegment.

6. Find the coordinates of D and E.
7. Verify that $\overline{DE}$ is parallel to $\overline{BC}$.
8. Verify that $DE = \frac{1}{2}BC$.

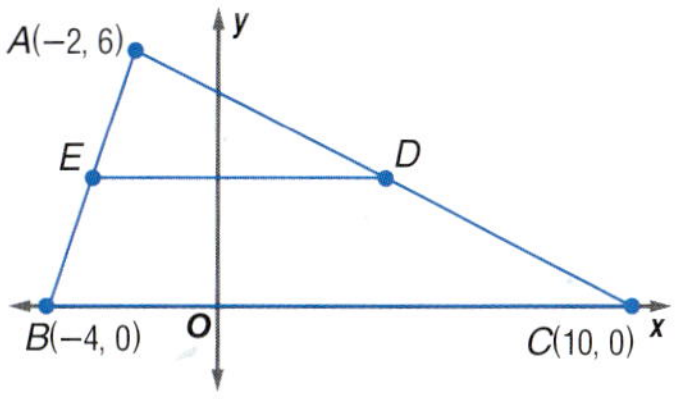

9. In $\triangle MQP$, $MP = 25$, $MN = 9$, $MR = 4.5$, and $MQ = 12.5$. Determine whether $\overline{RN} \parallel \overline{QP}$. Justify your answer.

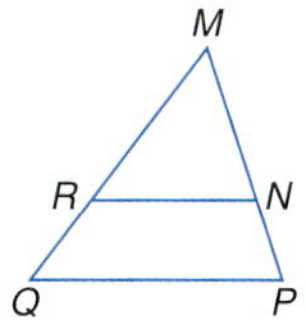

. (4, 3); (−3, 3)

. The slopes of $\overline{DE}$ and $\overline{BC}$ are both 0. So $\overline{DE} \parallel \overline{BC}$.

. $DE = 7$ and $BC = 14$; $DE = \frac{1}{2}BC$.

. Yes; $\frac{MN}{NP} = \frac{MR}{RQ} = \frac{9}{16}$, so $\overline{RN} \parallel \overline{QP}$.

About the Exercises...

Organization by Objective
- **Proportional Parts of Triangles:** 12–30, 36–37
- **Divide Segment Proportionally:** 31–35, 41

Odd/Even Assignments

Exercises 12–41 are structured so that students practice the same concepts whether they are assigned odd or even problems.

Alert! Exercises 40–42 require a compass and straightedge.

Assignment Guide

Basic: 15–25 odd, 29, 33–41, 44–47, 49–64 (optional: 48)

Average: 15–43 odd, 44–64

Advanced: 14–44 even, 46–61 (optional: 62–64)

Answers

1. Sample answer: If a line intersects two sides of a triangle and separates sides into corresponding segments of proportional lengths, then it is parallel to the third side.

2. Sample answer:

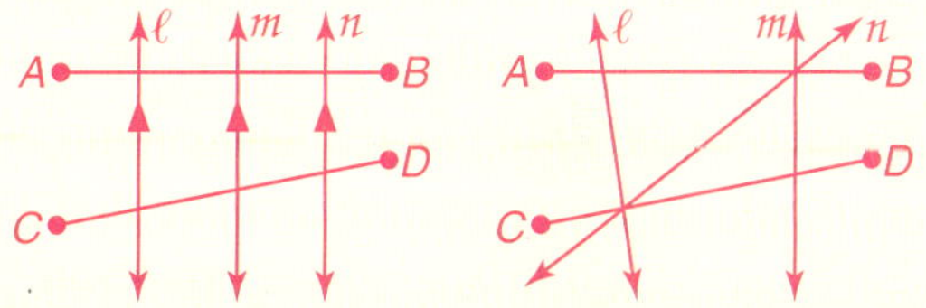

3. Given three or more parallel lines intersecting two transversals, Corollary 6.1 states that the parts of the transversals are proportional. Corollary 6.2 states that if the parts of one transversal are congruent, then the parts of every transversal are congruent.

Lesson 6-4 Parallel Lines and Proportional Parts 311

Study Guide and Intervention, p. 313 (shown) and p. 314

Proportional Parts of Triangles In any triangle, a line parallel to one side of a triangle separates the other two sides proportionally. The converse is also true.

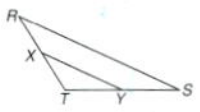

If X and Y are the midpoints of $\overline{RT}$ and $\overline{ST}$, then $\overline{XY}$ is a **midsegment** of the triangle. The Triangle Midsegment Theorem states that a midsegment is parallel to the third side and is half its length.

If $\overleftrightarrow{XY} \parallel \overleftrightarrow{RS}$, then $\frac{RX}{XT} = \frac{SY}{YT}$.

If $\frac{RX}{XT} = \frac{SY}{YT}$, then $\overleftrightarrow{XY} \parallel \overleftrightarrow{RS}$.

If $\overline{XY}$ is a midsegment, then $\overleftrightarrow{XY} \parallel \overleftrightarrow{RS}$ and $XY = \frac{1}{2}RS$.

Example 1 **In $\triangle ABC$, $\overline{EF} \parallel \overline{CB}$. Find x.**

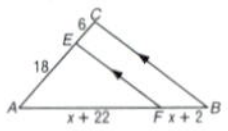

Since $\overline{EF} \parallel \overline{CB}$, $\frac{AF}{FB} = \frac{AE}{EC}$.

$$\frac{x+22}{x+2} = \frac{18}{6}$$
$$6x + 132 = 18x + 36$$
$$96 = 12x$$
$$8 = x$$

Example 2 **A triangle has vertices $D(3, 6)$, $E(-3, -2)$, and $F(7, -2)$. Midsegment $\overline{GH}$ is parallel to $\overline{EF}$. Find the length of $\overline{GH}$.**

$\overline{GH}$ is a midsegment, so its length is one-half that of $\overline{EF}$. Points E and F have the same y-coordinate, so $EF = 7 - (-3) = 10$. The length of midsegment $\overline{GH}$ is 5.

Exercises

Find x.

1. 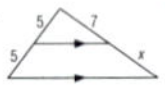

7

2. 

10

3. 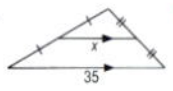

17.5

4. 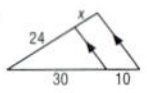

8

5. 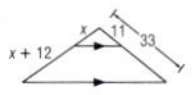

12

6. 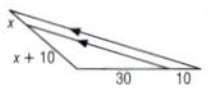

5

7. In Example 2, find the slope of $\overline{EF}$ and show that $\overline{EF} \parallel \overline{GH}$.
The endpoints of $\overline{GH}$ are at (0, 2) and (5, 2). The slope of $\overline{GH}$ is 0 and the slope of $\overline{EF}$ is 0. $\overline{GH}$ and $\overline{EF}$ are parallel.

Skills Practice, p. 315 and Practice, p. 316 (shown)

1. If $AD = 24$, $DB = 27$, and $EB = 18$, find CE.

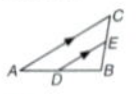

16

2. Find x, QT, and TR if $QT = x + 6$, $SR = 12$, $PS = 27$, and $TR = x - 4$.

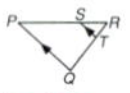

12; 18; 8

Determine whether $\overline{JK} \parallel \overline{NM}$.

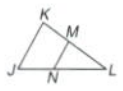

3. $JN = 18$, $JL = 30$, $KM = 21$, and $ML = 35$
no

4. $KM = 24$, $KL = 44$, and $NL = \frac{5}{6}JN$
yes

COORDINATE GEOMETRY **For Exercises 5 and 6, use the following information.**

Triangle EFG has vertices $E(-4, -1)$, $F(2, 5)$, and $G(2, -1)$. Point K is the midpoint of $\overline{EG}$ and H is the midpoint of $\overline{FG}$.

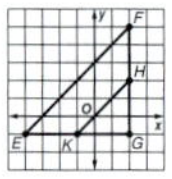

5. Show that $\overline{EF}$ is parallel to $\overline{KH}$.
The slope of $\overline{EF}$ is 1 and the slope of $\overline{KH}$ is 1, so $\overline{EF}$ is parallel to $\overline{KH}$.

6. Show that $KH = \frac{1}{2}EF$.
The length of $\overline{KH}$ is $3\sqrt{2}$, and the length of $\overline{EF}$ is $6\sqrt{2}$.

7. Find x and y.

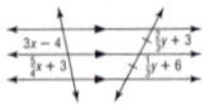

$x = 4$, $y = 9$

8. Find x and y.

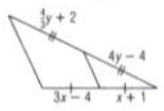

$x = \frac{5}{2}$, $y = \frac{9}{4}$

9. MAPS The distance from Wilmington to Ash Grove along Kendall is 820 feet and along Magnolia, 660 feet. If the distance between Beech and Ash Grove along Magnolia is 280 feet, what is the distance between the two streets along Kendall?

about 348 ft

Reading to Learn Mathematics, p. 317 ELL

Pre-Activity **How do city planners use geometry?**

Read the introduction to Lesson 6-4 at the top of page 307 in your textbook.

Use a geometric idea to explain why the distance between Chicago Avenue and Ontario Street is shorter along Michigan Avenue than along Lake Shore Drive. Sample answer: The shortest distance between two parallel lines is the perpendicular distance.

Reading the Lesson

1. Provide the missing words to complete the statement of each theorem. Then state the name of the theorem.
   a. If a line intersects two sides of a triangle and separates the sides into corresponding segments of ___proportional___ lengths, then the line is ___parallel___ to the third side. Converse of the Triangle Proportionality Theorem
   b. A midsegment of a triangle is ___parallel___ to one side of the triangle and its length is ___one-half___ the length of that side. Triangle Midsegment Theorem
   c. If a line is ___parallel___ to one side of a triangle and intersects the other two sides in ___two___ distinct points, then it separates these sides into ___segments___ of proportional length. Triangle Proportionality Theorem

2. Refer to the figure at the right.

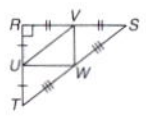

   a. Name the three midsegments of $\triangle RST$. $\overline{UV}$, $\overline{VW}$, $\overline{UW}$
   b. If $RS = 8$, $RU = 3$, and $TW = 5$, find the length of each of the midsegments. $UV = 5$, $VW = 3$, $UW = 4$
   c. What is the perimeter of $\triangle RST$? 24
   d. What is the perimeter of $\triangle UVW$? 12
   e. What are the perimeters of $\triangle RUV$, $\triangle SVW$, and $\triangle TUW$? 12; 12; 12
   f. How are the perimeters of each of the four small triangles related to the perimeter of the large triangle? The perimeter of each small triangle is one-half the perimeter of the large triangle.
   g. Would the relationship that you found in part f apply to any triangle in which the midpoints of the three sides are connected? yes

Helping You Remember

3. A good way to remember a new mathematical term is to relate it to other mathematical vocabulary that you already know. What is an easy way to remember the definition of *midsegment* using other geometric terms? Sample answer: A *midsegment* is a *segment* that connects the *midpoints* of two sides of a triangle.

---

10. In $\triangle ACE$, $ED = 8$, $DC = 20$, $BC = 25$, and $AB = 12$. Determine whether $\overline{DB} \parallel \overline{AE}$. **no**

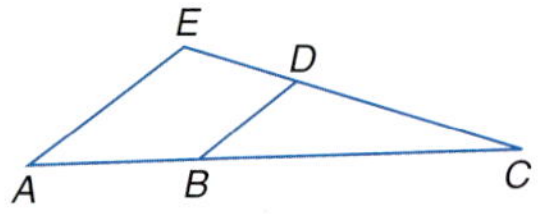

11. Find x and y. **$x = 2$; $y = 5$**

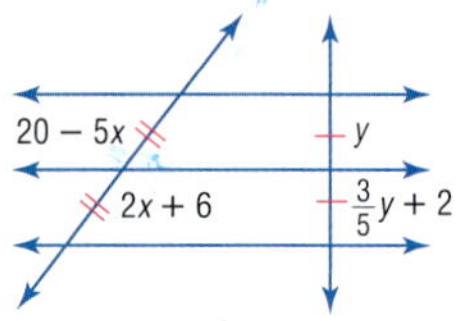

12. Find x and y. **$x = 18$; $y = 3$**

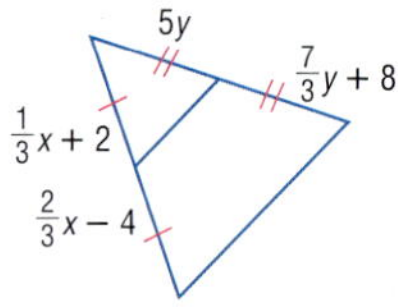

Application

13. **MAPS** The distance along Talbot Road from the Triangle Park entrance to the Walkthrough is 880 yards. The distance along Talbot Road from the Walkthrough to Clay Road is 1408 yards. The distance along Woodbury Avenue from the Walkthrough to Clay Road is 1760 yards. If the Walkthrough is parallel to Clay Road, find the distance from the entrance to the Walkthrough along Woodbury. **1100 yd**

★ indicates increased difficulty

Practice and Apply

Homework Help	
For Exercises	See Examples
14–19	1
20–26	2
27, 28	3
35–37, 43	4
33, 34	5

Extra Practice
See page 765.

For Exercises 14 and 15, refer to $\triangle XYZ$.

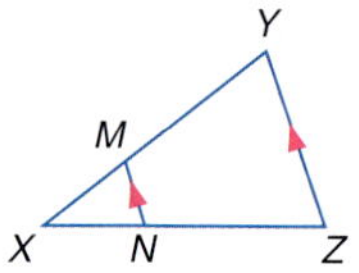

14. If $XM = 4$, $XN = 6$, and $NZ = 9$, find XY. **10**

15. If $XN = t - 2$, $NZ = t + 1$, $XM = 2$, and $XY = 10$, solve for t. **3**

16. If $DB = 24$, $AE = 3$, and $EC = 18$, find AD. **$AD = 4$**

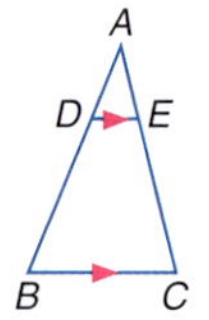

17. Find x and ED if $AE = 3$, $AB = 2$, $BC = 6$, and $ED = 2x - 3$. **$x = 6$, $ED =$**

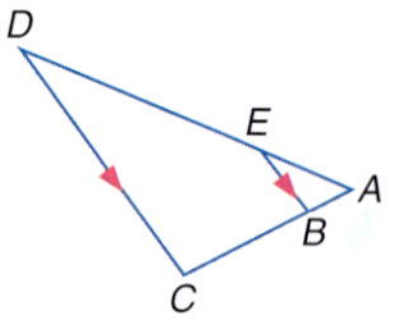

18. Find x, AC, and CD if $AC = x - 3$, $BE = 20$, $AB = 16$, and $CD = x + 5$.

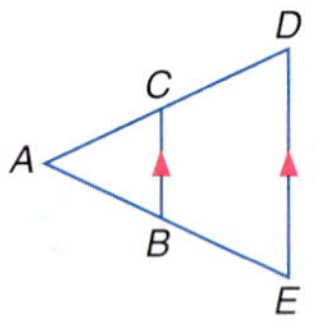

$x = 35$, $AC = 32$, $CD = 40$

19. Find BC, FE, CD, and DE if $AB = 6$, $AF = 8$, $BC = x$, $CD = y$, $DE = 2y - 3$, and $FE = x + \frac{10}{3}$.

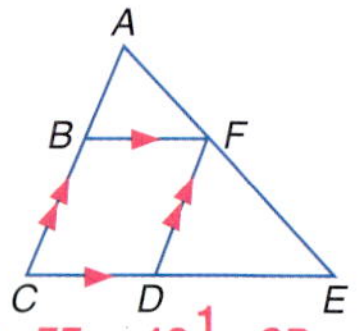

$BC = 10$, $FE = 13\frac{1}{3}$, $CD = 9$, $DE = 15$

Enrichment, p. 318

Parallel Lines and Congruent Parts

There is a theorem stating that if three parallel lines cut off congruent segments on one transversal, then they cut off congruent segments on any transversal. This can be shown for any number of parallel lines. The following drafting technique uses this fact to divide a segment into congruent parts.

$\overline{AB}$ is to be separated into five congruent parts. This can be done very accurately without using a ruler. All that is needed is a compass and a piece of notebook paper.

Step 1 Hold the corner of a piece of notebook paper at point A.

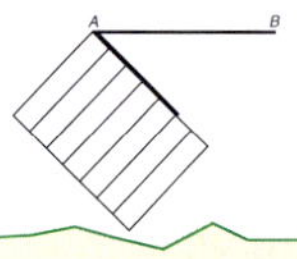

Find x so that $\overline{GJ} \parallel \overline{FK}$.

20. $GF = 12$, $HG = 6$, $HJ = 8$, $JK = x - 4$ **20**

21. $HJ = x - 5$, $JK = 15$, $FG = 18$, $HG = x - 4$ **10**

22. $GH = x + 3.5$, $HJ = x - 8.5$, $FH = 21$, $HK = 7$ **14.5**

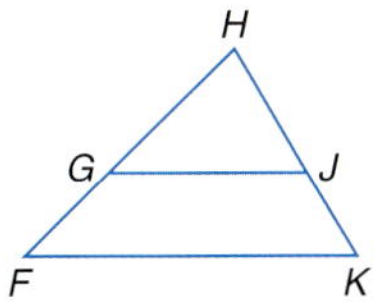

Determine whether $\overline{QT} \parallel \overline{RS}$. Justify your answer.

23–26. See margin for justifications.

23. $PR = 30$, $PQ = 9$, $PT = 12$, and $PS = 18$ **no**

24. $QR = 22$, $RP = 65$, and SP is 3 times TS. **no**

25. $TS = 8.6$, $PS = 12.9$, and PQ is half RQ. **yes**

26. $PQ = 34.88$, $RQ = 18.32$, $PS = 33.25$, and $TS = 11.45$ **yes**

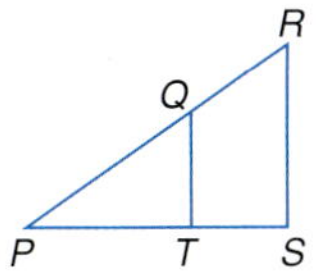

★ 27. Find the length of $\overline{BC}$ if $\overline{BC} \parallel \overline{DE}$ and $\overline{DE}$ is a midsegment of $\triangle ABC$. **$\sqrt{52}$**

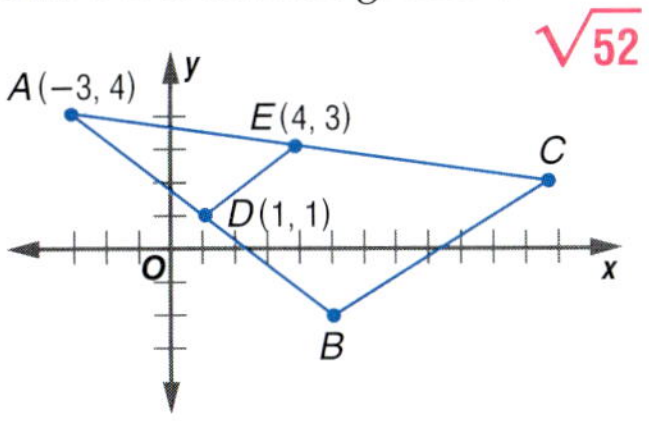

28. Show that $\overline{WM} \parallel \overline{TS}$ and determine whether $\overline{WM}$ is a midsegment.

28. The slopes of $\overline{TS}$ and $\overline{WM}$ are both −1 so $\overline{WM} \parallel \overline{TS}$; $\overline{WM}$ is not a midsegment because W and M are not midpoints of their respective sides.

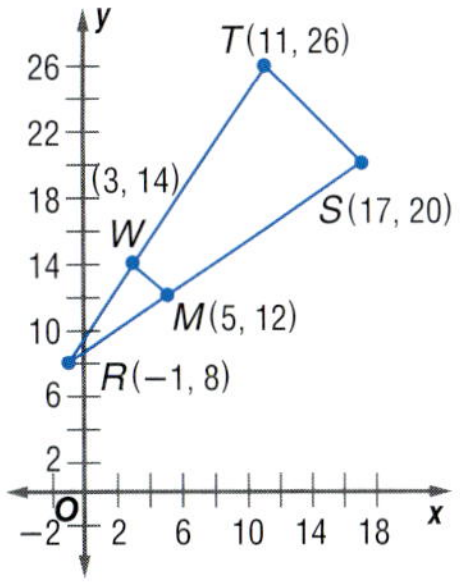

COORDINATE GEOMETRY **For Exercises 29 and 30, use the following information.**

Triangle ABC has vertices $A(-1, 6)$, $B(-4, -3)$, and $C(7, -5)$. $\overline{DE}$ is a midsegment.

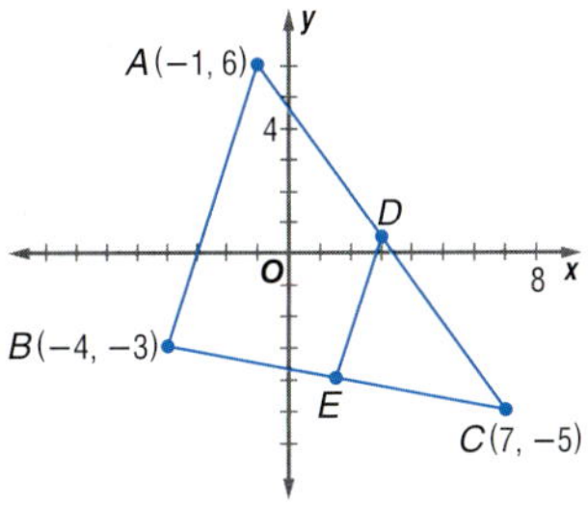

29. Verify that $\overline{DE}$ is parallel to $\overline{AB}$.

29. The endpoints of $\overline{DE}$ are $D\left(3, \frac{1}{2}\right)$ and $E\left(\frac{3}{2}, -4\right)$. Both $\overline{DE}$ and $\overline{AB}$ have slope of 3.

★ 30. Verify that $DE = \frac{1}{2}AB$. **$DE = \frac{\sqrt{90}}{2}$ and $AB = \sqrt{90}$**

★ 31. **COORDINATE GEOMETRY** Given $A(2, 12)$ and $B(5, 0)$, find the coordinates of P such that P separates $\overline{AB}$ into two parts with a ratio of 2 to 1. **(3, 8) or (4, 4)**

★ 32. **COORDINATE GEOMETRY** In $\triangle LMN$, $\overline{PR}$ divides $\overline{NL}$ and $\overline{MN}$ proportionally. If the vertices are $N(8, 20)$, $P(11, 16)$, and $R(3, 8)$ and $\frac{LP}{PN} = \frac{2}{1}$, find the coordinates of L and M. **$L(17, 8)$; $M(-7, -16)$**

ALGEBRA **Find x and y.**

33. **$x = 21$, $y = 15$**

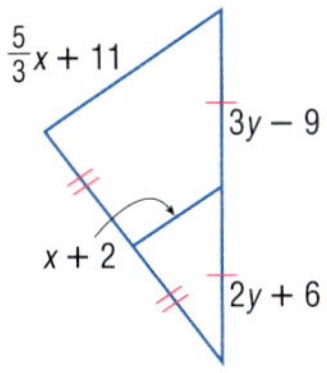

34. **$x = 1$, $y = \frac{3}{2}$**

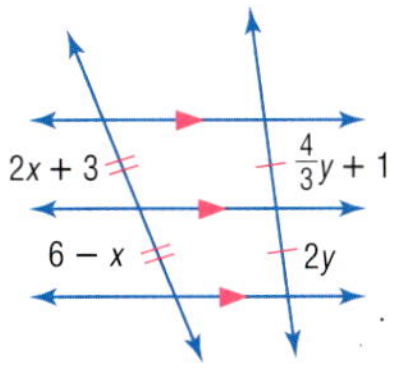

Answers

23. Segments are not proportional; $\frac{PQ}{QR} = \frac{3}{7}$, $\frac{PT}{TS} = 2$

24. Segments are not proportional; $\frac{PQ}{QR} = \frac{43}{22}$, $\frac{PT}{TS} = 2$

25. $\frac{PQ}{QR} = \frac{PT}{TS} = \frac{1}{2}$

26. $\frac{PQ}{QR} = \frac{PT}{TS} = \frac{436}{229}$

Lesson 6-4 Parallel Lines and Proportional Parts 313

Answers

38. Given: $\frac{DB}{AD} = \frac{EC}{AE}$

Prove: $\overline{DE} \parallel \overline{BC}$

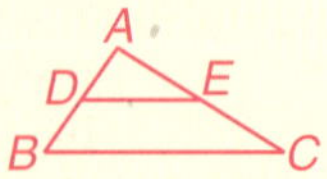

Statements (Reasons)

1. $\frac{DB}{AD} = \frac{EC}{AE}$ (Given)
2. $\frac{AD}{AD} + \frac{DB}{AD} = \frac{AE}{AE} + \frac{EC}{AE}$ (Addition Prop.)
3. $\frac{AD + DB}{AD} = \frac{AE + EC}{AE}$ (Subst.)
4. $AB = AD + DB$, $AC = AE + EC$ (Segment Addition Postulate)
5. $\frac{AB}{AD} = \frac{AC}{AE}$ (Substitution)
6. $\angle A \cong \angle A$ (Reflexive Prop.)
7. $\triangle ADE \sim \triangle ABC$ (SAS Similarity)
8. $\angle ADE \cong \angle ABC$ (Def. of $\sim$ polygons)
9. $\overline{DE} \parallel \overline{BC}$ (If corr. ∠s are ≅ then the lines are ∥.)

39. Given: D is the midpoint of $\overline{AB}$. E is the midpoint of $\overline{AC}$.

Prove: $\overline{DE} \parallel \overline{BC}$; $DE = \frac{1}{2}BC$

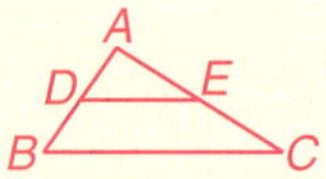

Statements (Reasons)

1. D is the midpoint of $\overline{AB}$; E is the midpoint of $\overline{AC}$. (Given)
2. $\overline{AD} \cong \overline{DB}$, $\overline{AE} \cong \overline{EC}$ (Midpoint Theorem)
3. $AD = DB$, $AE = EC$ (Def. of ≅ segments)
4. $AB = AD + DB$, $AC = AE + EC$ (Seg. Add. Post.)
5. $AB = AD + AD$, $AC = AE + AE$ (Substitution)
6. $AB = 2AD$, $AC = 2AE$ (Subst.)
7. $\frac{AB}{AD} = 2$, $\frac{AC}{AE} = 2$ (Div. Prop.)
8. $\frac{AB}{AD} = \frac{AC}{AE}$ (Transitive Prop.)
9. $\angle A \cong \angle A$ (Reflexive Prop.)
10. $\triangle ADE \sim \triangle ABC$ (SAS Similar.)
11. $\angle ADE \cong \angle ABC$ (Def. of $\sim$ polygons)
12. $\overline{DE} \parallel \overline{BC}$ (If corr. ∠s are ≅, the lines are parallel.)
13. $\frac{BC}{DE} = \frac{AB}{AD}$ (Def. of $\sim$ polygons)
14. $\frac{BC}{DE} = 2$ (Subst. Prop.)
15. $2DE = BC$ (Mult. Prop.)
16. $DE = \frac{1}{2}BC$ (Div. Prop.)

CONSTRUCTION **For Exercises 35–37, use the following information and drawing.**

Two poles, 30 feet and 50 feet tall, are 40 feet apart and perpendicular to the ground. The poles are supported by wires attached from the top of each pole to the bottom of the other, as in the figure. A coupling is placed at C where the two wires cross.

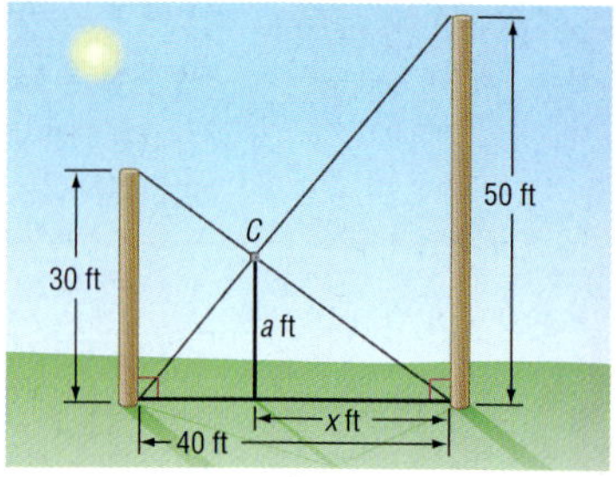

★ 35. Find x, the distance from C to the taller pole. **25 ft**

★ 36. How high above the ground is the coupling? **18.75 ft**

★ 37. How far down the wire from the smaller pole is the coupling? **18.75 ft**

PROOF **Write a two-column proof of each theorem.**

★ 38. Theorem 6.5 **See margin.**

★ 39. Theorem 6.6 **See margin.**

CONSTRUCTION **Construct each segment as directed.** **40–42. See margin.**

40. a segment 8 centimeters long, separated into three congruent segments

41. a segment separated into four congruent segments

★ 42. a segment separated into two segments in which their lengths have a ratio of 1 to 4

★ 43. **REAL ESTATE** In Lake Creek, the lots on which houses are to be built are laid out as shown. What is the lake frontage for each of the five lots if the total frontage is 135.6 meters? **$u = 24$; $w = 26.4$; $x = 30$; $y = 21.6$; $z = 33.6$**

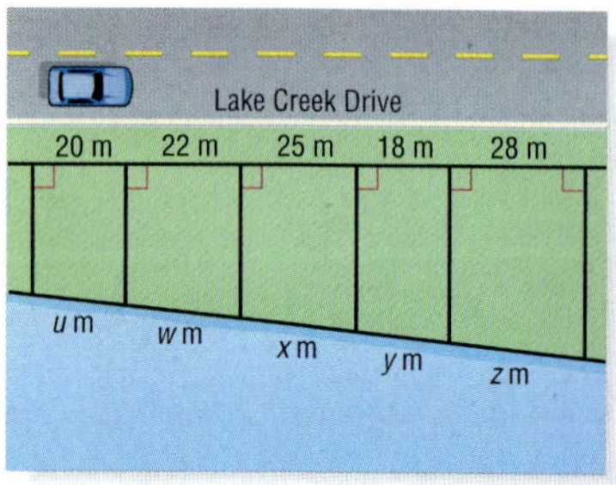

44. **CRITICAL THINKING** Copy the figure that accompanies Corollary 6.1 on page 309. Draw $\overline{DC}$. Let G be the intersection point of $\overline{DC}$ and $\overline{BE}$. Using that segment, explain how you could prove $\frac{AB}{BC} = \frac{DE}{EF}$.

44. Use Theorem 6.4 in $\triangle DCA$. $\overline{BG} \parallel \overline{AD}$, so $\frac{AB}{BC} = \frac{DG}{GC}$ and in $\triangle DCF$, $\frac{DG}{GC} = \frac{DE}{EF}$. Using the Transitive Property of Equality, $\frac{AB}{BC} = \frac{DE}{EF}$.

45. **WRITING IN MATH** Answer the question that was posed at the beginning of the lesson. **See margin.**

How do city planners use geometry?

Include the following in your answer:
- why maps are important to city planners, and
- what geometry facts a city planner needs to know to explain why the block between Chestnut and Pearson is longer on Lake Shore Drive than on Michigan Avenue.

Standardized Test Practice

★ 46. Find x. **B**

Ⓐ 16 Ⓑ 16.8

Ⓒ 24 Ⓓ 28.4

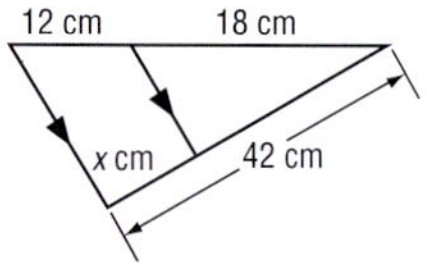

47. **GRID IN** The average of a and b is 18, and the ratio of a to b is 5 to 4. What is the value of $a - b$? **4**

314 Chapter 6 Proportions and Similarity

40. Figure not shown actual size.

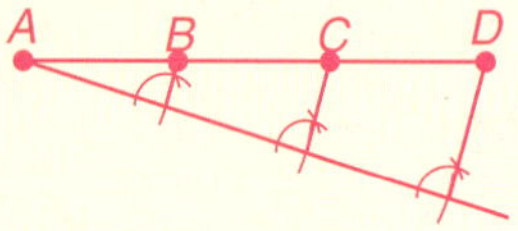

41.

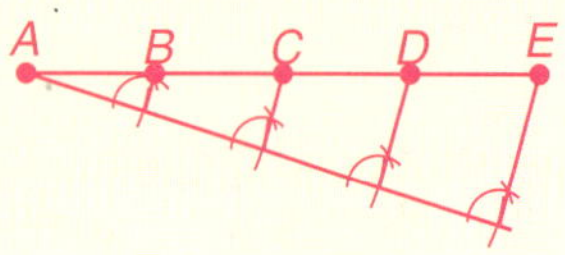

42.

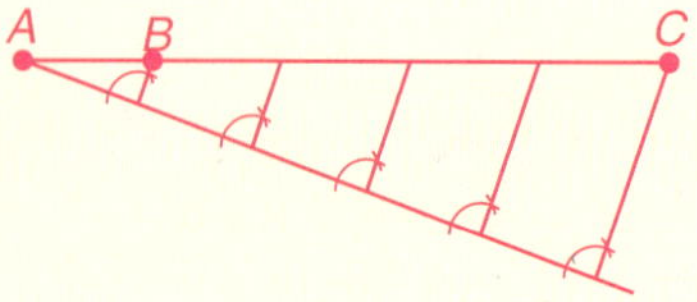

Extending the Lesson

48a. $\overline{EF} \parallel \overline{GH}$, $\overline{FG} \parallel \overline{EH}$, $\overline{EF} \cong \overline{GH}$, $\overline{FG} \cong \overline{EH}$

48. **MIDPOINTS IN POLYGONS** **Draw any quadrilateral *ABCD* on a coordinate plane. Points *E*, *F*, *G*, and *H* are midpoints of $\overline{AB}$, $\overline{BC}$, $\overline{CD}$, and $\overline{DA}$, respectively.**

    a. Connect the midpoints to form quadrilateral *EFGH*. Describe what you know about the sides of quadrilateral *EFGH*.

    b. Will the same reasoning work with five-sided polygons? Explain why or why not. **No, there will be an odd number of sides. It is not possible to pair opposite sides.**

Maintain Your Skills

Mixed Review

Determine whether each pair of triangles is similar. Justify your answer. *(Lesson 6-3)*

49.

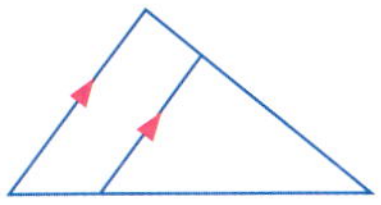

yes; AA

50.

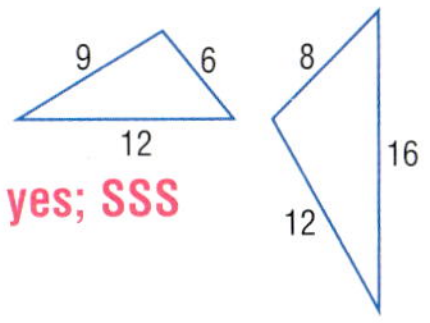

yes; SSS

51.

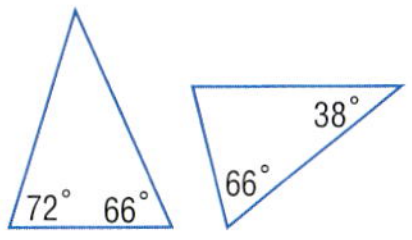

No; angles are not congruent.

Each pair of polygons is similar. Find *x* and *y*. *(Lesson 6-2)*

52.

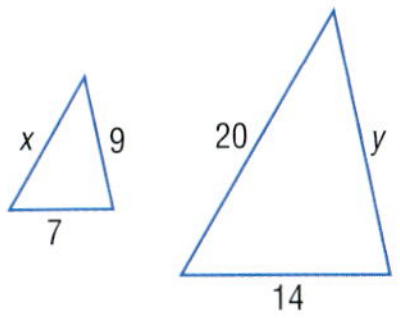

$x = 10$, $y = 18$

53.

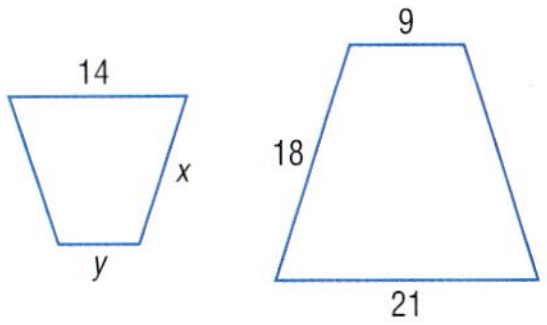

$x = 12$, $y = 6$

Determine the relationship between the measures of the given angles. *(Lesson 5-2)*

54. $\angle ADB$, $\angle ABD$ **$m\angle ADB > m\angle ABD$**

55. $\angle ABD$, $\angle BAD$ **$m\angle ABD > m\angle BAD$**

56. $\angle BCD$, $\angle CDB$ **$m\angle BCD < m\angle CDB$**

57. $\angle CBD$, $\angle BCD$ **$m\angle CBD > m\angle BCD$**

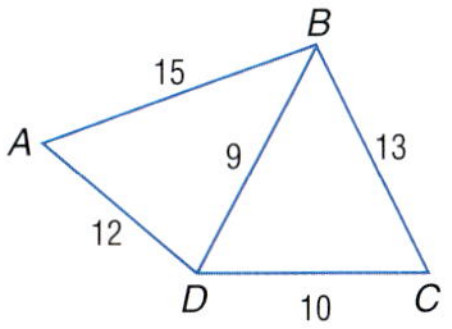

ARCHITECTURE **For Exercises 58 and 59, use the following information.**
The geodesic dome was developed by Buckminster Fuller in the 1940s as an energy-efficient building. The figure at the right shows the basic structure of one geodesic dome. *(Lesson 4-1)*

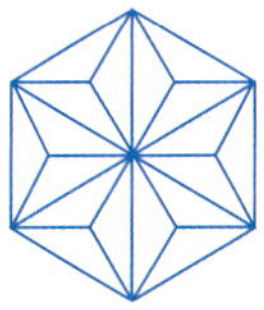

58. How many equilateral triangles are in the figure? **6**

59. How many obtuse triangles are in the figure? **18**

Determine the truth value of the following statement for each set of conditions.
If you have a fever, then you are sick. *(Lesson 2-3)*

60. You do not have a fever, and you are sick. **true**

61. You have a fever, and you are not sick. **false**

62. You do not have a fever, and you are not sick. **true**

63. You have a fever, and you are sick. **true**

Getting Ready for the Next Lesson

64–66. See margin.

PREREQUISITE SKILL **Write all the pairs of corresponding parts for each pair of congruent triangles.** *(To review **corresponding congruent parts**, see Lesson 4-3.)*

64. $\triangle ABC \cong \triangle DEF$ 65. $\triangle RST \cong \triangle XYZ$ 66. $\triangle PQR \cong \triangle KLM$

www.geometryonline.com/self_check_quiz/fcat

4 Assess

Open-Ended Assessment

Writing Have students explain the Triangle Proportionality Theorem using the triangle similarity properties.

Getting Ready for Lesson 6-5

Prerequisite Skill Students will learn about parts of similar triangles in Lesson 6-5. They will extend their knowledge of corresponding congruent parts to corresponding proportional parts of similar triangles. Use Exercises 64–66 to determine your students' familiarity with corresponding parts of congruent triangles.

Answers

64. $\angle A \cong \angle D$, $\angle B \cong \angle E$, $\angle C \cong \angle F$, $\overline{AB} \cong \overline{DE}$, $\overline{BC} \cong \overline{EF}$, $\overline{AC} \cong \overline{DF}$

65. $\angle R \cong \angle X$, $\angle S \cong \angle Y$, $\angle T \cong \angle Z$, $\overline{RS} \cong \overline{XY}$, $\overline{ST} \cong \overline{YZ}$, $\overline{RT} \cong \overline{XZ}$

66. $\angle P \cong \angle K$, $\angle Q \cong \angle L$, $\angle R \cong \angle M$, $\overline{PQ} \cong \overline{KL}$, $\overline{QR} \cong \overline{LM}$, $\overline{PR} \cong \overline{KM}$

Answer

45. Sample answer: City planners use maps in their work. Answers should include the following.

- City planners need to know geometry facts when developing zoning laws.
- A city planner would need to know that the shortest distance between two parallel lines is the perpendicular distance.

6-5 Lesson Notes

1 Focus

5-Minute Check Transparency 6-5 Use as a quiz or review of Lesson 6-4.

Mathematical Background notes are available for this lesson on p. 280D.

How is geometry related to photography?

Ask students:

- What is the size of the image on the film? **35 mm**
- What is the distance between the camera and the actual sculpture? **4 m**
- How is a photograph like a scale model? **Sample answer: The parts of an object in a photograph are in the same proportions as in the real object.**

6-5 Parts of Similar Triangles

Sunshine State Standards
MA.A.3.4.3, MA.B.1.4.1, MA.B.1.4.3, MA.C.1.4.1, MA.C.2.4.1, MA.C.3.4.1

What You'll Learn

- Recognize and use proportional relationships of corresponding perimeters of similar triangles.
- Recognize and use proportional relationships of corresponding angle bisectors, altitudes, and medians of similar triangles.

How is geometry related to photography?

The camera lens was 6.16 meters from this Dale Chihuly glass sculpture when this photograph was taken. The image on the film is 35 millimeters tall. Similar triangles enable us to find the height of the actual sculpture.

More About. . .

Dale Chihuly

Dale Chihuly (1941–), born in Tacoma, Washington, is widely recognized as one of the greatest glass artists in the world. His sculptures are made of hundreds of pieces of hand-blown glass that are assembled to resemble patterns in nature.

PERIMETERS Triangle ABC is similar to $\triangle DEF$ with a scale factor of 1:3. You can use variables and the scale factor to compare their perimeters. Let the measures of the sides of $\triangle ABC$ be a, b, and c. The measures of the corresponding sides of $\triangle DEF$ would be $3a$, $3b$, and $3c$.

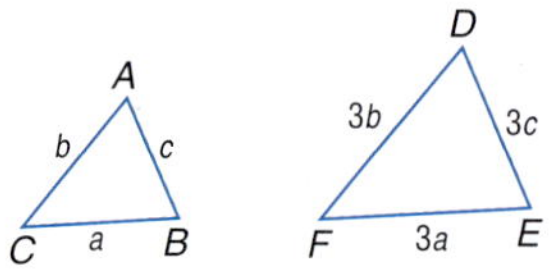

$$\frac{\text{perimeter of } \triangle ABC}{\text{perimeter of } \triangle DEF} = \frac{a+b+c}{3a+3b+3c}$$
$$= \frac{1(a+b+c)}{3(a+b+c)} \text{ or } \frac{1}{3}$$

The perimeters are in the same proportion as the side measures of the two similar figures. This suggests Theorem 6.7, the Proportional Perimeters Theorem.

Theorem 6.7

Proportional Perimeters Theorem If two triangles are similar, then the perimeters are proportional to the measures of corresponding sides.

You will prove Theorem 6.7 in Exercise 8.

Example 1 Perimeters of Similar Triangles

If $\triangle LMN \sim \triangle QRS$, $QR = 35$, $RS = 37$, $SQ = 12$, and $NL = 5$, find the perimeter of $\triangle LMN$.

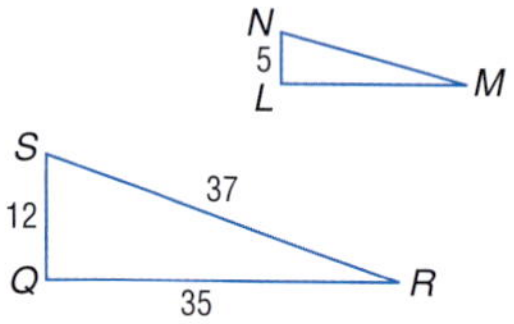

Let x represent the perimeter of $\triangle LMN$. The perimeter of $\triangle QRS = 35 + 37 + 12$ or 84.

$\frac{NL}{SQ} = \frac{\text{perimeter of } \triangle LMN}{\text{perimeter of } \triangle QRS}$	Proportional Perimeter Theorem
$\frac{5}{12} = \frac{x}{84}$	Substitution
$12x = 420$	Cross products
$x = 35$	Divide each side by 12.

The perimeter of $\triangle LMN$ is 35 units.

Resource Manager

Workbook and Reproducible Masters

Chapter 6 Resource Masters
- Study Guide and Intervention, pp. 319–320
- Skills Practice, p. 321
- Practice, p. 322
- Reading to Learn Mathematics, p. 323
- Enrichment, p. 324
- Assessment, p. 346

Prerequisite Skills Workbook, pp. 31–32

Transparencies

5-Minute Check Transparency 6-5
Answer Key Transparencies

Technology

Interactive Chalkboard

SPECIAL SEGMENTS OF SIMILAR TRIANGLES Think about a triangle drawn on a piece of paper being placed on a copy machine and either enlarged or reduced. The copy is similar to the original triangle. Now suppose you drew in special segments of a triangle, such as the altitudes, medians, or angle bisectors, on the original. When you enlarge or reduce that original triangle, all of those segments are enlarged or reduced at the same rate. This conjecture is formally stated in Theorems 6.8, 6.9, and 6.10.

Theorems — *Special Segments of Similar Triangles*

6.8 If two triangles are similar, then the measures of the corresponding altitudes are proportional to the measures of the corresponding sides.

Abbreviation: *~ △s have corr. altitudes proportional to the corr. sides.*

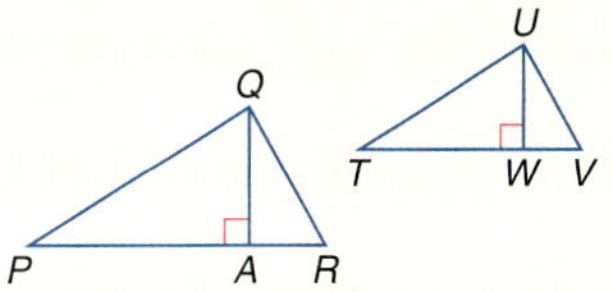

$$\frac{QA}{UW} = \frac{PR}{TV} = \frac{QR}{UV} = \frac{PQ}{TU}$$

6.9 If two triangles are similar, then the measures of the corresponding angle bisectors of the triangles are proportional to the measures of the corresponding sides.

Abbreviation: *~ △s have corr. ∠ bisectors proportional to the corr. sides.*

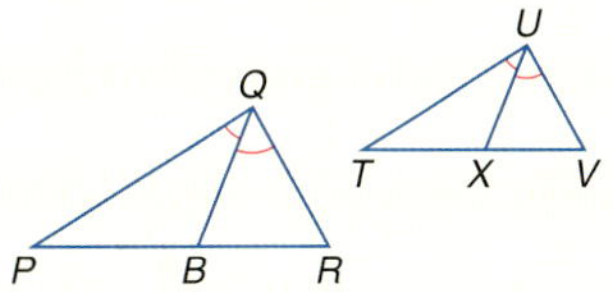

$$\frac{QB}{UX} = \frac{PR}{TV} = \frac{QR}{UV} = \frac{PQ}{TU}$$

6.10 If two triangles are similar, then the measures of the corresponding medians are proportional to the measures of the corresponding sides.

Abbreviation: *~ △s have corr. medians proportional to the corr. sides.*

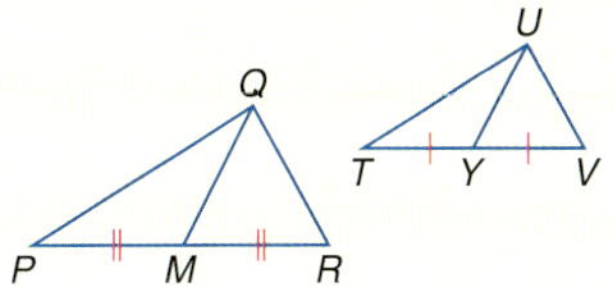

$$\frac{QM}{UY} = \frac{PR}{TV} = \frac{QR}{UV} = \frac{PQ}{TU}$$

You will prove Theorems 6.8 and 6.10 in Exercises 30 and 31, respectively.

Example 2 Write a Proof

Write a paragraph proof of Theorem 6.9.

Since the corresponding angles to be bisected are chosen at random, we need not prove this for every pair of bisectors.

Given: $\triangle RTS \sim \triangle EGF$

$\overline{TA}$ and $\overline{GB}$ are angle bisectors.

Prove: $\frac{TA}{GB} = \frac{RT}{EG}$

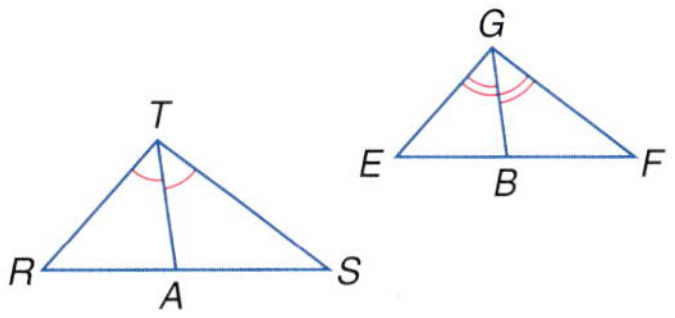

Paragraph Proof: Because corresponding angles of similar triangles are congruent, $\angle R \cong \angle E$ and $\angle RTS \cong \angle EGF$. Since $\angle RTS$ and $\angle EGF$ are bisected, we know that $\frac{1}{2}m\angle RTS = \frac{1}{2}m\angle EGF$ or $m\angle RTA = m\angle EGB$. This makes $\angle RTF \cong \angle EGB$ and $\triangle RTF \sim \triangle EGB$ by AA Similarity. Thus, $\frac{TA}{GB} = \frac{RT}{EG}$.

2 Teach

PERIMETERS

In-Class Example

Power Point®

Teaching Tip Students may not believe that if two triangles are similar, then the perimeters are proportional. Ask groups of students to cut out pairs of similar triangles on grid paper and measure each perimeter. Ask them to test if the perimeters have about the same proportion as the scale factor between the similar figures. Share the results with the class.

1 If $\triangle ABC \sim \triangle XYZ$, $AC = 32$, $AB = 16$, $BC = 16\sqrt{5}$, and $XY = 24$, find the perimeter of $\triangle XYZ$.

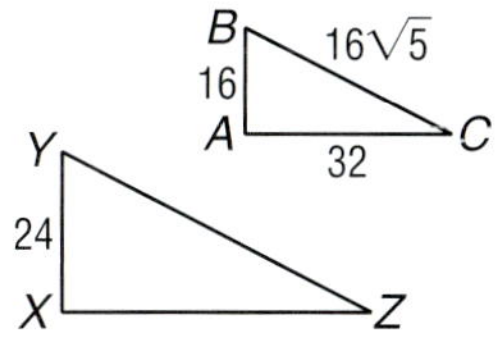

$72 + 24\sqrt{5}$

SPECIAL SEGMENTS OF SIMILAR TRIANGLES

In-Class Example

Power Point®

2 $\triangle ABC \sim \triangle MNO$ and $BC = \frac{1}{3}NO$. Find the ratio of the length of an altitude of $\triangle ABC$ to the length of an altitude of $\triangle MNO$. $\frac{1}{3}$

In-Class Examples

Power Point®

3 In the figure, $\triangle EFG \sim \triangle JKL$. $\overline{ED}$ is an altitude of $\triangle EFG$, and $\overline{JI}$ is an altitude of $\triangle JKL$. Find x if $EF = 36$, $ED = 18$, and $JK = 56$. **28**

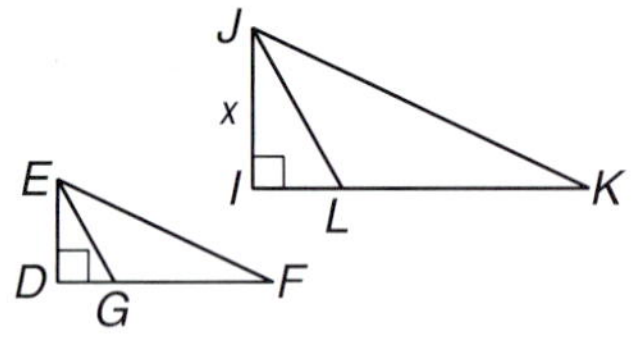

Teaching Tip You may want to ask students to conjecture what other segments related to triangles may have the same proportions as the corresponding sides of the triangles (angle bisectors and medians). Then ask them to test their conjectures.

4 The drawing below illustrates two poles supported by wires. $\triangle ABC \sim \triangle GED$. $\overline{AF} \cong \overline{CF}$ and $\overline{FG} \cong \overline{GC} \cong \overline{DC}$. Find the height of pole $\overline{EC}$. **15 ft**

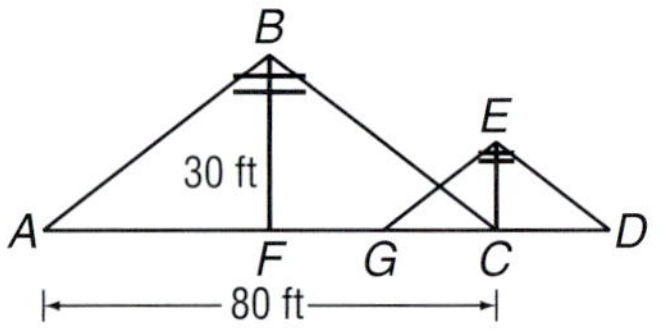

More About. . .

Photography The first consumer-oriented digital cameras were produced for sale in 1994 with a 640 × 480 pixel resolution. In 2002, a 3.3-megapixel camera could take a picture with 2048 × 1536 pixel resolution, which is a sharper picture than most computer monitors can display.

Source: www.howstuffworks.com

Example 3 Medians of Similar Triangles

In the figure, $\triangle ABC \sim \triangle DEF$. $\overline{BG}$ is a median of $\triangle ABC$, and $\overline{EH}$ is a median of $\triangle DEF$. Find EH if $BC = 30$, $BG = 15$, and $EF = 15$.

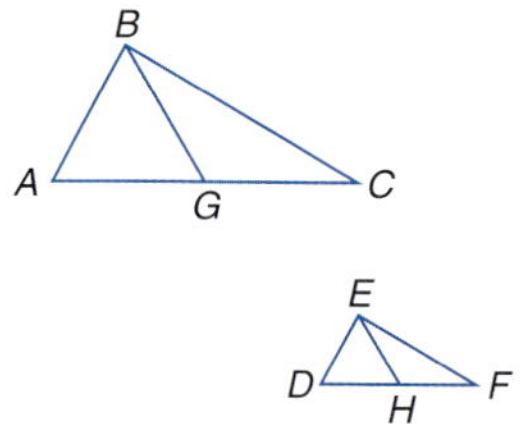

Let x represent EH.

$\frac{BG}{EH} = \frac{BC}{EF}$ Write a proportion.

$\frac{15}{x} = \frac{30}{15}$ $BG = 15$, $EH = x$, $BC = 30$, and $EF = 15$

$30x = 225$ Cross products

$x = 7.5$ Divide each side by 30.

Thus, $EH = 7.5$.

The theorems about the relationships of special segments in similar triangles can be used to solve real-life problems.

Example 4 Solve Problems with Similar Triangles

PHOTOGRAPHY **Refer to the application at the beginning of the lesson. The drawing below illustrates the position of the camera and the distance from the lens of the camera to the film. Find the height of the sculpture.**

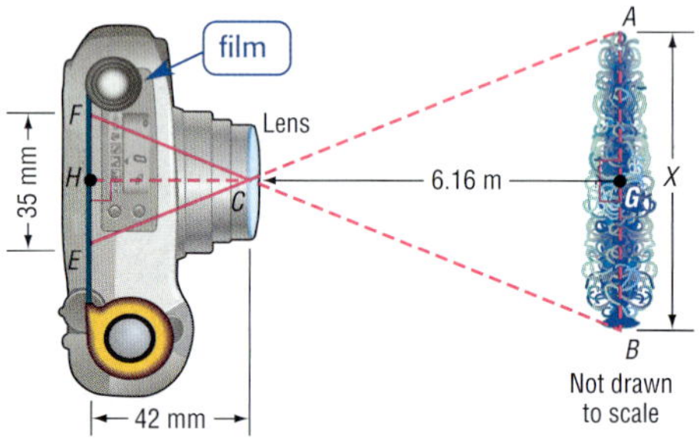

$\triangle ABC$ and $\triangle EFC$ are similar. The distance from the lens to the film in the camera is $CH = 42$ mm. $\overline{CG}$ and $\overline{CH}$ are altitudes of $\triangle ABC$ and $\triangle EFC$, respectively. If two triangles are similar, then the measures of the corresponding altitudes are proportional to the measures of the corresponding sides. This leads to the proportion $\frac{AB}{EF} = \frac{GC}{HC}$.

$\frac{AB}{EF} = \frac{GC}{HC}$ Write the proportion.

$\frac{x \text{ m}}{35 \text{ mm}} = \frac{6.16 \text{ m}}{42 \text{ mm}}$ $AB = x$ m, $EF = 35$ m, $GC = 6.16$ m, $HC = 42$ mm

$x \cdot 42 = 35(6.16)$ Cross products

$42x = 215.6$ Simplify.

$x \approx 5.13$ Divide each side by 42.

The sculpture is about 5.13 meters tall.

DAILY INTERVENTION

Differentiated Instruction

Auditory/Musical Ask students to call out the corresponding parts of given similar triangles as you use them in examples.

An angle bisector also divides the side of the triangle opposite the angle proportionally.

Theorem 6.11

Angle Bisector Theorem An angle bisector in a triangle separates the opposite side into segments that have the same ratio as the other two sides.

Example: $\frac{AD}{DB} = \frac{AC}{BC}$ ← segments with vertex A ← segments with vertex B

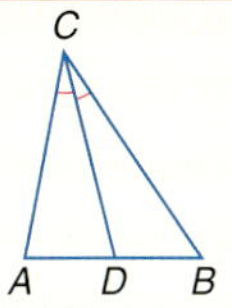

You will prove this theorem in Exercise 32.

Check for Understanding

Concept Check

1. **Explain** what must be true about $\triangle ABC$ and $\triangle MNQ$ before you can conclude that $\frac{AD}{MR} = \frac{BA}{NM}$.
$\triangle ABC \sim \triangle MNQ$ and $\overline{AD}$ and $\overline{MR}$ are altitudes, angle bisectors, or medians.

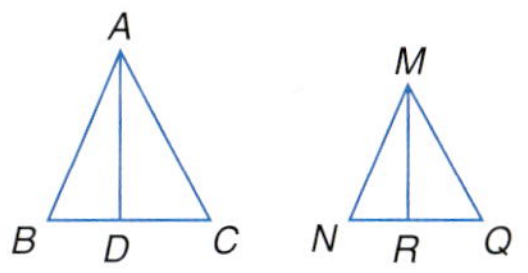

2. **OPEN ENDED** The perimeter of one triangle is 24 centimeters, and the perimeter of a second triangle is 36 centimeters. If the length of one side of the smaller triangle is 6, find possible lengths of the other sides of the triangles so that they are similar. Sample answer: 6, 8, 10 and 9, 12, 15

Guided Practice

GUIDED PRACTICE KEY	
Exercises	Examples
3–4	1
5–7	3
8	2
9	4

Find the perimeter of the given triangle.

3. $\triangle DEF$, if $\triangle ABC \sim \triangle DEF$, $AB = 5$, $BC = 6$, $AC = 7$, and $DE = 3$ 10.8

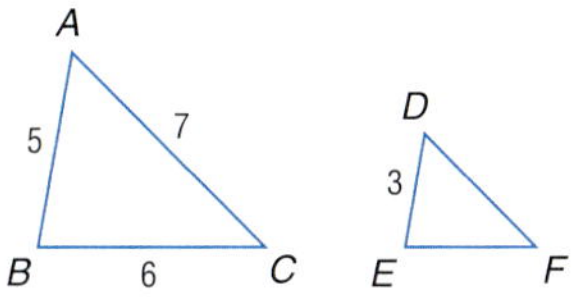

4. $\triangle WZX$, if $\triangle WZX \sim \triangle SRT$, $ST = 6$, $WX = 5$, and the perimeter of $\triangle SRT = 15$ 12.5

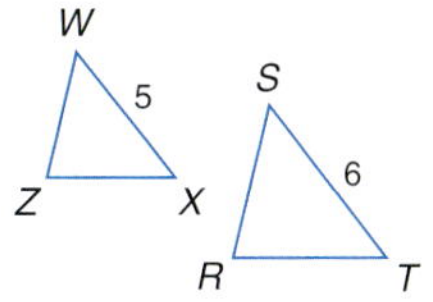

Find x.

5. 6

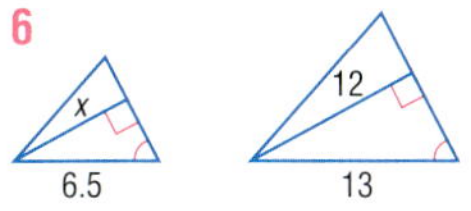

6. 15

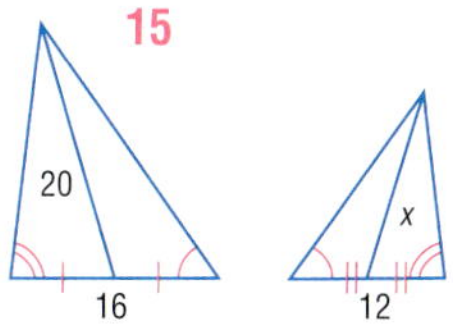

7. 6.75

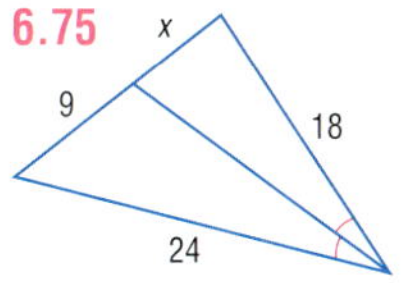

8. **PROOF** Write a paragraph proof of Theorem 6.7.

Given: $\triangle ABC \sim \triangle DEF$

$\frac{AB}{DE} = \frac{m}{n}$

Prove: $\frac{\text{perimeter of } \triangle ABC}{\text{perimeter of } \triangle DEF} = \frac{m}{n}$ See margin.

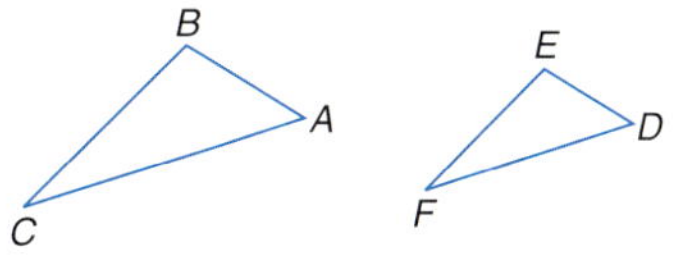

Application

9. **PHOTOGRAPHY** The distance from the film to the lens in a camera is 10 centimeters. The film image is 5 centimeters high. Tamika is 165 centimeters tall. How far should she be from the camera in order for the photographer to take a full-length picture? 330 cm or 3.3 m

3 Practice/Apply

Study Notebook

Have students—

- add the definitions/examples of the vocabulary terms to their Vocabulary Builder worksheets for Chapter 6.
- include a list of all parts related to similar triangles that have the same scale factor as the corresponding sides. This includes the altitudes, the medians, and the angle bisectors.
- include any other item(s) that they find helpful in mastering the skills in this lesson.

Answer

8. Given: $\triangle ABC \sim \triangle DEF$ and $\frac{AB}{DE} = \frac{m}{n}$

Prove: $\frac{\text{perimeter of } \triangle ABC}{\text{perimeter of } \triangle DEF} = \frac{m}{n}$

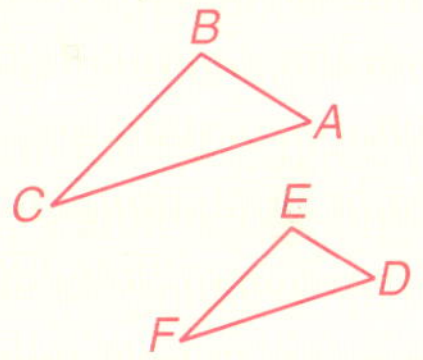

Proof: Because $\triangle ABC \sim \triangle DEF$, $\frac{AB}{DE} = \frac{BC}{EF} = \frac{AC}{DF}$. So $\frac{AB}{DE} = \frac{BC}{EF} = \frac{AC}{DF} = \frac{m}{n}$. Cross products yield $AB = DE\left(\frac{m}{n}\right)$, $BC = EF\left(\frac{m}{n}\right)$, and $AC = DF\left(\frac{m}{n}\right)$. Using substitution, the perimeter of $\triangle ABC = DE\left(\frac{m}{n}\right) + EF\left(\frac{m}{n}\right) + DF\left(\frac{m}{n}\right)$, or $\frac{m}{n}(DE + EF + DF)$. The ratio of the two perimeters $= \frac{\frac{m}{n}(DE + EF + DF)}{DE + EF + DF}$ or $\frac{m}{n}$.

DAILY INTERVENTION — Unlocking Misconceptions

A common error is to believe that the corresponding angles of two similar triangles have the same ratio as the corresponding sides. Emphasize that the corresponding angles of two similar triangles must be congruent.

About the Exercises...

Organization by Objective

- **Perimeters:** 10–15
- **Special Segments of Similar Triangles:** 16–28, 30–37

Odd/Even Assignments

Exercises 10–37 are structured so that students practice the same concepts whether they are assigned odd or even problems.

Assignment Guide

Basic: 11–19 odd, 23, 29–37 odd, 38–50

Average: 11–37 odd, 38–50

Advanced: 10–28 even, 29, 30–38 even, 39–47 (optional: 48–50)

All: Quiz 2 (1–10)

★ indicates increased difficulty

Practice and Apply

Homework Help

For Exercises	See Examples
10–15	1
16, 17, 28	4
18–27	3
30–37	2

Extra Practice
See page 766.

Find the perimeter of the given triangle.

10. $\triangle BCD$, if $\triangle BCD \sim \triangle FDE$, $CD = 12$, $FD = 5$, $FE = 4$, and $DE = 8$ **25.5**

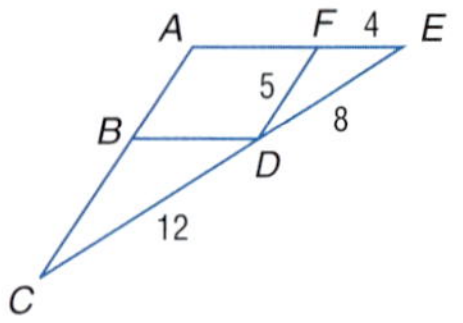

11. $\triangle ADF$, if $\triangle ADF \sim \triangle BCE$, $BC = 24$, $EB = 12$, $CE = 18$, and $DF = 21$ **63**

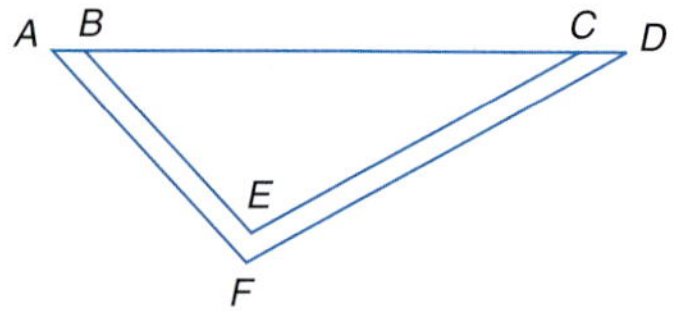

12. $\triangle CBH$, if $\triangle CBH \sim \triangle FEH$, $ADEG$ is a parallelogram, $CH = 7$, $FH = 10$, $FE = 11$, and $EH = 6$ **18.9**

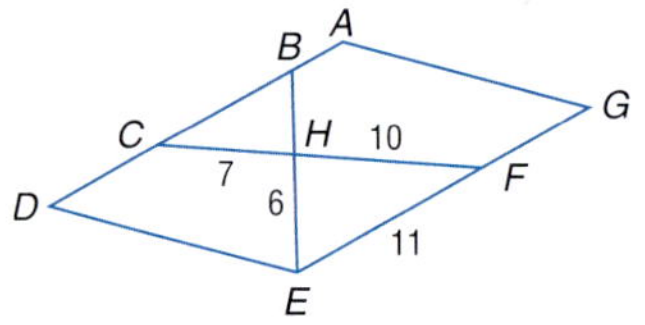

13. $\triangle DEF$, if $\triangle DEF \sim \triangle CBF$, perimeter of $\triangle CBF = 27$, $DF = 6$, and $FC = 8$ **20.25**

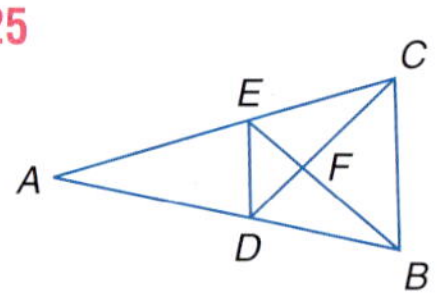

14. $\triangle ABC$, if $\triangle ABC \sim \triangle CBD$, $CD = 4$, $DB = 3$, and $CB = 5$ **20**

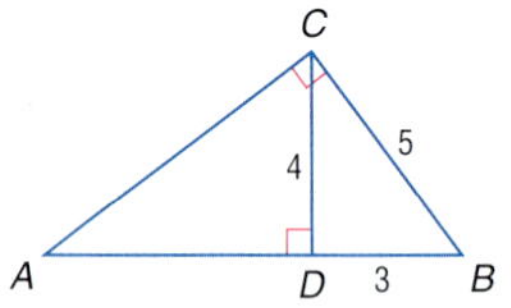

15. $\triangle ABC$, if $\triangle ABC \sim \triangle CBD$, $AD = 5$, $CD = 12$, and $BC = 31.2$ **78**

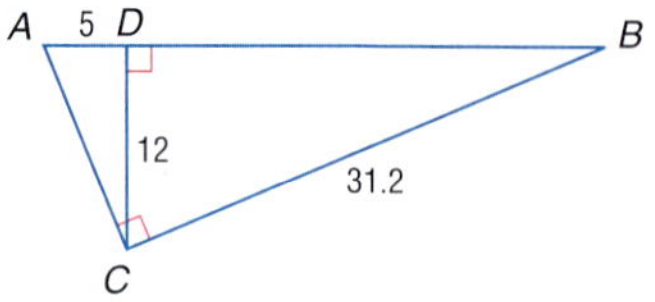

16. DESIGN Rosario wants to enlarge the dimensions of an 18-centimeter by 24-centimeter picture by 30%. She plans to line the inside edge of the frame with blue cord. The store only had 110 centimeters of blue cord in stock. Will this be enough to fit on the inside edge of the frame? Explain. **Yes; the enlarged picture will take approximately 109.2 cm of cord.**

17. PHYSICAL FITNESS A park has two similar triangular jogging paths as shown. The dimensions of the inner path are 300 meters, 350 meters, and 550 meters. The shortest side of the outer path is 600 meters. Will a jogger on the inner path run half as far as one on the outer path? Explain. **Yes; the perimeters are in the same ratio as the sides, $\frac{300}{600}$ or $\frac{1}{2}$.**

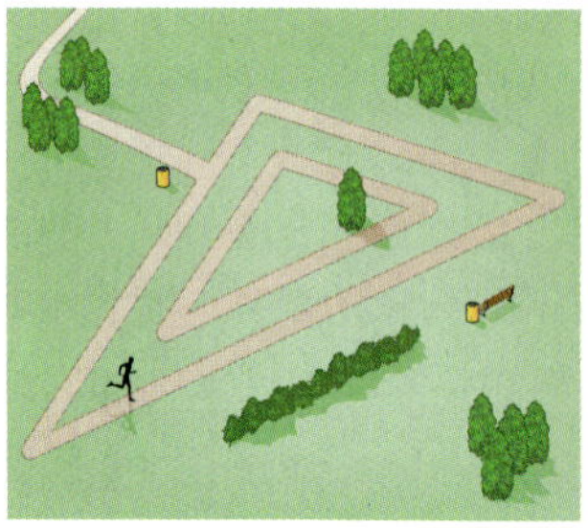

18. Find EG if $\triangle ACB \sim \triangle EGF$, $\overline{AD}$ is an altitude of $\triangle ACB$, $\overline{EH}$ is an altitude of $\triangle EGF$, $AC = 17$, $AD = 15$, and $EH = 7.5$. **8.5**

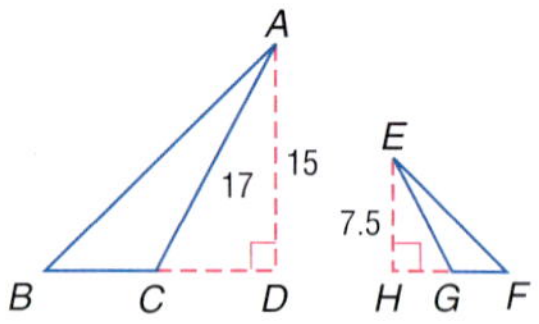

19. Find *EH* if $\triangle ABC \sim \triangle DEF$, $\overline{BG}$ is an altitude of $\triangle ABC$, $\overline{EH}$ is an altitude of $\triangle DEF$, $BG = 3$, $BF = 4$, $FC = 2$, and $CE = 1$. $\frac{3}{2}$

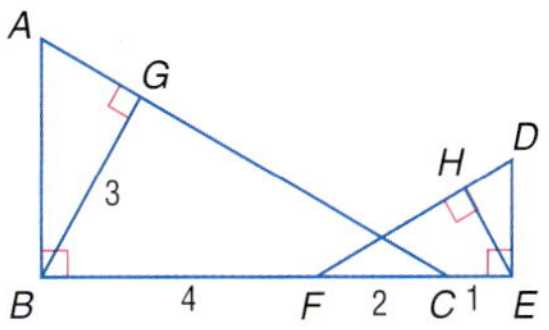

★ **20.** Find *FB* if $\overline{SA}$ and $\overline{FB}$ are altitudes and $\triangle RST \sim \triangle EFG$. 2

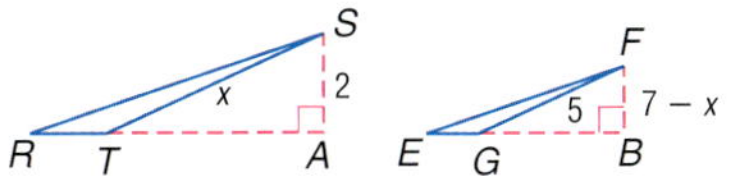

★ **21.** Find *DC* if $\overline{DG}$ and $\overline{JM}$ are altitudes and $\triangle KJL \sim \triangle EDC$. 4

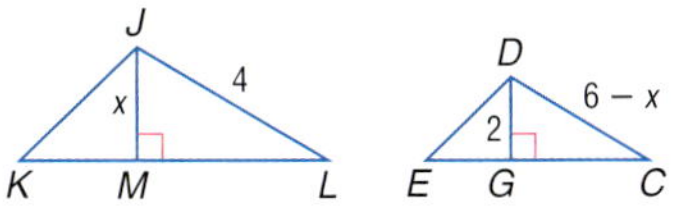

Find *x*.

22. 15.5

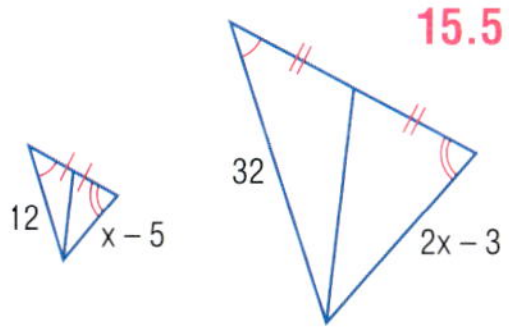

23. $11\frac{1}{5}$

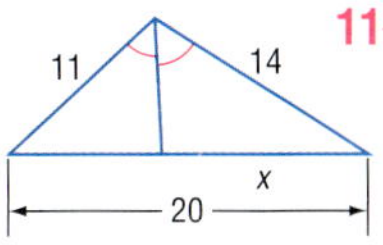

24. 6

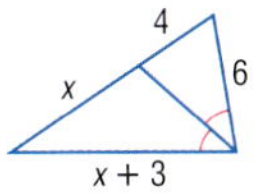

★ **25.** 6

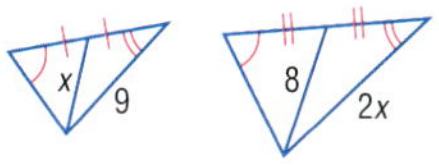

★ **26.** Find *UB* if $\triangle RST \sim \triangle UVW$, $\overline{TA}$ and $\overline{WB}$ are medians, $TA = 8$, $RA = 3$, $WB = 3x - 6$, and $UB = x + 2$. 36

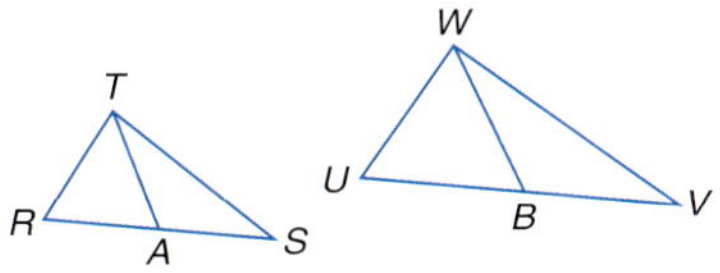

★ **27.** Find *CF* and *BD* if $\overline{BF}$ bisects $\angle ABC$ and $\overline{AC} \parallel \overline{ED}$, $BA = 6$, $BC = 7.5$, $AC = 9$, and $DE = 9$. 5, 13.5

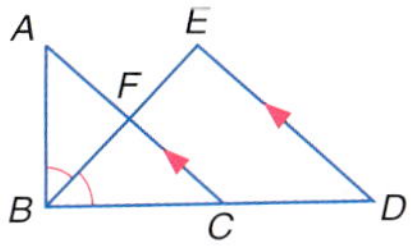

28. PHOTOGRAPHY One of the first cameras invented was called a *camera obscura*. Light entered an opening in the front, and an image was reflected in the back of the camera, upside down, forming similar triangles. If the image of the person on the back of the camera is 12 inches, the distance from the opening to the person is 7 feet, and the camera itself is 15 inches long, how tall is the person being photographed? about 5 ft 7 in.

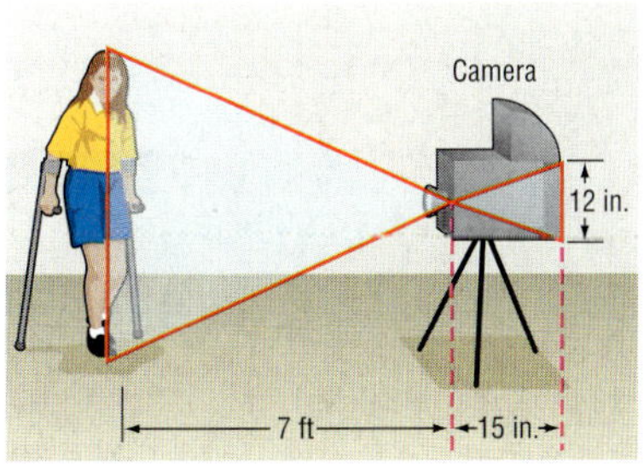

29. CRITICAL THINKING $\overline{CD}$ is an altitude to the hypotenuse $\overline{AB}$. Make a conjecture about *x*, *y*, and *z*. Justify your reasoning. $xy = z^2$; $\triangle ACD \sim \triangle CBD$ by AA Similarity. Thus, $\frac{CD}{BD} = \frac{AD}{CD}$ or $\frac{z}{y} = \frac{x}{z}$. The cross products yield $xy = z^2$.

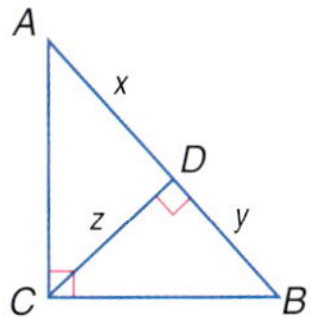

Study Guide and Intervention, p. 319 (shown) and p. 320

Perimeters If two triangles are similar, their perimeters have the same proportion as the corresponding sides.

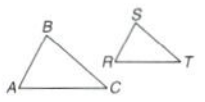

If $\triangle ABC \sim \triangle RST$, then

$$\frac{AB + BC + AC}{RS + ST + RT} = \frac{AB}{RS} = \frac{BC}{ST} = \frac{AC}{RT}.$$

Example **Use the diagram above with $\triangle ABC \sim \triangle RST$. If $AB = 24$ and $RS = 15$, find the ratio of their perimeters.**

Since $\triangle ABC \sim \triangle RST$, the ratio of the perimeters of $\triangle ABC$ and $\triangle RST$ is the same as the ratio of corresponding sides.

Therefore $\frac{\text{perimeter of } \triangle ABC}{\text{perimeter of } \triangle RST} = \frac{24}{15} = \frac{8}{5}$

Exercises

Each pair of triangles is similar. Find the perimeter of the indicated triangle.

1. $\triangle XYZ$

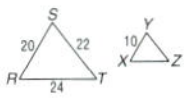

33

2. $\triangle BDE$

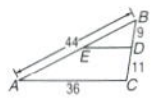

45

3. $\triangle ABC$

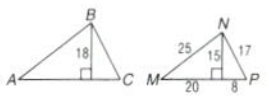

84

4. $\triangle XYZ$

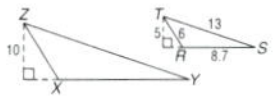

55.4

5. $\triangle ABC$

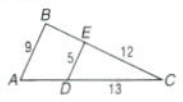

54

6. $\triangle RST$

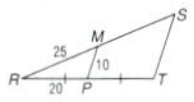

110

Skills Practice, p. 321 and Practice, p. 322 (shown)

Find the perimeter of the given triangle.

1. $\triangle DEF$, if $\triangle ABC \sim \triangle DEF$, $AB = 36$, $BC = 20$, $CA = 40$, and $DE = 35$

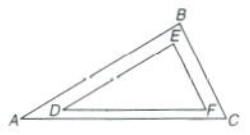

$93.\overline{3}$

2. $\triangle STU$, if $\triangle STU \sim \triangle KLM$, $KL = 12$, $LM = 31$, $MK = 32$, and $US = 28$

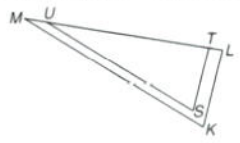

65.625

Use the given information to find each measure.

3. Find *PR* if $\triangle JKL \sim \triangle NPR$, $\overline{KM}$ is an altitude of $\triangle JKL$, $\overline{PT}$ is an altitude of $\triangle NPR$, $KL = 28$, $KM = 18$, and $PT = 15.75$.

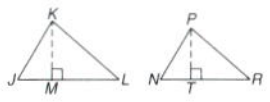

24.5

4. Find *ZY* if $\triangle STU \sim \triangle XYZ$, $\overline{UA}$ is an altitude of $\triangle STU$, $\overline{ZB}$ is an altitude of $\triangle XYZ$, $UT = 8.5$, $UA = 6$, and $ZB = 11.4$.

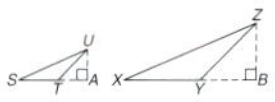

16.15

Find *x*.

5.

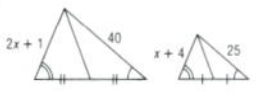

13.5

6.

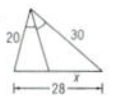

16.8

PHOTOGRAPHY For Exercises 7 and 8, use the following information.

Francine has a camera in which the distance from the lens to the film is 24 millimeters.

7. If Francine takes a full-length photograph of her friend from a distance of 3 meters and the height of her friend is 140 centimeters, what will be the height of the image on the film? (*Hint*: Convert to the same unit of measure.) 11.2 mm

8. Suppose the height of the image on the film of her friend is 15 millimeters. If Francine took a full-length shot, what was the distance between the camera and her friend? 2.24 m

Reading to Learn Mathematics, p. 323 ELL

Pre-Activity **How is geometry related to photography?**

Read the introduction to Lesson 6-5 at the top of page 316 in your textbook.

- How is similarity involved in the process of making a photographic print from a negative?

  Sample answer: The print is an enlarged version of the negative. The image on the print is similar to the image on the negative, so when you look at the print, you will see exactly the same things in the same proportions as on the negative, just larger.

- Why do photographers place their cameras on tripods?

  Sample answer: Three points determine a plane, so the three feet of the tripod are coplanar and the tripod will not wobble.

Reading the Lesson

1. In the figure, $\triangle RST \sim \triangle UVW$. Complete each proportion involving the lengths of segments in this figure by replacing the question mark. Then identify the definition or theorem from the list below that the completed proportion illustrates.

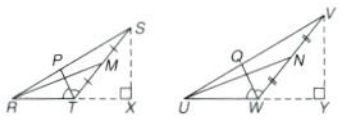

i. Definition of congruent polygons
ii. Definition of similar polygons
iii. Proportional Perimeters Theorem
iv. Angle Bisectors Theorem
v. Similar triangles have corresponding altitudes proportional to corresponding sides.
vi. Similar triangles have corresponding medians proportional to corresponding sides.
vii. Similar triangles have corresponding angle bisectors proportional to corresponding sides.

a. $\frac{RS + ST + TR}{?} = \frac{RS}{UV}$ $UV + VW + WU$; iii
b. $\frac{RT}{UW} = \frac{SX}{?}$ VY; v
c. $\frac{RM}{UN} = \frac{?}{VW}$ ST; vi
d. $\frac{RS}{UV} = \frac{ST}{?}$ VW; ii
e. $\frac{RP}{PS} = \frac{?}{ST}$ RT; iv
f. $\frac{UN}{?} = \frac{UW}{RT}$ RM; vi
g. $\frac{TP}{WQ} = \frac{RT}{?}$ UW; vii
h. $\frac{UW}{VW} = \frac{?}{QV}$ UQ; iv

Helping You Remember

2. A good way to remember a large amount of information is to remember key words. What key words will help you remember the features of similar triangles that are proportional to the lengths of the corresponding sides?

perimeters, altitudes, angle bisectors, medians

Enrichment, p. 324

Proportions for Similar Triangles

Recall that if a line crosses two sides of a triangle and is parallel to the third side, then the line separates the two sides that it crosses into segments of proportional lengths.

You can write many proportions by identifying similar triangles in the following diagram. In the diagram, $\overleftrightarrow{AM} \parallel \overleftrightarrow{BN}$, $\overleftrightarrow{AE} \parallel \overleftrightarrow{FL} \parallel \overleftrightarrow{MR}$, and $\overleftrightarrow{DQ} \parallel \overleftrightarrow{ER}$.

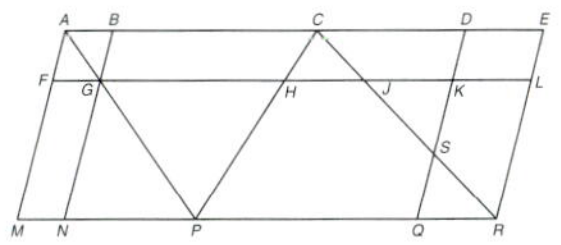

Answers

38. Sample answer: The geometry occurs inside the camera as the image is formed on the film. Answers should include the following.

- 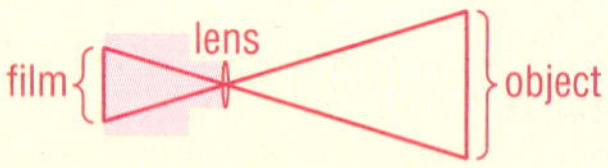

- The triangles are similar because the SAS Similarity Theorem holds. The congruent angles are the vertical angles and the corresponding sides are the congruent sides of the isosceles triangles.

PROOF **Write the indicated type of proof.** 30–37. pp. 339A–339B.

30. a paragraph proof of Theorem 6.8

31. a two-column proof of Theorem 6.10

32. a two-column proof of the Angle Bisector Theorem (Theorem 6.11)

Given: $\overline{CD}$ bisects $\angle ACB$

By construction, $\overline{AE} \parallel \overline{CD}$.

Prove: $\frac{AD}{AC} = \frac{BD}{BC}$

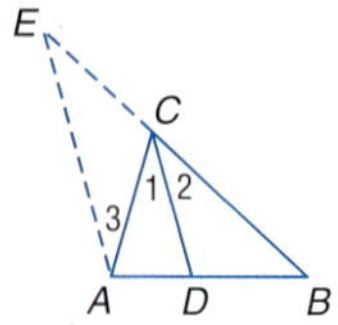

33. a paragraph proof

Given: $\triangle ABC \sim \triangle PQR$

$\overline{BD}$ is an altitude of $\triangle ABC$.

$\overline{QS}$ is an altitude of $\triangle PQR$.

Prove: $\frac{QP}{BA} = \frac{QS}{BD}$

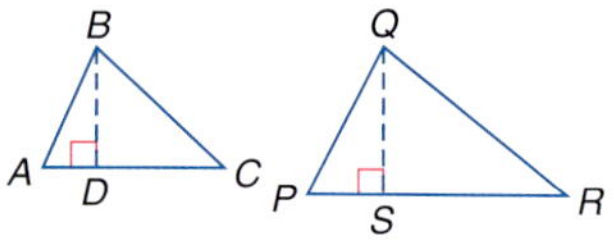

34. a flow proof

Given: $\angle C \cong \angle BDA$

Prove: $\frac{AC}{DA} = \frac{AD}{BA}$

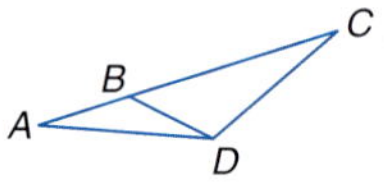

35. a two-column proof

Given: $\overline{JF}$ bisects $\angle EFG$.

$\overline{EH} \parallel \overline{FG}$, $\overline{EF} \parallel \overline{HG}$

Prove: $\frac{EK}{KF} = \frac{GJ}{JF}$

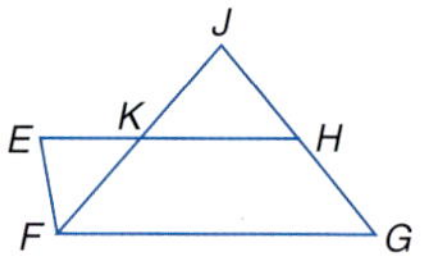

36. a two-column proof

Given: $\overline{RU}$ bisects $\angle SRT$;

$\overline{VU} \parallel \overline{RT}$.

Prove: $\frac{SV}{VR} = \frac{SR}{RT}$

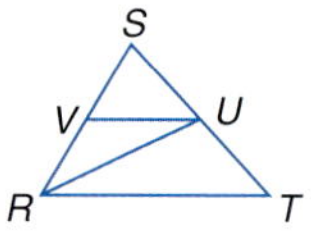

37. a flow proof

Given: $\triangle RST \sim \triangle ABC$; W and D are midpoints of $\overline{TS}$ and $\overline{CB}$.

Prove: $\triangle RWS \sim \triangle ADB$

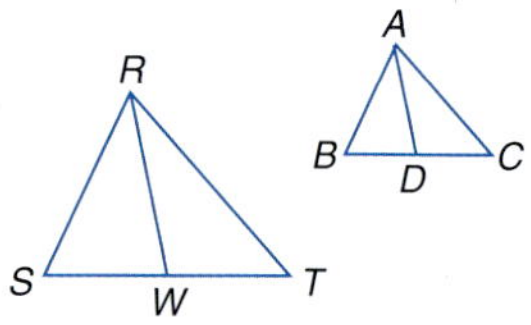

38. *WRITING IN MATH* Answer the question that was posed at the beginning of the lesson. See margin.

How is geometry related to photography?

Include the following in your answer:

- a sketch of how a camera works showing the image and the film, and
- why the two isosceles triangles are similar.

FCAT Practice
Standardized Test Practice

39. **GRID IN** Triangle ABC is similar to $\triangle DEF$. If $AC = 10.5$, $AB = 6.5$, and $DE = 8$, find DF. 12.9

40. **ALGEBRA** The sum of three numbers is 180. Two of the numbers are the same, and each of them is one-third of the greatest number. What is the least number? B

Ⓐ 30 Ⓑ 36 Ⓒ 45 Ⓓ 72

Maintain Your Skills

Mixed Review

Determine whether $\overline{MN} \parallel \overline{OP}$. Justify your answer. *(Lesson 6-4)*

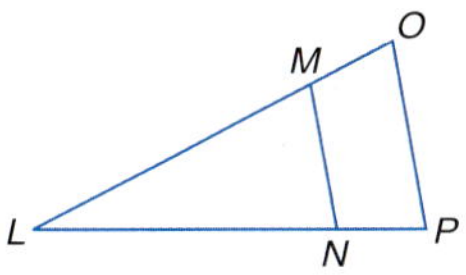

41. $LM = 7$, $LN = 9$, $LO = 14$, $LP = 16$
42. $LM = 6$, $MN = 4$, $LO = 9$, $OP = 6$
43. $LN = 12$, $NP = 4$, $LM = 15$, $MO = 5$

41. no; sides not proportional

42. not enough information to determine

43. yes; $\frac{LM}{MO} = \frac{LN}{NP}$

Identify the similar triangles. Find x and the measure(s) of the indicated side(s). *(Lesson 6-3)*

44. VW and WX $\triangle VZW \sim \triangle XYW$; $x = 6$, $VW = 12$, $WX = 10$

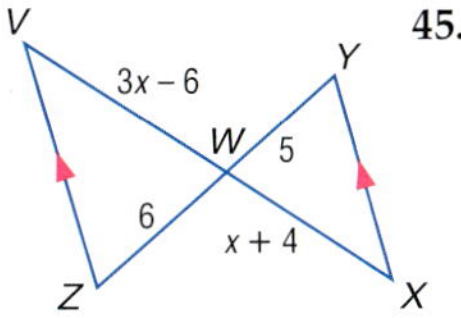

45. PQ $\triangle PQT \sim \triangle PRS$; $x = 7$, $PQ = 15$

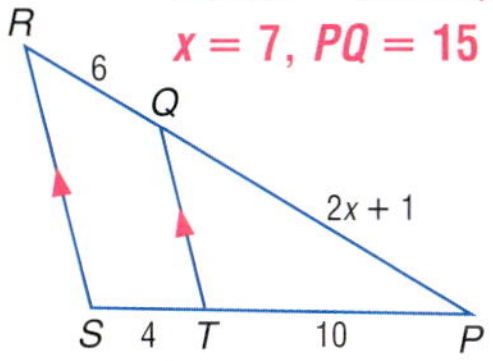

Write an equation in slope-intercept form for the line that satisfies the given conditions. *(Lesson 3-4)* 46. $y = x - 3$

46. x-intercept is 3, y-intercept is -3
47. $m = 2$, contains $(-1, -1)$ $y = 2x + 1$

Getting Ready for the Next Lesson

PREREQUISITE SKILL Name the next two numbers in each pattern.
*(To review **patterns**, see Lesson 2-1.)*

48. 5, 12, 19, 26, 33, … 40, 47
49. 10, 20, 40, 80, 160, … 320, 640
50. 0, 5, 4, 9, 8, 13, … 12, 17

Practice Quiz 2

Lessons 6-4 and 6-5

Refer to $\triangle ABC$. *(Lesson 6-4)*

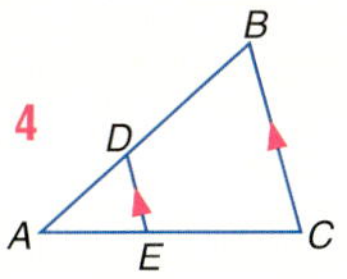

1. If $AD = 8$, $AE = 12$, and $EC = 18$, find AB. 20
2. If $AE = m - 2$, $EC = m + 4$, $AD = 4$, and $AB = 20$, find m. 4

3. no; sides not proportional 4. yes; $\frac{XV}{VY} = \frac{XW}{WZ}$

Determine whether $\overline{YZ} \parallel \overline{VW}$. Justify your answer. *(Lesson 6-4)*

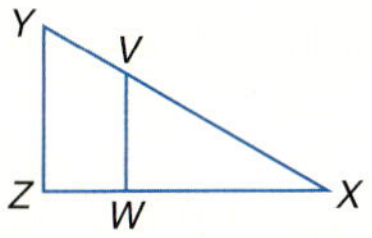

3. $XY = 30$, $XV = 9$, $XW = 12$, and $XZ = 18$
4. $XV = 34.88$, $VY = 18.32$, $XZ = 33.25$, and $WZ = 11.45$

Find the perimeter of the given triangle. *(Lesson 6-5)*

5. $\triangle DEF$ if $\triangle DEF \sim \triangle GFH$ 12.75

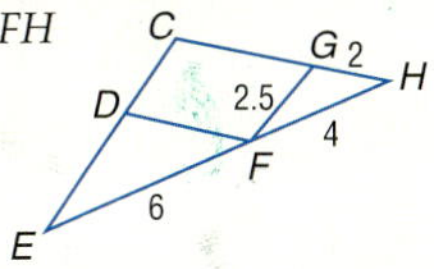

6. $\triangle RUW$ if $\triangle RUW \sim \triangle STV$, $ST = 24$, $VS = 12$, $VT = 18$, and $UW = 21$ 63

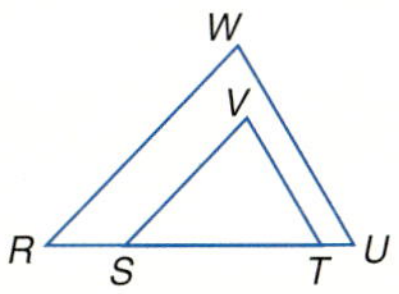

Find x. *(Lesson 6-5)*

7. 10.5

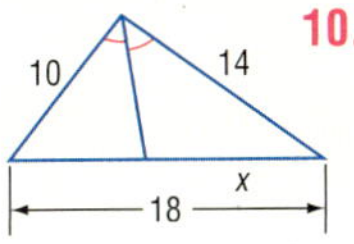

8. 6

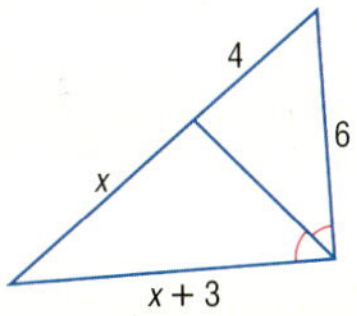

9. 5

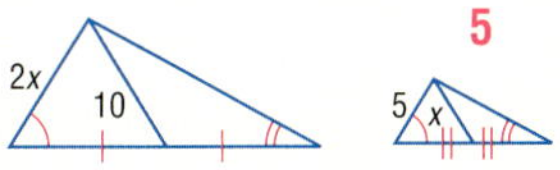

10. **LANDSCAPING** Paulo is designing two gardens shaped like similar triangles. One garden has a perimeter of 53.5 feet, and the longest side is 25 feet. He wants the second garden to have a perimeter of 32.1 feet. Find the length of the longest side of this garden. *(Lesson 6-5)* 15 ft

www.geometryonline.com/self_check_quiz/fcat

4 Assess

Open-Ended Assessment

Modeling Have students draw and label a triangle with perimeter 24 centimeters. Then ask them to draw and label a similar triangle with a scale factor of $\frac{2}{3}$. Ask them to find the perimeter. 16 cm

Getting Ready for Lesson 6-6

Prerequisite Skill Students will learn about fractals and self-similarity in Lesson 6-6. They will apply patterns and conjectures to finding the patterns in fractals. Use Exercises 48–50 to determine your students' familiarity with number patterns.

Assessment Options

Practice Quiz 2 The quiz provides students with a brief review of the concepts and skills in Lessons 6-4 and 6-5. Lesson numbers are given to the right of the exercises or instruction lines so students can review concepts not yet mastered.

Quiz (Lessons 6-4 and 6-5) is available on p. 346 of the *Chapter 6 Resource Masters*.

Geometry Activity

A Preview of Lesson 6-6

Sierpinski Triangle

Collect Data

Stage 0 On isometric dot paper, draw an equilateral triangle in which each side is 16 units long.

Stage 1 Connect the midpoints of each side to form another triangle. Shade the center triangle.

Stage 2 Repeat the process using the three nonshaded triangles. Connect the midpoints of each side to form other triangles.

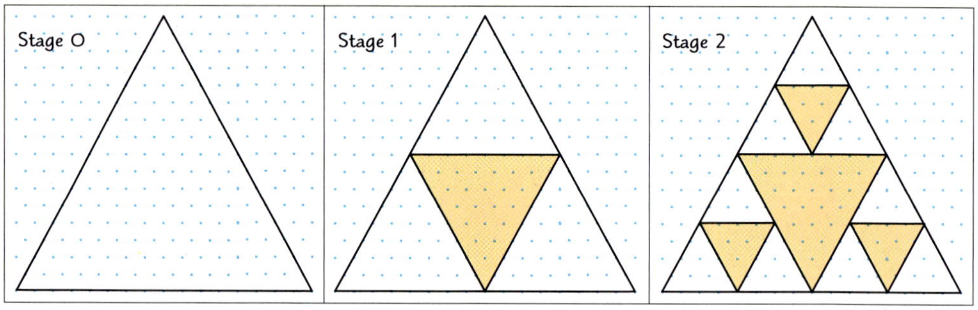

If you repeat this process indefinitely, the pattern that results is called the **Sierpinski Triangle.** Since this figure is created by repeating the same procedure over and over again, it is an example of a geometric shape called a *fractal*.

Analyze the Data

2. Stage 0: 48 units, Stage 1: 24 units, Stage 2: 12 units, Stage 3; 6 units, Stage 4: 3 units

1. Continue the process through Stage 4. How many nonshaded triangles do you have at Stage 4? **81**
2. What is the perimeter of a nonshaded triangle in Stage 0 through Stage 4?
3. If you continue the process indefinitely, describe what will happen to the perimeter of each nonshaded triangle. **The perimeter will approach zero.**
4. Study $\triangle DFM$ in Stage 2 of the Sierpenski Triangle shown at the right. Is this an equilateral triangle? Are $\triangle BCE$, $\triangle GHL$, or $\triangle IJN$ equilateral? **yes; yes**
5. Is $\triangle BCE \sim \triangle DFM$? Explain your answer. **yes, by AA Similarity**
6. How many Stage 1 Sierpinski triangles are there in Stage 2? **3**

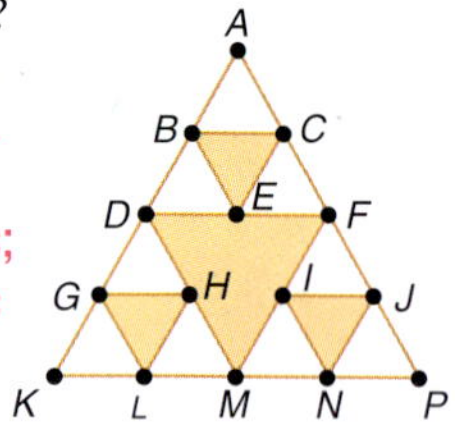

Make a Conjecture

7. How can three copies of a Stage 2 triangle be combined to form a Stage 3 triangle? **See p. 339B.**

8. Combine three copies of the Stage 4 Sierpinski triangle. Which stage of the Sierpinski Triangle is this? **Stage 5**
9. How many copies of the Stage 4 triangle would you need to make a Stage 6 triangle? **9**

Geometry Activity

A Preview of Lesson 6-6

Getting Started

Objective Draw a Sierpinski Triangle.

Materials isometric dot paper

Teach

- Discuss how each stage is determined so that students are clear about the process. Every time each side of the triangle is bisected to form one or more other triangles, it is one stage. Ask students to record the number of triangles at each stage so they can analyze the pattern in the numbers.
- You may want students to do this activity in groups of four. Ask one student to collect the data by drawing the triangles while another student analyzes the data by counting the triangles and forming a sequence from the count. Ask a third student to find the perimeter of each triangle as the activity continues, and the fourth student to generate a sequence of numbers representing the perimeters of the triangles. This sequence should help them see that the perimeter approaches zero as the number of triangles increases.

Assess

Ask students what stage triangle could be made if all students combined their triangles. Have students check their answers by making a giant class triangle.

Study Notebook

Ask students to summarize what they have learned about Sierpinski Triangles.

Resource Manager

Teaching Geometry with Manipulatives
- p. 109 (student recording sheet)
- p. 7 (isometric dot paper)

Glencoe Mathematics Classroom Manipulative Kit
- isometric dot grid stamp
- rulers

6-6 Fractals and Self-Similarity

Sunshine State Standards
MA.A.3.4.3, MA.A.4.4.1, MA.B.1.4.3, MA.C.1.4.1, MA.C.2.4.1, MA.C.3.4.1

Vocabulary
- iteration
- fractal
- self-similar

What You'll Learn
- Recognize and describe characteristics of fractals.
- Identify nongeometric iteration.

How is mathematics found in nature?

Patterns can be found in many objects in nature, including broccoli. If you take a piece of broccoli off the stalk, the small piece resembles the whole. This pattern of repeated shapes at different scales is part of fractal geometry.

CHARACTERISTICS OF FRACTALS Benoit Mandelbrot, a mathematician, coined the term *fractal* to describe things in nature that are irregular in shape, such as clouds, coastlines, or the growth of a tree. The patterns found in nature are analyzed and then recreated on a computer, where they can be studied more closely. These patterns are created using a process called **iteration**. Iteration is a process of repeating the same procedure over and over again. A **fractal** is a geometric figure that is created using iteration. The pattern's structure appears to go on infinitely.

Compare the pictures of a human circulatory system and the mouth of the Ganges in Bangladesh. Notice how the branches of the tributaries have the same pattern as the branching of the blood vessels.

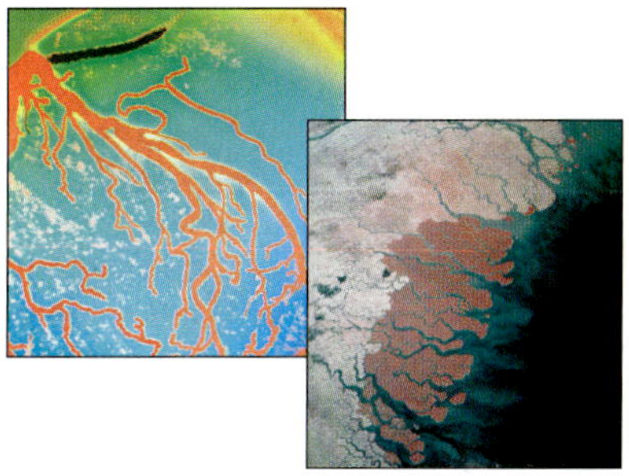

One characteristic of fractals is that they are **self similar**. That is, the smaller and smaller details of a shape have the same geometric characteristics as the original form.

WebQuest
By creating a Sierpinski Triangle, you can find a pattern in the area and perimeter of this well-known fractal. Visit www.geometryonline.com/webquest to continue work on your WebQuest project.

The Sierpinski Triangle is a fractal that is self-similar. Stage 1 is formed by drawing the midsegments of an equilateral triangle and shading in the triangle formed by them. Stage 2 repeats the process in the unshaded triangles. This process can continue indefinitely with each part still being similar to the original.

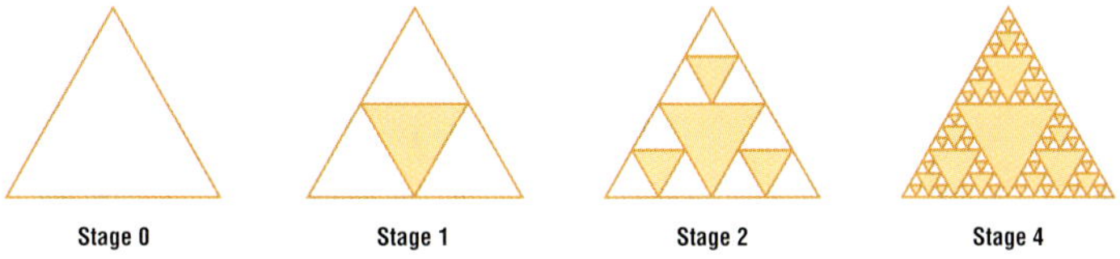

Study Tip
Sierpinski Triangle
The Sierpenski Triangle can also be shown without shading. The triangles that are used to continue the process are those with a vertex pointing upward.

The Sierpinski Triangle is said to be *strictly self-similar*, which means that any of its parts, no matter where they are located or what size is selected, contain a figure that is similar to the whole.

6-6 Lesson Notes

1 Focus

5-Minute Check Transparency 6-6 Use as a quiz or review of Lesson 6-5.

Mathematical Background notes are available for this lesson on p. 280D.

How is mathematics found in nature?

Ask students:
- How is broccoli an example of mathematics found in nature? **A small piece of broccoli resembles the whole; it has a pattern of repeated shapes.**
- What is another example of mathematics found in nature? **Sample answers: tree branches, cauliflower, branches of lightning**

Resource Manager

Workbook and Reproducible Masters

Chapter 6 Resource Masters
- Study Guide and Intervention, pp. 325–326
- Skills Practice, p. 327
- Practice, p. 328
- Reading to Learn Mathematics, p. 329
- Enrichment, p. 330
- Assessment, p. 346

Graphing Calculator and Computer Masters, p. 28
Teaching Geometry With Manipulatives Masters, pp. 1, 17

Transparencies
5-Minute Check Transparency 6-6
Answer Key Transparencies

Technology
Interactive Chalkboard

2 Teach

CHARACTERISTICS OF FRACTALS

In-Class Examples

PowerPoint®

1 $\triangle DEF$ below is found by connecting the midpoints of the sides of $\triangle ABC$. Prove that $\triangle ABC \sim \triangle DFE$.

Given: E, D, and F are midpoints of $\overline{AB}$, $\overline{BC}$, and $\overline{AC}$, respectively.

Prove: $\triangle ABC \sim \triangle DFE$

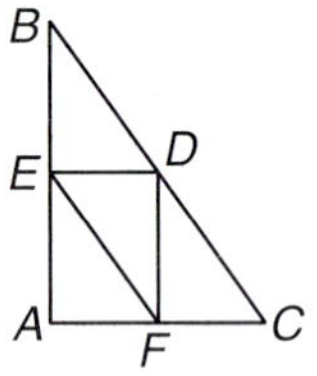

Proof:
Statements (Reasons)

1. E, D, and F are midpoints of $\overline{AB}$, $\overline{BC}$, and $\overline{AC}$, respectively. (Given)
2. $\overline{ED} \parallel \overline{AC}$ and $\overline{EF} \parallel \overline{BC}$ and $\overline{DF} \parallel \overline{BA}$ (△ Midsegment Thm.)
3. $\angle EFA \cong \angle DEF$ and $\angle DFC \cong \angle EDF$ (Alt. Int. ∠s Thm)
4. $\angle C \cong \angle EFA$ and $\angle A \cong \angle DFC$ (Corresponding ∠s Post.)
5. $\angle C \cong \angle DEF$ and $\angle A \cong \angle EDF$ (Trans. Prop.)
6. $\triangle ABC \sim \triangle DFE$ (AA Similarity)

2 Draw an equilateral triangle. Create a fractal by drawing another equilateral triangle within it and shading above or beneath the triangle that shares the horizontal side of the equilateral triangle. Stages 1 and 2 are shown.

Stage 1

Stage 2

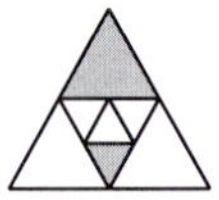

See students' work.

Example 1 Self-Similarity

Prove that a triangle formed in Stage 2 of a Sierpinski triangle is similar to the triangle in Stage 0.

The argument will be the same for any triangle in Stage 2, so we will use only $\triangle CGJ$ from Stage 2.

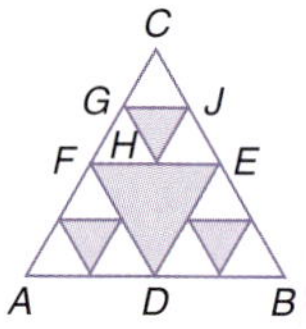

Given: $\triangle ABC$ is equilateral.
D, E, F, G, J, and H are midpoints of $\overline{AB}$, $\overline{BC}$, $\overline{CA}$, $\overline{FC}$, $\overline{CE}$, and $\overline{FE}$, respectively.

Prove: $\triangle CGJ \sim \triangle CAB$

Statements	Reasons
1. $\triangle ABC$ is equilateral; D, E, F are midpoints of $\overline{AB}$, $\overline{BC}$, $\overline{CA}$; G, J, and H are midpoints of $\overline{FC}$, $\overline{CE}$, $\overline{FE}$.	1. Given
2. $\overline{FE}$ is a midsegment of $\triangle CAB$; $\overline{GJ}$ is a midsegment of $\triangle CFE$.	2. Definition of a Triangle Midsegment
3. $\overline{FE} \parallel \overline{AB}$; $\overline{GJ} \parallel \overline{FE}$	3. Triangle Midsegment Theorem
4. $\overline{GJ} \parallel \overline{AB}$	4. Two segments parallel to the same segment are parallel.
5. $\angle CGJ \cong \angle CAB$	5. Corresponding ∠s Postulate
6. $\angle C \cong \angle C$	6. Reflexive Property
7. $\triangle CGJ \sim \triangle CAB$	7. AA Similarity

Thus, using the same reasoning, every triangle in Stage 2 is similar to the original triangle in Stage 0.

Study Tip

Look Back
To review **midsegment**, see Lesson 6-4.

You can generate many other fractal images using an iterative process.

Example 2 Create a Fractal

Draw a segment and trisect it. Create a fractal by replacing the middle third of the segment with two segments of the same length as the removed segment.

After the first geometric iteration, repeat the process on each of the four segments in Stage 1. Continue to repeat the process.

This fractal image is called a *Koch curve*.

Study Tip

Common Misconceptions
Not all repeated patterns are self-similar. The Koch curve is one such example when applied to a triangle.

If the first stage is an equilateral triangle, instead of a segment, this iteration will produce a fractal called *Koch's snowflake*.

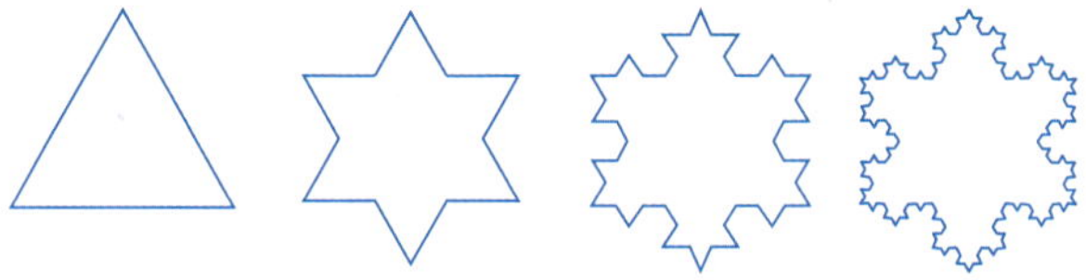

326 Chapter 6 Proportions and Similarity

NONGEOMETRIC ITERATION An iterative process does not always include manipulation of geometric shapes. Iterative processes can be translated into formulas or algebraic equations. These are called *recursive formulas*.

Study Tip

Recursion on the Graphing Calculator

To do recursion on a graphing calculator, store 2 as the value for X and press ENTER. Then enter X²:X² → X. Press ENTER for each iteration.

Example 3 Evaluate a Recursive Formula

Find the value of x^2, where x initially equals 2. Then use that value as the next x in the expression. Repeat the process four times and describe your observations.

The iterative process is to square the value repeatedly. Begin with $x = 2$. The value of x^2 becomes the next value for x.

x	2	4	16	256	65,536
x^2	4	16	256	65,536	4,294,967,296

The values grow greater with each iteration, approaching infinity.

Example 4 Find a Recursive Formula

PASCAL'S TRIANGLE ***Pascal's Triangle*** **is a numerical pattern in which each row begins and ends with 1 and all other terms in the row are the sum of the two numbers above it.**

a. Find a formula in terms of the row number for the sum of the values in any row in the Pascal's triangle.

To find the sum of the values in the tenth row, we can investigate a simpler problem. What is the sum of values in the first four rows of the triangle?

Row	Pascal's Triangle	Sum	Pattern
1	1	1	$2^0 = 2^{1-1}$
2	1 1	2	$2^1 = 2^{2-1}$
3	1 2 1	4	$2^2 = 2^{3-1}$
4	1 3 3 1	8	$2^3 = 2^{4-1}$
5	1 4 6 4 1	16	$2^4 = 2^{5-1}$

It appears that the sum of any row is a power of 2. The formula is 2 to a power that is one less than the row number: $A_n = 2^{n-1}$.

b. What is the sum of the values in the tenth row of Pascal's triangle?

The sum of the values in the tenth row will be 2^{10-1} or 512.

Example 5 Solve a Problem Using Iteration

BANKING **Felisa has \$2500 in a money market account that earns 3.2% interest. If the interest is compounded annually, find the balance of her account after 3 years.**

First, write an equation to find the balance after one year.

current balance + (current balance × interest rate) = new balance

$$2500 + (2500 \cdot 0.032) = 2580$$

$$2580 + (2580 \cdot 0.032) = 2662.56$$

$$2662.56 + (2662.56 \cdot 0.032) = 2747.76$$

After 3 years, Felisa will have \$2747.76 in her account.

www.geometryonline.com/extra_examples/fcat

NONGEOMETRIC ITERATION

In-Class Examples

Power Point®

Teaching Tip Point out that iterative processes are the basis for many computer operations that are algebraic in nature. In these examples, iteration is used to generate sequences.

3 Find the value of $3x^2 - 1$, where x initially equals 1. Then use that value as the next x in the expression. Repeat the process three more times and describe your observations. **2; 11, 362, 393,131; The x values increase with each iteration, approaching infinity.**

4 **a.** The diagram below represents the odd integers 1, 3, 5, and 7. Find a formula in terms of the row number for the sum of the values in the diagram. **$S_n = n^2$**

Row		Sum
1	•	1 = 1
2	• • •	1 + 3 = 4
3	• • • • •	4 + 5 = 9
4	• • • • • • •	9 + 7 = 16

b. What is the sum of the first 10 odd numbers? **100**

5 **BANKING** Joaquin has \$1500 in a savings account that earns 4.1% interest. If the interest is compounded annually, find the balance of his account after 4 years. **\$1761.55**

Concept Check

Ask students to describe how to find the next number in a sequence of numbers generated recursively if they are given the formula and the initial value. **Evaluate the formula using the initial value of the variable. Evaluate the formula again using the result of the first computation. Continue with these steps.**

DAILY INTERVENTION

Differentiated Instruction

Naturalist Students may not understand the concept of *self-similar* immediately. Consider bringing examples to class (like broccoli or cauliflower) to help students see examples. Also encourage students to study nature around them and take pictures to share about objects they think are self-similar.

3 Practice/Apply

Study Notebook

Have students—

- add the definitions/examples of the vocabulary terms to their Vocabulary Builder worksheets for Chapter 6.
- include an example of how to generate a fractal using an iterative process.
- include any other item(s) that they find helpful in mastering the skills in this lesson.

About the Exercises...

Organization by Objective

- **Characteristics of Fractals:** 11–13, 21–29
- **Nongeometric iteration:** 14–20, 25, 28, 30–40

Odd/Even Assignments

Exercises 11–40 are structured so that students practice the same concepts whether they are assigned odd or even problems.

Assignment Guide

Basic: 11–23 odd, 29–43 odd, 44, 46–58

Average: 11–43 odd, 44, 45–58

Advanced: 12–44 even, 46–58

Check for Understanding

Concept Check

1. **Describe** a *fractal* in your own words. Include characteristics of fractals in your answer. **Sample answer: irregular shape formed by iteration of self-similar shapes**
2. **Explain** why computers provide an efficient way to generate fractals.
3. **OPEN ENDED** Find an example of fractal geometry in nature, excluding those mentioned in the lesson. **Sample answer: icebergs, ferns, leaf veins**

2. They can accurately calculate thousands of iterations.

Guided Practice

GUIDED PRACTICE KEY	
Exercises	Examples
4	2
5	3
6	1
7–9	4
10	5

4. Stage 1: 2, Stage 2: 6, Stage 3: 14, Stage 4: 30; see margin for drawings.

For Exercises 4–6, use the following information.

A *fractal tree* can be drawn by making two new branches from the endpoint of each original branch, each one-third as long as the previous branch.

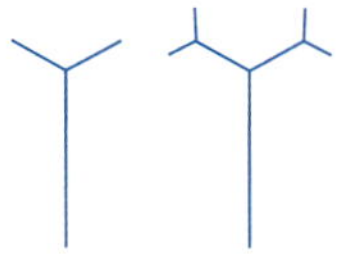

Stage 1 Stage 2

4. Draw Stages 3 and 4 of a fractal tree. How many total branches do you have in Stages 1 through 4? (Do not count the stems.)
5. Find a pattern to predict the number of branches at each stage. $A_n = 2(2^n - 1)$
6. Is a fractal tree strictly self-similar? Explain. **See margin.**

For Exercises 7–9, use a calculator. **7. 1.4142...; 1.1892..., 8. 1.0905...; approaches 1**

7. Find the square root of 2. Then find the square root of the result.
8. Find the square root of the result in Exercise 7. What would be the result after 100 repeats of taking the square root?
9. Determine whether this is an iterative process. Explain. **Yes, the procedure is repeated over and over again.**

Application

10. **BANKING** Jamir has $4000 in a savings account. The annual percent interest rate is 1.1%. Find the amount of money Jamir will have after the interest is compounded four times. **$4178.92**

★ indicates increased difficulty

Practice and Apply

Homework Help	
For Exercises	See Examples
11–13, 21–23	1, 2
14–20, 25, 28	4
24–29	2
30–37	3
38–40	5

Extra Practice
See page 766.

For Exercises 11–13, Stage 1 of a fractal is drawn on grid paper so that each side of the large square is 27 units long. The trisection points of the sides are connected to make 9 smaller squares with the middle square shaded. The shaded square is known as a *hole*. **11–12. See margin for drawings.**

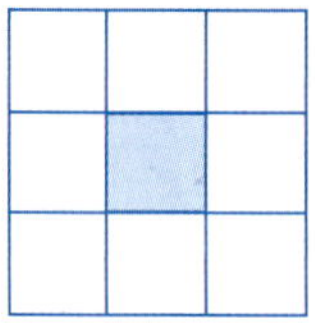

11. Copy Stage 1 on your paper. Then draw Stage 2 by repeating the Stage 1 process in each of the outer eight squares. How many holes are in this stage? **9 holes**
12. Draw Stage 3 by repeating the Stage 1 process in each unshaded square of Stage 2. How many holes are in Stage 3? **73 holes**
13. If you continue the process indefinitely, will the figure you obtain be strictly self-similar? Explain. **Yes, any part contains the same figure as the whole, 9 squares with the middle shaded.**
14. Count the number of dots in each arrangement. These numbers are called *triangular numbers*. The second triangular number is 3 because there are three dots in the array. How many dots will be in the seventh triangular number? **28**

Answers

4.

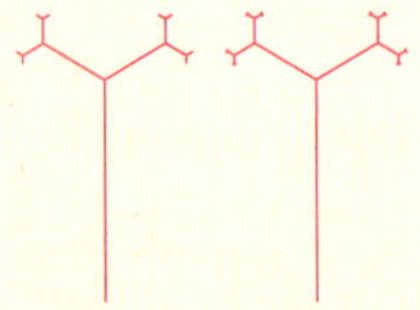

Stage 3 Stage 4

6. No, the base of the tree or segment of a branch without an end does not contain a replica of the entire tree.

11.

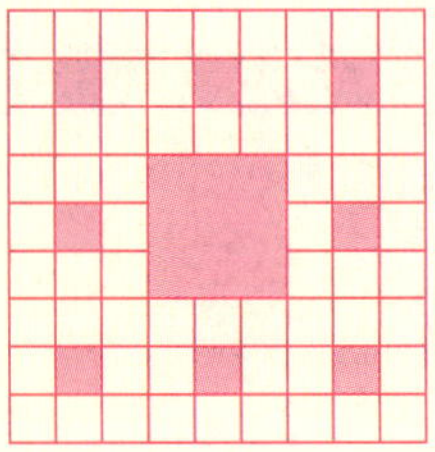

12.

15. 1, 3, 6, 10, 15...; Each difference is 1 more than the preceding difference.

16. The triangular numbers are the numbers in the diagonal.

More About...

Blaise Pascal

Pascal (1623–1662) did not discover Pascal's triangle, but it was named after him in honor of his work in 1653 called *Treatise on the Arithmetical Triangle*. The patterns in the triangle are also used in probability.

Source: *Great Moments in Mathematics/After 1650*

28. $P = 3\left(\frac{4}{3}\right)^n$; as the stages increase, the perimeter increases and approaches infinity.

30–33. See margin.

For Exercises 15–20, refer to Pascal's triangle on page 327. Look at the third diagonal from either side, starting at the top of the triangle.

15. Describe the pattern.

16. Explain how Pascal's triangle relates to the triangular numbers.

17. Generate eight rows of Pascal's triangle. Replace each of the even numbers with a 0 and each of the odd numbers with a 1. Color each 1 and leave the 0s uncolored. Describe the picture. The result is similar to a Stage 3 Sierpinski triangle.

18. Generate eight rows of Pascal's triangle. Divide each entry by 3. If the remainder is 1 or 2, shade the number cell black. If the remainder is 0, leave the cell unshaded. Describe the pattern that emerges. It is similar to a Stage 1 Sierpinski triangle.

19. Find the sum of the first 25 numbers in the outside diagonal of Pascal's triangle. 25

20. Find the sum of the first 50 numbers in the second diagonal. 1275

The three shaded interior triangles shown were made by trisecting the three sides of an equilateral triangle and connecting the points.

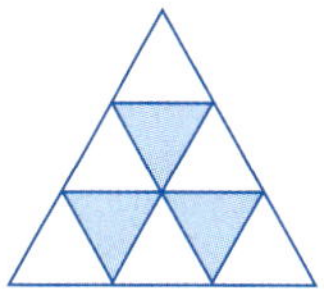

21. Prove that one of the nonshaded triangles is similar to the original triangle. See margin.

22. Repeat the iteration once more. See margin.

23. Is the new figure strictly self-similar? See margin.

24. How many nonshaded triangles are in Stages 1 and 2? Stage 1: 6, Stage 2: 36

Refer to the Koch Curve on page 326. 25. $A_n = 4^n$; 65,536

★25. What is a formula for the number of segments in terms of the stage number? Use your formula to predict the number of segments in Stage 8 of a Koch curve.

★26. If the length of the original segment is 1 unit, how long will the segments be in each of the first four stages? What will happen to the length of each segment as the number of stages continues to increase? See margin.

Refer to the Koch Snowflake on page 326. At Stage 1, the length of each side is 1 unit. 27. See margin.

★27. What is the perimeter at each of the first four stages of a Koch snowflake?

★28. What is a formula for the perimeter in terms of the stage number? Describe the perimeter as the number of stages continues to increase.

29. Write a paragraph proof to show that the triangles generated on the sides of a Koch Snowflake in Stage 1, are similar to the original triangle. See margin.

Find the value of each expression. Then use that value as the next x in the expression. Repeat the process four times, and describe your observations.

30. $\sqrt{x}$, where x initially equals 12

31. $\frac{1}{x}$, where x initially equals 5

32. $x^{\frac{1}{3}}$, where x initially equals 0.3

33. 2^x, where x initially equals 0

Find the first three iterates of each expression. 35. 0, −5, −10

34. $2x + 1$, x initially equals 1 3, 7, 15

35. $x - 5$, where x initially equals 5

36. $x^2 - 1$, x initially equals 2 3, 8, 63

37. $3(2 - x)$, where x initially equals 4 −6, 24, −66

38. **BANKING** Raini has a credit card balance of \$1250 and a monthly interest rate of 1.5%. If he makes payments of \$100 each month, what will the balance be after 3 months? \$1002.57

32. 0.6694..., 0.8747..., 0.9563..., 0.9852..., 0.9950...; the numbers converge to 1.

33. 1, 2, 4, 16, 65,536; the numbers approach positive infinity.

Answers

21. Given: $\triangle ABC$ is equilateral. $CD = \frac{1}{3}CB$ and $CE = \frac{1}{3}CA$

Prove: $\triangle CED \sim \triangle CAB$

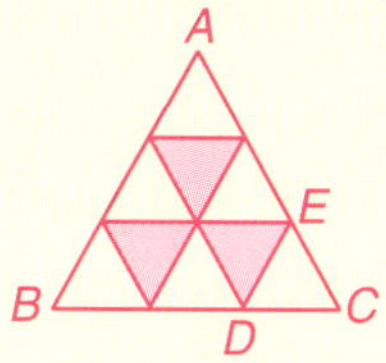

Statements (Reasons)

1. $\triangle ABC$ is equilateral. $CD = \frac{1}{3}CB$, $CE = \frac{1}{3}CA$. (Given)
2. $\overline{AC} \cong \overline{BC}$ (Def. of equilateral △)
3. $AC = BC$ (Def. of ≅ segments)
4. $\frac{1}{3}AC = \frac{1}{3}CB$ (Mult. Prop.)
5. $CD = CE$ (Substitution)
6. $\frac{CD}{CB} = \frac{CE}{CB}$ (Division Prop.)
7. $\frac{CD}{CB} = \frac{CE}{CA}$ (Substitution)
8. $\angle C \cong \angle C$ (Reflexive Prop.)
9. $\triangle CED \sim \triangle CAB$ (AA Similarity)

22.

23. Yes; the smaller and smaller details of the shape have the same geometric characteristics as the original form.

26. Stage 0: 1 unit, Stage 1: $\frac{1}{3}$ unit, Stage 2: $\frac{1}{9}$ unit, Stage 3: $\frac{1}{27}$ unit; as the stages increase, the length of the segments will approach zero.

27. Stage 0: 3 units, Stage 1: $3 \cdot \frac{4}{3}$ or 4 units, Stage 2: $3 \cdot \frac{4}{3} \cdot \frac{4}{3} = 3\left(\frac{4}{3}\right)^2$ or $5\frac{1}{3}$ units, Stage 3: $3\left(\frac{4}{3}\right)^3$ or $7\frac{1}{9}$ units

29. The original triangle and the new triangles are equilateral and thus, all of the angles are equal to 60. By AA Similarity, the triangles are similar.

30. 3.46410..., 1.8612..., 1.3642..., 1.168..., 1.0807...; the numbers converge to 1.

31. 0.2, 5, 0.2, 5, 0.2; the numbers alternate between 0.2 and 5.0.

WEATHER **For Exercises 39 and 40, use the following information.**
There are so many factors that affect the weather that it is difficult for meteorologists to make accurate long term predictions. Edward N. Lorenz called this dependence *the Butterfly Effect* and posed the question "Can the flap of a butterfly's wings in Brazil cause a tornado in Texas?"

39. Use a calculator to find the first ten iterates of $4x(1 - x)$ when x initially equals 0.200 and when the initial value is 0.201. Did the change in initial value affect the tenth value?

39. When $x = 0.200$: 0.64, 0.9216, 0.2890..., 0.8219..., 0.5854..., 0.9708..., 0.1133..., 0.4019..., 0.9615..., 0.1478...; when $x = 0.201$: 0.6423..., 0.9188..., 0.2981..., 0.8369..., 0.5458..., 0.9916..., 0.0333..., 0.1287..., 0.4487..., 0.9894.... Yes, the initial value affected the tenth value.

40. Why do you think this is called the Butterfly Effect? **A small difference in initial data can have a large effect in later data.**

41. **ART** Describe how artist Jean-Paul Agosti used iteration and self-similarity in his painting *Jardin du Creuset*. **Sample answer: The leaves in the tree and the branches of the trees are self-similar. These self-similar shapes are repeated throughout the painting.**

42. **NATURE** Some of these pictures are of real objects and others are fractal images of objects.
   a. Compare the pictures and identify those you think are of real objects.
   b. Describe the characteristics of fractals shown in the images.

42a. The flower and mountain are computer-generated; the feathers and moss are real. 42b. The fractals exhibit self-similarity and iteration.

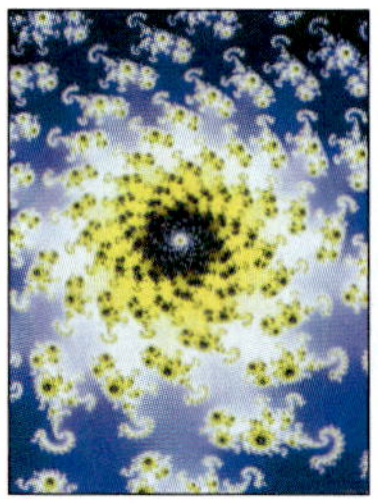
flower

mountain

feathers

moss

43. **RESEARCH** Use the Internet or other sources to find the names and pictures of the other fractals Waclaw Sierpinski developed. **See students' work.**

44. **CRITICAL THINKING** Draw a right triangle on grid paper with 6 and 8 units for the lengths of the perpendicular sides. Shade the triangle formed by the three midsegments. Repeat the process for each unshaded triangle. Find the perimeter of the shaded triangle in Stage 1. What is the total perimeter of all the shaded triangles in Stage 2? **12 units; 30 units**

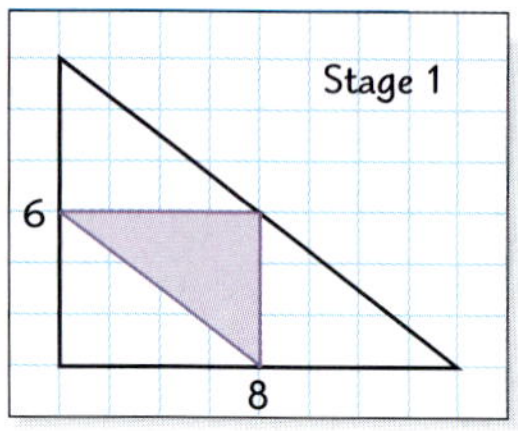

45. **WRITING IN MATH** Answer the question that was posed at the beginning of the lesson. **See margin.**

How is mathematics related to nature?

Include the following in your answer:
- explain why broccoli is an example of fractal geometry, and
- how scientists can use fractal geometry to better understand nature.

Study Guide and Intervention, p. 325 (shown) and p. 326

Characteristics of Fractals The act of repeating a process over and over, such as finding a third of a segment, then a third of the new segment, and so on, is called **iteration**. When the process of iteration is applied to some geometric figures, the results are called **fractals**. For objects such as fractals, when a portion of the object has the same shape or characteristics as the entire object, the object can be called **self-similar**.

Example **In the diagram at the right, notice that the details at each stage are similar to the details at Stage 1.**

Exercises

1. Follow the iteration process below to produce a fractal.

Stage 1
- Draw a square.
- Draw an isosceles right triangle on the top side of the square. Use the side of the square as the hypotenuse of the triangle.
- Draw a square on each leg of the right triangle.

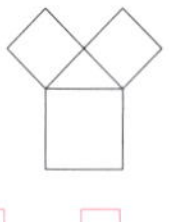

Stage 2
Repeat the steps in Stage 1, drawing an isosceles triangle and two small squares for each of the small squares from Stage 1.

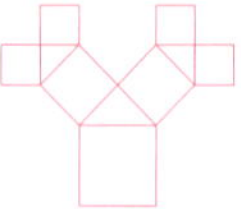

Stage 3
Repeat the steps in Stage 1 for each of the smallest squares in Stage 2.

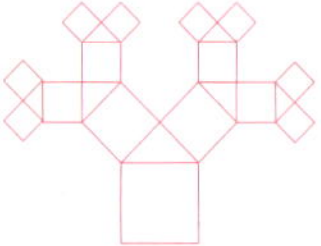

2. Is the figure produced in Stage 3 self-similar? yes

Skills Practice, p. 327 and Practice, p. 328 (shown)

An artist is designing a book cover and wants to show a copy of the book cover in the lower left corner of the cover. After Stages 1 and 2 of the design, he realizes that the design is developing into a fractal!

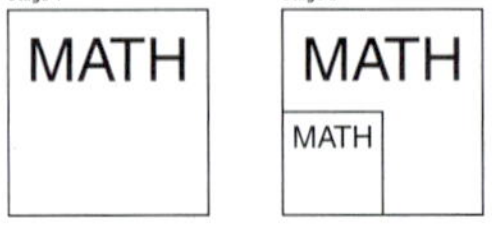

1. Draw Stages 3 and 4 of the book cover fractal.

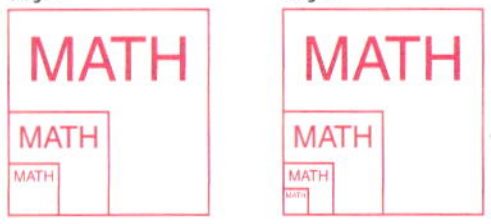

2. After infinitely many iterations, will the result be a self similar fractal? yes

3. On what part of the book cover should you focus your attention to be sure you can find a copy of the entire figure? the lower left corner

Find the value of each expression. Then, use that value as the next x in the expression. Repeat the process three times and describe your observations.

4. $2(x + 3)$, where x initially equals 0
   6, 18, 42, 90
   The difference between values doubles with each term.

5. $\frac{x}{2} - 3$, where x initially equals 10
   2, −2, −4, −5
   The difference between values is multiplied by $\frac{1}{2}$ with each term.

Find the first three iterates of each expression.

6. $\frac{x}{2} + 1$, where x initially equals 1
   1.5; 1.75; 1.875

7. $2x^2$, where x initially equals 2
   8; 128; 32,768

8. **HOUSING** The Andrews purchased a house for \$96,000. The real estate agent who sold the house said that comparable houses in the area appreciate at a rate of 4.5% per year. If this pattern continues, what will be the value of the house in three years? Round to the nearest whole number. \$109,552

Reading to Learn Mathematics, p. 329 ELL

Pre-Activity **How is mathematics found in nature?**

Read the introduction to Lesson 6-6 at the top of page 325 in your textbook.

Name two objects from nature other than broccoli in which a small piece resembles the whole. Sample answer: rings of a tree trunk; spiral shell

Reading the Lesson

1. Match each definition from the first column with a term from the second column. (Some words of phrases in the second column may be used more than once or not at all.)

Phrase	Term
a. a geometric figure that is created using iteration iii	i. similar
b. a pattern in which smaller and smaller details of a shape have the same geometric characteristics as the original shape iv	ii. iteration
c. the result of translating an iterative process into a formula or algebraic equation v	iii. fractal
d. the process of repeating the same procedure over and over again ii	iv. self-similar
	v. recursion formula
	vi. congruent

2. Refer to the three patterns below.

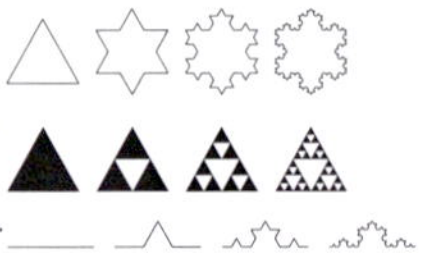

   a. Give the special name for each of the fractals you obtain by continuing the pattern without end. i. Koch snowflake; ii. Sierpinski triangle; iii. Koch curve
   b. Which of these fractals are self-similar? Sierpinski triangle, Koch curve
   c. Which of these fractals are strictly self-similar? Sierpinski triangle

Helping You Remember

3. A good way to remember a new mathematical term is to relate it to everyday English words. The word *fractal* is related to *fraction* and *fragment*. Use your dictionary to find at least one definition for each of these words that you think is related to the meaning of the word *fractal* and explain the connection. Sample answer: *Fraction*: a small part; a disconnected piece; *fragment*: a piece detached or broken off; something incomplete; in a fractal, the smaller pieces of the shape resemble the whole.

Enrichment, p. 330

The Möbius Strip

A Möbius strip is a special surface with only one side. It was discovered by August Ferdinand Möbius, a German astronomer and mathematician.

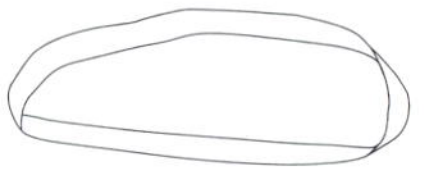

1. To make a Möbius strip, cut a strip of paper about 16 inches long and 1 inch wide. Mark the ends with the letters A, B, C, and D as shown below.

Twist th[illegible]er once, connectin[illegible] D and B to C. Tape the en[illegible]

330 Chapter 6 Proportions and Similarity

FCAT Practice

Standardized Test Practice

46. **GRID IN** A triangle has side lengths of 3 inches, 6 inches, and 8 inches. A similar triangle is 24 inches on one side. Find the maximum perimeter, in inches, of the second triangle. **136**

47. **ALGEBRA** A repair technician charges \$80 for the first thirty minutes of each house call plus \$2 for each additional minute. The repair technician charged a total of \$170 for a job. How many minutes did the repair technician work? **C**

Ⓐ 45 min Ⓑ 55 min Ⓒ 75 min Ⓓ 85 min

Maintain Your Skills

Mixed Review Find x. *(Lesson 6-5)*

48. **8**

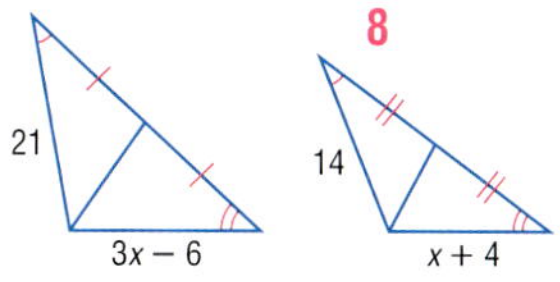

49. **$13\frac{3}{5}$**

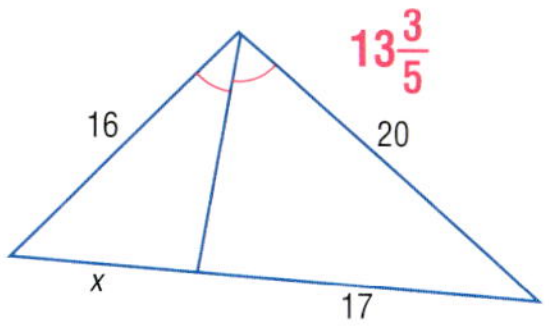

50. **4**

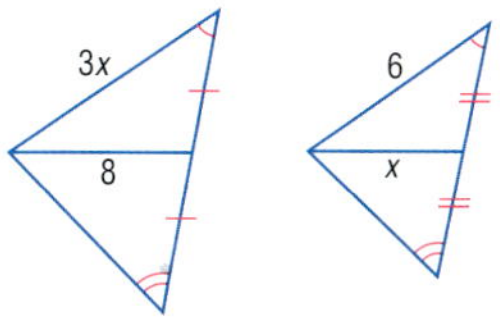

51. **7**

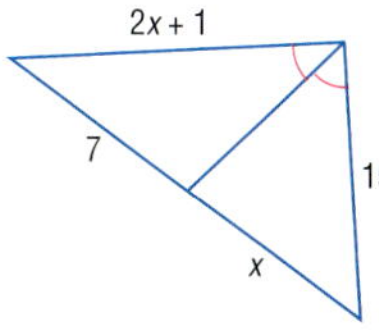

For Exercises 52–54, refer to $\triangle JKL$. *(Lesson 6-4)*

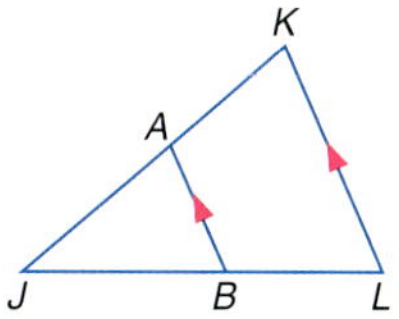

52. If $JL = 27$, $BL = 9$, and $JK = 18$, find JA. **12**

53. If $AB = 8$, $KL = 10$, and $JB = 13$, find JL. **$16\frac{1}{4}$**

54. If $JA = 25$, $AK = 10$, and $BL = 14$, find JB. **35**

55. **FOLKLORE** The Bermuda Triangle is an imaginary region located off the southeastern Atlantic coast of the United States. It is the subject of many stories about unexplained losses of ships, small boats, and aircraft. Use the vertex locations to name the angles in order from least measure to greatest measure. *(Lesson 5-4)* **Miami, Bermuda, San Juan**

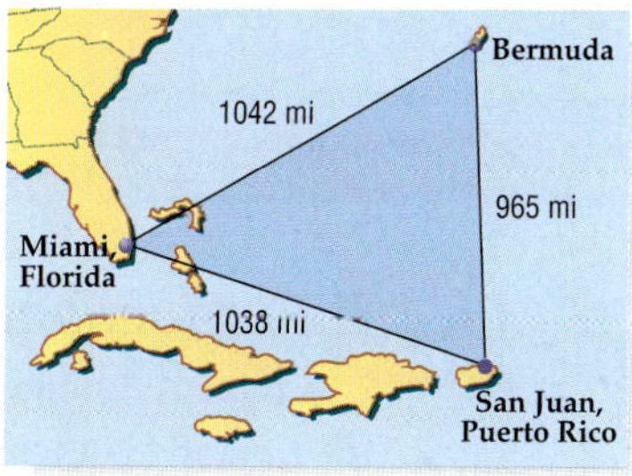

Find the length of each side of the polygon for the given perimeter. *(Lesson 1-6)*

56. $P = 60$ centimeters **12 cm**

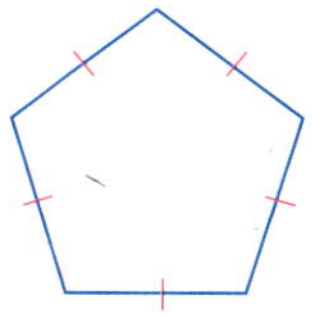

57. $P = 54$ feet

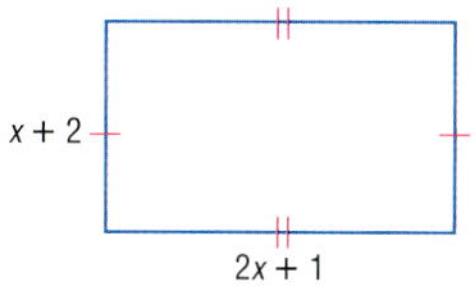

10 ft, 10 ft, 17 ft, 17 ft

58. $P = 57$ units

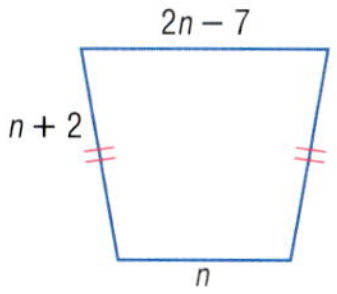

17, 14, 14, 12

www.geometryonline.com/self_check_quiz/fcat

4 Assess

Open-Ended Assessment

Modeling Ask students to create fractals of their own using construction materials or spaghetti.

Assessment Options

Quiz (Lesson 6-6) is available on p. 346 of the *Chapter 6 Resource Masters*.

Answers

45. Sample answer: Fractal geometry can be found in the repeating patterns of nature. Answers should include the following.
    - Broccoli is an example of fractal geometry because the shape of the florets is repeated throughout; one floret looks the same as the stalk.
    - Sample answer: Scientists can use fractals to study the human body, rivers, and tributaries, and to model how landscapes change over time.

Lesson 6-6 Fractals and Self-Similarity 331

Vocabulary and Concept Check

- This alphabetical list of vocabulary terms in Chapter 6 includes a page reference where each term was introduced.
- **Assessment** A vocabulary test/review for Chapter 6 is available on p. 344 of the *Chapter 6 Resource Masters*.

Lesson-by-Lesson Review

For each lesson,

- the main ideas are summarized,
- additional examples review concepts, and
- practice exercises are provided.

Vocabulary PuzzleMaker

ELL The Vocabulary PuzzleMaker software improves students' mathematics vocabulary using four puzzle formats—crossword, scramble, word search using a word list, and word search using clues. Students can work on a computer screen or from a printed handout.

MindJogger Videoquizzes

ELL MindJogger Videoquizzes provide an alternative review of concepts presented in this chapter. Students work in teams in a game show format to gain points for correct answers. The questions are presented in three rounds.

Round 1 Concepts (5 questions)
Round 2 Skills (4 questions)
Round 3 Problem Solving (4 questions)

Chapter 6 Study Guide and Review

Vocabulary and Concept Check

cross products (p. 283)	means (p. 283)	scale factor (p. 290)
extremes (p. 283)	midsegment (p. 308)	self-similar (p. 325)
fractal (p. 325)	proportion (p. 283)	similar polygons (p. 289)
iteration (p. 325)	ratio (p. 282)	

A complete list of postulates and theorems can be found on pages R1–R8.

Exercises **State whether each sentence is *true* or *false*. If false, replace the underlined expression to make a true sentence.**

1. A midsegment of a triangle is a segment whose endpoints are the midpoints of two sides of the triangle. **true**
2. Two polygons are similar if and only if their corresponding angles are congruent and the measures of the corresponding sides are congruent. **false, proportional**
3. If two angles of one triangle are congruent to two angles of another triangle, then the triangles are similar. **true**
4. If two triangles are similar, then the perimeters are proportional to the measures of the corresponding angles. **false, sides**
5. A fractal is a geometric figure that is created using *recursive formulas*. **false, iteration**
6. A midsegment of a triangle is parallel to one side of the triangle, and its length is twice the length of that side. **false, one-half**
7. For any numbers a and c and any nonzero numbers b and d, $\frac{a}{b} = \frac{c}{d}$ if and only if $ad = bc$. **true**
8. If two triangles are similar, then the measures of the corresponding angle bisectors of the triangle are proportional to the measures of the corresponding sides. **true**
9. If a line intersects two sides of a triangle and separates the sides into corresponding segments of proportional lengths, then the line is equal to one-half the length of the third side. **false, parallel to**

Lesson-by-Lesson Review

6-1 Proportions

See pages 282–287.

Concept Summary

- A ratio is a comparison of two quantities.
- A proportion is an equation stating that two ratios are equal.

Example Solve $\frac{z}{40} = \frac{5}{8}$.

$\frac{z}{40} = \frac{5}{8}$	Original proportion
$z \cdot 8 = 40(5)$	Cross products
$8z = 200$	Multiply.
$z = 25$	Divide each side by 8.

www.geometryonline.com/vocabulary_revie

FOLDABLES™ Study Organizer

For more information about Foldables, see *Teaching Mathematics with Foldables.*

Have students look through the chapter to make sure they have included notes and examples in their Foldables for each lesson of Chapter 6.

Encourage students to refer to their Foldables while completing the Study Guide and Review and to use them in preparing for the Chapter Test.

332 Chapter 6 Proportions and Similarity

Exercises **Solve each proportion.** *See Example 3 on page 284.*

10. $\frac{3}{4} = \frac{x}{12}$ **9**
11. $\frac{7}{3} = \frac{28}{z}$ **12**
12. $\frac{x+2}{5} = \frac{14}{10}$ **5**
13. $\frac{3}{7} = \frac{7}{y-3}$ $\frac{58}{3}$
14. $\frac{4-x}{3+x} = \frac{16}{25}$ $\frac{52}{41}$
15. $\frac{x-12}{6} = \frac{x+7}{-4}$ $\frac{3}{5}$

16. **BASEBALL** A player's slugging percentage is the ratio of the number of total bases from hits to the number of total at-bats. The ratio is converted to a decimal (rounded to three places) by dividing. If Alex Rodriguez of the Texas Rangers has 263 total bases in 416 at-bats, what is his slugging percentage? **0.632**

17. A 108-inch-long board is cut into two pieces that have lengths in the ratio 2:7. How long is each new piece? **24 in. and 84 in.**

6-2 Similar Polygons

See pages 289–297.

Concept Summary

- In similar polygons, corresponding angles are congruent, and corresponding sides are in proportion.
- The ratio of two corresponding sides in two similar polygons is the scale factor.

Example **Determine whether the pair of triangles is similar. Justify your answer.**

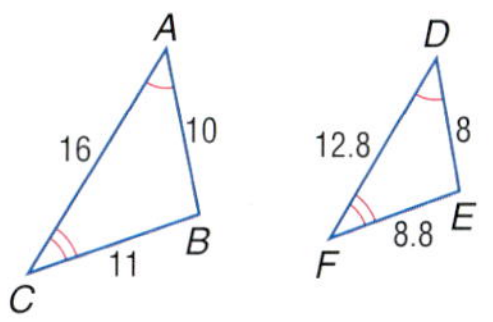

$\angle A \cong \angle D$ and $\angle C \cong \angle F$, so by the Third Angle Theorem, $\angle B \cong \angle E$. All of the corresponding angles are congruent.

Now, check to see if corresponding sides are in proportion.

$\frac{AB}{DE} = \frac{10}{8} = \frac{5}{4}$ or 1.25 $\quad \frac{BC}{EF} = \frac{11}{8.8} = \frac{5}{4}$ or 1.25 $\quad \frac{CA}{FD} = \frac{16}{12.8} = \frac{5}{4}$ or 1.25

The corresponding angles are congruent, and the ratios of the measures of the corresponding sides are equal, so $\triangle ABC \sim \triangle DEF$.

Exercises **Determine whether each pair of figures is similar. Justify your answer.** *See Example 1 on page 290.* **18–19. See margin.**

18.
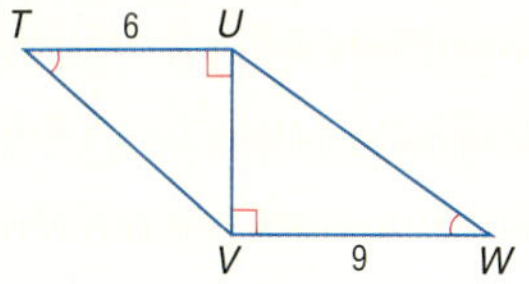

19.
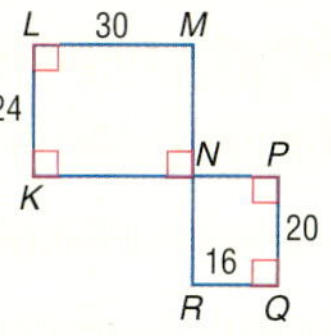

Answers

18. No, two of the corresponding sides are in a 2:3 ratio, while two others are equal in length.
19. Yes, these are rectangles, so all angles are congruent. Additionally, all sides are in a 3:2 ratio.

Answers

22. yes, $\triangle ABC \sim \triangle DFE$ by SAS Similarity
23. yes, $\triangle GHI \sim \triangle GJK$ by AA Similarity
24. No, the angles of the triangles are not congruent.

Each pair of polygons is similar. Write a similarity statement, and find x, the measures of the indicated sides, and the scale factor. *See Example 3 on page 291.*

20. $\overline{AB}$ and $\overline{AG}$

$ABCD \sim AEFG$; 6; $AB = 4$, $AG = 13.5$; $\frac{4}{9}$

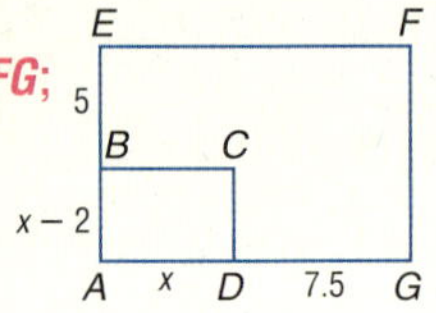

21. $\overline{PQ}$ and $\overline{QS}$

$\triangle PQT \sim \triangle RQS$; 0; $PQ = 6$; $QS = 3$; 1:2

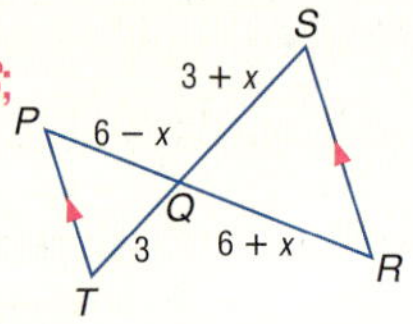

6-3 Similar Triangles

See pages 298–306.

Concept Summary

- AA, SSS, and SAS Similarity can all be used to prove triangles similar.
- Similarity of triangles is reflexive, symmetric, and transitive.

Example INDIRECT MEASUREMENT **Alonso wanted to determine the height of a tree on the corner of his block. He knew that a certain fence by the tree was 4 feet tall. At 3 P.M., he measured the shadow of the fence to be 2.5 feet tall. Then he measured the tree's shadow to be 11.3 feet. What is the height of the tree?**

Since the triangles formed are similar, a proportion can be written. Let x be the height of the tree.

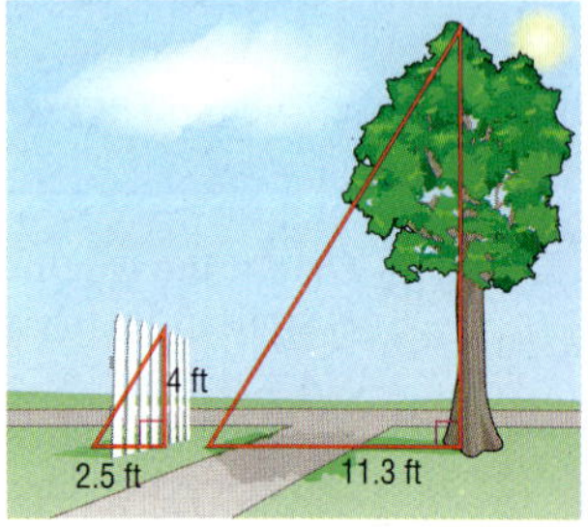

$$\frac{\text{height of the tree}}{\text{height of the fence}} = \frac{\text{tree shadow length}}{\text{fence shadow length}}$$

$\frac{x}{4} = \frac{11.3}{2.5}$ Substitution

$x \cdot 2.5 = 4(11.3)$ Cross products

$2.5x = 45.2$ Simplify.

$x = 18.08$ Divide each side by 2.5.

The height of the tree is 18.08 feet.

Exercises **Determine whether each pair of triangles is similar. Justify your answer.** *See Example 1 on page 299.* **22–24. See margin.**

22.

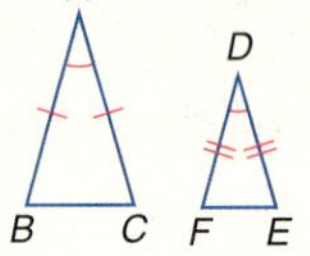

23.

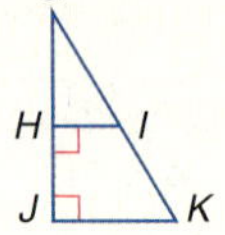

24.

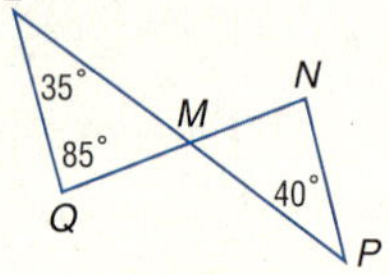

Identify the similar triangles. Find x. *See Example 2 on page 300.*

25. $\triangle ABC \sim \triangle DEC$, 4

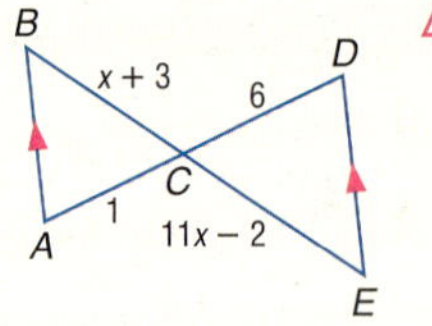

26. $\triangle RVT \sim \triangle UST$, 3

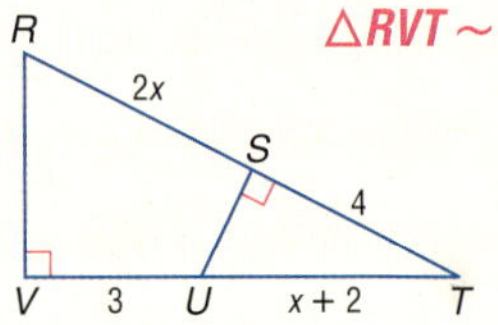

334 Chapter 6 Proportions and Similarity

6-4 Parallel Lines and Proportional Parts

See pages 307–315.

Concept Summary

- A segment that intersects two sides of a triangle and is parallel to the third side divides the two intersected sides in proportion.
- If two lines divide two segments in proportion, then the lines are parallel.

Example **In $\triangle TRS$, $TS = 12$. Determine whether $\overline{MN} \parallel \overline{SR}$.**

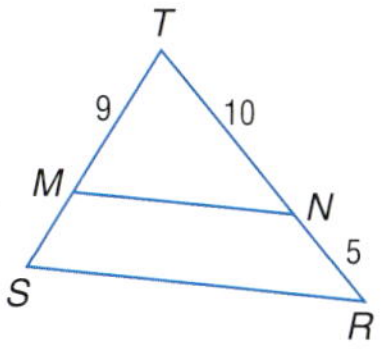

If $TS = 12$, then $MS = 12 - 9$ or 3. Compare the segment lengths to determine if the lines are parallel.

$\frac{TM}{MS} = \frac{9}{3} = 3$ $\quad$ $\frac{TN}{NR} = \frac{10}{5} = 2$

Because $\frac{TM}{MS} \neq \frac{TN}{NR}$, $\overline{MN} \nparallel \overline{SR}$.

Exercises **Determine whether $\overline{GL} \parallel \overline{HK}$. Justify your answer.**
See Example 2 on page 308. 27. no; lengths not proportional

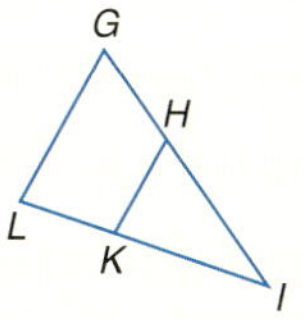

27. $IH = 21$, $HG = 14$, $LK = 9$, $KI = 15$

28. $GH = 10$, $HI = 35$, $IK = 28$, $IL = 36$ yes; $\frac{HI}{GH} = \frac{IK}{KL}$

29. yes; $\frac{HI}{GH} = \frac{IK}{KL}$ **29.** $GH = 11$, $HI = 22$, and IL is three times the length of $\overline{KL}$.

30. $LK = 6$, $KI = 18$, and IG is three times the length of $\overline{HI}$.
no; sides not proportional

Refer to the figure at the right. *See Example 1 on page 308.*

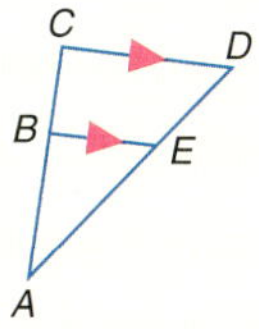

31. Find ED if $AB = 6$, $BC = 4$, and $AE = 9$. 6

32. Find AE if $AB = 12$, $AC = 16$, and $ED = 5$. 15

33. Find CD if $AE = 8$, $ED = 4$, and $BE = 6$. 9

34. Find BC if $BE = 24$, $CD = 32$, and $AB = 33$. 11

6-5 Parts of Similar Triangles

See pages 316–323.

Concept Summary

- Similar triangles have perimeters proportional to the corresponding sides.
- Corresponding angle bisectors, medians, and altitudes of similar triangles have lengths in the same ratio as corresponding sides.

Example **If $\overline{FB} \parallel \overline{EC}$, $\overline{AD}$ is an angle bisector of $\angle A$, $BF = 6$, $CE = 10$, and $AD = 5$, find AM.**

By AA Similarity using $\angle AFE \cong \angle ABF$ and $\angle A \cong \angle A$, $\triangle ABF \sim \triangle ACE$.

$\frac{AM}{AD} = \frac{BF}{CE}$ $\quad$ $\sim\triangle$s have angle bisectors in the same proportion as the corresponding sides.

$\frac{x}{5} = \frac{6}{10}$ $\quad$ $AD = 5$, $AF = 6$, $FE = 4$, $AM = x$

$10x = 30$ $\quad$ Cross products

$x = 3$ $\quad$ Divide each side by 10.

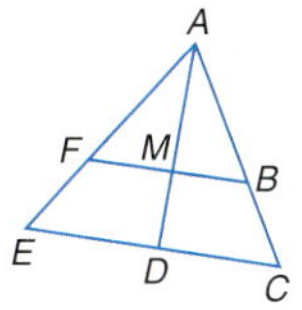

Thus, $AM = 3$.

Chapter 6 Study Guide and Review 335

Chapter 6 For More ...
• Extra Practice, see pages 764–766.
• Mixed Problem Solving, see page 787.

Answer

39. Stage 2 is not similar to Stage 1.

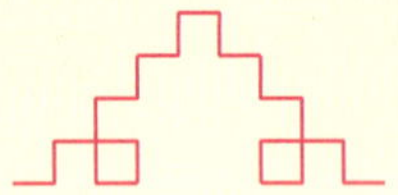

Answers (page 337)

7. pentagon *DCABE* ~ pentagon *DGIHF*; 2:3
8. $\triangle PQR \sim \triangle PST$; 2:3
9. $\triangle MAD \sim \triangle MCB$; 5:2
10. yes, by SSS Similarity
11. No, corresponding angles are not congruent.
12. yes, by AA Similarity

Exercises **Find the perimeter of the given triangle.** *See Example 1 on page 316.*

35. $\triangle DEF$ if $\triangle DEF \sim \triangle ABC$ 24

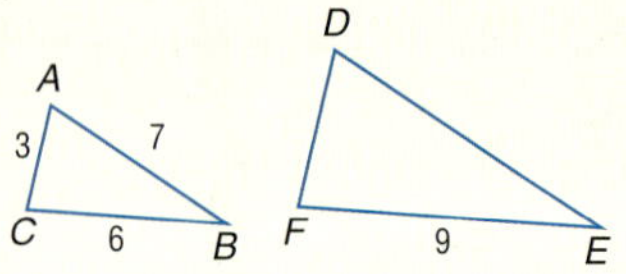

36. $\triangle QRS$ if $\triangle QRS \sim \triangle QTP$ 14

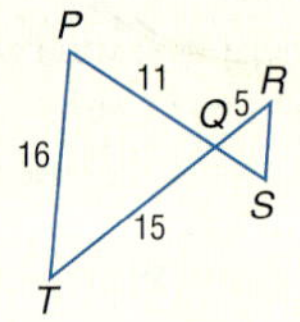

37. $\triangle CPD$ if the perimeter of $\triangle BPA$ is 12, $BM = \sqrt{13}$, and $CN = 3\sqrt{13}$ 36

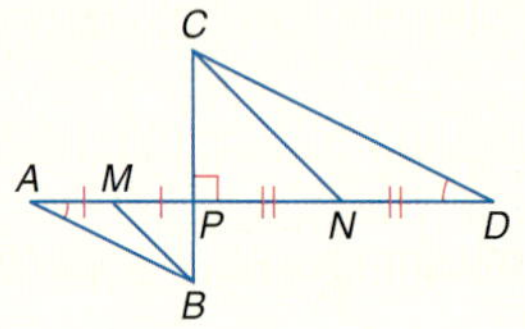

38. $\triangle PQR$, if $\triangle PQM \sim \triangle PRQ$ 78

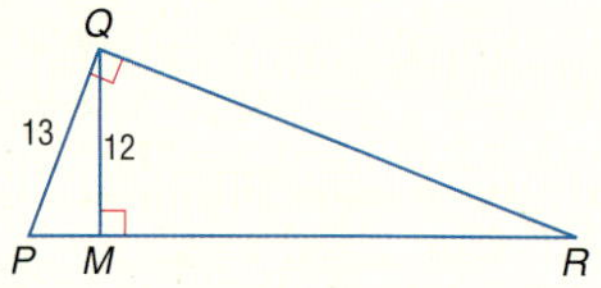

6-6 Fractals and Self-Similarity

See pages 325–331.

Concept Summary

- Iteration is the creation of a sequence by repetition of the same operation.
- A fractal is a geometric figure created by iteration.
- An iterative process involving algebraic equations is a recursive formula.

Example **Find the value of $\frac{x}{2} + 4$, where x initially equals −8. Use that value as the next x in the expression. Repeat the process five times and describe your observations.**

Make a table to organize each iteration.

Iteration	1	2	3	4	5	6
x	−8	0	4	6	7	7.5
$\frac{x}{2}+4$	0	4	6	7	7.5	7.75

The x values appear to get closer to the number 8 with each iteration.

Exercises **Draw Stage 2 of the fractal shown below. Determine whether Stage 2 is similar to Stage 1.** *See Example 2 on page 326.*

39. See margin.

40. 4, 60, 215,996 41. −8, −20, −56

Find the first three iterates of each expression. *See Example 3 on page 327.*

40. $x^3 - 4$, x initially equals 2

41. $3x + 4$, x initially equals −4

42. $\frac{1}{x}$, x initially equals 10 0.1, 10, 0.1

43. $\frac{x}{10} - 9$, x initially equals 30 −6, −9.6, −9.96

Chapter 6 Practice Test

Vocabulary and Concepts

Choose the answer that best matches each phrase.

1. an equation stating that two ratios are equal **b**
2. the ratio between corresponding sides of two similar figures **a**
3. the means multiplied together and the extremes multiplied together **c**

a. scale factor
b. proportion
c. cross products

Skills and Applications

Solve each proportion.

4. $\frac{x}{14} = \frac{1}{2}$ **7**
5. $\frac{4x}{3} = \frac{108}{x}$ **±9**
6. $\frac{k+2}{7} = \frac{k-2}{3}$ **5**

Each pair of polygons is similar. Write a similarity statement and find the scale factor.

7.
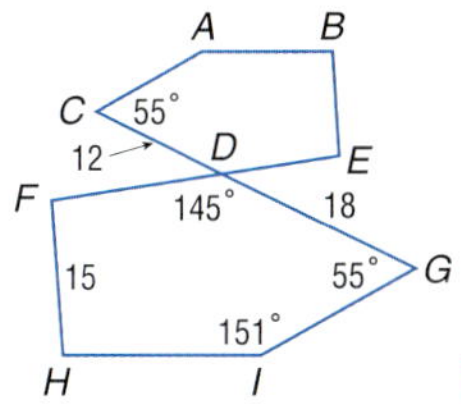

8.
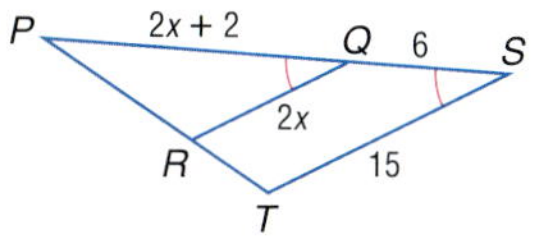

9.
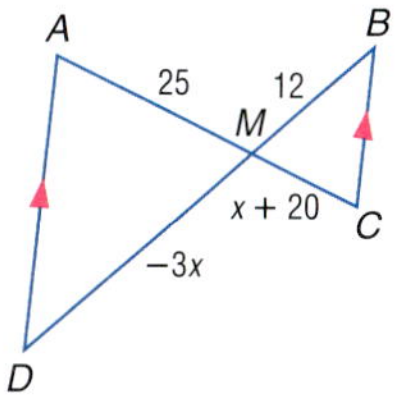

7–9. See margin.

Determine whether each pair of triangles is similar. Justify your answer. **10–12. See margin.**

10.
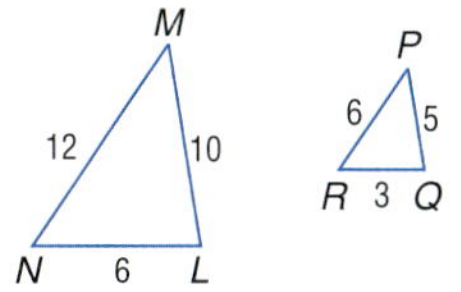

11.
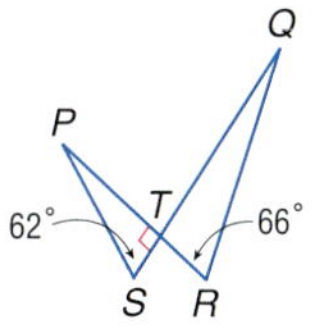

12.
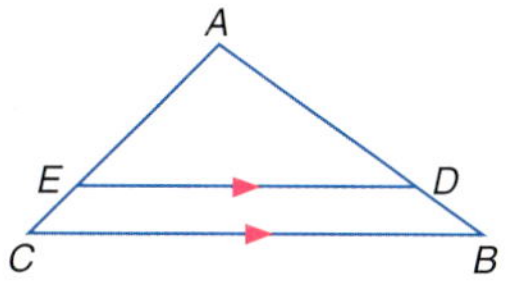

Refer to the figure at the right.

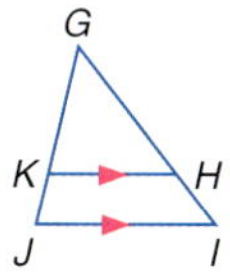

13. Find KJ if $GJ = 8$, $GH = 12$, and $HI = 4$. **2**
14. Find GK if $GI = 14$, $GH = 7$, and $KJ = 6$. **6**
15. Find GI if $GH = 9$, $GK = 6$, and $KJ = 4$. **15**

Find the perimeter of the given triangle.

16. $\triangle DEF$, if $\triangle DEF \sim \triangle ACB$ **42**

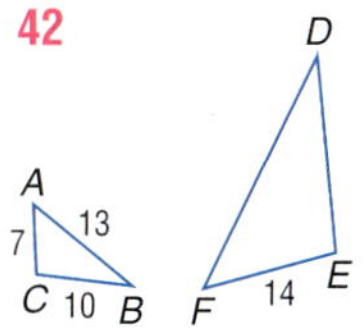

17. $\triangle ABC$ **30**

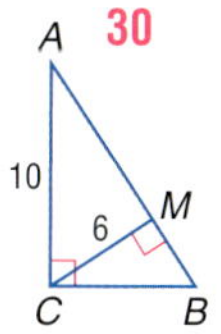

18. Find the first three iterates of $5x + 27$ when x initially equals -3. **12, 87, 462**
19. **BASKETBALL** Terry wants to measure the height of the top of the backboard of his basketball hoop. At 4:00, the shadow of a 4-foot fence is 20 inches, and the shadow of the backboard is 65 inches. What is the height of the top of the backboard? **13 ft or 156 in.**
20. **STANDARDIZED TEST PRACTICE** If a person's weekly salary is \$X and \$Y is saved, what part of the weekly salary is spent? **B**

Ⓐ $\frac{X}{Y}$ Ⓑ $\frac{X-Y}{X}$ Ⓒ $\frac{X-Y}{Y}$ Ⓓ $\frac{Y-X}{Y}$

www.geometryonline.com/chapter_test/fcat

Assessment Options

Vocabulary Test A vocabulary test/review for Chapter 6 can be found on p. 344 of the *Chapter 6 Resource Masters*.

Chapter Tests There are six Chapter 6 Tests and an Open-Ended Assessment task available in the *Chapter 6 Resource Masters*.

Chapter 6 Tests

Form	Type	Level	Pages
1	MC	basic	331–332
2A	MC	average	333–334
2B	MC	average	335–336
2C	FR	average	337–338
2D	FR	average	339–340
3	FR	advanced	341–342

MC = multiple-choice questions
FR = free-response questions

Open-Ended Assessment Performance tasks for Chapter 6 can be found on p. 343 of the *Chapter 6 Resource Masters*. A sample scoring rubric for these tasks appears on p. A25.

TestCheck and Worksheet Builder

This **networkable software** has three modules for assessment.

- **Worksheet Builder** to make worksheets and tests.
- **Student Module** to take tests on-screen.
- **Management System** to keep student records.

Portfolio Suggestion

Introduction Similarity, ratios, and proportions are the basis of many scale drawings and scale models.

Ask Students Ask students to do a scale drawing of a new park that includes a play area for children, a pond, and a picnic area for families. Have students add their scale drawings to their portfolios.

Chapter 6 Standardized Test Practice

These two pages contain practice questions in the various formats that can be found on the most frequently given standardized tests.

A practice answer sheet for these two pages can be found on p. A1 of the *Chapter 6 Resource Masters*.

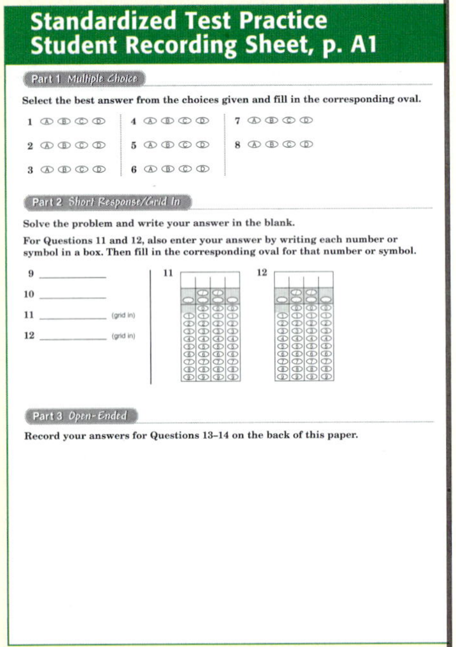

Standardized Test Practice
Student Recording Sheet, p. A1

Part 1 Multiple Choice

Select the best answer from the choices given and fill in the corresponding oval.

1 Ⓐ Ⓑ Ⓒ Ⓓ 4 Ⓐ Ⓑ Ⓒ Ⓓ 7 Ⓐ Ⓑ Ⓒ Ⓓ
2 Ⓐ Ⓑ Ⓒ Ⓓ 5 Ⓐ Ⓑ Ⓒ Ⓓ 8 Ⓐ Ⓑ Ⓒ Ⓓ
3 Ⓐ Ⓑ Ⓒ Ⓓ 6 Ⓐ Ⓑ Ⓒ Ⓓ

Part 2 Short Response/Grid In

Solve the problem and write your answer in the blank.

For Questions 11 and 12, also enter your answer by writing each number or symbol in a box. Then fill in the corresponding oval for that number or symbol.

9 ______
10 ______
11 ______ (grid in)
12 ______ (grid in)

Part 3 Open-Ended

Record your answers for Questions 13–14 on the back of this paper.

Additional Practice

See pp. 349–350 in the *Chapter 6 Resource Masters* for additional standardized test practice.

Chapter 6 Standardized Test Practice

FCAT Practice

Part 1 Multiple Choice

Record your answers on the answer sheet provided by your teacher or on a sheet of paper.

1. Which of the following is equivalent to $|-8 + 2|$? (Prerequisite Skill) **B**

Ⓐ 10 Ⓑ 6 Ⓒ -6 Ⓓ -10

2. Kip's family moved to a new house. He used a coordinate plane with units in miles to locate his new house and school in relation to his old house. What is the distance between his new house and school? (Lesson 1-3) **B**

Ⓐ 12 miles
Ⓑ $\sqrt{229}$ miles
Ⓒ 17 miles
Ⓓ $\sqrt{425}$ miles

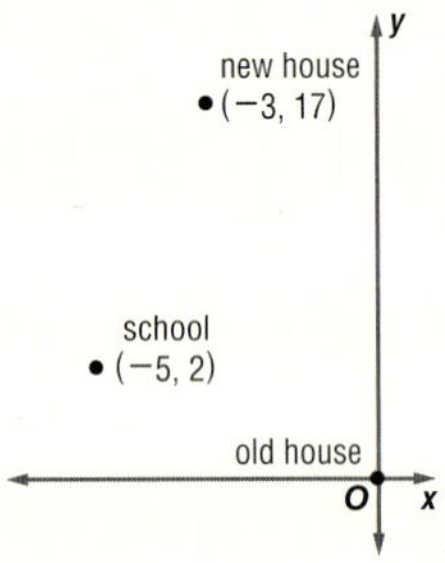

3. The diagonals of rectangle *ABCD* are $\overline{AC}$ and $\overline{BD}$. Hallie found that the distances from the point where the diagonals intersect to each vertex were the same. Which of the following conjectures could Hallie make? (Lesson 2-1) **A**

Ⓐ Diagonals of a rectangle are congruent.
Ⓑ Diagonals of a rectangle create equilateral triangles.
Ⓒ Diagonals of a rectangle intersect at more than one point.
Ⓓ Diagonals of a rectangle are congruent to the width.

4. If two sides of a triangular sail are congruent, which of the following terms *cannot* be used to describe the shape of the sail? (Lesson 4-1) **D**

Ⓐ acute Ⓑ equilateral
Ⓒ obtuse Ⓓ scalene

5. Miguel is using centimeter grid paper to make a scale drawing of his favorite car. Miguel's drawing is 11.25 centimeters wide. How many feet long is the actual car? (Lesson 6-1) **A**

Ⓐ 15.0 ft
Ⓑ 18.75 ft
Ⓒ 22.5 ft
Ⓓ 33.0 ft

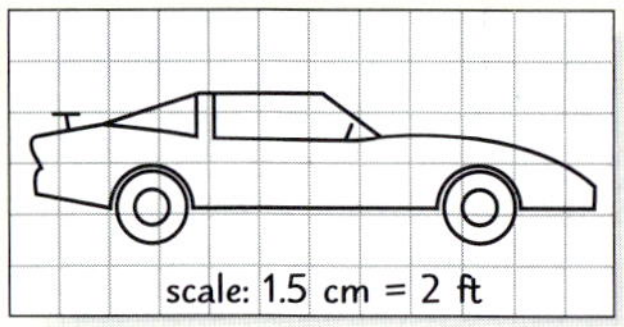

6. Joely builds a large corkboard for her room that is 45 inches tall and 63 inches wide. She wants to build a smaller corkboard with a similar shape for the kitchen. Which of the following could be the dimensions of that corkboard? (Lesson 6-2) **B**

Ⓐ 4 in. by 3 in.
Ⓑ 7 in. by 5 in.
Ⓒ 12 in. by 5 in.
Ⓓ 21 in. by 14 in.

7. If $\triangle PQR$ and $\triangle STU$ are similar, which of the following is a correct proportion? (Lesson 6-3) **C**

Ⓐ $\frac{s}{u} = \frac{t}{q}$
Ⓑ $\frac{s}{u} = \frac{p}{q}$
Ⓒ $\frac{s}{u} = \frac{p}{r}$
Ⓓ $\frac{s}{u} = \frac{r}{p}$

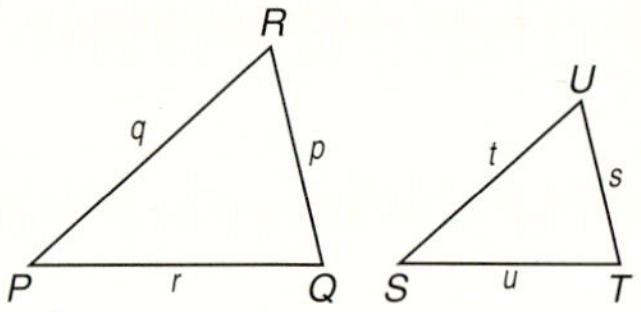

8. In $\triangle ABC$, *D* is the midpoint of $\overline{AB}$, and *E* is the midpoint of $\overline{AC}$. Which of the following is *not* true? (Lesson 6-4) **D**

Ⓐ $\frac{AD}{DB} = \frac{AE}{EC}$
Ⓑ $\overline{DE} \parallel \overline{BC}$
Ⓒ $\triangle ABC \sim \triangle ADE$
Ⓓ $\angle 1 \cong \angle 4$

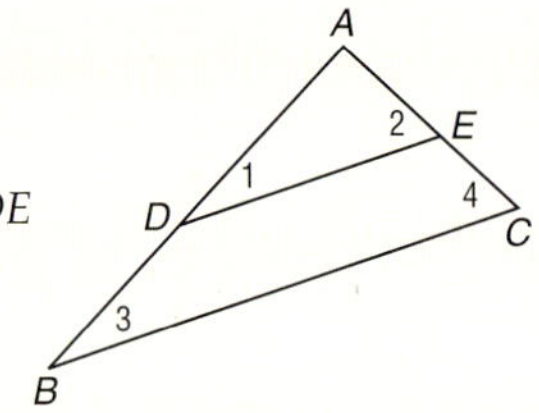

Log On for Test Practice

The Princeton Review offers additional test-taking tips and practice problems at their web site. Visit www.princetonreview.com or www.review.com

TestCheck and Worksheet Builder

Special banks of standardized test questions similar to those on the SAT, ACT, TIMSS 8, NAEP 8, and state proficiency tests can be found on this CD-ROM.

Aligned and verified by The Princeton Review

FCAT Practice

Part 2 Short Response/Grid In

Record your answers on the answer sheet provided by your teacher or on a sheet of paper.

9. During his presentation, Dante showed a picture of several types of balls used in sports. From this picture, he conjectured that all balls used in sports are spheres. Brianna then showed another ball. What is this type of example called? (Lesson 2-1)

Dante Brianna

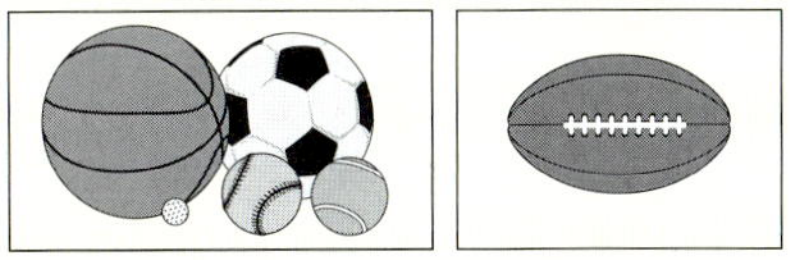

counterexample

10. What is the equation of a line with slope 3 that contains $A(2, 2)$? (Lesson 3-4)
$y = 3x - 4$

11. In $\triangle DEF$, P is the midpoint of $\overline{DE}$, and Q is the midpoint of side $\overline{DF}$. If $EF = 3x + 4$ and $PQ = 20$, what is the value of x? (Lesson 6-4)
12

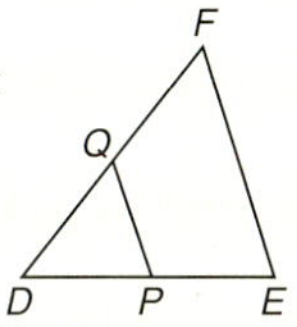

12. A city planner designs a triangular traffic median on Main Street to provide more green space in the downtown area. The planner builds a model so that the section of the median facing Main Street East measures 20 centimeters. What is the perimeter, in centimeters, of the model of the traffic median? (Lesson 6-5) 54.5

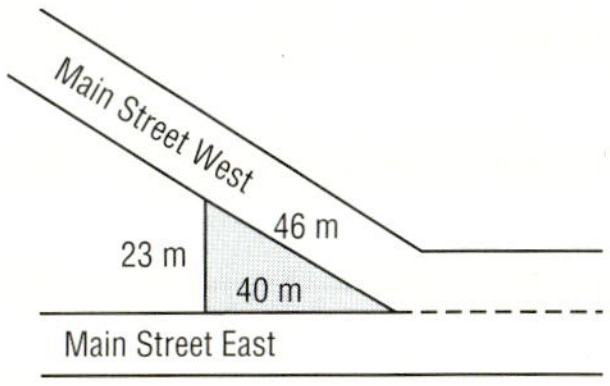

www.geometryonline.com/standardized_test/fcat

The Princeton Review **Test-Taking Tip**

Question 7 In similar triangles, corresponding angles are congruent and corresponding sides are proportional. When you set up a proportion, be sure that it compares corresponding sides. In Question 7, p corresponds to s, q corresponds to t, and r corresponds to u.

FCAT Practice

Part 3 Open Ended

Record your answers on a sheet of paper. Show your work.

13. A cable company charges a one-time connection fee plus a monthly flat rate as shown in the graph.

Reliable Cable Company

a. What is the slope of the line that joins the points on the graph? (Lesson 3-3) 30

b. Discuss what the value of the slope represents. (Lesson 3-3) See margin.

c. Write the equation of the line. (Lesson 3-4)
$y = 30x + 40$

d. If the company presents a special offer that lowers the monthly rate by \$5, how will the equation and graph change? (Lesson 3-4) $y = 25x + 40$; the graph will have a less steep slope.

14. In $\triangle ADE$, $\overline{BC}$ is equidistant from $\overline{DE}$.

a. Prove that $\triangle ABC \sim \triangle ADE$. (Lessons 6-3 and 6-4) See margin.

b. Suppose $AB = 3500$ feet, $BD = 1500$ feet, and $BC = 1400$ feet. Find DE. (Lesson 6-3)
2000 ft or 800 ft

Evaluating Open-Ended Assessment Questions

Open-Ended Assessment questions are graded by using a multilevel rubric that guides you in assessing a student's knowledge of a particular concept.

Goal for Exercise 13: Students analyze a line graph.

Goal for Exercise 14: Students prove two triangles are similar.

Sample Scoring Rubric: The following rubric is a sample scoring device. You may wish to add more detail to this sample to meet your individual scoring needs.

Score	Criteria
4	A correct solution that is supported by well-developed, accurate explanations
3	A generally correct solution, but may contain minor flaws in reasoning or computation
2	A partially correct interpretation and/or solution to the problem
1	A correct solution with no supporting evidence or explanation
0	An incorrect solution indicating no mathematical understanding of the concept or task, or no solution is given

Answers

13b. The slope represents the monthly flat rate, so the company charges a flat rate of \$30 per month.

14a. Using the Triangle Proportionality Theorem, if a line is parallel to one side of a triangle and intersects the other two sides in two distinct points, then it separates these sides into segments of proportional length, so $\frac{AE}{AC} = \frac{AD}{AB}$. Since the corresponding sides are proportional and the included angle, $\angle A$, is the same, by the SAS Similarity Theorem we know that the triangles are similar.

Page 301–306, Lesson 6-3

38.

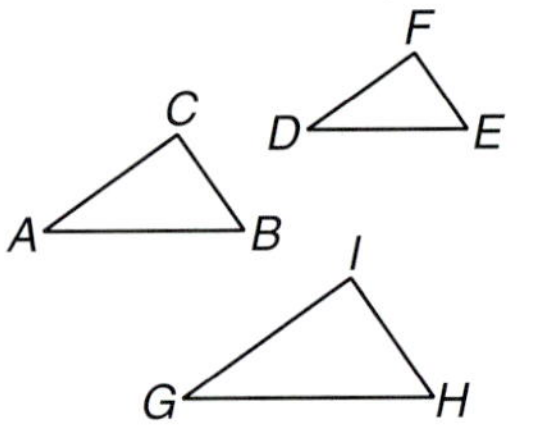

Reflexive Property of Similarity

Given: $\triangle ABC$

Prove: $\triangle ABC \sim \triangle ABC$

Proof:

Statements (Reasons)

1. $\triangle ABC$ (Given)
2. $\angle A \cong \angle A$, $\angle B \cong \angle B$ (Refl. Prop.)
3. $\triangle ABC \sim \triangle ABC$ (AA Similarity)

Symmetric Property of Similarity

Given: $\triangle ABC \sim \triangle DEF$

Prove: $\triangle DEF \sim \triangle ABC$

Proof:

Statements (Reasons)

1. $\triangle ABC \sim \triangle DEF$ (Given)
2. $\angle A \cong \angle D$, $\angle B \cong \angle E$ (Def. of $\sim$ polygons)
3. $\angle D \cong \angle A$, $\angle E \cong \angle B$ (Symmetric Prop.)
4. $\triangle DEF \sim \triangle ABC$ (AA Similarity)

Transitive Property of Similarity

Given: $\triangle ABC \sim \triangle DEF$ and $\triangle DEF \sim \triangle GHI$

Prove: $\triangle ABC \sim \triangle GHI$

Proof:

Statements (Reasons)

1. $\triangle ABC \sim \triangle DEF$, $\triangle DEF \sim \triangle GHI$ (Given)
2. $\angle A \cong \angle D$, $\angle B \cong \angle E$, $\angle D \cong \angle G$, $\angle E \cong \angle H$ (Def. of $\sim$ polygons)
3. $\angle A \cong \angle G$, $\angle B \cong \angle H$ (Trans. Prop.)
4. $\triangle ABC \sim \triangle GHI$ (AA Similarity)

Page 319–323, Lesson 6-5

30. Given: $\triangle ABC \sim \triangle PQR$

Prove: $\frac{BD}{QS} = \frac{BA}{QP}$

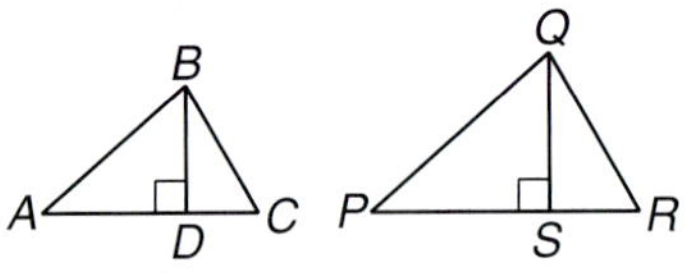

Proof: Since $\triangle ABC \sim \triangle PQR$, $\angle A \cong \angle P$. $\angle BDA \cong \angle QSP$ because they are both right angles created by the altitude drawn to the opposite side and all right angles are congruent. Thus $\triangle ABD \sim \triangle PQS$ by AA Similarity and $\frac{BD}{QS} = \frac{BA}{QP}$ by the definition of similar polygons.

31. Given: $\triangle ABC \sim \triangle RST$, $\overline{AD}$ is a median of $\triangle ABC$. $\overline{RU}$ is a median of $\triangle RST$.

Prove: $\frac{AD}{RU} = \frac{AB}{RS}$

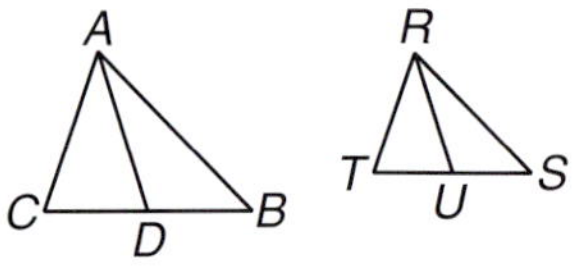

Proof:

Statements (Reasons)

1. $\triangle ABC \sim \triangle RST$; $\overline{AD}$ is a median of $\triangle ABC$; $\overline{RU}$ is a median of $\triangle RST$. (Given)
2. $CD = DB$; $TU = US$ (Def. of median)
3. $\frac{AB}{RS} = \frac{CB}{TS}$ (Def. of $\sim$ polygons)
4. $CB = CD + DB$; $TS = TU + US$ (Seg. Add. Post.)
5. $\frac{AB}{RS} = \frac{CD + DB}{TU + US}$ (Substitution)
6. $\frac{AB}{RS} = \frac{DB + DB}{US + US}$ or $\frac{2(DB)}{2(US)}$ (Substitution)
7. $\frac{AB}{RS} = \frac{DB}{US}$ (Substitution)
8. $\angle B \cong \angle S$ (Def. of $\sim$ polygons)
9. $\triangle ABD \sim \triangle RSU$ (SAS Similarity)
10. $\frac{AD}{RU} = \frac{AB}{RS}$ (Def. of $\sim$ polgyons)

32. Given: $\overline{CD}$ bisects $\angle ACB$. By construction $\overline{AE} \parallel \overline{CD}$.

Prove: $\frac{AD}{DB} = \frac{AC}{BC}$

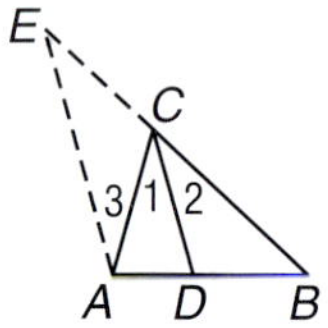

Proof:

Statements (Reasons)

1. $\overline{CD}$ bisects $\angle ACB$; By construction, $\overline{AE} \parallel \overline{CD}$. (Given)
2. $\frac{AD}{DB} = \frac{EC}{BC}$ (Triangle Proportionality Theorem)
3. $\angle 1 \cong \angle 2$ (Definition of Angle Bisector)
4. $\angle 3 \cong \angle 1$ (Alternate Interior Angle Theorem)
5. $\angle 2 \cong \angle E$ (Corresponding Angle Postulate)
6. $\angle 3 \cong \angle E$ (Transitive Prop.)
7. $\overline{EC} \cong \overline{AC}$ (Isosceles $\triangle$ Th.)
8. $EC = AC$ (Def. of congruent segments)
9. $\frac{AD}{DB} = \frac{AC}{BC}$ (Substitution)

33. Given: $\triangle ABC \sim \triangle PQR$, $\overline{BD}$ is an altitude of $\triangle ABC$. $\overline{QS}$ is an altitude of $\triangle PQR$.

Prove: $\frac{QP}{BA} = \frac{QS}{BD}$

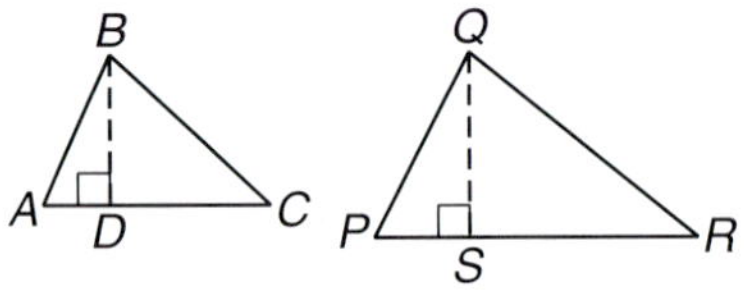

Proof: $\angle A \cong \angle P$ because of the definition of similar polygons. Since $\overline{BD}$ and $\overline{QS}$ are perpendicular to $\overline{AC}$ and $\overline{PR}$, $\angle BDA \cong \angle QSP$. So, $\triangle ABD \sim \triangle PQS$ by AA Similarity and $\frac{QP}{BA} = \frac{QS}{BD}$ by definition of similar polygons.

34. Given: $\angle C \cong \angle BDA$

Prove: $\frac{AC}{DA} = \frac{AD}{BA}$

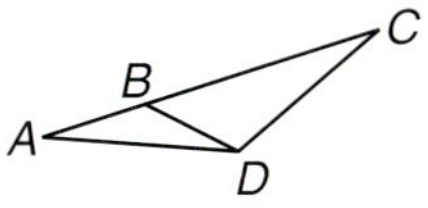

Proof:

$\angle C \cong \angle BDA$ (Given) → $\triangle ADB \sim \triangle ACD$

$\angle A \cong \angle A$ (Reflexive Prop.) → $\triangle ADB \sim \triangle ACD$

$\triangle ADB \sim \triangle ACD$ (AA Similarity) → $\frac{AC}{DA} = \frac{AD}{BA}$ (Def. of ~ polygons)

35. Given: $\overline{JF}$ bisects $\angle EFG$.
$\overline{EH} \parallel \overline{FG}$, $\overline{EF} \parallel \overline{HG}$

Prove: $\frac{EK}{KF} = \frac{GJ}{JF}$

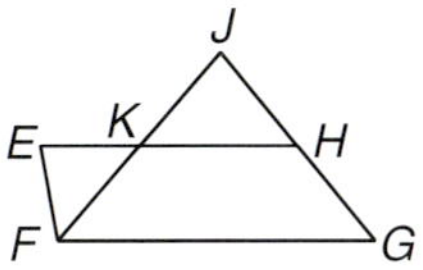

Proof:

Statements (Reasons)

1. $\overline{JF}$ bisects $\angle EFG$. $\overline{EH} \parallel \overline{FG}$, $\overline{EF} \parallel \overline{HG}$ (Given)
2. $\angle EFK \cong \angle KFG$ (Def. of $\angle$ bisector)
3. $\angle KFG \cong \angle JKH$ (Corresponding $\angle$s Postulate)
4. $\angle JKH \cong \angle EKF$ (Vertical $\angle$s are $\cong$.)
5. $\angle EFK \cong \angle EKF$ (Transitive Prop.)
6. $\angle FJH \cong \angle EFK$ (Alternate Interior $\angle$s Theorem)
7. $\angle FJH \cong \angle EKF$ (Transitive Prop.)
8. $\triangle EKF \sim \triangle GJF$ (AA Similarity)
9. $\frac{EK}{KF} = \frac{GJ}{JF}$ (Def. of ~ $\triangle$s)

36. Given: $\overline{RU}$ bisects $\angle SRT$.
$\overline{VU} \parallel \overline{RT}$

Prove: $\frac{SV}{VR} = \frac{SR}{RT}$

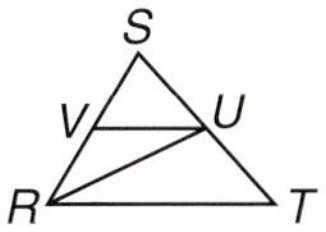

Proof:

Statements (Reasons)

1. $\overline{RU}$ bisects $\angle SRT$; $\overline{VU} \parallel \overline{RT}$ (Given)
2. $\angle S \cong \angle S$ (Reflexive Prop.)
3. $\angle SUV \cong \angle STR$ (Corresponding $\angle$s Postulate)
4. $\triangle SUV \sim \triangle STR$ (AA Similarity)
5. $\frac{SV}{VU} = \frac{SR}{RT}$ (Def. of ~ $\triangle$s)
6. $\angle URT \cong \angle VUR$ (Alternate Interior $\angle$s Theorem)
7. $\angle VRU \cong \angle URT$ (Def. of $\angle$ bisector)
8. $\angle VUR \cong \angle VRU$ (Transitive Prop.)
9. $\overline{VU} \cong \overline{VR}$ (If 2 $\angle$s of a $\triangle$ are $\cong$, the sides opp. these $\angle$s are $\cong$.)
10. $VU = VR$ (Def. of congruent segments)
11. $\frac{SV}{VR} = \frac{SR}{RT}$ (Substitution Prop.)

37. Given: $\triangle RST \sim \triangle ABC$,
W and D are midpoints of $\overline{TS}$ and $\overline{CB}$, respectively.

Prove: $\triangle RWS \sim \triangle ADB$

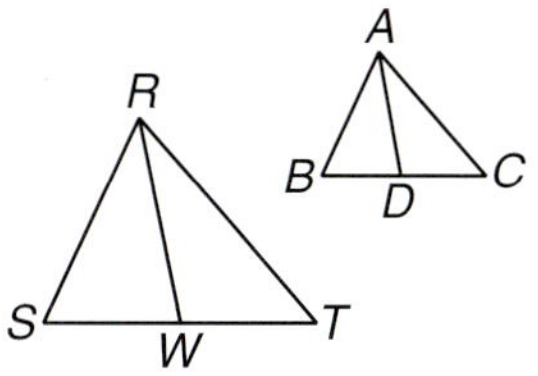

Proof:

$\triangle RST \sim \triangle ABC$ (Given) → $\angle S \cong \angle B$ (Def. of ~ polygons)

$\triangle RST \sim \triangle ABC$ (Given) → $\frac{RS}{AB} = \frac{TS}{CB}$ (Def. of ~ polygons)

W and D are midpoints. (Given) → $2WS = TS$, $2BD = CB$ (Def. of midpoint)

$\frac{RS}{AB} = \frac{TS}{CB}$ and $2WS = TS$, $2BD = CB$ → $\frac{RS}{AB} = \frac{2WS}{2BD}$ (Substitution)

$\frac{RS}{AB} = \frac{2WS}{2BD}$ → $\frac{RS}{AB} = \frac{WS}{BD}$ (Def. of Division)

$\frac{RS}{AB} = \frac{WS}{BD}$ and $\angle S \cong \angle B$ → $\triangle RWS \sim \triangle ABC$ (SAS Similarity)

Page 324, Geometry Activity

7.

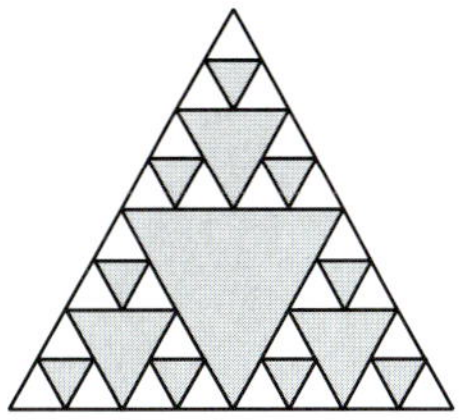

Chapter 7

Right Triangles and Trigonometry

Chapter Overview and Pacing

LESSON OBJECTIVES

	PACING (days)			
	Regular		Block	
	Basic/ Average	Advanced	Basic/ Average	Advanced
7-1 Geometric Mean *(pp. 342–348)* • Find the geometric mean between two numbers. • Solve problems involving relationships between parts of a right triangle and the altitude to its hypotenuse.	1	1	0.5	0.5
7-2 The Pythagorean Theorem and Its Converse *(pp. 349–356)* ***Preview:*** Physically explore side-length relationships in the Pythagorean Theorem. • Use the Pythagorean Theorem. • Use the converse of the Pythagorean Theorem.	2 (with 7-2 Preview)	1	1 (with 7-2 Preview)	0.5
7-3 Special Right Triangles *(pp. 357–363)* • Use properties of 45°-45°-90° triangles. • Use properties of 30°-60°-90° triangles.	2	2	1	1
7-4 Trigonometry *(pp. 364–370)* • Find trigonometric ratios using right triangles. • Solve problems using trigonometric ratios.	2	2	1	1
7-5 Angle of Elevation and Depression *(pp. 371–376)* • Solve problems involving angles of elevation. • Solve problems involving angles of depression.	1	2	0.5	1
7-6 The Law of Sines *(pp. 377–384)* • Use the Law of Sines to solve triangles. • Solve problems by using the Law of Sines. ***Follow-Up:*** Use geometry software to investigate the Law of Sines.	optional	2 (with 7-6 Follow-Up)	optional	1 (with 7-6 Follow-Up)
7-7 The Law of Cosines *(pp. 385–391)* • Use the Law of Cosines to solve triangles. • Solve problems by using the Law of Cosines. ***Follow-Up:*** Use trigonometric ratios to identify trigonometric identities.	optional	2	optional	1
Study Guide and Practice Test *(pp. 392–397)* **Standardized Test Practice** *(pp. 398–399)*	1	1	0.5	0.5
Chapter Assessment	1	1	0.5	0.5
TOTAL	**10**	**14**	**5**	**7**

Pacing suggestions for the entire year can be found on pages T20–T21.

Timesaving Tools
TeacherWorks™
All-In-One Planner and Resource Center
See pages T5 and T21.

Chapter Resource Manager

Chapter 7 Resource Masters: Study Guide and Intervention	Practice (Skills and Average)	Reading to Learn Mathematics	Enrichment	Assessment	Prerequisite Skills Workbook	Applications*	5-Minute Check Transparencies	Interactive Chalkboard	GeomPASS: Tutorial Plus (lessons)	Materials
351–352	353–354	355	356		11–12, 31–32		7-1	7-1		
357–358	359–360	361	362	407	11–12, 31–32	GCC 29 SC 13	7-2	7-2	13	(*Preview:* patty paper, ruler)
363–364	365–366	367	368		17–18		7-3	7-3	14	grid paper, scissors
369–370	371–372	373	374	407, 409	11–12	SC 14	7-4	7-4		paper, scissors, metric ruler, protractor
375–376	377–378	379	380				7-5	7-5	15	
381–382	383–384	385	386	408		GCC 30	7-6	7-6		
387–388	389–390	391	392	408	21–22, 25–26		7-7	7-7	16	
				393–406, 410–412						

**Key to Abbreviations:* GCC = Graphing Calculator and Computer Masters
SC = School-to-Career Masters

Chapter 7

Mathematical Connections and Background

Continuity of Instruction

Prior Knowledge

In previous courses, students evaluated radical expressions and equations involving fractions. The Pythagorean Theorem is reviewed in Chapter 1. In Chapter 6, students solve proportions. Chapter 4 introduces students to the Angle Sum Theorem.

This Chapter

This chapter provides students with an introduction to trigonometry. Students learn how to use the geometric mean to solve problems involving side length. They solve problems using the Pythagorean Theorem and its converse. Trigonometric ratios are defined and then used to solve right triangle problems. Students also use the Law of Sines and the Law of Cosines to solve non-right triangles.

Future Connections

Students gain a basic knowledge of trigonometric ratios from this chapter. This knowledge will help them in higher-level math courses, including pre-calculus and calculus. Architects, surveyors, and civil engineers use trigonometric ratios in their work.

7-1 Geometric Mean

The geometric mean between two numbers is the square root of their product. For two positive numbers a and b, the geometric mean is the positive number x where the proportions $a:x = x:b$ is true. This proportion can also be written $x = \sqrt{ab}$.

The geometric mean has a particular application to a right triangle. If an altitude is drawn from the vertex of the right angle of a right triangle to its hypotenuse, then the two triangles formed are similar to the given triangle and to each other. The measure of this altitude is the geometric mean between the measures of the two segments of the hypotenuse. Moreover, the measure of a leg of the triangle is the geometric mean between the measures of the hypotenuse and the segment of the hypotenuse adjacent to that leg.

7-2 The Pythagorean Theorem and Its Converse

The Pythagorean Theorem states that in a right triangle, the sum of the squares of the measures of the legs equals the square of the measure of the hypotenuse. The converse of the Pythagorean Theorem is useful in determining whether given measures are those of a right triangle. If the sum of the squares of the measures of two sides of a triangle equals the square of the measure of the longest side, then the triangle is a right triangle. A Pythagorean triple is a group of three whole numbers that satisfies the equation $a^2 + b^2 = c^2$, where c is the greatest number.

7-3 Special Right Triangles

A 45°-45°-90° triangle is the only isosceles right triangle. One of its special properties is that the hypotenuse is $\sqrt{2}$ times as long as a leg, so the ratio of the sides is $1:1:\sqrt{2}$. A 30°-60°-90° triangle also has special properties. The measures of the sides are x, $x\sqrt{3}$, and $2x$, giving the sides a ratio of $1:\sqrt{3}:2$. Knowing these properties can save you valuable time when you are solving problems involving special right triangles.

7-4 Trigonometry

A ratio of the lengths of the sides of a right triangle is called a trigonometric ratio. The three most common trigonometric ratios are sine, cosine, and tangent, abbreviated sin, cos, and tan. Sine of $\angle A$ is the measure of the leg opposite $\angle A$ divided by the measure of the hypotenuse. Cosine of $\angle A$ is the measure of the leg adjacent $\angle A$ divided by the measure of the hypotenuse. Tangent of $\angle A$ is the measure of the leg opposite the angle divided by the measure of the leg adjacent the angle. The value of this ratio does not depend on the size of the triangle or the measures of the sides.

Trigonometric ratios are used to find missing measures of a right triangle. You only need to know the measures of two sides or the measure of one side and one acute angle. The inverse of each trigonometric ratio yields the angle measure. The inverses are written $\sin^{-1}$, $\cos^{-1}$, and $\tan^{-1}$.

7-5 Angle of Elevation and Depression

An angle of elevation is the angle between the line of sight and the horizontal when an observer looks upward. An angle of depression is the angle between the line of sight and the horizontal when an observer looks downward. Trigonometric ratios can be used to solve problems involving angles of elevation and depression. Angles of elevation and depression to two different objects can be used to find the difference between those objects.

7-6 The Law of Sines

In trigonometry, the Law of Sines can be used to find missing measures of triangles that are not right triangles. Let $\triangle ABC$ be any triangle with a, b, and c representing the measures of the sides opposite the angles with measures A, B, and C, respectively. Then $\frac{\sin A}{a} = \frac{\sin B}{b} = \frac{\sin C}{c}$. The Law of Sines can be used to solve a triangle. This means finding the measure of every side and angle. The Law of Sines can be used to solve a triangle if you know the measures of two angles and any side of a triangle, or if you know the measures of two sides and an angle opposite one of these sides of the triangle.

There is one case in which using the Law of Sines to solve a triangle will yield an ambiguous solution. If you know the measures of two sides and an angle opposite one of the sides (SSA) and the angle is acute, it is possible to find two different triangles. One triangle will be an acute triangle, and the other will be obtuse.

7-7 The Law of Cosines

The Law of Cosines allows you to solve a triangle when the Law of Sines cannot be used. This occurs when you know the measures of two sides and the included angle (SAS) or three sides (SSS). Let $\triangle ABC$ be any triangle with a, b, and c representing the measures of sides opposite angles with measures A, B, and C, respectively. Then the following equations are true:

$$a^2 = b^2 + c^2 - 2bc \cos A$$
$$b^2 = a^2 + c^2 - 2ac \cos B$$
$$c^2 = a^2 + b^2 - 2ab \cos C$$

Chapter 7

DAILY INTERVENTION and Assessment

	Type	Student Edition	Teacher Resources	Technology/Internet
INTERVENTION	Ongoing	Prerequisite Skills, pp. 341, 348, 356, 363, 370, 376, 383 Practice Quiz 1, p. 363 Practice Quiz 2, p. 383	5-Minute Check Transparencies *Prerequisite Skills Workbook,* pp. 11–12, 17–18, 21–22, 25–26, 31–32 Quizzes, *CRM* pp. 407–408 Mid-Chapter Test, *CRM* p. 409 Study Guide and Intervention, *CRM* pp. 351–352, 357–358, 363–364, 369–370, 375–376, 381–382, 387–388	GeomPASS: Tutorial Plus www.geometryonline.com/self_check_quiz www.geometryonline.com/extra_examples
	Mixed Review	pp. 348, 356, 363, 370, 376, 383, 390	Cumulative Review, *CRM* p. 410	
	Error Analysis	Find the Error, pp. 348, 353, 380 Common Misconceptions, p. 372	Find the Error, *TWE* pp. 345, 353, 380 Unlocking Misconceptions, *TWE* p. 359 Tips for New Teachers, *TWE* p. 358	
	Standardized Test Practice	pp. 348, 356, 362, 370, 372, 373, 376, 382, 390, 392, 398, 399	*TWE* pp. 398–399 Standardized Test Practice, *CRM* pp. 411–412	Standardized Test Practice CD-ROM www.geometryonline.com/standardized_test
	Open-Ended Assessment	Writing in Math, pp. 348, 356, 362, 369, 376, 382, 389 Open Ended, pp. 345, 353, 360, 367, 373, 380, 387 Standardized Test, p. 399	Modeling: *TWE* pp. 356, 376 Speaking: *TWE* pp. 363, 370 Writing: *TWE* pp. 348, 383, 390 Open-Ended Assessment, *CRM* p. 405	
ASSESSMENT	Chapter Assessment	Study Guide, pp. 392–396 Practice Test, p. 397	Multiple-Choice Tests (Forms 1, 2A, 2B), *CRM* pp. 393–398 Free-Response Tests (Forms 2C, 2D, 3), *CRM* pp. 399–404 Vocabulary Test/Review, *CRM* p. 406	TestCheck and Worksheet Builder (see below) MindJogger Videoquizzes www.geometryonline.com/vocabulary_review www.geometryonline.com/chapter_test

Key to Abbreviations: TWE = Teacher Wraparound Edition; CRM = Chapter Resource Masters

Additional Intervention Resources

The Princeton Review's *Cracking the SAT & PSAT*

The Princeton Review's *Cracking the ACT*

TestCheck and Worksheet Builder

This **networkable** software has three modules for intervention and assessment flexibility:

- **Worksheet Builder** to make worksheet and tests
- **Student Module** to take tests on screen (optional)
- **Management System** to keep student records (optional)

Special banks are included for SAT, ACT, TIMSS, NAEP, and End-of-Course tests.

Intervention Technology

GeomPASS: Tutorial Plus CD-ROM offers a complete, self-paced geometry curriculum.

Geometry Lesson	GeomPASS Lesson
7-2	**13** *Solving Problems Using the Pythagorean Theorem*
7-3	**14** *Special Right Triangles*
7-5	**15** *Trigonometry I*
7-7	**16** *Trigonometry II*

Intervention at Home

Log on for student study help.

- For each lesson in the Student Edition, there are Extra Examples and Self-Check Quizzes.
  www.geometryonline.com/extra_examples
  www.geometryonline.com/self_check_quiz
- For chapter review, there is vocabulary review, test practice, and standardized test practice.
  www.geometryonline.com/vocabulary_review
  www.geometryonline.com/chapter_test
  www.geometryonline.com/standardized_test

For more information on Intervention and Assessment, see pp. T8–T11.

Reading and Writing in Mathematics

Glencoe Geometry provides numerous opportunities to incorporate reading and writing into the mathematics classroom.

Student Edition

- Foldables Study Organizer, p. 341
- Concept Check questions require students to verbalize and write about what they have learned in the lesson. (pp. 345, 353, 360, 367, 373, 380, 387)
- Writing in Math questions in every lesson, pp. 348, 356, 362, 369, 376, 382, 389
- Reading Study Tip, p. 364
- WebQuest, pp. 346, 390

Teacher Wraparound Edition

- Foldables Study Organizer, pp. 341, 392
- Study Notebook suggestions, pp. 345, 349, 353, 360, 367, 373, 380, 388
- Modeling activities, pp. 356, 376
- Speaking activities, pp. 363, 370
- Writing activities, pp. 348, 383, 390
- Differentiated Instruction (Verbal/Linguistic), p. 386
- ELL Resources, pp. 340, 346, 354, 361, 368, 374, 381, 386, 389, 392

Additional Resources

- Vocabulary Builder worksheets require students to define and give examples for key vocabulary terms as they progress through the chapter. (*Chapter 7 Resource Masters,* pp. vii-viii)
- Proof Builder helps students learn and understand theorems and postulates from the chapter. (*Chapter 7 Resource Masters,* pp. ix–x)
- Reading to Learn Mathematics master for each lesson (*Chapter 7 Resource Masters*, pp. 355, 361, 367, 373, 379, 385, 391)
- *Vocabulary PuzzleMaker* software creates crossword, jumble, and word search puzzles using vocabulary lists that you can customize.
- *Teaching Mathematics with Foldables* provides suggestions for promoting cognition and language.
- *Reading Strategies for the Mathematics Classroom*
- *WebQuest and Project Resources*

For more information on Reading and Writing in Mathematics, see pp. T6–T7.

Chapter 7 Notes

What You'll Learn

Have students read over the list of objectives and make a list of any words with which they are not familiar.

Why It's Important

Point out to students that this is only one of many reasons why each objective is important. Others are provided in the introduction to each lesson.

Lesson	NCTM Standards	Local Objectives
7-1	1, 3, 6, 8, 9, 10	
7-2 Preview	3, 6	
7-2	3, 6, 8, 9, 10	
7-3	3, 6, 8, 9, 10	
7-4	3, 6, 8, 9, 10	
7-5	3, 6, 8, 9, 10	
7-6	3, 6, 8, 9, 10	
7-6 Follow-Up	3, 6	
7-7	3, 6, 8, 9, 10	
7-7 Follow-Up	3, 6	

Key to NCTM Standards:

1=Number & Operations, 2=Algebra, 3=Geometry, 4=Measurement, 5=Data Analysis & Probability, 6=Problem Solving, 7=Reasoning & Proof, 8=Communication, 9=Connections, 10=Representation

Chapter 7 Right Triangles and Trigonometry

What You'll Learn

- **Lessons 7-1, 7-2, and 7-3** Solve problems using the geometric mean, the Pythagorean Theorem, and its converse.
- **Lessons 7-4 and 7-5** Use trigonometric ratios to solve right triangle problems.
- **Lessons 7-6 and 7-7** Solve triangles using the Law of Sines and the Law of Cosines.

Key Vocabulary

- geometric mean (p. 342)
- Pythagorean triple (p. 352)
- trigonometric ratio (p. 364)
- Law of Sines (p. 377)
- Law of Cosines (p. 385)

Why It's Important

Trigonometry is used to find the measures of the sides and angles of triangles. These ratios are frequently used in real-world applications such as architecture, aviation, and surveying. *You will learn how surveyors use trigonometry in Lesson 7-6.*

Vocabulary Builder

ELL

The Key Vocabulary list introduces students to some of the main vocabulary terms included in this chapter. For a more thorough vocabulary list with pronunciations of new words, give students the Vocabulary Builder worksheets found on pages vii and viii of the *Chapter 7 Resource Masters*. Encourage them to complete the definition of each term as they progress through the chapter. You may suggest that they add these sheets to their study notebooks for future reference when studying for the Chapter 7 test.

Getting Started

Prerequisite Skills To be successful in this chapter, you'll need to master these skills and be able to apply them in problem-solving situations. Review these skills before beginning Chapter 7.

For Lesson 7-1 — Proportions

Solve each proportion. Round to the nearest hundredth, if necessary. *(For review, see Lesson 6-1.)*

1. $\frac{3}{4} = \frac{12}{a}$ **16**
2. $\frac{c}{5} = \frac{8}{3}$ **13.33**
3. $\frac{e}{20} = \frac{6}{5} = \frac{f}{10}$ **$e = 24$, $f = 12$**
4. $\frac{4}{3} = \frac{6}{y} = \frac{1}{z}$ **$y = 4.5$, $z = 0.75$**

For Lesson 7-2 — Pythagorean Theorem

Find the measure of the hypotenuse of each right triangle having legs with the given measures. Round to the nearest hundredth, if necessary. *(For review, see Lesson 1-3.)*

5. 5 and 12 **13**
6. 6 and 8 **10**
7. 15 and 15 **21.21**
8. 14 and 27 **30.41**

For Lessons 7-3 and 7-4 — Radical Expressions

Simplify each expression. *(For review, see pages 744 and 745.)*

9. $\sqrt{8}$ **$2\sqrt{2}$**
10. $\sqrt{10^2 - 5^2}$ **$5\sqrt{3}$**
11. $\sqrt{39^2 - 36^2}$ **15**
12. $\frac{7}{\sqrt{2}}$ **$\frac{7\sqrt{2}}{2}$**

For Lessons 7-5 through 7-7 — Angle Sum Theorem

Find x. *(For review, see Lesson 4-2.)*

13. **98**

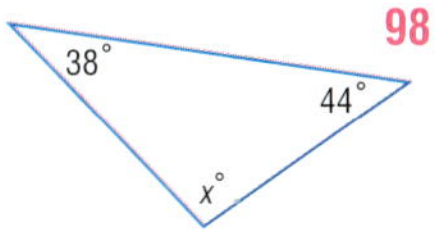

14. **115**

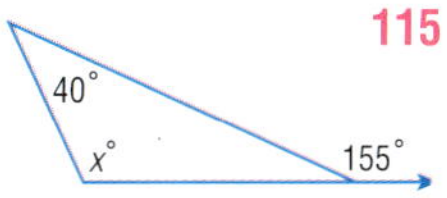

15. **23**

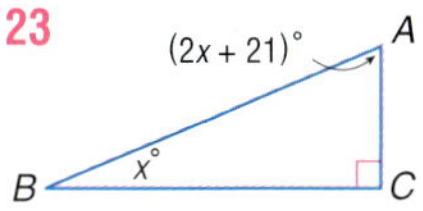

FOLDABLES™ Study Organizer

Right Triangles and Trigonometry Make this Foldable to record information from this chapter. Begin with seven sheets of grid paper.

Step 1 Fold

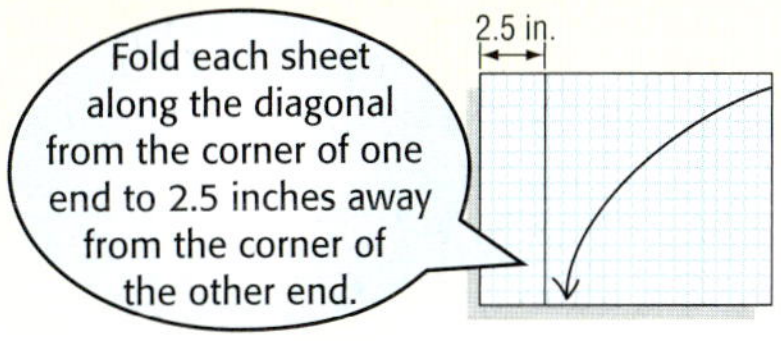

Step 2 Stack

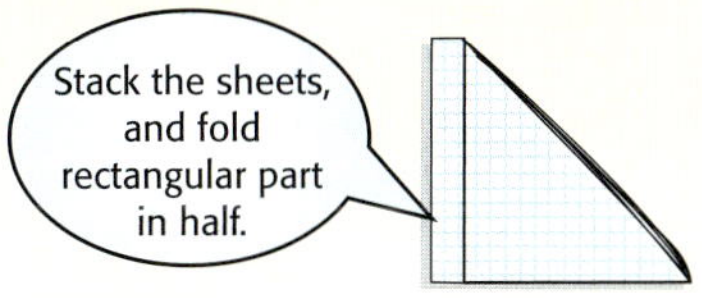

Step 3 Staple

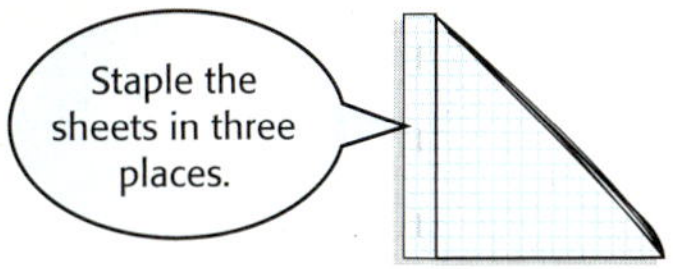

Step 4 Label

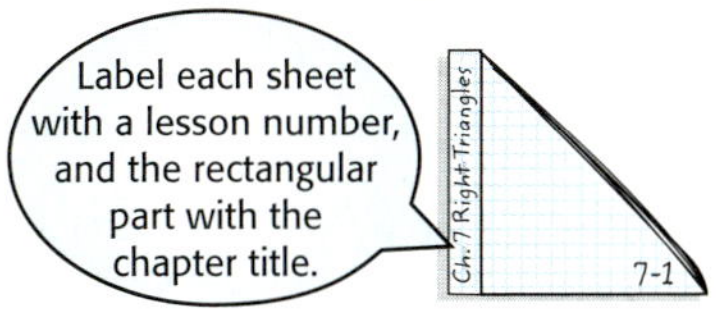

Reading and Writing As you read and study the chapter, write notes, define terms, and solve problems in your Foldable.

Getting Started

This section provides a review of the basic concepts needed before beginning Chapter 7. Page references are included for additional student help.

Additional review is provided in the *Prerequisite Skills Workbook,* pages 11–12, 17–18, 21–22, 25–26, 31–32.

Prerequisite Skills in the Getting Ready for the Next Lesson section at the end of each exercise set review a skill needed in the next lesson.

For Lesson	Prerequisite Skill
7-2	Using the Pythagorean Theorem, p. 348
7-3	Simplifying Radical Expressions, p. 356
7-4	Solving Equations That Involve Fractions, p. 363
7-5	Angles Formed by Parallel Lines and a Transversal, p. 370
7-6	Solving Proportions, p. 376
7-7	Evaluating Expressions, p. 383

FOLDABLES™ Study Organizer

For more information about Foldables, see *Teaching Mathematics with Foldables.*

Journals and Visuals Use this Foldable journal for writing about right triangles and trigonometry. After students make their triangle-shaped journal, have them label the front of each triangle to correspond to the seven lessons in this chapter. Under the tabs of their Foldable, students take notes, define terms, solve problems, and write examples of how laws are used. On the front of each section, ask students to design a visual (graph, diagram, picture, chart) that presents the information introduced in the lesson in a concise, easy-to-study format. Encourage students to label their visuals and write captions.

7-1 Lesson Notes

1 Focus

5-Minute Check Transparency 7-1 Use as a quiz or review of Chapter 6.

Mathematical Background notes are available for this lesson on p. 340C.

How can the geometric mean be used to view paintings?

Ask students:

- What happens if you view the painting at a distance that is less than the geometric mean of the two distances described above? **You are too close to the painting and cannot see all of it at once. Your eyes must move left, right, up, and down to see all the details of the painting.**
- Name some other scenarios where you would be best positioned at a geometric mean to view something. **watching a program on television or a movie in a theatre, viewing a sports event, seeing a production on a stage, using a computer, skimming a book or magazine, etc.**

7-1 Geometric Mean

Sunshine State Standards
MA.A.3.4.3, MA.B.1.4.1, MA.B.1.4.3, MA.B.2.4.1, MA.C.1.4.1, MA.C.2.4.1

What You'll Learn

- Find the geometric mean between two numbers.
- Solve problems involving relationships between parts of a right triangle and the altitude to its hypotenuse.

Vocabulary

- geometric mean

How can the geometric mean be used to view paintings?

When you look at a painting, you should stand at a distance that allows you to see all of the details in the painting. The distance that creates the best view is the geometric mean of the distance from the top of the painting to eye level and the distance from the bottom of the painting to eye level.

GEOMETRIC MEAN The **geometric mean** between two numbers is the positive square root of their product.

Study Tip

Means and Extremes
In the equation $x^2 = ab$, the two x's in x^2 represent the *means*, and a and b represent the *extremes* of the proportion.

Key Concept *Geometric Mean*

For two positive numbers a and b, the geometric mean is the positive number x where the proportion $a : x = x : b$ is true. This proportion can be written using fractions as $\frac{a}{x} = \frac{x}{b}$ or with cross products as $x^2 = ab$ or $x = \sqrt{ab}$.

Example 1 *Geometric Mean*

Find the geometric mean between each pair of numbers.

a. 4 and 9

Let x represent the geometric mean.

$\frac{4}{x} = \frac{x}{9}$ Definition of geometric mean

$x^2 = 36$ Cross products

$x = \sqrt{36}$ Take the positive square root of each side.

$x = 6$ Simplify.

b. 6 and 15

$\frac{6}{x} = \frac{x}{15}$ Definition of geometric mean

$x^2 = 90$ Cross products

$x = \sqrt{90}$ Take the positive square root of each side.

$x = 3\sqrt{10}$ Simplify.

$x \approx 9.5$ Use a calculator.

Resource Manager

Workbook and Reproducible Masters

Chapter 7 Resource Masters
- Study Guide and Intervention, pp. 351–352
- Skills Practice, p. 353
- Practice, p. 354
- Reading to Learn Mathematics, p. 355
- Enrichment, p. 356

Prerequisite Skills Workbook, pp. 11–12, 31–32

Transparencies

5-Minute Check Transparency 7-1
Answer Key Transparencies

Technology

Interactive Chalkboard

ALTITUDE OF A TRIANGLE Consider right triangle XYZ with altitude $\overline{WZ}$ drawn from the right angle Z to the hypotenuse $\overline{XY}$. A special relationship exists for the three right triangles, $\triangle XYZ$, $\triangle XZW$, and $\triangle ZYW$.

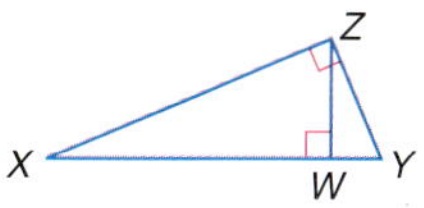

Geometry Software Investigation

Right Triangles Formed by the Altitude

Use The Geometer's Sketchpad to draw a right triangle XYZ with right angle Z. Draw the altitude $\overline{ZW}$ from the right angle to the hypotenuse. Explore the relationships among the three right triangles formed.

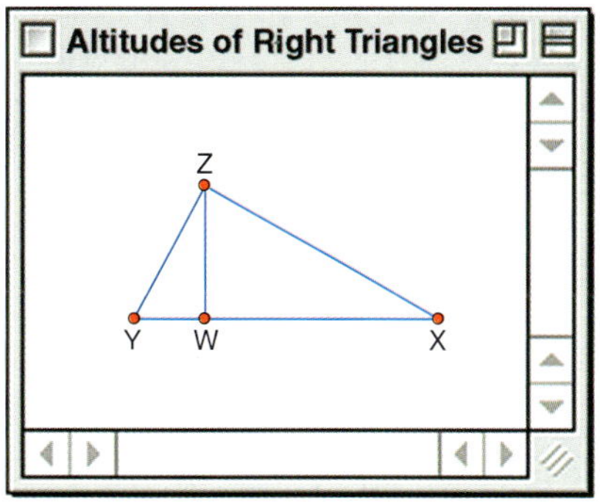

Think and Discuss

1. Find the measures of $\angle X$, $\angle XZY$, $\angle Y$, $\angle XWZ$, $\angle XZW$, $\angle YWZ$, and $\angle YZW$. **See students' work.**
2. What is the relationship between the measures of $\angle X$ and $\angle YZW$? What is the relationship between the measures of $\angle Y$ and $\angle XZW$? **They are equal.**
3. Drag point Z to another position. Describe the relationship between the measures of $\angle X$ and $\angle YZW$ and between the measures of $\angle Y$ and $\angle XZW$. **They are equal.**
4. Make a conjecture about $\triangle XYZ$, $\triangle XZW$, and $\triangle ZYW$. **They are similar.**

Study Tip

Altitudes of a Right Triangle
The altitude drawn to the hypotenuse originates from the right angle. The other two altitudes of a right triangle are the legs.

The results of the Geometry Software Investigation suggest the following theorem.

Theorem 7.1

If the altitude is drawn from the vertex of the right angle of a right triangle to its hypotenuse, then the two triangles formed are similar to the given triangle and to each other.

Example: $\triangle XYZ \sim \triangle XWY \sim \triangle YWZ$

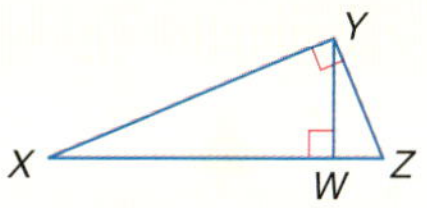

You will prove this theorem in Exercise 45.

By Theorem 7.1, since $\triangle XWY \sim \triangle YWZ$, the corresponding sides are proportional. Thus, $\frac{XW}{YW} = \frac{YW}{ZW}$. Notice that $\overline{XW}$ and $\overline{ZW}$ are segments of the hypotenuse of the largest triangle.

Theorem 7.2

The measure of an altitude drawn from the vertex of the right angle of a right triangle to its hypotenuse is the geometric mean between the measures of the two segments of the hypotenuse.

Example: YW is the geometric mean of XW and ZW.

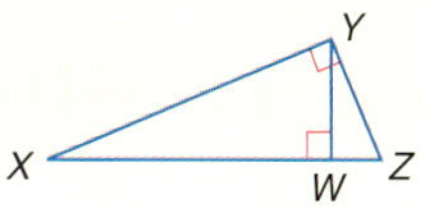

You will prove this theorem in Exercise 46.

www.geometryonline.com/extra_examples/fcat

Geometry Software Investigation

Students should recognize that if they were given just one angle measure other than the right angle, they would have enough information to find all the other angles represented in the figure. They should also see that given one segment length, they could use proportions to find any of the other segment lengths in the figure as well.

2 Teach

GEOMETRIC MEAN

In-Class Example Power Point®

1 Find the geometric mean between each pair of numbers.

a. 2 and 50 **10**

b. 25 and 7 **≈13.2**

ALTITUDE OF A TRIANGLE

In-Class Example Power Point®

Teaching Tip Remind students that they automatically discard the negative root when finding the altitude or geometric mean because these values represent lengths, and lengths cannot be negative.

2 In $\triangle ABC$, $BD = 6$ and $AD = 27$. Find CD.

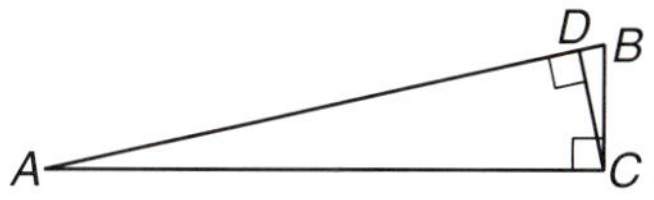

$CD \approx 12.7$

Interactive Chalkboard
PowerPoint® Presentations

This CD-ROM is a customizable Microsoft® PowerPoint® presentation that includes:

- Step-by-step, dynamic solutions of each In-Class Example from the Teacher Wraparound Edition
- Additional, Try These exercises for each example
- The 5-Minute Check Transparencies
- Hot links to Glencoe Online Study Tools

In-Class Examples

Power Point®

3 **KITES** Ms. Alspach is constructing a kite for her son. She has to arrange perpendicularly two support rods, the shorter of which is 27 inches long. If she has to place the short rod 7.25 inches from one end of the long rod in order to form two right triangles with the kite fabric, what is the length of the long rod?

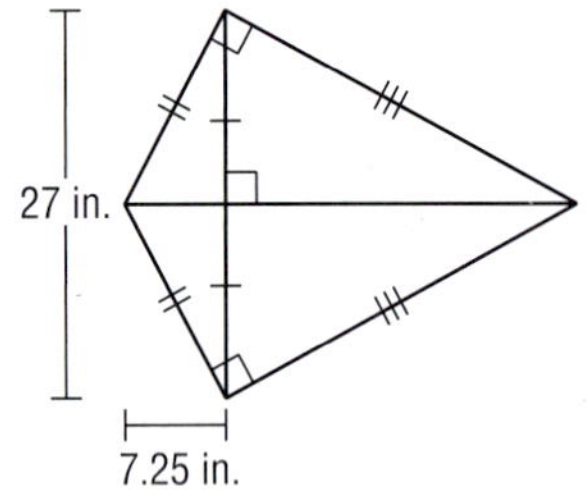

≈ 32.39 in.

4 Find c and d in $\triangle JKL$.

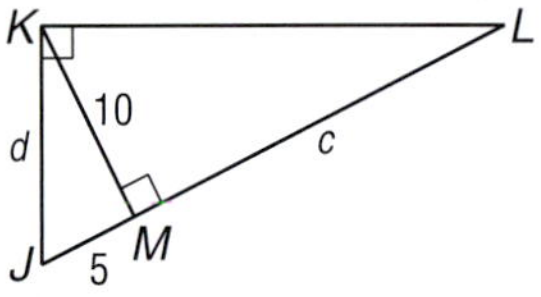

$c = 20$; $d \approx 11.2$

Example 2 Altitude and Segments of the Hypotenuse

In $\triangle PQR$, $RS = 3$ and $QS = 14$. Find PS.

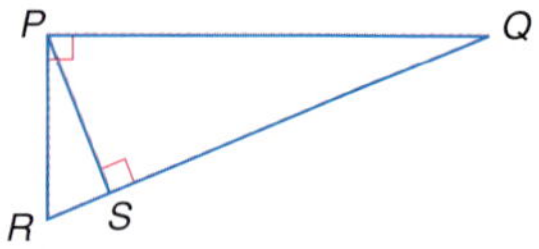

Let $x = PS$.

$\frac{RS}{PS} = \frac{PS}{QS}$

$\frac{3}{x} = \frac{x}{14}$ — $RS = 3$, $QS = 14$, and $PS = x$

$x^2 = 42$ — Cross products

$x = \sqrt{42}$ — Take the positive square root of each side.

$x \approx 6.5$ — Use a calculator.

PS is about 6.5.

Study Tip

Square Roots
Since these numbers represent measures, you can ignore the negative square root value.

Ratios in right triangles can be used to solve problems.

Example 3 Altitude and Length of the Hypotenuse

ARCHITECTURE **Mr. Martinez is designing a walkway that must pass over an elevated train. To find the height of the elevated train, he holds a carpenter's square at eye level and sights along the edges from the street to the top of the train. If Mr. Martinez's eye level is 5.5 feet above the street and he is 8.75 feet from the train, find the distance from the street to the top of the train. Round to the nearest tenth.**

Draw a diagram. Let $\overline{YX}$ be the altitude drawn from the right angle of $\triangle WYZ$.

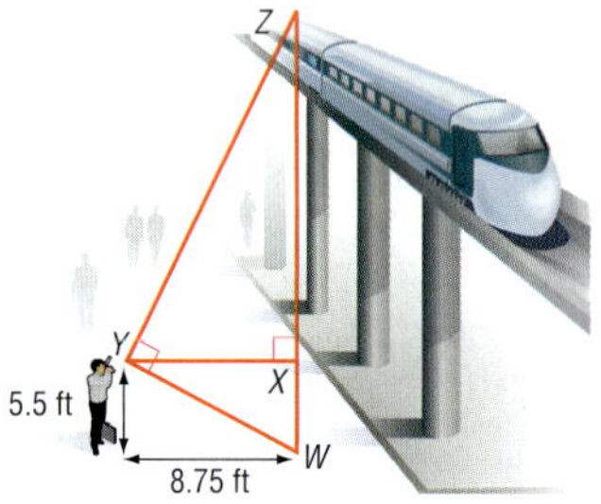

$\frac{WX}{YX} = \frac{YX}{ZX}$

$\frac{5.5}{8.75} = \frac{8.75}{ZX}$ — $WX = 5.5$ and $YX = 8.75$

$5.5ZX = 76.5625$ — Cross products

$ZX \approx 13.9$ — Divide each side by 5.5.

Mr. Martinez estimates that the elevated train is 5.5 + 13.9 or about 19.4 feet high.

The altitude to the hypotenuse of a right triangle determines another relationship between the segments.

Theorem 7.3

If the altitude is drawn from the vertex of the right angle of a right triangle to its hypotenuse, then the measure of a leg of the triangle is the geometric mean between the measures of the hypotenuse and the segment of the hypotenuse adjacent to that leg.

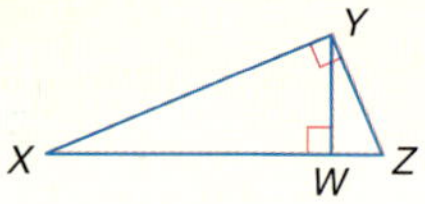

Example: $\frac{XZ}{XY} = \frac{XY}{XW}$ and $\frac{XZ}{YZ} = \frac{ZY}{WZ}$

You will prove Theorem 7.3 in Exercise 47.

DAILY INTERVENTION

Differentiated Instruction

Intrapersonal Allow students to sit quietly and explore similarities and differences between Theorem 7.2 and Theorem 7.3. Encourage students to use the examples in the book or create their own to reinforce the concepts outlined in these two theorems. Ask students to think and write about why the formulas for geometric mean work for a right triangle with an altitude drawn to its hypotenuse.

Example 4 Hypotenuse and Segment of Hypotenuse

Find x and y in $\triangle PQR$.

$\overline{PQ}$ and $\overline{RQ}$ are legs of right triangle PQR. Use Theorem 7.3 to write a proportion for each leg and then solve.

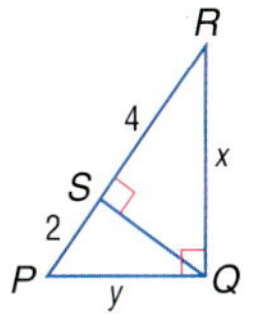

$\frac{PR}{PQ} = \frac{PQ}{PS}$

$\frac{6}{y} = \frac{y}{2}$ $PS = 2$, $PQ = y$, $PR = 6$

$y^2 = 12$ Cross products

$y = \sqrt{12}$ Take the square root.

$y = 2\sqrt{3}$ Simplify.

$y \approx 3.5$ Use a calculator.

$\frac{PR}{RQ} = \frac{RQ}{SR}$

$\frac{6}{x} = \frac{x}{4}$ $RS = 4$, $RQ = x$, $PR = 6$

$x^2 = 24$ Cross products

$x = \sqrt{24}$ Take the square root.

$x = 2\sqrt{6}$ Simplify.

$x \approx 4.9$ Use a calculator.

Study Tip

Simplifying Radicals
our teacher may ask you to simplify radicals, such as $\sqrt{12}$. Remember that $\sqrt{12} = \sqrt{4} \cdot \sqrt{3}$. Since $\sqrt{4} = 2$, $\sqrt{12} = 2\sqrt{3}$. For more practice simplifying radicals, see pages 744 and 745.

Check for Understanding

Concept Check

1. **OPEN ENDED** Find two numbers whose geometric mean is 12.

1\. Sample answer: 2 and 72

2. **Draw and label** a right triangle with an altitude drawn from the right angle. From your drawing, explain the meaning of *the hypotenuse and the segment of the hypotenuse adjacent to that leg* in Theorem 7.3. **See margin.**

3. **FIND THE ERROR** $\triangle RST$ is a right isosceles triangle. Holly and Ian are finding the measure of altitude $\overline{SU}$.

3\. Ian; his proportion shows that the altitude is the geometric mean of the two segments of the hypotenuse.

Holly	Ian
$\frac{RS}{SU} = \frac{SU}{RT}$	$\frac{RU}{SU} = \frac{SU}{UT}$
$\frac{9.9}{x} = \frac{x}{14}$	$\frac{7}{x} = \frac{x}{7}$
$x^2 = 138.6$	$x^2 = 49$
$x = \sqrt{138.6}$	$x = 7$
$x \approx 11.8$	

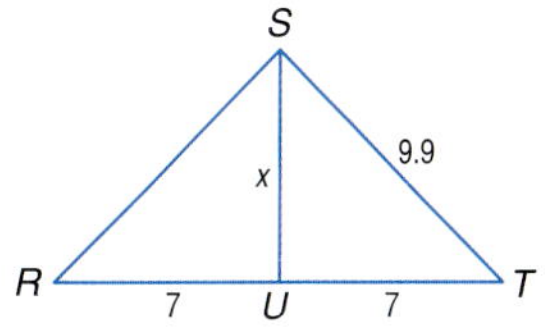

Who is correct? Explain your reasoning.

Guided Practice

Find the geometric mean between each pair of numbers.

4. 9 and 4 **6**
5. 36 and 49 **42**
6. 6 and 8 **$4\sqrt{3} \approx 6.9$**
7. $2\sqrt{2}$ and $3\sqrt{2}$ **$2\sqrt{3} \approx 3.5$**

GUIDED PRACTICE KEY	
Exercises	Examples
4–7	1
8, 9	2
10–12	3, 4

Find the measure of the altitude drawn to the hypotenuse.

8\.
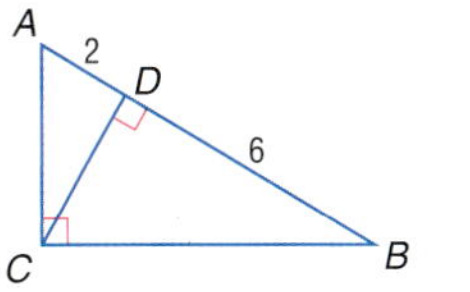

$2\sqrt{3} \approx 3.5$

9\.
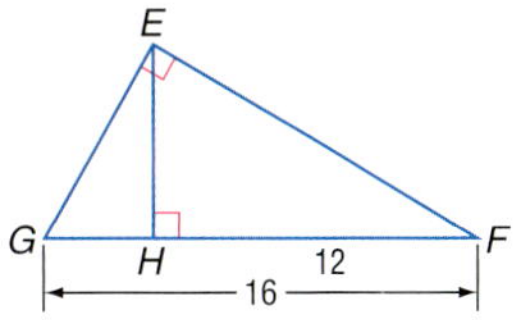

$4\sqrt{3} \approx 6.9$

3 Practice/Apply

Study Notebook

Have students—
- add the definitions/examples of the vocabulary terms to their Vocabulary Builder worksheets for Chapter 7.
- include any other item(s) that they find helpful in mastering the skills in this lesson.

DAILY INTERVENTION **FIND THE ERROR**

Ask students to name the hypotenuse of $\triangle RST$, and point out that Ian correctly found the geometric mean between the two segments of the *hypotenuse*, as the rule states in Theorem 7.2.

About the Exercises...

Organization by Objective
- **Geometric Mean:** 13–20
- **Altitude of a Triangle:** 21–32

Odd/Even Assignments

Exercises 13–32 are structured so that students practice the same concepts whether they are assigned odd or even problems.

Alert! Exercise 40 requires the Internet or other research materials.

Assignment Guide

Basic: 13–31 odd, 35–43 odd, 44, 48–66

Average: 13–43 odd, 44, 48–66

Advanced: 14–44 even, 45–63 (optional: 64–66)

Answer

2\. For leg $\overline{CB}$, $\overline{DB}$ is the segment of the hypotenuse that shares an endpoint. Thus, it is the adjacent segment. The same is true for leg $\overline{AC}$ and segment $\overline{AD}$.

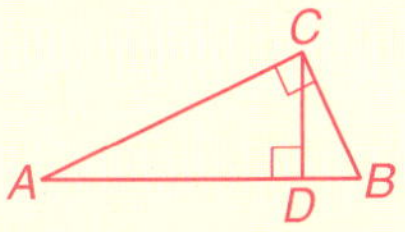

Study Guide and Intervention, p. 351 (shown) and p. 352

Geometric Mean The **geometric mean** between two numbers is the square root of their product. For two positive numbers a and b, the geometric mean of a and b is the positive number x in the proportion $\frac{a}{x} = \frac{x}{b}$. Cross multiplying gives $x^2 = ab$, so $x = \sqrt{ab}$.

Example **Find the geometric mean between each pair of numbers.**

a. 12 and 3

Let x represent the geometric mean.

$\frac{12}{x} = \frac{x}{3}$ Definition of geometric mean

$x^2 = 36$ Cross multiply.

$x = \sqrt{36}$ or 6 Take the square root of each side.

b. 8 and 4

Let x represent the geometric mean.

$\frac{8}{x} = \frac{x}{4}$

$x^2 = 32$

$x = \sqrt{32}$

≈ 5.7

Exercises

Find the geometric mean between each pair of numbers.

1. 4 and 4 4
2. 4 and 6 $\sqrt{24} \approx 4.9$
3. 6 and 9 $\sqrt{54} \approx 7.3$
4. $\frac{1}{2}$ and 2 1
5. $2\sqrt{3}$ and $3\sqrt{3}$ $\sqrt{18} \approx 4.2$
6. 4 and 25 10
7. $\sqrt{3}$ and $\sqrt{6}$ $18^{\frac{1}{4}} \approx 2.1$
8. 10 and 100 $\sqrt{1000} \approx 31.6$
9. $\frac{1}{2}$ and $\frac{1}{4}$ $\sqrt{\frac{1}{8}} \approx 0.4$
10. $\frac{2\sqrt{2}}{5}$ and $\frac{3\sqrt{2}}{5}$ $\sqrt{\frac{12}{25}} \approx 0.7$
11. 4 and 16 8
12. 3 and 24 $\sqrt{72} \approx 8.5$

The geometric mean and one extreme are given. Find the other extreme.

13. $\sqrt{24}$ is the geometric mean between a and b. Find b if $a = 2$. 12
14. $\sqrt{12}$ is the geometric mean between a and b. Find b if $a = 3$. 4

Determine whether each statement is *always*, *sometimes*, or *never* true.

15. The geometric mean of two positive numbers is greater than the average of the two numbers. never
16. If the geometric mean of two positive numbers is less than 1, then both of the numbers are less than 1. sometimes

Skills Practice, p. 353 and Practice, p. 354 (shown)

Find the geometric mean between each pair of numbers to the nearest tenth.

1. 8 and 12 $\sqrt{96} \approx 9.8$
2. $3\sqrt{7}$ and $6\sqrt{7}$ $\sqrt{126} \approx 11.2$
3. $\frac{4}{5}$ and 2 $\sqrt{\frac{8}{5}} \approx 1.3$

Find the measure of each altitude. State exact answers and answers to the nearest tenth.

4. 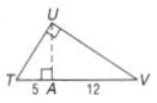

$\sqrt{60} \approx 7.7$

5. 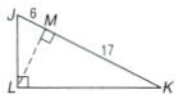

$\sqrt{102} \approx 10.1$

Find x, y, and z.

6. 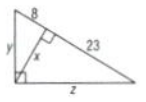

$\sqrt{184} \approx 13.6$; $\sqrt{248} \approx 15.7$; $\sqrt{713} \approx 26.7$

7. 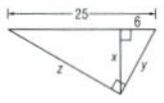

$\sqrt{114} \approx 10.7$; $\sqrt{150} \approx 12.2$; $\sqrt{475} \approx 21.8$

8. 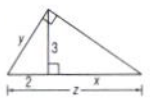

4.5; $\sqrt{13} \approx 3.6$; 6.5

9. 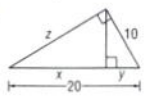

15; 5; $\sqrt{300} \approx 17.3$

10. **CIVIL ENGINEERING** An airport, a factory, and a shopping center are at the vertices of a right triangle formed by three highways. The airport and factory are 6.0 miles apart. Their distances from the shopping center are 3.6 miles and 4.8 miles, respectively. A service road will be constructed from the shopping center to the highway that connects the airport and factory. What is the shortest possible length for the service road? Round to the nearest hundredth. 2.88 mi

Reading to Learn Mathematics, p. 355 ELL

Pre-Activity **How can the geometric mean be used to view paintings?**

Read the introduction to Lesson 7-1 at the top of page 342 in your textbook.

- What is a disadvantage of standing too close to a painting? Sample answer: You don't get a good overall view.
- What is a disadvantage of standing too far from a painting? Sample answer: You can't see all the details in the painting.

Reading the Lesson

1. In the past, when you have seen the word *mean* in mathematics, it referred to the *average* or *arithmetic mean* of the two numbers.
   a. Complete the following by writing an algebraic expression in each blank.
   If a and b are two positive numbers, then the geometric mean between a and b is $\sqrt{ab}$ and their arithmetic mean is $\frac{a+b}{2}$.
   b. Explain in words, without using any mathematical symbols, the difference between the geometric mean and the algebraic mean. Sample answer: The geometric mean between two numbers is the square root of their product. The arithmetic mean of two numbers is half their sum.
2. Let r and s be two positive numbers. In which of the following equations is z equal to the geometric mean between r and s? A, C, D, F
   A. $\frac{s}{z} = \frac{z}{r}$ B. $\frac{r}{z} = \frac{s}{z}$ C. $s:z = z:r$ D. $\frac{r}{z} = \frac{z}{s}$ E. $\frac{z}{r} = \frac{z}{s}$ F. $\frac{z}{s} = \frac{r}{z}$
3. Supply the missing words or phrases to complete the statement of each theorem.
   a. The measure of the altitude drawn from the vertex of the right angle of a right triangle to its hypotenuse is the geometric mean between the measures of the two segments of the hypotenuse.
   b. If the altitude is drawn from the vertex of the right angle of a right triangle to its hypotenuse, then the measure of a leg of the triangle is the geometric mean between the measure of the hypotenuse and the segment of the hypotenuse adjacent to that leg.
   c. If the altitude is drawn from the vertex of the right angle of a right triangle to its hypotenuse, then the two triangles formed are similar to the given triangle and to each other.

Helping You Remember

4. A good way to remember a new mathematical concept is to relate it to something you already know. How can the meaning of *mean* in a proportion help you to remember how to find the geometric mean between two numbers? Sample answer: Write a proportion in which the two means are equal. This common mean is the geometric mean between the two extremes.

Find x and y.

10. $x = 2\sqrt{6}$; $y = \sqrt{33}$

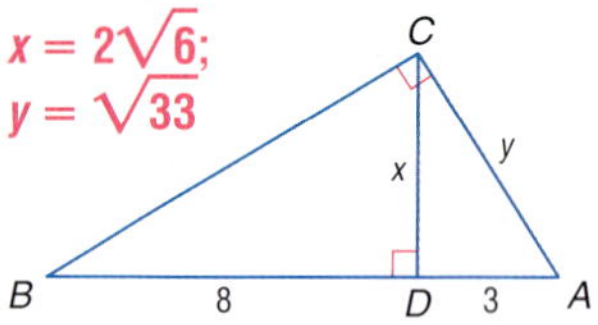

11. $x = 6$; $y = 4\sqrt{3}$

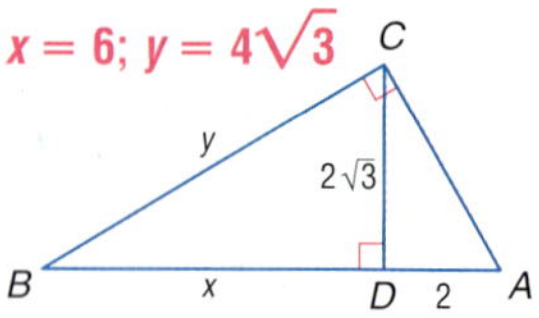

Application

12. **DANCES** Khaliah is making a banner for the dance committee. The banner is to be as high as the wall of the gymnasium. To find the height of the wall, Khaliah held a book up to her eyes so that the top and bottom of the wall were in line with the top edge and binding of the cover. If Khaliah's eye level is 5 feet off the ground and she is standing 12 feet from the wall, how high is the wall? 33.8 ft

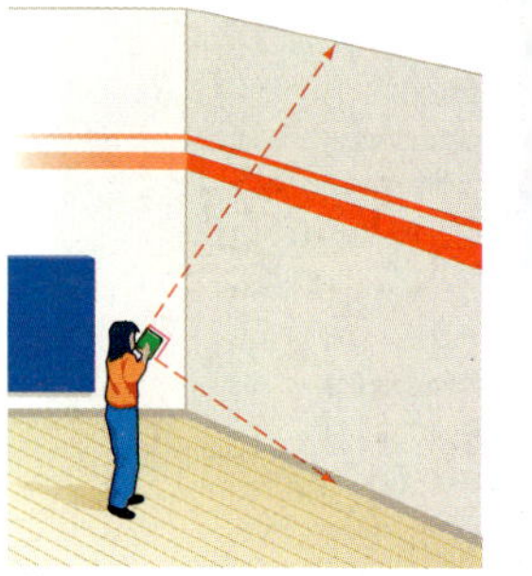

★ indicates increased difficulty

Practice and Apply

Homework Help	
For Exercises	**See Examples**
13–20	1
21–26	2
27–32	3, 4

Extra Practice
See page 766.

Find the geometric mean between each pair of numbers.

13. 5 and 6
14. 24 and 25
15. $\sqrt{45}$ and $\sqrt{80}$
16. $\sqrt{28}$ and $\sqrt{1372}$
17. $\frac{3}{5}$ and 1
18. $\frac{8\sqrt{3}}{5}$ and $\frac{6\sqrt{3}}{5}$
19. $\frac{2\sqrt{2}}{6}$ and $\frac{5\sqrt{2}}{6}$
20. $\frac{13}{7}$ and $\frac{5}{7}$

13. $\sqrt{30} \approx 5.5$
14. $10\sqrt{6} \approx 24.5$
15. $2\sqrt{15} \approx 7.7$
16. 14
17. $\frac{\sqrt{15}}{5} \approx 0.8$
18. $\frac{12}{5} = 2.4$
19. $\frac{\sqrt{5}}{3} \approx 0.7$
20. $\frac{\sqrt{65}}{7} \approx 1.2$

Find the measure of the altitude drawn to the hypotenuse.

21. $3\sqrt{5} \approx 6.7$

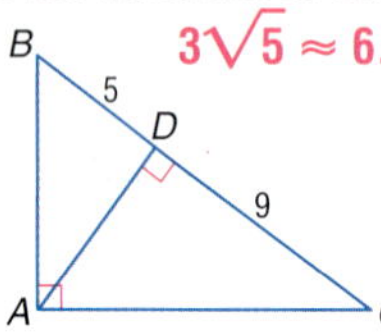

22. 12

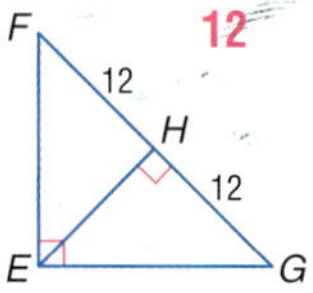

23. $8\sqrt{2} \approx 11.3$

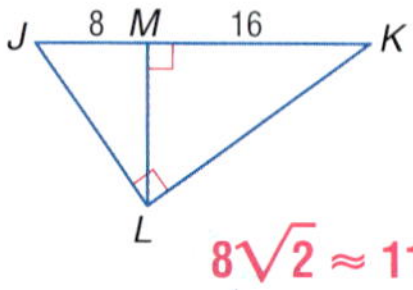

24. $\sqrt{147} \approx 12.1$

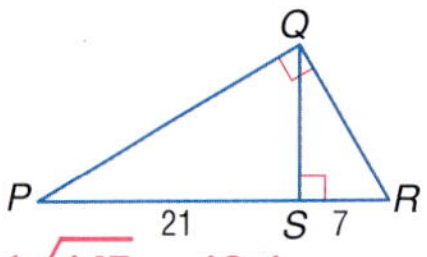

25. $\sqrt{26} \approx 5.1$

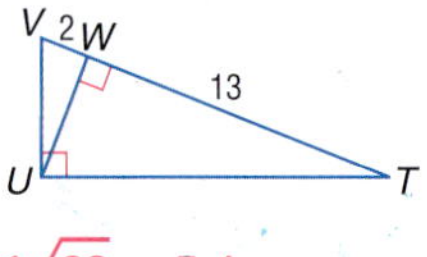

26. 5

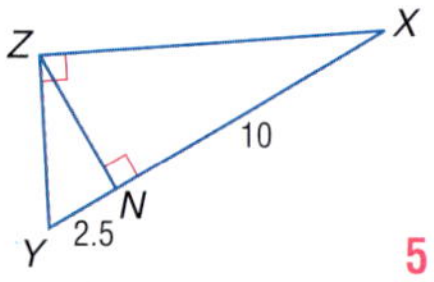

Find x, y, and z. 27–32. See margin.

27. 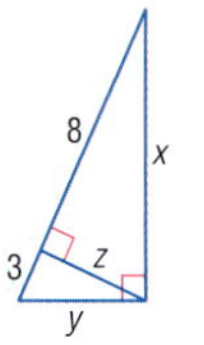

28. 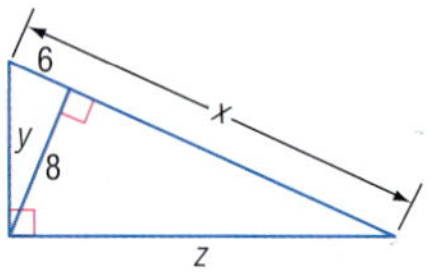

29. 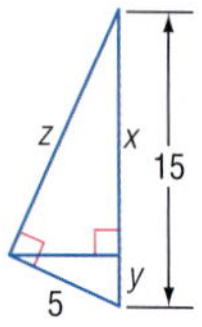

30. 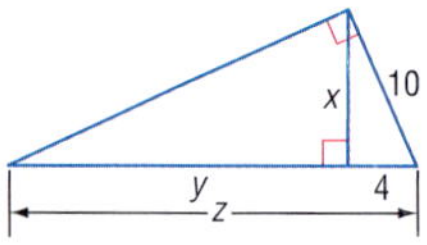

31. 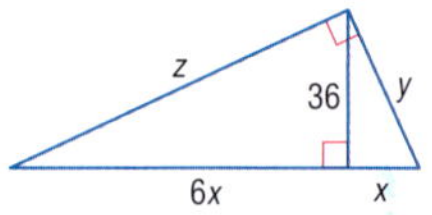

32. 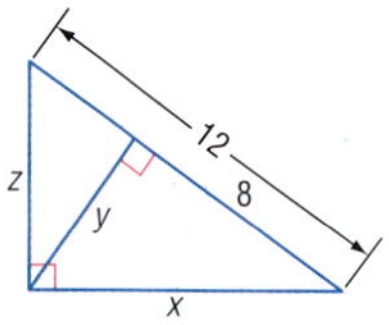

Enrichment, p. 356

Mathematics and Music

Pythagoras, a Greek philosopher who lived during the sixth century B.C., believed that all nature, beauty, and harmony could be expressed by whole-number relationships. Most people remember Pythagoras for his teachings about right triangles. (The sum of the squares of the legs equals the square of the hypotenuse.) But Pythagoras also discovered relationships between the musical notes of a scale. These relationships can be expressed as ratios.

C	D	E	F	G	A	B	C′
$\frac{1}{1}$	$\frac{8}{9}$	$\frac{4}{5}$	$\frac{3}{4}$	$\frac{2}{3}$	$\frac{3}{5}$	$\frac{8}{15}$	$\frac{1}{2}$

When you play a stringed instrument, you produce different notes by placing your finger on different places on a string. This is the result of changing the length of the vibrating part of the string.

The C string can be used to produce F by placing a finger $\frac{3}{4}$ of the way along the string.

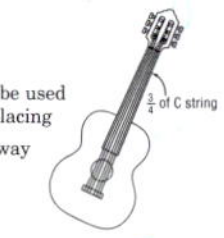

The geometric mean and one extreme are given. Find the other extreme.

★ 33. $\sqrt{17}$ is the geometric mean between a and b. Find b if $a = 7$. $\frac{17}{7}$

★ 34. $\sqrt{12}$ is the geometric mean between x and y. Find x if $y = \sqrt{3}$. $4\sqrt{3} \approx 6.9$

Determine whether each statement is *always*, *sometimes*, or *never* true.

35. The geometric mean for consecutive positive integers is the average of the two numbers. **never**

36. The geometric mean for two perfect squares is a positive integer. **always**

37. The geometric mean for two positive integers is another integer. **sometimes**

38. The measure of the altitude of a triangle is the geometric mean between the measures of the segments of the side it intersects. **sometimes**

More About . . .

Chambered Nautilus

One of the most common traits of the chambered nautilus is that it is a natural example of the golden ratio.

Source: www.infoplease.com

39. **BIOLOGY** The shape of the shell of a chambered nautilus can be modeled by a geometric mean. Consider the sequence of segments $\overline{OA}$, $\overline{OB}$, $\overline{OC}$, $\overline{OD}$, $\overline{OE}$, $\overline{OF}$, $\overline{OG}$, $\overline{OH}$, $\overline{OI}$, and $\overline{OJ}$. The length of each of these segments is the geometric mean between the lengths of the preceding segment and the succeeding segment. Explain this relationship. (*Hint:* Consider $\triangle FGH$.) **See margin.**

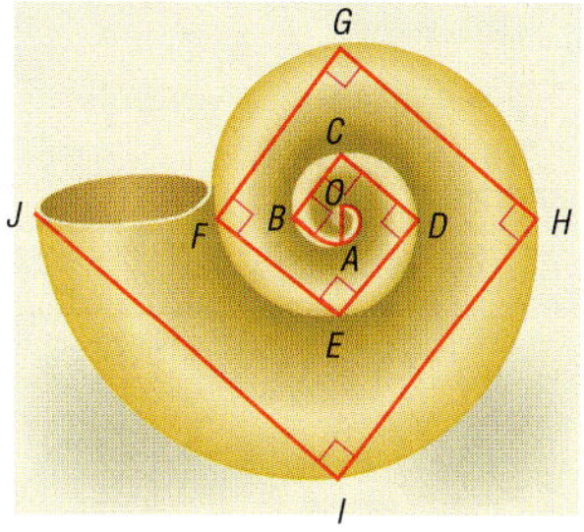

40. Sample answer: The golden ratio occurs when the geometric mean is approximately 1.62.

40. **RESEARCH** Refer to the information at the left. Use the Internet or other resource to write a brief description of the golden ratio.

41. **CONSTRUCTION** In the United States, most building codes limit the steepness of the slope of a roof to $\frac{4}{3}$, as shown at the right. A builder wants to put a support brace from point C perpendicular to $\overline{AP}$. Find the length of the brace. **2.4 yd**

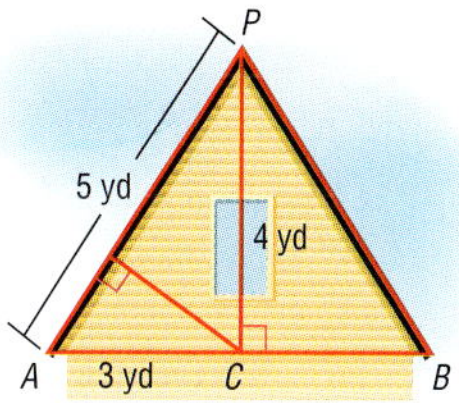

WebQuest

You can use geometric mean and the Quadratic Formula to discover the golden mean. Visit www.geometryonline.com/webquest to continue work on your WebQuest project.

SOCCER For Exercises 42 and 43, refer to the graphic.

42. Find the geometric mean between the number of players from Indiana and North Carolina. $\sqrt{70} \approx 8.4$

43. Are there two schools whose geometric mean is the same as the geometric mean between UCLA and Clemson? If so, which schools? **yes; Indiana and Virginia**

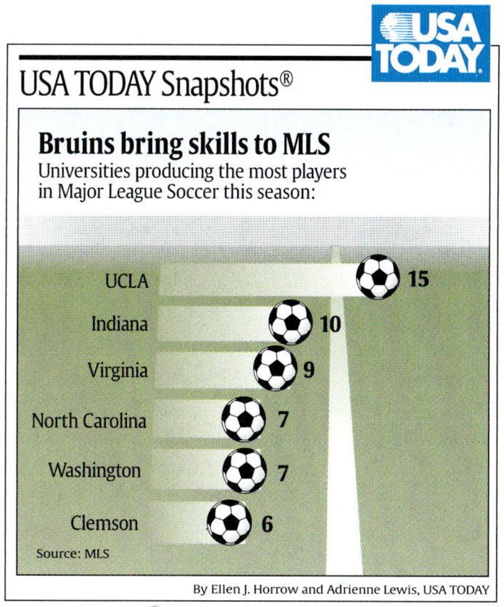

44. **CRITICAL THINKING** Find the exact value of DE, given $AD = 12$ and $BD = 4$. $2\sqrt{3}$

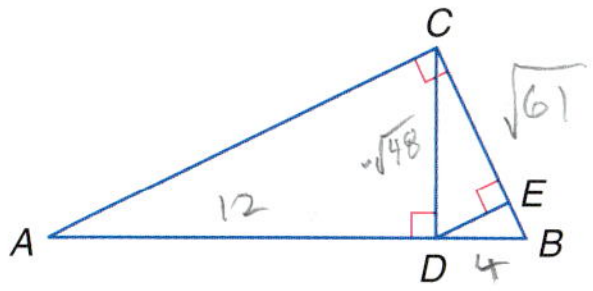

www.geometryonline.com/self_check_quiz/fcat

Answers

39. $\triangle FGH$ is a right triangle. $\overline{OG}$ is the altitude from the vertex of the right angle to the hypotenuse of that triangle. So, by Theorem 7.2, OG is the geometric mean between OF and OH, and so on.

P345 4, 6, P.346 14
22 24 26
28 30 32
Ec 34

Answers (p. 346)

27. $x = 2\sqrt{22} \approx 9.4$; $y = \sqrt{33} \approx 5.7$; $z = 2\sqrt{6} \approx 4.9$

28. $x = \frac{50}{3}$; $y = 10$; $z = \frac{40}{3}$

29. $x = \frac{40}{3}$; $y = \frac{5}{3}$; $z = 10\sqrt{2} \approx 14.1$

30. $x = 2\sqrt{21} \approx 9.2$; $y = 21$; $z = 25$

31. $x = 6\sqrt{6} \approx 14.7$; $y = 6\sqrt{42} \approx 38.9$; $z = 36\sqrt{7} \approx 95.2$

32. $x = 4\sqrt{6} \approx 9.8$; $y = 4\sqrt{2} \approx 5.7$; $z = 4\sqrt{3} \approx 6.9$

4 Assess

Open-Ended Assessment

Writing Draw and label several figures on the board, and ask students to write down the correct geometric mean proportions without working out the problem. Then have volunteers come to the board and write appropriate proportions that correspond with the figures, and allow seated students to check their work.

Getting Ready for Lesson 7-2

Prerequisite Skill Students will review the Pythagorean Theorem and learn about its converse in Lesson 7-2. Use Exercises 64–66 to determine your students' familiarity with using the Pythagorean Theorem.

Answers

48. Sample answer: The geometric mean can be used to help determine the optimum viewing distance. Answers should include the following.
    - If you are too far from a painting, you may not be able to see fine details. If you are too close, you may not be able to see the entire painting.
    - A curator can use the geometric mean to help determine how far from the painting the roping should be.

PROOF Write the specified type proof for each theorem. 45–47. See p. 399A.

45. two-column proof of Theorem 7.1
46. paragraph proof of Theorem 7.2
47. two-column proof of Theorem 7.3

48. **WRITING IN MATH** Answer the question that was posed at the beginning of the lesson. See margin.

How can the geometric mean be used to view paintings?

Include the following in your answer:
- an explanation of what happens when you are too far or too close to a painting, and
- an explanation of how the curator of a museum would determine where to place roping in front of paintings on display.

FCAT Practice
Standardized Test Practice

49. Find x and y. C

Ⓐ 4 and 6 Ⓑ 2.5 and 7.5
Ⓒ 3.6 and 6.4 Ⓓ 3 and 7

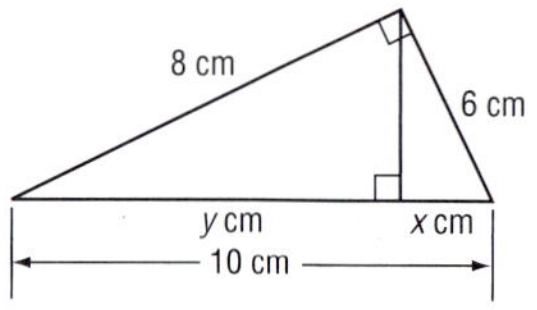

50. **ALGEBRA** Solve $5x^2 + 405 = 1125$. B

Ⓐ ± 15 Ⓑ ± 12
Ⓒ $\pm 4\sqrt{3}$ Ⓓ ± 4

Maintain Your Skills

Mixed Review

Find the first three iterations of each expression. *(Lesson 6-6)*

51. $x + 3$, where x initially equals 12 51. 15, 18, 21
52. $3x + 2$, where x initially equals 4 52. 14, 44, 134
53. $x^2 - 2$, where x initially equals 3 7, 47, 2207
54. $2(x - 3)$, where x initially equals 1 −4, −14, −34
55. The measures of the sides of a triangle are 20, 24, and 30. Find the measures of the segments formed where the bisector of the smallest angle meets the opposite side. *(Lesson 6-5)* $8\frac{8}{9}$, $11\frac{1}{9}$

Use the Exterior Angle Inequality Theorem to list all angles that satisfy the stated condition. *(Lesson 5-2)*

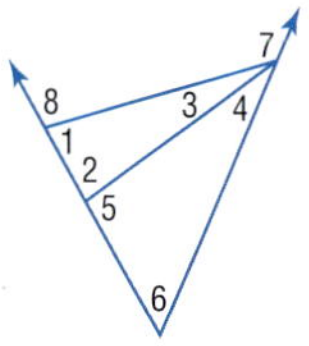

56. all angles with a measure less than $m\angle 8$ $\angle 6$, $\angle 4$, $\angle 2$, $\angle 3$
57. all angles with a measure greater than $m\angle 1$ $\angle 5$, $\angle 7$
58. all angles with a measure less than $m\angle 7$ $\angle 1$, $\angle 6$
59. all angles with a measure greater than $m\angle 6$ $\angle 2$, $\angle 7$, $\angle 8$

Write an equation in slope-intercept form for the line that satisfies the given conditions. *(Lesson 3-4)*

60. $m = 2$, y-intercept $= 4$ $y = 2x + 4$
61. x-intercept is 2, y-intercept $= -8$ $y = 4x - 8$
62. passes through $(2, 6)$ and $(-1, 0)$ $y = 2x + 2$
63. $m = -4$, passes through $(-2, -3)$ $y = -4x - 11$

Getting Ready for the Next Lesson

PREREQUISITE SKILL **Use the Pythagorean Theorem to find the length of the hypotenuse of each right triangle.**
(To review ***using the Pythagorean Theorem****, see Lesson 1-4.)*

64. 5 cm

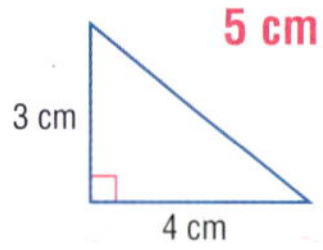

65. 13 ft

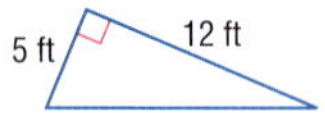

66. $\sqrt{34} \approx 5.8$ in.

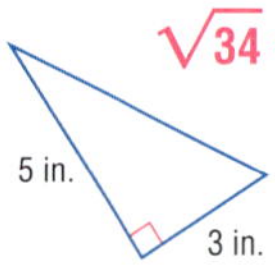

348 Chapter 7 Right Triangles and Trigonometry

Sunshine State Standards
MA.C.1.4.1, MA.B.1.4.1

The Pythagorean Theorem

In Chapter 1, you learned that the Pythagorean Theorem relates the measures of the legs and the hypotenuse of a right triangle. Ancient cultures used the Pythagorean Theorem before it was officially named in 1909.

Use square pieces of patty paper and algebra. Then you too can discover this relationship among the measures of the sides of a right triangle.

Activity

Use paper folding to develop the Pythagorean Theorem.

Step 1 On a piece of patty paper, make a mark along one side so that the two resulting segments are not congruent. Label one as a and the other as b.

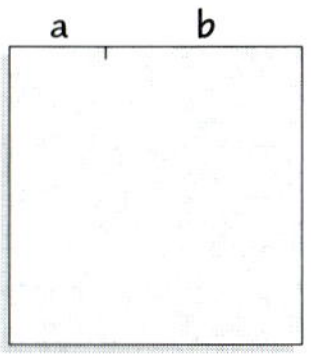

Step 2 Copy these measures on the other sides in the order shown at the right. Fold the paper to divide the square into four sections. Label the area of each section.

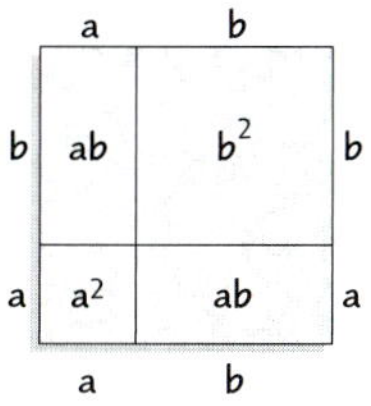

Step 3 On another sheet of patty paper, mark the same lengths a and b on the sides in the different pattern shown at the right.

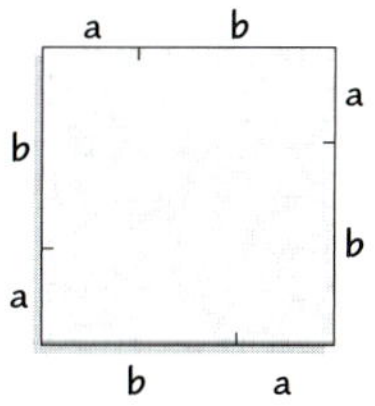

Step 4 Use your straightedge and pencil to connect the marks as shown at the right. Let c represent the length of each hypotenuse.

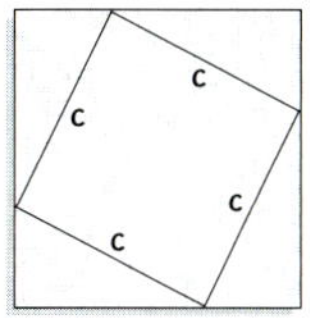

Step 5 Label the area of each section, which is $\frac{1}{2}ab$ for each triangle and c^2 for the square.

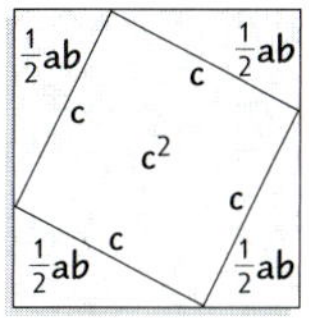

Step 6 Place the squares side by side and color the corresponding regions that have the same area. For example, $ab = \frac{1}{2}ab + \frac{1}{2}ab$.

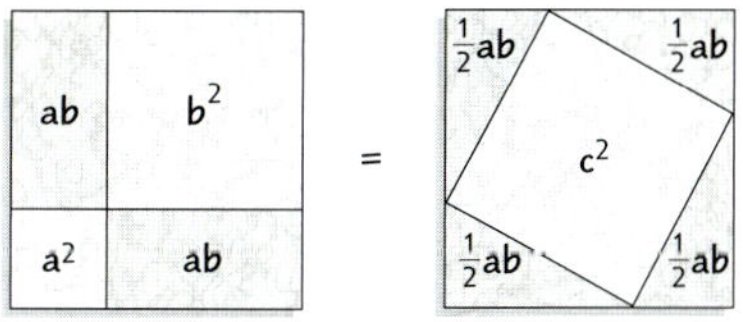

The parts that are not shaded tell us that $a^2 + b^2 = c^2$.

Model

1. Use a ruler to find actual measures for a, b, and c. Do these measures confirm that $a^2 + b^2 = c^2$? **yes**
2. Repeat the activity with different a and b values. What do you notice? **$a^2 + b^2 = c^2$**

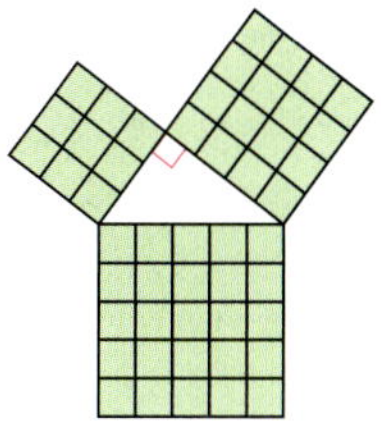

Analyze the model

3. **Explain** why the drawing at the right is an illustration of the Pythagorean Theorem. **Sample answer: The sum of the areas of the two smaller squares is equal to the area of the largest square.**

Resource Manager

Teaching Geometry with Manipulatives
- p. 115 (student recording sheet)
- p. 17 (ruler)

Glencoe Mathematics Classroom Manipulative Kit
- rulers

Geometry Activity

A Preview of Lesson 7-2

Getting Started

Objective Physically explore side-length relationships in the Pythagorean Theorem.

Materials patty paper, ruler, pencil

Teach

- For Step 2, have students use a ruler and a pencil to draw the lines that the creases make.
- Advise students that the key to this activity is making sure measures a and b are exactly the same on both sheets of paper. Without using a ruler, students may use one marked edge of the first paper to mark accurate lengths on the second paper.
- After Step 6, some students may still be skeptical that the shaded areas of the two pieces of patty paper are equal. Point out that the two pieces of paper are the same size, so they have the same area. Then ask students to cut out all the shaded triangles from the second piece of paper and arrange them so that they fit over the shaded areas on the first piece of paper.

Assess

Exercises 1 and 2 provide students with actual measurements they can use to confirm the Pythagorean Theorem. In **Exercise 3**, students prove that three lengths meet the criteria for the Pythagorean Theorem.

Study Notebook

Ask students to summarize what they have learned about the Pythagorean Theorem.

7-2 Lesson Notes

1 Focus

5-Minute Check Transparency 7-2 Use as a quiz or review of Lesson 7-1.

Mathematical Background notes are available for this lesson on p. 340C.

How are right triangles used to build suspension bridges?

Ask students:

- How can you reword the Pythagorean Theorem by substituting the names of the parts of the Talmadge Memorial Bridge for the legs and hypotenuse of the triangle? **The sum of the squares of the measure of the height of the tower from the roadway and length of the roadway from the tower to the end of the bridge equals the square of the measure of the cable.**
- You may notice that most suspension cables have some slack in them. In this case, does the actual length of the cable satisfy the criteria of the Pythagorean Theorem? Explain. **No; a suspension cable is cut to allow for expansion and contraction due to temperature changes, and when it has slack, its length actually exceeds the length required to satisfy the Pythagorean Theorem.**

7-2 The Pythagorean Theorem and Its Converse

Sunshine State Standards
MA.A.3.4.3, MA.B.1.4.1, MA.B.1.4.3, MA.B.2.4.1, MA.C.1.4.1, MA.C.2.4.1, MA.C.3.4.2

Vocabulary
- Pythagorean triple

What You'll Learn

- Use the Pythagorean Theorem.
- Use the converse of the Pythagorean Theorem.

How are right triangles used to build suspension bridges?

The Talmadge Memorial Bridge over the Savannah River has two soaring towers of suspension cables. Note the right triangles being formed by the roadway, the perpendicular tower, and the suspension cables. The Pythagorean Theorem can be used to find measures in any right triangle.

Study Tip

Look Back
To review **finding the hypotenuse of a right triangle,** see Lesson 1-3.

THE PYTHAGOREAN THEOREM In Lesson 1-3, you used the Pythagorean Theorem to find the distance between two points by finding the length of the hypotenuse when given the lengths of the two legs of a right triangle. You can also find the measure of any side of a right triangle given the other two measures.

Theorem 7.4

Pythagorean Theorem In a right triangle, the sum of the squares of the measures of the legs equals the square of the measure of the hypotenuse.

Symbols: $a^2 + b^2 = c^2$

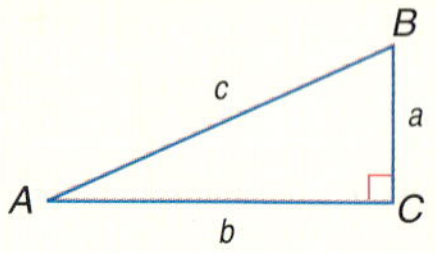

The geometric mean can be used to prove the Pythagorean Theorem.

Proof *Pythagorean Theorem*

Given: $\triangle ABC$ with right angle at C

Prove: $a^2 + b^2 = c^2$

Proof:

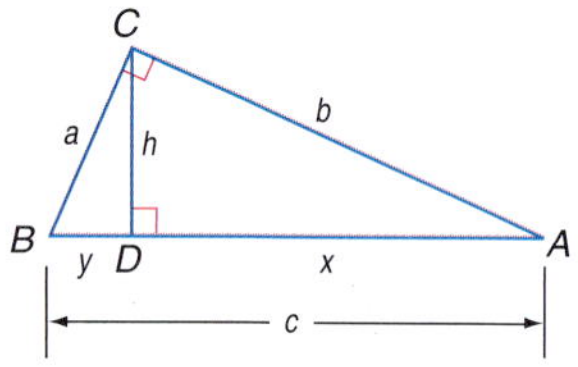

Draw right triangle ABC so C is the right angle. Then draw the altitude from C to $\overline{AB}$. Let $AB = c$, $AC = b$, $BC = a$, $AD = x$, $DB = y$, and $CD = h$.

Two geometric means now exist.

$$\frac{c}{a} = \frac{a}{y} \quad \text{and} \quad \frac{c}{b} = \frac{b}{x}$$

$$a^2 = cy \quad \text{and} \quad b^2 = cx \qquad \text{Cross products}$$

Add the equations.

$$a^2 + b^2 = cy + cx$$

$$a^2 + b^2 = c(y + x) \qquad \text{Factor.}$$

$$a^2 + b^2 = c^2 \qquad \text{Since } c = y + x \text{, substitute } c \text{ for } (y + x).$$

Resource Manager

Workbook and Reproducible Masters

Chapter 7 Resource Masters
- Study Guide and Intervention, pp. 357–358
- Skills Practice, p. 359
- Practice, p. 360
- Reading to Learn Mathematics, p. 361
- Enrichment, p. 362
- Assessment, p. 407

Graphing Calculator and Computer Masters, p. 29
School-to-Career Masters, p. 13
Prerequisite Skills Workbook, pp. 11–12, 31–32
Teaching Geometry With Manipulatives Masters, pp. 114, 116

Transparencies

5-Minute Check Transparency 7-2
Answer Key Transparencies

Technology

GeomPASS: Tutorial Plus, Lesson 13
Interactive Chalkboard

Aore About. . .

Maps

)ue to the curvature)f Earth, the distance etween two points is)ften expressed as *degree listance* using latitude nd longitude. This neasurement closely pproximates the listance on a plane.

ource: NASA

Example 1 Find the Length of the Hypotenuse

LONGITUDE AND LATITUDE **NASA Dryden is located at about 117 degrees longitude and 34 degrees latitude. NASA Ames is located at about 122 degrees longitude and 37 degrees latitude. Use the lines of longitude and latitude to find the degree distance to the nearest tenth between NASA Dryden and NASA Ames.**

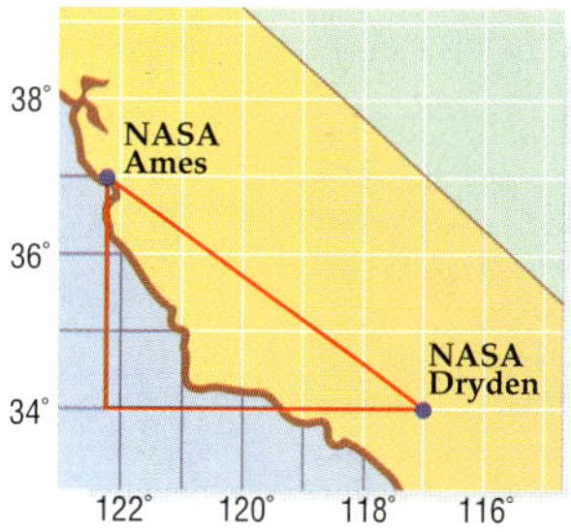

The change in longitude between the two locations is $|117 - 122|$ or 5 degrees. Let this distance be a.

The change in latitude is $|37 - 34|$ or 3 degrees latitude. Let this distance be b.

Use the Pythagorean Theorem to find the distance in degrees from NASA Dryden to NASA Ames, represented by c.

$a^2 + b^2 = c^2$	Pythagorean Theorem
$5^2 + 3^2 = c^2$	$a = 5, b = 3$
$25 + 9 = c^2$	Simplify.
$34 = c^2$	Add.
$\sqrt{34} = c$	Take the square root of each side.
$5.8 \approx c$	Use a calculator.

The degree distance between NASA Dryden and NASA Ames is about 5.8 degrees.

Example 2 Find the Length of a Leg

Find x.

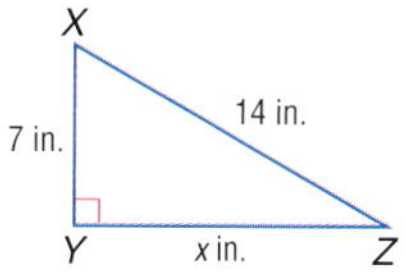

$(XY)^2 + (YZ)^2 = (XZ)^2$	Pythagorean Theorem
$7^2 + x^2 = 14^2$	$XY = 7, XZ = 14$
$49 + x^2 = 196$	Simplify.
$x^2 = 147$	Subtract 49 from each side.
$x = \sqrt{147}$	Take the square root of each side.
$x = 7\sqrt{3}$	Simplify.
$x \approx 12.1$	Use a calculator.

CONVERSE OF THE PYTHAGOREAN THEOREM The converse of the Pythagorean Theorem can help you determine whether three measures of the sides of a triangle are those of a right triangle.

Theorem 7.5

Converse of the Pythagorean Theorem If the sum of the squares of the measures of two sides of a triangle equals the square of the measure of the longest side, then the triangle is a right triangle.

Symbols: If $a^2 + b^2 = c^2$, then $\triangle ABC$ is a right triangle.

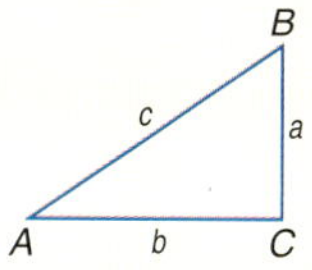

You will prove this theorem in Exercise 38.

www.geometryonline.com/extra_examples/fcat

2 Teach

THE PYTHAGOREAN THEOREM

In-Class Examples

Power Point®

1 **LONGITUDE AND LATITUDE** Carson City, Nevada, is located at about 120 degrees longitude and 39 degrees latitude. Use the lines of longitude and latitude to find the degree distance to the nearest tenth degree if you were to travel directly from NASA Ames to Carson City, Nevada. **about 2.8 degrees**

Teaching Tip Students can also rewrite the Pythagorean Theorem as $b^2 = c^2 - a^2$ or $a^2 = c^2 - b^2$ when they are using it to find the measure of either leg of a right triangle.

2 Find d. **$d \approx 5.2$**

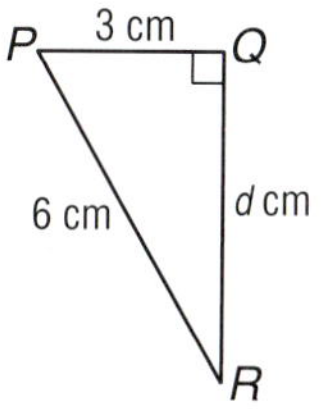

Building on Prior Knowledge

Students used the Distance Formula in Chapter 1 when they were learning how to calculate segment lengths. In this lesson, they use this formula to verify the side lengths of right triangles.

CONVERSE OF THE PYTHAGOREAN THEOREM

In-Class Examples

Power Point®

3 **COORDINATE GEOMETRY** Verify that $\triangle ABC$ is a right triangle.

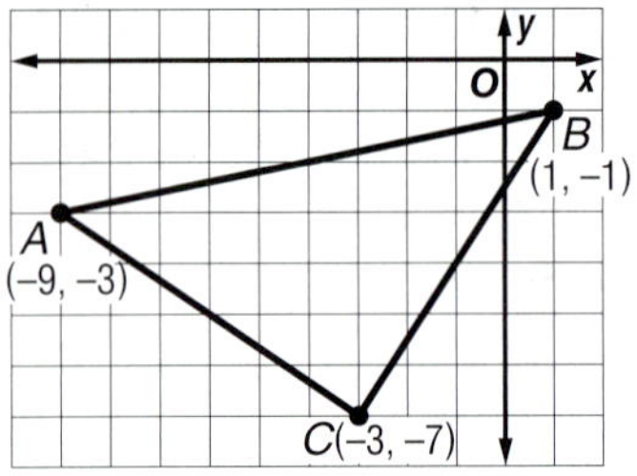

$AB = \sqrt{104}$; $BC = \sqrt{52}$; $AC = \sqrt{52}$; $\triangle ABC$ is a right triangle because $AC^2 + BC^2 = AB^2$.

4 Determine whether each set of measures are the sides of a right triangle. Then state whether they form a Pythagorean triple.

a. 9, 12, and 15
The segments form the sides of a right triangle, and the measures form a Pythagorean triple.

b. 21, 42, and 54
The segments do not form the sides of a right triangle, and the measures do not form a Pythagorean triple.

c. $4\sqrt{3}$, 4, and 8
The segments form the sides of a right triangle, but the measures do not form a Pythagorean triple.

Study Tip

Distance Formula
When using the Distance Formula, be sure to follow the order of operations carefully. Perform the operation inside the parentheses first, square each term, and then add.

Example 3 Verify a Triangle is a Right Triangle

COORDINATE GEOMETRY **Verify that ΔPQR is a right triangle.**

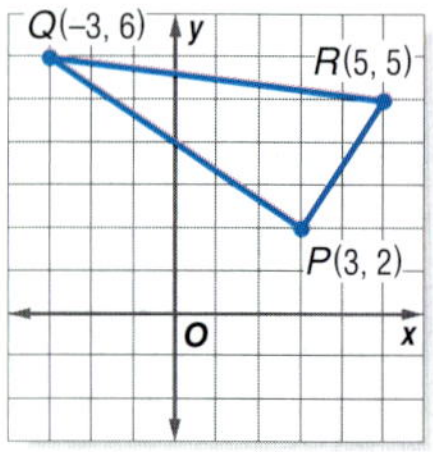

Use the Distance Formula to determine the lengths of the sides.

$PQ = \sqrt{(-3-3)^2 + (6-2)^2}$ $x_1 = 3, y_1 = 2, x_2 = -3, y_2 = 6$
$= \sqrt{(-6)^2 + 4^2}$ Subtract.
$= \sqrt{52}$ Simplify.

$QR = \sqrt{[5-(-3)]^2 + (5-6)^2}$ $x_1 = -3, y_1 = 6, x_2 = 5, y_2 = 5$
$= \sqrt{8^2 + (-1)^2}$ Subtract.
$= \sqrt{65}$ Simplify.

$PR = \sqrt{(5-3)^2 + (5-2)^2}$ $x_1 = 3, y_1 = 2, x_2 = 5, y_2 = 5$
$= \sqrt{2^2 + 3^2}$ Subtract.
$= \sqrt{13}$ Simplify.

By the converse of the Pythagorean Theorem, if the sum of the squares of the measures of two sides of a triangle equals the square of the measure of the longest side, then the triangle is a right triangle.

$PQ^2 + PR^2 = QR^2$ Converse of the Pythagorean Theorem
$(\sqrt{52})^2 + (\sqrt{13})^2 \stackrel{?}{=} (\sqrt{65})^2$ $PQ = \sqrt{52}, PR = \sqrt{13}, QR = \sqrt{65}$
$52 + 13 \stackrel{?}{=} 65$ Simplify.
$65 = 65$ Add.

Since the sum of the squares of two sides equals the square of the longest side, ΔPQR is a right triangle.

A **Pythagorean triple** is three whole numbers that satisfy the equation $a^2 + b^2 = c^2$, where c is the greatest number. One common Pythagorean triple is 3-4-5, in which the sides of a right triangle are in the ratio 3:4:5. If the measures of the sides of any right triangle are whole numbers, the measures form a Pythagorean triple.

Example 4 Pythagorean Triples

Determine whether each set of measures can be the sides of a right triangle. Then state whether they form a Pythagorean triple.

a. 8, 15, 16

Since the measure of the longest side is 16, 16 must be c, and a or b are 15 and 8.

$a^2 + b^2 = c^2$ Pythagorean Theorem
$8^2 + 15^2 \stackrel{?}{=} 16^2$ $a = 8, b = 15, c = 16$
$64 + 225 \stackrel{?}{=} 256$ Simplify.
$289 \neq 256$ Add.

Since $289 \neq 256$, segments with these measures cannot form a right triangle. Therefore, they do not form a Pythagorean triple.

DAILY INTERVENTION **Differentiated Instruction**

Visual/Spatial Explain that many artists use right triangles in their works because they are so appealing to the eye. Right triangles can serve as guidelines to draw mountains in the background or to create vanishing points and perspective. Ask students to construct one or more right triangles on a blank sheet of paper, find the lengths of the sides and then try to compose a picture using the triangle in an image or as a guide for an image. Examples could be a picture of a road vanishing in the distance or a house with right triangles as part of its roof.

b. 20, 48, and 52

$a^2 + b^2 = c^2$	Pythagorean Theorem
$20^2 + 48^2 \stackrel{?}{=} 52^2$	$a = 20, b = 48, c = 52$
$400 + 2304 \stackrel{?}{=} 2704$	Simplify.
$2704 = 2704$	Add.

These segments form the sides of a right triangle since they satisfy the Pythagorean Theorem. The measures are whole numbers and form a Pythagorean triple.

c. $\frac{\sqrt{3}}{5}, \frac{\sqrt{6}}{5}$, **and** $\frac{3}{5}$

$a^2 + b^2 = c^2$	Pythagorean Theorem
$\left(\frac{\sqrt{3}}{5}\right)^2 + \left(\frac{\sqrt{6}}{5}\right)^2 \stackrel{?}{=} \left(\frac{3}{5}\right)^2$	$a = \frac{\sqrt{3}}{5}, b = \frac{\sqrt{6}}{5}, c = \frac{3}{5}$
$\frac{3}{25} + \frac{6}{25} \stackrel{?}{=} \frac{9}{25}$	Simplify.
$\frac{9}{25} = \frac{9}{25}$	Add.

Since $\frac{9}{25} = \frac{9}{25}$, segments with these measures form a right triangle. However, the three numbers are not whole numbers. Therefore, they do not form a Pythagorean triple.

tudy Tip

omparing Numbers

you cannot quickly dentify the greatest umber, use a calculator find decimal values or each number and ompare.

Check for Understanding

Concept Check

1. **FIND THE ERROR** Maria and Colin are determining whether 5-12-13 is a Pythagorean triple.

Colin	Maria
$a^2 + b^2 = c^2$	$a^2 + b^2 = c^2$
$13^2 + 5^2 \stackrel{?}{=} 12^2$	$5^2 + 12^2 \stackrel{?}{=} 13^2$
$169 + 25 \stackrel{?}{=} 144$	$25 + 144 \stackrel{?}{=} 169$
$193 \neq 144$	$169 = 169$
no	yes

Who is correct? Explain your reasoning.

. Maria; Colin does ot have the longest side as the value of c.

2. **Explain** why a Pythagorean triple can represent the measures of the sides of a right triangle.

. Since the numbers n a Pythagorean triple satisfy the equation $^2 + b^2 = c^2$, they epresent the sides of right triangle by the onverse of the ythagorean Theorem.

3. **OPEN ENDED** Draw a pair of similar right triangles. List the corresponding sides, the corresponding angles, and the scale factor. Are the measures of the sides of each triangle a Pythagorean triple? See margin.

Guided Practice

GUIDED PRACTICE KEY	
Exercises	Examples
4, 5	2
6	1
7	3
8–10	4

Find x.

4. 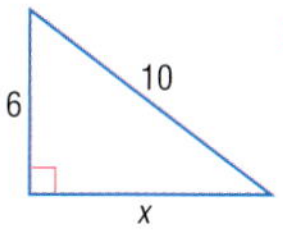

8

5. 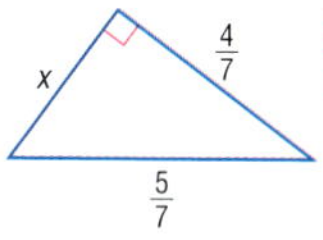

$\frac{3}{7}$

6. 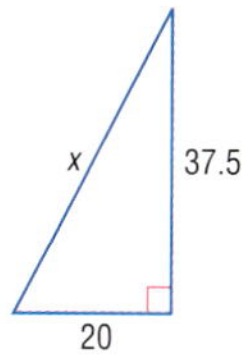

42.5

Answers

3. 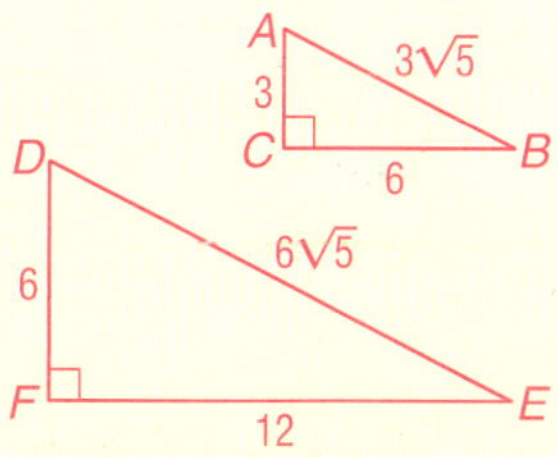

Sample answer: $\triangle ABC \sim \triangle DEF$, $\angle A \cong \angle D$, $\angle B \cong \angle E$, and $\angle C \cong \angle F$, $\overline{AB}$ corresponds to $\overline{DE}$, $\overline{BC}$ corresponds to $\overline{EF}$, and $\overline{AC}$ corresponds to $\overline{DF}$. The scale factor is $\frac{2}{1}$. No; the measures of the sides do not form a Pythagorean triple since $6\sqrt{5}$ and $3\sqrt{5}$ are not whole numbers.

3 Practice/Apply

Study Notebook

Have students—

- add the definitions/examples of the vocabulary terms to their Vocabulary Builder worksheets for Chapter 7.
- include any other item(s) that they find helpful in mastering the skills in this lesson.

DAILY INTERVENTION **FIND THE ERROR**

Point out that in all right triangles, the hypotenuse is always the longest side because it is opposite the 90° angle. Students have to make sure that the value they use to check for c is the largest value given. Also, tell students that the order of a and b does not matter. Maria could have used 12 for a and 5 for b, and she still would have been correct.

About the Exercises...

Organization by Objective

- **The Pythagorean Theorem:** 12–17
- **Converse of the Pythagorean Theorem:** 18–29

Odd/Even Assignments

Exercises 12–29 are structured so that students practice the same concepts whether they are assigned odd or even problems.

Alert! Exercises 48 and 49 require a graphing calculator.

Assignment Guide

Basic: 13–17 odd, 41, 43–70

Average: 13–43 odd, 44–70

Advanced: 12–44 even, 45–60 (optional: 61–70)

Study Guide and Intervention, p. 357 (shown) and p. 358

The Pythagorean Theorem In a right triangle, the sum of the squares of the measures of the legs equals the square of the measure of the hypotenuse.

$\triangle ABC$ is a right triangle, so $a^2 + b^2 = c^2$.

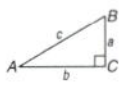

Example 1 **Prove the Pythagorean Theorem.**

With altitude $\overline{CD}$, each leg a and b is a geometric mean between hypotenuse c and the segment of the hypotenuse adjacent to that leg.

$\frac{c}{a} = \frac{a}{y}$ and $\frac{c}{b} = \frac{b}{x}$, so $a^2 = cy$ and $b^2 = cx$.

Add the two equations and substitute $c = y + x$ to get $a^2 + b^2 = cy + cx = c(y + x) = c^2$.

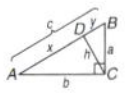

Example 2

a. Find *a*.

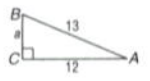

$a^2 + b^2 = c^2$	Pythagorean Theorem
$a^2 + 12^2 = 13^2$	$b = 12$, $c = 13$
$a^2 + 144 = 169$	Simplify.
$a^2 = 25$	Subtract.
$a = 5$	Take the square root of each side.

b. Find *c*.

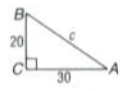

$a^2 + b^2 = c^2$	Pythagorean Theorem
$20^2 + 30^2 = c^2$	$a = 20$, $b = 30$
$400 + 900 = c^2$	Simplify.
$1300 = c^2$	Add.
$\sqrt{1300} = c$	Take the square root of each side.
$36.1 \approx c$	Use a calculator.

Exercises

Find *x*.

1. 

$\sqrt{18} \approx 4.2$

2. 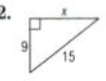

12

3. 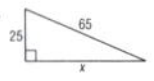

60

4. 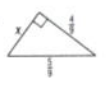

$\frac{3}{10}$

5. 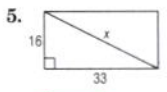

$\sqrt{1345} \approx 36.7$

6. 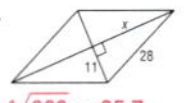

$\sqrt{663} \approx 25.7$

Skills Practice, p. 359 and Practice, p. 360 (shown)

Find *x*.

1. 

$\sqrt{698} \approx 26.4$

2. 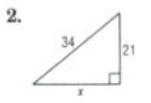

$\sqrt{715} \approx 26.7$

3. 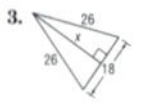

$\sqrt{595} \approx 24.4$

4. 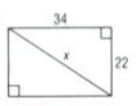

$\sqrt{1640} \approx 40.5$

5. 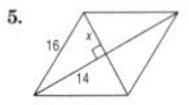

$\sqrt{60} \approx 7.7$

6. 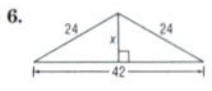

$\sqrt{135} \approx 11.6$

Determine whether $\triangle GHI$ is a right triangle for the given vertices. Explain.

7. $G(2, 7)$, $H(3, 6)$, $I(-4, -1)$
yes; $GH = \sqrt{2}$, $HI = \sqrt{98}$, $IG = \sqrt{100}$, $(\sqrt{2})^2 + (\sqrt{98})^2 = (\sqrt{100})^2$

8. $G(-6, 2)$, $H(1, 12)$, $I(-2, 1)$
no; $GH = \sqrt{149}$, $HI = \sqrt{130}$, $IG = \sqrt{17}$, $(\sqrt{130})^2 + (\sqrt{17})^2 \neq (\sqrt{149})^2$

9. $G(-2, 1)$, $H(3, -1)$, $I(-4, -4)$
yes; $GH = \sqrt{29}$, $HI = \sqrt{58}$, $IG = \sqrt{29}$, $(\sqrt{29})^2 + (\sqrt{29})^2 = (\sqrt{58})^2$

10. $G(-2, 4)$, $H(4, 1)$, $I(-1, -9)$
yes; $GH = \sqrt{45}$, $HI = \sqrt{125}$, $IG = \sqrt{170}$, $(\sqrt{45})^2 + (\sqrt{125})^2 = (\sqrt{170})^2$

Determine whether each set of measures can be the measures of the sides of a right triangle. Then state whether they form a Pythagorean triple.

11. 9, 40, 41 yes, yes
12. 7, 28, 29 no, no
13. 24, 32, 40 yes, yes
14. $\frac{9}{5}, \frac{12}{5}, 3$ yes, no
15. $\frac{1}{3}, \frac{2\sqrt{2}}{3}, 1$ yes, no
16. $\frac{\sqrt{4}}{7}, \frac{2\sqrt{3}}{7}, \frac{4}{7}$ yes, no

17. **CONSTRUCTION** The bottom end of a ramp at a warehouse is 10 feet from the base of the main dock and is 11 feet long. How high is the dock? about 4.6 ft high

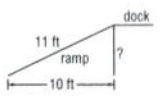

Reading to Learn Mathematics, p. 361 ELL

Pre-Activity **How are right triangles used to build suspension bridges?**

Read the introduction to Lesson 7-2 at the top of page 350 in your textbook.

Do the two right triangles shown in the drawing appear to be similar? Explain your reasoning. Sample answer: No; their sides are not proportional. In the smaller triangle, the longer leg is more than twice the length of the shorter leg, while in the larger triangle, the longer leg is less than twice the length of the shorter leg.

Reading the Lesson

1. Explain in your own words the difference between how the Pythagorean Theorem is used and how the Converse of the Pythagorean Theorem is used. Sample answer: The Pythagorean Theorem is used to find the third side of a right triangle if you know the lengths of any two of the sides. The converse is used to tell whether a triangle with three given side lengths is a right triangle.

2. Refer to the figure. For this figure, which statements are true? B, E, F, G

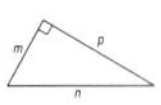

A. $m^2 + n^2 = p^2$
B. $n^2 = m^2 + p^2$
C. $m^2 = n^2 + p^2$
D. $m^2 = p^2 - n^2$
E. $p^2 = n^2 - m^2$
F. $n^2 - p^2 = m^2$
G. $n = \sqrt{m^2 + p^2}$
H. $p = \sqrt{m^2 - n^2}$

3. Is the following statement true or false?
A Pythagorean triple is any group of three numbers for which the sum of the squares of the smaller two numbers is equal to the square of the largest number. Explain your reasoning.
Sample answer: The statement is false because in a Pythagorean triple, all three numbers must be *whole* numbers.

4. If x, y, and z form a Pythagorean triple and k is a positive integer, which of the following groups of numbers are also Pythagorean triples? B, D
A. $3x, 4y, 5z$ B. $3x, 3y, 3z$ C. $-3x, -3y, -3z$ D. kx, ky, kz

Helping You Remember

5. Many students who studied geometry long ago remember the Pythagorean Theorem as the equation $a^2 + b^2 = c^2$, but cannot tell you what this equation means. A formula is useless if you don't know what it means and how to use it. How could you help someone who has forgotten the Pythagorean Theorem remember the meaning of the equation $a^2 + b^2 = c^2$?
Sample answer: Draw a right triangle. Label the lengths of the two legs as *a* and *b* and the length of the hypotenuse as *c*.

7. **COORDINATE GEOMETRY** Determine whether $\triangle JKL$ with vertices $J(-2, 2)$, $K(-1, 6)$, and $L(3, 5)$ is a right triangle. Explain.
yes; $JK = \sqrt{17}$, $KL = \sqrt{17}$, $JL = \sqrt{34}$; $(\sqrt{17})^2 + (\sqrt{17})^2 = (\sqrt{34})^2$

Determine whether each set of numbers can be the measures of the sides of a right triangle. Then state whether they form a Pythagorean triple.

8. 15, 36, 39 yes, yes
9. $\sqrt{40}$, 20, 21 no, no
10. $\sqrt{44}$, 8, $\sqrt{108}$ yes, no

Application

11. **COMPUTERS** Computer monitors are usually measured along the diagonal of the screen. A 19-inch monitor has a diagonal that measures 19 inches. If the height of the screen is 11.5 inches, how wide is the screen?
about 15.1 in.

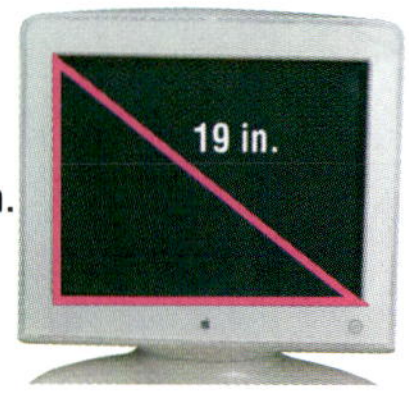

★ indicates increased difficulty

Practice and Apply

Homework Help

For Exercises	See Examples
14, 15	1
12, 13, 16, 17	2
18–21	3
22–29	4

Extra Practice
See page 767.

Find *x*. 12. $\sqrt{15} \approx 3.9$ 13. $4\sqrt{3} \approx 6.9$

12. 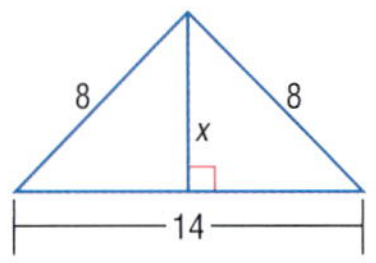

13. 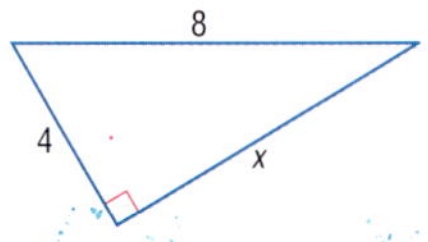

14. $4\sqrt{74} \approx 34.4$
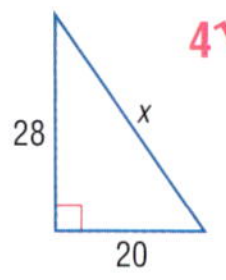

15. 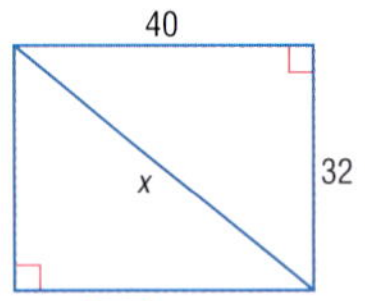

$8\sqrt{41} \approx 51.2$

16. 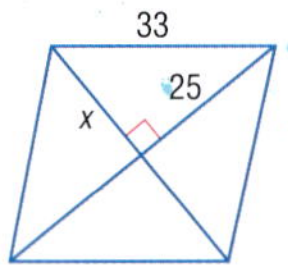

$4\sqrt{29} \approx 21.5$

17. 20
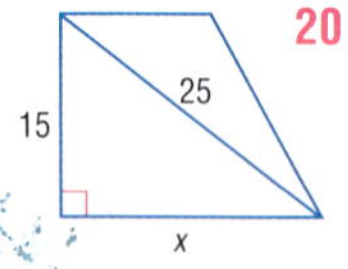

COORDINATE GEOMETRY **Determine whether $\triangle QRS$ is a right triangle for the given vertices. Explain.** 18–21. See margin for explanations.

18. $Q(1, 0)$, $R(1, 6)$, $S(9, 0)$ yes
19. $Q(3, 2)$, $R(0, 6)$, $S(6, 6)$ no
20. $Q(-4, 6)$, $R(2, 11)$, $S(4, -1)$ no
21. $Q(-9, -2)$, $R(-4, -4)$, $S(-6, -9)$ yes

Determine whether each set of numbers can be the measures of the sides of a right triangle. Then state whether they form a Pythagorean triple.

22. 8, 15, 17
23. 7, 24, 25
24. 20, 21, 31
25. 37, 12, 34
26. $\frac{1}{5}, \frac{1}{7}, \frac{\sqrt{74}}{35}$ yes, no
27. $\frac{\sqrt{3}}{2}, \frac{\sqrt{2}}{3}, \frac{35}{36}$ no, no
28. $\frac{3}{5}, \frac{4}{5}, 1$ yes, no
29. $\frac{6}{7}, \frac{8}{7}, \frac{10}{7}$ yes, no

22. yes, yes
23. yes, yes
24. no, no
25. no, no
31. 5-12-13
32. Sample answer: The triples are all multiples of the triple 5-12-13.
33. Sample answer: They consist of any number of similar triangles.
34. Yes; the measures of the sides are always multiples of 5, 12, and 13.
35a. 16-30-34; 24-45-51
35b. 18-80-82; 27-120-123
35c. 14-48-50; 21-72-75

For Exercises 30–35, use the table of Pythagorean triples.

a	b	c
5	12	13
10	24	? 26
15	? 36	39
? 20	48	52

30. Copy and complete the table.
31. A *primitive* Pythagorean triple is a Pythagorean triple with no common factors except 1. Name any primitive Pythagorean triples contained in the table.
32. Describe the pattern that relates these sets of Pythagorean triples.
33. These Pythagorean triples are called a *family*. Why do you think this is?
34. Are the triangles described by a family of Pythagorean triples similar? Explain.
35. For each Pythagorean triple, find two triples in the same family.
a. 8, 15, 17 b. 9, 40, 41 c. 7, 24, 25

Enrichment, p. 362

Converse of a Right Triangle Theorem

You have learned that the measure of the altitude from the vertex of the right angle of a right triangle to its hypotenuse is the geometric mean between the measures of the two segments of the hypotenuse. Is the converse of this theorem true? In order to find out, it will help to rewrite the original theorem in if-then form as follows.

If $\triangle ABQ$ is a right triangle with right angle at Q, then QP is the geometric mean between AP and PB, where P is between A and B and $\overline{QP}$ is perpendicular to $\overline{AB}$.

1. Write the converse of the if-then form of the theorem.

354 Chapter 7 Right Triangles and Trigonometry

GEOGRAPHY **For Exercises 36 and 37, use the following information.**

Denver is located at about 105 degrees longitude and 40 degrees latitude. San Francisco is located at about 122 degrees longitude and 38 degrees latitude. Las Vegas is located at about 115 degrees longitude and 36 degrees latitude. Using the lines of longitude and latitude, find each degree distance.

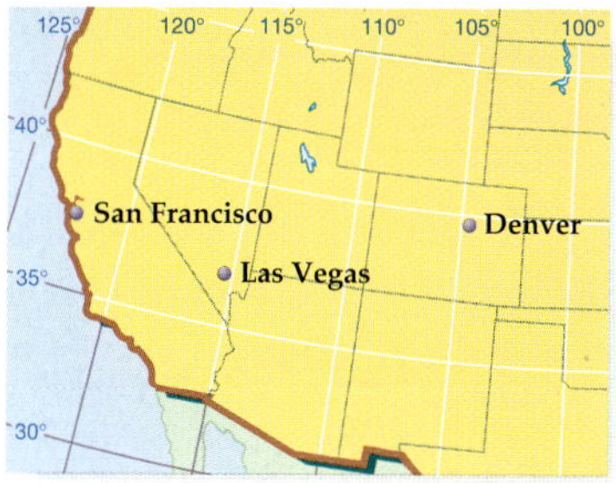

36. San Francisco to Denver ≈ **17.1 degrees**

37. Las Vegas to Denver ≈ **10.8 degrees**

38. **PROOF** Write a paragraph proof of Theorem 7.5. **See margin.**

★ 39. **PROOF** Use the Pythagorean Theorem and the figure at the right to prove the Distance Formula. **See margin.**

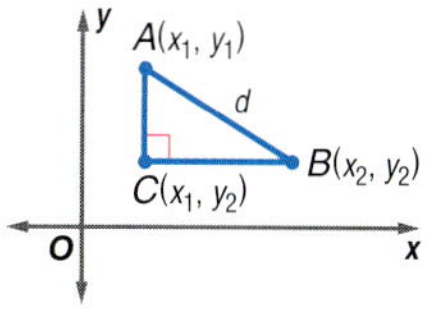

40. **PAINTING** A painter sets a ladder up to reach the bottom of a second-story window 16 feet above the ground. The base of the ladder is 12 feet from the house. While the painter mixes the paint, a neighbor's dog bumps the ladder, which moves the base 2 feet farther away from the house. How far up the side of the house does the ladder reach? **$2\sqrt{51}$ ft ≈ 14.3 ft**

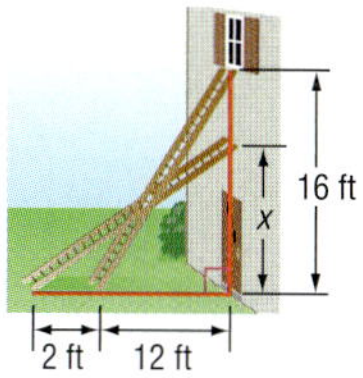

Career Choices

Military

All branches of the military use navigation. Some of the jobs using navigation include radar/sonar operators, boat operators, airplane navigators, and space operations officers.

Online Research

For information about a career in the military, visit: www.geometryonline.com/careers

41. **SAILING** The mast of a sailboat is supported by wires called *shrouds.* What is the total length of wire needed to form these shrouds? **about 76.53 ft**

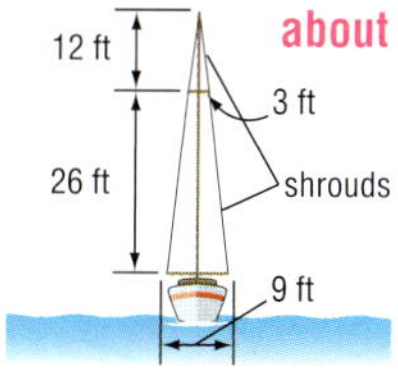

★ 42. **LANDSCAPING** Six congruent square stones are arranged in an L-shaped walkway through a garden. If $x = 15$ inches, then find the area of the L-shaped walkway. **270 in^2**

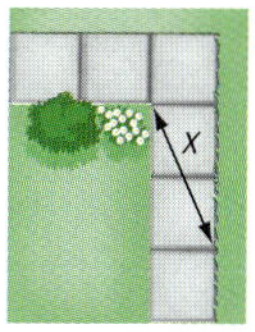

43. **NAVIGATION** A fishing trawler off the coast of Alaska was ordered by the U.S. Coast Guard to change course. They were to travel 6 miles west and then sail 12 miles south to miss a large iceberg before continuing on the original course. How many miles out of the way did the trawler travel? **about 4.6 mi**

44. **CRITICAL THINKING** The figure at the right is a rectangular prism with $AB = 8$, $BC = 6$, and $BF = 8$, and M is the midpoint of BD. Find BD and HM. How are EM, FM, and GM related to HM? **$BD = 10$, $HM = \sqrt{89} \approx 9.4$; $HM = EM = FM = GM$**

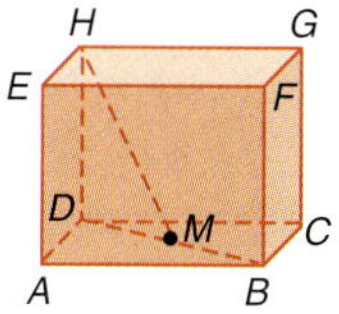

Answers

38. Given: $\triangle ABC$ with sides of measure a, b, and c, where $c^2 = a^2 + b^2$

Prove: $\triangle ABC$ is a right triangle.

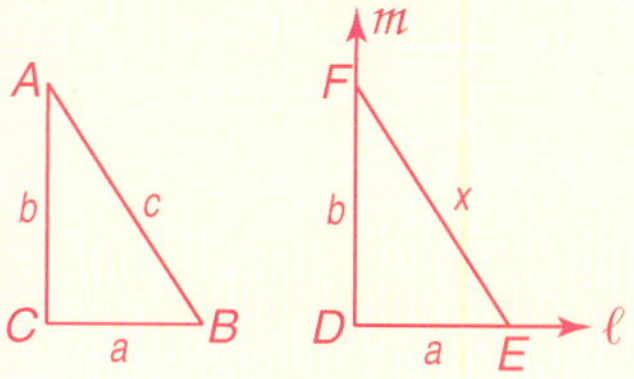

Proof: Draw $\overline{DE}$ on line ℓ with measure equal to a. At D, draw line $m \perp \overline{DE}$. Locate point F on m so that $DF = b$. Draw $\overline{FE}$ and call its measure x. Because $\triangle FED$ is a right triangle, $a^2 + b^2 = x^2$. But $a^2 + b^2 = c^2$, so $x^2 = c^2$ or $x = c$. Thus, $\triangle ABC \cong \triangle FED$ by SSS. This means $\angle C \cong \angle D$. Therefore, $\angle C$ must be a right angle, making $\triangle ABC$ a right triangle.

39. Given: $\triangle ABC$ with right angle at C, $AB = d$

Prove: $d = \sqrt{(x_2 - x_1)^2 + (y_2 - y_1)^2}$

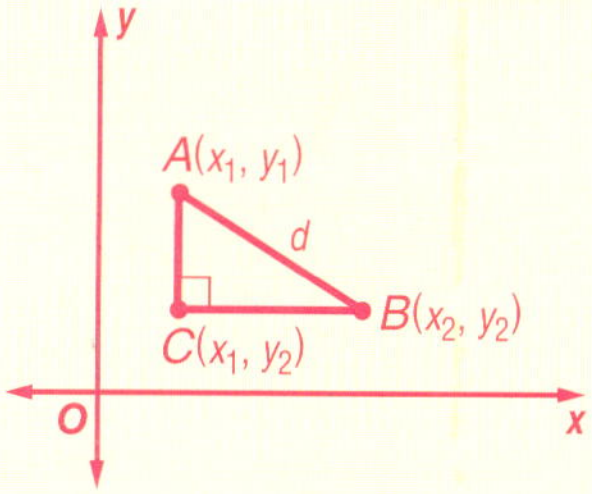

Proof:

Statements (Reasons)

1. $\triangle ABC$ with right angle at C, $AB = d$ (Given)
2. $(CB)^2 + (AC)^2 = (AB)^2$ (Pythagorean Theorem)
3. $|x_2 - x_1| = CB$; $|y_2 - y_1| = AC$ (Distance on a number line)
4. $|x_2 - x_1|^2 + |y_2 - y_1|^2 = d^2$ (Substitution)
5. $(x_2 - x_1)^2 + (y_2 - y_1)^2 = d^2$ (Substitution)
6. $\sqrt{(x_2 - x_1)^2 + (y_2 - y_1)^2} = d$ (Take the square root of each side.)
7. $d = \sqrt{(x_2 - x_1)^2 + (y_2 - y_1)^2}$ (Reflexive Property)

Answers (p. 354)

18. $QR = 6$, $RS = 10$, $QS = 8$; $6^2 + 8^2 = 10^2$
19. $QR = 5$, $RS = 6$, $QS = 5$; $5^2 + 5^2 \neq 6^2$
20. $QR = \sqrt{61}$, $RS = \sqrt{148}$, $QS = \sqrt{113}$; $(\sqrt{61})^2 + (\sqrt{113})^2 \neq (\sqrt{148})^2$
21. $QR = \sqrt{29}$, $RS = \sqrt{29}$, $QS = \sqrt{58}$; $(\sqrt{29})^2 + (\sqrt{29})^2 = (\sqrt{58})^2$

4 Assess

Open-Ended Assessment

Modeling Ask students to recall how right triangles are modeled in suspension bridges, and have them demonstrate how a right triangle could model real world objects such as sailboats, flags, houses, bookshelves, and so on.

Getting Ready for Lesson 7-3

Prerequisite Skill In Lesson 7-3, students will be working more extensively with radical expressions to find side lengths of special right triangles. Use Exercises 61–70 to determine your students' familiarity with simplifying radical expressions.

Assessment Options

Quiz (Lessons 7-1 and 7-2) is available on p. 407 of the *Chapter 7 Resource Masters*.

Answers

45. Sample answer: The road, the tower that is perpendicular to the road, and the cables form the right triangles. Answers should include the following.
   - Right triangles are formed by the bridge, the towers, and the cables.
   - The cable is the hypotenuse in each triangle.

---

45. **WRITING IN MATH** Answer the question that was posed at the beginning of the lesson. See margin.

How are right triangles used to build suspension bridges?

Include the following in your answer:
- the locations of the right triangles, and
- an explanation of which parts of the right triangle are formed by the cables.

FCAT Practice

Standardized Test Practice

46. In the figure, if $AE = 10$, what is the value of h? **A**

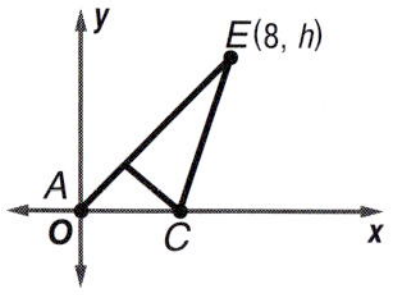

Ⓐ 6 Ⓑ 8
Ⓒ 10 Ⓓ 12

47. **ALGEBRA** If $x^2 + 36 = (9 - x)^2$, then find x. **C**

Ⓐ 6 Ⓑ no solution
Ⓒ 2.5 Ⓓ 10

Graphing Calculator

PROGRAMMING For Exercises 48 and 49, use the following information.

The TI-83 Plus program uses a procedure for finding *Pythagorean triples* that was developed by Euclid around 320 B.C. Run the program to generate a list of Pythagorean triples.

```
PROGRAM: PYTHTRIP
:For (X, 2, 6)
:For (Y, 1, 5)
:If X > Y
:Then
:int (X²−Y² + 0.5)→A
:2XY→B
:int (X²+Y²+0.5)→C
:If A > B
:Then
:Disp B,A,C
:Else
:Disp A,B,C
:End
:End
:Pause
:Disp " "
:End
:End
:Stop
```

48. List all the members of the 3-4-5 family that are generated by the program. **3-4-5, 6-8-10, 12-16-20, 24-32-40, 27-36-45**

49. A geometry student made the conjecture that if three whole numbers are a Pythagorean triple, then their product is divisible by 60. Does this conjecture hold true for each triple that is produced by the program? **yes**

Maintain Your Skills

Mixed Review

Find the geometric mean between each pair of numbers. *(Lesson 7-1)*

50. 3 and 12 **6**
51. 9 and 12 **$6\sqrt{3} \approx 10.4$**
52. 11 and 7 **$\sqrt{77} \approx 8.8$**
53. 6 and 9 **$3\sqrt{6} \approx 7.3$**
54. 2 and 7 **$\sqrt{14} \approx 3.7$**
55. 2 and 5 **$\sqrt{10} \approx 3.2$**

Find the value of each expression. Then use that value as the next x in the expression. Repeat the process and describe your observations. *(Lesson 6-6)*

56. $\sqrt{2x}$, where x initially equals 5
57. 3^x, where x initially equals 1
58. $x^{\frac{1}{2}}$, where x initially equals 4
59. $\frac{1}{x}$, where x initially equals 4

56. $\sqrt{10}$; it converges to 2.
57. 3; it approaches positive infinity.
58. 2; it converges to 1.
59. 0.25; it alternates between 0.25 and 4.

60. Determine whether the sides of a triangle could have the lengths 12, 13, and 25. Explain. *(Lesson 5-4)* **no; $12 + 13 \not> 25$**

Getting Ready for the Next Lesson

PREREQUISITE SKILL Simplify each expression by rationalizing the denominator.
*(To review **simplifying radical expressions**, see pages 744 and 745.)*

61. $\frac{7}{\sqrt{3}}$ **$\frac{7\sqrt{3}}{3}$**
62. $\frac{18}{\sqrt{2}}$ **$9\sqrt{2}$**
63. $\frac{\sqrt{14}}{\sqrt{2}}$ **$\sqrt{7}$**
64. $\frac{3\sqrt{11}}{\sqrt{3}}$ **$\sqrt{33}$**
65. $\frac{24}{\sqrt{2}}$ **$12\sqrt{2}$**
66. $\frac{12}{\sqrt{3}}$ **$4\sqrt{3}$**
67. $\frac{2\sqrt{6}}{\sqrt{3}}$ **$2\sqrt{2}$**
68. $\frac{15}{\sqrt{3}}$ **$5\sqrt{3}$**
69. $\frac{2}{\sqrt{8}}$ **$\frac{\sqrt{2}}{2}$**
70. $\frac{25}{\sqrt{10}}$ **$\frac{5\sqrt{10}}{2}$**

a must!

356 Chapter 7 Right Triangles and Trigonometry

7-3 Special Right Triangles

Sunshine State Standards
MA.A.3.4.3, MA.B.1.4.2, MA.B.2.4.1, MA.C.1.4.1, MA.C.3.4.1, MA.C.3.4.2

What You'll Learn

- Use properties of 45°-45°-90° triangles.
- Use properties of 30°-60°-90° triangles.

How is triangle tiling used in wallpaper design?

Triangle tiling is the process of laying copies of a single triangle next to each other to fill an area. One type of triangle tiling is *wallpaper tiling*. There are exactly 17 types of triangle tiles that can be used for wallpaper tiling.

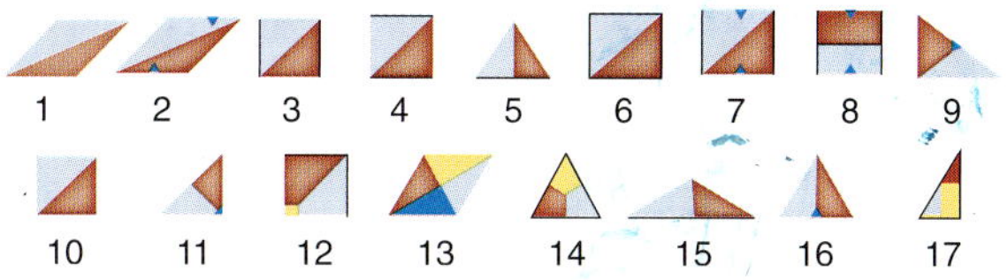

Tile 4 is made up of two 45°-45°-90° triangles that form a square. This tile is rotated to make the wallpaper design shown at the right

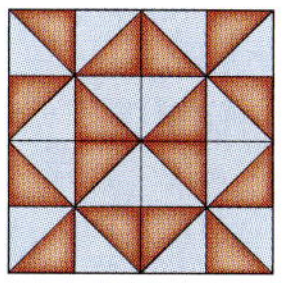

PROPERTIES OF 45°-45°-90° TRIANGLES Facts about 45°-45°-90° triangles are used to solve many geometry problems. The Pythagorean Theorem allows us to discover special relationships that exist among the sides of a 45°-45°-90° triangle.

Draw a diagonal of a square. The two triangles formed are isosceles right triangles. Let x represent the measure of each side and let d represent the measure of the hypotenuse.

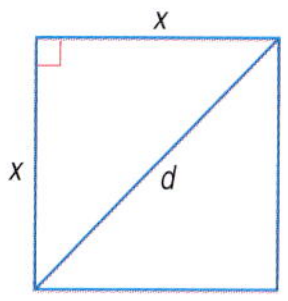

$d^2 = x^2 + x^2$	Pythagorean Theorem
$d^2 = 2x^2$	Add.
$d = \sqrt{2x^2}$	Take the positive square root of each side.
$d = \sqrt{2} \cdot \sqrt{x^2}$	Factor.
$d = x\sqrt{2}$	Simplify.

This algebraic proof verifies that the length of the hypotenuse of any 45°-45°-90° triangle is $\sqrt{2}$ times the length of its leg. The ratio of the sides is $1 : 1 : \sqrt{2}$.

Theorem 7.6

In a 45°-45°-90° triangle, the length of the hypotenuse is $\sqrt{2}$ times the length of a leg.

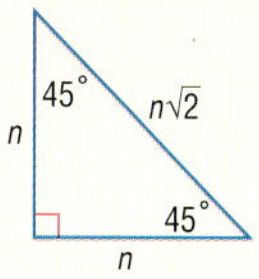

7-3 Lesson Notes

1 Focus

5-Minute Check Transparency 7-3 Use as a quiz or review of Lesson 7-2.

Mathematical Background notes are available for this lesson on p. 340C.

How is triangle tiling used in wallpaper design?

Ask students:

- How many of the wallpaper tiles are made up of right triangles? Which ones are they? **12 tiles; tiles 3, 4, 5, 6, 7, 10, 11, 12, 13, 15, 16, and 17**
- What percent of wallpaper tiling depends on right triangles? **12 out of 17 or about 71%**
- If the length of the hypotenuse of one of the triangles in tile 12 is $7\sqrt{2}$ centimeters, then how long is each leg of the triangle? **7 cm**

Resource Manager

Workbook and Reproducible Masters

Chapter 7 Resource Masters
- Study Guide and Intervention, pp. 363–364
- Skills Practice, p. 365
- Practice, p. 366
- Reading to Learn Mathematics, p. 367
- Enrichment, p. 368

Prerequisite Skills Workbook, pp. 17–18
Teaching Geometry With Manipulatives Masters, p. 1

Transparencies

5-Minute Check Transparency 7-3
Answer Key Transparencies

Technology

GeomPASS: Tutorial Plus, Lesson 14
Interactive Chalkboard

2 Teach

PROPERTIES OF 45°-45°-90° TRIANGLES

In-Class Examples Power Point®

1 **WALLPAPER TILING** Note that the figure in Example 1 can be divided into four equal square quadrants so that each square contains 8 triangles. What is the area of one of these squares if the hypotenuse of each 45°-45°-90° triangle measures $7\sqrt{2}$ millimeters? **196 mm²**

2 Find *a*.

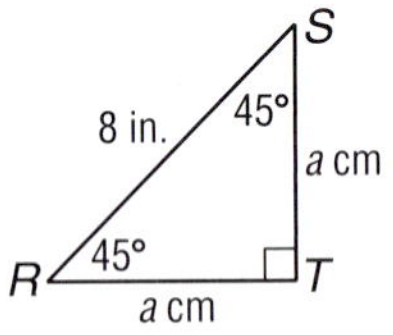

$a = 4\sqrt{2}$

Teaching Tip Point out to students that one corner of the wallpaper square consists of four 45°-45°-90° triangles, the combination of which forms a new larger 45°-45°-90° triangle. Explain how it is interesting that tiled groups of 4, 8, and 16 of the smaller triangles form larger 45°-45°-90° triangles.

Tips for New Teachers

Intervention Some students may have difficulty keeping the properties of 45°-45°-90° and 30°-60°-90° triangles straight. Suggest that they periodically reproduce the figures in Theorem 7.6 and Theorem 7.7 to keep these relationships fresh. Advise them, however, that they have the intuitive knowledge to derive either of these two figures by using the Pythagorean Theorem with a square or an equilateral triangle.

Example 1 Find the Measure of the Hypotenuse

WALLPAPER TILING Assume that the length of one of the legs of the 45°-45°-90° triangles in the wallpaper in the figure is 4 inches. What is the length of the diagonal of the entire wallpaper square?

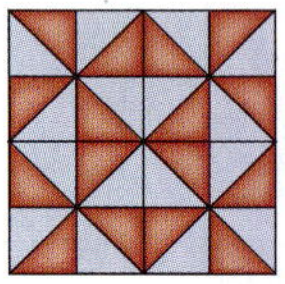

The length of each leg of the 45°-45°-90° triangle is 4 inches. The length of the hypotenuse is $\sqrt{2}$ times as long as a leg. The length of the hypotenuse of one of the triangles is $4\sqrt{2}$. There are four 45°-45°-90° triangles along the diagonal of the square. So, the length of the diagonal of the square is $4(4\sqrt{2})$ or $16\sqrt{2}$ inches.

Example 2 Find the Measure of the Legs

Find *x*.

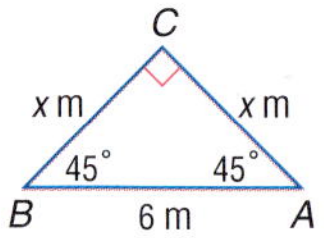

The length of the hypotenuse of a 45°-45°-90° triangle is $\sqrt{2}$ times the length of a leg of the triangle.

$AB = (AC)\sqrt{2}$

$6 = x\sqrt{2}$ — $AB = 6$, $AC = x$

$\frac{6}{\sqrt{2}} = x$ — Divide each side by $\sqrt{2}$.

$\frac{6}{\sqrt{2}} \cdot \frac{\sqrt{2}}{\sqrt{2}} = x$ — Rationalize the denominator.

$\frac{6\sqrt{2}}{2} = x$ — Multiply.

$3\sqrt{2} = x$ — Divide.

Study Tip

Rationalizing Denominators
To rationalize a denominator, multiply the fraction by 1 in the form of a radical over itself so that the product in the denominator is a rational number.

PROPERTIES OF 30°-60°-90° TRIANGLES There is also a special relationship among the measures of the sides of a 30°-60°-90° triangle.

When an altitude is drawn from any vertex of an equilateral triangle, two congruent 30°-60°-90° triangles are formed. $\overline{LM}$ and $\overline{KM}$ are congruent segments, so let $LM = x$ and $KM = x$. By the Segment Addition Postulate, $LM + KM = KL$. Thus, $KL = 2x$. Since $\triangle JKL$ is an equilateral triangle, $KL = JL = JK$. Therefore, $JL = 2x$ and $JK = 2x$.

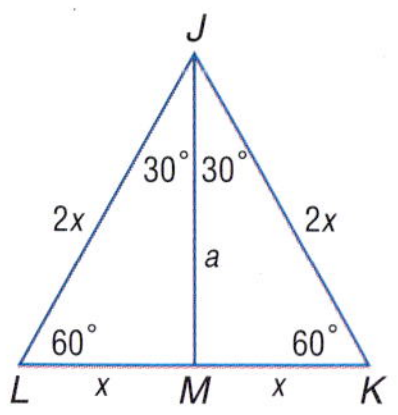

Let *a* represent the measure of the altitude. Use the Pythagorean Theorem to find *a*.

$(JM)^2 + (LM)^2 = (JL)^2$ — Pythagorean Theorem

$a^2 + x^2 = (2x)^2$ — $JM = a$, $LM = x$, $JL = 2x$

$a^2 + x^2 = 4x^2$ — Simplify.

$a^2 = 3x^2$ — Subtract x^2 from each side.

$a = \sqrt{3x^2}$ — Take the positive square root of each side.

$a = \sqrt{3} \cdot \sqrt{x^2}$ — Factor.

$a = x\sqrt{3}$ — Simplify.

So, in a 30°-60°-90° triangle, the measures of the sides are x, $x\sqrt{3}$, and $2x$. The ratio of the sides is $1:\sqrt{3}:2$.

DAILY INTERVENTION — **Differentiated Instruction**

Logical/Mathematical Suggest that students close their books and divide a piece of paper into two columns. At the top of one column, ask them to draw a square with diagonal *d* and side *x*. At the top of the other column, ask students to draw an equilateral triangle with one altitude and tell them to label the segment that the altitude divides with two *x*'s. Then have students systematically use the Pythagorean Theorem to figure out the side relationships of the 45°-45°-90° and 30°-60°-90° triangles in these two figures.

The relationship of the side measures leads to Theorem 7.7.

Theorem 7.7

In a 30°-60°-90° triangle, the length of the hypotenuse is twice the length of the shorter leg, and the length of the longer leg is $\sqrt{3}$ times the length of the shorter leg.

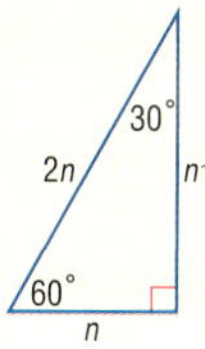

Study Tip

30°-60°-90° Triangle

The shorter leg is opposite the 30° angle, and the longer leg is opposite the 60° angle.

Example 3 30°-60°-90° Triangles

Find *AC*.

$\overline{AC}$ is the longer leg, $\overline{AB}$ is the shorter leg, and $\overline{BC}$ is the hypotenuse.

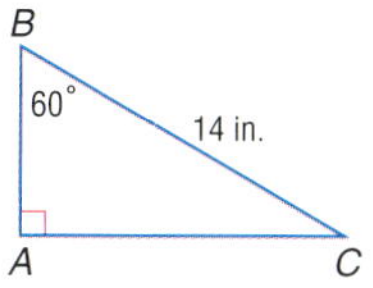

$AB = \frac{1}{2}(BC)$

$= \frac{1}{2}(14)$ or 7 $BC = 14$

$AC = \sqrt{3}(AB)$

$= \sqrt{3}(7)$ or $7\sqrt{3}$ $AB = 7$

Example 4 Special Triangles in a Coordinate Plane

COORDINATE GEOMETRY **Triangle *PCD* is a 30°-60°-90° triangle with right angle *C*. $\overline{CD}$ is the longer leg with endpoints *C*(3, 2) and *D*(9, 2). Locate point *P* in Quadrant I.**

Graph *C* and *D*. $\overline{CD}$ lies on a horizontal gridline of the coordinate plane. Since $\overline{PC}$ will be perpendicular to $\overline{CD}$, it lies on a vertical gridline. Find the length of $\overline{CD}$.

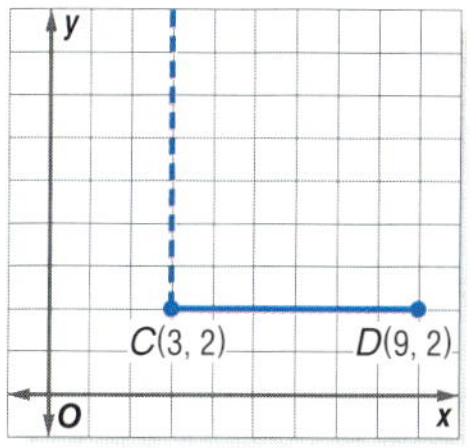

$CD = |9 - 3| = 6$

$\overline{CD}$ is the longer leg. $\overline{PC}$ is the shorter leg.

So, $CD = \sqrt{3}(PC)$. Use *CD* to find *PC*.

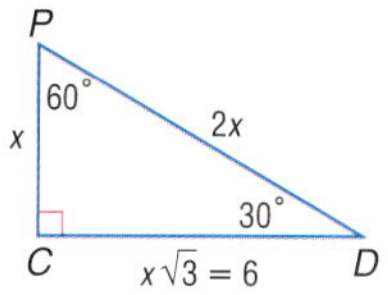

$CD = \sqrt{3}(PC)$

$6 = \sqrt{3}(PC)$ $CD = 6$

$\frac{6}{\sqrt{3}} = PC$ Divide each side by $\sqrt{3}$.

$\frac{6}{\sqrt{3}} \cdot \frac{\sqrt{3}}{\sqrt{3}} = PC$ Rationalize the denominator.

$\frac{6\sqrt{3}}{3} = PC$ Multiply.

$2\sqrt{3} = PC$ Simplify.

Point *P* has the same *x*-coordinate as *C*. *P* is located $2\sqrt{3}$ units above *C*. So, the coordinates of *P* are $(3, 2 + 2\sqrt{3})$ or about (3, 5.46).

www.geometryonline.com/extra_examples/fcat

PROPERTIES OF 30°-60°-90° TRIANGLES

In-Class Examples

Power Point®

3 Find *QR*.

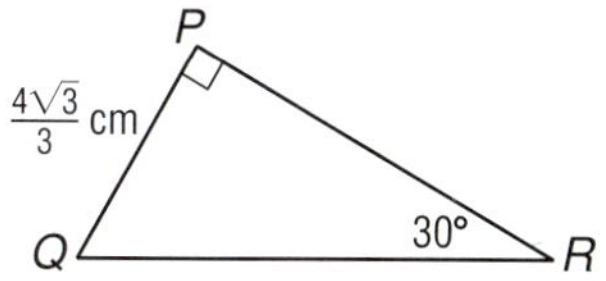

$\frac{8\sqrt{3}}{3}$ cm

4 $\triangle WXY$ is a 30°-60°-90° triangle with right angle *X* and $\overline{WX}$ as the longer leg. Graph points *X*(−2, 7) and *Y*(−7, 7), and locate point *W* in Quadrant III.

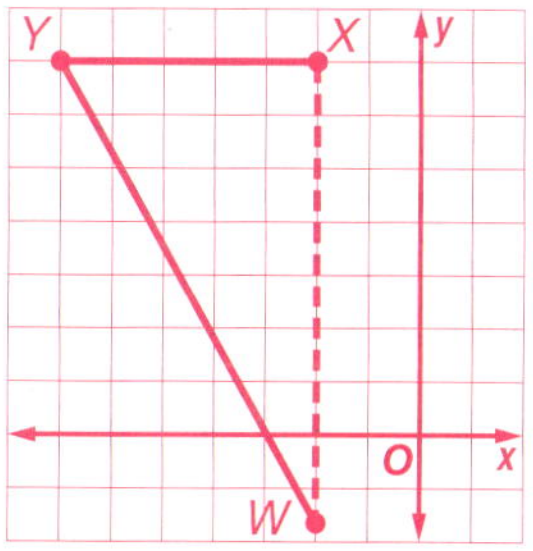

$W(-2, 7 - 5\sqrt{3})$

DAILY INTERVENTION

Unlocking Misconceptions

30°-60°-90° Triangles Point out that a common mistake is for students to quickly assume that the longer leg of a 30°-60°-90° triangle is twice the length of the shorter leg. Demonstrate for students that this cannot be so by drawing a 30°-60°-90° triangle on the board and extending the figure into an equilateral triangle. Explain that the shorter leg of the original right triangle is exactly half the distance of one side of the equilateral triangle. Tell students that the hypotenuse is also a side of the equilateral triangle, so this has to be the side that is twice the length of the shorter leg.

3 Practice/Apply

Study Notebook

Have students—

- add the definitions/examples of the vocabulary terms to their Vocabulary Builder worksheets for Chapter 7.
- include any other item(s) that they find helpful in mastering the skills in this lesson.

About the Exercises...

Organization by Objective

- **Properties of 45°-45°-90° Triangles:** 12, 13, 17, 22, 25
- **Properties of 30°-60°-90° Triangles:** 14–16, 18–21, 23–24, 27–31

Odd/Even Assignments

Exercises 12–31 are structured so that students practice the same concepts whether they are assigned odd or even problems.

Assignment Guide

Basic: 13–37 odd, 41–66

Average: 13–41 odd, 42–66

Advanced: 12–40 even, 41–58 (optional: 59–66)

All: Quiz 1 (1–5)

Check for Understanding

Concept Check

1. **OPEN ENDED** Draw a 45°-45°-90° triangle. Be sure to label the angles and the sides and to explain how you made the drawing. **1–2. See margin.**
2. **Explain** how to draw a 30°-60°-90° triangle with the shorter leg 2 centimeters long.
3. **Write** an equation to find the length of a rectangle that has a diagonal twice as long as its width. **The length of the rectangle is $\sqrt{3}$ times the width; $\ell = \sqrt{3}w$.**

Guided Practice

GUIDED PRACTICE KEY	
Exercises	Examples
4, 5, 11	1, 2
6–8	3
9, 10	4

Find x and y.

4. 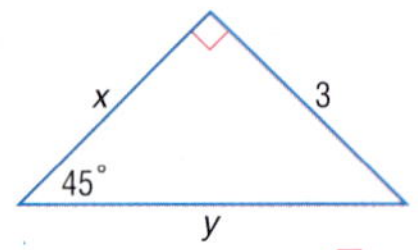

$x = 3$; $y = 3\sqrt{2}$

5. 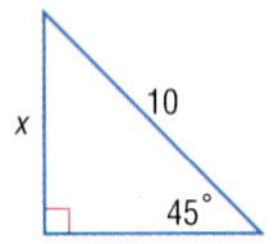

$x = 5\sqrt{2}$; $y = 5\sqrt{2}$

6. 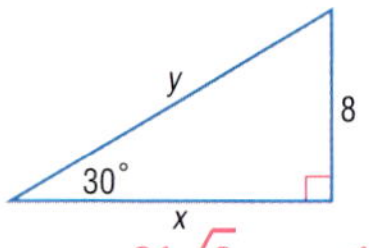

$x = 8\sqrt{3}$; $y = 16$

Find the missing measures.

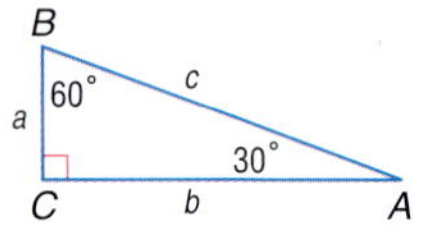

7. If $c = 8$, find a and b. **$a = 4$; $b = 4\sqrt{3}$**
8. If $b = 18$, find a and c. **$a = 6\sqrt{3}$; $c = 12\sqrt{3}$**

Triangle ABD is a 30°-60°-90° triangle with right angle B and with $\overline{AB}$ as the shorter leg. Graph A and B, and locate point D in Quadrant I. **9–10. See p. 399A.**

9. $A(8, 0)$, $B(8, 3)$
10. $A(6, 6)$, $B(2, 6)$

Application

11. **SOFTBALL** Find the distance from home plate to second base if the bases are 90 feet apart. **$90\sqrt{2}$ or about 127.28 ft**

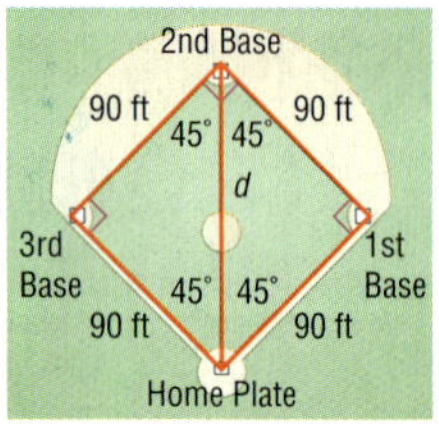

★ indicates increased difficulty

Practice and Apply

Homework Help	
For Exercises	See Examples
12, 13, 17, 22, 25	1, 2
14–16, 18–21, 23, 24	3
27–31	4

Extra Practice See page 767.

Find x and y.

12. 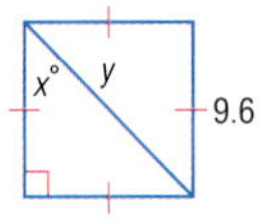

$x = 45$; $y = 9.6\sqrt{2}$

13. 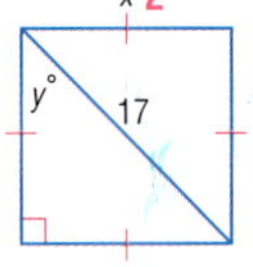

$x = \frac{17\sqrt{2}}{2}$; $y = 45$

14. 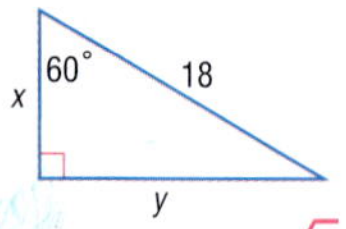

$x = 9$; $y = 9\sqrt{3}$

15. 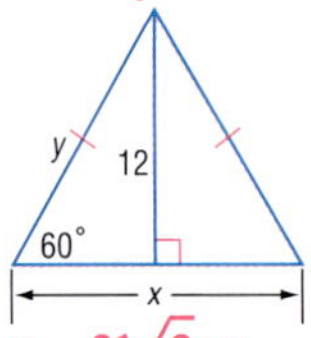

$x = 8\sqrt{3}$; $y = 8\sqrt{3}$

16. 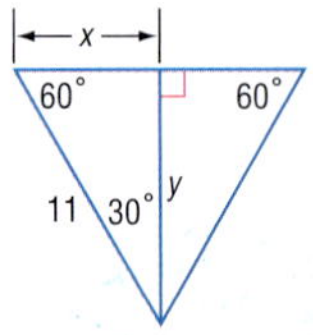

$x = 5.5$; $y = 5.5\sqrt{3}$

17. 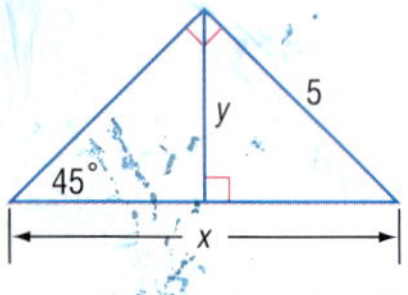

$x = 5\sqrt{2}$; $y = \frac{5\sqrt{2}}{2}$

For Exercises 18 and 19, use the figure at the right.

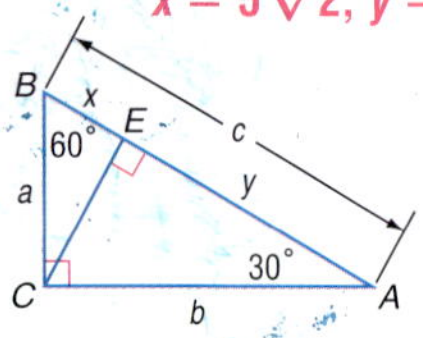

18. If $a = 10\sqrt{3}$, find CE and y. **$CE = 15$; $y = 15\sqrt{3}$**
19. If $x = 7\sqrt{3}$, find a, CE, y, and b. **$a = 14\sqrt{3}$; $CE = 21$; $y = 21\sqrt{3}$; $b = 42$**

Answers

1. Sample answer: Construct two perpendicular lines. Use a ruler to measure 3 cm from the point of intersection on the one ray. Use the compass to copy the 3 cm segment. Connect the two endpoints to form a 45°-45°-90° triangle with sides of 3 cm and a hypotenuse of $3\sqrt{2}$ cm.
2. Sample answer: Draw a line using a ruler. Then use a protractor to measure a 90° angle. On one ray mark a 2 cm length, and at that endpoint use the protractor to measure a 30° angle toward the other ray. Where this ray intersects the other ray should form a 60° angle, completing the 30°-60°-90° triangle with sides 2 cm, $2\sqrt{3}$ cm, and a hypotenuse of 4 cm.

20. The length of an altitude of an equilateral triangle is 12 feet. Find the length of a side of the triangle. $8\sqrt{3}$ ft ≈ 13.86 ft

21. The perimeter of an equilateral triangle is 45 centimeters. Find the length of an altitude of the triangle. $7.5\sqrt{3}$ cm ≈ 12.99 cm

22. The length of a diagonal of a square is $22\sqrt{2}$ millimeters. Find the perimeter of the square. 88 mm

23. The altitude of an equilateral triangle is 7.4 meters long. Find the perimeter of the triangle. $14.8\sqrt{3}$ m ≈ 25.63 m

24. The diagonals of a rectangle are 12 inches long and intersect at an angle of 60°. Find the perimeter of the rectangle. $12 + 12\sqrt{3}$ or about 32.78 in.

25. The sum of the squares of the measures of the sides of a square is 256. Find the measure of a diagonal of the square. $8\sqrt{2} \approx 11.31$

★ 26. Find x, y, z, and the perimeter of $ABCD$.
$x = 4$; $y = 4\sqrt{3}$; $z = 6$;
$24 + 4\sqrt{2} + 4\sqrt{3}$ or ≈ 36.59 units

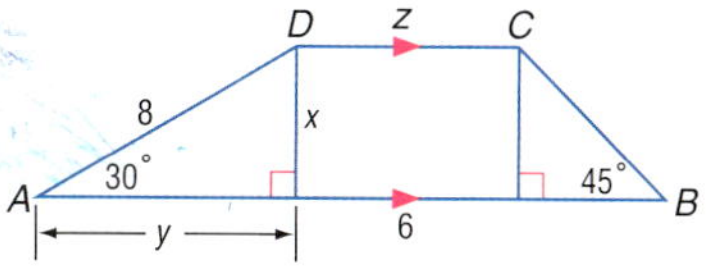

27. $\triangle PAB$ is a 45°-45°-90° triangle with right angle B. Find the coordinates of P in Quadrant I for $A(-3, 1)$ and $B(4, 1)$. (4, 8)

28. $\triangle PGH$ is a 45°-45°-90° triangle with $m\angle P = 90$. Find the coordinates of P in Quadrant I for $G(4, -1)$ and $H(4, 5)$. (1, 2), (7, 2)

29. $\triangle PCD$ is a 30°-60°-90° triangle with right angle C and $\overline{CD}$ the longer leg. Find the coordinates of P in Quadrant III for $C(-3, -6)$ and $D(-3, 7)$.

29. $\left(-3 - \frac{13\sqrt{3}}{3}, -6\right)$ or about $(-10.51, -6)$

★ 30. $\triangle PCD$ is a 30°-60°-90° triangle with $m\angle C = 30$ and hypotenuse $\overline{CD}$. Find the coordinates of P for $C(2, -5)$ and $D(10, -5)$ if P lies above $\overline{CD}$.
$(8, -5 + 2\sqrt{3})$

31. If $\overline{PQ} \parallel \overline{SR}$, use the figure to find a, b, c, and d.
$a = 3\sqrt{3}$, $b = 9$, $c = 3\sqrt{3}$, $d = 9$

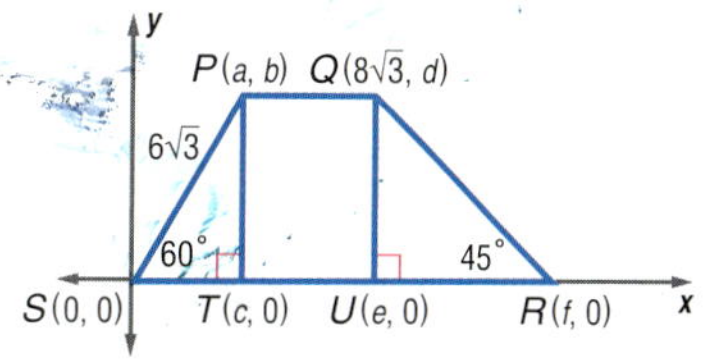

More About. . .

Triangle Tiling

Buildings in Federation Square in Melbourne, Australia, feature a tiling pattern called a pinwheel tiling. The sides of each right triangle are in the ratio $1 : 2 : \sqrt{5}$.

Source: www.federationsquare.com.au

TRIANGLE TILING **For Exercises 32–35, use the following information.**

Triangle tiling refers to the process of taking many copies of a single triangle and laying them next to each other to fill an area. For example, the pattern shown is composed of tiles like the one outlined.

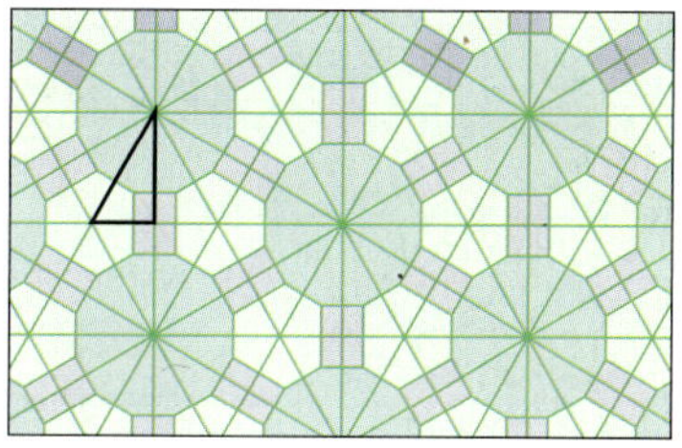

32. How many 30°-60°-90° triangles are used to create the basic circular pattern? 12

33. Which angle of the 30°-60°-90° triangle is being rotated to make the basic shape? 30° angle

34. Explain why there are no gaps in the basic pattern.

34. There are no gaps because when a 30° angle is rotated 12 times, it rotates 360°.

35. Use grid paper to cut out 30°-60°-90° triangles. Color the same pattern on each triangle. Create one basic figure that would be part of a wallpaper tiling. See margin.

Answer

35. Sample answer:

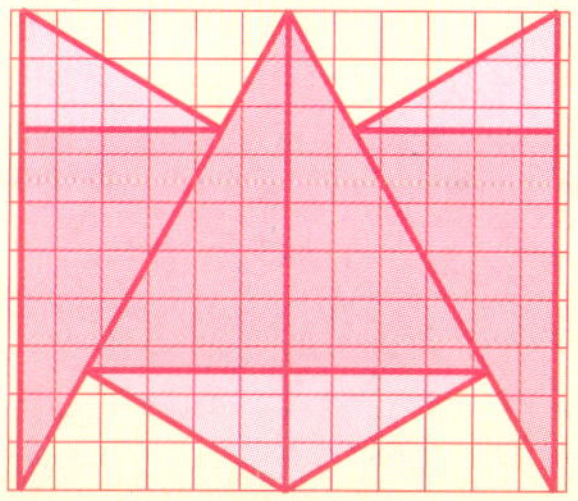

Enrichment, p. 368

Constructing Values of Square Roots

The diagram at the right shows a right isosceles triangle with two legs of length 1 inch. By the Pythagorean Theorem, the length of the hypotenuse is $\sqrt{2}$ inches. By constructing an adjacent right triangle with legs of $\sqrt{2}$ inches and 1 inch, you can create a segment of length $\sqrt{3}$.

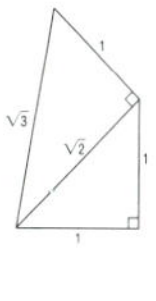

By continuing this process as shown below, you can construct a "wheel" of square roots. This wheel is called the "Wheel of Theodorus" after a Greek philosopher who lived about 400 B.C.

Continue constructing the wheel until you make a segment of length $\sqrt{18}$.

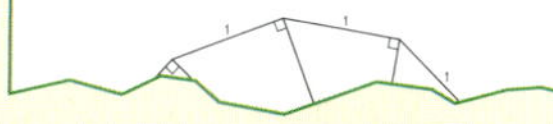

Study Guide and Intervention, p. 363 (shown) and p. 364

Properties of 45°-45°-90° Triangles The sides of a 45°-45°-90° right triangle have a special relationship.

Example 1 **If the leg of a 45°-45°-90° right triangle is x units, show that the hypotenuse is $x\sqrt{2}$ units.**

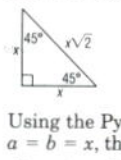

Using the Pythagorean Theorem with $a = b = x$, then

$$c^2 = a^2 + b^2 = x^2 + x^2 = 2x^2$$
$$c = \sqrt{2x^2} = x\sqrt{2}$$

Example 2 **In a 45°-45°-90° right triangle the hypotenuse is $\sqrt{2}$ times the leg. If the hypotenuse is 6 units, find the length of each leg.**

The hypotenuse is $\sqrt{2}$ times the leg, so divide the length of the hypotenuse by $\sqrt{2}$.

$$a = \frac{6}{\sqrt{2}} = \frac{6\sqrt{2}}{\sqrt{2}\sqrt{2}} = \frac{6\sqrt{2}}{2} = 3\sqrt{2} \text{ units}$$

Exercises

Find x.

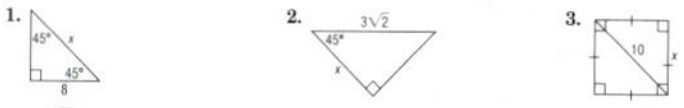

1. $8\sqrt{2} \approx 11.3$ 2. 3 3. $5\sqrt{2} \approx 7.1$

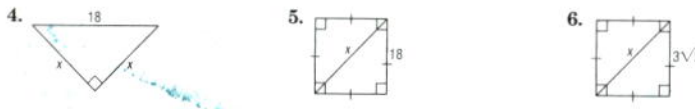

4. $9\sqrt{2} \approx 12.7$ 5. $18\sqrt{2} \approx 25.5$ 6. 6

7. Find the perimeter of a square with diagonal 12 centimeters. $24\sqrt{2} \approx 33.9$ cm

8. Find the diagonal of a square with perimeter 20 inches. $5\sqrt{2} \approx 7.1$ in.

9. Find the diagonal of a square with perimeter 28 meters. $7\sqrt{2} \approx 9.9$ m

Skills Practice, p. 365 and Practice, p. 366 (shown)

Find x and y.

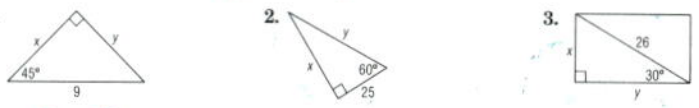

1. $\frac{9\sqrt{2}}{2}, \frac{9\sqrt{2}}{2}$ 2. $25\sqrt{3}$, 50 3. 13, $13\sqrt{3}$

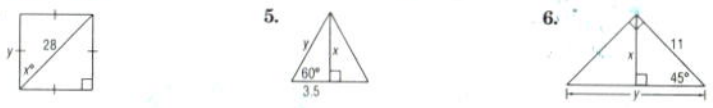

4. 45, $14\sqrt{2}$ 5. $3.5\sqrt{3}$, 7 6. $\frac{11\sqrt{2}}{2}$; $11\sqrt{2}$

For Exercises 7–9, use the figure at the right.

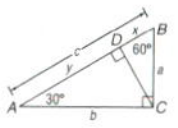

7. If $a = 4\sqrt{3}$, find b and c.
$b = 12$, $c = 8\sqrt{3}$

8. If $x = 3\sqrt{3}$, find a and CD.
$a = 6\sqrt{3}$, $CD = 9$

9. If $a = 4$, find CD, b, and y.
$CD = 2\sqrt{3}$, $b = 4\sqrt{3}$, $y = 6$

10. The perimeter of an equilateral triangle is 39 centimeters. Find the length of an altitude of the triangle.
$6.5\sqrt{3}$ in. or about 11.26 in.

11. $\triangle MIP$ is a 30°-60°-90° triangle with right angle at I, and $\overline{IP}$ the longer leg. Find the coordinates of M in Quadrant I for $I(3, 3)$ and $P(12, 3)$.
$(3, 3 + 3\sqrt{3})$ or about (3, 8.19)

12. $\triangle TJK$ is a 45°-45°-90° triangle with right angle at J. Find the coordinates of T in Quadrant II for $J(-2, -3)$ and $K(3, -3)$.
(−2, 2)

13. **BOTANICAL GARDENS** One of the displays at a botanical garden is an herb garden planted in the shape of a square. The square measures 6 yards on each side. Visitors can view the herbs from a diagonal pathway through the garden. How long is the pathway?
$6\sqrt{2}$ yd or about 8.48 yd

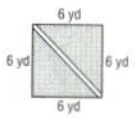

Reading to Learn Mathematics, p. 367 ELL

Pre-Activity **How is triangle tiling used in wallpaper design?**

Read the introduction to Lesson 7-3 at the top of page 357 in your textbook.

- How can you most completely describe the larger triangle and the two smaller triangles in tile 15? Sample answer: The larger triangle is an isosceles obtuse triangle. The two smaller triangles are congruent scalene right triangles.
- How can you most completely describe the larger triangle and the two smaller triangles in tile 16? (Include angle measures in describing all the triangles.) Sample answer: The larger triangle is equilateral, so each of its angle measures is 60. The two smaller triangles are congruent right triangles in which the angle measures are 30, 60, and 90.

Reading the Lesson

1. Supply the correct number or numbers to complete each statement.
   a. In a 45°-45°-90° triangle, to find the length of the hypotenuse, multiply the length of a leg by $\sqrt{2}$.
   b. In a 30°-60°-90° triangle, to find the length of the hypotenuse, multiply the length of the shorter leg by 2.
   c. In a 30°-60°-90° triangle, the longer leg is opposite the angle with a measure of 60.
   d. In a 30°-60°-90° triangle, to find the length of the longer leg, multiply the length of the shorter leg by $\sqrt{3}$.
   e. In an isosceles right triangle, each leg is opposite an angle with a measure of 45.
   f. In a 30°-60°-90° triangle, to find the length of the shorter leg, divide the length of the longer leg by $\sqrt{3}$.
   g. In 30°-60°-90° triangle, to find the length of the longer leg, divide the length of the hypotenuse by 2 and multiply the result by $\sqrt{3}$.
   h. To find the length of a side of a square, divide the length of the diagonal by $\sqrt{2}$.
2. Indicate whether each statement is *always*, *sometimes*, or *never* true.
   a. The lengths of the three sides of an isosceles triangle satisfy the Pythagorean Theorem. sometimes
   b. The lengths of the sides of a 30°-60°-90° triangle form a Pythagorean triple. never
   c. The lengths of all three sides of a 30°-60°-90° triangle are positive integers. never

Helping You Remember

3. Some students find it easier to remember mathematical concepts in terms of specific numbers rather than variables. How can you use specific numbers to help you remember the relationship between the lengths of the three sides in a 30°-60°-90° triangle? Sample answer: Draw a 30°-60°-90° triangle. Label the length of the shorter leg as 1. Then the length of the hypotenuse is 2, and the length of the longer leg is $\sqrt{3}$. Just remember: 1, 2, $\sqrt{3}$.

Answers

42. Sample answer: Congruent triangles of different color can be arranged to create patterns. Answers should include the following.
- 5, 9, 15, and 17; 3, 4, 6, 7, 10, 11, and 12
- Placing 45° angles next to one another forms 90° angles, which can be placed next to each other, leaving no holes.

36. Find x, y, and z. **$x = 4$; $y = 4\sqrt{3}$; $z = 2\sqrt{6}$**

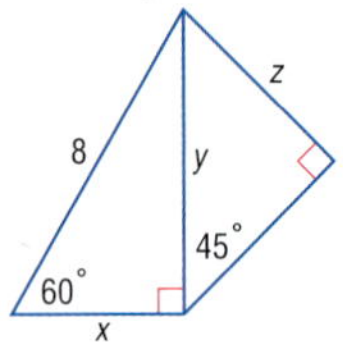

37. If $BD = 8\sqrt{3}$ and $m\angle DHB = 60$, find BH. **16**

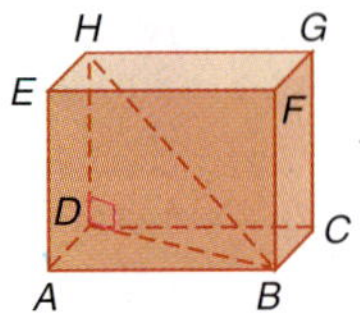

★ **38.** Each triangle in the figure is a 30°-60°-90° triangle. Find x. **2.25**

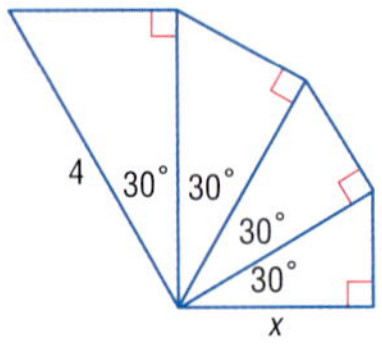

★ **39.** In regular hexagon $UVWXYZ$, each side is 12 centimeters long. Find WY. **$12\sqrt{3} \approx 20.78$ cm**

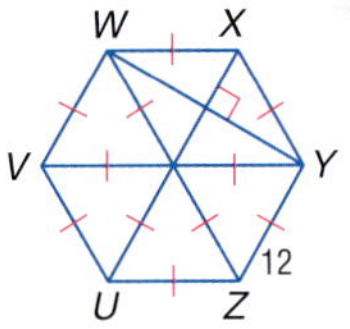

40. BASEBALL The diagram at the right shows some dimensions of Comiskey Park in Chicago, Illinois. $\overline{BD}$ is a segment from home plate to dead center field, and $\overline{AE}$ is a segment from the left field foul-ball pole to the right field foul-ball pole. If the center fielder is standing at C, how far is he from home plate? **$\frac{347\sqrt{2}}{2} \approx 245.4$ ft**

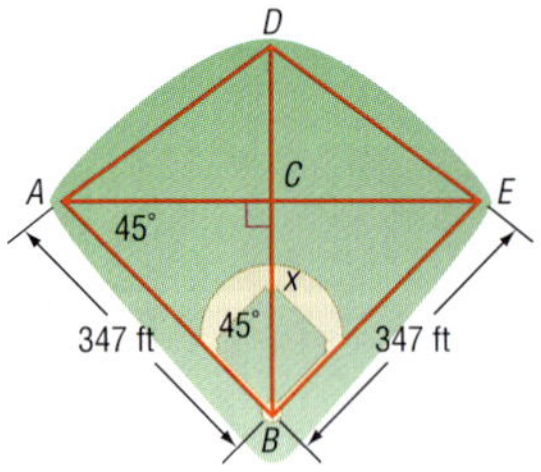

41. CRITICAL THINKING Given figure $ABCD$, with $\overline{AB} \parallel \overline{DC}$, $m\angle B = 60$, $m\angle D = 45$, $BC = 8$, and $AB = 24$, find the perimeter.
$52 + 4\sqrt{3} + 4\sqrt{6}$ units

42. WRITING IN MATH Answer the question that was posed at the beginning of the lesson. **See margin.**

How is triangle tiling used in wallpaper design?

Include the following in your answer:
- which of the numbered designs contain 30°-60°-90° triangles and which contain 45°-45°-90° triangles, and
- a reason why rotations of the basic design left no holes in the completed design.

FCAT Practice
Standardized Test Practice

43. In the right triangle, what is AB if $BC = 6$? **C**

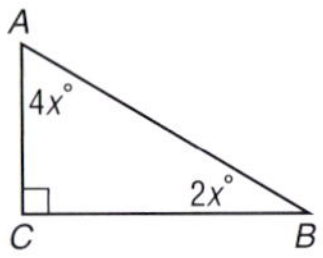

Ⓐ 12 units
Ⓑ $6\sqrt{2}$ units
Ⓒ $4\sqrt{3}$ units
Ⓓ $2\sqrt{3}$ units

44. SHORT RESPONSE For real numbers a and b, where $b \neq 0$, if $a \star b = \frac{a^2}{b^2}$, then $(3 \star 4)(5 \star 3) = $ __?__. **$\frac{25}{16}$**

362 Chapter 7 Right Triangles and Trigonometry

Maintain Your Skills

Mixed Review

Determine whether each set of measures can be the sides of a right triangle. Then state whether they form a Pythagorean triple. *(Lesson 7-2)*

45. 3, 4, 5 **yes, yes** **46.** 9, 40, 41 **yes, yes** **47.** 20, 21, 31 **no, no**

48. 20, 48, 52 **yes, yes** **49.** 7, 24, 25 **yes, yes** **50.** 12, 34, 37 **no, no**

Find x, y, and z. *(Lesson 7-1)*

51.

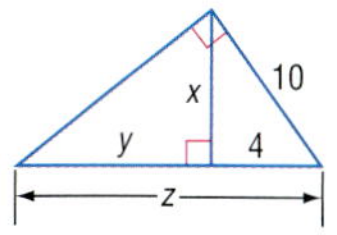

$2\sqrt{21} \approx 9.2$; 21; 25

52.

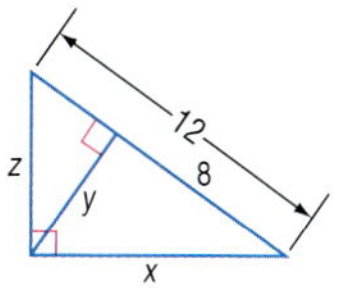

53.

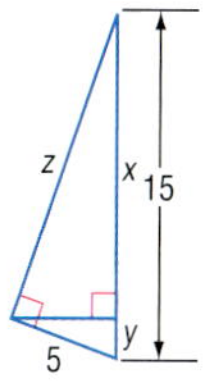

52. $4\sqrt{6} \approx 9.8$; $\sqrt{2} \approx 5.7$; $\sqrt{3} \approx 6.9$

53. $\frac{40}{3}$; $\frac{5}{3}$; $0\sqrt{2} \approx 14.1$

Write an inequality or equation relating each pair of angles. *(Lesson 5-5)*

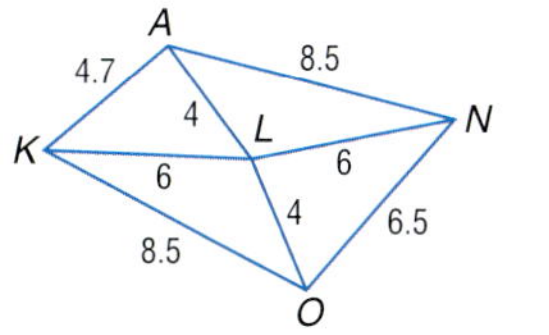

54. $m\angle ALK$, $m\angle ALN$ $m\angle ALK < m\angle ALN$

55. $m\angle ALK$, $m\angle NLO$ $m\angle ALK < m\angle NLO$

56. $m\angle OLK$, $m\angle NLO$ $m\angle OLK > m\angle NLO$

57. $m\angle KLO$, $m\angle ALN$ $m\angle KLO = m\angle ALN$

58. Determine whether $\triangle JKL$ with vertices $J(-3, 2)$, $K(-1, 5)$, and $L(4, 4)$ is congruent to $\triangle RST$ with vertices $R(-6, 6)$, $S(-4, 3)$, and $T(1, 4)$. Explain. *(Lesson 4-4)*

58. $JK = \sqrt{13}$, $KL = \sqrt{26}$, $JL = \sqrt{53}$, $RS = \sqrt{13}$, $ST = \sqrt{26}$, $RT = \sqrt{53}$; $\triangle JKL \cong \triangle RST$ by SSS

Getting Ready for the Next Lesson

PREREQUISITE SKILL **Solve each equation.**
*(To review **solving equations**, see pages 737 and 738.)*

59. $5 = \frac{x}{3}$ **15** **60.** $\frac{x}{9} = 0.14$ **1.26** **61.** $0.5 = \frac{10}{k}$ **20** **62.** $0.2 = \frac{13}{g}$ **65**

63. $\frac{7}{n} = 0.25$ **28** **64.** $9 = \frac{m}{0.8}$ **7.2** **65.** $\frac{24}{x} = 0.4$ **60** **66.** $\frac{35}{y} = 0.07$ **500**

Practice Quiz 1

Lessons 7-1 through 7-3

Find the measure of the altitude drawn to the hypotenuse. *(Lesson 7-1)*

1.

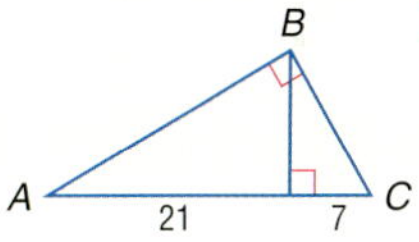

$7\sqrt{3} \approx 12.1$

2.

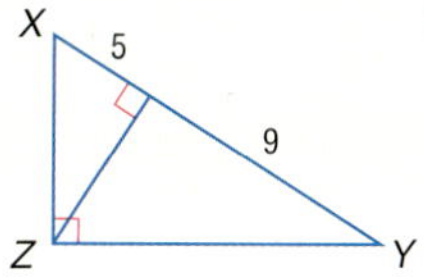

$3\sqrt{5} \approx 6.7$

3. Determine whether $\triangle ABC$ with vertices $A(2, 1)$, $B(4, 0)$, and $C(5, 7)$ is a right triangle. Explain. *(Lesson 7-2)* yes; $AB = \sqrt{5}$, $BC = \sqrt{50}$, $AC = \sqrt{45}$; $(\sqrt{5})^2 + (\sqrt{45})^2 = (\sqrt{50})^2$

Find x and y. *(Lesson 7-3)*

4.

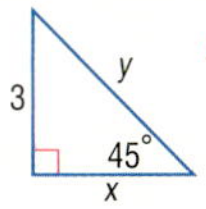

$x = 3$; $y = 3\sqrt{2}$

5.

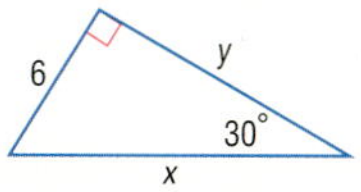

$x = 12$; $y = 6\sqrt{3}$

4 Assess

Open-Ended Assessment

Speaking Have students name the side relationships for 45°-45°-90° triangles and 30°-60°-90° triangles. For some examples and problems in the book, ask students to call out these side relationships using the numbers and variables in the figures.

Getting Ready for Lesson 7-4

Prerequisite Skill Students will learn about trigonometric ratios in Lesson 7-4. They will use fractions to solve equations and express side lengths. Use Exercises 59–66 to determine your students' familiarity with solving equations that involve fractions.

Assessment Options

Practice Quiz 1 The quiz provides students with a brief review of the concepts and skills in Lessons 7-1 through 7-3. Lesson numbers are given to the right of the exercises or instruction lines so students can review concepts not yet mastered.

7-4 Lesson Notes

1 Focus

5-Minute Check Transparency 7-4 Use as a quiz or review of Lesson 7-3.

Mathematical Background notes are available for this lesson on p. 340D.

How can surveyors determine angle measures?

Ask students:

- What side of a right triangle is represented by the theodolite's line of sight? **the hypotenuse**
- The Statue of Liberty is approximately 91.5 meters tall from the base of the pedestal to the tip of the torch. If a theodolite were placed 100 meters from the base of the pedestal, about how far would the theodolite be from the top of the torch? **about 135.5 m**

7-4 Trigonometry

Sunshine State Standards
MA.A.3.4.3, MA.B.2.4.1, MA.B.3.4.1, MA.C.3.4.1, MA.C.3.4.2

What You'll Learn

- Find trigonometric ratios using right triangles.
- Solve problems using trigonometric ratios.

Vocabulary

- trigonometry
- trigonometric ratio
- sine
- cosine
- tangent

How can surveyors determine angle measures?

The old surveyor's telescope shown at right is called a theodolite (thee AH duh lite). It is an optical instrument used to measure angles in surveying, navigation, and meteorology. It consists of a telescope fitted with a level and mounted on a tripod so that it is free to rotate about its vertical and horizontal axes. After measuring angles, surveyors apply trigonometry to calculate distance or height.

TEACHING TIP
In the expression sin *A*, *A* represents the measure of ∠*A* in degrees.

TRIGONOMETRIC RATIOS The word **trigonometry** comes from two Greek terms, *trigon*, meaning triangle, and *metron*, meaning measure. The study of trigonometry involves triangle measurement. A ratio of the lengths of sides of a right triangle is called a **trigonometric ratio**. The three most common trigonometric ratios are **sine**, **cosine**, and **tangent**.

Study Tip

Reading Math
SOH-CAH-TOA is a mnemonic device for learning the ratios for sine, cosine, and tangent using the first letter of each word in the ratios.

$\sin A = \frac{opp}{hyp}$

$\cos A = \frac{adj}{hyp}$

$\tan A = \frac{opp}{adj}$

Key Concept — Trigonometric Ratios

Words	Symbols	Models
sine of ∠*A* = $\frac{\text{measure of leg opposite } \angle A}{\text{measure of hypotenuse}}$ sine of ∠*B* = $\frac{\text{measure of leg opposite } \angle B}{\text{measure of hypotenuse}}$	$\sin A = \frac{BC}{AB}$ $\sin B = \frac{AC}{AB}$	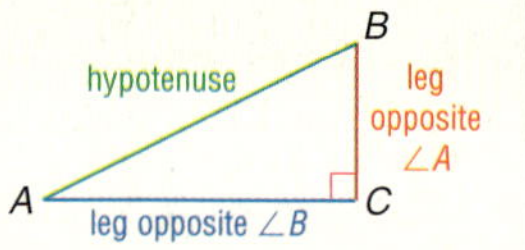
cosine of ∠*A* = $\frac{\text{measure of leg adjacent to } \angle A}{\text{measure of hypotenuse}}$ cosine of ∠*B* = $\frac{\text{measure of leg adjacent to } \angle B}{\text{measure of hypotenuse}}$	$\cos A = \frac{AC}{AB}$ $\cos B = \frac{BC}{AB}$	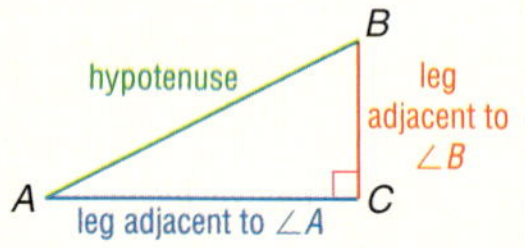
tangent of ∠*A* = $\frac{\text{measure of leg opposite } \angle A}{\text{measure of leg adjacent to } \angle A}$ tangent of ∠*B* = $\frac{\text{measure of leg opposite } \angle B}{\text{measure of leg adjacent to } \angle B}$	$\tan A = \frac{BC}{AC}$ $\tan B = \frac{AC}{BC}$	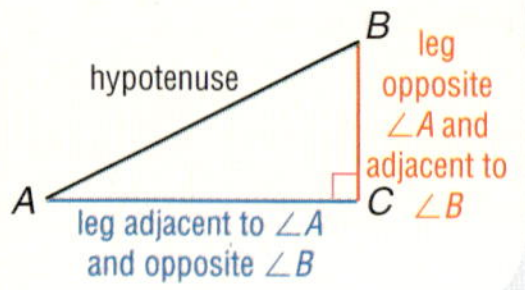

Resource Manager

Workbook and Reproducible Masters

Chapter 7 Resource Masters
- Study Guide and Intervention, pp. 369–370
- Skills Practice, p. 371
- Practice, p. 372
- Reading to Learn Mathematics, p. 373
- Enrichment, p. 374
- Assessment, pp. 407, 409

School-to-Career Masters, p. 14
Prerequisite Skills Workbook, pp. 11–12
Teaching Geometry With Manipulatives Masters, pp. 16, 17, 18, 117, 118

Transparencies

5-Minute Check Transparency 7-4
Answer Key Transparencies

Technology

Interactive Chalkboard

Trigonometric ratios are related to the acute angles of a right triangle, *not* the right angle.

Study Tip

Equivalent Ratios
Notice that the ratio $\frac{\text{leg opposite } \angle A}{\text{hypotenuse}}$ is the same as $\frac{\text{leg adjacent } \angle C}{\text{hypotenuse}}$. Thus, $\sin A = \cos C = \frac{a}{b}$. Likewise, $\cos A = \sin B = \frac{c}{b}$.

Geometry Activity

Trigonometric Ratios

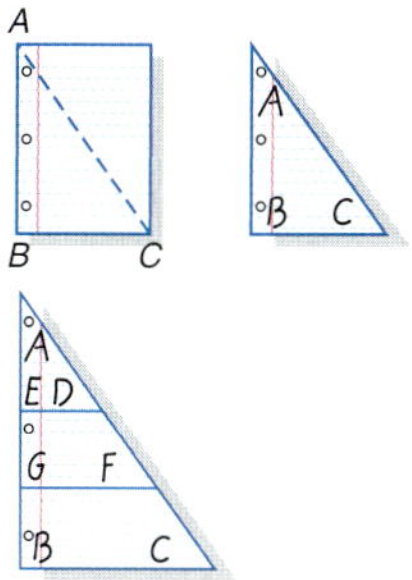

- Fold a rectangular piece of paper along a diagonal from A to C. Then cut along the fold to form right triangle ABC. Write the name of each angle on the inside of the triangle.
- Fold the triangle so that there are two segments perpendicular to $\overline{BA}$. Label points D, E, F, and G as shown. Use a ruler to measure $\overline{AC}$, $\overline{AB}$, $\overline{BC}$, $\overline{AF}$, $\overline{AG}$, $\overline{FG}$, $\overline{AD}$, $\overline{AE}$, and $\overline{DE}$ to the nearest millimeter.

Analyze

1. What is true of $\triangle AED$, $\triangle AGF$, and $\triangle ABC$? **They are similar triangles.**
2. Copy the table. Write the ratio of the side lengths for each trigonometric ratio. Then calculate a value for each ratio to the nearest ten-thousandth.

	In $\triangle AED$	In $\triangle AGF$	In $\triangle ABC$
sin A	$\frac{DE}{AD} \approx 0.6114$	$\frac{FG}{AF} \approx 0.6114$	$\frac{BC}{AC} \approx 0.6114$
cos A	$\frac{AE}{AD} \approx 0.7913$	$\frac{AG}{AF} \approx 0.7913$	$\frac{AB}{AC} \approx 0.7913$
tan A	$\frac{DE}{AE} \approx 0.7727$	$\frac{FG}{AG} \approx 0.7727$	$\frac{BC}{AB} \approx 0.7727$

3. Study the table. Write a sentence about the patterns you observe with the trigonometric ratios.
4. What is true about $m\angle A$ in each triangle? **$m\angle A$ is the same in all triangles.**

3. Sample answer: Regardless of the side lengths, the trigonometric ratio is the same when comparing angles in similar triangles.

As the Geometry Activity suggests, the value of a trigonometric ratio depends *only* on the measure of the angle. It does not depend on the size of the triangle.

Example 1 Find Sine, Cosine, and Tangent Ratios

Find sin R, cos R, tan R, sin S, cos S, and tan S. Express each ratio as a fraction and as a decimal.

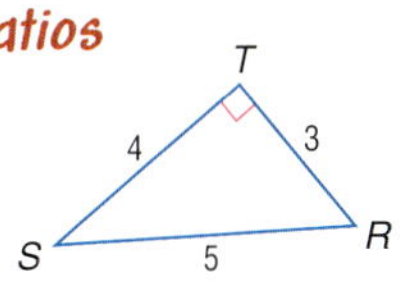

$\sin R = \frac{\text{opposite leg}}{\text{hypotenuse}} = \frac{ST}{RS} = \frac{4}{5}$ or 0.8

$\cos R = \frac{\text{adjacent leg}}{\text{hypotenuse}} = \frac{RT}{RS} = \frac{3}{5}$ or 0.6

$\tan R = \frac{\text{opposite leg}}{\text{adjacent leg}} = \frac{ST}{RT} = \frac{4}{3}$ or $1.\overline{3}$

$\sin S = \frac{\text{opposite leg}}{\text{hypotenuse}} = \frac{RT}{RS} = \frac{3}{5}$ or 0.6

$\cos S = \frac{\text{adjacent leg}}{\text{hypotenuse}} = \frac{ST}{RS} = \frac{4}{5}$ or 0.8

$\tan S = \frac{\text{opposite leg}}{\text{adjacent leg}} = \frac{RT}{ST} = \frac{3}{4}$ or 0.75

www.geometryonline.com/extra_examples/fcat

2 Teach

TRIGONOMETRIC RATIOS

Building on Prior Knowledge

In Lesson 7-1, students learned that they could use the Distance Formula and the Pythagorean Theorem to find lengths in a coordinate plane. They revisit these methods in this lesson and combine them with trigonometry to find the angles of a triangle in a coordinate plane.

In-Class Example Power Point®

Teaching Tip Assure the class that sine, cosine, and tangent are very useful and not as imposing as they first seem. Explain that if students regularly practice using these ratios to solve problems, they will become as familiar as squaring and taking the square root of numbers and are just as easy to key into a calculator.

1 Find sin L, cos L, tan L, sin N, cos N, and tan N. Express each ratio as a fraction and as a decimal.

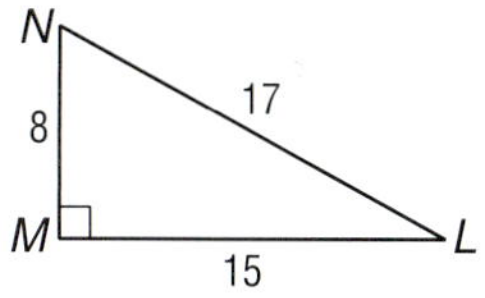

$\sin L = \frac{8}{17}$ or 0.47

$\cos L = \frac{15}{17}$ or 0.88

$\tan L = \frac{8}{15}$ or 0.53

$\sin N = \frac{15}{17}$ or 0.88

$\cos N = \frac{8}{17}$ or 0.47

$\tan N = \frac{15}{8}$ or 1.88

Geometry Activity

Materials: rectangular paper, scissors, metric ruler, pencil

- When students are folding the triangle to form two segments that are parallel to $\overline{BC}$, tell them that it does not matter how close or far apart these segments are, and encourage the class to come up with a variety of different lengths to demonstrate that this activity would work for any lengths.
- Students can also begin the activity with different sizes of rectangular paper.

Teaching Tip Encourage students to familiarize themselves with using their calculators to find cosines, sines and tangents and their inverses. Allow students to practice with easy values, such as $\cos 0° = 1$, $\sin 30° = 0.5$, $\cos^{-1} 0.5 = 60°$, etc., so they can be sure they are using their calculators correctly.

In-Class Example

Power Point®

2 Find each value to the nearest ten thousandth.

a. tan 56° **1.4826**

a. cos 89° **0.0175**

USE TRIGONOMETRIC RATIOS

In-Class Examples

Power Point®

3 **EXERCISING** A fitness trainer sets the incline on a treadmill to 7°. The walking surface is 5 feet long. Approximately how many inches did the trainer raise the end of the treadmill from the floor?

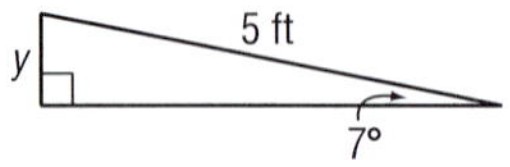

≈7.3 in.

4 **COORDINATE GEOMETRY** Find $m\angle X$ in right $\triangle XYZ$ for $X(-2, 8)$, $Y(-6, 4)$, and $Z(-3, 1)$.

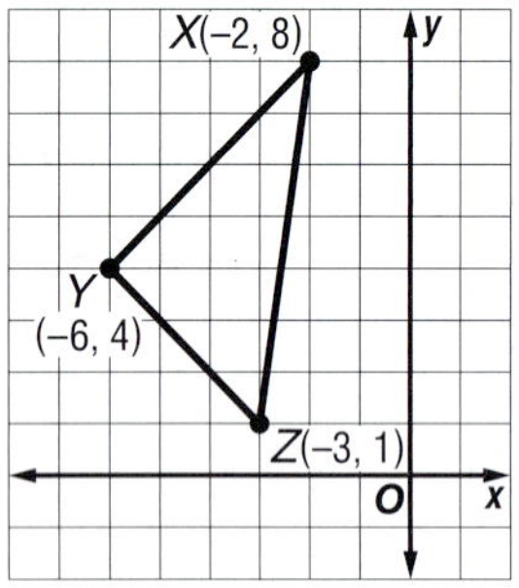

≈36.9

Study Tip

Graphing calculator

Be sure your calculator is in degree mode rather than radian mode. The SIN, COS, and TAN keys will calculate the value of the sine, cosine, and tangent of the angle measure. Your calculator may require you to input the angle *before* using the trigonometric key.

You can use a calculator to evaluate expressions involving trigonometric ratios.

Example 2 Evaluate Expressions

Use a calculator to find each value to the nearest ten thousandth.

a. cos 39°

KEYSTROKES: COS 39 ENTER

cos 39° ≈ 0.7771

b. sin 67°

KEYSTROKES: SIN 67 ENTER

sin 67° ≈ 0.9205

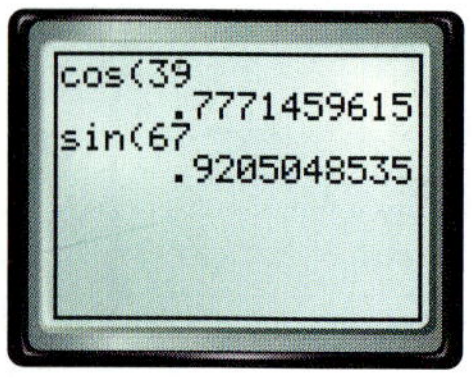

USE TRIGONOMETRIC RATIOS You can use trigonometric ratios to find the missing measures of a right triangle if you know the measures of two sides of a triangle or the measure of one side and one acute angle.

Example 3 Use Trigonometric Ratios to Find a Length

SURVEYING **Dakota is standing on the ground 97 yards from the base of a cliff. Using a theodolite, he noted that the angle formed by the ground and the line of sight to the top of the cliff was 56°. Find the height of the cliff to the nearest yard.**

Let x be the height of the cliff in yards.

$\tan 56° = \frac{x}{97}$ $\quad$ $\tan = \frac{\text{leg opposite}}{\text{leg adjacent}}$

$97 \tan 56° = x$ $\quad$ Multiply each side by 97.

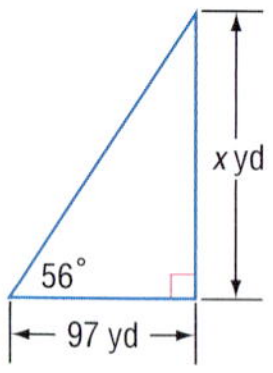

Use a calculator to find x.

KEYSTROKES: 97 TAN 56 ENTER 143.8084139

The cliff is about 144 yards high.

When solving equations like $3x = -27$, you use the inverse of multiplication to find x. In trigonometry, you can find the measure of the angle by using the inverse of sine, cosine, or tangent.

Given equation	To find the angle	Read as
$\sin A = x$	$A = \sin^{-1}(x)$	A equals *the inverse sine of x.*
$\cos A = y$	$A = \cos^{-1}(y)$	A equals *the inverse cosine of y.*
$\tan A = z$	$A = \tan^{-1}(z)$	A equals *the inverse tangent of z.*

Example 4 Use Trigonometric Ratios to Find an Angle Measure

COORDINATE GEOMETRY **Find $m\angle A$ in right triangle ABC for $A(1, 2)$, $B(6, 2)$, and $C(5, 4)$.**

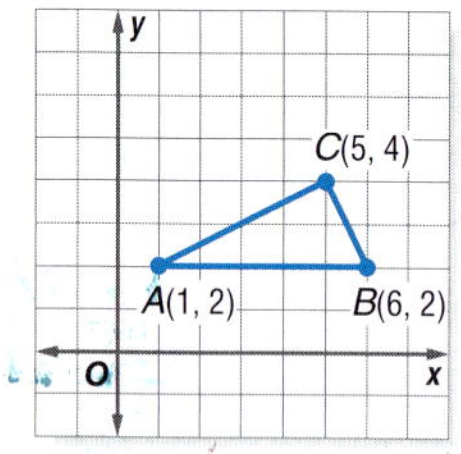

Explore You know the coordinates of the vertices of a right triangle and that $\angle C$ is the right angle. You need to find the measure of one of the angles.

Plan Use the Distance Formula to find the measure of each side. Then use one of the trigonometric ratios to write an equation. Use the inverse to find $m\angle A$.

DAILY INTERVENTION — Differentiated Instruction

Auditory/Musical The easiest way for auditory learners to remember the ratios for sine, cosine and tangent is for them to chant SOH-CAH-TOA. When introducing this mnemonic device to students, have them repeat it as a class a few times in rhythm. Point out that SOH and CAH each have one syllable because the "H" is silent, so students can remember that one "silent" hypotenuse is involved for the sine and cosine ratios. TOA has two syllables and involves the two legs for the tangent ratio.

Study Tip

Calculators

The second functions of the SIN, COS, and TAN keys are usually the inverses.

Solve

$AB = \sqrt{(6-1)^2 + (2-2)^2}$ $\quad BC = \sqrt{(5-6)^2 + (4-2)^2}$

$= \sqrt{25+0}$ or 5 $\quad = \sqrt{1+4}$ or $\sqrt{5}$

$AC = \sqrt{(5-1)^2 + (4-2)^2}$

$= \sqrt{16+4}$

$= \sqrt{20}$ or $2\sqrt{5}$

Use the cosine ratio.

$\cos A = \frac{AC}{AB}$ $\quad \cos = \frac{\text{leg adjacent}}{\text{hypotenuse}}$

$\cos A = \frac{2\sqrt{5}}{5}$ $\quad AC = 2\sqrt{5}$ and $AB = 5$

$A = \cos^{-1}\left(\frac{2\sqrt{5}}{5}\right)$ $\quad$ Solve for A.

Use a calculator to find $m\angle A$.

KEYSTROKES: 2nd [COS^{-1}] 2 2nd [$\sqrt{\ }$] 5) ÷ 5 ENTER

$m\angle A \approx 26.56505118$

The measure of $\angle A$ is about 26.6.

Examine Use the sine ratio to check the answer.

$\sin A = \frac{BC}{AB}$ $\quad \sin = \frac{\text{leg opposite}}{\text{hypotenuse}}$

$\sin A = \frac{\sqrt{5}}{5}$ $\quad BC = \sqrt{5}$ and $AB = 5$

KEYSTROKES: 2nd [SIN^{-1}] 2nd [$\sqrt{\ }$] 5) ÷ 5 ENTER

$m\angle A \approx 26.56505118$

The answer is correct.

Check for Understanding

Concept Check

1. The triangles are similar, so the ratios remain the same.

1. **Explain** why trigonometric ratios do not depend on the size of the right triangle.
2. **OPEN ENDED** Draw a right triangle and label the measures of one acute angle and the measure of the side opposite that angle. Then solve for the remaining measures. **2–4. See margin.**
3. **Compare and contrast** the sine, cosine, and tangent ratios.
4. **Explain** the difference between $\tan A = \frac{x}{y}$ and $\tan^{-1}\left(\frac{x}{y}\right) = A$.

Guided Practice

GUIDED PRACTICE KEY	
Exercises	Examples
5, 6	1
7–12	2
17	3
13–16	4

Use $\triangle ABC$ to find $\sin A$, $\cos A$, $\tan A$, $\sin B$, $\cos B$, and $\tan B$. Express each ratio as a fraction and as a decimal to the nearest hundredth. **5–6. See margin.**

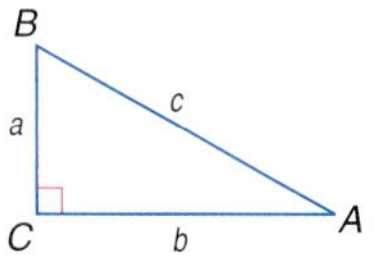

5. $a = 14$, $b = 48$, and $c = 50$
6. $a = 8$, $b = 15$, and $c = 17$

Use a calculator to find each value. Round to the nearest ten-thousandth.

7. sin 57° **0.8387**
8. cos 60° **0.5000**
9. cos 33° **0.8387**
10. tan 30° **0.5774**
11. tan 45° **1.0000**
12. sin 85° **0.9962**

3 Practice/Apply

Study Notebook

Have students—
- add the definitions/examples of the vocabulary terms to their Vocabulary Builder worksheets for Chapter 7.
- include the "SOH-CAH-TOA" mnemonic device and an example similar to Example 1 on p. 365 for quick reference.
- include any other item(s) that they find helpful in mastering the skills in this lesson.

About the Exercises...

Organization by Objective
- **Trigonometric Ratios:** 18–36
- **Use Trigonometric Ratios:** 37–48, 52–54

Odd/Even Assignments

Exercises 18–54 are structured so that students practice the same concepts whether they are assigned odd or even problems.

Alert! Exercises 22–27 require a scientific calculator.

Assignment Guide

Basic: 19–49 odd, 59, 61–64, 69–82 (optional: 65–68)

Average: 19–61 odd, 62–64, 69–82 (optional: 65–68)

Advanced: 18–60 even, 61–76 (optional: 77–82)

Answers

2. Sample answer:

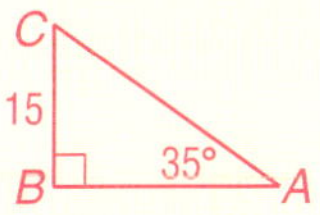

$m\angle B = 90$, $m\angle C = 55$, $b \approx 26.2$, $c \approx 21.4$

3. All three ratios involve two sides of a right triangle. The sine ratio is the measure of the opposite leg divided by the measure of the hypotenuse. The cosine ratio is the measure of the adjacent leg divided by the measure of the hypotenuse. The tangent ratio is the measure of the opposite leg divided by the measure of the adjacent leg.

4. The tan is the ratio of the measure of the opposite leg divided by the measure of the adjacent leg for a given angle in a right triangle. The tan^{-1} is the measure of the angle with a certain tangent ratio.

5. $\frac{14}{50} = 0.28$; $\frac{48}{50} = 0.96$; $\frac{14}{48} \approx 0.29$; $\frac{48}{50} = 0.96$; $\frac{14}{50} = 0.28$; $\frac{48}{14} \approx 3.43$

6. $\frac{8}{17} \approx 0.47$; $\frac{15}{17} \approx 0.88$; $\frac{8}{15} \approx 0.53$; $\frac{15}{17} \approx 0.88$; $\frac{8}{17} \approx 0.47$; $\frac{15}{8} \approx 1.88$

Find the measure of each angle to the nearest tenth of a degree.

13. $\tan A = 1.4176$ **$m\angle A \approx 54.8$**

14. $\sin B = 0.6307$ **$m\angle B \approx 39.1$**

COORDINATE GEOMETRY Find the measure of the angle to the nearest tenth in each right triangle ABC.

15. $\angle A$ in $\triangle ABC$, for $A(6, 0)$, $B(-4, 2)$, and $C(0, 6)$ **$m\angle A \approx 33.7$**

16. $\angle B$ in $\triangle ABC$, for $A(3, -3)$, $B(7, 5)$, and $C(7, -3)$ **$m\angle B \approx 26.6$**

Application

17. **SURVEYING** Maureen is standing on horizontal ground level with the base of the CN Tower in Toronto, Ontario. The angle formed by the ground and the line segment from her position to the top of the tower is 31.2°. She knows that the height of the tower to the top of the antennae is about 1815 feet. Find her distance from the CN Tower to the nearest foot. **2997 ft**

★ indicates increased difficulty

Practice and Apply

Homework Help

For Exercises	See Examples
18–21, 28–36	1
22–27	2
43–48	3
37–42, 52–54	4

Extra Practice See page 767.

Use $\triangle PQR$ with right angle R to find $\sin P$, $\cos P$, $\tan P$, $\sin Q$, $\cos Q$, and $\tan Q$. Express each ratio as a fraction, and as a decimal to the nearest hundredth. **18–21. See margin.**

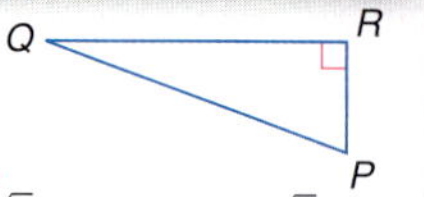

18. $p = 12$, $q = 35$, and $r = 37$

19. $p = \sqrt{6}$, $q = 2\sqrt{3}$, and $r = 3\sqrt{2}$

20. $p = \frac{3}{2}$, $q = \frac{3\sqrt{3}}{2}$, and $r = 3$

21. $p = 2\sqrt{3}$, $q = \sqrt{15}$, and $r = 3\sqrt{3}$

Use a calculator to find each value. Round to the nearest ten-thousandth.

22. $\sin 6°$ **0.1045**

23. $\tan 42.8°$ **0.9260**

24. $\cos 77°$ **0.2250**

25. $\sin 85.9°$ **0.9974**

26. $\tan 12.7°$ **0.2254**

27. $\cos 22.5°$ **0.9239**

Use the figure to find each trigonometric ratio. Express answers as a fraction and as a decimal rounded to the nearest ten-thousandth.

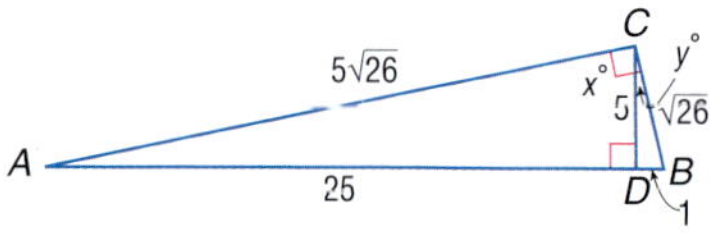

28. $\sin A$

29. $\tan B$

30. $\cos A$

31. $\sin x°$

32. $\cos x°$

33. $\tan A$

34. $\cos B$

35. $\sin y°$

36. $\tan x°$

28. $\frac{\sqrt{26}}{26} \approx 0.1961$

29. $\frac{5}{1} = 5.0000$

30. $\frac{5\sqrt{26}}{26} \approx 0.9806$

31. $\frac{5\sqrt{26}}{26} \approx 0.9806$

32. $\frac{\sqrt{26}}{26} \approx 0.1961$

33. $\frac{1}{5} = 0.2000$

34. $\frac{\sqrt{26}}{26} \approx 0.1961$

35. $\frac{\sqrt{26}}{26} \approx 0.1961$

36. $\frac{5}{1} = 5.0000$

Find the measure of each angle to the nearest tenth of a degree.

37. $\sin B = 0.7245$ **46.4**

38. $\cos C = 0.2493$ **75.6**

39. $\tan E = 9.4618$ **84.0**

40. $\sin A = 0.4567$ **27.2**

41. $\cos D = 0.1212$ **83.0**

42. $\tan F = 0.4279$ **23.2**

Find x. Round to the nearest tenth.

43. **8.5**

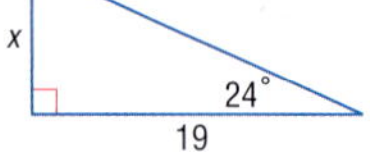

44. **44.9**

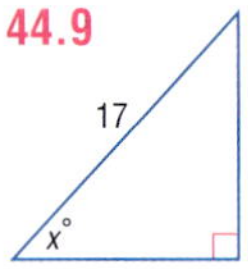

45. **28.2**

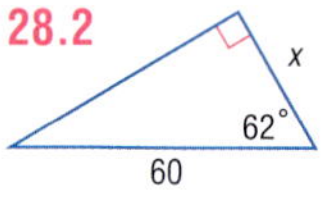

46. **29.1**

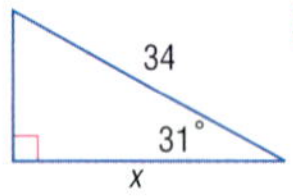

47. **22.6**

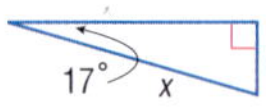

48. **39.8**

Study Guide and Intervention, p. 369 (shown) and p. 370

Trigonometric Ratios The ratio of the lengths of two sides of a right triangle is called a **trigonometric ratio**. The three most common ratios are **sine**, **cosine**, and **tangent**, which are abbreviated *sin*, *cos*, and *tan*, respectively.

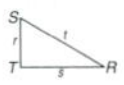

$\sin R = \frac{\text{leg opposite } \angle R}{\text{hypotenuse}} = \frac{r}{t}$ $\cos R = \frac{\text{leg adjacent to } \angle R}{\text{hypotenuse}} = \frac{s}{t}$ $\tan R = \frac{\text{leg opposite } \angle R}{\text{leg adjacent to } \angle R} = \frac{r}{s}$

Example **Find sin A, cos A, and tan A. Express each ratio as a decimal to the nearest thousandth.**

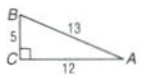

$\sin A = \frac{\text{opposite leg}}{\text{hypotenuse}} = \frac{BC}{AB} = \frac{5}{13} \approx 0.385$

$\cos A = \frac{\text{adjacent leg}}{\text{hypotenuse}} = \frac{AC}{AB} = \frac{12}{13} \approx 0.923$

$\tan A = \frac{\text{opposite leg}}{\text{adjacent leg}} = \frac{BC}{AC} = \frac{5}{12} \approx 0.417$

Exercises

Find the indicated trigonometric ratio as a fraction and as a decimal. If necessary, round to the nearest ten-thousandth.

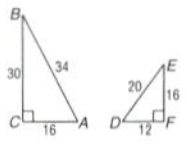

1. $\sin A$ $\frac{15}{17}$; 0.8824

2. $\tan B$ $\frac{8}{15}$; 0.5333

3. $\cos A$ $\frac{8}{17}$; 0.4706

4. $\cos B$ $\frac{15}{17}$; 0.8824

5. $\sin D$ $\frac{4}{5}$; 0.8

6. $\tan E$ $\frac{3}{4}$; 0.75

7. $\cos E$ $\frac{4}{5}$; 0.8

8. $\cos D$ $\frac{3}{5}$; 0.6

Skills Practice, p. 371 and Practice, p. 372 (shown)

Use $\triangle LMN$ to find sin L, cos L, tan L, sin M, cos M, and tan M. Express each ratio as a fraction and as a decimal to the nearest hundredth.

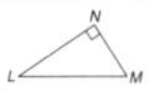

1. $\ell = 15$, $m = 36$, $n = 39$

$\sin L = \frac{15}{39} \approx 0.38$; $\cos L = \frac{36}{39} \approx 0.92$; $\tan L = \frac{15}{36} \approx 0.42$; $\sin M = \frac{36}{39} \approx 0.92$; $\cos M = \frac{15}{39} \approx 0.38$; $\tan M = \frac{36}{15} = 2.4$

2. $\ell = 12$, $m = 12\sqrt{3}$, $n = 24$

$\sin L = \frac{12}{24} = 0.50$; $\cos L = \frac{12\sqrt{3}}{24} \approx 0.87$; $\tan L = \frac{12}{12\sqrt{3}} \approx 0.58$; $\sin M = \frac{12\sqrt{3}}{24} \approx 0.87$; $\cos M = \frac{12}{24} = 0.50$; $\tan M = \frac{12\sqrt{3}}{12} \approx 1.73$

Use a calculator to find each value. Round to the nearest ten-thousandth.

3. sin 92.4 0.9991

4. tan 27.5 0.5206

5. cos 64.8 0.4258

Use the figure to find each trigonometric ratio. Express answers as a fraction and as a decimal rounded to the nearest ten-thousandth.

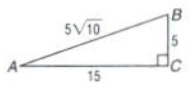

6. $\cos A$ $\frac{3\sqrt{10}}{10} \approx 0.9487$

7. $\tan B$ $\frac{3}{1} = 3.0000$

8. $\sin A$ $\frac{\sqrt{10}}{10} \approx 0.3162$

Find the measure of each acute angle to the nearest tenth of a degree.

9. $\sin B = 0.7823$ 51.5

10. $\tan A = 0.2356$ 13.3

11. $\cos R = 0.6401$ 50.2

Find x. Round to the nearest tenth.

12. 64.4

13. 18.1

14. 24.2

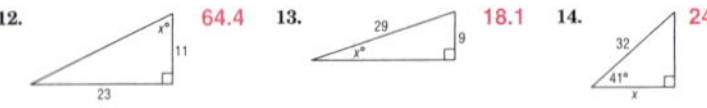

15. **GEOGRAPHY** Diego used a theodolite to map a region of land for his class in geomorphology. To determine the elevation of a vertical rock formation, he measured the distance from the base of the formation to his position and the angle between the ground and the line of sight to the top of the formation. The distance was 43 meters and the angle was 36 degrees. What is the height of the formation to the nearest meter? 31 m

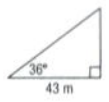

Reading to Learn Mathematics, p. 373 ELL

Pre-Activity **How can surveyors determine angle measures?**

Read the introduction to Lesson 7-4 at the top of page 364 in your textbook.

- Why is it important to determine the relative positions accurately in navigation? (Give two possible reasons.) Sample answers: (1) To avoid collisions between ships, and (2) to prevent ships from losing their bearings and getting lost at sea.
- What does *calibrated* mean? Sample answer: marked precisely to permit accurate measurements to be made

Reading the Lesson

1. Refer to the figure. Write a ratio using the side lengths in the figure to represent each of the following trigonometric ratios.

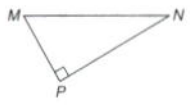

A. $\sin N$ $\frac{MP}{MN}$

B. $\cos N$ $\frac{NP}{MN}$

C. $\tan N$ $\frac{MP}{NP}$

D. $\tan M$ $\frac{NP}{MP}$

E. $\sin M$ $\frac{NP}{MN}$

F. $\cos M$ $\frac{MP}{MN}$

2. Assume that you enter each of the expressions in the list on the left into your calculator. Match each of these expressions with a description from the list on the right to tell what you are finding when you enter this expression.

a. sin 20 v b. cos 20 ii c. $\sin^{-1} 0.8$ vi d. $\tan^{-1} 0.8$ iii e. tan 20 iv f. $\cos^{-1} 0.8$ i	i. the degree measure of an acute angle whose cosine is 0.8 ii. the ratio of the length of the leg adjacent to the 20° angle to the length of hypotenuse in a 20°-70°-90° triangle iii. the degree measure of an acute angle in a right triangle for which the ratio of the length of the opposite leg to the length of the adjacent leg is 0.8 iv. the ratio of the length of the leg opposite the 20° angle to the length of the leg adjacent to it in a 20°-70°-90° triangle v. the ratio of the length of the leg opposite the 20° angle to the length of hypotenuse in a 20°-70°-90° triangle vi. the degree measure of an acute angle in a right triangle for which the ratio of the length of the opposite leg to the length of the hypotenuse is 0.8

Helping You Remember

3. How can the *co* in *cosine* help you to remember the relationship between the sines and cosines of the two acute angles of a right triangle?

Sample answer: The *co* in *cosine* comes from *complement*, as in *complementary* angles. The cosine of an acute angle is equal to the sine of its complement.

Enrichment, p. 374

Sine and Cosine of Angles

The following diagram can be used to obtain approximate values for the sine and cosine of angles from 0° to 90°. The radius of the circle is 1. So, the sine and cosine values can be read directly from the vertical and horizontal axes.

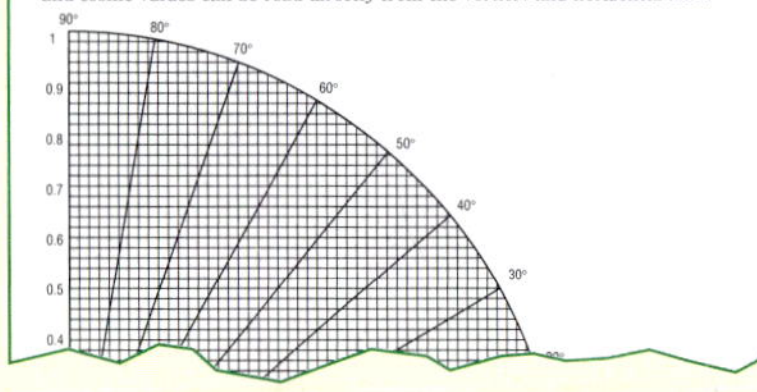

368 Chapter 7 Right Triangles and Trigonometry

49. **AVIATION** A plane is one mile above sea level when it begins to climb at a constant angle of 3° for the next 60 ground miles. About how far above sea level is the plane after its climb? **4.1 mi**

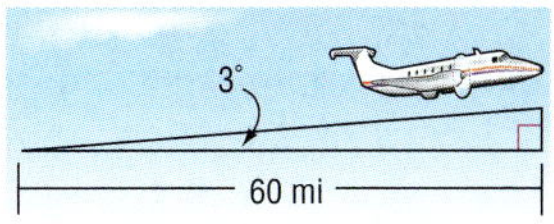

SAFETY **For Exercises 50 and 51, use the following information.**
To guard against a fall, a ladder should make an angle of 75° or less with the ground.

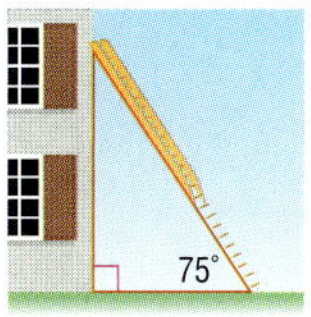

50. What is the maximum height that a 20-foot ladder can reach safely? **19.32 ft**

51. How far from the building is the base of the ladder at the maximum height? **5.18 ft**

COORDINATE GEOMETRY **Find the measure of each angle to the nearest tenth in each right triangle.**

52. $\angle J$ in $\triangle JCL$ for $J(2, 2)$, $C(2, -2)$, and $L(7, -2)$ **about 51.3**

53. $\angle C$ in $\triangle BCD$ for $B(-1, -5)$, $C(-6, -5)$, and $D(-1, 2)$ **about 54.5**

54. $\angle X$ in $\triangle XYZ$ for $X(-5, 0)$, $Y(7, 0)$, and $Z(0, \sqrt{35})$ **49.8**

★55. Find the perimeter of $\triangle ABC$ if $m\angle A = 35$, $m\angle C = 90$, and $AB = 20$ inches. **about 47.9 in.**

Find x and y. Round to the nearest tenth.

★56.

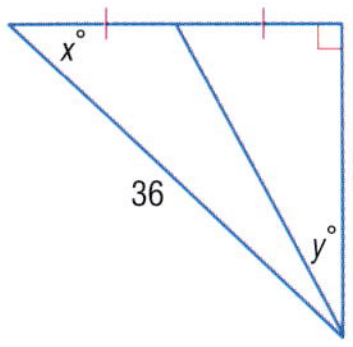

★57.

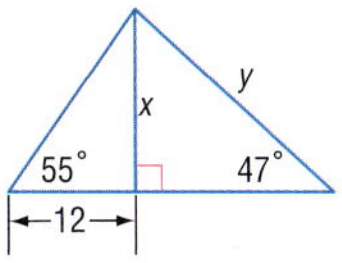

★58.

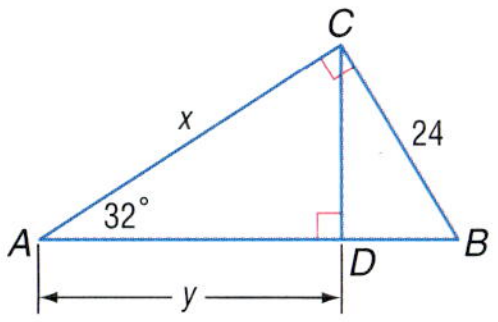

56. $x = 41.8$; $y = 29.2$
57. $x = 17.1$; $y = 23.4$
58. $x = 38.4$; $y = 32.6$
59. about 272,837 astronomical units
60. The stellar parallax would be too small.
62. See margin.

ASTRONOMY **For Exercises 59 and 60, use the following information.**
One way to find the distance between the sun and a relatively close star is to determine the angles of sight for the star exactly six months apart. Half the measure formed by these two angles of sight is called the *stellar parallax*. Distances in space are sometimes measured in *astronomical units*. An astronomical unit is equal to the average distance between Earth and the sun.

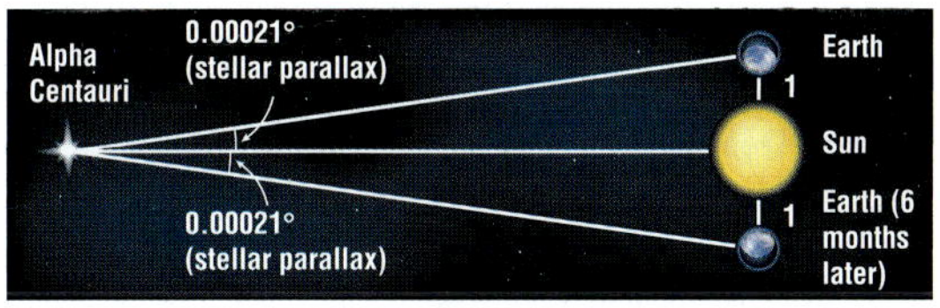

59. Find the distance between Alpha Centauri and the sun.

60. Make a conjecture as to why this method is used only for close stars.

61. **CRITICAL THINKING** Use the figure at the right to find sin $x°$. $\frac{2\sqrt{2}}{5}$

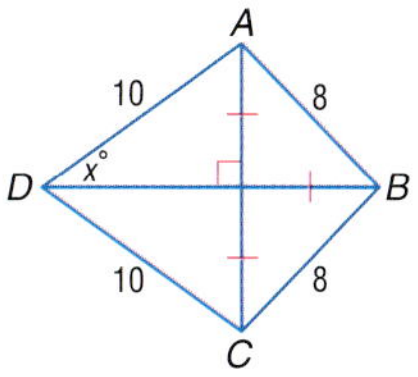

Exercise 61

62. **WRITING IN MATH** Answer the question that was posed at the beginning of the lesson.

How do surveyors determine angle measures?

Include the following in your answer:
- where theodolites are used, and
- the kind of information one obtains from a theodolite.

More About. . .

Astronomy

The stellar parallax is one of several methods of triangulation used to determine the distance of stars from the sun. Another method is trigonometric parallax, which measures the displacement of a nearby star relative to a more distant star.

Source: www.infoplease.com

Answer

62. Sample answer: Surveyors use a theodolite to measure angles to determine distances and heights. Answers should include the following.
- **Theodolites are used in surveying, navigation, and meteorology. They are used to measure angles.**
- **The angle measures from two points, which are a fixed distance apart, to a third point.**

www.geometryonline.com/self_check_quiz/fcat

Answers (p. 368)

18. $\frac{12}{37} \approx 0.32$; $\frac{35}{37} \approx 0.95$; $\frac{12}{35} \approx 0.34$; $\frac{35}{37} \approx 0.95$; $\frac{12}{37} \approx 0.32$; $\frac{35}{12} \approx 2.92$

19. $\frac{\sqrt{3}}{3} \approx 0.58$; $\frac{\sqrt{6}}{3} \approx 0.82$; $\frac{\sqrt{2}}{2} \approx 0.71$; $\frac{\sqrt{6}}{3} \approx 0.82$; $\frac{\sqrt{3}}{3} \approx 0.58$; $\sqrt{2} \approx 1.41$

20. $\frac{1}{2} = 0.5$; $\frac{\sqrt{3}}{2} \approx 0.87$; $\frac{\sqrt{3}}{3} \approx 0.58$; $\frac{\sqrt{3}}{2} \approx 0.87$; $\frac{1}{2} = 0.5$; $\sqrt{3} \approx 1.73$

21. $\frac{2}{3} \approx 0.67$; $\frac{\sqrt{5}}{3} \approx 0.75$; $\frac{2\sqrt{5}}{5} \approx 0.89$; $\frac{\sqrt{5}}{3} \approx 0.75$; $\frac{2}{3} \approx 0.67$; $\frac{\sqrt{5}}{2} \approx 1.12$

4 Assess

Open-Ended Assessment

Speaking Draw and label the sides and angles of several right triangles on the board, and call on volunteers to name the sides they would use to find the sine, cosine, and tangent of the angles. Encourage students to use the terms *opposite*, *adjacent*, and *hypotenuse* when naming the sides.

Getting Ready for Lesson 7-5

Prerequisite Skill Students will learn about angles of elevation and depression in Lesson 7-5. They will use angle relationships and trigonometry to find distances and angles. Use Exercises 77–82 to determine your students' familiarity with determining angle relationships among angles formed by parallel lines and a transversal.

Assessment Options

Quiz (Lessons 7-3 and 7-4) is available on p. 407 of the *Chapter 7 Resource Masters*.

Mid-Chapter Test (Lessons 7-1 through 7-4) is available on p. 409 of the *Chapter 7 Resource Masters*.

Answers

65. $\csc A = \frac{5}{3}$; $\sec A = \frac{5}{4}$; $\cot A = \frac{4}{3}$; $\csc B = \frac{5}{4}$; $\sec B = \frac{5}{3}$; $\cot B = \frac{3}{4}$

66. $\csc A = \frac{13}{12}$; $\sec A = \frac{13}{5}$; $\cot A = \frac{5}{12}$; $\csc B = \frac{13}{5}$; $\sec B = \frac{13}{12}$; $\cot B = \frac{12}{5}$

67. $\csc A = 2$; $\sec A = \frac{2\sqrt{3}}{3}$; $\cot A = \sqrt{3}$; $\csc B = \frac{2\sqrt{3}}{3}$; $\sec B = 2$; $\cot B = \frac{\sqrt{3}}{3}$

68. $\csc A = \sqrt{2}$; $\sec A = \sqrt{2}$; $\cot A = 1$; $\csc B = \sqrt{2}$; $\sec B = \sqrt{2}$; $\cot B = 1$

FCAT Practice

Standardized Test Practice

63. Find cos C. **C**

Ⓐ $\frac{3}{5}$ Ⓑ $\frac{3}{4}$ Ⓒ $\frac{4}{5}$ Ⓓ $\frac{5}{4}$

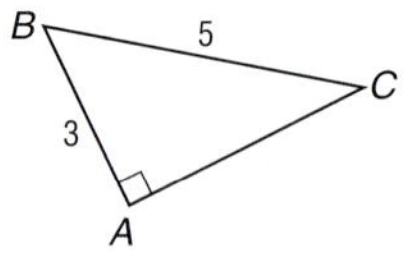

64. **ALGEBRA** If $x^2 = 15^2 + 24^2 - 15(24)$, find x. **B**

Ⓐ 20.8 Ⓑ 21 Ⓒ 12 Ⓓ 9

Extending the Lesson

Each of the basic trigonometric ratios has a reciprocal ratio. The reciprocals of the sine, cosine, and tangent are called the *cosecant*, *secant*, and the *cotangent*, respectively.

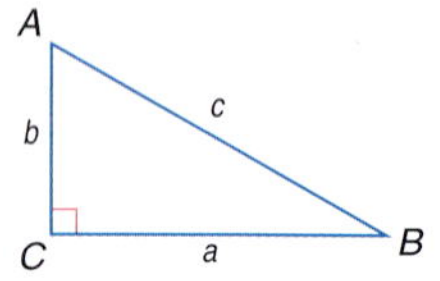

Reciprocal	Trigonometric Ratio	Abbreviation	Definition
$\frac{1}{\sin A}$	cosecant of $\angle A$	csc A	$\frac{\text{measure of the hypotenuse}}{\text{measure of the leg opposite } \angle A} = \frac{c}{a}$
$\frac{1}{\cos A}$	secant of $\angle A$	sec A	$\frac{\text{measure of the hypotenuse}}{\text{measure of the leg adjacent } \angle A} = \frac{c}{b}$
$\frac{1}{\tan A}$	cotangent of $\angle A$	cot A	$\frac{\text{measure of the leg adjacent } \angle A}{\text{measure of the leg opposite } \angle A} = \frac{b}{a}$

Use $\triangle ABC$ to find csc A, sec A, cot A, csc B, sec B, and cot B. Express each ratio as a fraction or as a radical in simplest form. **65–68. See margin.**

65. $a = 3$, $b = 4$, and $c = 5$

66. $a = 12$, $b = 5$, and $c = 13$

67. $a = 4$, $b = 4\sqrt{3}$, and $c = 8$

68. $a = 2\sqrt{2}$, $b = 2\sqrt{2}$, and $c = 4$

Maintain Your Skills

Mixed Review

Find each measure. *(Lesson 7-3)*

69. If $a = 4$, find b and c. $b = 4\sqrt{3}$, $c = 8$

70. If $b = 3$, find a and c. $a = \sqrt{3}$, $c = 2\sqrt{3}$

71. If $c = 5$, find a and b. $a = 2.5$, $b = 2.5\sqrt{3}$

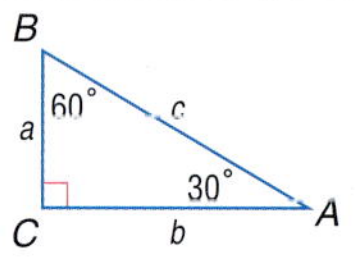

Determine whether each set of measures can be the sides of a right triangle. Then state whether they form a Pythagorean triple. *(Lesson 7-2)*

72. 4, 5, 6 **no, no**

73. 5, 12, 13 **yes, yes**

74. 9, 12, 15 **yes, yes**

75. 8, 12, 16 **no, no**

76. **TELEVISION** During a 30-minute television program, the ratio of minutes of commercials to minutes of the actual show is 4 : 11. How many minutes are spent on commercials? *(Lesson 6-1)* **8 min**

Getting Ready for the Next Lesson

PREREQUISITE SKILL **Find each angle measure if $h \parallel k$.** *(To review **angles formed by parallel lines and a transversal**, see Lesson 3-2.)*

77. $m\angle 15$ **117**

78. $m\angle 7$ **30**

79. $m\angle 3$ **150**

80. $m\angle 12$ **63**

81. $m\angle 11$ **63**

82. $m\angle 4$ **150**

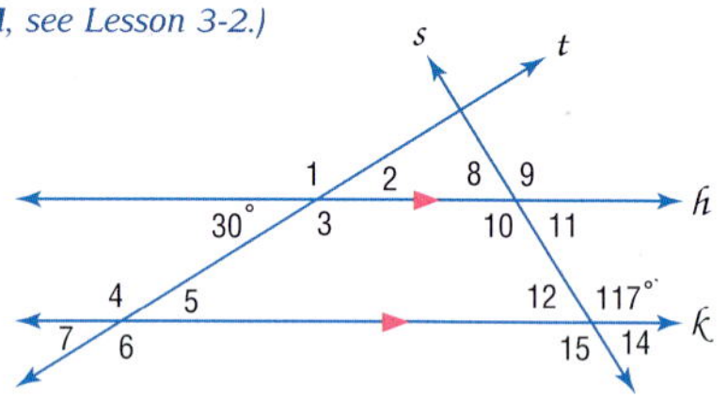

7-5 Angles of Elevation and Depression

Sunshine State Standards
MA.A.3.4.3, MA.B.2.4.1, MA.B.3.4.1, MA.C.3.4.1, MA.C.3.4.2

Vocabulary
- angle of elevation
- angle of depression

What You'll Learn
- Solve problems involving angles of elevation.
- Solve problems involving angles of depression.

How do airline pilots use angles of elevation and depression?

A pilot is getting ready to take off from Mountain Valley airport. She looks up at the peak of a mountain immediately in front of her. The pilot must estimate the speed needed and the angle formed by a line along the runway and a line from the plane to the peak of the mountain to clear the mountain.

ANGLES OF ELEVATION

An **angle of elevation** is the angle between the line of sight and the horizontal when an observer looks upward.

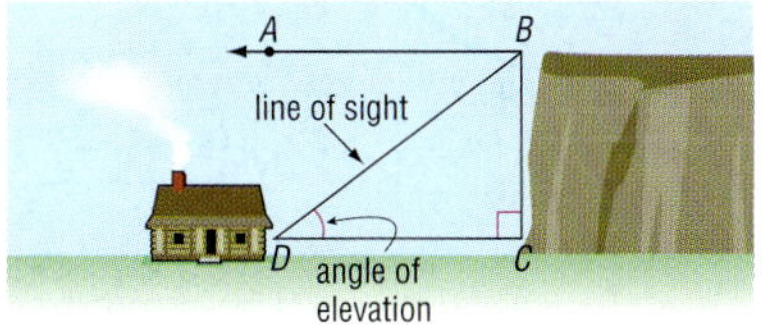

Example 1 Angle of Elevation

AVIATION **The peak of Goose Bay Mountain is 400 meters higher than the end of a local airstrip. The peak rises above a point 2025 meters from the end of the airstrip. A plane takes off from the end of the runway in the direction of the mountain at an angle that is kept constant until the peak has been cleared. If the pilot wants to clear the mountain by 50 meters, what should the angle of elevation be for the takeoff to the nearest tenth of a degree?**

Make a drawing.

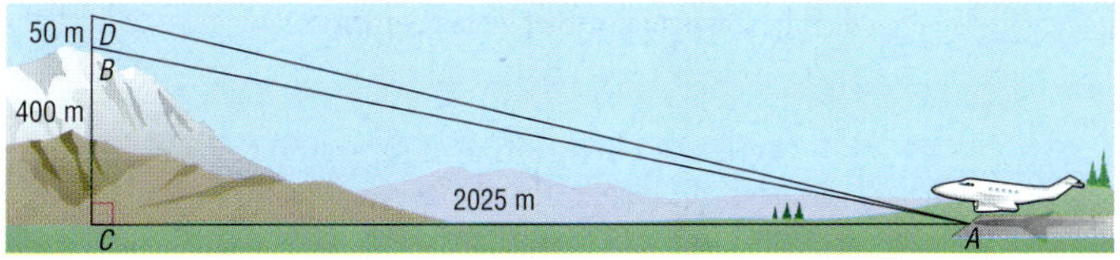

Since CB is 400 meters and BD is 50 meters, CD is 450 meters. Let x represent $m\angle DAC$.

$\tan x° = \frac{CD}{AC}$ $\tan = \frac{\text{opposite}}{\text{adjacent}}$

$\tan x° = \frac{450}{2025}$ $CD = 450, AC = 2025$

$x = \tan^{-1}\left(\frac{450}{2025}\right)$ Solve for x.

$x \approx 12.5$ Use a calculator.

The angle of elevation for the takeoff should be more than 12.5°.

7-5 Lesson Notes

1 Focus

5-Minute Check Transparency 7-5 Use as a quiz or review of Lesson 7-4.

Mathematical Background notes are available for this lesson on p. 340D.

How do airline pilots use angles of elevation and depression?

Ask students:

- In the right triangle formed by the pilot's scenario, what triangle side does the height of the mountain represent? What does the line along the runway represent? **the side opposite the unknown angle, the side adjacent to the unknown angle**
- If the pilot is going to calculate the angle that she needs to clear the mountain given the height of the mountain and the horizontal distance from her takeoff point to the base of the mountain, what trigonometric ratio would she use? **tangent**

Resource Manager

Workbook and Reproducible Masters

Chapter 7 Resource Masters
- Study Guide and Intervention, pp. 375–376
- Skills Practice, p. 377
- Practice, p. 378
- Reading to Learn Mathematics, p. 379
- Enrichment, p. 380

Teaching Geometry With Manipulatives Masters, p.18

Transparencies
5-Minute Check Transparency 7-5
Real-World Transparency 7
Answer Key Transparencies

Technology
GeomPASS: Tutorial Plus, Lesson 15
Interactive Chalkboard

2 Teach

ANGLE OF ELEVATION

In-Class Example

PowerPoint®

1 **CIRCUS ACTS** At the circus, a person in the audience at ground level watches the high-wire routine. A 5-foot-6-inch tall acrobat is standing on a platform that is 25 feet off the ground. How far is the audience member from the base of the platform, if the angle of elevation from the audience member's line of sight to the top of the acrobat is 27°? **about 60 ft**

ANGLE OF DEPRESSION

In-Class Examples

PowerPoint®

Teaching Tip Remind students to draw upon their previous knowledge of angle relationships formed by two parallel lines and a transversal. Students may want to review these relationships.

2 A wheelchair ramp is 3 meters long and inclines at 6°. Find the height of the ramp to the nearest tenth centimeter. **about 31.4 cm**

3 Vernon is on the top deck of a cruise ship and observes two dolphins following each other directly away from the ship in a straight line. Vernon's position is 154 meters above sea level, and the angles of depression to the two dolphins are 35° and 36°. Find the distance between the two dolphins to the nearest meter. **about 8 m**

ANGLES OF DEPRESSION An **angle of depression** is the angle between the line of sight when an observer looks downward, and the horizontal.

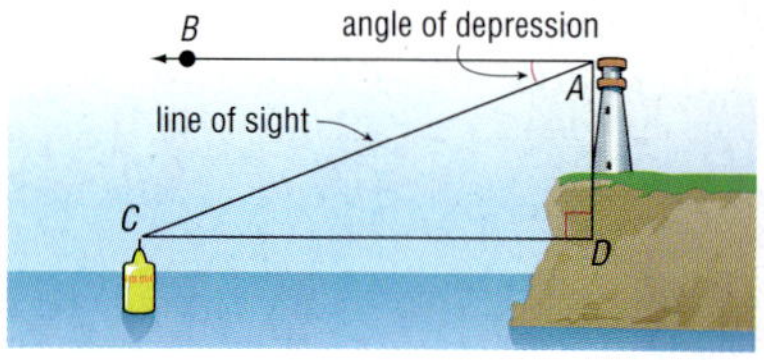

FCAT Practice

Standardized Test Practice

A B C D

Example 2 Angle of Depression

Short-Response Test Item

The tailgate of a moving van is 3.5 feet above the ground. A loading ramp is attached to the rear of the van at an incline of 10°. Find the length of the ramp to the nearest tenth foot.

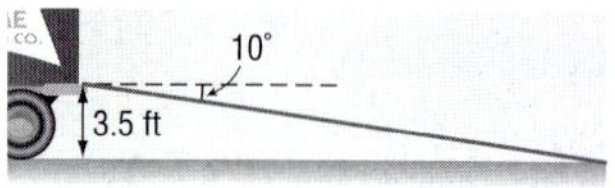

The Princeton Review

Test-Taking Tip

Draw a Figure It may be helpful to draw a figure using only information necessary to solve a problem.

Read the Test Item

The angle of depression between the ramp and the horizontal is 10°. Use trigonometry to find the length of the ramp.

Solve the Test Item

Method 1

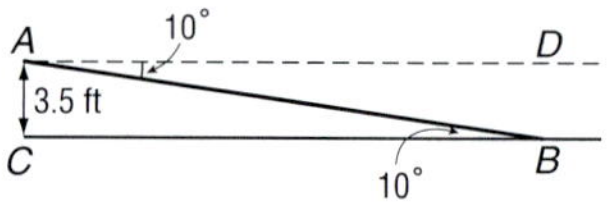

The ground and the horizontal level with the back of the van are parallel. Therefore, $m\angle DAB = m\angle ABC$ since they are alternate interior angles.

$$\sin 10° = \frac{3.5}{AB}$$

$$AB \sin 10° = 3.5$$

$$AB = \frac{3.5}{\sin 10°}$$

$$AB \approx 20.2$$

The ramp is about 20.2 feet long.

Method 2

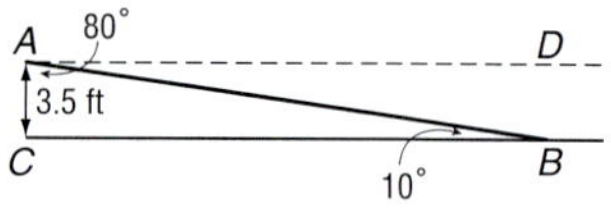

The horizontal line from the back of the van and the segment from the ground to the back of the van are perpendicular. So, $\angle DAB$ and $\angle BAC$ are complementary angles. Therefore, $m\angle BAC = 90 - 10$ or 80.

$$\cos 80° = \frac{3.5}{AB}$$

$$AB \cos 80° = 3.5$$

$$AB = \frac{3.5}{\cos 80°}$$

$$AB \approx 20.2$$

Study Tip

Common Misconception

The angle of depression is often not an angle of the triangle, but the complement to an angle of the triangle. In $\triangle DBC$, the angle of depression is $\angle BCE$, not $\angle DCB$.

Angles of elevation or depression to two different objects can be used to find the distance between those objects.

Example 3 Indirect Measurement

Olivia is in a lighthouse on a cliff. She observes two sailboats due east of the lighthouse. The angles of depression to the two boats are 33° and 57°. Find the distance between the two sailboats to the nearest foot.

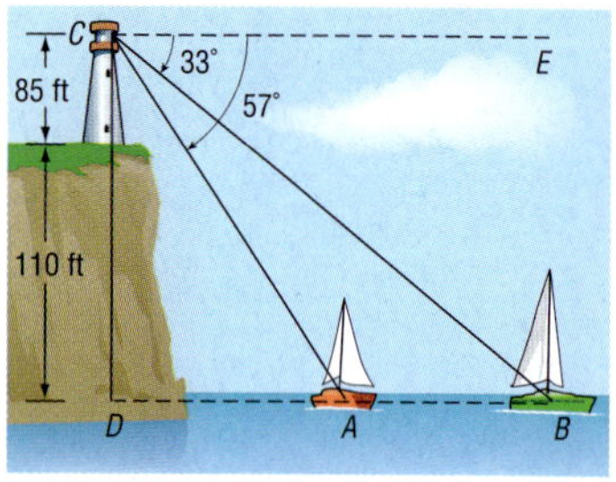

$\triangle CDA$ and $\triangle CDB$ are right triangles, and $CD = 110 + 85$ or 195. The distance between the boats is AB or $BD - AD$. Use the right triangles to find these two lengths.

DAILY INTERVENTION

Differentiated Instruction

Kinesthetic Using a meterstick and a calculator, groups of students can find angles of elevation and depression for different objects in the classroom. Groups can measure one person's eye level from the floor, and the topmost height of a wall clock from the floor. The person stands 5 feet away from the clock, and the group calculates the angle of elevation from the person's line of sight to the top of the object. Repeat for items placed on the floor, and include variations like having the person stand on a chair, or placing two objects on the floor a certain distance from each other.

372 Chapter 7 Right Triangles and Trigonometry

Study Tip

Look Back
To review **alternate interior angles**, see Lesson 3-2.

Because $\overline{CE}$ and $\overline{DB}$ are horizontal lines, they are parallel. Thus, $\angle ECB \cong \angle CBD$ and $\angle ECA \cong \angle CAD$ because they are alternate interior angles. This means that $m\angle CBD = 33$ and $m\angle CAD = 57$.

$\tan 33° = \frac{195}{DB}$ — $\tan = \frac{\text{opposite}}{\text{adjacent}}$; $m\angle CBD = 33$

$DB \tan 33° = 195$ — Multiply each side by DB.

$DB = \frac{195}{\tan 33°}$ — Divide each side by tan 33°.

$DB \approx 300.27$ — Use a calculator.

$\tan 57° = \frac{195}{DA}$ — $\tan = \frac{\text{opposite}}{\text{adjacent}}$; $m\angle CAD = 57$

$DA \tan 57° = 195$ — Multiply each side by DA.

$DA = \frac{195}{\tan 57°}$ — Divide each side by tan 57°.

$DA \approx 126.63$ — Use a calculator.

The distance between the boats is $DB - DA$.

$DB - DA \approx 300.27 - 126.63$ or about 174 feet.

Check for Understanding

Concept Check

1. **OPEN ENDED** Find a real-life example of an angle of depression. Draw a diagram and identify the angle of depression. **1–2. See margin.**

2. **Explain** why an angle of elevation is given that name.

3. **Name** the angles of depression and elevation in the figure. **The angle of depression is $\angle FPB$ and the angle of elevation is $\angle TBP$.**

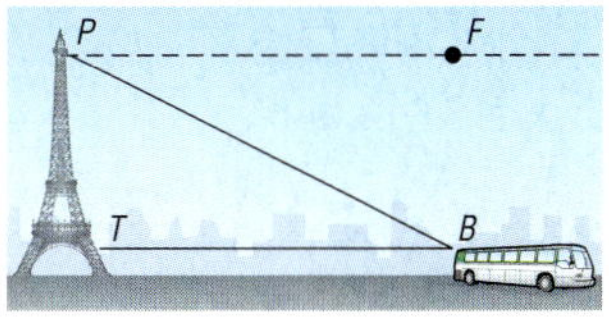

Guided Practice

GUIDED PRACTICE KEY	
Exercises	Examples
4, 5	1
6	2
7	3

4. **AVIATION** A pilot is flying at 10,000 feet and wants to take the plane up to 20,000 feet over the next 50 miles. What should be his angle of elevation to the nearest tenth? (*Hint:* There are 5280 feet in a mile.) **about 2.2°**

5. **SHADOWS** Find the angle of elevation of the sun when a 7.6-meter flagpole casts a 18.2-meter shadow. Round to the nearest tenth of a degree. **22.7°**

6. **SALVAGE** A salvage ship uses sonar to determine that the angle of depression to a wreck on the ocean floor is 13.25°. The depth chart shows that the ocean floor is 40 meters below the surface. How far must a diver lowered from the salvage ship walk along the ocean floor to reach the wreck? **about 169.9 m**

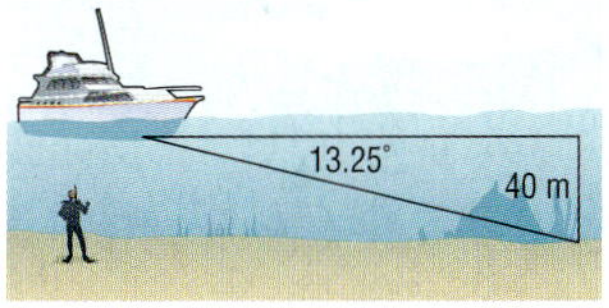

FCAT Practice Standardized Test Practice (A) (B) (C) (D)

7. **SHORT RESPONSE** From the top of a 150-foot high tower, an air traffic controller observes an airplane on the runway. To the nearest foot, how far from the base of the tower is the airplane? **706 ft**

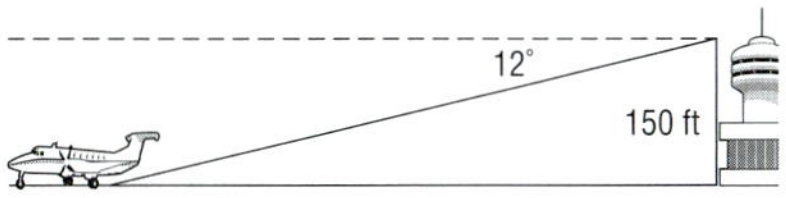

www.geometryonline.com/extra_examples/fcat

3 Practice/Apply

Study Notebook

Have students—

- add the definitions/examples of the vocabulary terms to their Vocabulary Builder worksheets for Chapter 7.
- include any other item(s) that they find helpful in mastering the skills in this lesson.

About the Exercises...

Organization by Objective

- **Angles of Elevation:** 12, 14–18
- **Angles of Depression:** 8–11, 13, 19–20

Odd/Even Assignments

Exercises 8–20 are structured so that students practice the same concepts whether they are assigned odd or even problems.

Alert! Exercise 23 requires the Internet or other research materials.

Assignment Guide

Basic: 9–23 odd, 26–48

Average: 9–25 odd, 26–48

Advanced: 8–26 even, 27–40 (optional: 41–48)

Answers

1. Sample answer: $\angle ABC$

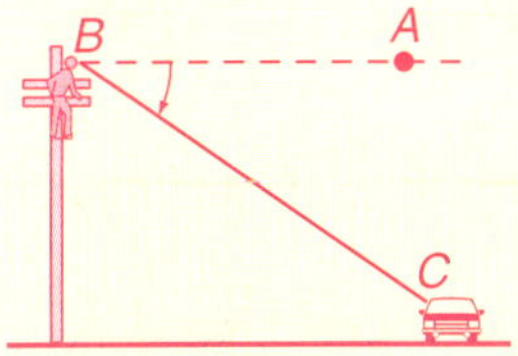

2. Sample answer: An angle of elevation is called that because the angle formed by a horizontal line and a segment joining two endpoints rises above the horizontal line.

Study Guide and Intervention, p. 375 (shown) and p. 376

Angles of Elevation Many real-world problems that involve looking up to an object can be described in terms of an **angle of elevation**, which is the angle between an observer's line of sight and a horizontal line.

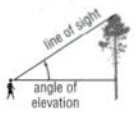

Example **The angle of elevation from point A to the top of a cliff is 34°. If point A is 1000 feet from the base of the cliff, how high is the cliff?**

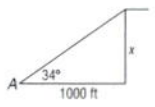

Let x = the height of the cliff.

$\tan 34° = \frac{x}{1000}$ $\tan = \frac{\text{opposite}}{\text{adjacent}}$

$1000(\tan 34°) = x$ Multiply each side by 1000.

$674.5 = x$ Use a calculator.

The height of the cliff is about 674.5 feet.

Exercises

Solve each problem. Round measures of segments to the nearest whole number and angles to the nearest degree.

1. The angle of elevation from point A to the top of a hill is 49°. If point A is 400 feet from the base of the hill, how high is the hill?
   460 ft

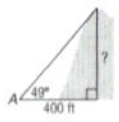

2. Find the angle of elevation of the sun when a 12.5-meter-tall telephone pole casts a 18-meter-long shadow.
   35°

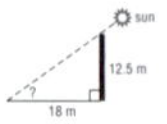

3. A ladder leaning against a building makes an angle of 78° with the ground. The foot of the ladder is 5 feet from the building. How long is the ladder?
   24 ft

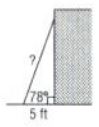

4. A person whose eyes are 5 feet above the ground is standing on the runway of an airport 100 feet from the control tower. That person observes an air traffic controller at the window of the 132-foot tower. What is the angle of elevation?
   52°

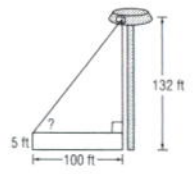

Skills Practice, p. 377 and Practice, p. 378 (shown)

Name the angle of depression or angle of elevation in each figure.

1. 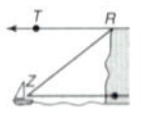

   $\angle TRZ$; $\angle YZR$

2. 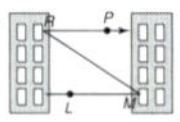

   $\angle PRM$; $\angle LMR$

3. **WATER TOWERS** A student can see a water tower from the closest point of the soccer field at San Lobos High School. The edge of the soccer field is about 110 feet from the water tower and the water tower stands at a height of 32.5 feet. What is the angle of elevation if the eye level of the student viewing the tower from the edge of the soccer field is 6 feet above the ground? Round to the nearest tenth degree.
   about 13.5°

4. **CONSTRUCTION** A roofer props a ladder against a wall so that the top of the ladder reaches a 30-foot roof that needs repair. If the angle of elevation from the bottom of the ladder to the roof is 55°, how far is the ladder from the base of the wall? Round your answer to the nearest foot.
   about 21 ft

5. **TOWN ORDINANCES** The town of Belmont restricts the height of flagpoles to 25 feet on any property. Lindsay wants to determine whether her school is in compliance with the regulation. Her eye level is 5.5 feet from the ground and she stands 36 feet from the flagpole. If the angle of elevation is about 25°, what is the height of the flagpole to the nearest tenth foot?
   about 22.3 ft

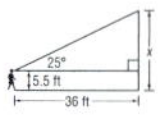

6. **GEOGRAPHY** Stephan is standing on a mesa at the Painted Desert. The elevation of the mesa is about 1380 meters and Stephan's eye level is 1.8 meters above ground. If Stephan can see a band of multicolored shale at the bottom and the angle of depression is 29°, about how far is the band of shale from his eyes? Round to the nearest meter.
   about 2850 m

7. **INDIRECT MEASUREMENT** Mr. Dominguez is standing on a 40-foot ocean bluff near his home. He can see his two dogs on the beach below. If his line of sight is 6 feet above the ground and the angles of depression to his dogs are 34° and 48°, how far apart are the dogs to the nearest foot?
   about 27 ft

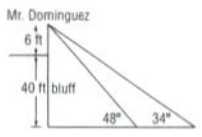

Reading to Learn Mathematics, p. 379 ELL

Pre-Activity **How do airline pilots use angles of elevation and depression?**

Read the introduction to Lesson 7-5 at the top of page 371 in your textbook.

What does the angle measure tell the pilot? Sample answer: how steep her ascent must be to clear the peak

Reading the Lesson

1. Refer to the figure. The two observers are looking at one another. Select the correct choice for each question.

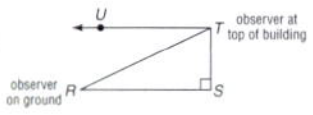

   a. What is the line of sight? iii
      (i) line RS (ii) line ST (iii) line RT (iv) line TU
   b. What is the angle of elevation? ii
      (i) $\angle RST$ (ii) $\angle SRT$ (iii) $\angle RTS$ (iv) $\angle UTR$
   c. What is the angle of depression? iv
      (i) $\angle RST$ (ii) $\angle SRT$ (iii) $\angle RTS$ (iv) $\angle UTR$
   d. How are the angle of elevation and the angle of depression related? ii
      (i) They are complementary.
      (ii) They are congruent.
      (iii) They are supplementary.
      (iv) The angle of elevation is larger than the angle of depression.
   e. Which postulate or theorem that you learned in Chapter 3 supports your answer for part c? iv
      (i) Corresponding Angles Postulate
      (ii) Alternate Exterior Angles Theorem
      (iii) Consecutive Interior Angles Theorem
      (iv) Alternate Interior Angles Theorem

2. A student says that the angle of elevation from his eye to the top of a flagpole is 135°. What is wrong with the student's statement?
   An angle of elevation cannot be obtuse.

Helping You Remember

3. A good way to remember something is to explain it to someone else. Suppose a classmate finds it difficult to distinguish between angles of elevation and angles of depression. What are some hints you can give her to help her get it right every time? Sample answers: (1) The angle of depression and the angle of elevation are both measured between the horizontal and the line of sight. (2) The angle of depression is always congruent to the angle of elevation in the same diagram. (3) Associate the word *elevation* with the word *up* and the word *depression* with the word *down*.

★ indicates increased difficulty

Practice and Apply

Homework Help

For Exercises	See Examples
12, 14–18	1
9–11	2
8, 13, 19, 20	3

Extra Practice
See page 768.

8. **BOATING** Two boats are observed by a parasailer 75 meters above a lake. The angles of depression are 12.5° and 7°. How far apart are the boats? about 273 m

9. **GOLF** A golfer is standing at the tee, looking up to the green on a hill. If the tee is 36 yards lower than the green and the angle of elevation from the tee to the hole is 12°, find the distance from the tee to the hole. about 173.2 yd

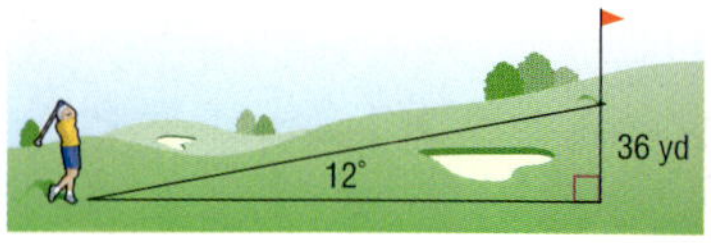

10. **AVIATION** After flying at an altitude of 500 meters, a helicopter starts to descend when its ground distance from the landing pad is 11 kilometers. What is the angle of depression for this part of the flight? about 2.6°

11. **SLEDDING** A sledding run is 300 yards long with a vertical drop of 27.6 yards. Find the angle of depression of the run. about 5.3°

12. **RAILROADS** The Monongahela Incline overlooks the city of Pittsburgh, Pennsylvania. Refer to the information at the left to determine the incline of the railway. about 35.6°

13. **AMUSEMENT PARKS** From the top of a roller coaster, 60 yards above the ground, a rider looks down and sees the merry-go-round and the Ferris wheel. If the angles of depression are 11° and 8° respectively, how far apart are the merry-go-round and the Ferris wheel? about 118.2 yd

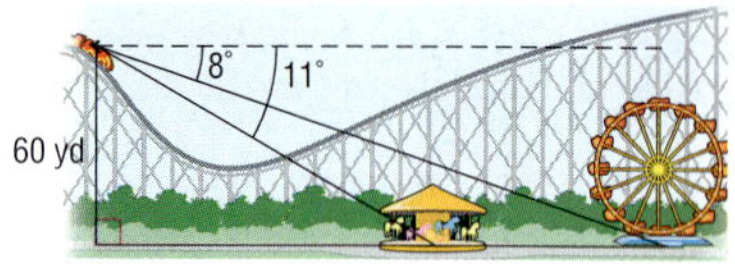

CIVIL ENGINEERING **For Exercises 14 and 15, use the following information.**
The percent grade of a highway is the ratio of the vertical rise or fall over a given horizontal distance. The ratio is expressed as a percent to the nearest whole number. Suppose a highway has a vertical rise of 140 feet for every 2000 feet of horizontal distance.

14. Calculate the percent grade of the highway. 7%
15. Find the angle of elevation that the highway makes with the horizontal. about 4°

16. **SKIING** A ski run has an angle of elevation of 24.4° and a vertical drop of 1100 feet. To the nearest foot, how long is the ski run? 2663 ft

GEYSERS **For Exercises 17 and 18, use the following information.**
Kirk visits Yellowstone Park and Old Faithful on a perfect day. His eyes are 6 feet from the ground, and the geyser can reach heights ranging from 90 feet to 184 feet.

17. If Kirk stands 200 feet from the geyser and the eruption rises 175 feet in the air, what is the angle of elevation to the top of the spray to the nearest tenth? about 40.2°
18. In the afternoon, Kirk returns and observes the geyser's spray reach a height of 123 feet when the angle of elevation is 37°. How far from the geyser is Kirk standing to the nearest tenth of a foot? about 155.3 ft

★ 19. **BIRDWATCHING** Two observers are 200 feet apart, in line with a tree containing a bird's nest. The angles of elevation to the bird's nest are 30° and 60°. How far is each observer from the base of the tree? 100 ft, 300 ft

More About. . .

Railroads

The Monongahela Incline is 635 feet long with a vertical rise of 369.39 feet. It was built at a cost of $50,000 and opened on May 28, 1870. It is still used by commuters to and from Pittsburgh.

Source: www.portauthority.org

Enrichment, p. 380

Reading Mathematics

The three most common trigonometric ratios are **sine**, **cosine**, and **tangent**. Three other ratios are the **cosecant**, **secant**, and **cotangent**. The chart below shows abbreviations and definitions for all six ratios. Refer to the triangle at the right.

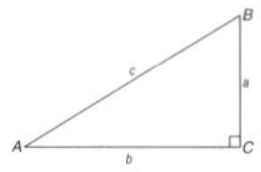

Abbreviation	Read as:	Ratio
sin A	the sine of $\angle A$	$\frac{\text{leg opposite } \angle A}{\text{hypotenuse}} = \frac{a}{c}$
cos A	the cosine of $\angle A$	$\frac{\text{leg adjacent to } \angle A}{\text{hypotenuse}} = \frac{b}{c}$
tan A	the tangent of $\angle A$	$\frac{\text{leg opposite } \angle A}{\text{leg adjacent to } \angle A} = \frac{a}{b}$
csc A	the cosecant of $\angle A$	$\frac{\text{hypotenuse}}{\text{leg opposite } \angle A} = \frac{c}{a}$
sec A	the secant of $\angle A$	$\frac{\text{hypotenuse}}{\text{leg adjacent to } \angle A} = \frac{c}{b}$

374 Chapter 7 Right Triangles and Trigonometry

20. **METEOROLOGY** The altitude of the base of a cloud formation is called the *ceiling*. To find the ceiling one night, a meteorologist directed a spotlight vertically at the clouds. Using a theodolite placed 83 meters from the spotlight and 1.5 meters above the ground, he found the angle of elevation to be 62.7°. How high was the ceiling? **about 162.3 m**

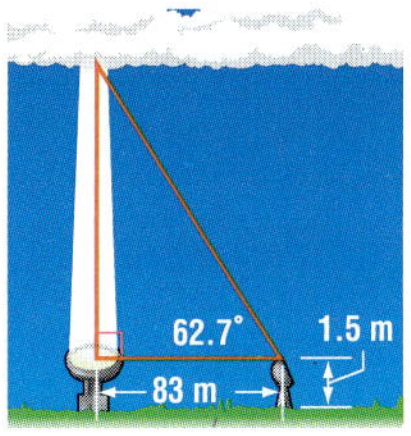

MEDICINE **For Exercises 21–23, use the following information.**
A doctor is using a treadmill to assess the strength of a patient's heart. At the beginning of the exam, the 48-inch long treadmill is set at an incline of 10°.

21. How far off the horizontal is the raised end of the treadmill at the beginning of the exam? **about 8.3 in.**

22. During one stage of the exam, the end of the treadmill is 10 inches above the horizontal. What is the incline of the treadmill to the nearest degree? **12°**

23. Suppose the exam is divided into five stages and the incline of the treadmill is increased 2° for each stage. Does the end of the treadmill rise the same distance between each stage? **no**

More About. . .

Travel

Ayers Rock is the largest monolith, a type of rock formation, in the world. It is approximately 3.6 kilometers long and 2 kilometers wide. The rock is believed to be the tip of a mountain, two thirds of which is underground.

Source: www.atn.com.au

★ 24. **TRAVEL** Ulura or Ayers Rock is a sacred place for Aborigines of the western desert of Australia. Kwan-Yong uses a theodolite to measure the angle of elevation from the ground to the top of the rock to be 15.85°. He walks half a kilometer closer and measures the angle of elevation to be 25.6°. How high is Ayers Rock to the nearest meter? **about 348 m**

★ 25. **AEROSPACE** On July 20, 1969, Neil Armstrong became the first human to walk on the moon. During this mission, the lunar lander *Eagle* traveled aboard *Apollo 11*. Before sending *Eagle* to the surface of the moon, *Apollo 11* orbited the moon three miles above the surface. At one point in the orbit, the onboard guidance system measured the angles of depression to the far and near edges of a large crater. The angles measured 16° and 29°, respectively. Find the distance across the crater. **about 5.1 mi**

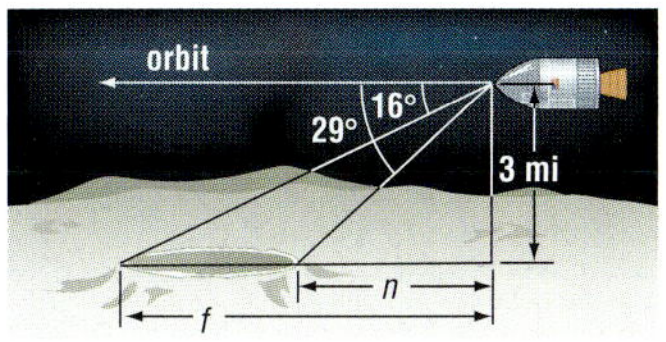

Online Research **Data Update** Use the Internet to determine the angle of depression formed by someone aboard the international space station looking down to your community. Visit www.geometryonline.com/data_update to learn more.

26. **CRITICAL THINKING** Two weather observation stations are 7 miles apart. A weather balloon is located between the stations. From Station 1, the angle of elevation to the weather balloon is 33°. From Station 2, the angle of elevation to the balloon is 52°. Find the altitude of the balloon to the nearest tenth of a mile. **about 3.0 mi**

27. **WRITING IN MATH** Answer the question that was posed at the beginning of the lesson. **See margin.**

How do airline pilots use angles of elevation and depression?

Include the following in your answer:
- when pilots use angles of elevation or depression, and
- the difference between angles of elevation and depression.

www.geometryonline.com/self_check_quiz/fcat

Answer

27. Answers should include the following.
- Pilots use angles of elevation when they are ascending and angles of depression when descending.
- Angles of elevation are formed when a person looks upward and angles of depression are formed when a person looks downward.

Teacher to Teacher

Susan M. Parece — Plymouth South High School, Plymouth, MA

I like to make a clinometer (also called a hypsometer) and have my students try to measure objects in the schoolyard, such as height of the building, height of the flagpole, and height of the lights. (Directions can be found at www.globe.org.uk/activities/toolkit/tollkit.pdf.)

4 Assess

Open-Ended Assessment

Modeling Using a ruler, have students model different angles of elevation and depression and name real-world situations that the angle depicted could be modeling.

Getting Ready for Lesson 7-6

Prerequisite Skill Students will learn about the Law of Sines in Lesson 7-6. They will use proportions to solve for side lengths and angle measures. Use Exercises 41–48 to determine your students' familiarity with solving proportions.

FCAT Practice

Standardized Test Practice

28. The top of a signal tower is 120 meters above sea level. The angle of depression from the top of tower to a passing ship is 25°. How many meters from the foot of the tower is the ship? **B**

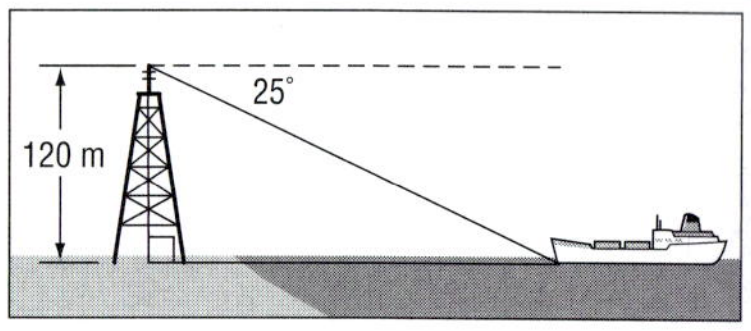

(A) 283.9 m (B) 257.3 m
(C) 132.4 m (D) 56 m

29. **ALGEBRA** If $\frac{y}{28} = \frac{x}{16}$, then find x when $y = \frac{1}{2}$. **A**

(A) $\frac{2}{7}$ (B) $\frac{4}{7}$ (C) $\frac{7}{4}$ (D) $3\frac{1}{2}$

Maintain Your Skills

Mixed Review

Find the measure of each angle to the nearest tenth of a degree. *(Lesson 7-4)*

30. $\cos A = 0.6717$ **47.8**
31. $\sin B = 0.5127$ **30.8**
32. $\tan C = 2.1758$ **65.3**
33. $\cos D = 0.3421$ **70.0**
34. $\sin E = 0.1455$ **8.4**
35. $\tan F = 0.3541$ **19.5**

Find x and y. *(Lesson 7-3)*

36.

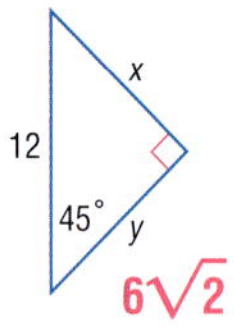

$6\sqrt{2}$; $6\sqrt{2}$

37.

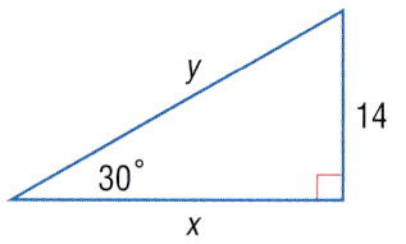

$14\sqrt{3}$; 28

38.

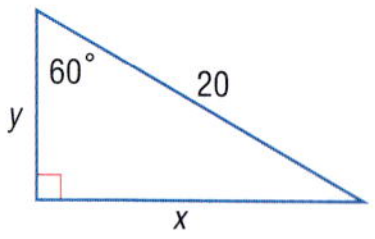

$10\sqrt{3}$; 10

39. **HOBBIES** A twin-engine airplane used for medium-range flights has a length of 78 meters and a wingspan of 90 meters. If a scale model is made with a wingspan of 36 centimeters, find its length. *(Lesson 6-2)* **31.2 cm**

40. Copy and complete the flow proof. *(Lesson 4-6)*

Given: $\angle 5 \cong \angle 6$
$\overline{FR} \cong \overline{GS}$

Prove: $\angle 4 \cong \angle 3$

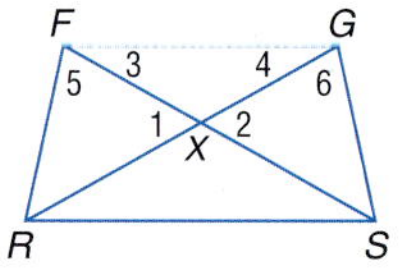

Proof:

- $\angle 5 \cong \angle 6$ — Given
- $\overline{FR} \cong \overline{GS}$ — Given
- a. ___?___ — Vert. ∠s are ≅. **$\angle 1 \cong \angle 2$**

→ $\triangle FXR \cong \triangle GXS$ — b. ___?___ **AAS**

→ c. ___?___ **$\overline{FX} \cong \overline{GX}$** — d. ___?___ **CPCTC**

→ e. ___?___ **$\angle 4 \cong \angle 3$** — f. ___?___ **Isos. △ Th.**

Getting Ready for the Next Lesson

PREREQUISITE SKILL **Solve each proportion.**
*(To review **solving proportions**, see Lesson 6-1.)*

41. $\frac{x}{6} = \frac{35}{42}$ **5**
42. $\frac{3}{x} = \frac{5}{45}$ **27**
43. $\frac{12}{17} = \frac{24}{x}$ **34**
44. $\frac{24}{36} = \frac{x}{15}$ **10**
45. $\frac{12}{13} = \frac{48}{x}$ **52**
46. $\frac{x}{18} = \frac{5}{8}$ **11.25**
47. $\frac{28}{15} = \frac{7}{x}$ **3.75**
48. $\frac{x}{40} = \frac{3}{26}$ **$\frac{60}{13}$**

376 Chapter 7 Right Triangles and Trigonometry

7-6 The Law of Sines

Sunshine State Standards
MA.A.3.4.3, MA.B.2.4.1, MA.B.3.4.1, MA.C.3.4.1, MA.C.3.4.2

Vocabulary

- Law of Sines
- solving a triangle

What You'll Learn

- Use the Law of Sines to solve triangles.
- Solve problems by using the Law of Sines.

How are triangles used in radio astronomy?

The Very Large Array (VLA), one of the world's premier astronomical radio observatories, consists of 27 radio antennas in a Y-shaped configuration on the Plains of San Agustin in New Mexico. Astronomers use the VLA to make pictures from the radio waves emitted by astronomical objects. Construction of the antennas is supported by a variety of triangles, many of which are not right triangles.

THE LAW OF SINES In trigonometry, the **Law of Sines** can be used to find missing parts of triangles that are not right triangles.

Tip

Angles
also values for
A, and tan *A*
= 90°. Values of
for these angles
und using the
etric functions
alculator.

Key Concept — *Law of Sines*

Let $\triangle ABC$ be any triangle with a, b, and c representing the measures of the sides opposite the angles with measures A, B, and C, respectively. Then

$$\frac{\sin A}{a} = \frac{\sin B}{b} = \frac{\sin C}{c}.$$

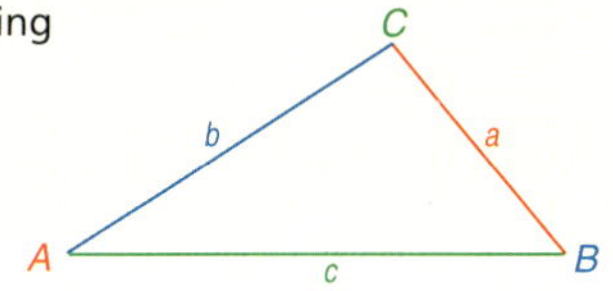

Proof — *Law of Sines*

$\triangle ABC$ is a triangle with an altitude from C that intersects $\overline{AB}$ at D. Let h represent the measure of $\overline{CD}$. Since $\triangle ADC$ and $\triangle BDC$ are right triangles, we can find $\sin A$ and $\sin B$.

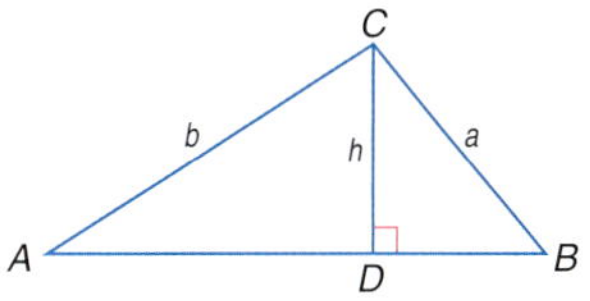

$\sin A = \frac{h}{b}$	$\sin B = \frac{h}{a}$	Definition of sine
$b \sin A = h$	$a \sin B = h$	Cross products

$b \sin A = a \sin B$	Substitution
$\frac{\sin A}{a} = \frac{\sin B}{b}$	Divide each side by ab.

The proof can be completed by using a similar technique with the other altitudes to show that $\frac{\sin A}{a} = \frac{\sin C}{c}$ and $\frac{\sin B}{b} = \frac{\sin C}{c}$.

7-6 Lesson Notes

1 Focus

5-Minute Check Transparency 7-6 Use as a quiz or review of Lesson 7-5.

Mathematical Background notes are available for this lesson on p. 340D.

How are triangles used in radio astronomy?

Ask students:

- Why is it called radio astronomy? **because they are analyzing radio signals**
- What are some examples of astronomical objects? **Sample answers: stars, planets, quasars, brown dwarfs, galaxies**

Resource Manager

Workbook and Reproducible Masters

Chapter 7 Resource Masters

- Study Guide and Intervention, pp. 381–382
- Skills Practice, p. 383
- Practice, p. 384
- Reading to Learn Mathematics, p. 385
- Enrichment, p. 386
- Assessment, p. 408

Graphing Calculator and Computer Masters, p. 30

Transparencies

5-Minute Check Transparency 7-6
Answer Key Transparencies

Technology

Interactive Chalkboard
Multimedia Applications: Virtual Activities

2 Teach

THE LAW OF SINES

In-Class Examples

Power Point®

1 **a.** Find p. Round to the nearest tenth.

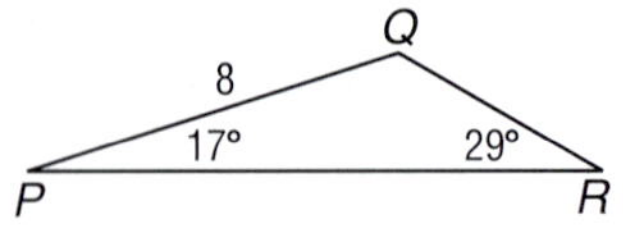

4.8

b. Find $m\angle L$ to the nearest degree in $\triangle LMN$ if $n = 7$, $\ell = 9$, and $m\angle N = 43$.

61

Teaching Tip To save a step, students can also write $a \sin B = b \sin A$ and then substitute values to solve.

2 **a.** Solve $\triangle DEF$ if $m\angle D = 112$, $m\angle F = 8$, and $f = 2$.

$d \approx 13.2$; $e \approx 12.4$; $m\angle E = 60$

b. Solve $\triangle IJK$ if $m\angle J = 32$, $i = 30$, and $j = 16$.

$k \approx 27.3$; $m\angle I \approx 83.5$; $m\angle K \approx 64.5$

Study Tip

Rounding

If you round before the final answer, your results may differ from results in which rounding was not done until the final answer.

Example 1 Use the Law of Sines

a. Find b. Round to the nearest tenth.

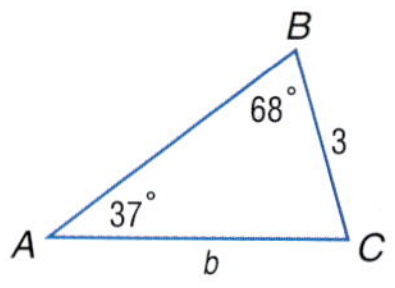

Use the Law of Sines to write a proportion.

$$\frac{\sin A}{a} = \frac{\sin B}{b} \quad \text{Law of Sines}$$

$$\frac{\sin 37°}{3} = \frac{\sin 68°}{b} \quad m\angle A = 37,\ a = 3,\ m\angle B = 68$$

$$b \sin 37° = 3 \sin 68° \quad \text{Cross products}$$

$$b = \frac{3 \sin 68°}{\sin 37°} \quad \text{Divide each side by } \sin 37°.$$

$$b \approx 4.6 \quad \text{Use a calculator.}$$

b. Find $m\angle Z$ to the nearest degree in $\triangle XYZ$ if $y = 17$, $z = 14$, and $m\angle Y = 92$.

Write a proportion relating $\angle Y$, $\angle Z$, y, and z.

$$\frac{\sin Y}{y} = \frac{\sin Z}{z} \quad \text{Law of Sines}$$

$$\frac{\sin 92°}{17} = \frac{\sin Z}{14} \quad m\angle Y = 92,\ y = 17,\ z = 14$$

$$14 \sin 92° = 17 \sin Z \quad \text{Cross products}$$

$$\frac{14 \sin 92°}{17} = \sin Z \quad \text{Divide each side by 17.}$$

$$\sin^{-1}\left(\frac{14 \sin 92°}{17}\right) = Z \quad \text{Solve for } Z.$$

$$55° \approx Z \quad \text{Use a calculator.}$$

So, $m\angle Z \approx 55$.

The Law of Sines can be used to solve a triangle. **Solving a triangle** means finding the measures of all of the angles and sides of a triangle.

Study Tip

Look Back

To review the **Angle Sum Theorem**, see Lesson 4-2.

Example 2 Solve Triangles

a. Solve $\triangle ABC$ if $m\angle A = 33$, $m\angle B = 47$, and $b = 14$. Round angle measures to the nearest degree and side measures to the nearest tenth.

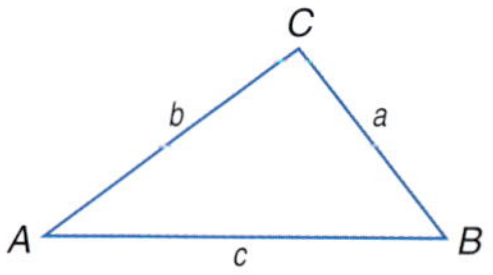

We know the measures of two angles of the triangle. Use the Angle Sum Theorem to find $m\angle C$.

$$m\angle A + m\angle B + m\angle C = 180 \quad \text{Angle Sum Theorem}$$

$$33 + 47 + m\angle C = 180 \quad m\angle A = 33,\ m\angle B = 47$$

$$80 + m\angle C = 180 \quad \text{Add.}$$

$$m\angle C = 100 \quad \text{Subtract 80 from each side.}$$

Since we know $m\angle B$ and b, use proportions involving $\frac{\sin B}{b}$.

To find a:		To find c:
$\frac{\sin B}{b} = \frac{\sin A}{a}$	Law of Sines	$\frac{\sin B}{b} = \frac{\sin C}{c}$
$\frac{\sin 47°}{14} = \frac{\sin 33°}{a}$	Substitute.	$\frac{\sin 47°}{14} = \frac{\sin 100°}{c}$
$a \sin 47° = 14 \sin 33°$	Cross products	$c \sin 47° = 14 \sin 100°$
$a = \frac{14 \sin 33°}{\sin 47°}$	Divide each side by sin 47°.	$c = \frac{14 \sin 100°}{\sin 47°}$
$a \approx 10.4$	Use a calculator.	$c \approx 18.9$

Therefore, $m\angle C = 100$, $a \approx 10.4$, and $c \approx 18.9$.

DAILY INTERVENTION

Differentiated Instruction

Interpersonal Separate the class into groups of three students to work through the exercises. One group member can select the exercise. The second member sets up the equation for the Law of Sines and fills in the values. The third person uses a calculator to solve the problem. Group members can rotate tasks so each member participates in each responsibility.

378 Chapter 7 Right Triangles and Trigonometry

Study Tip

An Equivalent Proportion

The Law of Sines may also be written as $\frac{a}{\sin A} = \frac{b}{\sin B} = \frac{c}{\sin C}$. You may wish to use this form when finding the length of a side.

b. Solve $\triangle ABC$ if $m\angle C = 98$, $b = 14$, and $c = 20$. Round angle measures to the nearest degree and side measures to the nearest tenth.

We know the measures of two sides and an angle opposite one of the sides.

$\frac{\sin B}{b} = \frac{\sin C}{c}$	Law of Sines
$\frac{\sin B}{14} = \frac{\sin 98°}{20}$	$m\angle C = 98$, $b = 14$, and $c = 20$
$20 \sin B = 14 \sin 98°$	Cross products
$\sin B = \frac{14 \sin 98°}{20}$	Divide each side by 20.
$B = \sin^{-1}\left(\frac{14 \sin 98°}{20}\right)$	Solve for B.
$B \approx 44°$	Use a calculator.
$m\angle A + m\angle B + m\angle C = 180$	Angle Sum Theorem
$m\angle A + 44 + 98 = 180$	$m\angle B = 44$ and $m\angle C = 98$
$m\angle A + 142 = 180$	Add.
$m\angle A = 38$	Subtract 142 from each side.
$\frac{\sin A}{a} = \frac{\sin C}{c}$	Law of Sines
$\frac{\sin 38°}{a} = \frac{\sin 98°}{20}$	$m\angle A = 38$, $m\angle C = 98$, and $c = 20$
$20 \sin 38° = a \sin 98°$	Cross products
$\frac{20 \sin 38°}{\sin 98°} = a$	Divide each side by sin 98°.
$12.4 \approx a$	Use a calculator.

Therefore, $A \approx 38°$, $B \approx 44°$, and $a \approx 12.4$.

USE THE LAW OF SINES TO SOLVE PROBLEMS The Law of Sines is very useful in solving direct and indirect measurement applications.

Example 3 Indirect Measurement

When the angle of elevation to the sun is 62°, a telephone pole tilted at an angle of 7° from the vertical casts a shadow of 30 feet long on the ground. Find the length of the telephone pole to the nearest tenth of a foot.

Draw a diagram.

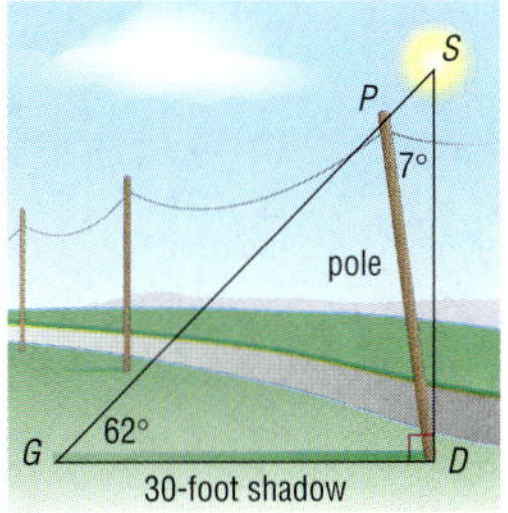

Draw $\overline{SD} \perp \overline{GD}$. Then find the $m\angle GDP$ and $m\angle GPD$.

$m\angle GDP = 90 - 7$ or 83

$m\angle GPD + 62 + 83 = 180$ or $m\angle GPD = 35$

Since you know the measures of two angles of the triangle, $m\angle GDP$ and $m\angle GPD$, and the length of a side opposite one of the angles ($\overline{GD}$ is opposite $\angle GPD$) you can use the Law of Sines to find the length of the pole.

(continued on the next page)

USE THE LAW OF SINES TO SOLVE PROBLEMS

In-Class Examples PowerPoint®

3 In Example 3, what happens to the length of the shadow later in the day when the angle of elevation decreases to 33°? Assume that the length and position of the telephone pole are still the same. **The length of the shadow increases to about 76.2 feet.**

Lesson 7-6 The Law of Sines 379

3 Practice/Apply

Study Notebook

Have students—

- add the definitions/examples of the vocabulary terms to their Vocabulary Builder worksheets for Chapter 7.
- include one completely worked example similar to Example 2 on pp. 378–379.
- include any other item(s) that they find helpful in mastering the skills in this lesson.

DAILY INTERVENTION

FIND THE ERROR Explain that if $\angle F$ in the figure had been 90°, then Makayla and Felipe would have found the same value for d. Students should recognize that they *could* use the Law of Sines to solve right triangles, but trigonometric ratios are more efficient.

About the Exercises...

Organization by Objective

- **The Law of Sines:** 16–29
- **Use the Law of Sines to Solve Problems:** 30–38

Odd/Even Assignments

Exercises 16–38 are structured so that students practice the same concepts whether they are assigned odd or even problems.

Assignment Guide

Basic: 17–39 odd, 40–58

Average: 17–39 odd, 40–58

Advanced: 16–40 even, 41–52 (optional: 53–58)

All: Quiz 2 (1–5)

$\frac{PD}{\sin \angle DGP} = \frac{GD}{\sin \angle GPD}$ Law of Sines

$\frac{PD}{\sin 62°} = \frac{30}{\sin 35°}$ $m\angle DGP = 62$, $m\angle GPD = 35$, and $GD = 30$

$PD \sin 35° = 30 \sin 62°$ Cross products

$PD = \frac{30 \sin 62°}{\sin 35°}$ Divide each side by sin 35°.

$PD \approx 46.2$ Use a calculator.

The telephone pole is about 46.2 feet long.

Concept Summary — Law of Sines

The Law of Sines can be used to solve a triangle in the following cases.

Case 1 You know the measures of two angles and any side of a triangle. (AAS or ASA)

Case 2 You know the measures of two sides and an angle opposite one of these sides of the triangle. (SSA)

Check for Understanding

Concept Check

1. **FIND THE ERROR** Makayla and Felipe are trying to find d in $\triangle DEF$.

Makayla: $\sin 59° = \frac{d}{12}$

Felipe: $\frac{\sin 59°}{d} = \frac{\sin 48°}{12}$

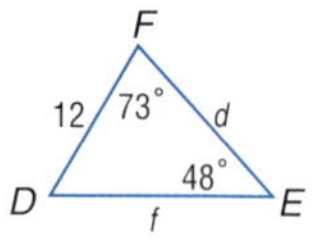

Who is correct? Explain your reasoning.

1. Felipe; Makayla is using the definition of the sine ratio for a right triangle, but this is not a right triangle.

2. **OPEN ENDED** Draw an acute triangle and label the measures of two angles and the length of one side. Explain how to solve the triangle.
3. **Compare** the two cases for the Law of Sines. *2–3. See margin.*

Guided Practice

GUIDED PRACTICE KEY	
Exercises	Examples
4–7	1
8–13	2
14	3

Find each measure using the given measures of $\triangle XYZ$. Round angle measures to the nearest degree and side measures to the nearest tenth.

4. Find y if $x = 3$, $m\angle X = 37$, and $m\angle Y = 68$. *4.6*
5. Find x if $y = 12.1$, $m\angle X = 57$, and $m\angle Z = 72$. *13.1*
6. Find $m\angle Y$ if $y = 7$, $z = 11$, and $m\angle Z = 37$. *23*
7. Find $m\angle Z$ if $y = 17$, $z = 14$, and $m\angle Y = 92$. *55*

Solve each $\triangle PQR$ described below. Round angle measures to the nearest degree and side measures to the nearest tenth.

8. $m\angle R = 66$, $m\angle Q = 59$, $p = 72$
9. $p = 32$, $r = 11$, $m\angle P = 105$
10. $m\angle P = 33$, $m\angle R = 58$, $q = 22$
11. $p = 28$, $q = 22$, $m\angle P = 120$
12. $m\angle P = 50$, $m\angle Q = 65$, $p = 12$
13. $q = 17.2$, $r = 9.8$, $m\angle Q = 110.7$

8. $m\angle P = 55$, $q \approx 75.3$, $r \approx 80.3$
9. $m\angle R \approx 19$, $m\angle Q \approx 56$, $q \approx 27.5$
10. $m\angle Q = 89$, $p \approx 12.0$, $r \approx 18.7$
11. $m\angle Q \approx 43$, $m\angle R \approx 17$, $r \approx 9.5$
12. $m\angle R = 65$, $q \approx 14.2$, $r \approx 14.2$
13. $m\angle P \approx 37$, $p \approx 11.1$; $m\angle R \approx 32$

14. Find the perimeter of parallelogram $ABCD$ to the nearest tenth. *34.6 units*

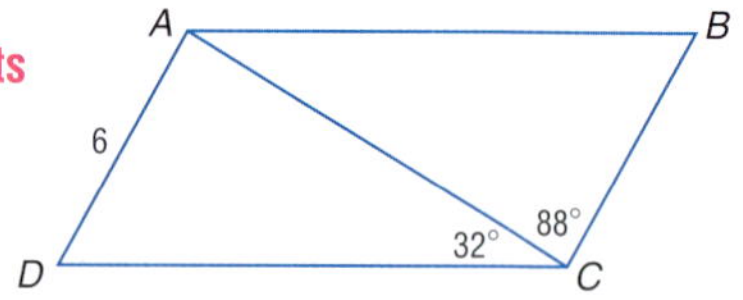

Answer

2. Sample answer: Let $m\angle D = 65$, $m\angle E = 73$, and $d = 15$. Then $\frac{\sin 65°}{15}$ is the fixed ratio or scale factor for the Law of Sines extended proportion. The length of e is found by using $\frac{\sin 65°}{15} = \frac{\sin 73°}{e}$. The $m\angle F$ is found by evaluating $180 - (m\angle D + m\angle E)$. In this problem $m\angle F = 42$. The length of f is found by using $\frac{\sin 65°}{15} = \frac{\sin 42°}{f}$.

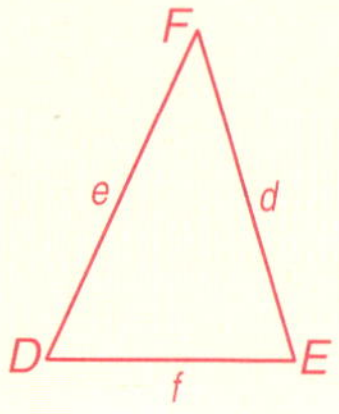

Application **15. SURVEYING** To find the distance between two points A and B that are on opposite sides of a river, a surveyor measures the distance to point C on the same side of the river as point A. The distance from A to C is 240 feet. He then measures the angle from A to B as 62° and measures the angle from C to B as 55°. Find the distance from A to B. **about 237.8 ft**

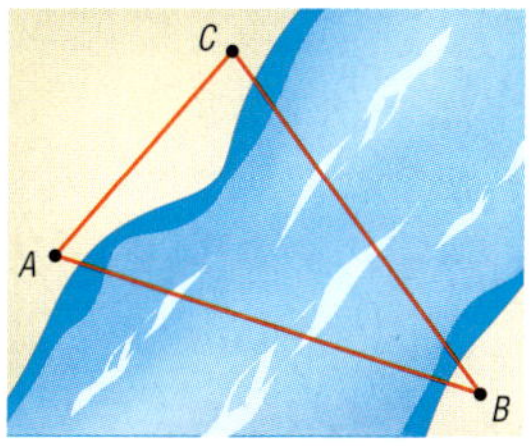

Practice and Apply

Homework Help

For Exercises	See Examples
16–21	1
22–29	2
30–38	3

Extra Practice
See page 768.

Find each measure using the given measures of $\triangle KLM$. Round angle measures to the nearest degree and side measures to the nearest tenth.

16. If $m\angle L = 45$, $m\angle K = 63$, and $\ell = 22$, find k. **27.7**
17. If $k = 3.2$, $m\angle L = 52$, and $m\angle K = 70$, find ℓ. **2.7**
18. If $m = 10.5$, $k = 18.2$, and $m\angle K = 73$, find $m\angle M$. **33**
19. If $k = 10$, $m = 4.8$, and $m\angle K = 96$, find $m\angle M$. **29**
20. If $m\angle L = 88$, $m\angle K = 31$, and $m = 5.4$, find ℓ. **6.2**
21. If $m\angle M = 59$, $\ell = 8.3$, and $m = 14.8$, find $m\angle L$. **29**

Solve each $\triangle WXY$ described below. Round measures to the nearest tenth.

22. $m\angle Y = 71$, $y = 7.4$, $m\angle X = 41$
23. $x = 10.3$, $y = 23.7$, $m\angle Y = 96$
24. $m\angle X = 25$, $m\angle W = 52$, $y = 15.6$
25. $m\angle Y = 112$, $x = 20$, $y = 56$
26. $m\angle W = 38$, $m\angle Y = 115$, $w = 8.5$
27. $m\angle W = 36$, $m\angle Y = 62$, $w = 3.1$
28. $w = 30$, $y = 9.5$, $m\angle W = 107$
29. $x = 16$, $w = 21$, $m\angle W = 88$

22. $m\angle W = 68$, $w \approx 7.3$, $x \approx 5.1$
23. $m\angle X \approx 25.6$, $m\angle W \approx 58.4$, $w \approx 20.3$
24. $m\angle Y = 103$, $w \approx 12.6$, $x \approx 6.8$
25. $m\angle X \approx 19.3$, $m\angle W \approx 48.7$, $w \approx 45.4$
26. $m\angle X = 27$, $x \approx 6.3$, $y \approx 12.5$
27. $m\angle X = 82$, $x \approx 5.2$, $y \approx 4.7$
28. $m\angle Y \approx 17.6$, $m\angle X \approx 55.4$, $x \approx 25.8$
29. $m\angle X \approx 49.6$, $m\angle Y \approx 42.4$, $y \approx 14.2$

30. An isosceles triangle has a base of 46 centimeters and a vertex angle of 44°. Find the perimeter. **about 168.8 cm**

31. Find the perimeter of quadrilateral $ABCD$ to the nearest tenth. **56.9 units**

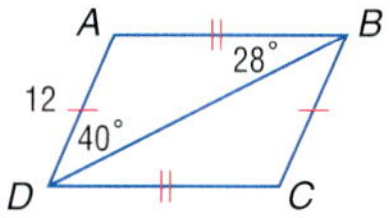

32. **GARDENING** Elena is planning a triangular garden. She wants to build a fence around the garden to keep out the deer. The length of one side of the garden is 26 feet. If the angles at the end of this side are 78° and 44°, find the length of fence needed to enclose the garden. **about 77.3 ft**

33. **AVIATION** Two radar stations that are 20 miles apart located a plane at the same time. The first station indicated that the position of the plane made an angle of 43° with the line between the stations. The second station indicated that it made an angle of 48° with the same line. How far is each station from the plane? **about 14.9 mi, about 13.6 mi**

34. **SURVEYING** Maria Lopez is a surveyor who must determine the distance across a section of the Rio Grande Gorge in New Mexico. Standing on one side of the ridge, she measures the angle formed by the edge of the ridge and the line of sight to a tree on the other side of the ridge. She then walks along the ridge 315 feet, passing the tree and measures the angle formed by the edge of the ridge and the new line of sight to the same tree. If the first angle is 80° and the second angle is 85°, find the distance across the gorge. **about 1194 ft**

Answers (page 380)

3. In one case you need the measures of two sides and the measure of an angle opposite one of the sides. In the other case you need the measure of two angles and the measure of a side.

Enrichment, p. 386

Identities

An **identity** is an equation that is true for all values of the variable for which both sides are defined. One way to verify an identity is to use a right triangle and the definitions for trigonometric functions.

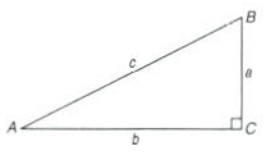

Example 1 **Verify that $(\sin A)^2 + (\cos A)^2 = 1$ is an identity.**

$$(\sin A)^2 + (\cos A)^2 = \left(\frac{a}{c}\right)^2 + \left(\frac{b}{c}\right)^2$$
$$= \frac{a^2 + b^2}{c} = \frac{c^2}{c^2} = 1$$

To check whether an equation *may* be an identity, you can test several values. However, since you cannot test all values, you cannot be *certain* that the equation is an identity.

Example 2 **Test $\sin 2x = 2 \sin x \cos x$ to see if it could be an identity.**

...lculator to...

Study Guide and Intervention, p. 381 (shown) and p. 382

The Law of Sines In any triangle, there is a special relationship between the angles of the triangle and the lengths of the sides opposite the angles.

Law of Sines	$\frac{\sin A}{a} = \frac{\sin B}{b} = \frac{\sin C}{c}$

Example 1 **In $\triangle ABC$, find b.**

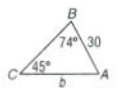

$\frac{\sin C}{c} = \frac{\sin B}{b}$ Law of Sines

$\frac{\sin 45°}{30} = \frac{\sin 74°}{b}$ $m\angle C = 45$, $c = 30$, $m\angle B = 74$

$b \sin 45° = 30 \sin 74°$ Cross multiply.

$b = \frac{30 \sin 74°}{\sin 45°}$ Divide each side by sin 45°.

$b \approx 40.8$ Use a calculator.

Example 2 **In $\triangle DEF$, find $m\angle D$.**

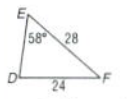

$\frac{\sin D}{d} = \frac{\sin E}{e}$ Law of Sines

$\frac{\sin D}{28} = \frac{\sin 58°}{24}$ $d = 28$, $m\angle E = 58$, $e = 24$

$24 \sin D = 28 \sin 58°$ Cross multiply.

$\sin D = \frac{28 \sin 58°}{24}$ Divide each side by 24.

$D = \sin^{-1} \frac{28 \sin 58°}{24}$ Use the inverse sine.

$D \approx 81.6°$ Use a calculator.

Exercises

Find each measure using the given measures of $\triangle ABC$. Round angle measures to the nearest degree and side measures to the nearest tenth.

1. If $c = 12$, $m\angle A = 80$, and $m\angle C = 40$, find a.
18.4
2. If $b = 20$, $c = 26$, and $m\angle C = 52$, find $m\angle B$.
37
3. If $a = 18$, $c = 16$, and $m\angle A = 84$, find $m\angle C$.
62
4. If $a = 25$, $m\angle A = 72$, and $m\angle B = 17$, find b.
7.7
5. If $b = 12$, $m\angle A = 89$, and $m\angle B = 80$, find a.
12.2
6. If $a = 30$, $c = 20$, and $m\angle A = 60$, find $m\angle C$.
35

Skills Practice, p. 383 and Practice, p. 384 (shown)

Find each measure using the given measures from $\triangle EFG$. Round angle measures to the nearest tenth degree and side measures to the nearest tenth.

1. If $m\angle G = 14$, $m\angle E = 67$, and $e = 14$, find g. **3.7**
2. If $e = 12.7$, $m\angle E = 42$, and $m\angle F = 61$, find f. **16.6**
3. If $g = 14$, $f = 5.8$, and $m\angle G = 83$, find $m\angle F$. **24.3**
4. If $e = 19.1$, $m\angle G = 34$, and $m\angle E = 56$, find g. **12.9**
5. If $f = 9.6$, $g = 27.4$, and $m\angle G = 43$, find $m\angle F$. **13.8**

Solve each $\triangle STU$ described below. Round measures to the nearest tenth.

6. $m\angle T = 85$, $s = 4.3$, $t = 8.2$ **$m\angle S \approx 31.5$, $m\angle U \approx 63.5$, $u \approx 7.4$**
7. $s = 40$, $u = 12$, $m\angle S = 37$ **$m\angle T \approx 132.6$, $m\angle U \approx 10.4$, $t \approx 48.9$**
8. $m\angle U = 37$, $t = 2.3$, $m\angle T = 17$ **$m\angle S \approx 126$, $s \approx 6.4$, $u \approx 4.7$**
9. $m\angle S = 62$, $m\angle U = 59$, $s = 17.8$ **$m\angle T \approx 59$, $t \approx 17.3$, $u \approx 17.3$**
10. $t = 28.4$, $u = 21.7$, $m\angle T = 66$ **$m\angle S \approx 69.7$, $m\angle U \approx 44.3$, $s \approx 29.2$**
11. $m\angle S = 89$, $s = 15.3$, $t = 14$ **$m\angle T \approx 66.2$, $m\angle U \approx 24.8$, $u \approx 6.4$**
12. $m\angle T = 98$, $m\angle U = 74$, $u = 9.6$ **$m\angle S \approx 8$, $s \approx 1.4$, $t \approx 9.9$**
13. $t = 11.8$, $m\angle S = 84$, $m\angle T = 47$ **$m\angle U = 49$, $s \approx 16.0$, $u \approx 12.2$**

14. **INDIRECT MEASUREMENT** To find the distance from the edge of the lake to the tree on the island in the lake, Hannah set up a triangular configuration as shown in the diagram. The distance from location A to location B is 85 meters. The measures of the angles at A and B are 51° and 83°, respectively. What is the distance from the edge of the lake at B to the tree on the island at C?
about 91.8 m

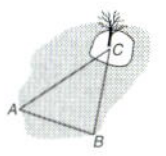

Reading to Learn Mathematics, p. 385 ELL

Pre-Activity **How are triangles used in radio astronomy?**

Read the introduction to Lesson 7-6 at the top of page 377 in your textbook.

Why might several antennas be better than one single antenna when studying distant objects? **Sample answer: Observing an object from only one position often does not provide enough information to calculate things such as the distance from the observer to the object.**

Reading the Lesson

1. Refer to the figure. According to the Law of Sines, which of the following are correct statements? **A, F**

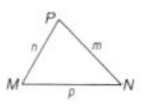

A. $\frac{m}{\sin M} = \frac{n}{\sin N} = \frac{p}{\sin P}$
B. $\frac{\sin m}{M} = \frac{\sin n}{N} = \frac{\sin p}{P}$
C. $\frac{\cos M}{m} = \frac{\cos N}{n} = \frac{\cos P}{p}$
D. $\frac{\sin M}{m} + \frac{\sin N}{n} = \frac{\sin P}{p}$
E. $(\sin M)^2 + (\sin N)^2 = (\sin P)^2$
F. $\frac{\sin P}{p} = \frac{\sin M}{m} = \frac{\sin N}{n}$

2. State whether each of the following statements is *true* or *false*. If the statement is false, explain why.
   a. The Law of Sines applies to all triangles. **true**
   b. The Pythagorean Theorem applies to all triangles. **False; sample answer: It only applies to right triangles.**
   c. If you are given the length of one side of a triangle and the measures of any two angles, you can use the Law of Sines to find the lengths of the other two sides. **true**
   d. If you know the measures of two angles of a triangle, you should use the Law of Sines to find the measure of the third angle. **False; sample answer: You should use the Angle Sum Theorem.**
   e. A friend tells you that in triangle RST, $m\angle R = 132$, $r = 24$ centimeters, and $s = 31$ centimeters. Can you use the Law of Sines to solve the triangle? Explain. **No; sample answer: In any triangle, the longest side is opposite the largest angle. Because a triangle can have only one obtuse angle, $\angle R$ must be the largest angle, but $s > r$, so it is impossible to have a triangle with the given measures.**

Helping You Remember

3. Many students remember mathematical equations and formulas better if they can state them in words. State the Law of Sines in your own words without using variables or mathematical symbols.
Sample answer: In any triangle, the ratio of the sine of an angle to the length of the opposite side is the same for all three angles.

Answers

40. Yes; in right $\triangle ABC$ $\frac{\sin A}{a} = \frac{\sin C}{c}$ where C is the right angle, $\sin A = \frac{a \sin C}{c}$. Since $m\angle C = 90$, then $\sin A = \frac{a \sin 90°}{c}$. Since the $\sin 90° = 1$, then $\sin A = \frac{a}{c}$, which is the definition of the sine ratio.

41. Sample answer: Triangles are used to determine distances in space. Answers should include the following.
   - The VLA is one of the world's premier astronomical radio observatories. It is used to make pictures from the radio waves emitted by astronomical objects.
   - Triangles are used in the construction of the antennas.

35. **REAL ESTATE** A house is built on a triangular plot of land. Two sides of the plot are 160 feet long, and they meet at an angle of 85°. If a fence is to be placed along the perimeter of the property, how much fencing material is needed? about 536 ft

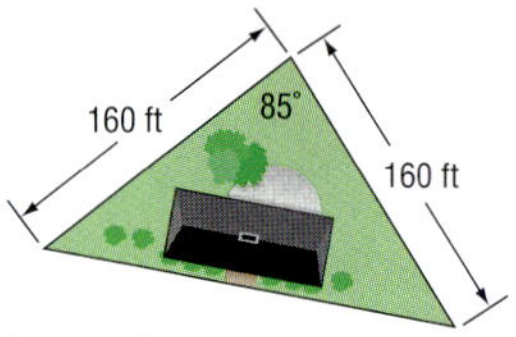

MIRRORS **For Exercises 36 and 37, use the following information.**
Kayla, Jenna, and Paige live in a town nicknamed "Perpendicular City" because the planners and builders took great care to have all the streets oriented north-south or east-west. The three of them play a game where they signal each other using mirrors. Kayla and Jenna signal each other from a distance of 1433 meters. Jenna turns 27° to signal Paige. Kayla rotates 40° to signal Paige.

36. To the nearest tenth of a meter, how far apart are Kayla and Paige? about 706.8 m

37. To the nearest tenth of a meter, how far apart are Jenna and Paige? about 1000.7 m

More About. . .

Aviation

From January 2001 to May 2001, Polly Vacher flew over 29,000 miles to become the first woman to fly the smallest single-engined aircraft around the world via Australia and the Pacific.

Source: www.worldwings.org

AVIATION **For Exercises 38 and 39, use the following information.**
Keisha Jefferson is flying a small plane due west. To avoid the jet stream, she must change her course. She turns the plane 27° to the south and flies 60 miles. Then she makes a turn of 124° heads back to her original course. 38. about 56.2 mi

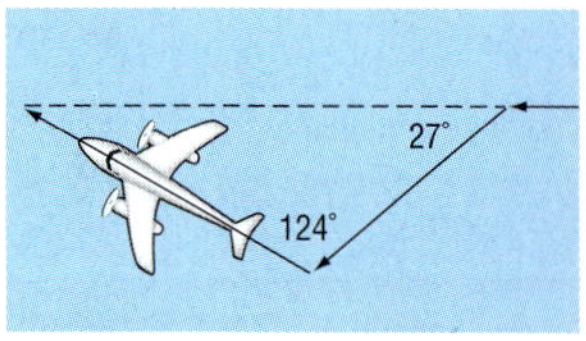

38. How far must she fly after the second turn to return to the original course?

39. How many miles did she add to the flight by changing course? about 13.6 mi

40. **CRITICAL THINKING** Does the Law of Sines apply to the acute angles of a right triangle? Explain your answer. See margin.

41. **WRITING IN MATH** Answer the question that was posed at the beginning of the lesson. See margin.

    How are triangles used in radio astronomy?

    Include the following in your answer:
    - a description of what the VLA is and the purpose of the VLA, and
    - the purpose of the triangles in the VLA.

FCAT Practice
Standardized Test Practice
A B C D

42. **SHORT RESPONSE** In $\triangle XYZ$, if $x = 12$, $m\angle X = 48$, and $m\angle Y = 112$, solve the triangle to the nearest tenth. $m\angle Z = 20$, $y = 15.0$, and $z = 5.5$

43. **ALGEBRA** The table below shows the customer ratings of five restaurants in the *Metro City Guide to Restaurants*. The rating scale is from 1, the worst, to 10, the best. Which of the five restaurants has the best average rating? A

Restaurant	Food	Decor	Service	Value
Del Blanco's	7	9	4	7
Aquavent	8	9	4	6
Le Circus	10	8	3	5
Sushi Mambo	7	7	5	6
Metropolis Grill	9	8	7	7

(A) Metropolis Grill
(B) Le Circus
(C) Aquavent
(D) Del Blanco's

382 Chapter 7 Right Triangles and Trigonometry

Maintain Your Skills

Mixed Review

ARCHITECTURE **For Exercises 44 and 45, use the following information.**
Mr. Martinez is an architect who designs houses so that the windows receive minimum sun in the summer and maximum sun in the winter. For Columbus, Ohio, the angle of elevation of the sun at noon on the longest day is 73.5° and on the shortest day is 26.5°. Suppose a house is designed with a south-facing window that is 6 feet tall. The top of the window is to be installed 1 foot below the overhang. *(Lesson 7-5)*

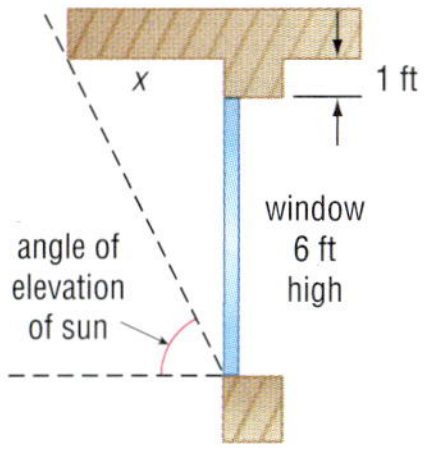

44. How long should the architect make the overhang so that the window gets no direct sunlight at noon on the longest day? **about 2.07 ft**

45. Using the overhang from Exercise 44, how much of the window will get direct sunlight at noon on the shortest day? **about 5.97 ft**

Use $\triangle JKL$ to find sin *J*, cos *J*, tan *J*, sin *L*, cos *L*, and tan *L*. Express each ratio as a fraction and as a decimal to the nearest hundredth. *(Lesson 7-4)* **46–49. See margin.**

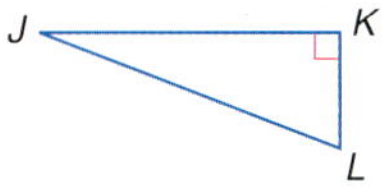

46. $j = 8, k = 17, \ell = 15$
47. $j = 20, k = 29, \ell = 21$
48. $j = 12, k = 24, \ell = 12\sqrt{3}$
49. $j = 7\sqrt{2}, k = 14, \ell = 7\sqrt{2}$

If $\overline{KH}$ is parallel to $\overline{JI}$, find the measure of each angle. *(Lesson 4-2)*

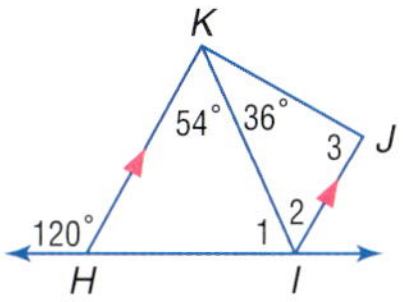

50. $\angle 1$ **66**
51. $\angle 2$ **54**
52. $\angle 3$ **90**

Getting Ready for the Next Lesson

PREREQUISITE SKILL Evaluate $\frac{c^2 - a^2 - b^2}{-2ab}$ for the given values of *a*, *b*, and *c*.
*(To review **evaluating expressions**, see page 736.)* **53.** $\frac{13}{112}$ **54.** $\frac{61}{72}$ **55.** $-\frac{11}{80}$

53. $a = 7, b = 8, c = 10$
54. $a = 4, b = 9, c = 6$
55. $a = 5, b = 8, c = 10$
56. $a = 16, b = 4, c = 13$ $\frac{103}{128}$
57. $a = 3, b = 10, c = 9$ $\frac{7}{15}$
58. $a = 5, b = 7, c = 11$ $-\frac{47}{70}$

Practice Quiz 2 — *Lessons 7-4 through 7-6*

Find *x* to the nearest tenth. *(Lesson 7-4)*

1. **58.0**

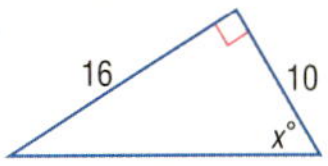

2. **9.3**

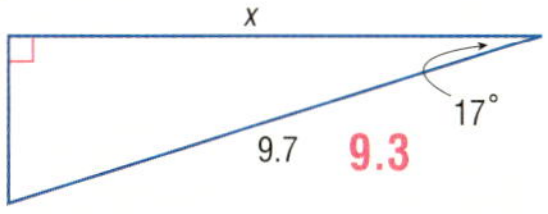

3. **53.2**

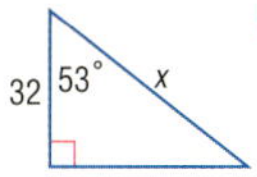

4. **COMMUNICATIONS** To secure a 500-foot radio tower against high winds, guy wires are attached to the tower 5 feet from the top. The wires form a 15° angle with the tower. Find the distance from the centerline of the tower to the anchor point of the wires. *(Lesson 7-5)* **132.6 ft**

5. Solve $\triangle DEF$. *(Lesson 7-6)*
$m\angle D \approx 41$
$m\angle E \approx 57$
$e \approx 10.2$

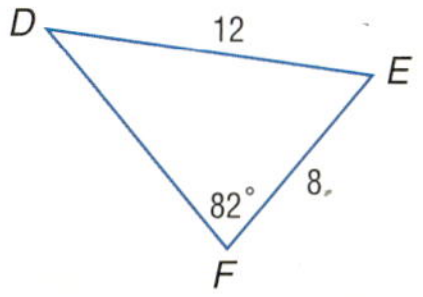

www.geometryonline.com/self_check_quiz/fcat

4 Assess

Open-Ended Assessment

Writing Ask students to draw three triangular figures on a single sheet of paper and label and assign values to them in correspondence with the three types of triangles that can be solved using the Law of Sines (AAS, ASA and SSA). For example, one triangle would have two angle measures, one adjacent side measure and one adjacent side variable. Students can use the examples in the book if necessary. Then tell them to set up the appropriate proportions to find the unknown values.

Getting Ready for Lesson 7-7

Prerequisite Skill Students will learn about the Law of Cosines in Lesson 7-7. They will solve problems using multivariable equations. Use Exercises 53–58 to determine your students' familiarity with evaluating multivariable expressions.

Assessment Options

Practice Quiz 2 This quiz provides students with a brief review of the concepts and skills in Lessons 7-4 through 7-6. Lesson numbers are given to the right of the exercises or instruction lines so students can review concepts not yet mastered.

Quiz (Lessons 7-5 and 7-6) is available on p. 408 of the *Chapter 7 Resource Masters*.

Answers

46. $\frac{8}{17} \approx 0.47; \frac{15}{17} \approx 0.88; \frac{8}{15} \approx 0.53; \frac{15}{17} \approx 0.88; \frac{8}{17} \approx 0.47; \frac{15}{8} \approx 1.88$

47. $\frac{20}{29} \approx 0.69; \frac{21}{29} \approx 0.72; \frac{20}{21} \approx 0.95; \frac{21}{29} \approx 0.72; \frac{20}{29} \approx 0.69; \frac{21}{20} = 1.05$

48. $\frac{1}{2} = 0.50; \frac{\sqrt{3}}{2} \approx 0.87; \frac{\sqrt{3}}{3} \approx 0.58; \frac{\sqrt{3}}{2} \approx 0.87; \frac{1}{2} = 0.50; \sqrt{3} \approx 1.73$

49. $\frac{\sqrt{2}}{2} \approx 0.71; \frac{\sqrt{2}}{2} \approx 0.71; 1.00; \frac{\sqrt{2}}{2} \approx 0.71; \frac{\sqrt{2}}{2} \approx 0.71; 1.00$

Geometry Software Investigation

A Follow-Up of Lesson 7-6

Getting Started

The Ambiguous Case of the Law of Sines As a variation, students can label the points D_1 and D_2 where the circle intersects $\overrightarrow{AD}$ and construct two radii $\overline{BD_1}$ and $\overline{BD_2}$. Then they can find and display all of the measures on the screen at one time.

Teach

- When students are comparing results, point out that the two possible measures for angle D are supplementary, and discuss why this is the case.
- Explain to students that if circle B intersects $\overrightarrow{AC}$ at just one point, then $\overrightarrow{AC}$ is tangent to circle B, and $\triangle ABD$ is a right triangle with right angle D.
- If students become concerned about the possibility of finding two solutions for a SSA problem, assure them that the problems will either be designed to yield one answer or set up in a way so students will know which solution is appropriate. Sometimes it might be appropriate for students to find both solutions.

Assess

Exercises 1–5 After working through the exercises, students will have demonstrated numerous examples of the ambiguous case of the Law of Sines and discussed the possibilities of finding one solution or no solution.

Geometry Software Investigation

A Follow-Up of Lesson 7-6

The Ambiguous Case of the Law of Sines

In Lesson 7-6, you learned that you could solve a triangle using the Law of Sines if you know the measures of two angles and any side of the triangle (AAS or ASA). You can also solve a triangle by the Law of Sines if you know the measures of two sides and an angle opposite one of the sides (SSA). When you use SSA to solve a triangle, and the given angle is acute, sometimes it is possible to find two different triangles. You can use The Geometer's Sketchpad to explore this case, called the **ambiguous case**, of the Law of Sines.

Model

Step 1 Construct $\overline{AB}$ and $\overrightarrow{AC}$. Construct a circle whose center is B so that it intersects $\overrightarrow{AC}$ at two points. Then, construct any radius $\overline{BD}$.

Step 2 Find the measures of $\overline{BD}$, $\overline{AB}$, and $\angle A$.

Step 3 Use the rotate tool to move D so that it lies on one of the intersection points of circle B and $\overrightarrow{AC}$. In $\triangle ABD$, find the measures of $\angle ABD$, $\angle BDA$, and $\overline{AD}$.

Step 4 Using the rotate tool, move D to the other intersection point of circle B and $\overrightarrow{AC}$.

Step 5 Note the measures of $\angle ABD$, $\angle BDA$, and $\overline{AD}$ in $\triangle ABD$.

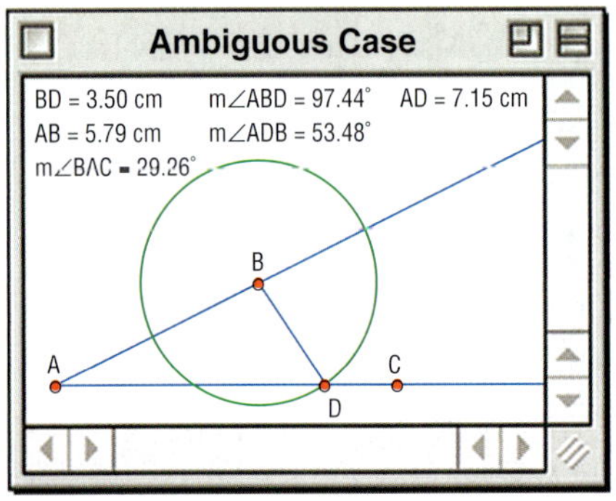

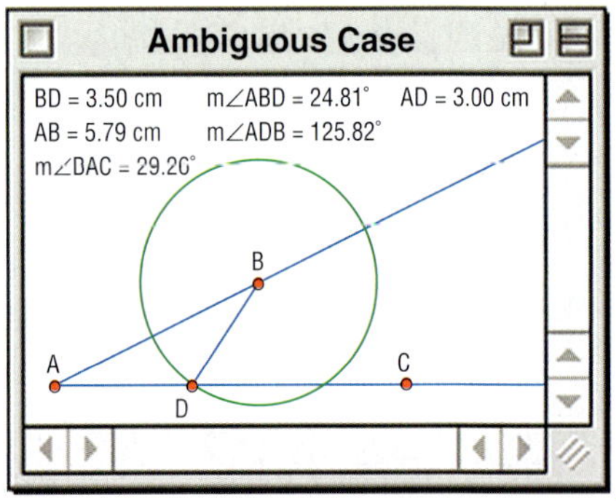

Analyze

1. Which measures are the same in both triangles? ***BD, AB, m∠A***
2. Repeat the activity using different measures for $\angle A$, $\overline{BD}$, and $\overline{AB}$. How do the results compare to the earlier results? **Sample answer: There are two different triangles.**

Make a Conjecture **3–5. See margin.**

3. Compare your results with those of your classmates. How do the results compare?
4. What would have to be true about circle B in order for there to be one unique solution? Test your conjecture by repeating the activity.
5. Is it possible, given the measures of $\overline{BD}$, $\overline{AB}$, and $\angle A$, to have no solution? Test your conjecture and explain.

Answers

3. **Sample answer: The results are the same. In each case, two triangles are possible.**
4. **Sample answer: Circle *B* intersects $\overrightarrow{AC}$ at only one point. See students' work.**
5. **Yes; sample answer: There is no solution if circle *B* does not intersect $\overrightarrow{AC}$.**

384 Chapter 7 Right Triangles and Trigonometry

7-7 The Law of Cosines

Sunshine State Standards
MA.A.3.4.3, MA.B.2.4.1, MA.B.3.4.1, MA.C.3.4.1, MA.C.3.4.2

Vocabulary
- Law of Cosines

What You'll Learn
- Use the Law of Cosines to solve triangles.
- Solve problems by using the Law of Cosines.

How are triangles used in building design?

The Chicago Metropolitan Correctional Center is a 27-story triangular federal detention center. The cells are arranged around a lounge-like common area. The architect found that a triangular floor plan allowed for the maximum number of cells to be most efficiently centered around the lounge.

THE LAW OF COSINES Suppose you know the lengths of the sides of the triangular building and want to solve the triangle. The **Law of Cosines** allows us to solve a triangle when the Law of Sines cannot be used.

Study Tip

Side and Angle
Note that the letter of the side length on the left-hand side of each equation corresponds to the angle measure used with the cosine.

Key Concept — *Law of Cosines*

Let $\triangle ABC$ be any triangle with a, b, and c representing the measures of sides opposite angles A, B, and C, respectively. Then the following equations are true.

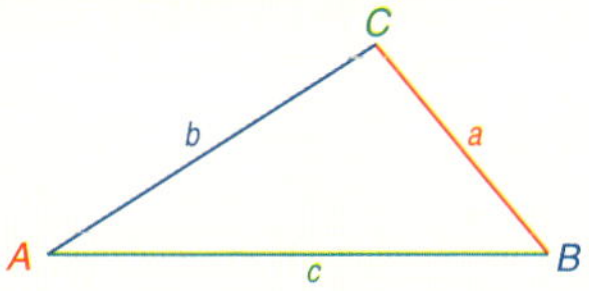

$$a^2 = b^2 + c^2 - 2bc \cos A$$

$$b^2 = a^2 + c^2 - 2ac \cos B$$

$$c^2 = a^2 + b^2 - 2ab \cos C$$

The Law of Cosines can be used to find missing measures in a triangle if you know the measures of two sides and their included angle.

Example 1 *Two Sides and the Included Angle*

Find a if $c = 8$, $b = 10$, and $m\angle A = 60$.

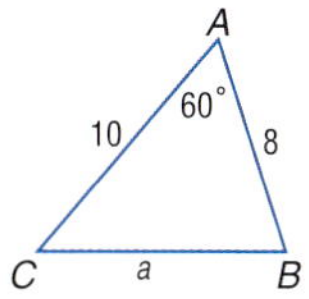

Use the Law of Cosines since the measures of two sides and the included are known.

$a^2 = b^2 + c^2 - 2bc \cos A$	Law of Cosines
$a^2 = 10^2 + 8^2 - 2(10)(8) \cos 60°$	$b = 10$, $c = 8$, and $m\angle A = 60$
$a^2 = 164 - 160 \cos 60°$	Simplify.
$a = \sqrt{164 - 160 \cos 60°}$	Take the square root of each side.
$a \approx 9.2$	Use a calculator.

7-7 Lesson Notes

1 Focus

5-Minute Check Transparency 7-7 Use as a quiz or review of Lesson 7-6.

Mathematical Background notes are available for this lesson on p. 340D.

How are triangles used in building design?

Ask students:

- What advantages might the triangular design offer the correction center employees? **The residents would be more confined and easier to monitor.**
- How might the triangular design of the correction facility be more cost efficient? **A triangular building may utilize less space than a rectangular one, so fewer materials would be needed, and the building could have more compact electrical and plumbing systems.**

Resource Manager

Workbook and Reproducible Masters

Chapter 7 Resource Masters
- Study Guide and Intervention, pp. 387–388
- Skills Practice, p. 389
- Practice, p. 390
- Reading to Learn Mathematics, p. 391
- Enrichment, p. 392
- Assessment, p. 408

Prerequisite Skills Workbook, pp. 21–22, 25–26

Teaching Geometry With Manipulatives Masters, p. 121

Transparencies
5-Minute Check Transparency 7-7
Answer Key Transparencies

Technology
GeomPASS: Tutorial Plus, Lesson 16
Interactive Chalkboard

2 Teach

THE LAW OF COSINES

In-Class Examples

Power Point®

1 Find x if $y = 11$, $z = 25$, and $m\angle X = 45$.

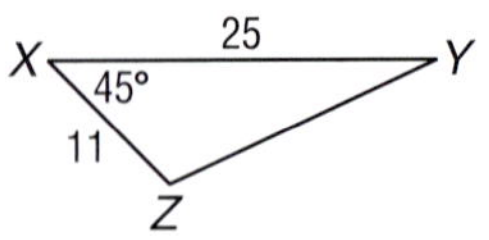

≈ 18.9

Teaching Tip In Example 2, show students how the Law of Cosines can be rewritten to provide an alternate form: $\cos R = \dfrac{q^2 + s^2 - r^2}{2qs}$.

2 Find $m\angle L$.

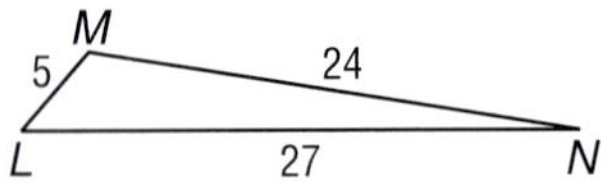

≈ 48.8

USE THE LAW OF COSINES TO SOLVE PROBLEMS

In-Class Example

Power Point®

Teaching Tip In Example 3, students can store the actual value of k in their calculator. Then if they use the stored value to find L, they will get 79.19°. Advise students that answers will vary when using rounded values instead of more exact values.

3 Determine whether the Law of Sines or the Law of Cosines should be used first to solve $\triangle DEF$. Then solve $\triangle DEF$. Round angle measures to the nearest degree and side measures to the nearest tenth.

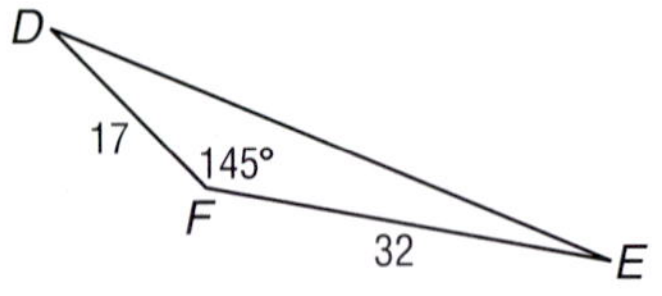

Cosines, $f \approx 46.9$; $m\angle D \approx 23$; $m\angle E \approx 12$

You can also use the Law of Cosines to find the measures of angles of a triangle when you know the measures of the three sides.

Example 2 Three Sides

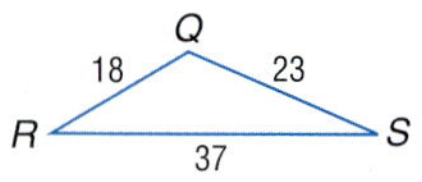

Find $m\angle R$.

$r^2 = q^2 + s^2 - 2qs \cos R$	Law of Cosines
$23^2 = 37^2 + 18^2 - 2(37)(18) \cos R$	$r = 23$, $q = 37$, $s = 18$
$529 = 1693 - 1332 \cos R$	Simplify.
$-1164 = -1332 \cos R$	Subtract 1693 from each side.
$\dfrac{-1164}{-1332} = \cos R$	Divide each side by −1332.
$R = \cos^{-1}\left(\dfrac{1164}{1332}\right)$	Solve for R.
$R \approx 29.1°$	Use a calculator.

USE THE LAW OF COSINES TO SOLVE PROBLEMS Most problems can be solved using more than one method. Choosing the most efficient way to solve a problem is sometimes not obvious.

When solving right triangles, you can use sine, cosine, or tangent ratios. When solving other triangles, you can use the Law of Sines or the Law of Cosines. You must decide how to solve each problem depending on the given information.

Example 3 Select a Strategy

Solve $\triangle KLM$. Round angle measure to the nearest degree and side measure to the nearest tenth.

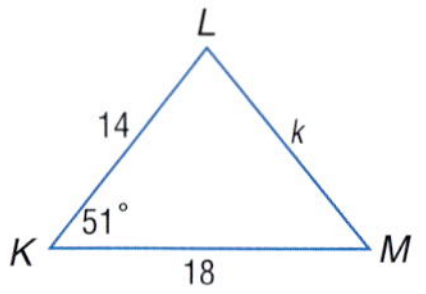

We do not know whether $\triangle KLM$ is a right triangle, so we must use the Law of Cosines or the Law of Sines. We know the measures of two sides and the included angle. This is SAS, so use the Law of Cosines.

$k^2 = \ell^2 + m^2 - 2\ell m \cos K$	Law of Cosines
$k^2 = 18^2 + 14^2 - 2(18)(14) \cos 51°$	$\ell = 18$, $m = 14$, and $m\angle K = 51$
$k = \sqrt{18^2 + 14^2 - 2(18)(14) \cos 51°}$	Take the square root of each side.
$k \approx 14.2$	Use a calculator.

Next, we can find $m\angle L$ or $m\angle M$. If we decide to find $m\angle L$, we can use either the Law of Sines or the Law of Cosines to find this value. In this case, we will use the Law of Sines.

$\dfrac{\sin L}{\ell} = \dfrac{\sin K}{k}$	Law of Sines
$\dfrac{\sin L}{18} \approx \dfrac{\sin 51°}{14.2}$	$\ell = 18$, $k \approx 14.2$, and $m\angle K = 51$
$14.2 \sin L \approx 18 \sin 51°$	Cross products
$\sin L \approx \dfrac{18 \sin 51°}{14.2}$	Divide each side by 14.2.
$L \approx \sin^{-1}\left(\dfrac{18 \sin 51°}{14.2}\right)$	Take the inverse sine of each side.
$L \approx 80°$	Use a calculator.

Study Tip

Law of Cosines
If you use the Law of Cosines to find another measure, your answer may differ slightly from one found using the Law of Sines. This is due to rounding.

TEACHING TIP

Help students learn that when using a calculator, they can skip some steps. For example,

$L = \sin^{-1}\left(\dfrac{18 \sin 51°}{14.2}\right) \approx 80.1.$

DAILY INTERVENTION — Differentiated Instruction — ELL

Verbal/Linguistic Have students rewrite the equations for the Law of Cosines in their own words without using variables. Then they can describe scenarios for which the Law of Cosines is the more useful. Finally, they can close their books, draw and label a triangular figure, and try to reproduce the equations for the Law of Cosines by using only their written explanations and descriptions.

Use the Angle Sum Theorem to find $m\angle M$.

$m\angle K + m\angle L + m\angle M = 180$ Angle Sum Theorem

$51 + 80 + m\angle M \approx 180$ $m\angle K = 51$ and $m\angle L \approx 80$

$m\angle M \approx 49$ Subtract 131 from each side.

Therefore, $k \approx 14.2$, $m\angle K \approx 80$, and $m\angle M \approx 49$.

Example 4 Use Law of Cosines to Solve Problems

REAL ESTATE Ms. Jenkins is buying some property that is shaped like quadrilateral *ABCD*. Find the perimeter of the property.

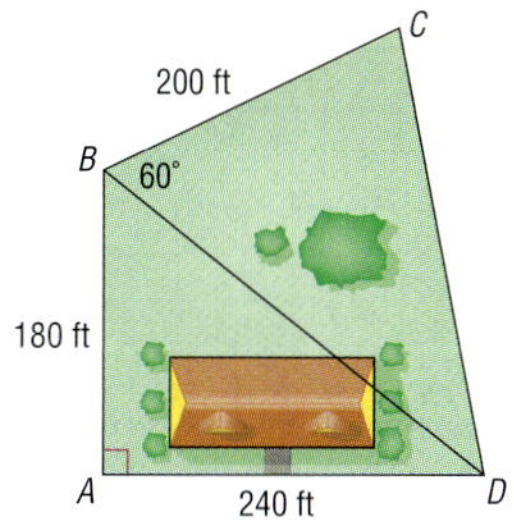

Use the Pythagorean Theorem to find BD in $\triangle ABD$.

$(AB)^2 + (AD)^2 = (BD)^2$ Pythagorean Theorem

$180^2 + 240^2 = (BD)^2$ $AB = 180$, $AD = 240$

$90{,}000 = (BD)^2$ Simplify.

$300 = BD$ Take the square root of each side.

Next, use the Law of Cosines to find CD in $\triangle BCD$.

$(CD)^2 = (BC)^2 + (BD)^2 - 2(BC)(BD)\cos\angle CBD$ Law of Cosines

$(CD)^2 = 200^2 + 300^2 - 2(200)(300)\cos 60°$ $BC = 200$, $BD = 300$, $m\angle CBD = 60$

$(CD)^2 = 130{,}000 - 120{,}000\cos 60°$ Simplify.

$CD = \sqrt{130{,}000 - 120{,}000\cos 60°}$ Take the square root of each side.

$CD \approx 264.6$ Use a calculator.

The perimeter of the property is 180 + 200 + 264.6 + 240 or about 884.6 feet.

Check for Understanding

Concept Check

1. **OPEN ENDED** Draw and label one acute and one obtuse triangle, illustrating when you can use the Law of Cosines to find the missing measures.
2. **Explain** when you should use the Law of Sines or the Law of Cosines to solve a triangle. **1–2. See margin.**
3. **Find a counterexample** for the following statement.
The Law of Cosines can be used to find the length of a missing side in any triangle.

3. If two angles and one side are given, then the Law of Cosines cannot be used.

Guided Practice

GUIDED PRACTICE KEY	
Exercises	Examples
4–5	1
6–7, 10	2
8–9	3

In $\triangle BCD$, given the following measures, find the measure of the missing side.

4. $c = \sqrt{2}$, $d = 5$, $m\angle B = 45$ **4.1**
5. $b = 107$, $c = 94$, $m\angle D = 105$ **159.7**

In $\triangle RST$, given the lengths of the sides, find the measure of the stated angle to the nearest degree.

6. $r = 33$, $s = 65$, $t = 56$; $m\angle S$ **90**
7. $r = 2.2$, $s = 1.3$, $t = 1.6$; $m\angle R$ **98**

Solve each triangle using the given information. Round angle measure to the nearest degree and side measure to the nearest tenth.

8. $\triangle XYZ$: $x = 5$, $y = 10$, $z = 13$
9. $\triangle KLM$: $k = 20$, $m = 24$, $m\angle L = 47$ $\ell \approx 17.9$; $m\angle K \approx 55$; $m\angle M \approx 78$

8. $m\angle X \approx 20$; $m\angle Y \approx 43$; $m\angle Z \approx 117$

www.geometryonline.com/extra_examples/fcat

In-Class Example

Power Point®

4 **AIRCRAFT** From the diagram of the plane shown, determine the approximate exterior perimeter of each wing. Round to the nearest tenth meter.

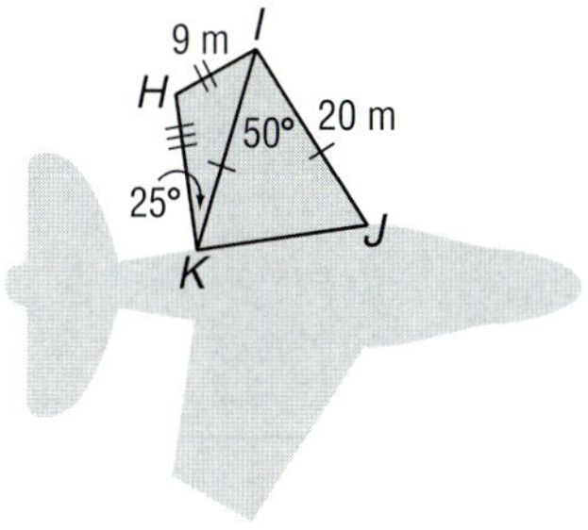

67.1 m

Concept Check

Have students compose a chart in which they include all of the types of triangles they could solve using the Pythagorean Theorem, trigonometric ratios, Law of Sines, and Law of Cosines. They can also include a column listing the formulas and/or ratios and a column for the limitations of each method. Students should recognize that the Pythagorean Theorem is the most limited but the most efficient, and the Law of Cosines is the most diverse but very involved and time-consuming. They may want to form a habit of checking for the most efficient way to solve a problem, keeping all these methods in mind.

Answers

1. Sample answer: Use the Law of Cosines when you have all three sides given (SSS) or two sides and the included angle (SAS).

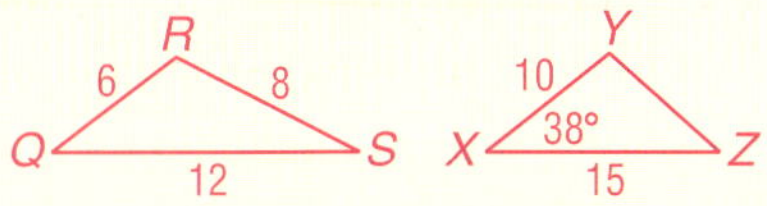

2. If you have all three sides (SSS) or two sides and the included angle (SAS) given, then use the Law of Cosines. If two angles and one side (ASA or AAS) or two sides with angle opposite one of the sides (SSA) are given, then use the Law of Sines.

3 Practice/Apply

Study Notebook

Have students—
- add the definitions/examples of the vocabulary terms to their Vocabulary Builder worksheets for Chapter 7.
- include an example each for using the Law of Cosines with SAS and with SSS.
- include any other item(s) that they find helpful in mastering the skills in this lesson.

About the Exercises...

Organization by Objective
- **Solve Triangles:** 11–18, 38
- **Use the Law of Cosines to Solve Problems:** 22–37, 39–41

Odd/Even Assignments
Exercises 11–41 are structured so that students practice the same concepts whether they are assigned odd or even problems.

Assignment Guide
Basic: 11–37 odd, 41, 43–59
Average: 11–43 odd, 44–59
Advanced: 12–44 even, 46–56 (optional: 57–59)

Application

10. **CRAFTS** Jamie, age 25, is creating a logo for herself and two cousins, ages 10 and 5. She is using a quarter (25 cents), a dime (10 cents), and a nickel (5 cents) to represent their ages. She will hold the coins together by soldering a triangular piece of silver wire so that the three vertices of the triangle lie at the centers of the three circular coins. The diameter of the quarter is 24 millimeters, the diameter of the nickel is 22 millimeters, and the diameter of a dime is 10 millimeters. Find the measures of the three angles in the triangle. $m\angle Q = 44.1$; $m\angle D = 88.3$; $m\angle N = 47.6$

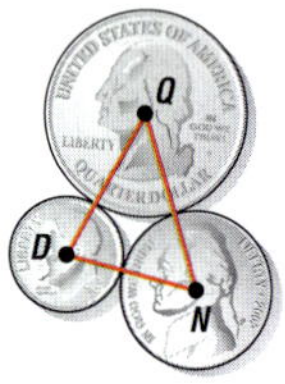

★ indicates increased difficulty

Practice and Apply

Homework Help	
For Exercises	See Examples
11–14	1
15–18, 38	2
22–37, 39–41	3

Extra Practice
See page 768.

In $\triangle TUV$, given the following measures, find the measure of the missing side.

11. $t = 9.1, v = 8.3, m\angle U = 32$ $u \approx 4.9$
12. $t = 11, u = 17, m\angle V = 78$ $v \approx 18.2$
13. $u = 11, v = 17, m\angle T = 105$ $t \approx 22.5$
14. $v = 11, u = 17, m\angle T = 59$ $t \approx 14.7$

In $\triangle EFG$, given the lengths of the sides, find the measure of the stated angle to the nearest degree.

15. $e = 9.1, f = 8.3, g = 16.7; m\angle F$ 16
16. $e = 14, f = 19, g = 32; m\angle E$ 12
17. $e = 325, f = 198, g = 208; m\angle F$ 36
18. $e = 21.9, f = 18.9, g = 10; m\angle G$ 27

Solve each triangle using the given information. Round angle measures to the nearest degree and side measures to the nearest tenth. 19–25. See margin.

19.
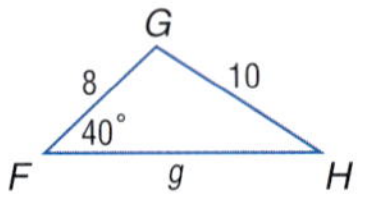

20.
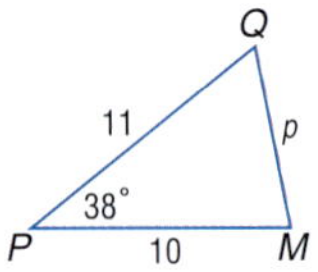

21.
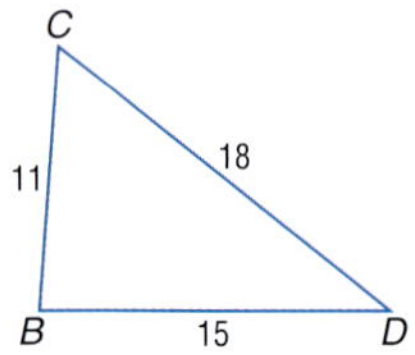

22. $\triangle ABC$: $m\angle A = 42, m\angle C = 77, c = 6$
23. $\triangle ABC$: $a = 10.3, b = 9.5, m\angle C = 37$
24. $\triangle ABC$: $a = 15, b = 19, c = 28$
25. $\triangle ABC$: $m\angle A = 53, m\angle C = 28, c = 14.9$

26. **KITES** Beth is building a kite like the one at the right. If $\overline{AB}$ is 5 feet long, $\overline{BC}$ is 8 feet long, and $\overline{BD}$ is $7\frac{2}{3}$ feet long, find the measure of the angle between the short sides and the angle between the long sides to the nearest degree. 100; 57

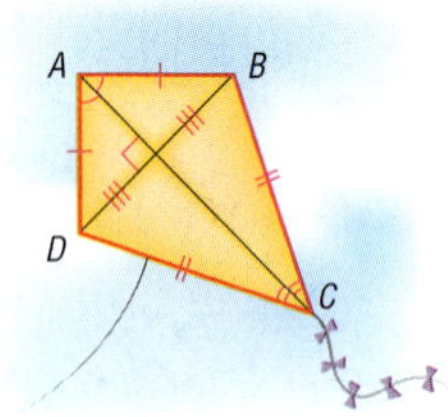

Solve each $\triangle LMN$ described below. Round measures to the nearest tenth.

27–38. See margin.

27. $m = 44, \ell = 54, m\angle L = 23$
28. $m = 18, \ell = 24, n = 30$
29. $m = 19, n = 28, m\angle L = 49$
30. $m\angle M = 46, m\angle L = 55, n = 16$
31. $m = 256, \ell = 423, n = 288$
32. $m\angle M = 55, \ell = 6.3, n = 6.7$
33. $m\angle M = 27, \ell = 5, n = 10$
34. $n = 17, m = 20, \ell = 14$
35. $\ell = 14, n = 21, m\angle M = 60$
36. $\ell = 14, m = 15, n = 16$
37. $m\angle L = 51, \ell = 40, n = 35$
38. $\ell = 10, m = 11, n = 12$

Answers

19. $m\angle H \approx 31$; $m\angle G \approx 109$; $g \approx 14.7$
20. $p \approx 6.9$; $m\angle M \approx 79$; $m\angle Q \approx 63$
21. $m\angle B \approx 86$; $m\angle C \approx 56$; $m\angle D \approx 38$
22. $m\angle B = 61$; $b \approx 5.4$; $a \approx 4.1$
23. $c \approx 6.3$; $m\angle A \approx 80$; $m\angle B \approx 63$
24. $m\angle A \approx 30$; $m\angle B \approx 39$; $m\angle C \approx 111$
25. $m\angle B = 99$; $b \approx 31.3$; $a \approx 25.3$
27. $m\angle M \approx 18.6$; $m\angle N \approx 138.4$; $n \approx 91.8$
28. $m\angle L \approx 53.2$; $m\angle M \approx 36.9$; $m\angle N \approx 89.9$
29. $\ell \approx 21.1$; $m\angle M \approx 42.8$; $m\angle N \approx 88.2$
30. $m\angle N = 79$; $\ell \approx 13.4$; $m \approx 11.7$
31. $m\angle L \approx 101.9$; $m\angle M \approx 36.3$; $m\angle N \approx 41.8$
32. $m \approx 6.0$; $m\angle L \approx 59.3$; $m\angle N \approx 65.7$
33. $m \approx 6.0$; $m\angle L \approx 22.2$; $m\angle N \approx 130.8$
34. $m\angle L \approx 43.5$; $m\angle M \approx 79.5$; $m\angle N \approx 57.0$
35. $m \approx 18.5$; $m\angle L \approx 40.9$; $m\angle N \approx 79.1$
36. $m\angle L \approx 53.6$; $m\angle M \approx 59.6$; $m\angle N \approx 66.8$
37. $m\angle N \approx 42.8$; $m\angle M \approx 86.2$; $m \approx 51.4$
38. $m\angle L \approx 51.3$; $m\angle M \approx 59.1$; $m\angle N \approx 69.6$

388 Chapter 7 Right Triangles and Trigonometry

★ **39.** In quadrilateral $ABCD$, $AC = 188$, $BD = 214$, $m\angle BPC = 70$, and P is the midpoint of $\overline{AC}$ and $\overline{BD}$. Find the perimeter of quadrilateral $ABCD$. **561.2 units**

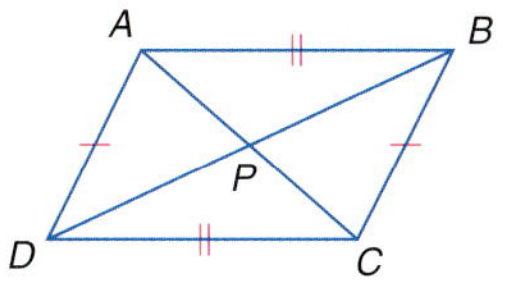

★ **40.** In quadrilateral $PQRS$, $PQ = 721$, $QR = 547$, $RS = 593$, $PS = 756$, and $m\angle P = 58$. Find QS, $m\angle PQS$, and $m\angle R$.
$QS \approx 716.7$; $m\angle PQS \approx 63.5$; $m\angle R \approx 77.8$

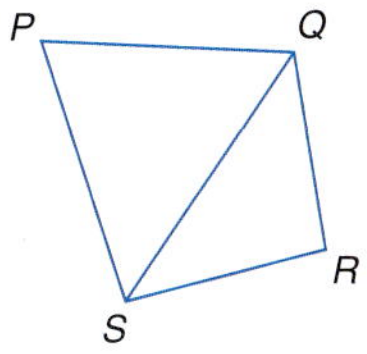

More About. . .

Buildings

The Swissôtel in Chicago, Illinois, is built in the shape of a triangular prism. The lengths of the sides of the triangle are 180 feet, 186 feet, and 174 feet.

Source: Swissôtel

41. BUILDINGS Refer to the information at the left. Find the measures of the angles of the triangular building to the nearest tenth. **59.8, 63.4, 56.8**

42. SOCCER Carlos and Adam are playing soccer. Carlos is standing 40 feet from one post of the goal and 50 feet from the other post. Adam is standing 30 feet from one post of the goal and 22 feet from the other post. If the goal is 24 feet wide, which player has a greater angle to make a shot on goal? **Adam; 52.3°**

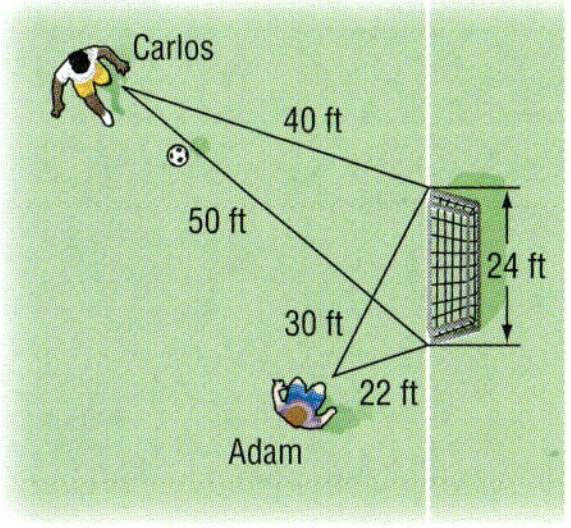

43. PROOF Justify each statement for the derivation of the Law of Cosines.

Given: $\overline{AD}$ is an altitude of $\triangle ABC$.

Prove: $c^2 = a^2 + b^2 - 2ab \cos C$

Proof:

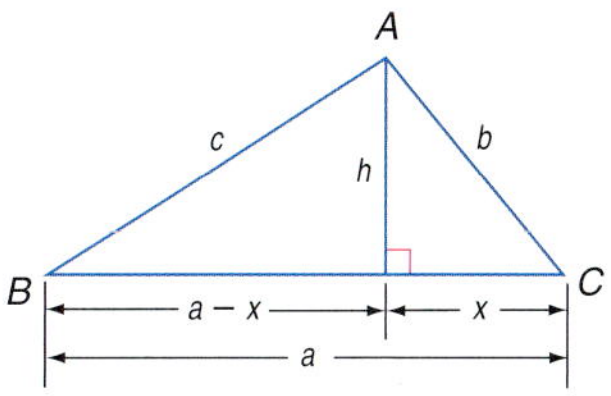

Statement	Reasons
a. $c^2 = (a - x)^2 + h^2$	a. __?__ **Pythagorean Theorem**
b. $c^2 = a^2 - 2ax + x^2 + h^2$	b. __?__ **Substitution**
c. $x^2 + h^2 = b^2$	c. __?__ **Pythagorean Theorem**
d. $c^2 = a^2 - 2ax + b^2$	d. __?__ **Substitution**
e. $\cos C = \frac{x}{b}$	e. __?__ **Def. of cosine**
f. $b \cos C = x$	f. __?__ **Cross products**
g. $c^2 = a^2 - 2a(b \cos C) + b^2$	g. __?__ **Substitution**
h. $c^2 = a^2 + b^2 - 2ab \cos C$	h. __?__ **Commutative Property**

44. CRITICAL THINKING Graph $A(-6, -8)$, $B(10, -4)$, $C(6, 8)$, and $D(5, 11)$ on a coordinate plane. Find the measure of interior angle ABC and the measure of exterior angle DCA. **$m\angle ABC \approx 85.6$; $m\angle DCA \approx 124.7$**

45. WRITING IN MATH Answer the question that was posed at the beginning of the lesson. **See margin.**

How are triangles used in building construction?

Include the following in your answer:
- why the building was triangular instead of rectangular, and
- why the Law of Sines could not be used to solve the triangle.

www.geometryonline.com/self_check_quiz/fcat

Answers

45. Sample answer: Triangles are used to build supports, walls, and foundations. Answers should include the following.
- **The triangular building was more efficient with the cells around the edge.**
- **The Law of Sines requires two angles and a side or two sides and an angle opposite one of those sides.**

Enrichment, p. 392

Spherical Triangles

Spherical trigonometry is an extension of plane trigonometry. Figures are drawn on the surface of a sphere. Arcs of great circles correspond to line segments in the plane. The arcs of three great circles intersecting on a sphere form a spherical triangle. Angles have the same measure as the tangent lines drawn to each great circle at the vertex. Since the sides are arcs, they too can be measured in degrees.

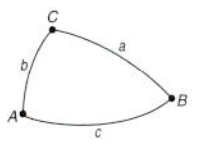

The sum of the sides of a spherical triangle is less than 360°.
The sum of the angles is greater than 180° and less than 540°.
The Law of Sines for spherical triangles is as follows.

$$\frac{\sin a}{\sin A} = \frac{\sin b}{\sin B} = \frac{\sin c}{\sin C}$$

There is also a Law of Cosines for spherical triangles.

$$\cos a = \cos b \cos c + \sin b \sin c \cos A$$
$$\cos b = \cos a \cos c + \sin a \sin c \cos B$$
$$\cos c = \cos a \cos b + \sin a \sin b \cos C$$

Study Guide and Intervention, p. 387 (shown) and p. 388

The Law of Cosines Another relationship between the sides and angles of any triangle is called the **Law of Cosines**. You can use the Law of Cosines if you know three sides of a triangle or if you know two sides and the included angle of a triangle.

Law of Cosines	Let $\triangle ABC$ be any triangle with a, b, and c representing the measures of the sides opposite the angles with measures A, B, and C, respectively. Then the following equations are true. $a^2 = b^2 + c^2 - 2bc \cos A$ $\quad b^2 = a^2 + c^2 - 2ac \cos B$ $\quad c^2 = a^2 + b^2 - 2ab \cos C$

Example 1 **In $\triangle ABC$, find c.**

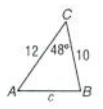

$c^2 = a^2 + b^2 - 2ab \cos C$ — Law of Cosines
$c^2 = 12^2 + 10^2 - 2(12)(10)\cos 48°$ — $a = 12$, $b = 10$, $m\angle C = 48$
$c = \sqrt{12^2 + 10^2 - 2(12)(10)\cos 48°}$ — Take the square root of each side.
$c \approx 9.1$ — Use a calculator.

Example 2 **In $\triangle ABC$, find $m\angle A$.**

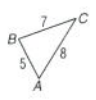

$a^2 = b^2 + c^2 - 2bc \cos A$ — Law of Cosines
$7^2 = 5^2 + 8^2 - 2(5)(8) \cos A$ — $a = 7$, $b = 5$, $c = 8$
$49 = 25 + 64 - 80 \cos A$ — Multiply.
$-40 = -80 \cos A$ — Subtract 89 from each side.
$\frac{1}{2} = \cos A$ — Divide each side by −80.
$\cos^{-1}\frac{1}{2} = A$ — Use the inverse cosine.
$60° = A$ — Use a calculator.

Exercises

Find each measure using the given measures from $\triangle ABC$. Round angle measures to the nearest degree and side measures to the nearest tenth.

1. If $b = 14$, $c = 12$, and $m\angle A = 62$, find a. **13.5**
2. If $a = 11$, $b = 10$, and $c = 12$, find $m\angle B$. **51**
3. If $a = 24$, $b = 18$, and $c = 16$, find $m\angle C$. **42**
4. If $a = 20$, $c = 25$, and $m\angle B = 82$, find b. **29.8**
5. If $b = 18$, $c = 28$, and $m\angle A = 59$, find a. **24.3**
6. If $a = 15$, $b = 19$, and $c = 15$, find $m\angle C$. **51**

Skills Practice, p. 389 and Practice, p. 390 (shown)

In $\triangle JKL$, given the following measures, find the measure of the missing side.

1. $j = 1.3$, $k = 10$, $m\angle L = 77$ **$\ell \approx 9.8$**
2. $j = 9.6$, $\ell = 1.7$, $m\angle K = 43$ **$k \approx 8.4$**
3. $j = 11$, $k = 7$, $m\angle L = 63$ **$\ell \approx 10.0$**
4. $k = 4.7$, $\ell = 5.2$, $m\angle J = 112$ **$j \approx 8.2$**

In $\triangle MNQ$, given the lengths of the sides, find the measure of the stated angle to the nearest tenth.

5. $m = 17$, $n = 23$, $q = 25$; $m\angle Q$ **75.7**
6. $m = 24$, $n = 28$, $q = 34$; $m\angle M$ **44.2**
7. $m = 12.9$, $n = 18$, $q = 20.5$; $m\angle N$ **60.2**
8. $m = 23$, $n = 30.1$, $q = 42$; $m\angle Q$ **103.7**

Determine whether the Law of Sines or the Law of Cosines should be used first to solve $\triangle ABC$. Then sole each triangle. Round angle measures to the nearest degree and side measure to the nearest tenth.

9. $a = 13$, $b = 18$, $c = 19$
Cosines; $m\angle A \approx 41$; $m\angle B \approx 65$; $m\angle C \approx 74$
10. $a = 6$, $b = 19$, $m\angle C = 38$
Cosines; $m\angle A \approx 15$; $m\angle B \approx 127$; $c \approx 14.7$
11. $a = 17$, $b = 22$, $m\angle B = 49$
Sines; $m\angle A \approx 36$; $m\angle C \approx 95$; $c \approx 29.0$
12. $a = 15.5$, $b = 18$, $m\angle C = 72$
Cosines; $m\angle A \approx 48$; $m\angle B \approx 60$; $c \approx 19.8$

Solve each $\triangle FGH$ described below. Round measures to the nearest tenth.

13. $m\angle F = 54$, $f = 12.5$, $g = 11$ **$m\angle G \approx 45.4$, $m\angle H \approx 80.6$, $h \approx 15.2$**
14. $f = 20$, $g = 23$, $m\angle H = 47$ **$m\angle F \approx 57.4$, $m\angle G \approx 75.6$, $h \approx 17.4$**
15. $f = 15.8$, $g = 11$, $h = 14$ **$m\angle F \approx 77.4$, $m\angle G \approx 42.8$, $m\angle H \approx 59.8$**
16. $f = 36$, $h = 30$, $m\angle G = 54$ **$m\angle F \approx 73.1$, $m\angle H \approx 52.9$, $g \approx 30.4$**

17. **REAL ESTATE** The Esposito family purchased a triangular plot of land on which they plan to build a barn and corral. The lengths of the sides of the plot are 320 feet, 286 feet, and 305 feet. What are the measures of the angles formed on each side of the property?
65.5, 54.4, 60.1

Reading to Learn Mathematics, p. 391 ELL

Pre-Activity **How are triangles used in building design?**

Read the introduction to Lesson 7-7 at the top of page 385 in your textbook.

What could be a disadvantage of a triangular room? **Sample answer: Furniture will not fit in the corners.**

Reading the Lesson

1. Refer to the figure. According to the Law of Cosines, which statements are correct for $\triangle DEF$? **B, E, H**

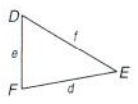

A. $d^2 = e^2 + f^2 - ef \cos D$
B. $e^2 = d^2 + f^2 - 2df \cos E$
C. $d^2 = e^2 + f^2 + 2ef \cos D$
D. $f^2 = d^2 + e^2 - 2ef \cos F$
E. $f^2 = d^2 + e^2 - 2de \cos F$
F. $d^2 = e^2 + f^2$
G. $\frac{\sin D}{d} = \frac{\sin E}{e} = \frac{\sin F}{f}$
H. $d = \sqrt{e^2 + f^2 - 2ef \cos D}$

2. Each of the following describes three given parts of a triangle. In each case, indicate whether you would use the Law of Sines or the Law of Cosines first in solving a triangle with those given parts. (In some cases, only one of the two laws would be used in solving the triangle.)
   a. SSS **Law of Cosines**
   b. ASA **Law of Sines**
   c. AAS **Law of Sines**
   d. SAS **Law of Cosines**
   e. SSA **Law of Sines**

3. Indicate whether each statement is *true* or *false*. If the statement is false, explain why.
   a. The Law of Cosines applies to right triangles. **true**
   b. The Pythagorean Theorem applies to acute triangles. **False; sample answer: It only applies to right triangles.**
   c. The Law of Cosines is used to find the third side of a triangle when you are given the measures of two sides and the nonincluded angle. **False; sample answer: It is used when you are given the measures of two sides and the included angle.**
   d. The Law of Cosines can be used to solve a triangle in which the measures of the three sides are 5 centimeters, 8 centimeters, and 15 centimeters. **False; sample answer: 5 + 8 < 15, so, by the Triangle Inequality Theorem, no such triangle exists.**

Helping You Remember

4. A good way to remember a new mathematical formula is to relate it to one you already know. The Law of Cosines looks somewhat like the Pythagorean Theorem. Both formulas must be true for a right triangle. How can that be? **cos 90 = 0, so in a right triangle, where the included angle is the right angle, the Law of Cosines becomes the Pythagorean Theorem.**

4 Assess

Open-Ended Assessment

Writing Select some exercises to reproduce on the board for the class. Have students write the equations necessary to solve for different angles and sides and fill in the values without solving the problems. Volunteers can write their equations on the board.

Assessment Options

Quiz (Lessons 7-7) is available on p. 408 of the *Chapter 7 Resource Masters.*

Answers

55. Given: $\triangle JFM \sim \triangle EFB$
$\triangle LFM \sim \triangle GFB$

Prove: $\triangle JFL \sim \triangle EFG$

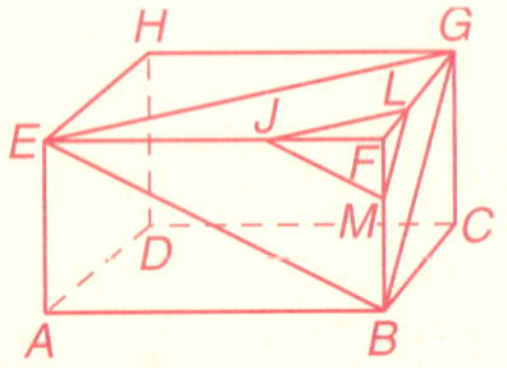

Proof: Since $\triangle JFM \sim \triangle EFB$ and $\triangle LFM \sim \triangle GFB$, then by definition of similar triangles, $\frac{JF}{EF} = \frac{MF}{BF}$ and $\frac{MF}{BF} = \frac{LF}{GF}$. By the Transitive Property of Equality, $\frac{JF}{EF} = \frac{LF}{GF}$. $\angle F \cong \angle F$ by the Reflexive Property of Congruence. Then, by SAS Similarity, $\triangle JFL \sim \triangle EFG$.

56. Given: $\overline{JM} \parallel \overline{EB}$
$\overline{LM} \parallel \overline{GB}$

Prove: $\overline{JL} \parallel \overline{EG}$

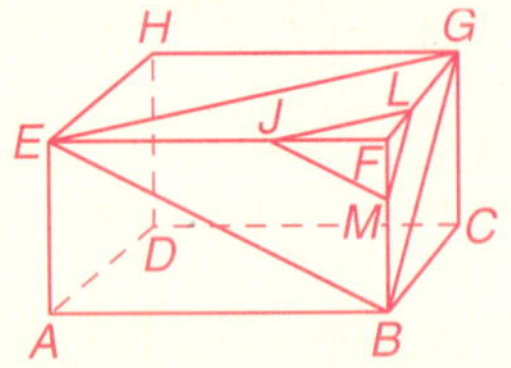

Proof: Since $\overline{JM} \parallel \overline{EB}$ and $\overline{LM} \parallel \overline{GB}$, then $\angle MJF \cong \angle BEF$ and $\angle FML \cong \angle FBG$ because if two parallel lines are cut by a transversal, corresponding angles are congruent. $\angle EFB \cong \angle EFB$ and $\angle BFG \cong \angle BFG$ by the Reflexive Property of Congruence. Then $\triangle EFB \sim \triangle JFM$ and $\triangle FBG \sim \triangle FML$ by AA Similarity. Then $\frac{JF}{EF} = \frac{MF}{BF}$, $\frac{MF}{BF} = \frac{LF}{GF}$, by the definition of similar triangles. $\frac{JF}{EF} = \frac{LF}{GF}$ by the Transitive Property of Equality, and $\angle EFG \cong \angle EFG$ by the Reflexive Property of Congruence. Thus, $\triangle JFL \sim \triangle EFG$ by SAS Similarity and $\angle FJL \cong \angle FEG$ by the definition of similar triangles. $\overline{JL} \parallel \overline{EG}$ because if two lines are cut by a tranversal so that the corresponding angles are congruent, then the lines are parallel.

FCAT Practice
Standardized Test Practice

46. For $\triangle DEF$, find d to the nearest tenth if $e = 12$, $f = 15$, and $m\angle D = 75$. **B**

Ⓐ 18.9 Ⓑ 16.6 Ⓒ 15.4 Ⓓ 9.8

47. **ALGEBRA** Ms. LaHue earns a monthly base salary of \$1280 plus a commission of 12.5% of her total monthly sales. At the end of one month, Ms. LaHue earned \$4455. What were her total sales for the month? **C**

Ⓐ \$3175 Ⓑ \$10,240 Ⓒ \$25,400 Ⓓ \$35,640

Maintain Your Skills

Mixed Review **Find each measure using the given measures from $\triangle XYZ$. Round angle measure to the nearest degree and side measure to the nearest tenth.** *(Lesson 7-6)*

48. If $y = 4.7$, $m\angle X = 22$, and $m\angle Y = 49$, find x. **2.3**

49. If $y = 10$, $x = 14$, and $m\angle X = 50$, find $m\angle Y$. **33**

50. **SURVEYING** A surveyor is 100 meters from a building and finds that the angle of elevation to the top of the building is 23°. If the surveyor's eye level is 1.55 meters above the ground, find the height of the building. *(Lesson 7-5)* **about 44.0 m**

For Exercises 51–54, determine whether $\overline{AB} \parallel \overline{CD}$. *(Lesson 6-4)*

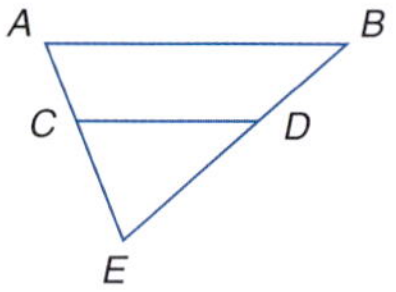

51. $AC = 8.4$, $BD = 6.3$, $DE = 4.5$, and $CE = 6$ **yes**

52. $AC = 7$, $BD = 10.5$, $BE = 22.5$, and $AE = 15$ **yes**

53. $AB = 8$, $AE = 9$, $CD = 4$, and $CE = 4$ **no**

54. $AB = 5.4$, $BE = 18$, $CD = 3$, and $DE = 10$ **yes**

Use the figure at the right to write a paragraph proof. *(Lesson 6-3)*

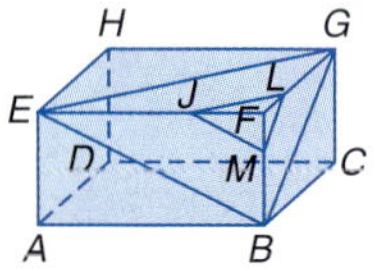

55. **Given:** $\triangle JFM \sim \triangle EFB$
$\triangle LFM \sim \triangle GFB$

Prove: $\triangle JFL \sim \triangle EFG$ **See margin.**

56. **Given:** $\overline{JM} \parallel \overline{EB}$
$\overline{LM} \parallel \overline{GB}$

Prove: $\overline{JL} \parallel \overline{EG}$ **See margin.**

COORDINATE GEOMETRY **The vertices of $\triangle XYZ$ are $X(8, 0)$, $Y(-4, 8)$, and $Z(0, 12)$. Find the coordinates of the points of concurrency of $\triangle XYZ$ to the nearest tenth.** *(Lesson 5-1)*

57. orthocenter **(−1.6, 9.6)** 58. centroid **(1.3, 6.7)** 59. circumcenter **(2.8, 5.2)**

WebQuest Internet Project

Who is Behind This Geometry Idea Anyway?

It's time to complete your project. Use the information and data you have gathered about your research topic, two mathematicians, and a geometry problem to prepare a portfolio or Web page. Be sure to include illustrations and/or tables in the presentation.

www.geometryonline.com/webquest

Geometry Activity

A Follow-Up of Lesson 7-7

Trigonometric Identities

In algebra, the equation $2(x + 2) = 2x + 4$ is called an *identity* because the equation is true for all values of x. There are equations involving trigonometric ratios that are true for all values of the angle measure. These are called **trigonometric identities**.

In the figure, $P(x, y)$ is in Quadrant I. The Greek letter theta (pronounced THAY tuh) θ, represents the measure of the angle formed by the x-axis and $\overrightarrow{OP}$. Triangle POR is a right triangle. Let r represent the length of the hypotenuse. Then the following are true.

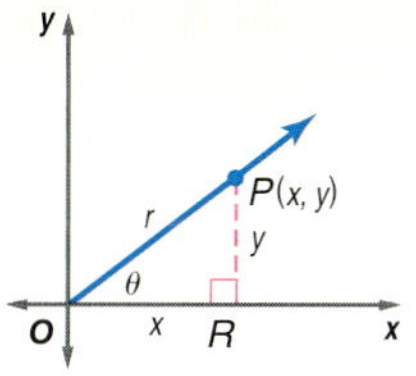

$\sin\theta = \frac{y}{r}$ $\quad\cos\theta = \frac{x}{r}$ $\quad\tan\theta = \frac{y}{x}$

$\csc\theta = \frac{r}{y}$ $\quad\sec\theta = \frac{r}{x}$ $\quad\cot\theta = \frac{x}{y}$

Notice that $\frac{1}{\sin\theta} = \frac{1}{\frac{y}{r}}$ $\qquad \frac{1}{\frac{y}{r}} = 1 \div \frac{y}{r}$ $\qquad 1 \div \frac{y}{r} = 1 \cdot \frac{r}{y}$ or $\frac{r}{y}$ $\qquad \frac{r}{y} = \csc\theta.$

So, $\frac{1}{\sin\theta} = \csc\theta$. This is known as one of the **reciprocal identities**.

Activity

Verify that $\cos^2\theta + \sin^2\theta = 1$.

The expression $\cos^2\theta$ means $(\cos\theta)^2$. To verify an identity, work on only one side of the equation and use what you know to show how that side is equivalent to the other side.

$\cos^2\theta + \sin^2\theta \stackrel{?}{=} 1$	Original equation
$\left(\frac{x}{r}\right)^2 + \left(\frac{y}{r}\right)^2 \stackrel{?}{=} 1$	Substitute.
$\frac{x^2}{r^2} + \frac{y^2}{r^2} \stackrel{?}{=} 1$	Simplify.
$\frac{x^2 + y^2}{r^2} \stackrel{?}{=} 1$	Combine fractions with like denominators.
$\frac{r^2}{r^2} \stackrel{?}{=} 1$	Pythagorean Theorem: $x^2 + y^2 = r^2$
$1 = 1 \checkmark$	Simplify.

Since $1 = 1$, $\cos^2\theta + \sin^2\theta = 1$.

Analyze 1–2. See margin.

1. The identity $\cos^2\theta + \sin^2\theta = 1$ is known as a **Pythagorean identity**. Why do you think the word *Pythagorean* is used to name this?
2. Find two more reciprocal identities involving $\frac{1}{\cos\theta}$ and $\frac{1}{\tan\theta}$.

Verify each identity. 3–6. See p. 399A.

3. $\frac{\sin\theta}{\cos\theta} = \tan\theta$
4. $\cot\theta = \frac{\cos\theta}{\sin\theta}$
5. $\tan^2\theta + 1 = \sec^2\theta$
6. $\cot^2\theta + 1 = \csc^2\theta$

Geometry Activity

A Follow-Up of Lesson 7-7

Getting Started

Trigonometric Identities
Explain to students that this investigation provides them with the tools necessary for deriving trigonometric identities, which they will later use to solve problems and prove theorems.

Teach

- Have students write an equation for the Law of Cosines and substitute the symbols in the figure, using θ as the angle.
- Ask students to set up a proportion with the Law of Sines, using θ and 90° for the angles and then solving for $\sin\theta$.
- Encourage students to become comfortable with verifying the identities, and advise them that these skills will be useful when they are faced with more complex problems and proofs.

Assess

Exercises 1–6 allow students to systematically practice and become more comfortable with finding equivalent trigonometric expressions and verifying trigonometric identities. Some students may need to strengthen their algebra skills before working through these exercises.

Answers

1. Sample answer: It is of the form $a^2 + b^2 = c^2$, where $c = 1$.
2. $\frac{1}{\cos\theta} = \sec\theta$; $\frac{1}{\tan\theta} = \cot\theta$

Chapter 7 Study Guide and Review

Vocabulary and Concept Check

- This alphabetical list of vocabulary terms in Chapter 7 includes a page reference where each term was introduced.
- **Assessment** A vocabulary test/review for Chapter 7 is available on p. 406 of the *Chapter 7 Resource Masters*.

Lesson-by-Lesson Review

For each lesson,

- the main ideas are summarized,
- additional examples review concepts, and
- practice exercises are provided.

Vocabulary PuzzleMaker

ELL The Vocabulary PuzzleMaker software improves students' mathematics vocabulary using four puzzle formats—crossword, scramble, word search using a word list, and word search using clues. Students can work on a computer screen or from a printed handout.

MindJogger Videoquizzes

ELL MindJogger Videoquizzes provide an alternative review of concepts presented in this chapter. Students work in teams in a game show format to gain points for correct answers. The questions are presented in three rounds.

Round 1 Concepts (5 questions)
Round 2 Skills (4 questions)
Round 3 Problem Solving (4 questions)

Chapter 7 Study Guide and Review

Vocabulary and Concept Check

ambiguous case (p. 384)	geometric mean (p. 342)	Pythagorean triple (p. 352)	tangent (p. 364)
angle of depression (p. 372)	Law of Cosines (p. 385)	reciprocal identities (p. 391)	trigonometric identity (p. 3
angle of elevation (p. 371)	Law of Sines (p. 377)	sine (p. 364)	trigonometric ratio (p. 364)
cosine (p. 364)	Pythagorean identity (p. 391)	solving a triangle (p. 378)	trigonometry (p. 364)

A complete list of postulates and theorems can be found on pages R1–R8.

Exercises **State whether each statement is *true* or *false*. If false, replace the underlined word or words to make a true sentence.**

1. The Law of Sines can be applied if you know the measures of two sides and an angle <u>opposite</u> one of these sides of the triangle. **true**
2. The tangent of $\angle A$ is the measure of the leg <u>adjacent</u> to $\angle A$ divided by the measure of the leg <u>opposite</u> $\angle A$. **false; opposite; adjacent**
3. In <u>any</u> triangle, the sum of the squares of the measures of the legs equals the square of the measure of the hypotenuse. **false; a right**
4. An angle of <u>elevation</u> is the angle between the line of sight and the horizontal when an observer looks upward. **true**
5. The geometric mean between two numbers is the positive square root of their <u>product</u>. **true**
6. In a <u>30°-60°-90°</u> triangle, two of the sides will have the same length. **false; 45°-45°-90°**
7. Looking at a city while flying in a plane is an example that uses angle of <u>elevation</u>. **false; depress**

Lesson-by-Lesson Review

7-1 Geometric Mean

See pages 342–348.

Concept Summary

- The geometric mean of two numbers is the square root of their product.
- You can use the geometric mean to find the altitude of a right triangle.

Examples

1 **Find the geometric mean between 10 and 30.**

$\frac{10}{x} = \frac{x}{30}$ Definition of geometric mean

$x^2 = 300$ Cross products

$x = \sqrt{300}$ or $10\sqrt{3}$ Take the square root of each side.

2 **Find NG in $\triangle TGR$.**

The measure of the altitude is the geometric mean between the measures of the two hypotenuse segments.

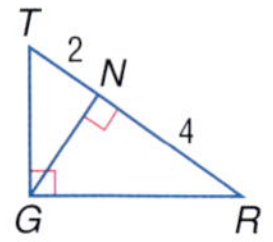

$\frac{TN}{GN} = \frac{GN}{RN}$ Definition of geometric mean

$\frac{2}{GN} = \frac{GN}{4}$ $TN = 2$, $RN = 4$

$8 = (GN)^2$ Cross products

$\sqrt{8}$ or $2\sqrt{2} = GN$ Take the square root of each side.

www.geometryonline.com/vocabulary_revie

FOLDABLES™ Study Organizer

For more information about Foldables, see *Teaching Mathematics with Foldables.*

Have students look through the chapter to make sure they have included notes and examples in their Foldables for each lesson of Chapter 7.

Encourage students to refer to their Foldables while completing the Study Guide and Review and to use them in preparing for the Chapter Test.

Exercises **Find the geometric mean between each pair of numbers.**
See Example 1 on page 342.

8. 4 and 16 **8** **9.** 4 and 81 **18** **10.** 20 and 35 **26.5** **11.** 18 and 44 **28.1**

12. In $\triangle PQR$, $PS = 8$, and $QS = 14$. Find RS. *See Example 2 on page 344.*
$4\sqrt{7} \approx 10.6$

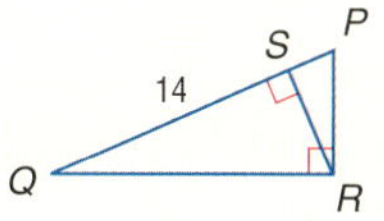

7-2 The Pythagorean Theorem and Its Converse

See pages 350–356.

Concept Summary

- The Pythagorean Theorem can be used to find the measures of the sides of a right triangle.
- If the measures of the sides of a triangle form a Pythagorean triple, then the triangle is a right triangle.

Example **Find *k*.**

$a^2 + (LK)^2 = (JL)^2$	Pythagorean Theorem
$a^2 + 8^2 = 13^2$	$LK = 8$ and $JL = 13$
$a^2 + 64 = 169$	Simplify.
$a^2 = 105$	Subtract 64 from each side.
$a = \sqrt{105}$	Take the square root of each side.
$a \approx 10.2$	Use a calculator.

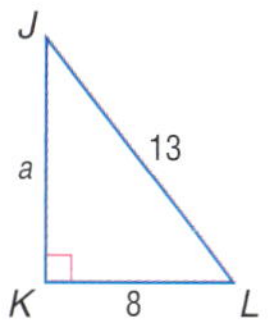

Exercises **Find *x*.** *See Example 2 on page 351.*

13. **25**

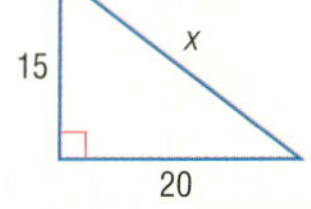

14. $\frac{12}{17}$

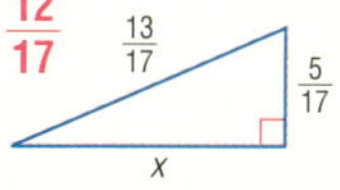

15. $4\sqrt{17} \approx 16.5$

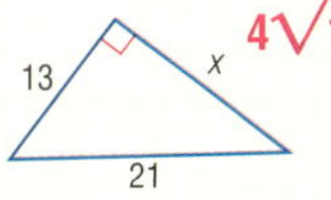

7-3 Special Right Triangles

See pages 357–363.

Concept Summary

- The measure of the hypotenuse of a 45°-45°-90° triangle is $\sqrt{2}$ times the length of the legs of the triangle. The measures of the sides are x, x, and $x\sqrt{2}$.
- In a 30°-60°-90° triangle, the measures of the sides are x, $x\sqrt{3}$, and $2x$.

Examples **1** **Find *x*.**

The measure of the shorter leg $\overline{XZ}$ of $\triangle XYZ$ is half the measure of the hypotenuse $\overline{XY}$. Therefore, $XZ = \frac{1}{2}(26)$ or 13. The measure of the longer leg is $\sqrt{3}$ times the measure of the shorter leg. So, $x = 13\sqrt{3}$.

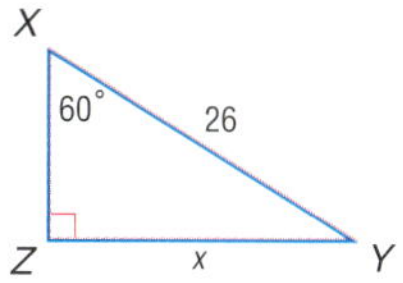

Chapter 7 Study Guide and Review 393

Answers

21. $\frac{3}{5} = 0.60; \frac{4}{5} = 0.80; \frac{3}{4} = 0.75;$
$\frac{4}{5} = 0.80; \frac{3}{5} = 0.60; \frac{4}{3} \approx 1.33$

22. $\frac{7}{25} = 0.28; \frac{24}{25} = 0.96; \frac{7}{24} \approx 0.29;$
$\frac{24}{25} = 0.96; \frac{7}{25} = 0.28; \frac{24}{7} \approx 3.43$

2 Find x.

The measure of the hypotenuse of a 45°-45°-90° triangle is $\sqrt{2}$ times the length of a leg of the triangle.

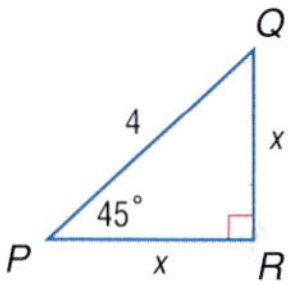

$x\sqrt{2} = 4$

$x = \frac{4}{\sqrt{2}}$

$x = \frac{4}{\sqrt{2}} \cdot \frac{\sqrt{2}}{\sqrt{2}}$ or $2\sqrt{2}$

Exercises **Find x and y.** *See Examples 1 and 3 on pages 358 and 359.*

16.

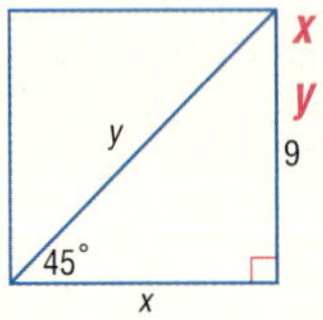

$x = 9$; $y = 9\sqrt{2}$

17.

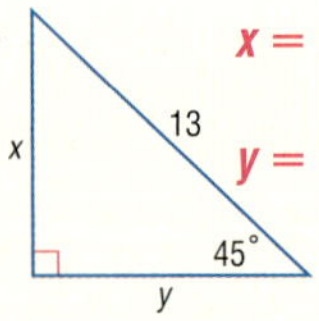

$x = \frac{13\sqrt{2}}{2}$; $y = \frac{13\sqrt{2}}{2}$

18.

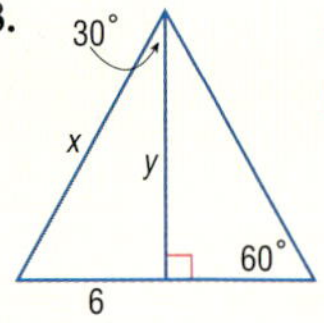

$x = 12$; $y = 6\sqrt{3}$

For Exercises 19 and 20, use the figure at the right.
See Example 3 on page 359.

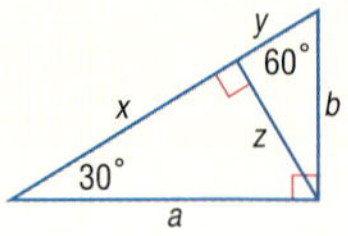

19. If $y = 18$, find z and a. $z = 18\sqrt{3}$, $a = 36\sqrt{3}$

20. If $x = 14$, find a, z, b, and y.
$a = \frac{28\sqrt{3}}{3}$, $z = \frac{14\sqrt{3}}{3}$, $b = \frac{28}{3}$, $y = \frac{14}{3}$

7-4 Trigonometry

See pages 364–370.

Concept Summary

- Trigonometric ratios can be used to find measures in right triangles.

Example **Find sin A, cos A, and tan A. Express each ratio as a fraction and as a decimal.**

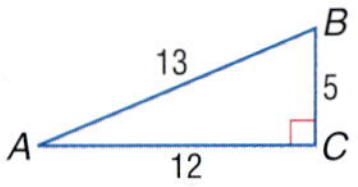

$\sin A = \frac{\text{opposite leg}}{\text{hypotenuse}} = \frac{BC}{AB} = \frac{5}{13}$ or about 0.38

$\cos A = \frac{\text{adjacent leg}}{\text{hypotenuse}} = \frac{AC}{AB} = \frac{12}{13}$ or about 0.92

$\tan A = \frac{\text{opposite leg}}{\text{adjacent leg}} = \frac{BC}{AC} = \frac{5}{12}$ or about 0.42

Exercises **Use $\triangle FGH$ to find sin F, cos F, tan F, sin G, cos G, and tan G. Express each ratio as a fraction and as a decimal to the nearest hundredth.** *See Example 1 on page 365.*

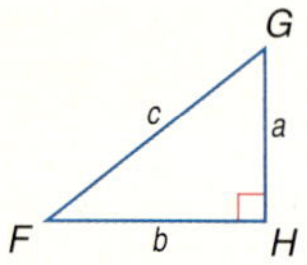

21. $a = 9, b = 12, c = 15$ **22.** $a = 7, b = 24, c = 25$

21–22. See margin.

Find the measure of each angle to the nearest tenth of a degree.
See Example 4 on pages 366 and 367.

23. $\sin P = 0.4522$ **26.9** **24.** $\cos Q = 0.1673$ **80.4** **25.** $\tan R = 0.9324$ **43.0**

394 Chapter 7 Right Triangles and Trigonometry

Study Guide and Review

7-5 Angles of Elevation and Depression

See pages 371–376.

Concept Summary

- Trigonometry can be used to solve problems related to angles of elevation and depression.

Example **A store has a ramp near its front entrance. The ramp measures 12 feet, and has a height of 3 feet. What is the angle of elevation?**

Make a drawing.

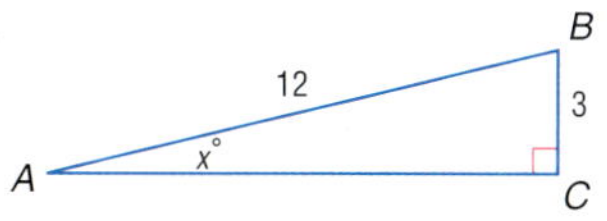

Let x represent $m\angle BAC$.

$\sin x° = \frac{BC}{AB}$ — $\sin x = \frac{\text{opposite}}{\text{hypotenuse}}$

$\sin x° = \frac{3}{12}$ — $BC = 3$ and $AB = 12$

$x = \sin^{-1}\left(\frac{3}{12}\right)$ — Find the inverse.

$x \approx 14.5$ — Use a calculator.

The angle of elevation for the ramp is about 14.5°.

Exercises **Determine the angles of elevation or depression in each situation.**

See Examples 1 and 2 on pages 371 and 372.

26. An airplane must clear a 60-foot pole at the end of a runway 500 yards long. **≈ 2.3°**

27. An escalator descends 100 feet for each horizontal distance of 240 feet. **≈ 22.6°**

28. A hot-air balloon descends 50 feet for every 1000 feet traveled horizontally. **≈ 2.9°**

29. **DAYLIGHT** At a certain time of the day, the angle of elevation of the sun is 44°. Find the length of a shadow cast by a building that is 30 yards high. **≈ 31.1 yd**

30. **RAILROADS** A railroad track rises 30 feet for every 400 feet of track. What is the measure of the angle of elevation of the track? **≈ 4.3°**

7-6 The Law of Sines

See pages 377–383.

Concept Summary

- To find the measures of a triangle by using the Law of Sines, you must either know the measures of two angles and any side (AAS or ASA), or two sides and an angle opposite one of these sides (SSA) of the triangle.
- To solve a triangle means to find the measures of all sides and angles.

Example **Solve $\triangle XYZ$ if $m\angle X = 32$, $m\angle Y = 61$, and $y = 15$. Round angle measures to the nearest degree and side measures to the nearest tenth.**

Find the measure of $\angle Z$.

$m\angle X + m\angle Y + m\angle Z = 180$ — Angle Sum Theorem

$32 + 61 + m\angle Z = 180$ — $m\angle X = 32$ and $m\angle Y = 61$

$93 + m\angle Z = 180$ — Add.

$m\angle Z = 87$ — Subtract 93 from each side.

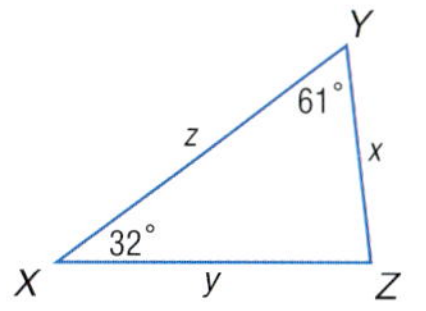

(continued on the next page)

Chapter 7 Study Guide and Review 395

Chapter 7 For More ...
• Extra Practice, see pages 766-768.
• Mixed Problem Solving, see page 788.

Answers

33. $m\angle B \approx 41$, $m\angle C \approx 75$, $c \approx 16.1$
34. $m\angle B = 58$, $b \approx 11.1$, $a \approx 10.7$
35. $m\angle B \approx 61$, $m\angle C \approx 90$, $c \approx 9.9$
36. $m\angle C = 87$, $a \approx 10.0$, $c \approx 20.6$
39. $a \approx 17.0$, $m\angle B \approx 43$, $m\angle C \approx 73$
40. $m\angle B \approx 38$, $m\angle A \approx 89$, $a \approx 8.4$

Since we know $m\angle Y$ and y, use proportions involving sin Y and y.

To find x:		To find z:
$\frac{\sin Y}{y} = \frac{\sin X}{x}$	Law of Sines	$\frac{\sin Y}{y} = \frac{\sin Z}{z}$
$\frac{\sin 61°}{15} = \frac{\sin 32°}{x}$	Substitute.	$\frac{\sin 61°}{15} = \frac{\sin 87°}{z}$
$x \sin 61° = 15 \sin 32°$	Cross products	$z \sin 61° = 15 \sin 87°$
$x = \frac{15 \sin 32°}{\sin 61°}$	Divide.	$z = \frac{15 \sin 87°}{\sin 61°}$
$x \approx 9.1$	Use a calculator.	$z \approx 17.1$

Exercises **Find each measure using the given measures of $\triangle FGH$. Round angle measures to the nearest degree and side measures to the nearest tenth.** *See Example 1 on page 378.*

31. Find f if $g = 16$, $m\angle G = 48$, and $m\angle F = 82$. **21.3**
32. Find $m\angle H$ if $h = 10.5$, $g = 13$, and $m\angle G = 65$. **47**

Solve each $\triangle ABC$ described below. Round angle measures to the nearest degree and side measures to the nearest tenth. *See Example 2 on pages 378 and 379.*

33. $a = 15, b = 11, m\angle A = 64$
34. $c = 12, m\angle C = 67, m\angle A = 55$
35. $m\angle A = 29, a = 4.8, b = 8.7$
36. $m\angle A = 29, m\angle B = 64, b = 18.5$

33–36. See margin.

7-7 The Law of Cosines

See pages 385–390.

Concept Summary

- The Law of Cosines can be used to solve triangles when you know the measures of two sides and the included angle (SAS) or the measures of the three sides (SSS).

Example **Find a if $b = 23$, $c = 19$, and $m\angle A = 54$.**

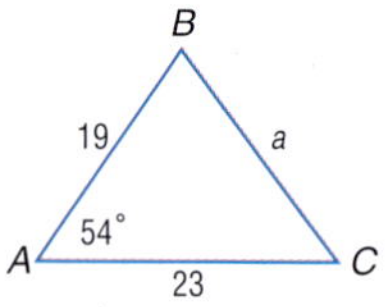

Since the measures of two sides and the included angle are known, use the Law of Cosines.

$a^2 = b + c^2 - 2bc \cos A$	Law of Cosines
$a^2 = 23^2 + 19^2 - 2(23)(19) \cos 54°$	$b = 23$, $c = 19$, and $m\angle A = 54$
$a = \sqrt{23^2 + 19^2 - 2(23)(19) \cos 54°}$	Take the square root of each side.
$a \approx 19.4$	Use a calculator.

Exercises **In $\triangle XYZ$, given the following measures, find the measure of the missing side.** *See Example 1 on page 385.*

37. $x = 7.6, y = 5.4, m\angle Z = 51$ **$z \approx 5.9$**
38. $x = 21, m\angle Y = 73, z = 16$ **$y \approx 22.4$**

Solve each triangle using the given information. Round angle measure to the nearest degree and side measure to the nearest tenth. *See Example 3 on pages 386 and 387.* **39–40. See margin.**

39. $c = 18, b = 13, m\angle A = 64$
40. $b = 5.2, m\angle C = 53, c = 6.7$

Answers (page 397)

2. Yes; two perfect squares can be written as $a \cdot a$ and $b \cdot b$. Multiplied together, we have $a \cdot a \cdot b \cdot b$. Taking the square root, we have ab, which is rational.

20. $m\angle A \approx 59$, $m\angle B \approx 76$, $c \approx 12.4$
21. $m\angle A \approx 56$, $m\angle C \approx 76$, $c \approx 14.2$
22. $m\angle A \approx 51$, $m\angle B \approx 72$, $m\angle C \approx 57$

Chapter 7 Practice Test

Vocabulary and Concepts

1. **State** the Law of Cosines for $\triangle ABC$ used to find $m\angle C$. $c^2 = a^2 + b^2 - 2ab \cos C$
2. **Determine** whether the geometric mean of two perfect squares will always be rational. Explain.
3. **Give** an example of side measures of a 30°-60°-90° triangle. Sample answer: 2, $2\sqrt{3}$, 4   2. See margin.

Skills and Applications

Find the geometric mean between each pair of numbers.

4. 7 and 63 21
5. 6 and 24 12
6. 10 and 50 $10\sqrt{5}$

Find the missing measures.

7. 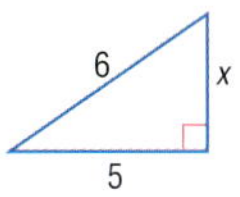

$\sqrt{11} \approx 3.32$

8. 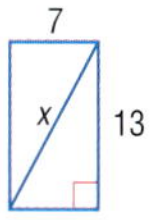

$\sqrt{218} \approx 14.8$

9. 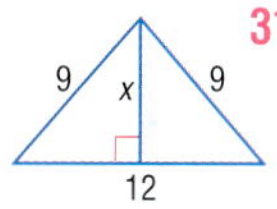

$3\sqrt{5} \approx 6.7$

10. 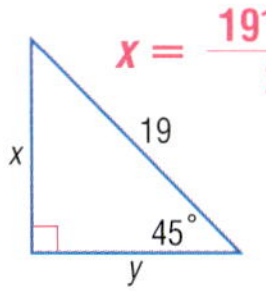

$x = \frac{19\sqrt{2}}{2}$; $y = \frac{19\sqrt{2}}{2}$

11. 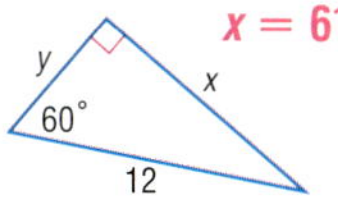

$x = 6\sqrt{3}$, $y = 6$

12. 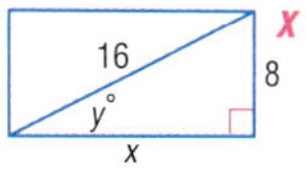

$x = 8\sqrt{3}$, $y = 30$

Use the figure to find each trigonometric ratio. Express answers as a fraction.

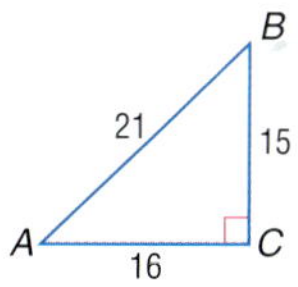

13. $\cos B$ $\frac{5}{7}$
14. $\tan A$ $\frac{15}{16}$
15. $\sin A$ $\frac{5}{7}$

Find each measure using the given measures from $\triangle FGH$. Round to the nearest tenth.

16. Find g if $m\angle F = 59$, $f = 13$, and $m\angle G = 71$. 14.3
17. Find $m\angle H$ if $m\angle F = 52$, $f = 10$, and $h = 12.5$. 80.1
18. Find f if $g = 15$, $h = 13$, and $m\angle F = 48$. 11.5
19. Find h if $f = 13.7$, $g = 16.8$, and $m\angle H = 71$. 17.9

Solve each triangle. Round each angle measure to the nearest degree and each side measure to the nearest tenth. 20–22. See margin.

20. $a = 15$, $b = 17$, $m\angle C = 45$
21. $a = 12.2$, $b = 10.9$, $m\angle B = 48$
22. $a = 19$, $b = 23.2$, $c = 21$

23. **TRAVEL** From an airplane, Janara looked down to see a city. If she looked down at an angle of 9° and the airplane was half a mile above the ground, what was the horizontal distance to the city? 3.2 mi

24. **CIVIL ENGINEERING** A section of freeway has a steady incline of 10°. If the horizontal distance from the beginning of the incline to the end is 5 miles, how high does the incline reach? 0.9 mi

25. **STANDARDIZED TEST PRACTICE** Find $\tan X$. D

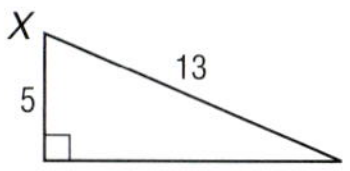

(A) $\frac{5}{12}$ (B) $\frac{12}{13}$ (C) $\frac{17}{12}$ (D) $\frac{12}{5}$

www.geometryonline.com/chapter_test/fcat

Portfolio Suggestion

Introduction After working through a chapter of very detailed material, it can help to determine strengths and weaknesses.

Ask Students List all the methods and formulas for solving triangles that you learned about in this chapter. Select a favorite method to write about and explain why you are most comfortable with this particular method. Write about your least favorite method and include possible ideas and activities to help you master it. Place this in your portfolio.

Chapter 7 Practice Test

Assessment Options

Vocabulary Test A vocabulary test/review for Chapter 7 can be found on p. 406 of the *Chapter 7 Resource Masters*.

Chapter Tests There are six Chapter 7 Tests and an Open-Ended Assessment task available in the *Chapter 7 Resource Masters*.

Chapter 7 Tests			
Form	**Type**	**Level**	**Pages**
1	MC	basic	393–394
2A	MC	average	395–396
2B	MC	average	397–398
2C	FR	average	399–400
2D	FR	average	401–402
3	FR	advanced	403–404

MC = multiple-choice questions
FR = free-response questions

Open-Ended Assessment Performance tasks for Chapter 7 can be found on p. 405 of the *Chapter 7 Resource Masters*. A sample scoring rubric for these tasks appears on p. A28.

Unit 2 Test A unit test/review can be found on pp. 413–414 of the *Chapter 7 Resource Masters*.

First Semester Test A test for Chapters 1–7 can be found on pp. 415–416 of the *Chapter 7 Resource Masters*.

TestCheck and Worksheet Builder

This **networkable software** has three modules for assessment.

- **Worksheet Builder** to make worksheets and tests.
- **Student Module** to take tests on-screen.
- **Management System** to keep student records.

Chapter 7 Standardized Test Practice

These two pages contain practice questions in the various formats that can be found on the most frequently given standardized tests.

A practice answer sheet for these two pages can be found on p. A1 of the *Chapter 7 Resource Masters*.

Standardized Test Practice Student Recording Sheet, p. A1

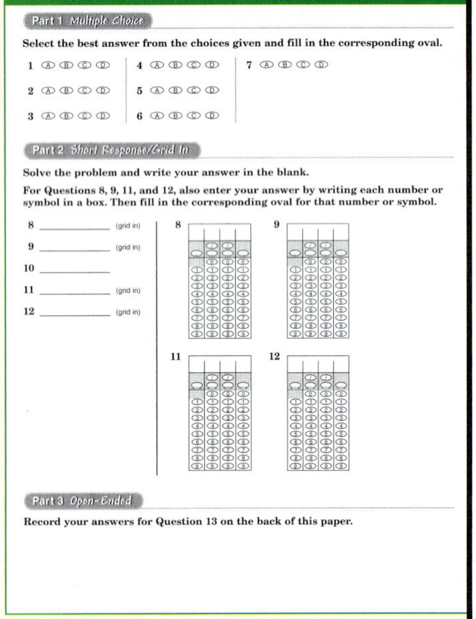
Part 1 Multiple Choice

Select the best answer from the choices given and fill in the corresponding oval.

1 Ⓐ Ⓑ Ⓒ Ⓓ 4 Ⓐ Ⓑ Ⓒ Ⓓ 7 Ⓐ Ⓑ Ⓒ Ⓓ

2 Ⓐ Ⓑ Ⓒ Ⓓ 5 Ⓐ Ⓑ Ⓒ Ⓓ

3 Ⓐ Ⓑ Ⓒ Ⓓ 6 Ⓐ Ⓑ Ⓒ Ⓓ

Part 2 Short Response/Grid In

Solve the problem and write your answer in the blank.

For Questions 8, 9, 11, and 12, also enter your answer by writing each number or symbol in a box. Then fill in the corresponding oval for that number or symbol.

8 ________ (grid in)

9 ________ (grid in)

10 ________

11 ________ (grid in)

12 ________ (grid in)

Part 3 Open-Ended

Record your answers for Question 13 on the back of this paper.

Additional Practice

See pp. 411–412 in the *Chapter 7 Resource Masters* for additional standardized test practice.

Chapter 7 Standardized Test Practice

FCAT Practice

Part 1 Multiple Choice

Record your answers on the answer sheet provided by your teacher or on a sheet of paper.

1. If $\angle 4$ and $\angle 3$ are supplementary, which reason could you use as the first step in proving that $\angle 1$ and $\angle 2$ are supplementary? (Lesson 2-7) C

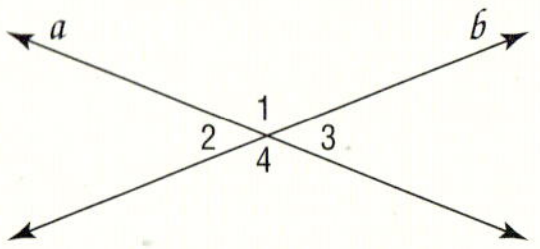

Ⓐ Definition of similar angles

Ⓑ Definition of perpendicular lines

Ⓒ Definition of a vertical angle

Ⓓ Division Property

2. In $\triangle ABC$, $\overline{BD}$ is a median. If $AD = 3x + 5$ and $CD = 5x - 1$, find AC. (Lesson 5-1) D

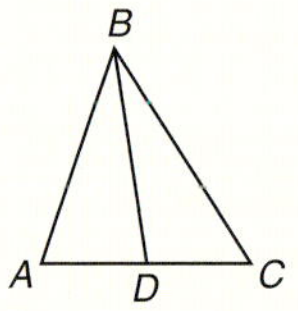

Ⓐ 3 Ⓑ 11

Ⓒ 14 Ⓓ 28

3. If pentagons $ABCDE$ and $PQRST$ are similar, find SR. (Lesson 6-2) B

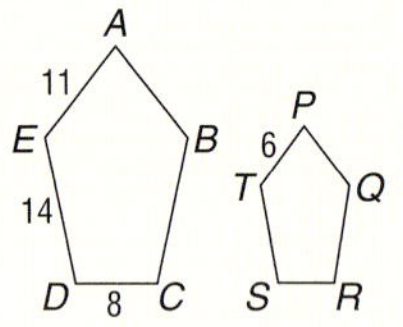

Ⓐ $14\frac{2}{3}$ Ⓑ $4\frac{4}{11}$

Ⓒ 3 Ⓓ $1\frac{5}{6}$

4. In $\triangle ABC$, $\overline{CD}$ is an altitude and $m\angle ACB = 90°$. If $AD = 12$ and $BD = 3$, find AC to the nearest tenth. (Lesson 7-1) C

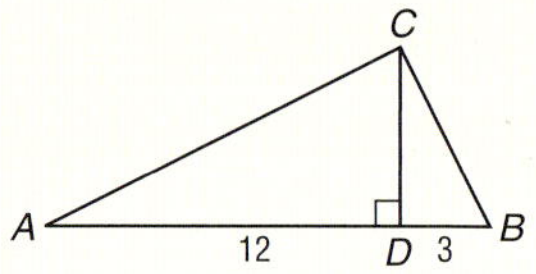

Ⓐ 6.5 Ⓑ 9.0 Ⓒ 13.4 Ⓓ 15.0

5. What is the length of $\overline{RT}$? (Lesson 7-3) B

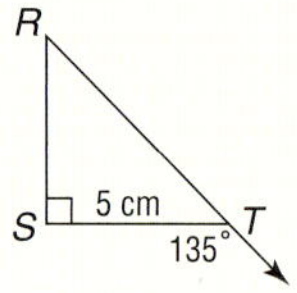

Ⓐ 5 cm Ⓑ $5\sqrt{2}$ cm

Ⓒ $5\sqrt{3}$ cm Ⓓ 10 cm

6. An earthquake damaged a communication tower. As a result, the top of the tower broke off at a point 60 feet above the base. If the fallen portion of the tower made a 36° angle with the ground, what was the approximate height of the original tower? (Lesson 7-4) D

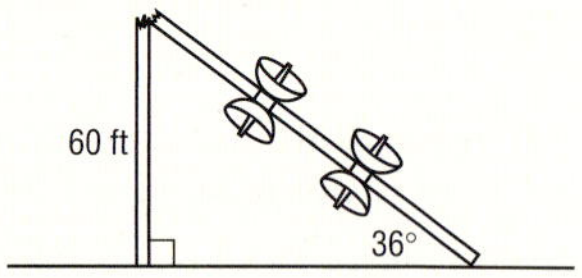

Ⓐ 35 ft Ⓑ 95 ft

Ⓒ 102 ft Ⓓ 162 ft

7. Miraku drew a map showing Regina's house, Steve's apartment, and Trina's workplace. The three locations formed $\triangle RST$, where $m\angle R = 34$, $r = 14$, and $s = 21$. Which could be $m\angle S$? (Lesson 7-6) C

Ⓐ 15 Ⓑ 22 Ⓒ 57 Ⓓ 84

398 Chapter 7 Right Triangles and Trigonometry

The Princeton Review

***Log On* for Test Practice**

The Princeton Review offers additional test-taking tips and practice problems at their web site. Visit www.princetonreview.com or www.review.com

TestCheck and Worksheet Builder

Special banks of standardized test questions similar to those on the SAT, ACT, TIMSS 8, NAEP 8, and state proficiency tests can be found on this CD-ROM.

Aligned and verified by The Princeton Review

FCAT Practice

Part 2 Short Response/Grid In

Record your answers on the answer sheet provided by your teacher or on a sheet of paper.

8. Find $m\angle ABC$ if $m\angle CDA = 61$. (Lesson 1-6) **140**

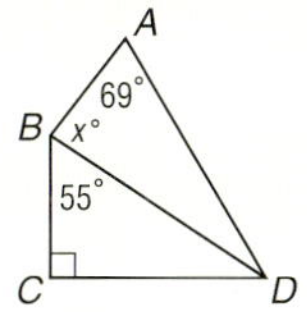

For Questions 9 and 10, refer to the graph.

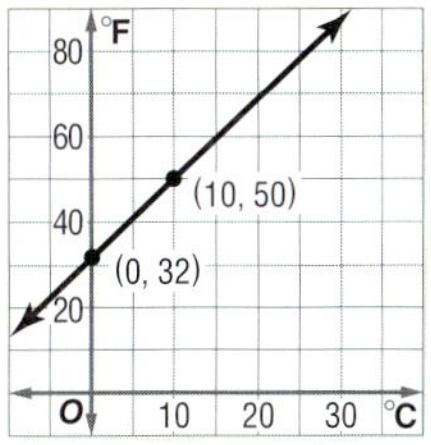

9. At the International Science Fair, a Canadian student recorded temperatures in degrees Celsius. A student from the United States recorded the same temperatures in degrees Fahrenheit. They used their data to plot a graph of Celsius versus Fahrenheit. What is the slope of their graph? (Lesson 3-3) $\frac{9}{5}$

10. Students used the equation of the line for the temperature graph of Celsius versus Fahrenheit to convert degrees Celsius to degrees Fahrenheit. If the line goes through points (0, 32) and (10, 50), what equation can the students use to convert degrees Celsius to degrees Fahrenheit? (Lesson 3-4)

$y = \frac{9}{5}x + 32$

11. $\triangle TUV$ and $\triangle XYZ$ are similar. Calculate the ratio $\frac{YZ}{UV}$. (Lesson 6-3) $\frac{3}{5}$

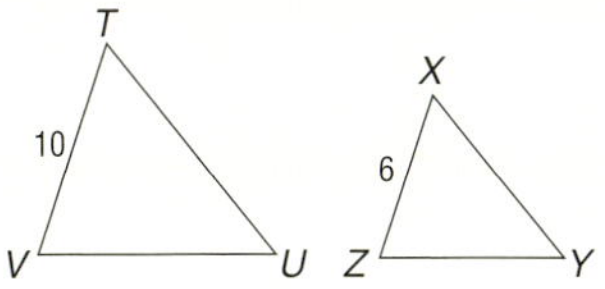

The Princeton Review Test-Taking Tip

Questions 6, 7, and 12 If a standardized test question involves trigonometric functions, draw a diagram that represents the problem and use a calculator (if allowed) or the table of trigonometric relationships provided with the test to help you find the answer.

12. Dee is parasailing at the ocean. The angle of depression from her line of sight to the boat is 41°. If the cable attaching Dee to the boat is 500 feet long, how many feet is Dee above the water? (Lesson 7-5) **328**

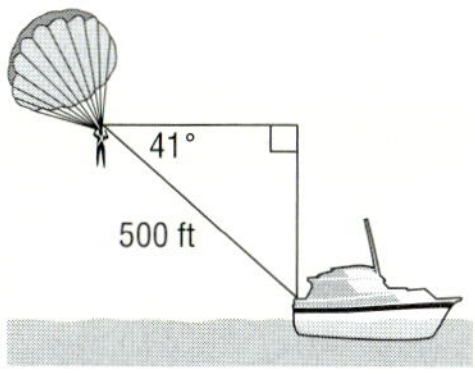

FCAT Practice

Part 3 Open Ended

Record your answers on a sheet of paper. Show your work.

13. Toby, Rani, and Sasha are practicing for a double Dutch rope-jumping tournament. Toby and Rani are standing at points T and R and are turning the ropes. Sasha is standing at S, equidistant from both Toby and Rani. Sasha will jump into the middle of the turning rope to point X. Prove that when Sasha jumps into the rope, she will be at the midpoint between Toby and Rani. (Lessons 4-5 and 4-6) **See margin.**

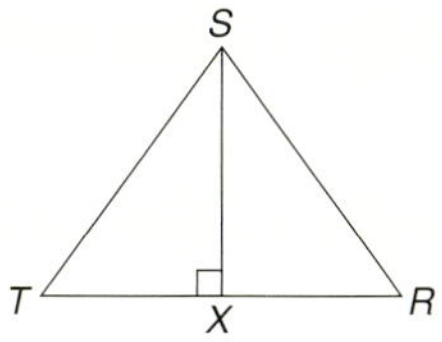

www.geometryonline.com/standardized_test/fcat

Evaluating Open-Ended Assessment Questions

Open-Ended Assessment questions are graded by using a multilevel rubric that guides you in assessing a student's knowledge of a particular concept.

Goal: Prove that a point is a midpoint.

Sample Scoring Rubric: The following rubric is a sample scoring device. You may wish to add more detail to this sample to meet your individual scoring needs.

Score	Criteria
4	A correct solution that is supported by well-developed, accurate explanations
3	A generally correct solution, but may contain minor flaws in reasoning or computation
2	A partially correct interpretation and/or solution to the problem
1	A correct solution with no supporting evidence or explanation
0	An incorrect solution indicating no mathematical understanding of the concept or task, or no solution is given

Answers

13. Since Sasha is equidistant from Toby and Rani, $\overline{ST}$ and $\overline{SR}$ are congruent and $\triangle STR$ is an isosceles triangle. According to the Isosceles Triangle Theorem, $\angle T$ and $\angle R$ are also congruent. $\overline{SX}$ is perpendicular to $\overline{TR}$, so $\angle SXT$ and $\angle SXR$ are both right angles and congruent. Two corresponding angles and the corresponding nonincluded sides are congruent (AAS Theorem), so $\triangle STX$ and $\triangle SRX$ are congruent triangles. Since these triangles are congruent, the corresponding sides $\overline{TX}$ and $\overline{RX}$ are congruent and have equal length; therefore when Sasha is jumping at Point X she will be at the midpoint between Toby and Rani.

Pages 345–348, Lesson 7-1

45. Given: $\angle PQR$ is a right angle.
$\overline{QS}$ is an altitude of $\triangle PQR$.

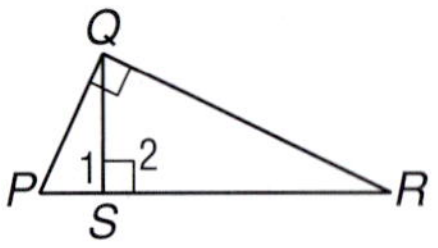

Prove: $\triangle PSQ \sim \triangle PQR$
$\triangle PQR \sim \triangle QSR$
$\triangle PSQ \sim \triangle QSR$

Proof:

Statements (Reasons)

1. $\angle PQR$ is a right angle; $\overline{QS}$ is an altitude of $\triangle PQR$. (Given)
2. $\overline{QS} \perp \overline{RP}$ (Definition of altitude)
3. $\angle 1$ and $\angle 2$ are right angles. (Definition of $\perp$ lines)
4. $\angle 1 \cong \angle PQR$; $\angle 2 \cong \angle PQR$ (All right $\angle$s are $\cong$.)
5. $\angle P \cong \angle P$; $\angle R \cong \angle R$ (Congruence of $\angle$s is reflexive.)
6. $\triangle PSQ \sim \triangle PQR$; $\triangle PQR \sim \triangle QSR$ (AA Similarity; Statements 4 and 5)
7. $\triangle PSQ \sim \triangle QSR$ (Similarity of triangles is transitive.)

46. Given: $\angle ADC$ is a right angle.
$\overline{DB}$ is an altitude of $\triangle ADC$.

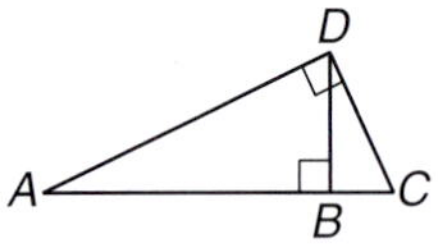

Prove: $\frac{AB}{DB} = \frac{DB}{CB}$

Proof: It is given that $\angle ADC$ is a right angle and $\overline{DB}$ is an altitude of $\triangle ADC$. $\triangle ADC$ is a right triangle by the definition of a right triangle. Therefore, $\triangle ADB \sim \triangle DCB$, because if the altitude is drawn from the vertex of the right angle to the hypotenuse of a right triangle, then the two triangles formed are similar to the given triangle and to each other. So $\frac{AB}{DB} = \frac{DB}{CB}$ by definition of similar polygons.

47. Given: $\angle ADC$ is a right angle.
$\overline{DB}$ is an altitude of $\triangle ADC$.

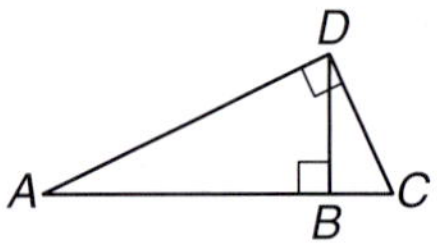

Prove: $\frac{AB}{AD} = \frac{AD}{AC}$; $\frac{BC}{DC} = \frac{DC}{AC}$

Proof:

Statements (Reasons)

1. $\angle ADC$ is a right angle; $\overline{DB}$ is an altitude of $\triangle ADC$. (Given)
2. $\triangle ADC$ is a right triangle. (Definition of right triangle)
3. $\triangle ABD \sim \triangle ADC$; $\triangle DBC \sim \triangle ADC$ (If the altitude is drawn from the vertex of the rt. $\angle$ to the hypotenuse of a rt. $\triangle$, then the 2 $\triangle$s formed are similar to the given $\triangle$ and to each other.)
4. $\frac{AB}{AD} = \frac{AD}{AC}$; $\frac{BC}{DC} = \frac{DC}{AC}$ (Definition of similar polygons)

Pages 360–363, Lesson 7-3

9.

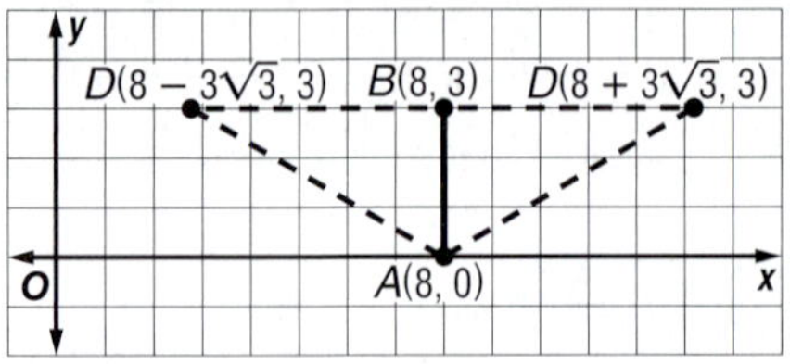

10.

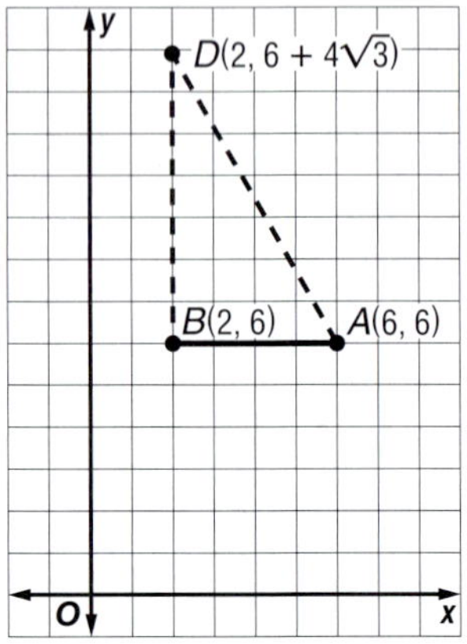

Page 391, Geometry Activity

3.

$\frac{\sin\theta}{\cos\theta} \stackrel{?}{=} \tan\theta$	Original equation
$\frac{\frac{y}{r}}{\frac{x}{r}} \stackrel{?}{=} \frac{y}{x}$	$\sin\theta = \frac{y}{r}$, $\cos\theta = \frac{x}{r}$, $\tan\theta = \frac{y}{x}$
$\frac{y}{r} \cdot \frac{r}{x} \stackrel{?}{=} \frac{y}{x}$	Multiply by the reciprocal of $\frac{x}{r}$.
$\frac{y}{x} = \frac{y}{x}\checkmark$	Multiply.

4.

$\frac{\cos\theta}{\sin\theta} \stackrel{?}{=} \cot\theta$	Original equation
$\frac{\frac{x}{r}}{\frac{y}{r}} \stackrel{?}{=} \frac{x}{y}$	$\sin\theta = \frac{y}{r}$, $\cos\theta = \frac{x}{r}$, $\cot\theta = \frac{x}{y}$
$\frac{x}{r} \cdot \frac{r}{y} \stackrel{?}{=} \frac{x}{y}$	Multiply by the reciprocal of $\frac{y}{x}$.
$\frac{x}{y} = \frac{x}{y}\checkmark$	Multiply.

5.

$\tan^2\theta + 1 \stackrel{?}{=} \sec^2\theta$	Original equation
$\left(\frac{y}{x}\right)^2 + 1 \stackrel{?}{=} \left(\frac{r}{x}\right)^2$	$\tan\theta = \frac{y}{x}$, $\sec\theta = \frac{r}{x}$
$\frac{y^2}{x^2} + 1 \stackrel{?}{=} \frac{r^2}{x^2}$	Evaluate exponents.
$x^2\left(\frac{y^2}{x^2} + 1\right) \stackrel{?}{=} x^2 \cdot \frac{r^2}{x^2}$	Multiply each side by x^2.
$y^2 + x^2 \stackrel{?}{=} r^2$	Simplify.
$r^2 = r^2\checkmark$	Substitution; $y^2 + x^2 = r^2$

6.

$\cot^2\theta + 1 \stackrel{?}{=} \csc^2\theta$	Original equation
$\left(\frac{x}{y}\right)^2 + 1 \stackrel{?}{=} \left(\frac{r}{y}\right)^2$	$\cot\theta = \frac{x}{y}$, $\sec\theta = \frac{r}{y}$
$\frac{x^2}{y^2} + 1 \stackrel{?}{=} \frac{r^2}{y^2}$	Evaluate exponents.
$y^2\left(\frac{x^2}{y^2} + 1\right) \stackrel{?}{=} y^2 \cdot \frac{r^2}{y^2}$	Multiply each side by y^2.
$x^2 + y^2 \stackrel{?}{=} r^2$	Simplify.
$r^2 = r^2\checkmark$	Substitution; $x^2 + y^2 = r^2$

Notes

UNIT 3

Notes

Introduction

In this unit students focus on quadrilaterals, transformations, and circles. They learn the properties of the various quadrilaterals and continue to use coordinate proofs to prove theorems.

Students explore reflections, translations, rotations, and dilations. Vectors and transformations with matrices are introduced.

Students learn the special properties of circles, including the form of their equations. They also learn about inscribed and circumscribed polygons, tangents, and secants.

Assessment Options

Unit 3 Test Pages 609–610 of the *Chapter 10 Resource Masters* may be used as a test or review for Unit 3. This assessment contains both multiple-choice and short answer items.

TestCheck and Worksheet Builder

This CD-ROM can be used to create additional unit tests and review worksheets.

UNIT 3

Quadrilaterals and Circles

Two-dimensional shapes such as quadrilaterals and circles can be used to describe and model the world around us. In this unit, you will learn about the properties of quadrilaterals and circles and how these two-dimensional figures can be transformed.

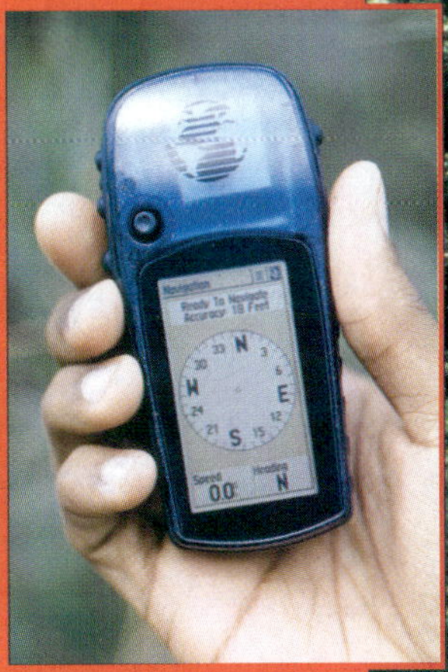

400 Unit 3 Quadrilaterals and Circles

WebQuest Internet Project

"Geocaching" Sends Folks on a Scavenger Hunt

Source: *USA TODAY*, July 26, 2001

"N42 DEGREES 02.054 W88 DEGREES 12.329 – Forget the poison ivy and needle-sharp brambles.

Dave April is a man on a mission. Clutching a palm-size Global Positioning System (GPS) receiver in one hand and a computer printout with latitude and longitude coordinates in the other, the 37-year-old software developer trudges doggedly through a suburban Chicago forest preserve, intent on finding a geek's version of buried treasure." Geocaching is one of the many new ways that people are spending their leisure time. In this project, you will use quadrilaterals, circles, and geometric transformations to give clues for a treasure hunt.

Log on to www.geometryonline.com/webquest. Begin your WebQuest by reading the Task.

Then continue working on your WebQuest as you study Unit 3.

Lesson	8-6	9-1	10-1
Page	444	469	527

Teaching Suggestions

Have students study the USA TODAY Snapshot.

- Ask them what the most popular leisure-time activity is. **reading**
- Have volunteers share their favorite leisure-time activities.
- Tell students that this WebQuest will send them on a scavenger hunt of their own.

Additional USA TODAY Snapshots appearing in Unit 3:

Chapter 8 Large companies have increased using the Internet to find employees (p. 411)

Chapter 9 Kids would rather be smart (p. 474)

Chapter 10 Majority microwave leftovers (p. 531)

WebQuest Internet Project

A WebQuest is an online project in which students do research on the Internet, gather data, and make presentations using word processing, graphing, page-making, or presentation software. In each chapter, students advance to the next step in their WebQuest. At the end of Chapter 10, the project culminates with a presentation of their findings.

Teaching notes and sample answers are available in the *WebQuest and Project Resources.*

Chapter 8

Quadrilaterals

Chapter Overview and Pacing

LESSON OBJECTIVES	PACING (days) Regular: Basic/Average	Regular: Advanced	Block: Basic/Average	Block: Advanced
8-1 Angles of Polygons *(pp. 404–410)* • Find the sum of the measures of the interior angles of a polygon. • Find the sum of the measures of the exterior angles of a polygon. ***Follow-Up:*** Use a spreadsheet to find interior and exterior angle measurements of a regular polygon.	1	2 (with 8-1 Follow-Up)	0.5	1 (with 8-1 Follow-Up)
8-2 Parallelograms *(pp. 411–416)* • Recognize and apply properties of the sides and angles of parallelograms. • Recognize and apply properties of the diagonals of parallelograms.	2	2	1	1
8-3 Tests for Parallelograms *(pp. 417–423)* • Recognize the conditions that ensure a quadrilateral is a parallelogram. • Prove that a set of points forms a parallelogram in the coordinate plane.	2	2	1	1
8-4 Rectangles *(pp. 424–430)* • Recognize and apply properties of rectangles. • Determine whether parallelograms are rectangles.	1	1	0.5	0.5
8-5 Rhombi and Squares *(pp. 431–438)* • Recognize and apply the properties of rhombi. • Recognize and apply the properties of squares. ***Follow-Up:*** Construct a kite.	2 (with 8-5 Follow-Up)	2 (with 8-5 Follow-Up)	1 (with 8-5 Follow-Up)	1 (with 8-5 Follow-Up)
8-6 Trapezoids *(pp. 439–445)* • Recognize and apply the properties of trapezoids. • Solve problems involving the medians of trapezoids.	2	1	1	0.5
8-7 Coordinate Proof with Quadrilaterals *(pp. 447–451)* • Position and label quadrilaterals for use in coordinate proofs. • Prove theorems using coordinate proofs.	2	2	1	1
Study Guide and Practice Test *(pp. 452–457)* **Standardized Test Practice** *(pp. 458–459)*	1	1	0.5	0.5
Chapter Assessment	1	1	0.5	0.5
TOTAL	**14**	**14**	**7**	**7**

Pacing suggestions for the entire year can be found on pages T20–T21.

Timesaving Tools
TeacherWorks™
All-In-One Planner and Resource Center
See pages T5 and T21.

Chapter Resource Manager

Chapter 8 Resource Masters										
Study Guide and Intervention	Practice (Skills and Average)	Reading to Learn Mathematics	Enrichment	Assessment	Prerequisite Skills Workbook	Applications*	5-Minute Check Transparencies	Interactive Chalkboard	GeomPASS: Tutorial Plus (lessons)	Materials
417–418	419–420	421	422		5–6		8-1	8-1		straightedge, protractor
423–424	425–426	427	428	473	81–82		8-2	8-2		patty paper, ruler
429–430	431–432	433	434			SC 15	8-3	8-3		straw, scissors, pipe cleaners, protractors
435–436	437–438	439	440	473, 475		GCC 31, 32	8-4	8-4		straightedge, compass
441–442	443–444	445	446		41–42		8-5	8-5		straightedge, compass, ruler, protractor (*Follow-Up:* compass)
447–448	449–450	451	452	474		SC 16	8-6	8-6	17	ruler, compass
453–454	455–456	457	458	474			8-7	8-7		
				459–472, 476–478						

**Key to Abbreviations:* GCC = Graphing Calculator and Computer Masters
SC = School-to-Career Masters

Chapter 8

Mathematical Connections and Background

Continuity of Instruction

Prior Knowledge

In Chapter 1, students used the Distance Formula. In Chapter 3, they found the slope of a line and proved that lines are perpendicular. They also identified the types of angles formed when a transversal intersects a pair of parallel lines. Chapter 4 challenged students to find the measure of a missing angle.

This Chapter

In this chapter, students explore quadrilaterals. They begin by investigating the interior and exterior angles of polygons. Then students learn to recognize and apply the properties of parallelograms. Students' knowledge of parallelograms is extended as they explore rectangles, rhombi, and squares and their special properties. Trapezoids are also explored. Finally, students position quadrilaterals on the coordinate plane for use in coordinate proofs.

Future Connections

In Chapter 11, students find the area of quadrilaterals. In Chapter 12, they again extend their knowledge of quadrilaterals by finding the surface area of prisms. Understanding the properties of quadrilaterals is essential to success in engineering, architecture, or design.

8-1 Angles of Polygons

The Interior Angle Sum Theorem states that if a convex polygon has n sides and S is the sum of the measures of its interior angles, then $S = 180(n - 2)$. This equation can also be used to find the measure of each interior angle in a regular polygon. Moreover, it can be used to find the number of sides in a polygon if the sum of the interior angle measures is known.

The sum of the exterior angles of a convex polygon is always 360, no matter the number of sides. This is called the Exterior Angle Sum Theorem.

8-2 Parallelograms

A parallelogram is a quadrilateral with both pairs of opposite sides parallel. Parallelograms have several special properties that help to define them. First, opposite sides of a parallelogram are congruent, and opposite angles of a parallelogram are congruent. Second, consecutive angles in a parallelogram are supplementary. Third, if a parallelogram has one right angle, it has four right angles. Finally, the diagonals of a parallelogram bisect each other, and each diagonal separates the parallelogram into two congruent triangles.

8-3 Tests for Parallelograms

In addition to the basic definition of a parallelogram as having opposite sides parallel, there are other tests to determine whether a quadrilateral is a parallelogram. If both pairs of opposite sides of a quadrilateral are congruent, then the quadrilateral is a parallelogram. If both pairs of opposite angles of a quadrilateral are congruent, then it is a parallelogram. If the diagonals of a quadrilateral bisect each other, the quadrilateral is a parallelogram. If one pair of opposite sides of a quadrilateral is both parallel *and* congruent, then it is a parallelogram.

A quadrilateral needs to pass only one of these five tests to be proved a parallelogram. All of the properties of a parallelogram do not need to be proved.

If a quadrilateral is graphed on the coordinate plane, you can use the Distance Formula and the Slope Formula to determine if it is a parallelogram. The Slope Formula is used to determine whether opposite sides are parallel. The Distance Formula is used to test opposite sides for congruence.

8-4 Rectangles

A rectangle is a quadrilateral with four right angles. Since both pairs of opposite sides are congruent, a rectangle has all the properties of a parallelogram. A rectangle has special properties of its own as well. For example, the diagonals of a rectangle are congruent. Congruent diagonals, in fact, can be used to prove that a parallelogram is a rectangle.

If a quadrilateral is graphed on a coordinate plane, the Slope Formula can be used to find out whether consecutive sides are perpendicular. If they are, then the quadrilateral is a rectangle. The Distance Formula can also be used to prove that a quadrilateral is a rectangle. After proving that the figure is a parallelogram, you can use the Distance Formula to calculate the measures of the diagonals. If the diagonals are congruent, then the parallelogram is a rectangle.

8-5 Rhombi and Squares

A rhombus is a quadrilateral with all four sides congruent. Since opposite sides are congruent, the rhombus is a parallelogram. Therefore, all the properties of parallelograms can be applied to rhombi. Rhombi also have special properties of their own. The diagonals of a rhombus are perpendicular. The converse of this theorem also holds true: If the diagonals of a parallelogram are perpendicular, then the parallelogram is a rhombus. Each diagonal of a rhombus bisects a pair of opposite angles.

If a quadrilateral is both a rhombus and a rectangle, then it is a square. A square is extremely specialized, having all the properties of a parallelogram, a rectangle, and a rhombus. It is important to note that while a square is a rhombus, a rhombus is not necessarily a square.

Coordinate geometry can be used to prove whether a parallelogram is a rhombus, a rectangle, a square, or none of those.

8-6 Trapezoids

A trapezoid is a quadrilateral with exactly one pair of parallel sides. The parallel sides are called *bases*, and the nonparallel sides are called *legs*. The base and one of the legs form the base angles. If the legs are congruent, then the trapezoid is an isosceles trapezoid. Both pairs of base angles of an isosceles trapezoid are congruent. The diagonals of an isosceles trapezoid are also congruent.

The segment that joins the midpoints of the legs of a trapezoid is the median. The median of a trapezoid is parallel to the bases, and its measure is one-half the sum of the measures of the bases. This is true for all trapezoids, not only isosceles trapezoids.

8-7 Coordinate Proof with Quadrilaterals

A coordinate proof is easier to write if the quadrilateral is correctly placed on the coordinate plane. Quadrilaterals should be placed so that the coordinates of the vertices are as simple as possible. In general, this means placing the quadrilateral in the first quadrant with one of its vertices at the origin. The base of the quadrilateral should run along the x-axis.

Once a figure has been placed on the coordinate plane, theorems can be proved using the Slope, Midpoint, and Distance Formulas.

Chapter 8

DAILY INTERVENTION and Assessment

	Type	Student Edition	Teacher Resources	Technology/Internet
INTERVENTION	Ongoing	Prerequisite Skills, pp. 403, 409, 416, 423, 430, 437, 445 Practice Quiz 1, p. 423 Practice Quiz 2, p. 445	5-Minute Check Transparencies *Prerequisite Skills Workbook*, pp. 5–6, 41–42, 81–82 Quizzes, *CRM* pp. 473–474 Mid-Chapter Test, *CRM* p. 475 Study Guide and Intervention, *CRM* pp. 417–418, 423–424, 429–430, 435–436, 441–442, 447–448, 453–454	GeomPASS: Tutorial Plus www.geometryonline.com/self_check_quiz www.geometryonline.com/extra_examples
	Mixed Review	pp. 409, 416, 423, 430, 437, 445, 451	Cumulative Review, *CRM* p. 476	
	Error Analysis	Find the Error, pp. 420, 427 Common Misconceptions, p. 419	Find the Error, *TWE* pp. 421, 427 Unlocking Misconceptions, *TWE* p. 434 Tips for New Teachers, *TWE* p. 406	
ASSESSMENT	Standardized Test Practice	pp. 409, 413, 414, 416, 423, 430, 437, 444, 445, 451, 457, 458, 459	*TWE* pp. 458–459 Standardized Test Practice, *CRM* pp. 477–478	Standardized Test Practice CD-ROM www.geometryonline.com/standardized_test
	Open-Ended Assessment	Writing in Math, pp. 409, 416, 422, 430, 436, 444, 451 Open Ended, pp. 407, 414, 420, 427, 434, 442, 449 Standardized Test, p. 459	Modeling: *TWE* pp. 423, 437 Speaking: *TWE* pp. 409, 445 Writing: *TWE* pp. 416, 430, 451 Open-Ended Assessment, *CRM* p. 471	
	Chapter Assessment	Study Guide, pp. 452–456 Practice Test, p. 457	Multiple-Choice Tests (Forms 1, 2A, 2B), *CRM* pp. 459–464 Free-Response Tests (Forms 2C, 2D, 3), *CRM* pp. 465–470 Vocabulary Test/Review, *CRM* p. 472	TestCheck and Worksheet Builder (see below) MindJogger Videoquizzes www.geometryonline.com/vocabulary_review www.geometryonline.com/chapter_test

Key to Abbreviations: TWE = Teacher Wraparound Edition; CRM = Chapter Resource Masters

Additional Intervention Resources

The Princeton Review's *Cracking the SAT & PSAT*

The Princeton Review's *Cracking the ACT*

TestCheck and Worksheet Builder

This **networkable** software has three modules for intervention and assessment flexibility:

- **Worksheet Builder** to make worksheet and tests
- **Student Module** to take tests on screen (optional)
- **Management System** to keep student records (optional)

Special banks are included for SAT, ACT, TIMSS, NAEP, and End-of-Course tests.

Intervention Technology

GeomPASS: Tutorial Plus CD-ROM offers a complete, self-paced geometry curriculum.

Geometry Lesson	GeomPASS Lesson
8-6	**17** *Quadrilaterals*

Intervention at Home

Log on for student study help.

- For each lesson in the Student Edition, there are Extra Examples and Self-Check Quizzes.
  www.geometryonline.com/extra_examples
  www.geometryonline.com/self_check_quiz
- For chapter review, there is vocabulary review, test practice, and standardized test practice.
  www.geometryonline.com/vocabulary_review
  www.geometryonline.com/chapter_test
  www.geometryonline.com/standardized_test

For more information on Intervention and Assessment, see pp. T8–T11.

Reading and Writing in Mathematics

Glencoe Geometry provides numerous opportunities to incorporate reading and writing into the mathematics classroom.

Student Edition

- Foldables Study Organizer, p. 403
- Concept Check questions require students to verbalize and write about what they have learned in the lesson. (pp. 407, 414, 420, 427, 434, 442, 449)
- Reading Mathematics, p. 446
- Writing in Math questions in every lesson, pp. 409, 416, 422, 430, 436, 444, 451
- Reading Study Tip, pp. 411, 432
- WebQuest, p. 444

Teacher Wraparound Edition

- Foldables Study Organizer, pp. 403, 452
- Study Notebook suggestions, pp. 404, 414, 421, 427, 434, 438, 442, 446, 449
- Modeling activities, pp. 423, 437
- Speaking activities, pp. 409, 445
- Writing activities, pp. 416, 430, 451
- ELL Resources, pp. 402, 408, 415, 422, 428, 435, 443, 446, 450, 452

Additional Resources

- Vocabulary Builder worksheets require students to define and give examples for key vocabulary terms as they progress through the chapter. (*Chapter 8 Resource Masters,* pp. vii-viii)
- Proof Builder helps students learn and understand theorems and postulates from the chapter. (*Chapter 8 Resource Masters,* pp. ix–x)
- Reading to Learn Mathematics master for each lesson (*Chapter 8 Resource Masters*, pp. 421, 427, 433, 439, 445, 451, 457)
- *Vocabulary PuzzleMaker* software creates crossword, jumble, and word search puzzles using vocabulary lists that you can customize.
- *Teaching Mathematics with Foldables* provides suggestions for promoting cognition and language.
- *Reading Strategies for the Mathematics Classroom*
- *WebQuest and Project Resources*

For more information on Reading and Writing in Mathematics, see pp. T6–T7.

Chapter 8 Notes

What You'll Learn

Have students read over the list of objectives and make a list of any words with which they are not familiar.

Why It's Important

Point out to students that this is only one of many reasons why each objective is important. Others are provided in the introduction to each lesson.

Chapter 8 Quadrilaterals

What You'll Learn

- **Lesson 8-1** Investigate interior and exterior angles of polygons.
- **Lessons 8-2 and 8-3** Recognize and apply the properties of parallelograms.
- **Lessons 8-4 through 8-6** Recognize and apply the properties of rectangles, rhombi, squares, and trapezoids.
- **Lesson 8-7** Position quadrilaterals for use in coordinate proof.

Key Vocabulary

- parallelogram (p. 411)
- rectangle (p. 424)
- rhombus (p. 431)
- square (p. 432)
- trapezoid (p. 439)

Why It's Important

Several different geometric shapes are examples of quadrilaterals. These shapes each have individual characteristics. A rectangle is a type of quadrilateral. Tennis courts are rectangles, and the properties of the rectangular court are used in the game. *You will learn more about tennis courts in Lesson 8-4.*

Lesson	NCTM Standards	Local Objectives
8-1	2, 3, 6, 8, 9, 10	
8-1 Follow-Up	2, 3, 6	
8-2	2, 3, 6, 8, 9, 10	
8-3	2, 3, 6, 7, 8, 9, 10	
8-4	2, 3, 6, 8, 9, 10	
8-5	2, 3, 6, 7, 8, 9, 10	
8-5 Follow-Up	3, 6	
8-6	2, 3, 6, 7, 8, 9, 10	
8-7	2, 3, 6, 7, 8, 9, 10	

Key to NCTM Standards:

1=Number & Operations, 2=Algebra, 3=Geometry, 4=Measurement, 5=Data Analysis & Probability, 6=Problem Solving, 7=Reasoning & Proof, 8=Communication, 9=Connections, 10=Representation

Vocabulary Builder

ELL

The Key Vocabulary list introduces students to some of the main vocabulary terms included in this chapter. For a more thorough vocabulary list with pronunciations of new words, give students the Vocabulary Builder worksheets found on pages vii and viii of the *Chapter 8 Resource Masters*. Encourage them to complete the definition of each term as they progress through the chapter. You may suggest that they add these sheets to their study notebooks for future reference when studying for the Chapter 8 test.

Getting Started

Prerequisite Skills To be successful in this chapter, you'll need to master these skills and be able to apply them in problem-solving situations. Review these skills before beginning Chapter 8.

For Lesson 8-1 — Exterior Angles of Triangles

Find x for each figure. *(For review, see Lesson 4-2.)*

1. 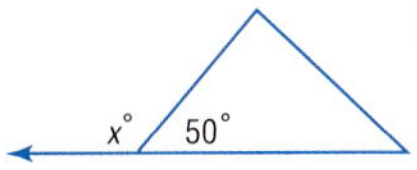

130

2. 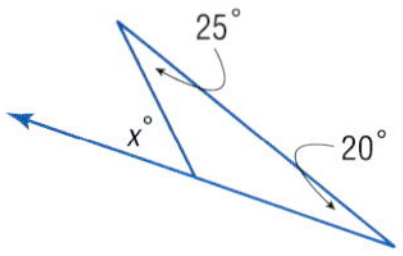

45

3. 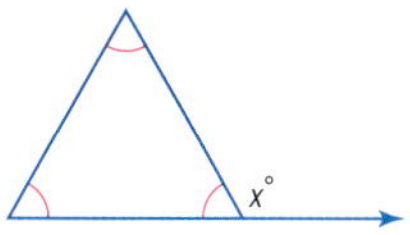

120

4. $-\frac{7}{5}, \frac{5}{7}$; perpendicular 5. $\frac{1}{6}, -6$; perpendicular

For Lessons 8-4 and 8-5 — Perpendicular Lines

Find the slopes of $\overline{RS}$ and $\overline{TS}$ for the given points, R, T, and S. Determine whether $\overline{RS}$ and $\overline{TS}$ are *perpendicular* or *not perpendicular.* *(For review, see Lesson 3-6.)*

4. $R(4, 3)$, $S(-1, 10)$, $T(13, 20)$

5. $R(-9, 6)$, $S(3, 8)$, $T(1, 20)$

6. $R(-6, -1)$, $S(5, 3)$, $T(2, 5)$ $\frac{4}{11}, -\frac{2}{3}$; not perpendicular

7. $R(-6, 4)$, $S(-3, 8)$, $T(5, 2)$ $\frac{4}{3}, -\frac{3}{4}$; perpendicular

For Lesson 8-7 — Slope

Write an expression for the slope of a segment given the coordinates of the endpoints. *(For review, see Lesson 3-3.)*

8. $\left(\frac{c}{2}, \frac{d}{2}\right)$, $(-c, d)$ $-\frac{d}{3c}$

9. $(0, a)$, $(b, 0)$ $-\frac{a}{b}$

10. $(-a, c)$, $(-c, a)$ 1

FOLDABLES™ Study Organizer

Quadrilaterals Make this Foldable to help you organize information about quadrilaterals. Begin with a sheet of notebook paper.

Step 1 Fold

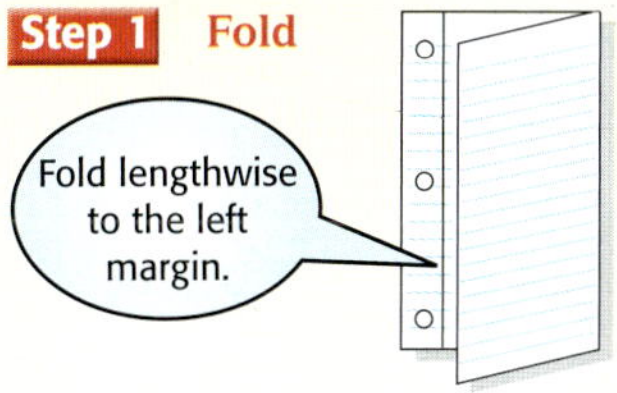

Step 2 Cut

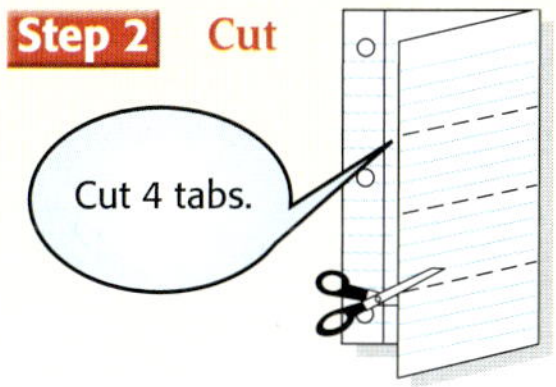

Step 3 Label

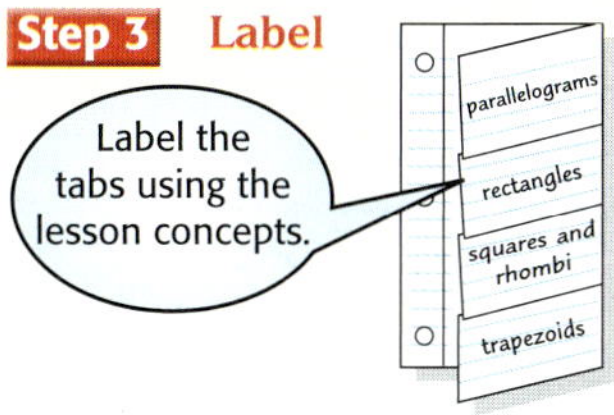

Reading and Writing As you read and study the chapter, use your Foldable to take notes, define terms, and record concepts about quadrilaterals.

Getting Started

This section provides a review of the basic concepts needed before beginning Chapter 8. Page references are included for additional student help.

Additional review is provided in the *Prerequisite Skills Workbook,* pages 5–6, 41–42, 81–82.

Prerequisite Skills in the Getting Ready for the Next Lesson section at the end of each exercise set review a skill needed in the next lesson.

For Lesson	Prerequisite Skill
8-2	Angles formed by parallel lines and a transversal, p. 409
8-3	Slope, p. 416
8-4	Using slope to determine perpendicularity, p. 423
8-5	Distance Formula, p. 430
8-6	Solving equations, p. 437
8-7	Slope, p. 445

FOLDABLES™ Study Organizer

For more information about Foldables, see *Teaching Mathematics with Foldables.*

Organization of Data for Comparing and Contrasting Use this Foldable to organize data about quadrilaterals. After students make their Foldable, have them label the tabs as illustrated. Students can use their Foldable to take notes, define terms, record concepts, and apply properties of quadrilaterals. Use the data recorded to compare and contrast the four quadrilaterals studied. For example, how are parallelograms and rectangles similar? different?

8-1 Lesson Notes

1 Focus

5-Minute Check Transparency 8-1 Use as a quiz or review of Chapter 7.

Mathematical Background notes are available for this lesson on p. 402C.

How does a scallop shell illustrate the angles of polygons?

Ask students:

- How many diagonals can be drawn from one vertex of a 12-sided polygon? **9**
- What other polygons are represented in nature? **Sample answer: Honeycombs resemble hexagons.**

8-1 Angles of Polygons

Sunshine State Standards
MA.A.3.4.3, MA.B.1.4.2

What You'll Learn

- Find the sum of the measures of the interior angles of a polygon.
- Find the sum of the measures of the exterior angles of a polygon.

How does a scallop shell illustrate the angles of polygons?

This scallop shell resembles a 12-sided polygon with diagonals drawn from one of the vertices. A **diagonal** of a polygon is a segment that connects any two nonconsecutive vertices. For example, $\overline{AB}$ is one of the diagonals of this polygon.

Vocabulary

- diagonal

SUM OF MEASURES OF INTERIOR ANGLES

Polygons with more than three sides have diagonals. The polygons below show all of the possible diagonals drawn from one vertex.

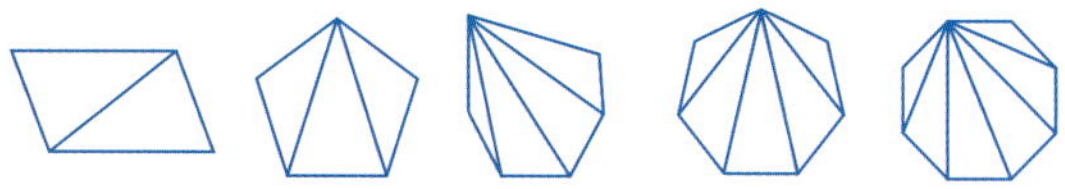

Study Tip

Look Back
To review the **sum of the measures of the angles of a triangle**, see Lesson 4-2.

In each case, the polygon is separated into triangles. Each angle of the polygon is made up of one or more angles of triangles. The sum of the measures of the angles of each polygon can be found by adding the measures of the angles of the triangles. Since the sum of the measures of the angles in a triangle is 180, we can easily find this sum. Make a table to find the sum of the angle measures for several convex polygons.

Convex Polygon	Number of Sides	Number of Triangle	Sum of Angle Measures
triangle	3	1	$(1 \cdot 180)$ or 180
quadrilateral	4	2	$(2 \cdot 180)$ or 360
pentagon	5	3	$(3 \cdot 180)$ or 540
hexagon	6	4	$(4 \cdot 180)$ or 720
heptagon	7	5	$(5 \cdot 180)$ or 900
octagon	8	6	$(6 \cdot 180)$ or 1080

Look for a pattern in the sum of the angle measures. In each case, the sum of the angle measures is 2 less than the number of sides in the polygon times 180. So in an n-gon, the sum of the angle measures will be $(n - 2)180$ or $180(n - 2)$.

Theorem 8.1

Interior Angle Sum Theorem If a convex polygon has n sides and S is the sum of the measures of its interior angles, then $S = 180(n - 2)$.

Example:

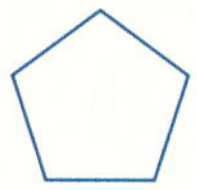

$n = 5$
$S = 180(n - 2)$
$= 180(5 - 2)$ or 540

Resource Manager

Workbook and Reproducible Masters

Chapter 8 Resource Masters

- Study Guide and Intervention, pp. 417–418
- Skills Practice, p. 419
- Practice, p. 420
- Reading to Learn Mathematics, p. 421
- Enrichment, p. 422

Prerequisite Skills Workbook, pp. 5–6
Teaching Geometry With Manipulatives Masters, pp. 16, 127, 128

Transparencies

5-Minute Check Transparency 8-1
Answer Key Transparencies

Technology

Interactive Chalkboard

Example 1 Interior Angles of Regular Polygons

CHEMISTRY **The benzene molecule, C_6H_6, consists of six carbon atoms in a regular hexagonal pattern with a hydrogen atom attached to each carbon atom. Find the sum of the measures of the interior angles of the hexagon.**

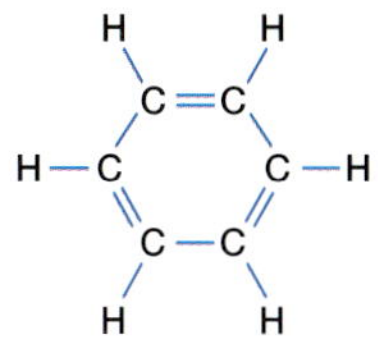

Since the molecule is a convex polygon, we can use the Interior Angle Sum Theorem.

$S = 180(n - 2)$ Interior Angle Sum Theorem

$= 180(6 - 2)$ $n = 6$

$= 180(4)$ or 720 Simplify.

The sum of the measures of the interior angles is 720.

The Interior Angle Sum Theorem can also be used to find the number of sides in a regular polygon if you are given the measure of one interior angle.

Example 2 Sides of a Polygon

The measure of an interior angle of a regular polygon is 108. Find the number of sides in the polygon.

Use the Interior Angle Sum Theorem to write an equation to solve for n, the number of sides.

$S = 180(n - 2)$ Interior Angle Sum Theorem

$(108)n = 180(n - 2)$ $S = 108n$

$108n = 180n - 360$ Distributive Property

$0 = 72n - 360$ Subtract $108n$ from each side.

$360 = 72n$ Add 360 to each side.

$5 = n$ Divide each side by 72.

The polygon has 5 sides.

In Example 2, the Interior Angle Sum Theorem was applied to a regular polygon. In Example 3, we will apply this theorem to a quadrilateral that is not a regular polygon.

Example 3 Interior Angles

ALGEBRA **Find the measure of each interior angle.**

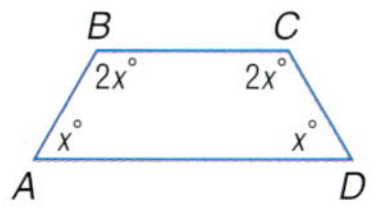

Since $n = 4$, the sum of the measures of the interior angles is $180(4 - 2)$ or 360. Write an equation to express the sum of the measures of the interior angles of the polygon.

$360 = m\angle A + m\angle B + m\angle C + m\angle D$ Sum of measures of angles

$360 = x + 2x + 2x + x$ Substitution

$360 = 6x$ Combine like terms.

$60 = x$ Divide each side by 6.

Use the value of x to find the measure of each angle.

$m\angle A = 60$, $m\angle B = 2 \cdot 60$ or 120, $m\angle C = 2 \cdot 60$ or 120, and $m\angle D = 60$.

www.geometryonline.com/extra_examples/fcat

Teacher to Teacher

Monique Siedschlag, Thoreau High School Thoreau, NM

I ask students "If the sum of the measures of the angles in a triangle is 180, what would a quadrilateral angle measure sum be? pentagon angle measure sum?" Then I pass out pre-cut polygons (4 to 9 sides) and have students measure the angles. The students soon discover that the sum is not 180, as many thought it would be. We list the results and students start to see a pattern. I then have them create a workable formula. Many do come up with the Interior Angle Sum Theorem.

2 Teach

SUM OF MEASURES OF INTERIOR ANGLES

Teaching Tip The proof of the Interior Angle Sum Theorem uses induction, which is not covered in this book.

In-Class Examples

Power Point®

Teaching Tip Point out that while a polygon can be made up of any number of sides, the Interior Angle Sum Theorem applies to *convex* polygons. That means that a segment connecting two points in the interior of the polygon is entirely contained within the polygon. Their diagonals separate the polygon into countable triangles.

1 **ARCHITECTURE** A mall is designed so that five walkways meet at a food court that is in the shape of a regular pentagon. Find the sum of the measures of the interior angles of the pentagon. **540**

Teaching Tip In Example 2, point out that you could not use this method to find the number of sides if the polygon was not regular.

2 The measure of an interior angle of a regular polygon is 135. Find the number of sides in the polygon. **8**

3 Find the measure of each interior angle.

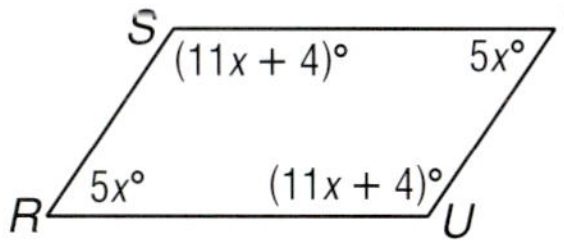

$m\angle R = m\angle T = 55$;
$m\angle S = m\angle U = 125$

SUM OF MEASURES OF EXTERIOR ANGLES

Tips for New Teachers

Intervention The Geometry Activity gives students experience with discovering patterns by collecting and analyzing data. Students should recognize that in a polygon, the number of sides equals the number of exterior angles.

In-Class Example

Power Point®

4 Find the measures of an exterior angle and an interior angle of convex regular nonagon *ABCDEFGHJ*.

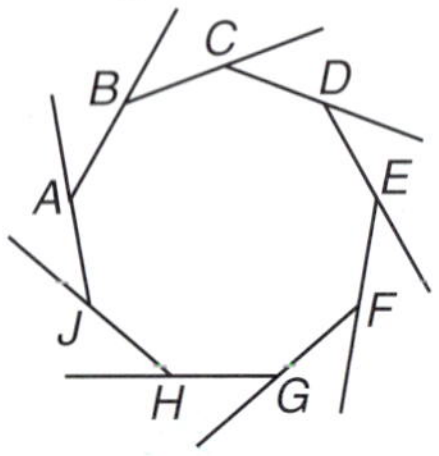

40; 140

Interactive Chalkboard

PowerPoint® Presentations

This CD-ROM is a customizable Microsoft® PowerPoint® presentation that includes:

- Step-by-step, dynamic solutions of each In-Class Example from the Teacher Wraparound Edition
- Additional, Try These exercises for each example
- The 5-Minute Check Transparencies
- Hot links to Glencoe Online Study Tools

SUM OF MEASURES OF EXTERIOR ANGLES

The Interior Angle Sum Theorem relates the interior angles of a convex polygon to the number of sides. Is there a relationship among the exterior angles of a convex polygon?

Study Tip

Look Back
To review **exterior angles**, see Lesson 4-2.

Geometry Activity

Sum of the Exterior Angles of a Polygon

Collect Data

- Draw a triangle, a convex quadrilateral, a convex pentagon, a convex hexagon, and a convex heptagon.
- Extend the sides of each polygon to form exactly one exterior angle at each vertex.
- Use a protractor to measure each exterior angle of each polygon and record it on your drawing.

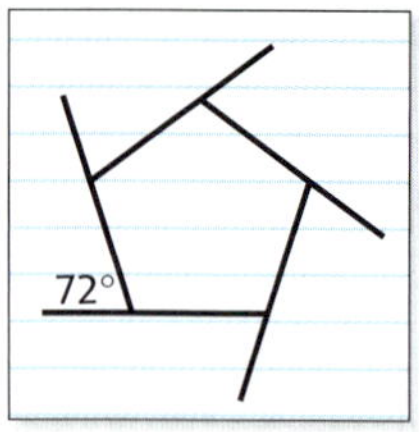

Analyze the Data

1. Copy and complete the table.

Polygon	triangle	quadrilateral	pentagon	hexagon	heptagon
number of exterior angles	3	4	5	6	7
sum of measure of exterior angles	360	360	360	360	360

2. What conjecture can you make? **The sum of the measures of exterior angles is 360.**

The Geometry Activity suggests Theorem 8.2.

Theorem 8.2

Exterior Angle Sum Theorem If a polygon is convex, then the sum of the measures of the exterior angles, one at each vertex, is 360.

Example:

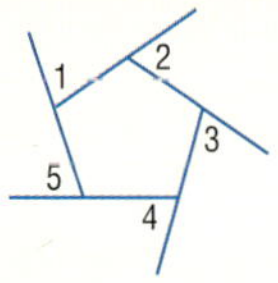

$m\angle 1 + m\angle 2 + m\angle 3 + m\angle 4 + m\angle 5 = 360$

You will prove Theorem 8.2 in Exercise 42.

Example 4 Exterior Angles

Find the measures of an exterior angle and an interior angle of convex regular octagon *ABCDEFGH*.

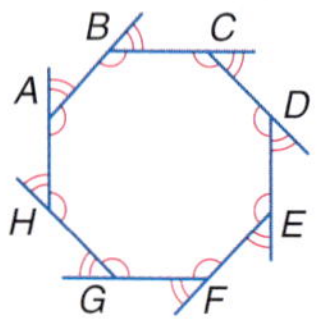

At each vertex, extend a side to form one exterior angle.
The sum of the measures of the exterior angles is 360.
A convex regular octagon has 8 congruent exterior angles.

$8n = 360$ n = measure of each exterior angle

$n = 45$ Divide each side by 8.

The measure of each exterior angle is 45. Since each exterior angle and its corresponding interior angle form a linear pair, the measure of the interior angle is 180 − 45 or 135.

Geometry Activity

Materials: straightedge, protractor

It may help students see the relationship between the interior angles and the exterior angles if you have them add a row to the bottom of the table in the activity. Have it include the measures of the exterior angles of the polygon if the polygon is regular.

Check for Understanding

Concept Check

1. **Explain** why the Interior Angle Sum Theorem and the Exterior Angle Sum Theorem only apply to convex polygons. **1–2. See margin.**
2. **Determine** whether the Interior Angle Sum Theorem and the Exterior Angle Sum Theorem apply to polygons that are not regular. Explain.
3. **OPEN ENDED** Draw a regular convex polygon and a convex polygon that is not regular with the same number of sides. Find the sum of the interior angles for each. **See p. 459A.**

Guided Practice

GUIDED PRACTICE KEY	
Exercises	Examples
4–5, 12	1
6–7	2
8–9	3
10–11	4

Find the sum of the measures of the interior angles of each convex polygon.

4. pentagon **540**
5. dodecagon **1800**

The measure of an interior angle of a regular polygon is given. Find the number of sides in each polygon.

6. 60 **3**
7. 90 **4**

ALGEBRA **Find the measure of each interior angle.**

8. 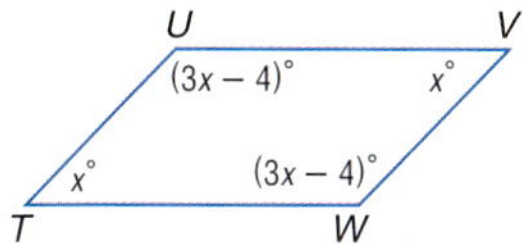

9. 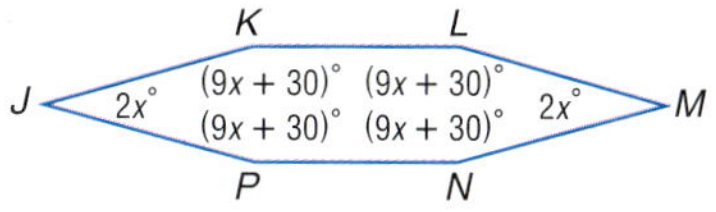

8. $m\angle T = m\angle V = 46$, $m\angle U = m\angle W = 134$

9. $m\angle J = m\angle M = 30$, $m\angle K = m\angle L = m\angle P = m\angle N = 165$

Find the measures of an exterior angle and an interior angle given the number of sides of each regular polygon.

10. 6 **60, 120**
11. 18 **20, 160**

Application

12. **AQUARIUMS** The regular polygon at the right is the base of a fish tank. Find the sum of the measures of the interior angles of the pentagon. **540**

★ indicates increased difficulty

Practice and Apply

Homework Help	
For Exercises	See Examples
13–20	1
21–26	2
27–34	3
35–44	4

Extra Practice
See page 769.

Find the sum of the measures of the interior angles of each convex polygon.

13. 32-gon **5400**
14. 18-gon **2880**
15. 19-gon **3060**
16. 27-gon **4500**
★ 17. $4y$-gon **$360(2y - 1)$**
★ 18. $2x$-gon **$360(x - 1)$**

19. **GARDENING** Carlotta is designing a garden for her backyard. She wants a flower bed shaped like a regular octagon. Find the sum of the measures of the interior angles of the octagon. **1080**

20. **GAZEBOS** A company is building regular hexagonal gazebos. Find the sum of the measures of the interior angles of the hexagon. **720**

The measure of an interior angle of a regular polygon is given. Find the number of sides in each polygon.

21. 140 **9**
22. 170 **36**
23. 160 **18**
24. 165 **24**
25. $157\frac{1}{2}$ **16**
26. $176\frac{2}{5}$ **100**

3 Practice/Apply

Study Notebook

Have students—

- add the definitions/examples of the vocabulary terms to their Vocabulary Builder worksheets for Chapter 8.
- include a table that shows the sum of the measures of the interior angles for polygons with 3, 4, 5, 6, 7, and 8 sides as shown on p. 404.
- include any other item(s) that they find helpful in mastering the skills in this lesson.

About the Exercises...

Organization by Objective

- **Sum of Measures of Interior Angles:** 13–34
- **Sum of Measures of Exterior Angles:** 35–44

Odd/Even Assignments

Exercises 13–44 are structured so that students practice the same concepts whether they are assigned odd or even problems.

Assignment Guide

Basic: 13–15 odd, 19–31 odd, 35–45 odd, 46–65

Average: 13–45 odd, 46–65

Advanced: 14–44 even, 45–61 (optional: 62–65)

Answers

1. A concave polygon has at least one obtuse angle, which means the sum will be different from the formula.
2. Yes; an irregular polygon can be separated by the diagonals into triangles so the theorems apply.

DAILY INTERVENTION

Differentiated Instruction

Logical Students should reason that as the number of sides in a regular convex polygon increases, the measure of each interior angle increases and the measure of each exterior angle decreases. For example, a regular polygon with 360 sides will have an interior angle of 179 and an exterior angle of 1. Make the connection that as the number sides increases without bound, there is a *limit* to the measure that each interior angle can reach, which happens when the length of the side becomes infinitely small so that the polygon looks like a circle and the exterior angles disappear.

Study Guide and Intervention, p. 417 (shown) and p. 418

Sum of Measures of Interior Angles The segments that connect the nonconsecutive sides of a polygon are called **diagonals**. Drawing all of the diagonals from one vertex of an ***n*-gon** separates the polygon into $n - 2$ triangles. The sum of the measures of the interior angles of the polygon can be found by adding the measures of the interior angles of those $n - 2$ triangles.

Interior Angle Sum Theorem	If a convex polygon has n sides, and S is the sum of the measures of its interior angles, then $S = 180(n - 2)$.

Example 1 **A convex polygon has 13 sides. Find the sum of the measures of the interior angles.**

$S = 180(n - 2)$
$= 180(13 - 2)$
$= 180(11)$
$= 1980$

Example 2 **The measure of an interior angle of a regular polygon is 120. Find the number of sides.**

The number of sides is n, so the sum of the measures of the interior angles is $120n$.

$S = 180(n - 2)$
$120n = 180(n - 2)$
$120n = 180n - 360$
$-60n = -360$
$n = 6$

Exercises

Find the sum of the measures of the interior angles of each convex polygon.

1. 10-gon **1440**
2. 16-gon **2520**
3. 30-gon **5040**
4. 8-gon **1080**
5. 12-gon **1800**
6. $3x$-gon **$180(3x - 2)$**

The measure of an interior angle of a regular polygon is given. Find the number of sides in each polygon.

7. 150 **12**
8. 160 **18**
9. 175 **72**
10. 165 **24**
11. 168.75 **32**
12. 135 **8**

13. Find x. **20**

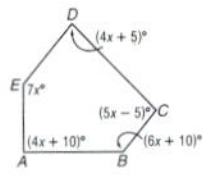

Skills Practice, p. 419 and Practice, p. 420 (shown)

Find the sum of the measures of the interior angles of each convex polygon.

1. 11-gon **1620**
2. 14-gon **2160**
3. 17-gon **2700**

The measure of an interior angle of a regular polygon is given. Find the number of sides in each polygon.

4. 144 **10**
5. 156 **15**
6. 160 **18**

Find the measure of each interior angle using the given information.

7.

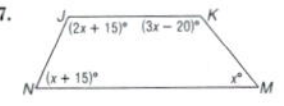

$m\angle J = 115$, $m\angle K = 130$, $m\angle M = 50$, $m\angle N = 65$

8. quadrilateral $RSTU$ with $m\angle R = 6x - 4$, $m\angle S = 2x + 8$

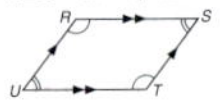

$m\angle R = 128$, $m\angle S = 52$, $m\angle T = 128$, $m\angle U = 52$

Find the measures of an interior angle and an exterior angle for each regular polygon. Round to the nearest tenth if necessary.

9. 16-gon **157.5, 22.5**
10. 24-gon **165, 15**
11. 30-gon **168, 12**

Find the measures of an interior angle and an exterior angle given the number of sides of each regular polygon. Round to the nearest tenth if necessary.

12. 14 **154.3, 25.7**
13. 22 **163.6, 16.4**
14. 40 **171, 9**

15. CRYSTALLOGRAPHY Crystals are classified according to seven crystal systems. The basis of the classification is the shapes of the faces of the crystal. Turquoise belongs to the triclinic system. Each of the six faces of turquoise is in the shape of a parallelogram. Find the sum of the measures of the interior angles of one such face. **360**

Reading to Learn Mathematics, p. 421 ELL

Pre-Activity **How does a scallop shell illustrate the angles of polygons?**

Read the introduction to Lesson 8-1 at the top of page 404 in your textbook.

- How many diagonals of the scallop shell shown in your textbook can be drawn from vertex A? **9**
- How many diagonals can be drawn from one vertex of an n-gon? Explain your reasoning. **$n - 3$; Sample answer: An n-gon has n vertices, but diagonals cannot be drawn from a vertex to itself or to either of the two adjacent vertices. Therefore, the number of diagonals from one vertex is 3 less than the number of vertices.**

Reading the Lesson

1. Write an expression that describes each of the following quantities for a regular n-gon. If the expression applies to regular polygons only, write *regular*. If it applies to all convex polygons, write *all*.
   a. the sum of the measures of the interior angles **$180(n - 2)$; all**
   b. the measure of each interior angle **$\frac{180(n-2)}{n}$; regular**
   c. the sum of the measures of the exterior angles (one at each vertex) **360; all**
   d. the measure of each exterior angle **$\frac{360}{n}$; regular**
2. Give the measure of an interior angle and the measure of an exterior angle of each polygon.
   a. equilateral triangle **60; 120**
   b. regular hexagon **120; 60**
   c. square **90; 90**
   d. regular octagon **135; 45**
3. Underline the correct word or phrase to form a true statement about regular polygons.
   a. As the number of sides increases, the sum of the measures of the interior angles (increases/decreases/stays the same).
   b. As the number of sides increases, the measure of each interior angle (increases/decreases/stays the same).
   c. As the number of sides increases, the sum of the measures of the exterior angles (increases/decreases/stays the same).
   d. As the number of sides increases, the measure of each exterior angle (increases/decreases/stays the same).
   e. If a regular polygon has more than four sides, each interior angle will be a(n) (acute/right/obtuse) angle, and each exterior angle will be a(n) (acute/right/obtuse) angle.

Helping You Remember

4. A good way to remember a new mathematical idea or formula is to relate it to something you already know. How can you use your knowledge of the Angle Sum Theorem (for a triangle) to help you remember the Interior Angle Sum Theorem? **Sample answer: The sum of the measures of the (interior) angles of a triangle is 180. Each time another side is added, the polygon can be subdivided into one more triangle, so 180 is added to the interior angle sum.**

27. $m\angle M = 30$, $m\angle P = 120$, $m\angle Q = 60$, $m\angle R = 150$

28. $m\angle E = 102$, $m\angle F = 122$, $m\angle G = 107$, $m\angle H = 97$, $m\angle J = 112$

29. $m\angle M = 60$, $m\angle N = 120$, $m\angle P = 60$, $m\angle Q = 120$

30. $m\angle T = 72$, $m\angle W = 72$, $m\angle Y = 108$, $m\angle Z = 108$

31. 105, 110, 120, 130, 135, 140, 160, 170, 180, 190

32. $m\angle A = 186$, $m\angle B = 137$, $m\angle C = 40$, $m\angle D = 54$, $m\angle E = 123$

ALGEBRA **Find the measure of each interior angle using the given information.**

27.

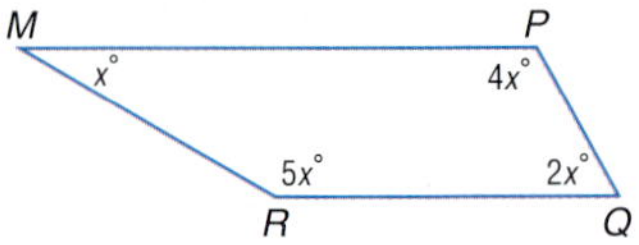

28.

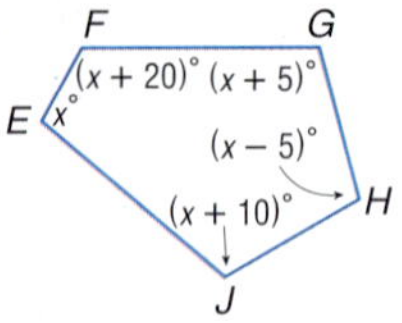

29. parallelogram $MNPQ$ with $m\angle M = 10x$ and $m\angle N = 20x$

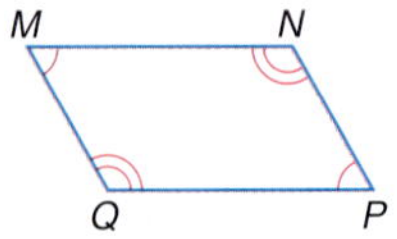

30. isosceles trapezoid $TWYZ$ with $\angle Z \cong \angle Y$, $m\angle Z = 30x$, $\angle T \cong \angle W$, and $m\angle T = 20x$

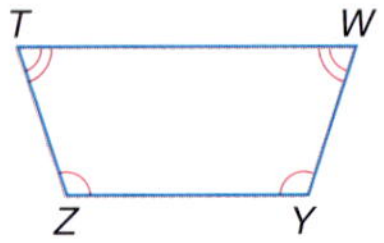

31. decagon in which the measures of the interior angles are $x + 5$, $x + 10$, $x + 20$, $x + 30$, $x + 35$, $x + 40$, $x + 60$, $x + 70$, $x + 80$, and $x + 90$

32. polygon $ABCDE$ with $m\angle A = 6x$, $m\angle B = 4x + 13$, $m\angle C = x + 9$, $m\angle D = 2x - 8$, and $m\angle E = 4x - 1$

★ 33. quadrilateral in which the measures of the angles are consecutive multiples of x **Sample answer: 36, 72, 108, 144**

★ 34. quadrilateral in which the measure of each consecutive angle increases by 10 **75, 85, 95, 105**

Find the measures of each exterior angle and each interior angle for each regular polygon.

35. decagon **36, 144**
36. hexagon **60, 120**
37. nonagon **40, 140**
38. octagon **45, 135**

Find the measures of an interior angle and an exterior angle given the number of sides of each regular polygon. Round to the nearest tenth if necessary.

39. 11 **147.3, 32.7**
40. 7 **128.6, 51.4**
41. 12 **150, 30**

★ 42. **PROOF** Use algebra to prove the Exterior Angle Sum Theorem. **See margin.**

43. **ARCHITECTURE** The Pentagon building in Washington, D.C., was designed to resemble a regular pentagon. Find the measure of an interior angle and an exterior angle of the courtyard. **108, 72**

44. **ARCHITECTURE** Compare the dome to the architectural elements on each side of the dome. Are the interior and exterior angles the same? Find the measures of the interior and exterior angles. **yes; 135, 45**

45. **CRITICAL THINKING** Two formulas can be used to find the measure of an interior angle of a regular polygon: $s = \frac{180(n-2)}{n}$ and $s = 180 - \frac{360}{n}$. Show that these are equivalent. **See margin.**

More About. . .

Architecture

Thomas Jefferson's home, Monticello, features a dome on an octagonal base. The architectural elements on either side of the dome were based on a regular octagon.

Source: www.monticello.org

408 Chapter 8 Quadrilaterals

Enrichment, p. 422

Tangrams

The tangram puzzle is composed of seven pieces that form a square, as shown at the right. This puzzle has been a popular amusement for Chinese students for hundreds and perhaps thousands of years.

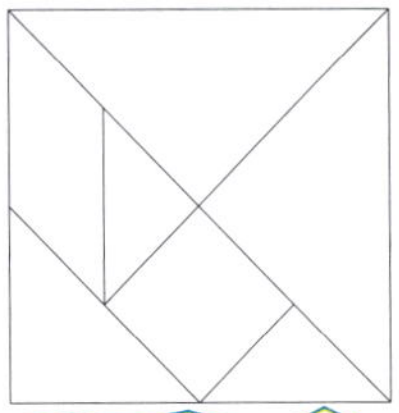

46. **WRITING IN MATH** Answer the question that was posed at the beginning of the lesson. See margin.

How does a scallop shell illustrate the angles of polygons?

Include the following in your answer:

- explain how triangles are related to the Interior Angle Sum Theorem, and
- describe how to find the measure of an exterior angle of a polygon.

FCAT Practice
Standardized Test Practice
Ⓐ Ⓑ Ⓒ Ⓓ

47. A regular pentagon and a square share a mutual vertex X. The sides $\overline{XY}$ and $\overline{XZ}$ are sides of a third regular polygon with a vertex at X. How many sides does this polygon have? **B**

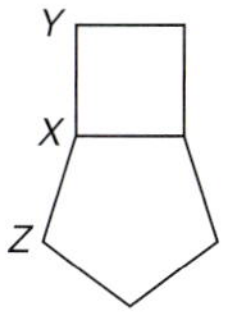

Ⓐ 19 Ⓑ 20

Ⓒ 28 Ⓓ 32

48. **GRID IN** If $6x + 3y = 48$ and $\frac{9y}{2x} = 9$, then $x = ?$ **4**

Maintain Your Skills

Mixed Review

In $\triangle ABC$, given the lengths of the sides, find the measure of the given angle to the nearest tenth. *(Lesson 7-7)*

49. $a = 6, b = 9, c = 11; m\angle C$ **92.1**
50. $a = 15.5, b = 23.6, c = 25.1; m\angle B$ **66.3**
51. $a = 47, b = 53, c = 56; m\angle A$ **51.0**
52. $a = 12, b = 14, c = 16; m\angle C$ **75.5**

Solve each $\triangle FGH$ described below. Round angle measures to the nearest degree and side measures to the nearest tenth. *(Lesson 7-6)*

53. $f = 15, g = 17, m\angle F = 54$
54. $m\angle F = 47, m\angle H = 78, g = 31$
55. $m\angle G = 56, m\angle H = 67, g = 63$
56. $g = 30.7, h = 32.4, m\angle G = 65$

53. $m\angle G \approx 66$, $m\angle H \approx 60$, $h \approx 16.1$
54. $m\angle G = 55$, $f \approx 27.7$, $h \approx 37.0$
55. $m\angle F = 57$, $f \approx 63.7$, $h \approx 70.0$
56. $m\angle H \approx 73$, $m\angle F \approx 42$, $f \approx 22.7$

57. **PROOF** Write a two-column proof. *(Lesson 4-5)* See p. 459A.

Given: $\overline{JL} \parallel \overline{KM}$, $\overline{JK} \parallel \overline{LM}$

Prove: $\triangle JKL \cong \triangle MLK$

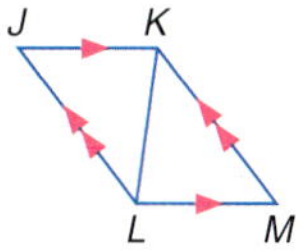

State the transversal that forms each pair of angles. Then identify the special name for the angle pair. *(Lesson 3-1)*

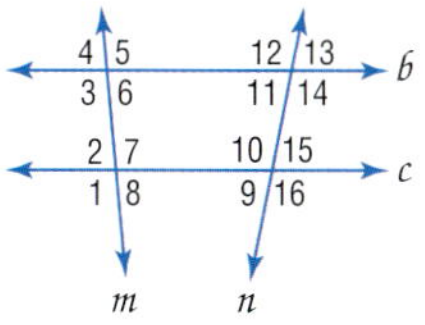

58. $\angle 3$ and $\angle 11$ *b*; **corr.**
59. $\angle 6$ and $\angle 7$ *m*; **cons. int.**
60. $\angle 8$ and $\angle 10$ *c*; **alt. int.**
61. $\angle 12$ and $\angle 16$ *n*; **alt. ext.**

Getting Ready for the Next Lesson

PREREQUISITE SKILL **In the figure, $\overline{AB} \parallel \overline{DC}$ and $\overline{AD} \parallel \overline{BC}$. Name all pairs of angles for each type indicated.** *(To review **angles formed by parallel lines and a transversal**, see Lesson 3-1.)*

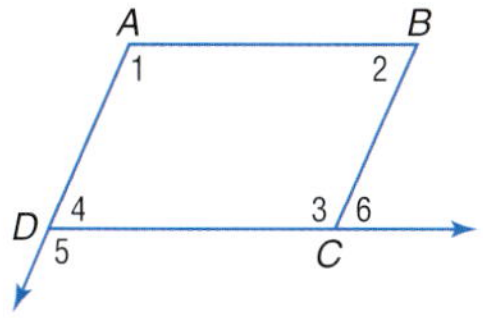

62. consecutive interior angles
63. alternate interior angles $\angle 3$ and $\angle 5$, $\angle 2$ and $\angle 6$
64. corresponding angles $\angle 1$ and $\angle 5$, $\angle 4$ and $\angle 6$
65. alternate exterior angles **none**

62. $\angle 1$ and $\angle 4$, $\angle 1$ and $\angle 2$, $\angle 2$ and $\angle 3$, $\angle 3$ and $\angle 4$

www.geometryonline.com/self_check_quiz/fcat

4 Assess

Open-Ended Assessment

Speaking Have students describe how to find the sum of the measures of the interior angles of a polygon.

Getting Ready for Lesson 8-2

Prerequisite Skill Students will learn about parallelograms in Lesson 8-2. They will use the angles formed by parallel lines and a transversal to find the angles in a quadrilateral with parallel sides. Use Exercises 62–65 to determine your students' familiarity with the angles formed by parallel lines and a transversal.

Answer

46. Sample answer: The outline of a scallop shell is a convex polygon that is not regular. The lines in the shell resemble diagonals drawn from one vertex of a polygon. These diagonals separate the polygon into triangles. Answers should include the following.

- The Interior Angle Sum Theorem is derived from the pattern between the number of sides in a polygon and the number of triangles. The formula is the product of the sum of the measures of the angles in a triangle, 180, and the number of triangles the polygon contains.
- The exterior angle and the interior angle of a polygon are a linear pair. So, the measure of an exterior angle is the difference between 180 and the measure of the interior angle.

Answers (p. 408)

42. Consider the sum of the measures of the exterior angles, N, for an n-gon.

N = sum of measures of linear pairs − sum of measures of interior angles

$= 180n - 180(n - 2)$

$= 180n - 180n + 360$

$= 360$

So, the sum of the exterior angle measures is 360 for any convex polygon.

45. $\frac{180(n-2)}{n} = \frac{180n - 360}{n}$

$= \frac{180n}{n} - \frac{360}{n}$

$= 180 - \frac{360}{n}$

Spreadsheet Investigation

A Follow-Up of Lesson 8-1

Getting Started

Students may use the "Fill-down" feature of a spreadsheet to enter the data in columns one and two. For example, if the cell reference for 3 sides is A2, then enter "=A2 + 1" in cell A3. Then highlight cell A3 and drag to "fill down" the column with the same pattern. Students should notice that they can find the sum of the interior angles for a many-sided polygon this way, including a polygon with 110 sides (Exercise 7).

Teach

- You may want students to do this activity in pairs, especially if there is limited availability of computers. Ask one student to enter the data into the computer while another student reads the information to be entered. Ask the pairs of students to discuss which formulas should be used for Exercises 1–2.

Assess

The answers for **Exercises 5–7** will be correct only if students entered their formulas correctly into the spreadsheet.

Spreadsheet Investigation

A Follow-Up of Lesson 8-1

Angles of Polygons

It is possible to find the interior and exterior measurements along with the sum of the interior angles of any regular polygon with *n* number of sides using a spreadsheet.

Example

Design a spreadsheet using the following steps.

- Label the columns as shown in the spreadsheet below.
- Enter the digits 3–10 in the first column.
- The number of triangles formed by diagonals from the same vertex in a polygon is 2 less than the number of sides. Write a formula for Cell B2 to subtract 2 from each number in Cell A2.
- Enter a formula for Cell C2 so the spreadsheet will find the sum of the measures of the interior angles. Remember that the formula is $S = (n - 2)180$.
- Continue to enter formulas so that the indicated computation is performed. Then, copy each formula through Row 9. The final spreadsheet will appear as below.

Polygons and Angles

	A	B	C	D	E	F	G
1	Number of Sides	Number of Triangles	Sum of Measures of Interior Angles	Measure of Each Interior Angle	Measure of Each Exterior Angle	Sum of Measures of Exterior Angles	
2	3	1	180	60	120	360	
3	4	2	360	90	90	360	
4	5	3	540	108	72	360	
5	6	4	720	120	60	360	
6	7	5	900	128.57	51.43	360	
7	8	6	1080	135	45	360	
8	9	7	1260	140	40	360	
9	10	8	1440	144	36	360	
10							

Sheet1 / Sheet2 / Sheet3

Exercises

1. Write the formula to find the measure of each interior angle in the polygon. **=C2/A2**
2. Write the formula to find the sum of the measures of the exterior angles. **=A2*E2**
3. What is the measure of each interior angle if the number of sides is 1? 2? **–180, 0**
4. Is it possible to have values of 1 and 2 for the number of sides? Explain. **No, a polygon is a closed figure formed by coplanar segments.**

For Exercises 5–8, use the spreadsheet.

5. How many triangles are in a polygon with 15 sides? **13**
6. Find the measure of the exterior angle of a polygon with 15 sides. **24**
7. Find the measure of the interior angle of a polygon with 110 sides. **176.7**
8. If the measure of the exterior angles is 0, find the measure of the interior angles. Is this possible? Explain. **Each interior angle measures 180. This is not possible for a polygon.**

8-2 Parallelograms

What You'll Learn

- Recognize and apply properties of the sides and angles of parallelograms.
- Recognize and apply properties of the diagonals of parallelograms.

Sunshine State Standards
MA.A.3.4.3, MA.C.1.4.1, MA.C.2.4.1, MA.C.3.4.2

Vocabulary
- parallelogram

How are parallelograms used to represent data?

The graphic shows the percent of Global 500 companies that use the Internet to find potential employees. The top surfaces of the wedges of cheese are all polygons with a similar shape. However, the size of the polygon changes to reflect the data. What polygon is this?

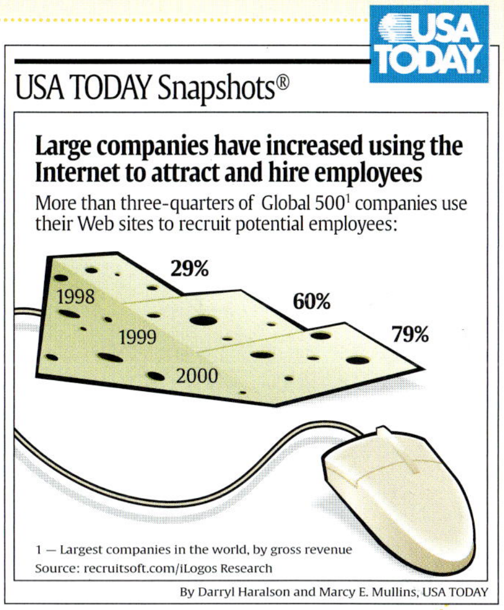

SIDES AND ANGLES OF PARALLELOGRAMS A quadrilateral with parallel opposite sides is called a **parallelogram**.

Study Tip

Reading Math
Recall that the matching arrow marks on the segments mean that the sides are parallel.

Key Concept — Parallelogram

- **Words** A parallelogram is a quadrilateral with both pairs of opposite sides parallel.
- **Symbols** ▱$ABCD$
- **Example**

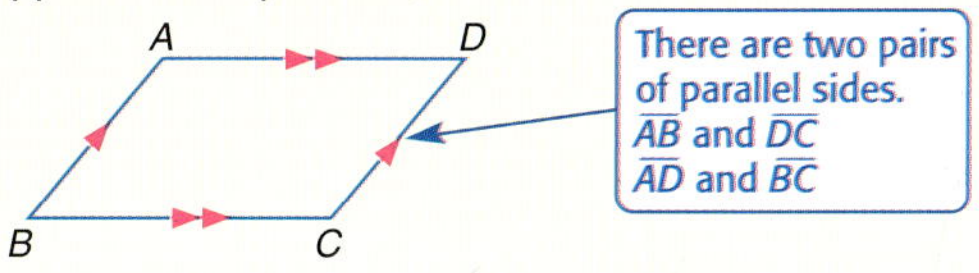

This activity will help you make conjectures about the sides and angles of a parallelogram.

Geometry Activity

Properties of Parallelograms

Make a model

Step 1 Construct two sets of intersecting parallel lines on patty paper. Label the vertices $FGHJ$.

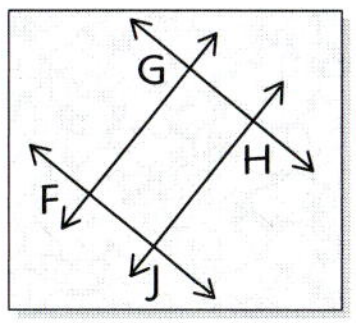

(continued on the next page)

8-2 Lesson Notes

1 Focus

5-Minute Check Transparency 8-2 Use as a quiz or review of Lesson 8-1.

Mathematical Background notes are available for this lesson on p. 402C.

How are parallelograms used to represent data?

Ask students:

- Why are parallelograms the shape used in this graphic? **A wedge can be sliced into parallelograms.**
- Why use cheese in this graphic? **Sample answer: A "mouse" likes cheese.**
- How would you describe the increase in Internet use to attract and hire employees in these companies? **Sample answer: it has more than doubled from 1998 to 2000.**

Resource Manager

Workbook and Reproducible Masters

Chapter 8 Resource Masters
- Study Guide and Intervention, pp. 423–424
- Skills Practice, p. 425
- Practice, p. 426
- Reading to Learn Mathematics, p. 427
- Enrichment, p. 428
- Assessment, p. 473

Prerequisite Skills Workbook, pp. 81–82
Teaching Geometry With Manipulatives Masters, pp. 8, 129

Transparencies

5-Minute Check Transparency 8-2
Answer Key Transparencies

Technology

Interactive Chalkboard

2 Teach

SIDES AND ANGLES OF PARALLELOGRAMS

In-Class Example

Power Point®

1 Prove that if a parallelogram has two consecutive sides congruent, it has four sides congruent.

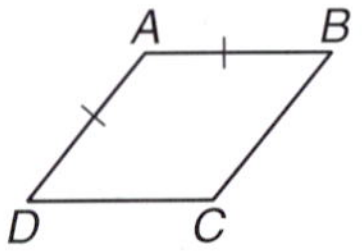

Given: ▱$ABCD$; $\overline{AD} \cong \overline{AB}$
Prove: $\overline{AD} \cong \overline{AB} \cong \overline{BC} \cong \overline{CD}$

Statements (Reasons)
1. ▱$ABCD$ (Given)
2. $\overline{AD} \cong \overline{AB}$ (Given)
3. $\overline{CD} \cong \overline{AB}$, $\overline{BC} \cong \overline{AD}$ (Opposite sides of a ▱ are ≅.)
4. $\overline{AD} \cong \overline{AB} \cong \overline{BC} \cong \overline{CD}$ (Transitive property)

1. $\overline{FG} \cong \overline{HJ} \cong \overline{PQ} \cong \overline{RS}$, $\overline{FJ} \cong \overline{GH} \cong \overline{PS} \cong \overline{QR}$
2. $\angle F \cong \angle P \cong \angle H \cong \angle R \cong$, $\angle J \cong \angle G \cong \angle Q \cong \angle S$
3. Opposite angles are congruent; consecutive angles are supplementary.

Step 2 Trace $FGHJ$. Label the second parallelogram $PQRS$ so $\angle F$ and $\angle P$ are congruent.

Step 3 Rotate ▱$PQRS$ on ▱$FGHJ$ to compare sides and angles.

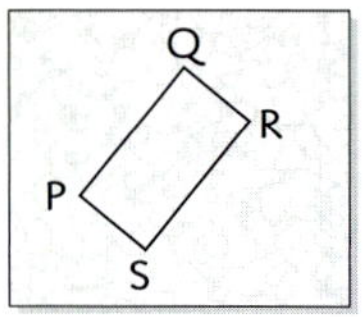

Analyze
1. List all of the segments that are congruent.
2. List all of the angles that are congruent.
3. Describe the angle relationships you observed.

The Geometry Activity leads to four properties of parallelograms.

Key Concept — *Properties of Parallelograms*

	Theorem	Example
8.3	Opposite sides of a parallelogram are congruent. **Abbreviation:** *Opp. sides of ▱ are ≅.*	$\overline{AB} \cong \overline{DC}$ $\overline{AD} \cong \overline{BC}$
8.4	Opposite angles in a parallelogram are congruent. **Abbreviation:** *Opp. ∠s of ▱ are ≅.*	$\angle A \cong \angle C$ $\angle B \cong \angle D$
8.5	Consecutive angles in a parallelogram are supplementary. **Abbreviation:** *Cons. ∠s in ▱ are suppl.*	$m\angle A + m\angle B = 180$ $m\angle B + m\angle C = 180$ $m\angle C + m\angle D = 180$ $m\angle D + m\angle A = 180$
8.6	If a parallelogram has one right angle, it has four right angles. **Abbreviation:** *If ▱ has 1 rt. ∠, it has 4 rt. ∠s.*	$m\angle G = 90$ $m\angle H = 90$ $m\angle J = 90$ $m\angle K = 90$

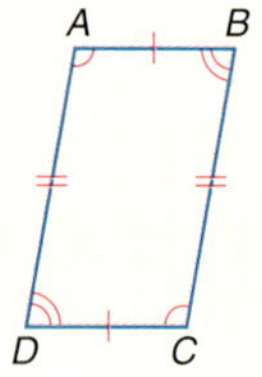

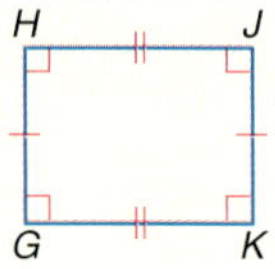

You will prove Theorems 8.3, 8.5, and 8.6 in Exercises 41, 42, and 43, respectively.

Study Tip

Including a Figure
Theorems are presented in general terms. In a proof, you must include a drawing so that you can refer to segments and angles specifically.

Example 1 *Proof of Theorem 8.4*

Write a two-column proof of Theorem 8.4.

Given: ▱$ABCD$
Prove: $\angle A \cong \angle C$
$\angle D \cong \angle B$

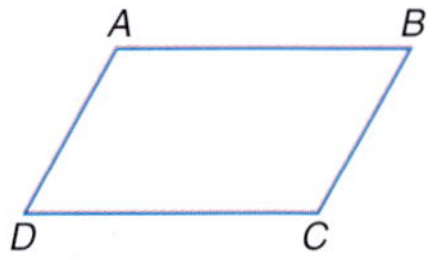

Proof:

Statements	Reasons
1. ▱$ABCD$	1. Given
2. $\overline{AB} \parallel \overline{DC}$, $\overline{AD} \parallel \overline{BC}$	2. Definition of parallelogram
3. $\angle A$ and $\angle D$ are supplementary. $\angle D$ and $\angle C$ are supplementary. $\angle C$ and $\angle B$ are supplementary.	3. If parallel lines are cut by a transversal, consecutive interior angles are supplementary.
4. $\angle A \cong \angle C$ $\angle D \cong \angle B$	4. Supplements of the same angles are congruent.

Geometry Activity

Materials: ruler

Ask students to recall what they know about the angles formed by parallel lines and transversals before you do this activity.

Example 2 Properties of Parallelograms

ALGEBRA Quadrilateral *LMNP* is a parallelogram. Find $m\angle PLM$, $m\angle LMN$, and d.

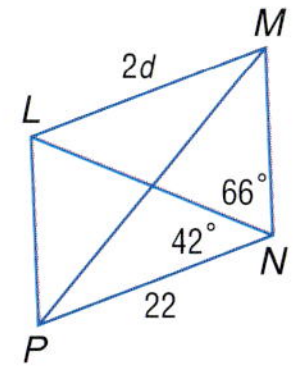

$m\angle MNP = 66 + 42$ or 108 Angle Addition Theorem

$\angle PLM \cong \angle MNP$ Opp. ∠s of ▱ are ≅.
$m\angle PLM = m\angle MNP$ Definition of congruent angles
$m\angle PLM = 108$ Substitution

$m\angle PLM + m\angle LMN = 180$ Cons. ∠s of ▱ are suppl.
$108 + m\angle LMN = 180$ Substitution
$m\angle LMN = 72$ Subtract 108 from each side.

$\overline{LM} \cong \overline{PN}$ Opp. sides of ▱ are ≅.
$LM = PN$ Definition of congruent segments
$2d = 22$ Substitution
$d = 11$ Substitution

DIAGONALS OF PARALLELOGRAMS

In parallelogram *JKLM*, $\overline{JL}$ and $\overline{KM}$ are diagonals. Theorem 8.7 states the relationship between diagonals of a parallelogram.

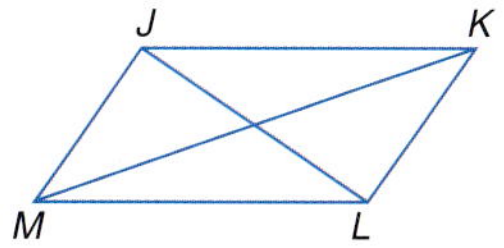

Theorem 8.7

The diagonals of a parallelogram bisect each other.

Abbreviation: *Diag. of ▱ bisect each other.*

Example: $\overline{RQ} \cong \overline{QT}$ and $\overline{SQ} \cong \overline{QU}$

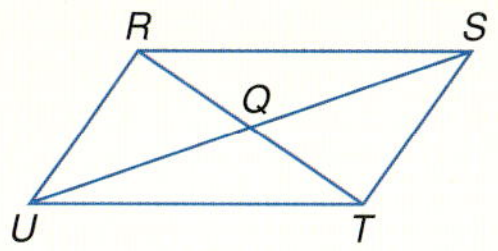

You will prove Theorem 8.7 in Exercise 44.

FCAT Practice
Standardized Test Practice
A B C D

Example 3 Diagonals of a Parallelogram

Multiple-Choice Test Item

What are the coordinates of the intersection of the diagonals of parallelogram *ABCD* with vertices *A*(2, 5), *B*(6, 6), *C*(4, 0), and *D*(0, –1)?

(A) (4, 2) (B) (4.5, 2) (C) $\left(\frac{7}{6}, \frac{-5}{2}\right)$ (D) (3, 2.5)

Read the Test Item

Since the diagonals of a parallelogram bisect each other, the intersection point is the midpoint of $\overline{AC}$ and $\overline{BD}$.

Solve the Test Item

Find the midpoint of $\overline{AC}$.

$$\left(\frac{x_1 + x_2}{2}, \frac{y_1 + y_2}{2}\right) = \left(\frac{2 + 4}{2}, \frac{5 + 0}{2}\right) \quad \text{Midpoint Formula}$$
$$= (3, 2.5)$$

The coordinates of the intersection of the diagonals of parallelogram *ABCD* are (3, 2.5). The answer is D.

The Princeton Review

Test-Taking Tip

Check Answers Always check your answer. To check the answer to this problem, find the coordinates of the midpoint of $\overline{BD}$.

In-Class Example

Power Point®

2 *RSTU* is a parallelogram. Find $m\angle URT$, $m\angle RST$, and y.

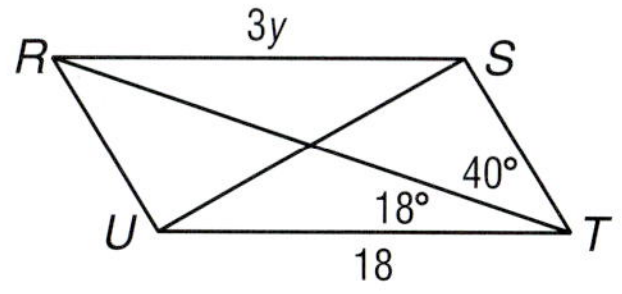

40; 122, 6

Building on Prior Knowledge

In Chapter 3, students learned that if two parallel lines are cut by a transversal, the alternate interior angles are congruent. In this lesson, emphasize that the diagonals of a parallelogram are transversals, and therefore the alternate interior angles are congruent.

DIAGONALS OF PARALLELOGRAMS

Teaching Tip Make sure students understand that if diagonals bisect each other, it implies that they have the same midpoint. It does not imply that all four segments created by the intersection of the diagonals are congruent.

In-Class Example

Power Point®

3 What are the coordinates of the intersection of the diagonals of parallelogram *MNPR*, with vertices *M*(−3, 0), *N*(−1, 3), *P*(5, 4), and *R*(3, 1)? **C**

A (2, 4) **B** $\left(\frac{9}{2}, \frac{5}{2}\right)$

C (1, 2) **D** $\left(-2, \frac{3}{2}\right)$

DAILY INTERVENTION

Differentiated Instruction

Visual/Spatial Stress that in some parallelograms, the diagonals appear to bisect the opposite angles, but this is not a property of parallelograms. Caution students not to assume that angles are bisected. In Lesson 8-5, students will study rhombi and squares. The diagonals do bisect the opposite angles in these parallelograms.

3 Practice/Apply

Study Notebook

Have students—
- add the definitions/examples of the vocabulary terms to their Vocabulary Builder worksheets for Chapter 8.
- include the parallelogram theorems studied in this lesson.
- include any other item(s) that they find helpful in mastering the skills in this lesson.

About the Exercises...

Organization by Objective
- **Sides and Angles of Parallelograms:** 16–33, 41–47
- **Diagonals of Parallelograms:** 34–40

Odd/Even Assignments
Exercises 16–47 are structured so that students practice the same concepts whether they are assigned odd or even problems.

Assignment Guide

Basic: 17–33 odd, 37–47 odd, 48–63

Average: 17–47 odd, 48–63

Advanced: 16–46 even, 48–60 (optional: 61–63)

Theorem 8.8 describes another characteristic of the diagonals of a parallelogram.

Theorem 8.8

Each diagonal of a parallelogram separates the parallelogram into two congruent triangles.

Abbreviation: *Diag. separates ▱ into 2 ≅ △s.*

Example: $\triangle ACD \cong \triangle CAB$

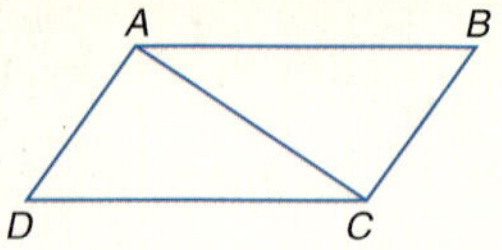

You will prove Theorem 8.8 in Exercise 45.

Check for Understanding

Concept Check

1. **Describe** the characteristics of the sides and angles of a parallelogram.
2. **Describe** the properties of the diagonals of a parallelogram.
3. **OPEN ENDED** Draw a parallelogram with one side twice as long as another side. **1–3. See margin.**

Guided Practice

GUIDED PRACTICE KEY	
Exercises	Examples
4–12	2
13–14	1
15	3

Complete each statement about ▱*QRST*. Justify your answer.

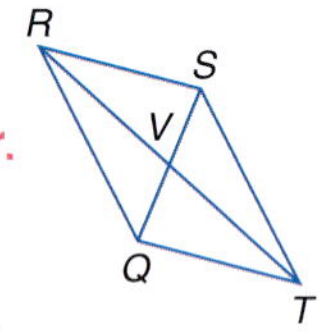

4. $\overline{SV} \cong$ ___?___ **$\overline{VQ}$; Diag. of ▱ bisect each other.**
5. $\triangle VRS \cong$ ___?___
6. $\angle TSR$ is supplementary to ___?___. **$\angle STQ$ and $\angle SRQ$; Consec. ∡ in ▱ are suppl.**

5. $\triangle VTQ$, SSS; diag. bisect each other and opp. sides of ▱ are ≅.

Use ▱*JKLM* to find each measure or value if $JK = 2b + 3$ and $JM = 3a$.

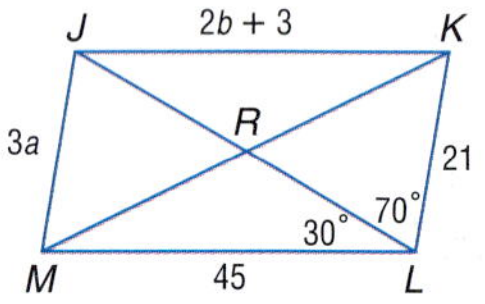

7. $m\angle MJK$ **100**
8. $m\angle JML$ **80**
9. $m\angle JKL$ **80**
10. $m\angle KJL$ **30**
11. a **7**
12. b **21**

PROOF **Write the indicated type of proof.** **13–14. See p. 459A.**

13. two-column

Given: ▱*VZRQ* and ▱*WQST*
Prove: $\angle Z \cong \angle T$

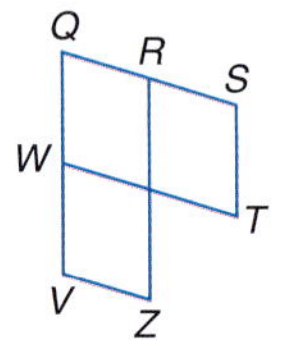

14. paragraph

Given: ▱*XYRZ*, $\overline{WZ} \cong \overline{WS}$
Prove: $\angle XYR \cong \angle S$

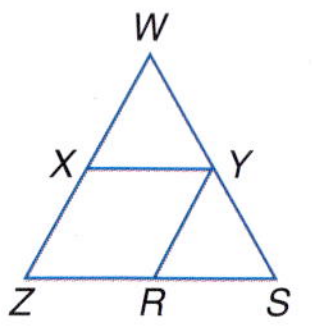

FCAT Practice
Standardized Test Practice

15. **MULTIPLE CHOICE** Find the coordinates of the intersection of the diagonals of parallelogram *GHJK* with vertices $G(-3, 4)$, $H(1, 1)$, $J(3, -5)$, and $K(-1, -2)$. **C**

Ⓐ (0, 0.5) Ⓑ (6, −1) Ⓒ (0, −0.5) Ⓓ (5, 0)

Answers

1. Opposite sides are congruent; opposite angles are congruent; consecutive angles are supplementary; and if there is one right angle, there are four right angles.
2. Diagonals bisect each other; each diagonal forms two congruent triangles.
3. Sample answer:

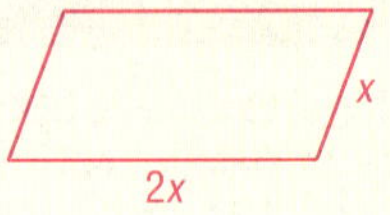

★ indicates increased difficulty

Practice and Apply

Homework Help

For Exercises	See Examples
16–33	2
34–40	3
41–47	1

Extra Practice
See page 769.

Complete each statement about ▱ABCD. Justify your answer. 16–21. See margin for justifications.

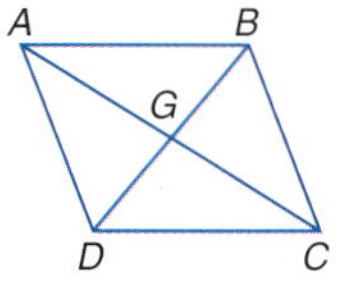

16. $\angle DAB \cong$ __?__ $\angle BCD$
17. $\angle ABD \cong$ __?__ $\angle CDB$
18. $\overline{AB} \parallel$ __?__ $\overline{DC}$
19. $\overline{BG} \cong$ __?__ $\overline{GD}$
20. $\triangle ABD \cong$ __?__ $\triangle CDB$
21. $\angle ACD \cong$ __?__ $\angle BAC$

ALGEBRA **Use ▱MNPR to find each measure or value.**

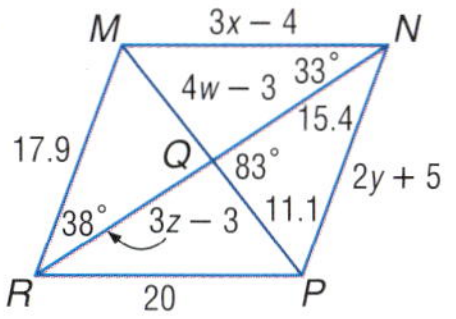

22. $m\angle MNP$ 71
23. $m\angle NRP$ 33
24. $m\angle RNP$ 38
25. $m\angle RMN$ 109
26. $m\angle MQN$ 97
27. $m\angle MQR$ 83
28. x 8
29. y 6.45
30. w 3.5
31. z 6.1

DRAWING **For Exercises 32 and 33, use the following information.**
The frame of a pantograph is a parallelogram.

32. Find x and EG if $EJ = 2x + 1$ and $JG = 3x$. $x = 1$, $EG = 6$
33. Find y and FH if $HJ = \frac{1}{2}y + 2$ and $JF = y - \frac{1}{2}$. $y = 5$, $FH = 9$

34. **DESIGN** The chest of drawers shown at the right is called *Side 2*. It was designed by Shiro Kuramata. Describe the properties of parallelograms the artist used to place each drawer pull.

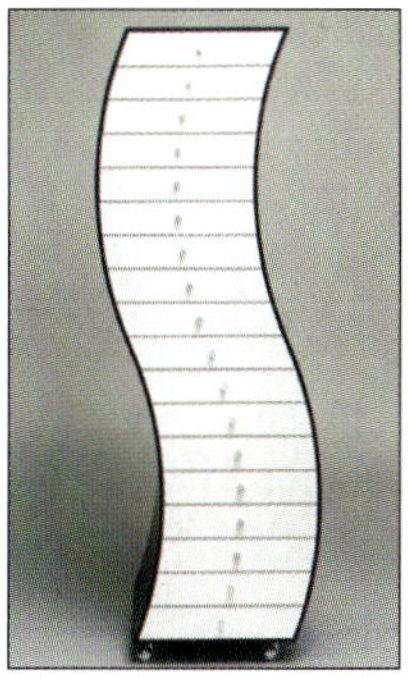

★35. **ALGEBRA** Parallelogram $ABCD$ has diagonals $\overline{AC}$ and $\overline{DB}$ that intersect at point P. If $AP = 3a + 18$, $AC = 12a$, $PB = a + 2b$, and $PD = 3b + 1$, find a, b, and DB. $a = 6$, $b = 5$, $DB = 32$

★36. **ALGEBRA** In parallelogram $ABCD$, $AB = 2x + 5$, $m\angle BAC = 2y$, $m\angle B = 120$, $m\angle CAD = 21$, and $CD = 21$. Find x and y. $x = 8$, $y = 19.5$

COORDINATE GEOMETRY **For Exercises 37–39, refer to ▱EFGH.**

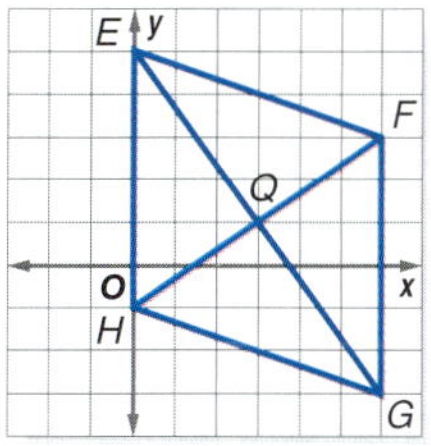

37. Use the Distance Formula to verify that the diagonals bisect each other.
38. Determine whether the diagonals of this parallelogram are congruent. no
39. Find the slopes of $\overline{EH}$ and $\overline{EF}$. Are the consecutive sides perpendicular? Explain.

40. Determine the relationship among $ACBX$, $ABYC$, and $ABCZ$ if $\triangle XYZ$ is equilateral and A, B, and C are midpoints of $\overline{XZ}$, $\overline{XY}$, and $\overline{ZY}$, respectively. See p. 459A.

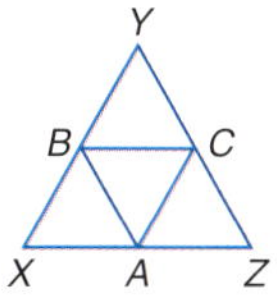

PROOF **Write the indicated type of proof.** 41–45. See p. 459A.

41. two-column proof of Theorem 8.3
42. two-column proof of Theorem 8.5
43. paragraph proof of Theorem 8.6
44. paragraph proof of Theorem 8.7
45. two-column proof of Theorem 8.8

www.geometryonline.com/self_check_quiz/fcat

More About. . .

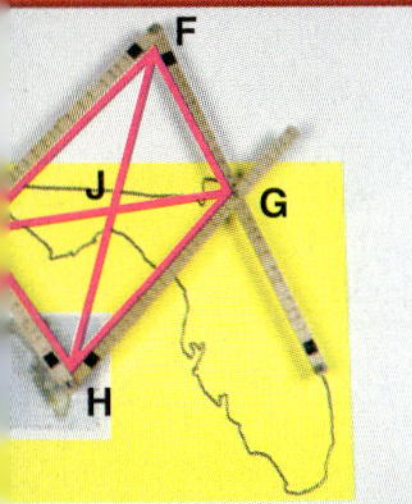

Drawing

The pantograph was used as a primitive copy machine. The device makes an exact replica as the user traces over a figure.

Source: www.infoplease.com

34. Since the diagonals of a ▱ bisect each other, the drawer pulls are at the intersection point of the diagonals.

37. $EQ = 5$, $QG = 5$, $HQ = \sqrt{13}$, $QF = \sqrt{13}$

39. Slope of $\overline{EH}$ is undefined, slope of $\overline{EF} = -\frac{1}{3}$; no, the slopes of the sides are not negative reciprocals of each other.

Answers

16. Opp. ⦞ of ▱ are ≅.
17. Alt. int. ⦞ are ≅.
18. Opp. sides of ▱ are ∥.
19. Diag. of ▱ bisect each other.
20. Diag. of ▱ separates ▱ into 2 ≅ △s.
21. Alt. int. ⦞ are ≅.

Enrichment, p. 428

Tessellations

A **tessellation** is a tiling pattern made of polygons. The pattern can be extended so that the polygonal tiles cover the plane completely with no gaps. A checkerboard and a honeycomb pattern are examples of tessellations. Sometimes the same polygon can make more than one tessellation pattern. Both patterns below can be formed from an isosceles triangle.

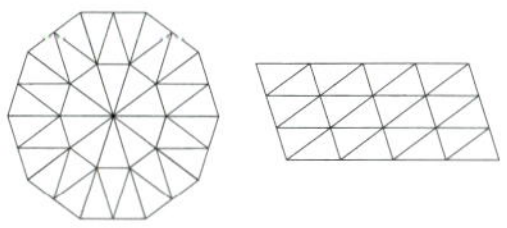

Study Guide and Intervention, p. 423 (shown) and p. 424

Sides and Angles of Parallelograms A quadrilateral with both pairs of opposite sides parallel is a **parallelogram**. Here are four important properties of parallelograms.

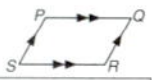

	If *PQRS* is a parallelogram, then
The opposite sides of a parallelogram are congruent.	$\overline{PQ} \cong \overline{SR}$ and $\overline{PS} \cong \overline{QR}$
The opposite angles of a parallelogram are congruent.	$\angle P \cong \angle R$ and $\angle S \cong \angle Q$
The consecutive angles of a parallelogram are supplementary.	$\angle P$ and $\angle S$ are supplementary; $\angle S$ and $\angle R$ are supplementary; $\angle R$ and $\angle Q$ are supplementary; $\angle Q$ and $\angle P$ are supplementary.
If a parallelogram has one right angle, then it has four right angles.	If $m\angle P = 90$, then $m\angle Q = 90$, $m\angle R = 90$, and $m\angle S = 90$.

Example **If *ABCD* is a parallelogram, find *a* and *b*.**

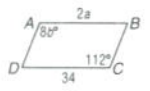

$\overline{AB}$ and $\overline{CD}$ are opposite sides, so $\overline{AB} \cong \overline{CD}$.
$2a = 34$
$a = 17$

$\angle A$ and $\angle C$ are opposite angles, so $\angle A \cong \angle C$.
$8b = 112$
$b = 14$

Exercises

Find *x* and *y* in each parallelogram.

1. 

$x = 30$; $y = 22.5$

2. 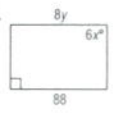

$x = 15$; $y = 11$

3. 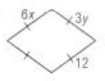
$x = 2$; $y = 4$

4. 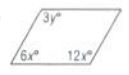
$x = 10$; $y = 40$

5. 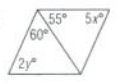
$x = 13$; $y = 32.5$

6. 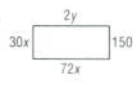

$x = 5$; $y = 180$

Skills Practice, p. 425 and Practice, p. 426 (shown)

Complete each statement about ▱*LMNP*. Justify your answer.

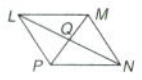

1. $\overline{LQ} \cong$ __?__ $\overline{NQ}$; diag. of ▱ bisect each other.
2. $\angle LMN \cong$ __?__ $\angle NPL$; opp. ⦞ of ▱ are ≅.
3. $\triangle LMP \cong$ __?__ $\triangle NPM$; diag. of ▱ separates ▱ into 2 ≅ △s.
4. $\angle NPL$ is supplementary to __?__. $\angle MNP$ or $\angle PLM$; cons. ⦞ in ▱ are suppl.
5. $\overline{LM} \cong$ __?__ $\overline{NP}$; opp. sides of ▱ are ≅.

ALGEBRA **Use ▱*RSTU* to find each measure or value.**

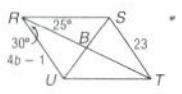

6. $m\angle RST$ = 125
7. $m\angle STU$ = 55
8. $m\angle TUR$ = 125
9. b = 6

COORDINATE GEOMETRY **Find the coordinates of the intersection of the diagonals of parallelogram *PRYZ* given each set of vertices.**

10. $P(2, 5)$, $R(3, 3)$, $Y(-2, -3)$, $Z(-3, -1)$
(0, 1)
11. $P(2, 3)$, $R(1, -2)$, $Y(-5, -7)$, $Z(-4, -2)$
(−1.5, −2)

12. **PROOF** Write a paragraph proof of the following.
Given: ▱*PRST* and ▱*PQVU*
Prove: $\angle V \cong \angle S$

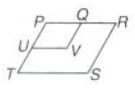

Proof: We are given ▱*PRST* and ▱*PQVU*. Since opposite angles of a parallelogram are congruent, $\angle P \cong \angle V$ and $\angle P \cong \angle S$. Since congruence of angles is transitive, $\angle V \cong \angle S$ by the Transitive Property of Congruence.

13. **CONSTRUCTION** Mr. Rodriquez used the parallelogram at the right to design a herringbone pattern for a paving stone. He will use the paving stone for a sidewalk. If $m\angle 1$ is 130, find $m\angle 2$, $m\angle 3$, and $m\angle 4$.
50, 130, 50

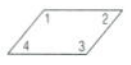

Reading to Learn Mathematics, p. 427 ELL

Pre-Activity **How are parallelograms used to represent data?**

Read the introduction to Lesson 8-2 at the top of page 411 in your textbook.

- What is the name of the shape of the top surface of each wedge of cheese? parallelogram
- Are the three polygons shown in the drawing similar polygons? Explain your reasoning. No; sample answer: Their sides are not proportional.

Reading the Lesson

1. Underline words or phrases that can complete the following sentences to make statements that are always true. (There may be more than one correct choice for some of the sentences.)
   a. Opposite sides of a parallelogram are (congruent/perpendicular/parallel).
   b. Consecutive angles of a parallelogram are (complementary/supplementary/congruent).
   c. A diagonal of a parallelogram divides the parallelogram into two (acute/right/obtuse/congruent) triangles.
   d. Opposite angles of a parallelogram are (complementary/supplementary/congruent).
   e. The diagonals of a parallelogram (bisect each other/are perpendicular/are congruent).
   f. If a parallelogram has one right angle, then all of its other angles are (acute/right/obtuse) angles.
2. Let $ABCD$ be a parallelogram with $AB \neq BC$ and with no right angles.
   a. Sketch a parallelogram that matches the description above and draw diagonal $\overline{BD}$.
   Sample answer:

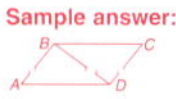

   In parts **b–f**, complete each sentence.
   b. $\overline{AB} \parallel$ $\overline{CD}$ and $\overline{AD} \parallel$ $\overline{BC}$.
   c. $\overline{AB} \cong$ $\overline{CD}$ and $\overline{BC} \cong$ $\overline{AD}$.
   d. $\angle A \cong$ $\angle C$ and $\angle ABC \cong$ $\angle CDA$.
   e. $\angle ADB \cong$ $\angle CBD$ because these two angles are alternate interior angles formed by the two parallel lines $\overleftrightarrow{AD}$ and $\overleftrightarrow{BC}$ and the transversal $\overleftrightarrow{BD}$.
   f. $\triangle ABD \cong$ $\triangle CDB$.

Helping You Remember

3. A good way to remember new theorems in geometry is to relate them to theorems you learned earlier. Name a theorem about parallel lines that can be used to remember the theorem that says, "If a parallelogram has one right angle, it has four right angles."
Perpendicular Transversal Theorem

4 Assess

Open-Ended Assessment

Writing Ask students to list all the properties of parallelograms they have learned.

Getting Ready for Lesson 8-3

Prerequisite Skill Students will learn about the tests for parallelograms in Lesson 8-3. They will use slope to prove that opposite sides of a quadrilateral are parallel. Use Exercises 61–63 to determine your students' familiarity with finding slope.

Assessment Options

Quiz (Lessons 8-1 and 8-2) is available on p. 473 of the *Chapter 8 Resource Masters*.

Answers

46. Given: $DGHK$ is a parallelogram.
$\overline{FH} \perp \overline{GD}$
$\overline{DJ} \perp \overline{HK}$
Prove: $\triangle DJK \cong \triangle HFG$

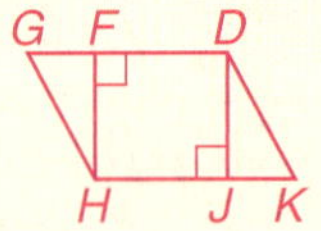

Proof:
Statements (Reasons)

1. $DGHK$ is a parallelogram; $\overline{FH} \perp \overline{GD}$, $\overline{DJ} \perp \overline{HK}$. (Given)
2. $\angle G \cong \angle K$ (Opp. ∡ of ▱ ≅.)
3. $\overline{GH} \cong \overline{DK}$ (Opp. sides of ▱ ≅.)
4. $\angle HFG$ and $\angle DJK$ are rt. ∡. (⊥ lines form four rt. ∡.)
5. $\triangle HFG$ and $\triangle DJK$ are rt. △s. (Def. of rt. △)
6. $\triangle HFG \cong \triangle DJK$ (HA)

47. Given: ▱$BCGH$, $\overline{HD} \cong \overline{FD}$
Prove: $\angle F \cong \angle GCB$
Proof:

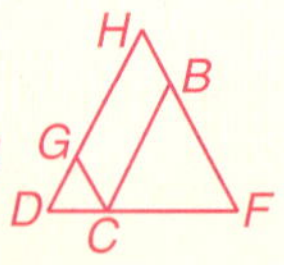

Statements (Reasons)

1. ▱$BCGH$, $\overline{HD} \cong \overline{FD}$ (Given)
2. $\angle F \cong \angle H$ (Isosceles △ Thm.)
3. $\angle H \cong \angle GCB$ (Opp. ∡ of ▱ ≅.)
4. $\angle F \cong \angle GCB$ (Congruence of ∡ is transitive.)

PROOF Write a two-column proof. **46–47. See margin.**

46. Given: ▱$DGHK$, $\overline{FH} \perp \overline{GD}$, $\overline{DJ} \perp \overline{HK}$
Prove: $\triangle DJK \cong \triangle HFG$

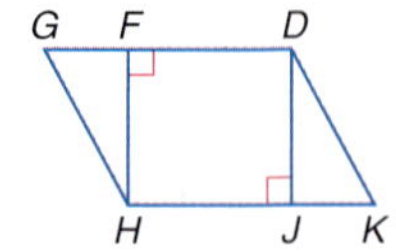

47. Given: ▱$BCGH$, $\overline{HD} \cong \overline{FD}$
Prove: $\angle F \cong \angle GCB$

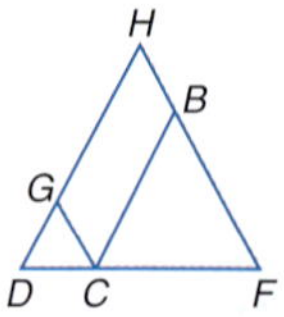

48. **CRITICAL THINKING** Find the ratio of MS to SP, given that $MNPQ$ is a parallelogram with $MR = \frac{1}{4} MN$. $\frac{1}{2}$

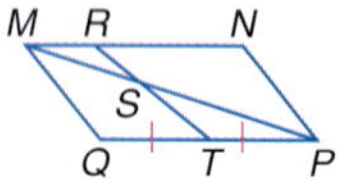

49. **WRITING IN MATH** Answer the question that was posed at the beginning of the lesson. **See margin.**

How are parallelograms used to represent data?

Include the following in your answer:
- properties of parallelograms, and
- a display of the data in the graphic with a different parallelogram.

FCAT Practice
Standardized Test Practice
Ⓐ Ⓑ Ⓒ Ⓓ

50. **SHORT RESPONSE** Two consecutive angles of a parallelogram measure $(3x + 42)°$ and $(9x - 18)°$. Find the measures of the angles. **81, 99**

51. **ALGEBRA** The perimeter of the rectangle $ABCD$ is equal to p and $x = \frac{y}{5}$. What is the value of y in terms of p? **B**

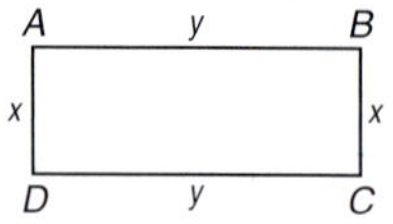

Ⓐ $\frac{p}{3}$ Ⓑ $\frac{5p}{12}$ Ⓒ $\frac{5p}{8}$ Ⓓ $\frac{5p}{6}$

Maintain Your Skills

Mixed Review **Find the sum of the measures of the interior angles of each convex polygon.** *(Lesson 8-1)*

52. 14-gon **2160** 53. 22-gon **3600** 54. 17-gon **2700** 55. 36-gon **6120**

Determine whether the *Law of Sines* or the *Law of Cosines* should be used to solve each triangle. Then solve each triangle. Round to the nearest tenth. *(Lesson 7-7)*

56. Cosines; $a \approx 8.8$, $m\angle B \approx 56.8$, $m\angle C \approx 81.2$
57. Sines; $m\angle C \approx 69.9$, $m\angle A \approx 53.1$, $a \approx 11.9$
58. Cosines; $c \approx 28.4$, $m\angle A \approx 46.3$, $m\angle B \approx 55.7$

56.
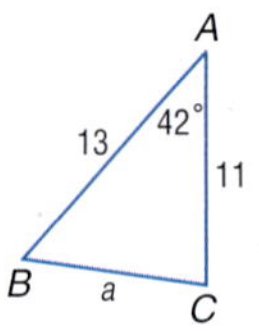

57.
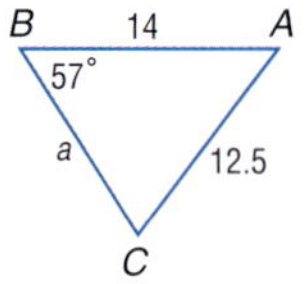

58.
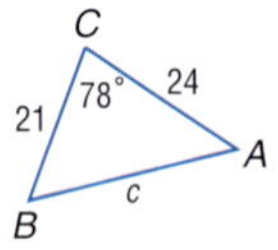

Use Pascal's Triangle for Exercises 59 and 60. *(Lesson 6-6)* **59. 30**

59. Find the sum of the first 30 numbers in the outside diagonal of Pascal's triangle.
60. Find the sum of the first 70 numbers in the second diagonal. **2485**

Getting Ready for the Next Lesson **PREREQUISITE SKILL The vertices of a quadrilateral are $A(-5, -2)$, $B(-2, 5)$, $C(2, -2)$, and $D(-1, -9)$. Determine whether each segment is a side or a diagonal of the quadrilateral, and find the slope of each segment.**
*(To review **slope**, see Lesson 3-3.)*

61. $\overline{AB}$ **side, $\frac{7}{3}$** 62. $\overline{BD}$ **diagonal, –14** 63. $\overline{CD}$ **side, $\frac{7}{3}$**

416 Chapter 8 Quadrilaterals

49. Sample answer: The graphic uses the illustration of wedges shaped like parallelograms to display the data. Answers should include the following.
- The opposite sides are parallel and congruent, the opposite angles are congruent, and the consecutive angles are supplementary.
- Sample answer:

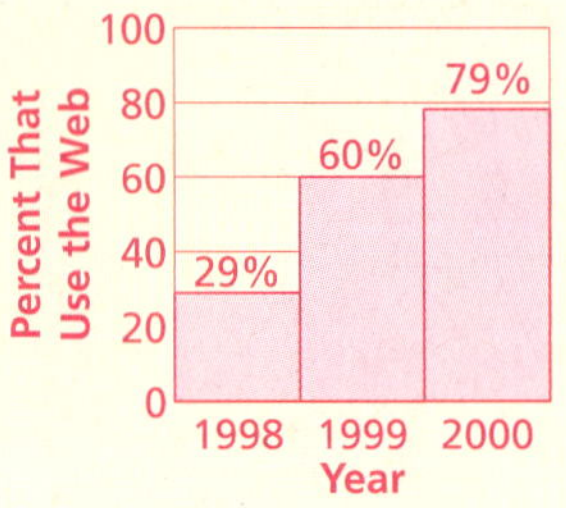

8-3 Tests for Parallelograms

unshine State Standards
IA.A.3.4.3, MA.C.1.4.1, MA.C.2.4.1,
IA.C.3.4.2

What You'll Learn

- Recognize the conditions that ensure a quadrilateral is a parallelogram.
- Prove that a set of points forms a parallelogram in the coordinate plane.

How are parallelograms used in architecture?

The roof of the covered bridge appears to be a parallelogram. Each pair of opposite sides looks like they are the same length. How can we know for sure if this shape is really a parallelogram?

CONDITIONS FOR A PARALLELOGRAM By definition, the opposite sides of a parallelogram are parallel. So, if a quadrilateral has each pair of opposite sides parallel it is a parallelogram. Other tests can be used to determine if a quadrilateral is a parallelogram.

Geometry Activity

Testing for a Parallelogram

Model

- Cut two straws to one length and two other straws to a different length.
- Connect the straws by inserting a pipe cleaner in one end of each size of straw to form a quadrilateral like the one shown at the right.
- Shift the sides to form quadrilaterals of different shapes.

Analyze

1. Measure the distance between the opposite sides of the quadrilateral in at least three places. Repeat this process for several figures. What can you conclude about opposite sides? They appear to be parallel.
2. Classify the quadrilaterals that you formed. parallelograms
3. Compare the measures of pairs of opposite sides. They are equal.
4. Measure the four angles in several of the quadrilaterals. What relationships do you find?

Make a Conjecture

5. What conditions are necessary to verify that a quadrilateral is a parallelogram?

. Opposite angles are ongruent, and onsecutive angles re supplementary.

. Opposite sides are arallel and ongruent, opposite ngles are congruent, r consecutive angles re supplementary.

8-3 Lesson Notes

1 Focus

5-Minute Check Transparency 8-3 Use as a quiz or review of Lesson 8-2.

Mathematical Background notes are available for this lesson on p. 402C.

How are parallelograms used in architecture?

Ask students:

- Which geometric shape is used most often in the construction of the roof of a house? parallelogram
- Why is this shape used for roofs? Sample answer: It is easy to build or construct.
- What is another structure that uses this shape in its construction? Sample answer: doors and doorways

Resource Manager

Workbook and Reproducible Masters

Chapter 8 Resource Masters

- Study Guide and Intervention, pp. 429–430
- Skills Practice, p. 431
- Practice, p. 432
- Reading to Learn Mathematics, p. 433
- Enrichment, p. 434

School-to-Career Masters, p. 15

Teaching Geometry With Manipulatives Masters, pp. 2, 16, 130, 132

Transparencies

5-Minute Check Transparency 8-3
Answer Key Transparencies

Technology

Interactive Chalkboard

2 Teach

CONDITIONS FOR A PARALLELOGRAM

Teaching Tip Note that Theorems 8.9, 8.10, and 8.11 are converses of the theorems in Lesson 8-2.

In-Class Examples

Power Point®

Teaching Tip Emphasize that the theorems on this page are used in proofs. In this lesson, students learn that to prove that a quadrilateral is a parallelogram, they must establish that one of these theorems applies to the quadrilateral.

1 Write a paragraph proof of the statement: *If a diagonal of a quadrilateral divides the quadrilateral into two congruent triangles, then the quadrilateral is a parallelogram.*

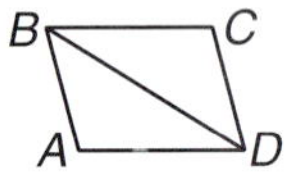

Given: $\triangle ABD \cong \triangle CDB$
Prove: $ABCD$ is a parallelogram.

Since $\triangle ABD \cong \triangle CDB$, $\overline{AB} \cong \overline{CD}$, and $\overline{BC} \cong \overline{DA}$ by CPCTC. Therefore $ABCD$ is a parallelogram because if a quadrilateral has both pairs of opposite sides congruent, it is a parallelogram.

2 Some of the shapes in this Bavarian crest appear to be parallelograms. Describe the information needed to determine whether the shapes are parallelograms.

If both pairs of opposite sides are the same length or if one pair of opposite sides is congruent and parallel, the quadrilateral is a parallelogram. If both pairs of opposite angles are congruent or if the diagonals bisect each other, the quadrilateral is a parallelogram.

Key Concept — Proving Parallelograms

Theorem	Example
8.9 If both pairs of opposite sides of a quadrilateral are congruent, then the quadrilateral is a parallelogram. **Abbreviation:** *If both pairs of opp. sides are* $\cong$*, then quad. is* $\square$*.*	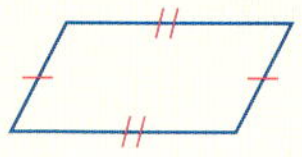
8.10 If both pairs of opposite angles of a quadrilateral are congruent, then the quadrilateral is a parallelogram. **Abbreviation:** *If both pairs of opp.* $\angle$s *are* $\cong$*, then quad. is* $\square$*.*	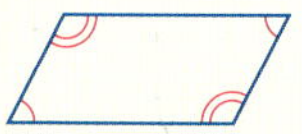
8.11 If the diagonals of a quadrilateral bisect each other, then the quadrilateral is a parallelogram. **Abbreviation:** *If diag. bisect each other, then quad. is* $\square$*.*	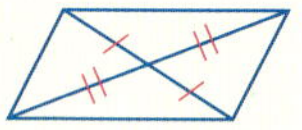
8.12 If one pair of opposite sides of a quadrilateral is both parallel and congruent, then the quadrilateral is a parallelogram. **Abbreviation:** *If one pair of opp. sides is* $\parallel$ *and* $\cong$*, then the quad. is a* $\square$*.*	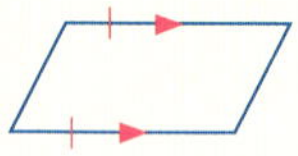

You will prove Theorems 8.9, 8.11, and 8.12 in Exercises 39, 40, and 41, respectively.

Example 1 Write a Proof

PROOF Write a paragraph proof for Theorem 8.10

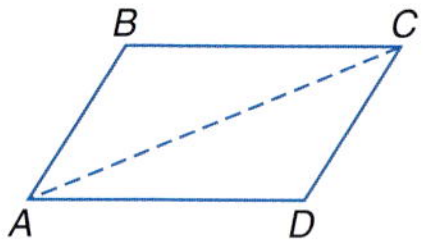

Given: $\angle A \cong \angle C$, $\angle B \cong \angle D$

Prove: $ABCD$ is a parallelogram.

Paragraph Proof:

Because two points determine a line, we can draw $\overline{AC}$. We now have two triangles. We know the sum of the angle measures of a triangle is 180, so the sum of the angle measures of two triangles is 360. Therefore, $m\angle A + m\angle B + m\angle C + m\angle D = 360$.

Since $\angle A \cong \angle C$ and $\angle B \cong \angle D$, $m\angle A = m\angle C$ and $m\angle B = m\angle D$. Substitute to find that $m\angle A + m\angle A + m\angle B + m\angle B = 360$, or $2(m\angle A) + 2(m\angle B) = 360$. Dividing each side of the equation by 2 yields $m\angle A + m\angle B = 180$. This means that consecutive angles are supplementary and $\overline{AD} \parallel \overline{BC}$.

Likewise, $2m\angle A + 2m\angle D = 360$, or $m\angle A + m\angle D = 180$. These consecutive supplementary angles verify that $\overline{AB} \parallel \overline{DC}$. Opposite sides are parallel, so $ABCD$ is a parallelogram.

More About. . .

Art

Ellsworth Kelly created *Sculpture for a Large Wall* in 1957. The sculpture is made of 104 aluminum panels. The piece is over 65 feet long, 11 feet high, and 2 feet deep.

Source: www.moma.org

Example 2 Properties of Parallelograms

ART **Some panels in the sculpture appear to be parallelograms. Describe the information needed to determine whether these panels are parallelograms.**

A panel is a parallelogram if both pairs of opposite sides are congruent, or if one pair of opposite sides is congruent and parallel. If the diagonals bisect each other, or if both pairs of opposite angles are congruent, then the panel is a parallelogram.

Geometry Activity

Materials: straws, scissors, pipe cleaners, protractors

You may want to do this activity in groups of four students. Ask one student to cut the straws. Ask a second student in the group to construct the parallelogram and a third student to measure the opposite sides. Ask the fourth student to measure the opposite angles. Then ask students to rotate roles and repeat the activity with different length straws.

Example 3 Properties of Parallelograms

Determine whether the quadrilateral is a parallelogram. Justify your answer.

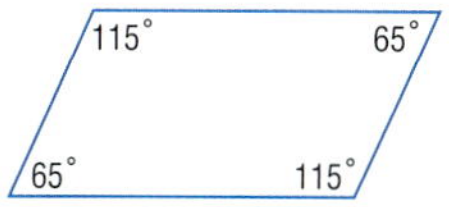

Each pair of opposite angles have the same measure. Therefore, they are congruent. If both pairs of opposite angles are congruent, the quadrilateral is a parallelogram.

A quadrilateral is a parallelogram if any one of the following is true.

Concept Summary — Tests for a Parallelogram

1. Both pairs of opposite sides are parallel. (Definition)
2. Both pairs of opposite sides are congruent. (Theorem 8.9)
3. Both pairs of opposite angles are congruent. (Theorem 8.10)
4. Diagonals bisect each other. (Theorem 8.11)
5. A pair of opposite sides is both parallel and congruent. (Theorem 8.12)

Study Tip

Common Misconceptions

If a quadrilateral meets one of the five tests, it is a parallelogram. All of the properties of parallelograms need not be shown.

Example 4 Find Measures

ALGEBRA **Find x and y so that each quadrilateral is a parallelogram.**

a.

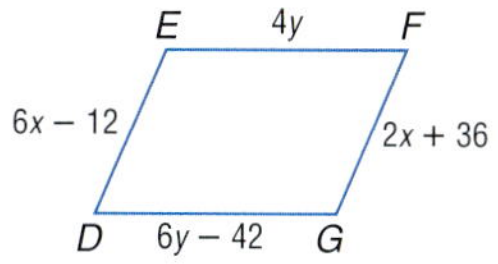

Opposite sides of a parallelogram are congruent.

$\overline{EF} \cong \overline{DG}$	Opp. sides of ▱ are ≅.	$\overline{DE} \cong \overline{FG}$	Opp. sides of ▱ are ≅.
$EF = DG$	Def. of ≅ segments	$DE = FG$	Def. of ≅ segments
$4y = 6y - 42$	Substitution	$6x - 12 = 2x + 36$	Substitution
$-2y = -42$	Subtract $6y$.	$4x = 48$	Subtract $2x$ and add 12.
$y = 21$	Divide by −2.	$x = 12$	Divide by 4.

So, when x is 12 and y is 21, $DEFG$ is a parallelogram.

b.

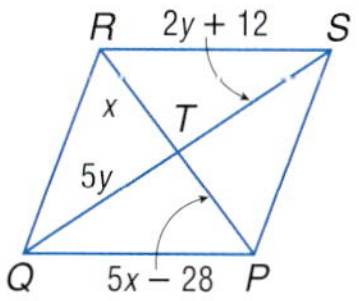

Diagonals in a parallelogram bisect each other.

$\overline{QT} \cong \overline{TS}$	Opp. sides of ▱ are ≅.	$\overline{RT} \cong \overline{TP}$	Opp. sides of ▱ are ≅.
$QT = TS$	Def. of ≅ segments	$RT = TP$	Def. of ≅ segments
$5y = 2y + 12$	Substitution	$x = 5x - 28$	Substitution
$3y = 12$	Subtract $2y$.	$-4x = -28$	Subtract $5x$.
$y = 4$	Divide by 3.	$x = 7$	Divide by −4.

$PQRS$ is a parallelogram when $x = 7$ and $y = 4$.

www.geometryonline.com/extra_examples/fcat

Building on Prior Knowledge

In Lesson 3-3, students learned how to find the slope of parallel lines. In this lesson, they will use slope to show that a quadrilateral in the coordinate plane is or is not a parallelogram.

In-Class Examples

Power Point®

3 Determine whether the quadrilateral is a parallelogram. Justify your answer.

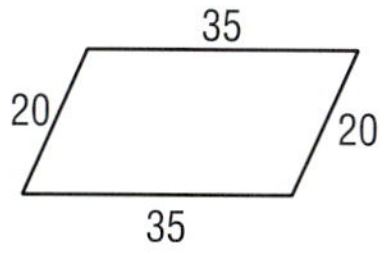

Each pair of opposite sides have the same measure. Therefore they are congruent. If both pairs of opposite sides are congruent, the quadrilateral is a parallelogram.

4 Find x and y so that each quadrilateral is a parallelogram.

a.

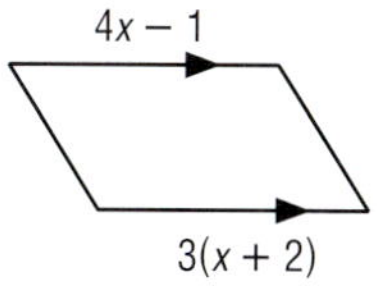

$x = 7$

b.

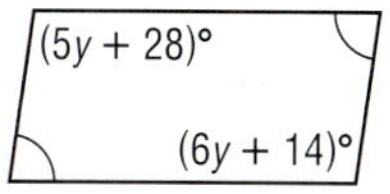

$y = 14$

DAILY INTERVENTION — Differentiated Instruction

Intrapersonal Have students choose a partner. Ask one student to draw a parallelogram in the plane. Then ask the partner to prove that the quadrilateral is a parallelogram. Next switch roles and do the activity again.

PARALLELOGRAMS ON THE COORDINATE PLANE

In-Class Example PowerPoint®

5 COORDINATE GEOMETRY Determine whether the figure with the given vertices is a parallelogram. Use the method indicated.

a. $A(-3, 0)$, $B(-1, 3)$, $C(3, 2)$, $D(1, -1)$; Slope Formula

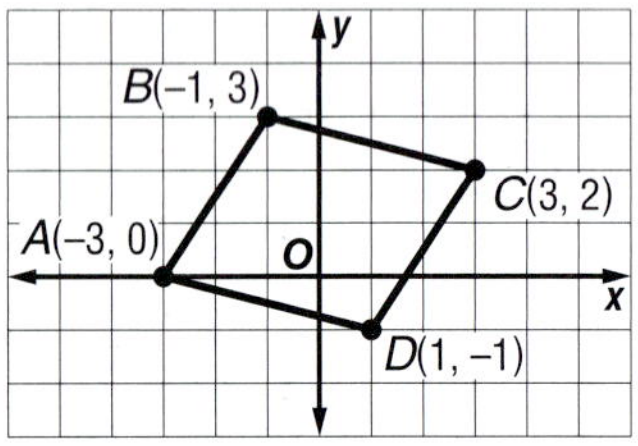

Slope of $\overline{AB} = \frac{3}{2}$; slope of $\overline{CD} = \frac{3}{2}$; slope of $\overline{AD} = -\frac{1}{4}$; slope of $\overline{BC} = -\frac{1}{4}$. Since opposite sides have the same slope, $\overline{AB} \parallel \overline{CD}$ and $\overline{AD} \parallel \overline{BC}$. Therefore, $ABCD$ is a parallelogram.

b. $P(-3, -1)$, $Q(-1, 3)$, $R(3, 1)$, $S(1, -3)$; Distance and Slope Formulas

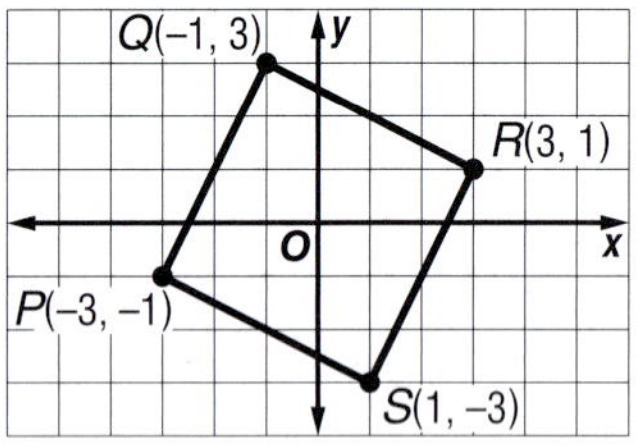

$PS = \sqrt{20}$; $QR = \sqrt{20}$; therefore $PS = QR$, and $\overline{PS} \cong \overline{QR}$. Slope of $\overline{PS} = -\frac{1}{2}$; slope of $\overline{QR} = -\frac{1}{2}$. Since opposite sides have the same slope, $\overline{PS} \parallel \overline{QR}$. Since one pair of opposite sides is congruent and parallel, $PQRS$ is a parallelogram.

Study Tip

Coordinate Geometry The Midpoint Formula can also be used to show that a quadrilateral is a parallelogram by Theorem 8.11.

PARALLELOGRAMS ON THE COORDINATE PLANE We can use the Distance Formula and the Slope Formula to determine if a quadrilateral is a parallelogram in the coordinate plane.

Example 5 Use Slope and Distance

COORDINATE GEOMETRY Determine whether the figure with the given vertices is a parallelogram. Use the method indicated.

a. $A(3, 3)$, $B(8, 2)$, $C(6, -1)$, $D(1, 0)$; Slope Formula

If the opposite sides of a quadrilateral are parallel, then it is a parallelogram.

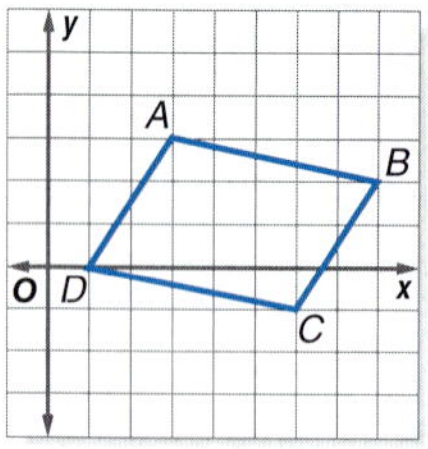

slope of $\overline{AB} = \frac{2-3}{8-3}$ or $\frac{-1}{5}$ slope of $\overline{DC} = \frac{-1-0}{6-1}$ or $\frac{-1}{5}$

slope of $\overline{AD} = \frac{3-0}{3-1}$ or $\frac{3}{2}$ slope of $\overline{BC} = \frac{-1-2}{6-8}$ or $\frac{3}{2}$

Since opposite sides have the same slope, $\overline{AB} \parallel \overline{DC}$ and $\overline{AD} \parallel \overline{BC}$. Therefore, $ABCD$ is a parallelogram by definition.

b. $P(5, 3)$, $Q(1, -5)$, $R(-6, -1)$, $S(-2, 7)$; Distance and Slope Formulas

First use the Distance Formula to determine whether the opposite sides are congruent.

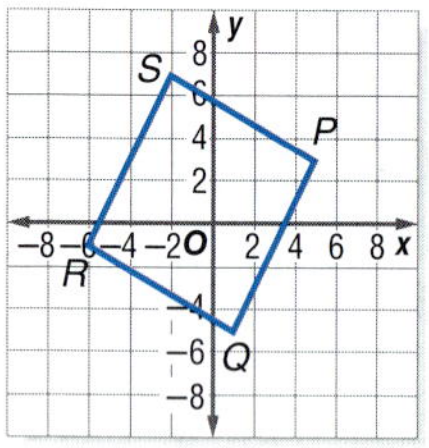

$$PS = \sqrt{[5-(-2)]^2 + (3-7)^2}$$
$$= \sqrt{7^2 + (-4)^2} \text{ or } \sqrt{65}$$

$$QR = \sqrt{[1-(-6)]^2 + [-5-(-1)]^2}$$
$$= \sqrt{7^2 + (-4)^2} \text{ or } \sqrt{65}$$

Since $PS = QR$, $\overline{PS} \cong \overline{QR}$.

Next, use the Slope Formula to determine whether $\overline{PS} \parallel \overline{QR}$.

slope of $\overline{PS} = \frac{3-7}{5-(-2)}$ or $-\frac{4}{7}$ slope of $\overline{QR} = \frac{-5-(-1)}{1-(-6)}$ or $-\frac{4}{7}$

$\overline{PS}$ and $\overline{QR}$ have the same slope, so they are parallel. Since one pair of opposite sides is congruent and parallel, $PQRS$ is a parallelogram.

Check for Understanding

Concept Check

1. List and describe four tests for parallelograms. **1–2. See margin.**
2. **OPEN ENDED** Draw a parallelogram. Label the congruent angles.
3. **FIND THE ERROR** Carter and Shaniqua are describing ways to show that a quadrilateral is a parallelogram.

Carter	Shaniqua
A quadrilateral is a parallelogram if one pair of opposite sides is congruent and one pair of opposite sides is parallel.	A quadrilateral is a parallelogram if one pair of opposite sides is congruent and parallel.

Who is correct? Explain your reasoning. **Shaniqua; Carter's description could result in a shape that is not a parallelogram.**

Answers

1. Both pairs of opposite sides are congruent; both pairs of opposite angles are congruent; diagonals bisect each other; one pair of opposite sides is parallel and congruent.

2. Sample answer:

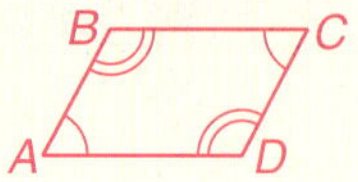

Guided Practice

GUIDED PRACTICE KEY	
Exercises	Examples
4–5	3
6–7	4
8–10	5
11	1
12	2

Determine whether each quadrilateral is a parallelogram. Justify your answer.

4. 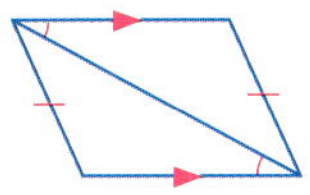
No; one pair of opp. sides is not parallel and congruent.

5. 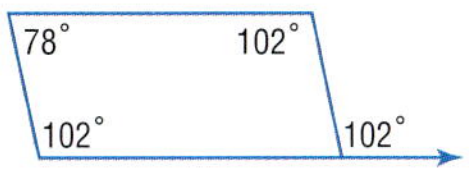

Yes; each pair of opp. ∠s is ≅.

ALGEBRA **Find x and y so that each quadrilateral is a parallelogram.**

6. 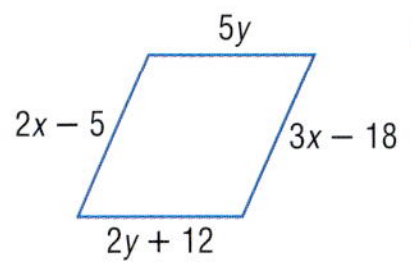

$x = 13$, $y = 4$

7. 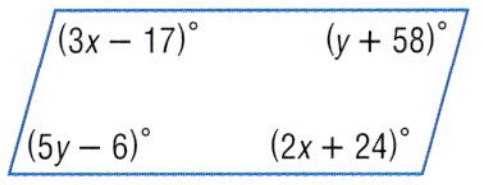

$x = 41$, $y = 16$

COORDINATE GEOMETRY **Determine whether the figure with the given vertices is a parallelogram. Use the method indicated.**

8. $B(0, 0)$, $C(4, 1)$, $D(6, 5)$, $E(2, 4)$; Slope Formula **yes**
9. $A(-4, 0)$, $B(3, 1)$, $C(1, 4)$, $D(-6, 3)$; Distance and Slope Formulas **yes**
10. $E(-4, -3)$, $F(4, -1)$, $G(2, 3)$, $H(-6, 2)$; Midpoint Formula **no**

11. **PROOF** Write a two-column proof to prove that $PQRS$ is a parallelogram given that $\overline{PT} \cong \overline{TR}$ and $\angle TSP \cong \angle TQR$. **See pp. 459A–459B.**

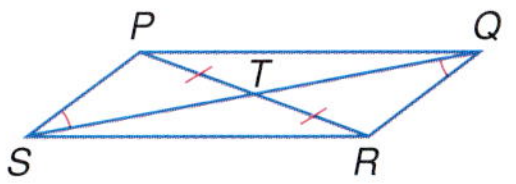

Application

12. **TANGRAMS** A tangram set consists of seven pieces: a small square, two small congruent right triangles, two large congruent right triangles, a medium-sized right triangle, and a quadrilateral. How can you determine the shape of the quadrilateral? Explain. **See margin.**

★ indicates increased difficulty

Practice and Apply

Homework Help	
For Exercises	See Examples
13–18	3
19–24	4
25–36	5
37–38	2
39–42	1

Extra Practice
See page 769.

Determine whether each quadrilateral is a parallelogram. Justify your answer.

13. 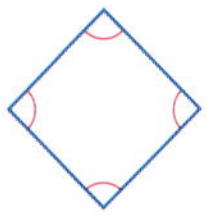
yes

14. 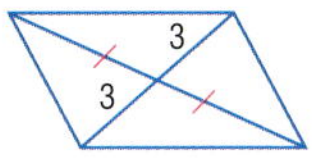

yes

15. **yes** 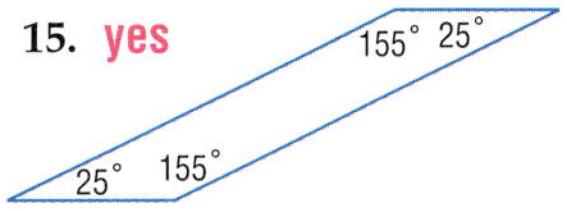

16. 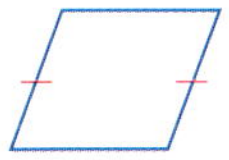
no

17. 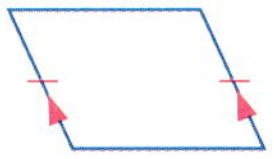
yes

18. 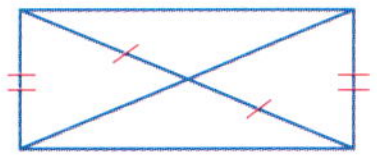
no

ALGEBRA **Find x and y so that each quadrilateral is a parallelogram.**

19. 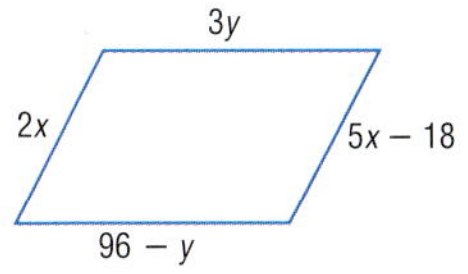

20. 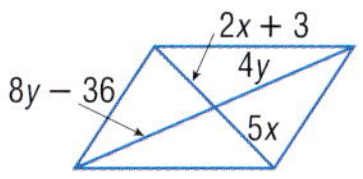

21. 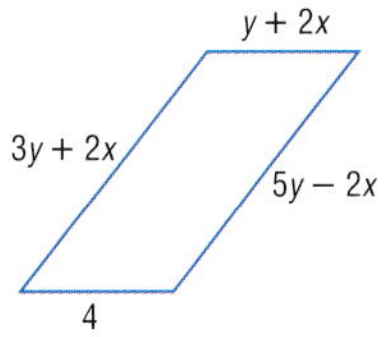

22. 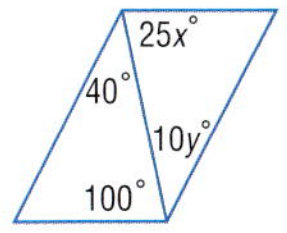

★23. 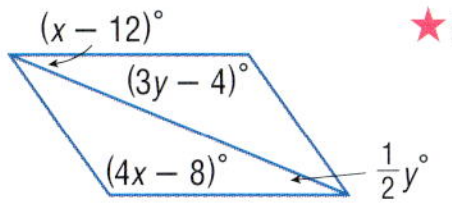

★24. 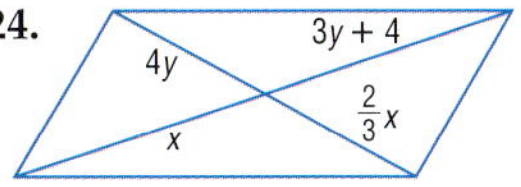

13–18. See margin for justifications.
19. $x = 6$, $y = 24$
20. $x = 1$, $y = 9$
21. $x = 1$, $y = 2$
22. $x = 4$, $y = 4$
23. $x = 34$, $y = 44$
24. $x = 8$, $y = 1\frac{1}{3}$

3 Practice/Apply

Study Notebook

Have students—
- add the definitions/examples of the vocabulary terms to their Vocabulary Builder worksheets for Chapter 8.
- write the four theorems to prove that a quadrilateral is a parallelogram in their notes. Ask them to include an example of each.
- include any other item(s) that they find helpful in mastering the skills in this lesson.

DAILY INTERVENTION

FIND THE ERROR
In Exercise 3, caution students that the same pair of opposite sides needs to be congruent and parallel in order to use the theorem that applies. Otherwise the quadrilateral could be an isosceles trapezoid (studied in Lesson 8-6).

About the Exercises...

Organization by Objective
- **Conditions for a Parallelogram:** 13–24, 37–42
- **Parallelograms on the Coordinate Plane:** 25–36

Odd/Even Assignments
Exercises 13–42 are structured so that students practice the same concepts whether they are assigned odd or even problems.

Assignment Guide

Basic: 13–21 odd, 25–31 odd, 37–43 odd, 44–63

Average: 13–43 odd, 44–63

Advanced: 14–42 even, 43–59 (optional: 60–63)

All: Quiz 1 (1–5)

Answers

12. If one pair of opposite sides are congruent and parallel, the quadrilateral is a parallelogram.
13. Each pair of opposite angles is congruent.
14. The diagonals bisect each other.
15. Opposite angles are congruent.
16. None of the tests for parallelograms are fulfilled.
17. One pair of opposite sides is parallel and congruent.
18. None of the tests for parallelograms are fulfilled.

Lesson 8-3 Tests for Parallelograms 421

COORDINATE GEOMETRY **Determine whether a figure with the given vertices is a parallelogram. Use the method indicated.**

25. $B(-6, -3)$, $C(2, -3)$, $E(4, 4)$, $G(-4, 4)$; Slope Formula **yes**
26. $Q(-3, -6)$, $R(2, 2)$, $S(-1, 6)$, $T(-5, 2)$; Slope Formula **no**
27. $A(-5, -4)$, $B(3, -2)$, $C(4, 4)$, $D(-4, 2)$; Distance Formula **yes**
28. $W(-6, -5)$, $X(-1, -4)$, $Y(0, -1)$, $Z(-5, -2)$; Midpoint Formula **yes**
29. $G(-2, 8)$, $H(4, 4)$, $J(6, -3)$, $K(-1, -7)$; Distance and Slope Formulas **no**
30. $H(5, 6)$, $J(9, 0)$, $K(8, -5)$, $L(3, -2)$; Distance Formula **no**
31. $S(-1, 9)$, $T(3, 8)$, $V(6, 2)$, $W(2, 3)$; Midpoint Formula **yes**
32. $C(-7, 3)$, $D(-3, 2)$, $F(0, -4)$, $G(-4, -3)$; Distance and Slope Formulas **yes**

★ 33. Quadrilateral *MNPR* has vertices $M(-6, 6)$, $N(-1, -1)$, $P(-2, -4)$, and $R(-5, -2)$. Determine how to move one vertex to make *MNPR* a parallelogram. **Move *M* to $(-4, 1)$, *N* to $(-3, 4)$, *P* to $(0, -9)$, or *R* to $(-7, 3)$.**

★ 34. Quadrilateral *QSTW* has vertices $Q(-3, 3)$, $S(4, 1)$, $T(-1, -2)$, and $W(-5, -1)$. Determine how to move one vertex to make *QSTW* a parallelogram. **Move *Q* to $(0, 2)$, *S* to $(1, 2)$, *T* to $(2, -3)$, or *W* to $(-8, 0)$.**

COORDINATE GEOMETRY **The coordinates of three of the vertices of a parallelogram are given. Find the possible coordinates for the fourth vertex.**

★ 35. $A(1, 4)$, $B(7, 5)$, and $C(4, -1)$. **$(-2, -2)$, $(4, 10)$, or $(10, 0)$**

★ 36. $Q(-2, 2)$, $R(1, 1)$, and $S(-1, -1)$. **$(2, -2)$, $(-4, 0)$, or $(0, 4)$**

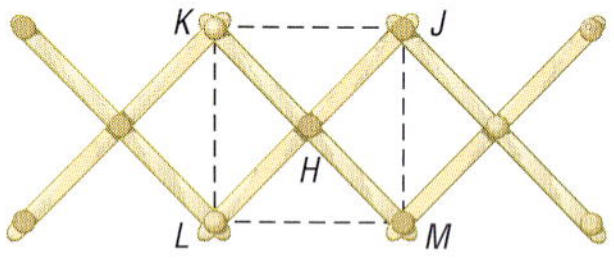

37. **STORAGE** Songan purchased an expandable hat rack that has 11 pegs. In the figure, *H* is the midpoint of $\overline{KM}$ and $\overline{JL}$. What type of figure is *JKLM*? Explain. **Parallelogram; $\overline{KM}$ and $\overline{JL}$ are diagonals that bisect each other.**

38. **METEOROLOGY** To show the center of a storm, television stations superimpose a "watchbox" over the weather map. Describe how you know that the watchbox is a parallelogram. **See margin.**

Online Research Data Update Each hurricane is assigned a name as the storm develops. What is the name of the most recent hurricane or tropical storm in the Atlantic or Pacific Oceans? Visit www.geometryonline.com/data_update to learn more.

PROOF Write a two-column proof of each theorem. **39–41. See p. 459B.**

39. Theorem 8.9
40. Theorem 8.11
41. Theorem 8.12

★ 42. Li-Cheng claims she invented a new geometry theorem. *A diagonal of a parallelogram bisects its angles.* Determine whether this theorem is true. Find an example or counterexample. **See margin.**

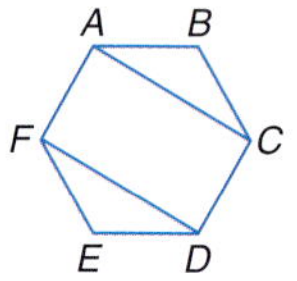

43. **CRITICAL THINKING** Write a proof to prove that *FDCA* is a parallelogram if *ABCDEF* is a regular hexagon. **See margin.**

44. **WRITING IN MATH** Answer the question that was posed at the beginning of the lesson. **See margin.**

How are parallelograms used in architecture?

Include the following in your answer:
- the information needed to prove that the roof of the covered bridge is a parallelogram, and
- another example of parallelograms used in architecture.

Career Choices

Atmospheric Scientist

Atmospheric scientists, or meteorologists, study weather patterns. They can work for private companies, the Federal Government or television stations.

Online Research For information about a career as an atmospheric scientist, visit: www.geometryonline.com/careers

Study Guide and Intervention, p. 429 (shown) and p. 430

Conditions for a Parallelogram There are many ways to establish that a quadrilateral is a parallelogram.

If:	If:
both pairs of opposite sides are parallel,	$\overline{AB} \parallel \overline{DC}$ and $\overline{AD} \parallel \overline{BC}$,
both pairs of opposite sides are congruent,	$\overline{AB} \cong \overline{DC}$ and $\overline{AD} \cong \overline{BC}$,
both pairs of opposite angles are congruent,	$\angle ABC \cong \angle ADC$ and $\angle DAB \cong \angle BCD$,
the diagonals bisect each other,	$\overline{AE} \cong \overline{CE}$ and $\overline{DE} \cong \overline{BE}$,
one pair of opposite sides is congruent and parallel,	$\overline{AB} \parallel \overline{CD}$ and $\overline{AB} \cong \overline{CD}$, or $\overline{AD} \parallel \overline{BC}$ and $\overline{AD} \cong \overline{BC}$,
then: the figure is a parallelogram.	**then:** *ABCD* is a parallelogram.

Example **Find x and y so that *FGHJ* is a parallelogram.**

FGHJ is a parallelogram if the lengths of the opposite sides are equal.

$6x + 3 = 15$, $6x = 12$, $x = 2$

$4x - 2y = 2$, $4(2) - 2y = 2$, $8 - 2y = 2$, $-2y = -6$, $y = 3$

Exercises

Find x and y so that each quadrilateral is a parallelogram.

1. $x = 7$; $y = 4$
2. $x = 5$; $y = 25$
3. $x = 31$; $y = 5$
4. $x = 5$; $y = 3$
5. $x = 15$; $y = 9$
6. $x = 30$; $y = 15$

Skills Practice, p. 431 and Practice, p. 432 (shown)

Determine whether each quadrilateral is a parallelogram. Justify your answer.

1. Yes; the diagonals bisect each other.
2. No; none of the tests for parallelograms is fulfilled.
3. Yes; both pairs of opposite angles are congruent.
4. No; none of the tests for parallelograms is fulfilled.

COORDINATE GEOMETRY **Determine whether a figure with the given vertices is a parallelogram. Use the method indicated.**

5. $P(-5, 1)$, $S(-2, 2)$, $F(-1, -3)$, $T(2, -2)$; Slope Formula
no

6. $R(-2, 5)$, $O(1, 3)$, $M(-3, -4)$, $Y(-6, -2)$; Distance and Slope Formula
yes

ALGEBRA **Find x and y so that each quadrilateral is a parallelogram.**

7. $x = 20$, $y = 12$
8. $x = -6$, $y = 13$
9. $x = -3$, $y = 2$
10. $x = -2$, $y = -5$

11. **TILE DESIGN** The pattern shown in the figure is to consist of congruent parallelograms. How can the designer be certain that the shapes are parallelograms?
Sample answer: Confirm that both pairs of opposite ∠s are ≅.

Reading to Learn Mathematics, p. 433 ELL

Pre-Activity **How are parallelograms used in architecture?**

Read the introduction to Lesson 8-3 at the top of page 417 in your textbook.

Make two observations about the angles in the roof of the covered bridge.
Sample answer: Opposite angles appear congruent. Consecutive angles appear supplementary.

Reading the Lesson

1. Which of the following conditions guarantee that a quadrilateral is a parallelogram? B, D, G, H, J, L
   - A. Two sides are parallel.
   - B. Both pairs of opposite sides are congruent.
   - C. The diagonals are perpendicular.
   - D. A pair of opposite sides is both parallel and congruent.
   - E. There are two right angles.
   - F. The sum of the measures of the interior angles is 360.
   - G. All four sides are congruent.
   - H. Both pairs of opposite angles are congruent.
   - I. Two angles are acute and the other two angles are obtuse.
   - J. The diagonals bisect each other.
   - K. The diagonals are congruent.
   - L. All four angles are right angles.

2. Determine whether there is enough given information to know that each figure is a parallelogram. If so, state the definition or theorem that justifies your conclusion.
   - a. no
   - b. Yes; if the diagonals bisect each other, then the quadrilateral is a parallelogram.
   - c. Yes; if both pairs of opposite sides are ≅, then quadrilateral is ▱.
   - d. no

Helping You Remember

3. A good way to remember a large number of mathematical ideas is to think of them in groups. How can you state the conditions as one group about the *sides* of quadrilaterals that guarantee that the quadrilateral is a parallelogram? Sample answer: both pairs of opposite sides parallel, both pairs of opposite sides congruent, or one pair of opposite sides both parallel and congruent

Enrichment, p. 434

Tests for Parallelograms

By definition, a quadrilateral is a parallelogram *if and only if* both pairs of opposite sides are parallel. What conditions other than both pairs of opposite sides parallel will guarantee that a quadrilateral is a parallelogram? In this activity, several possibilities will be investigated by drawing quadrilaterals to satisfy certain conditions. Remember that any test that seems to work is not guaranteed to work unless it can be formally proven.

Complete.

1. Draw a quadrilateral with one pair of opposite sides congruent. Must it be a parallelogram? no
2. Draw a quadrilateral with both pairs of opposite sides congruent. Must it be a parallelogram? yes

Answers

38. If both pairs of opposite sides are parallel and congruent, then the watchbox is a parallelogram.

422 Chapter 8 Quadrilaterals

FCAT Practice

Standardized Test Practice

45. A parallelogram has vertices at $(-2, 2)$, $(1, -6)$, and $(8, 2)$. Which ordered pair could represent the fourth vertex? **B**

Ⓐ $(5, 6)$ Ⓑ $(11, -6)$ Ⓒ $(14, 3)$ Ⓓ $(8, -8)$

46. **ALGEBRA** Find the distance between $X(5, 7)$ and $Y(-3, -4)$. **C**

Ⓐ $\sqrt{19}$ Ⓑ $3\sqrt{15}$ Ⓒ $\sqrt{185}$ Ⓓ $5\sqrt{29}$

Maintain Your Skills

Mixed Review Use ▱*NQRM* to find each measure or value. *(Lesson 8-2)*

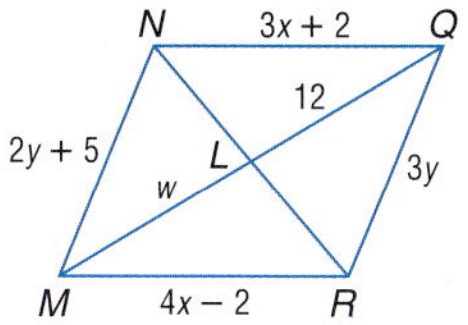

47. w **12**
48. x **4**
49. NQ **14 units**
50. QR **15 units**

The measure of an interior angle of a regular polygon is given. Find the number of sides in each polygon. *(Lesson 8-1)*

51. 135 **8**
52. 144 **10**
53. 168 **30**
54. 162 **20**
55. 175 **72**
56. 175.5 **80**

Find x and y. *(Lesson 7-3)*

57. **45, $12\sqrt{2}$**

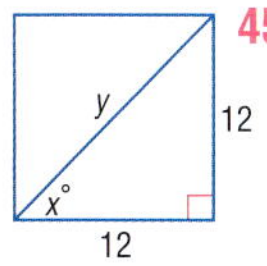

58. **$10\sqrt{3}$, 30**

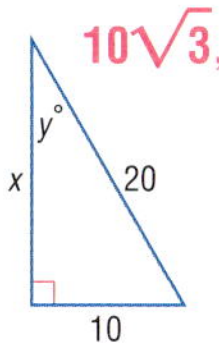

59. **$16\sqrt{3}$, 16**

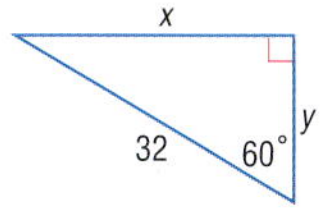

Getting Ready for the Next Lesson

PREREQUISITE SKILL Use slope to determine whether $\overline{AB}$ and $\overline{BC}$ are *perpendicular* or *not perpendicular*. *(To review **slope and perpendicularity**, see Lesson 3-3.)*

60. $A(2, 5)$, $B(6, 3)$, $C(8, 7)$ **$-\frac{1}{2}$, 2; ⊥**
61. $A(-1, 2)$, $B(0, 7)$, $C(4, 1)$ **61. 5, $-\frac{3}{2}$; not ⊥**
62. $A(0, 4)$, $B(5, 7)$, $C(8, 3)$ **$\frac{3}{5}$, $-\frac{4}{3}$; not ⊥**
63. $A(-2, -5)$, $B(1, -3)$, $C(-1, 0)$ **$\frac{2}{3}$, $-\frac{3}{2}$; ⊥**

Practice Quiz 1

Lessons 8-1 through 8-3

1. The measure of an interior angle of a regular polygon is $147\frac{3}{11}$. Find the number of sides in the polygon. *(Lesson 8-1)* **11**

Use ▱*WXYZ* to find each measure. *(Lesson 8-2)*

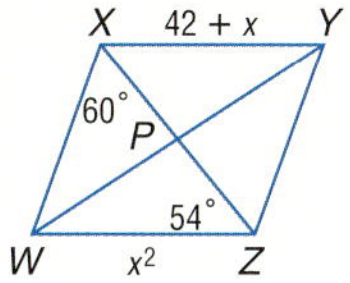

2. $WZ =$ __?__. **36 or 49**
3. $m\angle XYZ =$ __?__. **66**

ALGEBRA **Find x and y so that each quadrilateral is a parallelogram.** *(Lesson 8-3)*

4. **$x = 14$, $y = 31$**

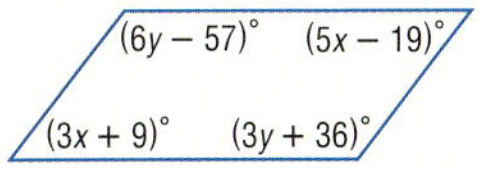

5. **$x = 8$, $y = 6$**

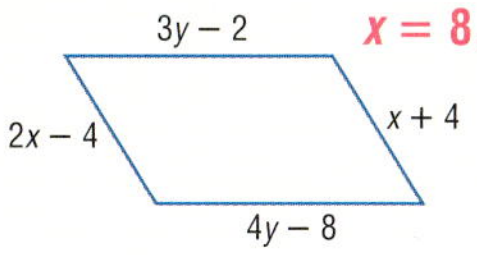

www.geometryonline.com/self_check_quiz/fcat

4 Assess

Open-Ended Assessment

Modeling Ask students to demonstrate how to "build" a parallelogram from straws of various lengths.

Getting Ready for Lesson 8-4

Prerequisite Skill Students will learn about rectangles in Lesson 8-4. They will find the slopes of adjacent sides of a quadrilateral to determine if they are perpendicular. Use Exercises 60–63 to determine your students' familiarity with finding slope.

Assessment Options

Practice Quiz 1 The quiz provides students with a brief review of the concepts and skills in Lessons 8-1 through 8-3. Lesson numbers are given to the right of the exercises or instruction lines so students can review concepts not yet mastered.

Answers

42. This theorem is not true. *ABCD* is a parallelogram with diagonal $\overline{BD}$, $\angle ABD \not\cong \angle CBD$.

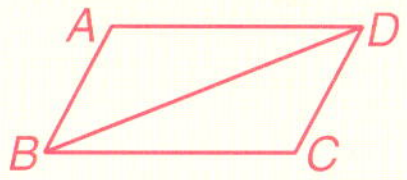

43. Given: *ABCDEF* is a regular hexagon.
Prove: *FDCA* is a parallelogram.

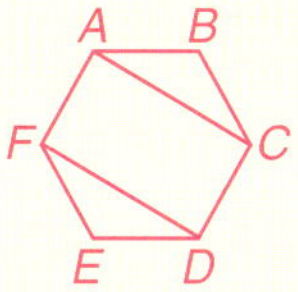

Proof:

Statements (Reasons)

1. *ABCDEF* is a regular hexagon. (Given)
2. $\overline{AB} \cong \overline{DE}$, $\overline{BC} \cong \overline{EF}$, $\angle E \cong \angle B$, $\overline{FA} \cong \overline{CD}$ (Def. regular hexagon)
3. $\triangle ABC \cong \triangle DEF$ (SAS)
4. $\overline{AC} \cong \overline{DF}$ (CPCTC)
5. *FDCA* is a ▱. (If both pairs of opp. sides are ≅, then the quad. is a ▱.)

44. Sample answer: The roofs of some covered bridges are parallelograms. The opposite sides are congruent and parallel. Answers should include the following.
- We need to know the length of the sides, or the measures of the angles formed.
- Sample answer: windows or tiles

8-4 Lesson Notes

1 Focus

5-Minute Check Transparency 8-4 Use as a quiz or review of Lesson 8-3.

Mathematical Background notes are available for this lesson on p. 402D.

How are rectangles used in tennis?

Ask students:

- What kind of lines frequently are used to make up tennis courts? **parallel and perpendicular lines**
- What other sports playing fields include parallel and perpendicular lines? **Sample answers: football, soccer, basketball**

8-4 Rectangles

Sunshine State Standards
MA.A.3.4.3, MA.B.3.4.1, MA.C.1.4.1, MA.C.2.4.1, MA.C.3.4.2

Vocabulary

- rectangle

What You'll Learn

- Recognize and apply properties of rectangles.
- Determine whether parallelograms are rectangles.

How are rectangles used in tennis?

Many sports are played on fields marked by parallel lines. A tennis court has parallel lines at half-court for each player. Parallel lines divide the court for singles and doubles play. The service box is marked by perpendicular lines.

Study Tip

Rectangles and Parallelograms
A rectangle is a parallelogram, but a parallelogram is not necessarily a rectangle.

PROPERTIES OF RECTANGLES A **rectangle** is a quadrilateral with four right angles. Since both pairs of opposite angles are congruent, it follows that it is a special type of parallelogram. Thus, a rectangle has all the properties of a parallelogram. Because the right angles make a rectangle a rigid figure, the diagonals are also congruent.

Theorem 8.13

If a parallelogram is a rectangle, then the diagonals are congruent.

Abbreviation: *If ▱ is rectangle, diag. are ≅.*

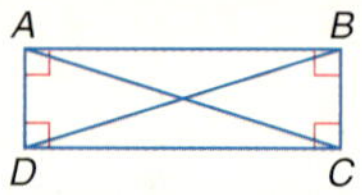

$\overline{AC} \cong \overline{BD}$

You will prove Theorem 8.13 in Exercise 40.

If a quadrilateral is a rectangle, then the following properties are true.

Key Concept — Rectangle

Words A rectangle is a quadrilateral with four right angles.

Properties	Examples
1. Opposite sides are congruent and parallel.	$\overline{AB} \cong \overline{DC}$, $\overline{BC} \cong \overline{AD}$; $\overline{AB} \parallel \overline{DC}$, $\overline{BC} \parallel \overline{AD}$
2. Opposite angles are congruent.	$\angle A \cong \angle C$, $\angle B \cong \angle D$
3. Consecutive angles are supplementary.	$m\angle A + m\angle B = 180$, $m\angle B + m\angle C = 180$, $m\angle C + m\angle D = 180$, $m\angle D + m\angle A = 180$
4. Diagonals are congruent and bisect each other.	$\overline{AC}$ and $\overline{BD}$ bisect each other. $\overline{AC} \cong \overline{BD}$
5. All four angles are right angles.	$m\angle DAB = m\angle BCD = m\angle ABC = m\angle ADC = 90$

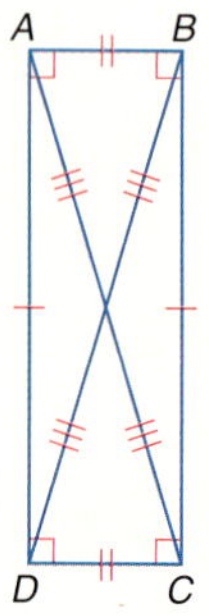

Resource Manager

Workbook and Reproducible Masters

Chapter 8 Resource Masters
- Study Guide and Intervention, pp. 435–436
- Skills Practice, p. 437
- Practice, p. 438
- Reading to Learn Mathematics, p. 439
- Enrichment, p. 440
- Assessment, pp. 473, 475

Graphing Calculator and Computer Masters, pp. 31, 32
Teaching Geometry With Manipulatives Masters, pp. 2, 8, 133

Transparencies

5-Minute Check Transparency 8-4
Real-World Transparency 8
Answer Key Transparencies

Technology

Interactive Chalkboard

Example 1 Diagonals of a Rectangle

ALGEBRA **Quadrilateral $MNOP$ is a rectangle. If $MO = 6x + 14$ and $PN = 9x + 5$, find x.**

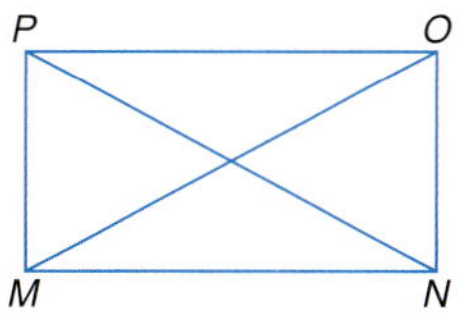

The diagonals of a rectangle are congruent, so $\overline{MO} \cong \overline{PN}$.

$\overline{MO} \cong \overline{PN}$	Diagonals of a rectangle are ≅.
$MO = PN$	Definition of congruent segments
$6x + 14 = 9x + 5$	Substitution
$14 = 3x + 5$	Subtract $6x$ from each side.
$9 = 3x$	Subtract 5 from each side.
$3 = x$	Divide each side by 3.

Rectangles can be constructed using perpendicular lines.

Construction

Rectangle

1. Use a straightedge to draw line ℓ. Label a point P on ℓ. Place the point at P and locate point Q on ℓ. Now construct lines perpendicular to ℓ through P and through Q. Label them m and n.

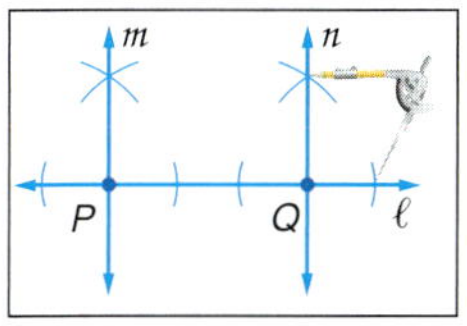

2. Place the compass point at P and mark off a segment on m. Using the same compass setting, place the compass at Q and mark a segment on n. Label these points R and S. Draw $\overline{RS}$.

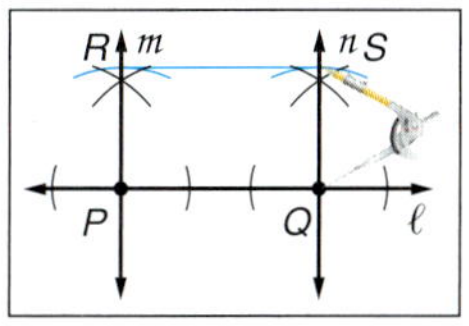

3. Locate the compass setting that represents PR and compare to the setting for QS. The measures should be the same.

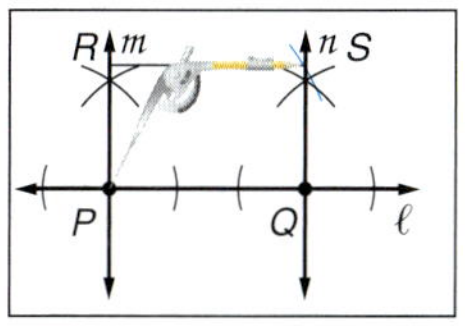

Study Tip

Look Back
To review **constructing perpendicular lines through a point**, see Lesson 3-6.

Example 2 Angles of a Rectangle

ALGEBRA **Quadrilateral $ABCD$ is a rectangle.**

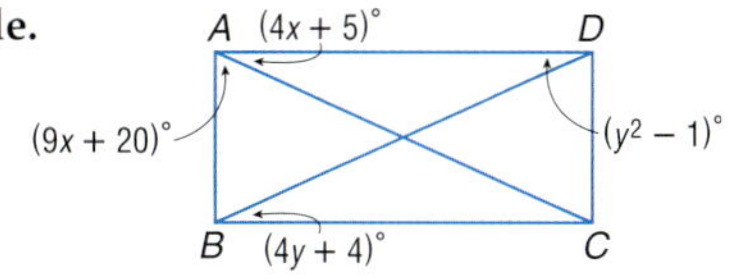

a. Find x.

$\angle DAB$ is a right angle, so $m\angle DAB = 90$.

$m\angle DAC + m\angle BAC = m\angle DAB$	Angle Addition Theorem
$4x + 5 + 9x + 20 = 90$	Substitution
$13x + 25 = 90$	Simplify.
$13x = 65$	Subtract 25 from each side.
$x = 5$	Divide each side by 13.

www.geometryonline.com/extra_examples/fcat

2 Teach

PROPERTIES OF RECTANGLES

In-Class Examples Power Point®

Teaching Tip In Example 1, remind students that all other properties of parallelograms apply to rectangles as well. Since the diagonals of a parallelogram bisect each other, students could also show that $\frac{1}{2}MO = \frac{1}{2}PN$.

1 Quadrilateral $RSTU$ is a rectangle. If $RT = 6x + 4$ and $SU = 7x - 4$, find x.

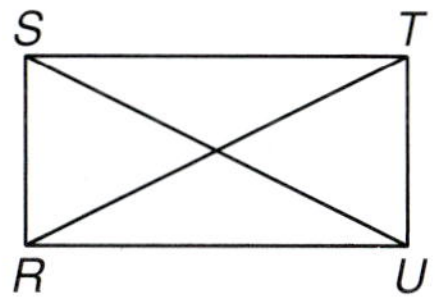

8

2 Quadrilateral $LMNP$ is a rectangle.

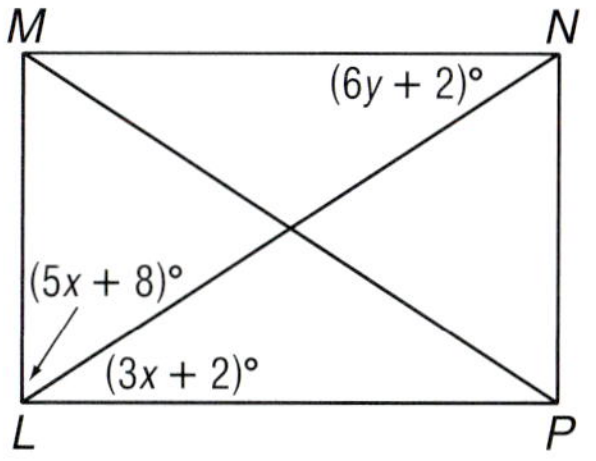

a. Find x. 10

b. Find y. 5

PROVE THAT PARALLELOGRAMS ARE RECTANGLES

Teaching Tip Watch for students who think that any quadrilateral with congruent diagonals is a rectangle. Isosceles trapezoids have congruent diagonals. Remind them that if the diagonals of a *parallelogram* are congruent, then the parallelogram is a rectangle.

In-Class Examples

Power Point®

3 Kyle is building a barn for his horse. He measures the diagonals of the door opening to make sure they bisect each other and they are congruent. How does he know that the measure of each corner is 90?

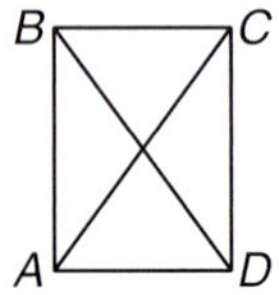

$\overline{AC}$ and $\overline{BD}$ are congruent diagonals. A parallelogram with congruent diagonals is a rectangle.

4 Quadrilateral *ABCD* has vertices $A(-2, 1)$, $B(4, 3)$, $C(5, 0)$, and $D(-1, -2)$. Determine whether *ABCD* is a rectangle using the Slope formula.

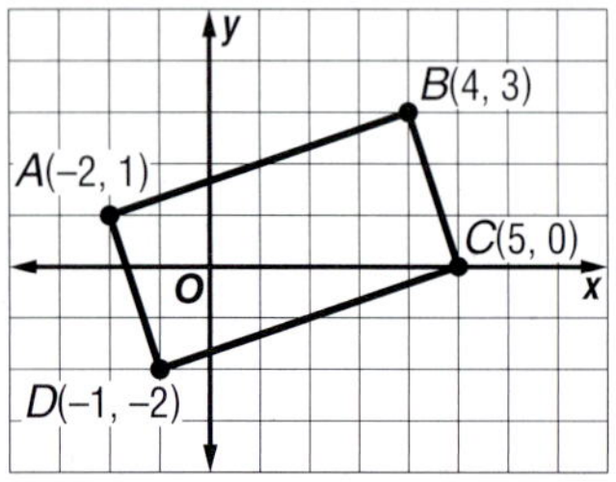

slope of $\overline{AB} = \frac{1}{3}$, slope of $\overline{AD} = -3$, slope of $\overline{BC} = -3$, slope of $\overline{CD} = \frac{1}{3}$. The product of consecutive slopes is -1 so $\overline{AB} \perp \overline{BC}$, $\overline{BC} \perp \overline{CD}$, $\overline{CD} \perp \overline{AD}$, and $\overline{AD} \perp \overline{AB}$. The perpendicular segments create 4 right angles. Therefore *ABCD* is a rectangle.

b. Find *y*.

Since a rectangle is a parallelogram, opposite sides are parallel. So, alternate interior angles are congruent.

$\angle ADB \cong \angle CBD$	Alternate Interior Angles Theorem
$m\angle ADB = m\angle CBD$	Definition of $\cong$ angles
$y^2 - 1 = 4y + 4$	Substitution
$y^2 - 4y - 5 = 0$	Subtract 4*y* and 4 from each side.
$(y - 5)(y + 1) = 0$	Factor.

$y - 5 = 0$ $\quad$ $y + 1 = 0$

$y = 5$ $\quad$ $y = -1$ $\quad$ Disregard $y = -1$ because it yields angle measures of 0.

PROVE THAT PARALLELOGRAMS ARE RECTANGLES

The converse of Theorem 8.13 is also true.

Theorem 8.14

If the diagonals of a parallelogram are congruent, then the parallelogram is a rectangle.

Abbreviation: *If diagonals of ▱ are ≅, ▱ is a rectangle.*

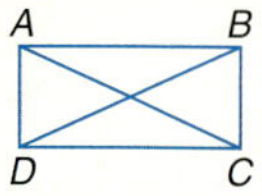

$\overline{AC} \cong \overline{BD}$

You will prove Theorem 8.14 in Exercise 41.

Example 3 Diagonals of a Parallelogram

WINDOWS **Trent is building a tree house for his younger brother. He has measured the window opening to be sure that the opposite sides are congruent. He measures the diagonals to make sure that they are congruent. This is called *squaring* the frame. How does he know that the corners are 90° angles?**

First draw a diagram and label the vertices. We know that $\overline{WX} \cong \overline{ZY}$, $\overline{XY} \cong \overline{WZ}$, and $\overline{WY} \cong \overline{XZ}$.

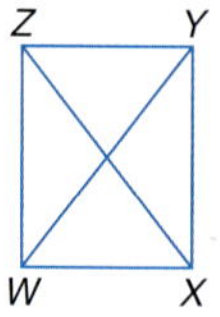

Because $\overline{WX} \cong \overline{ZY}$ and $\overline{XY} \cong \overline{WZ}$, *WXYZ* is a parallelogram.

$\overline{XZ}$ and $\overline{WY}$ are diagonals and they are congruent. A parallelogram with congruent diagonals is a rectangle. So, the corners are 90° angles.

More About. . .

Windows

It is important to square the window frame because over time the opening may have become "out-of-square." If the window is not properly situated in the framed opening, air and moisture can leak through cracks.

Source: www.supersealwindows.com/guide/measurement

Example 4 Rectangle on a Coordinate Plane

COORDINATE GEOMETRY **Quadrilateral *FGHJ* has vertices $F(-4, -1)$, $G(-2, -5)$, $H(4, -2)$, and $J(2, 2)$. Determine whether *FGHJ* is a rectangle.**

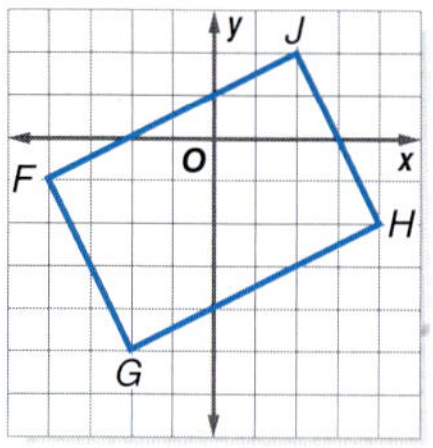

Method 1: Use the Slope Formula, $m = \frac{y_2 - y_1}{x_2 - x_1}$, to see if consecutive sides are perpendicular.

slope of $\overline{FJ} = \frac{2 - (-1)}{2 - (-4)}$ or $\frac{1}{2}$

426 Chapter 8 Quadrilaterals

DAILY INTERVENTION

Differentiated Instruction

Kinesthetic Have your students use string, masking tape, and a tiled floor to mark off congruent diagonals that intersect at their midpoint. Use the string to show that a rectangle is the polygon with those diagonals.

slope of $\overline{GH} = \frac{-2 - (-5)}{4 - (-2)}$ or $\frac{1}{2}$

slope of $\overline{FG} = \frac{-5 - (-1)}{-2 - (-4)}$ or -2

slope of $\overline{JH} = \frac{2 - (-2)}{2 - 4}$ or -2

Because $\overline{FJ} \parallel \overline{GH}$ and $\overline{FG} \parallel \overline{JH}$, quadrilateral $FGHJ$ is a parallelogram.

The product of the slopes of consecutive sides is -1. This means that $\overline{FJ} \perp \overline{FG}$, $\overline{FJ} \perp \overline{JH}$, $\overline{JH} \perp \overline{GH}$, and $\overline{FG} \perp \overline{GH}$. The perpendicular segments create four right angles. Therefore, by definition $FGHJ$ is a rectangle.

Method 2: Use the Distance Formula, $d = \sqrt{(x_2 - x_1)^2 + (y_2 - y_1)^2}$, to determine whether opposite sides are congruent.

First, we must show that quadrilateral $FGHJ$ is a parallelogram.

$$FJ = \sqrt{(-4-2)^2 + (-1-2)^2} = \sqrt{36+9} = \sqrt{45}$$

$$GH = \sqrt{(-2-4)^2 + [-5-(-2)]^2} = \sqrt{36+9} = \sqrt{45}$$

$$FG = \sqrt{[-4-(-2)]^2 + [-1-(-5)]^2} = \sqrt{4+16} = \sqrt{20}$$

$$JH = \sqrt{(2-4)^2 + [2-(-2)]^2} = \sqrt{4+16} = \sqrt{20}$$

Since each pair of opposite sides of the quadrilateral have the same measure, they are congruent. Quadrilateral $FGHJ$ is a parallelogram.

$$FH = \sqrt{(-4-4)^2 + [-1-(-2)]^2} = \sqrt{64+1} = \sqrt{65}$$

$$GJ = \sqrt{(-2-2)^2 + (-5-2)^2} = \sqrt{16+49} = \sqrt{65}$$

The length of each diagonal is $\sqrt{65}$. Since the diagonals are congruent, $FGHJ$ is a rectangle by Theorem 8.14.

Check for Understanding

Concept Check

1. How can you determine whether a parallelogram is a rectangle? **1–2. See margin.**
2. **OPEN ENDED** Draw two congruent right triangles with a common hypotenuse. Do the legs form a rectangle?
3. **FIND THE ERROR** McKenna and Consuelo are defining a rectangle for an assignment.

McKenna	Consuelo
A rectangle is a parallelogram with one right angle.	A rectangle has a pair of parallel opposite sides and a right angle.

Who is correct? Explain. **McKenna; Consuelo's definition is correct if one pair of opposite sides is parallel and congruent.**

3 Practice/Apply

Study Notebook

Have students—

- add the definitions/examples of the vocabulary terms to their Vocabulary Builder worksheets for Chapter 8.
- include the properties of a rectangle.
- include any other item(s) that they find helpful in mastering the skills in this lesson.

DAILY INTERVENTION

FIND THE ERROR Exercise 3 gives students the opportunity for critical thinking. This new definition is valid. Students should recognize that they can prove that one right angle in a parallelogram is sufficient to prove it a rectangle.

About the Exercises…

Organization by Objective

- **Properties of Rectangles:** 10–24, 36–37
- **Prove That Parallelograms Are Rectangles:** 25–35, 38–46

Odd/Even Assignments

Exercises 10–46 are structured so that students practice the same concepts whether they are assigned odd or even problems.

Alert! Exercise 37 requires the Internet or other research materials.

Assignment Guide

Basic: 11–41 odd, 44, 45, 47, 48–63

Average: 11–43 odd, 44, 45, 47, 48–63

Advanced: 10–48 even, 49–60 (optional: 61–63)

Answers

1. If consecutive sides are perpendicular or diagonals are congruent, then the parallelogram is a rectangle.
2. Sample answer: sometimes

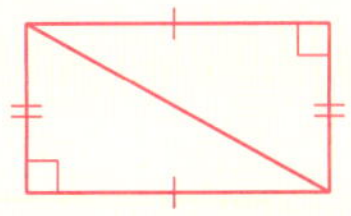

rectangle

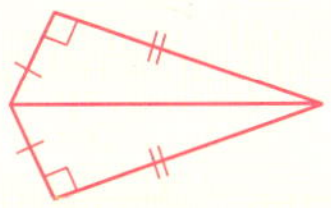

not rectangle

Study Guide and Intervention, p. 435 (shown) and p. 436

Properties of Rectangles A **rectangle** is a quadrilateral with four right angles. Here are the properties of rectangles.

A rectangle has all the properties of a parallelogram.

- Opposite sides are parallel.
- Opposite angles are congruent.
- Opposite sides are congruent.
- Consecutive angles are supplementary.
- The diagonals bisect each other.

Also:

- All four angles are right angles. $\angle UTS$, $\angle TSR$, $\angle SRU$, and $\angle RUT$ are right angles.
- The diagonals are congruent. $\overline{TR} \cong \overline{US}$

Example 1 **In rectangle *RSTU* above, $US = 6x + 3$ and $RT = 7x - 2$. Find *x*.**

The diagonals of a rectangle bisect each other, so $US = RT$.

$6x + 3 = 7x - 2$
$3 = x - 2$
$5 = x$

Example 2 **In rectangle *RSTU* above, $m\angle STR = 8x + 3$ and $m\angle UTR = 16x - 9$. Find $m\angle STR$.**

$\angle UTS$ is a right angle, so $m\angle STR + m\angle UTR = 90$.

$8x + 3 + 16x - 9 = 90$
$24x - 6 = 90$
$24x = 96$
$x = 4$

$m\angle STR = 8x + 3 = 8(4) + 3$ or 35

Exercises

***ABCD* is a rectangle.**

1. If $AE = 36$ and $CE = 2x - 4$, find *x*. 20
2. If $BE = 6y + 2$ and $CE = 4y + 6$, find *y*. 2
3. If $BC = 24$ and $AD = 5y - 1$, find *y*. 5
4. If $m\angle BEA = 62$, find $m\angle BAC$. 59
5. If $m\angle AED = 12x$ and $m\angle BEC = 10x + 20$, find $m\angle AED$. 120
6. If $BD = 8y - 4$ and $AC = 7y + 3$, find *BD*. 52
7. If $m\angle DBC = 10x$ and $m\angle ACB = 4x^2 - 6$, find $m\angle ACB$. 30
8. If $AB = 6y$ and $BC = 8y$, find *BD* in terms of *y*. 10*y*
9. In rectangle *MNOP*, $m\angle 1 = 40$. Find the measure of each numbered angle.
   $m\angle 2 = 40$; $m\angle 3 = 50$; $m\angle 4 = 50$; $m\angle 5 = 80$; $m\angle 6 = 100$

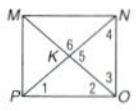

Skills Practice, p. 437 and Practice, p. 438 (shown)

ALGEBRA *RSTU* is a rectangle.

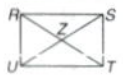

1. If $UZ = x + 21$ and $ZS = 3x - 15$, find *US*. 78
2. If $RZ = 3x + 8$ and $ZS = 6x - 28$, find *UZ*. 44
3. If $RT = 5x + 8$ and $RZ = 4x + 1$, find *ZT*. 9
4. If $m\angle SUT = 3x + 6$ and $m\angle RUS = 5x - 4$, find $m\angle SUT$. 39
5. If $m\angle SRT = x^2 + 9$ and $m\angle UTR = 2x + 44$, find *x*. −5 or 7
6. If $m\angle RSU = x^2 - 1$ and $m\angle TUS = 3x + 9$, find $m\angle RSU$. 24 or 3

***GHJK* is a rectangle. Find each measure if $m\angle 1 = 37$.**

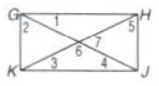

7. $m\angle 2$ 53
8. $m\angle 3$ 37
9. $m\angle 4$ 37
10. $m\angle 5$ 53
11. $m\angle 6$ 106
12. $m\angle 7$ 74

COORDINATE GEOMETRY Determine whether *BGHL* is a rectangle given each set of vertices. Justify your answer.

13. $B(-4, 3)$, $G(-2, 4)$, $H(1, -2)$, $L(-1, -3)$
    Yes; sample answer: Opposite sides are parallel and consecutive sides are perpendicular.
14. $B(-4, 5)$, $G(6, 0)$, $H(3, -6)$, $L(-7, -1)$
    Yes; sample answer: Opposite sides are congruent and diagonals are congruent.
15. $B(0, 5)$, $G(4, 7)$, $H(5, 4)$, $L(1, 2)$
    No; sample answer: Diagonals are not congruent.
16. **LANDSCAPING** Huntington Park officials approved a rectangular plot of land for a Japanese Zen garden. Is it sufficient to know that opposite sides of the garden plot are congruent and parallel to determine that the garden plot is rectangular? Explain.
    No; if you only know that opposite sides are congruent and parallel, the most you can conclude is that the plot is a parallelogram.

Reading to Learn Mathematics, p. 439 ELL

Pre-Activity How are rectangles used in tennis?

Read the introduction to Lesson 8-4 at the top of page 424 in your textbook.

Are the singles court and doubles court similar rectangles? Explain your answer. No; sample answer: All of their angles are congruent, but their sides are not proportional. The doubles court is wider in relation to its length.

Reading the Lesson

1. Determine whether each sentence is *always*, *sometimes*, or *never* true.
   a. If a quadrilateral has four congruent angles, it is a rectangle. always
   b. If consecutive angles of a quadrilateral are supplementary, then the quadrilateral is a rectangle. sometimes
   c. The diagonals of a rectangle bisect each other. always
   d. If the diagonals of a quadrilateral bisect each other, the quadrilateral is a rectangle. sometimes
   e. Consecutive angles of a rectangle are complementary. never
   f. Consecutive angles of a rectangle are congruent. always
   g. If the diagonals of a quadrilateral are congruent, the quadrilateral is a rectangle. sometimes
   h. A diagonal of a rectangle bisects two of its angles. sometimes
   i. A diagonal of a rectangle divides the rectangle into two congruent right triangles. always
   j. If the diagonals of a quadrilateral bisect each other and are congruent, the quadrilateral is a rectangle. always
   k. If a parallelogram has one right angle, it is a rectangle. always
   l. If a parallelogram has four congruent sides, it is a rectangle. sometimes
2. *ABCD* is a rectangle with $AD > AB$.
   Name each of the following in this figure.

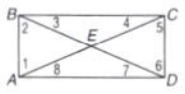

   a. all segments that are congruent to $\overline{BE}$ $\overline{AE}$, $\overline{CE}$, $\overline{DE}$
   b. all angles congruent to $\angle 1$ $\angle 2$, $\angle 5$, $\angle 6$
   c. all angles congruent to $\angle 7$ $\angle 3$, $\angle 4$, $\angle 8$
   d. two pairs of congruent triangles $\triangle AED \cong \triangle BEC$, $\triangle AEB \cong \triangle DEC$, $\triangle BCD \cong \triangle DAB$, $\triangle ABC \cong \triangle CDA$

Helping You Remember

3. It is easier to remember a large number of geometric relationships and theorems if you are able to combine some of them. How can you combine the two theorems about diagonals that you studied in this lesson? Sample answer: A parallelogram is a rectangle if and only if its diagonals are congruent.

Guided Practice

GUIDED PRACTICE KEY	
Exercises	Examples
4–5	1
6–7	2
8	4
9	3

4. **ALGEBRA** *ABCD* is a rectangle. If $AC = 30 - x$ and $BD = 4x - 60$, find *x*. 18

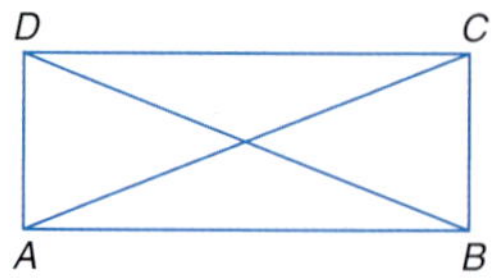

5. **ALGEBRA** *MNQR* is a rectangle. If $NR = 2x + 10$ and $NP = 2x - 30$, find *MP*. 40

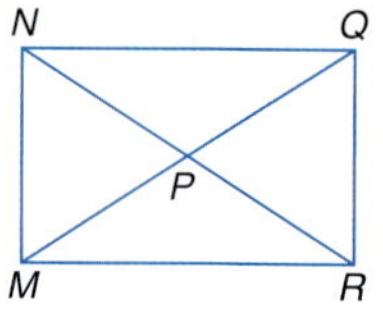

ALGEBRA Quadrilateral *QRST* is a rectangle. Find each measure or value.

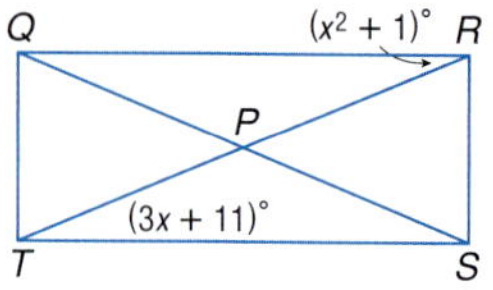

6. *x* 5 or −2
7. $m\angle RPS$ 52 or 10

8. **COORDINATE GEOMETRY** Quadrilateral *EFGH* has vertices $E(-4, -3)$, $F(3, -1)$, $G(2, 3)$, and $H(-5, 1)$. Determine whether *EFGH* is a rectangle.
   No; the lengths of the diagonals are not equal: $EG = \sqrt{72}$ and $HF = \sqrt{68}$.

Application

9. **FRAMING** Mrs. Walker has a rectangular picture that is 12 inches by 48 inches. Because this is not a standard size, a special frame must be built. What can the framer do to guarantee that the frame is a rectangle? Justify your reasoning.
   Make sure that the angles measure 90 or that the diagonals are congruent.

★ indicates increased difficulty

Practice and Apply

Homework Help	
For Exercises	**See Examples**
10–15, 36–37	1
16–24	2
25–26, 35, 38–46	3
27–34	4

Extra Practice
See page 770.

ALGEBRA Quadrilateral *JKMN* is a rectangle.

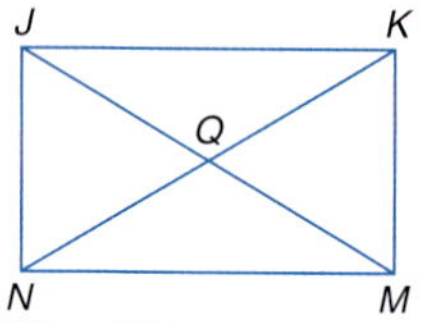

10. If $NQ = 5x - 3$ and $QM = 4x + 6$, find *NK*. 84
11. If $NQ = 2x + 3$ and $QK = 5x - 9$, find *JQ*. 11
12. If $NM = 8x - 14$ and $JK = x^2 + 1$, find *JK*. 10 or 26
13. If $m\angle NJM = 2x - 3$ and $m\angle KJM = x + 5$, find *x*. $29\frac{1}{3}$
14. If $m\angle NKM = x^2 + 4$ and $m\angle KNM = x + 30$, find $m\angle JKN$. 22 or 37
15. If $m\angle JKN = 2x^2 + 2$ and $m\angle NKM = 14x$, find *x*. 4

***WXYZ* is a rectangle. Find each measure if $m\angle 1 = 30$.**

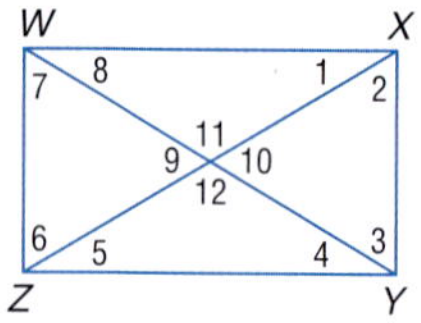

16. $m\angle 1$ 30
17. $m\angle 2$ 60
18. $m\angle 3$ 60
19. $m\angle 4$ 30
20. $m\angle 5$ 30
21. $m\angle 6$ 60
22. $m\angle 7$ 60
23. $m\angle 8$ 30
24. $m\angle 9$ 60

25. **PATIOS** A contractor has been hired to pour a rectangular concrete patio. How can he be sure that the frame in which to pour the concrete is rectangular?
    Measure the opposite sides and the diagonals to make sure they are congruent.

26. **TELEVISION** Television screens are measured on the diagonal. What is the measure of the diagonal of this screen?
    about 42 in.

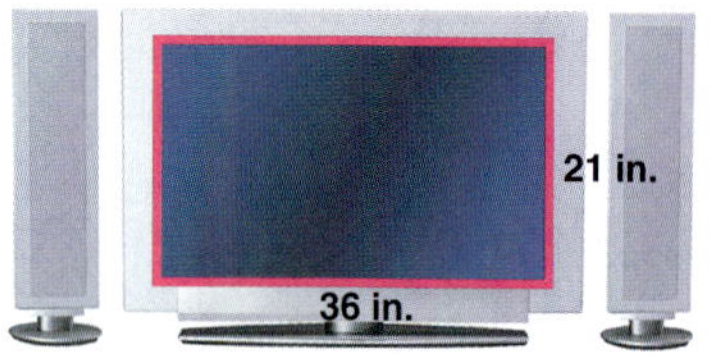

Enrichment, p. 440

Counting Squares and Rectangles

Each puzzle below contains many squares and/or rectangles. Count them carefully. You may want to label each region so you can list all possibilities.

Example **How many rectangles are in the figure at the right?**

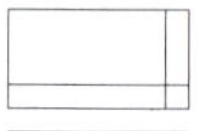

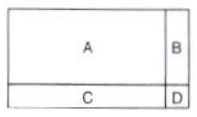

Label each small region with a letter. Then list each rectangle by writing the letters of regions it contains.

A, B, C, D, AB, CD, AC, BD, ABCD

There are 9 rectangles.

...es are in ea...

COORDINATE GEOMETRY Determine whether *DFGH* is a rectangle given each set of vertices. Justify your answer.

27. $D(9, -1), F(9, 5), G(-6, 5), H(-6, 1)$ No; $\overline{DH}$ and $\overline{FG}$ are not parallel.
28. $D(6, 2), F(8, -1), G(10, 6), H(12, 3)$ No; this is not a quadrilateral.
29. $D(-4, -3), F(-5, 8), G(6, 9), H(7, -2)$ Yes; opp. sides are ∥, diag. are ≅.

COORDINATE GEOMETRY The vertices of *WXYZ* are $W(2, 4)$, $X(-2, 0)$, $Y(-1, -7)$, and $Z(9, 3)$.

30. Find *WY* and *XZ*. $WY = \sqrt{130}$; $XZ = \sqrt{130}$
31. Find the coordinates of the midpoints of $\overline{WY}$ and $\overline{XZ}$. $\left(\frac{1}{2}, -\frac{3}{2}\right), \left(\frac{7}{2}, \frac{3}{2}\right)$
32. Is *WXYZ* a rectangle? Explain. No; the midpoints of the diagonals are not the same, so the diagonals do not bisect each other.

COORDINATE GEOMETRY The vertices of parallelogram *ABCD* are $A(-4, -4)$, $B(2, -1)$, $C(0, 3)$, and $D(-6, 0)$.

33. Determine whether *ABCD* is a rectangle. Yes; consec. sides are ⊥.
34. If *ABCD* is a rectangle and *E, F, G,* and *H* are midpoints of its sides, what can you conclude about *EFGH*? It is a parallelogram with all sides congruent.
35. **MINIATURE GOLF** The windmill section of a miniature golf course will be a rectangle 10 feet long and 6 feet wide. Suppose the contractor placed stakes and strings to mark the boundaries with the corners at *A, B, C* and *D*. The contractor measured $\overline{BD}$ and $\overline{AC}$ and found that $AC > BD$. Describe where to move the stakes *L* and *K* to make *ABCD* a rectangle. Explain. Move *L* and *K* until the length of the diagonals is the same.

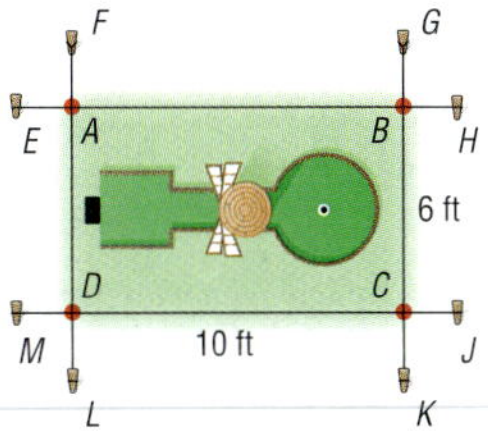

More About. . .

Golden Rectangles

The Parthenon in ancient Greece is an example of how the golden rectangle was applied to architecture. The ratio of the length to the height is the golden ratio.

Source: www.enc.org

GOLDEN RECTANGLES For Exercises 36 and 37, use the following information. Many artists have used *golden rectangles* in their work. In a golden rectangle, the ratio of the length to the width is about 1.618. This ratio is known as the *golden ratio.*

36. A rectangle has dimensions of 19.42 feet and 12.01 feet. Determine if the rectangle is a golden rectangle. Then find the length of the diagonal.
37. **RESEARCH** Use the Internet or other sources to find examples of golden rectangles. See students' work.
38. What are the minimal requirements to justify that a parallelogram is a rectangle? diagonals are congruent or one right angle
39. Draw a counterexample to the statement *If the diagonals are congruent, the quadrilateral is a rectangle.* See margin.

PROOF Write a two-column proof. 40–41. See margin.

40. Theorem 8.13
41. Theorem 8.14

★ 42. **Given:** *PQST* is a rectangle. $\overline{QR} \cong \overline{VT}$
Prove: $\overline{PR} \cong \overline{VS}$

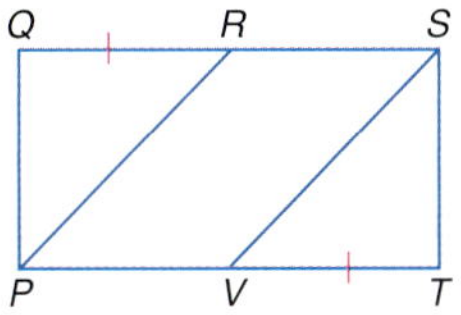

★ 43. **Given:** *DEAC* and *FEAB* are rectangles. $\angle GKH \cong \angle JHK$; $\overline{GJ}$ and $\overline{HK}$ intersect at *L*.
Prove: *GHJK* is a parallelogram.

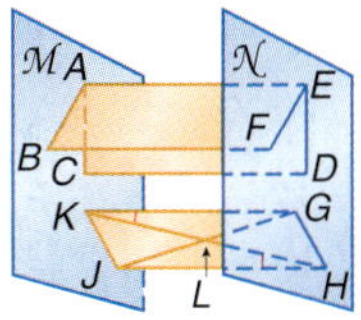

44. **CRITICAL THINKING** Using four of the twelve points as corners, how many rectangles can be drawn? 20

36. Yes, the ratio of sides is 1.617; 22.83 ft.

42–43. See p. 459B.

www.geometryonline.com/self_check_quiz/fcat

40. **Given:** *WXYZ* is a rectangle with diagonals $\overline{WY}$ and $\overline{XZ}$.
Prove: $\overline{WY} \cong \overline{XZ}$

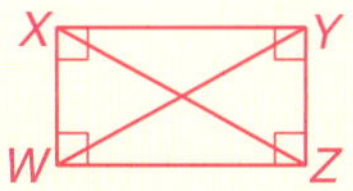

Proof:

1. *WXYZ* is a rectangle with diagonals $\overline{WY}$ and $\overline{XZ}$. (Given)
2. $\overline{WX} \cong \overline{ZY}$ (Opp. sides of ▱ are ≅.)
3. $\overline{WZ} \cong \overline{WZ}$ (Reflexive Property)
4. ∠*XWZ* and ∠*YZW* are right angles. (Def. of rectangle)
5. ∠*XWZ* ≅ ∠*YZW* (All right ∠s are ≅.)
6. △*XWZ* ≅ △*YZW* (SAS)
7. $\overline{WY} \cong \overline{XZ}$ (CPCTC)

41. **Given:** $\overline{WX} \cong \overline{YZ}$, $\overline{XY} \cong \overline{WZ}$, and $\overline{WY} \cong \overline{XZ}$
Prove: *WXYZ* is a rectangle.

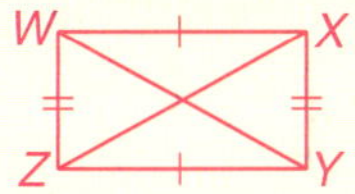

Proof:

1. $\overline{WX} \cong \overline{YZ}$, $\overline{XY} \cong \overline{WZ}$, and $\overline{WY} \cong \overline{XZ}$ (Given)
2. $\overline{WX} \cong \overline{WX}$ (Reflexive Prop.)
3. △*WZX* ≅ △*XYW* (SSS)
4. ∠*ZWX* ≅ ∠*YXW* (CPCTC)
5. $m\angle ZWX = m\angle YXW$ (Def. of ≅)
6. *WXYZ* is a parallelogram. (If both pairs of opp. sides are ≅, then quad. is ▱.)
7. ∠*ZWX* and ∠*YXW* are supplementary. (Cons. ∠s of ▱ are suppl.)
8. $m\angle ZWX + m\angle YXW = 180$ (Def of suppl.)
9. ∠*ZWX* and ∠*YXW* are right angles. (If 2 ∠s are ≅ and suppl., each ∠ is a rt. ∠.)
10. ∠*WZY* and ∠*XYZ* are right angles. (If ▱ has 1 rt. ∠, it has 4 rt. ∠s.)
11. *WXYZ* is a rectangle. (Def. of rectangle)

Answers

39. Sample answer: $\overline{AC} \cong \overline{BD}$ but *ABCD* is not a rectangle.

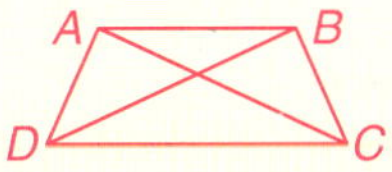

4 Assess

Open-Ended Assessment

Writing Have students explain how to prove that a quadrilateral with congruent diagonals is a rectangle.

Getting Ready for Lesson 8-5

Prerequisite Skill Students will learn about the properties of rhombi and squares in Lesson 8-5. They will use the Distance Formula to find information about rhombi and squares. Use Exercises 61–63 to determine your students' familiarity with the Distance Formula.

Assessment Options

Quiz (Lessons 8-3 and 8-4) is available on p. 473 of the *Chapter 8 Resource Masters*.

Mid-Chapter Test (Lessons 8-1 through 8-4) is available on p. 475 of the *Chapter 8 Resource Masters*.

Answers

48. Sample answer: The tennis court is divided into rectangular sections. The players use the rectangles to establish the playing area. Answers should include the following.
- Not counting overlap, there are 5 rectangles on each side of a tennis court.
- Measure each diagonal to make sure they are the same length and measure each angle to make sure they measure 90.

SPHERICAL GEOMETRY The figure shows a *Saccheri quadrilateral* on a sphere. Note that it has four sides with $\overline{CT} \perp \overline{TR}$, $\overline{AR} \perp \overline{TR}$, and $\overline{CT} \cong \overline{AR}$.

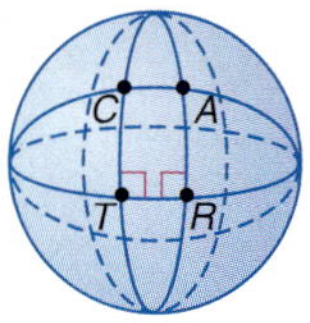

45. Is $\overline{CT}$ parallel to $\overline{AR}$? Explain.
46. How does AC compare to TR? **$AC < TR$**
47. Can a rectangle exist in spherical geometry? Explain.

45. No; there are no parallel lines in spherical geometry.
47. No; the sides are not parallel.

48. **WRITING IN MATH** Answer the question that was posed at the beginning of the lesson. **See margin.**

How are rectangles used in tennis?

Include the following in your answer:
- the number of rectangles on one side of a tennis court, and
- a method to ensure the lines on the court are parallel

FCAT Practice
Standardized Test Practice

49. In the figure, $\overline{AB} \parallel \overline{CE}$. If $DA = 6$, what is DB? **A**

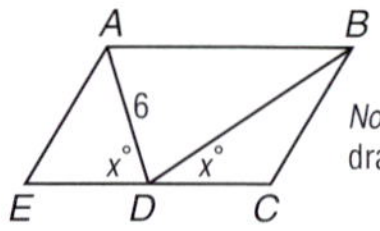

Note: Figure not drawn to scale

Ⓐ 6 Ⓑ 7
Ⓒ 8 Ⓓ 9

50. **ALGEBRA** A rectangular playground is surrounded by an 80-foot long fence. One side of the playground is 10 feet longer than the other. Which of the following equations could be used to find s, the shorter side of the playground? **D**

Ⓐ $10s + s = 80$ Ⓑ $4s + 10 = 80$
Ⓒ $s(s + 10) = 80$ Ⓓ $2(s + 10) + 2s = 80$

Maintain Your Skills

Mixed Review

51. **TEXTILE ARTS** The Navajo people are well known for their skill in weaving. The design at the right, known as the Eye-Dazzler, became popular with Navajo weavers in the 1880s. How many parallelograms, not including rectangles, are in the pattern? *(Lesson 8-3)* **31**

For Exercises 52–57, use ▱$ABCD$. Find each measure or value. *(Lesson 8-2)*

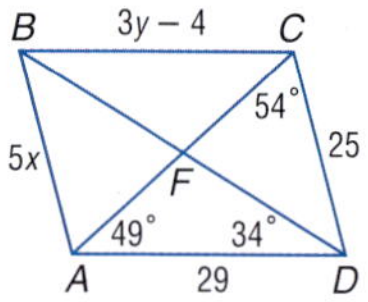

52. $m\angle AFD$ **97**
53. $m\angle CDF$ **43**
54. $m\angle FBC$ **34**
55. $m\angle BCF$ **49**
56. y **11**
57. x **5**

Find the measure of the altitude drawn to the hypotenuse. *(Lesson 7-2)*

58. **$\sqrt{612} \approx 24.7$**

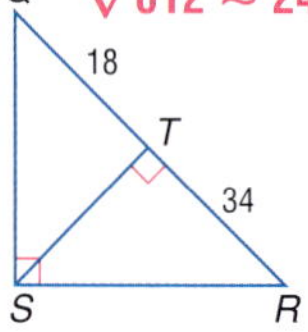

59.

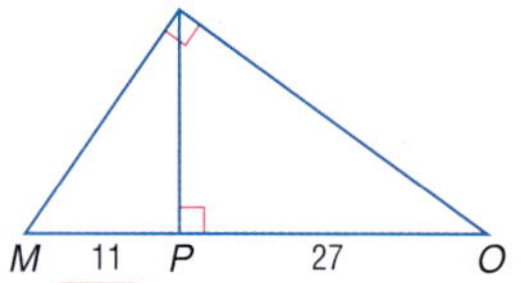

$\sqrt{297} \approx 17.2$

60. **$\sqrt{336} \approx 18.3$**

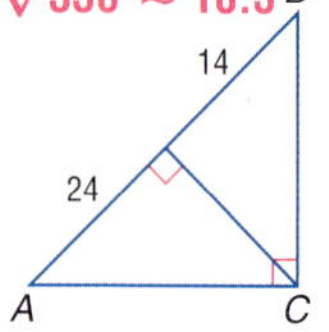

Getting Ready for the Next Lesson

PREREQUISITE SKILL **Find the distance between each pair of points.**
(To review the ***Distance Formula****, see Lesson 1-4.)*

61. $(1, -2), (-3, 1)$ **5**
62. $(-5, 9), (5, 12)$ **$\sqrt{109} \approx 10.4$**
63. $(1, 4), (22, 24)$ **29**

430 Chapter 8 Quadrilaterals

8-5 Rhombi and Squares

Sunshine State Standards
MA.A.3.4.3, MA.B.3.4.1, MA.C.1.4.1, MA.C.2.4.1, MA.C.3.4.2

Vocabulary
- rhombus
- square

What You'll Learn
- Recognize and apply the properties of rhombi.
- Recognize and apply the properties of squares.

How can you ride a bicycle with square wheels?

Most bicycles have round wheels, but Professor Stan Wagon at Macalester College in St. Paul, Minnesota, developed one with square wheels. There are two front wheels so the rider can balance without turning the handlebars. Riding over a specially curved road means there is a smooth ride.

PROPERTIES OF RHOMBI A square is a special type of parallelogram called a rhombus. A **rhombus** is a quadrilateral with all four sides congruent. Since opposite sides are congruent, the rhombus is a parallelogram, and all of the properties of parallelograms can be applied to rhombi. There are three other characteristics of rhombi described in the following theorems.

Key Concept — *Rhombus*

	Theorem	Example
8.15	The diagonals of a rhombus are perpendicular.	$\overline{AC} \perp \overline{BD}$
8.16	If the diagonals of a parallelogram are perpendicular, then the parallelogram is a rhombus. (Converse of Theorem 8.15)	If $\overline{BD} \perp \overline{AC}$, then $\square ABCD$ is a rhombus.
8.17	Each diagonal of a rhombus bisects a pair of opposite angles.	$\angle DAC \cong \angle BAC \cong \angle DCA \cong \angle BCA$ $\angle ABD \cong \angle CBD \cong \angle ADB \cong \angle CDB$

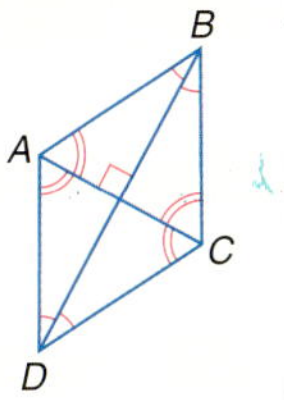

You will prove Theorems 8.16 and 8.17 in Exercises 35 and 36, respectively.

Study Tip

Proof
Since a rhombus has four congruent sides, one diagonal separates the rhombus into two congruent isosceles triangles. Drawing two diagonals separates the rhombus into four congruent right triangles.

Example 1 Proof of Theorem 8.15

Given: $PQRS$ is a rhombus.

Prove: $\overline{PR} \perp \overline{SQ}$

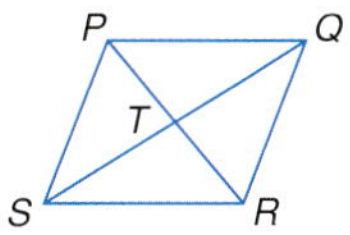

Proof:
By the definition of a rhombus, $\overline{PQ} \cong \overline{QR} \cong \overline{RS} \cong \overline{PS}$. A rhombus is a parallelogram and the diagonals of a parallelogram bisect each other, so $\overline{QS}$ bisects $\overline{PR}$ at T. Thus, $\overline{PT} \cong \overline{RT}$. $\overline{QT} \cong \overline{QT}$ because congruence of segments is reflexive. Thus, $\triangle PQT \cong \triangle RQT$ by SSS. $\angle QTP \cong \angle QTR$ by CPCTC. $\angle QTP$ and $\angle QTR$ also form a linear pair. Two congruent angles that form a linear pair are right angles. $\angle QTP$ is a right angle, so $\overline{PR} \perp \overline{SQ}$ by the definition of perpendicular lines.

8-5 Lesson Notes

1 Focus

5-Minute Check Transparency 8-5 Use as a quiz or review of Lesson 8-4.

Mathematical Background notes are available for this lesson on p. 402D.

How can you ride a bicycle with square wheels?

Ask students:
- Where is the frame of the bicycle attached to the square wheels? **where the diagonals of the square intersect**
- How can Mr. Wagon get a smooth ride with this bicycle? **He rides over a specially curved road.**
- How does the curved road produce a smooth ride? **Accept all reasonable answers.**

Resource Manager

Workbook and Reproducible Masters

Chapter 8 Resource Masters
- Study Guide and Intervention, pp. 441–442
- Skills Practice, p. 443
- Practice, p. 444
- Reading to Learn Mathematics, p. 445
- Enrichment, p. 446

Prerequisite Skills Workbook, pp. 41–42
Teaching Geometry With Manipulatives Masters, pp. 8, 16, 134, 135

Transparencies
5-Minute Check Transparency 8-5
Answer Key Transparencies

Technology
Interactive Chalkboard

2 Teach

PROPERTIES OF RHOMBI

In-Class Examples

Power Point®

Teaching Tip Students may not believe that a rhombus has perpendicular diagonals. Ask groups of students to cut out 4 congruent right triangles. Make sure each group's triangles are unique to that group. Ask them to join them so they share the common right angle vertex. They should form a rhombus. Share the results with the class.

1 **Given:** $BCDE$ is a rhombus, $m\angle AEB = \frac{1}{2}m\angle BCD$, $\overline{AE} \cong \overline{CE}$.

Prove: $\triangle ABE \cong \triangle CBE$

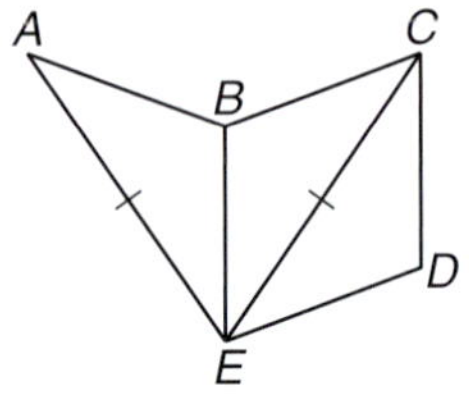

Because opposite angles of a rhombus are congruent and the diagonals of a rhombus bisect the angles, $\angle BEC \cong \angle DEC \cong \angle DCE \cong \angle BCE$ and $m\angle BEC = m\angle DEC = m\angle DCE = m\angle BCE$. Because $\overline{EC}$ bisects $\angle BCD$, $m\angle BCE = \frac{1}{2}m\angle BCD$. By substitution, $m\angle BCE = m\angle AEB$ and thus, $\angle BCE \cong \angle AEB$. $\overline{BE} \cong \overline{BE}$ by the Reflexive Property and it is given that $\overline{AE} \cong \overline{CE}$. Therefore $\triangle ABE \cong \triangle CBE$ by SAS.

2 **MEASURES OF A RHOMBUS** Use rhombus $LMNP$ and the given information to find each value.

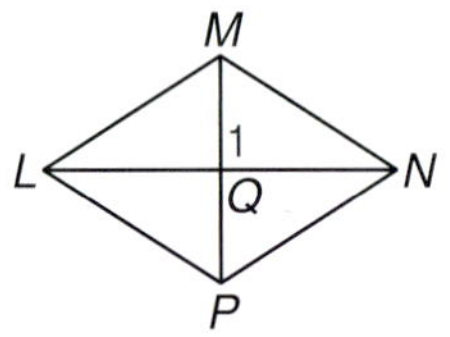

a. Find y if $m\angle 1 = y^2 - 54$. 12 or −12

b. Find $m\angle PNL$ if $m\angle MLP = 64$. 32

Study Tip

Reading Math The plural form of rhombus is *rhombi*, pronounced ROM-bye.

Example 2 Measures of a Rhombus

ALGEBRA **Use rhombus $QRST$ and the given information to find the value of each variable.**

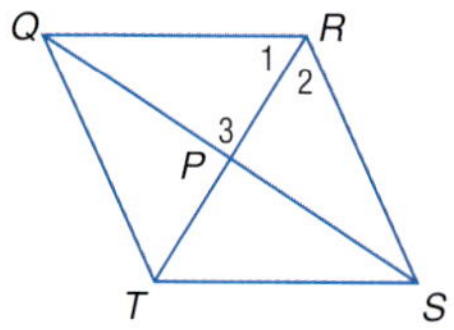

a. Find y if $m\angle 3 = y^2 - 31$.

$m\angle 3 = 90$ The diagonals of a rhombus are perpendicular.

$y^2 - 31 = 90$ Substitution

$y^2 = 121$ Add 31 to each side.

$y = \pm 11$ Take the square root of each side.

The value of y can be 11 or −11.

b. Find $m\angle TQS$ if $m\angle RST = 56$.

$m\angle TQR = m\angle RST$ Opposite angles are congruent.

$m\angle TQR = 56$ Substitution

The diagonals of a rhombus bisect the angles. So, $m\angle TQS$ is $\frac{1}{2}(56)$ or 28.

PROPERTIES OF SQUARES If a quadrilateral is both a rhombus and a rectangle, then it is a **square**. All of the properties of parallelograms and rectangles can be applied to squares.

Example 3 Squares

COORDINATE GEOMETRY **Determine whether parallelogram $ABCD$ is a *rhombus*, a *rectangle*, or a *square*. List all that apply. Explain.**

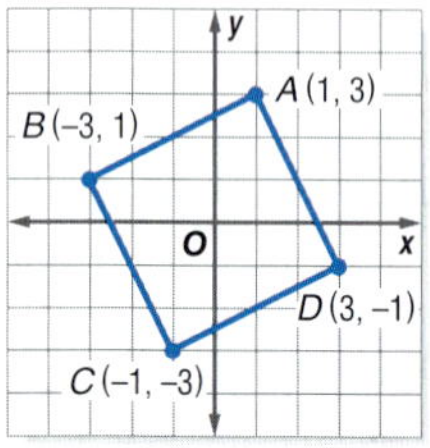

Explore Plot the vertices on a coordinate plane.

Plan If the diagonals are perpendicular, then $ABCD$ is either a rhombus or a square. The diagonals of a rectangle are congruent. If the diagonals are congruent and perpendicular, then $ABCD$ is a square.

Solve Use the Distance Formula to compare the lengths of the diagonals.

$$DB = \sqrt{[3-(-3)]^2 + (-1-1)^2} = \sqrt{36+4} = \sqrt{40}$$

$$AC = \sqrt{(1+1)^2 + (3+3)^2} = \sqrt{4+36} = \sqrt{40}$$

Use slope to determine whether the diagonals are perpendicular.

slope of $\overline{DB} = \frac{1-(-1)}{-3-3}$ or $-\frac{1}{3}$ slope of $\overline{AC} = \frac{-3-3}{-1-1}$ or 3

Since the slope of $\overline{AC}$ is the negative reciprocal of the slope of $\overline{DB}$, the diagonals are perpendicular. The lengths of $\overline{DB}$ and $\overline{AC}$ are the same so the diagonals are congruent. $ABCD$ is a rhombus, a rectangle, and a square.

Examine You can verify that $ABCD$ is a square by finding the measure and slope of each side. All four sides are congruent and consecutive sides are perpendicular.

432 Chapter 8 Quadrilaterals

Construction

Rhombus

1. Draw any segment $\overline{AD}$. Place the compass point at A, open to the width of AD, and draw an arc above $\overline{AD}$.

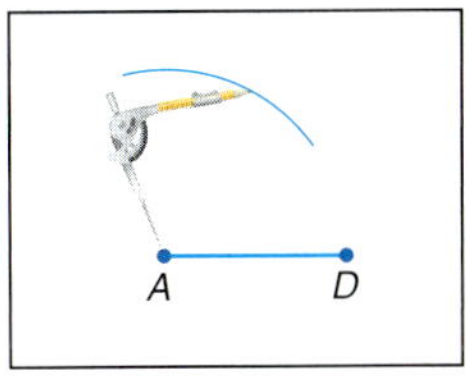

2. Label any point on the arc as B. Using the same setting, place the compass at B, and draw an arc to the right of B.

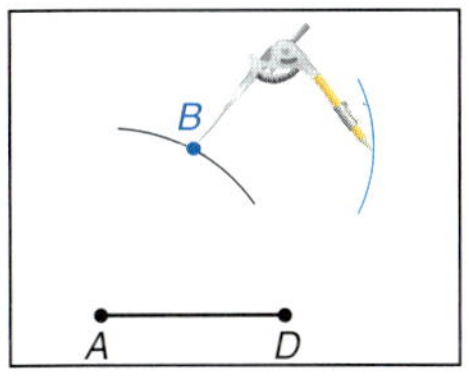

3. Then place the compass at point D, and draw an arc to intersect the arc drawn from point B. Label the point of intersection C.

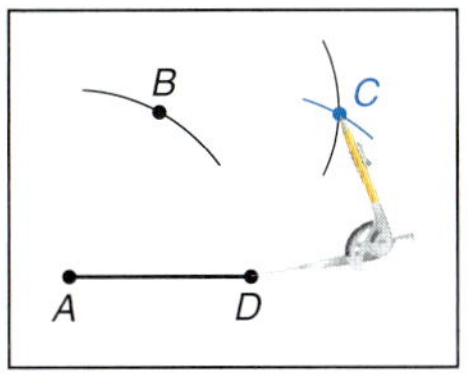

4. Use a straightedge to draw $\overline{AB}$, $\overline{BC}$, and $\overline{CD}$.

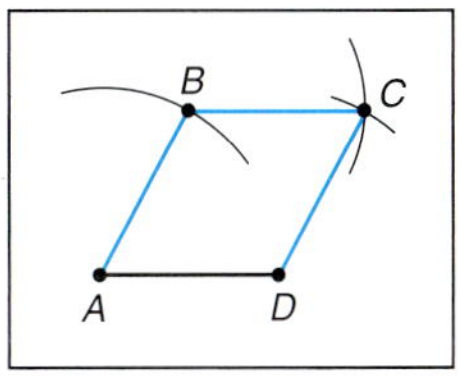

Conclusion: Since all of the sides are congruent, quadrilateral $ABCD$ is a rhombus.

PROPERTIES OF SQUARES

In-Class Examples

Power Point®

3 Determine whether parallelogram $ABCD$ is a *rhombus*, a *rectangle*, or a *square* for $A(-2, -1)$, $B(-1, 3)$, $C(3, 2)$, and $D(2, -2)$. List all that apply.

$AC = \sqrt{34}$; $BD = \sqrt{34}$; slope of $\overline{AC} = \frac{3}{5}$; slope of $\overline{BD} = -\frac{5}{3}$. Since the slope of $\overline{AC}$ is the negative reciprocal of the slope of $\overline{BD}$, the diagonals are perpendicular. The lengths of $\overline{AC}$ and $\overline{BD}$ are the same. $ABCD$ is a rhombus, a rectangle, and a square.

4 A square table has four legs that are 2 feet apart. The table is placed over an umbrella stand so that the hole in the center of the table lines up with the hole in the stand. How far away from a leg is the center of the hole? **about 1.4 ft**

Example 4 Diagonals of a Square

BASEBALL The infield of a baseball diamond is a square, as shown at the right. Is the pitcher's mound located in the center of the infield? Explain.

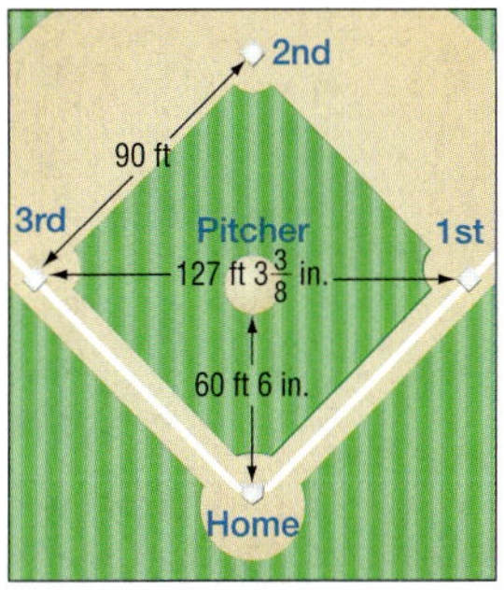

Since a square is a parallelogram, the diagonals bisect each other. Since a square is a rhombus, the diagonals are congruent. Therefore, the distance from first base to third base is equal to the distance between home plate and second base. Thus, the distance from home plate to the center of the infield is 127 feet $3\frac{3}{8}$ inches divided by 2 or 63 feet $7\frac{11}{16}$ inches. This distance is longer than the distance from home plate to the pitcher's mound so the pitcher's mound is not located in the center of the field. It is about 3 feet closer to home.

If a quadrilateral is a rhombus or a square, then the following properties are true.

Study Tip

Square and Rhombus

A square is a rhombus, but a rhombus is not necessarily a square.

Concept Summary — Properties of Rhombi and Squares

Rhombi

1. A rhombus has all the properties of a parallelogram.
2. All sides are congruent.
3. Diagonals are perpendicular.
4. Diagonals bisect the angles of the rhombus.

Squares

1. A square has all the properties of a parallelogram.
2. A square has all the properties of a rectangle.
3. A square has all the properties of a rhombus.

www.geometryonline.com/extra_examples/fcat

DAILY INTERVENTION

Differentiated Instruction

Auditory/Musical Ask students to name the similarities and the differences between rhombi, rectangles, and squares.

Lesson 8-5 Rhombi and Squares 433

3 Practice/Apply

Study Notebook

Have students—

- add the definitions/examples of the vocabulary terms to their Vocabulary Builder worksheets for Chapter 8.
- include a list of properties of rhombi and squares.
- include any other item(s) that they find helpful in mastering the skills in this lesson.

About the Exercises...

Organization by Objective

- **Properties of Rhombi:** 12–19, 37–42
- **Properties of Squares:** 20–36

Odd/Even Assignments

Exercises 12–43 are structured so that students practice the same concepts whether they are assigned odd or even problems.

Assignment Guide

Basic: 13–33 odd, 39, 41, 43–67

Average: 13–43 odd, 44–67

Advanced: 12–44 even, 45–63 (optional: 64–67)

Answers

1. Sample answer:

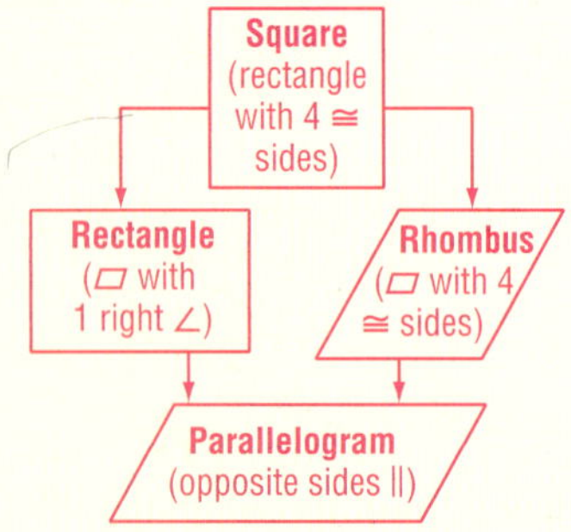

2. Sample answer:

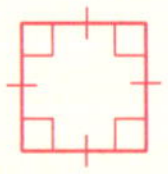

Check for Understanding

Concept Check

1. **Draw a diagram** to demonstrate the relationship among parallelograms, rectangles, rhombi, and squares. **1–2. See margin.**

2. **OPEN ENDED** Draw a quadrilateral that has the characteristics of a rectangle, a rhombus, and a square.

3. **Explain** the difference between a square and a rectangle. **A square is a rectangle with all sides congruent.**

Guided Practice

GUIDED PRACTICE KEY	
Exercises	Examples
4–7	2
8–9	3
10	1
11	4

ALGEBRA **In rhombus *ABCD*, $AB = 2x + 3$ and $BC = 5x$.**

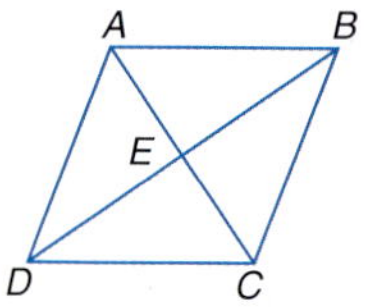

4. Find x. **1**
5. Find AD. **5**
6. Find $m\angle AEB$. **90**
7. Find $m\angle BCD$ if $m\angle ABC = 83.2$. **96.8**

COORDINATE GEOMETRY **Given each set of vertices, determine whether ▱*MNPQ* is a *rhombus*, a *rectangle*, or a *square*. List all that apply. Explain your reasoning.**

8. $M(0, 3)$, $N(-3, 0)$, $P(0, -3)$, $Q(3, 0)$ **8. rectangle, rhombus, square; consecutive sides are perpendicular; all sides are congruent.**
9. $M(-4, 0)$, $N(-3, 3)$, $P(2, 2)$, $Q(1, -1)$ **None; the diagonals are not congruent or perpendicular.**

10. **PROOF** Write a two-column proof. **See p. 459B.**
Given: $\triangle KGH$, $\triangle HJK$, $\triangle GHJ$, and $\triangle JKG$ are isosceles.
Prove: $GHJK$ is a rhombus.

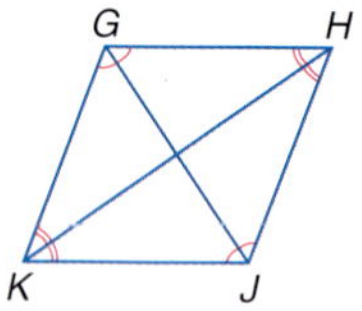

Application

11. **REMODELING** The Steiner family is remodeling their kitchen. Each side of the floor measures 10 feet. What other measurements should be made to determine whether the floor is a square? **If the measure of each angle is 90 or if the diagonals are congruent, then the floor is a square.**

★ indicates increased difficulty

Practice and Apply

Homework Help	
For Exercises	See Examples
12–19	2
20–23	3
24–36	4
37–42	1

Extra Practice See page 770.

In rhombus *ABCD*, $m\angle DAB = 2m\angle ADC$ and $CB = 6$.

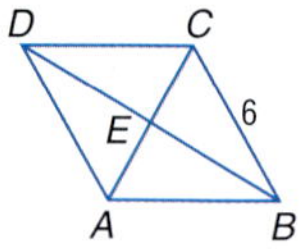

12. Find $m\angle ACD$. **60**
13. Find $m\angle DAB$. **120**
14. Find DA. **6**
15. Find $m\angle ADB$. **30**

ALGEBRA **Use rhombus *XYZW* with $m\angle WYZ = 53$, $VW = 3$, $XV = 2a - 2$, and $ZV = \frac{5a + 1}{4}$.**

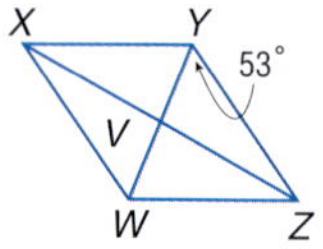

16. Find $m\angle YZV$. **37**
17. Find $m\angle XYW$. **53**
18. Find XZ. **8**
19. Find XW. **5**

COORDINATE GEOMETRY **Given each set of vertices, determine whether ▱*EFGH* is a *rhombus*, a *rectangle*, or a *square*. List all that apply. Explain your reasoning.**

20. $E(1, 10)$, $F(-4, 0)$, $G(7, 2)$, $H(12, 12)$ **Rhombus; the diagonals are perpendicular.**
21. $E(-7, 3)$, $F(-2, 3)$, $G(1, 7)$, $H(-4, 7)$ **Rhombus; the diagonals are perpendicular.**
22. $E(1, 5)$, $F(6, 5)$, $G(6, 10)$, $H(1, 10)$ **22. Square, rectangle, rhombus; all sides are congruent and perpendicular.**
23. $E(-2, -1)$, $F(-4, 3)$, $G(1, 5)$, $H(3, 1)$ **None; the diagonals are not congruent or perpendicular.**

DAILY INTERVENTION

Unlocking Misconceptions

A common error is to believe that all rectangles are squares and that all squares are rectangles. It is true that all squares are rectangles, but the converse is not true. Ask students to verify this by giving several counterexamples.

CONSTRUCTION Construct each figure using a compass and ruler.

24. a square with one side 3 centimeters long **24–25. See margin.**

25. a square with a diagonal 5 centimeters long

Use the Venn diagram to determine whether each statement is *always*, *sometimes*, or *never* true.

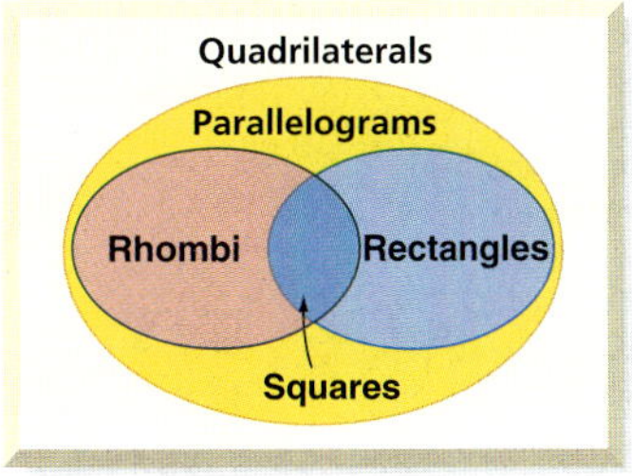

26. A parallelogram is a square. **sometimes**

27. A square is a rhombus. **always**

28. A rectangle is a parallelogram. **always**

29. A rhombus is a rectangle. **sometimes**

30. A rhombus is a square. **sometimes**

31. A square is a rectangle. **always**

More About. . .

Design

The plant stand is constructed from painted wood and metal. The overall dimensions are $36\frac{1}{2}$ inches tall by $15\frac{3}{4}$ inches wide.

Source: www.metmuseum.org

32. **DESIGN** Otto Prutscher designed the plant stand at the left in 1903. The base is a square, and the base of each of the five boxes is also a square. Suppose each smaller box is one half as wide as the base. Use the information at the left to find the dimensions of the base of one of the smaller boxes. **$7\frac{7}{8}$ in. by $7\frac{7}{8}$ in.**

33. **PERIMETER** The diagonals of a rhombus are 12 centimeters and 16 centimeters long. Find the perimeter of the rhombus. **40 cm**

34. **ART** This piece of art is Dorthea Rockburne's *Egyptian Painting: Scribe*. The diagram shows three of the shapes shown in the piece. Use a ruler or a protractor to determine which type of quadrilateral is represented by each figure. ***ABCD* is a rhombus; *EFGH* and *JKLM* are congruent squares.**

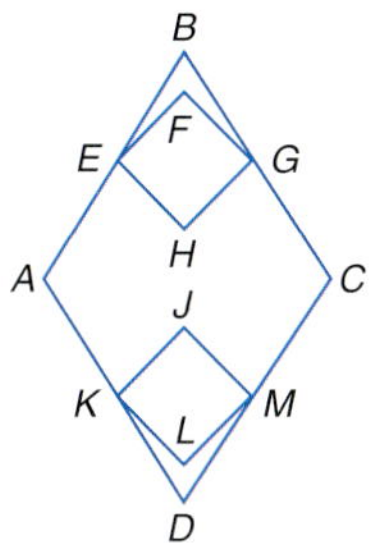

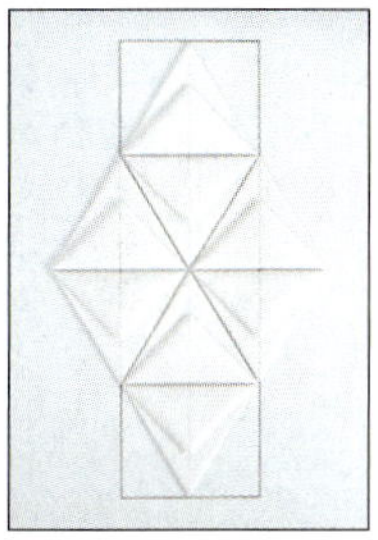

PROOF **Write a paragraph proof for each theorem.** **35–36. See p. 459C.**

★ 35. Theorem 8.16

★ 36. Theorem 8.17

SQUASH **For Exercises 37 and 38, use the diagram of the court for squash, a game similar to racquetball and tennis.**

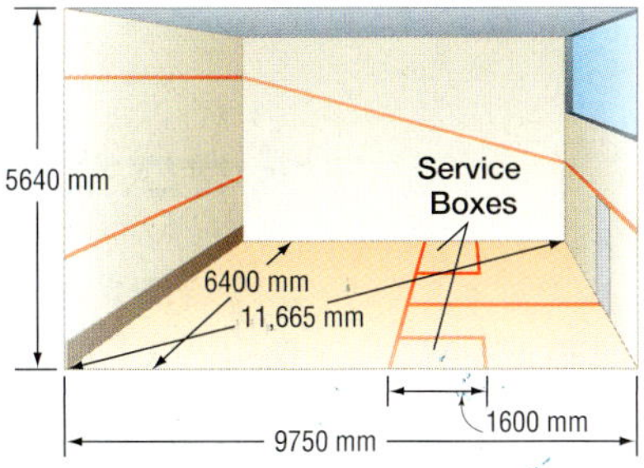

★ 37. The diagram labels the diagonal as 11,665 millimeters. Is this correct? Explain. **No; it is about 11,662.9 mm.**

★ 38. The service boxes are squares. Find the length of the diagonal. **about 2263 mm or 2.263 m**

Answers

24. Sample answer:

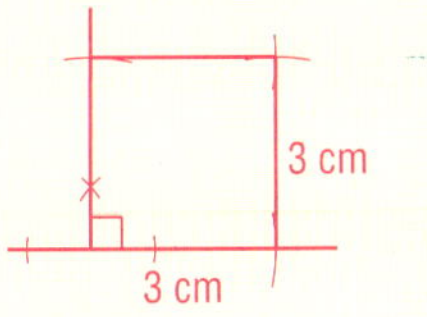

25. Sample answer:

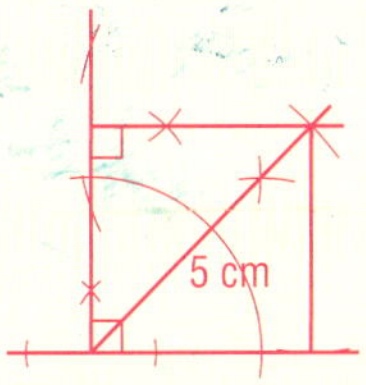

Enrichment, p. 446

Creating Pythagorean Puzzles

By drawing two squares and cutting them in a certain way, you can make a puzzle that demonstrates the Pythagorean Theorem. A sample puzzle is shown. You can create your own puzzle by following the instructions below.

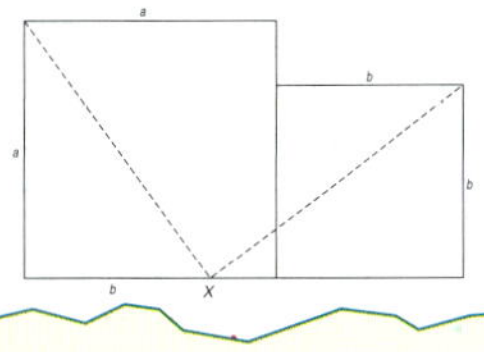

Study Guide and Intervention, p. 441 (shown) and p. 442

Properties of Rhombi A **rhombus** is a quadrilateral with four congruent sides. Opposite sides are congruent, so a rhombus is also a parallelogram and has all of the properties of a parallelogram. Rhombi also have the following properties.

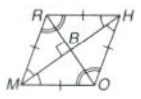

The diagonals are perpendicular.	$\overline{MH} \perp \overline{RO}$
Each diagonal bisects a pair of opposite angles.	$\overline{MH}$ bisects $\angle RMO$ and $\angle RHO$. $\overline{RO}$ bisects $\angle MRH$ and $\angle MOH$.
If the diagonals of a parallelogram are perpendicular, then the figure is a rhombus.	If *RHOM* is a parallelogram and $\overline{RO} \perp \overline{MH}$, then *RHOM* is a rhombus.

Example **In rhombus *ABCD*, $m\angle BAC = 32$. Find the measure of each numbered angle.**

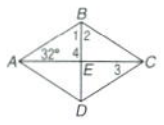

ABCD is a rhombus, so the diagonals are perpendicular and $\triangle ABE$ is a right triangle. Thus $m\angle 4 = 90$ and $m\angle 1 = 90 - 32$ or 58. The diagonals in a rhombus bisect the vertex angles, so $m\angle 1 = m\angle 2$. Thus, $m\angle 2 = 58$.

A rhombus is a parallelogram, so the opposite sides are parallel. $\angle BAC$ and $\angle 3$ are alternate interior angles for parallel lines, so $m\angle 3 = 32$.

Exercises

***ABCD* is a rhombus.**

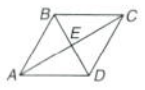

1. If $m\angle ABD = 60$, find $m\angle BDC$. **60**
2. If $AE = 8$, find AC. **16**
3. If $AB = 26$ and $BD = 20$, find AE. **24**
4. Find $m\angle CEB$. **90**
5. If $m\angle CBD = 58$, find $m\angle ACB$. **32**
6. If $AE = 3x - 1$ and $AC = 16$, find x. **3**
7. If $m\angle CDB = 6y$ and $m\angle ACB = 2y + 10$, find y. **10**
8. If $AD = 2x + 4$ and $CD = 4x - 4$, find x. **4**
9. a. What is the midpoint of $\overline{FH}$? **(5, 5)**
   b. What is the midpoint of $\overline{GJ}$? **(5, 5)**
   c. What kind of figure is *FGHJ*? Explain. ***FGHJ* is a parallelogram because the diagonals bisect each other.**
   d. What is the slope of $\overline{FH}$? **$\frac{1}{2}$**
   e. What is the slope of $\overline{GJ}$? **-2**
   f. Based on parts **c**, **d**, and **e**, what kind of figure is *FGHJ*? Explain. ***FGHJ* is a parallelogram with perpendicular diagonals, so it is a rhombus.**

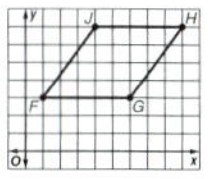

Skills Practice, p. 443 and Practice, p. 444 (shown)

Use rhombus *PRYZ* with $RK = 4y + 1$, $ZK = 7y - 14$, $PK = 3x - 1$, and $YK = 2x + 6$.

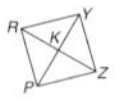

1. Find *PY*. **40**
2. Find *RZ*. **42**
3. Find *RY*. **29**
4. Find $m\angle YKZ$. **90**

Use rhombus *MNPQ* with $PQ = 3\sqrt{2}$, $PA = 4x - 1$, and $AM = 9x - 6$.

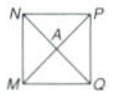

5. Find *AQ*. **3**
6. Find $m\angle APQ$. **45**
7. Find $m\angle MNP$. **90**
8. Find *PM*. **6**

COORDINATE GEOMETRY **Given each set of vertices, determine whether ▱*BEFG* is a *rhombus*, a *rectangle*, or a *square*. List all that apply. Explain your reasoning.**

9. $B(-9, 1)$, $E(2, 3)$, $F(12, -2)$, $G(1, -4)$
   Rhombus; all sides are congruent and the diagonals are perpendicular, but not congruent.

10. $B(1, 3)$, $E(7, -3)$, $F(1, -9)$, $G(-5, -3)$
   Rhombus, rectangle, square; all sides are congruent and the diagonals are perpendicular and congruent.

11. $B(-4, -5)$, $E(1, -5)$, $F(-7, -1)$, $G(-2, -1)$
   None; two of the opposite sides are not congruent.

12. **TESSELATIONS** The figure is an example of a tessellation. Use a ruler or protractor to measure the shapes and then name the quadrilaterals used to form the figure.
   The figure consists of 6 congruent rhombi.

Reading to Learn Mathematics, p. 445 ELL

Pre-Activity **How can you ride a bicycle with square wheels?**

Read the introduction to Lesson 8-5 at the top of page 431 in your textbook.

If you draw a diagonal on the surface of one of the square wheels shown in the picture in your textbook, how can you describe the two triangles that are formed? **Sample answer: two congruent isosceles right triangles**

Reading the Lesson

1. Sketch each of the following. **Sample answers are given.**
   a. a quadrilateral with perpendicular diagonals that is not a rhombus

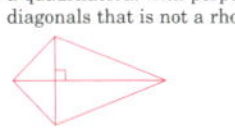

   b. a quadrilateral with congruent diagonals that is not a rectangle

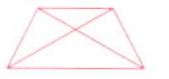

   c. a quadrilateral whose diagonals are perpendicular and bisect each other, but are not congruent

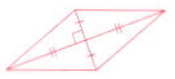

2. List all of the following special quadrilaterals that have each listed property: *parallelogram, rectangle, rhombus, square*.
   a. The diagonals are congruent. **rectangle, square**
   b. Opposite sides are congruent. **parallelogram, rectangle, rhombus, square**
   c. The diagonals are perpendicular. **rhombus, square**
   d. Consecutive angles are supplementary. **parallelogram, rectangle, rhombus, square**
   e. The quadrilateral is equilateral. **rhombus, square**
   f. The quadrilateral is equiangular. **rectangle, square**
   g. The diagonals are perpendicular and congruent. **square**
   h. A pair of opposite sides is both parallel and congruent. **parallelogram, rectangle, rhombus, square**
3. What is the common name for a regular quadrilateral? Explain your answer. **Square; sample answer: All four sides of a square are congruent and all four angles are congruent.**

Helping You Remember

4. A good way to remember something is to explain it to someone else. Suppose that your classmate Luis is having trouble remembering which of the properties he has learned in this chapter apply to squares. How can you help him? **Sample answer: A square is a parallelogram that is both a rectangle and a rhombus, so all properties of parallelograms, rectangles, and rhombi apply to squares.**

More About. . .

Flags

The state of Ohio has the only state flag in the United States that is not rectangular.

Source: *World Almanac*

39. **FLAGS** Study the flags shown below. Use a ruler and protractor to determine if any of the flags contain parallelograms, rectangles, rhombi, or squares. **See margin.**

Denmark

St. Vincent and The Grenadines

Trinidad and Tobago

PROOF Write a two-column proof. **40–43. See margin.**

40. **Given:** $\triangle WZY \cong \triangle WXY$, $\triangle WZY$ and $\triangle XYZ$ are isosceles.
Prove: $WXYZ$ is a rhombus.

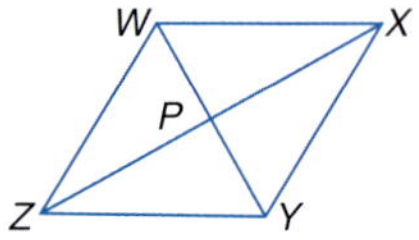

41. **Given:** $\triangle TPX \cong \triangle QPX \cong \triangle QRX \cong \triangle TRX$
Prove: $TPQR$ is a rhombus.

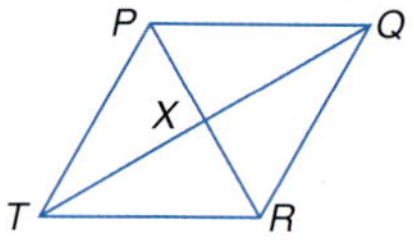

42. **Given:** $\triangle LGK \cong \triangle MJK$
$GHJK$ is a parallelogram.
Prove: $GHJK$ is a rhombus.

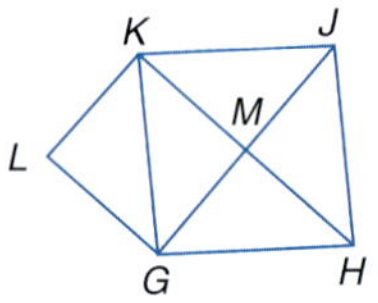

43. **Given:** $QRST$ and $QRTV$ are rhombi.
Prove: $\triangle QRT$ is equilateral.

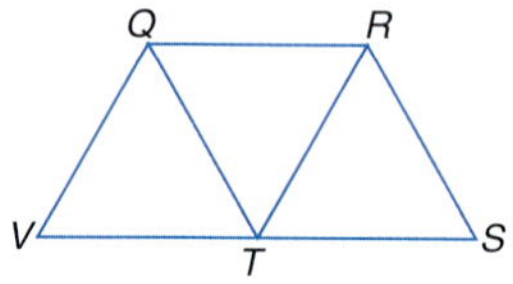

44. **CRITICAL THINKING** The pattern at the right is a series of rhombi that continue to form a hexagon that increases in size. Copy and complete the table.

Hexagon	Number of rhombi
1	3
2	12
3	27
4	48
5	**75**
6	**108**
x	$3x^2$

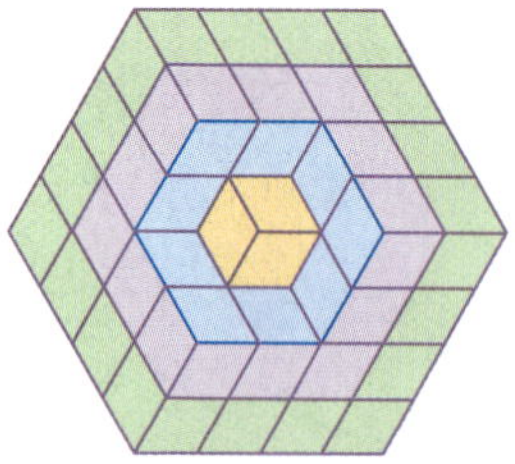

45. **WRITING IN MATH** Answer the question that was posed at the beginning of the lesson. **See margin.**

How can you ride a bicycle with square wheels?

Include the following in your answer:

- difference between squares and rhombi, and
- how nonsquare rhombus-shaped wheels would work with the curved road.

Answers

39. The flag of Denmark contains four red rectangles. The flag of St. Vincent and the Grenadines contains a blue rectangle, a green rectangle, a yellow rectangle, a blue and yellow rectangle, a yellow and green rectangle, and three green rhombi. The flag of Trinidad and Tobago contains two white parallelograms and one black parallelogram.

40. Given: $\triangle WZY \cong \triangle WXY$
$\triangle WZY$ and $\triangle XYZ$ are isosceles.
Prove: $WXYZ$ is a rhombus.

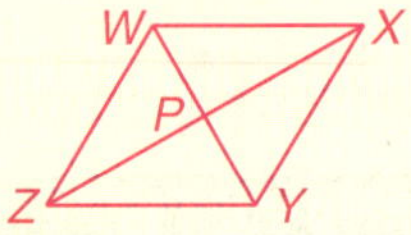

Proof:

Statements (Reasons)

1. $\triangle WZY \cong \triangle WXY$; $\triangle WZY$ and $\triangle XYZ$ are isosceles. (Given)
2. $\overline{WZ} \cong \overline{WX}$, $\overline{ZY} \cong \overline{XY}$ (CPCTC)
3. $\overline{WZ} \cong \overline{ZY}$, $\overline{WX} \cong \overline{XY}$ (Def. of isosceles $\triangle$)
4. $\overline{WZ} \cong \overline{WX} \cong \overline{ZY} \cong \overline{XY}$ (Substitution Property)
5. $WXYZ$ is a rhombus. (Def. of rhombus)

41. Given: $\triangle TPX \cong \triangle QPX \cong \triangle QRX \cong \triangle TRX$
Prove: $TPQR$ is a rhombus.

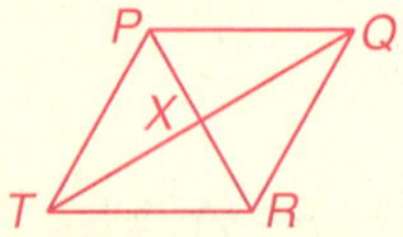

Proof:

Statements (Reasons)

1. $\triangle TPX \cong \triangle QPX \cong \triangle QRX \cong \triangle TRX$ (Given)
2. $\overline{TP} \cong \overline{PQ} \cong \overline{QR} \cong \overline{TR}$ (CPCTC)
3. $TPQR$ is a rhombus. (Def. of rhombus)

42. Given: $\triangle LGK \cong \triangle MJK$
$GHJK$ is a parallelogram.
Prove: $GHJK$ is a rhombus.

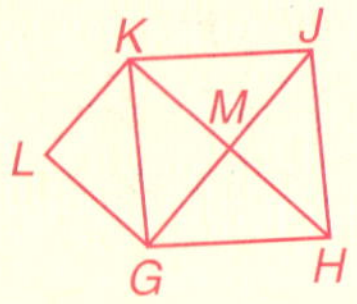

Proof:

Statements (Reasons)

1. $\triangle LGK \cong \triangle MJK$; $GHJK$ is a parallelogram. (Given)
2. $\overline{KG} \cong \overline{KJ}$ (CPCTC)
3. $\overline{KJ} \cong \overline{GH}$, $\overline{KG} \cong \overline{JH}$ (Opp. sides of ▱ are $\cong$.)
4. $\overline{KG} \cong \overline{JH} \cong \overline{GH} \cong \overline{JK}$ (Substitution Property)
5. $GHJK$ is a rhombus. (Def. of rhombus)

436 Chapter 8 Quadrilaterals

FCAT Practice

Standardized Test Practice

46. Points A, B, C, and D are on a square. The area of the square is 36 square units. Which of the following statements is true? **B**

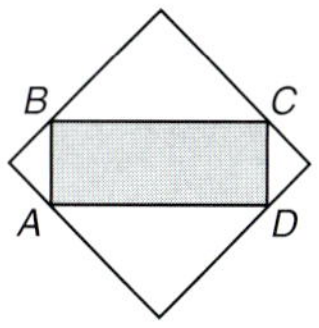

Ⓐ The perimeter of rectangle $ABCD$ is greater than 24 units.

Ⓑ The perimeter of rectangle $ABCD$ is less than 24 units.

Ⓒ The perimeter of rectangle $ABCD$ is equal to 24 units.

Ⓓ The perimeter of rectangle $ABCD$ cannot be determined from the information given.

47. **ALGEBRA** For all integers $x \neq 2$, let $<x> = \frac{1+x}{x-2}$. Which of the following has the greatest value? **C**

Ⓐ <0> Ⓑ <1> Ⓒ <3> Ⓓ <4>

Maintain Your Skills

Mixed Review

ALGEBRA **Use rectangle $LMNP$, parallelogram $LKMJ$, and the given information to solve each problem.** *(Lesson 8-4)*

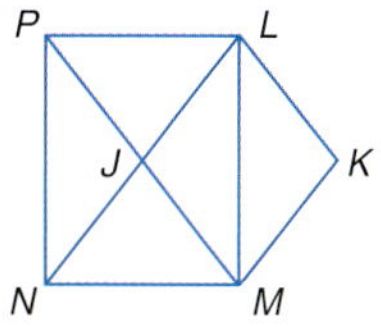

48. If $LN = 10$, $LJ = 2x + 1$, and $PJ = 3x - 1$, find x. **2**
49. If $m\angle PLK = 110$, find $m\angle LKM$. **140**
50. If $m\angle MJN = 35$, find $m\angle MPN$. **17.5**
51. If $MK = 6x$, $KL = 3x + 2y$, and $JN = 14 - x$, find x and y. **$x = 2$, $y = 3$**
52. If $m\angle LMP = m\angle PMN$, find $m\angle PJL$. **90**

COORDINATE GEOMETRY **Determine whether the points are the vertices of a parallelogram. Use the method indicated.** *(Lesson 8-3)*

53. $P(0, 2)$, $Q(6, 4)$, $R(4, 0)$, $S(-2, -2)$; Distance Formula **yes**
54. $F(1, -1)$, $G(-4, 1)$, $H(-3, 4)$, $J(2, 1)$; Distance Formula **no**
55. $K(-3, -7)$, $L(3, 2)$, $M(1, 7)$, $N(-3, 1)$; Slope Formula **no**
56. $A(-4, -1)$, $B(-2, -5)$, $C(1, 7)$, $D(3, 3)$; Slope Formula **yes**

Refer to $\triangle PQS$. *(Lesson 6-4)*

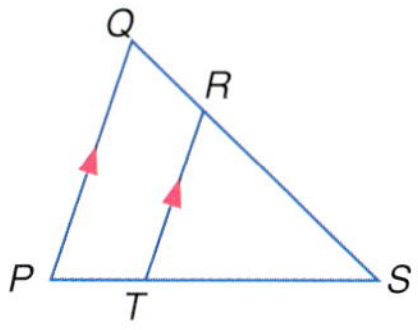

57. If $RT = 16$, $QP = 24$, and $ST = 9$, find PS. **13.5**
58. If $PT = y - 3$, $PS = y + 2$, $RS = 12$, and $QS = 16$, solve for y. **$4\frac{2}{3}$**
59. If $RT = 15$, $QP = 21$, and $PT = 8$, find TS. **20**

Refer to the figure. *(Lesson 4-6)*

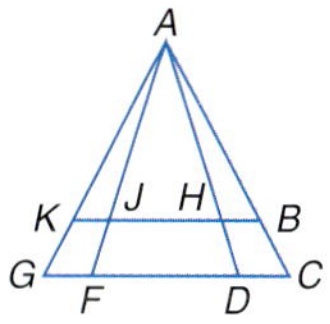

60. If $\overline{AG} \cong \overline{AC}$, name two congruent angles. **$\angle AGC \cong \angle ACG$**
61. If $\overline{AJ} \cong \overline{AH}$, name two congruent angles. **$\angle AJH \cong \angle AHJ$**
62. If $\angle AFD \cong \angle ADF$, name two congruent segments. **$\overline{AF} \cong \overline{AD}$**
63. If $\angle AKB \cong \angle ABK$, name two congruent segments. **$\overline{AK} \cong \overline{AB}$**

Getting Ready for the Next Lesson

PREREQUISITE SKILL **Solve each equation.**

(To review ***solving equations****, see pages 737 and 738.)*

64. $\frac{1}{2}(8x - 6x - 7) = 5$ **8.5**
65. $\frac{1}{2}(7x + 3x + 1) = 12.5$ **2.4**
66. $\frac{1}{2}(4x + 6 + 2x + 13) = 15.5$ **2**
67. $\frac{1}{2}(7x - 2 + 3x + 3) = 25.5$ **5**

www.geometryonline.com/self_check_quiz/fcat

4 Assess

Open-Ended Assessment

Modeling Have students construct a square by connecting the vertices of two congruent straws that intersect at their midpoints. Ask them if they can construct a rhombus from the same two straws. Ask them to describe the rhombus.

Getting Ready for Lesson 8-6

Prerequisite Skill Students will learn about trapezoids in Lesson 8-6. They will solve equations to find the measures of angles and sides of trapezoids. Use Exercises 64–67 to determine your students' familiarity with solving equations.

Answers

43. Given: $QRST$ and $QRTV$ are rhombi.
Prove: $\triangle QRT$ is equilateral.

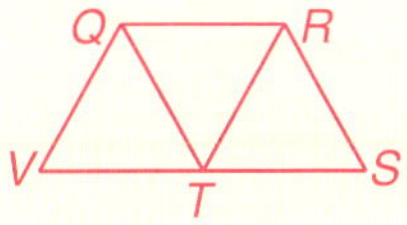

Proof:

Statements (Reasons)

1. $QRST$ and $QRTV$ are rhombi. (Given)
2. $\overline{QV} \cong \overline{VT} \cong \overline{TR} \cong \overline{QR}$, $\overline{QR} \cong \overline{TS} \cong \overline{RS} \cong \overline{QT}$ (Def. of rhombus)
3. $\overline{TR} \cong \overline{QR} \cong \overline{QT}$ (Transitive Property)
4. $\triangle QRT$ is equilateral. (Def. of equilateral triangle)

45. Sample answer: You can ride a bicycle with square wheels over a curved road. Answers should include the following.
- Rhombi and squares both have all four sides congruent, but the diagonals of a square are congruent. A square has four right angles and rhombi have each pair of opposite angles congruent, but not all angles are necessarily congruent.
- Sample answer: Since the angles of a rhombus are not all congruent, riding over the same road would not be smooth.

Geometry Activity

A Follow-Up of Lesson 8-5

Getting Started

Objective Construct a kite

Materials compass

Discuss why the compass setting has to be increased (or decreased) when drawing the second pair of sides for the kite. (If the compass setting is the same, the figure will have four congruent sides and be a rhombus.)

Teach

You may want students to do this activity in groups of four. Ask one student to construct the kite. Ask another student to measure each of the angles in the kite and make a conjecture. Ask a third student to measure the distance from the vertices of the kite to the intersection of the diagonals of the kite and make a conjecture. Ask the fourth student to make conjectures about the triangles formed by the diagonals of the kite.

Assess

As students complete **Exercise 6**, list their conjectures on the board and discuss as a class.

Geometry Activity

A Follow-Up to Lesson 8-5

Kites

A **kite** is a quadrilateral with exactly two distinct pairs of adjacent congruent sides. In kite *ABCD*, diagonal $\overline{BD}$ separates the kite into two congruent triangles. Diagonal $\overline{AC}$ separates the kite into two noncongruent isosceles triangles.

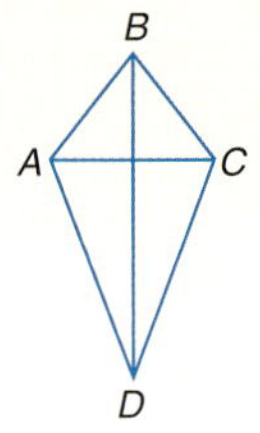

Activity

Construct a kite *QRST*.

1. Draw $\overline{RT}$.

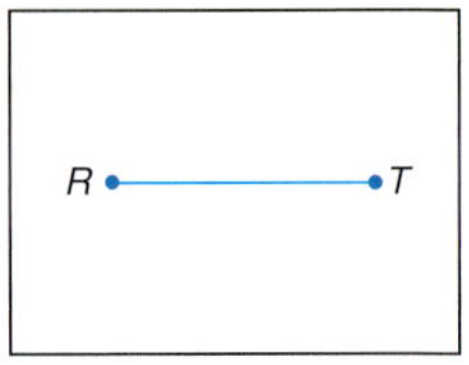

2. Choose a compass setting greater than $\frac{1}{2}$ $\overline{RT}$. Place the compass at point *R* and draw an arc above $\overline{RT}$. Then without changing the compass setting, move the compass to point *T* and draw an arc that intersects the first one. Label the intersection point *Q*. Increase the compass setting. Place the compass at *R* and draw an arc below $\overline{RT}$. Then, without changing the compass setting, draw an arc from point *T* to intersect the other arc. Label the intersection point *S*.

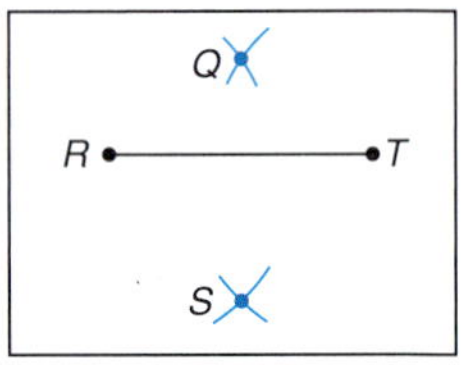

3. Draw *QRST*.

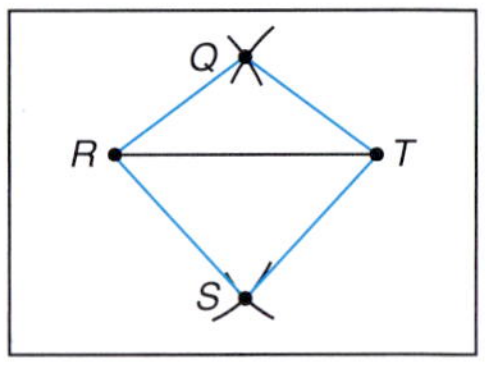

Model

1. Draw $\overline{QS}$ in kite *QRST*. Use a protractor to measure the angles formed by the intersection of $\overline{QS}$ and $\overline{RT}$. **The diagonals intersect at a right angle.**
2. Measure the interior angles of kite *QRST*. Are any congruent? **$\angle QRS \cong \angle QTS$**
3. Label the intersection of $\overline{QS}$ and $\overline{RT}$ as point *N*. Find the lengths of $\overline{QN}$, $\overline{NS}$, $\overline{TN}$, and $\overline{NR}$. How are they related? **See students' work; $\overline{NR} \cong \overline{TN}$, but $\overline{QN} \ncong \overline{NS}$.**
4. How many pairs of congruent triangles can be found in kite *QRST*?
5. Construct another kite *JKLM*. Repeat Exercises 1–4. **See p. 459C.**

4. 3 pairs: $\triangle QRN \cong \triangle QTN$, $\triangle RNS \cong \triangle TNS$, $\triangle QRS \cong \triangle QTS$

Analyze

6. Use your observations and measurements of kites *QRST* and *JKLM* to make conjectures about the angles, sides, and diagonals of kites. **See p. 459C.**

Resource Manager

Teaching Geometry with Manipulatives
- p. 136 (student recording sheet)

Glencoe Mathematics Classroom Manipulative Kit
- compass

Study Notebook

Ask students to summarize what they have learned about the properties of a kite.

8-6 Trapezoids

What You'll Learn

- Recognize and apply the properties of trapezoids.
- Solve problems involving the medians of trapezoids.

Sunshine State Standards
MA.A.3.4.3, MA.C.1.4.1, MA.C.2.4.1, MA.C.3.4.2

Vocabulary

- trapezoid
- isosceles trapezoid
- median

How are trapezoids used in architecture?

The Washington Monument in Washington, D.C., is an obelisk made of white marble. The width of the base is longer than the width at the top. Each face of the monument is an example of a trapezoid.

PROPERTIES OF TRAPEZOIDS A **trapezoid** is a quadrilateral with exactly one pair of parallel sides. The parallel sides are called *bases*. The *base angles* are formed by a base and one of the legs. The nonparallel sides are called *legs*.

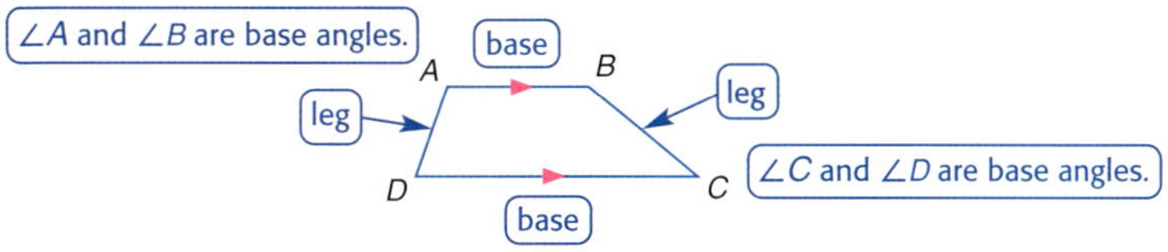

If the legs are congruent, then the trapezoid is an **isosceles trapezoid**. Theorems 8.18 and 8.19 describe two characteristics of isosceles trapezoids.

Study Tip

Isosceles Trapezoid
If you extend the legs of an isosceles trapezoid until they meet, you will have an isosceles triangle. Recall that the base angles of an isosceles triangle are congruent.

Theorems

8.18 Both pairs of base angles of an isosceles trapezoid are congruent.

8.19 The diagonals of an isosceles trapezoid are congruent.

Example:
$\angle DAB \cong \angle CBA$
$\angle ADC \cong \angle BCD$

$\overline{AC} \cong \overline{BD}$

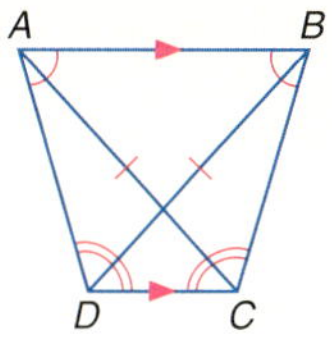

You will prove Theorem 8.18 in Exercise 36.

Example 1 Proof of Theorem 8.19

Write a flow proof of Theorem 8.19.

Given: *MNOP* is an isosceles trapezoid.

Prove: $\overline{MO} \cong \overline{NP}$

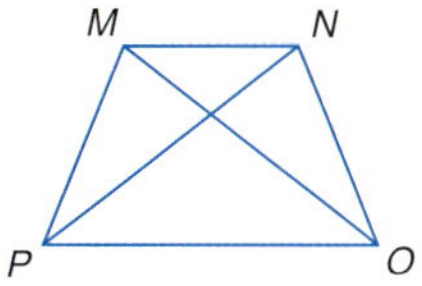

Proof:

- *MNOP* is an isosceles trapezoid. (Given)
  - → $\overline{MP} \cong \overline{NO}$ (Def. of isos. trapezoid)
  - → $\angle MPO \cong \angle NOP$ (Base ∠s of isos. trap. are ≅.)
- $\overline{PO} \cong \overline{PO}$ (Reflexive Property)
- → $\triangle MPO \cong \triangle NOP$ (SAS)
- → $\overline{MO} \cong \overline{NP}$ (CPCTC)

8-6 Lesson Notes

1 Focus

5-Minute Check Transparency 8-6 Use as a quiz or review of Lesson 8-5.

Mathematical Background notes are available for this lesson on p. 402D.

How are trapezoids used in architecture?

Ask students:

- What is the shape of the base of the obelisk? **square**
- How many sides does the base of the obelisk have? **4**
- How is a trapezoid different from other shapes you have studied? **Accept all reasonable answers.**

Resource Manager

Workbook and Reproducible Masters

Chapter 8 Resource Masters

- Study Guide and Intervention, pp. 447–448
- Skills Practice, p. 449
- Practice, p. 450
- Reading to Learn Mathematics, p. 451
- Enrichment, p. 452
- Assessment, p. 474

School-to-Career Masters, p. 16
Teaching Geometry With Manipulatives Masters, pp. 8, 137

Transparencies

5-Minute Check Transparency 8-6
Answer Key Transparencies

Technology

GeomPASS: Tutorial Plus, Lesson 17
Interactive Chalkboard

2 Teach

PROPERTIES OF TRAPEZOIDS

In-Class Examples

Power Point®

1 Write a flow proof.

Given: $KLMN$ is an isosceles trapezoid.

Prove: $\angle LKM \cong \angle MNL$

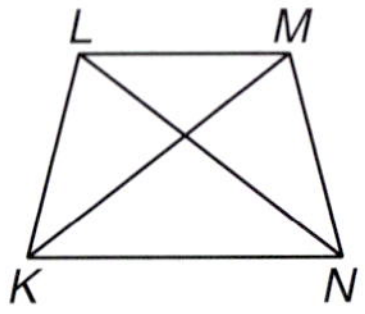

See bottom of page for flow proof.

2 The top of this work station appears to be two adjacent trapezoids. Use a ruler and protractor to determine if they are isosceles trapezoids.

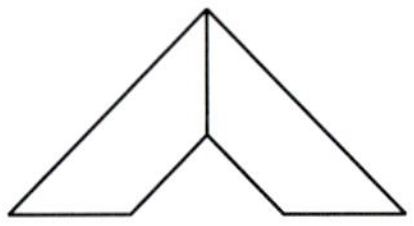

yes

3 $ABCD$ is a quadrilateral with vertices $A(5, 1)$, $B(-3, -1)$, $C(-2, 3)$, and $D(2, 4)$.

a. Verify that $ABCD$ is a trapezoid.

Slope of $\overline{AB} = \frac{1}{4}$; slope of $\overline{CD} = \frac{1}{4}$; slope of $\overline{AD} = -1$; slope of $\overline{BC} = 4$. Since $\overline{AB}$ and $\overline{CD}$ have the same slope, $\overline{AB} \parallel \overline{CD}$. Exactly one pair of opposite sides is parallel. Therefore, $ABCD$ is a trapezoid.

b. Determine whether $ABCD$ is an isosceles trapezoid. Explain.

$BC = \sqrt{17}$, and $AD = \sqrt{18}$; since the legs are not congruent, $ABCD$ is not an isosceles trapezoid.

More About . . .

Art

Barnett Newman designed this piece to be 50% larger. This piece was built for an exhibition in Japan but it could not be built as large as the artist wanted because of size limitations on cargo from New York to Japan.

Source: www.sfmoma.org

Example 2 Identify Isoceles Trapezoids

ART **The sculpture pictured is *Zim Zum I* by Barnett Newman. The walls are connected at right angles. In perspective, the rectangular panels appear to be trapezoids. Use a ruler and protractor to determine if the images of the front panels are isosceles trapezoids. Explain.**

The panel on the left is an isosceles trapezoid. The bases are parallel and are different lengths. The legs are not parallel and they are the same length.

The panel on the right is not an isosceles trapezoid. Each side is a different length.

Example 3 Identify Trapezoids

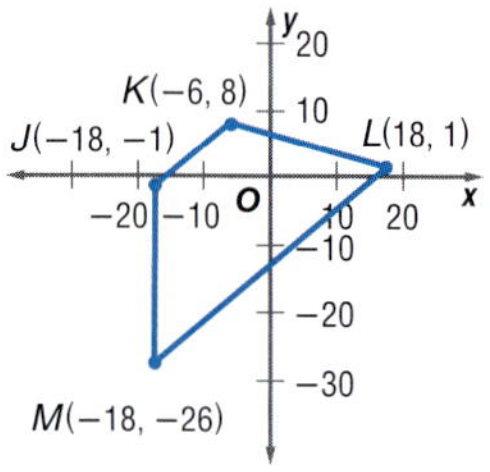

COORDINATE GEOMETRY **$JKLM$ is a quadrilateral with vertices $J(-18, -1)$, $K(-6, 8)$, $L(18, 1)$, and $M(-18, -26)$.**

a. Verify that $JKLM$ is a trapezoid.

A quadrilateral is a trapezoid if exactly one pair of opposite sides are parallel. Use the Slope Formula.

$$\text{slope of } \overline{JK} = \frac{-1-8}{-18-(-6)} = \frac{-9}{-12} \text{ or } \frac{3}{4} \qquad \text{slope of } \overline{ML} = \frac{1-(-26)}{18-(-18)} = \frac{27}{36} \text{ or } \frac{3}{4}$$

$$\text{slope of } \overline{JM} = \frac{-1-(-26)}{-18-(-18)} = \frac{25}{0} \text{ or undefined} \qquad \text{slope of } \overline{KL} = \frac{1-8}{18-(-6)} = \frac{-7}{24}$$

Exactly one pair of opposite sides are parallel, $\overline{JK}$ and $\overline{ML}$. So, $JKLM$ is a trapezoid.

b. Determine whether $JKLM$ is an isosceles trapezoid. Explain.

First use the Distance Formula to show that the legs are congruent.

$$JM = \sqrt{[-18-(-18)]^2 + [-1-(-26)]^2} = \sqrt{0 + 625} = \sqrt{625} \text{ or } 25$$

$$KL = \sqrt{(-6-18)^2 + (8-1)^2} = \sqrt{576 + 49} = \sqrt{625} \text{ or } 25$$

Since the legs are congruent, $JKLM$ is an isosceles trapezoid.

MEDIANS OF TRAPEZOIDS

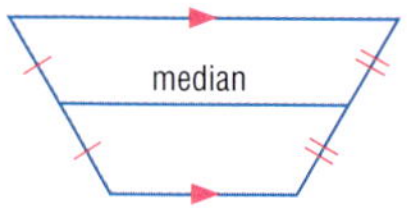

The segment that joins midpoints of the legs of a trapezoid is the **median**. The median of a trapezoid can also be called a *midsegment*. Recall from Lesson 6-4 that the midsegment of a triangle is the segment joining the midpoints of two sides. The median of a trapezoid has the same properties as the midsegment of a triangle. You can construct the median of a trapezoid using a compass and a straightedge.

Proof:

$KLMN$ is an isosceles trapezoid. (Given) → $\overline{KM} \cong \overline{NL}$ (Diag. of an isos. trap. are ≅.)

$KLMN$ is an isosceles trapezoid. (Given) → $\overline{KL} \cong \overline{NM}$ (Def. of isos. trap.)

$\overline{LM} \cong \overline{ML}$ (Reflexive Prop.)

$\overline{KM} \cong \overline{NL}$, $\overline{KL} \cong \overline{NM}$, $\overline{LM} \cong \overline{ML}$ → $\triangle KLM \cong \triangle NML$ (SSS) → $\angle LKM \cong \angle MNL$ (CPCTC)

440 Chapter 8 Quadrilaterals

Study Tip

Reading Math

The word *median* means *middle*. The median of a trapezoid is the segment that is parallel to and equidistant from each base.

Geometry Activity

Median of a Trapezoid

Model

1. Draw trapezoid $WXYZ$ with legs $\overline{XY}$ and $\overline{WZ}$.
2. Construct the perpendicular bisectors of $\overline{XY}$ and $\overline{WZ}$. Label the midpoints M and N.
3. Draw $\overline{MN}$.

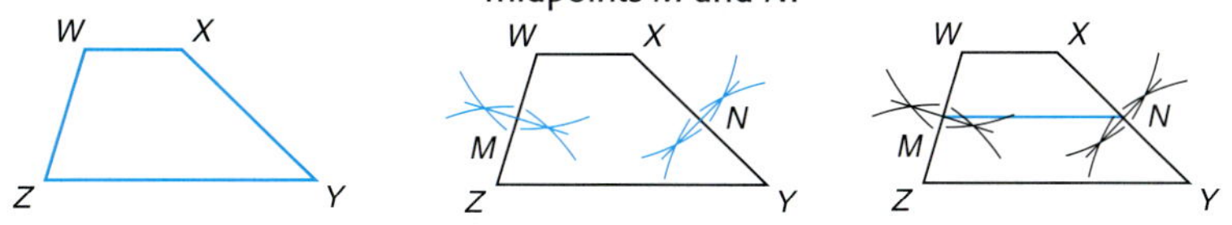

Analyze

1. Measure $\overline{WX}$, $\overline{ZY}$, and $\overline{MN}$ to the nearest millimeter. **See students' work.**
2. Make a conjecture based on your observations. **$MN = \frac{1}{2}(WX + ZY)$**

The results of the Geometry Activity suggest Theorem 8.20.

Theorem 8.20

The median of a trapezoid is parallel to the bases, and its measure is one-half the sum of the measures of the bases.

Example: $EF = \frac{1}{2}(AB + DC)$

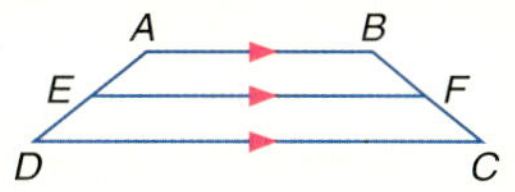

Example 4 Median of a Trapezoid

ALGEBRA **$QRST$ is an isosceles trapezoid with median $\overline{XY}$.**

a. Find TS if $QR = 22$ and $XY = 15$.

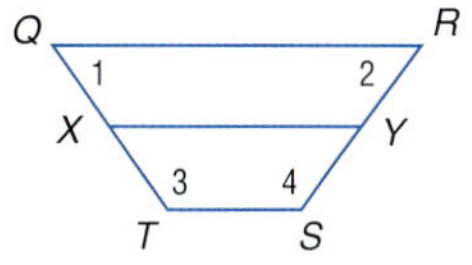

$XY = \frac{1}{2}(QR + TS)$	Theorem 8.20
$15 = \frac{1}{2}(22 + TS)$	Substitution
$30 = 22 + TS$	Multiply each side by 2.
$8 = TS$	Subtract 22 from each side.

b. Find $m\angle 1$, $m\angle 2$, $m\angle 3$, and $m\angle 4$ if $m\angle 1 = 4a - 10$ and $m\angle 3 = 3a + 32.5$.

Since $\overline{QR} \parallel \overline{TS}$, $\angle 1$ and $\angle 3$ are supplementary. Because this is an isosceles trapezoid, $\angle 1 \cong \angle 2$ and $\angle 3 \cong \angle 4$.

$m\angle 1 + m\angle 3 = 180$	Consecutive Interior Angles Theorem
$4a - 10 + 3a + 32.5 = 180$	Substitution
$7a + 22.5 = 180$	Combine like terms.
$7a = 157.5$	Subtract 22.5 from each side.
$a = 22.5$	Divide each side by 7.

If $a = 22.5$, then $m\angle 1 = 80$ and $m\angle 3 = 100$.
Because $\angle 1 \cong \angle 2$ and $\angle 3 \cong \angle 4$, $m\angle 2 = 80$ and $m\angle 4 = 100$.

www.geometryonline.com/extra_examples/fcat

MEDIANS OF TRAPEZOIDS

In-Class Examples

Power Point®

Teaching Tip You may want to stress the differences between a trapezoid and a parallelogram. A trapezoid has exactly one pair of opposite sides parallel, while a parallelogram has exactly two. Caution students that they should not assume all trapezoids are isosceles.

4 $DEFG$ is an isosceles trapezoid with median $\overline{MN}$.

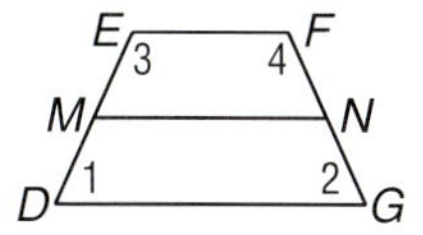

a. Find DG if $EF = 20$ and $MN = 30$. **40**

b. Find $m\angle 1$, $m\angle 2$, $m\angle 3$, and $m\angle 4$ if $m\angle 1 = 3x + 5$ and $m\angle 3 = 6x - 5$. **$m\angle 1 = 65$, $m\angle 2 = 65$, $m\angle 3 = 115$, and $m\angle 4 = 115$**

Concept Check

Ask students to describe the properties of an isosceles trapezoid.
Exactly one pair of opposite sides is parallel; the legs are congruent; each pair of base angles is congruent; the diagonals are congruent.

Geometry Activity

Materials: ruler, compass

Group students in pairs. Have each group draw a differently shaped trapezoid, or provide pairs of different trapezoids for them to use.

3 Practice/Apply

Study Notebook

Have students—
- add the definitions/examples of the vocabulary terms to their Vocabulary Builder worksheets for Chapter 8.
- include an example of a trapezoid with the bases, legs, and median labeled.
- include any other item(s) that they find helpful in mastering the skills in this lesson.

About the Exercises...

Organization by Objective
- **Properties of Trapezoids:** 9–12, 20–38
- **Medians of Trapezoids:** 13–19, 39

Odd/Even Assignments

Exercises 9–38 are structured so that students practice the same concepts whether they are assigned odd or even problems.

Assignment Guide

Basic: 9–17 odd, 21–33 odd, 37, 39–61

Average: 9–39 odd, 40–61

Advanced: 10–38 even, 39–55 (optional: 56–61)

Answers

9a. $\overline{AD} \parallel \overline{BC}$, $\overline{CD} \nparallel \overline{AB}$

9b. not isosceles, $AB = \sqrt{17}$ and $CD = 5$

10a. $\overline{KJ} \parallel \overline{GH}$, $\overline{KG} \nparallel \overline{JH}$

10b. not isosceles, $GK = 5$, $JH = \sqrt{26}$

11a. $\overline{DC} \parallel \overline{FE}$, $\overline{DE} \nparallel \overline{FC}$

11b. isosceles, $DE = \sqrt{50}$, $CF = \sqrt{50}$

12a. $\overline{QR} \parallel \overline{TS}$, $\overline{QT} \nparallel \overline{RS}$

12b. isosceles, $QT = \sqrt{26}$, $RS = \sqrt{26}$

Check for Understanding

Concept Check

1. **List** the minimum requirements to show that a quadrilateral is a trapezoid. 1. Exactly one pair of opposite sides is parallel.
2. **Make a chart** comparing the characteristics of the diagonals of a trapezoid, a rectangle, a square, and a rhombus. (*Hint:* Use the types of quadrilaterals as column headings and the properties of diagonals as row headings.)
3. **OPEN ENDED** Draw an isosceles trapezoid and a trapezoid that is not isosceles. Draw the median for each. Is the median parallel to the bases in both trapezoids? 2–3. See p. 459C.

Guided Practice

GUIDED PRACTICE KEY

Exercises	Examples
4–5	3
6	1
7	4
8	2

COORDINATE GEOMETRY ***QRST*** **is a quadrilateral with vertices** ***Q*****(−3, 2),** ***R*****(−1, 6),** ***S*****(4, 6),** ***T*****(6, 2).**

4. Verify that *QRST* is a trapezoid. $\overline{RS} \parallel \overline{QT}$ and $\overline{QR} \nparallel \overline{ST}$, *QRST* is a trapezoid.
5. Determine whether *QRST* is an isosceles trapezoid. Explain. isosceles, $QR = \sqrt{20}$, $ST = \sqrt{20}$
6. **PROOF** *CDFG* is an isosceles trapezoid with bases $\overline{CD}$ and $\overline{FG}$. Write a flow proof to prove $\angle DGF \cong \angle CFG$. See p. 459C.

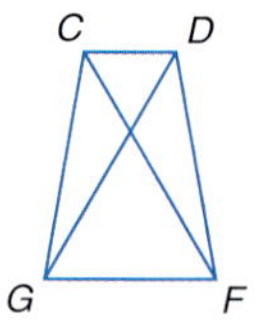

7. **ALGEBRA** *EFGH* is an isosceles trapezoid with bases $\overline{EF}$ and $\overline{GH}$ and median $\overline{YZ}$. If $EF = 3x + 8$, $HG = 4x - 10$, and $YZ = 13$, find x. 4

Application

8. **PHOTOGRAPHY** Photographs can show a building in a perspective that makes it appear to be a different shape. Identify the types of quadrilaterals in the photograph. trapezoids, parallelograms

★ indicates increased difficulty

Practice and Apply

Homework Help

For Exercises	See Examples
9–12, 23–32	3
13–19, 39	4
20–22, 38	2
33–37	1

Extra Practice
See page 770.

COORDINATE GEOMETRY For each quadrilateral whose vertices are given,
a. verify that the quadrilateral is a trapezoid, and
b. determine whether the figure is an isosceles trapezoid. 9–12. See margin.

9. $A(-3, 3)$, $B(-4, -1)$, $C(5, -1)$, $D(2, 3)$
10. $G(-5, -4)$, $H(5, 4)$, $J(0, 5)$, $K(-5, 1)$
11. $C(-1, 1)$, $D(-5, -3)$, $E(-4, -10)$, $F(6, 0)$
12. $Q(-12, 1)$, $R(-9, 4)$, $S(-4, 3)$, $T(-11, -4)$

ALGEBRA Find the missing measure(s) for the given trapezoid.

13. For trapezoid *DEGH*, *X* and *Y* are midpoints of the legs. Find *DE*. 8

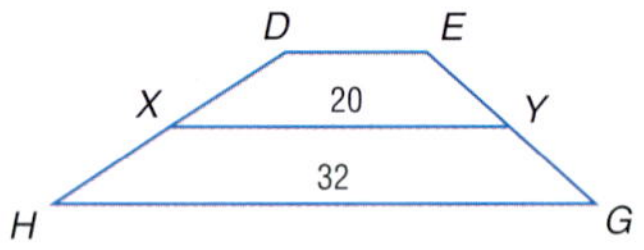

14. For trapezoid *RSTV*, *A* and *B* are midpoints of the legs. Find *VT*. 4

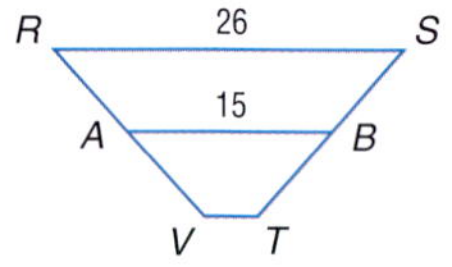

DAILY INTERVENTION

Differentiated Instruction

Naturalist Challenge students to think of where they have seen trapezoids in nature. Encourage students to research shapes of cells, scales, and crystal faces for examples.

15. For isosceles trapezoid $XYZW$, find the length of the median, $m\angle W$, and $m\angle Z$. **14, 110, 110**

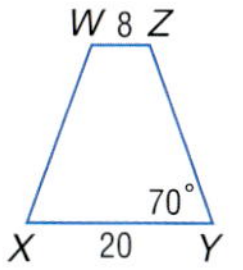

16. For trapezoid $QRST$, A and B are midpoints of the legs. Find AB, $m\angle Q$, and $m\angle S$. **16, 60, 135**

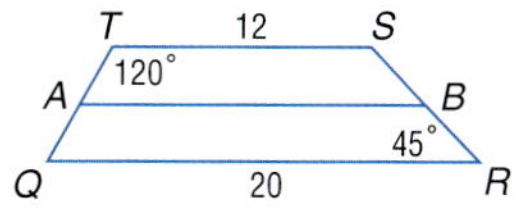

For Exercises 17 and 18, use trapezoid $QRST$.

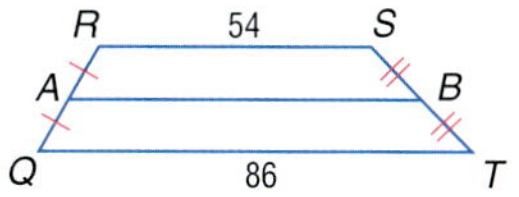

17. Let $\overline{GH}$ be the median of $RSBA$. Find GH. **62**

18. Let $\overline{JK}$ be the median of $ABTQ$. Find JK. **78**

★ 19. **ALGEBRA** $JKLM$ is a trapezoid with $\overline{JK} \parallel \overline{LM}$ and median $\overline{RP}$. Find RP if $JK = 2(x + 3)$, $RP = 5 + x$, and $ML = \frac{1}{2}x - 1$. **15**

20. **DESIGN** The bagua is a tool used in Feng Shui design. This bagua consists of two regular octagons centered around a yin-yang symbol. How can you determine the type of quadrilaterals in the bagua? **See margin.**

21. **SEWING** Madison is making a valance for a window treatment. She is using striped fabric cut on the bias, or diagonal, to create a chevron pattern. Identify the polygons formed in the fabric. **Sample answer: triangles, quadrilaterals, trapezoids, hexagons**

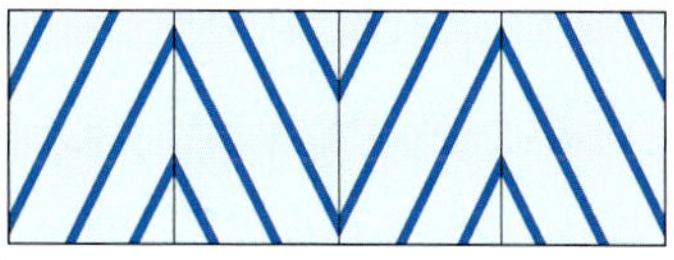

COORDINATE GEOMETRY **Determine whether each figure is a *trapezoid*, a *parallelogram*, a *square*, a *rhombus*, or a *quadrilateral*. Choose the most specific term. Explain.**

22.

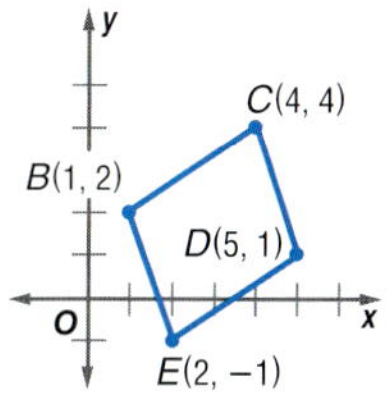

parallelogram, opp. sides ∥, no rt. ∠, no cons. sides ≅

23.

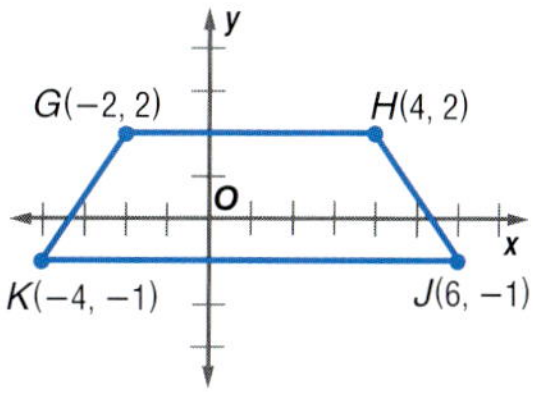

24.

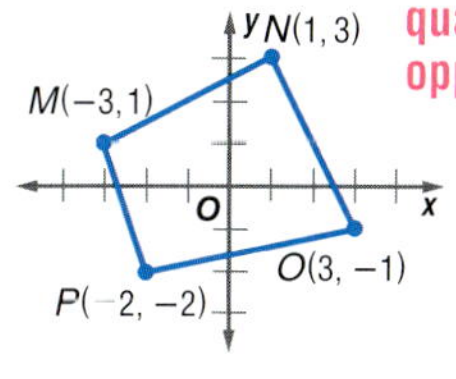

quadrilateral, opp. sides not ∥

25.

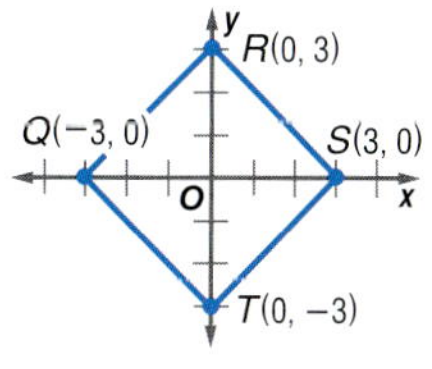

COORDINATE GEOMETRY **For Exercises 26–28, refer to quadrilateral $PQRS$.**

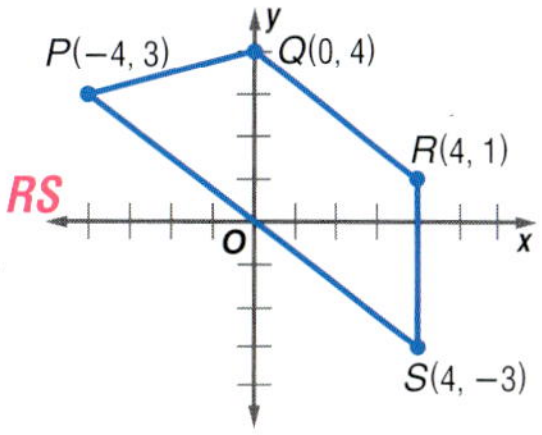

26. Determine whether the figure is a trapezoid. If so, is it isosceles? Explain. **$\overline{QR} \parallel \overline{PS}$, not isosceles, $PQ \neq RS$**

27. Find the coordinates of the midpoints of $\overline{PQ}$ and $\overline{RS}$, and label them A and B. **$A(-2, 3.5)$, $B(4, -1)$**

28. Find AB without using the Distance Formula. **7.5**

More About. . .

Design

Feng Shui is an ancient Chinese theory of design. The goal is to create spaces that enhance creativity and balance.

Source: www.fengshui2000.com

23. trapezoid, exactly one pair opp. sides ∥

25. square, all sides ≅, consecutive sides ⊥

Answers

20. Since the two octagons are regular polygons with the same center, the quadrilaterals are trapezoids with one pair of opposite sides parallel.

Enrichment, p. 452

Quadrilaterals in Construction

Quadrilaterals are often used in construction work.

1. The diagram at the right represents a roof frame and shows many quadrilaterals. Find the following shapes in the diagram and shade in their edges. **See students' work.**
   a. isosceles triangle
   b. scalene triangle
   c. rectangle
   d. rhombus
   e. trapezoid (not isosceles)
   f. isosceles trapezoid

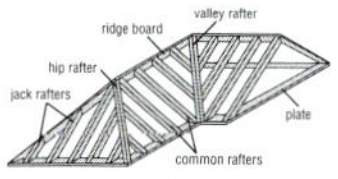

Roof Frame

Study Guide and Intervention, p. 447 (shown) and p. 448

Properties of Trapezoids A **trapezoid** is a quadrilateral with exactly one pair of parallel sides. The parallel sides are called **bases** and the nonparallel sides are called **legs**. If the legs are congruent, the trapezoid is an **isosceles trapezoid**. In an isosceles trapezoid both pairs of **base angles** are congruent.

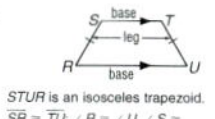

STUR is an isosceles trapezoid. $\overline{SR} \cong \overline{TU}$; $\angle R \cong \angle U$, $\angle S \cong \angle T$

Example

The vertices of $ABCD$ are $A(-3, -1)$, $B(-1, 3)$, $C(2, 3)$, and $D(4, -1)$. Verify that $ABCD$ is a trapezoid.

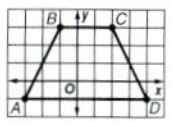

slope of $\overline{AB} = \frac{3 - (-1)}{-1 - (-3)} = \frac{4}{2} = 2$

slope of $\overline{AD} = \frac{-1 - (-1)}{4 - (-3)} = \frac{0}{7} = 0$

slope of $\overline{BC} = \frac{3 - 3}{2 - (-1)} = \frac{0}{3} = 0$

slope of $\overline{CD} = \frac{-1 - 3}{4 - 2} = \frac{-4}{2} = -2$

$AB = \sqrt{(-3 - (-1))^2 + (-1 - 3)^2} = \sqrt{4 + 16} = \sqrt{20} = 2\sqrt{5}$

$CD = \sqrt{(2 - 4)^2 + (3 - (-1))^2} = \sqrt{4 + 16} = \sqrt{20} = 2\sqrt{5}$

Exactly two sides are parallel, $\overline{AD}$ and $\overline{BC}$, so $ABCD$ is a trapezoid. $AB = CD$, so $ABCD$ is an isosceles trapezoid.

Exercises

In Exercises 1–3, determine whether $ABCD$ is a trapezoid. If so, determine whether it is an isosceles trapezoid. Explain.

1. $A(-1, 1)$, $B(2, 1)$, $C(3, -2)$, and $D(2, -2)$
   Slope of $\overline{AB} = 0$, slope of $\overline{DC} = 0$, slope of $\overline{AD} = -1$, slope of $\overline{BC} = -3$. Exactly two sides are parallel, so $ABCD$ is a trapezoid. $AD = 3\sqrt{2}$ and $BC = \sqrt{10}$; $AD \neq BC$, so $ABCD$ is not isosceles.

2. $A(3, -3)$, $B(-3, -3)$, $C(-2, 3)$, and $D(2, 3)$
   Slope of $\overline{AB} = 0$, slope of $\overline{DC} = 0$, slope of $\overline{BC} = 6$, slope of $\overline{AD} = -6$. Exactly 2 sides are ∥, so $ABCD$ is a trapezoid. $BC = \sqrt{37}$ and $AD = \sqrt{37}$; $BC = AD$, so $ABCD$ is isosceles.

3. $A(1, -4)$, $B(-3, -3)$, $C(-2, 3)$, and $D(2, 2)$
   Slope of $\overline{AB} = -\frac{1}{4}$, slope of $\overline{DC} = -\frac{1}{4}$, slope of $\overline{BC} = 6$, slope of $\overline{AD} = 6$. Both pairs of opposite sides are parallel, so $ABCD$ is not a trapezoid.

4. The vertices of an isosceles trapezoid are $R(-2, 2)$, $S(2, 2)$, $T(4, -1)$, and $U(-4, -1)$. Verify that the diagonals are congruent.
   $RT = \sqrt{((4 - (-2))^2 + (-1 - 2)^2)} = \sqrt{45}$
   $SU = \sqrt{((-4 - (2))^2 + (-1 - 2)^2)} = \sqrt{45}$

Skills Practice, p. 449 and Practice, p. 450 (shown)

COORDINATE GEOMETRY **$RSTU$ is a quadrilateral with vertices $R(-3, -3)$, $S(5, 1)$, $T(10, -2)$, $U(-4, -9)$.**

1. Verify that $RSTU$ is a trapezoid.
   $\overline{RS} \parallel \overline{TU}$

2. Determine whether $RSTU$ is an isosceles trapezoid. Explain.
   not isosceles; $RU = \sqrt{37}$ and $ST = \sqrt{34}$

COORDINATE GEOMETRY **$BGHJ$ is a quadrilateral with vertices $B(-9, 1)$, $G(2, 3)$, $H(12, -2)$, $J(-10, -6)$.**

3. Verify that $BGHJ$ is a trapezoid.
   $\overline{BG} \parallel \overline{HJ}$

4. Determine whether $BGHJ$ is an isosceles trapezoid. Explain.
   not isosceles; $BJ = \sqrt{50}$ and $GH = \sqrt{125}$

ALGEBRA **Find the missing measure(s) for the given trapezoid.**

5. For trapezoid $CDEF$, V and Y are midpoints of the legs. Find CD. **38**

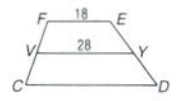

6. For trapezoid $WRLP$, B and C are midpoints of the legs. Find LP. **18**

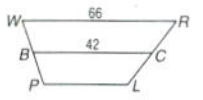

7. For trapezoid $FGHI$, K and M are midpoints of the legs. Find FI, $m\angle F$, and $m\angle I$. **51, 40, 55**

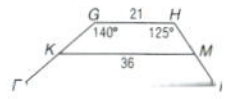

8. For isosceles trapezoid $TVZY$, find the length of the median, $m\angle T$, and $m\angle Z$. **19.5, 60, 120**

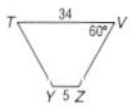

9. **CONSTRUCTION** A set of stairs leading to the entrance of a building is designed in the shape of an isosceles trapezoid with the longer base at the bottom of the stairs and the shorter base at the top. If the bottom of the stairs is 21 feet wide and the top is 14, find the width of the stairs halfway to the top. **17.5 ft**

10. **DESK TOPS** A carpenter needs to replace several trapezoid-shaped desktops in a classroom. The carpenter knows the lengths of both bases of the desktop. What other measurements, if any, does the carpenter need?
    Sample answer: the measures of the base angles

Reading to Learn Mathematics, p. 451 ELL

Pre-Activity **How are trapezoids used in architecture?**

Read the introduction to Lesson 8-6 at the top of page 439 in your textbook.

How might trapezoids be used in the interior design of a home?
Sample answer: floor tiles for a kitchen or bathroom

Reading the Lesson

1. In the figure at the right, $EFGH$ is a trapezoid, I is the midpoint of $\overline{FE}$, and J is the midpoint of $\overline{GH}$. Identify each of the following segments or angles in the figure.

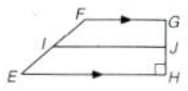

   a. the bases of trapezoid $EFGH$ $\overline{FG}$, $\overline{EH}$
   b. the two pairs of base angles of trapezoid $EFGH$ $\angle E$ and $\angle H$; $\angle F$ and $\angle G$
   c. the legs of trapezoid $EFGH$ $\overline{FE}$, $\overline{GH}$
   d. the median of trapezoid $EFGH$ $\overline{IJ}$

2. Determine whether each statement is *true* or *false*. If the statement is false, explain why.
   a. A trapezoid is a special kind of parallelogram. False; sample answer: A parallelogram has two pairs of parallel sides, but a trapezoid has only one pair of parallel sides.
   b. The diagonals of a trapezoid are congruent. False; sample answer: This is only true for isosceles trapezoids.
   c. The median of a trapezoid is parallel to the legs. False; sample answer: The median is parallel to the bases.
   d. The length of the median of a trapezoid is the average of the length of the bases. true
   e. A trapezoid has three medians. False; sample answer: A trapezoid has only one median.
   f. The bases of an isosceles trapezoid are congruent. False: sample answer: The legs are congruent.
   g. An isosceles trapezoid has two pairs of congruent angles. true
   h. The median of an isosceles trapezoid divides the trapezoid into two smaller isosceles trapezoids. true

Helping You Remember

3. A good way to remember a new geometric theorem is to relate it to one you already know. Name and state in words a theorem about triangles that is similar to the theorem in this lesson about the median of a trapezoid. Triangle Midsegment Theorem; a midsegment of a triangle is parallel to one side of the triangle, and its length is one-half the length of that side.

Answers

36. Given: $ABCD$ is an isosceles trapezoid.
$\overline{BC} \parallel \overline{AD}$
$\overline{AB} \cong \overline{CD}$
Prove: $\angle A \cong \angle D$
$\angle ABC \cong \angle DCB$

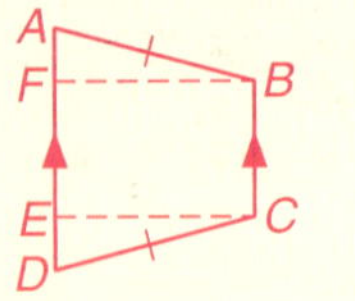

Proof: Draw auxiliary segments so that $\overline{BF} \perp \overline{AD}$ and $\overline{CE} \perp \overline{AD}$. Since $\overline{BC} \parallel \overline{AD}$ and parallel lines are everywhere equidistant, $\overline{BF} \cong \overline{CE}$. Perpendicular lines form right angles, so $\angle BFA$ and $\angle CED$ are right angles. $\triangle BFA$ and $\triangle CED$ are right triangles by definition. Therefore, $\triangle BFA \cong \triangle CED$ by HL. $\angle A \cong \angle D$ by CPCTC. Since $\angle CBF$ and $\angle BCE$ are right angles and all right angles are congruent, $\angle CBF \cong \angle BCE$. $\angle ABF \cong \angle DCE$ by CPCTC. So, $\angle ABC \cong \angle DCB$ by angle addition.

37. Sample answer:

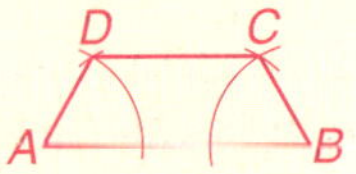

38. Sample answer:

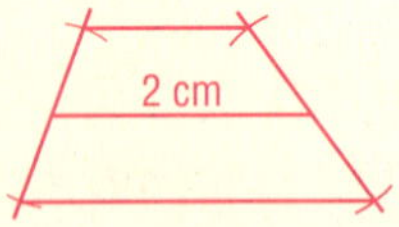

40. It is not possible. Since pairs of base angles of an isosceles trapezoid are congruent, if two angles are right, all four angles will be right. Then the quadrilateral would be a rectangle, not a trapezoid.

41. Sample answer: Trapezoids are used in monuments as well as other buildings. Answers should include the following.
- Trapezoids have exactly one pair of opposite sides parallel.
- Trapezoids can be used as window panes.

---

29. $\overline{DG} \parallel \overline{EF}$, $\overline{DE} \not\parallel \overline{GF}$ not isosceles, $DE \neq GF$

WebQuest

You can use a map of Seattle to locate and draw a quadrilateral that will help you begin to find the hidden treasure. Visit www.geometryonline.com/webquest to continue work on your WebQuest project.

COORDINATE GEOMETRY For Exercises 29–31, refer to quadrilateral $DEFG$.

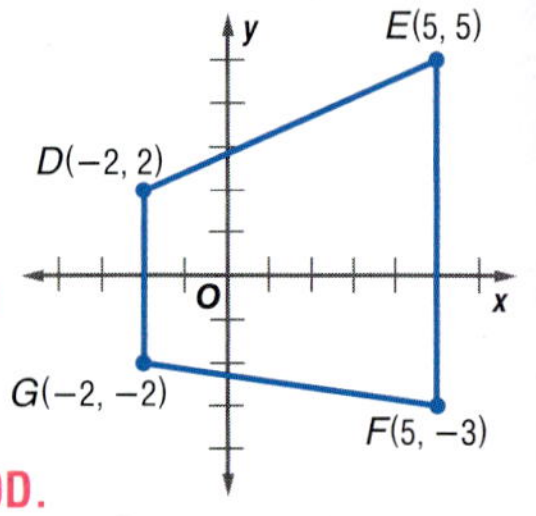

29. Determine whether the figure is a trapezoid. If so, is it isosceles? Explain.

30. Find the coordinates of the midpoints of $\overline{DE}$ and $\overline{GF}$, and label them W and V. $W(1.5, 3.5)$, $V(1.5, -2.5)$

31. Find WV without using the Distance Formula. $WV = 6$

PROOF Write a flow proof. 32–35. See pp. 459C–459D.

32. **Given:** $\overline{HJ} \parallel \overline{GK}$, $\triangle HGK \cong \triangle JKG$, $\overline{HG} \not\parallel \overline{JK}$
Prove: $GHJK$ is an isosceles trapezoid.

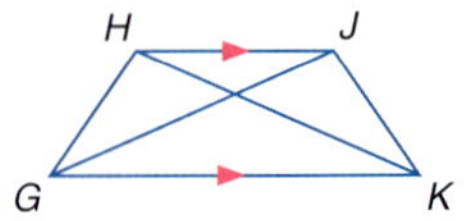

33. **Given:** $\triangle TZX \cong \triangle YXZ$, $\overline{WX} \not\parallel \overline{ZY}$
Prove: $XYZW$ is a trapezoid.

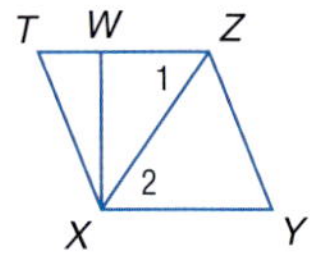

34. **Given:** $ZYXP$ is an isosceles trapezoid.
Prove: $\triangle PWX$ is isosceles.

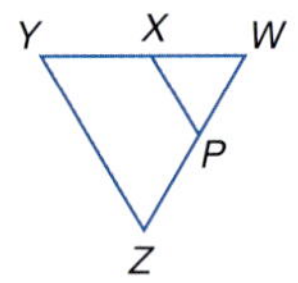

★ 35. **Given:** E and C are midpoints of $\overline{AD}$ and $\overline{DB}$. $\overline{AD} \cong \overline{DB}$
Prove: $ABCE$ is an isosceles trapezoid.

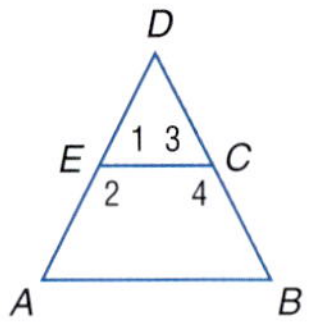

★ 36. Write a paragraph proof of Theorem 8.18. See margin.

CONSTRUCTION Use a compass and ruler to construct each figure.

37. an isosceles trapezoid 37–38. See margin.

★ 38. trapezoid with a median 2 centimeters long

39. **CRITICAL THINKING** In $RSTV$, $RS = 6$, $VT = 3$, and RX is twice the length of XV. Find XY. 4

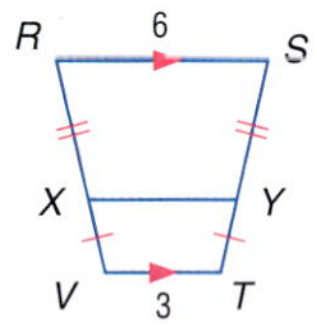

40. **CRITICAL THINKING** Is it possible for an isosceles trapezoid to have two right base angles? Explain. See margin.

41. **WRITING IN MATH** Answer the question that was posed at the beginning of the lesson. See margin.

How are trapezoids used in architecture?

Include the following in your answer:
- the characteristics of a trapezoid, and
- other examples of trapezoids in architecture.

FCAT Practice
Standardized Test Practice

42. **SHORT RESPONSE** What type of quadrilateral is $WXYZ$? Justify your answer.
Trapezoid; one pair of opp. sides is parallel.

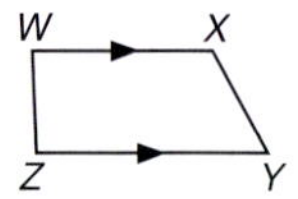

43. **ALGEBRA** In the figure, which point lies within the shaded region? **B**

Ⓐ (−2, 4) Ⓑ (−1, 3)

Ⓒ (1, −3) Ⓓ (2, −4)

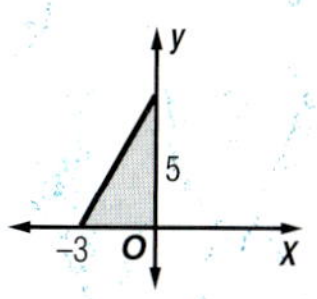

Maintain Your Skills

Mixed Review

ALGEBRA **In rhombus *LMPQ*, $m\angle QLM = 2x^2 - 10$, $m\angle QPM = 8x$, and $MP = 10$.** *(Lesson 8-5)*

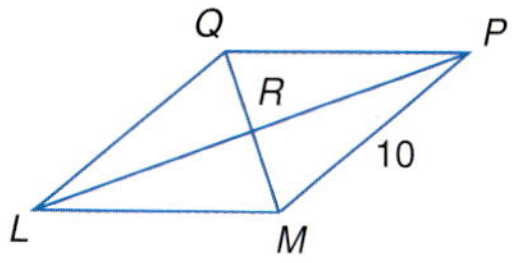

44. Find $m\angle LPQ$. **20**
45. Find QL. **10**
46. Find $m\angle LQP$. **140**
47. Find $m\angle LQM$. **70**
48. Find the perimeter of *LMPQ*. **40**

COORDINATE GEOMETRY **For Exercises 49–51, refer to quadrilateral *RSTV*.** *(Lesson 8-4)*

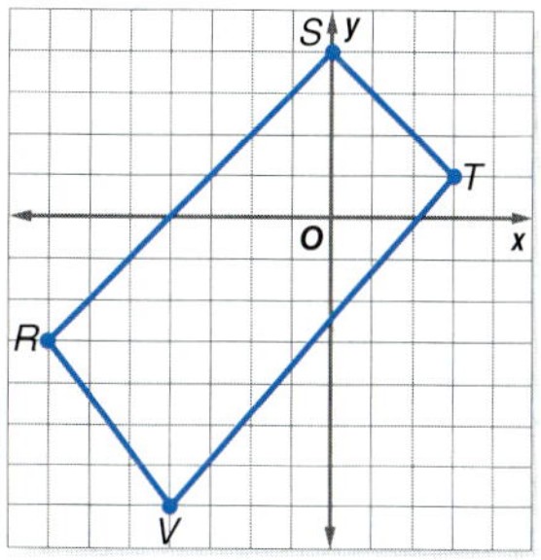

49. Find *RS* and *TV*. **$RS = 7\sqrt{2}$, $TV = \sqrt{113}$**
50. Find the coordinates of the midpoints of $\overline{RT}$ and $\overline{SV}$. **50. (−2, −1); $\left(-2, -\frac{3}{2}\right)$**
51. Is *RSTV* a rectangle? Explain. **No; opposite sides are not congruent and the diagonals do not bisect each other.**

Solve each proportion. *(Lesson 6-1)*

52. $\frac{16}{38} = \frac{24}{y}$ **57**
53. $\frac{y}{6} = \frac{17}{30}$ **$\frac{17}{5}$**
54. $\frac{5}{y+4} = \frac{20}{28}$ **3**
55. $\frac{2y}{9} = \frac{52}{36}$ **$\frac{13}{2}$**

Getting Ready for the Next Lesson

PREREQUISITE SKILL **Write an expression for the slope of the segment given the coordinates of the endpoints.** *(To review **slope**, see Lesson 3-3.)*

56. $(0, a), (-a, 2a)$ **−1**
57. $(-a, b), (a, b)$ **0**
58. $(c, c), (c, d)$ **undefined**
59. $(a, -b), (2a, b)$ **$\frac{2b}{a}$**
60. $(3a, 2b), (b, -a)$ **$\frac{a+2b}{3a-b}$**
61. $(b, c), (-b, -c)$ **$\frac{c}{b}$**

Practice Quiz 2 — *Lessons 8-4 through 8-6*

Quadrilateral *ABCD* is a rectangle. *(Lesson 8-4)*

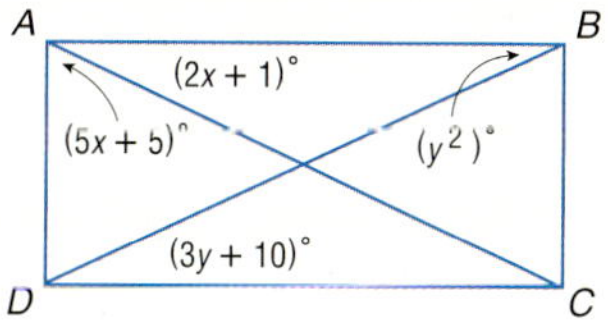

1. Find *x*. **12**
2. Find *y*. **5**

3. **COORDINATE GEOMETRY** Determine whether *MNPQ* is a *rhombus*, a *rectangle*, or a *square* for $M(-5, -3)$, $N(-2, 3)$, $P(1, -3)$, and $Q(-2, -9)$. List all that apply. Explain. *(Lesson 8-5)* **rhombus, opp. sides ∥, diag. ⊥, consec. sides not ⊥**

For trapezoid *TRSV*, *M* and *N* are midpoints of the legs. *(Lesson 8-6)*

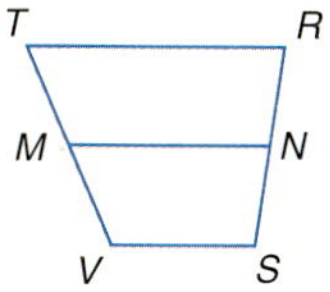

4. If $VS = 21$ and $TR = 44$, find *MN*. **32.5**
5. If $TR = 32$ and $MN = 25$, find *VS*. **18**

www.geometryonline.com/self_check_quiz/fcat

4 Assess

Open-Ended Assessment

Speaking Ask students to describe each type of quadrilateral. Ask them the difference between parallelograms, trapezoids, and kites. Also ask them to differentiate rectangles, rhombi, and squares.

Getting Ready for Lesson 8-7

Prerequisite Skill Students will learn about coordinate proofs in Lesson 8-7. They will find the slopes of segments on coordinate planes. Use Exercises 56–61 to determine your students' familiarity with finding slopes on coordinate planes.

Assessment Options

Practice Quiz 2 The quiz provides students with a brief review of the concepts and skills in Lessons 8-4 through 8-6. Lesson numbers are given to the right of the exercises or instruction lines so students can review concepts not yet mastered.

Quiz (Lessons 8-5 and 8-6) is available on p. 474 of the *Chapter 8 Resource Masters.*

Reading Mathematics

Getting Started

You may want to give groups of students the categories detailed in the hierarchy on the student page as cutouts in an envelope before referring to the textbook. Ask the groups to agree on the arrangement of the hierarchy, then use the textbook to check their result. Each group should be able to justify the order they decide on.

Teach

Hierarchy of Polygons In this activity, students will learn that each class is linked to the class above it in the hierarchy. For example, every square is a rectangle, but not every rectangle is a square. Also, rectangles, rhombi, and squares are all parallelograms. Make sure students understand that they should follow the direction of the arrows on the chart.

Assess

Study Notebook

Ask students to summarize what they have learned about quadrilaterals, parallelograms, trapezoids, and kites in their notebooks.

ELL English Language Learners may benefit from writing key concepts from this activity in their Study Notebooks in their native language and then in English.

Reading Mathematics

Hierarchy of Polygons

A *hierarchy* is a ranking of classes or sets of things. Examples of some classes of polygons are rectangles, rhombi, trapezoids, parallelograms, squares, and quadrilaterals. These classes are arranged in the hierarchy below.

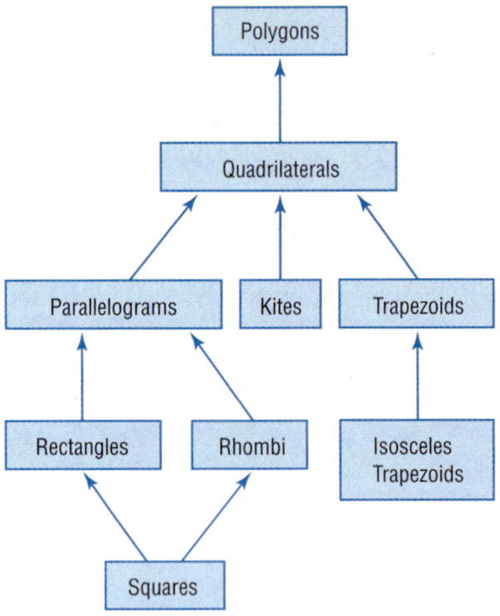

Use the following information to help read the hierarchy diagram.

- The class that is the broadest is listed first, followed by the other classes in order. For example, *polygons* is the broadest class in the hierarchy diagram above, and *squares* is a very specific class.
- Each class is contained within any class linked above it in the hierarchy. For example, *all* squares are also rhombi, rectangles, parallelograms, quadrilaterals, and polygons. However, an isosceles trapezoid is not a square or a kite.
- Some, but not all, elements of each class are contained within lower classes in the hierarchy. For example, some trapezoids are isosceles trapezoids, and some rectangles are squares.

Reading to Learn

Refer to the hierarchy diagram at the right. Write *true*, *false*, or *not enough information* for each statement.

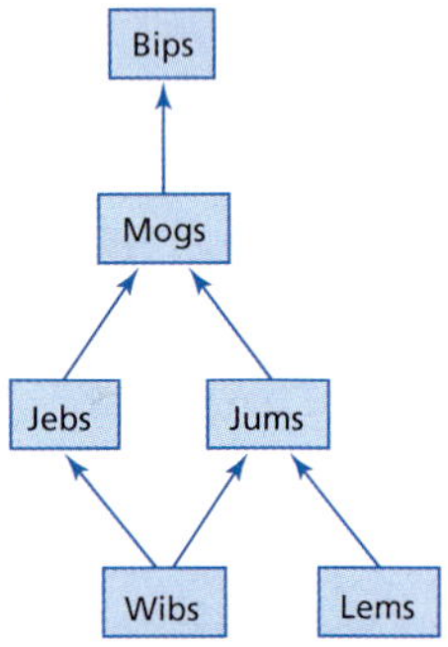

1. All mogs are jums. **false**
2. Some jebs are jums. **false**
3. All lems are jums. **true**
4. Some wibs are jums. **true**
5. All mogs are bips. **true**
6. Draw a hierarchy diagram to show these classes: equilateral triangles, polygons, isosceles triangles, triangles, and scalene triangles. **See margin.**

Answer

6.

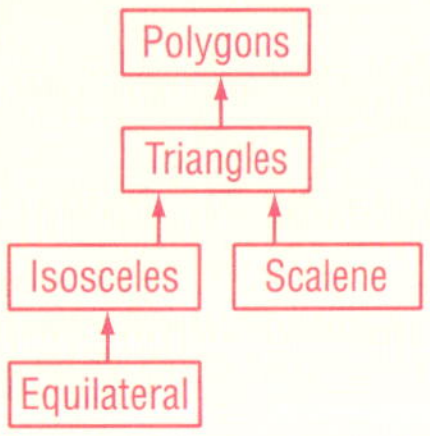

8-7 Coordinate Proof With Quadrilaterals

Sunshine State Standards
MA.B.3.4.1, MA.C.1.4.1, MA.C.3.4.2

What You'll Learn

- Position and label quadrilaterals for use in coordinate proofs.
- Prove theorems using coordinate proofs.

How can you use a coordinate plane to prove theorems about quadrilaterals?

In Chapter 4, you learned that variable coordinates can be assigned to the vertices of triangles. Then the Distance and Midpoint Formulas and coordinate proofs were used to prove theorems. The same is true for quadrilaterals.

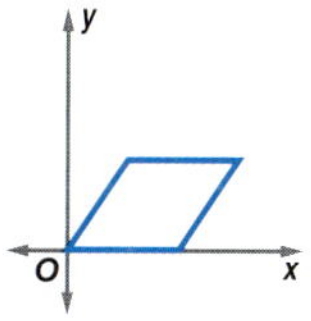

POSITION FIGURES The first step to using a coordinate proof is to place the figure on the coordinate plane. The placement of the figure can simplify the steps of the proof.

Study Tip

Look Back
To review **placing a figure on a coordinate plane**, see Lesson 4-7.

Example 1 Positioning a Square

Position and label a square with sides a units long on the coordinate plane.

- Let A, B, C, and D be vertices of a square with sides a units long.
- Place the square with vertex A at the origin, $\overline{AB}$ along the positive x-axis, and $\overline{AD}$ along the y-axis. Label the vertices A, B, C, and D.
- The y-coordinate of B is 0 because the vertex is on the x-axis. Since the side length is a, the x-coordinate is a.
- D is on the y-axis so the x-coordinate is 0. The y-coordinate is $0 + a$ or a.
- The x-coordinate of C is also a. The y-coordinate is $0 + a$ or a because the side $\overline{BC}$ is a units long.

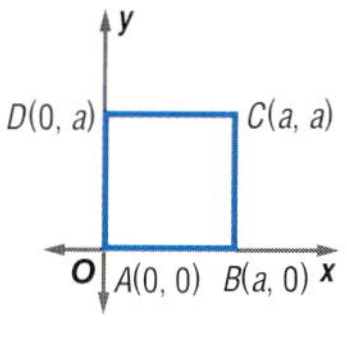

Some examples of quadrilaterals placed on the coordinate plane are given below. Notice how the figures have been placed so the coordinates of the vertices are as simple as possible.

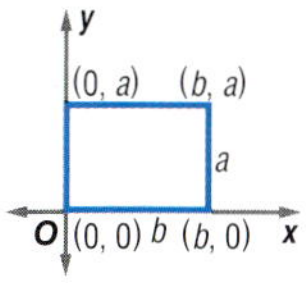

rectangle

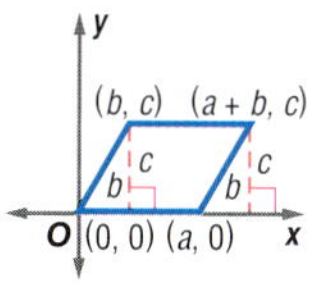

parallelogram

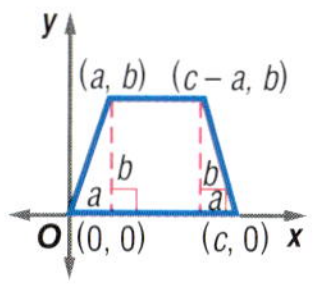

isosceles trapezoid

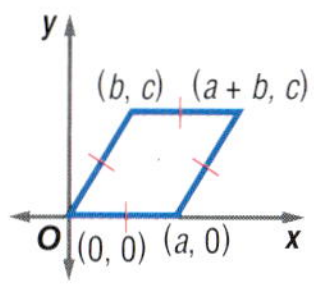

rhombus

8-7 Lesson Notes

1 Focus

5-Minute Check Transparency 8-7 Use as a quiz or review of Lesson 8-6.

Mathematical Background notes are available for this lesson on p. 402D.

How can you use a coordinate plane to prove theorems about quadrilaterals?

Ask students:

- What is a convenient location to place a square on the coordinate plane? with one vertex at (0, 0) and one side along the *x*-axis
- What is usually the best position to place a figure on the coordinate plane? Sample answer: with one side parallel to an axis and one vertex at (0, 0)
- What formulas have been used in the past in coordinate proofs? Distance Formula; Slope Formula; Midpoint Formula

Resource Manager

Workbook and Reproducible Masters

Chapter 8 Resource Masters

- Study Guide and Intervention, pp. 453–454
- Skills Practice, p. 455
- Practice, p. 456
- Reading to Learn Mathematics, p. 457
- Enrichment, p. 458
- Assessment, p. 474

Teaching Geometry With Manipulatives Masters, pp. 2, 8, 138

Transparencies

5-Minute Check Transparency 8-7
Answer Key Transparencies

Technology

Interactive Chalkboard
Multimedia Applications: Virtual Activities

2 Teach

POSITION FIGURES

In-Class Examples

Power Point®

1 POSITIONING A RECTANGLE Position and label a rectangle with sides a and b units long on the coordinate plane.

Sample answer:

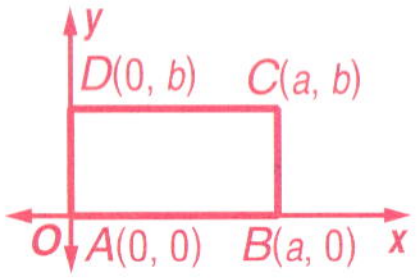

2 Name the missing coordinates for the isosceles trapezoid. **D(b, c)**

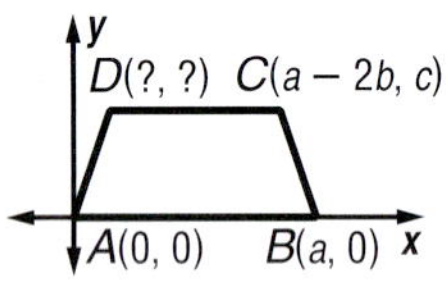

3 Place a rhombus on the coordinate plane. Label the midpoints of the sides M, N, P, and Q. Write a coordinate proof to prove that $MNPQ$ is a rectangle.

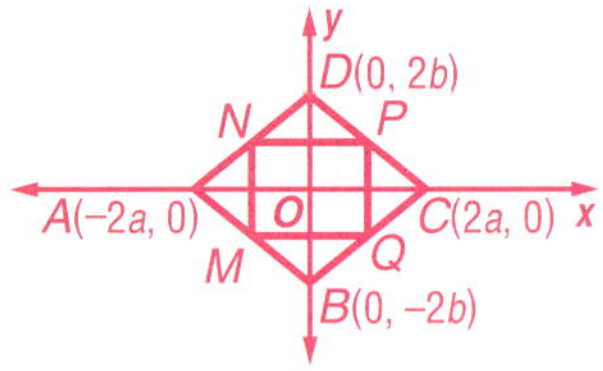

Place rhombus $ABCD$ on the coordinate plane so that the origin is the midpoint of the diagonals and the diagonals are on the axes, as shown.

Given: $ABCD$ is a rhombus as labeled. M, N, P, Q are midpoints.

Prove: $MNPQ$ is a rectangle.

The coordinates of M are $(-a, -b)$; the coordinates of N are $(-a, b)$; the coordinates of P are (a, b); the coordinates of Q are $(a, -b)$. Slope of $\overline{MQ} = 0$; slope of $\overline{QP}$ = undefined; slope of $\overline{PN} = 0$; slope of $\overline{NM}$ = undefined. A segment with slope 0 is perpendicular to a segment with undefined slope. Therefore, consecutive sides of this quadrilateral are perpendicular. Since consecutive sides are perpendicular, $MNPQ$ is, by definition, a rectangle.

Example 2 Find Missing Coordinates

Name the missing coordinates for the parallelogram.

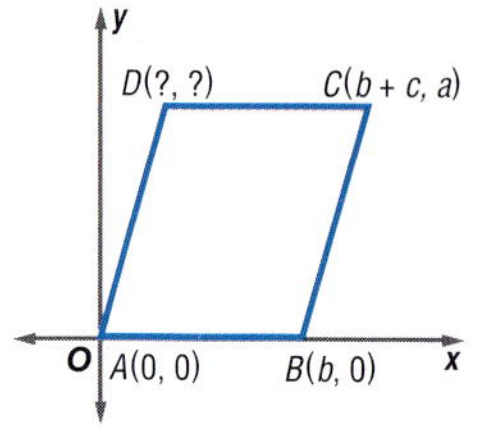

Opposite sides of a parallelogram are congruent and parallel. So, the y-coordinate of D is a.

The length of $\overline{AB}$ is b, and the length of $\overline{DC}$ is b. So, the x-coordinate of D is $(b + c) - b$ or c.

The coordinates of D are (c, a).

PROVE THEOREMS Once a figure has been placed on the coordinate plane, we can prove theorems using the Slope, Midpoint, and Distance Formulas.

Geometry Software Investigation

Quadrilaterals

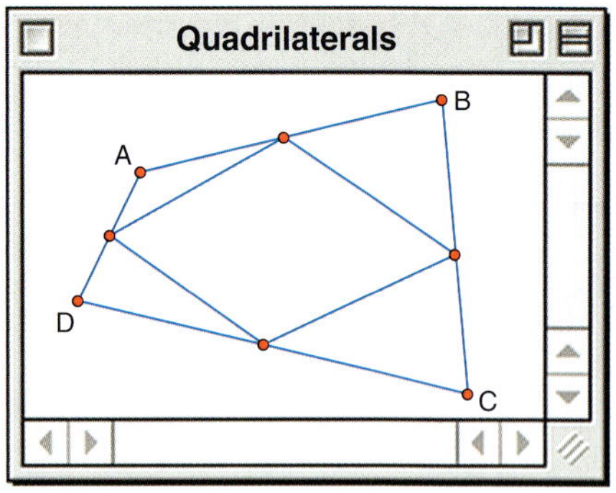

Model

- Use The Geometer's Sketchpad to draw a quadrilateral $ABCD$ with no two sides parallel or congruent.
- Construct the midpoints of each side.
- Draw the quadrilateral formed by the midpoints of the segments.

Analyze

1. Measure each side of the quadrilateral determined by the midpoints of $ABCD$. **See students' work.**
2. What type of quadrilateral is formed by the midpoints? Justify your answer. **Parallelogram; the opposite sides are congruent.**

In this activity, you discover that the quadrilateral formed from the midpoints of any quadrilateral is a parallelogram. You will prove this in Exercise 22.

Study Tip

Problem Solving

To prove that a quadrilateral is a square, you can also show that all sides are congruent and that the diagonals bisect each other.

Example 3 Coordinate Proof

Place a square on a coordinate plane. Label the midpoints of the sides, M, N, P, and Q. Write a coordinate proof to prove that $MNPQ$ is a square.

The first step is to position a square on the coordinate plane. Label the vertices to make computations as simple as possible.

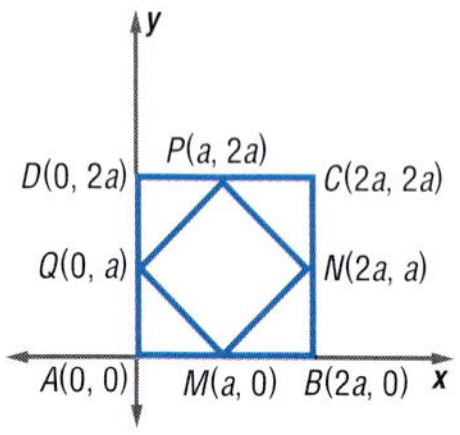

Given: $ABCD$ is a square. M, N, P, and Q are midpoints.

Prove: $MNPQ$ is a square.

Proof:

By the Midpoint Formula, the coordinates of M, N, P, and Q are as follows.

$$M\left(\frac{2a+0}{2}, \frac{0+0}{2}\right) = (a, 0) \qquad N\left(\frac{2a+2a}{2}, \frac{2a+0}{2}\right) = (2a, a)$$

$$P\left(\frac{0+2a}{2}, \frac{2a+2a}{2}\right) = (a, 2a) \qquad Q\left(\frac{0+0}{2}, \frac{0+2a}{2}\right) = (0, a)$$

Geometry Software Investigation

- Have students drag one or more of the vertices to create a different quadrilateral. Analyze the new quadrilateral.
- This activity can be repeated with different quadrilaterals. Investigate the quadrilateral formed by the midpoints of the sides of a square, a rhombus, or a trapezoid.

Find the slopes of $\overline{QP}$, $\overline{MN}$, $\overline{QM}$, and $\overline{PN}$.

slope of $\overline{QP} = \frac{2a - a}{a - 0}$ or 1 slope of $\overline{MN} = \frac{a - 0}{2a - a}$ or 1

slope of $\overline{QM} = \frac{a - 0}{0 - a}$ or -1 slope of $\overline{PN} = \frac{2a - a}{a - 2a}$ or -1

Each pair of opposite sides is parallel, so they have the same slope. Consecutive sides form right angles because their slopes are negative reciprocals.

Use the Distance Formula to find the length of $\overline{QP}$ and $\overline{QM}$.

$$QP = \sqrt{(a - 0)^2 + (2a - a)^2} = \sqrt{a^2 + a^2} = \sqrt{2a^2} \text{ or } a\sqrt{2}$$

$$QM = \sqrt{(a - 0)^2 + (0 - a)^2} = \sqrt{a^2 + a^2} = \sqrt{2a^2} \text{ or } a\sqrt{2}$$

MNPQ is a square because each pair of opposite sides is parallel, and consecutive sides form right angles and are congruent.

Example 4 *Properties of Quadrilaterals*

PARKING **Write a coordinate proof to prove that the sides of the parking space are parallel.**

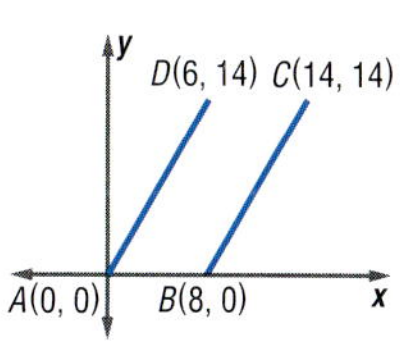

Given: $A(0, 0)$, $B(8, 0)$, $C(14, 14)$, $D(6, 14)$

Prove: $\overline{AD} \parallel \overline{BC}$

Proof:

slope of $\overline{AD} = \frac{14 - 0}{6 - 0}$ or $\frac{7}{3}$

slope of $\overline{BC} = \frac{14 - 0}{14 - 8}$ or $\frac{7}{3}$

Since $\overline{AD}$ and $\overline{BC}$ have the same slope, they are parallel.

Check for Understanding

Concept Check

1. **Explain** how to position a quadrilateral to simplify the steps of the proof.
2. **OPEN ENDED** Position and label a trapezoid with two vertices on the *y*-axis.
1–2. See p. 459D.

Guided Practice

GUIDED PRACTICE KEY	
Exercises	Examples
3	1
4–5	2
6–7	3
8	4

Position and label the quadrilateral on the coordinate plane.

3. rectangle with length a units and height $a + b$ units **See p. 459D.**

Name the missing coordinates for each quadrilateral.

4. **(a, a)**

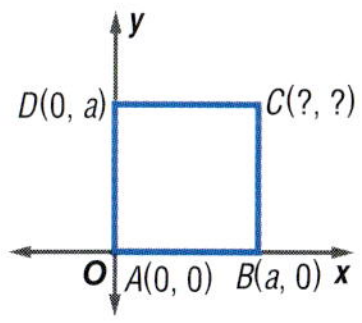

5. **(c, b)**

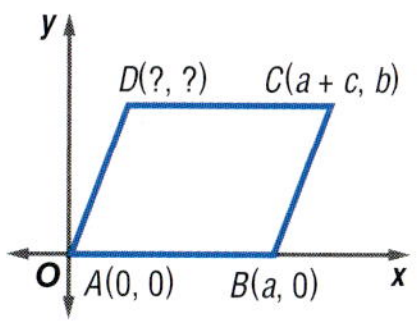

Write a coordinate proof for each statement. **6–7. See p. 459D.**

6. The diagonals of a parallelogram bisect each other.
7. The diagonals of a square are perpendicular.

www.geometryonline.com/extra_examples/fcat

PROVE THEOREMS

In-Class Example

Power Point®

4 Write a coordinate proof to prove that the supports of a platform lift are parallel.

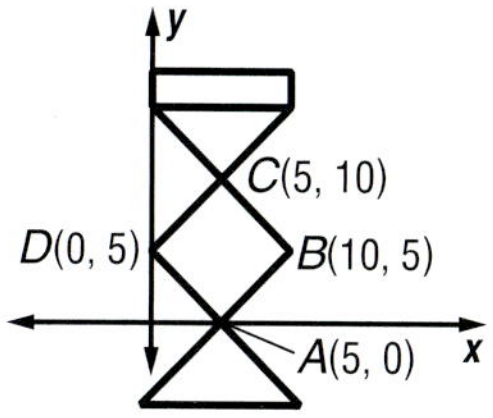

Given: $A(5, 0)$, $B(10, 5)$, $C(5, 10)$, and $D(0, 5)$.

Prove: $\overline{AB} \parallel \overline{CD}$

Slope of $\overline{AB} = 1$. Slope of $\overline{CD} = 1$. Since $\overline{AB}$ and $\overline{CD}$ have the same slope, they are parallel.

3 Practice/Apply

Study Notebook

Have students—

- add the definitions/examples of the vocabulary terms to their Vocabulary Builder worksheets for Chapter 8.
- include any other item(s) that they find helpful in mastering the skills in this lesson.

About the Exercises...

Organization by Objective

- **Position Figures:** 9–16, 23
- **Prove Theorems:** 17–22, 24–26

Odd/Even Assignments

Exercises 9–22 are structured so that students practice the same concepts whether they are assigned odd or even problems.

Assignment Guide

Basic: 9–19 odd, 23, 25, 27–39

Average: 9–27 odd, 28–39

Advanced: 10–22 even, 23, 24, 26, 27–39

DAILY INTERVENTION

Differentiated Instruction

Interpersonal Students may benefit from going on a short walk around the school campus with other geometry students to photograph or sketch examples of quadrilaterals in the real world. As they discover quadrilaterals, ask them to discuss how they can separate the quadrilaterals into the classes studied in this chapter.

Study Guide and Intervention, p. 453 (shown) and p. 454

Position Figures Coordinate proofs use properties of lines and segments to prove geometric properties. The first step in writing a coordinate proof is to place the figure on the coordinate plane in a convenient way. Use the following guidelines for placing a figure on the coordinate plane.

1. Use the origin as a vertex, so one set of coordinates is (0, 0), or use the origin as the center of the figure.
2. Place at least one side of the quadrilateral on an axis so you will have some zero coordinates.
3. Try to keep the quadrilateral in the first quadrant so you will have positive coordinates.
4. Use coordinates that make the computations as easy as possible. For example, use even numbers if you are going to be finding midpoints.

Example **Position and label a rectangle with sides a and b units long on the coordinate plane.**

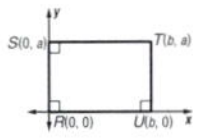

- Place one vertex at the origin for R, so one vertex is $R(0, 0)$.
- Place side $\overline{RU}$ along the x-axis and side $\overline{RS}$ along the y-axis, with the rectangle in the first quadrant.
- The sides are a and b units, so label two vertices $S(0, a)$ and $U(b, 0)$.
- Vertex T is b units right and a units up, so the fourth vertex is $T(b, a)$.

Exercises

Name the missing coordinates for each quadrilateral.

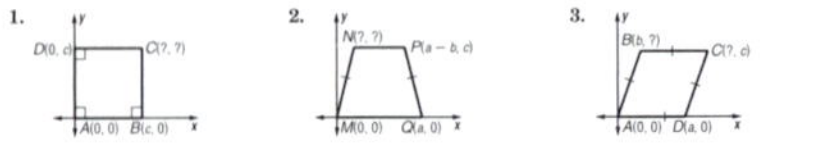

1. $C(c, c)$ 2. $N(b, c)$ 3. $B(b, c)$; $C(a + b, c)$

Position and label each quadrilateral on the coordinate plane. Sample answers are given.

4. square $STUV$ with side s units
5. parallelogram $PQRS$ with congruent diagonals
6. rectangle $ABCD$ with length twice the width

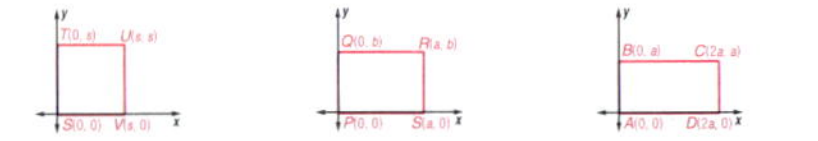

Skills Practice, p. 455 and Practice, p. 456 (shown)

Position and label each quadrilateral on the coordinate plane.

1. parallelogram with side length b units and height a units
2. isosceles trapezoid with height b units, bases $2c - a$ units and $2c + a$ units

Name the missing coordinates for each quadrilateral.

3. parallelogram
4. isosceles trapezoid

$K(2b + c, a)$, $L(c, a)$ $X(-a, 0)$, $Y(-b, c)$

Position and label the figure on the coordinate plane. Then write a coordinate proof for the following.

5. The opposite sides of a parallelogram are congruent.

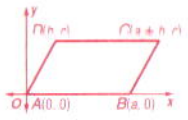

Given: $ABCD$ is a parallelogram.
Prove: $\overline{AB} \cong \overline{CD}$, $\overline{AD} \cong \overline{BC}$
Proof:
$AB = \sqrt{(a - 0)^2 + (0 - 0)^2} = \sqrt{a^2}$ or a
$CD = \sqrt{[(a + b) - b]^2 + (c - c)^2} = \sqrt{a^2}$ or a
$AD = \sqrt{(b - 0)^2 + (c - 0)^2} = \sqrt{b^2 + c^2}$
$AB = \sqrt{[(a + b) - a]^2 + (c - 0)^2} = \sqrt{b^2 + c^2}$
$AB = CD$ and $AD = BC$, so $\overline{AB} \cong \overline{CD}$ and $\overline{AD} \cong \overline{BC}$

6. **THEATER** A stage is in the shape of a trapezoid. Write a coordinate proof to prove that $\overline{TR}$ and $\overline{SF}$ are parallel.

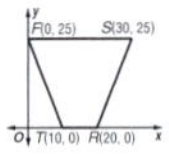

Given: $T(10, 0)$, $R(20, 0)$, $S(30, 25)$, $F(0, 25)$
Prove: $\overline{TR} \parallel \overline{SF}$
Proof: The slope of $\overline{TR} = \frac{0 - 0}{20 - 10} = \frac{0}{10}$ and the slope of $SF = \frac{25 - 25}{30 - 0} = \frac{0}{30}$. Since $\overline{TR}$ and $\overline{SF}$ both have a slope of 0, $\overline{TR} \parallel \overline{SF}$.

Reading to Learn Mathematics, p. 457 ELL

Pre-Activity **How can you use a coordinate plane to prove theorems about quadrilaterals?**

Read the introduction to Lesson 8-7 at the top of page 447 in your textbook.

What special kinds of quadrilaterals can be placed on a coordinate system so that two sides of the quadrilateral lie along the axes?
rectangles and squares

Reading the Lesson

1. Find the missing coordinates in each figure. Then write the coordinates of the four vertices of the quadrilateral.
   a. isosceles trapezoid
   b. parallelogram

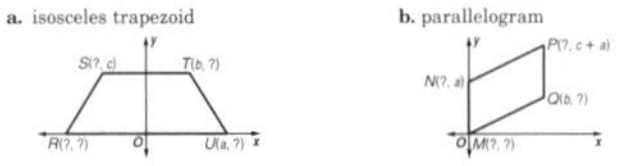

$R(-a, 0)$, $S(-b, c)$, $T(b, c)$, $U(a, 0)$ $M(0, 0)$, $N(0, a)$, $P(b, c + a)$, $Q(b, c)$

2. Refer to quadrilateral $EFGH$.

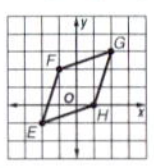

   a. Find the slope of each side. slope $\overline{EF} = 3$, slope $\overline{FG} = \frac{1}{3}$, slope $\overline{GH} = 3$, slope $\overline{HE} = \frac{1}{3}$
   b. Find the length of each side. $EF = \sqrt{10}$, $FG = \sqrt{10}$, $GH = \sqrt{10}$, $HE = \sqrt{10}$
   c. Find the slope of each diagonal. slope $\overline{EG} = 1$, slope $\overline{FH} = -1$
   d. Find the length of each diagonal. $EG = \sqrt{32}$, $FH = \sqrt{8}$
   e. What can you conclude about the sides of $EFGH$? All four sides are congruent and both pairs of opposite sides are parallel.
   f. What can you conclude about the diagonals of $EFGH$? The diagonals are perpendicular and they are not congruent.
   g. Classify $EFGH$ as a *parallelogram*, a *rhombus*, or a *square*. Choose the most specific term. Explain how your results from parts a-f support your conclusion. *EFGH* is a rhombus. All four sides are congruent and the diagonals are perpendicular. Since the diagonals are not congruent, *EFGH* is not a square.

Helping You Remember

3. What is an easy way to remember how best to draw a diagram that will help you devise a coordinate proof? Sample answer: A key point in the coordinate plane is the origin. The everyday meaning of *origin* is *place where something begins*. So look to see if there is a good way to begin by placing a vertex of the figure at the origin.

Application

8. **STATES** The state of Tennessee can be separated into two shapes that resemble quadrilaterals. Write a coordinate proof to prove that $DEFG$ is a trapezoid. All measures are approximate and given in kilometers. See p. 459D.

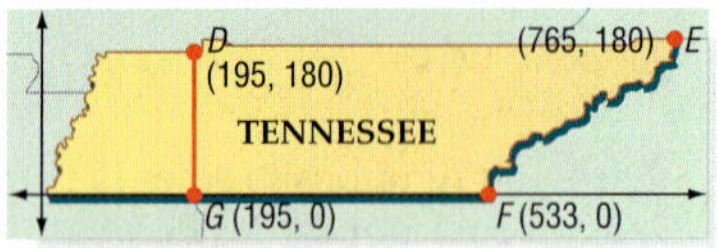

★ indicates increased difficulty

Practice and Apply

Homework Help	
For Exercises	**See Examples**
9–10	1
11–16, 23	2
17–22	3
24–26	4

Extra Practice See page 771.

Position and label each quadrilateral on the coordinate plane. 9–10. See margin.

9. isosceles trapezoid with height c units, bases a units and $a + 2b$ units
10. parallelogram with side length c units and height b units

Name the missing coordinates for each parallelogram or trapezoid.

11. 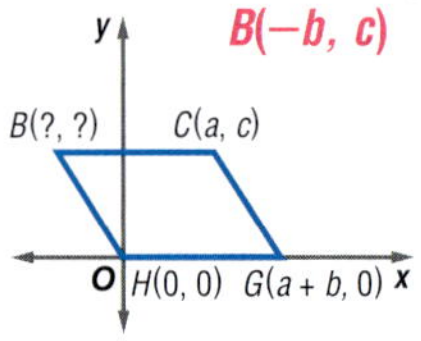

$B(-b, c)$

12. 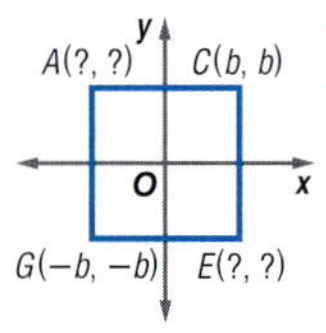

$A(-b, b)$, $E(b, -b)$

13. 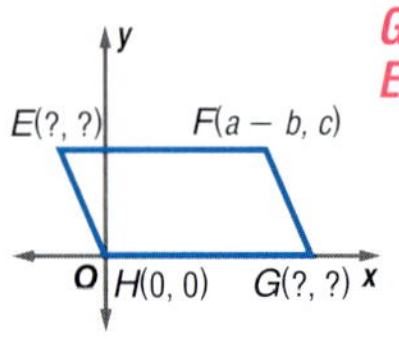

$G(a, 0)$, $E(-b, c)$

14. 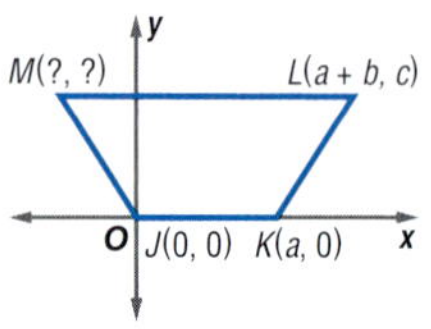

$M(-b, c)$

15. 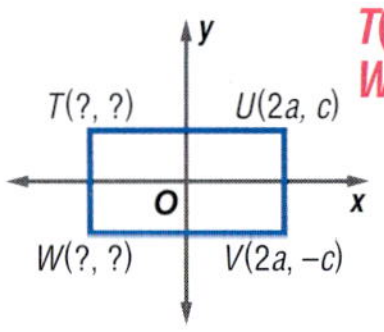

$T(-2a, c)$, $W(-2a, -c)$

16. 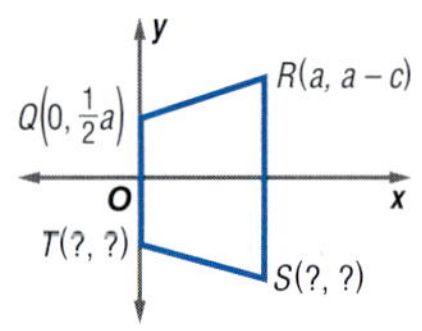

$T\left(0, -\frac{1}{2}a\right)$, $S(a, -a + c)$

Position and label each figure on the coordinate plane. Then write a coordinate proof for each of the following. 17–22. See pp. 459D–459E.

17. The diagonals of a rectangle are congruent.
18. If the diagonals of a parallelogram are congruent, then it is a rectangle.
19. The diagonals of an isosceles trapezoid are congruent.
20. The median of an isosceles trapezoid is parallel to the bases.
★21. The segments joining the midpoints of the sides of a rectangle form a rhombus.
★22. The segments joining the midpoints of the sides of a quadrilateral form a parallelogram.

★23. **CRITICAL THINKING** A has coordinates $(0, 0)$, and B has coordinates (a, b). Find the coordinates of C and D so $ABCD$ is an isosceles trapezoid. Sample answer: $C(a + c, b)$, $D(2a + c, 0)$

450 Chapter 8 Quadrilaterals

Enrichment, p. 458

Coordinate Proofs

An important part of planning a coordinate proof is correctly placing and labeling the geometric figure on the coordinate plane.

Example **Draw a diagram you would use in a coordinate proof of the following theorem.**

The median of a trapezoid is parallel to the bases of the trapezoid.

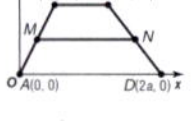

In the diagram, note that one vertex, A, is placed at the origin. Also, the coordinates of B, C, and D use $2a$, $2b$, and $2c$ in order to avoid the use of fractions when finding the coordinates of the midpoints, M and N.

When doing coordinate proofs, the following strategies may be helpful.

1. If you are asked to prove that segments are parallel or perpendicular, use slopes.
2. If you are asked to prove that segments are congruent or have related measures, use the distance formula.
3. If you are asked to prove that a segment is bisected, use the midpoint formula.

More About. . .

Architecture

The tower is also sinking. In 1838, the foundation was excavated to reveal the bases of the columns.

Source: www.torre.duomo.pisa.it

ARCHITECTURE **For Exercises 24–26, use the following information.**

The Leaning Tower of Pisa is approximately 60 meters tall, from base to belfry. The tower leans about 5.5° so the top level is 4.5 meters over the first level.

24. Position and label the tower on a coordinate plane. **24–26. See margin.**
25. Is it possible to write a coordinate proof to prove that the sides of the tower are parallel? Explain.
26. From the given information, what conclusion can be drawn?
27. **WRITING IN MATH** Answer the question that was posed at the beginning of the lesson. **See margin.**

How is the coordinate plane used in proofs?

Include the following in your answer:
- guidelines for placing a figure on a coordinate grid, and
- an example of a theorem from this chapter that could be proved using the coordinate plane.

Standardized Test Practice

28. In the figure, *ABCD* is a parallelogram. What are the coordinates of point *D*? **D**

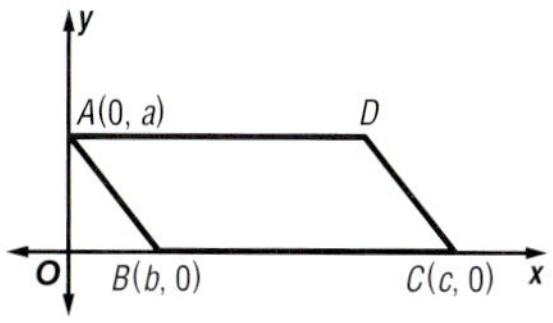

Ⓐ $(a, c + b)$ Ⓑ $(c + b, a)$
Ⓒ $(b - c, a)$ Ⓓ $(c - b, a)$

29. **ALGEBRA** If $p = -5$, then $5 - p^2 - p =$ __?__. **A**

Ⓐ −15 Ⓑ −5
Ⓒ 10 Ⓓ 30

Maintain Your Skills

Mixed Review

30. **PROOF** Write a two-column proof. *(Lesson 8-6)*

Given: *MNOP* is a trapezoid with bases $\overline{MN}$ and $\overline{OP}$. $\overline{MN} \cong \overline{QO}$

Prove: *MNOQ* is a parallelogram. **See margin.**

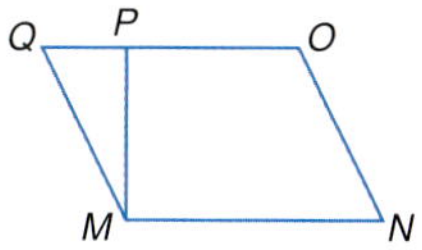

JKLM is a rectangle. MLPR is a rhombus. ∠JMK ≅ ∠RMP, m∠JMK = 55, and m∠MRP = 70. *(Lesson 8-5)*

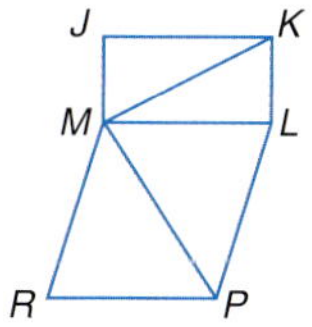

31. Find $m\angle MPR$. **55**
32. Find $m\angle KML$. **35**
33. Find $m\angle KLP$. **160**

Find the geometric mean between each pair of numbers. *(Lesson 7-1)*

34. 7 and 14 $\sqrt{98} \approx 9.9$
35. $2\sqrt{5}$ and $6\sqrt{5}$ $\sqrt{60} \approx 7.7$

Write an expression relating the given pair of angle measures. *(Lesson 5-5)*

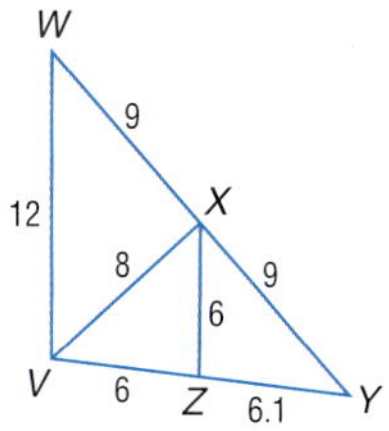

36. $m\angle WVX, m\angle VXY$ $m\angle WVX < m\angle VXY$
37. $m\angle XVZ, m\angle VXZ$ $m\angle XVZ = m\angle VXZ$
38. $m\angle XYV, m\angle VXY$ $m\angle XYV < m\angle VXY$
39. $m\angle XZY, m\angle ZXY$ $m\angle XZY > m\angle ZXY$

www.geometryonline.com/self_check_quiz

4 Assess

Open-Ended Assessment

Writing Ask students to describe how to position each type of quadrilateral on the coordinate plane so that the coordinates of the vertices are as simple as possible.

Assessment Options

Quiz (Lesson 8-7) is available on p. 474 of the *Chapter 8 Resource Masters.*

27. Sample answer: The coordinate plane is used in coordinate proofs. The Distance Formula, Midpoint Formula and Slope Formula are used to prove theorems. Answers should include the following.
- Place the figure so one of the vertices is at the origin. Place at least one side of the figure on the positive *x*-axis. Keep the figure in the first quadrant if possible and use coordinates that will simplify calculations.
- Sample answer: Theorem 8.3 Opposite sides of a parallelogram are congruent.

30. Given: *MNOP* is a trapezoid with bases $\overline{MN}$ and $\overline{OP}$. $\overline{MN} \cong \overline{QO}$

Prove: *MNOQ* is a parallelogram.

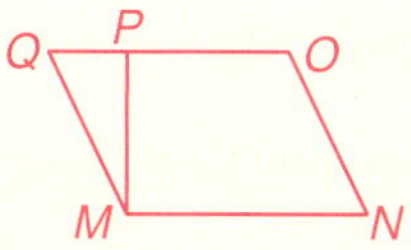

Proof:

Statements (Reasons)

1. *MNOP* is a trapezoid with bases $\overline{MN}$ and $\overline{OP}$; $\overline{MN} \cong \overline{QO}$ (Given)
2. $\overline{OP} \parallel \overline{MN}$ (Def. of a trapezoid.)
3. *MNOQ* is a parallelogram. (If one pair of opp. sides are ∥ and ≅, the quad. is a ▱.)

Answers

24.

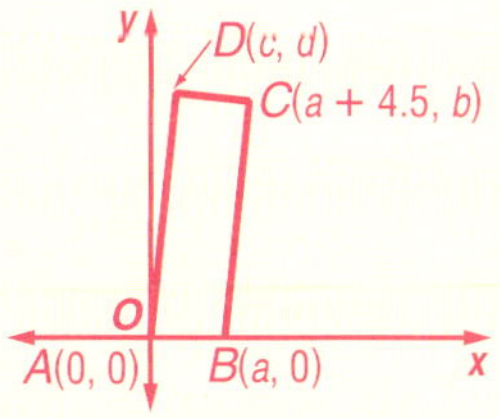

25. No, there is not enough information given to prove that the sides of the tower are parallel.
26. From the information given, we can approximate the height from the ground to the top level of the tower.

Chapter 8 Study Guide and Review

Vocabulary and Concept Check

- This alphabetical list of vocabulary terms in Chapter 8 includes a page reference where each term was introduced.
- **Assessment** A vocabulary test/review for Chapter 8 is available on p. 472 of the *Chapter 8 Resource Masters*.

Lesson-by-Lesson Review

For each lesson,
- the main ideas are summarized,
- additional examples review concepts, and
- practice exercises are provided.

Vocabulary PuzzleMaker

ELL The Vocabulary PuzzleMaker software improves students' mathematics vocabulary using four puzzle formats—crossword, scramble, word search using a word list, and word search using clues. Students can work on a computer screen or from a printed handout.

MindJogger Videoquizzes

ELL MindJogger Videoquizzes provide an alternative review of concepts presented in this chapter. Students work in teams in a game show format to gain points for correct answers. The questions are presented in three rounds.

Round 1 Concepts (5 questions)
Round 2 Skills (4 questions)
Round 3 Problem Solving (4 questions)

Chapter 8 Study Guide and Review

Vocabulary and Concept Check

diagonal (p. 404)	median (p. 440)	rhombus (p. 431)
isosceles trapezoid (p. 439)	parallelogram (p. 411)	square (p. 432)
kite (p. 438)	rectangle (p. 424)	trapezoid (p. 439)

A complete list of postulates and theorems can be found on pages R1–R8.

Exercises **State whether each sentence is *true* or *false*. If false, replace the underlined term to make a true sentence.**

1. The diagonals of a rhombus are perpendicular. **true**
2. All squares are rectangles. **true**
3. If a parallelogram is a rhombus, then the diagonals are congruent. **false; rectangle**
4. Every parallelogram is a quadrilateral. **true**
5. A(n) rhombus is a quadrilateral with exactly one pair of parallel sides. **false; trapezoid**
6. Each diagonal of a rectangle bisects a pair of opposite angles. **false; rhombus**
7. If a quadrilateral is both a rhombus and a rectangle, then it is a square. **true**
8. Both pairs of base angles in a(n) isosceles trapezoid are congruent. **true**

Lesson-by-Lesson Review

8-1 Angles of Polygons

See pages 404–409.

Concept Summary

- If a convex polygon has n sides and the sum of the measures of its interior angles is S, then $S = 180(n - 2)$.
- The sum of the measures of the exterior angles of a convex polygon is 360.

Example **Find the measure of an interior angle of a regular decagon.**

$S = 180(n - 2)$ Interior Angle Sum Theorem

$= 180(10 - 2)$ $n = 10$

$= 180(8)$ or 1440 Simplify.

The measure of each interior angle is $1440 \div 10$, or 144.

Exercises **Find the measure of each interior angle of a regular polygon given the number of sides.** *See Example 1 on page 405.*

9. 6 **120** 10. 15 **156** 11. 4 **90** 12. 20 **162**

ALGEBRA **Find the measure of each interior angle.** *See Example 3 on page 405.*

13.

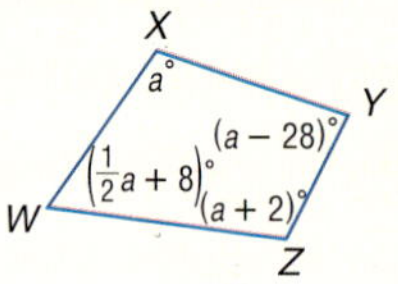

$m\angle W = 62$, $m\angle X = 108$, $m\angle Y = 80$, $m\angle Z = 110$

14.

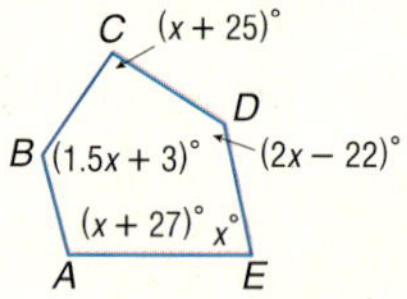

$m\angle A = 105$, $m\angle B = 120$, $m\angle C = 103$, $m\angle D = 134$, $m\angle E = 78$

www.geometryonline.com/vocabulary_revie

FOLDABLES™ Study Organizer

For more information about Foldables, see *Teaching Mathematics with Foldables.*

Have students look through the chapter to make sure they have included notes and examples in their Foldables for each lesson of Chapter 8.

Encourage students to refer to their Foldables while completing the Study Guide and Review and to use them in preparing for the Chapter Test.

452 Chapter 8 Quadrilaterals

8-2 Parallelograms

See pages 411–416.

Concept Summary

- In a parallelogram, opposite sides are parallel and congruent, opposite angles are congruent, and consecutive angles are supplementary.
- The diagonals of a parallelogram bisect each other.

Example **WXYZ is a parallelogram. Find $m\angle YZW$ and $m\angle XWZ$.**

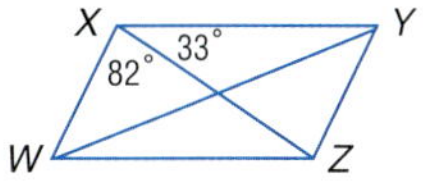

$m\angle YZW = m\angle WXY$	Opp. ∠s of ▱ are ≅.
$m\angle YZW = 82 + 33 \text{ or } 115$	$m\angle WXY = m\angle WXZ + m\angle YXZ$
$m\angle XWZ + m\angle WXY = 180$	Cons. ∠s in ▱ are suppl.
$m\angle XWZ + (82 + 33) = 180$	$m\angle WXY = m\angle WXZ + m\angle YXZ$
$m\angle XWZ + 115 = 180$	Simplify.
$m\angle XWZ = 65$	Subtract 115 from each side.

Exercises **Use ▱ABCD to find each measure.**

See Example 2 on page 413.

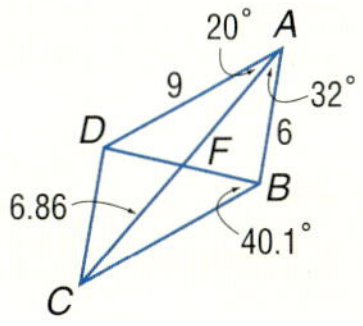

15. $m\angle BCD$ **52**
16. AF **6.86**
17. $m\angle BDC$ **87.9**
18. BC **9**
19. CD **6**
20. $m\angle ADC$ **128**

8-3 Tests for Parallelograms

See pages 417–423.

Concept Summary

A quadrilateral is a parallelogram if any one of the following is true.

- Both pairs of opposite sides are parallel and congruent.
- Both pairs of opposite angles are congruent.
- Diagonals bisect each other.
- A pair of opposite sides is both parallel and congruent.

Example COORDINATE GEOMETRY **Determine whether the figure with vertices $A(-5, 3)$, $B(-1, 5)$, $C(6, 1)$, and $D(2, -1)$ is a parallelogram. Use the Distance and Slope Formulas.**

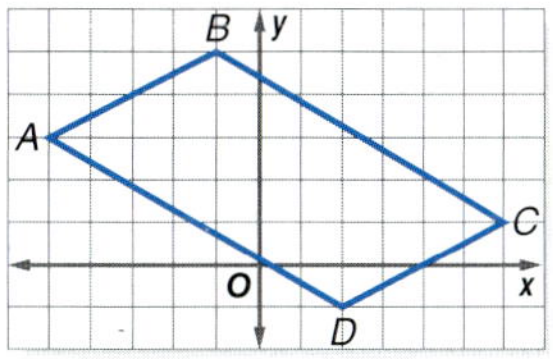

$$AB = \sqrt{[-5-(-1)]^2 + (3-5)^2}$$
$$= \sqrt{(-4)^2 + (-2)^2} \text{ or } \sqrt{20}$$
$$CD = \sqrt{(6-2)^2 + [1-(-1)]^2}$$
$$= \sqrt{4^2 + 2^2} \text{ or } \sqrt{20}$$

Since $AB = CD$, $\overline{AB} \cong \overline{CD}$.

slope of $\overline{AB} = \frac{5-3}{-1-(-5)}$ or $\frac{1}{2}$ slope of $\overline{CD} = \frac{-1-1}{2-6}$ or $\frac{1}{2}$

$\overline{AB}$ and $\overline{CD}$ have the same slope, so they are parallel. Since one pair of opposite sides is congruent and parallel, $ABCD$ is a parallelogram.

Chapter 8 Study Guide and Review 453

Exercises **Determine whether the figure with the given vertices is a parallelogram. Use the method indicated.** *See Example 5 on page 420.*

21. $A(-2, 5)$, $B(4, 4)$, $C(6, -3)$, $D(-1, -2)$; Distance Formula **no**
22. $H(0, 4)$, $J(-4, 6)$, $K(5, 6)$, $L(9, 4)$; Midpoint Formula **yes**
23. $S(-2, -1)$, $T(2, 5)$, $V(-10, 13)$, $W(-14, 7)$; Slope Formula **yes**

8-4 Rectangles

See pages 424–430.

Concept Summary

- A rectangle is a quadrilateral with four right angles and congruent diagonals.
- If the diagonals of a parallelogram are congruent, then the parallelogram is a rectangle.

Example

Quadrilateral *KLMN* is a rectangle. If $PL = x^2 - 1$ and $PM = 4x + 11$, find x.

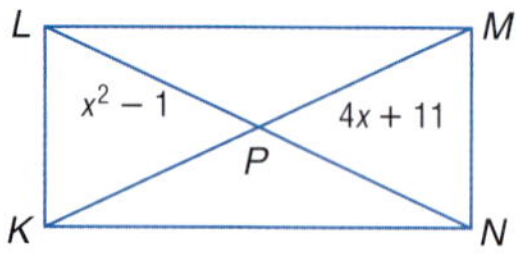

The diagonals of a rectangle are congruent and bisect each other, so $\overline{PL} \cong \overline{PM}$.

$\overline{PL} \cong \overline{PM}$	Diag. are ≅ and bisect each other.
$PL = PM$	Def. of ≅ angles
$x^2 - 1 = 4x + 11$	Substitution
$x^2 - 1 - 4x = 11$	Subtract 4x from each side.
$x^2 - 4x - 12 = 0$	Subtract 11 from each side.
$(x + 2)(x - 6) = 0$	Factor.

$x + 2 = 0$ $\quad$ $x - 6 = 0$

$x = -2$ $\quad$ $x = 6$

The value of x is -2 or 6.

Exercises ***ABCD* is a rectangle.**
See Examples 1 and 2 on pages 425 and 426.

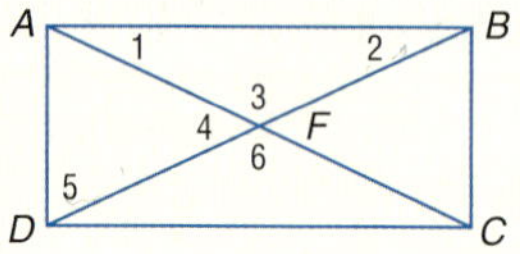

24. If $AC = 9x - 1$ and $AF = 2x + 7$, find AF. **13**
25. If $m\angle 1 = 12x + 4$ and $m\angle 2 = 16x - 12$, find $m\angle 2$. **52**
26. If $CF = 4x + 1$ and $DF = x + 13$, find x. **4**
27. If $m\angle 2 = 70 - 4x$ and $m\angle 5 = 18x - 8$, find $m\angle 5$. **28**

COORDINATE GEOMETRY **Determine whether *RSTV* is a rectangle given each set of vertices. Justify your answer.** *See Example 4 on pages 426 and 427.*

28. $R(-3, -5)$, $S(0, -5)$, $T(3, 4)$, $V(0, 4)$ **No, consec. sides are not perpendicular.**
29. $R(0, 0)$, $S(6, 3)$, $T(4, 7)$, $V(-2, 4)$ **Yes, opp. sides are parallel and diag. are congruent.**

454 Chapter 8 Quadrilaterals

8-5 Rhombi and Squares

See pages 431–437.

Concept Summary

- A rhombus is a quadrilateral with each side congruent, diagonals that are perpendicular, and each diagonal bisecting a pair of opposite angles.
- A quadrilateral that is both a rhombus and a rectangle is a square.

Example **Use rhombus *JKLM* to find $m\angle JMK$ and $m\angle KJM$.**

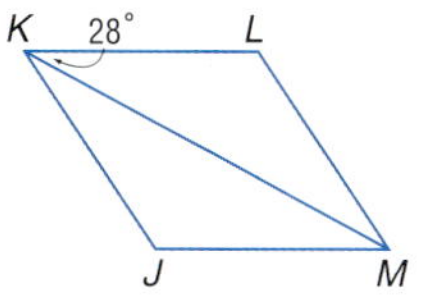

The opposite sides of a rhombus are parallel, so $\overline{KL} \parallel \overline{JM}$. $\angle JMK \cong \angle LKM$ because alternate interior angles are congruent.

$m\angle JMK = m\angle LKM$ Definition of congruence

$= 28$ Substitution

The diagonals of a rhombus bisect the angles, so $\angle JKM \cong \angle LKM$.

$m\angle KJM + m\angle JKL = 180$ Cons. $\angle$s in ▱ are suppl.

$m\angle KJM + (m\angle JKM + m\angle LKM) = 180$ $m\angle JKL = m\angle JKM + m\angle LKM$

$m\angle KJM + (28 + 28) = 180$ Substitution

$m\angle KJM + 56 = 180$ Add.

$m\angle KJM = 124$ Subtract 56 from each side.

Exercises **Use rhombus *ABCD* with $m\angle 1 = 2x + 20$, $m\angle 2 = 5x - 4$, $AC = 15$, and $m\angle 3 = y^2 + 26$.** *See Example 2 on page 432.*

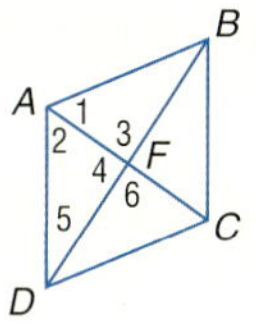

30. Find x. **8**

31. Find AF. **7.5**

32. Find y. **8 or −8**

8-6 Trapezoids

See pages 439–445.

Concept Summary

- In an isosceles trapezoid, both pairs of base angles are congruent and the diagonals are congruent.
- The median of a trapezoid is parallel to the bases, and its measure is one-half the sum of the measures of the bases.

Example **$RSTV$ is a trapezoid with bases $\overline{RV}$ and $\overline{ST}$ and median $\overline{MN}$. Find x if $MN = 60$, $ST = 4x - 1$, and $RV = 6x + 11$.**

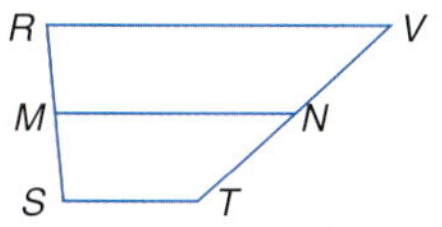

$MN = \frac{1}{2}(ST + RV)$

$60 = \frac{1}{2}[(4x - 1) + (6x + 11)]$ Substitution

$120 = 4x - 1 + 6x + 11$ Multiply each side by 2.

$120 = 10x + 10$ Simplify.

$110 = 10x$ Subtract 10 from each side.

$11 = x$ Divide each side by 10.

Chapter 8 Study Guide and Review 455

Exercises **Find the missing value for the given trapezoid.**
See Example 4 on page 441.

33. For isosceles trapezoid $ABCD$, X and Y are midpoints of the legs. Find $m\angle XBC$ if $m\angle ADY = 78$. **102**

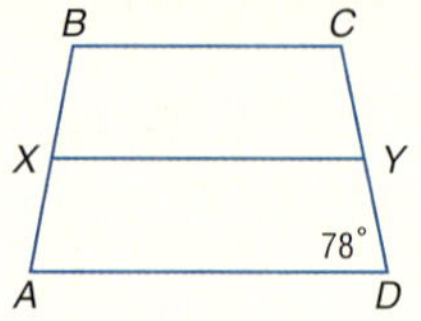

34. For trapezoid $JKLM$, A and B are midpoints of the legs. If $AB = 57$ and $KL = 21$, find JM. **93**

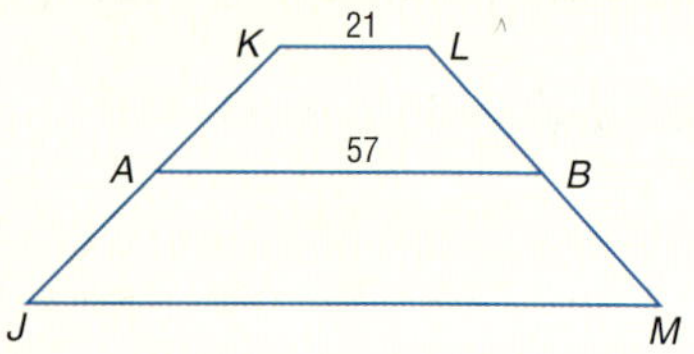

8-7 Coordinate Proof with Quadrilaterals

See pages 447–451.

Concept Summary

- Position a quadrilateral so that a vertex is at the origin and at least one side lies along an axis.

Example **Position and label rhombus $RSTV$ on the coordinate plane. Then write a coordinate proof to prove that each pair of opposite sides is parallel.**

First, draw rhombus $RSTV$ on the coordinate plane. Label the coordinates of the vertices.

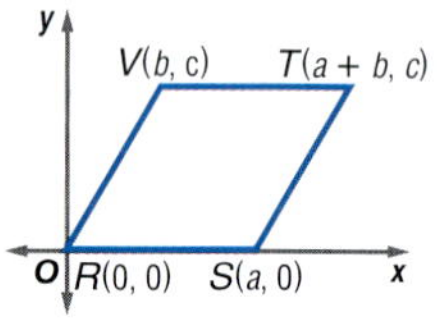

Given: $RSTV$ is a rhombus.

Prove: $\overline{RV} \parallel \overline{ST}$, $\overline{RS} \parallel \overline{VT}$

Proof:

slope of $\overline{RV} = \frac{c-0}{b-0}$ or $\frac{c}{b}$ slope of $\overline{ST} = \frac{c-0}{(a+b)-a}$ or $\frac{c}{b}$

slope of $\overline{RS} = \frac{0-0}{a-0}$ or 0 slope of $\overline{VT} = \frac{c-c}{(a+b)-b}$ or 0

$\overline{RV}$ and $\overline{ST}$ have the same slope. So $\overline{RV} \parallel \overline{ST}$. $\overline{RS}$ and $\overline{VT}$ have the same slope, and $\overline{RS} \parallel \overline{VT}$.

Exercises **Position and label each figure on the coordinate plane. Then write a coordinate proof for each of the following.** *See Example 3 on pages 448 and 449.*

35. The diagonals of a square are perpendicular. **35–36. See margin.**
36. A diagonal separates a parallelogram into two congruent triangles.

Name the missing coordinates for each quadrilateral. *See Example 2 on page 448.*

37. **$P(3a, c)$**

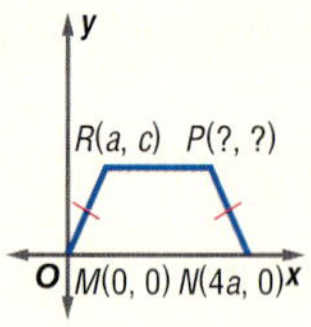

38. **$U(a + b, c)$**

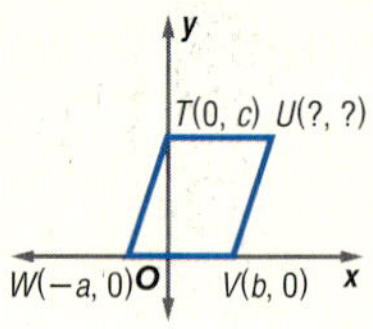

Answers

35. Given: $ABCD$ is a square.
Prove: $\overline{AC} \perp \overline{BD}$

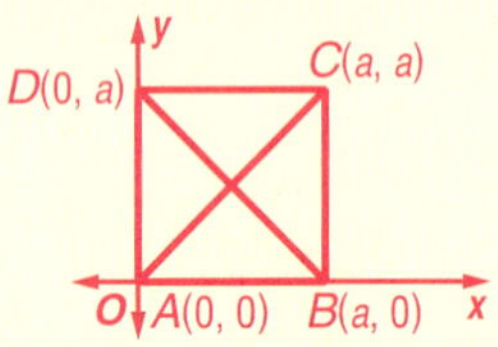

Proof:

Slope of $\overline{AC} = \frac{a-0}{a-0}$ or 1

Slope of $\overline{BD} = \frac{a-0}{0-a}$ or -1

The slope of $\overline{AC}$ is the negative reciprocal of the slope of $\overline{BD}$. Therefore, $\overline{AC} \perp \overline{BD}$.

36. Given: $ABCD$ is a parallelogram.
Prove: $\triangle ABC \cong \triangle CDA$

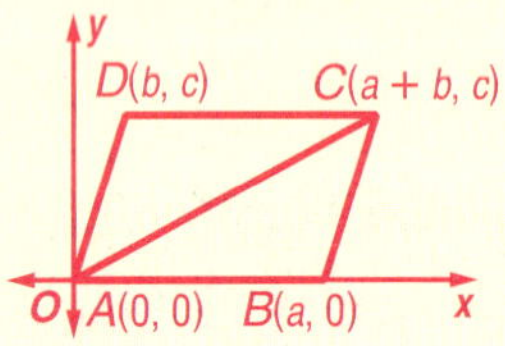

Proof:

$AB = \sqrt{(a-0)^2 + (0-0)^2}$
$= \sqrt{a^2 + 0^2}$ or a

$DC = \sqrt{[(a+b)-b]^2 + (c-c)^2}$
$= \sqrt{a^2 + 0^2}$ or a

$AD = \sqrt{(b-0)^2 + (c-0)^2}$
$= \sqrt{b^2 + c^2}$

$BC = \sqrt{[(a+b)-a]^2 + (c-0)^2}$
$= \sqrt{b^2 + c^2}$

AB and DC have the same measure, so $\overline{AB} \cong \overline{DC}$. AD and BC have the same measure, so $\overline{AD} \cong \overline{BC}$. $\overline{AC} \cong \overline{AC}$ by the Reflexive Property. Therefore, $\triangle ABC \cong \triangle CDA$ by SSS.

Answers (page 457)

2.

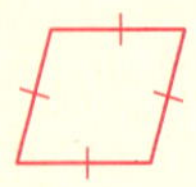

3.

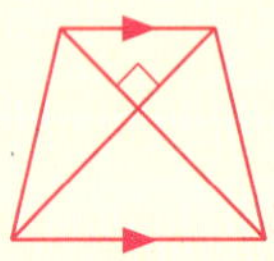

4. $\overline{FG}$; opp. sides of ▱ are ≅.
5. $\angle HGJ$; alt. int. ∠s are ≅.
6. $\angle FGH$; opp. ∠s of ▱ are ≅.
7. $\overline{FK}$; opp. sides of ▱ are ∥.
14. Rectangle, rhombus, square; all sides are ≅, cons. sides are ⊥.
15. Rhombus; all sides are ≅, diag. are ⊥.

Chapter 8 Practice Test

Vocabulary and Concepts

Determine whether each conditional is *true* or *false*. If false, draw a counterexample.

1. If a quadrilateral has four right angles, then it is a rectangle. **true**
2. If a quadrilateral has all four sides congruent, then it is a square. **False; see margin.**
3. If the diagonals of a quadrilateral are perpendicular, then it is a rhombus. **False; see margin.**

Skills and Applications

Complete each statement about ▱$FGHK$. Justify your answer. **4–7. See margin.**

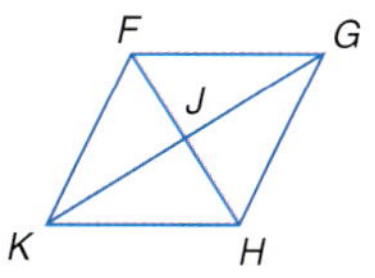

4. $\overline{HK} \cong$ __?__.
5. $\angle FKJ \cong$ __?__.
6. $\angle FKH \cong$ __?__.
7. $\overline{GH} \parallel$ __?__.

Determine whether the figure with the given vertices is a parallelogram. Justify your answer. **8. Yes, diag. bisect each other.** **9. Yes, opp. sides are ≅ and ∥.**

8. $A(4, 3), B(6, 0), C(4, -8), D(2, -5)$
9. $S(-2, 6), T(2, 11), V(3, 8), W(-1, 3)$
10. $F(7, -3), G(4, -2), H(6, 4), J(12, 2)$ **No, opp. sides are not ≅.**
11. $W(-4, 2), X(-3, 6), Y(2, 7), Z(1, 3)$ **Yes, both pairs of opp. sides are ≅.**

ALGEBRA ***QRST* is a rectangle.**

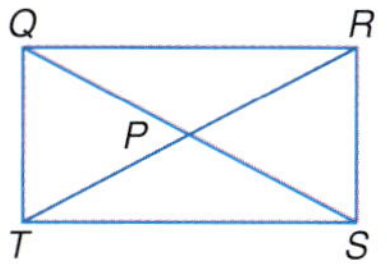

12. If $QP = 3x + 11$ and $PS = 4x + 8$, find QS. **40**
13. If $m\angle QTR = 2x^2 - 7$ and $m\angle SRT = x^2 + 18$, find $m\angle QTR$. **43**

COORDINATE GEOMETRY **Determine whether ▱$ABCD$ is a *rhombus*, a *rectangle*, or a *square*. List all that apply. Explain your reasoning.** **14–15. See margin.**

14. $A(12, 0), B(6, -6), C(0, 0), D(6, 6)$
15. $A(-2, 4), B(5, 6), C(12, 4), D(5, 2)$

Name the missing coordinates for each quadrilateral.

16. 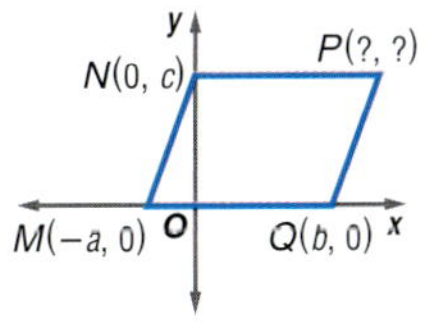

$P(a + b, c)$

17. 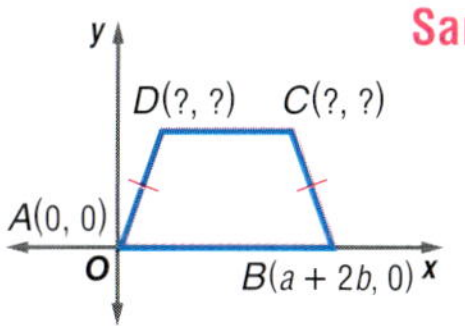

Sample answer: $C(a + b, c)$, $D(b, c)$

18. Position and label an isosceles trapezoid on the coordinate plane. Write a coordinate proof to prove that the median is parallel to each base. **See p. 459F.**

19. **SAILING** Many large sailboats have a *keel* to keep the boat stable in high winds. A keel is shaped like a trapezoid with its top and bottom parallel. If the root chord is 9.8 feet and the tip chord is 7.4 feet, find the length of the mid-chord. **8.6 ft**

20. **STANDARDIZED TEST PRACTICE** The measure of an interior angle of a regular polygon is 108. Find the number of sides. **C**
    Ⓐ 8 Ⓑ 6 Ⓒ 5 Ⓓ 3

www.geometryonline.com/chapter_test

Assessment Options

Vocabulary Test A vocabulary test/review for Chapter 8 can be found on p. 472 of the *Chapter 8 Resource Masters*.

Chapter Tests There are six Chapter 8 Tests and an Open-Ended Assessment task available in the *Chapter 8 Resource Masters*.

Chapter 8 Tests

Form	Type	Level	Pages
1	MC	basic	459–460
2A	MC	average	461–462
2B	MC	average	463–464
2C	FR	average	465–466
2D	FR	average	467–468
3	FR	advanced	469–470

MC = multiple-choice questions
FR = free-response questions

Open-Ended Assessment
Performance tasks for Chapter 8 can be found on p. 471 of the *Chapter 8 Resource Masters*. A sample scoring rubric for these tasks appears on p. A28.

TestCheck and Worksheet Builder

This **networkable software** has three modules for assessment.

- **Worksheet Builder** to make worksheets and tests.
- **Student Module** to take tests on-screen.
- **Management System** to keep student records.

Portfolio Suggestion

Introduction Rectangles, parallelograms, and other shapes are used in architecture.

Ask Students Ask students to design the façade of a house that incorporates as many of the shapes and concepts from this chapter as possible. For each shape, have students write a sentence describing a feature of the shape that is useful in this context. Have students add their designs and descriptions to their portfolios.

Chapter 8 Standardized Test Practice

These two pages contain practice questions in the various formats that can be found on the most frequently given standardized tests.

A practice answer sheet for these two pages can be found on p. A1 of the *Chapter 8 Resource Masters*.

Standardized Test Practice Student Recording Sheet, p. A1

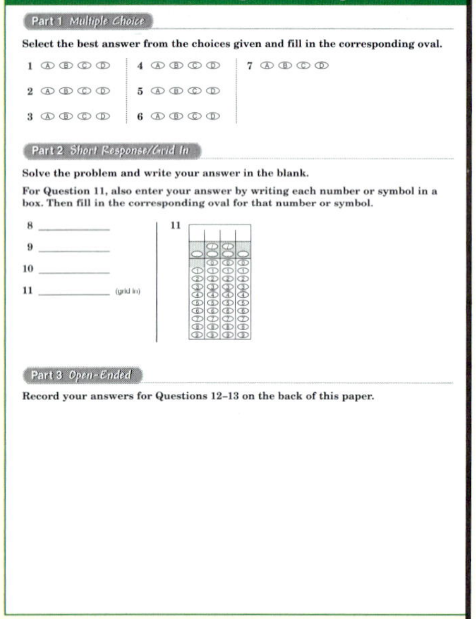
Part 1 Multiple Choice

Select the best answer from the choices given and fill in the corresponding oval.

1 Ⓐ Ⓑ Ⓒ Ⓓ 4 Ⓐ Ⓑ Ⓒ Ⓓ 7 Ⓐ Ⓑ Ⓒ Ⓓ
2 Ⓐ Ⓑ Ⓒ Ⓓ 5 Ⓐ Ⓑ Ⓒ Ⓓ
3 Ⓐ Ⓑ Ⓒ Ⓓ 6 Ⓐ Ⓑ Ⓒ Ⓓ

Part 2 Short Response/Grid In

Solve the problem and write your answer in the blank.

For Question 11, also enter your answer by writing each number or symbol in a box. Then fill in the corresponding oval for that number or symbol.

8 ________
9 ________
10 ________
11 ________ (grid in)

Part 3 Open-Ended

Record your answers for Questions 12–13 on the back of this paper.

Additional Practice

See pp. 477–478 in the *Chapter 8 Resource Masters* for additional standardized test practice.

Chapter 8 Standardized Test Practice

Part 1 Multiple Choice

Record your answers on the answer sheet provided by your teacher or on a sheet of paper.

1. A trucking company wants to purchase a ramp to use when loading heavy objects onto a truck. The closest that the truck can get to the loading area is 5 meters. The height from the ground to the bed of the truck is 3 meters. To the nearest meter, what should the length of the ramp be? (Lesson 1-3) C

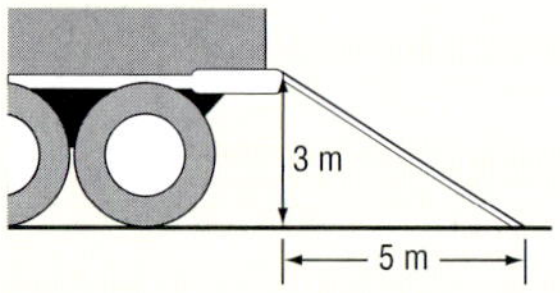

Ⓐ 4 m Ⓑ 5 m
Ⓒ 6 m Ⓓ 7 m

2. Which of the following is the contrapositive of the statement below? (Lesson 2-3) D

If an astronaut is in orbit, then he or she is weightless.

Ⓐ If an astronaut is weightless, then he or she is in orbit.

Ⓑ If an astronaut is not in orbit, then he or she is not weightless.

Ⓒ If an astronaut is on Earth, then he or she is weightless.

Ⓓ If an astronaut is not weightless, then he or she is not in orbit.

3. Rectangle *QRST* measures 7 centimeters long and 4 centimeters wide. Which of the following could be the dimensions of a rectangle similar to rectangle *QRST*? (Lesson 6-2) B

Ⓐ 28 cm by 14 cm

Ⓑ 21 cm by 12 cm

Ⓒ 14 cm by 4 cm

Ⓓ 7 cm by 8 cm

4. A 24 foot ladder, leaning against a house, forms a 60° angle with the ground. How far up the side of the house does the ladder reach? (Lesson 7-3) C

Ⓐ 12 ft

Ⓑ $12\sqrt{2}$ ft

Ⓒ $12\sqrt{3}$ ft

Ⓓ 20 ft

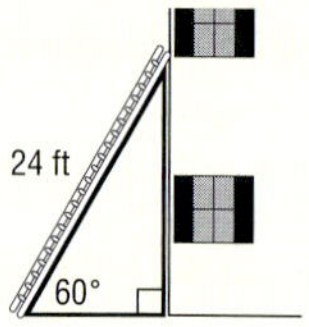

5. In rectangle *JKLM* shown below, $\overline{JL}$ and $\overline{MK}$ are diagonals. If $JL = 2x + 5$ and $KM = 4x - 11$, what is x? (Lesson 8-4) B

Ⓐ 10

Ⓑ 8

Ⓒ 6

Ⓓ 5

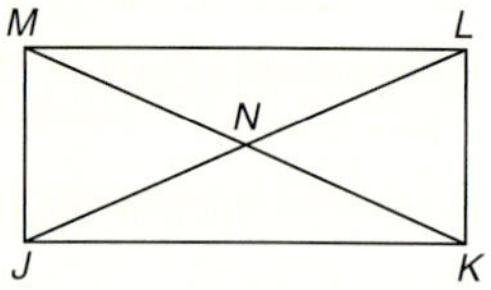

6. Joaquin bought a set of stencils for his younger sister. One of the stencils is a quadrilateral with perpendicular diagonals that bisect each other, but are **not** congruent. What kind of quadrilateral is this piece? (Lesson 8-5) C

Ⓐ square Ⓑ rectangle
Ⓒ rhombus Ⓓ trapezoid

7. In the diagram below, *ABCD* is a trapezoid with diagonals $\overline{AC}$ and $\overline{BD}$ intersecting at point *E*.

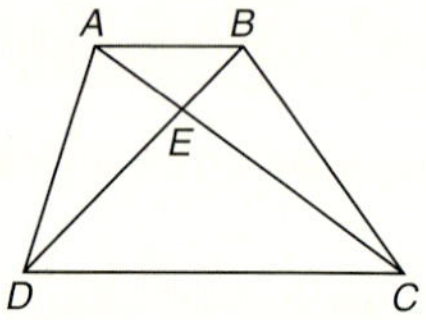

Which statement is true? (Lesson 8-6) A

Ⓐ $\overline{AB}$ is parallel to $\overline{CD}$.

Ⓑ $\angle ADC$ is congruent to $\angle BCD$.

Ⓒ $\overline{CE}$ is congruent to $\overline{DE}$.

Ⓓ $\overline{AC}$ and $\overline{BD}$ bisect each other.

The Princeton Review *Log On* for Test Practice

The Princeton Review offers additional test-taking tips and practice problems at their web site. Visit www.princetonreview.com or www.review.com

TestCheck and Worksheet Builder

Special banks of standardized test questions similar to those on the SAT, ACT, TIMSS 8, NAEP 8, and state proficiency tests can be found on this CD-ROM.

458 Chapter 8 Quadrilaterals

Aligned and verified by The Princeton Review

Part 2 Short Response/Grid In

Record your answers on the answer sheet provided by your teacher or on a sheet of paper.

8. At what point does the graph of $y = -4x + 5$ cross the x-axis on a coordinate plane? (Prerequisite Skill) $\left(\frac{5}{4}, 0\right)$

9. Candace and Julio are planning to see a movie together. They decide to meet at the house that is closer to the theater. From the locations shown on the diagram, whose house is closer to the theater? (Lesson 5-3)
Julio's house

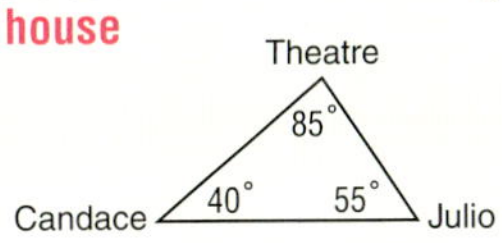

10. In the diagram, $\overline{CE}$ is the mast of a sailboat with sail $\triangle ABC$.

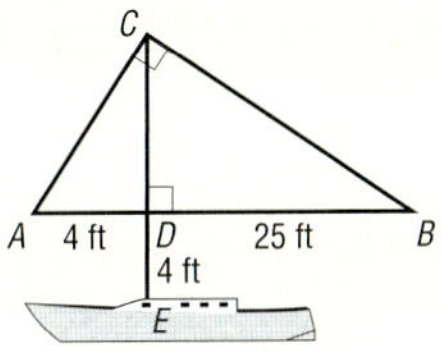

Marcia wants to calculate the length, in feet, of the mast. Write an equation in which the geometric mean is represented by x.
(Lesson 7-1) $\frac{4}{x} = \frac{x}{25}$

11. $\overline{AC}$ is a diagonal of rhombus $ABCD$. If $m\angle CDE$ is 116, what is $m\angle ACD$?
(Lesson 8-4) **58**

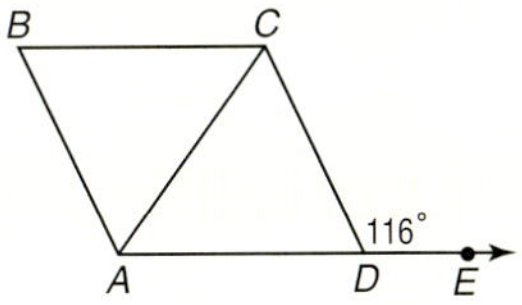

www.geometryonline.com/standardized_test

The Princeton Review Test-Taking Tip

Question 10 Read the question carefully to check that you answered the question that was asked. In Question 10, you are asked to write an equation, not to find the length of the mast.

Part 3 Open Ended

Record your answers on a sheet of paper. Show your work.

12. On the tenth hole of a golf course, a sand trap is located right before the green at point M. Matt is standing 126 yards away from the green at point N. Quintashia is standing 120 yards away from the beginning of the sand trap at point Q.

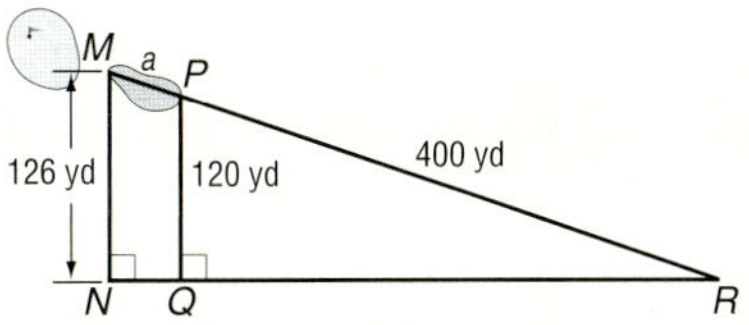

a. Explain why $\triangle MNR$ is similar to $\triangle PQR$.
(Lesson 6-3) **See margin.**

b. Write and solve a proportion to find the distance across the sand trap, a.
(Lesson 6-3) $\frac{400 + a}{126} = \frac{400}{120}$; $a = 20$ yd

13. Quadrilateral $ABCD$ has vertices with coordinates: $A(0, 0)$, $B(a, 0)$, $C(a + b, c)$, and $D(b, c)$.

a. Position and label $ABCD$ on the coordinate plane. Prove that $ABCD$ is a parallelogram. (Lesson 8-2 and 8-7)
See p. 459F.

b. If $a^2 = b^2 + c^2$, what can you determine about the slopes of the diagonals $\overline{AC}$ and $\overline{BD}$? (Lesson 8-7) **See p. 459F.**

c. What kind of parallelogram is $ABCD$?
(Lesson 8-7) **Since the diagonals are perpendicular, $ABCD$ is a rhombus.**

Evaluating Open-Ended Assessment Questions

Open-Ended Assessment questions are graded by using a multilevel rubric that guides you in assessing a student's knowledge of a particular concept.

Goal: In Exercise 12, students use properties to find a distance. In Exercise 13, students use a coordinate proof to draw conclusions about a quadrilateral.

Sample Scoring Rubric: The following rubric is a sample scoring device. You may wish to add more detail to this sample to meet your individual scoring needs.

Score	Criteria
4	A correct solution that is supported by well-developed, accurate explanations
3	A generally correct solution, but may contain minor flaws in reasoning or computation
2	A partially correct interpretation and/or solution to the problem
1	A correct solution with no supporting evidence or explanation
0	An incorrect solution indicating no mathematical understanding of the concept or task, or no solution is given

Answers

12a. $\angle MNR$ and $\angle PQR$ are both right angles and all right angles are congruent, so $\angle MNR \cong \angle PQR$. Since congruence of angles is reflexive, $\angle R \cong \angle R$. $\triangle MNR$ is similar to $\triangle PQR$ because two angles are congruent (AA Similarity).

Chapter 8 Standardized Test Practice 459

Pages 407–409, Lesson 8-1

3. Sample answer: regular quadrilateral, 360° 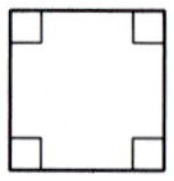 ;

quadrilateral that is not regular, 360°

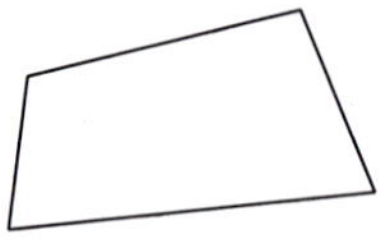

57. Given: $\overline{JL} \parallel \overline{KM}$, $\overline{JK} \parallel \overline{LM}$

Prove: $\triangle JKL \cong \triangle MLK$

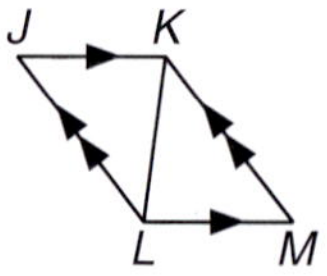

Proof:

Statements (Reasons)

1. $\overline{JL} \parallel \overline{KM}$, $\overline{JK} \parallel \overline{LM}$ (Given)
2. $\angle MKL \cong \angle JLK$, $\angle JKL \cong \angle MLK$ (Alt. int. $\angle$s are $\cong$.)
3. $\overline{KL} \cong \overline{KL}$ (Reflexive Property)
4. $\triangle JKL \cong \triangle MLK$ (ASA)

Pages 414–416, Lesson 8-2

13. Given: ▱$VZRQ$ and ▱$WQST$

Prove: $\angle Z \cong \angle T$

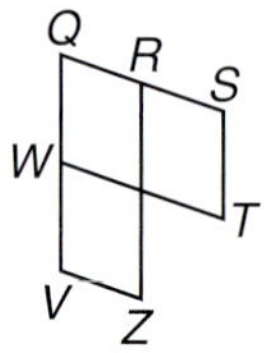

Proof:

Statements (Reasons)

1. ▱$VZRQ$ and ▱$WQST$ (Given)
2. $\angle Z \cong \angle Q$, $\angle Q \cong \angle T$ (Opp. $\angle$s of a ▱ are $\cong$.)
3. $\angle Z \cong \angle T$ (Transitive Prop.)

14. Given: ▱$XYRZ$, $\overline{WZ} \cong \overline{WS}$

Prove: $\angle XYR \cong \angle S$

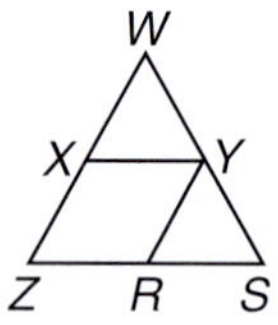

Proof: Opposite angles of a parallelogram are congruent, so $\angle Z \cong \angle XYR$. By the Isosceles Triangle Theorem, since $\overline{WZ} \cong \overline{WS}$, $\angle Z \cong \angle S$. By the Transitive Property, $\angle XYR \cong \angle S$.

40. They are all congruent parallelograms. Since A, B and C are midpoints, $\overline{AC}$, $\overline{AB}$, and $\overline{BC}$ are midsegments. The midsegment is parallel to the third side and equal to half the length of the third side. So, each pair of opposite sides of $ACBX$, $ABYC$, and $ABCZ$ are parallel.

41. Given: ▱$PQRS$

Prove: $\overline{PQ} \cong \overline{RS}$
$\overline{QR} \cong \overline{SP}$

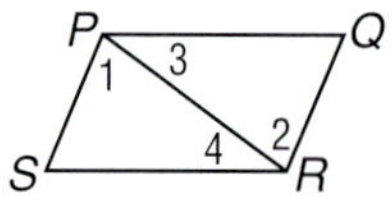

Proof:

Statements (Reasons)

1. ▱$PQRS$ (Given)
2. Draw an auxiliary segment $\overline{PR}$ and label angles 1, 2, 3, and 4 as shown. (Diagonal of $PQRS$)
3. $\overline{PQ} \parallel \overline{SR}$, $\overline{PS} \parallel \overline{QR}$ (Opp. sides of ▱ are $\parallel$.)
4. $\angle 1 \cong \angle 2$, and $\angle 3 \cong \angle 4$ (Alt. int. $\angle$s are $\cong$.)
5. $\overline{PR} \cong \overline{PR}$ (Reflexive Prop.)
6. $\triangle QPR \cong \triangle SRP$ (ASA)
7. $\overline{PQ} \cong \overline{RS}$ and $\overline{QR} \cong \overline{SP}$ (CPCTC)

42. Given: ▱$GKLM$

Prove: $\angle G$ and $\angle K$ are supplementary.
$\angle K$ and $\angle L$ are supplementary.
$\angle L$ and $\angle M$ are supplementary.
$\angle M$ and $\angle G$ are supplementary.

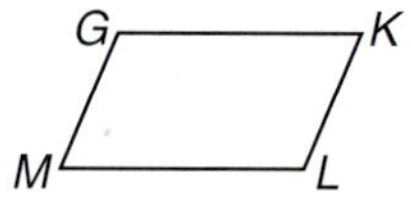

Proof:

Statements (Reasons)

1. ▱$GKLM$ (Given)
2. $\overline{GK} \parallel \overline{ML}$, $\overline{GM} \parallel \overline{KL}$ (Opp. sides of ▱ are $\parallel$.)
3. $\angle G$ and $\angle K$ are supplementary, $\angle K$ and $\angle L$ are supplementary, $\angle L$ and $\angle M$ are supplementary, $\angle M$ and $\angle G$ are supplementary. (Cons. int. $\angle$s are suppl.)

43. Given: ▱$MNPQ$
$\angle M$ is a right angle.

Prove: $\angle N$, $\angle P$ and $\angle Q$ are right angles.

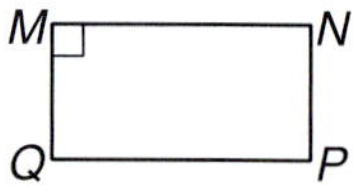

Proof: By definition of a parallelogram, $\overline{MN} \parallel \overline{QP}$. Since $\angle M$ is a right angle, $\overline{MQ} \perp \overline{MN}$. By the Perpendicular Transversal Theorem, $\overline{MQ} \perp \overline{QP}$. $\angle Q$ is a right angle, because perpendicular lines form a right angle. $\angle N \cong \angle Q$ and $\angle M \cong \angle P$ because opposite angles in a parallelogram are congruent. $\angle P$ and $\angle N$ are right angles, since all right angles are congruent.

44. Given: $ACDE$ is a parallelogram.

Prove: $\overline{EC}$ bisects $\overline{AD}$.
$\overline{AD}$ bisects $\overline{EC}$.

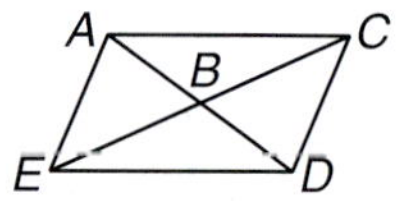

Proof: It is given that $ACDE$ is a parallelogram. Since opposite sides of a parallelogram are congruent, $\overline{EA} \cong \overline{DC}$. By definition of a parallelogram, $\overline{EA} \parallel \overline{DC}$. $\angle AEB \cong \angle DCB$ and $\angle EAB \cong \angle CDB$ because alternate interior angles are congruent. $\triangle EBA \cong \triangle CBD$ by ASA. $\overline{EB} \cong \overline{BC}$ and $\overline{AB} \cong \overline{BD}$ by CPCTC. By the definition of segment bisector, $\overline{EC}$ bisects $\overline{AD}$ and $\overline{AD}$ bisects $\overline{EC}$.

45. Given: ▱$WXYZ$

Prove: $\triangle WXZ \cong \triangle YZX$

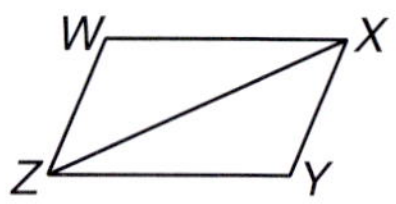

Proof:

Statements (Reasons)

1. ▱$WXYZ$ (Given)
2. $\overline{WX} \cong \overline{ZY}$, $\overline{WZ} \cong \overline{XY}$ (Opp. sides of ▱ are $\cong$.)
3. $\angle ZWX \cong \angle XYZ$ (Opp. $\angle$s of ▱ are $\cong$.)
4. $\triangle WXZ \cong \triangle YZX$ (SAS)

Pages 420–423, Lesson 8-3

11. Given: $\overline{PT} \cong \overline{TR}$
$\angle TSP \cong \angle TQR$

Prove: $PQRS$ is a parallelogram.

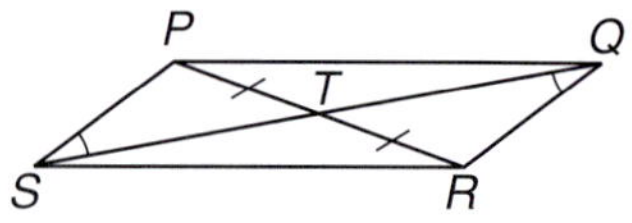

Proof:

Statements (Reasons)

1. $\overline{PT} \cong \overline{TR}$, $\angle TSP \cong \angle TQR$ (Given)
2. $\angle PTS \cong \angle RTQ$ (Vertical $\angle$s are $\cong$.)
3. $\triangle PTS \cong \triangle RTQ$ (AAS)
4. $\overline{PS} \cong \overline{QR}$ (CPCTC)
5. $\overline{PS} \parallel \overline{QR}$ (If alt. int. $\angle$s are $\cong$, lines are $\parallel$.)
6. *PQRS* is a parallelogram. (If one pair of opp. sides is $\parallel$ and $\cong$, then the quad. is a ▱.)

39. Given: $\overline{AD} \cong \overline{BC}$
$\overline{AB} \cong \overline{DC}$
Prove: *ABCD* is a parallelogram.

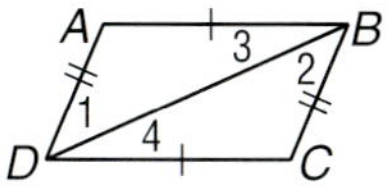

Proof:

Statements (Reasons)

1. $\overline{AD} \cong \overline{BC}$, $\overline{AB} \cong \overline{DC}$ (Given)
2. Draw $\overline{DB}$. (Two points determine a line.)
3. $\overline{DB} \cong \overline{DB}$ (Reflexive Property)
4. $\triangle ABD \cong \triangle CDB$ (SSS)
5. $\angle 1 \cong \angle 2$, $\angle 3 \cong \angle 4$ (CPCTC)
6. $\overline{AD} \parallel \overline{BC}$, $\overline{AB} \parallel \overline{DC}$ (If alt. int. $\angle$s are $\cong$, lines are $\parallel$.)
7. *ABCD* is a parallelogram. (Def. of parallelogram)

40. Given: $\overline{AE} \cong \overline{EC}$, $\overline{DE} \cong \overline{EB}$
Prove: *ABCD* is a parallelogram.

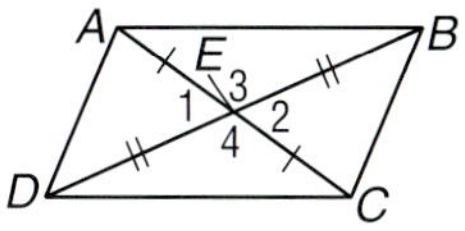

Proof:

Statements (Reasons)

1. $\overline{AE} \cong \overline{EC}$, $\overline{DE} \cong \overline{EB}$ (Given)
2. $\angle 1 \cong \angle 2$, $\angle 3 \cong \angle 4$ (Vertical $\angle$s are $\cong$.)
3. $\triangle ABE \cong \triangle CDE$, $\triangle ADE \cong \triangle CBE$ (SAS)
4. $\overline{AB} \cong \overline{DC}$, $\overline{AD} \cong \overline{BC}$ (CPCTC)
5. *ABCD* is a parallelogram. (If both pairs of opp. sides are $\cong$, then quad is ▱.)

41. Given: $\overline{AB} \cong \overline{DC}$
$\overline{AB} \parallel \overline{DC}$
Prove: *ABCD* is a parallelogram.

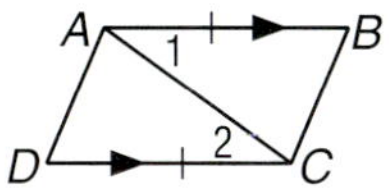

Proof:

Statements (Reasons)

1. $\overline{AB} \cong \overline{DC}$, $\overline{AB} \parallel \overline{DC}$ (Given)
2. Draw $\overline{AC}$ (Two points determine a line.)
3. $\angle 1 \cong \angle 2$ (If two lines are $\parallel$, then alt. int. $\angle$s are $\cong$.)
4. $\overline{AC} \cong \overline{AC}$ (Reflexive Property)
5. $\triangle ABC \cong \triangle CDA$ (SAS)
6. $\overline{AD} \cong \overline{BC}$ (CPCTC)
7. *ABCD* is a parallelogram. (If both pairs of opp. sides are $\cong$, then the quad. is ▱.)

Pages 427–430, Lesson 8-4

42. Given: *PQST* is a rectangle.
$\overline{QR} \cong \overline{VT}$
Prove: $\overline{PR} \cong \overline{VS}$

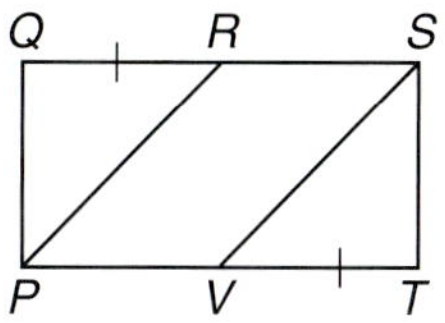

Proof:

Statements (Reasons)

1. *PQST* is a rectangle; $\overline{QR} \cong \overline{VT}$. (Given)
2. *PQST* is a parallelogram. (Def. of rectangle)
3. $\overline{TS} \cong \overline{PQ}$ (Opp. sides of ▱ are $\cong$.)
4. $\angle T$ and $\angle Q$ are rt. $\angle$s. (Definition of rectangle)
5. $\angle T \cong \angle Q$ (All rt. $\angle$s are $\cong$.)
6. $\triangle RPQ \cong \triangle VST$ (SAS)
7. $\overline{PR} \cong \overline{VS}$ (CPCTC)

43. Given: *DEAC* and *FEAB* are rectangles.
$\angle GKH \cong \angle JHK$
$\overline{GJ}$ and $\overline{HK}$ intersect at *L*.
Prove: *GHJK* is a parallelogram.

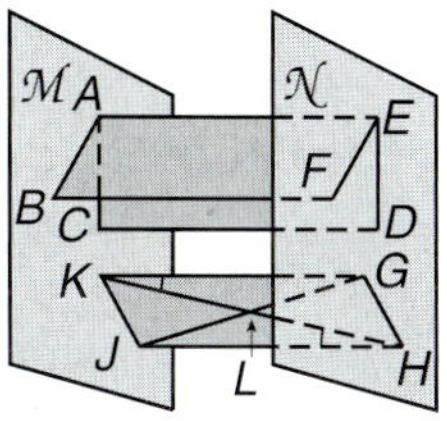

Proof:

Statements (Reasons)

1. *DEAC* and *FEAB* are rectangles; $\angle GKH \cong \angle JHK$; $\overline{GJ}$ and $\overline{HK}$ intersect at *L*. (Given)
2. $\overline{DE} \parallel \overline{AC}$ and $\overline{FE} \parallel \overline{AB}$ (Def. of parallelogram)
3. plane 𝒩 $\parallel$ plane ℳ (Def. of parallel plane)
4. *G*, *J*, *H*, *K*, *L* are in the same plane. (Def. of intersecting lines)
5. $\overline{GH} \parallel \overline{KJ}$ (Def. of parallel lines)
6. $\overline{GK} \parallel \overline{HJ}$ (Alt. int. $\angle$s are $\cong$)
7. *GHJK* is a parallelogram. (Def. of parallelogram)

Pages 434–437, Lesson 8-5

10. Given: $\triangle KGH$, $\triangle HJK$, $\triangle GHJ$, and $\triangle JKG$ are isosceles.
Prove: *GHJK* is a rhombus.

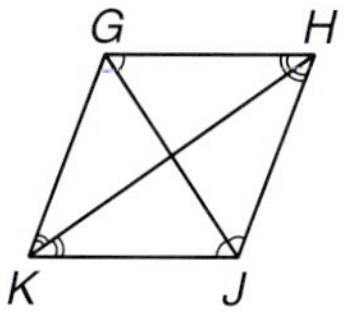

Proof:

Statements (Reasons)

1. $\triangle KGH$, $\triangle HJK$, $\triangle GHJ$, and $\triangle JKG$ are isosceles. (Given)
2. $\overline{KG} \cong \overline{GH}$, $\overline{HJ} \cong \overline{KJ}$, $\overline{GH} \cong \overline{HJ}$, $\overline{KG} \cong \overline{KJ}$ (Def. of isosceles $\triangle$)
3. $\overline{KG} \cong \overline{HJ}$, $\overline{GH} \cong \overline{KJ}$ (Transitive Property)
4. $\overline{KG} \cong \overline{GH}$, $\overline{HJ} \cong \overline{KJ}$ (Substitution)
5. *GHJK* is a rhombus. (Def. of rhombus)

35. Given: $ABCD$ is a parallelogram.
$\overline{AC} \perp \overline{BD}$

Prove: $ABCD$ is a rhombus.

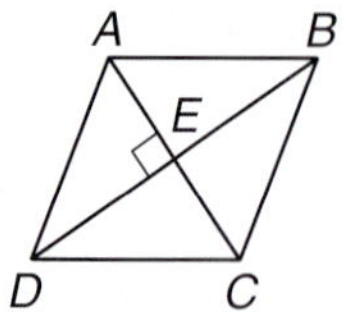

Proof: We are given that $ABCD$ is a parallelogram. The diagonals of a parallelogram bisect each other, so $\overline{AE} \cong \overline{EC}$. $\overline{BE} \cong \overline{BE}$ because congruence of segments is reflexive. We are also given that $\overline{AC} \perp \overline{BD}$. Thus, $\angle AEB$ and $\angle BEC$ are right angles by the definition of perpendicular lines. Then $\angle AEB \cong \angle BEC$ because all right angles are congruent. Therefore, $\triangle AEB \cong \triangle CEB$ by SAS. $\overline{AB} \cong \overline{CB}$ by CPCTC. Opposite sides of parallelograms are congruent, so $\overline{AB} \cong \overline{CD}$ and $\overline{BC} \cong \overline{AD}$. Then since congruence of segments is transitive, $\overline{AB} \cong \overline{CD} \cong \overline{BC} \cong \overline{AD}$. All four sides of $ABCD$ are congruent, so $ABCD$ is a rhombus by definition.

36. Given: $ABCD$ is a rhombus.

Prove: Each diagonal bisects a pair of opposite angles.

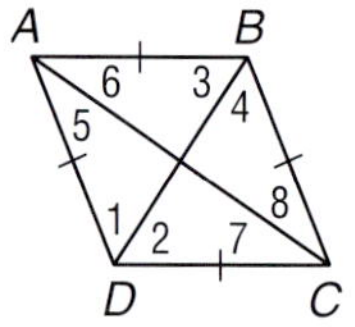

Proof: We are given that $ABCD$ is a rhombus. By definition of rhombus, $ABCD$ is a parallelogram. Opposite angles of a parallelogram are congruent, so $\angle ABC \cong \angle ADC$ and $\angle BAD \cong \angle BCD$. $\overline{AB} \cong \overline{BC} \cong \overline{CD} \cong \overline{DA}$ because all sides of a rhombus are congruent. $\triangle ABC \cong \triangle ADC$ by SAS. $\angle 5 \cong \angle 6$ and $\angle 7 \cong \angle 8$ by CPCTC. $\triangle BAD \cong \triangle BCD$ by SAS. $\angle 1 \cong \angle 2$ and $\angle 3 \cong \angle 4$ by CPCTC. By definition of angle bisector, each diagonal bisects a pair of opposite angles.

Page 438, Geometry Activity

5.

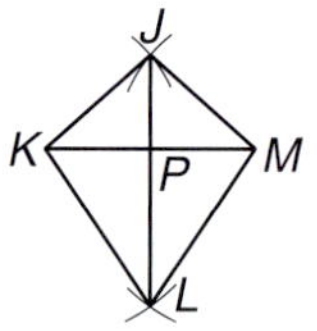

The diagonals intersect in a right angle; $\angle JKL \cong \angle JML$; $\overline{KP} \cong \overline{PM}$, $\overline{JP} \not\cong \overline{PL}$; 3 pairs: $\triangle JPK \cong \triangle JPM$, $\triangle KPL \cong \triangle MPL$, $\triangle JKL \cong \triangle JML$.

6. One pair of opposite angles is congruent. The diagonals are perpendicular. The longer diagonal bisects the shorter diagonal. The short sides are congruent and the long sides are congruent.

Pages 442–445, Lesson 8-6

2.

Properties	Trapezoid	Rectangle	Square	Rhombus
diagonals are congruent	only isosceles	yes	yes	no
diagonals are perpendicular	no	no	yes	yes
diagonals bisect each other	no	yes	yes	yes
diagonals bisect angles	no	no	yes	yes

3. Sample answer: The median of a trapezoid is parallel to both bases.

trapezoid   isosceles trapezoid

6. Given: $CDFG$ is an isosceles trapezoid with bases $\overline{CD}$ and $\overline{GF}$.

Prove: $\angle DGF \cong \angle CFG$

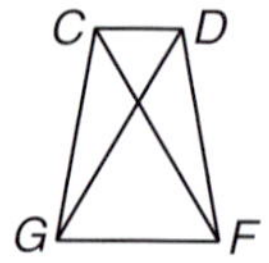

Proof:

- $CDFG$ is an isosceles trapezoid with bases $\overline{CD}$ and $\overline{GF}$. — Given
- $\overline{CG} \cong \overline{DF}$ — Def. isos. trap.
- $\overline{CF} \cong \overline{DG}$ — Diag. of isos. trap. are $\cong$.
- $\overline{GF} \cong \overline{GF}$ — Reflex. Prop.
- $\triangle CGF \cong \triangle DFG$ — SSS
- $\angle DGF \cong \angle CFG$ — CPCTC

32. Given: $\overline{HJ} \parallel \overline{GK}$, $\overline{HG} \not\parallel \overline{JK}$
$\triangle HGK \cong \triangle JKG$

Prove: $GHJK$ is an isosceles trapezoid.

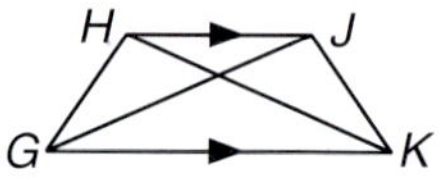

Proof:

- $\overline{HJ} \parallel \overline{GK}$, $\overline{HG} \not\parallel \overline{JK}$ — Given
- $\triangle HGK \cong \triangle JKG$ — Given
- $GHJK$ is a trapezoid. — Def. of trapezoid
- $\overline{HG} \cong \overline{JK}$ — CPCTC
- $GHJK$ is an isosceles trapezoid. — Def. of isosceles trapezoid

33. Given: $\triangle TZX \cong \triangle YXZ$, $\overline{WX} \not\parallel \overline{ZY}$

Prove: $XYZW$ is a trapezoid.

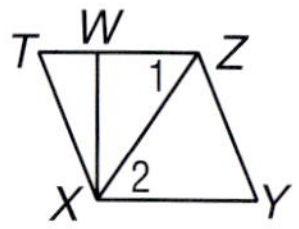

Proof:

- $\triangle TZX \cong \triangle YXZ$ — Given
- $\angle 1 \cong \angle 2$ — CPCTC
- $\overline{WZ} \parallel \overline{XY}$ — If alt. int. ∠s are $\cong$, then the lines are $\parallel$.
- $\overline{WX} \not\parallel \overline{ZY}$ — Given
- $XYZW$ is a trapezoid. — Def. of trapezoid

34. Given: *ZYXP* is an isosceles trapezoid.
Prove: $\triangle PWX$ is isosceles.

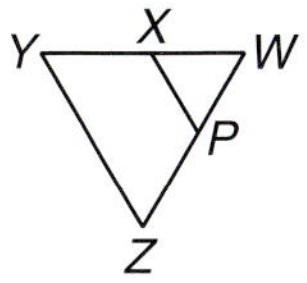

Proof:

- *ZYXP* is an isosceles trapezoid. (Given)
- $\angle Y \cong \angle Z$ (Base ∠s of an isos. trap. are ≅.)
- $\overline{WY} \cong \overline{WZ}$ (If 2 ∠s of a △ are ≅, the sides opp. are ≅.)
- $\overline{XY} \cong \overline{PZ}$ (Def. of isos. trap.)
- $WX + XY = WY$, $WP + PZ = WZ$ (Segment Addition)
- $WY = WZ$ (Def. of ≅)
- $XY = PZ$ (Def. of ≅)
- $WX + XY = WP + PZ$ (Substitution)
- $WX = WP$ (Subtraction)
- $\overline{WX} \cong \overline{WP}$ (Def. of ≅)
- $\triangle PWX$ is isosceles. (Def. of isos. △)

35. Given: *E* and *C* are midpoints of $\overline{AD}$ and $\overline{DB}$. $\overline{AD} \cong \overline{DB}$
Prove: *ABCE* is an isosceles trapezoid.

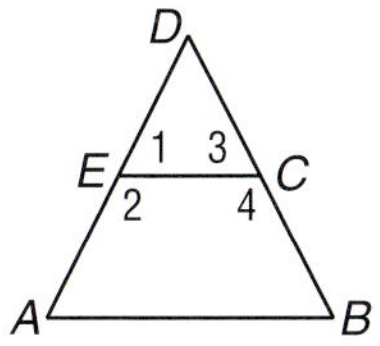

Proof:

- *E* and *C* are midpoints of $\overline{AD}$ and $\overline{DB}$. (Given)
- $\overline{AD} \cong \overline{DB}$ (Given)
- $\overline{EC} \parallel \overline{AB}$ (A segment joining the midpoints of two sides of a triangle is parallel to the third side.)
- $\frac{1}{2}AD = \frac{1}{2}DB$ (Def. of Midpt.)
- $AE = BC$ (Substitution)
- $\overline{AE} \cong \overline{BC}$ (Def. of ≅)
- *ABCE* is an isos. trapezoid. (Def. of isos. trapezoid)

Pages 449–451, Lesson 8-7

1. Place one vertex at the origin and position the figure so another vertex lies on the positive *x*-axis.
2. Sample answer:

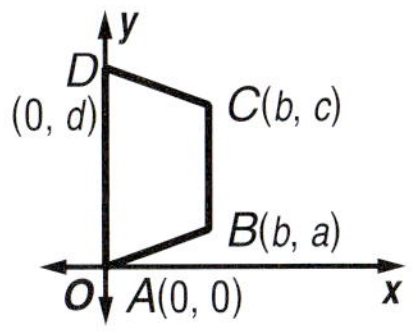

3.

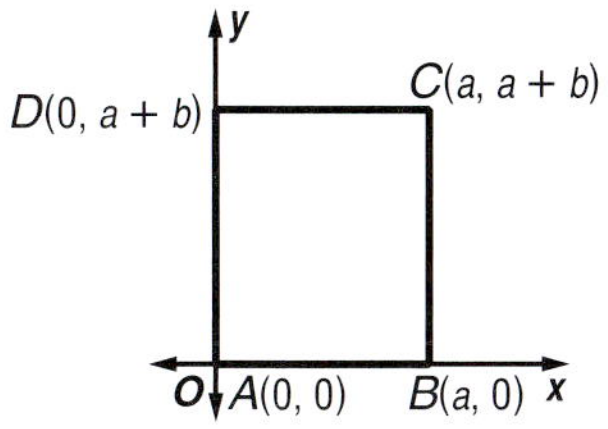

6. Given: *ABCD* is a parallelogram.
Prove: $\overline{AC}$ and $\overline{DB}$ bisect each other.

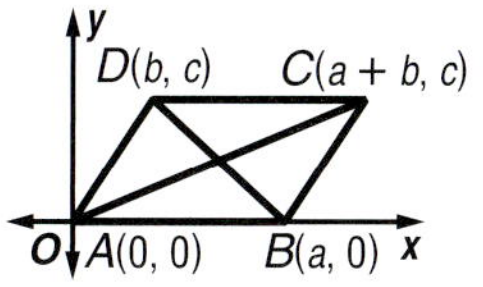

Proof:

The midpoint of $\overline{AC} = \left(\frac{0 + (a + b)}{2}, \frac{0 + c}{2}\right)$

$= \left(\frac{a + b}{2}, \frac{c}{2}\right)$

The midpoint of $\overline{DB} = \left(\frac{a + b}{2}, \frac{0 + c}{2}\right)$

$= \left(\frac{a + b}{2}, \frac{c}{2}\right)$

$\overline{AC}$ and $\overline{DB}$ bisect each other.

7. Given: *ABCD* is a square.
Prove: $\overline{AC} \perp \overline{DB}$

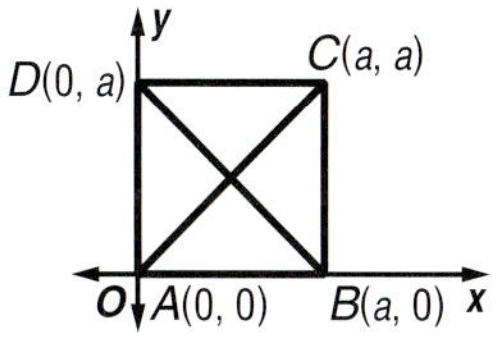

Proof:

Slope of $\overline{DB} = \frac{0 - a}{a - 0}$ or -1

Slope of $\overline{AC} = \frac{0 - a}{0 - a}$ or 1

The slope of $\overline{AC}$ is the negative reciprocal of the slope of $\overline{DB}$, so they are perpendicular.

8. Given: *D*(195, 180), *E*(765, 180), *F*(533, 0), *G*(195, 0)
Prove: *DEFG* is a trapezoid.

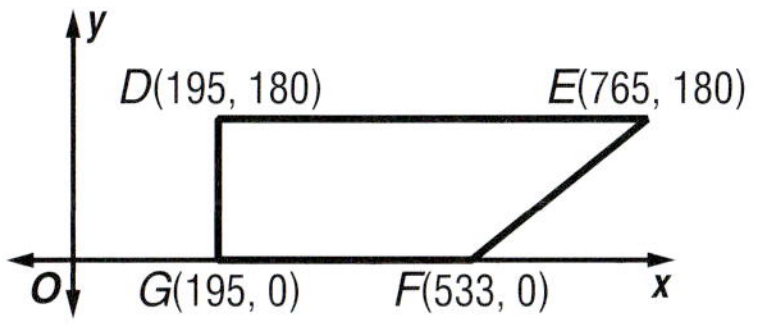

Proof:

Slope of $\overline{DE} = \frac{180 - 180}{765 - 195}$ or 0

Slope of $\overline{GF} = \frac{0 - 0}{533 - 195}$ or 0

Slope of $\overline{EF} = \frac{180 - 0}{765 - 533}$ or $\frac{45}{58}$

Slope of $\overline{DG} = \frac{180 - 0}{195 - 195}$ or undefined

$\overline{DE}$ and $\overline{GF}$ have the same slope, so exactly one pair of opposite sides are parallel. Therefore, *DEFG* is a trapezoid.

9.

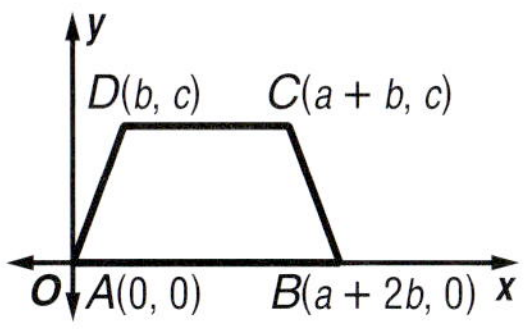

10.

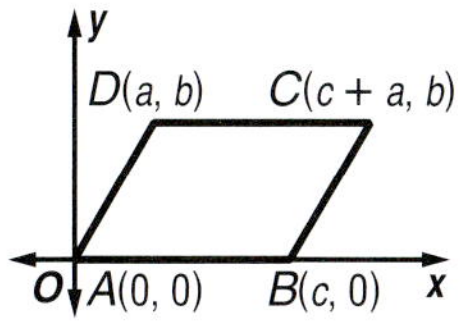

17. Given: *ABCD* is a rectangle.
Prove: $\overline{AC} \cong \overline{DB}$

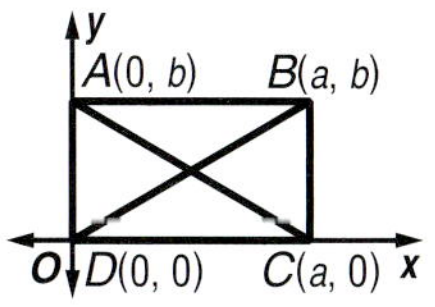

Proof: Use the Distance Formula to find $AC = \sqrt{a^2 + b^2}$ and $BD = \sqrt{a^2 + b^2}$. $\overline{AC}$ and $\overline{DB}$ have the same length, so they are congruent.

18. Given: $\square ABCD$ and $\overline{AC} \cong \overline{BD}$
Prove: $\square ABCD$ is a rectangle.

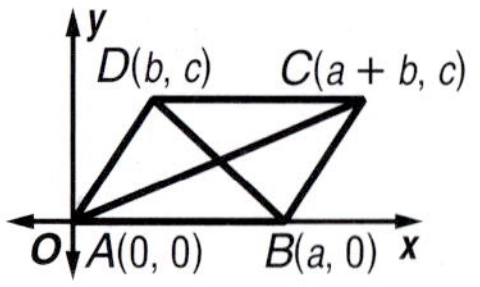

Proof:

$AC = \sqrt{(a + b - 0)^2 + (c - 0)^2}$

$BD = \sqrt{(b - a)^2 + (c - 0)^2}$

But $AC = BD$ and

$$\sqrt{(a + b - 0)^2 + (c - 0)^2} = \sqrt{(b - a)^2 + (c - 0)^2}$$
$$(a + b - 0)^2 + (c - 0)^2 = (b - a)^2 + (c - 0)^2$$
$$(a + b)^2 + c^2 = (b - a)^2 + c^2$$
$$a^2 + 2ab + b^2 + c^2 = b^2 - 2ab + a^2 + c^2$$
$$2ab = -2ab$$
$$4ab = 0$$
$$a = 0 \text{ or } b = 0$$

Because A and B are different points, $a \neq 0$. Then $b = 0$. The slope of $\overline{AD}$ is undefined and the slope of $\overline{AB} = 0$. Thus $\overline{AD} \perp \overline{AB}$. $\angle DAB$ is a right angle and $ABCD$ is a rectangle.

19. Given: isosceles trapezoid $ABCD$ with $\overline{AD} \cong \overline{BC}$
Prove: $\overline{BD} \cong \overline{AC}$

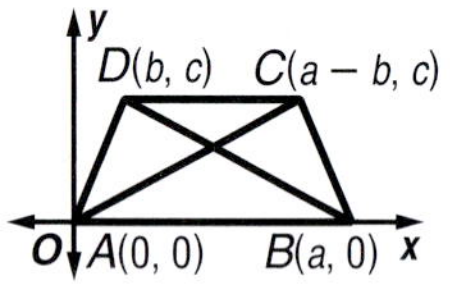

Proof:

$BD = \sqrt{(a - b)^2 + (0 - c)^2} = \sqrt{(a - b)^2 + c^2}$

$AC = \sqrt{((a - b) - 0)^2 + (c - 0)^2} = \sqrt{(a - b)^2 + c^2}$

$BD = AC$ and $\overline{BD} \cong \overline{AC}$

20. Given: $ABCD$ is an isosceles trapezoid with median $\overline{XY}$.
Prove: $\overline{XY} \parallel \overline{AB}$ and $\overline{XY} \parallel \overline{DC}$

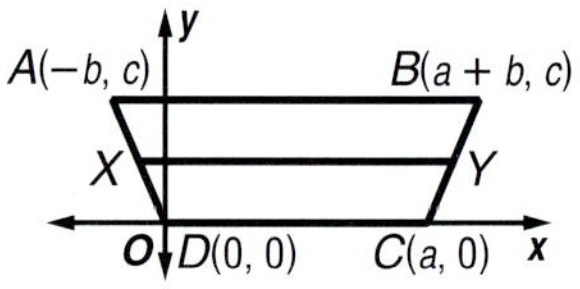

Proof: The midpoint of $\overline{AD}$ is X. The coordinates are $\left(\frac{-b}{2}, \frac{c}{2}\right)$. The midpoint of $\overline{BC}$ is $Y\left(\frac{2a + b}{2}, \frac{c}{2}\right)$. The slope of $\overline{AB} = 0$, the slope of $\overline{XY} = 0$ and the slope of $\overline{DC} = 0$. Thus, $\overline{XY} \parallel \overline{AB}$ and $\overline{XY} \parallel \overline{DC}$.

21. Given: $ABCD$ is a rectangle. Q, R, S, and T are midpoints of their respective sides.
Prove: $QRST$ is a rhombus.

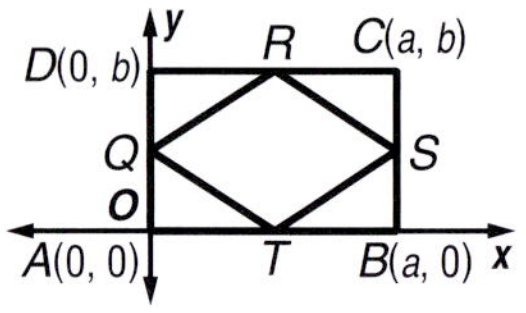

Proof:

Midpoint Q is $\left(\frac{0 + 0}{2}, \frac{b + 0}{2}\right)$ or $\left(0, \frac{b}{2}\right)$.

Midpoint R is $\left(\frac{a + 0}{2}, \frac{b + b}{2}\right)$ or $\left(\frac{a}{2}, \frac{2b}{2}\right)$ or $\left(\frac{a}{2}, b\right)$.

Midpoint S is $\left(\frac{a + a}{2}, \frac{b + 0}{2}\right)$ or $\left(\frac{2a}{2}, \frac{b}{2}\right)$ or $\left(a, \frac{b}{2}\right)$.

Midpoint T is $\left(\frac{a + 0}{2}, \frac{0 + 0}{2}\right)$ or $\left(\frac{a}{2}, 0\right)$.

$QR = \sqrt{\left(\frac{a}{2} - 0\right)^2 + \left(b - \frac{b}{2}\right)^2} = \sqrt{\left(\frac{a}{2}\right)^2 + \left(\frac{b}{2}\right)^2}$

$RS = \sqrt{\left(a - \frac{a}{2}\right)^2 + \left(\frac{b}{2} - b\right)^2}$

$= \sqrt{\left(\frac{a}{2}\right)^2 + \left(-\frac{b}{2}\right)^2}$ or $\sqrt{\left(\frac{a}{2}\right)^2 + \left(\frac{b}{2}\right)^2}$

$ST = \sqrt{\left(a - \frac{a}{2}\right)^2 + \left(\frac{b}{2} - 0\right)^2} = \sqrt{\left(\frac{a}{2}\right)^2 + \left(\frac{b}{2}\right)^2}$

$QT = \sqrt{\left(\frac{a}{2} - 0\right)^2 + \left(0 - \frac{b}{2}\right)^2}$

$= \sqrt{\left(\frac{a}{2}\right)^2 + \left(-\frac{b}{2}\right)^2}$ or $\sqrt{\left(\frac{a}{2}\right)^2 + \left(\frac{b}{2}\right)^2}$

$QR = RS = ST = QT$

$\overline{QR} \cong \overline{RS} \cong \overline{ST} \cong \overline{QT}$

$QRST$ is a rhombus.

22. Given: $RSTV$ is a quadrilateral. A, B, C, and D are midpoints of sides $\overline{RS}$, $\overline{ST}$, $\overline{TV}$, and $\overline{VR}$, respectively.
Prove: $ABCD$ is a parallelogram.

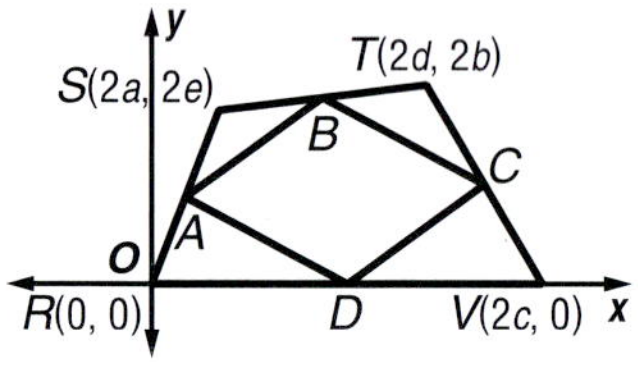

Proof: Place quadrilateral $RSTV$ on the coordinate plane and label coordinates as shown. (Using coordinates that are multiples of 2 will make the computation easier.) By the Midpoint Formula, the coordinates of A, B, C, and D are $A\left(\frac{2a}{2}, \frac{2e}{2}\right) = (a, e)$; $B\left(\frac{2d + 2a}{2}, \frac{2e + 2b}{2}\right) = (d + a, e + b)$; $C\left(\frac{2d + 2c}{2}, \frac{2b}{2}\right) = (d + c, b)$; and $D\left(\frac{2c}{2}, \frac{0}{2}\right) = (c, 0)$.

Find the slopes of $\overline{AB}$ and $\overline{DC}$.

Slope of $\overline{AB}$

$m = \frac{y_2 - y_1}{x_2 - x_1}$

$= \frac{(e + b) - e}{(d + a) - a}$

$= \frac{b}{d}$

Slope of $\overline{DC}$

$m = \frac{y_2 - y_1}{x_2 - x_1}$

$= \frac{0 - b}{c - (d + c)}$

$= \frac{-b}{-d}$ or $\frac{b}{d}$

The slopes of $\overline{AB}$ and $\overline{DC}$ are the same so the segments are parallel.

Use the Distance Formula to find AB and DC.

$AB = \sqrt{((d + a) - a)^2 + ((e + b) - e)^2}$

$= \sqrt{d^2 + b^2}$

$DC = \sqrt{((d + c) - c)^2 + (b - 0)^2}$

$= \sqrt{d^2 + b^2}$

Thus, $\overline{AB} \cong \overline{DC}$. Therefore, $ABCD$ is a parallelogram because if one pair of opposite sides of a quadrilateral are both parallel and congruent, then the quadrilateral is a parallelogram.

Page 457, Chapter 8 Practice Test

18. Given: isosceles trapezoid $WXYZ$ with median $\overline{ST}$

Prove: $\overline{WX} \parallel \overline{ST} \parallel \overline{YZ}$

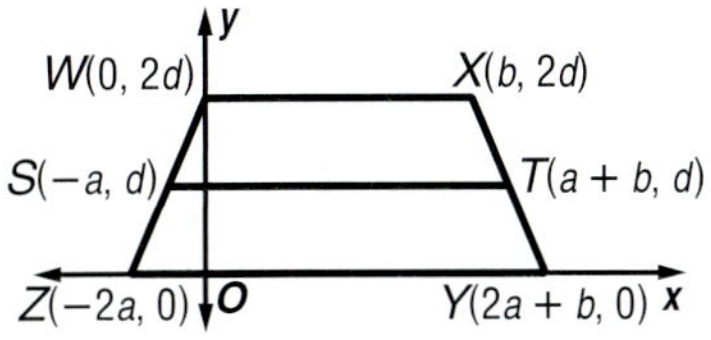

Proof: To prove lines parallel, show their slopes equal.

The slope of WX is $\frac{2d - 2d}{b - 0}$ or 0.

The slope of ST is $\frac{d - d}{(a + b) - (-a)}$ or 0.

The slope of YZ is $\frac{0 - 0}{(2a + b) - (-2a)}$ or 0.

Since WX, ST, and YZ all have zero slope, they are parallel.

Pages 458–459, Chapter 8 Standardized Test Practice

13a. Given: quadrilateral $ABCD$

Prove: $ABCD$ is a parallelogram

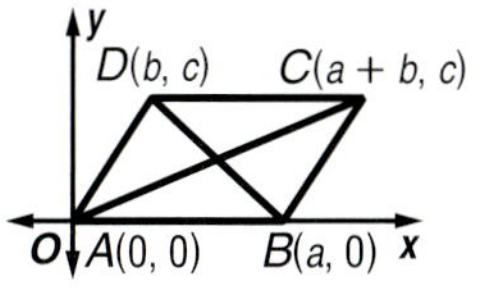

Proof: The slope of $\overline{AD}$ is $\frac{c - 0}{b - 0}$ or $\frac{c}{b}$. The slope of $\overline{BC}$ is $\frac{c - 0}{a + b - a}$ or $\frac{c}{b}$. $\overline{AD}$ and $\overline{BC}$ have the same slope so they are parallel.

$AD = \sqrt{(b - 0)^2 + (c - 0)^2} = \sqrt{b^2 + c^2}$.

$BC = \sqrt{(a + b - a)^2 + (c - 0)^2} = \sqrt{b^2 + c^2}$.

Since one pair of opposite sides are parallel and congruent, $ABCD$ is a parallelogram.

13b. The slope of $\overline{AC}$ is $\frac{c - 0}{a + b - 0}$ or $\frac{c}{a + b}$. The slope of $\overline{BD}$ is $\frac{c - 0}{b - a}$ or $\frac{c}{b - a}$. The product of the slopes is $\frac{c}{a + b} \times \frac{c}{b - a} = \frac{c^2}{b^2 - a^2}$. Since $c^2 = a^2 - b^2$, the product of the slopes is $\frac{a^2 - b^2}{b^2 - a^2}$ or -1, so the diagonals of $ABCD$ are perpendicular.

Chapter 9

Transformations

Chapter Overview and Pacing

LESSON OBJECTIVES

	PACING (days)			
	Regular		Block	
	Basic/ Average	Advanced	Basic/ Average	Advanced
9-1 Reflections *(pp. 462–469)* ***Preview:*** Identify transformations. • Draw reflected images. • Recognize and draw lines of symmetry and points of symmetry.	2 (with 9-1 Preview)	1	1 (with 9-1 Preview)	0.5
9-2 Translations *(pp. 470–475)* • Draw translated images using coordinates. • Draw translated images by using repeated reflections.	1	1	0.5	0.5
9-3 Rotations *(pp. 476–482)* • Draw rotated images using the angle of rotation. • Identify figures with rotational symmetry.	2	2	1	1
9-4 Tessellations *(pp. 483–489)* • Identify regular tessellations. • Create tessellations with specific attributes. ***Follow-Up:*** Make tessellations using a translation and a rotation.	2 (with 9-4 Follow-Up)	2 (with 9-4 Follow-Up)	1 (with 9-4 Follow-Up)	1 (with 9-4 Follow-Up)
9-5 Dilations *(pp. 490–497)* • Determine whether a dilation is an enlargement, a reduction, or a congruence transformation. • Determine the scale factor for a given dilation.	2	2	1	1
9-6 Vectors *(pp. 498–505)* • Find magnitudes and directions of vectors. • Perform translations with vectors.	optional	2	optional	1
9-7 Transformations with Matrices *(pp. 506–511)* • Use matrices to determine the coordinates of translations and dilations. • Use matrices to determine the coordinates of reflections and rotations.	optional	2	optional	1
Study Guide and Practice Test *(pp. 512–517)* **Standardized Test Practice** *(pp. 518–519)*	1	1	1	0.5
Chapter Assessment	1	1	0.5	0.5
TOTAL	**11**	**14**	**6**	**7**

Pacing suggestions for the entire year can be found on pages T20–T21.

Chapter Resource Manager

Timesaving Tools
TeacherWorks™
All-In-One Planner and Resource Center
See pages T5 and T21.

Chapter 9 Resource Masters: Study Guide and Intervention	Practice (Skills and Average)	Reading to Learn Mathematics	Enrichment	Assessment	Prerequisite Skills Workbook	Applications*	5-Minute Check Transparencies	Interactive Chalkboard	GeomPASS: Tutorial Plus (lessons)	Materials
479–480	481–482	483	484		5–6	SC 17	9-1	9-1		grid paper, ruler
485–486	487–488	489	490	535	1–2	GCC 33	9-2	9-2		grid paper, ruler, protractor
491–492	493–494	495	496				9-3	9-3		compass, protractor, grid paper, ruler
497–498	499–500	501	502	535, 537			9-4	9-4		paper, pattern blocks, grid paper, straightedge (*Follow-Up:* paper, pencil)
503–504	505–506	507	508		19–20, 43–44		9-5	9-5	18	grid paper, ruler, compass, protractor
509–510	511–512	513	514	536	5–6	SC 18	9-6	9-6		grid paper, straightedge
515–516	517–518	519	520	536			9-7	9-7		grid paper, straightedge
				521–534, 538–540						

**Key to Abbreviations:* GCC = Graphing Calculator and Computer Masters
SC = School-to-Career Masters

Chapter 9

Mathematical Connections and Background

Continuity of Instruction

Prior Knowledge

In previous courses students learned to graph points on the coordinate plane. They also found the product of two matrices. Chapter 6 introduced students to scale factors and similar polygons. In Chapter 7 students solved trigonometric ratios.

This Chapter

In this chapter, students explore the different types of transformations: reflections, translations, rotations, and dilations. They learn to identify, draw, and recognize figures that have been transformed. Students also identify and create different types of tessellations. Vectors are introduced in this chapter as well. Students find the magnitude and direction of vectors and perform operations on vectors. Students use matrices to perform transformations on the coordinate plane.

Future Connections

The knowledge about transformations, tessellations, and vectors that students gain while studying this chapter will be important to them in future mathematics courses and in many careers that they might choose.

9-1 Reflections

A reflection is a transformation representing a flip of a figure. Figures may be reflected in a point, a line, or a plane. A reflected image is always congruent to its preimage. In other words, a reflection is a congruence transformation or an isometry. Reflections can occur in the coordinate plane, allowing you to assign coordinates to each point in the image and preimage.

Figures that are commonly called symmetrical have a line of reflection, or line of symmetry. Some of these figures have a point that is a common point of reflection for all points on that figure. This common point of reflection is called a *point of symmetry*.

9-2 Translations

A translation is a transformation that moves all points of a figure the same distance in the same direction. Translations on the coordinate plane can be drawn if you know the direction and how far the figure is moving horizontally and/or vertically. One way to translate a figure in the coordinate plane is simply to count units on the x-axis and on the y-axis, much as you count for slope. Another way to find a translation is to perform a reflection in the first of two parallel lines and then reflect the image in the other parallel line. Successive reflections in parallel lines are called a *composition of reflections*.

Because a translation can be performed by two reflections, all translations are isometries or congruence transformations. Therefore, a translated figure is congruent to its preimage.

9-3 Rotations

A rotation is a transformation that turns every point of a preimage through a specified angle and direction about a fixed point. The fixed point is called the *center of rotation*. The angle of rotation is the angle formed by a point on the preimage, the center of rotation, and the corresponding point on the rotated image. A rotation exhibits all the properties of isometries, including preservation of distance and angle measure.

A rotation can be performed using a protractor to measure the angle of rotation and a compass to mark the new points. Another way to perform a rotation is to reflect a figure successively in two intersecting lines. If a figure is rotated by successive reflection, there is a relationship between the angle of rotation and the angle formed by the intersecting lines. In a given rotation, if A is the preimage, A'' is the image, and P is the center of rotation, then the measure of the angle of rotation

$\angle APA''$ is twice the measure of the acute or right angle formed by the intersecting lines of reflection. This means that reflecting an image successively in two perpendicular lines results in a 180° rotation.

Some objects have rotational symmetry. If a figure can be rotated less than 360° about a point so that the image and the preimage are indistinguishable, then the figure has rotational symmetry. Rotational symmetry can be described in terms of *order and magnitude*. The order is the number of times a figure can be rotated less than 360° and produce an image indistinguishable from the original. The magnitude is the measure of this rotation, or 360° divided by the order.

9-4 Tessellations

A tessellation is a pattern that covers a plane by transforming the same figure or set of figures so that there are no overlapping or empty spaces. In a tessellation, the sum of the measures of the angles of the polygons surrounding a point at a vertex is always 360. A regular tessellation is a tessellation formed by only one type of regular polygon, such as a hexagon. Tessellations containing the same arrangement of shapes and angles at each vertex are called *uniform*. A uniform tessellation formed by using two or more regular polygons is called a *semi-regular tessellation*.

9-5 Dilations

A dilation is a transformation that changes the size of a figure. The new figure may be smaller or larger than the original by a scale factor. If the scale factor is 1, then the dilation is a congruence transformation. If the scale factor is not 1, then the dilation is a similarity transformation. This means that the new figure and the original are similar. It is important to note that a negative scale factor does not result in a negative measure. It simply means that the new image falls on the opposite side of the center than the preimage. The center of a dilation is always its own image.

In the coordinate plane, you can use the scale factor to determine the coordinate of the image of dilations centered at the origin. If $P(x, y)$ is the preimage of a dilation centered at the origin with a scale factor r, then the image is $P'(rx, ry)$. To determine the scale factor of a dilation on the coordinate plane, divide the image length by the preimage length.

9-6 Vectors

A vector is a quantity that has both magnitude, or length, and direction. It is represented by a directed segment. A vector in standard position has its initial point at the origin, though a vector can be drawn anywhere in the coordinate plane. To write a vector as an ordered pair, find the change in the x values and the change in the y values from the tip to the tail of the directed segment. This ordered pair is called the component form of the vector and is written with pointed brackets: $\langle$change in x, change in $y\rangle$.

You can use the Distance Formula to find the magnitude of a vector. The direction of a vector is the measure of the angle that the vector forms with the positive x-axis or any other horizontal line. You can use trigonometric ratios to find the direction of a vector.

Vectors can be used to describe translations. Combining vectors allows you to perform a composition of translations. To combine vectors, add their corresponding components. The sum of two vectors is called the *resultant*. You can also multiply a vector by a constant, called a scalar, that will change the magnitude of the vector but not affect its direction. Multiplying a vector by a scalar is called *scalar multiplication*.

9-7 Transformations with Matrices

A vector can be represented by a column matrix. Likewise, polygons can be represented by placing all of the coordinates of the vertices into one matrix, called a *vertex matrix*. You can easily find the coordinates of a translated figure by adding the x and y value changes to the vertex matrix. A reflection matrix can be used to multiply the vertex matrix of a figure to find the coordinates of the image. Matrices can also be used to determine the vertices of a figure's image by rotation using a rotational matrix. Reflection and rotational matrices always have only two rows and two columns, no matter the number of columns in the vertex matrix.

Chapter 9

DAILY INTERVENTION and Assessment

	Type	Student Edition	Teacher Resources	Technology/Internet
INTERVENTION	Ongoing	Prerequisite Skills, pp. 461, 469, 475, 482, 488, 497, 505 Practice Quiz 1, p. 482 Practice Quiz 2, p. 497	5-Minute Check Transparencies *Prerequisite Skills Workbook*, pp. 1–2, 5–6, 19–20, 43–44 Quizzes, *CRM* pp. 535–536 Mid-Chapter Test, *CRM* p. 537 Study Guide and Intervention, *CRM* pp. 479–480, 485–486, 491–492, 497–498, 503–504, 509–510, 515–516	GeomPASS: Tutorial Plus www.geometryonline.com/self_check_quiz www.geometryonline.com/extra_examples
	Mixed Review	pp. 469, 475, 482, 488, 497, 505, 511	Cumulative Review, *CRM* p. 538	
	Error Analysis	Find the Error, pp. 472, 493 Common Misconceptions, pp. 478, 498	Find the Error, *TWE* pp. 473, 493 Unlocking Misconceptions, *TWE* p. 466 Tips for New Teachers, *TWE* p. 501	
	Standardized Test Practice	pp. 469, 474, 481, 488, 493, 494, 496, 505, 511, 517, 518, 519	*TWE* pp. 518–519 Standardized Test Practice, *CRM* pp. 539–540	Standardized Test Practice CD-ROM www.geometryonline.com/standardized_test
	Open-Ended Assessment	Writing in Math, pp. 469, 474, 481, 487, 496, 505, 511 Open Ended, pp. 467, 472, 478, 485, 493, 502, 508 Standardized Test, p. 519	Modeling: *TWE* pp. 482, 497 Speaking: *TWE* pp. 475, 505 Writing: *TWE* pp. 469, 488, 511 Open-Ended Assessment, *CRM* p. 533	
ASSESSMENT	Chapter Assessment	Study Guide, pp. 512–516 Practice Test, p. 517	Multiple-Choice Tests (Forms 1, 2A, 2B), *CRM* pp. 512–526 Free-Response Tests (Forms 2C, 2D, 3), *CRM* pp. 527–532 Vocabulary Test/Review, *CRM* p. 534	TestCheck and Worksheet Builder (see below) MindJogger Videoquizzes www.geometryonline.com/vocabulary_review www.geometryonline.com/chapter_test

Key to Abbreviations: TWE = Teacher Wraparound Edition; CRM = Chapter Resource Masters

Additional Intervention Resources

The Princeton Review's *Cracking the SAT & PSAT*

The Princeton Review's *Cracking the ACT*

TestCheck and Worksheet Builder

This **networkable** software has three modules for intervention and assessment flexibility:

- **Worksheet Builder** to make worksheet and tests
- **Student Module** to take tests on screen (optional)
- **Management System** to keep student records (optional)

Special banks are included for SAT, ACT, TIMSS, NAEP, and End-of-Course tests.

Intervention Technology

GeomPASS: Tutorial Plus CD-ROM offers a complete, self-paced geometry curriculum.

Geometry Lesson	GeomPASS Lesson
9-5	**18** Transformations

Intervention at Home

Log on for student study help.

- For each lesson in the Student Edition, there are Extra Examples and Self-Check Quizzes.
  www.geometryonline.com/extra_examples
  www.geometryonline.com/self_check_quiz
- For chapter review, there is vocabulary review, test practice, and standardized test practice.
  www.geometryonline.com/vocabulary_review
  www.geometryonline.com/chapter_test
  www.geometryonline.com/standardized_test

For more information on Intervention and Assessment, see pp. T8–T11.

Reading and Writing in Mathematics

Glencoe Geometry provides numerous opportunities to incorporate reading and writing into the mathematics classroom.

Student Edition

- Foldables Study Organizer, p. 461
- Concept Check questions require students to verbalize and write about what they have learned in the lesson. (pp. 467, 472, 478, 485, 493, 502, 508)
- Writing in Math questions in every lesson, pp. 469, 474, 481, 487, 496, 505, 511
- Reading Study Tip, pp. 463, 464, 470, 483
- WebQuest, p. 469

Teacher Wraparound Edition

- Foldables Study Organizer, pp. 461, 512
- Study Notebook suggestions, pp. 462, 467, 472, 479, 486, 489, 493, 502, 509
- Modeling activities, pp. 482, 497
- Speaking activities, pp. 475, 505
- Writing activities, pp. 469, 488, 511
- ELL Resources, pp. 460, 468, 474, 480, 487, 496, 503, 510, 512

Additional Resources

- Vocabulary Builder worksheets require students to define and give examples for key vocabulary terms as they progress through the chapter. (*Chapter 9 Resource Masters*, pp. vii-viii)
- Proof Builder helps students learn and understand theorems and postulates from the chapter. (*Chapter 9 Resource Masters*, pp. ix–x)
- Reading to Learn Mathematics master for each lesson (*Chapter 9 Resource Masters*, pp. 483, 489, 495, 501, 507, 513, 519)
- *Vocabulary PuzzleMaker* software creates crossword, jumble, and word search puzzles using vocabulary lists that you can customize.
- *Teaching Mathematics with Foldables* provides suggestions for promoting cognition and language.
- *Reading Strategies for the Mathematics Classroom*
- *WebQuest and Project Resources*

For more information on Reading and Writing in Mathematics, see pp. T6–T7.

Chapter 9 Notes

What You'll Learn

Have students read over the list of objectives and make a list of any words with which they are not familiar.

Why It's Important

Point out to students that this is only one of many reasons why each objective is important. Others are provided in the introduction to each lesson.

Lesson	NCTM Standards	Local Objectives
9-1 Preview	3	
9-1	2, 3, 6, 8, 9, 10	
9-2	2, 3, 6, 8, 9, 10	
9-3	2, 3, 6, 8, 9, 10	
9-4	2, 3, 6, 8, 9, 10	
9-4 Follow-Up	3	
9-5	2, 3, 6, 8, 9, 10	
9-6	2, 3, 6, 8, 9, 10	
9-7	2, 3, 6, 8, 9, 10	

Key to NCTM Standards:

1=Number & Operations, 2=Algebra, 3=Geometry, 4=Measurement, 5=Data Analysis & Probability, 6=Problem Solving, 7=Reasoning & Proof, 8=Communication, 9=Connections, 10=Representation

Chapter 9 Transformations

What You'll Learn

- **Lesson 9-1, 9-2, 9-3, and 9-5** Name, draw, and recognize figures that have been reflected, translated, rotated, or dilated.
- **Lesson 9-4** Identify and create different types of tessellations.
- **Lesson 9-6** Find the magnitude and direction of vectors and perform operations on vectors.
- **Lesson 9-7** Use matrices to perform transformations on the coordinate plane.

Key Vocabulary

- reflection (p. 463)
- translation (p. 470)
- rotation (p. 476)
- tessellation (p. 483)
- dilation (p. 490)
- vector (p. 498)

Why It's Important

Transformations, lines of symmetry, and tessellations can be seen in artwork, nature, interior design, quilts, amusement parks, and marching band performances. These geometric procedures and characteristics make objects more visually pleasing. *You will learn how mosaics are created by using transformations in Lesson 9-2.*

460 Chapter 9 Transformations

Vocabulary Builder

ELL

The Key Vocabulary list introduces students to some of the main vocabulary terms included in this chapter. For a more thorough vocabulary list with pronunciations of new words, give students the Vocabulary Builder worksheets found on pages vii and viii of the *Chapter 9 Resource Masters*. Encourage them to complete the definition of each term as they progress through the chapter. You may suggest that they add these sheets to their study notebooks for future reference when studying for the Chapter 9 test.

Getting Started

Prerequisite Skills To be successful in this chapter, you'll need to master these skills and be able to apply them in problem-solving situations. Review these skills before beginning Chapter 9.

For Lessons 9-1 through 9-5 — Graph Points

Graph each pair of points. *(For review, see pages 728 and 729.)* **1–6. See p. 519A.**

1. $A(1, 3), B(-1, 3)$
2. $C(-3, 2), D(-3, -2)$
3. $E(-2, 1), F(-1, -2)$
4. $G(2, 5), H(5, -2)$
5. $J(-7, 10), K(-6, 7)$
6. $L(3, -2), M(6, -4)$

For Lesson 9-6 — Distance and Slope

Find $m\angle A$. Round to the nearest tenth. *(For review, see Lesson 7-4.)*

7. $\tan A = \frac{3}{4}$ **36.9**
8. $\tan A = \frac{5}{8}$ **32.0**
9. $\sin A = \frac{2}{3}$ **41.8**
10. $\sin A = \frac{4}{5}$ **53.1**
11. $\cos A = \frac{9}{12}$ **41.4**
12. $\cos A = \frac{15}{17}$ **28.1**

For Lesson 9-7 — Multiply Matrices

Find each product. *(For review, see pages 752 and 753.)*

13. $\begin{bmatrix} 0 & 1 \\ 1 & -1 \end{bmatrix} \cdot \begin{bmatrix} 5 & 4 \\ -5 & -1 \end{bmatrix}$ $\begin{bmatrix} -5 & -1 \\ 10 & 5 \end{bmatrix}$
14. $\begin{bmatrix} -1 & 0 \\ 1 & 1 \end{bmatrix} \cdot \begin{bmatrix} 0 & -2 \\ -2 & 3 \end{bmatrix}$ $\begin{bmatrix} 0 & 2 \\ -2 & 1 \end{bmatrix}$
15. $\begin{bmatrix} 0 & 1 \\ -1 & 0 \end{bmatrix} \cdot \begin{bmatrix} -3 & 4 & 5 \\ -2 & -5 & 1 \end{bmatrix}$ $\begin{bmatrix} -2 & -5 & 1 \\ 3 & -4 & -5 \end{bmatrix}$
16. $\begin{bmatrix} -1 & 0 \\ 0 & -1 \end{bmatrix} \cdot \begin{bmatrix} -1 & -3 & -3 & 2 \\ 3 & -1 & -2 & 1 \end{bmatrix}$ $\begin{bmatrix} 1 & 3 & 3 & -2 \\ -3 & 1 & 2 & -1 \end{bmatrix}$

FOLDABLES™ Study Organizer

Transformations Make this Foldable to help you organize the types of transformations. Begin with one sheet of notebook paper.

Step 1 Fold

Step 2 Cut

Step 3 Label

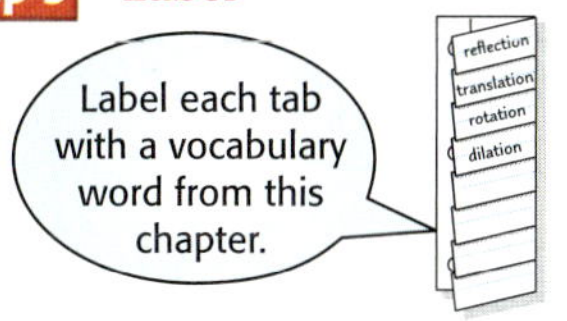

Reading and Writing As you read and study the chapter, use each page to write notes and examples of transformations, tessellations, and vectors on the coordinate plane.

Getting Started

This section provides a review of the basic concepts needed before beginning Chapter 9. Page references are included for additional student help.

Additional review is provided in the *Prerequisite Skills Workbook*, pages 1–2, 5–6, 19–20, 43–44.

Prerequisite Skills in the Getting Ready for the Next Lesson section at the end of each exercise set review a skill needed in the next lesson.

For Lesson	Prerequisite Skill
9-2	Distance Formula, p. 469
9-3	Drawing Angles, p. 475
9-4	Solving Problems by Making a Table, p. 482
9-5	Similar Polygons, p. 488
9-6	Finding Angles Using Inverses of Trigonometric Ratios, p. 497
9-7	Operations with Matrices, p. 505

FOLDABLES™ Study Organizer

For more information about Foldables, see *Teaching Mathematics with Foldables.*

Vocabulary and Writing Definitions Use this Foldable to help students organize information they learn about transformations and to give them practice writing concise definitions in their own words. Have students write a word or concept on the front of each tab and definitions on the back. Some examples of terms they can use are: transformations, reflections, symmetry, translations, vectors, magnitudes, rotation images, angle of rotation, dilations, and similarity transformations. Under the tabs, to the right of each definition, ask students to produce their own example of the concepts presented.

Geometry Activity

A Preview of Lesson 9-1

Getting Started

Tell students that the transformations they will be learning about are very easy to remember by name because the mathematical terms and their corresponding transformations are very similar to the literal definitions of the words.

Objective Students will learn to identify translation, reflection, rotation, and dilation transformations.

Materials
pencil, paper

Teach

- Tell students that if the object is symmetric, as in the reflection example, it can appear to be rotated when it is reflected and vice versa.
- Explain that sometimes two or more transformations can be applied to the same figure, as with the dilation example, which has been translated and dilated. Explain and demonstrate how a shape can be dilated without a translation.

Assess

Exercises 1–10 offer examples for students to identify transformations. Students make a conjecture about transformations in **Exercise 11**.

Study Notebook

Ask students to summarize what they have learned about transformations.

Geometry Activity
A Preview of Lesson 9-1

Transformations

In a plane, you can slide, flip, turn, enlarge, or reduce figures to create new figures. These corresponding figures are frequently designed into wallpaper borders, mosaics, and artwork. Each figure that you see will correspond to another figure. These corresponding figures are formed using transformations.

A **transformation** maps an initial figure, called a preimage, onto a final figure, called an image. Below are some of the types of transformations. The red lines show some corresponding points.

translation
A figure can be slid in any direction.

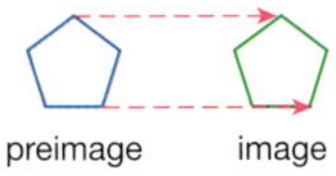

reflection
A figure can be flipped over a line.

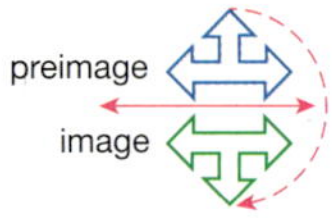

rotation
A figure can be turned around a point.

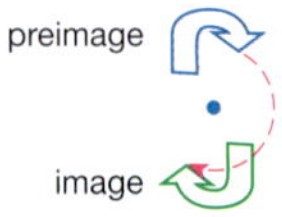

dilation
A figure can be enlarged or reduced.

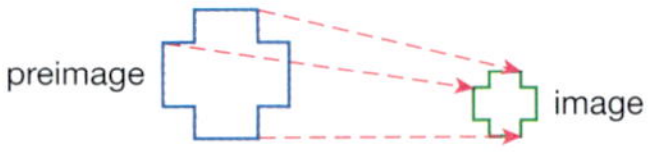

Exercises **Identify the following transformations. The blue figure is the preimage.**

1. 
rotation

2. 
dilation

3. 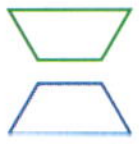
reflection or rotation

4. 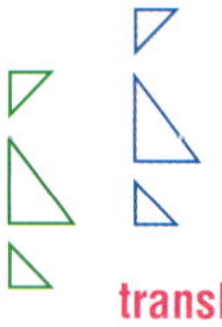
translation

5. 
dilation

6. 
reflection

7. 
translation

8. 
reflection or rotation

9. 
reflection

10. 
reflection

Make a Conjecture

11. An *isometry* is a transformation in which the resulting image is congruent to the preimage. Which transformations are isometries? rotation, reflection, and translation

Resource Manager

Teaching Geometry with Manipulatives

- p. 145 (student recording sheet)

9-1 Reflections

What You'll Learn

- Draw reflected images.
- Recognize and draw lines of symmetry and points of symmetry.

Vocabulary

- reflection
- line of reflection
- isometry
- line of symmetry
- point of symmetry

Where are reflections found in nature?

On a clear, bright day glacial-fed lakes can provide vivid reflections of the surrounding vistas. Note that each point above the water line has a corresponding point in the image in the lake. The distance that a point lies above the water line appears the same as the distance its image lies below the water.

DRAW REFLECTIONS A **reflection** is a transformation representing a flip of a figure. Figures may be reflected in a point, a line, or a plane.

The figure shows a reflection of *ABCDE* in line *m*. Note that the segment connecting a point and its image is perpendicular to line *m* and is bisected by line *m*. Line *m* is called the **line of reflection** for *ABCDE* and its image *A′B′C′D′E′*. Because *E* lies on the line of reflection, its preimage and image are the same point.

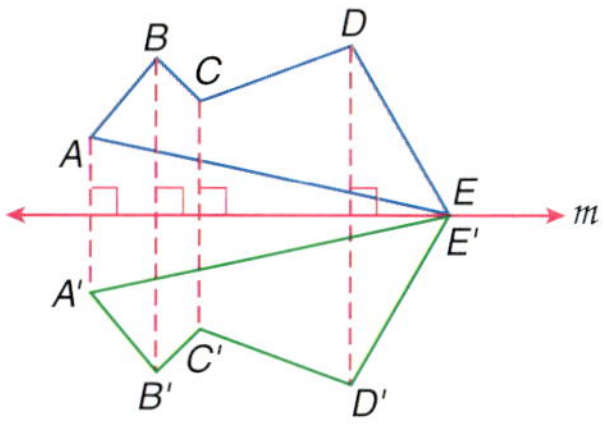

Study Tip

Reading Math

A′, A″, A‴, and so on name corresponding points for one or more transformations.

It is possible to reflect a preimage in a point. In the figure below, polygon *UVWXYZ* is reflected in point *P*.

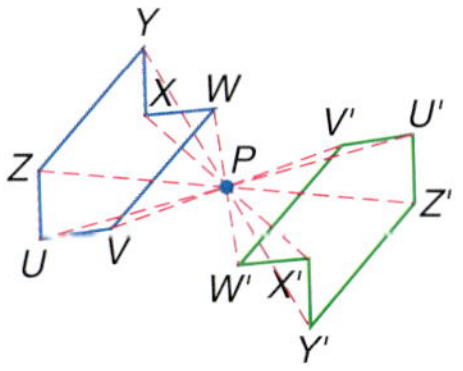

Note that *P* is the midpoint of each segment connecting a point with its image.

$\overline{UP} \cong \overline{PU'}$, $\overline{VP} \cong \overline{PV'}$,
$\overline{WP} \cong \overline{PW'}$, $\overline{XP} \cong \overline{PX'}$,
$\overline{YP} \cong \overline{PY'}$, $\overline{ZP} \cong \overline{PZ'}$

Study Tip

Look Back

To review **congruence transformations**, see Lesson 4-3.

When reflecting a figure in a line or in a point, the image is congruent to the preimage. Thus, a reflection is a *congruence transformation*, or an **isometry**. That is, reflections preserve distance, angle measure, betweenness of points, and collinearity. In the figure above, polygon *UVWXYZ* ≅ polygon *U′V′W′X′Y′Z′*.

Corresponding Sides	Corresponding Angles
$\overline{UV} \cong \overline{U'V'}$	$\angle UVW \cong \angle U'V'W'$
$\overline{VW} \cong \overline{V'W'}$	$\angle VWX \cong \angle V'W'X'$
$\overline{WX} \cong \overline{W'X'}$	$\angle WXY \cong \angle W'X'Y'$
$\overline{XY} \cong \overline{X'Y'}$	$\angle XYZ \cong \angle X'Y'Z'$
$\overline{YZ} \cong \overline{Y'Z'}$	$\angle YZU \cong \angle Y'Z'U'$
$\overline{UZ} \cong \overline{U'Z'}$	$\angle ZUV \cong \angle Z'U'V'$

9-1 Lesson Notes

1 Focus

5-Minute Check Transparency 9-1 Use as a quiz or review of Chapter 8.

Mathematical Background notes are available for this lesson on p. 460C.

Where are reflections found in nature?

Ask students:

- Is the entire top portion of the picture reflected in the water? Why? **No; the reflection depends on the vantage point of the observer.**
- Would you say that the image has horizontal symmetry or vertical symmetry? Why? **Horizontal symmetry, because it is reflected from top to bottom.**

2 Teach

DRAW REFLECTIONS

In-Class Example Power Point®

1 Construct the reflected image of quadrilateral *WXYZ* in line *p*.

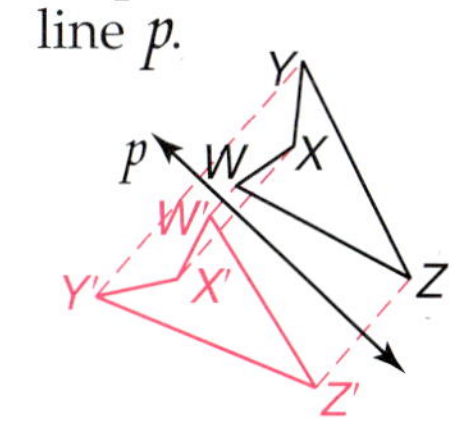

Resource Manager

Workbook and Reproducible Masters

Chapter 9 Resource Masters
- Study Guide and Intervention, pp. 479–480
- Skills Practice, p. 481
- Practice, p. 482
- Reading to Learn Mathematics, p. 483
- Enrichment, p. 484

School-to-Career Masters, p. 17
Prerequisite Skills Workbook, pp. 5–6
Teaching Geometry With Manipulatives Masters, pp. 1, 17, 146

Transparencies

5-Minute Check Transparency 9-1
Answer Key Transparencies

Technology

Interactive Chalkboard

In-Class Examples

Power Point®

2 **COORDINATE GEOMETRY** Quadrilateral *ABCD* has vertices $A(1, 1)$, $B(3, 2)$, $C(4, -1)$, and $D(2, -3)$. Graph *ABCD* and its image under reflection in the x-axis. Compare the coordinates of each vertex with the coordinates of its image.

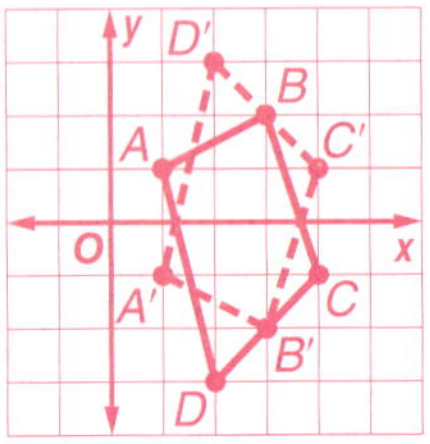

$A(1, 1) \rightarrow A'(1, -1)$, $B(3, 2) \rightarrow B'(3, -2)$, $C(4, -1) \rightarrow C'(4, 1)$, $D(2, -3) \rightarrow D'(2, 3)$. The x-coordinates stay the same, but the y-coordinates are opposite.

3 **COORDINATE GEOMETRY** Suppose quadrilateral *ABCD* from In-Class Example 2 is reflected in the y-axis. Graph *ABCD* and its image under reflection in the y-axis. Compare the coordinates of each vertex with the coordinates of its image.

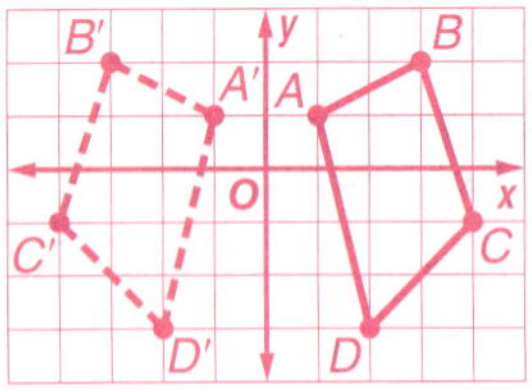

$A(1, 1) \rightarrow A'(-1, 1)$, $B(3, 2) \rightarrow B'(-3, 2)$, $C(4, -1) \rightarrow C'(-4, -1)$, $D(2, -3) \rightarrow D'(-2, -3)$. The y-coordinates stay the same, but the x-coordinates are opposite.

TEACHING TIP

The line of reflection is sometimes called a *mirror line.*

Example 1 Reflecting a Figure in a Line

Draw the reflected image of quadrilateral *DEFG* in line m.

Step 1 Since D is on line m, D is its own reflection. Draw segments perpendicular to line m from E, F, and G.

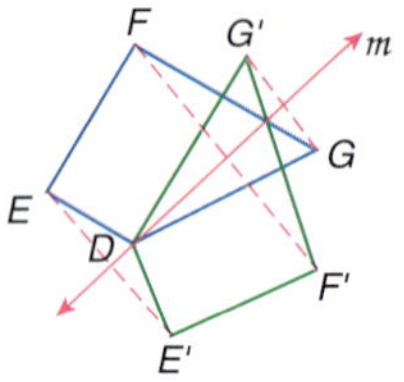

Step 2 Locate E', F', and G' so that line m is the perpendicular bisector of $\overline{EE'}$, $\overline{FF'}$, and $\overline{GG'}$. Points E', F', and G' are the respective images of E, F, and G.

Step 3 Connect vertices D, E', F', and G'.

Since points D, E', F', and G' are the images of points D, E, F, and G under reflection in line m, then quadrilateral $DE'F'G'$ is the reflection of quadrilateral *DEFG* in line m.

Reflections can also occur in the coordinate plane.

Study Tip

Reading Mathematics

The expression $K(2, -4) \rightarrow K'(2, 4)$ can be read as "point K is mapped to new location K'." This means that point K' in the image corresponds to point K in the preimage.

Example 2 Reflection in the x-axis

COORDINATE GEOMETRY **Quadrilateral *KLMN* has vertices $K(2, -4)$, $L(-1, 3)$, $M(-4, 2)$, and $N(-3, -4)$. Graph *KLMN* and its image under reflection in the x-axis. Compare the coordinates of each vertex with the coordinates of its image.**

Use the vertical grid lines to find a corresponding point for each vertex so that the x-axis is equidistant from each vertex and its image.

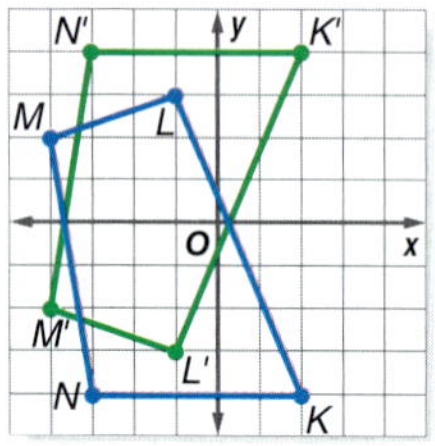

$K(2, -4) \rightarrow K'(2, 4)$ $\qquad$ $L(-1, 3) \rightarrow L'(-1, -3)$

$M(-4, 2) \rightarrow M'(-4, -2)$ $\qquad$ $N(-3, -4) \rightarrow N'(-3, 4)$

Plot the reflected vertices and connect to form the image $K'L'M'N'$. The x-coordinates stay the same, but the y-coordinates are opposite. That is, $(a, b) \rightarrow (a, -b)$.

Example 3 Reflection in the y-axis

COORDINATE GEOMETRY **Suppose quadrilateral *KLMN* from Example 2 is reflected in the y-axis. Graph *KLMN* and its image under reflection in the y-axis. Compare the coordinates of each vertex with the coordinates of its image.**

Use the horizontal grid lines to find a corresponding point for each vertex so that the y-axis is equidistant from each vertex and its image.

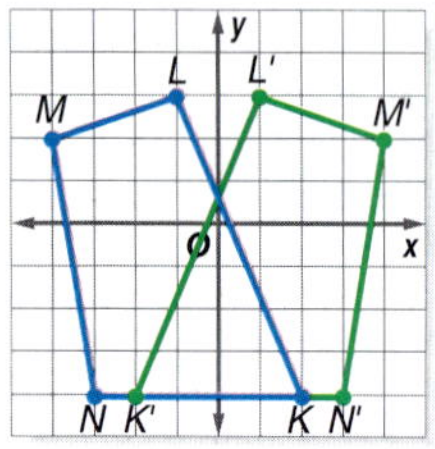

$K(2, -4) \rightarrow K'(-2, -4)$ $\qquad$ $L(-1, 3) \rightarrow L'(1, 3)$

$M(-4, 2) \rightarrow M'(4, 2)$ $\qquad$ $N(-3, -4) \rightarrow N'(3, -4)$

Plot the reflected vertices and connect to form the image $K'L'M'N'$. The x-coordinates are opposite and the y-coordinates are the same. That is, $(a, b) \rightarrow (-a, b)$.

Interactive Chalkboard
PowerPoint® Presentations

This CD-ROM is a customizable Microsoft® PowerPoint® presentation that includes:

- Step-by-step, dynamic solutions of each In-Class Example from the Teacher Wraparound Edition
- Additional, Try These exercises for each example
- The 5-Minute Check Transparencies
- Hot links to Glencoe Online Study Tools

Example 4 Reflection in the Origin

COORDINATE GEOMETRY **Suppose quadrilateral *KLMN* from Example 2 is reflected in the origin. Graph *KLMN* and its image under reflection in the origin. Compare the coordinates of each vertex with the coordinates of its image.**

Since $\overline{KK'}$ passes through the origin, use the horizontal and vertical distances from K to the origin to find the coordinates of K'. From K to the origin is 4 units up and 2 units left. K' is located by repeating that pattern from the origin. Four units up and 2 units left yields $K'(-2, 4)$.

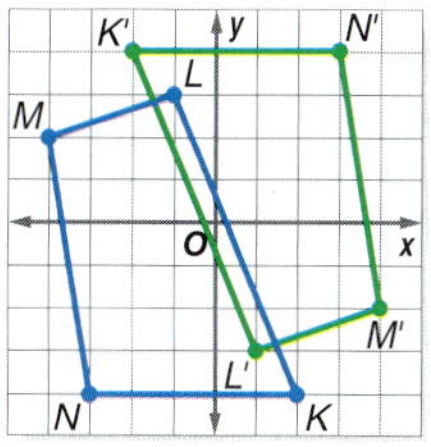

$K(2, -4) \rightarrow K'(-2, 4)$ $L(-1, 3) \rightarrow L'(1, -3)$

$M(-4, 2) \rightarrow M'(4, -2)$ $N(-3, -4) \rightarrow N'(3, 4)$

Plot the reflected vertices and connect to form the image $K'L'M'N'$. Comparing coordinates shows that $(a, b) \rightarrow (-a, -b)$.

Example 5 Reflection in the Line $y = x$

COORDINATE GEOMETRY **Suppose quadrilateral *KLMN* from Example 2 is reflected in the line $y = x$. Graph *KLMN* and its image under reflection in the line $y = x$. Compare the coordinates of each vertex with the coordinates of its image.**

The slope of $y = x$ is 1. $\overline{KK'}$ is perpendicular to $y = x$, so its slope is -1. From K to the line $y = x$, move up three units and left three units. From the line $y = x$ move up three units and left three units to $K'(-4, 2)$.

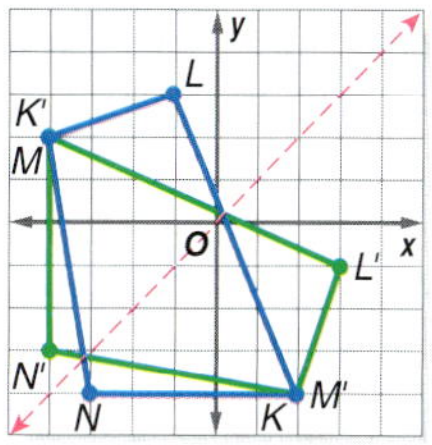

$K(2, -4) \rightarrow K'(-4, 2)$ $L(-1, 3) \rightarrow L'(3, -1)$

$M(-4, 2) \rightarrow M'(2, -4)$ $N(-3, -4) \rightarrow N'(-4, -3)$

Plot the reflected vertices and connect to form the image $K'L'M'N'$. Comparing coordinates shows that $(a, b) \rightarrow (b, a)$.

Concept Summary — Reflections in the Coordinate Plane

Reflection	x-axis	y-axis	origin	$y = x$
Preimage to Image	$(a, b) \rightarrow (a, -b)$	$(a, b) \rightarrow (-a, b)$	$(a, b) \rightarrow (-a, -b)$	$(a, b) \rightarrow (b, a)$
How to find coordinates	Multiply the y-coordinate by -1.	Multiply the x-coordinate by -1.	Multiply both coordinates by -1.	Interchange the x- and y-coordinates.
Example	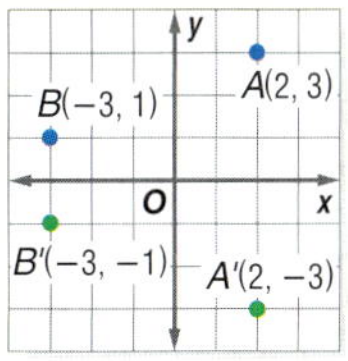 B(−3, 1), A(2, 3), B'(−3, −1), A'(2, −3)	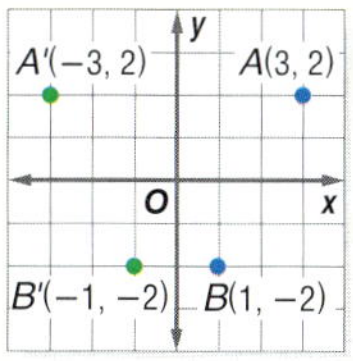 A'(−3, 2), A(3, 2), B'(−1, −2), B(1, −2)	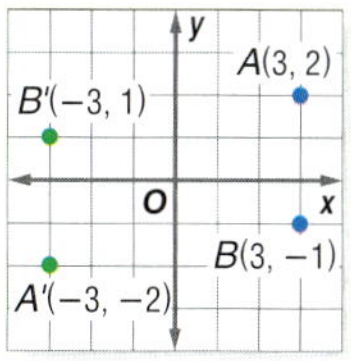 B'(−3, 1), A(3, 2), B(3, −1), A'(−3, −2)	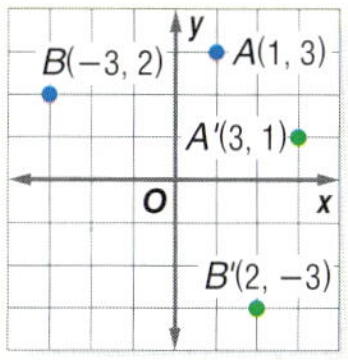 B(−3, 2), A(1, 3), A'(3, 1), B'(2, −3)

www.geometryonline.com/extra_examples

DAILY INTERVENTION — Differentiated Instruction

Naturalist Allow the class to discuss examples of reflections in nature and in everyday objects that they use. Students can explain where lines of reflection and lines of symmetry are in objects and decide whether they have point symmetry. Natural examples could be leaves, flowers, fruits, vegetables, animals, eggs, etc. Everyday objects could be pencils, paper, cars, compact discs, clothing, etc.

In-Class Examples

Power Point®

4 **COORDINATE GEOMETRY** Suppose quadrilateral *ABCD* with $A(1, 2)$, $B(3, 5)$, $C(4, -3)$, and $D(2, -5)$ is reflected in the origin. Graph *ABCD* and its image under reflection in the origin. Compare the coordinates of each vertex with the coordinates of its image.

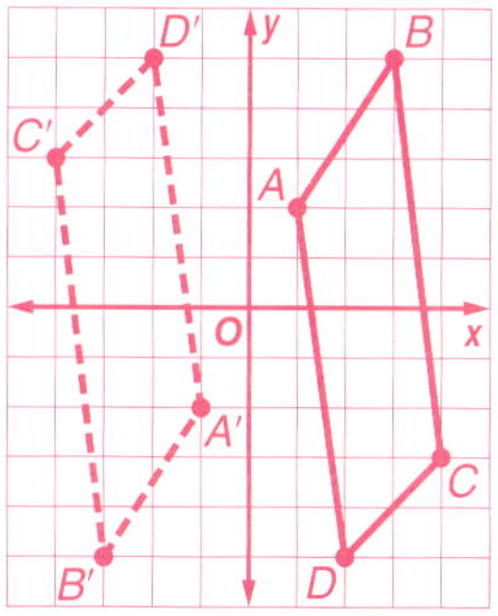

$A(1, 2) \rightarrow A'(-1, -2)$, $B(3, 5) \rightarrow B'(-3, -5)$, $C(4, -3) \rightarrow C'(-4, 3)$, $D(2, -5) \rightarrow D'(-2, 5)$; comparing coordinates shows that $(a, b) \rightarrow (-a, -b)$.

5 **COORDINATE GEOMETRY** Suppose quadrilateral *ABCD* from In-Class Example 4 is reflected in the line $y = x$. Graph *ABCD* and its image under reflection in the line $y = x$. Compare the coordinates of each vertex with the coordinates of its image.

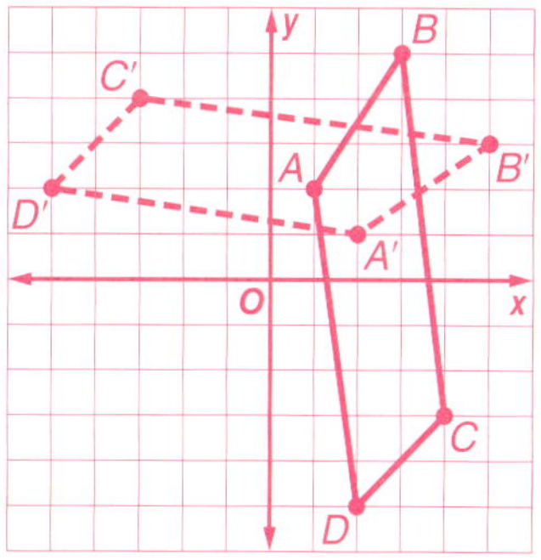

$A(1, 2) \rightarrow A'(2, 1)$, $B(3, 5) \rightarrow B'(5, 3)$, $C(4, -3) \rightarrow C'(-3, 4)$, $D(2, -5) \rightarrow D'(-5, 2)$; comparing coordinates shows that $(a, b) \rightarrow (b, a)$.

Building on Prior Knowledge

Students learned about wallpaper tiling in Lesson 7-3. In this chapter, they learn how to perform the transformations used in tiling and design.

In-Class Example

Power Point®

Teaching Tip Explain that in real-world situations, people must factor in other physical conditions, such as friction, force, and gravity, but geometric reflections play a very important role in many sports, such as golf, billiards, tennis, table tennis, and hockey.

6 **TABLE TENNIS** During a game of table tennis, Dipa decides that she wants to hit the ball so that it strikes her side of the table and then just clears the net. Describe how she should hit the ball using reflections.

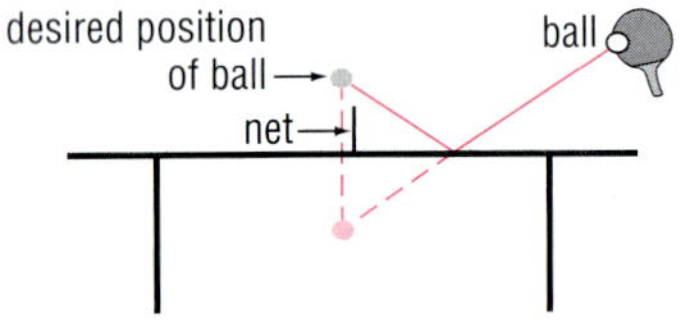

She should mentally reflect the desired position of the ball in the line of the table and aim toward the reflected image under the table.

LINES AND POINTS OF SYMMETRY

In-Class Example

Power Point®

7 Determine how many lines of symmetry a regular pentagon has. Then determine whether a regular pentagon has point symmetry. 5; no

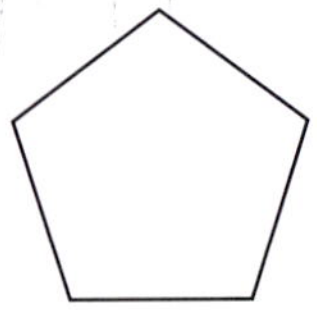

Example 6 Use Reflections

GOLF **Adeel and Natalie are playing miniature golf. Adeel says that he read how to use reflections to help make a hole-in-one on most miniature golf holes. Describe how he should putt the ball to make a hole-in-one.**

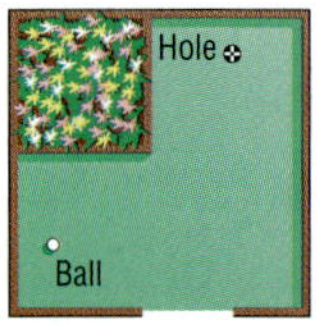

If Adeel tries to putt the ball directly to the hole, he will strike the border as indicated by the blue line. So, he can mentally reflect the hole in the line that contains the right border. If he putts the ball at the reflected image of the hole, the ball will strike the border, and it will rebound on a path toward the hole.

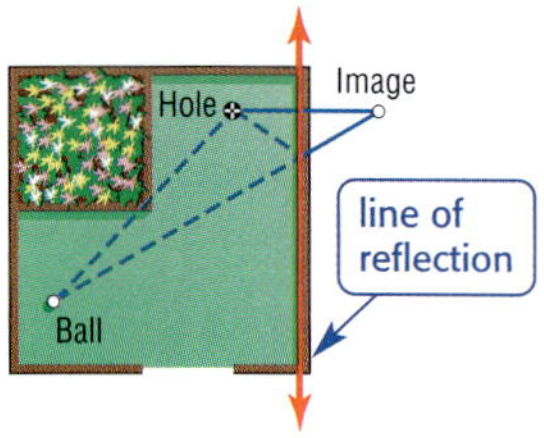

Study Tip

A Point of Symmetry
A point of symmetry is the intersection of all the segments joining a preimage to an image.

LINES AND POINTS OF SYMMETRY Some figures can be folded so that the two halves match exactly. The fold is a line of reflection called a **line of symmetry**. For some figures, a point can be found that is a common point of reflection for all points on a figure. This common point of reflection is called a **point of symmetry**.

Lines of Symmetry

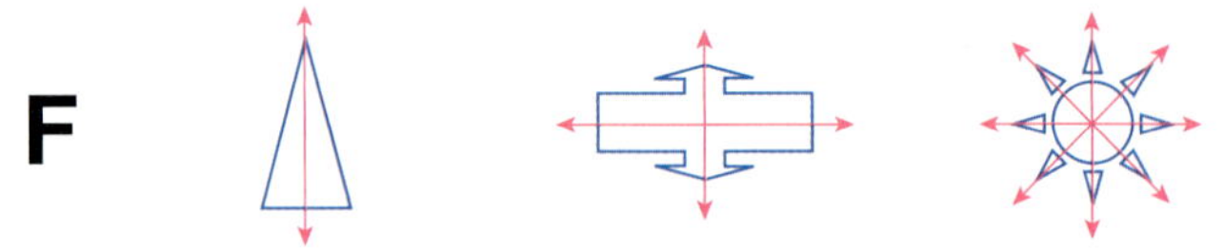

Points of Symmetry

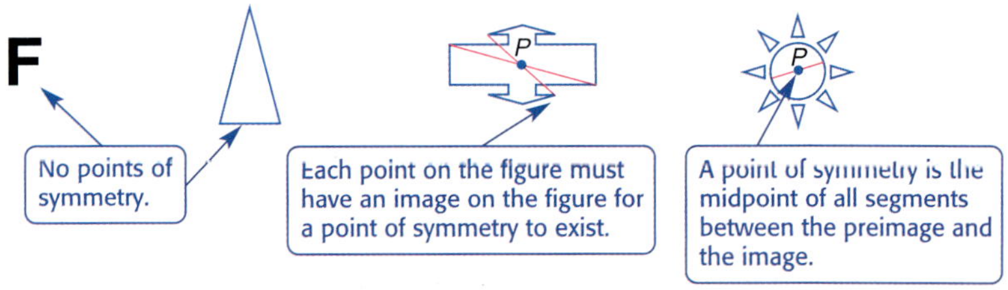

Example 7 Draw Lines of Symmetry

Determine how many lines of symmetry a square has. Then determine whether a square has point symmetry.

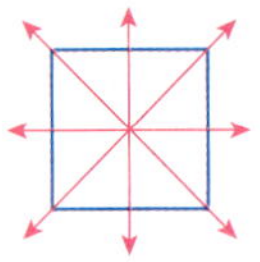

A square has four lines of symmetry.

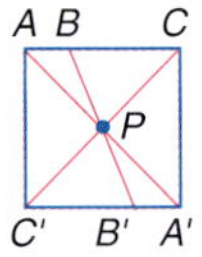

A square has point symmetry. P is the point of symmetry such that $AP = PA'$, $BP = PB'$, $CP = PC'$, and so on.

DAILY INTERVENTION — Unlocking Misconceptions

Reflections Explain that a very common mistake is to multiply the x-coordinate by -1 to reflect in the x-axis, and multiply the y-coordinate by -1 to reflect in the y-axis. Students can remember that reflecting in the x-axis means that the x-coordinate stays the same but the y-coordinate changes. Reflecting in the y-axis means that the y-coordinate stays the same but the x-coordinate changes. They can also remember that a reflection in the x-axis flips a figure up or down, affecting the y-coordinate, and a reflection in the y-axis flips a figure left or right, affecting the x-coordinate.

Check for Understanding

Concept Check

1–3. See margin.

1. **Find a counterexample** to disprove the statement *Two or more lines of symmetry for a plane figure intersect in a point of symmetry..*
2. **OPEN ENDED** Draw a figure on the coordinate plane and then reflect it in the line $y = x$. Label the coordinates of the preimage and the image.
3. **Identify** four properties that are preserved in reflections.

Guided Practice

GUIDED PRACTICE KEY	
Exercises	Examples
4	1
5–7, 11–13	7
8	2
10	3
11	4
12	5

4. Copy the figure at the right. Draw its reflected image in line m.

Determine how many lines of symmetry each figure has. Then determine whether the figure has point symmetry.

5.  2; yes
6. 3; no 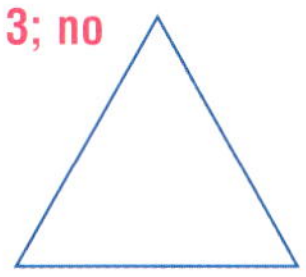
7.  6; yes

COORDINATE GEOMETRY **Graph each figure and its image under the given reflection.** 8–11. See p. 519A.

8. $\overline{AB}$ with endpoints $A(2, 4)$ and $B(-3, -3)$ reflected in the x-axis
9. $\triangle ABC$ with vertices $A(-1, 4)$, $B(4, -2)$, and $C(0, -3)$ reflected in the y-axis
10. $\triangle DEF$ with vertices $D(-1, -3)$, $E(3, -2)$, and $F(1, 1)$ reflected in the origin
11. $\square GHIJ$ with vertices $G(-1, 2)$, $H(2, 3)$, $I(6, 1)$, and $J(3, 0)$ reflected in the line $y = x$

Application

NATURE **Determine how many lines of symmetry each object has. Then determine whether each object has point symmetry.**

12.  1; no
13.  4; yes
14.  1; no

★ indicates increased difficulty

Practice and Apply

Homework Help

For Exercises	See Examples
15–26, 38, 39	1
29, 34, 40 41	2
30, 33, 41	3
27, 28	4
31, 32, 40	5
42	6
35–37, 44–47	7

Extra Practice
See page 771.

Refer to the figure at the right. Name the image of each figure under a reflection in:

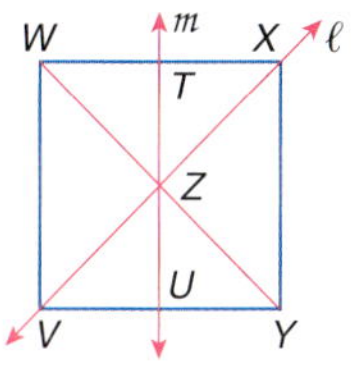

line ℓ	line m	point Z
15. $\overline{WX}$ $\overline{YX}$	18. T T	21. U T
16. $\overline{WZ}$ $\overline{YZ}$	19. $\overline{UY}$ $\overline{UV}$	22. $\angle TXZ$ $\angle UVZ$
17. $\angle XZY$ $\angle XZW$	20. $\triangle YVW$ $\triangle VYX$	23. $\triangle YUZ$ $\triangle WTZ$

Copy each figure. Draw the image of each figure under a reflection in line ℓ.

24. 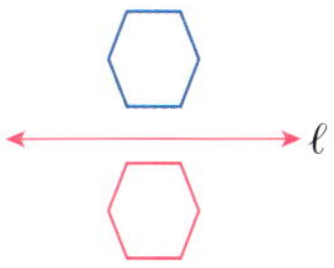

25. 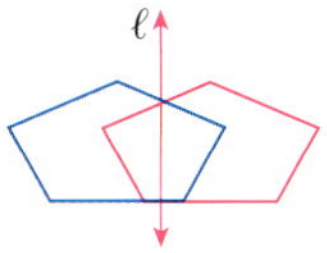

26. 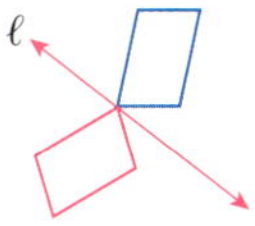

3 Practice/Apply

Study Notebook

Have students—
- add the definitions/examples of the vocabulary terms to their Vocabulary Builder worksheets for Chapter 9.
- include any other item(s) that they find helpful in mastering the skills in this lesson.

About the Exercises...

Organization by Objective
- **Draw Reflections:** 15–34, 38–42
- **Lines and Points of Symmetry:** 35–37, 44–47

Odd/Even Assignments
Exercises 15–41, 44–47 are structured so that students practice the same concepts whether they are assigned odd or even problems.

Assignment Guide

Basic: 15–31 odd, 35, 37, 43–47 odd, 49–62

Average: 15–47 odd, 49–62

Advanced: 16–42 even, 43, 44–48 even, 49–58 (optional: 59–62)

Answers

1. Sample Answer: The centroid of an equilateral triangle is not a point of symmetry.
2. Sample answer: $W(-3, 1)$, $X(-2, 3)$, $Y(3, 3)$ and $Z(3, 1)$ with reflected image $W'(1, -3)$, $X'(3, -2)$, $Y'(3, 3)$ and $Z'(1, 3)$

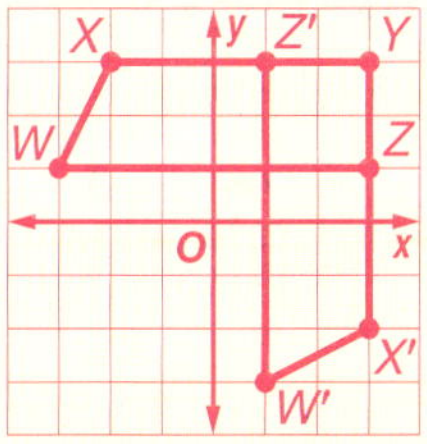

3. angle measure, betweenness of points, collinearity, distance

Study Guide and Intervention, p. 479 (shown) and p. 480

Draw Reflections The transformation called a **reflection** is a flip of a figure in a point, a line, or a plane. The new figure is the **image** and the original figure is the **preimage**. The preimage and image are congruent, so a reflection is a **congruence transformation** or **isometry**.

Example 1 **Construct the image of quadrilateral *ABCD* under a reflection in line *m*.**

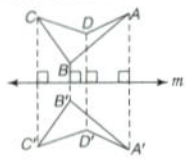

Draw a perpendicular from each vertex of the quadrilateral to *m*. Find vertices *A'*, *B'*, *C'*, and *D'* that are the same distance from *m* on the other side of *m*. The image is *A'B'C'D'*.

Example 2 **Quadrilateral *DEFG* has vertices $D(-2, 3)$, $E(4, 4)$, $F(3, -2)$, and $G(-3, -1)$. Find the image under reflection in the x-axis.**

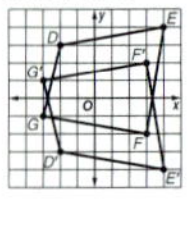

To find an image for a reflection in the x-axis, use the same x-coordinate and multiply the y-coordinate by -1. In symbols, $(a, b) \rightarrow (a, -b)$. The new coordinates are $D'(-2, -3)$, $E'(4, -4)$, $F'(3, 2)$, and $G'(-3, 1)$. The image is $D'E'F'G'$.

In Example 2, the notation $(a, b) \rightarrow (a, -b)$ represents a reflection in the x-axis. Here are three other common reflections in the coordinate plane.

- in the y-axis: $(a, b) \rightarrow (-a, b)$
- in the line $y = x$: $(a, b) \rightarrow (b, a)$
- in the origin: $(a, b) \rightarrow (-a, -b)$

Exercises

Draw the image of each figure under a reflection in line *m*.

1. 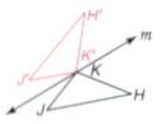

2. 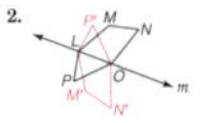

3. 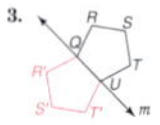

Graph each figure and its image under the given reflection.

4. $\triangle DEF$ with $D(-2, -1)$, $E(-1, 3)$, $F(3, -1)$ in the x-axis

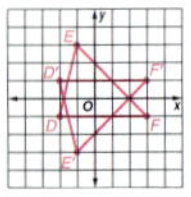

5. $ABCD$ with $A(1, 4)$, $B(3, 2)$, $C(2, -2)$, $D(-3, 1)$ in the y-axis

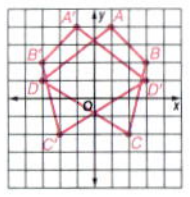

Skills Practice, p. 481 and Practice, p. 482 (shown)

Draw the image of each figure under a reflection in line ℓ.

1. 

2. 

COORDINATE GEOMETRY **Graph each figure and its image under the given reflection.**

3. quadrilateral $ABCD$ with vertices $A(-3, 3)$, $B(1, 4)$, $C(4, 0)$, and $D(-3, -3)$ in the origin

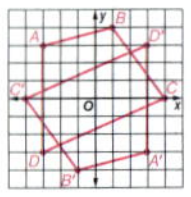

4. $\triangle FGH$ with vertices $F(-3, -1)$, $G(0, 4)$, and $H(3, -1)$ in the line $y = x$

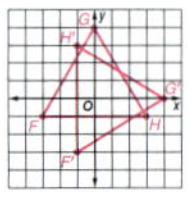

5. rectangle $QRST$ with vertices $Q(-3, 2)$, $R(-1, 4)$, $S(2, 1)$, and $T(0, -1)$ in the x-axis

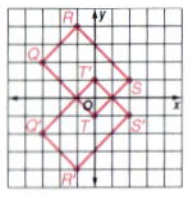

6. trapezoid $HIJK$ with vertices $H(-2, 5)$, $I(2, 5)$, $J(-4, -1)$, and $K(-4, 3)$ in the y-axis

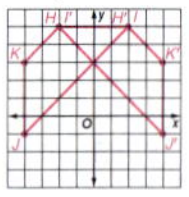

ROAD SIGNS **Determine how many lines of symmetry each sign has. Then determine whether the sign has point symmetry.**

7. 
0; yes

8. 
1; no

9. 
4; yes

Reading to Learn Mathematics, p. 483 ELL

Pre-Activity **Where are reflections found in nature?**

Read the introduction to Lesson 9-1 at the top of page 463 in your textbook.

Suppose you draw a line segment connecting a point at the peak of a mountain to its image in the lake. Where will the midpoint of this segment fall? on the boundary line between the shore and the surface of the lake

Reading the Lesson

1. Draw the reflected image for each reflection described below.
   a. reflection of trapezoid $ABCD$ in the line n. Label the image of $ABCD$ as $A'B'C'D'$.

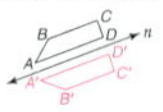

   b. reflection of $\triangle RST$ in point P. Label the image of RST as $R'S'T'$.

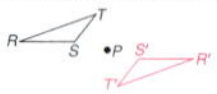

   c. reflection of pentagon $ABCDE$ in the origin. Label the image of $ABCDE$ as $A'B'C'D'E'$.

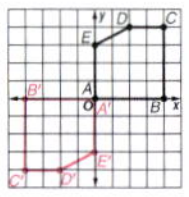

2. Determine the image of the given point under the indicated reflection.
   a. $(4, 6)$; reflection in the y-axis $(-4, 6)$
   b. $(-3, 5)$; reflection in the x-axis $(-3, -5)$
   c. $(-8, -2)$; reflection in the line $y = x$ $(-2, -8)$
   d. $(9, -3)$; reflection in the origin $(-9, 3)$
3. Determine the number of lines of symmetry for each figure described below. Then determine whether the figure has point symmetry and indicate this by writing *yes* or *no*.
   a. a square 4; yes
   b. an isosceles triangle (not equilateral) 1; no
   c. a regular hexagon 6; yes
   d. an isosceles trapezoid 1; no
   e. a rectangle (not a square) 2; yes
   f. the letter E 1; no

Helping You Remember

4. A good way to remember a new geometric term is to relate the word or its parts to geometric terms you already know. Look up the origins of the two parts of the word *isometry* in your dictionary. Explain the meaning of each part and give a term you already know that shares the origin of that part. Sample answer: The first part comes from *isos*, which means *equal*, as in *isosceles*. The second part comes from *metron*, which means *measure*, as in *geometry*.

COORDINATE GEOMETRY **Graph each figure and its image under the given reflection. 27–34. See p. 519A.**

27. rectangle $MNPQ$ with vertices $M(2, 3)$, $N(2, -3)$, $P(-2, -3)$, and $Q(-2, 3)$ in the origin
28. quadrilateral $GHIJ$ with vertices $G(-2, -2)$, $H(2, 0)$, $I(3, 3)$, and $J(-2, 4)$ in the origin
29. square $QRST$ with vertices $Q(-1, 4)$, $R(2, 5)$, $S(3, 2)$, and $T(0, 1)$ in the x-axis
30. trapezoid with vertices $D(4, 0)$, $E(-2, 4)$, $F(-2, -1)$, and $G(4, -3)$ in the y-axis
31. $\triangle BCD$ with vertices $B(5, 0)$, $C(-2, 4)$, and $D(-2, -1)$ in the line $y = x$
32. $\triangle KLM$ with vertices $K(4, 0)$, $L(-2, 4)$, and $M(-2, 1)$ in the line $y = 2$

★ 33. The reflected image of $\triangle FGH$ has vertices $F'(1, 4)$, $G'(4, 2)$, and $H'(3, -2)$. Describe the reflection in the y-axis.

★ 34. The reflected image of $\triangle XYZ$ has vertices $X'(1, 4)$, $Y'(2, 2)$, and $Z'(-2, -3)$. Describe the reflection in the line $x = -1$.

Determine how many lines of symmetry each figure has. Then determine whether the figure has point symmetry.

35. 
2; yes

36. 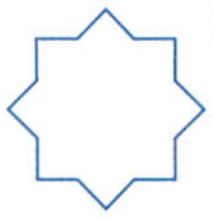
8; yes

37. 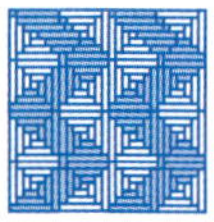
1; no

Copy each figure and then reflect the figure in line *m* first and then reflect that image in line *n*. Compare the preimage with the final image.

★ 38. 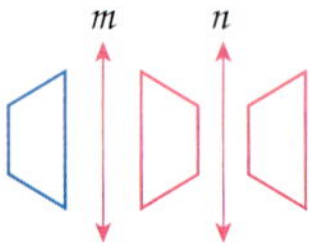

same shape and the same orientation

★ 39. 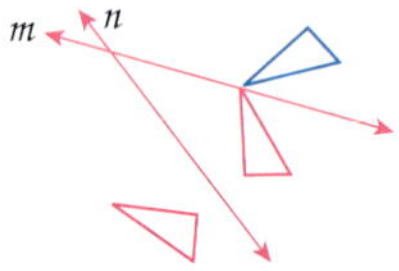

same shape, but turned or rotated

More About. . .

Billiards

The game in its present form was popular in the early 1800s, but games similar to billiards appeared as early as the 14th century. There are three types of billiards: carom billiards, pocket billiards (pool), and snooker.

Source: www.infoplease.com

★ 40. **COORDINATE GEOMETRY** Square $DEFG$ with vertices $D(-1, 4)$, $E(2, 8)$, $F(6, 5)$, and $G(3, 1)$ is reflected first in the x-axis, then in the line $y = x$. Find the coordinates of $D''E''F''G''$. $D''(-4, -1)$, $E''(-8, 2)$, $F''(-5, 6)$, and $G''(-1, 3)$

★ 41. **COORDINATE GEOMETRY** Triangle ABC has been reflected in the x-axis, then the y-axis, then the origin. The result has coordinates $A'''(4, 7)$, $B'''(10, -3)$, and $C'''(-6, -8)$. Find the coordinates of A, B, and C. $A(4, 7)$, $B(10, -3)$, and $C(-6, -8)$

42. **BILLIARDS** Tonya is playing billiards. She wants to pocket the eight ball in the lower right pocket using the white cue ball. Copy the diagram and sketch the path the eight ball must travel after being struck by the cue ball.

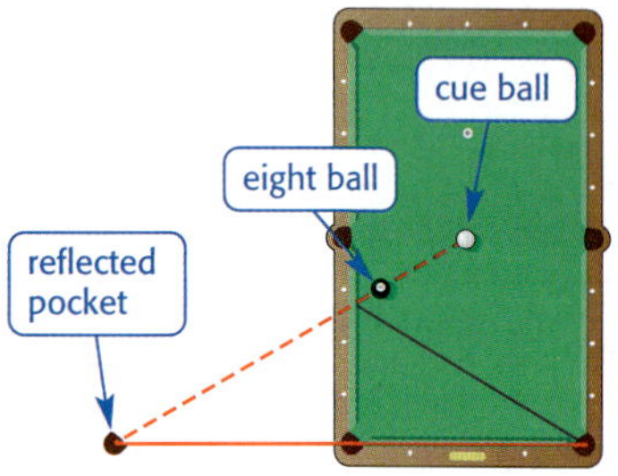

43. **CRITICAL THINKING** Show that the image of a point upon reflection in the origin is the same image obtained when reflecting a point in the x-axis and then the y-axis. See margin.

Enrichment, p. 484

Reflections in the Coordinate Plane

Study the diagram at the right. It shows how the triangle *ABC* is mapped onto triangle *XYZ* by the transformation $(x, y) \rightarrow (-x + 6, y)$. Notice that $\triangle XYZ$ is the reflection image with respect to the vertical line with equation $x = 3$.

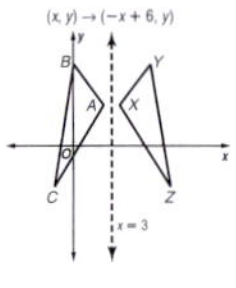

1. Prove that the vertical line with equation $x = 3$ is the perpendicular bisector of the segment with endpoints (x, y) and $(-x + 6, y)$. (Hint: Use the midpoint formula.)

Midpoint $= \left(\frac{x + (-x + 6)}{2}, \frac{y + y}{2}\right)$ or $(3, y)$

The segment joining (x, y) and $(-x + 6, y)$ is horizontal and hence is perpendicular to the vertical line.

Answer

43. Consider point (a, b). Upon reflection in the origin, its image is $(-a, -b)$. Upon reflection in the x-axis and then the y-axis, its image is $(a, -b) \Rightarrow (-a, -b)$. The images are the same.

WebQuest

A reflection of the quadrilateral on the map will help you come closer to locating the hidden treasure. Visit www.geometryonline.com/webquest to continue work on your WebQuest project.

DIAMONDS **For Exercises 44–47, use the following information.**
Diamond jewelers offer a variety of cuts. For each top view, identify any lines or points of symmetry. **44–47. See margin.**

44. round cut

45. pear cut

46. heart cut

47. emerald cut

48. WRITING IN MATH Answer the question that was posed at the beginning of the lesson. **See margin.**

Where are reflections found in nature?

Include the following in your answer:
- three examples in nature having line symmetry, and
- an explanation of how the distance from each point above the water line relates to the image in the water.

Standardized Test Practice

49. The image of $A(-2, 5)$ under a reflection is $A'(2, -5)$. Which reflection or group of reflections was used? **D**

I. reflected in the x-axis **II.** reflected in the y-axis **III.** reflected in the origin

Ⓐ I or III Ⓑ II and III Ⓒ I and II Ⓓ I and II, or III

50. **ALGEBRA** If $a \star c = 2a + b + 2c$, find $a \star c$ when $a = 25$, $b = 18$, and $c = 45$. **B**

Ⓐ 176 Ⓑ 158 Ⓒ 133 Ⓓ 88

Maintain Your Skills

Mixed Review

Write a coordinate proof for each of the following. *(Lesson 8-7)* **51–52. See pp. 519A–519B.**

51. The segments joining the midpoints of the opposite sides of a quadrilateral bisect each other.

52. The segments joining the midpoints of the sides of an isosceles trapezoid form a rhombus.

Refer to trapezoid *ACDF*. *(Lesson 8-6)*

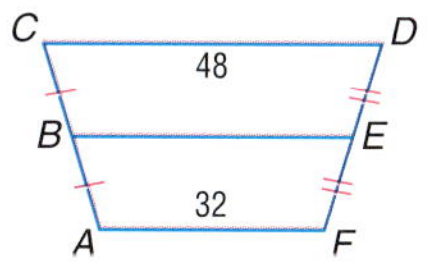

53. Find BE. **40**

54. Let $\overline{XY}$ be the median of $BCDE$. Find XY. **44**

55. Let $\overline{WZ}$ be the median of $ABEF$. Find WZ. **36**

Solve each $\triangle FGH$ described below. Round angle measures to the nearest degree and side measures to the nearest tenth. *(Lesson 7-6)*

56. $m\angle G = 53$, $m\angle H = 71$, $f = 48$ **$m\angle F = 56$, $g \approx 46.2$, $h \approx 54.7$**

57. $g = 21$, $m\angle G = 45$, $m\angle F = 59$ **$f \approx 25.5$, $m\angle H = 76$, $h \approx 28.8$**

58. $h = 13.2$, $m\angle F = 106$, $f = 14.5$ **$m\angle H \approx 61$, $m\angle G \approx 13$, $g \approx 3.4$**

Getting Ready for the Next Lesson

PREREQUISITE SKILL **Find the exact length of each side of quadrilateral *EFGH*.**
(To review the ***Distance Formula****, see Lesson 3-3.)*

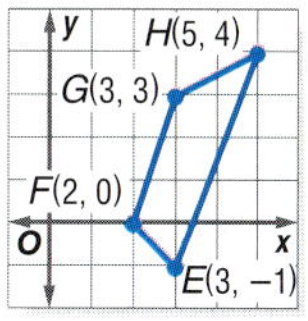

59. $\overline{EF}$ **$\sqrt{2}$**

60. $\overline{FG}$ **$\sqrt{10}$**

61. $\overline{GH}$ **$\sqrt{5}$**

62. $\overline{HE}$ **$\sqrt{29}$**

www.geometryonline.com/self_check_quiz

4 Assess

Open-Ended Assessment

Writing Students can practice drawing reflections with or without coordinate grids. They can use tracing paper, draw a figure on one half, and then fold and trace the figure on the other half. Pairs of students can also use grid paper to plot and draw figures and reflections. One student can plot a simple figure with three vertices and draw and label the point or line in which the other student will draw the reflection.

Getting Ready for Lesson 9-2

Prerequisite Skill Students will learn about translations using coordinates in Lesson 9-2. They will apply their knowledge of horizontal and vertical movement on a coordinate plane. Use Exercises 59–62 to determine your students' familiarity with finding segment length on a coordinate plane.

Answers

44. numerous lines of symmetry, including vertical and horizontal lines of symmetry; point of symmetry at the center

45. vertical line of symmetry

46. vertical line of symmetry

47. vertical, horizontal lines of symmetry; point of symmetry at the center

48. Sample answer: Reflections of the surrounding vistas can be seen in bodies of water. Answers should include the following.
- Three examples of line of symmetry in nature are the water's edge in a lake, the line through the middle of a pin oak leaf, and the lines of a four leaf clover.
- Each point above the water has a corresponding point in the image in the lake. The distance of a point above the water appears the same as the distance of the image below the water.

9-2 Lesson Notes

1 Focus

5-Minute Check Transparency 9-2 Use as a quiz or review of Lesson 9-1.

Mathematical Background notes are available for this lesson on p. 460C.

How are translations used in a marching band show?

Ask students:

- When band members hold their instruments in place tightly and move very precisely, how does this affect the overall appearance of the show? **Sample answer: Everything appears very uniform and pleasing to the eye.**
- Name some other forms of entertainment or art that use translated movements similar to those of the band members. **Sample answers: dance, interpretive movement, mime, animation**

9-2 Translations

What You'll Learn

- Draw translated images using coordinates.
- Draw translated images by using repeated reflections.

Vocabulary

- translation
- composition
- glide reflection

How are translations used in a marching band show?

The sights and pageantry of a marching band performance can add to the excitement of a sporting event. Band members dedicate a lot of time and energy to learning the music as well as the movements required for a performance.

The movements of each band member as they progress through the show are examples of *translations*.

Study Tip

Reading Math
A translation is also called a slide, a shift, or a glide.

TRANSLATIONS USING COORDINATES A **translation** is a transformation that moves all points of a figure the same distance in the same direction. Translations on the coordinate plane can be drawn if you know the direction and how far the figure is moving horizontally and/or vertically. For the fixed values of a and b, a translation moves every point $P(x, y)$ of a plane figure to an image $P'(x + a, y + b)$. One way to symbolize a transformation is to write $(x, y) \to (x + a, y + b)$.

In the figure, quadrilateral $DEFG$ has been translated 5 units to the left and three units down. This can be written as $(x, y) \to (x - 5, y - 3)$.

$D(1, 2) \to D'(1 - 5, 2 - 3)$ or $D'(-4, -1)$
$E(3, 1) \to E'(3 - 5, 1 - 3)$ or $E'(-2, -2)$
$F(4, -1) \to F'(4 - 5, -1 - 3)$ or $F'(-1, -4)$
$G(2, 0) \to G'(2 - 5, 0 - 3)$ or $G'(-3, -3)$

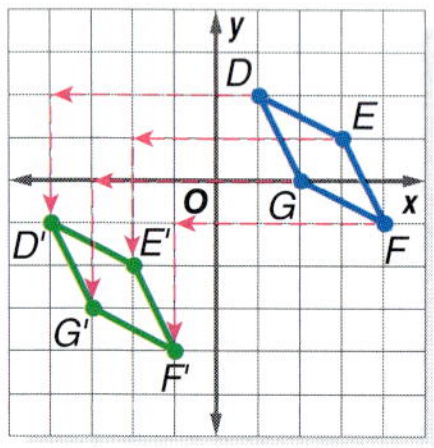

Example 1 Translations in the Coordinate Plane

Rectangle $PQRS$ has vertices $P(-3, 5)$, $Q(-4, 2)$, $R(3, 0)$, and $S(4, 3)$. Graph $PQRS$ and its image for the translation $(x, y) \to (x + 8, y - 5)$.

This translation moved every point of the preimage 8 units right and 5 units down.

$P(-3, 5) \to P'(-3 + 8, 5 - 5)$ or $P'(5, 0)$
$Q(-4, 2) \to Q'(-4 + 8, 2 - 5)$ or $Q'(4, -3)$
$R(3, 0) \to R'(3 + 8, 0 - 5)$ or $R'(11, -5)$
$S(4, 3) \to S'(4 + 8, 3 - 5)$ or $S'(12, -2)$

Plot the translated vertices and connect to form rectangle $P'Q'R'S'$.

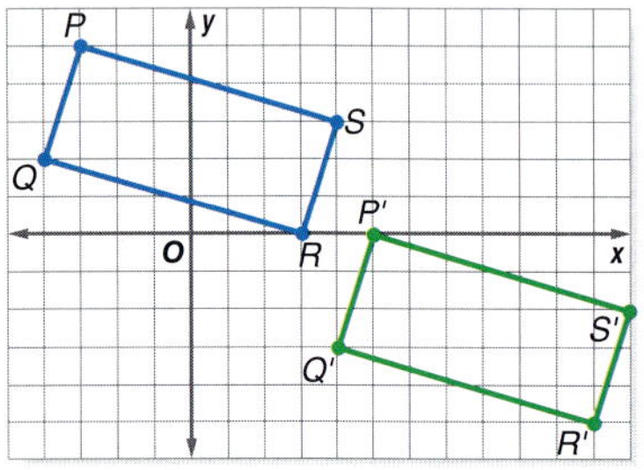

Study Tip

Counting Method
Each image point can be located using the same procedure as finding points on a line using the slope.

Resource Manager

Workbook and Reproducible Masters

Chapter 9 Resource Masters
- Study Guide and Intervention, pp. 485–486
- Skills Practice, p. 487
- Practice, p. 488
- Reading to Learn Mathematics, p. 489
- Enrichment, p. 490
- Assessment, p. 535

Graphing Calculator and Computer Masters, p. 33
Prerequisite Skills Workbook, pp. 1–2
Teaching Geometry With Manipulatives Masters, pp. 1, 16, 17, 147, 149, 151

Transparencies

5-Minute Check Transparency 9-2
Real-World Transparency 9
Answer Key Transparencies

Technology

Interactive Chalkboard

More About...

Animation

Animation began as a series of hand-drawn cells, each one with a subtle change made in the drawing. When the drawings were displayed in rapid succession, it visually created movement.

Example 2 Repeated Translations

ANIMATION **Computers are often used to create animation. The graph shows repeated translations that result in animation of the star. Find the translation that moves star 1 to star 2 and the translation that moves star 4 to star 5.**

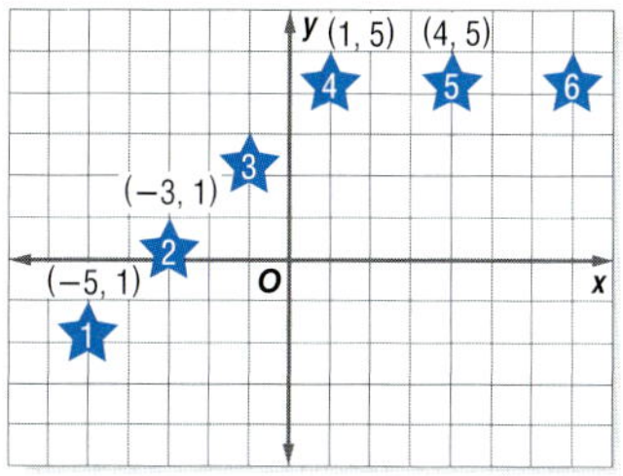

To find the translation from star 1 to star 2, use the coordinates at the top of each star. Use the coordinates $(-5, -1)$ and $(-3, 1)$ in the formula.

$(x, y) \rightarrow (x + a, y + b)$

$(-5, -1) \rightarrow (-3, 1)$

$x + a = -3$		$y + b = 1$	
$-5 + a = -3$	$x = -5$	$-1 + b = 1$	$y = -1$
$a = 2$	Add 5 to each side.	$b = 2$	Add 1 to each side.

The translation is $(x, y) \rightarrow (x + 2, y + 2)$.

Use the coordinates $(1, 5)$ and $(4, 5)$ to find the translation from star 4 to star 5.

$(x, y) \rightarrow (x + a, y + b)$

$(1, 5) \rightarrow (4, 5)$

$x + a = 4$		$y + b = 5$	
$1 + a = 4$	$x = 1$	$5 + b = 5$	$y = 5$
$a = 3$	Subtract 1 from each side.	$b = 0$	Subtract 5 from each side.

The translation is $(x, y) \rightarrow (x + 3, y)$ from star 4 to star 5 and from star 5 to star 6.

TRANSLATIONS BY REPEATED REFLECTIONS Another way to find a translation is to perform a reflection in the first of two parallel lines and then reflect the image in the other parallel line. Each successive transformation is called a **composition**.

Example 3 Find a Translation Using Reflections

In the figure, lines m and n are parallel. Determine whether the red figure is a translation image of the blue preimage, quadrilateral $ABCD$.

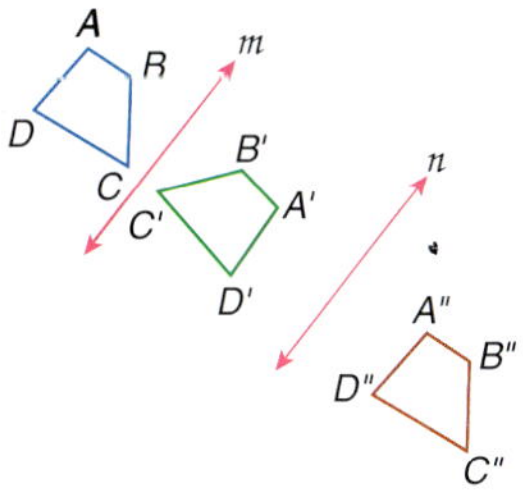

Reflect quadrilateral $ABCD$ in line m. The result is the green image, quadrilateral $A'B'C'D'$. Then reflect the green image, quadrilateral $A'B'C'D'$ in line n. The red image, quadrilateral $A''B''C''D''$, has the same orientation as quadrilateral $ABCD$.

Quadrilateral $A''B''C''D''$ is the translation image of quadrilateral $ABCD$.

Since translations are compositions of two reflections, all translations are isometries. Thus, all properties preserved by reflections are preserved by translations. These properties include betweenness of points, collinearity, and angle and distance measure.

www.geometryonline.com/extra_examples/fcat

2 Teach

TRANSLATIONS USING COORDINATES

In-Class Examples

Power Point®

Teaching Tip When translating the rectangle, students can either use the counting method first and then check their work algebraically or vice versa.

1 **COORDINATE GEOMETRY** Parallelogram $TUVW$ has vertices $T(-1, 4)$, $U(2, 5)$, $V(4, 3)$, and $W(1, 2)$. Graph $TUVW$ and its image for the translation $(x, y) \rightarrow (x - 4, y - 5)$.

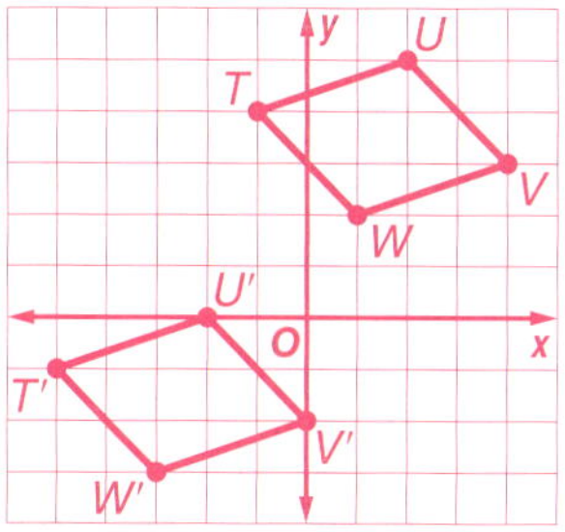

2 **ANIMATION** The graph shows repeated translations that result in the animation of the raindrop. Find the translation that moves raindrop 2 to raindrop 3 and then the translation that moves raindrop 3 to raindrop 4.

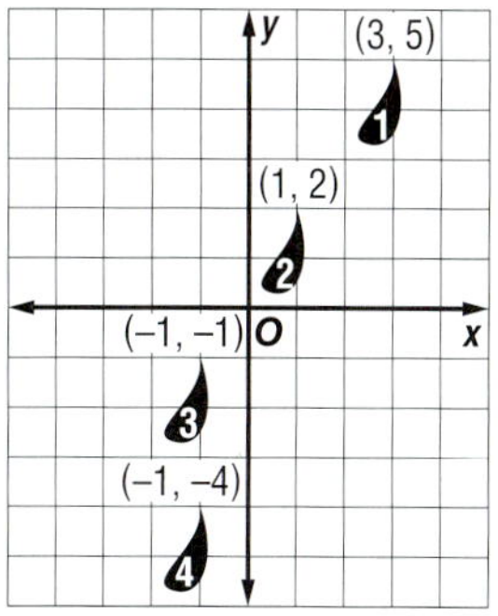

$(x, y) \rightarrow (x - 2, y - 3)$; $(x, y) \rightarrow (x, y - 3)$

DAILY INTERVENTION

Differentiated Instruction

Kinesthetic Create three or four large coordinate grids using poster board and a black permanent marker. Provide several laminated shapes, including rectangles, hexagons, pentagons, trapezoids, etc. Groups of students can practice physically translating shapes on the grids. Students can use examples of translations in the lesson or create their own.

TRANSLATIONS BY REPEATED REFLECTIONS

In-Class Example

Power Point®

Teaching Tip Explain that a translation by a composition of reflections only works with two parallel lines or two lines that are both perpendicular to the first line.

3 In the figure, lines *p* and *q* are parallel. Determine whether the red figure is a translation image of the blue preimage, quadrilateral *EFGH*.

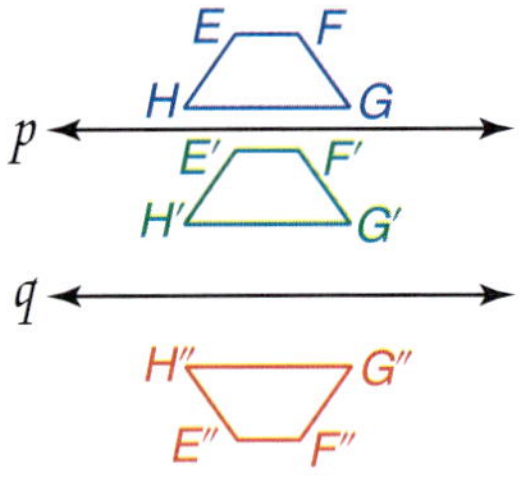

Quadrilateral *E″F″G″H″* is not a translation image of quadrilateral *EFGH*.

Teaching Tip Tell students that the shorter the distance is between translated figures, the more fluid an animation appears. Discuss how some cartoon animation companies might cut costs by translating images quicker using longer distances and how full-length animated movies may tend to use shorter translated distances for a more realistic effect.

3 Practice/Apply

Study Notebook

Have students—

- add the definitions/examples of the vocabulary terms to their Vocabulary Builder worksheets for Chapter 9.
- include any other item(s) that they find helpful in mastering the skills in this lesson.

Check for Understanding

Concept Check

1–2. See margin.

1. **OPEN ENDED** Choose integer coordinates for any two points *A* and *B*. Then describe how you could count to find the translation of point *A* to point *B*.

2. **Explain** which properties are preserved in a translation and why they are preserved.

3. Allie; counting from the point (−2, 1) to (1, −1) is right 3 and down 2 to the image. The reflections would be too far to the right. The image would be reversed as well.

3. **FIND THE ERROR** Allie and Tyrone are describing the transformation in the drawing.

Allie	Tyrone
This is a translation right 3 units and down 2 units.	This is a reflection in the y-axis and then the x-axis.

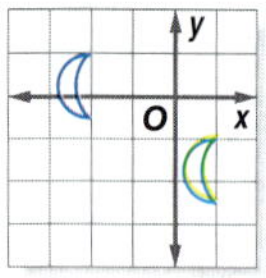

Who is correct? Explain your reasoning.

Guided Practice

GUIDED PRACTICE KEY	
Exercises	Examples
4, 5	3
6, 7	1
8	2

In each figure, *m* ∥ *n*. Determine whether the red figure is a translation image of the blue figure. Write *yes* or *no*. Explain your answer.

4. 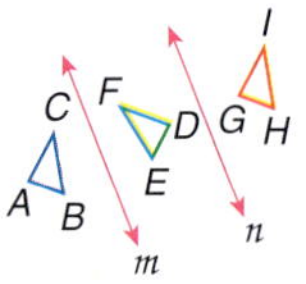

Yes; △*GHI* is a translation of △*ABC*.

5. 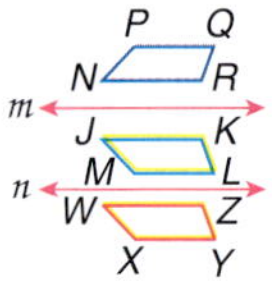

No; quadrilateral *WXYZ* is oriented differently than quadrilateral *NPQR*.

Application

COORDINATE GEOMETRY **Graph each figure and its image under the given translation.** 6–7. See p. 519B.

6. $\overline{DE}$ with endpoints $D(-3, -4)$ and $E(4, 2)$ under the translation $(x, y) \rightarrow (x + 1, y + 3)$

7. $\triangle KLM$ with vertices $K(5, -2)$, $L(-3, -1)$, and $M(0, 5)$ under the translation $(x, y) \rightarrow (x - 3, y - 4)$

8. **ANIMATION** Find the translations that move the hexagon on the coordinate plane in the order given.
$1 \rightarrow 2 = (x, y) \rightarrow (x, y + 3)$
$2 \rightarrow 3 = (x, y) \rightarrow (x + 4, y)$
$3 \rightarrow 4 = (x, y) \rightarrow (x + 4, y)$

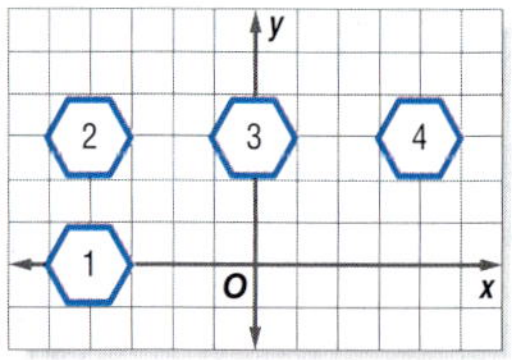

★ indicates increased difficulty

Practice and Apply

In each figure, *a* ∥ *b*. Determine whether the red figure is a translation image of the blue figure. Write *yes* or *no*. Explain your answer. 9–11. See margin.

9. 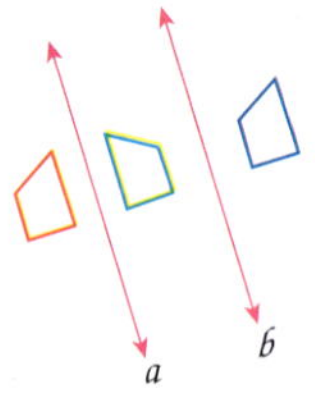

10. 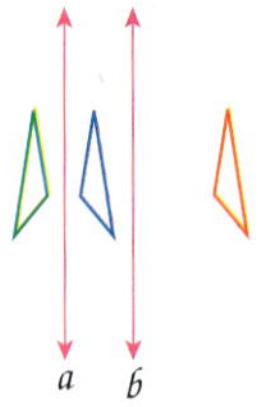

11. 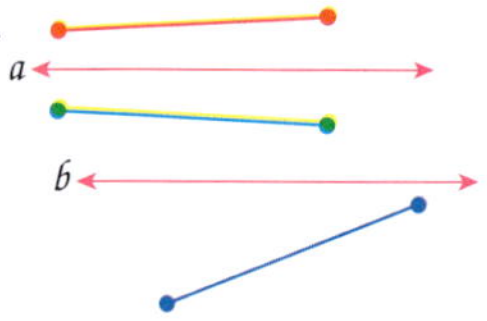

Answers

1. Sample answer: $A(3, 5)$ and $B(-4, 7)$; start at 3, count to the left to −4, which is 7 units to the left or −7. Then count up 2 units from 5 to 7 or +2. The translation from *A* to *B* is $(x, y) \rightarrow (x - 7, y + 2)$.

2. The properties that are preserved include betweenness of points, collinearity, and angle and distance measure. Since translations are composites of two reflections, all translations are isometries. Thus, all properties preserved by reflections are preserved by translations.

Homework Help

For Exercises	See Examples
9–14, 28, 29	3
15–20, 24–27	1
19, 21–23	2

Extra Practice
See page 771.

12–14. See margin.

12.

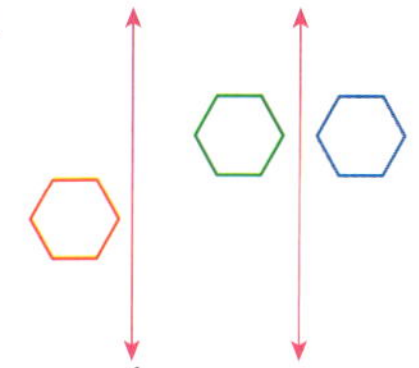

13.

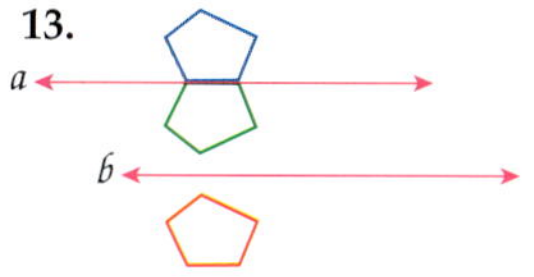

14.

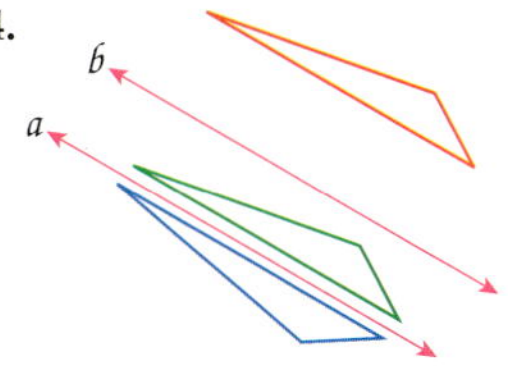

COORDINATE GEOMETRY Graph each figure and its image under the given translation. 15–20. See p. 519B.

15. $\overline{PQ}$ with endpoints $P(2, -4)$ and $Q(4, 2)$ under the translation left 3 units and up 4 units

16. $\overline{AB}$ with endpoints $A(-3, 7)$ and $B(-6, -6)$ under the translation 4 units to the right and down 2 units

17. $\triangle MJP$ with vertices $M(-2, -2)$, $J(-5, 2)$, and $P(0, 4)$ under the translation $(x, y) \rightarrow (x + 1, y + 4)$

18. $\triangle EFG$ with vertices $E(0, -4)$, $F(-4, -4)$, and $G(0, 2)$ under the translation $(x, y) \rightarrow (x + 2, y - 1)$

19. quadrilateral $PQRS$ with vertices $P(1, 4)$, $Q(-1, 4)$, $R(-2, -4)$, and $S(2, -4)$ under the translation $(x, y) \rightarrow (x - 5, y + 3)$

20. pentagon $VWXYZ$ with vertices $V(-3, 0)$, $W(-3, 2)$, $X(-2, 3)$, $Y(0, 2)$, and $Z(-1, 0)$ under the translation $(x, y) \rightarrow (x + 4, y - 3)$

21. **CHESS** The bishop shown in square f8 can only move diagonally along dark squares. If the bishop is in c1 after two moves, describe the translation. left 3 squares and down 7 squares

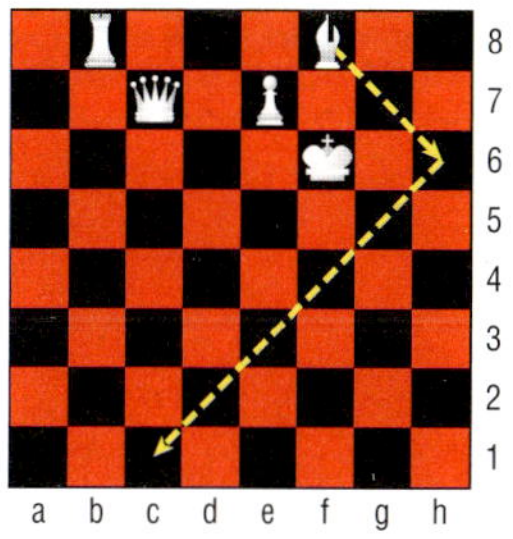

22. **RESEARCH** Use the Internet or other resource to write a possible translation for each chess piece for a single move. See margin.

MOSAICS For Exercises 23–25, use the following information.
The mosaic tiling shown on the right is a thirteenth-century Roman inlaid marble tiling. Suppose this pattern is a floor design where the length of a side of the small white equilateral triangle is 12 inches. All triangles and hexagons are regular. Describe the translations in inches represented by each line.

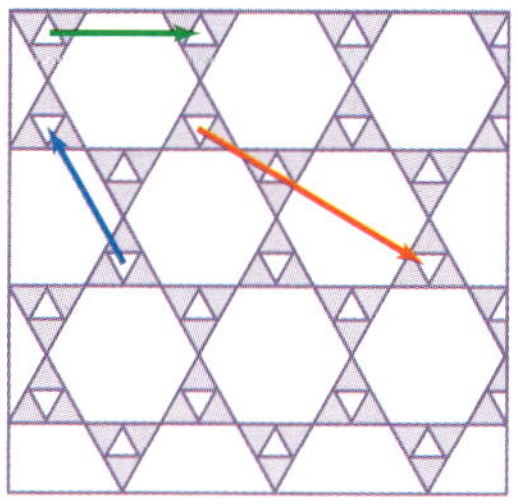

★ 23. green line 48 in. right

★ 24. blue line $24\sqrt{3}$ in. up, 24 in. left

★ 25. red line 72 in. right, $24\sqrt{3}$ in. down

26. **CRITICAL THINKING** Triangle TWY has vertices $T(3, -7)$, $W(7, -4)$, and $Y(9, -8)$. Triangle BDG has vertices $B(3, 3)$, $D(7, 6)$, and $G(9, 2)$. If $\triangle BDG$ is the translation image of $\triangle TWY$ with respect to two parallel lines, find the equations that represent two possible parallel lines. Sample answer: $y = 1$ and $y = -4$

DAILY INTERVENTION **FIND THE ERROR** Students must remember to check the orientation of the translated figure and examine how it relates to the original figure. Explain that a reflection changes the orientation of the original figure, while a translation does not.

About the Exercises...

Organization by Objective

- **Translations Using Coordinates:** 15–27, 30
- **Translations by Repeated Reflections:** 9–14, 28, 29

Odd/Even Assignments
Exercises 9–20, 27–30 are structured so that students practice the same concepts whether they are assigned odd or even problems.

Alert! Exercise 22 requires the Internet or other research materials.

Assignment Guide

Basic: 9–21 odd, 26, 27, 31, 33, 38–59 (optional: 35–37)

Average: 9–25 odd, 26, 27–33 odd, 38–59 (optional: 35–37)

Advanced: 10–34 even, 35–53 (optional: 54–59)

Answers

12. No; it is a reflection followed by a translation.
13. Yes; it is one reflection after another with respect to the two parallel lines.
14. No; it is a reflection followed by a translation.
22. Sample answers: pawn: up two squares; rook: left four squares; knight: down two squares, right 1 square; bishop: up three squares, right three squares; queen: up five squares; king: right 1 square

Answers (page 472)

9. Yes; it is one reflection after another with respect to the two parallel lines.
10. No; it is a reflection followed by a translation.
11. No; it is a reflection followed a rotation.

Study Guide and Intervention, p. 485 (shown) and p. 486

Translations Using Coordinates A transformation called a **translation** slides a figure in a given direction. In the coordinate plane, a translation moves every preimage point $P(x, y)$ to an image point $P(x + a, y + b)$ for fixed values a and b. In words, a translation shifts a figure a units horizontally and b units vertically; in symbols, $(x, y) \rightarrow (x + a, y + b)$.

Example **Rectangle *RECT* has vertices $R(-2, -1)$, $E(-2, 2)$, $C(3, 2)$, and $T(3, -1)$. Graph *RECT* and its image for the translation $(x, y) \rightarrow (x + 2, y - 1)$.**

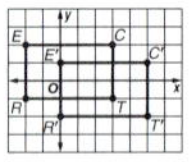

The translation moves every point of the preimage right 2 units and down 1 unit.

$(x, y) \rightarrow (x + 2, y - 1)$
$R(-2, -1) \rightarrow R'(-2 + 2, -1 - 1)$ or $R'(0, -2)$
$E(-2, 2) \rightarrow E'(-2 + 2, 2 - 1)$ or $E'(0, 1)$
$C(3, 2) \rightarrow C'(3 + 2, 2 - 1)$ or $C'(5, 1)$
$T(3, -1) \rightarrow T'(3 + 2, -1 - 1)$ or $T'(5, -2)$

Exercises

Graph each figure and its image under the given translation.

1. $\overline{PQ}$ with endpoints $P(-1, 3)$ and $Q(2, 2)$ under the translation left 2 units and up 1 unit

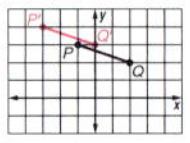

2. $\triangle PQR$ with vertices $P(-2, -4)$, $Q(-1, 2)$, and $R(2, 1)$ under the translation right 2 units and down 2 units

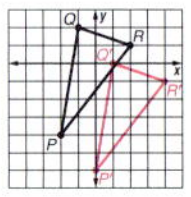

3. square $SQUR$ with vertices $S(0, 2)$, $Q(3, 1)$, $U(2, -2)$, and $R(-1, -1)$ under the translation right 3 units and up 1 unit

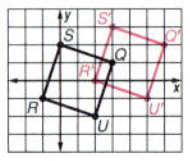

Skills Practice, p. 487 and Practice, p. 488 (shown)

In each figure, $c \parallel d$. Determine whether figure 3 is a translation image of figure 1. Write *yes* or *no*. Explain your answer.

1. 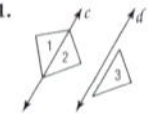

No; it is a translation of a reflection and is oriented differently than figure 1.

2. 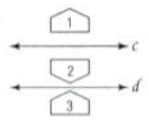

Yes; it is a reflection of a reflection with respect to the parallel lines.

COORDINATE GEOMETRY **Graph each figure and its image under the given translation.**

3. quadrilateral $TUWX$ with vertices $T(-1, 1)$, $U(4, 2)$, $W(1, 5)$, and $X(-1, 3)$ under the translation $(x, y) \rightarrow (x - 2, y - 4)$

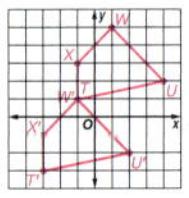

4. pentagon $DEFGH$ with vertices $D(-1, -2)$, $E(2, -1)$, $F(5, -2)$, $G(4, -4)$, $H(1, -4)$ under the translation $(x, y) \rightarrow (x - 1, y + 5)$

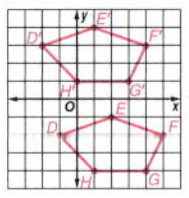

ANIMATION **Find the translation that moves the figure on the coordinate plane.**

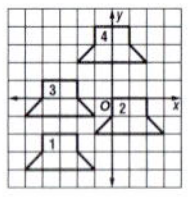

5. figure 1 → figure 2
$(x + 4, y + 2)$

6. figure 2 → figure 3
$(x - 4, y + 1)$

7. figure 3 → figure 4
$(x + 3, y + 3)$

Reading to Learn Mathematics, p. 489 ELL

Pre-Activity **How are translations used in a marching band show?**

Read the introduction to Lesson 9-2 at the top of page 470 in your textbook.

How do band directors get the marching band to maintain the shape of the figure they originally formed? Sample answer: The band practices marching in unison, with everyone making identical moves at the same time.

Reading the Lesson

1. Underline the correct word or phrase to form a true statement.
   a. All reflections and translations are (opposites/isometries/equivalent).
   b. The preimage and image of a figure under a reflection in a line have (the same orientation/opposite orientations).
   c. The preimage and image of a figure under a translation have (the same orientation/opposite orientations).
   d. The result of successive reflections over two parallel lines is a (reflection/rotation/translation).
   e. Collinearity (is/is not) preserved by translations.
   f. The translation $(x, y) \rightarrow (x + a, y + b)$ shifts every point a units (horizontally/vertically) and y units (horizontally/vertically).

2. Find the image of each preimage under the indicated translation.
   a. (x, y); 5 units right and 3 units up $(x + 5, y + 3)$
   b. (x, y); 2 units left and 4 units down $(x - 2, y - 4)$
   c. (x, y); 1 unit left and 6 units up $(x - 1, y + 6)$
   d. (x, y); 7 units right $(x + 7, y)$
   e. $(4, -3)$; 3 units up $(4, 0)$
   f. $(-5, 6)$; 3 units right and 2 units down $(-2, 4)$
   g. $(-7, 5)$; 7 units right and 5 units down $(0, 0)$
   h. $(-9, -2)$; 12 units right and 6 units down $(3, -8)$

3. $\triangle RST$ has vertices $R(-3, 3)$, $S(0, -2)$, and $T(2, 1)$. Graph $\triangle RST$ and its image $\triangle R'S'T'$ under the translation $(x, y) \rightarrow (x + 3, y - 2)$. List the coordinates of the vertices of the image.
$R'(0, 1)$, $S'(3, -4)$, $T'(5, -1)$

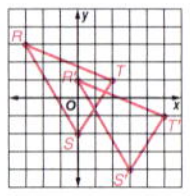

Helping You Remember

4. A good way to remember a new mathematical term is to relate it to an everyday meaning of the same word. How is the meaning of *translation* in geometry related to the idea of *translation* from one language to another? Sample answer: When you translate from one language to another, you *carry over* the meaning from one language to another. When you translate a geometric figure, you *carry over* the figure from one position to another without changing its basic

Study Tip

Coordinate Translations

To check your graph, trace the original image and cut it out. Place it on the original graph and, keeping the same orientation, move the copy the correct number of spaces up/down and right/left. Does the copy match your image?

COORDINATE GEOMETRY **Graph each figure and the image under the given translation.** 27–30. See p. 519B for figures.

27. $\triangle PQR$ with vertices $P(-3, -2)$, $Q(-1, 4)$, and $R(2, -2)$ under the translation $(x, y) \rightarrow (x + 2, y - 4)$

★ 28. $\triangle RST$ with vertices $R(-4, -1)$, $S(-1, 3)$, and $T(-1, 1)$ reflected in $y = 2$ and then reflected in $y = -2$

★ 29. Under $(x, y) \rightarrow (x - 4, y + 5)$, $\triangle ABC$ has translated vertices $A'(-8, 5)$, $B'(2, 7)$, and $C'(3, 1)$. Find the coordinates of A, B, and C. $A(-4, 0)$, $B(6, 2)$, $C(7, -4)$

★ 30. Triangle FGH is translated to $\triangle MNP$. Given $F(3, 9)$, $G(-1, 4)$, $M(4, 2)$, and $P(6, -3)$, find the coordinates of H and N. Then write the coordinate form of the translation. $H(5, 4)$, $N(0, -3)$; $(x, y) \rightarrow (x + 1, y - 7)$

STUDENTS **For Exercises 31–33, refer to the graphic at the right. Each bar of the graph is made up of a boy-girl-boy unit.**

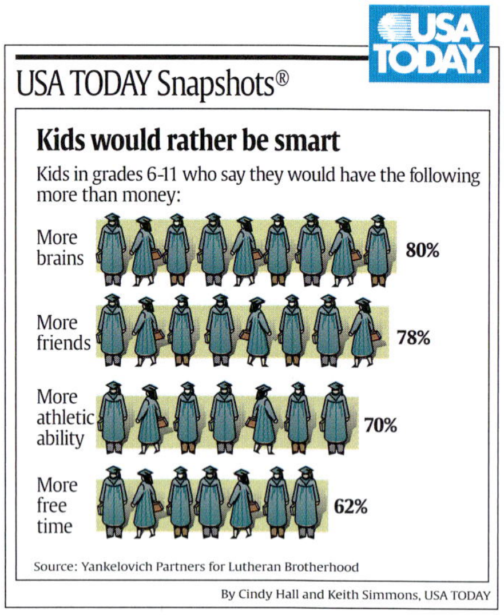

31. Which categories show a boy-girl-boy unit that is translated within the bar?

31. more brains; more free time

32. Which categories show a boy-girl-boy unit that is reflected within the bar?

32. more friends; more athletic ability

33. Does each person shown represent the same percent? Explain. No; the percent per figure is different in each category.

Online Research

Data Update How much allowance do teens receive? Visit www.geometryonline.com/data_update to learn more.

34. WRITING IN MATH Answer the question that was posed at the beginning of the lesson. See margin.

How are translations used in a marching band show?

Include the following in your answer:
- the types of movements by band members that are translations, and
- a description of a simple pattern for a band member.

Extending the Lesson

GLIDE REFLECTION **For Exercises 35–37, use the following information.**

A **glide reflection** is a composition of a translation and a reflection in a line parallel to the direction of the translation. 35–37. See margin.

35. Is a glide reflection an *isometry*? Explain.

36. Triangle DEF has vertices $D(4, 3)$, $E(2, -2)$ and $F(0, 1)$. Sketch the image of a glide reflection composed of the translation $(x, y) \rightarrow (x, y - 2)$ and a reflection in the y-axis.

37. Triangle ABC has vertices $A(-3, -2)$, $B(-1, -3)$ and $C(2, -1)$. Sketch the image of a glide reflection composed of the translation $(x, y) \rightarrow (x + 3, y)$ and a reflection in $y = 1$.

Enrichment, p. 490

Translations in The Coordinate Plane

You can use algebraic descriptions of reflections to show that the composite of two reflections with respect to parallel lines is a translation (that is, a slide).

1. Suppose a and b are two different real numbers. Let S and T be the following reflections.

   $S: (x, y) \rightarrow (-x + 2a, y)$
   $T: (x, y) \rightarrow (-x + 2b, y)$

   S is reflection with respect to the line with equation $x = a$, and T is reflection with respect to the line with equation $x = b$.

   a. Find an algebraic description (similar to those above for S and T) to describe the composite transformation "S followed by T."
   $(x, y) \rightarrow (x + 2(b - a), y)$

FCAT Practice

Standardized Test Practice

38. Triangle XYZ with vertices $X(5, 4)$, $Y(3, -1)$, and $Z(0, 2)$ is translated so that X' is at $(3, 1)$. State the coordinates of Y' and Z'. **C**

(A) $Y'(5, 2)$ and $Z'(2, 5)$ (B) $Y'(0, -3)$ and $Z'(-3, 0)$

(C) $Y'(1, -4)$ and $Z'(-2, -1)$ (D) $Y'(11, 4)$ and $Z'(8, 6)$

39. **ALGEBRA** Find the slope of a line through $P(-2, 5)$ and $T(2, -1)$. **A**

(A) $-\frac{3}{2}$ (B) -1 (C) 0 (D) $\frac{2}{3}$

Maintain Your Skills

Mixed Review Copy each figure. Draw the reflected image of each figure in line m. *(Lesson 9-1)*

40. 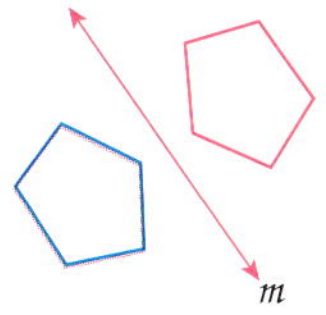

41. 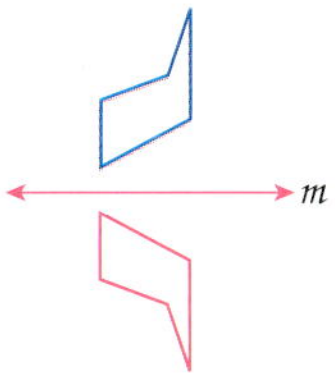

42. 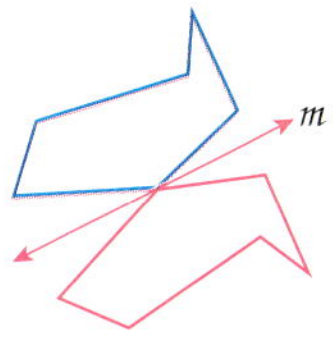

Name the missing coordinates for each quadrilateral. *(Lesson 8-7)*

43. $QRST$ is an isosceles trapezoid. **$Q(a - b, c)$, $T(0, 0)$**

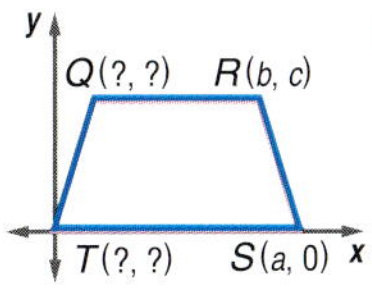

44. $ABCD$ is a parallelogram. **$B(a, b)$, $D(d, 0)$**

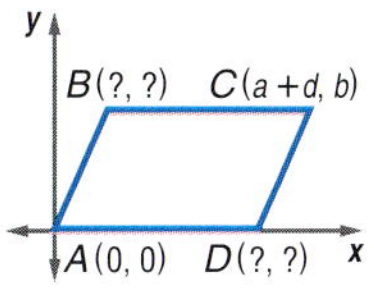

45. **LANDSCAPING** Juanna is a landscaper. She wishes to determine the height of a tree. Holding a drafter's 45° triangle so that one leg is horizontal, she sights the top of the tree along the hypotenuse as shown at the right. If she is 6 yards from the tree and her eyes are 5 feet from the ground, find the height of the tree. *(Lesson 7-3)* **23 ft**

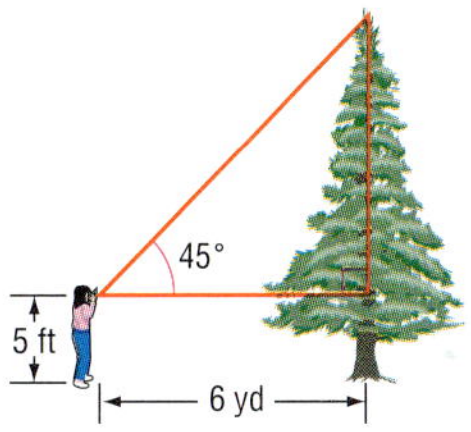

State the assumption you would make to start an indirect proof of each statement. *(Lesson 5-3)*

46. Every shopper that walks through the door is greeted by a salesperson.

47. If you get a job, you have filled out an application.

48. If $4y + 17 \leq 41$, $y \leq 6$. **$y > 6$**

49. If two lines are cut by a transversal and a pair of alternate interior angles are congruent, then the two lines are parallel. **The two lines are not parallel.**

46. A certain shopper is not greeted by a salesperson when he walks through the door.

47. You did not fill out an application.

Find the distance between each pair of parallel lines. *(Lesson 3-6)*

50. $x = -2$, $x = 5$ **7**

51. $y = -6$, $y = -1$ **5**

52. $y = 2x + 3$, $y = 2x - 7$ **$2\sqrt{5}$**

53. $y = x + 2$, $y = x - 4$ **$3\sqrt{2}$**

Getting Ready for the Next Lesson **PREREQUISITE SKILL** Use a protractor and draw an angle for each of the following degree measures. *(To review **drawing angles**, see Lesson 1-4.)*

54–59. See p. 519B.

54. 30 55. 45 56. 52 57. 60 58. 105 59. 150

www.geometryonline.com/self_check_quiz/fcat

4 Assess

Open-Ended Assessment

Speaking Create or draw selected translations from the book on the board and call on students to describe the transformations aloud.

Getting Ready for Lesson 9-3

Prerequisite Skill Students will learn about rotations in Lesson 9-3. They will rotate objects through specific angles. Use Exercises 54–59 to determine your students' familiarity with drawing angles.

Assessment Options

Quiz (Lessons 9-1 and 9-2) is available on p. 535 of the *Chapter 9 Resource Masters*.

35. Translations and reflections preserve the congruences of segments and angles. The composition of the two transformations will preserve both congruences. Therefore, a glide reflection is an isometry.

36. 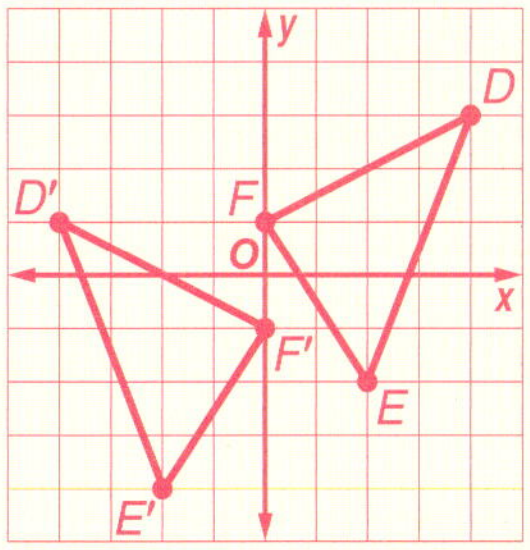

37. 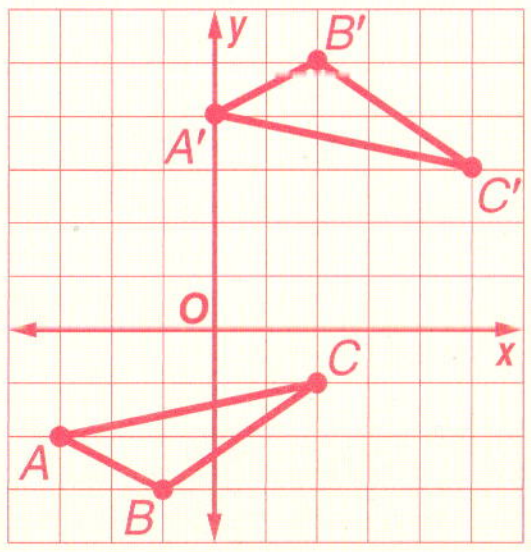

Answers (page 474)

34. Sample answer: Every time a band member takes a step, he or she moves a fixed amount in a certain direction. This is a translation. Answers should include the following.

- When a band member takes a step forward, backward, left, right, or on a diagonal without turning, this is a translation.
- To move in a rectangular pattern, the band member starting at (0,0) could move to (0, 5). Moving from (0, 5) to (4, 5), from (4, 5) to (4, 0) and from (4, 0) back to the origin, the band member would have completed the rectangle.

9-3 Lesson Notes

1 Focus

5-Minute Check Transparency 9-3 Use as a quiz or review of Lesson 9-2.

Mathematical Background notes are available for this lesson on p. 460C.

How do some amusement rides illustrate rotations?

Ask students:

- Some Tilt-A-Whirl cars have wheels in the center. As a rider turns the wheel, the car spins about the wheel. What must be true of the wheel in order for this to happen? **It must be stationary, thereby representing a fixed point.**
- Are you more likely to feel the dizzy and disoriented effects of the ride by a spinning car or a stationary car making its way around the circular track? **a spinning car**

9-3 Rotations

Sunshine State Standards
MA.A.3.4.3, MA.B.1.4.3, MA.C.2.4.1, MA.C.3.4.2

Vocabulary

- rotation
- center of rotation
- angle of rotation
- rotational symmetry
- direct isometry
- indirect isometry

What You'll Learn

- Draw rotated images using the angle of rotation.
- Identify figures with rotational symmetry.

How do some amusement rides illustrate rotations?

In 1926, Herbert Sellner invented the Tilt-A-Whirl. Today, no carnival is considered complete without these cars that send riders tipping and spinning as they make their way around a circular track. Manufactured in Faribault, Minnesota, the Tilt-A-Whirl provides an example of rotation.

Study Tip

Turns
A rotation, sometimes called a turn, is generally measured as a counterclockwise turn. A half-turn is 180° and a full turn is 360°.

DRAW ROTATIONS A **rotation** is a transformation that turns every point of a preimage through a specified angle and direction about a fixed point. The fixed point is called the **center of rotation**.

In the figure, R is the center of rotation for the preimage $ABCD$. The measures of angles ARA', BRB', CRC', and DRD' are equal. Any point P on the preimage of $ABCD$ has an image P' on $A'B'C'D'$ such that the measure of $\angle PRP'$ is a constant measure. This is called the **angle of rotation**.

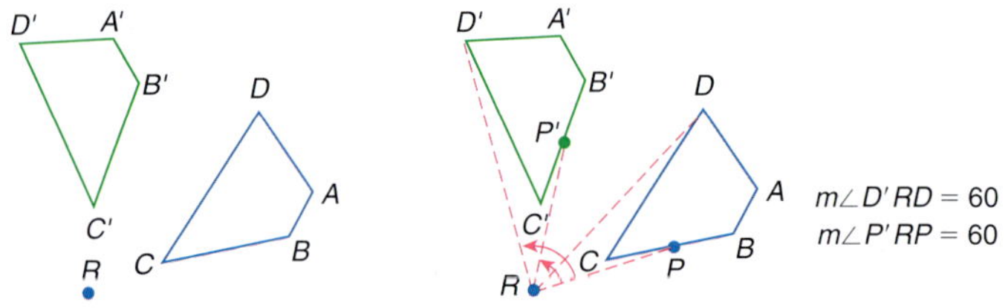

A rotation exhibits all of the properties of isometries, including preservation of distance and angle measure. Therefore, it is an isometry.

Example 1 Draw a Rotation

Triangle ABC has vertices $A(2, 3)$, $B(6, 3)$, and $C(5, 5)$. Draw the image of $\triangle ABC$ under a rotation of 60° counterclockwise about the origin.

- First graph $\triangle ABC$.
- Draw a segment from the origin O, to point A.
- Use a protractor to measure a 60° angle counterclockwise with $\overline{OA}$ as one side.
- Draw $\overrightarrow{OR}$.
- Use a compass to copy $\overline{OA}$ onto $\overrightarrow{OR}$. Name the segment $\overline{OA'}$

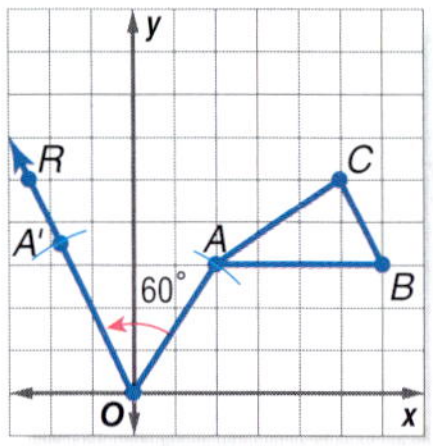

Resource Manager

Workbook and Reproducible Masters

Chapter 9 Resource Masters
- Study Guide and Intervention, pp. 491–492
- Skills Practice, p. 493
- Practice, p. 494
- Reading to Learn Mathematics, p. 495
- Enrichment, p. 496

Teaching Geometry With Manipulatives Masters, pp. 1, 16, 17, 152

Transparencies
5-Minute Check Transparency 9-3
Answer Key Transparencies

Technology
Interactive Chalkboard

- Repeat with points B and C.
$\triangle A'B'C'$ is the image of $\triangle ABC$ under a $60°$ counterclockwise rotation about the origin.

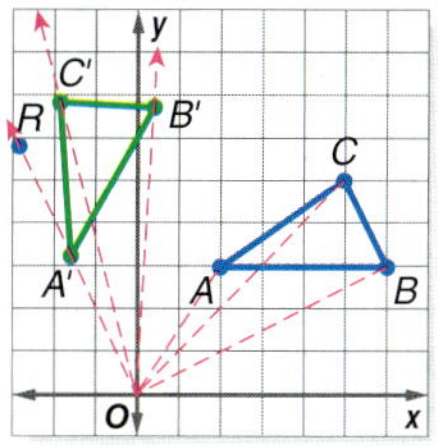

Another way to perform rotations is by reflecting a figure successively in two intersecting lines. Reflecting a figure once and then reflecting the image in a second line is another example of a composition of reflections.

Geometry Software Investigation

Reflections in Intersecting Lines

Construct a Figure

- Use The Geometer's Sketchpad to construct scalene triangle ABC.
- Construct lines m and n so that they intersect outside $\triangle ABC$.
- Label the point of intersection P.

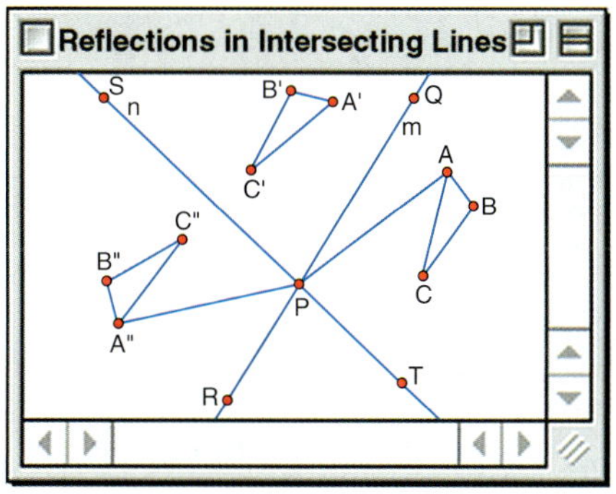

Analyze

1. Reflect $\triangle ABC$ in line m. Then, reflect $\triangle A'B'C'$ in line n.
2. Describe the transformation of $\triangle ABC$ to $\triangle A''B''C''$.
3. Measure the acute angle formed by m and n. **See students' work.**
4. Construct a segment from A to P and from A'' to P. Find the measure of the angle of rotation, $\angle APA''$.
5. Find $m\angle BPB''$ and $m\angle CPC''$. **See students' work. The angle measures should be the same as $m\angle APA''$ in Exercise 4.**

Make a Conjecture

6. What is the relationship between the measures of the angles of rotation and the measure of the acute angle formed by m and n?

1. See students' figures.
2. The transformation is a rotation about P.
4. See students' work. The angle measure should be twice the measure of the acute angle formed by the intersecting lines.
6. Sample answer: The measure of the angle of rotation is twice the measure of the acute angle formed by the intersecting lines.

When rotating a figure by reflecting it in two intersecting lines, there is a relationship between the angle of rotation and the angle formed by the intersecting lines.

Key Concept

Postulate 9.1 In a given rotation, if A is the preimage, A'' is the image, and P is the center of rotation, then the measure of the angle of rotation $\angle APA''$ is twice the measure of the acute or right angle formed by the intersecting lines of reflection.

Corollary 9.1 Reflecting an image successively in two perpendicular lines results in a $180°$ rotation.

www.geometryonline.com/extra_examples/fcat

2 Teach

DRAW ROTATIONS

In-Class Example Power Point®

Teaching Tip Students can also practice drawing and measuring the congruent segments with a metric ruler instead of copying the segments with a compass.

1 Triangle DEF has vertices $D(-2, -1)$, $E(-1, 1)$, and $F(1, -1)$. Draw the image of $\triangle DEF$ under a rotation of $115°$ clockwise about the point $G(-4, -2)$.

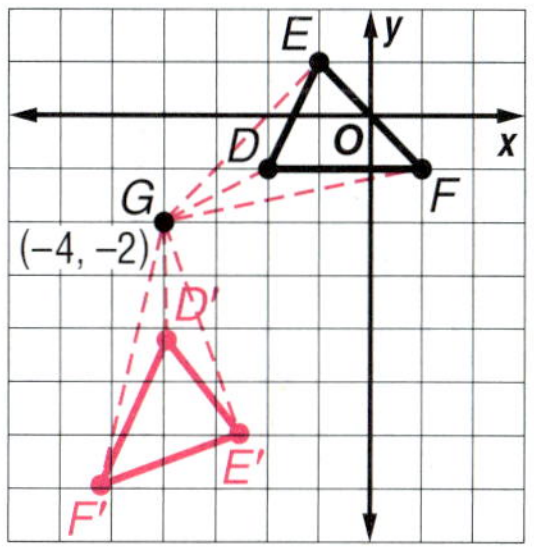

Geometry Software Investigation

Materials: geometry software

- Students' triangles do not have to match exactly the one in the text. However, it should be small enough that it can be reflected twice and still be visible.
- Students may need to adjust their sketches after performing the reflections in order for all three triangles to appear on the same screen.
- Encourage students to draw a scalene triangle so they can see for themselves how the triangle images are oriented with respect to the original triangle.

In-Class Example

Power Point®

2 Find the image of parallelogram *WXYZ* under reflections in line *p* and then line *q*.

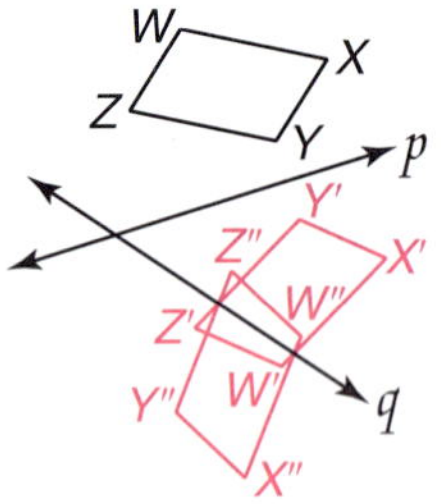

ROTATIONAL SYMMETRY

Teaching Tip When you introduce rotational symmetry, you can also label the vertices of the pentagon in each rotation, so students can visualize the concept better. Explain that it is important to determine the correct order because the magnitude is dependent on this value. Ask students to compare the order to the number of vertices in the figure.

In-Class Example

Power Point®

3 **QUILTS** Using the quilt by Judy Mathieson in Example 3, identify the order and magnitude of the symmetry in each part of the quilt.

a. medium star directly to the left of the large star in the center of the quilt **16; 22.5°**

b. tiny star above the medium-sized star in part *a* **8; 45°**

Answer

4.

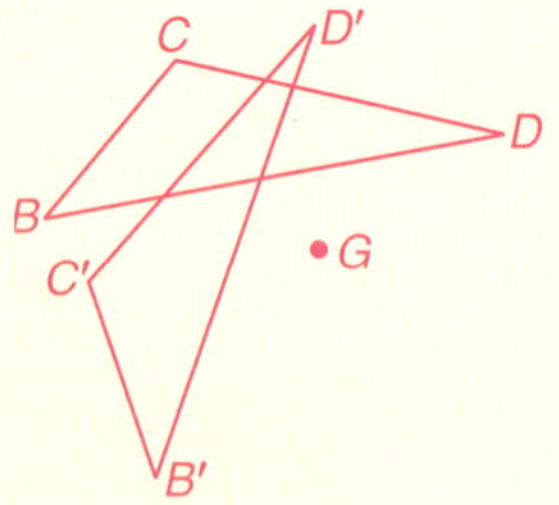

Study Tip

Common Misconception The order in which you reflect a figure in two nonperpendicular intersecting lines produces rotations of the same degree measure, but one is clockwise and the other is counterclockwise.

Example 2 *Reflections in Intersecting Lines*

Find the image of rectangle *DEFG* under reflections in line *m* and then line *n*.

First reflect rectangle *DEFG* in line *m*. Then label the image *D'E'F'G'*.

Next, reflect the image in line *n*.

Rectangle *D"E"F"G"* is the image of rectangle *DEFG* under reflections in lines *m* and *n*. How can you transform *DEFG* directly to *D"E"F"G"* by using a rotation?

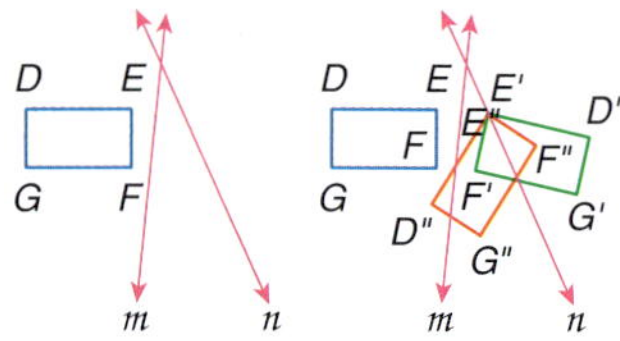

ROTATIONAL SYMMETRY Some objects have rotational symmetry. If a figure can be rotated less than 360 degrees about a point so that the image and the preimage are indistinguishable, then the figure has **rotational symmetry**.

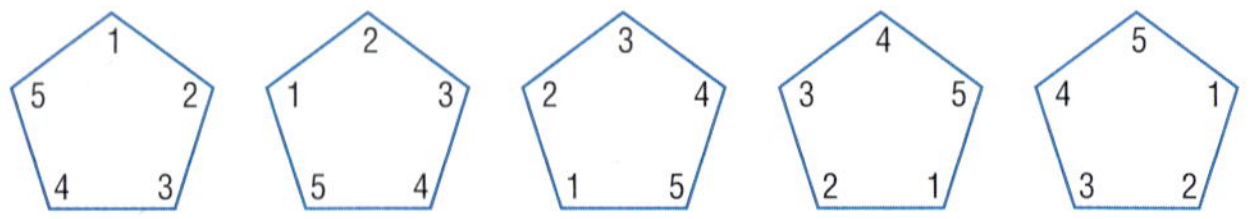

In the figure, the pentagon has rotational symmetry of *order* 5 because there are 5 rotations of less than 360° (including 0 degrees) that produce an image indistinguishable from the original. The rotational symmetry has a *magnitude* of 72° because 360 degrees divided by the order, in this case 5, produces the magnitude of the symmetry.

Example 3 *Identifying Rotational Symmetry*

QUILTS **One example of rotational symmetry artwork is quilt patterns. A quilt made by Judy Mathieson of Sebastopol, California, won the Masters Award for Contemporary Craftsmanship at the International Quilt Festival in 1999. Identify the order and magnitude of the symmetry in each part of the quilt.**

a. large star in center of quilt

The large star in the center of the quilt has rotational symmetry of order 20 and magnitude of 18°.

b. entire quilt

The entire quilt has rotational symmetry of order 4 and magnitude of 90°.

Check for Understanding

Concept Check **2–3. See p. 519C.**

1. **OPEN ENDED** Draw a figure on the coordinate plane in Quadrant I. Rotate the figure clockwise 90 degrees about the origin. Then rotate the figure 90 degrees counterclockwise. Describe the results using the coordinates. **See p. 519B.**
2. **Explain** two techniques that can be used to rotate a figure.
3. **Compare and contrast** translations and rotations.

Guided Practice

4. Copy $\triangle BCD$ and rotate the triangle 60° counterclockwise about point *G*. **See margin.**

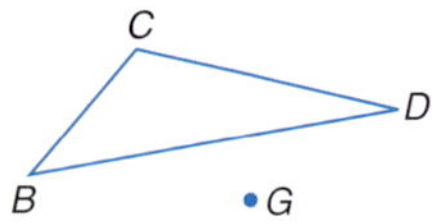

DAILY INTERVENTION

Differentiated Instruction

Logical/Mathematical Tell students to develop a system for rotating images. First, they should read the problem and locate or plot the figure for visual recognition. Next, they can reread the problem to determine if the problem requires an angle rotation or a composition of reflections. They should also carefully note the specifications, especially the direction of the rotation. Finally, they can apply the rotation. Students can use a system similar to this, or you can encourage them to create their own.

GUIDED PRACTICE KEY	
Exercises	Examples
4, 7, 8	1
5, 6	2
9–11	3

Copy each figure. Use a composition of reflections to find the rotation image with respect to lines ℓ and m.

5.

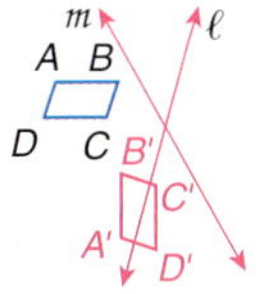

6.

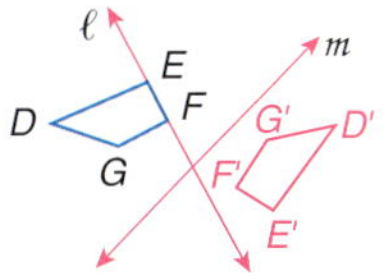

7. $\overline{XY}$ has endpoints $X(-5, 8)$ and $Y(0, 3)$. Draw the image of $\overline{XY}$ under a rotation of 45° clockwise about the origin. See p. 519C.

8. $\triangle PQR$ has vertices $P(-1, 8)$, $Q(4, -2)$, and $R(-7, -4)$. Draw the image of $\triangle PQR$ under a rotation of 90° counterclockwise about the origin. See p. 519C.

9. Identify the order and magnitude of the rotational symmetry in a regular hexagon.

9. order 6; magnitude 60°

10. Identify the order and magnitude of the rotational symmetry in a regular octagon. order 8; magnitude 45°

Application

11. **FANS** The blades of a fan exhibit rotational symmetry. Identify the order and magnitude of the symmetry of the blades of each fan in the pictures. See margin.

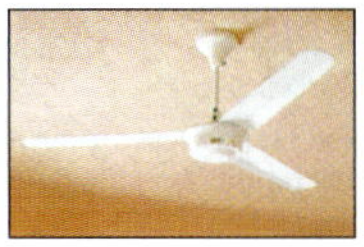

★ indicates increased difficulty

Practice and Apply

Homework Help	
For Exercises	See Examples
12–15, 27	1
19–26	2
16–28, 28	3

Extra Practice
See page 772.

12. Copy pentagon $BCDEF$. Then rotate the pentagon 110° counterclockwise about point R.

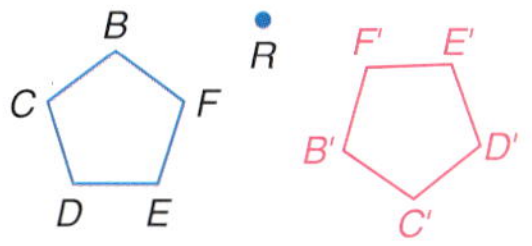

13. Copy $\triangle MNP$. Then rotate the triangle 180° counterclockwise around point Q.

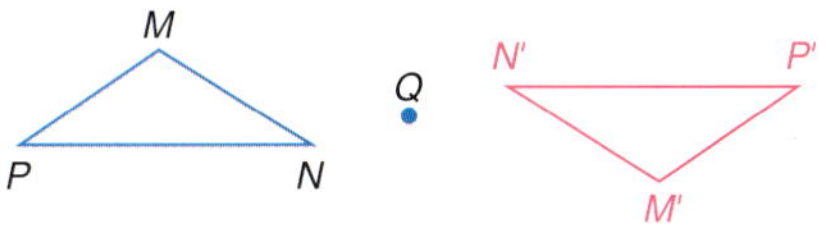

COORDINATE GEOMETRY **Draw the rotation image of each figure 90° in the given direction about the center point and label the coordinates.** 14–15. See margin.

14. $\triangle XYZ$ with vertices $X(0, -1)$, $Y(3,1)$, and $Z(1, 5)$ counterclockwise about the point $P(-1, 1)$

15. $\triangle RST$ with vertices $R(0, 1)$, $S(5,1)$, and $T(2, 5)$ clockwise about the point $P(-2, 5)$

RECREATION **For Exercises 16–18, use the following information.**
A Ferris wheel's motion is an example of a rotation. The Ferris wheel shown has 20 cars.

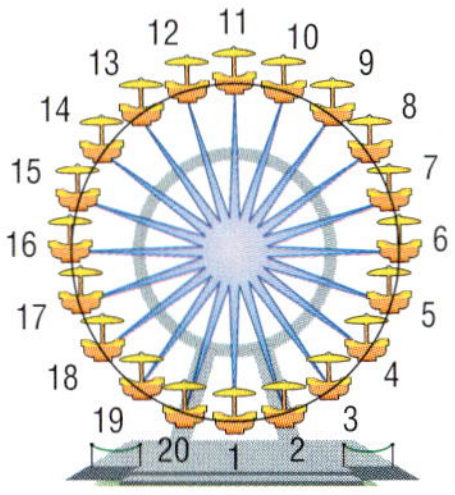

16. Identify the order and magnitude of the symmetry of a 20-seat Ferris wheel. order 20 and magnitude 18°

17. What is the measure of the angle of rotation if seat 1 of a 20-seat Ferris wheel is moved to the seat 5 position? 72°

18. If seat 1 of a 20-seat Ferris wheel is rotated 144°, find the original seat number of the position it now occupies. seat 9

3 Practice/Apply

Study Notebook

Have students—
- add the definitions/examples of the vocabulary terms to their Vocabulary Builder worksheets for Chapter 9.
- include any other item(s) that they find helpful in mastering the skills in this lesson.

About the Exercises...

Organization by Objective
- **Identify and Draw Rotations:** 12–15, 19–27
- **Rotational Symmetry:** 16–18, 28

Odd/Even Assignments
Exercises 12–28 are structured so that students practice the same concepts whether they are assigned odd or even problems.

Assignment Guide

Basic: 13–25 odd, 29–33 odd, 34, 35, 37–39, 41, 42, 47–67 (optional: 43–46)

Average: 13–33 odd, 34, 35, 37–39, 41, 42, 47–67 (optional: 43–46)

Advanced: 12–32 even, 33–61 (optional: 62–67)

All: Quiz 1: 1–5

Answers

11. order 5 and magnitude 72°; order 4 and magnitude 90°; order 3 and magnitude 120°

14.

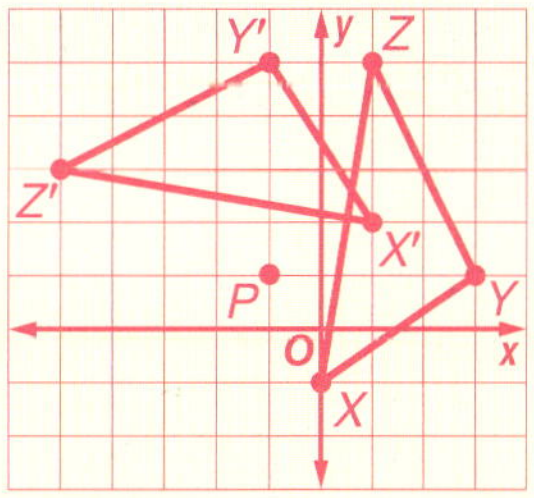

15.

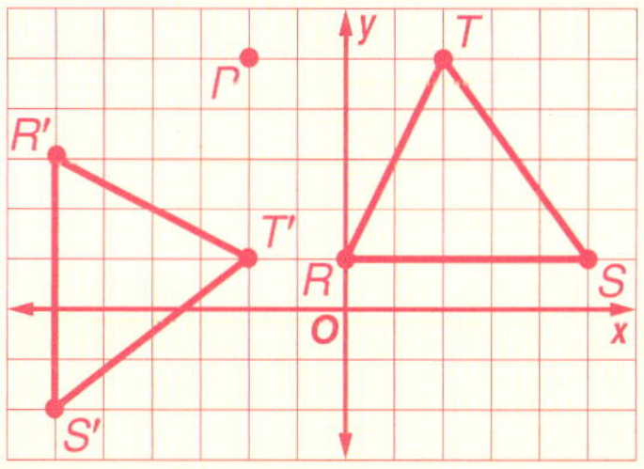

Copy each figure. Use a composition of reflections to find the rotation image with respect to lines m and t.

19. 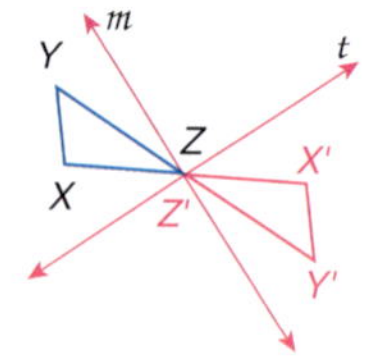

20. 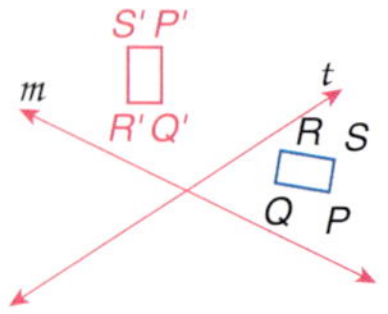

21. 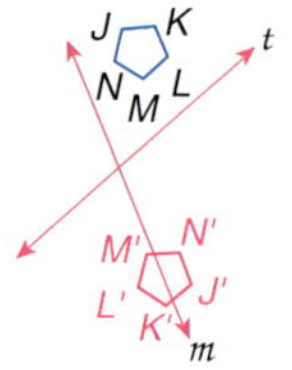

COORDINATE GEOMETRY Draw the rotation image of each triangle by reflecting the triangles in the given lines. State the coordinates of the rotation image and the angle of rotation. **22–24. See margin for figures.**

22. $\triangle TUV$ with vertices $T(4, 0)$, $U(2, 3)$, and $V(1, 2)$, reflected in the y-axis and then the x-axis **$T''(-4, 0)$, $U''(-2, -3)$, and $V''(-1, -2)$; 180°**

23. $\triangle KLM$ with vertices $K(5, 0)$, $L(2, 4)$, and $M(-2, 4)$, reflected in the line $y = x$ and then the x-axis **$K''(0, -5)$, $L''(4, -2)$, and $M''(4, 2)$; 90° clockwise**

24. $\triangle XYZ$ with vertices $X(5, 0)$, $Y(3, 4)$, and $Z(-3, 4)$, reflected in the line $y = -x$ and then the line $y = x$ **$X''(-5, 0)$, $Y''(-3, -4)$, and $Z''(3, -4)$; 180°**

25. **COORDINATE GEOMETRY** The point at (2, 0) is rotated 30° counterclockwise about the origin. Find the exact coordinates of its image. **$(\sqrt{3}, 1)$**

26. **MUSIC** A five-disc CD changer rotates as each CD is played. Identify the magnitude of the rotational symmetry as the changer moves from one CD to another. **72°**

More About. . .

Music

Some manufacturers now make CD changers that are referred to as CD jukeboxes. These can hold hundreds of CDs, which may be more than a person's entire CD collection.

Source: www.usatoday.com

Determine whether the indicated composition of reflections is a rotation. Explain.

★ 27. 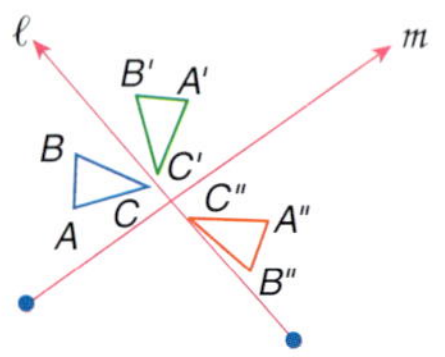

★ 28. 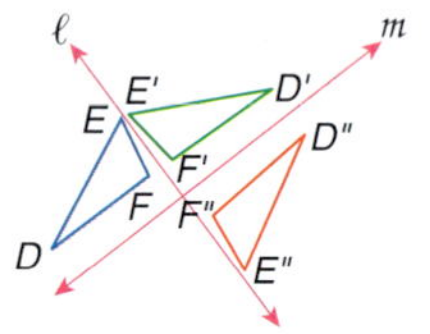

27. Yes; it is a proper successive reflection with respect to the two intersecting lines.

28. Yes; it is a proper successive reflection with respect to the two intersecting lines.

AMUSEMENT RIDES For each ride, determine whether the rider undergoes a rotation. Write *yes* or *no*.

29. spinning teacups **yes**

30. scrambler **yes**

31. roller coaster loop **no**

32. **ALPHABET** Which capital letters of the alphabet produce the same letter after being rotated 180°? **H, I, N, O, S, X, Z**

33. **CRITICAL THINKING** In $\triangle ABC$, $m\angle BAC = 40$. Triangle $AB'C$ is the image of $\triangle ABC$ under reflection and $\triangle AB'C'$ is the image of $\triangle AB'C$ under reflection. How many such reflections would be necessary to map $\triangle ABC$ onto itself? **9**

34. **CRITICAL THINKING** If a rotation is performed on a coordinate plane, what angles of rotation would make the rotations easier? Explain. **See margin.**

Study Guide and Intervention, p. 491 (shown) and p. 492

Draw Rotations A transformation called a **rotation** turns a figure through a specified angle about a fixed point called the **center of rotation**. To find the image of a rotation, one way is to use a protractor. Another way is to reflect a figure twice, in two intersecting lines.

Example 1 **$\triangle ABC$ has vertices $A(2, 1)$, $B(3, 4)$, and $C(5, 1)$. Draw the image of $\triangle ABC$ under a rotation of 90° counterclockwise about the origin.**

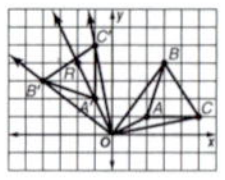

- First draw $\triangle ABC$. Then draw a segment from O, the origin, to point A.
- Use a protractor to measure 90° counterclockwise with $\overline{OA}$ as one side.
- Draw $\overrightarrow{OR}$.
- Use a compass to copy $\overline{OA}$ onto $\overrightarrow{OR}$. Name the segment $\overline{OA'}$.
- Repeat with segments from the origin to points B and C.

Example 2 **Find the image of $\triangle ABC$ under reflection in lines m and n.**

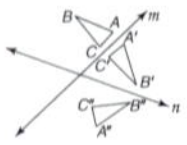

First reflect $\triangle ABC$ in line m. Label the image $\triangle A'B'C'$.
Reflect $\triangle A'B'C'$ in line n. Label the image $\triangle A''B''C''$.
$\triangle A''B''C''$ is a rotation of $\triangle ABC$. The center of rotation is the intersection of lines m and n. The angle of rotation is twice the measure of the acute angle formed by m and n.

Exercises

Draw the rotation image of each figure 90° in the given direction about the center point and label the coordinates.

1. $\overline{PQ}$ with endpoints $P(-1, -2)$ and $Q(1, 3)$ counterclockwise about the origin

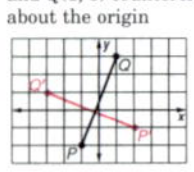

2. $\triangle PQR$ with vertices $P(-2, -3)$, $Q(2, -1)$, and $R(3, 2)$ clockwise about the point $T(1, 1)$

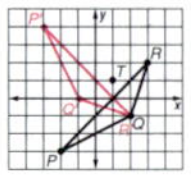

Find the rotation image of each figure by reflecting it in line m and then in line n.

3. 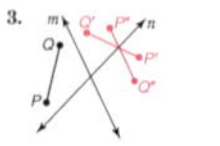

4. 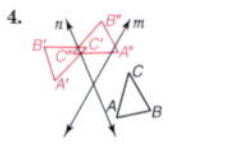

Skills Practice, p. 493 and Practice, p. 494 (shown)

Rotate each figure about point R under the given angle of rotation and the given direction. Label the vertices of the rotation image.

1. 80° counterclockwise

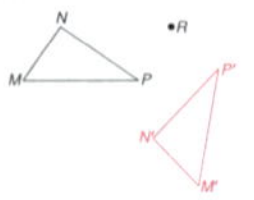

2. 100° clockwise

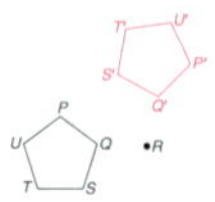

COORDINATE GEOMETRY **Draw the rotation image of each figure 90° in the given direction about the center point and label the coordinates.**

3. $\triangle RST$ with vertices $R(-3, 3)$, $S(2, 4)$, and $T(1, 2)$ clockwise about the point $P(1, 0)$

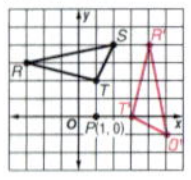

4. $\triangle HJK$ with vertices $H(3, 1)$, $J(3, -3)$, and $K(-3, -4)$ counterclockwise about the point $P(-1, -1)$

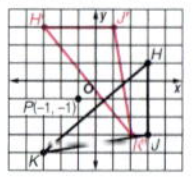

Use a composition of reflections to find the rotation image with respect to lines p and s. Then find the angle of rotation for each image.

5. 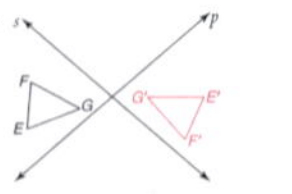

160° counterclockwise

6. 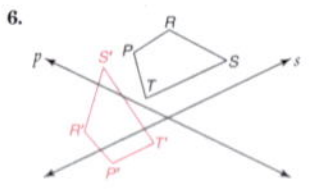

100° counterclockwise

7. STEAMBOATS A paddle wheel on a steamboat is driven by a steam engine and moves from one paddle to the next to propel the boat through the water. If a paddle wheel consists of 18 evenly spaced paddles, identify the order and magnitude of its rotational symmetry. order 18 and magnitude 20°

Reading to Learn Mathematics, p. 495 ELL

Pre-Activity **How do some amusement rides illustrate rotations?**

Read the introduction to Lesson 9-3 at the top of page 476 in your textbook.

What are two ways that each car rotates?

Each car spins around its own center and each car rotates around a point in the center of the circular track.

Reading the Lesson

1. List all of the following types of transformations that satisfy each description: *reflection, translation, rotation*.
   a. The transformation is an isometry. reflection, translation, rotation
   b. The transformation preserves the orientation of a figure. translation, rotation
   c. The transformation is the composite of successive reflections over two intersecting lines. rotation
   d. The transformation is the composite of successive reflections over two parallel lines. translation
   e. A specific transformation is defined by a fixed point and a specified angle. rotation
   f. A specific transformation is defined by a fixed point, a fixed line, or a fixed plane. reflection
   g. A specific transformation is defined by $(x, y) \rightarrow (x + a, x + b)$, for fixed values of a and b. translation
   h. The transformation is also called a slide. translation
   i. The transformation is also called a flip. reflection
   j. The transformation is also called a turn. rotation

2. Determine the order and magnitude of the rotational symmetry for each figure.

a. 

order 3; magnitude 120°

b. 

order 2; magnitude 180°

c. 

order 5; magnitude 72°

d. 

order 8; magnitude 45°

Helping You Remember

3. What is an easy way to remember the order and magnitude of the rotational symmetry of a regular polygon?

Sample answer: The order is the same as the number of sides. To find the magnitude, divide 360 by the number of sides.

480 Chapter 9 Transformations

Enrichment, p. 496

Finding the Center of Rotation

Suppose you are told that $\triangle X'Y'Z'$ is the rotation image of $\triangle XYZ$, but you are not told where the center of rotation is nor the measure of the angle of rotation. Can you find them? Yes, you can. Connect two pairs of corresponding vertices with segments. In the figure, the segments YY' and ZZ' are used. Draw the perpendicular bisectors, ℓ and m, of these segments. The point C where ℓ and m intersect is the center of rotation.

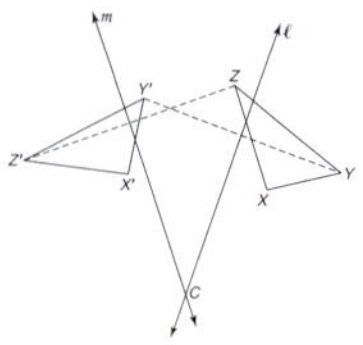

1. How can you find the measure of the angle of rotation in the figure above?

($\angle YCY'$ or ...

35. **COORDINATE GEOMETRY** Quadrilateral $QRST$ is rotated 90° clockwise about the origin. Describe the transformation using coordinate notation. $(x, y) \rightarrow (y, -x)$

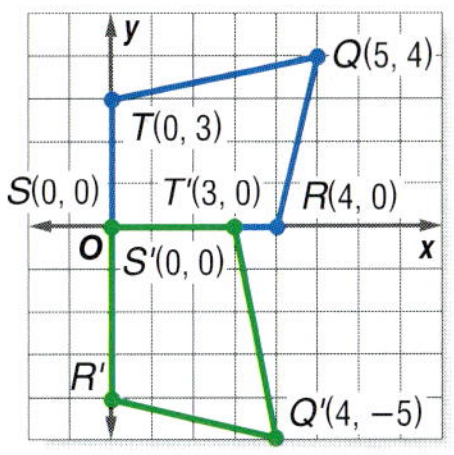

★ 36. Triangle FGH is rotated 80° clockwise and then rotated 150° counterclockwise about the origin. To what counterclockwise rotation about the origin is this equivalent? See margin.

CRITICAL THINKING **For Exercises 37–39, use the following information.**
Points that do not change position under a transformation are called *invariant points*. For each of the following transformations, identify any invariant points.

37. reflection in a line any point on the line of reflection

38. a rotation of $x°(0 < x < 360)$ about point P the center of rotation

39. $(x, y) \rightarrow (x + a, y + b)$, where a and b are not 0 no invariant points

40. **WRITING IN MATH** Answer the question that was posed at the beginning of the lesson. See margin.

How do amusement rides exemplify rotations?

Include the following in your answer:
- a description of how the Tilt-A-Whirl actually rotates two ways, and
- other amusement rides that use rotation.

FCAT Practice
Standardized Test Practice

41. In the figure, describe the rotation that moves triangle 1 to triangle 2. B

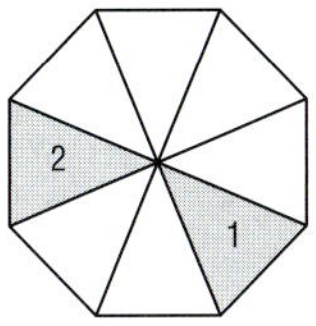

(A) 180° clockwise
(B) 135° clockwise
(C) 135° counterclockwise
(D) 90° counterclockwise

42. **ALGEBRA** Suppose x is $\frac{2}{5}$ of y and y is $\frac{1}{3}$ of z. If $x = 6$, then $z = ?$ D

(A) $\frac{4}{5}$ (B) $\frac{18}{5}$ (C) 5 (D) 45

Extending the Lesson

A **direct isometry** is one in which the image of a figure is found by moving the figure intact within the plane. An **indirect isometry** cannot be performed by maintaining the orientation of the points as in a direct isometry.

43. Copy and complete the table below. Determine whether each transformation preserves the given property. Write *yes* or *no*.

Transformation	angle measure	betweenness of points	orientation	collinearity	distance measure
reflection	yes	yes	no	yes	yes
translation	yes	yes	yes	yes	yes
rotation	yes	yes	yes	yes	yes

Identify each type of transformation as a direct isometry or an indirect isometry.

44. reflection indirect 45. translation direct 46. rotation direct

www.geometryonline.com/self_check_quiz/fcat

Answers (page 480)

22.

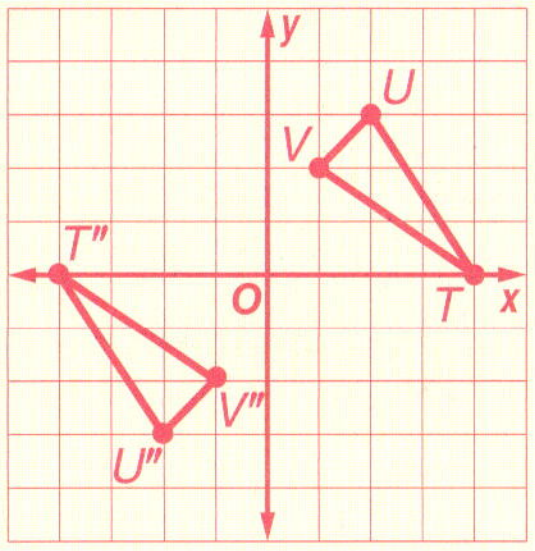

23.

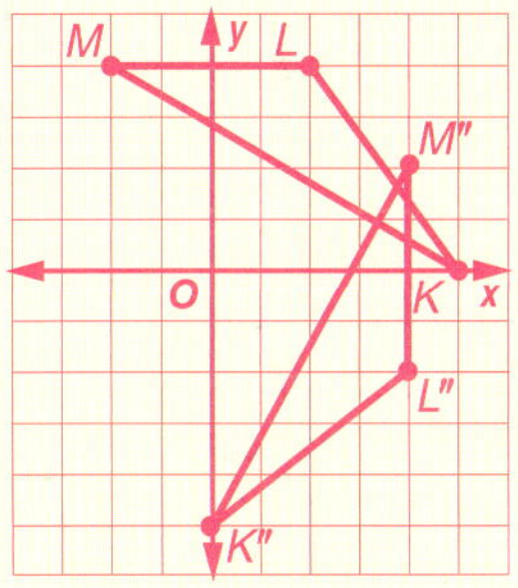

24.

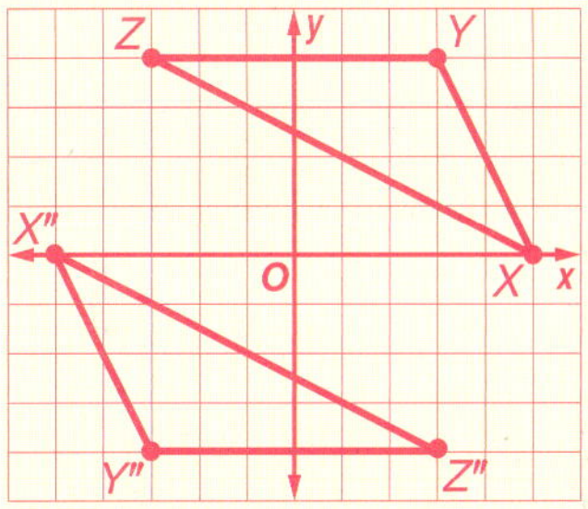

34. Angles of rotation with measures of 90 or 180 would be easier on a coordinate plane because of the grids used in graphing.

Answers

36. The 80° clockwise rotation and then 150° counterclockwise rotation about the origin is equivalent to a 70° counterclockwise rotation.

40. The Tilt-A-Whirl sends riders tipping and spinning on a circular track. Answers should include the following.
- The Tilt-A-Whirl shows rotation in two ways. The cars rotate about the center of the ride as the cars go around the track. Each car rotates around a pivot point in the car.
- Answers will vary but the Scrambler, Teacups, and many kiddie rides use rotation.

Lesson 9-3 Rotations 481

4 Assess

Open-Ended Assessment

Modeling Students can use a cork board marked as a grid, pushpins, laminated shapes, and string or yarn to model rotations. They can start by placing a shape on the board and affixing lengths of string or yarn to each vertex. Then they can choose a center of rotation and use a pushpin to secure the loose ends of each string to this point. Finally, they can determine an angle of rotation, use a protractor to measure the angle, and slide the shape to its new rotated position.

Getting Ready for Lesson 9-4

Prerequisite Skill Students will learn about tessellations in Lesson 9-4. They will apply algebraic concepts to model and analyze tessellations. Use Exercises 62–67 to determine your students' familiarity with solving problems by making a table.

Assessment Options

Practice Quiz 1 The quiz provides students with a brief review of the concepts and skills in Lessons 9-1 through 9-3. Lesson numbers are given to the right of the exercises or instruction lines so students can review concepts not yet mastered.

Maintain Your Skills

Mixed Review

47–49. See margin.

In each figure, $a \parallel b$. Determine whether the blue figure is a translation image of the red figure. Write *yes* or *no*. Explain your answer. *(Lesson 9-2)*

47.

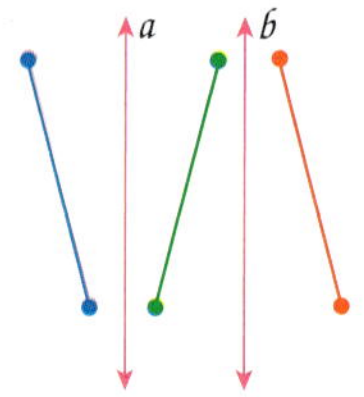

48.

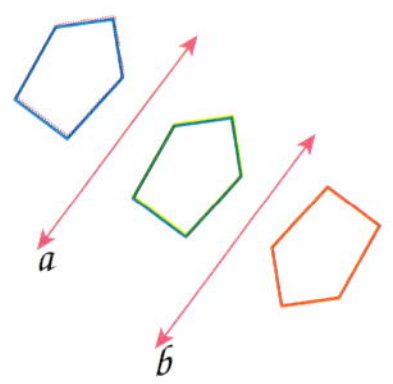

49.

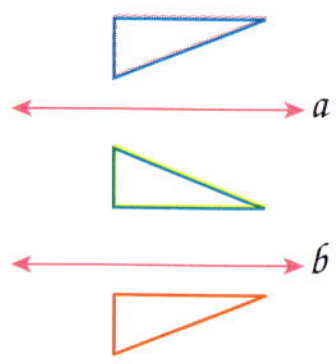

For the figure to the right, name the reflected image of each image. *(Lesson 9-1)*

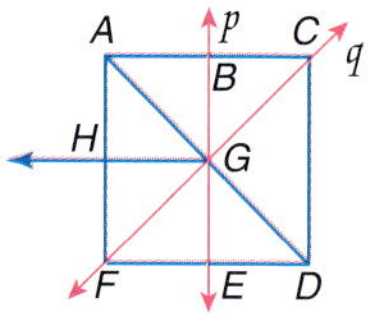

50. $\overline{AG}$ in line p $\overline{CG}$
51. F in point G C
52. $\overline{GE}$ in line q $\overline{GH}$
53. $\angle CGD$ in line p $\angle AGF$

Complete each statement about ▱$PQRS$. Justify your answer. *(Lesson 8-2)*

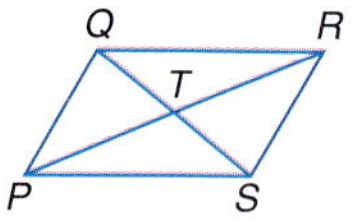

54. $\overline{PS}$; opp. sides ∥
55. $\overline{TR}$; diagonals bisect each other

54. $\overline{QR} \parallel$ __?__
55. $\overline{PT} \cong$ __?__
56. $\angle SQR \cong$ __?__ $\angle QSP$; alt. int. ∠s ≅
57. $\angle QPS \cong$ __?__ $\angle QRS$; opp. ∠s ≅

58. **SURVEYING** A surveyor is 100 meters from a building and finds that the angle of elevation to the top of the building is 23°. If the surveyor's eye level is 1.55 meters above the ground, find the height of the building. *(Lesson 7-5)* about 44.0 m

Determine whether the given measures can be the lengths of the sides of a triangle. Write *yes* or *no*. *(Lesson 5-4)*

59. 6, 8, 16 no
60. 12, 17, 20 yes
61. 22, 23, 37 yes

Getting Ready for the Next Lesson

62–67. See margin.

PREREQUISITE SKILL **Find whole number values for each variable so each equation is true.** *(To review **solving problems by making a table**, see pages 737–738.)*

62. $180a = 360$
63. $180a + 90b = 360$
64. $135a + 45b = 360$
65. $120a + 30b = 360$
66. $180a + 60b = 360$
67. $180a + 30b = 360$

Practice Quiz 1 — Lesson 9-1 through 9-3

Graph each figure and the image in the given reflection. *(Lesson 9-1)* 1–2. See p. 519C.

1. $\triangle DEF$ with vertices $D(-1, 1)$, $E(1, 4)$ and $F(3, 2)$ in the origin
2. quadrilateral $ABCD$ with vertices $A(0, 2)$, $B(2, 2)$, $C(3, 0)$, and $D(-1, 1)$ in the line $y = x$

Graph each figure and the image under the given translation. *(Lesson 9-2)* 3–4. See p. 519C.

3. $\overline{PQ}$ with endpoints $P(1, -4)$ and $Q(4, -1)$ under the translation left 3 units and up 4 units
4. $\triangle KLM$ with vertices $K(-2, 0)$, $L(-4, 2)$, and $M(0, 4)$ under the translation $(x, y) \rightarrow (x + 1, y - 4)$

5. Identify the order and magnitude of the symmetry of a 36-horse carousel. *(Lesson 9-3)* order 36; magnitude 10°

Answers

47. Yes; it is one reflection after another with respect to the two parallel lines.
48. No; it is a rotation followed by a reflection with respect to a line.
49. Yes; it is one reflection after another with respect to the two parallel lines.
62. 2
63. (0, 4), (1, 2), (2, 0)
64. (0, 8), (1, 5), (2, 2)
65. (0, 12), (1, 8), (2, 4), (3, 0)
66. (0, 6), (1, 3), (2, 0)
67. (0, 12), (1, 6), (2, 0)

9-4 Tessellations

Sunshine State Standards
MA.C.2.4.1

Vocabulary
- tessellation
- regular tessellation
- uniform
- semi-regular tessellation

What You'll Learn
- Identify regular tessellations.
- Create tessellations with specific attributes.

How are tessellations used in art?

M.C. Escher (1898–1972) was a Dutch graphic artist famous for repeating geometric patterns. He was also well known for his spatial illusions, impossible buildings, and techniques in wood-cutting and lithography. In the picture, figures can be reduced to basic regular polygons. Equilateral triangles and regular hexagons are prominent in the repeating patterns.

Symmetry Drawing E103,
M.C. Escher

REGULAR TESSELLATIONS Reflections, translations, and rotations can be used to create patterns using polygons. A pattern that covers a plane by transforming the same figure or set of figures so that there are no overlapping or empty spaces is called a **tessellation**.

In a tessellation, the sum of the measures of the angles of the polygons surrounding a point (at a vertex) is 360.

You can use what you know about angle measures in regular polygons to help determine which polygons tessellate.

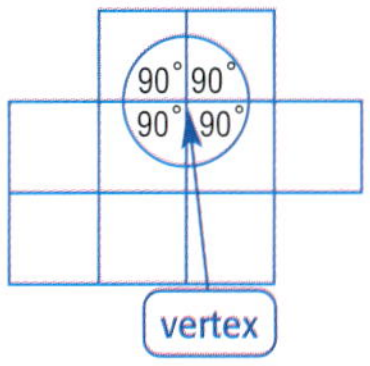

Study Tip

Reading Math
The word *tessellation* comes from the Latin word *tessera* which means "a square tablet." These small square pieces of stone or tile were used in mosaics.

Geometry Activity

Tessellations of Regular Polygons

Model and Analyze

- Study a set of pattern blocks to determine which shapes are regular.
- Make a tessellation with each type of regular polygon.

1. Which shapes in the set are regular?
2. Write an expression showing the sum of the angles at each vertex of the tessellation. 6(60) = 360; 4(90) = 360; 3(120) = 360
3. Copy and complete the table below.

Regular Polygon	triangle	square	pentagon	hexagon	heptagon	octagon
Measure of One Interior Angle	60	90	108	120	128.57	135
Does it tessellate?	yes	yes	no	yes	no	no

Make a Conjecture

4. What must be true of the angle measure of a regular polygon for a regular tessellation to occur?

1. equilateral triangle, square, and hexagon
4. If a regular polygon has an interior angle with a measure that is a factor of 360, then the polygon will tessellate the plane.

9-4 Lesson Notes

1 Focus

5-Minute Check Transparency 9-4 Use as a quiz or review of Lesson 9-3.

Mathematical Background notes are available for this lesson on p. 460D.

How are tessellations used in art?

Ask students:

- What transformation has been applied to the smallest equilateral triangles in the picture? reflection
- Do the regular hexagons have points of symmetry? yes
- How many different patterns are repeated in the picture? one

Resource Manager

Workbook and Reproducible Masters

Chapter 9 Resource Masters
- Study Guide and Intervention, pp. 497–498
- Skills Practice, p. 499
- Practice, p. 500
- Reading to Learn Mathematics, p. 501
- Enrichment, p. 502
- Assessment, pp. 535, 537

Teaching Geometry With Manipulatives Masters, pp. 1, 15, 17, 154

Transparencies
5-Minute Check Transparency 9-4
Answer Key Transparencies

Technology
Interactive Chalkboard

2 Teach

REGULAR TESSELLATIONS

In-Class Example

Power Point®

1 Determine whether a regular 16-gon tessellates the plane. Explain. **No; 157.5 is not a factor of 360.**

Study Tip

Look Back

To review finding the measure of an interior angle of a regular polygon, see Lesson 8-1.

The tessellations you formed in the Geometry Activity are regular tessellations. A **regular tessellation** is a tessellation formed by only one type of regular polygon. In the activity, you found that if a regular polygon has an interior angle with a measure that is a factor of 360, then the polygon will tessellate the plane.

Example 1 Regular Polygons

Determine whether a regular 24-gon tessellates the plane. Explain.

Let $\angle 1$ represent one interior angle of a regular 24-gon.

$m\angle 1 = \frac{180(n-2)}{n}$ Interior Angle Formula

$= \frac{180(24-2)}{24}$ Substitution

$= 165$ Simplify.

Since 165 is not a factor of 360, a 24-gon will not tessellate the plane.

TESSELLATIONS WITH SPECIFIC ATTRIBUTES A tessellation pattern can contain any type of polygon. Tessellations containing the same arrangement of shapes and angles at each vertex are called **uniform**.

uniform

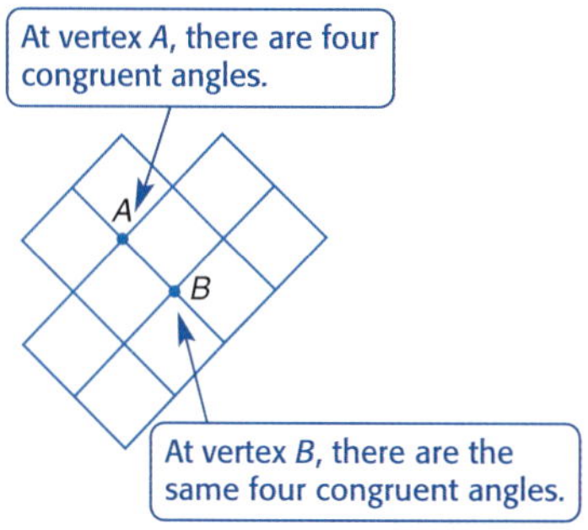

not uniform

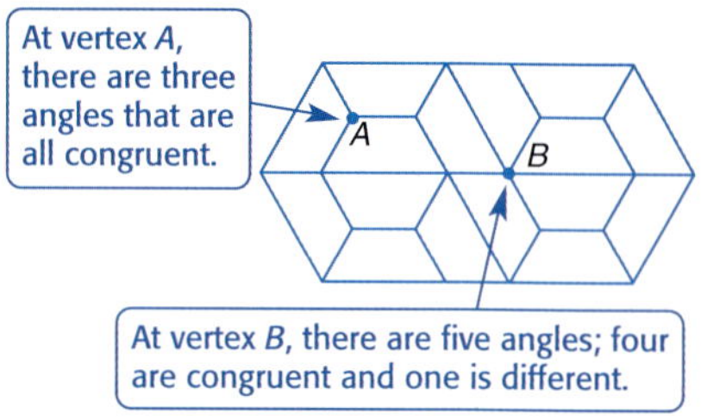

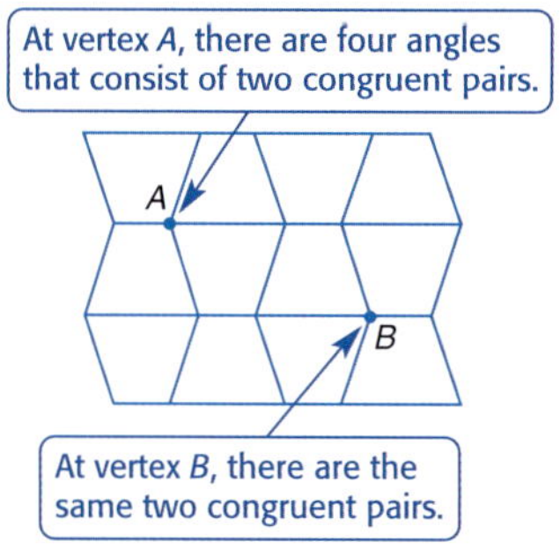

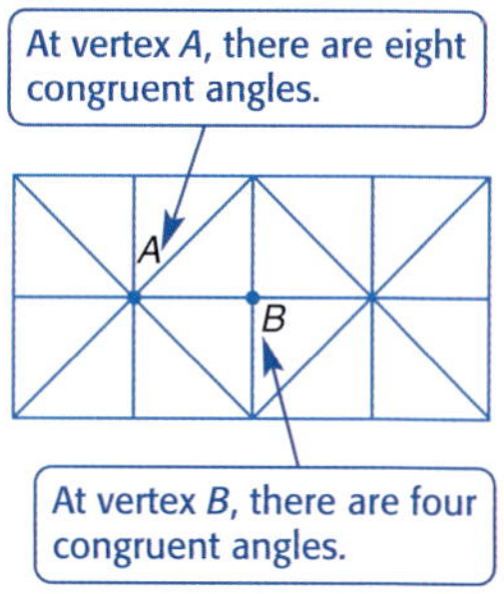

Tessellations can be formed using more than one type of polygon. A uniform tessellation formed using two or more regular polygons is called a **semi-regular tessellation**.

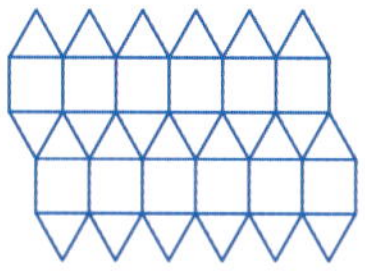

Geometry Activity

Materials: pencil, paper, pattern blocks

- Students could also include a row in the chart titled "Is the measure of one interior angle divisible by 360?"
- Explain that since tessellations are made with regular polygons, reflections, translations, and rotations would not distort the original figure. Students can also explore rotations with their tessellations and find that the original polygon needs to be rotated by increments divisible by the measure of one interior angle.

Study Tip

Drawing
When creating your own tessellation, it is sometimes helpful to complete one pattern piece, cut it out, and trace it for the other units.

Example 2 Semi-Regular Tessellation

Determine whether a semi-regular tessellation can be created from regular hexagons and equilateral triangles, all having sides 1 unit long.

Method 1 Make a model.
Two semi-regular models are shown. You will notice that the spaces at each vertex can be filled in with equilateral triangles. Model 1 has two hexagons and two equilateral triangles arranged in an alternating pattern around each vertex. Model 2 has one hexagon and four equilateral triangles at each vertex.

Model 1

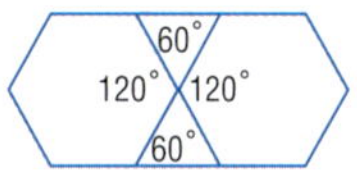

Model 2

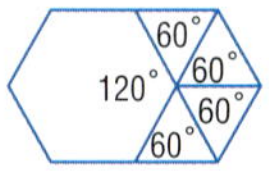

Method 2 Solve algebraically.
Each interior angle of a regular hexagon measures $\frac{180(6-2)}{6}$ or 120°.

Each angle of an equilateral triangle measures 60°. Find whole-number values for h and t so that $120h + 60t = 360$.

Let $h = 1$.

$120(1) + 60t = 360$	Substitution	$120(2) + 60t = 360$
$120 + 60t = 360$	Simplify.	$240 + 60t = 360$
$60t = 240$	Subtract from each side.	$60t = 120$
$t = 4$	Divide each side by 60.	$t = 2$

Let $h = 2$.

When $h = 1$ and $t = 4$, there is one hexagon with four equilateral triangles at each vertex. (Model 2)

When $h = 2$ and $t = 2$, there are two hexagons and two equilateral triangles. (Model 1)

Note if $h = 0$ and $t = 6$ or $h = 3$ and $t = 0$, then the tessellations are regular because there would be only one regular polygon.

Example 3 Classify Tessellations

FLOORING **Tile flooring comes in many shapes and patterns. Determine whether the pattern is a tessellation. If so, describe it as *uniform, not uniform, regular,* or *semi-regular.***

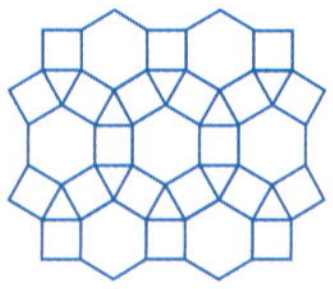

The pattern is a tessellation because at the different vertices the sum of the angles is 360°.

The tessellation is uniform because at each vertex there are two squares, a triangle, and a hexagon arranged in the same order. The tessellation is also semi-regular since more than one regular polygon is used.

Check for Understanding

Concept Check
1–3. See margin.

1. **Compare and contrast** a semi-regular tessellation and a uniform tessellation.
2. **OPEN ENDED** Use these pattern blocks that are 1 unit long on each side to create a tessellation.

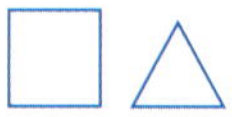

3. **Explain** why the tessellation is *not* a regular tessellation.

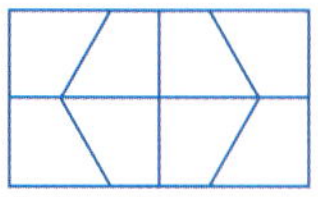

www.geometryonline.com/extra_examples/fcat

TESSELLATIONS WITH SPECIFIC ATTRIBUTES

In-Class Examples

Power Point®

Teaching Tip Students may find it easier to solve this type of problem algebraically first and then use their findings to make the model(s).

2 Determine whether a semi-regular tessellation can be created from regular nonagons and squares, all having sides 1 unit long. Explain. **No; there are no whole-number values for n and s so that $140n + 90s = 360$.**

3 **STAINED GLASS** Stained glass is a very popular design selection for church and cathedral windows. It is also fashionable to use stained glass for lampshades, decorative clocks and residential windows. Determine whether the pattern is a tessellation. If so, describe it as *uniform*, *regular*, *semi-regular*, or *not uniform*.

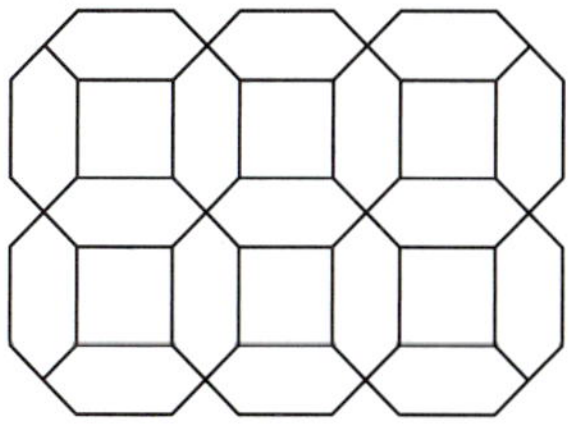

tessellation, not uniform

Answers

1. Semi-regular tessellations contain two or more regular polygons, but uniform tessellations can be any combination of shapes.
2. Sample answer:

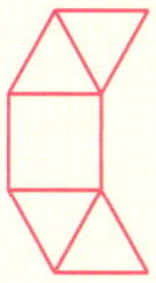

3. The figure used in the tessellation appears to be a trapezoid, which is not a regular polygon. Thus, the tessellation cannot be regular.

DAILY INTERVENTION — **Differentiated Instruction**

Visual/Spatial Provide several cardboard or laminated shapes to students. Include regular unit shapes and irregular shapes that would tessellate with each other or the regular ones. Ask students to select one or more shapes that will tessellate and copy the shape(s) repeatedly on a blank sheet of paper. Then ask them to color their designs and encourage students to create patterns or perspective with their colors. They can also classify their designs as uniform, regular, semi-regular, or not uniform.

3 Practice/Apply

Study Notebook

Have students—
- add the definitions/examples of the vocabulary terms to their Vocabulary Builder worksheets for Chapter 9.
- include an algebraic example of how to determine whether a regular polygon tessellates and an algebraic example of how to determine whether semi-regular tessellations can be created from regular polygons.
- include any other item(s) that they find helpful in mastering the skills in this lesson.

About the Exercises...

Organization by Objective
- **Regular Tessellations:** 11–16
- **Tessellations with Specific Attributes:** 17–28

Odd/Even Assignments
Exercises 11–28 are structured so that students practice the same concepts whether they are assigned odd or even problems.

Assignment Guide

Basic: 11–37 odd, 38–65
Average: 11–37 odd, 38–65
Advanced: 12–38 even, 40–61 (optional: 62–65)

Guided Practice

GUIDED PRACTICE KEY	
Exercises	Examples
4, 5	1
6, 7	2
8–10	3

Determine whether each regular polygon tessellates the plane. Explain.

4. decagon
no; measure of interior angle = 144

5. 30-gon
no; measure of interior angle = 168

Determine whether a semi-regular tessellation can be created from each set of figures. Assume that each figure has a side length of 1 unit.

6. a square and a triangle **yes**

7. an octagon and a square **yes**

Determine whether each pattern is a tessellation. If so, describe it as *uniform, not uniform, regular,* or *semi-regular.*

8. 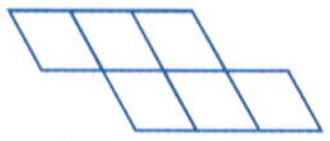 yes; uniform

9. 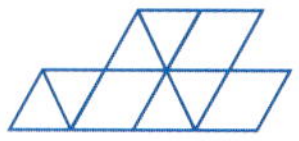 yes; not uniform

Application

10. **QUILTING** The "Postage Stamp" pattern can be used in quilting. Explain why this is a tessellation and what kind it is. **Each "postage stamp" is a square that has been tessellated and 90 is a factor of 360. It is a regular tessellation since only one polygon is used.**

Practice and Apply

Homework Help	
For Exercises	See Examples
11–16	1
17–20	2
21–28	3

Extra Practice
See page 772.

Determine whether each regular polygon tessellates the plane. Explain.

11. nonagon
12. hexagon
13. equilateral triangle
14. dodecagon
15. 23-gon
16. 36-gon

11. no; measure of interior angle = 140
12. yes; measure of interior angle = 120
13. yes; measure of interior angle = 60
14. no; measure of interior angle = 150
15. no; measure of interior angle ≈ 164.3
16. no; measure of interior angle = 170

Determine whether a semi-regular tessellation can be created from each set of figures. Assume that each figure has a side length of 1 unit.

17. regular octagons and non-square rhombi **no**
18. regular dodecagons and equilateral triangles **yes**
19. regular dodecagons, squares, and equilateral triangles **yes**
20. regular heptagons, squares, and equilateral triangles **no**

Determine whether each figure tessellates the plane. If so, describe the tessellation as *uniform, not uniform, regular,* or *semi-regular.*

21. parallelogram **yes; uniform**

22. kite **yes; uniform**

23. quadrilateral **yes; not uniform**

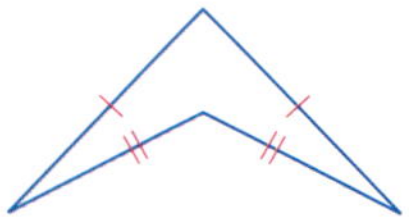

24. pentagon and square **no**

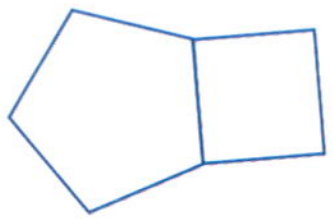

Determine whether each pattern is a tessellation. If so, describe it as *uniform, not uniform, regular,* or *semi-regular.*

25. 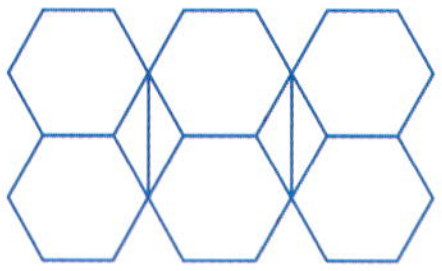 **yes; not uniform**

26. 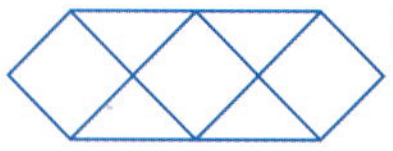 **yes; not uniform**

Teacher to Teacher

Liza Allen — Conway High School West, Conway, AR

Bring a quilt or have students bring quilts to class to find the transformations described in this chapter in the quilt patterns. Quilt patterns often use rotations, translations, and dilations as well as being examples of tessellating a plane.

Career Choices

Bricklayer

ricklayers arrange and ecure bricks, concrete locks, tiles, and other uilding materials to onstruct or repair walls, artitions, arches, fireplaces, nd other structures. They ust understand how the ieces tessellate to omplete the structure pecifications.

Online Research
or information about career as a bricklayer, isit: www.geometryonline.om/careers

27. 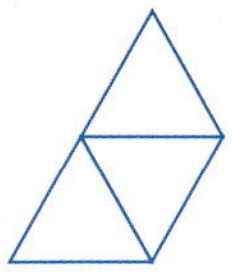 yes; uniform, regular

28. 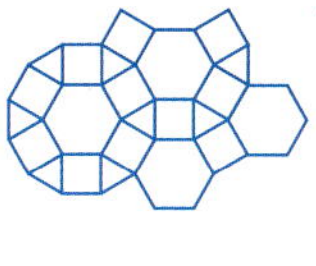 yes; uniform, semi-regular

29. **BRICKWORK** In the picture, suppose the side of the octagon is the same length as the side of the square. What kind of tessellation is this? semi-regular, uniform

Determine whether each statement is *always, sometimes,* or *never* true. Justify your answers.

30. Any triangle will tessellate the plane. always
31. Semi-regular tessellations are not uniform. never
32. Uniform tessellations are semi-regular. sometimes
33. Every quadrilateral will tessellate the plane. always
34. Regular 16-gons will tessellate the plane. never

30–34. See p. 519C for justifications.

INTERIOR DESIGN **For Exercises 35 and 36, use the following information.**
Kele's family is tiling the entry floor with the tiles in the pattern shown.

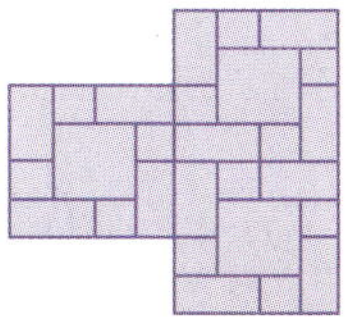

35. Determine whether the pattern is a tessellation. yes
36. Is the tessellation *uniform, regular,* or *semi-regular*? none of these

37. **BEES** A honeycomb is composed of hexagonal cells made of wax in which bees store honey. Determine whether this pattern is a tessellation. If so, describe it as *uniform, not uniform, regular,* or *semi-regular.* uniform, regular

38. **CRITICAL THINKING** What could be the measures of the interior angles in a pentagon that tessellate a plane? Is this tessellation regular? Is it uniform?

8. Sample answers: The measures are 90°, 0°, 90°, 135°, and 35°. The tessellation s not regular or niform.

39. **WRITING IN MATH** Answer the question that was posed at the beginning of the lesson. See margin.

How are tessellations used in art?
Include the following in your answer:
- how equilateral triangles and regular hexagons form a tessellation, and
- other geometric figures that can be found in the picture.

FCAT Practice
Standardized Test Practice
Ⓐ Ⓑ Ⓒ Ⓓ

40. Find the measure of an interior angle of a regular nonagon. C

Ⓐ 150 Ⓑ 147 Ⓒ 140 Ⓓ 115

41. **ALGEBRA** Evaluate $\frac{360(x-2)}{2x} - \frac{180}{x}$ if $x = 12$. A

Ⓐ 135 Ⓑ 150 Ⓒ 160 Ⓓ 225

www.geometryonline.com/self_check_quiz/fcat

Answers

39. Sample answer: Tessellations can be used in art to create abstract art. Answers should include the following.
- The equilateral triangles are arranged to form hexagons, which are arranged adjacent to one another.
- Sample answers: kites, trapezoids, isosceles triangles

Study Guide and Intervention, p. 497 (shown) and p. 498

Regular Tessellations A pattern that covers a plane with repeating copies of one or more figures so that there are no overlapping or empty spaces is a **tessellation**. A **regular tessellation** uses only one type of regular polygon. In a tessellation, the sum of the measures of the angles of the polygons surrounding a vertex is 360. If a regular polygon has an interior angle that is a factor of 360, then the polygon will tessellate.

regular tessellation | tessellation | Copies of a regular hexagon can form a tessellation. | Copies of a regular pentagon cannot form a tessellation.

Example **Determine whether a regular 16-gon tessellates the plane. Explain.**
If $m\angle 1$ is the measure of one interior angle of a regular polygon, then a formula for $m\angle 1$ is $m\angle 1 = \frac{180(n-2)}{n}$. Use the formula with $n = 16$.

$$m\angle 1 = \frac{180(n-2)}{n} = \frac{180(16-2)}{16} = 157.5$$

The value 157.5 is not a factor of 360, so the 16-gon will not tessellate.

Exercises

Determine whether each polygon tessellates the plane. If so, draw a sample figure.

1. scalene right triangle yes
2. isosceles trapezoid yes

Determine whether each regular polygon tessellates the plane. Explain.

3. square
Yes; the measure of each interior angle is 90, and 90 is a factor of 360.
4. 20-gon
No; the measure of each interior angle is 162, and 162 is not a factor of 360.
5. septagon
No; the measure of each interior angle is 128.6, and 128.6 is not a factor of 360.
6. 15-gon
No; the measure of each interior angle is 156, and 156 is not a factor of 360.
7. octagon
No; the measure of each interior angle is 135, and 135 is not a factor of 360.
8. pentagon
No; the measure of each interior angle is 108, and 108 is not a factor of 360.

Skills Practice, p. 499 and Practice, p. 500 (shown)

Determine whether each regular polygon tessellates the plane. Explain.

1. 22-gon
no, interior angle ≈ 163.6
2. 40-gon
no, interior angle = 171

Determine whether a semi-regular tessellation can be created from each set of figures. Assume each figure has a side length of 1 unit.

3. regular pentagons and regular decagons
no (will not tessellate the plane)
4. regular dodecagons, regular hexagons, and squares
yes

Determine whether each polygon tessellates the plane. If so, describe the tessellation as *uniform, not uniform, regular,* or *semi-regular.*

5. kite
6. octagon and decagon

5. yes; uniform
6. no

Determine whether each pattern is a tessellation. If so, describe it as *uniform, not uniform, regular,* or *semi-regular.*

7.
8.

7. yes; uniform, semi-regular
8. yes, not uniform

FLOOR TILES **For Exercises 9 and 10, use the following information.**
Mr. Martinez chose the pattern of tile shown to retile his kitchen floor.

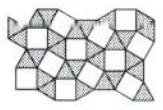

9. Determine whether the pattern is a tessellation. Explain
Yes; since there are 2 squares and 3 triangles at each vertex, the sum of the angles at the vertices is 360°.
10. Is the pattern uniform, regular, or semi-regular?
uniform and semi-regular

Reading to Learn Mathematics, p. 501 ELL

Pre-Activity **How are tessellations used in art?**

Read the introduction to Lesson 9-4 at the top of page 483 in your textbook.

- In the pattern shown in the picture in your textbook, how many small equilateral triangles make up one regular hexagon? 6
- In this pattern, how many fish make up one equilateral triangle? $1\frac{1}{2}$

Reading the Lesson

1. Underline the correct word, phrase, or number to form a true statement.
   a. A tessellation is a pattern that covers a plane with the same figure or set of figures so that there are no (congruent angles/overlapping or empty spaces/right angles).
   b. A tessellation that uses only one type of regular polygon is called a (uniform/regular/semi-regular) tessellation.
   c. The sum of the measures of the angles at any vertex in any tessellation is (90/180/360).
   d. A tessellation that contains the same arrangement of shapes and angles at every vertex is called a (uniform/regular/semi-regular) tessellation.
   e. In a regular tessellation made up of hexagons, there are (3/4/6) hexagons meeting at each vertex, and the measure of each of the angles at any vertex is (60/90/120).
   f. A uniform tessellation formed using two or more regular polygons is called a (rotational/regular/semi-regular) tessellation.
   g. In a regular tessellation made up of triangles, there are (3/4/6) triangles meeting at each vertex, and the measure of each of the angles at any vertex is (30/60/120).
   h. If a regular tessellation is made up of quadrilaterals, all of the quadrilaterals must be congruent (rectangles/parallelograms/squares/trapezoids).

2. Write all of the following words that describe each tessellation: *uniform, non-uniform, regular, semi-regular.*

a. non-uniform
b. uniform, semi-regular

Helping You Remember

3. Often the everyday meanings of a word can help you to remember its mathematical meaning. Look up *uniform* in your dictionary. How can its everyday meanings help you to remember the meaning of a *uniform* tessellation? Sample answer: Three similar meanings of *uniform* are *unvarying, consistent, and identical.* In a uniform tessellation, the arrangement of shapes and angles at each vertex is the same. Each of the three everyday meanings describes this situation.

Enrichment, p. 502

Polygonal Numbers

Certain numbers related to regular polygons are called **polygonal numbers**. The chart shows several triangular, square, and pentagonal numbers. The **rank** of a polygon number is the number of dots on each "side" of the outer polygon. For example, the pentagonal number 22 has a rank of 4.

Polygonal numbers can be described with formulas. For example, a triangular number T of rank r can be described by $T = \frac{r(r+1)}{2}$.

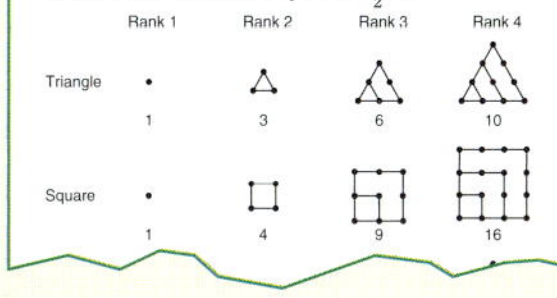

4 Assess

Open-Ended Assessment

Writing Ask students to compose a page with an example each of a uniform, regular, semi-regular, and non-uniform tessellation with a brief explanation of why each classification applies to its corresponding example. Students can select examples from the Practice and Apply section of the book, or you can encourage them to create their own.

Getting Ready for Lesson 9-5

Prerequisite Skill Students will learn about dilations in Lesson 9-5. They will apply concepts of similar polygons. Use Exercises 62–65 to determine your students' familiarity with similar polygons.

Assessment Options

Quiz (Lessons 9-3 and 9-4) is available on p. 535 of the *Chapter 9 Resource Masters*.

Mid-Chapter Test (Lessons 9-1 through 9-4) is available on p. 537 of the *Chapter 9 Resource Masters*.

Answers

42.

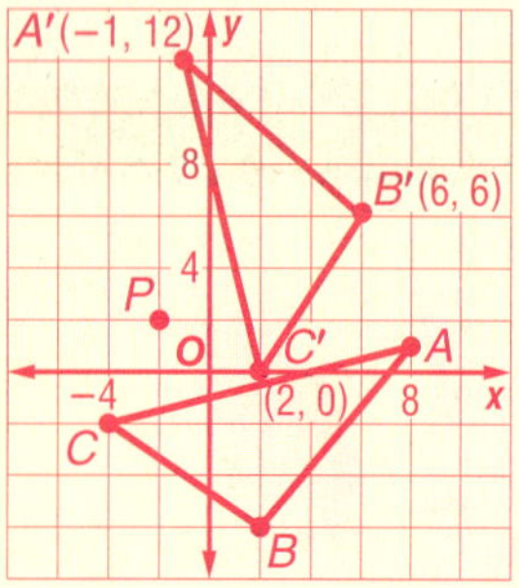

43.

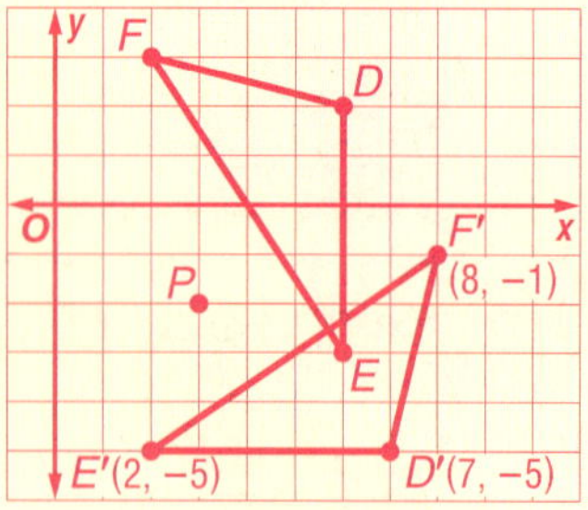

Maintain Your Skills

Mixed Review

COORDINATE GEOMETRY **Draw the rotation image of each figure 90° in the given direction about the center point and label the coordinates.** *(Lesson 9-3)*

42–45. See margin.

42. $\triangle ABC$ with $A(8, 1)$, $B(2, -6)$, and $C(-4, -2)$ counterclockwise about $P(-2, 2)$

43. $\triangle DEF$ with $D(6, 2)$, $E(6, -3)$, and $F(2, 3)$ clockwise about $P(3, -2)$

44. $\square GHIJ$ with $G(-1, 2)$, $H(-3, -3)$, $I(-5, -6)$, and $J(-3, -1)$ counterclockwise about $P(-2, -3)$

45. rectangle $KLMN$ with $K(-3, -5)$, $L(3, 3)$, $M(7, 0)$, and $N(1, -8)$ counterclockwise about $P(-2, 0)$

46. REMODELING The diagram at the right shows the floor plan of Justin's kitchen. Each square on the diagram represents a 3 foot by 3 foot area. While remodeling his kitchen, Justin moved his refrigerator from square A to square B. Describe the move. *(Lesson 9-2)*

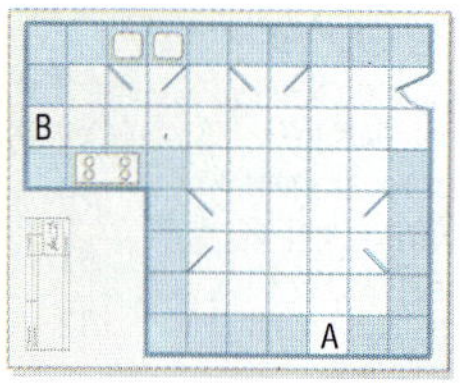

46. The move is a translation 15 feet out from the wall and 21 feet to the left, then a rotation of 90° clockwise.

ALGEBRA **Find x and y so that each quadrilateral is a parallelogram.** *(Lesson 8-3)*

47.

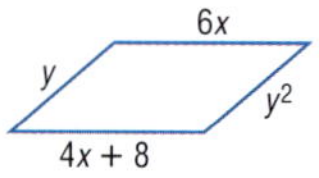

$x = 4$, $y = 1$

48.

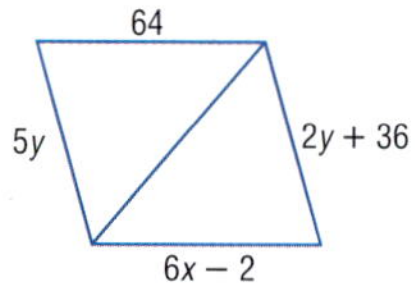

$x = 11$, $y = 12$

49.

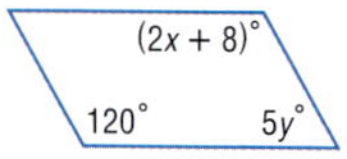

$x = 56$, $y = 12$

Determine whether each set of numbers can be the measures of the sides of a right triangle. Then state whether they form a Pythagorean triple. *(Lesson 7-2)*

50. 12, 16, 20 **yes, yes**

51. 9, 10, 15 **no, no**

52. 2.5, 6, 6.5 **yes, no**

53. $14, 14\sqrt{3}, 28$ **yes, no**

54. 14, 48, 50 **yes, yes**

55. $\frac{1}{2}, \frac{1}{3}, \frac{1}{4}$ **no, no**

Points A, B, and C are the midpoints of $\overline{DF}$, $\overline{DE}$, and $\overline{EF}$, respectively. *(Lesson 6-4)*

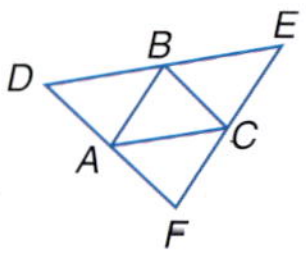

56. If $BC = 11$, $AC = 13$, and $AB = 15$, find the perimeter of $\triangle DEF$. **78**

57. If $DE = 18$, $DA = 10$, and $FC = 7$, find AB, BC, and AC. **$AB = 7$, $BC = 10$, $AC = 9$**

COORDINATE GEOMETRY **For Exercises 58–61, use the following information.**
The vertices of quadrilateral $PQRS$ are $P(5, 2)$, $Q(1, 6)$, $R(-3, 2)$, and $S(1, -2)$. *(Lesson 3-3)*

58. Show that the opposite sides of quadrilateral $PQRS$ are parallel.

59. Show that the adjacent sides of quadrilateral $PQRS$ are perpendicular

60. Determine the length of each side of quadrilateral $PQRS$.

61. What type of figure is quadrilateral $PQRS$? **square**

58. slope of $\overline{PQ} = -1$, slope of $\overline{RS} = -1$, slope of $\overline{QR} = 1$, slope of $\overline{PS} = 1$,
59. $-1(1) = -1$ and $-1(1) = -1$
60. $PQ = \sqrt{32}$, $QR = \sqrt{32}$, $RS = \sqrt{32}$, $PS = \sqrt{32}$

Getting Ready for the Next Lesson

PREREQUISITE SKILL **If quadrilateral $ABCD \sim$ quadrilateral $WXYZ$, find each of the following.** *(To review **similar polygons**, see Lesson 6-2.)*

62. scale factor of $ABCD$ to $WXYZ$ $\frac{2}{3}$

63. XY **15**

64. YZ **15**

65. WZ **22.5**

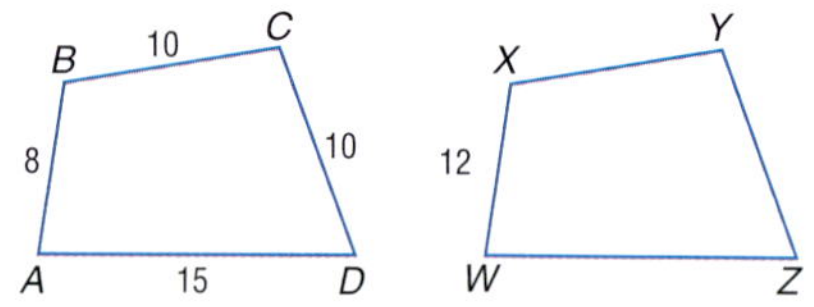

44.

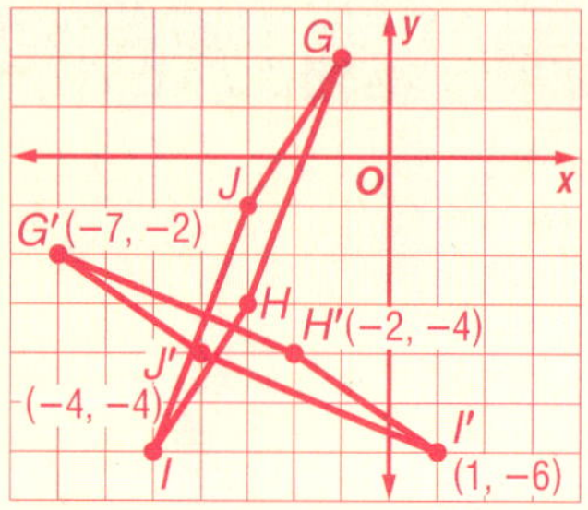

45.

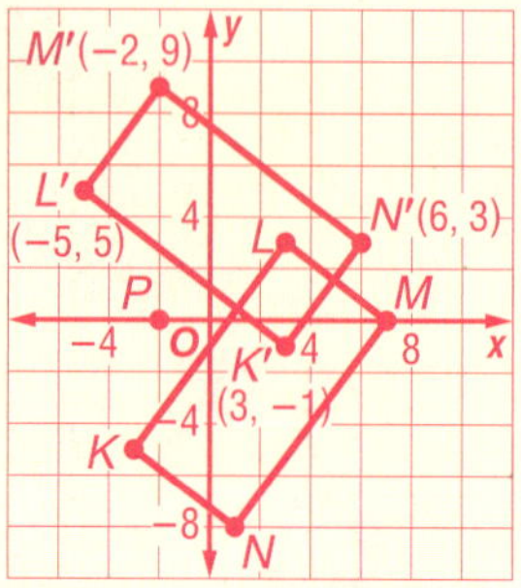

Tessellations and Transformations

Activity 1 **Make a tessellation using a translation.**

Step 1 Start by drawing a square. Then copy the figure shown below.

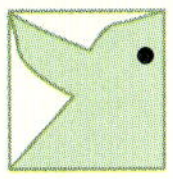

Step 2 Translate the shape on the top side to the bottom side.

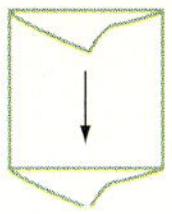

Step 3 Translate the figure on the left side and the dot to the right side to complete the pattern.

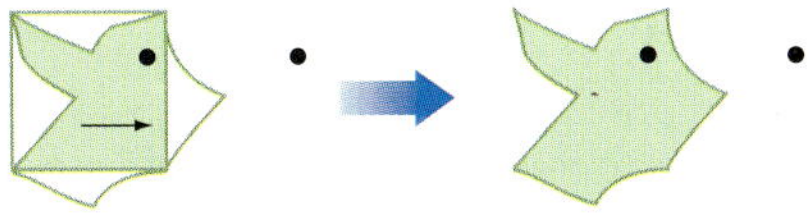

Step 4 Repeat this pattern on a tessellation of squares.

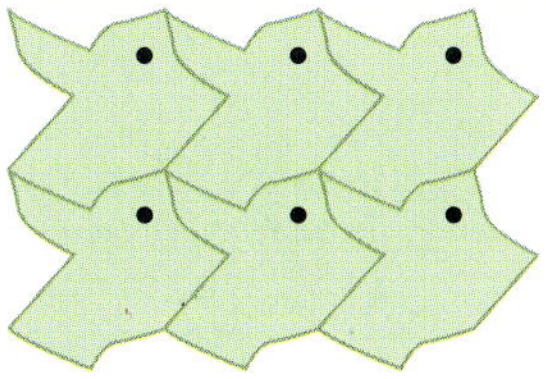

Activity 2 **Make a tessellation using a rotation.**

Step 1 Start by drawing an equilateral triangle. Then draw a trapezoid inside the right side of the triangle.

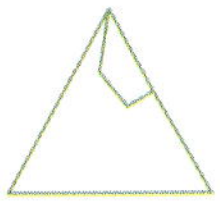

Step 2 Rotate the trapezoid so you can copy the change on the side indicated.

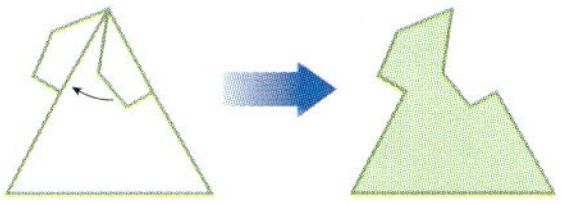

Step 3 Repeat this pattern on a tessellation of equilateral triangles. Alternating colors can be used to best show the tessellation.

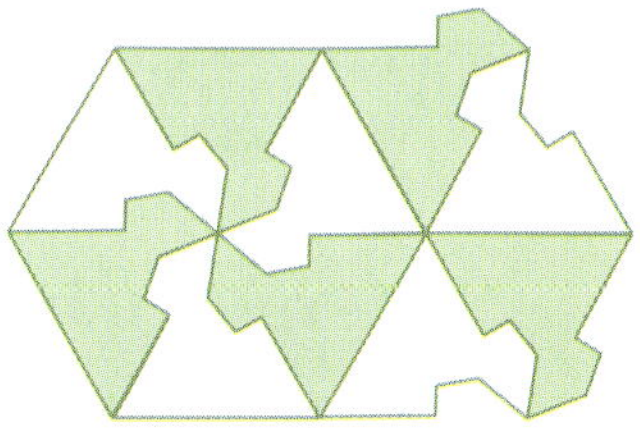

Model and Analyze 1–5. See p. 519C.

1. Is the area of the square in Step 1 of Activity 1 the same as the area of the new shape in Step 2? Explain.
2. Describe how you would create the unit for the pattern shown at the right.

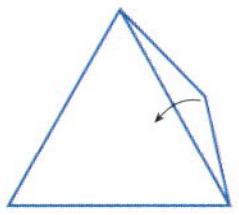

Make a tessellation for each pattern described. Use a tessellation of two rows of three squares as your base.

3. 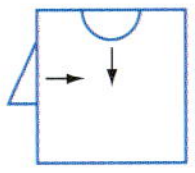
4. 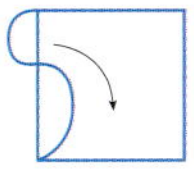
5. 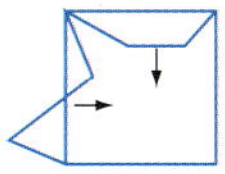

Geometry Activity

A Follow-Up of Lesson 9-4

Getting Started

Objective Students will make tessellations using a translation and a rotation.

Materials
paper, pencil

Teach

- Explain that students will start with a regular shape and then create a patterned distortion of the original shape. Students should note that they are starting with shapes from Lesson 9-4 proven to tessellate.
- Assign squares of different sizes to students. Some can start with a square 2 cm by 2 cm. Others can start with a square 5 cm by 5 cm. Repeat for the equilateral triangles.

Assess

Students analyze unit figures in **Exercises 1–2**. Students repeat the activity in **Exercises 3–5**.

Study Notebook

Ask students to summarize what they have learned about tessellations and transformations.

Resource Manager

Teaching Geometry with Manipulatives
- p. 155 (student recording sheet)

9-5 Lesson Notes

1 Focus

5-Minute Check Transparency 9-5 Use as a quiz or review of Lesson 9-4.

Mathematical Background notes are available for this lesson on p. 460D.

How do you use dilations when you use a computer?

Ask students:

- In many programs, you can hold the shift key down to keep from distorting the original figure while you scale it. When you do this, what is the relationship between the original figure and the transformed figure? **They are similar.**
- How many scaling boxes usually appear on the frame of a selected figure in a word processing program? **Sample answer: 6**

9-5 Dilations

Sunshine State Standards
MA.A.3.4.3, MA.B.1.4.3, MA.C.2.4.1, MA.C.3.4.1, MA.C.3.4.2

What You'll Learn

- Determine whether a dilation is an enlargement, a reduction, or a congruence transformation.
- Determine the scale factor for a given dilation.

Vocabulary

- dilation
- similarity transformation

How do you use dilations when you use a computer?

Have you ever tried to paste an image into a word processing document and the picture was too large? Many word processing programs allow you to scale the size of the picture so that you can fit it in your document. Scaling a picture is an example of a dilation.

Study Tip

Scale Factor
When discussing dilations, scale factor has the same meaning as with proportions. The letter r usually represents the scale factor.

CLASSIFY DILATIONS All of the transformations you have studied so far in this chapter produce images that are congruent to the original figure. A dilation is another type of transformation. A **dilation** is a transformation that may change the size of a figure. A dilation requires a center point and a scale factor. The figures below show how dilations can result in a larger figure and a smaller figure than the original.

$r = 2$

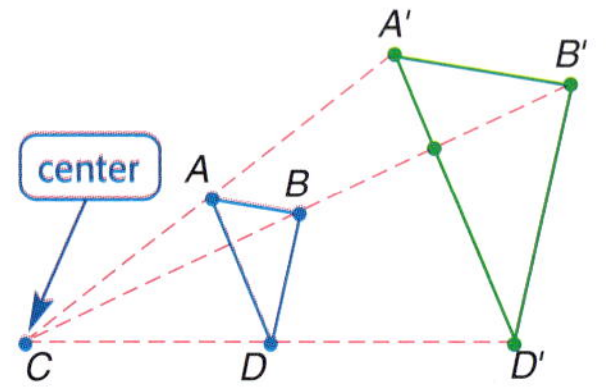

Triangle $A'B'D'$ is a dilation of $\triangle ABD$.

$CA' = 2(CA)$

$CB' = 2(CB)$

$CD' = 2(CD)$

$\triangle A'B'D'$ is larger than $\triangle ABD$.

$r = \frac{1}{3}$

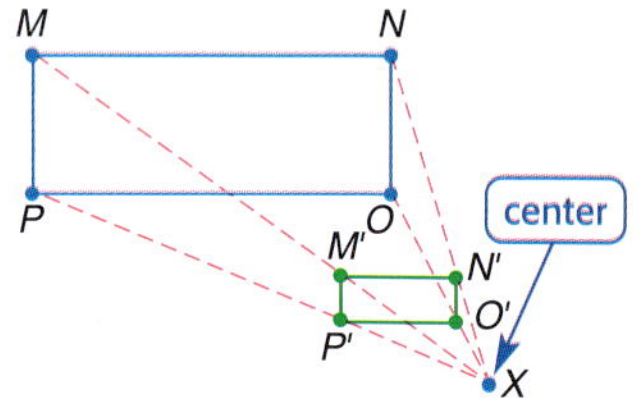

Rectangle $M'N'O'P'$ is a dilation of rectangle $MNOP$.

$XM' = \frac{1}{3}(XM)$ $\quad XN' = \frac{1}{3}(XN)$

$XO' = \frac{1}{3}(XO)$ $\quad XP' = \frac{1}{3}(XP)$

Rectangle $M'N'O'P'$ is smaller than rectangle $MNOP$.

The value of r determines whether the dilation is an enlargement or a reduction.

Key Concept *Dilation*

If $|r| > 1$, the dilation is an enlargement.

If $0 < |r| < 1$, the dilation is a reduction.

If $|r| = 1$, the dilation is a congruence transformation.

Resource Manager

Workbook and Reproducible Masters

Chapter 9 Resource Masters
- Study Guide and Intervention, pp. 503–504
- Skills Practice, p. 505
- Practice, p. 506
- Reading to Learn Mathematics, p. 507
- Enrichment, p. 508

Prerequisite Skills Workbook, pp. 19–20, 43–44

Teaching Geometry With Manipulatives Masters, pp. 1, 16, 17

Transparencies

5-Minute Check Transparency 9-5
Answer Key Transparencies

Technology

GeomPASS: Tutorial Plus, Lesson 18
Interactive Chalkboard

Study Tip

Isometry Dilation
A dilation with a scale factor of 1 produces an image that coincides with the preimage. The two are congruent.

As you can see in the figures on the previous page, dilation preserves angle measure, betweenness of points, and collinearity, but does not preserve distance. That is, dilations produce similar figures. Therefore, a dilation is a **similarity transformation**.

This means that $\triangle ABD \sim \triangle A'B'D'$ and $\square MNOP \sim \square M'N'O'P'$. This implies that $\frac{A'B'}{AB} = \frac{B'D'}{BD} = \frac{A'D'}{AD}$ and $\frac{M'N'}{MN} = \frac{N'O'}{NO} = \frac{O'P'}{OP} = \frac{M'P'}{MP}$. The ratios of measures of the corresponding parts is equal to the absolute value scale factor of the dilation, $|r|$. So, $|r|$ determines the size of the image as compared to the size of the preimage.

Theorem 9.1

If a dilation with center C and a scale factor of r transforms A to E and B to D, then $ED = |r|(AB)$.

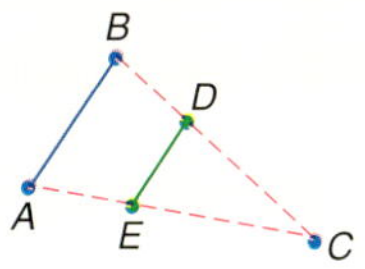

You will prove Theorem 9.1 in Exercise 41.

Example 1 Determine Measures Under Dilations

Find the measure of the dilation image $\overline{A'B'}$ or the preimage $\overline{AB}$ using the given scale factor.

a. $AB = 12$, $r = -2$

$A'B' = |r|(AB)$

$A'B' = 2(12)$ — $|r| = 2$, $AB = 12$

$A'B' = 24$ — Multiply.

b. $A'B' = 36$, $r = \frac{1}{4}$

$A'B' = |r|(AB)$

$36 = \frac{1}{4}(AB)$ — $A'B' = 36$, $|r| = \frac{1}{4}$

$144 = AB$ — Multiply each side by 4.

When the scale factor is negative, the image falls on the opposite side of the center than the preimage.

Key Concept — Dilations

If $r > 0$, P' lies on $\overrightarrow{CP}$, and $CP' = r \cdot CP$.

If $r < 0$, P' lies on $\overrightarrow{CP'}$ the ray opposite $\overrightarrow{CP}$, and $CP' = |r| \cdot CP$.

The center of a dilation is always its own image.

Example 2 Draw a Dilation

Draw the dilation image of $\triangle JKL$ with center C and $r = -\frac{1}{2}$.

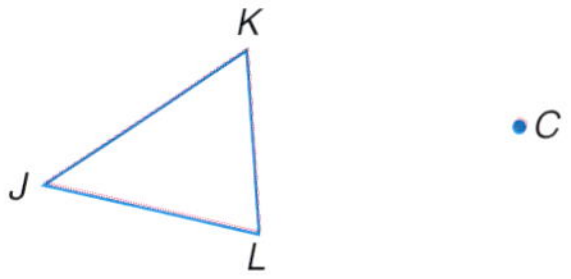

Since $0 < |r| < 1$, the dilation is a reduction of $\triangle JKL$.

Draw $\overline{CJ}$, $\overline{CK}$, and $\overline{CL}$. Since r is negative, J', K', and L' will lie on $\overrightarrow{CJ'}$, $\overrightarrow{CK'}$, and $\overrightarrow{CL'}$, respectively. Locate J', K', and L' so that $CJ' = \frac{1}{2}(CJ)$, $CK' = \frac{1}{2}(CK)$, and $CL' = \frac{1}{2}(CL)$. Draw $\triangle J'K'L'$.

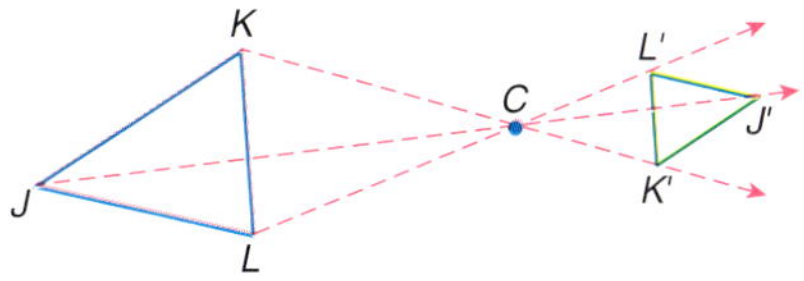

2 Teach

CLASSIFY DILATIONS

In-Class Examples — Power Point®

Teaching Tip Explain that C can also be located inside the original figure, in which case the sides of the dilated image would radiate outward from C keeping the same orientation as the original if r is positive and reflecting in C if r is negative.

1 Find the measure of the dilation image or the preimage of $\overline{CD}$ using the given scale factor.

a. $CD = 15$, $r = 3$ **45**

b. $C'D' = 7$, $r = -\frac{2}{3}$ **10.5**

2 Draw the dilation image of trapezoid $PQRS$ with center C and $r = -3$.

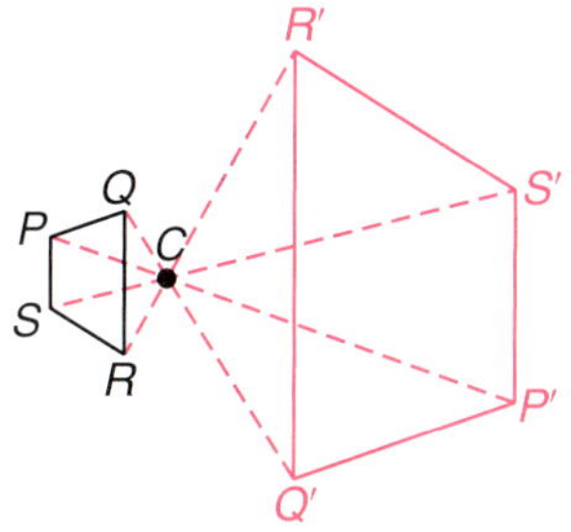

Concept Check

Have students plot a simple figure on grid paper and apply a reflection, a translation, a rotation, and a dilation to the figure, color-coding each figure with a label referring to the type of transformation applied. Then, ask students to plot the same figure elsewhere on the paper or on a new sheet and apply the same transformations in reverse order, starting with the dilation and ending with the reflection. Tell students to note the differences between the two procedures. As an extension, ask them to transform the same figure twice using two different orders of the four transformations so that the image ends up in the same position and orientation both times.

DAILY INTERVENTION — Differentiated Instruction

Auditory/Musical Students can relate dilations to music by how loud or soft a sound is. They can remember that a harmonica's sound is magnified or dilated with a scale factor of r greater than 1 when a great force is used to create a musical note, and the sound is much softer when the same note is produced with half the force. They can also correlate drawing a breath through the harmonica with a negative scale factor and exhaling into the harmonica with a positive scale factor.

In-Class Example

Power Point®

3 **COORDINATE GEOMETRY** Trapezoid *EFGH* has vertices $E(-8, 4)$, $F(-4, 8)$, $G(8, 4)$, and $H(-4, -8)$. Find the image of trapezoid *EFGH* after a dilation centered at the origin with a scale factor of $\frac{1}{4}$. Sketch the preimage and the image.

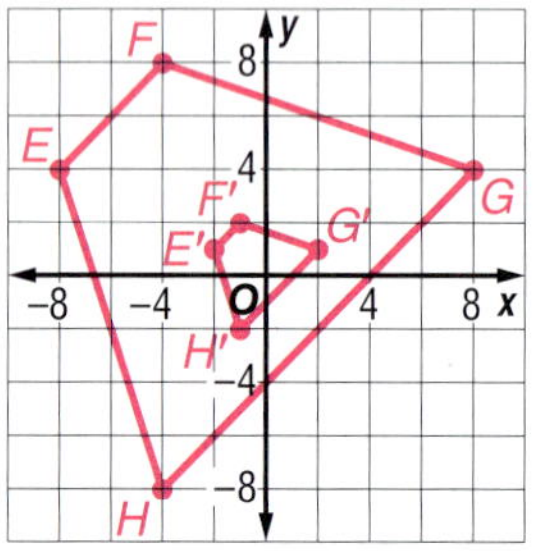

$E'(-2, 1)$, $F'(-1, 2)$, $G'(2, 1)$, $H'(-1, -2)$

IDENTIFY THE SCALE FACTOR

In-Class Example

Power Point®

4 Determine the scale factor used for each dilation with center *C*. Determine whether the dilation is an *enlargement*, *reduction*, or *congruence transformation*.

a.

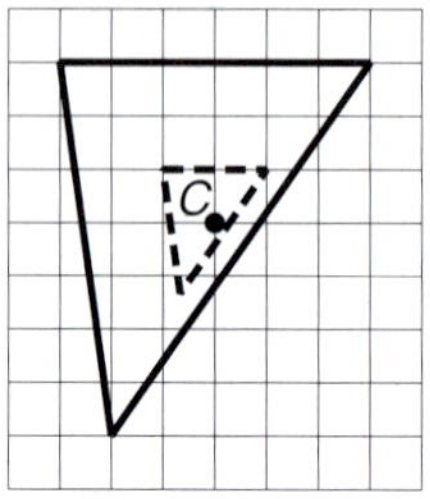

$\frac{1}{3}$; reduction

b.

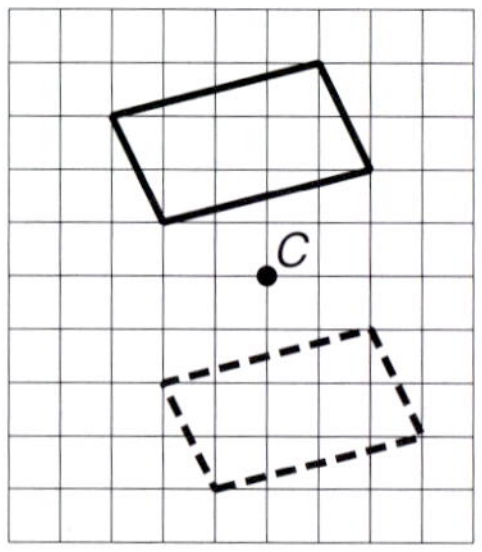

−1; congruence transformation

In the coordinate plane, you can use the scale factor to determine the coordinates of the image of dilations centered at the origin.

Theorem 9.2

If $P(x, y)$ is the preimage of a dilation centered at the origin with a scale factor r, then the image is $P'(rx, ry)$.

Example 3 Dilations in the Coordinate Plane

COORDINATE GEOMETRY **Triangle *ABC* has vertices $A(7, 10)$, $B(4, -6)$, and $C(-2, 3)$. Find the image of $\triangle ABC$ after a dilation centered at the origin with a scale factor of 2. Sketch the preimage and the image.**

Preimage (x, y)	Image $(2x, 2y)$
$A(7, 10)$	$A'(14, 20)$
$B(4, -6)$	$B'(8, -12)$
$C(-2, 3)$	$C'(-4, 6)$

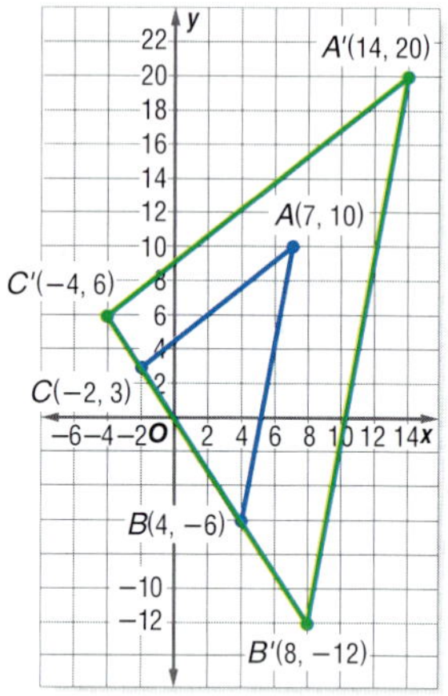

IDENTIFY THE SCALE FACTOR In Chapter 6, you found scale factors of similar figures. If you know the measurement of a figure and its dilated image, you can determine the scale factor.

Study Tip

Look Back
To review **scale factor**, see Lesson 6-2.

Example 4 Identify Scale Factor

Determine the scale factor for each dilation with center *C*. Then determine whether the dilation is an *enlargement*, *reduction*, or *congruence transformation*.

a.

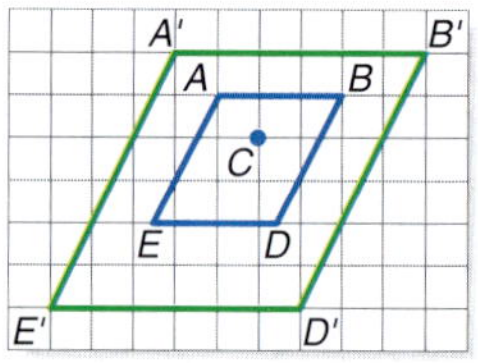

$$\text{scale factor} = \frac{\text{image length}}{\text{preimage length}}$$

$$= \frac{6 \text{ units}}{3 \text{ units}} \quad \begin{array}{l}\leftarrow \text{image length} \\ \leftarrow \text{preimage length}\end{array}$$

$$= 2 \quad \text{Simplify.}$$

Since the scale factor is greater than 1, the dilation is an enlargement.

b.

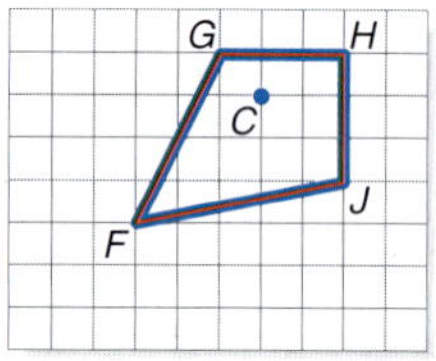

$$\text{scale factor} = \frac{\text{image length}}{\text{preimage length}}$$

$$= \frac{4 \text{ units}}{4 \text{ units}} \quad \begin{array}{l}\leftarrow \text{image length} \\ \leftarrow \text{preimage length}\end{array}$$

$$= 1 \quad \text{Simplify.}$$

Since the scale factor is 1, the dilation is a congruence transformation.

492 Chapter 9 Transformations

FCAT Practice

Standardized Test Practice

Example 5 Scale Drawing

Multiple-Choice Test Item

Jacob wants to make a scale drawing of a painting in an art museum. The painting is 4 feet wide and 8 feet long. Jacob decides on a dilation reduction factor of $\frac{1}{6}$. What size paper will he need to make a complete sketch?

Ⓐ $8\frac{1}{2}$ in. by 11 in. Ⓑ 9 in. by 12 in. Ⓒ 11 in. by 14 in. Ⓓ 11 in. by 17 in.

Read the Test Item

The painting's dimensions are given in feet, and the paper choices are in inches. You need to convert from feet to inches in the problem.

Solve the Test Item

Step 1 Convert feet to inches.

4 feet = 4(12) or 48 inches

8 feet = 8(12) or 96 inches

Step 2 Use the scale factor to find the image dimensions.

$w = \frac{1}{6}(48)$ or 8 $\qquad \ell = \frac{1}{6}(96)$ or 16

Step 3 The dimensions of the image are 8 inches by 16 inches. Choice D is the only size paper on which the scale drawing will fit.

The Princeton Review

Test-Taking Tip

Compare Measurements Compare the measurements given in the problem to those in the answers. These answers are in inches, so convert feet to inches before using the scale factor. It may make calculations easier.

In-Class Examples

Power Point®

5 Sharetta built a frame for a photograph that is 20 centimeters by 25 centimeters. The frame measures 400 millimeters by 500 millimeters. Which scale factor did she use? **A**

A 2 **B** 3

C $\frac{1}{2}$ **D** $\frac{1}{3}$

3 Practice/Apply

Study Notebook

Have students—

- add the definitions/examples of the vocabulary terms to their Vocabulary Builder worksheets for Chapter 9.
- include an example of an enlargement, a reduction, and a congruence transformation and a brief description of how to determine and/or apply scale factors.
- include any other item(s) that they find helpful in mastering the skills in this lesson.

Check for Understanding

Concept Check

1–2. See margin.

1. Find a counterexample to disprove the statement *All dilations are isometries.*

2. **OPEN ENDED** Draw a figure on the coordinate plane. Then show a dilation of the figure that is a reduction and a dilation of the figure that is an enlargement.

3. **FIND THE ERROR** Desiree and Trey are trying to describe the effect of a negative r value for a dilation of quadrilateral $WXYZ$.

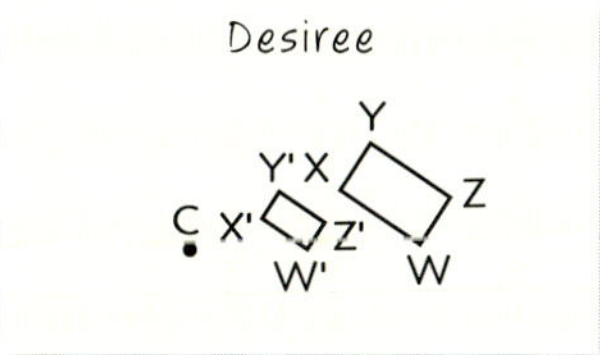

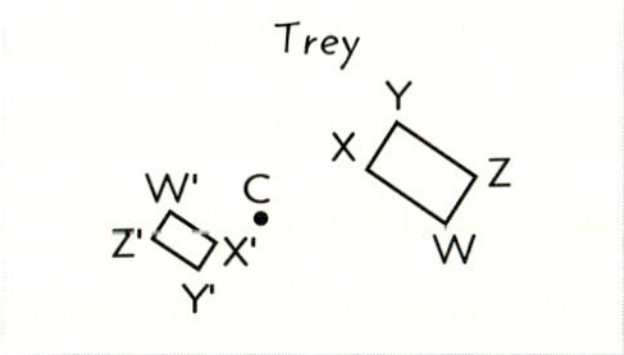

Who is correct? Explain your reasoning. **Trey; Desiree found the image using a positive scale factor.**

GUIDED PRACTICE KEY	
Exercises	Examples
4–6	1
7, 8	2
9, 10	3
11, 12	4
13	5

Guided Practice

4–6. See margin.

Draw the dilation image of each figure with center C and the given scale factor.

4. $r = 4$

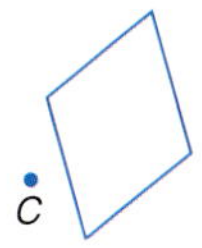

5. $r = \frac{1}{5}$

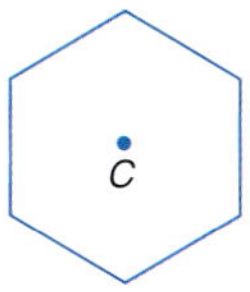

6. $r = -2$

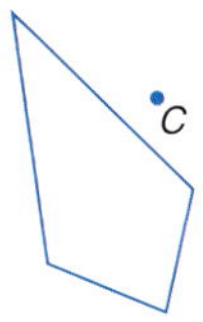

DAILY INTERVENTION

FIND THE ERROR Students should see that in this example, C would be between the image and preimage with the negative r value.

Answers

1. Dilations only preserve length if the scale factor is 1 or −1. So for any other scale factor, length is not preserved and the dilation is not an isometry.

2. Sample answer:

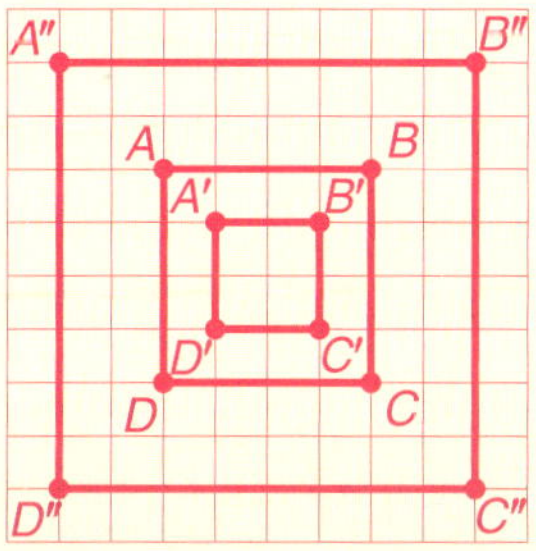

4.

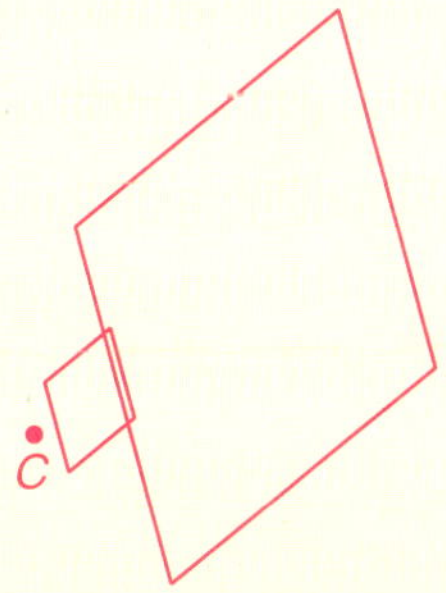

5.

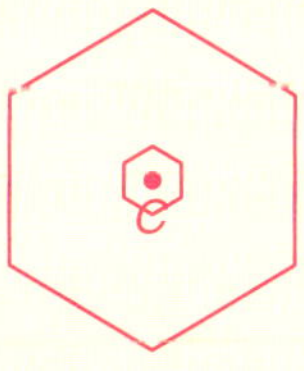

6.

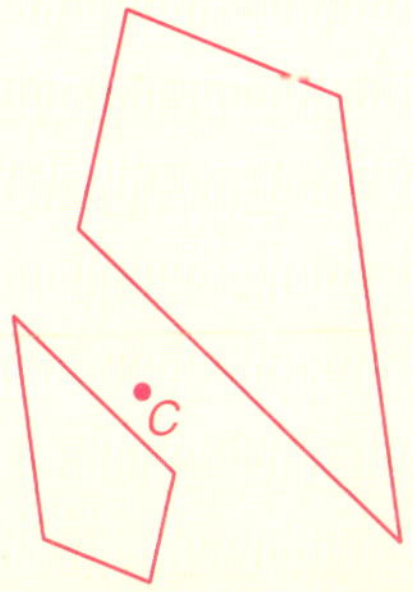

About the Exercises...

Organization by Objective
- **Classify Dilations:** 14–29
- **Identify the Scale Factor:** 30–40

Odd/Even Assignments
Exercises 14–43 are structured so that students practice the same concepts whether they are assigned odd or even problems.

Assignment Guide
Basic: 15–37 odd, 41, 45, 47, 49, 52–66

Average: 15–51 odd, 52–66

Advanced: 14–50 even, 52–63 (optional: 64–66)

All: Quiz 2 (1–5)

Answers
9.

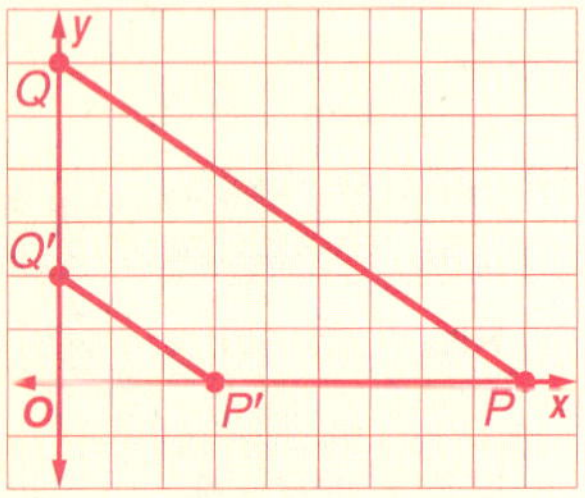

10.

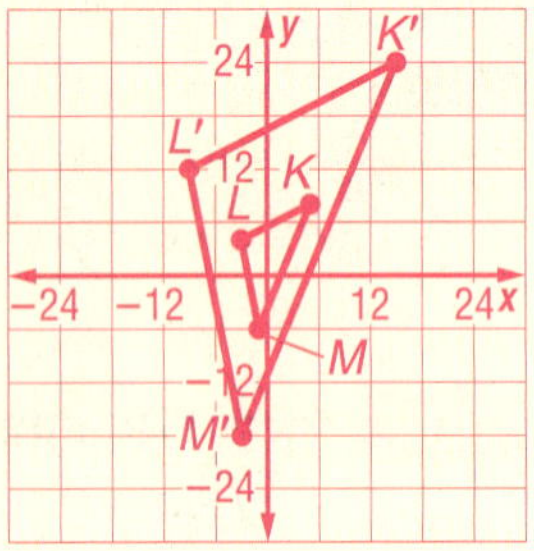

14.

15.

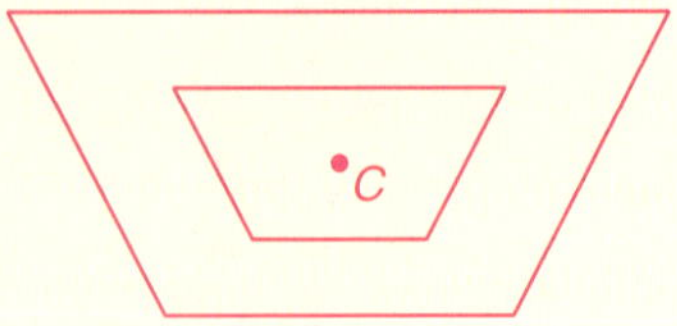

16.

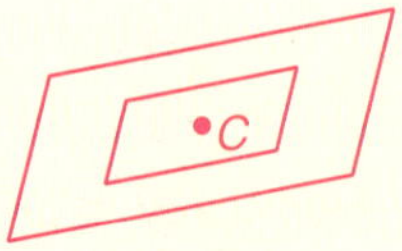

17.

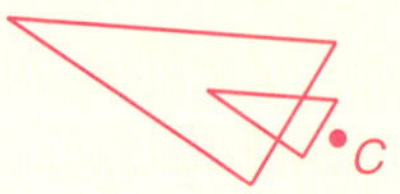

18.

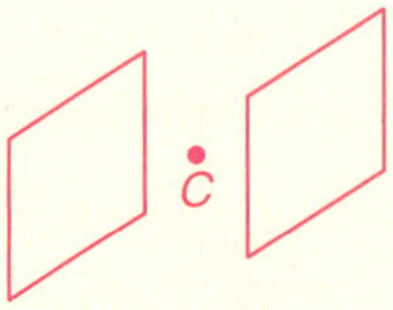

19.

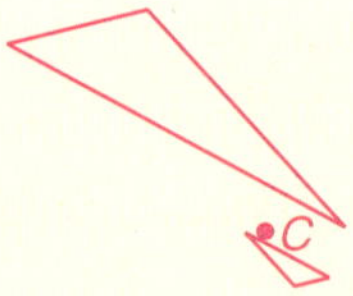

Find the measure of the dilation image $\overline{A'B'}$ or the preimage $\overline{AB}$ using the given scale factor.

7. $AB = 3, r = 4$ **$A'B' = 12$**

8. $A'B' = 8, r = -\frac{2}{5}$ **$AB = 20$**

9–10. See margin.

9. $\overline{PQ}$ has endpoints $P(9, 0)$ and $Q(0, 6)$. Find the image of $\overline{PQ}$ after a dilation centered at the origin with a scale factor $r = \frac{1}{3}$. Sketch the preimage and the image.

10. Triangle KLM has vertices $K(5, 8)$, $L(-3, 4)$, and $M(-1, -6)$. Find the image of $\triangle KLM$ after a dilation centered at the origin with scale factor of 3. Sketch the preimage and the image.

Determine the scale factor for each dilation with center C. Determine whether the dilation is an *enlargement*, *reduction*, or *congruence transformation*.

11.

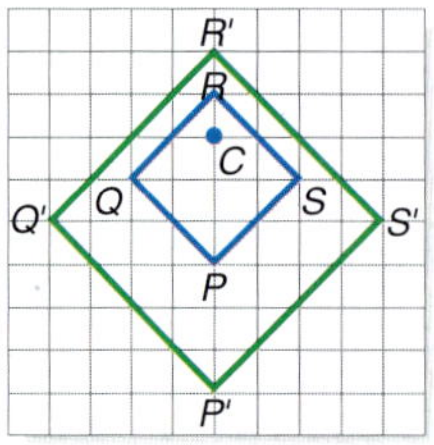

$r = 2$; enlargement

12.

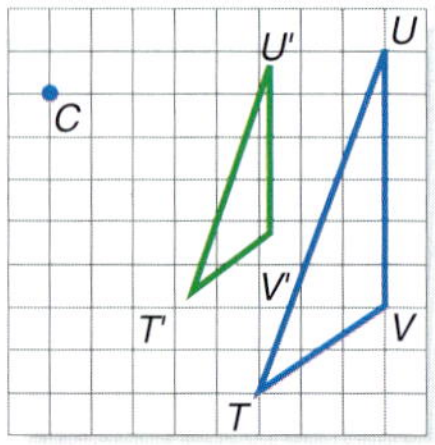

$r = \frac{2}{3}$; reduction

FCAT Practice
Standardized Test Practice

13. Alexis made a scale drawing of the plan for her spring garden. It will be a rectangle measuring 18 feet by 12 feet. On the scaled version, it measures 8 inches on the longer sides. What is the measure of each of the shorter sides? **C**

Ⓐ $6\frac{1}{2}$ in. Ⓑ $5\frac{1}{2}$ in. Ⓒ $5\frac{1}{3}$ in. Ⓓ $4\frac{3}{5}$ in.

★ indicates increased difficulty

Practice and Apply

Homework Help

For Exercises	See Examples
14–19	1
20–25	2
26–29	3
32–37	4
30–31, 38–40	5

Extra Practice
See page 772.

14–19. See margin.

Draw the dilation image of each figure with center C and the given scale factor.

14. $r = 3$

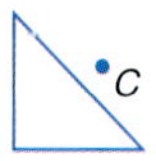

15. $r = 2$

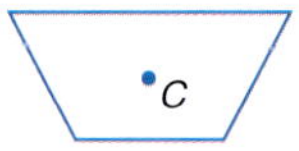

16. $r = \frac{1}{2}$

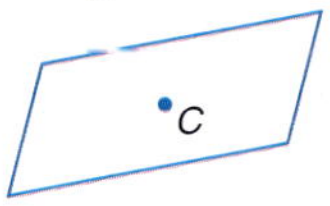

17. $r = \frac{2}{5}$

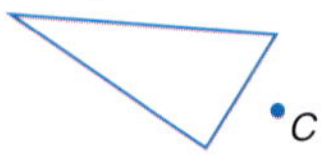

18. $r = -1$

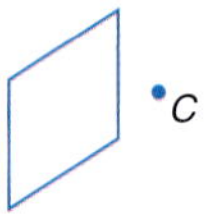

19. $r = -\frac{1}{4}$

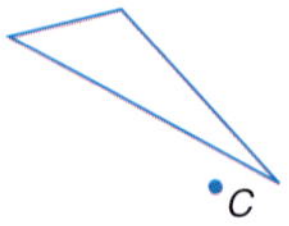

Find the measure of the dilation image $\overline{S'T'}$ or the preimage $\overline{ST}$ using the given scale factor.

20. $ST = 6, r = -1$ **$S'T' = 6$**

21. $ST = \frac{4}{5}, r = \frac{3}{4}$ **$S'T' = \frac{3}{5}$**

22. $S'T' = 12, r = \frac{2}{3}$ **$ST = 18$**

23. $S'T' = \frac{12}{5}, r = -\frac{3}{5}$ **$ST = 4$**

24. $ST = 32, r = -\frac{5}{4}$ **$S'T' = 40$**

25. $ST = 2.25, r = 0.4$ **$S'T' = 0.9$**

COORDINATE GEOMETRY Find the image of each polygon, given the vertices, after a dilation centered at the origin with a scale factor of 2. Then graph a dilation centered at the origin with a scale factor of $\frac{1}{2}$. **26–29. See margin.**

26. $F(3, 4)$, $G(6, 10)$, $H(-3, 5)$
27. $X(1, -2)$, $Y(4, -3)$, $Z(6, -1)$
28. $P(1, 2)$, $Q(3, 3)$, $R(3, 5)$, $S(1, 4)$
29. $K(4, 2)$, $L(-4, 6)$, $M(-6, -8)$, $N(6, -10)$

Determine the scale factor for each dilation with center *C*. Determine whether the dilation is an *enlargement*, *reduction*, or *congruence transformation*.

30. 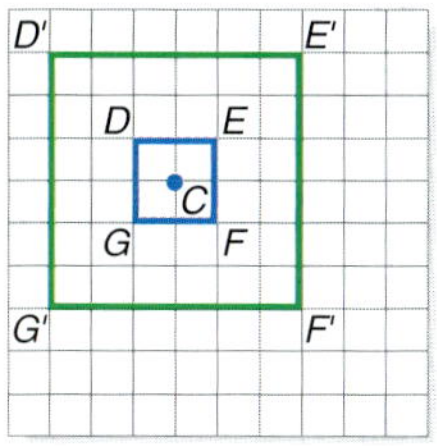

3; enlargement

31. 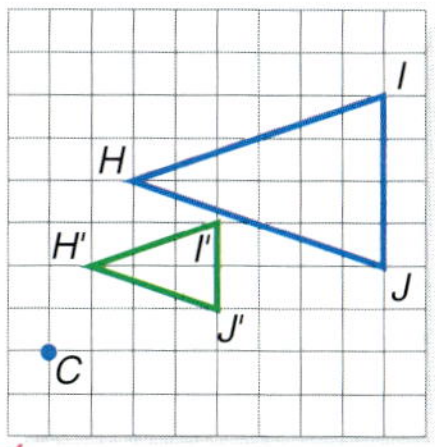

$\frac{1}{2}$; reduction

32. 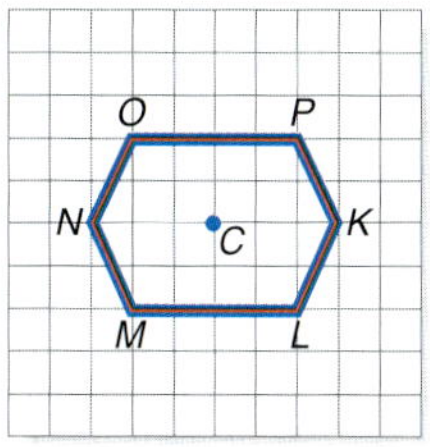

1; congruence transformation

33. 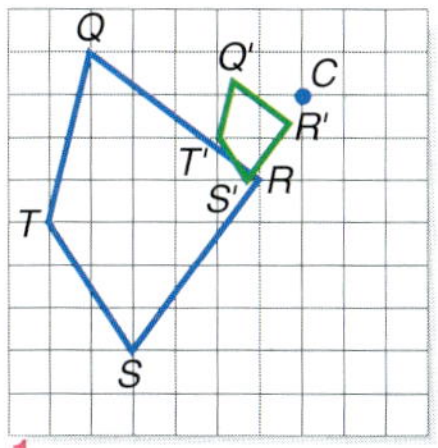

$\frac{1}{3}$; reduction

34. 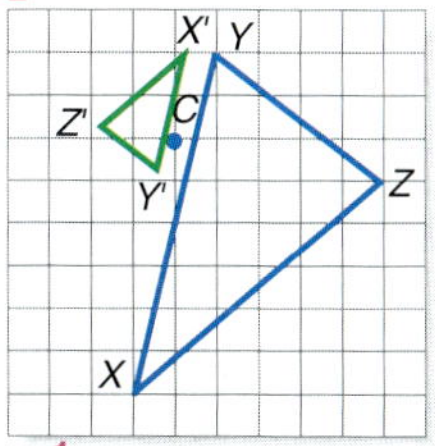

$-\frac{1}{4}$; reduction

35. 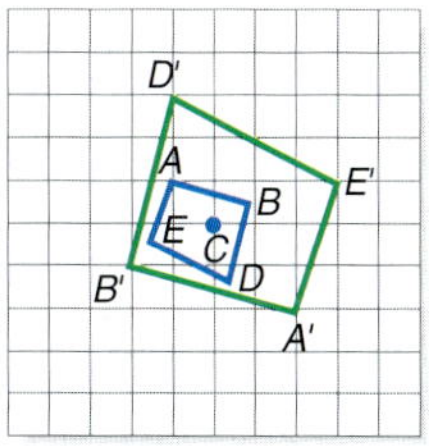

−2; enlargement

More About. . .

Airplanes

The SR-71 Blackbird is 107 feet 5 inches long with a wingspan of 55 feet 7 inches and can fly at speeds over 2200 miles per hour. It can fly nonstop from Los Angeles to Washington, D.C., in just over an hour, while a standard commercial jet takes about five hours to complete the trip.

Source: NASA

36. **AIRPLANES** Etay is building a model of the SR-71 Blackbird. If the wingspan of his model is 14 inches, what is the approximate scale factor of the model? **about $\frac{1}{48}$**

PHOTOCOPY **For Exercises 37 and 38, refer to the following information.**
A 10-inch by 14-inch rectangular design is being reduced on a photocopier by a factor of 75%.

37. What are the new dimensions of the design? **7.5 in. by 10.5 in.**
38. How has the area of the preimage changed? **It is $\frac{9}{16}$ of the original.**

For Exercises 39 and 40, use the following information.
A dilation on a rectangle has a scale factor of 4.

★ 39. What is the effect of the dilation on the perimeter of the rectangle? **39. The perimeter is four times the original perimeter.**
★ 40. What is the effect of the dilation on the area of the rectangle? **The area is 16 times the original area.**

41. **PROOF** Write a paragraph proof of Theorem 9.1. **See margin.**

★ 42. Triangle *ABC* has vertices $A(12, 4)$, $B(4, 8)$, and $C(8, -8)$. After two successive dilations centered at the origin with the same scale factor, the final image has vertices $A''(3, 1)$, $B''(1, 2)$, and $C''(2, -2)$. Determine the scale factor *r* of each dilation from $\triangle ABC$ to $\triangle A''B''C''$. **$\frac{1}{2}$**

★ 43. Segment *XY* has endpoints $X(4, 2)$ and $Y(0, 5)$. After a dilation, the image has endpoints of $X'(7, 17)$ and $Y'(15, 11)$. Find the absolute value of the scale factor. **2**

Answers

26. 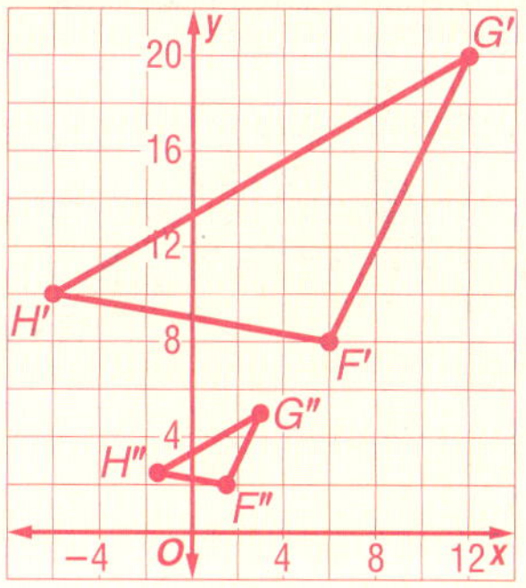

27. 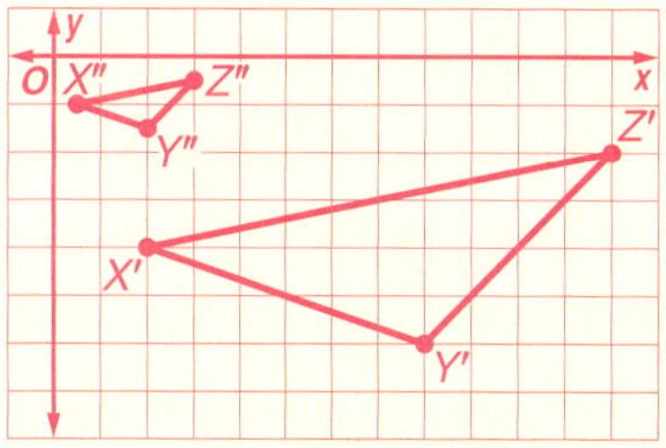

28. 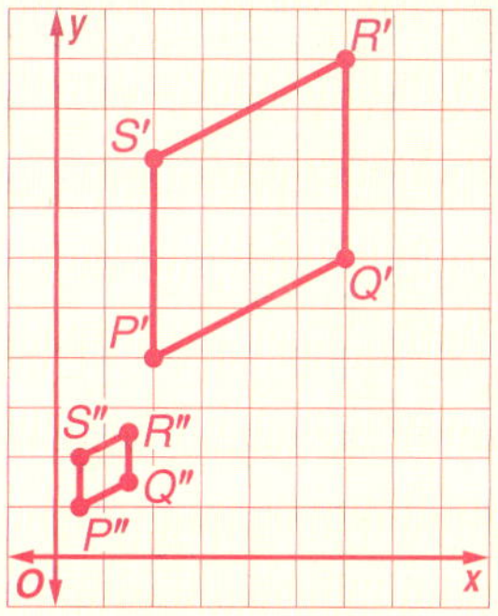

29. 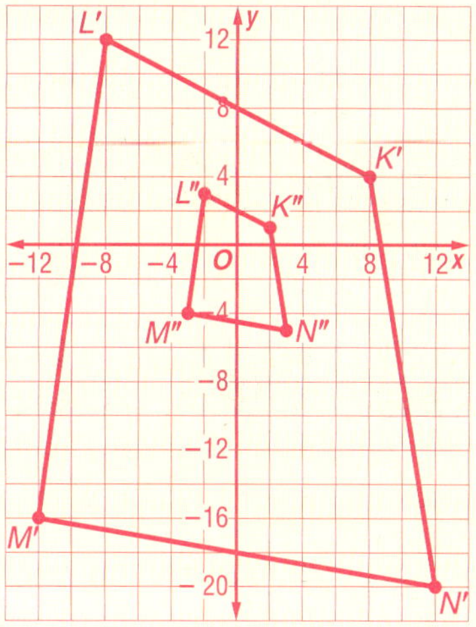

41. Given: dilation with center *C* and scale factor *r*
Prove: $ED = r(AB)$

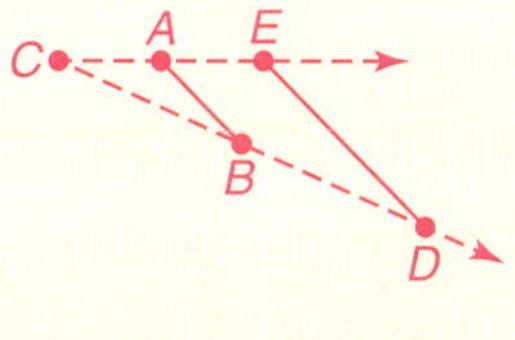

Proof: $CE = r(CA)$ and $CD = r(CB)$ by the definition of a dilation. $\frac{CE}{CA} = r$ and $\frac{CD}{CB} = r$. So, $\frac{CE}{CA} = \frac{CD}{CB}$ by substitution. $\angle ACB \cong \angle ECD$, since congruence of angles is reflexive. Therefore, by SAS Similarity, $\triangle ACB$ is similar to $\triangle ECD$. The corresponding sides of similar triangles are proportional, so $\frac{ED}{AB} = \frac{CE}{CA}$. We know that $\frac{CE}{CA} = r$, so $\frac{ED}{AB} = r$ by substitution. Therefore, $ED = r(AB)$ by the Multiplication Property of Equality.

DIGITAL PHOTOGRAPHY **For Exercises 44–46, use the following information.**
Dinah is editing a digital photograph that is 640 pixels wide and 480 pixels high on her monitor.

44. If Dinah zooms the image on her monitor 150%, what are the dimensions of the image? **960 pixels by 720 pixels**

45. $\frac{1}{20}$

45. Suppose that Dinah wishes to use the photograph on a web page and wants the image to be 32 pixels wide. What scale factor should she use to reduce the image?

46. Dinah resizes the photograph so that it is 600 pixels high. What scale factor did she use? $\frac{5}{4}$

47. **DESKTOP PUBLISHING** Grace is creating a template for her class newsletter. She has a photograph that is 10 centimeters by 12 centimeters, but the maximum space available for the photograph is 6 centimeters by 8 centimeters. She wants the photograph to be as large as possible on the page. When she uses a scanner to save the photograph, at what percent of the original photograph's size should she save the image file? **60%**

For Exercises 48–50, use quadrilateral *ABCD*.

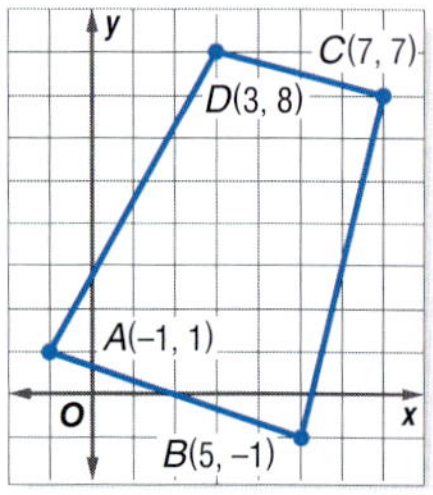

48. about 26.8 units
49. See p. 519C.
50. About 53.5 units; the perimeter of quadrilateral $A'B'C'D'$ is twice the perimeter of quadrilateral $ABCD$.

48. Find the perimeter of quadrilateral $ABCD$.

49. Graph the image of quadrilateral $ABCD$ after a dilation centered at the origin with scale factor -2.

50. Find the perimeter of quadrilateral $A'B'C'D'$ and compare it to the perimeter of quadrilateral $ABCD$.

★ 51. Triangle TUV has vertices $T(6, -5)$, $U(3, -8)$, and $V(-1, -2)$. Find the coordinates of the final image of triangle TUV after a reflection in the x-axis, a translation with $(x, y) \rightarrow (x + 4, y - 1)$, and a dilation centered at the origin with a scale factor of $\frac{1}{3}$. Sketch the preimage and the image. See p. 519C.

52. **CRITICAL THINKING** In order to perform a dilation not centered at the origin, you must first translate all of the points so the center is the origin, dilate the figure, and translate the points back. Consider a triangle with vertices $G(3, 5)$, $H(7, -4)$, and $I(-1, 0)$. State the coordinates of the vertices of the image after a dilation centered at (3, 5) with a scale factor of 2. $G'(3, 5)$, $H'(11, -13)$, $I'(-5, -5)$

53. **WRITING IN MATH** Answer the question that was posed at the beginning of the lesson. See p. 519C.

How do you use dilations when you use a computer?

Include the following in your answer:
- how a "cut and paste" in word processing may be an example of a dilation, and
- other examples of dilations when using a computer.

FCAT Practice

Standardized Test Practice

54. The figure shows two regular pentagons. Find the perimeter of the larger pentagon. **B**

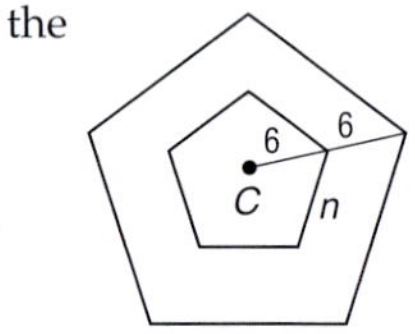

Ⓐ $5n$ Ⓑ $10n$

Ⓒ $15n$ Ⓓ $60n$

55. **ALGEBRA** What is the slope of a line perpendicular to the line given by the equation $3x + 5y = 12$? **A**

Ⓐ $\frac{5}{3}$ Ⓑ $\frac{3}{5}$ Ⓒ $-\frac{3}{5}$ Ⓓ $-\frac{5}{3}$

Study Guide and Intervention, p. 503 (shown) and p. 504

Classify Dilations A **dilation** is a transformation in which the image may be a different size than the preimage. A dilation requires a center point and a scale factor, r.

Let r represent the scale factor of a dilation.
If $|r| > 1$, then the dilation is an enlargement.
If $|r| = 1$, then the dilation is a congruence transformation.
If $0 < |r| < 1$, then the dilation is a reduction.

Example **Draw the dilation image of $\triangle ABC$ with center O and $r = 2$.**
Draw $\overrightarrow{OA}$, $\overrightarrow{OB}$, and $\overrightarrow{OC}$. Label points A', B', and C' so that $OA' = 2(OA)$, $OB' = 2(OB)$, and $OC' = 2(OC)$. $\triangle A'B'C'$ is a dilation of $\triangle ABC$.

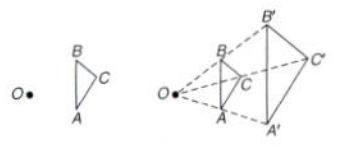

Exercises

Draw the dilation image of each figure with center C and the given scale factor. Describe each transformation as an *enlargement*, *congruence*, or *reduction*.

1. $r = 2$ enlargement

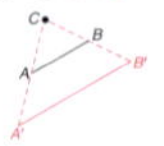

2. $r = \frac{1}{2}$ reduction

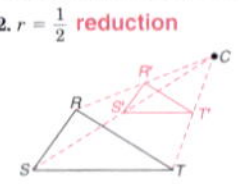

3. $r = 1$ congruence

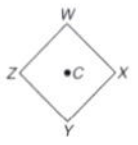

4. $r = 3$ enlargement

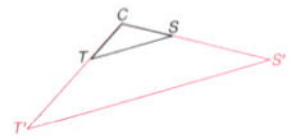

5. $r = \frac{2}{3}$ reduction

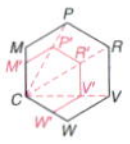

6. $r = 1$ congruence

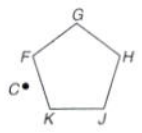

Skills Practice, p. 505 and Practice, p. 506 (shown)

Draw the dilation image of each figure with center C and the given scale factor.

1. $r = \frac{3}{2}$

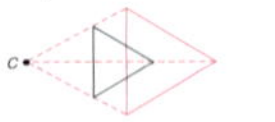

2. $r = \frac{2}{3}$

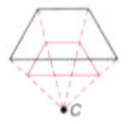

Find the measure of the dilation image $\overline{A'T'}$ or of the preimage $\overline{AT}$ using the given scale factor.

3. $AT = 15$, $r = \frac{3}{5}$ $A'T' = 9$

4. $AT = 30$, $r = -\frac{1}{6}$ $A'T' = 5$

5. $A'T' = 12$, $r = \frac{4}{3}$ $AT = 9$

COORDINATE GEOMETRY **Find the image of each polygon, given the vertices, after a dilation centered at the origin with a scale factor of 2. Then graph a dilation centered at the origin with a scale factor of $\frac{1}{2}$.**

6. $A(1, 1)$, $C(2, 3)$, $D(4, 2)$, $E(3, 1)$

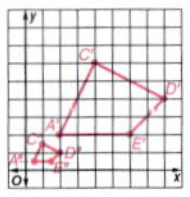

7. $Q(-1, -1)$, $R(0, 2)$, $S(2, 1)$

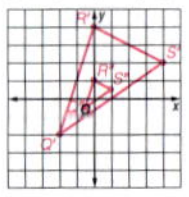

Determine the scale factor for each dilation with center C. Determine whether the dilation is an *enlargement*, *reduction*, or *congruence transformation*. The dotted figure is the dilation image.

8. 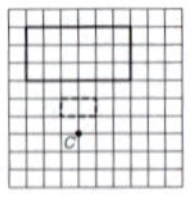 $\frac{1}{3}$; reduction

9. 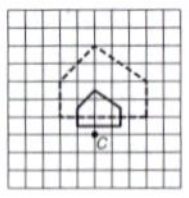 2; enlargement

10. PHOTOGRAPHY Estebe enlarged a 4-inch by 6-inch photograph by a factor of $\frac{5}{2}$. What are the new dimensions of the photograph? 10 in. by 15 in.

Reading to Learn Mathematics, p. 507 ELL

Pre-Activity **How do you use dilations when you use a computer?**

Read the introduction to Lesson 9-5 at the top of page 490 in your textbook.

In addition to the example given in your textbook, give two everyday examples of scaling an object, one that makes the object larger and another that makes it smaller. Sample answer: larger: enlarging a photograph; smaller: making a scale model of a building

Reading the Lesson

1. Each of the values of r given below represents the scale factor for a dilation. In each case, determine whether the dilation is an *enlargement*, a *reduction*, or a *congruence transformation*.
   a. $r = 3$ enlargement
   b. $r = 0.5$ reduction
   c. $r = -0.75$ reduction
   d. $r = -1$ congruence transformation
   e. $r = \frac{2}{3}$ reduction
   f. $r = -\frac{3}{2}$ enlargement
   g. $r = -1.01$ enlargement
   h. $r = 0.999$ reduction

2. Determine whether each sentence is *always*, *sometimes*, or *never* true. If the sentence is not always true, explain why. For explanations, sample answers are given.
   a. A dilation requires a center point and a scale factor. always
   b. A dilation changes the size of a figure. Sometimes; if the dilation is a congruence transformation, the size of the figure is unchanged.
   c. A dilation changes the shape of a figure. Never; all dilations produce similar figures, and similar figures have the same shape.
   d. The image of a figure under a dilation lies on the opposite side of the center from the preimage. Sometimes; this is only true when the scale factor is negative.
   e. A similarity transformation is a congruence transformation. Sometimes; this is true only when the scale factor is 1 or −1.
   f. The center of a dilation is its own image. always
   g. A dilation is an isometry. Sometimes; this is true only when the dilation is a congruence transformation.
   h. The scale factor for a dilation is a positive number. Sometimes; the scale factor can be any positive or negative number.
   i. Dilations produce similar figures. always

Helping You Remember

3. A good way to remember something is to explain it to someone else. Suppose that your classmate Lydia is having trouble understanding the relationship between *similarity transformations* and *congruence transformations*. How can you explain this to her? Sample answer: A congruence transformation preserves the size and shape of a figure, that is, the image is congruent to the preimage. A similarity transformation preserves the shape of a figure, that is, the image is similar to the preimage. A similarity transformation is a congruence transformation if the scale factor is 1 or −1.

Enrichment, p. 508

Similar Circles

You may be surprised to learn that two noncongruent circles that lie in the same plane and have no common interior points can be mapped one onto the other by more than one dilation.

1. Here is diagram that suggests one way to map a smaller circle onto a larger one using a dilation. The circles are given. The lines suggest how to find the center for the dilation. Describe how the center is found. Use segments in the diagram to name the scale factor.

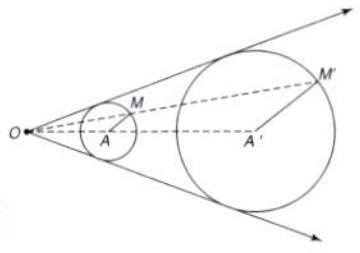

Draw the two common external tangents. The point where they intersect is the center of a dilation that maps ⊙A onto ⊙A'; scale factor $= \frac{A'M'}{AM}$.

496 Chapter 9 Transformations

Maintain Your Skills

Mixed Review

Determine whether a semi-regular tessellation can be created from each figure. Assume that each figure is regular and has a side length of 1 unit. *(Lesson 9-4)*

56. a triangle and a pentagon **no**

57. an octagon and a hexagon **no**

58. a square and a triangle **yes**

59. a hexagon and a dodecagon **no**

COORDINATE GEOMETRY **Draw the rotation image of each figure 90° in the given direction about the center point and label the vertices.** *(Lesson 9-3)*

60–61. See p. 519D.

60. $\triangle ABC$ with $A(7, -1)$, $B(5, 0)$, and $C(1, 6)$ counterclockwise about $P(-1, 4)$

61. $\square DEFG$ with $D(-4, -2)$, $E(-3, 3)$, $F(3, 1)$, and $G(2, -4)$ clockwise about $P(-4, -6)$

62. Yes; if the diagonals are congruent, it is a rectangle.

62. CONSTRUCTION The Vanamans are building an addition to their house. Ms. Vanaman is cutting an opening for a new window. If she measures to see that the opposite sides are the same length and that the diagonal measures are the same, can Ms. Vanaman be sure that the window opening is rectangular? Explain. *(Lesson 8-4)*

63. Given: $\angle J \cong \angle L$ **See margin.**

B is the midpoint of $\overline{JL}$.

Prove: $\triangle JHB \cong \triangle LCB$

(Lesson 4-4)

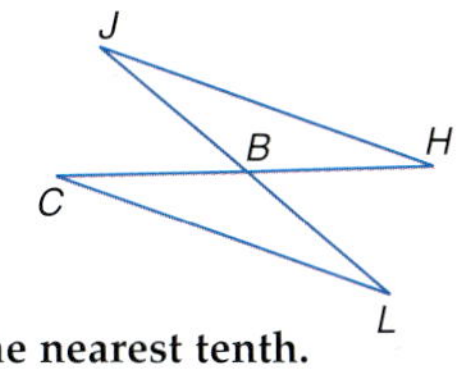

Getting Ready for the Next Lesson

PREREQUISITE SKILL **Find $m\angle A$ to the nearest tenth.**

(To review ***finding angles using inverses of trigonometric ratios****, see Lesson 7-3.)*

64. **33.7**

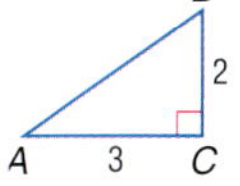

65. **76.0**

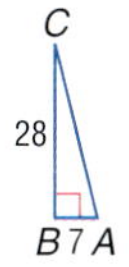

66. **51.3**

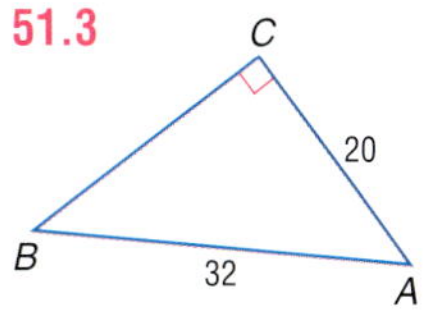

Practice Quiz 2

Lessons 9-4 and 9-5

Determine whether each pattern is a tessellation. If so, describe it as ***uniform, regular, semi-regular,*** **or** ***not uniform.*** *(Lesson 9-4)*

1. **yes; uniform; semi-regular**

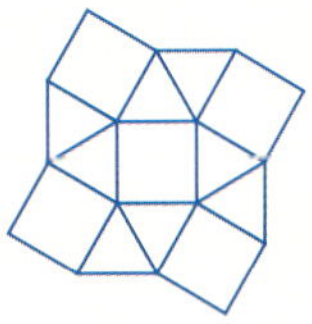

2. **yes; uniform**

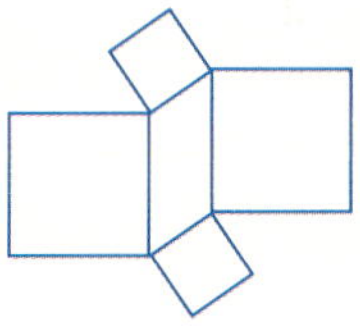

Draw the dilation image of each figure with center C and given scale factor. *(Lesson 9-5)* **3–4. See margin.**

3. $r = \frac{3}{4}$

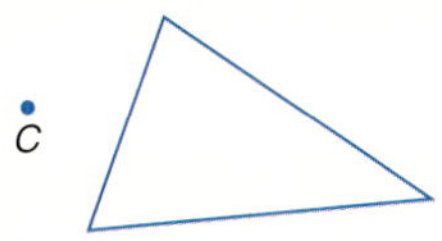

4. $r = -2$

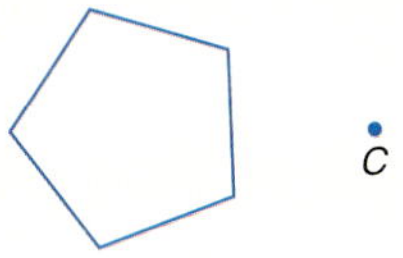

5. $A'(-5, -1)$, $B'\left(-\frac{1}{2}, -3\right)$, $C'(2, -2)$; **See margin for graph.**

5. Triangle ABC has vertices $A(10, 2)$, $B(1, 6)$, and $C(-4, 4)$. Find the image of $\triangle ABC$ after a dilation centered at the origin with scale factor of $-\frac{1}{2}$. Sketch the preimage and the image. *(Lesson 9-5)*

www.geometryonline.com/self_check_quiz/fcat

4 Assess

Open-Ended Assessment

Modeling Students can measure an object in the room and apply a dilation to reduce the object and make a model with clay, poster board, and/or construction paper. Examples could be the chalkboard, a wall clock, a particular shape or display on a bulletin board.

Getting Ready for Lesson 9-6

Prerequisite Skill Students will learn about vectors in Lesson 9-6. They will use trigonometric ratios and their inverses to find the direction of vectors. Use Exercises 64–66 to determine your students' familiarity with finding angles using inverses of trigonometric ratios.

Assessment Options

Practice Quiz 2 The quiz provides students with a brief review of the concepts and skills in Lessons 9-4 and 9-5. Lesson numbers are given to the right of the exercises or instruction lines so students can review concepts not yet mastered.

Answer

63. Given: $\angle J \cong \angle L$

B is the midpoint of $\overline{JL}$.

Prove: $\triangle JHB \cong \triangle LCB$

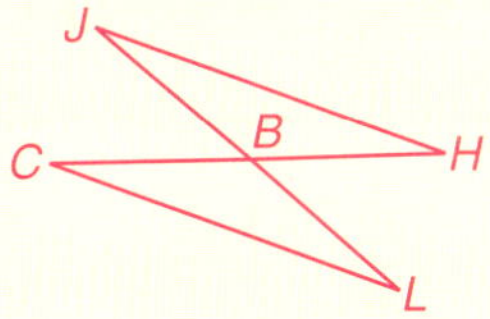

Proof: It is known that $\angle J \cong \angle L$. Since B is the midpoint of $\overline{JL}$, $\overline{JB} \cong \overline{LB}$ by the Midpoint Theorem. $\angle JBH \cong \angle LBC$ because vertical angles are congruent. Thus, $\triangle JHB \cong \triangle LCB$ by ASA.

Answers (Practice Quiz 2)

3.

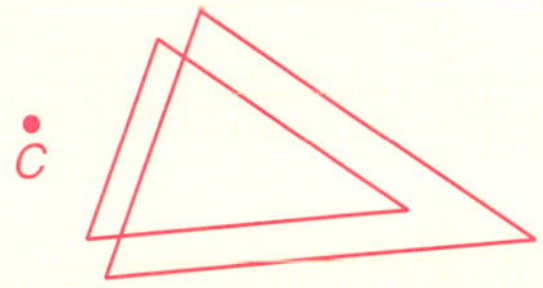

4.

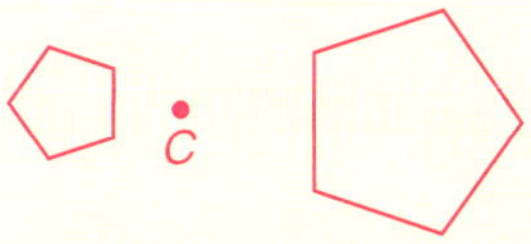

5.

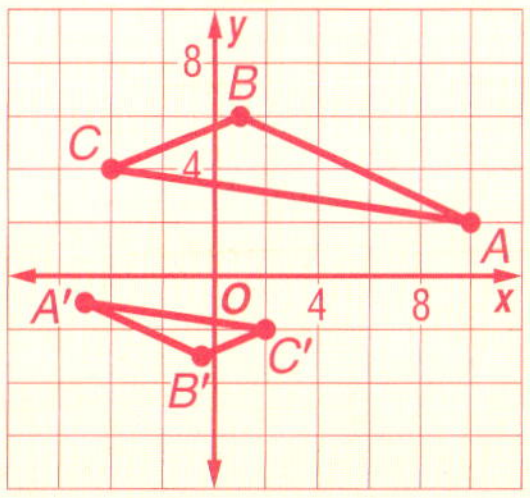

9-6 Lesson Notes

1 Focus

5-Minute Check Transparency 9-6 Use as a quiz or review of Lesson 9-5.

Mathematical Background notes are available for this lesson on p. 460D.

How do vectors help a pilot plan a flight?

Ask students:

- Why do you think commercial pilots have to submit flight plans? **Sample answer: Because the safety of the passengers and the crew depends on mathematically correct flight plans.**
- How might the speed and direction of strong winds directed toward a plane affect the speed and direction of the plane? **Strong winds directed toward the plane might slow the plane or push it in a direction other than its intended course.**

9-6 Vectors

Sunshine State Standards
MA.A.3.4.3, MA.B.3.4.1, MA.C.3.4.1, MA.C.3.4.2

Vocabulary

- vector
- magnitude
- direction
- standard position
- component form
- equal vectors
- parallel vectors
- resultant
- scalar
- scalar multiplication

What You'll Learn

- Find magnitudes and directions of vectors.
- Perform translations with vectors.

How do vectors help a pilot plan a flight?

Commercial pilots must submit flight plans prior to departure. These flight plans take into account the speed and direction of the plane as well as the speed and direction of the wind.

MAGNITUDE AND DIRECTION The speed and direction of a plane and the wind can be represented by vectors.

Key Concept — Vectors

- **Words** A **vector** is a quantity that has both **magnitude**, or length, and **direction**, and is represented by a directed segment.
- **Symbols** $\vec{v}$
  $\overrightarrow{AB}$, where A is the initial point and B is the endpoint

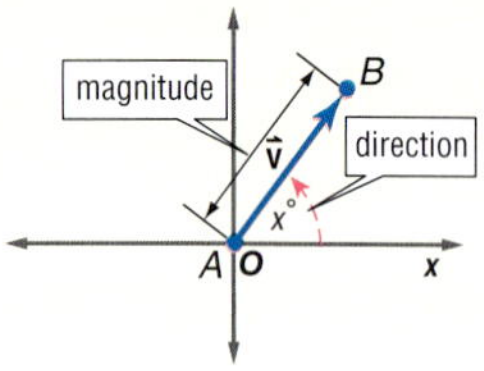

A vector in **standard position** has its initial point at the origin. In the diagram, $\overrightarrow{CD}$ is in standard position and can be represented by the ordered pair $\langle 4, 2\rangle$.

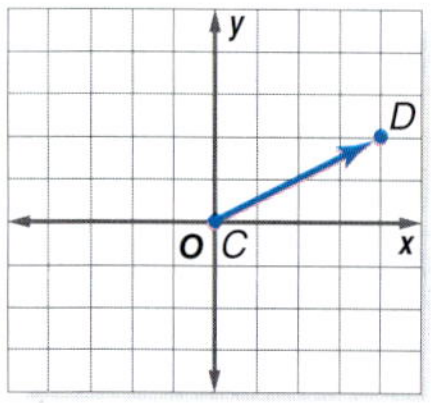

A vector can also be drawn anywhere in the coordinate plane. To write such a vector as an ordered pair, find the change in the x values and the change in y values, $\langle$change in x, change in $y\rangle$, from the tip to the tail of the directed segment. The ordered pair representation of a vector is called the **component form** of the vector.

Study Tip

Common Misconception
The notation for a vector from C to D, $\overrightarrow{CD}$, is similar to the notation for a ray from C to D, $\overrightarrow{CD}$. Be sure to use the correct arrow above the letters when writing each.

Example 1 Write Vectors in Component Form

Write the component form of $\overrightarrow{EF}$.

Find the change in x-values and the corresponding change in y-values.

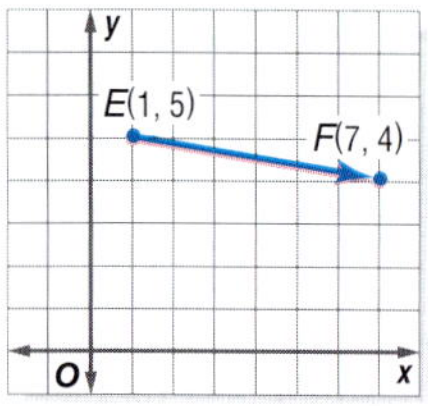

$\overrightarrow{EF} = \langle x_2 - x_1, y_2 - y_1\rangle$ Component form of vector

$= \langle 7 - 1, 4 - 5\rangle$ $x_1 = 1, y_1 = 5, x_2 = 7, y_2 = 4$

$= \langle 6, -1\rangle$ Simplify.

Because the magnitude and direction of a vector are not changed by translation, the vector $\langle 6, -1\rangle$ represents the same vector as $\overrightarrow{EF}$.

Resource Manager

Workbook and Reproducible Masters

Chapter 9 Resource Masters
- Study Guide and Intervention, pp. 509–510
- Skills Practice, p. 511
- Practice, p. 512
- Reading to Learn Mathematics, p. 513
- Enrichment, p. 514
- Assessment, p. 536

School-to-Career Masters, p. 18
Prerequisite Skills Workbook, pp. 5–6
Teaching Geometry With Manipulatives Masters, pp. 1, 17, 156

Transparencies

5-Minute Check Transparency 9-6
Answer Key Transparencies

Technology

Interactive Chalkboard
Multimedia Applications: Virtual Activities

The Distance Formula can be used to find the magnitude of a vector. The symbol for the magnitude of $\overrightarrow{AB}$ is $|\overrightarrow{AB}|$. The *direction* of a vector is the measure of the angle that the vector forms with the positive x-axis or any other horizontal line. You can use the trigonometric ratios to find the direction of a vector.

Example 2 Magnitude and Direction of a Vector

Find the magnitude and direction of $\overrightarrow{PQ}$ for $P(3, 8)$ and $Q(-4, 2)$.

Find the magnitude.

$|\overrightarrow{PQ}| = \sqrt{(x_2 - x_1)^2 + (y_2 - y_1)^2}$ Distance Formula

$= \sqrt{(-4 - 3)^2 + (2 - 8)^2}$ $x_1 = 3, y_1 = 8, x_2 = -4, y_2 = 2$

$= \sqrt{85}$ Simplify.

≈ 9.2 Use a calculator.

Graph $\overrightarrow{PQ}$ to determine how to find the direction. Draw a right triangle that has $\overrightarrow{PQ}$ as its hypotenuse and an acute angle at P.

$\tan P = \frac{y_2 - y_1}{x_2 - x_1}$ $\tan = \frac{\text{length of opposite side}}{\text{length of adjacent side}}$

$= \frac{2 - 8}{-4 - 3}$ Substitution

$= \frac{6}{7}$ Simplify.

$m\angle P = \tan^{-1}\frac{6}{7}$

≈ 40.6 Use a calculator.

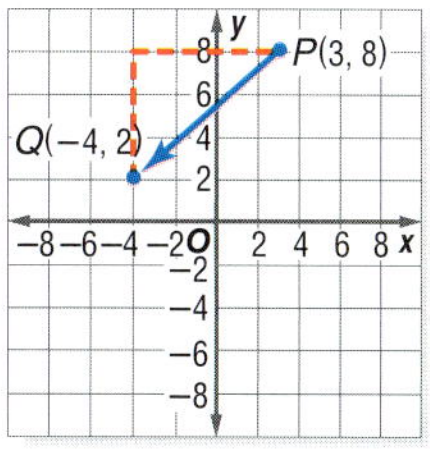

A vector in standard position that is equal to $\overrightarrow{PQ}$ lies in the third quadrant and forms an angle with the negative x-axis that has a measure equal to $m\angle P$. The x-axis is a straight angle whose measure is 180. So, the direction of $\overrightarrow{PQ}$ is $m\angle P + 180$ or about 220.6°.

Thus, $\overrightarrow{PQ}$ has a magnitude of about 9.2 units and a direction of about 220.6°.

In Example 1, we stated that a vector in standard position with magnitude of 9.2 units and a direction of 220.6° was equal to $\overrightarrow{PQ}$. This leads to a definition of equal vectors.

Key Concept

Equal Vectors Two vectors are equal if and only if they have the same magnitude and direction.

Example $\vec{v} = \vec{z}$

Nonexample $\vec{v} \neq \vec{u}$, $\vec{w} \neq \vec{y}$

Parallel Vectors Two vectors are parallel if and only if they have the same or opposite direction.

Example $\vec{v} \parallel \vec{w} \parallel \vec{y} \parallel \vec{z}$

Nonexample $\vec{v} \nparallel \vec{x}$

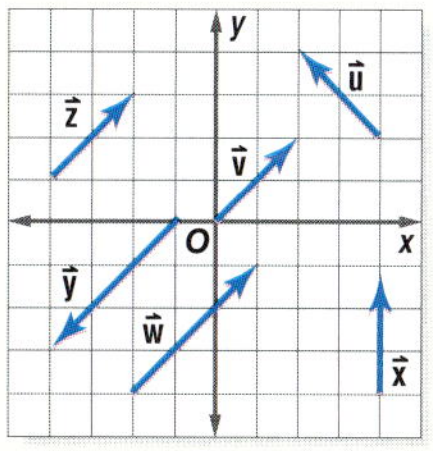

Study Tip

Using Slope Even though slope is not associated with vectors, you can use the concept of slope to determine if vectors are parallel before actually comparing their directions.

www.geometryonline.com/extra_examples/fcat

2 Teach

MAGNITUDE AND DIRECTION

In-Class Examples

Power Point®

1 Write the component form of $\overrightarrow{AB}$.

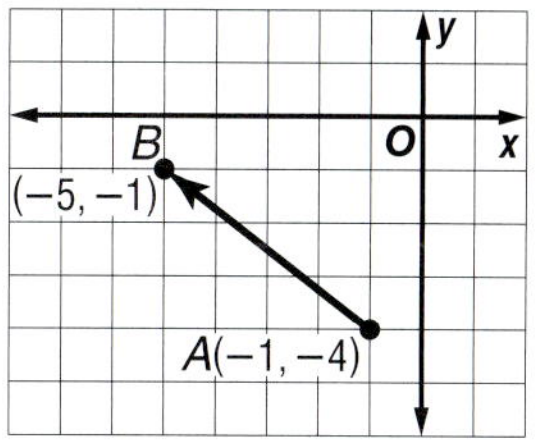

$\langle -4, 3 \rangle$

Teaching Tip When appropriate, students can also use the Pythagorean Theorem to find the magnitude of the resultant vector.

2 Find the magnitude and direction of $\overrightarrow{ST}$ for $S(-3, -2)$ and $T(4, -7)$.

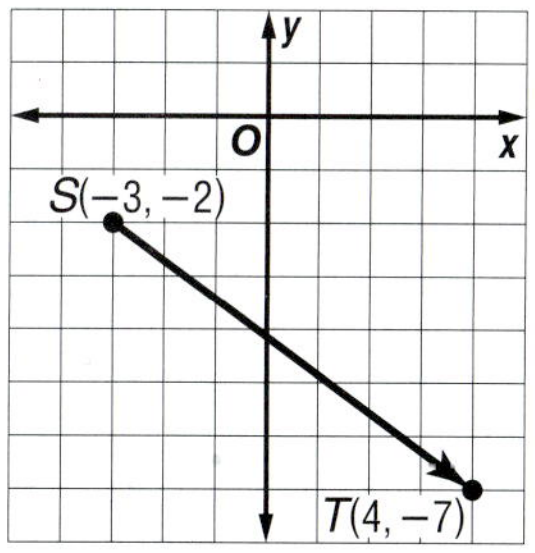

≈ 8.6; $\approx 324.5°$

DAILY INTERVENTION

Differentiated Instruction

Intrapersonal Ask students to read over the key concepts in this lesson and create and plot on one sheet of grid paper their own examples of equal vectors, parallel vectors, and a vector that is multiplied by a constant. Tell students to label the magnitude and direction of each vector in each example, label the vectors in component form, and highlight comparable values. Students can then create an example each of a translation with vectors and vector addition on another sheet of paper.

TRANSLATIONS WITH VECTORS

In-Class Examples

3 Graph the image of quadrilateral $HJLK$ with vertices $H(-4, 4)$, $J(-2, 4)$, $L(-1, 2)$, and $K(-3, 1)$ under the translation $\vec{v} = \langle 5, -5 \rangle$.

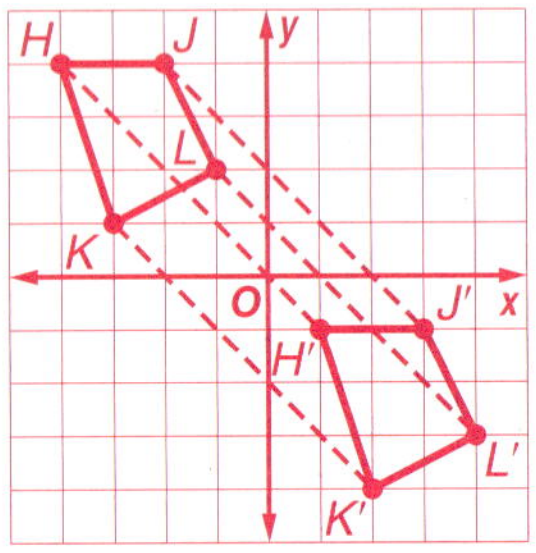

4 Graph the image of $\triangle EFG$ with vertices $E(1, -3)$, $F(3, -1)$, and $G(4, -4)$ under the translation by $\vec{a} = \langle -4, 2 \rangle$ and $\vec{b} = \langle 2, 3 \rangle$.

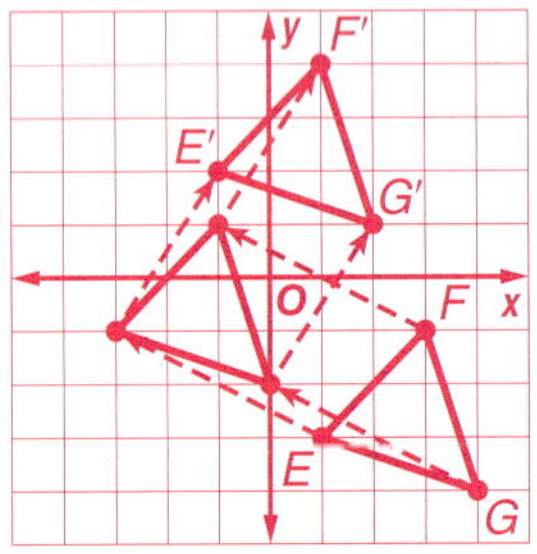

TRANSLATIONS WITH VECTORS Vectors can be used to describe translations.

Example 3 Translations with Vectors

Graph the image of $\triangle ABC$ with vertices $A(-3, -1)$, $B(-1, -2)$, and $C(-3, -3)$ under the translation $\vec{v} = \langle 4, 3 \rangle$.

First, graph $\triangle ABC$. Next, translate each vertex by $\vec{v}$, 4 units right and 3 units up. Connect the vertices to form $\triangle A'B'C'$.

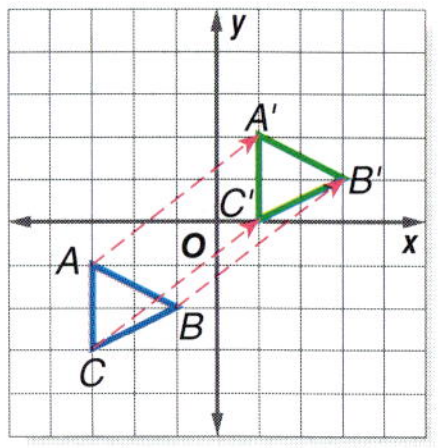

Vectors can be combined to perform a composition of translations by adding the vectors. To add vectors, add their corresponding components. The sum of two vectors is called the **resultant**.

Key Concept — Vector Addition

- **Words** To add two vectors, add the corresponding components.
- **Symbols** If $\vec{a} = \langle a_1, a_2 \rangle$ and $= \vec{b}\ \langle b_1, b_2 \rangle$, then $\vec{a} + \vec{b} = \langle a_1 + b_1, a_2 + b_2 \rangle$, and $\vec{b} + \vec{a} = \langle b_1 + a_1, b_2 + a_2 \rangle$.
- **Model**

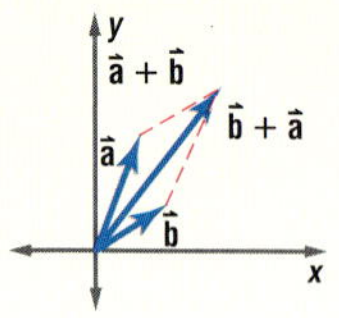

Example 4 Add Vectors

Graph the image of ▱$QRST$ with vertices $Q(-4, 4)$, $R(-1, 4)$, $S(-2, 2)$, and $T(-5, 2)$ under the translation $\vec{m} = \langle 5, -1 \rangle$ and $\vec{n} = \langle -2, -6 \rangle$.

Graph ▱$QRST$.

Method 1 Translate two times.
Translate ▱$QRST$ by $\vec{m}$. Then translate the image of ▱$QRST$ by $\vec{n}$.

Translate each vertex 5 units right and 1 unit down.

Then translate each vertex 2 units left and 6 units down.

Label the image ▱$Q'R'S'T'$.

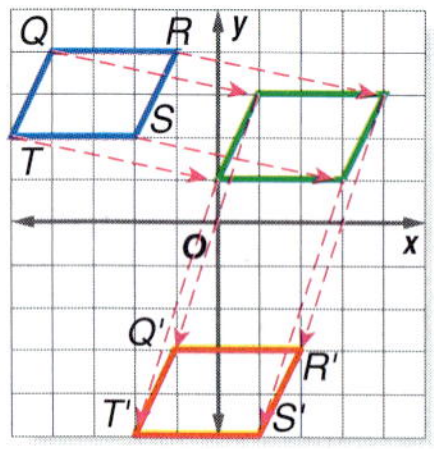

Method 2 Find the resultant, and then translate.
Add $\vec{m}$ and $\vec{n}$.

$\vec{m} + \vec{n} = \langle 5 - 2, -1 - 6 \rangle$

$= \langle 3, -7 \rangle$

Translate each vertex 3 units right and 7 units down.

Notice that the vertices for the image are the same for either method.

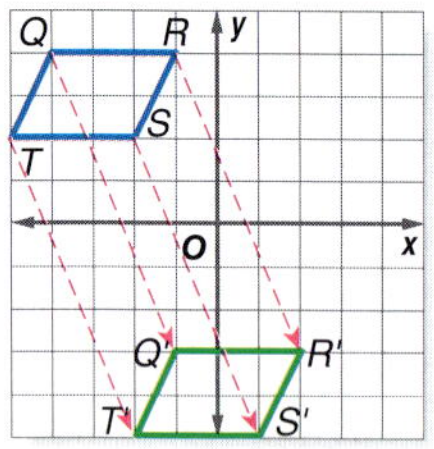

500 Chapter 9 Transformations

Geometry Activity

Comparing Magnitude and Components of Vectors

Model and Analyze

- Draw $\vec{a}$ in standard position.
- Draw $\vec{b}$ in standard position with the same direction as $\vec{a}$, but with a magnitude twice the magnitude of $\vec{a}$.

1. Write $\vec{a}$ and $\vec{b}$ in component form. See students' work.
2. What do you notice about the components of $\vec{a}$ and $\vec{b}$?
3. Draw $\vec{b}$ so that its magnitude is three times that of $\vec{a}$. How do the components of $\vec{a}$ and $\vec{b}$ compare? The components of $\vec{b}$ are three times the components of $\vec{a}$.

Make a Conjecture

4. Describe the vector magnitude and direction of a vector $\langle x, y\rangle$ after the components are multiplied by n.

. The components f $\vec{b}$ are twice the omponents of $\vec{a}$.

. Sample answer: he magnitude is n mes greater than the agnitude of $\langle x, y\rangle$, nd the direction is e same.

In the Geometry Activity, you found that a vector can be multiplied by a positive constant, called a **scalar**, that will change the magnitude of the vector, but not affect its direction. Multiplying a vector by a positive scalar is called **scalar multiplication**.

Key Concept — *Scalar Multiplication*

- **Words** To multiply a vector by a scalar multiply each component by the scalar.
- **Symbols** If $\vec{a} = \langle a_1, a_2\rangle$ has a magnitude $|\vec{a}|$ and direction d, then $n\vec{a} = n\langle a_1, a_2\rangle = \langle na_1, na_2\rangle$, where n is a positive real number, the magnitude is $|n\vec{a}|$, and its direction is d.
- **Model**

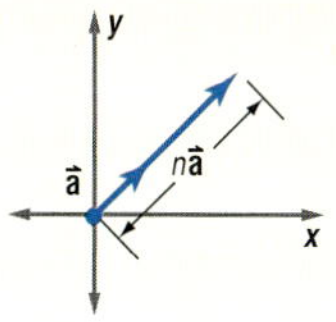

Example 5 Solve Problems Using Vectors

AVIATION Refer to the application at the beginning of the lesson.

a. Suppose a pilot begins a flight along a path due north flying at 250 miles per hour. If the wind is blowing due east at 20 miles per hour, what is the resultant velocity and direction of the plane?

The initial path of the plane is due north, so a vector representing the path lies on the positive y-axis 250 units long. The wind is blowing due east, so a vector representing the wind will be parallel to the positive x-axis 20 units long. The resultant path can be represented by a vector from the initial point of the vector representing the plane to the terminal point of the vector representing the wind.

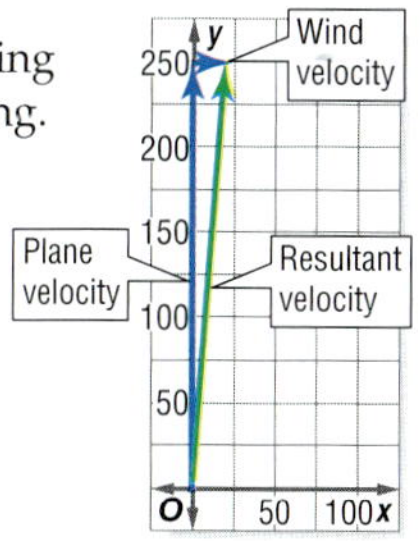

Use the Pythagorean Theorem.

$c^2 = a^2 + b^2$ — Pythagorean Theorem

$c^2 = 250^2 + 20^2$ — $a = 250, b = 20$

$c^2 = 62{,}900$ — Simplify.

$c = \sqrt{62{,}900}$ — Take the square root of each side.

$c \approx 250.8$

The resultant speed of the plane is about 250.8 miles per hour.

(continued on the next page)

More About. . .

Aviation

A tailwind will allow a lane to arrive faster than nticipated without the ailwind. A headwind will ause the plane to take nore time to travel than vithout the headwind.

Tips for New Teachers

At first, students may find it difficult to relate real-world values for velocity, path, and direction to vectors. They might also be confused about where to put these values in the tangent ratio. Before students work these types of problems, you may want to ask students how they would draw a vector to represent something moving due south, how they would draw a "force" vector that acts on the original vector and what the resultant vector means in terms of velocity and direction.

In-Class Example — Power Point®

5 CANOEING Suppose a person is canoeing due east across a river at 4 miles per hour.

a. If the river is flowing south at 3 miles an hour, what are the resultant direction and speed of the canoe? Resultant direction is about 36.9° south of due east; resultant speed is 5 mi/h.

b. If the current reduces to half of its original speed, what are the resultant direction and speed of the canoe? Resultant direction is about 20.6° south of due east; resultant speed is about 4.3 mi/h.

Geometry Activity

Materials: grid paper, pencil, straightedge

- Students can use a protractor to measure the direction of the vectors in the activity and note that this value stays the same throughout the activity.
- Repeat the activity for a specific direction in a different quadrant, and allow students to compare results.
- Ask students whether multiplying a vector by n changes its magnitude, direction, or both.

3 Practice/Apply

Study Notebook

Have students—

- add the definitions/examples of the vocabulary terms to their Vocabulary Builder worksheets for Chapter 9.
- include an example of solving a real-world problem using vectors.
- include any other item(s) that they find helpful in mastering the skills in this lesson.

About the Exercises...

Organization by Objective

- **Magnitude and Direction:** 15–36
- **Translations with Vectors:** 37–57

Odd/Even Assignments

Exercises 15–54 are structured so that students practice the same concepts whether they are assigned odd or even problems.

Assignment Guide

Basic: 15–55 odd, 58–78

Average: 15–57 odd, 58–78

Advanced: 16–58 even, 60–72 (optional: 73–78)

Study Tip

Reading Math

The Greek letter theta, θ, is used to represent the unknown measure of an angle.

Use the tangent ratio to find the direction of the plane.

$\tan\theta = \frac{20}{250}$ side opposite = 20, side adjacent = 250

$\theta = \tan^{-1}\frac{20}{250}$ Solve for θ.

$\theta \approx 4.6$ Use a calculator.

The resultant direction of the plane is about 4.6° east of due north. Therefore, the resultant vector is 250.8 miles per hour at 4.6° east of due north.

b. If the wind velocity doubles, what is the resultant path and velocity of the plane?

Use scalar multiplication to find the magnitude of the vector for wind velocity.

$n|\vec{a}| = 2|20|$ Magnitude of $n\vec{a}$; $n = 2$, $|\vec{a}| = 20$

$= 2(20)$ or 40 Simplify.

Next, use the Pythagorean Theorem to find the magnitude of the resultant vector.

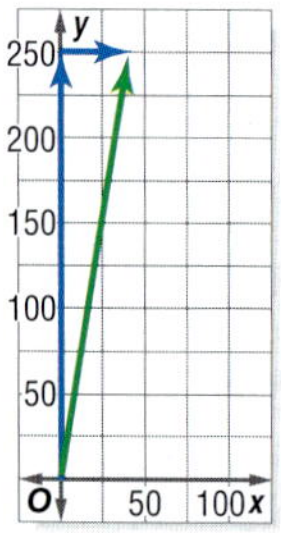

$c^2 = a^2 + b^2$ Pythagorean Theorem

$c^2 = 250^2 + 40^2$ $a = 250$, $b = 40$

$c^2 = 64{,}100$ Simplify.

$c = \sqrt{64{,}100}$ Take the square root of each side.

$c \approx 253.2$

Then, use the tangent ratio to find the direction of the plane.

$\tan\theta = \frac{40}{250}$ side opposite = 40, side adjacent = 250

$\theta = \tan^{-1}\frac{40}{250}$ Solve for θ.

$\theta \approx 9.1$ Use a calculator.

If the wind velocity doubles, the plane flies along a path approximately 9.1° east of due north at about 253.2 miles per hour.

Check for Understanding

Concept Check

1–3. See margin.

1. **OPEN ENDED** Draw a pair of vectors on a coordinate plane. Label each vector in component form and then find their sum.
2. **Compare and contrast** equal vectors and parallel vectors.
3. **Discuss** the similarity of using vectors to translate a figure and using an ordered pair.

Guided Practice

GUIDED PRACTICE KEY

Exercises	Examples
4, 5	1
6–8	2
9, 10	3
11	4
12–14	5

Write the component form of each vector.

4. 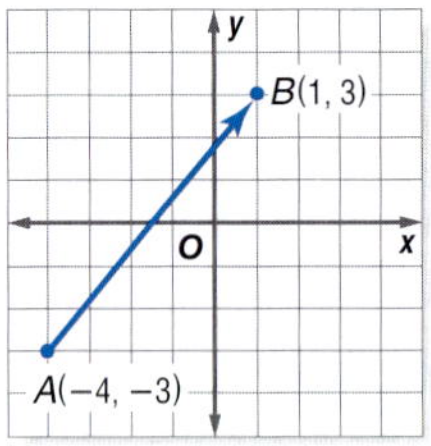

⟨5, 6⟩

5. 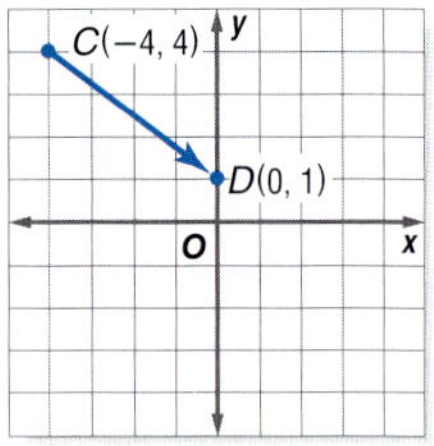

⟨4, −3⟩

Answers

1. Sample answer; ⟨7, 7⟩

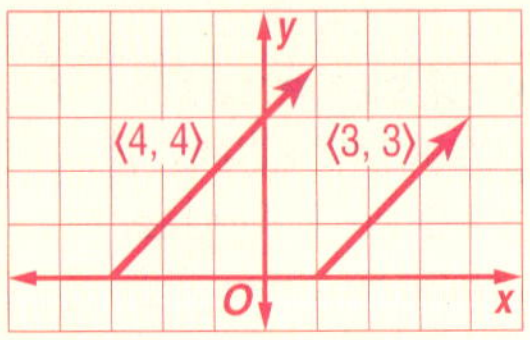

2. Two equal vectors must have the same magnitude and direction. Parallel vectors have the same direction or opposite directions. The magnitude of parallel vectors can be different.

3. Sample answer: Using a vector to translate a figure is the same as using an ordered pair because a vector has horizontal and vertical components, each of which can be represented by one coordinate of an ordered pair.

Find the magnitude and direction of $\overline{AB}$ for the given coordinates.

7. $2\sqrt{13} \approx 7.2, \approx 213.7°$

6. $A(2, 7), B(-3, 3)$ $\sqrt{41} \approx 6.4, \approx 218.7°$ 7. $A(-6, 0), B(-12, -4)$

8. What are the magnitude and direction of $\vec{v} = \langle 8, -15 \rangle$? $17, \approx 298.1°$

Graph the image of each figure under a translation by the given vector.

9–10. See p. 519D.

9. $\triangle JKL$ with vertices $J(2, -1), K(-7, -2), L(-2, 8)$; $\vec{t} = \langle -1, 9 \rangle$

10. trapezoid $PQRS$ with vertices $P(1, 2), Q(7, 3), R(15, 1), S(3, -1)$; $\vec{u} = \langle 3, -3 \rangle$

11. Graph the image of $\square WXYZ$ with vertices $W(6, -6), X(3, -8), Y(-4, -4)$, and $Z(-1, -2)$ under the translation by $\vec{e} = \langle -1, 6 \rangle$ and $\vec{f} = \langle 8, -5 \rangle$. See p. 519D.

Application

Find the magnitude and direction of each resultant for the given vectors.

12. $\vec{g} = \langle 4, 0 \rangle, \vec{h} = \langle 0, 6 \rangle$ $2\sqrt{13} \approx 7.2, \approx 56.3°$

13. $\vec{t} = \langle 0, -9 \rangle, \vec{u} = \langle 12, -9 \rangle$ $6\sqrt{13} \approx 21.6, \approx 303.7°$

14. **BOATING** Raphael sails his boat due east at a rate of 10 knots. If there is a current of 3 knots moving 30° south of east, what is the resultant speed and direction of the boat? about 12.7 knots, about 6.8° south of due east

★ indicates increased difficulty

Practice and Apply

Homework Help

For Exercises	See Examples
15–20	1
21–36	2
37–42	3
43–46	4
47–57	5

Extra Practice
See page 773.

Write the component form of each vector.

15.

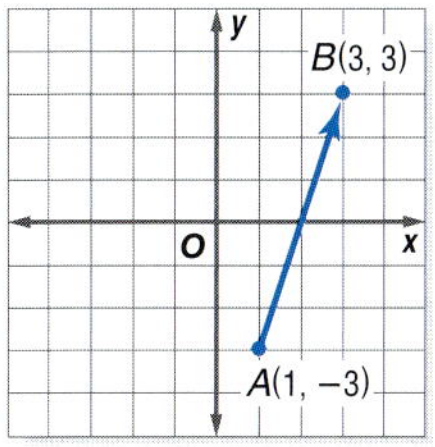

$\langle 2, 6 \rangle$

16.

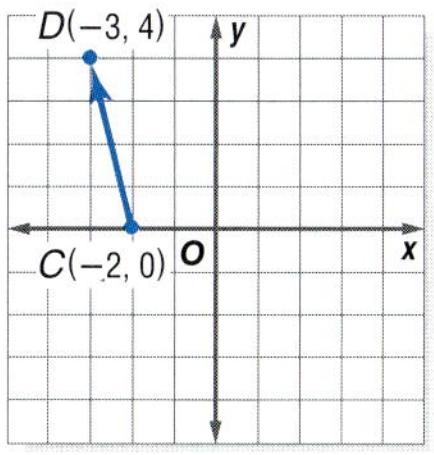

$\langle -1, 4 \rangle$

17.

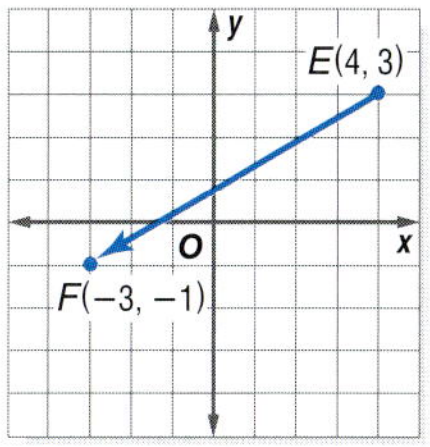

$\langle -7, -4 \rangle$

18.

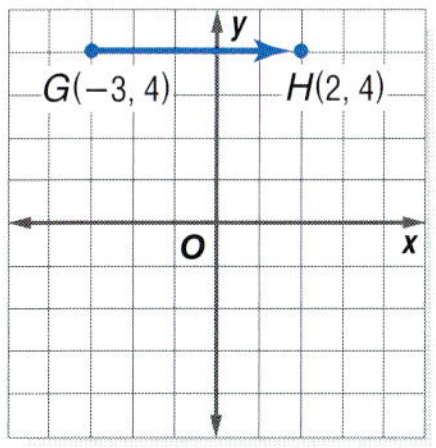

$\langle 5, 0 \rangle$

19.

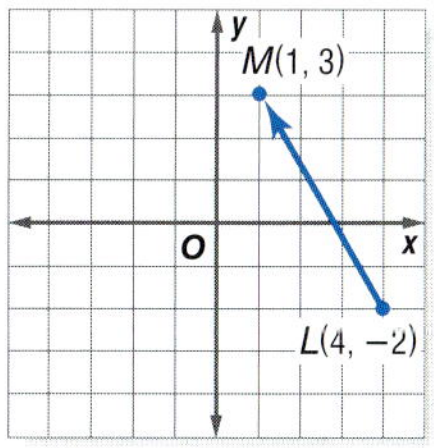

$\langle -3, 5 \rangle$

20.

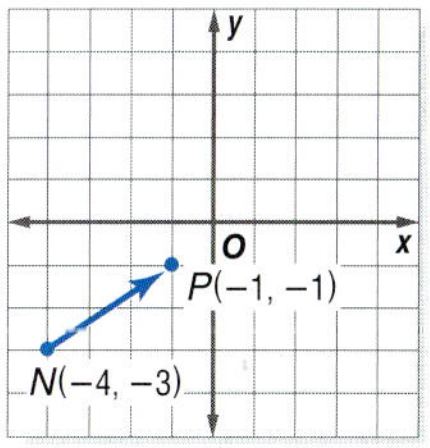

$\langle 3, 2 \rangle$

Find the magnitude and direction of $\overline{CD}$ for the given coordinates. Round to the nearest tenth. 22. $4\sqrt{2} \approx 5.7, 45°$ 23. $2\sqrt{5} \approx 4.5, 296.6°$

21. $C(4, 2), D(9, 2)$ $5, 0°$
22. $C(-2, 1), D(2, 5)$
23. $C(-5, 10), D(-3, 6)$
24. $C(0, -7), D(-2, -4)$ $\sqrt{13} \approx 3.6, 123.7°$
25. $C(-8, -7), D(6, 0)$ $7\sqrt{5} \approx 15.7, 26.6°$
26. $C(10, -3), D(-2, -2)$ $\sqrt{145} \approx 12.0, 175.2°$

27. What are the magnitude and direction of $\vec{t} = \langle 7, 24 \rangle$? $25, \approx 73.7°$

28. What are the magnitude and direction of $\vec{u} = \langle -12, 15 \rangle$? $3\sqrt{41} \approx 19.2, \approx 128.7°$

29. What are the magnitude and direction of $\vec{v} = \langle -25, -20 \rangle$? $5\sqrt{41} \approx 32.0, \approx 218.7°$

30. What are the magnitude and direction of $\vec{w} = \langle 36, -15 \rangle$? $39, \approx 337.4°$

Study Guide and Intervention, p. 509 (shown) and p. 510

Magnitude and Direction A vector is a directed segment representing a quantity that has both **magnitude**, or length, and **direction**. For example, the speed and direction of an airplane can be represented by a vector. In symbols, a vector is written as $\overrightarrow{AB}$, where A is the initial point and B is the endpoint, or as $\vec{v}$.

A vector in **standard position** has its initial point at (0, 0) and can be represented by the ordered pair for point B. The vector at the right can be expressed as $\vec{v} = \langle 5, 3 \rangle$.

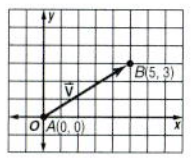

You can use the Distance Formula to find the magnitude $|\overline{AB}|$ of a vector. You can describe the direction of a vector by measuring the angle that the vector forms with the positive x-axis or with any other horizontal line.

Example Find the magnitude and direction of $\overline{AB}$ for $A(5, 2)$ and $B(8, 7)$.

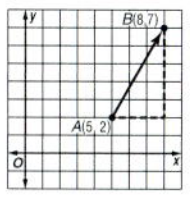

Find the magnitude.

$|AB| = \sqrt{(x_2 - x_1)^2 + (y_2 - y_1)^2}$
$= \sqrt{(8 - 5)^2 + (7 - 2)^2}$
$= \sqrt{34}$ or about 5.8 units

To find the direction, use the tangent ratio.

$\tan A = \frac{5}{3}$ The tangent ratio is opposite over adjacent.
$m\angle A \approx 59.0$ Use a calculator.

The magnitude of the vector is about 5.8 units and its direction is 59°.

Exercises

Find the magnitude and direction of $\overline{AB}$ for the given coordinates. Round to the nearest tenth.

1. $A(3, 1), B(-2, 3)$ 5.4; 158.2°
2. $A(0, 0), B(-2, 1)$ 2.2; 153.4°
3. $A(0, 1), B(3, 5)$ 5; 53.1°
4. $A(-2, 2), B(3, 1)$ 5.1; 348.7°
5. $A(3, 4), B(0, 0)$ 5; 233.1°
6. $A(4, 2), B(0, 3)$ 4.1; 166.0°

Skills Practice, p. 511 and Practice, p. 512 (shown)

Write the component form of each vector.

1.

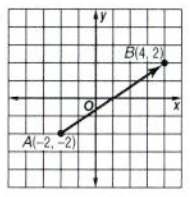

$\langle 6, 4 \rangle$

2.

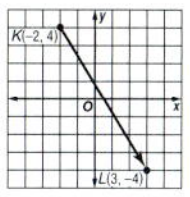

$\langle 5, -8 \rangle$

Find the magnitude and direction of $\overline{FG}$ for the given coordinates. Round to the nearest tenth.

3. $F(-8, -5), G(-2, 7)$ $6\sqrt{5} \approx 13.4, 63.4°$
4. $F(-4, 1), G(5, -6)$ $\sqrt{130} \approx 11.4, 322.1°$

Graph the image of each figure under a translation by the given vector(s).

5. $\triangle QRT$ with vertices $Q(-1, 1), R(1, 4), T(5, 1)$; $\vec{s} = \langle -2, -5 \rangle$
6. trapezoid with vertices $J(-4, -1), K(0, -1), L(-1, -3), M(-2, -3)$; $\vec{c} = \langle 5, 4 \rangle$ and $\vec{d} = \langle -2, 1 \rangle$

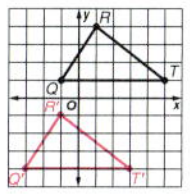

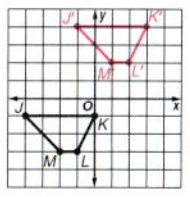

Find the magnitude and direction of each resultant for the given vectors.

7. $\vec{a} = \langle -6, 4 \rangle, \vec{b} = \langle 4, 6 \rangle$ $2\sqrt{26} \approx 10.2, 101.3°$
8. $\vec{c} = \langle -4, -5 \rangle, \vec{f} = \langle -1, 3 \rangle$ $\sqrt{29} \approx 5.4, 201.8°$

AVIATION For Exercises 9 and 10, use the following information.
A jet begins a flight along a path due north at 300 miles per hour. A wind is blowing due west at 30 miles per hour.

9. Find the resultant velocity of the plane. about 301.5 mph
10. Find the resultant direction of the plane. about 5.7° west of due north

Reading to Learn Mathematics, p. 513 ELL

Pre-Activity How do vectors help a pilot plan a flight?

Read the introduction to Lesson 9-6 at the top of page 498 in your textbook.

Why do pilots often head their planes in a slightly different direction from their destination? Sample answer: to compensate for the effect of the wind so that the result will be that the plane will actually fly in the correct direction

Reading the Lesson

1. Supply the missing words or phrases to complete the following sentences.
   a. A vector is a directed segment representing a quantity that has both magnitude and direction.
   b. The length of a vector is called its magnitude.
   c. Two vectors are parallel if and only if they have the same or opposite direction.
   d. A vector is in standard position if it is drawn with initial point at the origin.
   e. Two vectors are equal if and only if they have the same magnitude and the same direction.
   f. The sum of two vectors is called the resultant.
   g. A vector is written in component form if it is expressed as an ordered pair.
   h. The process of multiplying a vector by a constant is called scalar multiplication.
2. Write each vector described below in component form.
   a. a vector in standard position with endpoint (a, b) $\langle a, b \rangle$
   b. a vector with initial point (a, b) and endpoint (c, d) $\langle c - a, d - b \rangle$
   c. a vector in standard position with endpoint $(-3, 5)$ $\langle -3, 5 \rangle$
   d. a vector with initial point $(2, -3)$ and endpoint $(6, -8)$ $\langle 4, -5 \rangle$
   e. $\vec{a} + \vec{b}$ if $\vec{a} = \langle -3, 5 \rangle$ and $\vec{b} = \langle 6, -4 \rangle$ $\langle 3, 1 \rangle$
   f. $5\vec{u}$ if $\vec{u} = \langle 8, -6 \rangle$ $\langle 40, -30 \rangle$
   g. $-\frac{1}{3}\vec{v}$ if $\vec{v} = \langle -15, 24 \rangle$ $\langle 5, -8 \rangle$
   h. $0.5\vec{u} + 1.5\vec{v}$ if $\vec{u} = \langle 10, -10 \rangle$ and $\vec{v} = \langle -8, 6 \rangle$ $\langle -7, 4 \rangle$

Helping You Remember

3. A good way to remember a new mathematical term is to relate it to a term you already know. You learned about *scale factors* when you studied similarity and dilations. How is the idea of a *scalar* related to *scale factors*? Sample answer: A *scalar* is the term used for a constant (a specific real number) when working with vectors. A vector has both magnitude and direction, while a scalar is just a magnitude. Multiplying a vector by a positive scalar changes the magnitude of the vector, but not the direction, so it represents a change in scale.

Enrichment, p. 514

Reading Mathematics

Many quantities in nature can be thought of as vectors. The science of physics involves many vector quantities. In reading about applications of mathematics, ask yourself whether the quantities involve only magnitude or both magnitude and direction. The first kind of quantity is called **scalar**. The second kind is a **vector**.

Classify each of the following. Write *scalar* or *vector*.

1. the mass of a book scalar
2. a car traveling north at 55 mph vector
3. a balloon rising 24 feet per minute vector
... scalar

Answers

37.

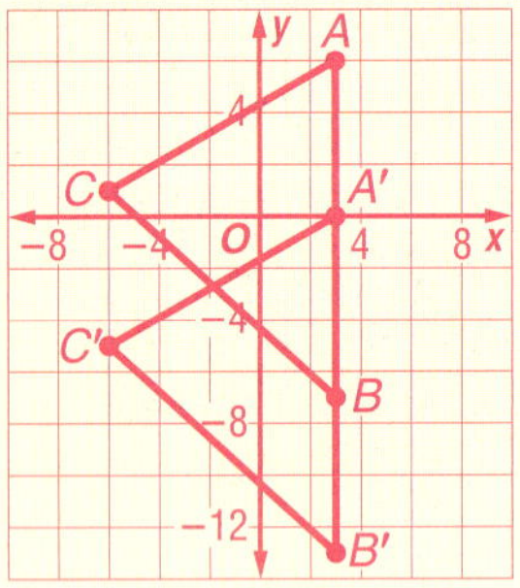

38.

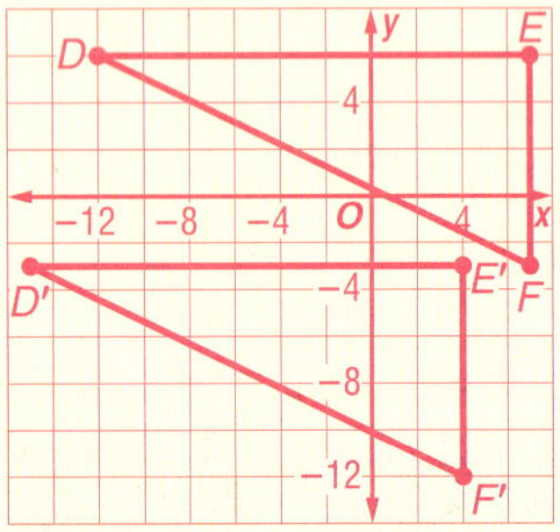

39.

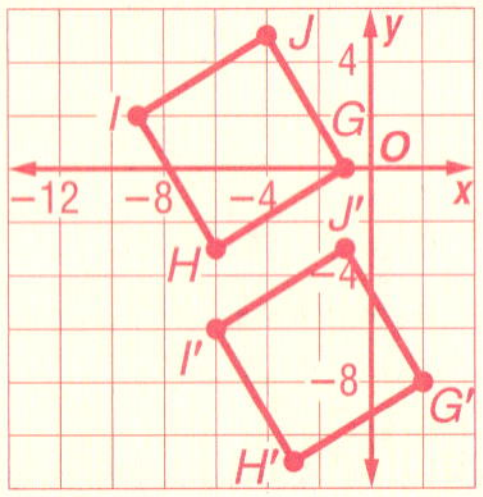

40.

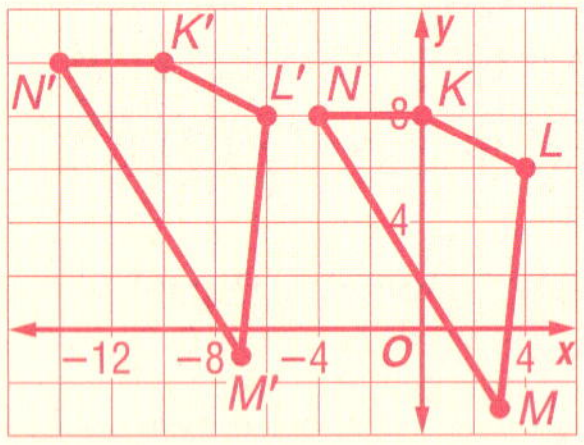

41.

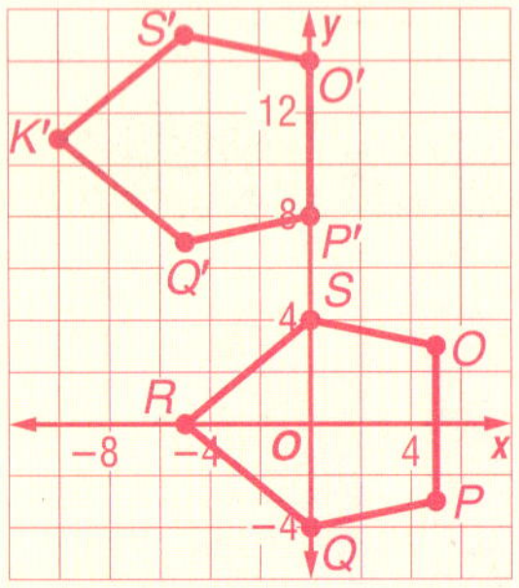

42.

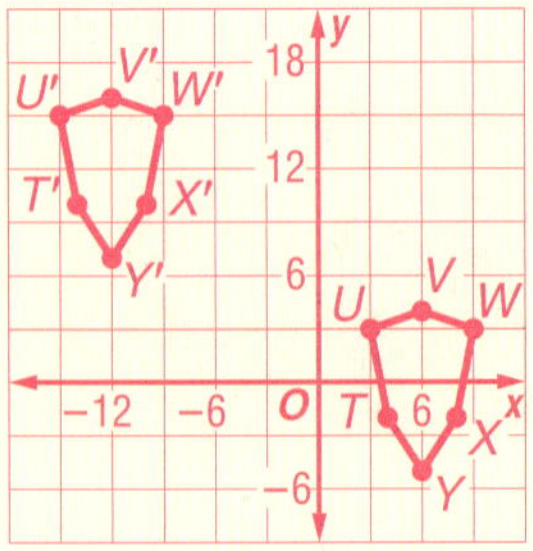

Find the magnitude and direction of $\overrightarrow{MN}$ for the given coordinates. Round to the nearest tenth.

31. $6\sqrt{2} \approx 8.5$, $135.0°$
32. $2\sqrt{13} \approx 7.2$, $146.3°$
33. $4\sqrt{10} \approx 12.6$, $198.4°$
34. $\sqrt{274} \approx 16.6$, $295.0°$
35. $2\sqrt{122} \approx 22.1$, $275.2°$
36. $2\sqrt{5} \approx 4.5$, $243.4°$
37–42. See margin.

31. $M(-3, 3)$, $N(-9, 9)$ **32.** $M(8, 1)$, $N(2, 5)$ **33.** $M(0, 2)$, $N(-12, -2)$
34. $M(-1, 7)$, $N(6, -8)$ **35.** $M(-1, 10)$, $N(1, -12)$ **36.** $M(-4, 0)$, $N(-6, -4)$

Graph the image of each figure under a translation by the given vector.

37. $\triangle ABC$ with vertices $A(3, 6)$, $B(3, -7)$, $C(-6, 1)$; $\vec{a} = \langle 0, -6 \rangle$
38. $\triangle DEF$ with vertices $D(-12, 6)$, $E(7, 6)$, $F(7, -3)$; $\vec{b} = \langle -3, -9 \rangle$
39. square $GHIJ$ with vertices $G(-1, 0)$, $H(-6, -3)$, $I(-9, 2)$, $J(-4, 5)$; $\vec{c} = \langle 3, -8 \rangle$
40. quadrilateral $KLMN$ with vertices $K(0, 8)$, $L(4, 6)$, $M(3, -3)$, $N(-4, 8)$; $\vec{x} = \langle -10, 2 \rangle$
41. pentagon $OPQRS$ with vertices $O(5, 3)$, $P(5, -3)$, $Q(0, -4)$, $R(-5, 0)$, $S(0, 4)$; $\vec{y} = \langle -5, 11 \rangle$
42. hexagon $TUVWXY$ with vertices $T(4, -2)$, $U(3, 3)$, $V(6, 4)$, $W(9, 3)$, $X(8, -2)$, $Y(6, -5)$; $\vec{z} = \langle -18, 12 \rangle$

Graph the image of each figure under a translation by the given vectors.

43–46. See p. 519D.

43. $\square ABCD$ with vertices $A(-1, -6)$, $B(4, -8)$, $C(-3, -11)$, $D(-8, -9)$; $\vec{p} = \langle 11, 6 \rangle$, $\vec{q} = \langle -9, -3 \rangle$
44. $\triangle XYZ$ with vertices $X(3, -5)$, $Y(9, 4)$, $Z(12, -2)$; $\vec{p} = \langle 2, 2 \rangle$, $\vec{q} = \langle -4, -7 \rangle$
45. quadrilateral $EFGH$ with vertices $E(-7, -2)$, $F(-3, 8)$, $G(4, 15)$, $H(5, -1)$; $\vec{p} = \langle -6, 10 \rangle$, $\vec{q} = \langle 1, -8 \rangle$
46. pentagon $STUVW$ with vertices $S(1, 4)$, $T(3, 8)$, $U(6, 8)$, $V(6, 6)$, $W(4, 4)$; $\vec{p} = \langle -4, 5 \rangle$, $\vec{q} = \langle 12, 11 \rangle$

48. $8\sqrt{2} \approx 11.3$, $225°$
50. $12\sqrt{2} \approx 17.0$, $45°$
51. $2\sqrt{5} \approx 4.5$, $26.6°$
52. $\sqrt{145} \approx 12.0$, $265.2°$

Find the magnitude and direction of each resultant for the given vectors.

47. $\vec{a} = \langle 5, 0 \rangle$, $\vec{b} = \langle 0, 12 \rangle$ 13, 67.4°
48. $\vec{c} = \langle 0, -8 \rangle$, $\vec{d} = \langle -8, 0 \rangle$
49. $\vec{e} = \langle -4, 0 \rangle$, $\vec{f} = \langle 7, -4 \rangle$ 5, 306.9°
50. $\vec{u} = \langle 12, 6 \rangle$, $\vec{v} = \langle 0, 6 \rangle$
51. $\vec{w} = \langle 5, 6 \rangle$, $\vec{x} = \langle -1, -4 \rangle$
52. $\vec{y} = \langle 9, -10 \rangle$, $\vec{z} = \langle -10, -2 \rangle$

More About . . .

Rivers

The Congo River is one of the fastest rivers in the world. It has no dry season, because it has tributaries both north and south of the Equator. The river flows so quickly that it doesn't form a delta where it ends in the Atlantic like most rivers do when they enter an ocean.

Source: Compton's Encyclopedia

53. SHIPPING A freighter has to go around an oil spill in the Pacific Ocean. The captain sails due east for 35 miles. Then he turns the ship and heads due south for 28 miles. What is the distance and direction of the ship from its original point of course correction? about 44.8 mi; about 38.7° south of due east

54. RIVERS Suppose a section of the New River in West Virginia has a current of 2 miles per hour. If a swimmer can swim at a rate of 4.5 miles per hour, how does the current in the New River affect the speed and direction of the swimmer as she tries to swim directly across the river? The swimmer is traveling about 4.9 mph at an angle of 24°.

AVIATION **For Exercises 55–57, use the following information.**

A jet is flying northwest, and its velocity is represented by $\langle -450, 450 \rangle$ miles per hour. The wind is from the west, and its velocity is represented by $\langle 100, 0 \rangle$ miles per hour.

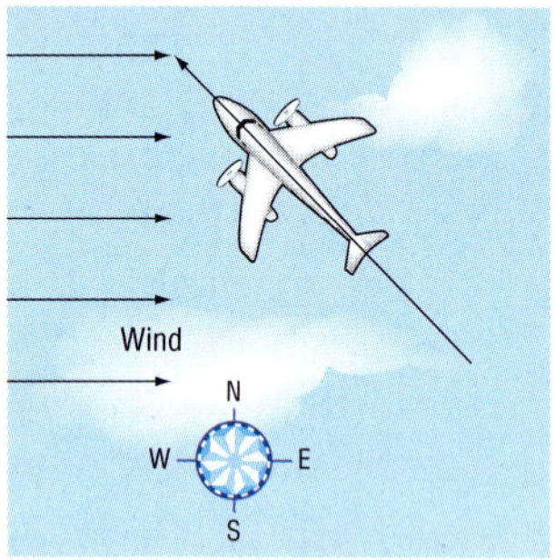

55. Find the resultant vector for the jet in component form. $\langle -350, 450 \rangle$ mph

56. Find the magnitude of the resultant. 570.1 mph

★ **57.** Find the direction of the resultant. 52.1° north of due west

58. Sample answer: $\langle 1, 0\rangle$, $\left\langle -\frac{1}{2}, \frac{\sqrt{3}}{2}\right\rangle$, $\left\langle -\frac{1}{2}, -\frac{\sqrt{3}}{2}\right\rangle$

58. **CRITICAL THINKING** If two vectors have opposite directions but the same magnitude, the resultant is $\langle 0, 0\rangle$ when they are added. Find three vectors of equal magnitude, each with its tail at the origin, the sum of which is $\langle 0, 0\rangle$.

59. **WRITING IN MATH** Answer the question that was posed at the beginning of the lesson. See margin.

How do vectors help a pilot plan a flight?

Include the following in your answer:
- an explanation of how a wind from the west affects the overall velocity of a plane traveling east, and
- an explanation as to why planes traveling from Hawaii to the continental U.S. take less time than planes traveling from the continental U.S. to Hawaii.

FCAT Practice

Standardized Test Practice

60. If $\vec{q} = \langle 5, 10\rangle$ and $\vec{r} = \langle 3, 5\rangle$, find the magnitude for the sum of these two vectors. B

(A) 23 (B) 17 (C) 7 (D) $\sqrt{29}$

61. **ALGEBRA** If $5^b = 125$, then find $4^b \times 3$. D

(A) 48 (B) 64 (C) 144 (D) 192

Maintain Your Skills

Mixed Review

Find the measure of the dilation image $\overline{A'B'}$ or the preimage of $\overline{AB}$ using the given scale factor. *(Lesson 9-5)*

62. $AB = 8, r = 2$ $A'B' = 16$

63. $AB = 12, r = \frac{1}{2}$ $A'B' = 6$

64. $A'B' = 15, r = 3$ $AB = 5$

65. $A'B' = 12, r = \frac{1}{4}$ $AB = 48$

Determine whether each pattern is a tessellation. If so, describe it as *uniform*, *not uniform*, *regular*, or *semi-regular*. *(Lesson 9-4)*

66.

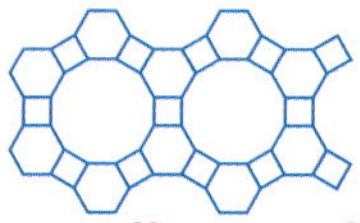

yes; uniform; semi-regular

67.

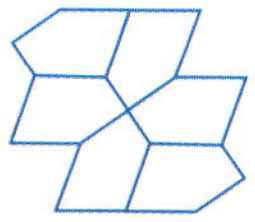

yes; not uniform

ALGEBRA **Use rhombus $WXYZ$ with $m\angle XYZ = 5m\angle WZY$ and $YZ = 12$.** *(Lesson 8-5)*

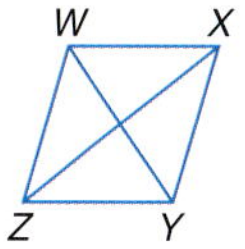

68. Find $m\angle XYZ$. 150

69. Find WX. 12

70. Find $m\angle XZY$. 15

71. Find $m\angle WXY$. 30

72. Each side of a rhombus is 30 centimeters long. One diagonal makes a 25° angle with a side. What is the length of each diagonal to the nearest tenth? *(Lesson 7-4)* 25.4 cm, 54.4 cm

Getting Ready for the Next Lesson

PREREQUISITE SKILL **Perform the indicated operation.**

*(To review **operations with matrices**, see pages 752–753.)*

73. $\begin{bmatrix} -5 & 5 \\ -3 & -2 \end{bmatrix} + \begin{bmatrix} 1 & -8 \\ -7 & 6 \end{bmatrix}$ $\begin{bmatrix} -4 & -3 \\ -10 & 4 \end{bmatrix}$

74. $\begin{bmatrix} -2 & 2 & -2 \\ -7 & -2 & -5 \end{bmatrix} + \begin{bmatrix} -8 & -8 & -8 \\ 1 & 1 & 1 \end{bmatrix}$

74. $\begin{bmatrix} -10 & -6 & -10 \\ -6 & -1 & -4 \end{bmatrix}$

75. $3\begin{bmatrix} -9 & -5 & -1 \\ 9 & 1 & 5 \end{bmatrix}$ $\begin{bmatrix} -27 & -15 & -3 \\ 27 & 3 & 15 \end{bmatrix}$

76. $\frac{1}{2}\begin{bmatrix} -4 & -5 & 0 & 2 \\ 4 & 4 & 6 & 0 \end{bmatrix}$ $\begin{bmatrix} -2 & -2.5 & 0 & 1 \\ 2 & 2 & 3 & 0 \end{bmatrix}$

77. $\begin{bmatrix} -4 & -4 \\ 2 & 2 \end{bmatrix} + 2\begin{bmatrix} 8 & 4 \\ -3 & -7 \end{bmatrix}$ $\begin{bmatrix} 12 & 4 \\ -4 & -12 \end{bmatrix}$

78. $\begin{bmatrix} 1 & -1 \\ -1 & 1 \end{bmatrix} \cdot \begin{bmatrix} 2 & -3 \\ -2 & -4 \end{bmatrix}$ $\begin{bmatrix} 4 & 1 \\ -4 & -1 \end{bmatrix}$

www.geometryonline.com/self_check_quiz/fcat

4 Assess

Open-Ended Assessment

Speaking Students can practice naming vector parts and explaining how vectors apply to real-world situations. Select examples from the book or create examples on a handout, and call on students to use the vocabulary terms and concepts of this lesson to analyze the problems aloud in class.

Getting Ready for Lesson 9-7

Prerequisite Skill Students will learn about transformations with matrices in Lesson 9-7. They will use matrices to determine coordinates of transformations. Use Exercises 73–78 to determine your students' familiarity with operations with matrices.

Assessment Options

Quiz (Lessons 9-5 and 9-6) is available on p. 536 of the *Chapter 9 Resource Masters.*

Answers

59. Quantities such as velocity are vectors. The velocity of the wind and the velocity of the plane together factor into the overall flight plan. Answers should include the following.
- A wind from the west would add to the velocity contributed by the plane resulting in an overall velocity with a larger magnitude.
- When traveling east, the prevailing winds add to the velocity of the plane. When traveling west, they detract from it.

9-7 Lesson Notes

1 Focus

5-Minute Check Transparency 9-7 Use as a quiz or review of Lesson 9-6.

Mathematical Background notes are available for this lesson on p. 460D.

How can matrices be used to make movies?

Ask students:

- Which types of movies are likely to use special effects? **science fiction movies, spy movies, horror movies, movies with superheroes, animated movies, and so on.**
- A special effect where a regular man becomes a giant would utilize which type of transformation? **dilation**

9-7 Transformations with Matrices

Sunshine State Standards
MA.C.2.4.1

What You'll Learn

- Use matrices to determine the coordinates of translations and dilations.
- Use matrices to determine the coordinates of reflections and rotations.

Vocabulary

- column matrix
- vertex matrix
- translation matrix
- reflection matrix
- rotation matrix

How can matrices be used to make movies?

Many movie directors use computers to create special effects that cannot be easily created in real life. A special effect is often a simple image that is enhanced using transformations. Complex images can be broken down into simple polygons, which are moved and resized using matrices to define new vertices for the polygons.

TRANSLATIONS AND DILATIONS In Lesson 9-6, you learned that a vector can be represented by the ordered pair $\langle x, y \rangle$. A vector can also be represented by a **column matrix** $\begin{bmatrix} x \\ y \end{bmatrix}$. Likewise, polygons can be represented by placing all of the column matrices of the coordinates of the vertices into one matrix, called a **vertex matrix**.

Triangle PQR with vertices $P(3, 5)$, $Q(1, -2)$, and $R(-4, 4)$ can be represented by the vertex matrix at the right.

$$\triangle PQR = \begin{bmatrix} 3 & 1 & -4 \\ 5 & -2 & 4 \end{bmatrix} \begin{matrix} \leftarrow \text{x-coordinates} \\ \leftarrow \text{y-coordinates} \end{matrix}$$

(columns labeled P, Q, R)

Like vectors, matrices can be used to perform translations. You can use matrix addition and a **translation matrix** to find the coordinates of a translated figure.

Study Tip

Translation Matrix
A translation matrix contains the same number of rows and columns as the vertex matrix of a figure.

Example 1 Translate a Figure

Use a matrix to find the coordinates of the vertices of the image of ▱$ABCD$ with $A(3, 2)$, $B(1, -3)$, $C(-3, -1)$, and $D(-1, 4)$ under the translation $(x, y) \rightarrow (x + 5, y - 3)$.

Write the vertex matrix for ▱$ABCD$. $\begin{bmatrix} 3 & 1 & -3 & -1 \\ 2 & -3 & -1 & 4 \end{bmatrix}$

To translate the figure 5 units to the right, add 5 to each x-coordinate. To translate the figure 3 units down, add -3 to each y-coordinate. This can be done by adding the translation matrix $\begin{bmatrix} 5 & 5 & 5 & 5 \\ -3 & -3 & -3 & -3 \end{bmatrix}$ to the vertex matrix of ▱$ABCD$.

Vertex Matrix of ▱$ABCD$ + Translation Matrix = Vertex Matrix of ▱$A'B'C'D'$

$$\begin{bmatrix} 3 & 1 & -3 & -1 \\ 2 & -3 & -1 & 4 \end{bmatrix} + \begin{bmatrix} 5 & 5 & 5 & 5 \\ -3 & -3 & -3 & -3 \end{bmatrix} = \begin{bmatrix} 8 & 6 & 2 & 4 \\ -1 & -6 & -4 & 1 \end{bmatrix}$$

The coordinates of ▱$A'B'C'D'$ are $A'(8, -1)$, $B'(6, -6)$, $C'(2, -4)$, and $D'(4, 1)$.

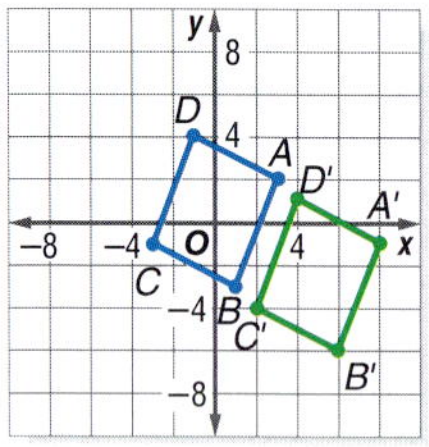

Resource Manager

Workbook and Reproducible Masters

Chapter 9 Resource Masters
- Study Guide and Intervention, pp. 515–516
- Skills Practice, p. 517
- Practice, p. 518
- Reading to Learn Mathematics, p. 519
- Enrichment, p. 520
- Assessment, p. 536

Teaching Geometry With Manipulatives Masters, pp. 1, 17

Transparencies
5-Minute Check Transparency 9-7
Answer Key Transparencies

Technology
Interactive Chalkboard

Scalars can be used with matrices to perform dilations.

Example 2 Dilate a Figure

Triangle FGH has vertices $F(-3, 1)$, $G(-1, 2)$, and $H(1, -1)$. Use scalar multiplication to dilate $\triangle FGH$ centered at the origin so that its perimeter is 3 times the original perimeter.

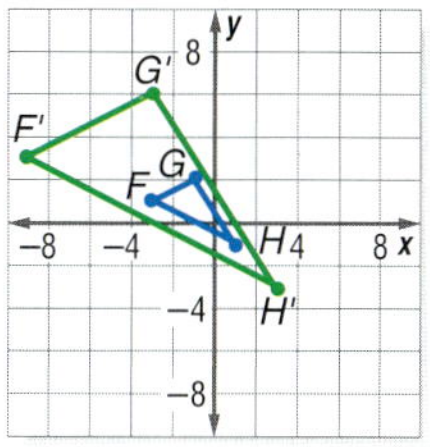

If the perimeter of a figure is 3 times the original perimeter, then the lengths of the sides of the figure will be 3 times the measures of the original lengths. Multiply the vertex matrix by a scale factor of 3.

$$3\begin{bmatrix}-3 & -1 & 1\\ 1 & 2 & -1\end{bmatrix} = \begin{bmatrix}-9 & -3 & 3\\ 3 & 6 & -3\end{bmatrix}$$

The coordinates of the vertices of $\triangle F'G'H'$ are $F'(-9, 3)$, $G'(-3, 6)$, and $H'(3, -3)$.

REFLECTIONS AND ROTATIONS A **reflection matrix** can be used to multiply the vertex matrix of a figure to find the coordinates of the image. The matrices used for four common reflections are shown below.

Study Tip

Reflection Matrices
The matrices used for reflections and rotations always have two rows and two columns no matter what the number of columns in the vertex matrix.

Concept Summary — *Reflection Matrices*

For a reflection in the:	x-axis	y-axis	origin	line $y = x$
Multiply the vertex matrix on the left by:	$\begin{bmatrix}1 & 0\\ 0 & -1\end{bmatrix}$	$\begin{bmatrix}-1 & 0\\ 0 & 1\end{bmatrix}$	$\begin{bmatrix}-1 & 0\\ 0 & -1\end{bmatrix}$	$\begin{bmatrix}0 & 1\\ 1 & 0\end{bmatrix}$
The product of the reflection matrix and the vertex matrix $\begin{bmatrix}a & b\\ c & d\end{bmatrix}$ would be:	$\begin{bmatrix}a & b\\ -c & -d\end{bmatrix}$	$\begin{bmatrix}-a & -b\\ c & d\end{bmatrix}$	$\begin{bmatrix}-a & -b\\ -c & -d\end{bmatrix}$	$\begin{bmatrix}c & d\\ a & b\end{bmatrix}$

Example 3 Reflections

Use a matrix to find the coordinates of the vertices of the image of $\overline{TU}$ with $T(-4, -4)$ and $U(3, 2)$ after a reflection in the x-axis.

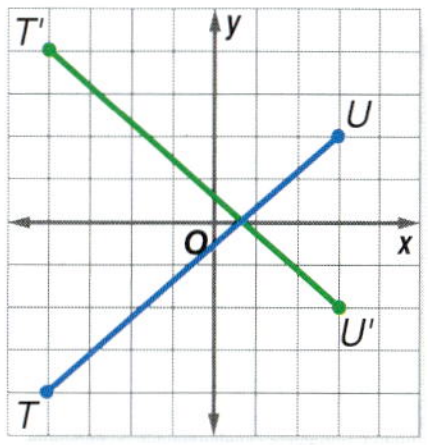

Write the ordered pairs as a vertex matrix. Then multiply the vertex matrix by the reflection matrix for the x-axis.

$$\begin{bmatrix}1 & 0\\ 0 & -1\end{bmatrix} \cdot \begin{bmatrix}-4 & 3\\ -4 & 2\end{bmatrix} = \begin{bmatrix}-4 & 3\\ 4 & -2\end{bmatrix}$$

The coordinates of the vertices of $\overline{T'U'}$ are $T'(-4, 4)$ and $U'(3, -2)$. Graph $\overline{TU}$ and $\overline{T'U'}$.

Matrices can also be used to determine the vertices of a figure's image by rotation using a **rotation matrix**. Commonly used rotation matrices are summarized on the next page.

www.geometryonline.com/extra_examples/fcat

2 Teach

Teaching Tip Students can remember that whether the coordinates are represented by a coordinate pair or by a column matrix, the x is always first, leftmost in the coordinate pair and topmost in the column matrix.

TRANSLATIONS AND DILATIONS

In-Class Examples Power Point®

1 Use a matrix to find the coordinates of the vertices of the image of quadrilateral $EFGH$ with $E(5, 3)$, $F(2, -6)$, $G(-5, -5)$, and $H(-3, 4)$ under the translation $(x, y) \rightarrow (x - 3, y + 6)$. $E'(2, 9)$, $F'(-1, 0)$, $G'(-8, 1)$, and $H'(-6, 10)$

2 Triangle ABC has vertices $A(4, -8)$, $B(-12, -8)$, and $C(-24, 12)$. Use scalar multiplication to find the vertices of $\triangle ABC$ so that its perimeter is $\frac{1}{4}$ the original perimeter. $A'(1, -2)$, $B'(-3, -2)$, and $C'(-6, 3)$

REFLECTIONS AND ROTATIONS

In-Class Example Power Point®

Teaching Tip Students can check their matrix multiplication by reflecting the preimage in the x-axis using $(a, b) \rightarrow (a, -b)$.

3 Use a matrix to find the coordinates of the end points of the image of $\overline{FG}$ with $F(-3, 1)$ and $G(0, 4)$ after a reflection in the y-axis. $F'(3, 1)$ and $G'(0, 4)$

In-Class Example

Power Point®

4 Triangle ABC has vertices $A(3, 1)$, $B(5, 2)$, and $C(3, 4)$.

a. Use a matrix to find the coordinates of the image under a 270° counterclockwise rotation about the origin.
$A'(1, -3)$, $B'(2, -5)$, $C'(4, -3)$

b. What are the coordinates of the image if $\triangle ABC$ is reflected in the y-axis before a 90° counterclockwise rotation about the origin?
$A'(-1, -3)$, $B'(-2, -5)$, $C'(-4, -3)$

Answers

2. Sample answer: The format used to represent the transformation is different in each method, but the result is the same. Positive values move a figure up or right, and negative values move a figure down or left.

3. Sample answer:
$$\begin{bmatrix} -2 & -2 & -2 & -2 \\ -1 & -1 & -1 & -1 \end{bmatrix}$$

Study Tip

Graphing Calculator
It may be helpful to store the reflection and rotation matrices in your calculator to avoid reentering them.

Concept Summary — Rotation Matrices

For a counterclockwise rotation about the origin of:	90°	180°	270°
Multiply the vertex matrix on the left by:	$\begin{bmatrix} 0 & -1 \\ 1 & 0 \end{bmatrix}$	$\begin{bmatrix} -1 & 0 \\ 0 & -1 \end{bmatrix}$	$\begin{bmatrix} 0 & 1 \\ -1 & 0 \end{bmatrix}$
The product of the rotation matrix and the vertext matrix $\begin{bmatrix} a & b \\ c & d \end{bmatrix}$ would be:	$\begin{bmatrix} -c & -d \\ a & b \end{bmatrix}$	$\begin{bmatrix} -a & -b \\ -c & -d \end{bmatrix}$	$\begin{bmatrix} c & d \\ -a & -b \end{bmatrix}$

Example 4 Use Rotations

COMPUTERS **A software designer is creating a screen saver by transforming various patterns like the one at the right.**

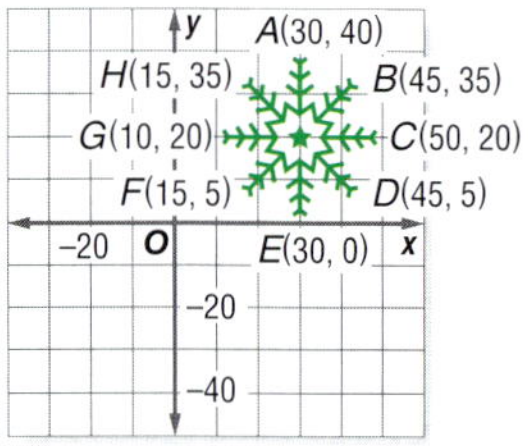

a. Write a vertex matrix for the figure.

One possible vertex matrix is
$$\begin{bmatrix} 30 & 45 & 50 & 45 & 30 & 15 & 10 & 15 \\ 40 & 35 & 20 & 5 & 0 & 5 & 20 & 35 \end{bmatrix}.$$

b. Use a matrix to find the coordinates of the image under a 90° counterclockwise rotation about the origin.

Enter the rotation matrix in your calculator as matrix A and enter the vertex matrix as matrix B. Then multiply.

KEYSTROKES: [2nd] [MATRX] 1 [2nd] [MATRX] 2 [ENTER]

$$AB = \begin{bmatrix} -40 & -35 & -20 & -5 & 0 & -5 & -20 & -35 \\ 30 & 45 & 50 & 45 & 30 & 15 & 10 & 15 \end{bmatrix}$$

The coordinates of the vertices of the figure are $A'(-40, 30)$, $B'(-35, 45)$, $C'(-20, 50)$, $D'(-5, 45)$, $E'(0, 30)$, $F'(-5, 15)$, $G'(-20, 10)$, and $H'(-35, 15)$.

c. What are the coordinates of the image if the pattern is enlarged to twice its size before a 180° rotation about the origin?

Enter the rotation matrix as matrix C. A dilation is performed before the rotation. Multiply the matrices by 2.

KEYSTROKES: 2 [2nd] [MATRX] 3 [2nd] [MATRX] 2 [ENTER]

$$2AC = \begin{bmatrix} -60 & -90 & -100 & -90 & -60 & -30 & -20 & -30 \\ -80 & -70 & -40 & -10 & 0 & -10 & -40 & -70 \end{bmatrix}$$

The coordinates of the vertices of the figure are $A'(-60, -80)$, $B'(-90, -70)$, $C'(-100, -40)$, $D'(-90, -10)$, $E'(-60, 0)$, $F'(-30, -10)$, $G'(-20, -40)$, and $H'(-30, -70)$.

Check for Understanding

Concept Check

1. $\begin{bmatrix} 0 & 1 \\ 1 & 0 \end{bmatrix}$

2–3. See margin.

1. **Write** the reflection matrix for $\triangle ABC$ and its image $\triangle A'B'C'$ at the right.

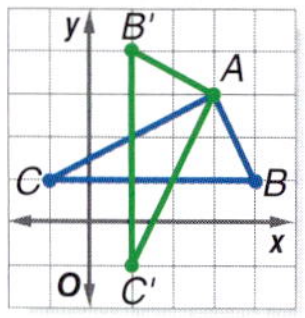

2. **Discuss** the similarities of using coordinates, vectors, and matrices to translate a figure.

3. **OPEN ENDED** Graph any ▱$PQRS$ on a coordinate grid. Then write a translation matrix that moves ▱$PQRS$ down and left on the grid.

508 Chapter 9 Transformations

Guided Practice

GUIDED PRACTICE KEY	
Exercises	Examples
4, 5	1
6, 7	2
8, 9	3
10–14	4

Use a matrix to find the coordinates of the vertices of the image of each figure under the given translation. **4. $A'(3, 3)$, $B'(1, -2)$, $C'(-2, 1)$**

4. $\triangle ABC$ with $A(5, 4)$, $B(3, -1)$, and $C(0, 2)$; $(x, y) \rightarrow (x - 2, y - 1)$
5. $\square DEFG$ with $D(-1, 3)$, $E(5, 3)$, $F(3, 0)$, and $G(-3, 0)$; $(x, y) \rightarrow (x, y + 6)$ **$D'(-1, 9)$, $E'(5, 9)$, $F'(3, 6)$, $G'(-3, 6)$**

Use scalar multiplication to find the coordinates of the vertices of each figure for a dilation centered at the origin with the given scale factor.

6. $\triangle XYZ$ with $X(3, 4)$, $Y(6, 10)$, and $Z(-3, 5)$; $r = 2$ **$X'(6, 8)$, $Y'(12, 20)$, $Z'(-6, 10)$**
7. $\square ABCD$ with $A(1, 2)$, $B(3, 3)$, $C(3, 5)$, and $D(1, 4)$; $r = -\frac{1}{4}$

7. $A'\left(-\frac{1}{4}, -\frac{1}{2}\right)$, $B'\left(-\frac{3}{4}, -\frac{3}{4}\right)$, $C'\left(-\frac{3}{4}, -\frac{5}{4}\right)$, $D'\left(-\frac{1}{4}, -1\right)$

Use a matrix to find the coordinates of the endpoints or vertices of the image of each figure under the given reflection.

8. $\overline{EF}$ with $E(-2, 4)$ and $F(5, 1)$; x-axis **$E'(-2, -4)$, $F'(5, -1)$**
9. quadrilateral $HIJK$ with $H(-5, 4)$, $I(-1, -1)$, $J(-3, -6)$, and $K(-7, -3)$; y-axis **$H'(5, 4)$, $I'(1, -1)$, $J'(3, -6)$, $K'(7, -3)$**

Use a matrix to find the coordinates of the endpoints or vertices of the image of each figure under the given rotation.

10. $\overline{LM}$ with $L(-2, 1)$ and $M(3, 5)$; 90° counterclockwise **$L'(-1, -2)$, $M'(-5, 3)$**
11. $\triangle PQR$ with $P(6, 3)$, $Q(6, 7)$, and $R(2, 7)$; 270° counterclockwise **$P'(3, -6)$, $Q'(7, -6)$, $R'(7, -2)$**
12. Use matrices to find the coordinates of the image of quadrilateral $STUV$ with $S(-4, 1)$, $T(-2, 2)$, $U(0, 1)$, and $V(-2, -2)$ after a dilation by a scale factor of 2 and a rotation 90° counterclockwise about the origin. **$S'(-2, -8)$, $T'(-4, -4)$, $U'(-2, 0)$, $V'(4, -4)$**

Application

LANDSCAPING **For Exercises 13 and 14, use the following information.**
A garden design is drawn on a coordinate grid. The original plan shows a rose bed with vertices at $(3, -1)$, $(7, -3)$, $(5, -7)$, and $(1, -5)$. Changes to the plan require that the rose bed's perimeter be half the original perimeter with respect to the origin, while the shape remains the same.

13. What are the new coordinates for the vertices of the rose bed?
14. If the center of the rose bed was originally located at $(4, -4)$, what will be the coordinates of the center after the changes have been made? **$(2, -2)$**

13. $(1.5, -0.5)$, $(3.5, -1.5)$, $(2.5, -3.5)$, $(0.5, -2.5)$

Practice and Apply

Homework Help

For Exercises	See Examples
15–18, 28, 31, 40, 41	1
19–22, 27, 32, 39, 42	2
23–26, 30, 33, 39, 41 29, 34,	3
35–38, 40, 42	4

Extra Practice
See page 773.

Use a matrix to find the coordinates of the endpoints or vertices of the image of each figure under the given translation. **15–18. See margin.**

15. $\overline{EF}$ with $E(-4, 1)$, and $F(-1, 3)$, $(x, y) \rightarrow (x - 2, y + 5)$
16. $\triangle JKL$ with $J(-3, 5)$, $K(4, 8)$, and $L(7, 5)$; $(x, y) \rightarrow (x - 3, y - 4)$
17. $\square MNOP$ with $M(-2, 7)$, $N(2, 9)$, $O(2, 7)$, and $P(-2, 5)$; $(x, y) \rightarrow (x + 3, y - 6)$
18. trapezoid $RSTU$ with $R(2, 3)$, $S(6, 2)$, $T(6, -1)$, and $U(-2, 1)$; $(x, y) \rightarrow (x - 6, y - 2)$

Use scalar multiplication to find the coordinates of the vertices of each figure for a dilation centered at the origin with the given scale factor. **19–22. See margin.**

19. $\triangle ABC$ with $A(6, 5)$, $B(4, 5)$, and $C(3, 7)$; $r = 2$
20. $\triangle DEF$ with $D(-1, 4)$, $E(0, 1)$, and $F(2, 3)$; $r = -\frac{1}{3}$
21. quadrilateral $GHIJ$ with $G(4, 2)$, $H(-4, 6)$, $I(-6, -8)$, and $J(6, -10)$; $r = -\frac{1}{2}$
22. pentagon $KLMNO$ with $K(1, -2)$, $L(3, -1)$, $M(6, -1)$, $N(4, -3)$, and $O(3, -3)$; $r = 4$

3 Practice/Apply

Study Notebook

Have students—
- add the definitions/examples of the vocabulary terms to their Vocabulary Builder worksheets for Chapter 9.
- include any other item(s) that they find helpful in mastering the skills in this lesson.

About the Exercises...

Organization by Objective
- **Translations and Dilations:** 15–20, 23–26, 29, 30, 35–38
- **Reflections and Rotations:** 21, 22, 27, 28, 31–42

Odd/Even Assignments
Exercises 15–42 are structured so that students practice the same concepts whether they are assigned odd or even problems.

Alert! Exercises 25, 26, 33, 34, 37–42 require a graphing calculator.

Assignment Guide

Basic: 15–23 odd, 27–31 odd, 35, 43, 45, 47, 49–58
Average: 15–47 odd, 49–58
Advanced: 16–46 even, 47–58

DAILY INTERVENTION

Differentiated Instruction

Interpersonal Groups of four students can work two questions from Exercises 35–38. Two students can write the matrix for the first transformation and the other two can write the second matrix. Then they can combine their efforts on a sheet of grid paper to plot and draw the two preimages and the final image.

Answers

15. $E'(-6, 6)$, $F'(-3, 8)$
16. $J'(-6, 1)$, $K'(1, 4)$, $L'(4, 1)$
17. $M'(1, 1)$, $N'(5, 3)$, $O'(5, 1)$, $P'(1, -1)$
18. $R'(-4, 1)$, $S'(0, 0)$, $T'(0, -3)$, $U'(-8, -1)$
19. $A'(12, 10)$, $B'(8, 10)$, $C'(6, 14)$
20. $D'\left(\frac{1}{3}, -\frac{4}{3}\right)$, $E'\left(0, -\frac{1}{3}\right)$, $F'\left(-\frac{2}{3}, -1\right)$
21. $G'(-2, -1)$, $H'(2, -3)$, $I'(3, 4)$, $J'(-3, 5)$
22. $K'(4, -8)$, $L'(12, -4)$, $M'(24, -4)$, $N'(16, -12)$, $O'(12, -12)$

Study Guide and Intervention, p. 515 (shown) and p. 516

Translations and Dilations A **vector** can be represented by the ordered pair $\langle x, y\rangle$ or by the **column matrix** $\begin{bmatrix} x \\ y \end{bmatrix}$. When the ordered pairs for all the vertices of a polygon are placed together, the resulting matrix is called the **vertex matrix** for the polygon.

For $\triangle ABC$ with $A(-2, 2)$, $B(2, 1)$, and $C(-1, -1)$, the vertex matrix for the triangle is $\begin{bmatrix} -2 & 2 & -1 \\ 2 & 1 & -1 \end{bmatrix}$.

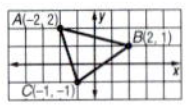

Example 1 **For $\triangle ABC$ above, use a matrix to find the coordinates of the vertices of the image of $\triangle ABC$ under the translation $(x, y) \to (x + 3, y - 1)$.**

To translate the figure 3 units to the right, add 3 to each x-coordinate. To translate the figure 1 unit down, add -1 to each y-coordinate.

Vertex Matrix of $\triangle ABC$		Translation Matrix		Vertex Matrix of $\triangle A'B'C'$
$\begin{bmatrix} -2 & 2 & -1 \\ 2 & 1 & -1 \end{bmatrix}$	+	$\begin{bmatrix} 3 & 3 & 3 \\ -1 & -1 & -1 \end{bmatrix}$	=	$\begin{bmatrix} 1 & 5 & 2 \\ 1 & 0 & -2 \end{bmatrix}$

The coordinates are $A'(1, 1)$, $B'(5, 0)$, and $C'(2, -2)$.

Example 2 **For $\triangle ABC$ above, use a matrix to find the coordinates of the vertices of the image of $\triangle ABC$ for a dilation centered at the origin with scale factor 3.**

Scale Factor		Vertex Matrix of $\triangle ABC$		Vertex Matrix of $\triangle A'B'C'$
3	·	$\begin{bmatrix} -2 & 2 & -1 \\ 2 & 1 & -1 \end{bmatrix}$	=	$\begin{bmatrix} -6 & 6 & -3 \\ 6 & 3 & -3 \end{bmatrix}$

The coordinates are $A'(-6, 6)$, $B'(6, 3)$, and $C'(-3, -3)$.

Exercises

Use a matrix to find the coordinates of the vertices of the image of each figure under the given translations or dilations.

1. $\triangle ABC$ with $A(3, 1)$, $B(-2, 4)$, $C(-2, -1)$; $(x, y) \to (x - 1, y + 2)$
   $A'(2, 3)$, $B'(-3, 6)$, $C'(-3, 1)$
2. parallelogram $RSTU$ with $R(-4, -2)$, $S(-3, 1)$, $T(3, 4)$, $U(2, 1)$; $(x, y) \to (x - 4, y - 3)$
   $R'(-8, -5)$, $S'(-7, -2)$, $T'(-1, 1)$, $U'(-2, -2)$
3. rectangle $PQRS$ with $P(4, 0)$, $Q(3, -3)$, $R(-3, -1)$, $S(-2, 2)$; $(x, y) \to (x - 2, y + 1)$
   $P'(2, 1)$, $Q'(1, -2)$, $R'(-5, 0)$, $S'(-4, 3)$
4. $\triangle ABC$ with $A(-2, -1)$, $B(-2, -3)$, $C(2, -1)$; dilation centered at the origin with scale factor 2
   $A'(-4, -2)$, $B'(-4, -6)$, $C'(4, -2)$
5. parallelogram $RSTU$ with $R(4, -2)$, $S(-4, -1)$, $T(-2, 3)$, $U(6, 2)$; dilation centered at the origin with scale factor 1.5
   $R'(6, -3)$, $S'(-6, -1.5)$, $T'(-3, 4.5)$, $U'(9, 3)$

Skills Practice, p. 517 and Practice, p. 518 (shown)

Use a matrix to find the coordinates of the vertices of the image of each figure under the given translations.

1. $\triangle KLM$ with $K(-7, -3)$, $L(4, 9)$, and $M(9, -6)$; $(x, y) \to (x - 7, y + 2)$
   $K'(-14, -1)$, $L'(-3, 11)$, $M'(2, -4)$
2. $\square ABCD$ with $A(-4, 3)$, $B(-2, 8)$, $C(3, 10)$, and $D(1, 5)$; $(x, y) \to (x + 3, y - 9)$
   $A'(-1, -6)$, $B'(1, -1)$, $C'(6, 1)$, $D'(4, -4)$

Use scalar multiplication to find the coordinates of the vertices of each figure for a dilation centered at the origin with the given scale factor.

3. quadrilateral $HIJK$ with $H(-2, 3)$, $I(2, 6)$, $J(8, 3)$, and $K(3, -4)$; $r = -\frac{1}{3}$
   $H'\left(\frac{2}{3}, -1\right)$, $I'\left(-\frac{2}{3}, -2\right)$, $J'\left(-\frac{8}{3}, -1\right)$, $K'\left(-1, \frac{4}{3}\right)$
4. pentagon $DEFGH$ with $D(-8, -4)$, $E(-8, 2)$, $F(2, 6)$, $G(8, 0)$, and $H(4, -6)$; $r = \frac{5}{4}$
   $D'(-10, -5)$, $E'\left(-10, \frac{5}{2}\right)$, $F'\left(\frac{5}{2}, \frac{15}{2}\right)$, $G'(10, 0)$, $H'\left(5, -\frac{15}{2}\right)$

Use a matrix to find the coordinates of the vertices of the image of each figure under the given reflection.

5. $\triangle QRS$ with $Q(-5, -4)$, $R(-1, -1)$, and $S(2, -6)$; x-axis
   $Q'(-5, 4)$, $R'(-1, 1)$, $S'(2, 6)$
6. quadrilateral $VXYZ$ with $V(-4, -2)$, $X(-3, 4)$, $Y(2, 1)$, and $Z(4, -3)$; $y = x$
   $V'(-2, -4)$, $X'(4, -3)$, $Y'(1, 2)$, $Z'(-3, 4)$

Use a matrix to find the coordinates of the vertices of the image of each figure under the given rotation.

7. $\square EFGH$ with $E(-5\ -4)$, $F(-3, -1)$, $G(5, -1)$, and $H(3, -4)$; 90° counterclockwise
   $E'(4, -5)$, $F'(1, -3)$, $G'(1, 5)$, $H'(4, 3)$
8. quadrilateral $PSTU$ with $P(-3, 5)$, $S(2, 6)$, $T(8, 1)$, and $U(-6, -4)$; 270° counterclockwise
   $P'(5, 3)$, $S'(6, -2)$, $T'(1, -8)$, $U'(-4, 6)$
9. **FORESTRY** A research botanist mapped a section of forested land on a coordinate grid to keep track of endangered plants in the region. The vertices of the map are $A(-2, 6)$, $B(9, 8)$, $C(14, 4)$, and $D(1, -1)$. After a month, the botanist has decided to decrease the research area to $\frac{3}{4}$ of its original size. If the center for the reduction is $O(0, 0)$,what are the coordinates of the new research area?
   $A\left(-\frac{3}{2}, \frac{9}{2}\right)$, $B\left(\frac{27}{4}, 6\right)$, $C\left(\frac{21}{2}, 3\right)$, $D\left(\frac{3}{4}, -\frac{3}{4}\right)$

Reading to Learn Mathematics, p. 519 ELL

Pre-Activity **How can matrices be used to make movies?**

Read the introduction to Lesson 9-7 at the top of page 506 in your textbook.

- What kind of transformation should be used to move a polygon? reflection, translation, or rotation
- What kind of transformation should be used to resize a polygon? dilation

Reading the Lesson

1. Write a vertex matrix for each figure.

   a. $\triangle ABC$

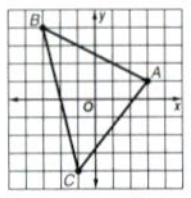

   $\begin{bmatrix} 3 & -3 & -1 \\ 1 & 4 & -4 \end{bmatrix}$

   b. parallelogram $ABCD$

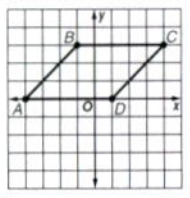

   $\begin{bmatrix} -4 & -1 & 4 & 1 \\ 0 & 3 & 3 & 0 \end{bmatrix}$

2. Match each transformation from the first column with the corresponding matrix from the second or third column. In each case, the vertex matrix for the preimage of a figure is multiplied on the left by one of the matrices below to obtain the image of the figure. All rotations listed are counterclockwise through the origin. (Some matrices may be used more than once or not at all.)

a. reflection over the y-axis vi	i. $\begin{bmatrix} 1 & 0 \\ 0 & 1 \end{bmatrix}$	v. $\begin{bmatrix} 0 & 1 \\ -1 & 0 \end{bmatrix}$
b. 90° rotation viii		
c. reflection over the line $y = x$ iii	ii. $\begin{bmatrix} -1 & 0 \\ 0 & -1 \end{bmatrix}$	vi. $\begin{bmatrix} -1 & 0 \\ 0 & 1 \end{bmatrix}$
d. 270° rotation v		
e. reflection over the origin ii	iii. $\begin{bmatrix} 0 & 1 \\ 1 & 0 \end{bmatrix}$	vii. $\begin{bmatrix} 1 & 0 \\ 0 & -1 \end{bmatrix}$
f. 180° rotation ii		
g. reflection over the x-axis vii	iv. $\begin{bmatrix} 0 & -1 \\ -1 & 0 \end{bmatrix}$	viii. $\begin{bmatrix} 0 & -1 \\ 1 & 0 \end{bmatrix}$
h. 360° rotation i		

Helping You Remember

3. How can you remember or quickly figure out the matrices for the transformations in Exercise 2? Visualize or sketch how the "unit points" (1, 0) on the x-axis and (0,1) on the y-axis are moved by the transformation. Write the ordered pair for the image points in a 2×2 matrix, with the coordinates of the image of the x-axis unit point in the first column and the coordinates of the image of the y-axis unit point in the second column.

Use a matrix to find the coordinates of the endpoints or vertices of the image of each figure under the given reflection. 23–26. See margin.

23. $\overline{XY}$ with $X(2, 2)$, and $Y(4, -1)$; y-axis
24. $\triangle ABC$ with $A(5, -3)$, $B(0, -5)$, and $C(-1, -3)$; $y = x$
★ 25. quadrilateral $DEFG$ with $D(-4, 5)$, $E(2, 6)$, $F(3, 1)$, and $G(-3, -4)$; x-axis
★ 26. quadrilateral $HIJK$ with $H(9, -1)$, $I(2, -6)$, $J(-4, -3)$, and $K(-2, 4)$; y-axis

Find the coordinates of the vertices of the image of $\triangle VWX$ under the given transformation. 27–30. See margin.

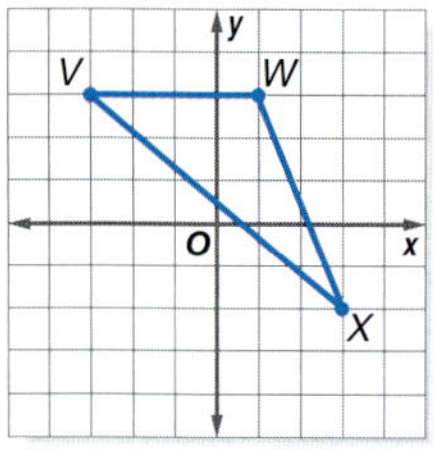

27. dilation by scale factor $\frac{2}{3}$
28. translation $(x, y) \to (x - 4, y - 1)$
29. rotation 90° counterclockwise about the origin
30. reflection in the line $y = x$

Find the coordinates of the vertices of the image of polygon $PQRST$ under the given transformation.

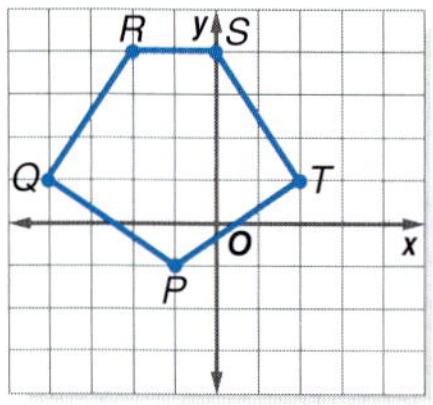

31. translation $(x, y) \to (x + 3, y - 2)$
32. dilation by scale factor -3
★ 33. reflection in the y-axis
★ 34. rotation 180° counterclockwise about the origin
    $P'(1, 1)$, $Q'(4, -1)$, $R'(2, -4)$, $S'(0, -4)$, $T'(-2, -1)$

31. $P'(2, -3)$, $Q'(-1, -1)$, $R'(1, 2)$, $S'(3, 2)$, $T'(5, -1)$
32. $P'(3, 3)$, $Q'(12, -3)$, $R'(6, -12)$, $S'(0, -12)$, $T'(-6, -3)$
33. $P'(1, -1)$, $Q'(4, 1)$, $R'(2, 4)$, $S'(0, 4)$, $T'(-2, 1)$

Use a matrix to find the coordinates of the endpoints or vertices of the image of each figure under the given rotation. 35–38. See margin.

35. $\overline{MN}$ with $M(12, 1)$ and $N(-3, 10)$; 90° counterclockwise
36. $\triangle PQR$ with $P(5, 1)$, $Q(1, 2)$, and $R(1, -4)$; 180° counterclockwise
★ 37. $\square STUV$ with $S(2, 1)$, $T(6, 1)$, $U(5, -3)$, and $V(1, -3)$; 90° counterclockwise
★ 38. pentagon $ABCDE$ with $A(-1, 1)$, $B(6, 0)$, $C(4, -8)$, $D(-4, -10)$, and $E(-5, -3)$; 270° counterclockwise

Find the coordinates of the image under the stated transformations. 39–42. See p. 519D.

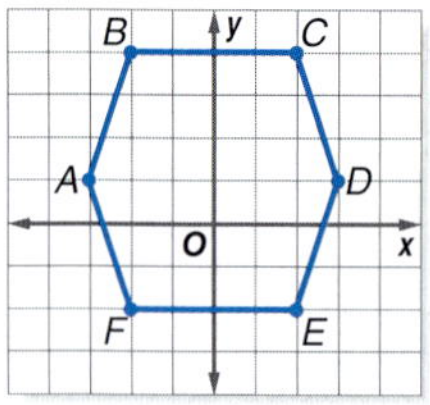

★ 39. dilation by scale factor $\frac{1}{3}$, then a reflection in the x-axis
★ 40. translation $(x, y) \to (x - 5, y + 2)$ then a rotation 90° counterclockwise about the origin
★ 41. reflection in the line $y = x$ then the translation $(x, y) \to (x + 1, y + 4)$
★ 42. rotation 180° counterclockwise about the origin, then a dilation by scale factor -2

PALEONTOLOGY **For Exercises 43 and 44, use the following information.** 43–44. See margin.
Paleontologists sometimes discover sets of fossilized dinosaur footprints like those shown at the right.

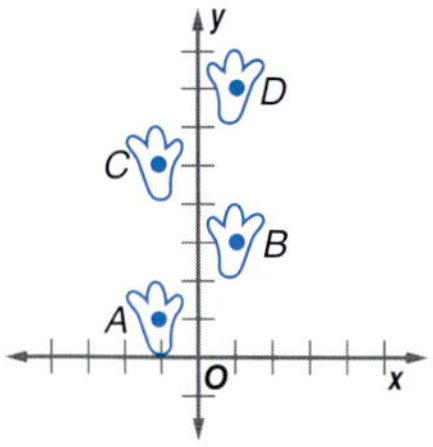

43. Describe the transformation combination shown.
44. Write two matrix operations that could be used to find the coordinates of point C.

Enrichment, p. 520

Vector Addition

Vectors are physical quantities with magnitude and direction. Force and velocity are two examples. We will investigate adding vector quantities. The sum of two vectors is called a **resultant vector** or just the resultant.

Example **Two separate forces, one measuring 20 units and the other measuring 40 units, act on an object. If the angle between the forces is 50°, find the magnitude and direction of the resultant force.**

First, the vectors must be rearranged by placing the tail of the 20-unit vector at the head of the 40-unit vector. Since these vectors are not perpendicular, the horizontal and vertical components of one of the vectors must be found. Using trigonometry, the horizontal component must be (20 cos 50°) units and the vertical component must be (20 sin 50°) units. Replacing the 20-unit vector with these components, we can now form two vectors perpendicular and use the Pythagorean Theorem to find the resultant.

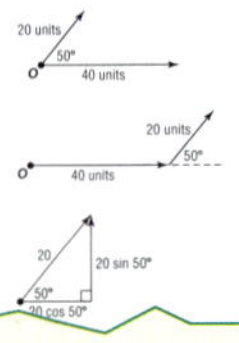

…os 50°)² + (20…

Answers

23. $X'(-2, 2)$, $Y'(-4, -1)$
24. $A'(-3, 5)$, $B'(-5, 0)$, $C'(-3, -1)$
25. $D'(-4, -5)$, $E'(2, -6)$, $F'(3, -1)$, $G'(-3, 4)$
26. $H'(-9, -1)$, $I'(-2, -6)$, $J'(4, -3)$, $K'(2, 4)$

CONSTRUCTION **For Exercises 45 and 46, use the following information.** House builders often use one set of blueprints for many projects. By changing the orientation of a floor plan, a builder can make several different looking houses.

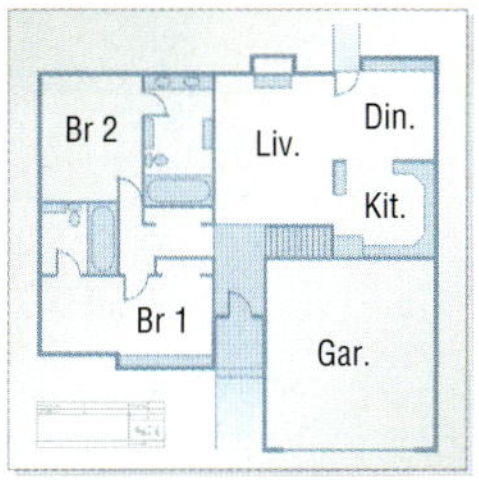

45. $\begin{bmatrix} -1 & 0 \\ 0 & 1 \end{bmatrix}$ 45. Write a transformation matrix that could be used to create a floor plan with the garage on the left.

46. $\begin{bmatrix} 0 & -1 \\ 1 & 0 \end{bmatrix}$ 46. If the current plan is of a house that faces south, write a transformation matrix that could be used to create a floor plan for a house that faces east.

47. $\begin{bmatrix} 0 & -1 \\ -1 & 0 \end{bmatrix}$ 47. **CRITICAL THINKING** Write a matrix to represent a reflection in the line $y = -x$.

48. **WRITING IN MATH** Answer the question that was posed at the beginning of the lesson. See margin.

How can matrices be used to make movies?

Include the following in your answer:

- an explanation of how transformations are used in movie production, and
- an everyday example of transformation that can be modeled using matrices.

FCAT Practice

Standardized Test Practice

49. $\begin{bmatrix} 0 & 1 \\ -1 & 0 \end{bmatrix}$ 49. **SHORT RESPONSE** Quadrilateral *ABCD* is rotated 90° clockwise about the origin. Write the transformation matrix.

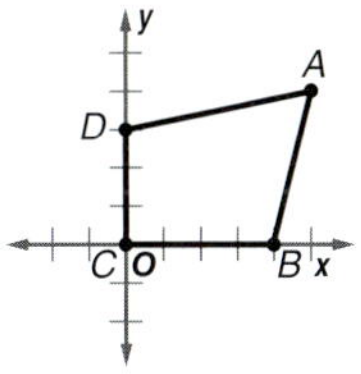

50. **ALGEBRA** A video store stocks 2500 different movie titles. If 26% of the titles are action movies and 14% are comedies, how many are neither action movies nor comedies? B

Ⓐ 1000 Ⓑ 1500

Ⓒ 1850 Ⓓ 2150

Maintain Your Skills

Mixed Review **Graph the image of each figure under a translation by the given vector.** *(Lesson 9-6)*

51–52. See margin.

51. $\triangle ABC$ with $A(-6, 1)$, $B(4, 8)$, and $C(1, -4)$; $\vec{v} = \langle -1, -5 \rangle$

52. quadrilateral *DEFG* with $D(3, -3)$, $E(1, 2)$, $F(8, -1)$, and $G(4, -6)$; $\vec{w} = \langle -7, 8 \rangle$

53. Determine the scale factor used for the dilation at the right, centered at *C*. Determine whether the dilation is an *enlargement*, *reduction*, or *congruence transformation*. *(Lesson 9-5)* $\frac{1}{2}$; reduction

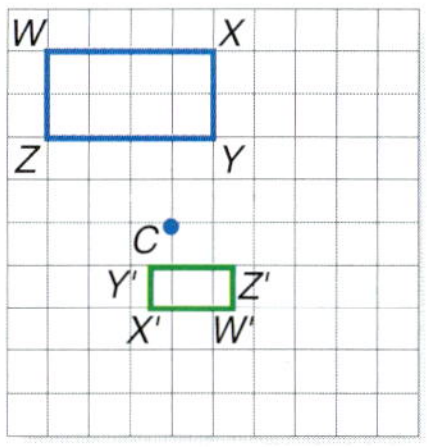

Find the measures of an exterior angle and an interior angle given the number of sides of a regular polygon. *(Lesson 8-1)*

54. 5 72, 108 55. 6 60, 120

56. 8 45, 135 57. 10 36, 144

58. **FORESTRY** To estimate the height of a tree, Lara sights the top of the tree in a mirror that is 34.5 meters from the tree. The mirror is on the ground and faces upward. Lara is standing 0.75 meter from the mirror, and the distance from her eyes to the ground is 1.75 meters. How tall is the tree? *(Lesson 6-3)* 80.5 m

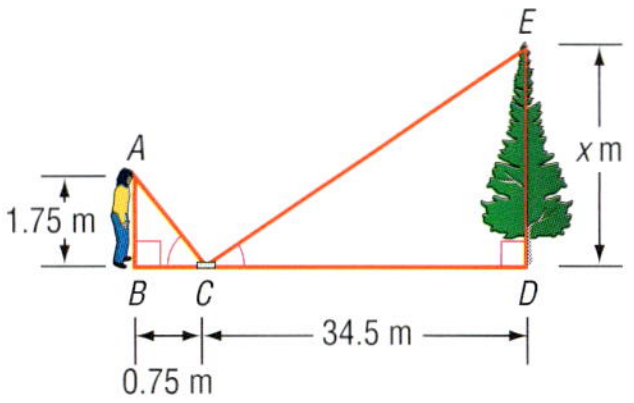

www.geometryonline.com/self_check_quiz/fcat

4 Assess

Open-Ended Assessment

Writing Select or create examples for students to practice writing matrices to perform transformations. Include at least two examples of each type of transformation. After you present the examples, ask students to write down the type of transformation first and then compose the matrix or matrices necessary to complete the problem without actually performing the calculations. Students can swap papers to check their work.

Assessment Options

Quiz (Lesson 9-7) is available on p. 536 of the *Chapter 9 Resource Masters.*

Answers

48. Matrices make it simpler for movie makers to move figures. Answers should include the following.

- By using a succession of matrix transformations, an object will move about in a scene.
- Sample answer: programming the animation in a screen saver

51.

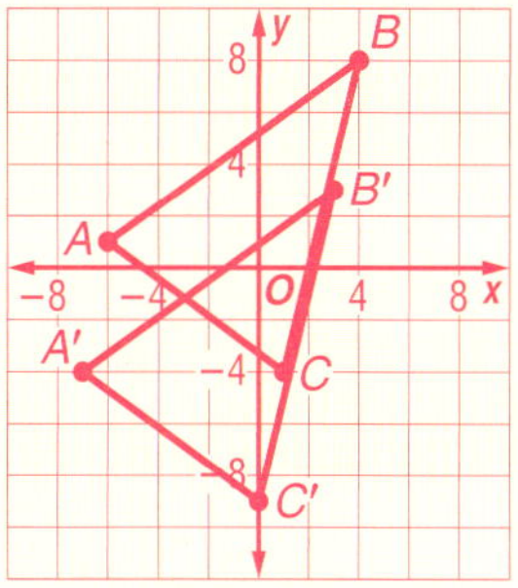

52.

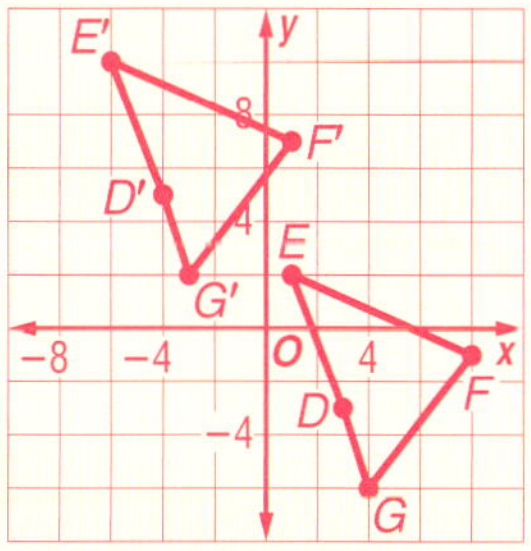

Answers (page 510)

27. $V'(-2, 2)$, $W'\left(\frac{2}{3}, 2\right)$, $X'\left(2, -\frac{4}{3}\right)$

28. $V'(-7, 2)$, $W'(-3, 2)$, $X'(-1, -3)$

29. $V'(-3, -3)$, $W'(-3, 1)$, $X'(2, 3)$

30. $V'(3, -3)$, $W'(3, 1)$, $X'(-2, 3)$

35. $M'(-1, 12)$, $N'(-10, -3)$

36. $P'(-5, -1)$, $Q'(-1, -2)$, $R'(-1, 4)$

37. $S'(-1, 2)$, $T'(-1, 6)$, $U'(3, 5)$, $V'(3, 1)$

38. $A'(1, 1)$, $B'(0, -6)$, $C'(-8, -4)$, $D'(-10, 4)$, $E'(-3, 5)$

43. Each footprint is reflected in the *y*-axis, then translated up two units.

44. $\begin{bmatrix} -1 & 0 \\ 0 & 1 \end{bmatrix} \cdot \begin{bmatrix} x \\ y \end{bmatrix} + \begin{bmatrix} 0 \\ 2 \end{bmatrix}$

Chapter 9 Study Guide and Review

Vocabulary and Concept Check

- angle of rotation (p. 476)
- center of rotation (p. 476)
- column matrix (p. 506)
- component form (p. 498)
- composition (p. 471)
- dilation (p. 490)
- direct isometry (p. 481)
- direction (p. 498)
- equal vectors (p. 499)
- glide reflection (p. 474)
- indirect isometry (p. 481)
- isometry (p. 463)
- line of reflection (p. 463)
- line of symmetry (p. 466)
- magnitude (p. 498)
- parallel vectors (p. 499)
- point of symmetry (p. 466)
- reflection (p. 463)
- reflection matrix (p. 507)
- regular tessellation (p. 484)
- resultant (p. 500)
- rotation (p. 476)
- rotation matrix (p. 507)
- rotational symmetry (p. 478)
- scalar (p. 501)
- scalar multiplication (p. 501)
- semi-regular tessellation (p. 484)
- similarity transformation (p. 491)
- standard position (p. 498)
- tessellation (p. 483)
- transformation (p. 462)
- translation (p. 470)
- translation matrix (p. 506)
- uniform (p. 484)
- vector (p. 498)
- vertex matrix (p. 506)

A complete list of postulates and theorems can be found on pages R1–R8.

Exercises **State whether each sentence is *true* or *false*. If false, replace the underlined word or phrase to make a true sentence.**

1. A dilation can change the distance between each point on the figure and the given *line of symmetry*. **false, center**
2. A tessellation is *uniform* if the same combination of shapes and angles is present at every vertex. **true**
3. Two vectors can be added easily if you know their *magnitude*. **false, component form**
4. Scalar multiplication affects only the *direction* of a vector. **false, magnitude**
5. In a rotation, the figure is turned about the *point of symmetry*. **false, center of rotation**
6. A *reflection* is a transformation determined by a figure and a line. **true**
7. A *congruence transformation* is the amount by which a figure is enlarged or reduced in a dilation. **false, scale factor**
8. A *scalar multiple* is the sum of two other vectors. **false, resultant vector**

Lesson-by-Lesson Review

9-1 Reflections

See pages 463–469.

Concept Summary

- A reflection in a line or a point is a congruence transformation.
- The line of symmetry in a figure is a line where the figure could be folded in half so that the two halves match exactly.

Example **Copy the figure. Draw the image of the figure under a reflection in line ℓ.**

The green triangle is the reflected image of the blue triangle.

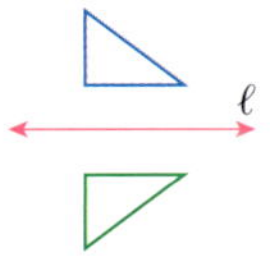

www.geometryonline.com/vocabulary_review

Vocabulary and Concept Check

- This alphabetical list of vocabulary terms in Chapter 9 includes a page reference where each term was introduced.
- **Assessment** A vocabulary test/review for Chapter 9 is available on p. 534 of the *Chapter 9 Resource Masters*.

Lesson-by-Lesson Review

For each lesson,

- the main ideas are summarized,
- additional examples review concepts, and
- practice exercises are provided.

Vocabulary PuzzleMaker

ELL The Vocabulary PuzzleMaker software improves students' mathematics vocabulary using four puzzle formats—crossword, scramble, word search using a word list, and word search using clues. Students can work on a computer screen or from a printed handout.

MindJogger Videoquizzes

ELL MindJogger Videoquizzes provide an alternative review of concepts presented in this chapter. Students work in teams in a game show format to gain points for correct answers. The questions are presented in three rounds.

Round 1 Concepts (5 questions)
Round 2 Skills (4 questions)
Round 3 Problem Solving (4 questions)

FOLDABLES™ Study Organizer

For more information about Foldables, see *Teaching Mathematics with Foldables.*

Have students look through the chapter to make sure they have included notes and examples in their Foldables for each lesson of Chapter 9.

Encourage students to refer to their Foldables while completing the Study Guide and Review and to use them in preparing for the Chapter Test.

512 Chapter 9 Transformations

Exercises **Graph each figure and its image under the given reflection.**
See Example 2 on page 464. **9–11. See margin.**

9. triangle ABC with $A(2, 1)$, $B(5, 1)$, and $C(2, 3)$ in the x-axis
10. parallelogram $WXYZ$ with $W(-4, 5)$, $X(-1, 5)$, $Y(-3, 3)$, and $Z(-6, 3)$ in the line $y = x$
11. rectangle $EFGH$ with $E(-4, -2)$, $F(0, -2)$, $G(0, -4)$, and $H(-4, -4)$ in the line $x = 1$

9-2 Translations

See pages 470–475.

Concept Summary

- A translation moves all points of a figure the same distance in the same direction.
- A translation can be represented as a composition of reflections.

Example **COORDINATE GEOMETRY** **Triangle ABC has vertices $A(2, 1)$, $B(4, -2)$, and $C(1, -4)$. Graph $\triangle ABC$ and its image for the translation $(x, y) \rightarrow (x - 5, y + 3)$.**

(x, y)	$(x - 5, y + 3)$
(2, 1)	(−3, 4)
(4, −2)	(−1, 1)
(1, −4)	(−4, −1)

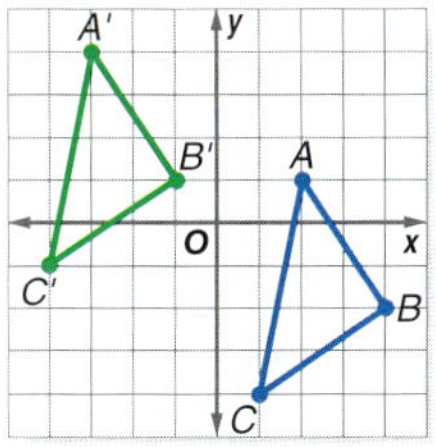

This translation moved every point of the preimage 5 units left and 3 units up.

Exercises **Graph each figure and the image under the given translation.**
See Example 1 on page 470. **12–14. See margin.**

12. quadrilateral $EFGH$ with $E(2, 2)$, $F(6, 2)$, $G(4, -2)$, $H(1, -1)$ under the translation $(x, y) \rightarrow (x - 4, y - 4)$
13. $\overline{ST}$ with endpoints $S(-3, -5)$, $T(-1, -1)$ under the translation $(x, y) \rightarrow (x + 2, y + 4)$
14. $\triangle XYZ$ with $X(2, 5)$, $Y(1, 1)$, $Z(5, 1)$ under the translation $(x, y) \rightarrow (x + 1, y - 3)$

9-3 Rotations

See pages 476–482.

Concept Summary

- A rotation turns each point in a figure through the same angle about a fixed point.
- An object has rotational symmetry when you can rotate it less than 360° and the preimage and image are indistinguishable.

Example **Identify the order and magnitude of the rotational symmetry in the figure.**

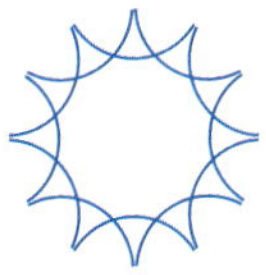

The figure has rotational symmetry of order 12 because there are 12 rotations of less than 360° (including 0°) that produce an image indistinguishable from the original.

The magnitude is 360° ÷ 12 or 30°.

Answers

9.

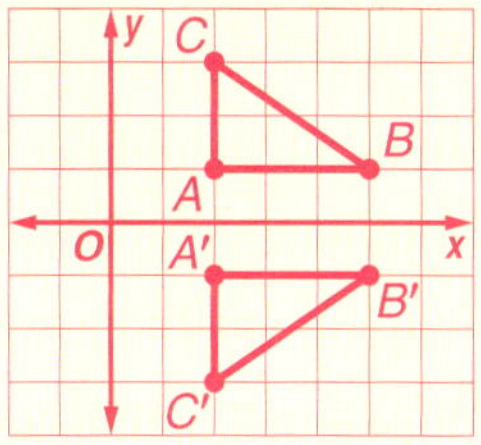

10.

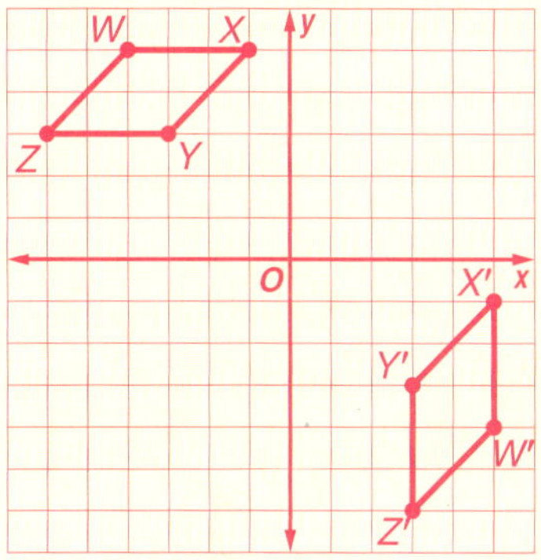

11.

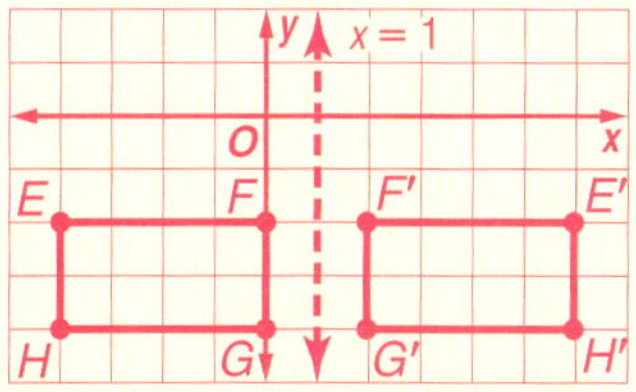

12.

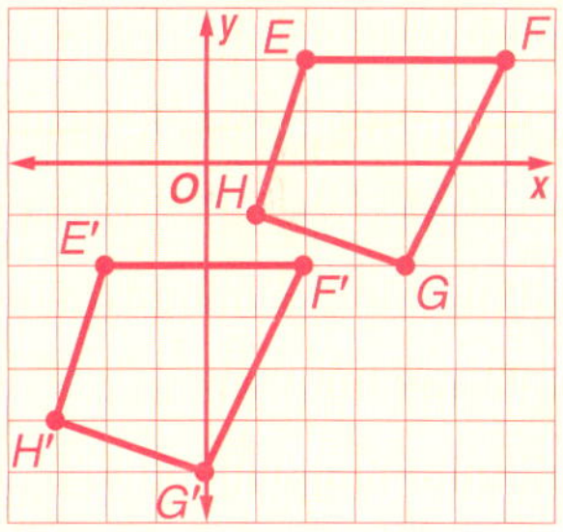

13.

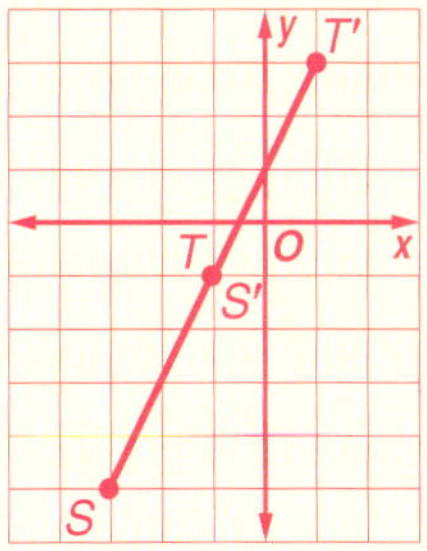

14.

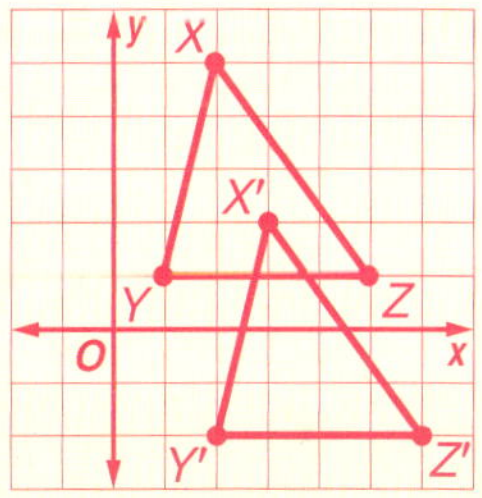

Answers

15.

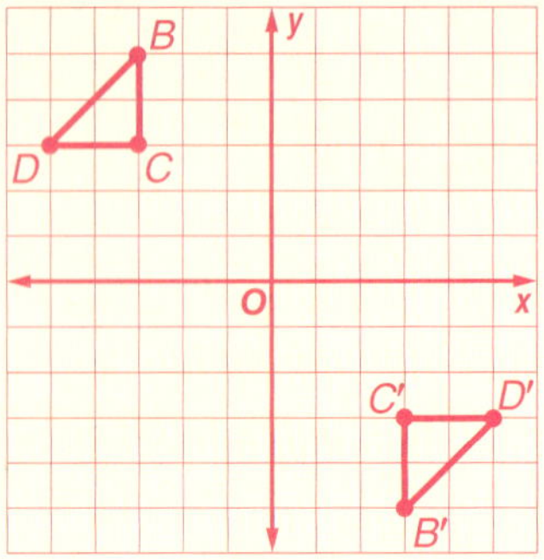

16.

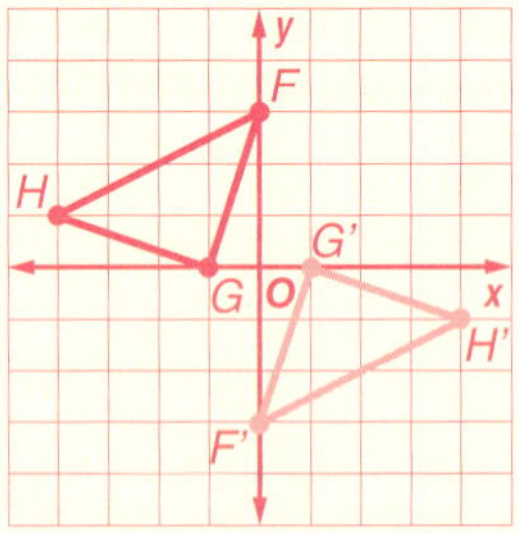

17.

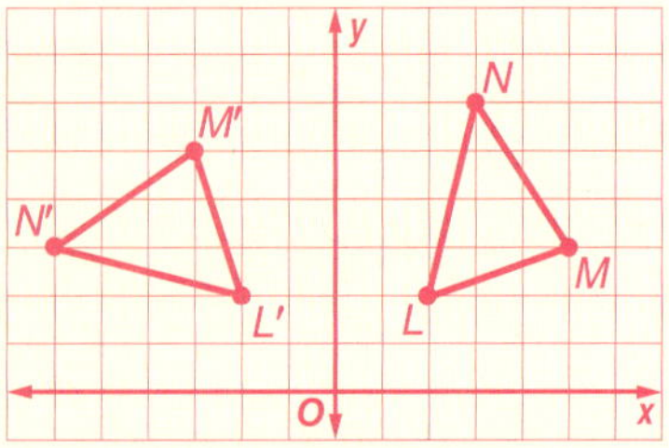

24. No; the measure of an interior angle is 108 which is not a factor of 360.

25. Yes; the measure of an interior angle is 60 which is a factor of 360.

26. No; the measure of an interior angle is 144 which is not a factor of 360.

Exercises **Draw the rotation image of each triangle by reflecting the triangles in the given lines. State the coordinates of the rotation image and the angle of rotation.** *See Example 2 on page 478.* 15–17. See margin for images.

15. $\triangle BCD$ with vertices $B(-3, 5)$, $C(-3, 3)$, and $D(-5, 3)$ reflected in the x-axis and then the y-axis $B'(3, -5)$, $C'(3, -3)$, $D'(5, -3)$; 180°

16. $\triangle FGH$ with vertices $F(0, 3)$, $G(-1, 0)$, $H(-4, 1)$ reflected in the line $y = x$ and then the line $y = -x$ $F'(0, -3)$, $G'(1, 0)$, $H'(4, -1)$; 180°

17. $\triangle LMN$ with vertices $L(2, 2)$, $M(5, 3)$, $N(3, 6)$ reflected in the line $y = -x$ and then the x-axis $L'(-2, 2)$, $M'(-3, 5)$, $N'(-6, 3)$; 90° counterclockwise

The figure at the right is a regular nonagon.
See Exercise 3 on page 478. 18. order 9 and magnitude 40°

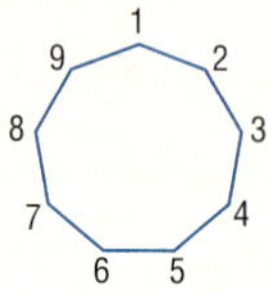

18. Identify the order and magnitude of the symmetry.

19. What is the measure of the angle of rotation if vertex 2 is moved counterclockwise to the current position of vertex 6? 200°

20. If vertex 5 is rotated 280° counterclockwise, find its new position. vertex 7

9-4 *Tessellations*

See pages 483–488.

Concept Summary

- A tessellation is a repetitious pattern that covers a plane without overlap.
- A regular tessellation contains the same combination of shapes and angles at every vertex.

Example **Classify the tessellation at the right.**

The tessellation is uniform, because at each vertex there are two squares and three equilateral triangles. Both the square and equilateral triangle are regular polygons.

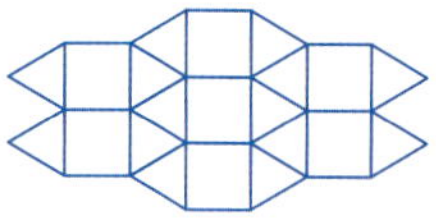

Since there is more than one regular polygon in the tessellation, it is a semi-regular tessellation.

Exercises **Determine whether each pattern is a tessellation. If so, describe it as** ***uniform, not uniform, regular,*** **or** ***semi-regular.*** *See Example 3 on page 485.*

21. 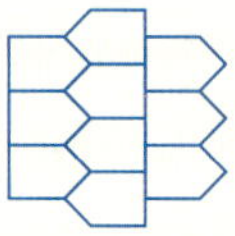 yes; not uniform

22. 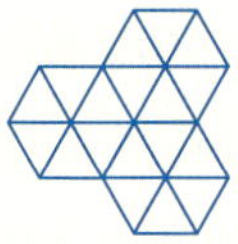 yes; uniform; regular

23.  yes; uniform

Determine whether each regular polygon will tessellate the plane. Explain.
See Example 1 on page 484. 24–26. See margin.

24. pentagon
25. triangle
26. decagon

9-5 Dilations

See pages 490–497.

Concept Summary

- Dilations can be enlargements, reductions, or congruence transformations.

Example **Triangle EFG has vertices $E(-4, -2)$, $F(-3, 2)$, and $G(1, 1)$. Find the image of $\triangle EFG$ after a dilation centered at the origin with a scale factor of $\frac{3}{2}$.**

Preimage (x, y)	Image $\left(\frac{3}{2}x, \frac{3}{2}y\right)$
$E(-4, -2)$	$E'(-6, -3)$
$F(-3, 2)$	$F'\left(-\frac{9}{2}, 3\right)$
$G(1, 1)$	$G'\left(\frac{3}{2}, \frac{3}{2}\right)$

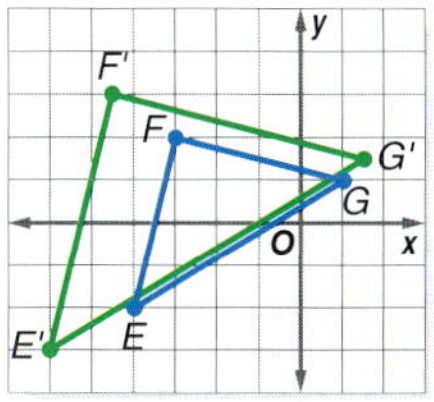

Exercises **Find the measure of the dilation image $\overline{C'D'}$ or preimage of $\overline{CD}$ using the given scale factor.** *See Example 1 on page 491.*

27. $CD = 8$, $r = 3$ **$C'D' = 24$**
28. $CD = \frac{2}{3}$, $r = -6$ **$C'D' = 4$**
29. $C'D' = 24$, $r = 6$ **$CD = 4$**
30. $C'D' = 60$, $r = \frac{10}{3}$ **$CD = 18$**
31. $CD = 12$, $r = -\frac{5}{6}$ **$C'D' = 10$**
32. $C'D' = \frac{55}{2}$, $r = \frac{5}{4}$ **$CD = 22$**

Find the image of each polygon, given the vertices, after a dilation centered at the origin with a scale factor of -2. *See Example 3 on page 492.*

33. $P(-1, 3)$, $Q(2, 2)$, $R(1, -1)$ **$P'(2, -6)$, $Q'(-4, -4)$, $R'(-2, 2)$**
34. $E(-3, 2)$, $F(1, 2)$, $G(1, -2)$, $H(-3, -2)$ **$E'(6, -4)$, $F'(-2, -4)$, $G'(-2, 4)$, $H'(6, 4)$**

9-6 Vectors

See pages 498–505.

Concept Summary

- A vector is a quantity with both magnitude and direction.
- Vectors can be used to translate figures on the coordinate plane.

Example **Find the magnitude and direction of $\overrightarrow{PQ}$ for $P(-8, 4)$ and $Q(6, 10)$.**

Find the magnitude.

$|\overrightarrow{PQ}| = \sqrt{(x_2 - x_1)^2 + (y_2 - y_1)^2}$ Distance Formula

$= \sqrt{(6 + 8)^2 + (10 - 4)^2}$ $x_1 = -8, y_1 = 4, x_2 = 6, y_2 = 10$

$= \sqrt{232}$ Simplify.

≈ 15.2 Use a calculator.

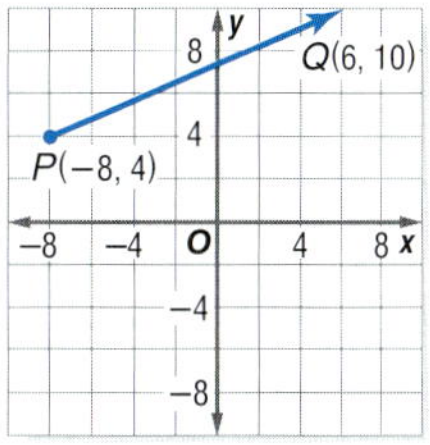

Find the direction.

$\tan P = \frac{y_2 - y_1}{x_2 - x_1}$ $\frac{\text{length of opposite side}}{\text{length of adjacent side}}$

$= \frac{10 - 4}{6 + 8}$ Substitution

$= \frac{6}{14}$ or $\frac{3}{7}$ Simplify.

$m\angle P = \tan^{-1} \frac{3}{7}$

≈ 23.2 Use a calculator.

Chapter 9 Study Guide and Review 515

Chapter 9 For More ...
• Extra Practice, see pages 771–773.
• Mixed Problem Solving, see page 790.

Exercises Write the component form of each vector. *See Example 1 on page 498.*

35.
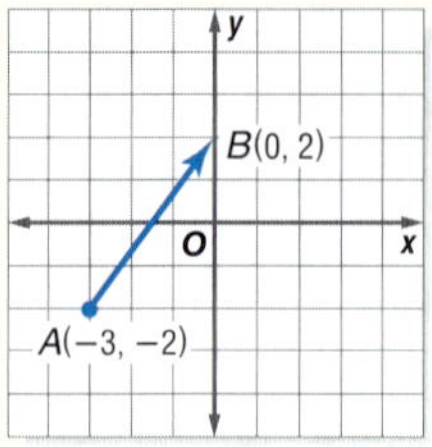

⟨3, 4⟩

36.
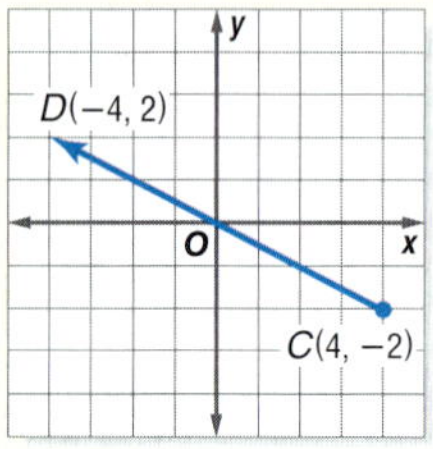

⟨−8, 4⟩

37.
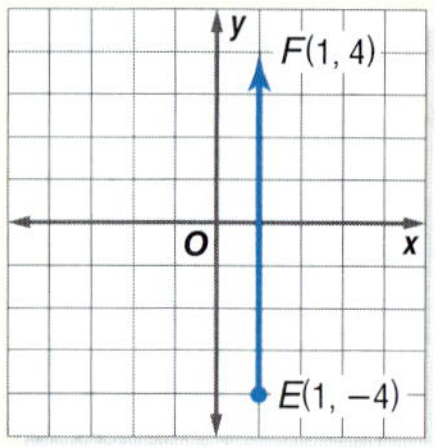

⟨0, 8⟩

Find the magnitude and direction of $\overrightarrow{AB}$ for the given coordinates. *See Example 2 on page 499.*

38. $A(-6, 4)$, $B(-9, -3)$ **≈ 7.6, ≈ 246.8°**
39. $A(8, 5)$, $B(-5, -2)$ **≈ 14.8, ≈ 208.3°**
40. $A(-14, 2)$, $B(15, -5)$ **≈ 29.8, ≈ 346.4°**
41. $A(16, 40)$, $B(-45, 0)$ **≈ 72.9, ≈ 213.3°**

9-7 Transformations with Matrices

See pages 506–511.

Concept Summary

- The vertices of a polygon can be represented by a vertex matrix.
- Matrix operations can be used to perform transformations.

Example

Use a matrix to find the coordinates of the vertices of the image of △ABC with A(1, −1), B(2, −4), C(7, −1) after a reflection in the y-axis.

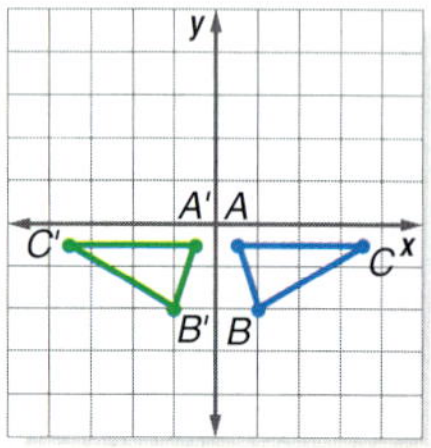

Write the ordered pairs in a vertex matrix. Then use a calculator to multiply the vertex matrix by the reflection matrix.

$$\begin{bmatrix} -1 & 0 \\ 0 & 1 \end{bmatrix} \cdot \begin{bmatrix} 1 & 2 & 7 \\ -1 & -4 & -1 \end{bmatrix} = \begin{bmatrix} -1 & -2 & -7 \\ -1 & -4 & -1 \end{bmatrix}$$

The coordinates of △A′B′C′ are A′(−1, −1), B′(−2, −4), and C′(−7, −1).

Exercises Use a matrix to find the coordinates of the vertices of the image after the stated transformation. *See Example 1 on page 506.*

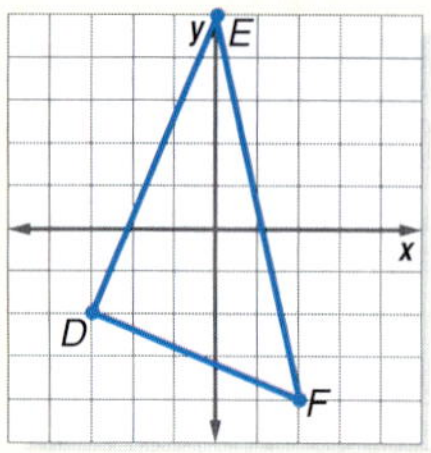

42. translation $(x, y) \rightarrow (x - 3, y - 6)$ **42–45. See margin.**
43. dilation by scale factor $\frac{4}{5}$
44. reflection in the line $y = x$
45. rotation 270° counterclockwise about the origin

Find the coordinates of the image after the stated transformations. *See Examples 2–4 on pages 507 and 508.* **46–47. See margin.**

46. △PQR with P(9, 2), Q(1, −1), and R(4, 5); $(x, y) \rightarrow (x + 2, y - 5)$, then a reflection in the x-axis
47. quadrilateral WXYZ with W(−8, 1), X(−2, 3), Y(−1, 0), and Z(−6, −3); a rotation 180° counterclockwise, then a dilation by scale factor −2

Answers

42. D′(−6, −8), E′(−3, −1), F′(−1, −10)
43. $D'\left(-\frac{12}{5}, -\frac{8}{5}\right)$, E′(0, 4), $F'\left(\frac{8}{5}, -\frac{16}{5}\right)$
44. D′(−2, −3), E′(5, 0), F′(−4, 2)
45. D′(−2, 3), E′(5, 0), F′(−4, −2)
46. P′(11, 3), Q′(3, 6), R′(6, 0)
47. W′(−16, 2), X′(−4, 6), Y′(−2, 0), Z′(−12, −6)

Answers (page 517)

7.
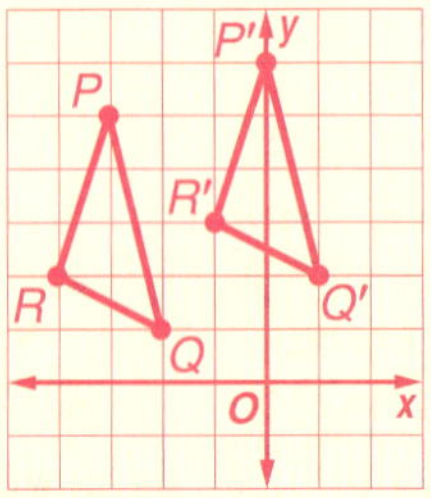

8.
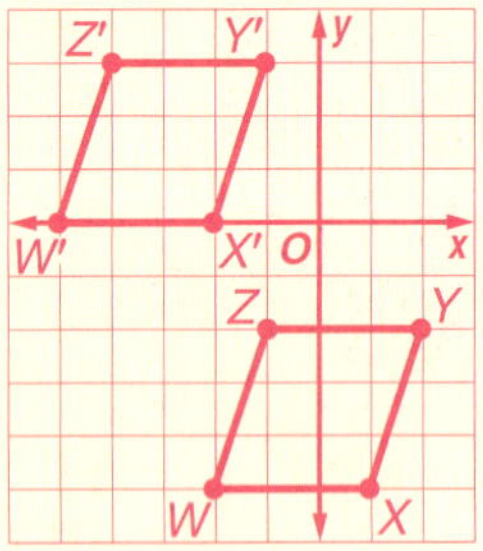

9.
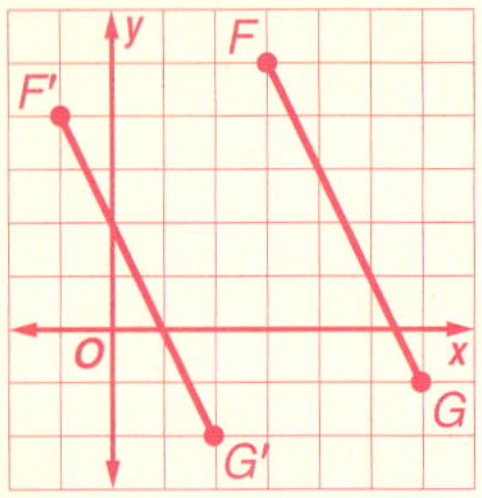

10. J′(1, 2), K′(3, 4), L′(−1, 4); 180°

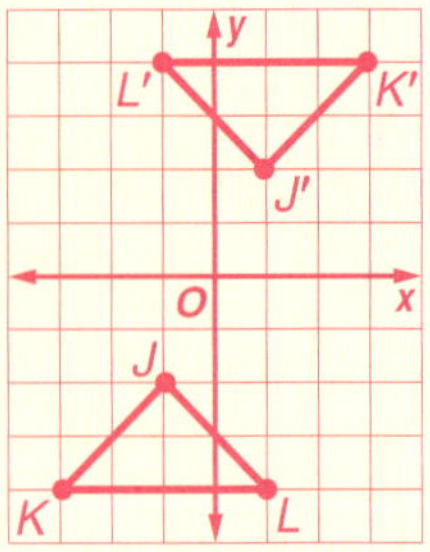

11. A′(3, 2), B′(1, −1), C′(−3, 1); 180°

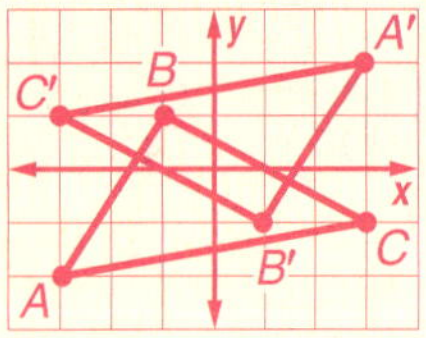

12. R′(6, −1), S′(1, −1), T′(−2, −3); 90° clockwise or 270° counterclockwise

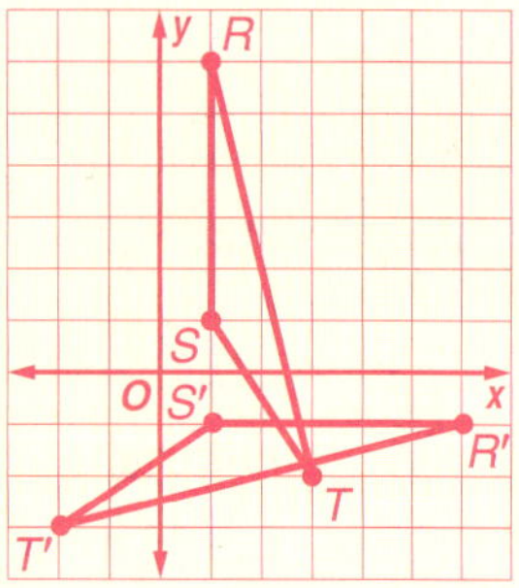

516 Chapter 9 Transformations

Chapter 9 Practice Test

Vocabulary and Concepts

Choose the correct term to complete each sentence.

1. If a dilation does not change the size of the object, then it is a(n) (*isometry*, *reflection*).
2. Tessellations with the same shapes and angles at each vertex are called (*uniform*, *regular*).
3. A vector multiplied by a (*vector*, *scalar*) results in another vector.

Skills and Applications

Name the reflected image of each figure under a reflection in line m.

4. A **E**
5. $\overline{BC}$ **$\overline{DC}$**
6. $\triangle DCE$ **$\triangle BCA$**

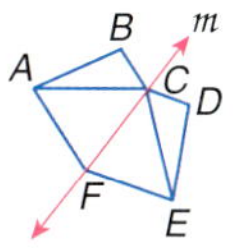

7–9. See margin.

COORDINATE GEOMETRY **Graph each figure and its image under the given translation.**

7. $\triangle PQR$ with $P(-3, 5)$, $Q(-2, 1)$, and $R(-4, 2)$ under the translation right 3 units and up 1 unit
8. Parallelogram $WXYZ$ with $W(-2, -5)$, $X(1, -5)$, $Y(2, -2)$, and $Z(-1, -2)$ under the translation up 5 units and left 3 units
9. $\overline{FG}$ with $F(3, 5)$ and $G(6, -1)$ under the translation $(x, y) \rightarrow (x - 4, y - 1)$

Draw the rotation image of each triangle by reflecting the triangles in the given lines. State the coordinates of the rotation image and the angle of rotation. **10–12. See margin.**

10. $\triangle JKL$ with $J(-1, -2)$, $K(-3, -4)$, $L(1, -4)$ reflected in the y-axis and then the x-axis
11. $\triangle ABC$ with $A(-3, -2)$, $B(-1, 1)$, $C(3, -1)$ reflected in the line $y = x$ and then the line $y = -x$
12. $\triangle RST$ with $R(1, 6)$, $S(1, 1)$, $T(3, -2)$ reflected in the y-axis and then the line $y = x$

Determine whether each pattern is a tessellation. If so, describe it as *uniform*, *not uniform*, *regular*, or *semi-regular*.

13. 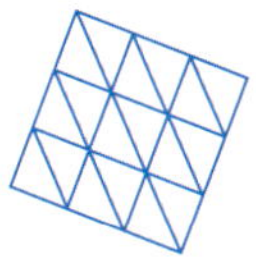 **yes; uniform**

14. 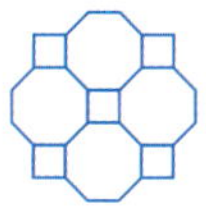 **yes; uniform; semi-regular**

15. 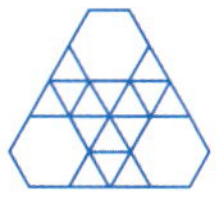 **yes; not uniform**

Find the measure of the dilation image $\overline{M'N'}$ or preimage of $\overline{MN}$ using the given scale factor.

16. $MN = 5, r = 4$ **$M'N' = 20$**
17. $MN = 8, r = \frac{1}{4}$ **$M'N' = 2$**
18. $M'N' = 36, r = 3$ **$MN = 12$**
19. $MN = 9, r = -\frac{1}{5}$ **$M'N' = \frac{9}{5}$**
20. $M'N' = 20, r = \frac{2}{3}$ **$MN = 30$**
21. $M'N' = \frac{29}{5}, r = -\frac{3}{5}$ **$MN = \frac{29}{3}$**

Find the magnitude and direction of each vector.

22. $\vec{v} = \langle -3, 2 \rangle$ **$\sqrt{13} \approx 3.6$, 146.3°**
23. $\vec{w} = \langle -6, -8 \rangle$ **10, 233.1°**

24. **TRAVEL** In trying to calculate how far she must travel for an appointment, Gunja measured the distance between Richmond, Virginia, and Charlotte, North Carolina, on a map. The distance on the map was 2.25 inches, and the scale factor was 1 inch equals 150 miles. How far must she travel? **337.5 mi**

25. **STANDARDIZED TEST PRACTICE** What reflections could be used to create the image (3, 4) from (3, −4)?
    I. reflection in the x-axis II. reflection in the y-axis III. reflection in the origin **A**
    (A) I only (B) III only (C) I and III (D) I and II

www.geometryonline.com/chapter_test/fcat

Chapter 9 Practice Test

Assessment Options

Vocabulary Test A vocabulary test/review for Chapter 9 can be found on p. 534 of the *Chapter 9 Resource Masters*.

Chapter Tests There are six Chapter 9 Tests and an Open-Ended Assessment task available in the *Chapter 9 Resource Masters*.

Chapter 9 Tests			
Form	**Type**	**Level**	**Pages**
1	MC	basic	521–522
2A	MC	average	523–524
2B	MC	average	525–526
2C	FR	average	527–528
2D	FR	average	529–530
3	FR	advanced	531–532

MC = multiple-choice questions
FR = free-response questions

Open-Ended Assessment
Performance tasks for Chapter 9 can be found on p. 533 of the *Chapter 9 Resource Masters*. A sample scoring rubric for these tasks appears on p. A28.

TestCheck and Worksheet Builder

This **networkable software** has three modules for assessment.

- **Worksheet Builder** to make worksheets and tests.
- **Student Module** to take tests on-screen.
- **Management System** to keep student records.

Portfolio Suggestion

Introduction Students may feel very confident that they understand all the major concepts presented in a chapter, but this may lead them to forget some of the important minor points.

Ask Students Ask students to search the chapter for one or two things that they had forgotten about or that may not have examined thoroughly while they were working through the chapter. Have students list the concept(s) on a sheet of paper with a brief definition or explanation and an example of a problem that uses the concept(s) and place this sheet in their portfolios.

Chapter 9 Standardized Test Practice

These two pages contain practice questions in the various formats that can be found on the most frequently given standardized tests.

A practice answer sheet for these two pages can be found on p. A1 of the *Chapter 9 Resource Masters*.

Standardized Test Practice Student Recording Sheet, p. A1

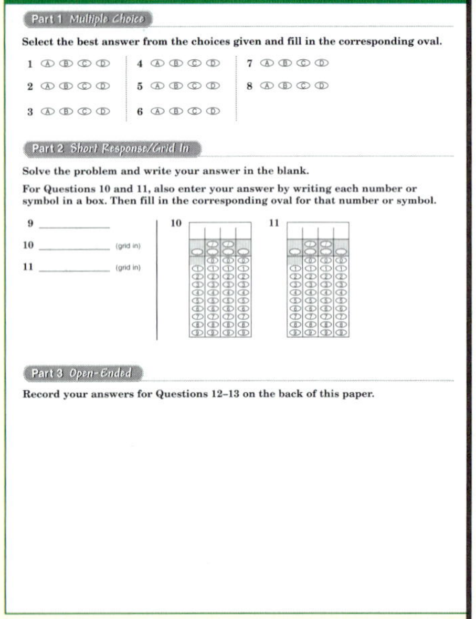
Part 1 Multiple Choice

Select the best answer from the choices given and fill in the corresponding oval.

1 Ⓐ Ⓑ Ⓒ Ⓓ 4 Ⓐ Ⓑ Ⓒ Ⓓ 7 Ⓐ Ⓑ Ⓒ Ⓓ
2 Ⓐ Ⓑ Ⓒ Ⓓ 5 Ⓐ Ⓑ Ⓒ Ⓓ 8 Ⓐ Ⓑ Ⓒ Ⓓ
3 Ⓐ Ⓑ Ⓒ Ⓓ 6 Ⓐ Ⓑ Ⓒ Ⓓ

Part 2 Short Response/Grid In

Solve the problem and write your answer in the blank.

For Questions 10 and 11, also enter your answer by writing each number or symbol in a box. Then fill in the corresponding oval for that number or symbol.

9 ______
10 ______ (grid in)
11 ______ (grid in)

Part 3 Open-Ended

Record your answers for Questions 12–13 on the back of this paper.

Additional Practice

See pp. 539–540 in the *Chapter 9 Resource Masters* for additional standardized test practice.

Chapter 9 Standardized Test Practice

FCAT Practice

Part 1 Multiple Choice

Record your answers on the answer sheet provided by your teacher or on a sheet of paper.

1. Ms. Lee told her students, "If you do not get enough rest, you will be tired. If you are tired, you will not be able to concentrate." Which of the following is a logical conclusion that could follow Ms. Lee's statements? (Lesson 2-4) D

Ⓐ If you get enough rest, you will be tired.

Ⓑ If you are tired, you will be able to concentrate.

Ⓒ If you do not get enough rest, you will be able to concentrate.

Ⓓ If you do not get enough rest, you will not be able to concentrate.

2. Which of the following statements is true? (Lesson 3-5) A

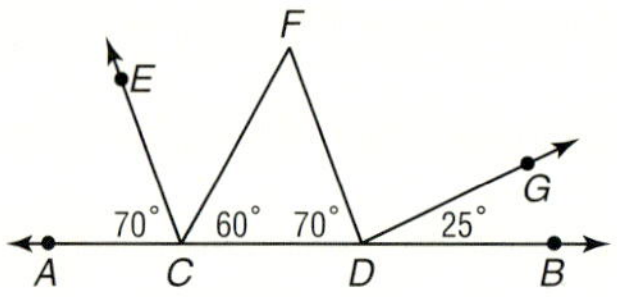

Ⓐ $\overline{CE} \parallel \overline{DF}$ Ⓑ $\overline{CF} \parallel \overline{DG}$

Ⓒ $\overline{CF} \cong \overline{DF}$ Ⓓ $\overline{CE} \cong \overline{DF}$

3. Which of the following would *not* prove that quadrilateral $QRST$ is a parallelogram? (Lesson 8-2) D

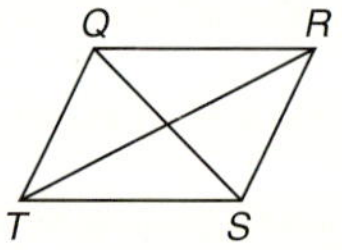

Ⓐ Both pairs of opposite angles are congruent.

Ⓑ Both pairs of opposite sides are parallel.

Ⓒ Diagonals bisect each other.

Ⓓ A pair of opposite sides is congruent.

4. If $Q(4, 2)$ is reflected in the y-axis, what will be the coordinates of Q'? (Lesson 9-1) B

Ⓐ $(-4, -2)$ Ⓑ $(-4, 2)$

Ⓒ $(2, -4)$ Ⓓ $(2, 4)$

5. Which of the following statements about the figures below is true? (Lesson 9-2) D

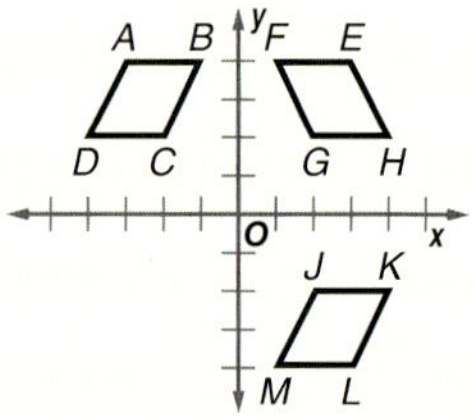

Ⓐ Parallelogram $JKLM$ is a reflection image of ▱$ABCD$.

Ⓑ Parallelogram $EFGH$ is a translation image of ▱$ABCD$.

Ⓒ Parallelogram $JKLM$ is a translation image of ▱$EFGH$.

Ⓓ Parallelogram $JKLM$ is a translation image of ▱$ABCD$.

6. Which of the following is **not** necessarily preserved in a congruence transformation? (Lesson 9-2) B

Ⓐ angle and distance measure

Ⓑ orientation

Ⓒ collinearity

Ⓓ betweenness of points

7. Which transformation is used to map $\triangle ABC$ to $\triangle A'B'C'$? (Lesson 9-3) A

Ⓐ rotation

Ⓑ reflection

Ⓒ dilation

Ⓓ translation

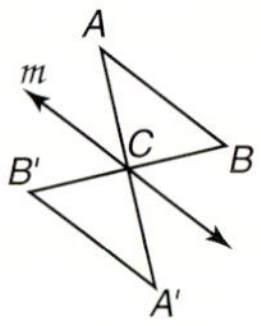

The Princeton Review

Log On **for Test Practice**

The Princeton Review offers additional test-taking tips and practice problems at their web site. Visit www.princetonreview.com or www.review.com

TestCheck and Worksheet Builder

Special banks of standardized test questions similar to those on the SAT, ACT, TIMSS 8, NAEP 8, and state proficiency tests can be found on this CD-ROM.

Aligned and verified by The Princeton Review

FCAT Practice

Part 2 Short Response/Grid In

Record your answers on the answer sheet provided by your teacher or on a sheet of paper.

8. A new logo was designed for GEO Company. The logo is shaped like a symmetrical hexagon. What are the coordinates of the missing vertex of the logo? (Lesson 1-1) **(1, −2)**

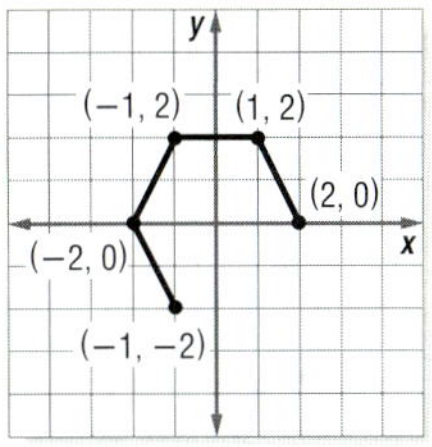

9. A soccer coach is having her players practice penalty kicks. She places two cones equidistant from the goal and asks the players to line up behind each cone. What is the value of x? (Lesson 4-6) **45**

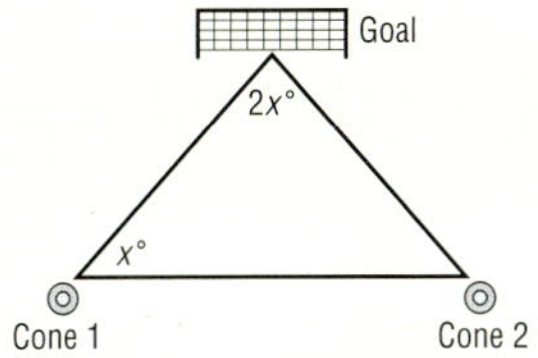

10. A steel cable, which supports a tram, needs to be replaced. To determine the length x of the cable currently in use, the engineer makes several measurements and draws the diagram below of two right triangles, $\triangle ABC$ and $\triangle EDC$. If $m\angle ACB = m\angle ECD$, what is the length x of the cable currently in use? Round the result to the nearest meter. (Lesson 6-3) **329m**

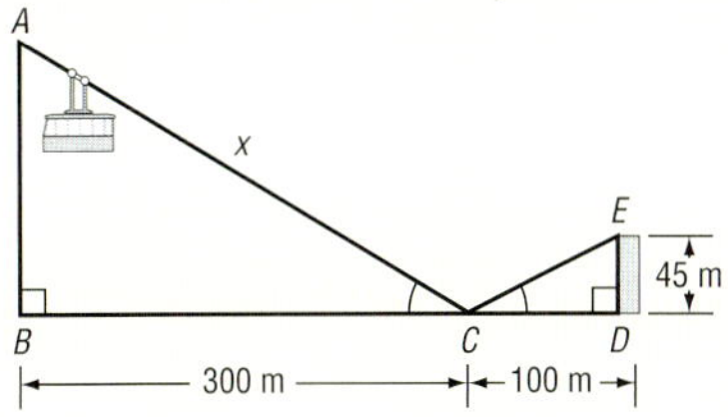

The Princeton Review Test-Taking Tip

Question 4 Always check your work for careless errors. To check your answer to this question, remember the following rule. In a reflection over the *x*-axis, the *x*-coordinate remains the same, and the *y*-coordinate changes its sign. In a reflection over the *y*-axis, the *y*-coordinate remains the same, and the *x*-coordinate changes its sign.

FCAT Practice

Part 3 Open Ended

Record your answers on a sheet of paper. Show your work.

11. Kelli drew the diagram below to show the front view of a circus tent. Prove that $\triangle ABD$ is congruent to $\triangle ACE$. (Lessons 4-5 and 4-6) **See p. 519D.**

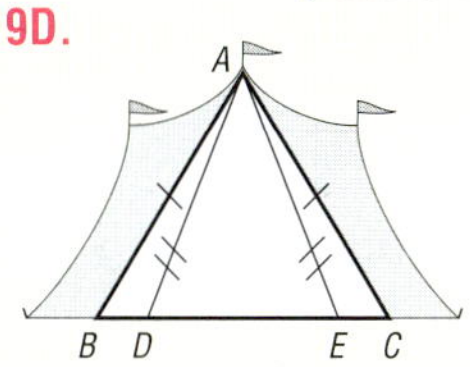

12. Paul is studying to become a landscape architect. He drew a map view of a park on the following vertices: $Q(2, 2)$, $R(-2, 4)$, $S(-3, -2)$, and $T(3, -4)$ **See margin.**

a. On a coordinate plane, graph quadrilateral $QRST$. (Prerequisite Skill)

b. Paul's original drawing appears small on his paper. His instructor says that he should dilate the image with the origin as center and a scale factor of 2. Graph and label the coordinates of the dilation image $Q'R'S'T'$. (Lesson 9-5)

c. Explain how Paul can determine the coordinates of the vertices of $Q'R'S'T'$ without using a coordinate plane. Use one of the vertices for a demonstration of your method. (Lesson 9-5)

d. Dilations are similarity transformations. What properties are preserved during an enlargement? reduction? congruence transformation? (Lesson 9-5)

www.geometryonline.com/standardized_test/fcat

Evaluating Open-Ended Assessment Questions

Open-Ended Assessment questions are graded by using a multilevel rubric that guides you in assessing a student's knowledge of a particular concept.

Goal: To prove triangle congruence and apply concepts of transformations to landscaping architecture.

Sample Scoring Rubric: The following rubric is a sample scoring device. You may wish to add more detail to this sample to meet your individual scoring needs.

Score	Criteria
4	A correct solution that is supported by well-developed, accurate explanations
3	A generally correct solution, but may contain minor flaws in reasoning or computation
2	A partially correct interpretation and/or solution to the problem
1	A correct solution with no supporting evidence or explanation
0	An incorrect solution indicating no mathematical understanding of the concept or task, or no solution is given

Answers

12a.

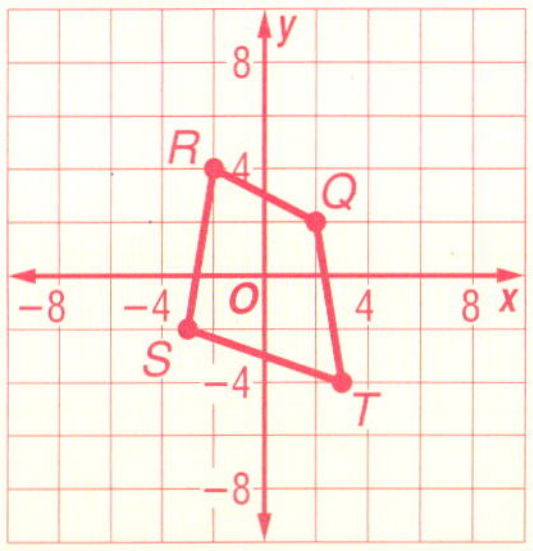

12b.

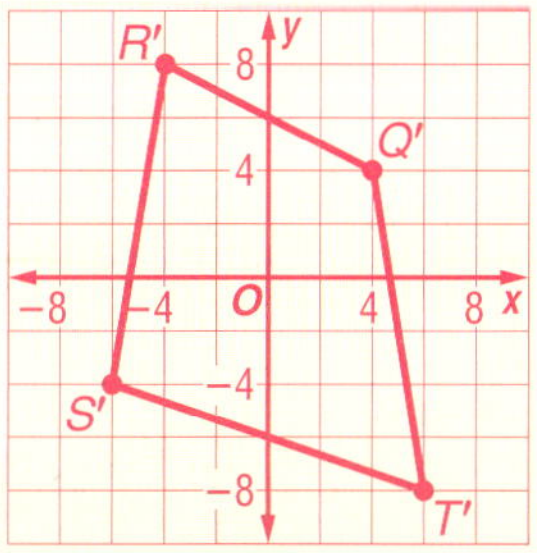

12c. Multiply the x- and y-coordinates of each vertex by the scale factor; $Q(2, 2)$ becomes $Q'(2 \times 2, 2 \times 2)$ or $Q'(4, 4)$.

12d. Sample answer: Enlargements and reductions preserve the shape of the figure. Congruence transformations preserve collinearity, betweenness of points, and angle and distance measures.

Page 461, Chapter 9 Getting Started

1.
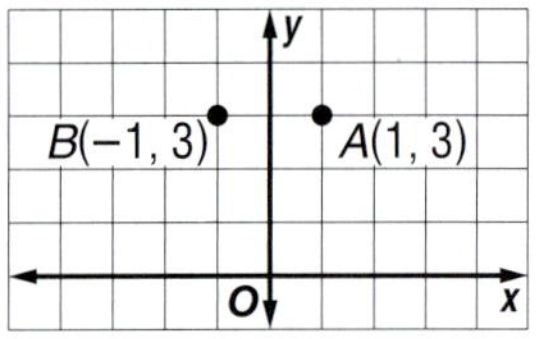

2.
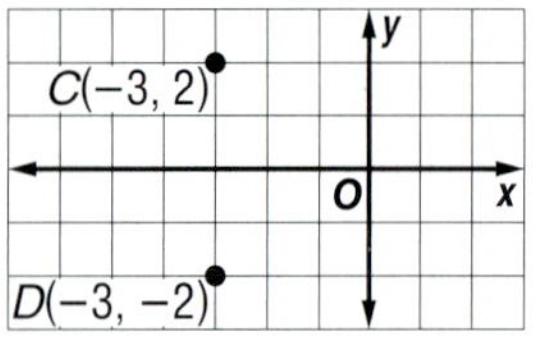

3.
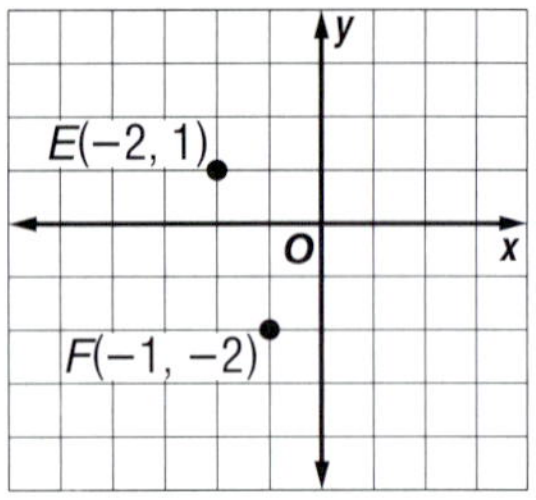

4.
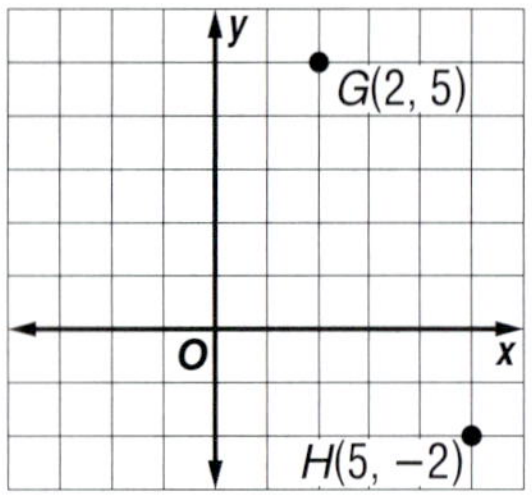

5.
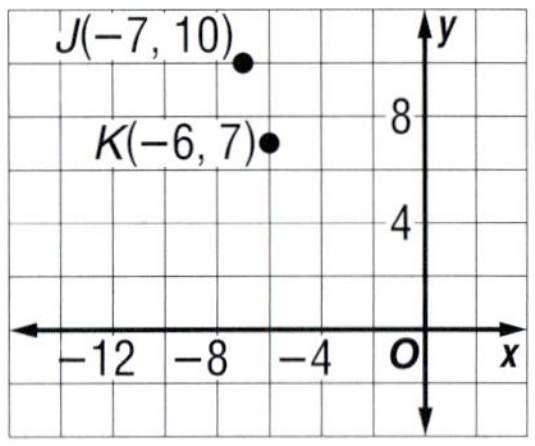

6.
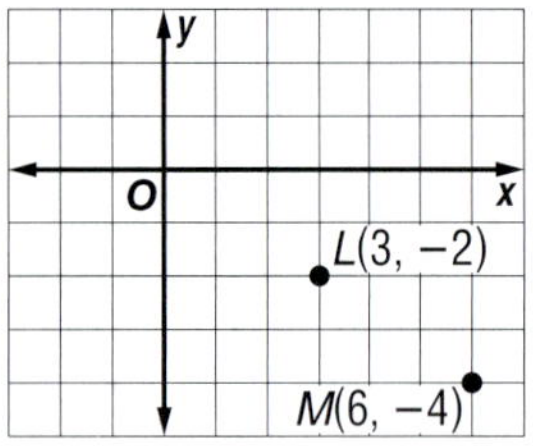

Pages 467–469, Lesson 9-1

8.
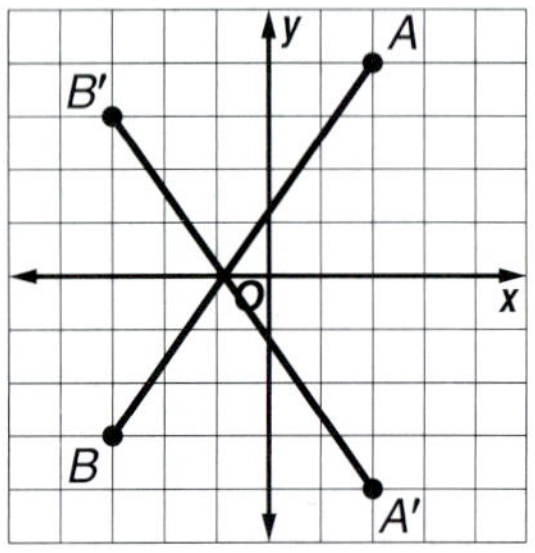

9.
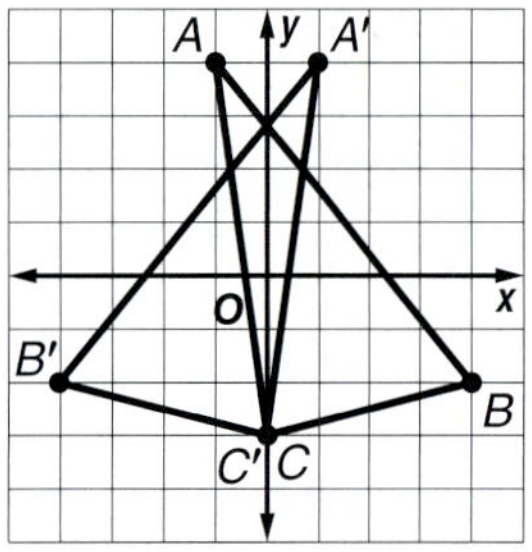

10.
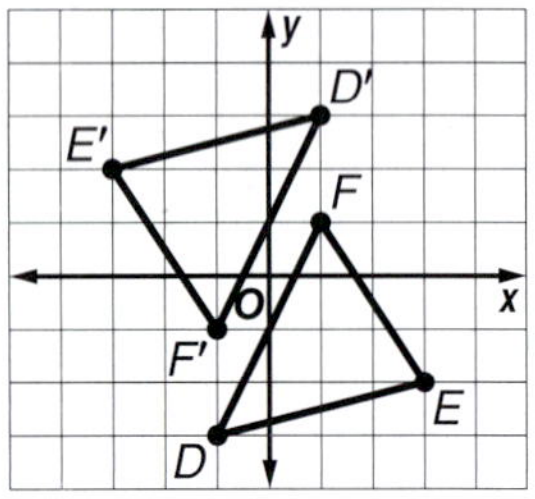

11.
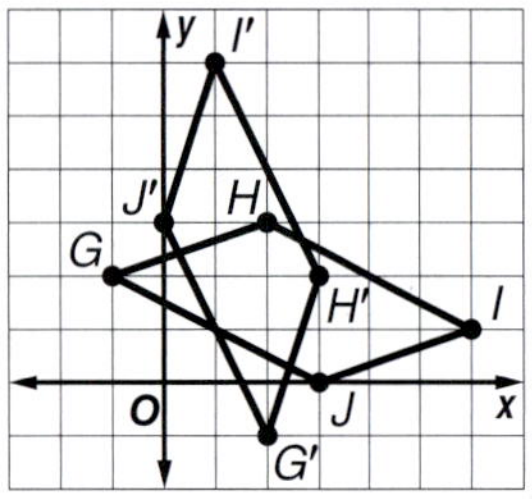

27.
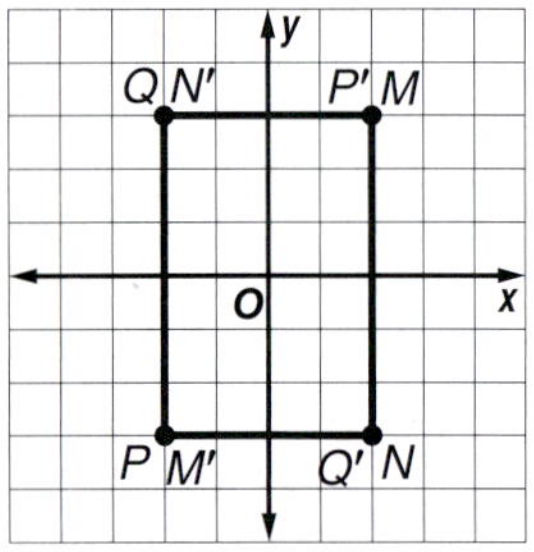

28.
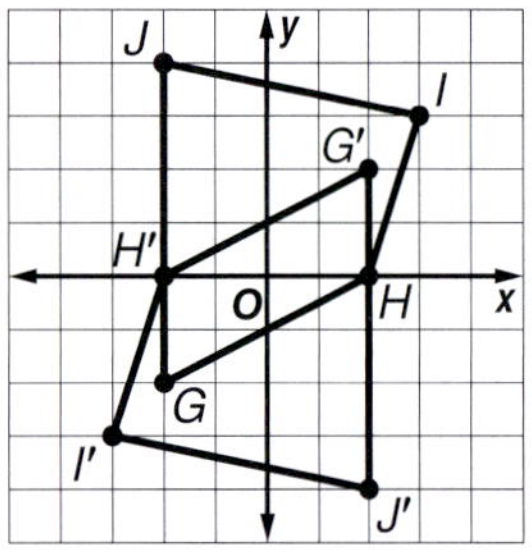

29.
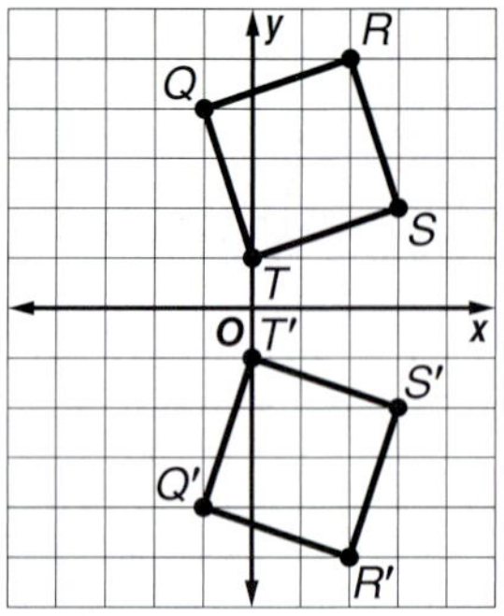

30.
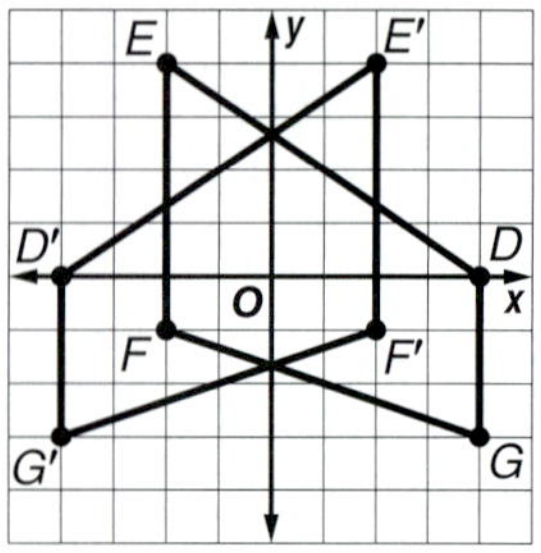

31.
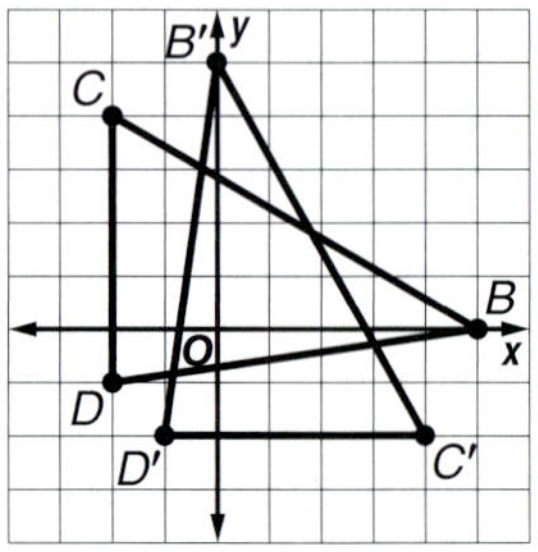

32.
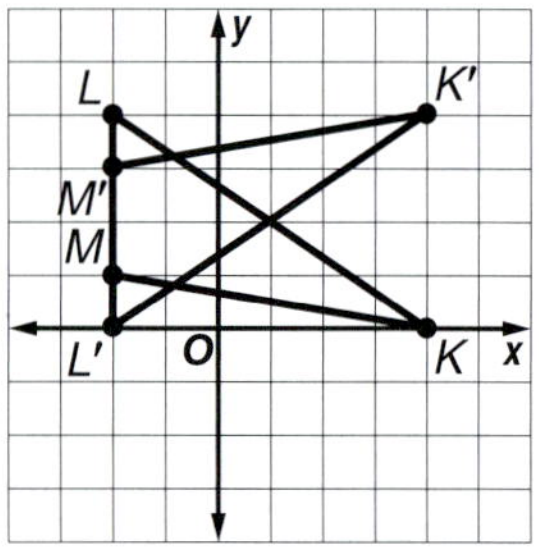

33.
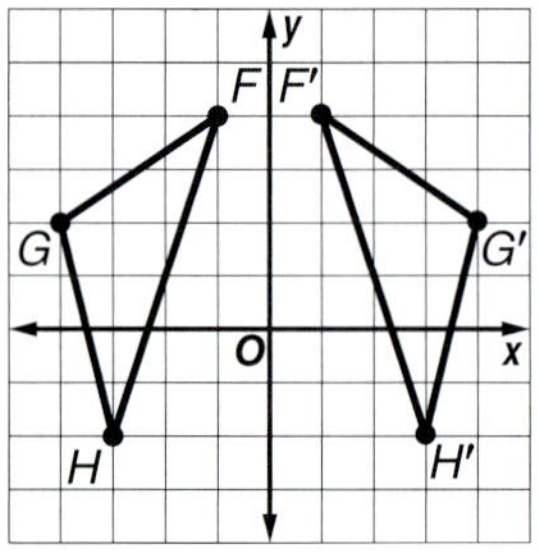

$(x, y) \rightarrow (-x, y)$

34.
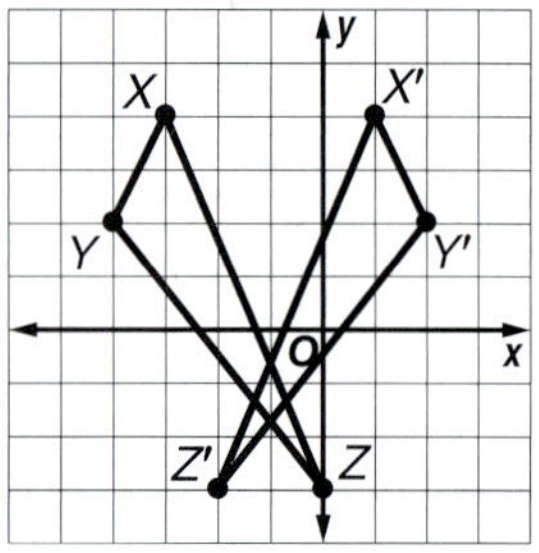

$(x, y) \rightarrow (-(x + 2), y)$

51. **Given:** Quadrilateral *LMNP*
X, *Y*, *Z*, and *W* are midpoints of their respective sides.
Prove: $\overline{YW}$ and $\overline{XZ}$ bisect each other.

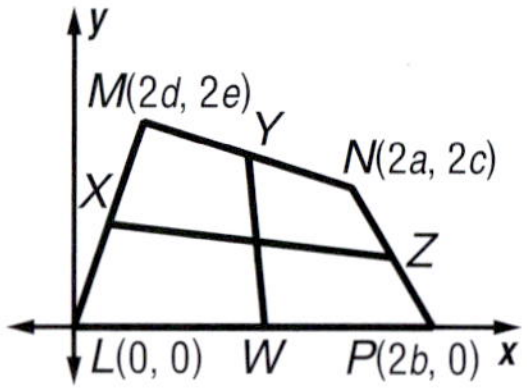

Proof:

Midpoint *Y* of $\overline{MN}$ is $\left(\frac{2d + 2a}{2}, \frac{2e + 2c}{2}\right)$ or $(d + a, e + c)$.

Midpoint *Z* of $\overline{NP}$ is $\left(\frac{2a + 2b}{2}, \frac{2c + 0}{2}\right)$ or $(a + b, c)$.

Midpoint *W* of $\overline{PL}$ is $\left(\frac{0 + 2b}{2}, \frac{0 + 0}{2}\right)$ or $(b, 0)$.

Midpoint *X* of $\overline{LM}$ is $\left(\frac{0 + 2d}{2}, \frac{0 + 2e}{2}\right)$ or (d, e).

Midpoint of $\overline{WY}$ is $\left(\frac{d + a + b}{2}, \frac{e + c + 0}{2}\right)$ or $\left(\frac{a + b + d}{2}, \frac{c + e}{2}\right)$.

Midpoint of $\overline{XZ}$ is $\left(\frac{d + a + b}{2}, \frac{e + c}{2}\right)$ or $\left(\frac{a + b + d}{2}, \frac{c + e}{2}\right)$.

The midpoints of $\overline{XZ}$ and $\overline{WY}$ are the same, so $\overline{XZ}$ and $\overline{WY}$ bisect each other.

52. **Given:** Isosceles trapezoid
$\overline{AD} \cong \overline{BC}$
H, *J*, *K*, and *G* are midpoints of their respective sides.
Prove: *GHJK* is a rhombus.

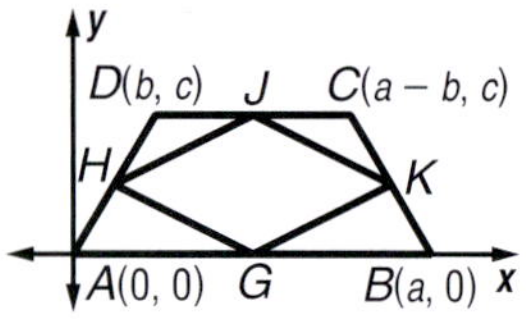

Proof:

Midpoint H of $\overline{AD}$ is $\left(\frac{b+0}{2}, \frac{c+0}{2}\right) = \left(\frac{b}{2}, \frac{c}{2}\right)$.

Midpoint J of $\overline{DC}$ is $\left(\frac{a-b+b}{2}, \frac{c+c}{2}\right) = \left(\frac{a}{2}, c\right)$.

Midpoint K of $\overline{CB}$ is $\left(\frac{a-b+a}{2}, \frac{c+0}{2}\right) = \left(\frac{2a-b}{2}, \frac{c}{2}\right)$.

Midpoint G of $\overline{DC}$ is $\left(\frac{a+0}{2}, \frac{0+0}{2}\right) = \left(\frac{a}{2}, 0\right)$.

$$HJ = \sqrt{\left(\frac{b}{2}-\frac{a}{2}\right)^2 + \left(\frac{c}{2}-c\right)^2} = \frac{\sqrt{b^2-2ab+a^2+c^2}}{2};$$

$$GK = \sqrt{\left(\frac{2a-b}{2}-\frac{a}{2}\right)^2 + \left(\frac{c}{2}-0\right)^2} = \frac{\sqrt{b^2-2ab+a^2+c^2}}{2};$$

$$HG = \sqrt{\left(\frac{b}{2}-\frac{a}{2}\right)^2 + \left(\frac{c}{2}-0\right)^2} = \frac{\sqrt{b^2-2ab+a^2+c^2}}{2};$$

$$KJ = \sqrt{\left(\frac{2a-b}{2}-\frac{a}{2}\right)^2 + \left(c-\frac{c}{2}\right)^2} = \frac{\sqrt{b^2-2ab+a^2+c^2}}{2};$$

$HJ = GK = HG = KJ$, so $\overline{HJ} \cong \overline{GK} \cong \overline{HG} \cong \overline{KJ}$ and $GHJK$ is a rhombus.

Pages 472–475, Lesson 9-2

6.
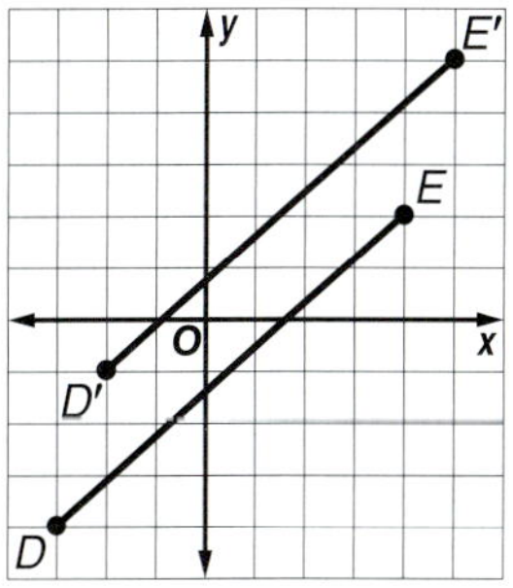

7.
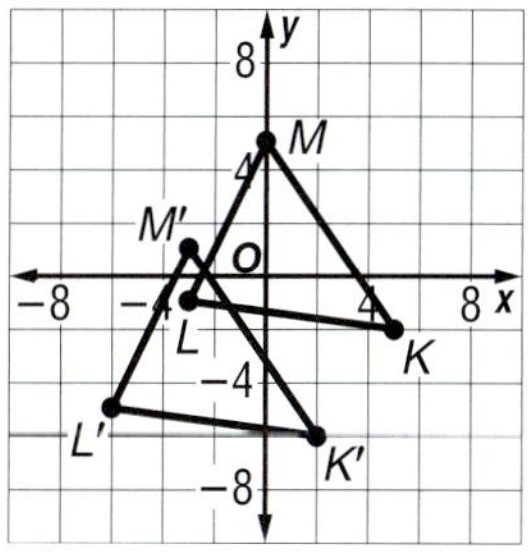

15.
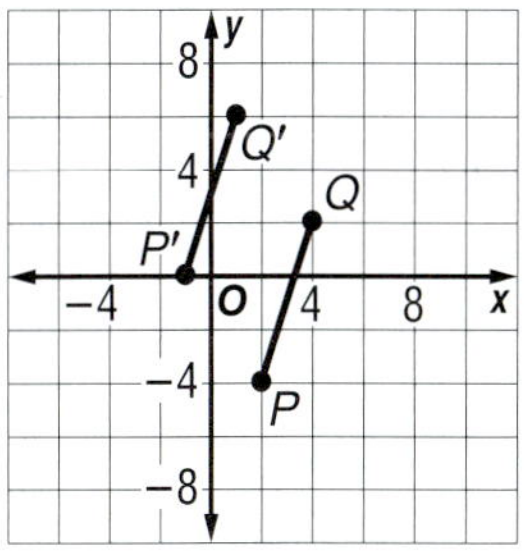

16.
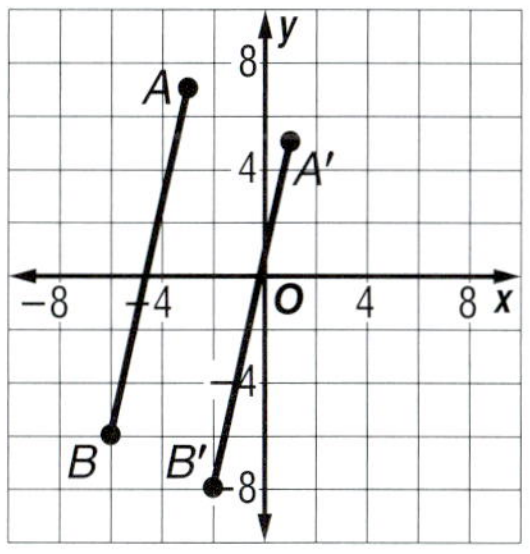

17.
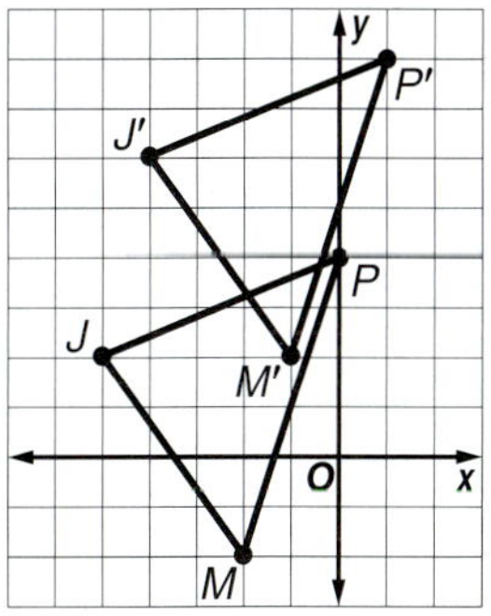

18.
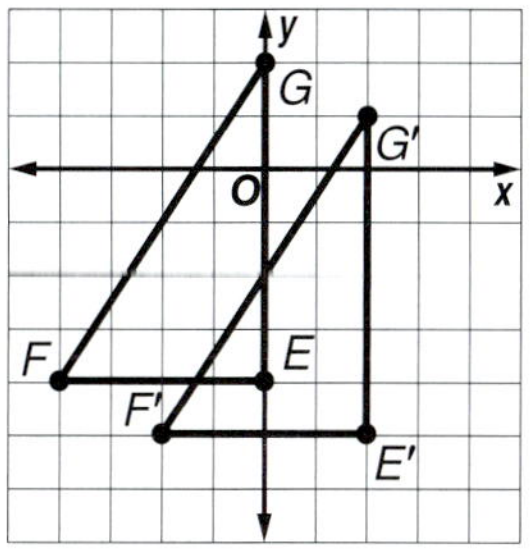

19.
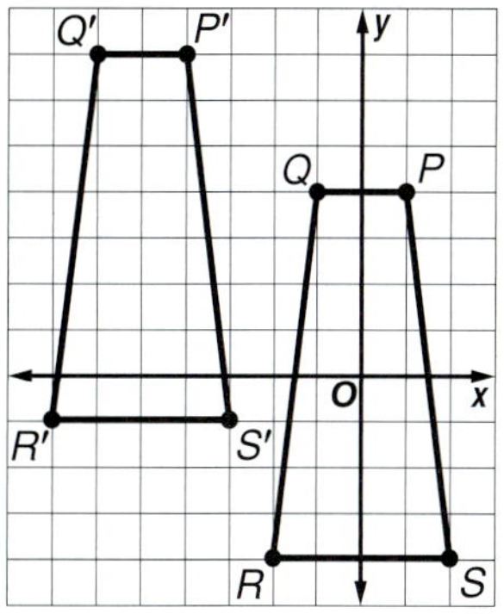

20.
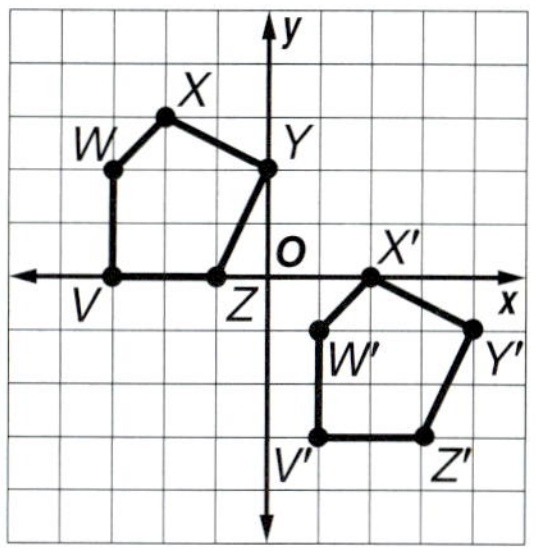

27.
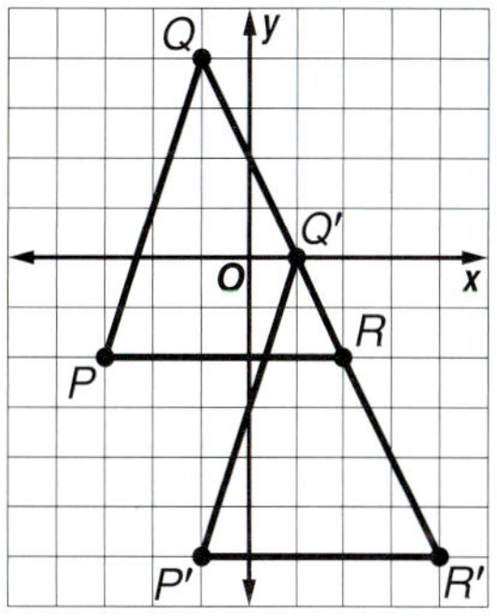

28.
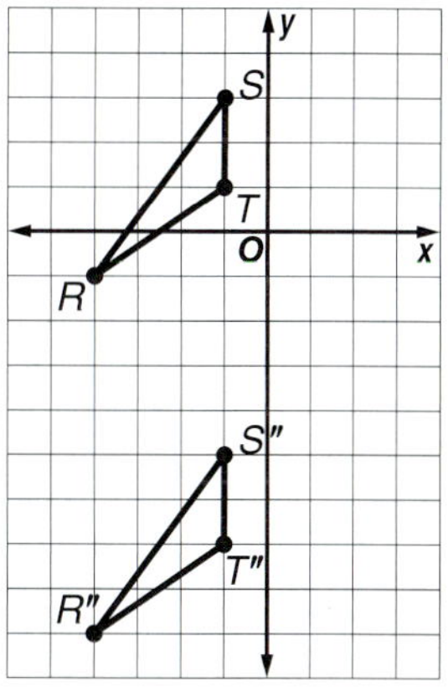

29.
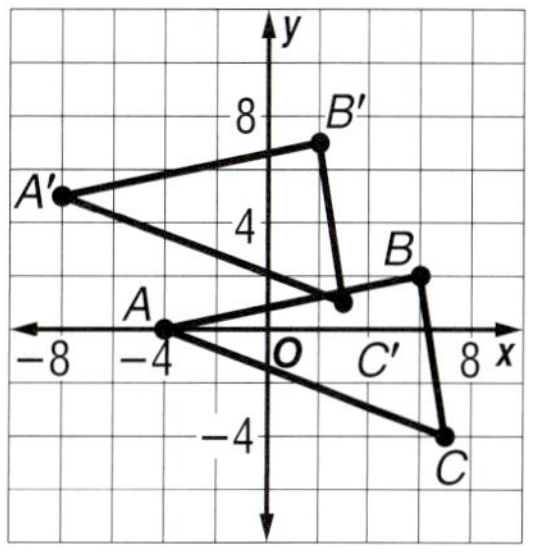

30.
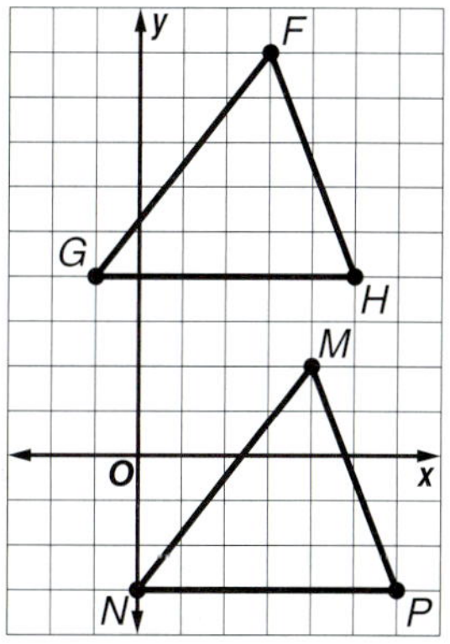

54.
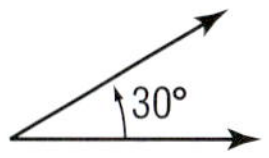

55.
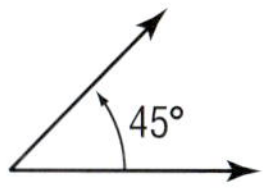

56.
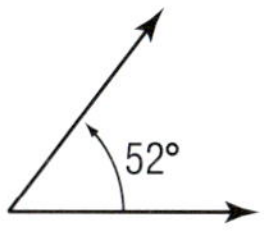

57.
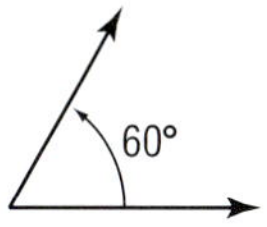

58.
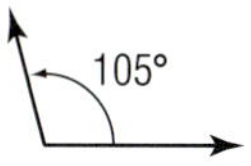

59.
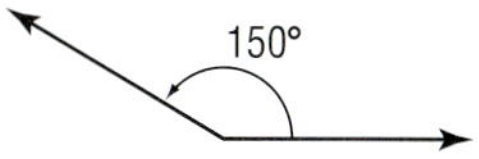

Pages 478–482, Lesson 9-3

1. clockwise
$(x, y) \rightarrow (y, -x)$

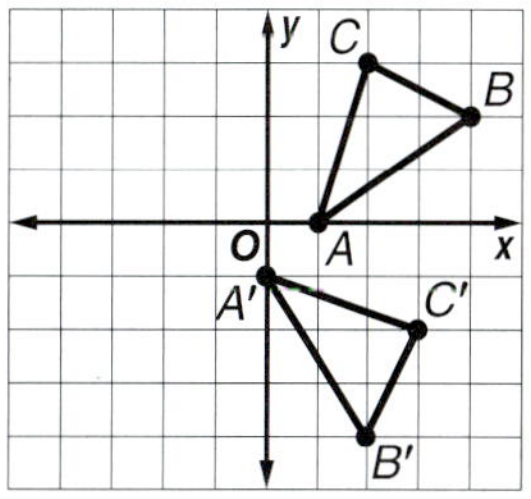

counterclockwise
$(x, y) \rightarrow (-y, x)$

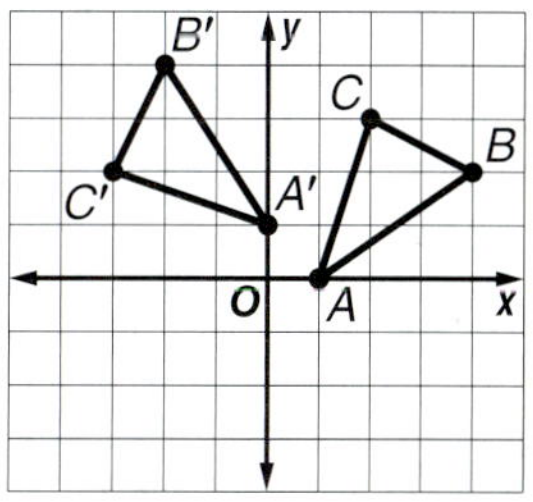

2. A rotation image can be found by reflecting the image in a line, then reflecting that image in a line that intersects the first. The second method is to rotate each point of the given figure using the angle of rotation.
3. Both translations and rotations are made up of two reflections. The difference is that translations reflect across parallel lines and rotations reflect across intersecting lines.

7.
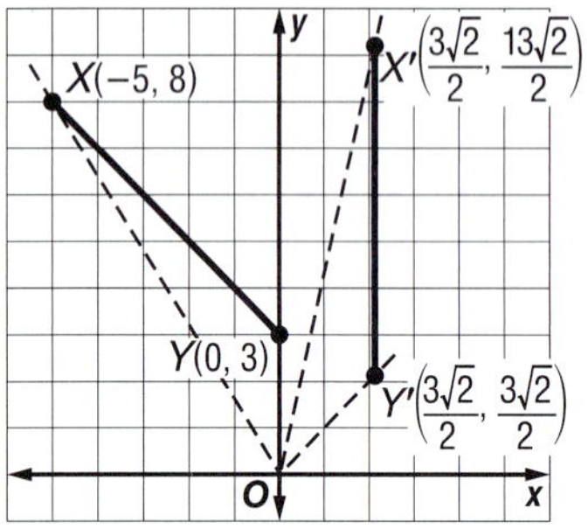

8.
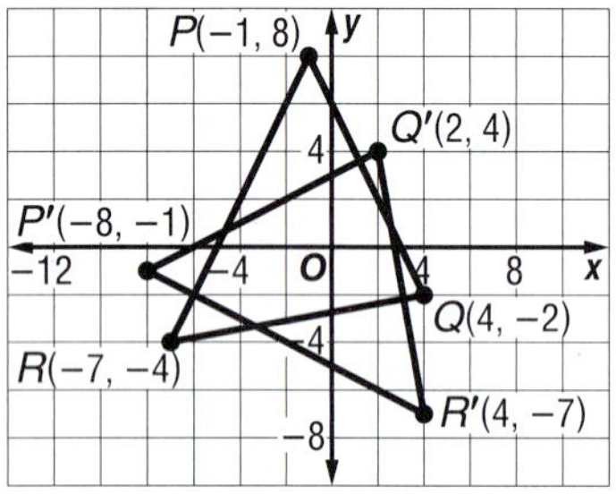

Page 482, Practice Quiz 1

1.
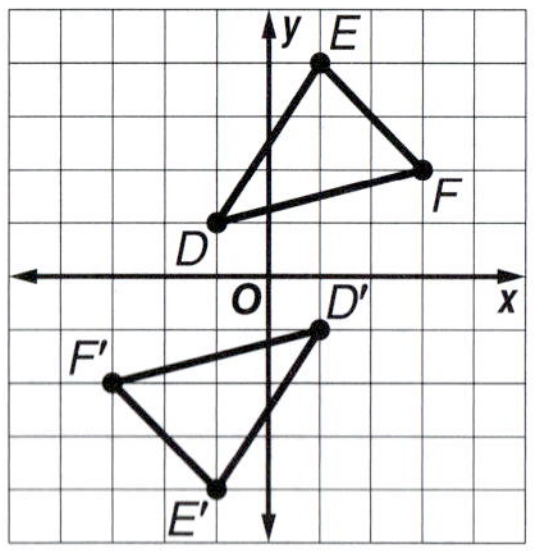

2.
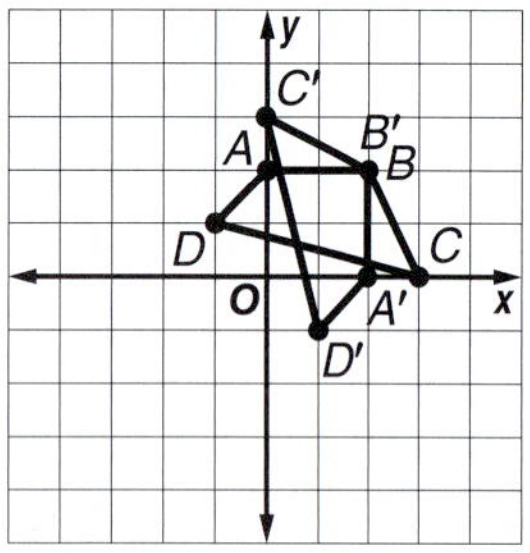

3.
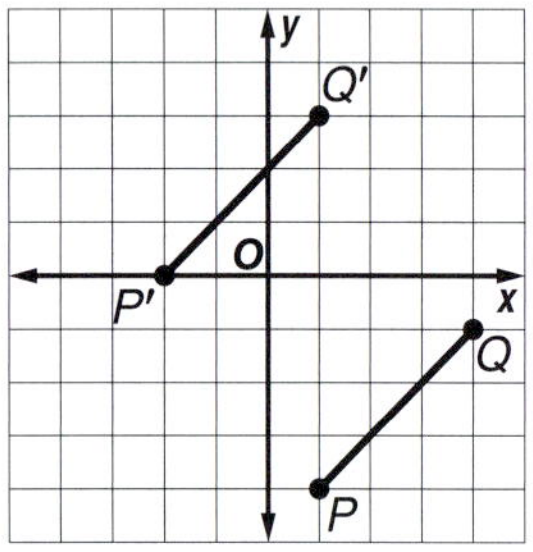

4.
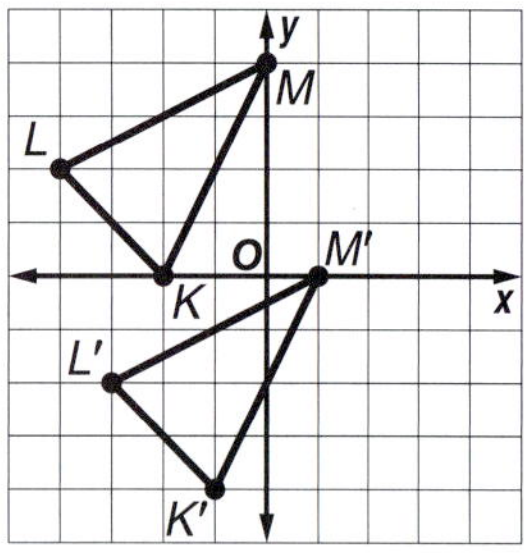

Pages 485–488, Lesson 9-4

30. The sum of the measures of the interior angles of a triangle is 180. If each angle is used twice at each vertex, the sum of the angles is 360.
31. Semi-regular tessellations have the same combination of shapes and angles at each vertex like uniform tessellations. The shapes for semi-regular tessellations are just regular.
32. When the combination of shapes are regular polygons, then the uniform tessellation becomes semi-regular.
33. The sum of the measures of the angles of a quadrilateral is 360. So if each angle of the quadrilateral is rotated at the vertex, then that equals 360 and the tessellation is possible.
34. The measure of an interior angle is 157.5, which is not a factor of 360.

Page 489, Geometry Activity

1. Yes; whatever space is taken out of the square is then added onto the outside of the square. The area does not change; only the shape changes.
2. Modify the bottom of the unit to be like the right side of the triangle. Erase the bottom and right original sides of the triangle.

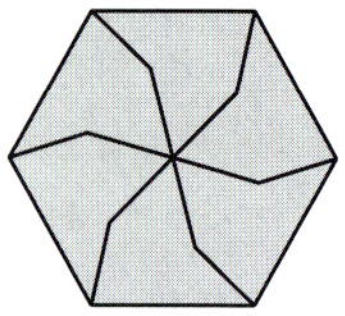

3.
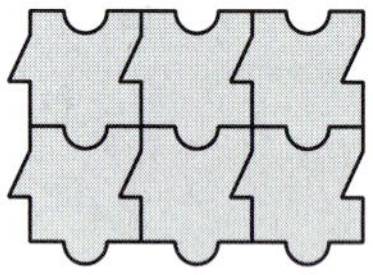

4.
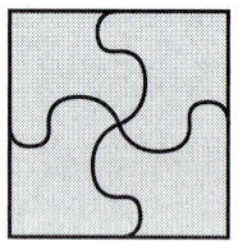

5.
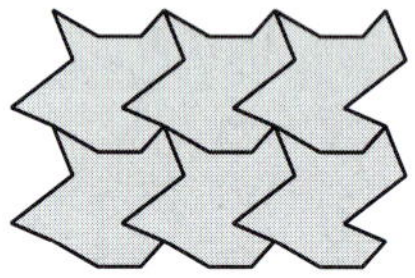

Pages 493–497, Lesson 9-5

49.
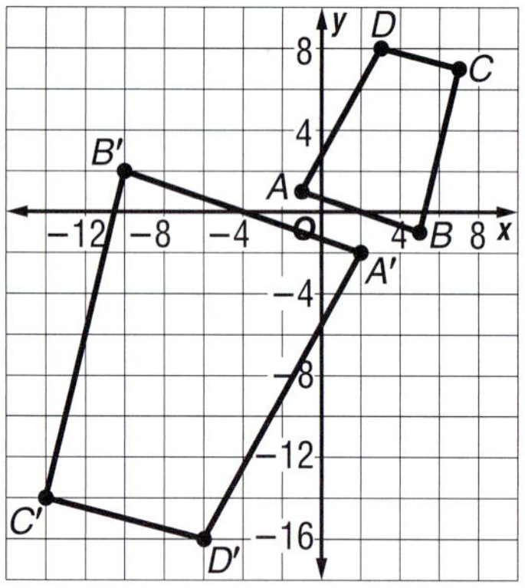

51.
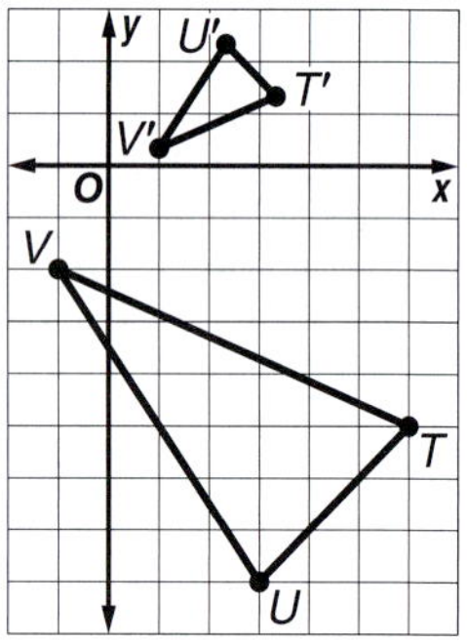

53. Sample answer: Yes; a cut and paste produces an image congruent to the original. Answers should include the following.
- Congruent figures are similar, so cutting and pasting is a similarity transformation.
- If you scale both horizontally and vertically by the same factor, you are creating a dilation.

60.

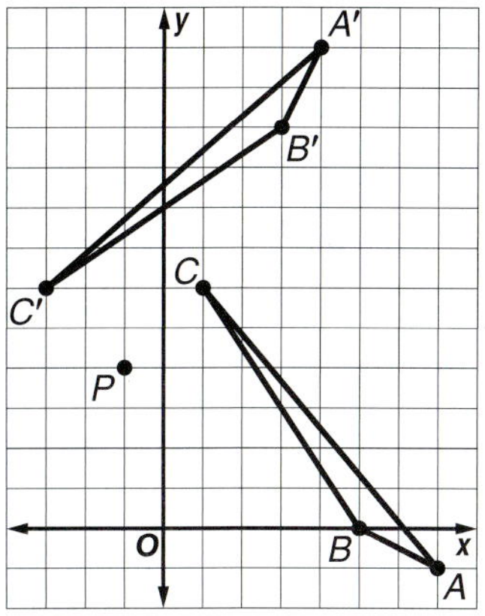

61.

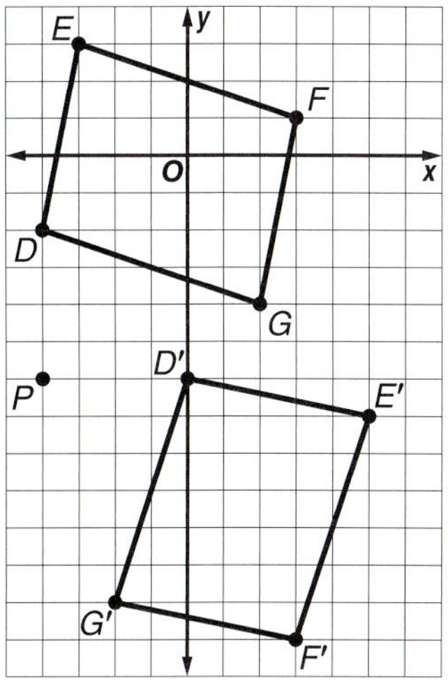

Pages 502–505, Lesson 9-6

9.

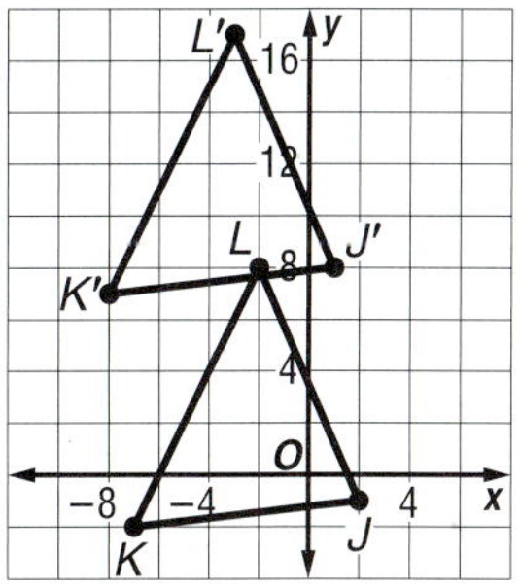

10.

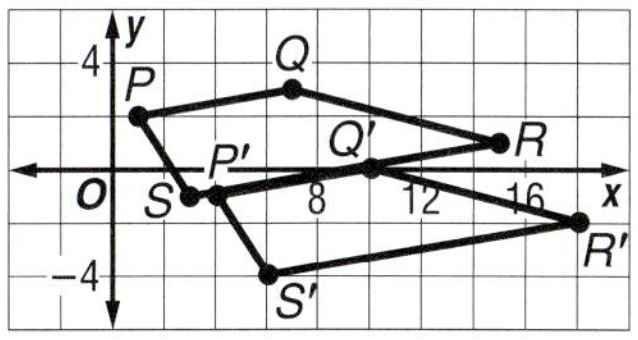

11.

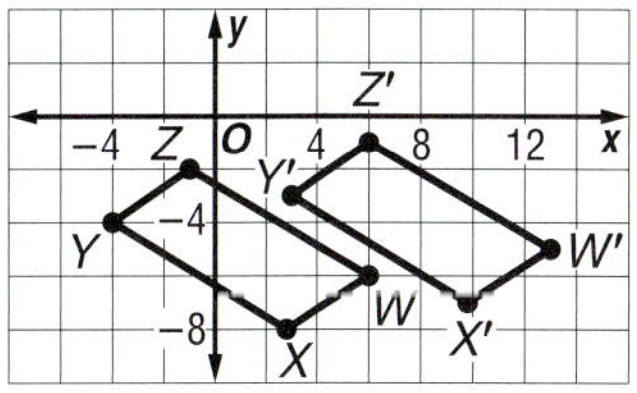

43.

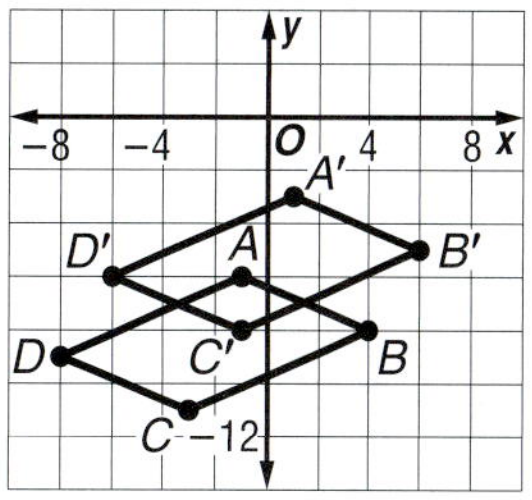

44.

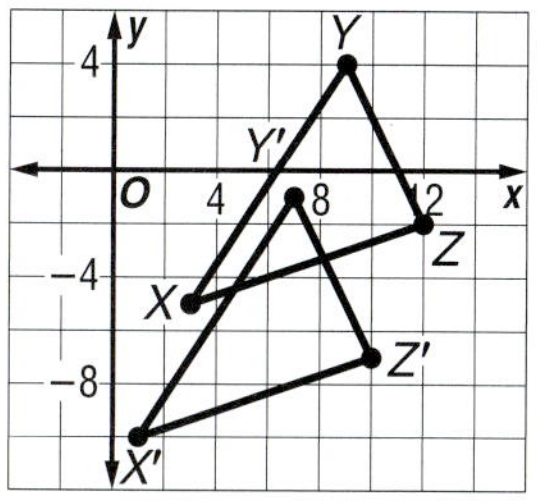

45.

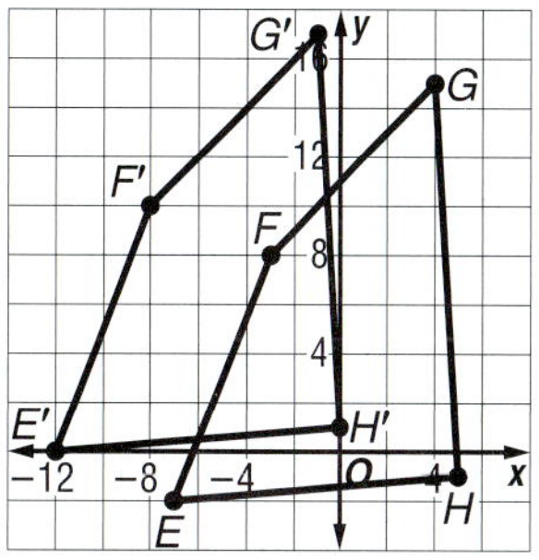

46.

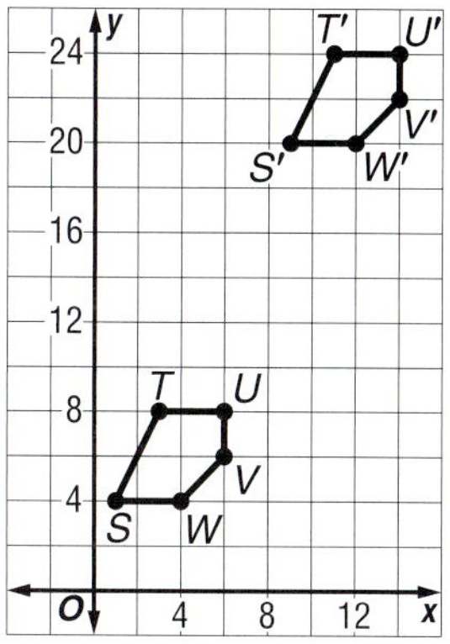

Pages 508–511, Lesson 9-7

39. $A'\left(-1, -\frac{1}{3}\right)$, $B'\left(-\frac{2}{3}, -\frac{4}{3}\right)$, $C'\left(\frac{2}{3}, -\frac{4}{3}\right)$, $D'\left(1, -\frac{1}{3}\right)$, $E'\left(\frac{2}{3}, \frac{2}{3}\right)$, $F'\left(-\frac{2}{3}, \frac{2}{3}\right)$

40. $A'(-3, -8)$, $B'(-6, -7)$, $C'(-6, -3)$, $D'(-3, -2)$, $E'(0, -3)$, $F'(0, -7)$

41. $A'(2, 1)$, $B'(5, 2)$, $C'(5, 6)$, $D'(2, 7)$, $E'(-1, 6)$, $F'(-1, 2)$

42. $A'(-6, 2)$, $B'(-4, 8)$, $C'(4, 8)$, $D'(6, 2)$, $E'(4, -4)$, $F'(-4, -4)$

Pages 518–519, Chapter 9 Standardized Test Practice

11. Given: $\overline{AB} \cong \overline{AC}$, $\overline{AD} \cong \overline{AE}$

Prove: $\triangle ABD \cong \triangle ACE$

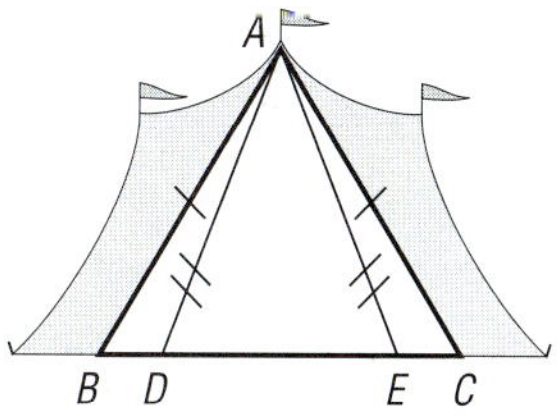

Proof:

Statements Reasons

1. $\overline{AB} \cong \overline{AC}$, $\overline{AD} \cong \overline{AE}$ (Given)
2. $\angle ABD \cong \angle ACE$; $\angle ADE \cong \angle AED$ (Isos. △ Thm.)
3. $\angle ADB$ and $\angle ADE$ are supplementary; $\angle AEC$ and $\angle AED$ are supplementary. (If 2 ∠s form a linear pair, then they are suppl.)
4. $\angle ADB \cong \angle AEC$ (∠s suppl. to ≅ ∠s are ≅.)
5. $\triangle ABD \cong \triangle ACE$ (AAS)

Chapter 10

Circles

Chapter Overview and Pacing

LESSON OBJECTIVES

LESSON OBJECTIVES	PACING (days) Regular Basic/Average	Regular Advanced	Block Basic/Average	Block Advanced
10-1 Circles and Circumference *(pp. 522–528)* • Identify and use parts of circles. • Solve problems involving the circumference of a circle.	1	1	0.5	0.5
10-2 Angles and Arcs *(pp. 529–535)* • Recognize major arcs, minor arcs, semicircles, and central angles and their measures. • Find arc length.	2	2	1	1
10-3 Arcs and Chords *(pp. 536–543)* • Recognize and use relationships between arcs and chords. • Recognize and use relationships between chords and diameters.	2	2	1	1
10-4 Inscribed Angles *(pp. 544–551)* • Find measures of inscribed angles. • Find measures of angles of inscribed polygons.	2	2	1	1
10-5 Tangents *(pp. 552–560)* • Use properties of tangents. • Solve problems involving circumscribed polygons. ***Follow-Up:*** To construct inscribed and circumscribed triangles.	2 (with 10-5 Follow-Up)	2 (with 10-5 Follow-Up)	1 (with 10-5 Follow-Up)	1 (with 10-5 Follow-Up)
10-6 Secants, Tangents, and Angle Measures *(pp. 561–568)* • Find measures of angles formed by lines intersecting on or inside a circle. • Find measures of angles formed by lines intersecting outside the circle.	2	2	1	1
10-7 Special Segments in a Circle *(pp. 569–574)* • Find measures of segments that intersect in the interior of a circle. • Find measures of segments that intersect in the exterior of a circle.	2	2	1	1
10-8 Equations of Circles *(pp. 575–580)* • Write the equation of a circle. • Graph a circle on the coordinate plane.	1	1	0.5	0.5
Study Guide and Practice Test *(pp. 581–587)* **Standardized Test Practice** *(pp. 588–589)*	1	1	0.5	0.5
Chapter Assessment	1	1	0.5	0.5
TOTAL	**16**	**16**	**8**	**8**

Pacing suggestions for the entire year can be found on pages T20–T21.

Timesaving Tools
TeacherWorks™
All-In-One Planner and Resource Center
See pages T5 and T21.

Chapter Resource Manager

Study Guide and Intervention	Practice (Skills and Average)	Reading to Learn Mathematics	Enrichment	Assessment	Prerequisite Skills Workbook	Applications*	5-Minute Check Transparencies	Interactive Chalkboard	GeomPASS: Tutorial Plus (lessons)	Materials
CHAPTER 10 RESOURCE MASTERS										
541–542	543–544	545	546		11–12, 23–24, 45–48	SC 19	10-1	10-1		
547–548	549–550	551	552	603	31–32, 61–64, 67–68, 71–72, 105–106, 109–110		10-2	10-2		compass, protractor
553–554	555–556	557	558				10-3	10-3		compass, patty paper, centimeter ruler, scissors, protractor
559–560	561–562	563	564	603, 605	41–42		10-4	10-4		compass, protractor, straightedge
565–566	567–568	569	570		15–16	GCC 35, 36 SC 20	10-5	10-5		compass, straightedge (*Follow-Up:* straightedge, compass, paper)
571–572	573–574	575	576	604	17–18		10-6	10-6		compass, straightedge
577–578	579–580	581	582		35–36, 51–52		10-7	10-7		compass, straightedge
583–584	585–586	587	588	604			10-8	10-8	19	grid paper, compass, straightedge
				589–602, 606–608						

**Key to Abbreviations:* GCC = Graphing Calculator and Computer Masters
SC = School-to-Career Masters

Chapter 10

Mathematical Connections and Background

Continuity of Instruction

Prior Knowledge

Students solved equations for a variable and used the Quadratic Formula in previous courses. In Chapter 4, students found the measures of the angles in isosceles triangles. In Chapter 7, they found the missing side length in a right triangle and used the Pythagorean Theorem to determine whether figures were right triangles.

This Chapter

This chapter focuses exclusively on circles and their special properties. A circle is a unique geometric shape in which the angles, arcs, and segments intersecting that circle have special relationships. In this chapter, students identify the parts of a circle and solve problems involving circumference. They find arc and angle measures and the measures of segments in a circle. In addition, students write the equation of a circle and graph circles in the coordinate plane.

Future Connections

Students will use their knowledge of circles to find the area of a circle in Chapter 11. They will also need to understand a circle to understand a sphere, which is introduced in Chapter 12.

10-1 Circles and Circumference

A circle is the locus of all points in a plane equidistant from a given point, which is the center of the circle. A circle is usually named by its center point. Any segment with endpoints on the circle is a chord of the circle. A chord that contains the center of the circle is a diameter of the circle. Any segment with endpoints that are the center and a point on the circle is a radius. All radii of a circle are congruent and all diameters are congruent.

The circumference of a circle is the distance around the circle. The ratio of the circumference to the diameter of a circle is always equal to π. For a circumference of C units and a diameter of d units or a radius of r units, $C = \pi d$ or $C = 2\pi r$.

10-2 Angles and Arcs

A central angle of a circle has the center of the circle as its vertex, and its sides are two radii of the circle. The sum of the measures of the central angles of a circle with no interior points in common is 360. A central angle separates the circle into two parts, each of which is an arc.

The measure of each arc is related to the measure of its central angle. A minor arc degree measure equals the measure of the central angle and is less than 180. A major arc degree measure equals 360 minus the measure of the minor arc and is greater than 180. A semicircle is also considered an arc and measures 180°. In the same or in congruent circles, two arcs are congruent if and only if their corresponding central angles are congruent.

In a circle graph, the central angles divide a circle into wedges, often expressed as percents. The size of the angle is proportional to the percent. By multiplying the percent by 360, you can determine the measure of the central angle. Another way to measure an arc is by its length. An arc is part of the circle, so the length of an arc is part of the circumference. The ratio of the arc degree measure to the measure of the whole circle is equal to the ratio of the arc length to the circumference. You can use these ratios to solve for arc length.

10-3 Arcs and Chords

The endpoints of a chord are also endpoints of an arc. Arcs and chords have a special relationship. In a circle or in congruent circles, two minor arcs are congruent if and only if their corresponding chords are congruent. In a circle or congruent circles, two chords are congruent if and only if they are equidistant from the center of the circle.

The chords of adjacent arcs can form a polygon. Such a polygon is said to be *inscribed* in the circle because all its vertices lie on the circle. The circle circumscribes the polygon.

Diameters that are perpendicular to chords create special segment and arc relationships. In a circle, if a diameter or radius is perpendicular to a chord, then it bisects the chord and its arc.

10-4 Inscribed Angles

An inscribed angle is an angle that has its vertex on the circle and its sides contained in chords of the circle. If an angle is inscribed in a circle, then the measure of the angle equals one-half of the measure of its intercepted arc (or the measure of the intercepted arc is twice the measure of the inscribed angle). If two inscribed angles of a circle (or congruent circles) intercept congruent arcs or the same arc, then the angles are congruent.

Inscribed polygons also have special properties. An inscribed triangle with a side that is a diameter is a special type of triangle. If an inscribed angle intercepts a semicircle, the angle is a right angle. If a quadrilateral is inscribed in a circle, then its opposite angles are supplementary.

10-5 Tangents

A tangent intersects a circle in exactly one point. This point is called the *point of tangency*. If a line is tangent to a circle, then it is perpendicular to the radius drawn to the point of tangency. The converse of that statement is also true: If a line is perpendicular to a radius of a circle at its endpoint on the circle, then the line is tangent to the circle.

More than one line can be tangent to the same circle. If two segments from the same exterior point are tangent to a circle, then they are congruent.

Circles can be inscribed in polygons, just as polygons can be inscribed in circles. If a circle is inscribed in a polygon, then every side of the polygon is tangent to the circle. You can use what you know about tangents to solve problems involving inscribed circles.

10-6 Secants, Tangents, and Angle Measures

A line that intersects a circle in exactly two points is called a *secant*. When two secants intersect inside a circle, the angles formed are related to the arcs they intercept. If two secants intersect in the interior of a circle, then the measure of an angle formed is one-half the sum of the measure of the arcs intercepted by the angle and its vertical angle.

A secant can also intersect a tangent at the point of tangency. If this occurs, then the measure of each angle formed is one-half the measure of its intercepted arc.

Secants and tangents can intersect outside a circle as well. If two secants, a tangent and a secant, or two tangents intersect in the exterior of a circle, then the measure of the angle formed is one-half the positive difference of the measures of the intercepted arcs.

10-7 Special Segments in a Circle

If two chords intersect in a circle, then the products of the measures of the segments of the chords are equal. You can also use intersecting chords to measure arcs.

If two secant segments are drawn to a circle from an exterior point, then the product of the measures of one secant segment and its external secant segment is equal to the product of the measures of the other secant segment and its external secant segment. This product can also be used if a tangent segment and a secant segment are drawn to a circle from an exterior point. In this case, the square of the measure of the tangent segment is equal to the product of the measures of the secant segment and its external secant segment.

10-8 Equations of Circles

An equation for a circle with center at (h, k) and radius of r units is $(x - h)^2 + (y - k)^2 = r^2$. You can analyze the equation of a circle to find information that will help you graph the circle on a coordinate plane. Once you know the coordinates of the center and the radius of a circle, you can graph the circle. In fact, if you know just three points on a circle, you can graph it and write its equation. By graphing the points as a triangle and constructing two perpendicular bisectors, you can locate the center of the circle. Then you can use the Distance Formula to calculate the radius. Finally, write an equation for the circle.

Chapter 10

DAILY INTERVENTION and Assessment

	Type	Student Edition	Teacher Resources	Technology/Internet
INTERVENTION	Ongoing	Prerequisite Skills, pp. 521, 528, 535, 543, 551, 558, 568, 574 Practice Quiz 1, p. 543 Practice Quiz 2, p. 568	5-Minute Check Transparencies *Prerequisite Skills Workbook,* pp. 11–12, 15–18, 23–24, 31–32, 35–36, 41–42, 45–48, 51–52, 61–64, 67–68, 71–72, 105–106, 109–110 Quizzes, *CRM* pp. 603–604 Mid-Chapter Test, *CRM* p. 605 Study Guide and Intervention, *CRM* pp. 541–542, 547–548, 553–554, 559–560, 565–566, 571–572, 577–578, 583–584	GeomPASS: Tutorial Plus www.geometryonline.com/self_check_quiz www.geometryonline.com/extra_examples
	Mixed Review	pp. 528, 535, 543, 551, 558, 568, 574, 580	Cumulative Review, *CRM* p. 606	
	Error Analysis	Find the Error, pp. 539, 571 Common Misconceptions, p. 555	Find the Error, *TWE* pp. 539, 571 Unlocking Misconceptions, *TWE* p. 532 Tips for New Teachers, *TWE* pp. 524, 562	
ASSESSMENT	Standardized Test Practice	pp. 525, 526, 528, 535, 543, 551, 558, 567, 574, 580, 587, 588, 589	*TWE* pp. 588–589 Standardized Test Practice, *CRM* pp. 607–608	Standardized Test Practice CD-ROM www.geometryonline.com/standardized_test
	Open-Ended Assessment	Writing in Math, pp. 527, 534, 542, 551, 558, 567, 574, 579 Open Ended, pp. 525, 532, 539, 548, 555, 564, 572, 577 Standardized Test, p. 589	Modeling: *TWE* pp. 551, 574 Speaking: *TWE* pp. 528, 568, 580 Writing: *TWE* pp. 535, 543, 558 Open-Ended Assessment, *CRM* p. 601	
	Chapter Assessment	Study Guide, pp. 581–586 Practice Test, p. 587	Multiple-Choice Tests (Forms 1, 2A, 2B), *CRM* pp. 589–594 Free-Response Tests (Forms 2C, 2D, 3), *CRM* pp. 595–600 Vocabulary Test/Review, *CRM* p. 602	TestCheck and Worksheet Builder (see below) MindJogger Videoquizzes www.geometryonline.com/vocabulary_review www.geometryonline.com/chapter_test

Key to Abbreviations: TWE = Teacher Wraparound Edition; CRM = Chapter Resource Masters

Additional Intervention Resources

The Princeton Review's *Cracking the SAT & PSAT*

The Princeton Review's *Cracking the ACT*

TestCheck and Worksheet Builder

This **networkable** software has three modules for intervention and assessment flexibility:

- **Worksheet Builder** to make worksheet and tests
- **Student Module** to take tests on screen (optional)
- **Management System** to keep student records (optional)

Special banks are included for SAT, ACT, TIMSS, NAEP, and End-of-Course tests.

Intervention Technology

GeomPASS: Tutorial Plus CD-ROM offers a complete, self-paced geometry curriculum.

Geometry Lesson	GeomPASS Lesson
10-8	**19** *Circle Relationships*

Intervention at Home

Log on for student study help.

- For each lesson in the Student Edition, there are Extra Examples and Self-Check Quizzes.
  www.geometryonline.com/extra_examples
  www.geometryonline.com/self_check_quiz
- For chapter review, there is vocabulary review, test practice, and standardized test practice.
  www.geometryonline.com/vocabulary_review
  www.geometryonline.com/chapter_test
  www.geometryonline.com/standardized_test

For more information on Intervention and Assessment, see pp. T8–T11.

Reading and Writing in Mathematics

Glencoe Geometry provides numerous opportunities to incorporate reading and writing into the mathematics classroom.

Student Edition

- Foldables Study Organizer, p. 521
- Concept Check questions require students to verbalize and write about what they have learned in the lesson. (pp. 525, 532, 539, 548, 555, 564, 571, 577)
- Writing in Math questions in every lesson, pp. 527, 534, 542, 551, 558, 567, 574, 579
- Reading Study Tip, pp. 522, 536
- WebQuest, pp. 527, 580

Teacher Wraparound Edition

- Foldables Study Organizer, pp. 521, 581
- Study Notebook suggestions, pp. 526, 533, 539, 548, 556, 560, 564, 571, 577
- Modeling activities, pp. 551, 574
- Speaking activities, pp. 528, 568, 580
- Writing activities, pp. 535, 543, 558
- Differentiated Instruction (Verbal/Linguistic), p. 525
- ELL Resources, pp. 520, 525, 527, 534, 541, 550, 557, 565, 573, 579, 581

Additional Resources

- Vocabulary Builder worksheets require students to define and give examples for key vocabulary terms as they progress through the chapter. (*Chapter 10 Resource Masters*, pp. vii-viii)
- Proof Builder helps students learn and understand theorems and postulates from the chapter. (*Chapter 10 Resource Masters*, pp. ix–x)
- Reading to Learn Mathematics master for each lesson (*Chapter 10 Resource Masters*, pp. 545, 551, 557, 563, 569, 575, 581, 587)
- *Vocabulary PuzzleMaker* software creates crossword, jumble, and word search puzzles using vocabulary lists that you can customize.
- *Teaching Mathematics with Foldables* provides suggestions for promoting cognition and language.
- *Reading Strategies for the Mathematics Classroom*
- *WebQuest and Project Resources*

For more information on Reading and Writing in Mathematics, see pp. T6–T7.

Chapter 10 Notes

What You'll Learn

Have students read over the list of objectives and make a list of any words with which they are not familiar.

Why It's Important

Point out to students that this is only one of many reasons why each objective is important. Others are provided in the introduction to each lesson.

Lesson	NCTM Standards	Local Objectives
10-1	3, 4, 6, 8, 9, 10	
10-2	3, 6, 8, 9, 10	
10-3	3, 4, 6, 8, 9, 10	
10-4	3, 6, 8, 9, 10	
10-5	3, 6, 8, 9, 10	
10-4 and 10-5 Follow-Up	3, 6	
10-6	3, 6, 8, 9, 10	
10-7	3, 4, 6, 8, 9, 10	
10-8	3, 4, 6, 8, 9, 10	

Key to NCTM Standards:

1=Number & Operations, 2=Algebra, 3=Geometry, 4=Measurement, 5=Data Analysis & Probability, 6=Problem Solving, 7=Reasoning & Proof, 8=Communication, 9=Connections, 10=Representation

Chapter 10 Circles

What You'll Learn

- **Lessons 10-1** Identify parts of a circle and solve problems involving circumference.
- **Lessons 10-2, 10-3, 10-4, and 10-6** Find arc and angle measures in a circle.
- **Lessons 10-5 and 10-7** Find measures of segments in a circle.
- **Lesson 10-8** Write the equation of a circle.

Key Vocabulary

- chord (p. 522)
- circumference (p. 523)
- arc (p. 530)
- tangent (p. 552)
- secant (p. 561)

Why It's Important

A circle is a unique geometric shape in which the angles, arcs, and segments intersecting that circle have special relationships. You can use a circle to describe a safety zone for fireworks, a location on Earth seen from space, and even a rainbow. *You will learn about angles of a circle when satellites send signals to Earth in Lesson 10-6.*

520 Chapter 10 Circles

Vocabulary Builder

ELL

The Key Vocabulary list introduces students to some of the main vocabulary terms included in this chapter. For a more thorough vocabulary list with pronunciations of new words, give students the Vocabulary Builder worksheets found on pages vii and viii of the *Chapter 10 Resource Masters*. Encourage them to complete the definition of each term as they progress through the chapter. You may suggest that they add these sheets to their study notebooks for future reference when studying for the Chapter 10 test.

Getting Started

Prerequisite Skills To be successful in this chapter, you'll need to master these skills and be able to apply them in problem-solving situations. Review these skills before beginning Chapter 10.

For Lesson 10-1 — Solve Equations

Solve each equation for the given variable. *(For review, see pages 737 and 738.)*

1. $\frac{4}{9}p = 72$ for p **162**
2. $6.3p = 15.75$ **2.5**
3. $3x + 12 = 8x$ for x **2.4**
4. $7(x + 2) = 3(x - 6)$ **−8**
5. $C = 2pr$ for r $r = \frac{C}{2p}$
6. $r = \frac{C}{6.28}$ for C $C = 6.28r$

For Lesson 10-5 — Pythagorean Theorem

Find x. Round to the nearest tenth if necessary. *(For review, see Lesson 7-2.)*

7. 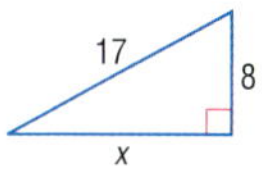

15

8. 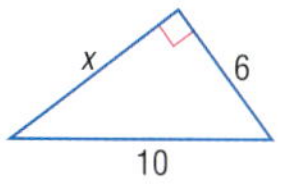

8

9. 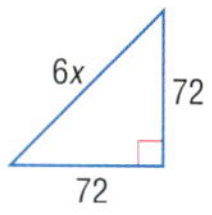

17.0

For Lesson 10-7 — Quadratic Formula

Solve each equation by using the Quadratic Formula. Round to the nearest tenth.

10. $x^2 - 4x = 10$ **5.7, −1.7**
11. $3x^2 - 2x - 4 = 0$ **1.5, −0.9**
12. $x^2 = x + 15$ **4.4, −3.4**
13. $2x^2 + x = 15$ **2.5, −3**

FOLDABLES™ Study Organizer

Circles Make this Foldable to help you organize information about circles. Begin with five sheets of plain $8\frac{1}{2}$" by 11" paper, and cut out five large circles that are the same size.

Step 1 Fold and Cut

Fold two of the circles in half and cut one-inch slits at each end of the folds.

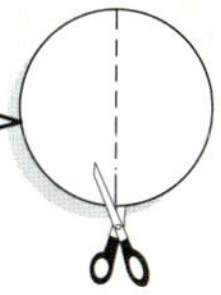

Step 2 Fold and Cut

Fold the remaining three circles in half and cut a slit in the middle of the fold.

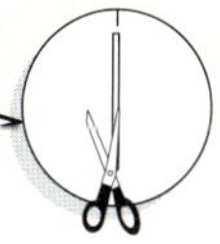

Step 3 Slide

Slide the two circles with slits on the ends through the large slit of the other circles.

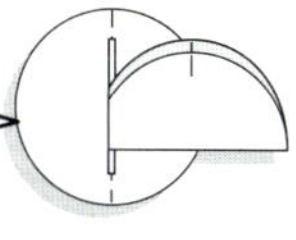

Step 4 Label

Label the cover with the title of the chapter and each sheet with a lesson number.

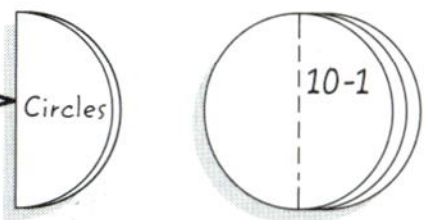

Reading and Writing As you read and study each lesson, take notes, and record concepts on the appropriate page of your Foldable.

Getting Started

This section provides a review of the basic concepts needed before beginning Chapter 10. Page references are included for additional student help.

Additional review is provided in the *Prerequisite Skills Workbook,* pages 11–12, 15–18, 23–24, 31–32, 35–36, 41–42, 45–48, 51–52, 61–64, 67–68, 71–72, 105–106, 109–110.

Prerequisite Skills in the Getting Ready for the Next Lesson section at the end of each exercise set review a skill needed in the next lesson.

For Lesson	Prerequisite Skill
10-2	Angle Addition, p. 528
10-3	Isosceles Triangles, p. 535
10-4	Solving Equations, p. 543
10-5	Pythagorean Theorem, p. 551
10-6	Solving Equations, p. 558
10-7	Solving Equations by Factoring, p. 568
10-8	Distance Formula, p. 574

FOLDABLES™ Study Organizer

Organization of Data and Expository Writing Use this Foldable for student writing about circles, angles, arcs, chords, tangents, secants, angle measurement, and equations. Students can use their Foldable to take notes, define terms, record concepts, use properties, and write and sketch examples. Ask students to write about circles in such a manner that someone who did not know what a circle was or understand how to solve problems using arcs and diameters will understand after reading what students have written.

For more information about Foldables, see *Teaching Mathematics with Foldables.*

Chapter 10 Circles 521

10-1 Lesson Notes

1 Focus

5-Minute Check Transparency 10-1 Use as a quiz or review of Chapter 9.

Mathematical Background notes are available for this lesson on p. 520C.

How far does a carousel animal travel in one rotation?

Ask students:

- Explain why an animal travels farther on the outside of the carousel than near the middle of the carousel. **The circumference of a circle with a large radius is greater than the circumference of a circle with a smaller radius.**
- Are there as many animals on the carousel in Wisconsin as there are degrees in a circle? Explain. **No; there are 100 degrees more in a circle than there are animals on the carousel.**

10-1 Circles and Circumference

Sunshine State Standards
MA.B.1.4.1

Vocabulary

- circle
- center
- chord
- radius
- diameter
- circumference
- pi (π)

What You'll Learn

- Identify and use parts of circles.
- Solve problems involving the circumference of a circle.

How far does a carousel animal travel in one rotation?

The largest carousel in the world still in operation is located in Spring Green, Wisconsin. It weighs 35 tons and contains 260 animals, none of which is a horse! The rim of the carousel base is a circle. The width, or diameter, of the circle is 80 feet. The distance that one of the animals on the outer edge travels can be determined by special segments in a circle.

Study Tip

Reading Mathematics
The plural of radius is *radii*, pronounced RAY-dee-eye. The term *radius* can mean a segment or the measure of that segment. This is also true of the term *diameter*.

PARTS OF CIRCLES A **circle** is the locus of all points in a plane equidistant from a given point called the **center** of the circle. A circle is usually named by its center point. The figure below shows circle C, which can be written as $\odot C$. Several special segments in circle C are also shown.

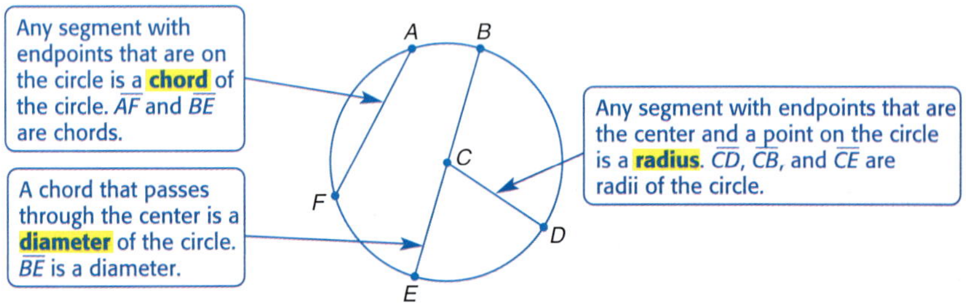

Note that diameter $\overline{BE}$ is made up of collinear radii $\overline{CB}$ and $\overline{CE}$.

Example 1 Identify Parts of a Circle

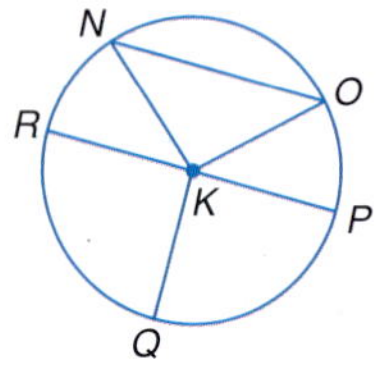

a. Name the circle.

The circle has its center at K, so it is named circle K, or $\odot K$.

In this textbook, the center of a circle will always be shown in the figure with a dot.

b. Name a radius of the circle.

Five radii are shown: $\overline{KN}$, $\overline{KO}$, $\overline{KP}$, $\overline{KQ}$, and $\overline{KR}$.

c. Name a chord of the circle.

Two chords are shown: $\overline{NO}$ and $\overline{RP}$.

d. Name a diameter of the circle.

$\overline{RP}$ is the only chord that goes through the center, so $\overline{RP}$ is a diameter.

Resource Manager

Workbook and Reproducible Masters

Chapter 10 Resource Masters
- Study Guide and Intervention, pp. 541–542
- Skills Practice, p. 543
- Practice, p. 544
- Reading to Learn Mathematics, p. 545
- Enrichment, p. 546

School-to-Career Masters, p. 19
Prerequisite Skills Workbook, pp. 11–12, 23–24, 45–48
Teaching Geometry With Manipulatives Masters, p. 161

Transparencies

5-Minute Check Transparency 10-1
Answer Key Transparencies

Technology

Interactive Chalkboard

Study Tip

Radii and Diameters
There are an infinite number of radii in each circle. Likewise, there are an infinite number of diameters.

By the definition of a circle, the distance from the center to any point on the circle is always the same. Therefore, all radii are congruent. A diameter is composed of two radii, so all diameters are congruent. The letters d and r are usually used to represent diameter and radius in formulas. So, $d = 2r$ and $r = \frac{d}{2}$ or $\frac{1}{2}d$.

Example 2 Find Radius and Diameter

Circle A has diameters $\overline{DF}$ and $\overline{PG}$.

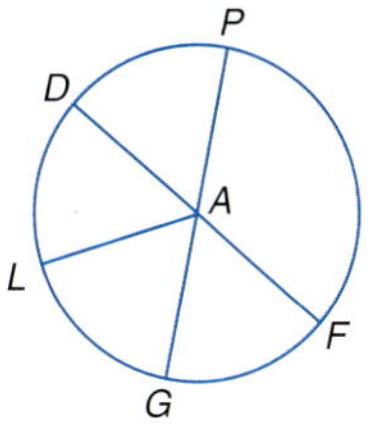

a. If $DF = 10$, find DA.

$r = \frac{1}{2}d$ Formula for radius

$r = \frac{1}{2}(10)$ or 5 Substitute and simplify.

b. If $PA = 7$, find PG.

$d = 2r$ Formula for diameter

$d = 2(7)$ or 14 Substitute and simplify.

c. If $AG = 12$, find LA.

Since all radii are congruent, $LA = AG$. So, $LA = 12$.

Circles can intersect. The segment connecting the centers of the two intersecting circles contains a radius of each circle.

Study Tip

Congruent Circles
The circles shown in Example 3 have different radii. They are *not* congruent circles. For two circles to be *congruent circles*, they must have congruent radii or congruent diameters.

Example 3 Find Measures in Intersecting Circles

The diameters of $\odot A$, $\odot B$, and $\odot C$ are 10 inches, 20 inches, and 14 inches, respectively.

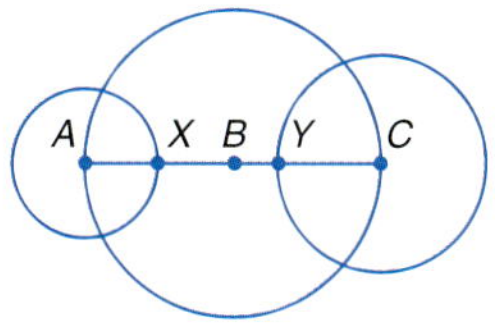

a. Find XB.

Since the diameter of $\odot A$ is 10, $AX = 5$.

Since the diameter of $\odot B$ is 20, $AB = 10$ and $BC = 10$.

$\overline{XB}$ is part of radius $\overline{AB}$.

$AX + XB = AB$ Segment Addition Postulate

$5 + XB = 10$ Substitution

$XB = 5$ Subtract 5 from each side.

b. Find BY.

$\overline{BY}$ is part of $\overline{BC}$.

Since the diameter of $\odot C$ is 14, $YC = 7$.

$BY + YC = BC$ Segment Addition Postulate

$BY + 7 = 10$ Substitution

$BY = 3$ Subtract 7 from each side.

CIRCUMFERENCE The **circumference** of a circle is the distance around the circle. Circumference is most often represented by the letter C.

www.geometryonline.com/extra_examples/fcat

2 Teach

PARTS OF CIRCLES

In-Class Examples

Power Point®

Teaching Tip Remind students that a diameter is a special chord of a circle because its endpoints are on the circle.

1

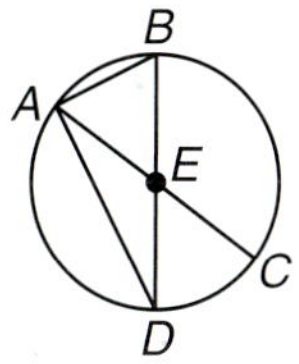

a. Name the circle. $\odot E$

b. Name a radius of the circle. $\overline{EB}$, $\overline{EA}$, $\overline{EC}$, or $\overline{ED}$

c. Name a chord of the circle. $\overline{AB}$, $\overline{AC}$, $\overline{BD}$, or $\overline{AD}$

d. Name a diameter of the circle. $\overline{AC}$ or $\overline{BD}$

2 Circle R has diameters $\overline{ST}$ and $\overline{QM}$.

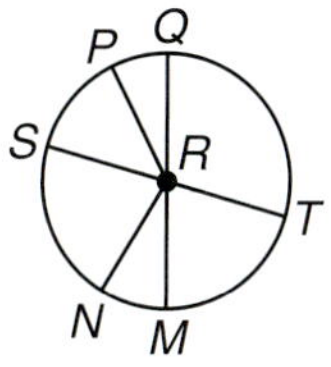

a. If $ST = 18$, find RS. 9

b. If $RM = 24$, find QM. 48

c. If $RN = 2$, find RP. 2

3 The diameters of $\odot X$, $\odot Y$, and $\odot Z$ are 22 millimeters, 16 millimeters, and 10 millimeters, respectively.

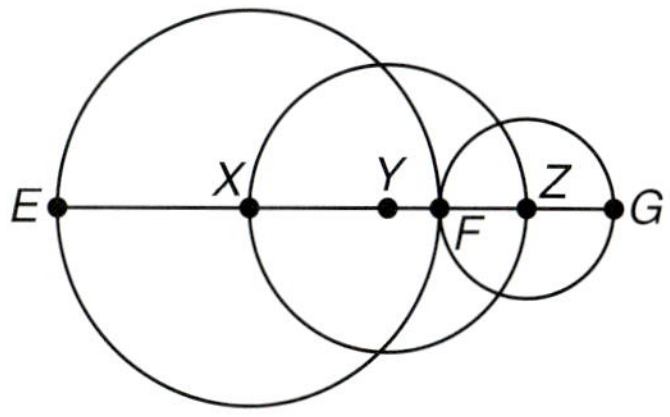

a. Find EZ. 27 mm

b. Find XF. 11 mm

CIRCUMFERENCE

In-Class Example

Power Point®

Teaching Tip Tell students that π can also be approximated by $\frac{22}{7}$ if students are using a nonscientific calculator that does not include π on its keyboard.

4 a. Find C if $r = 13$ inches. 26π or ≈ 81.68 in.

b. Find C if $d = 6$ millimeters. 6π or ≈ 18.85 mm

c. Find d and r to the nearest hundredth if $C = 65.4$ feet. $d \approx 20.82$ ft; $r \approx 10.41$ ft

Tips for New Teachers

If students wonder why they are given two formulas for the circumference of a circle when they already know that the diameter is twice the radius, tell them that the two formulas lead them to look closely at a question to determine if the problem gives a radius or a diameter. Explain that a common mistake is to erroneously calculate the circumference of a circle as πr.

Geometry Activity

Circumference Ratio

A special relationship exists between the circumference of a circle and its diameter.

Gather Data and Analyze

Collect ten round objects.

1. Measure the circumference and diameter of each object using a millimeter measuring tape. Record the measures in a table like the one at the right. See students' work.

Object	C	d	$\frac{C}{d}$
1			
2			
3			
⋮			
10			

2. Compute the value of $\frac{C}{d}$ to the nearest hundredth for each object. Record the result in the fourth column of the table. Each ratio should be near 3.1.

Make a Conjecture

3. What seems to be the relationship between the circumference and the diameter of the circle? $C \approx 3.14d$

Study Tip

Value of π
In this book, we will use a calculator to evaluate expressions involving π. If no calculator is available, 3.14 is a good estimate for π.

The Geometry Activity suggests that the circumference of any circle can be found by multiplying the diameter by a number slightly larger than 3. By definition, the ratio $\frac{C}{d}$ is an irrational number called **pi**, symbolized by the Greek letter π. Two formulas for the circumference can be derived using this definition.

$\frac{C}{d} = \pi$ Definition of pi

$C = \pi d$ Multiply each side by d.

$C = \pi d$

$C = \pi(2r)$ $d = 2r$

$C = 2\pi r$ Simplify.

Key Concept — *Circumference*

For a circumference of C units and a diameter of d units or a radius of r units, $C = \pi d$ or $C = 2\pi r$.

If you know the diameter or radius, you can find the circumference. Likewise, if you know the circumference, you can find the diameter or radius.

Example 4 Find Circumference, Diameter, and Radius

a. Find C if $r = 7$ centimeters.

$C = 2\pi r$ Circumference formula

$= 2\pi(7)$ Substitution

$= 14\pi$ or about 43.98 cm

b. Find C if $d = 12.5$ inches.

$C = \pi d$ Circumference formula

$= \pi(12.5)$ Substitution

$= 12.5\pi$ or 39.27 in.

c. Find d and r to the nearest hundredth if $C = 136.9$ meters.

$C = \pi d$ Circumference formula

$136.9 = \pi d$ Substitution

$\frac{136.9}{\pi} = d$ Divide each side by π.

$43.58 \approx d$ Use a calculator.

$d \approx 43.58$ m

$r = \frac{1}{2}d$ Radius formula

$\approx \frac{1}{2}(43.58)$ $d \approx 43.58$

≈ 21.79 m Use a calculator.

Geometry Activity

- You can provide students a handout with a blank 11-row by 4-column table.
- Ask students why they think they are measuring the objects in millimeters. Students should note that millimeters provide very accurate values for comparison in this activity.
- Point out that the relationship between the circumference and diameter of a circle is an extremely interesting concept that has been analyzed for centuries, and there are many books written just on this subject.

524 Chapter 10 Circles

You can also use other geometric figures to help you find the circumference of a circle.

FCAT Practice
Standardized Test Practice
(A) (B) (C) (D)

Example 5 Use Other Figures to Find Circumference

Multiple-Choice Test Item

Find the exact circumference of ⊙P.

(A) 13 cm
(B) 12π cm
(C) 40.84 cm
(D) 13π cm

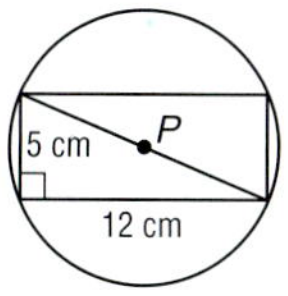

Read the Test Item

You are given a figure that involves a right triangle and a circle. You are asked to find the exact circumference of the circle.

Solve the Test Item

The diameter of the circle is the same as the hypotenuse of the right triangle.

$a^2 + b^2 = c^2$ Pythagorean Theorem

$5^2 + 12^2 = c^2$ Substitution

$169 = c^2$ Simplify.

$13 = c$ Take the square root of each side.

So the diameter of the circle is 13 centimeters.

$C = \pi d$ Circumference formula

$C = \pi(13)$ or 13π Substitution

Because we want the exact circumference, the answer is D.

The Princeton Review

Test-Taking Tip

Notice that the problem asks for an exact answer. Since you know that an exact circumference contains π, you can eliminate choices A and C.

In-Class Example

Power Point®

5 Find the exact circumference of ⊙K. **B**

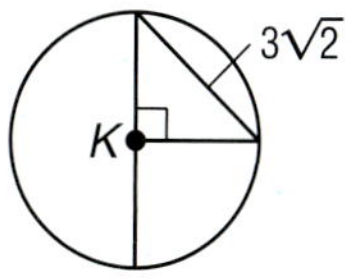

A $3\sqrt{2}\pi$ **B** 6π
C $6\sqrt{2}\pi$ **D** 12π

Answers

1. Sample answer: The value of π is calculated by dividing the circumference of a circle by the diameter.
3. Except for a diameter, two radii and a chord of a circle can form a triangle. The Triangle Inequality Theorem states that the sum of two sides has to be greater than the third. So, $2r$ has to be greater than the measure of any chord that is not a diameter, but $2r$ is the measure of the diameter. So the diameter has to be longer than any other chord of the circle.

Check for Understanding

Concept Check

1. **Describe** how the value of π can be calculated. See margin.
2. **Write** two equations that show how the diameter of a circle is related to the radius of a circle. $d = 2r$, $r = \frac{1}{2}d$
3. **OPEN ENDED** Explain why a diameter is the longest chord of a circle. See margin.

Guided Practice

GUIDED PRACTICE KEY	
Exercises	Examples
4–7	1
8, 9	2
10–12	3
13, 14	4
15	5

For Exercises 4–9, refer to the circle at the right. 6. $\overline{AB}$, $\overline{AC}$, or $\overline{BD}$

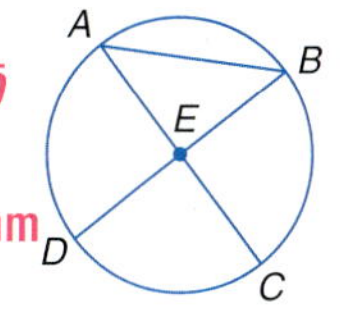

4. Name the circle. ⊙E
5. Name a radius. $\overline{EA}$, $\overline{EB}$, $\overline{EC}$, or $\overline{ED}$
6. Name a chord.
7. Name a diameter. $\overline{AC}$ or $\overline{BD}$
8. Suppose $BD = 12$ millimeters. Find the radius of the circle. 6 mm
9. Suppose $CE = 5.2$ inches. Find the diameter of the circle. 10.4 in.

Circle W has a radius of 4 units, ⊙Z has a radius of 7 units, and $XY = 2$. Find each measure.

10. YZ 5
11. IX 6
12. IC 20

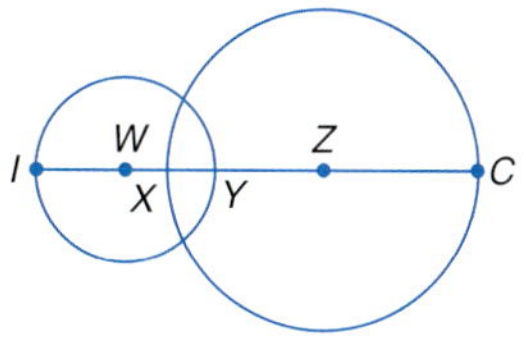

DAILY INTERVENTION

Differentiated Instruction ELL

Verbal/Linguistic Have students write about the parts of a circle and its circumference in their own words. They can write a paragraph that explains each vocabulary term and the relationship of the terms to each other, or they can list the terms and write a brief explanation and/or provide an example for each. Students can use these explanations for their study notebooks.

Lesson 10-1 Circles and Circumference 525

3 Practice/Apply

Study Notebook

Have students—
- add the definitions/examples of the vocabulary terms to their Vocabulary Builder worksheets for Chapter 10.
- include one circle that is labeled to demonstrate each vocabulary term in this lesson.
- include any other item(s) that they find helpful in mastering the skills in this lesson.

About the Exercises...

Organization by Objective
- **Parts of Circles:** 16–43
- **Circumference:** 48–52

Odd/Even Assignments
Exercises 16–55 are structured so that students practice the same concepts whether they are assigned odd or even problems.

Assignment Guide
Basic: 17–35 odd, 39, 41, 45–49 odd, 53–63 odd, 64, 66–80 (optional: 65)

Average: 17–61 odd, 63, 64, 66–80 (optional: 65)

Advanced: 16–60 even, 61–74 (optional: 75–80)

FCAT Practice
Standardized Test Practice

The radius, diameter, or circumference of a circle is given. Find the missing measures. Round to the nearest hundredth if necessary.

13. $r = 5$ m, $d =$ __?__, $C =$ __?__ **10 m, 31.42 m**

14. $C = 2368$ ft, $d =$ __?__, $r =$ __?__ **753.76 ft, 376.88 ft**

15. Find the exact circumference of the circle. **B**

Ⓐ 4.5π mm
Ⓑ 9π mm
Ⓒ 18π mm
Ⓓ 81π mm

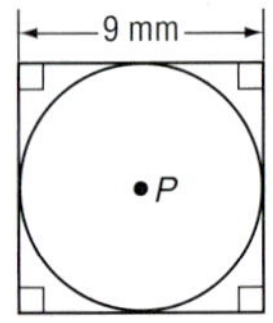

★ indicates increased difficulty

Practice and Apply

Homework Help	
For Exercises	See Examples
16–25	1
26–31	2
32–43	3
48–51	4
52	5

Extra Practice
See page 773.

For Exercises 16–20, refer to the circle at the right.

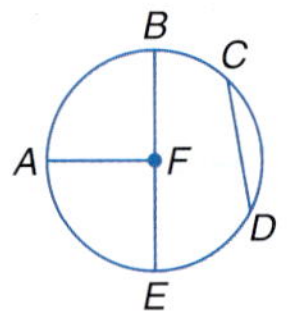

16. Name the circle. **⊙F**
17. Name a radius. **$\overline{FA}$, $\overline{FB}$, or $\overline{FE}$**
18. Name a chord. **$\overline{BE}$ or $\overline{CD}$**
19. Name a diameter. **$\overline{BE}$**
20. Name a radius not contained in a diameter. **$\overline{FA}$**

HISTORY **For Exercises 21–31, refer to the model of the Conestoga wagon wheel.**

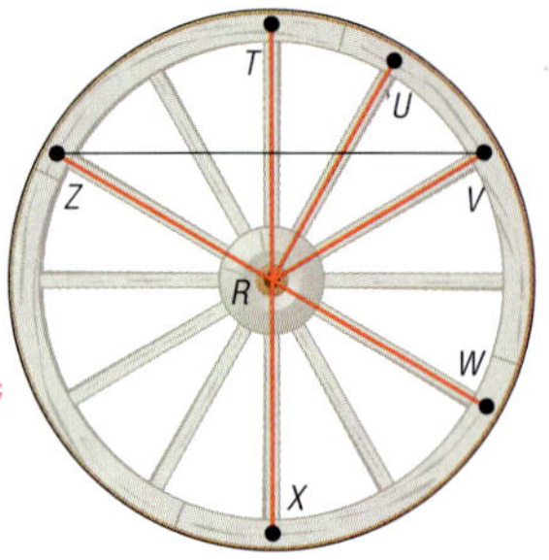

21. Name the circle. **⊙R**
22. Name a radius of the circle. **$\overline{RT}$, $\overline{RU}$, $\overline{RV}$, $\overline{RW}$, $\overline{RX}$, or $\overline{RZ}$**
23. Name a chord of the circle. **$\overline{ZV}$, $\overline{TX}$, or $\overline{WZ}$**
24. Name a diameter of the circle. **$\overline{TX}$ or $\overline{WZ}$**
25. Name a radius not contained in a diameter. **$\overline{RU}$, $\overline{RV}$**
26. Suppose the radius of the circle is 2 feet. Find the diameter. **4 ft**
27. The larger wheel of the wagon was often 5 or more feet tall. What is the radius of a 5-foot wheel? **2.5 ft**
28. If $TX = 120$ centimeters, find TR. **60 cm**
29. If $RZ = 32$ inches, find ZW. **64 in. or 5 ft 4 in.**
30. If $UR = 18$ inches, find RV. **18 in.**
31. If $XT = 1.2$ meters, find UR. **0.6 m**

More About. . .

History

In the early 1800s, thousands of people traveled west in Conestoga wagons. The spoke wheels enabled the wagons to carry cargo over rugged terrain. The number of spokes varied with the size of the wheel.

Source: Smithsonian Institute

The diameters of ⊙A, ⊙B, and ⊙C are 10, 30, and 10 units, respectively. Find each measure if $\overline{AZ} \cong \overline{CW}$ and $CW = 2$.

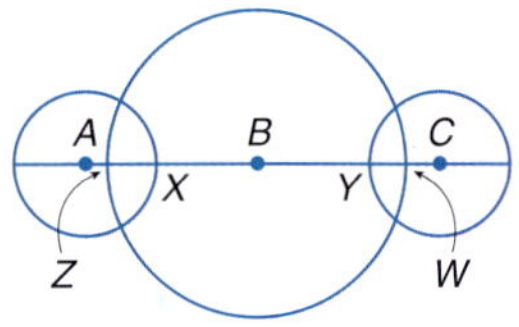

32. AZ **2**
33. ZX **3**
34. BX **12**
35. BY **12**
36. YW **3**
★37. AC **34**

Circles G, J, and K all intersect at L. If $GH = 10$, find each measure.

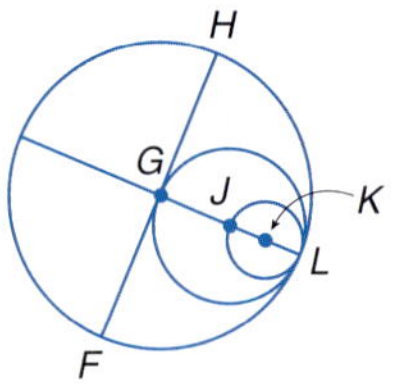

38. FG **10**
39. FH **20**
40. GL **10**
41. GJ **5**
42. JL **5**
★43. JK **2.5**

Answer

62. Sample answer: about 251.3 feet. Answers should include the following.
- The distance the animal travels is approximated by the circumference of the circle.
- The diameter for the circle on which the animal is located becomes 80 − 2 or 78. The circumference of this circle is 78π. Multiply by 22 to get a total distance of $22(78\pi)$ or 5391 feet. This is a little over a mile.

4. 14 mm, 43.98 mm
5. 13.4 cm, 84.19 cm
6. 26 mi, 13 mi
7. 24.32 m, 12.16 m
8. $6\frac{1}{4}$yd, 39.27 yd
9. $13\frac{1}{2}$ in., 42.41 in.

The radius, diameter, or circumference of a circle is given. Find the missing measures. Round to the nearest hundredth if necessary.

44. $r = 7$ mm, $d =$ _?_, $C =$ _?_
45. $d = 26.8$ cm, $r =$ _?_, $C =$ _?_
46. $C = 26\pi$ mi, $d =$ _?_, $r =$ _?_
47. $C = 76.4$ m, $d =$ _?_, $r =$ _?_
48. $d = 12\frac{1}{2}$yd, $r =$ _?_, $C =$ _?_
49. $r = 6\frac{3}{4}$in., $d =$ _?_, $C =$ _?_
50. $d = 2a$, $r =$ _?_, $C =$ _?_ a, $6.28a$
★51. $r = \frac{a}{6}$, $d =$ _?_, $C =$ _?_ $0.33a$, $1.05a$

Find the exact circumference of each circle.

52.

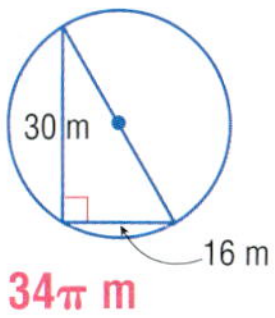

34π m

53.

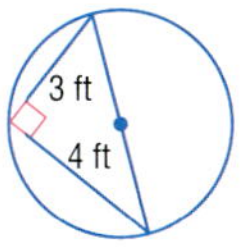

5π ft

54.

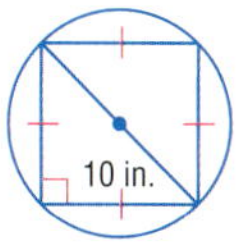

$10\pi\sqrt{2}$ in.

55.

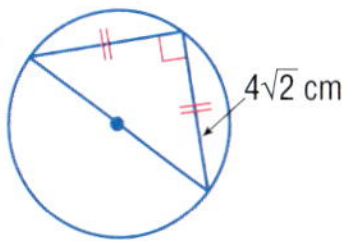

8π cm

6. 1; This description s the definition of a adius.

56. **PROBABILITY** Find the probability that a segment with endpoints that are the center of the circle and a point on the circle is a radius. Explain.

57. **PROBABILITY** Find the probability that a chord that does not contain the center of a circle is the longest chord of the circle.
0; The longest chord of a circle is the diameter, which contains the center.

FIREWORKS **For Exercises 58–60, use the following information.**

Every July 4th Boston puts on a gala with the Boston Pops Orchestra, followed by a huge fireworks display. The fireworks are shot from a barge in the river. There is an explosion circle inside which all of the fireworks will explode. Spectators sit outside a safety circle that is 800 feet from the center of the fireworks display.

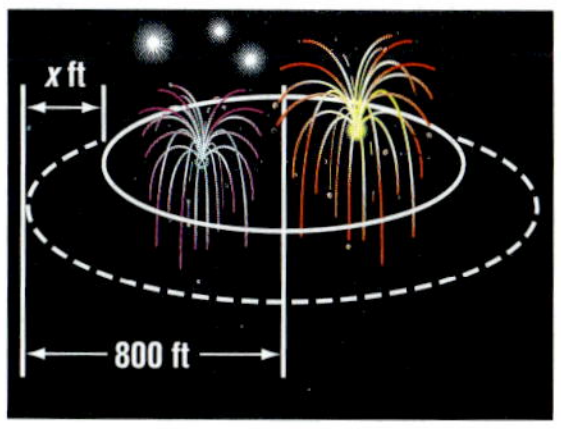

58. Find the approximate circumference of the safety circle. 5026.5 ft

59. If the safety circle is 200 to 300 feet farther from the center than the explosion circle, find the range of values for the radius of the explosion circle. 500–600 ft

60. Find the least and maximum circumference of the explosion circle to the nearest foot. 3142 ft; 3770 ft

WebQuest

Drawing a radius and circle on the map is the last clue to help you find the hidden treasure. Visit www.geometryonline.com/webquest to continue work on your WebQuest project.

Online Research **Data Update** Find the largest firework ever made. How does its dimension compare to the Boston display? Visit www.geometryonline.com/data_update to learn more.

61. **CRITICAL THINKING** In the figure, O is the center of the circle, and $x^2 + y^2 + p^2 + t^2 = 288$. What is the exact circumference of $\odot O$? 24π units

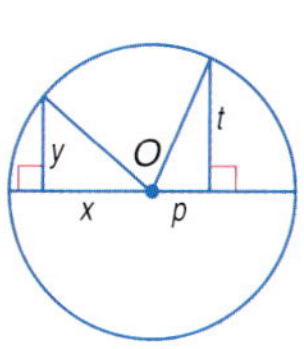

62. **WRITING IN MATH** Answer the question that was posed at the beginning of the lesson. See margin.

How far does a carousel animal travel in one rotation?

Include the following in your answer:
- a description of how the circumference of a circle relates to the distance traveled by the animal, and
- whether an animal located one foot from the outside edge of the carousel travels a mile when it makes 22 rotations for each ride.

www.geometryonline.com/self_check_quiz/fcat

Study Guide and Intervention, p. 541 (shown) and p. 542

Parts of Circles A **circle** consists of all points in a plane that are a given distance, called the **radius**, from a given point called the **center**.

A segment or line can intersect a circle in several ways.
- A segment with endpoints that are the center of the circle and a point of the circle is a **radius**.
- A segment with endpoints that lie on the circle is a **chord**.
- A chord that contains the circle's center is a **diameter**.

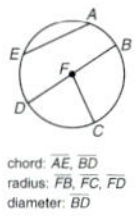

chord: $\overline{AE}$, $\overline{BD}$
radius: $\overline{FB}$, $\overline{FC}$, $\overline{FD}$
diameter: $\overline{BD}$

Example

a. **Name the circle.**
The name of the circle is $\odot O$.

b. **Name radii of the circle.**
$\overline{AO}$, $\overline{BO}$, $\overline{CO}$, and $\overline{DO}$ are radii.

c. **Name chords of the circle.**
$\overline{AB}$ and $\overline{CD}$ are chords.

d. **Name a diameter of the circle.**
$\overline{AB}$ is a diameter.

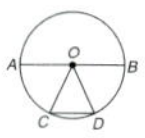

Exercises

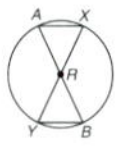

1. Name the circle. $\odot R$
2. Name radii of the circle. $\overline{RA}$, $\overline{RB}$, $\overline{RY}$, and $\overline{RX}$
3. Name chords of the circle. $\overline{BY}$, $\overline{AX}$, $\overline{AB}$, and $\overline{XY}$
4. Name diameters of the circle. $\overline{AB}$ and $\overline{XY}$
5. Find AR if AB is 18 millimeters. 9 mm
6. Find AR and AB if RY is 10 inches. $AR = 10$ in.; $AB = 20$ in.
7. Is $\overline{AB} \cong \overline{XY}$? Explain. Yes; all diameters of the same circle are congruent.

Skills Practice, p. 543 and Practice, p. 544 (shown)

For Exercises 1–5, refer to the circle.

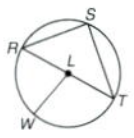

1. Name the circle.
$\odot L$
2. Name a radius.
$\overline{LR}$, $\overline{LT}$, or $\overline{LW}$
3. Name a chord.
$\overline{RT}$, $\overline{RS}$, or $\overline{ST}$
4. Name a diameter.
$\overline{RT}$
5. Name a radius not drawn as part of a diameter. $\overline{LW}$
6. Suppose the radius of the circle is 3.5 yards. Find the diameter. 7 yd
7. If $RT = 19$ meters, find LW. 9.5 m

The diameters of $\odot L$ and $\odot M$ are 20 and 13 units, respectively. Find each measure if $QR = 4$.

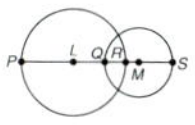

8. LQ
6
9. RM
2.5

The radius, diameter, or circumference of a circle is given. Find the missing measures to the nearest hundredth.

10. $r = 7.5$ mm
$d =$ 15 mm, $C \approx$ 47.12 mm
11. $C = 227.6$ yd
$d \approx$ 72.45 yd, $r \approx$ 36.22 yd

Find the exact circumference of each circle.

12.

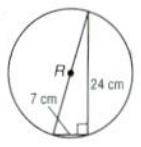

25π cm

13.

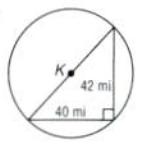

58π mi

SUNDIALS **For Exercises 14 and 15, use the following information.**
Herman purchased a sundial to use as the centerpiece for a garden. The diameter of the sundial is 9.5 inches.

14. Find the radius of the sundial. 4.75 in.
15. Find the circumference of the sundial to the nearest hundredth. 29.85 in.

Reading to Learn Mathematics, p. 545 ELL

Pre-Activity **How far does a carousel animal travel in one rotation?**

Read the introduction to Lesson 10-1 at the top of page 522 in your textbook.

How could you measure the approximate distance around the circular carousel using everyday measuring devices? Sample answer: Place a piece of string along the rim of the carousel. Cut off a length of string that covers the perimeter of the circle. Straighten the string and measure it with a yardstick.

Reading the Lesson

1. Refer to the figure.
a. Name the circle. $\odot Q$
b. Name four radii of the circle. $\overline{QP}$, $\overline{QR}$, $\overline{QS}$, and $\overline{QT}$
c. Name a diameter of the circle. $\overline{PR}$
d. Name two chords of the circle. $\overline{PR}$ and $\overline{ST}$

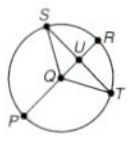

2. Match each description from the first column with the best term from the second column. (Some terms in the second column may be used more than once or not at all.)
a. a segment whose endpoints are on a circle iii
b. the set of all points in a plane that are the same distance from a given point iv
c. the distance between the center of a circle and any point on the circle i
d. a chord that passes through the center of a circle ii
e. a segment whose endpoints are the center and any point on a circle i
f. a chord made up of two collinear radii ii
g. the distance around a circle v

i. radius
ii. diameter
iii. chord
iv. circle
v. circumference

3. Which equations correctly express a relationship in a circle? A, D, G

A. $d = 2r$	B. $C = \pi r$	C. $C = 2d$	D. $d = \frac{C}{\pi}$
E. $r = \frac{d}{\pi}$	F. $C = r^2$	G. $C = 2\pi r$	H. $d = \frac{1}{2}r$

Helping You Remember

4. A good way to remember a new geometric term is to relate the word or its parts to geometric terms you already know. Look up the origins of the two parts of the word *diameter* in your dictionary. Explain the meaning of each part and give a term you already know that shares the origin of that part. Sample answer: The first part comes from *dia*, which means *across* or *through*, as in *diagonal*. The second part comes from *metron*, which means *measure*, as in *geometry*.

Enrichment, p. 546

The Four Color Problem

Mapmakers have long believed that only four colors are necessary to distinguish among any number of different countries on a plane map. Countries that meet only at a point may have the same color provided they do not have an actual border. The conjecture that four colors are sufficient for every conceivable plane map eventually attracted the attention of mathematicians and became known as the "four-color problem." Despite extraordinary efforts over many years to solve the problem, no definite answer was obtained until the 1980s. Four colors are indeed sufficient, and the proof was accomplished by making ingenious use of computers.

The following problems will help you appreciate some of the complexities of the four-color problem. For these "maps," assume that each closed region is a different country.

1. What is the minimum number of colors necessary for each map?
a.
b.
c.

4 Assess

Open-Ended Assessment

Speaking Students can practice the vocabulary terms in this lesson by describing selected circles and defining terms aloud. For example, find a circle in the lesson without values, and call on students to name its parts. Then ask students to state the values for the radius and circumference of the circle if the diameter is 10 units, 20 units, etc.

Getting Ready for Lesson 10-2

Prerequisite Skill Students will learn about angles and arcs in Lesson 10-2. They will use angle addition to find angle measures in circles. Use Exercises 75–80 to determine your students' familiarity with angle addition.

Answers

73. Given: $\overline{RQ}$ bisects $\angle SRT$.
Prove: $m\angle SQR > m\angle SRQ$

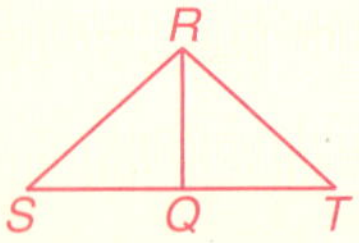

Proof:

Statements (Reasons)
1. $\overline{RQ}$ bisects $\angle SRT$. (Given)
2. $\angle SRQ \cong \angle QRT$ (Def. of ∠ bisector)
3. $m\angle SRQ = m\angle QRT$ (Def. of ≅ ∠s)
4. $m\angle SQR = m\angle T + m\angle QRT$ (Exterior Angle Theorem)
5. $m\angle SQR > m\angle QRT$ (Def. of Inequality)
6. $m\angle SQR > m\angle SRQ$ (Substitution)

FCAT Practice
Standardized Test Practice

63. **GRID IN** In the figure, the radius of circle A is twice the radius of circle B and four times the radius of circle C. If the sum of the circumferences of the three circles is 42π, find the measure of $\overline{AC}$. **27**

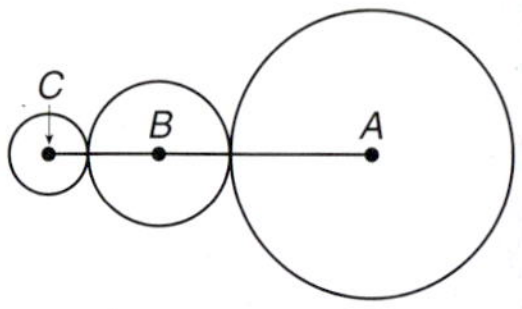

64. **ALGEBRA** There are k gallons of gasoline available to fill a tank. After d gallons have been pumped, what percent of gasoline, in terms of k and d, has been pumped? **A**

Ⓐ $\frac{100d}{k}\%$ Ⓑ $\frac{k}{100d}\%$ Ⓒ $\frac{100k}{d}\%$ Ⓓ $\frac{100k-d}{k}\%$

Extending the Lesson

65. **CONCENTRIC CIRCLES** Circles that have the same center, but different radii, are called *concentric circles*. Use the figure at the right to find the exact circumference of each circle. List the circumferences in order from least to greatest. **10π, 20π, 30π**

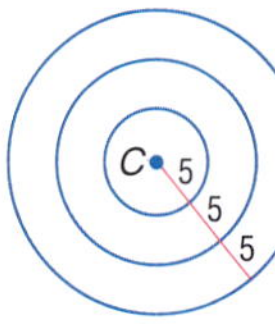

Maintain Your Skills

Mixed Review

Find the magnitude to the nearest tenth and direction to the nearest degree of each vector. *(Lesson 9-6)*

66. $\overrightarrow{AB} = \langle 1, 4\rangle$ **4.1; 76°**
67. $\vec{\mathbf{v}} = \langle 4, 9\rangle$ **9.8; 66°**
68. $\overrightarrow{AB}$ if $A(4, 2)$ and $B(7, 22)$ **20.2; 81°**
69. $\overrightarrow{CD}$ if $C(0, -20)$ and $D(40, 0)$ **44.7; 27°**

Find the measure of the dilation image of $\overline{AB}$ for each scale factor k. *(Lesson 9-5)*

70. $AB = 5, k = 6$ **30**
71. $AB = 16, k = 1.5$ **24**
72. $AB = \frac{2}{3}, k = -\frac{1}{2}$ **$\frac{1}{3}$**

73. **PROOF** Write a two-column proof. *(Lesson 5-3)*
Given: $\overline{RQ}$ bisects $\angle SRT$.
Prove: $m\angle SQR > m\angle SRQ$
See margin.

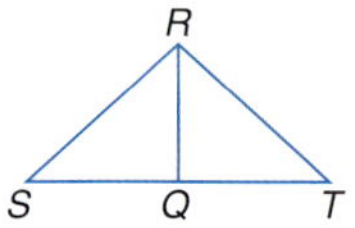

74. **COORDINATE GEOMETRY** Name the missing coordinates if $\triangle DEF$ is isosceles with vertex angle E. *(Lesson 4-3)* **(2a, 0)**

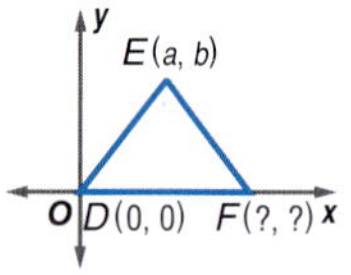

Getting Ready for the Next Lesson

PREREQUISITE SKILL Find x. *(To review **angle addition**, see Lesson 1-4.)*

75. **60**

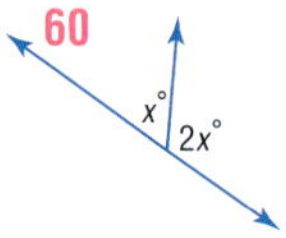

76. **18**

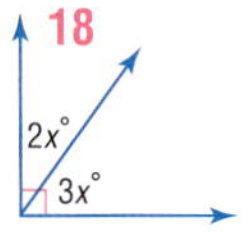

77. **30**

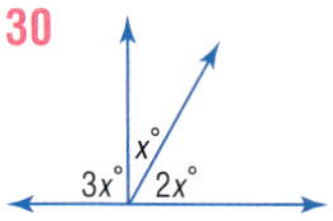

78. **22.5**

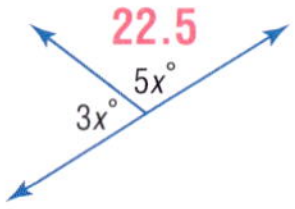

79. **30**

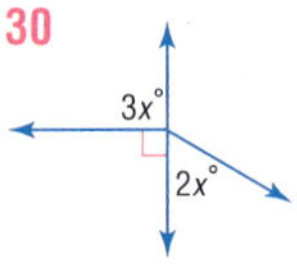

80. **120**

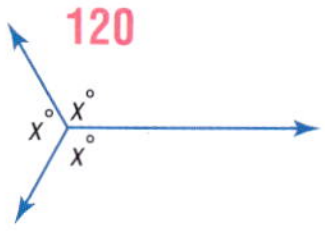

Teacher to Teacher

Kim A. Halvorson, DeSoto County High School — Arcadia, FL

My students are asked to decorate a T-shirt with a "pi" theme. Then they wear them on March 14 (3.14). The rest of the school (via morning announcements) is encouraged to ask the geometry students to discuss their shirts.

10-2 Angles and Arcs

Sunshine State Standards
MA.B.1.4.2

Vocabulary

- central angle
- arc
- minor arc
- major arc
- semicircle

What You'll Learn

- Recognize major arcs, minor arcs, semicircles, and central angles and their measures.
- Find arc length.

What kinds of angles do the hands on a clock form?

Most clocks on electronic devices are digital, showing the time as numerals. Analog clocks are often used in decorative furnishings and wrist watches. An analog clock has moving hands that indicate the hour, minute, and sometimes the second. This clock face is a circle. The three hands form three central angles of the circle.

ANGLES AND ARCS In Chapter 1, you learned that a degree is $\frac{1}{360}$ of the circular rotation about a point. This means that the sum of the measures of the angles about the center of the clock above is 360. Each of the angles formed by the clock hands is called a central angle. A **central angle** has the center of the circle as its vertex, and its sides contain two radii of the circle.

Key Concept — Sum of Central Angles

- **Words** The sum of the measures of the central angles of a circle with no interior points in common is 360.
- **Example** $m\angle 1 + m\angle 2 + m\angle 3 = 360$

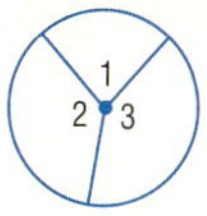

Example 1 Measures of Central Angles

ALGEBRA Refer to $\odot O$.

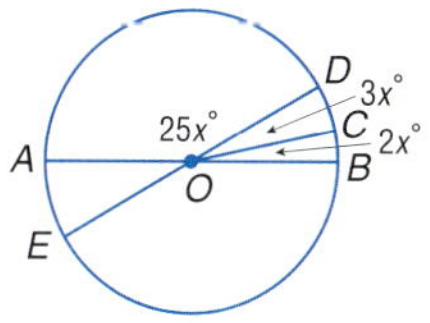

a. Find $m\angle AOD$.

$\angle AOD$ and $\angle DOB$ are a linear pair, and the angles of a linear pair are supplementary.

$m\angle AOD + m\angle DOB = 180$

$m\angle AOD + m\angle DOC + m\angle COB = 180$ — Angle Sum Theorem

$25x + 3x + 2x = 180$ — Substitution

$30x = 180$ — Simplify.

$x = 6$ — Divide each side by 60.

Use the value of x to find $m\angle AOD$.

$m\angle AOD = 25x$ — Given

$= 25(6)$ or 150 — Substitution

10-2 Lesson Notes

1 Focus

5-Minute Check Transparency 10-2 Use as a quiz or review of Lesson 10-1.

Mathematical Background notes are available for this lesson on p. 520C.

What kinds of angles do the hands on a clock form?

Ask students:

- Do the three angles on the clock appear to be acute, obtuse, or right angles? **2 acute, 1 obtuse**
- Why do you think the angles formed by the three hands are called central angles? **Because the three angles share the center of the circle as a vertex.**

Resource Manager

Workbook and Reproducible Masters

Chapter 10 Resource Masters

- Study Guide and Intervention, pp. 547–548
- Skills Practice, p. 549
- Practice, p. 550
- Reading to Learn Mathematics, p. 551
- Enrichment, p. 552
- Assessment, p. 603

Prerequisite Skills Workbook, pp. 31–32, 61–64, 67–68, 71–72, 105–106, 109–110

Teaching Geometry With Manipulatives Masters, p. 16

Transparencies

5-Minute Check Transparency 10-2
Answer Key Transparencies

Technology

Interactive Chalkboard

2 Teach

ANGLES AND ARCS

In-Class Example

Power Point®

1 **ALGEBRA** $\overline{RV}$ is a diameter of $\odot T$.

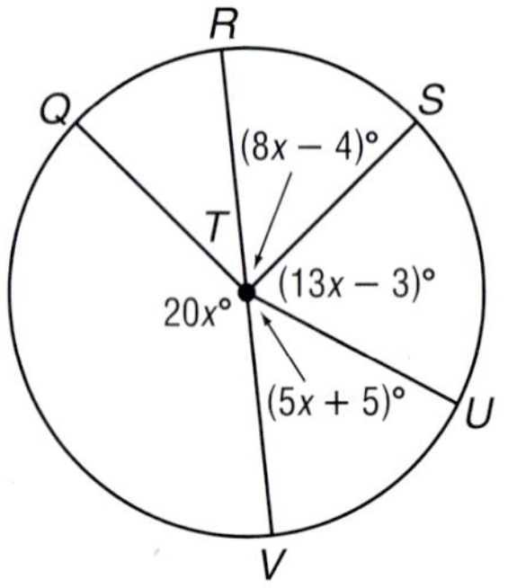

a. Find $m\angle RTS$. **52**

b. Find $m\angle QTR$. **40**

b. Find $m\angle AOE$.

$\angle AOE$ and $\angle AOD$ form a linear pair.

$m\angle AOE + m\angle AOD = 180$ Linear pairs are supplementary.

$m\angle AOE + 150 = 180$ Substitution

$m\angle AOE = 30$ Subtract 150 from each side.

A central angle separates the circle into two parts, each of which is an **arc**. The measure of each arc is related to the measure of its central angle.

Study Tip

Naming Arcs
Do not assume that because an arc is named by three letters that it is a semicircle or major arc. You can also correctly name a minor arc using three letters.

Key Concept — Arcs of a Circle

Type of Arc:	**minor arc**	**major arc**	**semicircle**
Example:	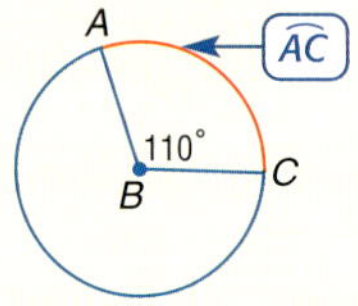	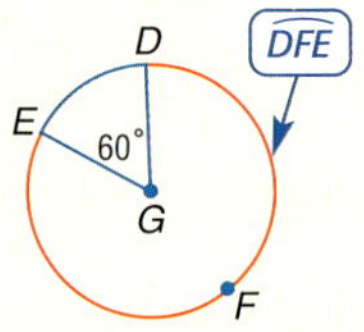	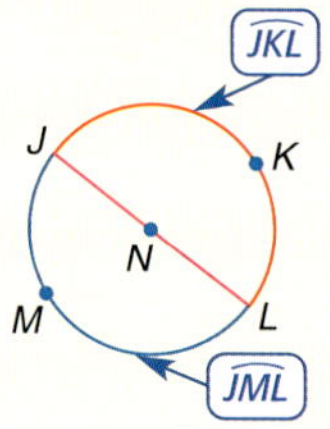
Named:	usually by the letters of the two endpoints $\widehat{AC}$	by the letters of the two endpoints and another point on the arc $\widehat{DFE}$	by the letters of the two endpoints and another point on the arc $\widehat{JML}$ and $\widehat{JKL}$
Arc Degree Measure Equals:	the measure of the central angle and is less than 180 $m\angle ABC = 110$, so $m\widehat{AC} = 110$	360 minus the measure of the minor arc and is greater than 180 $m\widehat{DFE} = 360 - m\widehat{DE}$ $m\widehat{DFE} = 360 - 60$ or 300	$360 \div 2$ or 180 $m\widehat{JML} = 180$ $m\widehat{JML} = 180$

Arcs with the same measure in the same circle or in congruent circles are congruent.

Theorem 10.1

In the same or in congruent circles, two arcs are congruent if and only if their corresponding central angles are congruent.

You will prove Theorem 10.1 in Exercise 54.

Arcs of a circle that have exactly one point in common are *adjacent arcs*. Like adjacent angles, the measures of adjacent arcs can be added.

Postulate 10.1

Arc Addition Postulate The measure of an arc formed by two adjacent arcs is the sum of the measures of the two arcs.

Example: In $\odot S$, $m\widehat{PQ} + m\widehat{QR} = m\widehat{PQR}$.

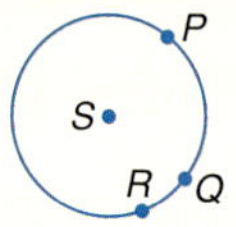

530 Chapter 10 Circles

Example 2 Measures of Arcs

In $\odot F$, $m\angle DFA = 50$ and $\overline{CF} \perp \overline{FB}$. Find each measure.

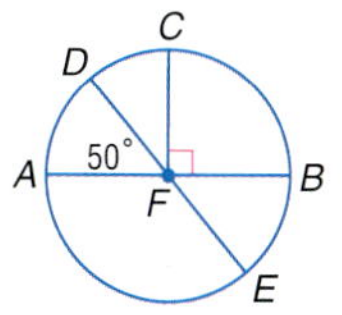

a. $m\overset{\frown}{BE}$

$\overset{\frown}{BE}$ is a minor arc, so $m\overset{\frown}{BE} = m\angle BFE$.

$\angle BFE$	$\cong \angle DFA$	Vertical angles are congruent.
$m\angle BFE$	$= m\angle DFA$	Definition of congruent angles
$m\overset{\frown}{BE}$	$= m\angle DFA$	Transitive Property
$m\overset{\frown}{BE}$	$= 50$	Substitution

b. $m\overset{\frown}{CBE}$

$\overset{\frown}{CBE}$ is composed of adjacent arcs, $\overset{\frown}{CB}$ and $\overset{\frown}{BE}$.

$m\overset{\frown}{CB}$	$= m\angle CFB$	
	$= 90$	$\angle CFB$ is a right angle.
$m\overset{\frown}{CBE}$	$= m\overset{\frown}{CB} + m\overset{\frown}{BE}$	Arc Addition Postulate
$m\overset{\frown}{CBE}$	$= 90 + 50$ or 140	Substitution

c. $m\overset{\frown}{ACE}$

One way to find $m\overset{\frown}{ACE}$ is by using $\overset{\frown}{ACB}$ and $\overset{\frown}{BE}$.

$\overset{\frown}{ACB}$ is a semicircle.

$m\overset{\frown}{ACE}$	$= m\overset{\frown}{ACB} + \overset{\frown}{BE}$	Arc Addition Postulate
$m\overset{\frown}{ACE}$	$= 180 + 50$ or 230	Substitution

In a circle graph, the central angles divide a circle into wedges to represent data, often expressed as a percent. The size of the angle is proportional to the percent.

Example 3 Circle Graphs

FOOD **Refer to the graphic.**

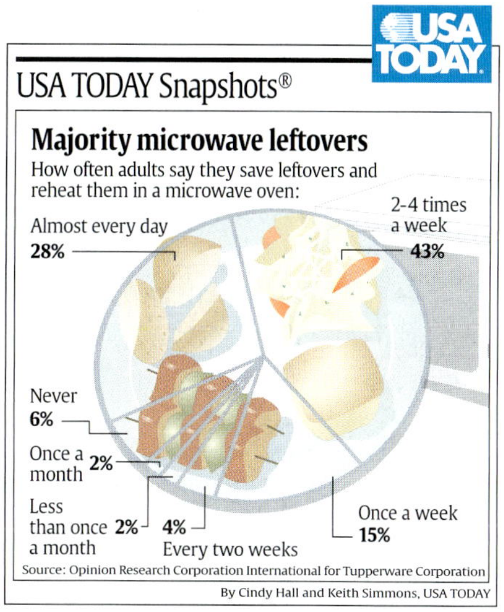

a. Find the measurement of the central angle for each category.

The sum of the percents is 100% and represents the whole. Use the percents to determine what part of the whole circle (360°) each central angle contains.

$2\%(360°) = 7.2°$

$6\%(360°) = 21.6°$

$28\%(360°) = 100.8°$

$43\%(360°) = 154.8°$

$15\%(360°) = 54°$

$4\%(360°) = 14.4°$

b. Use the categories to identify any arcs that are congruent.

The arcs for the wedges named *Once a month* and *Less than once a month* are congruent because they both represent 2% or 7.2° of the circle.

Log on for:
- Updated data
- More activities on circle graphs

www.geometryonline.com/usa_today

www.geometryonline.com/extra_examples/fcat

In-Class Examples

Power Point®

2 In $\odot P$, $m\angle NPM = 46$, $\overline{PL}$ bisects $\angle KPM$, and $\overline{OP} \perp \overline{KN}$. Find each measure.

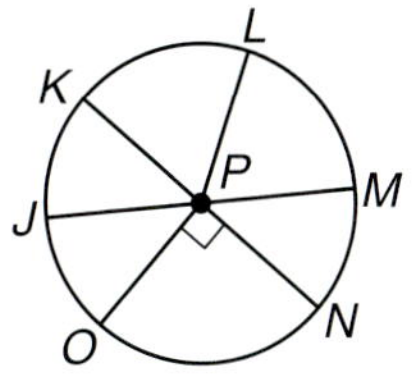

a. $m\overset{\frown}{OK}$ **90**

b. $m\overset{\frown}{LM}$ **67**

c. $m\overset{\frown}{JKO}$ **316**

3 **BICYCLES** This graph shows the percent of each type of bicycle sold in the United States in 2001.

Bicycles Bought Last Year (by type)

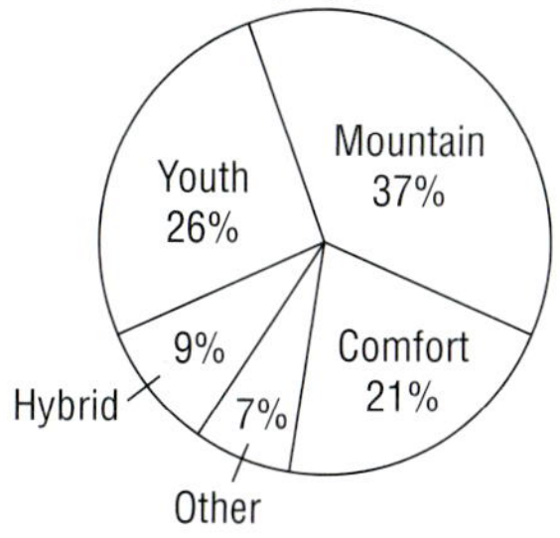

a. Find the measurement of the central angle representing each category. List them from least to greatest. **25.2°, 32.4°, 75.6°, 93.6°, 133.2°**

b. Is the arc for the wedge named *Youth* congruent to the arc for the combined wedges named *Other* and *Comfort*? **no**

DAILY INTERVENTION

Differentiated Instruction

Interpersonal Draw a circle segmented with different sizes of central angles. Shade each portion of the circle with a different color. Repeat for two other circles the same size, but with different central angles. Laminate the paper, cut out the circles, and separate each portion. Provide the cutouts to groups of students who can fit the pieces together to form the three circles, find the central angle measures, arc measures, circumferences and arc lengths. Groups can compare to check results and/or determine which group is the most efficient at finding all the correct information.

ARC LENGTH

Teaching Tip Tell students that they can set up a proportion to find an arc length because they are finding a *portion* of the circumference. Explain that this process is very similar to finding a percent of a whole.

In-Class Example

Power Point®

Teaching Tip If students want to see this problem another way, explain that 120 is $\frac{1}{3}$ of 360 as each arc length would be equal to $\frac{1}{3}$ of the total circumference. So, students can divide 30π by 3 and get the same answer.

4 In $\odot B$, $AC = 9$ and $m\angle ABD = 40$. Find the length of $\widehat{AD}$.

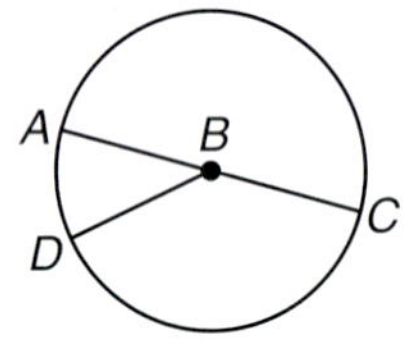

π units or about 3.14 units

Teaching Tip Have students construct a circle like the one in Example 4 and measure its radius. Have them use string to trace the circumference of the circle. Mark on the string the points that are the endpoints of the arc. After calculating the circumference, have them use a ruler to verify the arc length.

Answers

1. Sample answer: $\widehat{AB}$, $\widehat{BC}$, $\widehat{AC}$, $\widehat{ABC}$, $\widehat{BCA}$, $\widehat{CAB}$; $m\widehat{AB} = 110$, $m\widehat{BC} = 160$, $m\widehat{AC} = 90$, $m\widehat{ABC} = 270$, $m\widehat{BCA} = 250$, $m\widehat{CAB} = 200$

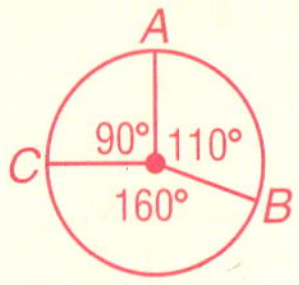

2. A diameter divides the circle into two congruent arcs. Without the third letter, it is impossible to know which semicircle is being referenced.
3. Sample answer: Concentric circles have the same center, but different radius measures; congruent circles usually have different centers but the same radius measure.

ARC LENGTH Another way to measure an arc is by its length. An arc is part of the circle, so the length of an arc is a part of the circumference.

Study Tip

Look Back
To review proportions, see Lesson 6-1.

Example 4 Arc Length

In $\odot P$, $PR = 15$ and $m\angle QPR = 120$. Find the length of $\widehat{QR}$.

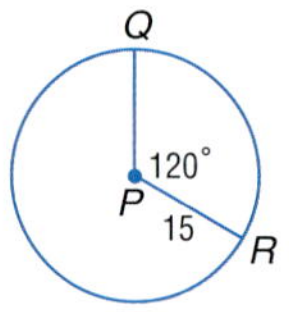

In $\odot P$, $r = 15$, so $C = 2\pi(15)$ or 30π and $m\widehat{QR} = m\angle QPR$ or 120. Write a proportion to compare each part to its whole.

degree measure of arc → $\frac{120}{360} = \frac{\ell}{30\pi}$ ← arc length
degree measure of whole circle → ← circumference

Now solve the proportion for ℓ.

$$\frac{120}{360} = \frac{\ell}{30\pi}$$

$$\frac{120}{360}(30\pi) = \ell \quad \text{Multiply each side by } 30\pi.$$

$$10\pi = \ell \quad \text{Simplify.}$$

The length of $\widehat{QR}$ is 10π units or about 31.42 units.

The proportion used to find the arc length in Example 4 can be adapted to find the arc length in any circle.

Key Concept — Arc Length

degree measure of arc → $\frac{A}{360} = \frac{\ell}{2\pi r}$ ← arc length
degree measure of whole circle → ← circumference

This can also be expressed as $\frac{A}{360} \cdot C = \ell$.

Check for Understanding

Concept Check

1. **OPEN-ENDED** Draw a circle and locate three points on the circle. Name all of the arcs determined by the three points and use a protractor to find the measure of each arc. **1–3. See margin.**
2. **Explain** why it is necessary to use three letters to name a semicircle.
3. **Describe** the difference between *concentric* circles and *congruent* circles.

Guided Practice

GUIDED PRACTICE KEY	
Exercises	Examples
4–7	1
8–11	2
12	3
13	4

ALGEBRA **Find each measure.**

4. $m\angle NCL$ **120**
5. $m\angle RCL$ **137**
6. $m\angle RCM$ **43**
7. $m\angle RCN$ **103**

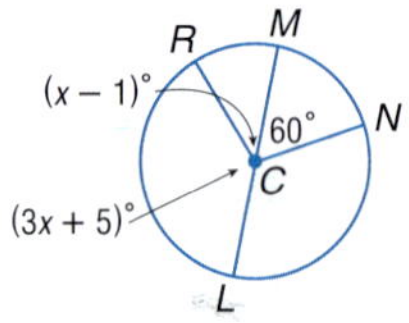

In $\odot A$, $m\angle EAD = 42$. Find each measure.

8. $m\widehat{BC}$ **42**
9. $m\widehat{CBE}$ **180**
10. $m\widehat{EDB}$ **222**
11. $m\widehat{CD}$ **138**

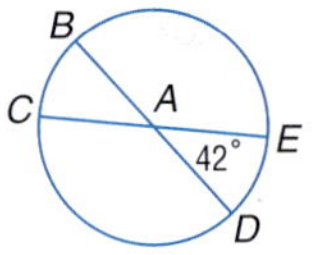

12. Points T and R lie on $\odot W$ so that $WR = 12$ and $m\angle TWR = 60$. Find the length of $\widehat{TR}$. **$4\pi \approx 12.57$ units**

DAILY INTERVENTION — Unlocking Misconceptions

Arcs Students may sometimes confuse the terms *arc measure* and *arc length*. Explain that they can remember that angles have degree measure, denoted $m\angle ABC$; similarly, arcs have degree measure, denoted $m\widehat{AC}$. Just as segment length is a distance along a line, arc length is a distance along a curve that you can actually follow or draw with a pencil. Point out that students should be careful to determine whether they need to find the *measure* or *length* of an arc.

532 Chapter 10 Circles

Application

13. **SURVEYS** The graph shows the results of a survey of 1400 chief financial officers who were asked how many hours they spend working on the weekend. Determine the measurement of each angle of the graph. Round to the nearest degree. **Sample answer: 25% = 90°, 23% = 83°, 28% = 101°, 22% = 79°, 2% = 7°**

Executives Working on the Weekend

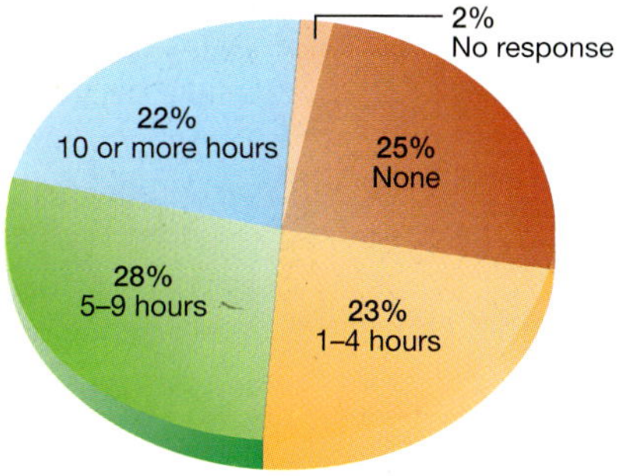

Source: Accountemps

★ indicates increased difficulty

Practice and Apply

Homework Help

For Exercises	See Examples
14–23	1
24–39	2
40–43	3
44–45	4

Extra Practice
See page 774.

Find each measure.

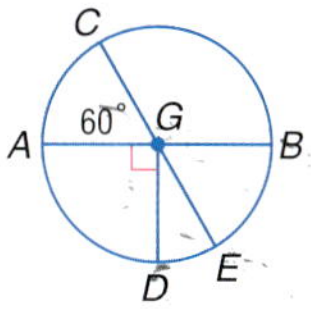

14. $m\angle CGB$ **120**
15. $m\angle BGE$ **60**
16. $m\angle AGD$ **90**
17. $m\angle DGE$ **30**
18. $m\angle CGD$ **150**
19. $m\angle AGE$ **120**

ALGEBRA Find each measure.

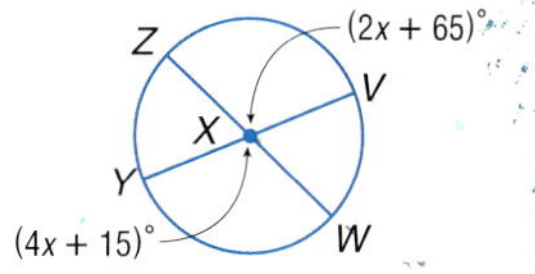

20. $m\angle ZXV$ **115**
21. $m\angle YXW$ **115**
22. $m\angle ZXY$ **65**
23. $m\angle VXW$ **65**

In $\odot O$, $\overline{EC}$ and $\overline{AB}$ are diameters, and $\angle BOD \cong \angle DOE \cong \angle EOF \cong \angle FOA$. Find each measure.

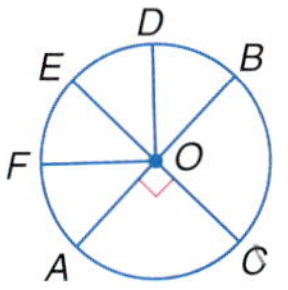

24. $m\widehat{BC}$ **90**
25. $m\widehat{AC}$ **90**
26. $m\widehat{AE}$ **90**
27. $m\widehat{EB}$ **90**
28. $m\widehat{ACB}$ **180**
29. $m\widehat{AD}$ **135**
30. $m\widehat{CBF}$ **225**
31. $m\widehat{ADC}$ **270**

ALGEBRA In $\odot Z$, $\angle WZX \cong \angle XZY$, $m\angle VZU = 4x$, $m\angle UZY = 2x + 24$, and $\overline{VY}$ and $\overline{WU}$ are diameters. Find each measure.

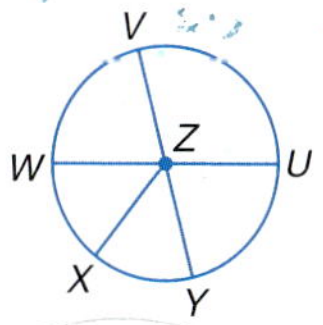

32. $m\widehat{UY}$ **76**
33. $m\widehat{WV}$ **76**
34. $m\widehat{WX}$ **52**
35. $m\widehat{XY}$ **52**
36. $m\widehat{WUY}$ **256**
37. $m\widehat{YVW}$ **256**
★ 38. $m\widehat{XVY}$ **308**
★ 39. $m\widehat{WUX}$ **308**

The diameter of $\odot C$ is 32 units long. Find the length of each arc for the given angle measure.

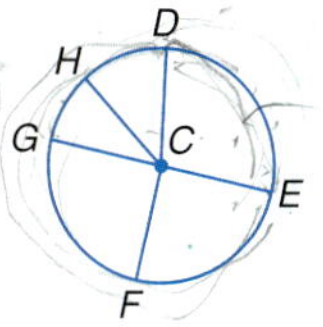

40. $\widehat{DE}$ if $m\angle DCE = 100$ **40. $\frac{188\pi}{9} \approx 27.93$ units**
41. $\widehat{DHE}$ if $m\angle DCE = 90$ **41. $24\pi \approx 75.40$ units**
42. $\widehat{HDF}$ if $m\angle HCF = 125$ **$\frac{188\pi}{9} \approx 65.62$ units**
43. $\widehat{HD}$ if $m\angle DCH = 45$ **$4\pi \approx 12.57$ units**

Study Guide and Intervention, p. 547 (shown) and p. 548

Angles and Arcs A **central angle** is an angle whose vertex is at the center of a circle and whose sides are radii. A central angle separates a circle into two arcs, a **major arc** and a **minor arc**.

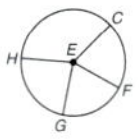

$\widehat{GF}$ is a minor arc.
$\widehat{CHG}$ is a major arc.
$\angle GEF$ is a central angle.

Here are some properties of central angles and arcs.

- The sum of the measures of the central angles of a circle with no interior points in common is 360. $m\angle HEC + m\angle CEF + m\angle FEG + m\angle GEH = 360$
- The measure of a minor arc equals the measure of its central angle. $m\widehat{CF} = m\angle CEF$
- The measure of a major arc is 360 minus the measure of the minor arc. $m\widehat{CGF} = 360 - m\widehat{CF}$
- Two arcs are congruent if and only if their corresponding central angles are congruent. $\widehat{CF} \cong \widehat{FG}$ if and only if $\angle CEF \cong \angle FEG$.
- The measure of an arc formed by two adjacent arcs is the sum of the measures of the two arcs. **(Arc Addition Postulate)** $m\widehat{CF} + m\widehat{FG} = m\widehat{CG}$

Example **In $\odot R$, $m\angle ARB = 42$ and $\overline{AC}$ is a diameter. Find $m\widehat{AB}$ and $m\widehat{ACB}$.**

$\angle ARB$ is a central angle and $m\angle ARB = 42$, so $m\widehat{AB} = 42$.
Thus $m\widehat{ACB} = 360 - 42$ or 318.

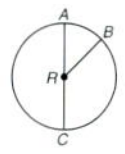

Exercises

Find each measure.

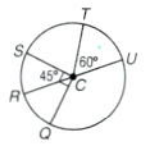

1. $m\angle SCT$ 75
2. $m\angle SCU$ 135
3. $m\angle SCQ$ 90
4. $m\angle QCT$ 165

If $m\angle BOA = 44$, find each measure.

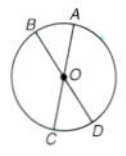

5. $m\widehat{BA}$ 44
6. $m\widehat{BC}$ 136
7. $m\widehat{CD}$ 44
8. $m\widehat{ACB}$ 316
9. $m\widehat{BCD}$ 180
10. $m\widehat{AD}$ 136

Skills Practice, p. 549 and Practice, p. 550 (shown)

ALGEBRA **In $\odot Q$, $\overline{AC}$ and $\overline{BD}$ are diameters. Find each measure.**

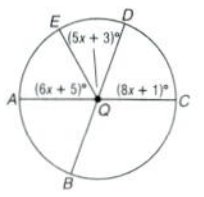

1. $m\angle AQE$ 59
2. $m\angle DQE$ 48
3. $m\angle CQD$ 73
4. $m\angle BQC$ 107
5. $m\angle CQE$ 121
6. $m\angle AQD$ 107

In $\odot P$, $m\angle GPH = 38$. Find each measure.

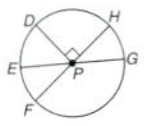

7. $m\widehat{EF}$ 38
8. $m\widehat{DE}$ 52
9. $m\widehat{FG}$ 142
10. $m\widehat{DHG}$ 128
11. $m\widehat{DFG}$ 232
12. $m\widehat{DGE}$ 308

The radius of $\odot Z$ is 13.5 units long. Find the length of each arc for the given angle measure.

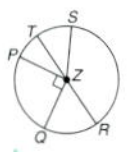

13. $\widehat{QPT}$ if $m\angle QZT = 120$
$9\pi \approx 28.27$ units
14. $\widehat{QR}$ if $m\angle QZR = 60$
$4.5\pi \approx 14.14$ units
15. $\widehat{PQR}$ if $m\angle PZR = 150$
$11.25\pi \approx 35.34$ units
16. $\widehat{QPS}$ if $m\angle QZS = 160$
$12\pi \approx 37.70$ units

HOMEWORK **For Exercises 17 and 18, refer to the table, which shows the number of hours students at Leland High School say they spend on homework each night.**

Homework	
Less than 1 hour	8%
1–2 hours	29%
2–3 hours	58%
3–4 hours	3%
Over 4 hours	2%

17. If you were to construct a circle graph of the data, how many degrees would be allotted to each category?
28.8°, 104.4°, 208.8°, 10.8°, 7.2°

18. Describe the arcs associated with each category.
The arc associated with 2–3 hours is a major arc; minor arcs are associated with the remaining categories.

Reading to Learn Mathematics, p. 551 ELL

Pre-Activity **What kinds of angles do the hands on a clock form?**

Read the introduction to Lesson 10-2 at the top of page 529 in your textbook.

- What is the measure of the angle formed by the hour hand and the minute hand of the clock at 5:00? 150
- What is the measure of the angle formed by the hour hand and the minute hand at 10:30? (Hint: How has each hand moved since 10:00?) 135

Reading the Lesson

1. Refer to $\odot P$. Indicate whether each statement is *true* or *false*.

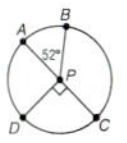

a. $\widehat{DAB}$ is a major arc. false
b. $\widehat{ADC}$ is a semicircle. true
c. $\widehat{AD} \cong \widehat{CD}$ true
d. $\widehat{DA}$ and $\widehat{AB}$ are adjacent arcs. true
e. $\angle BPC$ is an acute central angle. false
f. $\angle DPA$ and $\angle BPA$ are supplementary central angles. false

2. Refer to the figure in Exercise 1. Give each of the following arc measures.

a. $m\widehat{AB}$ 52
b. $m\widehat{CD}$ 90
c. $m\widehat{BC}$ 128
d. $m\widehat{ADC}$ 180
e. $m\widehat{DAB}$ 142
f. $m\widehat{DCB}$ 218
g. $m\widehat{DAC}$ 270
h. $m\widehat{BDA}$ 308

3. Underline the correct word or number to form a true statement.

a. The arc measure of a semicircle is (90/180/360).
b. Arcs of a circle that have exactly one point in common are (congruent/opposite/adjacent) arcs.
c. The measure of a major arc is greater than (0/90/180) and less than (90/180/360).
d. Suppose a set of central angles of a circle have interiors that do not overlap. If the angles and their interiors contain all points of the circle, then the sum of the measures of the central angles is (90/270/360).
e. The measure of an arc formed by two adjacent arcs is the (sum/difference/product) of the measures of the two arcs.
f. The measure of a minor arc is greater than (0/90/180) and less than (90/180/360).

Helping You Remember

4. A good way to remember something is to explain it to someone else. Suppose your classmate Luis does not like to work with proportions. What is a way that he can find the length of a minor arc of a circle without solving a proportion? Sample answer: Divide the measure of the central angle of the arc by 360 to form a fraction. Multiply this fraction by the circumference of the circle to find the length of the arc.

Enrichment, p. 552

Curves of Constant Width

A circle is called a curve of constant width because no matter how you turn it, the greatest distance across it is always the same. However, the circle is not the only figure with this property.

The figure at the right is called a Reuleaux triangle.

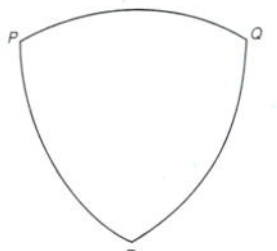

1. Use a metric ruler to find the distance from P to any point on the opposite side. 4.6 cm
2. Find the distance from Q to the opposite side. 4.6 cm
3. What is the distance from R to the opposite side? 4.6 cm

The Reuleaux triangle is made of three arcs. In the example shown, $\widehat{PQ}$ has center R, $\widehat{QR}$ has center P, and $\widehat{PR}$ has center Q.

4. Trace the Reuleaux triangle above on a piece of paper and cut it out. Make a square with sides the length you found in Exercise 1. Show that you can turn the triangle inside the ... eping its sides ... with ...

3 Practice/Apply

Study Notebook

Have students—
- add the definitions/examples of the vocabulary terms to their Vocabulary Builder worksheets for Chapter 10.
- include any other item(s) that they find helpful in mastering the skills in this lesson.

About the Exercises...

Organization by Objective
- **Angles and Arcs:** 14–43
- **Arc Length:** 44–45

Odd/Even Assignments
Exercises 14–43 are structured so that students practice the same concepts whether they are assigned odd or even problems.

Alert! Exercise 46 requires a compass.

Assignment Guide

Basic: 15–37 odd, 41–55 odd, 57–76

Average: 15–55 odd, 57–76

Advanced: 14–50 even, 51, 52, 54, 55–70 (optional: 71–76)

ONLINE MUSIC For Exercises 44–46, refer to the table and use the following information.

A recent survey asked online users how many legally free music files they have collected. The results are shown in the table.

Free Music Downloads How many free music files have you collected?	
100 files or less	76%
101 to 500 files	16%
501 to 1000 files	5%
More than 1000 files	3%

Source: QuickTake.com

44. If you were to construct a circle graph of this information, how many degrees would be needed for each category?

44. Sample answer: 76% = 273°, 16% = 58°, 5% = 18°, 3% = 11°

45. Describe the kind of arc associated with each category.

45. The first category is a major arc, and the other three categories are minor arcs.

★ 46. Construct a circle graph for these data. See margin.

Determine whether each statement is *sometimes, always,* or *never* true.

47. The measure of a major arc is greater than 180. always

48. The central angle of a minor arc is an acute angle. sometimes

49. The sum of the measures of the central angles of a circle depends on the measure of the radius. never

50. The semicircles of two congruent circles are congruent. always

51. **CRITICAL THINKING** Central angles 1, 2, and 3 have measures in the ratio 2 : 3 : 4. Find the measure of each angle. $m\angle 1 = 80$, $m\angle 2 = 120$, $m\angle 3 = 160$

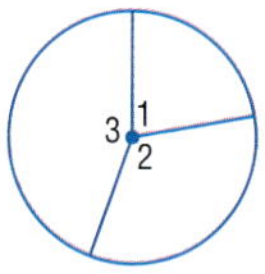

52. **CLOCKS** The hands of a clock form the same angle at various times of the day. For example, the angle formed at 2:00 is congruent to the angle formed at 10:00. If a clock has a diameter of 1 foot, what is the distance along the edge of the clock from the minute hand to the hour hand at 2:00? 2π in. ≈ 6.3 in.

More About. . .

Irrigation In the Great Plains of the United States, farmers use center-pivot irrigation systems to water crops. New low-energy spray systems water circles of land that are thousands of feet in diameter with minimal water loss to evaporation from the spray.
Source: U.S. Geological Survey

53. **IRRIGATION** Some irrigation systems spray water in a circular pattern. You can adjust the nozzle to spray in certain directions. The nozzle in the diagram is set so it does not spray on the house. If the spray has a radius of 12 feet, what is the approximate length of the arc that the spray creates? 56.5 ft

54. **PROOF** Write a proof of Theorem 10.1. See margin.

55. **CRITICAL THINKING** The circles at the right are concentric circles that both have point E as their center. If $m\angle 1 = 42$, determine whether $\overparen{AB} \cong \overparen{CD}$. Explain. No; the radii are not equal, so the proportional part of the circumferences would not be the same. Thus, the arcs would not be congruent.

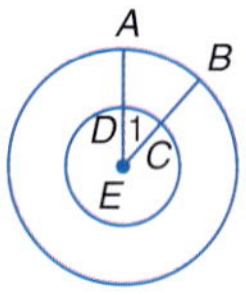

56. **WRITING IN MATH** Answer the question that was posed at the beginning of the lesson. See margin.

What kind of angles do the hands of a clock form?

Include the following in your answer:
- the kind of angle formed by the hands of a clock, and
- several times of day when these angles are congruent.

Answers

46.

54. Given: $\angle BAC \cong \angle DAE$
Prove: $\overparen{BC} \cong \overparen{DE}$

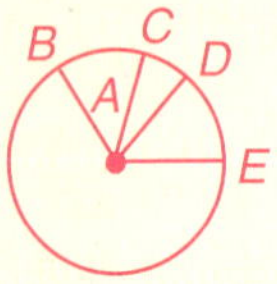

Proof:

Statements (Reasons)
1. $\angle BAC \cong \angle DAE$ (Given)
2. $m\angle BAC = m\angle DAE$ (Def. of ≅ ∠s)
3. $m\overparen{BC} = m\overparen{DE}$ (Def. of arc measure)
4. $\overparen{BC} \cong \overparen{DE}$ (Def. of ≅ arcs)

56. Sample answer: The hands of the clock form central angles. Answers should include the following.
- The hands form acute, right, and obtuse angles.
- Some times when the angles formed by the minute and hour hand are congruent are at 1:00 and 11:00, 2:00 and 10:00, 3:00 and 9:00, 4:00 and 8:00, and 5:00 and 7:00. They also form congruent angles at many other times of the day, such as 3:05 and 8:55.

FCAT Practice

Standardized Test Practice

Ⓐ Ⓑ Ⓒ Ⓓ

57. Compare the circumference of circle E with the perimeter of rectangle $ABCD$. Which statement is true? **B**

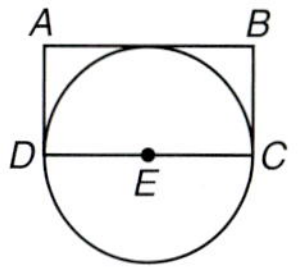

Ⓐ The perimeter of $ABCD$ is greater than the circumference of circle E.

Ⓑ The circumference of circle E is greater than the perimeter of $ABCD$.

Ⓒ The perimeter of $ABCD$ equals the circumference of circle E.

Ⓓ There is not enough information to determine this comparison.

58. **SHORT RESPONSE** A circle is divided into three central angles that have measures in the ratio 3 : 5 : 10. Find the measure of each angle. **60, 100, 200**

Maintain Your Skills

Mixed Review

The radius, diameter, or circumference of a circle is given. Find the missing measures. Round to the nearest hundredth if necessary. *(Lesson 10-1)*

59. 20; 62.83

60. 6.5; 40.84

59. $r = 10, d =$ ___?___ $, C =$ ___?___

60. $d = 13, r =$ ___?___ $, C =$ ___?___

61. $C = 28\pi, d =$ ___?___ $, r =$ ___?___ **28; 14**

62. $C = 75.4, d =$ ___?___ $, r =$ ___?___ **24.00; 12.00**

63. **SOCCER** Two soccer players kick the ball at the same time. One exerts a force of 72 newtons east. The other exerts a force of 45 newtons north. What are the magnitude to the nearest tenth and direction to the nearest degree of the resultant force on the soccer ball? *(Lesson 9-6)* **84.9 newtons, 32° north of due east**

ALGEBRA **Find x.** *(Lesson 6-5)*

64. **$8\frac{2}{11}$**

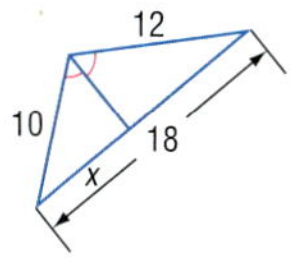

65. **36.68**

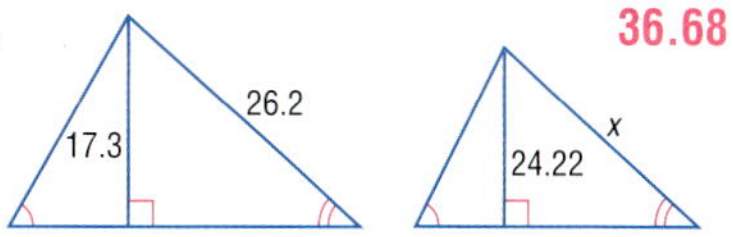

Find the exact distance between each point and line or pair of lines. *(Lesson 3-6)*

66. point $Q(6, -2)$ and the line with the equation $y - 7 = 0$ **9 units**

67. parallel lines with the equations $y = x + 3$ and $y = x - 4$ **$\sqrt{24.5}$**

68. Angle A has a measure of 57.5. Find the measures of the complement and supplement of $\angle A$. *(Lesson 2-8)* **32.5, 122.5**

Use the following statement for Exercises 69 and 70.
If ABC is a triangle, then ABC has three sides. *(Lesson 2-3)*

69. Write the converse of the statement. **If *ABC* has three sides, then *ABC* is a triangle.**

70. Determine the truth value of the statement and its converse. **Both are true.**

Getting Ready for the Next Lesson

PREREQUISITE SKILL **Find x.** *(To review **isosceles triangles**, see Lesson 4-6.)*

71. **42**

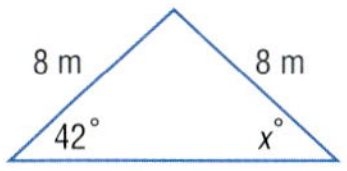

72. **75**

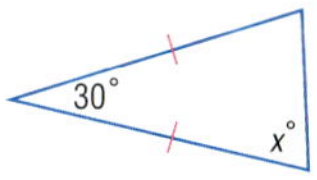

73. **100**

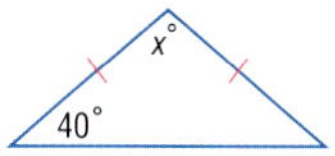

74. **45**

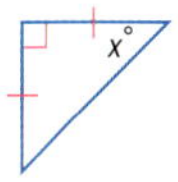

75. **36**

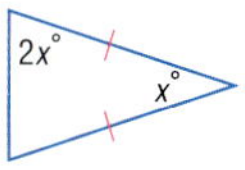

76. **60**

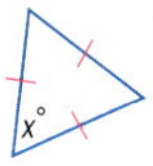

4 Assess

Open-Ended Assessment

Writing Provide examples on the board of circles marked with central angles, and have students take turns coming to the board and writing angle measure(s), arc measure(s), and arc length(s) for each example.

Getting Ready for Lesson 10-3

Prerequisite Skill Students will learn about arcs and chords in Lesson 10-3. They will use isosceles triangles to write proofs and to find arc measures and chord lengths. Use Exercises 71–76 to determine your students' familiarity with isosceles triangles.

Assessment Options

Quiz (Lessons 10-1 and 10-2) is available on p. 603 of the *Chapter 10 Resource Masters.*

10-3 Lesson Notes

1 Focus

5-Minute Check Transparency 10-3 Use as a quiz or review of Lesson 10-2.

Mathematical Background notes are available for this lesson on p. 520C.

How do the grooves in a Belgian waffle iron model segments in a circle?

Ask students:

- Excluding the diameter, how many chords can you count in the lower semicircle of the top heated plate on the waffle iron? **4**
- If the radius of the top heated plate measures 11 cm, then what is the circumference of the plate? **22π cm or about 69.12 cm**
- If one central angle of the waffle iron measures 90°, then what is the measure of its corresponding minor arc? **90**

10-3 Arcs and Chords

Sunshine State Standards
MA.B.1.4.2

Vocabulary

- inscribed
- circumscribed

What You'll Learn

- Recognize and use relationships between arcs and chords.
- Recognize and use relationships between chords and diameters.

How do the grooves in a Belgian waffle iron model segments in a circle?

Waffle irons have grooves in each heated plate that result in the waffle pattern when the batter is cooked. One model of a Belgian waffle iron is round, and each groove is a chord of the circle.

ARCS AND CHORDS The endpoints of a chord are also endpoints of an arc. If you trace the waffle pattern on patty paper and fold along the diameter, $\overline{AB}$ and $\overline{CD}$ match exactly, as well as $\widehat{AB}$ and $\widehat{CD}$. This suggests the following theorem.

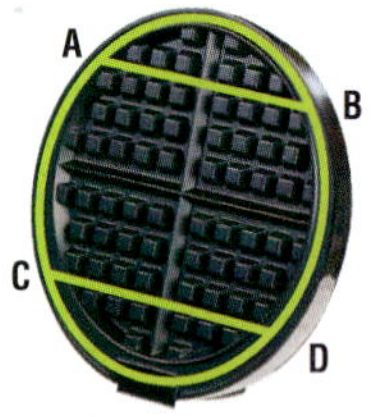

Study Tip

Reading Mathematics
Remember that the phrase *if and only if* means that the conclusion and the hypothesis can be switched and the statement is still true.

Theorem 10.2

In a circle or in congruent circles, two minor arcs are congruent if and only if their corresponding chords are congruent.

Abbreviations:
In ⊙, 2 minor arcs are ≅, corr. chords are ≅.
In ⊙, 2 chords are ≅, corr. minor arcs are ≅

Examples
If $\widehat{AB} \cong \widehat{CD}$, $\overline{AB} \cong \overline{CD}$.
If $\overline{AB} \cong \overline{CD}$, $\widehat{AB} \cong \widehat{CD}$.

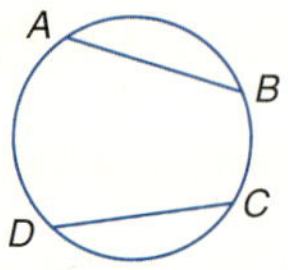

You will prove part 2 of Theorem 10.2 in Exercise 4.

Example 1 Prove Theorems

PROOF Theorem 10.2 (part 1)

Given: $\odot X$, $\widehat{UV} \cong \widehat{YW}$

Prove: $\overline{UV} \cong \overline{YW}$

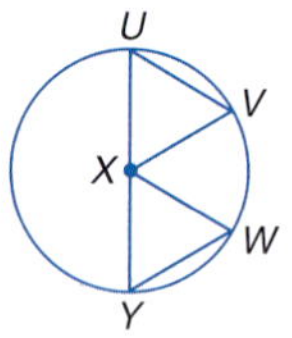

Proof:

Statements	Reasons
1. $\odot X$, $\widehat{UV} \cong \widehat{YW}$	1. Given
2. $\angle UXV \cong \angle WXY$	2. If arcs are ≅, their corresponding central ∠s are ≅.
3. $\overline{UX} \cong \overline{XV} \cong \overline{XW} \cong \overline{XY}$	3. All radii of a circle are congruent.
4. $\triangle UXV \cong \triangle WXY$	4. SAS
5. $\overline{UV} \cong \overline{YW}$	5. CPCTC

Resource Manager

Workbook and Reproducible Masters

Chapter 10 Resource Masters
- Study Guide and Intervention, pp. 553–554
- Skills Practice, p. 555
- Practice, p. 556
- Reading to Learn Mathematics, p. 557
- Enrichment, p. 558

Teaching Geometry With Manipulatives Masters, pp. 16, 17, 162, 163, 164

Transparencies

5-Minute Check Transparency 10-3
Real-World Transparency 10
Answer Key Transparencies

Technology

Interactive Chalkboard

Study Tip

Inscribed and Circumscribed A circle can also be inscribed in a polygon, so that the polygon is circumscribed about the circle. You will learn about this in Lesson 10-5.

The chords of adjacent arcs can form a polygon. Quadrilateral $ABCD$ is an **inscribed** polygon because all of its vertices lie on the circle. Circle E is **circumscribed** about the polygon because it contains all the vertices of the polygon.

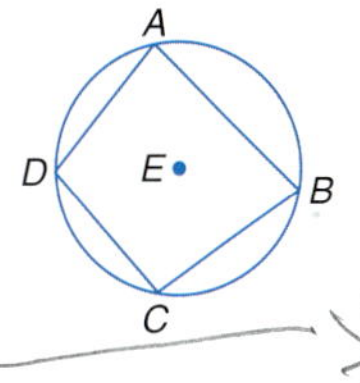

Example 2 Inscribed Polygons

SNOWFLAKES **The main veins of a snowflake create six congruent central angles. Determine whether the hexagon containing the flake is regular.**

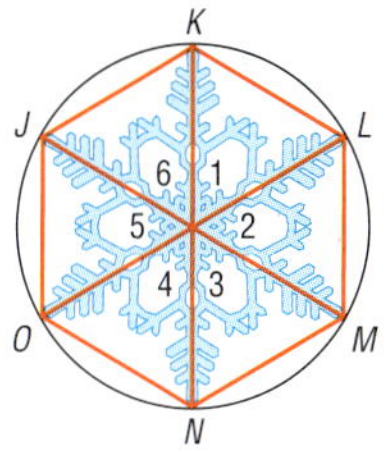

$\angle 1 \cong \angle 2 \cong \angle 3 \cong \angle 4 \cong \angle 5 \cong \angle 6$ Given

$\overset{\frown}{KL} \cong \overset{\frown}{LM} \cong \overset{\frown}{MN} \cong \overset{\frown}{NO} \cong \overset{\frown}{OJ} \cong \overset{\frown}{JK}$ If central $\angle$s are $\cong$, corresponding arcs are $\cong$.

$\overline{KL} \cong \overline{LM} \cong \overline{MN} \cong \overline{NO} \cong \overline{OJ} \cong \overline{JK}$ In $\odot$, 2 minor arcs $\cong$, corr. chords are $\cong$.

Because all the central angles are congruent, the measure of each angle is $360 \div 6$ or 60.

Let x be the measure of each base angle in the triangle containing $\overline{KL}$.

$m\angle 1 + x + x = 180$ Angle Sum Theorem

$60 + 2x = 180$ Substitution

$2x = 120$ Subtract 60 from each side.

$x = 60$ Divide each side by 2.

This applies to each triangle in the figure, so each angle of the hexagon is 2(60) or 120. Thus the hexagon has all sides congruent and all vertex angles congruent.

DIAMETERS AND CHORDS Diameters that are perpendicular to chords create special segment and arc relationships. Suppose you draw circle C and one of its chords $\overline{WX}$ on a piece of patty paper and fold the paper to construct the perpendicular bisector. You will find that the bisector also cuts $\overline{WX}$ in half and passes through the center of the circle, making it contain a diameter.

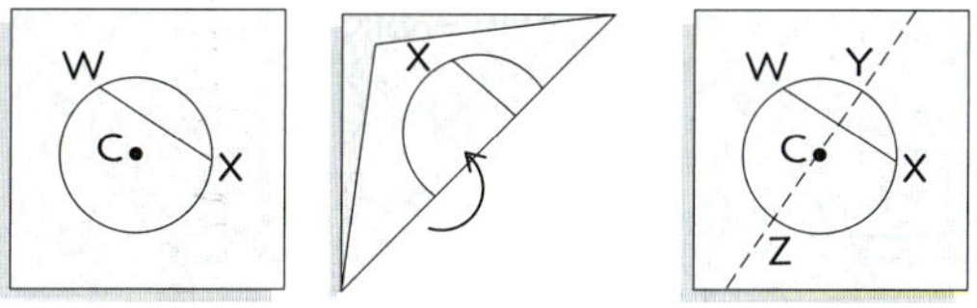

This is formally stated in the next theorem.

Theorem 10.3

In a circle, if a diameter (or radius) is perpendicular to a chord, then it bisects the chord and its arc.

Example: If $\overline{BA} \perp \overline{TV}$, then $\overline{UT} \cong \overline{UV}$ and $\overset{\frown}{AT} \cong \overset{\frown}{AV}$.

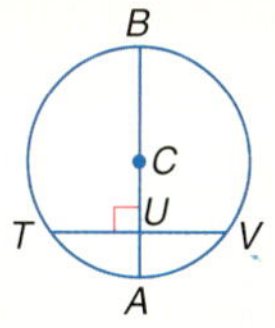

You will prove Theorem 10.3 in Exercise 36.

www.geometryonline.com/extra_examples/fcat

2 Teach

ARCS AND CHORDS

In-Class Examples Power Point®

1 **PROOF** Write a proof.

Given: $\overline{DE} \cong \overline{FG}$
$m\angle EBF = 24$
$\overset{\frown}{DFG}$ is a semicircle.

Prove: $m\angle FBG = 78$

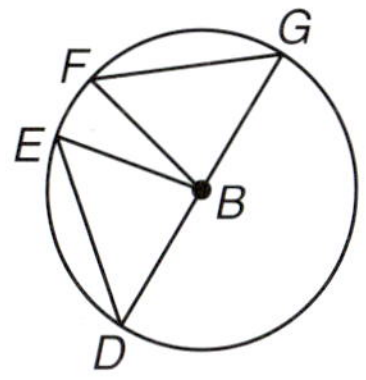

1. $\overline{DE} \cong \overline{FG}$; $m\angle EBF = 24$; $\overset{\frown}{DFG}$ is a semicircle. (Given)
2. $m\overset{\frown}{DFG} = 180$ (Def. of semicircle)
3. $\overset{\frown}{DE} \cong \overset{\frown}{FG}$ (In $\odot$, 2 chords are $\cong$, corr. minor arcs are $\cong$.)
4. $m\overset{\frown}{DE} = m\overset{\frown}{FG}$ (Def. of $\cong$ arcs)
5. $m\overset{\frown}{EF} = 24$ (Def. of arc measure)
6. $m\overset{\frown}{ED} + m\overset{\frown}{EF} + m\overset{\frown}{FG} = m\overset{\frown}{DFG}$ (Arc Addition Post.)
7. $m\overset{\frown}{FG} + 24 + m\overset{\frown}{FG} = 180$ (Substitution)
8. $2(m\overset{\frown}{FG}) = 156$ (Subtr. Prop. and simplify)
9. $m\overset{\frown}{FG} = 78$ (Div. Prop.)
10. $m\overset{\frown}{FG} = m\angle FBG$ (Def. of arc measure)
11. $m\angle FBG = 78$ (Substitution)

2 **TESSELLATIONS** The rotations of a tessellation can create twelve congruent central angles. Determine whether $\overline{PQ} \cong \overline{ST}$. yes

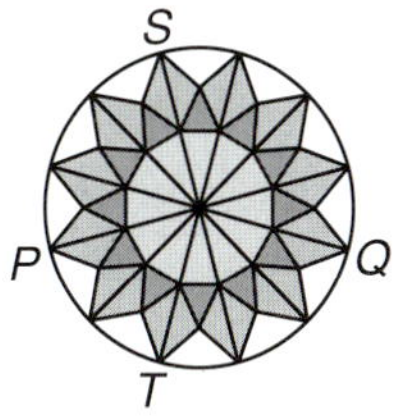

DAILY INTERVENTION **Differentiated Instruction**

Auditory/Musical Summarize the three major concepts of this lesson aloud for auditory students. Explain that they now know that congruent chords share endpoints with congruent arcs and vice versa; plus, the chords are only congruent if and only if they are equidistant from the center of the circle. They have also learned that a diameter perpendicularly bisects any chord of a circle.

DIAMETERS AND CHORDS

Building on Prior Knowledge

Students learned about the Pythagorean Theorem in Chapter 1. They determined how to prove triangle congruence in Chapter 4, and they learned about segment bisectors in Chapter 5. Students will apply all of these concepts in this lesson as they prove triangle congruence and find segment lengths in circles.

In-Class Example

Power Point®

Teaching Tip Remind students that they can add any known information to a figure to help them solve problems, as radius $\overline{OC}$ has been added to the figure in the example. Tell students to remember that angles, segment lengths, arcs, radii, and diameters all exist even if they are not drawn. Students need to be careful to follow geometric conditions/definitions when they add anything to a figure.

3 Circle W has a radius of 10 centimeters. Radius $\overline{WL}$ is perpendicular to chord $\overline{HK}$, which is 16 centimeters long.

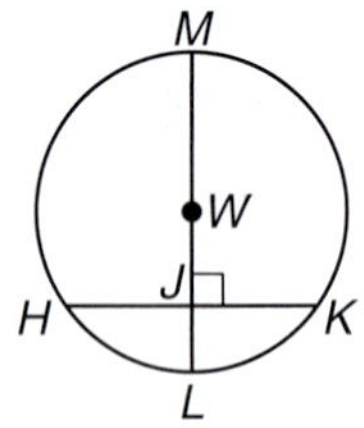

a. If $m\widehat{HL} = 53$, find $m\widehat{MK}$. **127**

b. Find JL. **4**

Example 3 Radius Perpendicular to a Chord

Circle O has a radius of 13 inches. Radius $\overline{OB}$ is perpendicular to chord $\overline{CD}$, which is 24 inches long.

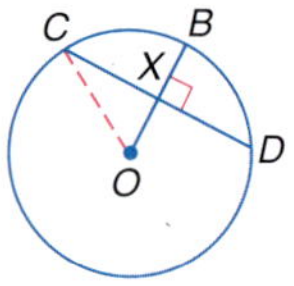

a. If $m\widehat{CD} = 134$, find $m\widehat{CB}$.

$\overline{OB}$ bisects $\widehat{CD}$, so $m\widehat{CB} = \frac{1}{2}m\widehat{CD}$.

$m\widehat{CB} = \frac{1}{2}m\widehat{CD}$	Definition of arc bisector
$m\widehat{CB} = \frac{1}{2}(134)$ or 67	$m\widehat{CD} = 134$

b. Find OX.

Draw radius $\overline{OC}$. $\triangle CXO$ is a right triangle.

$CO = 13$	$r = 13$
$\overline{OB}$ bisects $\overline{CD}$.	A radius perpendicular to a chord bisects it.
$CX = \frac{1}{2}(CD)$	Definition of segment bisector
$= \frac{1}{2}(24)$ or 12	$CD = 24$

Use the Pythagorean Theorem to find XO.

$(CX)^2 + (OX)^2 = (CO)^2$	Pythagorean Theorem
$12^2 + (OX)^2 = 13^2$	$CX = 12$, $CO = 13$
$144 + (OX)^2 = 169$	Simplify.
$(OX)^2 = 25$	Subtract 144 from each side.
$OX = 5$	Take the square root of each side.

In the next activity, you will discover another property of congruent chords.

Geometry Activity

Congruent Chords and Distance

Model

Step 1 Use a compass to draw a large circle on patty paper. Cut out the circle.

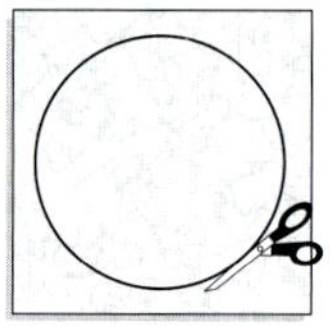

Step 2 Fold the circle in half.

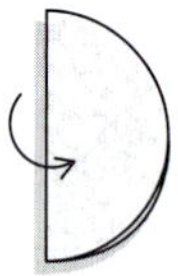

Step 3 Without opening the circle, fold the edge of the circle so it does not intersect the first fold.

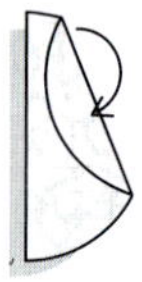

Step 4 Unfold the circle and label as shown.

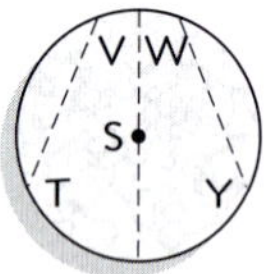

Step 5 Fold the circle, laying point V onto T to bisect the chord. Open the circle and fold again to bisect $\overline{WY}$. Label as shown.

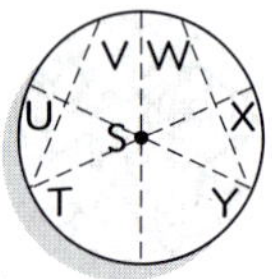

Analyze **1. $\overline{SU}$ and $\overline{SX}$ are perpendicular bisectors of $\overline{VT}$ and $\overline{WY}$, respectively.**

1. What is the relationship between $\overline{SU}$ and $\overline{VT}$? $\overline{SX}$ and $\overline{WY}$?

2. Use a centimeter ruler to measure $\overline{VT}$, $\overline{WY}$, $\overline{SU}$, and $\overline{SX}$. What do you find? **$VT = WY$, $SU = SX$**

3. Make a conjecture about the distance that two chords are from the center when they are congruent.

Sample answer: When the chords are congruent, they are equidistant from the center of the circle.

Geometry Activity

Materials: compass, patty paper, centimeter ruler

- Students can use a ruler to draw $\overline{VT}$, $\overline{WY}$, $\overline{ST}$, and $\overline{SY}$.
- To reinforce concepts, have students measure the central angles and determine if $\widehat{TV} \cong \widehat{WY}$.

The Geometry Activity suggests the following theorem.

Theorem 10.4

In a circle or in congruent circles, two chords are congruent if and only if they are equidistant from the center.

You will prove Theorem 10.4 in Exercises 37 and 38.

Example 4 Chords Equidistant from Center

Chords $\overline{AC}$ and $\overline{DF}$ are equidistant from the center. If the radius of $\odot G$ is 26, find AC and DE.

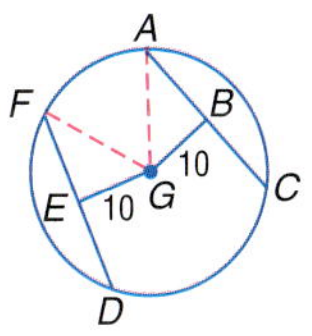

$\overline{AC}$ and $\overline{DF}$ are equidistant from G, so $\overline{AC} \cong \overline{DF}$.

Draw $\overline{AG}$ and $\overline{GF}$ to form two right triangles. Use the Pythagorean Theorem.

$(AB)^2 + (BG)^2 = (AG)^2$	Pythagorean Theorem
$(AB)^2 + 10^2 = 26^2$	$BG = 10$, $AG = 26$
$(AB)^2 + 100 = 676$	Simplify.
$(AB)^2 = 576$	Subtract 100 from each side.
$AB = 24$	Take the square root of each side.

$AB = \frac{1}{2}(AC)$, so $AC = 2(24)$ or 48.

$\overline{AC} \cong \overline{DF}$, so DF also equals 48. $DE = \frac{1}{2}DF$, so $DE = \frac{1}{2}(48)$ or 24.

Check for Understanding

Concept Check

1–2. See margin.

3. Tokei; to bisect the chord, it must be a diameter and be perpendicular.

1. Explain the difference between an inscribed polygon and a circumscribed circle.
2. **OPEN ENDED** Construct a circle and inscribe any polygon. Draw the radii to the vertices of the polygon and use a protractor to determine whether any sides of the polygon are congruent.
3. **FIND THE ERROR** Lucinda and Tokei are writing conclusions about the chords in $\odot F$. Who is correct? Explain your reasoning.

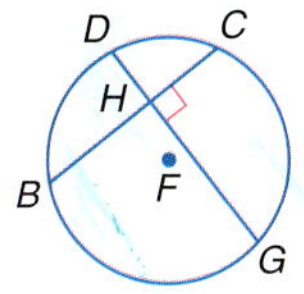

Lucinda	Tokei
Because $\overline{DG} \perp \overline{BC}$, $\angle DHB \cong \angle DHC \cong \angle CHG \cong \angle BHG$, and $\overline{DG}$ bisects $\overline{BC}$.	$\overline{DG} \perp \overline{BC}$, but $\overline{DG}$ does not bisect $\overline{BC}$ because it is not a diameter.

Guided Practice

GUIDED PRACTICE KEY	
Exercises	Examples
4	1
5–7	3
8–9	4
10	2

4. **PROOF** Prove part 2 of Theorem 10.2.

Given: $\odot X$, $\overline{UV} \cong \overline{WY}$

Prove: $\widehat{UV} \cong \widehat{WY}$ See margin.

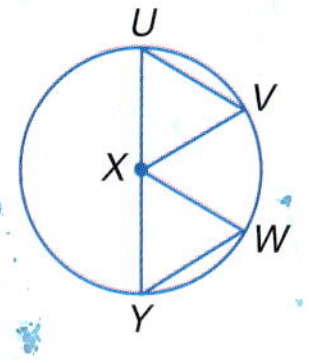

Circle O has a radius of 10, $AB = 10$, and $m\widehat{AB} = 60$. Find each measure.

5. $m\widehat{AY}$ 30
6. AX 5
7. OX $5\sqrt{3}$

In $\odot P$, $PD = 10$, $PQ = 10$, and $QE = 20$. Find each measure.

8. AB 40
9. PE $10\sqrt{5} \approx 22.36$

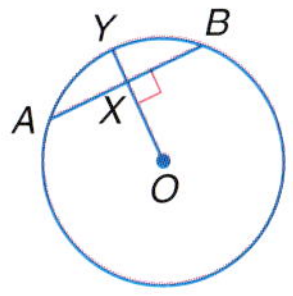

Exercises 5–7

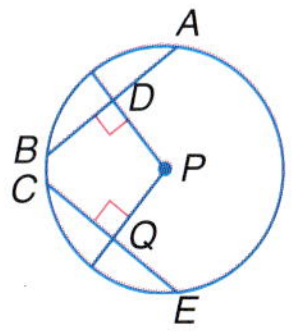

Exercises 8–9

In-Class Example

Power Point®

4 Chords $\overline{EF}$ and $\overline{GH}$ are equidistant from the center. If the radius of $\odot P$ is 15 and $EF = 24$, find PR and RH. 9; 12

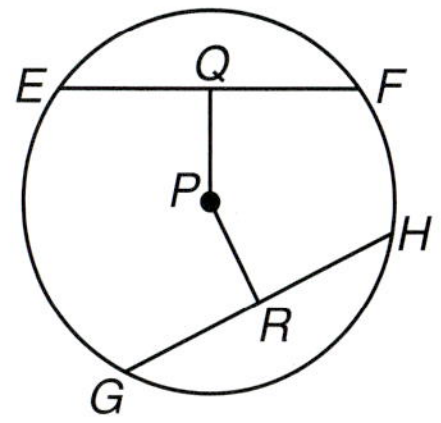

3 Practice/Apply

Study Notebook

Have students—

- add the definitions/examples of the vocabulary terms to their Vocabulary Builder worksheets for Chapter 10.
- include any other item(s) that they find helpful in mastering the skills in this lesson.

DAILY INTERVENTION

FIND THE ERROR Stress the importance of carefully analyzing figures to determine relationships. Point out that since a diameter or radius that is perpendicular to a chord does bisect the chord, no other segment perpendicular to the chord could bisect it.

Answers

1. Sample answer: An inscribed polygon has all vertices on the circle. A circumscribed circle means the circle is drawn around so that the polygon lies in its interior and all vertices lie on the circle.

2. Sample answer:

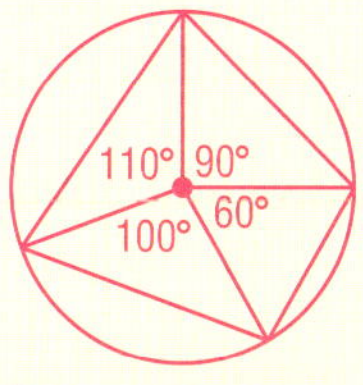

None of the sides are congruent.

4. Given: $\odot X$, $\overline{UV} \cong \overline{WY}$

Prove: $\widehat{UV} \cong \widehat{WY}$

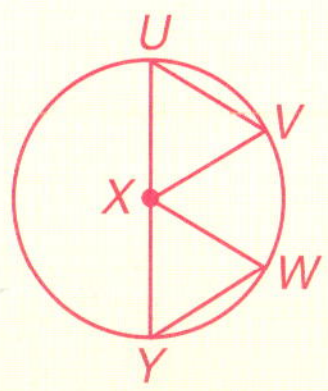

Proof: Because all radii are congruent, $\overline{XU} \cong \overline{XV} \cong \overline{XW} \cong \overline{XY}$. You are given that $\overline{UV} \cong \overline{WY}$, so $\triangle UVX \cong \triangle WYX$, by SSS. Thus, $\angle UXV \cong \angle WXY$ by CPCTC. Since the central angles have the same measure, their intercepted arcs have the same measure and are therefore, congruent. Thus, $\widehat{UV} \cong \widehat{WY}$.

About the Exercises...

Organization by Objective

- **Arcs and Chords:** 23–25, 36–38
- **Diameters and Chords:** 11–22, 26–33

Odd/Even Assignments

Exercises 11–33 are structured so that students practice the same concepts whether they are assigned odd or even problems.

Alert! Exercises 46–47 require a compass.

Assignment Guide

Basic: 11–19 odd, 23–35 odd, 39, 45–49 odd, 50–65

Average: 11–49 odd, 50–65

Advanced: 12–44 even, 45, 46, 48, 50, 52–59 (optional: 60–65)

All: Quiz 1 (1–10)

Answers (page 541)

40.

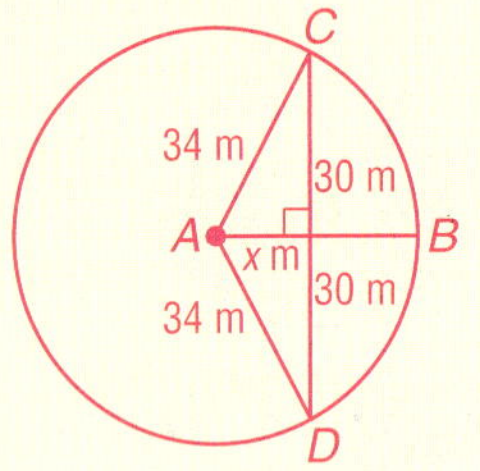

41.

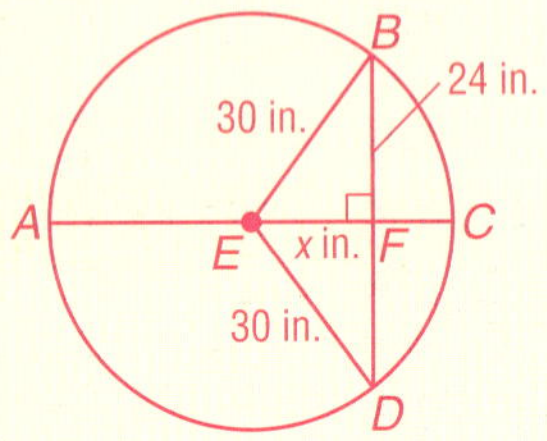

42.

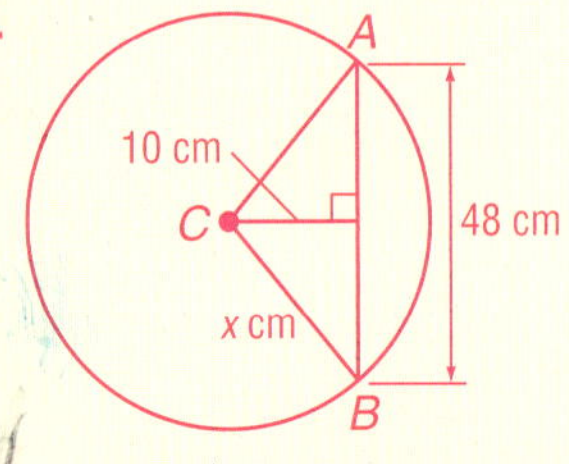

43.

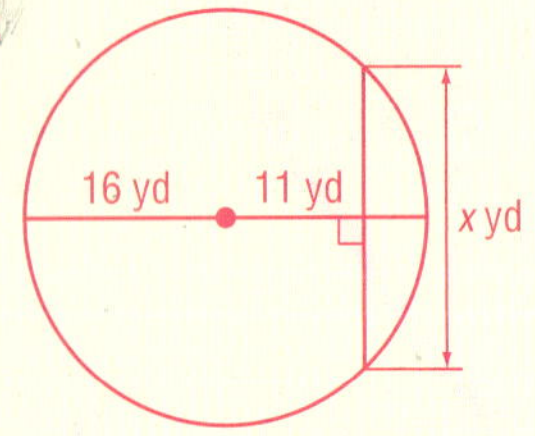

Application

10. **TRAFFIC SIGNS** A yield sign is an equilateral triangle. Find the measure of each arc of the circle circumscribed about the yield sign. **Each arc measures 120°.**

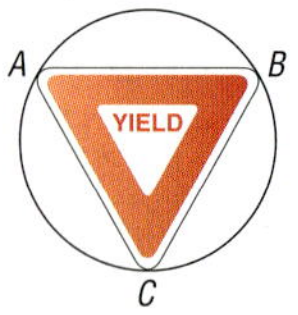

★ indicates increased difficulty

Practice and Apply

Homework Help

For Exercises	See Examples
11–22	3
23–25	2
26–33	4
36–38	1

Extra Practice
See page 774.

In $\odot X$, $AB = 30$, $CD = 30$, and $m\widehat{CZ} = 40$. Find each measure.

11. AM **15**
12. MB **15**
13. CN **15**
14. ND **15**
15. $m\widehat{DZ}$ **40**
16. $m\widehat{CD}$ **80**
17. $m\widehat{AB}$ **80**
18. $m\widehat{YB}$ **40**

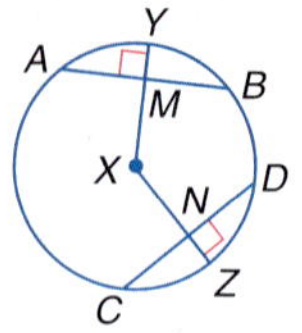

The radius of $\odot P$ is 5 and $PR = 3$. Find each measure.

19. QR **4**
20. QS **8**

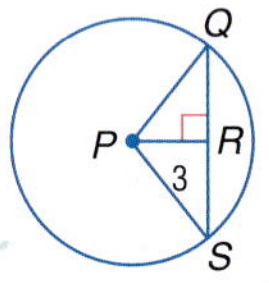

Exercises 19–20

In $\odot T$, $ZV = 1$, and $TW = 13$. Find each measure.

★ 21. XV **5**
22. XY **10**

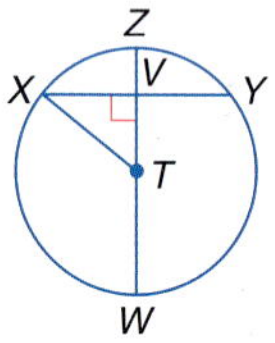

Exercises 21–22

TRAFFIC SIGNS Determine the measure of each arc of the circle circumscribed about the traffic sign.

23. regular octagon
24. square
25. rectangle

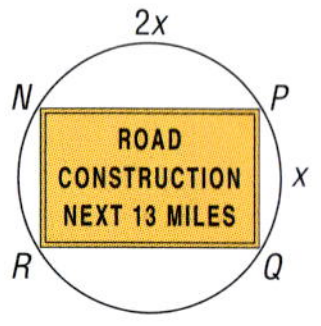

23. $m\widehat{AB} = m\widehat{BC} = m\widehat{CD} = m\widehat{DE} = m\widehat{EF} = m\widehat{FG} = m\widehat{GH} = m\widehat{HA} = 45$

24. $m\widehat{LM} = m\widehat{MJ} = m\widehat{JK} = m\widehat{KL} = 90$

25. $m\widehat{NP} = m\widehat{RQ} = 120$; $m\widehat{NR} = m\widehat{PQ} = 60$

In $\odot F$, $\overline{FH} \cong \overline{FL}$ and FK = 17. Find each measure.

26. LK **15**
27. KM **30**
28. JG **30**
29. JH **15**

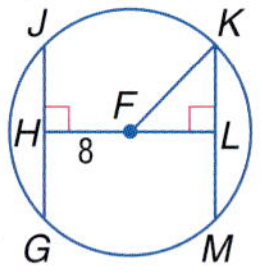

Exercises 26–29

In $\odot D$, $CF = 8$, $DE = FD$, and $DC = 10$. Find each measure.

30. FB **8**
31. BC **16**
32. AB **16**
33. ED **6**

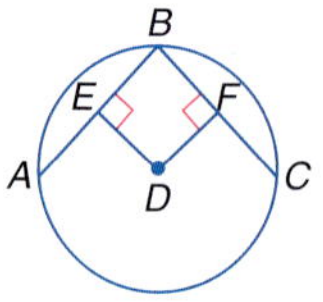

Exercises 30–33

34. **ALGEBRA** In $\odot Z$, $PZ = ZQ$, $XY = 4a - 5$, and $ST = -5a + 13$. Find SQ. **1.5**

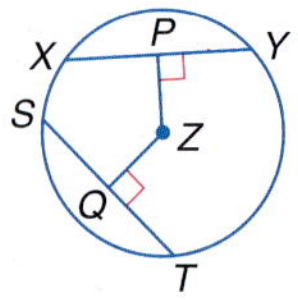

Exercise 34

35. **ALGEBRA** In $\odot B$, the diameter is 20 units long, and $m\angle ACE = 45$. Find x. $\sqrt{2} \approx 1.41$ **units**

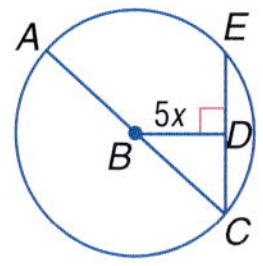

Exercise 35

More About. . .

Sayings

Many everyday sayings or expressions have a historical origin. The square peg comment is attributed to Sydney Smith (1771–1845), a witty British lecturer, who said "Trying to get those two together is like trying to put a square peg in a round hole."

36. **PROOF** Copy and complete the flow proof of Theorem 10.3.

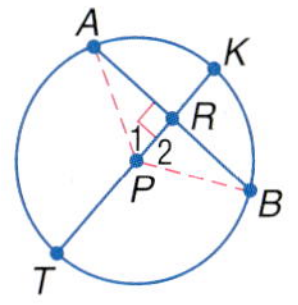

Given: $\odot P$, $\overline{AB} \perp \overline{TK}$

Prove: $\overline{AR} \cong \overline{BR}$, $\widehat{AK} \cong \widehat{BK}$

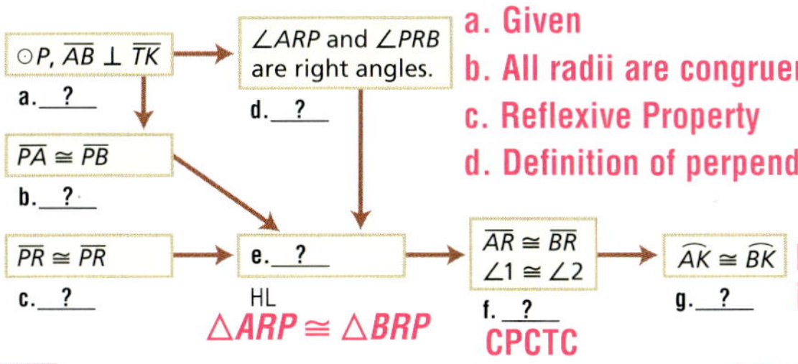

a. Given
b. All radii are congruent.
c. Reflexive Property
d. Definition of perpendicular lines

$\triangle ARP \cong \triangle BRP$ CPCTC   If central ∠s are ≅, intercepted arcs are ≅.

PROOF Write a proof for each part of Theorem 10.4. 37–38. See 589A.

★ 37. In a circle, if two chords are equidistant from the center, then they are congruent.

★ 38. In a circle, if two chords are congruent, then they are equidistant from the center.

39. **SAYINGS** An old adage states that "You can't fit a square peg in a round hole." Actually, you can, it just won't fill the hole. If a hole is 4 inches in diameter, what is the approximate width of the largest square peg that fits in the round hole? 2.82 in.

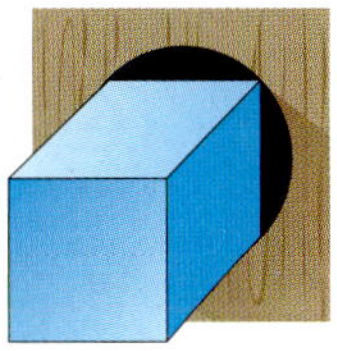

40–43. See margin for sample figures.

For Exercises 40–43, draw and label a figure. Then solve.

★ 40. The radius of a circle is 34 meters long, and a chord of the circle is 60 meters long. How far is the chord from the center of the circle? 16 m

★ 41. The diameter of a circle is 60 inches, and a chord of the circle is 48 inches long. How far is the chord from the center of the circle? 18 in.

★ 42. A chord of a circle is 48 centimeters long and is 10 centimeters from the center of the circle. Find the radius. 26 cm

★ 43. A diameter of a circle is 32 yards long. A chord is 11 yards from the center. How long is the chord? $2\sqrt{135} \approx 23.24$ yd

44. The line through the midpoint bisects the chord and is perpendicular to the chord, so the line is a diameter of the circle. Where two diameters meet would locate the center of the circle.

Study Tip

Finding the Center of a Circle

The process Mr. Ortega used can be done by construction and is often called *locating the center of a circle.*

44. **CARPENTRY** Mr. Ortega wants to drill a hole in the center of a round picnic table for an umbrella pole. To locate the center of the circle, he draws two chords of the circle and uses a ruler to find the midpoint for each chord. Then he uses a framing square to draw a line perpendicular to each chord at its midpoint. Explain how this process locates the center of the tabletop.

45. **CRITICAL THINKING** A diameter of $\odot P$ has endpoints A and B. Radius $\overline{PQ}$ is perpendicular to $\overline{AB}$. Chord $\overline{DE}$ bisects $\overline{PQ}$ and is parallel to $\overline{AB}$. Does $DE = \frac{1}{2}(AB)$? Explain.

45. Let r be the radius of $\odot P$. Draw radii to points D and E to create triangles. The length DE is $r\sqrt{3}$ and $AB = 2r$; $r\sqrt{3} \neq \frac{1}{2}(2r)$.

Enrichment, p. 558

Patterns from Chords

Some beautiful and interesting patterns result if you draw chords to connect evenly spaced points on a circle. On the circle shown below, 24 points have been marked to divide the circle into 24 equal parts. Numbers from 1 to 48 have been placed beside the points. Study the diagram to see exactly how this was done.

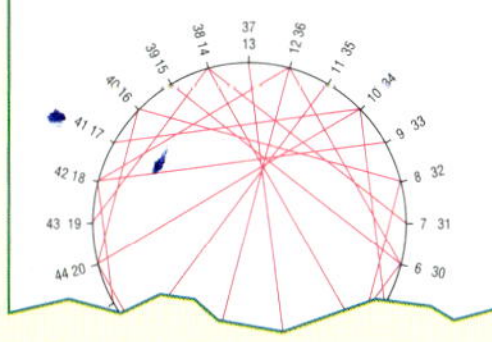

Study Guide and Intervention, p. 553 (shown) and p. 554

Arcs and Chords Points on a circle determine both chords and arcs. Several properties are related to points on a circle.

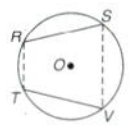

$\widehat{RS} \cong \widehat{TV}$ if and only if $\overline{RS} \cong \overline{TV}$.
$RSVT$ is inscribed in $\odot O$.
$\odot O$ is circumscribed about $RSVT$.

- In a circle or in congruent circles, two minor arcs are congruent if and only if their corresponding chords are congruent.
- If all the vertices of a polygon lie on a circle, the polygon is said to be **inscribed** in the circle and the circle is **circumscribed** about the polygon.

Example **Trapezoid $ABCD$ is inscribed in $\odot O$. If $\overline{AB} \cong \overline{BC} \cong \overline{CD}$ and $m\widehat{BC} = 50$, what is $m\widehat{APD}$?**

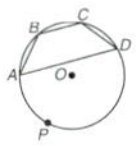

Chords $\overline{AB}$, $\overline{BC}$, and $\overline{CD}$ are congruent, so $\widehat{AB}$, $\widehat{BC}$, and $\widehat{CD}$ are congruent. $m\widehat{BC} = 50$, so $m\widehat{AB} + m\widehat{BC} + m\widehat{CD} = 50 + 50 + 50 = 150$. Then $m\widehat{APD} = 360 - 150$ or 210.

Exercises

Each regular polygon is inscribed in a circle. Determine the measure of each arc that corresponds to a side of the polygon.

1. hexagon 60
2. pentagon 72
3. triangle 120
4. square 90
5. octagon 45
6. 36-gon 10

Determine the measure of each arc of the circle circumscribed about the polygon.

7.

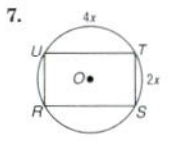

$m\widehat{UT} = m\widehat{RS} = 120$
$m\widehat{ST} = m\widehat{RU} = 60$

8.

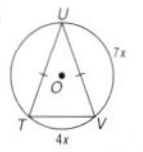

$m\widehat{UT} = m\widehat{UV} = 140$
$m\widehat{TV} = 80$

9.

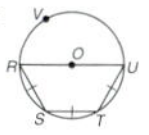

$m\widehat{RS} = m\widehat{ST} = m\widehat{TU} = 60$
$m\widehat{RVU} = 180$

Skills Practice, p. 555 and Practice, p. 556 (shown)

In $\odot E$, $m\widehat{HQ} = 48$, $HI = JK$, and $JR = 7.5$. Find each measure.

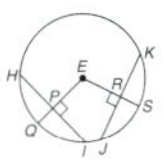

1. $m\widehat{HI}$ 96
2. $m\widehat{QI}$ 48
3. $m\widehat{JK}$ 96
4. HI 15
5. PI 7.5
6. JK 15

The radius of $\odot N$ is 18, $NK = 9$, and $m\widehat{DE} = 120$. Find each measure.

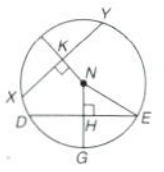

7. $m\widehat{GE}$ 60
8. $m\angle HNE$ 60
9. $m\angle HEN$ 30
10. HN 9

The radius of $\odot O = 32$, $\widehat{PQ} \cong \widehat{RS}$, and $PQ = 56$. Find each measure.

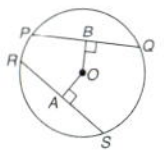

11. PB 28
14. BQ 28
12. OB $4\sqrt{15} \approx 15.49$
16. RS 56

13. **MANDALAS** The base figure in a mandala design is a nine-pointed star. Find the measure of each arc of the circle circumscribed about the star. Each arc measures 40°.

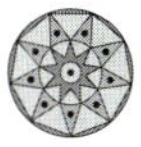

Reading to Learn Mathematics, p. 557   ELL

Pre-Activity **How do the grooves in a Belgian waffle iron model segments in a circle?**

Read the introduction to Lesson 10-3 at the top of page 536 in your textbook.

What do you observe about any two of the grooves in the waffle iron shown in the picture in your textbook? They are either parallel or perpendicular.

Reading the Lesson

1. Supply the missing words or phrases to form true statements.
   a. In a circle, if a radius is ___perpendicular___ to a chord, then it bisects the chord and its ___arc___.
   b. In a circle or in ___congruent___ circles, two ___minor arcs___ are congruent if and only if their corresponding chords are congruent.
   c. In a circle or in ___congruent___ circles, two chords are congruent if they are ___equidistant___ from the center.
   d. A polygon is inscribed in a circle if all of its ___vertices___ lie on the circle.
   e. All of the sides of an inscribed polygon are ___chords___ of the circle.

2. If $\odot P$ has a diameter 40 centimeters long, and $AC = FD = 24$ centimeters, find each measure.

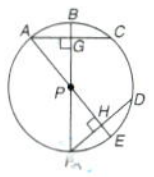

   a. PA 20 cm   b. AG 12 cm
   c. PE 20 cm   d. PH 16 cm
   e. HE 4 cm   f. FG 36 cm

3. In $\odot Q$, $RS = VW$ and $m\widehat{RS} = 70$. Find each measure.

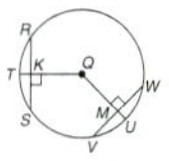

   a. $m\widehat{RT}$ 35   b. $m\widehat{ST}$ 35
   c. $m\widehat{VW}$ 70   d. $m\widehat{VU}$ 35

4. Find the measure of each arc of a circle that is circumscribed about the polygon.
   a. an equilateral triangle 120   b. a regular pentagon 72
   c. a regular hexagon 60   d. a regular decagon 36
   e. a regular dodecagon 30   f. a regular n-gon $\frac{360}{n}$

Helping You Remember

5. Some students have trouble distinguishing between *inscribed* and *circumscribed* figures. What is an easy way to remember which is which? Sample answer: The *in*scribed figure is *in*side the circle.

Answers

46. The chords and the radii of the circle are congruent by construction. Thus, all triangles formed by these segments are equilateral triangles. That means each angle of the hexagon measures 120°, making all angles of the hexagon congruent and all sides congruent.

47. The six arcs making up the circle are congruent because the chords intercepting them were congruent by construction. Each of the three chords drawn intercept two of the congruent chords. Thus, the three larger arcs are congruent. So, the three chords are congruent, making this an equilateral triangle.

51. Sample answer: The grooves of a waffle iron are chords of the circle. The ones that pass horizontally and vertically through the center are diameters. Answers should include the following.

- If you know the measure of the radius and the distance the chord is from the center, you can use the Pythagorean Theorem to find the length of half of the chord and then multiply by 2.
- There are four grooves on either side of the diameter, so each groove is about 1 in. from the center. In the figure, $EF = 2$ and $EB = 4$ because the radius is half the diameter. Using the Pythagorean Theorem, you find that $FB \approx 3.464$ in. so $AB \approx 6.93$ in. Approximate lengths for other chords are 5.29 in. and 7.75 in., but exactly 8 in. for the diameter.

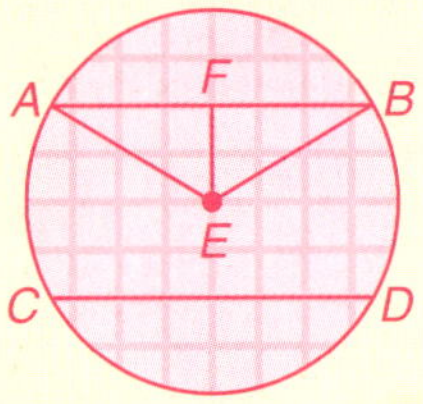

CONSTRUCTION **Use the following steps for each construction in Exercises 46 and 47.**

1. Construct a circle, and place a point on the circle.

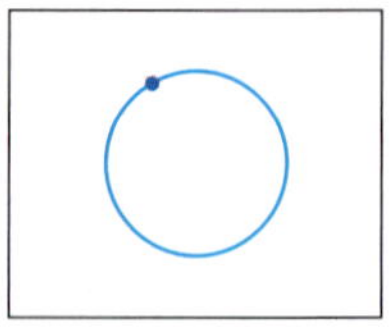

2. Using the same radius, place the compass on the point and draw a small arc to intercept the circle.

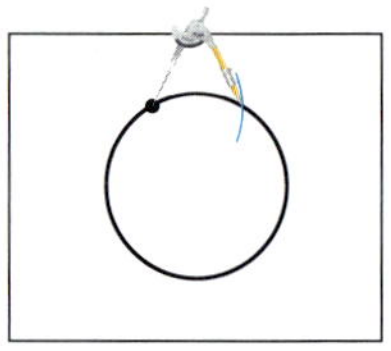

3. Using the same radius, place the compass on the intersection and draw another small arc to intercept the circle.

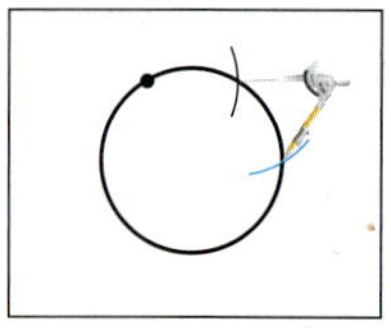

4. Continue the process in Step 3 until you return to the original point.

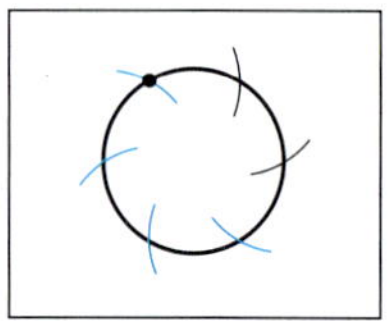

46–47. See margin for verifications.

46. Connect the intersections with chords of the circle. What type of figure is formed? Verify you conjecture. inscribed regular hexagon

47. Repeat the construction. Connect every other intersection with chords of the circle. What type of figure is formed? Verify your conjecture. inscribed equilateral triangle

COMPUTERS **For Exercises 48 and 49, use the following information.**
The hard drive of a computer contains platters divided into tracks, which are defined by concentric circles, and sectors, which are defined by radii of the circles.

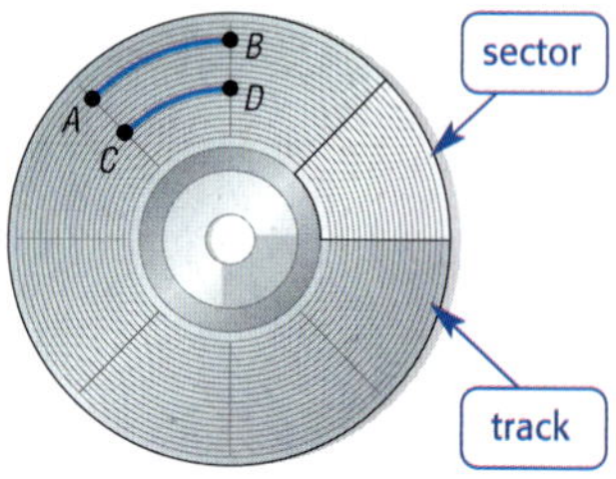

48. In the diagram of a hard drive platter at the right, what is the relationship between $m\widehat{AB}$ and $m\widehat{CD}$? $m\widehat{AB} = m\widehat{CD}$

49. Are $\widehat{AB}$ and $\widehat{CD}$ congruent? Explain. No; congruent arcs must be in the same circle or congruent circles, but these are in concentric circles.

50. $\overline{AB} \cong \overline{CD}$; in the smaller circle, $\overline{OX} \cong \overline{OY}$ because they are radii. This means that in the larger circle, $\overline{AB}$ and $\overline{CD}$ are equidistant from the center, making them congruent chords.

50. **CRITICAL THINKING** The figure shows two concentric circles with $\overline{OX} \perp \overline{AB}$ and $\overline{OY} \perp \overline{CD}$. Write a statement relating $\overline{AB}$ and $\overline{CD}$. Verify your reasoning.

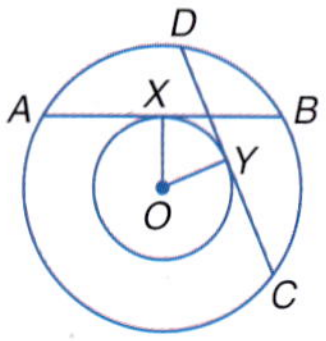

51. **WRITING IN MATH** Answer the question that was posed at the beginning of the lesson. See margin.

How do the grooves in a Belgian waffle iron model segments in a circle?

Include the following in your answer:

- a description of how you might find the length of a groove without directly measuring it, and
- a sketch with measurements for a waffle iron that is 8 inches wide.

542 Chapter 10 Circles

FCAT Practice

Standardized Test Practice

A B C D

52. Refer to the figure. Which of the following statements is true? **C**
I. $\overline{DB}$ bisects $\overline{AC}$. **II.** $\overline{AC}$ bisects $\overline{DB}$. **III.** $OA = OC$
Ⓐ I and II Ⓑ II and III
Ⓒ I and III Ⓓ I, II, and III

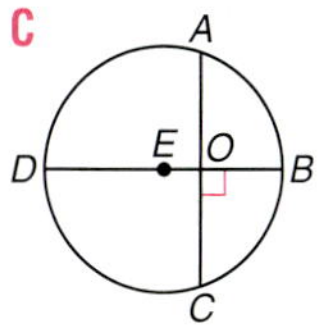

53. **SHORT RESPONSE** According to the 2000 census, the population of Bridgeworth was 204 thousand, and the population of Sutterly was 216 thousand. If the population of each city increased by exactly 20% ten years later, how many more people will live in Sutterly than in Bridgeworth in 2010? **14,400**

Maintain Your Skills

Mixed Review

In ⊙S, $m\angle TSR = 42$. Find each measure. *(Lesson 10-2)*

54. $m\widehat{KT}$ **138** 55. $m\widehat{ERT}$ **180** 56. $m\widehat{KRT}$ **222**

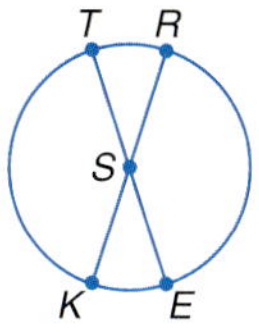

Exercises 54–56

Refer to ⊙M. *(Lesson 10-1)*

57. Name a chord that is not a diameter. $\overline{SU}$
58. If $MD = 7$, find RI. **14**
59. Name congruent segments in ⊙M.
$\overline{RM}$, $\overline{AM}$, $\overline{DM}$, $\overline{IM}$

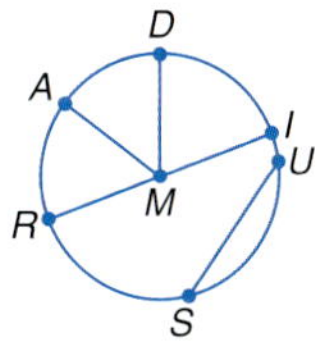

Exercises 57–59

Getting Ready for the Next Lesson

PREREQUISITE SKILL Solve each equation.
*(To review **solving equations**, see pages 742 and 743.)*

60. $\frac{1}{2}x = 120$ **240** 61. $\frac{1}{2}x = 25$ **50** 62. $2x = \frac{1}{2}(45 + 35)$ **20**

63. $3x = \frac{1}{2}(120 - 60)$ **10** 64. $45 = \frac{1}{2}(4x + 30)$ **15** 65. $90 = \frac{1}{2}(6x + 3x)$ **20**

Practice Quiz 1 — *Lessons 10-1 through 10-3*

PETS For Exercises 1–6, refer to the front circular edge of the hamster wheel shown at the right. *(Lessons 10-1 and 10-2)*

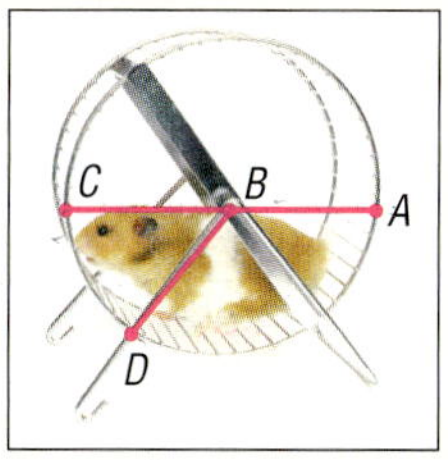

1. Name three radii of the wheel. $\overline{BC}$, $\overline{BD}$, $\overline{BA}$
2. If $BD = 3x$ and $CB = 7x - 3$, find AC. **4.5**
3. If $m\angle CBD = 85$, find $m\widehat{AD}$. **95**
4. If $r = 3$ inches, find the circumference of circle B to the nearest tenth of an inch. **18.8 in.**
5. There are 40 equally-spaced rungs on the wheel. What is the degree measure of an arc connecting two consecutive rungs? **9**
6. What is the length of $\widehat{CAD}$ to the nearest tenth if $m\angle ABD = 150$ and $r = 3$? **17.3 units**

Find each measure. *(Lesson 10-3)*

7. $m\angle CAM$ **28** 8. $m\widehat{ES}$ **100** 9. SC **21** 10. x **24**

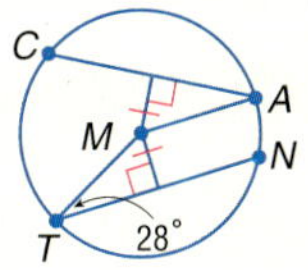

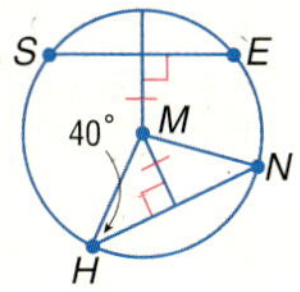

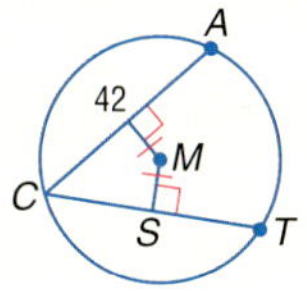

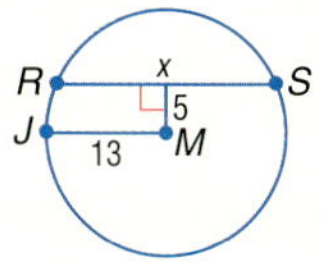

www.geometryonline.com/self_check_quiz/fcat

4 Assess

Open-Ended Assessment

Writing Have students draw three circles on a sheet of paper. Tell students to draw a chord anywhere on the first circle and then construct and label a perpendicular bisector for this chord. For the second circle, have students draw two segments extending from the center of the circle so that the chords perpendicular to these segments are congruent. Ask students to place two chords on the third circle so that their corresponding arcs are congruent. Tell students to write the rule they used from the lesson underneath each figure.

Getting Ready for Lesson 10-4

Prerequisite Skill Students will learn about inscribed angles in Lesson 10-4. They will apply concepts of solving equations to find the measures of inscribed angles. Use Exercises 60–65 to determine your students' familiarity with solving equations.

Assessment Options

Practice Quiz 1 The quiz provides students with a brief review of the concepts and skills in Lessons 10-1 through 10-3. Lesson numbers are given to the right of the exercises or instruction lines so students can review concepts not yet mastered.

10-4 Lesson Notes

1 Focus

5-Minute Check Transparency 10-4 Use as a quiz or review of Lesson 10-3 .

Mathematical Background notes are available for this lesson on p. 520D.

How is a socket like an inscribed polygon?

Ask students:

- Why do you think socket wrenches and nuts have a hexagonal shape? **Because a hexagon offers good strength and leverage while still distributing the force applied by the user evenly.**
- What are the advantages of the wrench mechanism to which the hexagonal cylinder is attached? **This mechanism gives more leverage to the user than just the hexagon itself and allows more maneuverability and flexibility.**

10-4 Inscribed Angles

What You'll Learn

- Find measures of inscribed angles.
- Find measures of angles of inscribed polygons.

How is a socket like an inscribed polygon?

A socket is a tool that comes in varying diameters. It is used to tighten or unscrew nuts. The "hole" in the socket is a hexagon cast in a metal cylinder, which attaches to a wrench mechanism for leverage.

Vocabulary

- intercepted

INSCRIBED ANGLES In Lesson 10-3, you learned that a polygon that has its vertices on a circle is called an inscribed polygon. Likewise, an *inscribed angle* is an angle that has its vertex on the circle and its sides contained in chords of the circle.

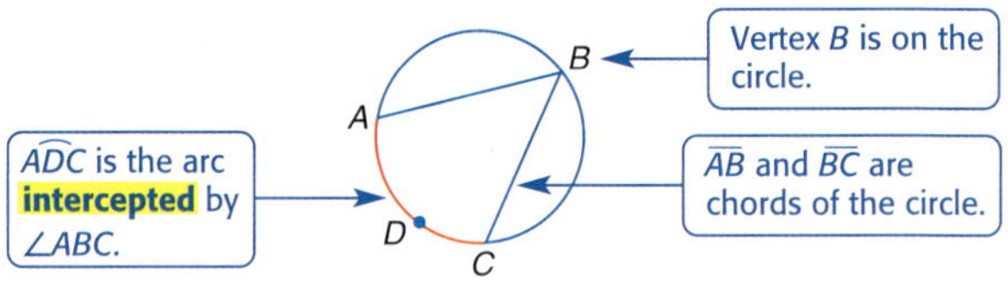

Geometry Activity

Measure of Inscribed Angles

Model

- Use a compass to draw a circle and label the center W.
- Draw an inscribed angle and label it XYZ.
- Draw $\overline{WX}$ and $\overline{WZ}$.

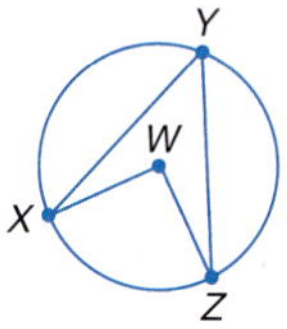

Analyze

1. Measure $\angle XYZ$ and $\angle XWZ$. **See students' work.**
2. Find $m\widehat{XZ}$ and compare it with $m\angle XYZ$. $m\widehat{XZ} = 2(m\angle XYZ)$
3. **Make a conjecture** about the relationship of the measure of an inscribed angle and the measure of its intercepted arc.

3. The measure of an inscribed angle is one-half the measure of its intercepted arc.

This activity suggests the following theorem.

Theorem 10.5

Inscribed Angle Theorem If an angle is inscribed in a circle, then the measure of the angle equals one-half the measure of its intercepted arc (or the measure of the intercepted arc is twice the measure of the inscribed angle).

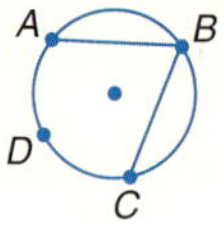

Example: $m\angle ABC = \frac{1}{2}(m\widehat{ADC})$ or $2(m\angle ABC) = m\widehat{ADC}$

Resource Manager

Workbook and Reproducible Masters

Chapter 10 Resource Masters
- Study Guide and Intervention, pp. 559–560
- Skills Practice, p. 561
- Practice, p. 562
- Reading to Learn Mathematics, p. 563
- Enrichment, p. 564
- Assessment, pp. 603, 605

Prerequisite Skills Workbook, pp. 41–42
Teaching Geometry With Manipulatives Masters, pp. 16, 17, 165, 166, 167

Transparencies

5-Minute Check Transparency 10-4
Answer Key Transparencies

Technology

Interactive Chalkboard

To prove Theorem 10.5, you must consider three cases.

	Case 1	Case 2	Case 3
Model of Angle Inscribed in ⊙O	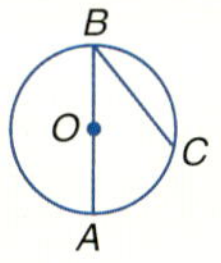	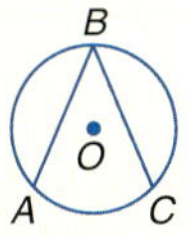	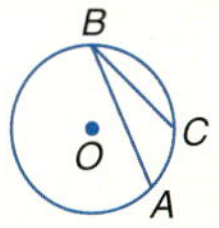
Location of center of circle	on a side of the angle	in the interior of the angle	in the exterior of the angle

Proof Theorem 10.5 (Case 1)

Given: $\angle ABC$ inscribed in $\odot D$ and $\overline{AB}$ is a diameter.

Prove: $m\angle ABC = \frac{1}{2}m\widehat{AC}$

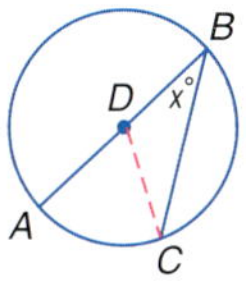

Draw $\overline{DC}$ and let $m\angle B = x$.

Proof:

Since $\overline{DB}$ and $\overline{DC}$ are congruent radii, $\triangle BDC$ is isosceles and $\angle B \cong \angle C$. Thus, $m\angle B = m\angle C = x$. By the Exterior Angle Theorem, $m\angle ADC = m\angle B + m\angle C$. So $m\angle ADC = 2x$. From the definition of arc measure, we know that $m\widehat{AC} = m\angle ADC$ or $2x$. Comparing $m\widehat{AC}$ and $m\angle ABC$, we see that $m\widehat{AC} = 2(m\angle ABC)$ or that $m\angle ABC = \frac{1}{2}m\widehat{AC}$.

You will prove Cases 2 and 3 of Theorem 10.5 in Exercises 35 and 36.

Study Tip

Using Variables

You can also assign a variable to an unknown measure. So, if you let $m\widehat{AD} = x$, the second equation becomes $140 + 100 + x + x = 360$, or $240 + 2x = 360$. This last equation may seem simpler to solve.

Example 1 Measures of Inscribed Angles

In $\odot O$, $m\widehat{AB} = 140$, $m\widehat{BC} = 100$, and $m\widehat{AD} = m\widehat{DC}$. Find the measures of the numbered angles.

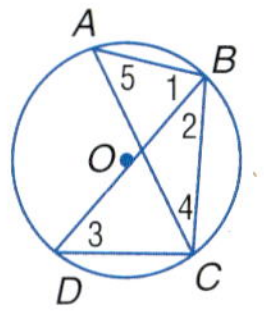

First determine $m\widehat{DC}$ and $m\widehat{AD}$.

$m\widehat{AB} + m\widehat{BC} + m\widehat{DC} + m\widehat{AD} = 360$ Arc Addition Theorem

$140 + 100 + m\widehat{DC} + m\widehat{DC} = 360$ $m\widehat{AB} = 140$, $m\widehat{BC} = 100$, $m\widehat{DC} = m\widehat{AD}$

$240 + 2(m\widehat{DC}) = 360$ Simplify.

$2(m\widehat{DC}) = 120$ Subtract 240 from each side.

$m\widehat{DC} = 60$ Divide each side by 2.

So, $m\widehat{DC} = 60$ and $m\widehat{AD} = 60$.

$m\angle 1 = \frac{1}{2}m\widehat{AD}$
$= \frac{1}{2}(60)$ or 30

$m\angle 2 = \frac{1}{2}m\widehat{DC}$
$= \frac{1}{2}(60)$ or 30

$m\angle 3 = \frac{1}{2}m\widehat{BC}$
$= \frac{1}{2}(100)$ or 50

$m\angle 4 = \frac{1}{2}m\widehat{AB}$
$= \frac{1}{2}(140)$ or 70

$m\angle 5 = \frac{1}{2}m\widehat{BC}$
$= \frac{1}{2}(100)$ or 50

www.geometryonline.com/extra_examples/fcat

2 Teach

INSCRIBED ANGLES

In-Class Example Power Point®

1 In $\odot F$, $m\widehat{WX} = 20$, $m\widehat{XY} = 40$, $m\widehat{UZ} = 108$, and $m\widehat{UW} = m\widehat{YZ}$. Find the measures of the numbered angles.

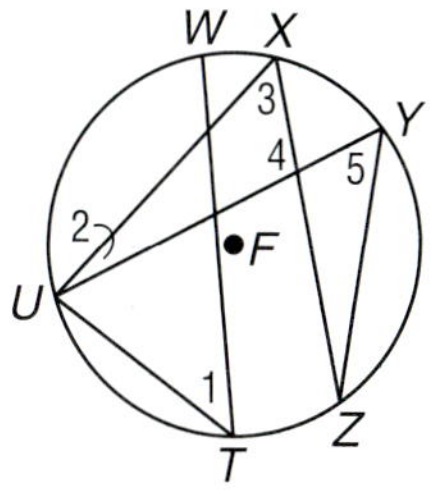

$m\angle 1 = 48$; $m\angle 2 = 20$; $m\angle 3 = 54$; $m\angle 4 = 106$; $m\angle 5 = 54$

Geometry Activity

Materials: compass, protractor, straightedge

Have some students start with an acute inscribed angle, some with a right inscribed angle, and some with an obtuse inscribed angle and compare results.

In-Class Examples

Power Point®

2 **Given:** $\odot C$ with $\overline{QR} \cong \overline{GF}$ and $\overline{JK} \cong \overline{HG}$

Prove: $\triangle PJK \cong \triangle EHG$

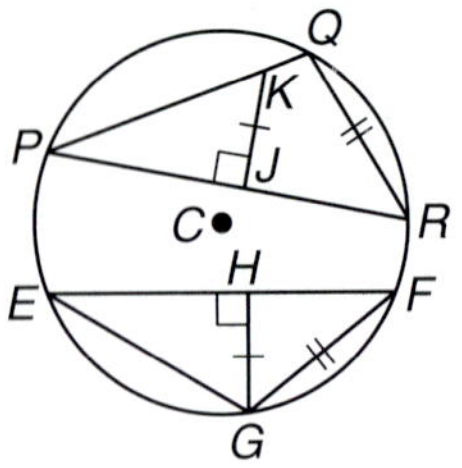

Statement (Reason)

1. $\overline{QR} \cong \overline{GF}$ and $\overline{JK} \cong \overline{HG}$ (Given)
2. $\widehat{QR} \cong \widehat{GF}$ (If two chords are ≅, corr. minor arcs are ≅.)
3. $\angle GEF$ intercepts $\widehat{FG}$; $\angle QPR$ intercepts $\widehat{QR}$. (Def. of intercepted arc)
4. $\angle GEF \cong \angle QPR$ (Inscribed ∠s of ≅ arcs are ≅.)
5. $\angle PJK \cong \angle EHG$ (Right ∠s are congruent.)
6. $\triangle PJK \cong \triangle EHG$ (AAS)

3 **PROBABILITY** Points M and N are on a circle so that $m\widehat{MN} = 72$. Suppose point L is randomly located on the same circle so that it does not coincide with M or N. What is the probability that $m\angle MLN = 144$? $\frac{1}{5}$

In Example 1, note that ∠3 and ∠5 intercept the same arc and are congruent.

Theorem 10.6

If two inscribed angles of a circle (or congruent circles) intercept congruent arcs or the same arc, then the angles are congruent.

Abbreviations:

Inscribed ∠s of ≅ arcs are ≅.

Inscribed ∠s of same arc are ≅.

Examples:

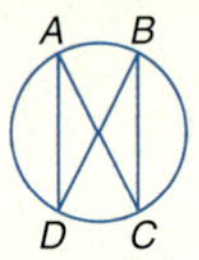

$\angle DAC \cong \angle DBC$

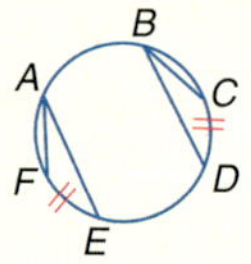

$\angle FAE \cong \angle CBD$

You will prove Theorem 10.6 in Exercise 37.

Example 2 Proofs with Inscribed Angles

Given: $\odot P$ with $\overline{CD} \cong \overline{AB}$

Prove: $\triangle AXB \cong \triangle CXD$

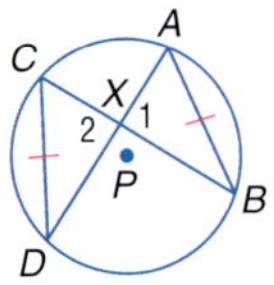

Proof:

Statements	Reasons
1. $\angle DAB$ intercepts $\widehat{DB}$. $\angle DCB$ intercepts $\widehat{DB}$.	1. Definition of intercepted arc
2. $\angle DAB \cong \angle DCB$	2. Inscribed ∠s of same arc are ≅.
3. $\angle 1 \cong \angle 2$	3. Vertical ∠s are ≅.
4. $\overline{CD} \cong \overline{AB}$	4. Given
5. $\triangle AXB \cong \triangle CXD$	5. AAS

You can also use the measure of an inscribed angle to determine probability of a point lying on an arc.

Example 3 Inscribed Arcs and Probability

PROBABILITY **Points A and B are on a circle so that $m\widehat{AB} = 60$. Suppose point D is randomly located on the same circle so that it does not coincide with A or B. What is the probability that $m\angle ADB = 30$?**

Since the angle measure is half the arc measure, inscribed $\angle ADB$ must intercept $\widehat{AB}$, so D must lie on major arc AB. Draw a figure and label any information you know.

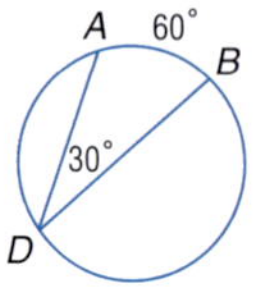

$$m\widehat{BDA} = 360 - m\widehat{AB}$$
$$= 360 - 60 \text{ or } 300$$

Since $\angle ADB$ must intercept $\widehat{AB}$, the probability that $m\angle ADB = 30$ is the same as the probability of D being contained in $\widehat{BDA}$.

The probability that D is located on $\widehat{ADB}$ is $\frac{5}{6}$. So, the probability that $m\angle ADB = 30$ is also $\frac{5}{6}$.

Study Tip

Eliminate the Possibilities
Think about what would be true if D was on minor arc AB. Then $\angle ADB$ would intercept the major arc. Thus, $m\angle ADB$ would be half of 300 or 150. This is not the desired angle measure in the problem, so you can eliminate the possibility that D can lie on $\widehat{AB}$.

DAILY INTERVENTION

Differentiated Instruction

Intrapersonal Select or provide examples that cover each concept in the lesson so that students can sit quietly and work on them at their desks. Ask students to make a note if a particular type of problem gives them difficulty. Encourage students to reread and use the examples and theorems in the book to help them work and understand the problems.

Study Tip

Inscribed Polygons
Remember that for a polygon to be an inscribed polygon, *all* of its vertices must lie on the circle.

ANGLES OF INSCRIBED POLYGONS

An inscribed triangle with a side that is a diameter is a special type of triangle.

Theorem 10.7

If an inscribed angle intercepts a semicircle, the angle is a right angle.

Example: $\overparen{ADC}$ is a semicircle, so $m\angle ABC = 90$.

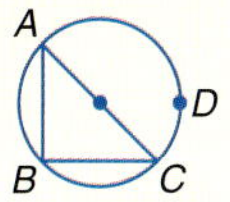

You will prove Theorem 10.7 in Exercise 38.

Example 4 Angles of an Inscribed Triangle

ALGEBRA Triangles ABD and ADE are inscribed in $\odot F$ with $\overparen{AB} \cong \overparen{BD}$. Find the measure of each numbered angle if $m\angle 1 = 12x - 8$ and $m\angle 2 = 3x + 8$.

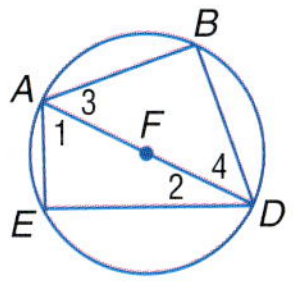

$\angle AED$ is a right angle because $\overparen{AED}$ is a semicircle.

$m\angle 1 + m\angle 2 + m\angle AED = 180$	Angle Sum Theorem
$(12x - 8) + (3x + 8) + 90 = 180$	$m\angle 1 = 12x - 8$, $m\angle 2 = 3x + 8$, $m\angle AED = 90$
$15x + 90 = 180$	Simplify.
$15x = 90$	Subtract 90 from each side.
$x = 6$	Divide each side by 15.

Use the value of x to find the measures of $\angle 1$ and $\angle 2$.

$m\angle 1 = 12x - 8$	Given	$m\angle 2 = 3x + 8$	Given
$= 12(6) - 8$	$x = 6$	$= 3(6) + 8$	$x = 6$
$= 64$	Simplify.	$= 26$	Simplify.

Angle ABD is a right angle because it intercepts a semicircle.

Because $\overparen{AB} \cong \overparen{BD}$, $\overline{AB} \cong \overline{BD}$, which leads to $\angle 3 \cong \angle 4$. Thus, $m\angle 3 = m\angle 4$.

$m\angle 3 + m\angle 4 + m\angle ABD = 180$	Angle Sum Theorem
$m\angle 3 + m\angle 3 + 90 = 180$	$m\angle 3 = m\angle 4$, $m\angle ABD = 90$
$2(m\angle 3) + 90 = 180$	Simplify.
$2(m\angle 3) = 90$	Subtract 90 from each side.
$m\angle 3 = 45$	Divide each side by 2.

Since $m\angle 3 = m\angle 4$, $m\angle 4 = 45$.

Example 5 Angles of an Inscribed Quadrilateral

Quadrilateral $ABCD$ is inscribed in $\odot P$. If $m\angle B = 80$ and $m\angle C = 40$, find $m\angle A$ and $m\angle D$.

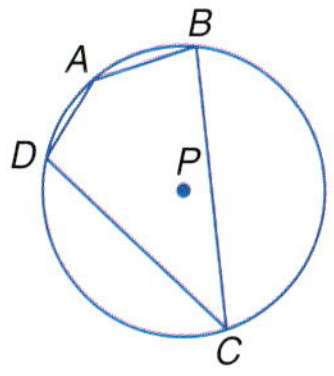

Draw a sketch of this situation.

To find $m\angle A$, we need to know $m\overparen{BCD}$.

To find $m\overparen{BCD}$, first find $m\overparen{DAB}$.

$m\overparen{DAB} = 2(m\angle C)$	Inscribed Angle Theorem
$= 2(40)$ or 80	$m\angle C = 40$

(continued on the next page)

ANGLES OF INSCRIBED POLYGONS

In-Class Examples

Power Point®

4 **ALGEBRA** Triangles TVU and TSU are inscribed in $\odot P$ with $\overparen{VU} \cong \overparen{SU}$. Find the measure of each numbered angle if $m\angle 2 = x + 9$ and $m\angle 4 = 2x + 6$.

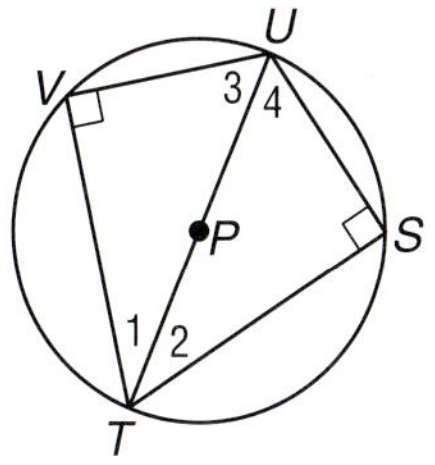

$m\angle 1 = 34$; $m\angle 2 = 34$; $m\angle 3 = 56$; $m\angle 4 = 56$

Teaching Tip Students can also remember that the sum of the interior angles of a quadrilateral is 360 and subtract the first three angle measures from 360 to find the fourth angle measure.

5 Quadrilateral $QRST$ is inscribed in $\odot M$. If $m\angle Q = 87$ and $m\angle R = 102$, find $m\angle S$ and $m\angle T$. **93; 78**

Lesson 10-4 Inscribed Angles **547**

3 Practice/Apply

Study Notebook

Have students—
- add the definitions/examples of the vocabulary terms to their Vocabulary Builder worksheets for Chapter 10.
- include a definition/example for an inscribed angle.
- include any other item(s) that they find helpful in mastering the skills in this lesson.

Answers

1. Sample answer:

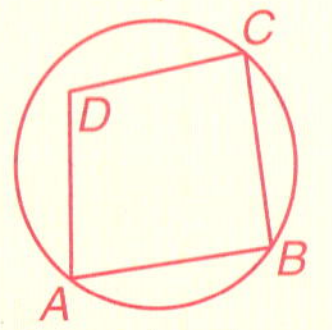

2. The measures of an inscribed angle and a central angle for the same intercepted arc can be calculated using the measure of the arc. However, the measure of the central angle equals the measure of the arc, while the measure of the inscribed angle is half the measure of the arc.

4. Given: Quadrilateral $ABCD$ is inscribed in $\odot P$.
$m\angle C = \frac{1}{2} m\angle B$
Prove: $m\widehat{CDA} = 2(m\widehat{DAB})$

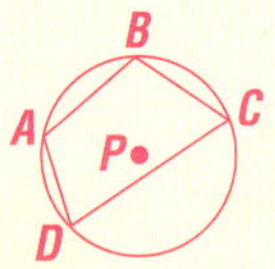

Proof: Given $m\angle C = \frac{1}{2}(m\angle B)$ means that $m\angle B = 2(m\angle C)$. Since $m\angle B = \frac{1}{2}(m\widehat{CDA})$ and $m\angle C = \frac{1}{2}(m\widehat{DAB})$, the equation becomes $\frac{1}{2}(m\widehat{CDA}) = 2[\frac{1}{2}(m\widehat{DAB})]$. Multiplying each side by 2 results in $m\widehat{CDA} = 2(m\widehat{DAB})$.

$m\widehat{BCD} + m\widehat{DAB} = 360$	Sum of angles in circle = 360
$m\widehat{BCD} + 80 = 360$	$m\widehat{DAB} = 80$
$m\widehat{BCD} = 280$	Subtract 80 from each side.
$m\widehat{BCD} = 2(m\angle A)$	Inscribed Angle Theorem
$280 = 2(m\angle A)$	Substitution
$140 = m\angle A$	Divide each side by 2.

To find $m\angle D$, we need to know $m\widehat{ABC}$, but first we must find $m\widehat{ADC}$.

$m\widehat{ADC} = 2(m\angle B)$	Inscribed Angle Theorem
$m\widehat{ADC} = 2(80)$ or 160	$m\angle B = 80$
$m\widehat{ABC} + m\widehat{ADC} = 360$	Sum of angles in circle = 360
$m\widehat{ABC} + 160 = 360$	$m\widehat{ADC} = 160$
$m\widehat{ABC} = 200$	Subtract 160 from each side.
$m\widehat{ABC} = 2(m\angle D)$	Inscribed Angle Theorem
$200 = 2(m\angle D)$	Substitution
$100 = m\angle D$	Divide each side by 2.

In Example 5, note that the opposite angles of the quadrilateral are supplementary. This is stated in Theorem 10.8 and can be verified by considering that the arcs intercepted by opposite angles of an inscribed quadrilateral form a circle.

Theorem 10.8

If a quadrilateral is inscribed in a circle, then its opposite angles are supplementary.

Example:
Quadrilateral $ABCD$ is inscribed in $\odot P$.
$\angle A$ and $\angle C$ are supplementary.
$\angle B$ and $\angle D$ are supplementary.

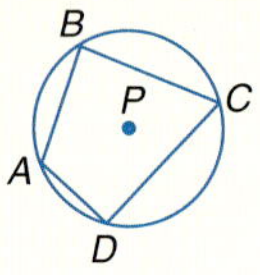

You will prove this theorem in Exercise 39.

Check for Understanding

Concept Check

1. **OPEN ENDED** Draw a counterexample of an inscribed trapezoid. If possible, include at least one angle that is an inscribed angle. **1–2. See margin.**
2. **Compare and contrast** an inscribed angle and a central angle that intercepts the same arc.

Guided Practice

GUIDED PRACTICE KEY	
Exercises	Examples
3	1
4	2
5	4
6	5
7	3

3. In $\odot R$, $m\widehat{MN} = 120$ and $m\widehat{MQ} = 60$. Find the measure of each numbered angle.
$m\angle 1 = 30$, $m\angle 2 = 60$, $m\angle 3 = 60$, $m\angle 4 = 30$, $m\angle 5 = 30$, $m\angle 6 = 60$, $m\angle 7 = 60$, $m\angle 8 = 30$

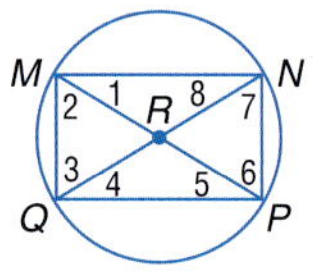

4. **PROOF** Write a paragraph proof. **See margin.**
Given: Quadrilateral $ABCD$ is inscribed in $\odot P$.
$m\angle C = \frac{1}{2} m\angle B$
Prove: $m\widehat{CDA} = 2(m\widehat{DAB})$

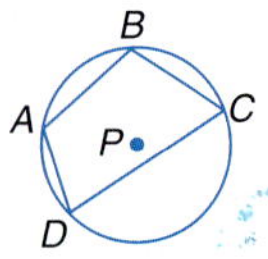

5. $m\angle 1 = 35$, $m\angle 2 = 55$, $m\angle 3 = 39$, $m\angle 4 = 39$

5. **ALGEBRA** In $\odot A$ at the right, $\widehat{PQ} \cong \widehat{RS}$. Find the measure of each numbered angle if $m\angle 1 = 6x + 11$, $m\angle 2 = 9x + 19$, $m\angle 3 = 4y - 25$, and $m\angle 4 = 3y - 9$.

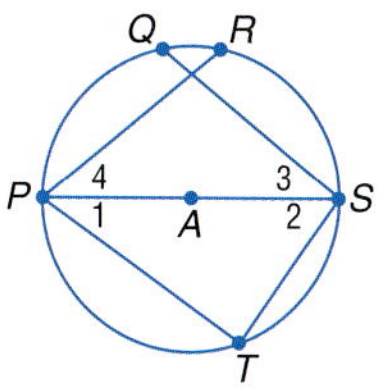

6. Suppose quadrilateral $VWXY$ is inscribed in $\odot C$. If $m\angle X = 28$ and $m\angle W = 110$, find $m\angle V$ and $m\angle Y$. **152, 70**

Application

7. **PROBABILITY** Points X and Y are endpoints of a diameter of $\odot W$. Point Z is another point on the circle. Find the probability that $\angle XZY$ is a right angle. **1**

★ indicates increased difficulty

Practice and Apply

Homework Help

For Exercises	See Examples
8–10	1
11–12, 35–39	2
13–17	4
18–21, 26–29	5
31–34	3

Extra Practice
See page 774.

Find the measure of each numbered angle for each figure.

8. $\overline{PQ} \cong \overline{RQ}$, $m\widehat{PS} = 45$, and $m\widehat{SR} = 75$

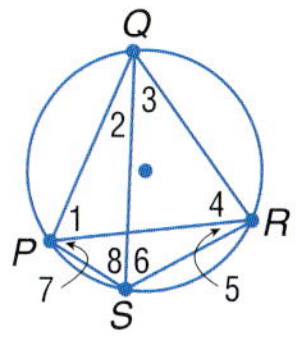

9. $m\angle BDC = 25$, $m\widehat{AB} = 120$, and $m\widehat{CD} = 130$

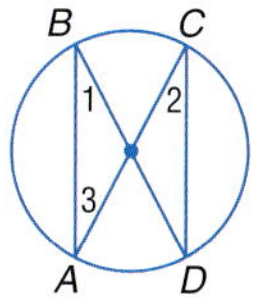

$m\angle 1 = m\angle 2 = 30$, $m\angle 3 = 25$

10. $m\widehat{XZ} = 100$, $\overline{XY} \perp \overline{ST}$, and $\overline{ZW} \perp \overline{ST}$

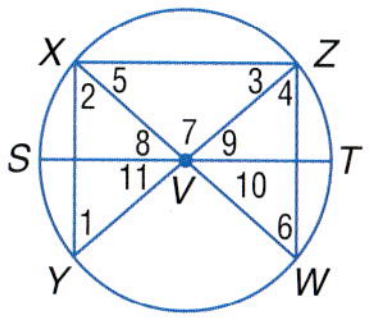

8. $m\angle 1 = 60$, $m\angle 2 = 22.5$, $m\angle 3 = 37.5$, $m\angle 4 = 60$, $m\angle 5 = 22.5$, $m\angle 6 = 60$, $m\angle 7 = 37.5$, $m\angle 8 = 60$

10. $m\angle 1 = m\angle 2 = 50$, $m\angle 3 = 40$, $m\angle 4 = 50$, $m\angle 5 = 40$, $m\angle 6 = 50$, $m\angle 7 = 100$, $m\angle 8 = m\angle 9 = m\angle 10 = m\angle 11 = 40$

PROOF **Write a two-column proof.** **11–12. See margin.**

11. **Given:** $\widehat{AB} \cong \widehat{DE}$, $\widehat{AC} \cong \widehat{CE}$
Prove: $\triangle ABC \cong \triangle EDC$

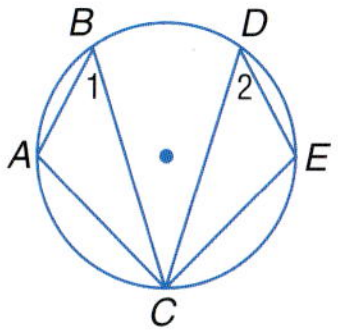

12. **Given:** $\odot P$
Prove: $\triangle AXB \sim \triangle CXD$

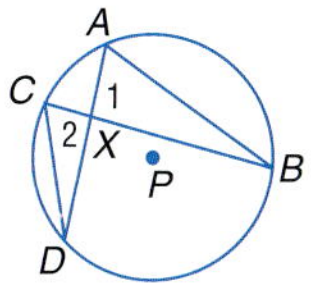

ALGEBRA **Find the measure of each numbered angle for each figure.**

13. $m\angle 1 = x$, $m\angle 2 = 2x - 13$

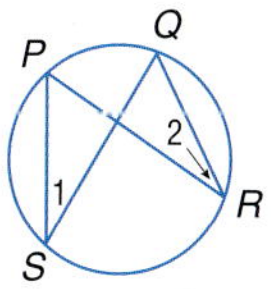

$m\angle 1 = m\angle 2 = 13$

14. $m\widehat{AB} = 120$

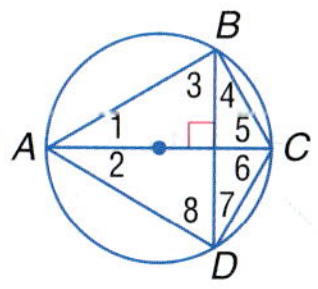

$m\angle 1 = m\angle 2 = 30$, $m\angle 3 = 60$, $m\angle 4 = 30$, $m\angle 5 = m\angle 6 = 60$, $m\angle 7 = 30$, $m\angle 8 = 60$

15. $m\angle R = \frac{1}{3}x + 5$, $m\angle K = \frac{1}{2}x$

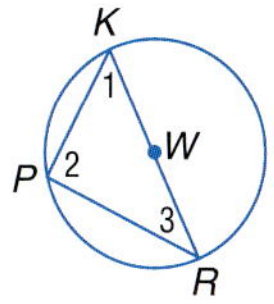

$m\angle 1 = 51$, $m\angle 2 = 90$, $m\angle 3 = 39$

16. $PQRS$ is a rhombus inscribed in a circle. Find $m\angle QRP$ and $m\widehat{SP}$. **45; 90**

★ 17. In $\odot D$, $\overline{DE} \cong \overline{EC}$, $m\widehat{CF} = 60$, and $\overline{DE} \perp \overline{EC}$. Find $m\angle 4$, $m\angle 5$, and $m\widehat{AF}$. **45, 30, 120**

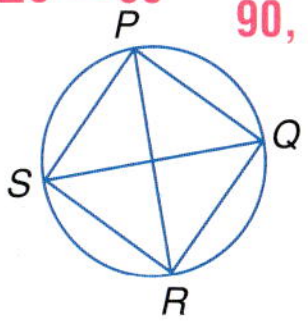

Exercise 16

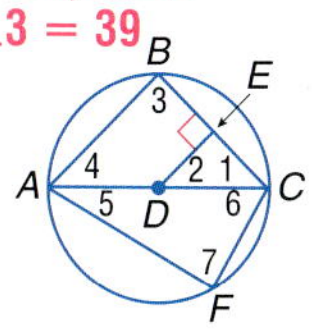

Exercise 17

12. Given: $\odot P$
Prove: $\triangle AXB \sim \triangle CXD$

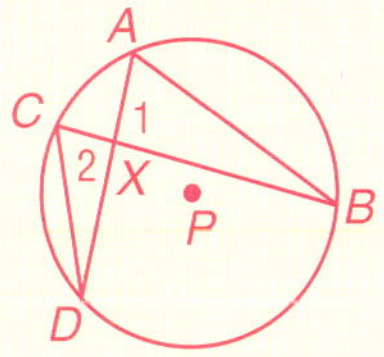

Proof:

Statements	(Reasons)
1. $\odot P$	(Given)
2. $\angle A \cong \angle C$	(Inscribed ∠s intercepting same arc are ≅.)
3. $\angle 1 \cong \angle 2$	(Vertical ∠s are ≅.)
4. $\triangle AXB \sim \triangle CXD$	(AA Similarity)

About the Exercises...

Organization by Objective

- **Inscribed Angles:** 8–12, 31–39
- **Angles of Inscribed Polygons:** 13–21, 26–29

Odd/Even Assignments

Exercises 8–15, 22–29 are structured so that students practice the same concepts whether they are assigned odd or even problems.

Assignment Guide

Basic: 9–15 odd, 19–27 odd, 31–43 odd, 44–58

Average: 9–43 odd, 44–58

Advanced: 8–42 even, 44–55 (optional: 56–58)

Answers

11. Given: $\widehat{AB} \cong \widehat{DE}$, $\widehat{AC} \cong \widehat{CE}$
Prove: $\triangle ABC \cong \triangle EDC$

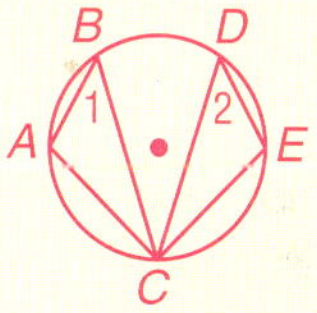

Proof:

Statements	(Reasons)
1. $\widehat{AB} \cong \widehat{DE}$, $\widehat{AC} \cong \widehat{CE}$	(Given)
2. $m\widehat{AB} = m\widehat{DE}$, $m\widehat{AC} = m\widehat{CE}$	(Def. of ≅ arcs)
3. $\frac{1}{2}m\widehat{AB} = \frac{1}{2}m\widehat{DE}$, $\frac{1}{2}m\widehat{AC} = \frac{1}{2}m\widehat{CE}$	(Mult. Prop.)
4. $m\angle ACB = \frac{1}{2}m\widehat{AB}$, $m\angle ECD = \frac{1}{2}m\widehat{DE}$, $m\angle 1 = \frac{1}{2}m\widehat{AC}$, $m\angle 2 = \frac{1}{2}m\widehat{CE}$	(Inscribed ∠ Theorem)
5. $m\angle ACB = m\angle ECD$, $m\angle 1 = m\angle 2$	(Substitution)
6. $\angle ACB \cong \angle ECD$, $\angle 1 \cong \angle 2$	(Def. of ≅ ∠s)
7. $\overline{AB} \cong \overline{DE}$	(≅ arcs have ≅ chords.)
8. $\triangle ABC \cong \triangle EDC$	(AAS)

18. Quadrilateral $WRTZ$ is inscribed in a circle. If $m\angle W = 45$ and $m\angle R = 100$, find $m\angle T$ and $m\angle Z$. **135, 80**

19. Trapezoid $ABCD$ is inscribed in a circle. If $m\angle A = 60$, find $m\angle B$, $m\angle C$, and $m\angle D$. **$m\angle B = 120$, $m\angle C = 120$, $m\angle D = 60$**

20. Rectangle $PDQT$ is inscribed in a circle. What can you conclude about $\overline{PQ}$? **Sample answer: $\overline{PQ}$ is a diagonal of $PDQT$ and a diameter of the circle.**

21. Square $EDFG$ is inscribed in a circle. What can you conclude about $\overline{EF}$? **Sample answer: $\overline{EF}$ is a diameter of the circle and a diagonal and angle bisector of $EDFG$.**

Equilateral pentagon $PQRST$ is inscribed in $\odot U$. Find each measure.

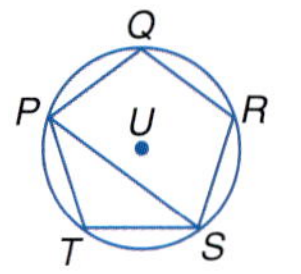

22. $m\widehat{QR}$ **72**
23. $m\angle PSR$ **72**
24. $m\angle PQR$ **108**
25. $m\widehat{PTS}$ **144**

Quadrilateral $ABCD$ is inscribed in $\odot Z$ such that $m\angle BZA = 104$, $m\widehat{CB} = 94$, and $\overline{AB} \parallel \overline{DC}$. Find each measure.

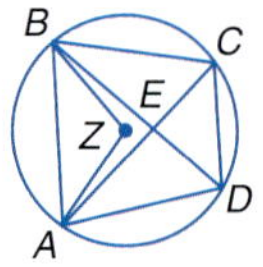

26. $m\widehat{BA}$ **104**
27. $m\widehat{ADC}$ **162**
★ 28. $m\angle BDA$ **52**
★ 29. $m\angle ZAC$ **9**

30. **SCHOOL RINGS** Some designs of class rings involve adding gold or silver to the surface of the round stone. The design at the right includes two inscribed angles. If $m\angle ABC = 50$ and $m\widehat{DBF} = 128$, find $m\widehat{AC}$ and $m\angle DEF$. **100, 64**

More About. . .

School Rings

Many companies that sell school rings also offer schools and individuals the option to design their own ring.

PROBABILITY **For Exercises 31–34, use the following information.**
Point T is randomly selected on $\odot C$ so that it does not coincide with points P, Q, R, or S. $\overline{SQ}$ is a diameter of $\odot C$.

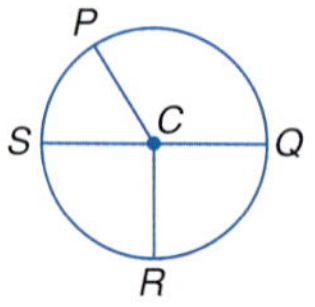

31. Find the probability that $m\angle PTS = 20$ if $m\widehat{PS} = 40$. **$\frac{8}{9}$**
32. Find the probability that $m\angle PTR = 55$ if $m\widehat{PSR} = 110$. **$\frac{25}{36}$**
33. Find the probability that $m\angle STQ = 90$. **1**
34. Find the probability that $m\angle PTQ = 180$. **0**

PROOF **Write the indicated proof for each theorem.** **35–39. See pp. 589A–589B.**

35. two-column proof:
Case 2 of Theorem 10.5
Given: T lies inside $\angle PRQ$.
$\overline{RK}$ is a diameter of $\odot T$.
Prove: $m\angle PRQ = \frac{1}{2}m\widehat{PKQ}$

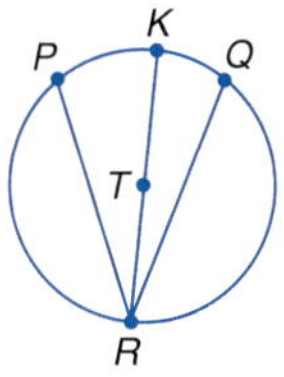

36. two-column proof:
Case 3 of Theorem 10.5
Given: T lies outside $\angle PRQ$.
$\overline{RK}$ is a diameter of $\odot T$.
Prove: $m\angle PRQ = \frac{1}{2}m\widehat{PQ}$

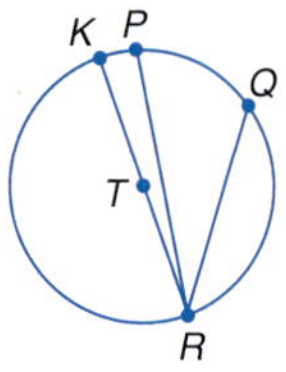

37. two-column proof:
Theorem 10.6

38. paragraph proof:
Theorem 10.7

39. paragraph proof:
Theorem 10.8

Study Guide and Intervention, p. 559 (shown) and p. 560

Inscribed Angles An **inscribed angle** is an angle whose vertex is on a circle and whose sides contain chords of the circle. In $\odot G$, inscribed $\angle DEF$ **intercepts** $\widehat{DF}$.

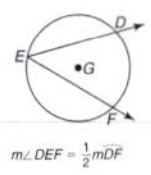

Inscribed Angle Theorem	If an angle is inscribed in a circle, then the measure of the angle equals one-half the measure of its intercepted arc.

$m\angle DEF = \frac{1}{2}m\widehat{DF}$

Example **In $\odot G$ above, $m\widehat{DF} = 90$. Find $m\angle DEF$.**
$\angle DEF$ is an inscribed angle so its measure is half of the intercepted arc.

$$m\angle DEF = \frac{1}{2}m\widehat{DF} = \frac{1}{2}(90) \text{ or } 45$$

Exercises

Use $\odot P$ for Exercises 1–10. In $\odot P$, $\overline{RS} \parallel \overline{TV}$ and $\overline{RT} \cong \overline{SV}$.

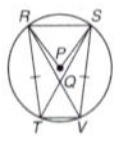

1. Name the intercepted arc for $\angle RTS$. $\widehat{RS}$
2. Name an inscribed angle that intercepts $\widehat{SV}$. $\angle SRV$ or $\angle STV$

In $\odot P$, $m\widehat{SV} = 120$ and $m\angle RPS = 76$. Find each measure.

3. $m\angle PRS$ 52
4. $m\widehat{RSV}$ 196
5. $m\widehat{RT}$ 120
6. $m\angle RVT$ 60
7. $m\angle QRS$ 60
8. $m\angle STV$ 60
9. $m\widehat{TV}$ 44
10. $m\angle SVT$ 98

Skills Practice, p. 561 and Practice, p. 562 (shown)

In $\odot B$, $m\widehat{WX} = 104$, $m\widehat{WZ} = 88$, and $m\angle ZWY = 26$. Find the measure of each angle.

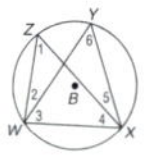

1. $m\angle 1$ 52
2. $m\angle 2$ 26
3. $m\angle 3$ 58
4. $m\angle 4$ 44
5. $m\angle 5$ 26
6. $m\angle 6$ 52

ALGEBRA **Find the measure of each numbered angle.**

7. $m\angle 1 = 5x + 2$, $m\angle 2 = 2x - 3$, $m\angle 3 = 7y - 1$, $m\angle 4 = 2y + 10$

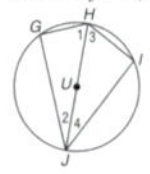

$m\angle 1 = 67$, $m\angle 2 = 23$
$m\angle 3 = 62$, $m\angle 4 = 28$

8. $m\angle 1 = 4x - 7$, $m\angle 2 = 2x + 11$, $m\angle 3 = 5y - 14$, $m\angle 4 = 3y + 8$

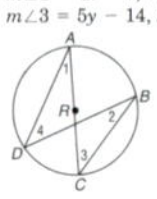

$m\angle 1 = 29$, $m\angle 2 = 29$
$m\angle 3 = 41$, $m\angle 4 = 41$

Quadrilateral $EFGH$ is inscribed in $\odot N$ such that $m\widehat{FG} = 97$, $m\widehat{GH} = 117$, and $m\widehat{EHG} = 164$. Find each measure.

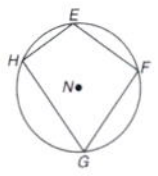

9. $m\angle E$ 107
10. $m\angle F$ 82
11. $m\angle G$ 73
12. $m\angle H$ 98

13. **PROBABILITY** In $\odot V$, point C is randomly located so that it does not coincide with points R or S. If $m\widehat{RS} = 140$, what is the probability that $m\angle RCS = 70$? $\frac{11}{18}$

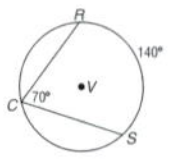

Reading to Learn Mathematics, p. 563 ELL

Pre-Activity **How is a socket like an inscribed polygon?**

Read the introduction to Lesson 10-4 at the top of page 544 in your textbook.

- Why do you think regular hexagons are used rather than squares for the "hole" in a socket? Sample answer: If a square were used, the points might be too sharp for the tool to work smoothly.
- Why do you think regular hexagons are used rather than regular polygons with more sides? Sample answer: If there are too many sides, the polygon would be too close to a circle, so the wrench might slip.

Reading the Lesson

1. Underline the correct word or phrase to form a true statement.
   a. An angle whose vertex is on a circle and whose sides contain chords of the circle is called a(n) (central/inscribed/circumscribed) angle.
   b. Every inscribed angle that intercepts a semicircle is a(n) (acute/right/obtuse) angle.
   c. The opposite angles of an inscribed quadrilateral are (congruent/complementary/supplementary).
   d. An inscribed angle that intercepts a major arc is a(n) (acute/right/obtuse) angle.
   e. Two inscribed angles of a circle that intercept the same arc are (congruent/complementary/supplementary).
   f. If a triangle is inscribed in a circle and one of the sides of the triangle is a diameter of the circle, the diameter is (the longest side of an acute triangle/a leg of an isosceles triangle/the hypotenuse of a right triangle).

2. Refer to the figure. Find each measure.

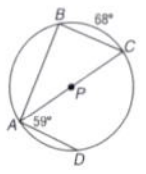

   a. $m\angle ABC$ 90
   b. $m\widehat{CD}$ 118
   c. $m\widehat{AD}$ 62
   d. $m\angle BAC$ 34
   e. $m\angle BCA$ 56
   f. $m\widehat{AB}$ 112
   g. $m\widehat{BCD}$ 186
   h. $m\widehat{BDA}$ 248

Helping You Remember

3. A good way to remember a geometric relationship is to visualize it. Describe how you could make a sketch that would help you remember the relationship between the measure of an inscribed angle and the measure of its intercepted arc. Sample answer: Draw a diameter of the circle to divide it into two semicircles. Inscribe an angle in one of the semicircles; this angle will intercept the other semicircle. From your sketch, you can see that the inscribed angle is a right angle. The measure of the semicircle arc is 180, so the measure of the inscribed angle is half the measure of its intercepted arc.

Enrichment, p. 564

Formulas for Regular Polygons

Suppose a regular polygon of n sides is inscribed in a circle of radius r. The figure shows one of the isosceles triangles formed by joining the endpoints of one side of the polygon to the center C of the circle. In the figure, s is the length of each side of the regular polygon, and a is the length of the segment from C perpendicular to $\overline{AB}$.

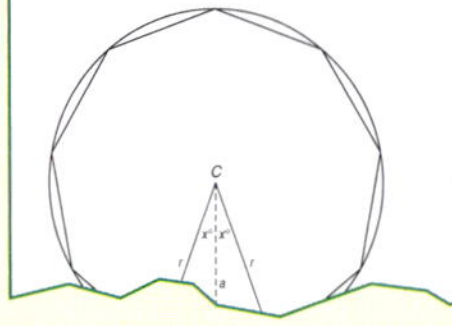

41–43. See p. 589B for explanations.

41. isosceles right triangle

STAINED GLASS In the stained glass window design, all of the small arcs around the circle are congruent. Suppose the center of the circle is point O.

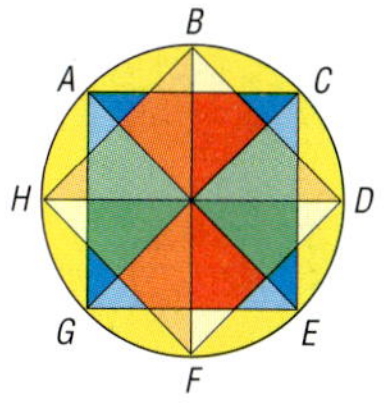

40. What is the measure of each of the small arcs? **45**
41. What kind of figure is $\triangle AOC$? Explain.
42. What kind of figure is quadrilateral $BDFH$? Explain. **square**
43. What kind of figure is quadrilateral $ACEG$? Explain. **square**

44. **CRITICAL THINKING** A trapezoid $ABCD$ is inscribed in $\odot O$. Explain how you can verify that $ABCD$ must be an isosceles trapezoid. **See margin.**

45. **WRITING IN MATH** Answer the question that was posed at the beginning of the lesson. **See margin.**

How is a socket like an inscribed polygon?

Include the following in your answer:
- a definition of an inscribed polygon, and
- the side length of a regular hexagon inscribed in a circle $\frac{3}{4}$ inch wide.

FCAT Practice
Standardized Test Practice

46. What is the ratio of the measure of $\angle ACB$ to the measure of $\angle AOB$? **C**

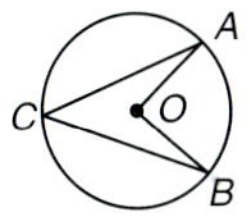

(A) 1 : 1 (B) 2 : 1
(C) 1 : 2 (D) not enough information

47. **GRID IN** The daily newspaper always follows a particular format. Each even-numbered page contains six articles, and each odd-numbered page contains seven articles. If today's paper has 36 pages, how many articles does it contain? **234**

Maintain Your Skills

Mixed Review

Find each measure. *(Lesson 10-3)*

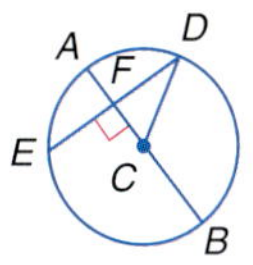

48. If $AB = 60$ and $DE = 48$, find CF. **18**
49. If $AB = 32$ and $FC = 11$, find FE.
50. If $DE = 60$ and $FC = 16$, find AB. **68**

49. $\sqrt{135} \approx 11.62$

Points Q and R lie on $\odot P$. Find the length of $\overset{\frown}{QR}$ for the given radius and angle measure. *(Lesson 10-2)*

51. $PR = 12$, and $m\angle QPR = 60$ **4π units**
52. $m\angle QPR = 90$, $PR = 16$ **8π units**

Complete each sentence with *sometimes*, *always*, or *never*. *(Lesson 4-1)*

53. Equilateral triangles are __?__ isosceles. **always**
54. Acute triangles are __?__ equilateral. **sometimes**
55. Obtuse triangles are __?__ scalene. **sometimes**

Getting Ready for the Next Lesson

PREREQUISITE SKILL Determine whether each figure is a right triangle.
*(To review the **Pythagorean Theorem**, see Lesson 7-2.)*

56. **no**

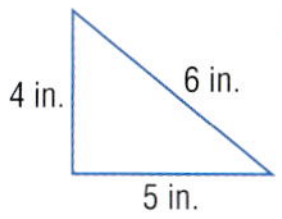

57. **no**

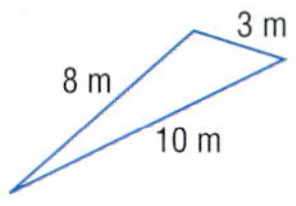

58. **yes**

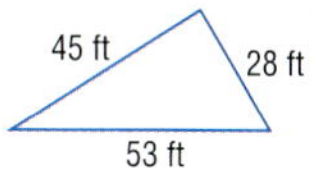

www.geometryonline.com/self_check_quiz/fcat

4 Assess

Open-Ended Assessment

Modeling Use a cork board, pushpins, a cutout circle, and a flexible rubber band to model inscribed angles. Place the circle on the cork board and put two pushpins on the circle to represent the points of an intercepted arc. Wrap the rubber band around the pins and use a pencil to drag the rubber band to the opposite end of the circle to represent the vertex of the inscribed angle. Students can move the pencil along the circle and use a protractor to note that the measure of the angle stays the same.

Getting Ready for Lesson 10-5

Prerequisite Skill Students will learn about tangents in Lesson 10-5. They will apply the Pythagorean Theorem to determine if segments are tangents and find lengths. Use Exercises 56–58 to determine your students' familiarity with the Pythagorean Theorem.

Assessment Options

Quiz (Lessons 10-3 and 10-4) is available on p. 603 of the *Chapter 10 Resource Masters*.

Mid-Chapter Test (Lessons 10-1 through 10-4) is available on p. 605 of the *Chapter 11 Resource Masters*.

Answers

44. Use the properties of trapezoids and inscribed quadrilaterals to verify that $ABCD$ is isosceles.

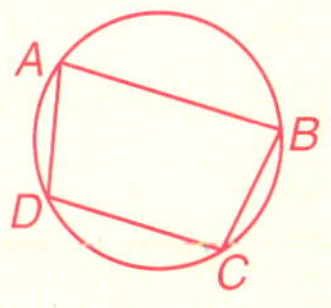

$m\angle A + m\angle D = 180$ (same side interior angles = 180)
$m\angle A + m\angle C = 180$ (opposite angles of inscribed quadrilaterals = 180)
$m\angle A + m\angle D = m\angle A + m\angle C$ (Substitution)
$m\angle D = m\angle C$ (Subtraction Property)
$\angle D \cong \angle C$ (Def. of $\cong$ $\triangle$s)
Trapezoid $ABCD$ is isosceles because the base angles are congruent.

45. Sample answer: The socket is similar to an inscribed polygon because the vertices of the hexagon can be placed on a circle that is concentric with the outer circle of the socket. Answers should include the following.
- An inscribed polygon is one in which all of its vertices are points on a circle.
- The side of the regular hexagon inscribed in a circle $\frac{3}{4}$ inch wide is $\frac{3}{8}$ inch.

10-5 Lesson Notes

1 Focus

5-Minute Check Transparency 10-5 Use as a quiz or review of Lesson 10-4.

Mathematical Background notes are available for this lesson on p. 520D.

How are tangents related to track and field events?

Ask students:

- What does *tangent* mean? Where have you heard the word used? **Accept all reasonable answers that suggest tangent means touching.**
- What other situations can be modeled by a circle and a tangent? **Sample answers: fishing line unrolling from a spool, the string of a yo-yo, a line of paint being applied to a wall by a roller**

10-5 Tangents

Sunshine State Standards MA.C.2.4.1, NC.C.1.4.1, NC.C.2.4.1, NC.C.2.4.1, NC.C.1.4.1, NC.C.2.4.1, NC.C.2.4.1

Vocabulary

- tangent
- point of tangency

What You'll Learn

- Use properties of tangents.
- Solve problems involving circumscribed polygons.

How are tangents related to track and field events?

In July 2001, Yipsi Moreno of Cuba won her first major title in the hammer throw at the World Athletic Championships in Edmonton, Alberta, Canada, with a throw of 70.65 meters. The hammer is a metal ball, usually weighing 16 pounds, attached to a steel wire at the end of which is a grip. The ball is spun around by the thrower and then released, with the greatest distance thrown winning the event.

TANGENTS The figure models the hammer throw event. Circle A represents the circular area containing the spinning thrower. Ray BC represents the path the hammer takes when released. $\overrightarrow{BC}$ is **tangent** to $\odot A$, because the line containing $\overrightarrow{BC}$ intersects the circle in exactly one point. This point is called the **point of tangency**.

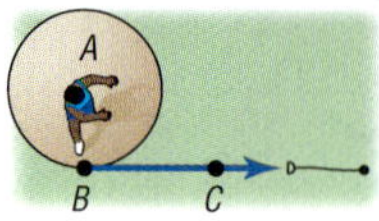

Geometry Software Investigation

Tangents and Radii

Model

- Use The Geometer's Sketchpad to draw a circle with center W. Then draw a segment tangent to $\odot W$. Label the point of tangency as X.
- Choose another point on the tangent and name it Y. Draw $\overline{WY}$.

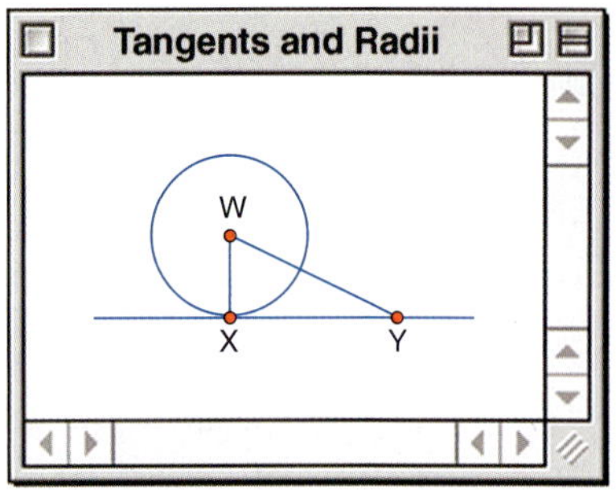

Think and Discuss

1. What is $\overline{WX}$ in relation to the circle? **radius**
2. Measure $\overline{WY}$ and $\overline{WX}$. Write a statement to relate WX and WY. **$WX < WY$**
3. Move point Y along the tangent. How does the location of Y affect the statement you wrote in Exercise 2? **It doesn't, unless Y and X coincide.**
4. Measure $\angle WXY$. What conclusion can you make? **$\overline{WX} \perp \overline{XY}$**
5. **Make a conjecture** about the shortest distance from the center of the circle to a tangent of the circle.

Study Tip

Tangent Lines
All of the theorems applying to tangent lines also apply to parts of the line that are tangent to the circle.

5. Sample answer: The shortest distance from the center of a circle to the tangent is the radius of the circle, which is perpendicular to the tangent.

This investigation suggests an indirect proof of Theorem 10.9.

Resource Manager

Workbook and Reproducible Masters

Chapter 10 Resource Masters

- Study Guide and Intervention, pp. 565–566
- Skills Practice, p. 567
- Practice, p. 568
- Reading to Learn Mathematics, p. 569
- Enrichment, p. 570

Graphing Calculator and Computer Masters, pp. 35, 36
School-to-Career Masters, p. 20
Prerequisite Skills Workbook, pp. 15–16
Teaching Geometry With Manipulatives Masters, pp. 17, 170, 171, 173

Transparencies

5-Minute Check Transparency 10-5
Answer Key Transparencies

Technology

Interactive Chalkboard

Theorem 10.9

If a line is tangent to a circle, then it is perpendicular to the radius drawn to the point of tangency.

Example: If $\overleftrightarrow{RT}$ is a tangent, $\overline{OR} \perp \overleftrightarrow{RT}$.

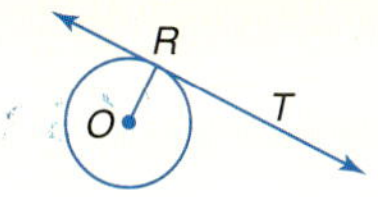

Example 1 Find Lengths

ALGEBRA $\overline{ED}$ **is tangent to** $\odot F$ **at point** E**. Find** x**.**

Because the radius is perpendicular to the tangent at the point of tangency, $\overline{EF} \perp \overline{DE}$. This makes $\angle DEF$ a right angle and $\triangle DEF$ a right triangle. Use the Pythagorean Theorem to find x.

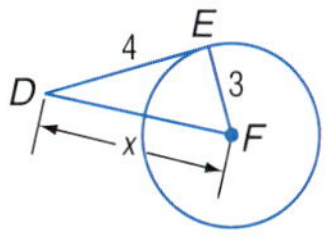

$(EF)^2 + (DE)^2 = (DF)^2$ Pythagorean Theorem

$3^2 + 4^2 = x^2$ $EF = 3$, $DE = 4$, $DF = x$

$25 = x^2$ Simplify.

$\pm 5 = x$ Take the square root of each side.

Because x is the length of $\overline{DF}$, ignore the negative result. Thus, $x = 5$.

The converse of Theorem 10.9 is also true.

Theorem 10.10

If a line is perpendicular to a radius of a circle at its endpoint on the circle, then the line is tangent to the circle.

Example: If $\overline{OR} \perp \overleftrightarrow{RT}$, $\overleftrightarrow{RT}$ is a tangent.

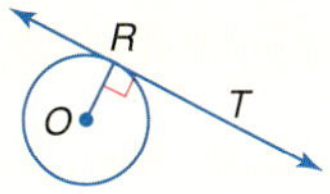

You will prove this theorem in Exercise 22.

Study Tip

Identifying Tangents
Never assume that a segment is tangent to a circle by appearance unless told otherwise. The figure must either have a right angle symbol or include the measurements that confirm a right angle.

Example 2 Identify Tangents

a. Determine whether $\overline{MN}$ **is tangent to** $\odot L$**.**

First determine whether $\triangle LMN$ is a right triangle by using the converse of the Pythagorean Theorem.

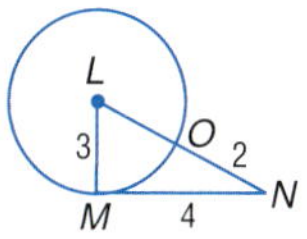

$(LM)^2 + (MN)^2 \stackrel{?}{=} (LN)^2$ Converse of Pythagorean Theorem

$3^2 + 4^2 \stackrel{?}{=} 5^2$ $LM = 3$, $MN = 4$, $LN = 3 + 2$ or 5

$25 = 25$ ✓ Simplify.

Because the converse of the Pythagorean Theorem is true, $\triangle LMN$ is a right triangle and $\angle LMN$ is a right angle. Thus, $\overline{LM} \perp \overline{MN}$, making $\overline{MN}$ a tangent to $\odot L$.

b. Determine whether $\overline{PQ}$ **is tangent to** $\odot R$**.**

Since $RQ = RS$, $RP = 4 + 4$ or 8 units.

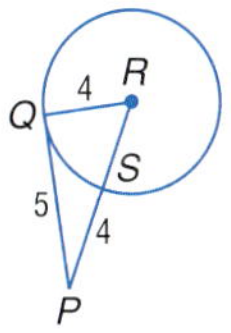

$(RQ)^2 + (PQ)^2 \stackrel{?}{=} (RP)^2$ Converse of Pythagorean Theorem

$4^2 + 5^2 \stackrel{?}{=} 8^2$ $RQ = 4$, $PQ = 5$, $RP = 8$

$41 \neq 64$ Simplify.

Because the converse of the Pythagorean Theorem did not prove true in this case, $\triangle RQP$ is not a right triangle.

So, $\overline{PQ}$ is not tangent to $\odot R$.

www.geometryonline.com/extra_examples/fcat

2 Teach

TANGENTS

In-Class Examples Power Point®

Teaching Tip Explain that even though a tangent intersects a circle, there is never any part of a tangent contained inside a circle. The only point that the tangent and the circle have in common is the point of intersection.

1 **ALGEBRA** $\overline{RS}$ is tangent to $\odot Q$ at point R. Find y. **24**

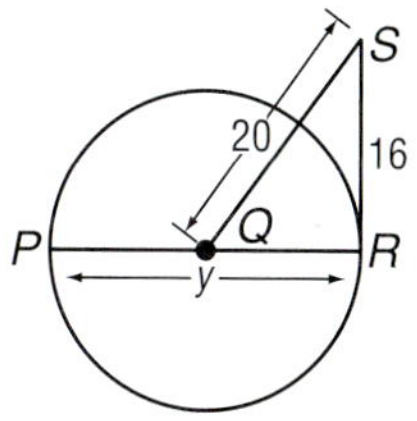

2 **a.** Determine whether $\overline{BC}$ is tangent to $\odot A$. **no**

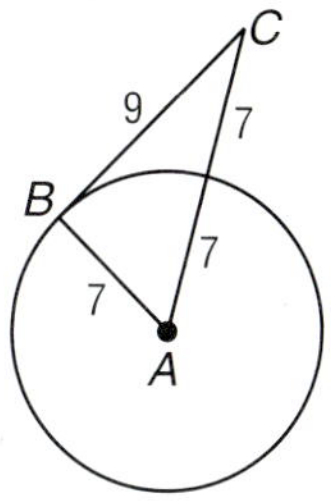

b. Determine whether $\overline{WE}$ is tangent to $\odot D$. **yes**

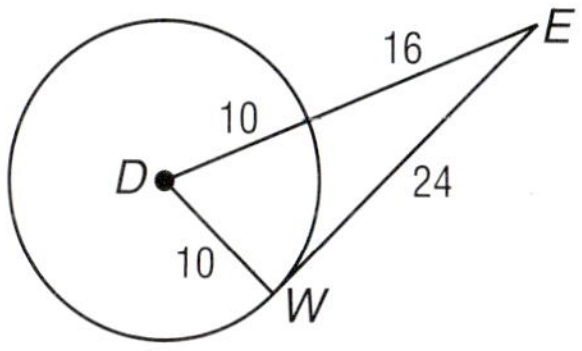

Geometry Software Investigation

Tangents and Radii Tell students that since a radius is perpendicular to a tangent at the point of tangency, the diameter containing that radius is also perpendicular to the tangent at the same point. Students can also repeat the activity for other points of tangency. They can start with a new circle, or you can ask students where they could place another tangent on $\odot W$ that is perpendicular to $\overline{WY}$.

In-Class Example

Power Point®

3 **ALGEBRA** Find x. Assume that segments that appear tangent to circles are tangent.

1

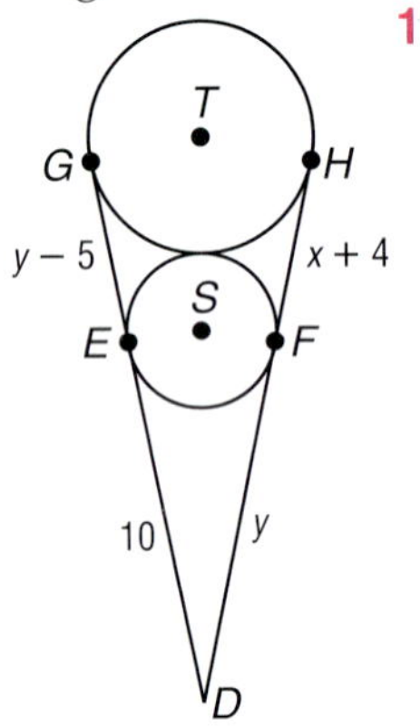

More than one line can be tangent to the same circle. In the figure, $\overline{AB}$ and $\overline{BC}$ are tangent to $\odot D$. So, $(AB)^2 + (AD)^2 = (DB)^2$ and $(BC)^2 + (CD)^2 = (DB)^2$.

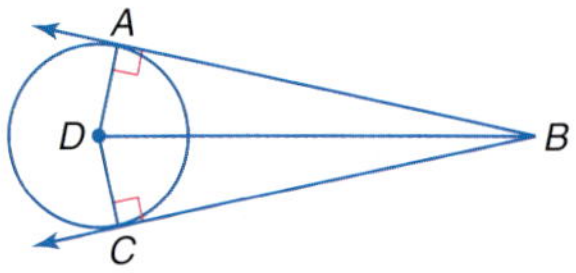

$(AB)^2 + (AD)^2 = (BC)^2 + (CD)^2$	Substitution
$(AB)^2 + (AD)^2 = (BC)^2 + (AD)^2$	$AD = CD$
$(AB)^2 = (BC)^2$	Subtract $(AD)^2$ from each side.
$AB = BC$	Take the square root of each side.

The last statement implies that $\overline{AB} \cong \overline{BC}$. This is a proof of Theorem 10.10.

Theorem 10.11

If two segments from the same exterior point are tangent to a circle, then they are congruent.

Example: $\overline{AB} \cong \overline{AC}$

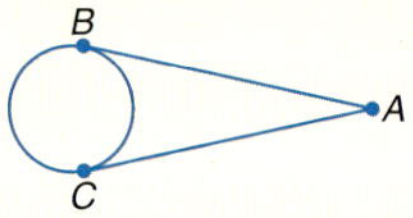

You will prove this theorem in Exercise 27.

Example 3 Solve a Problem Involving Tangents

ALGEBRA **Find x. Assume that segments that appear tangent to circles are tangent.**

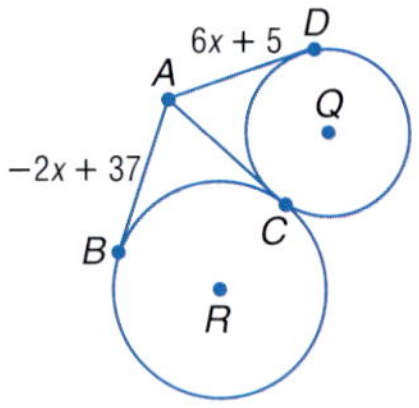

$\overline{AD}$ and $\overline{AC}$ are drawn from the same exterior point and are tangent to $\odot Q$, so $\overline{AD} \cong \overline{AC}$. $\overline{AC}$ and $\overline{AB}$ are drawn from the same exterior point and are tangent to $\odot R$, so $\overline{AC} \cong \overline{AB}$. By the Transitive Property, $\overline{AD} \cong \overline{AB}$.

$AD = AB$	Definition of congruent segments
$6x + 5 = -2x + 37$	Substitution
$8x + 5 = 37$	Add $2x$ to each side.
$8x = 32$	Subtract 5 from each side.
$x = 4$	Divide each side by 8.

Construction

Line Tangent to a Circle Through a Point Exterior to the Circle

1. Construct a circle. Label the center C. Draw a point outside $\odot C$. Then draw $\overline{CA}$.
2. Construct the perpendicular bisector of $\overline{CA}$ and label it line ℓ. Label the intersection of ℓ and $\overline{CA}$ as point X.
3. Construct circle X with radius $\overline{XC}$. Label the points where the circles intersect as D and E.
4. Draw $\overleftrightarrow{AD}$. $\triangle ADC$ is inscribed in a semicircle. So $\angle ADC$ is a right angle, and $\overleftrightarrow{AD}$ is a tangent.

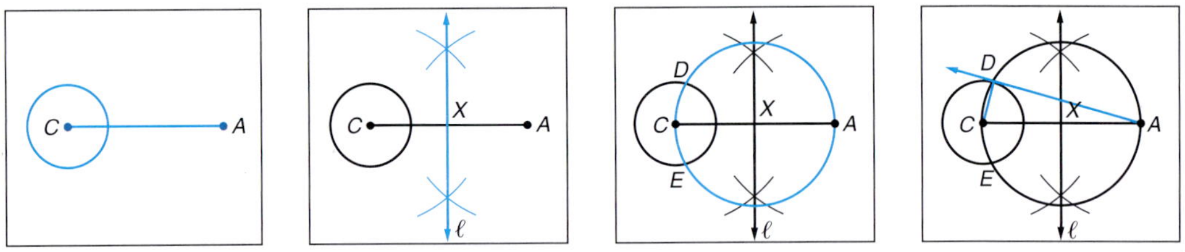

You will construct a line tangent to a circle through a point on the circle in Exercise 21.

DAILY INTERVENTION

Differentiated Instruction

Kinesthetic Have students place a compact disc on a sheet of blank paper and trace around it. Then have students place two metric rulers beside the disc to model two tangents. Have students arrange the rulers so that they form two tangents that intersect on the sheet. Then have students lightly draw the two tangents from the circle to the point of intersection and measure the distance from the circle to the point of intersection to confirm that the measures are equal.

CIRCUMSCRIBED POLYGONS In Lesson 10-3, you learned that circles can be circumscribed about a polygon. Likewise, polygons can be circumscribed about a circle, or the circle is inscribed in the polygon. Notice that the vertices of the polygon *do not* lie on the circle, but every side of the polygon is tangent to the circle.

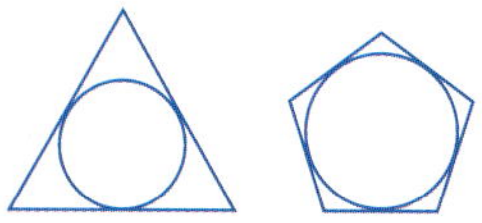

Polygons are circumscribed.

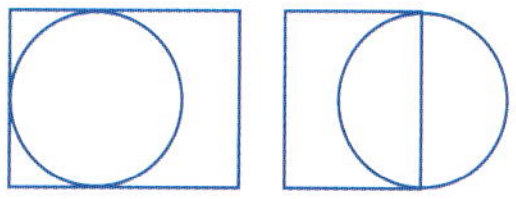

Polygons are *not* circumscribed.

Study Tip

Common Misconceptions
Just because the circle is tangent to one or more of the sides of a polygon does not mean that the polygon is circumscribed about the circle, as shown in the second pair of figures.

Example 4 Triangles Circumscribed About a Circle

Triangle ADC is circumscribed about $\odot O$. Find the perimeter of $\triangle ADC$ if $EC = DE + AF$.

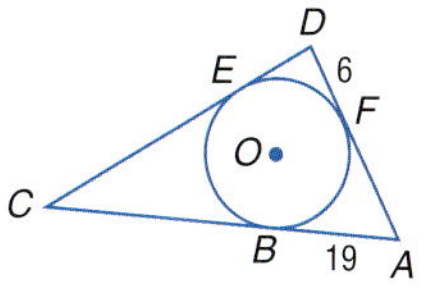

Use Theorem 10.10 to determine the equal measures.
$AB = AF = 19$, $FD = DE = 6$, and $EC = CB$.
We are given that $EC = DE + AF$, so $EC = 6 + 19$ or 25.

$P = AB + BC + EC + DE + FD + AF$ Definition of perimeter

$= 19 + 25 + 25 + 6 + 6 + 19$ or 100 Substitution

The perimeter of $\triangle ADC$ is 100 units.

Check for Understanding

Concept Check

1. **Determine** the number of tangents that can be drawn to a circle for each point. Explain your reasoning. See margin for reasoning.
   a. containing a point outside the circle two
   b. containing a point inside the circle none
   c. containing a point on the circle one
2. Write an argument to support or provide a counterexample to the statement *If two lines are tangent to the same circle, they intersect.*
3. **OPEN ENDED** Draw an example of a circumscribed polygon and an example of an inscribed polygon. See margin.

2. If the lines are tangent at the endpoints of a diameter, they are parallel and thus, not intersecting.

Guided Practice

For Exercises 4 and 5, use the figure at the right.

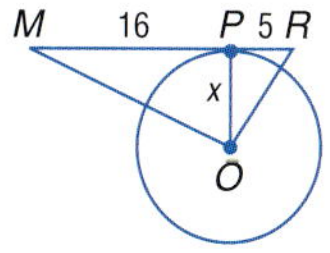

4. Tangent $\overline{MP}$ is drawn to $\odot O$. Find x if $MO = 20$. 12
5. If $RO = 13$, determine whether $\overline{PR}$ is tangent to $\odot O$. Yes; $5^2 + 12^2 = 13^2$.
6. Rhombus $ABCD$ is circumscribed about $\odot P$ and has a perimeter of 32. Find x. 5

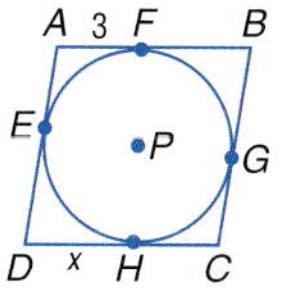

GUIDED PRACTICE KEY

Exercises	Examples
4	1
5	2
6	4
7	3

Application

7. **AGRICULTURE** A pivot-circle irrigation system waters part of a fenced square field. If the spray extends to a distance of 72 feet, what is the total length of the fence around the field? 576 ft

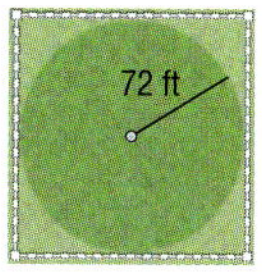

CIRCUMSCRIBED POLYGONS

In-Class Example PowerPoint®

4 Triangle HJK is circumscribed about $\odot G$. Find the perimeter of $\triangle HJK$ if $NK = JL + 29$.

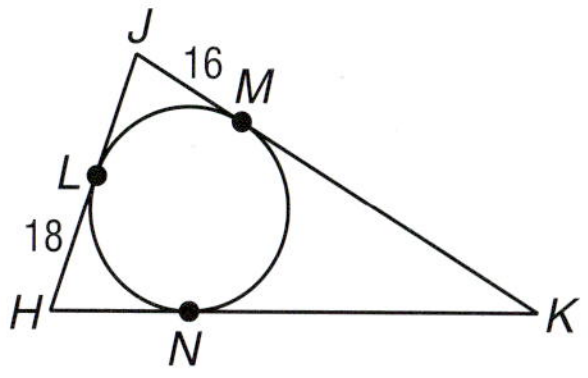

158 units

Answers

1a. From any point outside the circle, you can draw only two tangents.

1b. A line containing a point inside the circle would intersect the circle in two points. A tangent can only intersect a circle in one point.

1c. Since a tangent intersects a circle in exactly one point, there is one tangent containing a point on the circle.

3. Sample answer:
polygon circumscribed about a circle:

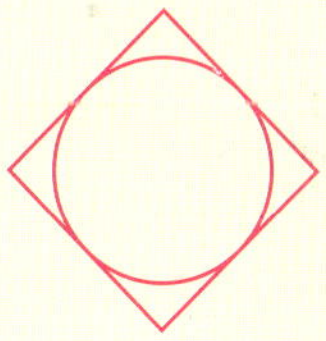

polygon inscribed in a circle:

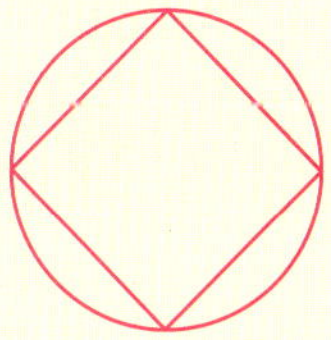

Lesson 10-5 Tangents 555

3 Practice/Apply

Study Notebook

Have students—

- add the definitions/examples of the vocabulary terms to their Vocabulary Builder worksheets for Chapter 10.
- include an example of a circumscribed polygon and an example that demonstrates how to use tangents to find segment lengths.
- include any other item(s) that they find helpful in mastering the skills in this lesson.

About the Exercises...

Organization by Objective

- **Tangents:** 8–20
- **Circumscribed Polygons:** 17–18

Odd/Even Assignments

Exercises 8–20, 23–26 are structured so that students practice the same concepts whether they are assigned odd or even problems.

Alert! Exercise 21 requires a compass and straightedge.

Assignment Guide

Basic: 9–17 odd, 21–31 odd, 32–34 (optional: 35–36), 37–45

Average: 9–31 odd, 32–34 (optional: 35–36), 37–45

Advanced: 8–30 even, 31–41 (optional: 42–45)

★ indicates increased difficulty

Practice and Apply

Homework Help

For Exercises	See Examples
8–11	2
12–20	1, 3
17, 18, 23–26	4

Extra Practice
See page 775.

Determine whether each segment is tangent to the given circle.

8. $\overline{BC}$ yes

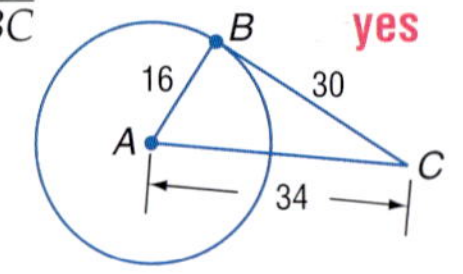

9. $\overline{DE}$ no

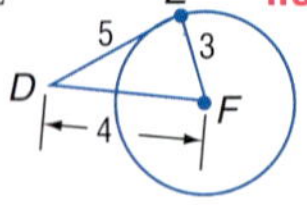

10. $\overline{GH}$ no

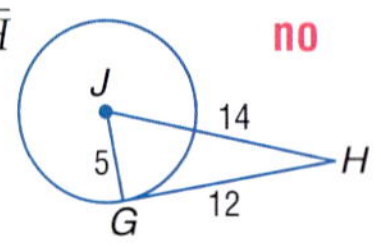

11. $\overline{KL}$ yes

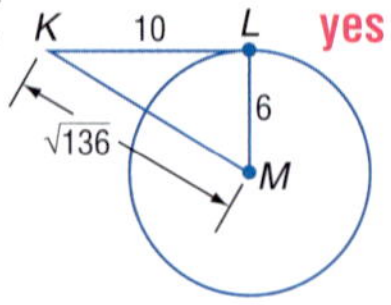

Find x. Assume that segments that appear to be tangent are tangent.

12. 8

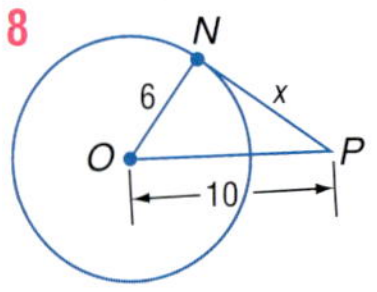

13. 16

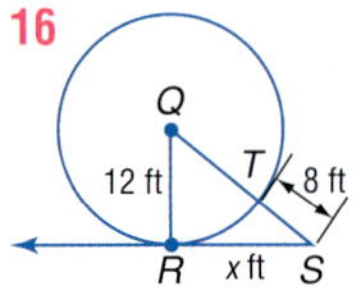

14. $\sqrt{193}$

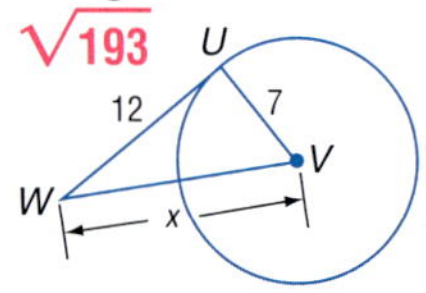

15. 15

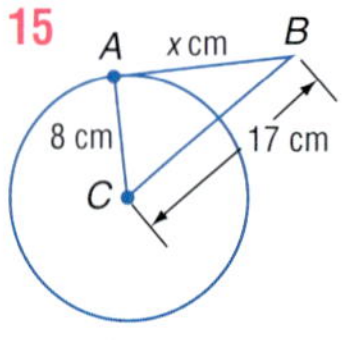

16. 16

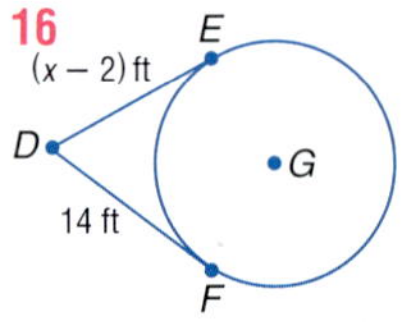

17. 3

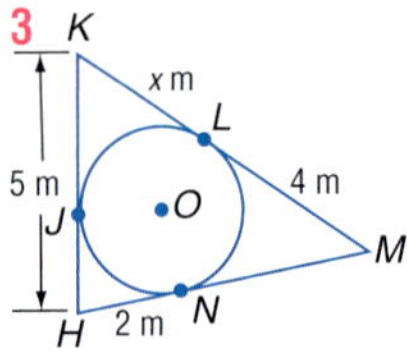

18. 10

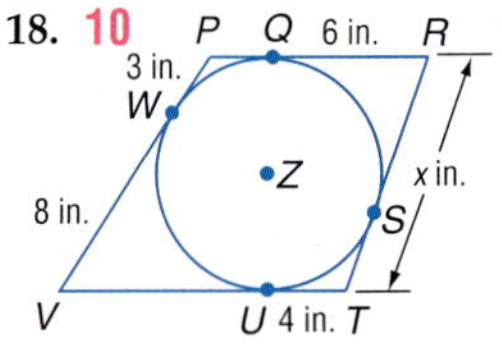

★ 19. 30

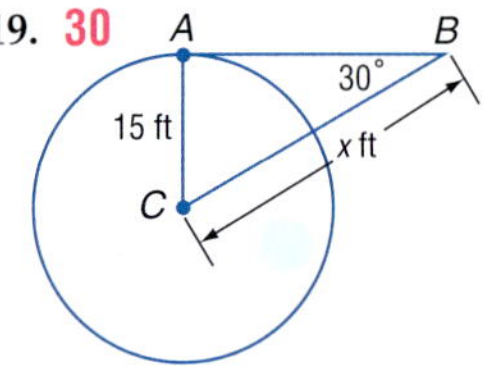

20. ★ $4\frac{2}{3}$

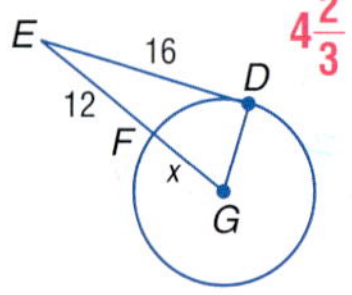

Study Tip

Look Back
To review constructing perpendiculars to a line, see Lesson 3-6.

21. **CONSTRUCTION** Construct a line tangent to a circle through a point on the circle following these steps. See students' work.
- Construct a circle with center T.
- Locate a point P on $\odot T$ and draw $\overrightarrow{TP}$.
- Construct a perpendicular to $\overrightarrow{TP}$ through point P.

22. **PROOF** Write an indirect proof of Theorem 10.10 by assuming that ℓ is not tangent to $\odot A$.
Given: $\ell \perp \overline{AB}$, $\overline{AB}$ is a radius of $\odot A$.
Prove: Line ℓ is tangent to $\odot A$. See margin.

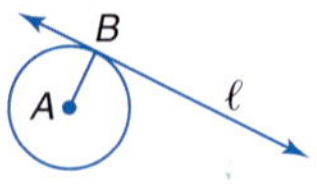

Answer

22. Given: $\ell \perp \overline{AB}$
$\overline{AB}$ is a radius of $\odot A$.
Prove: ℓ is tangent to $\odot A$.

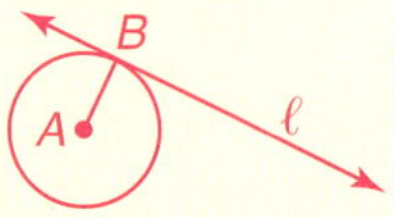

Proof: Assume ℓ is not tangent to $\odot A$. Since ℓ intersects $\odot A$ at B, it must intersect the circle in another place. Call this point C. Then $AB = AC$. But if $\overline{AB} \perp \ell$, then $\overline{AB}$ must be the shortest segment from A to ℓ. If $AB = AC$, then $\overline{AC}$ is the shortest segment from A to ℓ. Since B and C are two different points on ℓ, this is a contradiction. Therefore, ℓ is tangent to $\odot A$.

Find the perimeter of each polygon for the given information.

23. 60 units

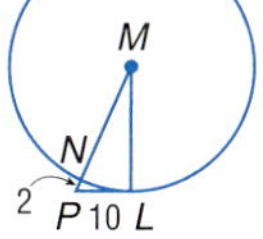

24. $ST = 18$, radius of $\odot P = 5$ 58.5 units

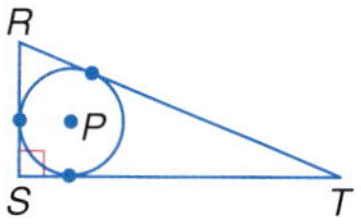

25. $BY = CZ = AX = 2.5$ diameter of $\odot G = 5$
$15\sqrt{3}$ units

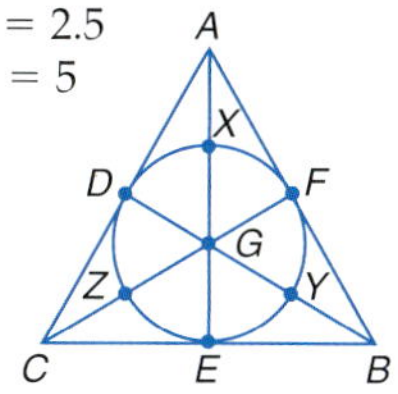

★26. $CF = 6(3 - x)$, $DB = 12y - 4$ 36 units

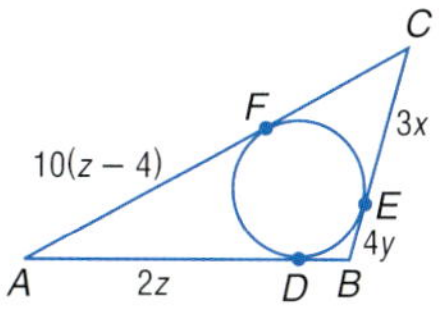

27. **PROOF** Write a two-column proof to show that if two segments from the same exterior point are tangent to a circle, then they are congruent. (Theorem 10.11) See p. 589B.

28. **PHOTOGRAPHY** The film in a 35-mm camera unrolls from a cylinder, travels across an opening for exposure, and then is forwarded into another circular chamber as each photograph is taken. The roll of film has a diameter of 25 millimeters, and the distance from the center of the roll to the intake of the chamber is 100 millimeters. To the nearest millimeter, how much of the film would be exposed if the camera were opened before the roll had been totally used? 99 mm

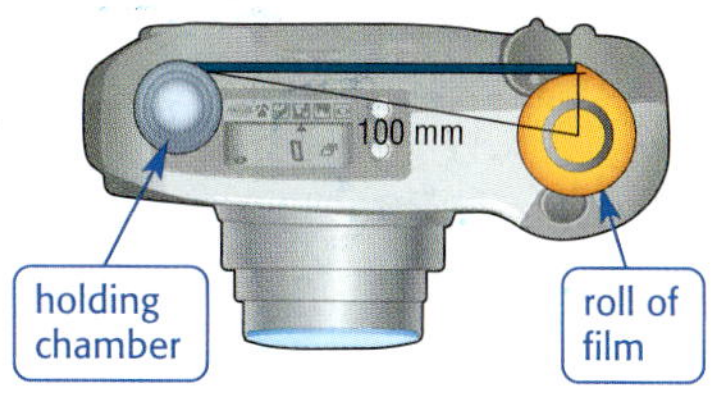

More About. . .

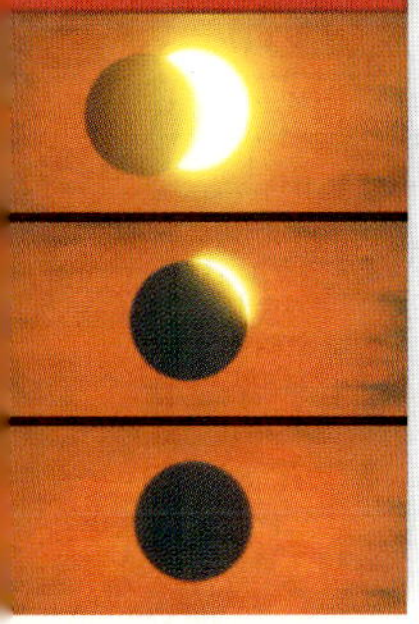

Astronomy

During the 20th century, there were 78 total solar eclipses, but only 15 of these affected parts of the United States. The next total solar eclipse visible in the U.S. will be in 2017.

Source: *World Almanac*

ASTRONOMY **For Exercises 29 and 30, use the following information.**
A solar eclipse occurs when the moon blocks the sun's rays from hitting Earth. Some areas of the world will experience a total eclipse, others a partial eclipse, and some no eclipse at all, as shown in the diagram below.

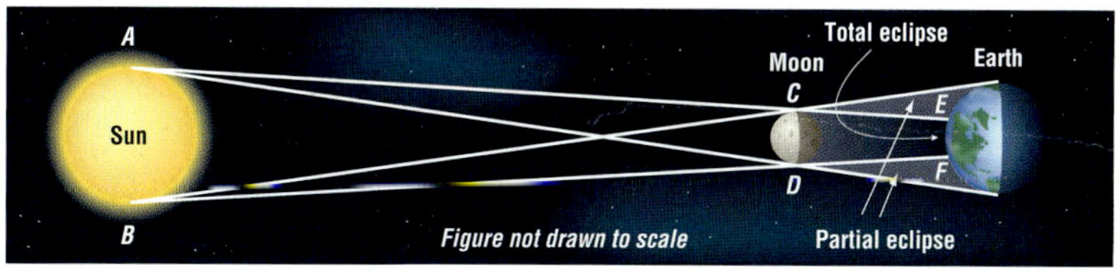

29. The blue section denotes a total eclipse on that part of Earth. Which tangents define the blue area? $\overline{AE}$ and $\overline{BF}$

30. The pink areas denote the portion of Earth that will have a partial eclipse. Which tangents define the northern and southern boundaries of the partial eclipse? $\overline{AD}$ and $\overline{BC}$

31. **CRITICAL THINKING** Find the measure of tangent $\overline{GN}$. Explain your reasoning. See p. 589B.

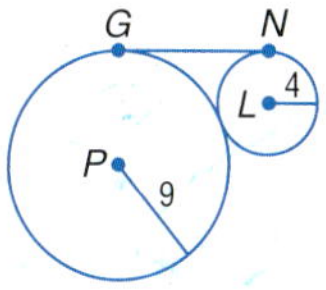

www.geometryonline.com/self_check_quiz/fcat

Study Guide and Intervention, p. 565 (shown) and p. 566

Tangents A tangent to a circle intersects the circle in exactly one point, called the **point of tangency**. There are three important relationships involving tangents.

- If a line is tangent to a circle, then it is perpendicular to the radius drawn to the point of tangency.
- If a line is perpendicular to a radius of a circle at its endpoint on the circle, then the line is a tangent to the circle.
- If two segments from the same exterior point are tangent to a circle, then they are congruent.

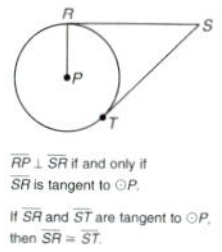

$\overline{RP} \perp \overline{SR}$ if and only if $\overline{SR}$ is tangent to $\odot P$.

If $\overline{SR}$ and $\overline{ST}$ are tangent to $\odot P$, then $\overline{SR} \cong \overline{ST}$.

Example $\overline{AB}$ **is tangent to** $\odot C$**. Find** x**.**

$\overline{AB}$ is tangent to $\odot C$, so $\overline{AB}$ is perpendicular to radius $\overline{BC}$. $\overline{CD}$ is a radius, so $CD = 8$ and $AC = 9 + 8$ or 17. Use the Pythagorean Theorem with right $\triangle ABC$.

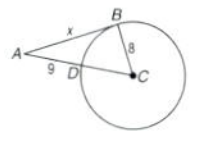

$(AB)^2 + (BC)^2 = (AC)^2$ Pythagorean Theorem
$x^2 + 8^2 = 17^2$ Substitution
$x^2 + 64 = 289$ Multiply.
$x^2 = 225$ Subtract 64 from each side.
$x = 15$ Take the square root of each side.

Exercises

Find x**. Assume that segments that appear to be tangent are tangent.**

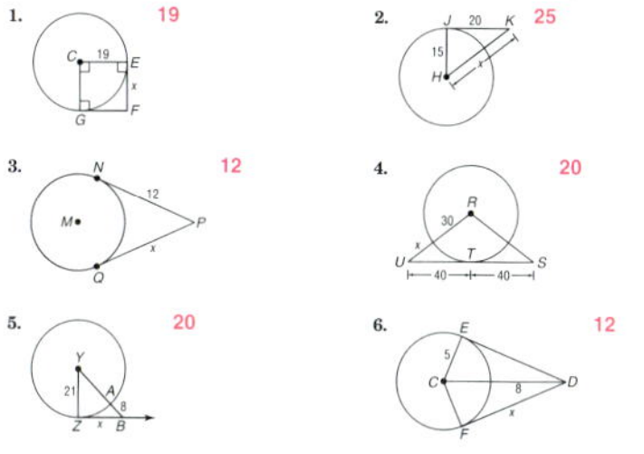

1. 19
2. 25
3. 12
4. 20
5. 20
6. 12

Skills Practice, p. 567 and Practice, p. 568 (shown)

Determine whether each segment is tangent to the given circle.

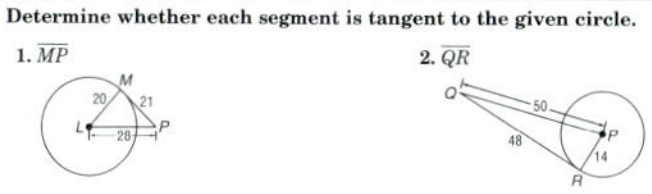

1. $\overline{MP}$ no
2. $\overline{QR}$ yes

Find x**. Assume that segments that appear to be tangent are tangent.**

3. 2
4. $5\sqrt{13}$

Find the perimeter of each polygon for the given information. Assume that segments that appear to be tangent are tangent.

5. $CD = 52$, $CU = 18$, $TB = 12$ 128 units
6. $KG = 32$, $HG = 56$ 154 units

CLOCKS **For Exercises 7 and 8, use the following information.**
The design shown in the figure is that of a circular clock face inscribed in a triangular base. AF and FC are equal.

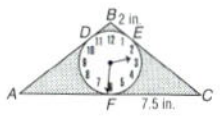

7. Find AB. 9.5 in.

8. Find the perimeter of the clock. 34 in.

Reading to Learn Mathematics, p. 569 ELL

Pre-Activity **How are tangents related to track and field events?**

Read the introduction to Lesson 10-5 at the top of page 552 in your textbook.

How is the hammer throw event related to the mathematical concept of a tangent line?
Sample answer: When the hammer is released, its initial path is a good approximation of a tangent line to the circular path around which it was traveling just before it was released.

Reading the Lesson

1. Refer to the figure. Name each of the following in the figure.

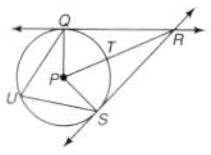

a. two lines that are tangent to $\odot P$ $\overleftrightarrow{RQ}$ and $\overleftrightarrow{RS}$
b. two points of tangency Q, S
c. two chords of the circle $\overline{UQ}$ and $\overline{US}$
d. three radii of the circle $\overline{PQ}$, $\overline{PS}$, and $\overline{PT}$
e. two right angles $\angle PQR$ and $\angle PSR$
f. two congruent right triangles $\triangle PQR$ and $\triangle PSR$
g. the hypotenuse or hypotenuses in the two congruent right triangles $\overline{PR}$
h. two congruent central angles $\angle QPT$ and $\angle SPT$
i. two congruent minor arcs $\overset{\frown}{QT}$ and $\overset{\frown}{ST}$
j. an inscribed angle $\angle QUS$

2. Explain the difference between an *inscribed polygon* and a *circumscribed polygon*. Use the words *vertex* and *tangent* in your explanation.
Sample answer: If a polygon is *inscribed* in a circle, every vertex of the polygon lies on the circle. If a polygon is *circumscribed* about a circle, every side of the polygon is tangent to the circle.

Helping You Remember

3. A good way to remember a mathematical term is to relate it to a word or expression that is used in a nonmathematical way. Sometimes a word or expression used in English is derived from a mathematical term. What does it mean to "go off on a tangent," and how is this meaning related to the geometric idea of a *tangent* line?
Sample answer: To "go off on a tangent" means to suddenly change the subject when you are talking or writing. You can visualize this as being like a tangent line "going off" from a circle as you go farther from the point of tangency.

Enrichment, p. 570

Tangent Circles

Two circles in the same plane are **tangent circles** if they have exactly one point in common. Tangent circles with no common interior points are **externally tangent**. If tangent circles have common interior points, then they are **internally tangent**. Three or more circles are **mutually tangent** if each pair of them are tangent.

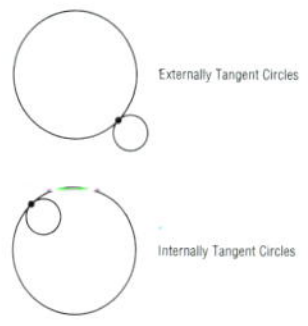

1. Make sketches to show all possible positions of three mutually tangent circles.

Lesson 10-5 Tangents 557

4 Assess

Open-Ended Assessment

Writing Provide an example on the board with a triangle formed by a tangent, a radius, and the line from the center of the circle to a point on the tangent. Assign lengths to the figure and ask students to write the equation necessary to solve the problem. Have a volunteer write his or her equation on the board, and allow students to check their work. Repeat for the other concepts presented in this lesson.

Getting Ready for Lesson 10-6

Prerequisite Skill Students will learn about secants, tangents, and angle measures in Lesson 10-6. They will apply concepts of solving equations to writing proofs and finding values. Use Exercises 42–45 to determine your students' familiarity with solving equations.

Answers

32. Sample answer: Many of the field events have the athlete moving in a circular motion and releasing an object (discus, hammer, shot). The movement of the athlete models a circle and the path of the released object models a tangent. Answers should include the following.
    - The arm of the thrower, the handle, the wire, and hammer form the radius defining the circle when the hammer is spun around. The tangent is the path of the hammer when it is released.
    - The distance the hammer was from the athlete was about 70.68 meters.

41. Sample answer:
Given: *ABCD* is a rectangle. *E* is the midpoint of $\overline{AB}$.
Prove: $\triangle CED$ is isosceles.

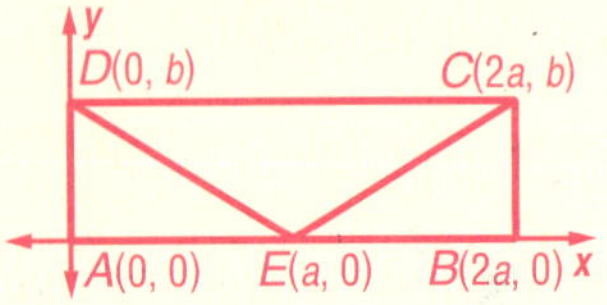

Proof: Let the coordinates of *E* be $(a, 0)$. Since *E* is the midpoint and is halfway between *A* and *B*, the coordinates of *B* will be $(2a, 0)$. Let the coordinates of *D* be $(0, b)$. The coordinates of *C* will be $(2a, b)$ because it is on the same horizontal as *D* and the same vertical as *B*.

$ED = \sqrt{(a-0)^2 + (0-b)^2} = \sqrt{a^2 + b^2}$

$EC = \sqrt{(a-2a)^2 + (0-b)^2} = \sqrt{a^2 + b^2}$

Since $ED = EC$, $\overline{ED} \cong \overline{EC}$. $\triangle DEC$ has two congruent sides, so it is isosceles.

32. **WRITING IN MATH** Answer the question that was posed at the beginning of the lesson. See margin.

How are tangents related to track and field events?

Include the following in your answer:
- how the hammer throw models a tangent, and
- the distance the hammer landed from the athlete if the wire and handle are 1.2 meters long and the athlete's arm is 0.8 meter long.

FCAT Practice
Standardized Test Practice
Ⓐ Ⓑ Ⓒ Ⓓ

33. **GRID IN** $\overline{AB}$, $\overline{BC}$, $\overline{CD}$, and $\overline{AD}$ are tangent to a circle. If $AB = 19$, $BC = 6$, and $CD = 14$, find AD. 27

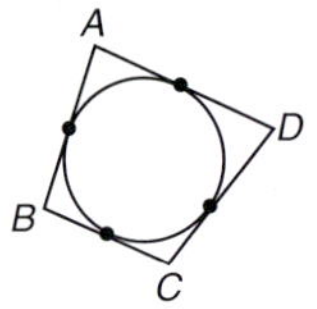

34. **ALGEBRA** Find the mean of all of the numbers from 1 to 1000 that end in 2. B

Ⓐ 496 Ⓑ 497
Ⓒ 498 Ⓓ 500

Extending the Lesson

A line that is tangent to two circles in the same plane is called a *common tangent.*

Common internal tangents intersect the segment connecting the centers.	***Common external tangents*** do not intersect the segment connecting the centers.
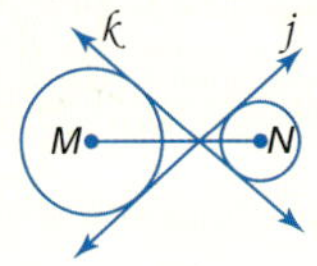 Lines *k* and *j* are common internal tangents.	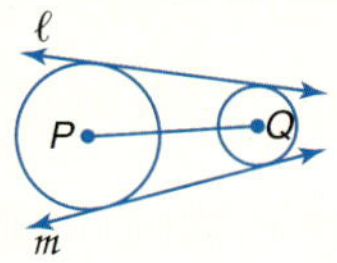 Lines ℓ and *m* are common external tangents.

Refer to the diagram of the eclipse on page 557.

35. Name two common internal tangents. $\overleftrightarrow{AD}$ and $\overleftrightarrow{BC}$

36. Name two common external tangents. $\overleftrightarrow{AE}$ and $\overleftrightarrow{BF}$

Maintain Your Skills

Mixed Review

37. **LOGOS** Circles are often used in logos for commercial products. The logo at the right shows two inscribed angles and two central angles. If $\widehat{AC} \cong \widehat{BD}$, $m\widehat{AF} = 90$, $m\widehat{FE} = 45$, and $m\widehat{ED} = 90$, find $m\angle AFC$ and $m\angle BED$. *(Lesson 10-4)* 45, 45

Find each measure. *(Lesson 10-3)*

38. x

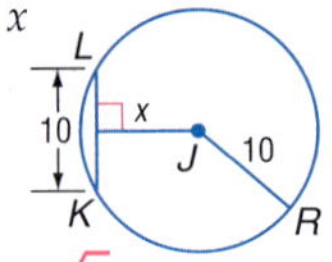

$5\sqrt{3} \approx 8.7$

39. BC 4

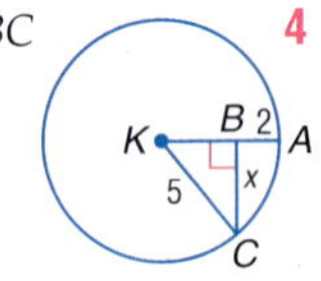

40. AP $8\sqrt{3} \approx 13.9$

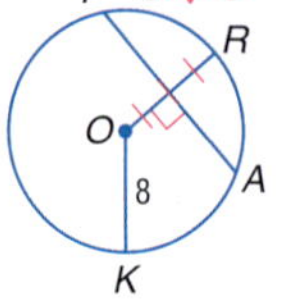

41. **PROOF** Write a coordinate proof to show that if *E* is the midpoint of $\overline{AB}$ in rectangle *ABCD*, then $\triangle CED$ is isosceles. *(Lesson 8-7)* See margin.

Getting Ready for the Next Lesson

PREREQUISITE SKILL Solve each equation.
*(To review **solving equations**, see pages 737 and 738.)*

42. $x + 3 = \frac{1}{2}[(4x + 6) - 10]$ 5

43. $2x - 5 = \frac{1}{2}[(3x + 16) - 20]$ 6

44. $2x + 4 = \frac{1}{2}[(x + 20) - 10]$ $\frac{2}{3}$

45. $x + 3 = \frac{1}{2}[(4x + 10) - 45]$ 20.5

Geometry Activity

A Follow-Up of Lessons 10-4 and 10-5

Inscribed and Circumscribed Triangles

In Lesson 5-1, you learned that there are special points of concurrency in a triangle. Two of these will be used in these activities.

- The *incenter* is the point at which the angle bisectors meet. It is equidistant from the sides of the triangle.
- The *circumcenter* is the point at which the perpendicular bisectors of the sides intersect. It is equidistant from the vertices of the triangle.

Activity 1

Construct a circle inscribed in a triangle. *The triangle is circumscribed about the circle.*

1. Draw a triangle and label its vertices A, B, and C. Construct two angle bisectors of the triangle to locate the incenter. Label it D.

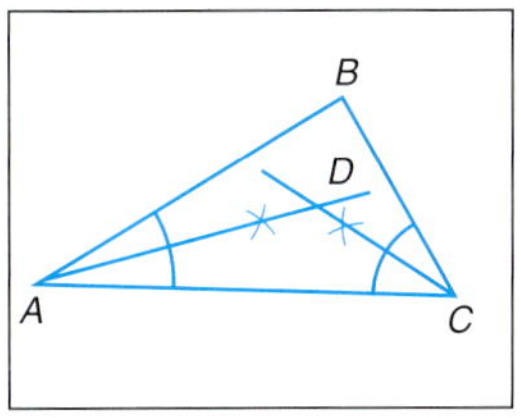

2. Construct a segment perpendicular to a side of $\triangle ABC$ through the incenter. Label the intersection E.

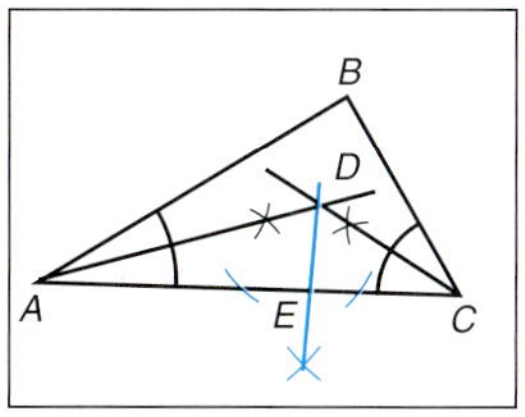

3. Use the compass to measure DE. Then put the point of the compass on D, and draw a circle with that radius.

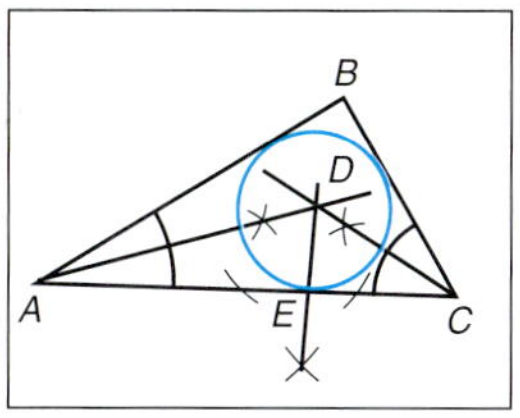

Activity 2

Construct a circle through any three noncollinear points.
This construction may be referred to as circumscribing a circle about a triangle.

1. Draw a triangle and label its vertices A, B, and C. Construct perpendicular bisectors of two sides of the triangle to locate the circumcenter. Label it D.

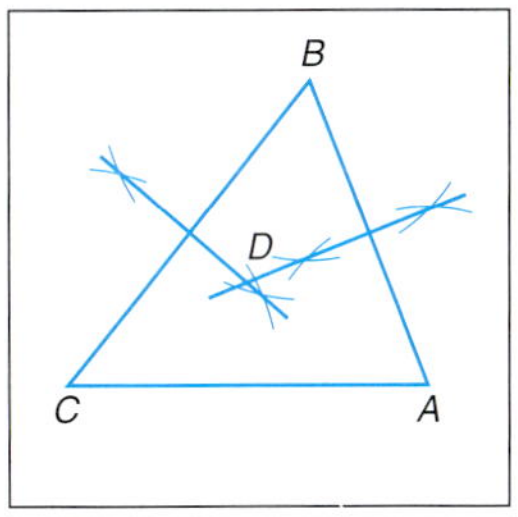

2. Use the compass to measure the distance from the circumcenter D to any of the three vertices.

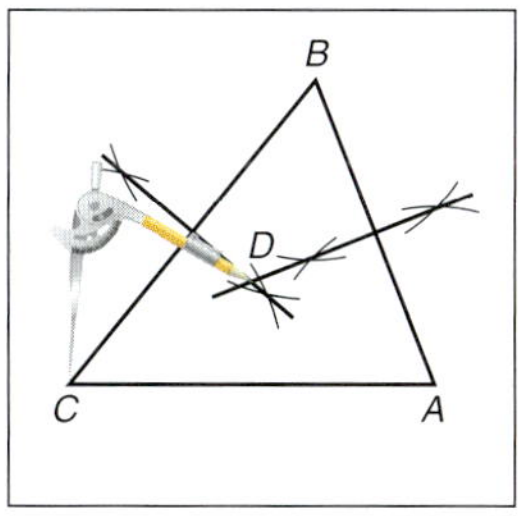

3. Using that setting, place the compass point at D, and draw a circle about the triangle.

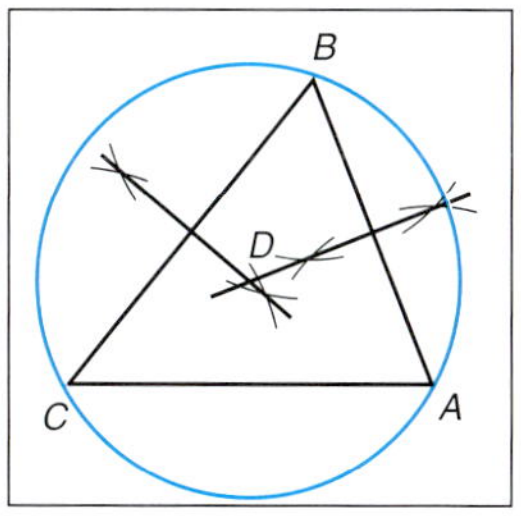

(continued on the next page)

Resource Manager

Teaching Geometry with Manipulatives

- p. 174 (student recording sheet)

Glencoe Mathematics Classroom Manipulative Kit

- compasses
- rulers

Geometry Activity

A Follow-Up of Lessons 10-4 and 10-5

Getting Started

Explain that students will use the incenter of a triangle to construct a circle so that the triangle is circumscribed about the circle, and they will use the circumcenter of a triangle to construct a circle in which the triangle is inscribed. They will also learn how to construct an equilateral triangle circumscribed about a circle.

Objective To construct inscribed and circumscribed triangles.

Materials
straightedge, compass, pencil, paper

Teach

- For Activity 1, have students draw an acute triangle. Have them use an acute triangle or obtuse triangle for Activity 2.
- In Activity 1, remind students that they only need two angle bisectors to locate the incenter because by definition, students know that the third angle bisector would pass through the same point.
- Explain that students must construct the angle bisectors and perpendicular bisectors for these activities because they need very accurate positions for the incenter and circumcenter in order to complete the activities successfully.
- In Activity 3, tell students that the radius is the setting needed to construct six congruent arcs in a circle.

Assess

Exercises 1–3 lead students to repeat the activities for different types of triangles and practice the constructions. Students draw upon their knowledge of circles and use **Exercises 4–10** to analyze their constructions and form a conjecture about the terms *incenter* and *circumcenter*.

Study Notebook

Ask students to summarize what they have learned about inscribed and circumscribed triangles and the terms incenter and circumcenter.

Geometry Activity

For the next activity, refer to the construction of an inscribed regular hexagon on page 542.

Activity 3

Construct an equilateral triangle circumscribed about a circle.

1. Construct a circle and divide it into six congruent arcs.

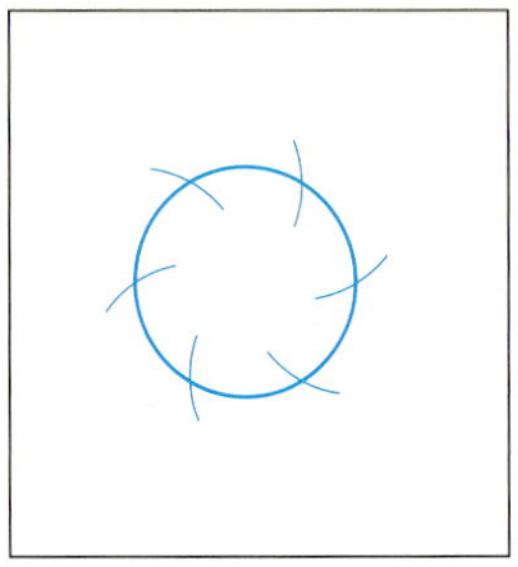

2. Place a point at every other arc. Draw rays from the center through these points.

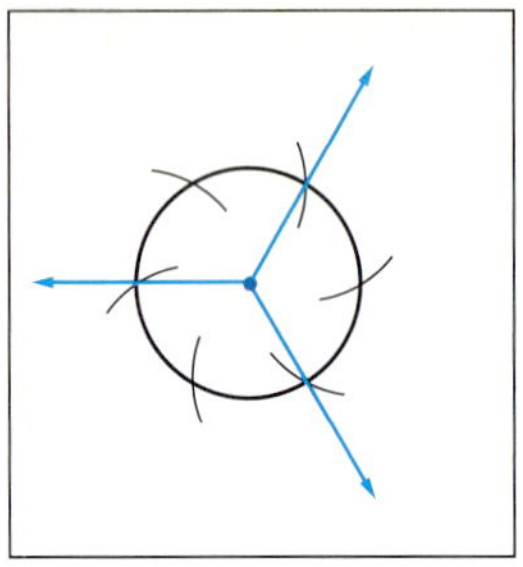

3. Construct a line perpendicular to each of the rays through the points.

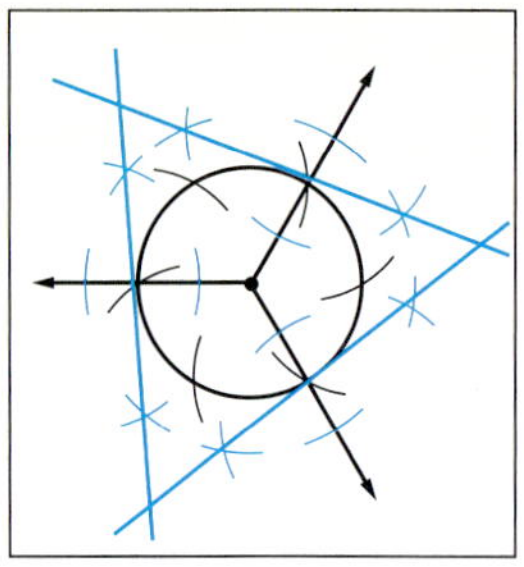

Model

1. Draw an obtuse triangle and inscribe a circle in it. **1–3. See students' work.**

2. Draw a right triangle and circumscribe a circle about it.

3. Draw a circle of any size and circumscribe an equilateral triangle about it.

Analyze

Refer to Activity 1. **4–5. See margin.**

4. Why do you only have to construct the perpendicular to one side of the triangle?

5. How can you use the Incenter Theorem to explain why this construction is valid?

Refer to Activity 2. **6–7. See margin.**

6. Why do you only have to measure the distance from the circumcenter to any one vertex?

7. How can you use the Circumcenter Theorem to explain why this construction is valid?

Refer to Activity 3.

8. What is the measure of each of the six congruent arcs? **60**

9. Write a convincing argument as to why the lines constructed in Step 3 form an equilateral triangle. **See margin.**

10. Why do you think the terms *incenter* and *circumcenter* are good choices for the points they define? **See margin.**

Answers

4. The incenter is equidistant from each side. The perpendicular to one side should be the same length as it is to the other two sides.

5. The incenter is equidistant from all the sides. The radius of the circle is perpendicular to the tangent sides and all radii are congruent, matching the distance from the incenter to the sides.

6. The circumcenter is equidistant from all three vertices, so the distance from the circumcenter to one vertex is the same as the distance to each of the others.

7. The circumcenter is equidistant from the vertices and all of the vertices must lie on the circle. So, this distance is the radius of the circle containing the vertices.

9. Suppose all six radii are drawn. Each central angle measures 60°. Thus, six 30°-60°-90° triangles are formed. Each triangle has a side which is a radius r units long. Using 30°-60°-90° side ratios, the segment tangent to the circle has length $r\sqrt{3}$, making each side of the circumscribed triangle $2r\sqrt{3}$. If all three sides have the same measure, then the triangle is equilateral.

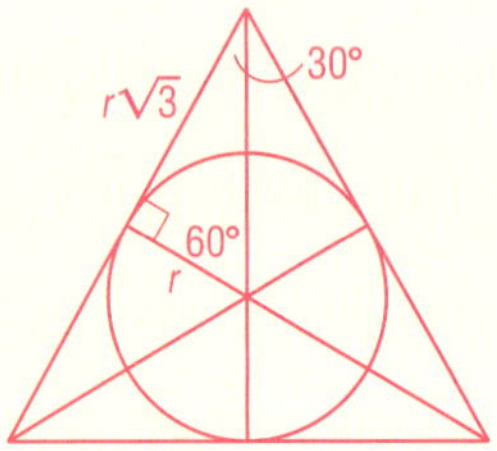

10. The incenter is the point from which you can construct a circle "in" the triangle. Circum means *around*. So the circumcenter is the point from which you can construct a circle "around" the triangle.

10-6 Secants, Tangents, and Angle Measures

What You'll Learn

- Find measures of angles formed by lines intersecting on or inside a circle.
- Find measures of angles formed by lines intersecting outside the circle.

Vocabulary
- secant

How is a rainbow formed by segments of a circle?

Droplets of water in the air refract or bend sunlight as it passes through them, creating a rainbow. The various angles of refraction result in an arch of colors. In the figure, the sunlight from point S enters the raindrop at B and is bent. The light proceeds to the back of the raindrop, and is reflected at C to leave the raindrop at point D heading to Earth. Angle F represents the measure of how the resulting ray of light deviates from its original path.

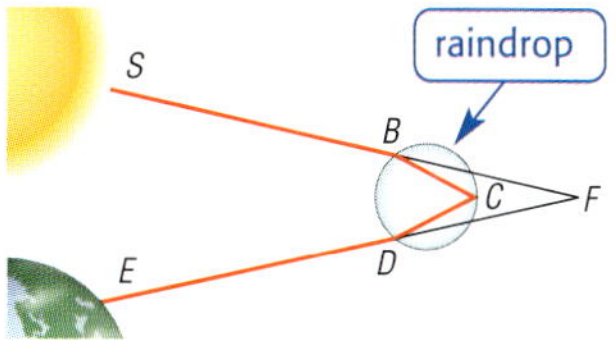

INTERSECTIONS ON OR INSIDE A CIRCLE A line that intersects a circle in exactly two points is called a **secant**. In the figure above, $\overline{SF}$ and $\overline{EF}$ are secants of the circle. When two secants intersect inside a circle, the angles formed are related to the arcs they intercept.

Theorem 10.12

If two secants intersect in the interior of a circle, then the measure of an angle formed is one-half the sum of the measure of the arcs intercepted by the angle and its vertical angle.

Examples: $m\angle 1 = \frac{1}{2}(m\widehat{AC} + m\widehat{BD})$

$m\angle 2 = \frac{1}{2}(m\widehat{AD} + m\widehat{BC})$

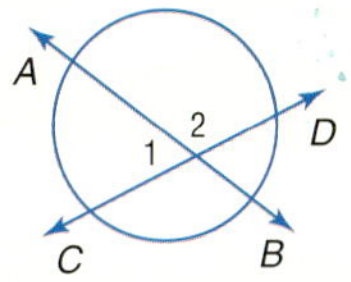

Proof *Theorem 10.12*

Given: secants $\overleftrightarrow{RT}$ and $\overleftrightarrow{SU}$

Prove: $m\angle 1 = \frac{1}{2}(m\widehat{ST} + m\widehat{RU})$

Draw $\overline{RS}$. Label $\angle TRS$ as $\angle 2$ and $\angle USR$ as $\angle 3$.

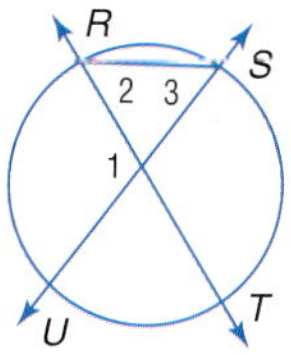

Proof:

Statements	Reasons
1. $m\angle 1 = m\angle 2 + m\angle 3$	1. Exterior Angle Theorem
2. $m\angle 2 = \frac{1}{2}m\widehat{ST}$, $m\angle 3 = \frac{1}{2}m\widehat{RU}$	2. The measure of inscribed $\angle$ = half the measure of the intercepted arc.
3. $m\angle 1 = \frac{1}{2}m\widehat{ST} + \frac{1}{2}m\widehat{RU}$	3. Substitution
4. $m\angle 1 = \frac{1}{2}(m\widehat{ST} + m\widehat{RU})$	4. Distributive Property

10-6 Lesson Notes

1 Focus

5-Minute Check Transparency 10-6 Use as a quiz or review of Lesson 10-5.

Mathematical Background notes are available for this lesson on p. 520D.

How **is a rainbow formed by segments of a circle?**

Ask students:

- If you were to connect B and D in the figure, what would you have? **a triangle inscribed in the circle**
- Name some situations that allow you to see a rainbow formed by segments of a circle. **Sample answers: a rainy, misty day when the sun is low in the sky; a bright sunny day when you are spraying a mist of water from a hose**

Resource Manager

Workbook and Reproducible Masters

Chapter 10 Resource Masters
- Study Guide and Intervention, pp. 571–572
- Skills Practice, p. 573
- Practice, p. 574
- Reading to Learn Mathematics, p. 575
- Enrichment, p. 576
- Assessment, p. 604

Prerequisite Skills Workbook, pp. 17–18

Teaching Geometry With Manipulatives Masters, p. 17

Transparencies

5-Minute Check Transparency 10-6

Answer Key Transparencies

Technology

Interactive Chalkboard

2 Teach

INTERSECTIONS ON OR INSIDE A CIRCLE

In-Class Examples

Power Point®

1 Find $m\angle 4$ if $m\widehat{FG} = 88$ and $m\widehat{EH} = 76$. **98**

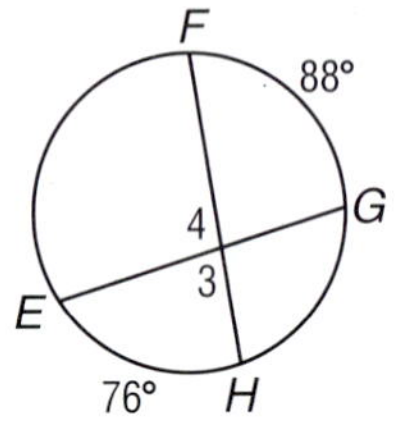

2 Find $m\angle RPS$ if $m\widehat{PT} = 114$ and $m\widehat{TS} = 136$. **55**

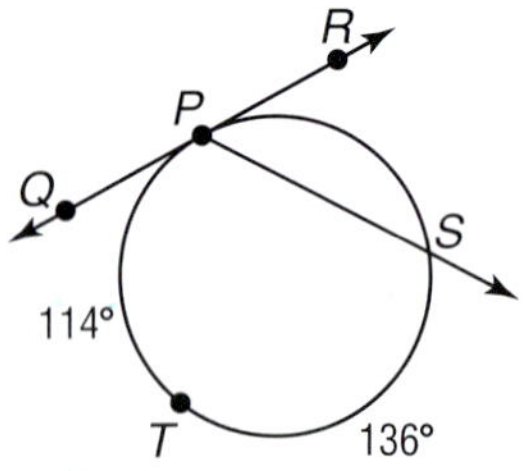

Tips for New Teachers Some students may ask you what the difference is between chords and secants and why there are two names for something that intersects a circle at two points. You may want to review how segments are parts of lines and explain that chords are segments of secants, which are lines that intersect circles. Tell students that every chord lies on a secant and that every secant contains a chord.

Example 1 Secant-Secant Angle

Find $m\angle 2$ if $m\widehat{BC} = 30$ and $m\widehat{AD} = 20$.

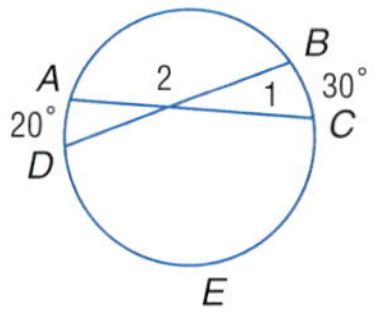

Method 1

$m\angle 1 = \frac{1}{2}(m\widehat{BC} + m\widehat{AD})$

$= \frac{1}{2}(30 + 20)$ or 25 Substitution

$m\angle 2 = 180 - m\angle 1$

$= 180 - 25$ or 155

Method 2

$m\angle 2 = \frac{1}{2}(m\widehat{AB} + m\widehat{DEC})$

Find $m\widehat{AB} + m\widehat{DEC}$.

$m\widehat{AB} + m\widehat{DEC} = 360 - (m\widehat{BC} + m\widehat{AD})$

$= 360 - (30 + 20)$

$= 360 - 50$ or 310

$m\angle 2 = \frac{1}{2}(m\widehat{AB} + m\widehat{DEC})$

$= \frac{1}{2}(310)$ or 155

A secant can also intersect a tangent at the point of tangency. Angle ABC intercepts $\widehat{BC}$, and $\angle DBC$ intercepts $\widehat{BEC}$. Each angle formed has a measure half that of the arc it intercepts.

$m\angle ABC = \frac{1}{2}m\widehat{BC}$ $\quad m\angle DBC = \frac{1}{2}m\widehat{BEC}$

This is stated formally in Theorem 10.13.

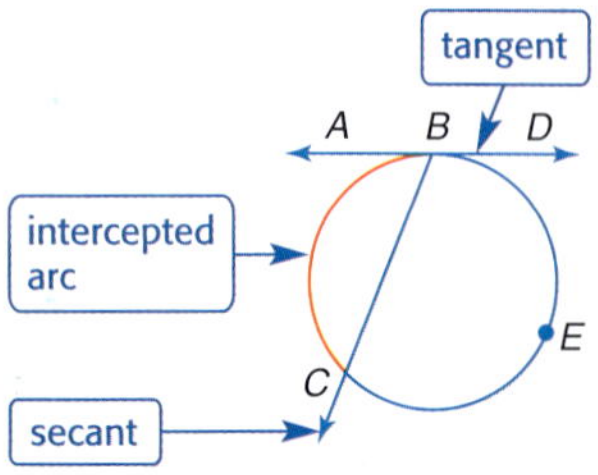

Theorem 10.13

If a secant and a tangent intersect at the point of tangency, then the measure of each angle formed is one-half the measure of its intercepted arc.

You will prove this theorem in Exercise 43.

Example 2 Secant-Tangent Angle

Find $m\angle ABC$ if $m\widehat{AB} = 102$.

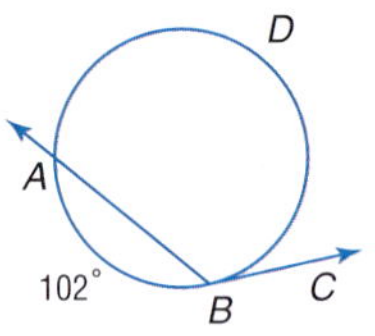

$m\widehat{ADB} = 360 - m\widehat{AB}$

$= 360 - 102$ or 258

$m\angle ABC = \frac{1}{2}m\widehat{ADC}$

$= \frac{1}{2}(258)$ or 129

DAILY INTERVENTION **Differentiated Instruction**

Naturalist Explain that the relationships presented in this chapter are naturally occurring relationships that have been mathematically defined and explained. Tell students that scientists from all fields can use these relationships to examine everything from raindrops and soap bubbles to cells and microorganisms.

INTERSECTIONS OUTSIDE A CIRCLE Secants and tangents can also meet outside a circle. The measure of the angle formed also involves half of the measures of the arcs they intercept.

Study Tip

Absolute Value
The measure of each $\angle A$ can also be expressed as one-half the absolute value of the difference of the arc measures. In this way, the order of the arc measures does not affect the outcome of the calculation.

Theorem 10.14

If two secants, a secant and a tangent, or two tangents intersect in the exterior of a circle, then the measure of the angle formed is one-half the positive difference of the measures of the intercepted arcs.

Two Secants	Secant-Tangent	Two Tangents
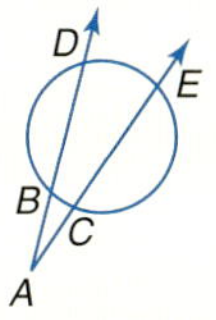	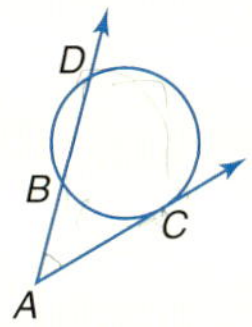	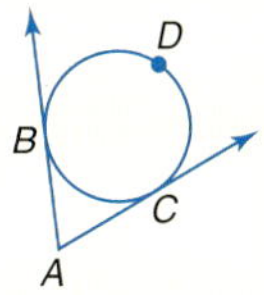
$m\angle A = \frac{1}{2}(m\widehat{DE} - m\widehat{BC})$	$m\angle A = \frac{1}{2}(m\widehat{DC} - m\widehat{BC})$	$m\angle A = \frac{1}{2}(m\widehat{BDC} - m\widehat{BC})$

You will prove this theorem in Exercise 40.

Example 3 Secant-Secant Angle

Find x.

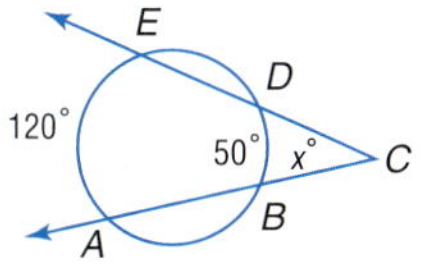

$m\angle C = \frac{1}{2}(m\widehat{EA} - m\widehat{DB})$

$x = \frac{1}{2}(120 - 50)$ Substitution

$x = \frac{1}{2}(70)$ or 35 Simplify.

Example 4 Tangent-Tangent Angle

SATELLITES **Suppose a geostationary satellite S orbits about 35,000 kilometers above Earth rotating so that it appears to hover directly over the equator. Use the figure to determine the arc measure on the equator visible to this geostationary satellite.**

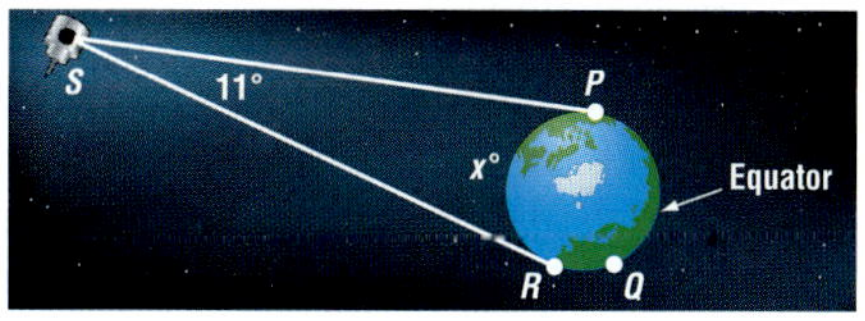

$\widehat{PR}$ represents the arc along the equator visible to the satellite S. If $x = m\widehat{PR}$, then $m\widehat{PQR} = 360 - x$. Use the measure of the given angle to find $m\widehat{PR}$.

$m\angle S = \frac{1}{2}(m\widehat{PQR} - m\widehat{PR})$

$11 = \frac{1}{2}[(360 - x) - x]$ Substitution

$22 = 360 - 2x$ Multiply each side by 2 and simplify.

$-338 = -2x$ Subtract 360 from each side.

$169 = x$ Divide each side by -2.

The measure of the arc on Earth visible to the satellite is 169.

www.geometryonline.com/extra_examples/fcat

INTERSECTIONS OUTSIDE A CIRCLE

In-Class Examples Power Point®

3 Find x. 17

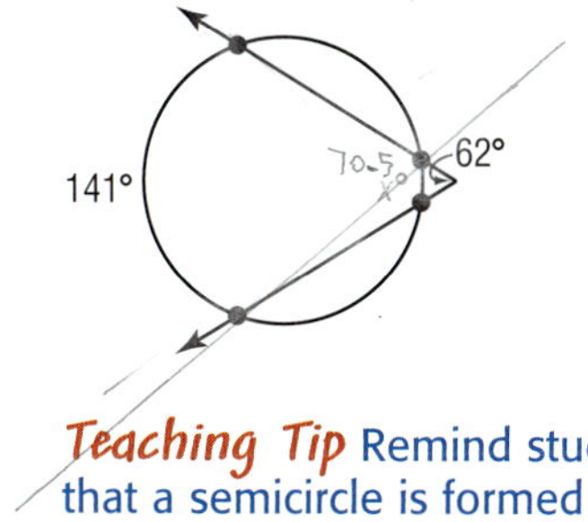

Teaching Tip Remind students that a semicircle is formed when a chord passes through the center of the circle to make a diameter. Explain that students should always examine a figure for known information that might not be labeled.

4 **JEWELRY** A jeweler wants to craft a pendant with the shape shown. Use the figure to determine the measure of the arc at the bottom of the pendant. 220

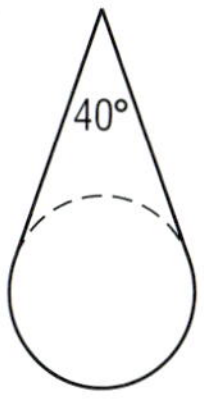

Lesson 10-6 Secants, Tangents, and Angle Measures 563

In-Class Example

Power Point®

5 Find x. 25

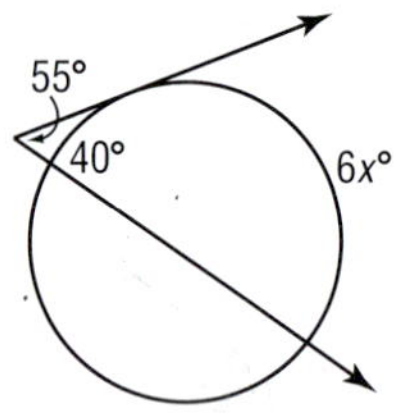

3 Practice/Apply

Study Notebook

Have students—
- add the definitions/examples of the vocabulary terms to their Vocabulary Builder worksheets for Chapter 10.
- include examples to demonstrate the relationships described in the theorems introduced in this lesson.
- include any other item(s) that they find helpful in mastering the skills in this lesson.

About the Exercises...

Organization by Objective
- **Intersections on or Inside a Circle:** 12–19
- **Intersections Outside a Circle:** 21–32

Odd/Even Assignments

Exercises 12–32 are structured so that students practice the same concepts whether they are assigned odd or even problems.

Assignment Guide

Basic: 13, 17–29 odd, 33–45 odd, 46–59

Average: 13–45 odd, 46–59

Advanced: 12–40 even, 41, 42, 44, 45–56 (optional: 57–59)

All: Quiz 2 (1–5)

Example 5 Secant-Tangent Angle

Find x.

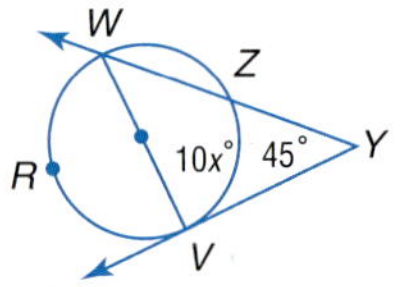

$\widehat{WRV}$ is a semicircle because $\overline{WV}$ is a diameter.

So, $m\widehat{WRV} = 180$.

$m\angle Y = \frac{1}{2}(m\widehat{WV} - m\widehat{ZV})$

$45 = \frac{1}{2}(180 - 10x)$ Substitution

$90 = 180 - 10x$ Multiply each side by 2.

$-90 = -10x$ Subtract 180 from each side.

$9 = x$ Divide each side by -10.

Check for Understanding

Concept Check

1. **Describe** the difference between a secant and a tangent. **1–2. See margin.**
2. **OPEN ENDED** Draw a circle and one of its diameters. Call the diameter $\overline{AC}$. Draw a line tangent to the circle at A. What type of angle is formed by the tangent and the diameter? Explain.

Guided Practice

GUIDED PRACTICE KEY	
Exercises	Examples
3	1
4, 8	2
5	3
6	4
7	5

Find each measure.

3. $m\angle 1$ **138**

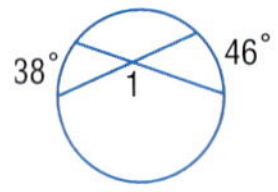

4. $m\angle 2$ **130**

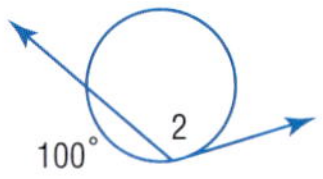

Find x.

5. **20**

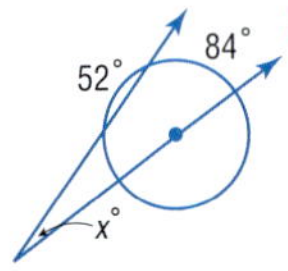

6. **22**

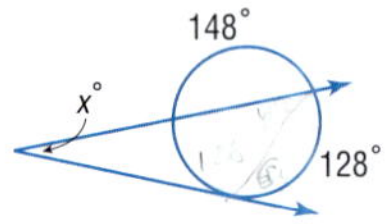

7. **235**

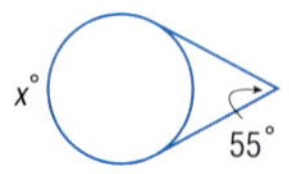

Application

CIRCUS For Exercises 8–11, refer to the figure and the information below.

One of the acrobatic acts in the circus requires the artist to balance on a board that is placed on a round drum as shown at the right. Find each measure if $\overline{SA} \parallel \overline{LK}$, $m\angle SLK = 78$, and $m\widehat{SA} = 46$.

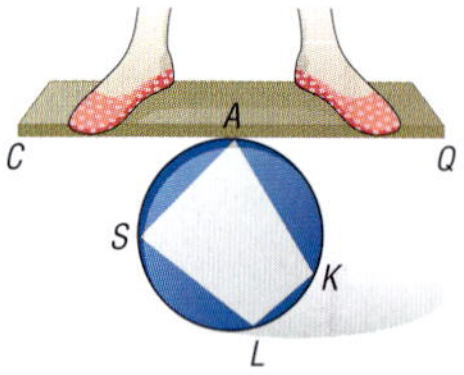

8. $m\angle CAS$ **23**
9. $m\angle QAK$ **55**
10. $m\widehat{KL}$ **94**
11. $m\widehat{SL}$ **110**

★ indicates increased difficulty

Practice and Apply

Find each measure.

12. $m\angle 3$ **110**

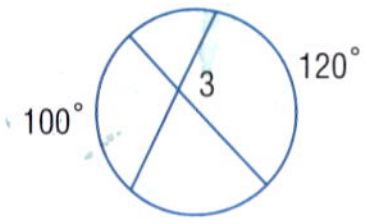

13. $m\angle 4$ **60**

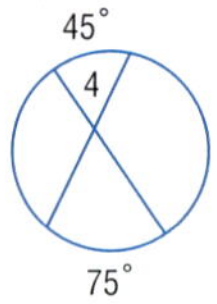

14. $m\angle 5$ **50**

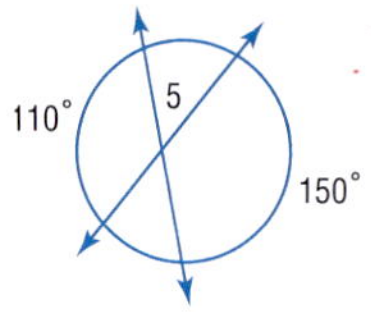

Answers

1. Sample answer: A tangent intersects the circle in only one point and no part of the tangent is in the interior of the circle. A secant intersects the circle in two points and some of its points do lie in the interior of the circle.

2. Sample answer: $\angle TAC$ is a right angle; There are two reasons: (1) If the point of tangency is the endpoint of a diameter, then the tangent is perpendicular to the diameter at that point. (2) The arc intercepted by the secant (diameter) and the tangent is a semicircle. Thus the measure of the angle is half of 180 or 90.

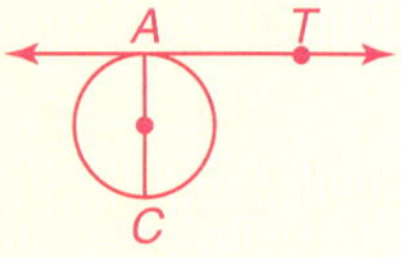

Homework Help	
For Exercises	See Examples
12–15	1
16–19	2
21–24, 31	3
25–28, 32	5
29–30	4

Extra Practice
See page 775.

★ 15. $m\angle 6$ **110**

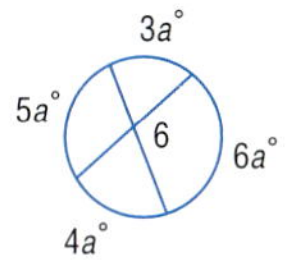

16. $m\angle 7$ **98**

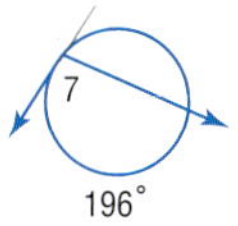

17. $m\angle 8$ **90**

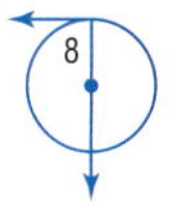

18. $m\angle 9$ **120**

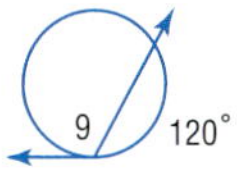

19. $m\angle 10$ **50**

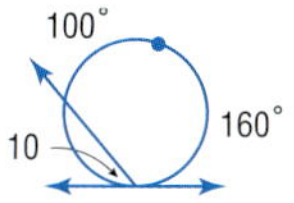

20. $m\widehat{AC}$ **58**

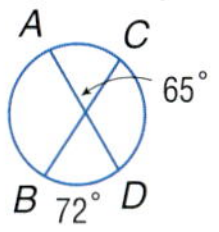

Find x. Assume that any segment that appears to be tangent is tangent.

21. **30**

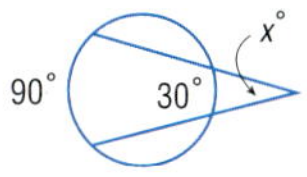

22. **5**

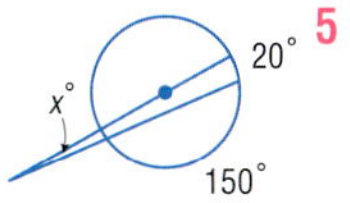

23. **8**

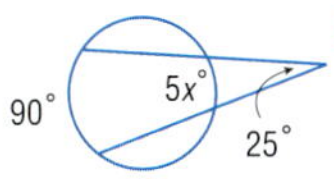

24. **13**

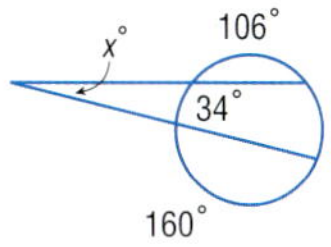

25. **4**

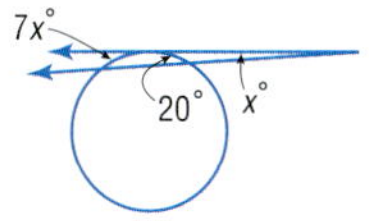

26. **5**

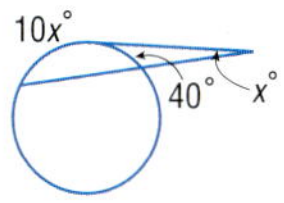

27. **25**

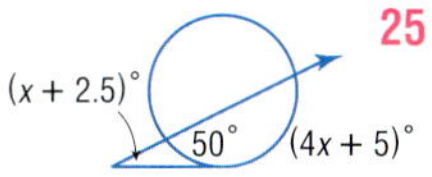

★ 28. **9**

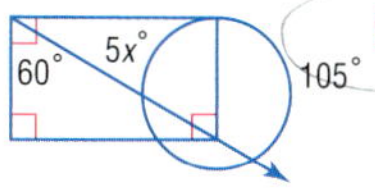

29. **130**

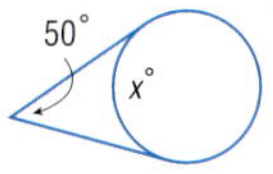

30. **210**

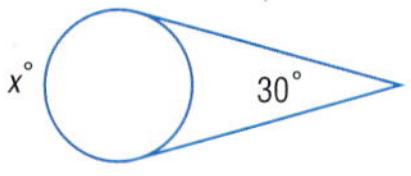

★ 31. **10**

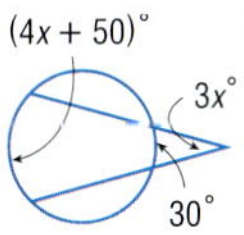

★ 32. **8**

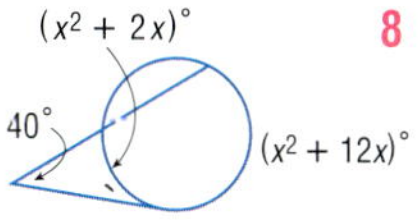

33. **WEAVING** Once yarn is woven from wool fibers, it is often dyed and then threaded along a path of pulleys to dry. One set of pulleys is shown below. Note that the yarn appears to intersect itself at C, but in reality it does not. Use the information from the diagram to find $m\widehat{BH}$. **141**

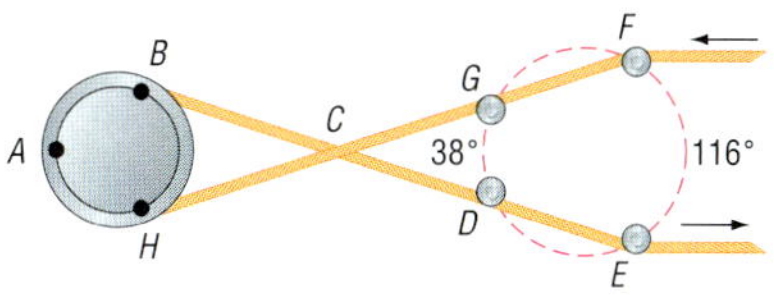

Study Guide and Intervention, p. 571 (shown) and p. 572

Intersections On or Inside a Circle A line that intersects a circle in exactly two points is called a **secant**. The measures of angles formed by secants and tangents are related to intercepted arcs.

- If two secants intersect in the interior of a circle, then the measure of the angle formed is one-half the sum of the measure of the arcs intercepted by the angle and its vertical angle.
- If a secant and a tangent intersect at the point of tangency, then the measure of each angle formed is one-half the measure of its intercepted arc.

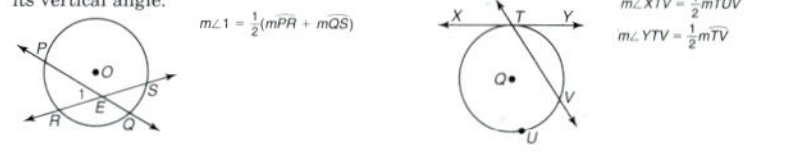

Example 1 **Find x.**

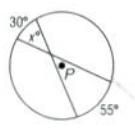

The two secants intersect inside the circle, so x is equal to one-half the sum of the measures of the arcs intercepted by the angle and its vertical angle.

$x = \frac{1}{2}(30 + 55)$
$= \frac{1}{2}(85)$
$= 42.5$

Example 2 **Find y.**

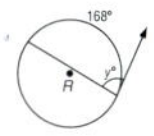

The secant and the tangent intersect at the point of tangency, so the measure the angle is one-half the measure of its intercepted arc.

$y = \frac{1}{2}(168)$
$= 84$

Exercises

Find each measure.

1. $m\angle 1$ 46
2. $m\angle 2$ 46
3. $m\angle 3$ 110
4. $m\angle 4$ 30
5. $m\angle 5$ 70
6. $m\angle 6$ 100

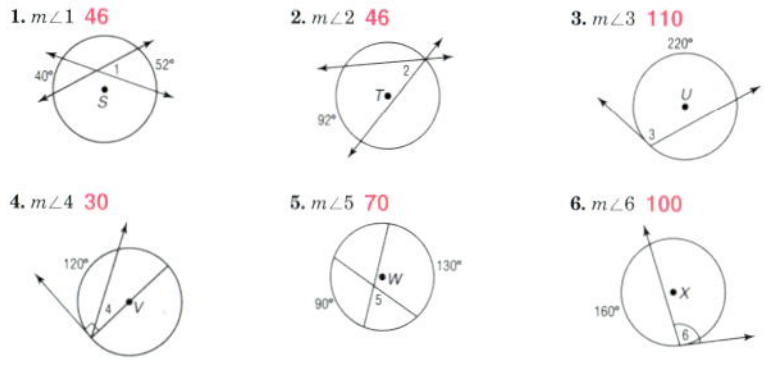

Skills Practice, p. 573 and Practice, p. 574 (shown)

Find each measure.

1. $m\angle 1$ 79
2. $m\angle 2$ 113
3. $m\angle 3$ 72

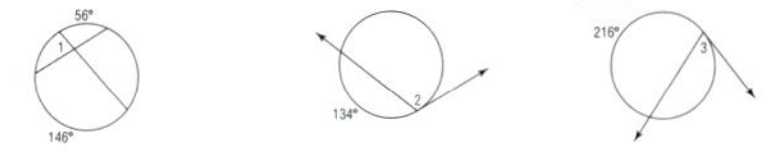

Find x. Assume that any segment that appears to be tangent is tangent.

7. 31
8. 14.5
9. 60

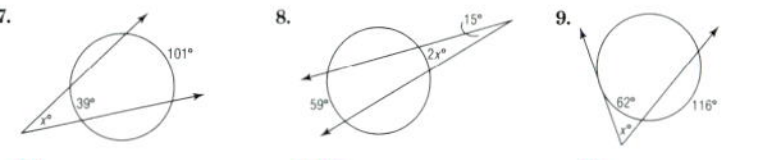

10. 21
11. 128
12. 217

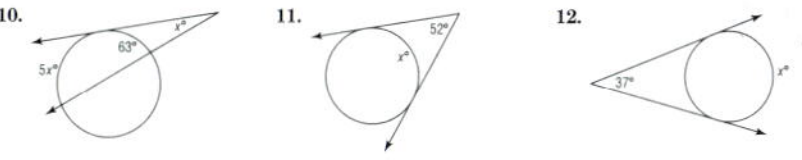

9. **RECREATION** In a game of kickball, Rickie has to kick the ball through a semicircular goal to score. If $m\widehat{XZ} = 58$ and the $m\widehat{XY} = 122$, at what angle must Rickie kick the ball to score? Explain.
Rickie must kick the ball at an angle less than 32° since the measure of the angle from the ground that a tangent would make with the goal post is 32°.

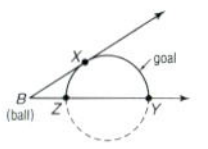

Reading to Learn Mathematics, p. 575 ELL

Pre-Activity **How is a rainbow formed by segments of a circle?**

Read the introduction to Lesson 10-6 at the top of page 561 in your textbook.

- How would you describe $\angle C$ in the figure in your textbook? Sample answer: $\angle C$ is an inscribed angle in the circle that represents the raindrop.
- When you see a rainbow, where is the sun in relation to the circle of which the rainbow is an arc? Sample answer: behind you and opposite the center of the circle

Reading the Lesson

1. Underline the correct word to form a true statement.
   a. A line can intersect a circle in at most (one/two/three) points.
   b. A line that intersects a circle in exactly two points is called a (tangent/secant/radius).
   c. A line that intersects a circle in exactly one point is called a (tangent/secant/radius).
   d. Every secant of a circle contains a (radius/tangent/chord).
2. Determine whether each statement is *always*, *sometimes*, or *never* true.
   a. A secant of a circle passes through the center of the circle. sometimes
   b. A tangent to a circle passes through the center of the circle. never
   c. A secant-secant angle is a central angle of the circle. sometimes
   d. A vertex of a secant-tangent angle is a point on the circle. sometimes
   e. A secant-tangent angle passes through the center of the circle. sometimes
   f. The vertex of a tangent-tangent angle is a point on the circle. never
   g. If one side of a secant-tangent angle passes through the center of the circle, the angle is a right angle. always
   h. The measure of a secant-secant angle is one-half the positive difference of the measures of its intercepted arcs. sometimes
   i. The sum of the measures of the arcs intercepted by a tangent-tangent angle is 360. always
   j. The two arcs intercepted by a tangent-tangent angle are congruent. never

Helping You Remember

4. Some students have trouble remembering the difference between a *secant* and a *tangent*. What is an easy way to remember which is which?
Sample answer: A secant cuts a circle, while a tangent just touches it at one point. You can associate *tangent* with *touches* because they both start with *t*. Then associate *secant* with *cuts*.

Enrichment, p. 576

Orbiting Bodies

The path of the Earth's orbit around the sun is elliptical. However, it is often viewed as circular.

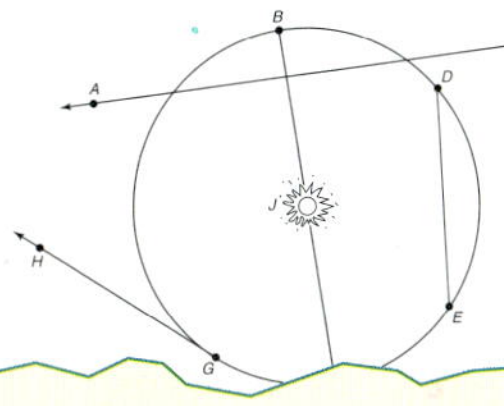

More About. . .

Landmarks

Stonehenge is located in southern England near Salisbury. In its final form, Stonehenge included 30 upright stones about 18 feet tall by 7 feet thick.

Source: *World Book Encyclopedia*

Find each measure if $m\widehat{FE} = 118$, $m\widehat{AB} = 108$, $m\angle EGB = 52$, and $m\angle EFB = 30$.

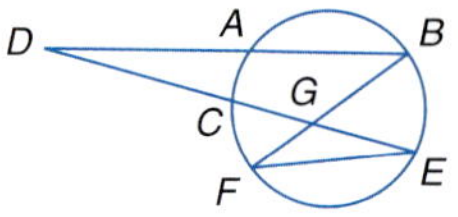

34. $m\widehat{AC}$ 30

35. $m\widehat{CF}$ 44

36. $m\angle EDB$ 15

LANDMARKS **For Exercises 37–39, use the following information.** Stonehenge is a British landmark made of huge stones arranged in a circular pattern that reflects the movements of Earth and the moon. The diagram shows that the angle formed by the north/south axis and the line aligned from the station stone to the northmost moonrise position measures 23.5°.

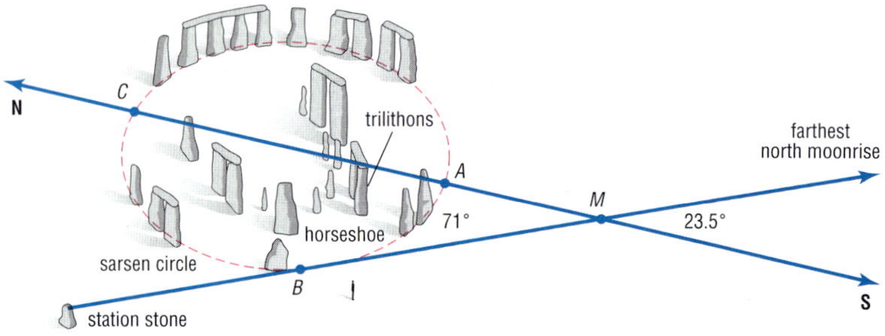

37. Find $m\widehat{BC}$. 118

38. Is $\widehat{ABC}$ a semicircle? Explain. No, its measure is 189.

39. If the circle measures about 100 feet across, approximately how far would you walk around the circle from point B to point C? about 103 ft

40. **PROOF** Write a two-column proof of Theorem 10.14. Consider each case.

a. Case 1: Two Secants See margin.

Given: $\overleftrightarrow{AC}$ and $\overleftrightarrow{AT}$ are secants to the circle.

Prove: $m\angle CAT = \frac{1}{2}(m\widehat{CT} - m\widehat{BR})$

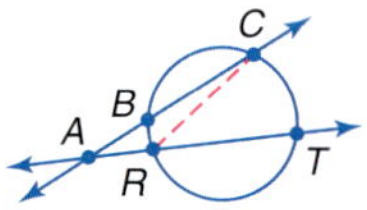

b. Case 2: Secant and a Tangent See margin.

Given: $\overleftrightarrow{DG}$ is a tangent to the circle. $\overleftrightarrow{DF}$ is a secant to the circle.

Prove: $m\angle FDG = \frac{1}{2}(m\widehat{FG} - m\widehat{GE})$

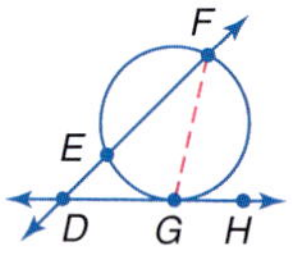

c. Case 3: Two Tangents See margin.

Given: $\overleftrightarrow{HI}$ and $\overleftrightarrow{HJ}$ are tangents to the circle.

Prove: $m\angle IHJ = \frac{1}{2}(m\widehat{IXJ} - m\widehat{IJ})$

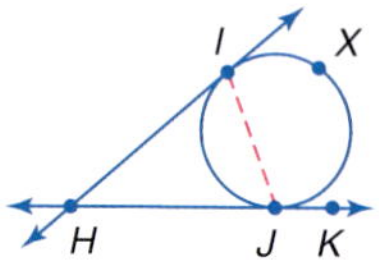

41. **CRITICAL THINKING** Circle E is inscribed in rhombus $ABCD$. The diagonals of the rhombus are 10 centimeters and 24 centimeters long. To the nearest tenth centimeter, how long is the radius of circle E? (*Hint:* Draw an altitude from E.) 4.6cm

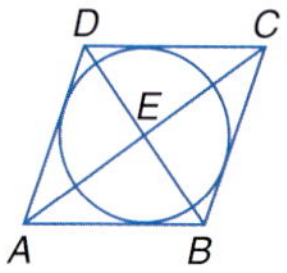

Answers

40a. Given: $\overleftrightarrow{AC}$ and $\overleftrightarrow{AT}$ are secants to the circle.

Prove: $m\angle CAT = \frac{1}{2}(m\widehat{CT} - m\widehat{BR})$

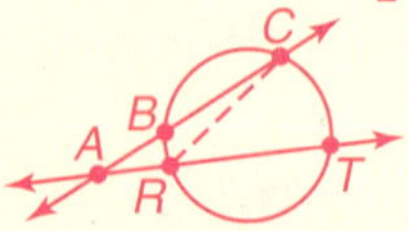

Statements (Reasons)

1. $\overleftrightarrow{AC}$ and $\overleftrightarrow{AT}$ are secants to the circle. (Given)
2. $m\angle CRT = \frac{1}{2}m\widehat{CT}$, $m\angle ACR = \frac{1}{2}m\widehat{BR}$ (The meas. of an inscribed ∠ = $\frac{1}{2}$ the meas. of its intercepted arc.)
3. $m\angle CRT = m\angle ACR + m\angle CAT$ (Exterior ∠ Theorem)
4. $\frac{1}{2}m\widehat{CT} = \frac{1}{2}m\widehat{BR} + m\angle CAT$ (Substitution)
5. $\frac{1}{2}m\widehat{CT} - \frac{1}{2}m\widehat{BR} = m\angle CAT$ (Subtraction Prop.)
6. $\frac{1}{2}(m\widehat{CT} - m\widehat{BR}) = m\angle CAT$ (Distributive Prop.)

40b. Given: $\overleftrightarrow{DG}$ is a tangent to the circle. $\overleftrightarrow{DF}$ is a secant to the circle.

Prove: $m\angle FDG = \frac{1}{2}(m\widehat{FG} - m\widehat{GE})$

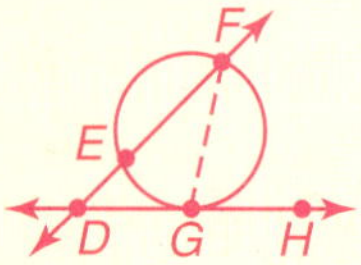

Statements (Reasons)

1. $\overleftrightarrow{DG}$ is a tangent to the circle. $\overleftrightarrow{DF}$ is a secant to the circle. (Given)
2. $m\angle DFG = \frac{1}{2}m\widehat{GE}$, $m\angle FGH = \frac{1}{2}m\widehat{FG}$ (The meas. of an inscribed ∠ = $\frac{1}{2}$ the meas. of its intercepted arc.)
3. $m\angle FGH = m\angle DFG + m\angle FDG$ (Exterior ∠ Theorem)
4. $\frac{1}{2}m\widehat{FG} = \frac{1}{2}m\widehat{GE} + m\angle FDG$ (Substitution)
5. $\frac{1}{2}m\widehat{FG} - \frac{1}{2}m\widehat{GE} = m\angle FDG$ (Subtraction Prop.)
6. $\frac{1}{2}(m\widehat{FG} - m\widehat{GE}) = m\angle FDG$ (Distributive Prop.)

40c. Given: $\overleftrightarrow{HI}$ and $\overleftrightarrow{HJ}$ are tangents to the circle.

Prove: $m\angle IHJ = \frac{1}{2}(m\widehat{IXJ} - m\widehat{IJ})$

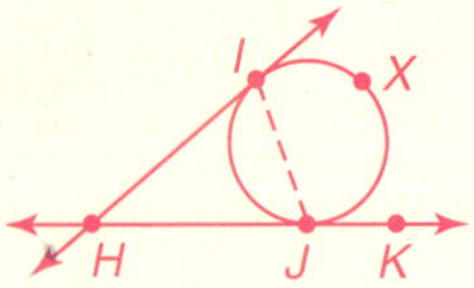

Statements (Reasons)

1. $\overleftrightarrow{HI}$ and $\overleftrightarrow{HJ}$ are tangents to the circle. (Given)
2. $m\angle IJK = \frac{1}{2}m\widehat{IXJ}$, $m\angle HIJ = \frac{1}{2}m\widehat{IJ}$ (The measure of an inscribed ∠ = $\frac{1}{2}$ the measure of its intercepted arc.)
3. $m\angle IJK = m\angle HIJ + m\angle IHJ$ (Ext. ∠ Th.)
4. $\frac{1}{2}m\widehat{IXJ} = \frac{1}{2}m\widehat{IJ} + m\angle IHJ$ (Substitution)
5. $\frac{1}{2}m\widehat{IXJ} - \frac{1}{2}m\widehat{IJ} = m\angle IHJ$ (Subtr. Prop.)
6. $\frac{1}{2}(m\widehat{IXJ} - m\widehat{IJ}) = m\angle IHJ$ (Distrib. Prop.)

566 Chapter 10 Circles

42. **TELECOMMUNICATION** The signal from a telecommunication tower follows a ray that has its endpoint on the tower and is tangent to Earth. Suppose a tower is located at sea level as shown in the figure. Determine the measure of the arc intercepted by the two tangents. **93.5**

Note: Art not drawn to scale

43. **PROOF** Write a paragraph proof of Theorem 10.13

a. **Given:** $\overleftrightarrow{AB}$ is a tangent of $\odot O$.
$\overrightarrow{AC}$ is a secant of $\odot O$.
$\angle CAB$ is acute.

Prove: $m\angle CAB = \frac{1}{2}m\overset{\frown}{CA}$

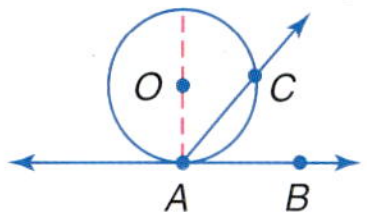

b. Prove Theorem 10.13 if the angle in part **a** is obtuse. **See margin.**

44. **SATELLITES** A satellite is orbiting so that it maintains a constant altitude above the equator. The camera on the satellite can detect an arc of 6000 kilometers on Earth's surface. This arc measures 54°. What is the measure of the angle of view of the camera located on the satellite? **126**

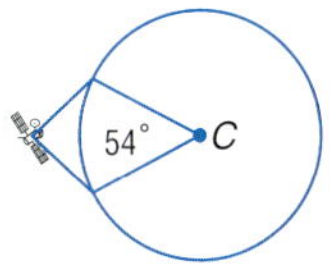

45. **CRITICAL THINKING** In the figure, $\angle 3$ is a central angle. List the numbered angles in order from greatest measure to least measure. Explain your reasoning.

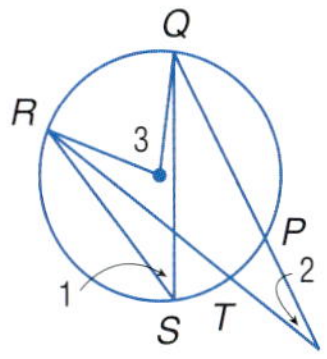

46. **WRITING IN MATH** Answer the question that was posed at the beginning of the lesson. **See margin.**

How is a rainbow formed by segments of a circle?

Include the following in your answer:
- the types of segments represented in the figure on page 561, and
- how you would calculate the angle representing how the light deviates from its original path.

FCAT Practice
Standardized Test Practice

47. What is the measure of $\angle B$ if $m\angle A = 10$? **A**

Ⓐ 30 Ⓑ 35
Ⓒ 47.5 Ⓓ 90

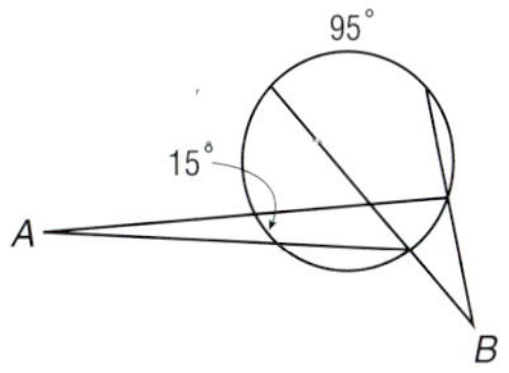

48. **ALGEBRA** Which of the following sets of data can be represented by a linear equation? **C**

Ⓐ

x	y
1	2
2	4
3	8
4	16

Ⓑ

x	y
1	4
2	2
3	2
4	4

Ⓒ

x	y
2	2
4	3
6	4
8	5

Ⓓ

x	y
1	1
3	9
5	25
7	49

45. $\angle 3, \angle 1, \angle 2$; $m\angle 3 = m\overset{\frown}{RQ}$, $m\angle 1 = \frac{1}{2}m\overset{\frown}{RQ}$ so $m\angle 3 > m\angle 1$, $m\angle 2 = \frac{1}{2}(m\overset{\frown}{RQ} - m\overset{\frown}{TP}) = \frac{1}{2}m\overset{\frown}{RQ} - \frac{1}{2}m\overset{\frown}{TP}$, which is less than $\frac{1}{2}m\overset{\frown}{RQ}$, so $m\angle 2 < m\angle 1$.

46. Sample answer: Each raindrop refracts light from the sun and sends the beam to Earth. The raindrop is actually spherical, but the angle of the light is an inscribed angle from the bent rays. Answers should include the following.
- $\angle C$ is an inscribed angle and $\angle F$ is a secant-secant angle.
- The measure of $\angle F$ can be calculated by finding the positive difference between $m\overset{\frown}{BD}$ and the measure of the small intercepted arc containing point C.

Answers

43a. Given: $\overleftrightarrow{AB}$ is a tangent to $\odot O$. $\overrightarrow{AC}$ is a secant to $\odot O$. $\angle CAB$ is acute.
Prove: $m\angle CAB = \frac{1}{2}m\overset{\frown}{CA}$

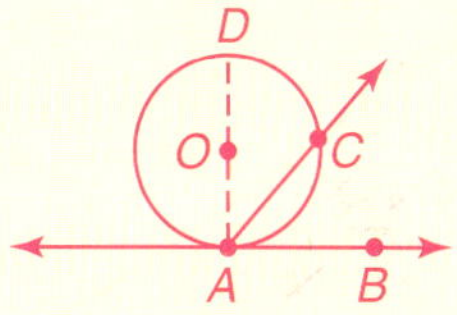

Proof: $\angle DAB$ is a right $\angle$ with measure 90, and $\overset{\frown}{DCA}$ is a semicircle with measure 180, since if a line is tangent to a $\odot$, it is $\perp$ to the radius at the point of tangency. Since $\angle CAB$ is acute, C is in the interior of $\angle DAB$, so by the Angle and Arc Addition Postulates, $m\angle DAB = m\angle DAC + m\angle CAB$ and $m\overset{\frown}{DCA} = m\overset{\frown}{DC} + m\overset{\frown}{CA}$. By substitution, $90 = m\angle DAC + m\angle CAB$ and $180 = m\overset{\frown}{DC} + m\overset{\frown}{CA}$. So, $90 = \frac{1}{2}m\overset{\frown}{DC} + \frac{1}{2}m\overset{\frown}{CA}$ by Division Prop., and $m\angle DAC + m\angle CAB = \frac{1}{2}m\overset{\frown}{DC} + \frac{1}{2}m\overset{\frown}{CA}$ by substitution. $m\angle DAC = \frac{1}{2}m\overset{\frown}{DC}$ since $\angle DAC$ is inscribed, so substitution yields $\frac{1}{2}m\overset{\frown}{DC} + m\angle CAB = \frac{1}{2}m\overset{\frown}{DC} + \frac{1}{2}m\overset{\frown}{CA}$. By Subtraction Prop., $m\angle CAB = \frac{1}{2}m\overset{\frown}{CA}$.

43b. Given: $\overleftrightarrow{AB}$ is a tangent to $\odot O$. $\overrightarrow{AC}$ is a secant to $\odot O$. $\angle CAB$ is obtuse.
Prove: $m\angle CAB = \frac{1}{2}m\overset{\frown}{CDA}$

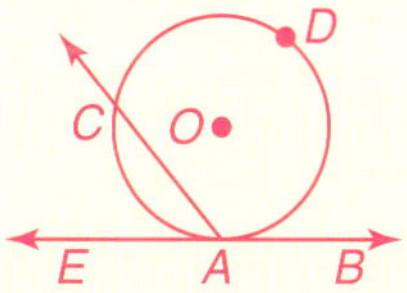

Proof: $\angle CAB$ and $\angle CAE$ form a linear pair, so $m\angle CAB + m\angle CAE = 180$. Since $\angle CAB$ is obtuse, $\angle CAE$ is acute and Case 1 applies, so $m\angle CAE = \frac{1}{2}m\overset{\frown}{CA}$. $m\overset{\frown}{CA} + m\overset{\frown}{CDA} = 360$, so $\frac{1}{2}m\overset{\frown}{CA} + \frac{1}{2}m\overset{\frown}{CDA} = 180$ by Division Prop., and $m\angle CAE + \frac{1}{2}m\overset{\frown}{CDA} = 180$ by substitution. By the Transitive Prop., $m\angle CAB + m\angle CAE = m\angle CAE + \frac{1}{2}m\overset{\frown}{CDA}$, so by Subtraction Prop., $m\angle CAB = \frac{1}{2}m\overset{\frown}{CDA}$.

4 Assess

Open-Ended Assessment

Speaking Select examples and ask students to call out the names of the segments in the figure. Then call on volunteers to explain how they would find missing angle measures or arc lengths.

Getting Ready for Lesson 10-7

Prerequisite Skill Students will learn about special segments in a circle in Lesson 10-7. They will apply solving quadratic equations to find values for segments that intersect in the interior and exterior of a circle. Use Exercises 57–59 to determine your students' familiarity with solving quadratic equations by factoring.

Assessment Options

Practice Quiz 2 The quiz provides students with a brief review of the concepts and skills in Lessons 10-4 through 10-6. Lesson numbers are given to the right of the exercises or instruction lines so students can review concepts not yet mastered.

Quiz (Lessons 10-5 and 10-6) is available on p. 604 of the *Chapter 10 Resource Masters*.

Answer

56. Given: $\overline{AC} \cong \overline{BF}$
Prove: $AB = CF$

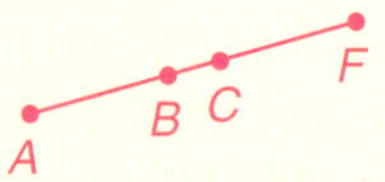

Proof: By definition of congruent segments, $AC = BF$. Using the Segment Addition Postulate, we know that $AC = AB + BC$ and $BF = BC + CF$. Since $AC = BF$, this means that $AB + BC = BC + CF$. If BC is subtracted from each side of this equation, the result is $AB = CF$.

Maintain Your Skills

Mixed Review

Find x. Assume that segments that appear to be tangent are tangent. *(Lesson 10-5)*

49. 16

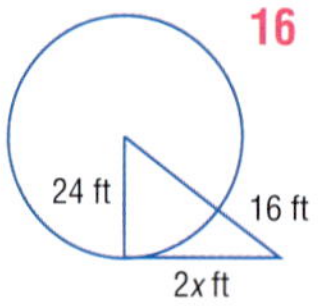

50. 4

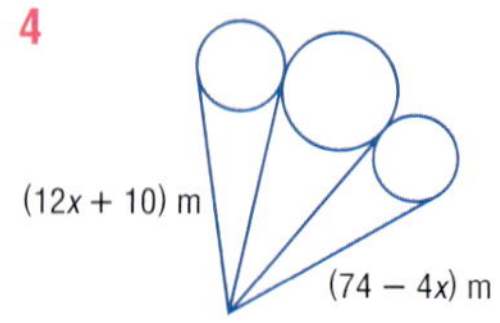

In $\odot P$, $m\widehat{EN} = 66$ and $m\angle GPM = 89$. Find each measure. *(Lesson 10-4)*

51. $m\angle EGN$ 33
52. $m\angle GME$ 57
53. $m\angle GNM$ 44.5

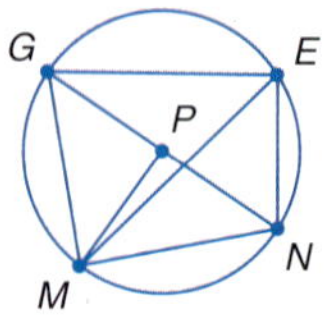

RAMPS **Use the following information for Exercises 54 and 55.**
The *Americans with Disabilities Act* (ADA), which went into effect in 1990, requires that wheelchair ramps have at least a 12-inch run for each rise of 1 inch. *(Lesson 3-3)*

54. Determine the slope represented by this requirement. $\frac{1}{12}$
55. The maximum length the law allows for a ramp is 30 feet. How many inches tall is the highest point of this ramp? 30 in.

56. **PROOF** Write a paragraph proof to show that $AB = CF$ if $\overline{AC} \cong \overline{BF}$. *(Lesson 2-5)* See margin.

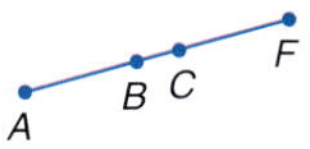

Getting Ready for the Next Lesson

PREREQUISITE SKILL **Solve each equation by factoring.**
*(To review **solving equations by factoring**, see pages 750 and 751.)*

57. $x^2 + 6x - 40 = 0$ 4, −10
58. $2x^2 + 7x - 30 = 0$ $-6, 2\frac{1}{2}$
59. $3x^2 - 24x + 45 = 0$ 3, 5

Practice Quiz 2 — Lessons 10-4 through 10-6

1. **AMUSEMENT RIDES** A Ferris wheel is shown at the right. If the distances between the seat axles are the same, what is the measure of an angle formed by the braces attaching consecutive seats? *(Lesson 10-4)* 67.5

2. Find the measure of each numbered angle. *(Lesson 10-4)* $m\angle 1 = m\angle 2 = 34$

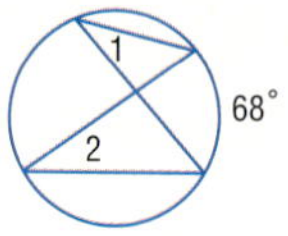

Find x. Assume that any segment that appears to be tangent is tangent. *(Lessons 10-5 and 10-6)*

3. 12

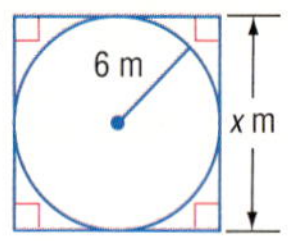

4. 13

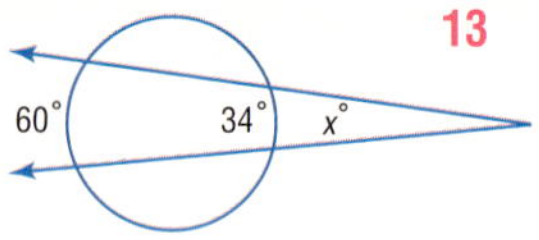

5. 115.5

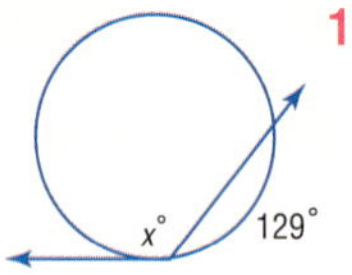

10-7 Special Segments in a Circle

What You'll Learn

- Find measures of segments that intersect in the interior of a circle.
- Find measures of segments that intersect in the exterior of a circle.

How are lengths of intersecting chords related?

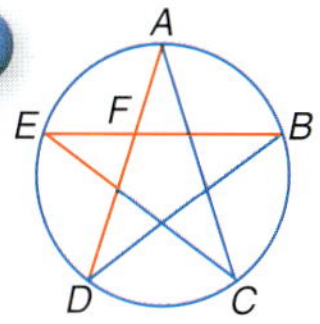

The star is inscribed in a circle. It was formed by intersecting chords. Segments AD and EB are two of those chords. When two chords intersect, four smaller segments are defined.

SEGMENTS INTERSECTING INSIDE A CIRCLE In Lesson 10-2, you learned how to find lengths of parts of a chord that is intersected by the perpendicular diameter. But how do you find lengths for other intersecting chords?

Geometry Activity

Intersecting Chords

Make A Model

- Draw a circle and two intersecting chords.
- Name the chords $\overline{PQ}$ and $\overline{RS}$ intersecting at T.
- Draw $\overline{PS}$ and $\overline{RQ}$.

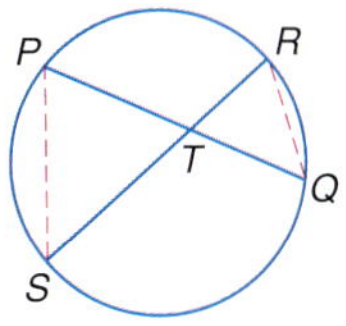

Analyze

1. Name pairs of congruent angles. Explain your reasoning.
2. How are $\triangle PTS$ and $\triangle RTQ$ related? Why? similar by AA Similarity
3. **Make a conjecture** about the relationship of $\overline{PT}$, $\overline{TQ}$, $\overline{RT}$, and $\overline{ST}$.

1. $\angle PTS \cong \angle RTQ$ (Vertical $\angle$s are $\cong$.); $\angle P \cong \angle R$ ($\angle$s intercepting same arc are $\cong$.); $\angle S \cong \angle Q$ ($\angle$s intercepting same arc are $\cong$.)

3. $\frac{PT}{RT} = \frac{ST}{TQ}$ or $PT \cdot TQ = RT \cdot ST$

The results of the Geometry Activity suggest a proof for Theorem 10.15.

Theorem 10.15

If two chords intersect in a circle, then the products of the measures of the segments of the chords are equal.

Example: $AE \cdot EC = BE \cdot ED$

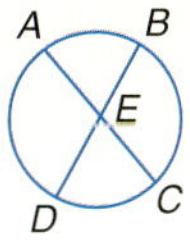

You will prove Theorem 10.15 in Exercise 21.

Example 1 Intersection of Two Chords

Find x.

$AE \cdot EB = CE \cdot ED$

$x \cdot 6 = 3 \cdot 4$	Substitution
$6x = 12$	Multiply.
$x = 2$	Divide each side by 6.

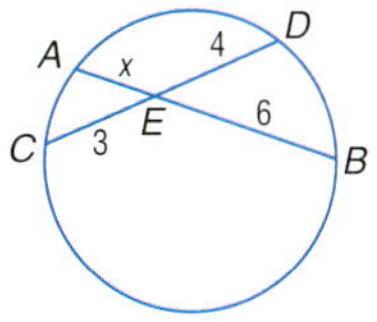

10-7 Lesson Notes

1 Focus

5-Minute Check Transparency 10-7 Use as a quiz or review of Lesson 10-6.

Mathematical Background notes are available for this lesson on p. 520D.

How are lengths of intersecting chords related?

Ask students:

- Name the segments that are defined by the intersection of $\overline{AD}$ and $\overline{EB}$ in the figure. $\overline{AF}$, $\overline{FD}$, $\overline{EF}$, and $\overline{FB}$
- Name the inscribed angles in the figure. $\angle A$, $\angle B$, $\angle C$, $\angle D$, and $\angle E$
- Is $\triangle DFB$ inscribed in the circle? Why or why not? No; because F is not a point on the circle.

Resource Manager

Workbook and Reproducible Masters

Chapter 10 Resource Masters

- Study Guide and Intervention, pp. 577–578
- Skills Practice, p. 579
- Practice, p. 580
- Reading to Learn Mathematics, p. 581
- Enrichment, p. 582

Prerequisite Skills Workbook, pp. 35–36, 51–52

Teaching Geometry With Manipulatives Masters, p. 17

Transparencies

5-Minute Check Transparency 10-7

Answer Key Transparencies

Technology

Interactive Chalkboard

Multimedia Applications: Virtual Activities

2 Teach

SEGMENTS INTERSECTING INSIDE A CIRCLE

In-Class Examples Power Point®

1 Find x. **13.5**

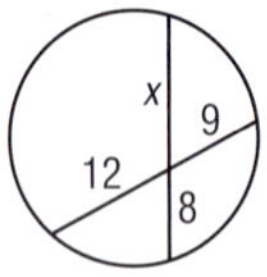

Teaching Tip Remind students that a diameter can be drawn to bisect any chord of a circle.

2 **BIOLOGY** Biologists often examine organisms under microscopes. The circle represents the field of view under the microscope with a diameter of 2 mm. Determine the length of the organism if it is located 0.25 mm from the bottom of the field of view. Round to the nearest hundredth. **0.66 mm**

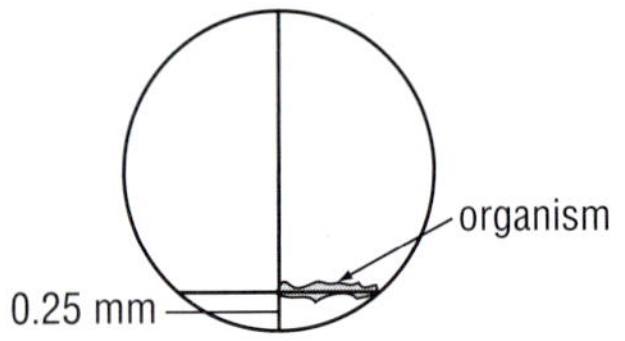

SEGMENTS INTERSECTING OUTSIDE A CIRCLE

In-Class Example Power Point®

3 Find x if $EF = 10$, $EH = 8$, and $FG = 24$. **34.5**

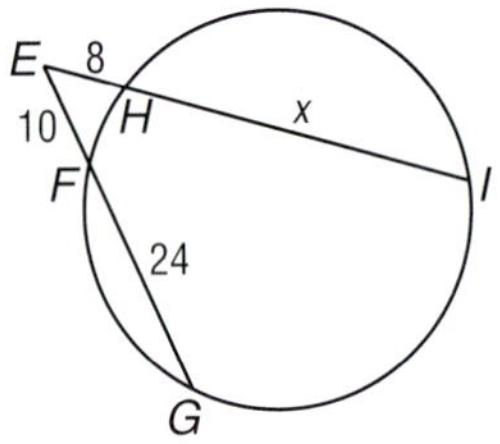

Intersecting chords can also be used to measure arcs.

Example 2 Solve Problems

TUNNELS **Tunnels are constructed to allow roadways to pass through mountains. What is the radius of the circle containing the arc if the opening is not a semicircle?**

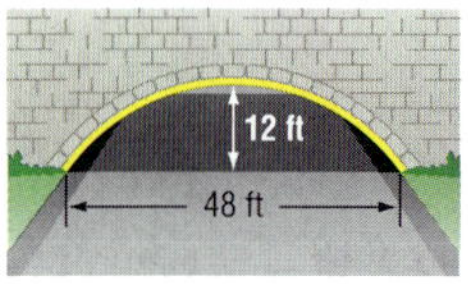

Draw a model using a circle. Let x represent the unknown measure of the segment of diameter $\overline{AB}$. Use the products of the lengths of the intersecting chords to find the length of the diameter.

$AE \cdot EB = DE \cdot EC$	Segment products
$12x = 24 \cdot 24$	Substitution
$x = 48$	Divide each side by 12.
$AB = AE + EB$	Segment Addition Postulate
$AB = 12 + 48$ or 60	Substitution and addition

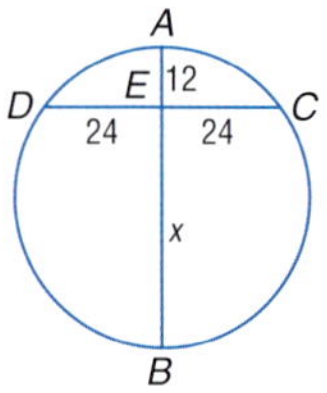

Since the diameter is 60, $r = 30$.

SEGMENTS INTERSECTING OUTSIDE A CIRCLE Nonparallel chords of a circle can be extended to form secants that intersect in the exterior of a circle. The special relationship among secant segments excludes the chord.

Study Tip

Helping You Remember

To remember this concept, the wording of Theorem 10.16 can be simplified by saying that each side of the equation is the product of the exterior part and the whole segment.

Theorem 10.16

If two secant segments are drawn to a circle from an exterior point, then the product of the measures of one secant segment and its external secant segment is equal to the product of the measures of the other secant segment and its external secant segment.

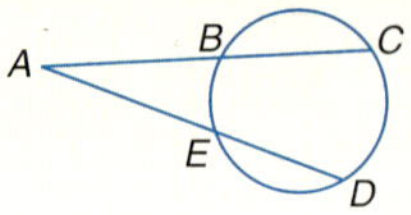

Example: $AB \cdot AC = AE \cdot AD$

You will prove this theorem in Exercise 30.

Example 3 Intersection of Two Secants

Find RS if $PQ = 12$, $QR = 2$, and $TS = 3$.

Let $RS = x$.

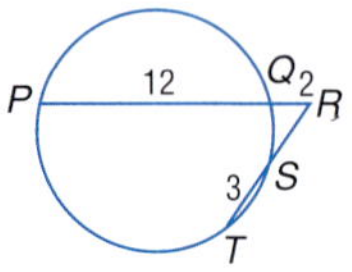

$QR \cdot PR = RS \cdot RT$	Secant Segment Products
$2 \cdot (12 + 2) = x \cdot (x + 3)$	Substitution
$28 = x^2 + 3x$	Distributive Property
$0 = x^2 + 3x - 28$	Subtract 28 from each side.
$0 = (x + 7)(x - 4)$	Factor.

$x + 7 = 0$ $\quad$ $x - 4 = 0$

$x = -7$ $\quad$ $x = 4$ $\quad$ Disregard negative value.

570 Chapter 10 Circles

Geometry Activity

Materials: compass, straightedge

- Tell students to draw chords that are not congruent and that do not intersect at the center of the circle.
- Students can use a protractor to verify that the triangles are similar, and they can determine the scale factor by which the triangles are related.

The same secant segment product can be used with a secant segment and a tangent. In this case, the tangent is both the exterior part and the whole segment. This is stated in Theorem 10.17.

Theorem 10.17

If a tangent segment and a secant segment are drawn to a circle from an exterior point, then the square of the measure of the tangent segment is equal to the product of the measures of the secant segment and its external secant segment.

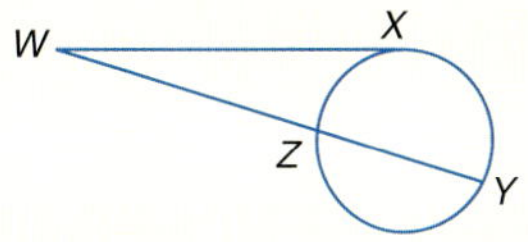

Example: $WX \cdot WX = WZ \cdot WY$

You will prove this theorem in Exercise 31.

Example 4 Intersection of a Secant and a Tangent

Find x. Assume that segments that appear to be tangent are tangent.

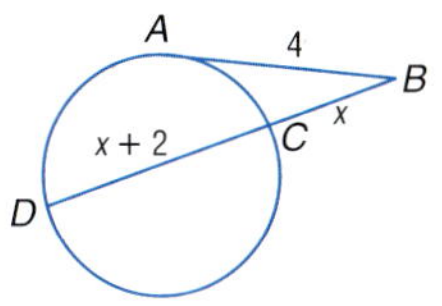

$(AB)^2 = BC \cdot BD$

$4^2 = x(x + x + 2)$

$16 = x(2x + 2)$

$16 = 2x^2 + 2x$

$0 = 2x^2 + 2x - 16$

$0 = x^2 + x - 8$

This expression is not factorable. Use the Quadratic Formula.

$x = \frac{-b \pm \sqrt{b^2 - 4ac}}{2a}$ Quadratic Formula

$= \frac{-1 \pm \sqrt{1^2 - 4(1)(-8)}}{2(1)}$ $a = 1, b = 1, c = -8$

$= \frac{-1 + \sqrt{33}}{2}$ or $x = \frac{-1 - \sqrt{33}}{2}$ Disregard the negative solution.

≈ 2.37 Use a calculator.

Check for Understanding

Concept Check

1. Show how the products for secant segments are similar to the products for a tangent and a secant segment. **See margin.**

2. **FIND THE ERROR** Becky and Latisha are writing products to find x. Who is correct? Explain your reasoning.

2. Latisha; the length of the tangent segment squared equals the product of the exterior secant segment and the entire secant, not the interior secant segment.

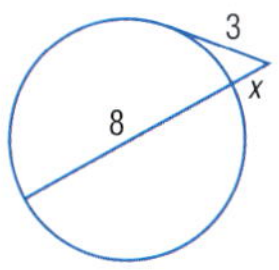

Becky	Latisha
$3^2 = x \cdot 8$	$3^2 = x(x + 8)$
$9 = 8x$	$9 = x^2 + 8x$
$\frac{9}{8} = x$	$0 = x^2 + 8x - 9$
	$0 = (x + 9)(x - 1)$
	$x = 1$

www.geometryonline.com/extra_examples/fcat

In-Class Example

Power Point®

4 Find x. Assume that segments that appear to be tangent are tangent. 8

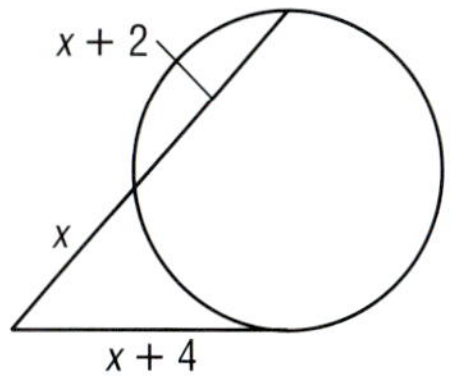

3 Practice/Apply

Study Notebook

Have students—
- add the definitions/examples of the vocabulary terms to their Vocabulary Builder worksheets for Chapter 10.
- include any other item(s) that they find helpful in mastering the skills in this lesson.

DAILY INTERVENTION

FIND THE ERROR Point out to students that in this lesson they do not use the entire length of the interior secant segment in any of the relationships. Explain that in each case with exterior secant segments, they do use the entire length of the exterior secant segment.

Answer

1. Sample answer: The product equation for secant segments equates the product of exterior segment measure and the whole segment measure for each secant. In the case of secant-tangent, the product involving the tangent segment becomes (measure of tangent segment)^2 because the exterior segment and the whole segment are the same segment.

DAILY INTERVENTION

Differentiated Instruction

Visual/Spatial Students can use their spatial skills to remember the concepts of this lesson. Tell students to visualize the similar triangles that would be made by the chords, secants, and/or tangents in the figures to help them determine how to set up their products.

Lesson 10-7 Special Segments in a Circle 571

About the Exercises...

Organization by Objective

- **Segments Intersecting Inside a Circle:** 8–11, 20, 27
- **Segments Intersecting Outside a Circle:** 12–19

Odd/Even Assignments

Exercises 8–19, 22–28 are structured so that students practice the same concepts whether they are assigned odd or even problems.

Assignment Guide

Basic: 9, 13–31 odd, 32–48

Average: 9–31 odd, 32–48

Advanced: 8–30 even, 32–45 (optional: 46–48)

Concept Check

Have students create and label a three-column chart with an example of each segment relationship described in this lesson, color code the parts that are equal, and write each relationship in algebraic form.

Answer

3. Sample answer:

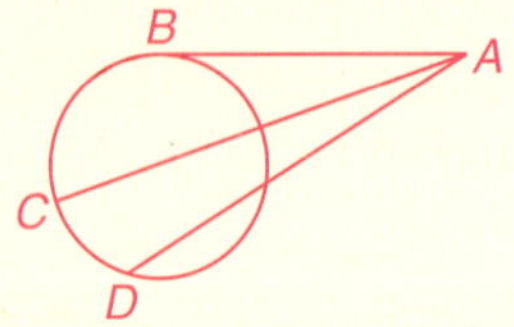

3. **OPEN ENDED** Draw a circle with two secant segments and one tangent segment that intersect at the same point. **See margin.**

Guided Practice

GUIDED PRACTICE KEY	
Exercises	Examples
4	1
5	4
6	3
7	2

Find x. Round to the nearest tenth if necessary. Assume that segments that appear to be tangent are tangent.

4. **2**

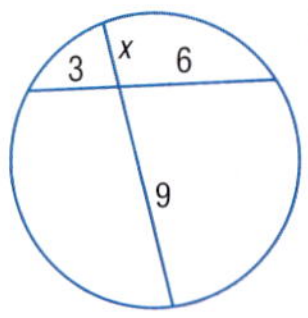

5. **28.1**

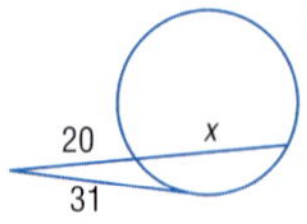

6. **2**

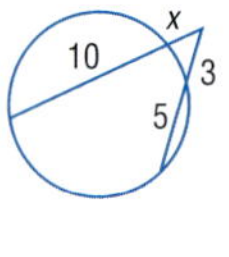

Application

7. **HISTORY** The Roman Coliseum has many "entrances" in the shape of a door with an arched top. The ratio of the arch width to the arch height is 7:3. Find the ratio of the arch width to the radius of the circle that contains the arch. **≈7:3.54**

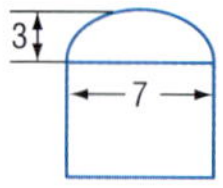

★ indicates increased difficulty

Practice and Apply

Homework Help	
For Exercises	See Examples
8–11	1
12–15	4
16–19	3
20, 27	2

Extra Practice
See page 775.

Find x. Round to the nearest tenth if necessary. Assume that segments that appear to be tangent are tangent.

8. **10**

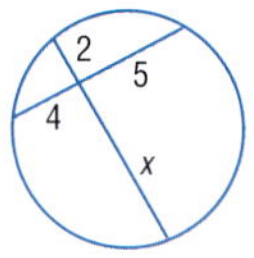

9. **4**

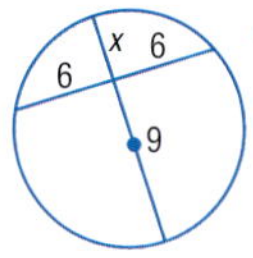

10. **4.7**

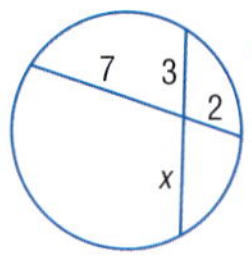

★11. **2**

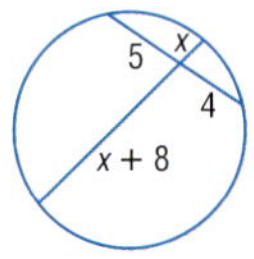

12. **6**

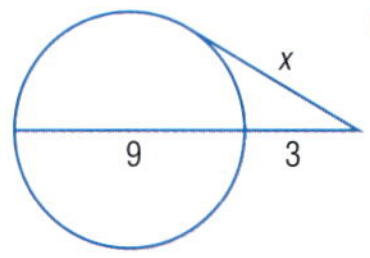

13. **6**

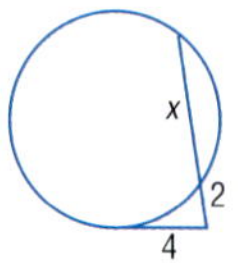

14. **8**

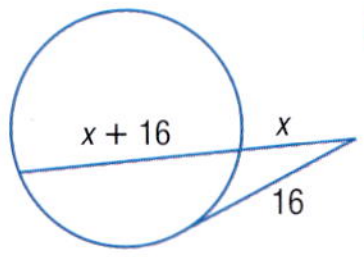

15. **3.2**

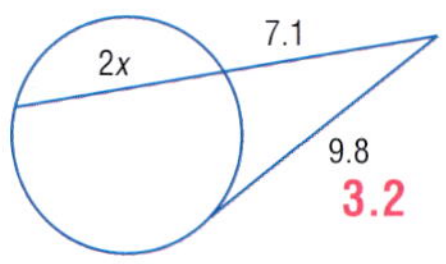

16. **5**

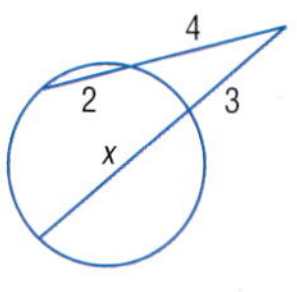

17. **4**

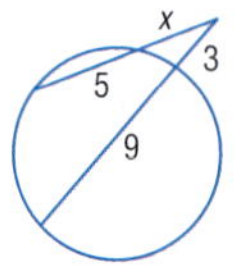

18. **5**

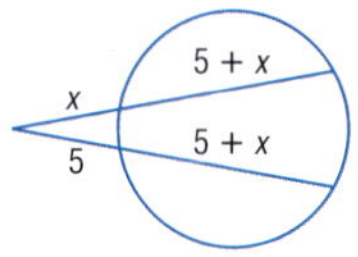

19. **5.6**

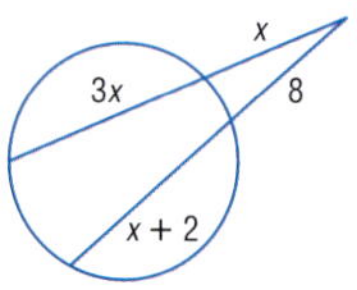

20. **KNOBS** If you remove a knob from a kitchen appliance, you may notice that the hole is not completely round. Suppose the flat edge is 4 millimeters long and the distance from the curved edge to the flat edge is about 4.25 millimeters. Find the radius of the circle containing the hole. **about 2.6 mm**

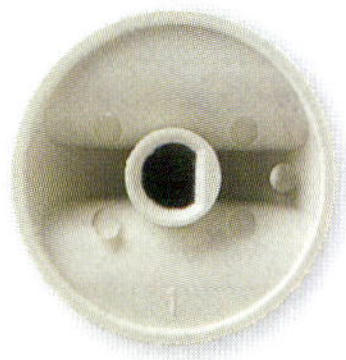

572 Chapter 10 Circles

21. **PROOF** Copy and complete the proof of Theorem 10.15.

Given: $\overline{WY}$ and $\overline{ZX}$ intersect at T.

Prove: $WT \cdot TY = ZT \cdot TX$

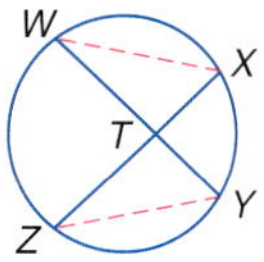

Statements	Reasons
a. $\angle W \cong \angle Z$, $\angle X \cong \angle Y$	a. __?__ Inscribed angles that intercept the same arc are congruent.
b. __?__ $\triangle WXT \sim \triangle ZYT$	b. AA Similarity
c. $\frac{WT}{ZT} = \frac{TX}{TY}$	c. __?__ Definition of similar triangles
d. __?__ $WT \cdot TY = ZT \cdot TX$	d. Cross products

Find each variable. Round to the nearest tenth, if necessary.

22. 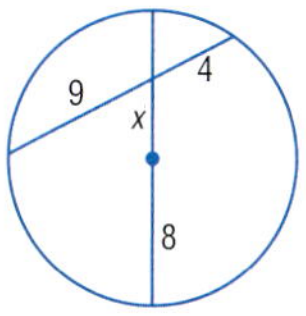

5.3

23. 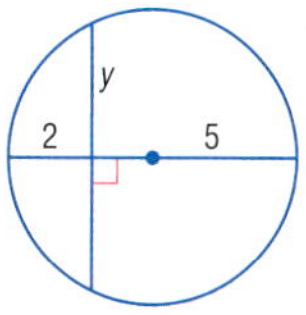

4

24. 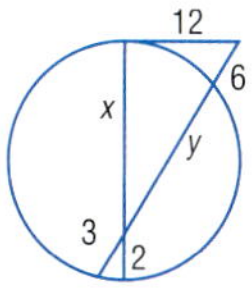

$y = 15$, $x = 22.5$

25. 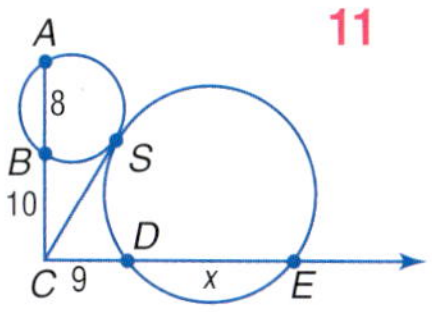

11

26. 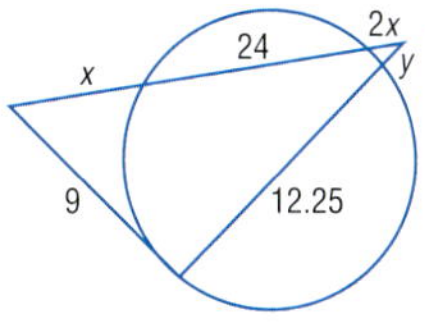

$x = 3$, $y = 8.6$

27. 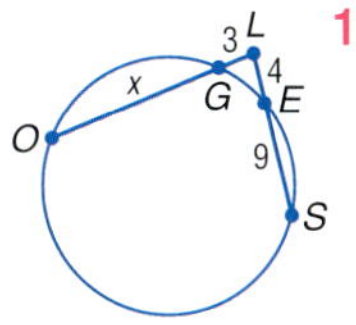

14.3

28. 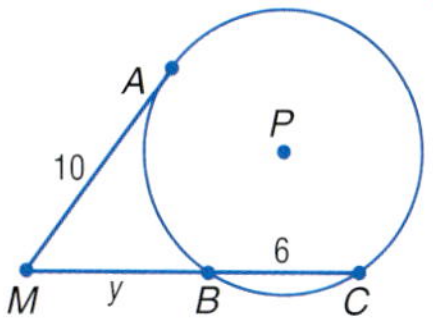

7.4

Career Choices

Construction Worker

Construction workers must know how to measure and fit shapes together to make a sound building that will last for years to come. These workers also must master using machines to cut wood and metal to certain specifications that are based on geometry.

Online Research For more information about a career as a construction worker, visit: www.geometryonline.com/careers

29. **CONSTRUCTION** An arch over a courtroom door is 60 centimeters high and 200 centimeters wide. Find the radius of the circle containing the arc of the arch. 113.$\overline{3}$ cm

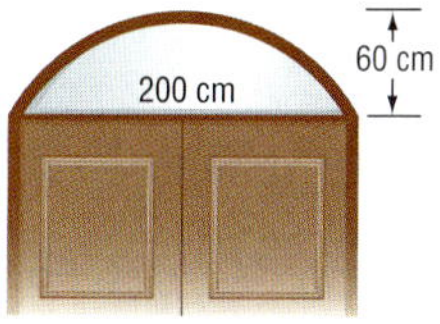

30–31. See p. 589B.

30. **PROOF** Write a two-column proof of Theorem 10.16.

Given: secants $\overline{EC}$ and $\overline{EB}$

Prove: $EA \cdot EC = ED \cdot EB$

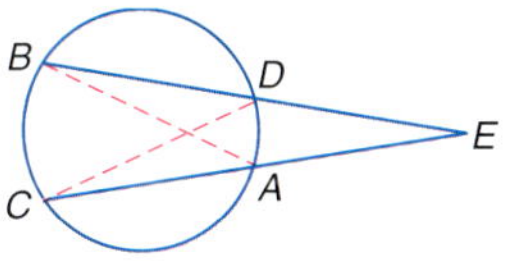

31. **PROOF** Write a two-column proof of Theorem 10.17.

Given: tangent $\overline{RS}$, secant $\overline{SU}$

Prove: $(RS)^2 = ST \cdot SU$

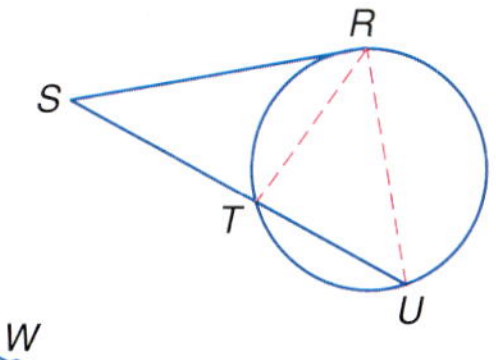

32. **CRITICAL THINKING** In the figure, Y is the midpoint of $\overline{XZ}$. Find WX in tems of XY. Explain your reasoning.

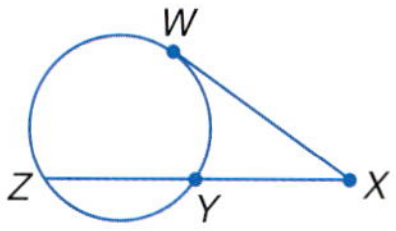

$WX = \sqrt{2} \cdot XY$; see margin for explanation.

Answer

32. $ZY = XY$

$(WX)^2 = XY \cdot XZ$

$(WX)^2 = XY(XY + ZY)$

$(WX)^2 = XY(2XY)$

$(WX)^2 = 2(XY)^2$

$WX = \sqrt{2(XY)^2}$

$WX = \sqrt{2} \cdot XY$

Enrichment, p. 582

The Nine-Point Circle

The figure below illustrates a surprising fact about triangles and circles. Given any $\triangle ABC$, there is a circle that contains all of the following nine points:

(1) the midpoints K, L, and M of the sides of $\triangle ABC$

(2) the points X, Y, and Z, where $\overline{AX}$, $\overline{BY}$, and $\overline{CZ}$ are the altitudes of $\triangle ABC$

(3) the points R, S, and T which are the midpoints of the segments $\overline{AH}$, $\overline{BH}$, and $\overline{CH}$ that join the vertices of $\triangle ABC$ to the point H where the lines containing the altitudes intersect.

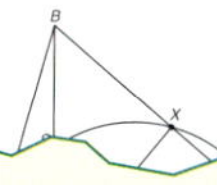

Study Guide and Intervention, p. 577 (shown) and p. 578

Segments Intersecting Inside a Circle If two chords intersect in a circle, then the products of the measures of the chords are equal.

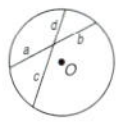

$a \cdot b = c \cdot d$

Example **Find x.**

The two chords intersect inside the circle, so the products $AB \cdot BC$ and $EB \cdot BD$ are equal.

$AB \cdot BC = EB \cdot BD$

$6 \cdot x = 8 \cdot 3$	Substitution
$6x = 24$	Simplify.
$x = 4$	Divide each side by 6.

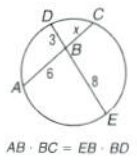

$AB \cdot BC = EB \cdot BD$

Exercises

Find x to the nearest tenth.

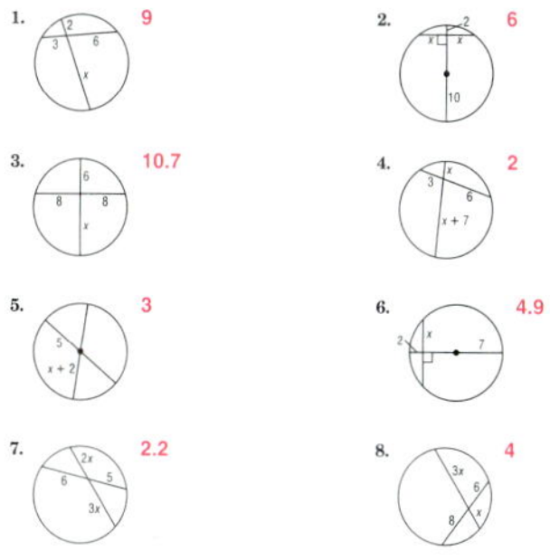

1. 9
2. 6
3. 10.7
4. 2
5. 3
6. 4.9
7. 2.2
8. 4

Skills Practice, p. 579 and Practice, p. 580 (shown)

Find x to the nearest tenth. Assume that segments that appear to be tangent are tangent.

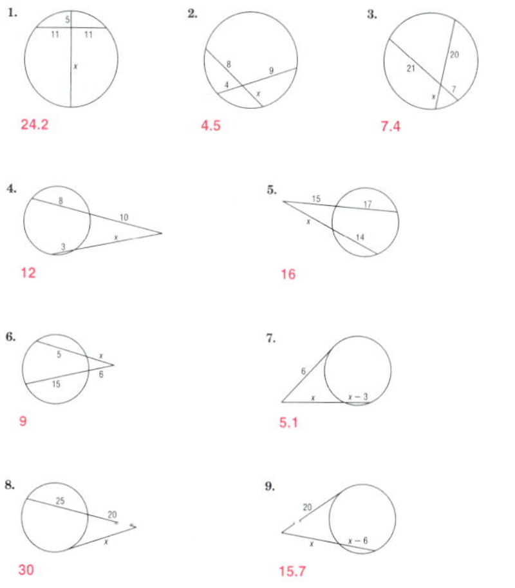

1. 24.2
2. 4.5
3. 7.4
4. 12
5. 16
6. 9
7. 5.1
8. 30
9. 15.7

10. **CONSTRUCTION** An arch over an apartment entrance is 3 feet high and 9 feet wide. Find the radius of the circle containing the arc of the arch. 4.875 ft

Reading to Learn Mathematics, p. 581 ELL

Pre-Activity **How are lengths of intersecting chords related?**

Read the introduction to Lesson 10-7 at the top of page 569 in your textbook.

- What kinds of angles of the circle are formed at the points of the star? inscribed angles
- What is the sum of the measures of the five angles of the star? 180

Reading the Lesson

1. Refer to $\odot O$. Name each of the following.

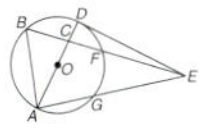

a. a diameter $\overline{AD}$

b. a chord that is not a diameter $\overline{AB}$, $\overline{BF}$, or $\overline{AG}$

c. two chords that intersect in the interior of the circle $\overline{AD}$ and $\overline{BF}$

d. an exterior point E

e. two secant segments that intersect in the exterior of the circle $\overline{EA}$ and $\overline{EB}$

f. a tangent segment $\overline{ED}$

g. a right angle $\angle ADE$

h. an external secant segment $\overline{EF}$ or $\overline{EG}$

i. a secant-tangent angle with vertex on the circle $\angle ADE$

j. an inscribed angle $\angle BAD$, $\angle DAG$, $\angle BAG$, or $\angle ABF$

2. Supply the missing length to complete each equation.

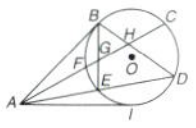

a. $BH \cdot HD = FH \cdot$ __HC__ b. $AC \cdot AF = AD \cdot$ __AE__

c. $AD \cdot AE = AB \cdot$ __AB__ d. $AB =$ __AI__

e. $AF \cdot AC = ($__AI or AB__$)^2$ f. $EG \cdot$ __GB__ $= FG \cdot GC$

Helping You Remember

3. Some students find it easier to remember geometric theorems if they restate them in their own words. Restate Theorem 10.16 in a way that you find easier to remember.

Sample answer: Suppose you draw a secant to a circle through a point *A* outside the circle. Multiply the distances from point *A* to the points where the secant intersects the circle. The corresponding product will be the same for any other secant through point *A* to the same circle.

4 Assess

Open-Ended Assessment

Modeling Provide students with three or four cutout circles, thin masking tape, and two or three sheets of construction paper. Have students arbitrarily model segments intersecting inside and outside a circle with the tape and the cutouts. Use the sheets of construction paper for points outside the circle. Students can then use a metric ruler to measure segments and test the theorems in this lesson.

Getting Ready for Lesson 10-8

Prerequisite Skill Students will learn about equations of circles in Lesson 10-8. They will apply the Distance Formula to write equations for circles and graph circles. Use Exercises 46–48 to determine your students' familiarity with the Distance Formula.

Answer

33. Sample answer: The product of the parts of one intersecting chord equals the product of the parts of the other chord. Answers should include the following.
- $\overline{AF}$, $\overline{FD}$, $\overline{EF}$, $\overline{FB}$
- $AF \cdot FD = EF \cdot FB$

33. **WRITING IN MATH** Answer the question that was posed at the beginning of the lesson. See margin.

How are the lengths of intersecting chords related?

Include the following in your answer:
- the segments formed by intersecting segments, $\overline{AD}$ and $\overline{EB}$, and
- the relationship among these segments.

FCAT Practice
Standardized Test Practice

34. Find two possible values for x from the information in the figure. D

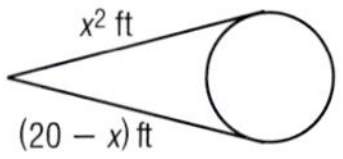

Ⓐ −4, −5 Ⓑ −4, 5
Ⓒ 4, 5 Ⓓ 4, −5

35. **ALGEBRA** Mr. Rodriguez can wash his car in 15 minutes, while his son Marcus takes twice as long to do the same job. If they work together, how long will it take them to wash the car? C

Ⓐ 5 min Ⓑ 7.5 min Ⓒ 10 min Ⓓ 22.5 min

Maintain Your Skills

Mixed Review Find the measure of each numbered angle. Assume that segments that appear tangent are tangent. *(Lesson 10-6)*

36. 129

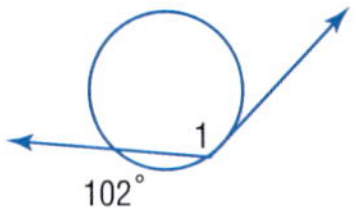

37. 157.5

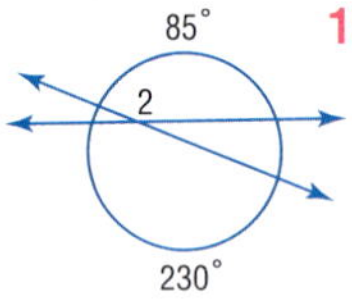

38. 26

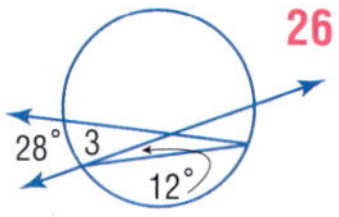

Find x. Assume that segments that appear to be tangent are tangent. *(Lesson 10-5)*

39. 7

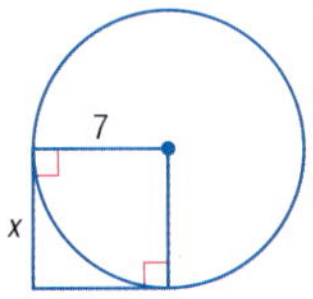

40. 8

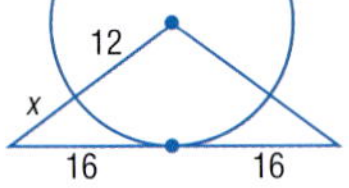

41. 36

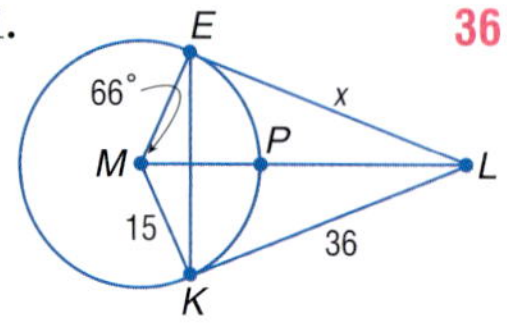

42. **INDIRECT MEASUREMENT** Joseph Blackarrow is measuring the width of a stream on his land to build a bridge over it. He picks out a rock across the stream as landmark A and places a stone on his side as point B. Then he measures 5 feet at a right angle from $\overline{AB}$ and marks this C. From C, he sights a line to point A on the other side of the stream and measures the angle to be about 67°. How far is it across the stream rounded to the nearest whole foot? *(Lesson 7-5)* 12 ft

Classify each triangle by its sides and by its angles. *(Lesson 4-1)*

43.

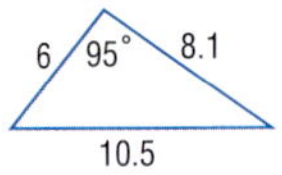

scalene, obtuse

44.

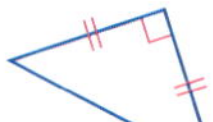

isosceles, right

45.

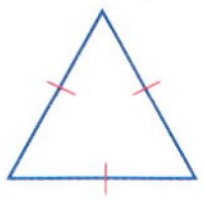

equilateral, acute, or equiangular

Getting Ready for the Next Lesson

PREREQUISITE SKILL Find the distance between each pair of points.
*(To review the **Distance Formula**, see Lesson 1-3.)*

46. $C(-2, 7)$, $D(10, 12)$ 13
47. $E(1, 7)$, $F(3, 4)$ $\sqrt{13}$
48. $G(9, -4)$, $H(15, -2)$ $\sqrt{40}$

574 Chapter 10 Circles

10-8 Equations of Circles

What You'll Learn
- Write the equation of a circle.
- Graph a circle on the coordinate plane.

What kind of equations describes the ripples of a splash?

When a rock enters the water, ripples move out from the center forming concentric circles. If the rock is assigned coordinates, each ripple can be modeled by an equation of a circle.

EQUATION OF A CIRCLE The fact that a circle is the *locus* of points in a plane equidistant from a given point creates an equation for any circle.

Suppose the center is at (3, 2) and the radius is 4. The radius is the distance from the center. Let $P(x, y)$ be the endpoint of any radius.

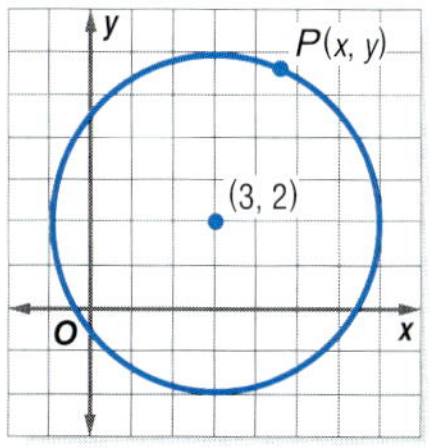

$d = \sqrt{(x_2 - x_1)^2 + (y_2 - y_1)^2}$	Distance Formula
$4 = \sqrt{(x - 3)^2 + (y - 2)^2}$	$d = 4, (x_1, y_1) = (3, 2)$
$16 = (x - 3)^2 + (y - 2)^2$	Square each side.

Applying this same procedure to an unknown center (h, k) and radius r yields a general equation for any circle.

Key Concept — *Standard Equation of a Circle*

An equation for a circle with center at (h, k) and radius of r units is $(x - h)^2 + (y - k)^2 = r^2$.

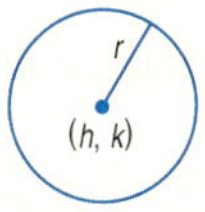

Study Tip

Equation of Circles
Note that the equation of a circle is kept in the form shown above. The terms being squared are not expanded.

Example 1 *Equation of a Circle*

Write an equation for each circle.

a. center at (−2, 4), $d = 4$

If $d = 4$, $r = 2$.

$(x - h)^2 + (y - k)^2 = r^2$	Equation of a circle
$[x - (-2)]^2 + [y - 4]^2 = 2^2$	$(h, k) = (-2, 4), r = 2$
$(x + 2)^2 + (y - 4)^2 = 4$	Simplify.

b. center at origin, $r = 3$

$(x - h)^2 + (y - k)^2 = r^2$	Equation of a circle
$(x - 0)^2 + (y - 0)^2 = 3^2$	$(h, k) = (0, 0), r = 3$
$x^2 + y^2 = 9$	Simplify.

10-8 Lesson Notes

1 Focus

5-Minute Check Transparency 10-8 Use as a quiz or review of Lesson 10-7.

Mathematical Background notes are available for this lesson on p. 520D.

What kind of equations describes the ripples of a splash?

Ask students:
- In order to cause water ripples to form concentric circles, what has to happen? **Something must break the surface tension of the water, creating the force that causes the ripples, like the rock in the example.**
- If the rock is thrown with a greater force, would you see fewer circles or more circles? **more circles**

Resource Manager

Workbook and Reproducible Masters

Chapter 10 Resource Masters
- Study Guide and Intervention, pp. 583–584
- Skills Practice, p. 585
- Practice, p. 586
- Reading to Learn Mathematics, p. 587
- Enrichment, p. 588
- Assessment, p. 604

Teaching Geometry With Manipulatives Masters, pp. 1, 17

Transparencies

5-Minute Check Transparency 10-8
Answer Key Transparencies

Technology

GeomPASS: Tutorial Plus, Lesson 19
Interactive Chalkboard
Multimedia Applications: Virtual Activities

2 Teach

EQUATION OF A CIRCLE

In-Class Examples

Power Point®

1 Write an equation for each circle.

a. center at (3, −3), $d = 12$
$(x - 3)^2 + (y + 3)^2 = 36$

b. center at (−12, −1), $r = 8$
$(x + 12)^2 + (y + 1)^2 = 64$

Teaching Tip Students should note that the two tangent lines have slopes that indicate they are perpendicular to each other. Students should also remember that a radius is the shortest distance from the tangent to the center of a circle.

2 A circle with a diameter of 10 has its center in the first quadrant. The lines $y = -3$ and $x = -1$ are tangent to the circle. Write an equation of the circle.
$(x - 4)^2 + (y - 2)^2 = 25$

GRAPH CIRCLES

In-Class Example

Power Point®

3 **a.** Graph $(x - 2)^2 + (y + 3)^2 = 4$.

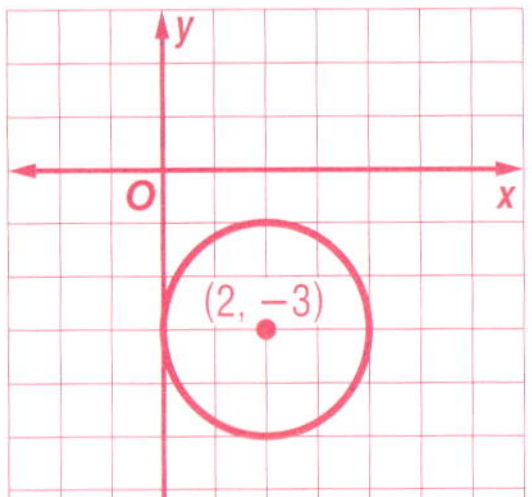

b. Graph $(x - 3)^2 + y^2 = 16$.

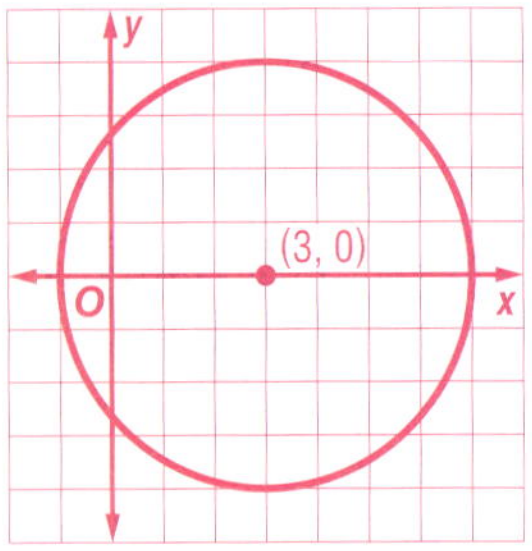

Other information about a circle can be used to find the equation of the circle.

Example 2 *Use Characteristics of Circles*

A circle with a diameter of 14 has its center in the third quadrant. The lines $y = -1$ and $x = 4$ are tangent to the circle. Write an equation of the circle.

Sketch a drawing of the two tangent lines.

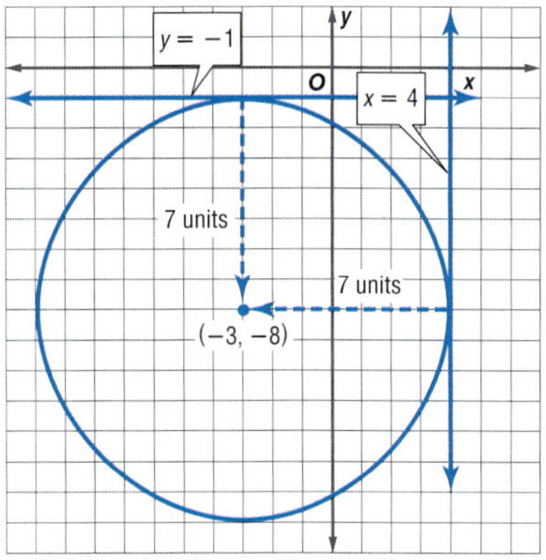

Since $d = 14$, $r = 7$. The line $x = 4$ is perpendicular to a radius. Since $x = 4$ is a vertical line, the radius lies on a horizontal line. Count 7 units to the left from $x = 4$. Find the value of h.

$$h = 4 - 7 \text{ or } -3$$

Likewise, the radius perpendicular to the line $y = -1$ lies on a vertical line. The value of k is 7 units down from −1.

$$k = -1 - 7 \text{ or } -8$$

The center is at (−3, −8), and the radius is 7. An equation for the circle is $(x + 3)^2 + (y + 8)^2 = 49$.

GRAPH CIRCLES You can analyze the equation of a circle to find information that will help you graph the circle on a coordinate plane.

Study Tip

Graphing Calculator
You can also use the center and radius to graph a circle on a graphing calculator. Select a suitable window that contains the center of the circle. For a TI-83 Plus, press ZOOM 5. Then use **9: Circle (** on the **Draw** menu. Put in the coordinates of the center and then the radius so that the screen shows "Circle (−2, 3, 4)". Then press ENTER.

Example 3 *Graph a Circle*

a. Graph $(x + 2)^2 + (y - 3)^2 = 16$.

Compare each expression in the equation to the standard form.

$(x - h)^2 = (x + 2)^2$ $\quad\quad$ $(y - k)^2 = (y - 3)^2$
$x - h = x + 2$ $\quad\quad$ $y - k = y - 3$
$-h = 2$ $\quad\quad$ $-k = -3$
$h = -2$ $\quad\quad$ $k = 3$

$r^2 = 16$, so $r = 4$.

The center is at (−2, 3), and the radius is 4. Graph the center. Use a compass set at a width of 4 grid squares to draw the circle.

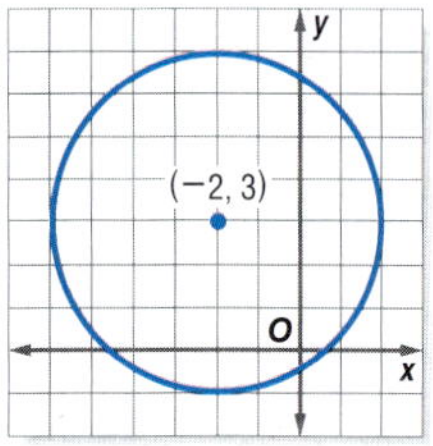

b. Graph $x^2 + y^2 = 9$.

Write the equation in standard form.

$(x - 0)^2 + (y - 0)^2 = 3^2$

The center is at (0, 0), and the radius is 3. Draw a circle with radius 3, centered at the origin.

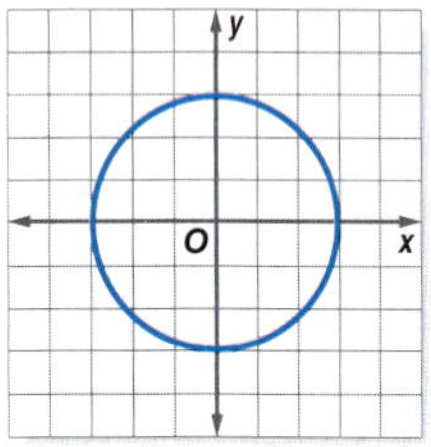

If you know three points on the circle, you can find the center and radius of the circle and write its equation.

DAILY INTERVENTION — **Differentiated Instruction**

Logical/Mathematical Explain that students will rely heavily on their geometric knowledge and reasoning skills to solve the problems in this lesson. Allow students to explain how to explore and collaborate as they work through examples and exercises. Encourage students to recall definitions, concepts, and theorems to help explain why they use certain methods to solve problems.

Example 4 A Circle Through Three Points

CELL PHONES Cell phones work by the transfer of phone signals from one tower to another via a satellite. Cell phone companies try to locate towers so that they service multiple communities. Suppose three large metropolitan areas are modeled by the points $A(4, 4)$, $B(0, -12)$, and $C(-4, 6)$, and each unit equals 100 miles. Determine the location of a tower equidistant from all three cities, and write an equation for the circle.

Explore You are given three points that lie on a circle.

Plan Graph $\triangle ABC$. Construct the perpendicular bisectors of two sides to locate the center, which is the location of the tower. Find the length of a radius. Use the center and radius to write an equation.

Solve Graph $\triangle ABC$ and construct the perpendicular bisectors of two sides. The center appears to be at $(-2, -3)$. This is the location of the tower.

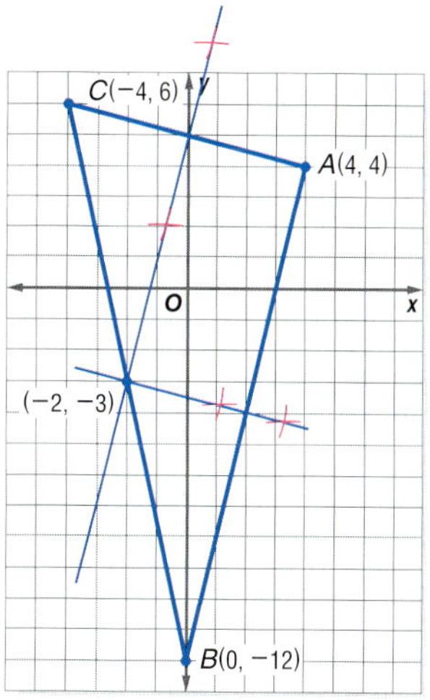

Find r by using the Distance Formula with the center and any of the three points.

$$r = \sqrt{[-2 - 4]^2 + [-3 - 4]^2}$$
$$= \sqrt{85}$$

Write an equation.

$$[x - (-2)]^2 + [y - (-3)]^2 = \left(\sqrt{85}\right)^2$$
$$(x + 2)^2 + (y + 3)^2 = 85$$

Study Tip

Look Back
To review **writing equations of lines**, see Lesson 5-4.

Examine You can verify the location of the center by finding the equations of the two bisectors and solving a system of equations. You can verify the radius by finding the distance between the center and another of the three points on the circle.

Check for Understanding

Concept Check

1. **OPEN ENDED** Draw an obtuse triangle on a coordinate plane and construct the circle that circumscribes it. **1–2. See margin.**
2. **Explain** how the definition of a circle leads to its equation.

Guided Practice

GUIDED PRACTICE KEY	
Exercises	Examples
3, 4, 8	1
5	2
6	3
7	3, 4

Write an equation for each circle.

3. center at $(-3, 5)$, $r = 10$ $(x + 3)^2 + (y - 5)^2 = 100$
4. center at origin, $r = \sqrt{7}$ $x^2 + y^2 = 7$
5. diameter with endpoints at $(2, 7)$ and $(-6, 15)$ $(x + 2)^2 + (y - 11)^2 = 32$

Graph each equation. **6–7. See margin.**

6. $(x + 5)^2 + (y - 2)^2 = 9$
7. $(x - 3)^2 + y^2 = 16$

8. Write an equation of a circle that contains $M(-2, -2)$, $N(2, -2)$, and $Q(2, 2)$. Then graph the circle. $x^2 + y^2 = 8$; **See margin for graph.**

www.geometryonline.com/extra_examples/fcat

In-Class Example Power Point®

4 **ELECTRICITY** Strategically located substations are extremely important in the transmission and distribution of a power company's electric supply. Suppose three substations are modeled by the points $D(3, 6)$, $E(-1, 0)$, and $F(3, -4)$. Determine the location of a town equidistant from all three substations, and write an equation for the circle.
$(4, 1)$; $(x - 4)^2 + (y - 1)^2 = 26$

3 Practice/Apply

Study Notebook

Have students—
- add the definitions/examples of the vocabulary terms to their Vocabulary Builder worksheets for Chapter 10.
- include examples of how to write the equation for a circle and how to graph a circle given various information.
- include any other item(s) that they find helpful in mastering the skills in this lesson.

Answers

1. Sample answer:

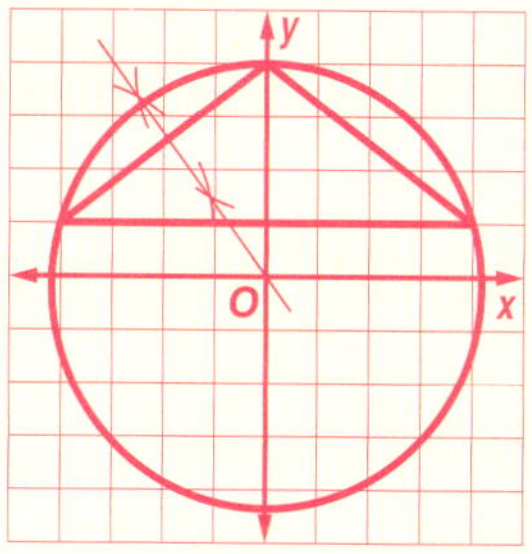

2. A circle is the locus of all points in a plane (coordinate plane) a given distance (the radius) from a given point (the center). The equation of a circle is written from knowing the location of the given point and the radius.

6.

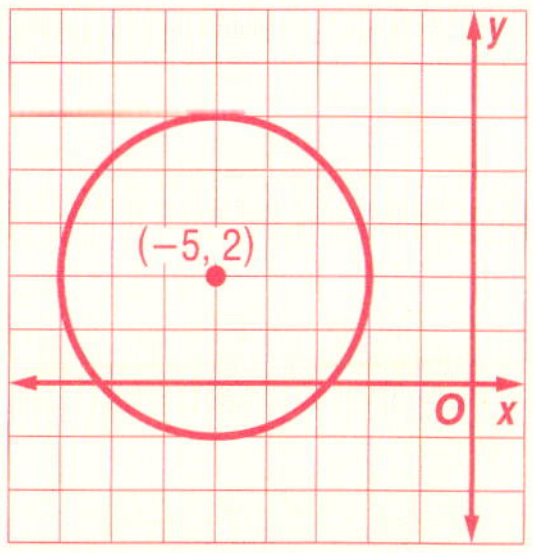

7.

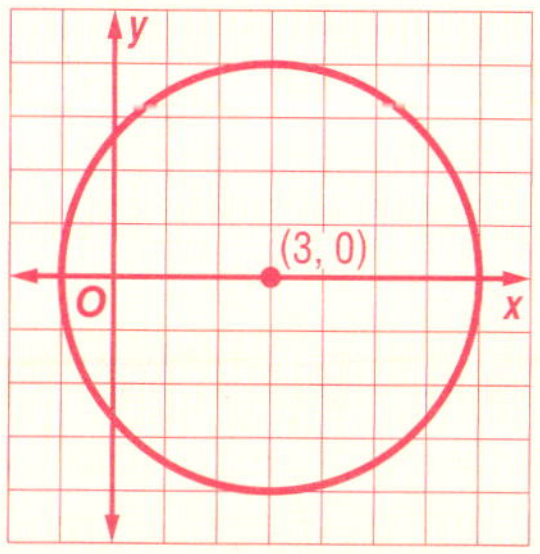

8.

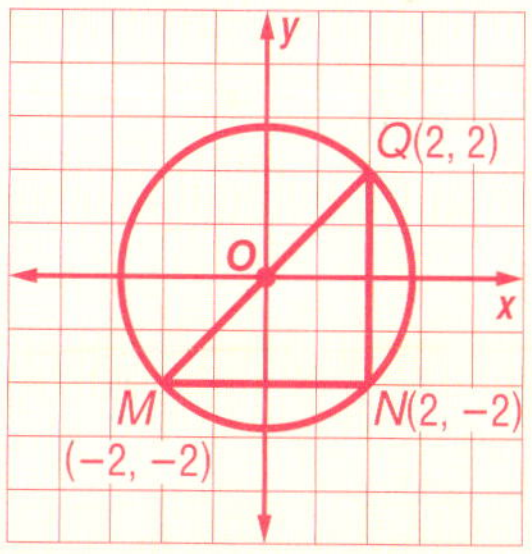

About the Exercises...

Organization by Objective

- **Equation of a Circle:** 10–23
- **Graph Circles:** 24–31

Odd/Even Assignments

Exercises 10–33 are structured so that students practice the same concepts whether they are assigned odd or even problems.

Assignment Guide

Basic: 11, 15–21 odd, 25–37 odd, 41, 43, 45–56

Average: 11–41 odd, 43, 45–56

Advanced: 10–42 even, 43–56

Application

9. **WEATHER** Meteorologists track severe storms using Doppler radar. A polar grid is used to measure distances as the storms progress. If the center of the radar screen is the origin and each ring is 10 miles farther from the center, what is the equation of the fourth ring? $x^2 + y^2 = 1600$

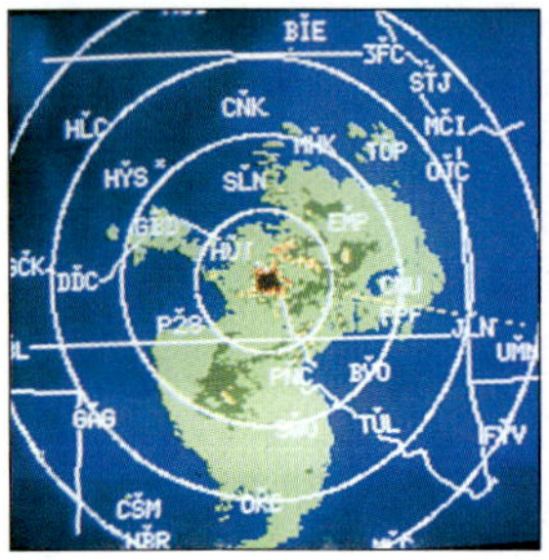

★ indicates increased difficulty

Practice and Apply

Homework Help

For Exercises	See Examples
10–17	1
18–23	2
24–29	3
30–31	4

Extra Practice
See page 776.

Write an equation for each circle.

10. center at origin, $r = 3$ $x^2 + y^2 = 9$
11. center at $(-2, -8)$, $r = 5$
12. center at $(1, -4)$, $r = \sqrt{17}$
★ 13. center at $(0, 0)$, $d = 12$
14. center at $(5, 10)$, $r = 7$
15. center at $(0, 5)$, $d = 20$
16. center at $(-8, 8)$, $d = 16$
17. center at $(-3, -10)$, $d = 24$
18. a circle with center at $(-3, 6)$ and a radius with endpoint at $(0, 6)$
19. a circle with a diameter that has endpoints at $(2, -2)$ and $(-2, 2)$ $x^2 + y^2 = 8$
20. a circle with a diameter that has endpoints at $(-7, -2)$ and $(-15, 6)$
21. a circle with center at $(-2, 1)$ and a radius with endpoint at $(1, 0)$

★ 22. a circle with $d = 12$ and a center translated 18 units left and 7 units down from the origin $(x + 18)^2 + (y + 7)^2 = 36$

★ 23. a circle with its center in quadrant I, radius of 5 units, and tangents $x = 2$ and $y = 3$ $(x - 7)^2 + (y - 8)^2 = 25$

Graph each equation. 24–29. See margin.

24. $x^2 + y^2 = 25$
25. $x^2 + y^2 = 36$
26. $x^2 + y^2 - 1 = 0$
27. $x^2 + y^2 - 49 = 0$
28. $(x - 2)^2 + (y - 1)^2 = 4$
29. $(x + 1)^2 + (y + 2)^2 = 9$

Write an equation of the circle containing each set of points. Copy and complete the graph of the circle.

30.

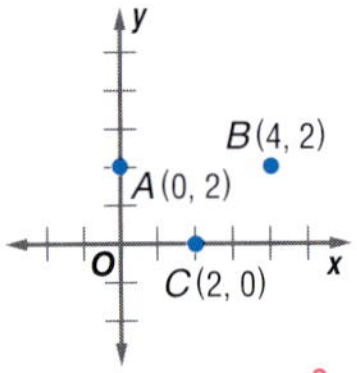

$(x - 2)^2 + (y - 2)^2 = 4$

31.

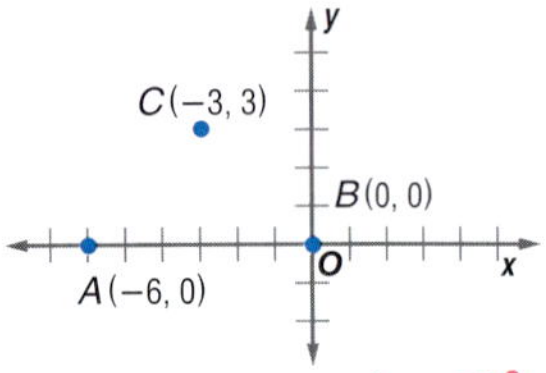

$(x + 3)^2 + y^2 = 9$

32. Find the radius of a circle with equation $(x - 2)^2 + (y - 2)^2 = r^2$ that contains the point at $(2, 5)$. 3

33. Find the radius of a circle with equation $(x - 5)^2 + (y - 3)^2 = r^2$ that contains the point at $(5, 1)$. 2

34. Refer to the Examine part of Example 4. Verify the coordinates of the center by solving a system of equations that represent the perpendicular bisectors. See p. 589B.

Answers

11. $(x + 2)^2 + (y + 8)^2 = 25$

12. $(x - 1)^2 + (y + 4)^2 = 17$

13. $x^2 + y^2 = 36$

14. $(x - 5)^2 + (y - 10)^2 = 49$

15. $x^2 + (y - 5)^2 = 100$

16. $(x + 8)^2 + (y - 8)^2 = 64$

17. $(x + 3)^2 + (y + 10)^2 = 144$

18. $(x + 3)^2 + (y - 6)^2 = 9$

20. $(x + 11)^2 + (y - 2)^2 = 32$

21. $(x + 2)^2 + (y - 1)^2 = 10$

24.

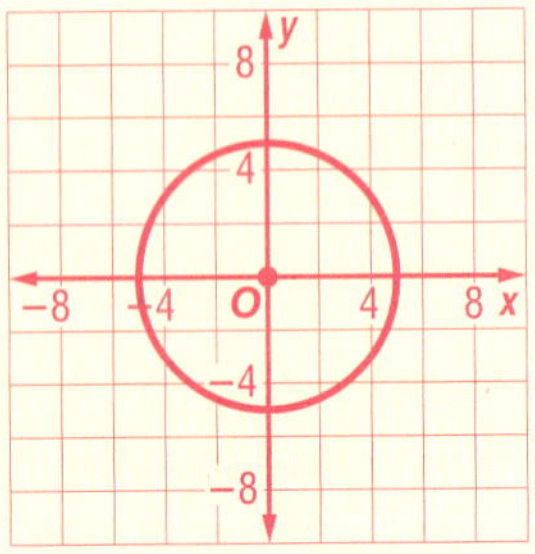

25.

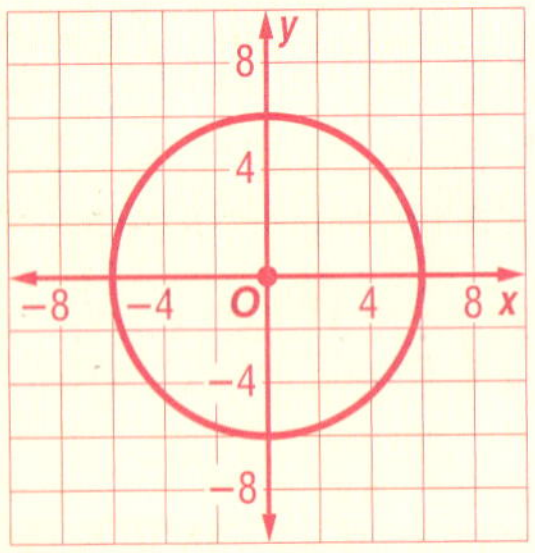

26.

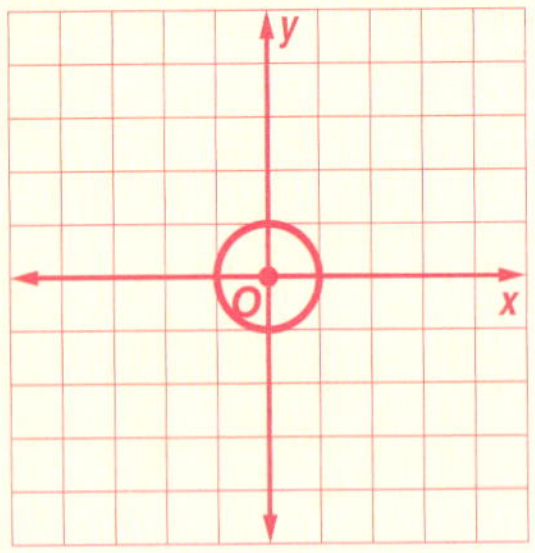

27.

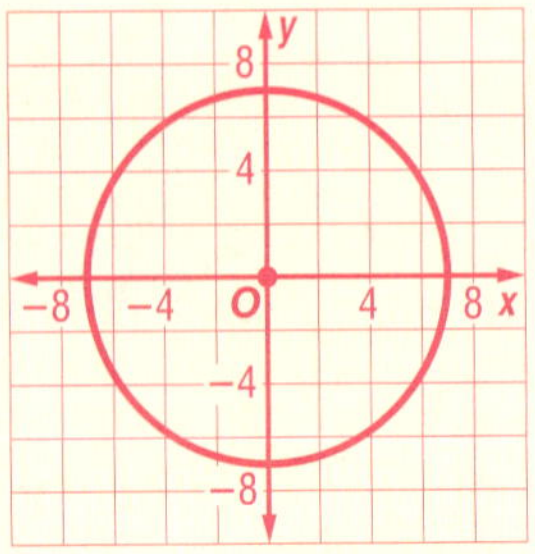

28.

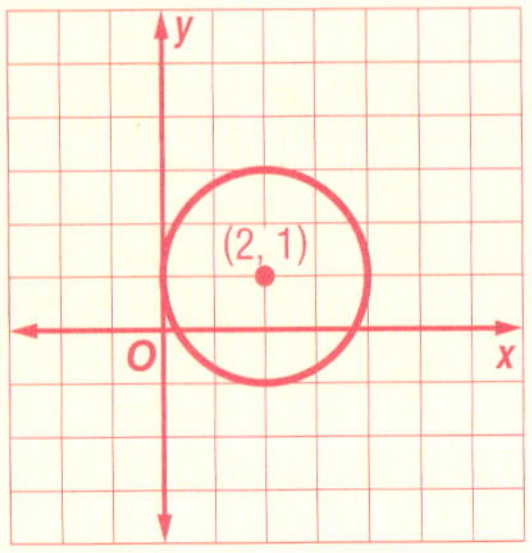

29.

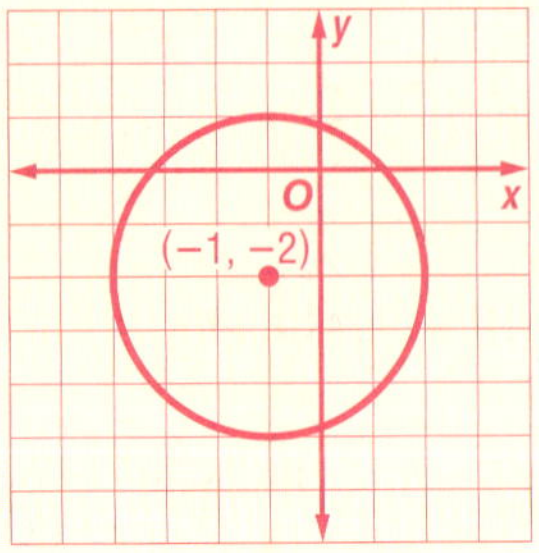

AERODYNAMICS **For Exercises 35–37, use the following information.**
The graph shows cross sections of spherical sound waves produced by a supersonic airplane. When the radius of the wave is 1 unit, the plane is 2 units from the origin. A wave of radius 3 occurs when the plane is 6 units from the center.

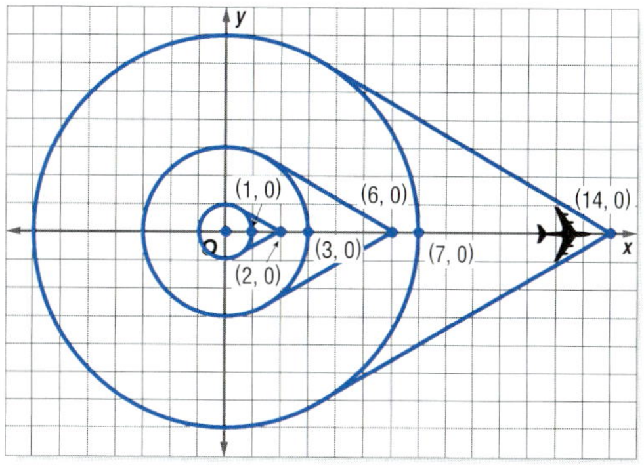

35. $x^2 + y^2 = 49$

35. Write the equation of the circle when the plane is 14 units from the center.
36. What type of circles are modeled by the cross sections? **concentric circles**
37. What is the radius of the circle for a plane 26 units from the center? **13**

★ 38. The equation of a circle is $(x - 6)^2 + (y + 2)^2 = 36$. Determine whether the line $y = 2x - 2$ is a secant, a tangent, or neither of the circle. Explain. **See margin.**

★ 39. The equation of a circle is $x^2 - 4x + y^2 + 8y = 16$. Find the center and radius of the circle. **(2, −4); $r = 6$**

40. **WEATHER** The geographic center of Tennessee is near Murfreesboro. The closest Doppler weather radar is in Nashville. If Murfreesboro is designated as the origin, then Nashville has coordinates (−58, 55), where each unit is one mile. If the radar has a radius of 80 miles, write an equation for the circle that represents the radar coverage from Nashville. $(x + 58)^2 + (y - 55)^2 = 6400$

41. **RESEARCH** Use the Internet or other materials to find the closest Doppler radar to your home. Write an equation of the circle for the radar coverage if your home is the center. **See students' work.**

More About. . .

Space Travel

The Apollo program was designed to successfully land a man on the moon. The first landing was July 20, 1969. There were a total of six landings on the moon during 1969–1972.

Source: www.infoplease.com

42. **SPACE TRAVEL** Apollo 8 was the first manned spacecraft to orbit the moon at an average altitude of 185 kilometers above the moon's surface. Determine an equation to model a single circular orbit of the Apollo 8 command module if the radius of the moon is 1740 kilometers. Let the center of the moon be at the origin. $x^2 + y^2 = 3{,}705{,}625$

43. **CRITICAL THINKING** Determine the coordinates of any intersection point of the graphs of each pair of equations.
a. $x^2 + y^2 = 9$, $y = x + 3$ **(0, 3) or (−3, 0)**
b. $x^2 + y^2 = 25$, $x^2 + y^2 = 9$ **none**
c. $(x + 3)^2 + y^2 = 9$, $(x - 3)^2 + y^2 = 9$ **(0, 0)**

44. **WRITING IN MATH** Answer the question that was posed at the beginning of the lesson. **See margin.**

What kind of equations describe the ripples of a splash?

Include the following in your answer:
- the general form of the equation of a circle, and
- the equations of five ripples if each ripple is 3 inches farther from the center.

www.geometryonline.com/self_check_quiz/fcat

Answers

38. secant, because it intersects the circle at (0, −2) and (2.4, 2.8)

44. Sample answer: Equations of concentric circles; answers should include the following.
- $(x - h)^2 + (y - k)^2 = r^2$
- $x^2 + y^2 = 9$, $x^2 + y^2 = 36$, $x^2 + y^2 = 81$, $x^2 + y^2 = 144$, $x^2 + y^2 = 225$

Enrichment, p. 588

Equations of Circles and Tangents

Recall that the circle whose radius is r and whose center has coordinates (h, k) is the graph of $(x - h)^2 + (y - k)^2 = r^2$. You can use this idea and what you know about circles and tangents to find an equation of the circle that has a given center and is tangent to a given line.

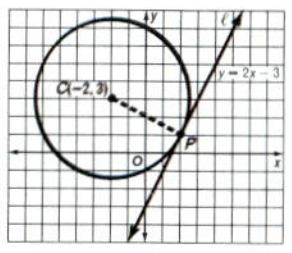

Use the following steps to find an equation for the circle that has center $C(-2, 3)$ and is tangent to the graph $y = 2x - 3$. Refer to the figure.

1. State the slope of the line ℓ that has equation $y = 2x - 3$.
2

Study Guide and Intervention, p. 583 (shown) and p. 584

Equation of a Circle A **circle** is the locus of points in a plane equidistant from a given point. You can use this definition to write an equation of a circle.

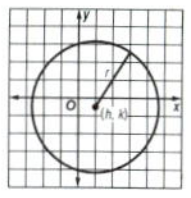

Standard Equation of a Circle	An equation for a circle with center at (h, k) and a radius of r units is $(x - h)^2 + (y - k)^2 = r^2$.

Example **Write an equation for a circle with center (−1, 3) and radius 6.**
Use the formula $(x - h)^2 + (y - k)^2 = r^2$ with $h = -1$, $k = 3$, and $r = 6$.

$(x - h)^2 + (y - k)^2 = r^2$ Equation of a circle
$(x - (-1))^2 + (y - 3)^2 = 6^2$ Substitution
$(x + 1)^2 + (y - 3)^2 = 36$ Simplify.

Exercises

Write an equation for each circle.

1. center at (0, 0), $r = 8$
$x^2 + y^2 = 64$
2. center at (−2, 3), $r = 5$
$(x + 2)^2 + (y - 3)^2 = 25$
3. center at (2, −4), $r = 1$
$(x - 2)^2 + (y + 4)^2 = 1$
4. center at (−1, −4), $r = 2$
$(x + 1)^2 + (y + 4)^2 = 4$
5. center at (−2, −6), diameter = 8
$(x + 2)^2 + (y + 6)^2 = 16$
6. center at $\left(-\frac{1}{2}, \frac{1}{4}\right)$, $r = \sqrt{3}$
$\left(x + \frac{1}{2}\right)^2 + \left(y - \frac{1}{4}\right)^2 = 3$
7. center at the origin, diameter = 4
$x^2 + y^2 = 4$
8. center at $\left(1, -\frac{5}{8}\right)$, $r = \sqrt{5}$
$(x - 1)^2 + \left(y + \frac{5}{8}\right)^2 = 5$
9. Find the center and radius of a circle with equation $x^2 + y^2 = 20$.
center (0, 0); radius $2\sqrt{5}$
10. Find the center and radius of a circle with equation $(x + 4)^2 + (y + 3)^2 = 16$.
center (−4, −3); radius 4

Skills Practice, p. 585 and Practice, p. 586 (shown)

Write an equation for each circle.

1. center at origin, $r = 7$
$x^2 + y^2 = 49$
2. center at (0, 0), $d = 18$
$x^2 + y^2 = 81$
3. center at (−7, 11), $r = 8$
$(x + 7)^2 + (y - 11)^2 = 64$
4. center at (12, −9), $d = 22$
$(x - 12)^2 + (y + 9)^2 = 121$
5. center at (−6, −4), $r = \sqrt{5}$
$(x + 6)^2 + (y + 4)^2 = 5$
6. center at (3, 0), $d = 28$
$(x - 3)^2 + y^2 = 196$
7. a circle with center at (−5, 3) and a radius with endpoint (2, 3)
$(x + 5)^2 + (y - 3)^2 = 49$
8. a circle whose diameter has endpoints (4, 6) and (−2, 6)
$(x - 1)^2 + (y - 6)^2 = 9$

Graph each equation.

9. $x^2 + y^2 = 4$

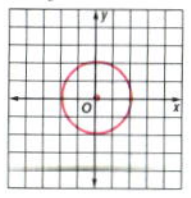

10. $(x + 3)^2 + (y - 3)^2 = 9$

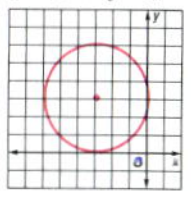

11. **EARTHQUAKES** When an earthquake strikes, it releases seismic waves that travel in concentric circles from the epicenter of the earthquake. Seismograph stations monitor seismic activity and record the intensity and duration of earthquakes. Suppose a station determines that the epicenter of an earthquake is located about 50 kilometers from the station. If the station is located at the origin, write an equation for the circle that represents a possible epicenter of the earthquake. $x^2 + y^2 = 2500$

Reading to Learn Mathematics, p. 587 ELL

Pre-Activity **What kind of equations describe the ripples of a splash?**

Read the introduction to Lesson 10-8 at the top of page 575 in your textbook.

In a series of concentric circles, what is the same about all the circles, and what is different? Sample answer: They all have the same center, but different radii.

Reading the Lesson

1. Identify the center and radius of each circle.
a. $(x - 2)^2 + (y - 3)^2 = 16$ (2, 3); 4
b. $(x + 1)^2 + (y + 5)^2 = 9$ (−1, −5); 3
c. $x^2 + y^2 = 49$ (0, 0); 7
d. $(x - 8)^2 + (y + 1)^2 = 36$ (8, −1); 6
e. $x^2 + (y - 10)^2 = 144$ (0, 10); 12
f. $(x + 3)^2 + y^2 = 5$ (−3, 0); $\sqrt{5}$

2. Write an equation for each circle.
a. center at origin, $r = 8$ $x^2 + y^2 = 64$
b. center at (3, 9), $r = 1$ $(x - 3)^2 + (y - 9)^2 = 1$
c. center at (−5, −6), $r = 10$ $(x + 5)^2 + (y + 6)^2 = 100$
d. center at (0, −7), $r = 7$ $x^2 + (y + 7)^2 = 49$
e. center at (12, 0), $d = 12$ $(x - 12)^2 + y^2 = 36$
f. center at (−4, 8), $d = 22$ $(x + 4)^2 + (y - 8)^2 = 121$
g. center at (4.5, −3.5), $r = 1.5$ $(x - 4.5)^2 + (y + 3.5)^2 = 2.25$
h. center at (0, 0), $r = \sqrt{13}$ $x^2 + y^2 = 13$

3. Write an equation for each circle.

a.
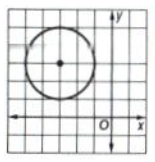

$(x + 3)^2 + (y - 3)^2 = 1$

b.
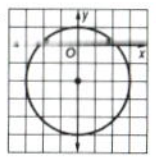

$x^2 + (y + 2)^2 = 9$

c.
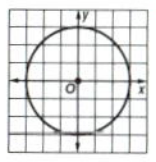

$x^2 + y^2 = 9$

d.
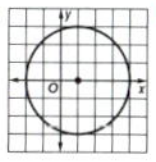

$(x - 1)^2 + y^2 = 9$

Helping You Remember

4. A good way to remember a new mathematical formula or equation is to relate it to one you already know. How can you use the Distance Formula to help you remember the standard equation of a circle? Sample answer: Use the Distance Formula to find the distance between the center (h, k) and a general point (x, y) on the circle. Square each side to obtain the standard equation of a circle.

Lesson 10-8 Equations of Circles 579

4 Assess

Open-Ended Assessment

Speaking Allow pairs of students to quiz each other with selected questions from the Practice and Apply section. Let students take turns calling out the equations of circles, naming the centers of circles, and stating the lengths of radii.

Assessment Options

Quiz (Lessons 10-7 and 10-8) is available on p. 604 of the *Chapter 10 Resource Masters*.

FCAT Practice
Standardized Test Practice

45. Which of the following is an equation of a circle with center at $(-2, 7)$ and a diameter of 18? **B**

Ⓐ $x^2 + y^2 - 4x + 14y + 53 = 324$ Ⓑ $x^2 + y^2 + 4x - 14y + 53 = 81$

Ⓒ $x^2 + y^2 - 4x + 14y + 53 = 18$ Ⓓ $x^2 + y^2 + 4x - 14y + 53 = 3$

46. ALGEBRA Jordan opened a one-gallon container of milk and poured one pint of milk into his glass. What is the fractional part of one gallon left in the container? **D**

Ⓐ $\frac{1}{8}$ Ⓑ $\frac{1}{2}$ Ⓒ $\frac{3}{4}$ Ⓓ $\frac{7}{8}$

Maintain Your Skills

Mixed Review **Find each measure if $EX = 24$ and $DE = 7$.** *(Lesson 10-7)*

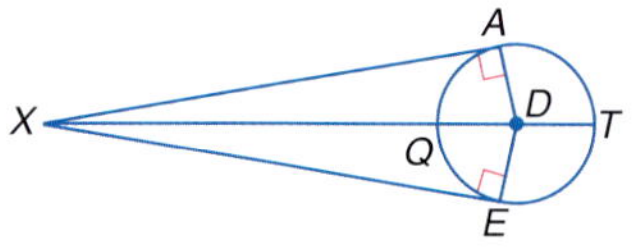

47. AX **24** **48.** DX **25**

49. QX **18** **50.** TX **32**

Find x. *(Lesson 10-6)*

51. **59**

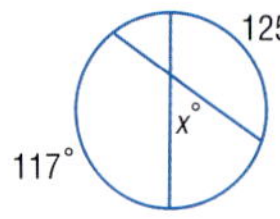

52. **35**

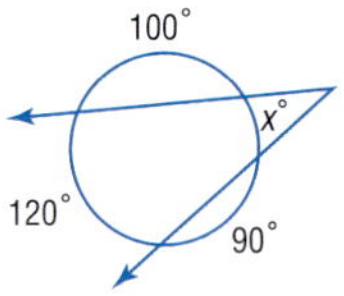

53. **20**

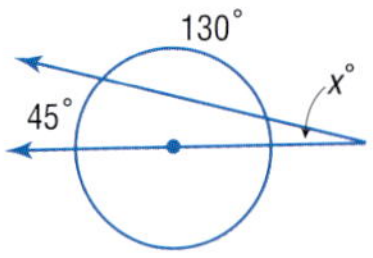

For Exercises 54 and 55, use the following information.
Triangle ABC has vertices $A(-3, 2)$, $B(4, -1)$, and $C(0, -4)$.

54. What are the coordinates of the image after moving $\triangle ABC$ 3 units left and 4 units up? *(Lesson 9-2)* **(−6, 6), (1, 3), (−3, 0)**

55. What are the coordinates of the image of $\triangle ABC$ after a reflection in the y-axis? *(Lesson 9-1)* **(3, 2), (−4, −1), (0, −4)**

56. CRAFTS For a Father's Day present, a kindergarten class is making foam plaques. The edge of each plaque is covered with felt ribbon all the way around with 1 inch overlap. There are 25 children in the class. How much ribbon does the teacher need to buy for all 25 children to complete this craft? *(Lesson 1-6)* **1125 in. or 31.25 yd**

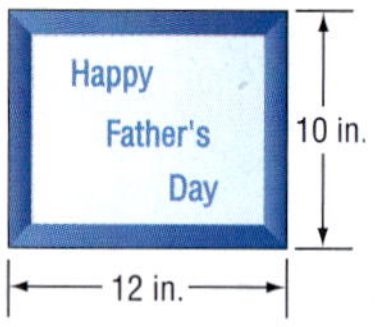

WebQuest Internet Project

"Geocaching" Sends Folks on a Scavenger Hunt

It's time to complete your project. Use the information and data you have gathered about designing a treasure hunt to prepare a portfolio or Web page. Be sure to include illustrations and/or tables in the presentation.

www.geometryonline.com/webquest

580 Chapter 10 Circles

Chapter 10

Study Guide and Review

Vocabulary and Concept Check

arc (p. 530)	circumference (p. 523)	major arc (p. 530)	radius (p. 522)
center (p. 522)	circumscribed (p. 537)	minor arc (p. 530)	secant (p. 561)
central angle (p. 529)	diameter (p. 522)	pi (π) (p. 524)	semicircle (p. 530)
chord (p. 522)	inscribed (p. 537)	point of tangency (p. 552)	tangent (p. 552)
circle (p. 522)	intercepted (p. 544)		

A complete list of postulates and theorems can be found on pages R1–R8.

Exercises **Choose the letter of the term that best matches each phrase.**

1. arcs of a circle that have exactly one point in common a
2. a line that intersects a circle in exactly one point j
3. an angle with a vertex that is on the circle and with sides containing chords of the circle h
4. a line that intersects a circle in exactly two points i
5. an angle with a vertex that is at the center of the circle b
6. arcs that have the same measure f
7. the distance around a circle d
8. circles that have the same radius g
9. a segment that has its endpoints on the circle c
10. circles that have different radii, but the same center e

a. adjacent arcs
b. central angle
c. chord
d. circumference
e. concentric circles
f. congruent arcs
g. congruent circles
h. inscribed angle
i. secant
j. tangent

Lesson-by-Lesson Review

10-1 Circles and Circumference

See pages 522–528.

Concept Summary

- The diameter of a circle is twice the radius.
- The circumference C of a circle with diameter d or a radius of r can be written in the form $C = \pi d$ or $C = 2\pi r$.

Example **Find r to the nearest hundredth if $C = 76.2$ feet.**

$C = 2\pi r$	Circumference formula
$76.2 = 2\pi r$	Substitution
$\frac{76.2}{2\pi} = r$	Divide each side by 2π.
$12.13 \approx r$	Use a calculator.

11. 7.5 in.; 47.12 in. 12. 12.8 m, 40.21 m
13. 10.82 yd; 21.65 yd 14. 26 cm; 163.36 cm

Exercises **The radius, diameter, or circumference of a circle is given. Find the missing measures. Round to the nearest hundredth if necessary.** *See Example 4 on page 524.*

11. $d = 15$ in., $r =$ __?__, $C =$ __?__
12. $r = 6.4$ m, $d =$ __?__, $C =$ __?__
13. $C = 68$ yd, $r =$ __?__, $d =$ __?__
14. $d = 52$ cm, $r =$ __?__, $C =$ __?__
15. $C = 138$ ft, $r =$ __?__, $d =$ __?__
16. $r = 11$ mm, $d =$ __?__, $C =$ __?__

21.96 ft; 43.93 ft 22 mm; 69.12 mm

www.geometryonline.com/vocabulary_review

Chapter 10

Study Guide and Review

Vocabulary and Concept Check

- This alphabetical list of vocabulary terms in Chapter 10 includes a page reference where each term was introduced.
- **Assessment** A vocabulary test/review for Chapter 10 is available on p. 602 of the *Chapter 10 Resource Masters*.

Lesson-by-Lesson Review

For each lesson,

- the main ideas are summarized,
- additional examples review concepts, and
- practice exercises are provided.

Vocabulary PuzzleMaker

ELL The Vocabulary PuzzleMaker software improves students' mathematics vocabulary using four puzzle formats—crossword, scramble, word search using a word list, and word search using clues. Students can work on a computer screen or from a printed handout.

MindJogger Videoquizzes

ELL MindJogger Videoquizzes provide an alternative review of concepts presented in this chapter. Students work in teams in a game show format to gain points for correct answers. The questions are presented in three rounds.

Round 1 Concepts (5 questions)
Round 2 Skills (4 questions)
Round 3 Problem Solving (4 questions)

FOLDABLES™ Study Organizer

For more information about Foldables, see *Teaching Mathematics with Foldables.*

Have students look through the chapter to make sure they have included notes and examples in their Foldables for each lesson of Chapter 10.

Encourage students to refer to their Foldables while completing the Study Guide and Review and to use them in preparing for the Chapter Test.

10-2 Angles and Arcs

See pages 529–535.

Concept Summary

- The sum of the measures of the central angles of a circle with no interior points in common is 360.
- The measure of each arc is related to the measure of its central angle.
- The length of an arc is proportional to the length of the circumference.

Examples **In $\odot P$, $m\angle MPL = 65$ and $\overline{NP} \perp \overline{PL}$.**

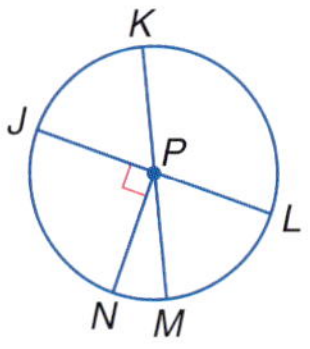

1 Find $m\widehat{NM}$.

$\widehat{NM}$ is a minor arc, so $m\widehat{NM} = m\angle NPM$.

$\angle JPN$ is a right angle and $m\angle MPL = 65$, so $m\angle NPM = 25$.

$m\widehat{NM} = 25$

2 Find $m\widehat{NJK}$.

$\widehat{NJK}$ is composed of adjacent arcs, $\widehat{NJ}$ and $\widehat{JK}$. $\angle MPL \cong \angle JPK$, so $m\angle JPK = 65$.

$m\widehat{NJ} = m\angle NPJ$ or 90	$\angle NPJ$ is a right angle
$m\widehat{NJK} = m\widehat{NJ} + m\widehat{JK}$	Arc Addition Postulate
$m\widehat{NJK} = 90 + 65$ or 155	Substitution

Exercises **Find each measure.**
See Example 1 on page 529.

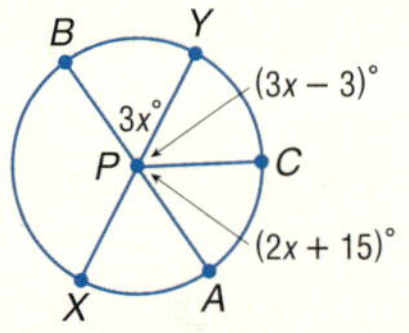

17. $m\widehat{YC}$ 60
18. $m\widehat{BC}$ 123
19. $m\widehat{BX}$ 117
20. $m\widehat{BCA}$ 180

In $\odot G$, $m\angle AGB = 30$ and $\overline{CG} \perp \overline{GD}$. Find each measure. *See Example 2 on page 531.*

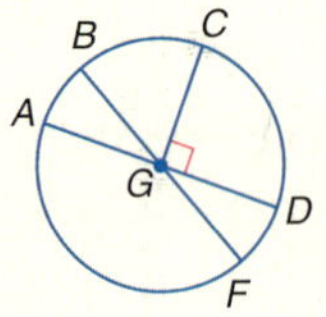

21. $m\widehat{AB}$ 30
22. $m\widehat{BC}$ 60
23. $m\widehat{FD}$ 30
24. $m\widehat{CDF}$ 120
25. $m\widehat{BCD}$ 150
26. $m\widehat{FAB}$ 180

Find the length of the indicated arc in each $\odot I$. *See Example 4 on page 532.*

27. $\widehat{DG}$ if $m\angle DGI = 24$ and $r = 6$ $\frac{22}{5}\pi$

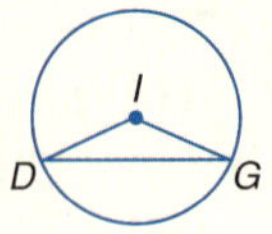

28. $\widehat{WN}$ if $\triangle IWN$ is equilateral and $WN = 5$ $\frac{5}{3}\pi$

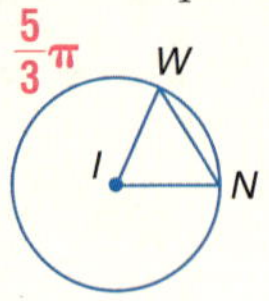

582 Chapter 10 Circles

Study Guide and Review

10-3 Arcs and Chords

See pages 536–543.

Concept Summary

- The endpoints of a chord are also the endpoints of an arc.
- Diameters perpendicular to chords bisect chords and intercepted arcs.

Examples

Circle L has a radius of 32 centimeters. $\overline{LH} \perp \overline{GJ}$, and $GJ = 40$ centimeters. Find LK.

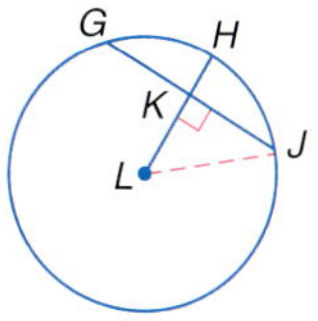

Draw radius $\overline{LJ}$. $LJ = 32$ and $\triangle LKJ$ is a right triangle.

$\overline{LH}$ bisects $\overline{GJ}$, since they are perpendicular.

$KJ = \frac{1}{2}(GJ)$ Definition of segment bisector

$= \frac{1}{2}(40)$ or 20 $GJ = 40$, and simplify.

Use the Pythagorean Theorem to find LK.

$(LK)^2 + (KJ)^2 = (LJ)^2$ Pythagorean Theorem

$(LK)^2 + 20^2 = 32^2$ $KJ = 20$, $LJ = 32$

$(LK)^2 + 400 = 1024$ Simplify.

$(LK)^2 = 624$ Subtract 400 from each side.

$LK = \sqrt{624}$ Take the square root of each side.

$LK \approx 24.98$ Use a calculator.

Exercises **In $\odot R$, $SU = 20$, $YW = 20$, and $m\widehat{YX} = 45$. Find each measure.** *See Example 3 on page 538.*

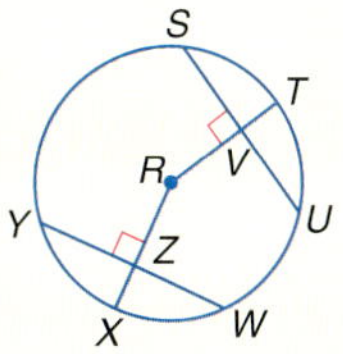

29. SV 10
30. WZ 10
31. UV 10
32. $m\widehat{YW}$ 90
33. $m\widehat{ST}$ 45
34. $m\widehat{SU}$ 90

10-4 Inscribed Angles

See pages 544–551.

Concept Summary

- The measure of the inscribed angle is half the measure of its intercepted arc.
- The angles of inscribed polygons can be found by using arc measures.

Example

ALGEBRA **Triangles FGH and FHJ are inscribed in $\odot K$ with $\widehat{FG} \cong \widehat{FJ}$. Find x if $m\angle 1 = 6x - 5$, and $m\angle 2 = 7x + 4$.**

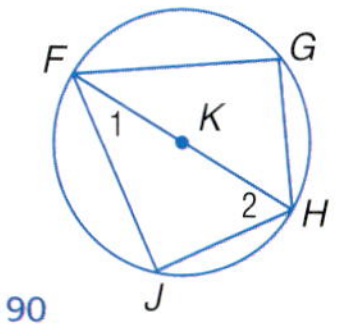

FJH is a right angle because $\widehat{FJH}$ is a semicircle.

$m\angle 1 + m\angle 2 + m\angle FJH = 180$ Angle Sum Theorem

$(6x - 5) + (7x + 4) + 90 = 180$ $m\angle 1 = 6x - 5$, $m\angle 2 = 7x + 4$, $m\angle FJH = 90$

$13x + 89 = 180$ Simplify.

$x = 7$ Solve for x.

Chapter 10 Study Guide and Review 583

Exercises **Find the measure of each numbered angle.**
See Example 1 on page 545.

35. 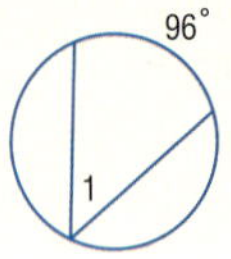

48

36. 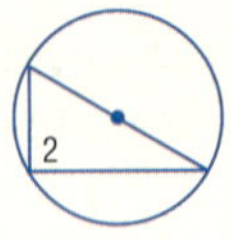

90

37. 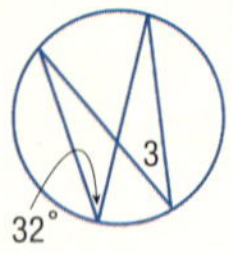

32

Find the measure of each numbered angle for each situation given.
See Example 4 on page 547.

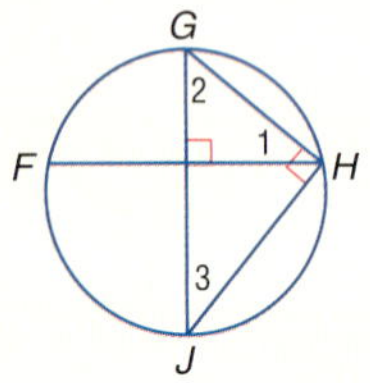

38. $m\widehat{GH} = 78$ $m\angle 1 = m\angle 3 = 39$, $m\angle 2 = 51$

39. $m\angle 2 = 2x$, $m\angle 3 = x$ $m\angle 1 = m\angle 3 = 30$, $m\angle 2 = 60$

40. $m\widehat{JH} = 114$ $m\angle 2 = 57$, $m\angle 3 = m\angle 1 = 33$

10-5 *Tangents*

See pages 552–558.

Concept Summary

- A line that is tangent to a circle intersects the circle in exactly one point.
- A tangent is perpendicular to a radius of a circle.
- Two segments tangent to a circle from the same exterior point are congruent.

Example **ALGEBRA** **Given that the perimeter of $\triangle ABC = 25$, find x. Assume that segments that appear tangent to circles are tangent.**

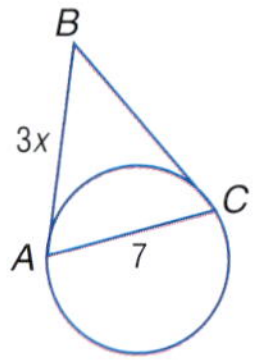

In the figure, $\overline{AB}$ and $\overline{AC}$ are drawn from the same exterior point and are tangent to $\odot Q$. So $\overline{AB} \cong \overline{AC}$.

The perimeter of the triangle, $AB + BC + AC$, is 25.

$AB + BC + AC = 25$	Definition of perimeter
$3x + 3x + 7 = 25$	$AB = BC = 3x$, $AC = 7$
$6x + 7 = 25$	Simplify.
$6x = 18$	Subtract 7 from each side.
$x = 3$	Divide each side by 6.

Exercises **Find x. Assume that segments that appear to be tangent are tangent.**
See Example 3 on page 554.

41. 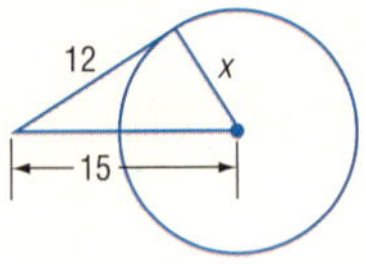

9

42. 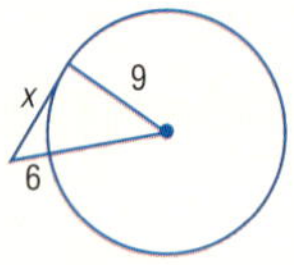

12

43. 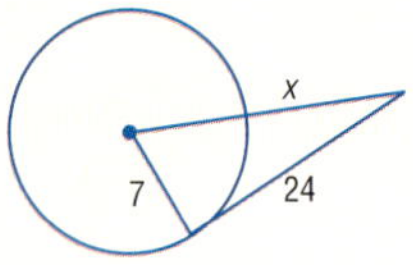

18

584 Chapter 10 Circles

10-6 Secants, Tangents, and Angle Measures

See pages 561–568.

Concept Summary

- The measure of an angle formed by two secant lines is half the positive difference of its intercepted arcs.
- The measure of an angle formed by a secant and tangent line is half its intercepted arc.

Example **Find x.**

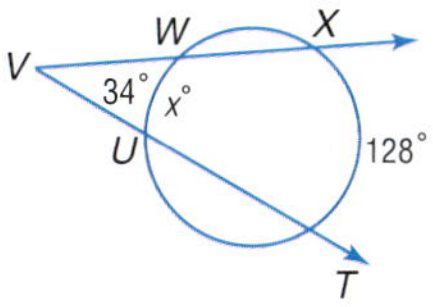

$m\angle V = \frac{1}{2}\left(m\widehat{XT} - m\widehat{WU}\right)$

$34 = \frac{1}{2}(128 - x)$ Substitution

$-30 = -\frac{1}{2}x$ Simplify.

$x = 60$ Multiply each side by -2.

Exercises **Find x.** *See Example 3 on page 563.*

44. 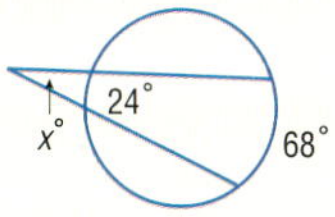

22

45. 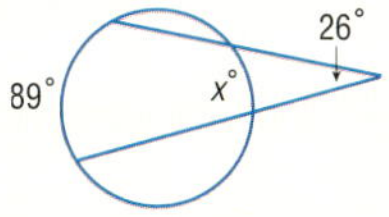

37

46. 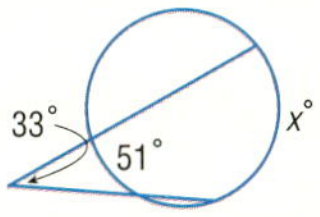

117

10-7 Special Segments in a Circle

See pages 569–574.

Concept Summary

- The lengths of intersecting chords in a circle can be found by using the products of the measures of the segments.
- The secant segment product also applies to segments that intersect outside the circle, and to a secant segment and a tangent.

Example **Find a, if $FG = 18$, $GH = 42$, and $FK = 15$.**

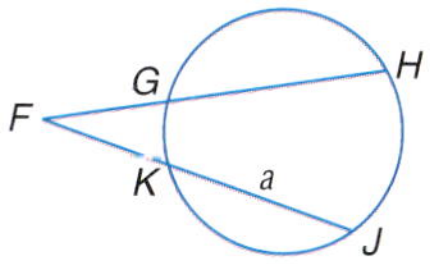

Let $KJ = a$.

$FK \cdot FJ = FG \cdot FH$ Secant Segment Products

$15(a + 15) = 18(18 + 42)$ Substitution

$15a + 225 = 1080$ Distributive Property

$15a = 855$ Subtract 225 from each side.

$a = 57$ Divide each side by 15.

Exercises **Find x to the nearest tenth. Assume that segments that appear to be tangent are tangent.** *See Examples 3 and 4 on pages 570 and 571.*

47. 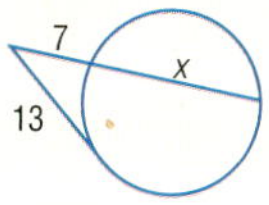

17.1

48. 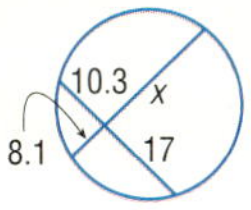

21.6

49. 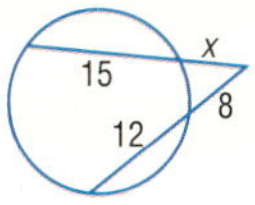

7.2

(10.3)(17) = (8.1) x

Chapter 10 Study Guide and Review 585

Chapter 10 For More ...
• Extra Practice, see pages 773–776.
• Mixed Problem Solving, see page 791.

10-8 Equations of Circles

See pages 575–580.

Concept Summary

- The coordinates of the center of a circle (h, k) and its radius r can be used to write an equation for the circle in the form $(x - h)^2 + (y - k)^2 = r^2$.
- A circle can be graphed on a coordinate plane by using the equation written in standard form.
- A circle can be graphed through any three noncollinear points on the coordinate plane.

Examples

1 Write an equation of a circle with center (−1, 4) and radius 3.

Since the center is at $(-1, 4)$ and the radius is 3, $h = -1$, $k = 4$, and $r = 3$.

$(x - h)^2 + (y - k)^2 = r^2$ Equation of a circle

$[x - (-1)]^2 + (y - 4)^2 = 3^2$ $h = -1$, $k = 4$, and $r = 3$

$(x + 1)^2 + (y - 4)^2 = 9$ Simplify.

2 Graph $(x - 2)^2 + (y + 3)^2 = 6.25$.

Identify the values of h, k, and r by writing the equation in standard form.

$(x - 2)^2 + (y + 3)^2 = 6.25$

$(x - 2)^2 + [y - (-3)]^2 = 2.5^2$

$h = 2$, $k = -3$, and $r = 2.5$

Graph the center $(2, -3)$ and use a compass to construct a circle with radius 2.5 units.

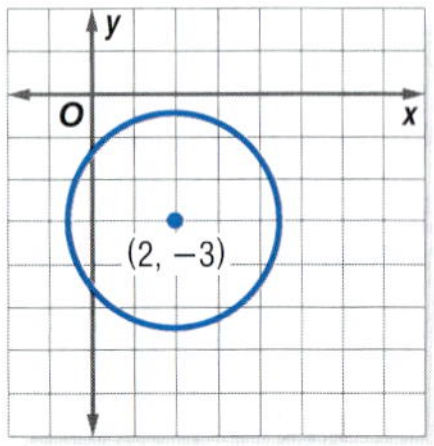

Exercises **Write an equation for each circle.** *See Examples 1 and 2 on pages 575 and 576.*

50. center at $(0, 0)$, $r = \sqrt{5}$ $x^2 + y^2 = 5$

51. center at $(-4, 8)$, $d = 6$ $(x + 4)^2 + (y - 8)^2 = 9$

52. diameter with endpoints at $(0, -4)$ and $(8, -4)$ $(x - 4)^2 + (y + 4)^2 = 16$

53. center at $(-1, 4)$ and is tangent to $x = 1$ $(x + 1)^2 + (y - 4)^2 = 4$

Graph each equation. *See Example 3 on page 576.*

54. $x^2 + y^2 = 2.25$

55. $(x - 4)^2 + (y + 1)^2 = 9$

54–55. See margin.

For Exercises 56 and 57, use the following information.

A circle graphed on a coordinate plane contains $A(0, 6)$, $B(6, 0)$, and $C(6, 6)$.

See Example 4 on page 577.

56. Write an equation of the circle. $(x - 3)^2 + (y - 3)^2 = 18$

57. Graph the circle. **See margin.**

Answers

54.

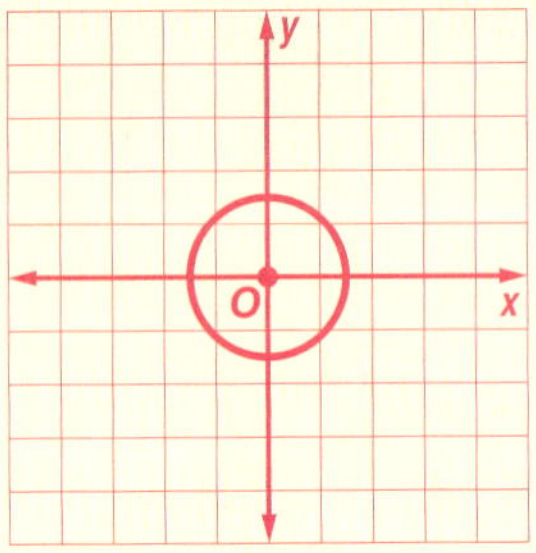

55.

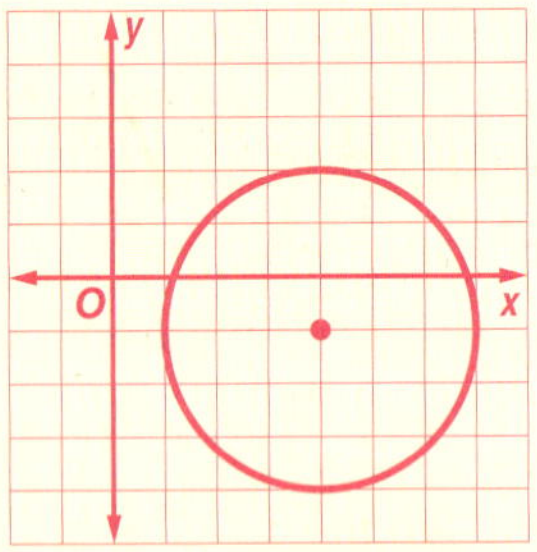

57.

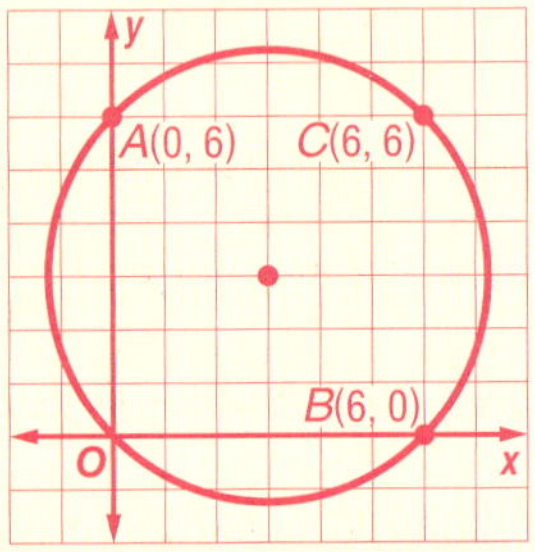

Answers (page 587)

1. Sample answer: A chord is a segment that has its endpoints on a circle. A secant contains a chord and is a line that intersects a circle in two points. A tangent is a line that intersects a circle in exactly one point and no point of the tangent lies in the interior of the circle.
2. Find the midpoint of the diameter using the Midpoint Formula with the coordinates of the diameter's endpoints.

22.

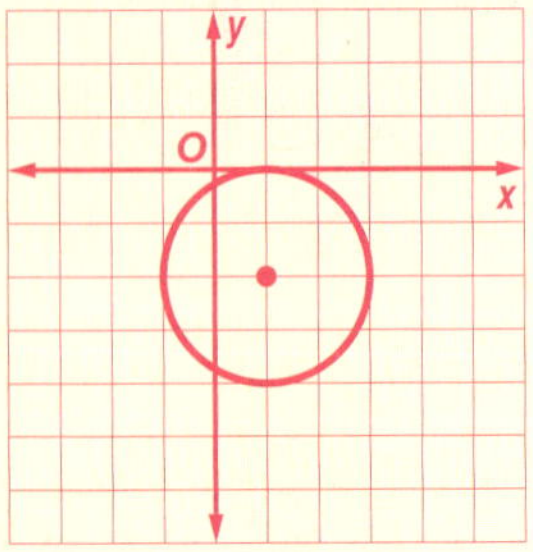

23. Sample answer:

Given: ⊙X with diameters $\overline{RS}$ and $\overline{TV}$

Prove: $\widehat{RT} \cong \widehat{VS}$

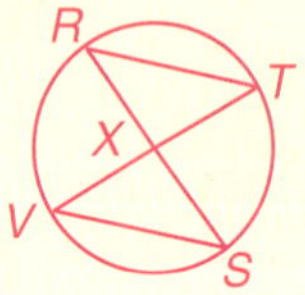

Proof:

Statements (Reasons)

1. ⊙X with diameters $\overline{RS}$ and $\overline{TV}$ (Given)
2. $\angle RXT \cong \angle VXS$ (Vertical ∠s are ≅.)
3. $m\angle RXT = m\angle VXS$ (Def. of ≅ ∠s)
4. $m\widehat{RT} = m\angle RXT$, $m\widehat{VS} = m\angle XIS$ (Measure of arc equals measure of its central angle.)
5. $m\widehat{RT} = m\widehat{VS}$ (Substitution)
6. $\widehat{RT} \cong \widehat{VS}$ (Def. of ≅ arcs)

Chapter 10 Practice Test

Vocabulary and Concepts

1. **Describe** the differences among a tangent, a secant, and a chord of a circle. **1–2. See margin.**
2. **Explain** how to find the center of a circle given the coordinates of the endpoints of a diameter.

Skills and Applications

3. Determine the radius of a circle with circumference 25π units. Round to the nearest tenth. **12.5 units**

For Questions 4–11, refer to ⊙N.

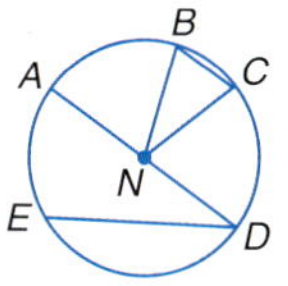

4. Name the radii of ⊙N. $\overline{NA}$, $\overline{NB}$, $\overline{NC}$, $\overline{ND}$
5. If $AD = 24$, find CN. **12**
6. Is $ED > AD$? Explain. **No; diameters are the longest chords of a circle.**
7. If AN is 5 meters long, find the exact circumference of ⊙N. **10π m**
8. If $m\angle BNC = 20$, find $m\widehat{BC}$. **20**
9. If $m\widehat{BC} = 30$ and $\widehat{AB} \cong \widehat{CD}$, find $m\widehat{AB}$. **75**
10. If $\widehat{BE} \cong \widehat{ED}$ and $m\widehat{ED} = 120$, find $m\widehat{BE}$. **120**
11. If $m\widehat{AE} = 75$, find $m\angle ADE$. **37.5**

Find x. Assume that segments that appear to be tangent are tangent.

12. **15**

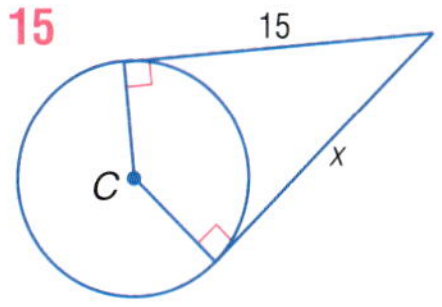

13. **4**

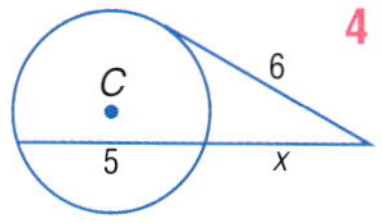

14. **9.6**

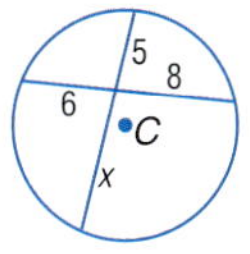

15. **4**

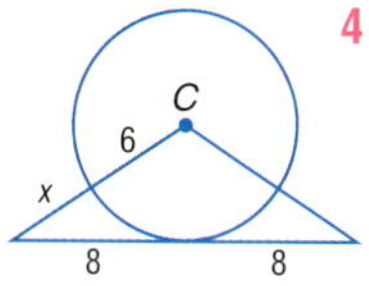

16. **3.8**

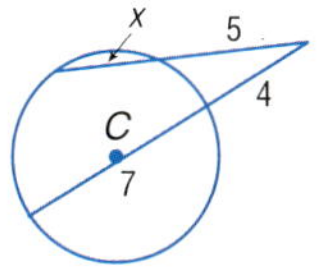

17. **145**

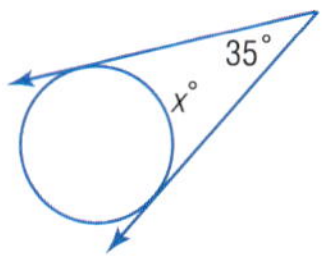

18. **10**

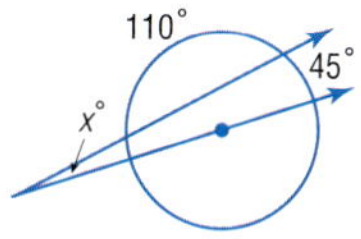

19. **50**

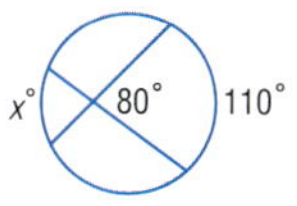

20. **AMUSEMENT RIDES** Suppose a Ferris wheel is 50 feet wide. Approximately how far does a rider travel in one rotation of the wheel? **157 ft**
21. Write an equation of a circle with center at $(-2, 5)$ and a diameter of 50. **$(x + 2)^2 + (y - 5)^2 = 625$**
22. Graph $(x - 1)^2 + (y + 2)^2 = 4$. **See margin.**
23. **PROOF** Write a two-column proof.
    Given: ⊙X with diameters $\overline{RS}$ and $\overline{TV}$
    Prove: $\widehat{RT} \cong \widehat{VS}$
    See margin.

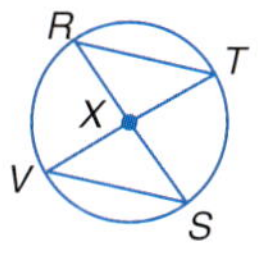

24. **CRAFTS** Takita is making bookends out of circular wood pieces as shown at the right. What is the height of the cut piece of wood? **about 7.1 in.**

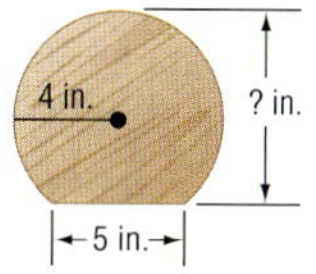

25. **STANDARDIZED TEST PRACTICE** Circle C has radius r and $ABCD$ is a rectangle. Find DB. **A**
    (A) r (B) $r\frac{\sqrt{2}}{2}$ (C) $r\sqrt{3}$ (D) $r\frac{\sqrt{3}}{2}$

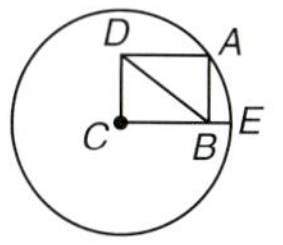

Portfolio Suggestion

Introduction After completing a chapter containing several concepts, students might benefit from going back and categorizing the concepts they found easy or challenging.

Ask Students Label two sheets of paper "Chapter 10—Concepts I Already Knew" and "Chapter 10—Concepts I Learned." Go back through each lesson and note the concepts in the lesson. Then categorize them on their pieces of paper. You can either write the name of the concept, explain it in your own words, or draw an example. Place these sheets in your portfolio.

Chapter 10 Practice Test

Assessment Options

Vocabulary Test A vocabulary test/review for Chapter 10 can be found on p. 602 of the *Chapter 10 Resource Masters*.

Chapter Tests There are six Chapter 10 Tests and an Open-Ended Assessment task available in the *Chapter 10 Resource Masters*.

Chapter 10 Tests			
Form	**Type**	**Level**	**Pages**
1	MC	basic	589–590
2A	MC	average	591–592
2B	MC	average	593–594
2C	FR	average	595–596
2D	FR	average	597–598
3	FR	advanced	599–600

MC = multiple-choice questions
FR = free-response questions

Open-Ended Assessment Performance tasks for Chapter 10 can be found on p. 601 of the *Chapter 10 Resource Masters*. A sample scoring rubric for these tasks appears on p. A31.

Unit 3 Test A unit test/review can be found on pp. 609–610 of the *Chapter 10 Resource Masters*.

TestCheck and Worksheet Builder

This **networkable software** has three modules for assessment.

- **Worksheet Builder** to make worksheets and tests.
- **Student Module** to take tests on-screen.
- **Management System** to keep student records.

Chapter 10 Standardized Test Practice

These two pages contain practice questions in the various formats that can be found on the most frequently given standardized tests.

A practice answer sheet for these two pages can be found on p. A1 of the *Chapter 10 Resource Masters*.

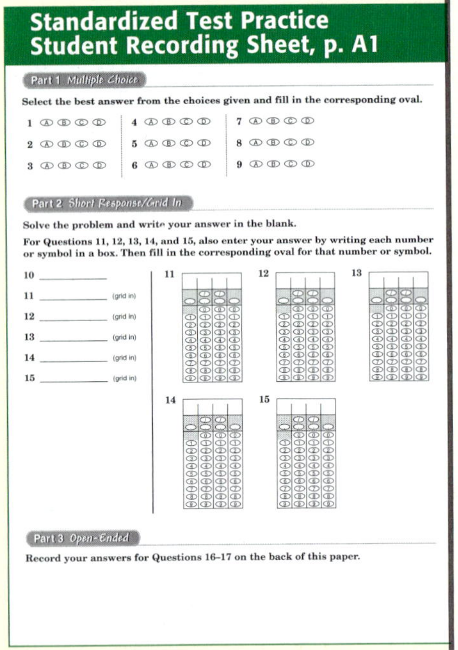
Standardized Test Practice
Student Recording Sheet, p. A1

Part 1 Multiple Choice

Select the best answer from the choices given and fill in the corresponding oval.

1 Ⓐ Ⓑ Ⓒ Ⓓ 4 Ⓐ Ⓑ Ⓒ Ⓓ 7 Ⓐ Ⓑ Ⓒ Ⓓ
2 Ⓐ Ⓑ Ⓒ Ⓓ 5 Ⓐ Ⓑ Ⓒ Ⓓ 8 Ⓐ Ⓑ Ⓒ Ⓓ
3 Ⓐ Ⓑ Ⓒ Ⓓ 6 Ⓐ Ⓑ Ⓒ Ⓓ 9 Ⓐ Ⓑ Ⓒ Ⓓ

Part 2 Short Response/Grid In

Solve the problem and write your answer in the blank.

For Questions 11, 12, 13, 14, and 15, also enter your answer by writing each number or symbol in a box. Then fill in the corresponding oval for that number or symbol.

10 ______
11 ______ (grid in)
12 ______ (grid in)
13 ______ (grid in)
14 ______ (grid in)
15 ______ (grid in)

Part 3 Open-Ended

Record your answers for Questions 16–17 on the back of this paper.

Additional Practice

See pp. 607–608 in the *Chapter 10 Resource Masters* for additional standardized test practice.

Chapter 10 Standardized Test Practice

FCAT Practice

Part 1 Multiple Choice

Record your answer on the answer sheet provided by your teacher or on a sheet of paper.

1. Which of the following shows the graph of $3y = 6x - 9$? (Prerequisite Skill) **A**

Ⓐ
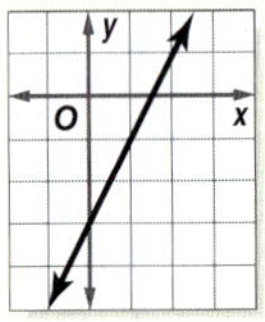

Ⓑ
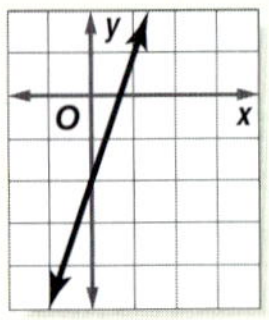

Ⓒ
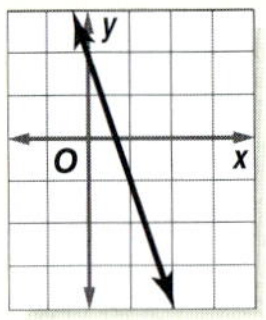

Ⓓ
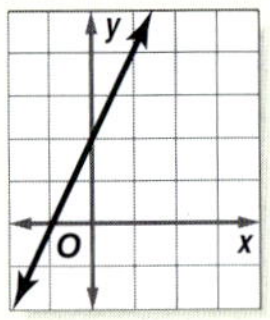

2. In Hyde Park, Main Street and Third Avenue do not meet at right angles. Use the figure below to determine the measure of $\angle 1$ if $m\angle 1 = 6x - 5$ and $m\angle 2 = 3x + 13$. (Lesson 1-5) **C**

Ⓐ 6
Ⓑ 18
Ⓒ 31
Ⓓ 36

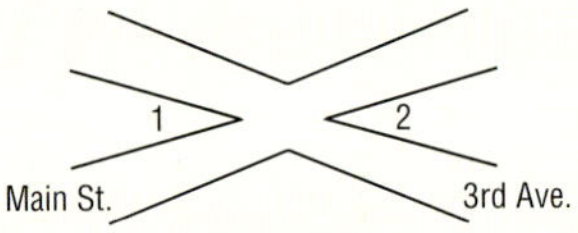

3. Part of a proof is shown below. What is the reason to justify Step b? (Lesson 2-5) **A**

Given: $4x + \frac{4}{3} = 12$ **Prove:** $x = \frac{8}{3}$

Statements	Reasons
a. $4x + \frac{4}{3} = 12$	**a.** Given
b. $3\left(4x + \frac{4}{3}\right) = 3(12)$	**b.** ?

Ⓐ Multiplication Property
Ⓑ Distributive Property
Ⓒ Cross products
Ⓓ none of the above

4. If an equilateral triangle has a perimeter of $(2x + 9)$ miles and one side of the triangle measures $(x + 2)$ miles, how long (in miles) is the side of the triangle? (Lesson 4-1) **B**

Ⓐ 3 Ⓑ 5 Ⓒ 9 Ⓓ 15

5. A pep team is holding up cards to spell out the school name. What symmetry does the card shown below have? (Lesson 9-1) **A**

Ⓐ only line symmetry
Ⓑ only point symmetry
Ⓒ both line and point symmetry
Ⓓ neither line nor point symmetry

Use the figure below for Questions 6 and 7.

6. In circle *F*, which are chords? (Lesson 10-1) **D**

Ⓐ $\overline{AD}$ and $\overline{EF}$
Ⓑ $\overline{AF}$ and $\overline{BC}$
Ⓒ $\overline{EF}$, $\overline{DF}$, and $\overline{AF}$
Ⓓ $\overline{AD}$ and $\overline{BC}$

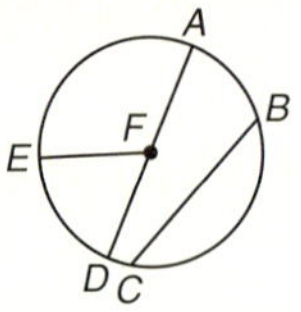

7. In circle *F*, what is the measure of $\widehat{EA}$ if $m\angle DFE$ is 36? (Lesson 10-2) **C**

Ⓐ 54 Ⓑ 104 Ⓒ 144 Ⓓ 324

8. Which statement is false? (Lesson 10-3) **C**

Ⓐ Two chords that are equidistant from the center of a circle are congruent.
Ⓑ A diameter of a circle that is perpendicular to a chord bisects the chord and its arc.
Ⓒ The measure of a major arc is the measure of its central angle.
Ⓓ Minor arcs in the same circle are congruent if their corresponding chords are congruent.

9. Which of the segments described could be a secant of a circle? (Lesson 10-6) **D**

Ⓐ intersects exactly one point on a circle
Ⓑ has its endpoints on a circle
Ⓒ one endpoint at the center of the circle
Ⓓ intersects exactly two points on a circle

The Princeton Review

Log On for Test Practice

The Princeton Review offers additional test-taking tips and practice problems at their web site. Visit www.princetonreview.com or www.review.com

TestCheck and Worksheet Builder

Special banks of standardized test questions similar to those on the SAT, ACT, TIMSS 8, NAEP 8, and state proficiency tests can be found on this CD-ROM.

Aligned and verified by The Princeton Review

FCAT Practice

Part 2 Short Response/Grid In

Record your answers on the answer sheet provided by your teacher or on a sheet of paper.

10. What is the shortest side of quadrilateral *DEFG*? (Lesson 5-3) $\overline{FG}$

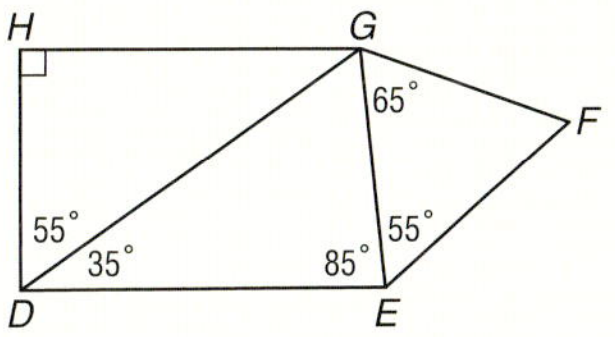

11. An architect designed a house and a garage that are similar in shape. How many feet long is $\overline{ST}$? (Lesson 6-2) **20**

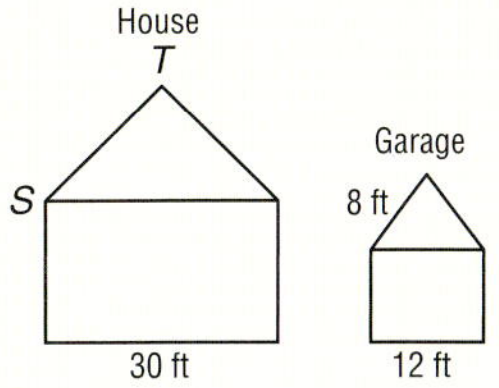

12. Two triangles are drawn on a coordinate grid. One has vertices at (0, 1), (0, 7), and (6, 4). The other has vertices at (7, 7), (10, 7), and (8.5, 10). What scale factor can be used to compare the smaller triangle to the larger? (Lesson 9-5) **2**

Use the figure below for Questions 13–15.

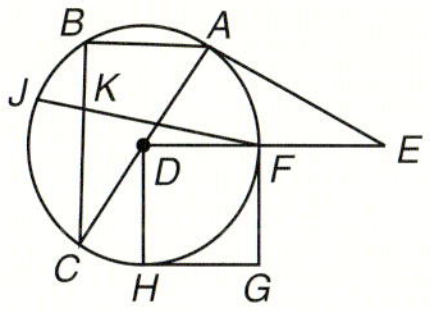

13. Point *D* is the center of the circle. What is $m\angle ABC$? (Lesson 10-4) **90**

14. $\overline{AE}$ is a tangent. If $AD = 12$ and $FE = 18$, how long is $\overline{AE}$ to the nearest tenth unit? (Lesson 10-5) **27.5**

15. Chords $\overline{JF}$ and $\overline{BC}$ intersect at *K*. If $BK = 8$, $KC = 12$, and $KF = 16$, find *JK*. (Lesson 10-7) **6**

www.geometryonline.com/standardized_test/fcat

The Princeton Review Test-Taking Tip

Question 4 If a question does not provide you with a figure that represents the problem, draw one yourself. By drawing and labeling an equilateral triangle with all the information you know, the problem becomes more understandable.

FCAT Practice

Part 3 Open Ended

Record your answers on a sheet of paper. Show your work.

16. The Johnson County High School flag is shown below. Points have been added for reference.

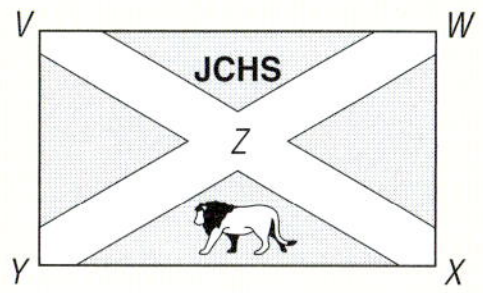

a. Which diagonal segments would have to be congruent for *VWXY* to be a rectangle? (Lesson 8-3) $\overline{VX}$ **and** $\overline{WY}$

b. Suppose the length of rectangle *VWXY* is 2 more than 3 times the width and the perimeter is 164 inches. What are the dimensions of the flag? (Lesson 1-6) **20 by 62**

17. The segment with endpoints $A(1, -2)$ and $B(1, 6)$ is the diameter of a circle.

a. Graph the points and draw the circle. (Lesson 10-1) **See margin.**

b. What is the center of the circle? (Lesson 10-1) **(1, 2)**

c. What is the length of the radius? (Lesson 10-8) **4**

d. What is the circumference of the circle? (Lesson 10-8) **8π units**

e. What is the equation of the circle? (Lesson 10-8) $(x - 1)^2 + (y - 2)^2 = 16$

Evaluating Open-Ended Assessment Questions

Open-Ended Assessment questions are graded by using a multilevel rubric that guides you in assessing a student's knowledge of a particular concept.

Goal for Question 16: Determine the relationship between segment lengths for the flag to be a parallelogram, and find the dimensions of the flag.

Goal for Question 17: Using a segment in a coordinate plane as a diameter, write an equation for a circle and find its center, radius, circumference.

Sample Scoring Rubric: The following rubric is a sample scoring device. You may wish to add more detail to this sample to meet your individual scoring needs.

Score	Criteria
4	A correct solution that is supported by well-developed, accurate explanations
3	A generally correct solution, but may contain minor flaws in reasoning or computation
2	A partially correct interpretation and/or solution to the problem
1	A correct solution with no supporting evidence or explanation
0	An incorrect solution indicating no mathematical understanding of the concept or task, or no solution is given

Answer

17a.

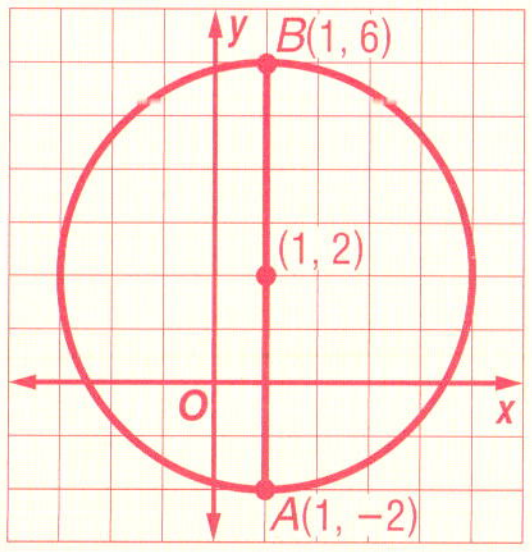

Pages 539–543, Lesson 10-3

37. Given: $\odot O$, $\overline{OS} \perp \overline{RT}$, $\overline{OV} \perp \overline{UW}$, $\overline{OS} \cong \overline{OV}$

Prove: $\overline{RT} \cong \overline{UW}$

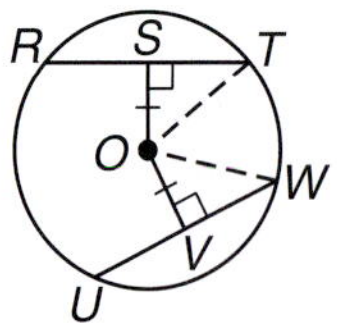

Proof:

Statements (Reasons)

1. $\overline{OT} \cong \overline{OW}$ (All radii of a $\odot$ are $\cong$.)
2. $\overline{OS} \perp \overline{RT}$, $\overline{OV} \perp \overline{UW}$, $\overline{OS} \cong \overline{OV}$ (Given)
3. $\angle OST$, $\angle OVW$ are right angles. (Def. of $\perp$ lines)
4. $\triangle STO \cong \triangle VWO$ (HL)
5. $\overline{ST} \cong \overline{VW}$ (CPCTC)
6. $ST = VW$ (Definition of $\cong$ segments)
7. $2(ST) = 2(VW)$ (Multiplication Property)
8. $\overline{OS}$ bisects $\overline{RT}$; $\overline{OV}$ bisects $\overline{UW}$. (Radius $\perp$ to a chord bisects the chord.)
9. $RT = 2(ST)$, $UW = 2(VW)$ (Def. of seg. bisector)
10. $RT = UW$ (Substitution)
11. $\overline{RT} \cong \overline{UW}$ (Definition of $\cong$ segments)

38. Given: $\odot O$, $\overline{MN} \cong \overline{PQ}$, $\overline{ON}$ and $\overline{OQ}$ are radii. $\overline{OA} \perp \overline{MN}$; $\overline{OB} \perp \overline{PQ}$

Prove: $\overline{OA} \cong \overline{OB}$

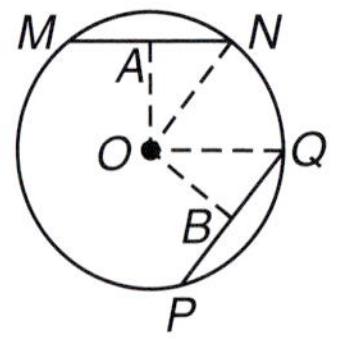

Proof:

Statements (Reasons)

1. $\odot O$, $\overline{MN} \cong \overline{PQ}$, $\overline{ON}$ and $\overline{OQ}$ are radii, $\overline{OA} \perp \overline{MN}$, $\overline{OB} \perp \overline{PQ}$ (Given)
2. $\overline{OA}$ bisects $\overline{MN}$; $\overline{OB}$ bisects $\overline{PQ}$. ($\overline{OA}$ and $\overline{OB}$ are contained in radii. A radius $\perp$ to a chord bisects the chord.)
3. $AN = \frac{1}{2}MN$; $BQ = \frac{1}{2}PQ$ (Def. of bisector)
4. $MN = PQ$ (Def. of $\cong$ segments)
5. $\frac{1}{2}MN = \frac{1}{2}PQ$ (Mult. Prop.)
6. $AN = BQ$ (Substitution)
7. $\overline{AN} \cong \overline{BQ}$ (Def. of $\cong$ segments)
8. $\overline{ON} \cong \overline{OQ}$ (All radii of a circle are $\cong$.)
9. $\triangle AON \cong \triangle BOQ$ (HL)
10. $\overline{OA} \cong \overline{OB}$ (CPCTC)

Pages 548–551 Lesson 10-4

35. Given: T lies inside $\angle PRQ$. $\overline{RK}$ is a diameter of $\odot T$.

Prove: $m\angle PRQ = \frac{1}{2}m\overset{\frown}{PKQ}$

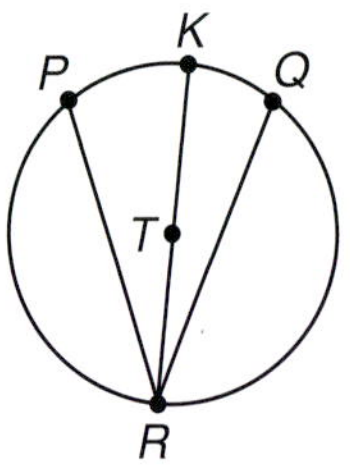

Proof:

Statements (Reasons)

1. $m\angle PRQ = m\angle PRK + m\angle KRQ$ ($\angle$ Addition Th.)
2. $m\overset{\frown}{PKQ} = m\overset{\frown}{PK} + m\overset{\frown}{KQ}$ (Arc Addition Theorem)
3. $\frac{1}{2}m\overset{\frown}{PKQ} = \frac{1}{2}m\overset{\frown}{PK} + \frac{1}{2}m\overset{\frown}{KQ}$ (Multiplication Prop.)
4. $m\angle PRK = \frac{1}{2}m\overset{\frown}{PK}$, $m\angle KRQ = \frac{1}{2}m\overset{\frown}{KQ}$ (The measure of an inscribed $\angle$ whose side is a diameter is half the measure of the intercepted arc (Case 1).)
5. $\frac{1}{2}m\overset{\frown}{PKQ} = m\angle PRK + m\angle KRQ$ (Subst. (Steps 3, 4))
6. $\frac{1}{2}m\overset{\frown}{PKQ} = m\angle PRQ$ (Substitution (Steps 5, 1))

36. Given: T lies outside $\angle PRQ$. $\overline{RK}$ is a diameter of $\odot T$.

Prove: $m\angle PRQ = \frac{1}{2}m\overset{\frown}{PQ}$

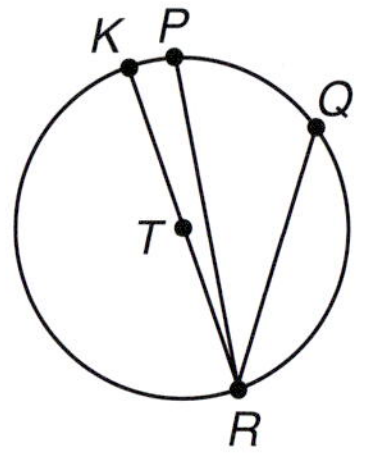

Proof:

Statements (Reasons)

1. $m\angle PRQ = m\angle KRQ - m\angle PRK$ (Angle Addition Theorem, Subtraction Property)
2. $m\overset{\frown}{PQ} = m\overset{\frown}{KQ} - m\overset{\frown}{KP}$ (Arc Addition Theorem, Subtraction Property)
3. $\frac{1}{2}m\overset{\frown}{PQ}, = \frac{1}{2}(m\overset{\frown}{KQ} - m\overset{\frown}{KP})$ (Division Property)
4. $m\angle PRK = \frac{1}{2}m\overset{\frown}{KP}$, $m\angle KRQ = \frac{1}{2}m\overset{\frown}{KQ}$ (The measure of an inscribed $\angle$ whose side is a diameter is half the measure of the intercepted arc (Case 1).)
5. $m\angle PRQ = \frac{1}{2}m\overset{\frown}{KQ} - \frac{1}{2}m\overset{\frown}{KP}$ (Subst. (Steps 1, 4))
6. $m\angle PRQ = \frac{1}{2}(m\overset{\frown}{KQ} - m\overset{\frown}{KP})$ (Distributive Property)
7. $m\angle PRQ = \frac{1}{2}m\overset{\frown}{PQ}$ (Substitution (Steps 6, 3))

37. Given: inscribed $\angle MLN$ and $\angle CED$ $\overset{\frown}{CD} \cong \overset{\frown}{MN}$

Prove: $\angle CED \cong \angle MLN$

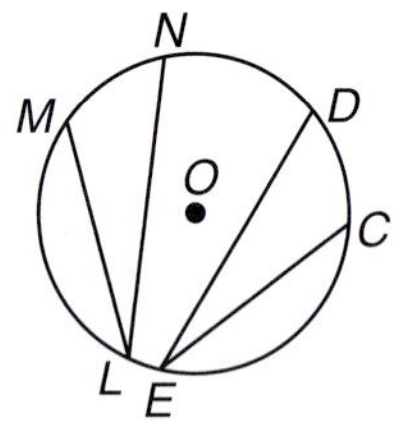

Proof:

Statements (Reasons)

1. $\angle MLN$ and $\angle CED$ are inscribed; $\overset{\frown}{CD} \cong \overset{\frown}{MN}$ (Given)
2. $m\angle MLN = \frac{1}{2}m\overset{\frown}{MN}$; $m\angle CED = \frac{1}{2}m\overset{\frown}{CD}$ (Measure of an inscribed $\angle$ = half measure of intercepted arc.)
3. $m\overset{\frown}{CD} = m\overset{\frown}{MN}$ (Def. of $\cong$ arcs)
4. $\frac{1}{2}m\overset{\frown}{CD} = \frac{1}{2}m\overset{\frown}{MN}$ (Mult. Prop.)
5. $m\angle CED = m\angle MLN$ (Substitution)
6. $\angle CED \cong \angle MLN$ (Def. of $\cong$ $\angle$s)

38. Given: $\overset{\frown}{PQR}$ is a semicircle.

Prove: $\angle PQR$ is a right angle.

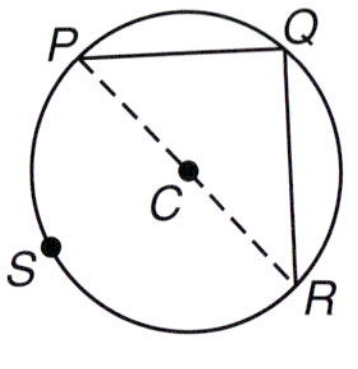

Proof: Since $\overset{\frown}{PQR}$ is a semicircle, $\overset{\frown}{PSR}$ is also a semicircle and $m\overset{\frown}{PSR} = 180$. $\angle PQR$ is an inscribed angle, and $m\angle PQR = \frac{1}{2}(m\overset{\frown}{PSR})$ or 90, making $\angle PQR$ a right angle.

39. Given: quadrilateral $ABCD$ inscribed in $\odot O$

Prove: $\angle A$ and $\angle C$ are supplementary. $\angle B$ and $\angle D$ are supplementary.

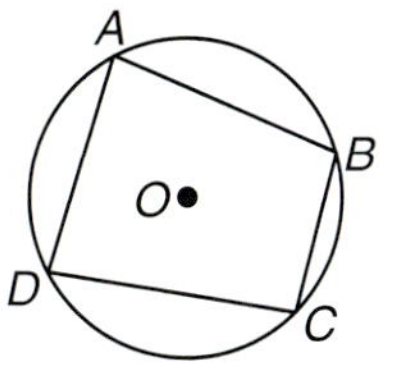

Proof: By arc addition and the definitions of arc measure and the sum of central angles, $m\widehat{DCB} + m\widehat{DAB} = 360$. Since $m\angle C = \frac{1}{2}m\widehat{DAB}$ and $m\angle A = \frac{1}{2}m\widehat{DCB}$, $m\angle C + m\angle A = \frac{1}{2}(m\widehat{DCB} + m\widehat{DAB})$, but $m\widehat{DCB} + m\widehat{DAB} = 360$, so $m\angle C + m\angle A = \frac{1}{2}(360)$ or 180. This makes $\angle C$ and $\angle A$ supplementary. Because the sum of the measures of the interior angles of a quadrilateral is 360, $m\angle A + m\angle C + m\angle B + m\angle D = 360$. But $m\angle A + m\angle C = 180$, so $m\angle B + m\angle D = 180$, making them supplementary also.

41. Sides are congruent radii making it isosceles and $\angle AOC$ is a central angle for an arc of 90°, making it a right angle.

42. Each angle intercepts a semicircle, making them 90° angles. Each side is a chord of congruent arcs, so the chords are congruent.

43. Each angle intercepts a semicircle, making them 90° angles. Each side is a chord of congruent arcs, so the chords are congruent.

Pages 555–558, Lesson 10-5

27. Given: $\overline{AB}$ is tangent to $\odot X$ at B. $\overline{AC}$ is tangent to $\odot X$ at C.

Prove: $\overline{AB} \cong \overline{AC}$

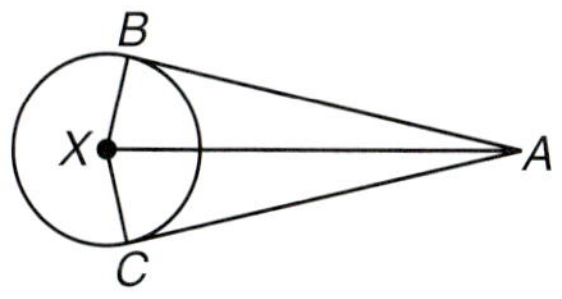

Proof:

Statements (Reasons)

1. $\overline{AB}$ is tangent to $\odot X$ at B, $\overline{AC}$ is tangent to $\odot X$ at C. (Given)
2. Draw $\overline{BX}$, $\overline{CX}$, and $\overline{AX}$. (Through any two points, there is one line.)
3. $\overline{AB} \perp \overline{BX}$, $\overline{AC} \perp \overline{CX}$ (Line tangent to a circle is $\perp$ to the radius at the pt. of tangency.)
4. $\angle ABX$ and $\angle ACX$ are right angles. (Def. of $\perp$ lines)
5. $\overline{BX} \cong \overline{CX}$ (All radii of a circle are $\cong$.)
6. $\overline{AX} \cong \overline{AX}$ (Reflexive Prop.)
7. $\triangle ABX \cong \triangle ACX$ (HL)
8. $\overline{AB} \cong \overline{AC}$ (CPCTC)

31. 12; Draw $\overline{PG}$, $\overline{NL}$, and $\overline{PL}$. Construct $\overline{LQ} \perp \overline{GP}$, thus $LQGN$ is a rectangle. $GQ = NL = 4$, so $QP = 5$. Using the Pythagorean Theorem, $(QP)^2 + (QL)^2 = (PL)^2$. So, $QL = 12$. Since $GN = QL$, $GN = 12$.

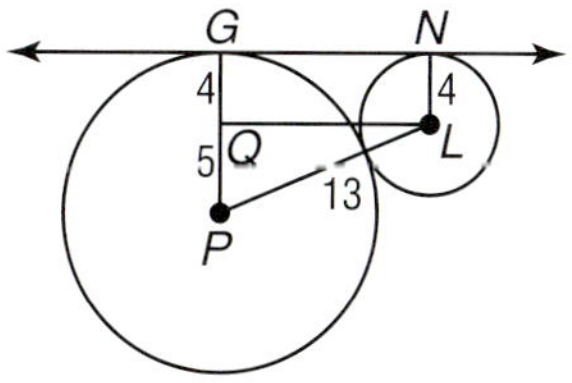

Pages 571–574, Lesson 10-7

30. Given: $\overline{EC}$ and $\overline{EB}$ are secant segments.

Prove: $EA \cdot EC = ED \cdot EB$

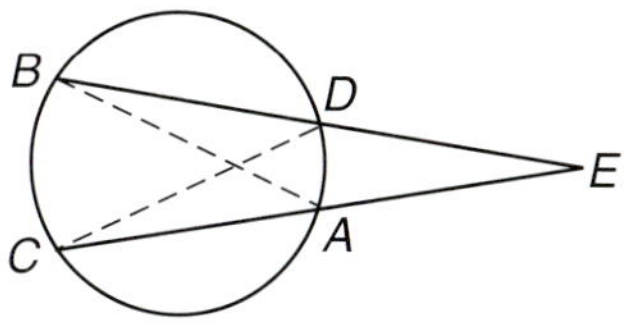

Proof:

Statements (Reasons)

1. $\overline{EC}$ and $\overline{EB}$ are secant segments. (Given)
2. $\angle DEC \cong \angle AEB$ (They name the same angle. (Reflexive Prop.))
3. $\angle ECD \cong \angle EBA$ (Inscribed $\measuredangle$ that intercept the same arc are $\cong$.)
4. $\triangle ABE \sim \triangle DCE$ (AA Similarity)
5. $\frac{EA}{ED} = \frac{EB}{EC}$ (Definition of similar triangles)
6. $EA \cdot EC = ED \cdot EB$ (Cross Products)

31. Given: tangent $\overline{RS}$ and secant $\overline{US}$

Prove: $(RS)^2 = US \cdot TS$

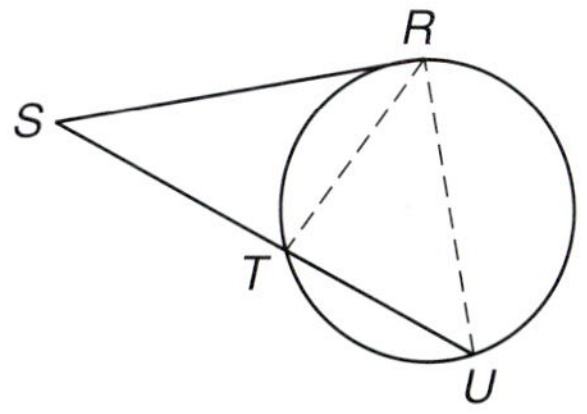

Proof:

Statements (Reasons)

1. tangent $\overline{RS}$ and secant $\overline{US}$ (Given)
2. $m\angle RUT = \frac{1}{2}m\widehat{RT}$ (The measure of an inscribed angle equals half the measure of its intercepted arc.)
3. $m\angle SRT = \frac{1}{2}m\widehat{RT}$ (The measure of an angle formed by a secant and a tangent equals half the measure of its intercepted arc.)
4. $m\angle RUT = m\angle SRT$ (Substitution)
5. $\angle RUT \cong \angle SRT$ (Definition of congruent angles)
6. $\angle S \cong \angle S$ (Reflexive Prop.)
7. $\triangle SUR \sim \triangle SRT$ (AA Similarity)
8. $\frac{RS}{US} = \frac{TS}{RS}$ (Definition of similar triangles)
9. $(RS)^2 = US \cdot TS$ (Cross Products)

Pages 577–580, Lesson 10-8

34. The slope of $\overline{AC}$ is $-\frac{1}{4}$, so the slope of its bisector is 4. The midpoint of $\overline{AC}$ is (0, 5). Use the slope and the midpoint to write an equation for the bisector of $\overline{AC}$: $y = 4x + 5$. The slope of $\overline{BC}$ is $-\frac{9}{2}$, so the slope of its bisector is $\frac{2}{9}$. The midpoint of $\overline{BC}$ is (−2, −3). Use the slope and the midpoint to write an equation for the bisector of $\overline{BC}$: $y = \frac{2}{9}x - \frac{23}{9}$. Solving the system of equations, $y = 4x + 5$ and $y = \frac{2}{9}x - \frac{23}{9}$, yields (−2, −3), which is the circumcenter. Let (−2, −3) be D, then $DA = DB = DC = \sqrt{85}$.

UNIT 4 Notes

Introduction

In this unit students learn to calculate measures in two and three dimensions: area, surface area, and volume. They find the areas of triangles and several types of quadrilaterals in addition to regular polygons, circles, and irregular figures. Students make models of three-dimensional figures and find surface area. Geometric probability is also explored.

Three-dimensional figures are investigated further as students learn to find volume. They also identify congruent or similar solids and graph solids in space.

About the Photographs The large photograph is the Korean War Veterans Memorial located in Washington, D.C., across the reflecting pool from the Vietnam Veterans Memorial. The smaller photo is a sketch of the National World War II Memorial also being constructed in Washington, D.C., scheduled to open in spring 2004.

Assessment Options

Unit 4 Test Pages 773–774 of the *Chapter 13 Resource Masters* may be used as a test or review for Unit 4. This assessment contains both multiple-choice and short answer items.

TestCheck and Worksheet Builder

This CD-ROM can be used to create additional unit tests and review worksheets.

UNIT 4

Area and Volume

Area and volume can be used to analyze real-world situations. In this unit, you will learn about formulas used to find the areas of two-dimensional figures and the surface areas and volumes of three-dimensional figures.

WebQuest Internet Project

Town With Major D-Day Losses Gets Memorial

Source: *USA TODAY,* May 27, 2001

"BEDFORD, Va. For years, World War II was a sore subject that many families in this small farming community avoided. 'We lost so many men,' said Boyd Wilson, 79, who joined Virginia's 116th National Guard before it was sent to war. 'It was just painful.' The war hit Bedford harder than perhaps any other small town in America, taking 19 of its sons, fathers and brothers in the opening moments of the Allied invasion of Normandy. Within a week, 23 of Bedford's 35 soldiers were dead. It was the highest per capita loss for any U.S. community." In this project, you will use scale drawings, surface area, and volume to design a memorial to honor war veterans.

Log on to www.geometryonline.com/webquest. Begin your WebQuest by reading the Task.

Continue working on your WebQuest as you study Unit 4.

Lesson	11-4	12-5	13-3
Page	618	662	703

Teaching Suggestions

Have students study the USA TODAY Snapshot.

- Ask them which two wars cost about the same. **World War I and the Vietnam War**
- Discuss with students why World War II cost so much more than the other U.S. conflicts.
- Point out to students that there is often a contest to choose a design for a memorial. The choice of Maya Ying Lin's design for the Vietnam Veterans Memorial Wall surprised many people because the design was so simple. Today it is one of the most admired and visited memorials in the nation.

Additional USA TODAY Snapshots appearing in Unit 4:

Chapter 11 Who's licensed to fly (p. 614)

Chapter 12 College kids plug in for fun (p. 653)

Chapter 13 Every day is mom's day (p. 705)

WebQuest Internet Project

A WebQuest is an online project in which students do research on the Internet, gather data, and make presentations using word processing, graphing, page-making, or presentation software. In each chapter, students advance to the next step in their WebQuest. At the end of Chapter 13, the project culminates with a presentation of their findings.

Teaching notes and sample answers are available in the *WebQuest and Project Resources.*

Chapter 11

Areas of Polygons and Circles

Chapter Overview and Pacing

LESSON OBJECTIVES	PACING (days)			
	Regular		Block	
	Basic/ Average	Advanced	Basic/ Average	Advanced
11-1 Areas of Parallelograms *(pp. 595–600)* • Find perimeters and areas of parallelograms. • Determine whether points on a coordinate plane define a parallelogram.	1	1	0.5	0.5
11-2 Areas of Triangles, Trapezoids, and Rhombi *(pp. 601–609)* • Find areas of triangles. • Find areas of trapezoids and rhombi.	2	2	1	1
11-3 Areas of Regular Polygons and Circles *(pp. 610–616)* • Find areas of regular polygons. • Find areas of circles.	2	2	1	1
11-4 Areas of Irregular Figures *(pp. 617–621)* • Find areas of irregular figures. • Find areas of irregular figures on the coordinate plane.	2	2	1	1
11-5 Geometric Probability *(pp. 622–627)* • Solve problems involving geometric probability. • Solve problems involving sectors and segments of circles.	1	1	0.5	0.5
Study Guide and Practice Test *(pp. 628–631)* **Standardized Test Practice** *(p. 632–633)*	1	1	0.5	0.5
Chapter Assessment	1	1	0.5	0.5
TOTAL	10	10	5	5

Pacing suggestions for the entire year can be found on pages T20–T21.

Timesaving Tools
TeacherWorks™
All-In-One Planner and Resource Center
See pages T5 and T21.

Chapter Resource Manager

Chapter 11 Resource Masters: Study Guide and Intervention	Practice (Skills and Average)	Reading to Learn Mathematics	Enrichment	Assessment	Prerequisite Skills Workbook	Applications*	5-Minute Check Transparencies	Interactive Chalkboard	GeomPASS: Tutorial Plus (lessons)	Materials
611–612	613–614	615	616			GCC 37, 38 SC 21	11-1	11-1		straightedge, grid paper
617–618	619–620	621	622	655			11-2	11-2		calculator, grid paper, straightedge
623–624	625–626	627	628	655, 657	43–44		11-3	11-3		
629–630	631–632	633	634	656		SC 22	11-4	11-4	20	grid paper, straightedge
635–636	637–638	639	640	656	27–28, 107–108		11-5	11-5		
				641–654, 658–660						

**Key to Abbreviations:* GCC = Graphing Calculator and Computer Masters
SC = School-to-Career Masters

Chapter 11

Mathematical Connections and Background

Continuity of Instruction

Prior Knowledge

Students found the area of a rectangle and evaluated expressions containing variables in previous courses. In Chapter 7, they worked with triangles, finding the height of a triangle and solving for the missing side of a right triangle. They also used trigonometric ratios in that chapter to solve triangles.

This Chapter

In this chapter, students find areas of parallelograms, rhombi, trapezoids, and triangles. They identify the apothem of a regular polygon and use that measure to find the areas of regular polygons. They also find the areas of irregular figures, circles, and sectors and segments of circles. Students determine geometric probability, which is a probability that involves a geometric measure.

Future Connections

In Chapter 12, students use the formulas for finding area to calculate the surface area of prisms and other three-dimensional figures. In Chapter 13, students will extend their understanding of area when they find the volumes of prisms, spheres, and other figures.

11-1 Areas of Parallelograms

A parallelogram is a quadrilateral with both pairs of opposite sides parallel. Any side of a parallelogram can be called a base. For each base, there is a corresponding altitude that is perpendicular to the base. The altitude corresponds to the height of the parallelogram. If a parallelogram has an area of A square units, a base of b units, and a height of h units, then $A = bh$.

To find the area of a quadrilateral on the coordinate plane, you must first determine whether the figure is a parallelogram. You can use the formula for slope to determine whether opposite sides are parallel. Then you find the measures of the base and height and use those to calculate the area.

11-2 Areas of Triangles, Trapezoids, and Rhombi

The formula for the area of a triangle is related to the formula for the area of a parallelogram or rectangle. It is $A = \frac{1}{2}bh$. This formula, in turn, yields the formulas for the areas of trapezoids and rhombi.

If a trapezoid has an area of A square units, bases of b_1 units and b_2 units, and a height of h units, then $A = \frac{1}{2}h(b_1 + b_2)$. If a rhombus has an area of A square units and diagonals of d_1 and d_2 units, then $A = \frac{1}{2}d_1d_2$.

11-3 Areas of Regular Polygons and Circles

A regular polygon can be divided into congruent isosceles triangles by drawing a line from each vertex to the center of the polygon. The altitude of one of these triangles is called an *apothem*. The area of the polygon can be determined by adding the areas of the triangles. If a polygon has a side of s units and an apothem of a units, then the area of one of these triangles is $\frac{1}{2}sa$. By multiplying this formula by the number of sides and substituting P for the formula for perimeter contained in the result, you will find that the formula for the area of a regular polygon is $A = \frac{1}{2}Pa$.

The area of a circle cannot be found without the value known as π. If a circle has an area of A square units and a radius of r units, then $A = \pi r^2$. You can use the properties of circles and regular polygons to find the areas of inscribed and circumscribed polygons.

11-4 Areas of Irregular Figures

An irregular figure is a figure that cannot be classified into the specific shapes that you have studied. To find the areas of irregular figures, separate the figures into shapes for which you can find the area. The area of the irregular figure is the sum of the areas of these separate shapes.

The formula for the area of a regular polygon does not apply to an irregular polygon. To find the area of an irregular polygon, separate the polygon into known figures.

11-5 Geometric Probability

Probability that involves a geometric measure such as length or area is called geometric probability. You can find the probability that a point lies in part of a figure by comparing the area of the part to the area of the whole figure. If a point in region A is chosen at random, then the probability $P(B)$ that the point is in region B, which is in the interior of region A, is $P(B) = \frac{\text{area of region } B}{\text{area of region } A}$. When determining the geometric probability with targets, assume that the object lands within the target area. You should also assume that it is equally likely that the object will land anywhere in the region.

Sometimes you need to know the area of a sector of a circle to find a geometric probability. A sector is a region of a circle bounded by a central angle and its intercepted arc. If a sector of a circle has an area of A square units, a central angle measuring $N°$, and a radius of r units, then $A = \frac{N}{360}\pi r^2$. The region of a circle bounded by an arc and a chord is called a *segment* of a circle. To find the area of a segment, subtract the area of the triangle formed by the radii and the chord from the area of the sector containing the segment.

Chapter 11

DAILY INTERVENTION and Assessment

INTERVENTION

ASSESSMENT

Type	Student Edition	Teacher Resources	Technology/Internet
Ongoing	Prerequisite Skills, pp. 593, 600, 609, 616, 621 Practice Quiz 1, p. 609 Practice Quiz 2, p. 621	5-Minute Check Transparencies *Prerequisite Skills Workbook*, pp. 27–28, 43–44, 107–108 Quizzes, *CRM* pp. 655–656 Mid-Chapter Test, *CRM* p. 657 Study Guide and Intervention, *CRM* pp. 611–612, 617–618, 623–624, 629–630, 635–636	GeomPASS: Tutorial Plus www.geometryonline.com/self_check_quiz www.geometryonline.com/extra_examples
Mixed Review	pp. 600, 609, 616, 621, 627	Cumulative Review, *CRM* p. 658	
Error Analysis	Find the Error, pp. 605, 625 Common Misconceptions, p. 623	Find the Error, *TWE* pp. 605, 624 Unlocking Misconceptions, *TWE* p. 625 Tips for New Teachers, *TWE* pp. 596, 602	
Standardized Test Practice	pp. 600, 608, 616, 621, 622, 625, 627, 631, 632, 633	*TWE* pp. 632–633 Standardized Test Practice, *CRM* pp. 659–660	Standardized Test Practice CD-ROM www.geometryonline.com/standardized_test
Open-Ended Assessment	Writing in Math, pp. 600, 608, 616, 620, 627 Open Ended, pp. 598, 605, 613, 619, 625 Standardized Test, p. 633	Modeling: *TWE* pp. 616, 627 Speaking: *TWE* p. 600 Writing: *TWE* pp. 609, 621 Open-Ended Assessment, *CRM* p. 653	
Chapter Assessment	Study Guide, pp. 628–630 Practice Test, p. 631	Multiple-Choice Tests (Forms 1, 2A, 2B), *CRM* pp. 641–646 Free-Response Tests (Forms 2C, 2D, 3), *CRM* pp. 647–652 Vocabulary Test/Review, *CRM* p. 654	TestCheck and Worksheet Builder (see below) MindJogger Videoquizzes www.geometryonline.com/vocabulary_review www.geometryonline.com/chapter_test

Key to Abbreviations: TWE = Teacher Wraparound Edition; CRM = Chapter Resource Masters

Additional Intervention Resources

The Princeton Review's *Cracking the SAT & PSAT*

The Princeton Review's *Cracking the ACT*

TestCheck and Worksheet Builder

This **networkable** software has three modules for intervention and assessment flexibility:

- **Worksheet Builder** to make worksheet and tests
- **Student Module** to take tests on screen (optional)
- **Management System** to keep student records (optional)

Special banks are included for SAT, ACT, TIMSS, NAEP, and End-of-Course tests.

Intervention Technology

GeomPASS: Tutorial Plus CD-ROM offers a complete, self-paced geometry curriculum.

Geometry Lesson	GeomPASS Lesson
11-4	20 *Exploring Polygons and Area*

Intervention at Home

Log on for student study help.

- For each lesson in the Student Edition, there are Extra Examples and Self-Check Quizzes.
  www.geometryonline.com/extra_examples
  www.geometryonline.com/self_check_quiz
- For chapter review, there is vocabulary review, test practice, and standardized test practice.
  www.geometryonline.com/vocabulary_review
  www.geometryonline.com/chapter_test
  www.geometryonline.com/standardized_test

For more information on Intervention and Assessment, see pp. T8–T11.

Reading and Writing in Mathematics

Glencoe Geometry provides numerous opportunities to incorporate reading and writing into the mathematics classroom.

Student Edition

- Foldables Study Organizer, p. 593
- Concept Check questions require students to verbalize and write about what they have learned in the lesson. (pp. 598, 605, 613, 619, 625)
- Reading Mathematics, p. 594
- Writing in Math questions in every lesson, pp. 600, 608, 616, 620, 627
- Reading Study Tip, p. 617
- WebQuest, p. 618

Teacher Wraparound Edition

- Foldables Study Organizer, pp. 592, 628
- Study Notebook suggestions, pp. 594, 598, 605, 613, 619, 624
- Modeling activities, pp. 616, 627
- Speaking activities, p. 600
- Writing activities, pp. 609, 621
- ELL Resources, pp. 592, 594, 599, 606, 614, 620, 626, 628

Additional Resources

- Vocabulary Builder worksheets require students to define and give examples for key vocabulary terms as they progress through the chapter. (*Chapter 11 Resource Masters,* pp. vii-viii)
- Proof Builder helps students learn and understand theorems and postulates from the chapter. (*Chapter 11 Resource Masters,* pp. ix–x)
- Reading to Learn Mathematics master for each lesson (*Chapter 11 Resource Masters*, pp. 615, 621, 627, 633, 639)
- *Vocabulary PuzzleMaker* software creates crossword, jumble, and word search puzzles using vocabulary lists that you can customize.
- *Teaching Mathematics with Foldables* provides suggestions for promoting cognition and language.
- *Reading Strategies for the Mathematics Classroom*
- *WebQuest and Project Resources*

For more information on Reading and Writing in Mathematics, see pp. T6–T7.

Chapter 11 Notes

What You'll Learn

Have students read over the list of objectives and make a list of any words with which they are not familiar.

Why It's Important

Point out to students that this is only one of many reasons why each objective is important. Others are provided in the introduction to each lesson.

Lesson	NCTM Standards	Local Objectives
11-1	3, 6, 8, 9, 10	
11-2	3, 6, 8, 9, 10	
11-3	3, 6, 8, 9, 10	
11-4	3, 6, 8, 9, 10	
11-5	3, 5, 6, 8, 9, 10	

Key to NCTM Standards:

1=Number & Operations, 2=Algebra, 3=Geometry, 4=Measurement, 5=Data Analysis & Probability, 6=Problem Solving, 7=Reasoning & Proof, 8=Communication, 9=Connections, 10=Representation

Chapter 11 Areas of Polygons and Circles

What You'll Learn

- **Lessons 11-1, 11-2, and 11-3** Find areas of parallelograms, triangles, rhombi, trapezoids, regular polygons, and circles.
- **Lesson 11-4** Find areas of irregular figures.
- **Lesson 11-5** Find geometric probability and areas of sectors and segments of circles.

Key Vocabulary

- apothem (p. 610)
- irregular figure (p. 617)
- geometric probability (p. 622)
- sector (p. 623)
- segment (p. 624)

Why It's Important

Skydivers use geometric probability when they attempt to land on a target marked on the ground. They can determine the chances of landing in the center of the target. *You will learn about skydiving in Lesson 11-5.*

Vocabulary Builder

ELL

The Key Vocabulary list introduces students to some of the main vocabulary terms included in this chapter. For a more thorough vocabulary list with pronunciations of new words, give students the Vocabulary Builder worksheets found on pages vii and viii of the *Chapter 11 Resource Masters*. Encourage them to complete the definition of each term as they progress through the chapter. You may suggest that they add these sheets to their study notebooks for future reference when studying for the Chapter 11 test.

Getting Started

Prerequisite Skills To be successful in this chapter, you'll need to master these skills and be able to apply them in problem-solving situations. Review these skills before beginning Chapter 11.

For Lesson 11-1 — Area of a Rectangle

The area and width of a rectangle are given. Find the length of the rectangle.
(For review, see pages 732–733.)

1. $A = 150, w = 15$ 10 **2.** $A = 38, w = 19$ 2 **3.** $A = 21.16, w = 4.6$ 4.6

4. $A = 2000, w = 32$ 62.5 **5.** $A = 450, w = 25$ 18 **6.** $A = 256, w = 20$ 12.8

For Lessons 11-2 and 11-4 — Evaluate a Given Expression

Evaluate each expression if $a = 6$, $b = 8$, $c = 10$, and $d = 11$. *(For review, see page 736.)*

7. $\frac{1}{2}a(b + c)$ 54 **8.** $\frac{1}{2}ab$ 24 **9.** $\frac{1}{2}(2b + c)$ 13

10. $\frac{1}{2}d(a + c)$ 88 **11.** $\frac{1}{2}(b + c)$ 9 **12.** $\frac{1}{2}cd$ 55

For Lesson 11-3 — Height of a Triangle

Find h in each triangle. *(For review, see Lesson 7-3.)*

13. 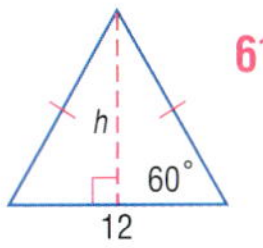

$6\sqrt{3}$

14. 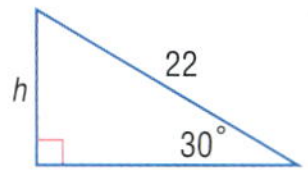

11

15. 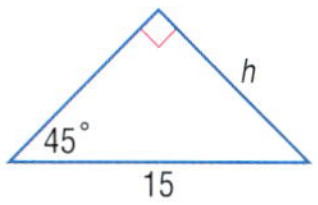

$\frac{15\sqrt{2}}{2}$

FOLDABLES™ Study Organizer

Areas of Polygons and Circles Make this Foldable to help you organize your notes about areas of polygons and circles. Begin with five sheets of notebook paper.

Step 1 Stack

Stack 4 of the 5 sheets of notebook paper as illustrated.

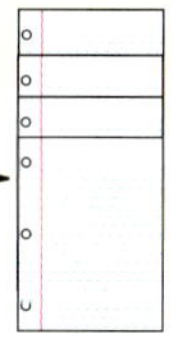

Step 2 Cut

Cut in about 1 inch along the heading line on the top sheet of paper.

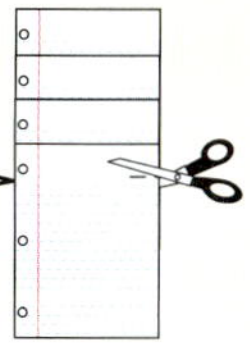

Step 3 Cut

Cut the margins off along the right edge.

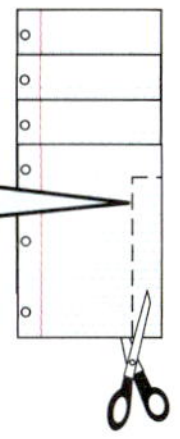

Step 4 Stack

Stack in order of cuts, placing uncut fifth sheet at the back. Label tabs as shown.

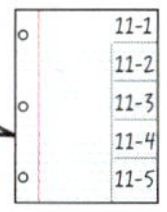

Reading and Writing As you read and study the chapter, take notes and record examples of areas of polygons and circles.

Getting Started

This section provides a review of the basic concepts needed before beginning Chapter 11. Page references are included for additional student help.

Additional review is provided in the *Prerequisite Skills Workbook,* pages 27–28, 43–44, 107–108.

Prerequisite Skills in the Getting Ready for the Next Lesson section at the end of each exercise set review a skill needed in the next lesson.

For Lesson	Prerequisite Skill
11-2	Evaluating expressions, p. 600
11-3	Trigonometric ratios in right triangles, p. 609
11-4	Special right triangles, p. 616

FOLDABLES™ Study Organizer

For more information about Foldables, see *Teaching Mathematics with Foldables.*

Summarizing Use this Foldable for student writing about polygons and area. After students make their Foldable, have them label the side tabs to correspond to the five lessons in this chapter. Students use their Foldable to take notes, define terms, record concepts, solve problems, and explain how to find areas. At the end of each lesson, ask students to write a summary of the lesson, or write in their own words what the lesson was about. Summaries are useful for condensing data.

Getting Started

Before referring to the Student Edition, you may want to give groups of students the prefixes listed in the table on p. 594 as cutouts in an envelope. Ask groups to make lists of everyday words and their meanings that start with that prefix. Then ask different groups to compare their lists.

Teach

Prefixes In this activity, students will learn that knowing the prefix or meaning of the root of a word can help them learn vocabulary. Encourage students to make their lists of everyday words as extensive as possible. Use the Internet or a dictionary to expand the lists.

Assess

Study Notebook

Ask students to summarize what they have learned about using prefixes or roots of words to learn vocabulary in their notebooks.

ELL English Language Learners may benefit from writing key concepts from this activity in their Study Notebooks in their native language and then in English.

Reading Mathematics

Prefixes

Many of the words used in mathematics use the same prefixes as other everyday words. Understanding the meaning of the prefixes can help you understand the terminology better.

Prefix	Meaning	Everyday Words	Meaning
bi-	2	bicycle	a 2-wheeled vehicle
		bipartisan	involving members of 2 political parties
tri-	3	triangle	closed figure with 3 sides
		tricycle	a 3-wheeled vehicle
		triplet	one of 3 children born at the same time
quad-	4	quadrilateral	closed figure with 4 sides
		quadriceps	muscles with 4 parts
		quadruple	four times as many
penta-	5	pentagon	closed figure with 5 sides
		pentathlon	athletic contest with 5 events
hexa-	6	hexagon	closed figure with 6 sides
hept-	7	heptagon	closed figure with 7 sides
oct-	8	octagon	closed figure with 8 sides
		octopus	animal with 8 legs
dec-	10	decagon	closed figure with 10 sides
		decade	a period of 10 years
		decathlon	athletic contest with 10 events

Several pairs of words in the chart have different prefixes, but the same root word. *Pentathlon* and *decathlon* are both athletic contests. *Heptagon* and *octagon* are both closed figures. Knowing the meaning of the root of the term as well as the prefix can help you learn vocabulary.

Reading to Learn

Use a dictionary to find the meanings of the prefix and root for each term. Then write a definition of the term. **1–6. See margin.**

1. bisector **2.** polygon **3.** equilateral

4. concentric **5.** circumscribe **6.** collinear

7. **RESEARCH** Use a dictionary to find the meanings of the prefix and root of *circumference*. ***circum-*** **around, about; *ferre-* to carry**

8. **RESEARCH** Use a dictionary or the Internet to find as many words as you can with the prefix *poly-* and the definition of each. **See margin.**

Answers

1. *bi-* 2, *sector-* a subdivision or region; divide into 2 regions
2. *poly-* many, *gon-* closed figure; closed figure with many sides
3. *equi-* equal, *lateral-* sides; having sides of equal length
4. *co-* together, *centr-* center; circles with a common center
5. *circum-* around, *scribe-* write; to write around (a geometrical figure)
6. *co-* together, *linear-* line; together on the same line
8. Sample answers: polychromatic—multicolored, polymer—a chemical compound composed of a repeating structural unit, polysyllabic—a word with more than three syllables

11-1 Areas of Parallelograms

What You'll Learn

- Find perimeters and areas of parallelograms.
- Determine whether points on a coordinate plane define a parallelogram.

Sunshine State Standards
MA.A.3.4.3, MA.A.4.4.1, MA.B.1.4.1, MA.B.3.4.1, MA.C.3.4.2

How is area related to garden design?

This composition of square-cut granite and moss was designed by Shigemori Mirei in Kyoto, Japan. How could you determine how much granite was used in this garden?

AREAS OF PARALLELOGRAMS Recall that a *parallelogram* is a quadrilateral with both pairs of opposite sides parallel. Any side of a parallelogram can be called a base. For each base, there is a corresponding altitude that is perpendicular to the base.

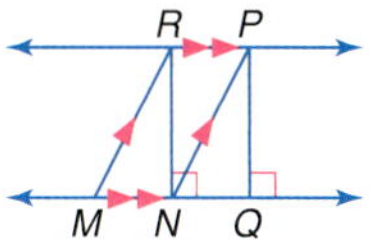

In ▱$MNPR$, if $\overline{MN}$ is the base, $\overline{RN}$ and $\overline{PQ}$ are altitudes. The length of an altitude is called the *height* of the parallelogram. If $\overline{MR}$ is the base, then the altitudes are $\overline{PT}$ and $\overline{NS}$.

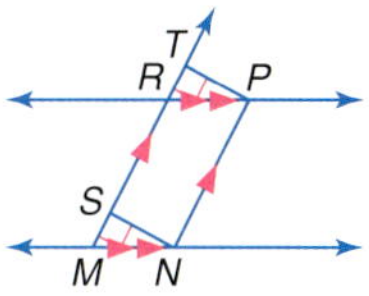

Geometry Activity

Area of a Parallelogram

Model

- Draw parallelogram $ABCD$ on grid paper. Label the vertices on the interior of the angles with letters A, B, C, and D.
- Fold ▱$ABCD$ so that A lies on B and C lies on D, forming a rectangle.

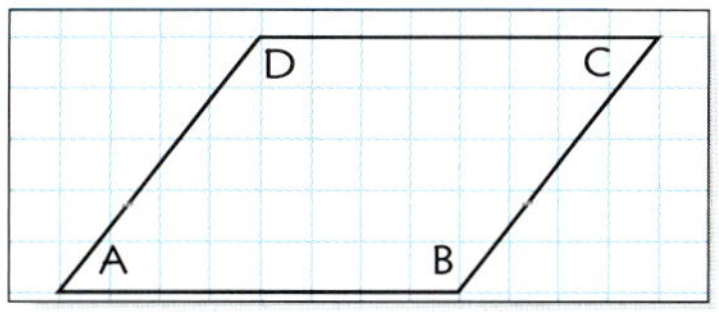

Analyze

1. What is the area of the rectangle? 20 units^2
2. How many rectangles form the parallelogram? 2
3. What is the area of the parallelogram? 40 units^2
4. How do the base and altitude of the parallelogram relate to the length and width of the rectangle?
5. **Make a conjecture** Use what you observed to write a formula for the area of a parallelogram. $A = bh$

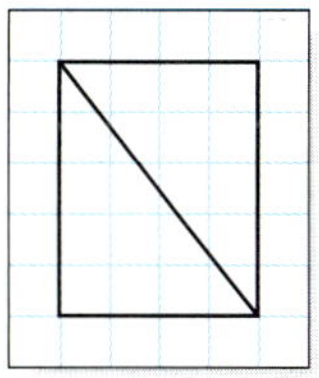

4. The base of the parallelogram is twice the length of the rectangle. The altitude of the parallelogram is the same length as the width of the rectangle.

11-1 Lesson Notes

1 Focus

5-Minute Check Transparency 11-1 Use as a quiz or review of Chapter 10.

Mathematical Background notes are available for this lesson on p. 592C.

How is area related to garden design?

Ask students:

- What kind of polygon is used in the garden design? square
- How could grid paper help model this garden design? Sample answer: You could shade squares of the grid paper to represent the moss.
- What other polygons are used in garden design? Sample answer: triangles and rectangles
- What is another real-world example of special garden designs? Sample answer: a rose garden made in the form of a geometric design

Resource Manager

Workbook and Reproducible Masters

Chapter 11 Resource Masters

- Study Guide and Intervention, pp. 611–612
- Skills Practice, p. 613
- Practice, p. 614
- Reading to Learn Mathematics, p. 615
- Enrichment, p. 616

Graphing Calculator and Computer Masters, pp. 37, 38
School-to-Career Masters, p. 21
Teaching Geometry With Manipulatives Masters, pp. 1, 180

Transparencies

5-Minute Check Transparency 11-1
Answer Key Transparencies

Technology

Interactive Chalkboard

2 Teach

AREAS OF PARALLELOGRAMS

Tips for New Teachers

Intervention Point out that many different parallelograms can be drawn with the same altitude, the same two parallel segments as the bases, and thus the same area. Use a geoboard or similar modeling device to show different parallelograms with these same characteristics, but different slants.

In-Class Examples

Power Point®

Teaching Tip Point out that each parallelogram has two altitudes. Ask students to sketch the other altitude in this figure.

1 Find the area and perimeter of ▱*RSTU*.

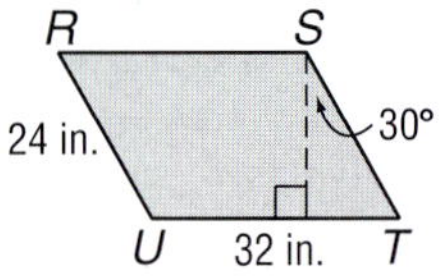

area = $384\sqrt{3}$ or about 665.1 in^2; perimeter = 112 in.

Teaching Tip In Example 2, point out that another way to find the square yardage of the rooms they need to recarpet is to find the total area of the large rectangle and subtract the area of the noncarpeted section.

2 The Kanes are planning to sod some parts of their yard. Find the number of square yards of grass needed. about 2111 yd^2

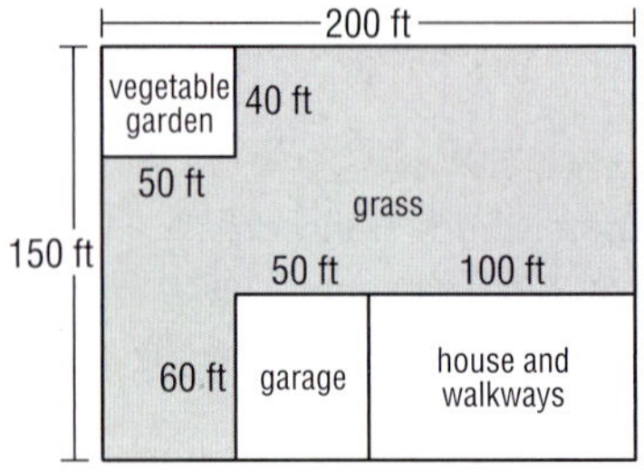

The Geometry Activity leads to the formula for the area of a parallelogram.

Study Tip

Units

Length is measured in linear units, and area is measured in square units.

Key Concept — Area of a Parallelogram

If a parallelogram has an area of A square units, a base of b units, and a height of h units, then $A = bh$.

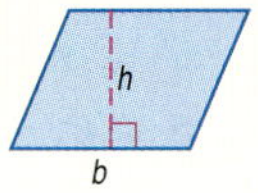

Example 1 Perimeter and Area of a Parallelogram

Find the perimeter and area of ▱*TRVW*.

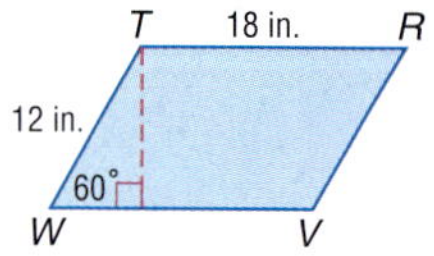

Base and Side: Each pair of opposite sides of a parallelogram has the same measure. Each base is 18 inches long, and each side is 12 inches long.

Study Tip

Look Back

To review **perimeter of polygons**, see Lesson 1-6.

Perimeter: The perimeter of a polygon is the sum of the measures of its sides. So, the perimeter of ▱*TRVW* is 2(18) + 2(12) or 60 inches.

Height: Use a 30°-60°-90° triangle to find the height. Recall that if the measure of the leg opposite the 30° angle is x, then the length of the hypotenuse is $2x$, and the length of the leg opposite the 60° angle is $x\sqrt{3}$.

$12 = 2x$ Substitute 12 for the hypotenuse.

$6 = x$ Divide each side by 2.

So, the height of the parallelogram is $x\sqrt{3}$ or $6\sqrt{3}$ inches.

Area:

$A = bh$ Area of a parallelogram

$= 18\left(6\sqrt{3}\right)$ $b = 18$, $h = 6\sqrt{3}$

$= 108\sqrt{3}$ or about 187.1

The perimeter of ▱*TRVW* is 60 inches, and the area is about 187.1 square inches.

Example 2 Use Area to Solve a Real-World Problem

INTERIOR DESIGN **The Waroners are planning to recarpet part of the first floor of their house. Find the amount of carpeting needed to cover the living room, den, and hall.**

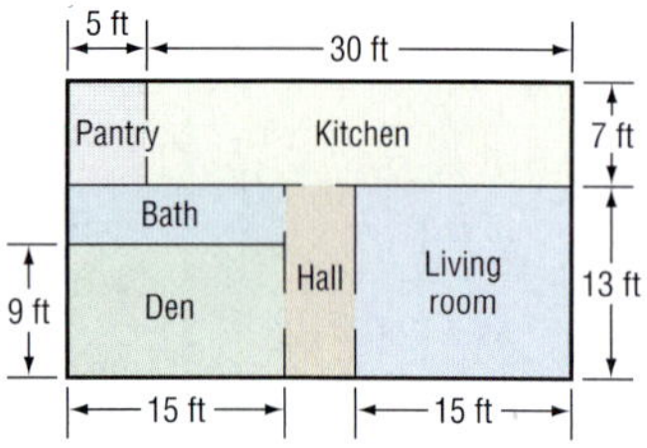

To estimate how much they can spend on carpeting, they need to find the square yardage of each room.

Living Room: $w = 13$ ft, $\ell = 15$ ft

Den: $w = 9$ ft, $\ell = 15$ ft

Hall: It is the same width as the living room, so $w = 13$. The total length of the house is 35 feet. So, $\ell = 35 - 15 - 15$ or 5 feet.

Area

Living Room	Den	Hall
$A = \ell w$	$A = \ell w$	$A = \ell w$
$= 13 \cdot 15$	$= 9 \cdot 15$	$= 5 \cdot 13$
$= 195$ ft^2	$= 135$ ft^2	$= 65$ ft^2

Geometry Activity

Materials: grid paper

To help students see the relationship between the area of a parallelogram and the area of a rectangle, have students cut out several different rectangles. Ask them to cut the rectangle on the diagonal to form two triangles. Then reposition the triangles to form a parallelogram. The area of the parallelogram is the same as the area of the rectangle.

The total area is 195 + 135 + 65 or 395 square feet. There are 9 square feet in one square yard, so divide by 9 to convert from square feet to square yards.

$$395 \text{ ft}^2 \div \frac{9 \text{ ft}^2}{1 \text{ yd}^2} = 395 \text{ ft}^2 \times \frac{1 \text{ yd}^2}{9 \text{ ft}^2}$$
$$\approx 43.9 \text{ yd}^2$$

Therefore, 44 square yards of carpeting are needed to cover these areas.

PARALLELOGRAMS ON THE COORDINATE PLANE Recall the properties of quadrilaterals that you studied in Chapter 8. Using these properties as well as the formula for slope and the Distance Formula, you can find the areas of quadrilaterals on the coordinate plane.

Study Tip

Look Back
To review **properties of parallelograms**, **rectangles**, and **squares**, see Lessons 8-3, 8-4, and 8-5.

Example 3 Area on the Coordinate Plane

COORDINATE GEOMETRY **The vertices of a quadrilateral are $A(-4, -3)$, $B(2, -3)$, $C(4, -6)$, and $D(-2, -6)$.**

a. Determine whether the quadrilateral is a *square*, a *rectangle*, or a *parallelogram*.

First graph each point and draw the quadrilateral. Then determine the slope of each side.

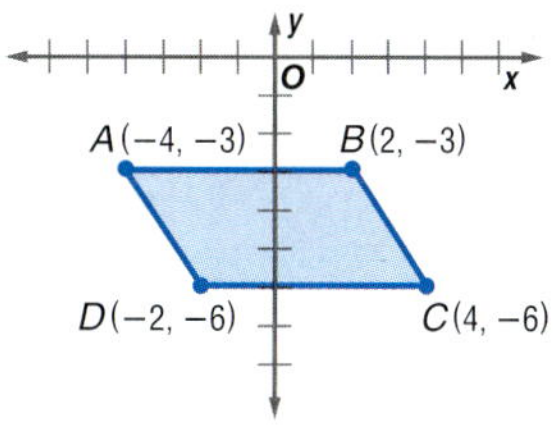

slope of $\overline{AB} = \frac{-3-(-3)}{-4-2}$

$= \frac{0}{-6}$ or 0

slope of $\overline{CD} = \frac{-6-(-6)}{4-(-2)}$

$= \frac{0}{6}$ or 0

slope of $\overline{BC} = \frac{-3-(-6)}{2-4}$

$= \frac{3}{-2}$

slope of $\overline{AD} = \frac{-3-(-6)}{-4-(-2)}$

$= \frac{3}{-2}$

Opposite sides have the same slope, so they are parallel. $ABCD$ is a parallelogram. The slopes of the consecutive sides are *not* negative reciprocals of each other, so the sides are not perpendicular. Thus, the parallelogram is neither a square nor a rectangle.

b. Find the area of quadrilateral $ABCD$.

Base: $\overline{CD}$ is parallel to the x-axis, so subtract the x-coordinates of the endpoints to find the length: $CD = |4-(-2)|$ or 6.

Height: Since $\overline{AB}$ and $\overline{CD}$ are horizontal segments, the distance between them, or the height, can be measured on any vertical segment. Reading from the graph, the height is 3.

$A = bh$ Area formula

$= 6(3)$ $b = 6, h = 3$

$= 18$ Simplify.

The area of ▱$ABCD$ is 18 square units.

DAILY INTERVENTION **Differentiated Instruction**

Logical For a parallelogram graphed on the coordinate plane, if two sides have slope 0, students should reason that an altitude of the parallelogram can be found easily by counting the number of grid units between the sides.

PARALLELOGRAMS ON THE COORDINATE PLANE

In-Class Example Power Point®

3 The vertices of a quadrilateral are at $A(-2, 3)$, $B(4, 1)$, $C(3, -2)$, and $D(-3, 0)$.

a. Determine whether the quadrilateral is a *square*, a *rectangle*, or a *parallelogram*. **rectangle**

b. Find the area of quadrilateral $ABCD$. **20 units^2**

Interactive Chalkboard
PowerPoint® Presentations

This CD-ROM is a customizable Microsoft® PowerPoint® presentation that includes:

- Step-by-step, dynamic solutions of each In-Class Example from the Teacher Wraparound Edition
- Additional, Try These exercises for each example
- The 5-Minute Check Transparencies
- Hot links to Glencoe Online Study Tools

3 Practice/Apply

Study Notebook

Have students—

- add the definitions/examples of the vocabulary terms to their Vocabulary Builder worksheets for Chapter 11.
- include the formula for the areas of a rectangle, square, and parallelogram with base *b* and height *h*.
- include any other item(s) that they find helpful in mastering the skills in this lesson.

About the Exercises...

Organization by Objective

- **Areas of Parallelograms:** 9–17, 27, 28, 31
- **Parallelograms on the Coordinate Plane:** 20–25

Odd/Even Assignments

Exercises 9–31 are structured so that students practice the same concepts whether they are assigned odd or even problems.

Assignment Guide

Basic: 9–17 odd, 21–29 odd, 35–53

Average: 9–35 odd, 36–53

Advanced: 10–34 even, 35–49 (optional: 50–53)

Answer

1. The area of a rectangle is the product of the length and the width. The area of a parallelogram is the product of the base and the height. For both quadrilaterals, the measure of the length of one side is multiplied by the length of the altitude.

Check for Understanding

Concept Check

1. **Compare and contrast** finding the area of a rectangle and the area of a parallelogram. See margin.
2. **OPEN ENDED** Make and label a scale drawing of your bedroom. Then find its area in square yards. See students' work.

Guided Practice

GUIDED PRACTICE KEY	
Exercises	Examples
3–5	1
6–7	3
8	2

Find the perimeter and area of each parallelogram. Round to the nearest tenth if necessary.

3. 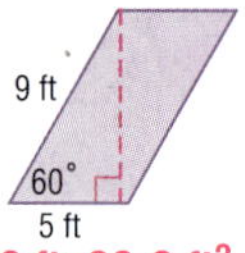

28 ft; 39.0 ft^2

4. 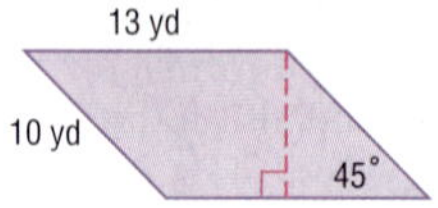

46 yd; 91.9 yd^2

5. 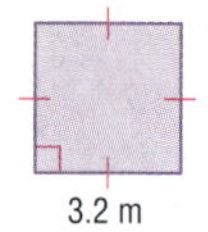

12.8 m; 10.2 m^2

Given the coordinates of the vertices of quadrilateral *TVXY*, determine whether it is a *square*, a *rectangle*, or a *parallelogram*. Then find the area of *TVXY*.

6. $T(0, 0)$, $V(2, 6)$, $X(6, 6)$, $Y(4, 0)$ parallelogram, 24 units^2
7. $T(10, 16)$, $V(2, 18)$, $X(-3, -2)$, $Y(5, -4)$ rectangle, 170 units^2

Application

8. **DESIGN** Mr. Kang is planning to stain his deck. To know how much stain to buy, he needs to find the area of the deck. What is the area? 1170 ft^2

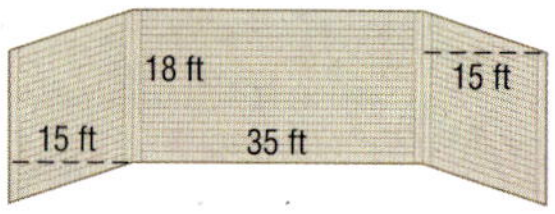

★ indicates increased difficulty

Practice and Apply

Homework Help	
For Exercises	See Examples
9–14	1
15–17, 27, 28, 31	2
20–25	3

Extra Practice See page 776.

Find the perimeter and area of each parallelogram. Round to the nearest tenth if necessary.

9. 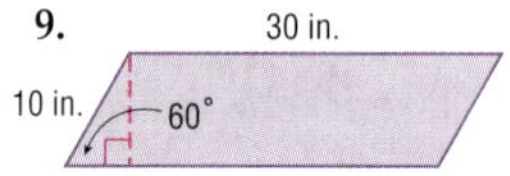

80 in.; 259.8 in^2

10. 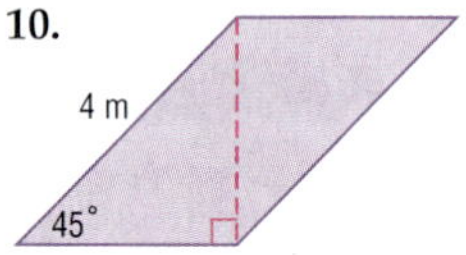

13.7 m; 8 m^2

11. 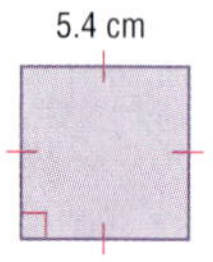

21.6 cm; 29.2 cm^2

12. 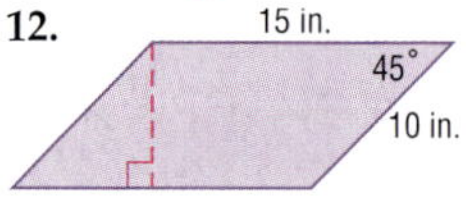

50 in.; 106.1 in^2

13. 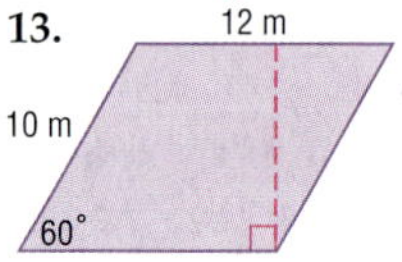

44 m; 103.9 m^2

14. 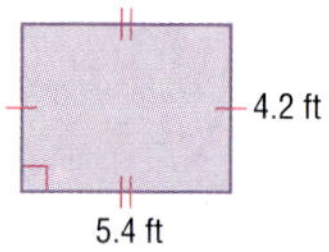

19.2 ft; 22.7 ft^2

Find the area of each shaded region. Round to the nearest tenth if necessary.

15. 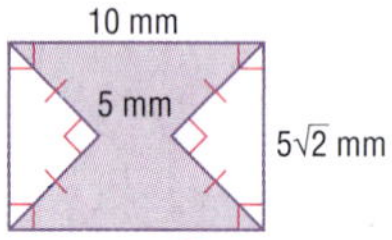

45.7 mm^2

16. 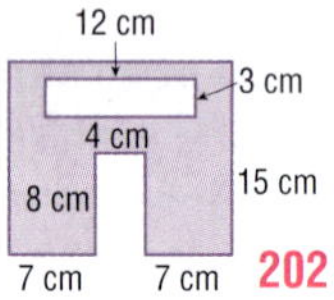

202 cm^2

17. 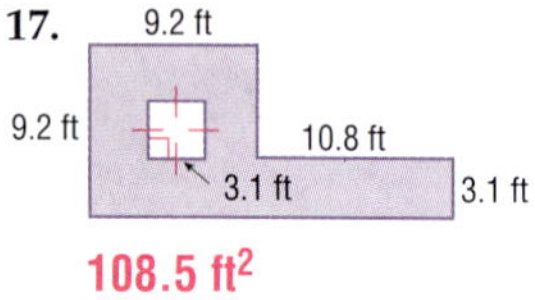

108.5 ft^2

Find the height and base of each parallelogram given its area.

★18. 100 square units

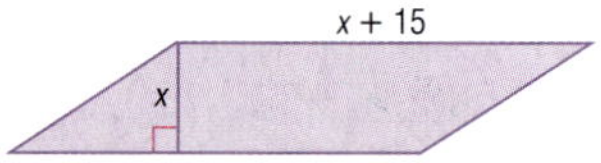

$h = 5$ units, $b = 20$ units

★19. 2000 square units

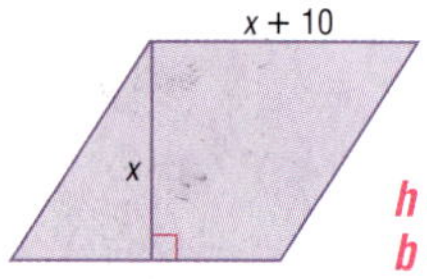

$h = 40$ units, $b = 50$ units

598 Chapter 11 Areas of Polygons and Circles

COORDINATE GEOMETRY **Given the coordinates of the vertices of a quadrilateral, determine whether it is a *square*, a *rectangle*, or a *parallelogram*. Then find the area of the quadrilateral.**

20. parallelogram, 20 units^2

21. parallelogram, 56 units^2

22. parallelogram, 50 units^2

23. parallelogram, 64 units^2

20. $A(0, 0)$, $B(4, 0)$, $C(5, 5)$, $D(1, 5)$

21. $E(-5, -3)$, $F(3, -3)$, $G(5, 4)$, $H(-3, 4)$

22. $J(-1, -4)$, $K(4, -4)$, $L(6, 6)$, $M(1, 6)$

23. $N(-6, 2)$, $O(2, 2)$, $P(4, -6)$, $Q(-4, -6)$

24. $R(-2, 4)$, $S(8, 4)$, $T(8, -3)$, $U(-2, -3)$ rectangle, 70 units^2

25. $V(1, 10)$, $W(4, 8)$, $X(2, 5)$, $Y(-1, 7)$ square, 13 units^2

26. **INTERIOR DESIGN** The Bessos are planning to have new carpet installed in their guest bedroom, family room, and hallway. Find the number of square yards of carpet they should order. 119 yd^2

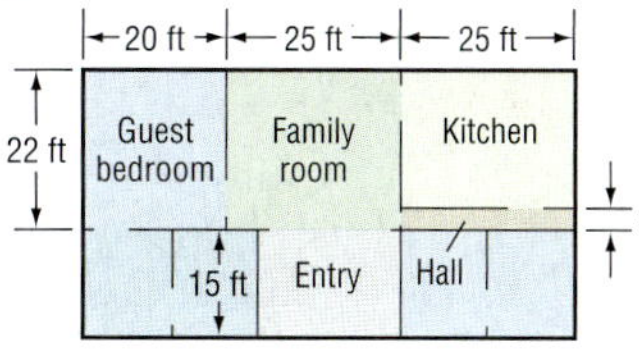

Find the area of each figure.

27.

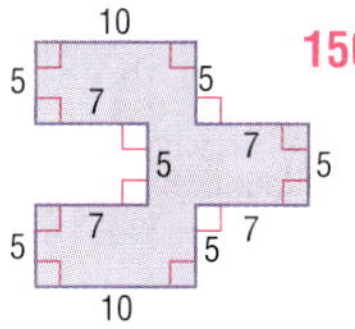

150 units^2

28.

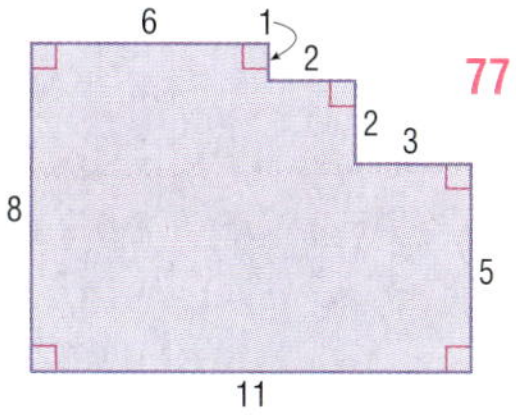

77 units^2

ART For Exercises 29 and 30, use the following information.

A *triptych* painting is a series of three pieces with a similar theme displayed together. Suppose the center panel is a 12-inch square and the panels on either side are 12 inches by 5 inches. The panels are 2 inches apart with a 3 inch wide border around the edges.

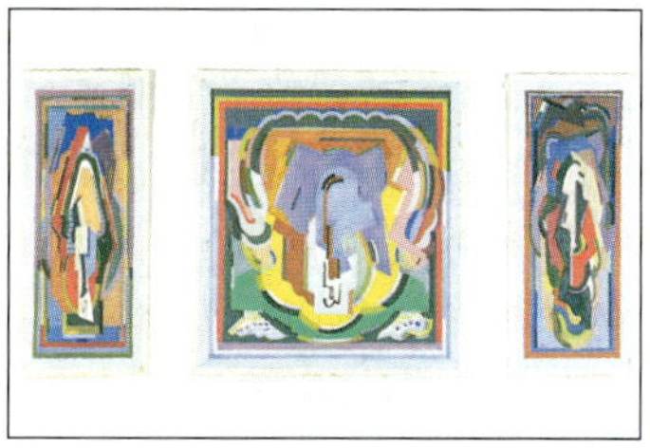

Study for a Triptych, by Albert Gleizes

29. Determine whether the triptych will fit a 45-inch by 20-inch frame. Explain.

30. Find the area of the artwork. 576 in^2

29. Yes; the dimensions are 32 in. by 18 in.

More About. . .

Art

A triptych originally referred to a Roman writing tablet with three panels that were hinged together.

Source: www.artlex.com

31. **CROSSWALKS** A crosswalk with two stripes each 52 feet long is at a 60° angle to the curb. The width of the crosswalk at the curb is 16 feet. Find the perpendicular distance between the stripes of the crosswalk. ≈ 13.9 ft

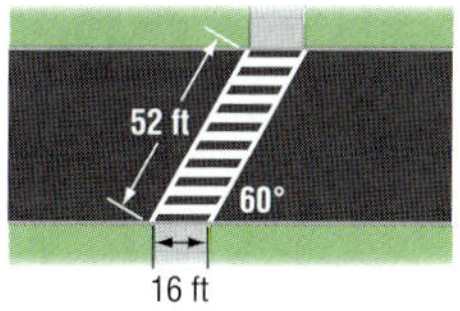

VARYING DIMENSIONS For Exercises 32–34, use the following information.

A parallelogram has a base of 8 meters, sides of 11 meters, and a height of 10 meters.

32. Find the perimeter and area of the parallelogram. 38 m, 80 m^2

★ **33.** Suppose the dimensions of the parallelogram were divided in half. Find the perimeter and the area.

★ **34.** Compare the perimeter and area of the parallelogram in Exercise 33 with the original.

33. The perimeter is 19 m, half of 38 m. The area is 20 m^2.

34. The new perimeter is half of the original. The new area is one half squared or one fourth the area of the original parallelogram.

35. **CRITICAL THINKING** A piece of twine 48 inches long is cut into two lengths. Each length is then used to form a square. The sum of the areas of the two squares is 74 square inches. Find the length of each side of the smaller square and the larger square. 5 in., 7 in.

www.geometryonline.com/self_check_quiz/fcat

Enrichment, p. 616

Area of a Parallelogram

You can prove some interesting results using the formula you have proved for the area of a parallelogram by drawing auxiliary lines to form congruent regions. Consider the top parallelogram shown at the right. In the figure, d is the length of the diagonal $\overline{BD}$, and k is the length of the perpendicular segment from A to $\overline{BD}$. Now consider the second figure, which shows the same parallelogram with a number of auxiliary perpendiculars added. Use what you know about perpendicular lines, parallel lines, and congruent triangles to answer the following.

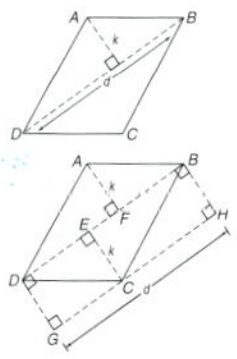

1. What kind of figure is $DBHG$?
rectangle

Study Guide and Intervention, p. 611 (shown) and p. 612

Areas of Parallelograms A parallelogram is a quadrilateral with both pairs of opposite sides parallel. Any side of a parallelogram can be called a **base**. Each base has a corresponding **altitude**, and the length of the altitude is the **height** of the parallelogram. The area of a parallelogram is the product of the base and the height.

Area of a Parallelogram	If a parallelogram has an area of A square units, a base of b units, and a height of h units, then $A = bh$.

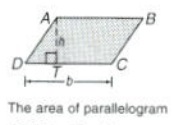

The area of parallelogram $ABCD$ is $CD \cdot AT$.

Example **Find the area of parallelogram *EFGH*.**

$A = bh$ Area of a parallelogram
$= 30(18)$ $b = 30$, $h = 18$
$= 540$ Multiply.

The area is 540 square meters.

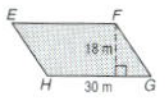

Exercises

Find the area of each parallelogram.

1. 288 ft^2
2. $288\sqrt{3}$ in^2
3. 2.56 cm^2

Find the area of each shaded region.

4. $WXYZ$ and $ABCD$ are rectangles.
5. All angles are right angles.
6. $EFGH$ and $NOPQ$ are rectangles; $JKLM$ is a square.

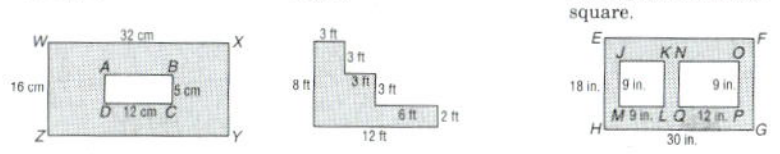

4. 452 cm^2 5. 51 ft^2 6. 351 in^2

7. The area of a parallelogram is 3.36 square feet. The base is 2.8 feet. If the measures of the base and height are each doubled, find the area of the resulting parallelogram. 13.44 ft^2

8. A rectangle is 4 meters longer than it is wide. The area of the rectangle is 252 square meters. Find the length. 18 m

Skills Practice, p. 613 and Practice, p. 614 (shown)

Find the perimeter and area of each parallelogram. Round to the nearest tenth if necessary.

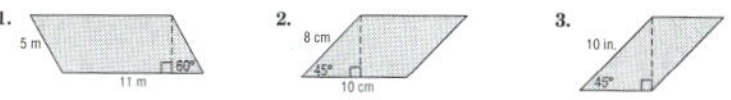

1. 32 m, 47.6 m^2
2. 36 cm, 56.6 cm^2
3. 34.1 in., 50 in^2

Find the area of each figure.

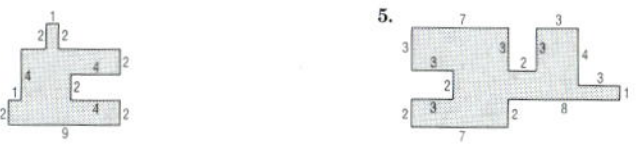

4. 44 units^2
5. 65 units^2

COORDINATE GEOMETRY **Given the coordinates of the vertices of a quadrilateral, determine whether it is a *square*, a *rectangle*, or a *parallelogram*. Then find the area of the quadrilateral.**

6. $C(-4, -1)$, $D(-4, 2)$, $F(1, 2)$, $G(1, -1)$ rectangle, 15 units^2

7. $W(2, 2)$, $X(1, -2)$, $Y(-2, -2)$, $Z(-1, 2)$ parallelogram, 12 units^2

8. $M(0, 4)$, $N(4, 6)$, $O(6, 2)$, $P(2, 0)$ square, 20 units^2

9. $P(-5, 2)$, $Q(4, 2)$, $R(5, 5)$, $S(-4, 5)$ parallelogram, 27 units^2

FRAMING **For Exercises 10–12, use the following information.**

A rectangular poster measures 42 inches by 26 inches. A frame shop fitted the poster with a half-inch mat border.

10. Find the area of the poster. 1092 in^2

11. Find the area of the mat border. 69 in^2

12. Suppose the wall is marked where the poster will hang. The marked area includes an additional 12-inch space around the poster and frame. Find the total wall area that has been marked for the poster. 3417 in^2

Reading to Learn Mathematics, p. 615 ELL

Pre-Activity **How is area related to garden design?**

Read the introduction to Lesson 11-1 at the top of page 595 in your textbook.

How could you describe the pattern you see in the picture of the garden so that someone who doesn't have the picture will know what it looks like?

Sample answer: There are squares of granite and squares of moss of the same size placed in a checkerboard design.

Reading the Lesson

1. Which expression gives the area of the parallelogram? (Hint: There can be more than one correct response.) B, D, E, G

A. ab	B. cb	C. ed
D. af	E. ce	F. cd
G. df	H. bf	I. cf

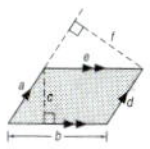

2. Refer to the figure. Determine whether each statement is *true* or *false*. If the statement is false, explain why.

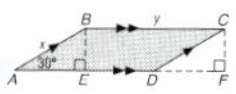

a. $\overline{AB}$ is an altitude of the parallelogram. False; $\overline{AB}$ is not perpendicular to any side of the parallelogram.

b. $\overline{CD}$ is a base of parallelogram $ABCD$.
true

c. The perimeter of $ABCD$ is $(2x + 2y)$ units^2. False; perimeter is measured in linear units, not square units. The perimeter is $(2x + 2y)$ units.

d. $BE = CF$
true

e. $BE = \frac{\sqrt{3}}{2}x$ False; $\overline{BE}$ is opposite the 30° angle in a 30°-60°-90° triangle, so $BE = \frac{1}{2}x$ or $\frac{x}{2}$.

f. The area of $ABCD$ is $2xy$ units^2. False; since $BE = \frac{x}{2}$, the area of $ABCD$ is $\frac{xy}{2}$ units^2.

Helping You Remember

3. A good way to remember a new formula in geometry is to relate it to a formula you already know. How can you use the formula for the area of a rectangle to help you remember the formula for the area of a parallelogram?

Sample answer: To find the area of a rectangle, you multiply the lengths of two segments that are perpendicular to each other. To find the area of a parallelogram, you do the same thing, but the height is not necessarily one of the sides.

4 Assess

Open-Ended Assessment

Speaking Have students describe how to find the area of a parallelogram in the coordinate plane.

Getting Ready for Lesson 11-2

Prerequisite Skill Students will learn about the areas of triangles, rhombi, and trapezoids in Lesson 11-2. They will substitute values and evaluate expressions to find areas. Use Exercises 50–53 to determine your students' familiarity with evaluating expressions.

Answer

36. Sample answer: Area is used when designing a garden to find the total amount of materials needed. Answers should include the following.
   - Find the area of one square and multiply by the number of squares in the garden.
   - Knowing the area is useful when planning a stone walkway or fencing in flowers or vegetables.

46.

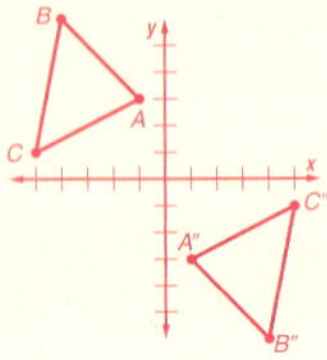

47.

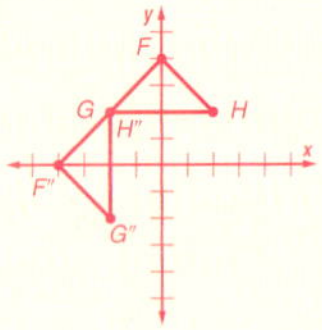

48.

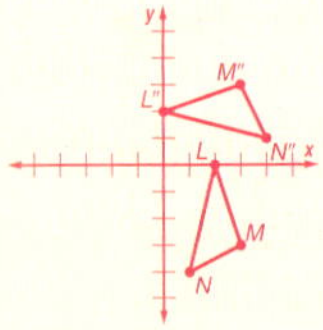

36. **WRITING IN MATH** Answer the question that was posed at the beginning of the lesson. See margin.

How is area related to garden design?

Include the following in your answer:
- how to determine the total area of granite squares, and
- other uses for area.

FCAT Practice *Standardized Test Practice*

37. What is the area of ▱$ABCD$? C

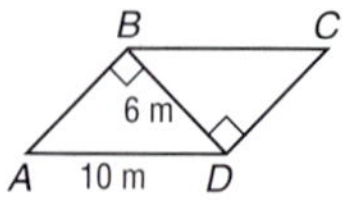

Ⓐ 24 m^2 Ⓑ 30 m^2 Ⓒ 48 m^2 Ⓓ 60 m^2

38. **ALGEBRA** Which statement is correct? D

Ⓐ $x^2 > (x - 1)^2$ Ⓒ $x^2 < (x - 1)^2$

Ⓑ $x^2 = (x - 1)^2$ Ⓓ The relationship cannot be determined.

Maintain Your Skills

Mixed Review **Determine the coordinates of the center and the measure of the radius for each circle with the given equation.** *(Lesson 10-8)*

39. $(x - 5)^2 + (y - 2)^2 = 49$ (5, 2), $r = 7$

40. $(x + 3)^2 + (y + 9)^2 - 81 = 0$ (−3, −9), $r = 9$

41. $\left(x + \frac{2}{3}\right)^2 + \left(y - \frac{1}{9}\right)^2 - \frac{4}{9} = 0$ $\left(-\frac{2}{3}, \frac{1}{9}\right)$, $r = \frac{2}{3}$

42. $(x - 2.8)^2 + (y + 7.6)^2 = 34.81$ (2.8, −7.6), $r = 5.9$

Find x. Assume that segments that appear to be tangent are tangent. *(Lesson 10-7)*

43. 32

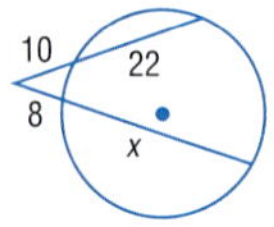

44. 6

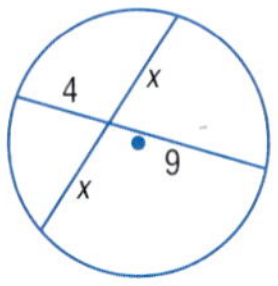

45. 21

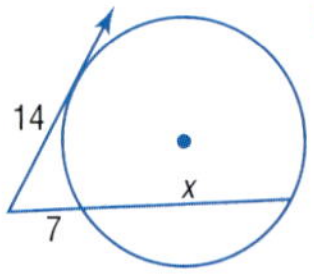

COORDINATE GEOMETRY **Draw the rotation image of each triangle by reflecting the triangles in the given lines. State the coordinates of the rotation image and the angle of rotation.** *(Lesson 9-3)*

46. △ABC with vertices $A(-1, 3)$, $B(-4, 6)$, and $C(-5, 1)$, reflected in the y-axis and then the x-axis 46–48. See margin for figures.

47. △FGH with vertices $F(0, 4)$, $G(-2, 2)$, and $H(2, 2)$, reflected in $y = x$ and then the y-axis

48. △LMN with vertices $L(2, 0)$, $M(3, -3)$, and $N(1, -4)$, reflected in the y-axis and then the line $y = -x$

46. $A''(1, -3)$, $B''(4, -6)$, $C''(5, -1)$; 180°
47. $F''(-4, 0)$, $G''(-2, -2)$, $H''(-2, 2)$; 90° counterclockwise
48. $L''(0, 2)$, $M''(3, 3)$, $N''(4, 1)$; 90° counterclockwise

49. **BIKES** Nate is making a ramp for bike jumps. The ramp support forms a right angle. The base is 12 feet long, and the height is 5 feet. What length of plywood does Nate need for the ramp? *(Lesson 7-2)* 13 ft

Getting Ready for the Next Lesson **PREREQUISITE SKILL** **Evaluate each expression if $w = 8$, $x = 4$, $y = 2$, and $z = 5$.**
*(To review **evaluating expressions**, see page 736.)*

50. $\frac{1}{2}(7y)$ 7 51. $\frac{1}{2}wx$ 16 52. $\frac{1}{2}z(x + y)$ 15 53. $\frac{1}{2}x(y + w)$ 20

11-2 Areas of Triangles, Trapezoids, and Rhombi

Sunshine State Standards
MA.A.3.4.3, MA.A.4.4.1, MA.B.1.4.1, MA.B.3.4.1

What You'll Learn

- Find areas of triangles.
- Find areas of trapezoids and rhombi.

How is the area of a triangle related to beach umbrellas?

Umbrellas can protect you from rain, wind, and sun. The umbrella shown at the right is made of triangular panels. To cover the umbrella frame with canvas panels, you need to know the area of each panel.

AREAS OF TRIANGLES You have learned how to find the areas of squares, rectangles, and parallelograms. The formula for the area of a triangle is related to these formulas.

Geometry Activity

Area of a Triangle

Model

You can determine the area of a triangle by using the area of a rectangle.

- Draw a triangle on grid paper so that one edge is along a horizontal line. Label the vertices on the interior of the angles of the triangle as *A*, *B*, and *C*.
- Draw a line perpendicular to $\overline{AC}$ through *A*.
- Draw a line perpendicular to $\overline{AC}$ through *C*.
- Draw a line parallel to $\overline{AC}$ through *B*.
- Label the points of intersection of the lines drawn as *D* and *E* as shown.
- Find the area of rectangle *ACDE* in square units.
- Cut out rectangle *ACDE*. Then cut out $\triangle ABC$. Place the two smaller pieces over $\triangle ABC$ to completely cover the triangle.

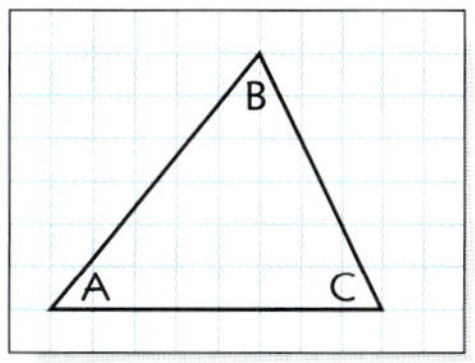

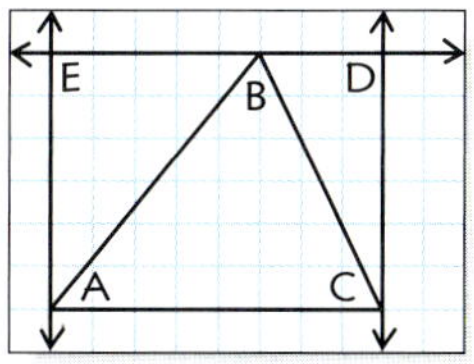

Analyze 1. Together the two smaller triangles are the same size as $\triangle ABC$.

1. What do you observe about the two smaller triangles and $\triangle ABC$?
2. What fraction of rectangle *ACDE* is $\triangle ABC$? $\frac{1}{2}$
3. Derive a formula that could be used to find the area of $\triangle ABC$. $A = \frac{1}{2}bh$

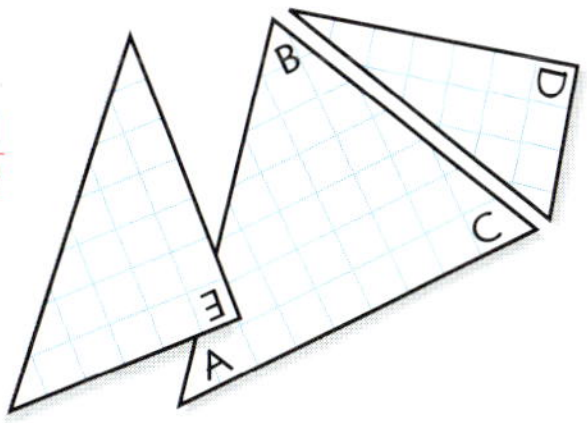

11-2 Lesson Notes

1 Focus

5-Minute Check Transparency 11-2 Use as a quiz or review of Lesson 11-1.

Mathematical Background notes are available for this lesson on p. 592C.

How is the area of a triangle related to beach umbrellas?

Ask students:

- What is the shape of the umbrella fabric when it is flat? a many-sided polygon
- Why are triangles the shape used to make umbrellas? Sample answer: They fit together at one vertex.
- Give another example of an item made from a triangular piece of fabric. Sample answer: a sail

Resource Manager

Workbook and Reproducible Masters

Chapter 11 Resource Masters

- Study Guide and Intervention, pp. 617–618
- Skills Practice, p. 619
- Practice, p. 620
- Reading to Learn Mathematics, p. 621
- Enrichment, p. 622
- Assessment, p. 655

Teaching Geometry With Manipulatives Masters, pp. 1, 181, 182

Transparencies

5-Minute Check Transparency 11-2
Answer Key Transparencies

Technology

Interactive Chalkboard

2 Teach

AREAS OF TRIANGLE

Tips for New Teachers

Intervention Help students understand the relationship between the area of a triangle and the area of a parallelogram or rectangle by showing them a model. Cut a piece of 8.5 × 11 paper in half along the diagonal to demonstrate that the area of a triangle is one-half the area of a rectangle. Then cut a right triangle from an end of another sheet of 8.5 × 11 paper so that it has the same height as the original paper. Form a parallelogram from the sheet by sliding it to the other side of the sheet. Then cut it in half along the diagonal. The area of a triangle is half the area of this corresponding parallelogram.

In-Class Example Power Point®

1 Find the area of quadrilateral $ABCD$ if $AC = 35$, $BF = 18$, and $DE = 10$.

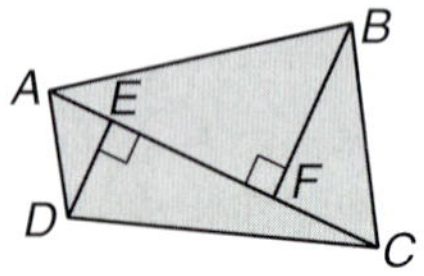

490 units²

The Geometry Activity suggests the formula for finding the area of a triangle.

Study Tip

Look Back
To review **the height and altitude of a triangle**, see Lesson 5-1.

Key Concept — *Area of a Triangle*

If a triangle has an area of A square units, a base of b units, and a corresponding height of h units, then $A = \frac{1}{2}bh$.

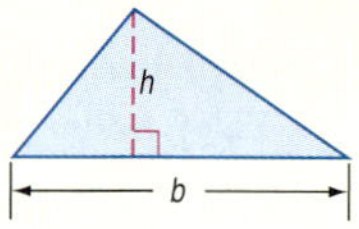

Example 1 *Areas of Triangles*

Find the area of quadrilateral $XYZW$ if $XZ = 39$, $HW = 20$, and $YG = 21$.

The area of the quadrilateral is equal to the sum of the areas of $\triangle XWZ$ and $\triangle XYZ$.

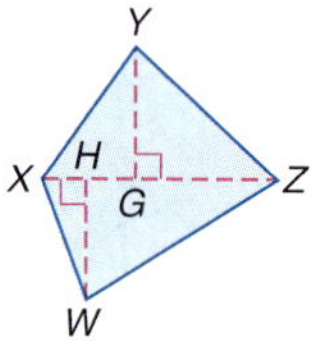

area of $XYZW$ = area of $\triangle XYZ$ + area of $\triangle XWZ$

$= \frac{1}{2}bh_1 + \frac{1}{2}bh_2$

$= \frac{1}{2}(39)(21) + \frac{1}{2}(39)(20)$ Substitution

$= 409.5 + 390$ Simplify.

$= 799.5$

The area of quadrilateral $XYZW$ is 799.5 square units.

AREAS OF TRAPEZOIDS AND RHOMBI

The formulas for the areas of trapezoids and rhombi are related to the formula for the area of a triangle.

Trapezoid $MNPQ$ has diagonal $\overline{QN}$ with parallel bases $\overline{MN}$ and $\overline{PQ}$. Therefore, the altitude h from vertex Q to the extension of base $\overline{MN}$ is the same length as the altitude from vertex N to the base $\overline{QP}$. Since the area of the trapezoid is the area of two nonoverlapping parts, we can write the following equation.

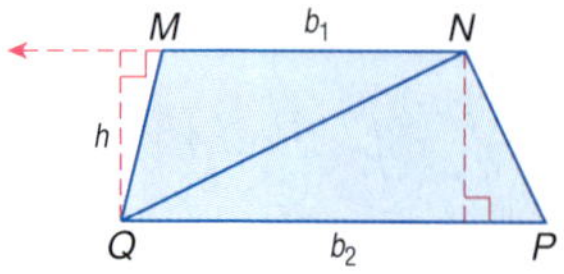

area of trapezoid $MNPQ$ = area of $\triangle MNQ$ + area of $\triangle NPQ$

$A = \frac{1}{2}(b_1)h + \frac{1}{2}(b_2)h$ Let the area be A, MN be b_1, and QP be b_2.

$A = \frac{1}{2}(b_1 + b_2)h$ Factor.

$A = \frac{1}{2}h(b_1 + b_2)$ Commutative Property

This is the formula for the area of any trapezoid.

Key Concept — *Area of a Trapezoid*

If a trapezoid has an area of A square units, bases of b_1 units and b_2 units, and a height of h units, then $A = \frac{1}{2}h(b_1 + b_2)$.

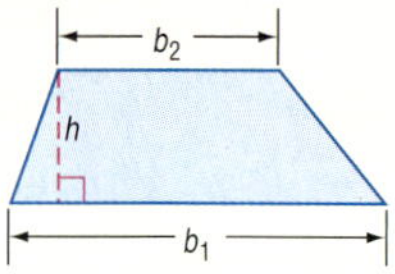

Geometry Activity

Materials: grid paper

- Ask students to recall what they know about perpendiculars and the altitudes of triangles before you do this activity.
- Some students may seem uninterested because they already know the formula for the area of a triangle. Ask them if they can demonstrate why the area formula works. In this activity, they will explore why the formula works and derive the formula for themselves.

602 Chapter 11 Areas of Polygons and Circles

Example 2 Area of a Trapezoid on the Coordinate Plane

COORDINATE GEOMETRY **Find the area of trapezoid *TVWZ* with vertices $T(-3, 4)$, $V(3, 4)$, $W(6, -1)$, and $Z(-5, -1)$.**

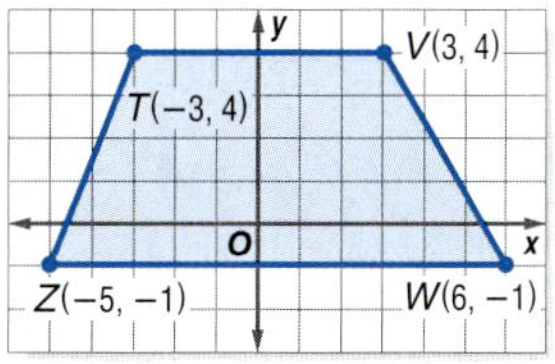

Bases: Since $\overline{TV}$ and $\overline{ZW}$ are horizontal, find their length by subtracting the x-coordinates of their endpoints.

$TV = |-3 - 3|$ $\quad$ $ZW = |-5 - 6|$

$= |-6|$ or 6 $\quad$ $= |-11|$ or 11

Height: Because the bases are horizontal segments, the distance between them can be measured on a vertical line. That is, subtract the y-coordinates.

$h = |4 - (-1)|$ or 5

Area: $A = \frac{1}{2}h(b_1 + b_2)$ Area of a trapezoid

$= \frac{1}{2}(5)(6 + 11)$ $h = 5, b_1 = 6, b_2 = 11$

$= 42.5$ Simplify.

The area of trapezoid *TVWZ* is 42.5 square units.

The formula for the area of a triangle can also be used to derive the formula for the area of a rhombus.

Key Concept — Area of a Rhombus

If a rhombus has an area of A square units and diagonals of d_1 and d_2 units, then $A = \frac{1}{2}d_1d_2$.

Example: $A = \frac{1}{2}(AC)(BD)$

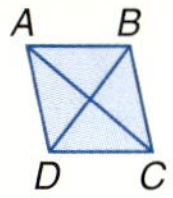

You will derive this formula in Exercise 46.

Study Tip

Area of a Rhombus
Because a rhombus is also a parallelogram, you can also use the formula $A = bh$ to determine the area.

Example 3 Area of a Rhombus on the Coordinate Plane

COORDINATE GEOMETRY **Find the area of rhombus *EFGH* with vertices at $E(-1, 3)$, $F(2, 7)$, $G(5, 3)$, and $H(2, -1)$.**

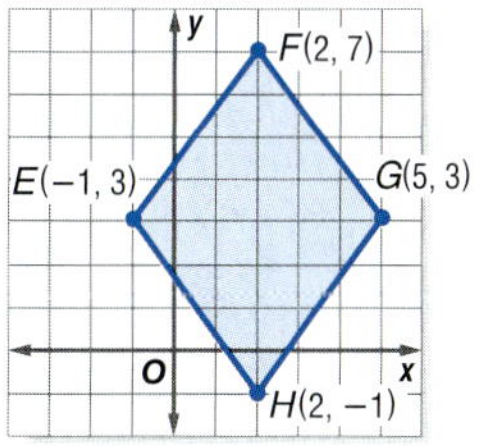

Explore To find the area of the rhombus, we need to know the lengths of each diagonal.

Plan Use coordinate geometry to find the length of each diagonal. Use the formula to find the area of rhombus *EFGH*.

Solve Let $\overline{EG}$ be d_1 and $\overline{FH}$ be d_2.

Subtract the x-coordinates of E and G to find that d_1 is 6.
Subtract the y-coordinates of F and H to find that d_2 is 8.

$A = \frac{1}{2}d_1d_2$ Area of a rhombus

$= \frac{1}{2}(6)(8)$ or 24 $\quad d_1 = 6, d_2 = 8$

Examine The area of rhombus *EFGH* is 24 square units.

If you know all but one measure in a quadrilateral, you can solve for the missing measure using the appropriate area formula.

www.geometryonline.com/extra_examples/fcat

AREAS OF TRAPEZOIDS AND RHOMBI

In-Class Examples

Power Point®

Teaching Tip Make sure students understand that the formula for the area of a trapezoid is derived from adding the areas of the two nonoverlapping triangles formed by one diagonal of the trapezoid. The height of the trapezoid is the same as the height of *each* triangle.

2 Find the area of trapezoid *RSTU* with vertices $R(4, 2)$, $S(6, -1)$, $T(-2, -1)$, and $U(-1, 2)$. **19.5 units²**

3 Find the area of rhombus *MNPR* with vertices at $M(0, 1)$, $N(4, 2)$, $P(3, -2)$, and $R(-1, -3)$. **15 units²**

Building on Prior Knowledge

In Chapter 8, students learned that if a parallelogram has all sides congruent, it is a rhombus. In this lesson, emphasize that the diagonals of a rhombus are used to find its area instead of the base and height.

DAILY INTERVENTION

Differentiated Instruction

Visual/Spatial Stress that in some triangles, one side seems to be an altitude. This is not true unless the triangle is a right triangle, in which the legs are perpendicular. Students should not assume that angles are right angles unless they are clearly marked.

Lesson 11-2 Areas of Triangles, Trapezoids, and Rhombi 603

In-Class Examples

Power Point®

4 **a.** Rhombus *RSTU* has an area of 64 square inches. Find *US* if *RT* = 8 inches.

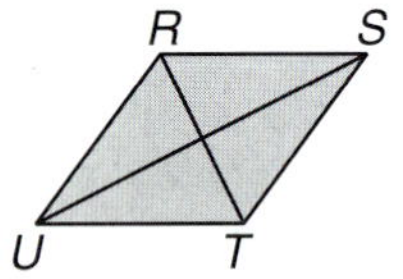

16 in.

b. Trapezoid *DEFG* has an area of 120 square feet. Find the height of *DEFG*.

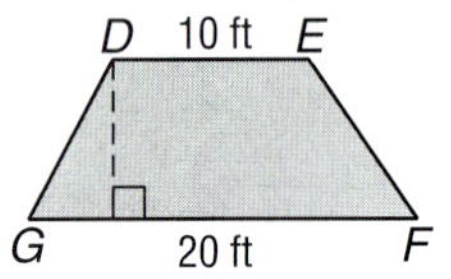

8 ft

5 **STAINED GLASS** This stained glass window is composed of 8 congruent trapezoidal shapes. The total area of the design is 72 square feet. Each trapezoid has bases of 3 and 6 feet. Find the height of each trapezoid.

2 ft

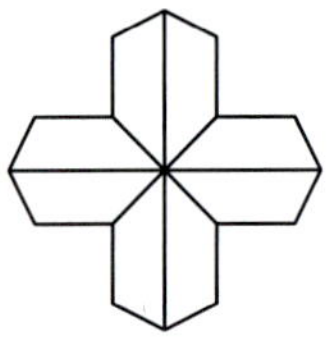

Example 4 Algebra: Find Missing Measures

a. Rhombus *WXYZ* has an area of 100 square meters. Find *WY* if *XZ* = 10 meters.

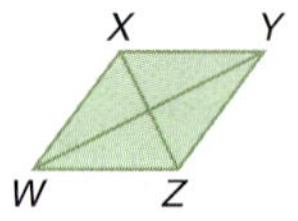

Use the formula for the area of a rhombus and solve for d_2.

$A = \frac{1}{2}d_1d_2$

$100 = \frac{1}{2}(10)(d_2)$

$100 = 5d_2$

$20 = d_2$

WY is 20 meters long.

b. Trapezoid *PQRS* has an area of 250 square inches. Find the height of *PQRS*.

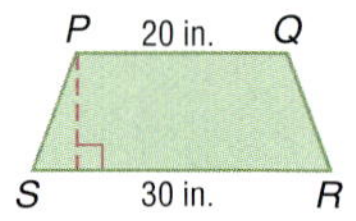

Use the formula for the area of a trapezoid and solve for h.

$A = \frac{1}{2}h(b_1 + b_2)$

$250 = \frac{1}{2}h(20 + 30)$

$250 = \frac{1}{2}(50)h$

$250 = 25h$

$10 = h$

The height of trapezoid *PQRS* is 10 inches.

Since the dimensions of congruent figures are equal, the areas of congruent figures are also equal.

Postulate 11.1

Congruent figures have equal areas.

Study Tip

Look Back

To review **the properties of rhombi and trapezoids**, see Lessons 8-5 and 8-6.

Example 5 Area of Congruent Figures

QUILTING **This quilt block is composed of twelve congruent rhombi arranged in a regular hexagon. The height of the hexagon is 8 inches. If the total area of the rhombi is 48 square inches, find the lengths of each diagonal and the area of one rhombus.**

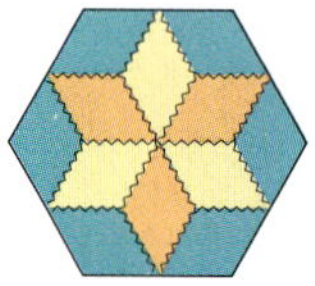

First, find the area of one rhombus. From Postulate 11.1, the area of each rhombus is the same. So, the area of each rhombus is 48 ÷ 12 or 4 square inches.

Next, find the length of one diagonal. The height of the hexagon is equal to the sum of the long diagonals of two rhombi. Since the rhombi are congruent, the long diagonals must be congruent. So, the long diagonal is equal to 8 ÷ 2, or 4 inches.

Use the area formula to find the length of the other diagonal.

$A = \frac{1}{2}d_1d_2$	Area of a rhombus
$4 = \frac{1}{2}(4)\,d_2$	$A = 4, d_1 = 4$
$2 = d_2$	Solve for d_2.

Each rhombus in the pattern has an area of 4 square inches and diagonals 4 inches and 2 inches long.

Check for Understanding

Concept Check

2. Kiku; she simplified the formula properly by adding the terms in the parentheses before multiplying.

1. **OPEN ENDED** Draw an isosceles trapezoid that contains at least one isosceles triangle. See margin.

2. **FIND THE ERROR** Robert and Kiku are finding the area of trapezoid $JKLM$.

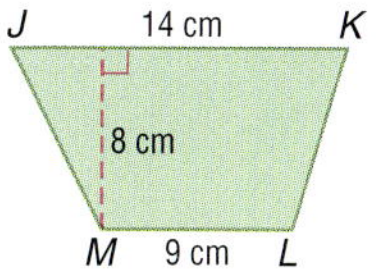

Robert	Kiku
$A = \frac{1}{2}(8)(14 + 9)$	$A = \frac{1}{2}(8)(14 + 9)$
$= \frac{1}{2}(8)(14) + 9$	$= \frac{1}{2}(8)(23)$
$= 56 + 9$	$= 4(23)$
$= 65 \text{ cm}^2$	$= 92 \text{ cm}^2$

Who is correct? Explain your reasoning.

3. **Determine** whether it is *always, sometimes,* or *never* true that rhombi with the same area have the same diagonal lengths. Explain your reasoning. Sometimes; two rhombi can have different corresponding diagonal lengths and have the same area.

Guided Practice

GUIDED PRACTICE KEY	
Exercises	Examples
5, 7	1
6, 8	2
4, 9	3
10, 11	4
12	5

Find the area of each quadrilateral.

4. 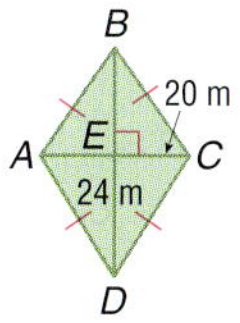

240 m^2

5. 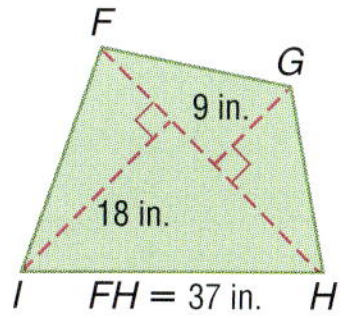

499.5 in^2

6. 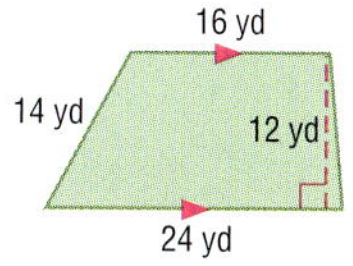

240 yd^2

COORDINATE GEOMETRY **Find the area of each figure given the coordinates of the vertices.**

7. $\triangle ABC$ with $A(2, -3)$, $B(-5, -3)$, and $C(-1, 3)$ 21 $units^2$
8. trapezoid $FGHJ$ with $F(-1, 8)$, $G(5, 8)$, $H(3, 4)$, and $J(1, 4)$ 16 $units^2$
9. rhombus $LMPQ$ with $L(-4, 3)$, $M(-2, 4)$, $P(0, 3)$, and $Q(-2, 2)$ 4 $units^2$

ALGEBRA **Find the missing measure for each quadrilateral.**

10. Trapezoid $NOPQ$ has an area of 250 square inches. Find the height of $NOPQ$. 10 in.

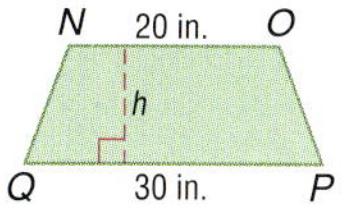

11. Rhombus $RSTU$ has an area of 675 square meters. Find SU. 45 m

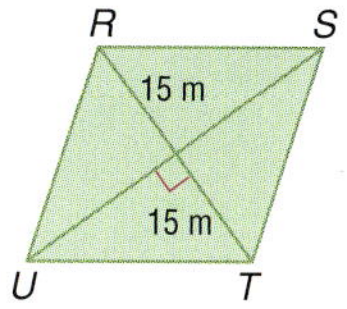

Application

12. **INTERIOR DESIGN** Jacques is designing a window hanging composed of 13 congruent rhombi. The total width of the window hanging is 15 inches, and the total area is $82\frac{7}{8}$ square inches. Find the length of each diagonal and the area of one rhombus.
$4\frac{1}{4}$ in., 3 in., $6\frac{3}{8}$ in^2

Answer

1. Sample answer:

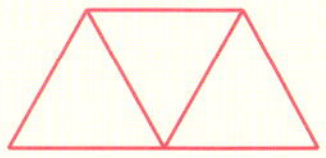

3 Practice/Apply

Study Notebook

Have students—
- add the definitions/examples of the vocabulary terms to their Vocabulary Builder worksheets for Chapter 11.
- include the formulas and examples of the area of a triangle, rhombus, and trapezoid.
- include any other item(s) that they find helpful in mastering the skills in this lesson.

DAILY INTERVENTION

FIND THE ERROR In Exercise 2, caution students that the order of operations applies to all mathematical expressions, including area formulas. Kiku did the problem correctly by adding the terms in parentheses before multiplying by the other terms.

About the Exercises...

Organization by Objective
- **Areas of Triangles:** 13, 14, 19–21
- **Areas of Trapezoids and Rhombi:** 15–18, 22–44

Odd/Even Assignments
Exercises 13–44 are structured so that students practice the same concepts whether they are assigned odd or even problems.

Assignment Guide

Basic: 13–43 odd, 47–57 odd, 58–60, 65–76 (optional: 61–64)
Average: 13–57 odd, 58–60, 65–76 (optional: 61–64)
Advanced: 14–56 even, 57–73 (optional: 74–76)
All: Quiz 1 (1–5)

Study Guide and Intervention, p. 617 (shown) and p. 618

Areas of Triangles The area of a triangle is half the area of a rectangle with the same base and height as the triangle.

If a triangle has an area of A square units, a base of b units, and a corresponding height of h units, then $A = \frac{1}{2}bh$.

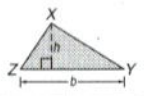

Example Find the area of the triangle.

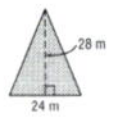

$A = \frac{1}{2}bh$ Area of a triangle

$= \frac{1}{2}(24)(28)$ $b = 24$, $h = 28$

$= 336$ Multiply.

The area is 336 square meters.

Exercises

Find the area of each figure.

1. 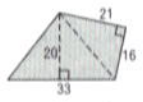 498 units²
2. 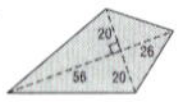 1640 units²
3.  25 units²
4. 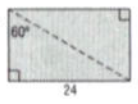 332.6 units²
5. 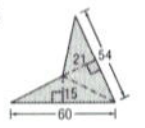 1017 units²
6. 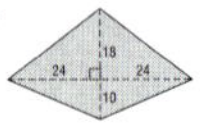 672 units²
7. The area of a triangle is 72 square inches. If the height is 8 inches, find the length of the base. 18 in.
8. A right triangle has a perimeter of 36 meters, a hypotenuse of 15 meters, and a leg of 9 meters. Find the area of the triangle. 54 m²

Skills Practice, p. 619 and Practice, p. 620 (shown)

Find the area of each figure. Round to the nearest tenth if necessary.

1. 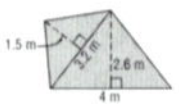 7.6 m²
2. 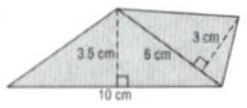 26.5 cm²
3. 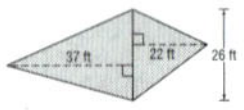 767 ft²

Find the area of each quadrilateral given the coordinates of the vertices.

4. trapezoid $ABCD$ $A(-7, 1)$, $B(-4, 4)$, $C(-4, -6)$, $D(-7, -3)$ 21 units²
5. rhombus $LMNO$ $L(6, 8)$, $M(14, 4)$, $N(6, 0)$, $O(-2, 4)$ 64 units²

Find the missing measure for each figure.

6. Trapezoid $WXYZ$ has an area of 13.75 square meters. Find WX. 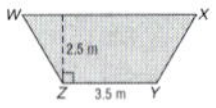 7.5 m
7. Triangle PRS has an area of 68 square yards. If the height of $\triangle PRS$ is 8 yards, find the base. 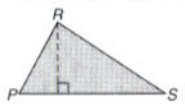 17 yd

DESIGN For Exercises 8 and 9, use the following information.
Mr. Hagarty used 16 congruent rhombi-shaped tiles to design the midsection of the backsplash area above a kitchen sink. The length of the design is 27 inches and the total area is 108 square inches.

8. Find the area of one rhombus. $6\frac{3}{4}$ in²
9. Find the length of each diagonal. $4\frac{1}{2}$ in., 3 in.

Reading to Learn Mathematics, p. 621 ELL

Pre-Activity How is the area of a triangle related to beach umbrellas?

Read the introduction to Lesson 11-2 at the top of page 601 in your textbook.

Classify the polygons in the panels of the beach umbrella. Isosceles triangles and isosceles trapezoids

Reading the Lesson

1. Match each area formula from the first column with the corresponding polygon in the second column.
   - a. $A = \ell w$ vi — i. triangle
   - b. $A = \frac{1}{2}d_1d_2$ iv — ii. parallelogram
   - c. $A = s^2$ v — iii. trapezoid
   - d. $A = \frac{1}{2}h(b_1 + b_2)$ iii — iv. rhombus
   - e. $A = \frac{1}{2}bh$ i — v. square
   - f. $A = bh$ ii — vi. rectangle
2. Determine whether each statement is *always*, *sometimes*, or *never* true. In each case, explain your reasoning. For explanations, sample answers are given.
   - a. The area of a square is half the product of its diagonals. Always; a square is a rhombus, so you can use the rhombus formula.
   - b. The area of a triangle is half the product of two of its sides. Sometimes; this is true only for a right triangle.
   - c. You can find the area of a rectangle by multiplying base times height. Always; a rectangle is a parallelogram, so you can use the parallelogram formula. If the length of a rectangle is used as the base, then the width is the height.
   - d. You can find the area of a rectangle by multiplying the lengths of any two of its sides. Sometimes; this is true only for a square. Otherwise, you must use two *consecutive* sides, not any two sides.
   - e. The area of a trapezoid is the product of its height and the sum of the bases. Never; the area is one-half the product of its height and the sum of the bases.
   - f. The square of the length of a side of a square is equal to half the product of its diagonals. Always; a square is a rhombus, so the formulas for a square and a rhombus must give the same answer whenever the rhombus is a square.

Helping You Remember

3. A good way to remember a new geometric formula is to state it in words. Write a short sentence that tells how to find the area of a trapezoid in a way that is easy to remember. Sample answer: Average the lengths of the bases and multiply by the height.

★ indicates increased difficulty

Practice and Apply

Homework Help

For Exercises	See Examples
13, 14, 19–21	1
15, 16, 22–25	2
17, 18, 26–29	3
30–35, 40–44	4
36–39	5

Extra Practice See page 776.

Find the area of each figure. Round to the nearest tenth if necessary.

13. 12.4 cm²
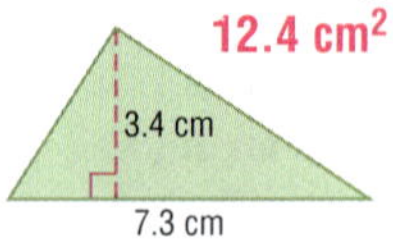

14. 35.7 ft²
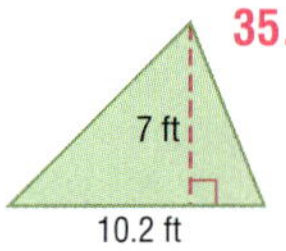

15. 95 km²
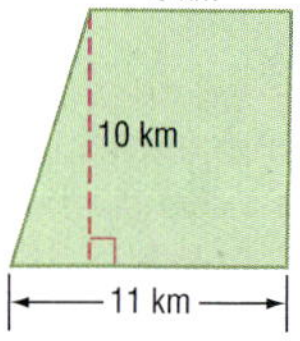

16. 96.5 yd²
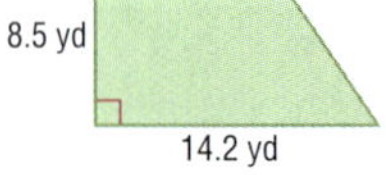

17. 1200 ft²
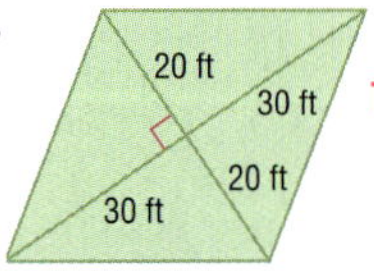

18. 408 cm²
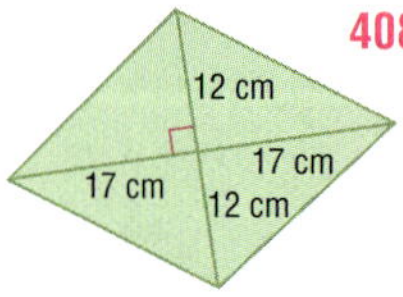

19. 50 m²
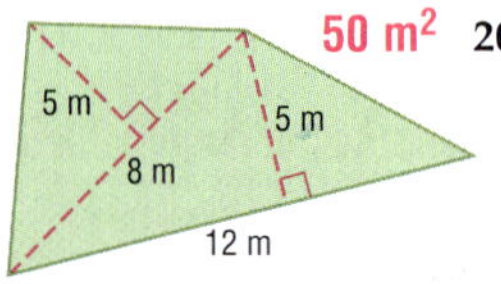

20. 99 in²
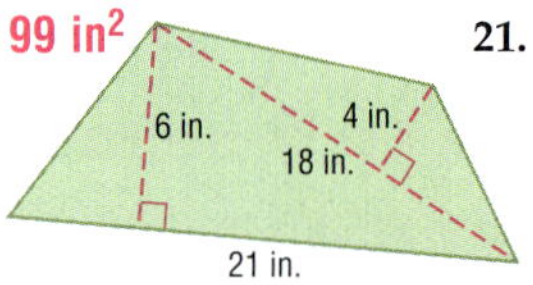

21. 129.9 mm²
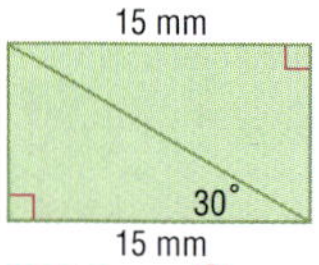

COORDINATE GEOMETRY Find the area of trapezoid $PQRT$ given the coordinates of the vertices. 23. 55 units²

22. $P(0, 3)$, $Q(3, 7)$, $R(5, 7)$, $T(6, 3)$ 16 units²
23. $P(-4, -5)$, $Q(-2, -5)$, $R(4, 6)$, $T(-4, 6)$
24. $P(-3, 8)$, $Q(6, 8)$, $R(6, 2)$, $T(1, 2)$ 42 units²
25. $P(-6, 3)$, $Q(1, 3)$, $R(-2, -2)$, $T(-4, -2)$ 22.5 units²

COORDINATE GEOMETRY Find the area of rhombus $JKLM$ given the coordinates of the vertices. 26. 30 units² 27. 20 units²

26. $J(2, 1)$, $K(7, 4)$, $L(12, 1)$, $M(7, -2)$
27. $J(-1, 2)$, $K(1, 7)$, $L(3, 2)$, $M(1, -3)$
28. $J(-1, -4)$, $K(2, 2)$, $L(5, -4)$, $M(2, -10)$ 36 units²
29. $J(2, 4)$, $K(6, 6)$, $L(10, 4)$, $M(6, 2)$ 16 units²

ALGEBRA Find the missing measure for each figure.

30. Trapezoid $ABCD$ has an area of 750 square meters. Find the height of $ABCD$. 25 m
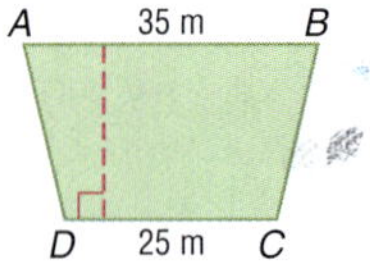

31. Trapezoid $GHJK$ has an area of 188.35 square feet. If HJ is 16.5 feet, find GK. 26.8 ft
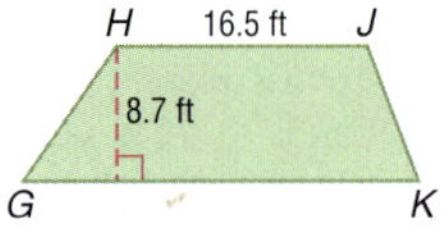

32. Rhombus $MNPQ$ has an area of 375 square inches. If MP is 25 inches, find NQ. 30 in.
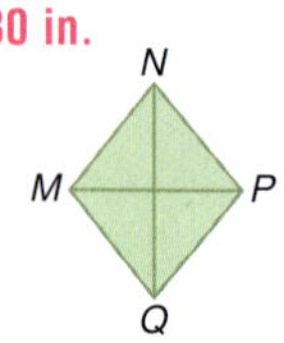

33. Rhombus $QRST$ has an area of 137.9 square meters. If RT is 12.2 meters, find QS. 22.6 m
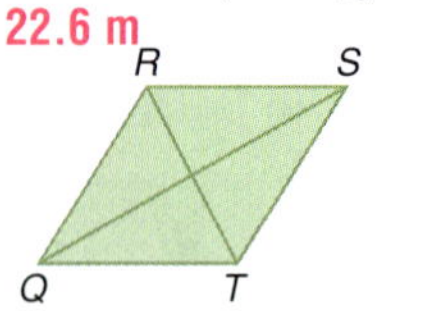

34. Triangle WXY has an area of 248 square inches. Find the length of the base. 31 in.
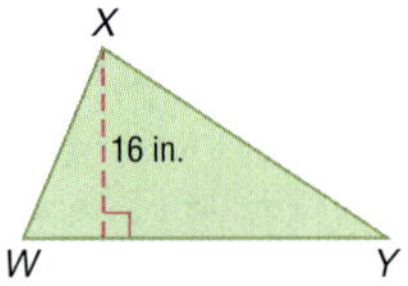

35. Triangle PQS has an area of 300 square centimeters. Find the height. 20 cm
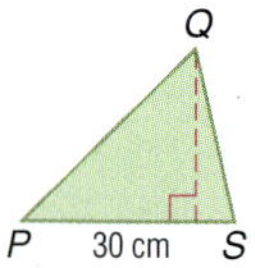

Enrichment, p. 622

Areas of Similar Triangles

You have learned that if two triangles are similar, the ratio of the lengths of corresponding altitudes is equal to the ratio of the lengths of a pair of corresponding sides. However, there is a different relationship between the areas of the two triangles.

Theorem If two triangles are similar, the ratio of their areas is the square of the ratio of the lengths of a pair of corresponding sides.

Triangle II is k times larger than Triangle I. Thus, its base is k times as large as that of Triangle I and its height is k times as large as that of Triangle I.

$\frac{\text{side of }\triangle II}{\text{side of }\triangle I} = \frac{kb}{b}$ or $\frac{k}{1}$

$\frac{\text{area of }\triangle II}{\text{area of }\triangle I} = \frac{\frac{1}{2}k^2bh}{\frac{1}{2}bh}$ or $\frac{k^2}{1}$

Triangle I area $\triangle I = \frac{1}{2}bh$

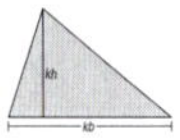

Triangle II area $\triangle II = \frac{1}{2}(kb)(kh)$ $= \frac{1}{2}k^2bh$

GARDENS For Exercises 36 and 37, use the following information.

Keisha designed a garden that is shaped like two congruent rhombi. She wants the long diagonals lined with a stone walkway. The total area of the garden is 150 square feet, and the shorter diagonals are each 12 feet long.

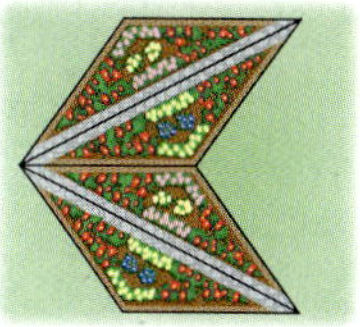

36. Find the length of each stone walkway. **12.5 ft**

37. Find the length of each side of the garden. **about 8.7 ft**

Career Choices

Real Estate Agent

Real estate agents must have a broad knowledge of the neighborhoods in their community. About two-thirds of real estate agents are self-employed. Success is dependent on selling properties.

Source: ww.bls.gov

REAL ESTATE For Exercises 38 and 39, use the following information.

The map shows the layout and dimensions of several lot parcels in Linworth Village. Suppose Lots 35 and 12 are trapezoids.

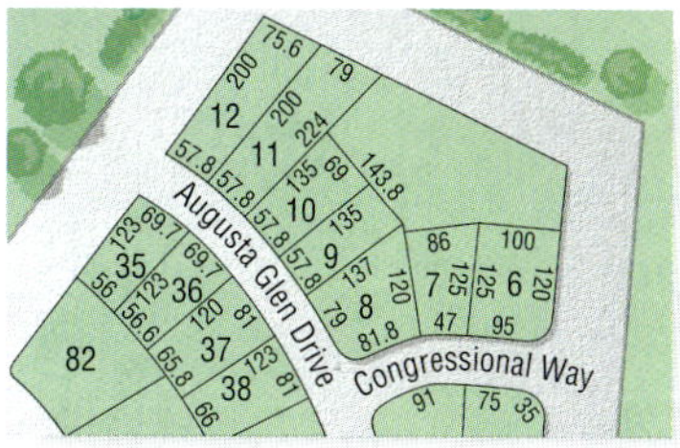

38. If the height of Lot 35 is 122.81 feet, find the area of this lot. **7718.6 ft^2**

39. If the height of Lot 12 is 199.8 feet, find the area of this lot. **13,326.7 ft^2**

Online Research **Data Update** Use the Internet or other resource to find the median price of homes in the United States. How does this compare to the median price of homes in your community? Visit www.geometryonline.com/data_update to learn more.

Find the area of each figure.

40. rhombus with a perimeter of 20 meters and a diagonal of 8 meters **24 m^2**

41. rhombus with a perimeter of 52 inches and a diagonal of 24 inches **120 in^2**

42. isosceles trapezoid with a perimeter of 52 yards; the measure of one base is 10 yards greater than the other base, the measure of each leg is 3 yards less than twice the length of the shorter base **156 yd^2**

43. equilateral triangle with a perimeter of 15 inches **≈10.8 in^2**

★ 44. scalene triangle with sides that measure 34.0 meters, 81.6 meters, and 88.4 meters. **1387.2 m^2**

★ 45. Find the area of $\triangle JKM$. **21 ft^2**

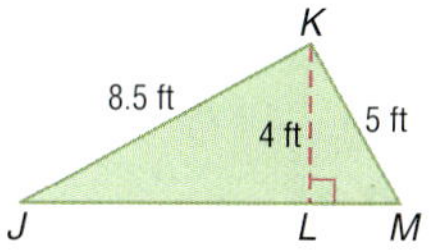

46. Derive the formula for the area of a rhombus using the formula for the area of a triangle. **See margin.**

47. Determine whether the statement *Two triangles that have the same area also have the same perimeter* is true or false. Give an example or counterexample. **See margin.**

48–49. See margin.

Each pair of figures is similar. Find the area and perimeter of each figure. Describe how changing the dimensions affects the perimeter and area.

48.

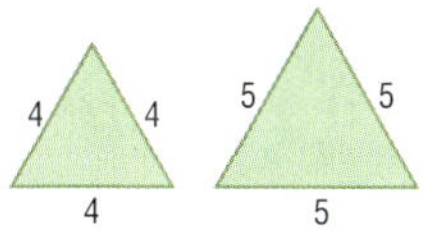

49.

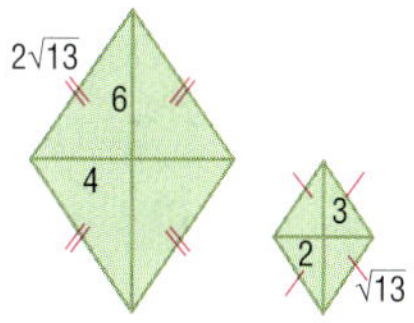

50. **RECREATION** Becky wants to cover a kite frame with decorative paper. If the length of one diagonal is 20 inches and the other diagonal measures 25 inches, find the area of the surface of the kite. **250 in^2**

Answers

46. A rhombus is made up of two congruent triangles. Using d_1 and d_2 instead of b and h, its area in reference to $A = \frac{1}{2}bh$ is $2\left[\frac{1}{2}(d_1)\left(\frac{1}{2}d_2\right)\right]$ or $\frac{1}{2}d_1d_2$.

47. False; Sample answer:

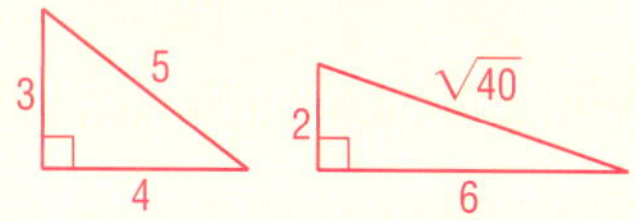

The area for each of these right triangles is 6 square units. The perimeter of one triangle is 12 and the perimeter of the other is $8 + \sqrt{40}$ or about 14.3.

48. area ≈ 6.9, area ≈ 10.8; perimeter = 12, perimeter = 15; scale factor and ratio of perimeters is $\frac{5}{4}$, ratio of areas is $\left(\frac{5}{4}\right)^2$

49. area = 12, area = 3, perimeter = $8\sqrt{13}$, perimeter = $4\sqrt{13}$; scale factor and ratio of perimeters = $\frac{1}{2}$, ratio of areas = $\left(\frac{1}{2}\right)^2$

Answers

58. Sample answer: Umbrellas have triangular panels of fabric or nylon. In order to make the panels to fit the umbrella frame, the area of the triangles is needed. Answers should include the following.

- Find the area of a triangle by multiplying the base and the height and dividing by two.
- Rhombi are composed of two congruent isosceles triangles, and trapezoids are composed of two triangles and a rectangle.

SIMILAR FIGURES **For Exercises 51–56, use the following information.**

Triangle *ABC* is similar to triangle *DEF*.

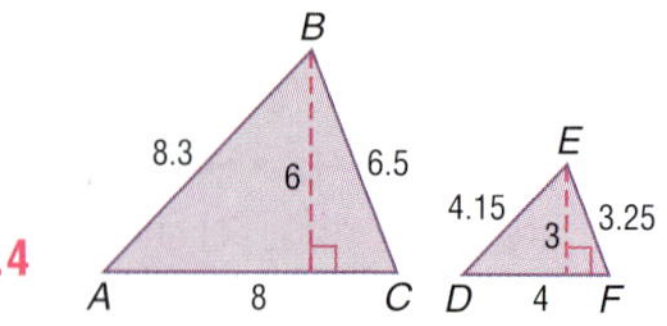

51. Find the scale factor. $\frac{2}{1}$
52. Find the perimeter of each triangle. 22.8; 11.4
53. Compare the ratio of the perimeters of the triangles to the scale factor.

53. The ratio is the same.

54. Find the area of each triangle. 24, 6
55. Compare the ratio of the areas of the triangles to the scale factor.

55. 4:1; The ratio of the areas is the square of the scale factor.

56. Compare the ratio of the areas of the triangles to the ratio of the perimeters of the triangles. The ratio of the areas is the square of the ratio of the perimeters.

57. **CRITICAL THINKING** In the figure, the vertices of quadrilateral *ABCD* intersect square *EFGH* and divide its sides into segments with measures that have a ratio of 1:2. Find the area of *ABCD*. Describe the relationship between the areas of *ABCD* and *EFGH*. 45 ft^2; The ratio of the areas is 5:9.

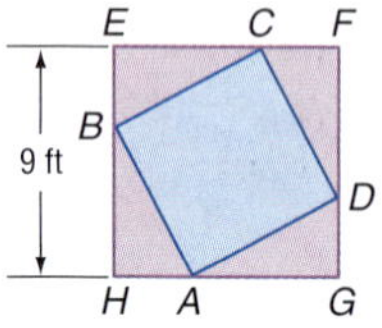

58. **WRITING IN MATH** Answer the question that was posed at the beginning of the lesson. See margin.

How is the area of a triangle related to beach umbrellas?

Include the following in your answer:
- how to find the area of a triangle, and
- how the area of a triangle can help you find the areas of rhombi and trapezoids.

FCAT Practice
Standardized Test Practice
A B C D

59. In the figure, if point *B* lies on the perpendicular bisector of $\overline{AC}$, what is the area of $\triangle ABC$? B

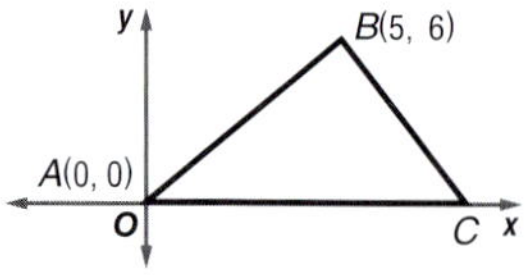

(A) 15 units^2 (B) 30 units^2

(C) 50 units^2 (D) 1602 units^2

60. **ALGEBRA** What are the solutions of the equation $(2x - 7)(x + 10) = 0$? D

(A) -3.5 and 10 (B) 7 and -10 (C) $\frac{2}{7}$ and -10 (D) 3.5 and -10

Extending the Lesson

Trigonometric Ratios and the Areas of Triangles

The area of any triangle can be found given the measures of two sides of the triangle and the measure of the included angle. Suppose we are given $AC = 15$, $BC = 8$, and $m\angle C = 60$. To find the height of the triangle, use the sine ratio, $\sin C = \frac{h}{BC}$. Then use the value of h in the formula for the area of a triangle. So, the area is $\frac{1}{2}(15)(8 \sin 60°)$ or 52.0 square meters.

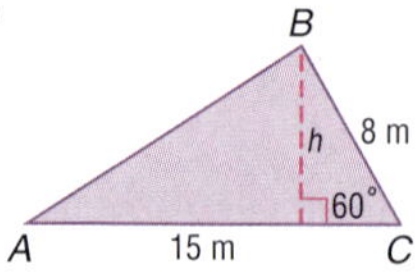

61. Derive a formula to find the area of any triangle, given the measures of two sides of the triangle and their included angle. area $= \frac{1}{2}ab \sin C$

Find the area of each triangle. Round to the nearest hundredth.

62.

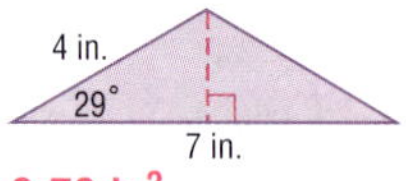

6.79 in^2

63.

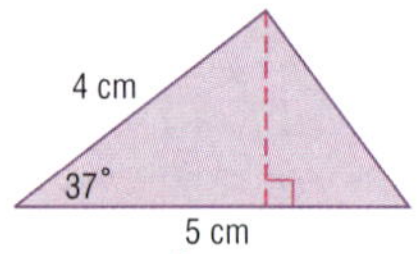

6.02 cm^2

64.

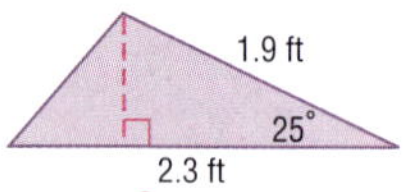

0.92 ft^2

608 Chapter 11 Areas of Polygons and Circles

Maintain Your Skills

Mixed Review

Find the area of each figure. Round to the nearest tenth. *(Lesson 11-1)*

65.

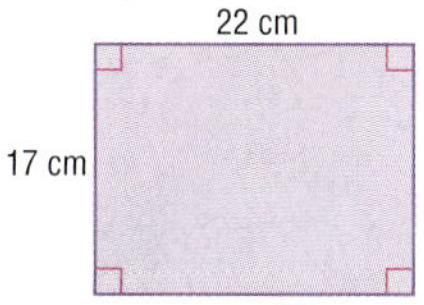

374 cm^2

66.

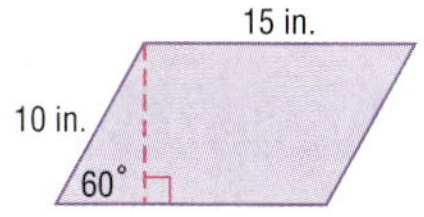

129.9 in^2

67.

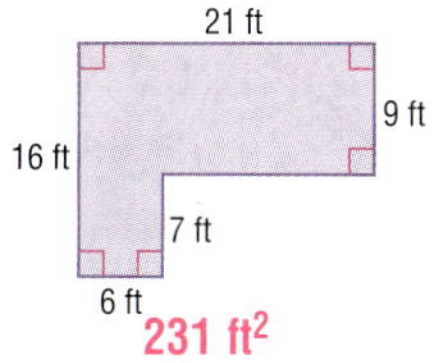

231 ft^2

Write an equation of circle *R* based on the given information. *(Lesson 10-8)*

68. center: $R(1, 2)$
radius: 7

69. center: $R\left(-4, \frac{1}{2}\right)$
radius: $\frac{11}{2}$

70. center: $R(-1.3, 5.6)$
radius: 3.5

8. $(x - 1)^2 +$ $y - 2)^2 = 49$

9. $(x + 4)^2 +$ $y - \frac{1}{2})^2 = \frac{121}{4}$

0. $(x + 1.3)^2 +$ $y - 5.6)^2 = 12.25$

71. CRAFTS Andria created a pattern to appliqué flowers onto a quilt by first drawing a regular pentagon that was 3.5 inches long on each side. Then she added a semicircle onto each side of the pentagon to create the appearance of five petals. How many inches of gold trim does she need to edge 10 flowers? *(Lesson 10-1)* 275 in.

Given the magnitude and direction of a vector, find the component form with values rounded to the nearest tenth. *(Lesson 9-6)* 72. $\langle 123.3, 57.5 \rangle$

72. magnitude of 136 at a direction of 25 degrees with the positive *x*-axis

73. magnitude of 280 at a direction of 52 degrees with the positive *x*-axis
$\langle 172.4, 220.6 \rangle$

Getting Ready for the Next Lesson

PREREQUISITE SKILL **Find *x*. Round to the nearest tenth.**
*(To review **trigonometric ratios in right triangles**, see Lesson 7-4.)*

74. 44.0

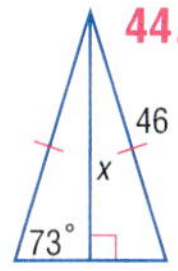

75. 20.1

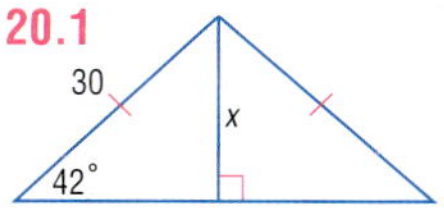

76. 4.8

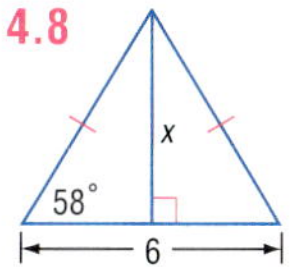

Practice Quiz 1

Lessons 11-1 and 11-2

The coordinates of the vertices of quadrilateral *JKLM* are *J*(−8, 4), *K*(−4, 0), *L*(0, 4), and *M*(−4, 8). *(Lesson 11-1)*

1. Determine whether *JKLM* is a *square*, a *rectangle*, or a *parallelogram*. square
2. Find the area of *JKLM*. 32 $units^2$

Find the area of each trapezoid. *(Lesson 11-2)*

3.

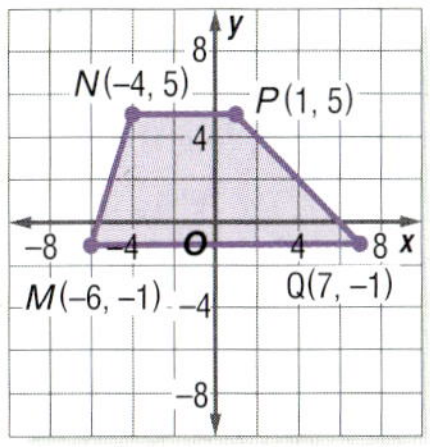

54 $units^2$

4.

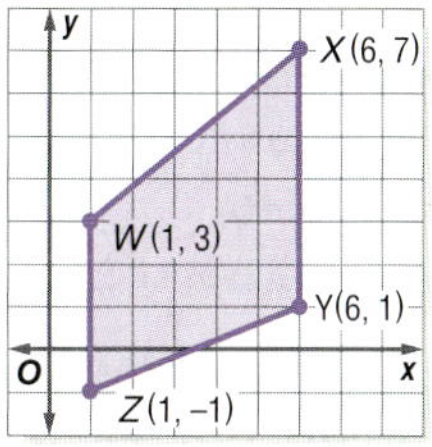

25 $units^2$

5. The area of a rhombus is 546 square yards. If d_1 is 26 yards long, find the length of d_2. *(Lesson 11-2)* 42 yd

www.geometryonline.com/self_check_quiz/fcat

4 Assess

Open-Ended Assessment

Writing Ask students to list all the formulas for area they have learned so far in this chapter. Ask them to include a labeled diagram corresponding to each formula.

Getting Ready for Lesson 11-3

Prerequisite Skill Students will use right triangle trigonometry in Lesson 11-3 to find the altitude of a regular polygon so that they can calculate the area of the polygon. Use Exercises 74–76 to determine your students' familiarity with trigonometric ratios in right triangles.

Assessment Options

Practice Quiz 1 The quiz provides students with a brief review of the concepts and skills in Lessons 11-1 and 11-2. Lesson numbers are given to the right of the exercises or instruction lines so students can review concepts not yet mastered.

Quiz (Lessons 11-1 and 11-2) is available on p. 655 of the *Chapter 11 Resource Masters*.

Teacher to Teacher

Sarah L. Waldrop, Forestview High School — Gastonia, NC

We play MATHO as a review. I give students (on the overhead) about 30-36 answers to be placed on their MATHO sheet. This sheet is arranged like bingo—with a free space. They fill in any 24 answers. After their card is filled in, I ask questions with the given answer. When they have MATHO they get a prize (candy, points on a quiz, etc.).

11-3 Lesson Notes

1 Focus

5-Minute Check Transparency 11-3 Use as a quiz or review of Lesson 11-2.

Mathematical Background notes are available for this lesson on p. 592C.

How can you find the area of a polygon?

Ask students:

- Which geometric shape is used most often in the construction of a gazebo? **a regular hexagon**
- Why is this shape used for gazebos? **Sample answer: nice shape; large area**
- What is another structure that uses this shape in its construction? **Sample answer: barns**
- How is this shape used in nature? **Sample answer: honeycombs**

11-3 Areas of Regular Polygons and Circles

Sunshine State Standards
MA.A.3.4.3, MA.B.1.4.1, MA.B.3.4.1

What You'll Learn

- Find areas of regular polygons.
- Find areas of circles.

Vocabulary

- apothem

How can you find the area of a polygon?

The foundations of most gazebos are shaped like regular hexagons. Suppose the owners of this gazebo would like to install tile on the floor. If tiles are sold in square feet, how can they find out the actual area of tiles needed to cover the floor?

Study Tip

Look Back
To review **apothem**, see Lesson 1-4.

AREAS OF REGULAR POLYGONS In regular hexagon $ABCDEF$ inscribed in circle G, $\overline{GA}$ and $\overline{GF}$ are radii from the center of the circle G to two vertices of the hexagon. $\overline{GH}$ is drawn from the center of the regular polygon perpendicular to a side of the polygon. This segment is called an **apothem**.

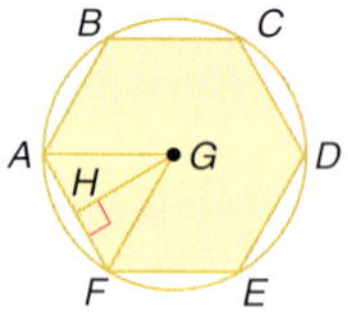

Triangle GFA is an isosceles triangle, since the radii are congruent. If all of the radii were drawn, they would separate the hexagon into 6 nonoverlapping congruent isosceles triangles.

The area of the hexagon can be determined by adding the areas of the triangles. Since $\overline{GH}$ is perpendicular to $\overline{AF}$, it is an altitude of $\triangle AGF$. Let a represent the length of $\overline{GH}$ and let s represent the length of a side of the hexagon.

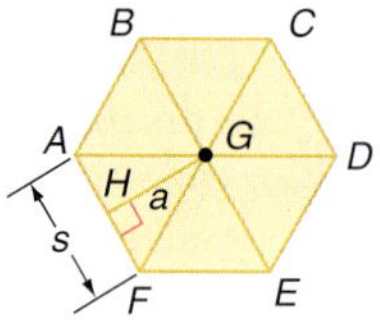

$$\text{Area of } \triangle AGF = \frac{1}{2}bh$$
$$= \frac{1}{2}sa$$

The area of one triangle is $\frac{1}{2}sa$ square units. So the area of the hexagon is $6\left(\frac{1}{2}sa\right)$ square units. Notice that the perimeter P of the hexagon is $6s$ units. We can substitute P for $6s$ in the area formula. So, $A = 6\left(\frac{1}{2}sa\right)$ becomes $A = \frac{1}{2}Pa$. This formula can be used for the area of any regular polygon.

Key Concept *Area of a Regular Polygon*

If a regular polygon has an area of A square units, a perimeter of P units, and an apothem of a units, then $A = \frac{1}{2}Pa$.

Resource Manager

Workbook and Reproducible Masters

Chapter 11 Resource Masters

- Study Guide and Intervention, pp. 623–624
- Skills Practice, p. 625
- Practice, p. 626
- Reading to Learn Mathematics, p. 627
- Enrichment, p. 628
- Assessment, pp. 655, 657

Prerequisite Skills Workbook, pp. 43–44

Teaching Geometry With Manipulatives Masters, pp. 1, 18, 184, 185, 186, 189

Transparencies

5-Minute Check Transparency 11-3
Answer Key Transparencies

Technology

Interactive Chalkboard

Example 1 Area of a Regular Polygon

Find the area of a regular pentagon with a perimeter of 40 centimeters.

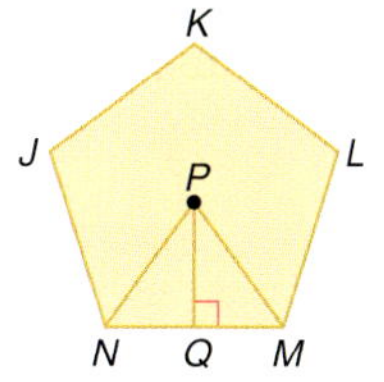

Apotem: The central angles of a regular pentagon are all congruent. Therefore, the measure of each angle is $\frac{360}{5}$ or 72. $\overline{PQ}$ is an apothem of pentagon *JKLMN*. It bisects $\angle NPM$ and is a perpendicular bisector of $\overline{NM}$. So, $m\angle MPQ = \frac{1}{2}(72)$ or 36. Since the perimeter is 40 centimeters, each side is 8 centimeters and $QM = 4$ centimeters. Write a trigonometric ratio to find the length of $\overline{PQ}$.

$\tan \angle MPQ = \frac{QM}{PQ}$ — $\tan \theta = \frac{\text{length of opposite side}}{\text{length of adjacent side}}$

$\tan 36° = \frac{4}{PQ}$ — $m\angle MPQ = 36$, $QM = 4$

$(PQ) \tan 36° = 4$ — Multiply each side by *PQ*.

$PQ = \frac{4}{\tan 36°}$ — Divide each side by tan 36°.

$PQ \approx 5.5$ — Use a calculator.

Area: $A = \frac{1}{2}Pa$ — Area of a regular polygon

$\approx \frac{1}{2}(40)(5.5)$ — $P = 40$, $a \approx 5.5$

≈ 110 — Simplify.

So, the area of the pentagon is about 110 square centimeters.

AREAS OF CIRCLES You can use a calculator to help derive the formula for the area of a circle from the areas of regular polygons.

Geometry Activity

Area of a Circle

Collect Data

Suppose each regular polygon is inscribed in a circle of radius *r*.

1. Copy and complete the following table. Round to the nearest hundredth.

Inscribed Polygon						
Number of Sides	3	5	8	10	20	50
Measure of a Side	1.73*r*	1.18*r*	0.77*r*	0.62*r*	0.31*r*	0.126*r*
Measure of Apothem	0.5*r*	0.81*r*	0.92*r*	0.95*r*	0.99*r*	0.998*r*
Area	$1.30r^2$	$2.39r^2$	$2.83r^2$	$2.95r^2$	$3.07r^2$	$3.14r^2$

Analyze the Data

2. What happens to the appearance of the polygon as the number of sides increases? The polygon appears to be a circle.
3. What happens to the areas as the number of sides increases?
4. **Make a conjecture** about the formula for the area of a circle.

Study Tip

Problem Solving
There is another method for finding the apothem of a regular polygon. You can use the Interior Angle Sum Theorem to find $m\angle PMQ$ and then write a trigonometric ratio to find *PQ*.

3. The areas of the polygons approach the area of the circle.
4. The formula for the area of a circle is πr^2 or about $3.14r^2$.

www.geometryonline.com/extra_examples/fcat

2 Teach

Building on Prior Knowledge

In Chapter 7, students learned how to use trigonometry to solve a triangle. In this lesson, they will use trigonometry to find the apothem of a regular polygon so that they can apply the area formula for a regular polygon. The area is one-half the product of the apothem and the perimeter.

Tips for New Teachers

Rounding Rounding during computation may result in a final answer that differs from the ones given in the Teacher Wraparound Edition.

AREAS OF REGULAR POLYGONS

In-Class Example Power Point®

Teaching Tip Students may have difficulty finding the apothem of a regular hexagon. Emphasize the construction of a 30°-60°-90° triangle from the center to a side of the hexagon. Then use the side relationships to find the apothem.

1. Find the area of a regular pentagon with a perimeter of 90 meters. about 557 m^2

Geometry Activity

Materials: calculator

- Ask why the radius is left as a variable in this data. (The table can be developed for a circle with any radius *r*.
- Also point out that the measure of a side and measure of the apothem of an equilateral triangle is found using 30°-60°-90° triangle relationships. Trigonometric ratios are used to determine the measures of a side and an apothem for each polygon.

AREAS OF CIRCLES

In-Class Examples

Power Point®

2 An outdoor accessories company manufactures circular covers for outdoor umbrellas. If the cover is 8 inches longer than the umbrella on each side, find the area of the cover in square yards. **4.7 yd²**

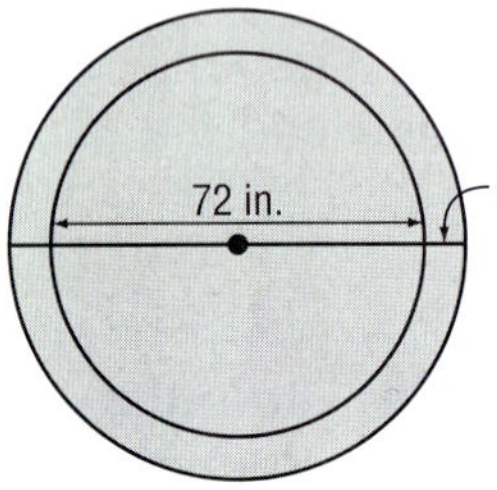

3 Find the area of the shaded region. Assume that the triangle is equilateral. Round to the nearest tenth.

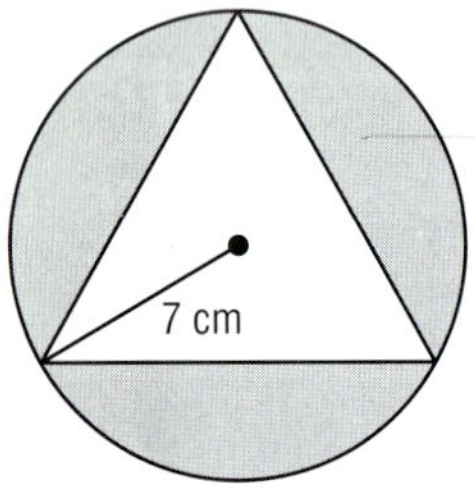

90.3 cm²

You can see from the Geometry Activity that the more sides a regular polygon has, the more closely it resembles a circle.

Key Concept — *Area of a Circle*

If a circle has an area of A square units and a radius of r units, then $A = \pi r^2$.

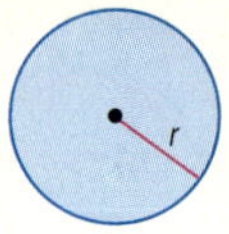

Study Tip

Square Yards
A square yard measures 36 inches by 36 inches or 1296 square inches.

Example 2 Use Area of a Circle to Solve a Real-World Problem

SEWING **A caterer has a 48-inch diameter table that is 34 inches tall. She wants a tablecloth that will touch the floor. Find the area of the tablecloth in square yards.**

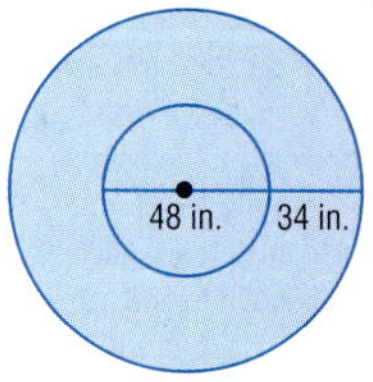

The diameter of the table is 48 inches, and the tablecloth must extend 34 inches in each direction. So the diameter of the tablecloth is 34 + 48 + 34 or 116 inches. Divide by 2 to find that the radius is 58 inches.

$A = \pi r^2$	Area of a circle
$= \pi(58)^2$	Substitution
$\approx 10{,}568.3$	Use a calculator.

The area of the tablecloth is 10,568.3 square inches. To convert to square yards, divide by 1296. The area of the tablecloth is 8.2 square yards to the nearest tenth.

Study Tip

Look Back
To review **inscribed and circumscribed polygons,** see Lesson 10-4.

You can use the properties of circles and regular polygons to find the areas of inscribed and circumscribed polygons.

Example 3 Area of an Inscribed Polygon

Find the area of the shaded region. Assume that the triangle is equilateral.

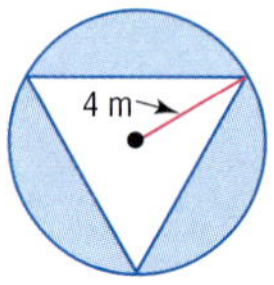

The area of the shaded region is the difference between the area of the circle and the area of the triangle. First, find the area of the circle.

$A = \pi r^2$	Area of a circle
$= \pi(4)^2$	Substitution
≈ 50.3	Use a calculator.

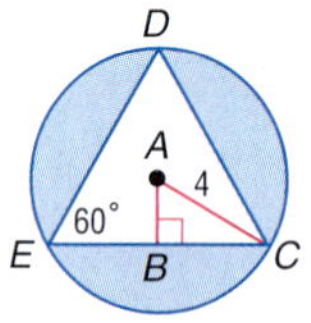

To find the area of the triangle, use properties of 30°-60°-90° triangles. First, find the length of the base. The hypotenuse of $\triangle ABC$ is 4, so BC is $2\sqrt{3}$. Since $EC = 2(BC)$, $EC = 4\sqrt{3}$.

Next, find the height of the triangle, DB. Since $m\angle DCB$ is 60, $DB = 2\sqrt{3}(\sqrt{3})$ or 6.

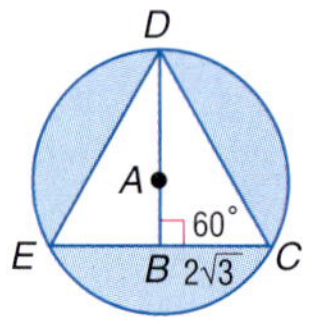

Use the formula to find the area of the triangle.

$A = \frac{1}{2}bh$	Area of a triangle
$= \frac{1}{2}(4\sqrt{3})(6)$	$b = 4\sqrt{3}$, $h = 6$
≈ 20.8	Use a calculator.

The area of the shaded region is 50.3 − 20.8 or 29.5 square meters to the nearest tenth.

DAILY INTERVENTION — **Differentiated Instruction**

Interpersonal Have students choose a partner. Ask one student to draw a polygon and give the radius of the circumscribed circle. Then ask the partner to find the area by first finding the apothem. Next switch roles and do the activity again. If students have difficulty finding the apothem of the polygon, ask them to discuss how right triangle trigonometry may be used.

612 Chapter 11 Areas of Polygons and Circles

Check for Understanding

Concept Check

1. **Explain** how to derive the formula for the area of a regular polygon. See margin.
2. **OPEN ENDED** Describe a method for finding the base or height of a right triangle given one acute angle and the length of one side. See margin.

Guided Practice

GUIDED PRACTICE KEY	
Exercises	**Examples**
3–4	1
5–6	3
7	2

Find the area of each polygon. Round to the nearest tenth.

3. a regular hexagon with a perimeter of 42 yards $127.3\ \text{yd}^2$
4. a regular nonagon with a perimeter of 108 meters $890.2\ \text{m}^2$

Find the area of each shaded region. Assume that all polygons that appear to be regular are regular. Round to the nearest tenth.

5. 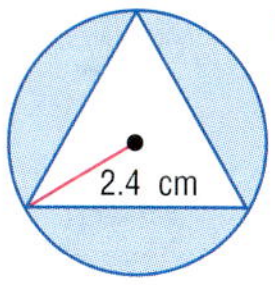

$10.6\ \text{cm}^2$

6. 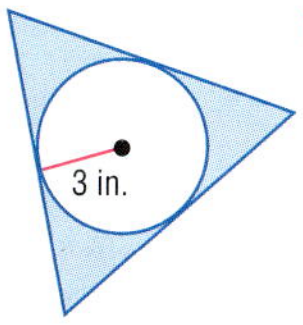

$18.5\ \text{in}^2$

Application

7. **UPHOLSTERY** Tyra wants to cover the cushions of her papasan chair with a different fabric. If there are seven circular cushions that are the same size with a diameter of 12 inches, around a center cushion with a diameter of 20 inches, find the area of fabric in square yards that she will need to cover both sides of the cushions. Allow an extra 3 inches of fabric around each cushion. about $3.6\ \text{yd}^2$

★ indicates increased difficulty

Practice and Apply

Homework Help

For Exercises	See Examples
8–13, 26, 27	1
14–23, 37–42	3
24, 25, 28–31	2

Extra Practice See page 777.

Find the area of each polygon. Round to the nearest tenth.

$2s^2(1+\sqrt{2})$ area of octagon

8. a regular octagon with a perimeter of 72 inches $391.1\ \text{in}^2$
9. a square with a perimeter of $84\sqrt{2}$ meters $882\ \text{m}^2$
10. a square with apothem length of 12 centimeters $576\ \text{cm}^2$
11. a regular hexagon with apothem length of 24 inches $1995.3\ \text{in}^2$
12. a regular triangle with side length of 15.5 inches $104.0\ \text{in}^2$
13. a regular octagon with side length of 10 kilometers $482.8\ \text{km}^2$

Find the area of each shaded region. Assume that all polygons that appear to be regular are regular. Round to the nearest tenth.

14. 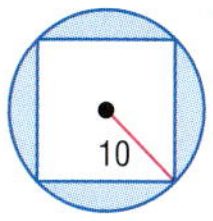

$114.2\ \text{units}^2$

15. 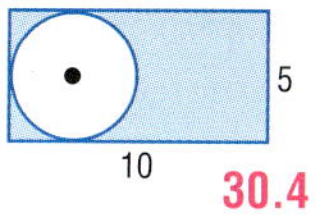

$30.4\ \text{units}^2$

16. 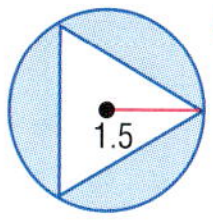

$4.1\ \text{units}^2$

17. 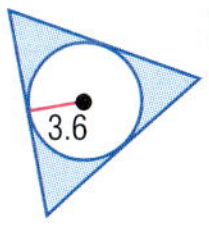

$26.6\ \text{units}^2$

18. 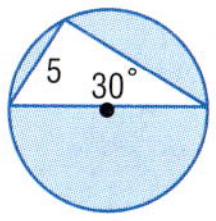

$56.9\ \text{units}^2$

19. 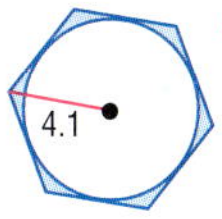

$4.1\ \text{units}^2$

20. 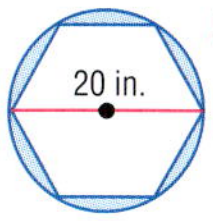

$54.4\ \text{in}^2$

★ 21. 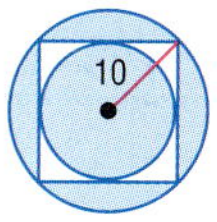

$271.2\ \text{units}^2$

★ 22. 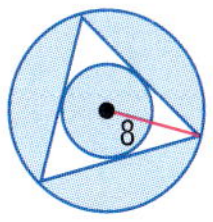

$168.2\ \text{units}^2$

Answers

1. Sample answer: Separate a hexagon inscribed in a circle into six congruent nonoverlapping isosceles triangles. The area of one triangle is one-half the product of one side of the hexagon and the apothem of the hexagon. The area of the hexagon is $6\left(\frac{1}{2}sa\right)$. The perimeter of the hexagon is $6s$, so the formula is $\frac{1}{2}Pa$.
2. Sample answer: Use the given angle measure, the given side length, and trigonometric ratios to find the missing lengths.

Concept Check

Ask students to describe how to find the area of an inscribed polygon. First find the apothem, or the distance from the center of the polygon perpendicular to the midpoint of the opposite side. Then find half of the product of the apothem and the perimeter.

3 Practice/Apply

Study Notebook

Have students—

- add the definitions/examples of the vocabulary terms to their Vocabulary Builder worksheets for Chapter 11.
- write the formula for the area of a regular polygon and the area of a circle. Ask them to include an example of each.
- include any other item(s) that they find helpful in mastering the skills in this lesson.

About the Exercises...

Organization by Objective

- **Areas of Regular Polygons:** 8–13, 26, 27
- **Areas of Circles:** 14–25, 28–31, 37–42

Odd/Even Assignments

Exercises 8–42 are structured so that students practice the same concepts whether they are assigned odd or even problems.

Assignment Guide

Basic: 9–19 odd, 23–37 odd, 38, 45–55 odd, 56–72

Average: 9–37 odd, 38, 39–55 odd, 56–72

Advanced: 8–54 even, 55–68 (optional: 69–72)

Study Guide and Intervention, p. 623 (shown) and p. 624

Areas of Regular Polygons In a regular polygon, the segment drawn from the center of the polygon perpendicular to the opposite side is called the **apothem**. In the figure at the right, $\overline{AP}$ is the apothem and $\overline{AR}$ is the radius of the circumscribed circle.

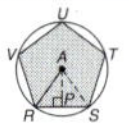

Area of a Regular Polygon	If a regular polygon has an area of A square units, a perimeter of P units, and an apothem of a units, then $A = \frac{1}{2}Pa$.

Example 1 **Verify the formula $A = \frac{1}{2}Pa$ for the regular pentagon above.**

For $\triangle RAS$, the area is $A = \frac{1}{2}bh = \frac{1}{2}(RS)(AP)$. So the area of the pentagon is $A = 5\left(\frac{1}{2}\right)(RS)(AP)$. Substituting P for $5RS$ and substituting a for AP, then $A = \frac{1}{2}Pa$.

Example 2 **Find the area of regular pentagon $RSTUV$ above if its perimeter is 60 centimeters.**

First find the apothem.

The measure of central angle RAS is $\frac{360}{5}$ or 72. Therefore $m\angle RAP = 36$. The perimeter is 60, so $RS = 12$ and $RP = 6$.

$$\tan \angle RAP = \frac{RP}{AP}$$
$$\tan 36° = \frac{6}{AP}$$
$$AP = \frac{6}{\tan 36°} \approx 8.26$$

So $A = \frac{1}{2}Pa = \frac{1}{2}60(8.26)$ or 247.7.

The area is about 248 square centimeters.

Exercises

Find the area of each regular polygon. Round to the nearest tenth.

1. 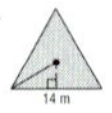

84.9 m^2

2. 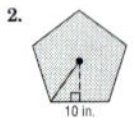

172.0 in^2

3. 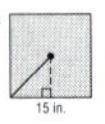

225 in^2

4. 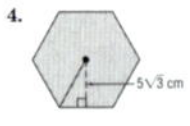

259.8 cm^2

5. 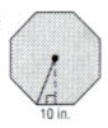

482.8 in^2

6. 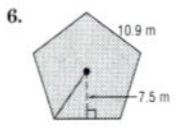

204.4 m^2

Skills Practice, p. 625 and Practice, p. 626 (shown)

Find the area of each regular polygon. Round to the nearest tenth.

1. a nonagon with a perimeter of 117 millimeters
1044.7 mm^2

2. an octagon with a perimeter of 96 yards
695.3 yd^2

Find the area of each circle. Round to the nearest tenth.

3. a circle with a diameter of 26 feet
530.9 ft^2

4. a circle with a circumference of 88 kilometers
616.2 km^2

Find the area of each shaded region. Assume that all polygons are regular. Round to the nearest tenth.

5. 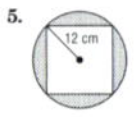

164.4 cm^2

6. 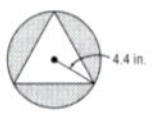

35.7 in^2

7. 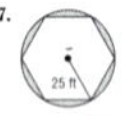

339.7 ft^2

8. 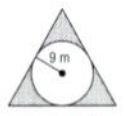

166.4 m^2

DISPLAYS **For Exercises 9 and 10, use the following information.**

A display case in a jewelry store has a base in the shape of a regular octagon. The length of each side of the base is 10 inches. The owners of the store plan to cover the base in black velvet.

9. Find the area of the base of the display case.
about 482.8 in^2

10. Find the number of square yards of fabric needed to cover the base.
about 0.37 yd^2

Reading to Learn Mathematics, p. 627 ELL

Pre-Activity **How can you find the area of a polygon?**

Read the introduction to Lesson 11-3 at the top of page 610 in your textbook.

How can you find the area of a regular hexagon without a new area formula? Sample answer: Divide the hexagon into six congruent equilateral triangles. Use properties of the 30°-60°-90° triangle to find the area of one of these triangles and multiply the result by 6.

Reading the Lesson

1. $ABCDEF$ and $RSTUV$ are regular polygons. Name each of the following in one of the figures.

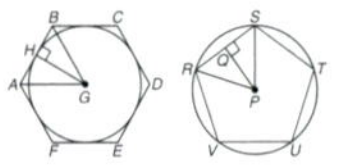

a. a circumscribed polygon hexagon $ABCDEF$
b. an inscribed polygon pentagon $RSTUV$
c. an apothem of a regular hexagon $\overline{GH}$
d. an isosceles triangle $\triangle PRS$ or $\triangle GAB$
e. a 30°-60°-90° triangle $\triangle GAH$ or $\triangle GBH$
f. a central angle with a measure of 72 $\angle RPS$

2. Refer to the figures in Exercise 1. Match each item in the first column with an expression in the second column.

a. perimeter of $ABCDEF$ v — i. $\pi(PS)^2$
b. circumference of circle G viii — ii. $2\pi(PR)$
c. perimeter of $RSTUV$ vii — iii. $\frac{5}{2}(RS)(PQ)$
d. area of circle G vi — iv. $3(AB)(HG)$
e. area of $RSTUV$ iii — v. $6(CD)$
f. area of $ABCDEF$ iv — vi. $\pi(GH)^2$
g. area of circle P i — vii. $5(UV)$
h. circumference of circle P ii — viii. $2\pi(GH)$

3. Explain in your own words how to find the area of a circle if you know the circumference. Sample answer: Divide the circumference by 2π to find the radius. Then square the radius and multiply by π to find the area.

Helping You Remember

4. A good way to remember something is to explain it to someone else. Suppose your classmate Joelle is having trouble remembering which formula is for circumference and which is for area. How can you help her? Sample answer: Circumference is measured in linear units, while area is measured in square units, so the formula containing r^2 must be the one for area.

23. **ALGEBRA** A circle is inscribed in a square, which is circumscribed by another circle. If the diagonal of the square is $2x$, find the ratio of the area of the large circle to the area of the small circle. **2 : 1**

24. **CAKE** A bakery sells single-layer mini-cakes that are 3 inches in diameter for $4 each. They also have a 9-inch cake for $15. If both cakes are the same thickness, which option gives you more cake for the money, nine mini-cakes or one 9-inch cake? Explain.

25. **PIZZA** A pizza shop sells 8-inch pizzas for $5 and 16-inch pizzas for $10. Which would give you more pizza, two 8-inch pizzas or one 16-inch pizza? Explain.

24. The total area is equal. Nine mini-cakes are the same size as one 9-inch cake, but nine mini-cakes cost 9 · $4 or $36 while the 9-inch cake is only $15.

25. One 16-inch pizza; the area of the 16-inch pizza is greater than the area of two 8-inch pizzas, so you get more pizza for the same price.

COORDINATE GEOMETRY **The coordinates of the vertices of a regular polygon are given. Find the area of each polygon to the nearest tenth.**

26. $T(0, 0), U(-7, -7), V(0, -14), W(7, -7)$ **98 units²**

27. $G(-12, 0), H(0, 4\sqrt{3}), J(0, -4\sqrt{3})$ **83.1 units²**

28. $J(5, 0), K\left(\frac{5\sqrt{2}}{2}, -\frac{5\sqrt{2}}{2}\right), L(0, -5), M\left(-\frac{5\sqrt{2}}{2}, -\frac{5\sqrt{2}}{2}\right), N(-5, 0), P\left(-\frac{5\sqrt{2}}{2}, \frac{5\sqrt{2}}{2}\right), Q(0, 5), R\left(\frac{5\sqrt{2}}{2}, \frac{5\sqrt{2}}{2}\right)$ **70.7 units²**

29. $A(-2\sqrt{2}, 2\sqrt{2}), B(0, 4), C(2\sqrt{2}, 2\sqrt{2}), D(4, 0), E(2\sqrt{2}, -2\sqrt{2}), F(0, -4), G(-2\sqrt{2}, -2\sqrt{2}), H(-4, 0)$ **45.3 units²**

Find the area of each circle. Round to the nearest tenth.

30. $C = 34\pi$ **907.9 units²**

31. $C = 17\pi$ **227.0 units²**

32. $C = 54.8$ **239.0 units²**

33. $C = 91.4$ **664.8 units²**

SWIMMING POOL **For Exercises 34 and 35, use the following information.**

The area of a circular pool is approximately 7850 square feet. The owner wants to replace the tiling at the edge of the pool.

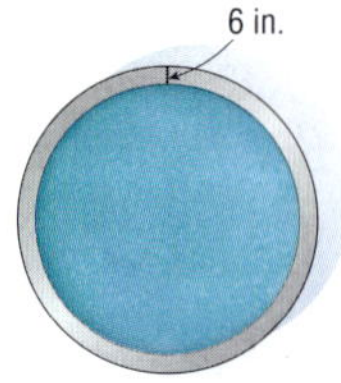

34. The edging is 6 inches wide, so she plans to use 6-inch square tiles to form a continuous inner edge. How many tiles will she need to purchase? **629 tiles**

35. Once the square tiles are in place around the pool, there will be extra space between the tiles. What shape of tile will best fill this space? How many tiles of this shape should she purchase? **triangles; 629 tiles**

AVIATION **For Exercises 36–38, refer to the circle graph.**

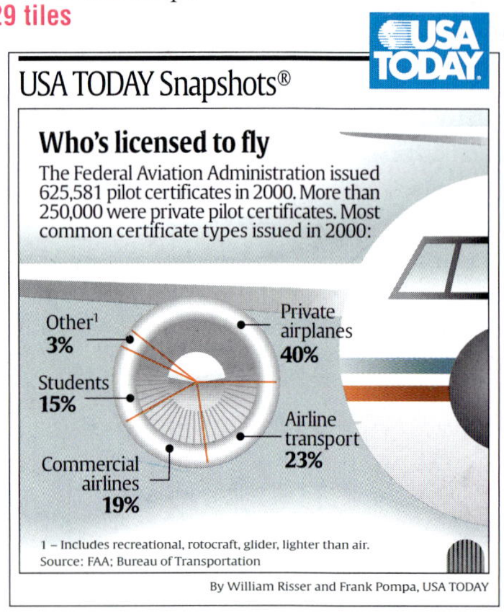

36. Suppose the radius of the circle on the graph is 1.3 centimeters. Find the area of the circle on the graph. **≈ 5.3 cm²**

37. Francesca wants to use this circle graph for a presentation. She wants the circle to use as much space on a 22″ by 28″ sheet of poster board as possible. Find the area of the circle. **≈ 380.1 in²**

38. **CRITICAL THINKING** Make a conjecture about how you could determine the area of the region representing the pilots who are certified to fly private airplanes.

38. Sample answer: Multiply the total area by 40%.

Study Tip

Look Back

To review **circle graphs**, see Lesson 10-2.

Enrichment, p. 628

Areas of Inscribed Polygons

A protractor can be used to inscribe a regular polygon in a circle. Follow the steps below to inscribe a regular nonagon in $\odot N$.

Step 1 Find the degree measure of each of the nine congruent arcs. 40

Step 2 Draw 9 radii to form 9 angles with the measure you found in Step 1. The radii will intersect the circle in 9 points.

Step 3 Connect the nine points to form the nonagon.

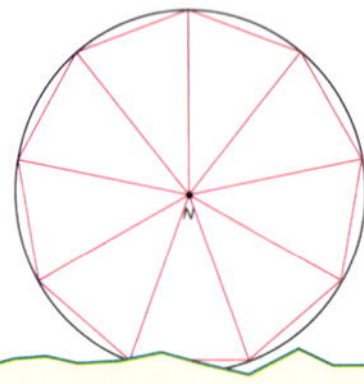

1. Find the length of one side of the nonagon to the nearest tenth of a centimeter. What is the perimeter of the nonagon? 2.5 cm, $P = 22.5$ cm

Find the area of each shaded region. Round to the nearest tenth.

★ 39.

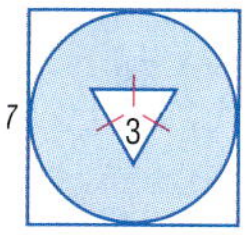

34.6 units^2

★ 40.

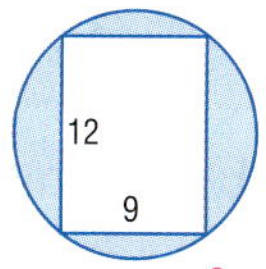

68.7 units^2

★ 41.

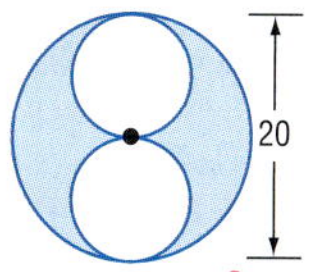

157.1 units^2

★ 42.

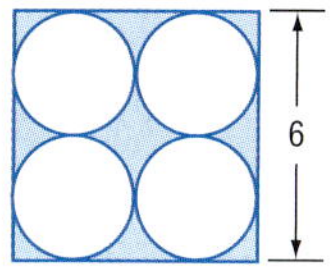

7.7 units^2

★ 43.

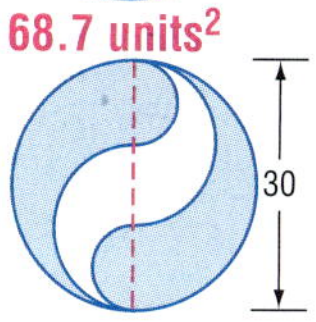

471.2 units^2

★ 44.

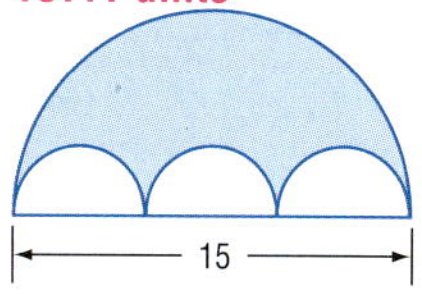

58.9 units^2

GARDENS For Exercises 45–47, use the following information.
The Elizabeth Park Rose Garden in Hartford, Connecticut, was designed with a gazebo surrounded by two concentric rose garden plots. Wide paths emanate from the center, dividing the garden into square and circular sections.

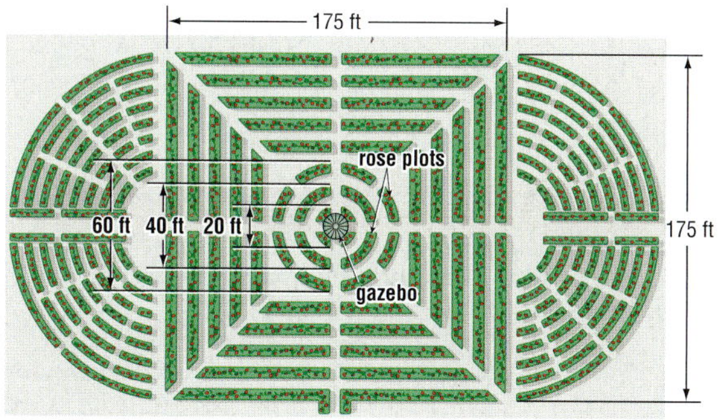

45. Find the area and perimeter of the entire Rose Garden. Round to the nearest tenth. 54,677.8 ft^2; 899.8 ft

46. What is the total of the circumferences of the three concentric circles formed by the gazebo and the two circular rose garden plots? (Ignore the width of the rose plots and the width of the paths.) $120\pi \approx 377.0$ ft

47. Each rose plot has a width of 5 feet. What is the area of the path between the outer two complete circles of rose garden plots? $225\pi \approx 706.9$ ft^2

More About...

Architecture

This structure is a *pagoda*. Pagodas are characterized by having several hexagonal or octagonal stories each topped with a curved roof. In this temple, the sizes of the floors are in the ratio 1 : 3 : 5 : 7.

Source: www.infoplease.com

48. **ARCHITECTURE** The Anraku-ji Temple in Japan is composed of four octagonal floors of different sizes that are separated by four octagonal roofs of different sizes. Refer to the information at the left. Determine whether the areas of each of the four floors are in the same ratio as their sizes. Explain. See margin.

SIMILAR FIGURES For Exercises 49–54, use the following information.
Polygons *FGHJK* and *VWXUZ* are similar regular pentagons.

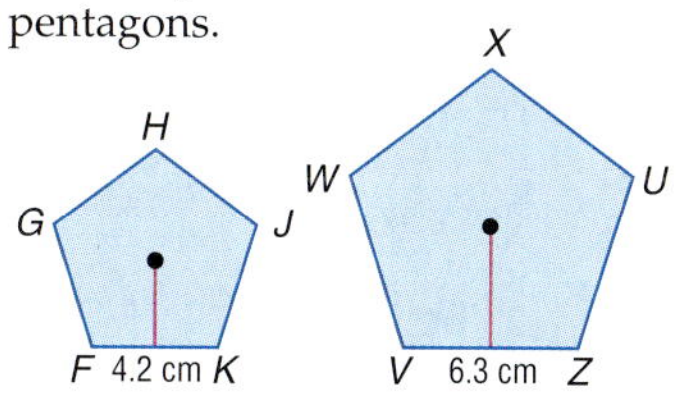

49. Find the scale factor. 2 : 3

50. Find the perimeter of each pentagon.

51. Compare the ratio of the perimeters of the pentagons to the scale factor.

52. Find the area of each pentagon.

53. Compare the ratio of the areas of the pentagons to the scale factor.

54. Compare the ratio of the areas of the pentagons to the ratio of the perimeters of the pentagons. The ratio of the areas is the square of the ratio of the perimeters.

50. 21 cm, 31.5 cm
51. The ratio is the same.
52. ≈ 30.35 cm^2, ≈ 68.29 cm^2
53. The ratio of the areas is the square of the scale factor.

www.geometryonline.com/self_check_quiz/fcat

Answer

48. No; the areas of the floors will increase by the squares of 1, 3, 5, and 7, or 1, 9, 25, and 49. The ratio of area is the square of the scale factor.

$\pi r^2 - 2\left[\frac{1}{2}\pi r^2 - \frac{1}{2}\pi r^2\right]$

Lesson 11-3 Areas of Regular Polygons and Circles 615

4 Assess

Open-Ended Assessment

Modeling Ask students to demonstrate how to construct a circle with radius 6 feet circumscribed about an equilateral triangle on the floor of the classroom. Ask them to use the tiles in the floor to demonstrate how to estimate the area of the region outside the triangle but inside the circle. Then use the area formula to calculate it exactly.

Getting Ready for Lesson 11-4

Prerequisite Skill Students will learn about areas of irregular figures in Lesson 11-4. They will use the properties of special right triangles to break the region into known shapes so that they can find the area. Use Exercises 67–70 to determine your students' familiarity with special right triangles.

Quiz (Lesson 11-3) is available on p. 655 of the *Chapter 11 Resource Masters*.

Mid-Chapter Test (Lessons 11-1 through 11-3) is available on p. 657 of the *Chapter 11 Resource Masters*.

Answers

56. Sample answer: You can find the areas of regular polygons by finding the product of the perimeter and the apothem and then multiplying by one half. Answers should include the following.
   - We need to know the length of each side and the length of the apothem.
   - One method is to divide the area of the floor by the area of each tile. Since the floor is hexagonal and not rectangular, tiles of different shapes will need to be ordered to cover the floor.

55. **CRITICAL THINKING** A circle inscribes one regular hexagon and circumscribes another. If the radius of the circle is 10 units long, find the ratio of the area of the smaller hexagon to the area of the larger hexagon. **3 to 4**

56. **WRITING IN MATH** Answer the question that was posed at the beginning of the lesson. **See margin.**

How can you find the area of a polygon?

Include the following in your answer:
- information needed about the gazebo floor to find the area, and
- how to find the number of tiles needed to cover the floor.

FCAT Practice Standardized Test Practice

57. A square is inscribed in a circle of area 18π square units. Find the length of a side of the square. **B**

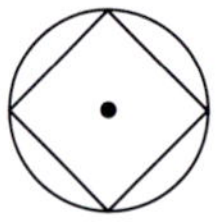

Ⓐ 3 units Ⓑ 6 units
Ⓒ $3\sqrt{2}$ units Ⓓ $6\sqrt{2}$ units

58. **ALGEBRA** The average of x numbers is 15. If the sum of the x numbers is 90, what is the value of x? **B**

Ⓐ 5 Ⓑ 6 Ⓒ 8 Ⓓ 15

Maintain Your Skills

Mixed Review Find the area of each quadrilateral. *(Lesson 11-2)*

59. 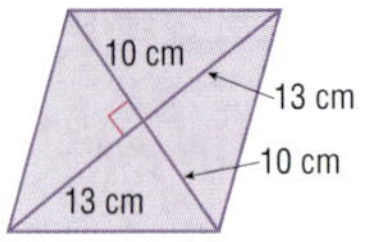

260 cm^2

60. 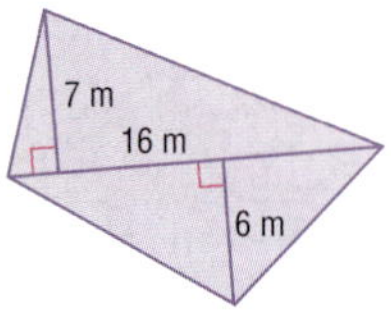

104 m^2

61. 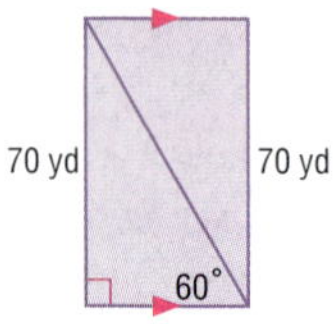

≈ 2829.0 yd^2

COORDINATE GEOMETRY Given the coordinates of the vertices of a quadrilateral, determine whether it is a *square*, a *rectangle*, or a *parallelogram*. Then find the area of the quadrilateral. *(Lesson 11-1)*

62. $A(-3, 2)$, $B(4, 2)$, $C(2, -1)$, $D(-5, -1)$ **parallelogram; 21 units^2**
63. $F(4, 1)$, $G(4, -5)$, $H(-2, -5)$, $J(-2, 1)$ **square; 36 units^2**
64. $K(-1, -3)$, $L(-2, 5)$, $M(1, 5)$, $N(2, -3)$ **parallelogram; 24 units^2**
65. $P(5, -7)$, $Q(-1, -7)$, $R(-1, -2)$, $S(5, -2)$ **rectangle; 30 units^2**

Refer to trapezoid *CDFG* with median $\overline{HE}$. *(Lesson 8-6)*

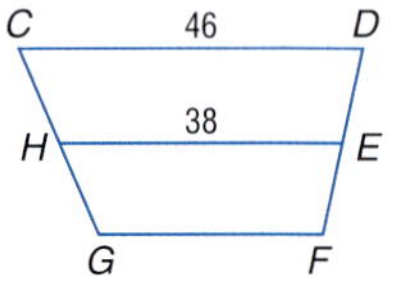

66. Find *GF*. **30**
67. Let $\overline{WX}$ be the median of *CDEH*. Find *WX*. **42**
68. Let $\overline{YZ}$ be the median of *HEFG*. Find *YZ*. **34**

Getting Ready for the Next Lesson

PREREQUISITE SKILL Find *h*. *(To review **special right triangles**, see Lesson 7-3.)*

69. **6**

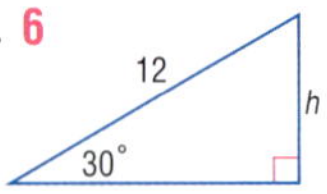

70. **$15\sqrt{3}$**

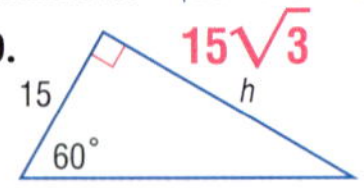

71. **$4\sqrt{2}$**

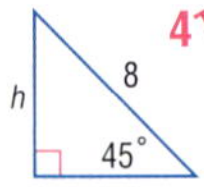

72. **$\frac{21\sqrt{2}}{2}$**

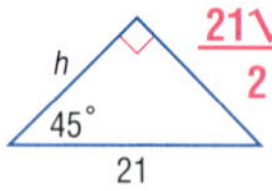

11-4 Areas of Irregular Figures

What You'll Learn

- Find areas of irregular figures.
- Find areas of irregular figures on the coordinate plane.

How do windsurfers use area?

The sail for a windsurf board cannot be classified as a triangle or a parallelogram. However, it can be separated into figures that can be identified, such as trapezoids and a triangle.

Vocabulary

- irregular figure
- irregular polygon

Study Tip

Reading Math
Irregular figures are also called *composite figures* because the regions can be separated into smaller regions.

IRREGULAR FIGURES An **irregular figure** is a figure that cannot be classified into the specific shapes that we have studied. To find areas of irregular figures, separate the figure into shapes of which we can find the area. Draw auxiliary lines as necessary. The sum of the areas of each is the area of the figure.

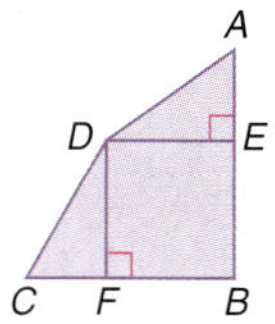

Auxiliary lines are drawn in quadrilateral $ABCD$. $\overline{DE}$ and $\overline{DF}$ separate the figure into $\triangle ADE$, $\triangle CDF$, and rectangle $BEDF$.

Postulate 11.2

The area of a region is the sum of all of its nonoverlapping parts.

Example 1 Area of an Irregular Figure

Find the area of the figure.

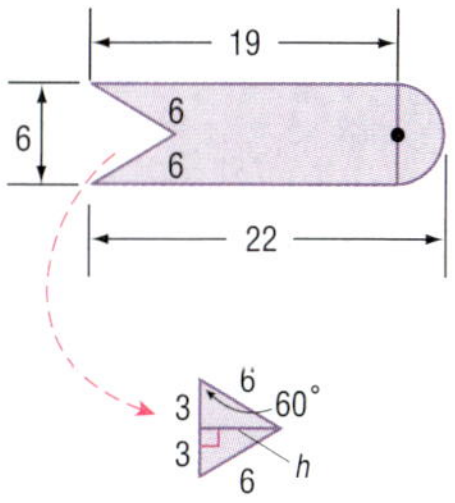

The figure can be separated into a rectangle with dimensions 6 units by 19 units, an equilateral triangle with sides each measuring 6 units, and a semicircle with a radius of 3 units.

Use 30°-60°-90° relationships to find that the height of the triangle is $3\sqrt{3}$.

area of irregular figure = area of rectangle − area of triangle + area of semicircle

$= \ell w - \frac{1}{2}bh + \frac{1}{2}\pi r^2$	Area formulas
$= 19 \cdot 6 - \frac{1}{2}(6)(3\sqrt{3}) + \frac{1}{2}\pi(3^2)$	Substitution
$= 114 - 9\sqrt{3} + \frac{9}{2}\pi$	Simplify.
≈ 112.5	Use a calculator.

The area of the irregular figure is 112.5 square units to the nearest tenth.

www.geometryonline.com/extra_examples/fcat

11-4 Lesson Notes

1 Focus

5-Minute Check Transparency 11-4 Use as a quiz or review of Lesson 11-3.

Mathematical Background notes are available for this lesson on p. 592D.

How do windsurfers use area?

Ask students:

- What kind of shapes make up the shape of a windsurf sail? **triangles and trapezoids**
- Where are these same shapes also used? **Sample answer: building design**

Resource Manager

Workbook and Reproducible Masters

Chapter 11 Resource Masters

- Study Guide and Intervention, pp. 629–630
- Skills Practice, p. 631
- Practice, p. 632
- Reading to Learn Mathematics, p. 633
- Enrichment, p. 634
- Assessment, p. 656

School-to-Career Masters, p. 22

Teaching Geometry With Manipulatives Masters, p. 1

Transparencies

5-Minute Check Transparency 11-4
Answer Key Transparencies

Technology

GeomPASS: Tutorial Plus, Lesson 20
Interactive Chalkboard
Multimedia Applications: Virtual Activities

2 Teach

IRREGULAR FIGURES

In-Class Examples

Power Point®

Teaching Tip Remind students that they may break up an irregular figure in different ways if they want to find the area. Suggest that they plan ahead to find the combination of shapes that are easiest to work with.

1 Find the area of the figure in square feet. Round to the nearest tenth if necessary. **953.1 ft²**

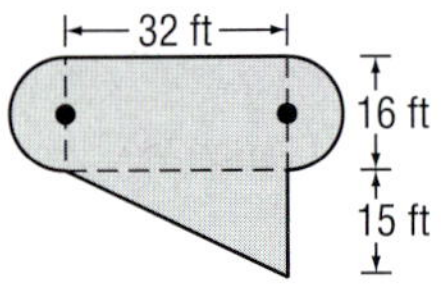

2 A rectangular rose garden is centered in a border of lawn. Find the area of the lawn around the garden in square feet. **8500 ft²**

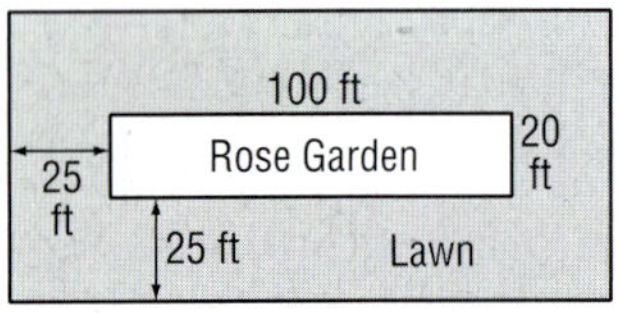

IRREGULAR FIGURES ON THE COORDINATE PLANE

In-Class Example

Power Point®

Teaching Tip Watch for students who think that the area formula for a regular polygon applies to any polygon. Stress that to find the area of a polygon that is not regular, you may need to divide the polygon into shapes with known formulas. Show students both types of polygons.

3 Find the area of polygon *MNPQR*. **44.5 units²**

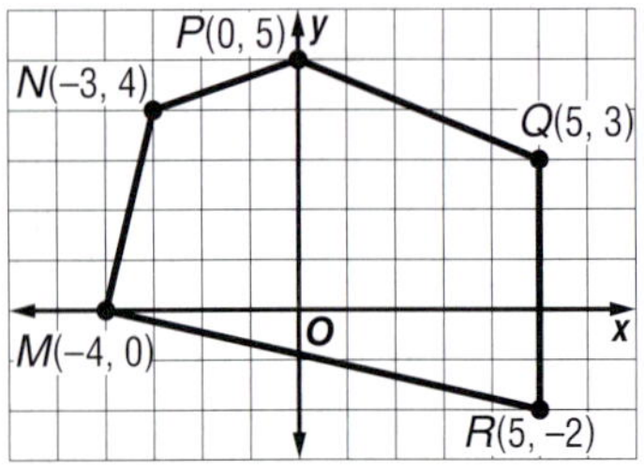

WebQuest

Identifying the polygons forming a region such as a tessellation will help you determine the type of tessellation. Visit www.geometryonline.com/webquest to continue work on your WebQuest project.

Example 2 Find the Area of an Irregular Figure to Solve a Problem

FURNITURE **Melissa's dining room table has hardwood around the outside. Find the area of wood around the edge of the table.**

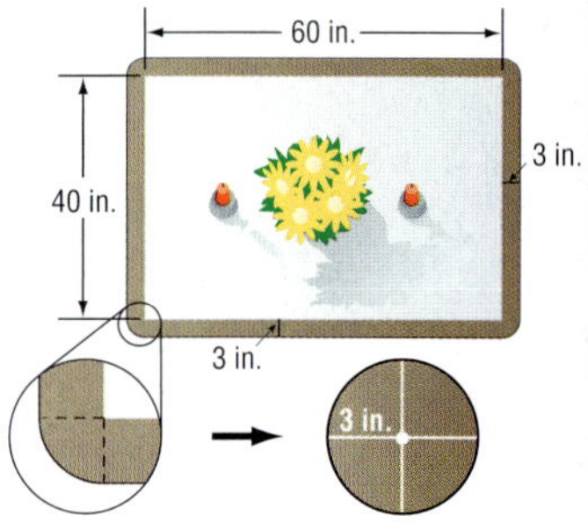

First, draw auxiliary lines to separate the figure into regions. The table can be separated into four rectangles and four corners.

The four corners of the table form a circle with radius 3 inches.

area of wood edge = area of rectangles + area of circle

$= 2\ell w + 2\ell w + \pi r^2$	Area formulas
$= 2(3)(60) + 2(3)(40) + \pi(3^2)$	Substitution
$= 360 + 240 + 9\pi$	Simplify.
≈ 628.3	Use a calculator.

The area of the wood edge of the table is 628.3 square inches to the nearest tenth.

IRREGULAR FIGURES ON THE COORDINATE PLANE The formula for the area of a regular polygon does not apply to an **irregular polygon**, a polygon that is not regular. To find the area of an irregular polygon on the coordinate plane, separate the polygon into known figures.

Study Tip

Estimation
Estimate the area of the simple closed curves by counting the unit squares. Use the estimate to determine if your answer is reasonable.

Example 3 Coordinate Plane

COORDINATE GEOMETRY **Find the area of polygon *RSTUV*.**

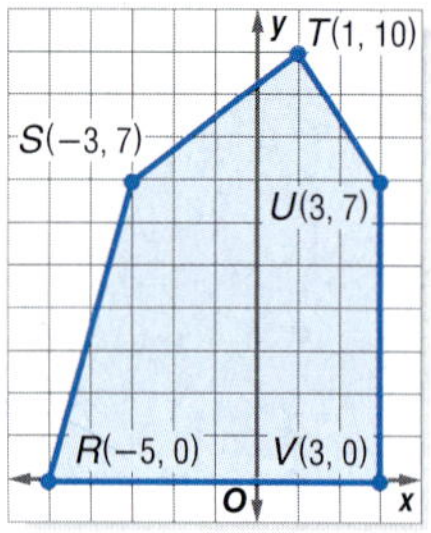

First, separate the figure into regions. Draw an auxiliary line from *S* to *U*. This divides the figure into triangle *STU* and trapezoid *RSUV*.

Find the difference between x-coordinates to find the length of the base of the triangle and the lengths of the bases of the trapezoid. Find the difference between y-coordinates to find the heights of the triangle and trapezoid.

area of *RSTUV* = area of $\triangle STU$ + area of trapezoid *RSUV*

$= \frac{1}{2}bh + \frac{1}{2}h(b_1 + b_2)$	Area formulas
$= \frac{1}{2}(6)(3) + \frac{1}{2}(7)(8 + 6)$	Substitution
$= 58$	Simplify.

The area of *RSTUV* is 58 square units.

618 Chapter 11 Areas of Polygons and Circles

DAILY INTERVENTION

Differentiated Instruction

Kinesthetic Have your students use string, masking tape, and a tiled floor to mark off irregular shapes on the floor. Ask them to estimate the area by counting squares and then verify the estimate by calculating the sum of the individual parts.

Check for Understanding

Concept Check

1. **OPEN ENDED** Sketch an irregular figure on a coordinate plane and find its area.
2. **Describe** the difference between an irregular figure and an irregular polygon. **1–2. See margin.**

Guided Practice

GUIDED PRACTICE KEY	
Exercises	Examples
3–4	1
5–6	3
7	2

Find the area of each figure. Round to the nearest tenth if necessary.

3. 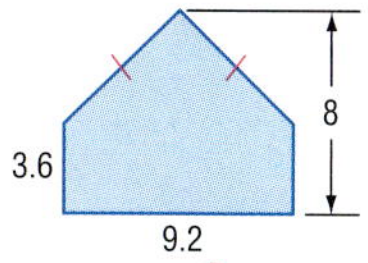

53.4 units^2

4. 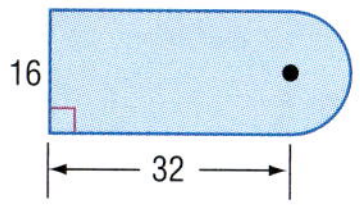

612.5 units^2

COORDINATE GEOMETRY **Find the area of each figure.**

5. 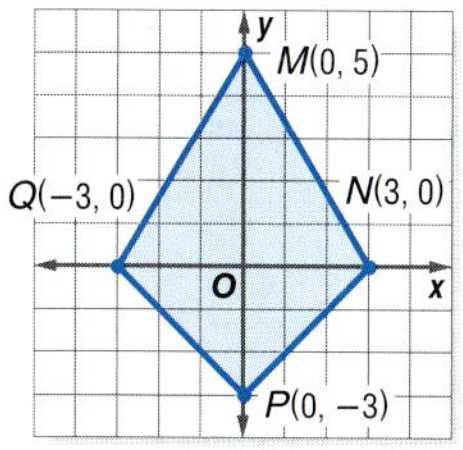

24 units^2

6. 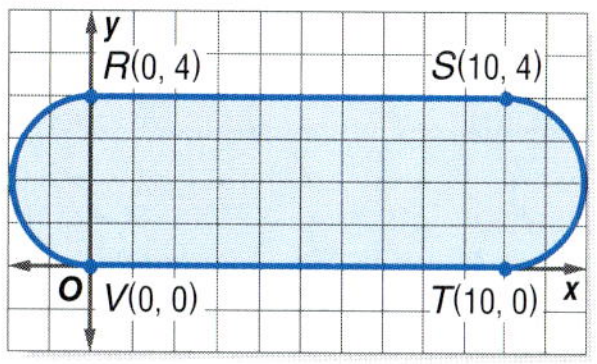

52.6 units^2

Application

7. **GATES** The Roths have a series of interlocking gates to form a play area for their baby. Find the area enclosed by the wall and gates. **1247.4 in^2**

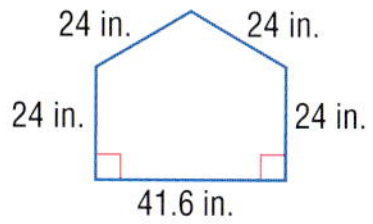

Practice and Apply

Homework Help	
For Exercises	See Examples
8–13	1
14, 15, 23–27	2
16–22	3

Extra Practice
See page 777.

Find the area of each figure. Round to the nearest tenth if necessary.

8. 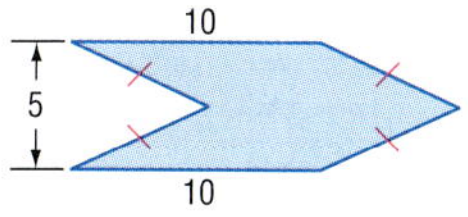

50 units^2

9. 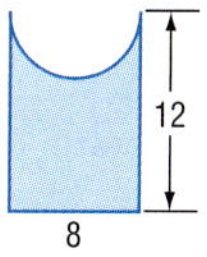

70.9 units^2

10. 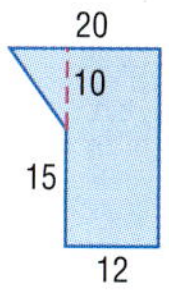

340 units^2

11. 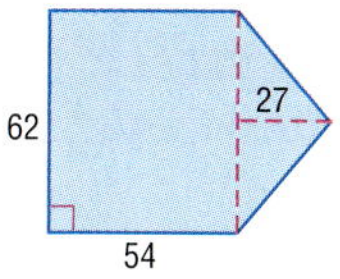

4185 units^2

12. 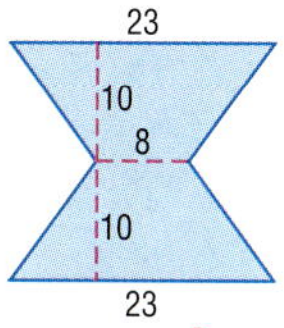

310 units^2

13. 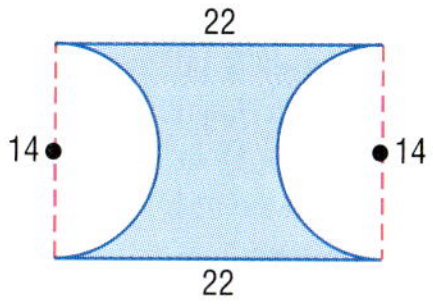

154.1 units^2

WINDOWS **For Exercises 14 and 15, use the following information.**

Mr. Cortez needs to replace this window in his house. The window panes are rectangles and sectors.

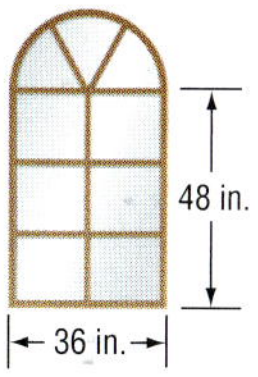

14. Find the perimeter of the window. **188.5 in.**
15. Find the area of the window. **2236.9 in^2**

3 Practice/Apply

Study Notebook

Have students—
- add the definitions/examples of the vocabulary terms to their Vocabulary Builder worksheets for Chapter 11.
- include an example of finding the area of an irregular shape that is divided into known figures.
- include any other item(s) that they find helpful in mastering the skills in this lesson.

About the Exercises...

Organization by Objective
- **Irregular Figures:** 8–15, 23–27
- **Irregular Figures on the Coordinate Plane:** 16–22

Odd/Even Assignments
Exercises 8–22 are structured so that students practice the same concepts whether they are assigned odd or even problems.

Alert! Exercise 24 requires the Internet or other research materials.

Assignment Guide

Basic: 9–27 odd, 28–42
Average: 9–27 odd, 28–42
Advanced: 8–28 even, 29–38 (optional: 39–42)
All: Quiz 2 (1–5)

Answers

1. Sample answer: ≈18.3 units^2

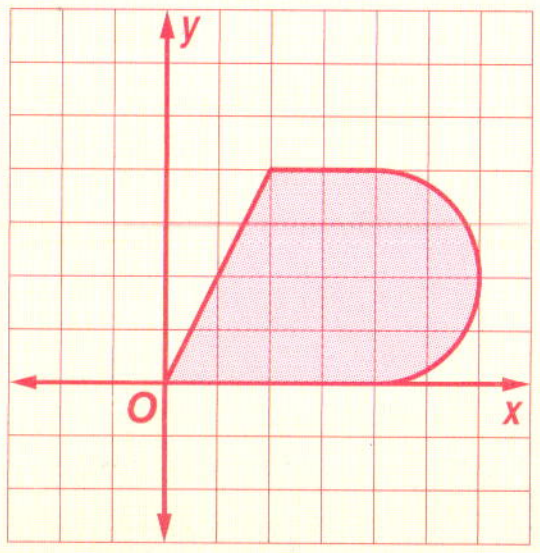

2. An irregular polygon is a polygon in which all sides are not congruent. If a shape can be separated into semicircles or smaller circular regions, it is an irregular figure.

Study Guide and Intervention, p. 629 (shown) and p. 630

Irregular Figures An **irregular** figure is one that cannot be classified as one of the previously-studied shapes. To find the area of an irregular figure, break it into familiar shapes. Find the area of each shape and add the areas.

Example 1 **Find the area of the irregular figure.**

50 ft, 30 ft

The figure is a rectangle minus one half of a circle. The radius of the circle is one half of 30 or 15.

$A = lw - \frac{1}{2}\pi r^2$
$\approx 50(30) - 0.5(3.14)(15)^2$
$= 1146.6$ or about 1147 ft²

Example 2 **Find the area of the shaded region.**

5 cm

The dimensions of the rectangle are 10 centimeters and 30 centimeters. The area of the shaded region is

$(10)(30) - 3\pi(5^2) = 300 - 75\pi$
≈ 64.4 cm²

Exercises

Find the area of each figure. Round to the nearest tenth if necessary.

1. 34 ft, 15 ft — 598.4 ft²
2. 40 in., 24 in., 24 in. — 960 in²
3. 14 cm, 10 cm, 38 cm, 40 cm — 466 cm²
4. 22 cm, 42 cm — 704 cm²
5. 64 m, 20 m, 40 m, 20 m — 1920 m²
6. 35 yd, 15 yd — 262.5 yd²
7. Refer to Example 2 above. Draw the largest possible square inside each of the three circles. What is the total area of the three squares? 150 cm²

Skills Practice, p. 631 and Practice, p. 632 (shown)

Find the area of each figure. Round to the nearest tenth if necessary.

1. 20, 20 — 400 units²
2. 38, 22, 22 — 869.6 units²
3. 9, 7, 23 — 143.8 units²
4. 20, 13, 30, 13 — 952.4 units²
5. D(2, 5), E(4, 4), B(−1, 3), C(1, 4), A(−1, 0), F(4, 0) — 20.5 units²
6. Q(2, 5), P(−2, 3), R(4, 3), T(−2, 1), S(2, 0) — 22 units²

LANDSCAPING For Exercises 7 and 8, use the following information.
One of the displays at a botanical garden is a koi pond with a walkway around it. The figure shows the dimensions of the pond and the walkway.

7 ft, 15 ft, 13 ft, 35 ft

7. Find the area of the pond to the nearest tenth. 129.5 ft²

8. Find the area of the walkway to the nearest tenth. 572.2 ft²

Reading to Learn Mathematics, p. 633 ELL

Pre-Activity **How do windsurfers use area?**

Read the introduction to Lesson 11-4 at the top of page 617 in your textbook.

How do you think the areas of the figures outlined in the picture of the sail are related? Sample answer: The areas get smaller as you move further up the sail. The area of the triangle is smaller than the area of any of the trapezoids.

Reading the Lesson

1. Use dashed segments to show how each figure can be subdivided into figures for which you have learned area formulas. Name the smaller figures that you have formed as specifically as possible and indicate whether any of them are congruent to each other. Sample answers are given.
   a. rectangle and isosceles triangle
   b. square and two congruent isosceles triangles
   c. rectangle and two congruent semicircles

2. In the figure, B is the midpoint of $\widehat{ABC}$. Complete the following steps to derive a formula for the area of the shaded region in terms of the radius r of the circle.

The area of circle P is πr^2.

$m\angle ABC =$ 90 because Sample answer: It is an inscribed angle that intercepts a semicircle.

$m\widehat{AB} = m\widehat{BC}$ because Sample answer: B is the midpoint of $\widehat{ABC}$ (definition of midpoint).

$\overline{AB} \cong \overline{BC}$ because Sample answer: If two minor arcs of a circle are congruent, their corresponding chords are congruent.

Therefore, $\triangle ABC$ is a(n) isosceles right or 45°-45°-90° triangle.

$AC = 2r$, so $AB = \frac{2r}{\sqrt{2}}$ or $r\sqrt{2}$ and $BC = \frac{2r}{\sqrt{2}}$ or $r\sqrt{2}$.

The area of $\triangle ABC$ is $\frac{1}{2} \cdot r\sqrt{2} \cdot r\sqrt{2} = r^2$.

Therefore, the area of the shaded region is given by

$A = \frac{1}{2}\pi r^2 - r^2 = \left(\frac{\pi}{2} - 1\right)r^2$.

Helping You Remember

3. Rolando is having trouble remembering when to subtract an area when finding the area of an irregular figure. How can you help him remember? Sample answer: Subtract when there is an indentation, or a hole in the figure.

COORDINATE GEOMETRY Find the area of each figure. Round to the nearest tenth if necessary.

16.
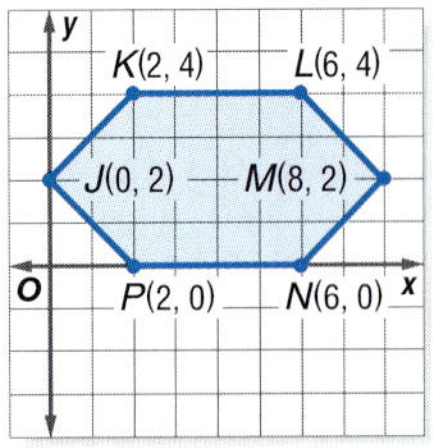

24 units²

17.
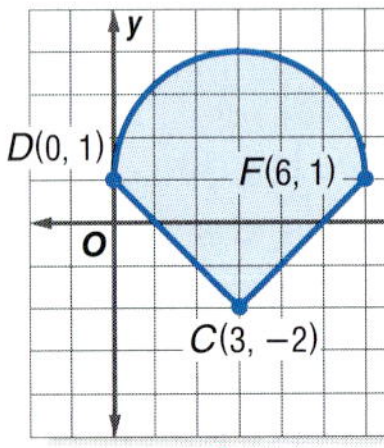

23.1 units²

18.
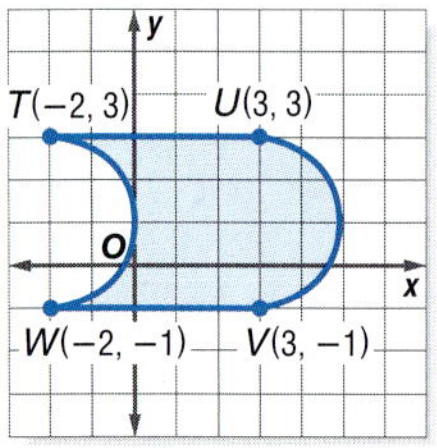

20 units²

COORDINATE GEOMETRY The vertices of an irregular figure are given. Find the area of each figure.

19. $M(-4, 0)$, $N(0, 3)$, $P(5, 3)$, $Q(5, 0)$ 21 units²
20. $T(-4, -2)$, $U(-2, 2)$, $V(3, 4)$, $W(3, -2)$ 29 units²
21. $G(-3, -1)$, $H(-3, 1)$, $I(2, 4)$, $J(5, -1)$, $K(1, -3)$ 33 units²
22. $P(-8, 7)$, $Q(3, 7)$, $R(3, -2)$, $S(-1, 3)$, $T(-11, 1)$ 67 units²

23. **GEOGRAPHY** Estimate the area of the state of Alabama. Each square on the grid represents 2500 square miles. Sample answer: 57,500 mi²

24. **RESEARCH** Find a map of your state or a state of your choice. Estimate the area. Then use the Internet or other source to check the accuracy of your estimate. See students' work.

More About. . .

Calculus

Twenty percent of students in Japan take a calculus class in high school. In Korea and Switzerland, nearly all high school students take a course in calculus. In the U.S., only 1% of students take a calculus course in high school.

Source: NCTM

CALCULUS **For Exercises 25–27, use the following information.** 26. See margin.

The irregular region under the curve has been separated into rectangles of equal width.

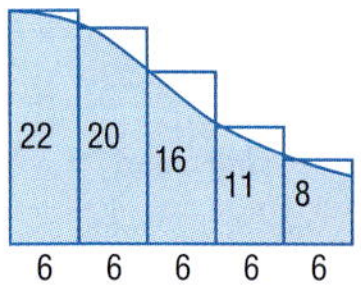

25. Use the rectangles to approximate the area of the region. 462
26. Analyze the estimate. Do you think the actual area is larger or smaller than your estimate? Explain.
27. How could the irregular region be separated to give an estimate of the area that is more accurate? Sample answer: Reduce the width of each rectangle.

28. **CRITICAL THINKING** Find the ratio of the area of $\triangle ABC$ to the area of square $BCDE$. $\frac{\sqrt{3}}{4} : 1$

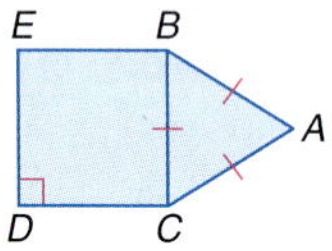

29. **WRITING IN MATH** Answer the question that was posed at the beginning of the lesson. See margin.

How do windsurfers use area?

Include the following in your answer:
- describe how to find the area of the sail, and
- another example of an irregular figure.

Enrichment, p. 634

Aerial Surveyors and Area

Many land regions have irregular shapes. Aerial surveyors often use coordinates when finding areas of such regions. The coordinate method described in the steps below can be used to find the area of *any* polygonal region. Study how this method is used to find the area of the region at the right.

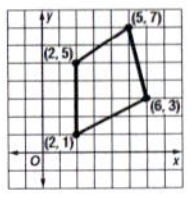

Step 1 List the ordered pairs for the vertices in counter-clockwise order, repeating the first ordered pair at the bottom of the list.

Step 2 Find D, the sum of the downward diagonal products (from left to right).
$D = (5 \cdot 5) + (2 \cdot 1) + (2 \cdot 3) + (6 \cdot 7)$
$= 25 + 2 + 6 + 42$ or 75

(5, 7)
(2, 5)
(2, 1)
(6, 3)

Step 3 Find U, the sum of the upward diagonal products (from left to right).
U … 7) + (2 · 5) + (6 · 1) + (5 · 3) …

Answer

26. The actual area of the irregular region should be smaller than the estimate. The rectangles drawn are larger than the region.

FCAT Practice

Standardized Test Practice

30. In the figure consisting of squares A, B, and C, $JK = 2KL$ and $KL = 2LM$. If the perimeter of the figure is 66 units, what is the area? **B**

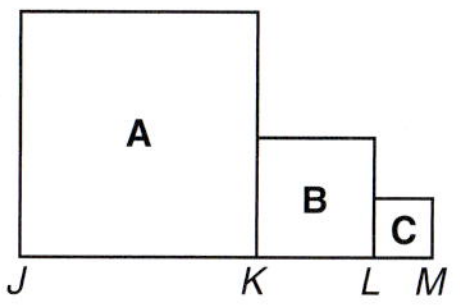

Ⓐ 117 units^2 Ⓑ 189 units^2

Ⓒ 224 units^2 Ⓓ 258 units^2

31. **ALGEBRA** For all integers n, $\boxed{n} = n^2$ if n is odd and $\boxed{n} = \sqrt{n}$ if n is even. What is the value of $\boxed{16} + \boxed{9}$? **C**

Ⓐ 7 Ⓑ 25 Ⓒ 85 Ⓓ 97

Maintain Your Skills

Mixed Review

Find the area of each shaded region. Assume that all polygons are regular unless otherwise stated. Round to the nearest tenth. *(Lesson 11-3)*

32. 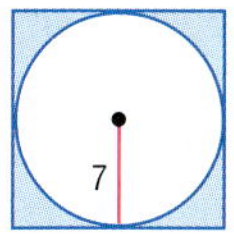

42.1 units^2

33. 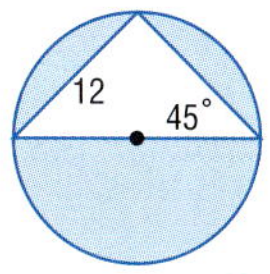

154.2 units^2

34. 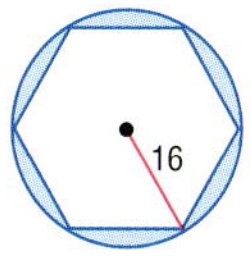

139.1 units^2

Find the area of each figure. Round to the nearest tenth if necessary. *(Lesson 11-2)*

35. equilateral triangle with perimeter of 57 feet **156.3 ft^2**

36. rhombus with a perimeter of 40 yards and a diagonal of 12 yards **96 yd^2**

37. isosceles trapezoid with a perimeter of 90 meters if the longer base is 5 meters less than twice as long as the other base, and each leg is 3 meters less than the shorter base **≈ 384.0**

38. **COORDINATE GEOMETRY** The point (6, 0) is rotated 45° clockwise about the origin. Find the exact coordinates of its image. *(Lesson 9-3)* $(3\sqrt{2}, -3\sqrt{2})$

Getting Ready for the Next Lesson

BASIC SKILL **Express each fraction as a decimal to the nearest hundredth.**

39. $\frac{5}{8}$ **0.63** 40. $\frac{13}{16}$ **0.81** 41. $\frac{9}{47}$ **0.19** 42. $\frac{10}{21}$ **0.48**

Practice Quiz 2

Lessons 11-3 and 11-4

Find the area of each polygon. Round to the nearest tenth. *(Lesson 11-3)*

1. regular hexagon with apothem length of 14 millimeters **679.0 mm^2**
2. regular octagon with a perimeter of 72 inches **391.1 in^2**

Find the area of each shaded region. Assume that all polygons are regular. Round to the nearest tenth. *(Lesson 11-3)*

3. 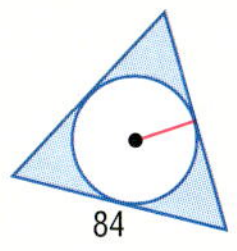

1208.1 units^2

4. 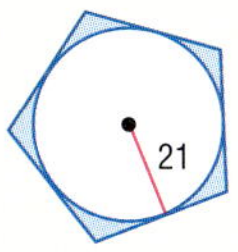

216.6 units^2

5. **COORDINATE GEOMETRY** Find the area of $CDGHJ$ with vertices $C(-3, -2)$, $D(1, 3)$, $G(5, 5)$, $H(8, 3)$, and $J(5, -2)$. *(Lesson 11-4)* **44.5 units^2**

www.geometryonline.com/self_check_quiz/fcat

4 Assess

Open-Ended Assessment

Writing Have students explain how to find the area of an irregular figure.

Getting Ready for Lesson 11-5

Prerequisite Skill Students will learn about geometric probability in Lesson 11-5. They will write fractions as decimals and percents to express probabilities involving geometric figures. Use Exercises 39–42 to determine your students' familiarity with writing fractions as decimals and percents.

Assessment Options

Practice Quiz 2 The quiz provides students with a brief review of the concepts and skills in Lessons 11-3 and 11-4. Lesson numbers are given to the right of the exercises or instruction lines so students can review concepts not yet mastered.

Quiz (Lesson 11-4) is available on p. 656 of the *Chapter 11 Resource Masters.*

Answer

29. Sample answer: Windsurfers use the area of the sail to catch the wind and stay afloat on the water. Answers should include the following.
   - To find the area of the sail, separate it into shapes. Then find the area of each shape. The sum of areas is the area of the sail.
   - Sample answer: Surfboards and sailboards are also irregular figures.

11-5 Lesson Notes

1 Focus

5-Minute Check Transparency 11-5 Use as a quiz or review of Lesson 11-4.

Mathematical Background notes are available for this lesson on p. 592D.

How can geometric probability help you win a game of darts?

Ask students:

- What shapes are used in a dart board? **circles and trapezoids**
- How can you get the most points playing darts? **Throw the darts into the smaller circular regions.**
- How are concentric circles used in the real-world? **Sample answer: a roof landing pad for a helicopter**

11-5 Geometric Probability

Sunshine State Standards
MA.A.3.4.3, MA.A.4.4.1, MA.B.3.4.1, MA.C.3.4.1

What You'll Learn

- Solve problems involving geometric probability.
- Solve problems involving sectors and segments of circles.

How can geometric probability help you win a game of darts?

To win at darts, you have to throw a dart at either the center or the part of the dartboard that earns the most points. In games, probability can sometimes be used to determine chances of winning. Probability that involves a geometric measure such as length or area is called **geometric probability**.

Vocabulary

- geometric probability
- sector
- segment

Study Tip

Look Back
To review **probability with line segments**, see page 20.

GEOMETRIC PROBABILITY In Chapter 1, you learned that the probability that a point lies on a part of a segment can be found by comparing the length of the part to the length of the whole segment. Similarly, you can find the probability that a point lies in a part of a two-dimensional figure by comparing the area of the part to the area of the whole figure.

Key Concept — *Probability and Area*

If a point in region A is chosen at random, then the probability $P(B)$ that the point is in region B, which is in the interior of region A, is

$$P(B) = \frac{\text{area of region } B}{\text{area of region } A}.$$

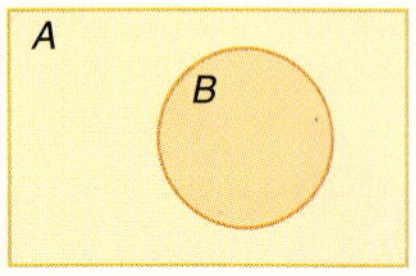

When determining geometric probability with targets, we assume

- that the object lands within the target area, and
- it is equally likely that the object will land anywhere in the region.

FCAT Practice
Standardized Test Practice
A B C D

The Princeton Review
Test-Taking Tip
Analyze the Question
If allowed, circle or underline key words as you read the test question.

Example 1 Probability with Area

Grid In Test Item

A square game board has black and white stripes of equal width as shown. What is the chance that a dart thrown at the board will land on a white stripe?

Read the Test Item

You want to find the probability of landing on a white stripe, not a black stripe.

Resource Manager

Workbook and Reproducible Masters

Chapter 11 Resource Masters
- Study Guide and Intervention, pp. 635–636
- Skills Practice, p. 637
- Practice, p. 638
- Reading to Learn Mathematics, p. 639
- Enrichment, p. 640
- Assessment, p. 656

Prerequisite Skills Workbook, pp. 27–28, 107–108

Teaching Geometry With Manipulatives Masters, pp. 1, 18

Transparencies

5-Minute Check Transparency 11-5
Real-World Transparency 11
Answer Key Transparencies

Technology

Interactive Chalkboard

Solve the Test Item

We need to divide the area of the white stripes by the total area of the game board. Extend the sides of each stripe. This separates the square into 36 small unit squares.

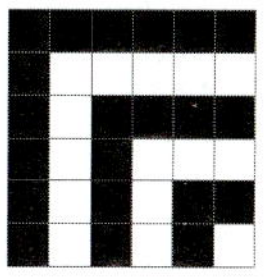

The white stripes have an area of 15 square units. The total area is 36 square units.

The probability of tossing a chip onto the white stripes is $\frac{15}{36}$ or $\frac{5}{12}$.

Fill in the Grid

Write $\frac{5}{12}$ as 5/12 in the top row of the grid-in.

Then shade in the appropriate bubble under each entry.

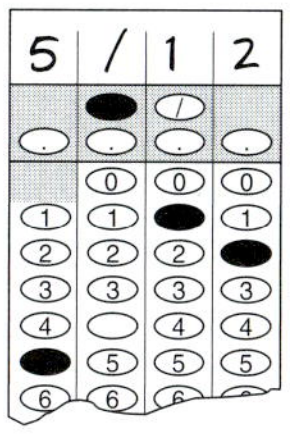

SECTORS AND SEGMENTS OF CIRCLES

Sometimes you need to know the area of a sector of a circle in order to find a geometric probability. A **sector** of a circle is a region of a circle bounded by a central angle and its intercepted arc.

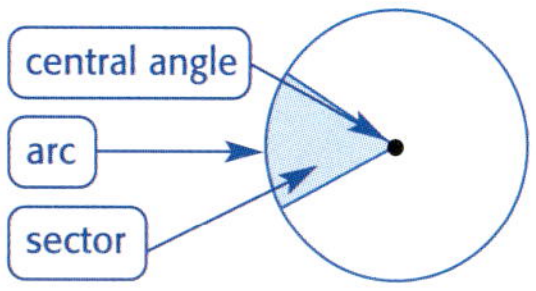

Key Concept — *Area of a Sector*

If a sector of a circle has an area of A square units, a central angle measuring $N°$, and a radius of r units, then $A = \frac{N}{360}\pi r^2$.

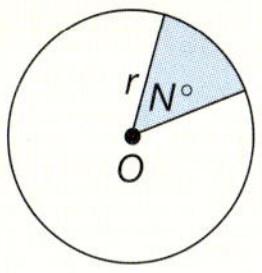

Study Tip

Common Misconceptions

The probability of an event can be expressed as a decimal or a fraction. These numbers are also sometimes represented by a percent.

Example 2 *Probability with Sectors*

a. Find the area of the blue sector.

Use the formula to find the area of the sector.

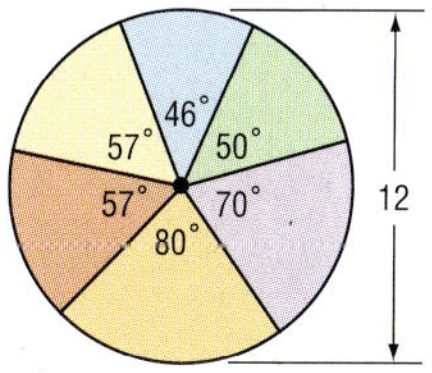

$A = \frac{N}{360}\pi r^2$ — Area of a sector

$= \frac{46}{360}\pi(6^2)$ — $N = 46, r = 6$

$= 4.6\pi$ — Simplify.

b. Find the probability that a point chosen at random lies in the blue region.

To find the probability, divide the area of the sector by the area of the circle. The area of the circle is πr^2 with a radius of 6.

$P(\text{blue}) = \frac{\text{area of sector}}{\text{area of circle}}$ — Geometric probability formula

$= \frac{4.6\pi}{\pi \cdot 6^2}$ — Area of sector = 4.6π, area of circle = $\pi \cdot 6^2$

≈ 0.13 — Use a calculator.

The probability that a random point is in the blue sector is about 0.13 or 13%.

www.geometryonline.com/extra_examples/fcat

2 Teach

GEOMETRIC PROBABILITY

In-Class Example — Power Point®

Teaching Tip Some students may not believe that you can discuss probability in relation to geometric figures. Point out that many games already use the notion of probability to determine possible outcomes for the game. Ask volunteers for examples of games that use geometric probability.

1 **GRID IN** A game board consists of a circle inscribed in a square. What is the chance that a dart thrown at the board will land in the shaded area? **0.215**

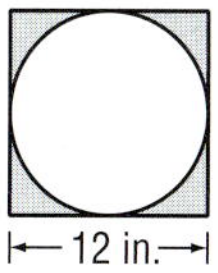

SECTORS AND SEGMENTS OF CIRCLES

In-Class Examples — Power Point®

2

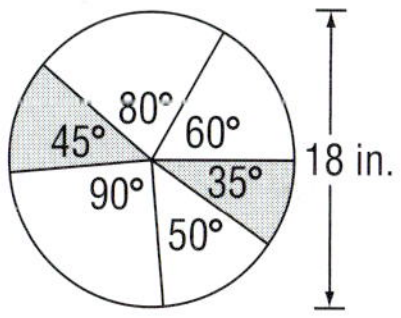

a. Find the total area of the shaded sectors. **≈ 56.5 in^2**

b. Find the probability that a point chosen at random lies in the shaded region. **$\frac{2}{9} \approx 0.22$**

3 A regular hexagon is inscribed in a circle with a diameter of 12.

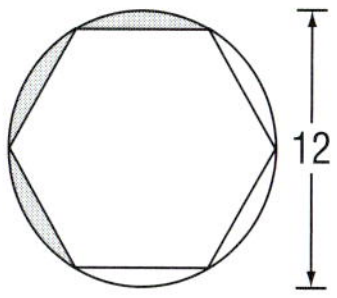

a. Find the area of the shaded regions. **≈ 9.78 units^2**

b. Find the probability that a point chosen at random lies in the shaded regions. **0.087 or 8.7%**

Lesson 11-5 Geometric Probability 623

3 Practice/Apply

Study Notebook

Have students—

- add the definitions/examples of the vocabulary terms to their Vocabulary Builder worksheets for Chapter 11.
- include the formula for finding geometric probability.
- include any other item(s) that they find helpful in mastering the skills in this lesson.

DAILY INTERVENTION

FIND THE ERROR In Exercise 3, caution students that this equation can be set up in two ways. Rachel added the degree measures of each green sector first. Another method is to find the probability for each green sector and then add. If Taimi had multiplied the second term by $\pi(5^2)$, her method would be correct.

The region of a circle bounded by an arc and a chord is called a **segment** of a circle. To find the area of a segment, subtract the area of the triangle formed by the radii and the chord from the area of the sector containing the segment.

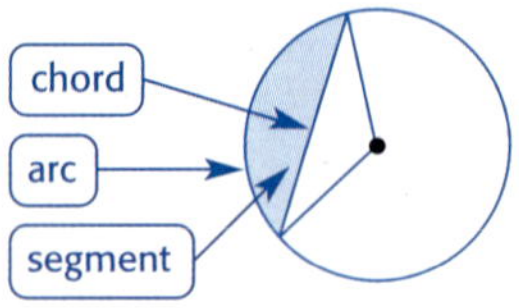

Example 3 Probability with Segments

A regular hexagon is inscribed in a circle with a diameter of 14.

a. Find the area of the red segment.

Area of the sector:

$A = \frac{N}{360}\pi r^2$ — Area of a sector

$= \frac{60}{360}\pi(7^2)$ — $N = 60$, $r = 7$

$= \frac{49}{6}\pi$ — Simplify.

≈ 25.66 — Use a calculator.

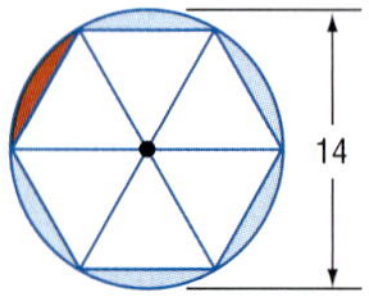

Area of the triangle:

Since the hexagon was inscribed in the circle, the triangle is equilateral, with each side 7 units long. Use properties of 30°-60°-90° triangles to find the apothem. The value of x is 3.5, the apothem is $x\sqrt{3}$ or $3.5\sqrt{3}$ which is approximately 6.06.

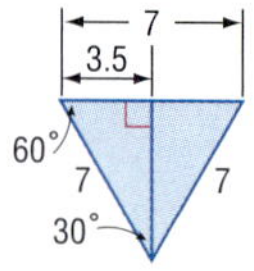

Next, use the formula for the area of a triangle.

$A = \frac{1}{2}bh$ — Area of a triangle

$= \frac{1}{2}(7)(6.06)$ — $b = 7$, $h = 6.06$

≈ 21.22 — Simplify.

Area of the segment:

area of segment = area of sector − area of triangle

$\approx 25.66 - 21.22$ — Substitution

≈ 4.44 — Simplify.

b. Find the probability that a point chosen at random lies in the red region.

Divide the area of the sector by the area of the circle to find the probability. First, find the area of the circle. The radius is 7, so the area is $\pi(7^2)$ or about 153.94 square units.

$P(\text{blue}) = \frac{\text{area of segment}}{\text{area of circle}}$

$\approx \frac{4.44}{153.94}$

≈ 0.03

The probability that a random point is on the red segment is about 0.03 or 3%.

DAILY INTERVENTION

Differentiated Instruction

Auditory/Musical Ask students to name the similarities between probability and geometric probability. **The similarity is that the probability is still found by dividing the favorable outcomes by the total number of outcomes. In geometric probability, the favorable outcome is an area or a length, and the total number of outcomes is an area or a length.**

Check for Understanding

Concept Check

1. **Explain** how to find the area of a sector of a circle.

2. **OPEN ENDED** List three games that involve geometric probability.

3. **FIND THE ERROR** Rachel and Taimi are finding the probability that a point chosen at random lies in the green region.

Rachel	Taimi
$A = \frac{N}{360}\pi r^2$	$A = \frac{N}{360}\pi r^2$
$= \frac{59+62}{360}\pi(5^2)$	$= \frac{59}{360}\pi(5^2) + \frac{62}{360}$
≈ 26.4	≈ 13.0
$P(\text{green}) \approx \frac{26.4}{25\pi} \approx 0.34$	$P(\text{green}) \approx \frac{13.0}{25\pi} \approx 0.17$

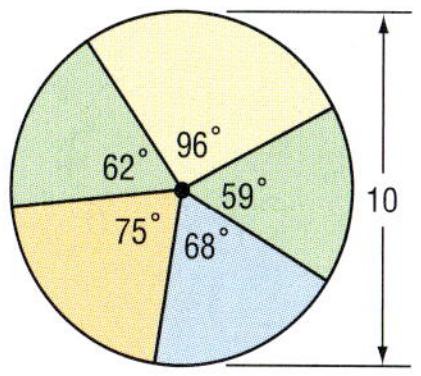

Who is correct? Explain your answer. **Rachel; Taimi did not multiply $\frac{62}{360}$ by the area of the circle.**

1. **Multiply the measure of the central angle of the sector by the area of the circle and then divide the product by 360°.**
2. **Sample answer: darts, archery, shuffleboard**

Guided Practice

GUIDED PRACTICE KEY	
Exercises	**Examples**
4	2
5	3
6	1

Find the area of the blue region. Then find the probability that a point chosen at random will be in the blue region.

4. 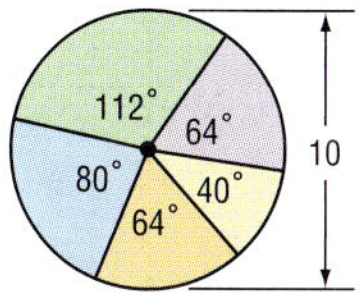

≈ 17.5 units^2, ≈ 0.22

5. 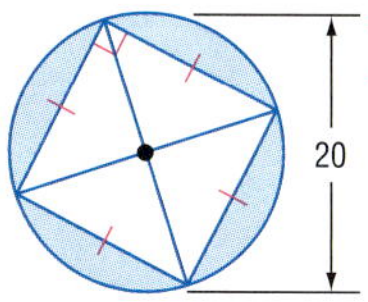

≈ 114.2 units^2, ≈ 0.36

FCAT Practice
Standardized Test Practice

6. What is the chance that a point chosen at random lies in the shaded region? **$\frac{3}{5}$ or 0.6**

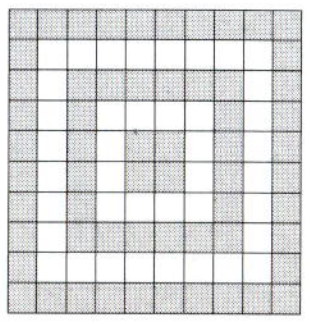

Practice and Apply

Homework Help

For Exercises	See Examples
7–9, 16, 24–30	1
10–15, 20–23	2
17–19	3

Extra Practice
See page 777.

Find the probability that a point chosen at random lies in the shaded region.

7. 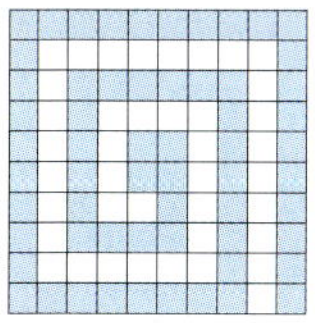
0.60

8. 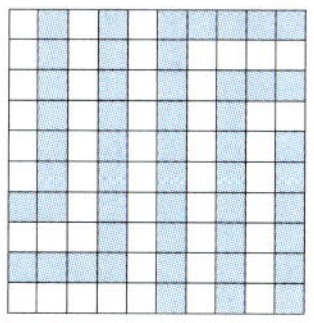
0.50

9. 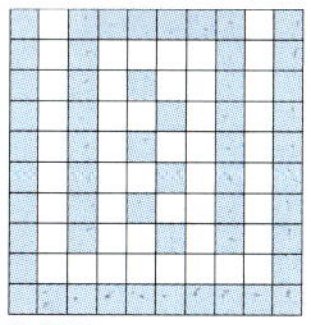
0.54

Find the area of the indicated sector. Then find the probability of spinning the color indicated if the diameter of each spinner is 15 centimeters.

10. blue **≈ 35.3 units^2, 0.20**

11. pink

12. purple

11. **≈ 58.9 units^2, 0.$\overline{3}$**
12. **≈ 66.3 units^2, 0.375**

About the Exercises...

Organization by Objective
- **Geometric Probability:** 7–9, 16, 24–30
- **Sectors and Segments of Circles:** 10–15, 17–23

Odd/Even Assignments
Exercises 7–30 are structured so that students practice the same concepts whether they are assigned odd or even problems.

Assignment Guide

Basic: 7–31 odd, 32–45
Average: 7–31 odd, 32–45
Advanced: 8–30 even, 31–45

DAILY INTERVENTION

Unlocking Misconceptions

A common error is to believe that you can estimate a geometric probability by looking at the figure. Remind students to be careful that they calculate all areas by using formulas, and then compare the calculated probability and the estimate.

Lesson 11-5 Geometric Probability 625

Study Guide and Intervention, p. 635 (shown) and p. 636

Geometric Probability The probability that a point in a figure will lie in a particular part of the figure can be calculated by dividing the area of the part of the figure by the area of the entire figure. The quotient is called the **geometric probability** for the part of the figure.

If a point in region A is chosen at random, then the probability $P(B)$ that the point is in region B, which is in the interior of region A, is

$P(B) = \frac{\text{area of region } B}{\text{area of region } A}$.

Example **Darts are thrown at a circular dartboard. If a dart hits the board, what is the probability that the dart lands in the bull's-eye?**

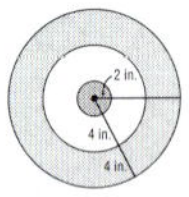

Area of bull's-eye: $A = \pi(2)^2$ or 4π

Area of entire dartboard: $A = \pi(10)^2$ or 100π

The probability of landing in the bull's-eye is

$\frac{\text{area of bull's-eye}}{\text{area of dartboard}} = \frac{4\pi}{100\pi}$

$= \frac{1}{25}$ or 0.04.

Exercises

Find the probability that a point chosen at random lies in the shaded region. Round to the nearest hundredth if necessary.

1. 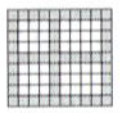 0.53

2. 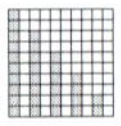 0.3

3. 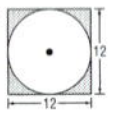

0.21

4. 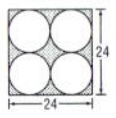

0.21

5. 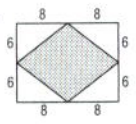

0.5

6. 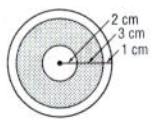

0.58

Skills Practice, p. 637 and Practice, p. 638 (shown)

Find the probability that a point chosen at random lies in the shaded region.

1. 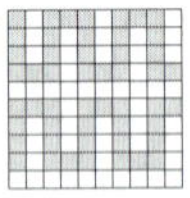 0.50

2. 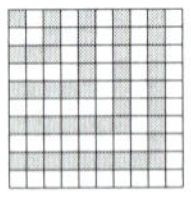 0.45

Find the area of the indicated sector. Then find the probability of spinning the color indicated if the diameter of the spinner is 9 meters.

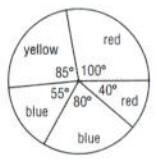

3. red 24.7 m^2, 0.39

4. blue 23.9 m^2, 0.38

5. yellow 15.0 m^2, 0.24

Find the area of the shaded region. Then find the probability that a point chosen at random is in the shaded region. Assume that all inscribed polygons are regular.

6. 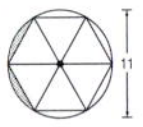

5.5 $units^2$, 0.058

7. 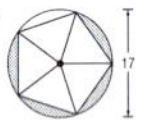

44.2 $units^2$, 0.195

8. **ARCHERY** A target consists of four concentric rings. The radius of the center circle is 4 inches, and the circles are spaced 2 inches apart. Find the probability that an arrow shot at random by an inexperienced archer will land in a shaded region. 0.44

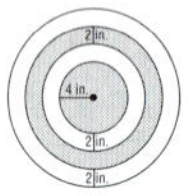

Reading to Learn Mathematics, p. 639 ELL

Pre-Activity **How can geometric probability help you win a game of darts?**

Read the introduction to Lesson 11-5 at the top of page 622 in your textbook.

To find the probability of winning at darts, would you use geometric probability to compare areas or lengths? areas

Reading the Lesson

1. Explain the difference between a sector of a circle and a segment of a circle Sample answer: A sector of a circle is bounded by a central angle and its intercept arc, while a segment is bounded by an arc and a chord.

2. Suppose you are playing a game of darts with a target like the one shown at the right. If your dart lands inside equilateral $\triangle UVW$, you get a point. Assume that every dart will land on the target. The radius of the circle is 1. Complete the following steps to figure out the probability of getting a point.

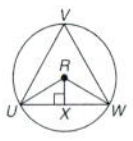

The area of circle R is $\underline{\pi}$.

$\triangle URW$ is a(n) __isosceles__ triangle because $\overline{RU}$ and $\overline{RW}$ are __radii__ of the same __circle__.

$\angle URW$ is a(n) __central__ angle of the circle, and $m\angle URW =$ __120__.

$m\angle RUX =$ __30__ and $m\angle RWX =$ __30__.

The angle measures in $\triangle RUX$ are __30__, __60__, and __90__.

$\overline{RU}$ is a __radius__ of the circle, so $RU =$ __1__.

$\overline{RX}$ is the leg of $\triangle RUX$ opposite the __30°__ angle, so $RX = \frac{1}{2}$.

Also, $\overline{UX}$ is the leg of $\triangle RUX$ opposite the __60°__ angle, so $UX = \frac{\sqrt{3}}{2}$.

$UW = \sqrt{3}$, so the area of $\triangle URW$ is $\frac{1}{2} \cdot \sqrt{3} \cdot \frac{1}{2} = \frac{\sqrt{3}}{4}$.

Then, the area of $\triangle UVW = 3 \cdot \frac{\sqrt{3}}{4} = \frac{3\sqrt{3}}{4}$.

Therefore, the probability that the dart will fall inside the triangle is the ratio of $\frac{3\sqrt{3}}{4}$ to π, which is approximately __0.413__ (to the nearest thousandth).

Helping You Remember

3. Many students find it difficult to remember a large number of geometric formulas. How can you use the formula for the area of a circle to find the area of a sector of a circle without having to learn a new formula? Sample answer: First use $A = \pi r^2$ to find the area of the circle. Then use the measure of the central angle to find out what fraction of the circle the sector is. Multiply the area of the circle by this fraction and you will have the area of the sector.

14. $\approx$ 84.9 $units^2$, $\approx$ 0.48

15. $\approx$ 74.6 $units^2$, $\approx$ 0.42

Find the area of the indicated sector. Then find the probability of choosing the color indicated if the diameter of each spinner is 15 centimeters.

13. red $\approx$ 19.6 $units^2$, $0.\overline{1}$

14. green

15. yellow

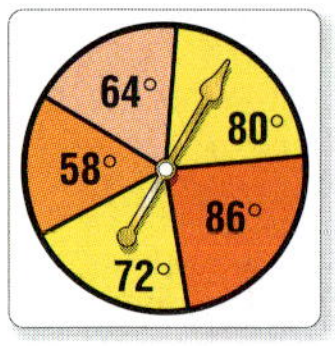

16. **PARACHUTES** A skydiver must land on a target of three concentric circles. The diameter of the center circle is 2 yards, and the circles are spaced 1 yard apart. Find the probability that she will land on the shaded area. $\frac{2}{3}$

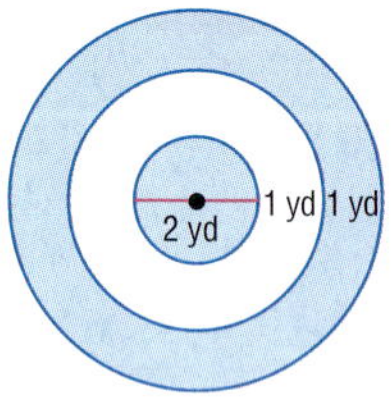

Find the area of the shaded region. Then find the probability that a point chosen at random is in the shaded region. Assume all inscribed polygons are regular.

17. 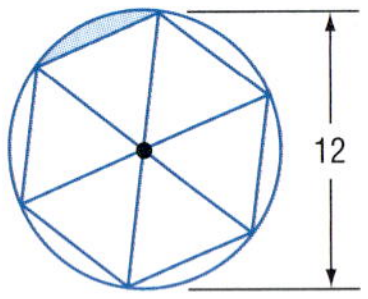

$\approx$ 3.3 $units^2$, $\approx$ 0.03

18. 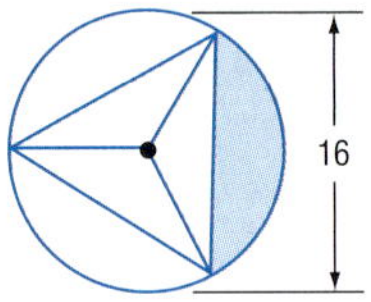

$\approx$ 39.3 $units^2$, $\approx$ 0.20

19. 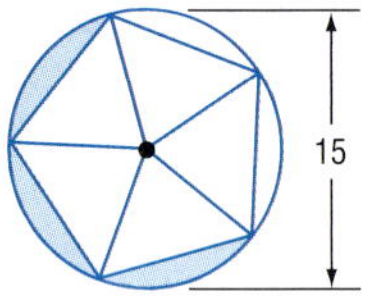

$\approx$ 25.8 $units^2$, $\approx$ 0.15

SURVEYS **For Exercises 20–23, use the following information.**

A survey was taken at a high school, and the results were put in a circle graph. The students were asked to list their favorite colors. The measurement of each central angle is shown. If a person is chosen at random from the school, find the probability of each response.

What's Your Favorite Color?

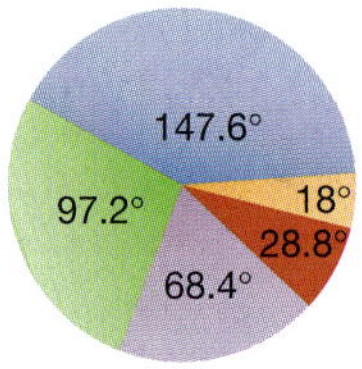

20. Favorite color is red. 0.08

21. Favorite color is blue or green. 0.68

22. Favorite color is *not* red or blue. 0.51

23. Favorite color is *not* orange or green. 0.68

More About. . .

Tennis

In tennis, the linesman determines whether the hit ball is in or out. The umpire may only overrule the linesman if he or she immediately thinks the call was wrong without a doubt and never as a result of a player's protest.

Source: www.usta.com

TENNIS **For Exercises 24 and 25, use the following information.**

A tennis court has stripes dividing it into rectangular regions. For singles play, the inbound region is defined by segments $\overline{AB}$ and $\overline{CD}$. The doubles court is bound by the segments $\overline{EF}$ and $\overline{GH}$.

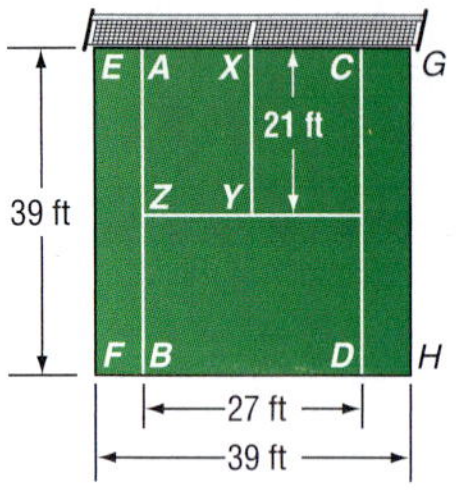

24. Find the probability that a ball in a singles game will land inside the court, but out of bounds. 0.31

25. When serving, the ball must land within $AXYZ$, the service box. Find the probability that a ball will land in the service box, relative to the court. 0.19

Enrichment, p. 640

Polygon Probability

Each problem on this page involves one or more regular polygons. To find the probability of a point chosen at random being in the shaded region, you need to find the ratio of the shaded area to the total area. If you wish, you may substitute numbers for the variables.

Find the probability that a point chosen at random in each figure is in the shaded region. Assume polygons that appear to be regular are regular. Round your answer to the nearest hundredth.

1. 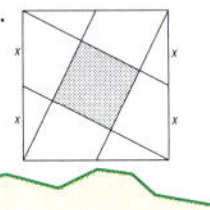

2. 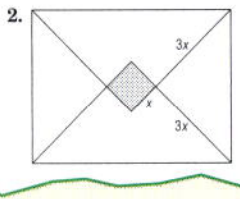

626 Chapter 11 Areas of Polygons and Circles

DARTS For Exercises 26–30, use the following information.

Each sector of the dartboard has congruent central angles. Find the probability that the dart will land on the indicated color. The diameter of the center circle is 2 units.

26. black **0.29**　　27. white **0.29**　　28. red **0.43**

29. Point values are assigned to each color. Should any of the colors have the same point value? Explain. **See margin.**

30. Which color should have the lowest point value? Explain. **See margin.**

31a–b. **See margin.**

31. **CRITICAL THINKING** Study each spinner in Exercises 13–15.
    a. Are the chances of landing on each color equal? Explain.
    b. Would this be considered a fair spinner to use in a game? Explain.

32. **WRITING IN MATH** Answer the question that was posed at the beginning of the lesson. **See margin.**

    How can geometric probability help you win a game of darts?

    Include the following in your answer:
    - an explanation of how to find the geometric probability of landing on a red sector, and
    - an explanation of how to find the geometric probability of landing in the center circle.

FCAT Practice
Standardized Test Practice
(A) (B) (C) (D)

33. One side of a square is a diameter of a circle. The length of one side of the square is 5 feet. What is the probability that a point chosen at random is in the shaded region? **C**

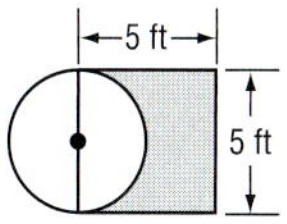

(A) 0.08　　(B) 0.22　　(C) 0.44　　(D) 0.77

34. **ALGEBRA** If $4y = 16$, then $12 \div y =$ **C**

(A) 1.　　(B) 2.　　(C) 3.　　(D) 4.

Maintain Your Skills

Mixed Review

Find the area of each figure. Round to the nearest tenth if necessary. *(Lesson 11-4)*

35. **1050 units^2**

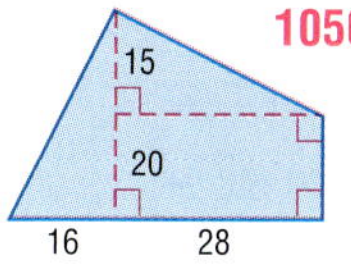

36. **82.9 ft^2**

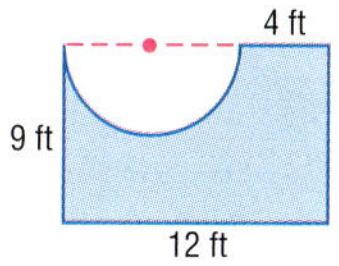

Find the area of each polygon. *(Lesson 11-3)*

37. a regular triangle with a perimeter of 48 feet **110.9 ft^2**

38. a square with a side length of 21 centimeters **441 cm^2**

39. a regular hexagon with an apothem length of 8 inches **221.7 in^2**

ALGEBRA **Find the measure of each angle on $\odot F$ with diameter $\overline{AC}$.** *(Lesson 10-2)*

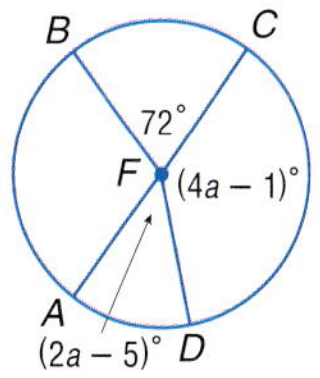

40. $\angle AFB$　　41. $\angle CFD$　　42. $\angle AFD$　　43. $\angle DFB$

40. **108**
41. **123**
42. **57**
43. **165**

Find the length of the third side of a triangle given the measures of two sides and the included angle of the triangle. Round to the nearest tenth. *(Lesson 7-7)*

44. $m = 6.8$, $n = 11.1$, $m\angle P = 57$ **$p = 9.3$**

45. $f = 32$, $h = 29$, $m\angle G = 41$ **$g = 21.5$**

www.geometryonline.com/self_check_quiz/fcat

4 Assess

Open-Ended Assessment

Modeling Have students design a gameboard. Ask them to draw a shaded region in polygons or squares whose area can be found. Then calculate the area of the shaded region and the corresponding probability.

Assessment Options

Quiz (Lesson 11-5) is available on p. 656 of the *Chapter 11 Resource Masters.*

Answers

29. The chances of landing on a black or white sector are the same, so they should have the same point value.

30. Of the three colors, there is the highest probability of landing on red, so red should have a lower point value than white or black.

31a. No; each colored sector has a different central angle.

31b. No; there is not an equal chance of landing on each color.

32. Sample answer: Geometric probability can help you determine the chance of a dart landing on the bullseye or high scoring sector. Answers should include the following.
    - Find the area of the circles containing the red sector. Divide the difference by the area of the larger circle.
    - Find the area of the center circle and divide by the area of the largest circle on the board.

Lesson 11-5 Geometric Probability 627

Chapter 11 Study Guide and Review

Vocabulary and Concept Check

- This alphabetical list of vocabulary terms in Chapter 11 includes a page reference where each term was introduced.
- **Assessment** A vocabulary test/review for Chapter 11 is available on p. 654 of the *Chapter 11 Resource Masters*.

Lesson-by-Lesson Review

For each lesson,
- the main ideas are summarized,
- additional examples review concepts, and
- practice exercises are provided.

Vocabulary PuzzleMaker

ELL The Vocabulary PuzzleMaker software improves students' mathematics vocabulary using four puzzle formats—crossword, scramble, word search using a word list, and word search using clues. Students can work on a computer screen or from a printed handout.

MindJogger Videoquizzes

ELL MindJogger Videoquizzes provide an alternative review of concepts presented in this chapter. Students work in teams in a game show format to gain points for correct answers. The questions are presented in three rounds.

Round 1 Concepts (5 questions)
Round 2 Skills (4 questions)
Round 3 Problem Solving (4 questions)

Chapter 11 Study Guide and Review

Vocabulary and Concept Check

apothem (p. 610)	irregular figure (p. 617)	sector (p. 623)
geometric probability (p. 622)	irregular polygon (p. 618)	segment (p. 624)

A complete list of postulates and theorems can be found on pages R1–R8.

Exercises Choose the formula to find the area of each shaded figure.

1. c 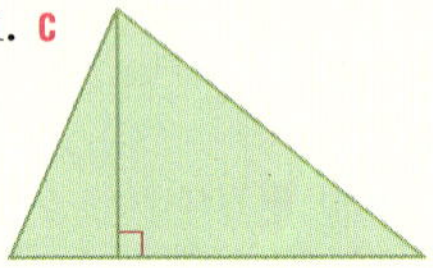
2. e 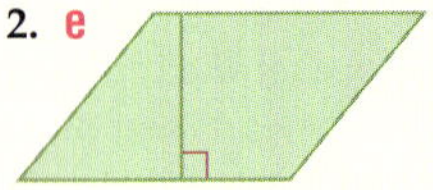
3. a 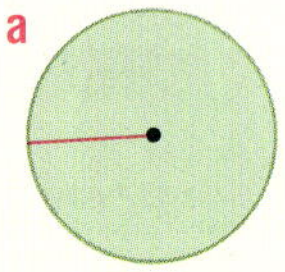
4. f 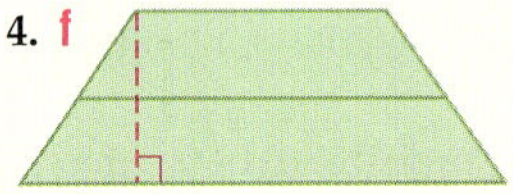
5. b 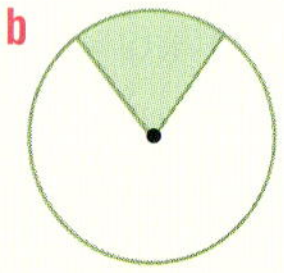
6. d 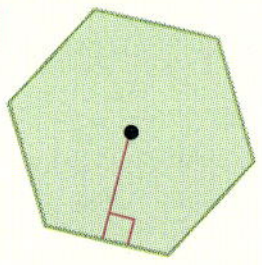

a. $A = \pi r^2$
b. $A = \frac{N}{360}\pi r^2$
c. $A = \frac{1}{2}bh$
d. $A = \frac{1}{2}Pa$
e. $A = bh$
f. $A = \frac{1}{2}h(b_1 + b_2)$

Lesson-by-Lesson Review

11-1 Area of Parallelograms

See pages 595–600.

Concept Summary

- The area of a parallelogram is the product of the base and the height.

Example **Find the area of $\square GHJK$.**

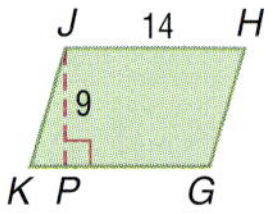

The area of a parallelogram is given by the formula $A = bh$.

$A = bh$ — Area of a parallelogram
$= 14(9)$ or 126 — $b = 14, h = 9$

The area of the parallelogram is 126 square units.

Exercises Find the perimeter and area of each parallelogram. *See Example 1 on page 596.*

7. 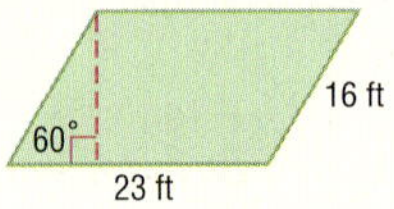

78 ft, ≈ 318.7 ft^2

8. 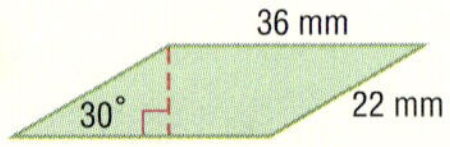

116 mm, 396 mm^2

COORDINATE GEOMETRY Given the coordinates of the vertices of a quadrilateral, determine whether it is a *square*, a *rectangle*, or a *parallelogram*. Then find the area of the quadrilateral. *See Example 3 on page 597.*

9. $A(-6, 1), B(1, 1), C(1, -6), D(-6, -6)$ square; 49 units^2
10. $E(7, -2), F(1, -2), G(2, 2), H(8, 2)$ parallelogram; 24 units^2
11. $J(-1, -4), K(-5, 0), L(-5, 5), M(-1, 1)$ parallelogram; 20 units^2
12. $P(-7, -1), Q(-3, 3), R(-1, 1), S(-5, -3)$ rectangle; 16 units^2

www.geometryonline.com/vocabulary_review

FOLDABLES™ Study Organizer

For more information about Foldables, see *Teaching Mathematics with Foldables.*

Have students look through the chapter to make sure they have included notes and examples in their Foldables for each lesson of Chapter 11.

Encourage students to refer to their Foldables while completing the Study Guide and Review and to use them in preparing for the Chapter Test.

Study Guide and Review

11-2 Areas of Triangles, Rhombi, and Trapezoids

See pages 601–609.

Concept Summary

- The formula for the area of a triangle can be used to find the areas of many different figures.
- Congruent figures have equal areas.

Example **Trapezoid *MNPQ* has an area of 360 square feet. Find the length of $\overline{MN}$.**

$A = \frac{1}{2}h(b_1 + b_2)$ Area of a trapezoid

$360 = \frac{1}{2}(18)(b_1 + 26)$ $A = 360, h = 18, b_2 = 26$

$360 = 9b_1 + 234$ Multiply.

$14 = b_1$ Solve for b_1.

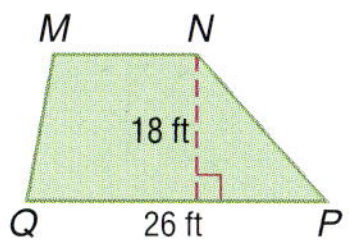

The length of $\overline{MN}$ is 14 feet.

Exercises **Find the missing measure for each quadrilateral.** *See Example 4 on page 604.*

13. Triangle *CDE* has an area of 336 square inches. Find *CE*. **28 in.**
14. Trapezoid *GHJK* has an area of 75 square meters. Find the height. **5 m**

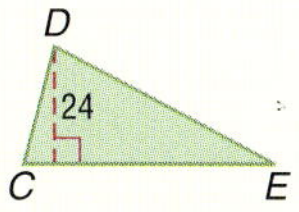

Exercise 13

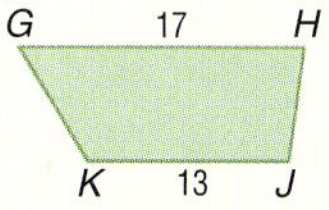

Exercise 14

11-3 Areas of Regular Polygons and Circles

See pages 610–616.

Concept Summary

- A regular *n*-gon is made up of *n* congruent isosceles triangles.
- The area of a circle of radius *r* units is πr^2 square units.

Example **Find the area of a regular hexagon with a perimeter of 72 feet.**

Since the perimeter is 72 feet, the measure of each side is 12 feet. The central angle of a hexagon is 60°. Use the properties of 30°-60°-90° triangles to find that the apothem is $6\sqrt{3}$ feet.

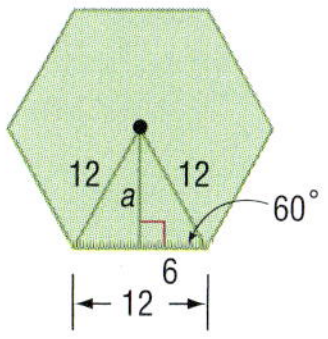

$A = \frac{1}{2}Pa$ Area of a regular polygon

$= \frac{1}{2}(72)(6\sqrt{3})$ $P = 72, a = 6\sqrt{3}$

$= 216\sqrt{3}$ Simplify.

≈ 374.1

The area of the regular hexagon is 374.1 square feet to the nearest tenth.

Exercises **Find the area of each polygon. Round to the nearest tenth.**
See Example 1 on page 611.

15. a regular pentagon with perimeter of 100 inches **688.2 in^2**
16. a regular decagon with side length of 12 millimeters **1108.0 mm^2**

Chapter 11 Study Guide and Review 629

Chapter 11 For More ...
• Extra Practice, see pages 776–777.
• Mixed Problem Solving, see page 792.

11-4 Areas of Irregular Figures

See pages 617–621.

Concept Summary

- The area of an irregular figure is the sum of the areas of its nonoverlapping parts.

Example

Find the area of the figure.

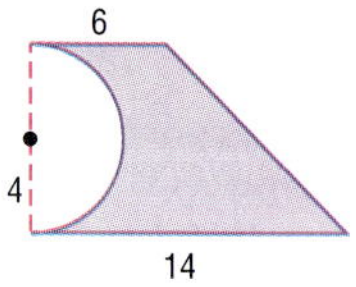

Separate the figure into a rectangle and a triangle.

area of irregular figure = area of rectangle − area of semicircle + area of triangle

$= \ell w - \frac{1}{2}\pi r^2 + \frac{1}{2}bh$ Area formulas

$= (6)(8) - \frac{1}{2}\pi(4^2) + \frac{1}{2}(8)(8)$ Substitution

$= 48 - 8\pi + 32$ or about 54.9 Simplify.

The area of the irregular figure is 54.9 square units to the nearest tenth.

Exercises **Find the area of each figure to the nearest tenth.** *See Example 1 on page 617.*

17. 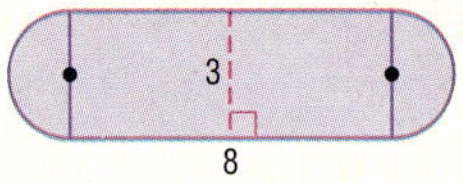

31.1 units^2

18. 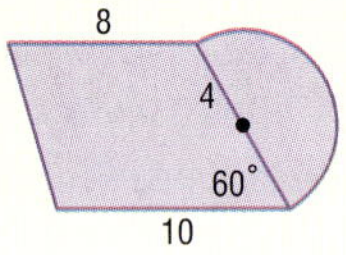

87.5 units^2

11-5 Geometric Probability

See pages 622–627.

Concept Summary

- To find a geometric probability, divide the area of a part of a figure by the total area.

Example

Find the probability that a point chosen at random will be in the blue sector.

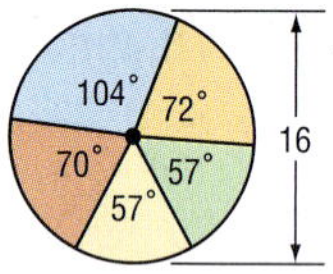

First find the area of the blue sector.

$A = \frac{N}{360}\pi r^2$ Area of a sector

$= \frac{104}{360}\pi(8^2)$ or about 58.08 Substitute and simplify.

To find the probability, divide the area of the sector by the area of the circle.

$P(\text{blue}) = \frac{\text{area of sector}}{\text{area of circle}}$ Geometric probability formula

$= \frac{58.08}{\pi 8^2}$ or about 0.29 The probability is about 0.29 or 29%.

Exercises **Find the probability that a point chosen at random will be in the sector of the given color.** *See Example 2 on page 623.*

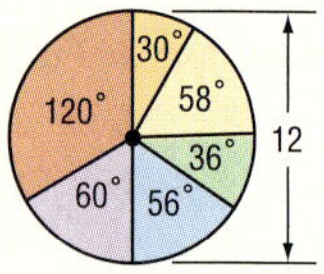

19. red $0.\overline{3}$

20. purple or green ≈ 0.27

630 Chapter 11 Areas of Polygons and Circles

Practice Test

Vocabulary and Concepts

Choose the letter of the correct area formula for each figure.

1. regular polygon **a**
2. trapezoid **c**
3. triangle **b**

a. $A = \frac{1}{2}Pa$

b. $A = \frac{1}{2}bh$

c. $A = \frac{1}{2}h(b_1 + b_2)$

Skills and Applications

COORDINATE GEOMETRY Given the coordinates of the vertices of a quadrilateral, determine whether it is a *square*, a *rectangle*, or a *parallelogram*. Then find the area of the quadrilateral.

4. $R(-6, 8)$, $S(-1, 5)$, $T(-1, 1)$, $U(-6, 4)$ **parallelogram, 20 units^2**
5. $R(7, -1)$, $S(9, 3)$, $T(5, 5)$, $U(3, 1)$ **square, 20 units^2**
6. $R(2, 0)$, $S(4, 5)$, $T(7, 5)$, $U(5, 0)$ **parallelogram, 15 units^2**
7. $R(3, -6)$, $S(9, 3)$, $T(12, 1)$, $U(6, -8)$ **rectangle, 39 units^2**

Find the area of each figure. Round to the nearest tenth if necessary.

8. **261 m^2**

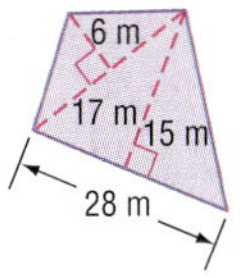

9. **855 yd^2**

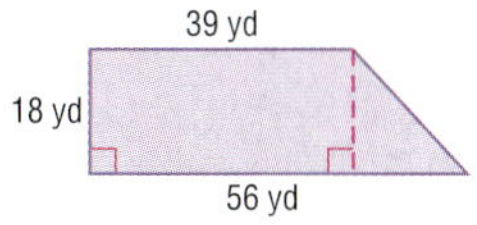

10. **814 cm^2**

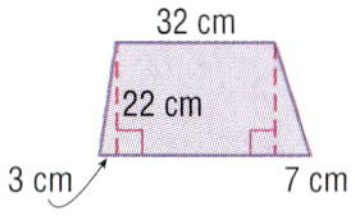

11. a regular octagon with apothem length of 3 ft **29.8 ft^2**
12. a regular pentagon with a perimeter of 115 cm **910.1 cm^2**

Each spinner has a diameter of 12 inches. Find the probability of spinning the indicated color.

13. red **0.28**

14. orange **0.24**

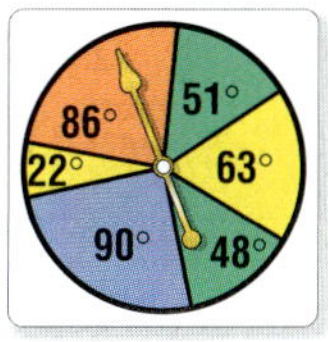

15. green **0.38**

Find the area of each figure. Round to the nearest tenth.

16. **474 units^2**

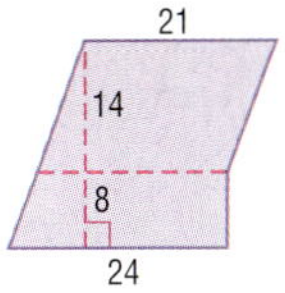

17. **91.2 units^2**

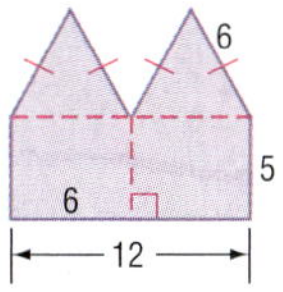

18. **87.5 units^2**

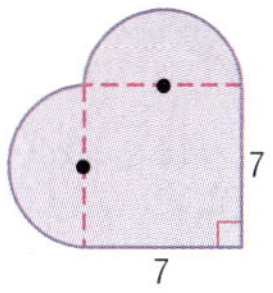

19. **SOCCER BALLS** The surface of a soccer ball is made of a pattern of regular pentagons and hexagons. If each hexagon on a soccer ball has a perimeter of 9 inches, what is the area of a hexagon? **5.8 in^2**

20. **STANDARDIZED TEST PRACTICE** What is the area of a quadrilateral with vertices at $(-3, -1)$, $(-1, 4)$, $(7, 4)$, and $(5, -1)$? **D**

Ⓐ 50 units^2 Ⓑ 45 units^2 Ⓒ $8\sqrt{29}$ units^2 Ⓓ 40 units^2

www.geometryonline.com/chapter_test/fcat

Chapter 11 Practice Test

Assessment Options

Vocabulary Test A vocabulary test/review for Chapter 11 can be found on p. 654 of the *Chapter 11 Resource Masters*.

Chapter Tests There are six Chapter 11 Tests and an Open-Ended Assessment task available in the *Chapter 11 Resource Masters*.

Chapter 11 Tests			
Form	**Type**	**Level**	**Pages**
1	MC	basic	641–642
2A	MC	average	643–644
2B	MC	average	645–646
2C	FR	average	647–648
2D	FR	average	649–650
3	FR	advanced	651–652

MC = multiple-choice questions
FR = free-response questions

Open-Ended Assessment
Performance tasks for Chapter 11 can be found on p. 653 of the *Chapter 11 Resource Masters*. A sample scoring rubric for these tasks appears on p. A22.

TestCheck and Worksheet Builder

This **networkable software** has three modules for assessment.

- **Worksheet Builder** to make worksheets and tests.
- **Student Module** to take tests on-screen.
- **Management System** to keep student records.

Portfolio Suggestion

Introduction Areas of irregular figures are used in architecture.

Ask Students Ask students to design the floor plan of an elaborate house or garden. Challenge them to include rooms of many different shapes. Have them find the area of each room, showing their calculations. Have students add their designs and area calculations to their portfolios.

Chapter 11 Standardized Test Practice

These two pages contain practice questions in the various formats that can be found on the most frequently given standardized tests.

A practice answer sheet for these two pages can be found on p. A1 of the *Chapter 11 Resource Masters*.

Standardized Test Practice Student Recording Sheet, p. A1

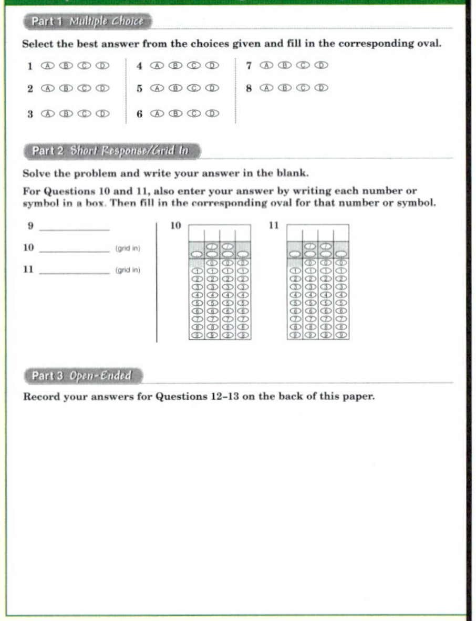

Part 1 Multiple Choice

Select the best answer from the choices given and fill in the corresponding oval.

1 Ⓐ Ⓑ Ⓒ Ⓓ 4 Ⓐ Ⓑ Ⓒ Ⓓ 7 Ⓐ Ⓑ Ⓒ Ⓓ

2 Ⓐ Ⓑ Ⓒ Ⓓ 5 Ⓐ Ⓑ Ⓒ Ⓓ 8 Ⓐ Ⓑ Ⓒ Ⓓ

3 Ⓐ Ⓑ Ⓒ Ⓓ 6 Ⓐ Ⓑ Ⓒ Ⓓ

Part 2 Short Response/Grid In

Solve the problem and write your answer in the blank.

For Questions 10 and 11, also enter your answer by writing each number or symbol in a box. Then fill in the corresponding oval for that number or symbol.

9 ______

10 ______ (grid in)

11 ______ (grid in)

Part 3 Open-Ended

Record your answers for Questions 12–13 on the back of this paper.

Additional Practice

See pp. 659–660 in the *Chapter 11 Resource Masters* for additional standardized test practice.

Chapter 11 Standardized Test Practice

FCAT Practice

Part 1 Multiple Choice

Record your answers on the answer sheet provided by your teacher or on a sheet of paper.

1. Solve $3\left(\frac{2x-4}{-6}\right) = 18$. (Prerequisite Skill) **B**

Ⓐ −19 Ⓑ −16 Ⓒ 4 Ⓓ 12

2. Sam rode his bike along the path from the library to baseball practice. What type of angle did he form during the ride? (Lesson 1-5) **B**

Ⓐ straight

Ⓑ obtuse

Ⓒ acute

Ⓓ right

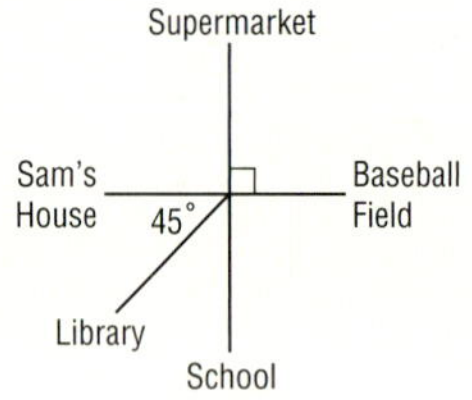

3. What is the logical conclusion of these statements?

If you exercise, you will maintain better health.
If you maintain better health, you will live longer.
(Lesson 2-4) **A**

Ⓐ If you exercise, you will live longer.

Ⓑ If you do not exercise, you will not live longer.

Ⓒ If you do not exercise, you will not maintain better health.

Ⓓ If you maintain better health, you will not live longer.

4. Which segments are parallel? (Lesson 3-5) **C**

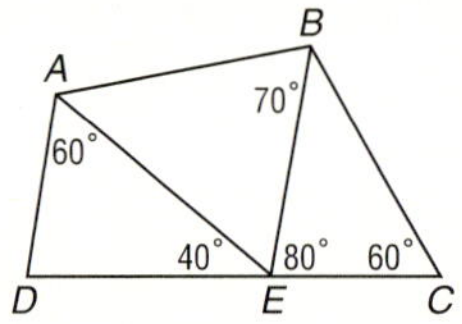

Ⓐ $\overline{AB}$ and $\overline{CD}$ Ⓑ $\overline{AD}$ and $\overline{BC}$

Ⓒ $\overline{AD}$ and $\overline{BE}$ Ⓓ $\overline{AE}$ and $\overline{BC}$

5. The front view of a pup tent resembles an isosceles triangle. The entrance to the tent is an angle bisector. The tent is secured by stakes. What is the distance between the two stakes? (Lesson 5-1) **D**

Ⓐ 3 ft

Ⓑ 4 ft

Ⓒ 5 ft

Ⓓ 6 ft

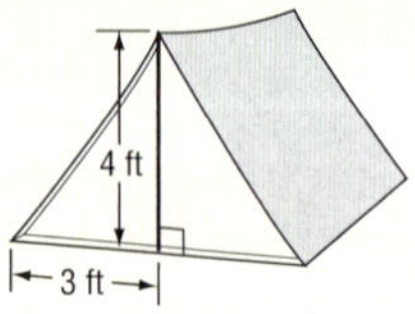

6. A carpenter is building steps leading to a hexagonal gazebo. The outside edges of the steps need to be cut at an angle. Find x. (Lesson 8-1) **D**

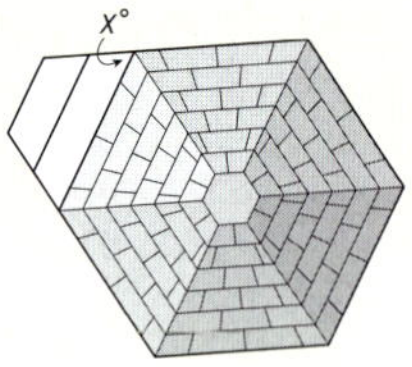

Ⓐ 180 Ⓑ 120 Ⓒ 72 Ⓓ 60

7. Which statement is *always* true? (Lesson 10-4) **A**

Ⓐ When an angle is inscribed in a circle, the angle's measure equals one-half of the measure of the intercepted arc.

Ⓑ In a circle, an inscribed quadrilateral will have consecutive angles that are supplementary.

Ⓒ In a circle, an inscribed angle that intercepts a semicircle is obtuse.

Ⓓ If two inscribed angles of a circle intercept congruent arcs, then the angles are complementary.

8. The apothem of a regular hexagon is 7.8 centimeters. If the length of each side is 9 centimeters, what is the area of the hexagon? (Lesson 11-3) **C**

Ⓐ 35.1 cm^2 Ⓑ 70.2 cm^2

Ⓒ 210.6 cm^2 Ⓓ 421.2 cm^2

The Princeton Review — ***Log On* for Test Practice**

The Princeton Review offers additional test-taking tips and practice problems at their web site. Visit www.princetonreview.com or www.review.com

TestCheck and Worksheet Builder

Special banks of standardized test questions similar to those on the SAT, ACT, TIMSS 8, NAEP 8, and state proficiency tests can be found on this CD-ROM.

Aligned and verified by The Princeton Review

FCAT Practice

Part 2 Short Response/Grid In

Record your answers on the answer sheet provided by your teacher or on a sheet of paper.

9. The post office is located halfway between the fire station and the library. What are the coordinates of the post office? (Lesson 1-3) **(2, 0)**

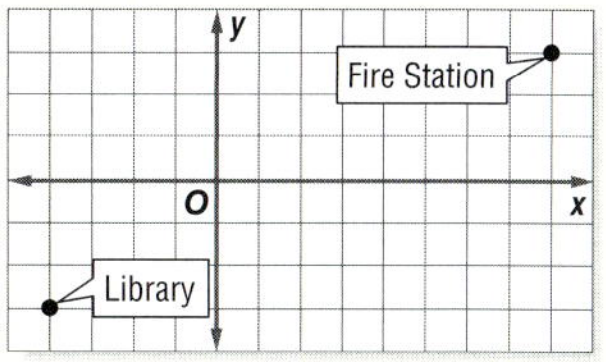

10. What is the slope of a line perpendicular to the line represented by the equation $3x - 6y = 12$? (Lesson 3-3) **-2**

11. $\triangle RST$ is a right triangle. Find $m\angle R$. (Lesson 4-2) **60**

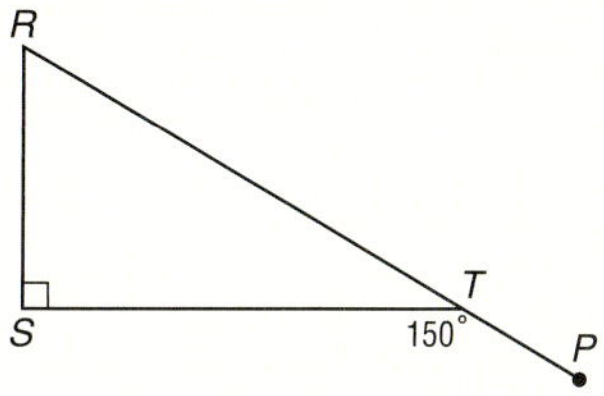

12. If $\angle A$ and $\angle E$ are congruent, find AB, the distance in feet across the pond. (Lesson 6-3) **140**

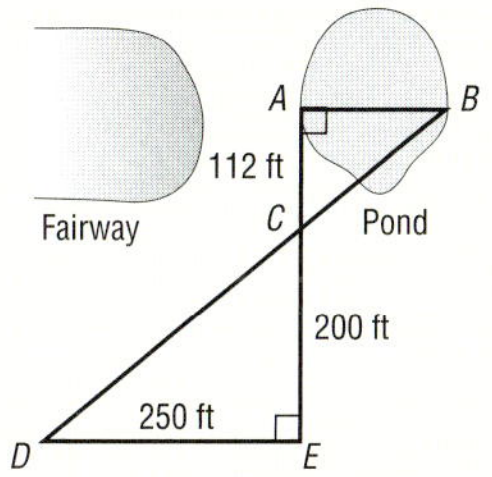

13. If point $J(6, -3)$ is translated 5 units up and then reflected over the y-axis, what will the new coordinates of J' be? (Lesson 9-2) **$(-6, 2)$**

www.geometryonline.com/standardized_test/fcat

The Princeton Review Test-Taking Tip

Question 4 Multi-Step Problems To find the pair of parallel lines, first you need to find the missing angle measures. Use the Angle Sum Theorem to find the measures of the angles in each triangle.

FCAT Practice

Part 3 Open Ended

Record your answers on a sheet of paper. Show your work.

14. Lori and her family are camping near a mountain. Their campground is in a clearing next to a stretch of forest.

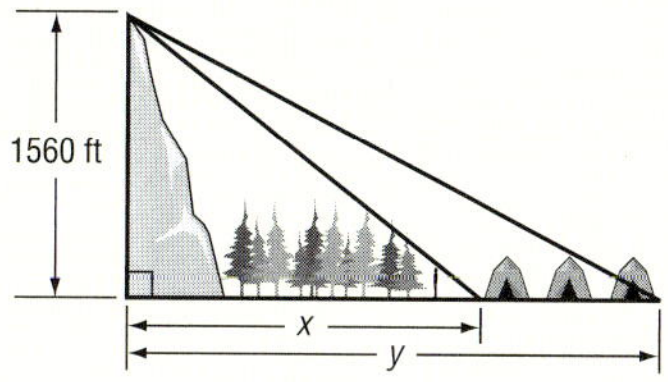

a. The angle of elevation from Lori's line of sight at the edge of the forest to the top of the mountain, is 38°. Find the distance x from the base of the mountain to the edge of the forest. Round to the nearest foot. (Lesson 7-5) **1997**

b. The angle of elevation from the far edge of the campground to the top of the mountain is 35°. Find the distance y from the base of the mountain to the far edge of the campground. Round to the nearest foot. (Lesson 7-5) **2228**

c. What is the width of the campground? Round to the nearest foot. (Lesson 7-5) **231**

15. Parallelogram $ABCD$ has vertices $A(0, 0)$, $B(3, 4)$, and $C(8, 4)$.

a. Find the possible coordinates for D. (Lesson 8-2) **(5, 0)**

b. Find the area of $ABCD$. (Lesson 11-1) **20 units^2**

Evaluating Open-Ended Assessment Questions

Open-Ended Assessment questions are graded by using a multilevel rubric that guides you in assessing a student's knowledge of a particular concept.

Goal: In Exercise 14, students use angles of elevation to find distances. In Exercise 15, students find the area of a parallelogram on a coordinate grid.

Sample Scoring Rubric: The following rubric is a sample scoring device. You may wish to add more detail to this sample to meet your individual scoring needs.

Score	Criteria
4	A correct solution that is supported by well-developed, accurate explanations
3	A generally correct solution, but may contain minor flaws in reasoning or computation
2	A partially correct interpretation and/or solution to the problem
1	A correct solution with no supporting evidence or explanation
0	An incorrect solution indicating no mathematical understanding of the concept or task, or no solution is given

Chapter 12

Surface Area

Chapter Overview and Pacing

LESSON OBJECTIVES

	PACING (days)			
	Regular		Block	
	Basic/ Average	Advanced	Basic/ Average	Advanced
12-1 Three-Dimensional Figures *(pp. 636–642)* • Use orthogonal drawings of three-dimensional figures to make models. • Identify and use three-dimensional figures.	1	1	0.5	0.5
12-2 Nets and Surface Area *(pp. 643–648)* • Draw two-dimensional models for three-dimensional figures. • Find surface area.	1	1	0.5	0.5
12-3 Surface Areas of Prisms *(pp. 649–654)* • Find lateral areas of prisms. • Find surface areas of prisms.	1	1	0.5	0.5
12-4 Surface Areas of Cylinders *(pp. 655–659)* • Find lateral areas of cylinders. • Find surface areas of cylinders.	1	1	0.5	0.5
12-5 Surface Areas of Pyramids *(pp. 660–665)* • Find lateral areas of regular pyramids. • Find surface areas of regular pyramids.	1	1	0.5	0.5
12-6 Surface Areas of Cones *(pp. 666–670)* • Find lateral areas of cones. • Find surface areas of cones.	1	1	0.5	0.5
12-7 Surface Areas of Spheres *(pp. 671–677)* • Recognize and define basic properties of spheres. • Find surface areas of spheres. ***Follow-Up:*** Find the locus of points a given distance from the endpoints of a segment.	2	2 (with 12-7 Follow-Up)	1	1 (with 12-7 Follow-Up)
Study Guide and Practice Test *(pp. 678–683)* **Standardized Test Practice** *(pp. 684–685)*	1	1	0.5	0.5
Chapter Assessment	1	1	0.5	0.5
TOTAL	**10**	**10**	**5**	**5**

Pacing suggestions for the entire year can be found on pages T20–T21.

Timesaving Tools
TeacherWorks™
All-In-One Planner and Resource Center
See pages T5 and T21.

Chapter Resource Manager

Chapter 12 Resource Masters

Study Guide and Intervention	Practice (Skills and Average)	Reading to Learn Mathematics	Enrichment	Assessment	Prerequisite Skills Workbook	Applications*	5-Minute Check Transparencies	Interactive Chalkboard	GeomPASS: Tutorial Plus (lessons)	Materials
661–662	663–664	665	666			GCC 39	12-1	12-1	21	isometric dot paper, straightedge
667–668	669–670	671	672	717			12-2	12-2		isometric dot paper, rectangular dot paper, straightedge
673–674	675–676	677	678			SC 23	12-3	12-3		straightedge, rectangular dot paper
679–680	681–682	683	684	717, 719		GCC 40	12-4	12-4		compass, straightedge, isometric dot paper
685–686	687–688	689	690			SC 24	12-5	12-5		straightedge
691–692	693–694	695	696	718			12-6	12-6		straightedge
697–698	699–700	701	702	718			12-7	12-7	22	compass, polystyrene ball, scissors, tape, glue (*Follow-Up:* ruler, compass)
				703–716, 720–722						

**Key to Abbreviations:* GCC = Graphing Calculator and Computer Masters
SC = School-to-Career Masters

Chapter 12

Mathematical Connections and Background

Continuity of Instruction

Prior Knowledge

In Chapter 3, students identified coplanar and noncoplanar lines as well as parallel and intersecting planes. Students found the circumference of a circle in Chapter 10. In Chapter 11, they found the areas of polygons and circles.

This Chapter

In this chapter, students begin their exploration of solids. The basic types of geometric figures are described and their characteristics are discussed. Students represent three-dimensional figures using orthogonal drawings, corner views, and nets. They find the lateral areas of prisms, cylinders, pyramids, and cones. They also find the surface areas of these figures. Students identify the parts of a sphere and find the surface areas of spheres and hemispheres.

Future Connections

In Chapter 13, students extend their knowledge of prisms, cylinders, pyramids, cones, and spheres by finding the volumes of those solids.

12-1 Three-Dimensional Figures

An orthogonal drawing shows the top, left, front, and right sides of an object, giving you a complete picture of that object. An orthogonal drawing can be used to render a corner view, or the view of a figure from a corner.

A solid with all flat surfaces that encloses a single region of space is called a polyhedron. Each flat surface, or face, is a polygon. A regular polyhedron has all congruent edges and all its faces are regular congruent polygons.

There are two basic types of polyhedra. A prism has two parallel congruent faces called bases. The other faces are parallelograms. A regular prism is a prism with bases that are regular polygons. A prism is named by the shape of its bases. The other type of polyhedron is a pyramid. All the faces of a pyramid (except for the base) intersect at one vertex. Pyramids are also named for their bases, which can be any polygon.

Not all solids are polyhedra. A cylinder is a solid with congruent circular bases in a pair of parallel planes. A cone has a circular base and a vertex. A sphere is a set of points in space that are a given distance from a given point.

12-2 Nets and Surface Area

Given its orthogonal drawing, you can use isometric dot paper to draw the corner view of a solid. You can also use isometric dot paper to draw two-dimensional models of geometric solids.

Another two-dimensional model of a geometric solid is a net. A net is the pattern for a three-dimensional figure. It can be folded into the shape of the solid without any overlap. Many nets can easily be drawn on rectangular dot paper. Nets are particularly useful as an aid to determining surface area. The surface area is the sum of the areas of each face of a solid. The net shows all the faces in two dimensions, making it easy to categorize each face and determine its area.

12-3 Surface Areas of Prisms

The bases of a prism are congruent faces in parallel planes. The faces that are not bases are called lateral faces. A segment perpendicular to the bases, with an endpoint in each plane, is the altitude of the prism. The height of the prism is the length of the altitude. A prism with lateral edges that are also altitudes is a right prism. If the lateral edges are not perpendicular to the bases, it is an oblique prism.

The lateral area of a prism is the sum of the areas of the lateral faces. If a right prism has a lateral area of L square units, a height of h units, and each base has a perimeter of P units, then $L = Ph$.

The surface area is the lateral area plus the area of the bases. If the surface area of a right prism is T square units, its height is h units, and each base has an area of B square units and a perimeter of P units, then $T = L + 2B$.

12-4 Surface Areas of Cylinders

A cylinder is a solid with bases that are congruent circles that lie in parallel planes. The axis of the cylinder is the segment whose endpoints are centers of those circles. If the axis is also the altitude, then the cylinder is called a *right cylinder*. Otherwise, it is an oblique cylinder.

The net of a cylinder is composed of two congruent circles and a rectangle. The area of this rectangle is the lateral area of the cylinder. The length of the rectangle is the same as the circumference of the base, $2\pi r$. If a right cylinder has a lateral area of L square units, a height of h units, and the bases have radii of r units, then $L = 2\pi rh$. If a right cylinder has a surface area of T square units, a height of h units, and the bases have radii of r units, then $T = 2\pi rh + 2\pi r^2$.

12-5 Surface Areas of Pyramids

A pyramid is a solid with faces that, except the base, intersect at one point called the *vertex*. The faces that intersect at the vertex are called *lateral faces* and form triangles. The altitude is the segment from the vertex perpendicular to the base.

If the base of a pyramid is a regular polygon and the segment with endpoints that are the center of the base and the vertex is perpendicular to the base, then the pyramid is called a *regular pyramid*. Regular pyramids have certain special characteristics. The altitude is the segment with endpoints that are the center of the base and the vertex. All the lateral faces are congruent isosceles triangles. The height of each lateral face is called the *slant height* of the pyramid. If a regular pyramid has a lateral area of L square units, a slant height of ℓ units, and its base has a perimeter of P units, then $L = \frac{1}{2}P\ell$. If a regular pyramid has a surface area of T square units, a slant height of ℓ units, and its base has a perimeter of P units and an area of B square units, then $T = \frac{1}{2}P\ell + B$.

12-6 Surface Areas of Cones

A cone with an axis that is also an altitude is a right cone. Otherwise, it is an oblique cone. The measure of any segment joining the vertex of a right cone to the edge of the circular base is called the slant height. The measure of the altitude is the height of the cone. The lateral region of a cone is a sector of a circle whose radius is equal to the slant height of the cone. If a right circular cone has a lateral area of L square units, a slant height of ℓ units, and the radius of the base is r units, then $L = \pi r\ell$. If a right circular cone has a surface area of T square units, a slant height of ℓ units, and the radius of the base is r units, then $T = \pi r\ell + \pi r^2$.

12-7 Surface Areas of Spheres

A sphere is the locus of all points that are a given distance from a given point called its *center*. A chord of a sphere is a segment with endpoints that are points on the sphere. A tangent to a sphere is a line that intersects the sphere in exactly one point. The intersection of a plane and a sphere can be a point or a circle. When a plane intersects a sphere so that it contains the center of the sphere, the intersection is called a *great circle*. Each great circle separates the sphere into two hemispheres. If a sphere has a surface area of T square units and a radius of r units, then $T = 4\pi r^2$.

Chapter 12

DAILY INTERVENTION and Assessment

	Type	Student Edition	Teacher Resources	Technology/Internet
INTERVENTION	Ongoing	Prerequisite Skills, pp. 635, 642, 648, 654, 659, 665, 670 Practice Quiz 1, p. 659 Practice Quiz 2, p. 670	5-Minute Check Transparencies Quizzes, *CRM* pp. 717–718 Mid-Chapter Test, *CRM* p. 719 Study Guide and Intervention, *CRM* pp. 661–662, 667–668, 673–674, 679–680, 685–686, 691–692, 697–698	GeomPASS: Tutorial Plus www.geometryonline.com/self_check_quiz www.geometryonline.com/extra_examples
	Mixed Review	pp. 642, 648, 654, 659, 665, 670, 676	Cumulative Review, *CRM* p. 720	
	Error Analysis	Find the Error, pp. 657, 674 Common Misconceptions, p. 638	Find the Error, *TWE* pp. 657, 674 Unlocking Misconceptions, *TWE* p. 663 Tips for New Teachers, *TWE* pp. 637, 644, 662, 672	
ASSESSMENT	Standardized Test Practice	pp. 642, 644, 646, 648, 653, 658, 664, 665, 670, 676, 683, 684, 685	*TWE* pp. 684–685 Standardized Test Practice, *CRM* pp. 721–722	Standardized Test Practice CD-ROM www.geometryonline.com/standardized_test
	Open-Ended Assessment	Writing in Math, pp. 641, 648, 653, 658, 664, 669, 676 Open Ended, pp. 639, 645, 651, 657, 663, 668, 674 Standardized Test, p. 685	Modeling: *TWE* pp. 654, 665 Speaking: *TWE* pp. 642, 676 Writing: *TWE* pp. 648, 659, 670 Open-Ended Assessment, *CRM* p. 715	
	Chapter Assessment	Study Guide, pp. 678–682 Practice Test, p. 683	Multiple-Choice Tests (Forms 1, 2A, 2B), *CRM* pp. 703–708 Free-Response Tests (Forms 2C, 2D, 3), *CRM* pp. 709–714 Vocabulary Test/Review, *CRM* p. 716	TestCheck and Worksheet Builder (see below) MindJogger Videoquizzes www.geometryonline.com/vocabulary_review www.geometryonline.com/chapter_test

Key to Abbreviations: TWE = Teacher Wraparound Edition; CRM = Chapter Resource Masters

Additional Intervention Resources

The Princeton Review's *Cracking the SAT & PSAT*

The Princeton Review's *Cracking the ACT*

TestCheck and Worksheet Builder

This **networkable** software has three modules for intervention and assessment flexibility:

- **Worksheet Builder** to make worksheet and tests
- **Student Module** to take tests on screen (optional)
- **Management System** to keep student records (optional)

Special banks are included for SAT, ACT, TIMSS, NAEP, and End-of-Course tests.

Intervention Technology

GeomPASS: Tutorial Plus CD-ROM offers a complete, self-paced geometry curriculum.

Geometry Lesson	GeomPASS Lesson
12-1	**21** Exploring Solids
12-7	**22** Surface Area of Solids

Intervention at Home

Log on for student study help.

- For each lesson in the Student Edition, there are Extra Examples and Self-Check Quizzes.
  www.geometryonline.com/extra_examples
  www.geometryonline.com/self_check_quiz
- For chapter review, there is vocabulary review, test practice, and standardized test practice.
  www.geometryonline.com/vocabulary_review
  www.geometryonline.com/chapter_test
  www.geometryonline.com/standardized_test

For more information on Intervention and Assessment, see pp. T8–T11.

Reading and Writing in Mathematics

Glencoe Geometry provides numerous opportunities to incorporate reading and writing into the mathematics classroom.

Student Edition

- Foldables Study Organizer, p. 635
- Concept Check questions require students to verbalize and write about what they have learned in the lesson. (pp. 639, 645, 651, 657, 663, 668, 674)
- Writing in Math questions in every lesson, pp. 641, 648, 653, 658, 664, 669, 676
- Reading Study Tip, pp. 637, 649, 666
- WebQuest, p. 662

Teacher Wraparound Edition

- Foldables Study Organizer, pp. 635, 678
- Study Notebook suggestions, pp. 640, 646, 652, 657, 663, 668, 674, 677
- Modeling activities, pp. 654, 665
- Speaking activities, pp. 642, 676
- Writing activities, pp. 648, 659, 670
- Differentiated Instruction (Verbal/Linguistic), p. 667
- ELL Resources, pp. 634, 641, 647, 653, 658, 664, 667, 669, 675, 678

Additional Resources

- Vocabulary Builder worksheets require students to define and give examples for key vocabulary terms as they progress through the chapter. (*Chapter 12 Resource Masters*, pp. vii-viii)
- Reading to Learn Mathematics master for each lesson (*Chapter 12 Resource Masters*, pp. 665, 671, 677, 683, 689, 695, 701)
- *Vocabulary PuzzleMaker* software creates crossword, jumble, and word search puzzles using vocabulary lists that you can customize.
- *Teaching Mathematics with Foldables* provides suggestions for promoting cognition and language.
- *Reading Strategies for the Mathematics Classroom*
- *WebQuest and Project Resources*

For more information on Reading and Writing in Mathematics, see pp. T6–T7.

Chapter 12 Notes

What You'll Learn

Have students read over the list of objectives and make a list of any words with which they are not familiar.

Why It's Important

Point out to students that this is only one of many reasons why each objective is important. Others are provided in the introduction to each lesson.

Lesson	NCTM Standards	Local Objectives
12-1	3, 6, 8, 9, 10	
12-2	3, 6, 8, 9, 10	
12-3	3, 6, 8, 9, 10	
12-4	3, 6, 8, 9, 10	
12-5	3, 6, 8, 9, 10	
12-6	3, 6, 8, 9, 10	
12-7	3, 6, 8, 9, 10	
12-7 Follow-Up	3, 8, 9, 10	

Key to NCTM Standards:

1=Number & Operations, 2=Algebra, 3=Geometry, 4=Measurement, 5=Data Analysis & Probability, 6=Problem Solving, 7=Reasoning & Proof, 8=Communication, 9=Connections, 10=Representation

Chapter 12 Surface Area

What You'll Learn

- **Lesson 12-1** Identify three-dimensional figures.
- **Lesson 12-2** Draw two-dimensional models for solids.
- **Lessons 12-3 through 12-6** Find the lateral areas and surface areas of prisms, cylinders, pyramids, and cones.
- **Lesson 12-7** Find the surface areas of spheres and hemispheres.

Key Vocabulary

- **polyhedron** (p. 637)
- **net** (p. 644)
- **surface area** (p. 644)
- **lateral area** (p. 649)

Why It's Important

Diamonds and other gems are cut to enhance the beauty of the stones. The stones are cut into regular geometric shapes. Each cut has a special name.
You will learn more about gemology in Lesson 12-1.

Vocabulary Builder

ELL

The Key Vocabulary list introduces students to some of the main vocabulary terms included in this chapter. For a more thorough vocabulary list with pronunciations of new words, give students the Vocabulary Builder worksheets found on pages vii and viii of the *Chapter 12 Resource Masters*. Encourage them to complete the definition of each term as they progress through the chapter. You may suggest that they add these sheets to their study notebooks for future reference when studying for the Chapter 12 test.

Getting Started

Prerequisite Skills To be successful in this chapter, you'll need to master these skills and be able to apply them in problem-solving situations. Review these skills before beginning Chapter 12.

For Lesson 12-1 — Parallel Lines and Planes

In the figure, $\overline{AC} \parallel \ell$. Determine whether each statement is *true*, *false*, or *cannot be determined*. (For review, see Lesson 3-1.)

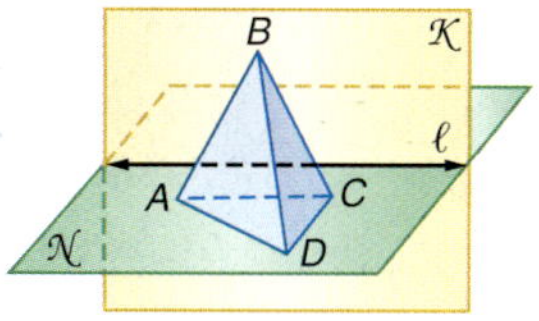

1. $\triangle ADC$ lies in plane $\mathcal{N}$. **true**
2. $\triangle ABC$ lies in plane $\mathcal{K}$. **false** **3. cannot be determined**
3. The line containing $\overline{AB}$ is parallel to plane $\mathcal{K}$.
4. The line containing $\overline{AC}$ lies in plane $\mathcal{K}$. **false**

For Lessons 12-3 and 12-5 — Areas of Triangles and Trapezoids

Find the area of each figure. Round to the nearest tenth if necessary. (For review, see Lesson 11-2.)

5. **384 ft²**

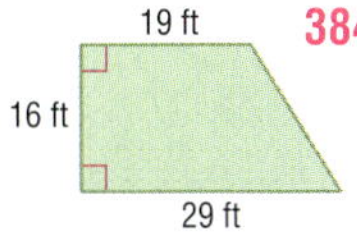

6. **305.5 mm²**

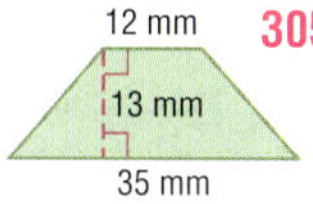

7. **1.8 m²**

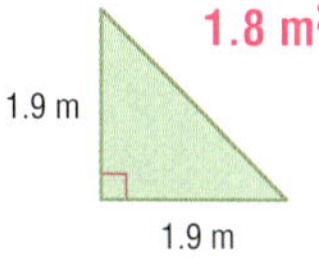

For Lessons 12-4, 12-6, and 12-7 — Area of Circles

Find the area of each circle with the given radius or diameter. Round to the nearest tenth. (For review, see Lesson 11-3.)

8. $d = 19.0$ cm **283.5 cm²**
9. $r = 1.5$ yd **7.1 yd²**
10. $d = 10.4$ m **84.9 m²**

FOLDABLES™ Study Organizer

Surface Area Make this Foldable to help you organize what you learn about surface area. Begin with a large sheet of paper, such as 11 inches × 17 inches.

Step 1 Fold Lengthwise

Step 2 Fold

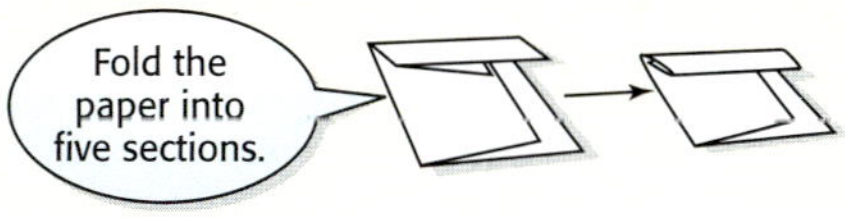

Step 3 Cut

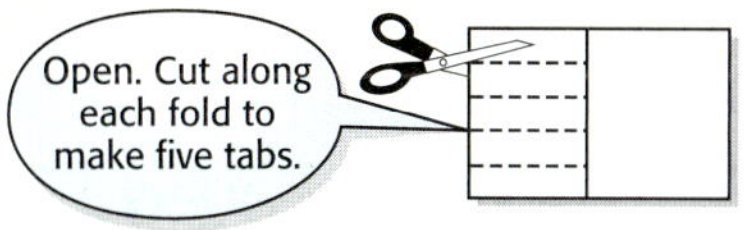

Step 4 Label

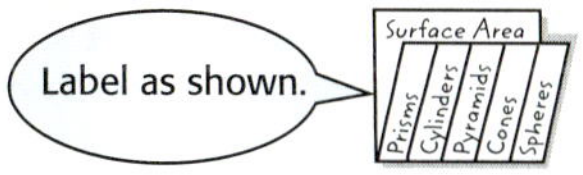

Reading and Writing As you read and study the chapter, define terms and write notes about surface area for each three-dimensional figure.

Getting Started

This section provides a review of the basic concepts needed before beginning Chapter 12. Page references are included for additional student help.

Prerequisite Skills in the Getting Ready for the Next Lesson section at the end of each exercise set review a skill needed in the next lesson.

For Lesson	Prerequisite Skill
12-2	Finding the area of a rectangle, p. 642
12-3	Finding areas of parallelograms, triangles, and trapezoids, p. 648
12-4	Finding the area of a circle, p. 654
12-5	Finding areas of triangles and trapezoids, p. 659
12-6	Pythagorean Theorem, p. 665
12-7	Finding the circumference of a circle, p. 670

FOLDABLES™ Study Organizer

For more information about Foldables, see *Teaching Mathematics with Foldables.*

Organization of Data with a Concept Map Use this Foldable concept map for student writing about five three-dimensional figures. Begin with the central chapter theme of *Surface Area* as the title, and have students write the names of the five shapes studied in this chapter: prisms, cylinders, pyramids, cones, and spheres. Students can use their Foldable to take notes, define terms, record concepts, and define basic properties. Foldable concept maps make great study guides because students view main ideas, recall what they know, and check their responses by looking under the tabs.

12-1 Lesson Notes

1 Focus

5-Minute Check Transparency 12-1 Use as a quiz or review of Chapter 11.

Mathematical Background notes are available for this lesson on p. 634C.

Why are drawings of three-dimensional structures valuable to archeologists?

Ask students:

- What three-dimensional structure is studied by archeologists? **pyramid**
- What other shapes are used in archeology or Egyptology? **Sample answer: triangles and rectangles**
- What is another real-world use of pyramids? **Sample answer: ornamental crystals**

12-1 Three-Dimensional Figures

Sunshine State Standards
MA.C.2.4.2 NC.C.1.4.1, NC.C.2.4.1, NC.C.2.4.1, NC.C.1.4.1, NC.C.2.4.1, NC.C.2.4.1

Vocabulary

- orthogonal drawing
- corner view
- perspective view
- polyhedron
- face
- edges
- prism
- bases
- regular prism
- pyramid
- regular polyhedron
- Platonic solids
- cylinder
- cone
- sphere
- cross section

What You'll Learn

- Use orthogonal drawings of three-dimensional figures to make models.
- Identify and use three-dimensional figures.

Why are drawings of three-dimensional structures valuable to archaeologists?

Archaeologists and Egyptologists continue to study the Great Pyramids of Egypt. Drawings of these three-dimensional structures are helpful in their study.

DRAWINGS OF THREE-DIMENSIONAL FIGURES If you see a three-dimensional object from only one viewpoint, you may not know its true shape. Here are four views of the pyramid of Menkaure in Giza, Egypt. The two-dimensional views of the top, left, front, and right sides of an object are called an **orthogonal drawing**.

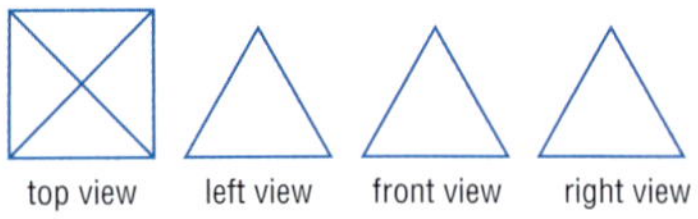

This sculpture is *Stacked Pyramid* by Jackie Ferrara. How can we show the stacks of blocks on each side of the piece in a two-dimensional drawing? Let the edge of each block represent a unit of length and use a dark segment to indicate a break in the surface.

Study Tip

Corner View Drawings
Use models to help you visualize the corner view of a solid.

The view of a figure from a corner is called the **corner view** or **perspective view**. You can use isometric dot paper to draw the corner view of a solid figure. One corner view of a cube is shown at the right.

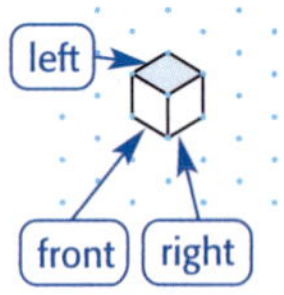

Example 1 Use Orthogonal Drawings

a. Draw the back view of a figure given its orthogonal drawing.

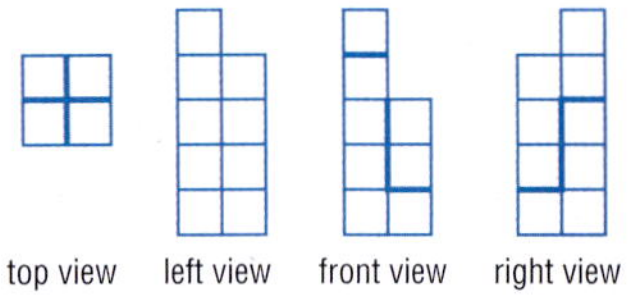

Use blocks to make a model. Then use your model to draw the back view.

- The top view indicates two rows and two columns of different heights.
- The front view indicates that the left side is 5 blocks high and the right side is 3 blocks high. The dark segments indicate breaks in the surface.

Resource Manager

Workbook and Reproducible Masters

Chapter 12 Resource Masters

- Study Guide and Intervention, pp. 661–662
- Skills Practice, p. 663
- Practice, p. 664
- Reading to Learn Mathematics, p. 665
- Enrichment, p. 666

Graphing Calculator and Computer Masters, p. 39

Teaching Geometry With Manipulatives Masters, pp. 7, 17

Transparencies

5-Minute Check Transparency 12-1
Answer Key Transparencies

Technology

GeomPASS: Tutorial Plus, Lesson 21
Interactive Chalkboard

- The right view indicates that the right front column is only one block high. The left front column is 4 blocks high. The right back column is 3 blocks high.
- Check the left side of your model. All of the blocks should be flush.
- Check to see that all views correspond to the model.

Now that your model is accurate, turn it around to the back and draw what you see. The blocks are flush, so no heavy segments are needed.

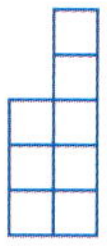

b. Draw the corner view of the figure.

Turn your model so you are looking at the corners of the blocks. The lowest columns should be in front so the differences in height between the columns is visible.

Connect the dots on the isometric dot paper to represent the edges of the solid. Shade the tops of each column.

Study Tip

Reading Math
The plural version of a polyhedron is either *polyhedrons* or *polyhedra.* Both forms are correct.

IDENTIFY THREE-DIMENSIONAL FIGURES

A solid with all flat surfaces that enclose a single region of space is called a **polyhedron**. Each flat surface, or **face**, is a polygon. The line segments where the faces intersect are called **edges**. Edges intersect at a point called a *vertex*.

A **prism** is a polyhedron with two parallel congruent faces called **bases**. The other faces are parallelograms. The intersection of three edges is a vertex. Prisms are named by the shape of their bases. A **regular prism** is a prism with bases that are regular polygons. A cube is an example of a regular prism. Some common prisms are shown below.

Name	Triangular Prism	Rectangular Prism	Pentagonal Prism
Model	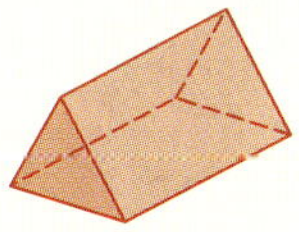	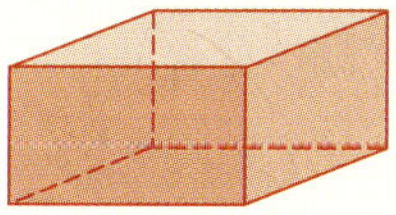	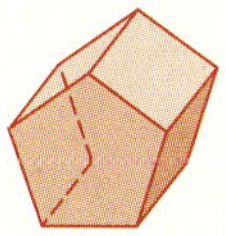
Shape of Base	triangle	rectangle	pentagon

More About. . .

Plato
Plato was a teacher of mathematics and philosophy. Around 387 B.C., he founded a school in Athens, Greece, called the "Academy."
Source: www.infoplease.com

A polyhedron with all faces (except for one) intersecting at one vertex is a **pyramid**. Pyramids are named for their bases, which can be any polygon.

A polyhedron is a **regular polyhedron** if all of its faces are regular congruent polygons and all of the edges are congruent.

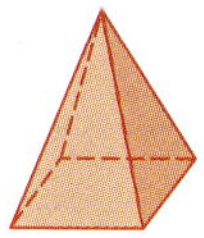

square pyramid

There are exactly five types of regular polyhedra. These are called the **Platonic Solids** because Plato described them extensively in his writings.

www.geometryonline.com/extra_examples/fcat

2 Teach

DRAWINGS OF THREE-DIMENSIONAL FIGURES

Tips for New Teachers
Introduce the lesson by having students name familiar three-dimensional figures and give examples of how they are used.

In-Class Example Power Point®

Teaching Tip Point out that an orthogonal drawing includes the top, front, and both side views of an object. Each view may be a different shape.

1

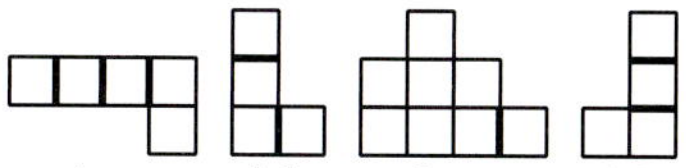

a. Draw the back view of a figure given its orthogonal drawing.

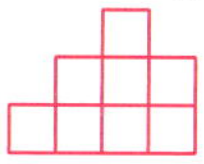

b. Draw the corner view of the figure.

Interactive Chalkboard
PowerPoint® Presentations

This CD-ROM is a customizable Microsoft® PowerPoint® presentation that includes:
- Step-by-step, dynamic solutions of each In-Class Example from the Teacher Wraparound Edition
- Additional, Try These exercises for each example
- The 5-Minute Check Transparencies
- Hot links to Glencoe Online Study Tools

IDENTIFY THREE-DIMENSIONAL FIGURES

In-Class Example Power Point®

Teaching Tip Emphasize that each prism may be shown resting on a face or a base. Point out that students must learn how to recognize these prisms regardless of their orientation.

2 Identify each solid. Name the bases, faces, edges, and vertices.

a.

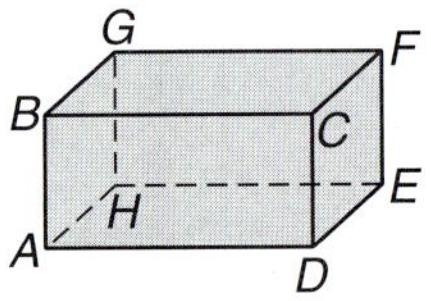

rectangular prism; bases: Sample answer: ▱*ADEH*, ▱*BCFG*, faces: ▱*ADEH*, ▱*BCFG*, ▱*ABGH*, ▱*CDEF*, ▱*ABCD*, ▱*EFGH*; edges: $\overline{AB}$, $\overline{CD}$, $\overline{EF}$, $\overline{GH}$, $\overline{AD}$, $\overline{BC}$, $\overline{FG}$, $\overline{EH}$, $\overline{BG}$, $\overline{AH}$, $\overline{CF}$, $\overline{DE}$; vertices: *A*, *B*, *C*, *D*, *E*, *F*, *G*, *H*

b.

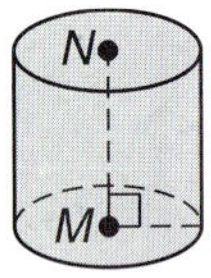

cylinder; bases: circle *N* and circle *M*

c.

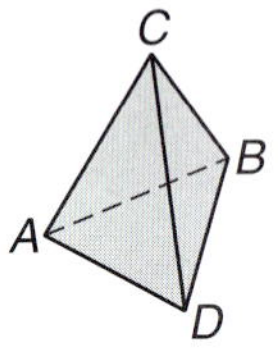

triangular pyramid; base: △*ABD*; faces: △*ABD*, △*ABC*, △*ACD*, △*BCD*; edges: $\overline{AD}$, $\overline{AC}$, $\overline{BD}$, $\overline{CD}$, $\overline{AB}$, $\overline{BC}$; vertices: *A*, *B*, *C*, *D*

Platonic Solids					
Name	**tetrahedron**	**hexahedron**	**octahedron**	**dodecahedron**	**icosahedron**
Model	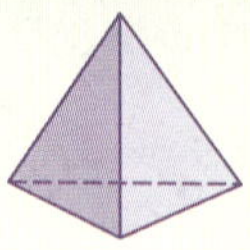	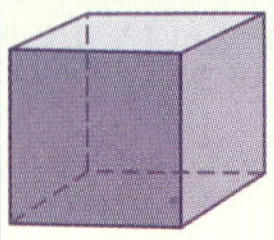	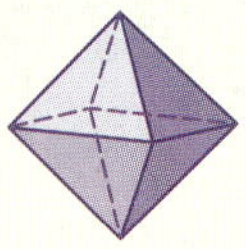	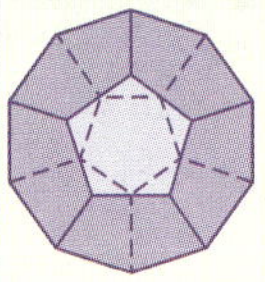	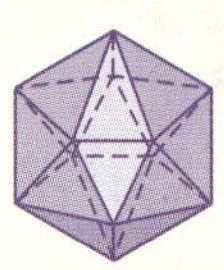
Number of Faces	4	6	8	12	20
Shape of Face	equilateral triangle	square	equilateral triangle	regular pentagon	equilateral triangle

There are solids that are *not* polyhedrons. All of the faces in each solid are not polygons. A **cylinder** is a solid with congruent circular bases in a pair of parallel planes. A **cone** has a circular base and a vertex. A **sphere** is a set of points in space that are a given distance from a given point.

Name	**cylinder**	**cone**	**sphere**
Model	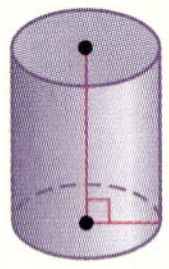	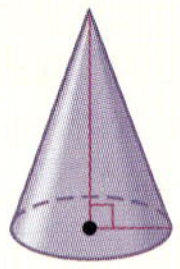	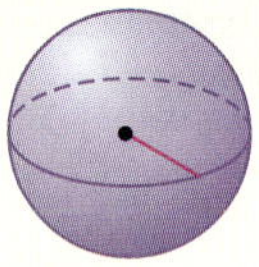

Study Tip

Common Misconception
Prisms can be oriented so the bases are not the top and bottom of the solid.

Example 2 *Identify Solids*

Identify each solid. Name the bases, faces, edges, and vertices.

a.

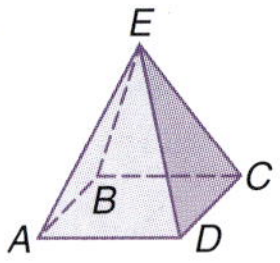

The base is a rectangle, and the other four faces meet in a point. So this solid is a rectangular pyramid.

Base: ▱*ABCD*

Faces: ▱*ABCD*, △*AED*, △*DEC*, △*CEB*, △*AEB*

Edges: $\overline{AB}$, $\overline{BC}$, $\overline{CD}$, $\overline{DA}$, $\overline{AE}$, $\overline{DE}$, $\overline{CE}$, $\overline{BE}$

Vertices: *A*, *B*, *C*, *D*, *E*

b.

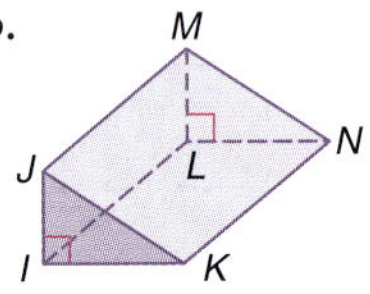

The bases are right triangles. So this is a triangular prism.

Bases: △*IJK*, △*LMN*

Faces: △*IJK*, △*LMN*, ▱*ILNK*, ▱*KJMN*, ▱*IJML*

Edges: $\overline{IL}$, $\overline{LN}$, $\overline{NK}$, $\overline{IK}$, $\overline{IJ}$, $\overline{LM}$, $\overline{JM}$, $\overline{MN}$, $\overline{JK}$

Vertices: *I*, *J*, *K*, *L*, *M*, *N*

c.

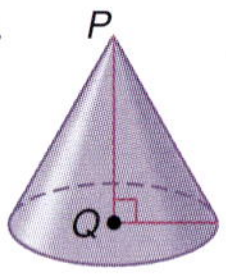

This solid has a circle for a base and a vertex. So it is a cone.

Base: ⊙*Q*

Vertex: *P*

DAILY INTERVENTION **Differentiated Instruction**

Logical Students should reason that while every pyramid has triangles for sides, the base can be any shape, including another triangle. The great pyramids of Egypt all have square bases, so they are called square pyramids. Pyramids with triangular, rectangular, or any other shaped base are named accordingly.

Interesting shapes occur when a plane intersects, or slices, a solid figure. If the plane is parallel to the base or bases of the solid, then the intersection of the plane and solid is called a **cross section** of the solid.

Study Tip

Modeling
You can model each of these cuts by using modeling clay for the cylinder and cutting the clay with dental floss.

Example 3 Slicing Three-Dimensional Figures

CARPENTRY **A carpenter purchased a section of a tree trunk. He wants to cut the trunk into a circle, an oval, and a rectangle. How could he cut the tree trunk to get each shape?**

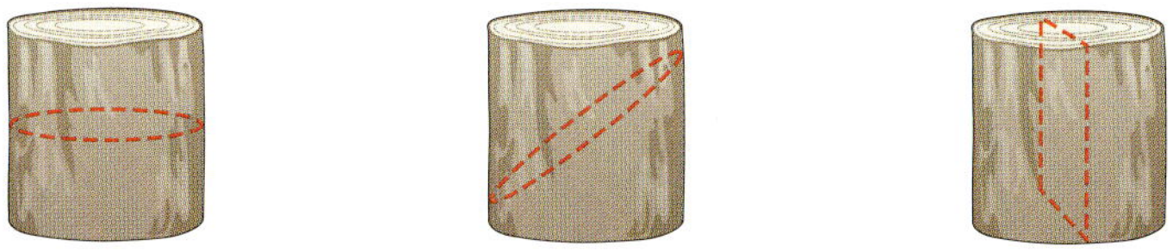

The tree trunk has a cylindrical shape. If the blade of the saw was placed parallel to the bases, the cross section would be a circle.

If the blade was placed at an angle to the bases of the tree trunk, the slice would be an oval shape, or an ellipse.

To cut a rectangle from the cylinder, place the blade perpendicular to the bases. The slice is a rectangle.

In-Class Example Power Point®

3 **BAKERY** A customer ordered a two-layer sheet cake. Describe the possible cross sections of the cake.

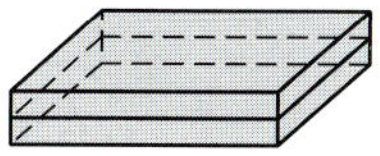

Sample answer: If the cake is cut horizontally, the cross section will be a rectangle. If the cake is cut vertically, the cross section will also be a rectangle.

Check for Understanding

Concept Check

1. **Explain** how the Platonic Solids are different from other polyhedra.
2. **Explain** the difference between a square pyramid and a square prism.
3. **OPEN ENDED** Draw a rectangular prism.

1–3. See margin.

Guided Practice

4. **Draw** the back view and corner view of a figure given its orthogonal drawing.

See margin.

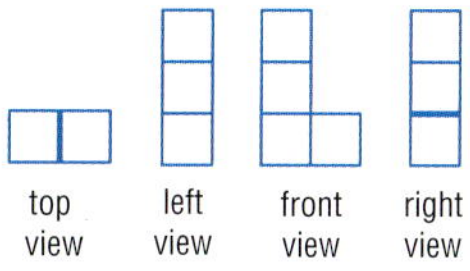

GUIDED PRACTICE KEY	
Exercises	Examples
4	1
5–7	2
8	3

Identify each solid. Name the bases, faces, edges, and vertices. 5–7. See margin.

5.

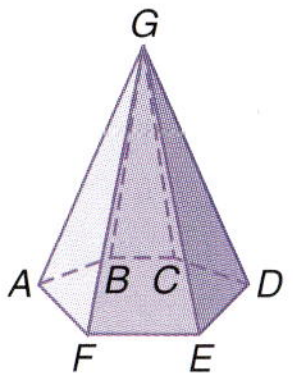

6.

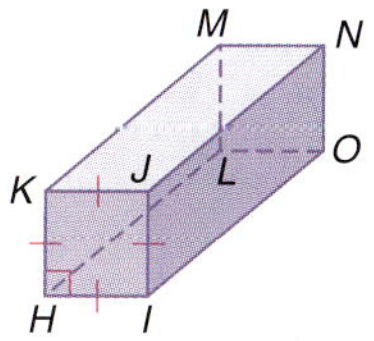

7.

Application

8. **DELICATESSEN** A slicer is used to cut whole pieces of meat and cheese for sandwiches. Suppose a customer wants slices of cheese that are round and slices that are rectangular. How can the cheese be placed on the slicer to get each shape? See margin.

4.

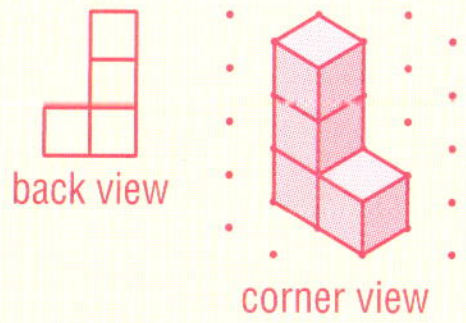

5. hexagonal pyramid; base: *ABCDEF*; faces: *ABCDEF*, △*AGF*, △*FGE*, △*EGD*, △*DGC*, △*CGB*, △*BGA*; edges: $\overline{AF}$, $\overline{FE}$, $\overline{ED}$, $\overline{DC}$, $\overline{CB}$, $\overline{BA}$, $\overline{AG}$, $\overline{FG}$, $\overline{EG}$, $\overline{DG}$, $\overline{CG}$, and $\overline{BG}$; vertices: *A*, *B*, *C*, *D*, *E*, *F*, and *G*

6. square prism; bases: Sample answer: ▱*KJIH*, ▱*MNOL*; faces: ▱*KJIH*, ▱*MNOL*, ▱*JNOI*, ▱*JKMN*, ▱*KHLM*, ▱*IHLO*; edges: $\overline{KH}$, $\overline{KJ}$, $\overline{JI}$, $\overline{IH}$, $\overline{JN}$, $\overline{IO}$, $\overline{HL}$, $\overline{KM}$, $\overline{MN}$, $\overline{ML}$, $\overline{NO}$, and $\overline{LO}$; vertices: *H*, *K*, *J*, *I*, *L*, *M*, *N*, and *O*

7. cylinder; bases: circles *P* and *Q*

8. To get round slices of cheese, slice the cheese parallel to the bases. To get rectangular slices, place the cheese on the slicer so the bases are perpendicular to the blade.

Answers

1. The Platonic solids are the five regular polyhedra. All of the faces are congruent, regular polygons. In other polyhedra, the bases are congruent parallel polygons, but the faces are not necessarily congruent.

2. Sample answers: In a square pyramid, the lateral faces are triangles. In a square prism, the faces are rectangles.

3. Sample answer:

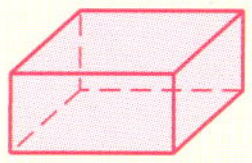

3 Practice/Apply

Study Notebook

Have students—
- add the definitions/examples of the vocabulary terms to their Vocabulary Builder worksheets for Chapter 12.
- include examples of orthogonal drawings.
- include any other item(s) that they find helpful in mastering the skills in this lesson.

About the Exercises...

Organization by Objective
- **Drawings of Three-Dimensional Figures:** 9–15, 23, 24
- **Identify Three-Dimensional Figures:** 16–22, 25–41

Odd/Even Assignments

Exercises 9–21 and 25–35 are structured so that students practice the same concepts whether they are assigned odd or even problems.

Assignment Guide

Basic: 9–29 odd, 37–41 odd, 42–45, 49–62 (optional: 46–48)

Average: 9–41 odd, 42–45, 49–62 (optional: 46–48)

Advanced: 10–40 even, 42–58 (optional: 59–62)

★ indicates increased difficulty

Practice and Apply

Homework Help

For Exercises	See Examples
9–15, 23, 24	1
16–22, 36–41	2
25–35	3

Extra Practice
See page 778.

Draw the back view and corner view of a figure given each orthogonal drawing.

9. 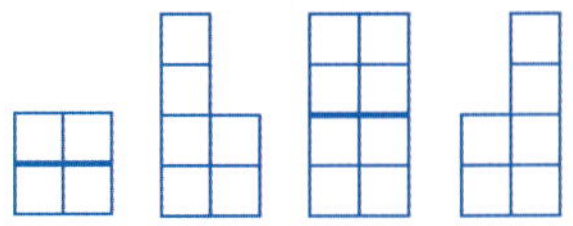

10. 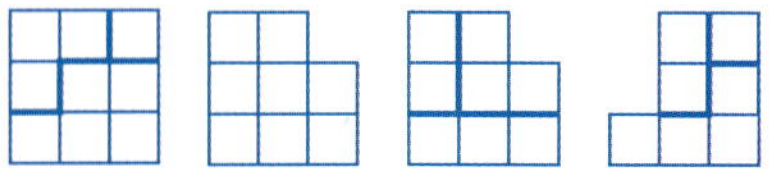

11. 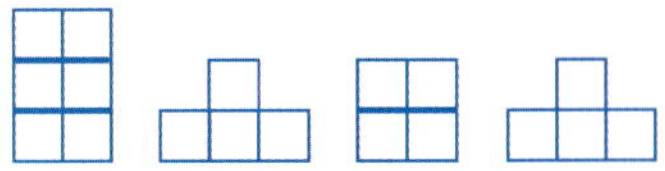

12. 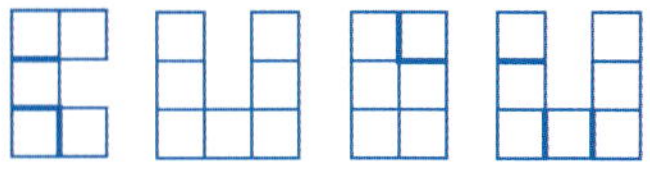

9–12. See margin.

Given the corner view of a figure, sketch the orthogonal drawing. 13–15. See p. 68

13. 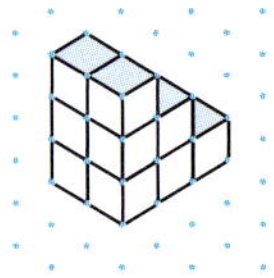

14. 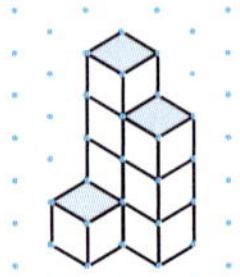

15. 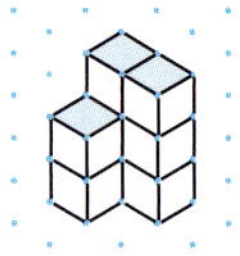

Identify each solid. Name the bases, faces, edges, and vertices.

16–18. See p. 685A.

16. 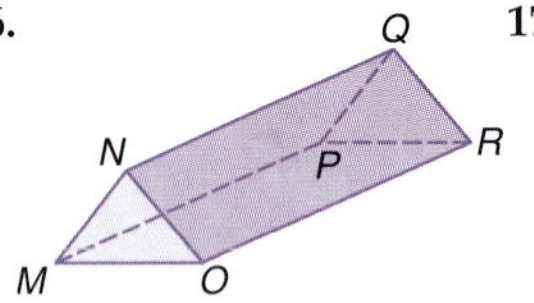

17. 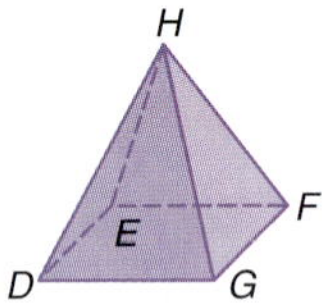

18. 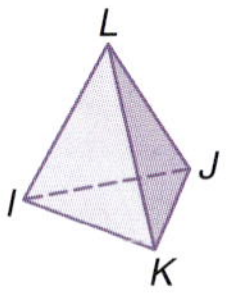

19. cylinder; bases: circles *S* and *T*

19. 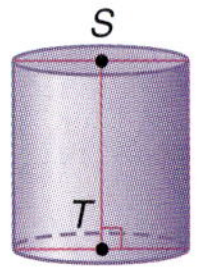

20. 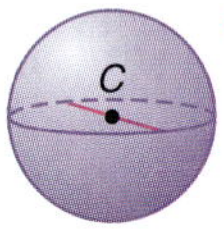

sphere

21. 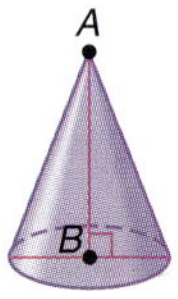

cone; base: circle *B*; vertex: *A*

22. **EULER'S FORMULA** The number of faces *F*, vertices *V*, and edges *E* of a polyhedron are related by Euler's (OY luhrz) formula: $F + V = E + 2$. Determine whether Euler's formula is true for each of the figures in Exercises 16–21. See margin.

SPEAKERS **For Exercises 23 and 24, use the following information.**
The top and front views of a speaker for a stereo system are shown.

23. Is it possible to determine the shape of the speaker? Explain.

23. No, not enough information is provided by the top and front views to determine the shape.

24. Describe possible shapes for the speaker. Draw the left and right views of one of the possible shapes. See p. 685A.

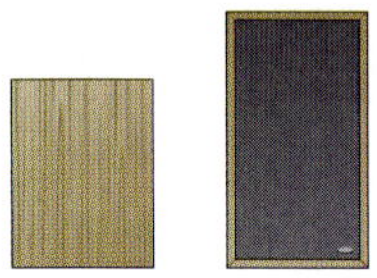

Top View Front View

Determine the shape resulting from each slice of the cone.

25. 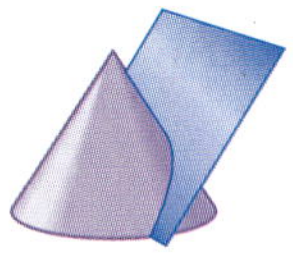
parabola

26. 
triangle

27. 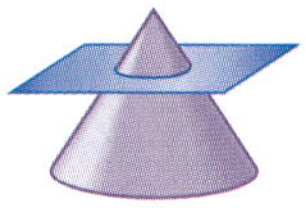
circle

Answers

9. 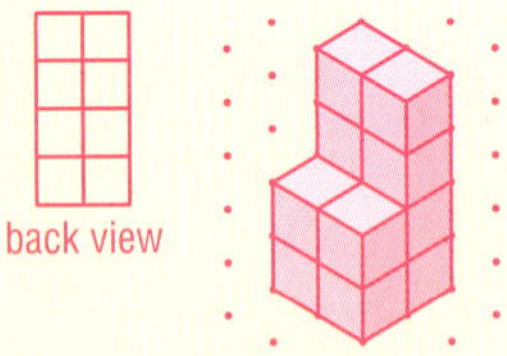

10. 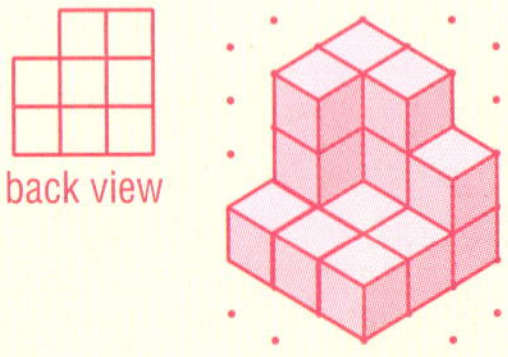

11. 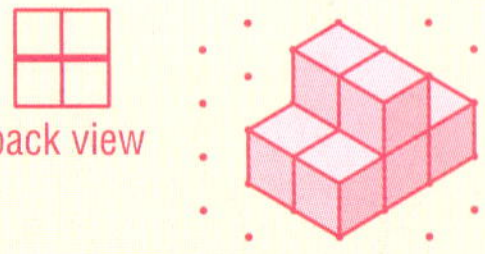

12. 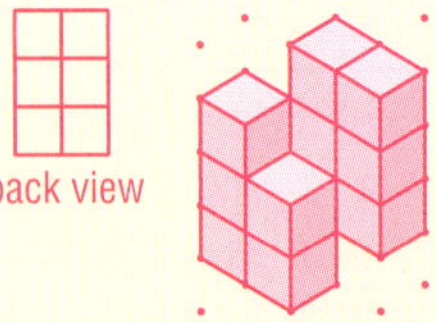

22. Ex. 16: yes, 5 + 6 = 9 + 2; Ex. 17: yes, 5 + 5 = 8 + 2; Ex. 18: yes, 4 + 4 = 6 + 2; Ex. 19–21: No, these figures are not polyhedrons so Euler's formula does not apply.

Determine the shape resulting from each slice of the rectangular prism.

28.

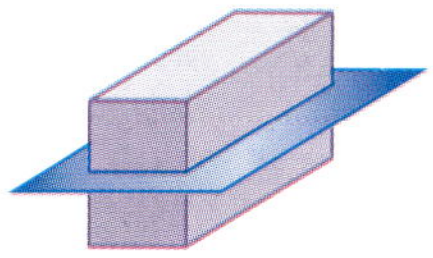

rectangle

29.

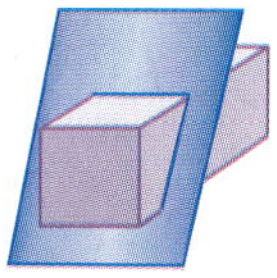

rectangle

30.

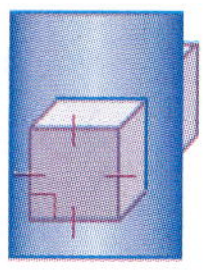

square

Draw a diagram and describe how a plane can slice a tetrahedron to form the following shapes. 31–33. See p. 685A.

★ 31. equilateral triangle ★ 32. isosceles triangle ★ 33. quadrilateral

Describe the solid that results if the number of sides of each base increases infinitely. The bases of each solid are regular polygons inscribed in a circle.

★ 34. pyramid cone ★ 35. prism cylinder

GEMOLOGY For Exercises 36–38, use the following information.

A well-cut diamond enhances the natural beauty of the stone. These cuts are called *facets*.

uncut

emerald cut

round cut

36. Describe the shapes seen in an uncut diamond. triangles, square or rectangle
37. What shapes are seen in the emerald-cut diamond? 37. rectangles, triangles, quadrilaterals
38. List the shapes seen in the round-cut diamond. octagon, triangles, quadrilaterals

For Exercises 39–41, use the following table.

Number of Faces	Prism	Pyramid
4	none	tetrahedron
5	a. ___?___ triangular	square or rectangular
6	b. ___?___	c. ___?___ pentagonal
7	pentagonal	d. ___?___ hexagonal
8	e. ___?___ hexagonal	heptagonal

39b. cube, rectangular, or hexahedron

39. Name the type of prism or pyramid that has the given number of faces.
40. Analyze the information in the table. Is there a pattern between the number of faces and the bases of the corresponding prisms and pyramids?
41. Is it possible to classify a polyhedron given only the number of faces? Explain.

40–41. See p. 685A.

42. **CRITICAL THINKING** Construct a Venn diagram that shows the relationship among polyhedra, Platonic solids, prisms, and pyramids. See margin.

43. **WRITING IN MATH** Answer the question that was posed at the beginning of the lesson. See p. 685A.

Why are drawings of three-dimensional structures valuable to archaeologists?

Include the following in your answer:
- types of two-dimensional models and drawings, and
- the views of a structure used to show three dimensions.

www.geometryonline.com/self_check_quiz/fcat

Answer

42.

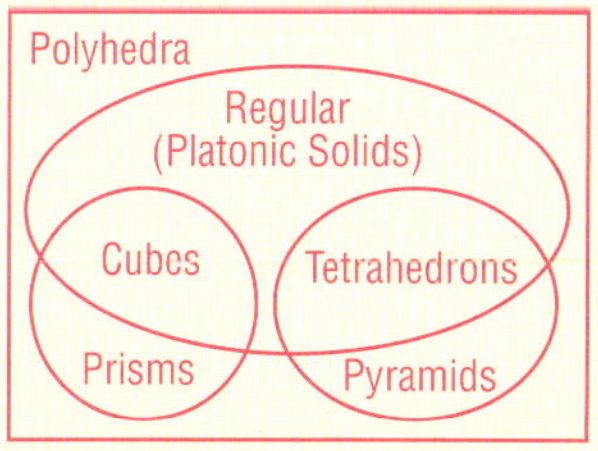

Study Guide and Intervention, p. 661 (shown) and p. 662

Drawings of Three-Dimensional Figures To work with a three-dimensional object, a useful skill is the ability to make an **orthogonal drawing**, which is a set of two-dimensional drawings of the different sides of the object. For a square pyramid, you would show the top view, the left view, the front view, and the right view.

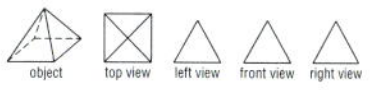

Example **Draw the back view of the figure given the orthogonal drawing.**

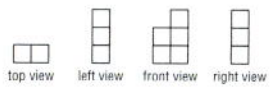

- The top view indicates two columns.
- The left view indicates that the height of figure is three blocks.
- The front view indicates that the columns have heights 2 and 3 blocks.
- The right view indicates that the height of the figure is three blocks.

Use blocks to make a model of the object. Then use your model to draw the back view. The back view indicates that the columns have heights 3 and 2 blocks.

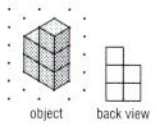

Exercises

Draw the back view and corner view of a figure given each orthogonal drawing.

1.

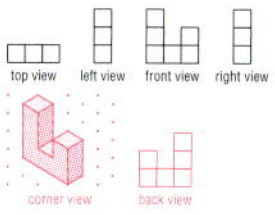

2.

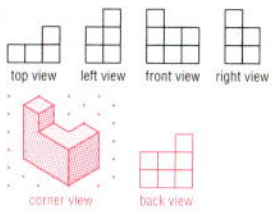

3.

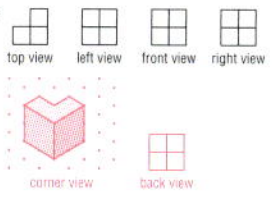

4.

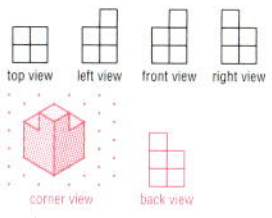

Skills Practice, p. 663 and Practice, p. 664 (shown)

Draw the back view and corner view of a figure given each orthogonal drawing.

1.

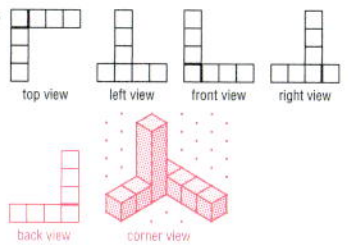

2.

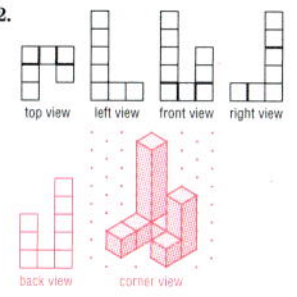

Identify each solid. Name the bases, faces, edges, and vertices.

3.

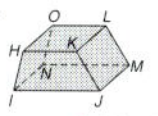

trapezoidal prism; bases: trapezoid *HIJK*, trapezoid *LMNO*; faces: trapezoid *HIJK*, trapezoid *LMNO*, ▱*HKLO*, ▱*HINO*, ▱*IJMN*, ▱*JKLM*; edges: $\overline{HI}$, $\overline{IJ}$, $\overline{JK}$, $\overline{HK}$, $\overline{HO}$, $\overline{ON}$, $\overline{IN}$, $\overline{KL}$, $\overline{LO}$, $\overline{LM}$, $\overline{JM}$, $\overline{MN}$; vertices: *H*, *I*, *J*, *K*, *L*, *M*, *N*, and *O*

4.

cylinder; bases: circle *T*, circle *Z*

5. **MINERALS** Pyrite, also known as fool's gold, can form crystals that are perfect cubes. Suppose a gemologist wants to cut a cube of pyrite to get a square and a rectangular face. What cuts should be made to get each of the shapes? Illustrate your answers.

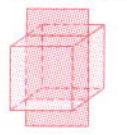

a cut parallel to the bases to get a square

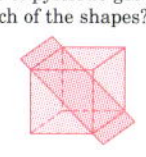

a cut through diagonally opposite top and bottom edges to get a rectangle

Reading to Learn Mathematics, p. 665 ELL

Pre-Activity **Why are drawings of three-dimensional structures valuable to archeologists?**

Read the introduction to Lesson 12-1 at the top of page 636 in your textbook.

Why do you think archeologists would want to know the details of the sizes and shapes of ancient three-dimensional structures? Sample answer: This information might provide clues about construction methods and tools and help archeologists trace advances in construction methods over time.

Reading the Lesson

1. Match each description from the first column with one of the terms from the second column. (Some of the terms may be used more than once or not at all.)

a. a polyhedron with two parallel congruent bases v	i. octahedron
b. the set of points in space that are a given distance from a given point viii	ii. face
c. a regular polyhedron with eight faces i	iii. icosahedron
d. a polyhedron that has all faces but one intersecting at one point xi	iv. edge
e. a line segment where two faces of a polyhedron intersect iv	v. prism
f. a solid with congruent circular bases in a pair of parallel planes vii	vi. dodecahedron
g. a regular polyhedron whose faces are squares x	vii. cylinder
h. a flat surface of a polyhedron ii	viii. sphere
	ix. cone
	x. hexahedron
	xi. pyramid
	xii. tetrahedron

2. Fill in the missing numbers, words, or phrases to complete each sentence.
   a. A triangular prism has __6__ vertices. It has __5__ faces: __2__ bases are congruent __triangles__, and __3__ faces are parallelograms.
   b. A regular octahedron has __6__ vertices and __8__ faces. Each face is a(n) __equilateral triangle__.
   c. A hexagonal prism has __12__ vertices. It has __8__ faces: __2__ of them are the bases, which are congruent __hexagons__, and the other __6__ faces are parallelograms.
   d. An octagonal pyramid has __9__ vertices and __9__ faces. The base is a(n) __octagon__, and the other __8__ faces are __triangles__.
   e. There are exactly __5__ types of regular polyhedra. These are called the __Platonic__ solids. A polyhedron whose faces are regular pentagons is a __dodecahedron__, which has __12__ faces.

Helping You Remember

3. A good way to remember the characteristics of geometric solids is to think about how different solids are alike. Name a way in which pyramids and cones are alike. Sample answer: They just have one base and the surfaces that are not the base meet at one point (the vertex).

Enrichment, p. 666

Drawing Solids on Isometric Dot Paper

Isometric dot paper is helpful for drawing solids. Remember to use dashed lines for hidden edges.

For each solid shown, draw another solid whose dimensions are twice as large.

1.

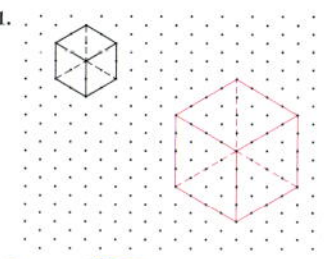

2.

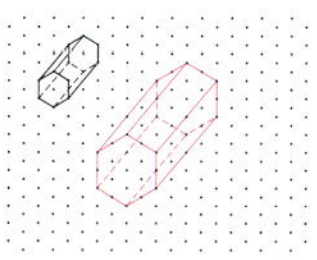

4 Assess

Open-Ended Assessment

Speaking Have students describe how to find the faces, edges, and vertices of a prism.

Getting Ready for Lesson 12-2

Prerequisite Skill Students will learn how to find the surface areas of solids in Lesson 12-2. They will make nets of triangles, circles, and rectangles and find the areas of the shapes. Use Exercises 59–62 to determine your students' familiarity with finding the area of a rectangle.

FCAT Practice
Standardized Test Practice

44. All of the following can be formed by the intersection of a cube and a plane *except*

Ⓐ a triangle. Ⓑ a rectangle. Ⓒ a point. Ⓓ a circle. **D**

45. ALGEBRA For which of the following values of x is $\frac{x^3}{x^4}$ the least? **D**

Ⓐ -4 Ⓑ -3 Ⓒ -2 Ⓓ -1

Extending the Lesson

SYMMETRY AND SOLIDS In a two-dimensional plane, figures are symmetric with respect to a line or a point. In three-dimensional space, solids are symmetric with respect to a plane. A square pyramid has four planes of symmetry. Two pass through the height and each vertex of the base, and two pass through the height and the center of each edge of the base.

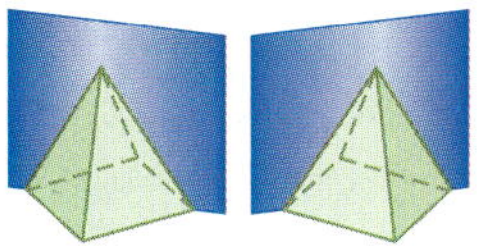

For each solid, determine the number of planes of symmetry.

46. tetrahedron **6** **47.** cylinder **infinite** **48.** sphere **infinite**

Maintain Your Skills

Mixed Review

SURVEYS **For Exercises 49–52, use the following information.**
The results of a restaurant survey are shown in the circle graph with the measurement of each central angle. Each customer was asked to choose a favorite entrée. If a customer is chosen at random, find the probability of each response. *(Lesson 11-5)*

Favorite Entrée

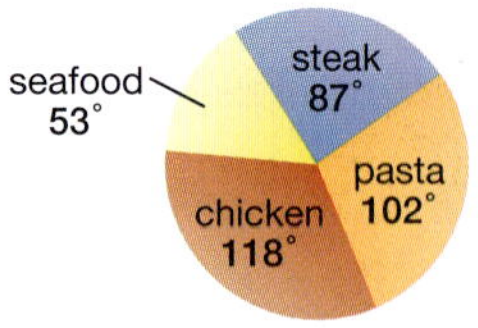

49. steak **0.242** **50.** not seafood **0.853**

51. either pasta or chicken **0.611** **52.** neither pasta nor steak **0.475**

COORDINATE GEOMETRY **The coordinates of the vertices of an irregular figure are given. Find the area of each figure.** *(Lesson 11-4)*

53. $A(1, 4), B(4, 1), C(1, -2), D(-3, 1)$ **21 units²**

54. $F(-2, -4), G(-2, -1), H(1, 1), J(4, 1), K(6, -4)$ **32 units²**

55. $L(-2, 2), M(0, 1), N(0, -2), P(-4, -2)$ **11 units²**

Find the perimeter and area of each parallelogram. Round to the nearest tenth if necessary. *(Lesson 11-1)*

56.

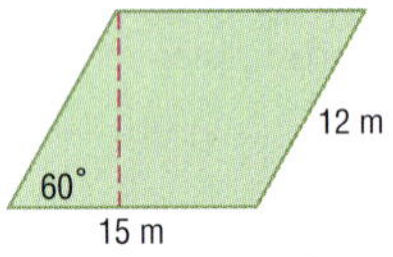

57.

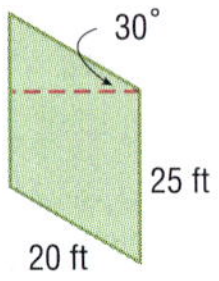

58.

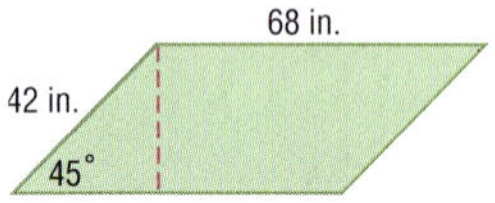

54 m, 155.9 m² **90 ft, 433.0 ft²** **220 in., 2019.5 in²**

Getting Ready for the Next Lesson

PREREQUISITE SKILL **Find the area of each rectangle. Round to the nearest tenth if necessary.** *(To review **finding the area of a rectangle**, see pages 732–733.)*

59.

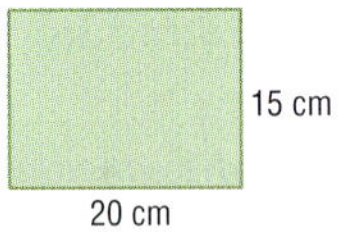

60.

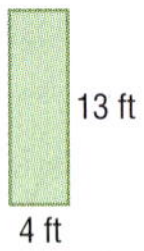

61.

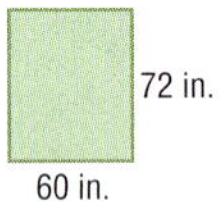

62.

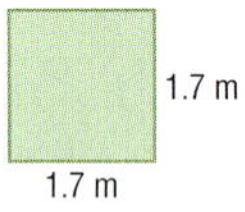

300 cm² **52 ft²** **4320 in²** **2.9 m²**

642 Chapter 12 Surface Area

12-2 Nets and Surface Area

What You'll Learn

- Draw two-dimensional models for three-dimensional figures.
- Find surface area.

Sunshine State Standards
MA.B.1.4.1

Vocabulary

- net
- surface area

Why is surface area important to car manufacturers?

Have you wondered why cars have evolved from boxy shapes to sleeker shapes with rounded edges? Car manufacturers use aerodynamics, or the study of wind resistance, and the shapes of surfaces to design cars that are faster and more efficient.

MODELS FOR THREE-DIMENSIONAL FIGURES You have used isometric dot paper to draw corner views of solids given the orthogonal view. In this lesson, isometric dot paper will be used to draw two-dimensional models of geometric solids.

Study Tip

Isometric Dot Paper
Note that right angles of the prism are 60° and 120° angles on isometric dot paper. This is to show perspective.

Example 1 Draw a Solid

Sketch a rectangular prism 2 units high, 5 units long, and 3 units wide using isometric dot paper.

Step 1 Draw the corner of the solid; 2 units down, 5 units to the left, and 3 units to the right.

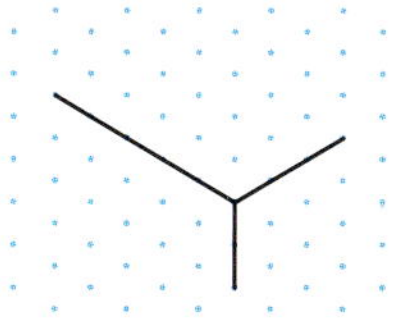

Step 2 Draw a parallelogram for the top of the solid.

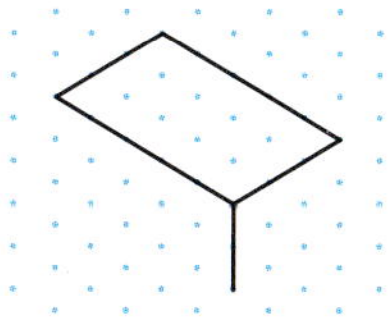

Step 3 Draw segments 2 units down from each vertex for the vertical edges.

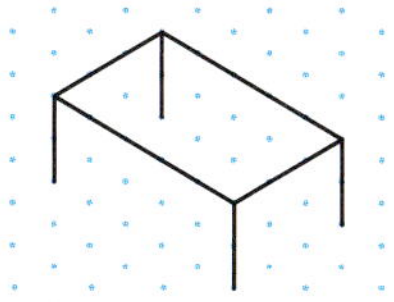

Step 4 Connect the corresponding vertices. Use dashed lines for the hidden edges. Shade the top of the solid.

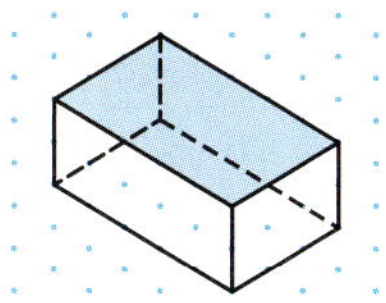

12-2 Lesson Notes

1 Focus

5-Minute Check Transparency 12-2 Use as a quiz or review of Lesson 12-1.

Mathematical Background notes are available for this lesson on p. 634C.

Why is surface area important to car manufacturers?

Ask students:

- Why have cars evolved from boxy shapes into their sleek designs? **so that they are faster and more efficient**
- Why is surface area important to designing a car? **Sample answer: The surface area is the outside of the car. It is the part that "shows off" the design.**

Resource Manager

Workbook and Reproducible Masters

Chapter 12 Resource Masters

- Study Guide and Intervention, pp. 667–668
- Skills Practice, p. 669
- Practice, p. 670
- Reading to Learn Mathematics, p. 671
- Enrichment, p. 672
- Assessment, p. 717

Teaching Geometry With Manipulatives Masters, pp. 6, 7, 9–13, 17, 194

Transparencies

5-Minute Check Transparency 12-2
Answer Key Transparencies

Technology

Interactive Chalkboard

2 Teach

MODELS FOR THREE-DIMENSIONAL FIGURES

Tips for New Teachers Watch for students who have difficulty visualizing three-dimensional models. Have models available for them to examine and manipulate. Have them practice drawing the faces of the solids by looking directly at a model of the solid.

In-Class Examples

Power Point®

Teaching Tip Point out that when three-dimensional figures are modeled using isometric dot paper, students should never need to draw a horizontal line.

1 Sketch a rectangular prism 4 units long, 3 units wide, and 2 units high using isometric dot paper.

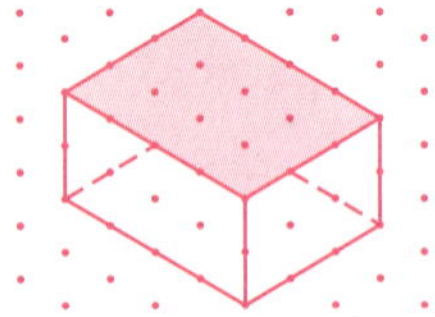

2 Which net could be folded into a triangular prism if folds are made only along dotted lines? **D**

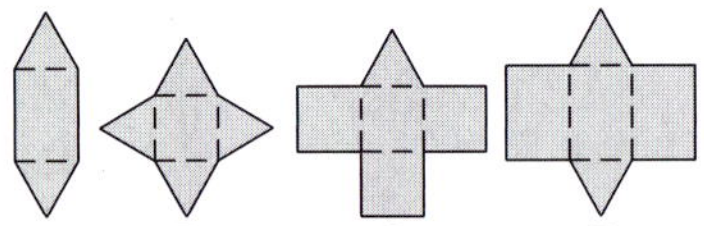

A B C D

If you cut a cardboard box at the edges and lay it flat, you will have a pattern, or **net**, for the three-dimensional solid. Nets can be made for most solid figures. This net is a pattern for the cube. It can be folded into the shape of the cube without any overlap.

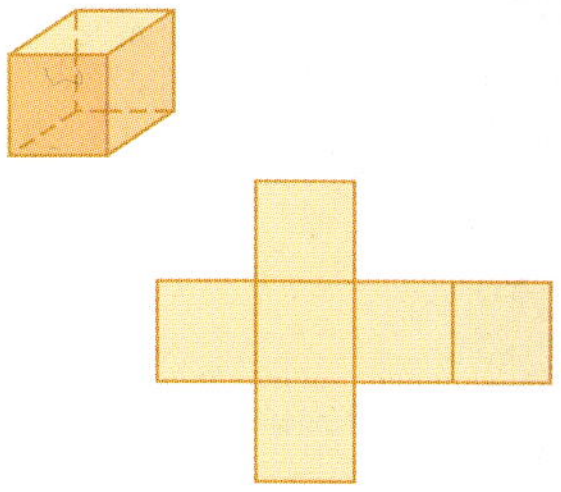

FCAT Practice

Standardized Test Practice Ⓐ Ⓑ Ⓒ Ⓓ

The Princeton Review

Test-Taking Tip

Nets One figure may have several different nets that represent the shape.

Example 2 Nets for a Solid

Multiple-Choice Test Item

Which net could be folded into a pyramid if folds are made only along the dotted lines?

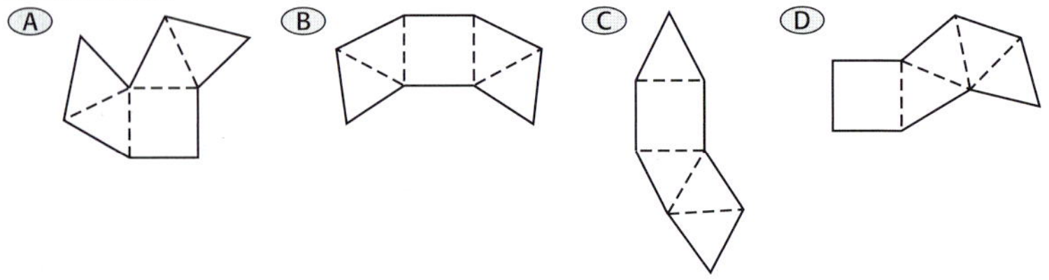

Read the Test Item

You are given four nets, only one of which can be folded into a pyramid.

Solve the Test Item

Each of the answer choices has one square and four triangles. So the square is the base of the pyramid. Each triangle in the sketch represents a face of the pyramid. The faces must meet at a point and cannot overlap. Analyze each answer choice carefully.

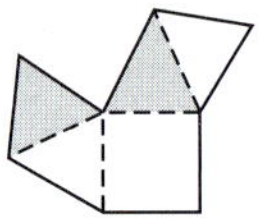

This net has overlapping triangles.

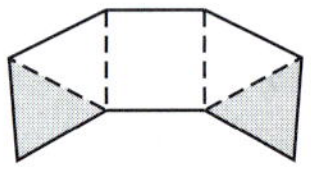

This net also has two triangles that overlap.

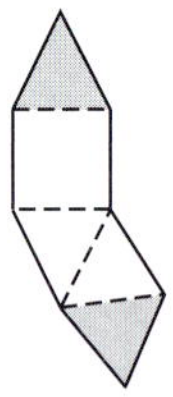

This also has overlapping triangles.

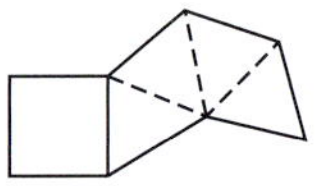

None of the triangles overlap. Each face of the pyramid is represented. This choice is correct.

The answer is D.

SURFACE AREA Nets are very useful in visualizing the polygons that make up the surface of the solid. A net for tetrahedron $QRST$ is shown at the right. The **surface area** is the sum of the areas of each face of the solid. Add the areas of $\triangle QRT$, $\triangle QTS$, $\triangle QRS$, and $\triangle RST$ to find the surface area of tetrahedron $QRST$.

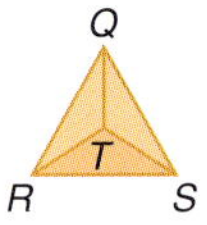

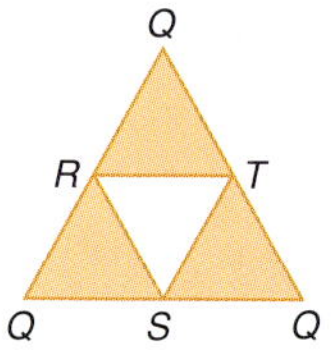

DAILY INTERVENTION

Differentiated Instruction

Visual/Spatial Use a pair of scissors and grid paper to show students how to create nets of solids. Draw the net on the grid paper, cut it out, and fold or glue it together to show students the solid.

Study Tip

Drawing Nets
It is helpful to let each unit of the dot paper represent one unit of measure. When the numbers are large, let each unit of dot paper represent two units of measure.

Example 3 Nets and Surface Area

a. Draw a net for the right triangular prism shown.

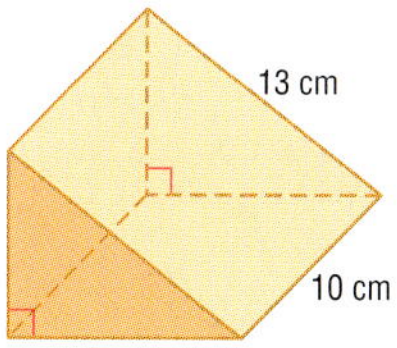

We need to know the height of the triangular base. Use the Pythagorean Theorem to find the height of the triangle.

$13^2 = 12^2 + h^2$ Pythagorean Theorem
$169 = 144 + h^2$ Simplify.
$25 = h^2$ Subtract 144 from each side.
$5 = h$ Take the square root of each side.

Use rectangular dot paper to draw a net. Let one unit on the dot paper represent 2 centimeters.

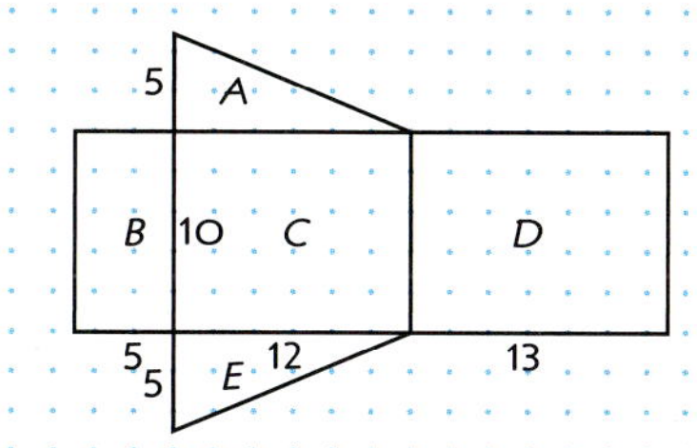

b. Use the net to find the surface area of the triangular prism.

To find the surface area of the prism, add the areas of the three rectangles and the two triangles.

Write an equation for the surface area.

$$\text{Surface area} = B + C + D + A + E$$
$$= 10 \cdot 5 + 10 \cdot 12 + 10 \cdot 13 + \frac{5 \cdot 12}{2} + \frac{5 \cdot 12}{2}$$
$$= 50 + 120 + 130 + 30 + 30 \text{ or } 360$$

The surface area of the right triangular prism is 360 square centimeters.

SURFACE AREA

Building on Prior Knowledge

In Chapter 11, students learned how to find the areas of polygons and circles. Point out that the surface area of a solid is the sum of the areas of each face of the solid.

In-Class Example

Power Point®

Teaching Tip Make sure students understand that rectangular dot paper is used to draw a net because each face can be represented by the squares of rectangular dot paper.

3

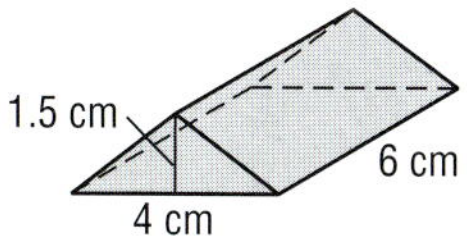

a. Draw a net for the triangular prism.

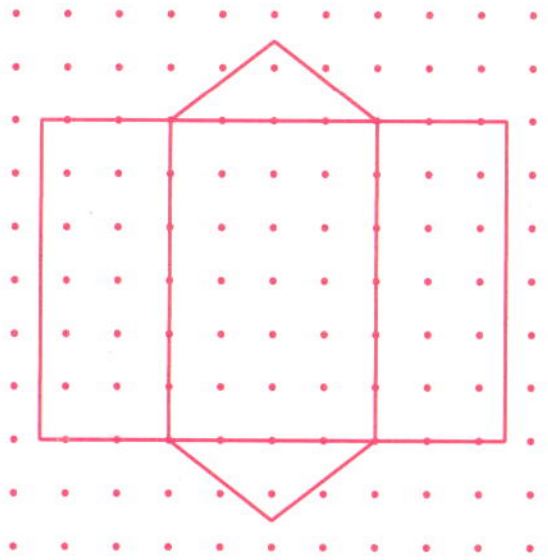

b. Use the net to find the surface area of the prism. **60 cm²**

Check for Understanding

Concept Check

1. **OPEN ENDED** Draw a net for a cube different from the one on page 644.
2. **Compare and contrast** isometric dot paper and rectangular dot paper. When is each type of paper useful?

1–2. See margin.

Guided Practice **Sketch each solid using isometric dot paper. 3–4. See margin.**

3. rectangular prism 4 units high, 2 units long, and 3 units wide
4. cube 2 units on each edge

GUIDED PRACTICE KEY

Exercises	Examples
3–4	1
5–7	3
8	2

For each solid, draw a net and find the surface area. 5–7. See p. 685A for nets.

5. 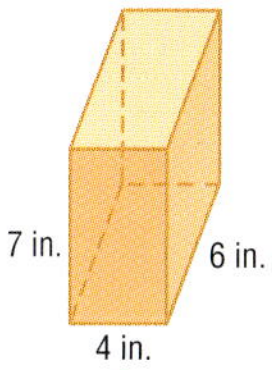

188 in²

6. 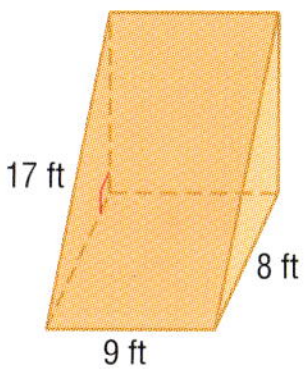

480 ft²

7. 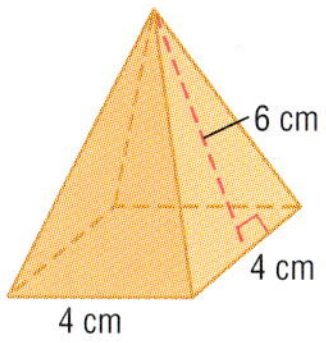

64 cm²

FCAT Practice
Standardized Test Practice

8. Which shape *cannot* be folded to make a pyramid? **C**

(A) 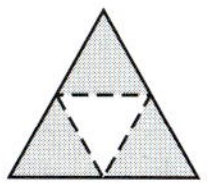
(B) 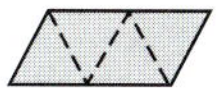
(C) 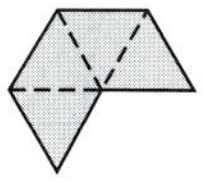
(D) 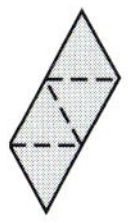

www.geometryonline.com/extra_examples/fcat

Answers

1. Sample answer:

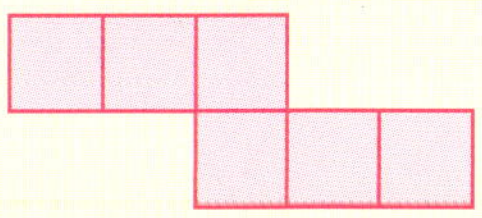

2. On isometric dot paper, the dots are arranged in triangles, which aid in drawing three-dimensional objects. On rectangular dot paper, the dots are arranged in squares, which aid in drawing the nets and orthogonal views of three-dimensional objects.

3. 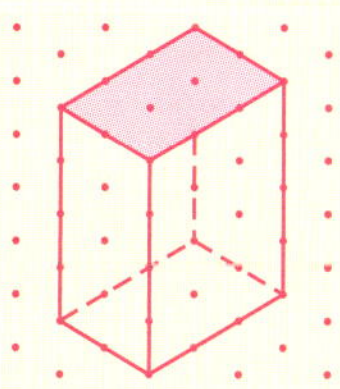

4. 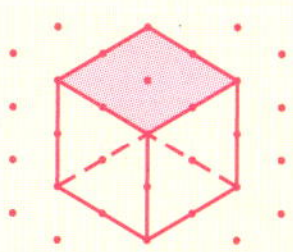

Lesson 12-2 Nets and Surface Area 645

3 Practice/Apply

Study Notebook

Have students—
- add the definitions/examples of the vocabulary terms to their Vocabulary Builder worksheets for Chapter 12.
- include sample nets for various solids.
- include any other item(s) that they find helpful in mastering the skills in this lesson.

About the Exercises...

Organization by Objective
- **Models for Three-Dimensional Figures:** 9–27, 35–38
- **Surface Area:** 28–34

Odd/Even Assignments

Exercises 9–23 and 25–34 are structured so that students practice the same concepts whether they are assigned odd or even problems.

Assignment Guide

Basic: 9–23 odd, 29, 33–39 odd, 40–53

Average: 9–39 odd, 40–53

Advanced: 10–38 even, 39–49 (optional: 50–53)

★ indicates increased difficulty

Practice and Apply

Homework Help

For Exercises	See Examples
9–14, 25–27	1
15–24, 35–38	2
28–34	3

Extra Practice See page 778.

Sketch each solid using isometric dot paper. 9–14. See pp. 685A–685B.

9. rectangular prism 3 units high, 4 units long, and 5 units wide
10. cube 5 units on each edge
11. cube 4 units on edge
12. rectangular prism 6 units high, 6 units long, and 3 units wide
13. triangular prism 4 units high, with bases that are right triangles with legs 5 units and 4 units long
14. triangular prism 2 units high, with bases that are right triangles with legs 3 units and 7 units long

For each solid, draw a net and find the surface area. Round to the nearest tenth if necessary. 15–23. See pp. 685B–685C for nets.

15. 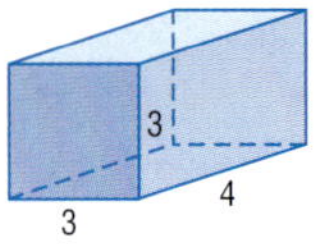

66 units^2

16. 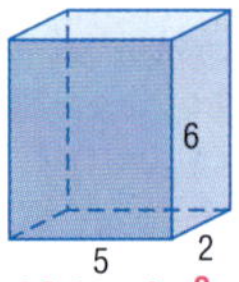

104 units^2

17. 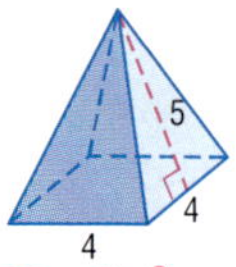

56 units^2

18. 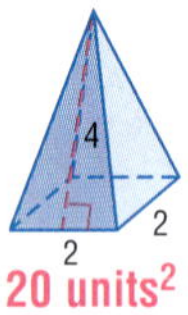

20 units^2

19. 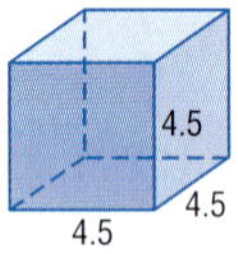

121.5 units^2

20. 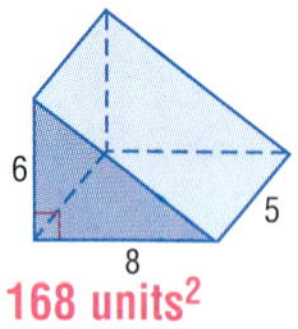

168 units^2

21. 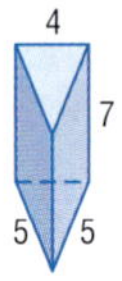

116.3 units^2

22. 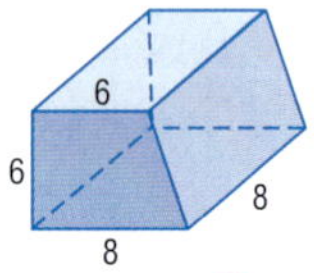

294.6 units^2

23. 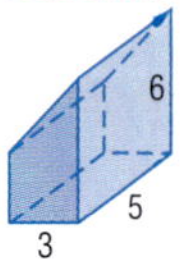

108.2 units^2

24. **FOOD** In 1999, Marks & Spencer, a British grocery store, created the biggest sandwich ever made. The tuna and cucumber sandwich was in the form of a triangular prism. Suppose each slice of bread was 8 inches thick. Draw a net of the sandwich, and find the surface area in square feet to the nearest tenth. 107.5 ft^2; See margin for net.

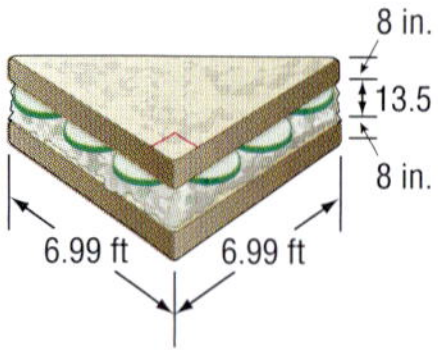

Online Research **Data Update** Are there records for other types of sandwiches? Visit www.geometryonline.com/data_update to learn more.

Given the net of a solid, use isometric dot paper to draw the solid. 25–27. See margin.

★ 25. 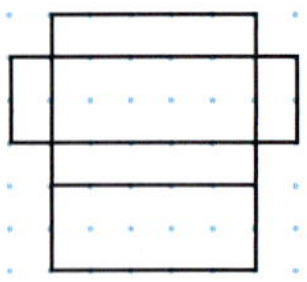

★ 26. 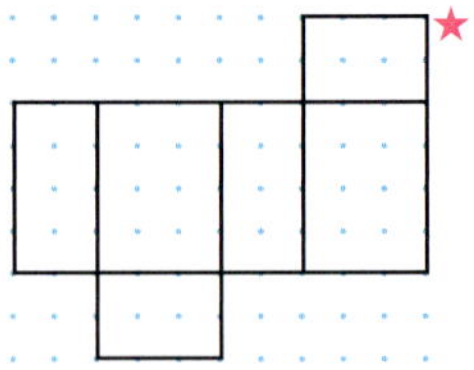

★ 27. 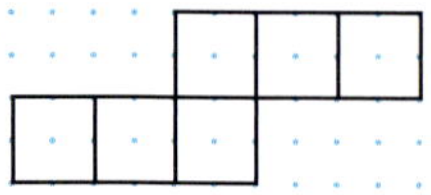

Answers

24. 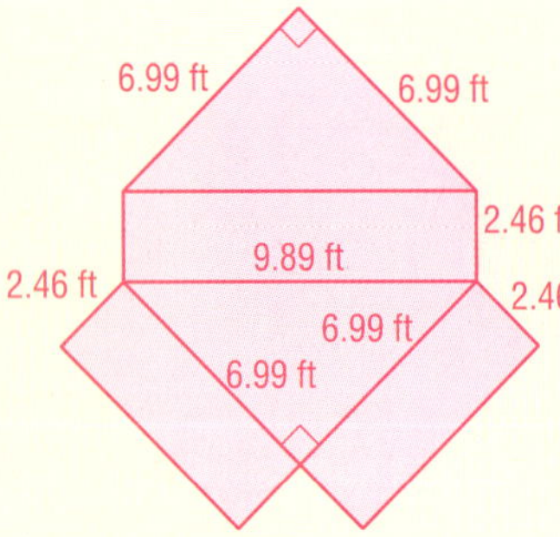

25. 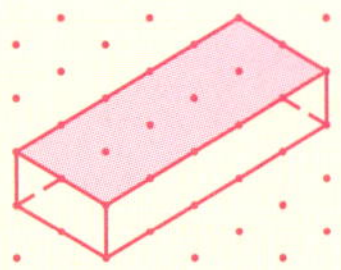

26. 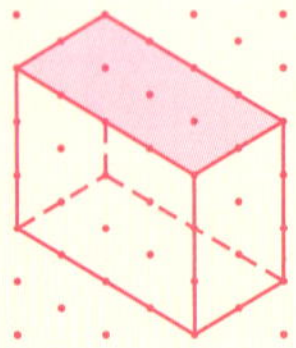

27. 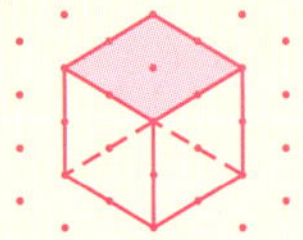

32. 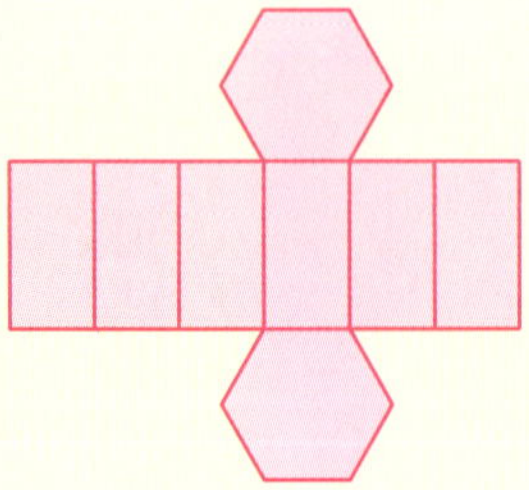

33. 

646 Chapter 12 Surface Area

Given each polyhedron, copy its net and label the remaining vertices.

28.–31.

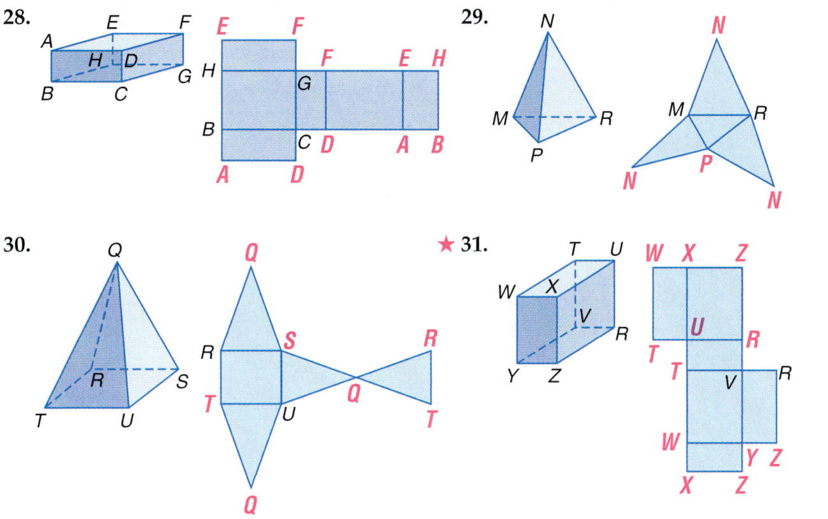

GEOLOGY For Exercises 32–34, use the following information.
Many minerals have a crystalline structure. The forms of three minerals are shown below. Draw a net of each crystal. **32–34. See margin.**

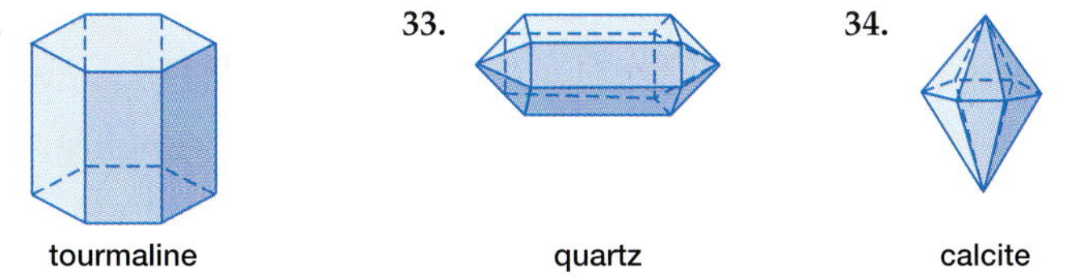

More About. . .

Geology

There are 32 different classes of crystals, each with a different type of symmetry.

Source: www.infoplease.com

VARYING DIMENSIONS For Exercises 35–38, use Figures A, B, and C.

35. Draw a net for each solid and find its surface area. **See p. 685C for nets.**

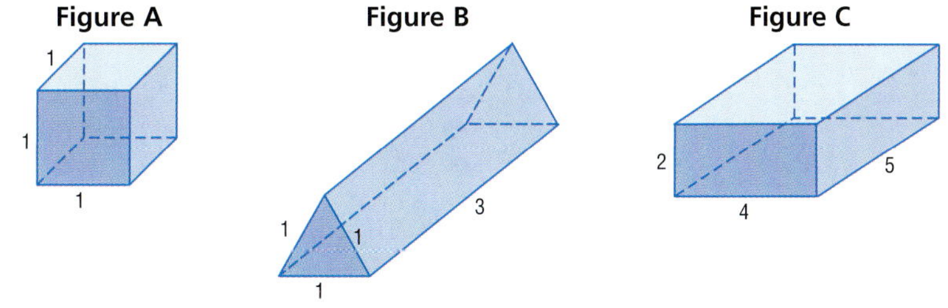

36. Double the dimensions of each figure. Find the surface areas.
37. How does the surface area change when the dimensions are doubled? Explain.
38. Make a conjecture about the surface area of a solid whose dimensions have been tripled. Check your conjecture by finding the surface area.

35. A, 6 units^2; B, $\left(9 + \frac{\sqrt{3}}{2}\right)$ units^2; C, 76 units^2
36. A, 24 units^2; B, $(36 + 2\sqrt{3})$ units^2; C, 76 units^2
37–38. See p. 685C.

39. **CRITICAL THINKING** Many board games use a standard die like the one shown. The sum of the number of dots on each pair of opposite faces is 7. Determine whether the net represents a standard die. Explain. **No; 5 and 3 are opposite faces; the sum is 8.**

www.geometryonline.com/self_check_quiz/fcat

34.

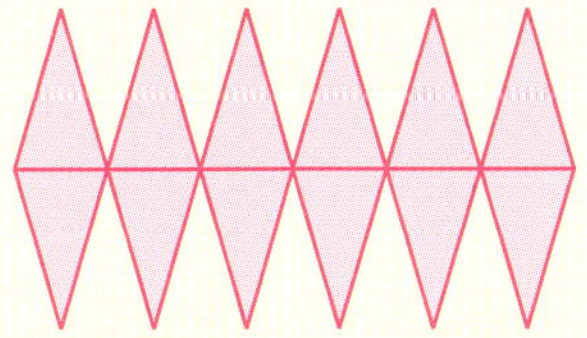

Enrichment, p. 672

Pyramids

1. On a sheet of paper, draw the figure at the right so that the measurements are accurate. Cut out the shape and fold it to make a pyramid.
See students' work.

2. Measure the height from the highest point straight down (perpendicular) to the base. What is the height of the pyramid?
12 cm

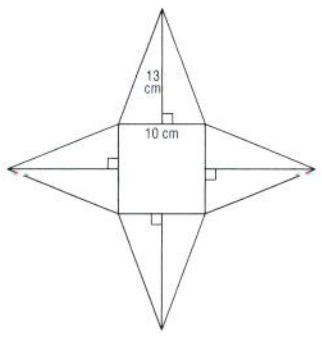

The drawing at the right shows the pyramid you made. ... following

Study Guide and Intervention, p. 667 (shown) and p. 668

Models for Three-Dimensional Figures One way to relate a three-dimensional figure and a two-dimensional drawing is to use isometric dot paper. Another way is to make a flat pattern, called a *net*, for the surfaces of a solid.

Example 1 Use isometric dot paper to sketch a triangular prism with 3-4-5 right triangles as bases and with a height of 3 units.
Step 1 Draw $\overline{AB}$ at 3 units and draw $\overline{AC}$ at 4 units.
Step 2 Draw $\overline{AD}$, $\overline{BE}$, and $\overline{CF}$, each at 3 units.
Step 3 Draw $\overline{BC}$ and $\triangle DEF$.

Example 2 Match the net at the right with one of the solids below.

The six squares of the net can be folded into a cube. The net represents solid c.

Exercises

Sketch each solid using isometric dot paper.
1. cube with edge 4
2. rectangular prism 1 unit high, 5 units long, and 4 units wide

Draw a net for each solid.
3.
4.
5.
6.

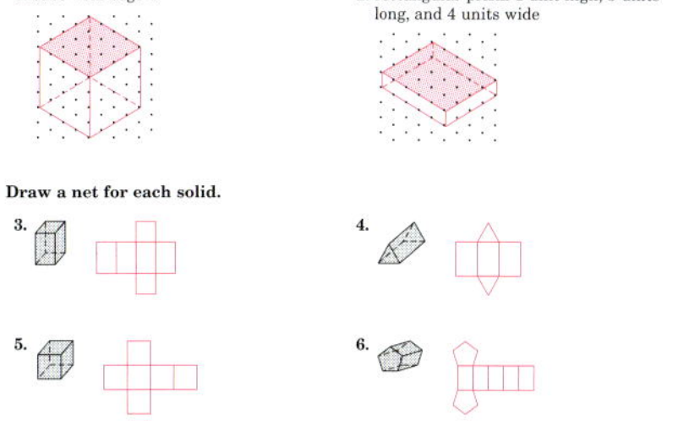

Skills Practice, p. 669 and Practice, p. 670 (shown)

Sketch each solid using isometric dot paper.
1. rectangular prism 3 units high, 3 units long, and 2 units wide
2. triangular prism 3 units high, whose bases are right triangles with legs 2 units and 4 units long

3. For the solid, draw a net and find the surface area.
Sample net: 342 units^2

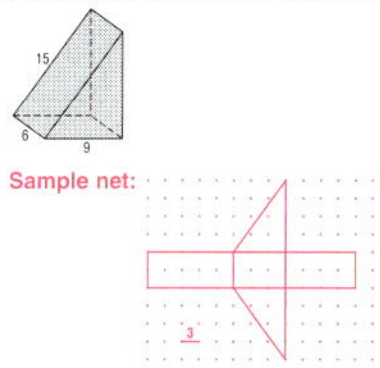

4. **SHIPPING** Rawanda needs to wrap a package to ship to her aunt. The rectangular package measures 2 inches high, 10 inches long, and 4 inches wide. Draw a net of the package. How much wrapping paper does Rawanda need to cover the package?
Sample net: 136 in^2

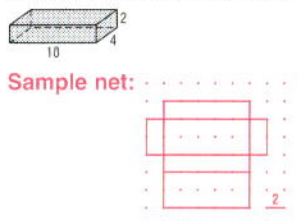

Reading to Learn Mathematics, p. 671 ELL

Pre-Activity Why is surface area important to car manufacturers?
Read the introduction to Lesson 12-2 at the top of page 643 in your textbook.
Suppose you are a passenger in a car that is moving at 50 miles per hour on the highway. If you hold your right hand just outside the right window, in what position does your hand encounter the greatest air resistance? palm of the hand open, with the air hitting the palm at a 90° angle

Reading the Lesson
1. Name the solid that can be formed by folding each net without any overlap.

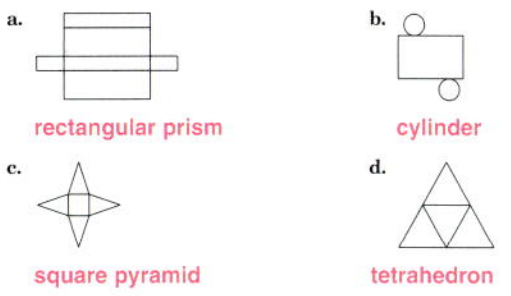

a. rectangular prism
b. cylinder
c. square pyramid
d. tetrahedron

2. Supply the missing number, word, or phrase to make a true statement. Be as specific as possible.
a. To find the surface area of a cube, add the areas of 6 congruent squares.
b. To find the surface area of a cylinder, add the areas of two congruent circles and one rectangle.
c. To find the surface area of a dodecahedron, add the areas of 12 congruent regular pentagons.
d. To find the surface area of an icosahedron, add the areas of 20 congruent equilateral triangles.

Helping You Remember
3. A good way to remember a mathematical concept is to think about how the concept relates to everyday life. How could you explain the concept of *surface area* using an everyday situation?
Sample answer: The surface area of a solid shape tells you how much wrapping paper you would need to wrap a present with that shape and size.

4 Assess

Open-Ended Assessment

Writing Ask students to define *net* and then describe how to find the surface area of a solid by using a net.

Getting Ready for Lesson 12-3

Prerequisite Skill Students will find the surface areas of prisms in Lesson 12-3. They use the areas of parallelograms, triangles, and other polygons to find the areas of prisms. Use Exercises 50–53 to determine your students' familiarity with finding the areas of parallelograms, triangles, and trapezoids.

Assessment Options

Quiz (Lessons 1 and 2) is available on p. 717 of the *Chapter 12 Resource Masters*.

Answers

40. Sample answer: Car manufacturers want their cars to be as fuel efficient as possible. If the car is designed so the front grill and windshield have a smaller surface area, the car meets less resistance from the wind. Answers should include the following.
   - A small compact car has less surface facing the wind than a larger truck, so smaller sedans tend to be more efficient than larger vehicles.
   - Of the two-dimensional models studied in this chapter, orthogonal drawings would be helpful to the designers.

40. WRITING IN MATH Answer the question that was posed at the beginning of the lesson. See margin.

Why is surface area important to car manufacturers?

Include the following in your answer:
- compare the surface area of a subcompact car and a large truck, and
- explain which two-dimensional models of cars would be helpful to designers.

FCAT Practice
Standardized Test Practice

41. Which shape could be folded into a rectangular prism if folds are made only along the dotted lines? C

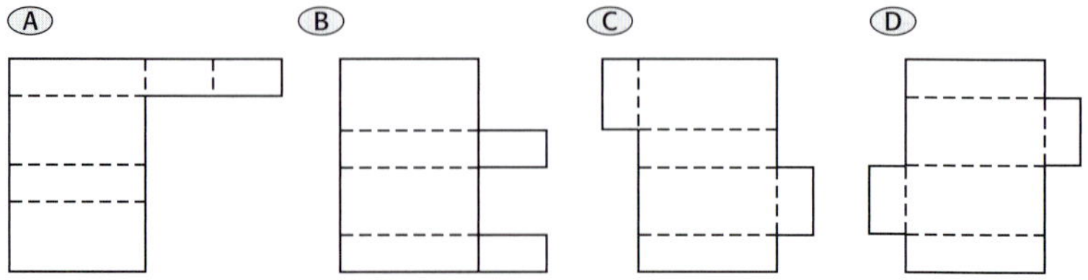

42. **ALGEBRA** What is the complete factorization of $16a^3 - 54b^3$? B

(A) $(2a - 3b)(4a^2 + 6ab + 9b^2)$ (B) $2(2a - 3b)(4a^2 + 6ab + 9b^2)$

(C) $2(2a - 3b)(4a^2 - 6ab + 9b^2)$ (D) $2(2a + 3b)(4a^2 + 6ab + 9b^2)$

Maintain Your Skills

Mixed Review **Determine the shape resulting from each slice of the triangular prism.** *(Lesson 12-1)*

43.

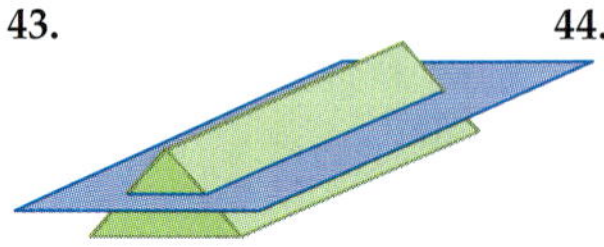

rectangle

44.

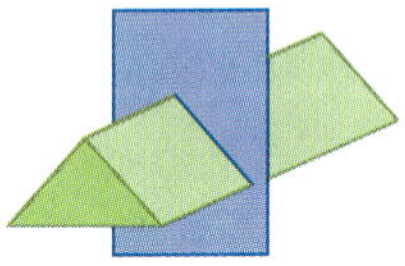

triangle

45.

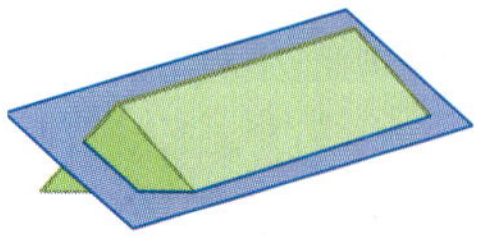

rectangle

46. **PROBABILITY** A rectangular garden is 100 feet long and 200 feet wide and includes a square flower bed that is 20 feet on each side. Find the probability that a butterfly in the garden is somewhere in the flower bed. *(Lesson 11-5)* 0.02

Equilateral hexagon *FGHJKL* is inscribed in ⊙*M*. Find each measure. *(Lesson 10-4)*

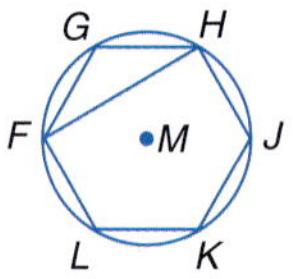

47. $m\angle FHJ$ 90

48. $m\overparen{LK}$ 60

49. $m\angle LFG$ 120

Getting Ready for the Next Lesson **PREREQUISITE SKILL** **Find the area of each figure.**
*(To review finding **areas of parallelograms, triangles, and trapezoids**, see Lessons 11-1 and 11-2.)*

50.

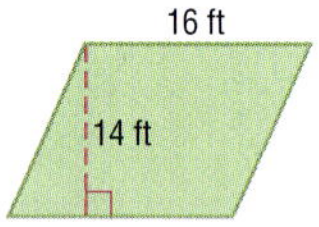

224 ft^2

51.

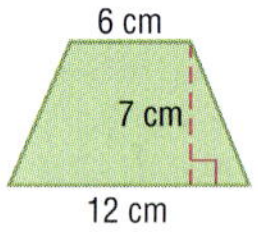

63 cm^2

52.

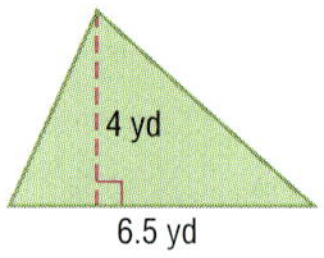

13 yd^2

53.

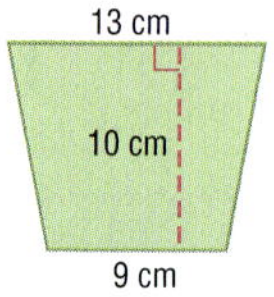

110 cm^2

12-3 Surface Areas of Prisms

Sunshine State Standards
MA.B.1.4.1

What You'll Learn

- Find lateral areas of prisms.
- Find surface areas of prisms.

How do brick masons know how many bricks to order for a project?

The owner of a house wants to build a new unattached brick garage. The sides of the garage will be brick. The brick mason has to estimate the number of bricks needed to complete the project.

Vocabulary

- lateral faces
- lateral edges
- right prism
- oblique prism
- lateral area

LATERAL AREAS OF PRISMS Most buildings are prisms or combinations of prisms. The garage shown above could be separated into a rectangular prism and a rectangular pyramid. Prisms have the following characteristics.

- The bases are congruent faces in parallel planes.
- The faces that are not bases are called **lateral faces**.
- The lateral faces intersect at the **lateral edges**. Lateral edges are parallel segments.
- A segment perpendicular to the bases, with an endpoint in each plane, is called an *altitude* of the prism. The height of a prism is the length of the altitude.
- A prism with lateral edges that are also altitudes is called a **right prism**. If the lateral edges are not perpendicular to the bases, it is an **oblique prism**.

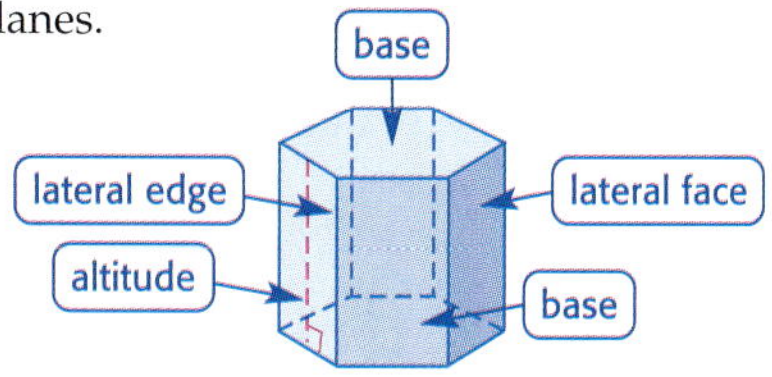

right hexagonal prism

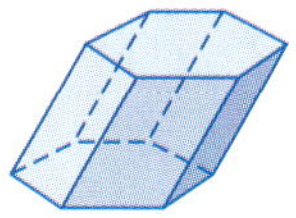
obliquc hcxagonal prism

Study Tip

Lateral Faces
The lateral faces of right prisms are always rectangles.

Study Tip

Reading Math
From this point in the text, you can assume that solids are right solids. If a solid is oblique, it will be clearly stated.

The **lateral area** L is the sum of the areas of the lateral faces.

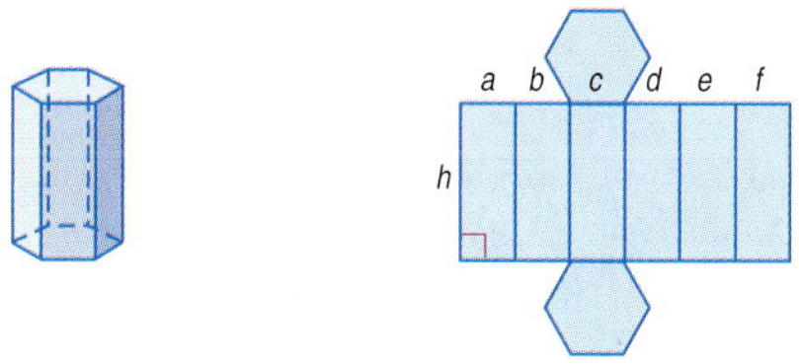

$$
\begin{aligned}
L &= ah + bh + ch + dh + eh + fh \\
&= h(a + b + c + d + e + f) \quad \text{Distributive Property} \\
&= Ph \quad P = a + b + c + d + e + f
\end{aligned}
$$

12-3 Lesson Notes

1 Focus

5-Minute Check Transparency 12-3 Use as a quiz or review of Lesson 12-2.

Mathematical Background notes are available for this lesson on p. 634C.

How do brick masons know how many bricks to order for a project?

Ask students:

- Which geometric solid is used most often to make bricks? **rectangular prism**
- How would you estimate the number of bricks needed to construct one side of the garage? **Sample answer: Multiply the number of bricks needed for the height by the number of bricks needed for the width.**

Resource Manager

Workbook and Reproducible Masters

Chapter 12 Resource Masters
- Study Guide and Intervention, pp. 673–674
- Skills Practice, p. 675
- Practice, p. 676
- Reading to Learn Mathematics, p. 677
- Enrichment, p. 678

School-to-Career Masters, p. 23
Teaching Geometry With Manipulatives Masters, pp. 6, 17

Transparencies

5-Minute Check Transparency 12-3
Answer Key Transparencies

Technology

Interactive Chalkboard

2 Teach

LATERAL AREAS OF PRISMS

In-Class Example Power Point®

Teaching Tip Point out that the formula to find the lateral area of a prism is applicable to both right and oblique prisms, and that the areas of the bases are not included in the lateral area. Emphasize that the height of an oblique prism is not the length of a lateral edge.

1 Find the lateral area of the regular hexagonal prism.

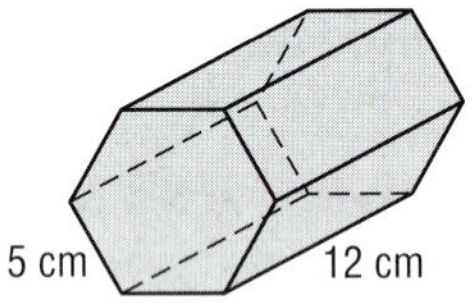

360 cm²

SURFACE AREAS OF PRISMS

In-Class Example Power Point®

2 Find the surface area of the square prism.

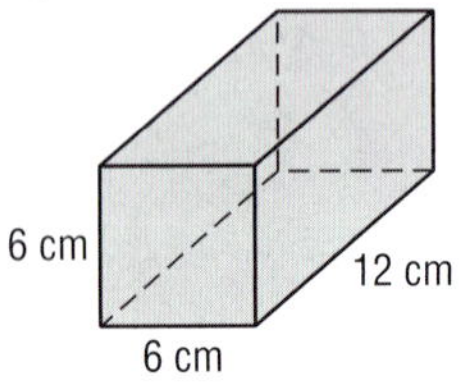

360 cm²

Key Concept — *Lateral Area of a Prism*

If a right prism has a lateral area of L square units, a height of h units, and each base has a perimeter of P units, then $L = Ph$.

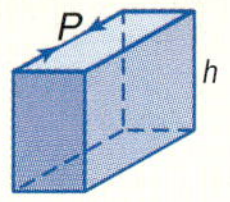

Example 1 Lateral Area of a Pentagonal Prism

Find the lateral area of the regular pentagonal prism.

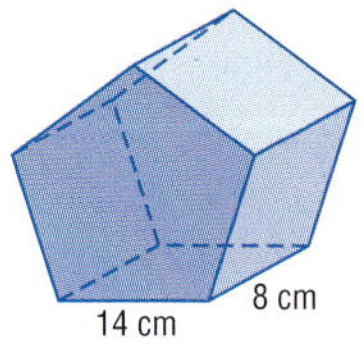

The bases are regular pentagons. So the perimeter of one base is 5(14) or 70 centimeters.

$L = Ph$	Lateral area of a prism
$= (70)(8)$	$P = 70$, $h = 8$
$= 560$	Multiply.

The lateral area is 560 square centimeters.

Study Tip

Right Prisms
The bases of a right prism are congruent, but the faces are not always congruent.

SURFACE AREAS OF PRISMS The surface area of a prism is the lateral area plus the areas of the bases. The bases are congruent, so the areas are equal.

Key Concept — *Surface Area of a Prism*

If the surface area of a right prism is T square units, its height is h units, and each base has an area of B square units and a perimeter of P units, then $T = L + 2B$.

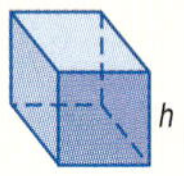

Example 2 Surface Area of a Triangular Prism

Find the surface area of the triangular prism.

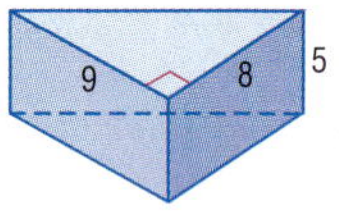

First, find the measure of the third side of the triangular base.

$c^2 = a^2 + b^2$	Pythagorean Theorem
$c^2 = 8^2 + 9^2$	Substitution
$c^2 = 145$	Simplify.
$c = \sqrt{145}$	Take the square root of each side.

$T = L + 2B$	Surface area of a prism
$= Ph + 2B$	$L = Ph$
$= \left(8 + 9 + \sqrt{145}\right)5 + 2\left[\frac{1}{2}(8 \cdot 9)\right]$	Substitution
≈ 217.2	Use a calculator.

The surface area is approximately 217.2 square units.

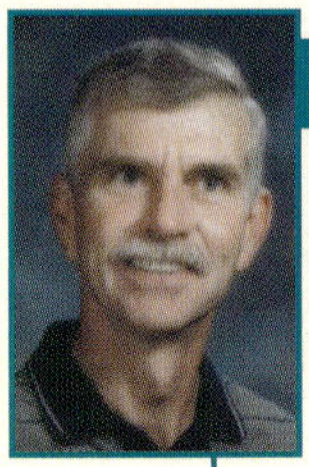

Teacher to Teacher

John R. Kennedy, Derby High School — Derby, KS

As a group activity, I set up 7–10 stations, each with a solid figure. Often the figure is an object in my room, such as a file cabinet (rectangular prism), overhead projector (trapezoidal prism), or trash basket (cylinder), or I bring in cones or a basketball. Students move from station to station, taking measurements and recording them so that they can figure surface area (and in Chapter 13, volume). They report their calculations along with their measurements, on large paper that we post up and then discuss.

Example 3 Use Surface Area to Solve a Problem

FURNITURE **Rick wants to have an ottoman reupholstered. Find the surface area that will be reupholstered.**

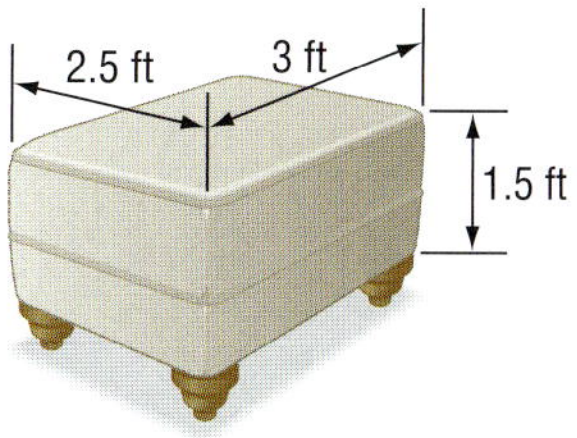

The ottoman is shaped like a rectangular prism. Since the bottom of the ottoman is not covered with fabric, find the lateral area and then add the area of one base. The perimeter of a base is $2(3) + 2(2.5)$ or 11 feet. The area of a base is $3(2.5)$ or 7.5 square feet.

$T = L + B$	Formula for surface area
$= (11)(1.5) + 7.5$	$P = 11$, $h = 1.5$, and $B = 7.5$
$= 24$	Simplify.

The total area that will be reupholstered is 24 square feet.

In-Class Example

Power Point®

3 **SCULPTURE** A solid block of marble will be used for a sculpture. If the block is 3 feet wide, 4 feet long, and $9\frac{1}{2}$ feet high, find the surface area of the block. **157 ft²**

Check for Understanding

Concept Check

1. **Explain** the difference between a right prism and an oblique prism.
2. **OPEN ENDED** Draw a prism and label the bases, lateral faces, and lateral edges.

1–2. See margin.

Guided Practice

Find the lateral area and surface area of each prism.

GUIDED PRACTICE KEY	
Exercises	Examples
3–4	1, 2
5	3

3. 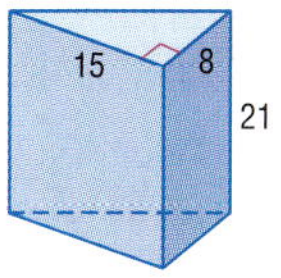

840 units², 960 units²

4. 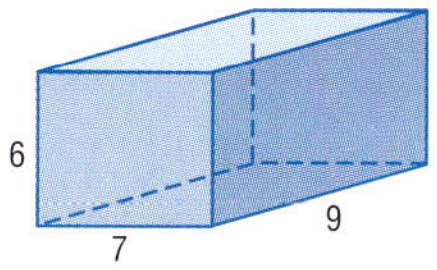

192 units² (7 × 9 base), 210 units² (6 × 9 base), 234 units² (6 × 7 base); 318 units²

Application

5. **PAINTING** Eva and Casey are planning to paint the walls and ceiling of their living room. The room is 20 feet long, 15 feet wide, and 12 feet high. Find the surface area to be painted. **1140 ft²**

★ indicates increased difficulty

Practice and Apply

Homework Help	
For Exercises	See Examples
6–11, 14, 15	1
12, 13, 16–21	2
22–36	3

Extra Practice
See page 778.

Find the lateral area of each prism or solid. Round to the nearest tenth if necessary.

6. 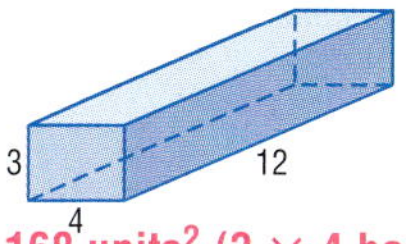

168 units² (3 × 4 base)
120 units² (3 × 12 base)
96 units² (4 × 12 base)

7. 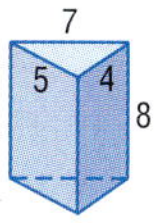

128 units²

8. 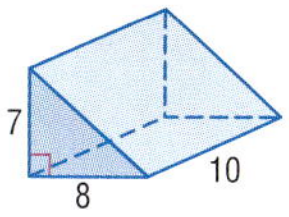

256.3 units²

9. 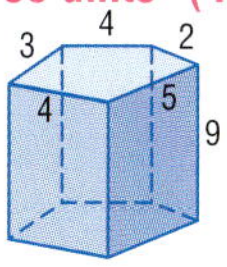

162 units²

★ 10. 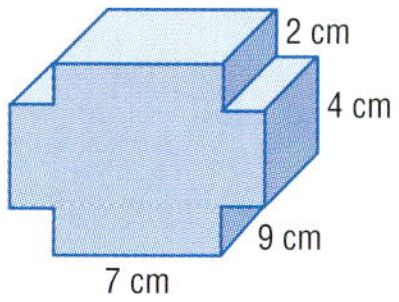

342 cm²

★ 11. 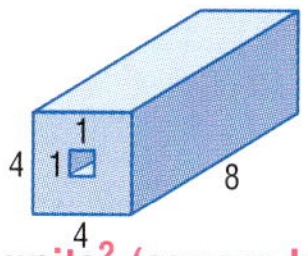

160 units² (square base)
126 units² (rectangular base)

www.geometryonline.com/extra_examples/fcat

Concept Check

Ask students to describe how to find the surface area of a prism. **First find the lateral area of the prism by multiplying the perimeter of the base times the height of the prism. Then find the area of the polygonal bases. Finally, add the lateral area to the area of the bases.**

Answers

1. **In a right prism a lateral edge is also an altitude. In an oblique prism, the lateral edges are not perpendicular to the bases.**
2. **Sample answer: bases: *ACHG*, *BDFE*; lateral faces: *ABDC*, *GEFH*, *BEGA*, *DFHC*; lateral edges: $\overline{BA}$, $\overline{EG}$, $\overline{FH}$, $\overline{DC}$**

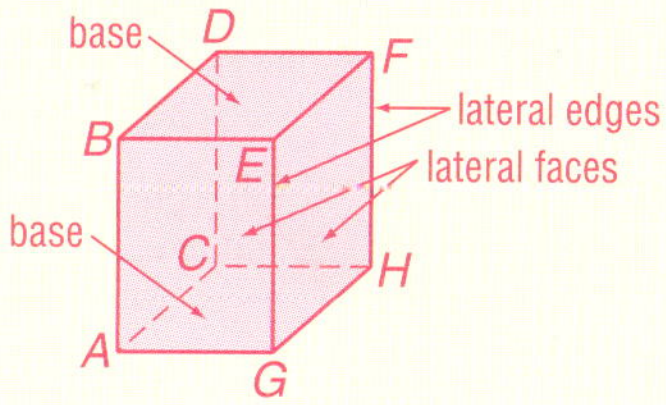

DAILY INTERVENTION

Differentiated Instruction

Intrapersonal Have students choose a partner. Ask one student to construct a prism using grid paper, tape, and scissors. Then ask the partner to find the lateral area and surface area of the prism. Next switch roles and do the activity again. If students have difficulty finding the area of the base of the prism, ask them to recall how to find the area of the polygon that is the base.

3 Practice/Apply

Study Notebook

Have students—
- add the definitions/examples of the vocabulary terms to their Vocabulary Builder worksheets for Chapter 12.
- write the formula for the lateral area of a prism and the surface area of a prism. Ask them to include an example of each.
- include any other item(s) that they find helpful in mastering the skills in this lesson.

About the Exercises...

Organization by Objective
- **Lateral Areas of Prisms:** 6–11, 14, 15
- **Surface Areas of Prisms:** 12, 13, 16–36

Odd/Even Assignments
Exercises 6–21 are structured so that students practice the same concepts whether they are assigned odd or even problems.

Alert! Exercise 43 requires the Internet or other research materials.

Assignment Guide

Basic: 7, 9, 17, 19, 23–37 odd, 38–40, 44–57 (optional: 41–43)

Average: 7–37 odd, 38–40, 44–57 (optional: 41–43)

Advanced: 6–36 even, 37–53 (optional: 54–57)

12. The surface area of a cube is 864 square inches. Find the length of the lateral edge of the cube. **12 in.**

★ 13. The surface area of a triangular prism is 540 square centimeters. The bases are right triangles with legs measuring 12 centimeters and 5 centimeters. Find the height. **16 cm**

★ 14. The lateral area of a rectangular prism is 156 square inches. What are the possible whole-number dimensions of the prism if the height is 13 inches? **See margin.**

★ 15. The lateral area of a rectangular prism is 96 square meters. What are the possible whole-number dimensions of the prism if the height is 4 meters? **See margin.**

Find the surface area of each prism. Round to the nearest tenth if necessary.

16.
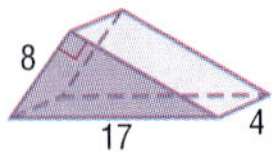

280 units^2

17.
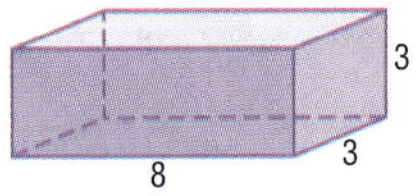

114 units^2

18.
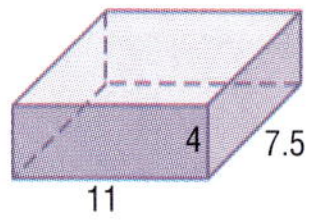

313 units^2

19.
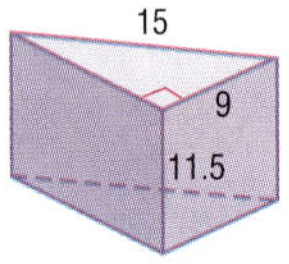

522 units^2

★ 20.
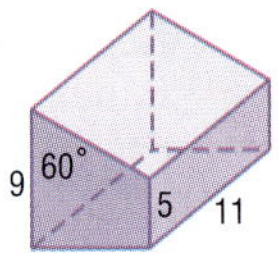

415.2 units^2

★ 21.
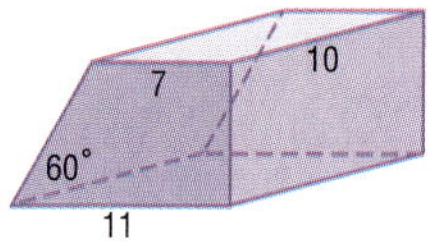

454.0 units^2

PAINTING **For Exercises 22–24, use the following information.**
A gallon of paint costs $16 and covers 400 square feet. Two coats of paint are recommended for even coverage. The room to be painted is 10 feet high, 15 feet long, and 15 feet wide. Only $1\frac{1}{2}$ gallons of paint are left to paint the room.

22. Is this enough paint for the walls of the room? Explain.

22. No, the walls are 600 ft^2; 1.5 gallons will only be enough for 1 coat.

23. How many gallons of paint are needed to paint the walls? **3 gallons for 2 coats**

24. How much would it cost to paint the walls and ceiling? **$80**

TOURISM **For Exercises 25–27, use the following information.**
The World's Only Corn Palace is located in Mitchell, South Dakota. The sides of the building are covered with huge murals made from corn and other grains.

25. Estimate the area of the Corn Palace to be covered if its base is 310 by 185 feet and it is 45 feet tall, not including the turrets. **44,550 ft^2**

26. Suppose a bushel of grain can cover 15 square feet. How many bushels of grain does it take to cover the Corn Palace? **2970 bushels**

27. Will the actual amount of grain needed be higher or lower than the estimate? Explain. **The actual amount needed will be higher because the area of the curved architectural elements appears to be greater than the area of the doors.**

More About. . .

Tourism

Tourists visit the Corn Palace each year to see the murals. New murals are created each year by local artists. It costs approximately $100,000 each year.

Source: www.cornpalace.org

★ 28. **GARDENING** This greenhouse is designed for a home gardener. The frame on the back of the greenhouse attaches to one wall of the house. The outside of the greenhouse is covered with tempered safety glass. Find the surface area of the glass covering the greenhouse. **≈ 75.4 ft^2**

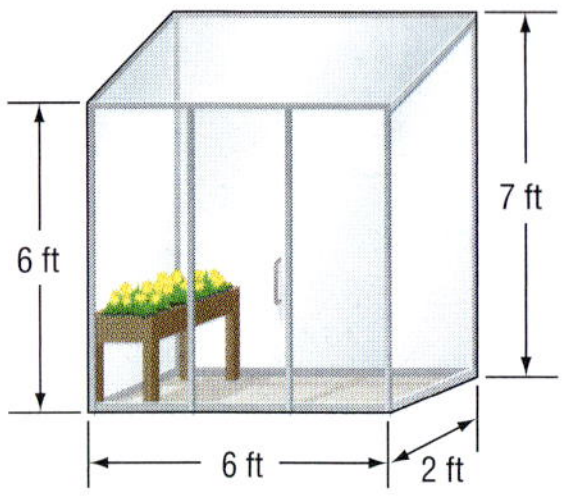

Answers

14. The perimeter of the base must be 12 inches. There are three rectangles with integer values for the dimensions that have a perimeter of 12. The dimensions of the base could be 5 × 1, 4 × 2, or 3 × 3.

15. The perimeter of the base must be 24 meters. There are six rectangles with integer values for the dimensions that have a perimeter of 24. The dimensions of the base could be 1 × 11, 2 × 10, 3 × 9, 4 × 8, 5 × 7, or 6 × 6.

29. base of A ≅ base of C; base of A ~ base of B; base of C ~ base of B
30. A:B = 1:2, B:C = 2:1, A:C = 1:1
31. A:B = 1:4, B:C = 4:1, A:C = 1:1
32. A:B = 1:4, B:C = 30:11, A:C = 15:22
33. A:B, because the heights of A and B are in the same ratio as perimeters of bases.

For Exercises 29–33, use prisms A, B, and C.

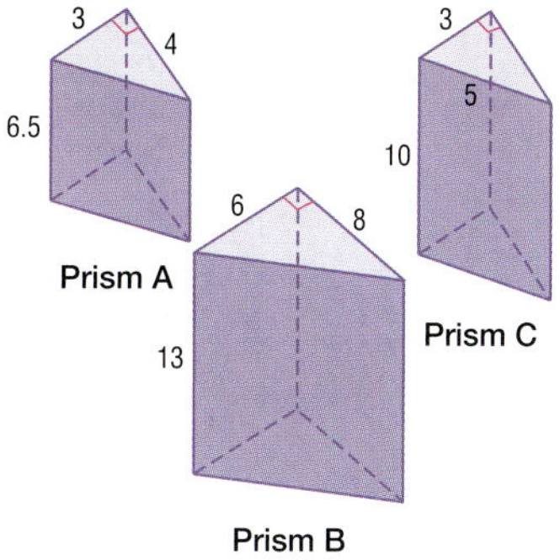

29. Compare the bases of each prism.
30. Write three ratios to compare the perimeters of the bases of the prisms.
31. Write three ratios to compare the areas of the bases of the prisms.
32. Write three ratios to compare the surface areas of the prisms.
33. Which pairs of prisms have the same ratio of base areas as ratio of surface areas? Why do you think this is so?

STATISTICS **For Exercises 34–36, use the graphic at the right.** Malik plans to build a three-dimensional model of the data from the graph.

- A rectangular prism will represent each category.
- Each prism will be 30 centimeters wide and 20 centimeters deep.
- The length of the prism for TV will be 84 centimeters.

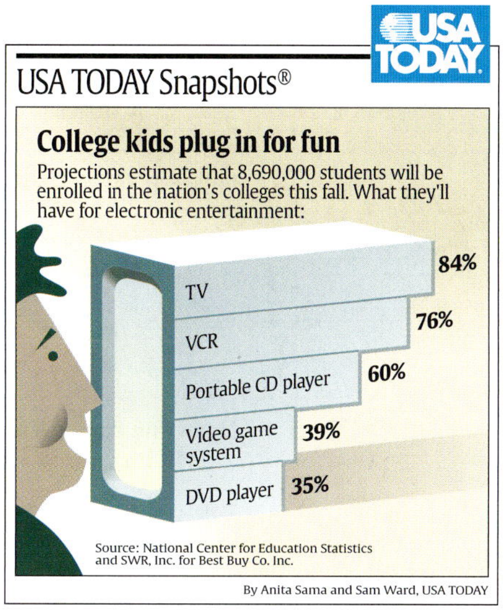

34. TV = 9600 cm^2, VCR = 8800 cm^2, CD = 7200 cm^2, video game system = 5100 cm^2, DVD = 4700 cm^2

34. Find the surface area of each prism that Malik builds.
35. Will the surface area of the finished product be the sum of the surface areas of each prism? Explain. See margin.
36. Find the total surface area of the finished model. **22,800 cm^2**

37. **CRITICAL THINKING** Suppose the lateral area of a right rectangular prism is 144 square centimeters. If the length is three times the width and the height is twice the width, find the surface area. **198 cm^2**

38. **WRITING IN MATH** Answer the question that was posed at the beginning of the lesson. See p. 685C.

How do brick masons know how many bricks to order for a project?

Include the following in your answer:
- how lateral area is used, and
- why overestimation is important in the process.

FCAT Practice
Standardized Test Practice (A) (B) (C) (D)

39. The surface area of a cube is 121.5 square meters. What is the length of each edge? **B**

(A) 4.05 m (B) 4.5 m (C) 4.95 m (D) 5 m

40. **ALGEBRA** For all $a \neq 4$, $\frac{a^2 - 16}{4a - 16} =$ __?__. **D**

(A) $a + 16$ (B) $a + 1$ (C) $\frac{a-4}{4}$ (D) $\frac{a+4}{4}$

www.geometryonline.com/self_check_quiz/fcat

Answers

35. No, the surface area of the finished product will be the sum of the lateral areas of each prism plus the area of the bases of the TV and DVD prisms. It will also include the area of the overhang between each prism, but not the area of the overlapping prisms.

Enrichment, p. 678

Cross Sections of Prisms

When a plane intersects a solid figure to form a two-dimensional figure, the results is called a **cross section**. The figure at the right shows a plane intersecting a cube. The cross section is a hexagon.

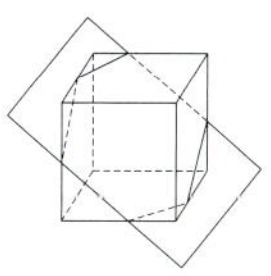

For each right prism, connect the labeled points in alphabetical order to show a cross section. Then identify the polygon.

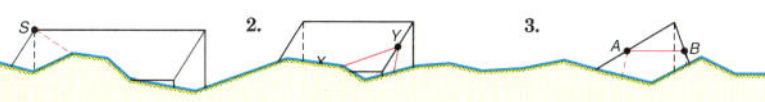

Study Guide and Intervention, p. 673 (shown) and p. 674

Lateral Areas of Prisms Here are some characteristics of prisms.

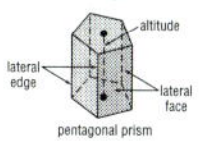

- The bases are parallel and congruent.
- The **lateral faces** are the faces that are not bases.
- The lateral faces intersect at **lateral edges**, which are parallel.
- The **altitude** of a prism is a segment that is perpendicular to the bases with an endpoint in each base.
- For a **right prism**, the lateral edges are perpendicular to the bases. Otherwise, the prism is **oblique**.

Lateral Area of a Prism	If a prism has a lateral area of L square units, a height of h units, and each base has a perimeter of P units, then $L = Ph$.

Example **Find the lateral area of the regular pentagonal prism above if each base has a perimeter of 75 centimeters and the altitude is 10 centimeters.**

$L = Ph$ Lateral area of a prism
$= 75(10)$ $P = 75, h = 10$
$= 750$ Multiply.

The lateral area is 750 square centimeters.

Exercises

Find the lateral area of each prism.

1. 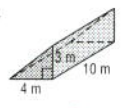

120 m^2

2. 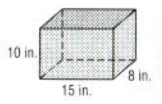

460 in^2 (8 in. × 15 in. base); 400 in^2 (10 in. × 15 in. base); 540 in^2 (10 in. × 8 in. base)

3. 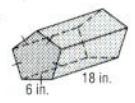

540 in^2

4. 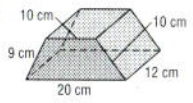

588 cm^2

5. 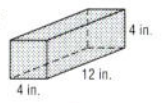

128 in^2 (rectangular base) 192 in^2 (square base)

6. 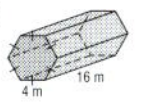

384 m^2

Skills Practice, p. 675 and Practice, p. 676 (shown)

Drawing Solids on Isometric Dot Paper

Isometric dot paper is helpful for drawing solids. Remember to use dashed lines for hidden edges.

For each solid shown, draw another solid whose dimensions are twice as large.

1. 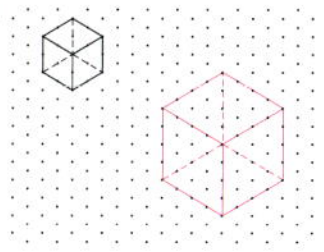

2. 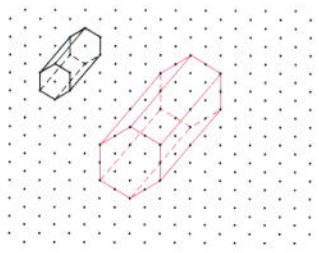

3. 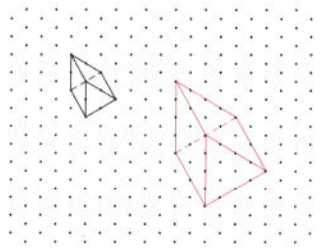

4. 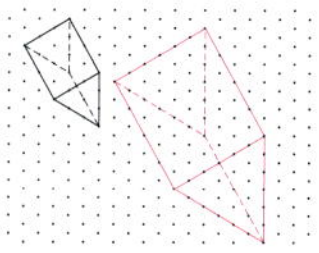

5. 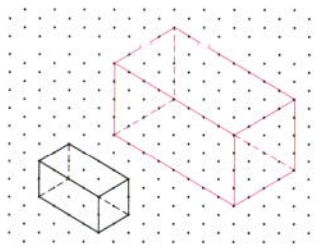

6. 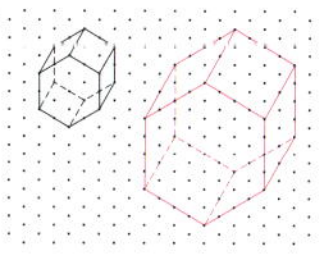

Reading to Learn Mathematics, p. 677 ELL

Pre-Activity **How do brick masons know how many bricks to order for a project?**

Read the introduction to Lesson 12-3 at the top of page 649 in your textbook.

How could the brick mason figure out approximately how many bricks are needed for the garage? Sample answer: Find the area of each wall of the garage and add these areas. Divide this total area by the area of one face of a brick, but take into account the amount of space for the mortar between the bricks.

Reading the Lesson

1. Determine whether each sentence is *always*, *sometimes*, or *never* true.
   a. A base of a prism is a face of the prism. always
   b. A face of a prism is a base of the prism. sometimes
   c. The lateral faces of a prism are rectangles. sometimes
   d. If a base of a prism has n vertices, then the prism has n faces. never
   e. If a base of a prism has n vertices, then the prism has n lateral edges. always
   f. In a right prism, the lateral edges are also altitudes. always
   g. The bases of a prism are congruent regular polygons. sometimes
   h. Any two lateral edges of a prism are perpendicular to each other. never
   i. In a rectangular prism, any pair of opposite faces can be called the bases. always
   j. All of the lateral faces of a prism are congruent to each other. sometimes

2. Explain the difference between the *lateral area* of a prism and the *surface area* of a prism. Your explanation should apply to both right and oblique prisms. Do not use any formulas in your explanation. Sample answer: The lateral area is the sum of the areas of lateral faces. The surface area is the sum of the areas of all the faces, including both the lateral faces and the bases.

3. Refer to the figure.

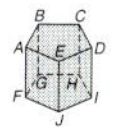

   a. Name this solid with as specific a name as possible. right pentagonal prism
   b. Name the bases of the solid. pentagons *ABCDE* and *FGHIJ*
   c. Name the lateral faces. rectangles *ABGF*, *BCHG*, *CDIH*, *DEJI*, and *EAFJ*
   d. Name the edges. $\overline{AF}$, $\overline{BG}$, $\overline{CH}$, $\overline{DI}$, $\overline{EJ}$, $\overline{AB}$, $\overline{BC}$, $\overline{CD}$, $\overline{DE}$, $\overline{EA}$, $\overline{FG}$, $\overline{GH}$, $\overline{HI}$, $\overline{IJ}$, $\overline{JF}$
   e. Name an altitude of the solid. $\overline{AF}$, $\overline{BG}$, $\overline{CH}$, $\overline{DI}$, or $\overline{EJ}$
   f. If a represents the area, P represents the perimeter of one of the bases, and $x = AF$, write an expression for the surface area of the solid that involves a, P, and x. $Px + 2a$

Helping You Remember

4. A good way to remember a new mathematical term is to relate it to an everyday use of the same word. How can the way the word *lateral* is used in sports help you remember the meaning of the *lateral area* of a solid? Sample answer: In football, a pass thrown to the *side* is called a *lateral pass*. In geometry, the *lateral area* is the "side area," or the sum of the areas of all the side surfaces of the solid (the surfaces other than the bases).

4 Assess

Open-Ended Assessment

Modeling Ask students to use grid paper to demonstrate how to construct a net for a rectangular solid and find the surface area.

Getting Ready for Lesson 12-4

Prerequisite Skill Students will learn about the surface areas of cylinders in Lesson 12-4. They will use the area of a circle to find the area of the base of a cylinder. Use Exercises 54–57 to determine your students' familiarity with finding the area of a circle.

Answers

47.

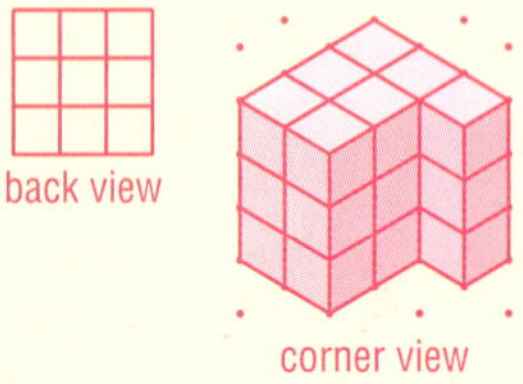

48.

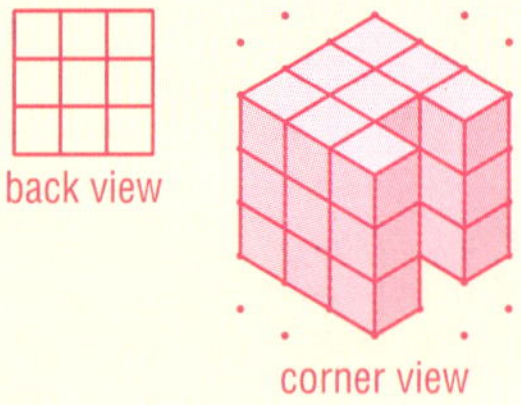

Extending the Lesson

OBLIQUE PRISMS The altitude of an oblique prism is not the length of a lateral edge. For an oblique rectangular prism, the bases are rectangles, two face are rectangles, and two faces are parallelograms. To find the lateral area and the surface area, apply the definitions of each.

Find the lateral area and surface area of each oblique prism.

41.

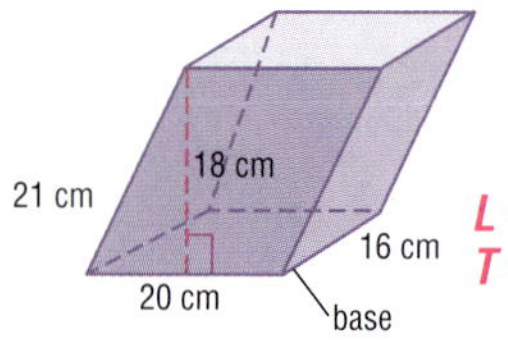

$L = 1416 \text{ cm}^2$, $T = 2056 \text{ cm}^2$

42.

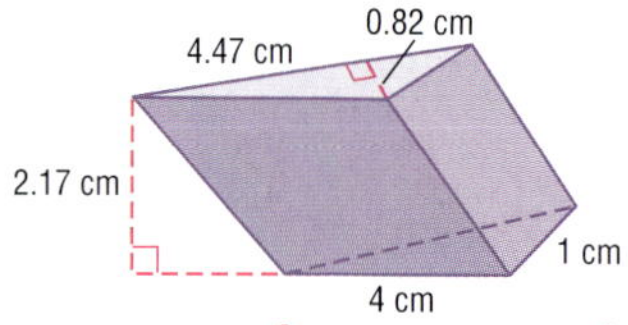

$L = 20.81 \text{ cm}^2$, $T = 24.48 \text{ cm}^2$

43. **RESEARCH** Use a dictionary to find the meaning of the term *oblique*. How is the everyday meaning related to the mathematical meaning? **See students' work.**

Maintain Your Skills

Mixed Review

44–46. See p. 685C for nets.

For each solid, draw a net and find the surface area. *(Lesson 12-2)*

44.

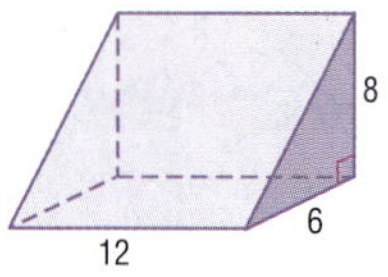

336 units^2

45.

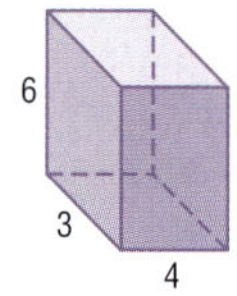

108 units^2

46.

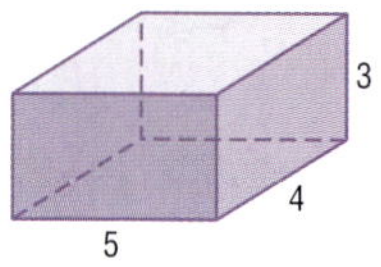

94 units^2

Draw the back view and corner view of the figure given the orthogonal drawing. *(Lesson 12-1)* **47–48. See margin.**

47.

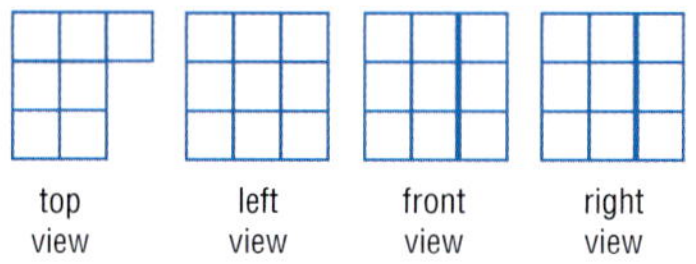

48.

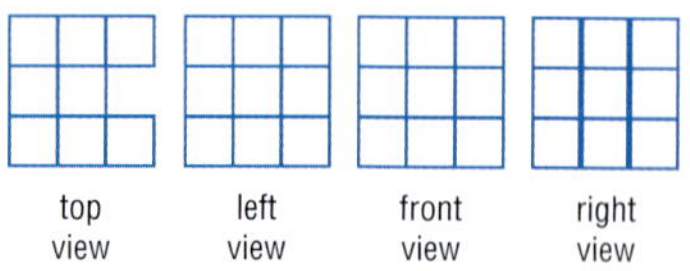

Circle Q has a radius of 24 units, $\odot R$ has a radius of 16 units, and $BC = 5$. Find each measure. *(Lesson 10-1)*

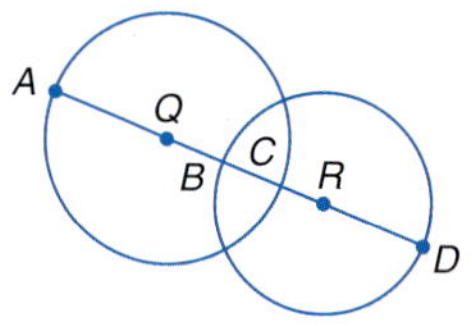

49. AB **43**

50. AD **75**

51. QR **35**

52. **NAVIGATION** An airplane is three miles above sea level when it begins to climb at a 3.5° angle. If this angle is constant, how far above sea level is the airplane after flying 50 miles? *(Lesson 7-4)* **≈ 6.1 mi**

53. **ART** Kiernan drew a sketch of a house. If the height of the house in her drawing was 5.5 inches and the actual height of the house was 33 feet, find the scale factor of the drawing. *(Lesson 6-1)* $\frac{1}{72}$

Getting Ready for the Next Lesson

PREREQUISITE SKILL **Find the area of each circle. Round to the nearest hundredth.** *(To review **finding the area of a circle**, see Lesson 11-3.)*

54.

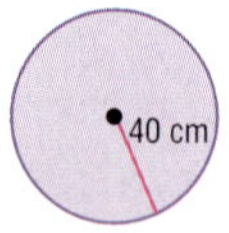

5026.55 cm^2

55.

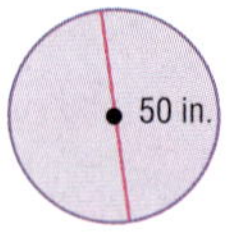

1963.50 in^2

56.

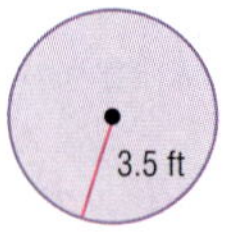

38.48 ft^2

57.

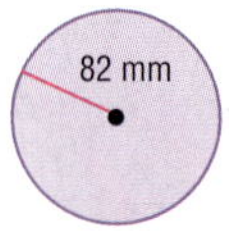

$21{,}124.07 \text{ mm}^2$

654 Chapter 12 Surface Area

12-4 Surface Areas of Cylinders

Sunshine State Standards
MA.B.1.4.1

What You'll Learn

- Find lateral areas of cylinders.
- Find surface areas of cylinders.

Vocabulary

- axis
- right cylinder
- oblique cylinder

How are cylinders used in extreme sports?

Extreme sports, such as in-line skating, biking, skateboarding, and snowboarding use a cylindrical-shaped ramp called a half-pipe. The half-pipe looks like half of a cylinder. Usually there is a flat section in the middle with sides almost 8 feet high. Near the top, the sides are almost vertical.

LATERAL AREAS OF CYLINDERS The **axis** of the cylinder is the segment with endpoints that are centers of the circular bases. If the axis is also the altitude, then the cylinder is called a **right cylinder**. Otherwise, the cylinder is an **oblique cylinder**.

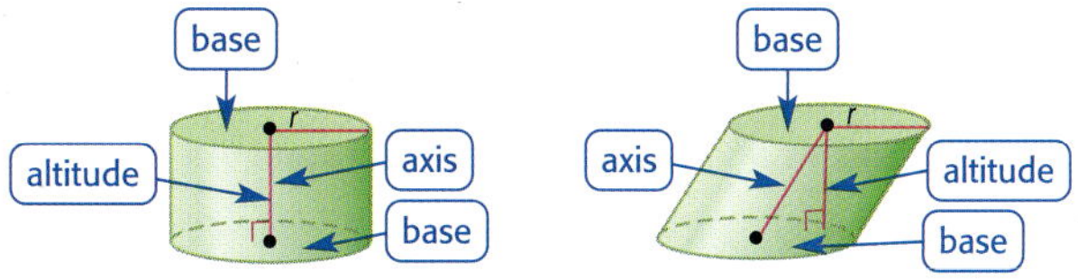

right cylinder
oblique cylinder

The net of a cylinder is composed of two congruent circles and a rectangle. The area of this rectangle is the lateral area. The length of the rectangle is the same as the circumference of the base, $2\pi r$. So, the lateral area of a right cylinder is $2\pi rh$.

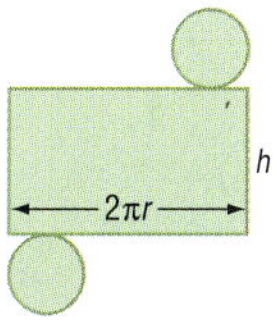

Study Tip

Formulas
An alternate formula for the lateral area of a cylinder is $L = \pi dh$, with πd as the circumference of a circle.

Key Concept — *Lateral Area of a Cylinder*

If a right cylinder has a lateral area of L square units, a height of h units, and the bases have radii of r units, then $L = 2\pi rh$.

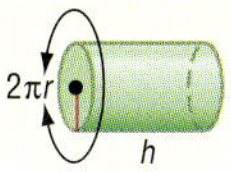

Example 1 *Lateral Area of a Cylinder*

MANUFACTURING An office has recycling barrels for cans and paper. The barrels are cylindrical with cardboard sides and plastic lids and bases. Each barrel is 3 feet tall, and the diameter is 30 inches. How many square feet of cardboard are used to make each barrel?

The cardboard section of the barrel represents the lateral area of the cylinder. If the diameter of the lid is 30 inches, then the radius is 15 inches. The height is 3 feet or 36 inches. Use the formula to find the lateral area. *(continued on the next page)*

12-4 Lesson Notes

1 Focus

5-Minute Check Transparency 12-4 Use as a quiz or review of Lesson 12-3.

Mathematical Background notes are available for this lesson on p. 634D.

How are cylinders used in extreme sports?

Ask students:

- What shape is a half-pipe? **cylindrical**
- Where are cylindrical shapes also used? **Sample answers: glasses, cups, buildings**
- Why is a cylinder a better shape for skating than a flat surface? **Sample answer: The athletes build enough momentum by speed skating down the half-pipe to skate up a near vertical surface.**

Resource Manager

Workbook and Reproducible Masters

Chapter 12 Resource Masters
- Study Guide and Intervention, pp. 679–680
- Skills Practice, p. 681
- Practice, p. 682
- Reading to Learn Mathematics, p. 683
- Enrichment, p. 684
- Assessment, pp. 717, 719

Graphing Calculator and Computer Masters, p. 40
Teaching Geometry With Manipulatives Masters, pp. 7, 17

Transparencies

5-Minute Check Transparency 12-4
Answer Key Transparencies

Technology

Interactive Chalkboard

2 Teach

LATERAL AREAS OF CYLINDERS

In-Class Example Power Point®

Teaching Tip Remind students that they may use a net to find the area. Suggest that they plan ahead to find the combination of shapes that make up the net.

1 A fruit juice can is cylindrical with aluminum sides and bases. Each can is 12 centimeters tall, and the diameter of the can is 6.3 centimeters. How many square centimeters of aluminum are used to make the sides of the can? about 237.5 cm^2

SURFACE AREAS OF CYLINDERS

In-Class Examples Power Point®

Teaching Tip Students should be able to determine that the curved side of a right cylinder is a rectangle if it is flattened. Cut apart a cardboard tube to show students the cylinder as a net. Point out that the width of the rectangle is the height of the cylinder, while the length of the rectangle is the circumference of the base of the cylinder.

2 Find the surface area of the cylinder.

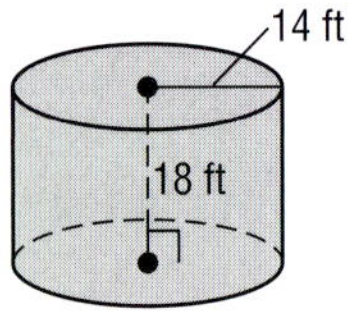

$\approx$ 2814.9 ft^2

3 Find the radius of the base of a right cylinder if the surface area is 528π square feet and the height is 10 feet. 12 ft

$L = 2\pi rh$ — Lateral area of a cylinder

$= 2\pi(15)(36)$ — $r = 15$, $h = 36$

≈ 3392.9 — Use a calculator.

Each barrel uses approximately 3393 square inches of cardboard. Because 144 square inches equal one square foot, there are 3393 ÷ 144 or about 23.6 square feet of cardboard per barrel.

SURFACE AREAS OF CYLINDERS To find the surface area of a cylinder, first find the lateral area and then add the areas of the bases. This leads to the formula for the surface area of a right cylinder.

Study Tip

Making Connections
The formula for the surface area of a right cylinder is like that of a prism, $T = L + 2B$.

Key Concept — *Surface Area of a Cylinder*

If a right cylinder has a surface area of T square units, a height of h units, and the bases have radii of r units, then $T = 2\pi rh + 2\pi r^2$.

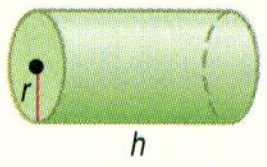

Example 2 Surface Area of a Cylinder

Find the surface area of the cylinder.

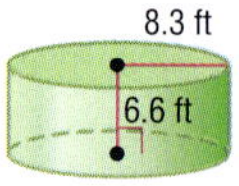

The radius of the base and the height of the cylinder are given. Substitute these values in the formula to find the surface area.

$T = 2\pi rh + 2\pi r^2$ — Surface area of a cylinder

$= 2\pi(8.3)(6.6) + 2\pi(8.3)^2$ — $r = 8.3$, $h = 6.6$

≈ 777.0 — Use a calculator.

The surface area is approximately 777.0 square feet.

Example 3 Find Missing Dimensions

Find the radius of the base of a right cylinder if the surface area is 128π square centimeters and the height is 12 centimeters.

Use the formula for surface area to write and solve an equation for the radius.

$T = 2\pi rh + 2\pi r^2$ — Surface area of a cylinder

$128\pi = 2\pi(12)r + 2\pi r^2$ — Substitution

$128\pi = 24\pi r + 2\pi r^2$ — Simplify.

$64 = 12r + r^2$ — Divide each side by 2π.

$0 = r^2 + 12r - 64$ — Subtract 64 from each side.

$0 = (r - 4)(r + 16)$ — Factor.

$r = 4$ or -16

Since the radius of a circle cannot have a negative value, -16 is eliminated. So, the radius of the base is 4 centimeters.

DAILY INTERVENTION — **Differentiated Instruction**

Kinesthetic Have your students use string, masking tape, and a tiled floor to mark off the net for a cylinder on the floor. Ask them to estimate the area by counting squares and then verify the area by calculating the sum of the individual parts in the net.

Check for Understanding

Concept Check

1. **Explain** how to find the surface area of a cylinder.

1. Multiply the circumference of the base by the height and add the area of each base.

2. **OPEN ENDED** Draw a net for a cylinder. See margin.

3. **FIND THE ERROR** Jamie and Dwayne are finding the surface area of a cylinder with one base removed.

Jamie	Dwayne
$T = 2\pi(4)(9) + \pi(4^2)$	$T = 2\pi(4)(9) + 2\pi(4^2)$
$T = 72\pi + 16\pi$	$T = 72\pi + 32\pi$
$T = 88\pi$ in^2	$T = 104\pi$ in^2

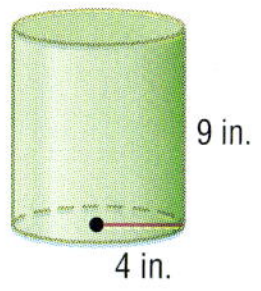

Who is correct? Explain. Jamie; since the cylinder has one base removed, the surface area will be the sum of the lateral area and one base.

Guided Practice

GUIDED PRACTICE KEY	
Exercises	Examples
4–5	2
6–7	3
8	1

4. Find the surface area of a cylinder with a radius of 4 feet and height of 6 feet. Round to the nearest tenth. 251.3 ft^2

5. Find the surface area of the cylinder. Round to the nearest tenth. 1520.5 m^2

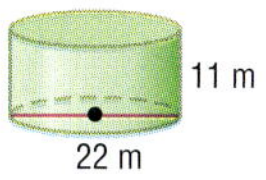

Find the radius of the base of each cylinder.

6. The surface area is 96π square centimeters, and the height is 8 centimeters. 4 cm

7. The surface area is 140π square feet, and the height is 9 feet. 5 ft

Application

8. **CONTESTS** Mrs. Fairway's class is collecting labels from soup cans to raise money for the school. The students collected labels from 3258 cans. If the cans are 4 inches high with a diameter of 2.5 inches, find the area of the labels that were collected. ≈ 102,353.1 in^2

★ indicates increased difficulty

Practice and Apply

Homework Help	
For Exercises	See Examples
9–16	2
17–20	3
21, 22, 24–25	1

Extra Practice See page 779.

Find the surface area of a cylinder with the given dimensions. Round to the nearest tenth.

9. $r = 13$ m, $h = 15.8$ m 2352.4 m^2
10. $d = 13.6$ ft, $h = 1.9$ ft 371.7 ft^2
11. $d = 14.2$ in., $h = 4.5$ in. 517.5 in^2
12. $r = 14$ mm, $h = 14$ mm 2463.0 mm^2

Find the surface area of each cylinder. Round to the nearest tenth.

13. 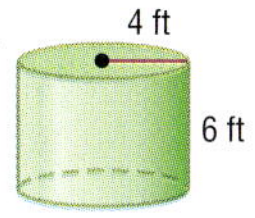

251.3 ft^2

14. 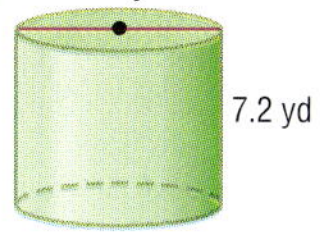

291.1 yd^2

15. 

30.0 cm^2

★16. 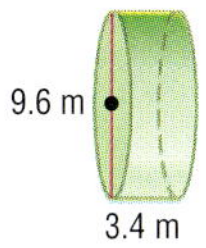

247.3 m^2

Find the radius of the base of each cylinder.

17. The surface area is 48π square centimeters, and the height is 5 centimeters.

17. 3 cm

18. The surface area is 340π square inches, and the height is 7 inches. 10 in.

19. The surface area is 320π square meters, and the height is 12 meters. 8 m

★ 20. The surface area is 425.1 square feet, and the height is 6.8 feet. ≈ 5.5 ft

3 Practice/Apply

Study Notebook

Have students—

- add the definitions/examples of the vocabulary terms to their Vocabulary Builder worksheets for Chapter 12.
- include an example of finding the lateral area and surface area of a cylinder.
- include any other item(s) that they find helpful in mastering the skills in this lesson.

DAILY INTERVENTION

FIND THE ERROR In Exercise 3, caution students that the surface area usually includes both bases of the cylinder. However, this cylinder only has one base, so there is no top, like a water glass.

About the Exercises...

Organization by Objective

- **Lateral Areas of Cylinders:** 21–22, 24–26
- **Surface Areas of Cylinders:** 9–20

Odd/Even Assignments

Exercises 9–20 are structured so that students practice the same concepts whether they are assigned odd or even problems.

Assignment Guide

Basic: 9–21 odd, 25–29, 32–44 (optional: 30–31)

Average: 9–25 odd, 26–29, 32–44 (optional: 30–31)

Advanced: 10–26 even, 27–41 (optional: 42–44)

All: Quiz 1 (1–5)

Answer

2. Sample answer:

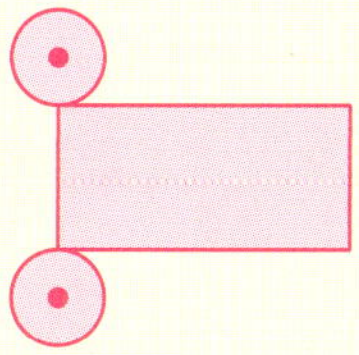

Study Guide and Intervention, p. 679 (shown) and p. 680

Lateral Areas of Cylinders A **cylinder** is a solid whose bases are congruent circles that lie in parallel planes. The **axis** of a cylinder is the segment whose endpoints are the centers of these circles. For a **right cylinder**, the axis and the altitude of the cylinder are equal. The lateral area of a right cylinder is the circumference of the cylinder multiplied by the height.

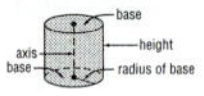

Lateral Area of a Cylinder	If a cylinder has a lateral area of L square units, a height of h units, and the bases have radii of r units, then $L = 2\pi rh$.

Example **Find the lateral area of the cylinder above if the radius of the base is 6 centimeters and the height is 14 centimeters.**

$L = 2\pi rh$ Lateral area of a cylinder
$= 2\pi(6)(14)$ Substitution
≈ 527.8 Simplify.

The lateral area is about 527.8 square centimeters.

Exercises

Find the lateral area of each cylinder. Round to the nearest tenth.

1. 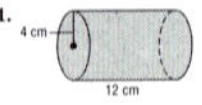

301.6 cm^2

2. 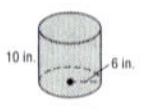

377.0 in^2

3. 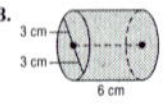

113.1 cm^2

4. 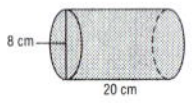

502.7 cm^2

5. 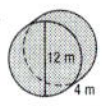

150.8 m^2

6. 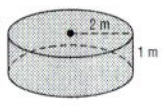

12.6 m^2

Skills Practice, p. 681 and Practice, p. 682 (shown)

Find the surface area of a cylinder with the given dimensions. Round to the nearest tenth.

1. $r = 8$ cm, $h = 9$ cm
854.5 cm^2

2. $r = 12$ in., $h = 14$ in.
1960.4 in^2

3. $d = 14$ mm, $h = 32$ mm
1715.3 mm^2

4. $d = 6$ yd, $h = 12$ yd
282.7 yd^2

5. $r = 2.5$ ft, $h = 7$ ft
149.2 ft^2

6. $d = 13$ m, $h = 20$ m
1082.3 m^2

Find the surface area of each cylinder. Round to the nearest tenth.

7. 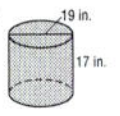

1581.8 in^2

8. 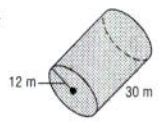

3166.7 m^2

Find the radius of the base of each right cylinder.

9. The surface area is 628.3 square millimeters, and the height is 15 millimeters.
5 mm

10. The surface area is 892.2 square feet, and the height is 4.2 feet.
10 ft

11. The surface area is 158.3 square inches, and the height is 5.4 inches.
3 in.

12. **KALEIDOSCOPES** Nathan built a kaleidoscope with a 20-centimeter barrel and a 5-centimeter diameter. He plans to cover the barrel with embossed paper of his own design. How many square centimeters of paper will it take to cover the barrel of the kaleidoscope?
about 314.2 cm^2

Reading to Learn Mathematics, p. 683 ELL

Pre-Activity **How are cylinders used in extreme sports?**

Read the introduction to Lesson 12-4 at the top of page 655 in your textbook.

Why is the surface area of a half-pipe more than half the surface area of a complete pipe?
Sample answer: The surface area of the half-pipe includes an added flat section in the middle.

Reading the Lesson

1. Underline the correct word or phrase to form a true statement.
   a. The bases of a cylinder are (rectangles/regular polygons/circles).
   b. The (axis/radius/diameter) of a cylinder is the segment whose endpoints are the centers of the bases.
   c. The net of a cylinder is composed of two congruent (rectangles/circles) and one (rectangle/semicircle).
   d. In a right cylinder, the axis of the cylinder is also a(n) (base/lateral edge/altitude).
   e. A cylinder that is not a right cylinder is called an (acute/obtuse/oblique) cylinder.

2. Match each description from the first column with an expression from the second column that represents its value.
   a. the lateral area of a right cylinder in which the radius of each base is x cm and the length of the axis is y cm v
   b. the surface area of a right prism with square bases in which the length of a side of a base is x cm and the length of a lateral edge is y cm i
   c. the surface area of a right cylinder in which the radius of a base is x cm and the height is y cm ii
   d. the surface area of regular hexahedron (cube) in which the length of each edge is x cm iv
   e. the lateral area of a triangular prism in which the bases are equilateral triangles with side length x cm and the height is y cm iii
   f. the surface area of a right cylinder in which the diameter of the base is x cm and the length of the axis is y cm vi

   i. $(2x^2 + 4xy)$ cm^2
   ii. $(2\pi xy + 2\pi x^2)$ cm^2
   iii. $3xy$ cm^2
   iv. $6x^2$ cm^2
   v. $2\pi xy$ cm^2
   vi. $\left(\frac{\pi x^2}{2} + \pi xy\right)$ cm^2

Helping You Remember

3. Often the best way to remember a mathematical formula is to think about where the different parts of the formula come from. How can you use this approach to remember the formula for the surface area of a cylinder? Sample answer: Think about the net for a cylinder. The two bases are circles, each with radius r, so the sum of their areas is $2\pi r^2$. The lateral surface is a rectangle whose dimensions are the circumference $2\pi r$ of the circular base and the height h of the cylinder, so the lateral area is $2\pi rh$. Add the lateral area and the area of the two bases to get the surface area of the cylinder: $2\pi rh + 2\pi r^2$.

21. **KITCHENS** Raul purchased a set of canisters with diameters of 5 inches and heights of 9 inches, 6 inches, and 3 inches. Make a conjecture about the relationship between the heights of the canisters and their lateral areas. Check your conjecture. The lateral areas will be in the ratio 3 : 2 : 1; 45π in^2, 30π in^2, 15π in^2.

22. **CAMPING** Campers can use a solar cooker to cook food. You can make a solar cooker from supplies you have on hand. The reflector in the cooker shown at the right is half of a cardboard cylinder covered with aluminum foil. The reflector is 18 inches long and has a diameter of $5\frac{1}{2}$ inches. How much aluminum foil was needed to cover the inside of the reflector? ≈ 179.3 in^2

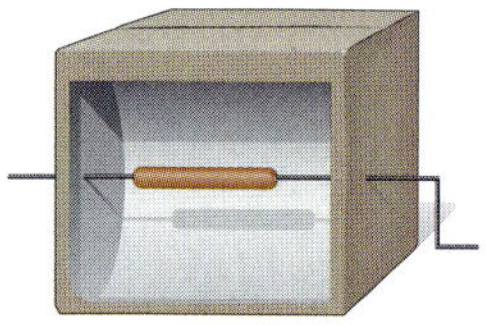

★ 23. Suppose the height of a right cylinder is tripled. Is the surface area or lateral area tripled? Explain. The lateral area is tripled. The surface area is increased, but not tripled.

Career Choices

Agricultural Engineer

Agricultural engineers use engineering techniques to develop equipment and methods that assist farmers in efficient agricultural production.

Source: www.bls.gov

AGRICULTURE **For Exercises 24 and 25, use the following information.**
The acid from the contents of a silo can weaken its concrete walls and seriously damage the silo's structure. So the inside of the silo must occasionally be resurfaced. The cost of the resurfacing is a function of the lateral area of the inside of the silo. 24. ≈ 204.2 m^2

24. Find the lateral area of a silo 13 meters tall with an interior diameter of 5 meters.

25. A second grain silo is 26 meters tall. If both silos have the same lateral area, find the radius of the second silo. 1.25 m

26. **CRITICAL THINKING** Some pencils are cylindrical, and others are hexagonal prisms. If the diameter of the cylinder is the same length as the longest diagonal of the hexagon, which has the greater surface area? Explain. Assume that each pencil is 11 inches long and unsharpened. See p. 685D.

27. WRITING IN MATH Answer the question that was posed at the beginning of the lesson. See margin.

How are cylinders used in extreme sports?

Include the following in your answer:
- how to find the lateral area of a semicylinder, and
- how to determine if the half-pipe ramp is a semicylinder.

FCAT Practice
Standardized Test Practice
(A) (B) (C) (D)

28. A cylinder has a height of 13.4 centimeters and a diameter of 8.2 centimeters. To the nearest tenth, what is the surface area of the cylinder? B
(A) 51.5 cm^2 (B) 450.8 cm^2 (C) 741.9 cm^2 (D) 1112.9 cm^2

29. **ALGEBRA** For the band concert, student tickets cost \$2 and adult tickets cost \$5. A total of 200 tickets were sold. If the total sales were more than \$500, what was the minimum number of adult tickets sold? C
(A) 30 (B) 33 (C) 34 (D) 40

Extending the Lesson

LOCUS A cylinder can be defined in terms of locus. The locus of points in space a given distance from a line is the lateral surface of a cylinder.

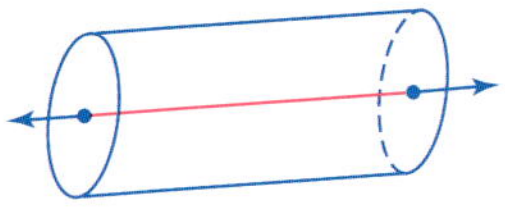

Draw a figure and describe the locus of all points in space that satisfy each set of conditions. 30–31. See margin.

30. 5 units from a given line

31. equidistant from two opposite vertices of a face of a cube

Enrichment, p. 684

Can-didly Speaking

Beverage companies have mathematically determined the ideal shape for a 12-ounce soft-drink can. Why won't any shape do? Some shapes are more economical than others. In order to hold 12 ounces of soft drink, an extremely skinny can would have to be very tall. The total amount of aluminum used for such a shape would be greater than the amount used for the conventional shape, thus costing the company more to make the skinny can. The radius r chosen determines the height h needed to hold 12 ounces of liquid. This also determines the amount of aluminum needed to make the can. Companies also have to keep in mind that the top of the can is three times thicker than the bottom and sides. Why? So you won't tear off the entire top when you open the can! The following formulas can be used to find the height and amount of aluminum a needed for a 12-ounce soft-drink can.

$h = \frac{17.89}{\pi r^2}$

$a = 0.02\pi\left[\frac{r^2 + 35.78}{r}\right]$ The values of r and h are measured in inches.

Answers

27. Sample answer: Extreme sports participants use a semicylinder for a ramp. Answers should include the following.
- To find the lateral area of a semicylinder like the half-pipe, multiply the height by the circumference of the base and then divide by 2.
- A half-pipe ramp is half of a cylinder if the ramp is an equal distance from the axis of the cylinder.

Maintain Your Skills

Mixed Review

32. 276 units^2(8 × 15 ase), 336 units^2 6 × 15 base), 420 nits^2 (8 × 6 base)
34. 312 units^2(8 × 18 ase), 384 units^2 6 × 18 base), 504 nits^2 (8 × 6 base)

Find the lateral area of each prism. *(Lesson 12-3)*

32.

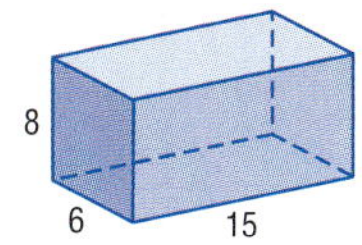

33.

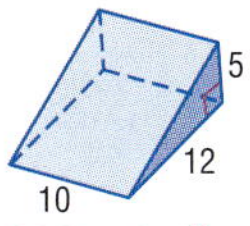

300 units^2

34.

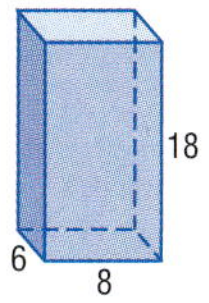

Given the net of a solid, use isometric dot paper to draw the solid. *(Lesson 12-2)*

35.

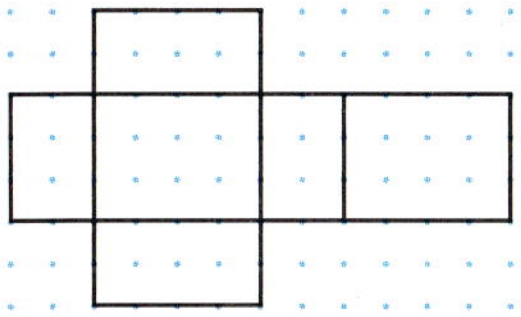

36.

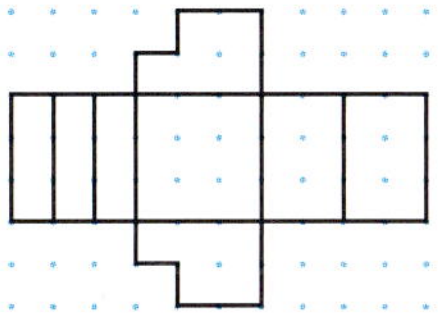

35–36. See margin.

Find x. Assume that segments that appear to be tangent are tangent. *(Lesson 10-5)*

37. 27

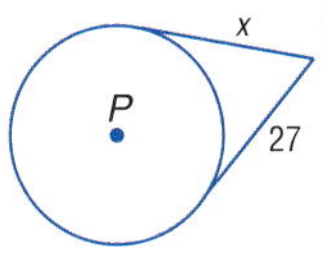

38. 8

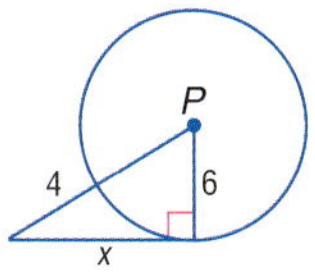

39. 8

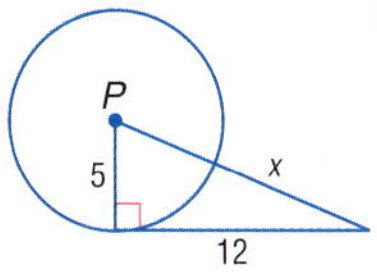

Solve each $\triangle ABC$ described below. Round to the nearest tenth if necessary. *(Lesson 7-7)*

40. $m\angle A = 54$, $b = 6.3$, $c = 7.1$
$m\angle B \approx 56.3$, $m\angle C \approx 69.7$, $a \approx 6.1$

41. $m\angle B = 47$, $m\angle C = 69$, $a = 15$
$m\angle A = 64$, $b \approx 12.2$, $c \approx 15.6$

Getting Ready for the Next Lesson

PREREQUISITE SKILL **Find the area of each figure.**
*(To review **finding areas of triangles and trapezoids**, see Lesson 11-2.)*

42.

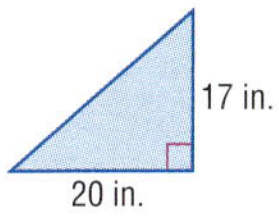

170 in^2

43.

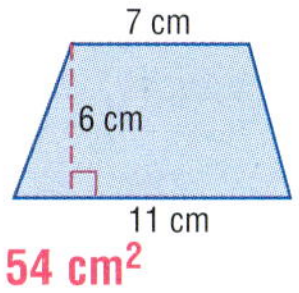

54 cm^2

44.

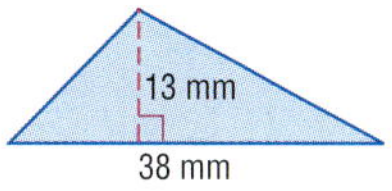

247 mm^2

Practice Quiz 1 — *Lessons 12-1 through 12-4*

1. Draw a corner view of the figure given the orthogonal drawing. *(Lesson 12-1)* **1–2. See margin.**

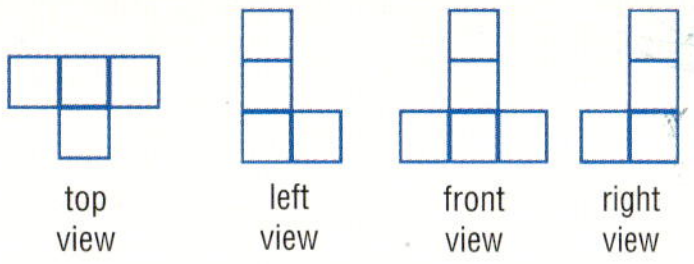

2. Sketch a rectangular prism 2 units wide, 3 units long, and 2 units high using isometric dot paper. *(Lesson 12-2)*
3. Find the lateral area of the prism. Round to the nearest tenth. *(Lesson 12-3)* **231.5 m^2**

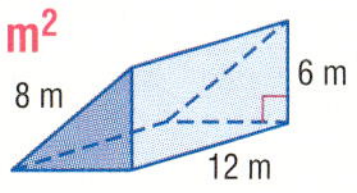

4. Find the surface area of the prism. Round to the nearest tenth. *(Lesson 12-4)* **263.2 m^2**
5. Find the radius of the base of a right cylinder if the surface area is 560 square feet and the height is 11 feet. Round to the nearest tenth. *(Lesson 12-4)* **5.4 ft**

www.geometryonline.com/self_check_quiz/fcat

4 Assess

Open-Ended Assessment

Writing Have students explain how to find the surface area of a right cylinder.

Getting Ready for Lesson 12-5

Prerequisite Skill Students will find the surface areas of pyramids in Lesson 12-5. They will use the formula for the area of a triangle to find the surface area of a pyramid. Use Exercises 42–44 to determine your students' familiarity with finding the areas of triangles and trapezoids.

Assessment Options

Practice Quiz 1 The quiz provides students with a brief review of the concepts and skills in Lessons 12-1 through 12-4. Lesson numbers are given to the right of the exercises or instruction lines so students can review concepts not yet mastered.

Quiz (Lessons 3 and 4) is available on p. 717 of the *Chapter 12 Resource Masters*.

Mid-Chapter Test (Lessons 12-1 through 12-4) is available on p. 719 of the *Chapter 12 Resource Masters*.

Answers (Practice Quiz)

1.

corner view

2.

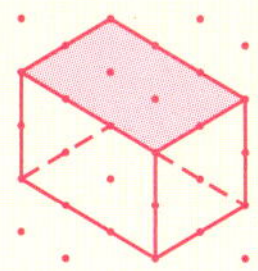

Answers

30. a cylinder with a radius of 5 units

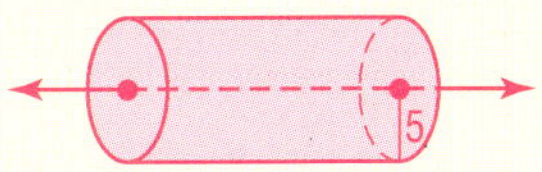

31. a plane perpendicular to the line containing the opposite vertices of the face of the cube

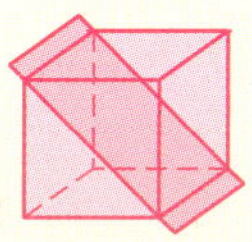

35.

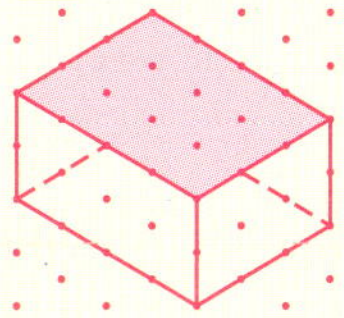

36.

12-5 Lesson Notes

1 Focus

5-Minute Check Transparency 12-5 Use as a quiz or review of Lesson 12-4.

Mathematical Background notes are available for this lesson on p. 634D.

How are pyramids used in architecture?

Ask students:

- From the picture, which shapes are used to make a pyramid? **triangles and squares**
- How is the shape of a pyramid used at the Louvre museum? **The entrance of the Louvre is in the shape of a giant glass pyramid.**
- How are pyramids used in the real-world? **Sample answer: as crowns for some buildings**

12-5 Surface Areas of Pyramids

Sunshine State Standards
MA.B.1.4.1

What You'll Learn

- Find lateral areas of regular pyramids.
- Find surface areas of regular pyramids.

Vocabulary

- regular pyramid
- slant height

How are pyramids used in architecture?

In 1989, a new entrance was completed in the courtyard of the Louvre museum in Paris, France. Visitors can enter the museum through a glass pyramid that stands 71 feet tall. The pyramid is glass with a structural system of steel rods and cables.

LATERAL AREAS OF REGULAR PYRAMIDS Pyramids have the following characteristics.

- All of the faces, except the base, intersect at one point called the *vertex*.
- The base is always a polygon.
- The faces that intersect at the vertex are called *lateral faces* and form triangles. The edges of the lateral faces that have the vertex as an endpoint are called *lateral edges*.
- The *altitude* is the segment from the vertex perpendicular to the base.

Study Tip

Right Pyramid
In a *right pyramid*, the altitude is the segment with endpoints that are the center of the base and the vertex. But the base is not always a regular polygon.

If the base of a pyramid is a regular polygon and the segment with endpoints that are the center of the base and the vertex is perpendicular to the base, then the pyramid is called a **regular pyramid**. They have specific characteristics. The altitude is the segment with endpoints that are the center of the base and the vertex. All of the lateral faces are congruent isosceles triangles. The height of each lateral face is called the **slant height** ℓ of the pyramid.

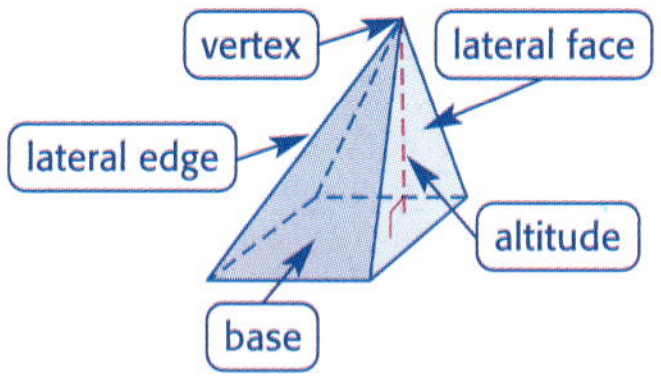

square pyramid

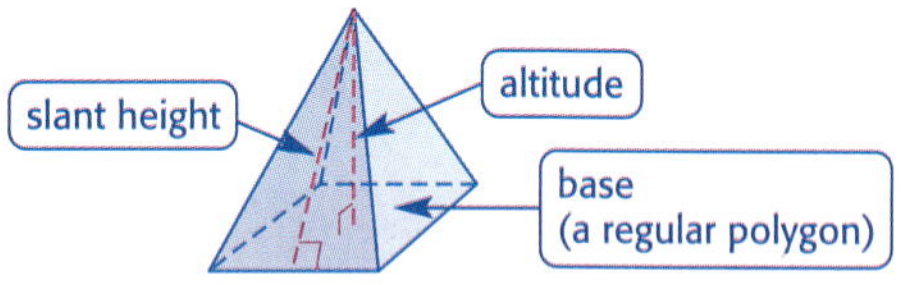

regular square pyramid

The figure below is a regular hexagonal pyramid. Its lateral area L can be found by adding the areas of all its congruent triangular faces as shown in its net.

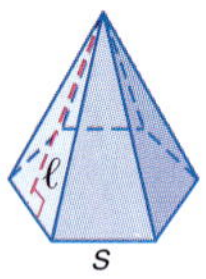

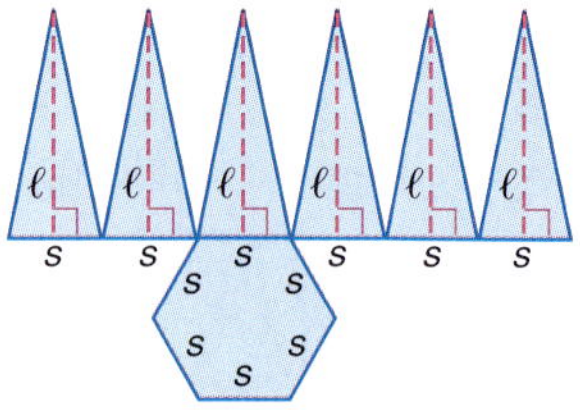

Resource Manager

Workbook and Reproducible Masters

Chapter 12 Resource Masters
- Study Guide and Intervention, pp. 685–686
- Skills Practice, p. 687
- Practice, p. 688
- Reading to Learn Mathematics, p. 689
- Enrichment, p. 690

School-to-Career Masters, p. 24
Teaching Geometry With Manipulatives Masters, p. 17

Transparencies

5-Minute Check Transparency 12-5
Answer Key Transparencies

Technology

Interactive Chalkboard

Area of the net

$L = \frac{1}{2}s\ell + \frac{1}{2}s\ell + \frac{1}{2}s\ell + \frac{1}{2}s\ell + \frac{1}{2}s\ell + \frac{1}{2}s\ell$ Sum of the areas of the lateral faces

$= \frac{1}{2}\ell(s + s + s + s + s + s)$ Distributive Property

$= \frac{1}{2}P\ell$ $P = s + s + s + s + s + s$

Key Concept *Lateral Area of a Regular Pyramid*

If a regular pyramid has a lateral area of L square units, a slant height of ℓ units, and its base has a perimeter of P units, then $L = \frac{1}{2}P\ell$.

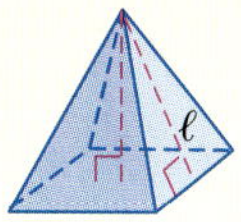

Example 1 Use Lateral Area to Solve a Problem

BIRDHOUSES The roof of a birdhouse is a regular hexagonal pyramid. The base of the pyramid has sides of 4 inches, and the slant height of the roof is 12 inches. If the roof is made of copper, find the amount of copper used for the roof.

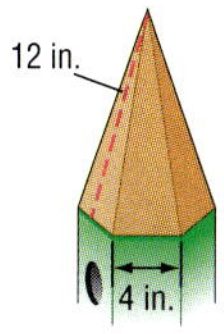

We need to find the lateral area of the hexagonal pyramid. The sides of the base measure 4 inches, so the perimeter is 6(4) or 24 inches.

$L = \frac{1}{2}P\ell$ Lateral area of a regular pyramid

$= \frac{1}{2}(24)(12)$ $P = 24, \ell = 12$

$= 144$ Multiply.

So, 144 square inches of copper are used to cover the roof of the birdhouse.

SURFACE AREAS OF REGULAR PYRAMIDS The surface area of a regular pyramid is the sum of the lateral area and the area of the base.

Key Concept *Surface Area of a Regular Pyramid*

If a regular pyramid has a surface area of T square units, a slant height of ℓ units, and its base has a perimeter of P units and an area of B square units, then $T = \frac{1}{2}P\ell + B$.

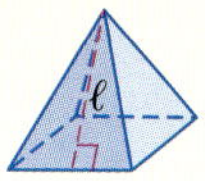

Example 2 Surface Area of a Square Pyramid

Find the surface area of the square pyramid.

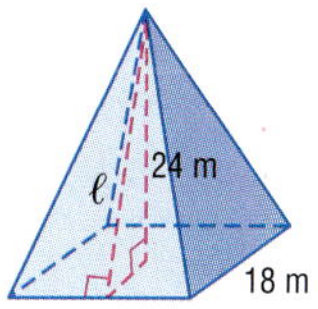

To find the surface area, first find the slant height of the pyramid. The slant height is the hypotenuse of a right triangle with legs that are the altitude and a segment with a length that is one-half the side measure of the base.

$c^2 = a^2 + b^2$ Pythagorean Theorem

$\ell^2 = 9^2 + 24^2$ $a = 9, b = 24, c = \ell$

$\ell = \sqrt{657}$ Simplify.

(continued on the next page)

Study Tip

Making Connections
The total surface area for a pyramid is $L + B$, because there is only one base to consider.

www.geometryonline.com/extra_examples/fcat

2 Teach

LATERAL AREAS OF REGULAR PYRAMIDS

In-Class Example PowerPoint®

Teaching Tip The difference between the slant height and the height of a pyramid may be confusing for some students. Point out that the slant height is the height of a *lateral* face, while the height is the altitude of the pyramid.

1 **CANDLES** A candle store offers a pyramidal candle that burns for 20 hours. The square base is 6 centimeters on a side and the slant height of the candle is 22 centimeters. Find the lateral area of the candle. **264 cm²**

SURFACE AREAS OF REGULAR PYRAMIDS

In-Class Example PowerPoint®

2 Find the surface area of the regular pyramid to the nearest tenth.

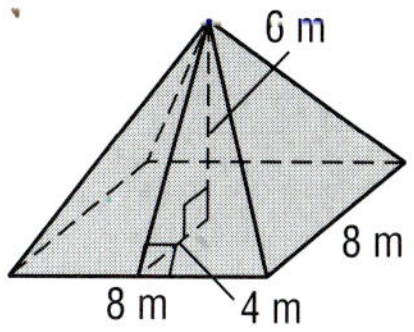

179.4 m²

Lesson 12-5 Surface Areas of Pyramids **661**

In-Class Example

Power Point®

3 Find the surface area of the regular pyramid. Round to the nearest tenth.

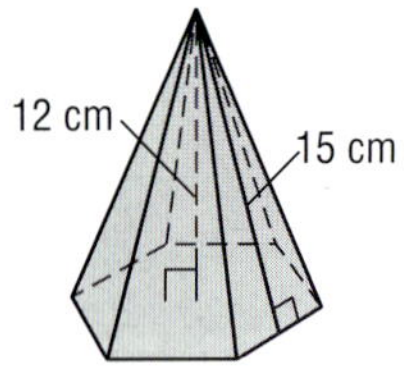

748.2 cm²

Tips for New Teachers

Rounding Rounding during computation may result in a final answer that differs from the ones given in the Teacher Wraparound Edition.

Answer

1. Sample answer:

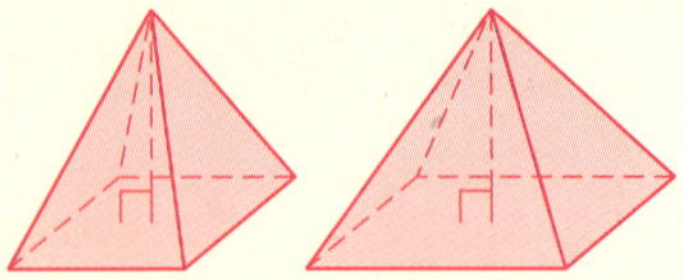

Study Tip

Look Back
To review **finding the areas of regular polygons,** see Lesson 11-3. To review **trigonometric ratios,** see Lesson 7-4.

Now find the surface area of a regular pyramid. The perimeter of the base is 4(18) or 72 meters, and the area of the base is 18^2 or 324 square meters.

$T = \frac{1}{2}P\ell + B$	Surface area of a regular pyramid
$T = \frac{1}{2}(72)\sqrt{657} + 324$	$P = 72$, $\ell = \sqrt{657}$, $B = 324$
$T \approx 1246.8$	Use a calculator.

The surface area is 1246.8 square meters to the nearest tenth.

Example 3 Surface Area of Pentagonal Pyramid

Find the surface area of the regular pyramid.

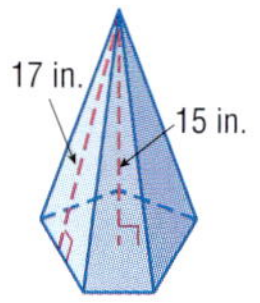

The altitude, slant height, and apothem form a right triangle. Use the Pythagorean Theorem to find the apothem. Let a represent the length of the apothem.

$c^2 = a^2 + b^2$	Pythagorean Theorem
$(17)^2 = a^2 + 15^2$	$b = 15$, $c = 17$
$8 = a$	Simplify.

Now find the length of the sides of the base. The central angle of the pentagon measures $\frac{360°}{5}$ or 72°. Let x represent the measure of the angle formed by a radius and the apothem. Then, $x = \frac{72}{2}$ or 36.

Use trigonometry to find the length of the sides.

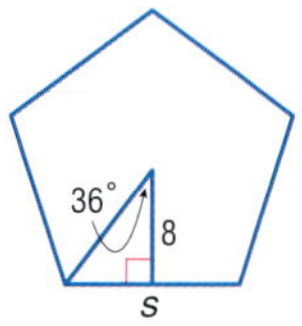

$\tan 36° = \frac{\frac{1}{2}s}{8}$	$\tan x° = \frac{\text{opposite}}{\text{adjacent}}$
$8(\tan 36°) = \frac{1}{2}s$	Multiply each side by 8.
$16(\tan 36°) = s$	Multiply each side by 2.
$11.6 \approx s$	Use a calculator.

Next, find the perimeter and area of the base.

$P = 5s$

$\approx 5(11.6)$ or 58

$B = \frac{1}{2}Pa$

$\approx \frac{1}{2}(58)(8)$ or 232

Finally, find the surface area.

$T = \frac{1}{2}P\ell + B$	Surface area of a regular pyramid
$\approx \frac{1}{2}(58)(17) + 232$	$P \approx 58$, $\ell = 17$, $B \approx 232$
≈ 726.5	Simplify.

The surface area is approximately 726.5 square inches.

WebQuest

Making a sketch of a pyramid can help you find its slant height, lateral area, and base area. Visit www.geometryonline.com/webquest to continue work on your WebQuest project.

DAILY INTERVENTION — Differentiated Instruction

Auditory/Musical Ask students to name the similarities and differences between finding the lateral area of a pyramid versus the lateral area of a prism. **The similarity is that both lateral areas use the perimeter of the base in their formula. The difference is that the lateral area of a pyramid is one-half the perimeter of the base times its slant height, while the lateral area of a prism is the perimeter of the base times its regular height.**

Check for Understanding

Concept Check

1–2. See margin.

1. **OPEN ENDED** Draw a regular pyramid and a pyramid that is not regular.
2. **Explain** whether a regular pyramid can also be a regular polyhedron.

Guided Practice

GUIDED PRACTICE KEY

Exercises	Examples
3–4	2
5	3
6	1

Find the surface area of each regular pyramid. Round to the nearest tenth if necessary.

3.
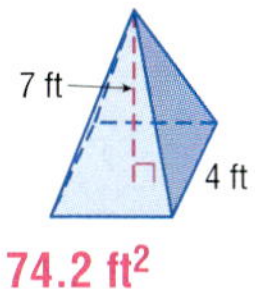

74.2 ft^2

4.
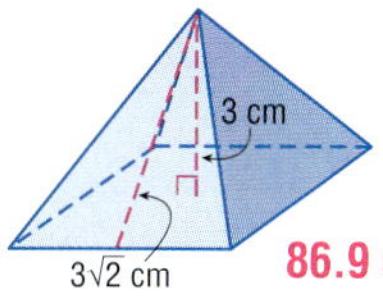

86.9 cm^2

5.
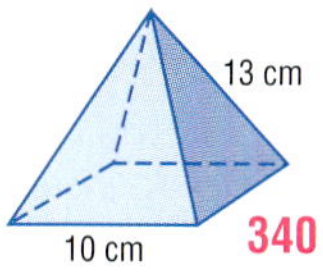

340 cm^2

Application

6. **DECORATIONS** Minowa purchased 3 decorative three-dimensional stars. Each star is composed of 6 congruent square pyramids with faces of paper and a base of cardboard. If the base is 2 inches on each side and the slant height is 4 inches, find the amount of paper used for one star. 16 in^2 per pyramid, 96 in^2 per star

★ indicates increased difficulty

Practice and Apply

Homework Help

For Exercises	See Examples
7, 9, 10, 14, 15	2
8, 11–13	3
16, 18–24	1

Extra Practice
See page 779.

Find the surface area of each regular pyramid. Round to the nearest tenth if necessary.

7.
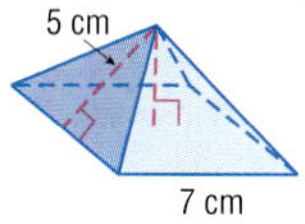

119 cm^2

8.
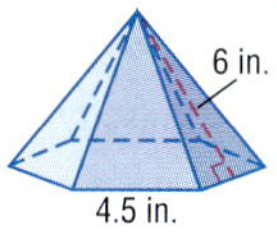

133.6 in^2

9.
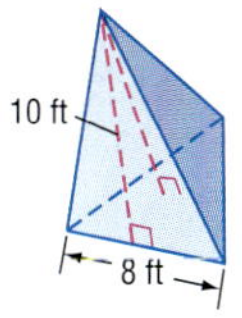

147.7 ft^2

10.
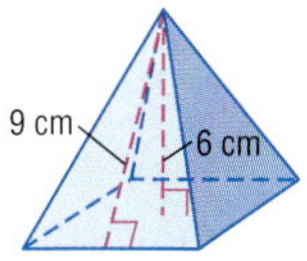

421.5 cm^2

11.
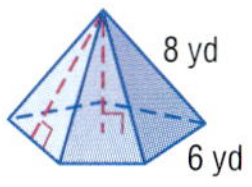

173.2 yd^2

12.
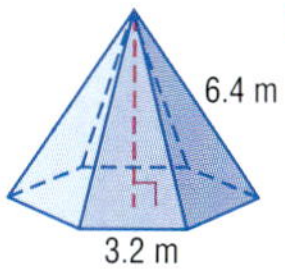

86.1 m^2

13.
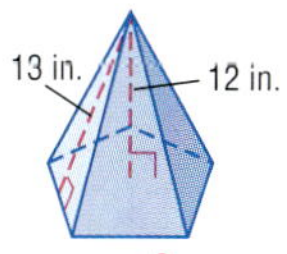

326.9 in^2

★ 14.
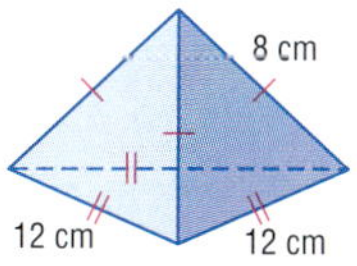

157.6 cm^2

★ 15.
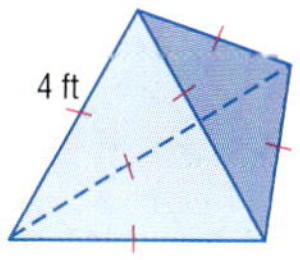

27.7 ft^2

16. **CONSTRUCTION** The roof on a building is a square pyramid with no base. If the altitude of the pyramid measures 5 feet and the slant height measures 20 feet, find the area of the roof. 1549.2 ft^2

★ 17. **PERFUME BOTTLES** Some perfumes are packaged in square pyramidal containers. The base of one bottle is 3 inches square, and the slant height is 4 inches. A second bottle has the same surface area, but the slant height is 6 inches long. Find the dimensions of the base of the second bottle. ≈ 2.3 inches on each side

DAILY INTERVENTION

Unlocking Misconceptions

A common error is to assume that the formulas in this lesson apply to any pyramid. Point out that these formulas apply only to *regular* pyramids. The lateral area of a non-regular pyramid would require finding the areas of triangles that are not congruent.

3 Practice/Apply

Study Notebook

Have students—

- add the definitions/examples of the vocabulary terms to their Vocabulary Builder worksheets for Chapter 12.
- include the formulas and examples of finding the lateral area and surface area of a regular pyramid.
- include any other item(s) that they find helpful in mastering the skills in this lesson.

About the Exercises...

Organization by Objective

- **Lateral Areas of Regular Pyramids:** 16, 18–24
- **Surface Areas of Regular Pyramids:** 7–15

Odd/Even Assignments

Exercises 7–15 are structured so that students practice the same concepts whether they are assigned odd or even problems.

Assignment Guide

Basic: 7–13 odd, 19–25 odd, 26–44

Average: 7–25 odd, 26–44

Advanced: 8–24 even, 25–41 (optional: 42–44)

Answer

2. A regular pyramid is only a regular polyhedron if all of the faces including the base are congruent regular polygons. Since the faces of a pyramid are triangles, the only regular pyramid that is also a regular polyhedron is a tetrahedron.

Lesson 12-5 Surface Areas of Pyramids 663

Study Guide and Intervention, p. 685 (shown) and p. 686

Lateral Areas of Regular Pyramids Here are some properties of pyramids.

- The base is a polygon.
- All of the faces, except the base, intersect in a common point known as the **vertex**.
- The faces that intersect at the vertex, which are called **lateral faces**, are triangles.

For a **regular pyramid**, the base is a regular polygon and the **slant height** is the height of each lateral face.

Lateral Area of a Regular Pyramid	If a regular pyramid has a lateral area of L square units, a slant height of ℓ units, and its base has a perimeter of P units, then $L = \frac{1}{2}P\ell$.

Example **The roof of a barn is a regular octagonal pyramid. The base of the pyramid has sides of 12 feet, and the slant height of the roof is 15 feet. Find the lateral area of the roof.**

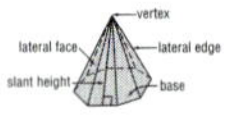

The perimeter of the base is 8(12) or 96 feet.

$L = \frac{1}{2}P\ell$ Lateral area of a pyramid

$= \frac{1}{2}(96)(15)$ $P = 96$, $\ell = 15$

$= 720$ Multiply.

The lateral area is 720 square feet.

Exercises

Find the lateral area of each regular pyramid. Round to the nearest tenth if necessary.

1. 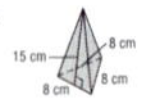

180 cm²

2. 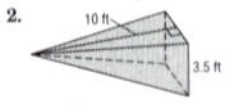

70 ft²

3. 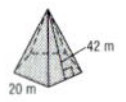

2100 m²

4. 

62.4 ft²

5. 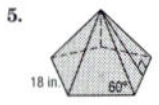

701.5 in²

6. 

216 yd²

Skills Practice, p. 687 and Practice, p. 688 (shown)

Find the surface area of each regular pyramid. Round to the nearest tenth if necessary.

1. 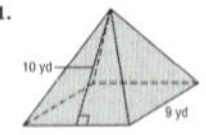

261 yd²

2. 

147.2 m²

3. 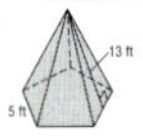

205.5 ft²

4. 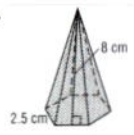

76.2 cm²

5. 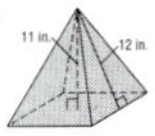

322.2 in²

6. 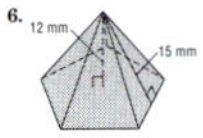

784.7 mm²

7. 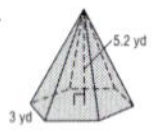

75.7 yd²

8. 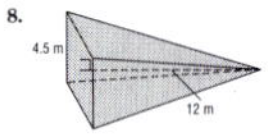

130.1 m²

9. **GAZEBOS** The roof of a gazebo is a regular octagonal pyramid. If the base of the pyramid has sides of 0.5 meters and the slant height of the roof is 1.9 meters, find the area of the roof. 3.8 m²

Reading to Learn Mathematics, p. 689 ELL

Pre-Activity **How are pyramids used in architecture?**

Read the introduction to Lesson 12-5 at the top of page 660 in your textbook.

Why do you think that the architect for the new entrance to the Louvre decided to use a pyramid rather than a rectangular prism?

Sample answer: The pyramid, with its sharp point at the vertex and its sloping sides, is more unusual and dramatic than a rectangular prism.

Reading the Lesson

1. In the figure, $ABCDE$ has congruent sides and congruent angles.

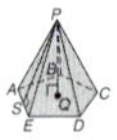

   a. Describe this pyramid with as specific a name as possible. regular pentagonal pyramid
   b. Use the figure to name the base of this pyramid. *ABCDE*
   c. Describe the base of the pyramid. regular pentagon
   d. Name the vertex of the pyramid. *P*
   e. Name the lateral faces of the pyramid. △*PAB*, △*PBC*, △*PCD*, △*PDE*, and △*PEA*
   f. Describe the lateral faces. five congruent isosceles triangles
   g. Name the lateral edges of the pyramid. $\overline{PA}$, $\overline{PB}$, $\overline{PC}$, $\overline{PD}$, and $\overline{PE}$
   h. Name the altitude of the pyramid. $\overline{PQ}$
   i. Write an expression for the height of the pyramid. *PQ*
   j. Write an expression for the slant height of the pyramid. *PS*

2. In a regular square pyramid, let s represent the side length of the base, h represent the height, a represent the apothem, and ℓ represent the slant height. Also, let L represent the lateral area and let T represent the surface area. Which of the following relationships are correct? A, D, E, F

   A. $s = 2a$ B. $a^2 + \ell^2 = h^2$ C. $L = 4\ell s$

   D. $h = \sqrt{\ell^2 - a^2}$ E. $\left(\frac{s}{2}\right)^2 + h^2 = \ell^2$ F. $T = s^2 + 2\ell s$

Helping You Remember

3. A good way to remember something is to explain it to someone else. Suppose that one of your classmates is having trouble remembering the difference between the *height* and the *slant height* of a regular pyramid. How can you explain this concept?

Sample answer: The *height* of the pyramid is the length of the perpendicular segment from the vertex to the center of the base. The *slant height* is the length of the perpendicular segment from the vertex of the pyramid to the midpoint of one of the sides of the base of the pyramid.

18. **STADIUMS** The Pyramid Arena in Memphis, Tennessee, is the third largest pyramid in the world. The base is 360,000 square feet, and the pyramid is 321 feet tall. Find the lateral area of the pyramid. (Assume that the base is a square). ≈ 527,237.2 ft²

19. **HOTELS** The Luxor Hotel in Las Vegas is a black glass pyramid. The base is a square with edges 646 feet long. The hotel is 350 feet tall. Find the area of the glass. ≈ 615,335.3 ft²

More About. . .

History

Egyptologists believe that the Great Pyramids of Egypt were originally covered with white limestone that has worn away or been removed.

Source: www.pbs.org

20. **HISTORY** Each side of the square base of Khafre's Pyramid is 214.5 meters. ★ The sides rise at an angle of about 53°. Find the lateral area of the pyramid. ≈ 76,452.5 m²

For Exercises 21–23, use the following information.

This solid is a composite of a cube and square pyramid. The base of the solid is the base of the cube. Find the indicated measurements for the solid.

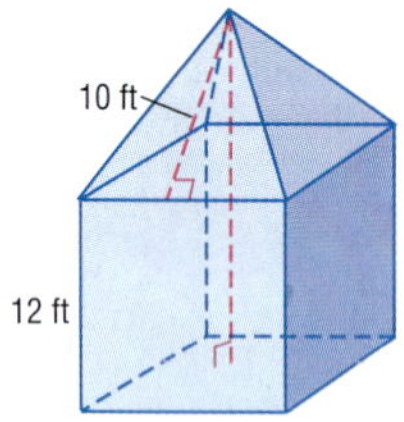

21. Find the height. 20 ft
22. Find the lateral area. 816 ft²
23. Find the surface area. 960 ft²

24. A *frustum* is the part of a solid that remains after the top portion has been cut by a plane parallel to the base. Find the lateral area of the frustum of a regular pyramid. 36 yd²

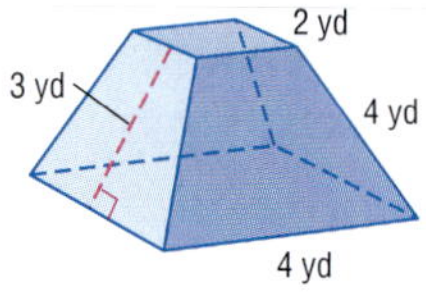

25. **CRITICAL THINKING** This square prism measures 1 inch on each side. The corner of the cube is cut off, or *truncated*. Does this change the surface area of the cube? Include the surface area of the original cube and that of the truncated cube in your answer. See margin.

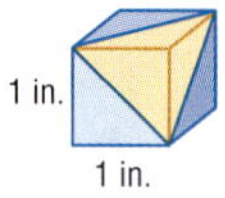

26. **WRITING IN MATH** Answer the question that was posed at the beginning of the lesson. See margin.

How are pyramids used in architecture?

Include the following in your answer:
- information needed to find the lateral area and surface area, and
- other examples of pyramidal shapes used in architecture.

FCAT Practice
Standardized Test Practice
Ⓐ Ⓑ Ⓒ Ⓓ

27. The base of a square pyramid has a perimeter of 20 centimeters, and the slant height is 10 centimeters. What is the surface area of the pyramid? D

Ⓐ 96.8 cm² Ⓑ 116 cm² Ⓒ 121.8 cm² Ⓓ 125 cm²

Enrichment, p. 690

Two Truncated Solids

To create a truncated solid, you could start with an ordinary solid and then cut off the corners. Another way to make such a shape is to use the patterns on this page.

The Truncated Octahedron

1. Two copies of the pattern at the right can be used to make a *truncated octahedron*, a solid with 6 square faces and 8 regular hexagonal faces.

   Each pattern makes half of the truncated octahedron. Attach adjacent faces using glue or tape to make a cup-shaped figure.

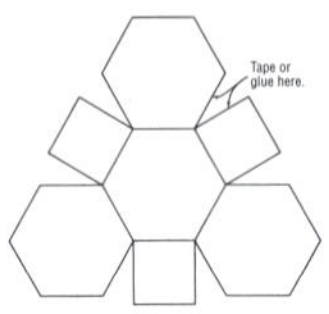

Answer

25. The surface area of the original cube is 6 square inches. The surface area of the truncated cube is approximately 5.37 square inches. Truncating the corner of the cube reduces the surface area by 0.63 square inch.

28. **ALGEBRA** If $x \otimes y = \frac{1}{x - y}$, what is the value of $\frac{1}{2} \otimes \frac{3}{4}$? **A**

Ⓐ -4 Ⓑ $-\frac{1}{4}$ Ⓒ $\frac{4}{5}$ Ⓓ $\frac{5}{4}$

Maintain Your Skills

Mixed Review

Find the surface area of each cylinder. Round to the nearest tenth. *(Lesson 12-4)*

29. 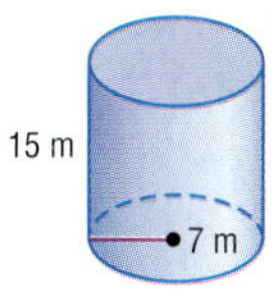

967.6 m²

30. 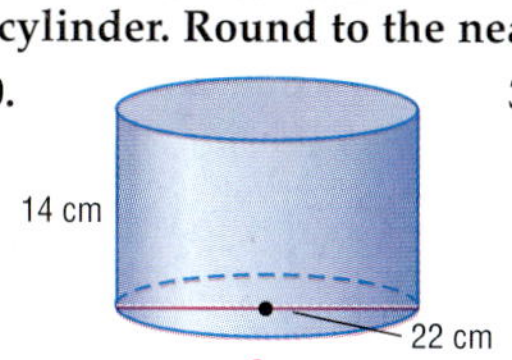

1727.9 cm²

31. 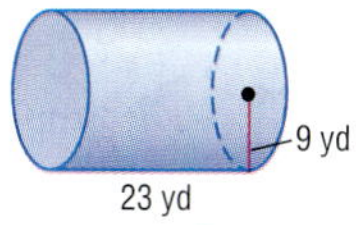

1809.6 yd²

32. **FOOD** Most cereals are packaged in cardboard boxes. If a box of cereal is 14 inches high, 6 inches wide, and 2.5 inches deep, find the surface area of the box. *(Lesson 12-3)* **268 in²**

Find the perimeter and area of each parallelogram. Round to the nearest tenth if necessary. *(Lesson 11-1)*

33. 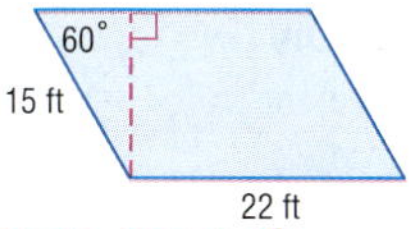

74 ft, 285.8 ft²

34. 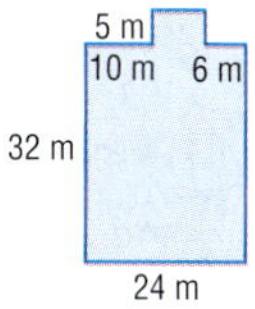

122 m, 808 m²

35. 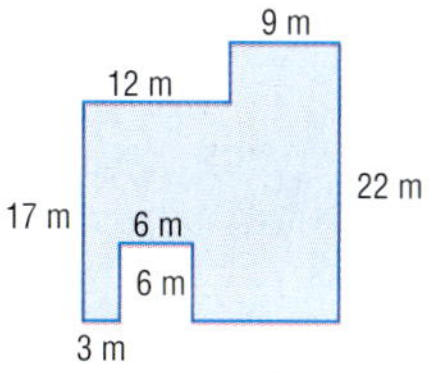

98 m, 366 m²

For Exercises 36–39, refer to the figure at the right. Name the reflected image of each figure. *(Lesson 9-1)*

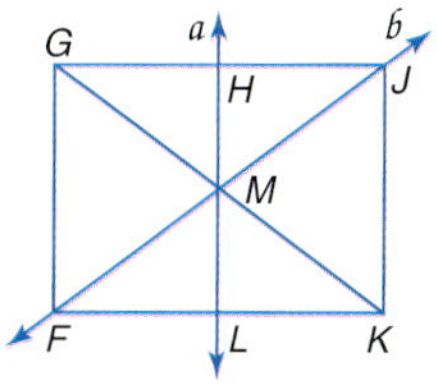

36. $\overline{FM}$ in line b **$\overline{FM}$**

37. $\overline{JK}$ in line a **$\overline{GF}$**

38. L in point M **H**

39. $\overline{GM}$ in line a **$\overline{JM}$**

Determine whether each statement is *true* or *false*. Explain. *(Lesson 8-3)*

40. If two pairs of consecutive sides of a quadrilateral are congruent, then the quadrilateral must be a parallelogram.

40. False; each pair of opposite sides must be congruent.

41. If all four sides of a quadrilateral are congruent, then the quadrilateral is a parallelogram. **True; each pair of opposite sides are congruent.**

Getting Ready for the Next Lesson

PREREQUISITE SKILL **Use the Pythagorean Theorem to solve for the missing length in each triangle. Round to the nearest tenth.**
*(To review the **Pythagorean Theorem**, see Lesson 7-2.)*

42. 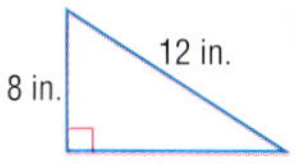

8.9 in.

43. 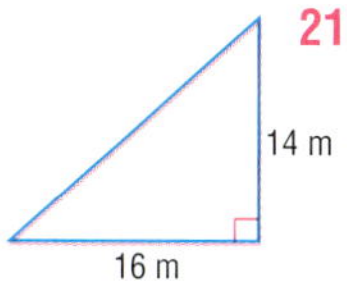

21.3 m

44. 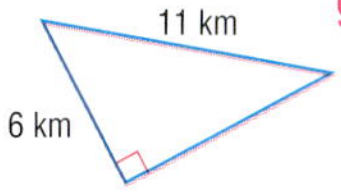

9.2 km

www.geometryonline.com/self_check_quiz/fcat

4 Assess

Open-Ended Assessment

Modeling Have students construct regular pyramids. Ask them to measure the slant height and the height of the pyramid, and then find the lateral area and surface area of their pyramid.

Getting Ready for Lesson 12-6

Prerequisite Skill Students will find the surface areas of cones in Lesson 12-6. They will use the Pythagorean Theorem to find the unknown height of a cone. Use Exercises 42–44 to determine your students' familiarity with using the Pythagorean Theorem.

Answer (page 664)

26. Sample answer: Pyramids are used as an alternative to rectangular prisms for the shapes of buildings. Answers should include the following.
    - We need to know the dimensions of the base and slant height to find the lateral area and surface area of a pyramid.
    - Sample answer: The roof of a gazebo is often a hexagonal pyramid.

12-6 Lesson Notes

1 Focus

5-Minute Check Transparency 12-6 Use as a quiz or review of Lesson 12-5.

Mathematical Background notes are available for this lesson on p. 634D.

How is the lateral area of a cone used to cover tepees?

Ask students:

- Which shape is used to make a tepee? **cone**
- What are some reasons this shape may be used to construct a living area? **Sample answer: A circle for a base is an efficient shape; the triangles formed by the sticks with the ground are a strong supportive shape.**
- How are cones used in the real-world? **Sample answer: containers for ice cream**

12-6 Surface Areas of Cones

Sunshine State Standards
MA.B.1.4.1

What You'll Learn

- Find lateral areas of cones.
- Find surface areas of cones.

Vocabulary

- circular cone
- right cone
- oblique cone

How is the lateral area of a cone used to cover tepees?

Native American tribes on the Great Plains typically lived in tepees, or tipis. Tent poles were arranged in a conical shape, and animal skins were stretched over the frame for shelter. The top of a tepee was left open for smoke to escape.

LATERAL AREAS OF CONES The shape of a tepee suggests a **circular cone**. Cones have the following characteristics.

- The base is a circle and the vertex is the point V.
- The *axis* is the segment with endpoints that are the vertex and the center of the base.
- The segment that has the vertex as one endpoint and is perpendicular to the base is called the *altitude* of the cone.

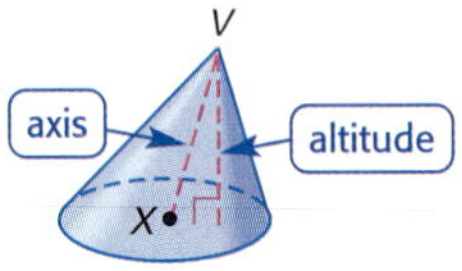

oblique cone

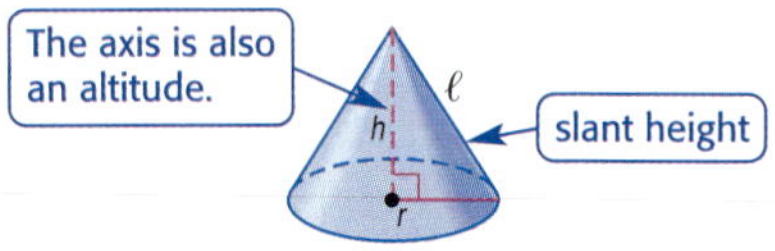

right cone

Study Tip

Reading Math
From this point in the text, you can assume that cones are right circular cones. If the cone is oblique, it will be clearly stated.

A cone with an axis that is also an altitude is called a **right cone**. Otherwise, it is called an **oblique cone**. The measure of any segment joining the vertex of a right cone to the edge of the circular base is called the *slant height*, ℓ. The measure of the altitude is the height h of the cone.

We can use the net for a cone to derive the formula for the lateral area of a cone. The lateral region of the cone is a sector of a circle with radius ℓ, the slant height of the cone. The arc length of the sector is the same as the circumference of the base, or $2\pi r$. The circumference of the circle containing the sector is $2\pi\ell$. The area of the sector is proportional to the area of the circle.

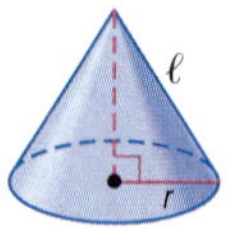

$$\frac{\text{area of sector}}{\text{area of circle}} = \frac{\text{measure of arc}}{\text{circumference of circle}} \quad \text{Write a proportion.}$$

$$\frac{\text{area of sector}}{\pi\ell^2} = \frac{2\pi r}{2\pi\ell} \quad \text{Substitution}$$

$$\text{area of sector} = \frac{(\pi\ell^2)(2\pi r)}{2\pi\ell} \quad \text{Multiply each side by } \pi\ell^2.$$

$$\text{area of sector} = \pi r\ell \quad \text{Simplify.}$$

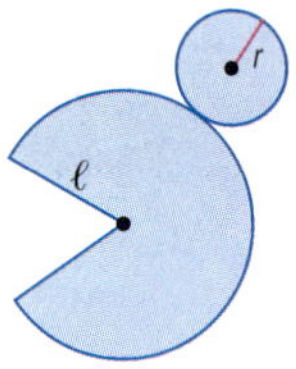

This derivation leads to the formula for the lateral area of a right circular cone.

Resource Manager

Workbook and Reproducible Masters

Chapter 12 Resource Masters
- Study Guide and Intervention, pp. 691–692
- Skills Practice, p. 693
- Practice, p. 694
- Reading to Learn Mathematics, p. 695
- Enrichment, p. 696
- Assessment, p. 718

Teaching Geometry With Manipulatives Masters, pp. 17, 196, 199

Transparencies

5-Minute Check Transparency 12-6
Answer Key Transparencies

Technology

Interactive Chalkboard

Key Concept — *Lateral Area of a Cone*

If a right circular cone has a lateral area of L square units, a slant height of ℓ units, and the radius of the base is r units, then $L = \pi r\ell$.

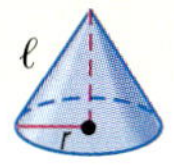

Example 1 *Lateral Area of a Cone*

LAMPS **Diego has a conical lamp shade with an altitude of 6 inches and a diameter of 12 inches. Find the lateral area of the lampshade.**

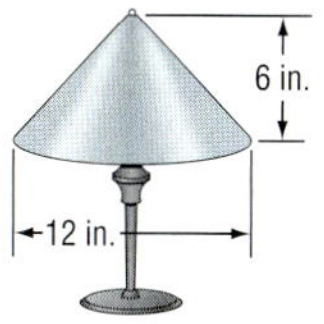

Explore We are given the altitude and the diameter of the base. We need to find the slant height of the cone.

Plan The radius of the base, height, and slant height form a right triangle. Use the Pythagorean Theorem to solve for the slant height. Then use the formula for the lateral area of a right circular cone.

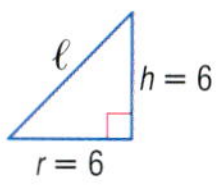

Solve Write an equation and solve for ℓ.

$\ell^2 = 6^2 + 6^2$ — Pythagorean Theorem

$\ell^2 = 72$ — Simplify.

$\ell = \sqrt{72}$ or $6\sqrt{2}$ — Take the square root of each side.

Next, use the formula for the lateral area of a right circular cone.

$L = \pi r\ell$ — Lateral area of a cone

$\approx \pi(6)(6\sqrt{2})$ — $r = 6$, $\ell = 6\sqrt{2}$

≈ 159.9 — Use a calculator.

The lateral area is approximately 159.9 square inches.

Examine Use estimation to check the reasonableness of this result. The lateral area is approximately $3 \cdot 6 \cdot 9$ or 162 square inches. Compared to the estimate, the answer is reasonable.

SURFACE AREAS OF CONES To find the surface area of a cone, add the area of the base to the lateral area.

Key Concept — *Surface Area of a Cone*

If a right circular cone has a surface area of T square units, a slant height of ℓ units, and the radius of the base is r units, then $T = \pi r\ell + \pi r^2$.

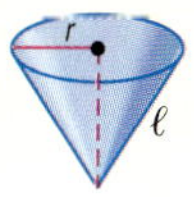

Example 2 *Surface Area of a Cone*

Find the surface area of the cone.

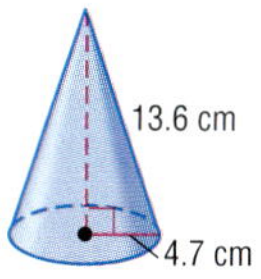

$T = \pi r\ell + \pi r^2$ — Surface area of a cone

$= \pi(4.7)(13.6) + \pi(4.7)^2$ — $r = 4.7$, $\ell = 13.6$

≈ 270.2 — Use a calculator.

The surface area is approximately 270.2 square centimeters.

Study Tip

Storing Values in Calculator Memory

A graphing calculator will let you store values for future use. For example, you could store the calculated value of ℓ by [√] 72 [STO▸] [ALPHA] [L]. To find the lateral area, use [2nd] [π] [X] 6 [X] [ALPHA] [L] [ENTER].

Study Tip

Making Connections

The surface area of a cone is like the surface area of a pyramid, $T = L + B$.

www.geometryonline.com/extra_examples/fcat

2 Teach

LATERAL AREAS OF CONES

In-Class Example PowerPoint®

Teaching Tip The difference between the slant height of a cone and the height of the cone may be confusing for some students. Point out that the *slant height* is the length from the vertex to the circle of the base, while the *height* is the altitude of the cone. The meanings of height and slant height are the same for pyramids and cones.

1 **ICE CREAM** A sugar cone has an altitude of 8 inches and a diameter of $2\frac{1}{2}$ inches. Find the lateral area of the sugar cone. ≈ 31.8 in^2

SURFACE AREAS OF CONES

In-Class Example PowerPoint®

2 Find the surface area of the cone. Round to the nearest tenth.

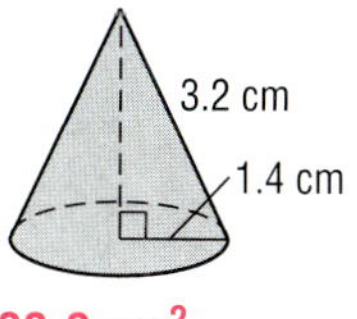

20.2 cm^2

DAILY INTERVENTION — **Differentiated Instruction** — **ELL**

Verbal/Linguistic Ask students to name the similarities and differences between finding the lateral area and surface area of a cone versus the lateral area and surface area of a regular pyramid. **The similarity is that both lateral areas use the perimeter of the base and the slant height in their formulas. Likewise, they both add on the area of the base to find the surface area. The difference is that the cone has a circular base and the pyramid has a polygonal base so the formulas for base area will differ.**

3 Practice/Apply

Study Notebook

Have students—

- add the definitions/examples of the vocabulary terms to their Vocabulary Builder worksheets for Chapter 12.
- include the formulas and examples of finding the lateral area and surface area of a cone.
- include any other item(s) that they find helpful in mastering the skills in this lesson.

About the Exercises...

Organization by Objective

- **Lateral Areas of Cones:** 22–30
- **Surface Areas of Cones:** 7–21, 31

Odd/Even Assignments

Exercises 7–21 are structured so that students practice the same concepts whether they are assigned odd or even problems.

Assignment Guide

Basic: 7–15 odd, 19–23 odd, 27–31 odd, 32–55

Average: 7–31 odd, 32–55

Advanced: 8–32 even, 33–49 (optional: 50–55)

All: Quiz 2 (1–5)

Answers

1. Sample answer:

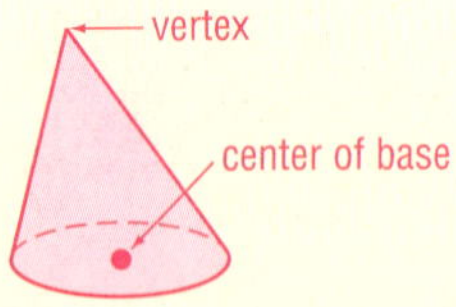

2. The formula for the lateral area is derived from the area of a sector of a circle. If the vertex of the cone is not the center of this circle, the formula is not valid.

Check for Understanding

Concept Check

1–2. See margin.

1. **OPEN ENDED** Draw an oblique cone. Mark the vertex and the center of the base.
2. **Explain** why the formula for the lateral area of a right circular cone does not apply to oblique cones.

Guided Practice

GUIDED PRACTICE KEY

Exercises	Examples
3–5	2
6	1

Find the surface area of each cone. Round to the nearest tenth.

3. 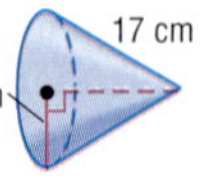

848.2 cm^2

4. 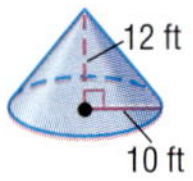

804.9 ft^2

5. 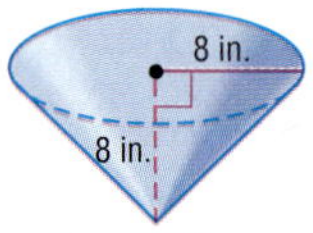

485.4 in^2

Application

6. **TOWERS** In 1921, Italian immigrant Simon Rodia bought a home in Los Angeles, California, and began building conical towers in his backyard. The structures are made of steel mesh and cement mortar. Suppose the height of one tower is 55 feet and the diameter of the base is 8.5 feet, find the lateral area of the tower. 736.5 ft^2

★ indicates increased difficulty

Practice and Apply

Homework Help

For Exercises	See Examples
7–21, 31	2
22–30	1

Extra Practice
See page 779.

Find the surface area of each cone. Round to the nearest tenth.

7. 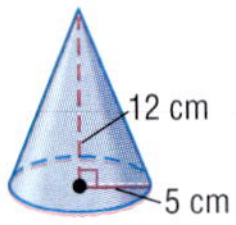

282.7 cm^2

8. 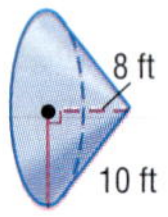

301.6 ft^2

9. 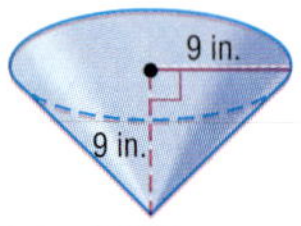

614.3 in^2

10. 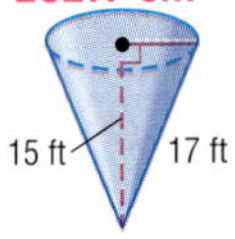

628.3 ft^2

11. 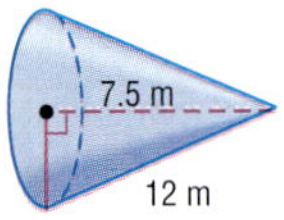

628.8 m^2

12. 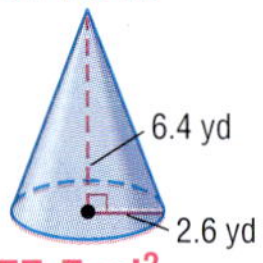

77.7 yd^2

For Exercises 13–16, round to the nearest tenth.

13. Find the surface area of the cone if the height is 16 inches and the slant height is 18 inches. 679.9 in^2
14. Find the surface area of the cone if the height is 8.7 meters and the slant height is 19.1 meters. 1928.6 m^2
15. The surface area of a cone is 1020 square meters and the radius is 14.5 meters. Find the slant height. 7.9 m
16. The surface area of a cone is 293.2 square feet and the radius is 6.1 feet. Find the slant height. 9.2 ft

Find the radius of a cone given the surface area and slant height. Round to the nearest tenth.

★17. $T = 359\ ft^2$, $\ell = 15$ ft 5.6 ft

★18. $T = 523\ m^2$, $\ell = 12.1$ m 8.2 m

Find the surface area of each solid. Round to the nearest tenth.

19. 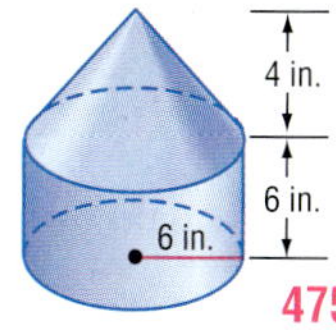

475.2 in^2

20. 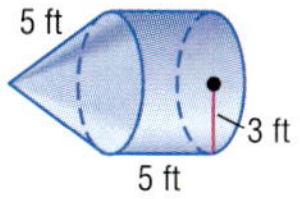

169.6 ft^2

21. 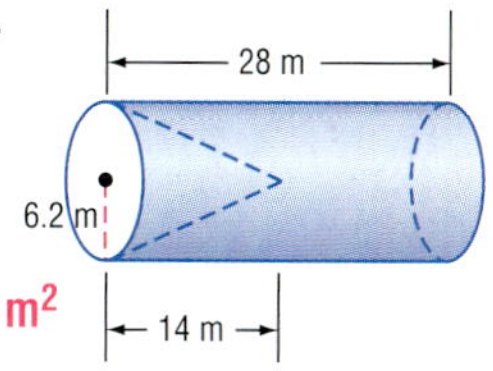

1509.8 m^2

More About. . .

Tepees

The Saamis Tepee in Medicine Hat, Alberta, Canada, is the world's largest tepee. The frame is made of steel, instead of wood. It weighs 200 tons.

Source: www.medicinehatchamber.com

22. **TEPEES** Find the area of canvas used to cover a tepee if the diameter of the base is 42 feet and the slant height is 47.9 feet. $\approx$ 3160.1 ft^2

23. **PARTY HATS** Shelley plans to make eight conical party hats for her niece's birthday party. She wants each hat to be 18 inches tall and the bases of each to be 22 inches in circumference. How much material will she use to make the hats? 1613.7 in^2

24. **WINTER STORMS** Many states use a cone structure to store salt used to melt snow on highways and roads. Find the lateral area of one of these cone structures if the building measures 24 feet tall and the diameter is 45 feet. 2325.4 ft^2

★ 25. **SPOTLIGHTS** A yellow-pink spotlight was positioned directly above a performer. If the surface area of the cone of light was approximately 500 square feet and the slant height was 20 feet, find the diameter of light on stage. $\approx$ 12 ft

The height of a cone is 7 inches, and the radius is 4 inches. Round final answers to the nearest ten-thousandth.

26. Find the lateral area of the cone using the store feature of a calculator.

26. 101.3133 in^2

27. Round the slant height to the nearest tenth and then calculate the lateral area of the cone. 8.1 in., 101.7876 in^2

28. Round the slant height to the nearest hundredth and then calculate the lateral area of the cone. 8.06 in., 101.2849 in^2

29. Compare the lateral areas for Exercises 26–28. Which is most accurate? Explain. See margin.

Determine whether each statement is *sometimes*, *always*, or *never* true. Explain.

30. If the diagonal of the base of a square pyramid is equal to the diameter of the base of a cone and the heights of both solids are equal, then the pyramid and cone have equal lateral areas. Never; the pyramid could be inscribed in the cone.

31. The ratio of the radii of the bases of two cones is equal to the ratio of the surface areas of the cone. Sometimes; only when the heights are in the same ratio as the radii of the bases.

32. **CRITICAL THINKING** If you were to move the vertex of a right cone down the axis toward the center of the base, explain what would happen to the lateral area and surface area of the cone.

32. As the altitude approaches zero, the slant height of the cone approaches the radius of the base. The lateral area approaches the area of the base. The surface area approaches twice the area of the base.

33. **WRITING IN MATH** Answer the question that was posed at the beginning of the lesson. See p. 685D.

How is the lateral area of a cone used to cover tepees?

Include the following in your answer:

- information needed to find the lateral area of the canvas covering, and
- how the open top of a tepee affects the lateral area of the canvas covering it.

Answer

29. Using the store feature on the calculator is the most accurate technique to find the lateral area. Rounding the slant height to either the tenths place or hundredths place changes the value of the slant height, which affects the final computation of the lateral area.

Study Guide and Intervention, p. 691 (shown) and p. 692

Lateral Areas of Cones Cones have the following properties.

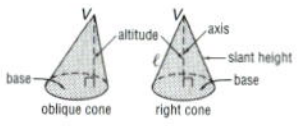

- A cone has one circular base and one vertex.
- The segment whose endpoints are the vertex and the center of the base is the **axis** of the cone.
- The segment that has one endpoint at the vertex, is perpendicular to the base, and has its other endpoint on the base is the **altitude** of the cone.
- For a **right cone** the axis is also the altitude, and any segment from the circumference of the base to the vertex is the **slant height** ℓ. If a cone is not a right cone, it is oblique.

Lateral Area of a Cone	If a cone has a lateral area of L square units, a slant height of ℓ units, and the radius of the base is r units, then $L = \pi r\ell$.

Example **Find the lateral area of a cone with slant height of 10 centimeters and a base with a radius of 6 centimeters.**

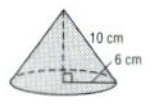

$L = \pi r\ell$ Lateral area of a cone
$= \pi(6)(10)$ $r = 6$, $\ell = 10$
≈ 188.5 Simplify.

The lateral area is about 188.5 square centimeters.

Exercises

Find lateral area of each circular cone. Round to the nearest tenth.

1. 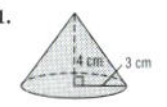

47.1 cm^2

2. 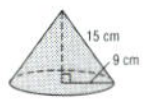

424.1 cm^2

3. 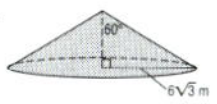

391.8 m^2

4. 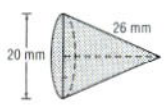

816.8 mm^2

5. 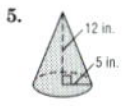

204.2 in^2

6. 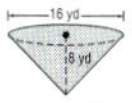

284.3 yd^2

Skills Practice, p. 693 and Practice, p. 694 (shown)

Find the surface area of each cone. Round to the nearest tenth if necessary.

1. 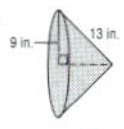

622.0 in^2

2. 

578.7 mm^2

3. 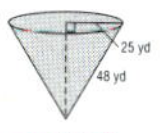

5733.4 yd^2

4. 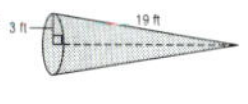

207.3 ft^2

5. 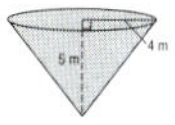

130.7 m^2

6. 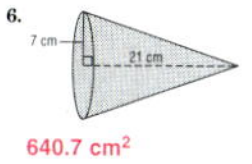

640.7 cm^2

7. Find the surface area of a cone if the height is 8 feet and the slant height is 10 feet. 301.6 ft^2

8. Find the surface area of a cone if the height is 14 centimeters and the slant height is 16.4 centimeters. 669.3 cm^2

9. Find the surface area of a cone if the height is 12 inches and the diameter is 27 inches. 1338.6 in^2

10. **HATS** Cuong bought a conical hat on a recent trip to central Vietnam. The basic frame of the hat is 16 hoops of bamboo that gradually diminish in size. The hat is covered in palm leaves. If the hat has a diameter of 50 centimeters and a slant height of 32 centimeters, what is the lateral area of the conical hat? about 2513.3 cm^2

Reading to Learn Mathematics, p. 695 ELL

Pre-Activity **How is the lateral area of a cone used to cover tepees?**

Read the introduction to Lesson 12-6 at the top of page 666 in your textbook.

If you wanted to build a tepee of a certain size, how would it help you to know the formula for the lateral area of a cone?

Sample answer: The formula would help you estimate how many animal skins you would need for the tepee.

Reading the Lesson

1. 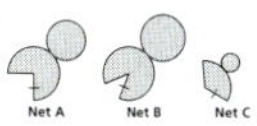

a. Which net will give the cone with the greatest lateral area? Net B

b. Which net will give the tallest cone? Net C

2. Refer to the figure at the right. Suppose you have removed the circular base of the cone and cut from V to A so that you can unroll the lateral surface onto a flat table.

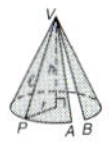

a. How can you be sure that the flattened-out piece is a sector of a circle? Sample answer: Before you unroll the lateral surface, all points on the bottom rim of the cone are ℓ units from V. After you flatten out the surface, those points are still ℓ units from V. This means that they will lie on a circle with center V and radius ℓ.

b. How do you know that the flattened-out piece is not a full circle? Sample answer: $r < \ell$

3. Suppose you have a right cone with radius r, diameter d, height h, and slant height ℓ. Which of the following relationships involving these lengths are correct? C, E

A. $r = 2d$ B. $r + h = \ell$ C. $r^2 + h^2 = \ell^2$

D. $r^2 + \ell^2 = h^2$ E. $r = \sqrt{\ell^2 - h^2}$ F. $h = \pm\sqrt{\ell^2 - r^2}$

Helping You Remember

4. One way to remember a new formula is to relate it to a formula you already know. Explain how the formulas for the lateral areas of a pyramid and a cone are similar. Sample answer: Both formulas say that the lateral area includes the distance around the base multiplied by the slant height.

Enrichment, p. 696

Cone Patterns

The pattern at the right is made from a circle. It can be folded to make a cone.

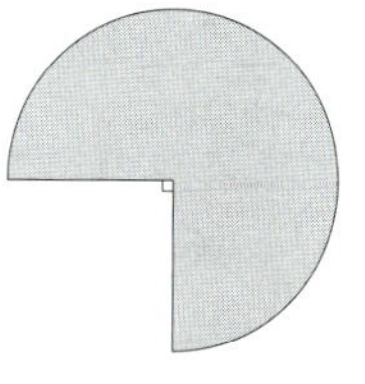

1. Measure the radius of the circle to the nearest centimeter. 4 cm

2. The pattern is what fraction of the complete circle? $\frac{3}{4}$

3. What is the circumference of the complete circle? 8π cm

4. How long is the circular arc that is the outside of the pattern? 6π cm

5. Cut out the pattern and tape it together to form a cone. See students' work.

6. Measure the diameter of the circular base of the cone. 6 cm

4 Assess

Open-Ended Assessment

Writing Have students describe how to find the lateral area and the surface area of a cone. Ask them to draw an example that includes the measure of the slant height and the height of the cone, and then find the lateral area and surface area of the cone.

Getting Ready for Lesson 12-7

Prerequisite Skill Students will find the surface areas of spheres in Lesson 12-7. They will use the circumference formula for a circle to find the radius of a sphere. Use Exercises 50–55 to determine your students' familiarity with finding the circumference of a circle given the radius or diameter.

Assessment Options

Practice Quiz 2 The quiz provides students with a brief review of the concepts and skills in Lessons 12-5 and 12-6. Lesson numbers are given to the right of the exercises or instruction lines so students can review concepts not yet mastered.

Quiz (Lessons 5 and 6) is available on p. 718 of the *Chapter 12 Resource Masters.*

FCAT Practice

Standardized Test Practice

(A) (B) (C) (D)

34. The lateral area of the cone is 91.5π square feet. What is the radius of the base? **B**

(A) 5.9 ft (B) 6.1 ft (C) 7.5 ft (D) 10 ft

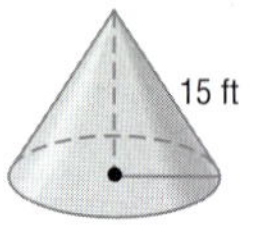

35. ALGEBRA Three times the first of three consecutive odd integers is 3 more than twice the third. Find the third integer. **D**

(A) 9 (B) 11 (C) 13 (D) 15

Maintain Your Skills

Mixed Review

36. ARCHITECTURE The Transamerica Tower in San Francisco is a regular pyramid with a square base that is 149 feet on each side and a height of 853 feet. Find its lateral area. *(Lesson 12-5)* **≈ 255,161.7 ft²**

Find the radius of the base of the right cylinder. Round to the nearest tenth. *(Lesson 12-4)*

37. The surface area is 563 square feet, and the height is 9.5 feet. **5.8 ft**

38. The surface area is 185 square meters, and the height is 11 meters. **2.2 m**

39. The surface area is 470 square yards, and the height is 6.5 yards. **6.0 yd**

40. The surface area is 951 square centimeters, and the height is 14 centimeters. **7.2 cm**

In ⊙M, $FL = 24$, $HJ = 48$, and $m\widehat{HP} = 45$. Find each measure. *(Lesson 10-3)*

41. FG **48** **42.** NJ **24**

43. HN **24** **44.** LG **24**

45. $m\widehat{PJ}$ **45** **46.** $m\widehat{HJ}$ **90**

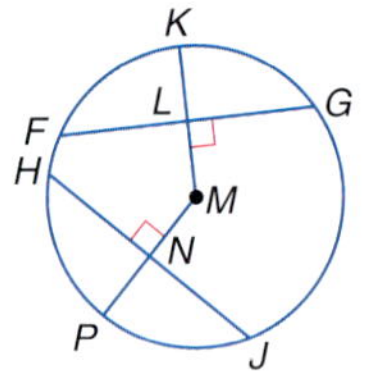

Find the geometric mean between each pair of numbers. *(Lesson 7-1)*

47. 7 and 63 **21** **48.** 8 and 18 **12** **49.** 16 and 44 **$8\sqrt{11} \approx 26.5$**

Getting Ready for the Next Lesson

PREREQUISITE SKILL **Find the circumference of each circle given the radius or the diameter. Round to the nearest tenth.**

(To review ***finding the circumference of a circle****, see Lesson 10-1.)*

50. $r = 6$ **37.7** **51.** $d = 8$ **25.1** **52.** $d = 18$ **56.5**

53. $r = 8.2$ **51.5** **54.** $d = 19.8$ **62.2** **55.** $r = 4.1$ **25.8**

Practice Quiz 2 — Lessons 12-5 and 12-6

Find the surface area of each solid. Round to the nearest tenth. *(Lessons 12-5 and 12-6)*

1.

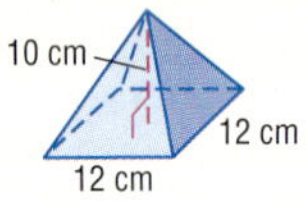

423.9 cm²

2.

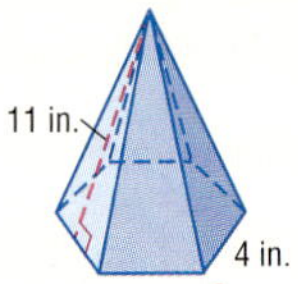

173.6 in²

3.

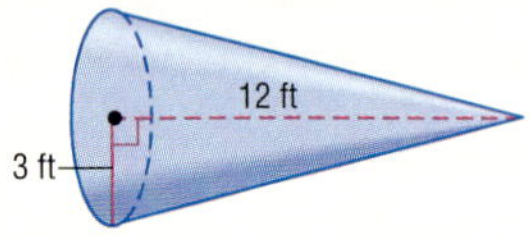

144.9 ft²

4. Find the surface area of a cone if the radius is 6 meters and the height is 2 meters. Round to the nearest tenth. *(Lesson 12-6)* **232.3 m²**

5. Find the slant height of a cone if the lateral area is 123 square inches and the radius is 10 inches. Round to the nearest tenth. *(Lesson 12-6)* **3.9 in.**

670 Chapter 12 Surface Area

12-7 Surface Areas of Spheres

What You'll Learn

- Recognize and define basic properties of spheres.
- Find surface areas of spheres.

How do manufacturers of sports equipment use the surface areas of spheres?

The sports equipment industry has grown significantly in recent years because people are engaging in more physical activities. Balls are used in many sports, such as golf, basketball, baseball, and soccer. Some are hollow, while others have solid inner cores. Each ball is identifiable by its design, texture, color, and size.

Vocabulary

- great circle
- hemisphere

PROPERTIES OF SPHERES To visualize a sphere, consider infinitely many congruent circles in space, all with the same point for their center. Considered together, these circles form a sphere. In space, a sphere is the locus of all points that are a given distance from a given point called its *center*.

There are several special segments and lines related to spheres.

- A segment with endpoints that are the center of the sphere and a point on the sphere is a *radius* of the sphere. In the figure, $\overline{DC}$, $\overline{DA}$, and $\overline{DB}$ are radii.
- A *chord* of a sphere is a segment with endpoints that are points on the sphere. In the figure, $\overline{GF}$ and $\overline{AB}$ are chords.
- A chord that contains the center of the sphere is a *diameter* of the sphere. In the figure, $\overline{AB}$ is a diameter.
- A *tangent* to a sphere is a line that intersects the sphere in exactly one point. In the figure, $\overleftrightarrow{JH}$ is tangent to the sphere at *E*.

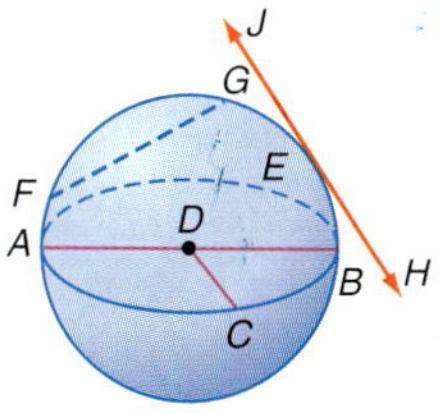

Study Tip

Circles and Spheres

The shortest distance between any two points on a sphere is the length of the arc of a great circle passing through those two points.

The intersection of a plane and a sphere can be a point or a circle. When a plane intersects a sphere so that it contains the center of the sphere, the intersection is called a **great circle**. A great circle has the same center as the sphere, and its radii are also radii of the sphere.

a point   a circle   a great circle

12-7 Lesson Notes

1 Focus

5-Minute Check Transparency 12-7 Use as a quiz or review of Lesson 12-6.

Mathematical Background notes are available for this lesson on p. 634D.

How do manufacturers of sports equipment use the surface areas of spheres?

Ask students:

- From the picture, which shapes are used to make balls for sports? **spheres**
- Why is this shape used for balls? **Sample answer: It is easy for the hand to hold and it allows the balls to bounce.**
- How could you approximate the surface area of a sphere? **Accept all reasonable answers.**

Resource Manager

Workbook and Reproducible Masters

Chapter 12 Resource Masters

- Study Guide and Intervention, pp. 697–698
- Skills Practice, p. 699
- Practice, p. 700
- Reading to Learn Mathematics, p. 701
- Enrichment, p. 702
- Assessment, p. 718

Teaching Geometry With Manipulatives Masters, pp. 200, 201, 202

Transparencies

5-Minute Check Transparency 12-7
Real-World Transparency 12
Answer Key Transparencies

Technology

GeomPASS: Tutorial Plus, Lesson 22
Interactive Chalkboard

2 Teach

PROPERTIES OF SPHERES

Tips for New Teachers Demonstrate the characteristics of spheres by using a ball. If possible, cut the ball in half and ask a student to point out the diameter, a chord, and the center. Use a pencil or pen as a tangent to the sphere.

In-Class Example

Power Point®

Teaching Tip Point out that the intersection of any plane with a sphere in its interior is a circle. with a radius less than or equal to the radius of the sphere.

1 In the figure, O is the center of the sphere, and plane $\mathcal{P}$ intersects the sphere in a circle R. If $OR = 6$ centimeters and $OS = 14$ centimeters, find RS.

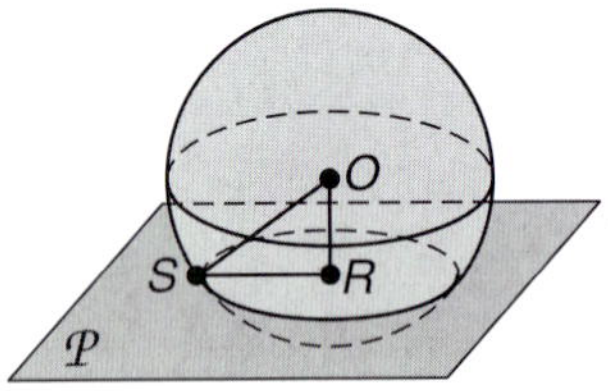

12.6 cm

Study Tip

Great Circles
A sphere has an infinite number of great circles.

Each great circle separates a sphere into two congruent halves, each called a **hemisphere**. Note that a hemisphere has a circular base.

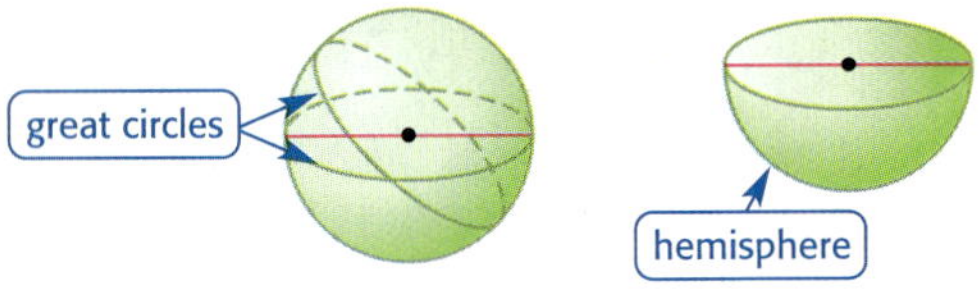

Example 1 Spheres and Circles

In the figure, O is the center of the sphere, and plane $\mathcal{R}$ intersects the sphere in circle A. If $AO = 3$ centimeters and $OB = 10$ centimeters, find AB.

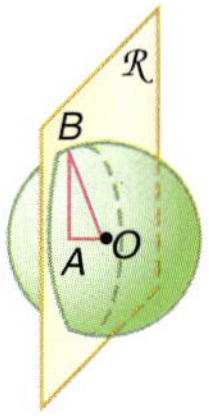

The radius of circle A is the segment $\overline{AB}$, B is a point on circle A and on sphere O. Use the Pythagorean Theorem for right triangle ABO to solve for AB.

$OB^2 = AB^2 + AO^2$	Pythagorean Theorem
$10^2 = AB^2 + 3^2$	$OB = 10$, $AO = 3$
$100 = AB^2 + 9$	Simplify.
$91 = AB^2$	Subtract 9 from each side.
$9.5 \approx AB$	Use a calculator.

AB is approximately 9.5 centimeters.

SURFACE AREAS OF SPHERES

You will investigate the surface area of a sphere in the geometry activity.

Geometry Activity

Surface Area of a Sphere

Model

- Cut a polystyrene ball along a great circle. Trace the great circle onto a piece of paper. Then cut out the circle.
- Fold the circle into eight sectors. Then unfold and cut the pieces apart. Tape the pieces back together in the pattern shown at the right.

- Use tape or glue to put the two pieces of the ball together. Tape the paper pattern to the sphere.

Analyze 1. $\frac{1}{4}$

1. Approximately what fraction of the surface of the sphere is covered by the pattern?
2. What is the area of the pattern in terms of r, the radius of the sphere? πr^2

Make a Conjecture

3. Make a conjecture about the formula for the surface area of a sphere.
The surface area of a sphere is 4 times the area of the great circle.

Geometry Activity

Materials: polystyrene ball, scissors

To help students see the relationship between the surface area of a sphere and the area of a great circle of the sphere, ask them to do the experiment with more than one size ball. In each case, the area of the great circle is about one fourth the area of the sphere. As an alternative, ask them to tightly wrap a ball in a square sheet of tissue paper and color the paper on the ball. Then ask them to unwrap the paper. About one-fourth of the paper should be colored.

The activity leads us to the formula for the surface area of a sphere.

Key Concept — *Surface Area of a Sphere*

If a sphere has a surface area of T square units and a radius of r units, then $T = 4\pi r^2$.

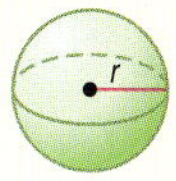

Example 2 Surface Area

a. Find the surface area of the sphere given the area of the great circle.

From the activity, we find that the surface area of a sphere is four times the area of the great circle.

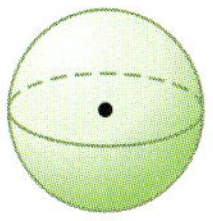

$A \approx 201.1$ in²

$T = 4\pi r^2$ Surface area of a sphere

$\approx 4(201.1)$ $\pi r^2 \approx 201.1$

≈ 804.4 Multiply.

The surface area is approximately 804.4 square inches.

b. Find the surface area of the hemisphere.

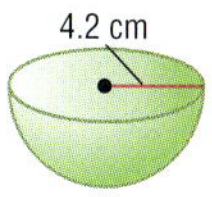

A hemisphere is half of a sphere. To find the surface area, find half of the surface area of the sphere and add the area of the great circle.

surface area $= \frac{1}{2}(4\pi r^2) + \pi r^2$ Surface area of a hemisphere

$= \frac{1}{2}[4\pi(4.2)^2] + \pi(4.2)^2$ Substitution

≈ 166.3 Use a calculator.

The surface area is approximately 166.3 square centimeters.

More About. . .

Baseball

A great circle of a standard baseball has a circumference between 9 and $9\frac{1}{4}$ inches.

Source: www.mlb.com

Example 3 Surface Area

BASEBALL **Find the surface area of a baseball with a circumference of 9 inches to determine how much leather is needed to cover the ball.**

First, find the radius of the sphere.

$C = 2\pi r$ Circumference of a circle

$9 = 2\pi r$ $C = 9$

$\frac{9}{2\pi} = r$ Divide each side by 2π.

$1.4 \approx r$ Use a calculator.

Next, find the surface area of a sphere.

$T = 4\pi r^2$ Surface area of a sphere

$\approx 4\pi(1.4)^2$ $r \approx 1.4$

≈ 25.8 Use a calculator.

The surface area is approximately 25.8 square inches.

www.geometryonline.com/extra_examples/fcat

2 Teach

SURFACE AREAS OF SPHERES

In-Class Examples Power Point®

2 a. Find the surface area of the sphere, given a great circle with an area of 907.9 square centimeters. **3631.6 cm²**

b. Find the surface area of a hemisphere with a radius of 3.8 inches. **136.1 in²**

3 Find the surface area of a ball with a circumference of 24 inches to determine how much leather is needed to make the ball. **183.3 in²**

DAILY INTERVENTION — **Differentiated Instruction**

Naturalist One way to compare moons is by their approximate diameters. Ask students to describe how to find the surface area of a moon. **To find the surface area, determine the radius from the diameter. Then substitute the radius into the formula 4π times the square of the radius.**

3 Practice/Apply

Study Notebook

Have students—
- add the definitions/examples of the vocabulary terms to their Vocabulary Builder worksheets for Chapter 12.
- include the formula and an example of finding the surface area of a sphere.
- include any other item(s) that they find helpful in mastering the skills in this lesson.

DAILY INTERVENTION

FIND THE ERROR In Exercise 2, caution students that the surface area includes the circular top of the hemisphere. It is a common error to exclude this surface when finding the area of a hemisphere because this circle is not part of the original sphere.

About the Exercises...

Organization by Objective
- **Properties of Spheres:** 10–15, 25–29
- **Surface Areas of Spheres:** 16–24, 30–40

Odd/Even Assignments

Exercises 10–15 and 17–29 are structured so that students practice the same concepts whether they are assigned odd or even problems.

Assignment Guide

Basic: 11–37 odd, 41–54

Average: 11–41 odd, 42–54

Advanced: 10–40 even, 41–54

Check for Understanding

Concept Check

1. **OPEN ENDED** Draw a sphere and a great circle. **See margin.**
2. **FIND THE ERROR** Loesha and Tim are finding the surface area of a hemisphere with a radius of 6 centimeters.

Loesha	Tim
$T = \frac{1}{2}(4\pi r^2)$	$T = \frac{1}{2}(4\pi r^2) + \pi r^2$
$T = 2\pi(6^2)$	$T = 2\pi(6^2) + \pi(6^2)$
$T = 72\pi$	$T = 72\pi + 36\pi$
	$T = 108\pi$

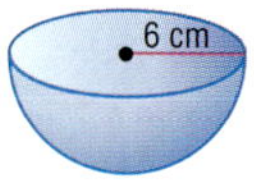

Who is correct? Explain. **Tim; the surface area of a hemisphere is half of the surface area of the sphere plus the area of the great circle.**

Guided Practice

GUIDED PRACTICE KEY	
Exercises	Examples
3–5	1
6–8	2
9	3

In the figure, A is the center of the sphere, and plane $\mathcal{M}$ intersects the sphere in circle C.

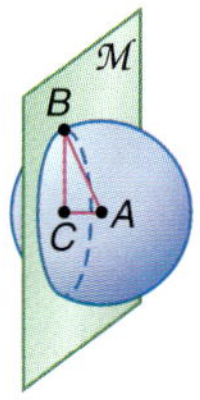

3. If $AC = 9$ and $BC = 12$, find AB. **15**
4. If the radius of the sphere is 15 units and the radius of the circle is 10 units, find AC. **≈ 11.2**
5. If Q is a point on $\odot C$ and $AB = 18$, find AQ. **18**

Find the surface area of each sphere or hemisphere. Round to the nearest tenth.

6. a sphere with radius 6.8 inches **581.1 in^2**
7. a hemisphere with the circumference of a great circle 8π centimeters **150.8 cm^2**
8. a sphere with the area of a great circle approximately 18.1 square meters **72.4 m^2**

Application

9. **BASKETBALL** An NCAA (National Collegiate Athletic Association) basketball has a radius of $4\frac{3}{4}$ inches. Find the surface area. **≈ 283.5 in^2**

★ indicates increased difficulty

Practice and Apply

Homework Help	
For Exercises	See Examples
10–15, 25–29	1
16, 30–38	3
17–24, 39, 40	2

Extra Practice See page 780.

In the figure, P is the center of the sphere, and plane $\mathcal{K}$ intersects the sphere in circle T.

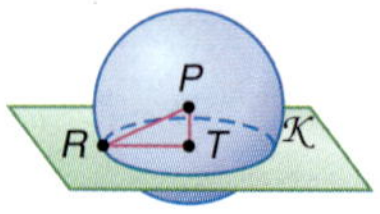

10. If $PT = 4$ and $RT = 3$, find PR. **5**
11. If $PT = 3$ and $RT = 8$, find PR. **≈ 8.5**
12. If the radius of the sphere is 13 units and the radius of $\odot T$ is 12 units, find PT. **5**
13. If the radius of the sphere is 17 units and the radius of $\odot T$ is 15 units, find PT. **8**
14. If X is a point on $\odot T$ and $PR = 9.4$, find PX. **9.4**
15. If Y is a point on $\odot T$ and $PR = 12.8$, find PY. **12.8**

16. **GRILLS** A hemispherical barbecue grill has two racks, one for the food and one for the charcoal. The food rack is a great circle of the grill and has a radius of 11 inches. The charcoal rack is 5 inches below the food rack. Find the difference in the areas of the two racks. **25π ≈ 78.5 in^2**

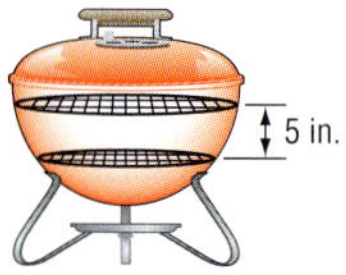

Answer

1. Sample answer:

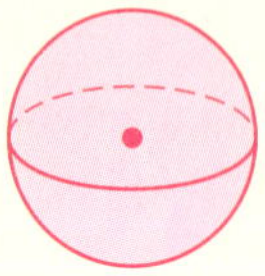

Find the surface area of each sphere or hemisphere. Round to the nearest tenth.

17.
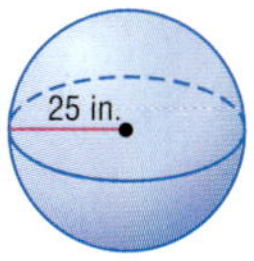

7854.0 in^2

18.

2642.1 cm^2

19.
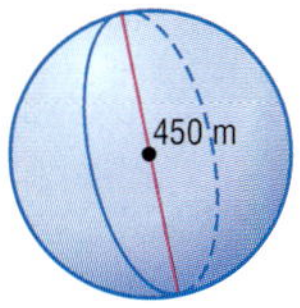

636,172.5 m^2

20.
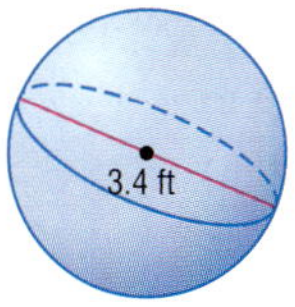

36.3 ft^2

21. Hemisphere: The circumference of a great circle is 40.8 inches. 397.4 in^2
22. Sphere: The circumference of a great circle is 30.2 feet. 290.3 ft^2
23. Sphere: The area of a great circle is 814.3 square meters. 3257.2 m^2
24. Hemisphere: The area of a great circle is 227.0 square kilometers. 681.0 km^2

Determine whether each statement is *true* or *false*. If false, give a counterexample.

25. The radii of a sphere are congruent to the radius of its great circle. true
26. In a sphere, two different great circles intersect in only one point. false
27. Two spheres with congruent radii can intersect in a circle. true
28. A sphere's longest chord will pass through the center of the circle. true
29. Two spheres can intersect in one point. true

26. See margin for counterexample.

EARTH For Exercises 30–32, use the following information.
The diameter of Earth is 7899.83 miles from the North Pole to the South Pole and 7926.41 miles from opposite points at the equator.

30. Approximate the surface area of Earth using each measure.
31. If the atmosphere of Earth extends to about 100 miles above the surface, find the surface area of the atmosphere surrounding Earth. Use the mean of the two diameters.
32. About 75% of Earth's surface is covered by water. Find the surface area of water on Earth, using the mean of the two diameters. $\approx$ 147,538,933.4 mi^2

30. pole to pole, 196,058,359.3 mi^2; equator, 197,379,906.2 mi^2

31. $\approx$ 206,788,161.4 mi^2

33. **IGLOOS** Use the information at the left to find the surface area of the living area if the diameter is 13 feet. 398.2 ft^2

34. Find the ratio of the surface area of two spheres if the radius of one is twice the radius of the second sphere. 4:1

35. Find the ratio of the radii of two spheres if the surface area of one is one half the surface area of the other. $\frac{\sqrt{2}}{2}:1$

36. Find the ratio of the surface areas of two spheres if the radius of one is three times the radius of the other. 9:1

More About. . .

Igloos

Igloos are made of hard-packed snow cut into blocks. The blocks are put together in a spiral that grows smaller at the top to form a hemispherical dome.

Source: www.infoplease.com

ASTRONOMY For Exercises 37 and 38, use the following information.
In 2002, NASA's Chandra X-Ray Observatory found two unusual neutron stars. These two stars are smaller than previously found neutron stars, but they have the mass of a larger neutron star, causing astronomers to think this star may not be made of neutrons, but a different form of matter.

37. Neutron stars have diameters from 12 to 20 miles in size. Find the range of the surface area.
38. One of the new stars has a diameter of 7 miles. Find the surface area of this star. $\approx$ 153.9 mi^2

37. The surface area can range from about 452.4 to about 1256.6 mi^2.

www.geometryonline.com/self_check_quiz/fcat

Answer

26. False, two great circles will intersect at two points.

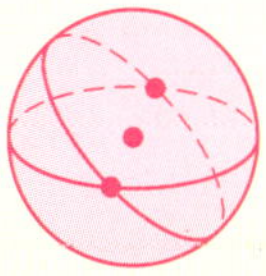

Enrichment, p. 702

Doubling Sizes

Consider what happens to surface area when the sides of a figure are doubled.

The sides of the large cube are twice the size of the sides of the small cube.

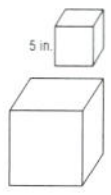

1. How long are the edges of the large cube? 6 in.
2. What is the surface area of the small cube? 54 in^2
3. What is the surface area of the large cube? 216 in^2
4. The surface area of the large cube is how many times greater than that of the small cube? 4 times

The radius of the large sphere at the right is twice the radius of the small sphere.

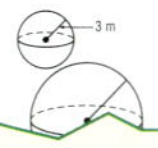

5. What is the surface area of the small sphere? 400π m^2
6. What is the surface area of the large sphere? 1600π m^2

Study Guide and Intervention, p. 697 (shown) and p. 698

Properties of Spheres A **sphere** is the locus of all points that are a given distance from a given point called its **center**.

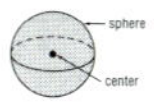

Here are some terms associated with a sphere.
- A **radius** is a segment whose endpoints are the center of the sphere and a point on the sphere.
- A **chord** is a segment whose endpoints are points on the sphere.
- A **diameter** is a chord that contains the sphere's center.
- A **tangent** is a line that intersects the sphere in exactly one point.
- A **great circle** is the intersection of a sphere and a plane that contains the center of the sphere.
- A **hemisphere** is one-half of a sphere. Each great circle of a sphere determines two hemispheres.

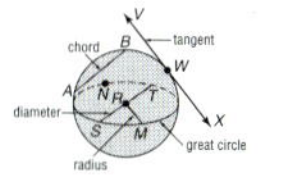

$\overline{RS}$ is a radius. $\overline{AB}$ is a chord. $\overline{ST}$ is a diameter. $\overleftrightarrow{VX}$ is a tangent. The circle that contains points S, M, T, and N is a great circle; it determines two hemispheres.

Example Determine the shapes you get when you intersect a plane with a sphere.

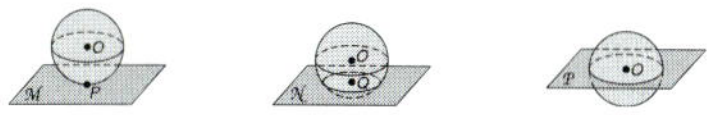

The intersection of plane $\mathcal{M}$ and sphere O is point P.
The intersection of plane $\mathcal{N}$ and sphere O is circle Q.
The intersection of plane $\mathcal{P}$ and sphere O is circle O.

A plane can intersect a sphere in a point, in a circle, or in a great circle.

Exercises

Describe each object as a model of a *circle*, a *sphere*, a *hemisphere*, or *none of these*.

1. a baseball sphere
2. a pancake circle
3. the Earth sphere
4. a kettle grill cover hemisphere
5. a basketball rim circle
6. cola can none of these

Determine whether each statement is *true* or *false*.

7. All lines intersecting a sphere are tangent to the sphere. false
8. Every plane that intersects a sphere makes a great circle. false
9. The eastern hemisphere of Earth is congruent to the western hemisphere. true
10. The diameter of a sphere is congruent to the diameter of a great circle. true

Skills Practice, p. 699 and Practice, p. 700 (shown)

In the figure, C is the center of the sphere, and plane $\mathcal{B}$ intersects the sphere in circle R. Round to the nearest tenth if necessary.

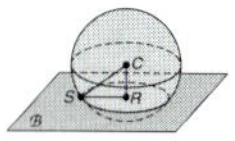

1. If $CR = 4$ and $SR = 14$, find CS. 14.6
2. If $CR = 7$ and $SR = 24$, find CS. 25
3. If the radius of the sphere is 28 units and the radius of $\odot R$ is 26 units, find CR. 10.4
4. If J is a point on $\odot R$ and $CS = 7.3$, find CJ. 7.3

Find the surface area of each sphere or hemisphere. Round to the nearest tenth.

5.
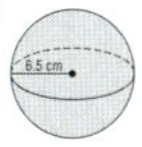

530.9 cm^2

6.

24,884.6 ft^2

7. a sphere with the area of a great circle 29.8 meters 119.2 m^2
8. a hemisphere with a radius of the great circle 8.4 inches 665.0 in^2
9. a hemisphere with the circumference of a great circle 18 millimeters 77.3 mm^2
10. **SPORTS** A standard size 5 soccer ball for ages 13 and older has a circumference of 27–28 inches. Suppose Breck is on a team that plays with a 28-inch soccer ball. Find the surface area of the ball. about 249.6 in^2

Reading to Learn Mathematics, p. 701 ELL

Pre-Activity **How do manufacturers of sports equipment use the surface areas of spheres?**

Read the introduction to Lesson 12-7 at the top of page 671 in your textbook.

How would knowing the formula for the surface area of a sphere help manufacturers of beach balls? Sample answer: The surface area of the ball tells the manufacturer how much material is needed to make each ball.

Reading the Lesson

1. In the figure, P is the center of the sphere. Name each of the following in the figure.

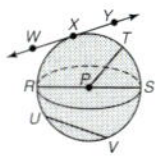

a. three radii of the sphere $\overline{PT}$, $\overline{PR}$, and $\overline{PS}$
b. a diameter of the sphere $\overline{RS}$
c. two chords of the sphere $\overline{RS}$ and $\overline{UV}$
d. a great circle of the sphere $\odot P$
e. a tangent to the sphere $\overleftrightarrow{WY}$
f. the point of tangency X

2. Determine whether each sentence is *sometimes*, *always*, or *never* true.
a. If a sphere and a plane intersect in more than one point, their intersection will be a great circle. sometimes
b. A great circle has the same center as the sphere. always
c. The endpoints of a radius of a sphere are two points on the sphere. never
d. A chord of a sphere is a diameter of the sphere. sometimes
e. A radius of a great circle is also a radius of the sphere. always

3. Match each surface area formula with the name of the appropriate solid.

a. $T = \pi r\ell + \pi r^2$ vi	i. regular pyramid
b. $T = Ph + 2B$ iv	ii. hemisphere
c. $T = 4\pi r^2$ v	iii. cylinder
d. $T = \frac{1}{2}P\ell + B$ i	iv. prism
e. $T = 2\pi rh + 2\pi r^2$ iii	v. sphere
f. $T = 3\pi r^2$ ii	vi. cone

Helping You Remember

4. Many students have trouble remembering all of the formulas they have learned in this chapter. What is an easy way to remember the formula for the surface area of a sphere? Sample answer: A sphere doesn't have any lateral faces or bases, so the expression in the formula for its surface area has just one term, $4\pi r^2$, rather than being the sum of the expressions for the lateral area and the area of the bases like the others.

4 Assess

Open-Ended Assessment

Speaking Ask students to describe how to find the surface area of a basketball that is sized by its circumference (usually 30 inches).

Assessment Options

Quiz (Lesson 7) is available on p. 718 of the *Chapter 12 Resource Masters*.

Answer

42. Sample answer: Sports equipment manufacturers use the surface area of spheres to determine the amount of material to cover the balls for different sports. Answers should include the following.
   - The surface area of a sphere is four times the area of the great circle of the sphere.
   - Racquetball and basketball are other sports that use balls.

Answer (page 677)

1.

The locus of all points in space at a specific distance from a given point is a sphere. Thus, for this problem, the locus of points is two spheres each with a radius of 5 units with centers that are endpoints of the given line segment.

39. The radius of the sphere is half the side of the cube.

40. The radius of the sphere is $\frac{x\sqrt{3}}{2}$, where x is the length of each edge of the cube.

★ **39.** A sphere is inscribed in a cube. Describe how the radius of the sphere is related to the dimensions of the cube.

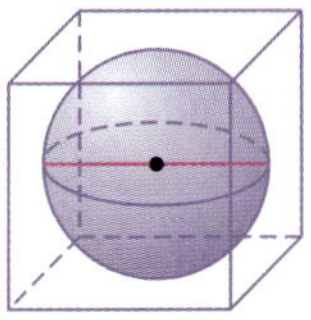

★ **40.** A sphere is circumscribed about a cube. Find the length of the radius of the sphere in terms of the dimensions of the cube.

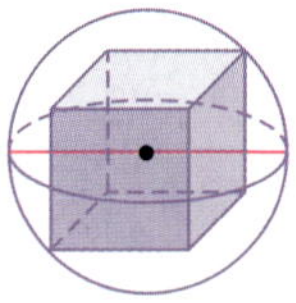

41. None; every line (great circle) that passes through X will also intersect g. All great circles intersect.

41. **CRITICAL THINKING** In spherical geometry, a plane is the surface of a sphere and a line is a great circle. How many lines exist that contain point X and do not intersect line g?

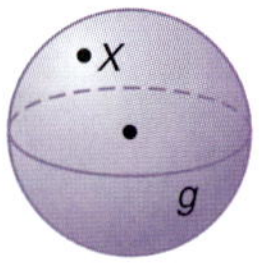

42. **WRITING IN MATH** Answer the question that was posed at the beginning of the lesson. See margin.

How do manufacturers of sports equipment use the surface area of spheres?

Include the following in your answer:
- how to find the surface area of a sphere, and
- other examples of sports that use spheres.

FCAT Practice
Standardized Test Practice

43. A rectangular solid that is 4 inches long, 5 inches high, and 7 inches wide is inscribed in a sphere. What is the radius of this sphere? A

Ⓐ 4.74 in. Ⓑ 5.66 in. Ⓒ 7.29 in. Ⓓ 9.49 in.

44. **ALGEBRA** Solve $\sqrt{x^2 + 7} - 2 = x - 1$. C

Ⓐ -3 Ⓑ $\frac{1}{3}$ Ⓒ 3 Ⓓ no solution

Maintain Your Skills

Mixed Review

Find the surface area of each cone. Round to the nearest tenth. *(Lesson 12-6)*

45. $h = 13$ inches, $\ell = 19$ inches 1430.3 in^2 **46.** $r = 7$ meters, $h = 10$ meters 422.4 m^2

47. $r = 4.2$ cm, $\ell = 15.1$ cm 254.7 cm^2 **48.** $d = 11.2$ ft, $h = 7.4$ ft 261.8 ft^2

Find the surface area of each regular pyramid. Round to the nearest tenth if necessary. *(Lesson 12-5)*

49. 969 yd^2

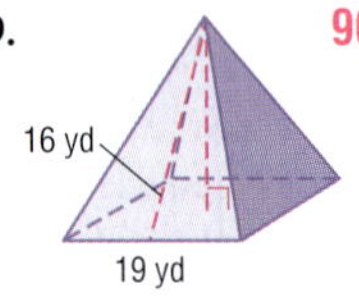

50. 487.6 ft^2

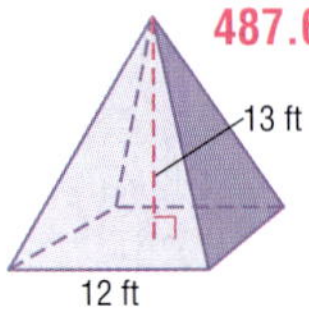

51. 649 cm^2

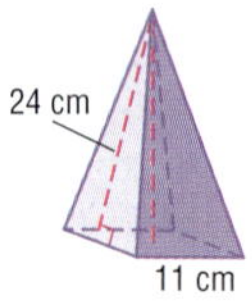

52. **RECREATION** Find the area of fabric needed to cover one side of a frisbee with a diameter of 9 inches. Allow an additional 3 inches around the frisbee. *(Lesson 11-3)* 176.7 in^2

Write an equation for each circle. *(Lesson 10-8)* 53. $(x + 2)^2 + (y - 7)^2 = 50$

53. a circle with center at $(-2, 7)$ and a radius with endpoint at $(3, 2)$

54. a diameter with endpoints at $(6, -8)$ and $(2, 5)$ $(x - 4)^2 + \left(y + \frac{3}{2}\right)^2 = \frac{185}{4}$

676 Chapter 12 Surface Area

Geometry Activity

A Follow-Up of Lesson 12-7

Locus and Spheres

Spheres are defined in terms of a locus of points in space. The definition of a sphere is the set of all points that are a given distance from a given point.

Activity 1

Find the locus of points a given distance from the endpoints of a segment.

Collect the Data

- Draw a given line segment with endpoints S and T.
- Create a set of points that are equidistant from S and a set of points that are equidistant from T.

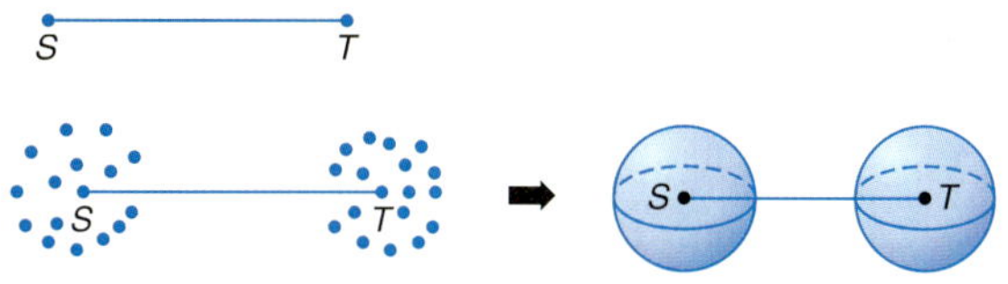

Analyze
3. Each sphere has a radius of 5 units and a diameter of 10 units.

1. Draw a figure and describe the locus of points in space that are 5 units from each endpoint of a given segment that is 25 units long. See margin.
2. Are the two spheres congruent? Yes, the radii are congruent.
3. What are the radius and diameter of each sphere?
4. Find the distance between the two spheres. The spheres are 15 units apart on the given segment.

Activity 2

Investigate spheres that intersect.

Find the locus of all points that are equidistant from the centers of two intersecting spheres with the same radius.

Collect the Data

- Draw a line segment.
- Draw congruent overlapping spheres, with the centers at the endpoints of the given line segment.

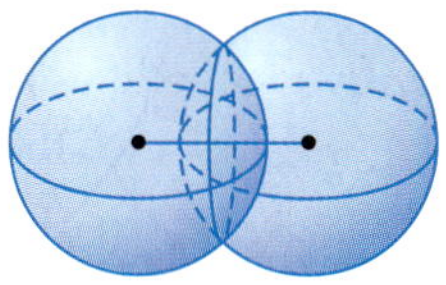

6. A circle is a locus of points on a plane.
7. The intersection is the set of all points equidistant from the midpoint of the given segment in the plane perpendicular to the given segment at its midpoint.

Analyze

5. What is the shape of the intersection of the spheres? circle
6. Can this be described as a locus of points in space or on a plane? Explain.
7. Describe the intersection as a locus.
8. **MINING** What is the locus of points that describes how particles will disperse in an explosion at ground level if the expected distance a particle could travel is 300 feet? a hemisphere with a radius of 300 ft

Resource Manager

Teaching Geometry with Manipulatives
- p. 17 (ruler)

Glencoe Mathematics Classroom Manipulative Kit
- compasses
- rulers

Geometry Activity

A Follow-Up of Lesson 12-7

Getting Started

Objective Find the locus of points a given distance from the endpoints of a segment.

Materials ruler, compass

Make sure students are clear that the points a given distance from the endpoint of a segment are points in space, and that they form the surface of a sphere. Model this by tracing out the points a given distance from the end of a pencil.

Teach

You may want students to do this activity in groups of four. Ask two students to do the experiments while the other students in the group answer the discussion questions. Then they can switch roles and repeat the activity.

Assess

Analyze Encourage students to discuss their answers to the questions and to explain their answers to others or to the class. This should help them visualize how to find a locus of points.

Study Notebook

Ask students to summarize what they have learned about the locus of points a given distance from the endpoints of a segment.

Chapter 12 Study Guide and Review

Vocabulary and Concept Check

- This alphabetical list of vocabulary terms in Chapter 12 includes a page reference where each term was introduced.
- **Assessment** A vocabulary test/review for Chapter 12 is available on p. 716 of the *Chapter 12 Resource Masters*.

Lesson-by-Lesson Review

For each lesson,

- the main ideas are summarized,
- additional examples review concepts, and
- practice exercises are provided.

Vocabulary PuzzleMaker

ELL The Vocabulary PuzzleMaker software improves students' mathematics vocabulary using four puzzle formats—crossword, scramble, word search using a word list, and word search using clues. Students can work on a computer screen or from a printed handout.

MindJogger Videoquizzes

ELL MindJogger Videoquizzes provide an alternative review of concepts presented in this chapter. Students work in teams in a game show format to gain points for correct answers. The questions are presented in three rounds.

Round 1 Concepts (5 questions)
Round 2 Skills (4 questions)
Round 3 Problem Solving (4 questions)

Chapter 12 Study Guide and Review

Vocabulary and Concept Check

axis (p. 655)	great circle (p. 671)	oblique prism (p. 649)	regular prism (p. 637)
bases (p. 637)	hemisphere (p. 672)	orthogonal drawing (p. 636)	regular pyramid (p. 660)
circular cone (p. 666)	lateral area (p. 649)	perspective view (p. 636)	right cone (p. 666)
cone (p. 638)	lateral edges (p. 649)	Platonic Solids (p. 637)	right cylinder (p. 655)
corner view (p. 636)	lateral faces (p. 649)	polyhedron (p. 637)	right prism (p. 649)
cross section (p. 639)	net (p. 644)	prism (p. 637)	slant height (p. 660)
cylinder (p. 638)	oblique cone (p. 666)	pyramid (p. 637)	sphere (p. 638)
edges (p. 637)	oblique cylinder (p. 655)	regular polyhedron (p. 637)	surface area (p. 644)
face (p. 637)			

A complete list of postulates and theorems can be found on pages R1–R8.

Exercises **Match each expression with the correct formula.**

1. lateral area of a prism d
2. surface area of a prism i
3. lateral area of a cylinder b
4. surface area of a cylinder h
5. lateral area of a regular pyramid a
6. surface area of a regular pyramid j
7. lateral area of a cone e
8. surface area of a cone g
9. surface area of a sphere c
10. surface area of a cube f

a. $L = \frac{1}{2}P\ell$	**f.** $T = 6s^2$
b. $L = 2\pi rh$	**g.** $T = \pi r\ell + \pi r^2$
c. $T = 4\pi r^2$	**h.** $T = 2\pi rh + 2\pi r^2$
d. $L = Ph$	**i.** $T = Ph + 2B$
e. $L = \pi r\ell$	**j.** $T = \frac{1}{2}P\ell + B$

Lesson-by-Lesson Review

12-1 Three-Dimensional Figures

See pages 636–642.

Concept Summary

- A solid can be determined from its orthogonal drawing.
- Solids can be classified by bases, faces, edges, and vertices.

Examples **Identify each solid. Name the bases, faces, edges, and vertices.**

a. 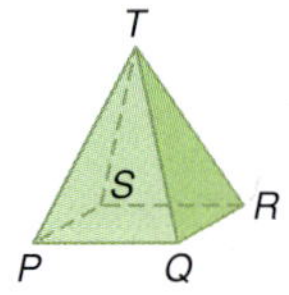

The base is a rectangle, and all of the lateral faces intersect at point T, so this solid is a rectangular pyramid.
Base: $\square PQRS$
Faces: $\triangle TPQ, \triangle TQR, \triangle TRS, \triangle TSP$
Edges: $\overline{PQ}, \overline{QR}, \overline{RS}, \overline{PS}, \overline{PT}, \overline{QT}, \overline{RT}, \overline{ST}$
Vertices: P, Q, R, S, T

b. 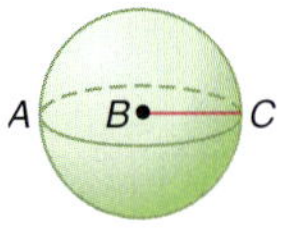

This solid has no bases, faces, or edges. It is a sphere.

www.geometryonline.com/vocabulary_review

FOLDABLES™ Study Organizer

For more information about Foldables, see *Teaching Mathematics with Foldables.*

Have students look through the chapter to make sure they have included notes and examples in their Foldables for each lesson of Chapter 12.

Encourage students to refer to their Foldables while completing the Study Guide and Review and to use them in preparing for the Chapter Test.

Exercises **Identify each solid. Name the bases, faces, edges, and vertices.**
See Example 2 on page 638. **11–13. See margin.**

11.

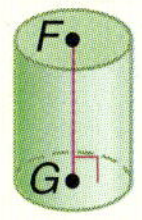

12.

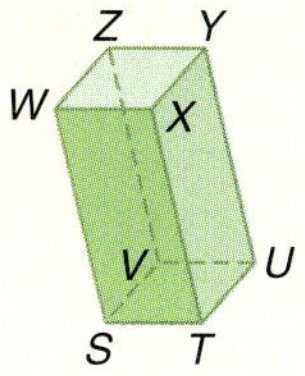

13.

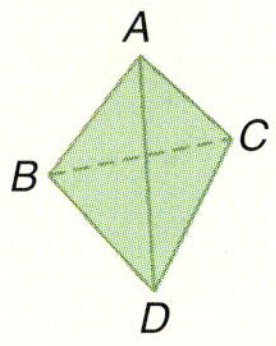

12-2 Nets and Surface Area

See pages 643–648.

Concept Summary

- Every three-dimensional solid can be represented by one or more two-dimensional nets.
- The area of the net of a solid is the same as the surface area of the solid.

Examples

Draw a net and find the surface area for the right rectangular prism shown.

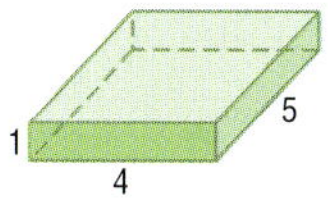

Use rectangular dot paper to draw a net. Since each face is a rectangle, opposite sides have the same measure.

To find the surface area of the prism, add the areas of the six rectangles.

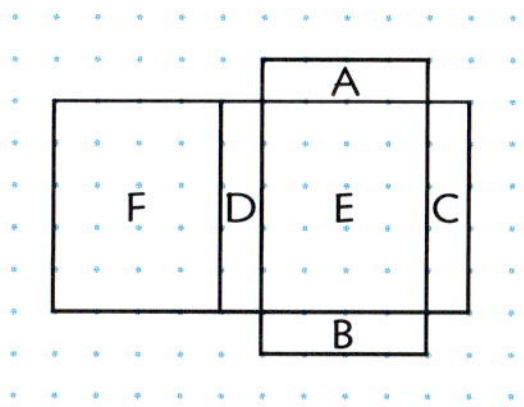

$$\text{Surface area} = A + B + C + D + E + F$$
$$= 4 \cdot 1 + 4 \cdot 1 + 5 \cdot 1 + 5 \cdot 1 + 4 \cdot 5 + 4 \cdot 5$$
$$= 4 + 4 + 5 + 5 + 20 + 20$$
$$= 58$$

The surface area is 58 square units.

Exercises **For each solid, draw a net and find the surface area.**
See Example 3 on page 645. **14–19. See p. 685D for nets.**

14. **84 units²**

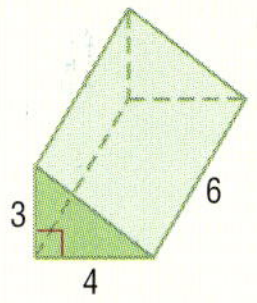

15. **340 units²**

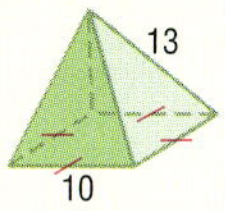

16. **96 units²**

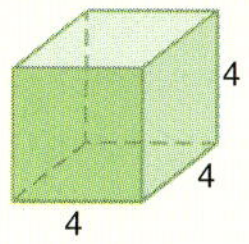

17.

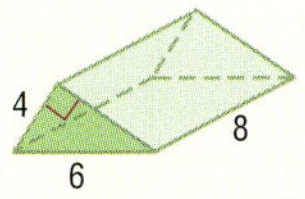

≈ 133.7 units²

18.

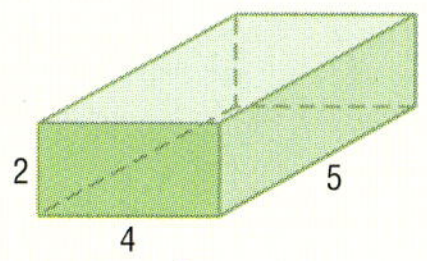

76 units²

19.

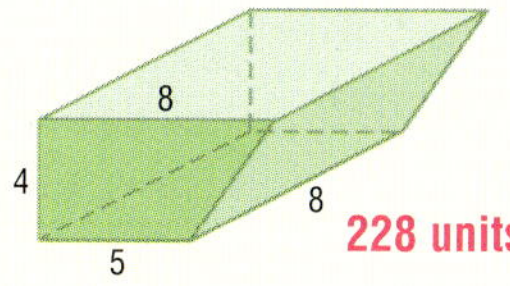

228 units²

Answers

11. cylinder; bases: $\odot F$ and $\odot G$
12. Sample answer: rectangular prism; bases: rectangle *WXYZ* and rectangle *STUV*; faces: rectangles *WXYZ*, *STUV*, *WXTS*, *XTUY*, *YUVZ*, and *WZVS*; edges: $\overline{WX}$, $\overline{XY}$, $\overline{YZ}$, $\overline{ZW}$, $\overline{ST}$, $\overline{TU}$, $\overline{UV}$, $\overline{VS}$, $\overline{WS}$, $\overline{XT}$, $\overline{YU}$, and $\overline{ZV}$; vertices: *S*, *T*, *U*, *V*, *W*, *X*, *Y*, and *Z*
13. Sample answer: triangular prism; base: $\triangle BCD$; faces: $\triangle ABC$, $\triangle ABD$, $\triangle ACD$, and $\triangle BCD$; edges: $\overline{AB}$, $\overline{BC}$, $\overline{AC}$, $\overline{AD}$, $\overline{BD}$, and $\overline{CD}$; vertices: *A*, *B*, *C*, and *D*

Chapter 12 Study Guide and Review 679

12-3 Surface Areas of Prisms

See pages 649–654.

Concept Summary

- The lateral faces of a prism are the faces that are not bases of the prism.
- The lateral surface area of a right prism is the perimeter of a base of the prism times the height of the prism.

Example **Find the lateral area of the regular hexagonal prism.**

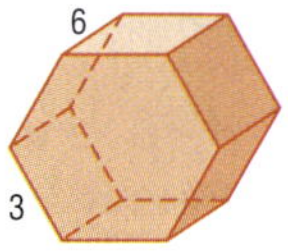

The bases are regular hexagons. So the perimeter of one base is 6(3) or 18. Substitute this value into the formula.

$L = Ph$ Lateral area of a prism

$= (18)(6)$ $P = 18, h = 6$

$= 108$ Multiply.

The lateral area is 108 square units.

Exercises **Find the lateral area of each prism.** *See Example 1 on page 650.*

20. 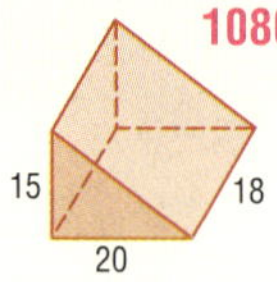

1080 units^2

21. 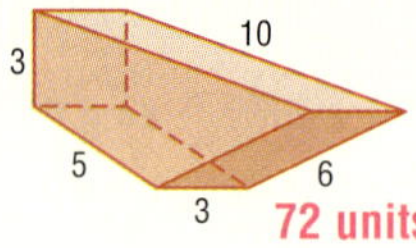

72 units^2

22. 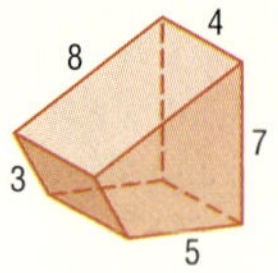

92 units^2

12-4 Surface Areas of Cylinders

See pages 655–659.

Concept Summary

- The lateral surface area of a cylinder is 2π multiplied by the product of the radius of a base of the cylinder and the height of the cylinder.
- The surface area of a cylinder is the lateral surface area plus the area of both circular bases.

Example **Find the surface area of a cylinder with a radius of 38 centimeters and a height of 123 centimeters.**

$T = 2\pi rh + 2\pi r^2$ Surface area of a cylinder

$= 2\pi(38)(123) + 2\pi(38)^2$ $r = 38, h = 123$

$\approx 38{,}440.5$ Use a calculator.

The surface area of the cylinder is approximately 38,440.5 square centimeters.

Exercises **Find the surface area of a cylinder with the given dimensions. Round to the nearest tenth.** *See Example 2 on page 656.*

23. $d = 4$ in., $h = 12$ in. 175.9 in^2

24. $r = 6$ ft, $h = 8$ ft 527.8 ft^2

25. $r = 4$ mm, $h = 58$ mm 1558.2 mm^2

26. $d = 4$ km, $h = 8$ km 125.7 km^2

680 Chapter 12 Surface Area

12-5 Surface Areas of Pyramids

See pages 660–665.

Concept Summary

- The slant height ℓ of a regular pyramid is the length of an altitude of a lateral face.
- The lateral area of a pyramid is $\frac{1}{2}P\ell$, where ℓ is the slant height of the pyramid and P is the perimeter of the base of the pyramid.

Example **Find the surface area of the regular pyramid.**

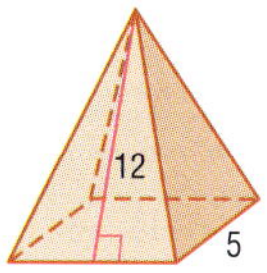

The perimeter of the base is 4(5) or 20 units, and the area of the base is 5^2 or 25 square units. Substitute these values into the formula for the surface area of a pyramid.

$T = \frac{1}{2}P\ell + B$ — Surface area of a regular pyramid

$= \frac{1}{2}(20)(12) + 25$ — $P = 20$, $\ell = 12$, $B = 25$

$= 145$ — Simplify.

The surface area is 145 square units.

Exercises **Find the surface area of each regular pyramid. Round to the nearest tenth if necessary.** *See Example 2 on pages 661 and 662.*

27. 304 units^2

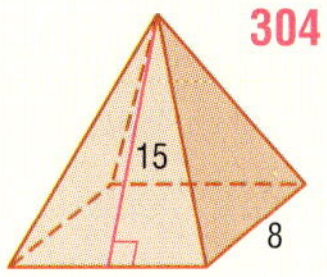

28. 472.0 units^2

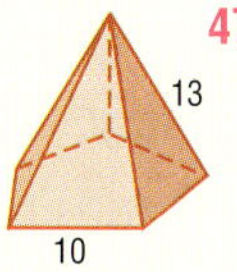

29. 33.3 units^2

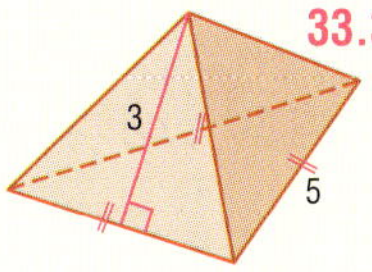

12-6 Surface Areas of Cones

See pages 666–670.

Concept Summary

- A cone is a solid with a circular base and a single vertex.
- The lateral area of a right cone is $\pi r\ell$, where ℓ is the slant height of the cone and r is the radius of the circular base.

Example **Find the surface area of the cone.**

Substitute the known values into the formula for the surface area of a right cylinder.

$T = \pi r\ell + \pi r^2$ — Surface area of a cone

$T = \pi(3)(12) + \pi(3)^2$ — $r = 3$, $\ell = 12$

$T \approx 141.4$ — Use a calculator.

The surface area is approximately 141.4 square meters.

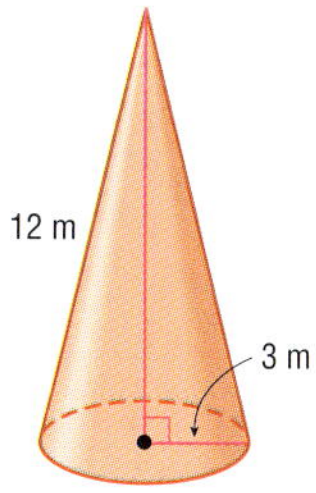

Chapter 12 Study Guide and Review 681

Chapter 12 For More ...
- Extra Practice, see pages 778–780.
- Mixed Problem Solving, see page 793.

Answers (page 683)

4. Sample answer: rectangular prism; bases: rectangles *PQRS* and *TUVW*; faces: rectangles *PQRS*, *TUVW*, *SPUT*, *QRWV*, *STWR*, and *PUVQ*; edges: $\overline{PS}$, $\overline{QR}$, $\overline{VW}$, $\overline{UT}$, $\overline{PU}$, $\overline{ST}$, $\overline{RW}$, $\overline{QV}$, $\overline{PQ}$, $\overline{SR}$, $\overline{UV}$, $\overline{TW}$; vertices: *P*, *Q*, *R*, *S*, *T*, *U*, *V*, and *W*
5. sphere
6. cone; base: ⊙*F*; vertex: *H*
7.

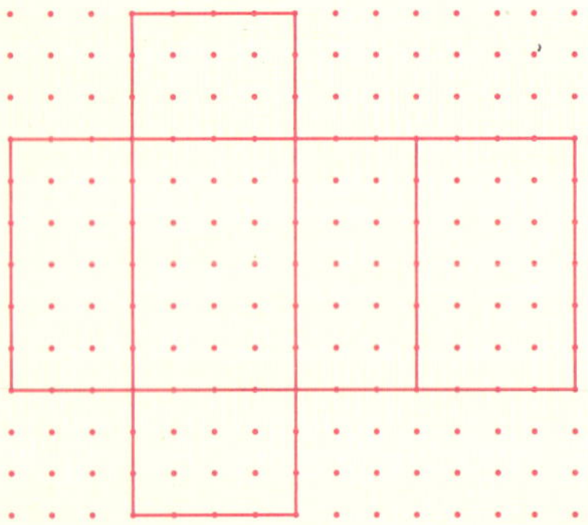

Exercises **Find the surface area of each cone. Round to the nearest tenth.**
See Example 2 on page 667.

30. 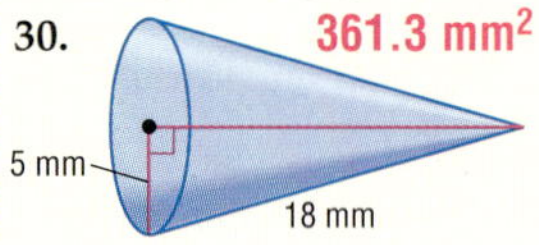

361.3 mm^2

31. 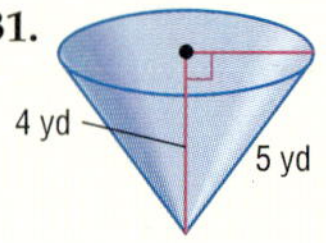

75.4 yd^2

32. 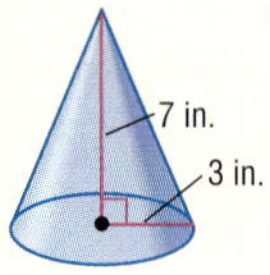

100.1 in^2

12-7 Surface Areas of Spheres

See pages 671–676.

Concept Summary

- The set of all points in space a given distance from one point is a sphere.
- The surface area of a sphere is $4\pi r^2$, where r is the radius of the sphere.

Examples

a. Find the surface area of a sphere with a diameter of 10 centimeters.

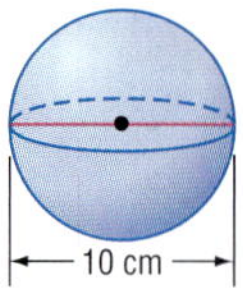

$T = 4\pi r^2$ Surface area of a sphere

$= 4\pi(5)^2$ $r = 5$

≈ 314.2 Use a calculator.

The surface area is approximately 314.2 square centimeters.

b. Find the surface area of a hemisphere with radius 6.3 inches.

To find the surface area of a hemisphere, add the area of the great circle to half of the surface area of the sphere.

surface area $= \frac{1}{2}(4\pi r^2) + \pi r^2$ Surface area of a hemisphere

$= \frac{1}{2}[4\pi(6.3)^2] + \pi(6.3)^2$ $r = 6.3$

≈ 374.1 Use a calculator.

The surface area is approximately 374.1 square inches.

Exercises **Find the surface area of each sphere or hemisphere. Round to the nearest tenth if necessary.** *See Example 2 on page 673.*

33. 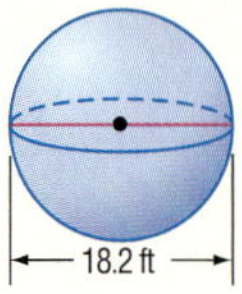

1040.6 ft^2

34. 

143.4 cm^2

35. Area of great circle = 121 mm^2
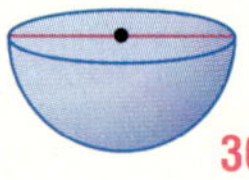
363 mm^2

36. Area of great circle = 218 in^2
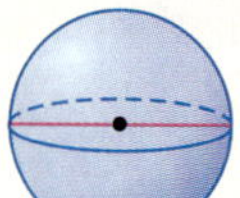
872 in^2

37. a hemisphere with radius 16 ft 2412.7 ft^2

38. a sphere with diameter 5 m 78.5 m^2

39. a sphere that has a great circle with an area of 220 ft^2 880 ft^2

40. a hemisphere that has a great circle with an area of 30 cm^2 90 cm^2

682 Chapter 12 Surface Area

Vocabulary and Concepts

Match each expression to the correct formula.

1. surface area of a prism **c**
2. surface area of a cylinder **a**
3. surface area of a regular pyramid **b**

a. $T = 2\pi rh + 2\pi r^2$

b. $T = \frac{1}{2}P\ell + B$

c. $T = Ph + 2B$

Skills and Applications

Identify each solid. Name the bases, faces, edges, and vertices. **4–6. See margin.**

4.
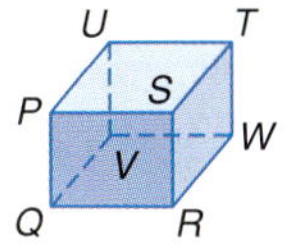

5.
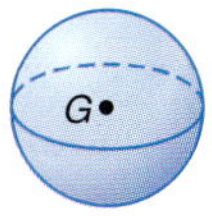

6.
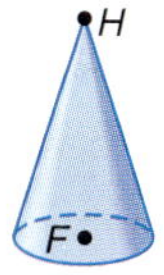

For each solid, draw a net and find the surface area. **7–8. See margin for nets.**

7.
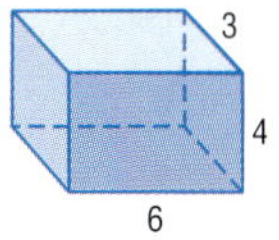

108 units^2

8.
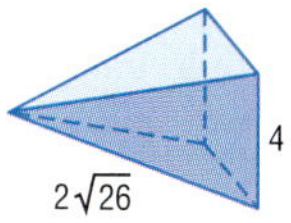

96 units^2

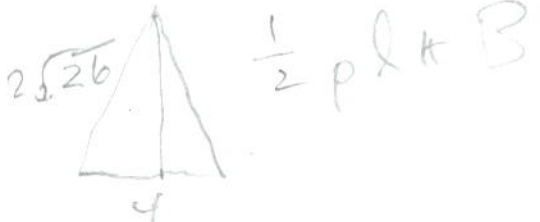

Find the lateral area of each prism.

9.
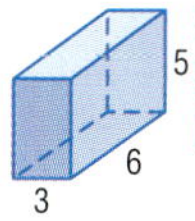

96 units^2 (3 × 5 base)
90 units^2 (3 × 6 base)
66 units^2 (6 × 5 base)

10.
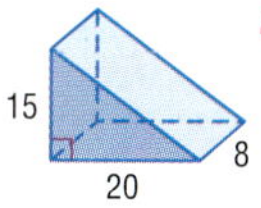

480 units^2

11.
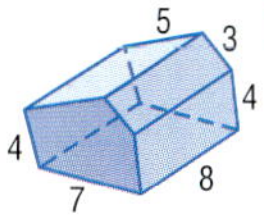

184 units^2

Find the surface area of a cylinder with the given dimensions. Round to the nearest tenth.

12. $r = 8$ ft, $h = 22$ ft **1508.0 ft^2**
13. $r = 3$ mm, $h = 2$ mm **94.2 mm^2**
14. $r = 78$ m, $h = 100$ m **87,235.7 m^2**

The figure at the right is a composite solid of a tetrahedron and a triangular prism. Find each measure in the solid. Round to the nearest tenth if necessary.

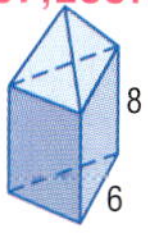

15. height **12.9 units**
16. lateral area **190.8 units^2**
17. surface area **206.4 units^2**

Find the surface area of each cone. Round to the nearest tenth.

18. $h = 24, r = 7$ **703.7 units^2**
19. $h = 3$ m, $\ell = 4$ m **55.2 units^2**
20. $r = 7, \ell = 12$ **417.8 units^2**

Find the surface area of each sphere. Round to the nearest tenth if necessary.

21. $r = 15$ in. **2827.4 in^2**
22. $d = 14$ m **615.8 m^2**
23. The area of a great circle of the sphere is 116 square feet. **464 ft^2**

24. **GARDENING** The surface of a greenhouse is covered with plastic or glass. Find the amount of plastic needed to cover the greenhouse shown. **1188 ft^2**

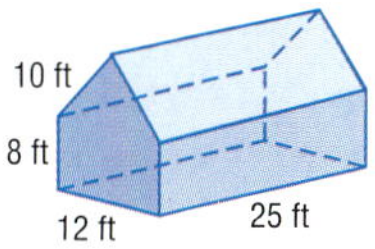

25. **STANDARDIZED TEST PRACTICE** A cube has a surface area of 150 square centimeters. What is the length of each edge? **D**

Ⓐ 25 cm Ⓑ 15 cm Ⓒ 12.5 cm Ⓓ 5 cm

www.geometryonline.com/chapter_test/fcat

Chapter 12 Practice Test

Assessment Options

Vocabulary Test A vocabulary test/review for Chapter 12 can be found on p. 716 of the *Chapter 12 Resource Masters*.

Chapter Tests There are six Chapter 12 Tests and an Open-Ended Assessment task available in the *Chapter 12 Resource Masters*.

Chapter 12 Tests			
Form	**Type**	**Level**	**Pages**
1	MC	basic	703–704
2A	MC	average	705–706
2B	MC	average	707–708
2C	FR	average	709–710
2D	FR	average	711–712
3	FR	advanced	713–714

MC = multiple-choice questions
FR = free-response questions

Open-Ended Assessment
Performance tasks for Chapter 12 can be found on p. 715 of the *Chapter 12 Resource Masters*. A sample scoring rubric for these tasks appears on p. A28.

TestCheck and Worksheet Builder

This **networkable software** has three modules for assessment.

- **Worksheet Builder** to make worksheets and tests.
- **Student Module** to take tests on-screen.
- **Management System** to keep student records.

Portfolio Suggestion

Introduction Surface area is a concept applied in the manufacturing of containers such as boxes and cans.

Ask Students Have students think of a product they would like their imaginary manufacturing company to produce. Students should design the three-dimensional container that their product would ship in, make a net of the design, and then calculate the surface area of the container. Have students add their designs and calculations to their portfolios.

Chapter 12 Standardized Test Practice

These two pages contain practice questions in the various formats that can be found on the most frequently given standardized tests.

A practice answer sheet for these two pages can be found on p. A1 of the *Chapter 12 Resource Masters*.

Standardized Test Practice Student Recording Sheet, p. A1

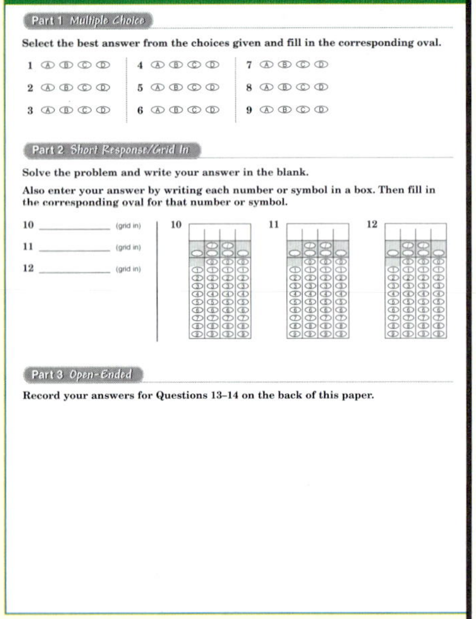
Part 1 Multiple Choice

Select the best answer from the choices given and fill in the corresponding oval.

1 Ⓐ Ⓑ Ⓒ Ⓓ 4 Ⓐ Ⓑ Ⓒ Ⓓ 7 Ⓐ Ⓑ Ⓒ Ⓓ

2 Ⓐ Ⓑ Ⓒ Ⓓ 5 Ⓐ Ⓑ Ⓒ Ⓓ 8 Ⓐ Ⓑ Ⓒ Ⓓ

3 Ⓐ Ⓑ Ⓒ Ⓓ 6 Ⓐ Ⓑ Ⓒ Ⓓ 9 Ⓐ Ⓑ Ⓒ Ⓓ

Part 2 Short Response/Grid In

Solve the problem and write your answer in the blank.

Also enter your answer by writing each number or symbol in a box. Then fill in the corresponding oval for that number or symbol.

10 ________ (grid in)

11 ________ (grid in)

12 ________ (grid in)

Part 3 Open-Ended

Record your answers for Questions 13–14 on the back of this paper.

Chapter 12 Standardized Test Practice

FCAT Practice

Part 1 Multiple Choice

Record your answers on the answer sheet provided by your teacher or on a sheet of paper.

1. A decorative strip of wood is used to bisect the center sector window panes. What is the measure of each angle formed when the pane is bisected? (Lesson 1-5) D

Ⓐ 12°

Ⓑ 25°

Ⓒ 50°

Ⓓ 59°

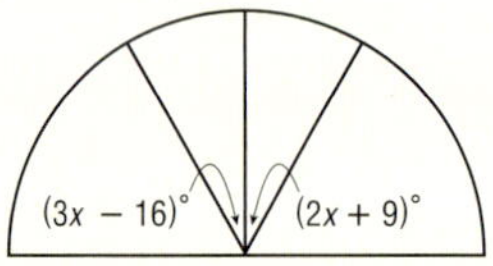

2. The coordinates of the endpoints of a segment are (0, 1) and (6, 9). A congruent segment has one endpoint at (10, 6). Which could be the coordinates of the other endpoint? (Lesson 4-3) D

Ⓐ (10, 15) Ⓑ (14, 14)

Ⓒ (16, 15) Ⓓ (16, 14)

3. Which piece of additional information is enough to prove that $\triangle ABC$ is similar to $\triangle DEF$? (Lesson 6-3) C

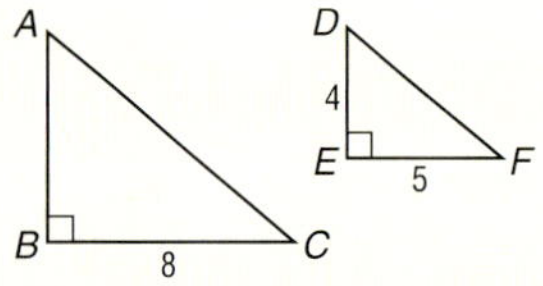

Ⓐ $\triangle ABC$ and $\triangle DEF$ are right triangles.

Ⓑ The length of $\overline{AB}$ is proportional to the length of $\overline{DE}$.

Ⓒ $\angle A$ is congruent to $\angle D$.

Ⓓ The length of $\overline{BC}$ is twice the length of $\overline{DE}$.

4. Which of the following lists the sides of $\triangle ABC$ in order from longest to shortest? (Lesson 5-3) B

Ⓐ $\overline{BC}, \overline{AB}, \overline{AC}$

Ⓑ $\overline{AC}, \overline{AB}, \overline{BC}$

Ⓒ $\overline{AC}, \overline{BC}, \overline{AB}$

Ⓓ $\overline{BC}, \overline{AC}, \overline{AB}$

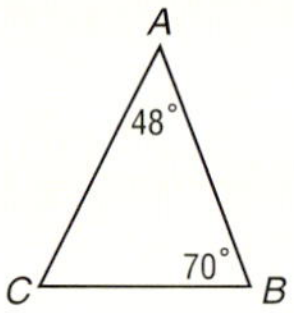

5. What is the approximate length of $\overline{AB}$? (Lesson 7-2) B

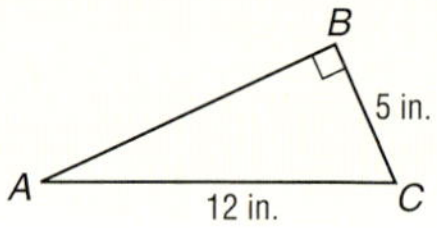

Ⓐ 8.9 in. Ⓑ 10.9 in.

Ⓒ 12 in. Ⓓ 13 in.

6. The diameter of circle *P* is 18 inches. What is the area of the shaded region? (Lesson 11-5) B

Ⓐ 27π

Ⓑ 54π

Ⓒ 81π

Ⓓ 216π

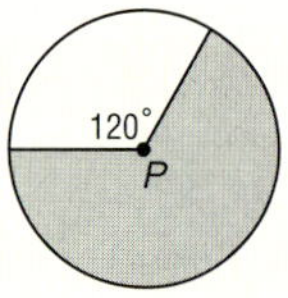

7. Which statement is false? (Lesson 12-1) C

Ⓐ A pyramid is a polyhedron.

Ⓑ The bases of a cylinder are in parallel planes.

Ⓒ All of the Platonic Solids are regular prisms.

Ⓓ A cone has a vertex.

8. Shelly bought a triangular prism at the science museum. The bases of the prism are equilateral triangles with side lengths of 2 centimeters. The height of the prism is 4 centimeters. What is the surface area of Shelly's prism to the nearest square centimeter? (Lesson 12-3) B

Ⓐ 16 cm² Ⓑ 27 cm²

Ⓒ 28 cm² Ⓓ 31 cm²

9. A spherical weather balloon has a diameter of 4 feet. What is the surface area of the balloon to the nearest square foot? (Lesson 12-7) A

Ⓐ 50 ft² Ⓑ 25 ft²

Ⓒ 16 ft² Ⓓ 13 ft²

The Princeton Review ***Log On*** **for Test Practice**

The Princeton Review offers additional test-taking tips and practice problems at their web site. Visit www.princetonreview.com or www.review.com

TestCheck and Worksheet Builder

Special banks of standardized test questions similar to those on the SAT, ACT, TIMSS 8, NAEP 8, and state proficiency tests can be found on this CD-ROM.

Aligned and verified by The Princeton Review

FCAT Practice

Part 2 Short Response/Grid In

Record your answers on the answer sheet provided by your teacher or on a sheet of paper.

10. Samantha wants to estimate the height of the tree.

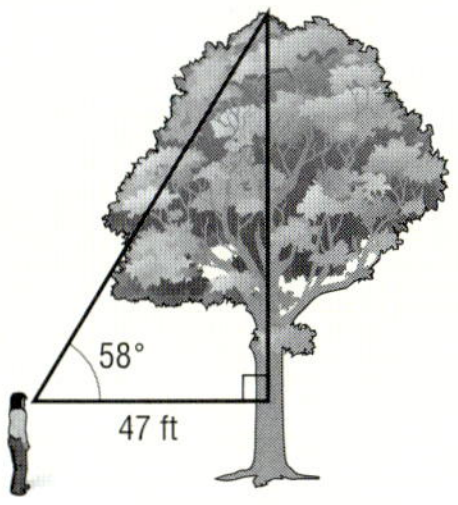

If Samantha is 5 feet tall, what is the height of the tree to the nearest foot? (Lesson 7-5) **80 ft**

11. If the base and height of the triangle are decreased by $2x$ units, what is the area of the resulting triangle? (Lesson 11-2) **$30x^2$ units^2**

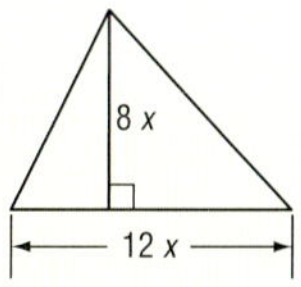

12. Mr. Jiliana built a wooden deck around half of his circular swimming pool. He needs to know the area of the deck so he can buy cans of stain. What is the area, to the nearest square foot, of the deck? (Lesson 11-4) **259**

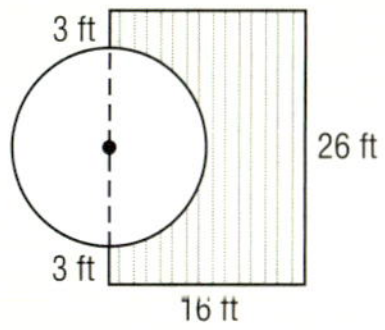

13. What is the surface area of this regular pentagonal pyramid to the nearest tenth of a square centimeter? (Lesson 12-5) **476.9**

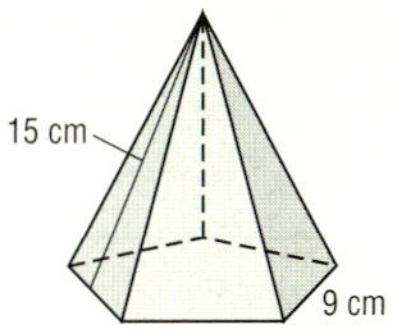

The Princeton Review

Test-Taking Tip

Question 2 Sometimes backsolving is the best technique to use to solve a multiple-choice test question. Check each of the given choices until you find the correct coordinates.

FCAT Practice

THINK SOLVE EXPLAIN

Part 3 Open Ended

Record your answers on a sheet of paper. Show your work.

14. A cylindrical pole is being used to support a large tent. The diameter of the base is 18 inches, and the height is 15 feet. (Lessons 12-2 and 12-4)

a. Draw a net of the cylinder and label the dimensions. **See margin.**

b. What is the lateral area of the pole to the nearest square foot? **71 ft^2**

c. What is the surface area of the pole to the nearest square foot? **74 ft^2**

15. Aliya is constructing a model of a rocket. She uses a right cylinder for the base and a right cone for the top. (Lessons 12-4 and 12-6)

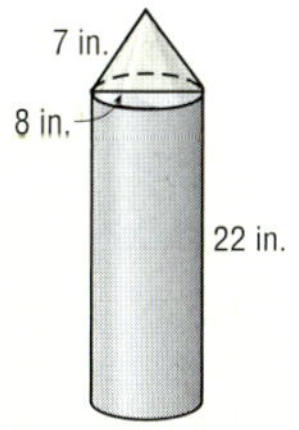

a. What is the surface area of the cone to the nearest square inch? **138 in^2**

b. What is the surface area of the cylinder to the nearest square inch? **653 in^2**

c. What is the surface area of the rocket, once it is assembled? Round to the nearest square inch. **691 in^2**

www.geometryonline.com/standardized_test/fcat

Evaluating Open-Ended Assessment Questions

Open-Ended Assessment questions are graded by using a multilevel rubric that guides you in assessing a student's knowledge of a particular concept.

Goal: Find the lateral area and surface area of cones and cylinders.

Sample Scoring Rubric: The following rubric is a sample scoring device. You may wish to add more detail to this sample to meet your individual scoring needs.

Score	Criteria
4	A correct solution that is supported by well-developed, accurate explanations
3	A generally correct solution, but may contain minor flaws in reasoning or computation
2	A partially correct interpretation and/or solution to the problem
1	A correct solution with no supporting evidence or explanation
0	An incorrect solution indicating no mathematical understanding of the concept or task, or no solution is given

Answer

14a. Sample answer:

Pages 639–642, Lesson 12-1

13.

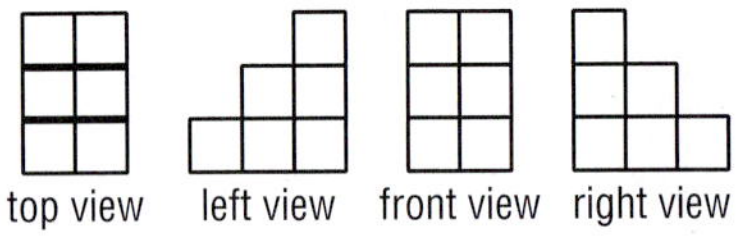

14.

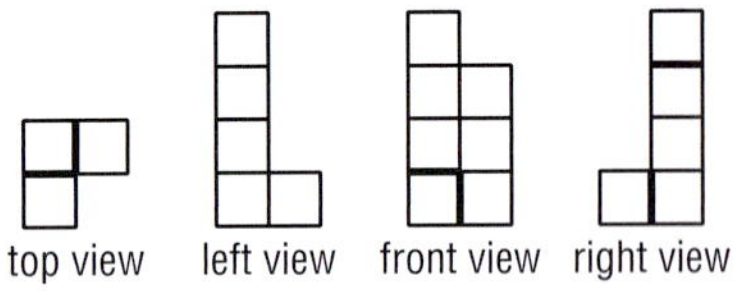

15.

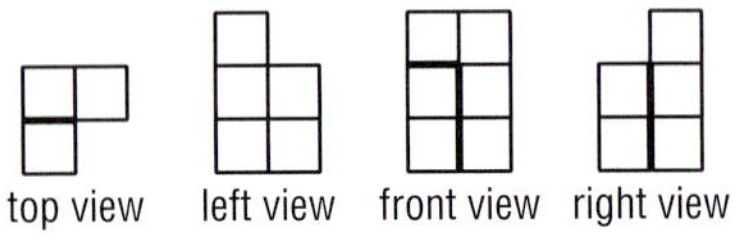

16. triangular prism; bases: $\triangle MNO$, $\triangle PQR$; faces: $\triangle MNO$, $\triangle PQR$, $OMPR$, $ONQR$, $PQNM$; edges: $\overline{MN}$, $\overline{NO}$, $\overline{OM}$, $\overline{PQ}$, $\overline{QR}$, $\overline{PR}$, $\overline{NQ}$, $\overline{MP}$, and $\overline{OR}$; vertices: M, N, O, P, Q, and R

17. rectangular pyramid; base: ▱$DEFG$; faces: ▱$DEFG$, $\triangle DHG$, $\triangle GHF$, $\triangle FHE$, $\triangle DHE$; edges: $\overline{DG}$, $\overline{GF}$, $\overline{FE}$, $\overline{ED}$, $\overline{DH}$, $\overline{EH}$, $\overline{FH}$, and $\overline{GH}$; vertices: D, E, F, G, and H

18. triangular pyramid; base: $\triangle IJK$; faces: $\triangle IJK$, $\triangle ILK$, $\triangle KLJ$, $\triangle ILJ$; edges: $\overline{IK}$, $\overline{KJ}$, $\overline{IJ}$, $\overline{IL}$, $\overline{KL}$, and $\overline{JL}$; vertices: I, K, J, and L

24. Sample answer: The speaker could be shaped like a rectangular prism, or the sides could be angled.

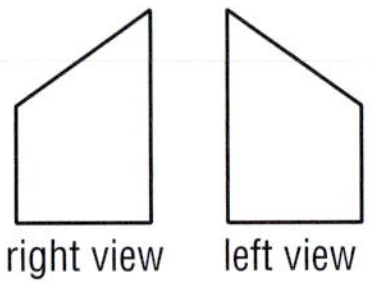

31. intersecting three faces and parallel to base

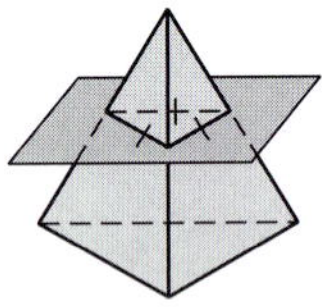

32. intersecting three faces and edges of base

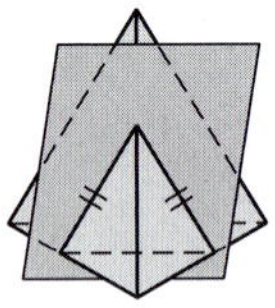

33. intersecting all four faces, not parallel to any face

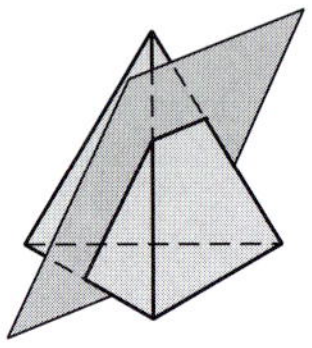

40. Yes, there is a pattern. The number of sides of the base of a prism is 2 less than the number of faces in the polyhedron. The number of sides of the base of a pyramid is 1 less than the number of faces.

41. No; the number of faces is not enough information to classify a polyhedron. A polyhedron with 6 faces could be a cube, rectangular prism, hexahedron, or a pentagonal pyramid. More information is needed to classify a polyhedron.

43. Sample answer: Archaeologists use two dimensional drawings to learn more about the structure they are studying. Egyptologists can compare two-dimensional drawings of the pyramids and note similarities and any differences. Answers should include the following.

- Viewpoint drawings and corner views are types of two-dimensional drawings that show three dimensions.
- To show three dimensions in a drawing, you need to know the views from the front, top, and each side.

Pages 645–648, Lesson 12-2

5.

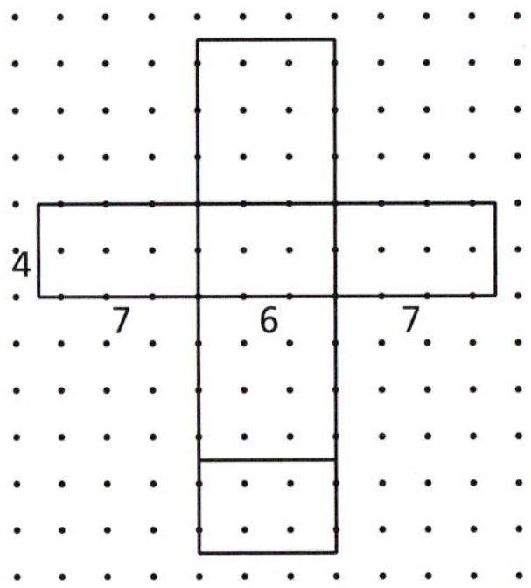

6.

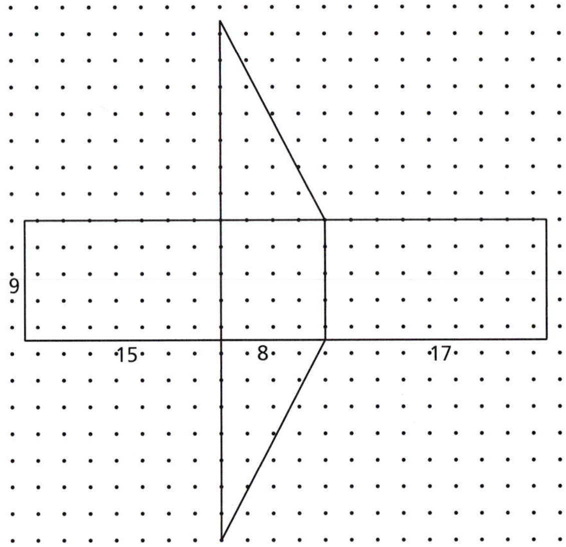

7.

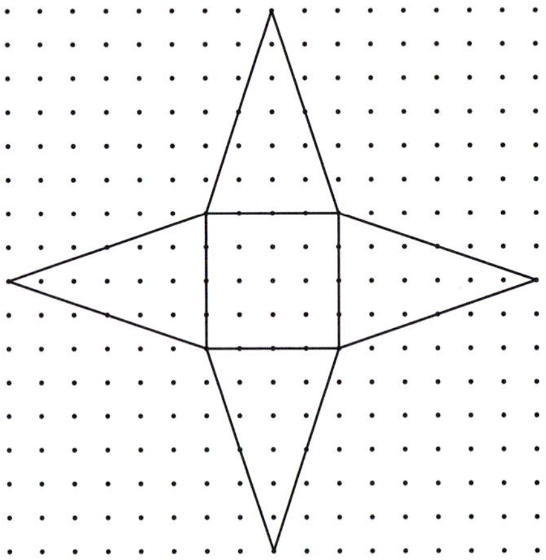

9.

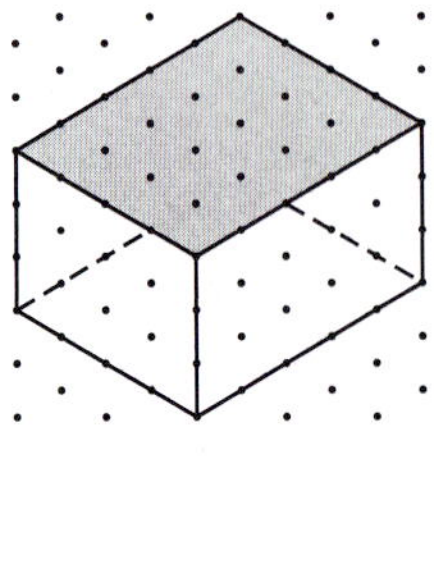

10.

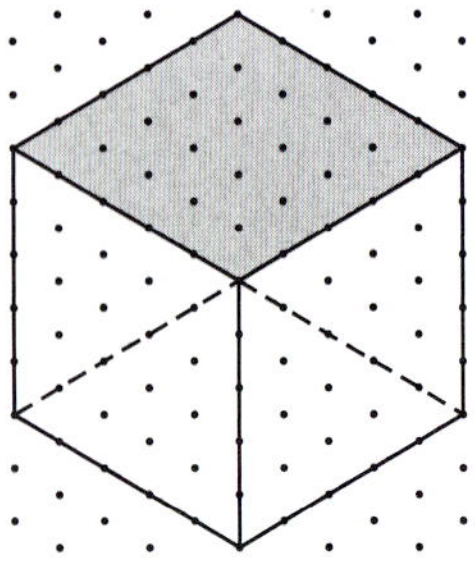

11.

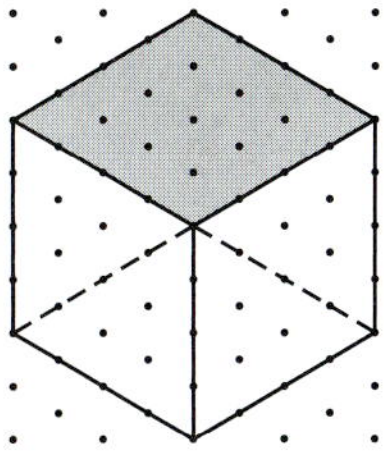

12.

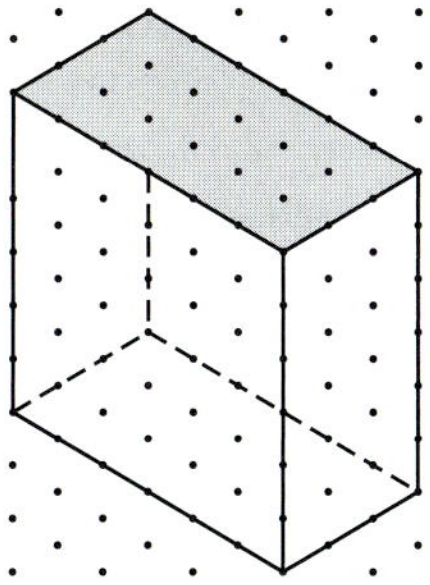

13.

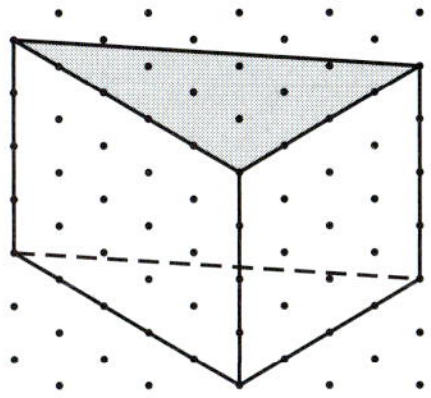

14.

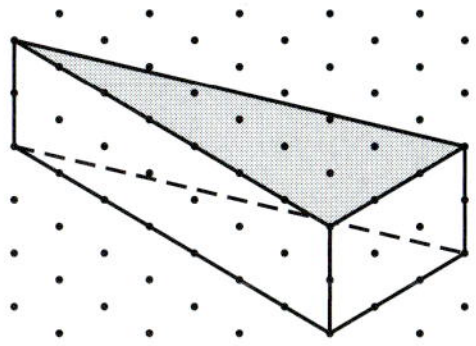

15.

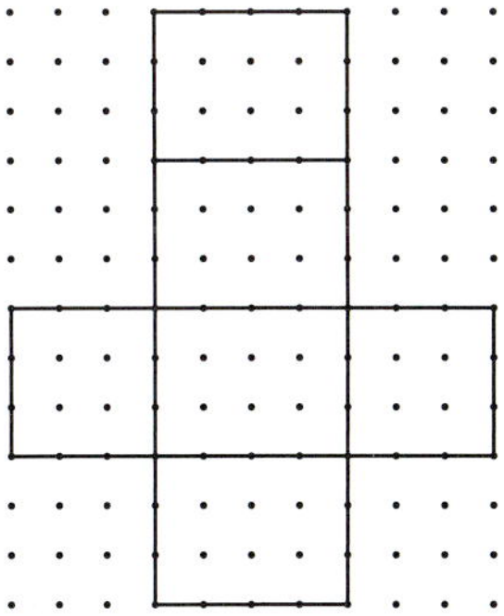

16.

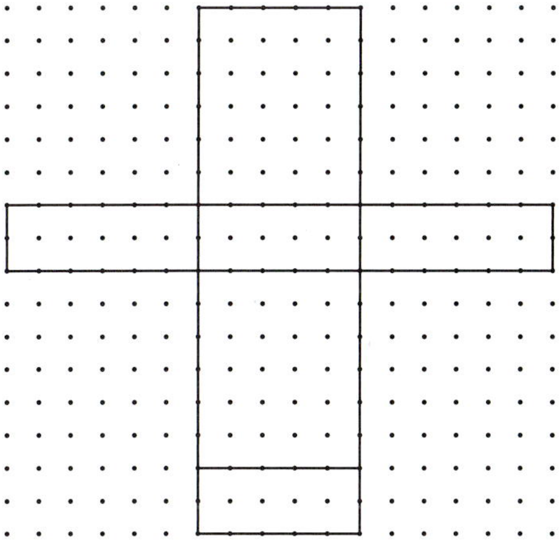

17.

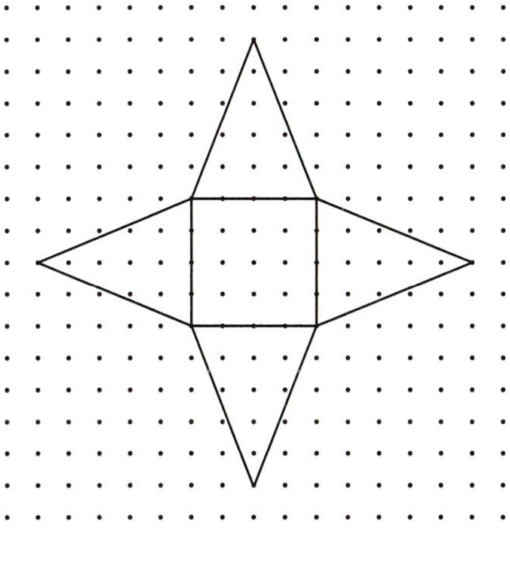

18.

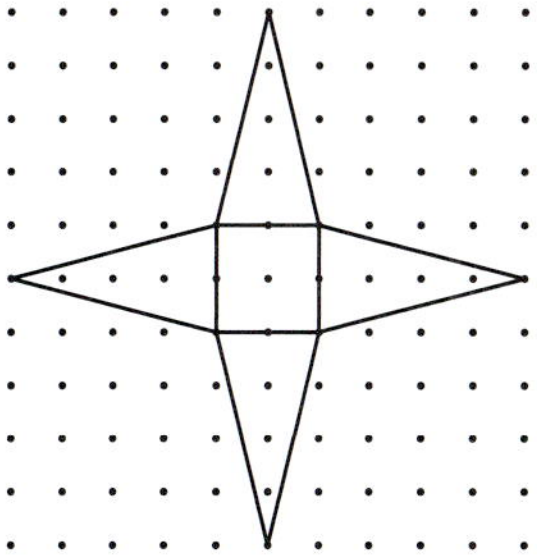

19.

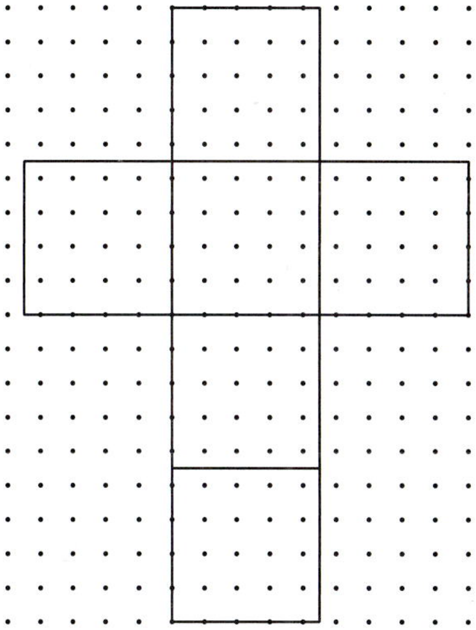

20.

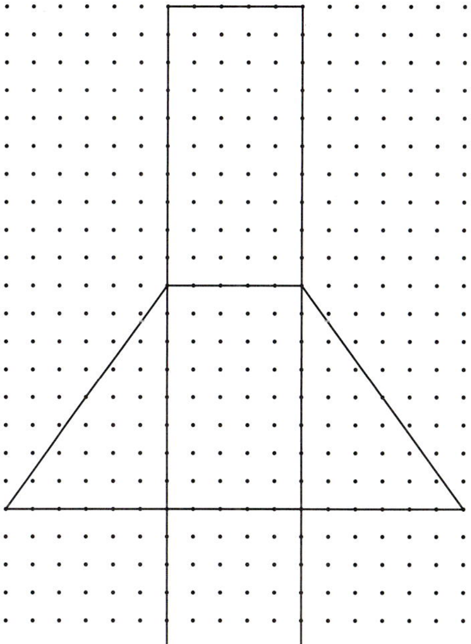

21.

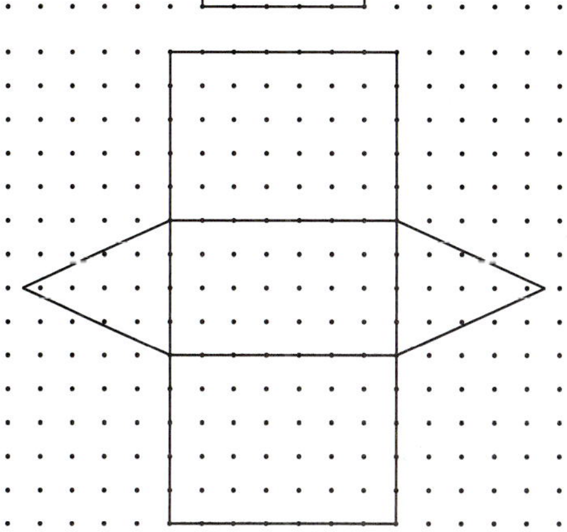

22.

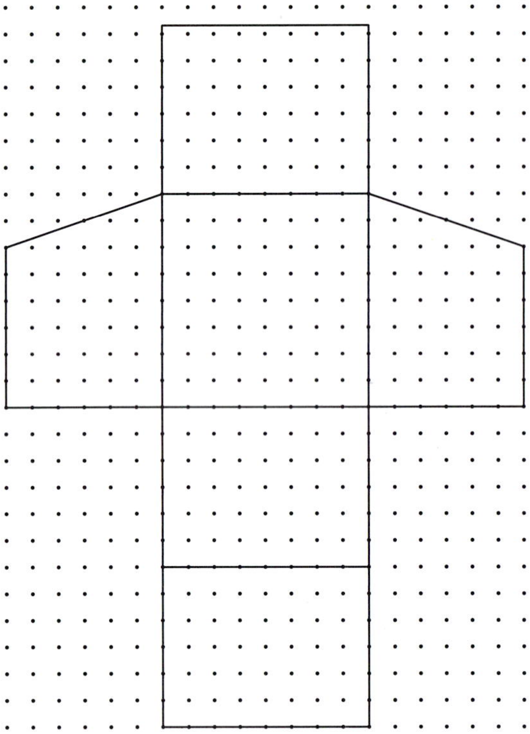

23.

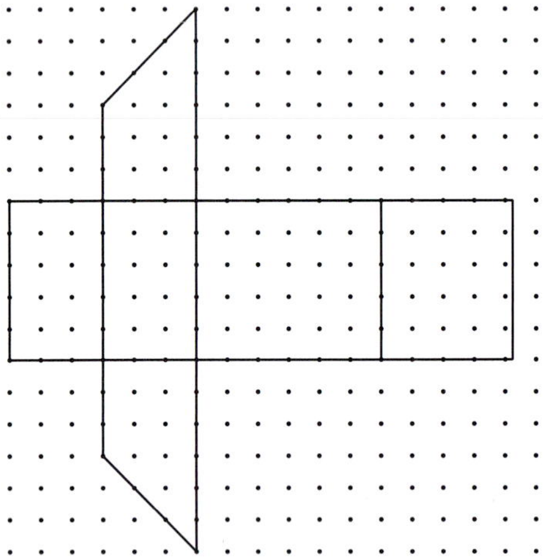

35.

A

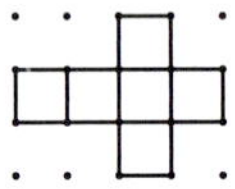

B

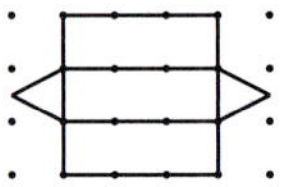

C

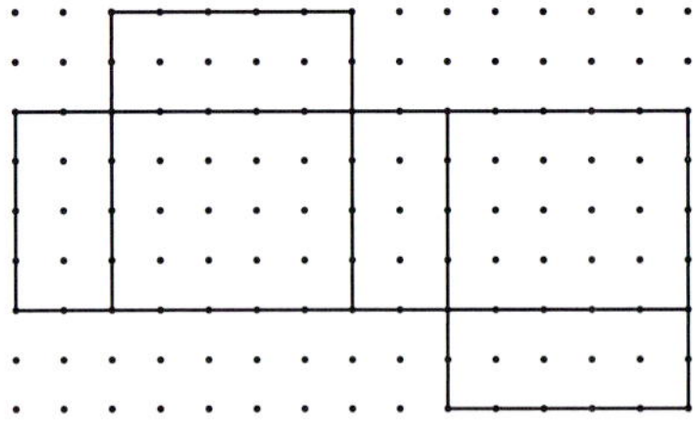

37. The surface area quadruples when the dimensions are doubled. For example, the surface area of the cube is $6(1^2)$ or 6 square units. When the dimensions are doubled the surface area is $6(2^2)$ or 24 square units.

38. When dimensions are tripled, the surface area will be nine times greater than the original surface area. For example, the surface area of the cube is $6(1^2)$ or 6 square units. The new surface area is $6(3^2)$ or 54 square units.

Pages 651–654, Lesson 12-3

38. Sample answer: Brick masons use the measurements of the structure and the measurements of the bricks to find the number of bricks that will be needed. Answers should include the following.

- The lateral area is important because the sides of the brick will show. Also, depending on the project, only the lateral area of the structure may be covered with brick.
- It is important to overestimate the number of bricks ordered in case some are damaged or the calculations were inaccurate.

44.

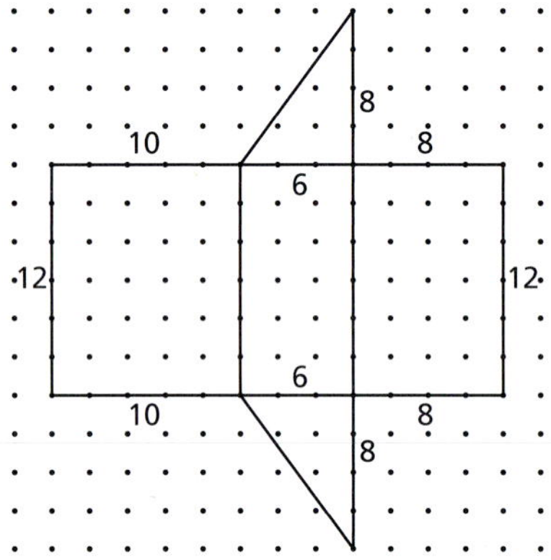

45.

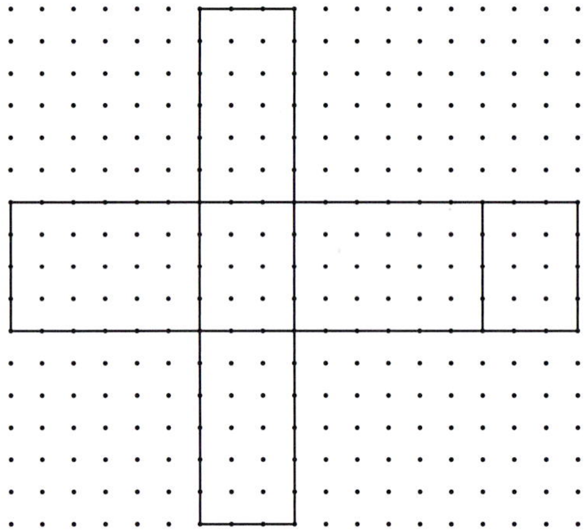

46.

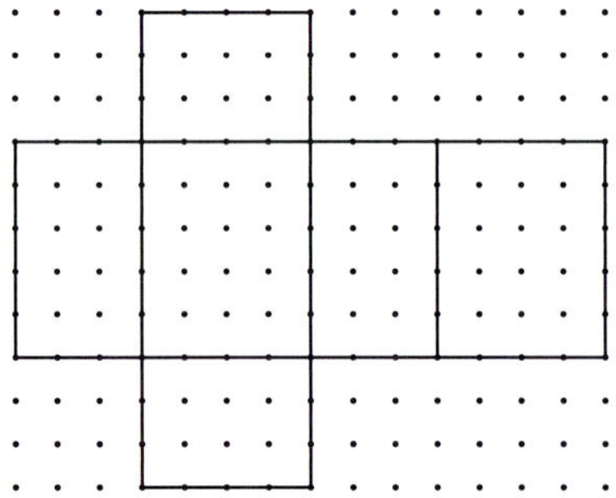

Pages 657–659, Lesson 12-4

26.

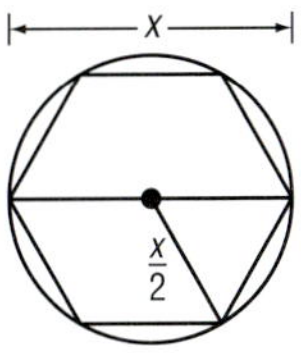

Let the diameter of the circle be x. A regular hexagon can be separated into 6 congruent nonoverlapping equilateral triangles. The sides of each triangle are $\frac{x}{2}$. The perimeter of the hexagon is $3x$. The lateral area of the hexagonal pencil is $33x$. The radius of the circle is also $\frac{x}{2}$. The circumference of the circle is $2\pi\left(\frac{x}{2}\right)$ or πx. The lateral area is approximately $34.6x$ square inches. The cylindrical pencil has the greater surface area.

Pages 668–670, Lesson 12-6

33. Sample answer: Tepees are conical shaped structures. Lateral area is used because the ground may not always be covered in circular canvas. Answers should include the following.

- We need to know the circumference of the base or the radius of the base and the slant height of the cone.
- The open top reduces the lateral area of canvas needed to cover the sides. To find the actual lateral area, subtract the lateral area of the conical opening from the lateral area of the structure.

Pages 678–682, Chapter 12 Study Guide and Review

14.

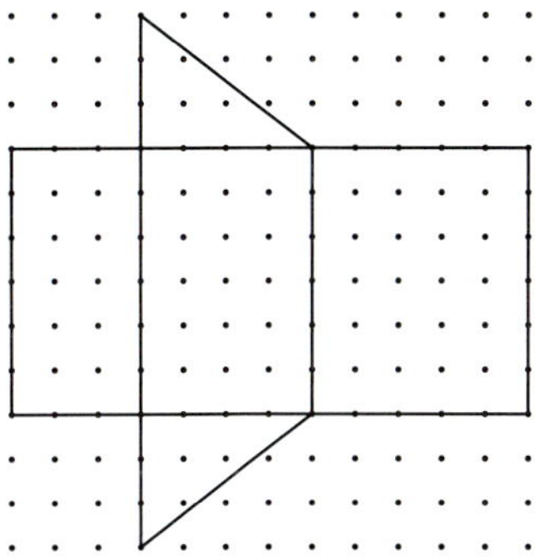

15.

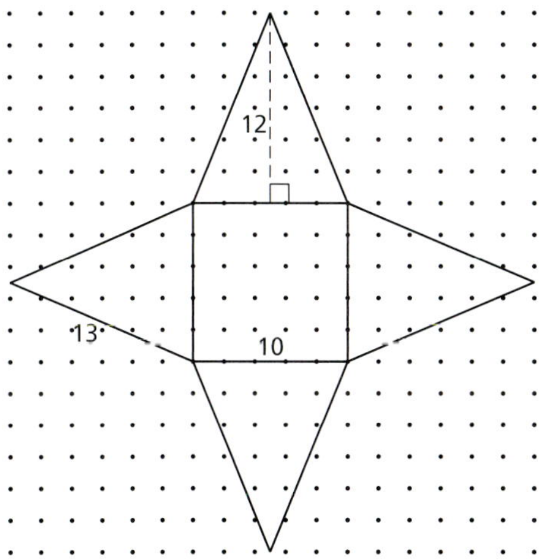

16.

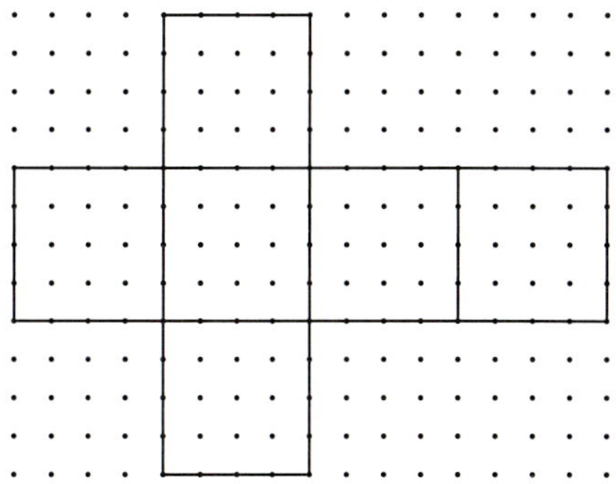

17.

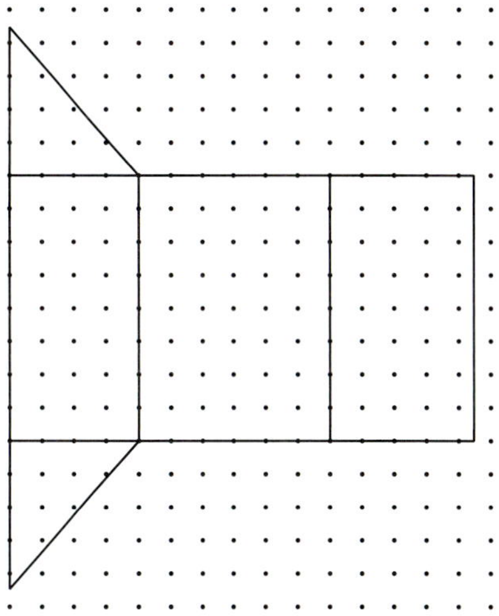

18.

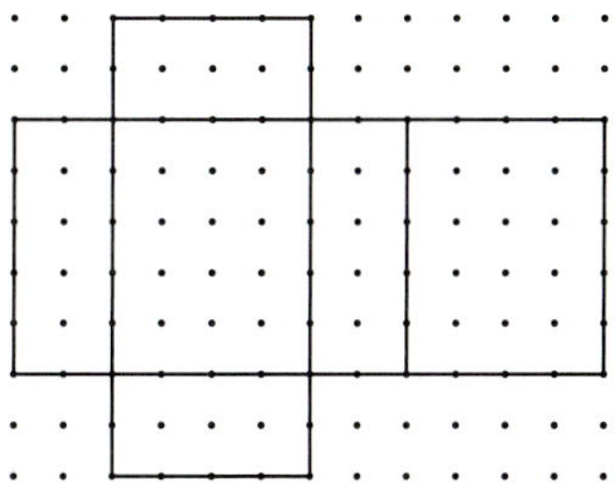

19.

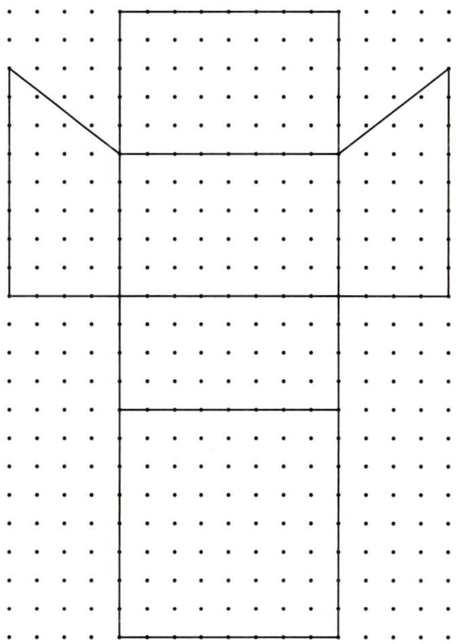

Chapter 13

Volume

Chapter Overview and Pacing

LESSON OBJECTIVES

	PACING (days)			
	Regular		Block	
	Basic/ Average	Advanced	Basic/ Average	Advanced
13-1 Volumes of Prisms and Cylinders *(pp. 688–695)* • Find volumes of prisms. • Find volumes of cylinders. ***Follow-Up:*** Use a spreadsheet to change the dimensions of a prism.	1	2 (with 13-1 Follow-Up)	0.5	1 (with 13-1 Follow-Up)
13-2 Volumes of Pyramids and Cones *(pp. 696–701)* • Find volumes of pyramids. • Find volumes of circular cones.	2	2	1	1
13-3 Volumes of Spheres *(pp. 702–706)* • Find volumes of spheres. • Solve problems involving volumes of spheres.	2	1	1	0.5
13-4 Congruent and Similar Solids *(pp. 707–713)* • Identify congruent or similar solids. • State the properties of similar solids.	2	2	1	1
13-5 Coordinates in Space *(pp. 714–719)* • Graph solids in space. • Use the Distance and Midpoint Formulas for points in space.	optional	2	optional	1
Study Guide and Practice Test *(pp. 720–723)* **Standardized Test Practice** *(p. 724–725)*	1	1	0.5	0.5
Chapter Assessment	1	1	0.5	0.5
TOTAL	**9**	**11**	**4.5**	**5.5**

Pacing suggestions for the entire year can be found on pages T20–T21.

Timesaving Tools
TeacherWorks™
All-In-One Planner and Resource Center
See pages T5 and T21.

Chapter Resource Manager

CHAPTER 13 RESOURCE MASTERS

Study Guide and Intervention	Practice (Skills and Average)	Reading to Learn Mathematics	Enrichment	Assessment	Prerequisite Skills Workbook	Applications*	5-Minute Check Transparencies	Interactive Chalkboard	GeomPASS: Tutorial Plus (lessons)	Materials
723–724	725–726	727	728		21–22, 97–100	SC 25	13-1	13-1		cubes
729–730	731–732	733	734	767		SC 26	13-2	13-2		card stock, straightedge, scissors, tape, rice
735–736	737–738	739	740	767, 769	37–38	GCC 41, 42	13-3	13-3	23	
741–742	743–744	745	746	768	97–98		13-4	13-4	24	computer with spreadsheet software
747–748	749–750	751	752	768	5–6		13-5	13-5	25	straightedge
				753–766, 770–772						

**Key to Abbreviations:* GCC = Graphing Calculator and Computer Masters
SC = School-to-Career Masters

Chapter 13

Mathematical Connections and Background

Continuity of Instruction

Prior Knowledge

Students evaluated and simplified exponential expressions in previous courses. In Chapter 1, students mastered the Midpoint Formula and graphed points in the coordinate plane. In Chapter 11, they found the area of regular polygons.

This Chapter

In this chapter, students explore volume. They find the volume of prisms, cylinders, pyramids, cones, and spheres. Students identify congruent and similar solids and state the properties of similar solids. The *z*-axis and *z*-coordinate are introduced so that students can graph solids in space. Students also use the Distance and Midpoint Formulas in space.

Future Connections

The knowledge about volume that students gain while studying this chapter will be important to them in future mathematics courses, in physics, and in many careers that they might choose.

13-1 Volumes of Prisms and Cylinders

The volume of a figure is the measure of the amount of space that it encloses. Volume is measured in cubic units. If a prism has a volume of V cubic units, a height of h units, and each base has an area of B square units, then $V = Bh$.

Like the volume of a prism, the volume of a cylinder is a product of the area of the base and the height. If a cylinder has a volume of V cubic units, a height of h units, and the bases have radii of r units, then $V = \pi r^2 h$.

The formulas for prisms and cylinders hold true for both right solids and oblique solids. Cavalieri's Principle, which applies to the volumes of all solids, states that if two solids have the same height and the same cross-sectional area at every level, then they have the same volume. This means that you would find the volume of an oblique cylinder just the same way you would find the volume for a right cylinder—with the formula $V = \pi r^2 h$.

13-2 Volumes of Pyramids and Cones

The formulas for the volumes of pyramids and circular cones are also based on the product of the area of the base and the height. If a pyramid has a volume of V cubic units, a height of h units, and a base with an area of B square units, then $V = \frac{1}{3}Bh$. This formula is derived from the formula for a prism.

To determine the formula for the volume of a cone, examine the formula for the volume of a cylinder. If a cone and a cylinder are the same height and their bases have the same area, the volume of the cylinder is three times the volume of the cone. If a right circular cone has a volume of V cubic units, a height of h units, and the base has a radius of r units, then $V = \frac{1}{3}Bh$ or $V = \frac{1}{3}\pi r^2 h$.

13-3 Volumes of Spheres

The formula for the volume of a sphere is derived from the formulas for the volume of a right pyramid and the surface area of a sphere. If a sphere has a volume of V cubic units and a radius of r units, then $V = \frac{4}{3}\pi r^3$. You can use this formula to determine the volume of a hemisphere. This formula is also useful if you are called on to compare the volume of a sphere and that of another solid.

13-4 Congruent and Similar Solids

Similar solids are solids that have exactly the same shape but not necessarily the same size. You can determine if two solids are similar by comparing the ratios of corresponding linear measurements. For example, you can compare the measures of the length, width, and height of two prisms. The ratio of the measures of two similar figures is called the *scale factor*. If the scale factor is 1, then the solids are congruent.

Congruent solids are exactly the same shape and exactly the same size. For two solids to be congruent, several conditions must be met. First, the corresponding angles are congruent. Second, the corresponding edges are congruent. Third, the corresponding faces are congruent. Finally, the volumes of the two solids are equal.

13-5 Coordinates in Space

To describe the location of a point on the coordinate plane, an ordered pair of coordinates is used. In space, each point requires three numbers, or coordinates, to describe its location because space has three dimensions. In addition to the x- and y-axes, there is the z-axis. These three axes are perpendicular to one another. A point in space is represented by an ordered triple of real numbers (x, y, z).

You can determine the distance between points in space just as you can determine the distance between points on a plane. Once again, the Pythagorean Theorem is used to derive the Distance Formula. The Distance Formula in space uses all three coordinates. Given two points $A(x_1, y_1, z_1)$ and $B(x_2, y_2, z_2)$ in space, the distance between A and B is given by the equation $d = \sqrt{(x_2 - x_1)^2 + (y_2 - y_1)^2 + (z_2 - z_1)^2}$. The Midpoint Formula can also be extended to three dimensions. Given two points $A(x_1, y_1, z_1)$ and $B(x_2, y_2, z_2)$ in space, the midpoint of $\overline{AB}$ is at $\left(\frac{x_1 + x_2}{2}, \frac{y_1 + y_2}{2}, \frac{z_1 + z_2}{2}\right)$.

Translations in space can be performed using a translation equation that adds the changes in the x-, y-, and z-coordinates to the original coordinates. You can use a matrix for a transformation in space such as a dilation. First, write a vertex matrix for the solid. Next, multiply each element of the vertex matrix by the scale factor to determine the coordinates of the dilated image.

Chapter 13

DAILY INTERVENTION and Assessment

	Type	Student Edition	Teacher Resources	Technology/Internet
INTERVENTION	Ongoing	Prerequisite Skills, pp. 687, 694, 701, 706, 713 Practice Quiz 1, p. 701 Practice Quiz 2, p. 713	5-Minute Check Transparencies *Prerequisite Skills Workbook*, pp. 5–6, 21–22, 37–38, 97–100 Quizzes, *CRM* pp. 767–768 Mid-Chapter Test, *CRM* p. 769 Study Guide and Intervention, *CRM* pp. 723–724, 729–730, 735–736, 741–742, 747–748	GeomPASS: Tutorial Plus www.geometryonline.com/self_check_quiz www.geometryonline.com/extra_examples
	Mixed Review	pp. 694, 701, 706, 713, 719	Cumulative Review, *CRM* p. 770	
	Error Analysis	Find the Error, pp. 691, 704 Common Misconceptions, p. 698	Find the Error, *TWE* pp. 691, 704 Unlocking Misconceptions, *TWE* p. 717 Tips for New Teachers, *TWE* pp. 689, 690	
	Standardized Test Practice	pp. 694, 701, 703, 704, 706, 713, 719, 723, 724, 725	*TWE* pp. 724–725 Standardized Test Practice, *CRM* pp. 771–772	Standardized Test Practice CD-ROM www.geometryonline.com/standardized_test
ASSESSMENT	Open-Ended Assessment	Writing in Math, pp. 693, 701, 706, 712, 719 Open Ended, pp. 691, 698, 710, 717 Standardized Test, p. 725	Modeling: *TWE* pp. 706, 719 Speaking: *TWE* p. 694 Writing: *TWE* pp. 701, 713 Open-Ended Assessment, *CRM* p. 765	
	Chapter Assessment	Study Guide, pp. 720–722 Practice Test, p. 723	Multiple-Choice Tests (Forms 1, 2A, 2B), *CRM* pp. 753–758 Free-Response Tests (Forms 2C, 2D, 3), *CRM* pp. 759–764 Vocabulary Test/Review, *CRM* p. 766	TestCheck and Worksheet Builder (see below) MindJogger Videoquizzes www.geometryonline.com/vocabulary_review www.geometryonline.com/chapter_test

Key to Abbreviations: TWE = Teacher Wraparound Edition; CRM = 5 Chapter Resource Masters

Additional Intervention Resources

The Princeton Review's *Cracking the SAT & PSAT*

The Princeton Review's *Cracking the ACT*

TestCheck and Worksheet Builder

This **networkable** software has three modules for intervention and assessment flexibility:

- **Worksheet Builder** to make worksheet and tests
- **Student Module** to take tests on screen (optional)
- **Management System** to keep student records (optional)

Special banks are included for SAT, ACT, TIMSS, NAEP, and End-of-Course tests.

Intervention Technology

GeomPASS: Tutorial Plus CD-ROM offers a complete, self-paced geometry curriculum.

Geometry Lesson	GeomPASS Lesson
13-3	**23** *Finding the Volume of Geometric Solids*
13-4	**24** *Similar and Congruent Solids*
13-5	**25** *Three-Dimensional Coordinate Systems*

Intervention at Home

Log on for student study help.

- For each lesson in the Student Edition, there are Extra Examples and Self-Check Quizzes.
  www.geometryonline.com/extra_examples
  www.geometryonline.com/self_check_quiz
- For chapter review, there is vocabulary review, test practice, and standardized test practice.
  www.geometryonline.com/vocabulary_review
  www.geometryonline.com/chapter_test
  www.geometryonline.com/standardized_test

For more information on Intervention and Assessment, see pp. T8–T11.

Reading and Writing in Mathematics

Glencoe Geometry provides numerous opportunities to incorporate reading and writing into the mathematics classroom.

Student Edition

- Foldables Study Organizer, p. 687
- Concept Check questions require students to verbalize and write about what they have learned in the lesson. (pp. 691, 698, 704, 710, 717)
- Writing in Math questions in every lesson, pp. 693, 701, 706, 713, 719
- Reading Study Tip, p. 714
- WebQuest, pp. 697, 719

Teacher Wraparound Edition

- Foldables Study Organizer, pp. 687, 720
- Study Notebook suggestions, pp. 691, 699, 704, 710, 717
- Modeling activities, pp. 706, 719
- Speaking activities, p. 694
- Writing activities, pp. 701, 713
- ELL Resources, pp. 686, 693, 700, 705, 711, 718, 720

Additional Resources

- Vocabulary Builder worksheets require students to define and give examples for key vocabulary terms as they progress through the chapter. (*Chapter 13 Resource Masters,* pp. vii-viii)
- Reading to Learn Mathematics master for each lesson (*Chapter 13 Resource Masters*, pp. 727, 733, 739, 745, 751)
- *Vocabulary PuzzleMaker* software creates crossword, jumble, and word search puzzles using vocabulary lists that you can customize.
- *Teaching Mathematics with Foldables* provides suggestions for promoting cognition and language.
- *Reading Strategies for the Mathematics Classroom*
- *WebQuest and Project Resources*

For more information on Reading and Writing in Mathematics, see pp. T6–T7.

Chapter 13 Notes

What You'll Learn

Have students read over the list of objectives and make a list of any words with which they are not familiar.

Why It's Important

Point out to students that this is only one of many reasons why each objective is important. Others are provided in the introduction to each lesson.

Lesson	NCTM Standards	Local Objectives
13-1	2, 3, 6, 8, 9, 10	
13-1 Follow-Up	2, 3, 6	
13-2	2, 3, 6, 8, 9, 10	
13-3	2, 3, 6, 8, 9, 10	
13-4	3, 6, 8, 9, 10	
13-5	3, 6, 8, 9, 10	

Key to NCTM Standards:

1=Number & Operations, 2=Algebra, 3=Geometry, 4=Measurement, 5=Data Analysis & Probability, 6=Problem Solving, 7=Reasoning & Proof, 8=Communication, 9=Connections, 10=Representation

Chapter 13 Volume

What You'll Learn

- **Lessons 13-1, 13-2, and 13-3** Find volumes of prisms, cylinders, pyramids, cones, and spheres.
- **Lesson 13-4** Identify congruent and similar solids, and state the properties of similar solids.
- **Lesson 13-5** Graph solids in space, and use the Distance and Midpoint Formulas in space.

Key Vocabulary

- volume (p. 688)
- similar solids (p. 707)
- congruent solids (p. 707)
- ordered triple (p. 714)

Why It's Important

Volcanoes like those found in Lassen Volcanic National Park, California, are often shaped like cones. By applying the formula for volume of a cone, you can find the amount of material in a volcano. *You will learn more about volcanoes in Lesson 13-2.*

Vocabulary Builder

ELL

The Key Vocabulary list introduces students to some of the main vocabulary terms included in this chapter. For a more thorough vocabulary list with pronunciations of new words, give students the Vocabulary Builder worksheets found on pages vii and viii of the *Chapter 13 Resource Masters*. Encourage them to complete the definition of each term as they progress through the chapter. You may suggest that they add these sheets to their study notebooks for future reference when studying for the Chapter 13 test.

Getting Started

Prerequisite Skills To be successful in this chapter, you'll need to master these skills and be able to apply them in problem-solving situations. Review these skills before beginning Chapter 13.

For Lesson 13-1 — Pythagorean Theorem

Find the value of the variable in each equation. *(For review, see Lesson 7-2.)*

1. $a^2 + 12^2 = 13^2$ **± 5**
2. $(4\sqrt{3})^2 + b^2 = 8^2$ **± 4**
3. $a^2 + a^2 = (3\sqrt{2})^2$ **± 3**
4. $b^2 + 3b^2 = 192$ **$\pm 4\sqrt{3}$**
5. $256 + 7^2 = c^2$ **$\pm\sqrt{305}$**
6. $144 + 12^2 = c^2$ **$\pm 12\sqrt{2}$**

For Lesson 13-2 — Area of Polygons

Find the area of each regular polygon. Round to the nearest tenth. *(For review, see Lesson 11-3.)*

7. hexagon with side length 7.2 cm **134.7 cm^2**
8. hexagon with side length 7 ft **127.3 ft^2**
9. octagon with side length 13.4 mm **867.0 mm^2**
10. octagon with side length 10 in. **482.8 in^2**

For Lesson 13-4 — Exponential Expressions

Simplify. *(For review, see page 746.)*

11. $(5b)^2$ **$25b^2$**
12. $\left(\frac{n}{4}\right)^2$ **$\frac{n^2}{16}$**
13. $\left(\frac{3x}{4y}\right)^2$ **$\frac{9x^2}{16y^2}$**
14. $\left(\frac{4y}{7}\right)^2$ **$\frac{16y^2}{49}$**

For Lesson 13-5 — Midpoint Formula

W is the midpoint of $\overline{AB}$. For each pair of points, find the coordinates of the third point. *(For review, see Lesson 1-3.)*

15. $A(0, -1)$, $B(-5, 4)$ **$W(-2.5, 1.5)$**
16. $A(5, 0)$, $B(-3, 6)$ **$W(1, 3)$**
17. $A(1, -1)$, $W(10, 10)$ **$B(19, 21)$**
18. $W(0, 0)$, $B(-2, 2)$ **$A(2, -2)$**

FOLDABLES™ Study Organizer

Volume Make this Foldable to help you organize information about volume. Begin with a sheet of $8\frac{1}{2}$-inch $\times$ 11-inch paper.

Step 1 Fold

Step 2 Fold and Label

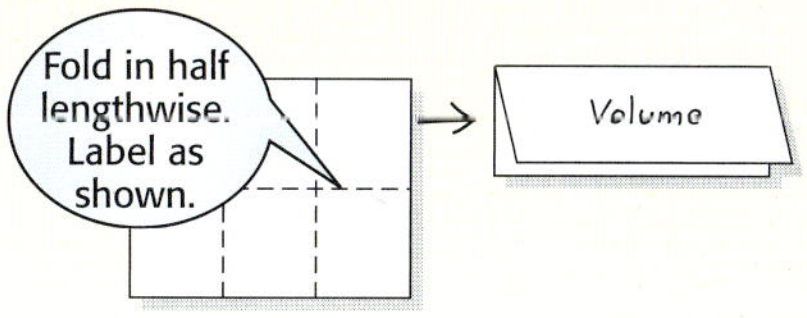

Step 3 Label

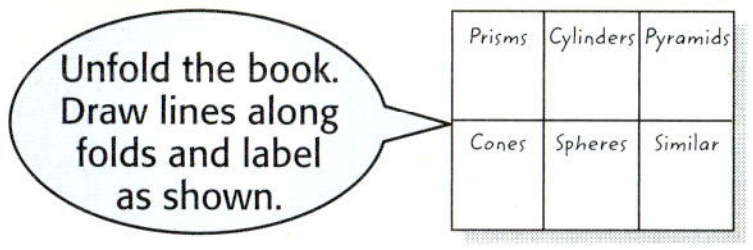

Reading and Writing As you read and study the chapter, write examples and notes about the volume of each solid and about similar solids.

Getting Started

This section provides a review of the basic concepts needed before beginning Chapter 13. Page references are included for additional student help.

Additional review is provided in the *Prerequisite Skills Workbook,* pages 5–6, 21–22, 37–38, 97–100.

Prerequisite Skills in the Getting Ready for the Next Lesson section at the end of each exercise set review a skill needed in the next lesson.

For Lesson	Prerequisite Skill
13-2	Finding the areas of polygons, p. 694
13-3	Evaluating Expressions, p. 701
13-4	Simplifying expressions involving exponents, p. 706
13-5	Graphs in the coordinate plane, p. 713

FOLDABLES™ Study Organizer

For more information about Foldables, see *Teaching Mathematics with Foldables.*

Organization of Data using a Table to Make Comparisons
Use this Foldable table for student writing about volume. After students make their Foldable table, have them label the columns and rows as illustrated. Students can use their table to take notes, define terms, record concepts, and explain how to find the volume of each of the five three-dimensional shapes. Use the data recorded to make comparisons. For example, how are pyramids and cones similar? different?

13-1 Lesson Notes

1 Focus

5-Minute Check Transparency 13-1 Use as a quiz or review of Chapter 12.

Mathematical Background notes are available for this lesson on p. 686C.

How is mathematics used in comics?

Ask students:

- What is the geometric topic referred to in the cartoon? **volume**
- Why is Shoe confused about volume? **He thinks the teacher is talking about volumes of books.**
- How is volume important in a real world context? **Sample answer: The volume can determine the cost of building a container or whether the container is the appropriate size to hold a given amount.**
- What is a figure for which you know how to calculate volume? **Sample answer: boxes**

13-1 Volumes of Prisms and Cylinders

Sunshine State Standards
MA.A.3.4.3, MA.B.1.4.1, MA.B.3.4.1

Vocabulary
- volume

What You'll Learn

- Find volumes of prisms.
- Find volumes of cylinders.

How is mathematics used in comics?

Creators of comics occasionally use mathematics.

SHOE

In the comic above, the teacher is getting ready to teach a geometry lesson on volume. Shoe seems to be confused about the meaning of volume.

VOLUMES OF PRISMS The **volume** of a figure is the measure of the amount of space that a figure encloses. Volume is measured in cubic units. You can create a rectangular prism from different views of the figure to investigate its volume.

Geometry Activity

Volume of a Rectangular Prism

Model

Use cubes to make a model of the solid with the given orthogonal drawing.

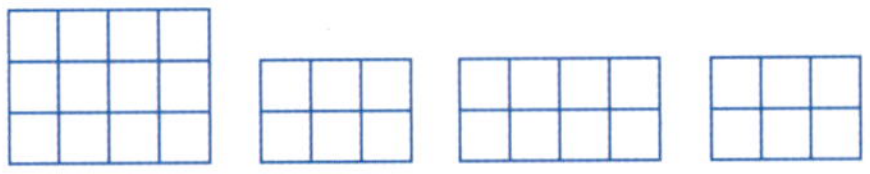

Analyze **2. 4 × 3 × 2 or 24** **4. See students' work.**

1. How many cubes make up the prism? **24**
2. Find the product of the length, width, and height of the prism.
3. Compare the number of cubes to the product of the length, width, and height. **They are the same.**
4. Repeat the activity with a prism of different dimensions.
5. **Make a conjecture** about the formula for the volume of a right rectangular prism. $V = \ell wh$

Resource Manager

Workbook and Reproducible Masters

Chapter 13 Resource Masters
- Study Guide and Intervention, pp. 723–724
- Skills Practice, p. 725
- Practice, p. 726
- Reading to Learn Mathematics, p. 727
- Enrichment, p. 728

School-to-Career Masters, p. 25
Prerequisite Skills Workbook, pp. 21–22, 97–100
Teaching Geometry With Manipulatives Masters, p. 206

Transparencies

5-Minute Check Transparency 13-1
Answer Key Transparencies

Technology

Interactive Chalkboard

Study Tip

Volume and Area
Area is two-dimensional, so it is measured in square units. Volume is three-dimensional, so it is measured in cubic units.

The Geometry activity leads to the formula for the volume of a prism.

Key Concept — *Volume of a Prism*

If a prism has a volume of V cubic units, a height of h units, and each base has an area of B square units, then $V = Bh$.

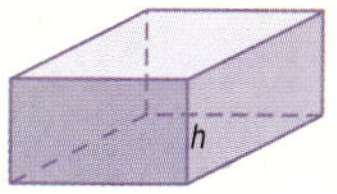

Area of base = B

Example 1 *Volume of a Triangular Prism*

Find the volume of the triangular prism.

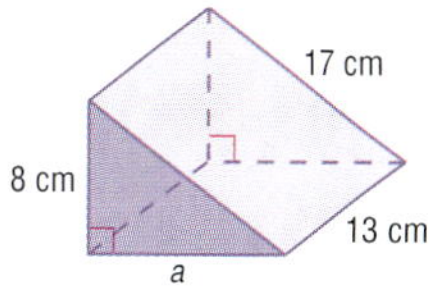

Use the Pythagorean Theorem to find the height of the base of the prism.

$a^2 + b^2 = c^2$ Pythagorean Theorem

$a^2 + 8^2 = 17^2$ $b = 8, c = 17$

$a^2 + 64 = 289$ Multiply.

$a^2 = 225$ Subtract 64 from each side.

$a = 15$ Take the square root of each side.

Next, find the volume of the prism.

$V = Bh$ Volume of a prism

$= \frac{1}{2}(8)(15)(13)$ $B = \frac{1}{2}(8)(15), h = 13$

$= 780$ Simplify.

The volume of the prism is 780 cubic centimeters.

The volume formula can be used to solve real-world problems.

More About. . .

Snow

Snow is made up mostly of air, but a "wet snow" contains a higher water content than usual. Wet sticky snow can be more than twice as heavy as dry fluffy snow.

Source: *Chicago Tribune*

Example 2 *Volume of a Rectangular Prism*

SNOW **The weight of wet snow is 0.575 times the volume of snow in cubic inches divided by 144. How many pounds of wet snow would a person shovel in a rectangular driveway 25 feet by 10 feet after 12 inches of snow have fallen?**

First, make a drawing.
Then convert feet to inches.
25 feet = 25 × 12 or 300 inches
10 feet = 10 × 12 or 120 inches

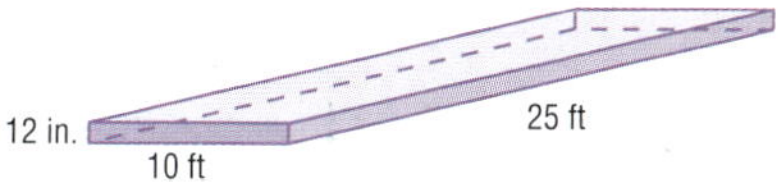

To find the pounds of wet snow shoveled, first find the volume of snow on the driveway.

$V = Bh$ Volume of a prism

$= 300(120)(12)$ $B = 300(120), h = 12$

$= 432{,}000$ The volume is 432,000 cubic inches.

Now multiply the volume by 0.575 and divide by 144.

$\frac{0.575(432{,}000)}{144} = 1725$ Simplify.

A person shoveling 12 inches of snow on a rectangular driveway 25 feet by 10 feet would shovel 1725 pounds of snow.

www.geometryonline.com/extra_examples/fcat

2 Teach

VOLUMES OF PRISMS

Tips for New Teachers

It may be helpful to use a deck of cards to help students see how the volume formula for a prism is related to finding the volume. The area of each card is like the area of the base of the prism. The volume of the deck of cards is found by stacking up the cards, or stacking the area of the base. Hence, the formula for the volume of a prism is the area of the base times the height. The cards are also useful to show oblique prisms.

In-Class Examples Power Point®

Teaching Tip Point out that the volume of *any* prism is the area of the base times the height.

1 Find the volume of the triangular prism.

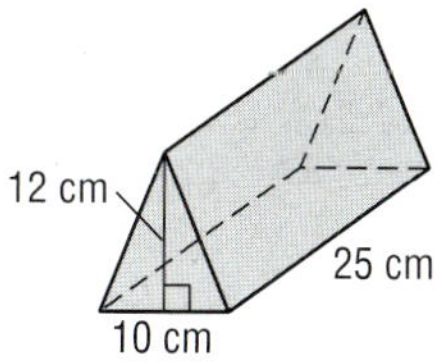

1500 cm³

2 The weight of water is 0.036 pounds times the volume of water in cubic inches. How many pounds of water would fit into a rectangular child's pool that is 12 inches deep, 3 feet wide, and 4 feet long?

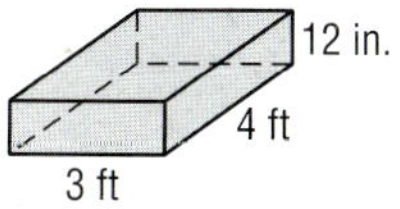

746.5 lb

Geometry Activity

Materials: cubes

It may be helpful to review how to do an orthogonal drawing from Chapter 12. To repeat the activity, ask students to create a table including the dimensions of different prisms. See if they can make conjectures about the way to find volume from their table values.

VOLUMES OF CYLINDERS

Tips for New Teachers

You may want to give students hands-on experience finding the volumes of cylinders by having them measure the actual volume of soda cans or soup cans. Ask them to discuss how their measured volume compares with the published volume on the can.

In-Class Example PowerPoint®

3 Find the volume of each cylinder to the nearest tenth.

a.

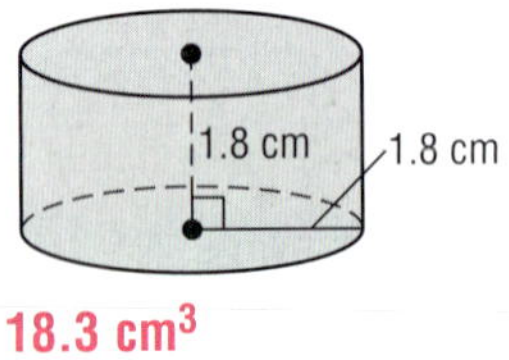

18.3 cm³

b.

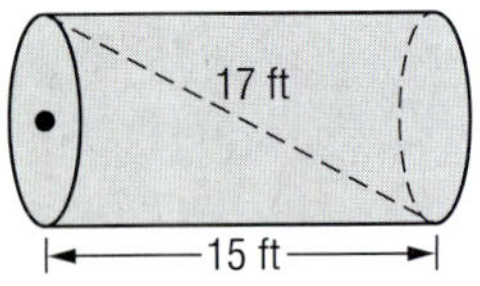

754.0 ft³

Interactive Chalkboard
PowerPoint® Presentations

This CD-ROM is a customizable Microsoft® PowerPoint® presentation that includes:

- Step-by-step, dynamic solutions of each In-Class Example from the Teacher Wraparound Edition
- Additional, Try These exercises for each example
- The 5-Minute Check Transparencies
- Hot links to Glencoe Online Study Tools

VOLUMES OF CYLINDERS Like the volume of a prism, the volume of a cylinder is the product of the area of the base and the height.

Key Concept — *Volume of a Cylinder*

If a cylinder has a volume of V cubic units, a height of h units, and the bases have radii of r units, then $V = Bh$ or $V = \pi r^2 h$.

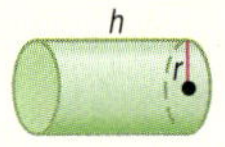

Area of base = πr^2

Study Tip

Square Roots
When you take the square root of each side of $h^2 = 144$, the result is actually ± 12. Since this is a measure, the negative value is discarded.

Example 3 *Volume of a Cylinder*

Find the volume of each cylinder.

a.

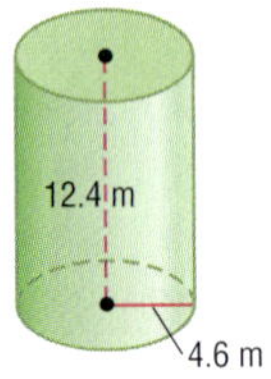

The height h is 12.4 meters, and the radius r is 4.6 meters.

$V = \pi r^2 h$	Volume of a cylinder
$= \pi(4.6^2)(12.4)$	$r = 4.6$, $h = 12.4$
≈ 824.3	Use a calculator.

The volume is approximately 824.3 cubic meters.

b.

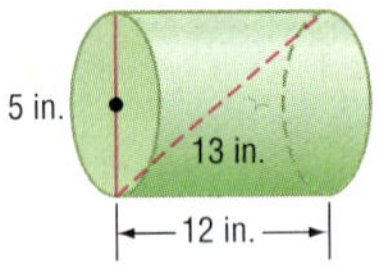

The diameter of the base, the diagonal, and the lateral edge of the cylinder form a right triangle. Use the Pythagorean Theorem to find the height.

$a^2 + b^2 = c^2$	Pythagorean Theorem
$h^2 + 5^2 = 13^2$	$a = h$, $b = 5$, and $c = 13$
$h^2 + 25 = 169$	Multiply.
$h^2 = 144$	Subtract 25 from each side.
$h = 12$	Take the square root of each side.

Now find the volume.

$V = \pi r^2 h$	Volume of a cylinder
$= \pi(2.5^2)(12)$	$r = 2.5$ and $h = 12$
≈ 235.6	Use a calculator.

The volume is approximately 235.6 cubic inches.

Study Tip

Look Back
To review **oblique solids**, see Lesson 12-3.

Thus far, we have only studied the volumes of right solids. Do the formulas for volume apply to oblique solids as well as right solids?

Study the two stacks of quarters. The stack on the left represents a right cylinder, and the stack on the right represents an oblique cylinder. Since each stack has the same number of coins, with each coin the same size and shape, the two cylinders must have the same volume. Cavalieri, an Italian mathematician of the seventeenth century, was credited with making this observation first.

DAILY INTERVENTION — **Differentiated Instruction**

Logical Students should reason that an oblique cylinder can also be likened to a stack of circles that has been shifted so that they make an oblique angle with the base. If a line segment is drawn connecting the center of each base (called the axis), then the axis is also an altitude for a right cylinder, but the axis is not an altitude for an oblique cylinder.

Study Tip

Cavalieri's Principle
This principle applies to the volumes of all solids.

Key Concept — *Cavalieri's Principle*

If two solids have the same height and the same cross-sectional area at every level, then they have the same volume.

If a cylinder has a base with an area of B square units and a height of h units, then its volume is Bh cubic units, whether it is right or oblique.

Example 4 Volume of an Oblique Cylinder

Find the volume of the oblique cylinder.

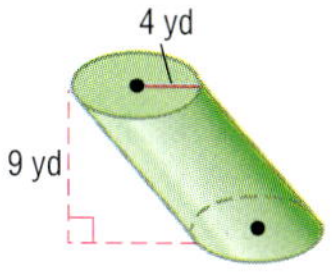

To find the volume, use the formula for a right cylinder.

$V = \pi r^2 h$ — Volume of a cylinder

$= \pi(4^2)(9)$ — $r = 4, h = 9$

≈ 452.4 — Use a calculator.

The volume is approximately 452.4 cubic yards.

Check for Understanding

Concept Check

1. **OPEN ENDED** List three objects that are cylinders and three that are prisms.

1. Sample answers: cans, roll of paper towels, and chalk; boxes, crystals, and buildings

2. **FIND THE ERROR** Che and Julia are trying to find the number of cubic feet in a cubic yard.

Che	Julia
$V = Bh$	$V = Bh$
$= 3 \times 3 \times 3$	$= 3 \times 3 \times 3$
$= 9$	$= 27$
There are 9 cubic feet in one cubic yard.	There are 27 cubic feet in one cubic yard.

Who is correct? Explain your reasoning. **Julia; Che did not multiply 3^3 correctly.**

Guided Practice

GUIDED PRACTICE KEY	
Exercises	Examples
3	1
4	3
5	4
6	2

Find the volume of each prism or cylinder. Round to the nearest tenth if necessary.

3. **288 cm^3**

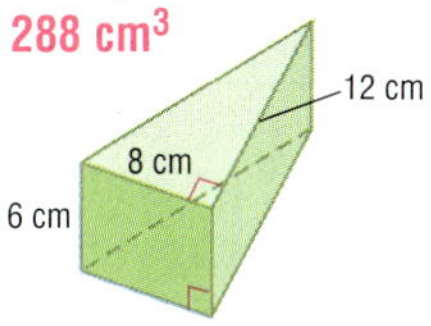

4. **754.0 in^3**

5. **3180.9 mm^3**

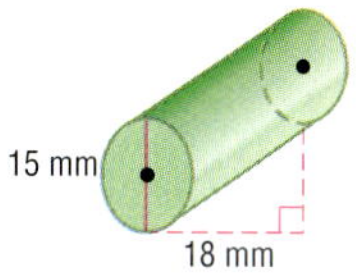

Application

6. **DIGITAL CAMERA** The world's most powerful digital camera is located in New Mexico at the Apache Point Observatory. It is surrounded by a rectangular prism made of aluminum that protects the camera from wind and unwanted light. If the prism is 12 feet long, 12 feet wide, and 14 feet high, find its volume to the nearest cubic foot. **2016 ft^3**

In-Class Example

Power Point®

4 Find the volume of the oblique cylinder to the nearest tenth.

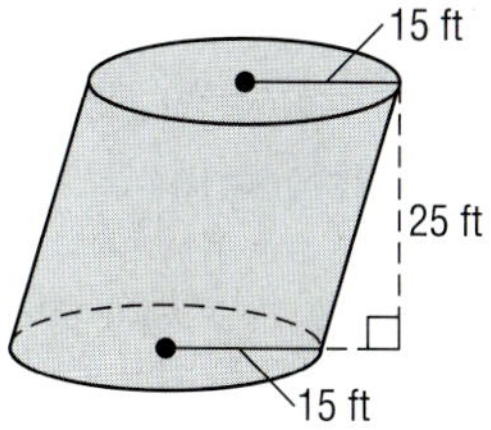

17,671.5 ft^3

3 Practice/Apply

Study Notebook

Have students—

- add the definitions/examples of the vocabulary terms to their Vocabulary Builder worksheets for Chapter 13.
- include a page that shows each volume formula with a sketch of the corresponding solid and an example of how to find the volume.
- include any other item(s) that they find helpful in mastering the skills in this lesson.

DAILY INTERVENTION **FIND THE ERROR**

In Exercise 2, caution students that Che added the three 3's instead of multiplying the three 3's. Since each side of a cube is 3 feet on a side, the correct calculation is done by Julia.

About the Exercises...

Organization by Objective

- **Volumes of Prisms:** 8, 9, 11, 12, 17, 19–20, 23, 25–30
- **Volumes of Cylinders:** 7, 10, 13–16, 18, 21, 22, 24

Odd/Even Assignments

Exercises 7–24 are structured so that students practice the same concepts whether they are assigned odd or even problems.

Assignment Guide

Basic: 7–21 odd, 25, 27, 31, 33–52

Average: 7–31 odd, 32–52

Advanced: 8–30 even, 31–48 (optional: 49–52)

★ indicates increased difficulty

Practice and Apply

Homework Help

For Exercises	See Examples
7, 10, 17	3
8, 11, 18	2
9, 12	1
13–16	4

Extra Practice
See page 780.

Find the volume of each prism or cylinder. Round to the nearest tenth if necessary.

7. 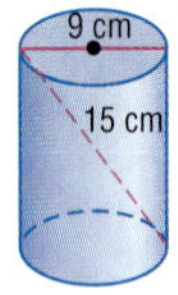

763.4 cm^3

8. 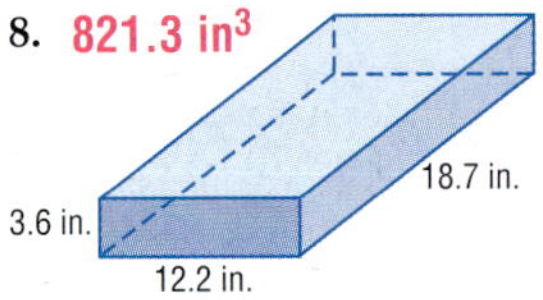

821.3 in^3

9. 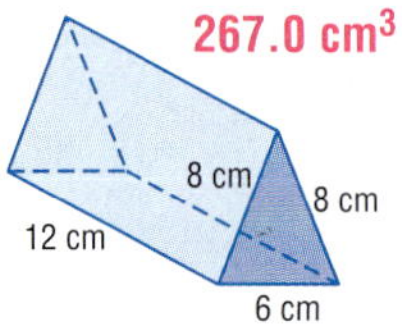

267.0 cm^3

10. 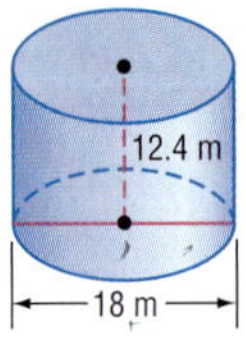

3155.4 m^3

11. 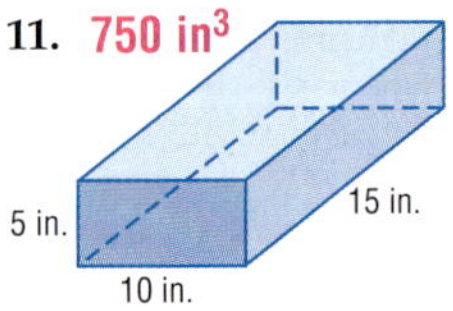

750 in^3

12. 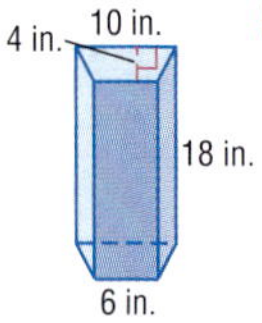

576 in^3

Find the volume of each oblique prism or cylinder. Round to the nearest tenth if necessary.

13. 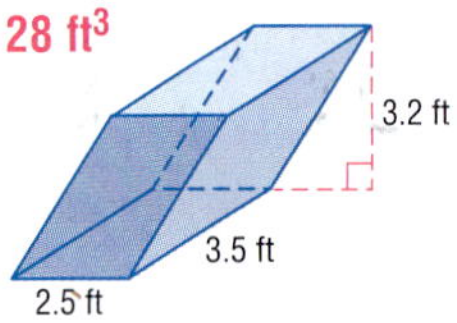

28 ft^3

14. 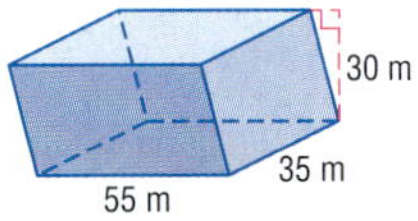

57,750 m^3

15. 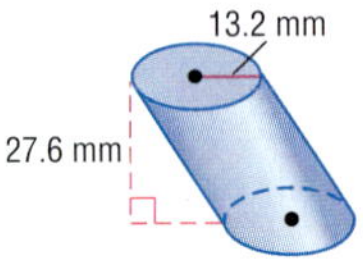

15,108.0 mm^3

16. 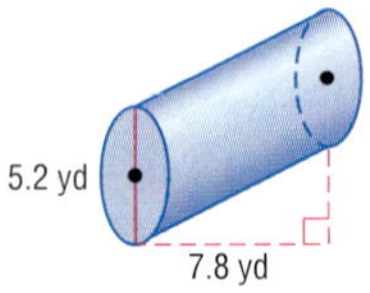

165.6 yd^3

17. The volume of a cylinder is 615.8 cubic meters, and its height is 4 meters. Find the diameter of the cylinder. ≈ 14 m

18. The volume of a right rectangular prism is 1152 cubic inches, and the area of each base is 64 square inches. Find the length of the lateral edge of the prism. 18 in.

Find the volume of the solid formed by each net. Round to the nearest tenth if necessary. 19. 24 $units^3$ 20. 2.5 $units^3$ 21. 48.5 mm^3

19. 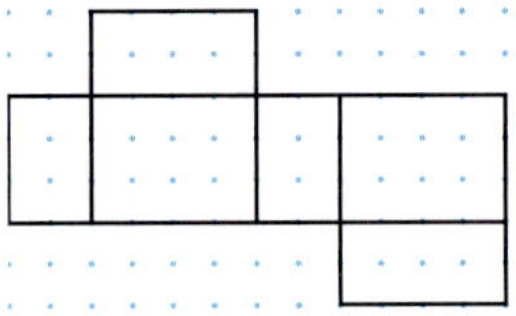

20. 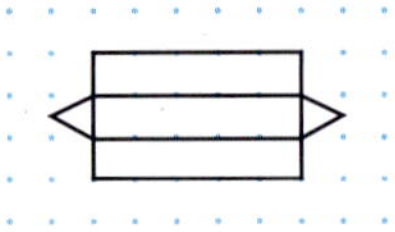

21. 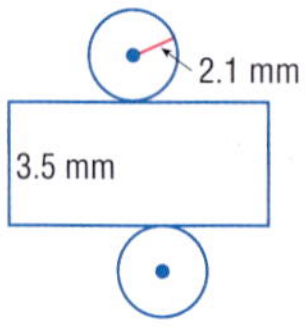

Find the volume of each solid. Round to the nearest tenth if necessary.

★ 22. 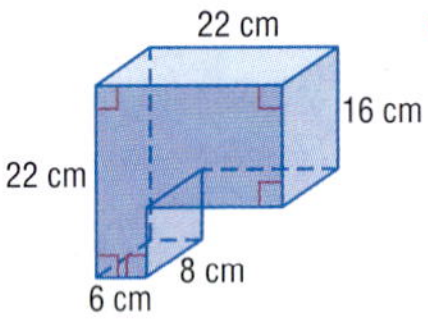

3104 cm^3

★ 23. 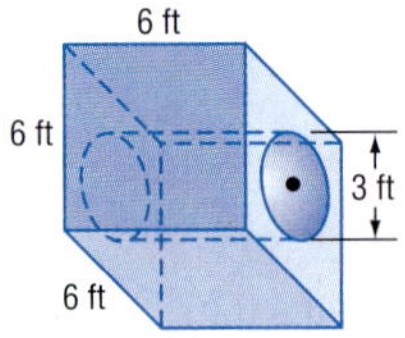

173.6 ft^3

★ 24. 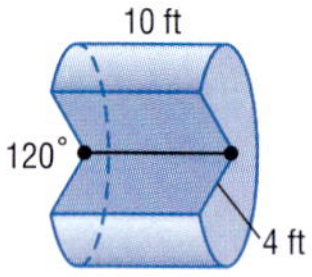

335.1 ft^3

692 Chapter 13 Volume

25. **MANUFACTURING** A can is 12 centimeters tall and has a diameter of 6.5 centimeters. It fits into a rubberized cylindrical holder that is 11.5 centimeters tall, including 1 centimeter, which is the thickness of the base of the holder. The thickness of the rim of the holder is 1 centimeter. What is the volume of the rubberized material that makes up the holder? $\approx$ 304.1 cm^3

26. **ARCHITECTURE** The Marina Towers in Chicago are cylindrical shaped buildings that are 586 feet tall. There is a 35-foot-diameter cylindrical core in the center of each tower. If the core extends 40 feet above the roof of the tower, find the volume of the core. $\approx$ 602,282.6 ft^3

27. **AQUARIUM** The New England Aquarium in Boston, Massachusetts, has one of the world's largest cylindrical tanks. The Giant Ocean tank holds approximately 200,000 gallons and is 23 feet deep. If it takes about $7\frac{1}{2}$ gallons of water to fill a cubic foot, what is the radius of the Giant Ocean Tank? about 19.2 ft

★ 28. **SWIMMING** A swimming pool is 50 meters long and 25 meters wide. The adjustable bottom of the pool can be up to 3 meters deep for competition and as shallow as 0.3 meter deep for recreation. The pool was filled to the recreational level, and then the floor was lowered to the competition level. If the volume of a liter of water is 0.001 cubic meter, how much water had to be added to fill the pool? 3,375,000 L

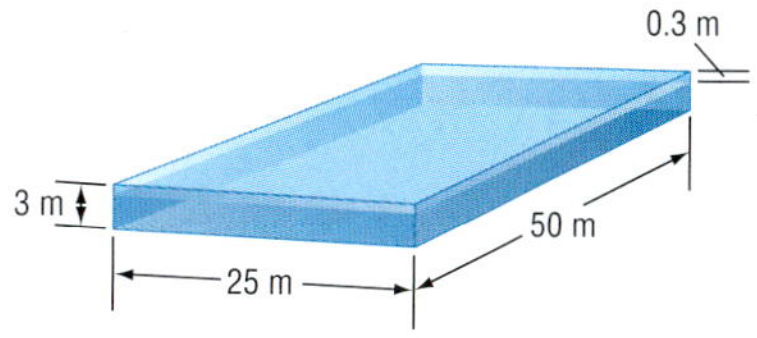

Career Choices

Machinist

Some machinists use computerized equipment to cut parts. The machinist determines how fast the part should be cut based on the type of metal and harmonic vibrations.

Source: *Occupational Outlook Handbook*

Online Research For information about a career as a machinist, visit: www.geometryonline.com/careers

ENGINEERING For Exercises 29 and 30, use the following information.

Machinists make parts for intricate pieces of equipment. Suppose a part has a regular hexagonal hole drilled in a brass block.

★ 29. Find the volume of the resulting part. $\approx$ 104,411.5 mm^3

★ 30. The *density* of a substance is its mass per unit volume. At room temperature, the density of brass is 8.0 grams per cubic centimeter. What is the mass of this block of brass? $\approx$ 835.3 g

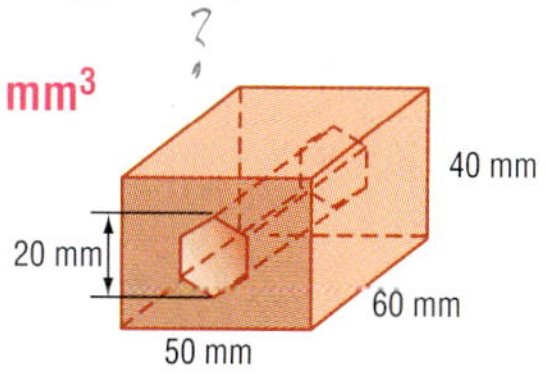

31. **CRITICAL THINKING** Find the volume of a regular pentagonal prism with a height of 5 feet and a perimeter of 20 feet. $\approx$ 137.6 ft^3

32. **WRITING IN MATH** Answer the question that was posed at the beginning of the lesson. See margin.

How is mathematics used in comics?

Include the following in your answer:
- the meaning of volume that Shoe has in the comic, and
- the mathematical meaning of volume.

www.geometryonline.com/self_check_quiz/fcat

Study Guide and Intervention, p. 723 (shown) and p. 724

Volumes of Prisms The measure of the amount of space that a three-dimensional figure encloses is the **volume** of the figure. Volume is measured in units such as cubic feet, cubic yards, or cubic meters. One cubic unit is the volume of a cube that measures one unit on each edge.

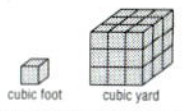

27 cubic feet = 1 cubic yard

Volume of a Prism	If a prism has a volume of V cubic units, a height of h units, and each base has an area of B square units, then $V = Bh$.

Example 1 Find the volume of the prism.

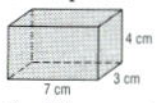

$V = Bh$ Formula for volume
$= (7)(3)(4)$ $B = (7)(3)$, $h = 4$
$= 84$ Multiply.

The volume of the prism is 84 cubic centimeters.

Example 2 Find the volume of the prism if the area of each base is 6.3 square feet.

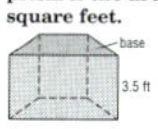

$V = Bh$ Formula for volume
$= (6.3)(3.5)$ $B = 6.3$, $h = 3.5$
$= 22.05$ Multiply.

The volume is 22.05 cubic feet.

Exercises

Find the volume of each prism. Round to the nearest tenth if necessary.

1. 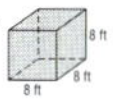

512 ft^3

2. 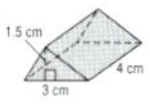

9 cm^3

3. 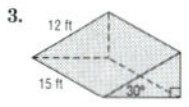

467.7 ft^3

4. 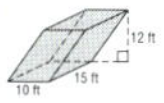

1800 ft^3

5. 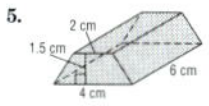

27 cm^3

6. 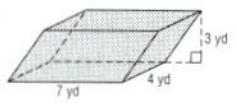

84 yd^3

Skills Practice, p. 725 and Practice, p. 726 (shown)

Find the volume of each prism or cylinder. Round to the nearest tenth if necessary.

1. 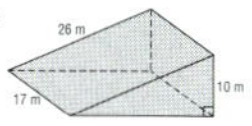

2040 m^3

2. 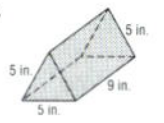

97.4 in^3

3. 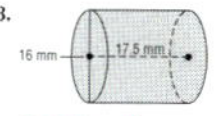

3518.6 mm^3

4. 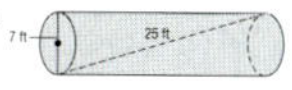

923.6 ft^3

5. 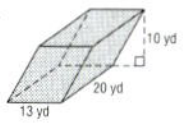

2600 yd^3

6. 

6031.9 cm^3

AQUARIUM **For Exercises 7–9, use the following information. Round answers to the nearest tenth.**

Mr. Gutierrez purchased a cylindrical aquarium for his office. The aquarium has a height of $25\frac{1}{2}$ inches and a radius of 21 inches.

7. What is the volume of the aquarium in cubic feet?
5.1 ft^3

8. If there are 7.48 gallons in a cubic foot, how many gallons of water does the aquarium hold?
38.2 gal

9. If a cubic foot of water weighs about 62.4 pounds, what is the weight of the water in the aquarium to the nearest five pounds?
2385 lb

Reading to Learn Mathematics, p. 727 ELL

Pre-Activity **How is mathematics used in comics?**

Read the introduction to Lesson 13-1 at the top of page 688 in your textbook.

In the cartoon, why was Shoe confused when the teacher said the class was going to discuss volumes? Sample answer: The teacher was referring to the mathematical meaning of volume, which is the amount of space enclosed by a three-dimensional figure. Shoe was thinking about volumes as books, which is a completely different meaning of the word.

Reading the Lesson

1. In each case, write a formula for the volume V of the solid in terms of the given variables.
   a. a rectangular box with length a, width b, and height c $V = abc$
   b. a rectangular box with square bases with side length x, and with height y $V = x^2y$
   c. a cube with edges of length e $V = e^3$
   d. a triangular prism whose bases are isosceles right triangles with legs of length x, and whose height is y $V = \frac{1}{2}x^2y$ or $V = \frac{x^2y}{2}$
   e. a prism whose bases are regular polygons with perimeter P and apothem a, and whose height is h $V = \frac{1}{2}aPh$ or $V = \frac{aPh}{2}$
   f. a cylinder whose bases each have radius r, and whose height is three times the radius of the bases $V = 3\pi r^3$
   g. a regular octagonal prism in which each base has sides of length s and apothem a, and whose height is t $V = 4ast$
   h. a cylinder with height h whose bases each have diameter d
   $V = \pi\left(\frac{d}{2}\right)^2 h$ or $V = \frac{\pi d^2 h}{4}$ or $V = \frac{1}{4}\pi d^2 h$
   i. an oblique cylinder whose bases have radius a and whose height is b $V = \pi a^2 b$
   j. a regular hexagonal prism whose bases have side length s, and whose height is h
   $V = \frac{3}{2}s^2h\sqrt{3}$

Helping You Remember

2. A good way to remember a mathematical concept is to explain it to someone else. Suppose that your younger sister, who is in eighth grade, is having trouble understanding why square units are used to measure area, but cubic units are needed to measure volume. How can you explain this to her in a way that will make it easy for her to understand and remember the correct units to use? Sample answer: Area measures the amount of space inside a two-dimensional figure, while volume measures the amount of space inside a three-dimensional figure. A two-dimensional figure can be covered with small squares, which represent square units, while a three-dimensional figure can be filled with small cubes, which represent cubic units.

Answers

32. Sample answer: Cartoonists use mathematical concepts or terms in comics because of the difficulty that many people have had with math. Answers should include the following.
- A volume means a book.
- In mathematics, volume refers to the amount of space that a figure encloses.

Enrichment, p. 728

Visible Surface Area

Use paper, scissors, and tape to make five cubes that have one-inch edges. Arrange the cubes to form each shape shown. Then find the volume and the visible surface area. In other words, do not include the area of surface covered by other cubes or by the table or desk.

1. 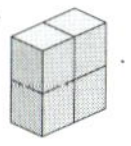
volume = 4 in^3
surface area = 14 in^2

2. 
volume = 4 in^3
surface area = 15 in^2

4 Assess

Open-Ended Assessment

Speaking Have students describe the similarities and differences between the volume of a cylinder and the volume of a prism. **They are similar because each volume is found by multiplying the area of the base times the height. They are different because the bases (and hence the formula for the area) are different. One is a circle and one is a polygon.**

Getting Ready for Lesson 13-2

Prerequisite Skill Students will learn about finding the volumes of pyramids and cones in Lesson 13-2. They will find the area of the base of the pyramid or cone before they find the volume. Use Exercises 49–52 to determine your students' familiarity with finding the area of a polygon.

FCAT Practice
Standardized Test Practice

33. A rectangular swimming pool has a volume of 16,320 cubic feet, a depth of 8 feet, and a length of 85 feet. What is the width of the swimming pool? **A**

    (A) 24 ft (B) 48 ft (C) 192 ft (D) 2040 ft

34. **ALGEBRA** Factor $\pi r^2h - 2\pi rh$ completely. **B**

    (A) $\pi(r^2h - 2rh)$ (B) $\pi rh(r - 2)$
    (C) $\pi rh(rh - 2rh)$ (D) $2\pi r(rh - h)$

Maintain Your Skills

Mixed Review

Find the surface area of each sphere. Round to the nearest tenth. *(Lesson 12-7)*

35. 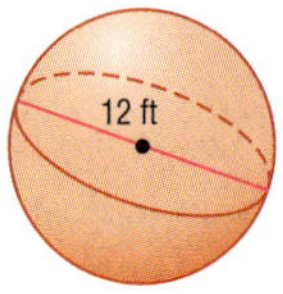

452.4 ft²

36. 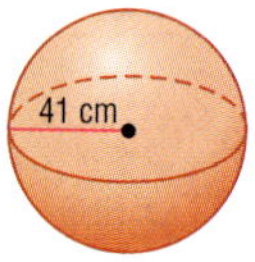

21,124.1 cm²

37. 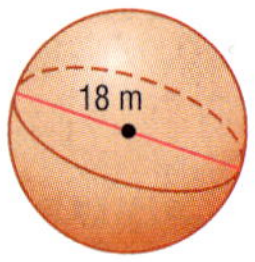

1017.9 m²

38. 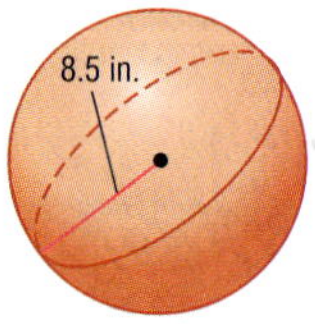

907.9 in²

Find the surface area of each cone. Round to the nearest tenth. *(Lesson 12-6)*

39. slant height = 11 m, radius = 6 m **320.4 m²**
40. diameter = 16 cm, slant height = 13.5 cm **540.4 cm²**
41. radius = 5 in., height = 12 in. **282.7 in²**
42. diameter = 14 in., height = 24 in. **703.7 in²**

43. **HOUSING** Martin lost a file at his home, and he only has time to search three of the rooms before he has to leave for work. If the shaded parts of his home will be searched, what is the probability that he finds his file? *(Lesson 11-5)* **0.42**

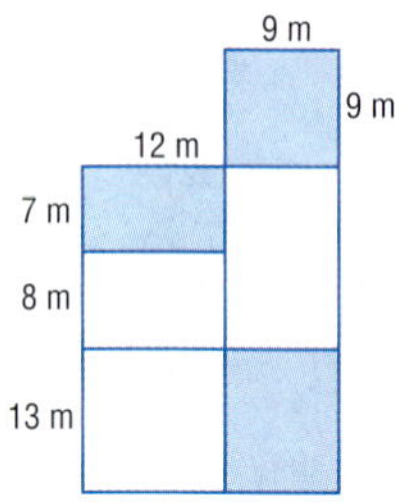

Find the area of each polygon. Round to the nearest tenth. *(Lesson 11-3)*

44. a regular hexagon with a perimeter of 156 inches **1756.3 in²**
45. a regular octagon with an apothem 7.5 meters long and a side 6.2 meters long **186 m²**

Find x to the nearest tenth. Assume that any segment that appears to be tangent is tangent. *(Lesson 10-7)*

46. **9.6**
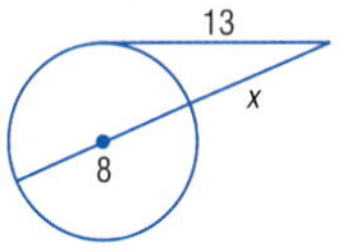

47. **8.8**
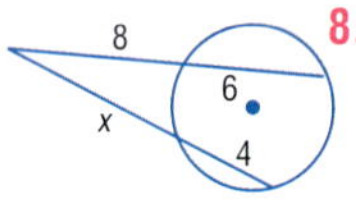

48. **16.1**
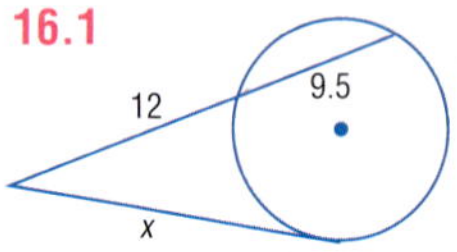

Getting Ready for the Next Lesson

PREREQUISITE SKILL Find the area of each polygon with given side length, s. Round to the nearest hundredth.
*(To review **finding areas of regular polygons**, see Lesson 11-3)* **50. 374.12 cm²**

49. equilateral triangle, $s = 7$ in. **21.22 in²**
50. regular hexagon, $s = 12$ cm
51. regular pentagon, $s = 6$ m **61.94 m²**
52. regular octagon, $s = 50$ ft **12,071.07 ft²**

694 Chapter 13 Volume

Spreadsheet Investigation

A Follow-Up of Lesson 13-1

Prisms

Changing the dimensions of a prism affects the surface area and the volume of the prism. You can investigate the changes by using a spreadsheet.

- Create a spreadsheet by entering the length of the rectangular prism in column B, the width in column C, and the height in column D.
- In cell E2, enter the formula for the total surface area.
- Copy the formula in cell E2 to the other cells in column E.
- Write a formula to find the volume of the prism. Enter the formula in cell F2.
- Copy the formula in cell F2 into the other cells in column F.

Prisms.xls

	A	B	C	D	E	F
1	Prism	ℓ	w	h	Surface Area	Volume
2	1	1	2	3	22	6
3	2					
4	3					
5	4					
6	5					
7						

Sheet1 / Sheet2

Use your spreadsheet to find the surface areas and volumes of prisms with the dimensions given in the table below.

Prism	Length	Width	Height	Surface Area	Volume
1	1	2	3	22	6
2	2	4	6	88	48
3	3	6	9	198	162
4	4	8	12	352	384
5	8	16	24	1408	3072

1. For each pair of prisms, the dimensions of the second prism are two times the dimensions of the first prism.
2. The surface areas of prisms 2, 4, and 5 are four times the respective surface areas of prisms 1, 2, and 4.
3. The volumes of prisms 2, 4, and 5 are eight times the respective volumes of prisms 1, 2, and 4.
4. When the dimensions are doubled, the surface area is multiplied by 4, or 2^2, and the volume is multiplied by 8, or 2^3.

Exercises

1. Compare the dimensions of prisms 1 and 2, prisms 2 and 4, and prisms 4 and 5.
2. Compare the surface areas of prisms 1 and 2, prisms 2 and 4, and prisms 4 and 5.
3. Compare the volumes of prisms 1 and 2, prisms 2 and 4, and prisms 4 and 5.
4. Write a statement about the change in the surface area and volume of a prism when the dimensions are doubled.

Spreadsheet Investigation

A Follow-Up of Lesson 13-1

Getting Started

Students may use the "Fill-down" feature of a spreadsheet to enter the data in 5 rows of columns one, two, and three. For example, if the cell reference for length is B2, then enter "=B2 + 1" in cell B3. Then highlight cell B3 and use the editing features of a spreadsheet to "fill down" the column with the same pattern. Likewise, if the cell reference for width is C2, then enter "=C2 + 2". Then fill down the column with the same pattern. Students should notice that they can find the surface area or volume of any prism by using spreadsheet features to copy formulas into columns.

Teach

You may want students to do this activity in pairs, especially if there is limited availability of computers. Ask one student to enter the data into the computer while another student directs the partner with the information to enter. Ask the pairs of students to discuss which formulas should be used for Exercises 1–4.

Assess

In **Exercises 1–4**, students use technology to confirm that when the dimensions of a prism are doubled, the surface area increases by a factor of 4, and the volume increases by a factor of 8.

Spreadsheet Investigation Prisms 695

13-2 Lesson Notes

1 Focus

5-Minute Check Transparency 13-2 Use as a quiz or review of Lesson 13-1.

Mathematical Background notes are available for this lesson on p. 686C.

Why do architects use geometry?

Ask students:

- Why are pyramids important to architects? **Many roofs of buildings are pyramids, so architects need to know the geometry of the pyramid.**
- What kind of pyramid is used for the building in this graphic? **a square pyramid**
- What other real-world examples of buildings use pyramids in their construction? **Sample answer: the Washington Monument**

13-2 Volumes of Pyramids and Cones

Sunshine State Standards
MA.A.3.4.3, MA.B.1.4.1, MA.B.3.4.1

What You'll Learn

- Find volumes of pyramids.
- Find volumes of cones.

Why do architects use geometry?

The Transamerica Pyramid is the tallest skyscraper in San Francisco. The 48-story building is a square pyramid. The building was designed to allow more light to reach the street.

VOLUMES OF PYRAMIDS The pyramid and the prism at the right share a base and have the same height. As you can see, the volume of the pyramid is less than the volume of the prism.

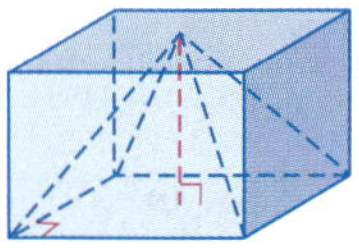

Geometry Activity

Investigating the Volume of a Pyramid

Activity

- Draw each net on card stock.
- Cut out the nets. Fold on the dashed lines.
- Tape the edges together to form models of the solids with one face removed.
- Estimate how much greater the volume of the prism is than the volume of the pyramid.
- Fill the pyramid with rice. Then pour this rice into the prism. Repeat until the prism is filled.

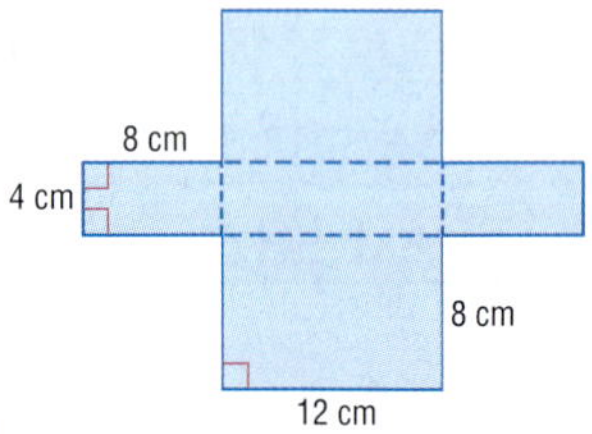

Analyze

1. How many pyramids of rice did it take to fill the prism? **3**
2. Compare the areas of the bases of the prism and pyramid. **The areas of the bases are the same.**
3. Compare the heights of the prism and the pyramid. **The heights are the same.**
4. Make a conjecture about the formula for the volume of a pyramid. $V = \frac{1}{3}Bh$

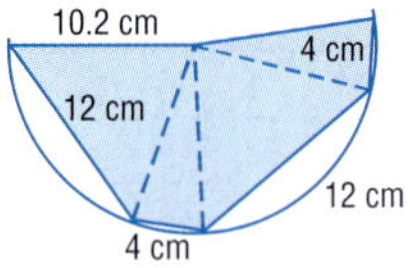

The Geometry Activity leads to the formula for the volume of a pyramid.

Key Concept — *Volume of a Pyramid*

If a pyramid has a volume of V cubic units, a height of h units, and a base with an area of B square units, then $V = \frac{1}{3}Bh$.

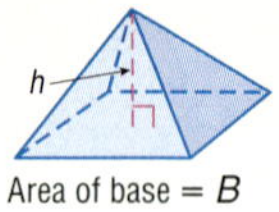

Resource Manager

Workbook and Reproducible Masters

Chapter 13 Resource Masters
- Study Guide and Intervention, pp. 729–730
- Skills Practice, p. 731
- Practice, p. 732
- Reading to Learn Mathematics, p. 733
- Enrichment, p. 734
- Assessment, p. 767

School-to-Career Masters, p. 26
Teaching Geometry With Manipulatives Masters, pp. 9, 207

Transparencies

5-Minute Check Transparency 13-2
Real-World Transparency 13
Answer Key Transparencies

Technology

Interactive Chalkboard
Multimedia Applications: Virtual Activities

Example 1 Volume of a Pyramid

NUTRITION **Travis is making a plaster model of the Food Guide Pyramid for a class presentation. The model is a square pyramid with a base edge of 12 inches and a height of 15 inches. Find the volume of plaster needed to make the model.**

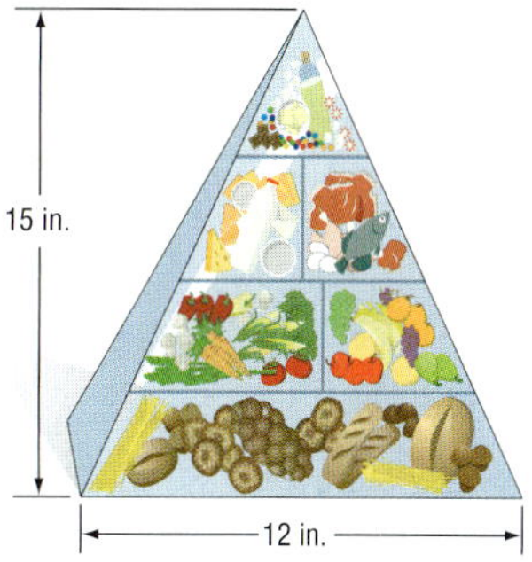

$V = \frac{1}{3}Bh$ Volume of a pyramid

$= \frac{1}{3}s^2h$ $B = s^2$

$= \frac{1}{3}(12^2)(15)$ $s = 12, h = 15$

$= 720$ Multiply.

Travis needs 720 cubic inches of plaster to make the model.

VOLUMES OF CONES The derivation of the formula for the volume of a cone is similar to that of a pyramid. If the areas of the bases of a cone and a cylinder are the same and if the heights are equal, then the volume of the cylinder is three times as much as the volume of the cone.

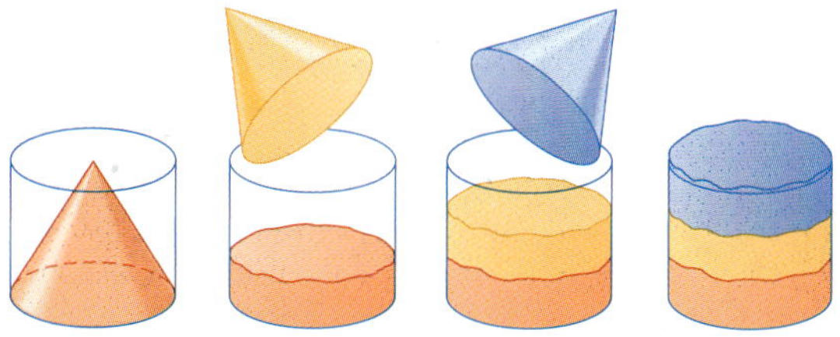

Key Concept — Volume of a Cone

If a right circular cone has a volume of V cubic units, a height of h units, and the base has a radius of r units, then $V = \frac{1}{3}Bh$ or $V = \frac{1}{3}\pi r^2 h$.

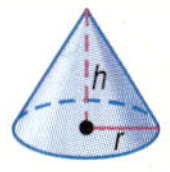

Area of base = πr^2

Example 2 Volumes of Cones

Find the volume of each cone.

a.

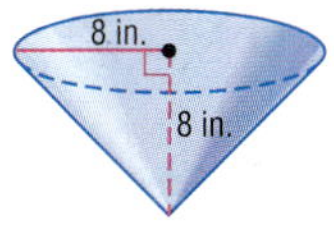

$V = \frac{1}{3}\pi r^2 h$ Volume of a cone

$= \frac{1}{3}\pi(8^2)(8)$ $r = 8, h = 8$

≈ 536.165 Use a calculator.

The volume of the cone is approximately 536.2 cubic inches.

www.geometryonline.com/extra_examples/fcat

Study Tip

Volume Formulas
Notice how the formulas for the volumes of a cone and pyramid are the same. However, the area of the base of a cone is πr^2.

2 Teach

VOLUMES OF PYRAMIDS

In-Class Example (PowerPoint®)

1 Teofilo has a solid clock that is in the shape of a square pyramid. The clock has a base of 3 inches and a height of 7 inches. Find the volume of the clock.

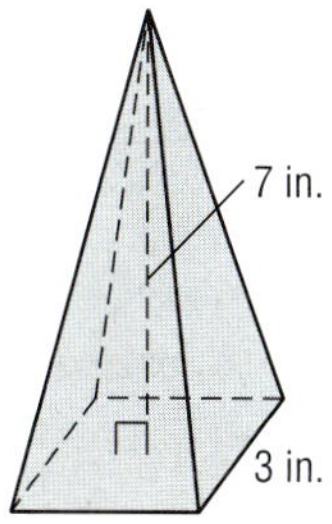

21 in^3

VOLUMES OF CONES

In-Class Example (PowerPoint®)

2 Find the volume of each cone to the nearest tenth.

a.

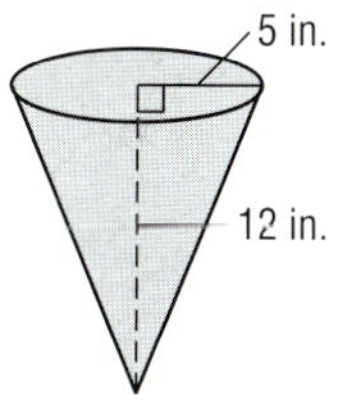

314.2 in^3

b.

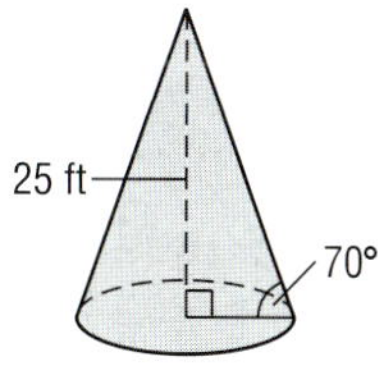

2167.6 ft^3

Geometry Activity

Materials: card stock

Ask students to recall what they remember about constructing a net for a prism. Alternatively, demonstrate in front of a class with a ready-made pyramid and prism to save time.

Building on Prior Knowledge

In Chapter 11, students found the area of polygons and circles. Stress that in this chapter, the volume formulas include the area formulas of the polygons or circles as bases of the solids.

In-Class Example

Power Point®

3 Find the volume of the oblique cone to the nearest tenth.

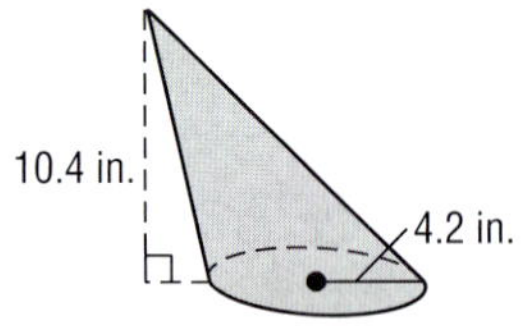

192.1 in^3

Answers

2. The volume of a pyramid is one-third the volume of a prism of the same height as the pyramid and with bases congruent to the bases of the pyramid.
3. Sample answer:

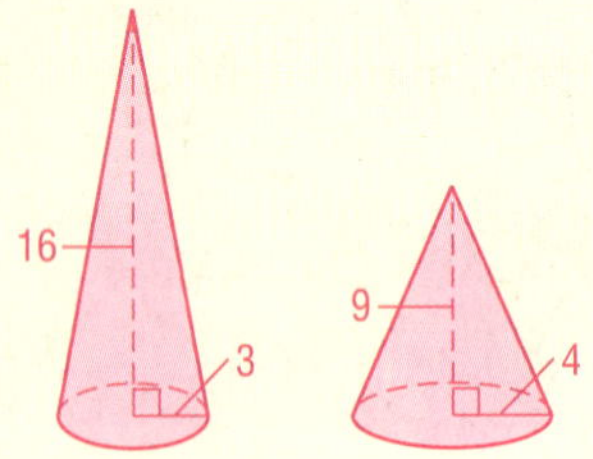

$V = \frac{1}{3}\pi(3^2)(16)$
$= 48\pi$

$V = \frac{1}{3}\pi(4^2)(9)$
$= 48\pi$

b.

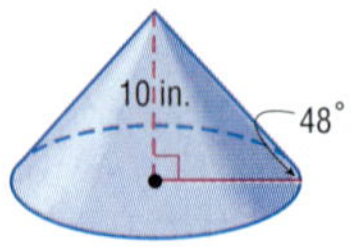

Use trigonometry to find the radius of the base.

$\tan A = \frac{\text{opposite}}{\text{adjacent}}$ Definition of tangent

$\tan 48° = \frac{10}{r}$ $A = 48°$, opposite $= 10$, and adjacent $= r$

$r = \frac{10}{\tan 48°}$ Solve for r.

$r \approx 9.0$ Use a calculator.

Now find the volume.

$V = \frac{1}{3}Bh$ Volume of a cone

$= \frac{1}{3}\pi r^2 h$ $B = \pi r^2$

$\approx \frac{1}{3}\pi(9^2)(10)$ $r \approx 9$, $h = 10$

≈ 848.992 Use a calculator.

The volume of the cone is approximately 849.0 cubic inches.

Study Tip

Common Misconceptions
The formula for the surface area of a cone only applies to right cones. However, a right cone and an oblique cone with the same radius and height have the same volume but different surface areas.

Recall that Cavalieri's Principle applies to all solids. So, the formula for the volume of an oblique cone is the same as that of a right cone.

Example 3 Volume of an Oblique Cone

Find the volume of the oblique cone.

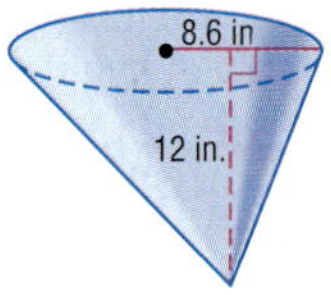

$V = \frac{1}{3}Bh$ Volume of a cone

$= \frac{1}{3}\pi r^2 h$ $B = \pi r^2$

$= \frac{1}{3}\pi(8.6^2)(12)$ $r = 8.6$, $h = 12$

≈ 929.4 Use a calculator.

The volume of the oblique cone is approximately 929.4 cubic inches.

Check for Understanding

Concept Check

1. **Describe** the effect on the volumes of a cone and a pyramid if the dimensions are doubled. **Each volume is 8 times as large as the original.**

2–3. See margin.

2. **Explain** how the volume of a pyramid is related to that of a prism with the same height and a base congruent to that of the pyramid.
3. **OPEN ENDED** Draw and label two cones with different dimensions, but with the same volume.

DAILY INTERVENTION — Differentiated Instruction

Visual/Spatial When you discuss cones and pyramids, show students that 3 cones or pyramids fit into a cylinder or rectangular prism with the same corresponding base and height by filling a cone or pyramid with water, rice, or beans and pouring it into the corresponding cylinder or pyramid.

698 Chapter 13 Volume

Guided Practice

GUIDED PRACTICE KEY	
Exercises	Examples
4, 7	1
5	2
6	3

Find the volume of each pyramid or cone. Round to the nearest tenth if necessary.

4. 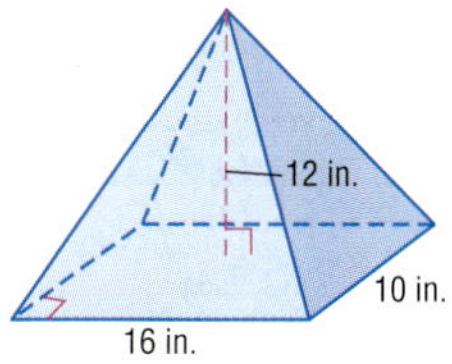

640 in^3

5. 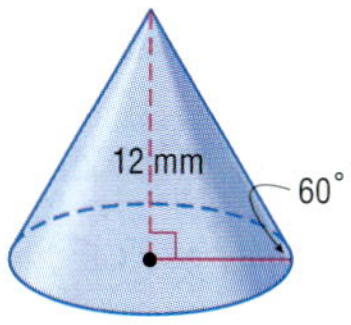

603.2 mm^3

6. 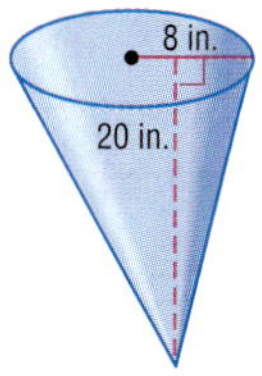

1340.4 ft^3

Application

7. **MECHANICAL ENGINEERING** The American Heritage Center at the University of Wyoming is a conical building. If the height is 77 feet, and the area of the base is about 38,000 square feet, find the volume of air that the heating and cooling systems would have to accommodate. Round to the nearest tenth. $975,333.3 \text{ ft}^3$

★ indicates increased difficulty

Practice and Apply

Homework Help	
For Exercises	See Examples
8–10, 17–18, 24–27, 30	1
11–13, 19–23, 28–29	2
14–16	3

Extra Practice
See page 780.

Find the volume of each pyramid or cone. Round to the nearest tenth if necessary.

8. 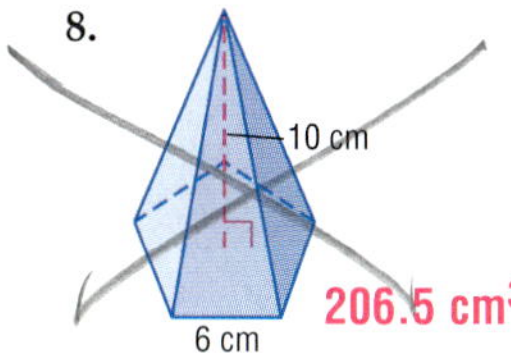

206.5 cm^3

9. 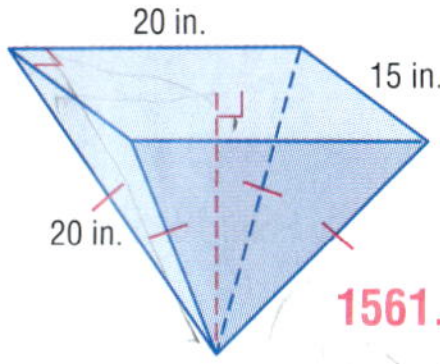

1561.2 ft^3

10. 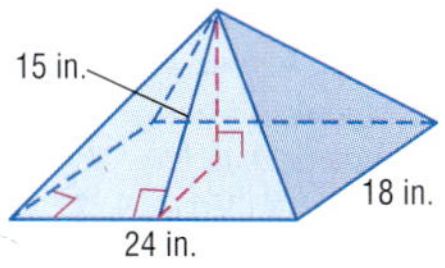

1728 in^3

11. 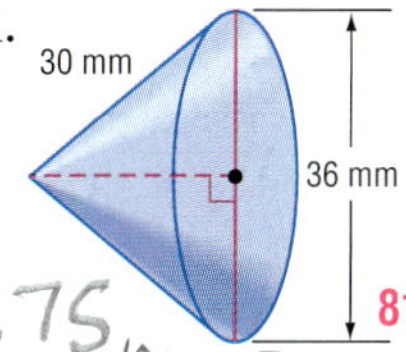

8143.0 mm^3

10178.75 mm3

12. 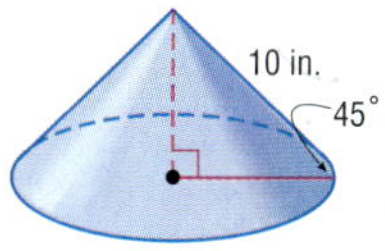

370.2 in^3

13. 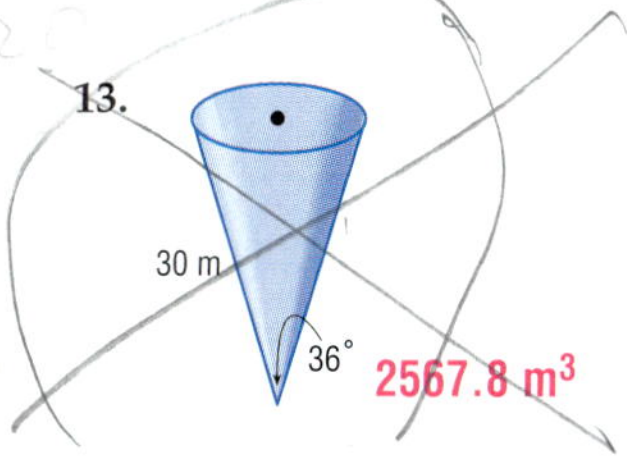

2567.8 m^3

14. 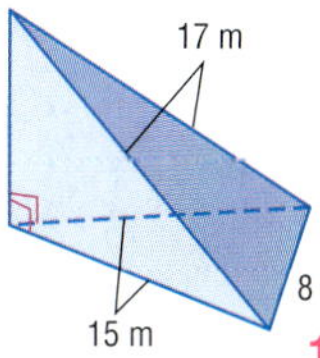

154.2 m^3

15. 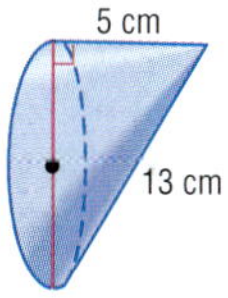

188.5 cm^3

16. 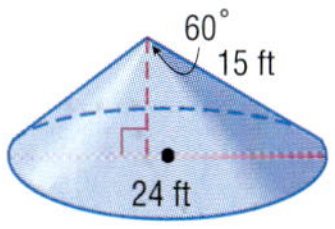

1131.0 ft^3

Find the volume of each solid. Round to the nearest tenth.

★ 17. 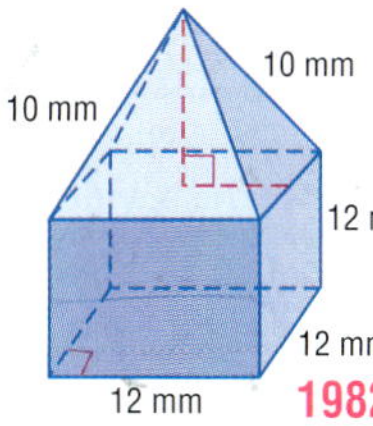

1982.0 mm^3

★ 18. 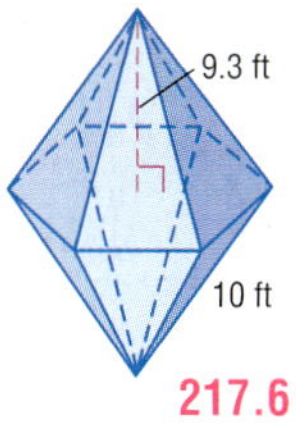

217.6 ft^3

★ 19. 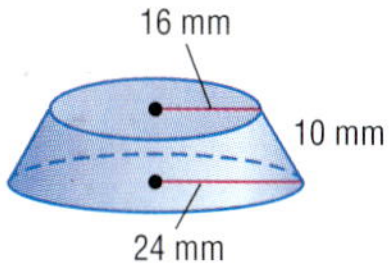

7640.4 cm^3

3 Practice/Apply

Study Notebook

Have students—

- add the definitions/examples of the vocabulary terms to their Vocabulary Builder worksheets for Chapter 13.
- include a review of the volume formulas for pyramids and cones and examples of finding the volume of each.
- include any other item(s) that they find helpful in mastering the skills in this lesson.

About the Exercises...

Organization by Objective

- **Volumes of Pyramids:** 8–10, 14, 17, 18
- **Volumes of Cones:** 11–13, 15, 16, 19

Odd/Even Assignments

Exercises 8–19 are structured so that students practice the same concepts whether they are assigned odd or even problems.

Assignment Guide

Basic: 9–15 odd, 21–27 odd, 31, 33–43

Average: 9–31 odd, 33–43

Advanced: 8–30 even, 31–40 (optional: 41–43)

All: Quiz 1 (1–5)

Teacher to Teacher

Nita Carpenter — Okemos High School, Okemos, MI

I introduce the formulas for volume of a pyramid or cone using transparent models and water. I show the students a model of a cube and a square pyramid that have the same base and height. I then ask them to tell me how many of the pyramids (when filled with water) would fit into the cube. I then fill the pyramid, pour it in, and let students change their guess if they want to. When finished it is very clear that the formula for volume of a pyramid is $V = \frac{Bh}{3}$, or $V = \frac{1}{3}Bh$.

Lesson 13-2 Volumes of Pyramids and Cones 699

VOLCANOES For Exercises 20–23, use the following information.

The slope of a volcano is the angle made by the side of the cone and a horizontal line. Find the volume of material in each volcano, assuming that it is a solid cone.

Volcano	Location	Type	Characteristics
Mauna Loa	Hawaii, United States	shield dome	4170 m tall, 103 km across at base
Mount Fuji	Honshu, Japan	composite	3776 m tall, slope of 9°
Paricutín	Michoacán, Mexico	cinder cone	410 m tall, 33° slope
Vesuvius	Campania, Italy	composite	22.3 km across at base, 1220 m tall

20. Mauna Loa $\approx 11{,}581.9\ km^3$
21. Mount Fuji $\approx 2247.5\ km^3$
22. Paricutin $\approx 171{,}137{,}610.4\ m^3$
23. Vesuvius $\approx 158.8\ km^3$
24. The shared base of the pyramids that make up the solid on the left is congruent to the base of the solid on the right. Write a ratio comparing the volumes of the solids. Explain your answer. See margin.

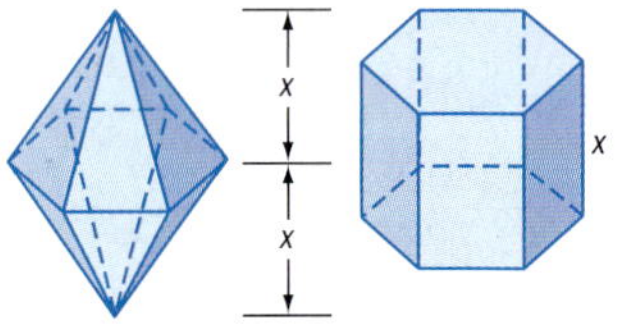

HISTORY For Exercises 25–27, use the following information.

The Great Pyramid of Khufu is a square pyramid. The lengths of the sides of the base are 755 feet. The original height was 481 feet. The current height is 449 feet.

25. Find the original volume of the pyramid. $\approx 91{,}394{,}008.3\ ft^3$
26. Find the present day volume of the pyramid. $\approx 85{,}313{,}741.7\ ft^3$
27. Compare the volumes of the pyramid. What volume of material has been lost? $\approx 6{,}080{,}266.7\ ft^3$

★ 28. **PROBABILITY** What is the probability of randomly choosing a point inside the cylinder, but not inside the cone that has the same base and height as the cylinder. $\frac{2}{3}$

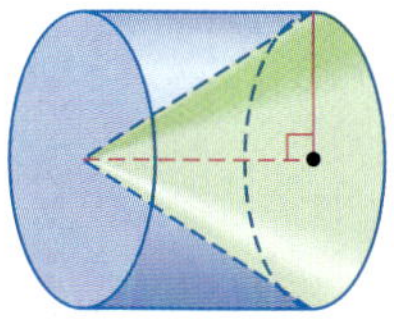

★ 29. A pyramid with a square base is next to a circular cone as shown at the right. The circular base is inscribed in the square. Isosceles $\triangle ABC$ is perpendicular to the base of the pyramid. DE is the slant height of the cone. Find the volume of the figure. $\approx 522.3\ units^3$

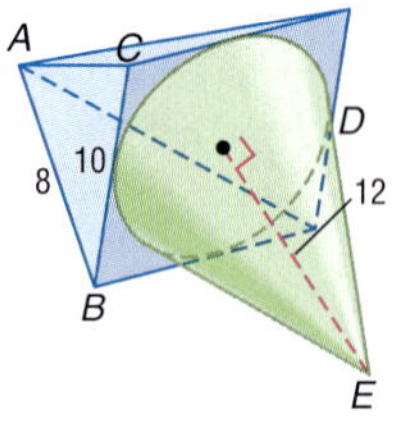

★ 30. **ARCHITECTURE** In an attempt to rid Florida's Lower Sugarloaf Key of mosquitoes, Richter Perky built a tower to attract bats. The Perky Bat Tower is a frustum of a pyramid with a square base. Each side of the base of the tower is 15 feet long, the top is a square with sides 8 feet long, and the tower is 35 feet tall. How many cubic feet of space does the tower supply for bats? (*Hint*: Draw the pyramid that contains the frustrum.) $\approx 4771.7\ ft^3$

31. **CRITICAL THINKING** Find the volume of a regular tetrahedron with one side measuring 12 inches. $\approx 203.6\ in^3$

More About. . .

Volcanoes

There are three major types of volcanoes: cinder cone, shield dome, and composite. Shield dome volcanoes are formed almost exclusively from molten lava. Composite volcanoes are formed from layers of molten lava and hardened chunks of lava.

Source: pubs.usgs.gov

Study Tip

Look Back

To review **frustum of a pyramid**, see Lesson 12-5.

Study Guide and Intervention, p. 729 (shown) and p. 730

Volumes of Pyramids This figure shows a prism and a pyramid that have the same base and the same height. It is clear that the volume of the pyramid is less than the volume of the prism. More specifically, the volume of the pyramid is one-third of the volume of the prism.

Volume of a Pyramid	If a pyramid has a volume of V cubic units, a height of h units, and a base with an area of B square units, then $V = \frac{1}{3}Bh$.

Example **Find the volume of the square pyramid.**

$V = \frac{1}{3}Bh$ Volume of a pyramid

$= \frac{1}{3}(8)(8)10$ $B = (8)(8)$, $h = 10$

≈ 213.3 Multiply.

The volume is about 213.3 cubic feet.

Exercises

Find the volume of each pyramid. Round to the nearest tenth if necessary.

1. $320\ ft^3$
2. $120\ ft^3$
3. $110.9\ cm^3$
4. $561.2\ ft^3$
5. $1200\ in^3$
6. $64\ yd^3$

Skills Practice, p. 731 and Practice, p. 732 (shown)

Find the volume of each pyramid or cone. Round to the nearest tenth if necessary.

1. $343.1\ yd^3$
2. $2395.8\ cm^3$
3. $1419.4\ ft^3$
4. $1104.6\ mm^3$
5. $132\ in^3$
6. $4688.3\ ft^3$

7. **CONSTRUCTION** Mr. Ganty built a conical storage shed. The base of the shed is 4 meters in diameter, and the height of the shed is 3.8 meters. What is the volume of the shed? about $15.9\ m^3$

8. **HISTORY** The start of the pyramid age began with King Zoser's pyramid, erected in the 27th century B.C. In its original state, it stood 62 meters high with a rectangular base that measured 140 meters by 118 meters. Find the volume of the original pyramid. about $341{,}413.3\ m^3$

Reading to Learn Mathematics, p. 733 ELL

Pre-Activity **How do architects use geometry?**

Read the introduction to Lesson 13-2 at the top of page 696 in your textbook.

In addition to reflecting more light, why do you think the architect of the Transamerica Pyramid may have designed the building as a square pyramid rather than a rectangular prism? Sample answer: The pyramid is more unusual and has a more dramatic appearance, so it attracts more attention. With the sharp point at the top, it seems to soar up into the sky.

Reading the Lesson

1. In each case, two solids are described. Determine whether the first solid or the second solid has the greater volume, or if the two solids have the same volume. (Answer by writing *first*, *second*, or *same*.)
   a. First solid: A rectangular prism with length x, width y, and height z
      Second solid: A rectangular prism with length $2x$, width y, height z second
   b. First solid: a rectangular prism that has a square base with side length x and that has height y
      Second solid: a square pyramid whose base has side length x and that has height y first
   c. First solid: a right cone whose base has radius x and that has height y
      Second solid: an oblique cone whose base has radius x and that has height y same
   d. First solid: a cone whose base has radius x, and whose height is y
      Second solid: a cylinder whose bases have radius x, and whose height is y second
   e. First solid: a cone whose base has radius x and whose height is y
      Second solid: a square pyramid whose base has side length x and whose height is y first
2. Supply the missing numbers to form true statements.
   a. If the length, width, and height of a rectangular box are all doubled, its volume will be multiplied by 8.
   b. If the radius of a cylinder is tripled and the height is unchanged, the volume will be multiplied by 9.
   c. In a square pyramid, if the side length of the base is multiplied by 1.5 and the height is doubled, the volume will be multiplied by 4.5.
   d. In a cone, if the radius of the base is tripled and the height is doubled, the volume will be multiplied by 18.
   e. In a cube, if the edge length is multiplied by 5, the volume will be multiplied by 125.

Helping You Remember

3. Many students find it easier to remember mathematical formulas if they can put them in words. Use words to describe in one sentence how to find the volume of any pyramid or cylinder. Sample answer: Multiply the area of the base by the height and divide by 3.

Enrichment, p. 734

Frustums

A **frustum** is a figure formed when a plane intersects a pyramid or cone so that the plane is parallel to the solid's base. The frustum is the part of the solid between the plane and the base. To find the volume of a frustum, the areas of both bases must be calculated and used in the formula

$$V = \frac{1}{3}h(B_1 + B_2 + \sqrt{B_1B_2}),$$

where h = height (perpendicular distance between the bases), B_1 = area of top base, and B_2 = area of bottom base.

Describe the shape of the bases of each frustum. Then find the volume. Round to the nearest tenth.

1. 13 cm, 6 cm
2. 3 in.

Answers

24. $\frac{2}{3}$; The volume of each pyramid that makes up the solid on the left is $\frac{1}{3}$ of the volume of the prism, so the total volume of the solid on the left is $\frac{1}{3} + \frac{1}{3}$ or $\frac{2}{3}$ of the volume of the prism.

32. **WRITING IN MATH** Answer the question that was posed at the beginning of the lesson. **See margin.**

How do architects use geometry?

Include the following in your answer:

- compare the available office space on the first floor and on the top floor, and
- explain how a pyramidal building allows more light to reach the street than a rectangular prism building.

FCAT Practice
Standardized Test Practice

33. Which of the following is the volume of the square pyramid if $b = 2h$? **B**

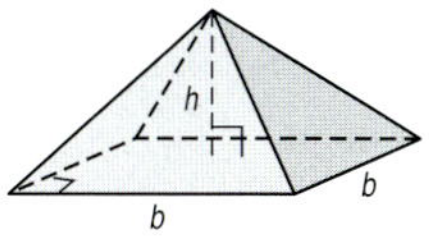

(A) $\frac{h^3}{3}$ (B) $\frac{4h^3}{3}$ (C) $4h^3$ (D) $\frac{8h^3 - 4h}{3}$

34. **ALGEBRA** If the volume of a right prism is represented by $x^3 \pm 9x$, find the factors that could represent the length, width, and height. **A**

(A) $x, x - 3, x + 3$ (B) $x, x - 9, x + 1$
(C) $x, x + 9, x \pm 1$ (D) $x, x - 3, x \pm 3$

Maintain Your Skills

Mixed Review

Find the volume of each prism or cylinder. Round to the nearest tenth if necessary. *(Lesson 13-1)*

35.
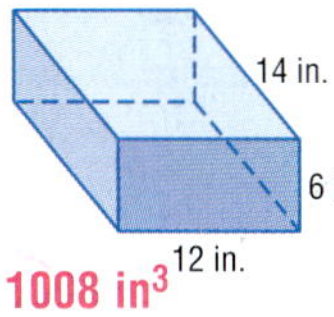

1008 in^3

36.
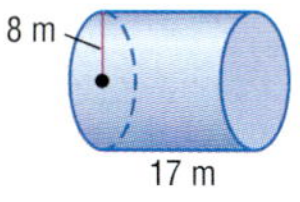

3418.1 m^3

37.
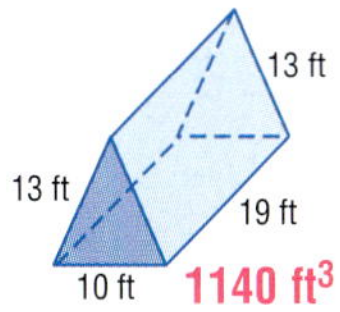

1140 ft^3

Find the surface area of each sphere. Round to the nearest tenth if necessary. *(Lesson 12-7)*

38. The circumference of a great circle is 86 centimeters. **2354.2 cm^2**

39. The area of a great circle is 64.5 square yards. **258 yd^2**

40. **BASEBALL** A baseball field has the shape of a rectangle with a corner cut out as shown at the right. What is the total area of the baseball field? *(Lesson 11-4)* **101,262.5 ft^2**

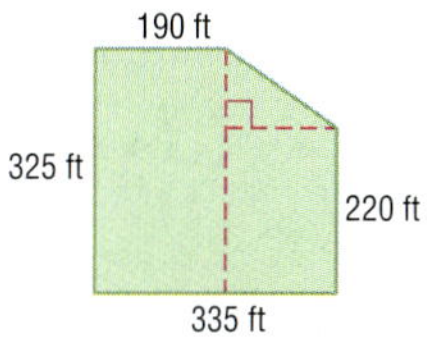

Getting Ready for the Next Lesson

PREREQUISITE SKILL **Evaluate each expression. Round to the nearest hundredth.**

*(To review **evaluating expressions**, see page 736.)*

41. $4\pi r^2, r = 3.4$ **145.27** 42. $\frac{4}{3}\pi r^3, r = 7$ **1436.76** 43. $4\pi r^2, r = 12$ **1809.56**

Practice Quiz 1

Lessons 13-1 and 13-2

1. **FOOD** A canister of oatmeal is 10 inches tall with a diameter of 4 inches. Find the maximum volume of oatmeal that the canister can hold to the nearest tenth. *(Lesson 13-1)* **125.7 in^3**

Find the volume of each solid. Round to the nearest tenth. *(Lessons 13-1 and 13-2)*

2.
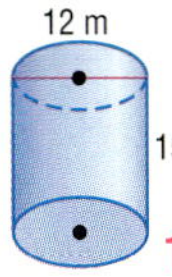

1696.5 m^3

3.
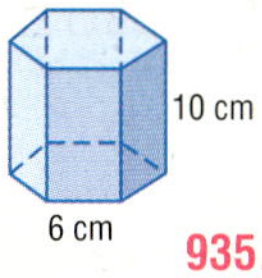

935.3 cm^3

4.
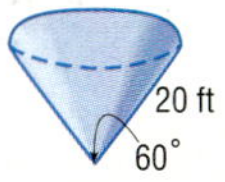

1813.8 ft^3

5.
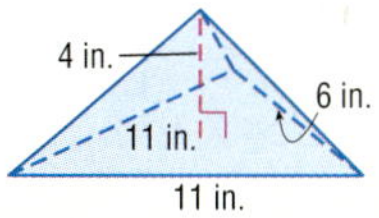

42.3 in^3

www.geometryonline.com/self_check_quiz/fcat

4 Assess

Open-Ended Assessment

Writing Ask students to list all the volume formulas they have learned thus far, along with an example of how to find each.

Getting Ready for Lesson 13-3

Prerequisite Skill Students will learn how to find volumes of spheres in Lesson 13-3. They will use their knowledge of how to evaluate an expression to evaluate the formula for the volume of a sphere. Use Exercises 41–43 to determine your students' familiarity with evaluating expressions containing π.

Assessment Options

Practice Quiz 1 The quiz provides students with a brief review of the concepts and skills in Lessons 13-1 and 13-2. Lesson numbers are given to the right of the exercises or instruction lines so students can review concepts not yet mastered.

Quiz (Lessons 1 and 2) is available on p. 767 of the *Chapter 13 Resource Masters*.

Answers

32. Sample answer: Architects use geometry to design buildings that meet the needs of their clients. Answers should include the following.
- The surface area at the top of a pyramid is much smaller than the surface area of the base. There is less office space at the top, than on the first floor.
- The silhouette of a pyramid-shaped building is smaller than the silhouette of a rectangular prism with the same height. If the light conditions are the same, the shadow cast by the pyramid is smaller than the shadow cast by the rectangular prism.

13-3 Lesson Notes

1 Focus

5-Minute Check Transparency 13-3 Use as a quiz or review of Lesson 13-2.

Mathematical Background notes are available for this lesson on p. 686D.

How can you find the volume of Earth?

Ask students:

- Why is the Earth used as an example for this lesson? **The Earth is approximately in the shape of a sphere.**
- How did Eratosthenes use the circumference of the Earth to find the volume? **He used the circumference to find the radius of the Earth. Then he used the radius to get the volume.**
- If you know the circumference, how do you find the radius? **Divide the circumference by two times pi.**

Building on Prior Knowledge

In Chapter 12, students learned how to find the surface area of a sphere using the formula $S = 4\pi r^2$. In this lesson, the formula for volume may be confusing because it looks the same to them. Write the formulas side by side and point out the differences between them.

13-3 Volumes of Spheres

Sunshine State Standards
MA.A.3.4.3, MA.B.1.4.1, MA.B.3.4.1

What You'll Learn

- Find volumes of spheres.
- Solve problems involving volumes of spheres.

How can you find the volume of Earth?

Eratosthenes was an ancient Greek mathematician who estimated the circumference of Earth. He assumed that Earth was a sphere and estimated that the circumference was about 40,000 kilometers. From the circumference, the radius of Earth can be calculated. Then the volume of Earth can be determined.

VOLUMES OF SPHERES You can relate finding a formula for the volume of a sphere to finding the volume of a right pyramid and the surface area of a sphere.

Suppose the space inside a sphere is separated into infinitely many pyramids, all with vertices located at the center of the sphere. Observe that the height of these pyramids is equal to the radius r of the sphere. The sum of the areas of all the pyramid bases equals the surface area of the sphere.

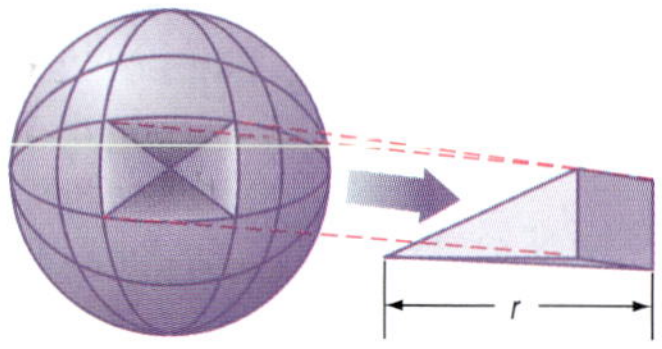

Each pyramid has a volume of $\frac{1}{3}Bh$, where B is the area of its base and h is its height. The volume of the sphere is equal to the sum of the volumes of all of the small pyramids.

$V = \frac{1}{3}B_1h_1 + \frac{1}{3}B_2h_2 + \frac{1}{3}B_3h_3 + \ldots + \frac{1}{3}B_nh_n$	Sum of the volumes of all of the pyramids
$= \frac{1}{3}B_1r + \frac{1}{3}B_2r + \frac{1}{3}B_3r + \ldots + \frac{1}{3}B_nr$	Replace h with r.
$= \frac{1}{3}r(B_1 + B_2 + B_3 + \ldots + B_n)$	Distributive Property
$= \frac{1}{3}r(4\pi r^2)$	Replace $B_1 + B_2 + B_3 + \ldots + B_n$ with $4\pi r^2$.
$= \frac{4}{3}\pi r^3$	Simplify.

Study Tip

Look Back
Recall that the surface area of a sphere, $4\pi r^2$, is equal to $B_1 + B_2 + B_3 + \ldots + B_n$. To review **surface area of a sphere**, see Lesson 12-7.

Key Concept — *Volume of a Sphere*

If a sphere has a volume of V cubic units and a radius of r units, then $V = \frac{4}{3}\pi r^3$.

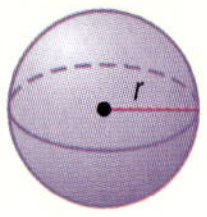

Resource Manager

Workbook and Reproducible Masters

Chapter 13 Resource Masters

- Study Guide and Intervention, pp. 735–736
- Skills Practice, p. 737
- Practice, p. 738
- Reading to Learn Mathematics, p. 739
- Enrichment, p. 740
- Assessment, pp. 767, 769

Graphing Calculator and Computer Masters, pp. 41, 42

Prerequisite Skills Workbook, pp. 37–38

Transparencies

5-Minute Check Transparency 13-3
Answer Key Transparencies

Technology

GeomPASS: Tutorial Plus, Lesson 23
Interactive Chalkboard

Example 1 Volumes of Spheres

Find the volume of each sphere.

a.

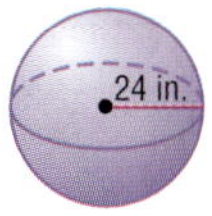

$V = \frac{4}{3}\pi r^3$ Volume of a sphere

$= \frac{4}{3}\pi(24^3)$ $r = 24$

$\approx 57{,}905.8 \text{ in}^3$ Use a calculator.

b. $C = 36$ cm

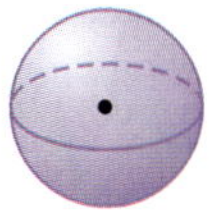

First find the radius of the sphere.

$C = 2\pi r$ Circumference of a circle

$36 = 2\pi r$ $C = 36$

$\frac{18}{\pi} = r$ Solve for r.

Now find the volume.

$V = \frac{4}{3}\pi r^3$ Volume of a sphere

$= \frac{4}{3}\pi\left(\frac{18}{\pi}\right)^3$ $r = \frac{18}{\pi}$

$\approx 787.9 \text{ cm}^3$ Use a calculator.

Example 2 Volume of a Hemisphere

Find the volume of the hemisphere.

The volume of a hemisphere is one-half the volume of the sphere.

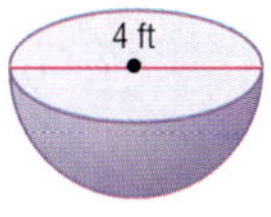

$V = \frac{1}{2}\left(\frac{4}{3}\pi r^3\right)$ Volume of a hemisphere

$= \frac{2}{3}\pi(2^3)$ $r = 2$

$\approx 16.8 \text{ ft}^3$ Use a calculator.

SOLVE PROBLEMS INVOLVING VOLUMES OF SPHERES Often spherical objects are contained in other solids. A comparison of the volumes is necessary to know if one object can be contained in the other.

Example 3 Volume Comparison

Short Response Test Item

Compare the volumes of the sphere and the cylinder. Determine which quantity is greater.

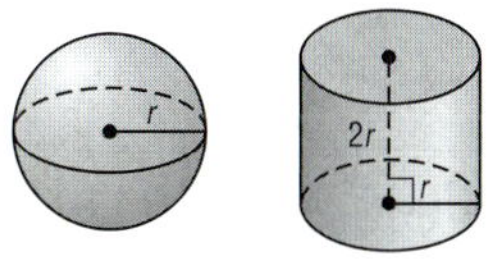

Read the Test Item

You are asked to compare the volumes of the sphere and the cylinder.

(continued on the next page)

Study Tip

Graphing Calculator

When using a graphing calculator to find the cube of a number, you can find the cubed symbol in the MATH menu, or you can use ^ followed by a 3 to cube a number.

WebQuest

The formulas for the surface area and volume of a sphere can help you find the surface area and volume of the hemisphere. Visit www.geometryonline.com/webquest to continue work on your WebQuest project.

FCAT Practice
Standardized Test Practice
A B C D

www.geometryonline.com/extra_examples/fcat

2 Teach

VOLUMES OF SPHERES

In-Class Examples

Power Point®

Teaching Tip Emphasize that the volume of a sphere is developed on this page by considering it as the sum of numerous right pyramids, as is shown in the diagram. Point out that the base of the pyramid has to be very small so that the square base is very close to being flat.

1 Find the volume of each sphere to the nearest tenth.

a.

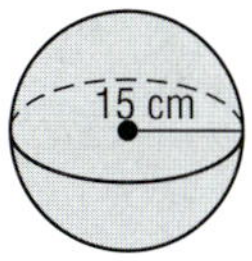

b.

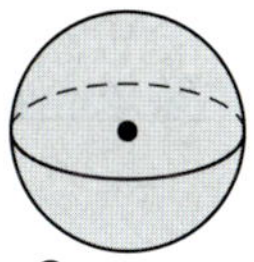

$C = 25$ cm

a. $14{,}137.2 \text{ cm}^3$ b. 263.9 cm^3

2 Find the volume of the hemisphere.

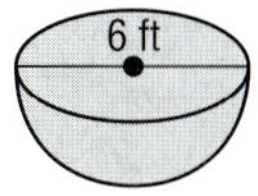

56.5 ft^3

SOLVE PROBLEMS INVOLVING VOLUMES OF SPHERES

In-Class Example

Power Point®

3 Compare the volumes of the sphere and the cylinder with the same radius and height as the radius of the sphere.

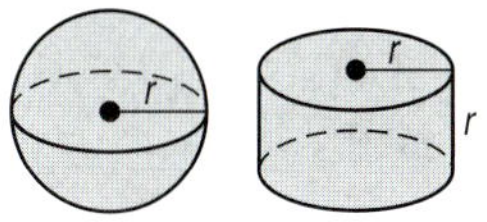

The sphere has $\frac{4}{3}$ the volume.

Concept Check

Ask students to explain how to find the volume of a sphere if you know the circumference. **Use the circumference to find the radius by solving the formula $C = 2\pi r$ for r. Then find the volume by substituting the radius into the formula $V = \frac{4}{3}\pi r^3$.**

DAILY INTERVENTION

Differentiated Instruction

Interpersonal Have students choose a partner. Ask one student to draw a sphere and label the radius or diameter. Then ask the partner to find the volume of the sphere. Next switch roles and do the activity again.

3 Practice/Apply

Study Notebook

Have students—
- add the definitions/examples of the vocabulary terms to their Vocabulary Builder worksheets for Chapter 13.
- write the formula for the volume of a sphere. Ask them to include an example of how to find the volume of a sphere and of a hemisphere.
- include any other item(s) that they find helpful in mastering the skills in this lesson.

DAILY INTERVENTION

FIND THE ERROR In Exercise 2, Winona made a very common mistake in the order of operations. Exponentiation comes before division, so 12^3 needs to be evaluated first.

About the Exercises...

Organization by Objective
- **Volumes of Spheres:** 9–18, 29–31
- **Solve Problems Involving Volumes of Spheres:** 19–28, 32

Odd/Even Assignments

Exercises 9–18 are structured so that students practice the same concepts whether they are assigned odd or even problems.

Assignment Guide

Basic: 9–29 odd, 33, 36–47

Average: 9–35 odd, 36–47

Advanced: 10–32 even, 33–43 (optional: 44–47)

The Princeton Review

Test-Taking Tip

Labeling When finding the volume for two solids, label your calculations with the type of solid.

Solve the Test Item

Volume of the sphere: $\frac{4}{3}\pi r^3$

Volume of the cylinder: $\pi r^2 h = \pi r^2(2r)$ $h = 2r$

$= 2\pi r^3$ Simplify.

Compare $2\pi r^3$ to $\frac{4}{3}\pi r^3$. Since 2 is greater than $\frac{4}{3}$, the volume of the cylinder is greater than the volume of the sphere.

Check for Understanding

Concept Check

1. **Explain** how to find the formula for the volume of a sphere. See margin.
2. **FIND THE ERROR** Winona and Kenji found the volume of a sphere with a radius of 12 centimeters.

Winona	Kenji
$V = \frac{4}{3}\pi(12)^3$	$V = \frac{4}{3}\pi(12)^3$
$= 4\pi(4)^3$	$= \frac{4}{3}\pi(1728)$
$= 256\pi\ cm^3$	$= 2304\pi\ cm^3$

Who is correct? Explain your reasoning.

2. Kenji; Winona divided the 12 by 3 before raising the result to the third power. Thus the order of operations was not followed correctly.

Guided Practice

GUIDED PRACTICE KEY	
Exercises	Examples
3–6	1
7	2
8	3

Find the volume of each sphere or hemisphere. Round to the nearest tenth.

3. The radius is 13 inches long. $9202.8\ in^3$
4. The diameter of the sphere is 12.5 centimeters. $1022.7\ cm^3$
5. 

$268.1\ in^3$
6. C = 18 cm 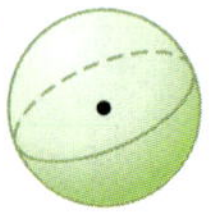
$98.5\ cm^3$
7. 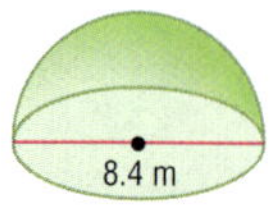

$155.2\ m^3$

FCAT Practice Standardized Test Practice

8. **SHORT RESPONSE** Compare the volumes of a sphere with a radius of 5 inches and a cone with a height of 20 inches and a base with a diameter of 10 inches.
Volumes are equal; $\frac{500}{3}\pi$.

★ indicates increased difficulty

Practice and Apply

Homework Help	
For Exercises	See Examples
9–19	1–2
20–22, 28–29, 32	3

Extra Practice See page 781.

Find the volume of each sphere or hemisphere. Round to the nearest tenth.

9. The radius of the sphere is 7.62 meters. $1853.3\ m^3$
10. The diameter of the sphere is 33 inches. $18{,}816.6\ in^3$
11. The diameter of the sphere is 18.4 feet. $3261.8\ ft^3$
12. The radius of the sphere is $\frac{\sqrt{3}}{2}$ centimeters. $2.7\ cm^3$
13. C = 24 in. 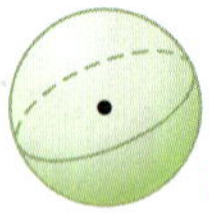
$233.4\ in^3$
14. 

$192{,}193.1\ mm^3$
15. 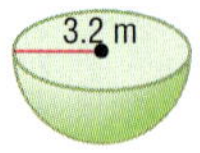

$68.6\ m^3$

Answer

1. The volume of a sphere was generated by adding the volumes of an infinite number of small pyramids. Each pyramid has its base on the surface of the sphere and its height from the base to the center of the sphere.

16. 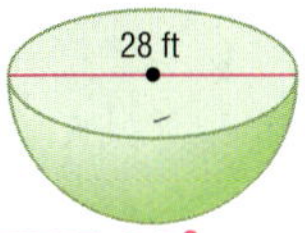

5747.0 ft³

17. 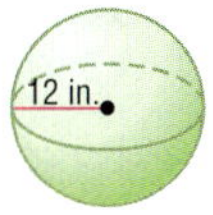

7238.2 in³

18. $C = 48$ cm **1867.6 cm³**

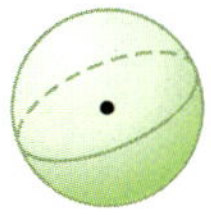

19. **ASTRONOMY** The diameter of the moon is 3476 kilometers. Find the volume of the moon. $\approx$ **21,990,642,871 km³**

20. **SPORTS** If a golf ball has a diameter of 4.3 centimeters and a tennis ball has a diameter of 6.9 centimeters, find the difference between the volumes of the two balls. $\approx$ **130.4 cm³**

FOOD **For Exercises 21 and 22, use the following information.**
Suppose a sugar cone is 10 centimeters deep and has a diameter of 4 centimeters. A spherical scoop of ice cream with a diameter of 4 centimeters rests on the top of the cone.

21. If all the ice cream melts into the cone, will the cone overflow? Explain.

22. If the cone does not overflow, what percent of the cone will be filled? **80%**

More About. . .

Food

On average, each person in the United States consumes 16.8 pounds of ice cream per year.

Source: *Statistical Abstract of the United States*

21. No, the volume of the cone is about 41.9 cm³; the volume of the ice cream is about 33.5 cm³.

FAMILY **For Exercises 23–26, use the following information.**
Suppose the bubble in the graphic is a sphere with a radius of 17 millimeters. **25. $\approx$ 1162.1 mm²**

23. What is the volume of the bubble? $\approx$ **20,579.5 mm³**

24. What is the volume of the portion of the bubble in which the kids had just the right amount of time with their mother? $\approx$ **12,141.9 mm³**

25. What is the surface area of that portion of the bubble in which the kids wish they could spend more time with their mother?

26. What is the area of the two-dimensional sector of the circle in which the kids wish they could spend less time with their mother? $\approx$ **81.7 mm²**

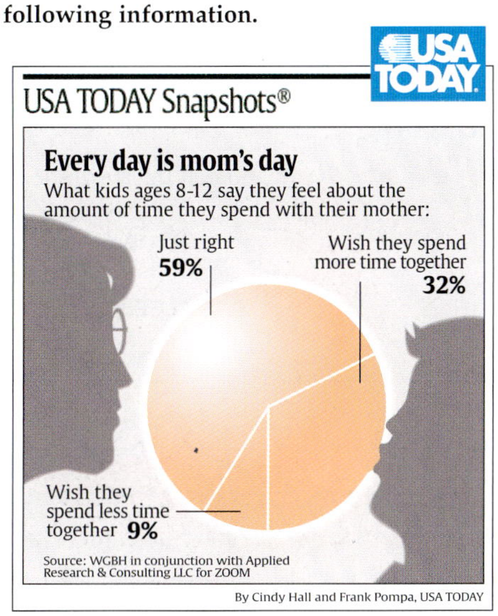

27. **PROBABILITY** Find the probability of choosing a point at random inside a sphere that has a radius of 6 centimeters and is inscribed in a cylinder. $\frac{2}{3}$

28. **TENNIS** Find the volume of the empty space in a cylindrical tube of three tennis balls. The diameter of each ball is about 2.5 inches. The cylinder is 2.5 inches in diameter and is 7.5 inches tall. $\approx$ **12.3 in³**

29. Find the volume of a sphere that is circumscribed about a cube with a volume of 216 cubic inches. $\approx$ **587.7 in³**

Find the volume of each sphere or hemisphere. Round to the nearest tenth.

★ 30. The surface area of a sphere is 784π square inches. **11,494.0 in³**

★ 31. A hemisphere has a surface area of 18.75π square meters. **32.7 m³**

Enrichment, p. 740

Spheres and Density

The **density** of a metal is a ratio of its mass to its volume. For example, the mass of aluminum is 2.7 grams per cubic centimeter. Here is a list of several metals and their densities.

Aluminum	2.7 g/cm³	Copper	8.96 g/cm³
Gold	19.32 g/cm³	Iron	7.874 g/cm³
Lead	11.35 g/cm³	Platinum	21.45 g/cm³
Silver	10.50 g/cm³		

To calculate the mass of a piece of metal, multiply volume by density.

Example **Find the mass of a silver ball that is 0.8 cm in diameter.**

$M = D \cdot V$

$= 10.5 \cdot \frac{4}{3}\pi(0.4)^3$

≈ 10.5 (0

Study Guide and Intervention, p. 735 (shown) and p. 736

Volumes of Spheres A sphere has one basic measurement, the length of its radius. If you know the radius of a sphere, you can calculate its volume.

Volume of a Sphere	If a sphere has a volume of V cubic units and a radius of r units, then $V = \frac{4}{3}\pi r^3$.

Example 1 **Find the volume of a sphere with radius 8 centimeters.**

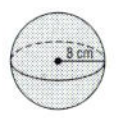

$V = \frac{4}{3}\pi r^3$ Volume of a sphere

$= \frac{4}{3}\pi(8)^3$ $r = 8$

≈ 2144.7 Simplify.

The volume is about 2144.7 cubic centimeters.

Example 2 **A sphere with radius 5 inches just fits inside a cylinder. What is the difference between the volume of the cylinder and the volume of the sphere? Round to the nearest cubic inch.**

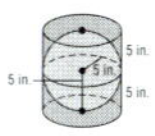

The base of the cylinder is 25π in² and the height is 10 in., so the volume of the cylinder is 250π in³. The volume of the sphere is $\frac{4}{3}\pi(5)^3$ or $\frac{500\pi}{3}$ in³. The difference in the volumes is $250\pi - \frac{500\pi}{3}$ or about 262 in³.

Exercises

Find the volume of each solid. Round to the nearest tenth.

1. 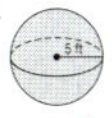

523.6 ft³

2. 

452.4 in³

3. 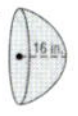

8578.6 in³

4. 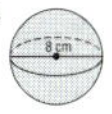

268.1 cm³

5. 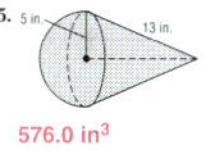

576.0 in³

6. 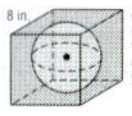

243.9 in³

7. A hemisphere with radius 16 centimeters just fits inside a rectangular prism. What is the difference between the volume of the prism and the volume of the hemisphere? Round to the nearest cubic centimeter. 7805 cm³

Skills Practice, p. 737 and Practice, p. 738 (shown)

Find the volume of each sphere or hemisphere. Round to the nearest tenth.

1. The radius of the sphere is 12.4 centimeters.
7986.4 cm³

2. The diameter of the sphere is 17 feet.
2572.4 ft³

3. The circumference of the sphere is 38 meters.
926.6 m³

4. The diameter of the hemisphere is 21 inches.
2424.5 in³

5. The circumference of the hemisphere is 18 millimeters.
49.2 mm³

6. 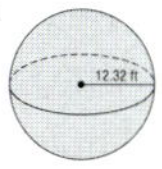

7832.9 ft³

7. 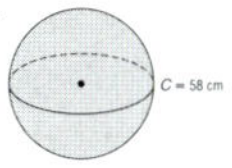

3294.8 cm³

8. 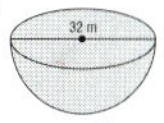

8578.6 m³

9. 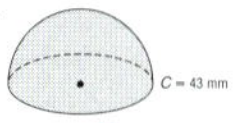

671.3 mm³

10. **PACKAGING** Amber plans to ship a mini-basketball she bought for her nephew. The circumference of the ball is 24 inches and the package she wants to ship it in is a rectangular box that measures 8 inches × 8 inches × 9 inches. Will the basketball fit in the box? Explain.
Yes; the diameter of the ball is about 7.64 in., so the ball will fit in the box.

Reading to Learn Mathematics, p. 739 ELL

Pre-Activity **How can you find the volume of Earth?**

Read the introduction to Lesson 13-3 at the top of page 702 in your textbook.

How would you estimate the radius of Earth based on Eratosthenes' estimate of its diameter? Sample answer: Use a calculator to divide 46,250 km by 2π.

Reading the Lesson

1. Name all solids from the following list for which each volume formula can be used: *prism, pyramid, cone, cylinder, sphere, hemisphere.*
   a. $V = Bh$ prism, cylinder
   b. $V = \frac{4}{3}\pi r^3$ sphere
   c. $V = \frac{1}{3}Bh$ pyramid, cone
   d. $V = \pi r^2 h$ cylinder
   e. $V = \frac{1}{3}\pi r^2 h$ cone
   f. $V = \frac{2}{3}\pi r^3$ hemisphere

2. Let r represent the radius and d represent the diameter of a sphere. Determine whether each formula below can be used to find the volume of a *sphere*, a *hemisphere*, or *neither*.
   a. $V = \frac{2\pi r^3}{3}$ hemisphere
   b. $V = \frac{1}{6}\pi d^3$ sphere
   c. $V = \frac{1}{3}\pi r^3$ neither
   d. $V = \frac{3}{4}\pi r^3$ neither
   e. $V = \frac{\pi d^3}{12}$ hemisphere
   f. $V = \frac{4}{3}\pi r^2 h$ neither

3. Compare the volumes of these three solids. Then complete the sentence below.

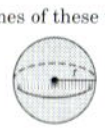
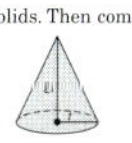

Of the three solids shown above, the sphere has the largest volume and the cone has the smallest volume.

Helping You Remember

4. A good way to remember something is to explain it to someone else. Suppose that your classmate Loretta knows that the expressions $\frac{4}{3}\pi r^3$ and $4\pi r^2$ are used in finding measurements related to spheres, but can't remember which one is used to find the surface area of a sphere and which one is used to find the volume. How can you help her to remember which is which? Sample answer: Look at the powers of r in the two expressions. The expression with r^3 will give a measurement in cubic units, so it is the expression for volume. The expression with r^2 will give a measurement in square units, so it is the expression for surface area.

4 Assess

Open-Ended Assessment

Modeling Ask students to measure the circumference of a basketball and use it to find the volume. **The circumference of a basketball is 30 inches. The volume is 455.9 in^3.**

Getting Ready for Lesson 13-4

Prerequisite Skill Students will learn about congruent and similar solids in Lesson 13-4. They will simplify expressions involving exponents. Use Exercises 44–47 to determine your students' familiarity with simplifying exponents.

Assessment Options

Quiz (Lesson 3) is available on p. 767 of the *Chapter 13 Resource Masters.*

Mid-Chapter Test (Lessons 13-1 through 13–3) is available on p. 769 of the *Chapter 13 Resource Masters.*

Answers

34. Sample answer: If a student knows the circumference of a sphere, then the volume can be found. Answers should include the following.
    - One needs to know the radius of the Earth.
    - The radius of Earth is about 6366.2 km and the volume is about 1.1×10^{12} km^3.

★ 32. **ARCHITECTURE** The Pantheon in Rome is able to contain a perfect sphere. The building is a cylinder 142 feet in diameter with a hemispherical domed roof. The total height is 142 feet. Find the volume of the interior of the Pantheon. **≈ 1,874,017.6 ft^3**

33. **CRITICAL THINKING** A vitamin capsule consists of a right cylinder with a hemisphere on each end. The capsule is 16 millimeters long and 4 millimeters thick. What is the volume of the capsule? **about 184 mm^3**

34. **WRITING IN MATH** Answer the question that was posed at the beginning of the lesson. **See margin.**

    How can you find the volume of Earth?

    Include the following in your answer:
    - the important dimension you must have to find the volume of Earth, and
    - the radius and volume of Earth from this estimate.

35. **RESEARCH** Use the Internet or other source to find the most current calculations for the volume of Earth. **See students' work.**

FCAT Practice *Standardized Test Practice*

36. If the radius of a sphere is increased from 3 units to 5 units, what percent would the volume of the smaller sphere be of the volume of the larger sphere? **A**

    (A) 21.6% (B) 40% (C) 60% (D) 463%

37. **ALGEBRA** Simplify $\frac{1}{2}(4\pi r^2) + \pi r^2 h + \frac{1}{2}(4\pi r^2)$. **A**

    (A) $\pi r^2(4 + h)$ (B) $4\pi r^2 h$ (C) $\pi r^2(9 + h)$ (D) $2\pi r^2(2 + h)$

Maintain Your Skills

Mixed Review **Find the volume of each cone. Round to the nearest tenth.** *(Lesson 13-2)*

38. height = 9.5 meters, radius = 6 meters **358.1 m^3**
39. height = 7 meters, diameter = 15 meters **412.3 m^3**

40. **REFRIGERATORS** A refrigerator has a volume of 25.9 cubic feet. If the interior height is 5.0 feet and the width is 2.4 feet, find the depth. *(Lesson 13-1)* **≈ 2.2 ft**

Write an equation for each circle. *(Lesson 10-8)*

41. center at (2, −1), $r = 8$ **$(x - 2)^2 + (y + 1)^2 = 64$**
42. center at (−4, −3), $r = \sqrt{19}$ **$(x + 4)^2 + (y + 3)^2 = 19$**
43. diameter with endpoints at (5, −4) and (−1, 6) **$(x - 2)^2 + (y - 1)^2 = 34$**

Getting Ready for the Next Lesson **PREREQUISITE SKILL** **Simplify.**

*(To review **simplifying expressions involving exponents**, see pages 746 and 747.)*

44. $(2a)^2$ **$4a^2$** 45. $(3x)^3$ **$27x^3$** 46. $\left(\frac{5a}{b}\right)^2$ **$\frac{25a^2}{b^2}$** 47. $\left(\frac{2k}{5}\right)^3$ **$\frac{8k^3}{125}$**

13-4 Congruent and Similar Solids

Sunshine State Standards
MA.A.3.4.3, MA.A.4.4.1, MA.B.1.4.1, MA.B.1.4.3, MA.B.3.4.1, MA.C.2.4.1, MA.C.3.4.1

Vocabulary
- similar solids
- congruent solids

What You'll Learn
- Identify congruent or similar solids.
- State the properties of similar solids.

How are similar solids applied to miniature collectibles?

People collect miniatures of race cars, farm equipment, and monuments such as the Statue of Liberty and the Washington Monument. The scale factors commonly used for miniatures include 1:16, 1:24, 1:32, 1:43, and 1:64. One of the smallest miniatures has a scale factor of 1:1000. If a car is 108 inches long, then a 1:24 scale model would be 108 ÷ 24 or 4.5 inches long.

CONGRUENT OR SIMILAR SOLIDS **Similar solids** are solids that have exactly the same shape but not necessarily the same size. You can determine if two solids are similar by comparing the ratios of corresponding linear measurements. In two similar polyhedra, all of the corresponding faces are similar, and all of the corresponding edges are proportional.

Study Tip
Spheres
All spheres are similar just like all circles are similar.

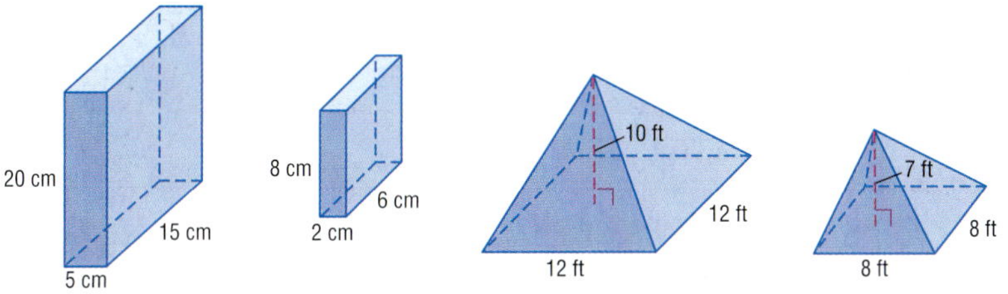

Similar Solids Nonsimilar Solids

In the similar solids above, $\frac{8}{20} = \frac{2}{5} = \frac{6}{15}$. Recall that the ratio of the measures is called the *scale factor*.

Study Tip
Look Back
To review **scale factor,** see lesson 6-2 and 9-5.

If the ratio of corresponding measurements of two solids is 1:1, then the solids are congruent. For two solids to be congruent, all of the following conditions must be met.

Key Concept *Congruent Solids*

Two solids are congruent if:
- the corresponding angles are congruent,
- the corresponding edges are congruent,
- the corresponding faces are congruent, and
- the volumes are equal.

Congruent solids are exactly the same shape and exactly the same size. They are a special case of similar solids. They have a scale factor of 1.

13-4 Lesson Notes

1 Focus

5-Minute Check Transparency 13-4 Use as a quiz or review of Lesson 13-3.

Mathematical Background notes are available for this lesson on p. 686D.

How are similar solids applied to miniature collectibles?

Ask students:
- How are scale factors used to determine the size of a miniature? **The scale factor is used to calculate the dimensions of the miniature.**
- The radius of the Earth is 4000 miles. If a scale model of the Earth at a museum is shown in a display with a scale factor of 1:2,000,000, what would be the radius of the model in feet? **about $10\frac{1}{2}$ ft**

Resource Manager

Workbook and Reproducible Masters

Chapter 13 Resource Masters
- Study Guide and Intervention, pp. 741–742
- Skills Practice, p. 743
- Practice, p. 744
- Reading to Learn Mathematics, p. 745
- Enrichment, p. 746
- Assessment, p. 768

Prerequisite Skills Workbook, pp. 97–98
Teaching Geometry With Manipulatives Masters, p. 208

Transparencies
5-Minute Check Transparency 13-4
Answer Key Transparencies

Technology
GeomPASS: Tutorial Plus, Lesson 24
Interactive Chalkboard

2 Teach

CONGRUENT OR SIMILAR SOLIDS

In-Class Example

Power Point®

Teaching Tip Point out that *all* corresponding dimensions of the two pyramids need to have the same scale factor for the pyramids to be similar.

1 Determine whether each pair of solids is *similar*, *congruent*, or *neither*.

a.

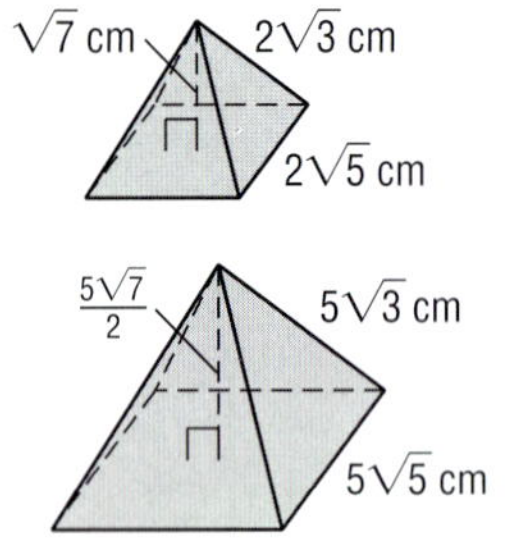

similar

b.

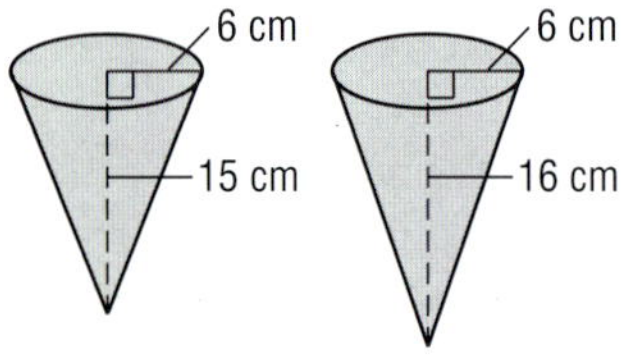

neither

Example 1 Similar and Congruent Solids

Determine whether each pair of solids are *similar*, *congruent*, or *neither*.

a. Find the ratios between the corresponding parts of the regular hexagonal pyramids.

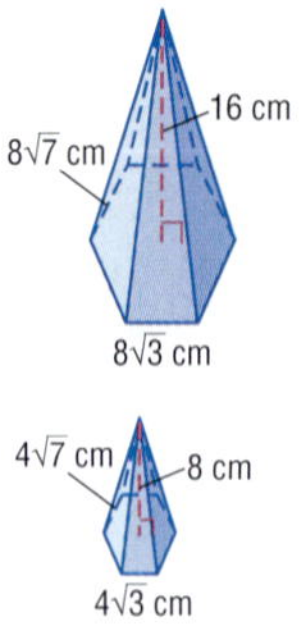

$\dfrac{\text{base edge of larger pyramid}}{\text{base edge of smaller pyramid}} = \dfrac{8\sqrt{3}}{4\sqrt{3}}$ Substitution

$= 2$ Simplify.

$\dfrac{\text{height of larger pyramid}}{\text{height of smaller pyramid}} = \dfrac{16}{8}$ Substitution

$= 2$ Simplify.

$\dfrac{\text{lateral edge of larger pyramid}}{\text{lateral edge of smaller pyramid}} = \dfrac{8\sqrt{7}}{4\sqrt{7}}$ Substitution

$= 2$ Simplify.

The ratios of the measures are equal, so we can conclude that the pyramids are similar. Since the scale factor is not 1, the solids are not congruent.

b. Compare the ratios between the corresponding parts of the cones.

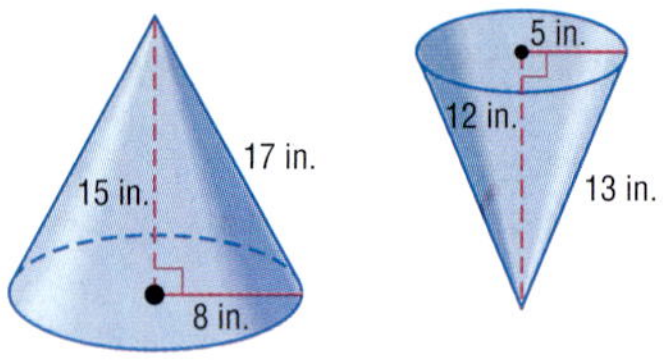

$\dfrac{\text{radius of larger cone}}{\text{radius of smaller cone}} = \dfrac{8}{5}$ Substitution

$\dfrac{\text{height of larger cone}}{\text{height of smaller cone}} = \dfrac{15}{12}$ Substitution

Since the ratios are not the same, there is no need to find the ratio of the slant heights. The cones are not similar.

Study Tip

Look Back
To review **regular polygonal pyramids**, see Lesson 12-1.

PROPERTIES OF SIMILAR SOLIDS You can investigate the relationships between similar solids using spreadsheets.

Spreadsheet Investigation

Explore Similar Solids

Collect the Data

Step 1 In Column A, enter the labels *length, width, height, surface area, volume, scale factor, ratios of surface area,* and the *ratios of volume.* Columns B, C, D, and E will be used for four similar prisms.

Step 2 Enter the formula for the surface area of the prism in cell B4. Copy the formula into the other cells in row 4.

Step 3 Write a similar formula to find the volume of the prism. Copy the formula in the cells in row 5.

Step 4 Enter the formula =C1/B1 in cell C6, enter =D1/B1 in cell D6, and so on. These formulas find the scale factor of prism B and each other solid.

Step 5 Type the formula =C4/B4 in cell C7, type =D4/B4 in cell D7, and so on. This formula will find the ratio of the surface area of prism B to the surface areas of each of the other prisms.

Geometry Activity

Materials: computer with spreadsheet software

It may be helpful to review the surface area formulas from Chapter 12. To repeat the activity, ask students to create a table including the prisms with different dimensions. See if they can make conjectures about the way to find similar solids from their table values.

Step 6 Write a formula for the ratio of the volume of prism C to the volume of prism B. Enter the formula in cell C8. Enter similar formulas in the cells in row 8.

Step 7 Use the spreadsheet to find the surface areas, volumes, and ratios for prisms with the dimensions given.

Similar Solids.xls

	A	B	C	D	E	F
1	length	1	2	3	4	
2	width	4	8	12	16	
3	height	6	12	18	24	
4	surface area	68	272	612	1088	
5	volume	24	192	648	1536	
6	scale factor		2	3	4	
7	ratios of surface area		4	9	16	
8	ratios of volume		8	27	64	
9						
10						

Sheet1 / Sheet2

Analyze

1. Compare the ratios in cells 6, 7, and 8 of columns C, D, and E. What do you observe?
2. Write a statement about the ratio of the surface areas of two solids if the scale factor is $a:b$. **The ratio of the surface areas is $a^2 : b^2$.**
3. Write a statement about the ratio of the volumes of two solids if the scale factor is $a:b$. **The ratio of the volumes is $a^3 : b^3$.**

1. If the number in row 6 is a, then row 7 contains a^2, and row 8 contains a^3.

The Spreadsheet Investigation suggests the following theorem.

Theorem 13.1

If two solids are similar with a scale factor of $a:b$, then the surface areas have a ratio of $a^2:b^2$, and the volumes have a ratio of $a^3:b^3$.

Example:

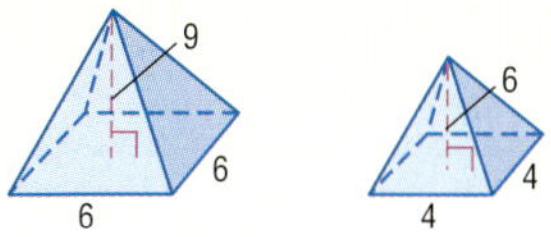

Scale factor 3:2

Ratio of surface areas $3^2:2^2$ or 9:4

Ratio of volumes $3^3:2^3$ or 27:8

Study Tip

Standardized Tests

Problems using Theorem 13.1 are frequently on standardized tests. Be sure to note that the solids must be similar for the theorem to apply.

Example 2 *Mirror Balls*

ENTERTAINMENT **Mirror balls are spheres that are covered with reflective tiles. One ball has a diameter of 4 inches, and another has a diameter of 20 inches.**

a. Find the scale factor of the two spheres.

Write the ratio of the corresponding measures of the spheres.

$$\frac{\text{diameter of the smaller sphere}}{\text{diameter of the larger sphere}} = \frac{4}{20} \quad \text{Substitution}$$

$$= \frac{1}{5} \quad \text{Simplify.}$$

The scale factor is 1:5.

www.geometryonline.com/extra_examples/fcat

PROPERTIES OF SIMILAR SOLIDS

In-Class Example

Power Point®

Teaching Tip Point out that Theorem 13.1 can be used with *any* similar solids, including prisms.

2 Softballs and baseballs are both used to play a game with a bat. A softball has a diameter of 3.8 inches, while a baseball has a diameter of about 2.9 inches.

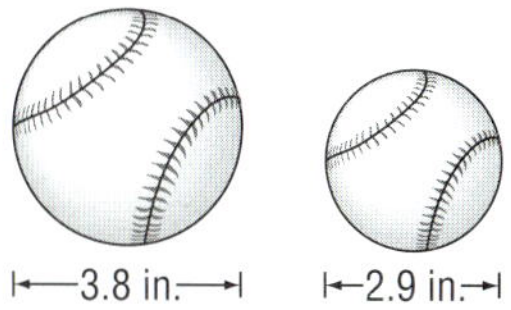

a. Find the scale factor of the two balls. **1.3:1**

b. Find the ratio of the surface area of the two balls. **1.7:1**

c. Find the ratio of the volumes of the two balls. **2.2:1**

Concept Check

Ask students to explain how to determine if two solids are similar or congruent. **Sample answer: Two solids are similar if they have exactly the same shape. They are congruent if the ratio of the corresponding measurements is 1:1. That is, the corresponding angles are congruent, the corresponding edges are congruent, and the corresponding faces are congruent.**

DAILY INTERVENTION

Differentiated Instruction

Kinesthetic Have your students use string and a ruler to determine whether two solids are similar. Then use the dimensions to compare their areas and volumes.

3 Practice/Apply

Study Notebook

Have students—
- add the definitions/examples of the vocabulary terms to their Vocabulary Builder worksheets for Chapter 13.
- include how to find the surface area and volume of two solids that are similar if the scale factor is known.
- include any other item(s) that they find helpful in mastering the skills in this lesson.

About the Exercises...

Organization by Objective
- **Congruent or Similar Solids:** 11–16
- **Properties of Similar Solids:** 18–23, 27–31

Odd/Even Assignments

Exercises 9–21 are structured so that students practice the same concepts whether they are assigned odd or even problems.

Assignment Guide

Basic: 11–37 odd, 40, 41, 43–59

Average: 11–39 odd, 40–59

Advanced: 12–38 even, 40–56 (optional: 57–59)

All: Quiz 2 (1–5)

b. Find the ratio of the surface areas of the two spheres.

If the scale factor is $a:b$, then the ratio of the surface areas is $a^2:b^2$.

$$\frac{\text{surface area of the smaller sphere}}{\text{surface area of the larger sphere}} = \frac{a^2}{b^2} \quad \text{Theorem 13.1}$$

$$= \frac{1^2}{5^2} \quad a = 1 \text{ and } b = 5$$

$$= \frac{1}{25} \quad \text{Simplify.}$$

The ratio of the surface areas is 1:25.

c. Find the ratio of the volumes of the two spheres.

If the scale factor is $a : b$, then the ratio of the volumes is $a^3:b^3$.

$$\frac{\text{volume of the smaller sphere}}{\text{volume of the larger sphere}} = \frac{a^3}{b^3} \quad \text{Theorem 13.1}$$

$$= \frac{1^3}{5^3} \quad a = 1 \text{ and } b = 5$$

$$= \frac{1}{125} \quad \text{Simplify.}$$

The ratio of the volumes of the two spheres is 1:125.

Check for Understanding

Concept Check

1–2. See margin.

1. **OPEN ENDED** Draw and label the dimensions of a pair of cones that are similar and a pair of cones that are neither similar nor congruent.
2. **Explain** the relationship between the surface areas of similar solids and volumes of similar solids.

Guided Practice

GUIDED PRACTICE KEY	
Exercises	Examples
3–4	1
5–10	2

Determine whether each pair of solids are *similar*, *congruent*, or *neither*.

3. 

congruent

4. 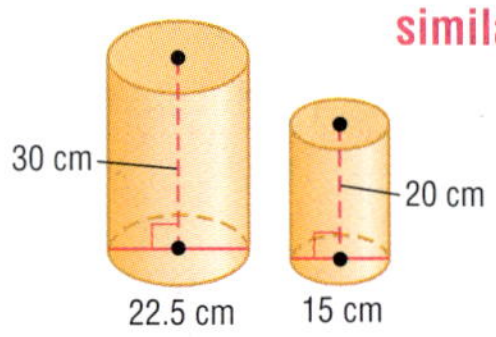

similar

For Exercises 5–7, refer to the pyramids on the right.

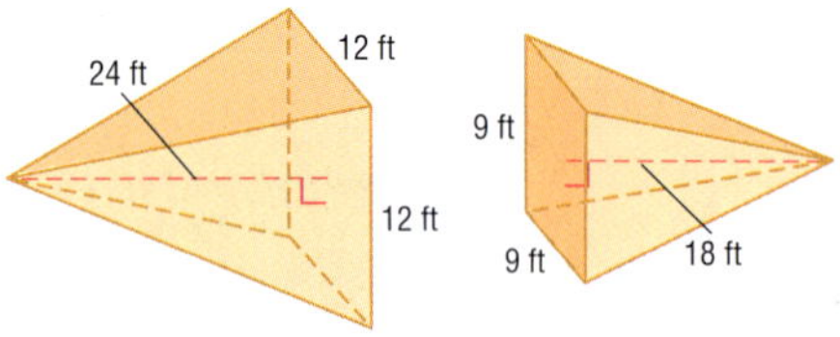

5. Find the scale factor of the two pyramids. $\frac{4}{3}$
6. Find the ratio of the surface areas of the two pyramids. $\frac{16}{9}$
7. Find the ratio of the volumes of the two pyramids. $\frac{64}{27}$

Application

LAWN ORNAMENTS **For Exercises 8–10, use the following information.**
There are two gazing balls in a garden. One has a diameter of 2 inches, and the other has a diameter of 16 inches.

8. Find the scale factor of the two gazing balls. 1:8
9. Determine the ratio of the surface areas of the two spheres. 1:64
10. What is the ratio of the volumes of the gazing balls? 1:512

Answers

1. Sample answer:

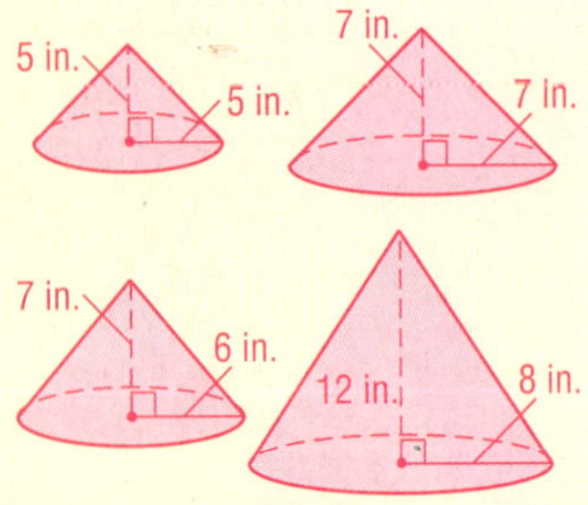

2. If two solids are similar with a scale factor of $a:b$, then the surface areas have a ratio of $a^2:b^2$ and the volumes have a ratio of $a^3:b^3$.

★ indicates increased difficulty

Practice and Apply

Homework Help

For Exercises	See Examples
11–16, 38, 39	1
17–37	2

Extra Practice
See page 781.

Determine whether each pair of solids are *similar, congruent,* or *neither.*

11. neither
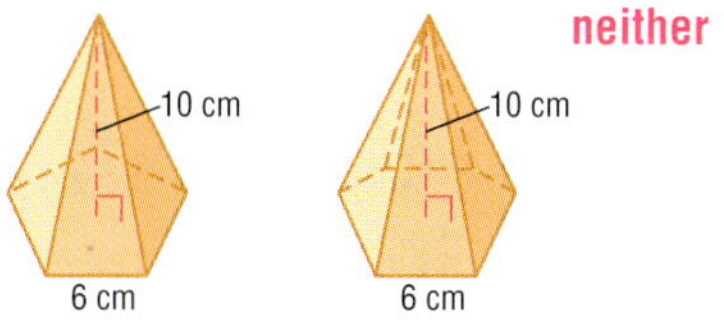

12. similar
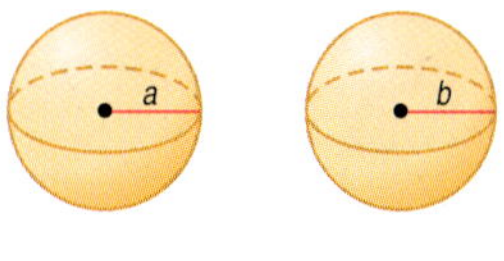

13. congruent
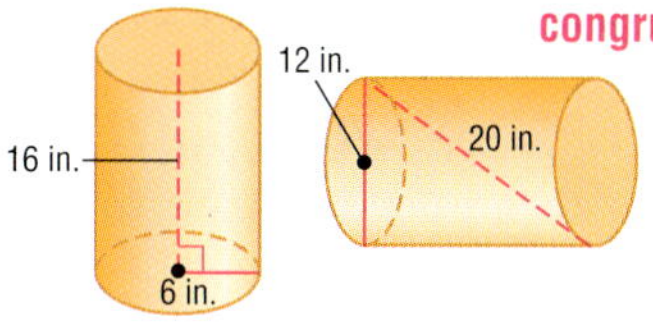

14. neither
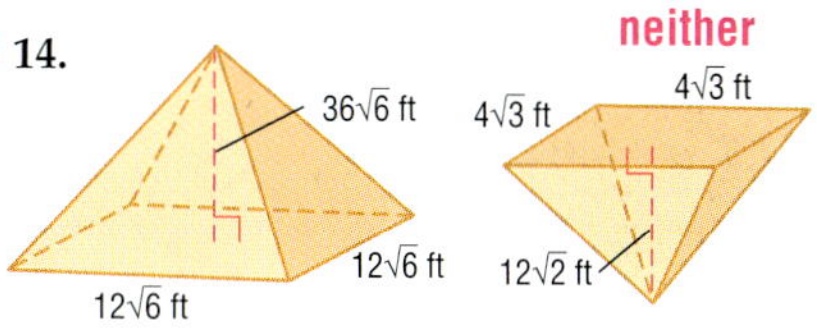

15. neither
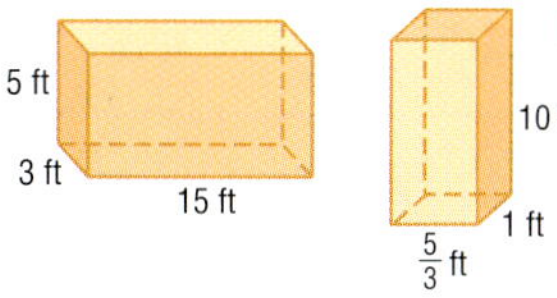

16. similar
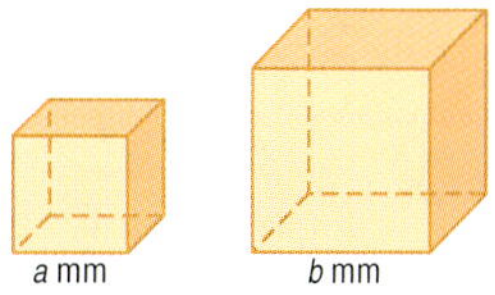

17. **ARCHITECTURE** To encourage recycling, the people of Rome, Italy, built a model of Basilica di San Pietro from empty beverage cans. The model was built to a 1:5 scale. The model measured 26 meters high, 49 meters wide, and 93 meters long. Find the dimensions of the actual Basilica di San Pietro. **130 m high, 245 m wide, and 465 m long**

Determine whether each statement is *sometimes, always,* or *never* true. Justify your answer. **18–23. See margin for justifications.**

18. Two spheres are similar. **always**
19. Congruent solids have equal surface areas. **always**
20. Similar solids have equal volumes. **sometimes**
21. A pyramid is similar to a cone. **never**
22. Cones and cylinders with the same height and base are similar. **never**
23. Nonsimilar solids have different surface areas. **sometimes**

More About. . .

Miniatures

The world's smallest car is a miniature version of a 1936 Model AA sedan called a DENSO Micro-Car. The scale factor is 1000 : 1.

Source: *The Guinness Book of Records*

MINIATURES **For Exercises 24–26, use the information at the left.**

24. If the door handle of the full-sized car is 15 centimeters long, how long is the door handle on the Micro-Car? **0.015 cm**
25. If the surface area of the Micro-Car is x square centimeters, what is the surface area of the full-sized car? **$1{,}000{,}000x$ cm^2**
26. If the scale factor was 1 : 18 instead of 1 : 1000, find the length of the miniature door handle. **≈ 0.83 cm**

For Exercises 27–30, refer to the two similar right prisms.

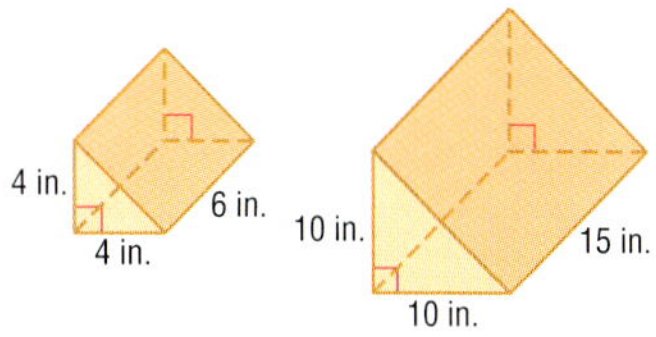

27. Find the ratio of the perimeters of the bases. $\frac{2}{5}$
28. What is the ratio of the surface areas? $\frac{4}{25}$
29. What is the ratio of the volumes? $\frac{8}{125}$
30. Suppose the volume of the smaller prism is 48 cubic inches. Find the volume of the larger prism. **750 in^3**

Answers

18. Spheres have only one measure to compare.
19. Congruent solids have equal dimensions.
20. If the solids have a scale factor of 1, the volumes will be equal.
21. Different types of solids cannot be similar.
22. Different types of solids cannot be similar.
23. Solids that are not similar can have the same surface area.

Enrichment, p. 746

Congruent and Similar Solids

Determine whether each pair of solids are *similar, congruent,* or *neither.*

1. 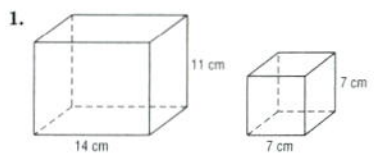

neither

2. 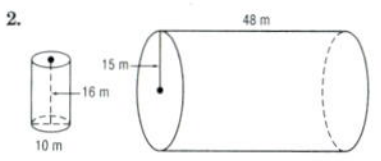

similar

3. 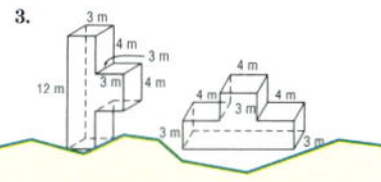

4. 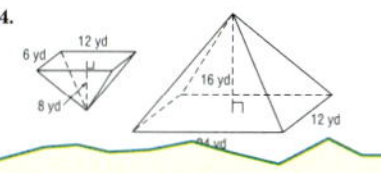

Study Guide and Intervention, p. 741 (shown) and p. 742

Congruent or Similar Solids If the corresponding angles and sides of two solids are congruent, then the solids are congruent. Also, the corresponding faces are congruent and their surface areas and volumes are equal. Solids that have the same shape but are different sizes are **similar**. You can determine whether two solids are similar by comparing the ratio, or **scale factor**, of corresponding linear measurements.

Example **Describe each pair of solids.**

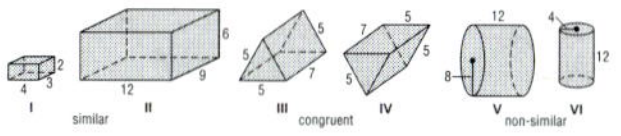

- Figures I and II are similar because the figures have the same shape. The ratio of each pair of corresponding sides is 1:3.
- Figures III and IV are congruent because they have the same shape and all corresponding measurements are the same.
- Figures V and VI are not congruent, and they are not similar because $\frac{4}{8} \neq \frac{12}{12}$.

Exercises

Determine whether each pair of solids are *similar, congruent,* or *neither.*

1. 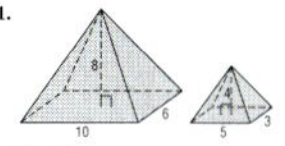
similar

2. 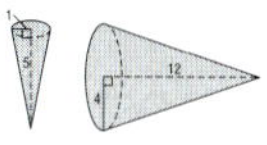
neither

3. 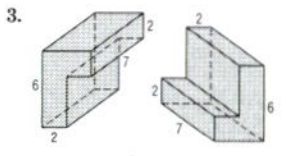
congruent

4. 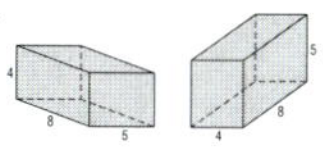
congruent

5. 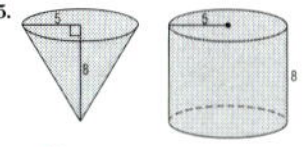
neither

6. 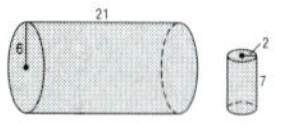
similar

Skills Practice, p. 743 and Practice, p. 744 (shown)

Determine whether each pair of solids are *similar, congruent,* or *neither.*

1. 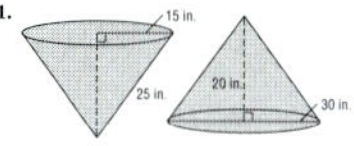

congruent

2. 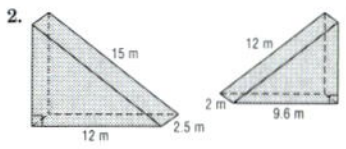

similar

3. 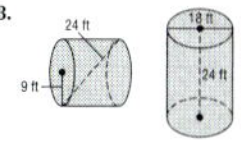

neither

4. 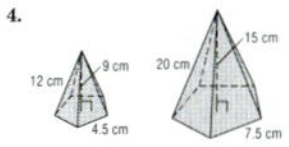

similar

For Exercises 5–8, refer to the two similar prisms.

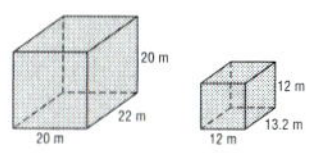

5. Find the scale factor of the two prisms. $\frac{5}{3}$
6. Find the ratio of the surface areas. $\frac{25}{9}$
7. Find the ratio of the volumes. $\frac{125}{27}$
8. Suppose the surface area of the larger prism is 2560 square meters. Find the surface area of the smaller prism. 921.6 m^2
9. **MINIATURES** Frank Lloyd Wright designed every aspect of the Imperial Hotel in Tokyo, including the chairs. The dimensions of a miniature Imperial Hotel chair are 6.25 inches × 3 inches × 2.5 inches. If the scale of the replica is 1:6, what are the dimensions of the original chair?
37.5 in. × 18 in. × 15 in.

Reading to Learn Mathematics, p. 745 — ELL

Pre-Activity **How are similar solids applied to miniature collectibles?**

Read the introduction to Lesson 13-4 at the top of page 707 in your textbook.

If you want to make a miniature with a scale factor of 1:64, how can you use the actual object to find the measurements you should use to construct the miniature? Sample answer: Take linear measurements of the actual object. Divide each measurement by 64 to find the corresponding measurement for the miniature.

Reading the Lesson

1. Determine whether each statement is *always, sometimes,* or *never* true.
   a. Two cubes are similar. always
   b. Two cones are similar. sometimes
   c. Two cylinders in which the height is twice the diameter are similar. always
   d. Two cylinders with the same volume are congruent. sometimes
   e. A prism with a square base and a square pyramid are similar. never
   f. Two rectangular prisms with equal surface areas are similar. sometimes
   g. Nonsimilar solids have different volumes. sometimes
   h. Two hemispheres with the same radius are congruent. always
2. Supply the missing ratios.
   a. If the ratio of the diameters of two spheres is 3:1, then the ratio of their surface areas is 9:1, and the ratio of their volumes is 27:1.
   b. If the ratio of the radii of two hemispheres is 2:5, then the ratio of their surface areas is 4:25, and the ratio of their volumes is 8:125.
   c. If two cones are similar and the ratio of their heights is $\frac{4}{3}$, then the ratio of their volumes is $\frac{64}{27}$, and the ratio of their surface areas is $\frac{16}{9}$.
   d. If two cylinders are similar and the ratio of their surface areas is 100:49, then the ratio of the radii of their bases is 10:7, and the ratio of their volumes is 1000:343.

Helping You Remember

3. A good way to remember a new mathematical concept is to relate it to something you already know. How can what you know about the units used to measure lengths, areas, and volumes help you to remember the theorem about the ratios of surface areas and volumes of similar solids? Sample answer: Lengths are measured in linear units, surface areas in square units, and volumes in cubic units. Take the scale factor, which is the ratio of linear measurements in the solids, and square it to get the ratio of their surface areas or cube it to get the ratio of their volumes.

Answers

40. Yes, both cones have congruent radii. If the heights are the same measure, the cones are congruent.

41. The volume of the cone on the right is equal to the sum of the volumes of the cones inside the cylinder. Justification: Call h the height of both solids. The volume of the cone on the right is $\frac{1}{3}\pi r^2 h$. If the height of one cone inside the cylinder is c, then the height of the other one is $h - c$. Therefore, the sum of the volumes of the two cones is: $\frac{1}{3}\pi r^2 c + \frac{1}{3}\pi r^2(h - c)$ or $\frac{1}{3}\pi r^2(c + h - c)$ or $\frac{1}{3}\pi r^2(h)$.

42. Sample answer: Scale factors relate the actual object to the miniatures. Answers should include the following.
   - The scale factors that are commonly used are 1:24, 1:32, 1:43, and 1:64.
   - The actual object is 108 in. long.

31. The diameters of two similar cones are in the ratio 5 to 6. If the volume of the smaller cone is 125π cubic centimeters and the diameter of the larger cone is 12 centimeters, what is the height of the larger cone? **18 cm**

32. **FESTIVALS** The world's largest circular pumpkin pie was made for the Circleville Pumpkin Show in Circleville, Ohio. The diameter was 5 feet. Most pies are 8 inches in diameter. If the pies are similar, what is the ratio of the volumes? $\frac{8}{3375}$

Online Research **Data Update** How many pies do Americans purchase in a year? Visit www.geometryonline.com/data_update to learn more.

More About. . .

Basketball

The National Collegiate Athletic Association (NCAA) states that the maximum circumference of a basketball for men is 30 inches. The maximum circumference of a women's basketball is 29 inches.

Source: www.ncaa.org

BASKETBALL **For Exercises 33–35, use the information at the left.** Find the indicated ratio of the smaller ball to the larger ball.

33. scale factor $\frac{29}{30}$
34. ratio of surface areas $\frac{841}{900}$
35. ratio of the volumes $\frac{24,389}{27,000}$

TOURISM **For Exercises 36 and 37, use the following information.**

Dale Ungerer, a farmer in Hawkeye, Iowa, constructed a gigantic ear of corn to attract tourists to his farm. The ear of corn is 32 feet long and has a circumference of 12 feet. Each "kernel" is a one-gallon milk jug with a volume of 231 cubic inches.

36. If a real ear of corn is 10 inches long, what is the scale factor between the gigantic ear of corn and the similar real ear of corn? $\frac{192}{5}$

37. Estimate the volume of a kernel of the real ear of corn. $\approx 0.004 \text{ in}^3$

For Exercises 38 and 39, use the following information.

When a cone is cut by a plane parallel to its base, a cone similar to the original is formed.

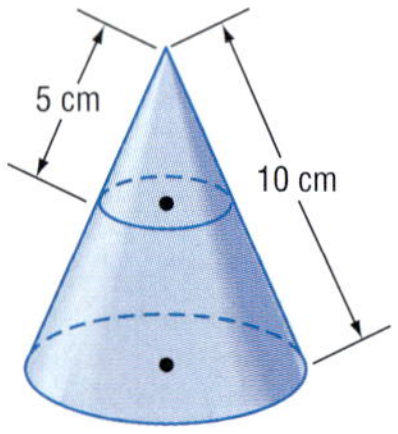

★38. What is the ratio of the volume of the frustum to that of the original cone? to the smaller cone? **7:8; 7:1**

★39. What is the ratio of the lateral area of the frustum to that of the original cone? to the smaller cone? **3:4; 3:1**

CRITICAL THINKING **For Exercises 40 and 41, refer to the figure.** **40–41. See margin.**

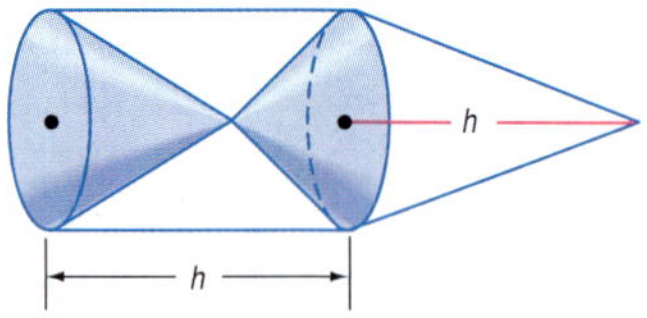

40. Is it possible for the two cones inside the cylinder to be congruent? Explain.

41. Is the volume of the cone on the right *equal to, greater than,* or *less than* the sum of the volume of the cones inside the cylinder? Explain.

42. **WRITING IN MATH** Answer the question that was posed at the beginning of the lesson. **See margin.**

How is the geometry of similar solids applied to miniature collectibles?

Include the following in your answer:
- the scale factors that are commonly used, and
- the answer to this question: If a miniature is 4.5 inches with a scale factor of 1:24, then how long is the actual object?

712 Chapter 13 Volume

FCAT Practice

Standardized Test Practice

43. If the ratio of the surface areas of two similar solids is 4:9, find the ratio of the volumes. **C**

Ⓐ 64:729 Ⓑ 2:3 Ⓒ 8:27 Ⓓ 16:81

44. **ALGEBRA** If $xyz = 4$ and $y^2z = 5$, then what is the value of $\frac{x}{y}$? **D**

Ⓐ 20.0 Ⓑ 2.0 Ⓒ 1.5 Ⓓ 0.8

Maintain Your Skills

Mixed Review

Find the volume of each sphere. Round to the nearest tenth. *(Lesson 13-3)*

45. diameter = 8 feet **268.1 ft^3**
46. radius = 9.5 meters **3591.4 m^3**
47. radius = 15.1 centimeters **14,421.8 cm^3**
48. diameter = 23 inches **6370.6 in^3**

Find the volume of each pyramid or cone. Round to the nearest tenth if necessary. *(Lesson 13-2)*

49. **323.3 in^3**

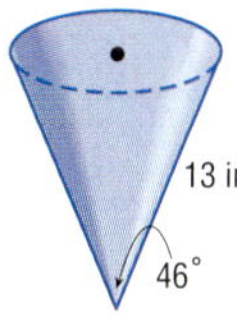

50. **385 m^3**

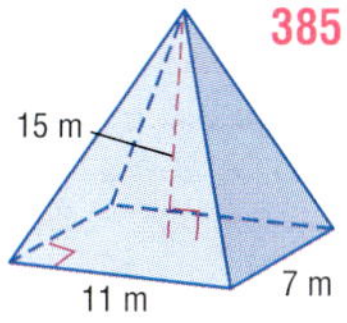

51. **2741.8 ft^3**

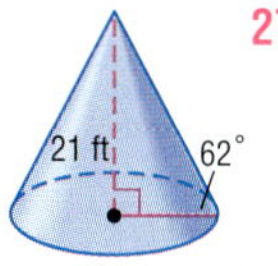

Find the radius of the base of each cylinder. Round to the nearest tenth. *(Lesson 12-4)* **52. 5.4 cm**

52. The surface area is 430 square centimeters, and the height is 7.4 centimeters.
53. The surface area is 224.7 square yards, and the height is 10 yards. **2.8 yd**

NAVIGATION **For Exercises 54–56, use the following information.**

As part of a scuba diving exercise, a 12-foot by 3-foot rectangular-shaped rowboat was sunk in a quarry. A boat takes a scuba diver to a random spot in the enclosed section of the quarry and anchors there so that the diver can search for the rowboat. *(Lesson 11-5)*

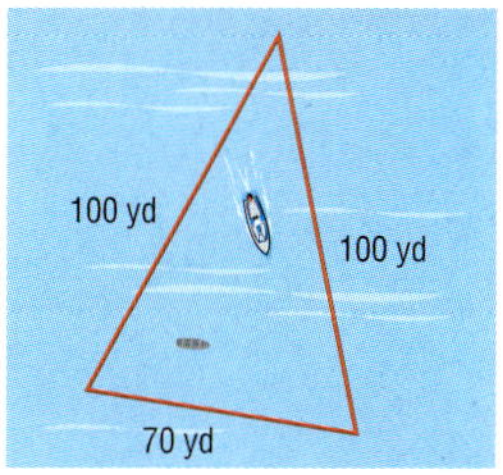

54. What is the approximate area of the enclosed section of the quarry? **≈ 3279 yd^2**
55. What is the area of the rowboat? **36 ft^2**
56. What is the probability that the boat will anchor over the sunken rowboat?

56. $\approx \frac{36}{29,511}$ or ≈ 0.0012

Getting Ready for the Next Lesson

PREREQUISITE SKILL **Determine whether the ordered pair is on the graph of the given equation. Write** ***yes*** **or** ***no.*** *(To review* ***graphs in the coordinate plane****, see Lesson 1-1.)*

57. $y = 3x + 5$, (4, 17) **yes**
58. $y = -4x + 1$, (−2, 9) **yes**
59. $y = 7x - 4$, (−1, 3) **no**

Practice Quiz 2

Lessons 13-3 and 13-4

Find the volume of each sphere. Round to the nearest tenth. *(Lesson 13-3)*

1. radius = 25.3 ft **67,834.4 ft^3**
2. diameter = 36.8 cm **26,094.1 cm^3**

The two square pyramids are similar. *(Lesson 13-4)*

3. Find the scale factor of the pyramids. **$\frac{7}{5}$**
4. What is the ratio of the surface areas? **$\frac{49}{25}$**
5. What is the ratio of the volumes? **$\frac{343}{125}$**

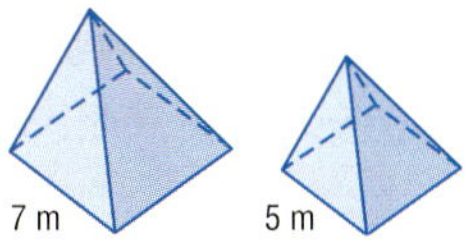

www.geometryonline.com/self_check_quiz/fcat

4 Assess

Open-Ended Assessment

Writing Have students explain how to tell if two solids are congruent. Two solids are congruent if the corresponding angles, edges, and faces are congruent, and if their volumes are congruent.

Getting Ready for Lesson 13-5

Prerequisite Skill Students will graph solids in space in Lesson 13-5. Students should be comfortable answering questions about graphs and points in the coordinate plane. Use Exercises 57–59 to determine your students' familiarity with graphs in the coordinate plane.

Assessment Options

Practice Quiz 2 The quiz provides students with a brief review of the concepts and skills in Lessons 13-3 and 13-4. Lesson numbers are given to the right of the exercises or instruction lines so students can review concepts not yet mastered.

Quiz (Lesson 4) is available on p. 768 of the *Chapter 13 Resource Masters*.

13-5 Lesson Notes

1 Focus

5-Minute Check Transparency 13-5 Use as a quiz or review of Lesson 13-4.

Mathematical Background notes are available for this lesson on p. 686D.

How is three-dimensional graphing used in computer animation?

Ask students:

- What is a computer mesh? **It is an outline that shows the size and shape of the image.**
- What is "rendering"? **adding color and texture to the image**
- How is computer animation used in the real world? **Sample answer: analyze sports, computer games, movies**

13-5 Coordinates in Space

Sunshine State Standards
MA.A.3.4.3, MA.A.4.4.1, MA.B.3.4.1, MA.C.3.4.2

Vocabulary
- ordered triple

What You'll Learn

- Graph solids in space.
- Use the Distance and Midpoint Formulas for points in space.

How is three-dimensional graphing used in computer animation?

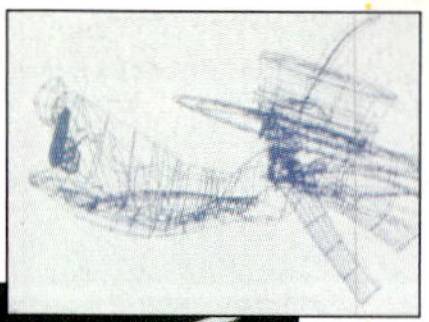

The initial step in computer animation is creating a three-dimensional image. A *mesh* is created first. This is an outline that shows the size and shape of the image. Then the image is *rendered*, adding color and texture. The image is animated using software. There is a way to describe the location of each point in the image.

Study Tip

Reading Math
When the three planes intersect to form the three dimensional coordinate system, eight regions are formed. These regions are called *octants*.

GRAPH SOLIDS IN SPACE To describe the location of a point on the coordinate plane, we use an ordered pair of two coordinates. In space, each point requires three numbers, or coordinates, to describe its location because space has three dimensions. In space, the x-, y-, and z-axes are perpendicular to each other.

A point in space is represented by an **ordered triple** of real numbers (x, y, z). In the figure at the right, the ordered triple (2, 3, 6) locates point P. Notice that a rectangular prism is used to show perspective.

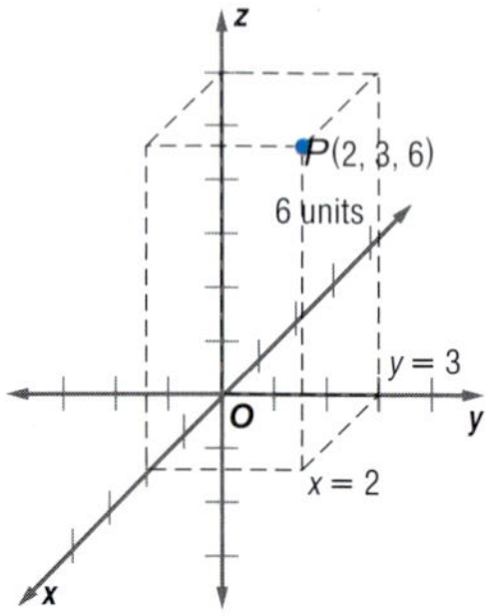

Study Tip

Drawing in Three Dimensions
Use the properties of a rectangular prism to correctly locate the z-coordinate. A is the vertex farthest from the origin.

Example 1 Graph a Rectangular Solid

Graph a rectangular solid that has $A(-4, 2, 4)$ and the origin as vertices. Label the coordinates of each vertex.

- Plot the x-coordinate first. Draw a segment from the origin 4 units in the negative direction.
- To plot the y-coordinate, draw a segment 2 units in the positive direction.
- Next, to plot the z-coordinate, draw a segment 4 units long in the positive direction.
- Label the coordinate A.
- Draw the rectangular prism and label each vertex.

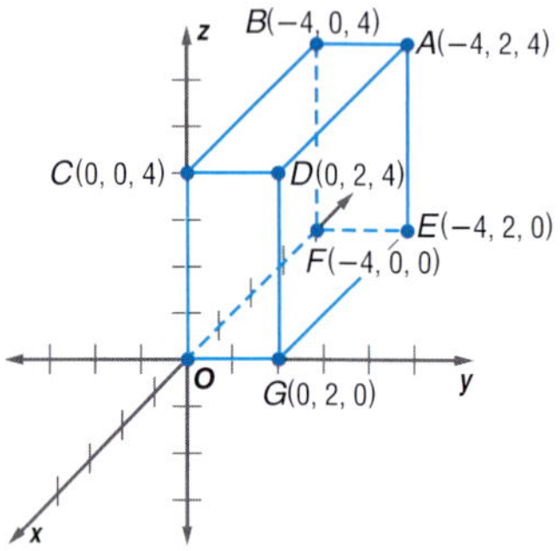

Resource Manager

Workbook and Reproducible Masters

Chapter 13 Resource Masters
- Study Guide and Intervention, pp. 747–748
- Skills Practice, p. 749
- Practice, p. 750
- Reading to Learn Mathematics, p. 751
- Enrichment, p. 752
- Assessment, p. 768

Prerequisite Skills Workbook, pp. 5–6
Teaching Geometry With Manipulatives Masters, p. 17

Transparencies

5-Minute Check Transparency 13-5
Answer Key Transparencies

Technology

GeomPASS: Tutorial Plus, Lesson 25
Interactive Chalkboard

DISTANCE AND MIDPOINT FORMULA Recall that the Distance Formula is derived from the Pythagorean Theorem. The Pythagorean Theorem can also be used to find the formula for the distance between two points in space.

Study Tip

Look Back
To review **Distance and Midpoint Formulas**, see Lesson 1-3.

Key Concept — *Distance Formula in Space*

Given two points $A(x_1, y_1, z_1)$ and $B(x_2, y_2, z_2)$ in space, the distance between A and B is given by the following equation.

$$d = \sqrt{(x_2 - x_1)^2 + (y_2 - y_1)^2 + (z_2 - z_1)^2}$$

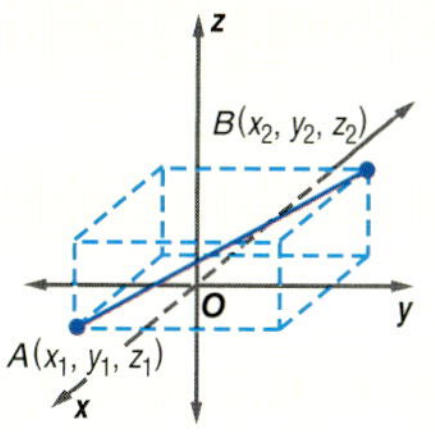

This formula is an extension of the Distance Formula in two dimensions. The Midpoint Formula can also be extended to the three-dimensions.

Key Concept — *Midpoint Formula in Space*

Given two points $A(x_1, y_1, z_1)$ and $B(x_2, y_2, z_2)$ in space, the midpoint of $\overline{AB}$ is at $\left(\frac{x_1 + x_2}{2}, \frac{y_1 + y_2}{2}, \frac{z_1 + z_2}{2}\right)$.

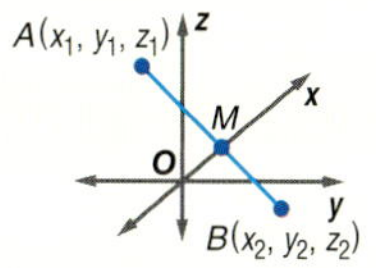

Example 2 Distance and Midpoint Formulas in Space

Study Tip

Look Back
To review **translations**, see Lesson 9-2.

a. Determine the distance between $T(6, 0, 0)$ and $Q(-2, 4, 2)$.

$TQ = \sqrt{(x_2 - x_1)^2 + (y_2 - y_1)^2 + (z_2 - z_1)^2}$ Distance Formula in Space

$= \sqrt{[6 - (-2)]^2 + (0 - 4)^2 + (0 - 2)^2}$ Substitution

$= \sqrt{84}$ or $2\sqrt{21}$ Simplify.

b. Determine the coordinates of the midpoint M of $\overline{TQ}$.

$M = \left(\frac{x_1 + x_2}{2}, \frac{y_1 + y_2}{2}, \frac{z_1 + z_2}{2}\right)$ Midpoint Formula in Space

$= \left(\frac{6 - 2}{2}, \frac{0 + 4}{2}, \frac{0 + 2}{2}\right)$ Substitution

$= (2, 2, 1)$ Simplify.

Example 3 Translating a Solid

ELEVATORS **Suppose an elevator is 5 feet wide, 6 feet deep, and 8 feet tall. Position the elevator on the ground floor at the origin of a three dimensional space. If the distance between the floors of a warehouse is 10 feet, write the coordinates of the vertices of the elevator after going up to the third floor.**

Explore Since the elevator is a rectangular prism, use positive values for x, y, and z. Write the coordinates of each corner. The points on the elevator will rise 10 feet for each floor. When the elevator ascends to the third floor, it will have traveled 20 feet.

(continued on the next page)

www.geometryonline.com/extra_examples/fcat

2 Teach

GRAPH SOLIDS IN SPACE

In-Class Example Power Point®

1 *Teaching Tip* Students may have difficulty graphing ordered triples. You may want to have them draw a prism first, then add the x-, y-, and z-axes, and then label the axes and vertices.

Graph a rectangular solid that contains the ordered triple $A(-3, 1, 2)$ and the origin as vertices. Label the coordinates of each vertex.

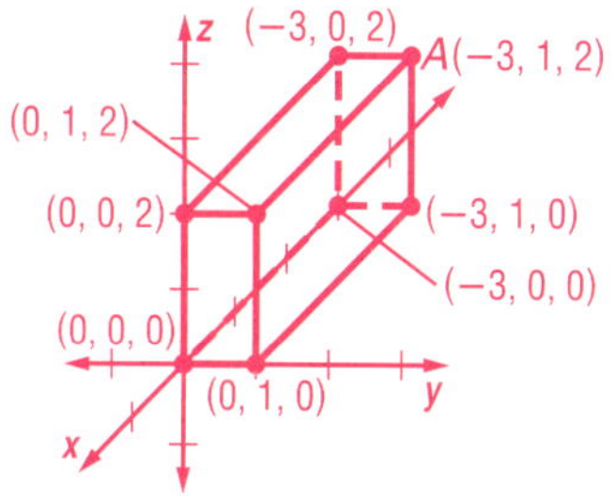

DISTANCE AND MIDPOINT FORMULAS

In-Class Examples Power Point®

2 **DISTANCE AND MIDPOINT FORMULAS IN SPACE**

a. Determine the distance between $F(4, 0, 0)$ and $G(-2, 3, -1)$. $\sqrt{46}$

b. Determine the midpoint M of $\overline{FG}$. $\left(1, \frac{3}{2}, -\frac{1}{2}\right)$

3 Suppose a two-story home is built with a bathroom on the first floor that is 9 feet wide, 6 feet deep, and 8 feet tall. Likewise, a bathroom on the second floor is directly above the one on the first floor and has the same dimensions. If the height of the second floor is 10 feet above the first floor, find the coordinates of each vertex of the rectangular prism that represents the second floor bathroom. **(0, 0, 10); (0, 9, 10); (6, 9, 10); (6, 0, 10); (0, 0, 18); (0, 9, 18); (6, 9, 18); (6, 0, 18)**

In-Class Example

Power Point®

4 Dilate the prism by a scale factor of $\frac{1}{2}$. Graph the image under the dilation.

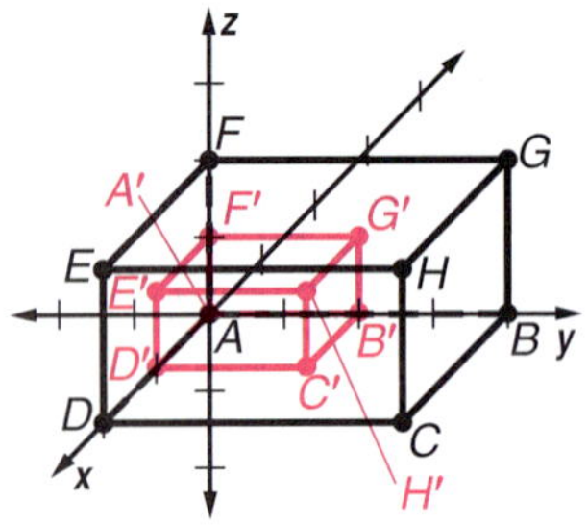

The coordinates of the vertices are $A'(0, 0, 0)$, $B'(0, 2, 0)$, $C'(1, 2, 0)$, $D'(1, 0, 0)$, $E'(1, 0, 1)$, $F'(0, 0, 1)$, $G'(0, 2, 1)$, $H'(1, 2, 1)$.

Answers (page 717)

1. The coordinate plane has 4 regions or quadrants with 4 possible combinations of signs for the ordered pairs. Three-dimensional space is the intersection of 3 planes that create 8 regions with 8 possible combinations of signs for the ordered triples.
2. Sample answer: Use the point at (2, 3, 4); $A(2, 3, 4)$, $B(2, 0, 4)$, $C(0, 0, 4)$, $D(0, 3, 4)$ $E(2, 3, 0)$, $F(2, 0, 0)$, $G(0, 0, 0)$, and $H(0, 3, 0)$.

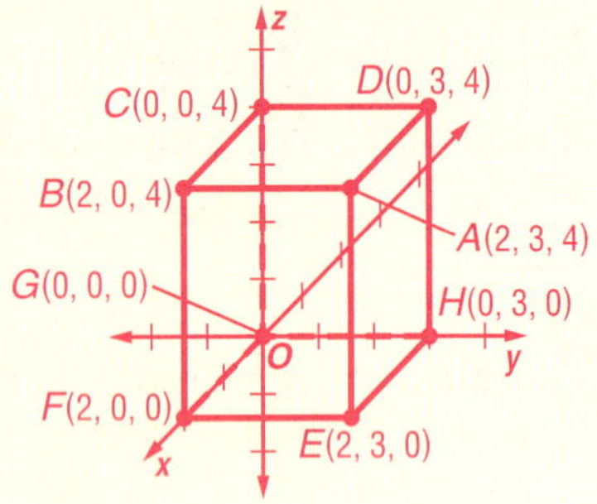

Plan Use the translation $(x, y, z) \rightarrow (x, y, z + 20)$ to find the coordinates of each vertex of the rectangular prism that represents the elevator.

Solve

Coordinates of the vertices, (x, y, z) Preimage	Translated coordinates, $(x, y, z + 20)$ Image
$J(0, 5, 8)$	$J'(0, 5, 28)$
$K(6, 5, 8)$	$K'(6, 5, 28)$
$L(6, 0, 8)$	$L'(6, 0, 28)$
$M(0, 0, 8)$	$M'(0, 0, 28)$
$N(6, 0, 0)$	$N'(6, 0, 20)$
$O(0, 0, 0)$	$O'(0, 0, 20)$
$P(0, 5, 0)$	$P'(0, 5, 20)$
$Q(6, 5, 0)$	$Q'(6, 5, 20)$

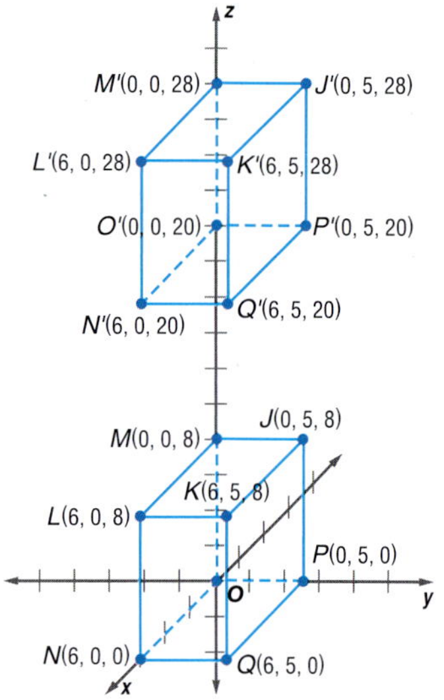

Examine Check that the distance between corresponding vertices is 20 feet.

Matrices can be used for transformations in space such as dilations.

Study Tip

Look Back

To review **matrices and transformations**, see Lesson 9-7.

Example 4 Dilation with Matrices

Dilate the prism by a scale factor of 2. Graph the image under the dilation.

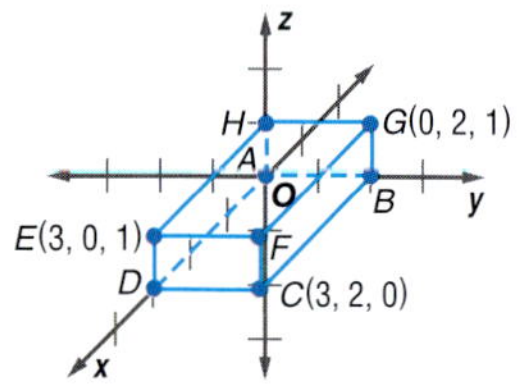

First, write a vertex matrix for the rectangular prism.

$$\begin{array}{c} \\ x \\ y \\ z \end{array}\begin{array}{c} \begin{array}{cccccccc} A & B & C & D & E & F & G & H \end{array} \\ \begin{bmatrix} 0 & 0 & 3 & 3 & 3 & 3 & 0 & 0 \\ 0 & 2 & 2 & 0 & 0 & 2 & 2 & 0 \\ 0 & 0 & 0 & 0 & 1 & 1 & 1 & 1 \end{bmatrix} \end{array}$$

Next, multiply each element of the vertex matrix by the scale factor, 2.

$$2\begin{array}{c} \begin{array}{cccccccc} A & B & C & D & E & F & G & H \end{array} \\ \begin{bmatrix} 0 & 0 & 3 & 3 & 3 & 3 & 0 & 0 \\ 0 & 2 & 2 & 0 & 0 & 2 & 2 & 0 \\ 0 & 0 & 0 & 0 & 1 & 1 & 1 & 1 \end{bmatrix} \end{array} = \begin{array}{c} \begin{array}{cccccccc} A' & B' & C' & D' & E' & F' & G' & H' \end{array} \\ \begin{bmatrix} 0 & 0 & 6 & 6 & 6 & 6 & 0 & 0 \\ 0 & 4 & 4 & 0 & 0 & 4 & 4 & 0 \\ 0 & 0 & 0 & 0 & 2 & 2 & 2 & 2 \end{bmatrix} \end{array}$$

The coordinates of the vertices of the dilated image are $A'(0, 0, 0)$, $B'(0, 4, 0)$, $C'(6, 4, 0)$, $D'(6, 0, 0)$, $E'(6, 0, 2)$, $F'(6, 4, 2)$, $G'(0, 4, 2)$, and $H'(0, 0, 2)$.

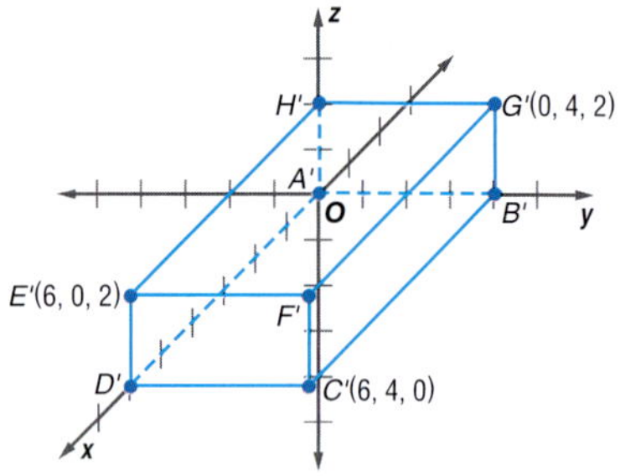

DAILY INTERVENTION

Differentiated Instruction

Auditory/Musical Ask students to describe how to plot three-dimensional coordinates in space.

Check for Understanding

Concept Check

1–2. See margin.

1. **Compare and contrast** the number of regions on the coordinate plane and in three-dimensional coordinate space.

2. **OPEN ENDED** Draw and label the vertices of a rectangular prism that has a volume of 24 cubic units.

3. **Find a counterexample** for the following statement.
Every rectangular prism will be congruent to its image from any type of transformation.
A dilation of a rectangular prism will provide a similar figure, but not a congruent one unless $r = 1$ or $r = -1$

Guided Practice

Graph a rectangular solid that contains the given point and the origin as vertices. Label the coordinates of each vertex. 4–5. See p. 725A.

GUIDED PRACTICE KEY	
Exercises	Examples
4–5	1
6–7	2
8	4
9	3

4. $A(2, 1, 5)$

5. $P(-1, 4, 2)$

Determine the distance between each pair of points. Then determine the coordinates of the midpoint M of the segment joining the pair of points.

6. $D(0, 0, 0)$ and $E(1, 5, 7)$ $\sqrt{75}; \left(\frac{1}{2}, \frac{5}{2}, \frac{7}{2}\right)$

7. $G(-3, -4, 6)$ and $H(5, -3, -5)$ $\sqrt{186}; \left(1, -\frac{7}{2}, \frac{1}{2}\right)$

8. The vertices of a rectangular prism are $M(0, 0, 0)$, $N(-3, 0, 0)$, $P(-3, 4, 0)$, $Q(0, 4, 0)$, $R(0, 0, 2)$, $S(0, 4, 2)$, $T(-3, 4, 2)$, and $V(-3, 0, 2)$. Dilate the prism by a scale factor of 2. Graph the image under the dilation. See margin.

Application

9. **STORAGE** A storage container is 12 feet deep, 8 feet wide, and 8 feet high. To allow the storage company to locate and identify the container, they assign ordered triples to the corners using positive x, y, and z values. If the container is stored 16 feet up and 48 feet back in the warehouse, find the ordered triples of the vertices describing the new location. Use the translation $(x, y, z) \rightarrow (x - 48, y, z + 16)$. See margin.

★ indicates increased difficulty

Practice and Apply

Homework Help	
For Exercises	See Examples
10–15	1
16–21, 32–34	2
22, 25–28, 31, 35	3
23–24, 29–30	4

Extra Practice
See page 781.

Graph a rectangular solid that contains the given point and the origin as vertices. Label the coordinates of each vertex. 10–15. See p. 725A.

10. $C(-2, 2, 2)$
11. $R(3, -4, 1)$
12. $P(4, 6, -3)$
13. $G(4, 1, -3)$
14. $K(-2, -4, -4)$
15. $W(-1, -3, -6)$

16. $KL = \sqrt{32}; (0, 0, 0)$ 17. $PQ = \sqrt{115}; \left(\frac{1}{2}, -\frac{7}{2}, \frac{7}{2}\right)$

Determine the distance between each pair of points. Then determine the coordinates of the midpoint M of the segment joining the pair of points.

16. $K(2, 2, 0)$ and $L(-2, -2, 0)$

17. $P(-2, -5, 8)$ and $Q(3, -2, -1)$

18. $F\left(\frac{3}{5}, 0, \frac{4}{5}\right)$ and $G(0, 3, 0)$

19. $G(1, -1, 6)$ and $H\left(\frac{1}{5}, -\frac{2}{5}, 2\right)$

20. $S(6\sqrt{3}, 4, 4\sqrt{2})$ and $T(4\sqrt{3}, 5, \sqrt{2})$
$ST = \sqrt{31}; \left(5\sqrt{3}, \frac{9}{2}, \frac{5\sqrt{2}}{2}\right)$

21. $B(\sqrt{3}, 2, 2\sqrt{2})$ and $C(-2\sqrt{3}, 4, 4\sqrt{2})$
$BC = \sqrt{39}; \left(\frac{-\sqrt{3}}{2}, 3, 3\sqrt{2}\right)$

18. $FG = \sqrt{10}; \left(\frac{3}{10}, \frac{3}{2}, \frac{2}{5}\right)$

19. $GH = \sqrt{17}; \left(\frac{3}{5}, -\frac{7}{10}, 4\right)$

22. **AVIATION** An airplane at an elevation of 2 miles is 50 miles east and 100 miles north of an airport. This location can be written as (50, 100, 2). A second airplane is at an elevation of 2.5 miles and is located 240 miles west and 140 miles north of the airport. This location can be written as (−240, 140, 2.5). Find the distance between the airplanes to the nearest tenth of a mile. 292.7 mi

DAILY INTERVENTION

Unlocking Misconceptions

A common error is to think that a point with coordinates such as (2, 0, 4) is on the y-axis because the y-coordinate is 0. Point out that this point is located on the xz-plane. A point on an axis has to have two zero coordinates.

3 Practice/Apply

Study Notebook

Have students—
- add the definitions/examples of the vocabulary terms to their Vocabulary Builder worksheets for Chapter 13.
- include an example graph of a rectangular solid.
- include any other item(s) that they find helpful in mastering the skills in this lesson.

About the Exercises...

Organization by Objective
- **Graph Solids in Space:** 10–15
- **Distance and Midpoint Formula:** 16–21

Odd/Even Assignments
Exercises 10–21, 25–30 are structured so that students practice the same concepts whether they are assigned odd or even problems.

Assignment Guide

Basic: 11–35 odd, 36–39, 42–47 (optional: 40–41)

Average: 11–35 odd, 36–39, 42–47 (optional: 40–41)

Advanced: 10–36 even, 38–47

Answers

8.

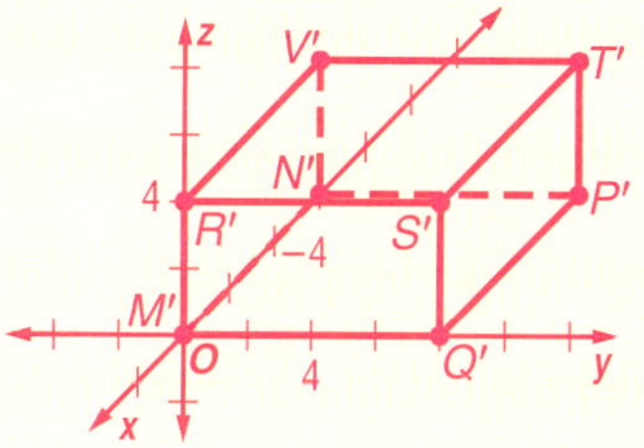

9. (12, 8, 8), (12, 0, 8), (0, 0, 8), (0, 8, 8), (12, 8, 0), (12, 0, 0), (0, 0, 0), and (0, 8, 0); (−36, 8, 24), (−36, 0, 24), (−48, 0, 24), (−48, 8, 24), (−36, 8, 16), (−36, 0, 16), (−48, 0, 16), and (−48, 8, 16)

Study Guide and Intervention, p. 747 (shown) and p. 748

Graph Solids in Space In space, you can describe the location of a point using an **ordered triple** of real numbers. The x-, y-, and z-axes are perpendicular to each other, and the coordinates for point P are the ordered triple $(-4, 6, 5)$. A rectangular prism can be drawn to show perspective.

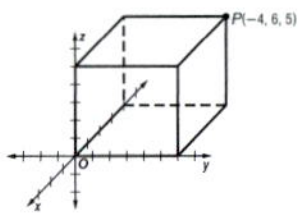

Example **Graph the rectangular solid that contains the ordered triple (2, 1, −2) and the origin. Label the coordinates of each vertex.**

- Plot the x-coordinate first. Draw a solid segment from the origin 2 units in the positive direction.
- Plot the y-coordinate next. Draw a solid segment 1 unit in the positive direction.
- Plot the z-coordinate next. Draw a solid segment 2 units in the negative direction.
- Draw the rectangular prism, using dotted lines for hidden edges of the prism.
- Label the coordinates of each vertex.

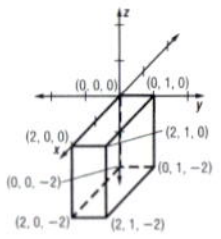

Exercises

Graph the rectangular solid that contains the given point and the origin as vertices. Label the coordinates of each vertex.

1. $A(2, 1, 3)$

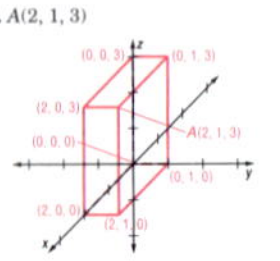

2. $G(-1, 2, 3)$

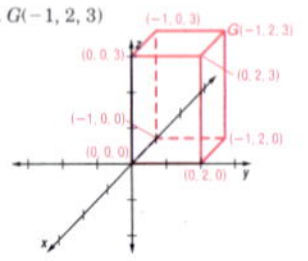

3. $P(-2, 1, -1)$

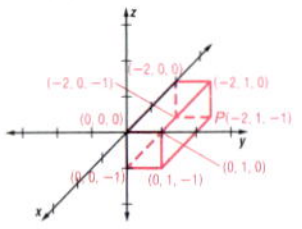

4. $T(-1, 3, 2)$

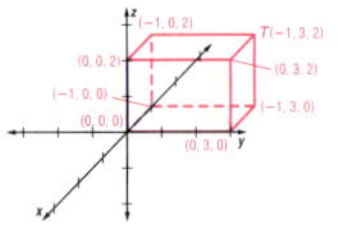

Skills Practice, p. 749 and Practice, p. 750 (shown)

Graph the rectangular solid that contains the given point and the origin as vertices. Label the coordinates of each vertex.

1. $E(4, 6, -2)$

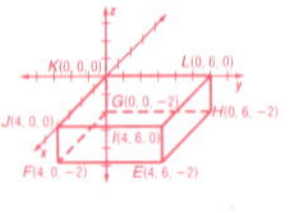

2. $R(-3, -5, 4)$

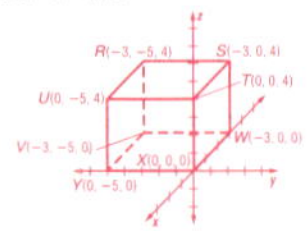

Determine the distance between each pair of points. Then determine the coordinates of the midpoint M of the segment joining the pair of points.

3. $Y(-5, 1, 2)$ and $Z(3, -3, 1)$ $YZ = 9; \left(-1, -1, \frac{3}{2}\right)$

4. $E(4, 2, 0)$ and $F(3, 2, -2)$ $EF = \sqrt{5}; \left(\frac{7}{2}, 2, -1\right)$

5. $B(-2, -2, -3)$ and $C(1, -3, 0)$ $BC = \sqrt{19}; \left(-\frac{1}{2}, -\frac{5}{2}, -\frac{3}{2}\right)$

6. $H(2, 0, -3)$ and $I(4, -1, 5)$ $HI = \sqrt{69}; \left(3, -\frac{1}{2}, 1\right)$

7. **ANIMATION** Derek wants to animate an image for his science presentation by moving it from one position to another. The mesh of the image is a rectangular prism with coordinates $A(-3, 2, 3)$, $B(-3, 0, 3)$, $C(0, 0, 3)$, $D(0, 2, 3)$, $E(-3, 2, 0)$, $F(-3, 0, 0)$, $G(0, 0, 0)$, and $H(0, 2, 0)$. Find the coordinates of the mesh after the translation $(x, y, z) \to (x - 7, y, z)$. Graph both the preimage and image of the mesh.
$A'(-10, 2, 3)$, $B'(-10, 0, 3)$, $C'(-7, 0, 3)$, $D'(-7, 2, 3)$, $E'(-10, 2, 0)$, $F'(-10, 0, 0)$, $G'(-7, 0, 0)$, $H'(-7, 2, 0)$

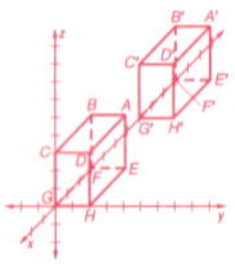

Reading to Learn Mathematics, p. 751 ELL

Pre-Activity **How is three-dimensional graphing used in computer animation?**

Read the introduction to Lesson 13-5 at the top of page 714 in your textbook.

Why would a mesh be created first? Sample answer: A mesh is an outline that the animator would use first like a sketch before rendering a final image.

Reading the Lesson

1. Refer to the figure. Match each point from the first column with its coordinates from the second column.

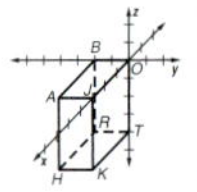

a. A iii	i. (3, 0, 0)
b. B vi	ii. (3, 0, −4)
c. O v	iii. (3, −2, 0)
d. J i	iv. (3, −2, −4)
e. H iv	v. (0, 0, 0)
f. K ii	vi. (0, −2, 0)
g. T viii	vii. (0, −2, −4)
h. R vii	viii. (0, 0, −4)

2. Which of the following expressions give the distance between the points at $(4, -1, -5)$ and $(-3, 2, -9)$? A, E, H

A. $\sqrt{7^2 + (-3)^2 + 4^2}$
B. $\sqrt{1^2 + 1^2 + (-14)^2}$
C. $\sqrt{2^2 + 2^2 + 4^2}$
D. $\left(\frac{1}{2}, \frac{1}{2}, -7\right)$
E. $\sqrt{(-3 - 4)^2 + (-1 - 2)^2 + (-9 + 5)^2}$
F. $\sqrt{24}$
G. $\sqrt{(-3 + 4)^2 + [2 + (-1)]^2 + [-9 + (-5)]^2}$
H. $\sqrt{74}$

Helping You Remember

3. A good way to remember new mathematical formulas is to relate them to ones you already know. How can you use your knowledge of the Distance and Midpoint Formulas in two dimensions to remember the formulas in three dimensions? Sample answer: Start with the formulas for two dimensions. Add a third term under the radical in the Distance Formula and a third coordinate in the Midpoint Formula that is just like the other two except that the variable is z rather than x or y.

More About. . .

Recreation

Modern hot-air balloons were developed by the Montgolfier brothers in 1783. The first passengers were a sheep, a chicken, and a duck. When the animals returned safely, humans boarded the balloon.

Source: www.howstuffworks.com

Dilate each prism by the given scale factor. Graph the image under the dilation. 23–24. See p. 725A.

23. scale factor of 3

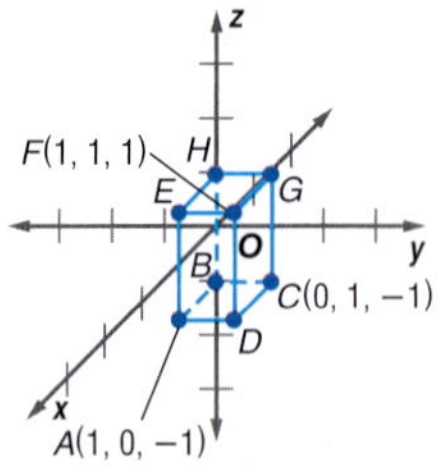

24. scale factor of 2

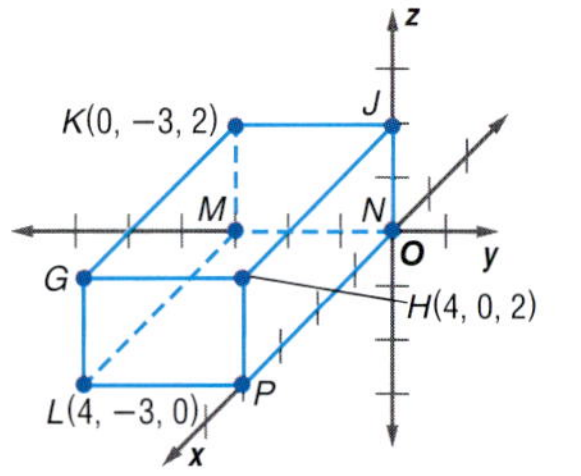

Consider a rectangular prism with the given coordinates. Find the coordinates of the vertices of the prism after the translation. 25–26. See margin.

25. $P(-2, -3, 3)$, $Q(-2, 0, 3)$, $R(0, 0, 3)$, $S(0, -3, 3)$ $T(-2, 0, 0)$, $U(-2, -3, 0)$, $V(0, -3, 0)$, and $W(0, 0, 0)$; $(x, y, z) \to (x + 2, y + 5, z - 5)$

26. $A(2, 0, 1)$, $B(2, 0, 0)$, $C(2, 1, 0)$, $D(2, 1, 1)$, $E(0, 0, 1)$, $F(0, 1, 1)$, $G(0, 1, 0)$, and $H(0, 0, 0)$; $(x, y, z) \to (x - 2, y + 1, z - 1)$.

Consider a cube with coordinates $A(3, 3, 3)$, $B(3, 0, 3)$, $C(0, 0, 3)$, $D(0, 3, 3)$, $E(3, 3, 0)$, $F(3, 0, 0)$, $G(0, 0, 0)$, and $H(0, 3, 0)$. Find the coordinates of the image under each transformation. Graph the preimage and the image. 27–30. See pp. 725A–725B.

27. Use the translation $(x, y, z) \to (x + 1, y + 2, z - 2)$.

28. Use the translation $(x, y, z) \to (x - 2, y - 3, z + 2)$.

29. Dilate the cube by a factor of 2. What is the volume of the image?

30. Dilate the cube by a factor of $\frac{1}{3}$. What is the ratio of the volumes for these two cubes?

31. **RECREATION** Two hot-air balloons take off from the same site. One hot-air balloon is 12 miles west and 12 miles south of the takeoff point and 0.4 mile above the ground. The other balloon is 4 miles west and 10 miles south of the takeoff site and 0.3 mile above the ground. Find the distance between the two balloons to the nearest tenth of a mile. **8.2 mi**

32. If $M(5, 1, 2)$ is the midpoint of segment $\overline{AB}$ and point A has coordinates $(2, 4, 7)$, then what are the coordinates of point B? **(8, −2, −3)**

33. The center of a sphere is at $(4, -2, 6)$, and the endpoint of a diameter is at $(8, 10, -2)$. What are the coordinates of the other endpoint of the diameter? **(0, −14, 14)**

34. Find the center and the radius of a sphere if the diameter has endpoints at $(-12, 10, 12)$ and $(14, -8, 2)$. **center (1, 1, 7); radius = $5\sqrt{11}$**

35. **GAMES** The object of a video game is to move a rectangular prism around to fit with other solids. The prism has moved to combine with the red L-shaped solid. Write the translation that moved the prism to the new location. **$(x, y, z) \to (x + 2, y + 3, z - 5)$**

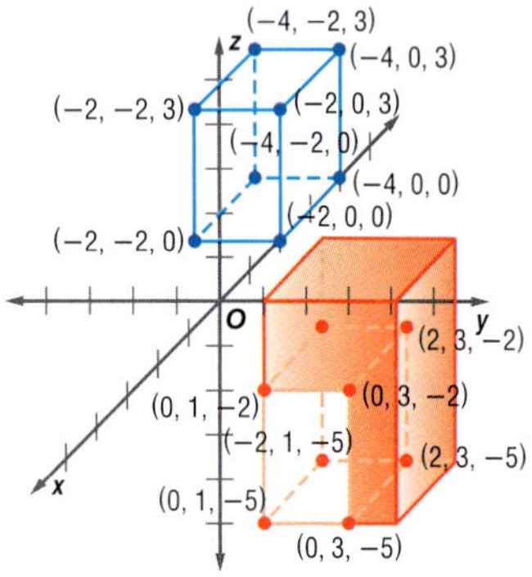

36. **CRITICAL THINKING** A sphere with its center at $(2, 4, 6)$ and a radius of 4 units is inscribed in a cube. Graph the cube and determine the coordinates of the vertices. **See margin.**

Enrichment, p. 752

Planes and Cylindrical Surfaces

Consider the points (x, y, z) in space whose coordinates satisfy the equation $z = 1$. Since x and y do not occur in the equation, any point with its z-coordinate equal to 1 has coordinates that satisfy the equation. These are the points in the plane 1 unit above the xy-plane. This plane is perpendicular to the z-axis at $(0, 0, 1)$.

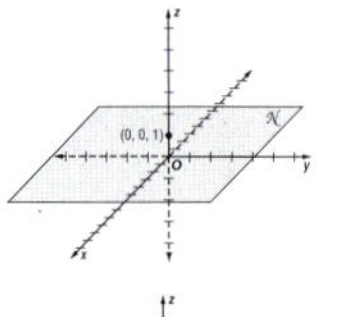

Next consider the points (x, y, z) whose coordinates satisfy $x^2 + y^2 = 16$. In the xy-plane, all points on the circle with center $(0, 0, 0)$ and radius 4 have coordinates that satisfy the equation. In the plane perpendicular to the z-axis at (0, … the points that sati… the …

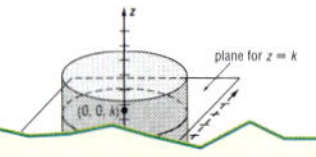

Answers

25. $P'(0, 2, -2)$, $Q'(0, 5, -2)$, $R'(2, 5, -2)$, $S'(2, 2, -2)$ $T'(0, 5, -5)$, $U'(0, 2, -5)$, $V'(2, 2, -5)$, and $W'(2, 5, -5)$

26. $A'(0, 1, 0)$, $B'(0, 1, -1)$, $C'(0, 2, -1)$, $D'(0, 2, 0)$, $E'(-2, 1, 0)$, $F'(-2, 2, 0)$, $G'(-2, 2, -1)$, and $H'(-2, 1, -1)$

37. **WRITING IN MATH** Answer the question that was posed at the beginning of the lesson. **See margin.**

How is three-dimensional graphing used in computer animation?

Include the following in your answer:

- the purpose of using an ordered triple, and
- why three-dimensional graphing is used instead of two-dimensional graphing.

FCAT Practice
Standardized Test Practice

38. The center of a sphere is at (4, −5, 3), and the endpoint of a diameter is at (5, −4, −2). What are the coordinates of the other endpoint of the diameter? **C**

Ⓐ (−1, −1, 5) Ⓑ $\left(-\frac{1}{2}, -\frac{1}{2}, \frac{5}{2}\right)$ Ⓒ (3, −6, 8) Ⓓ (13, −14, 4)

39. **ALGEBRA** Solve $\sqrt{x+1} = x - 1$. **B**

Ⓐ 0 and 3 Ⓑ 3 Ⓒ −2 and 1 Ⓓ −3 and 0

Extending the Lesson

LOCUS The locus of points in space with coordinates that satisfy the equation $y = 2x - 6$ is a plane perpendicular to the xy-plane whose intersection with the xy-plane is the graph of $y = 2x - 6$ in the xy-plane.

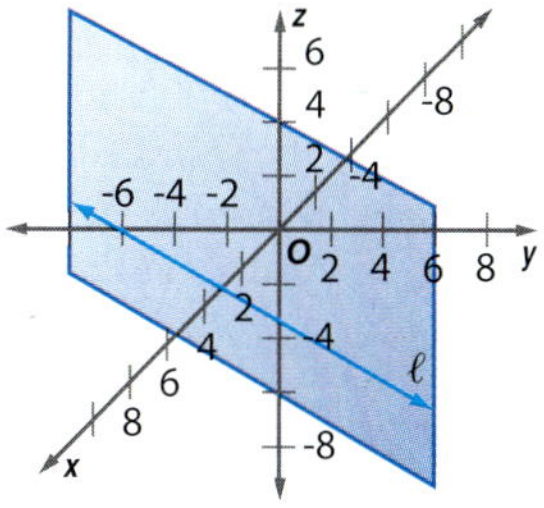

40–41. See margin.

40. Describe the locus of points in space that satisfy the equation $x + y = -5$.

41. Describe the locus of points in space that satisfy the equation $x + z = 4$.

Maintain Your Skills

Mixed Review **Determine whether each pair of solids are *similar, congruent,* or *neither.*** *(Lesson 13-4)*

42. **neither**

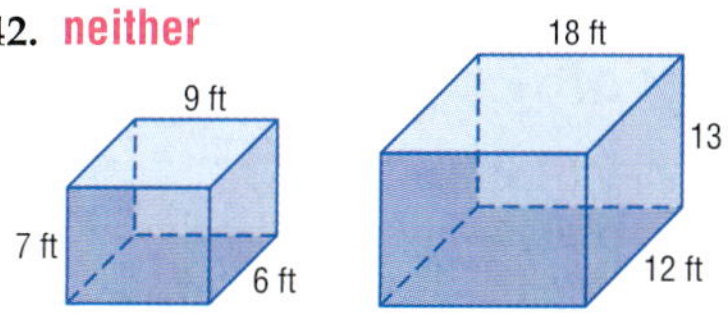

43. **similar**

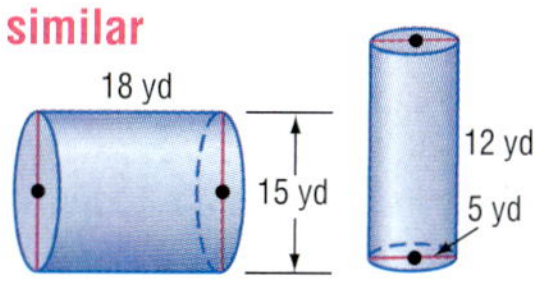

Find the volume of a sphere having the given radius or diameter. Round to the nearest tenth. *(Lesson 13-3)*

44. $r = 10$ cm **4188.8 cm³**

45. $d = 13$ yd **1150.3 yd³**

46. $r = 17.2$ m **21,314.4 m³**

47. $d = 29$ ft **12,770.1 ft³**

WebQuest Internet Project

Town With Major D-Day Losses Gets Memorial

It's time to complete your project. Use the information and data you have gathered about designing your memorial. Add some additional data or pictures to your portfolio or Web page. Be sure to include your scale drawings and calculations in the presentation.

www.geometryonline.com/webquest

www.geometryonline.com/self_check_quiz/fcat

4 Assess

Open-Ended Assessment

Modeling Have students construct and label a three-dimensional coordinate system with cardboard or with different colored wire or pipe cleaners. Then have them demonstrate locating points in the system.

Assessment Options

Quiz (Lesson 5) is available on p. 768 of the *Chapter 13 Resource Masters*.

Answers

37. Sample answer: Three-dimensional graphing is used in computer animation to render images and allow them to move realistically. Answers should include the following.
- Ordered triples are a method of locating and naming points in space. An ordered triple is unique to one point.
- Applying transformations to points in space would allow an animator to create realistic movement in animation.

40. The locus of points in space with coordinates that satisfy the equation $x + y = -5$ is a plane perpendicular to the xy-plane whose intersection with the xy-plane is the graph of $y = -x - 5$ in the xy-plane.

41. The locus of points in space with coordinates that satisfy the equation $x + z = 4$ is a plane perpendicular to the xz-plane whose intersection with the xz-plane is the graph of $z = -x + 4$ in the xz-plane.

Answer (page 718)

36. $A(-2, 0, 2)$, $B(6, 0, 2)$, $C(6, 8, 2)$, $D(-2, 8, 2)$, $E(-2, 8, 10)$, $F(6, 8, 10)$, $G(6, 0, 10)$, and $H(-2, 0, 10)$;

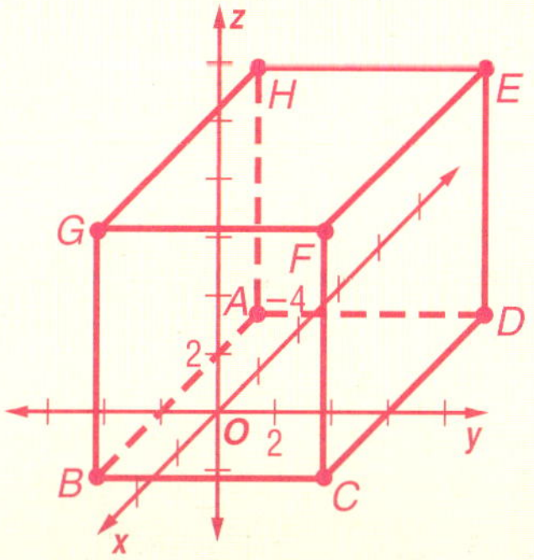

Chapter 13 Study Guide and Review

Vocabulary and Concept Check

- This alphabetical list of vocabulary terms in Chapter 13 includes a page reference where each term was introduced.
- **Assessment** A vocabulary test/review for Chapter 13 is available on p. 766 of the *Chapter 13 Resource Masters.*

Lesson-by-Lesson Review

For each lesson,

- the main ideas are summarized,
- additional examples review concepts, and
- practice exercises are provided.

Vocabulary PuzzleMaker

ELL The Vocabulary PuzzleMaker software improves students' mathematics vocabulary using four puzzle formats—crossword, scramble, word search using a word list, and word search using clues. Students can work on a computer screen or from a printed handout.

MindJogger Videoquizzes

ELL MindJogger Videoquizzes provide an alternative review of concepts presented in this chapter. Students work in teams in a game show format to gain points for correct answers. The questions are presented in three rounds.

Round 1 Concepts (5 questions)
Round 2 Skills (4 questions)
Round 3 Problem Solving (4 questions)

Chapter 13 Study Guide and Review

Vocabulary and Concept Check

congruent solids (p. 707) ordered triple (p. 714) similar solids (p. 707) volume (p. 688)

A complete list of postulates and theorems can be found on pages R1–R8.

Exercises **Complete each sentence with the correct italicized term.**

1. You can use $V = \frac{1}{3}Bh$ to find the volume of a (*prism*, *pyramid*).
2. (*Similar*, *Congruent*) solids always have the same volume.
3. Every point in space can be represented by (*an ordered triple*, *an ordered pair*).
4. $V = \pi r^2 h$ is the formula for the volume of a (*sphere*, *cylinder*).
5. In (*similar*, *congruent*) solids, if $a \neq b$ and $a : b$ is the ratio of the lengths of corresponding edges, then $a^3 : b^3$ is the ratio of the volumes.
6. The formula $V = Bh$ is used to find the volume of a (*prism*, *pyramid*).
7. To find the length of an edge of a pyramid, you can use (*the Distance Formula in Space*, *Cavalieri's Principle*).
8. You can use $V = \frac{4}{3}\pi r^3$ to find the volume of a (*cylinder*, *sphere*).
9. To find the volume of an oblique pyramid, you can use (*Cavalieri's Principle*, *the Distance Formula in Space*).
10. The formula $V = \frac{1}{3}Bh$ is used to find the volume of a (*cylinder*, *cone*).

Lesson-by-Lesson Review

13-1 Volumes of Prisms and Cylinders

See pages 688–694.

Concept Summary

- The volumes of prisms and cylinders are given by the formula $V = Bh$.
- Cavalieri's Principle states that if two objects have the same height and the same cross-sectional area, then they have the same volume.

Example **Find the volume of the cylinder.**

$V = \pi r^2 h$ Volume of a cylinder

$= \pi(12^2)(5)$ $r = 12$ and $h = 5$

≈ 2261.9 Use a calculator.

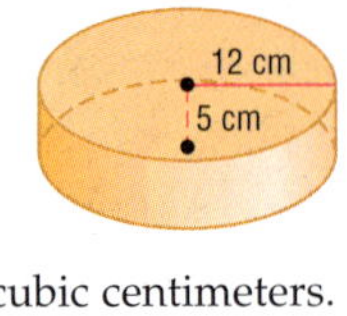

The volume is approximately 2261.9 cubic centimeters.

Exercises **Find the volume of each prism or cylinder. Round to the nearest tenth if necessary.** *See Examples 1 and 3 on pages 689 and 690.*

11.
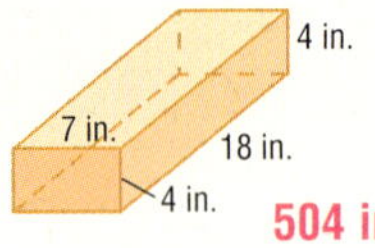

504 in^3

12.
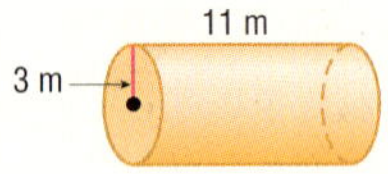

311.0 m^3

13.
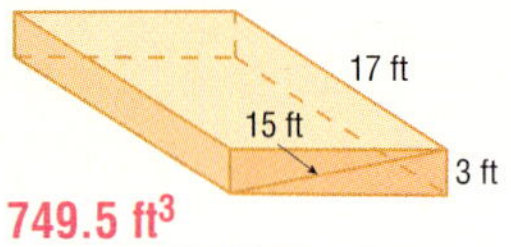

749.5 ft^3

www.geometryonline.com/vocabulary_review

FOLDABLES Study Organizer

For more information about Foldables, see *Teaching Mathematics with Foldables.*

Have students look through the chapter to make sure they have included notes and examples in their Foldables for each lesson of Chapter 13.

Encourage students to refer to their Foldables while completing the Study Guide and Review and to use them in preparing for the Chapter Test.

Study Guide and Review

13-2 Volumes of Pyramids and Cones

See pages 696–701.

Concept Summary

- The volume of a pyramid is given by the formula $V = \frac{1}{3}Bh$.
- The volume of a cone is given by the formula $V = \frac{1}{3}\pi r^2 h$.

Example **Find the volume of the square pyramid.**

$V = \frac{1}{3}Bh$ Volume of a pyramid

$= \frac{1}{3}(21^2)(19)$ $B = 21^2$ and $h = 19$

$= 2793$ Simplify.

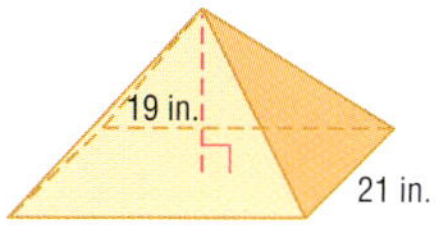

The volume of the pyramid is 2793 cubic inches.

Exercises **Find the volume of each pyramid or cone. Round to the nearest tenth.**

See Examples 1 and 2 on pages 697 and 698.

14. 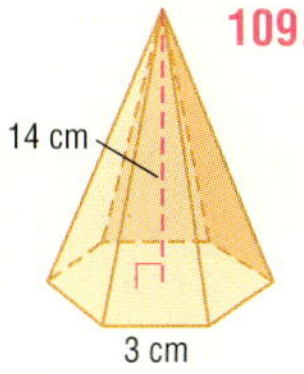

109.1 cm^3

15. 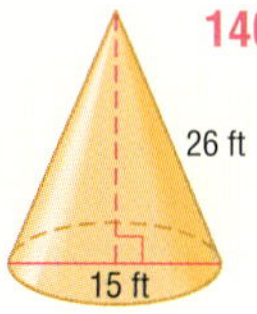

1466.4 ft^3

16. 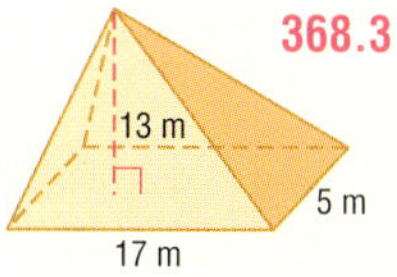

368.3 m^3

13-3 Volume of Spheres

See pages 702–706.

Concept Summary

- The volume of a sphere is given by the formula $V = \frac{4}{3}\pi r^3$.

Example **Find the volume of the sphere.**

$V = \frac{4}{3}\pi r^3$ Volume of a sphere

$= \frac{4}{3}\pi(5^3)$ $r = 5$

≈ 523.6 Use a calculator.

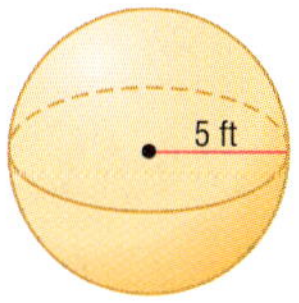

The volume of the sphere is about 523.6 cubic feet.

Exercises **Find the volume of each sphere. Round to the nearest tenth.**

See Example 1 on page 703.

17. The radius of the sphere is 2 feet. 33.5 ft^3

18. The diameter of the sphere is 4 feet. 33.5 ft^3

19. The circumference of the sphere is 65 millimeters. 4637.6 mm^3

20. The surface area of the sphere is 126 square centimeters. 133.0 cm^3

21. The area of a great circle of the sphere is 25π square units. 523.6 $units^3$

Chapter 13 Study Guide and Review 721

Answers

24. $AB = 10$; $(-1, -8, 1)$
25. $CD = \sqrt{58}$; $(-9, 5.5, 5.5)$
26. $EO = \sqrt{66}$; $(-2, 2.5, 2.5)$
27. $FG = \sqrt{422}$; $(1.5\sqrt{2}, 3\sqrt{7}, -3)$

Answers (page 723)

19. $\sqrt{34}$; $(0, -1.5, 2.5)$
20. $\sqrt{126}$; $(-0.5, 5, -2.5)$
21. $\sqrt{155}$; $(4.5, 2.5, -3.5)$
22. $\sqrt{86}$; $(-2.5, -1.5, -1)$
23. $2\sqrt{107}$; $(0, -2, 5)$
24. $\sqrt{323}$; $(2.5, -0.5, 5.5)$

13-4 Congruent and Similar Solids

See pages 707–713.

Concept Summary

- Similar solids have the same shape, but not necessarily the same size.
- Congruent solids are similar solids with a scale factor of 1.

Example **Determine whether the two cylinders are *congruent*, *similar*, or *neither*.**

$\frac{\text{diameter of larger cylinder}}{\text{diameter of smaller cylinder}} = \frac{6}{3}$ Substitution

$= 2$ Simplify.

$\frac{\text{height of larger cylinder}}{\text{height of smaller cylinder}} = \frac{15}{5}$ Substitution

$= 3$ Simplify.

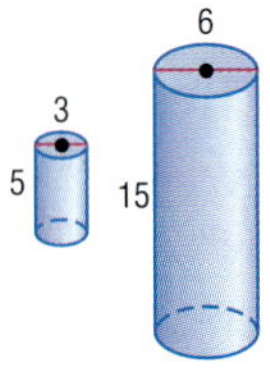

The ratios of the measures are not equal, so the cylinders are not similar.

Exercises **Determine whether the two solids are *congruent*, *similar*, or *neither*.**
See Example 1 on page 708.

22. **congruent**

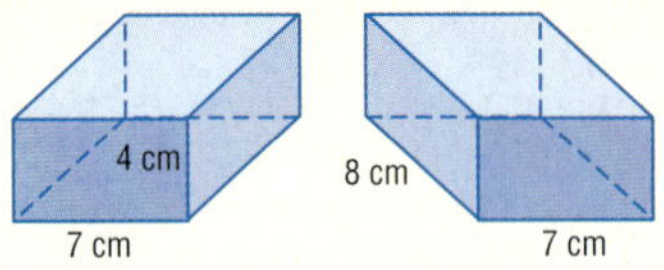

23. **similar**

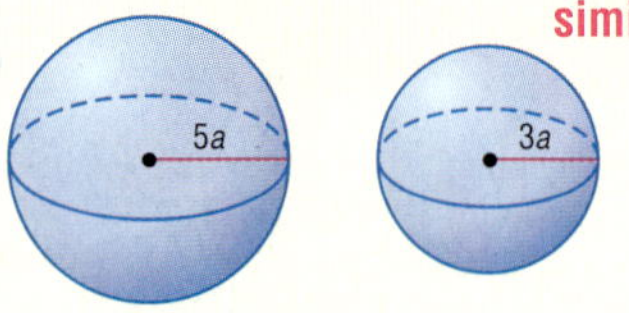

13-5 Coordinates in Space

See pages 714–719.

Concept Summary

- The Distance Formula in Space is $d = \sqrt{(x_2 - x_1)^2 + (y_2 - y_1)^2 + (z_2 - z_1)^2}$.
- Given $A(x_1, y_1, z_1)$ and $B(x_2, y_2, z_2)$, the midpoint of $\overline{AB}$ is at $\left(\frac{x_1 + x_2}{2}, \frac{y_1 + y_2}{2}, \frac{z_1 + z_2}{2}\right)$.

Example **Consider $\triangle ABC$ with vertices $A(13, 7, 10)$, $B(17, 18, 6)$, and $C(15, 10, 10)$. Find the length of the median from A to $\overline{BC}$ of ABC.**

$M = \left(\frac{17 + 15}{2}, \frac{18 + 10}{2}, \frac{6 + 10}{2}\right)$ Formula for the midpoint of $\overline{BC}$

$= (16, 14, 8)$ Simplify.

$\overline{AM}$ is the desired median, so AM is the length of the median.

$AM = \sqrt{(16 - 13)^2 + (14 - 7)^2 + (8 - 10)^2}$ or $\sqrt{62}$ Distance Formula in Space

Exercises **Determine the distance between each pair of points. Then determine the coordinates of the midpoint M of the segment joining the pair of points.**
See Example 2 on page 715. **24–27. See margin.**

24. $A(-5, -8, -2)$ and $B(3, -8, 4)$
25. $C(-9, 2, 4)$ and $D(-9, 9, 7)$
26. $E(-4, 5, 5)$ and the origin
27. $F(5\sqrt{2}, 3\sqrt{7}, 6)$ and $G(-2\sqrt{2}, 3\sqrt{7}, -12)$

722 Chapter 13 Volume

Chapter 13

Practice Test

Vocabulary and Concepts

Write the letter of the formula used to find the volume of each of the following figures.

1. right cylinder **b**
2. right pyramid **c**
3. sphere **a**

a. $V = \frac{4}{3}\pi r^3$
b. $V = \pi r^2 h$
c. $V = \frac{1}{3}Bh$

Skills and Applications

Find the volume of each solid. Round to the nearest tenth if necessary.

4. 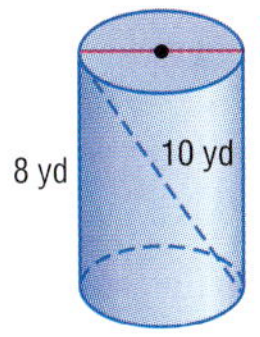

226.2 yd³

5. 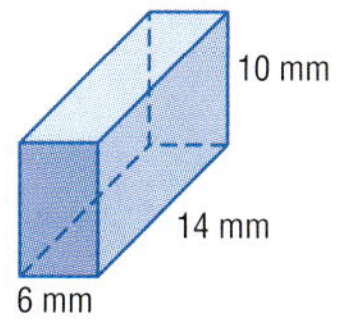

840 mm³

6. 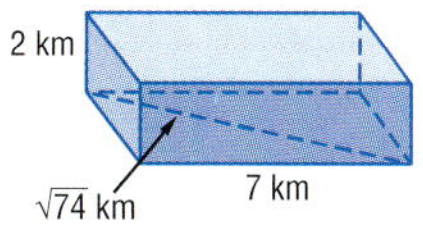

70 km³

7. 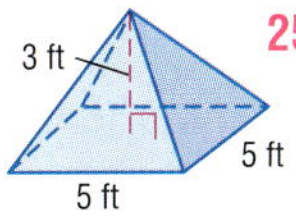

25 ft³

8. 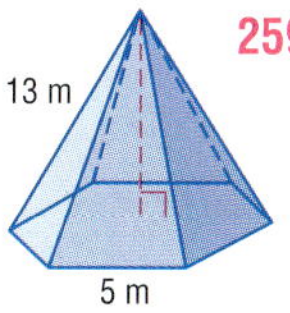

259.8 m³

9. 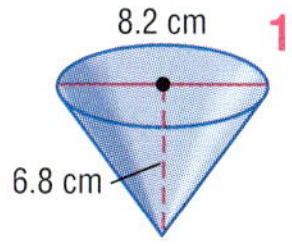

119.7 cm³

10. $C = 22\pi$ 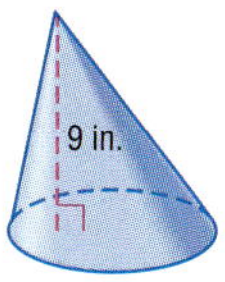

1140.4 in³

11. **SPORTS** The diving pool at the Georgia Tech Aquatic Center was used for the springboard and platform diving competitions of the 1996 Olympic Games. The pool is 78 feet long and 17 feet deep, and it is 110.3 feet from one corner on the surface of the pool to the opposite corner on the surface. If it takes about 7.5 gallons of water to fill one cubic foot of space, approximately how many gallons of water are needed to fill the diving pool? **775,588 gal**

Find the volume of each sphere. Round to the nearest tenth.

12. The radius has a length of 3 cm. **113.1 cm³**
13. The circumference of the sphere is 34 ft. **663.7 ft³**
14. The surface area of the sphere is 184 in². **234.7 in³**
15. The area of a great circle is 157 mm². **1479.8 mm³**

The two cylinders at the right are similar.

16. Find the ratio of the radii of the bases of the cylinders. **3 : 2**
17. What is the ratio of the surface areas? **9 : 4**
18. What is the ratio of the volumes? **27 : 8**

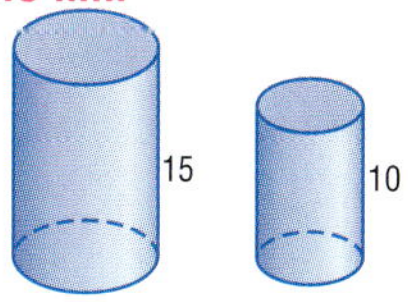

Determine the distance between each pair of points in space. Then determine the coordinates of the midpoint *M* of the segment joining the pair of points. **19–24. See margin.**

19. the origin and (0, −3, 5)
20. the origin and (−1, 10, −5)
21. the origin and (9, 5, −7)
22. (−2, 2, 2) and (−3, −5, −4)
23. (9, 3, 4) and (−9, −7, 6)
24. (8, −6, 1) and (−3, 5, 10)

25. **STANDARDIZED TEST PRACTICE** A rectangular prism has a volume of 360 cubic feet. If the prism has a length of 15 feet and a height of 2 feet, what is the width? **C**
Ⓐ 30 ft Ⓑ 24 ft Ⓒ 12 ft Ⓓ 7.5 ft

www.geometryonline.com/chapter_test/fcat

Portfolio Suggestion

Introduction A greenhouse is a closed structure that has air, temperature, and humidity control.

Ask Students Ask students to design a greenhouse that is a right rectangular prism with a pyramid for a roof. They should make the base of the pyramid the same shape as the base of the prism and include all dimensions. Ask them to find the volume of the greenhouse.

Chapter 13

Practice Test

Assessment Options

Vocabulary Test A vocabulary test/review for Chapter 13 can be found on p. 766 of the *Chapter 13 Resource Masters.*

Chapter Tests There are six Chapter 13 Tests and an Open-Ended Assessment task available in the *Chapter 13 Resource Masters.*

Chapter 13 Tests			
Form	**Type**	**Level**	**Pages**
1	MC	basic	753–754
2A	MC	average	755–756
2B	MC	average	757–758
2C	FR	average	759–760
2D	FR	average	761–762
3	FR	advanced	763–764

MC = multiple-choice questions
FR = free-response questions

Open-Ended Assessment Performance tasks for Chapter 13 can be found on p. 765 of the *Chapter 13 Resource Masters.* A sample scoring rubric for these tasks appears on p. A22.

Unit 4 Test A unit test/review can be found on pp. 773–774 of the *Chapter 13 Resource Masters.*

End-of-Year Tests A Second Semester Test for Chapters 8–13 and a Final Test for Chapters 1–13 can be found on pp. 775–784 of the *Chapter 13 Resource Masters.*

TestCheck and Worksheet Builder

This **networkable software** has three modules for assessment.

- **Worksheet Builder** to make worksheets and tests.
- **Student Module** to take tests on-screen.
- **Management System** to keep student records.

Chapter 13 Standardized Test Practice

These two pages contain practice questions in the various formats that can be found on the most frequently given standardized tests.

A practice answer sheet for these two pages can be found on p. A1 of the *Chapter 13 Resource Masters*.

Standardized Test Practice Student Recording Sheet, p. A1

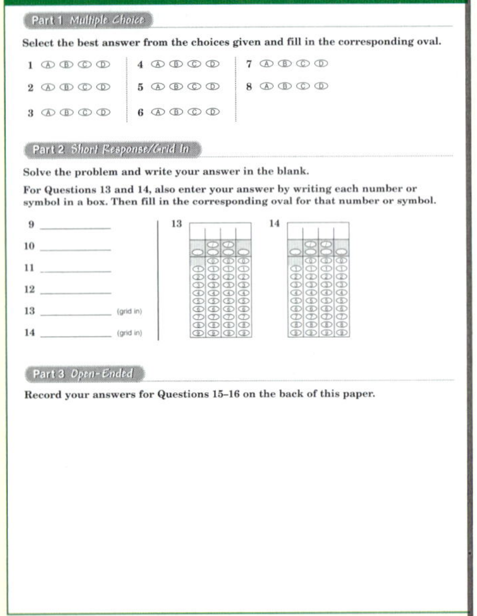
Part 1 Multiple Choice

Select the best answer from the choices given and fill in the corresponding oval.

1 Ⓐ Ⓑ Ⓒ Ⓓ 4 Ⓐ Ⓑ Ⓒ Ⓓ 7 Ⓐ Ⓑ Ⓒ Ⓓ
2 Ⓐ Ⓑ Ⓒ Ⓓ 5 Ⓐ Ⓑ Ⓒ Ⓓ 8 Ⓐ Ⓑ Ⓒ Ⓓ
3 Ⓐ Ⓑ Ⓒ Ⓓ 6 Ⓐ Ⓑ Ⓒ Ⓓ

Part 2 Short Response/Grid In

Solve the problem and write your answer in the blank.

For Questions 13 and 14, also enter your answer by writing each number or symbol in a box. Then fill in the corresponding oval for that number or symbol.

9 ______
10 ______
11 ______
12 ______
13 ______ (grid in)
14 ______ (grid in)

Part 3 Open-Ended

Record your answers for Questions 15–16 on the back of this paper.

Additional Practice

See pp. 771–772 in the *Chapter 13 Resource Masters* for additional standardized test practice.

Chapter 13 Standardized Test Practice

FCAT Practice

Part 1 Multiple Choice

Record your answers on the answer sheet provided by your teacher or on a sheet of paper.

1. *ABCD* is a rectangle. What is the relationship between $\angle ACD$ and $\angle ACB$? (Lesson 1-6) **A**

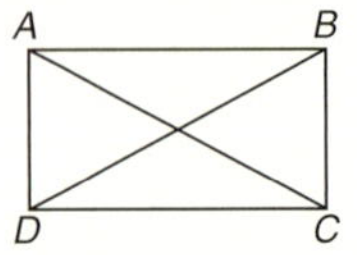

Ⓐ They are complementary angles.
Ⓑ They are perpendicular angles.
Ⓒ They are supplementary angles.
Ⓓ They are corresponding angles.

2. What is the measure of $\angle DEF$? (Lesson 4-2) **B**

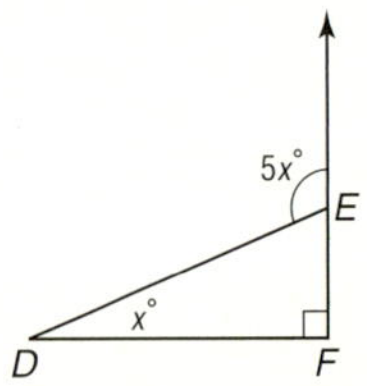

Ⓐ 22.5 Ⓑ 67.5 Ⓒ 112.5 Ⓓ 157.5

3. Two sides of a triangle measure 13 and 21 units. Which could be the measure of the third side? (Lesson 5-4) **C**

Ⓐ 5 Ⓑ 8 Ⓒ 21 Ⓓ 34

4. $\triangle QRS$ is similar to $\triangle TUV$. Which statement is true? (Lesson 6-3) **C**

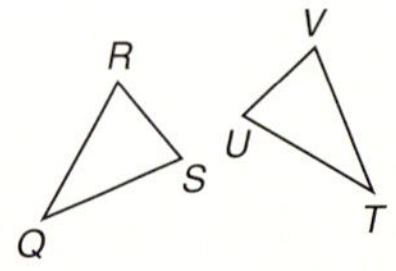

Ⓐ $m\angle Q = m\angle V$ Ⓑ $m\angle Q = m\angle S$
Ⓒ $m\angle Q = m\angle T$ Ⓓ $m\angle Q = m\angle U$

5. The wooden block shown must be able to slide onto a cylindrical rod. What is the volume of the block after the hole is drilled? Round to the nearest tenth. (Lesson 13-1) **B**

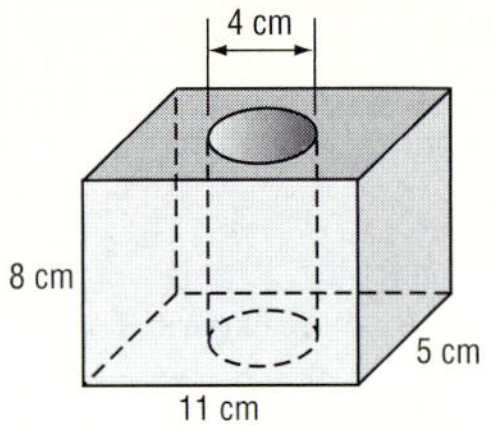

Ⓐ 100.5 cm^3 Ⓑ 339.5 cm^3
Ⓒ 402.0 cm^3 Ⓓ 440.0 cm^3

6. The circumference of a regulation soccer ball is 25 inches. What is the volume of the soccer ball to the nearest cubic inch? (Lesson 13-3) **B**

Ⓐ 94 in^3 Ⓑ 264 in^3
Ⓒ 333 in^3 Ⓓ 8177 in^3

7. If the two cylinders are similar, what is the volume of the larger cylinder to the nearest tenth of a cubic centimeter? (Lesson 13-4) **C**

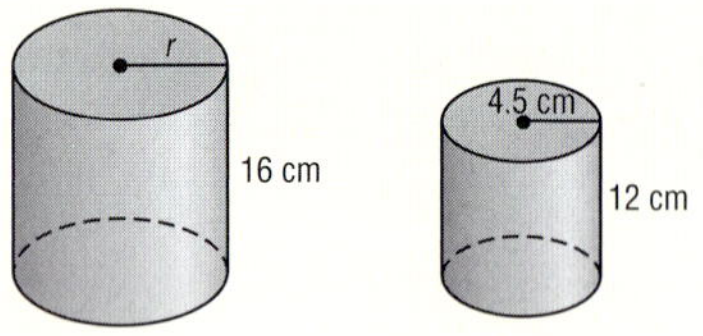

Ⓐ 730.0 cm^3 Ⓑ 1017.4 cm^3
Ⓒ 1809.6 cm^3 Ⓓ 2122.6 cm^3

8. The center of a sphere has coordinates (3, 1, 4). A point on the surface of the sphere has coordinates (9, −2, −2). What is the measure of the radius of the sphere? (Lesson 13-5) **D**

Ⓐ 7 Ⓑ $\sqrt{61}$
Ⓒ $\sqrt{73}$ Ⓓ 9

***Log On* for Test Practice**

The Princeton Review offers additional test-taking tips and practice problems at their web site. Visit www.princetonreview.com or www.review.com

TestCheck and Worksheet Builder

Special banks of standardized test questions similar to those on the SAT, ACT, TIMSS 8, NAEP 8, and state proficiency tests can be found on this CD-ROM.

FCAT Practice

Part 2 Short Response/Grid In

Record your answers on the answer sheet provided by your teacher or on a sheet of paper.

9. Find $\frac{12z^5 + 27z^2 - 6z}{3z}$. (Prerequisite Skill) $4z^4 + 9z - 2$

10. Sierra said, "If math is my favorite subject, then I like math." Carlos then said, "If I do not like math, then it is not my favorite subject." Carlos formed the __?__ of Sierra's statement. (Lesson 2-3) **contrapositive**

11. Describe the information needed about two triangles to prove that they are congruent by the SSS Postulate. (Lesson 4-1) **See margin.**

12. The figure is a regular octagon. Find x. (Lesson 8-1) **22.5**

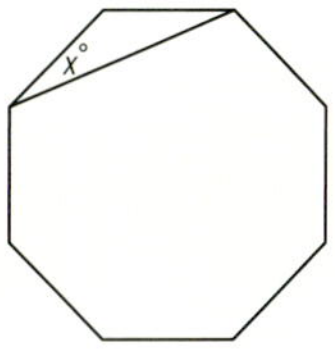

13. $ABCD$ is an isosceles trapezoid. What are the coordinates of A? (Lesson 8-7) $(-b, 0)$

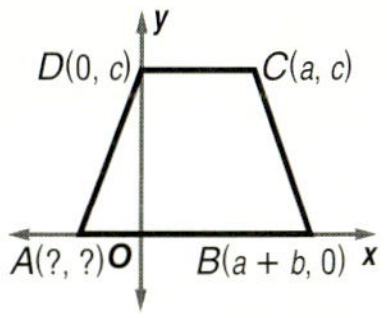

14. What is the volume of the cone? (Lesson 13-2) 600π cm^3

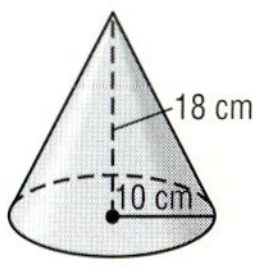

The Princeton Review Test-Taking Tip

Question 6 Sometimes more than one step is required to find the answer. In this question, you need to use the circumference formula, $C = 2\pi r$, to find the length of the radius. Then you can use the surface area formula, $T = 4\pi r^2$.

FCAT Practice

Part 3 Open Ended

Record your answers on a sheet of paper. Show your work.

15. A manufacturing company packages their product in the small cylindrical can shown in the diagram. During a promotion for the product, they doubled the height of the cans and sold them for the same price.

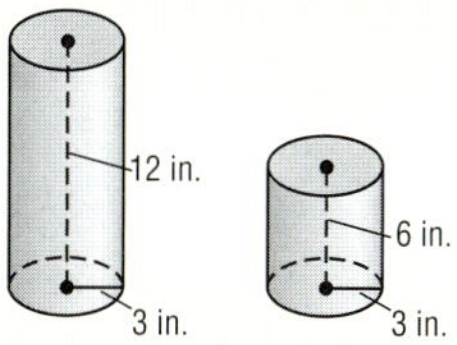

a. Find the surface area of each can. Explain the effect of doubling the height on the amount of material used to produce the can. (Lesson 12-4)

b. Find the volume of each can. Explain the effect doubling the height on the amount of product that can fit inside. (Lesson 13-1)
15a–15b. See margin.

16. Engineering students designed an enlarged external fuel tank for a space shuttle as part of an assignment.

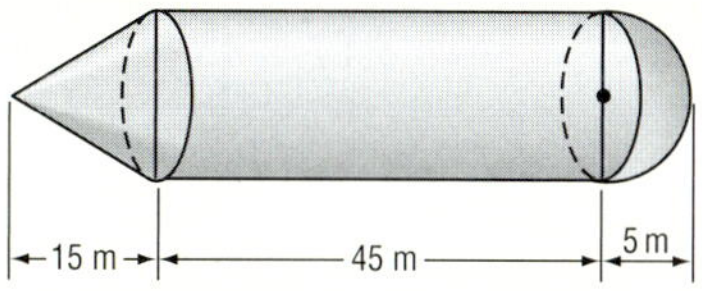

What is the volume of the entire fuel tank to the nearest cubic meter? Show your work. (Lesson 13-1) **4189 m^3**

www.geometryonline.com/standardized_test/fcat

Evaluating Open-Ended Assessment Questions

Open-Ended Assessment questions are graded by using a multilevel rubric that guides you in assessing a student's knowledge of a particular concept.

Goal: Explain the effect of changing a dimension of a solid on the surface area or volume of the solid.

Sample Scoring Rubric: The following rubric is a sample scoring device. You may wish to add more detail to this sample to meet your individual scoring needs.

Score	Criteria
4	A correct solution that is supported by well-developed, accurate explanations
3	A generally correct solution, but may contain minor flaws in reasoning or computation
2	A partially correct interpretation and/or solution to the problem
1	A correct solution with no supporting evidence or explanation
0	An incorrect solution indicating no mathematical understanding of the concept or task, or no solution is given

Answers

11. If the measures of the corresponding sides are the same, the triangles are congruent.

15a. The surface area of the small can is 54π in^2 and the surface area of the large can is 90π in^2. When the height is doubled, the lateral area of the cylinder is doubled, but the area of the bases remains the same. The surface area increases by a factor of $1\frac{2}{3}$ times.

15b. The volume of the small can is 54π in^3 and the volume of the larger can is 108π in^3. The volume increases by a factor of 2.

Chapter 13 Standardized Test Practice 725

Pages 717–719, Lesson 13-5

4.

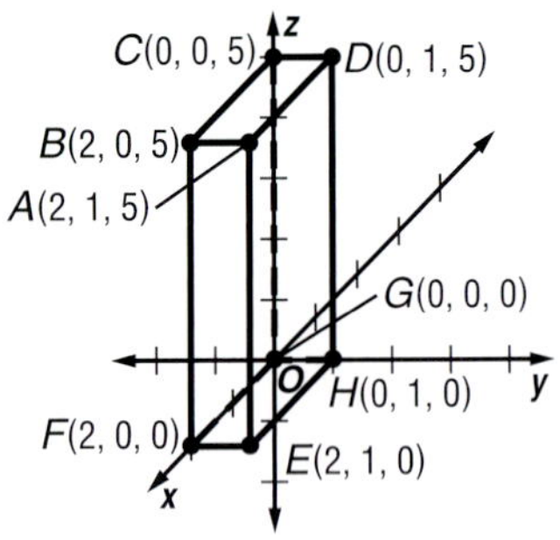

5.

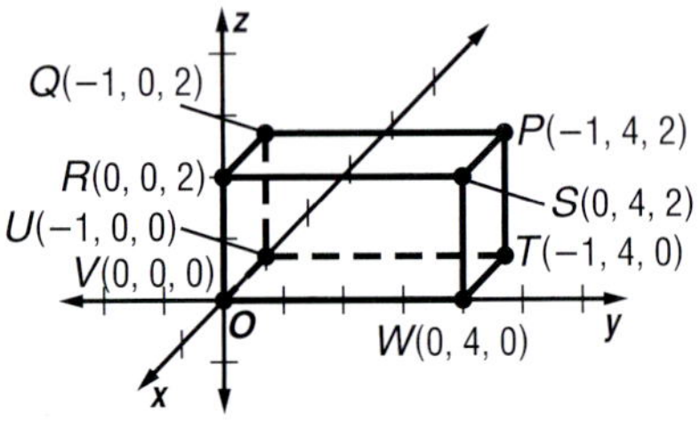

10.

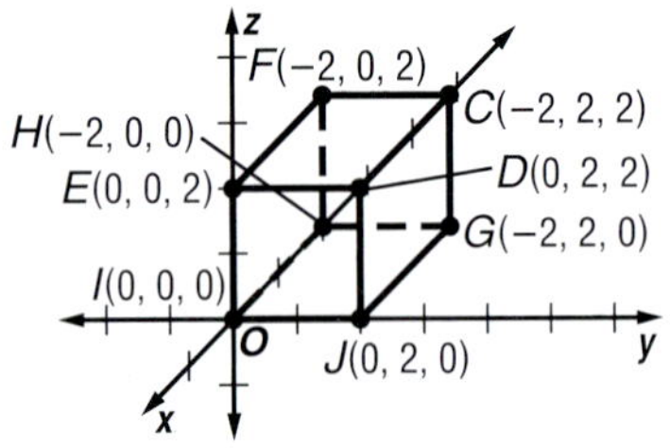

11.

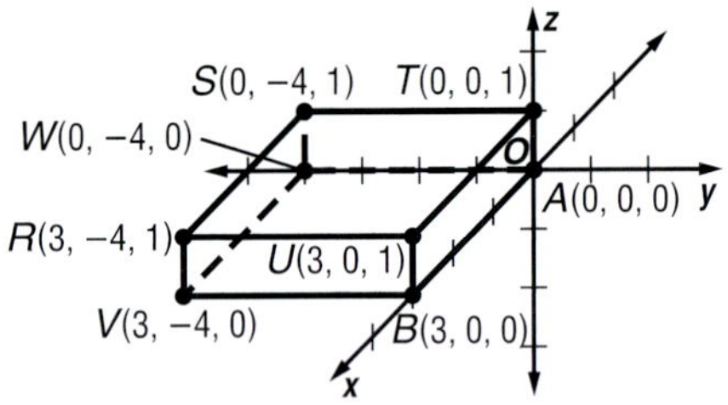

12.

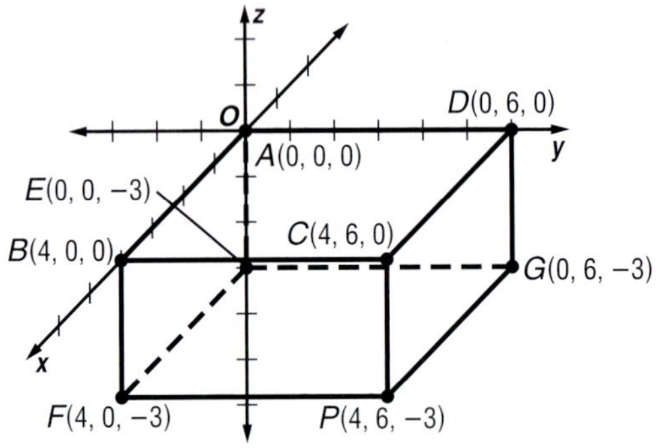

13.

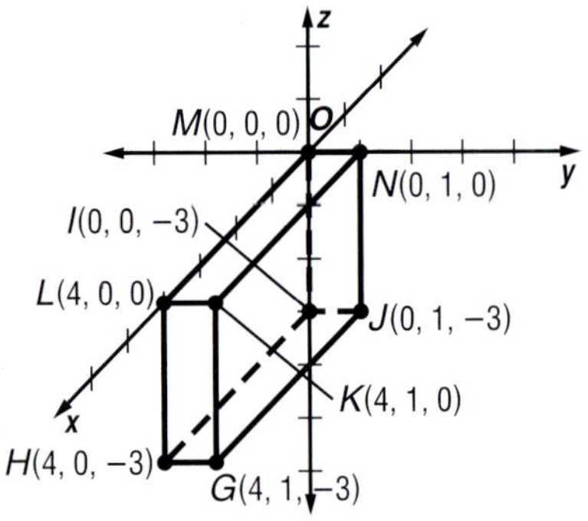

14.

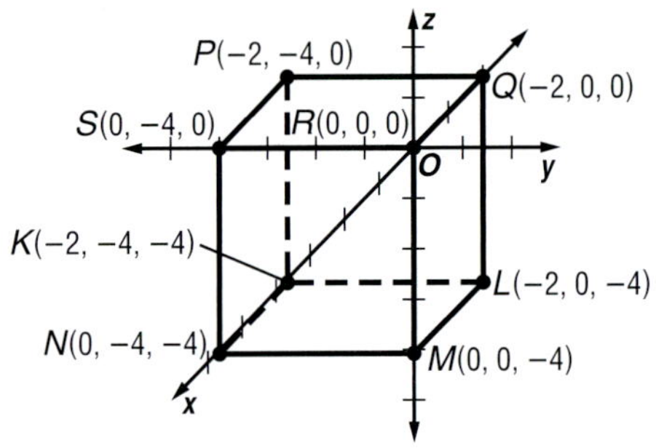

15.

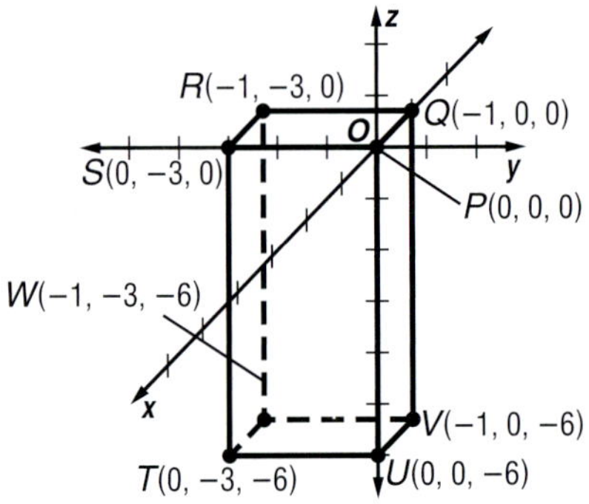

23.

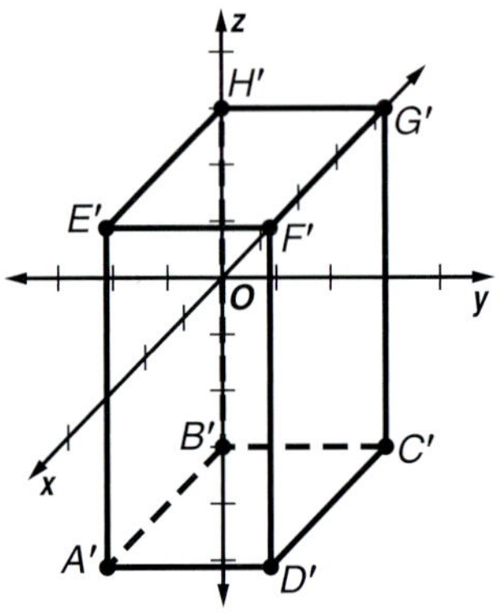

24.

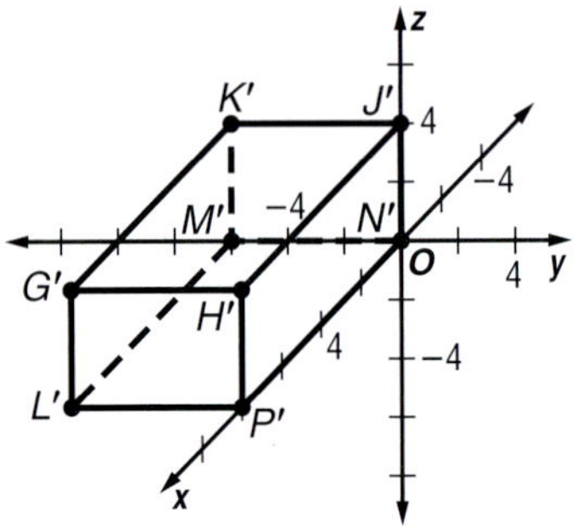

27. $A'(4, 5, 1)$, $B'(4, 2, 1)$, $C'(1, 2, 1)$, $D'(1, 5, 1)$, $E'(4, 5, -2)$, $F'(4, 2, -2)$, $G'(1, 2, -2)$, and $H'(1, 5, -2)$;

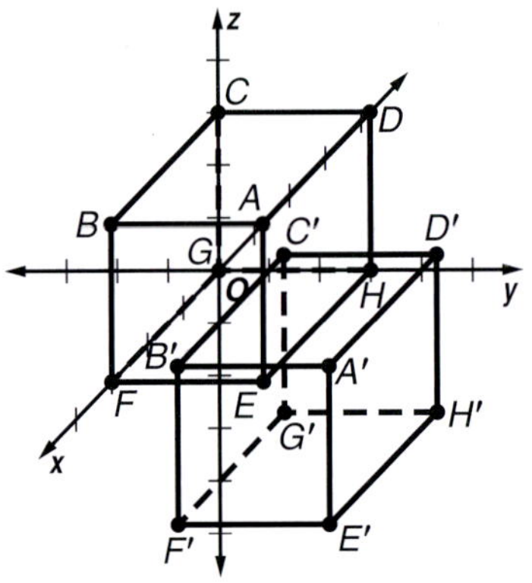

28. $A'(1, 0, 5)$, $B'(1, -3, 5)$, $C'(-2, -3, 5)$, $D'(-2, 0, 5)$, $E'(1, 0, 2)$, $F'(1, -3, 2)$, $G'(-2, -3, 2)$, and $H'(-2, 0, 2)$;

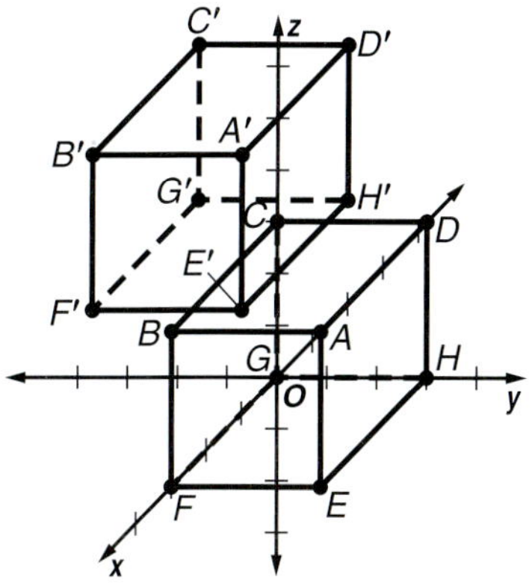

29. $A'(6, 6, 6)$, $B'(6, 0, 6)$, $C'(0, 0, 6)$, $D'(0, 6, 6)$, $E'(6, 6, 0)$, $F'(6, 0, 0)$, $G'(0, 0, 0)$, and $H'(0, 6, 0)$; $V = 216$ cubic units;

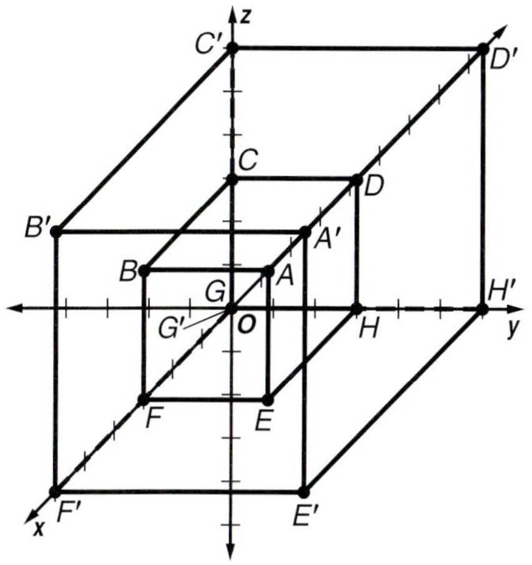

30. $A'(1, 1, 1)$, $B'(1, 0, 1)$, $C'(0, 0, 1)$, $D'(0, 1, 1)$, $E'(1, 1, 0)$, $F'(1, 0, 0)$, $G'(0, 0, 0)$, and $H'(0, 1, 0)$; $\frac{1}{27}$;

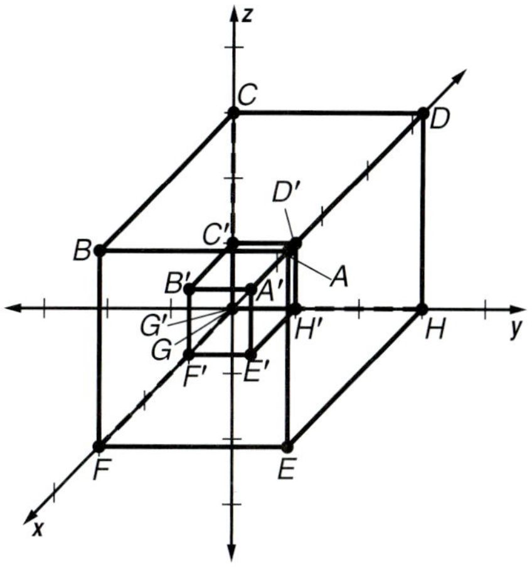

Additional Answers for Chapter 13

Student Handbook

Skills

Reference

Prerequisite Skills

Prerequisite Skills

Graphing Ordered Pairs

- Points in the coordinate plane are named by **ordered pairs** of the form (x, y). The first number, or **x-coordinate**, corresponds to a number on the x-axis. The second number, or **y-coordinate**, corresponds to a number on the y-axis.

Example 1 **Write the ordered pair for each point.**

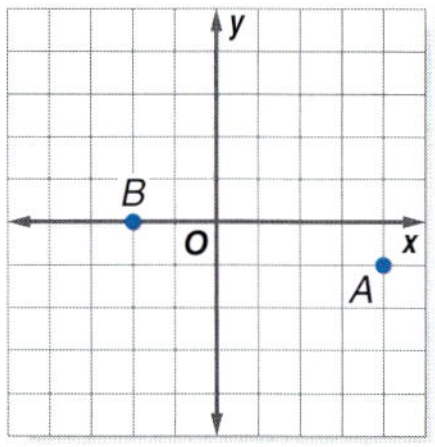

a. A
The x-coordinate is 4.
The y-coordinate is -1.
The ordered pair is $(4, -1)$.

b. B
The x-coordinate is -2.
The point lies on the x-axis, so its y-coordinate is 0.
The ordered pair is $(-2, 0)$.

- The x-axis and y-axis separate the coordinate plane into four regions, called **quadrants**. The point at which the axes intersect is called the **origin**. The axes and points on the axes are not located in any of the quadrants.

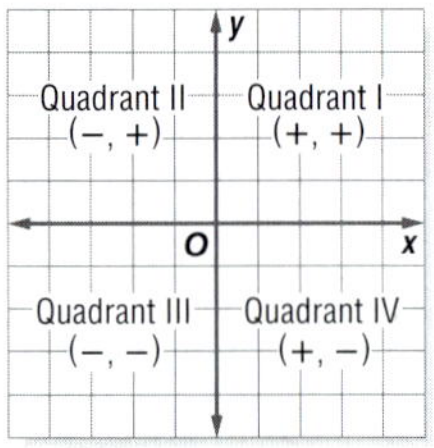

Example 2 **Graph and label each point on a coordinate plane. Name the quadrant in which each point is located.**

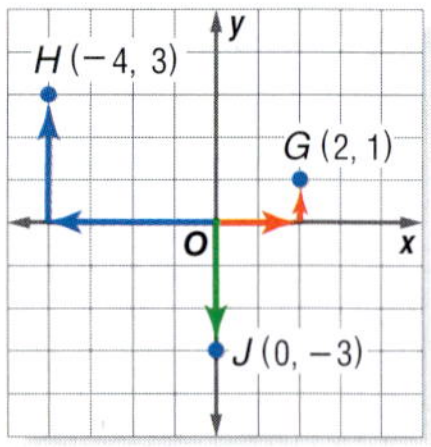

a. $G(2, 1)$
Start at the origin. Move 2 units right, since the x-coordinate is 2. Then move 1 unit up, since the y-coordinate is 1. Draw a dot, and label it G. Point $G(2, 1)$ is in Quadrant I.

b. $H(-4, 3)$
Start at the origin. Move 4 units left, since the x-coordinate is -4. Then move 3 units up, since the y-coordinate is 3. Draw a dot, and label it H. Point $H(-4, 3)$ is in Quadrant II.

c. $J(0, -3)$
Start at the origin. Since the x-coordinate is 0, the point lies on the y-axis. Move 3 units down, since the y-coordinate is -3. Draw a dot, and label it J. Because it is on one of the axes, point $J(0, -3)$ is not in any quadrant.

728 Prerequisite Skills

Example 3 Graph a polygon with vertices $A(-3, 3)$, $B(1, 3)$, $C(0, 1)$, and $D(-4, 1)$.

Graph the ordered pairs on a coordinate plane.
Connect each pair of consecutive points.
The polygon is a parallelogram.

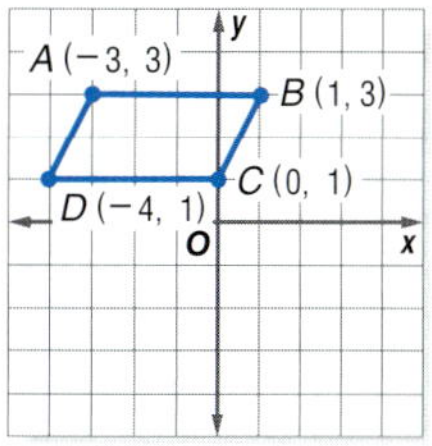

Example 4 Graph four points that satisfy the equation $y = 4 - x$.

Make a table.
Choose four values for x.
Evaluate each value of x for $4 - x$.

x	$4 - x$	y	(x, y)
0	4 − 0	4	(0, 4)
1	4 − 1	3	(1, 3)
2	4 − 2	2	(2, 2)
3	4 − 3	1	(3, 1)

Plot the points.

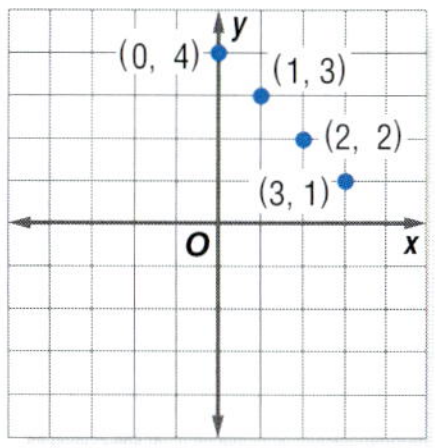

Exercises Write the ordered pair for each point shown at the right.

1. B **(−2, 3)** 2. C **(1, −1)** 3. D **(2, 2)**
4. E **(−3, −3)** 5. F **(−3, 1)** 6. G **(0, −3)**
7. H **(4, 1)** 8. I **(3, −2)** 9. J **(−1, −1)**
10. K **(1, 4)** 11. W **(3, 0)** 12. M **(−2, −4)**
13. N **(2, −4)** 14. P **(3, 3)** 15. Q **(−4, 2)**

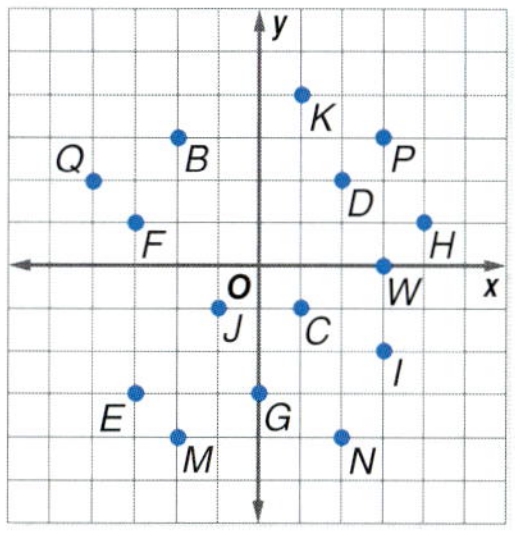

Graph and label each point on a coordinate plane. Name the quadrant in which each point is located. **16–31. See margin for graph.**

16. $M(-1, 3)$ **II** 17. $S(2, 0)$ **none** 18. $R(-3, -2)$ **III** 19. $P(1, -4)$ **IV**
20. $B(5, -1)$ **IV** 21. $D(3, 4)$ **I** 22. $T(2, 5)$ **I** 23. $L(-4, -3)$ **III**
24. $A(-2, 2)$ **II** 25. $N(4, 1)$ **I** 26. $H(-3, -1)$ **III** 27. $F(0, -2)$ **none**
28. $C(-3, 1)$ **II** 29. $E(1, 3)$ **I** 30. $G(3, 2)$ **I** 31. $I(3, -2)$ **IV**

Graph the following geometric figures. **32–35. See margin.**

32. a square with vertices $W(-3, 3)$, $X(-3, -1)$, $Y(1, 3)$, and $Z(1, -1)$
33. a polygon with vertices $J(4, 2)$, $K(1, -1)$, $L(-2, 2)$, and $M(1, 5)$
34. a triangle with vertices $F(2, 4)$, $G(-3, 2)$, and $H(-1, -3)$
35. a rectangle with vertices $P(-2, -1)$, $Q(4, -1)$, $R(-2, 1)$, and $S(4, 1)$

Graph four points that satisfy each equation. **36–39. See margin for sample answers.**

36. $y = 2x$ 37. $y = 1 + x$ 38. $y = 3x - 1$ 39. $y = 2 - x$

Prerequisite Skills

34.

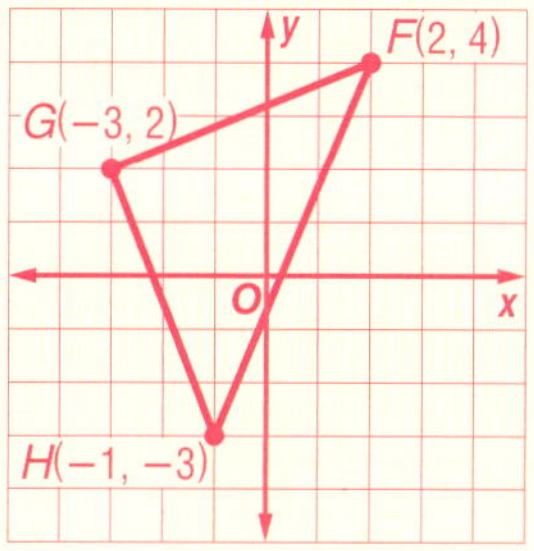

35.

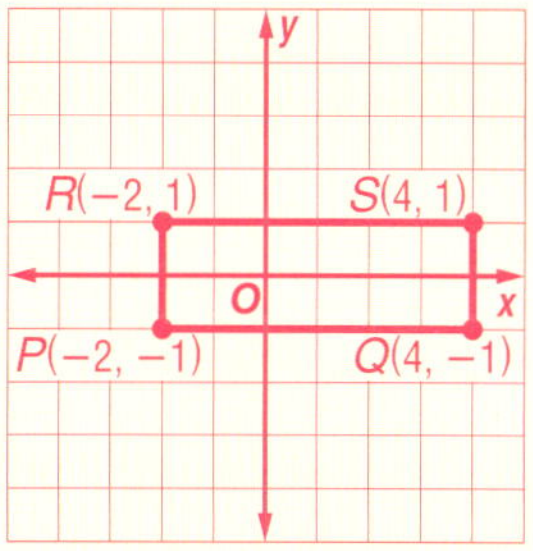

36.

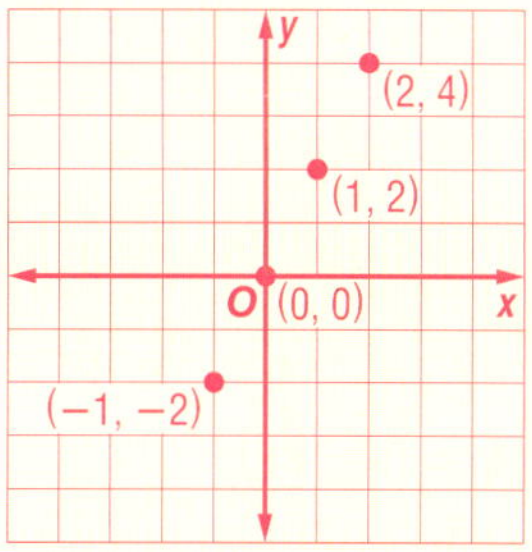

37.

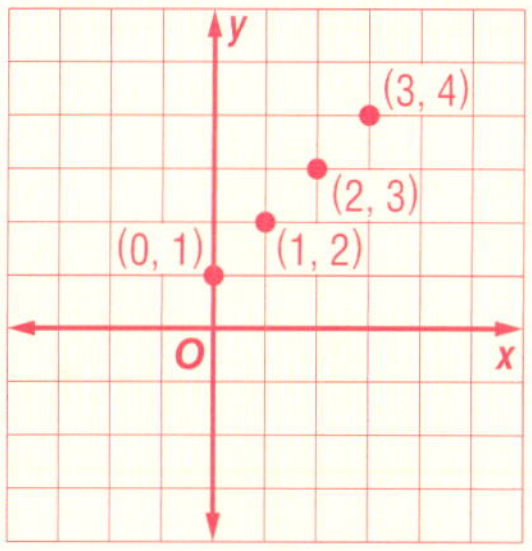

38.

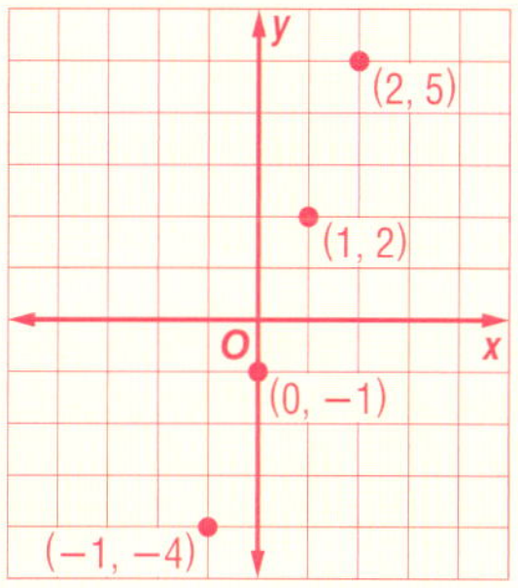

39.

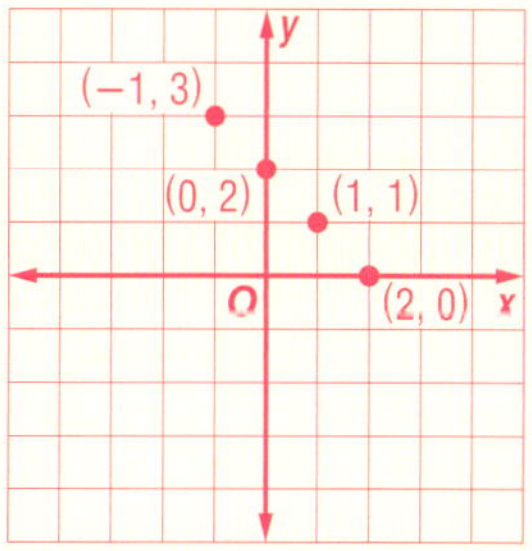

16–31.

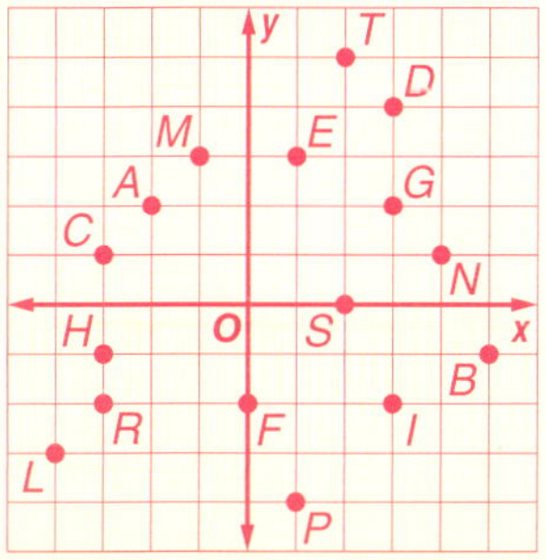

32.

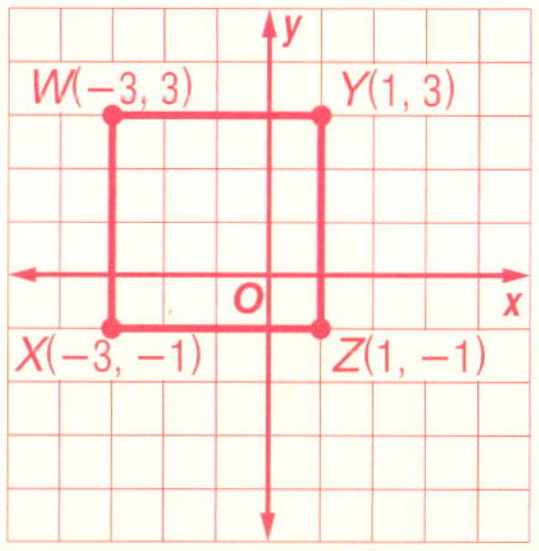

33.

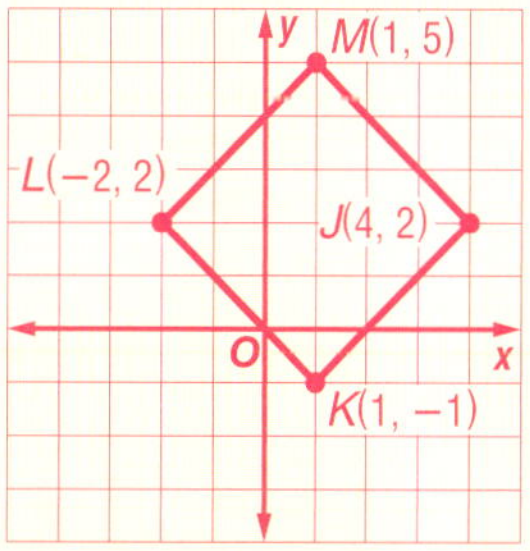

Prerequisite Skills

Changing Units of Measure within Systems

Metric Units of Length
1 kilometer (km) = 1000 meters (m)
1 m = 100 centimeters (cm)
1 cm = 10 millimeters (mm)

Customary Units of Length
1 foot (ft) = 12 inches (in.)
1 yard (yd) = 3 ft
1 mile (mi) = 5280 ft

- To convert from larger units to smaller units, multiply.
- To convert from smaller units to larger units, divide.

Example 1 **State which metric unit you would use to measure the length of your pen.**

Since a pen has a small length, the *centimeter* is the appropriate unit of measure.

Example 2 **Complete each sentence.**

a. 4.2 km = __?__ m

There are 1000 meters in a kilometer.

4.2 km × 1000 = 4200 m

b. 125 mm = __?__ cm

There are 10 millimeters in a centimeter.

125 mm ÷ 10 = 12.5 cm

c. 16 ft = __?__ in.

There are 12 inches in a foot.

16 ft × 12 = 192 in.

d. 39 ft = __?__ yd

There are 3 feet in a yard.

39 ft ÷ 3 = 13 yd

- Example 3 involves two-step conversions.

Example 3 **Complete each sentence.**

a. 17 mm = __?__ m

There are 100 centimeters in a meter. First change *millimeters* to *centimeters*.

17 mm = __?__ cm — smaller unit → larger unit

17 mm ÷ 10 = 1.7 cm — Since 10 mm = 1 cm, divide by 10.

Then change *centimeters* to *meters*.

1.7 cm = __?__ m — smaller unit → larger unit

1.7 cm ÷ 100 = 0.017 m — Since 100 cm = 1 m, divide by 100.

b. 6600 yd = __?__ mi

There are 5280 feet in one mile. First change *yards* to *feet*.

6600 yd = __?__ ft — larger unit → smaller unit

6600 yd × 3 = 19,800 ft — Since 3 ft = 1 yd, multiply by 3.

Then change *feet* to *miles*.

19,800 ft = __?__ mi — smaller unit → larger unit

19,800 ft ÷ 5280 = $3\frac{3}{4}$ or 3.75 mi — Since 5280 ft = 1 mi, divide by 5280.

Metric Units of Capacity
1 liter (L) = 1000 millimeters (mL)

Customary Units of Capacity	
1 cup (c) = 8 fluid ounces (fl oz)	1 quart (qt) = 2 pt
1 pint (pt) = 2 c	1 gallon (gal) = 4 qt

Example 4 **Complete each sentence.**

a. 3.7 L = __?__ mL

There are 1000 milliliters in a liter.

3.7 L × 1000 = 3700 mL

b. 16 qt = __?__ gal

There are 4 quarts in a gallon.

16 qt ÷ 4 = 4 gal

Prerequisite Skills

730 Prerequisite Skills

- Examples c and d involve two-step conversions.

c. 7 pt = __?__ fl oz

There are 8 fluid ounces in a cup.
First change *pints* to *cups*.

7 pt = __?__ c

7 pt × 2 = 14 c

Then change *cups* to *fluid ounces*.

14 c = __?__ fl oz

14 c × 8 = 112 fl oz

d. 4 gal = __?__ pt

There are 4 quarts in a gallon.
First change *gallons* to *quarts*.

4 gal = __?__ qt

4 gal × 4 = 16 qt

Then change *quarts* to *pints*.

16 qt = __?__ pt

16 qt × 2 = 32 pt

- The mass of an object is the amount of matter that it contains.

Metric Units of Mass
1 kilogram (kg) = 1000 grams (g)
1 g = 1000 milligrams (mg)

Customary Units of Weight
1 pound (lb) = 16 ounces (oz)
1 ton (T) = 2000 lb

Example 5 **Complete each sentence.**

a. 2300 mg = __?__ g

There are 1000 milligrams in a gram.

2300 mg ÷ 1000 = 2.3 g

b. 120 oz = __?__ lb

There are 16 ounces in a pound.

120 oz ÷ 16 = 7.5 lb

- Examples c and d involve two-step conversions.

c. 5.47 kg = __?__ mg

There are 1000 milligrams in a gram.
Change *kilograms* to *grams*.

5.47 kg = __?__ g

5.47 kg × 1000 = 5470 g

Then change *grams* to *milligrams*.

5470 g = __?__ mg

5470 g × 1000 = 5,470,000 mg

d. 5 T = __?__ oz

There are 16 ounces in a pound.
Change *tons* to *pounds*.

5 T = __?__ lb

5 T × 2000 = 10,000 lb

Then change *pounds* to *ounces*.

10,000 lb = __?__ oz

10,000 lb × 16 = 160,000 oz

Exercises

State which metric unit you would probably use to measure each item.

1. radius of a tennis ball cm
2. length of a notebook cm
3. mass of a textbook kg
4. mass of a beach ball g
5. width of a football field m
6. thickness of a penny mm
7. amount of liquid in a cup mL
8. amount of water in a bath tub L

Complete each sentence.

9. 120 in. = __?__ ft 10
10. 18 ft = __?__ yd 6
11. 10 km = __?__ m 10,000
12. 210 mm = __?__ cm 21
13. 180 mm = __?__ m 0.18
14. 3100 m = __?__ km 3.1
15. 90 in. = __?__ yd 2.5
16. 5280 yd = __?__ mi 3
17. 8 yd = __?__ ft 24
18. 0.62 km = __?__ m 620
19. 370 mL = __?__ L 0.370
20. 12 L = __?__ mL 12,000
21. 32 fl oz = __?__ c 4
22. 5 qt = __?__ c 20
23. 10 pt = __?__ qt 5
24. 48 c = __?__ gal 3
25. 4 gal = __?__ qt 16
26. 36 mg = __?__ g 0.036
27. 13 lb = __?__ oz 208
28. 130 g = __?__ kg 0.130
29. 9.05 kg = __?__ g 9050

Prerequisite Skills

Prerequisite Skills 731

Perimeter and Area of Rectangles and Squares

Perimeter is the distance around a figure whose sides are segments. Perimeter is measured in linear units.

Perimeter of a Rectangle	Perimeter of a Square
Words Multiply two times the sum of the length and width.	**Words** Multiply 4 times the length of a side.
Formula $P = 2(\ell + w)$	**Formula** $P = 4s$
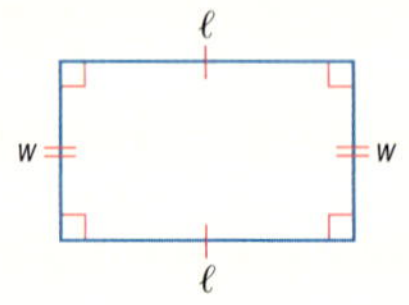	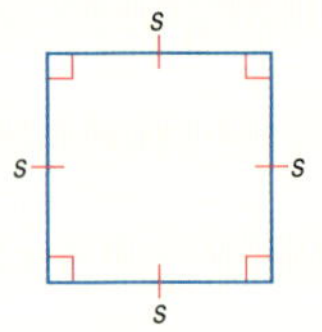

Area is the number of square units needed to cover a surface. Area is measured in square units.

Area of a Rectangle	Area of a Square
Words Multiply the length and width.	**Words** Square the length of a side.
Formula $A = \ell w$	**Formula** $A = s^2$
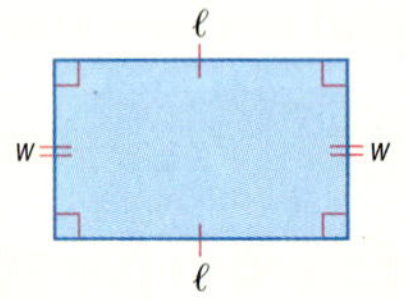	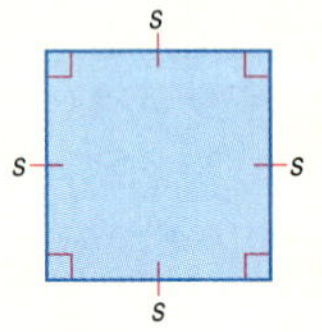

Example 1 **Find the perimeter and area of each rectangle.**

a.

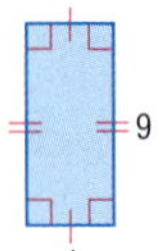

$P = 2(\ell + w)$ Perimeter formula

$= 2(4 + 9)$ Replace ℓ with 4 and w with 9.

$= 26$ Simplify.

$A = \ell w$ Area formula

$= 4 \cdot 9$ Replace ℓ with 4 and w with 9.

$= 36$ Multiply.

The perimeter is 26 units, and the area is 36 square units.

732 Prerequisite Skills

b. a rectangle with length 8 units and width 3 units.

$P = 2(\ell + w)$ Perimeter formula

$= 2(8 + 3)$ Replace ℓ with 8 and w with 3.

$= 22$ Simplify.

$A = \ell \cdot w$ Area formula

$= 8 \cdot 3$ Replace ℓ with 8 and w with 3.

$= 24$ Multiply

The perimeter is 22 units, and the area is 24 square units.

Example 2 **Find the perimeter and area of a square that has a side of length 14 feet.**

$P = 4s$ Perimeter formula

$= 4(14)$ $s = 14$

$= 56$ Multiply.

$A = s^2$ Area formula

$= 14^2$ $s = 14$

$= 196$ Multiply.

The perimeter is 56 feet, and the area is 196 square feet.

Exercises **Find the perimeter and area of each figure.**

1. 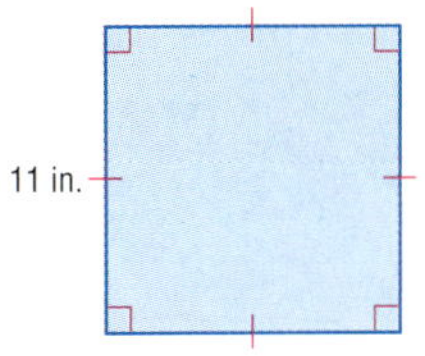

$P = 44$ in., $A = 121$ in^2

2. 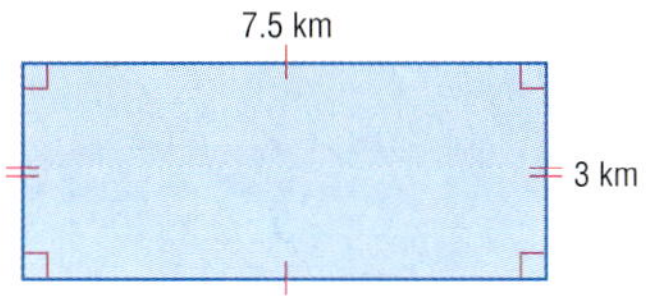

$P = 21$ km, $A = 22.5$ km^2

3. 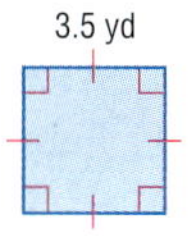

$P = 14$ yd, $A = 12.25$ yd^2

4. 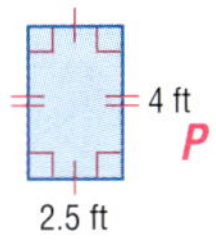

$P = 13$ ft, $A = 10$ ft^2

5. 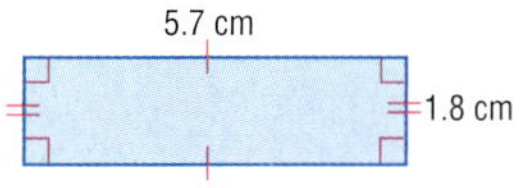

$P = 15$ cm, $A = 10.26$ cm^2

6. 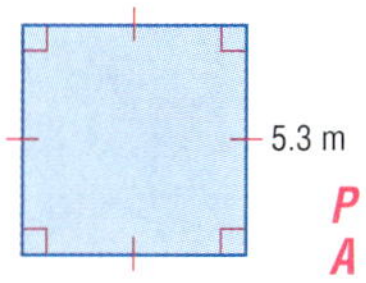

$P = 21.2$ m, $A = 28.09$ m^2

7. a rectangle with length 7 meters and width 11 meters $P = 36$ m, $A = 77$ m^2
8. a square with length 4.5 inches $P = 18$ in., $A = 20.25$ in^2
9. a rectangular sandbox with length 2.4 meters and width 1.6 meters $P = 8$ m, $A = 3.84$ m^2
10. a square with length 6.5 yards $P = 26$ yd, $A = 42.25$ yd^2
11. a square office with length 12 feet $P = 48$ ft, $A = 144$ ft^2
12. a rectangle with length 4.2 inches and width 15.7 inches $P = 39.8$ in., $A = 65.94$ in^2
13. a square with length 18 centimeters $P = 72$ cm, $A = 324$ cm^2
14. a rectangle with length 5.3 feet and width 7 feet $P = 24.6$ ft, $A = 37.1$ ft^2

15. **FENCING** Jansen purchased a lot that was 121 feet in width and 360 feet in length. If he wants to build a fence around the entire lot, how many feet of fence does he need? 962 ft

16. **CARPETING** Leonardo's bedroom is 10 feet wide and 11 feet long. If the carpet store has a remnant whose area is 105 square feet, could it be used to cover his bedroom floor? Explain. No, $10(11) = 110$ and $110 > 105$.

Prerequisite Skills

Prerequisite Skills 733

Prerequisite Skills

Prerequisite Skills

Operations with Integers

- The absolute value of any number n is its distance from zero on a number line and is written as $|n|$. Since distance cannot be less than zero, the absolute value of a number is always greater than or equal to zero.

Example 1 **Evaluate each expression.**

a. $|3|$

$|3| = 3$ Definition of absolute value

b. $|-7|$

$|-7| = 7$ Definition of absolute value

c. $|-4 + 2|$

$|-4 + 2| = |-2|$ $-4 + 2 = -2$

$= 2$ Simplify.

- To add integers with the same sign, add their absolute values. Give the result the same sign as the integers. To add integers with different signs, subtract their absolute values. Give the result the same sign as the integer with the greater absolute value.

Example 2 **Find each sum.**

a. $-3 + (-5)$ Both numbers are negative, so the sum is negative.

$-3 + (-5) = -8$ Add $|-3|$ and $|-5|$.

b. $-4 + 2$ The sum is negative because $|-4| > |2|$.

$-4 + 2 = -2$ Subtract $|2|$ from $|-4|$.

c. $6 + (-3)$ The sum is positive because $|6| > |-3|$.

$6 + (-3) = 3$ Subtract $|-3|$ from $|6|$.

d. $1 + 8$ Both numbers are positive, so the sum is positive

$1 + 8 = 9$ Add $|1|$ and $|8|$.

- To subtract an integer, add its additive inverse.

Example 3 **Find each difference.**

a. $4 - 7$

$4 - 7 = 4 + (-7)$ To subtract 7, add -7.

$= -3$

b. $2 - (-4)$

$2 - (-4) = 2 + 4$ To subtract -4, add 4.

$= 6$

- The product of two integers with different signs is negative. The product of two integers with the same sign is positive. Similarly, the quotient of two integers with different signs is negative, and the quotient of two integers with the same sign is positive.

734 Prerequisite Skills

Example 4 Find each product or quotient.

a. $4(-7)$ — The factors have different signs.

$4(-7) = -28$ — The product is negative.

b. $-64 \div (-8)$ — The dividend and divisor have the same sign.

$-64 \div (-8) = 8$ — The quotient is positive.

c. $-9(-6)$ — The factors have the same sign.

$-9(-6) = 54$ — The product is positive.

d. $-55 \div 5$ — The dividend and divisor have different signs.

$-55 \div 5 = -11$ — The quotient is negative.

e. $\frac{24}{-3}$ — The dividend and divisor have different signs.

$\frac{24}{-3} = -8$ — The quotient is negative.

- To evaluate expressions with absolute value, evaluate the absolute values first and then perform the operation.

Example 5 Evaluate each expression.

a. $|-3| - |5|$

$|-3| - |5| = 3 - 5$ — $|-3| = 3$, $|5| = 5$

$= -2$ — Simplify.

b. $|-5| + |-2|$

$|-5| + |-2| = 5 + 2$ — $|-5| = 5$, $|-2| = 2$

$= 7$ — Simplify.

Exercises Evaluate each absolute value.

1. $|-3|$ **3**
2. $|4|$ **4**
3. $|0|$ **0**
4. $|-5|$ **5**

Find each sum or difference.

5. $-4 - 5$ **−9**
6. $3 + 4$ **7**
7. $9 - 5$ **4**
8. $-2 - 5$ **−7**
9. $3 - 5$ **−2**
10. $-6 + 11$ **5**
11. $-4 + (-4)$ **−8**
12. $5 - 9$ **−4**
13. $-3 + 1$ **−2**
14. $-4 + (-2)$ **−6**
15. $2 - (-8)$ **10**
16. $7 + (-3)$ **4**
17. $-4 - (-2)$ **−2**
18. $3 - (-3)$ **6**
19. $3 + (-4)$ **−1**
20. $-3 - (-9)$ **6**

Evaluate each expression.

21. $|-4| - |6|$ **−2**
22. $|-7| + |-1|$ **8**
23. $|1| + |-2|$ **3**
24. $|2| - |-5|$ **−3**
25. $|-5 + 2|$ **3**
26. $|6 + 4|$ **10**
27. $|3 - 7|$ **4**
28. $|-3 - 3|$ **6**

Find each product or quotient.

29. $-36 \div 9$ **−4**
30. $-3(-7)$ **21**
31. $6(-4)$ **−24**
32. $-25 \div 5$ **−5**
33. $-6(-3)$ **18**
34. $7(-8)$ **−56**
35. $-40 \div (-5)$ **8**
36. $11(3)$ **33**
37. $44 \div (-4)$ **−11**
38. $-63 \div (-7)$ **9**
39. $6(5)$ **30**
40. $-7(12)$ **−84**
41. $-10(4)$ **−40**
42. $80 \div (-16)$ **−5**
43. $72 \div 9$ **8**
44. $39 \div 3$ **13**

Prerequisite Skills

Prerequisite Skills 735

Prerequisite Skills

Evaluating Algebraic Expressions

An expression is an algebraic expression if it contains sums and/or products of variables and numbers. To evaluate an algebraic expression, replace the variable or variables with known values, and then use the order of operations.

Order of Operations	
Step 1	Evaluate expressions inside grouping symbols.
Step 2	Evaluate all powers.
Step 3	Do all multiplications and/or divisions from left to right.
Step 4	Do all additions and/or subtractions from left to right.

Example 1 **Evaluate each expression.**

a. $x - 5 + y$ **if** $x = 15$ **and** $y = -7$

$x - 5 + y = 15 - 5 + (-7)$ $\quad$ $x = 15, y = -7$

$= 10 + (-7)$ $\quad$ Subtract 5 from 15.

$= 3$ $\quad$ Add.

b. $6ab^2$ **if** $a = -3$ **and** $b = 3$

$6ab^2 = 6(-3)(3)^2$ $\quad$ $a = -3, b = 3$

$= 6(-3)(9)$ $\quad$ $3^2 = 9$

$= (-18)(9)$ $\quad$ Multiply.

$= -162$ $\quad$ Multiply.

Example 2 **Evaluate each expression if** $m = -2$, $n = -4$, **and** $p = 5$.

a. $\dfrac{2m + n}{p - 3}$

The division bar is a grouping symbol. Evaluate the numerator and denominator before dividing.

$\dfrac{2m + n}{p - 3} = \dfrac{2(-2) + (-4)}{5 - 3}$ $\quad$ Replace m with -2, n with -4, and p with 5.

$= \dfrac{-4 - 4}{5 - 3}$ $\quad$ Multiply.

$= \dfrac{-8}{2}$ $\quad$ Subtract.

$= -4$ $\quad$ Simplify.

b. $-3(m^2 + 2n)$

$-3(m^2 + 2n) = -3[(-2)^2 + 2(-4)]$ $\quad$ Replace m with -2 and n with -4.

$= -3[4 + (-8)]$ $\quad$ Multiply.

$= -3(-4)$ $\quad$ Add.

$= 12$ $\quad$ Multiply.

Example 3 **Evaluate** $3|a - b| + 2|c - 5|$ **if** $a = -2$, $b = -4$, **and** $c = 3$.

$3|a - b| + 2|c - 5| = 3|-2 - (-4)| + 2|3 - 5|$ $\quad$ Substitute for a, b, and c.

$= 3|2| + 2|-2|$ $\quad$ Simplify.

$= 3(2) + 2(2)$ $\quad$ Find absolute values.

$= 10$ $\quad$ Simplify.

Exercises **Evaluate each expression if** $a = 2$, $b = -3$, $c = -1$, **and** $d = 4$.

1. $2a + c$ **3**
2. $\dfrac{bd}{2c}$ **6**
3. $\dfrac{2d - a}{b}$ **−2**
4. $3d - c$ **13**
5. $\dfrac{3b}{5a + c}$ **−1**
6. $5bc$ **15**
7. $2cd + 3ab$ **−26**
8. $\dfrac{c - 2d}{a}$ $-\dfrac{9}{2}$

Evaluate each expression if $x = 2$, $y = -3$, **and** $z = 1$.

9. $24 + |x - 4|$ **26**
10. $13 + |8 + y|$ **18**
11. $|5 - z| + 11$ **15**
12. $|2y - 15| + 7$ **28**
13. $|y| - 7$ **−4**
14. $11 - 7 + |-x|$ **6**
15. $|x| - |2z|$ **0**
16. $|z - y| + 6$ **10**

736 Prerequisite Skills

Solving Linear Equations

- If the same number is added to or subtracted from each side of an equation, the resulting equation is true.

Example 1 **Solve each equation.**

a. $x - 7 = 16$

$x - 7 = 16$	Original equation
$x - 7 + 7 = 16 + 7$	Add 7 to each side.
$x = 23$	Simplify.

b. $m + 12 = -5$

$m + 12 = -5$	Original equation
$m + 12 + (-12) = -5 + (-12)$	Add -12 to each side.
$m = -17$	Simplify.

c. $k + 31 = 10$

$k + 31 = 10$	Original equation
$k + 31 - 31 = 10 - 31$	Subtract 31 from each side.
$k = -21$	Simplify.

- If each side of an equation is multiplied or divided by the same number, the resulting equation is true.

Example 2 **Solve each equation.**

a. $4d = 36$

$4d = 36$	Original equation
$\frac{4d}{4} = \frac{36}{4}$	Divide each side by 4.
$x = 9$	Simplify.

b. $-\frac{t}{8} = -7$

$-\frac{t}{8} = -7$	Original equation.
$-8\left(-\frac{t}{8}\right) = -8(-7)$	Multiply each side by -8.
$t = 56$	Simplify.

c. $\frac{3}{5}x = -8$

$\frac{3}{5}x = -8$	Original equation.
$\frac{5}{3}\left(\frac{3}{5}x\right) = \frac{5}{3}(-8)$	Multiply each side by $\frac{5}{3}$.
$x = -\frac{40}{3}$	Simplify.

- To solve equations with more than one operation, often called *multi-step equations*, undo operations by working backward.

Example 3 **Solve each equation.**

a. $12 - m = 20$

$12 - m = 20$	Original equation
$12 - m - 12 = 20 - 12$	Subtract 12 from each side.
$-m = 8$	Simplify.
$m = -8$	Divide each side by -1.

Prerequisite Skills

Prerequisite Skills

Prerequisite Skills 737

Prerequisite Skills

b. $8q - 15 = 49$

$8q - 15 = 49$	Original equation
$8q - 15 + 15 = 49 + 15$	Add 15 to each side.
$8q = 64$	Simplify.
$\frac{8q}{8} = \frac{64}{8}$	Divide each side by 8.
$q = 8$	Simplify.

c. $12y + 8 = 6y - 5$

$12y + 8 = 6y - 5$	Original equation
$12y + 8 - 8 = 6y - 5 - 8$	Subtract 8 from each side.
$12y = 6y - 13$	Simplify.
$12y - 6y = 6y - 13 - 6y$	Subtract 6y from each side.
$6y = -13$	Simplify.
$\frac{6y}{6} = \frac{-13}{6}$	Divide each side by 6.
$y = -\frac{13}{6}$	Simplify.

- When solving equations that contain grouping symbols, first use the Distributive Property to remove the grouping symbols.

Example 4 Solve $3(x - 5) = 13$.

$3(x - 5) = 13$	Original equation
$3x - 15 = 13$	Distributive Property
$3x - 15 + 15 = 13 + 15$	Add 15 to each side.
$3x = 28$	Simplify.
$x = \frac{28}{3}$	Divide each side by 3.

Exercises Solve each equation.

1. $r + 11 = 3$ **−8**
2. $n + 7 = 13$ **6**
3. $d - 7 = 8$ **15**
4. $\frac{8}{5}a = -6$ **$-\frac{15}{4}$**
5. $-\frac{p}{12} = 6$ **−72**
6. $\frac{x}{4} = 8$ **32**
7. $\frac{12}{5}f = -18$ **$-\frac{15}{2}$**
8. $\frac{y}{7} = -11$ **−77**
9. $\frac{6}{7}y = 3$ **$\frac{7}{2}$**
10. $c - 14 = -11$ **3**
11. $t - 14 = -29$ **−15**
12. $p - 21 = 52$ **73**
13. $b + 2 = -5$ **−7**
14. $q + 10 = 22$ **12**
15. $-12q = 84$ **−7**
16. $5s = 30$ **6**
17. $5c - 7 = 8c - 4$ **−1**
18. $2\ell + 6 = 6\ell - 10$ **4**
19. $\frac{m}{10} + 15 = 21$ **60**
20. $-\frac{m}{8} + 7 = 5$ **16**
21. $8t + 1 = 3t - 19$ **−4**
22. $9n + 4 = 5n + 18$ **$\frac{7}{2}$**
23. $5c - 24 = -4$ **4**
24. $3n + 7 = 28$ **7**
25. $-2y + 17 = -13$ **15**
26. $-\frac{t}{13} - 2 = 3$ **−65**
27. $\frac{2}{9}x - 4 = \frac{2}{3}$ **21**
28. $9 - 4g = -15$ **6**
29. $-4 - p = -2$ **−2**
30. $21 - b = 11$ **10**
31. $-2(n + 7) = 15$ **$-\frac{29}{2}$**
32. $5(m - 1) = -25$ **−4**
33. $-8a - 11 = 37$ **−6**
34. $\frac{7}{4}q - 2 = -5$ **$-\frac{12}{7}$**
35. $2(5 - n) = 8$ **1**
36. $-3(d - 7) = 6$ **5**

Solving Inequalities in One Variable

Statements with **greater than** ($>$), **less than** ($<$), **greater than or equal to** ($\geq$), or **less than or equal to** ($\leq$) are **inequalities**.

- If any number is added or subtracted to each side of an inequality, the resulting inequality is true.

Example 1 **Solve each inequality.**

a. $x - 17 > 12$

$x - 17 > 12$	Original inequality
$x - 17 + 17 > 12 + 17$	Add 17 to each side.
$x > 29$	Simplify.

The solution set is $\{x \mid x > 29\}$.

b. $y + 11 \leq 5$

$y + 11 \leq 5$	Original inequality
$y + 11 - 11 \leq 5 - 11$	Subtract 11 from each side.
$y \leq -6$	Simplify.

The solution set is $\{y \mid y \leq -6\}$.

- If each side of an inequality is multiplied or divided by a positive number, the resulting inequality is true.

Example 2 **Solve each inequality.**

a. $\frac{t}{6} \geq 11$

$\frac{t}{6} \geq 11$	Original inequality
$(6)\frac{t}{6} \geq (6)11$	Multiply each side by 6.
$t \geq 66$	Simplify.

The solution set is $\{t \mid t > 66\}$.

b. $8p < 72$

$8p < 72$	Original inequality
$\frac{8p}{8} < \frac{72}{8}$	Divide each side by 8.
$p < 9$	Simplify.

The solution set is $\{p \mid p < 9\}$.

- If each side of an inequality is multiplied or divided by the same negative number, the direction of the inequality symbol must be *reversed* so that the resulting inequality is true.

Example 3 **Solve each inequality.**

a. $-5c > 30$

$-5c > 30$	Original inequality
$\frac{-5c}{-5} < \frac{30}{-5}$	Divide each side by -5. Change $>$ to $<$.
$c < -6$	Simplify.

The solution set is $\{c \mid c < -6\}$.

Prerequisite Skills

Prerequisite Skills 739

Page 741

1.
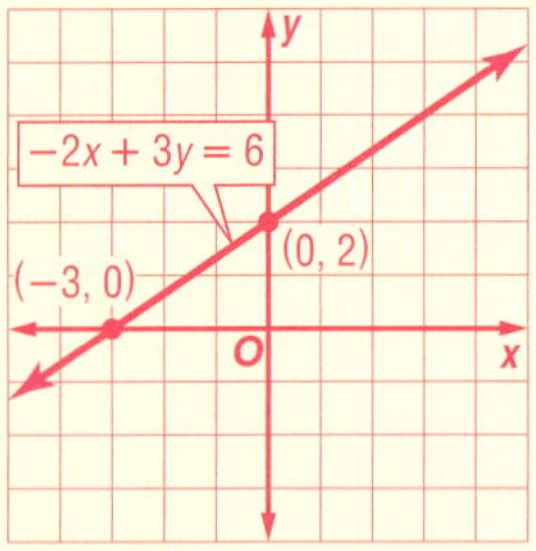

2.
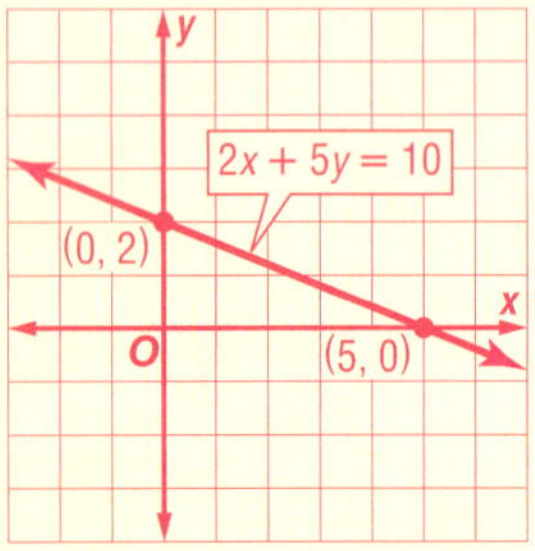

3.
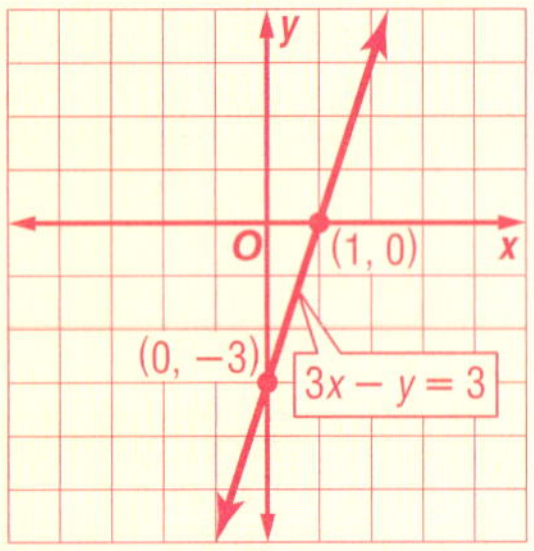

4.
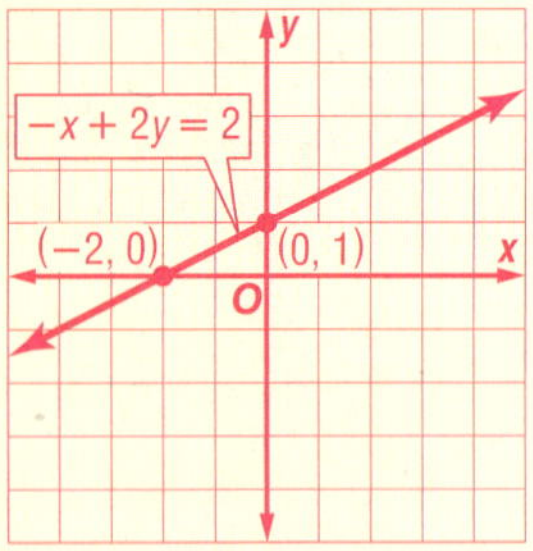

5.
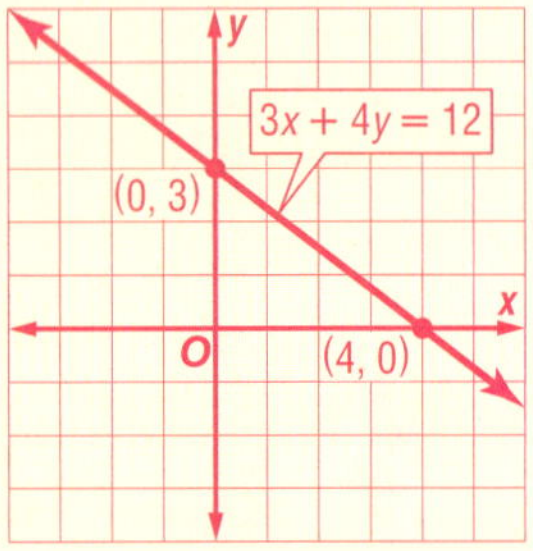

6.
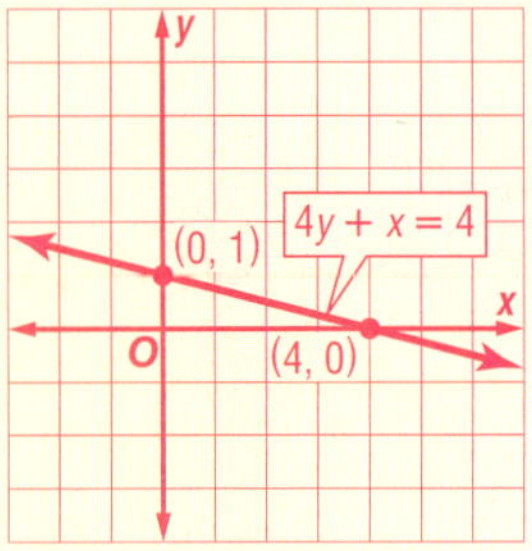

b. $-\frac{d}{13} \leq -4$

$-\frac{d}{13} \leq -4$	Original inequality
$(-13)(-\frac{d}{13}) \geq (-13)(-4)$	Multiply each side by −13. Change ≤ to ≥.
$d \geq 52$	Simplify.

The solution set is $\{d \mid d \geq 52\}$.

- Inequalities involving more than one operation can be solved by undoing the operations in the same way you would solve an equation with more than one operation.

Example 4 **Solve each inequality.**

a. $-6a + 13 < -7$

$-6a + 13 < -7$	Original inequality
$-6a + 13 - 13 < -7 - 13$	Subtract 13 from each side.
$-6a < -20$	Simplify.
$\frac{-6a}{-6} > \frac{-20}{-6}$	Divide each side by −6. Change < to >.
$a > \frac{10}{3}$	Simplify.

The solution set is $\left\{a \mid a > \frac{10}{3}\right\}$.

b. $4z + 7 \geq 8z - 1$

$4z + 7 \geq 8z - 1$	Original inequality.
$4z + 7 - 7 \geq 8z - 1 - 7$	Subtract 7 from each side.
$4z \geq 8z - 8$	Simplify.
$4z - 8z \geq 8z - 8 - 8z$	Subtract 8z from each side.
$-4z \geq -8$	Simplify.
$\frac{-4z}{-4} \leq \frac{-8}{-4}$	Divide each side by −4. Change ≥ to ≤ .
$z \leq 2$	Simplify.

The solution set is $\{z \mid z \leq 2\}$.

Exercises **Solve each inequality.**

1. $x - 7 < 6$ $\{x \mid x < 13\}$
2. $4c + 23 \leq -13$ $\{c \mid c \leq -9\}$
3. $-\frac{p}{5} \geq 14$ $\{p \mid p \leq -70\}$
4. $-\frac{a}{8} < 5$ $\{a \mid a > -40\}$
5. $\frac{t}{6} > -7$ $\{t \mid t > -42\}$
6. $\frac{a}{11} \leq 8$ $\{a \mid a \leq 88\}$
7. $d + 8 \leq 12$ $\{d \mid d \leq 4\}$
8. $m + 14 > 10$ $\{m \mid m > -4\}$
9. $2z - 9 < 7z + 1$ $\{z \mid z > -2\}$
10. $6t - 10 \geq 4t$ $\{t \mid t \geq 5\}$
11. $3z + 8 < 2$ $\{z \mid z < -2\}$
12. $a + 7 \geq -5$ $\{a \mid a \geq -12\}$
13. $m - 21 < 8$ $\{m \mid m < 29\}$
14. $x - 6 \geq 3$ $\{x \mid x \geq 9\}$
15. $-3b \leq 48$ $\{b \mid b \geq -16\}$
16. $4y < 20$ $\{y \mid y < 5\}$
17. $12k \geq -36$ $\{k \mid k \geq -3\}$
18. $-4h > 36$ $\{h \mid h < -9\}$
19. $\frac{2}{5}b - 6 \leq -2$ $\{b \mid b \leq 10\}$
20. $\frac{8}{3}t + 1 > -5$ $\left\{t \mid t > -\frac{9}{4}\right\}$
21. $7q + 3 \geq -4q + 25$ $\{q \mid q \geq 2\}$
22. $-3n - 8 > 2n + 7$ $\{n \mid n < -3\}$
23. $-3w + 1 \leq 8$ $\left\{w \mid w \geq -\frac{7}{3}\right\}$
24. $-\frac{4}{5}k - 17 > 11$ $\{k \mid k < -35\}$

7.
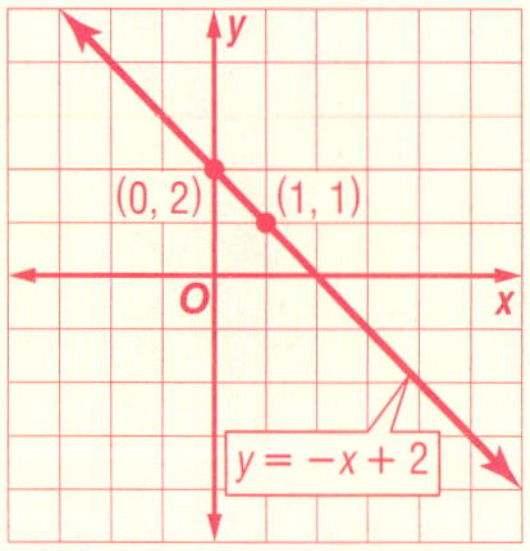

8.
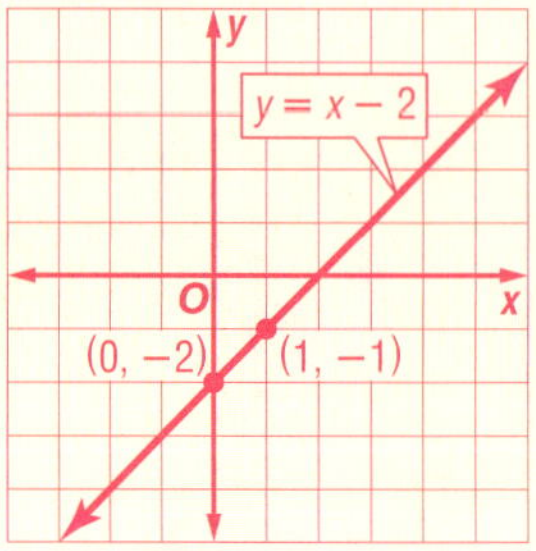

9.
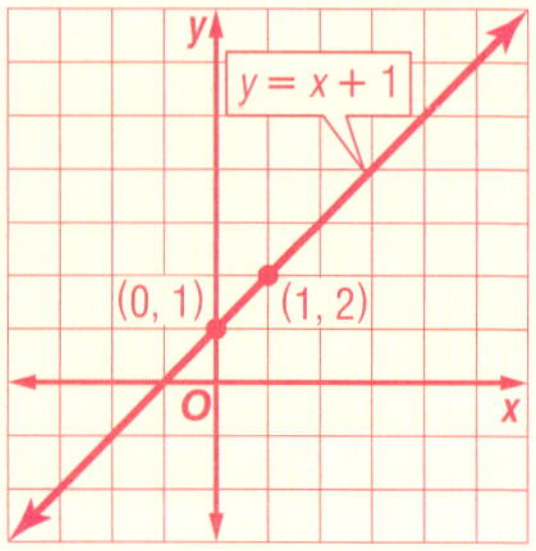

Graphing Using Intercepts and Slope

- The x-coordinate of the point at which a line crosses the x-axis is called the **x-intercept**. The y-coordinate of the point at which a line crosses the y-axis is called the **y-intercept**. Since two points determine a line, one method of graphing a linear equation is to find these intercepts.

Example 1 **Determine the x-intercept and y-intercept of $4x - 3y = 12$. Then graph the equation.**

To find the x-intercept, let $y = 0$.

$4x - 3y = 12$	Original equation
$4x - 3(0) = 12$	Replace y with 0.
$4x = 12$	Simplify.
$x = 3$	Divide each side by 4.

To find the y-intercept, let $x = 0$.

$4x - 3y = 12$	Original equation
$4(0) - 3y = 12$	Replace x with 0.
$-3y = 12$	Divide each side by -3.
$y = -4$	Simplify.

Put a point on the x-axis at 3 and a point on the y-axis at -4. Draw the line through the two points.

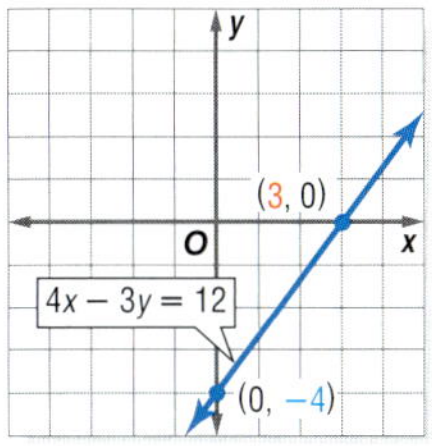

- A linear equation of the form $y = mx + b$ is in *slope-intercept* form, where m is the slope and b is the y-intercept. When an equation is written in this form, you can graph the equation quickly.

Example 2 **Graph $y = \frac{3}{4}x - 2$.**

Step 1 The y-intercept is -2. So, plot a point at $(0, -2)$.

Step 2 The slope is $\frac{3}{4}$. $\frac{\text{rise}}{\text{run}}$
From $(0, -2)$, move up 3 units and right 4 units. Plot a point.

Step 3 Draw a line connecting the points.

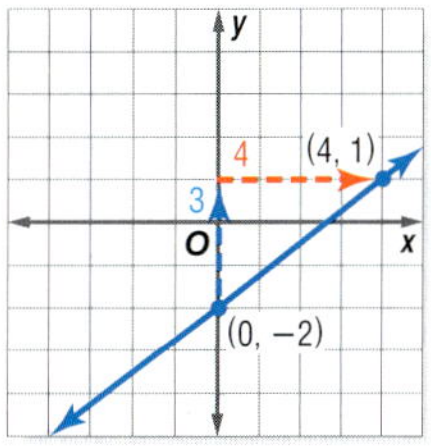

Exercises **Graph each equation using both intercepts.** **1–6. See margin.**

1. $-2x + 3y = 6$
2. $2x + 5y = 10$
3. $3x - y = 3$
4. $-x + 2y = 2$
5. $3x + 4y = 12$
6. $4y + x = 4$

Graph each equation using the slope and y-intercept. **7–12. See margin.**

7. $y = -x + 2$
8. $y = x - 2$
9. $y = x + 1$
10. $y = 3x - 1$
11. $y = -2x + 3$
12. $y = -3x - 1$

Graph each equation using either method. **13–21. See margin.**

13. $y = \frac{2}{3}x - 3$
14. $y = \frac{1}{2}x - 1$
15. $y = 2x - 2$
16. $-6x + y = 2$
17. $2y - x = -2$
18. $3x + 4y = -12$
19. $4x - 3y = 6$
20. $4x + y = 4$
21. $y = 2x - \frac{3}{2}$

13.

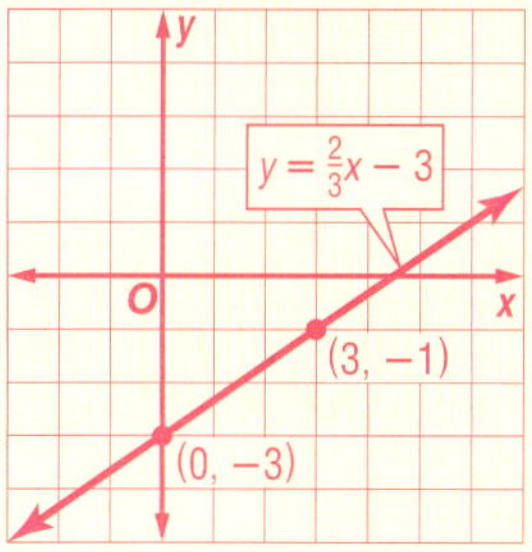

14.

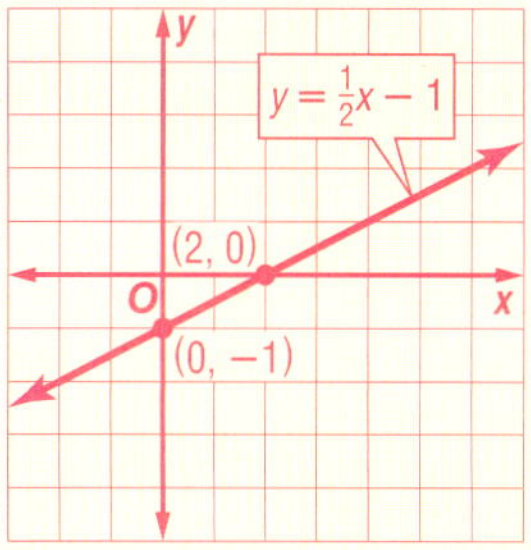

15.

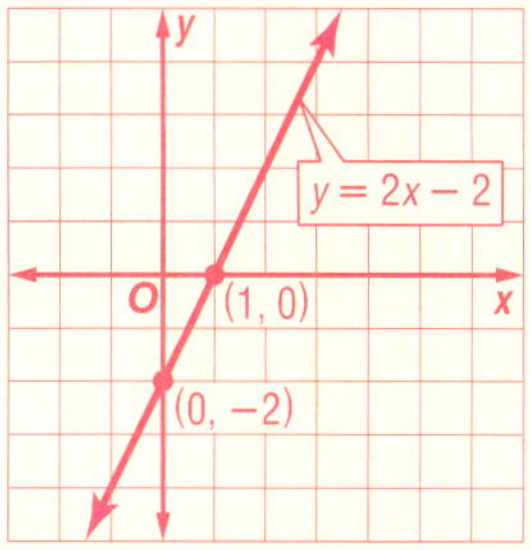

16.

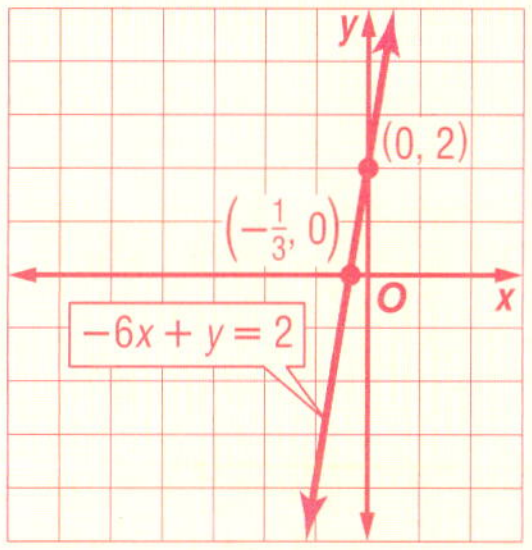

17.

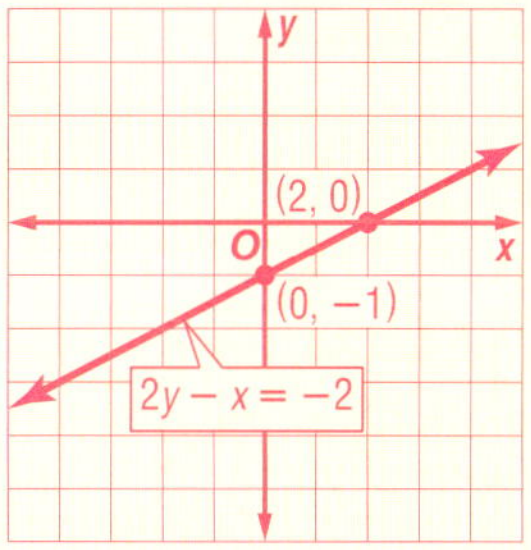

18.

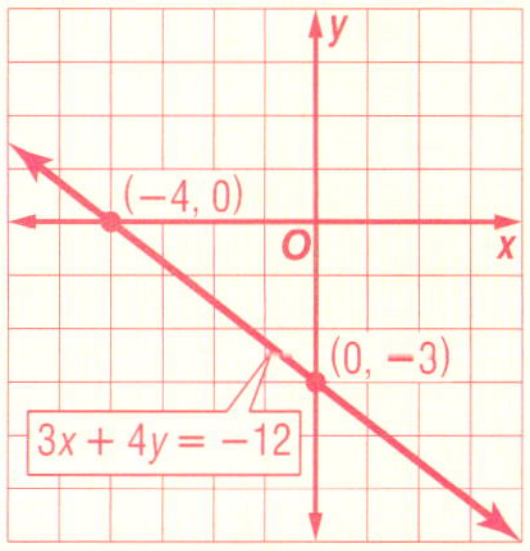

Answers continued on the following page.

10.

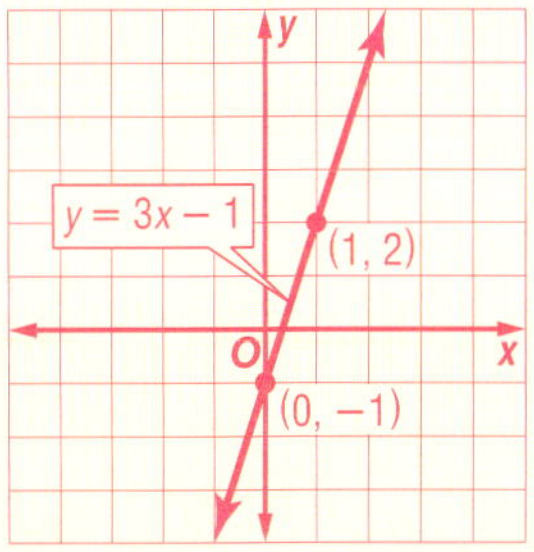

11.

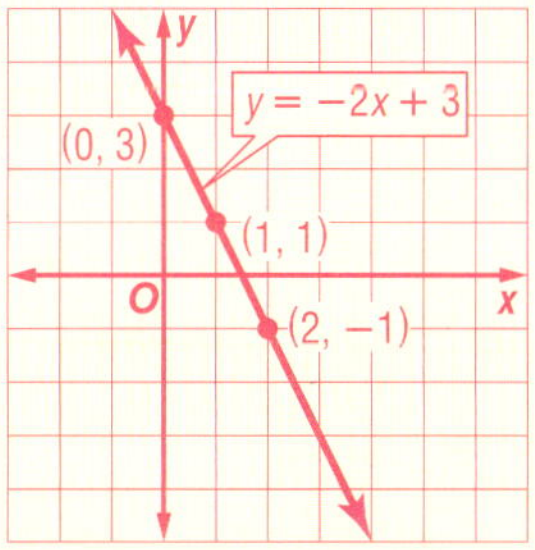

12.

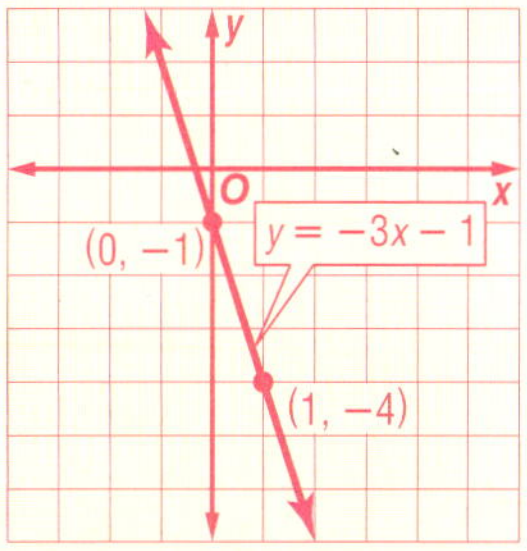

Page 741 (continued)

19.

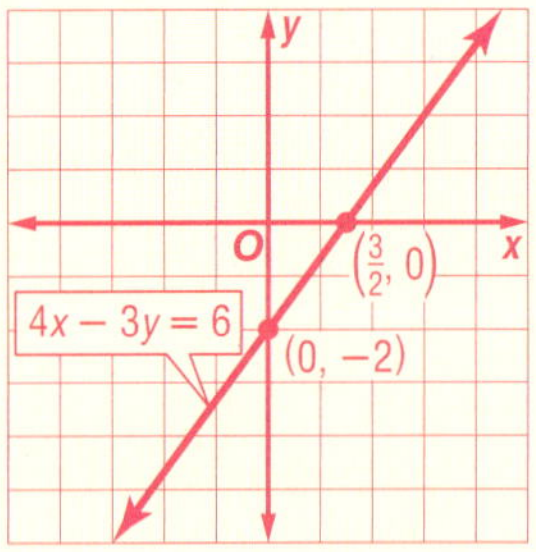

20.

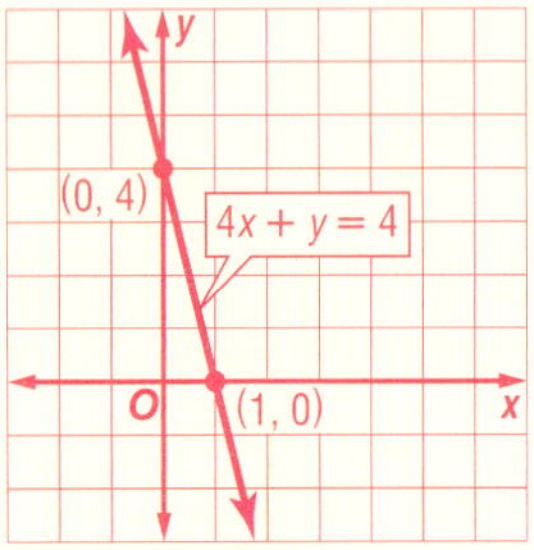

21.

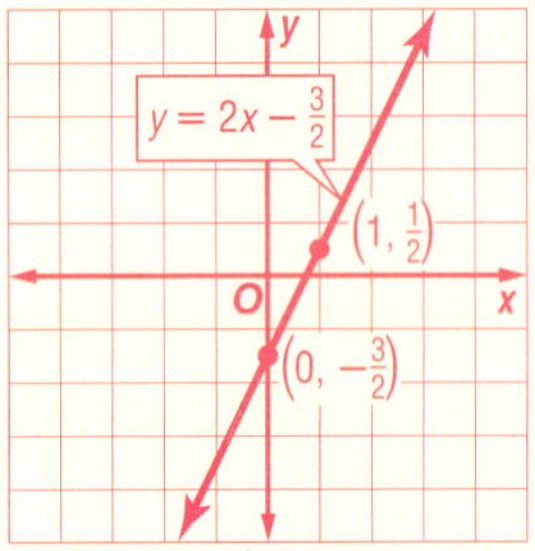

Solving Systems of Linear Equations

- Two or more equations that have common variables are called a **system of equations**. The solution of a system of equations in two variables is an ordered pair of numbers that satisfies both equations. A system of two linear equations can have zero, one, or an infinite number of solutions. There are three methods by which systems of equations can be solved: graphing, elimination, and substitution.

Example 1 **Solve each system of equations by graphing. Then determine whether each system has *no* solution, *one* solution, or *infinitely many* solutions.**

a. $y = -x + 3$
$y = 2x - 3$

The graphs appear to intersect at (2, 1). Check this estimate by replacing x with 2 and y with 1 in each equation.

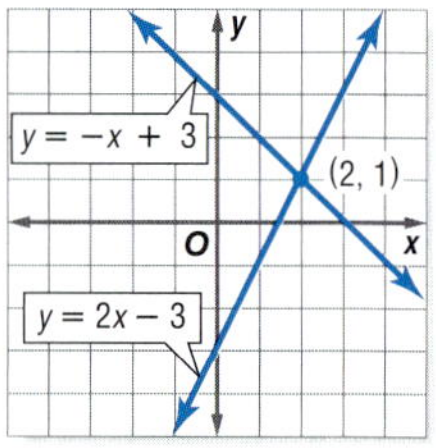

Check:

$y = -x + 3$	$y = 2x - 3$
$1 \stackrel{?}{=} -2 + 3$	$1 \stackrel{?}{=} 2(2) - 3$
$1 = 1$ ✓	$1 = 1$ ✓

The system has one solution at (2, 1).

b. $y - 2x = 6$
$3y - 6x = 9$

The graphs of the equations are parallel lines. Since they do not intersect, there are no solutions of this system of equations. Notice that the lines have the same slope but different y-intercepts. Equations with the same slope *and* the same y-intercepts have an infinite number of solutions.

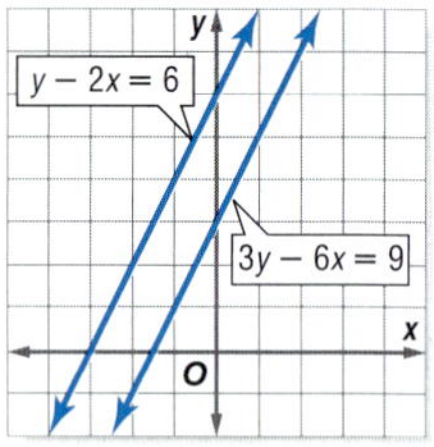

- It is difficult to determine the solution of a system when the two graphs intersect at noninteger values. There are algebraic methods by which an exact solution can be found. One such method is **substitution**.

Example 2 **Use substitution to solve the system of equations.**

$y = -4x$

$2y + 3x = 8$

Since $y = -4x$, substitute $-4x$ for y in the second equation.

$2y + 3x = 8$	Second equation
$2(-4x) + 3x = 3$	$y = -4x$
$-8x + 3x = 8$	Simplify.
$-5x = 8$	Combine like terms.
$\frac{-5x}{-5} = \frac{8}{-5}$	Divide each side by -5.
$x = -\frac{8}{5}$	Simplify.

Use $y = -4x$ to find the value of y.

$y = -4x$	First equation
$y = -4\left(-\frac{8}{5}\right)$	$x = -\frac{8}{5}$
$y = \frac{32}{5}$	Simplify.

The solution is $\left(-\frac{8}{5}, \frac{32}{5}\right)$.

- Sometimes adding or subtracting two equations together will eliminate one variable. Using this step to solve a system of equations is called **elimination**.

Example 3 **Use elimination to solve the system of equations.**

$3x + 5y = 7$

$4x + 2y = 0$

Either x or y can be eliminated. In this example, we will eliminate x.

$3x + 5y = 7$ Multiply by 4. $12x + 20y = 28$

$4x + 2y = 0$ Multiply by -3. $+ \; -12x - 6y = 0$

$14y = 28$ Add the equations.

$\frac{14y}{14} = \frac{28}{14}$ Divide each side by 14.

$y = 2$ Simplify.

Now substitute 2 for y in either equation to find the value of x.

$4x + 2y = 0$ Second equation

$4x + 2(2) = 0$ $y = 2$

$4x + 4 = 0$ Simplify.

$4x + 4 - 4 = 0 - 4$ Subtract 4 from each side.

$4x = -4$ Simplify.

$\frac{4x}{4} = \frac{-4}{4}$ Divide each side by 4.

$x = -1$ Simplify.

The solution is $(-1, 2)$.

Exercises

Solve by graphing.

1. $y = -x + 2$
   $y = -\frac{1}{2}x + 1$ **(2, 0)**
2. $y = 3x - 3$
   $y = x + 1$ **(2, 3)**
3. $y - 2x = 1$
   $2y - 4x = 1$ **no solution**
4. $2x - 4y = -2$
   $-6x + 12y = 6$ **infinitely many solutions**
5. $4x + 3y = 12$
   $3x - y = 9$ **(3, 0)**
6. $3y + x = -3$
   $y - 3x = -1$ **(0, −1)**

Solve by substitution.

7. $-5x + 3y = 12$
   $x + 2y = 8$ **(0, 4)**
8. $x - 4y = 22$
   $2x + 5y = -21$ **(2, −5)**
9. $y + 5x = -3$
   $3y - 2x = 8$ **(−1, 2)**
10. $y - 2x = 2$
    $7y + 4x = 23$ $\left(\frac{1}{2}, 3\right)$
11. $2x - 3y = -8$
    $x + 2y = 5$ **(−1, 2)**
12. $4x + 2y = 5$
    $3x - y = 10$ $\left(\frac{5}{2}, -\frac{5}{2}\right)$

Solve by elimination.

13. $-3x + y = 7$
    $3x + 2y = 2$ $\left(-\frac{4}{3}, 3\right)$
14. $3x + 4y = -1$
    $-9x - 4y = 13$ $\left(-2, \frac{5}{4}\right)$
15. $-4x + 5y = -11$
    $2x + 3y = 11$ **(4, 1)**
16. $6x - 5y = 1$
    $-2x + 9y = 7$ **(1, 1)**
17. $3x - 2y = 8$
    $5x - 3y = 16$ **(8, 8)**
18. $4x + 7y = -17$
    $3x + 2y = -3$ **(1, −3)**

Name an appropriate method to solve each system of equations. Then solve the system.

19. $4x - y = 11$
    $2x - 3y = 3$ **elimination or substitution, (3, 1)**
20. $4x + 6y = 3$
    $-10x - 15y = -4$ **elimination, no solution**
21. $3x - 2y = 6$
    $5x - 5y = 5$ **graphing, (4, 3)**
22. $3y + x = 3$
    $-2y + 5x = 15$ **elimination or substitution, (3, 0)**
23. $4x - 7y = 8$
    $-2x + 5y = -1$ **elimination,** $\left(\frac{11}{2}, 2\right)$
24. $x + 3y = 6$
    $4x - 2y = -32$ **elimination or substitution, (−6, 4)**

Prerequisite Skills

Prerequisite Skills 743

Prerequisite Skills

Square Roots and Simplifying Radicals

- A radical expression is an expression that contains a square root. The expression is in simplest form when the following three conditions have been met.
- No radicands have perfect square factors other than 1.
- No radicands contain fractions.
- No radicals appear in the denominator of a fraction.
- The **Product Property** states that for two numbers a and $b \geq 0$, $\sqrt{ab} = \sqrt{a} \cdot \sqrt{b}$.

Example 1 **Simplify.**

a. $\sqrt{45}$

$\sqrt{45} = \sqrt{3 \cdot 3 \cdot 5}$ — Prime factorization of 45

$= \sqrt{3^2} \cdot \sqrt{5}$ — Product Property of Square Roots

$= 3\sqrt{5}$ — Simplify.

b. $\sqrt{3} \cdot \sqrt{3}$

$\sqrt{3} \cdot \sqrt{3} = \sqrt{3 \cdot 3}$ — Product Property

$= \sqrt{9}$ or 3 — Simplify.

c. $\sqrt{6} \cdot \sqrt{15}$

$\sqrt{6} \cdot \sqrt{15} = \sqrt{6 \cdot 15}$ — Product Property

$= \sqrt{3 \cdot 2 \cdot 3 \cdot 5}$ — Prime factorization

$= \sqrt{3^2} \cdot \sqrt{10}$ — Product Property

$= 3\sqrt{10}$ — Simplify.

- For radical expressions in which the exponent of the variable inside the radical is *even* and the resulting simplified exponent is *odd*, you must use absolute value to ensure nonnegative results.

Example 2 $\sqrt{20x^3y^5z^6}$

$\sqrt{20x^3y^5z^6} = \sqrt{2^2 \cdot 5 \cdot x^3 \cdot y^5 \cdot z^6}$ — Prime factorization

$= \sqrt{2^2} \cdot \sqrt{5} \cdot \sqrt{x^3} \cdot \sqrt{y^5} \cdot \sqrt{z^6}$ — Product Property

$= 2 \cdot \sqrt{5} \cdot x \cdot \sqrt{x} \cdot y^2 \cdot \sqrt{y} \cdot |z^3|$ — Simplify.

$= 2xy^2|z^3|\sqrt{5xy}$ — Simplify.

- The **Quotient Property** states that for any numbers a and b, where $a \geq 0$ and $b \geq 0$, $\sqrt{\frac{a}{b}} = \frac{\sqrt{a}}{\sqrt{b}}$.

Example 3 **Simplify** $\sqrt{\frac{25}{16}}$.

$\sqrt{\frac{25}{16}} = \frac{\sqrt{25}}{\sqrt{16}}$ — Quotient Property

$= \frac{5}{4}$ — Simplify.

744 Prerequisite Skills

- Rationalizing the denominator of a radical expression is a method used to eliminate radicals from the denominator of a fraction. To rationalize the denominator, multiply the expression by a fraction equivalent to 1 such that the resulting denominator is a perfect square.

Example 4 **Simplify.**

a. $\dfrac{2}{\sqrt{3}}$

$\dfrac{2}{\sqrt{3}} = \dfrac{2}{\sqrt{3}} \cdot \dfrac{\sqrt{3}}{\sqrt{3}}$ Multiply by $\dfrac{\sqrt{3}}{\sqrt{3}}$.

$= \dfrac{2\sqrt{3}}{3}$ Simplify.

b. $\dfrac{\sqrt{13y}}{\sqrt{18}}$

$\dfrac{\sqrt{13y}}{\sqrt{18}} = \dfrac{\sqrt{13y}}{\sqrt{2 \cdot 3 \cdot 3}}$ Prime factorization

$= \dfrac{\sqrt{13y}}{3\sqrt{2}}$ Product Property

$= \dfrac{\sqrt{13y}}{3\sqrt{2}} \cdot \dfrac{\sqrt{2}}{\sqrt{2}}$ Multiply by $\dfrac{\sqrt{2}}{\sqrt{2}}$.

$= \dfrac{\sqrt{26y}}{6}$ Product Property

- Sometimes, conjugates are used to simplify radical expressions. Conjugates are binomials of the form $p\sqrt{q} + r\sqrt{s}$ and $p\sqrt{q} - r\sqrt{s}$.

Example 5 **Simplify** $\dfrac{3}{5 - \sqrt{2}}$.

$\dfrac{3}{5 - \sqrt{2}} = \dfrac{3}{5 - \sqrt{2}} \cdot \dfrac{5 + \sqrt{2}}{5 + \sqrt{2}}$ $\dfrac{5 + \sqrt{2}}{5 + \sqrt{2}} = 1$

$= \dfrac{3(5 + \sqrt{2})}{5^2 - (\sqrt{2})^2}$ $(a - b)(a + b) = a^2 - b^2$

$= \dfrac{15 + 3\sqrt{2}}{25 - 2}$ Multiply. $(\sqrt{2})^2 = 2$

$= \dfrac{15 + 3\sqrt{2}}{23}$ Simplify.

Exercises **Simplify.** 8. $2|a|b^2c^2\sqrt{14c}$ 19. $\dfrac{6\sqrt{5} + 3\sqrt{10}}{2}$

1. $\sqrt{32}$ $4\sqrt{2}$
2. $\sqrt{75}$ $5\sqrt{3}$
3. $\sqrt{50} \cdot \sqrt{10}$ $10\sqrt{5}$
4. $\sqrt{12} \cdot \sqrt{20}$ $4\sqrt{15}$
5. $\sqrt{6} \cdot \sqrt{6}$ 6
6. $\sqrt{16} \cdot \sqrt{25}$ 20
7. $\sqrt{98x^3y^6}$ $7x|y^3|\sqrt{2x}$
8. $\sqrt{56a^2b^4c^5}$
9. $\sqrt{\dfrac{81}{49}}$ $\dfrac{9}{7}$
10. $\sqrt{\dfrac{121}{16}}$ $\dfrac{11}{4}$
11. $\sqrt{\dfrac{63}{8}}$ $\dfrac{3\sqrt{14}}{4}$
12. $\sqrt{\dfrac{288}{147}}$ $\dfrac{4\sqrt{6}}{7}$
13. $\dfrac{\sqrt{10p^3}}{\sqrt{27}}$ $\dfrac{p\sqrt{30p}}{9}$
14. $\dfrac{\sqrt{108}}{\sqrt{2q^6}}$ $\dfrac{3\sqrt{6}}{|q^3|}$
15. $\dfrac{4}{5 - 2\sqrt{3}}$ $\dfrac{20 + 8\sqrt{3}}{13}$
16. $\dfrac{7\sqrt{3}}{5 - 2\sqrt{6}}$ $35\sqrt{3} + 42\sqrt{2}$
17. $\dfrac{3}{\sqrt{48}}$ $\dfrac{\sqrt{3}}{4}$
18. $\dfrac{\sqrt{24}}{\sqrt{125}}$ $\dfrac{2\sqrt{30}}{25}$
19. $\dfrac{3\sqrt{5}}{2 - \sqrt{2}}$
20. $\dfrac{3}{-2 + \sqrt{13}}$ $\dfrac{2 + \sqrt{13}}{3}$

Prerequisite Skills

Prerequisite Skills 745

Prerequisite Skills

Multiplying Polynomials

- The **Product of Powers** rule states that for any number a and all integers m and n, $a^m \cdot a^n = a^{m+n}$.

Example 1 **Simplify each expression.**

a. $(4p^5)(p^4)$

$(4p^5)(p^4) = (4)(1)(p^5 \cdot p^4)$ Commutative and Associative Properties

$= (4)(1)(p^{5+4})$ Product of powers

$= 4p^9$ Simplify.

b. $(3yz^5)(-9y^2z^2)$

$(3yz^5)(-9y^2z^2) = (3)(-9)(y \cdot y^2)(z^5 \cdot z^2)$ Commutative and Associative Properties

$= -27(y^{1+2})(z^{5+2})$ Product of powers

$= -27y^3z^7$ Simplify.

- The Distributive Property can be used to multiply a monomial by a polynomial.

Example 2 **Simplify $3x^3(-4x^2 + x - 5)$.**

$3x^3(-4x^2 + x - 5) = 3x^3(-4x^2) + 3x^3(x) - 3x^3(5)$ Distributive Property

$= -12x^5 + 3x^4 - 15x^3$ Multiply.

- To find the power of a power, multiply the exponents. This is called the **Power of a Power** rule.

Example 3 **Simplify each expression.**

a. $(-3x^2y^4)^3$

$(-3x^2y^4)^3 = (-3)^3(x^2)^3(y^4)^3$ Power of a product

$= -27x^6y^{12}$ Power of a power

b. $(xy)^3(-2x^4)^2$

$(xy)^3(-2x^4)^2 = x^3y^3(-2)^2(x^4)^2$ Power of a product

$= x^3y^3(4)x^8$ Power of a power

$= 4x^3 \cdot x^8 \cdot y^3$ Commutative Property

$= 4x^{11}y^3$ Product of powers

- To multiply two binomials, find the sum of the products of

F the *First* terms,
O the *Outer* terms,
I the *Inner* terms, and
L the *Last* terms.

Example 4 **Find each product.**

a. $(2x - 3)(x + 1)$

F O I L

$(2x - 3)(x + 1) = (2x)(x) + (2x)(1) + (-3)(x) + (-3)(1)$ FOIL method

$= 2x^2 + 2x - 3x - 3$ Multiply.

$= 2x^2 - x - 3$ Combine like terms.

b. $(x + 6)(x + 5)$

F O I L

$(x + 6)(x + 5) = (x)(x) + (x)(5) + (6)(x) + (6)(5)$ FOIL method

$= x^2 + 5x + 6x + 30$ Multiply.

$= x^2 + 11x + 30$ Combine like terms.

746 Prerequisite Skills

- The Distributive Property can be used to multiply any two polynomials.

Example 5 **Find $(3x - 2)(2x^2 + 7x - 4)$.**

$(3x - 2)(2x^2 + 7x - 4) = 3x(2x^2 + 7x - 4) - 2(2x^2 + 7x - 4)$		Distributive Property
$= 6x^3 + 21x^2 - 12x - 4x^2 - 14x + 8$		Distributive Property
$= 6x^3 + 17x^2 - 26x + 8$		Combine like terms.

- Three special products are: $(a + b)^2 = a^2 + 2ab + b^2$, $(a - b)^2 = a^2 - 2ab + b^2$, and $(a + b)(a - b) = a^2 - b^2$.

Example 6 **Find each product.**

a. $(2x - z)^2$

$(a - b)^2 = a^2 - 2ab + b^2$	Square of a difference
$(2x - z)^2 = (2x)^2 - 2(2x)(z) + (z)^2$	$a = 2x$ and $b = z$
$= 4x^2 - 4xz + z^2$	Simplify.

b. $(3x + 7)(3x - 7)$

$(a + b)(a - b) = a^2 - b^2$	Product of sum and difference
$(3x + 7)(3x - 7) = (3x)^2 - (7)^2$	$a = 3x$ and $b = 7$
$= 9x^2 - 49$	Simplify.

Exercises **Find each product.**

1. $(3q^2)(q^5)$ $3q^7$
2. $(5m)(4m^3)$ $20m^4$
3. $\left(\frac{9}{2}c\right)\left(8c^5\right)$ $36c^6$
4. $(n^6)(10n^2)$ $10n^8$
5. $(fg^8)(15f^2g)$ $15f^3g^9$
6. $(6j^4k^4)(j^2k)$ $6j^6k^5$
7. $(2ab^3)(4a^2b^2)$ $8a^3b^5$
8. $\left(\frac{8}{5}x^3y\right)\left(4x^3y^2\right)$ $\frac{32}{5}x^6y^3$
9. $-2q^2(q^2 + 3)$ $-2q^4 - 6q^2$
10. $5p(p - 18)$ $5p^2 - 90p$
11. $15c(-3c^2 + 2c + 5)$ $-45c^3 + 30c^2 + 75c$
12. $8x(-4x^2 - x + 11)$ $-32x^3 - 8x^2 + 88x$
13. $4m^2(-2m^2 + 7m - 5)$ $-8m^4 + 28m^3 - 20m^2$
14. $8y^2(5y^3 - 2y + 1)$ $40y^5 - 16y^3 + 8y^2$
15. $\left(\frac{3}{2}m^3n^2\right)^2$ $\frac{9}{4}m^6n^4$
16. $(-2c^3d^2)^2$ $4c^6d^4$
17. $(-5wx^5)^3$ $-125w^3x^{15}$
18. $(6a^5b)^3$ $216a^{15}b^3$
19. $(k^2\ell)^3(13k^2)^2$ $169k^{10}\ell^3$
20. $(-5w^3x^2)^2(2w^5)^2$ $100w^{16}x^4$
21. $(-7y^3z^2)(4y^2)^4$ $-1792y^{11}z^2$
22. $\left(\frac{1}{2}p^2q^2\right)^2\left(4pq^3\right)^3$ $16p^7q^{13}$
23. $(m - 1)(m - 4)$ $m^2 - 5m + 4$
24. $(s - 7)(s - 2)$ $s^2 - 9s + 14$
25. $(x - 3)(x + 4)$ $x^2 + x - 12$
26. $(a + 3)(a - 6)$ $a^2 - 3a - 18$
27. $(5d + 3)(d - 4)$ $5d^2 - 17d - 12$
28. $(q + 2)(3q + 5)$ $3q^2 + 11q + 10$
29. $(2q + 3)(5q + 2)$ $10q^2 + 19q + 6$
30. $(2a - 3)(2a - 5)$ $4a^2 - 16a + 15$
31. $(d + 1)(d - 1)$ $d^2 - 1$
32. $(4a - 3)(4a + 3)$ $16a^2 - 9$
33. $(s - 5)^2$ $s^2 - 10s + 25$
34. $(3f - g)^2$ $9f^2 - 6fg + g^2$
35. $(2r - 5)^2$ $4r^2 - 20r + 25$
36. $\left(t + \frac{8}{3}\right)^2$ $t^2 + \frac{16}{3}t + \frac{64}{9}$
37. $(x + 4)(x^2 - 5x - 2)$ $x^3 - x^2 - 22x - 8$
38. $(x - 2)(x^2 + 3x - 7)$ $x^3 + x^2 - 13x + 14$
39. $(3b - 2)(3b^2 + b + 1)$ $9b^3 - 3b^2 + b - 2$
40. $(2j + 7)(j^2 - 2j + 4)$ $2j^3 + 3j^2 - 6j + 28$

Prerequisite Skills

Prerequisite Skills 747

Prerequisite Skills

Dividing Polynomials

- The **Quotient of Powers** rule states that for any nonzero number a and all integers m and n, $\frac{a^m}{a^n} = a^{m-n}$.
- To find the power of a quotient, find the power of the numerator and the power of the denominator.

Example 1 **Simplify.**

a. $\frac{x^5y^8}{-xy^3}$

$\frac{x^5y^8}{-xy^3} = \left(\frac{x^5}{-x}\right)\left(\frac{y^8}{y^3}\right)$	Group powers that have the same base.
$= -(x^{5-1})(y^{8-3})$	Quotient of powers
$= -x^4y^5$	Simplify.

b. $\left(\frac{4z^3}{3}\right)^3$

$\left(\frac{4z^3}{3}\right)^3 = \frac{(4z^3)^3}{3^3}$	Power of a quotient
$= \frac{4^3(z^3)^3}{3^3}$	Power of a product
$= \frac{64z^9}{27}$	Power of a product

c. $\frac{w^{-2}x^4}{2w^{-5}}$

$\frac{w^{-2}x^4}{2w^{-5}} = \frac{1}{2}\left(\frac{w^{-2}}{w^{-5}}\right)x^4$	Group powers that have the same base.
$= \frac{1}{2}(w^{-2-(-5)})x^4$	Quotient of powers
$= \frac{1}{2}w^3x^4$	Simplify.

- You can divide a polynomial by a monomial by separating the terms of the numerator.

Example 2 **Simplify** $\frac{15x^3 - 3x^2 + 12x}{3x}$**.**

$\frac{15x^3 - 3x^2 + 12x}{3x} = \frac{15x^3}{3x} - \frac{3x^2}{3x} + \frac{12x}{3x}$	Divide each term by $3x$.
$= 5x^2 - x + 4$	Simplify.

- Division can sometimes be performed using factoring.

Example 3 **Find** $(n^2 - 8n - 9) \div (n - 9)$**.**

$(n^2 - 8n - 9) \div (n - 9) = \frac{n^2 - 8n - 9}{(n - 9)}$	Write as a rational expression.
$= \frac{(n-9)(n+1)}{(n-9)}$	Factor the numerator.
$= \frac{\cancel{(n-9)}(n+1)}{\cancel{(n-9)}}$	Divide by the GCF.
$= n + 1$	Simplify.

748 Prerequisite Skills

Prerequisite Skills

- When you cannot factor, you can use a long division process similar to the one you use in arithmetic.

Example 4 **Find $(n^3 - 4n^2 - 9) \div (n - 3)$.**

In this case, there is no n term, so you must rename the dividend using 0 as the coefficient of the missing term.

$(n^3 - 4n^2 + 9) \div (n - 3) = (n^3 - 4n^2 + 0n + 9) \div (n - 3)$

Divide the first term of the dividend, n^3, by the first term of the divisor, n.

$$n - 3\overline{)n^3 - 4n^2 + 0n + 12} \quad \text{quotient: } n^2 - n - 3$$

$(-)\ n^3 - 3n^2$	Multiply n^2 and $n - 3$.
$-n^2 + 0n$	Subtract and bring down $0n$.
$(-)-n^2 + 3n$	Multiply $-n$ and $n - 3$.
$-3n + 12$	Subtract and bring down 12.
$(-)-3n + 9$	Multiply -3 and $n - 3$.
3	Subtract.

Therefore, $(n^3 - 4n^2 + 9) \div (n - 3) = n^2 - n - 3 + \frac{3}{n - 3}$. Since the quotient has a nonzero remainder, $n - 3$ is not a factor of $n^3 - 4n^2 + 9$.

Exercises **Find each quotient.**

1. $\frac{a^2c^2}{2a}$ $\frac{ac^2}{2}$
2. $\frac{5q^5r^3}{q^2r^2}$ $5q^3r$
3. $\frac{b^2d^5}{8b^{-2}d^3}$ $\frac{b^4d^2}{8}$
4. $\frac{5p^{-3}x}{2p^{-7}}$ $\frac{5}{2}p^4x$
5. $\frac{3r^{-3}s^2t^4}{2r^2st^{-3}}$ $\frac{3st^7}{2r^5}$
6. $\frac{3x^3y^{-1}z^5}{xyz^2}$ $\frac{3x^2z^3}{y^2}$
7. $\left(\frac{w^4}{6}\right)^3$ $\frac{w^{12}}{216}$
8. $\left(\frac{-3q^2}{5}\right)^3$ $\frac{-27q^6}{125}$
9. $\left(\frac{-2y^2}{7}\right)^2$ $\frac{4y^4}{49}$
10. $\left(\frac{5m^2}{3}\right)^4$ $\frac{625m^8}{81}$
11. $\frac{4z^2 - 16z - 36}{4z}$ $z - 4 - \frac{9}{z}$
12. $(5d^2 + 8d - 20) \div 10d$ $\frac{d}{2} + \frac{4}{5} - \frac{2}{d}$
13. $(p^3 - 12p^2 + 3p + 8) \div 4p$ $\frac{p^2}{4} - 3p + \frac{3}{4} + \frac{2}{p}$
14. $(b^3 + 4b^2 + 10) \div 2b$ $\frac{b^2}{2} + 2b + \frac{5}{b}$
15. $\frac{a^3 - 6a^2 + 4a - 3}{a^2}$ $a - 6 + \frac{4}{a} - \frac{3}{a^2}$
16. $\frac{8x^2y - 10xy^2 + 6x^3}{2x^2}$ $4y - \frac{5y^2}{x} + 3x$
17. $\frac{s^2 - 2s - 8}{s - 4}$ $s + 2$
18. $(r^2 + 9r + 20) \div (r + 5)$ $r + 4$
19. $(t^2 - 7t + 12) \div (t - 3)$ $t - 4$
20. $(c^2 + 3c - 54) \div (c + 9)$ $c - 6$
21. $(2q^2 - 9q - 5) \div (q - 5)$ $2q + 1$
22. $\frac{3z^2 - 2z - 5}{z + 1}$ $3z - 5$
23. $\frac{(m^3 + 3m^2 - 5m + 1)}{m - 1}$ $m^2 + 4m - 1$
24. $(d^3 - 2d^2 + 4d + 24) \div (d + 2)$ $d^2 - 4d + 12$
25. $(2j^3 + 5j + 26) \div (j + 2)$ $2j^2 - 4j + 13$
26. $\frac{2x^3 + 3x^2 - 176}{x - 4}$ $2x^2 + 11x + 44$
27. $(x^2 + 6x - 3) \div (x + 4)$ $x + 2 - \frac{11}{x + 4}$
28. $\frac{h^3 + 2h^2 - 6h + 1}{h - 2}$ $h^2 + 4h + 2 + \frac{5}{h - 2}$

Prerequisite Skills 749

Prerequisite Skills

Factoring to Solve Equations

- Some polynomials can be factored using the Distributive Property.

Example 1 **Factor $5t^2 + 15t$.**

Find the greatest common factor (GCF) of $5t^2$ and $15t$.

$5t^2 = 5 \cdot t \cdot t,\ 15t = 3 \cdot 5 \cdot t$ — GCF: $5 \cdot t$ or $5t$

$5t^2 + 15t = 5t(t) + 5t(3)$ — Rewrite each term using the GCF.

$= 5t(t + 3)$ — Distributive Property

- To factor polynomials of the form $x^2 + bx + c$, find two integers m and n so that $mn = c$ and $m + n = b$. Then write $x^2 + bx + c$ using the pattern $(x + m)(x + n)$.

Example 2 **Factor each polynomial.**

a. $x^2 + 7x + 10$ — Both b and c are positive.

In this equation, b is 7 and c is 10. Find two numbers with a product of 10 and with a sum of 7.

Factors of 10	Sum of Factors
1, 10	11
2, 5	7

$x^2 + 7x + 10 = (x + m)(x + n)$ — The correct factors are 2 and 5.

$= (x + 2)(x + 5)$ — Write the pattern; $m = 2$ and $n = 5$.

b. $x^2 - 8x + 15$ — b is negative and c is positive.

In this equation, b is -8 and c is 15. This means that $m + n$ is negative and mn is positive. So m and n must both be negative.

Factors of 15	Sum of Factors
−1, −15	−16
−3, −5	−8

$x^2 - 8x + 15 = (x + m)(x + n)$ — The correct factors are -3 and -5.

$= (x - 3)(x - 5)$ — Write the pattern; $m = -3$ and $n = -5$.

- To factor polynomials of the form $ax^2 + bx + c$, find two integers m and n with a product equal to ac and with a sum equal to b. Write $ax^2 + bx + c$ using the pattern $ax^2 + mx + nx + c$. Then factor by grouping.

c. $5x^2 - 19x - 4$ — b is negative and c is negative.

In this equation, a is 5, b is -19, and c is -4.
Find two numbers with a product of -20 and with a sum of -19.

Factors of −20	Sum of Factors
−2, 10	8
2, −10	−8
−1, 20	19
1, −20	−19

The correct factors are 1 and -20.

$5x^2 - 19x - 4 = 5x^2 + mx + nx - 4$ — Write the pattern.

$= 5x^2 + x + (-20)x - 4$ — $m = 1$ and $n = -20$

$= (5x^2 + x) - (20x + 4)$ — Group terms with common factors.

$= x(5x + 1) - 4(5x + 1)$ — Factor the GCF from each group.

$= (x - 4)(5x + 1)$ — Distributive Property

750 Prerequisite Skills

- Here are some special products.

Perfect Square Trinomials

$a^2 + 2ab + b^2 = (a + b)(a + b)$
$= (a + b)^2$

$a^2 - 2ab + b^2 = (a - b)(a - b)$
$= (a - b)^2$

Difference of Squares

$a^2 - b^2 = (a + b)(a - b)$

Example 3 **Factor each polynomial.**

a. $9x^2 + 6x + 1$ ← The first and last terms are perfect squares, and the middle term is equal to $2(3x)(1)$.

$9x^2 + 6x + 1 = (3x)^2 + 2(3x)(1) + 1^2$ Write as $a^2 + 2ab + b^2$.
$= (3x + 1)^2$ Factor using the pattern.

b. $x^2 - 9 = 0$ ← This is a difference of squares.

$x^2 - 9 = x^2 - (3)^2$ Write in the form $a^2 - b^2$.
$= (x - 3)(x + 3)$ Factor the difference of squares.

- The binomial $x - a$ is a factor of the polynomial $f(x)$ if and only if $f(a) = 0$. Since 0 times any number is equal to zero, this implies that we can use factoring to solve equations.

Example 4 **Solve $x^2 - 5x + 4 = 0$ by factoring.**

Factor the polynomial. This expression is of the form $x^2 + bx + c$.

$x^2 - 5x + 4 = 0$ Original equation
$(x - 1)(x - 4) = 0$ Factor the polynomial.

If $ab = 0$, then $a = 0$, $b = 0$, or both equal 0. Let each factor equal 0.

$x - 1 = 0$ or $x - 4 = 0$
$x = 1$ $x = 4$

Exercises **Factor each polynomial.**

1. $u^2 - 12u$ $u(u - 12)$
2. $w^2 + 4w$ $w(w + 4)$
3. $7j^2 - 28j$ $7j(j - 4)$
4. $2g^2 + 24g$ $2g(g + 12)$
5. $6x^2 + 2x$ $2x(3x + 1)$
6. $5t^2 - 30t$ $5t(t - 6)$
7. $z^2 + 10z + 21$ $(z + 7)(z + 3)$
8. $n^2 + 8n + 15$ $(n + 3)(n + 5)$
9. $h^2 + 8h + 12$ $(h + 2)(h + 6)$
10. $x^2 + 14x + 48$ $(x + 6)(x + 8)$
11. $m^2 + 6m - 7$ $(m - 1)(m + 7)$
12. $b^2 + 2b - 24$ $(b - 4)(b + 6)$
13. $q^2 - 9q + 18$ $(q - 3)(q - 6)$
14. $p^2 - 5p + 6$ $(p - 2)(p - 3)$
15. $a^2 - 3a - 4$ $(a - 4)(a + 1)$
16. $k^2 - 4k - 32$ $(k - 8)(k + 4)$
17. $n^2 - 7n - 44$ $(n - 11)(n + 4)$
18. $y^2 - 3y - 88$ $(y - 11)(y + 8)$
19. $3z^2 + 4z - 4$ $(3z - 2)(z + 2)$
20. $2y^2 + 9y - 5$ $(2y - 1)(y + 5)$
21. $5x^2 + 7x + 2$ $(5x + 2)(x + 1)$
22. $3s^2 + 11s - 4$ $(3s - 1)(s + 4)$
23. $6r^2 - 5r + 1$ $(2r - 1)(3r - 1)$
24. $8a^2 + 15a - 2$ $(8a - 1)(a + 2)$
25. $w^2 - \frac{9}{4}$ $\left(w + \frac{3}{2}\right)\left(w - \frac{3}{2}\right)$
26. $c^2 - 64$ $(c - 8)(c + 8)$
27. $r^2 + 14r + 49$ $(r + 7)^2$
28. $b^2 + 18b + 81$ $(b + 9)^2$
29. $j^2 - 12j + 36$ $(j - 6)^2$
30. $4t^2 - 25$ $(2t - 5)(2t + 5)$

Solve each equation by factoring.

31. $10r^2 - 35r = 0$ $0, \frac{7}{2}$
32. $3x^2 + 15x = 0$ $0, -5$
33. $k^2 + 13k + 36 = 0$ $-4, -9$
34. $w^2 - 8w + 12 = 0$ $2, 6$
35. $c^2 - 5c - 14 = 0$ $-2, 7$
36. $z^2 - z - 42 = 0$ $-6, 7$
37. $2y^2 - 5y - 12 = 0$ $-\frac{3}{2}, 4$
38. $3b^2 - 4b - 15 = 0$ $-\frac{5}{3}, 3$
39. $t^2 + 12t + 36 = 0$ -6
40. $u^2 + 5u + \frac{25}{4} = 0$ $-\frac{5}{2}$
41. $q^2 - 8q + 16 = 0$ 4
42. $a^2 - 6a + 9 = 0$ 3

Prerequisite Skills

Prerequisite Skills 751

Prerequisite Skills

Operations with Matrices

- A **matrix** is a rectangular arrangement of numbers in rows and columns. Each entry in a matrix is called an **element**. A matrix is usually described by its **dimensions**, or the number of **rows** and **columns**, with the number of rows stated first.

- For example, matrix A has dimensions 3×2 and matrix B has dimensions 2×4.

$$\text{matrix } A = \begin{bmatrix} 6 & -2 \\ 0 & 5 \\ -4 & 10 \end{bmatrix} \qquad \text{matrix } B = \begin{bmatrix} 7 & -1 & -2 & 0 \\ 3 & 6 & -5 & 2 \end{bmatrix}$$

- If two matrices have the same dimensions, you can add or subtract them. To do this, add or subtract corresponding elements of the two matrices.

Example 1 **If $A = \begin{bmatrix} 12 & 7 & -3 \\ 0 & -1 & -6 \end{bmatrix}$, $B = \begin{bmatrix} -3 & 0 & 5 \\ 2 & 7 & -7 \end{bmatrix}$, and $C = \begin{bmatrix} 9 & 1 & -5 \\ 0 & -1 & 15 \end{bmatrix}$, find the sum and difference.**

a. $A + B$

$$A + B = \begin{bmatrix} 12 & 7 & -3 \\ 0 & -1 & -6 \end{bmatrix} + \begin{bmatrix} -3 & 0 & 5 \\ 2 & 7 & -7 \end{bmatrix} \quad \text{Substitution}$$

$$= \begin{bmatrix} 12 + (-3) & 7 + 0 & -3 + 5 \\ 0 + 2 & -1 + 7 & -6 + (-7) \end{bmatrix} \quad \text{Definition of matrix addition}$$

$$= \begin{bmatrix} 9 & 7 & 2 \\ 2 & 6 & -13 \end{bmatrix} \quad \text{Simplify.}$$

b. $B - C$

$$B - C = \begin{bmatrix} -3 & 0 & 5 \\ 2 & 7 & -7 \end{bmatrix} - \begin{bmatrix} 9 & 1 & -5 \\ 0 & -1 & 15 \end{bmatrix} \quad \text{Substitution}$$

$$= \begin{bmatrix} -3 - 9 & 0 - 1 & 5 - (-5) \\ 2 - 0 & 7 - (-1) & -7 - 15 \end{bmatrix} \quad \text{Definition of matrix subtraction}$$

$$= \begin{bmatrix} -12 & -1 & 10 \\ 2 & 8 & -22 \end{bmatrix} \quad \text{Simplify.}$$

- You can multiply any matrix by a constant called a *scalar*. This is called **scalar multiplication**. To perform scalar multiplication, multiply each element by the scalar.

Example 2 **If $D = \begin{bmatrix} -4 & 6 & -1 \\ 0 & 7 & 2 \\ -3 & -8 & -4 \end{bmatrix}$, find $2D$.**

$$2D = 2\begin{bmatrix} -4 & 6 & -1 \\ 0 & 7 & 2 \\ -3 & -8 & -4 \end{bmatrix} \quad \text{Substitution}$$

$$= \begin{bmatrix} 2(-4) & 2(6) & 2(-1) \\ 2(0) & 2(7) & 2(2) \\ 2(-3) & 2(-8) & 2(-4) \end{bmatrix} \quad \text{Definition of scalar multiplication}$$

$$= \begin{bmatrix} -8 & 12 & -2 \\ 0 & 14 & 4 \\ -6 & -16 & -8 \end{bmatrix} \quad \text{Simplify.}$$

752 Prerequisite Skills

- You can multiply two matrices if and only if the number of columns in the first matrix is equal to the number of rows in the second matrix. The product of two matrices is found by multiplying columns and rows. The entry in the first row and first column of AB, the resulting product, is found by multiplying corresponding elements in the first row of A and the first column of B and then adding.

Example 3 Find EF if $E = \begin{bmatrix} 3 & -2 \\ 0 & 6 \end{bmatrix}$ and $F = \begin{bmatrix} -1 & 5 \\ 6 & -3 \end{bmatrix}$.

$$EF = \begin{bmatrix} 3 & -2 \\ 0 & 6 \end{bmatrix} \cdot \begin{bmatrix} -1 & 5 \\ 6 & -3 \end{bmatrix}$$

Multiply the numbers in the first row of E by the numbers in the first column of F and add the products.

$$EF = \begin{bmatrix} 3 & -2 \\ 0 & 6 \end{bmatrix} \cdot \begin{bmatrix} -1 & 5 \\ 6 & -3 \end{bmatrix} = \begin{bmatrix} 3(-1) + (-2)(6) & \\ & \end{bmatrix}$$

Multiply the numbers in the first row of E by the numbers in the second column of F and add the products.

$$EF = \begin{bmatrix} 3 & -2 \\ 0 & 6 \end{bmatrix} \cdot \begin{bmatrix} -1 & 5 \\ 6 & -3 \end{bmatrix} = \begin{bmatrix} 3(-1) + (-2)(6) & 3(5) + (-2)(-3) \\ & \end{bmatrix}$$

Multiply the numbers in the second row of E by the numbers in the first column of F and add the products.

$$EF = \begin{bmatrix} 3 & -2 \\ 0 & 6 \end{bmatrix} \cdot \begin{bmatrix} -1 & 5 \\ 6 & -3 \end{bmatrix} = \begin{bmatrix} 3(-1) + (-2)(6) & 3(5) + (-2)(-3) \\ 0(-1) + 6(6) & \end{bmatrix}$$

Multiply the numbers in the second row of E by the numbers in the second column of F and add the products.

$$EF = \begin{bmatrix} 3 & -2 \\ 0 & 6 \end{bmatrix} \cdot \begin{bmatrix} -1 & 5 \\ 6 & -3 \end{bmatrix} = \begin{bmatrix} 3(-1) + (-2)(6) & 3(5) + (-2)(-3) \\ 0(-1) + 6(6) & 0(5) + 6(-3) \end{bmatrix}$$

Simplify the matrix.

$$\begin{bmatrix} 3(-1) + (-2)(6) & 3(5) + (-2)(-3) \\ 0(-1) + 6(6) & 0(5) + 6(-3) \end{bmatrix} = \begin{bmatrix} -15 & 21 \\ 36 & -18 \end{bmatrix}$$

Exercises If $A = \begin{bmatrix} 10 & -9 \\ 4 & -3 \\ -1 & 11 \end{bmatrix}$, $B = \begin{bmatrix} -1 & -3 \\ 2 & 8 \\ 7 & 6 \end{bmatrix}$, and $C = \begin{bmatrix} 8 & 0 \\ -2 & 2 \\ -10 & 6 \end{bmatrix}$, find each sum, difference, or product. **1–24. See margin.**

1. $A + B$
2. $B + C$
3. $A - C$
4. $C - B$
5. $3A$
6. $5B$
7. $-4C$
8. $\frac{1}{2}C$
9. $2A + C$
10. $A - 5C$
11. $\frac{1}{2}C + B$
12. $3A - 3B$

If $X = \begin{bmatrix} 2 & -8 \\ 10 & 4 \end{bmatrix}$, $Y = \begin{bmatrix} -1 & 0 \\ 6 & -5 \end{bmatrix}$, and $Z = \begin{bmatrix} 4 & -8 \\ -7 & 0 \end{bmatrix}$, find each sum, difference, or product.

13. $X + Z$
14. $Y + Z$
15. $X - Y$
16. $3Y$
17. $-6X$
18. $\frac{1}{2}X + Z$
19. $5Z - 2Y$
20. XY
21. YZ
22. XZ
23. $\frac{1}{2}(XZ)$
24. $XY + 2Z$

Prerequisite Skills

1. $\begin{bmatrix} 9 & -12 \\ 6 & 5 \\ 6 & 17 \end{bmatrix}$
2. $\begin{bmatrix} 7 & -3 \\ 0 & 10 \\ -3 & 12 \end{bmatrix}$
3. $\begin{bmatrix} 2 & -9 \\ 6 & -5 \\ 9 & 5 \end{bmatrix}$
4. $\begin{bmatrix} 9 & 3 \\ -4 & -6 \\ -17 & 0 \end{bmatrix}$
5. $\begin{bmatrix} 30 & -27 \\ 12 & -9 \\ -3 & 33 \end{bmatrix}$
6. $\begin{bmatrix} -5 & -15 \\ 10 & 40 \\ 35 & 30 \end{bmatrix}$
7. $\begin{bmatrix} -32 & 0 \\ 8 & -8 \\ 40 & -24 \end{bmatrix}$
8. $\begin{bmatrix} 4 & 0 \\ -1 & 1 \\ -5 & 3 \end{bmatrix}$
9. $\begin{bmatrix} 28 & -18 \\ 6 & -4 \\ -12 & 28 \end{bmatrix}$
10. $\begin{bmatrix} -30 & -9 \\ 14 & -13 \\ 49 & -19 \end{bmatrix}$
11. $\begin{bmatrix} 3 & -3 \\ 1 & 9 \\ 2 & 9 \end{bmatrix}$
12. $\begin{bmatrix} 33 & -18 \\ 6 & -33 \\ -24 & 15 \end{bmatrix}$
13. $\begin{bmatrix} 6 & -16 \\ 3 & 4 \end{bmatrix}$
14. $\begin{bmatrix} 3 & -8 \\ -1 & -5 \end{bmatrix}$
15. $\begin{bmatrix} 3 & -8 \\ 4 & 9 \end{bmatrix}$
16. $\begin{bmatrix} -3 & 0 \\ 18 & -15 \end{bmatrix}$
17. $\begin{bmatrix} -12 & 48 \\ -60 & -24 \end{bmatrix}$
18. $\begin{bmatrix} 5 & -12 \\ -2 & 2 \end{bmatrix}$
19. $\begin{bmatrix} 22 & -40 \\ -47 & 10 \end{bmatrix}$
20. $\begin{bmatrix} -50 & 40 \\ 14 & -20 \end{bmatrix}$
21. $\begin{bmatrix} -4 & 8 \\ 59 & -48 \end{bmatrix}$
22. $\begin{bmatrix} 64 & -16 \\ 12 & -80 \end{bmatrix}$
23. $\begin{bmatrix} 32 & -8 \\ 6 & -40 \end{bmatrix}$
24. $\begin{bmatrix} -42 & 24 \\ 0 & -20 \end{bmatrix}$

Prerequisite Skills 753

Lesson 1-1

8.

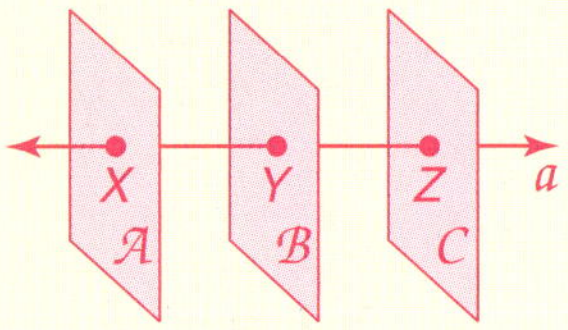

9.

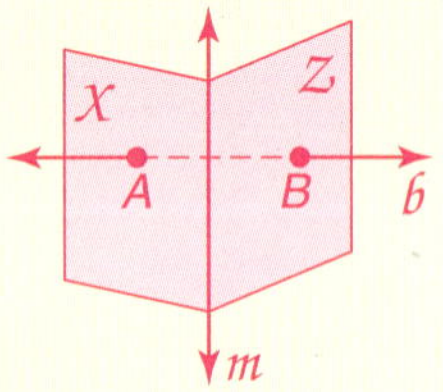

Extra Practice

Extra Practice

Answers not in back

Lesson 1-1

(pages 6–12)

For Exercises 1–7, refer to the figure.

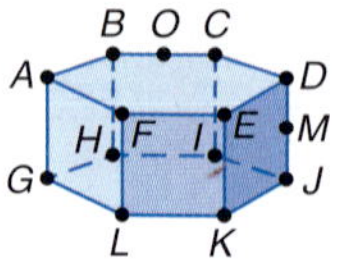

1. How many planes are shown in the figure? **8**
2. Name three collinear points. ***B*, *O*, *C* or *D*, *M*, *J***
3. Name all planes that contain point *G*. **planes *AFG*, *ABG*, and *GLK***
4. Name the intersection of plane *ABD* and plane *DJK*. $\overleftrightarrow{DE}$
5. Name two planes that do not intersect. **Sample answer: planes *ABD* and *GHJ***
6. Name a plane that contains $\overleftrightarrow{FK}$ and $\overleftrightarrow{EL}$. **plane *FEK***
7. Is the intersection of plane *ACD* and plane *EDJ* a point or a line? Explain. **A line; two planes intersect in a line, not a point.**

Draw and label a figure for each relationship. **8–9. See margin.**

8. Line *a* intersects planes 𝒜, ℬ, and *C* at three distinct points.
9. Planes *X* and *Z* intersect in line *m*. Line *b* intersects the two planes in two distinct points.

Extra Practice

Lesson 1-2

(pages 13–19)

Find the precision for each measurement. Explain its meaning. **2. 0.5 mm; 85.5 to 86.5 mm**

1. 42 in. **$\frac{1}{2}$ in.; $41\frac{1}{2}$ to $42\frac{1}{2}$ in.**
2. 86 mm
3. 251 cm **0.5 cm; 250.5 to 251.5 cm**
4. 33.5 in. **0.05 in.; 33.45 to 33.55 in.**
5. $5\frac{1}{4}$ ft **$\frac{1}{8}$ ft; $5\frac{1}{8}$ to $5\frac{3}{8}$ ft**
6. 89 m **0.5 m; 88.5 to 89.5 m**

Find the value of the variable and *BC* if *B* is between *A* and *C*.

7. $AB = 4x$, $BC = 5x$; $AB = 16$ **$x = 4$; $BC = 20$**
8. $AB = 17$, $BC = 3m$, $AC = 32$ **$m = 5$; $BC = 15$**
9. $AB = 9a$, $BC = 12a$, $AC = 42$ **$a = 2$; $BC = 24$**
10. $AB = 25$, $BC = 3b$, $AC = 7b + 13$ **$b = 3$; $BC = 9$**
11. $AB = 5n + 5$, $BC = 2n$; $AC = 54$ **$n = 7$; $BC = 14$**
12. $AB = 6c - 8$, $BC = 3c + 1$, $AC = 65$ **$c = 8$; $BC = 25$**

Lesson 1-3

(pages 21–27)

Use the Pythagorean Theorem to find the distance between each pair of points.

1. $A(0, 0)$, $B(-3, 4)$ **5**
2. $C(-1, 2)$, $N(5, 10)$ **10**
3. $X(-6, -2)$, $Z(6, 3)$ **13**
4. $M(-5, -8)$, $O(3, 7)$ **17**
5. $T(-10, 2)$, $R(6, -10)$ **20**
6. $F(5, -6)$, $N(-5, 6)$ **$\sqrt{244} \approx 15.6$**

Use the Distance Formula to find the distance between each pair of points.

7. $D(0, 0)$, $M(8, -7)$ **$\sqrt{113} \approx 10.6$**
8. $X(-1, 1)$, $Y(1, -1)$ **$\sqrt{8} \approx 2.8$**
9. $Z(-4, 0)$, $A(-3, 7)$ **$\sqrt{50} \approx 7.1$**
10. $K(6, 6)$, $D(-3, -3)$ **$\sqrt{162} \approx 12.7$**
11. $T(-1, 3)$, $N(0, 2)$ **$\sqrt{2} \approx 1.4$**
12. $S(7, 2)$, $E(-6, 7)$ **$\sqrt{194} \approx 13.9$**

Find the coordinates of the midpoint of a segment having the given endpoints.

13. $A(0, 0)$, $D(-2, -8)$ **$(-1, -4)$**
14. $D(-4, -3)$, $E(2, 2)$ **$(-1, -0.5)$**
15. $K(-4, -5)$, $M(5, 4)$ **$(0.5, -0.5)$**
16. $R(-10, 5)$, $S(8, 4)$ **$(-1, 4.5)$**
17. $B(2.8, -3.4)$, $Z(1.2, 5.6)$ **$(2, 1.1)$**
18. $D(-6.2, 7)$, $K(3.4, -4.8)$ **$(-1.4, 1.1)$**

Find the coordinates of the missing endpoint given that *B* is the midpoint of $\overline{AC}$.

19. $C(0, 0)$, $B(5, -6)$ **$(10, -12)$**
20. $C(-7, -4)$, $B(3, 5)$ **$(13, 14)$**
21. $C(8, -4)$, $B(-10, 2)$ **$(-28, 8)$**
22. $C(6, 8)$, $B(-3, 5)$ **$(-12, 2)$**
23. $C(6, -8)$, $B(3, -4)$ **$(0, 0)$**
24. $C(-2, -4)$, $B(0, 5)$ **$(2, 14)$**

754 Extra Practice

Lesson 1-4

(pages 29–36)

For Exercises 1–14, use the figure at the right. Name the vertex of each angle.

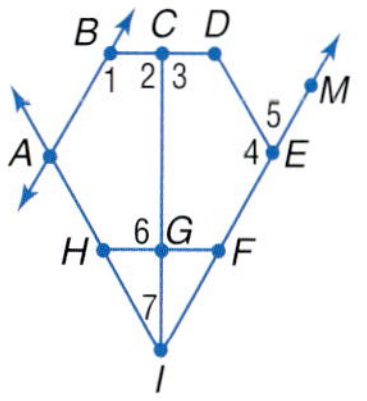

1. ∠1 **B**
2. ∠4 **E**
3. ∠6 **G**
4. ∠7 **I**

Name the sides of each angle.

5. ∠AIE $\overrightarrow{IA}$, $\overrightarrow{IE}$
6. ∠4 $\overrightarrow{ED}$, $\overrightarrow{EF}$
7. ∠6 $\overrightarrow{GC}$, $\overrightarrow{GH}$
8. ∠AHF $\overrightarrow{HA}$, $\overrightarrow{HF}$

Write another name for each angle.

9. ∠3 **∠DCG**
10. ∠DEF **∠4**
11. ∠2 **∠BCG**

Measure each angle and classify it as *right*, *acute*, or *obtuse*.

12. ∠ABC **120°, obtuse**
13. ∠CGF **90°, right**
14. ∠HIF **60°, acute**

Lesson 1-5

(pages 37–43)

For Exercises 1–7, refer to the figure.

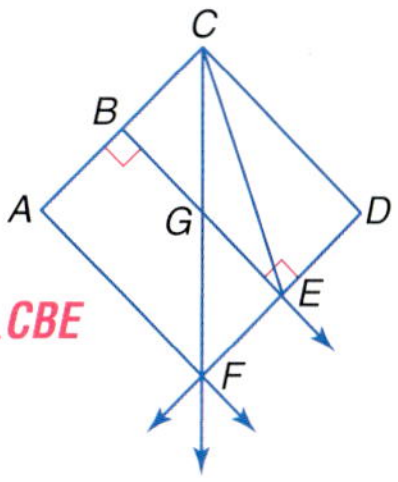

1. Name two acute vertical angles. **Sample answer: ∠BGC, ∠FGE**
2. Name two obtuse vertical angles. **Sample answer: ∠BGF, ∠CGE**
3. Name a pair of complementary adjacent angles. **Sample: ∠BEC, ∠CED**
4. Name a pair of supplementary adjacent angles. **Sample: ∠CEF, ∠CED**
5. Name a pair of congruent supplementary adjacent angles. **Sample: ∠ABE, ∠CBE**
6. If $m\angle BGC = 4x + 5$ and $m\angle FGE = 6x - 15$, find $m\angle BGF$. **135**
7. If $m\angle BCG = 5a + 5$, $m\angle GCE = 3a - 4$, and $m\angle ECD = 4a - 7$, find the value of a so that $\overline{AC} \perp \overline{CD}$. **8**
8. The measure of ∠A is nine less than the measure of ∠B. If ∠A and ∠B form a linear pair, what are their measures? **85.5, 94.5**
9. The measure of an angle's complement is 17 more than the measure of the angle. Find the measure of the angle and its complement. **36.5, 53.5**

Lesson 1-6

(pages 45–50)

Name each polygon by its number of sides. Classify it as *convex* or *concave* and *regular* or *irregular*. Then find the perimeter.

1.
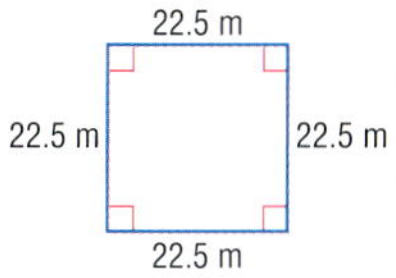

quadrilateral; convex; regular; 90 m

2.
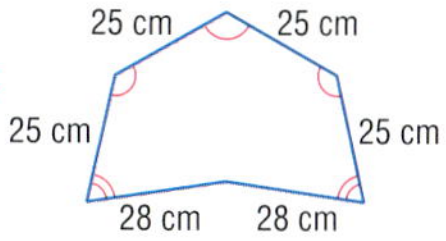

hexagon; concave; irregular; 156 cm

3.
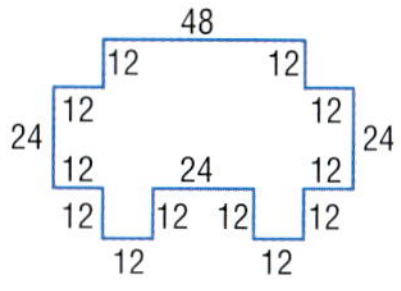

16-gon; concave; irregular; 264 in.

Find the perimeter of each polygon.

4. triangle with vertices at $X(3, 3)$, $Y(-2, 1)$, and $Z(1, -3)$ **≈ 16.7 units**
5. pentagon with vertices at $P(-2, 3)$, $E(-5, 0)$, $N(-2, -4)$, $T(2, -1)$, and $A(2, 2)$ **≈ 21.4 units**
6. hexagon with vertices at $H(0, 4)$, $E(-3, 2)$, $X(-3, -2)$, $G(0, -5)$, $O(5, -2)$, and $N(5, 2)$ **≈ 27.1 units**

Extra Practice

Extra Practice 755

Lesson 2-1

1.

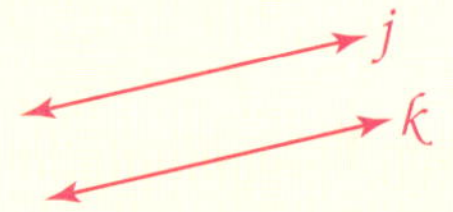

lines *j* and *k* do not intersect.

2.

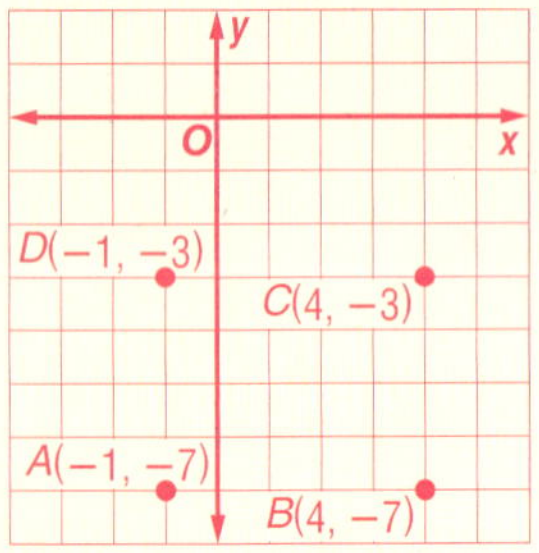

ABCD is a rectangle.

3.

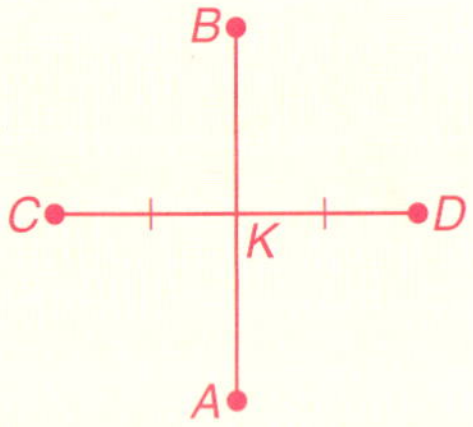

$CK = KD$

4.

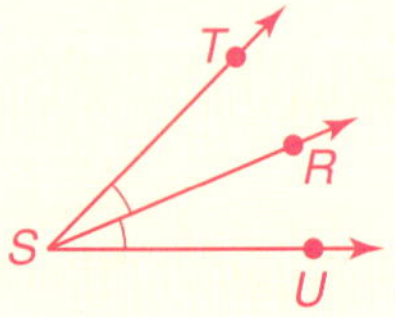

$\angle TSR \cong \angle RSU$

8.

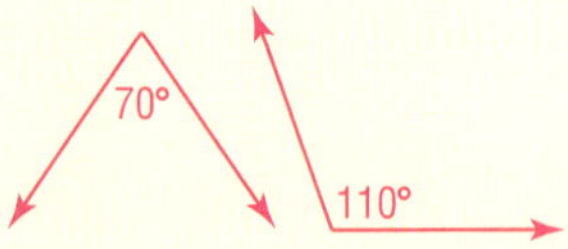

Lesson 2-2

1. $(-3)^2 = 9$ and a robin is a fish; false
2. $(-3)^2 = 9$ or a robin is a fish; true
3. $(-3)^2 = 9$ and an acute angle measures less than 90°; true
4. $(-3)^2 = 9$ or an acute angle measures less than 90°; true
5. $(-3)^2 \neq 9$ or a robin is a fish; false
6. $(-3)^2 = 9$ or an acute angle measures 90° or more; true
7. A robin is a fish and an acute angle measures less than 90°; false
8. $(-3)^2 = 9$ and a robin is a fish, or an acute angle measures less than 90°; true
9. $(-3)^2 \neq 9$ or an acute angle measures 90° or more; false

Lesson 2-1 (pages 62–66)

Make a conjecture based on the given information. Draw a figure to illustrate your conjecture. 1–4. See margin.

1. Lines *j* and *k* are parallel.
2. $A(-1, -7)$, $B(4, -7)$, $C(4, -3)$, $D(-1, -3)$
3. $\overline{AB}$ bisects $\overline{CD}$ at *K*.
4. $\overrightarrow{SR}$ is an angle bisector of $\angle TSU$.

Determine whether each conjecture is *true* or *false*. Give a counterexample for any false conjecture. 6. False; sample counterexample: $r = 0.5$

5. **Given:** *EFG* is an equilateral triangle.
   Conjecture: $EF = FG$ true
6. **Given:** *r* is a rational number.
   Conjecture: *r* is a whole number.
7. **Given:** *n* is a whole number.
   Conjecture: *n* is a rational number. true
8. **Given:** $\angle 1$ and $\angle 2$ are supplementary angles.
   Conjecture: $\angle 1$ and $\angle 2$ form a linear pair.
   False; see margin for counterexample.

Lesson 2-2 (pages 67–74)

Use the following statements to write a compound statement for each conjunction and disjunction. Then find its truth value. 1–9. See margin.

p: $(-3)^2 = 9$ *q*: A robin is a fish. *r*: An acute angle measures less than 90°.

1. *p* and *q*
2. *p* or *q*
3. *p* and *r*
4. *p* or *r*
5. $\sim p$ or *q*
6. *p* or $\sim r$
7. $q \wedge r$
8. $(p \wedge q) \vee r$
9. $\sim p \vee \sim r$

Copy and complete each truth table.

10.

p	q	$\sim q$	$p \vee \sim q$
T	T	F	T
T	F	T	T
F	T	F	F
F	F	T	T

11.

p	q	$\sim p$	$\sim q$	$\sim p \vee \sim q$
T	T	F	F	F
T	F	F	T	T
F	T	T	F	T
F	F	T	T	T

Lesson 2-3 (pages 75–80)

Identify the hypothesis and conclusion of each statement. 1–4. See margin.

1. If no sides of a triangle are equal, then it is a scalene triangle.
2. If it rains today, you will be wearing your raincoat.
3. If $6 - x = 11$, then $x = -5$.
4. If you are in college, you are at least 18 years old.

Write each statement in if-then form. 5–8. See margin.

5. The sum of the measures of two supplementary angles is 180.
6. A triangle with two congruent sides is an isosceles triangle.
7. Two lines that do not intersect are parallel lines.
8. A Saint Bernard is a dog.

Write the converse, inverse, and contrapositive of each conditional statement. Determine whether each related conditional is *true* or *false*. If a statement is false, find a counterexample. 9–12. See margin.

9. All triangles are polygons.
10. If two angles are congruent angles, then they have the same measure.
11. If three points lie on the same line, then they are collinear.
12. If $\overleftrightarrow{PQ}$ is a perpendicular bisector of $\overline{LM}$, then a right angle is formed.

756 Extra Practice

Lesson 2-3

1. H: no sides of a triangle are equal; C: it is a scalene triangle
2. H: it rains today; C: you will be wearing your raincoat
3. H: $6 - x = 11$; C: $x = -5$
4. H: you are in college; C: you are at least 18 years old
5. If two angles are supplementary, then the sum of their measures is 180.
6. If a triangle has two congruent sides, then it is an isosceles triangle.
7. If two lines do not intersect, then they are parallel lines.
8. If an animal is a Saint Bernard, then it is a dog.

Lesson 2-4

(pages 82–87)

Use the Law of Syllogism to determine whether a valid conclusion can be reached from each set of statements. If a valid conclusion is possible, write it. If not, write *no conclusion*.

1. (1) If it rains, then the field will be muddy. See margin.
   (2) If the field is muddy, then the game will be cancelled.
2. (1) If you read a book, then you enjoy reading.
   (2) If you are in the 10th grade, then you passed the 9th grade. no conclusion

Determine if statement (3) follows from statements (1) and (2) by the Law of Detachment or the Law of Syllogism. If it does, state which law was used. If it does not, write *invalid*.

3. (1) If it snows outside, you will wear your winter coat.
   (2) It is snowing outside.
   (3) You will wear your winter coat. yes; Law of Detachment
4. (1) Two complementary angles are both acute angles.
   (2) $\angle 1$ and $\angle 2$ are acute angles.
   (3) $\angle 1$ and $\angle 2$ are complementary angles. invalid

Lesson 2-5

(pages 89–93)

Determine whether the following statements are *always*, *sometimes*, or *never* true. Explain.

1. $\overrightarrow{RS}$ is perpendicular to $\overrightarrow{PS}$. Sometimes; $\overrightarrow{RS}$ and $\overrightarrow{PS}$ could intersect to form a 45° angle.
2. Three points will lie on one line. Sometimes; if they are collinear, then they lie on one line.
3. Points B and C are in plane $\mathcal{K}$. A line perpendicular to line BC is in plane $\mathcal{K}$. Sometimes; the line could lie in a plane perpendicular to plane $\mathcal{K}$.

For Exercises 4–7, use the figure at the right. In the figure, $\overleftrightarrow{EC}$ and $\overleftrightarrow{CD}$ are in plane $\mathcal{R}$, and F is on $\overleftrightarrow{CD}$. State the postulate that can be used to show each statement is true. 4–7. See margin.

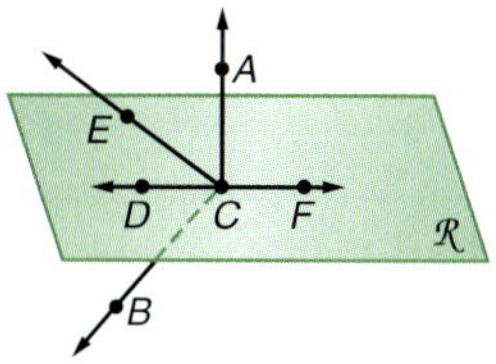

4. $\overleftrightarrow{DF}$ lies in plane $\mathcal{R}$.
5. E and C are collinear.
6. D, F, and E are coplanar.
7. E and F are collinear.

Lesson 2-6

(pages 94–100)

State the property that justifies each statement.

1. If $x - 5 = 6$, then $x = 11$. Addition Property
2. If $AB = CD$ and $CD = EF$, then $AB = EF$. Transitive Property
3. If $a - b = r$, then $r = a - b$. Symmetric Property
4. Copy and complete the following proof.

Given: $\frac{5x-1}{8} = 3$

Prove: $x = 5$

Proof:

Statements	Reasons
a. ? a. $\frac{5x-1}{8} = 3$	a. Given
b. ? b. $8\left(\frac{5x-1}{8}\right) = 8(3)$	b. Multiplication Prop.
c. $5x - 1 = 24$	c. ? Substitution
d. $5x = 25$	d. ? Addition Prop.
e. ? $x = 5$	e. Division Property

9. Converse: If a figure is a polygon, then it is a triangle; false; pentagons are polygons but are not triangles.
   Inverse: If a figure is not a triangle, then it is not a polygon; false; a hexagon is not a triangle, but it is a polygon.
   Contrapositive: If a figure is not a polygon, then it is not a triangle; true

10. Converse: If two angles have the same measure, then they are congruent angles; true
    Inverse: If two angles are not congruent angles, then they do not have the same measure; true
    Contrapositive: If two angles do not have the same measure, then they are not congruent angles; true

11. Converse: If three points are collinear, then they lie on the same line; true
    Inverse: If three points do not lie on the same line, then they are not collinear; true
    Contrapositive: If three points are not collinear, then they do not lie on the same line; true

12. Converse: If a right angle is formed by $\overleftrightarrow{PQ}$ and $\overline{LM}$, then $\overleftrightarrow{PQ}$ is a perpendicular bisector of $\overline{LM}$; false; $\overleftrightarrow{PQ}$ may not pass through the midpoint of $\overline{LM}$.
    Inverse: If $\overleftrightarrow{PQ}$ is not a perpendicular bisector of $\overline{LM}$, then a right angle is not formed; false; $\overleftrightarrow{PQ}$ could be perpendicular to $\overline{LM}$, without bisecting $\overline{LM}$.
    Contrapositive: If a right angle is not formed by $\overleftrightarrow{PQ}$ and $\overline{LM}$ then $\overleftrightarrow{PQ}$ is not a perpendicular bisector of $\overline{LM}$; true

Lesson 2-4

1. If it rains then the game will be cancelled.

Lesson 2-5

4. If two points lie in a plane, then the entire line containing those points lies in that plane.
5. Through any two points, there is exactly one line.
6. Through any three points not on the same line, there is exactly one plane.
7. Through any two points, there is exactly one line.

Extra Practice 757

Lesson 2-7

9. Given: $\overline{AB} \cong \overline{AF}$, $\overline{AF} \cong \overline{ED}$, $\overline{ED} \cong \overline{CD}$

Prove: $\overline{AB} \cong \overline{CD}$

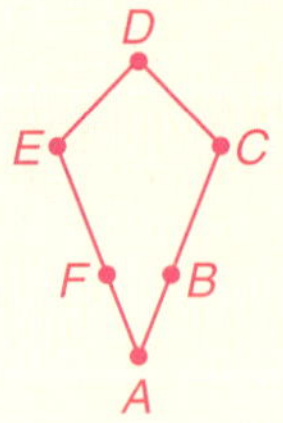

Proof:

Statements (Reasons)
1. $\overline{AB} \cong \overline{AF}$, $\overline{AF} \cong \overline{ED}$ (Given)
2. $\overline{AB} \cong \overline{ED}$ (Transitive)
3. $\overline{ED} \cong \overline{CD}$ (Given)
4. $\overline{AB} \cong \overline{CD}$ (Transitive)

10. Given: $AC = DF$, $AB = DE$

Prove: $BC = EF$

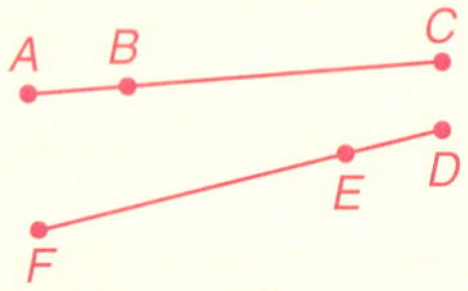

Proof:

Statements (Reasons)
1. $AC = AB + BC$ and $DF = DE + EF$ (Segment Addition Postulate)
2. $AC = DF$ (Given)
3. $AB + BC = DE + EF$ (Substitution)
4. $AB = DE$ (Given)
5. $BC = EF$ (Subtraction)

Lesson 2-7

(pages 101–106)

Justify each statement with a property of equality or a property of congruence.

1. If $CD = OP$, then $CD + GH = OP + GH$. Addition
2. If $\overline{MN} \cong \overline{PQ}$, then $\overline{PQ} \cong \overline{MN}$. Symmetric
3. If $\overline{TU} \cong \overline{JK}$ and $\overline{JK} \cong \overline{DF}$, then $\overline{TU} \cong \overline{DF}$. Transitive
4. If $AB = 10$ and $CD = 10$, then $AB = CD$. Substitution
5. $\overline{XB} \cong \overline{XB}$ Reflexive
6. If $GH = RS$, then $GH - VW = RS - VW$. Subtraction
7. If $EF = XY$, then $EF + KL = XY + KL$. Addition
8. If $\overline{JK} \cong \overline{XY}$ and $\overline{XY} \cong \overline{LM}$, then $\overline{JK} \cong \overline{LM}$. Transitive

Write a two-column proof. 9–10. See margin.

9. **Given:** $\overline{AB} \cong \overline{AF}$, $\overline{AF} \cong \overline{ED}$, $\overline{ED} \cong \overline{CD}$
   Prove: $\overline{AB} \cong \overline{CD}$

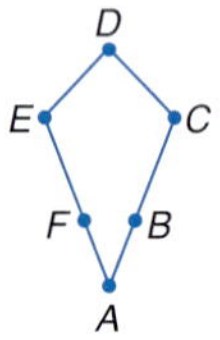

10. **Given:** $AC = DF$, $AB = DE$
    Prove: $BC = EF$

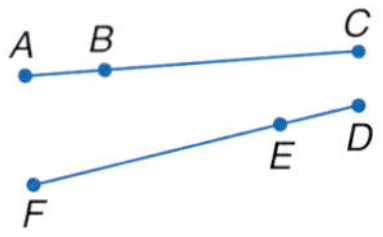

Lesson 2-8

(pages 107–114)

Find the measure of each numbered angle.

1. $m\angle 9 = 141 + x$
   $m\angle 10 = 25 + x$

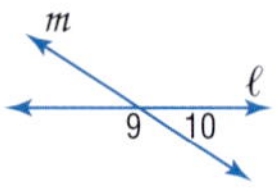

$m\angle 9 = 148$, $m\angle 10 = 32$

2. $m\angle 11 = x + 40$
   $m\angle 12 = x + 10$
   $m\angle 13 = 3x + 30$

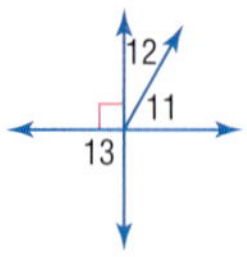

$m\angle 11 = 60$, $m\angle 12 = 30$, $m\angle 13 = 90$

3. $m\angle 14 = x + 25$
   $m\angle 15 = 4x + 50$
   $m\angle 16 = x + 45$

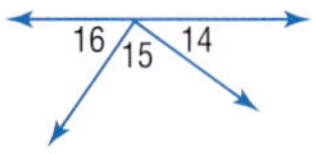

$m\angle 14 = 35$, $m\angle 15 = 90$, $m\angle 16 = 55$

Determine whether the following statements are *always*, *sometimes*, or *never* true.

4. Two angles that are complementary are congruent. sometimes
5. Two angles that form a linear pair are complementary. never
6. Two congruent angles are supplementary. sometimes
7. Perpendicular lines form four right angles. always
8. Two right angles are supplementary. always
9. Two lines intersect to form four right angles. sometimes

Lesson 3-1

(pages 126–131)

For Exercises 1–3, refer to the figure at the right.

1. Name all segments parallel to $\overline{AE}$. $\overline{LP}$
2. Name all planes intersecting plane BCN.
3. Name all segments skew to $\overline{DC}$.

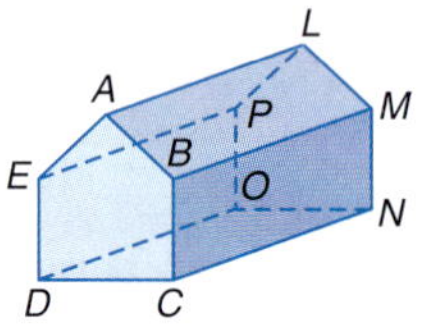

2. *ABM*, *OCN*, *ABC*, *LMN*, *AEP*
3. $\overline{BM}$, $\overline{AL}$, $\overline{EP}$, $\overline{OP}$, $\overline{PL}$, $\overline{LM}$, $\overline{MN}$

Identify each pair of angles as *alternate interior*, *alternate exterior*, *corresponding*, or *consecutive interior* angles.

4. $\angle 2$ and $\angle 5$ cons. int.
5. $\angle 9$ and $\angle 13$ corresponding
6. $\angle 12$ and $\angle 13$ alt. int.
7. $\angle 3$ and $\angle 6$ alt. ext.

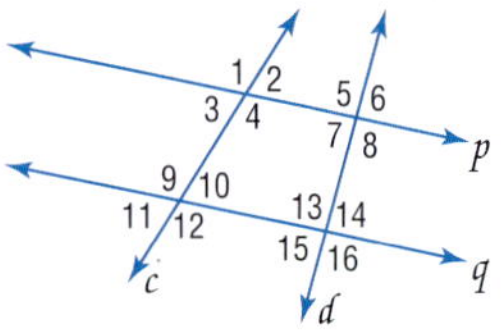

Lesson 3-2
(pages 133–138)

In the figure, $m\angle 5 = 72$ and $m\angle 9 = 102$. Find the measure of each angle.

1. $m\angle 1$ **102**
2. $m\angle 13$ **72**
3. $m\angle 4$ **102**
4. $m\angle 10$ **78**
5. $m\angle 7$ **108**
6. $m\angle 16$ **72**

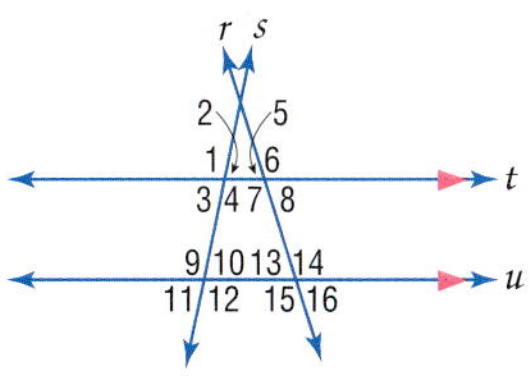

Find x and y in each figure.

7.

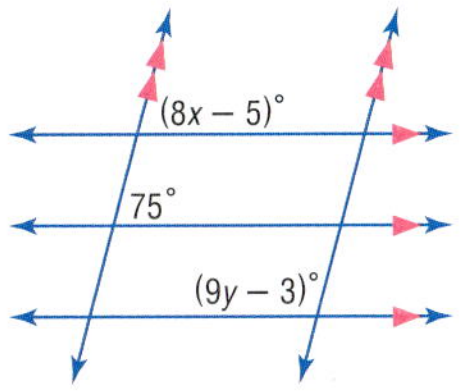

$x = 10$; $y = 12$

8.

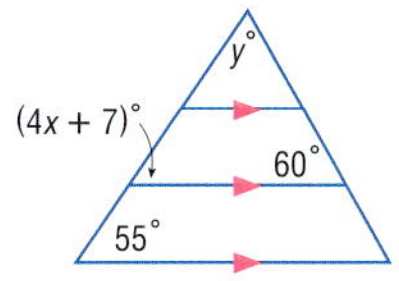

$x = 12$; $y = 65$

Lesson 3-3
(pages 139–144)

Find the slope of each line.

1. $\overleftrightarrow{RS}$ **$-\frac{4}{3}$**
2. $\overleftrightarrow{TU}$ **$\frac{1}{6}$**
3. $\overleftrightarrow{WV}$ **undefined**
4. $\overleftrightarrow{WR}$ **$-\frac{1}{5}$**
5. a line parallel to $\overleftrightarrow{TU}$ **$\frac{1}{6}$**
6. a line perpendicular to $\overleftrightarrow{WR}$ **5**
7. a line perpendicular to $\overleftrightarrow{WV}$ **0**

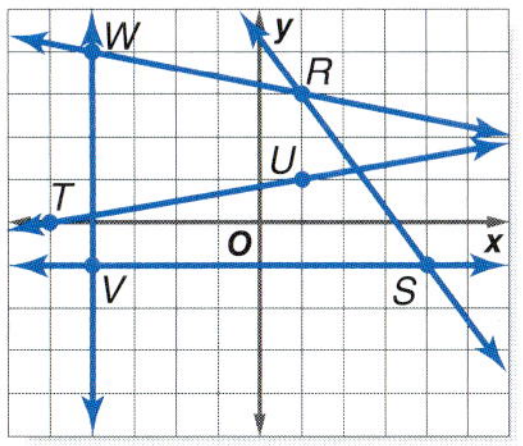

Determine whether $\overleftrightarrow{RS}$ and $\overleftrightarrow{TU}$ are *parallel, perpendicular,* or *neither*.

8. $R(3, 5)$, $S(5, 6)$, $T(-2, 0)$, $U(4, 3)$ **parallel**
9. $R(5, 11)$, $S(2, 2)$, $T(-1, 0)$, $U(2, 1)$ **neither**
10. $R(-1, 4)$, $S(-3, 7)$, $T(5, -1)$, $U(8, 1)$ **perpendicular**
11. $R(-2, 5)$, $S(-4, 1)$, $T(3, 3)$, $U(1, 5)$ **neither**

Lesson 3-4
(pages 145–150)

Write an equation in slope-intercept form of the line having the given slope and y-intercept.

1. $m = 1$, y-intercept: -5 **$y = x - 5$**
2. $m = -\frac{1}{2}$, y-intercept: $\frac{1}{2}$ **$y = -\frac{1}{2}x + \frac{1}{2}$**
3. $m = 3$, $b = -\frac{1}{4}$ **$y = 3x - \frac{1}{4}$**

Write an equation in point-slope form of the line having the given slope that contains the given point.

4. $m = 3$, $(-2, 4)$ **$y - 4 = 3(x + 2)$**
5. $m = -4$, $(0, 3)$ **$y - 3 = -4x$**
6. $m = \frac{2}{3}$, $(5, -7)$ **$y + 7 = \frac{2}{3}(x - 5)$**

For Exercises 7–14, use the graph at the right. Write an equation in slope-intercept form for each line.

7. p **$y = -2x + 1$**
8. q **$y = x - 3$**
9. r **$y = \frac{2}{3}x - 2$**
10. s **$y = -\frac{1}{3}x$**
11. parallel to line q, contains $(2, -5)$ **$y = x - 7$**
12. perpendicular to line r, contains $(0, 1)$ **$y = -\frac{3}{2}x + 1$**
13. parallel to line s, contains $(-2, -2)$ **$y = -\frac{1}{3}x - \frac{8}{3}$**
14. perpendicular to line p, contains $(0, 0)$ **$y = \frac{1}{2}x$**

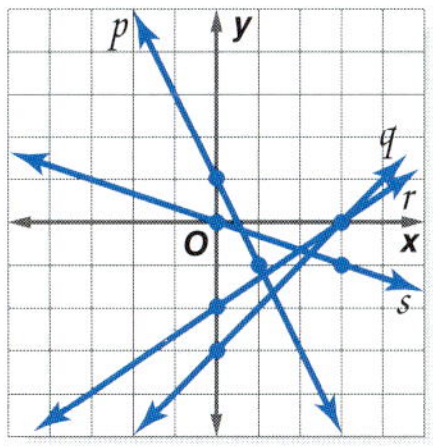

Extra Practice

Extra Practice

Lesson 3-5

1. $c \parallel d$; ≅ alternate exterior ∠s
2. none
3. $c \parallel d$; ≅ alternate interior ∠s
4. $c \parallel d$; supplementary consecutive interior ∠s

Lesson 3-6

7. $d = \frac{7\sqrt{2}}{2}$;

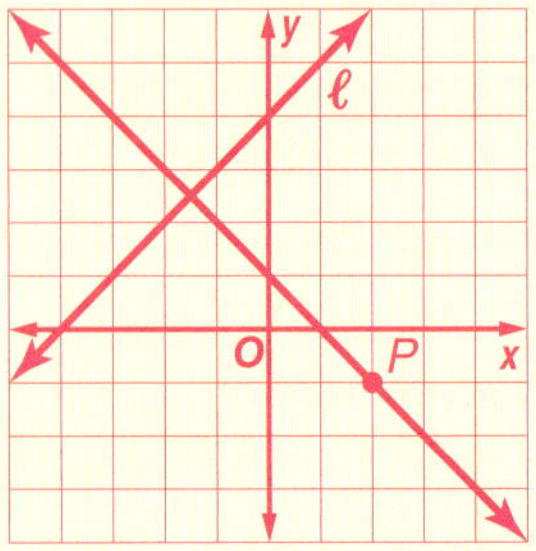

8. $d \approx 1.4$;

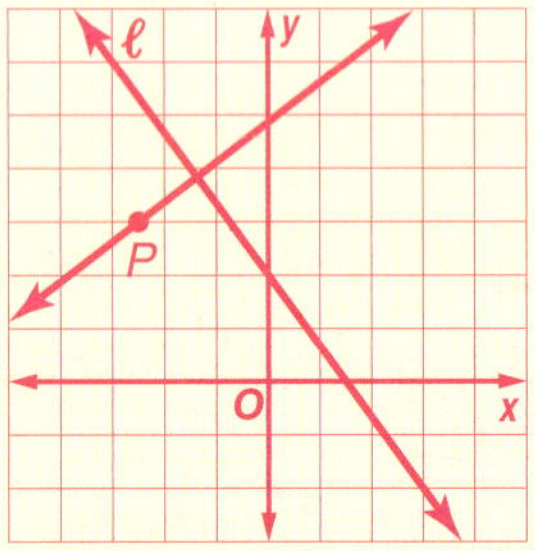

Extra Practice

Lesson 3-5 *(pages 151–157)*

Given the following information, determine which lines, if any, are parallel. State the postulate or theorem that justifies your answer.

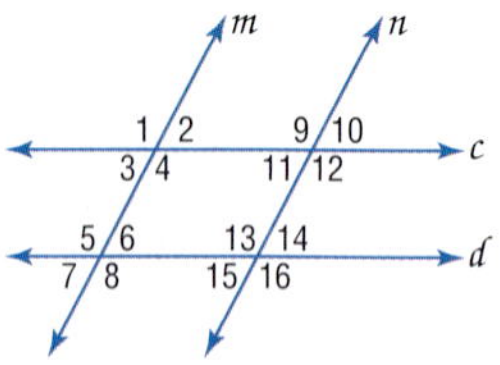

1. $\angle 9 \cong \angle 16$
2. $\angle 10 \cong \angle 16$
3. $\angle 12 \cong \angle 13$
4. $m\angle 12 + m\angle 14 = 180$ **1–4. See margin.**

Find x so that $r \parallel s$.

5. 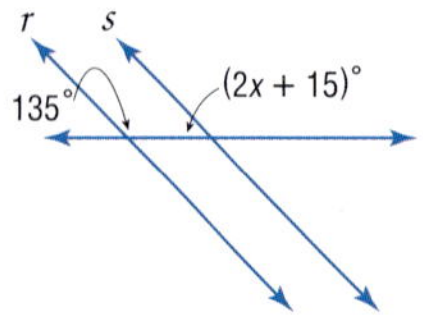

15

6. 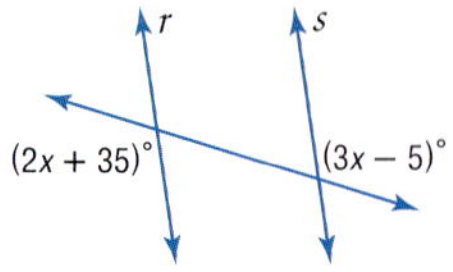

40

7. 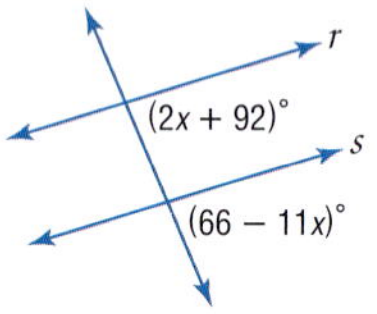

−2

Lesson 3-6 *(pages 159–164)*

Copy each figure. Draw the segment that represents the distance indicated.

1. P to $\overleftrightarrow{RS}$

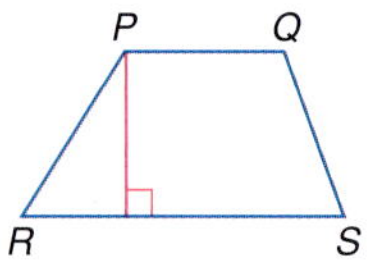

2. J to $\overleftrightarrow{KL}$

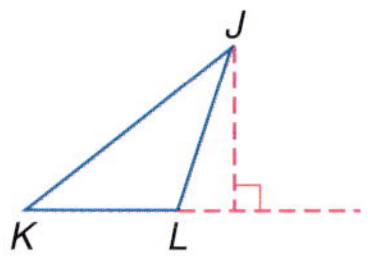

3. B to $\overleftrightarrow{FE}$

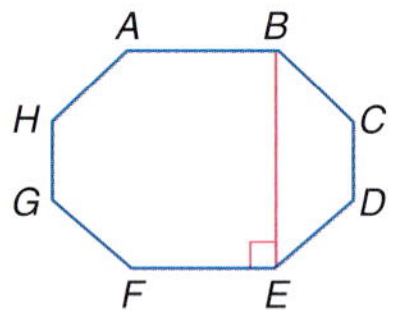

Find the distance between each pair of parallel lines.

4. $y = \frac{2}{3}x - 2$, $y = \frac{2}{3}x + \frac{1}{2}$ **≈ 2.08**
5. $y = 2x + 4$, $y - 2x = -5$ **≈ 4.02**
6. $x + 4y = -6$, $x + 4y = 4$ **≈ 2.43**

COORDINATE GEOMETRY Construct a line perpendicular to ℓ through P. Then find the distance from P to ℓ. **7–8. See margin.**

7. Line ℓ contains points (0, 4) and (−4, 0). Point P has coordinates (2, −1).
8. Line ℓ contains points (3, −2) and (0, 2). Point P has coordinates (−2.5, 3).

Lesson 4-1 *(pages 178–183)*

Use a protractor to classify each triangle as *acute*, *equiangular*, *obtuse*, or *right*.

1. 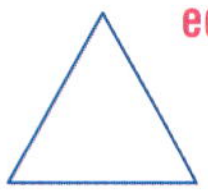 **equiangular**
2. 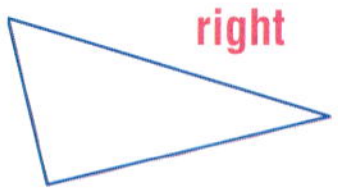 **right**
3. **obtuse** 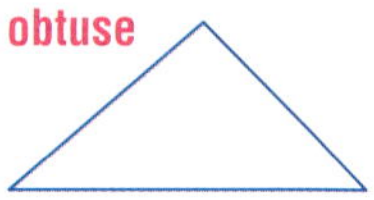

Identify the indicated type of triangles in the figure if $\overline{AB} \cong \overline{CD}$, $\overline{AD} \cong \overline{BC}$, $\overline{AE} \cong \overline{BE} \cong \overline{EC} \cong \overline{ED}$, and $m\angle BAD = m\angle ABC = m\angle BCD = m\angle ADC = 90$.

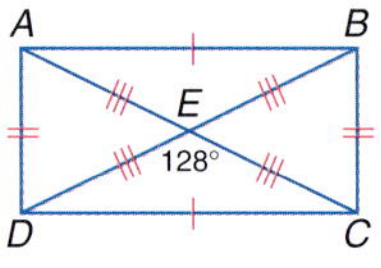

4. right
5. obtuse **△ABE, △CDE**
6. acute **△BEC, △AED**
7. isosceles **△ABE, △CDE, △BEC, △AED**

4. △DAB, △ABC, △BCD, △ADC

8. Find a and the measure of each side of equilateral triangle MNO if $MN = 5a$, $NO = 4a + 6$, and $MO = 7a - 12$. **$a = 6$; $MN = NO = MO = 30$**
9. Triangle TAC is an isosceles triangle with $\overline{TA} \cong \overline{AC}$. Find b, TA, AC, and TC if $TA = 3b + 1$, $AC = 4b - 11$, and $TC = 6b - 2$. **$b = 12$; $TA = AC = 37$, $TC = 70$**

Lesson 4-3

5. Given: $\triangle ANG \cong \triangle NGA$, $\triangle NGA \cong \triangle GAN$
Prove: $\triangle AGN$ is equilateral and equiangular.

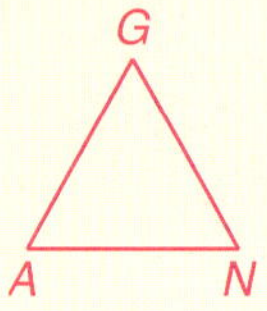

Proof:
Statements (Reasons)

1. $\triangle ANG \cong \triangle NGA$ (Given)
2. $\overline{AN} \cong \overline{NG}$, $\angle A \cong \angle N$ (CPCTC)
3. $\triangle NGA \cong \triangle GAN$ (Given)
4. $\overline{NG} \cong \overline{GA}$, $\angle N \cong \angle G$ (CPCTC)
5. $\overline{AN} \cong \overline{NG} \cong \overline{GA}$ (Transitive Property of ≅)
6. $\triangle AGN$ is equilateral. (Def. of equilateral △)
7. $\angle A \cong \angle N \cong \angle G$ (Transitive Property of ≅)
8. $\triangle AGN$ is equiangular. (Def. of equiangular △)

Lesson 4-2
(pages 185–191)

Find the measure of each angle.

1. $\angle 1$ **60** 2. $\angle 2$ **60**
3. $\angle 3$ **55** 4. $\angle 4$ **120**
5. $\angle 5$ **94** 6. $\angle 6$ **86**
7. $\angle 7$ **94** 8. $\angle 8$ **86**
9. $\angle 9$ **52** 10. $\angle 10$ **24**

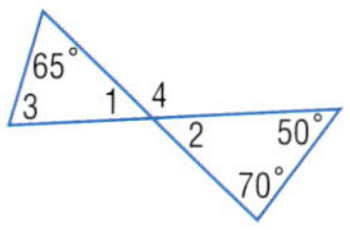

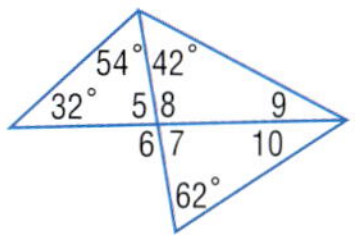

Lesson 4-3
(pages 192–198)

Identify the congruent triangles in each figure.

1. 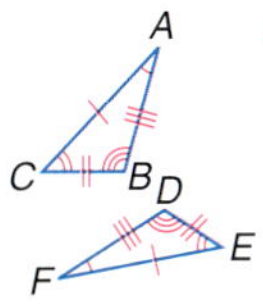

$\triangle ABC \cong \triangle FDE$

2. 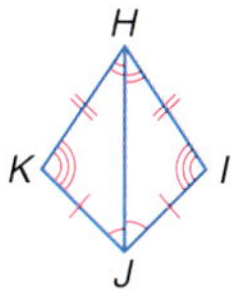

$\triangle JKH \cong \triangle JIH$

3. 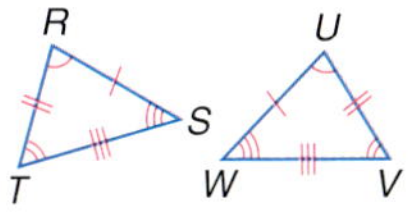

$\triangle RTS \cong \triangle UVW$

4. 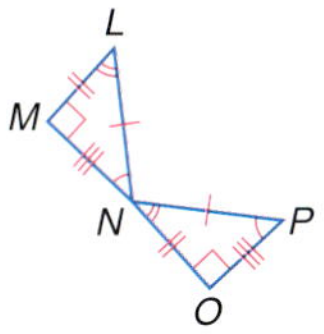

$\triangle LMN \cong \triangle NOP$

5. Write a two-column proof. **See margin.**
 Given: $\triangle ANG \cong \triangle NGA$
 $\triangle NGA \cong \triangle GAN$
 Prove: $\triangle AGN$ is equilateral and equiangular.

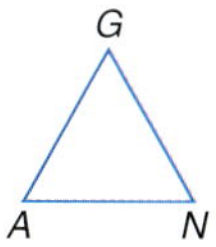

Lesson 4-4
(pages 200–206)

Determine whether $\triangle RST \cong \triangle JKL$ given the coordinates of the vertices. Explain.

1. $R(-6, 2)$, $S(-4, 4)$, $T(-2, 2)$, $J(6, -2)$, $K(4, -4)$, $L(2, -2)$ **Yes; see margin for explanation.**
2. $R(-6, 3)$, $S(-4, 7)$, $T(-2, 3)$, $J(2, 3)$, $K(5, 7)$, $L(6, 3)$ **No; see margin for explanation.**

Write a two-column proof. **3–4. See margin.**

3. **Given:** $\triangle GWN$ is equilateral.
 $\overline{WS} \cong \overline{WI}$
 $\angle SWG \cong \angle IWN$
 Prove: $\triangle SWG \cong \triangle IWN$

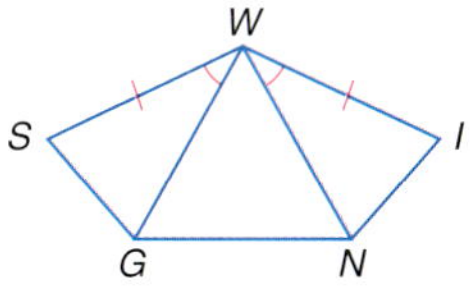

4. **Given:** $\triangle ANM \cong \triangle ANI$
 $\overline{DI} \cong \overline{OM}$
 $\overline{ND} \cong \overline{NO}$
 Prove: $\triangle DIN \cong \triangle OMN$

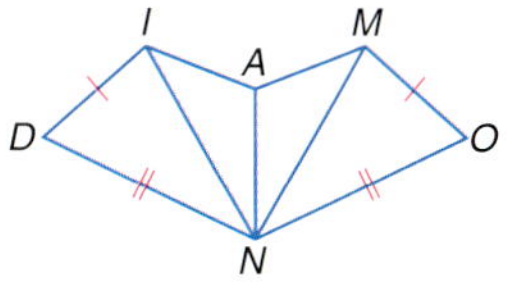

Lesson 4-4

1. $RS = \sqrt{(-6-(-4))^2 + (4-2)^2} = \sqrt{4+4}$ or $\sqrt{8}$
 $JK = \sqrt{(6-4)^2 + (-2-(-4))^2} = \sqrt{4+4}$ or $\sqrt{8}$
 $ST = \sqrt{(-4-(-2))^2 + (4-2)^2} = \sqrt{4+4}$ or $\sqrt{8}$
 $KL = \sqrt{(4-2)^2 + (-4-(-2))^2} = \sqrt{4+4}$ or $\sqrt{8}$
 $RT = \sqrt{(-6-(-2))^2 + (2-2)^2} = \sqrt{16}$ or 4
 $JL = \sqrt{(6-2)^2 + (-2-(-2))^2} = \sqrt{16}$ or 4
 $RS = JK$, $ST = KL$, and $RT = JL$. By definition of congruent segments, all corresponding segments are congruent. Therefore, $\triangle RST \cong \triangle JKL$.

2. $RS = \sqrt{(-6-(-4))^2 + (3-7)^2} = \sqrt{4+16}$ or $\sqrt{20}$
 $JK = \sqrt{(2-5)^2 + (3-7)^2} = \sqrt{9+16}$ or 5
 Since, $RS \neq JK$ the triangles are not congruent.

3. Given: $\triangle GWN$ is equilateral.
 $\overline{WS} \cong \overline{WI}$
 $\angle SWG \cong \angle IWN$
 Prove: $\triangle SWG \cong \triangle IWN$

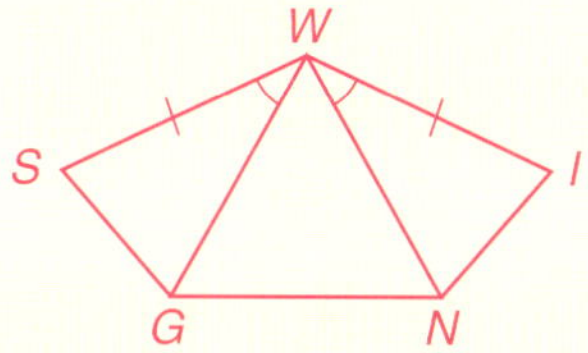

Proof:

Statements (Reasons)

1. $\triangle GWN$ is equilateral. (Given)
2. $\overline{WG} \cong \overline{WN}$ (Def. of equilateral triangle)
3. $\overline{WS} \cong \overline{WI}$ (Given)
4. $\angle SWG \cong \angle IWN$ (Given)
5. $\triangle SWG \cong \triangle IWN$ (SAS)

4. Given: $\triangle ANM \cong \triangle ANI$
 $\overline{DI} \cong \overline{OM}$
 $\overline{ND} \cong \overline{NO}$
 Prove: $\triangle DIN \cong \triangle OMN$

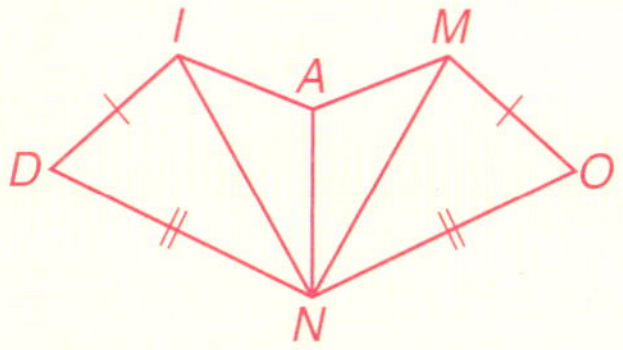

Proof:

Statements (Reasons)

1. $\triangle ANM \cong \triangle ANI$ (Given)
2. $\overline{IN} \cong \overline{MN}$ (CPCTC)
3. $\overline{DI} \cong \overline{OM}$ (Given)
4. $\overline{ND} \cong \overline{NO}$ (Given)
5. $\triangle DIN \cong \triangle OMN$ (SSS)

Extra Practice

Lesson 4-5

1. Given: $\triangle TEN$ is isosceles with base $\overline{TN}$. $\angle 1 \cong \angle 4$, $\angle T \cong \angle N$

   Prove: $\triangle TEC \cong \triangle NEA$

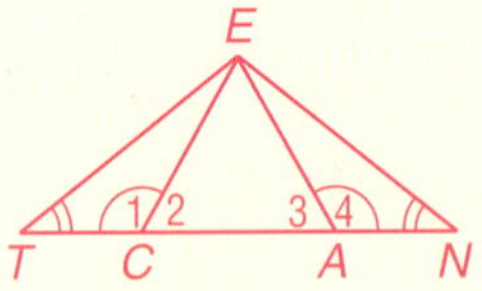

Proof: If $\triangle TEN$ is isosceles with base $\overline{TN}$, then $\overline{TE} \cong \overline{NE}$. Since $\angle 1 \cong \angle 4$ and $\angle T \cong \angle N$ are given, then $\triangle TEC \cong \triangle NEA$ by AAS.

2. Given: $\angle S \cong \angle W$, $\overline{SY} \cong \overline{YW}$

   Prove: $\overline{ST} \cong \overline{WV}$

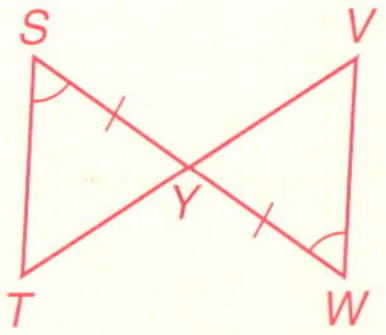

Proof: $\angle S \cong \angle W$ and $\overline{SY} \cong \overline{YW}$ are given and $\angle SYT \cong \angle WYV$ since vertical angles are congruent. Then $\triangle SYT \cong \triangle WYV$ by ASA and $\overline{ST} \cong \overline{WV}$ by CPCTC.

3. Given: $\angle 1 \cong \angle 2$, $\angle 3 \cong \angle 4$

   Prove: $\overline{PT} \cong \overline{LX}$

   Proof:

   - $\angle 1 \cong \angle 2$ (Given)
   - $\angle 3 \cong \angle 4$ (Given)
   - $\overline{TX} \cong \overline{TX}$ (Reflexive Prop.)
   - → $\triangle PXT \cong \triangle LTX$ (AAS)
   - → $\overline{PT} \cong \overline{LX}$ (CPCTC)

4. Given: $\overline{FP} \parallel \overline{ML}$, $\overline{FL} \parallel \overline{MP}$

   Prove: $\overline{MP} \cong \overline{FL}$

   Proof:

   - $\overline{FP} \parallel \overline{ML}$ (Given) → $\angle 3 \cong \angle 4$ (Alt. Int. ∠s Th.)
   - $\overline{FL} \parallel \overline{MP}$ (Given) → $\angle 1 \cong \angle 2$ (Alt. Int. ∠s Th.)
   - $\overline{PL} \cong \overline{PL}$ (Reflex. Prop.)
   - → $\triangle FLP \cong \triangle MPL$ (ASA)
   - → $\overline{MP} \cong \overline{FL}$ (CPCTC)

Lesson 4-5 (pages 207–213)

Write a paragraph proof. 1–2. See margin.

1. Given: $\triangle TEN$ is isosceles with base $\overline{TN}$. $\angle 1 \cong \angle 4$, $\angle T \cong \angle N$

Prove: $\triangle TEC \cong \triangle NEA$

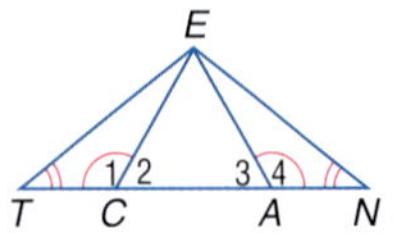

2. Given: $\angle S \cong \angle W$, $\overline{SY} \cong \overline{YW}$

Prove: $\overline{ST} \cong \overline{WV}$

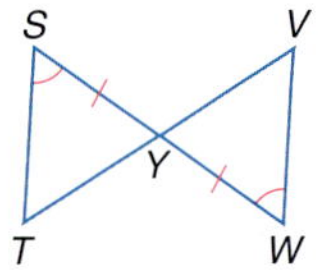

Write a flow proof. 3–4. See margin.

3. Given: $\angle 1 \cong \angle 2$, $\angle 3 \cong \angle 4$

Prove: $\overline{PT} \cong \overline{LX}$

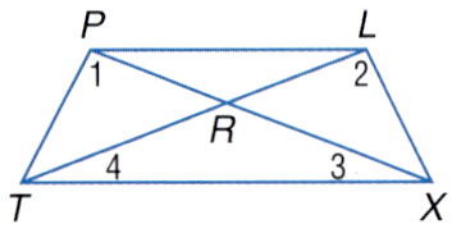

4. Given: $\overline{FP} \parallel \overline{ML}$, $\overline{FL} \parallel \overline{MP}$

Prove: $\overline{MP} \cong \overline{FL}$

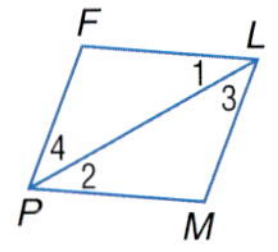

Lesson 4-6 (pages 216–221)

Refer to the figure for Exercises 1–6.

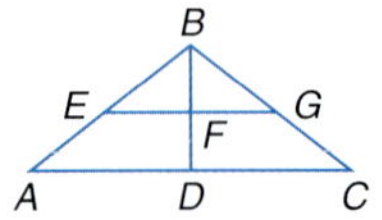

1. If $\overline{AD} \cong \overline{BD}$, name two congruent angles. $\angle DAB \cong \angle DBA$
2. If $\overline{BF} \cong \overline{FG}$, name two congruent angles. $\angle FBG \cong \angle FGB$
3. If $\overline{BE} \cong \overline{BG}$, name two congruent angles. $\angle BEF \cong \angle BGF$
4. If $\angle FBE \cong \angle FEB$, name two congruent segments. $\overline{FB} \cong \overline{FE}$
5. If $\angle BCA \cong \angle BAC$, name two congruent segments. $\overline{BA} \cong \overline{BC}$
6. If $\angle DBC \cong \angle BCD$, name two congruent segments. $\overline{BD} \cong \overline{CD}$

Lesson 4-7 (pages 222–226)

Position and label each triangle on the coordinate plane. 1–4. See margin for sample answers.

1. isosceles $\triangle ABC$ with base $\overline{BC}$ that is r units long
2. equilateral $\triangle XYZ$ with sides $4b$ units long
3. isosceles right $\triangle RST$ with hypotenuse $\overline{ST}$ and legs $(3 + a)$ units long
4. equilateral $\triangle CDE$ with base $\overline{DE}$ $\frac{1}{4}b$ units long.

Name the missing coordinates of each triangle.

5.

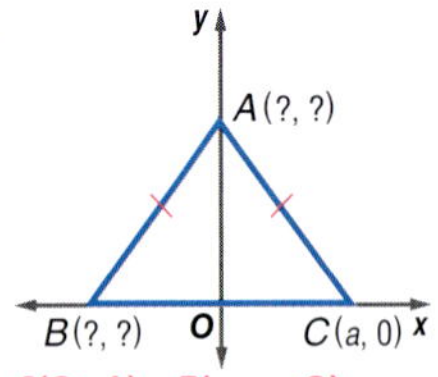

$A(0, b)$, $B(-a, 0)$

6.

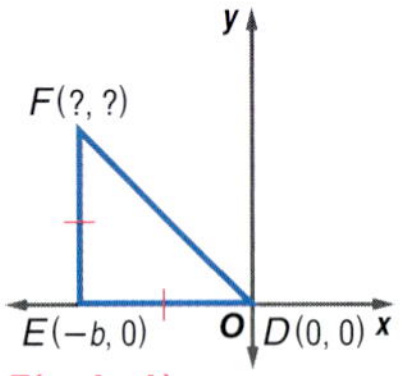

$F(-b, b)$

7.

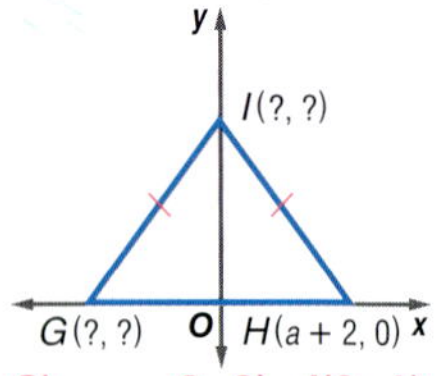

$G(-a - 2, 0)$, $I(0, b)$

Lesson 4-7

1.

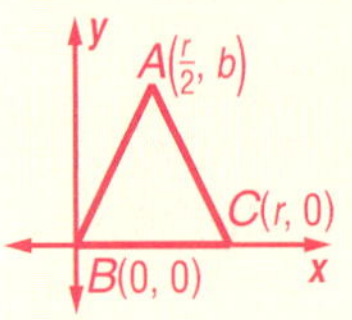

2.

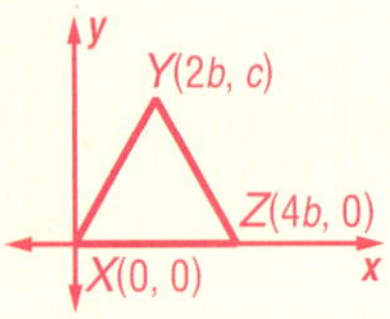

3.

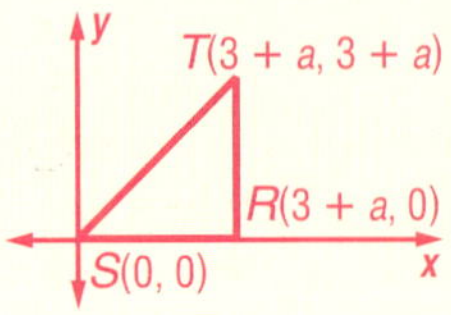

4.

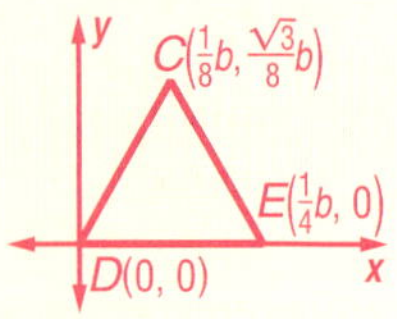

Lesson 5-1

(pages 238–246)

For Exercises 1–4, refer to the figures at the right.

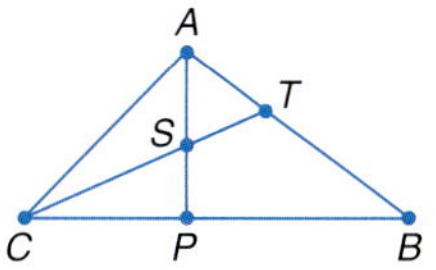

1. Suppose $CP = 7x - 1$ and $PB = 6x + 3$. If S is the circumcenter of $\triangle ABC$, find x and CP. **4; 27**
2. Suppose $m\angle ACT = 15a - 8$ and $m\angle ACB = 74$. If S is the incenter of $\triangle ABC$, find a and $m\angle ACT$. **3; 37**
3. Suppose $TO = 7b + 5$, $OR = 13b - 10$, and $TR = 18b$. If Z is the centroid of $\triangle TRS$, find b and TR. **2.5; 45**
4. Suppose $XR = 19n - 14$ and $ZR = 10n + 4$. If Z is the centroid of $\triangle TRS$, find n and ZR. **5; 54**

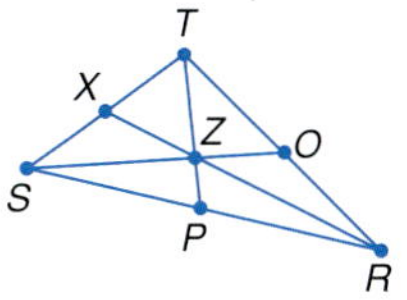

State whether each sentence is *always*, *sometimes*, or *never* true.

5. The circumcenter and incenter of a triangle are the same point. **sometimes**
6. The three altitudes of a triangle intersect at a point inside the triangle. **sometimes**
7. In an equilateral triangle, the circumcenter, incenter, and centroid are the same point. **always**
8. The incenter is inside of a triangle. **always**

Lesson 5-2

(pages 247–254)

Determine the relationship between the measures of the given angles.

1. $\angle TPS$, $\angle TSP$ **$m\angle TPS > m\angle TSP$**
2. $\angle PRZ$, $\angle ZPR$ **$m\angle PRZ > m\angle ZPR$**
3. $\angle SPZ$, $\angle SZP$ **$m\angle SPZ < m\angle SZP$**
4. $\angle SPR$, $\angle SRP$ **$m\angle SPR = m\angle SRP$**

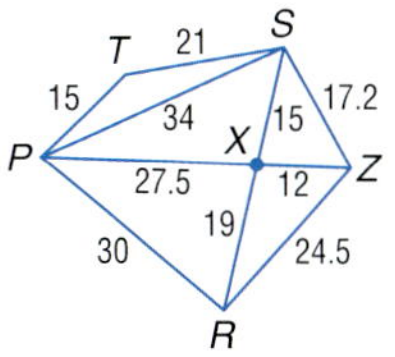

5. **Given:** $FH > FG$ **See margin.**
   Prove: $m\angle 1 > m\angle 2$

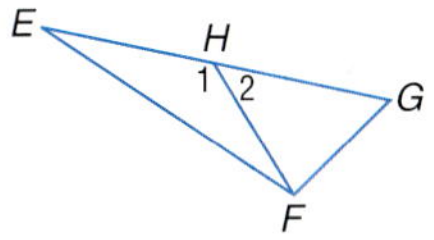

6. **Given:** $\overrightarrow{RQ}$ bisects $\angle SRT$. **See margin.**
   Prove: $m\angle SQR > m\angle SRQ$

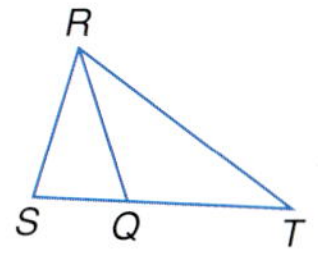

Lesson 5-3

(pages 255–260)

State the assumption you would make to start an indirect proof of each statement.

1. $\angle ABC \cong \angle XYZ$ **$\angle ABC \ncong \angle XYZ$**
2. An angle bisector of an equilateral triangle is also a median.
3. $\overrightarrow{RS}$ bisects $\angle ARC$ **$\overrightarrow{RS}$ does not bisect $\angle ARC$.**

2. An angle bisector of an equilateral triangle is not a median.

Write an indirect proof. **4–5. See margin.**

4. **Given:** $\angle AOY \cong \angle AOX$
   $\overline{XO} \ncong \overline{YO}$
   Prove: $\overrightarrow{AO}$ is not the angle bisector of $\angle XAY$.

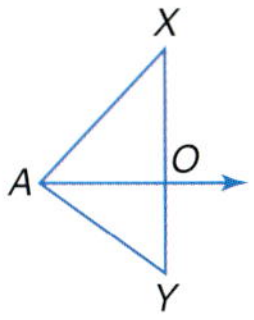

5. **Given:** $\triangle RUN$
   Prove: There can be no more than one right angle in $\triangle RUN$.

6. Given: $\overrightarrow{RQ}$ bisects $\angle SRT$.
Prove: $m\angle SQR > m\angle SRQ$

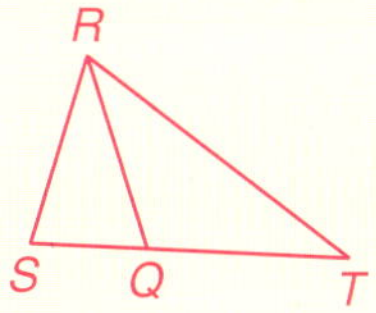

Proof:

Statements (Reasons)

1. $\overrightarrow{RQ}$ bisects $\angle SRT$ (Given)
2. $\angle SRQ \cong \angle QRT$ (Def. $\angle$ bisector)
3. $m\angle SRQ = m\angle QRT$ (Def. $\cong$ $\angle$s)
4. $m\angle SQR > m\angle QRT$ (Exterior Angle Inequality Theorem)
5. $m\angle SQR > m\angle SRQ$ (Subst.)

Lesson 5-3

4. Given: $\angle AOY \cong \angle AOX$, $\overline{XO} \ncong \overline{YO}$
Prove: $\overrightarrow{AO}$ is not the angle bisector of $\angle XAY$.

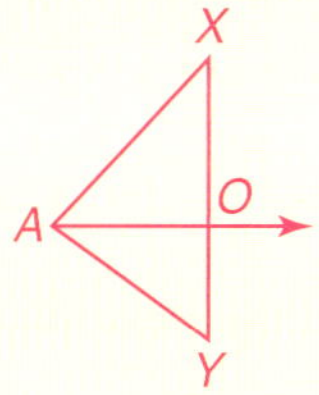

Proof:

Step 1: Assume $\overrightarrow{AO}$ is the angle bisector of $\angle XAY$.

Step 2: If $\overrightarrow{AO}$ is the angle bisector of $\angle XAY$, then $\angle XAO \cong \angle YAO$. $\angle AOY \cong \angle AOX$ by given and $\overline{AO} \cong \overline{AO}$ by reflexive. Then $\triangle XAO \cong \triangle YAO$ by ASA. $\overline{XO} \cong \overline{YO}$ by CPCTC.

Step 3: This conclusion contradicts the given fact $\overline{XO} \ncong \overline{YO}$. Thus, $\overrightarrow{AO}$ is not the angle bisector of $\angle XAY$.

5. Given: $\triangle RUN$
Prove: There can be no more than one right angle in $\triangle RUN$.

Proof:

Step 1: Assume $\triangle RUN$ has two right angles.

Step 2: By the Angle Sum Theorem, $m\angle R + m\angle U + m\angle N = 180$. If you substitute 90 for two of the $\angle$ measures, since the $\triangle$ has two right $\angle$s, then $90 + 90 + m\angle N = 180$. Then, $180 + m\angle N = 180$.

Step 3: This conclusion means that $m\angle N = 0$. This is not possible if $\triangle RUN$ is a $\triangle$. Thus, there can be no more than one right $\angle$ in $\triangle RUN$.

Lesson 5-2

5. Given: $FH > FG$
Prove: $m\angle 1 > m\angle 2$

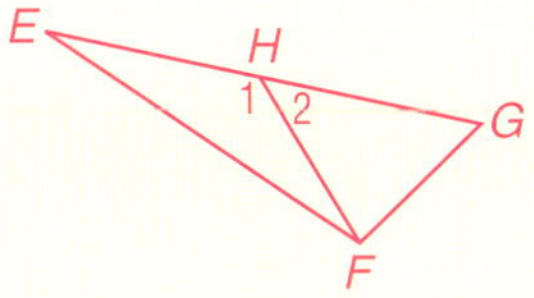

Proof:

Statements (Reasons)

1. $FH > FG$ (Given)
2. $m\angle FGH > m\angle 2$ (If one side of a $\triangle$ is longer than another, the $\angle$ opp. the longer side $>$ than the $\angle$ opp. the shorter side.)
3. $m\angle 1 > m\angle FGH$ (Exterior Angle Inequality Theorem)
4. $m\angle 1 > m\angle 2$ (Transitive Prop. of Inequality)

Extra Practice 763

Lesson 5-4

17. Given: $RS = RT$
Prove: $UV + VS > UT$

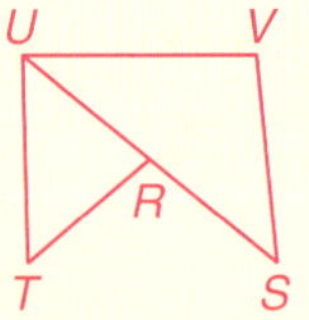

Proof:

Statements (Reasons)

1. $RS = RT$ (Given)
2. $UV + VS > US$ (Triangle Inequality Theorem)
3. $US = UR + RS$ (Segment Addition Postulate)
4. $UV + VS > UR + RS$ (Substitution)
5. $UV + VS > UR + RT$ (Substitution)
6. $UR + RT > UT$ (Triangle Inequality Theorem)
7. $UV + VS > UT$ (Transitive Property of Inequality)

18. Given: quadrilateral $ABCD$
Prove: $AD + CD + AB > BC$

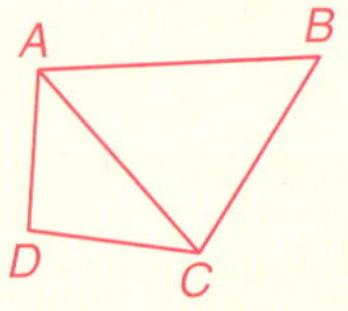

Proof:

Statements (Reasons)

1. quadrilateral $ABCD$ (Given)
2. Draw $\overline{AC}$. (Through any 2 pts. there is 1 line.)
3. $AD + CD > AC$; $AB + AC > BC$ (Triangle Inequality Theorem)
4. $AC > BC - AB$ (Subtraction Prop. of Inequality)
5. $AD + CD > BC - AB$ (Transitive Prop. of Inequality)
6. $AD + CD + AB > BC$ (Addition Prop. of Inequality)

Lesson 5-4 *(pages 261–266)*

Determine whether the given measures can be the lengths of the sides of a triangle. Write *yes* or *no*.

1. 2, 2, 6 no
2. 2, 3, 4 yes
3. 6, 8, 10 yes
4. 1, 1, 2 no
5. 15, 20, 30 yes
6. 1, 3, 5 no
7. 2.5, 3.5, 6.5 no
8. 0.3, 0.4, 0.5 yes

Find the range for the measure of the third side of a triangle given the measures of two sides.

9. 6 and 10 $4 < n < 16$
10. 2 and 5 $3 < n < 7$
11. 20 and 12 $8 < n < 32$
12. 8 and 8 $0 < n < 16$
13. 18 and 36 $18 < n < 54$
14. 32 and 34 $2 < n < 66$
15. 2 and 29 $27 < n < 31$
16. 80 and 25 $55 < n < 105$

Write a two-column proof. 17–18. See margin.

17. **Given:** $RS = RT$
Prove: $UV + VS > UT$

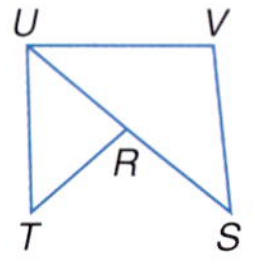

18. **Given:** quadrilateral $ABCD$
Prove: $AD + CD + AB > BC$

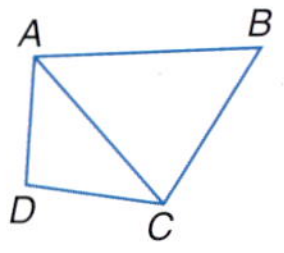

Lesson 5-5 *(pages 267–273)*

Write an inequality relating the given pair of angle or segment measures.

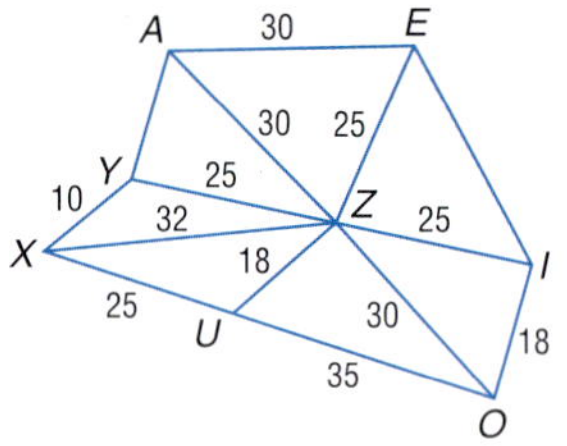

1. XZ, OZ $XZ > OZ$
2. $m\angle ZIO, m\angle ZUX$ $m\angle ZIO < m\angle ZUX$
3. $m\angle AEZ, m\angle AZE$ $m\angle AEZ = m\angle AZE$
4. IO, AE $IO < AE$
5. $m\angle AZE, m\angle IZO$ $m\angle AZE > m\angle IZO$

Write an inequality to describe the possible values of *x*.

6. $6.78 < x < 15.22$

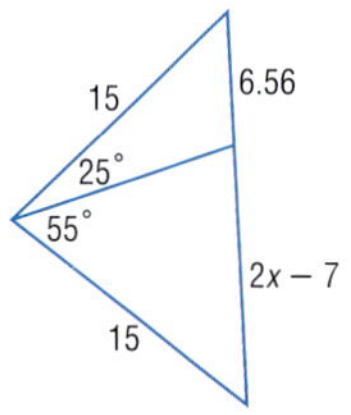

7. $4.2 < x < 10$

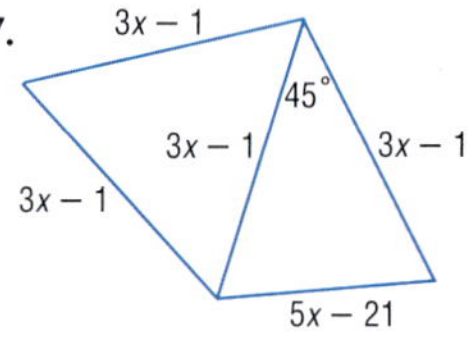

Lesson 6-1 *(pages 282–287)*

1. **ARCHITECTURE** The ratio of the height of a model of a house to the actual house is 1:63. If the width of the model is 16 inches, find the width of the actual house in feet. 84 ft

2. **CONSTRUCTION** A 64-inch long board is divided into lengths in the ratio 2:3. What are the two lengths into which the board is divided? 25.6 in., 38.4 in.

ALGEBRA **Solve each proportion.**

3. $\frac{x+4}{26} = -\frac{1}{3}$ $-\frac{38}{3}$
4. $\frac{3x+1}{14} = \frac{5}{7}$ 3
5. $\frac{x-3}{4} = \frac{x+1}{5}$ 19
6. $\frac{2x+2}{2x-1} = \frac{1}{3}$ $-\frac{7}{4}$

7. Find the measures of the sides of a triangle if the ratio of the measures of three sides of a triangle is 9:6:5, and its perimeter is 100 inches. 45 in., 30 in., 25 in.

8. Find the measures of the angles in a triangle if the ratio of the measures of the three angles is 13:16:21. 46.8, 57.6, 75.6

Lesson 6-2 *(pages 289–297)*

Determine whether each pair of figures is similar. Justify your answer. 1–2. See margin.

1.

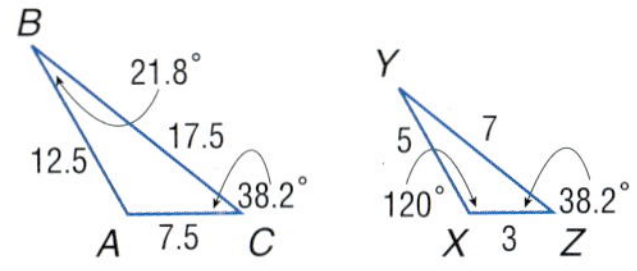

2.

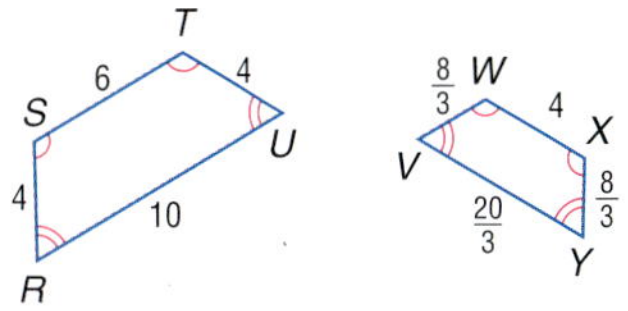

For Exercises 3 and 4, use △*RST* with vertices *R*(3, 6), *S*(1, 2), and *T*(3, −1). Explain. 3–4. See margin.

3. If the coordinates of each vertex are decreased by 3, describe the new figure. Is it similar to △*RST*?
4. If the coordinates of each vertex are multiplied by 0.5, describe the new figure. Is it similar to △*RST*?

Lesson 6-3 *(pages 298–306)*

Determine whether each pair of triangles is similar. Justify your answer. 1–2. See margin.

1.

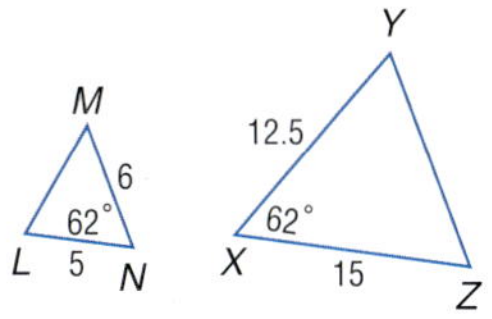

2.

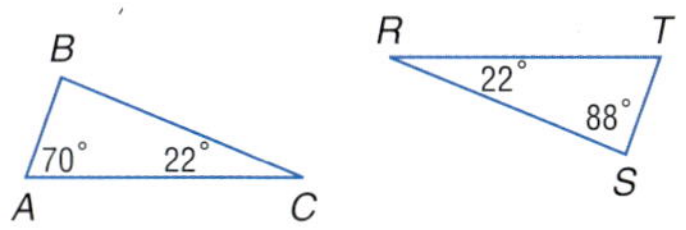

ALGEBRA **Identify the similar triangles. Find *x* and the measures of the indicated sides.**

3. *RT* and *SV*

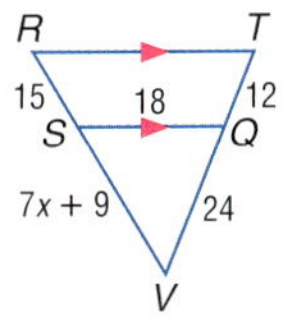

3–4. See margin.

4. *PN* and *MN*

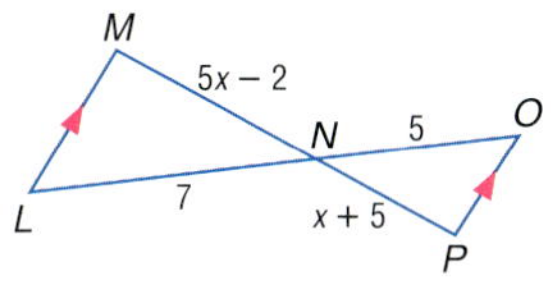

Lesson 6-4 *(pages 307–315)*

1. If $HI = 28$, $LH = 21$, and $LK = 8$, find *IJ*. $10\frac{2}{3}$

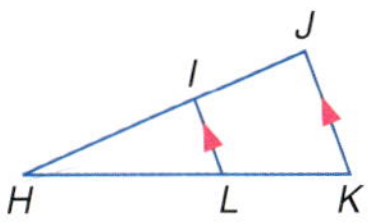

2. Find *x*, *AD*, *DR*, and *QR* if $AU = 15$, $QU = 25$, $AD = 3x + 6$, $DR = 8x - 2$, and $UD = 15$.

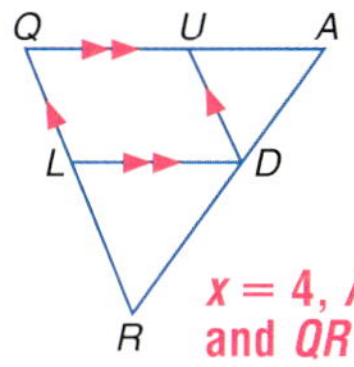

$x = 4$, $AD = 18$, $DR = 30$, and $QR = 40$

Find *x* so that $\overline{XY} \parallel \overline{LM}$.

3. $XL = 3$, $YM = 5$, $LD = 9$, $MD = x + 3$ 12
4. $YM = 3$, $LD = 3x + 1$, $XL = 4$, $MD = x + 7$ 5
5. $MD = 5x - 6$, $YM = 3$, $LD = 5x + 1$, $XL = 5$ 3.3

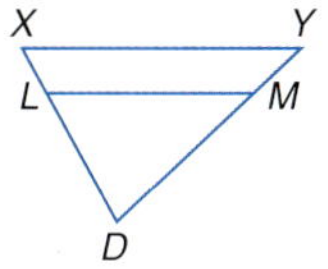

Lesson 6-2

1. $m\angle A = 180 - 21.8 - 38.2 = 120$, so $m\angle A = m\angle X$. Therefore $\angle A \cong \angle X$.
   $m\angle Y = 180 - 120 - 38.2 = 21.8$, so $m\angle Y = m\angle B$. Therefore $\angle Y \cong \angle B$.
   $m\angle C = m\angle Z$, therefore $\angle C \cong \angle Z$.
   All of the corresponding angles are congruent. Now determine whether corresponding sides are proportional.
   $\frac{AB}{XY} = \frac{12.5}{5} = 2.5$ $\frac{BC}{YZ} = \frac{17.5}{7} = 2.5$ $\frac{AC}{XZ} = \frac{7.5}{3} = 2.5$
   The ratios of the measures of the corresponding sides are equal, and the corresponding angles are congruent, so △*ABC* ~ △*XYZ*.
2. $\angle S \cong \angle W$, $\angle T \cong \angle X$, $\angle U \cong \angle Y$, $\angle R \cong \angle V$.
   All of the corresponding angles are congruent. Now determine whether corresponding sides are proportional.
   $\frac{RS}{VW} = \frac{4}{\frac{8}{3}} = 1.5$
   $\frac{ST}{WX} = \frac{6}{4} = 1.5$
   $\frac{TU}{XY} = \frac{4}{\frac{8}{3}} = 1.5$
   $\frac{RU}{VY} = \frac{10}{\frac{20}{3}} = 1.5$
   The ratios of the measures of the corresponding sides are equal, and the corresponding angles are congruent, so polygon *RSTU* ~ polygon *VWXY*.
3. Yes; the new triangle is congruent and similar to the original, but shifted to the left 3 units and down 3 units.
4. Yes; the new triangle is similar to the original, but the length of each side is one half the length of the corresponding sides of the original triangle.

Lesson 6-3

1. Yes; △*LNM* ~ △*YXZ*; SAS Similarity
2. Yes; △*ABC* ~ △*TSR*; AA Similarity.
3. △*RTV* ~ △*SQV*; $x = 3$; $RT = 27$; $SV = 30$
4. △*MNL* ~ △*PNO*; $x = 2.5$; $PN = 7.5$; $MN = 10.5$

Lesson 6-5

(pages 316–323)

Find the perimeter of each triangle.

1. $\triangle ABC$ if $\triangle ABC \sim \triangle DBE$, $AB = 17.5$, $BC = 15$, $BE = 6$, and $DE = 5$ **45**

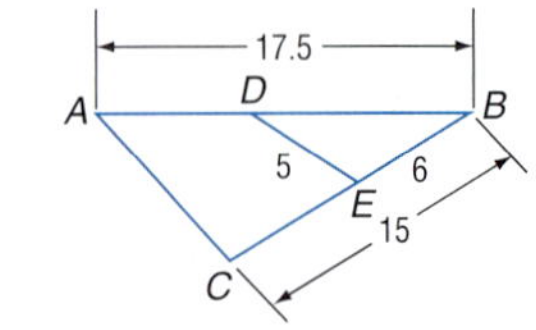

2. $\triangle RST$ if $\triangle RST \sim \triangle XYZ$, $RT = 12$, $XZ = 8$, and the perimeter of $\triangle XYZ = 22$ **33**

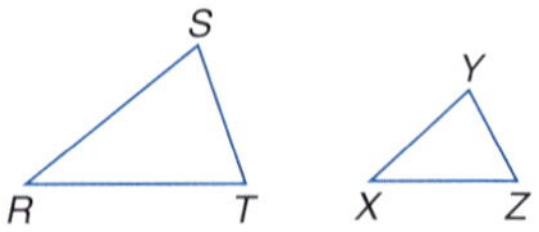

3. $\triangle LMN$ if $\triangle LMN \sim \triangle NXY$, $NX = 14$, $YX = 11$, $YN = 9$, and $LN = 27$ **102**

4. $\triangle GHI$ if $\triangle ABC \sim \triangle GHI$, $AB = 6$, $GH = 10$, and the perimeter of $\triangle ABC = 25$ **$41\frac{2}{3}$**

Lesson 6-6

(pages 325–331)

Stage 1 of a fractal is shown drawn on grid paper. Stage 1 is made by dividing a square into 4 congruent squares and shading the top left-hand square.

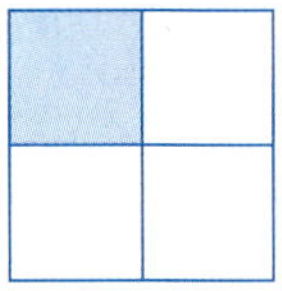

1. Draw Stage 2 by repeating the Stage 1 process in each of the 3 remaining unshaded squares. How many shaded squares are at this stage? **4**

2. Draw Stage 3 by repeating the Stage 1 process in each of the unshaded squares in Stage 2. How many shaded squares are at this stage? **13**

1–2. See margin for fractals.

Find the value of each expression. Then, use that value as the next x in the expression. Repeat the process and describe your observations. **3–6. See margin.**

3. $x^{\frac{1}{4}}$, where x initially equals 6
4. 4^x, where x initially equals 0.4
5. x^3, where x initially equals 0.5
6. 3^x, where x initially equals 10

Lesson 7-1

(pages 342–348)

Find the geometric mean between each pair of numbers. State exact answers and answers to the nearest tenth.

1. 8 and 12 **$4\sqrt{6} \approx 9.8$**
2. 15 and 20 **$10\sqrt{3} \approx 17.3$**
3. 1 and 2 **$\sqrt{2} \approx 1.4$**
4. 4 and 16 **8**
5. $3\sqrt{2}$ and $6\sqrt{2}$ **6**
6. $\frac{1}{2}$ and 10 **$\sqrt{5} \approx 2.2$**
7. $\frac{3}{8}$ and $\frac{1}{2}$ **$\frac{\sqrt{3}}{4} \approx 0.4$**
8. $\frac{\sqrt{2}}{2}$ and $\frac{3\sqrt{2}}{2}$ **$\frac{\sqrt{6}}{2} \approx 1.2$**
9. $\frac{1}{10}$ and $\frac{7}{10}$ **$\frac{\sqrt{7}}{10} \approx 0.3$**

Find the altitude of each triangle.

10.

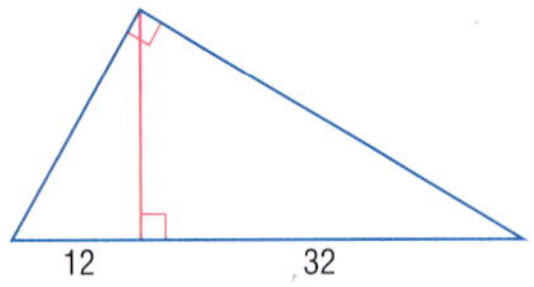

$8\sqrt{6} \approx 19.6$

11. **$4\sqrt{2} \approx 5.7$**

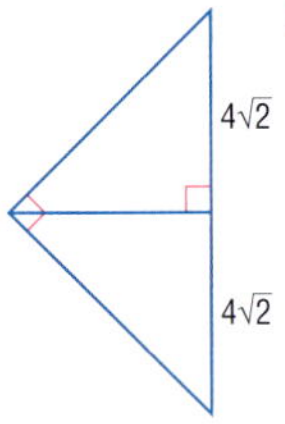

12. **$2\sqrt{42} \approx 13.0$**

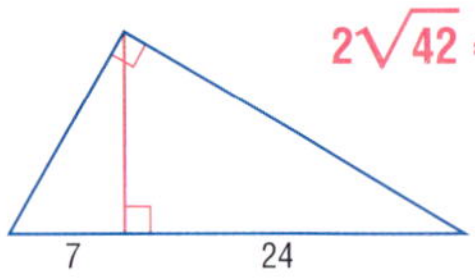

Extra Practice

Lesson 6-6

1.

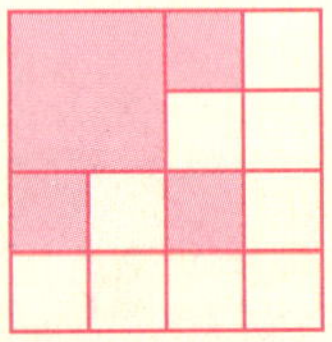

2.

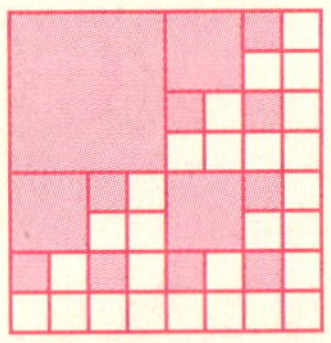

3. converges to 1
4. approaches positive infinity
5. converges to 0
6. approaches positive infinity

Lesson 7-2 *(pages 350–356)*

Determine whether $\triangle DEF$ is a right triangle for the given vertices. Explain. 1–4. See margin.

1. $D(0, 1), E(3, 2), F(2, 3)$
2. $D(-2, 2), E(3, -1), F(-4, -3)$
3. $D(2, -1), E(-2, -4), F(-4, -1)$
4. $D(1, 2), E(5, -2), F(-2, -1)$

Determine whether each set of measures are the sides of a right triangle. Then state whether they form a Pythagorean triple. 5–13. See margin.

5. 1, 1, 2
6. 21, 28, 35
7. 3, 5, 7
8. 2, 5, 7
9. 24, 45, 51
10. $\frac{1}{3}, \frac{5}{3}, \frac{\sqrt{26}}{3}$
11. $\frac{6}{11}, \frac{8}{11}, \frac{10}{11}$
12. $\frac{1}{2}, \frac{1}{2}, 1$
13. $\frac{\sqrt{6}}{3}, \frac{\sqrt{10}}{5}, \frac{\sqrt{240}}{15}$

Lesson 7-3 *(pages 357–363)*

Find the measures of x and y.

1. 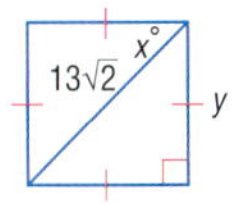

$x = 45, y = 13$

2. 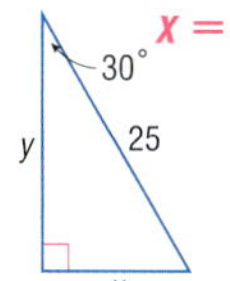

$x = 12.5, y = 12.5\sqrt{3}$

3. 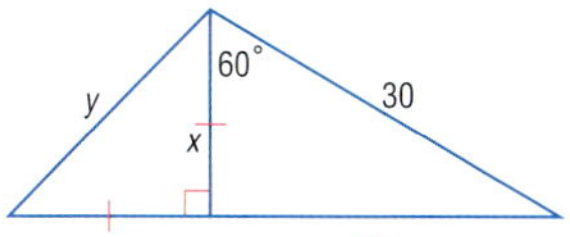

$x = 15, y = 15\sqrt{2}$

4. 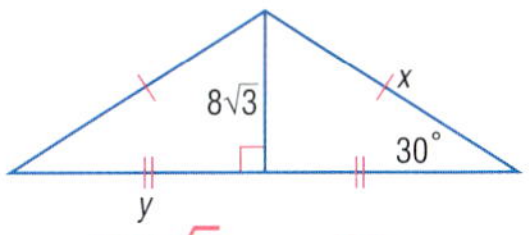

$x = 16\sqrt{3}, y = 24$

5. 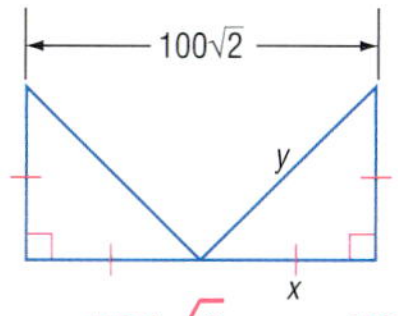

$x = 50\sqrt{2}, y = 100$

6. 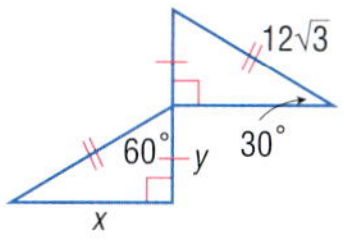

$x = 18, y = 6\sqrt{3}$

Lesson 7-4 *(pages 364–370)*

Use $\triangle MAN$ with right angle N to find $\sin M$, $\cos M$, $\tan M$, $\sin A$, $\cos A$, and $\tan A$. Express each ratio as a fraction, and as a decimal to the nearest hundredth. 1–4. See margin.

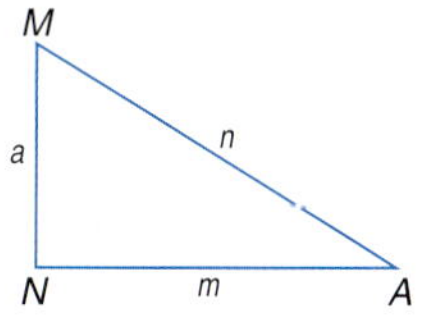

1. $m = 21, a = 28, n = 35$
2. $m = \sqrt{2}, a = \sqrt{3}, n = \sqrt{5}$
3. $m = \frac{\sqrt{2}}{2}, a = \frac{\sqrt{2}}{2}, n = 1$
4. $m = 3\sqrt{5}, a = 5\sqrt{3}, n = 2\sqrt{30}$

Find the measure of each angle to the nearest tenth of a degree.

5. $\cos A = 0.6293$ 51.0
6. $\sin B = 0.5664$ 34.5
7. $\tan C = 0.2665$ 14.9
8. $\sin D = 0.9352$ 69.3
9. $\tan M = 0.0808$ 4.6
10. $\cos R = 0.1097$ 83.7

Find x. Round to the nearest tenth.

11. 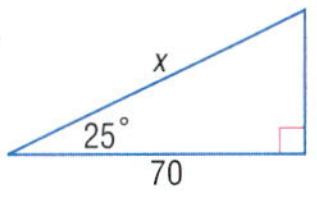

77.2

12. 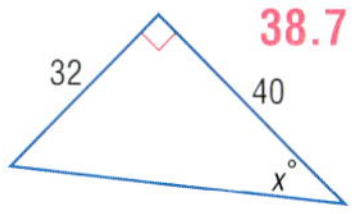

38.7

13. 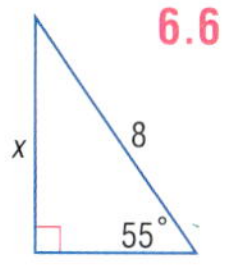

6.6

Lesson 7-2

1. yes; $DE = \sqrt{10}$, $EF = \sqrt{2}$, $DF = \sqrt{8}$; $(\sqrt{2})^2 + (\sqrt{8})^2 = (\sqrt{10})^2$
2. no; $DE = \sqrt{34}$, $EF = \sqrt{53}$, $DF = \sqrt{29}$; $(\sqrt{29})^2 + (\sqrt{34})^2 \neq (\sqrt{53})^2$
3. no; $DE = 5$, $EF = \sqrt{13}$, $DF = 6$; $(\sqrt{13})^2 + 5^2 \neq 6^2$
4. yes; $DE = \sqrt{32}$, $EF = \sqrt{50}$, $DF = \sqrt{18}$; $(\sqrt{18})^2 + (\sqrt{32})^2 = (\sqrt{50})^2$
5. no; no
6. yes; yes
7. no; no
8. no; no
9. yes, yes
10. yes; no
11. yes; no
12. no; no
13. yes; no

Lesson 7-4

1. $\frac{3}{5} = 0.60$; $\frac{4}{5} = 0.80$; $\frac{3}{4} = 0.75$; $\frac{4}{5} = 0.80$; $\frac{3}{5} = 0.60$; $\frac{4}{3} = 1.33$
2. $\frac{\sqrt{10}}{5} \approx 0.63$; $\frac{\sqrt{15}}{5} \approx 0.77$; $\frac{\sqrt{6}}{3} \approx 0.82$; $\frac{\sqrt{15}}{5} \approx 0.77$; $\frac{\sqrt{10}}{5} \approx 0.63$; $\frac{\sqrt{6}}{2} \approx 1.22$
3. $\frac{\sqrt{2}}{2} \approx 0.71$; $\frac{\sqrt{2}}{2} \approx 0.71$; 1.00; $\frac{\sqrt{2}}{2} \approx 0.71$; $\frac{\sqrt{2}}{2} \approx 0.71$; 1.00
4. $\frac{\sqrt{6}}{4} \approx 0.61$; $\frac{\sqrt{10}}{4} \approx 0.79$; $\frac{\sqrt{15}}{5} \approx 0.77$; $\frac{\sqrt{10}}{4} \approx 0.79$; $\frac{\sqrt{6}}{4} \approx 0.61$; $\frac{\sqrt{15}}{3} \approx 1.29$

Lesson 7-5

(pages 371–376)

1. **COMMUNICATIONS** A house is located below a hill that has a satellite dish. If $MN = 450$ feet and $RN = 120$ feet, what is the measure of the angle of elevation to the top of the hill? **about 14.9**

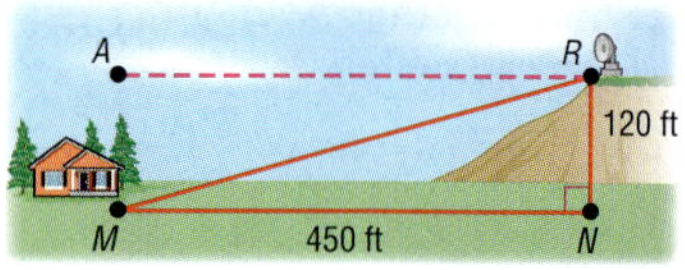

2. **AMUSEMENT PARKS** Mandy is at the top of the Mighty Screamer roller coaster. Her friend Bryn is at the bottom of the coaster waiting for the next ride. If the angle of depression from Mandy to Bryn is 26° and OL is 75 feet, what is the distance from L to C? **about 153.8 ft**

3. **SKIING** Mitchell is at the top of the Bridger Peak ski run. His brother Scott is looking up from the ski lodge at I. If the angle of elevation from Scott to Mitchell is 13° and the distance from K to I is 2000 ft, what is the length of the ski run SI? **about 2052.6 ft**

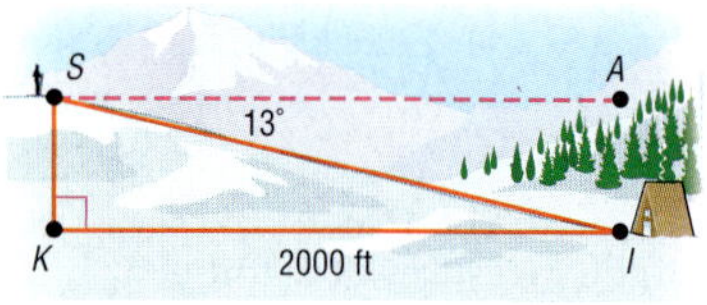

Lesson 7-6

(pages 377–383)

Find each measure using the given measures from $\triangle ANG$. Round angle measures to the nearest degree and side measures to the nearest tenth.

1. If $m\angle N = 32$, $m\angle A = 47$, and $n = 15$, find a. **20.7**
2. If $a = 10.5$, $m\angle N = 26$, $m\angle A = 75$, find n. **4.8**
3. If $n = 18.6$, $a = 20.5$, $m\angle A = 65$, find $m\angle N$. **55**
4. If $a = 57.8$, $n = 43.2$, $m\angle A = 33$, find $m\angle N$. **24**

Solve each $\triangle AKX$ described below. Round angle measures to the nearest degree and side measures to the nearest tenth.

5. $m\angle X = 62$, $a = 28.5$, $m\angle K = 33$ **$m\angle A = 85$, $x \approx 25.3$, $k \approx 15.6$**
6. $k = 3.6$, $x = 3.7$, $m\angle X = 55$ **$m\angle K \approx 53°$, $m\angle A \approx 72°$, $a \approx 4.3$**
7. $m\angle K = 35$, $m\angle A = 65$, $x = 50$ **$m\angle X = 80$, $a \approx 46.0$, $k \approx 29.1$**
8. $m\angle A = 122$, $m\angle X = 15$, $a = 33.2$ **$m\angle K = 43$, $k \approx 26.7$, $x \approx 10.1$**

Lesson 7-7

(pages 385–390)

In $\triangle CDE$, given the lengths of the sides, find the measure of the stated angle to the nearest tenth.

1. $c = 100$, $d = 125$, $e = 150$; $m\angle E$ **82.8**
2. $c = 5$, $d = 6$, $e = 9$; $m\angle C$ **31.6**
3. $c = 1.2$, $d = 3.5$, $e = 4$; $m\angle D$ **57.3**
4. $c = 42.5$, $d = 50$, $e = 81.3$; $m\angle E$ **122.8**

Solve each triangle using the given information. Round angle measures to the nearest degree and side measures to the nearest tenth.

5.

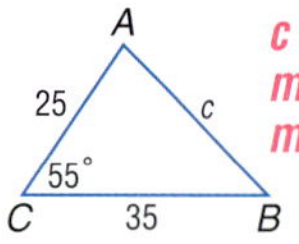

$c \approx 29.1$, $m\angle A \approx 80$, $m\angle B \approx 45$

6.

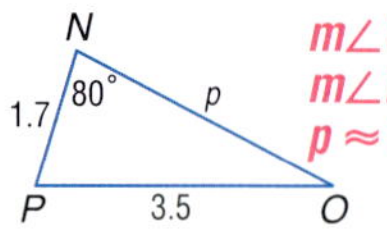

$m\angle O \approx 29$, $m\angle P \approx 71$, $p \approx 3.4$

7.

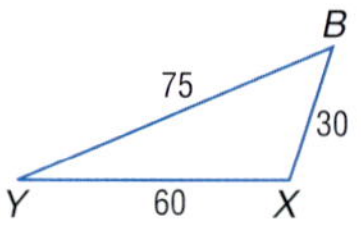

$m\angle B \approx 50$, $m\angle X \approx 108$, $m\angle Y \approx 22$

Extra Practice

768 Extra Practice

Lesson 8-1

(pages 404–409)

Find the sum of the measures of the interior angles of each convex polygon.

1. 25-gon **4140**
2. 30-gon **5040**
3. 22-gon **3600**
4. 17-gon **2700**
5. $5a$-gon **$180(5a - 2)$**
6. b-gon **$180(b - 2)$**

The measure of an interior angle of a regular polygon is given. Find the number of sides in each polygon.

7. 156 **15**
8. 168 **30**
9. 162 **20**

Find the measures of an interior angle and an exterior angle given the number of sides of a regular polygon. Round to the nearest tenth.

10. 15 **156, 24**
11. 13 **152.3, 27.7**
12. 42 **171.4, 8.6**

Lesson 8-2

(pages 411–416)

Complete each statement about ▱$RSTU$. Justify your answer.

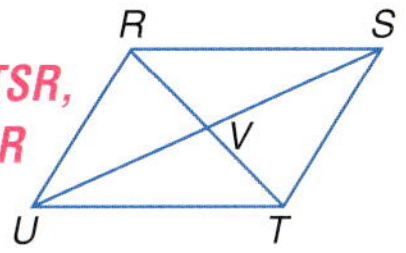

1. $\angle SRU \cong$ ___?___ **$\angle UTS$**
2. $\angle UTS$ is supplementary to ___?___ **$\angle TSR$, $\angle TUR$**
3. $\overline{RU} \parallel$ ___?___ **$\overline{ST}$**
4. $\overline{RU} \cong$ ___?___ **$\overline{ST}$**
5. $\triangle RST \cong$ ___?___ **$\triangle TUR$**
6. $\overline{SV} \cong$ ___?___ **$\overline{VU}$**

1–6. See margin for justification.

ALGEBRA Use ▱$ABCD$ to find each measure or value.

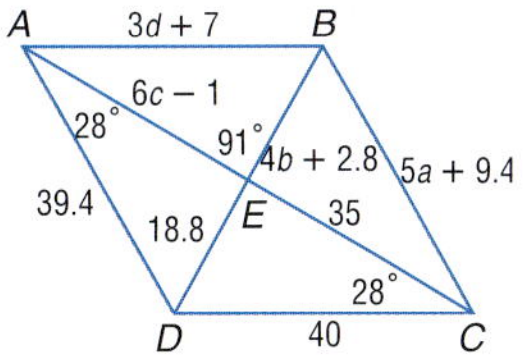

7. $m\angle BAE =$ ___?___ **28**
8. $m\angle BCE =$ ___?___ **28**
9. $m\angle BEC =$ ___?___ **89**
10. $m\angle CED =$ ___?___ **91**
11. $m\angle ABE =$ ___?___ **61**
12. $m\angle EBC =$ ___?___ **63**
13. $a =$ ___?___ **6**
14. $b =$ ___?___ **4**
15. $c =$ ___?___ **6**
16. $d =$ ___?___ **11**

Lesson 8-3

(pages 417–423)

Determine whether each quadrilateral is a parallelogram. Justify your answer. 1–3. See margin for justification.

1. 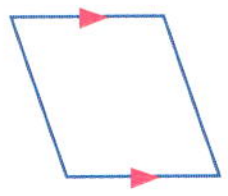
no

2. 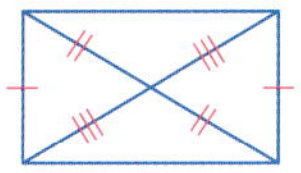
yes

3. 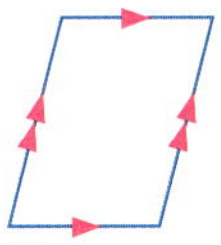
yes

ALGEBRA Find x and y so that each quadrilateral is a parallelogram.

4. 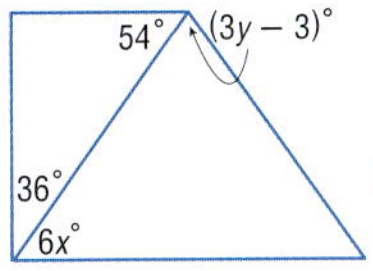

$x = 9$, $y = 13$

5. 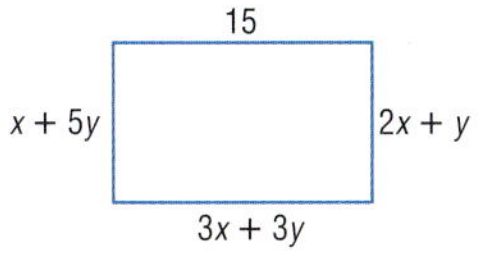

$x = 4$, $y = 1$

6. 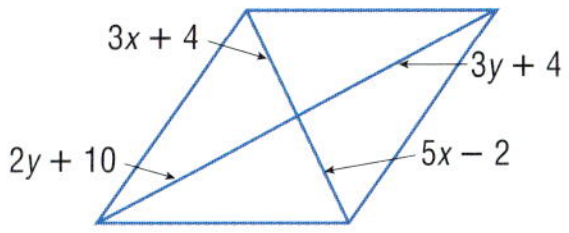

$x = 3$, $y = 6$

Determine whether a figure with the given vertices is a parallelogram. Use the method indicated.

7. $L(-3, 2)$, $M(5, 2)$, $N(3, -6)$, $O(-5, -6)$; Slope Formula **yes**
8. $W(-5, 6)$, $X(2, 5)$, $Y(-3, -4)$, $Z(-8, -2)$; Distance Formula **no**
9. $Q(-5, 4)$, $R(0, 6)$, $S(3, -1)$, $T(-2, -3)$; Midpoint Formula **yes**
10. $G(-5, 0)$, $H(-13, 5)$, $I(-10, 9)$, $J(-2, 4)$; Distance and Slope Formulas **yes**

Extra Practice

Lesson 8-2

1. $\angle UTS$; opp. ∠s of ▱ are ≅.
2. $\angle TSR$; cons. ∠s in ▱ are suppl.
3. Opp. sides of ▱ are parallel.
4. Opp. sides of ▱ are ≅.
5. Diag. of ▱ separates ▱ into 2 ≅ △s.
6. Diag. of ▱ bisect each other.

Lesson 8-3

1. Only one pair of sides is shown to be parallel.
2. If diag. bisect each other, then quad. is ▱.
3. If both pairs of opp. sides are ∥, then quad. is ▱.

Extra Practice

Extra Practice 769

Extra Practice

Lesson 8-6

1a. $\overline{AD} \parallel \overline{BC}$; $ABCD$ is a trapezoid.

1b. $\overline{AB} \cong \overline{CD}$; $ABCD$ is an isosceles trapezoid.

2a. $\overline{QR} \parallel \overline{ST}$; $QRST$ is a trapezoid.

2b. $\overline{QT} \not\cong \overline{RS}$; $QRST$ is not an isosceles trapezoid.

3a. $\overline{ON} \parallel \overline{LM}$; $LMNO$ is a trapezoid.

3b. $\overline{LO} \cong \overline{MN}$; $LMNO$ is an isosceles trapezoid.

4a. $\overline{WX} \parallel \overline{ZY}$; $WXYZ$ is a trapezoid.

4b. $\overline{WZ} \not\cong \overline{XY}$; $WXYZ$ is not an isosceles trapezoid.

Lesson 8-7

3. Given: $ABCD$ is a square.
   Prove: $\overline{AC} \cong \overline{BD}$

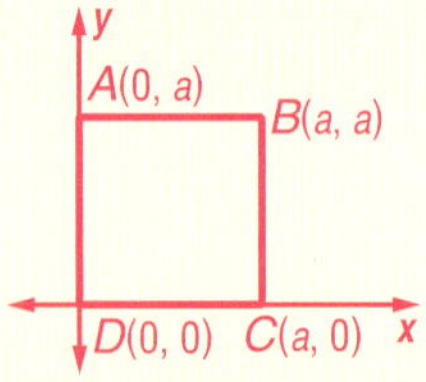

Proof:

$$AC = \sqrt{(a-0)^2 + (0-a)^2} = \sqrt{a^2 + a^2} = \sqrt{2a^2}$$

$$BD = \sqrt{(a-0)^2 + (a-0)^2} = \sqrt{a^2 + a^2} = \sqrt{2a^2}$$

$AC = BD$

$\overline{AC} \cong \overline{BD}$

4. Given: $EFGH$ is a quadrilateral.
   Prove: $EFGH$ is a rhombus.

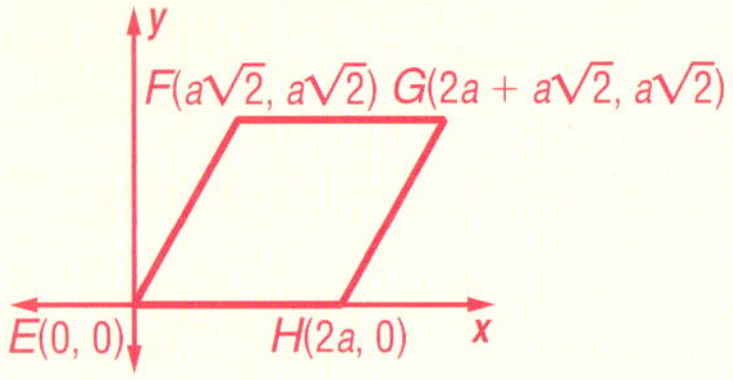

Proof:

$$EF = \sqrt{(a\sqrt{2}-0)^2 + (a\sqrt{2}-0)^2} = \sqrt{2a^2 + 2a^2} = \sqrt{4a^2} \text{ or } 2a$$

$$FG = \sqrt{((2a + a\sqrt{2}) - a\sqrt{2})^2 + (a\sqrt{2} - a\sqrt{2})^2} = \sqrt{(2a)^2 + 0^2} = \sqrt{4a^2} \text{ or } 2a$$

$$GH = \sqrt{((2a + a\sqrt{2}) - 2a)^2 + (a\sqrt{2} - 0)^2} = \sqrt{2a^2 + 2a^2} = \sqrt{4a^2} \text{ or } 2a$$

$$EH = \sqrt{(2a-0)^2 + (0-0)^2} = \sqrt{(2a)^2 + 0^2} = \sqrt{4a^2} \text{ or } 2a$$

$EF = FG = GH = EH$

$\overline{EF} \cong \overline{FG} \cong \overline{GH} \cong \overline{EH}$

Since all four sides are congruent, $EFGH$ is a rhombus.

Extra Practice

Lesson 8-4

(pages 424–430)

ALGEBRA **Refer to rectangle $QRST$.**

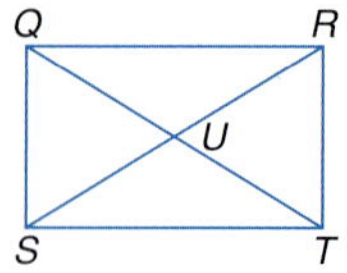

1. If $QU = 2x + 3$ and $UT = 4x - 9$, find SU. **15**
2. If $RU = 3x - 6$ and $UT = x + 9$, find RS. **33**
3. If $QS = 3x + 40$ and $RT = 16 - 3x$, find QS. **28**
4. If $m\angle STQ = 5x + 3$ and $m\angle RTQ = 3 - x$, find x. **21**
5. If $m\angle SRQ = x^2 + 6$ and $m\angle RST = 36 - x$, find $m\angle SRT$. **48 or 59**
6. If $m\angle TQR = x^2 + 16$ and $m\angle QTR = x + 32$, find $m\angle TQS$. **25 or 38**

Find each measure in rectangle $LMNO$ if $m\angle 5 = 38$.

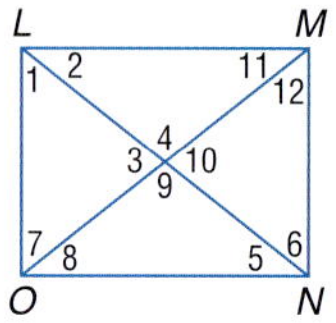

7. $m\angle 1$ **52**
8. $m\angle 2$ **38**
9. $m\angle 3$ **76**
10. $m\angle 4$ **104**
11. $m\angle 6$ **52**
12. $m\angle 7$ **52**
13. $m\angle 8$ **38**
14. $m\angle 9$ **104**
15. $m\angle 10$ **76**
16. $m\angle 11$ **38**
17. $m\angle 12$ **52**
18. $m\angle OLM$ **90**

Lesson 8-5

(pages 431–437)

In rhombus $QRST$, $m\angle QRS = m\angle TSR - 40$ and $TS = 15$.

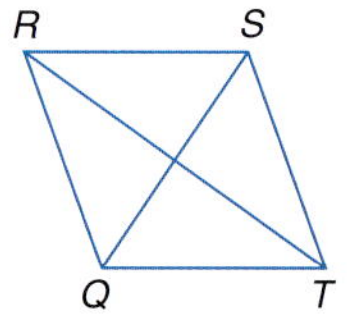

1. Find $m\angle TSQ$. **55**
2. Find $m\angle QRS$. **70**
3. Find $m\angle SRT$. **35**
4. Find QR. **15**

ALGEBRA **Use rhombus $ABCD$ with $AY = 6$, $DY = 3r + 3$, and $BY = \frac{10r - 4}{2}$.**

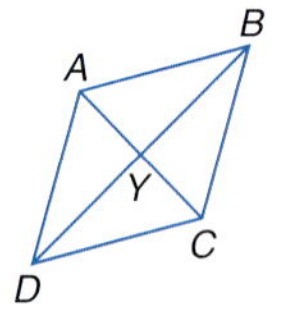

5. Find $m\angle ACB$. **60**
6. Find $m\angle ABD$. **30**
7. Find BY. **10.5**
8. Find AC. **12**

Lesson 8-6

(pages 439–445)

COORDINATE GEOMETRY **For each quadrilateral with the given vertices,**
a. verify that the quadrilateral is a trapezoid, and
b. determine whether the figure is an isosceles trapezoid. **1–4. See margin.**

1. $A(0, 9)$, $B(3, 4)$, $C(-5, 4)$, $D(-2, 9)$
2. $Q(1, 4)$, $R(4, 6)$, $S(10, 7)$, $T(1, 1)$
3. $L(1, 2)$, $M(4, -1)$, $N(3, -5)$, $O(-3, 1)$
4. $W(1, -2)$, $X(3, -1)$, $Y(7, -2)$, $Z(1, -5)$

5. For trapezoid $ABDC$, E and F are midpoints of the legs. Find CD. **18**

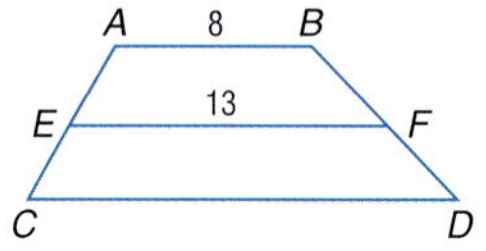

6. For trapezoid $LMNO$, P and Q are midpoints of the legs. Find PQ, $m\angle M$, and $m\angle O$. **19, 84, 144**

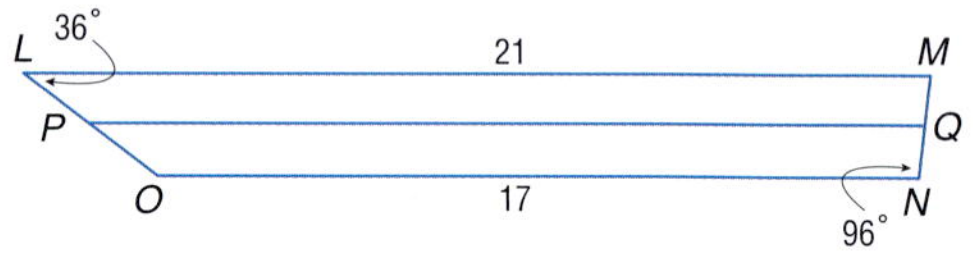

7. For isosceles trapezoid $QRST$, find the length of the median, $m\angle S$, and $m\angle R$. **18, 52, 128**

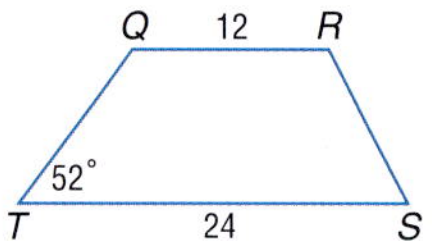

8. For trapezoid $XYZW$, A and B are midpoints of the legs. For trapezoid $XYBA$, C and D are midpoints of the legs. Find CD. **15**

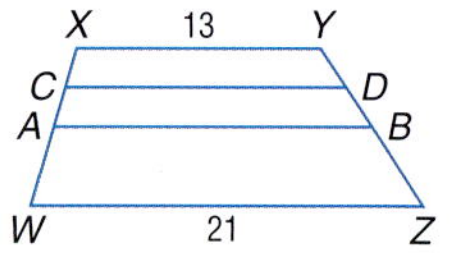

Lesson 8-7 (pages 447–451)

Name the missing coordinates for each quadrilateral.

1. isosceles trapezoid *ABCD*

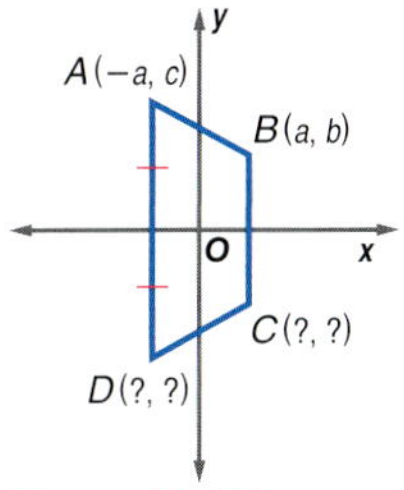

$C(a, -b)$, $D(-a, -c)$

2. rectangle *QRST*

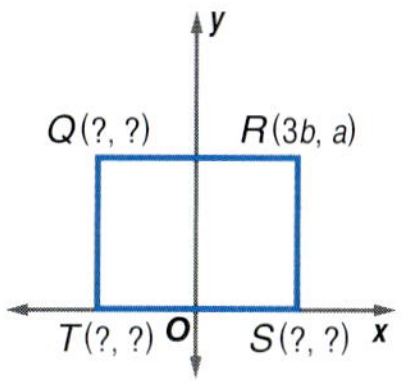

$S(3b, 0)$, $T(-3b, 0)$, $Q(-3b, a)$

Position and label each figure on the coordinate plane. Then write a coordinate proof for each of the following. 3–4. See margin.

3. The diagonals of a square are congruent.
4. Quadrilateral *EFGH* with vertices $E(0, 0)$, $F(a\sqrt{2}, a\sqrt{2})$, $G(2a + a\sqrt{2}, a\sqrt{2})$, and $H(2a, 0)$ is a rhombus.

Lesson 9-1 (pages 463–469)

COORDINATE GEOMETRY **Graph each figure and its image under the given reflection.**

1. $\triangle ABN$ with vertices $A(2, 2)$, $B(3, -2)$, and $N(-3, -1)$ in the x-axis 1–7. See margin.
2. rectangle *BARN* with vertices $B(3, 3)$, $A(3, -4)$, $R(-1, -4)$, and $N(-1, 3)$ in the line $y = x$
3. trapezoid *ZOID* with vertices $Z(2, 3)$, $O(2, -4)$, $I(-3, -3)$, and $D(-3, 1)$ in the origin
4. $\triangle PQR$ with vertices $P(-2, 1)$, $Q(2, -2)$, and $R(-3, -4)$ in the y-axis
5. square *BDFH* with vertices $B(-4, 4)$, $D(-1, 4)$, $F(-1, 1)$, and $H(-4, 1)$ in the origin
6. quadrilateral *QUAD* with vertices $Q(1, 3)$, $U(3, 1)$, $A(-1, 0)$, and $D(-3, 4)$ in the line $y = -1$
7. $\triangle CAB$ with vertices $C(0, 4)$, $A(1, -3)$, and $B(-4, 0)$ in the line $x = -2$

Lesson 9-2 (pages 470–475)

In each figure, $c \parallel d$. Determine whether the red figure is a translation image of the blue figure. Write *yes* or *no*. Explain your answer. 1–3. See margin.

1.

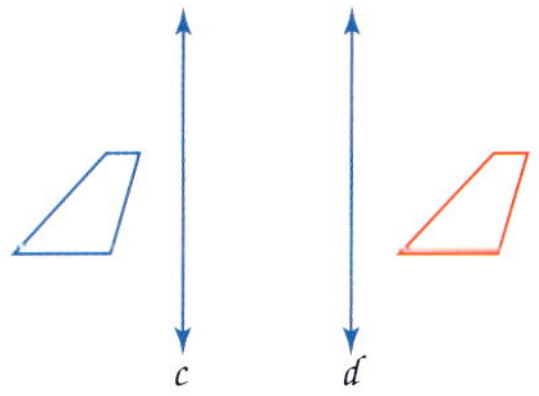

2.

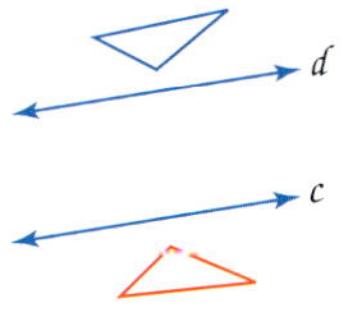

3.

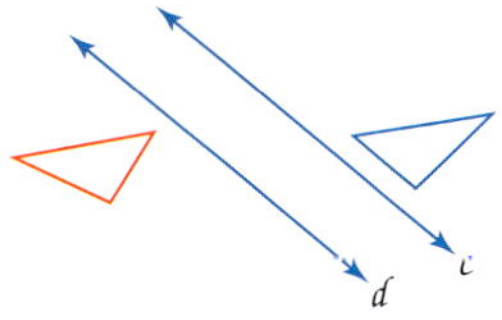

COORDINATE GEOMETRY **Graph each figure and its image under the given translation.** 4–8. See p. 781A.

4. $\overline{LM}$ with endpoints $L(2, 3)$ and $M(-4, 1)$ under the translation $(x, y) \to (x + 2, y + 1)$
5. $\triangle DEF$ with vertices $D(1, 2)$, $E(-2, 1)$, and $F(-3, -1)$ under the translation $(x, y) \to (x - 1, y - 3)$
6. quadrilateral *WXYZ* with vertices $W(1, 1)$, $X(-2, 3)$, $Y(-3, -2)$, and $Z(2, -2)$ under the translation $(x, y) \to (x + 1, y - 1)$
7. pentagon *ABCDE* with vertices $A(1, 3)$, $B(-1, 1)$, $C(-1, -2)$, $D(3, -2)$, and $E(3, 1)$ under the translation $(x, y) \to (x - 2, y + 3)$
8. $\triangle RST$ with vertices $R(-4, 3)$, $S(-2, -3)$, and $T(2, -1)$ under the translation $(x, y) \to (x + 3, y - 2)$

Extra Practice

Lesson 9-1

1.

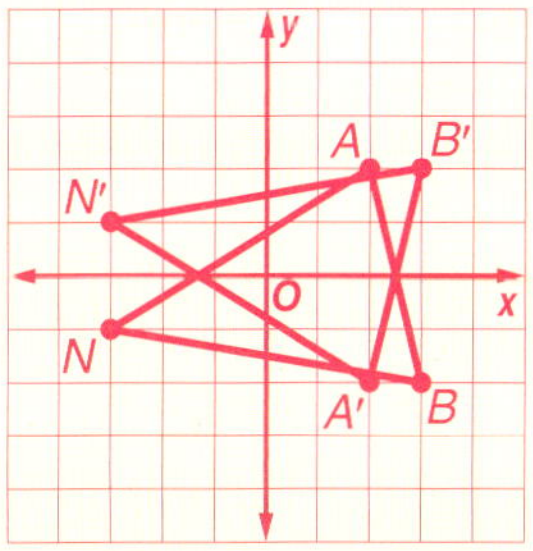

2.

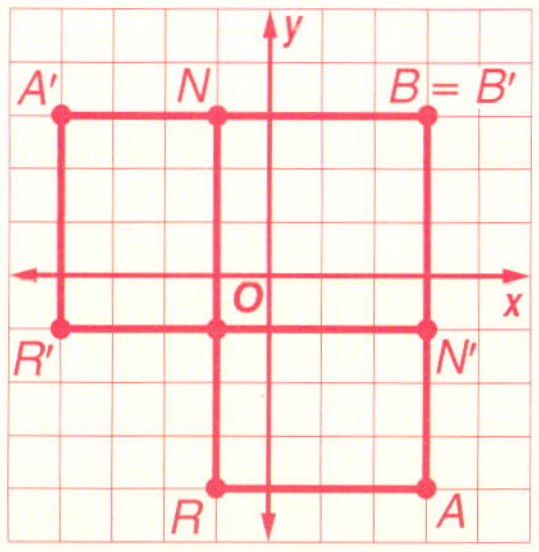

3.

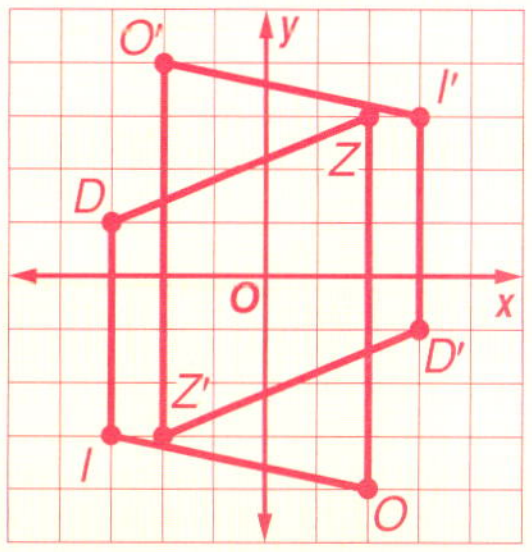

4.

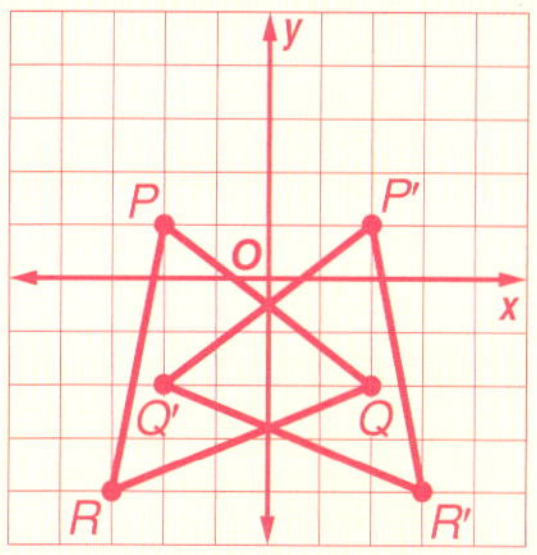

5.

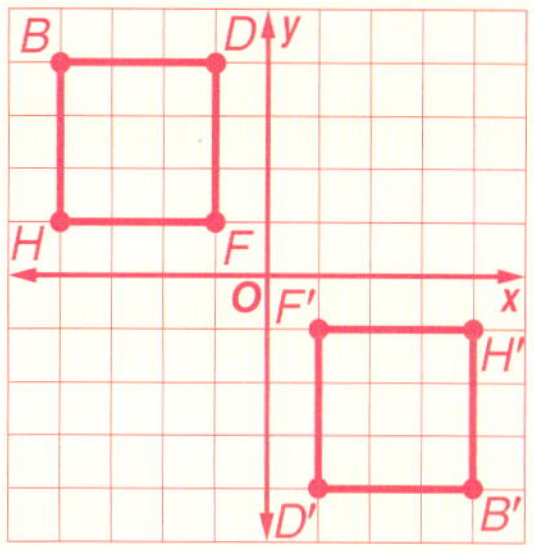

6.

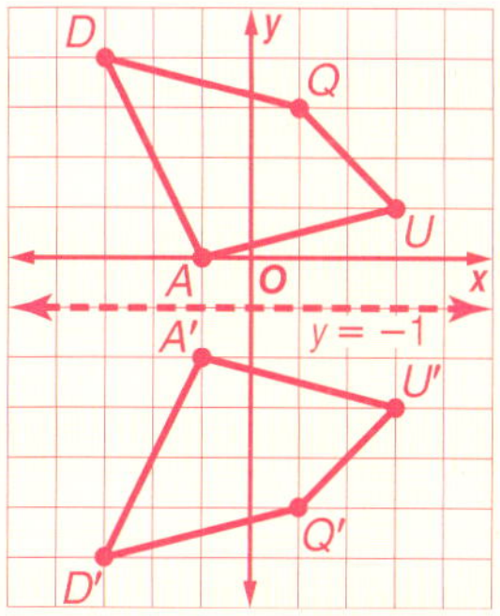

7.

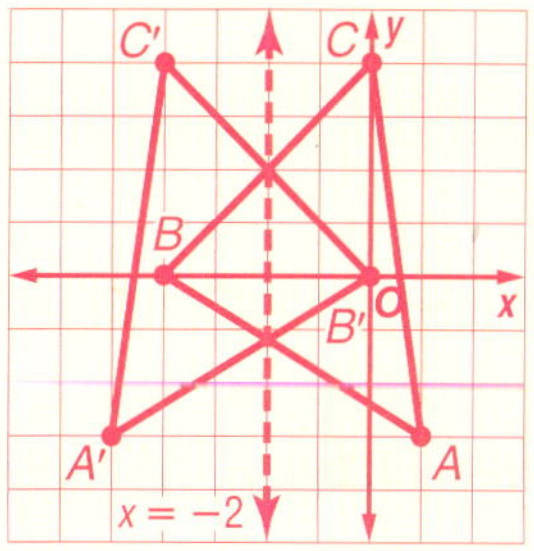

Lesson 9-2

1. Yes; it is one reflection after another with respect to the two parallel lines.
2. No; the figure has a different orientation.
3. No; it is not one reflection after another with respect to the two parallel lines.

Extra Practice

Lesson 9-3

1.
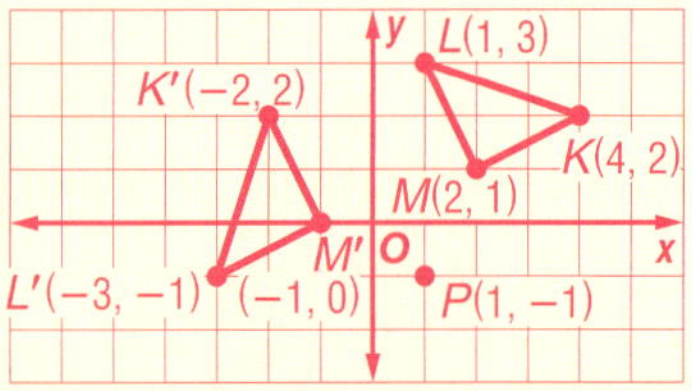

2.
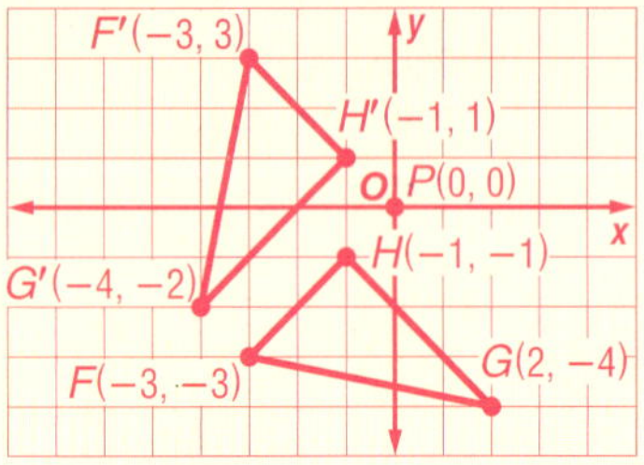

3.
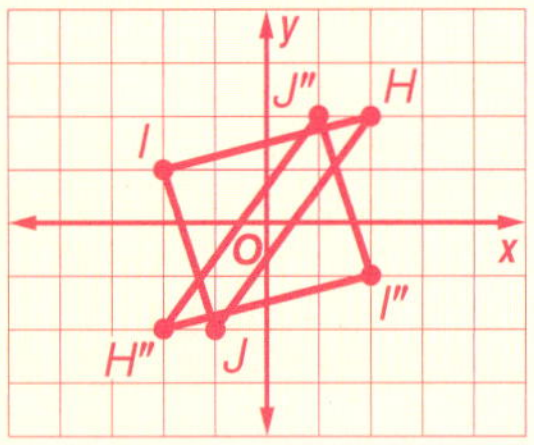

H″(−2, −2), I″(2, −1), and J″(1, 2); 180°

4.
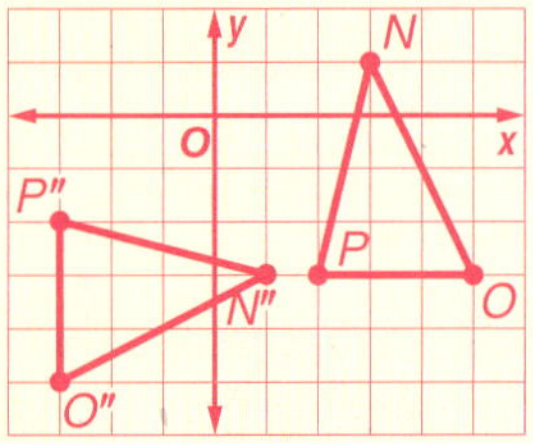

N″(1, −3), O″(−3, −5), and P″(−3, −2); 90° clockwise

5.
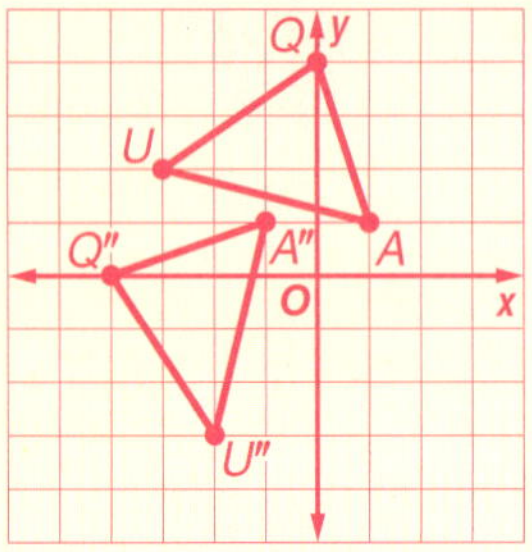

Q″(−4, 0), U″(−2, −3), and A″(−1, 1); 90° counterclockwise

6.
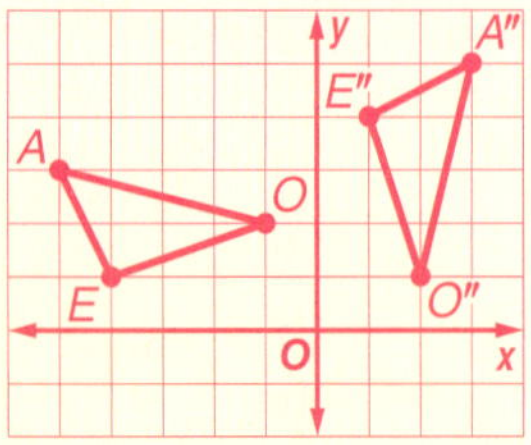

A″(3, 5), E″(1, 4), and O″(2, 1); 90° clockwise

Extra Practice

Lesson 9-3 *(pages 476–482)*

COORDINATE GEOMETRY Draw the rotation image of each figure 90° in the given direction about the center point and label the coordinates. **1–2. See margin.**

1. $\triangle KLM$ with vertices $K(4, 2)$, $L(1, 3)$, and $M(2, 1)$ counterclockwise about the point $P(1, -1)$
2. $\triangle FGH$ with vertices $F(-3, -3)$, $G(2, -4)$, and $H(-1, -1)$ clockwise about the point $P(0, 0)$

COORDINATE GEOMETRY Draw the rotational image of each triangle by reflecting the triangles in the given lines. State the coordinates of the rotation image and the angle of rotation. **3–6. See margin.**

3. $\triangle HIJ$ with vertices $H(2, 2)$, $I(-2, 1)$, and $J(-1, -2)$, reflected in the x-axis and then in the y-axis
4. $\triangle NOP$ with vertices $N(3, 1)$, $O(5, -3)$, and $P(2, -3)$, reflected in the y-axis and then in the line $y = x$
5. $\triangle QUA$ with vertices $Q(0, 4)$, $U(-3, 2)$, and $A(1, 1)$, reflected in the x-axis and then in the line $y = x$
6. $\triangle AEO$ with vertices $A(-5, 3)$, $E(-4, 1)$, and $O(-1, 2)$, reflected in the line $y = -x$ and then in the y-axis

Lesson 9-4 *(pages 483–488)*

Determine whether a semi-regular tessellation can be created from each set of figures. Assume each figure has a side length of 1 unit.

1. regular hexagons and squares **no**
2. squares and regular pentagons **no**
3. regular hexagons and regular octagons **no**

Determine whether each statement is *always, sometimes,* or *never* true.

4. Any right isosceles triangle can be used to form a uniform tessellation. **never**
5. A semi-regular tessellation is uniform. **always**
6. A polygon that is not regular can tessellate the plane. **sometimes**
7. If the measure of one interior angle of a regular polygon is greater than 120°, it cannot tessellate the plane. **always**

Lesson 9-5 *(pages 490–497)*

Find the measure of the dilation image or the preimage of $\overline{OM}$ with the given scale factor.

1. $OM = 1, r = -2$ **$O'M' = 2$**
2. $OM = 3, r = \frac{1}{3}$ **$O'M' = 1$**
3. $O'M' = \frac{3}{4}, r = 3$ **$OM = \frac{1}{4}$**
4. $OM = \frac{7}{8}, r = -\frac{5}{7}$ **$O'M' = \frac{5}{8}$**
5. $O'M' = 4, r = -\frac{2}{3}$ **$OM = 6$**
6. $O'M' = 4.5, r = -1.5$ **$OM = 3$**

COORDINATE GEOMETRY Find the image of each polygon, given the vertices, after a dilation centered at the origin with scale factor $r = 3$. Then graph a dilation with $r = \frac{1}{3}$. **7–10. See p. 781A.**

7. $T(1, 1)$, $R(-1, 2)$, $I(-2, 0)$
8. $E(2, 1)$, $I(3, -3)$, $O(-1, -2)$
9. $A(0, -1)$, $B(-1, 1)$, $C(0, 2)$, $D(1, 1)$
10. $B(1, 0)$, $D(2, 0)$, $F(3, -2)$, $H(0, -2)$

772 Extra Practice

Lesson 9-6

(pages 498–505)

Find the magnitude and direction of $\overrightarrow{XY}$ for the given coordinates. 1–6. See margin.

1. $X(1, 1)$, $Y(-2, 3)$
2. $X(-1, -1)$, $Y(2, 2)$
3. $X(-5, 4)$, $Y(-2, -3)$
4. $X(2, 1)$, $Y(-4, -4)$
5. $X(-2, -1)$, $Y(2, -2)$
6. $X(3, -1)$, $Y(-3, 1)$

Graph the image of each figure under a translation by the given vector. 7–9. See margin.

7. $\triangle HIJ$ with vertices $H(2, 3)$, $I(-4, 2)$, $J(-1, 1)$; $\vec{a} = \langle 1, 3 \rangle$
8. quadrilateral $RSTW$ with vertices $R(4, 0)$, $S(0, 1)$, $T(-2, -2)$, $W(3, -1)$; $\vec{x} = \langle -3, 4 \rangle$
9. pentagon $AEIOU$ with vertices $A(-1, 3)$, $E(2, 3)$, $I(2, 0)$, $O(-1, -2)$, $U(-3, 0)$; $\vec{b} = \langle -2, -1 \rangle$

10. $\sqrt{74} \approx 8.6$, 54.5° 11. $3\sqrt{5} \approx 6.7$, 116.6° 12. $\sqrt{41} \approx 6.4$, 321.3°

Find the magnitude and direction of each resultant for the given vectors.

10. $\vec{c} = \langle 2, 3 \rangle$, $\vec{d} = \langle 3, 4 \rangle$
11. $\vec{a} = \langle 1, 3 \rangle$, $\vec{b} = \langle -4, 3 \rangle$
12. $\vec{x} = \langle 1, 2 \rangle$, $\vec{y} = \langle 4, -6 \rangle$
13. $\vec{s} = \langle 2, 5 \rangle$, $\vec{t} = \langle -6, -8 \rangle$ 5, 216.9°
14. $\vec{m} = \langle 2, -3 \rangle$, $\vec{n} = \langle -2, 3 \rangle$ 0, 0°
15. $\vec{u} = \langle -7, 2 \rangle$, $\vec{v} = \langle 4, 1 \rangle$ $3\sqrt{2} \approx 4.2$, 135°

Lesson 9-7

(pages 506–511)

Find the coordinates of the image under the stated transformation. 1–4. See margin.

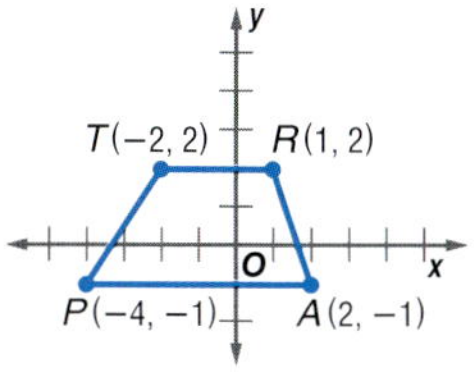

1. reflection in the x-axis
2. rotation 90° clockwise about the origin
3. translation $(x, y) \rightarrow (x - 4, y + 3)$
4. dilation by scale factor -4

Use a matrix to find the coordinates of the vertices of the image of each figure after the stated transformation. 5–10. See margin.

5. $\triangle DEF$ with $D(2, 4)$, $E(-2, -4)$, and $F(4, -6)$; dilation by a scale factor of 2.5
6. $\triangle RST$ with $R(3, 4)$, $S(-6, -2)$, and $T(5, -3)$; reflection in the x-axis
7. quadrilateral $CDEF$ with $C(1, 1)$, $D(-2, 5)$, $E(-2, 0)$, and $F(-1, -2)$; rotation of 90° counterclockwise
8. quadrilateral $WXYZ$ with $W(0, 4)$, $X(-5, 0)$, $Y(0, -3)$, and $Z(5, -2)$; translation $(x, y) \rightarrow (x + 1, y - 4)$
9. quadrilateral $JKLM$ with $J(-6, -2)$, $K(-2, -8)$, $L(4, -4)$, and $M(6, 6)$; dilation by a scale factor of $-\frac{1}{2}$
10. pentagon $ABCDE$ with $A(2, 2)$, $B(0, 4)$, $C(-3, 2)$, $D(-3, -4)$, and $E(2, -4)$; reflection in the line $y = x$

Lesson 10-1

(pages 522–528)

The radius, diameter, or circumference of a circle is given. Find the missing measures to the nearest hundredth.

1. $r = 18$ in., $d =$ _?_, $C =$ _?_ 36 in., 113.10 in.
2. $d = 34.2$ ft, $r =$ _?_, $C =$ _?_ 17.1 ft, 107.44 ft
3. $C = 12\pi$ m, $r =$ _?_, $d =$ _?_ 6 m, 12 m
4. $C = 84.8$ mi, $r =$ _?_, $d =$ _?_ 13.50 mi, 26.99 mi
5. $d = 8.7$ cm, $r =$ _?_, $C =$ _?_ 4.35 cm, 27.33 cm
6. $r = 3b$ in., $d =$ _?_, $C =$ _?_ $6b$ in., $18.85b$ in.

Find the exact circumference of each circle.

7.
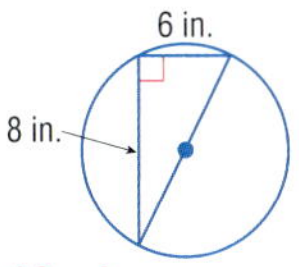

10π in.

8.
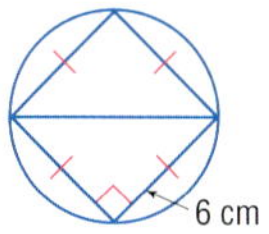

$6\sqrt{2}\pi$ cm

9.
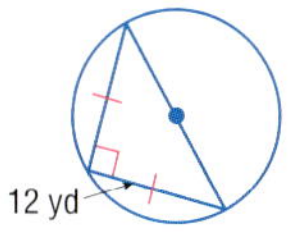

$12\sqrt{2}\pi$ yd

10.
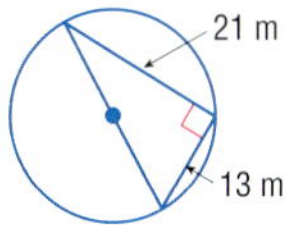

$\sqrt{610}\pi$ m

Extra Practice

Lesson 9-6

1. $\sqrt{13} \approx 3.6$, 146.3°
2. $3\sqrt{2} \approx 4.2$, 45°
3. $\sqrt{58} \approx 7.6$, 293.2°
4. $\sqrt{61} \approx 7.8$, 219.8°
5. $\sqrt{17} \approx 4.1$, 346.0°
6. $2\sqrt{10} \approx 6.3$, 161.6°
7.
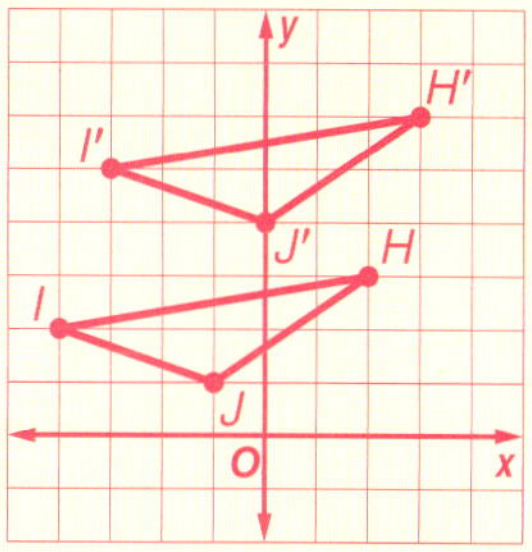

8.
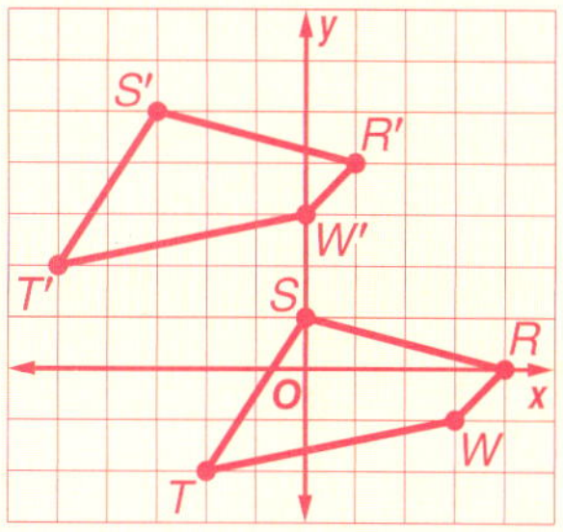

9.
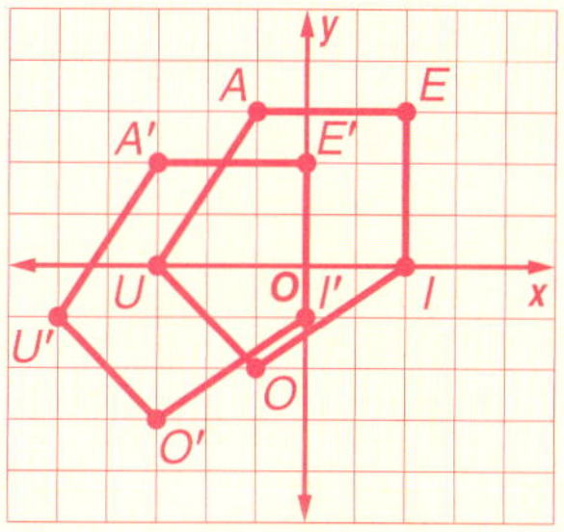

Lesson 9-7

1. $R'(1, -2)$, $T'(-2, -2)$, $P'(-4, 1)$, $A'(2, 1)$
2. $R'(2, -1)$, $T'(2, 2)$, $P'(-1, 4)$, $A'(-1, -2)$
3. $R'(-3, 5)$, $T'(-6, 5)$, $P'(-8, 2)$, $A'(-2, 2)$
4. $R'(-4, -8)$, $T'(8, -8)$, $P'(16, 4)$, $A'(-8, 4)$
5. $D'(5, 10)$, $E'(-5, -10)$, $F'(10, -15)$
6. $R'(3, -4)$, $S'(-6, 2)$, $T'(5, 3)$
7. $C'(-1, 1)$, $D'(-5, -2)$, $E'(0, -2)$, $F'(2, -1)$
8. $W'(1, 0)$, $X'(-4, -4)$, $Y'(1, -7)$, $Z'(6, -6)$
9. $J'(3, 1)$, $K'(1, 4)$, $L'(-2, 2)$, $M'(-3, -3)$
10. $A'(2, 2)$, $B'(4, 0)$, $C'(2, -3)$, $D'(-4, -3)$, $E'(-4, 2)$

Lesson 10-2

(pages 529–535)

Find each measure.

1. $m\angle GKI$ **90**
2. $m\angle LKJ$ **23**
3. $m\angle LKI$ **113**
4. $m\angle LKG$ **157**
5. $m\angle HKI$ **67**
6. $m\angle HKJ$ **157**

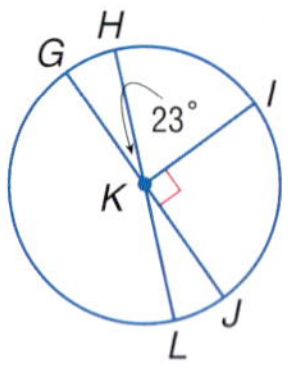

In $\odot X$, $\overline{WS}$, $\overline{VR}$, and $\overline{QT}$ are diameters, $m\angle WXV = 25$ and $m\angle VXU = 45$. Find each measure.

7. $m\widehat{QR}$ **90**
8. $m\widehat{QW}$ **65**
9. $m\widehat{TU}$ **45**
10. $m\widehat{WRV}$ **335**
11. $m\widehat{SV}$ **155**
12. $m\widehat{TRW}$ **245**

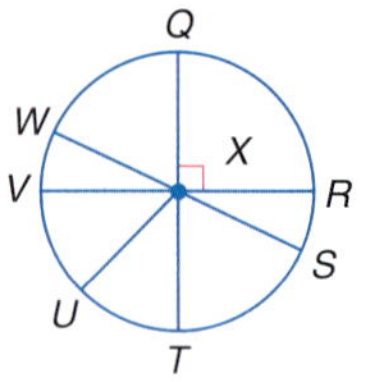

Lesson 10-3

(pages 536–543)

In $\odot S$, $HJ = 22$, $LG = 18$, $m\widehat{IJ} = 35$, and $m\widehat{LM} = 30$. Find each measure.

1. HR **11**
2. RJ **11**
3. LT **9**
4. TG **9**
5. $m\widehat{HJ}$ **70**
6. $m\widehat{LG}$ **60**
7. $m\widehat{MG}$ **30**
8. $m\widehat{HI}$ **35**

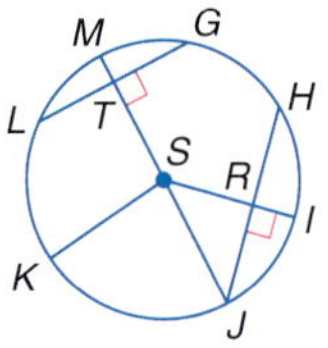

In $\odot R$, $CR = RF$, and $ED = 30$. Find each measure.

9. AB **30**
10. EF **15**
11. DF **15**
12. BC **15**

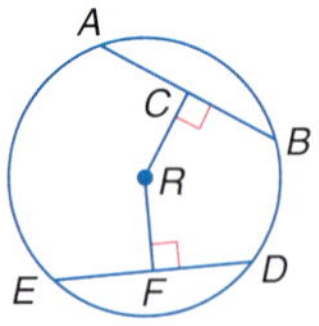

Lesson 10-4

(pages 544–551)

Find the measure of each numbered angle for each figure. **1–6. See margin.**

1. $m\widehat{AB} = 176$, and $m\widehat{BC} = 42$

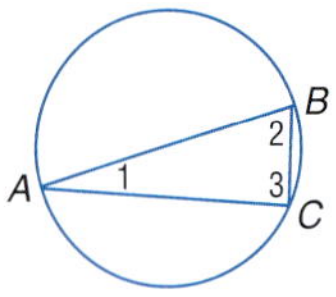

2. $\overline{WX} \cong \overline{ZY}$, and $m\widehat{ZW} = 120$

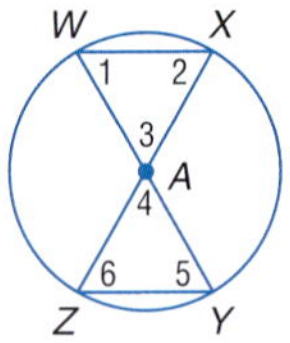

3. $m\widehat{QR} = 40$, and $m\widehat{TS} = 110$

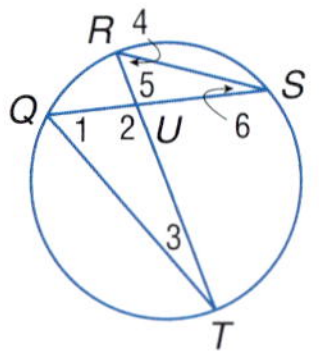

4. $\square ABCD$ is a rectangle, and $m\widehat{BC} = 70$.

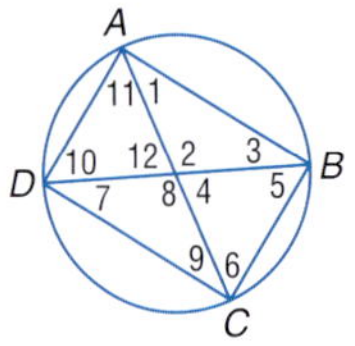

5. $m\widehat{TR} = 100$, and $\overline{SR} \perp \overline{QT}$

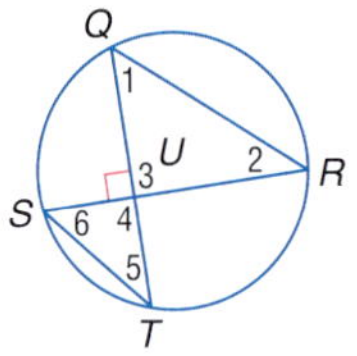

6. $m\widehat{UY} = m\widehat{XZ} = 56$ and $m\widehat{UV} = m\widehat{XW} = 56$

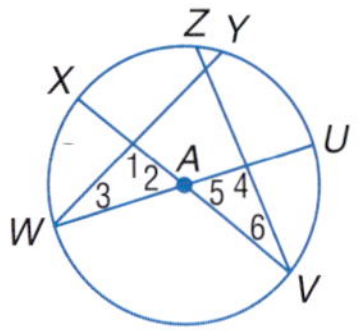

7. Rhombus $ABCD$ is inscribed in a circle. What can you conclude about $\overline{BD}$? **It is a diameter of the circle.**
8. Triangle RST is inscribed in a circle. If the measure of $\widehat{RS}$ is 170, what is the measure of $\angle T$? **85**

Extra Practice

Lesson 10-4

1. $m\angle 1 = 21$, $m\angle 2 = 71$, $m\angle 3 = 88$
2. $m\angle 1 = 60$, $m\angle 2 = 60$, $m\angle 3 = 60$, $m\angle 4 = 60$, $m\angle 5 = 60$, $m\angle 6 = 60$
3. $m\angle 1 = 55$, $m\angle 2 = 105$, $m\angle 3 = 20$, $m\angle 4 = 55$, $m\angle 5 = 105$, $m\angle 6 = 20$
4. $m\angle 1 = 35$, $m\angle 2 = 110$, $m\angle 3 = 35$, $m\angle 4 = 70$, $m\angle 5 = 55$, $m\angle 6 = 55$, $m\angle 7 = 35$, $m\angle 8 = 110$, $m\angle 9 = 35$, $m\angle 10 = 55$, $m\angle 11 = 55$, $m\angle 12 = 70$
5. $m\angle 1 = 50$, $m\angle 2 = 40$, $m\angle 3 = 90$, $m\angle 4 = 90$, $m\angle 5 = 40$, $m\angle 6 = 50$
6. $m\angle 1 = 96$, $m\angle 2 = 56$, $m\angle 3 = 28$, $m\angle 4 = 96$, $m\angle 5 = 56$, $m\angle 6 = 28$

Lesson 10-5

(pages 552–558)

Determine whether each segment is tangent to the given circle.

1. 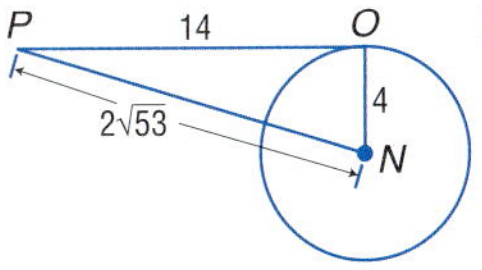

yes

2. 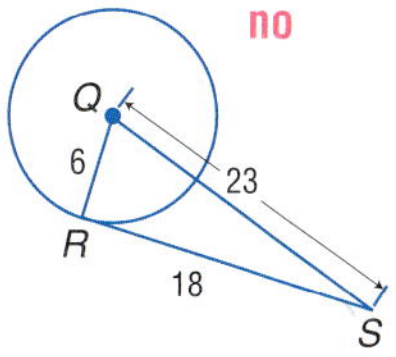

no

Find x. Assume that segments that appear to be tangent are tangent.

3. 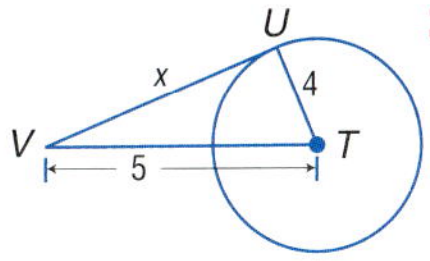

3

4. 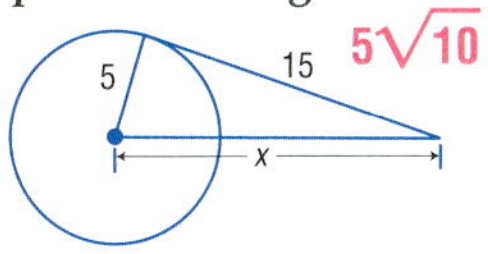

$5\sqrt{10}$

5. 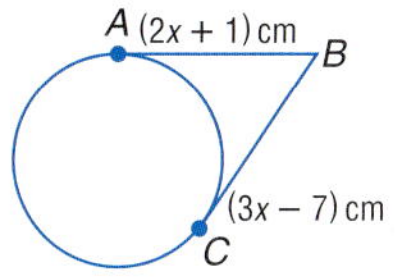

8

Lesson 10-6

(pages 561–568)

Find each measure.

1. $m\angle 5$ 75

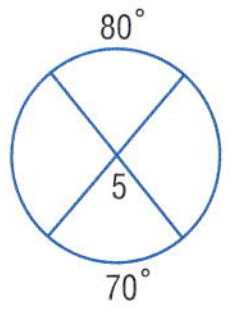

2. $m\angle 6$ 142.5

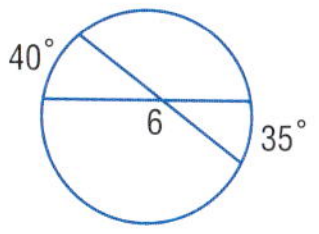

3. $m\angle 7$ 110

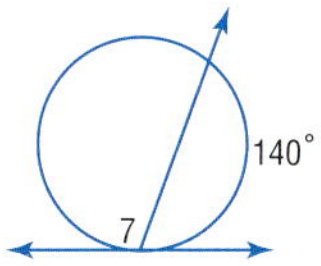

Find x. Assume that any segment that appears to be tangent is tangent.

4. 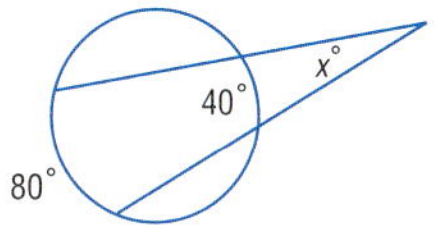

20

5. 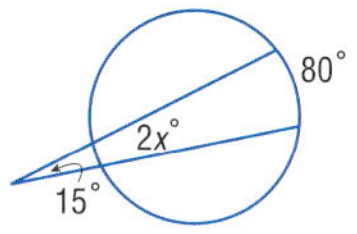

25

6. 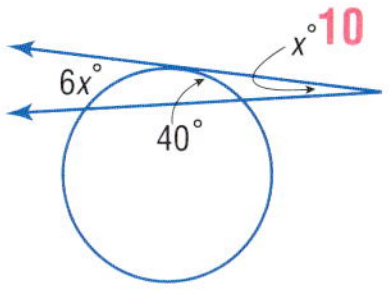

10

Lesson 10-7

(pages 569–574)

Find x. Assume that segments that appear to be tangent are tangent.

1. 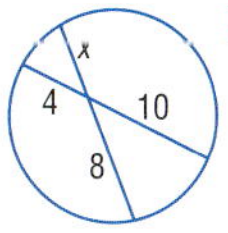

5

2. 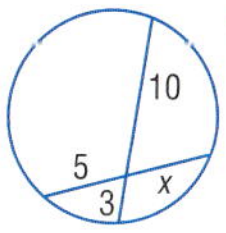

6

3. 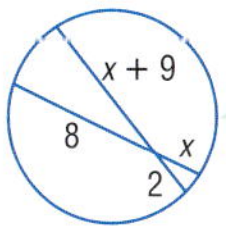

3

Find each variable to the nearest tenth.

4. 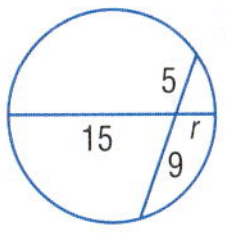

3.0

5. 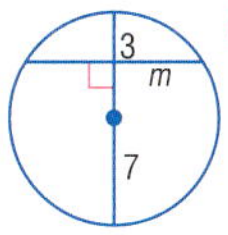

5.7

6. 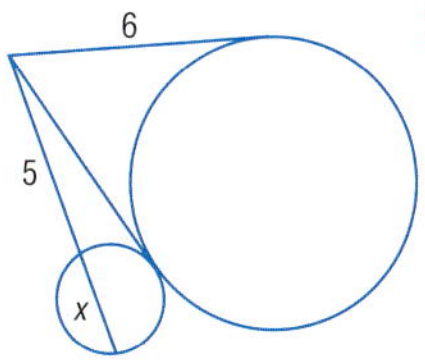

2.2

Extra Practice

Extra Practice

Extra Practice 775

Lesson 10-8

1. $(x-1)^2+(y+2)^2=4$
2. $x^2+y^2=16$
3. $(x+3)^2+(y+4)^2=11$
4. $(x-3)^2+(y+1)^2=9$
5. $(x-6)^2+(y-12)^2=49$
6. $(x-4)^2+y^2=16$
7. $(x-6)^2+(y+6)^2=121$
8. $(x+5)^2+(y-1)^2=1$

9.

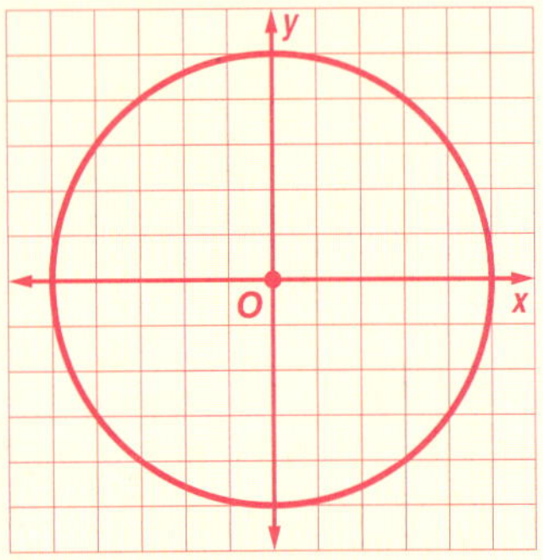

10.

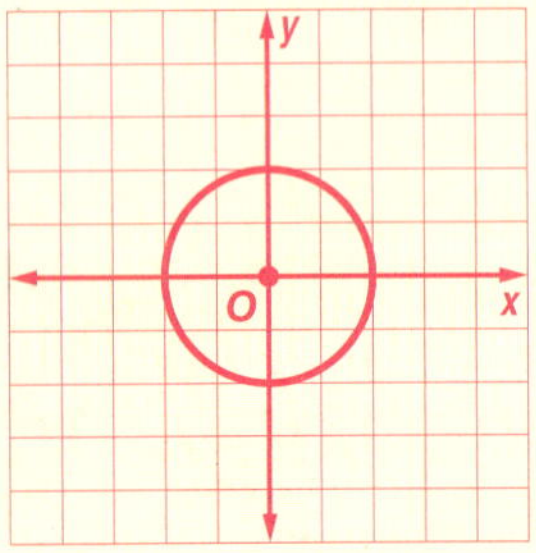

11.

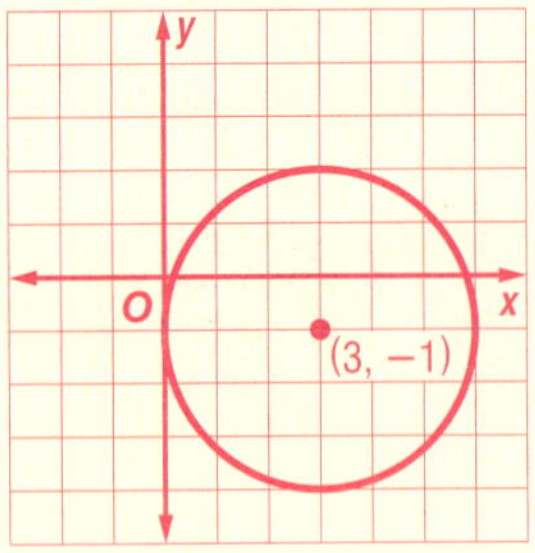

12.

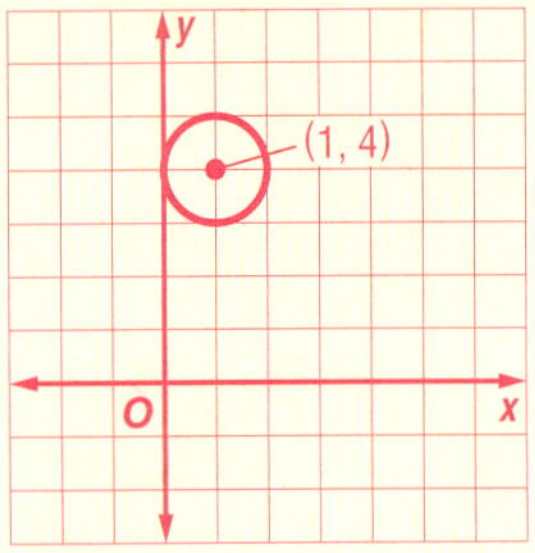

Lesson 10-8 *(pages 575–580)*

Write an equation for each circle. 1–8. See margin.

1. center at $(1, -2)$, $r = 2$
2. center at origin, $r = 4$
3. center at $(-3, -4)$, $r = \sqrt{11}$
4. center at $(3, -1)$, $d = 6$
5. center at $(6, 12)$, $r = 7$
6. center at $(4, 0)$, $d = 8$
7. center at $(6, -6)$, $d = 22$
8. center at $(-5, 1)$, $d = 2$

Graph each equation. 9–12. See margin.

9. $x^2 + y^2 = 25$
10. $x^2 + y^2 - 3 = 1$
11. $(x-3)^2 + (y+1)^2 = 9$
12. $(x-1)^2 + (y-4)^2 = 1$

13. Find the radius of a circle whose equation is $(x+3)^2 + (y-1)^2 = r^2$ and contains $(-2, 1)$. 1
14. Find the radius of a circle whose equation is $(x-4)^2 + (y-3)^2 = r^2$ and contains $(8, 3)$. 4

Lesson 11-1 *(pages 595–600)*

Find the area and perimeter of each parallelogram. Round to the nearest tenth if necessary.

1.

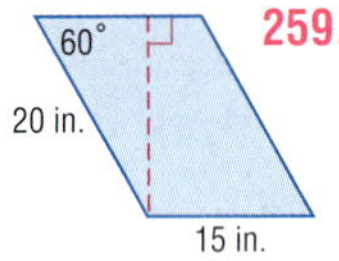

259.8 in², 70 in.

2.

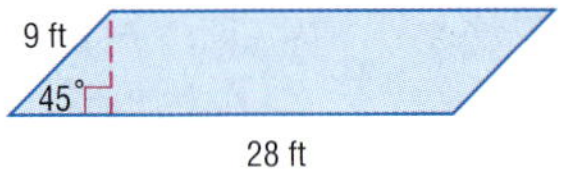

178.2 ft², 74 ft

3.

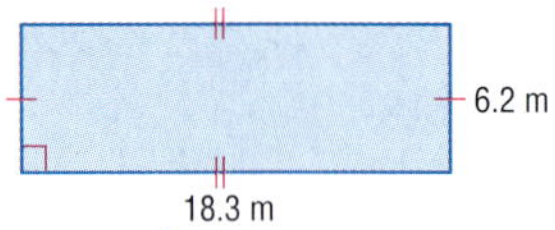

113.5 m², 49 m

COORDINATE GEOMETRY **Given the coordinates of the vertices of a quadrilateral, determine whether it is a *square*, a *rectangle*, or a *parallelogram*. Then find the area of the quadrilateral.**

4. $Q(-3, 3)$, $R(-1, 3)$, $S(-1, 1)$, $T(-3, 1)$ square, 4 units²
5. $A(-7, -6)$, $B(-2, -6)$, $C(-2, -3)$, $D(-7, -3)$ 5. rectangle, 15 units²
6. $L(5, 3)$, $M(8, 3)$, $N(9, 7)$, $O(6, 7)$ parallelogram, 12 units²
7. $W(-1, -2)$, $X(-1, 1)$, $Y(2, 1)$, $Z(2, -2)$ square, 9 units²

Lesson 11-2 *(pages 601–609)*

Find the area of each quadrilateral.

1.

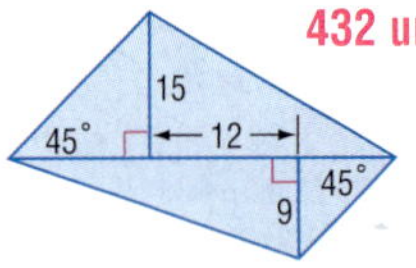

432 units²

2.

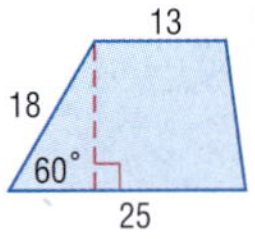

296.2 units²

3.

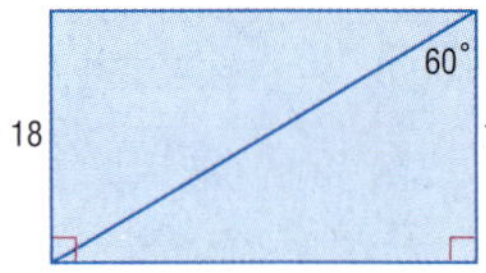

561.2 uni

COORDINATE GEOMETRY **Find the area of trapezoid *ABCD* given the coordinates of the vertices.**

4. $A(1, 1)$, $B(2, 3)$, $C(4, 3)$, $D(7, 1)$ 8 units²
5. $A(-2, 2)$, $B(2, 2)$, $C(7, -3)$, $D(-4, -3)$ 37.5 units²
6. $A(1, -1)$, $B(4, -1)$, $C(8, 5)$, $D(1, 5)$ 30 units²
7. $A(-2, 2)$, $B(4, 2)$, $C(3, -2)$, $D(1, -2)$ 16 units²

COORDINATE GEOMETRY **Find the area of rhombus *LMNO* given the coordinates of the vertices.**

8. $L(-3, 0)$, $M(1, -2)$, $N(-3, -4)$, $O(-7, -2)$ 16 units²
9. $L(-3, -2)$, $M(-4, 2)$, $N(-3, 6)$, $O(-2, 2)$ 8 units²
10. $L(-1, -4)$, $M(3, 4)$, $N(-1, 12)$, $O(-5, 4)$ 64 units²
11. $L(-2, -2)$, $M(4, 4)$, $N(10, -2)$, $O(4, -8)$ 72 units²

Lesson 11-3

(pages 610–616)

Find the area of each regular polygon. Round to the nearest tenth.

1. a square with perimeter 54 feet **182.3 ft^2**
2. a triangle with side length 9 inches **35.1 inches^2**
3. an octagon with side length 6 feet **173.8 ft^2**
4. a decagon with apothem length of 22 centimeters **1572.6 cm^2**

Find the area of each shaded region. Assume that all polygons that appear to be regular are regular. Round to the nearest tenth.

5. **66.3 cm^2**

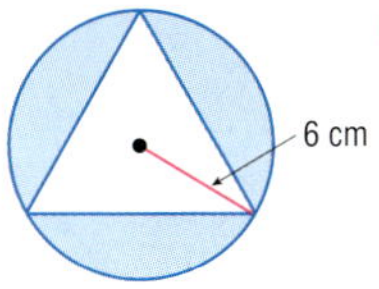

6. **61.7 ft^2**

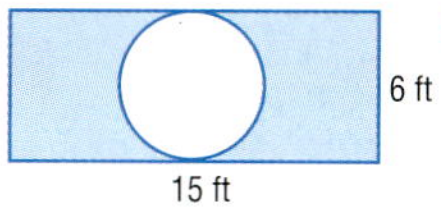

7. **37.4 in^2**

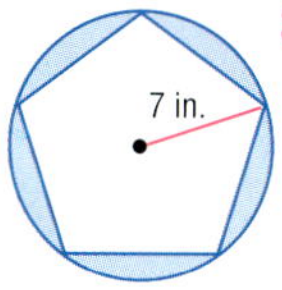

Lesson 11-4

(pages 617–621)

Find the area of each figure. Round to the nearest tenth if necessary.

1. **187.2 units^2**

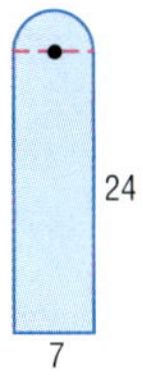

2. **420 units^2**

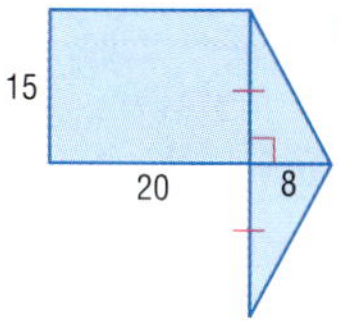

3. **88.3 units^2**

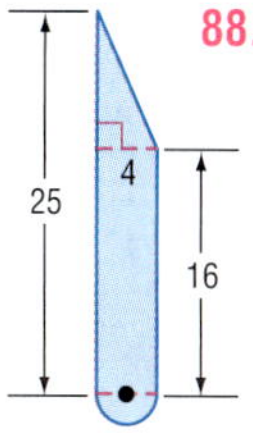

COORDINATE GEOMETRY The vertices of an irregular figure are given. Find the area of each figure.

4. $R(0, 5)$, $S(3, 3)$, $T(3, 0)$ **4.5 units^2**
5. $A(-5, -3)$, $B(-3, 0)$, $C(2, -1)$, $D(2, -3)$ **15.5 units^2**
6. $L(-1, 4)$, $M(3, 2)$, $N(3, -1)$, $O(-1, -2)$, $P(-3, 1)$ **24 units^2**

Lesson 11-5

(pages 622–627)

Find the total area of the sectors of the indicated color. Then find the probability of spinning the color indicated if the diameter of each spinner is 20 inches.

1. orange **≈ 62.8 in^2; 0.20**

2. blue **≈ 87.3 in^2; ≈ 0.28**

3. green **≈ 165.8 in^2; ≈ 0.53**

Find the area of the shaded region. Then find the probability that a point chosen at random is in the shaded region.

4.

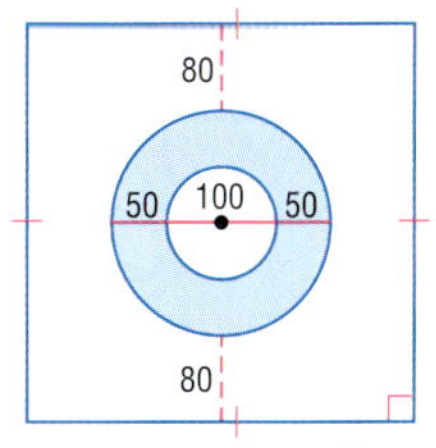

≈ 23,561.9 units^2; ≈ 0.18

5.

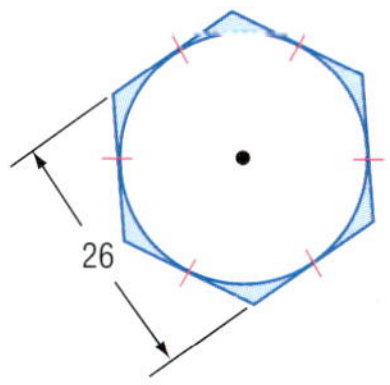

≈ 54.5 units^2; ≈ 0.09

6.

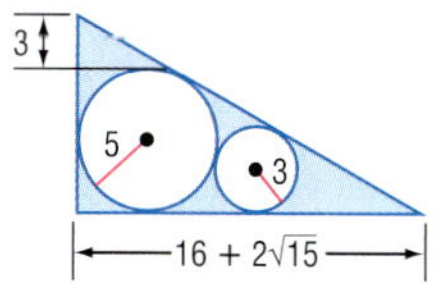

≈ 47.6 units^2; ≈ 0.31

Extra Practice

Extra Practice

Extra Practice 777

Lesson 12-1

1. 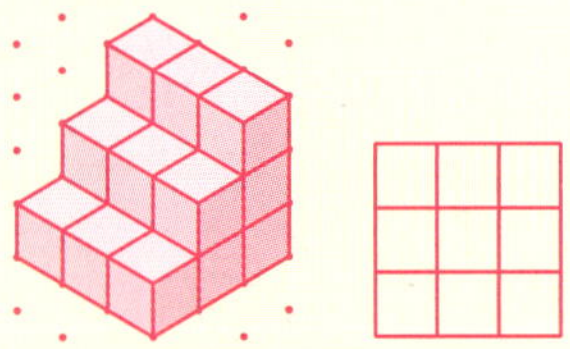

2. 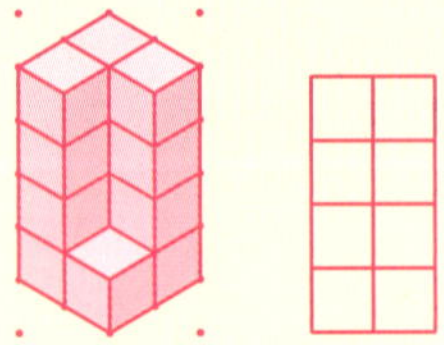

3. pentagonal pyramid; base: *OXNEP*; faces: *OXNEP*, △*PET*, △*ETN*, △*NTX*, △*XTO*, △*OTP*; edges: $\overline{PE}$, $\overline{EN}$, $\overline{NX}$, $\overline{XO}$, $\overline{OP}$, $\overline{TP}$, $\overline{TE}$, $\overline{TN}$, $\overline{TX}$, $\overline{TO}$; vertices: *T*, *P*, *E*, *N*, *X*, *O*
4. cone; base: circle *L*; vertex: *Z*
5. octagonal prism; bases: *ABCDEFGH*, *STUVWXYZ*; faces: *ABCDEFGH*, *STUVWXYZ*, *ABXY*, *BCWX*, *CDVW*, *DEUV*, *FEUT*, *GFTS*, *HGSZ*, *HAYZ*; edges: $\overline{AB}$, $\overline{BC}$, $\overline{CD}$, $\overline{DE}$, $\overline{EF}$, $\overline{FG}$, $\overline{GH}$, $\overline{AH}$, $\overline{ST}$, $\overline{TU}$, $\overline{UV}$, $\overline{VW}$, $\overline{WX}$, $\overline{XY}$, $\overline{YZ}$, $\overline{ZS}$, $\overline{AY}$, $\overline{BX}$, $\overline{CW}$, $\overline{DV}$, $\overline{EU}$, $\overline{FT}$, $\overline{GS}$, $\overline{HZ}$; vertices: *A*, *B*, *C*, *D*, *E*, *F*, *G*, *H*, *S*, *T*, *U*, *V*, *W*, *X*, *Y*, *Z*

Lesson 12-2

1. 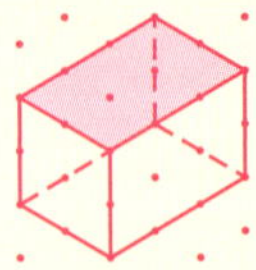

2. 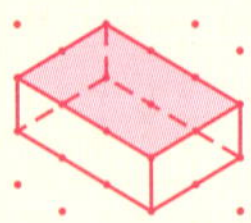

3. 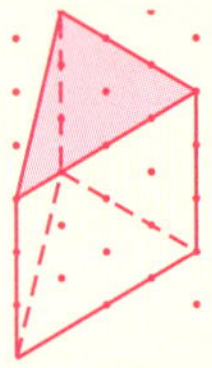

4. 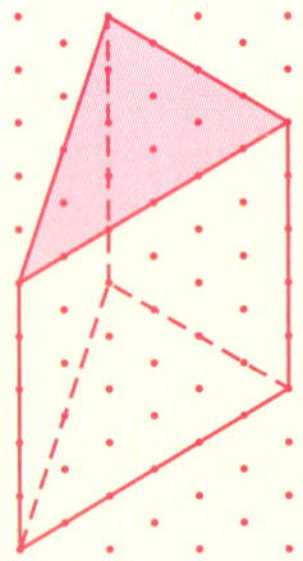

Extra Practice

Lesson 12-1 (pages 636–642)

Draw the back view and corner view of a figure given its orthogonal drawing. 1–2. See margin.

1. 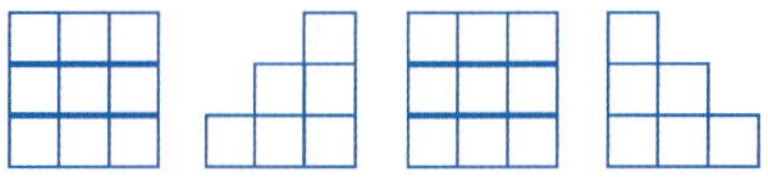

2. 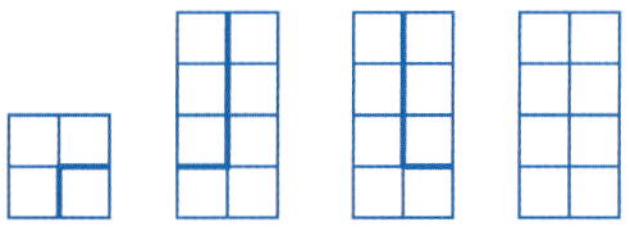

Identify each solid. Name the bases, faces, edges, and vertices. 3–5. See margin.

3. 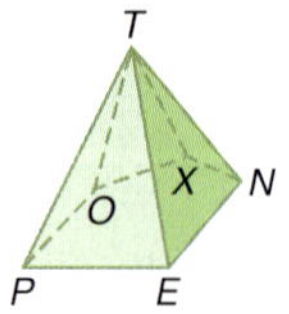

4. 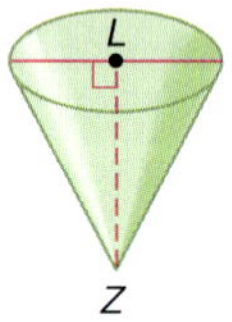

5. 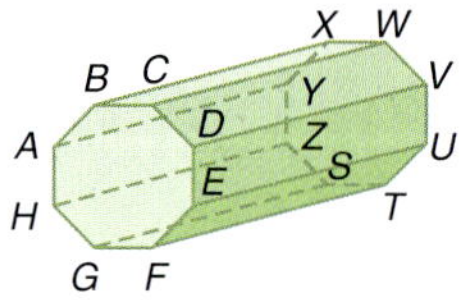

Lesson 12-2 (pages 643–648)

Sketch each solid using isometric dot paper. 1–4. See margin.

1. rectangular prism 2 units high, 3 units long, and 2 units wide
2. rectangular prism 1 unit high, 2 units long, and 3 units wide
3. triangular prism 3 units high with bases that are right triangles with legs 3 units and 4 units long
4. triangular prism 5 units high with bases that are right triangles with legs 4 units and 6 units long

5–7. See margin for nets.

For each solid, draw a net and find the surface area. Round to the nearest tenth if necessary.

5. 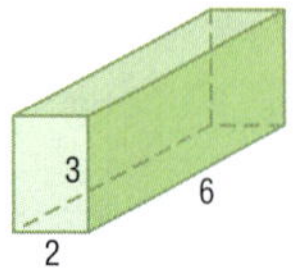

72 units^2

6. 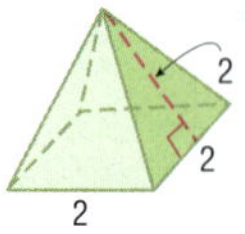

12 units^2

7. 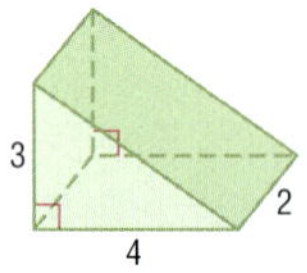

36 units^2

Lesson 12-3 (pages 649–654)

Find the lateral area and the surface area of each prism. Round to the nearest tenth if necessary.

1. 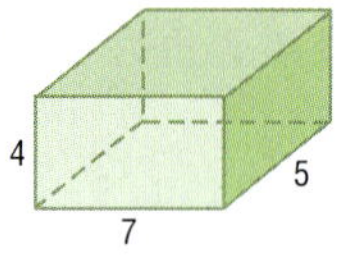

96 units^2; 166 units^2

2. 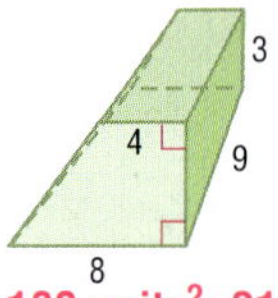

180 units^2; 216 units^2

3. 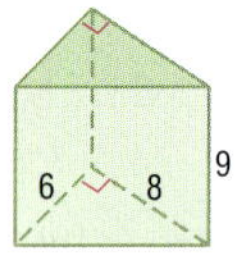

216 units^2; 264 units^2

4. 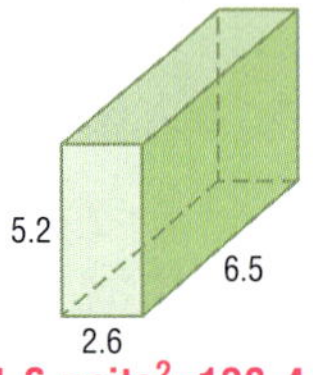

94.6 units^2; 128.4 units^2

5. 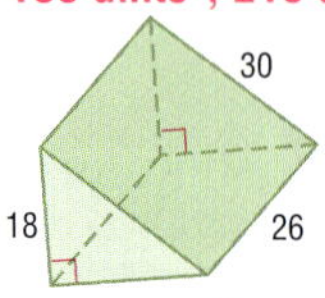

1872 units^2; 2304 units^2

6. 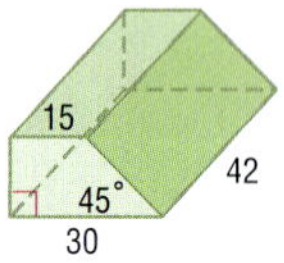

3411.0 units^2; 4086.0 units^2

7. The surface area of a right triangular prism is 228 square inches. The base is a right angle with legs measuring 6 inches and 8 inches. Find the height of the prism. 7.5 in.
8. The surface area of a right triangular prism with height 18 inches is 1,380 square inches. The base is a right angle with a leg measuring 15 inches and a hypotenuse of length 25 inches. Find the length of the other leg of the base. 20 in.

5. 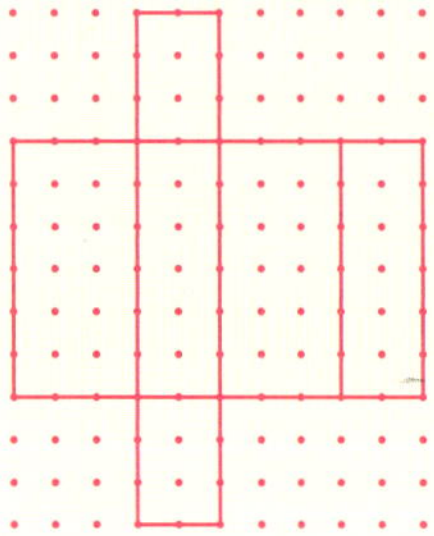

6. 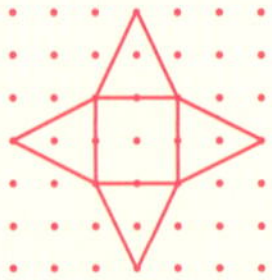

7. 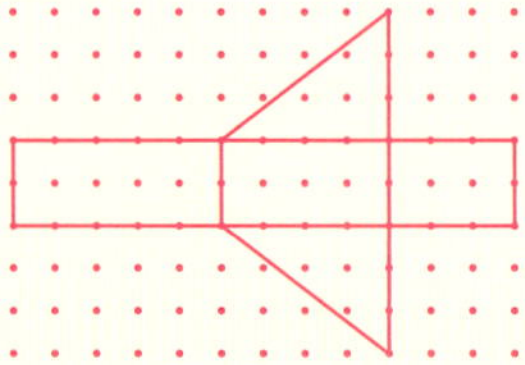

Lesson 12-4

(pages 655–659)

Find the surface area of a cylinder with the given dimensions. Round to the nearest tenth.

1. $r = 2$ ft, $h = 3.5$ ft **69.1 ft²**
2. $d = 15$ in., $h = 20$ in. **1295.9 in²**
3. $r = 3.7$ m, $h = 6.2$ m **230.2 m²**
4. $d = 19$ mm, $h = 32$ mm **2477.1 mm²**

Find the surface area of each cylinder. Round to the nearest tenth.

5. 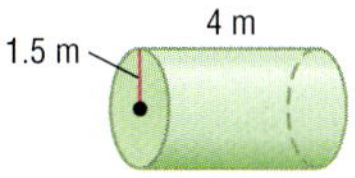

51.8 m²

6. 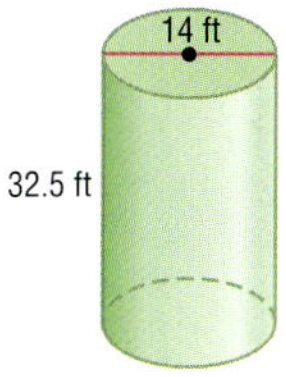

1737.3 ft²

7. 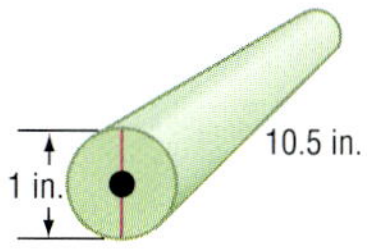

34.6 in²

8. 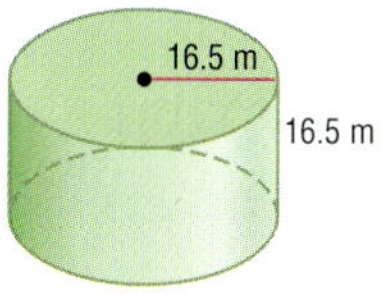

3421.2 m²

Lesson 12-5

(pages 660–665)

Find the surface area of each regular pyramid. Round to the nearest tenth.

1. 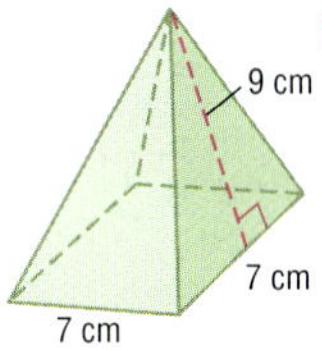

175 cm²

2. 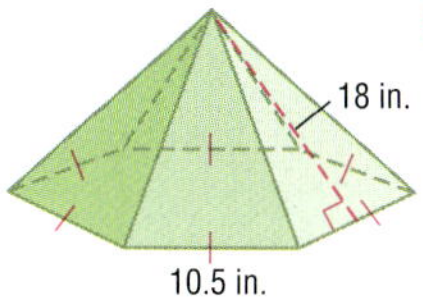

853.4 in²

3. 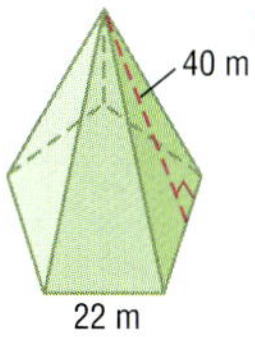

3032.7 m²

4. 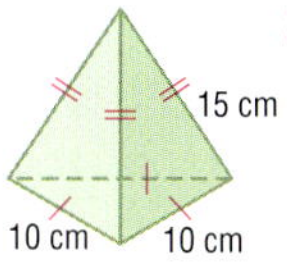

255.4 cm²

5. 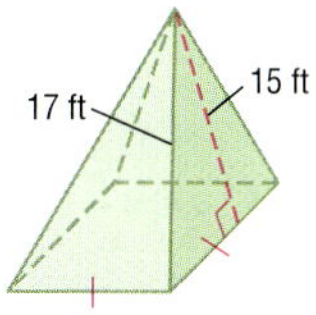

736 ft²

6. 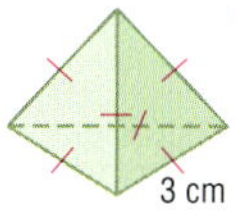

15.6 cm²

Lesson 12-6

(pages 666–670)

Find the surface area of each cone. Round to the nearest tenth.

1. 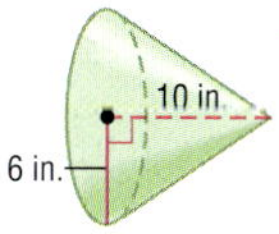

332.9 in²

2. 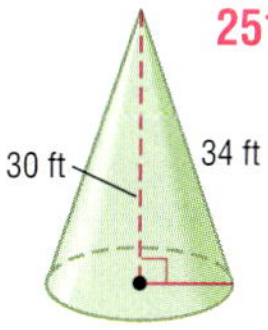

2513.3 ft²

3. 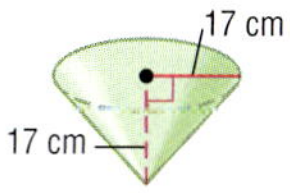

2191.9 cm²

4. 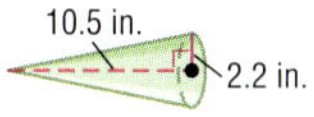

89.4 in²

5. 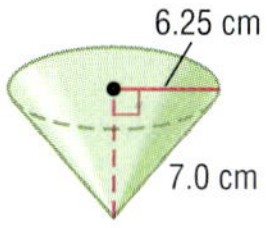

260.2 cm²

6. 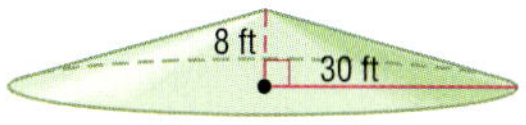

5753.7 ft²

7. Find the surface area of a cone if the height is 28 inches and the slant height is 40 inches. **6153.2 in²**

8. Find the surface area of a cone if the height is 7.5 centimeters and the radius is 2.5 centimeters. **81.7 cm²**

Extra Practice 779

Lesson 12-7
(pages 671–676)

Find the surface area of each sphere or hemisphere. Round to the nearest tenth.

1.
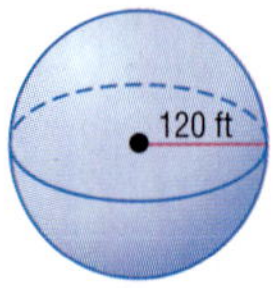

180,955.7 ft^2

2.
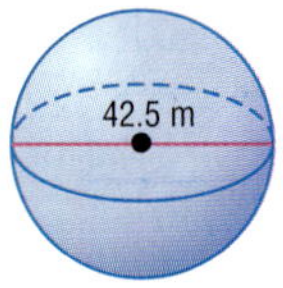

5674.5 m^2

3.

19,950,370.0 mi^2

4.
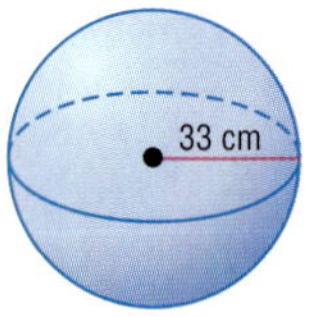

13,684.8 cm^2

5. a hemisphere with the circumference of a great circle 14.1 cm **47.5 cm^2**
6. a sphere with the circumference of a great circle 50.3 in. **805.4 in^2**
7. a sphere with the area of a great circle 98.5 m^2 **394 m^2**
8. a hemisphere with the circumference of a great circle 3.1 in. **2.3 in^2**
9. a hemisphere with the area of a great circle 31,415.9 ft^2 **94,247.7 ft^2**

Extra Practice

Lesson 13-1
(pages 688–694)

Find the volume of each prism or cylinder. Round to the nearest tenth if necessary.

1.
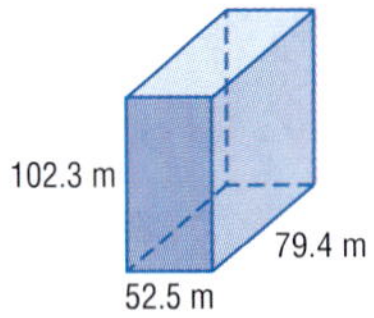

426,437.6 m^3

2.
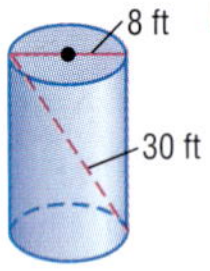

5102.4 ft^3

3.
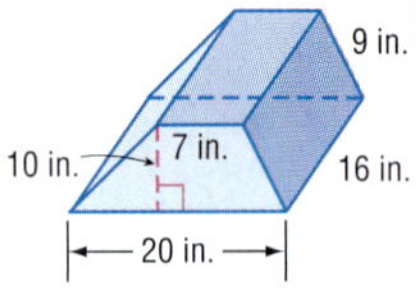

2160 in^3

Find the volume of each solid to the nearest tenth.

4.
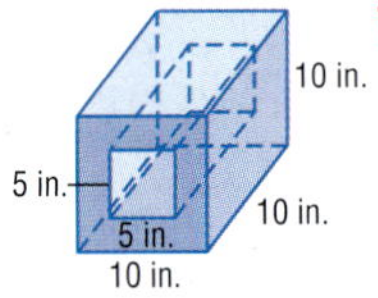

750 in^3

5.
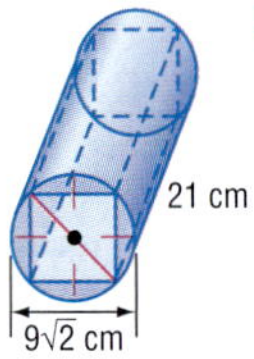

970.9 cm^3

6.
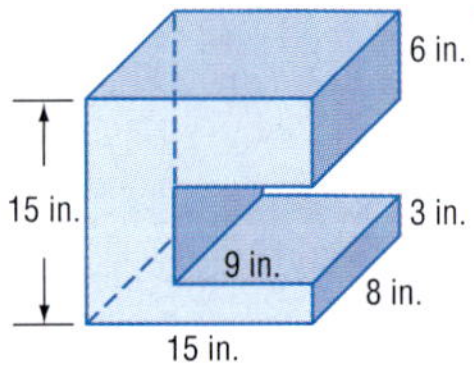

1368 in^3

Lesson 13-2
(pages 696–701)

Find the volume of each cone or pyramid. Round to the nearest tenth if necessary.

1.
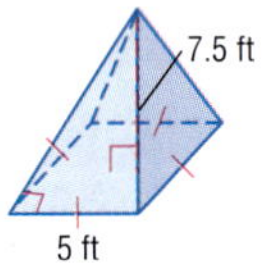

62.5 ft^3

2.
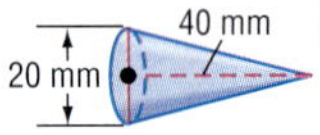

4188.8 mm^3

3.
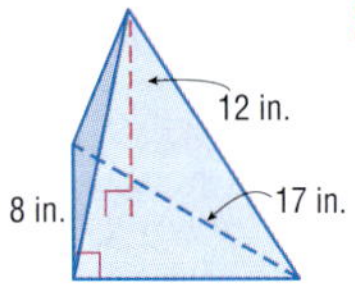

240 in^3

4.
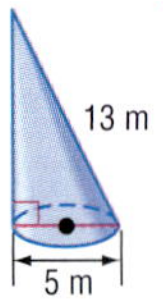

78.5 m^3

5.
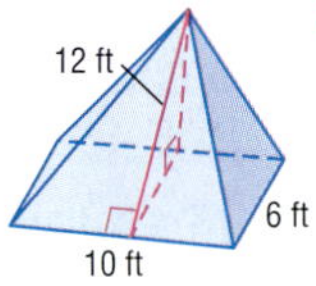

207.8 m^3

6.
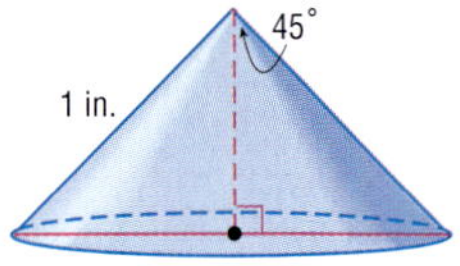

0.4 in^3

Extra Practice

780 Extra Practice

Lesson 13-3

(pages 702–706)

Find the volume of each sphere or hemisphere. Round to the nearest tenth.

1.
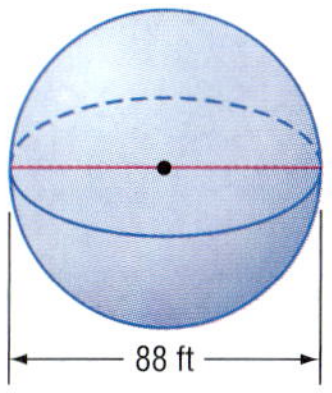

356,817.9 ft³

2. $C = 4$ m
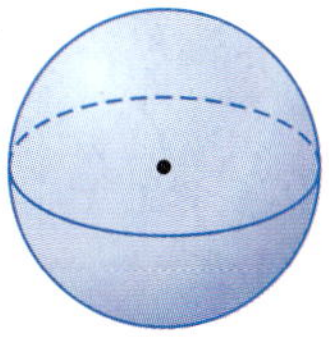

1.1 m³

3.
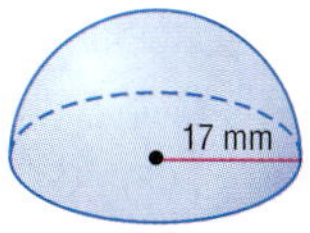

10,289.8 mm³

4. The diameter of the sphere is 3 cm. **14.1 cm³**
5. The radius of the hemisphere is $7\sqrt{2}$ m. **2031.9 m³**
6. The diameter of the hemisphere is 90 ft. **190,851.8 ft³**
7. The radius of the sphere is 0.5 in. **0.5 in³**

Lesson 13-4

(pages 707–713)

Determine whether each pair of solids are *similar, congruent,* or *neither.*

1.
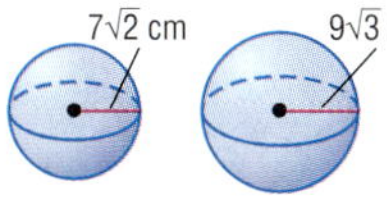

similar

2.
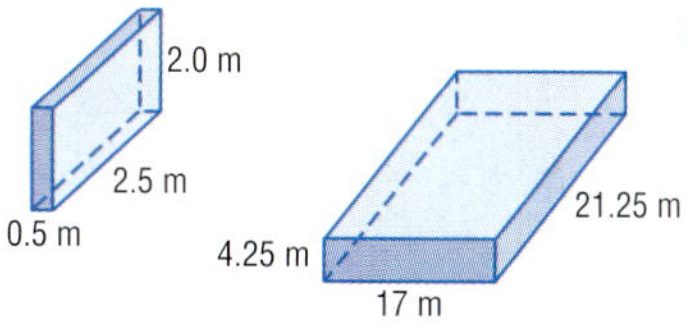

similar

3.
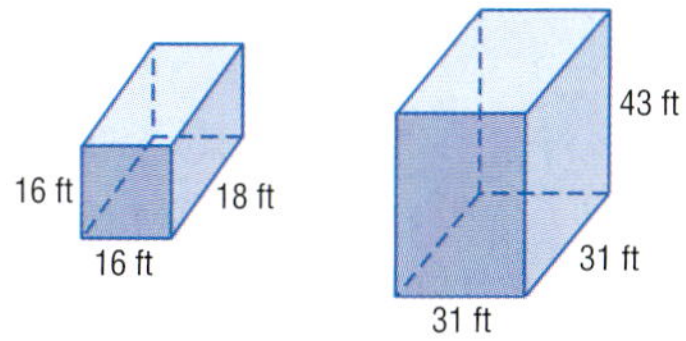

neither

4.
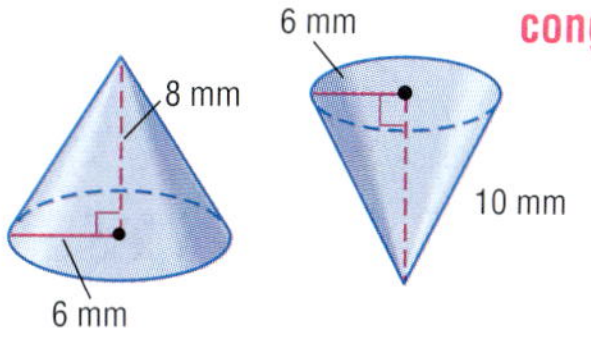

congruent

5.
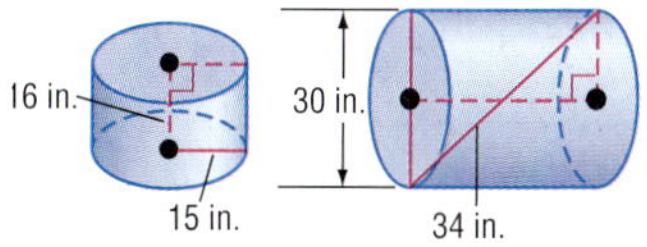

congruent

6.
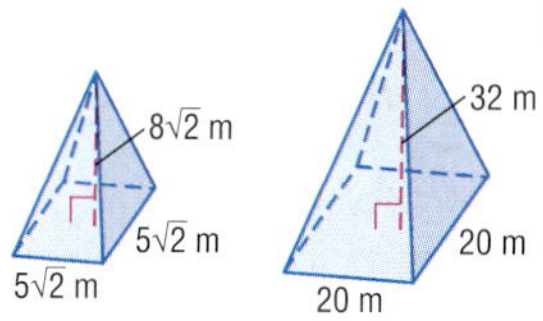

similar

Lesson 13-5

(pages 714–719)

Graph the rectangular solid that contains the given point and the origin. Label the coordinates of each vertex. **1–6. See margin.**

1. $A(3, -3, -3)$
2. $E(-1, 2, -3)$
3. $I(3, -1, 2)$
4. $Z(2, -1, 3)$
5. $Q(-4, -2, -4)$
6. $Y(-3, 1, -4)$

Determine the distance between each pair of points. Then determine the coordinates of the midpoint, *M*, of the segment joining the pair of points. **7–12. See margin.**

7. $A(-3, 3, 1)$ and $B(3, -3, -1)$
8. $O(2, -1, -3)$ and $P(-2, 4, -4)$
9. $D(0, -5, -3)$ and $E(0, 5, 3)$
10. $J(-1, 3, 5)$ and $K(3, -5, -3)$
11. $A(2, 1, 6)$ and $Z(-4, -5, -3)$
12. $S(-8, 3, -5)$ and $T(6, -1, 2)$

3.
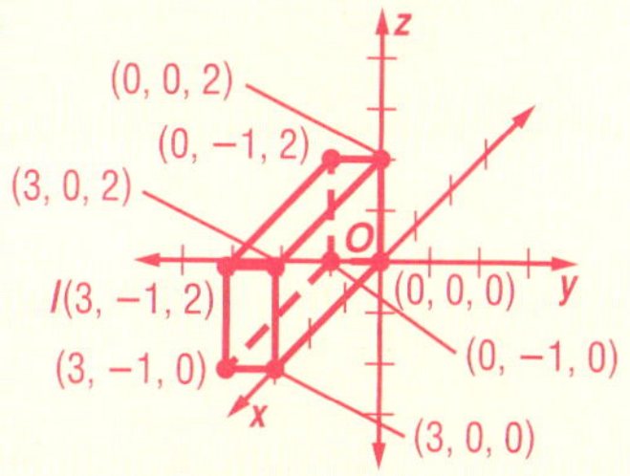

4.
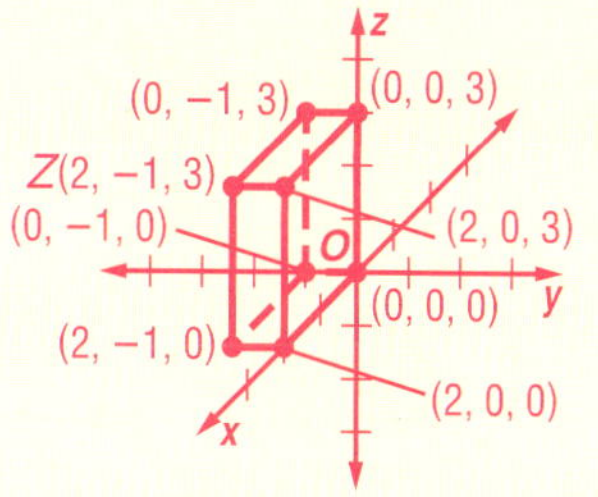

5.
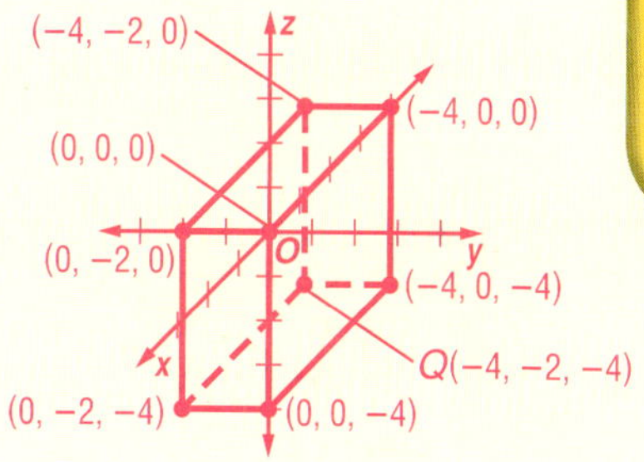

6.
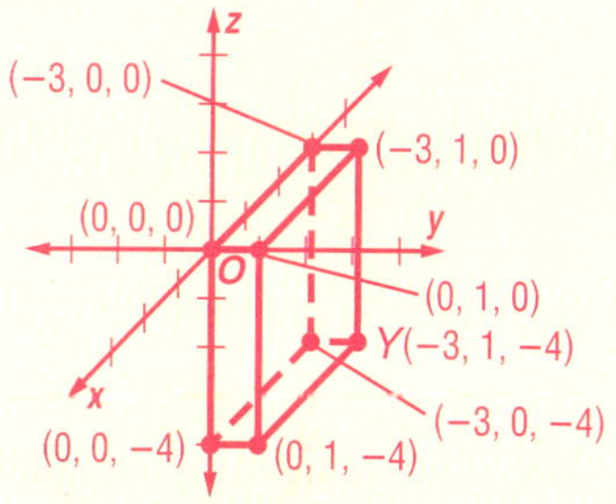

7. $AB = 2\sqrt{19}$; $(0, 0, 0)$
8. $OP = \sqrt{42}$; $(0, 1.5, -3.5)$
9. $DE = 2\sqrt{34}$; $(0, 0, 0)$
10. $JK = 12$; $(1, -1, 1)$
11. $AZ = 3\sqrt{17}$; $(-1, -2, 1.5)$
12. $ST = 3\sqrt{29}$; $(-1, 1, -1.5)$

Lesson 13-5

1.
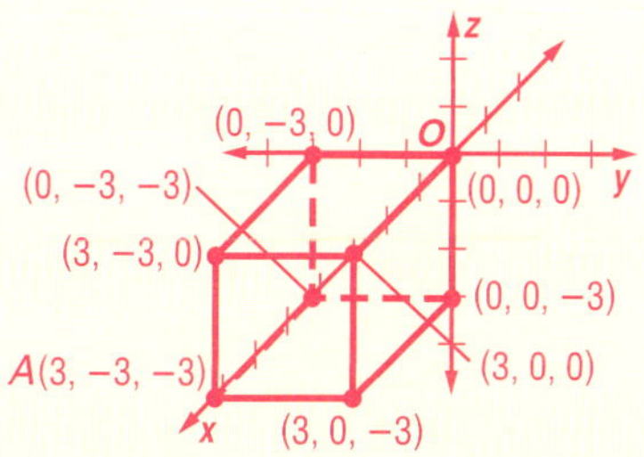

2.
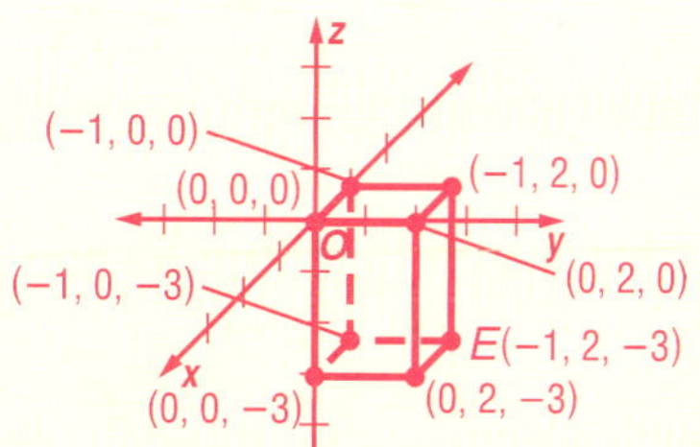

Extra Practice

Page 771, Lesson 9-2

4.

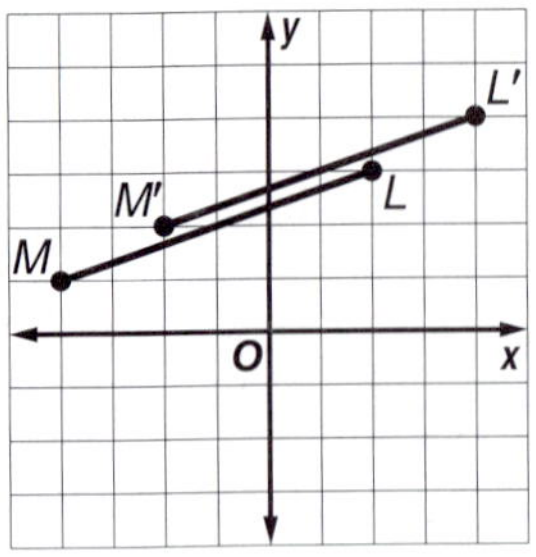

5.

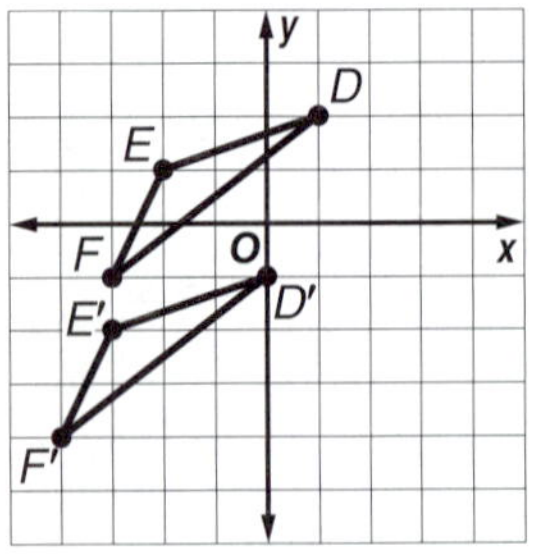

6.

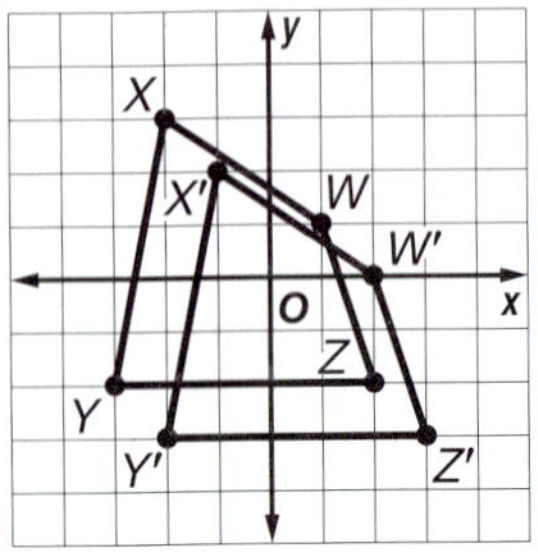

7.

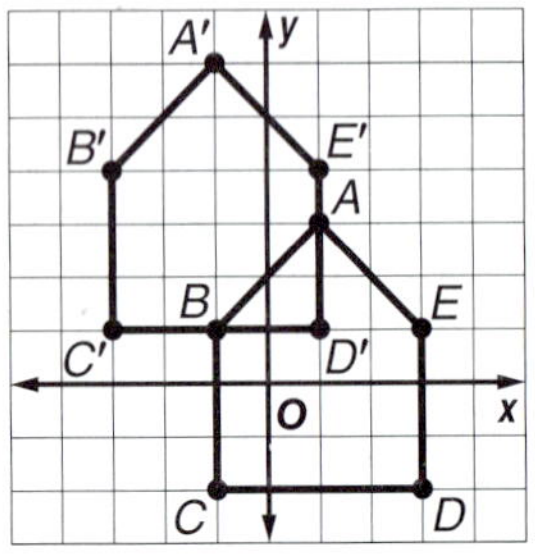

8.

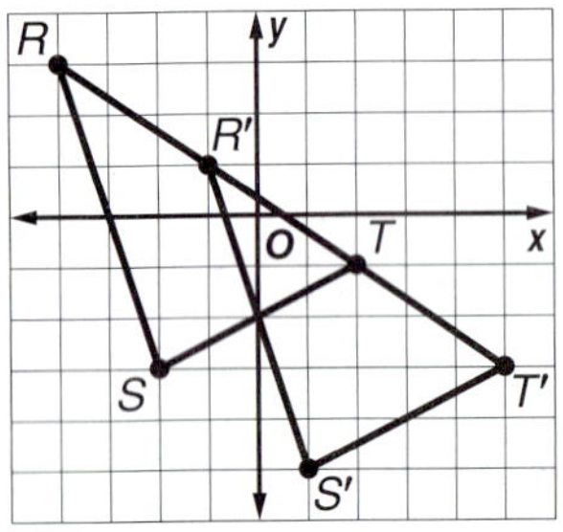

9.

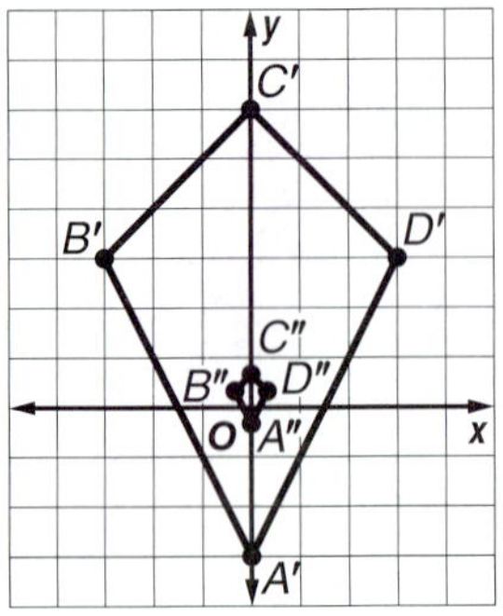

10.

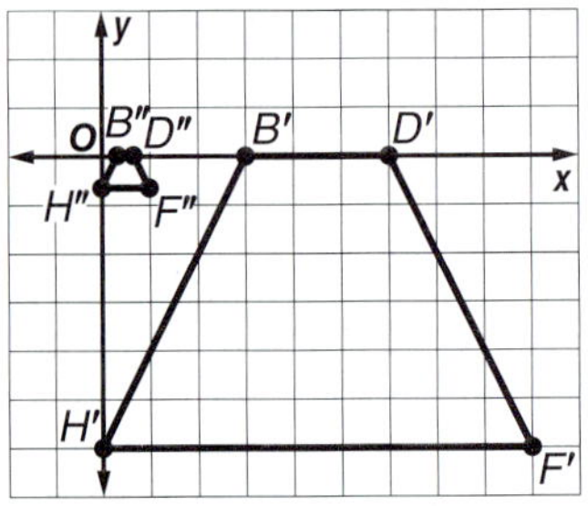

Page 772, Lesson 9-5

7.

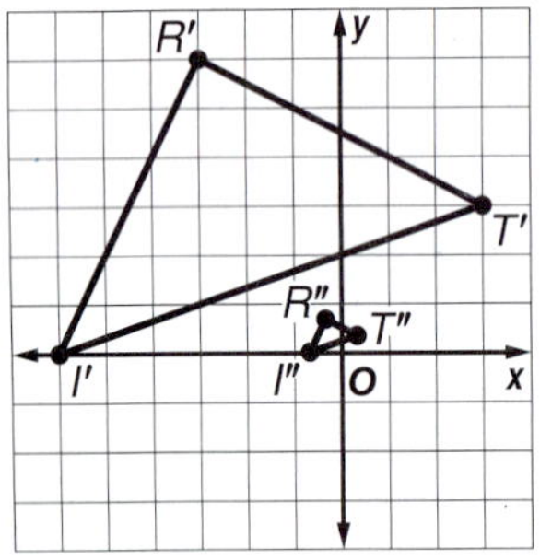

8.

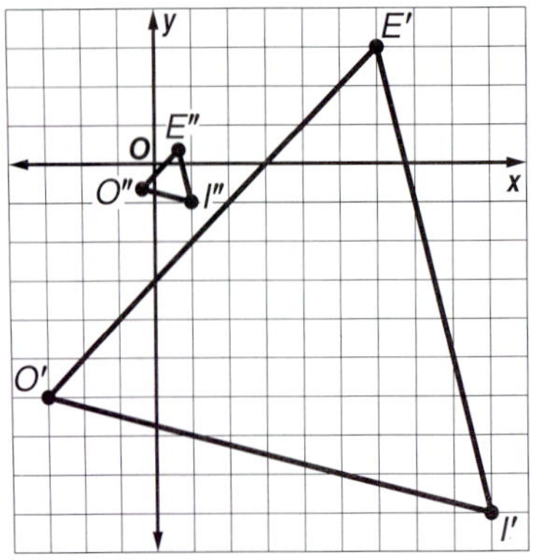

Notes

Mixed Problem Solving and Proof

Chapter 1 Points, Lines, Planes, and Angles (pages 4–59)

ARCHITECTURE **For Exercises 1–4, use the following information.**
The Burj Al Arab in Dubai, United Arab Emirates, is one of the world's tallest hotels. *(Lesson 1-1)*

1. Trace the outline of the building on your paper.
2. Label three different planes suggested by the outline.
3. Highlight three lines in your drawing that, when extended, do not intersect.
4. Label three points on your sketch. Determine if they are coplanar and collinear.

1–4. See margin.

SKYSCRAPERS **For Exercises 5–7, use the following information.** *(Lesson 1-2)*

Tallest Buildings in San Antonio, TX	
Name	**Height (ft)**
Tower of the Americas	622
Marriot Rivercenter	546
Weston Centre	444
Tower Life	404

Source: www.skyscrapers.com

5. What is the precision for the measures of the heights of the buildings? **0.5 ft**
6. What does the precision mean for the measure of the Tower of the Americas?
7. What is the difference in height between Weston Centre and Tower Life? **39–41 ft**

6. The height is between 621.5 and 622.5 ft.

PERIMETER **For Exercises 8–11, use the following information.** *(Lesson 1-3)* **10. 18.5 units**
The coordinates of the vertices of $\triangle ABC$ are $A(0, 6)$, $B(-6, -2)$, and $C(8, -4)$. Round to the nearest tenth.

8. Find the perimeter of $\triangle ABC$. **36.9 units**
9. Find the coordinates of the midpoints of each side of $\triangle ABC$. **$(-3, 2)$, $(1, -3)$, $(4, 1)$**
10. Suppose the midpoints are connected to form a triangle. Find the perimeter of this triangle.
11. Compare the perimeters of the two triangles. **See margin.**

12. **TRANSPORTATION** Mile markers are used to name the exits on Interstate 70 in Kansas. The exit for Hays is 3 miles farther than halfway between Exits 128 and 184. What is the exit number for the Hays exit? *(Lesson 1-3)* **159**

13. **ENTERTAINMENT** The Ferris wheel at the Navy Pier in Chicago has forty gondolas. What is the measure of an angle with a vertex that is the center of the wheel and with sides that are two consecutive spokes on the wheel? Assume that the gondolas are equally spaced. *(Lesson 1-4)* **9**

CONSTRUCTION **For Exercises 14–15, use the following information.**
A framer is installing a cathedral ceiling in a newly built home. A protractor and a plumb bob are used to check the angle at the joint between the ceiling and wall. The wall is vertical, so the angle between the vertical plumb line and the ceiling is the same as the angle between the wall and the ceiling. *(Lesson 1-5)*

14. How are $\angle ABC$ and $\angle CBD$ related?
15. If $m\angle ABC = 110$, what is $m\angle CBD$? **70**

14. They form a linear pair and are supplementary.

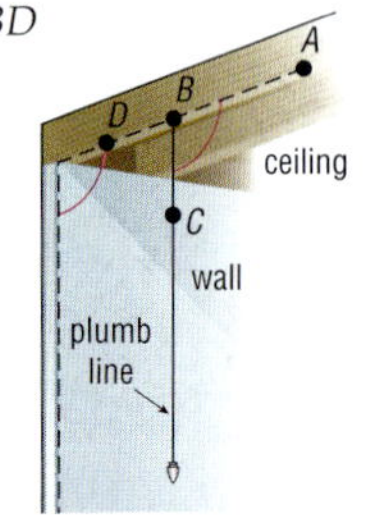

STRUCTURES **For Exercises 16–17, use the following information.** *(Lesson 1-6)*
The picture shows the Hongkong and Shanghai Bank located in Hong Kong, China.

16. Name five different polygons suggested by the picture.
17. Classify each polygon you identified as *convex* or *concave* and *regular* or *irregular*.

16–17. See margin.

Chapter 2

2. Sample answer: In 2010, California will have about 245 people per square mile. In 2010, Michigan will have about 185 people per square mile.
6. The Hatter is correct; Alice exchanged the hypothesis and conclusion.
8. then she should not accept it and should notify airline personnel immediately

Chapter 1

1–3. Sample answer:

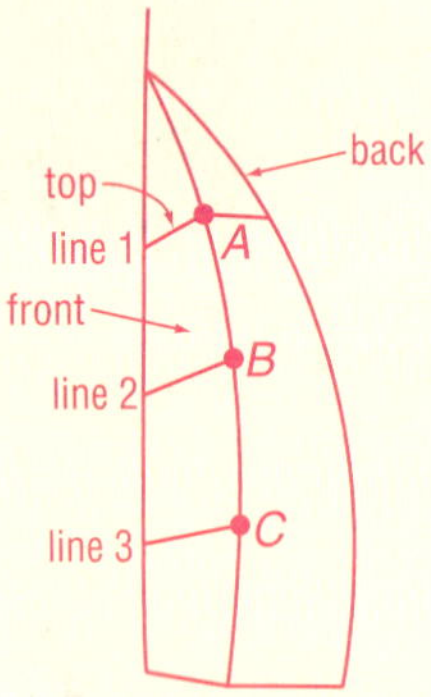

4. See figure for Exercises 1–3; points *A*, *B*, and *C* might be coplanar, but they are not collinear.
11. $\triangle ABC$ has a perimeter twice that of the smaller triangle.
16. Sample answer: isosceles triangle, rectangle, pentagon, hexagon, square
17. triangle: convex irregular; rectangle: convex irregular; pentagon: convex irregular; hexagon: concave irregular; square: convex regular

POPULATION For Exercises 1–2, use the table showing the population density for various states in 1960, 1980, and 2000. The figures represent the number of people per square mile. *(Lesson 2-1)*

State	1960	1980	2000
CA	100.4	151.4	217.2
CT	520.6	637.8	702.9
DE	225.2	307.6	401.0
HI	98.5	150.1	188.6
MI	137.7	162.6	175.0

Source: U.S. Census Bureau

1. Find a counterexample for the following statement. The population density for each state in the table increased by at least 30 during each 20-year period. **MI for both periods**
2. Write two conjectures for the year 2010. **See margin.**

STATES For Exercises 3–5, refer to the Venn diagram. *(Lesson 2-2)*

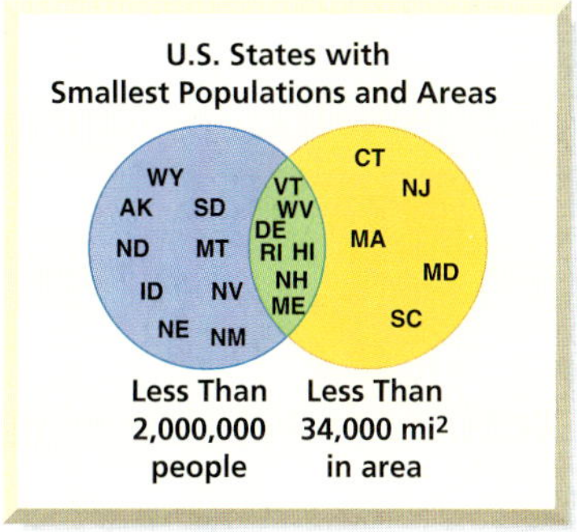

Source: *World Almanac*

3. How many states have less than 2,000,000 people? **16 states**
4. How many states have less than 34,000 square miles in area? **12 states**
5. How many states have less than 2,000,000 people and are less than 34,000 square miles in area? **7 states**

LITERATURE For Exercises 6–7, use the following quote from Lewis Carroll's *Alice's Adventures in Wonderland.* *(Lesson 2-3)*

"Then you should say what you mean," the March Hare went on.
"I do," Alice hastily replied; "at least—at least I mean what I say—that's the same thing, you know."
"Not the same thing a bit!" said the Hatter.

6. Who is correct? Explain. **See margin.**
7. How are the phrases *say what you mean* and *mean what you say* related? **They are converses of each other.**
8. **AIRLINE SAFETY** Airports in the United States post a sign stating *If any unknown person attempts to give you any items including luggage to transport on your flight, do not accept it and notify airline personnel immediately.* Write a valid conclusion to the hypothesis, *If a person Candace does not know attempts to give her an item to take on her flight, . . .* *(Lesson 2-4)* **See margin.**
9. **PROOF** Write a paragraph proof to show that $\overline{AB} \cong \overline{CD}$ if B is the midpoint of $\overline{AC}$ and C is the midpoint of $\overline{BD}$. *(Lesson 2-5)* **See margin.**

10. **CONSTRUCTION** Engineers consider the expansion and contraction of materials used in construction. The coefficient of linear expansion, k, is dependent on the change in length and the change in temperature and is found by the formula, $k = \frac{\Delta\ell}{\ell(T - t)}$. Solve this formula for T and justify each step. *(Lesson 2-6)* **See margin.**
11. **PROOF** Write a two-column proof. *(Lesson 2-7)*
 Given: $ABCD$ has 4 congruent sides.
 $DH = BF = AE$; $EH = FE$ **See margin.**
 Prove: $AB + BE + AE = AD + AH + DH$

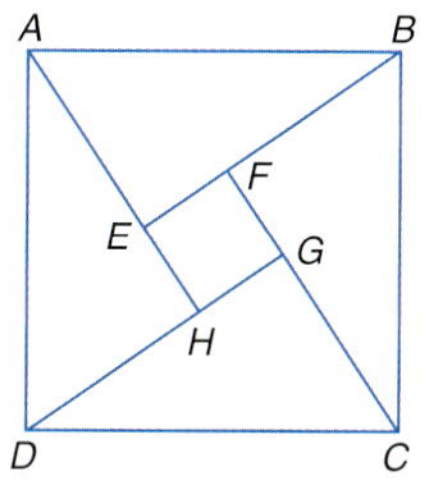

ILLUSIONS This drawing was created by German psychologist Wilhelm Wundt. *(Lesson 2-8)*

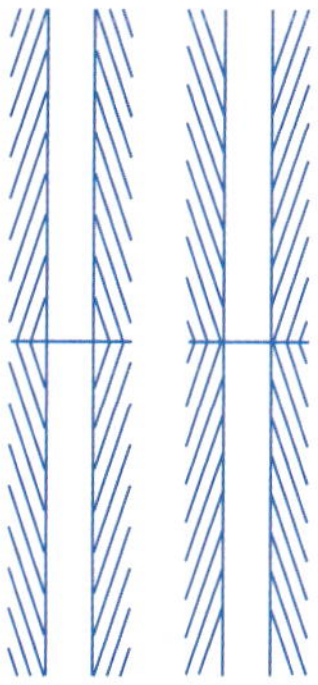

12. Describe the relationship between each pair of vertical lines. **12–13. See margin.**
13. A close-up of the angular lines is shown below. If $\angle 4 \cong \angle 2$, write a two-column proof to show that $\angle 3 \cong \angle 1$.

Mixed Problem Solving and Proof

10. Given: $k = \frac{\Delta\ell}{\ell(T - t)}$
Prove: $T = \frac{\Delta\ell}{k\ell} + t$
Proof:
Statements (Reasons)
1. $k = \frac{\Delta\ell}{\ell(T - t)}$ (Given)
2. $k(T - t) = \frac{\Delta\ell}{\ell}$ (Mult. Prop.)
3. $T - t = \frac{\Delta\ell}{k\ell}$ (Division Prop.)
4. $T = \frac{\Delta\ell}{k\ell} + t$ (Addition Prop.)

11. Given: $ABCD$ has 4 $\cong$ sides.
$DH = BF = AE$; $EH = FE$
Prove: $AB + BE + AE = AD + AH + DH$

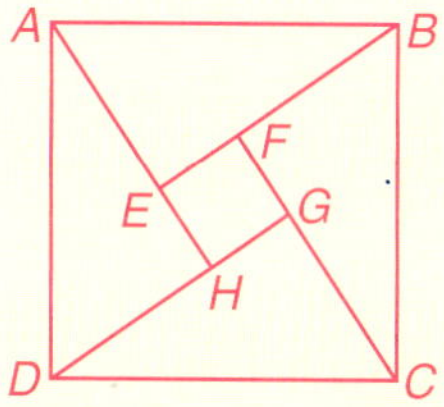

Proof:
Statements (Reasons)
1. $DH = BF = AE$; $EH = FE$ (Given)
2. $BE = BF + FE$; $AE + EH = AH$ (Segment Add. Prop.)
3. $BF + FE = AH$ (Substitution)
4. $BF + FE = AE + EH$ (Addition Prop.)
5. $BE = AH$ (Transitive Prop.)
6. $ABCD$ has 4 $\cong$ sides. (Given)
7. $AB = AD$ (Def. of $\cong$ segments)
8. $AB + BE = AD + AH$ (Addition Prop.)
9. $AB + BE + AE = AD + AH + DH$ (Addition Prop.)

12. The vertical lines are parallel. The first pair of vertical lines appear to curve inward, the second pair appear to curve outward.

13. Given: $\angle 4 \cong \angle 2$
Prove: $\angle 3 \cong \angle 1$

Proof:
Statements (Reasons)
1. $\angle 4 \cong \angle 2$ (Given)
2. $\angle 4$ and $\angle 3$ form a linear pair; $\angle 2$ and $\angle 1$ form a linear pair. (Def. of linear pair)
3. $\angle 4$ and $\angle 3$ are supplementary; $\angle 2$ and $\angle 1$ are supplementary. (Supplement Theorem)
4. $\angle 3 \cong \angle 1$ ($\angle$s suppl. to $\cong$ $\angle$s are $\cong$.)

9. Given: B is the midpoint of $\overline{AC}$ and C is the midpoint of $\overline{BD}$.
Prove: $\overline{AB} \cong \overline{CD}$

Proof: By the definition of midpoint, $AB = BC$ and $BC = CD$. By the Transitive Property, $AB = CD$. By definition of congruence, $\overline{AB} \cong \overline{CD}$.

Chapter 3

1. Alternate interior angles are congruent, so $\angle 1 \cong \angle 2$.

11. Given: $\overline{MQ} \parallel \overline{NP}$
$\angle 4 \cong \angle 3$
Prove: $\angle 1 \cong \angle 5$
Proof:
Statements (Reasons)
1. $\overline{MQ} \parallel \overline{NP}$; $\angle 4 \cong \angle 3$ (Given)
2. $\angle 3 \cong \angle 5$ (Alt. Int. $\angle$s Theorem)
3. $\angle 4 \cong \angle 5$ (Transitive Prop.)
4. $\angle 1 \cong \angle 4$ (Corres. $\angle$s Post.)
5. $\angle 1 \cong \angle 5$ (Transitive Prop.)

15. If two lines in a plane are cut by a transversal so that corresponding angles are congruent, then the lines are parallel.

16. Given: $\angle 1 \cong \angle 3$, $\overline{AB} \parallel \overline{DC}$
Prove: $\overline{BC} \parallel \overline{AD}$
Proof:
Statements (Reasons)
1. $\overline{AB} \parallel \overline{DC}$ (Given)
2. $\angle 1 \cong \angle 4$ (Alt. Int. $\angle$s Theorem)
3. $\angle 1 \cong \angle 3$ (Given)
4. $\angle 4 \cong \angle 3$ (Transitive Prop.)
5. $\overline{BC} \parallel \overline{AD}$ (If corr. $\angle$s are $\cong$, then lines are $\parallel$.)

17. The shortest distance is a perpendicular segment. You cannot walk this route because there are no streets that exactly follow this route and you cannot walk through or over buildings.

Chapter 4

1. The triangles appear to be scalene. One leg looks longer than the other leg.

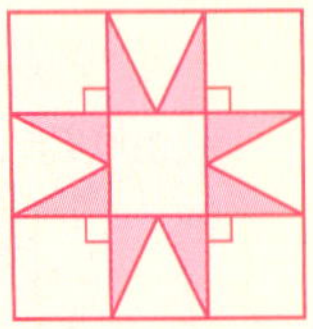

2. The triangles appear to be isosceles. Two of the sides appear to be the same length.

(Shading should be blue.)

Mixed Problem Solving and Proof

1. **OPTICAL ILLUSIONS** Lines ℓ and m are parallel, but appear to be bowed due to the transversals drawn through ℓ and m. Make a conjecture about the relationship between $\angle 1$ and $\angle 2$. *(Lesson 3-1)*

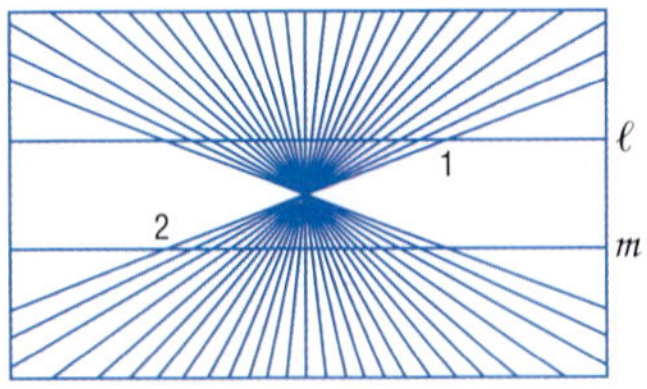

See margin.

ARCHITECTURE For Exercises 2–10, use the following information. The picture shows one of two towers of the Puerta de Europa in Madrid, Spain. Lines a, b, c, and d are parallel. The lines are cut by transversals e and f. If $m\angle 1 = m\angle 2 = 75$, find the measure of each angle. *(Lesson 3-2)*

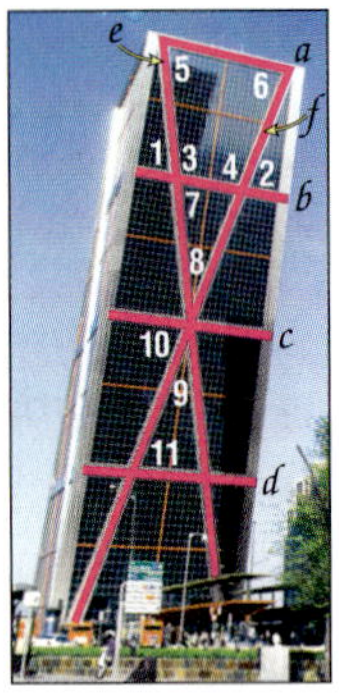

2. $\angle 3$ 105
3. $\angle 4$ 105
4. $\angle 5$ 75
5. $\angle 6$ 75
6. $\angle 7$ 75
7. $\angle 8$ 30
8. $\angle 9$ 30
9. $\angle 10$ 75
10. $\angle 11$ 75

11. **PROOF** Write a two-column proof. *(Lesson 3-2)*
Given: $\overline{MQ} \parallel \overline{NP}$ See margin.
$\angle 4 \cong \angle 3$
Prove: $\angle 1 \cong \angle 5$

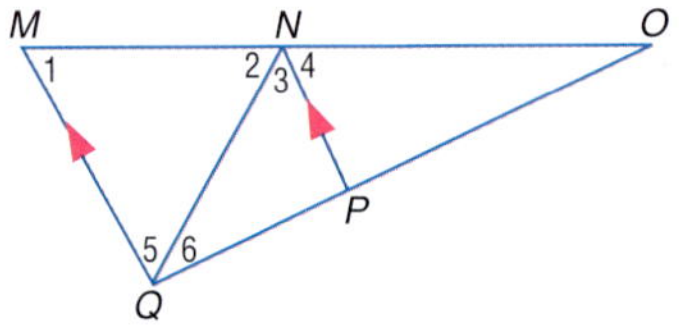

12. **EDUCATION** Between 1995 and 2000, the average cost for tuition and fees for American universities increased by an average rate of \$84.20 per year. In 2000, the average cost was \$2600. If costs increase at the same rate, what will the total average cost be in 2010? *(Lesson 3-3)* \$3442

RECREATION For Exercises 13 and 14, use the following information. *(Lesson 3-4)*
The Three Forks community swimming pool holds 74,800 gallons of water. At the end of the summer, the pool is drained and winterized.

13. If the pool drains at the rate of 1200 gallons per hour, write an equation to describe the number of gallons left after x hours. $y = 74{,}800 - 1200x$

14. How many hours will it take to drain the pool? $62\frac{1}{3}$ h

15. **CONSTRUCTION** An *engineer and carpenter square* is used to draw parallel line segments. Martin makes two cuts at an angle of 120° with the edge of the wood through points D and P. Explain why these cuts will be parallel. *(Lesson 3-5)* See margin.

16. **PROOF** Write a two-column proof. *(Lesson 3-5)*
Given: $\angle 1 \cong \angle 3$
$\overline{AB} \parallel \overline{DC}$
Prove: $\overline{BC} \parallel \overline{AD}$ See margin.

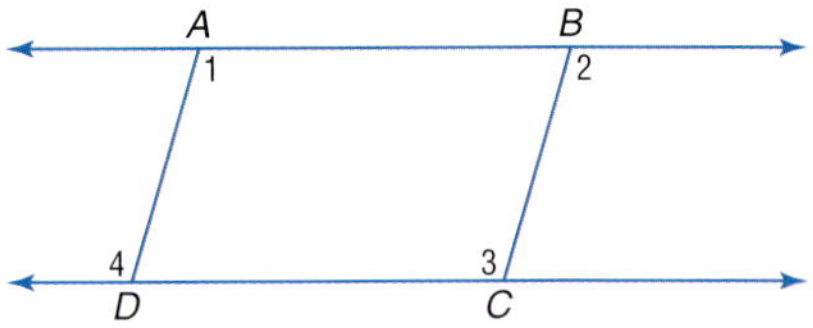

17. **CITIES** The map shows a portion of Seattle, Washington. Describe a segment that represents the shortest distance from the Bus Station to Denny Way. Can you walk the route indicated by your segment? Explain. *(Lesson 3-6)* See margin.

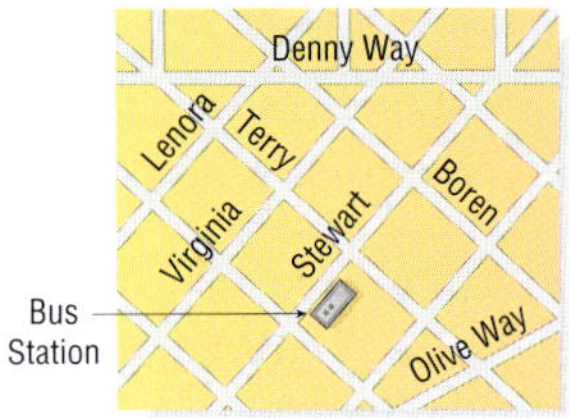

4. $\triangle BED \cong \triangle CFG$; $\triangle BJH \cong \triangle CKM$; $\triangle BPN \cong \triangle CQS$; $\triangle DIH \cong \triangle GLM$; $\triangle DON \cong \triangle GRS$

7. Given: $\overline{AC} \cong \overline{CI} \cong \overline{IG} \cong \overline{AG}$; $\overline{AI} \cong \overline{GC}$
Prove: $\triangle ACI \cong \triangle CAG$

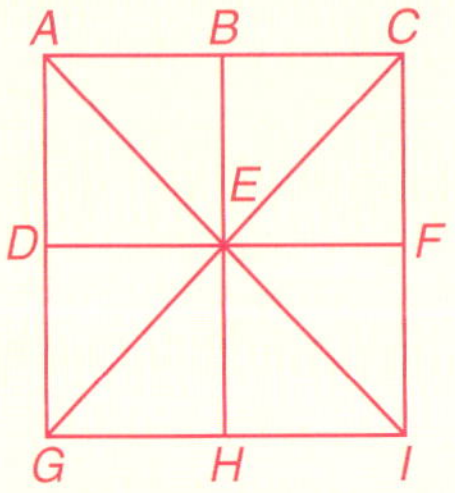

Proof:
$\overline{CI} \cong \overline{AG}$ (Given)
$\overline{AI} \cong \overline{GC}$ (Given)
$\overline{CA} \cong \overline{CA}$ (Reflexive Prop.)
→ $\triangle ACI \cong \triangle CAG$ (SSS)

QUILTING **For Exercises 1 and 2, trace the quilt pattern square below.** *(Lesson 4-1)*

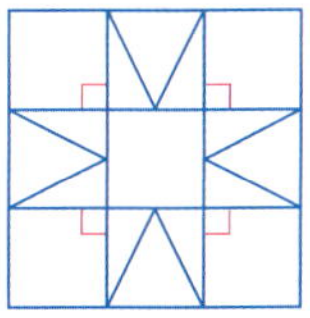

1. Shade all right triangles red. Do these triangles appear to be scalene or isosceles? Explain.
2. Shade all acute triangles blue. Do these triangles appear to be scalene, isoscles, or equilateral? Explain. **1–2. See margin.**

3. **ASTRONOMY** Leo is a constellation that represents a lion. Three of the brighter stars in the constellation form $\triangle LEO$. If the angles have measures as shown in the figure, find $m\angle OLE$. *(Lesson 4-2)* **66**

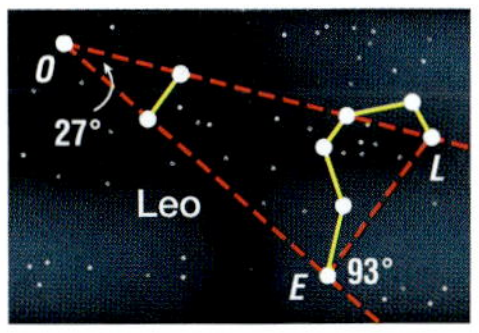

4. **ARCHITECTURE** The diagram shows an A-frame house with various points labeled. Assume that segments and angles that appear to be congruent in the diagram are congruent. Indicate which triangles are congruent. *(Lesson 4-3)* **See margin.**

RECREATION **For Exercises 5–7, use the following information.**
Tapatan is a game played in the Philippines on a square board, like the one shown at the top right. Players take turns placing each of their three pieces on a different point of intersection. After all the pieces have been played, the players take turns moving a piece along a line to another intersection. A piece cannot jump over another piece. A player who gets all their pieces in a straight line wins. Point E bisects all four line segments that pass through it. All sides are congruent, and the diagonals are congruent. Suppose a letter is assigned to each intersection. *(Lesson 4-4)*

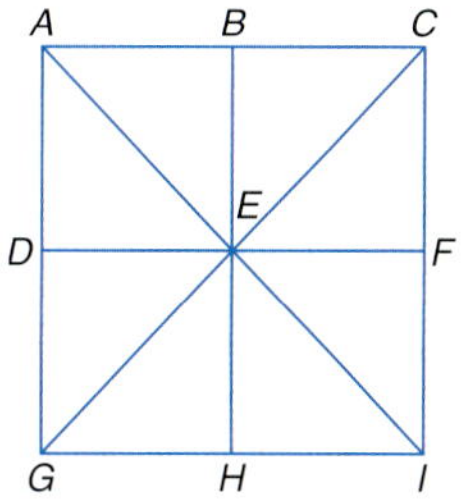

5. Is $\triangle GHE \cong \triangle CBE$? Explain. **yes; SAS**
6. Is $\triangle AEG \cong \triangle IEG$? Explain. **yes; SSS or SAS**
7. Write a flow proof to show that $\triangle ACI \cong \triangle CAG$. **See margin.**

8. **HISTORY** It is said that Thales determined the distance from the shore to the Greek ships by sighting the angle to the ship from a point P on the shore, walking to point Q, and then sighting the angle to the ship from Q. He then reproduced the angles on the other side of $\overline{PQ}$ and continued these lines until they intersected. Is this method valid? Explain. *(Lesson 4-5)* **See margin.**

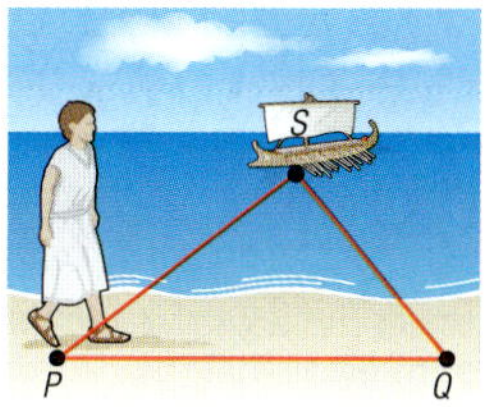

9. **PROOF** Write a two-column proof. *(Lesson 4-6)* **See margin.**

Given: $\overline{PH}$ bisects $\angle YHX$.
$\overline{PH} \perp \overline{YX}$

Prove: $\triangle YHX$ is an isosceles triangle.

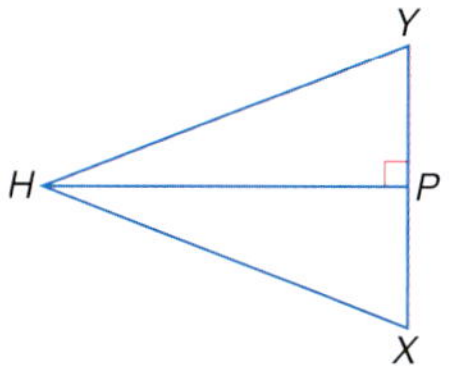

10. **PROOF** $\triangle ABC$ is a right isosceles triangle with hypotenuse $\overline{AB}$. M is the midpoint of $\overline{AB}$. Write a coordinate proof to show that $\overline{CM}$ is perpendicular to $\overline{AB}$. *(Lesson 4-7)* **See margin.**

8. Yes, the method is valid. Thales sighted $\angle SPQ$ and $\angle SQP$. He then constructed $\angle QPA$ congruent to $\angle SPQ$ and $\angle PQA$ congruent to $\angle SQP$. $\triangle SPQ$ and $\triangle APQ$ share the side $\overline{PQ}$. Since $\angle QPA \cong \angle SPQ$, $\angle PQA \cong \angle SQP$, and $\overline{PQ} \cong \overline{PQ}$, $\triangle SPQ \cong \triangle APQ$ by the ASA Postulate.

9. Given: $\overline{PH}$ bisects $\angle YHX$, $\overline{PH} \perp \overline{YX}$
Prove: $\triangle YHX$ is an isosceles triangle.

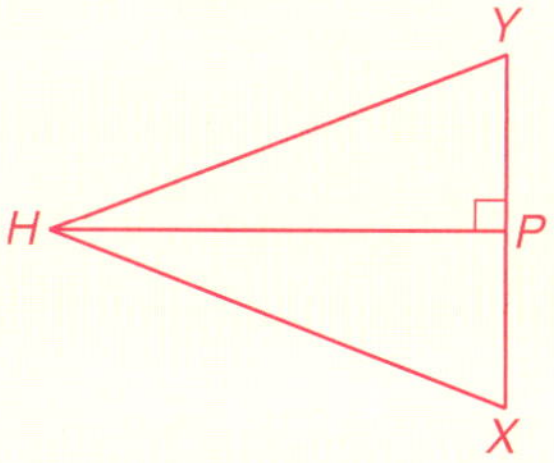

Proof:
Statements (Reasons)
1. **$\overline{PH}$ bisects $\angle YHX$. (Given)**
2. **$\angle YHP \cong \angle XHP$ (Def. of $\angle$ bisector)**
3. **$\overline{PH} \perp \overline{YX}$ (Given)**
4. **$\angle YPH$ and $\angle XPH$ are rt. $\angle$s (Def. of $\perp$ lines)**
5. **$\angle YPH \cong \angle XPH$ (All rt. $\angle$s are $\cong$.)**
6. **$\angle Y \cong \angle X$ (Third $\angle$ Th.)**
7. **$\overline{HX} \cong \overline{HY}$ (Conv. of Isos. $\triangle$ Th.)**
8. **$\triangle YHX$ is an isosceles triangle. (Def. of isos. $\triangle$)**

10. Given: $\triangle ABC$ is a right isosceles triangle. M is the midpoint of $\overline{AB}$.
Prove: $\overline{CM} \perp \overline{AB}$

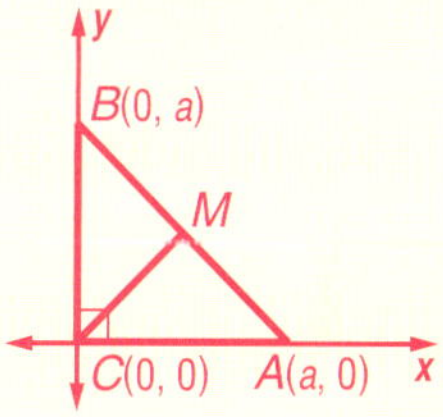

Proof: Place the triangle so that the vertices are $A(a, 0)$, $B(0, a)$, and $C(0, 0)$.
By the Midpoint Formula, the coordinates of M are $\left(\frac{0+a}{2}, \frac{a+0}{2}\right)$ or $\left(\frac{a}{2}, \frac{a}{2}\right)$.
Find the slopes of $\overline{AB}$ and $\overline{CM}$.

$$\text{Slope of } \overline{AB} = \frac{0-a}{a-0} = \frac{-a}{a} = -1$$

$$\text{Slope of } \overline{CM} = \frac{\frac{a}{2}-0}{\frac{a}{2}-0} = \frac{\frac{a}{2}}{\frac{a}{2}} = 1$$

The product of the slopes is -1, so $\overline{CM} \perp \overline{AB}$.

Mixed Problem Solving and Proof

Chapter 5

1.

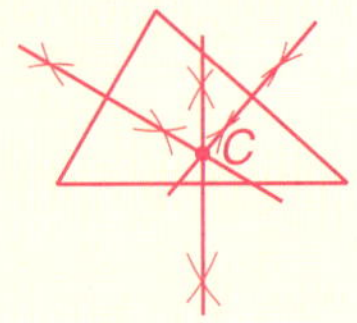

2.

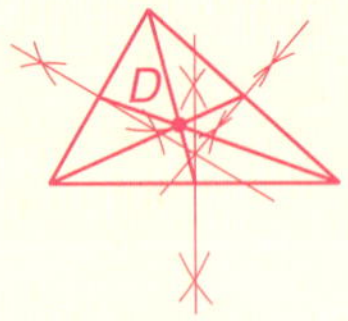

3.

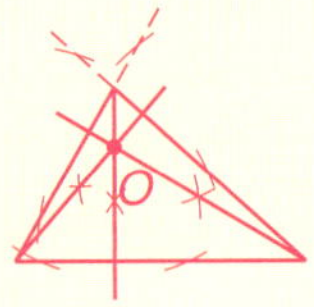

4.

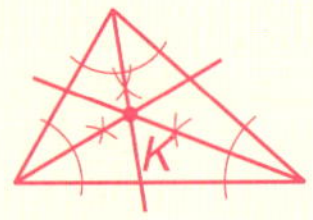

9. Given: $x + y > 634$
Prove: $x > 317$ or $y > 317$
Proof:
Step 1: Assume $x < 317$ and $y < 317$.
Step 2: $x + y < 634$
Step 3: This contradicts the fact that $x + y > 634$. Therefore, at least one of the legs was longer than 317 miles.

11. Given: $\angle ZST \cong \angle ZTS$
$\angle XRA \cong \angle XAR$
$TA = 2AX$
Prove: $2XR + AZ > SZ$

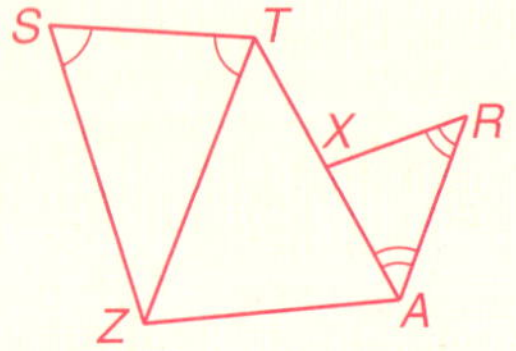

Proof:
Statements (Reasons)

1. $\angle ZST \cong \angle ZTS$ (Given)
2. $\overline{SZ} \cong \overline{TZ}$ (Isos. △ Th.)
3. $SZ = TZ$ (Def. of ≅)
4. $TA + AZ > TZ$ (△ Inequal. Th.)
5. $TA = 2AX$ (Given)
6. $2AX + AZ > TZ$ (Substitution)
7. $\angle XRA \cong \angle XAR$ (Given)
8. $\overline{XR} \cong \overline{XA}$ (Isos. △ Th.)
9. $XR = XA$ (Def. of ≅)
10. $2XR + AZ > TZ$ (Substitution)
11. $2XR + AZ > SZ$ (Substitution)

CONSTRUCTION **For Exercises 1–4, draw a large, acute scalene triangle. Use a compass and straightedge to make the required constructions.** *(Lesson 5-1)*

1. Find the circumcenter. Label it *C*.
2. Find the centroid of the triangle. Label it *D*.
3. Find the orthocenter. Label it *O*.
4. Find the incenter of the triangle. Label it *I*.

1–4. See margin.

RECREATION **For Exercises 5–7, use the following information.** *(Lesson 5-2)*
Kailey plans to fly over the route marked on the map of Oahu in Hawaii.

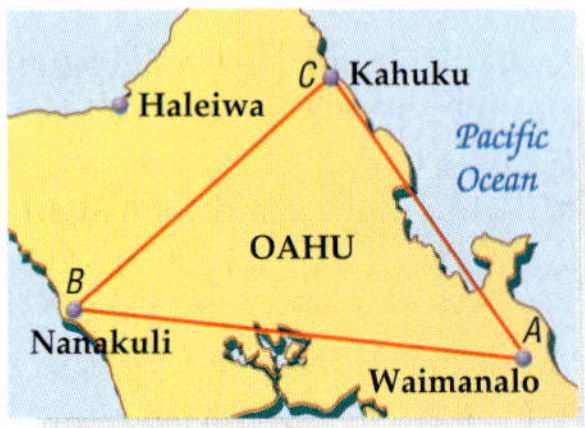

5. The measure of angle *A* is two degrees more than the measure of angle *B*. The measure of angle *C* is fourteen degrees less than twice the measure of angle *B*. What are the measures of the three angles? $m\angle A = 50$, $m\angle B = 48$, $m\angle C = 82$
6. Write the lengths of the legs of Kailey's trip in order from least to greatest. *AC, BC, BA*
7. The length of the entire trip is about 68 miles. The middle leg is 11 miles greater than one-half the length of the shortest leg. The longest leg is 12 miles greater than three-fourths of the shortest leg. What are the lengths of the legs of the trip? 20 mi, 21 mi, 27 mi
8. **LAW** A man is accused of comitting a crime. If the man is telling the truth when he says, "I work every Tuesday from 3:00 P.M. to 11:00 P.M.," what fact about the crime could be used to prove by indirect reasoning that the man was innocent? *(Lesson 5-3)* that the crime was committed on Tuesday between 3:00 P.M. and 11:00 P.M.

TRAVEL **For Exercises 9 and 10, use the following information.**
The total air distance to fly from Bozeman, Montana, to Salt Lake City, Utah, to Boise, Idaho is just over 634 miles.

9. Write an indirect proof to show that at least one of the legs of the trip is longer than 317 miles. *(Lesson 5-3)* See margin.
10. The air distance from Bozeman to Salt Lake City is 341 miles and the distance from Salt Lake to Boise is 294 miles. Find the range for the distance from Bozeman to Boise. *(Lesson 5-4)* $47 < n < 635$
11. **PROOF** Write a two-column proof.
Given: $\angle ZST \cong \angle ZTS$
$\angle XRA \cong \angle XAR$
$TA = 2AX$
Prove: $2XR + AZ > SZ$
(Lesson 5-4) See margin.

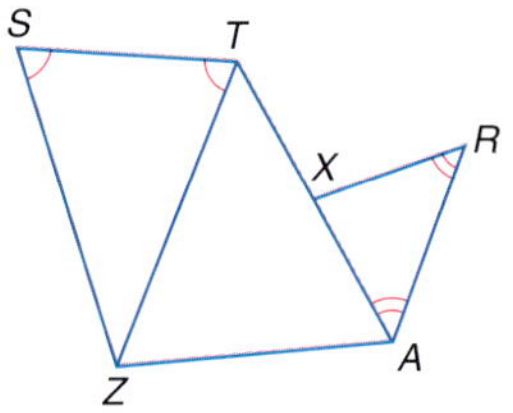

12. **GEOGRAPHY** The map shows a portion of Nevada. The distance from Tonopah to Round Mountain is the same as the distance from Tonopah to Warm Springs. The distance from Tonopah to Hawthorne is the same as the distance from Tonopah to Beatty. Use the angle measures to determine which distance is greater, Round Mountain to Hawthorne or Warm Springs to Beatty. *(Lesson 5-5)* Warm Springs to Beatty

13. **PROOF** Write a two-column proof. *(Lesson 5-5)*
Given: $\overline{DB}$ is a median of $\triangle ABC$.
$m\angle 1 > m\angle 2$
Prove: $m\angle C > m\angle A$
See margin.

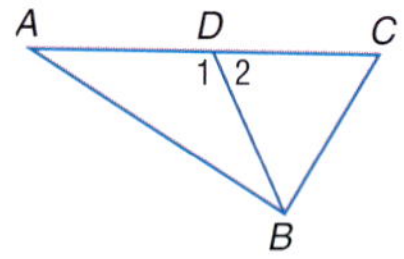

13. Given: $\overline{DB}$ is a median of $\triangle ABC$.
$m\angle 1 > m\angle 2$
Prove: $m\angle C > m\angle A$

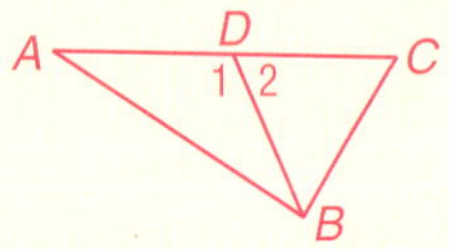

Proof:
Statements (Reasons)

1. $\overline{DB}$ is a median of $\triangle ABC$; $m\angle 1 > m\angle 2$ (Given)
2. *D* is the midpoint of $\overline{AC}$. (Def. of median)
3. $\overline{AD} \cong \overline{DC}$ (Midpoint Theorem)
4. $\overline{DB} \cong \overline{DB}$ (Reflexive Property)
5. $AB > BC$ (SAS Inequality)
6. $m\angle C > m\angle A$ (If one side of a △ is longer than another, the ∠ opp. the longer side > the ∠ opp. the shorter side.)

1. **TOYS** In 2000, \$34,554,900,000 was spent on toys in the U.S. The U.S. population in 2000 was 281,421,906, with 21.4% of the population 14 years and under. If all of the toys purchased in 2000 were for children 14 years and under, what was the average amount spent per child? *(Lesson 6-1)* about \$573.77

QUILTING For Exercises 2–4, use the following information. *(Lesson 6-2)*
Felicia found a pattern for a quilt square. The pattern measures three-quarters of an inch on a side. Felicia wants to make a quilt that is 77 inches by 110 inches when finished.

2. If Felicia wants to use only whole quilt squares, what is the greatest side length she can use for each square? 11 in.
3. How many quilt squares will she need for the quilt? 70 squares
4. By what scale factor will she need to increase the pattern for the quilt square? $\frac{44}{3}$

PROOF For Exercises 5 and 6, write a paragraph proof. *(Lesson 6-3)* 5–6. See margin.

5. Given: $\triangle WYX \sim \triangle QYR$, $\triangle ZYX \sim \triangle SYR$
   Prove: $\triangle WYZ \sim \triangle QYS$

6. Given: $\overline{WX} \parallel \overline{QR}$, $\overline{ZX} \parallel \overline{SR}$
   Prove: $\overline{WZ} \parallel \overline{QS}$

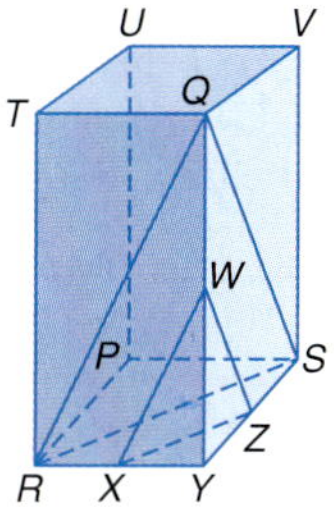

HISTORY For Exercises 7 and 8, use the following information. *(Lesson 6-4)*
In the fifteenth century, mathematicians and artists tried to construct the perfect letter. Damiano da Moile used a square as a frame to design the letter "A" as shown in the diagram. The thickness of the major stroke of the letter was to be $\frac{1}{12}$ of the height of the letter.

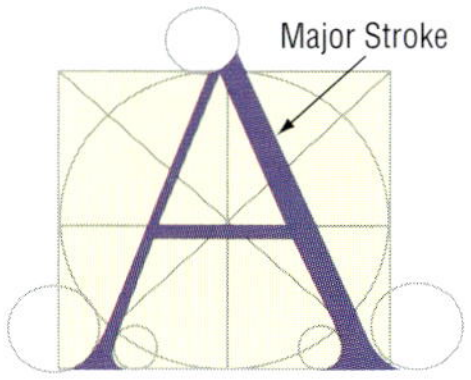

7. Explain why the bar through the middle of the A is half the length between the outside bottom corners of the sides of the letter. See margin.
8. If the letter were 3 centimeters tall, how wide would the major stroke of the A be? 0.25 cm

9. **PROOF** Write a two-column proof. *(Lesson 6-5)*
   Given: $\overline{WS}$ bisects $\angle RWT$. $\angle 1 \cong \angle 2$ See margin.
   Prove: $\frac{VW}{WT} = \frac{RS}{ST}$

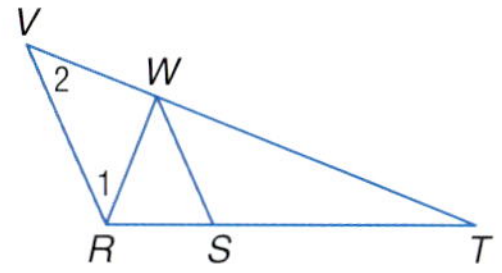

ART For Exercises 10 and 11, use the diagram of a square mosaic tile. $AB = BC = CD = \frac{1}{3}AD$ and $DE = EF = FG = \frac{1}{3}DG$. *(Lesson 6-5)*

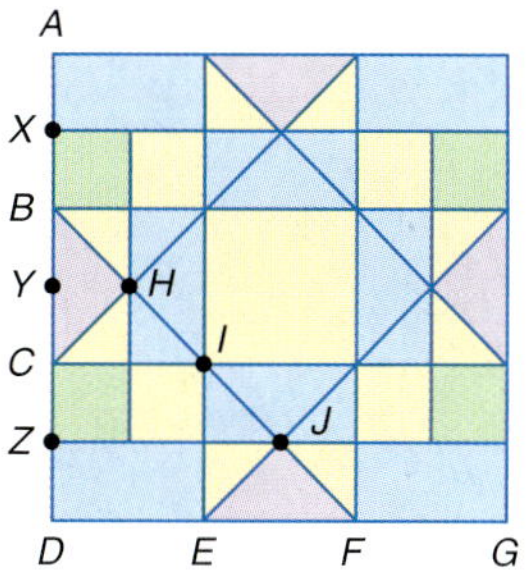

10. What is the ratio of the perimeter of $\triangle BDF$ to the perimeter of $\triangle BCI$? Explain.
11. Find two triangles such that the ratio of their perimeters is 2:3. Explain. 10–11. See margin.

12. **TRACK** A triangular track is laid out as shown. $\triangle RST \sim \triangle WVU$. If $UV = 500$ feet, $VW = 400$ feet, $UW = 300$ feet, and $ST = 1000$ feet, find the perimeter of $\triangle RST$. *(Lesson 6-5)* 2400 ft

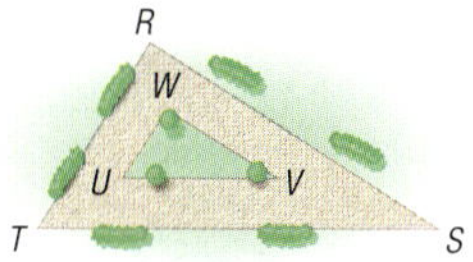

13. **BANKING** Ashante has \$5000 in a savings account with a yearly interest rate of 2.5%. The interest is compounded twice per year. What will be the amount in the savings account after 5 years? *(Lesson 6-6)* \$5661.35

Mixed Problem Solving and Proof

6. Given: $\overline{WX} \parallel \overline{QR}$, $\overline{ZX} \parallel \overline{SR}$
   Prove: $\overline{WZ} \parallel \overline{QS}$
   Proof: We are given that $\overline{WX} \parallel \overline{QR}$, $\overline{ZX} \parallel \overline{SR}$. By the Corresponding Angles Postulate, $\angle XWY \cong \angle RQY$ and $\angle YXZ \cong \angle YRS$. By the Reflexive Property, $\angle QYS \cong \angle QYS$, $\angle QYR \cong \angle QYR$ and $\angle RYS \cong \angle RYS$. $\triangle QYR \sim \triangle WYX$ and $\triangle YRS \sim \triangle YXZ$ by AA Similarity. By the definition of similar triangles, $\frac{WY}{QY} = \frac{YX}{YR}$ and $\frac{YX}{YR} = \frac{ZY}{SY}$. $\frac{WY}{QY} = \frac{ZY}{SY}$ by the Transitive Property. $\triangle WYZ \sim \triangle QYS$ by SAS Similarity. By the definition of similar triangles $\angle YWZ \cong \angle YQS$. $\overline{WZ} \parallel \overline{QS}$ by the Corresponding Angles Postulate.

7. The bar connects the midpoints of each leg of the letter and is parallel to the base. Therefore, the length of the bar is one-half the length of the base because a midsegment of a triangle is parallel to one side of the triangle, and its length is one-half the length of that side.

9. Given: $\overline{WS}$ bisects $\angle RWT$, $\angle 1 \cong \angle 2$
   Prove: $\frac{VW}{WT} = \frac{RS}{ST}$

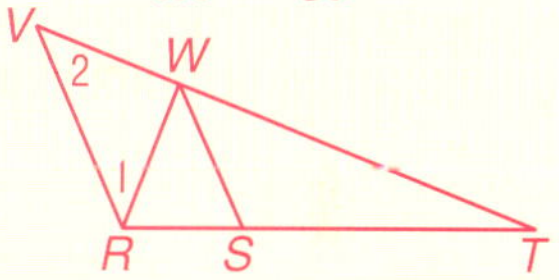

Proof:
Statements (Reasons)
1. $\overline{WS}$ bisects $\angle RWT$ (Given)
2. $\frac{RW}{WT} = \frac{RS}{ST}$ ($\angle$ Bisector Th.)
3. $\angle 1 \cong \angle 2$ (Given)
4. $\overline{RW} \cong \overline{VW}$ (Conv. of Isos. $\triangle$ Th.)
5. $RW = VW$ (Def. of $\cong$)
6. $\frac{VW}{WT} = \frac{RS}{ST}$ (Substitution)

10. Since $\triangle BDF \sim \triangle BCI$ and the ratio of side lengths is 2:1, the ratio of perimeters will be 2:1 by the Proportional Perimeters Theorem.

11. Sample answer: $\triangle BCI \sim \triangle BZJ$ and both are isosceles right triangles with a ratio of side length of 2:3. By the Proportional Perimeters Theorem, the ratio of their perimeters will be 2:3.

Chapter 6

5. Given: $\triangle WYX \sim \triangle QYR$, $\triangle ZYX \sim \triangle SYR$
   Prove: $\triangle WYZ \sim \triangle QYS$

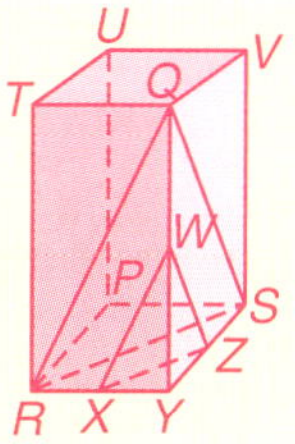

Proof: It is given that $\triangle WYX \sim \triangle QYR$ and $\triangle ZYX \sim \triangle SYR$. By definition of similar polygons we know that $\frac{WY}{QY} = \frac{YX}{YR}$ and $\frac{YX}{YR} = \frac{ZY}{SY}$. Then $\frac{WY}{QY} = \frac{ZY}{SY}$ by the Transitive Property. $\angle WYZ \cong \angle QYS$ because congruence of angles is reflexive. Therefore, $\triangle WYZ \sim \triangle QYS$ by SAS Similarity.

Chapter 7

1. Given: D is the midpoint of $\overline{BE}$, $\overline{BD}$ is an altitude of right triangle ABC

Prove: $\frac{AD}{DE} = \frac{DE}{DC}$

Proof:

Statements (Reasons)

1. $\overline{BD}$ is an altitude of right triangle ABC. (Given)
2. $\frac{AD}{DB} = \frac{DB}{DC}$ (The measure of an altitude drawn from the vertex of the right angle of a right triangle to its hypotenuse is the geometric mean between the measures of the two segments of the hypotenuse.)
3. D is the midpoint of $\overline{BE}$. (Given)
4. $DB = DE$ (Def. of midpoint)
5. $\frac{AD}{DE} = \frac{DE}{DC}$ (Substitution)

3. No; the measures do not satisfy the Pythagorean Theorem since $(2.7)^2 + (3.0)^2 \neq (5.3)^2$.

8. $AE \approx 339.4$ ft, $EB = 300$ ft, $CF \approx 134.2$ ft, $DF \approx 84.9$ ft

Chapter 8

3. Sample answer: Make sure that opposite sides are congruent or make sure that opposite angles are congruent.

4. Given: $\square ABCD$, $\overline{AE} \cong \overline{CF}$

Prove: Quadrilateral $EBFD$ is a $\square$.

Proof:

Statements (Reasons)

1. $\square ABCD$, $\overline{AE} \cong \overline{CF}$ (Given)
2. $\overline{AB} \cong \overline{DC}$ (Opp. sides of a $\square$ are $\cong$.)
3. $\angle A \cong \angle C$ (Opp. $\angle$s of a $\square$ are $\cong$.)
4. $\triangle BAE \cong \triangle DCF$ (SAS)
5. $\overline{EB} \cong \overline{DF}$, $\angle BEA \cong \angle DFC$ (CPCTC)
6. $\overline{BC} \parallel \overline{AD}$ (Def. of $\square$)
7. $\angle DFC \cong \angle FDE$ (Alt. Int. $\angle$s Th.)
8. $\angle BEA \cong \angle FDE$ (Trans. Prop.)
9. $\overline{EB} \parallel \overline{DF}$ (Corres. $\angle$s Post.)
10. Quadrilateral $EBFD$ is a $\square$. (If one pair of opp. sides is $\parallel$ and $\cong$, then the quad. is a $\square$.)

5. The legs are made so that they will bisect each other, so the quadrilateral formed by the ends of the legs is always a parallelogram. Therefore, the top of the stand is parallel to the floor.

Mixed Problem Solving and Proof

1. **PROOF** Write a two-column proof. *(Lesson 7-1)*

Given: D is the midpoint of $\overline{BE}$, $\overline{BD}$ is an altitude of right triangle $\triangle ABC$ See margin.

Prove: $\frac{AD}{DE} = \frac{DE}{DC}$

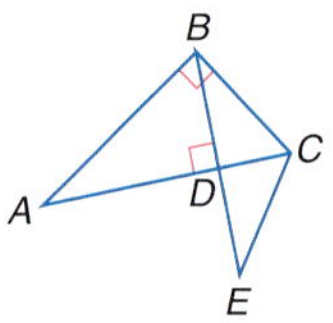

2. **AMUSEMENT PARKS** The map shows the locations of four rides at an amusement park. Find the length of the path from the roller coaster to the bumper boats. Round to the nearest tenth. *(Lesson 7-1)* 86.6 ft

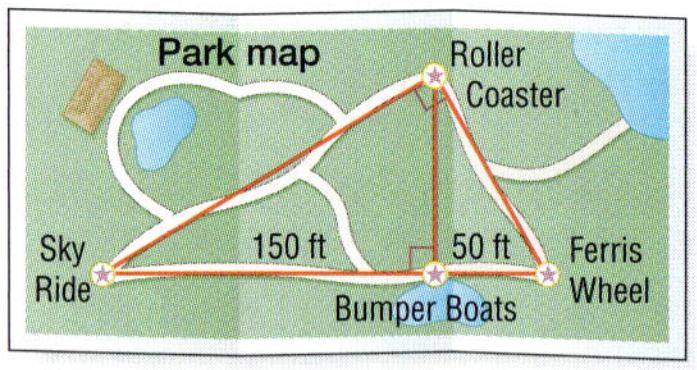

3. **CONSTRUCTION** Carlotta drew a diagram of a right triangular brace with side measures of 2.7 centimeters, 3.0 centimeters, and 5.3 centimeters. Is the diagram correct? Explain. *(Lesson 7-2)* See margin.

DESIGN For Exercises 4–5, use the following information. *(Lesson 7-3)*
Kwan designed the pinwheel. The blue triangles are congruent equilateral triangles each with an altitude of 4 inches. The red triangles are congruent isosceles right triangles. The hypotenuse of a red triangle is congruent to a side of a blue triangle.

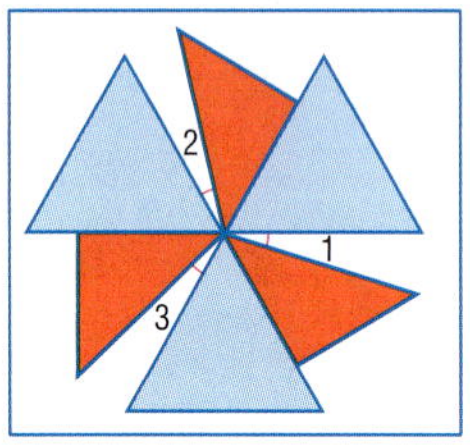

4. If angles 1, 2, and 3 are congruent, find the measure of each angle. 15°

5. Find the perimeter of the pinwheel. Round to the nearest inch. 55 in.

COMMUNICATION For Exercises 6–9, use the following information. *(Lesson 7-4)*
The diagram shows a radio tower secured by four pairs of guy wires that are equally spaced apart with $DX = 60$ feet. Round to the nearest tenth if necessary.

6. $\triangle AEX$, $\triangle AHX$, $\triangle DFX$, $\triangle DGX$, $\triangle AEH$, $\triangle CFG$, $\triangle BEH$, $\triangle DFG$

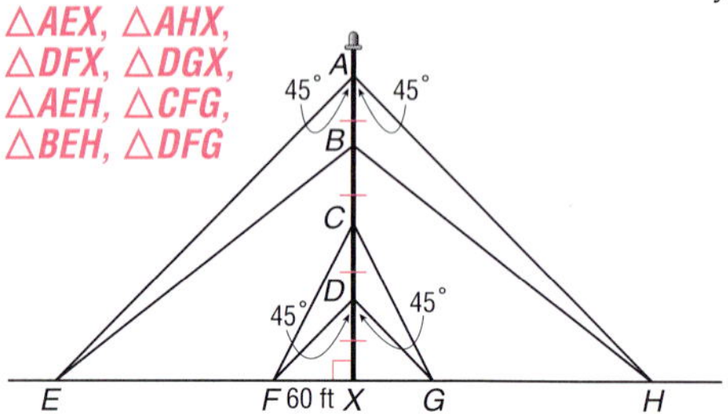

6. Name the isosceles triangles in the diagram.

7. Find $m\angle BEX$ and $m\angle CFX$. 36.9, 63.4

8. Find AE, EB, CF, and DF. See margin.

9. Find the total amount of wire used to support the tower. 1717 ft

10. **METEOROLOGY** A searchlight is 6500 feet from a weather station. If the angle of elevation to the spot of light on the clouds above the station is 47°, how high is the cloud ceiling? *(Lesson 7-5)* ≈ 6970 ft

GARDENING For Exercises 11 and 12, use the information below. *(Lesson 7-6)*
A flower bed at Magic City Rose Garden is in the shape of an obtuse scalene triangle with the shortest side measuring 7.5 feet. Another side measures 14 feet and the measure of the opposite angle is 103°. 11. 31°, 46°

11. Find the measures of the other angles of the triangle. Round to the nearest degree.

12. Find the perimeter of the garden. Round to the nearest tenth. 31.8 ft

13. **HOUSING** Mr. and Mrs. Abbott bought a lot at the end of a cul-de-sac. They want to build a fence on three sides of the lot, excluding $\overline{HE}$. To the nearest foot, how much fencing will they need to buy? *(Lesson 7-7)* 741 ft

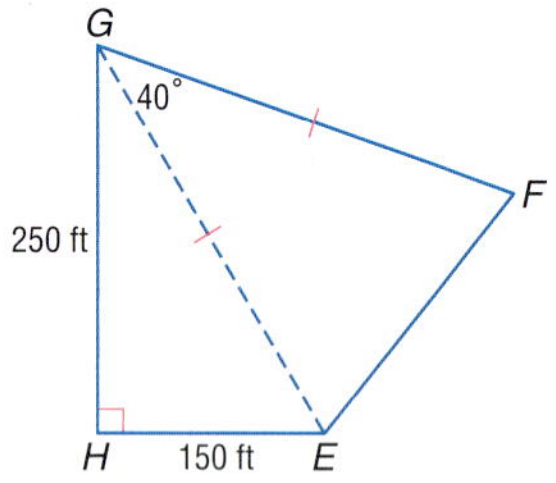

6. Given: $\square WXZY$, $\angle 1$ and $\angle 2$ are complementary.

Prove: $WXZY$ is a rectangle.

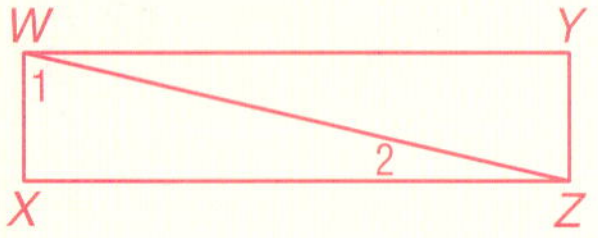

Proof:

Statements (Reasons)

1. $\square WXZY$, $\angle 1$ and $\angle 2$ are complementary (Given)
2. $m\angle 1 + m\angle 2 = 90$ (Def. of complementary $\angle$s)
3. $m\angle 1 + m\angle 2 + m\angle X = 180$ (Angle Sum Th.)
4. $90 + m\angle X = 180$ (Substitution)
5. $m\angle X = 90$ (Subtraction)
6. $\angle X \cong \angle Y$ (Opp. $\angle$s of a $\square$ are $\cong$.)
7. $m\angle Y = 90$ (Substitution)

ENGINEERING For Exercises 1–2, use the following information.
The London Eye in London, England, is the world's largest observation wheel. The ride has 32 enclosed capsules for riders. *(Lesson 8-1)*

1. Suppose each capsule is connected with a straight piece of metal forming a 32-gon. Find the sum of the measures of the interior angles. **5400**
2. What is the measure of one interior angle of the 32-gon? **168.75**

3. **QUILTING** The quilt square shown is called the Lone Star pattern. Describe two ways that the quilter could ensure that the pieces will fit properly. *(Lesson 8-2)* **See margin.**

4. **PROOF** Write a two-column proof. *(Lesson 8-3)*
Given: $\square ABCD$, $\overline{AE} \cong \overline{CF}$
Prove: Quadrilateral *EBFD* is a parallelogram.
See margin.

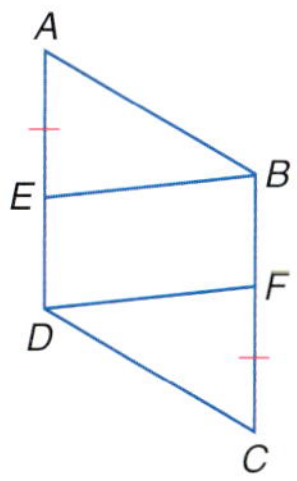

5. **MUSIC** Why will the keyboard stand shown always remain parallel to the floor? *(Lesson 8-3)*
See margin.

6. **PROOF** Write a two-column proof. *(Lesson 8-4)*
Given: $\square WXZY$, $\angle 1$ and $\angle 2$ are complementary.
Prove: *WXZY* is a rectangle. **See margin.**

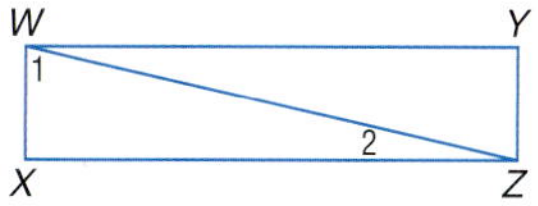

7. **PROOF** Write a paragraph proof. *(Lesson 8-4)*
Given: $\square KLMN$
Prove: *PQRS* is a rectangle. **See margin.**

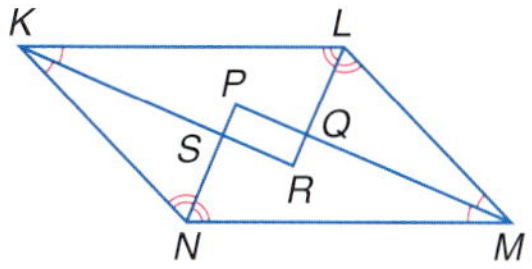

8. **CONSTRUCTION** Mr. Redwing is building a sandbox. He placed stakes at what he believes will be the four vertices of a square with a distance of 5 feet between each stake. How can he be sure that the sandbox will be a square? *(Lesson 8-5)* **See margin.**

DESIGN For Exercises 9 and 10, use the square floor tile design shown below. *(Lesson 8-6)*

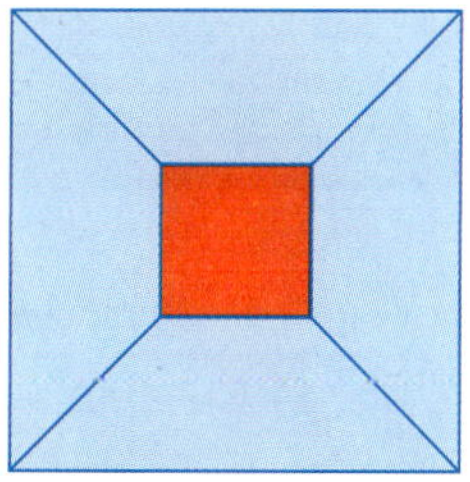

9. Explain how you know that the trapezoids in the design are isosceles. **See margin.**
10. The perimeter of the floor tile is 48 inches, and the perimeter of the interior red square is 16 inches. Find the perimeter of one trapezoid. **$16 + 8\sqrt{2}$ in. ≈ 27.3 in.**

11. **PROOF** Position a quadrilateral on the coordinate plane with vertices $Q(-a, 0)$, $R(a, 0)$, $S(b, c)$, and $T(-b, c)$. Prove that the quadrilateral is an isosceles trapezoid. *(Lesson 8-7)*
See margin

8. $\angle X$ and $\angle XWY$ are suppl., $\angle X$ and $\angle XZY$ are suppl. (Cons. ∠s in ▱ are suppl.)
9. $m\angle X + m\angle XWY = 180$, $m\angle X + m\angle XZY = 180$ (Def. of suppl. ∠s)
10. $90 + m\angle XWY = 180$, $90 + m\angle XZY = 180$ (Substitution)
11. $m\angle XWY = 90$, $m\angle XZY = 90$ (Subtraction)
12. $\angle X$, $\angle Y$, $\angle XWY$, and $\angle XZY$ are rt. ∠s (Def. rt. ∠)
13. *WXZY* is a rect. (Def. of rect.)

7. Given: $\square KLMN$
Prove: *PQRS* is a rectangle.
Proof: The diagram indicates that $\angle KNS \cong \angle SNM \cong \angle MLQ \cong \angle QLK$ and $\angle NKS \cong \angle SKL \cong \angle LMQ \cong \angle QMN$ in $\square KLMN$. Since $\triangle KLR$, $\triangle KNS$, $\triangle MLQ$, and $\triangle MNP$ all have two angles congruent, the third angles are congruent by the Third Angle Theorem. So $\angle QRS \cong \angle KSN \cong \angle MQL \cong \angle SPQ$. Since they are vertical angles, $\angle KSN \cong \angle PSR$ and $\angle MQL \cong \angle PQR$. Therefore, $\angle QRS \cong \angle PSR \cong \angle PQR \cong \angle SPQ$. *PQRS* is a parallelogram since if both pairs of opposite angles are congruent, the quadrilateral is a parallelogram. $\angle KSN$ and $\angle KSP$ form a linear pair and are therefore supplementary angles. $\angle KSP$ and $\angle PSR$ form a linear pair and are supplementary angles. Therefore, $\angle KSN$ and $\angle PSR$ are supplementary. Since they are also congruent, each is a right angle. If a parallelogram has one right angle, it has four right angles. Therefore, *PQRS* is a rectangle.

8. Sample answer: He should measure the angles at the vertices to see if they are 90 or he can check to see if the diagonals are congruent.

9. The legs of the trapezoids are part of the diagonals of the square. The diagonals of a square bisect opposite angles, so each base angle of a trapezoid measures 45°. One pair of sides is parallel and the base angles are congruent.

11. Given: Quadrilateral *QRST*
Prove: *QRST* is an isosceles trapezoid

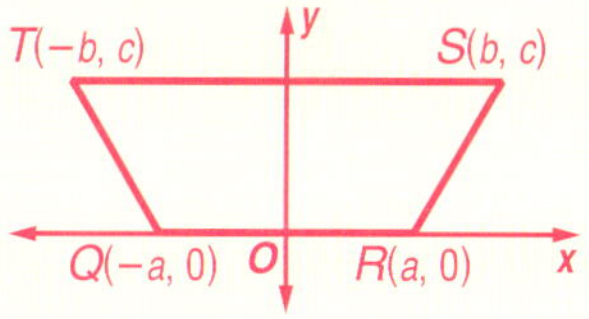

Proof:
$$TQ = \sqrt{(-b-(-a))^2 + (c-0)^2} = \sqrt{b^2 - 2ab + a^2 + c^2}$$
$$SR = \sqrt{(b-a)^2 + (c-0)^2} = \sqrt{b^2 - 2ab + a^2 + c^2}$$
Slope of $\overline{TS} = \frac{c-c}{b-(-b)} = \frac{0}{2b}$ or 0.
Slope of $\overline{QR} = \frac{0-0}{a-(-a)} = \frac{0}{2a}$ or 0.
Slope of $\overline{TQ} = \frac{c-0}{-b-(-a)}$ or $\frac{c}{-b+a}$.
Slope of $\overline{SR} = \frac{c-0}{b-a}$ or $\frac{c}{b-a}$.
Exactly one pair of opposite sides are parallel. The legs are congruent. *QRST* is an isosceles trapezoid.

Chapter 9

2. Sample answer: Look at the upper right-hand square containing two squares and four triangles. The blue triangles are reflections over a line representing the diagonal of the square. The purple pentagon is formed by reflecting a trapezoid over a line through the center of the square surrounding the pentagon. Any small pink square is a reflection of a small yellow square reflected over a diagonal of the larger square.

3. 50 mi;

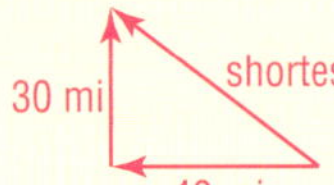

4. either 45° clockwise or 45° counterclockwise

5. either 45° clockwise or 45° counterclockwise

7. Yes; the measure of one interior angle is 90, which is a factor of 360. So, a square can tessellate the plane.

9.

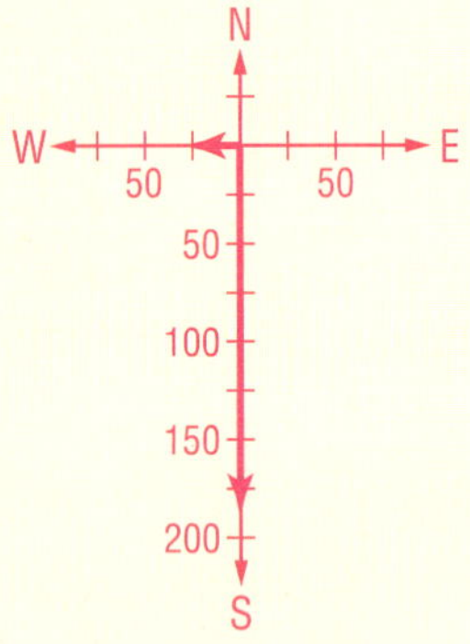

11. Sample answer: The matrix $\begin{bmatrix} -1 & 0 \\ 0 & 1 \end{bmatrix}$ will produce the vertices for a reflection of the figure in the y-axis. Then the matrix $\begin{bmatrix} 1 & 0 \\ 0 & -1 \end{bmatrix}$ will produce the vertices for a reflection of the second figure in the x-axis. This figure will be upside down.

12. The matrix $\begin{bmatrix} -1 & 0 \\ 0 & -1 \end{bmatrix}$ will produce the vertices for a 180° rotation about the origin. The figure will be upside down and in Quadrant III.

13. The matrix for Exercise 12 has the first row entries for the first matrix used in Exercise 11 and the second row entries for the second matrix used in 11.

Mixed Problem Solving and Proof

QUILTING **For Exercises 1 and 2, use the diagram of a quilt square.** *(Lesson 9-1)*

1. How many lines of symmetry are there for the entire quilt square? **4**

2. Consider different sections of the quilt square. Describe at least three different lines of reflection and the figures reflected in those lines. **See margin.**

3. **ENVIRONMENT** A cloud of dense gas and dust pours out of Surtsey, a volcanic island off the south coast of Iceland. If the cloud blows 40 miles north and then 30 miles east, make a sketch to show the translation of the smoke particles. Then find the distance of the shortest path that would take the particles to the same position. *(Lesson 9-2)* **See margin.**

ART **For Exercises 4–7, use the mosaic tile.**

4. Identify the order and magnitude of rotation that takes a yellow triangle to a blue triangle. *(Lesson 9-3)*

5. Identify the order and magnitude of rotation that takes a blue triangle to a yellow triangle. *(Lesson 9-3)* **4–5. See margin.**

6. Identify the magnitude of rotation that takes a trapezoid to a consecutive trapezoid. *(Lesson 9-3)* **90°**

7. Can the mosaic tile tessellate the plane? Explain. *(Lesson 9-4)* **See margin.**

8. **CRAFTS** Eduardo found a pattern for cross-stitch on the Internet. The pattern measures 2 inches by 3 inches. He would like to enlarge the piece to 4 inches by 6 inches. The copy machine available to him enlarges 150% or less by increments of whole number percents. Find two whole number percents by which he can consecutively enlarge the piece and get as close to the desired dimensions as possible without exceeding them. *(Lesson 9-5)*
Sample answer: 150% followed by 133%

AVIATION **For Exercises 9 and 10, use the following information.** *(Lesson 9-6)*
A small aircraft flies due south at an average speed of 190 miles per hour. The wind is blowing due west at 30 miles per hour.

9. Draw a diagram using vectors to represent this situation. **See margin.**

10. Find the resultant velocity and direction of the plane. **about 92.4 mph; about 9.0° west of due south**

GRAPHICS **For Exercises 11–14, use the graphic shown on the computer screen.** *(Lesson 9-7)*

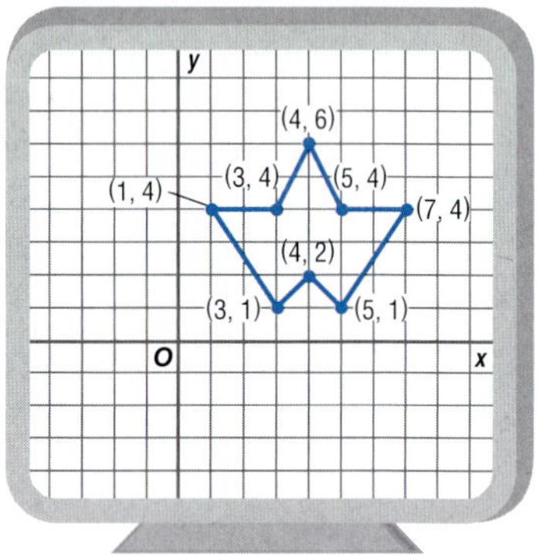

11–14. See margin.

11. Suppose you want the figure to move to Quadrant III but be upside down. Write two matrices that make this transformation, if they are applied consecutively.

12. Write one matrix that can be used to do the same transformation as in Exercise 11. What type of transformation is this?

13. Compare the two matrices in Exercise 11 to the matrix in Exercise 12. What do you notice?

14. Write the vertex matrix for the figure in Quadrant III and graph it on the coordinate plane.

14. $\begin{bmatrix} -4 & -5 & -7 & -5 & -4 & -3 & -1 & -3 \\ -6 & -4 & -4 & -1 & -2 & -1 & -4 & -4 \end{bmatrix}$;

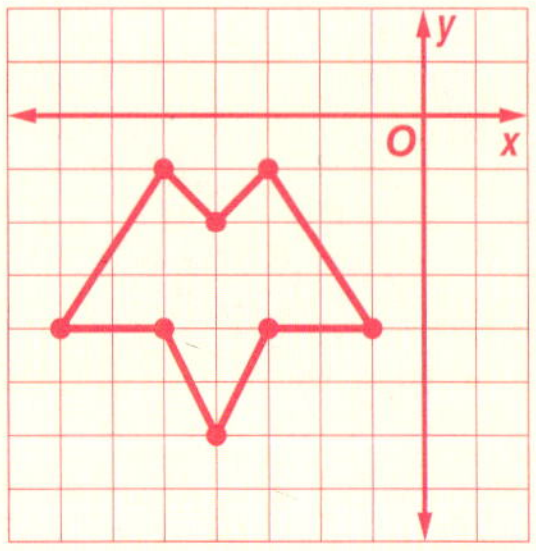

1. **CYCLING** A bicycle tire travels about 50.27 inches during one rotation of the wheel. What is the diameter of the tire? *(Lesson 10-1)* **about 16 in.**

SPACE **For Exercises 2–4, use the following information.** *(Lesson 10-2)*
School children were recently surveyed about what they believe to be the most important reason to explore Mars. They were given five choices and the table below shows the results.

Reason to Visit Mars	Number of Students
Learn about life beyond Earth	910
Learn more about Earth	234
Seek potential for human inhabitance	624
Use as a base for further exploration	364
Increase human knowledge	468

Source: *USA TODAY*

2. If you were to construct a circle graph of this data, how many degrees would be allotted to each category? **2–4. See margin.**
3. Describe the type of arc associated with each category.
4. Construct a circle graph for these data.

5. **CRAFTS** Yvonne uses wooden spheres to make paperweights to sell at craft shows. She cuts off a flat surface for each base. If the original sphere has a radius of 4 centimeters and the diameter of the flat surface is 6 centimeters, what is the height of the paperweight? *(Lesson 10-3)* **about 6.6 cm**

6. **PROOF** Write a two-column proof. *(Lesson 10-4)*
Given: $\widehat{MHT}$ is a semicircle.
$\overline{RH} \perp \overline{TM}$
Prove: $\frac{TR}{RH} = \frac{TH}{HM}$ **See margin.**

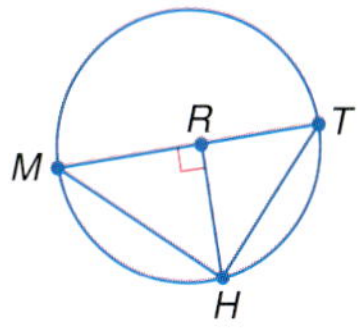

7. **PROOF** Write a paragraph proof. *(Lesson 10-5)*
Given: $\overline{GR}$ is tangent to $\odot D$ at G. **See margin.**
$\overline{AG} \cong \overline{DG}$
Prove: $\overline{AG}$ bisects $\overline{RD}$.

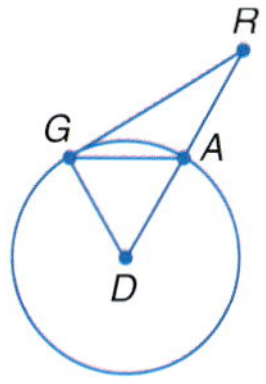

8. **METEOROLOGY** A rainbow is really a full circle with a center at a point in the sky directly opposite the Sun. The position of a rainbow varies according to the viewer's position, but its angular size, $\angle ABC$, is always 42°. If $m\widehat{CD} = 160$, find the measure of the visible part of the rainbow, $m\widehat{AC}$. *(Lesson 10-6)* **76**

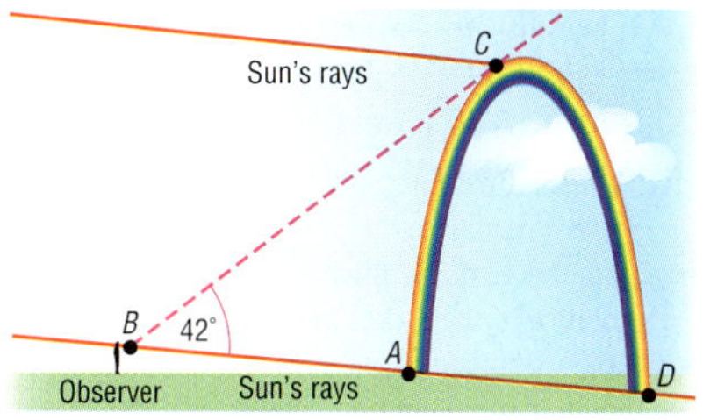

9. **CONSTRUCTION** An arch over an entrance is 100 centimeters wide and 30 centimeters high. Find the radius of the circle that contains the arch. *(Lesson 10-7)* **about 56.7 cm**

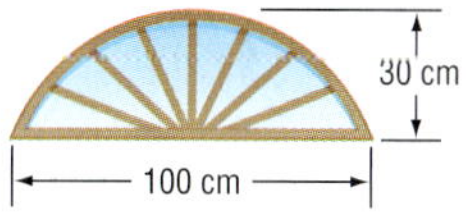

10. **SPACE** Objects that have been left behind in Earth's orbit from space missions are called "space junk." These objects are a hazard to current space missions and satellites. Eighty percent of space junk orbits Earth at a distance of 1,200 miles from the surface of Earth, which has a diameter of 7,926 miles. Write an equation to model the orbit of 80% of space junk with Earth's center at the origin. *(Lesson 10-8)*
$x^2 + y^2 = 26{,}656{,}569$

Mixed Problem Solving and Proof

6. Given: $\widehat{MHT}$ is a semicircle, $\overline{RH} \perp \overline{TM}$
Prove: $\frac{TR}{RH} = \frac{TH}{HM}$

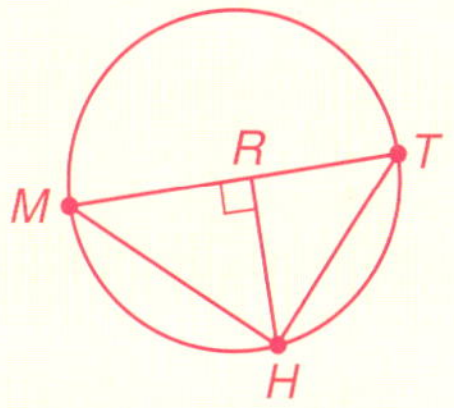

Proof:
Statements (Reasons)

1. $\widehat{MHT}$ is a semicircle, $\overline{RH} \perp \overline{TM}$ (Given)
2. $\angle THM$ is a rt. $\angle$. (If an inscribed $\angle$ intercepts a semicircle, the $\angle$ is a rt. $\angle$.)
3. $\angle TRH$ is a rt. $\angle$. (Def. $\perp$ lines)
4. $\angle THM \cong \angle TRH$ (All rt. $\angle$s are $\cong$.)
5. $\angle T \cong \angle T$ (Reflexive Prop.)
6. $\triangle TRH \sim \triangle THM$ (AA Sim.)
7. $\frac{TR}{RH} = \frac{TH}{HM}$ (Def. $\sim \triangle$s)

7. Given: $\overline{GR}$ is tangent to $\odot D$ at G. $\overline{AG} \cong \overline{DG}$
Prove: $\overline{AG}$ bisects $\overline{RD}$.

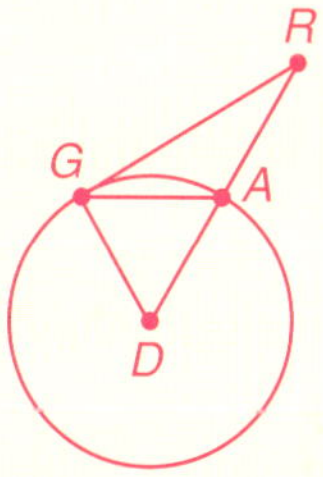

Proof: Since $\overline{DA}$ is a radius, $\overline{DG} \cong \overline{DA}$. Since $\overline{AG} \cong \overline{DG} \cong \overline{DA}$, $\triangle GDA$ is equilateral. Therefore, each angle has a measure of 60. Since $\overline{GR}$ is tangent to $\odot D$, $m\angle RGD = 90$. Since $m\angle AGD = 60$, then by the Angle Addition Postulate, $m\angle RGA = 30$. If $m\angle DAG = 60$, then $m\angle RAG = 120$. Then $m\angle R = 30$. Then, $\triangle RAG$ is isosceles, and $\overline{RA} \cong \overline{AG}$. By the Transitive Property, $\overline{RA} \cong \overline{DA}$. Therefore, $\overline{AG}$ bisects $\overline{RD}$.

Chapter 10

2. Learn about life beyond Earth: 126°; Learn more about Earth: 32.4°; Seek potential for human inhabitance: 86.4°; Use as a base for further exploration: 50.4°; Increase human knowledge: 64.8°
3. All of the categories are represented by minor arcs.

4.

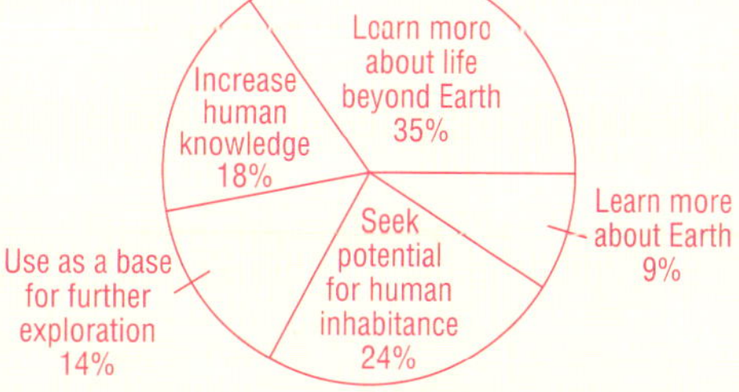

Chapter 11

9. The total for the black tiles is greater. For the red tiles, there are 4 hexagons and 5 squares for a perimeter of $2[4(4 + 2\sqrt{2}) + 5.4] = 72 + 16\sqrt{2}$ feet. For the black tiles, there are 8 squares and 8 triangles for a perimeter of $2[8.4 + 8(2 + \sqrt{2})] = 96 + 16\sqrt{2}$ feet.

Mixed Problem Solving and Proof

REMODELING **For Exercises 1–3, use the following information.**
The diagram shows the floor plan of the home that the Summers are buying. They want to replace the patio with a larger sunroom to increase their living space by one-third. *(Lesson 11-1)*

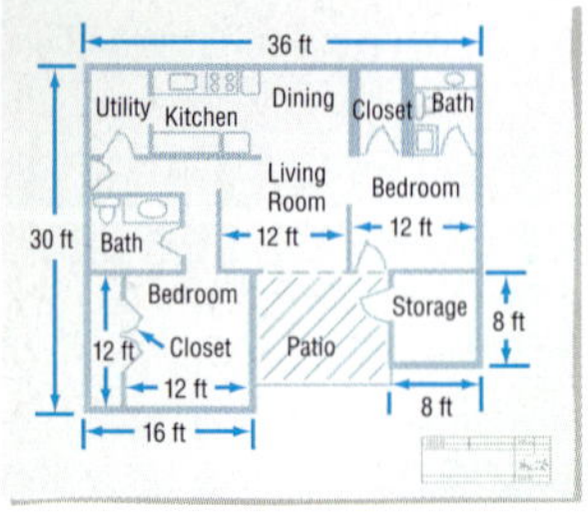

1. Excluding the patio and storage area, how many square feet of living area are in the current house? **840 ft²**
2. What area should be added to the house to increase the living area by one-third? **280 ft²**
3. The Summers want to connect the bedroom and storage area with the sunroom. What will be the dimensions of the sunroom? **12 ft by 23.3 ft**

HOME REPAIR **For Exercises 4 and 5, use the following information.**
Scott needs to replace the shingles on the roof of his house. The roof is composed of two large isosceles trapezoids, two smaller isosceles trapezoids, and a rectangle. Each trapezoid has the same height. *(Lesson 11-2)*

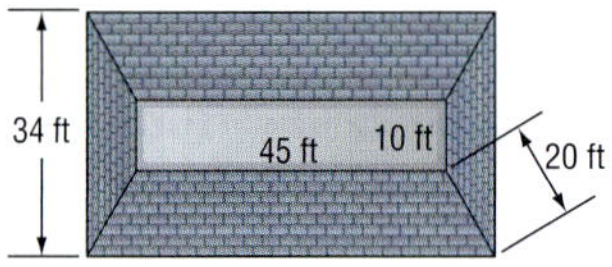

4. Find the height of the trapezoids. **16 ft**
5. Find the area of the roof covered by shingles. **2528 ft²**

6. **SPORTS** The Moore High School basketball team wants to paint their basketball court as shown. They want the center circle and the free throw areas painted blue. What is the area of the court that they will paint blue? *(Lesson 11-3)* **682.19 ft²**

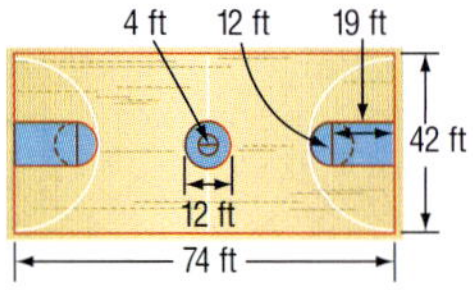

MUSEUMS **For Exercises 7–9, use the following information.**
The Hyalite Hills Museum plans to install the square mosaic pattern shown below in the entry hall. It is 10 feet on each side with each small black or red square tile measuring 2 feet on each side. *(Lesson 11-4)*

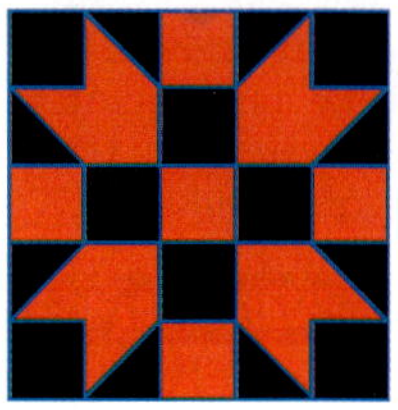

7. Find the area of black tiles. **48 ft²**
8. Find the area of red tiles. **52 ft²**
9. Which is greater, the total perimeter of the red tiles or the total perimeter of the black tiles? Explain. **See margin.**

10. **GAMES** If the dart lands on the target, find the probability that it lands in the blue region. *(Lesson 11-5)* **≈ 0.378**

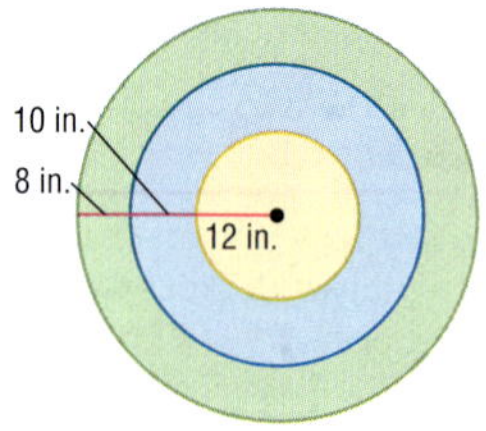

11. **ACCOMMODATIONS** The convention center in Washington D.C., lies in the northwest sector of the city between New York and Massachusetts Avenues, which intersect at a 130° angle. If the amount of hotel space is evenly distributed over an area with that intersection as the center and a radius of 1.5 miles, what is the probability that a vistor, randomly assigned to a hotel, will be housed in the sector containing the convention center? *(Lesson 11-5)* $\frac{13}{36}$ **or 36.1%**

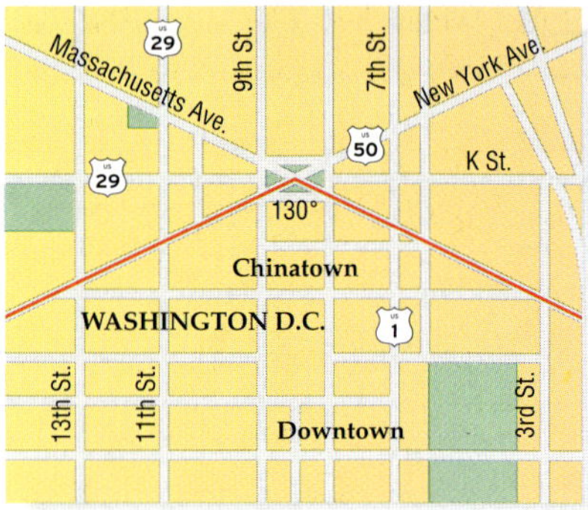

792 Mixed Problem Solving and Proof

1. **ARCHITECTURE** Sketch an orthogonal drawing of the Eiffel Tower. *(Lesson 12-1)* **See margin.**

2. **CONSTRUCTION** The roof shown below is a hip-and-valley style. Use the dimensions given to find the area of the roof that would need to be shingled. *(Lesson 12-2)* **2344.8 ft^2**

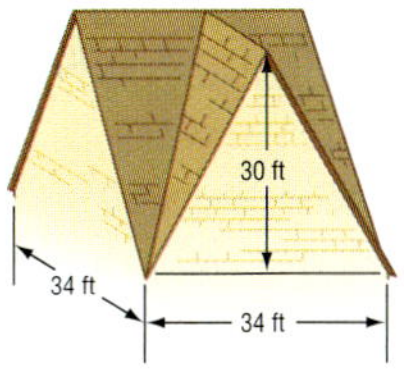

3. **AERONAUTICAL ENGINEERING** The surface area of the wing on an aircraft is used to determine a design factor known as wing loading. If the total weight of the aircraft and its load is w and the total surface area of its wings is s, then the formula for the wing loading factor, ℓ, is $\ell = \frac{w}{s}$. If the wing loading factor is exceeded, the pilot must either reduce the fuel load or remove passengers or cargo. Find the wing loading factor for a plane if it had a take-off weight of 750 pounds and the surface area of the wings was 532 square feet. *(Lesson 12-2)* **about 1.41**

4. **MANUFACTURING** Many baking pans are given a special nonstick coating. A rectangular cake pan is 9 inches by 13 inches by 2 inches deep. What is the area of the inside of the pan that needs to be coated? *(Lesson 12-3)* **205 in^2**

5. **COMMUNICATIONS** Coaxial cable is used to transmit long-distance telephone calls, cable television programming, and other communications. A typical coaxial cable contains 22 copper tubes and has a diameter of 3 inches. What is the lateral area of a coaxial cable that is 500 feet long? *(Lesson 12-4)* **about 392.7 ft^2**

COLLECTIONS For Exercises 6 and 7, use the following information.
Soledad collects unique salt-and-pepper shakers. She inherited a pair of tetrahedral shakers from her mother. *(Lesson 12-5)*

6. Each edge of a shaker measures 3 centimeters. Make a sketch of one shaker. **See margin.**

7. Find the total surface area of one shaker. **about 15.6 cm^2**

8. **FARMING** The picture below shows a combination hopper cone and bin used by farmers to store grain after harvest. The cone at the bottom of the bin allows the grain to be emptied more easily. Use the dimensions shown in the diagram to find the entire surface area of the bin with a conical top and bottom. Write the exact answer and the answer rounded to the nearest square foot. *(Lesson 12-6)*
$\pi(216 + 9\sqrt{106} + 81\sqrt{2}) \approx 1330$ ft^2

GEOGRAPHY For Exercises 9–11, use the following information.
Joaquin is buying Dennis a globe for his birthday. The globe has a diameter of 16 inches. *(Lesson 12-7)*

9. What is the surface area of the globe? **804.2 in^2**

10. If the diameter of Earth is 7926 miles, find the surface area of Earth. **197,359,487.5 mi^2**

11. The continent of Africa occupies about 11,700,000 square miles. How many square inches will be used to represent Africa on the globe? **about 47.7 in^2**

Chapter 12

1.

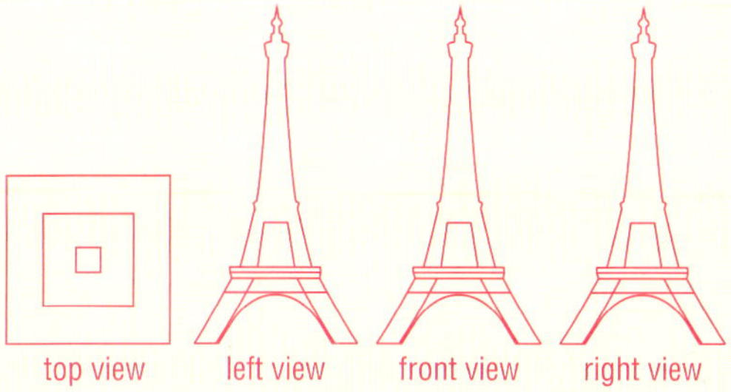

6.

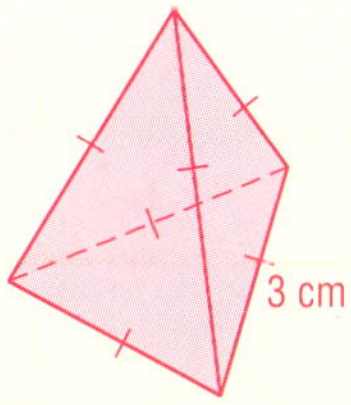

1. **METEOROLOGY** The TIROS weather satellites were a series of weather satellites, the first being launched on April 1, 1960. These satellites carried television and infrared cameras and were covered by solar cells. If the cylinder-shaped body of a TIROS had a diameter of 42 inches and a height of 19 inches, what was the volume available for carrying instruments and cameras? Round to the nearest tenth. *(Lesson 13-1)* **26,323.4 in^3**

2. **SPACECRAFT** The smallest manned spacecraft, used by astronauts for jobs outside the Space Shuttle, is the Manned Maneuvering Unit. It is 4 feet tall, 2 feet 8 inches wide, and 3 feet 8 inches deep. Find the volume of this spacecraft in cubic feet. Round to the nearest tenth. *(Lesson 13-1)* **39.1 ft^3**

3. **MUSIC** To play a concertina, you push and pull the end plates and press the keys. The air pressure causes vibrations of the metal reeds that make the notes. When fully expanded, the concertina is 36 inches from end to end. If the concertina is compressed, it is 7 inches from end to end. Find the volume of air in the instrument when it is fully expanded and when it is compressed. (*Hint:* Each endplate is a regular hexagonal prism and contains no air.) *(Lesson 13-1)* **2993.0 in^3; 280.6 in^3**

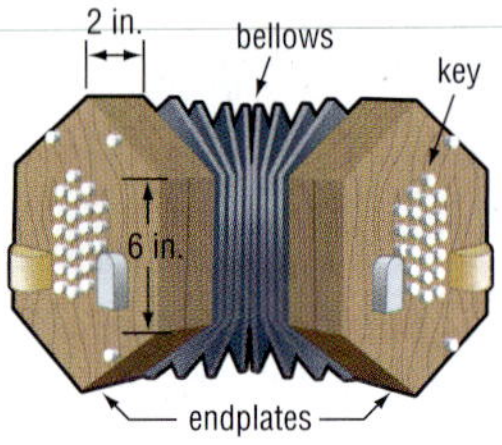

4. **ENGINEERING** The base of an oil drilling platform is made up of 24 concrete cylindrical cells. Twenty of the cells are used for oil storage. The pillars that support the platform deck rest on the four other cells. Find the total volume of the storage cells. *(Lesson 13-1)* **18,555,031.6 ft^3**

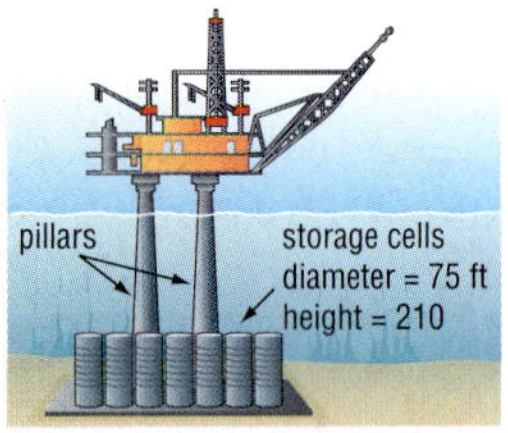

5. **HOME BUSINESS** Jodi has a home-based business selling homemade candies. She is designing a pyramid-shaped box for the candy. The base is a square measuring 14.5 centimeters on a side. The slant height of the pyramid is 16 centimeters. Find the volume of the box. Round to the nearest cubic centimeter. *(Lesson 13-2)* **1000 cm^3**

ENTERTAINMENT For Exercises 6–10, use the following information.
Some people think that the Spaceship Earth geosphere at Epcot® in Disney World resembles a golf ball. The building is a sphere measuring 165 feet in diameter. A typical golf ball has a diameter of approximately 1.5 inches.

6. Find the volume of Spaceship Earth. Round to the nearest cubic foot. *(Lesson 13-3)* **2,352,071 ft^3**

7. Find the volume of a golf ball. Round to the nearest tenth. *(Lesson 13-3)* **1.8 in^3**

8. What is the scale factor that compares Spaceship Earth to a golf ball? *(Lesson 13-4)* **1320 to 1**

9. What is the ratio of the volume of Spaceship Earth to the volume of a golf ball? *(Lesson 13-4)*

10. Suppose a six-foot-tall golfer plays golf with a 1.5 inch diameter golf ball. If the ratio between golfer and ball remains the same, how tall would a golfer need to be to use Spaceship Earth as a golf ball? *(Lesson 13-4)* **7920 ft tall**

9. 1320^3 to 1 or 2,299,968,000 to 1

ASTRONOMY For Exercises 11 and 12, use the following information.
A museum has set aside a children's room containing objects suspended from the ceiling to resemble planets and stars. Suppose an imaginary coordinate system is placed in the room with the center of the room at (0, 0, 0). Three particular stars are located at $S(-10, 5, 3)$, $T(3, -8, -1)$, and $R(-7, -4, -2)$, where the coordinates represent the distance in feet from the center of the room. *(Lesson 13-5)*

11. Find the distance between each pair of stars.

12. Which star is farthest from the center of the room? **the star located at *S***

11. $ST = \sqrt{354}$ ft, $TR = 3\sqrt{13}$ ft, $SR = \sqrt{115}$ ft

794 Mixed Problem Solving and Proof

Postulates, Theorems, and Corollaries

Chapter 2 Reasoning and Proof

Postulate 2.1 Through any two points, there is exactly one line. (p. 89)

Postulate 2.2 Through any three points not on the same line, there is exactly one plane. (p. 89)

Postulate 2.3 A line contains at least two points. (p. 90)

Postulate 2.4 A plane contains at least three points not on the same line. (p. 90)

Postulate 2.5 If two points lie in a plane, then the entire line containing those points lies in that plane. (p. 90)

Postulate 2.6 If two lines intersect, then their intersection is exactly one point. (p. 90)

Postulate 2.7 If two planes intersect, then their intersection is a line. (p. 90)

Theorem 2.1 **Midpoint Theorem** If M is the midpoint of $\overline{AB}$, then $\overline{AM} \cong \overline{MB}$. (p. 91)

Postulate 2.8 **Ruler Postulate** The points on any line or line segment can be paired with real numbers so that, given any two points A and B on a line, A corresponds to zero, and B corresponds to a positive real number. (p. 101)

Postulate 2.9 **Segment Addition Postulate** If B is between A and C, then $AB + BC = AC$. If $AB + BC = AC$, then B is between A and C. (p. 102)

Theorem 2.2 Congruence of segments is reflexive, symmetric, and transitive. (p. 102)

Postulate 2.10 **Protractor Postulate** Given $\overrightarrow{AB}$ and a number r between 0 and 180, there is exactly one ray with endpoint A, extending on either side of $\overrightarrow{AB}$, such that the measure of the angle formed is r. (p. 107)

Postulate 2.11 **Angle Addition Postulate** If R is in the interior of $\angle PQS$, then $m\angle PQR + m\angle RQS = m\angle PQS$. If $m\angle PQR + m\angle RQS = m\angle PQS$, then R is in the interior of $\angle PQS$. (p. 107)

Theorem 2.3 **Supplement Theorem** If two angles form a linear pair, then they are supplementary angles. (p. 108)

Theorem 2.4 **Complement Theorem** If the noncommon sides of two adjacent angles form a right angle, then the angles are complementary angles. (p. 108)

Theorem 2.5 Congruence of angles is reflexive, symmetric, and transitive. (p. 108)

Theorem 2.6 Angles supplementary to the same angle or to congruent angles are congruent. (p. 109) Abbreviation: $\angle$s suppl. to same $\angle$ or $\cong$ $\angle$s are $\cong$.

Theorem 2.7 Angles complementary to the same angle or to congruent angles are congruent. (p. 109) Abbreviation: $\angle$s compl. to same $\angle$ or $\cong$ $\angle$s are $\cong$.

Theorem 2.8 **Vertical Angle Theorem** If two angles are vertical angles, then they are congruent. (p. 110)

Theorem 2.9 Perpendicular lines intersect to form four right angles. (p. 110)

Theorem 2.10 All right angles are congruent. (p. 110)

Postulates, Theorems, and Corollaries

Theorem 2.11 Perpendicular lines form congruent adjacent angles. (p. 110)

Theorem 2.12 If two angles are congruent and supplementary, then each angle is a right angle. (p. 110)

Theorem 2.13 If two congruent angles form a linear pair, then they are right angles. (p. 110)

Chapter 3 Perpendicular and Parallel Lines

Postulate 3.1 **Corresponding Angles Postulate** If two parallel lines are cut by a transversal, then each pair of corresponding angles is congruent. (p. 133)

Theorem 3.1 **Alternate Interior Angles Theorem** If two parallel lines are cut by a transversal, then each pair of alternate interior angles is congruent. (p. 134)

Theorem 3.2 **Consecutive Interior Angles Theorem** If two parallel lines are cut by a transversal, then each pair of consecutive interior angles is supplementary. (p. 134)

Theorem 3.3 **Alternate Exterior Angles Theorem** If two parallel lines are cut by a transversal, then each pair of alternate exterior angles is congruent. (p. 134)

Theorem 3.4 **Perpendicular Transversal Theorem** In a plane, if a line is perpendicular to one of two parallel lines, then it is perpendicular to the other. (p. 134)

Postulate 3.2 Two nonvertical lines have the same slope if and only if they are parallel. (p. 141)

Postulate 3.3 Two nonvertical lines are perpendicular if and only if the product of their slopes is -1. (p. 141)

Postulate 3.4 If two lines in a plane are cut by a transversal so that corresponding angles are congruent, then the lines are parallel. (p. 151) Abbreviation: If corr. $\angle$s are $\cong$, lines are $\parallel$.

Postulate 3.5 **Parallel Postulate** If there is a line and a point not on the line, then there exists exactly one line through the point that is parallel to the given line. (p. 152)

Theorem 3.5 If two lines in a plane are cut by a transversal so that a pair of alternate exterior angles is congruent, then the two lines are parallel. (p. 152)
Abbreviation: If alt. ext. $\angle$s are $\cong$, then lines are $\parallel$.

Theorem 3.6 If two lines in a plane are cut by a transversal so that a pair of consecutive interior angles is supplementary, then the lines are parallel. (p. 152)
Abbreviation: If cons. int. $\angle$s are suppl., then lines are $\parallel$.

Theorem 3.7 If two lines in a plane are cut by a transversal so that a pair of alternate interior angles is congruent, then the lines are parallel. (p. 152)
Abbreviation: If alt. int. $\angle$s are $\cong$, then lines are $\parallel$.

Theorem 3.8 In a plane, if two lines are perpendicular to the same line, then they are parallel. (p. 152) Abbreviation: If 2 lines are $\perp$ to the same line, then lines are $\parallel$.

Theorem 3.9 In a plane, if two lines are each equidistant from a third line, then the two lines are parallel to each other. (p. 161)

Chapter 4 Congruent Triangles

Theorem 4.1 **Angle Sum Theorem** The sum of the measures of the angles of a triangle is 180. (p. 185)

Theorem 4.2 **Third Angle Theorem** If two angles of one triangle are congruent to two angles of a second triangle, then the third angles of the triangles are congruent. (p. 186)

Theorem 4.3 **Exterior Angle Theorem** The measure of an exterior angle of a triangle is equal to the sum of the measures of the two remote interior angles. (p. 186)

Corollary 4.1 The acute angles of a right triangle are complementary. (p. 188)

Corollary 4.2 There can be at most one right or obtuse angle in a triangle. (p. 188)

Theorem 4.4 Congruence of triangles is reflexive, symmetric, and transitive. (p. 193)

Postulate 4.1 **Side-Side-Side Congruence (SSS)** If the sides of one triangle are congruent to the sides of a second triangle, then the triangles are congruent. (p. 201)

Postulate 4.2 **Side-Angle-Side Congruence (SAS)** If two sides and the included angle of one triangle are congruent to two sides and the included angle of another triangle, then the triangles are congruent. (p. 202)

Postulate 4.3 **Angle-Side-Angle Congruence (ASA)** If two angles and the included side of one triangle are congruent to two angles and the included side of another triangle, the triangles are congruent. (p. 207)

Theorem 4.5 **Angle-Angle-Side Congruence (AAS)** If two angles and a nonincluded side of one triangle are congruent to the corresponding two angles and side of a second triangle, then the two triangles are congruent. (p. 208)

Theorem 4.6 **Leg-Leg Congruence (LL)** If the legs of one right triangle are congruent to the corresponding legs of another right triangle, then the triangles are congruent. (p. 214)

Theorem 4.7 **Hypotenuse-Angle Congruence (HA)** If the hypotenuse and acute angle of one right triangle are congruent to the hypotenuse and corresponding acute angle of another right triangle, then the two triangles are congruent. (p. 215)

Theorem 4.8 **Leg-Angle Congruence (LA)** If one leg and an acute angle of one right triangle are congruent to the corresponding leg and acute angle of another right triangle, then the triangles are congruent. (p. 215)

Postulate 4.4 **Hypotenuse-Leg Congruence (HL)** If the hypotenuse and a leg of one right triangle are congruent to the hypotenuse and corresponding leg of another right triangle, then the triangles are congruent. (p. 215)

Theorem 4.9 **Isosceles Triangle Theorem** If two sides of a triangle are congruent, then the angles opposite those sides are congruent. (p. 216)

Theorem 4.10 If two angles of a triangle are congruent, then the sides opposite those angles are congruent. (p. 218) Abbreviation: Conv. of Isos. $\triangle$Th.

Corollary 4.3 A triangle is equilateral if and only if it is equiangular. (p. 218)

Corollary 4.4 Each angle of an equilateral triangle measures 60°. (p. 218)

Chapter 5 Relationships in Triangles

Theorem 5.1 Any point on the perpendicular bisector of a segment is equidistant from the endpoints of the segment. (p. 238)

Theorem 5.2 Any point equidistant from the endpoints of a segment lies on the perpendicular bisector of the segment. (p. 238)

Theorem 5.3 **Circumcenter Theorem** The circumcenter of a triangle is equidistant from the vertices of the triangle. (p. 239)

Theorem 5.4 Any point on the angle bisector is equidistant from the sides of the angle. (p. 239)

Theorem 5.5 Any point equidistant from the sides of an angle lies on the angle bisector. (p. 239)

Theorem 5.6 **Incenter Theorem** The incenter of a triangle is equidistant from each side of the triangle. (p. 240)

Theorem 5.7 **Centroid Theorem** The centroid of a triangle is located two-thirds of the distance from a vertex to the midpoint of the side opposite the vertex on a median. (p. 240)

Theorem 5.8 **Exterior Angle Inequality Theorem** If an angle is an exterior angle of a triangle, then its measure is greater than the measure of either of its corresponding remote interior angles. (p. 248)

Theorem 5.9 If one side of a triangle is longer than another side, then the angle opposite the longer side has a greater measure than the angle opposite the shorter side. (p. 249)

Theorem 5.10 If one angle of a triangle has a greater measure than another angle, then the side opposite the greater angle is longer than the side opposite the lesser angle. (p. 250)

Theorem 5.11 **Triangle Inequality Theorem** The sum of the lengths of any two sides of a triangle is greater than the length of the third side. (p. 261)

Theorem 5.12 The perpendicular segment from a point to a line is the shortest segment from the point to the line. (p. 262)

Corollary 5.1 The perpendicular segment from a point to a plane is the shortest segment from the point to the plane. (p. 263)

Theorem 5.13 **SAS Inequality/Hinge Theorem** If two sides of a triangle are congruent to two sides of another triangle and the included angle in one triangle has a greater measure than the included angle in the other, then the third side of the first triangle is longer than the third side of the second triangle. (p. 267)

Theorem 5.14 **SSS Inequality** If two sides of a triangle are congruent to two sides of another triangle and the third side in one triangle is longer than the third side in the other, then the angle between the pair of congruent sides in the first triangle is greater than the corresponding angle in the second triangle. (p. 268)

Chapter 6 Proportions and Similarity

Postulate 6.1 **Angle-Angle (AA) Similarity** If the two angles of one triangle are congruent to two angles of another triangle, then the triangles are similar. (p. 298)

Theorem 6.1 **Side-Side-Side (SSS) Similarity** If the measures of the corresponding sides of two triangles are proportional, then the triangles are similar. (p. 299)

Theorem 6.2 **Side-Angle-Side (SAS) Similarity** If the measures of two sides of a triangle are proportional to the measures of two corresponding sides of another triangle and the included angles are congruent, then the triangles are similar. (p. 299)

Theorem 6.3 Similarity of triangles is reflexive, symmetric, and transitive. (p. 300)

Theorem 6.4 **Triangle Proportionality Theorem** If a line is parallel to one side of a triangle and intersects the other two sides in two distinct points, then it separates these sides into segments of proportional lengths. (p. 307)

Theorem 6.5 **Converse of the Triangle Proportionality Theorem** If a line intersects two sides of a triangle and separates the sides into corresponding segments of proportional lengths, then the line is parallel to the third side. (p. 308)

Theorem 6.6 **Triangle Midsegment Theorem** A midsegment of a triangle is parallel to one side of the triangle, and its length is one-half the length of that side. (p. 308)

Corollary 6.1 If three or more parallel lines intersect two transversals, then they cut off the transversals proportionally. (p. 309)

Corollary 6.2 If three or more parallel lines cut off congruent segments on one transversal, then they cut off congruent segments on every transversal. (p. 309)

Theorem 6.7 **Proportional Perimeters Theorem** If two triangles are similar, then the perimeters are proportional to the measures of corresponding sides. (p. 316)

Theorem 6.8 If two triangles are similar, then the measures of the corresponding altitudes are proportional to the measures of the corresponding sides. (p. 317)
Abbreviation: ~ △s have corr. altitudes proportional to the corr. sides.

Theorem 6.9 If two triangles are similar, then the measures of the corresponding angle bisectors of the triangles are proportional to the measures of the corresponding sides. (p. 317)
Abbreviation: ~ △s have corr. ∠ bisectors proportional to the corr. sides.

Theorem 6.10 If two triangles are similar, then the measures of the corresponding medians are proportional to the measures of the corresponding sides. (p. 317)
Abbreviation: ~ △s have corr. medians proportional to the corr. sides.

Theorem 6.11 **Angle Bisector Theorem** An angle bisector in a triangle separates the opposite side into segments that have the same ratio as the other two sides. (p. 319)

Chapter 7 Right Triangles and Trigonometry

Theorem 7.1 If the altitude is drawn from the vertex of the right angle of a right triangle to its hypotenuse, then the two triangles formed are similar to the given triangle and to each other. (p. 343)

Theorem 7.2 The measure of the altitude drawn from the vertex of the right angle of a right triangle to its hypotenuse is the geometric mean between the measures of the two segments of the hypotenuse. (p. 343)

Theorem 7.3 If the altitude is drawn from the vertex of the right angle of a right triangle to its hypotenuse, then the measure of a leg of the triangle is the geometric mean between the measures of the hypotenuse and the segment of the hypotenuse adjacent to that leg. (p. 344)

Theorem 7.4 **Pythagorean Theorem** In a right triangle, the sum of the squares of the measures of the legs equals the square of the measure of the hypotenuse. (p. 350)

Theorem 7.5 **Converse of the Pythagorean Theorem** If the sum of the squares of the measures of two sides of a triangle equals the square of the measure of the longest side, then the triangle is a right triangle. (p. 351)

Theorem 7.6 In a 45°-45°-90° triangle, the length of the hypotenuse is $\sqrt{2}$ times the length of a leg. (p. 357)

Theorem 7.7 In a $30°$-$60°$-$90°$ triangle, the length of the hypotenuse is twice the length of the shorter leg, and the length of the longer leg is $\sqrt{3}$ times the length of the shorter leg. (p. 359)

Chapter 8 Quadrilaterals

Theorem 8.1 **Interior Angle Sum Theorem** If a convex polygon has n sides and S is the sum of the measures of its interior angles, then $S = 180(n - 2)$. (p. 404)

Theorem 8.2 **Exterior Angle Sum Theorem** If a polygon is convex, then the sum of the measures of the exterior angles, one at each vertex, is 360. (p. 406)

Theorem 8.3 Opposite sides of a parallelogram are congruent. (p. 412)
Abbreviation: Opp. sides of ▱ are ≅.

Theorem 8.4 Opposite angles of a parallelogram are congruent. (p. 412)
Abbreviation: Opp. ∠s of ▱ are ≅.

Theorem 8.5 Consecutive angles in a parallelogram are supplementary. (p. 412)
Abbreviation: Cons. ∠s in ▱ are suppl.

Theorem 8.6 If a parallelogram has one right angle, it has four right angles. (p. 412)
Abbreviation: If ▱ has 1 rt. ∠, it has 4 rt. ∠s.

Theorem 8.7 The diagonals of a parallelogram bisect each other. (p. 413)
Abbreviation: Diag. of ▱ bisect each other.

Theorem 8.8 The diagonal of a parallelogram separates the parallelogram into two congruent triangles. (p. 414) Abbreviation: Diag. of ▱ separates ▱ into 2 ≅ △s.

Theorem 8.9 If both pairs of opposite sides of a quadrilateral are congruent, then the quadrilateral is a parallelogram. (p. 418) Abbreviation: If both pairs of opp. sides are ≅, then quad. is ▱.

Theorem 8.10 If both pairs of opposite angles of a quadrilateral are congruent, then the quadrilateral is a parallelogram. (p. 418) Abbreviation: If both pairs of opp. ∠s are ≅, then quad. is ▱.

Theorem 8.11 If the diagonals of a quadrilateral bisect each other, then the quadrilateral is a parallelogram. (p. 418) Abbreviation: If diag. bisect each other, then quad. is ▱.

Theorem 8.12 If one pair of opposite sides of a quadrilateral is both parallel and congruent, then the quadrilateral is a parallelogram. (p. 418)
Abbreviation: If one pair of opp. sides is ∥ and ≅, then the quad. is a ▱.

Theorem 8.13 If a parallelogram is a rectangle, then the diagonals are congruent. (p. 424)
Abbreviation: If ▱ is rectangle, diag. are ≅.

Theorem 8.14 If the diagonals of a parallelogram are congruent, then the parallelogram is a rectangle. (p. 426) Abbreviation: If diagonals of ▱ are ≅, ▱ is a rectangle.

Theorem 8.15 The diagonals of a rhombus are perpendicular. (p. 431)

Theorem 8.16 If the diagonals of a parallelogram are perpendicular, then the parallelogram is a rhombus. (p. 431)

Theorem 8.17 Each diagonal of a rhombus bisects a pair of opposite angles. (p. 431)

Theorem 8.18 Both pairs of base angles of an isosceles trapezoid are congruent. (p. 439)

Theorem 8.19 The diagonals of an isosceles trapezoid are congruent. (p. 439)

Theorem 8.20 The median of a trapezoid is parallel to the bases, and its measure is one-half the sum of the measures of the bases. (p. 441)

Chapter 9 Transformations

Postulate 9.1 In a given rotation, if A is the preimage, A' is the image, and P is the center of rotation, then the measure of the angle of rotation, $\angle APA'$ is twice the measure of the acute or right angle formed by the intersecting lines of reflection. (p. 477)

Corollary 9.1 Reflecting an image successively in two perpendicular lines results in a 180° rotation. (p. 477)

Theorem 9.1 If a dilation with center C and a scale factor of r transforms A to E and B to D, then $ED = |r|(AB)$. (p. 491)

Theorem 9.2 If $P(x, y)$ is the preimage of a dilation centered at the origin with a scale factor r, then the image is $P'(rx, ry)$. (p. 492)

Chapter 10 Circles

Theorem 10.1 Two arcs are congruent if and only if their corresponding central angles are congruent. (p. 530)

Postulate 10.1 **Arc Addition Postulate** The measure of an arc formed by two adjacent arcs is the sum of the measures of the two arcs. (p. 531)

Theorem 10.2 In a circle or in congruent circles, two minor arcs are congruent if and only if their corresponding chords are congruent. (p. 536)
Abbreviations: In ⊙, 2 minor arcs are ≅, *iff* corr. chords are ≅.
In ⊙, 2 chords are ≅, *iff* corr. minor arcs are ≅.

Theorem 10.3 In a circle, if a diameter (or radius) is perpendicular to a chord, then it bisects the chord and its arc. (p. 537)

Theorem 10.4 In a circle or in congruent circles, two chords are congruent if and only if they are equidistant from the center. (p. 539)

Theorem 10.5 If an angle is inscribed in a circle, then the measure of the angle equals one-half the measure of its intercepted arc (or the measure of the intercepted arc is twice the measure of the inscribed angle). (p. 544)

Theorem 10.6 If two inscribed angles of a circle (or congruent circles) intercept congruent arcs or the same arc, then the angles are congruent. (p. 546) Abbreviations: Inscribed ∠s of same arc are ≅. Inscribed ∠s of ≅ arcs are ≅.

Theorem 10.7 If an inscribed angle intercepts a semicircle, the angle is a right angle. (p. 547)

Theorem 10.8 If a quadrilateral is inscribed in a circle, then its opposite angles are supplementary. (p. 548)

Theorem 10.9 If a line is tangent to a circle, then it is perpendicular to the radius drawn to the point of tangency. (p. 553)

Theorem 10.10 If a line is perpendicular to a radius of a circle at its endpoint on the circle, then the line is a tangent to the circle. (p. 553)

Theorem 10.11 If two segments from the same exterior point are tangent to a circle, then they are congruent. (p. 554)

Theorem 10.12 If two secants intersect in the interior of a circle, then the measure of an angle formed is one-half the sum of the measure of the arcs intercepted by the angle and its vertical angle. (p. 561)

Theorem 10.13 If a secant and a tangent intersect at the point of tangency, then the measure of each angle formed is one-half the measure of its intercepted arc. (p. 562)

Theorem 10.14 If two secants, a secant and a tangent, or two tangents intersect in the exterior of a circle, then the measure of the angle formed is one-half the positive difference of the measures of the intercepted arcs. (p. 563)

Theorem 10.15 If two chords intersect in a circle, then the products of the measures of the segments of the chords are equal. (p. 569)

Theorem 10.16 If two secant segments are drawn to a circle from an exterior point, then the product of the measures of one secant segment and its external secant segment is equal to the product of the measures of the other secant segment and its external secant segment. (p. 570)

Theorem 10.17 If a tangent segment and a secant segment are drawn to a circle from an exterior point, then the square of the measure of the tangent segment is equal to the product of the measures of the secant segment and its external secant segment. (p. 571)

Chapter 11 Area of Polygons And Circles

Postulate 11.1 Congruent figures have equal areas. (p. 603)

Postulate 11.2 The area of a region is the sum of the areas of all of its nonoverlapping parts. (p. 619)

Chapter 13 Volume

Theorem 13.1 If two solids are similar with a scale factor of $a : b$, then the surface areas have a ratio of $a^2 : b^2$, and the volumes have a ratio of $a^3 : b^3$. (p. 709)

Glossary/Glosario

English | Español

A

acute angle (p. 30) An angle with a degree measure less than 90.

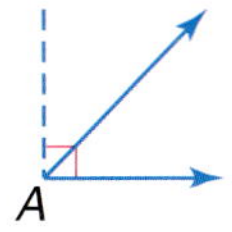

$0 < m\angle A < 90$

ángulo agudo Ángulo cuya medida en grados es menos de 90.

acute triangle (p. 178) A triangle in which all of the angles are acute angles.

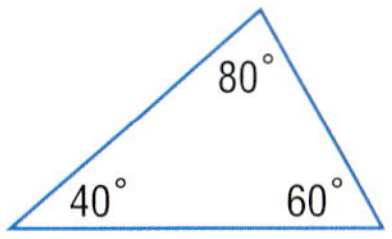

three acute angles
tres ángulos agudos

triángulo acutángulo Triángulo cuyos ángulos son todos agudos.

adjacent angles (p. 37) Two angles that lie in the same plane, have a common vertex and a common side, but no common interior points.

ángulos adyacentes Dos ángulos que yacen sobre el mismo plano, tienen el mismo vértice y un lado en común, pero ningún punto interior.

alternate exterior angles (p. 128) In the figure, transversal t intersects lines ℓ and m. $\angle 5$ and $\angle 3$, and $\angle 6$ and $\angle 4$ are alternate exterior angles.

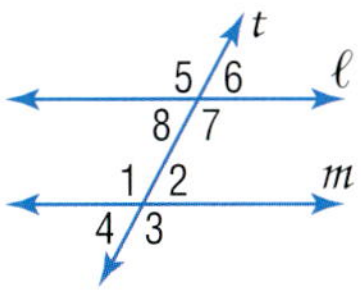

ángulos alternos externos En la figura, la transversal t interseca las rectas ℓ y m. $\angle 5$ y $\angle 3$, y $\angle 6$ y $\angle 4$ son ángulos alternos externos.

alternate interior angles (p. 128) In the figure above, transversal t intersects lines ℓ and m. $\angle 1$ and $\angle 7$, and $\angle 2$ and $\angle 8$ are alternate interior angles.

ángulos alternos internos En la figura anterior, la transversal t interseca las rectas ℓ y m. $\angle 1$ y $\angle 7$, y $\angle 2$ y $\angle 8$ son ángulos alternos internos .

altitude **1.** (p. 241) In a triangle, a segment from a vertex of the triangle to the line containing the opposite side and perpendicular to that side. **2.** (pp. 649, 655) In a prism or cylinder, a segment perpendicular to the bases with an endpoint in each plane. **3.** (pp. 660, 666) In a pyramid or cone, the segment that has the vertex as one endpoint and is perpendicular to the base.

altura **1.** En un triángulo, segmento trazado desde el vértice de un triángulo hasta el lado opuesto y que es perpendicular a dicho lado. **2.** El segmento perpendicular a las bases de prismas y cilindros que tiene un extremo en cada plano. **3.** El segmento que tiene un extremo en el vértice de pirámides y conos y que es perpendicular a la base.

ambiguous case of the Law of Sines (p. 384) Given the measures of two sides and a nonincluded angle, there exist two possible triangles.

caso ambiguo de la ley de los senos Dadas las medidas de dos lados y de un ángulo no incluido, existen dos triángulos posibles.

angle (p. 29) The intersection of two noncollinear rays at a common endpoint. The rays are called *sides* and the common endpoint is called the *vertex*.

ángulo La intersección de dos semirrectas no colineales en un punto común. Las semirrectas se llaman *lados* y el punto común se llama *vértice*.

angle bisector (p. 32) A ray that divides an angle into two congruent angles.

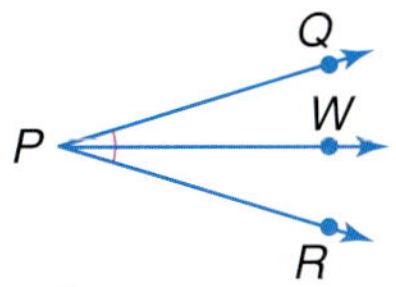

$\overrightarrow{PW}$ is the bisector of $\angle P$.
$\overrightarrow{PW}$ *es la bisectriz del* $\angle P$.

bisectriz de un ángulo Semirrecta que divide un ángulo en dos ángulos congruentes.

Glossary/Glosario

angle of depression (p. 372) The angle between the line of sight and the horizontal when an observer looks downward.

ángulo de depresión Ángulo formado por la horizontal y la línea de visión de un observador que mira hacia abajo.

angle of elevation (p. 371) The angle between the line of sight and the horizontal when an observer looks upward.

ángulo de elevación Ángulo formado por la horizontal y la línea de visión de un observador que mira hacia arriba.

angle of rotation (p. 476) The angle through which a preimage is rotated to form the image.

ángulo de rotación El ángulo a través del cual se rota una preimagen para formar la imagen.

apothem (p. 610) A segment that is drawn from the center of a regular polygon perpendicular to a side of the polygon.

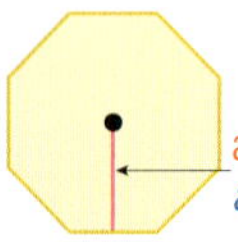

apotema Segmento perpendicular trazado desde el centro de un polígono regular hasta uno de sus lados.

arc (p. 530) A part of a circle that is defined by two endpoints.

arco Parte de un círculo definida por los dos extremos de una recta.

axis **1.** (p. 655) In a cylinder, the segment with endpoints that are the centers of the bases. **2.** (p. 666) In a cone, the segment with endpoints that are the vertex and the center of the base.

eje **1.** El segmento en un cilindro cuyos extremos forman el centro de las bases.
2. El segmento en un cono cuyos extremos forman el vértice y el centro de la base.

B

between (p. 14) For any two points A and B on a line, there is another point C between A and B if and only if A, B, and C are collinear and $AC + CB = AB$.

ubicado entre Para cualquier par de puntos A y B de una recta, existe un punto C ubicado entre A y B si y sólo si A, B y C son colineales y $AC + CB = AB$.

biconditional (p. 81) The conjunction of a conditional statement and its converse.

bicondicional La conjunción entre un enunciado condicional y su recíproco.

C

center of rotation (p. 476) A fixed point around which shapes move in a circular motion to a new position.

centro de rotación Punto fijo alrededor del cual gira una figura hasta alcanzar una posición determinada.

central angle (p. 529) An angle that intersects a circle in two points and has its vertex at the center of the circle.

ángulo central Ángulo que interseca un círculo en dos puntos y cuyo vértice se localiza en el centro del círculo.

centroid (p. 240) The point of concurrency of the medians of a triangle.

centroide Punto de intersección de las medianas de un triángulo.

chord **1.** (p. 522) For a given circle, a segment with endpoints that are on the circle.
2. (p. 671) For a given sphere, a segment with endpoints that are on the sphere.

cuerda **1.** Segmento cuyos extremos están en un círculo.
2. Segmento cuyos extremos están en una esfera.

circle (p. 522) The locus of all points in a plane equidistant from a given point called the *center* of the circle.

P is the center of the circle.
P es el centro del círculo.

círculo Lugar geométrico formado por el conjunto de puntos en un plano, equidistantes de un punto dado llamado *centro*.

Glossary/Glosario

circumcenter (p. 238) The point of concurrency of the perpendicular bisectors of a triangle.

circuncentro Punto de intersección de las mediatrices de un triángulo.

circumference (p. 523) The distance around a circle.

circunferencia Distancia alrededor de un círculo.

circumscribed (p. 537) A circle is circumscribed about a polygon if the circle contains all the vertices of the polygon.

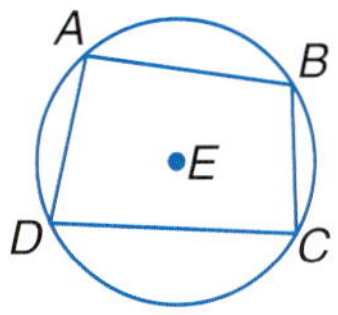

⊙*E* is circumscribed about quadrilateral *ABCD*.
⊙E está circunscrito al cuadrilátero ABCD.

circunscrito Un polígono está circunscrito a un círculo si todos sus vértices están contenidos en el círculo.

collinear (p. 6) Points that lie on the same line.

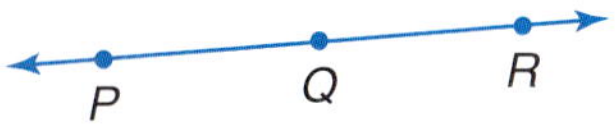

P, *Q*, and *R* are collinear.
P, Q y R son colineales.

colineal Puntos que yacen en la misma recta.

column matrix (p. 506) A matrix containing one column often used to represent an ordered pair or a vector, such as $\langle x, y\rangle = \begin{bmatrix} x \\ y \end{bmatrix}$.

matriz columna Matriz formada por una sola columna y que se usa para representar pares ordenados o vectores como, por ejemplo, $\langle x, y\rangle = \begin{bmatrix} x \\ y \end{bmatrix}$.

complementary angles (p. 39) Two angles with measures that have a sum of 90.

ángulos complementarios Dos ángulos cuya suma es igual a 90 grados.

component form (p. 498) A vector expressed as an ordered pair, ⟨change in x, change in y⟩.

componente Vector representado en forma de par ordenado, ⟨cambio en x, cambio en y⟩.

composition of reflections (p. 471) Successive reflections in parallel lines.

composición de reflexiones Reflexiones sucesivas en rectas paralelas.

compound statement (p. 67) A statement formed by joining two or more statements.

enunciado compuesto Enunciado formado por la unión de dos o más enunciados.

concave polygon (p. 45) A polygon for which there is a line containing a side of the polygon that also contains a point in the interior of the polygon.

polígono cóncavo Polígono para el cual existe una recta que contiene un lado del polígono y un punto interior del polígono.

conclusion (p. 75) In a conditional statement, the statement that immediately follows the word *then*.

conclusión Parte del enunciado condicional que está escrita después de la palabra *entonces*.

concurrent lines (p. 238) Three or more lines that intersect at a common point.

rectas concurrentes Tres o más rectas que se intersecan en un punto común.

conditional statement (p. 75) A statement that can be written in *if-then form*.

enunciado condicional Enunciado escrito en la forma *si-entonces*.

cone (p. 666) A solid with a circular base, a vertex not contained in the same plane as the base, and a lateral surface area composed of all points in the segments connecting the vertex to the edge of the base.

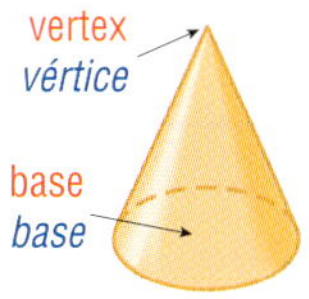

cono Sólido de base circular cuyo vértice no se localiza en el mismo plano que la base y cuya superficie lateral está formada por todos los segmentos que unen el vértice con los límites de la base.

Glossary/Glosario

congruence transformations (p. 194) A mapping for which a geometric figure and its image are congruent.

transformación de congruencia Transformación en un plano en la que la figura geométrica y su imagen son congruentes.

congruent (p. 15) Having the same measure.

congruente Que miden lo mismo.

congruent arcs (p. 530) Arcs of the same circle or congruent circles that have the same measure.

arcos congruentes Arcos de un mismo círculo, o de círculos congruentes, que tienen la misma medida.

congruent solids (p. 707) Two solids are congruent if all of the following conditions are met.
1. The corresponding angles are congruent.
2. Corresponding edges are congruent.
3. Corresponding faces are congruent.
4. The volumes are congruent.

sólidos congruentes Dos sólidos son congruentes si cumplen todas las siguientes condiciones:
1. Los ángulos correspondientes son congruentes.
2. Las aristas correspondientes son congruentes.
3. Las caras correspondientes son congruentes.
4. Los volúmenes son congruentes.

congruent triangles (p. 192) Triangles that have their corresponding parts congruent.

triángulos congruentes Triángulos cuyas partes correspondientes son congruentes.

conjecture (p. 62) An educated guess based on known information.

conjetura Juicio basado en información conocida.

conjunction (p. 68) A compound statement formed by joining two or more statements with the word *and*.

conjunción Enunciado compuesto que se obtiene al unir dos o más enunciados con la palabra *y*.

consecutive interior angles (p. 128) In the figure, transversal t intersects lines ℓ and m. There are two pairs of consecutive interior angles: $\angle 8$ and $\angle 1$, and $\angle 7$ and $\angle 2$.

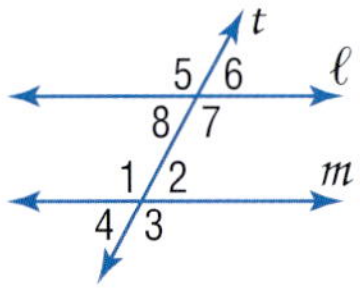

ángulos internos consecutivos En la figura, la transversal t interseca las rectas ℓ y m. La figura presenta dos pares de ángulos consecutivos internos: $\angle 8$ y $\angle 1$, y $\angle 7$ y $\angle 2$.

construction (p. 15) A method of creating geometric figures without the benefit of measuring tools. Generally, only a pencil, straightedge, and compass are used.

construcción Método para dibujar figuras geométricas sin el uso de instrumentos de medición. En general, sólo requiere de un lápiz, una regla sin escala y un compás.

contrapositive (p. 77) The statement formed by negating both the hypothesis and conclusion of the converse of a conditional statement.

antítesis Enunciado formado por la negación de la hipótesis y la conclusión del recíproco de un enunciado condicional dado.

converse (p. 77) The statement formed by exchanging the hypothesis and conclusion of a conditional statement.

recíproco Enunciado que se obtiene al intercambiar la hipótesis y la conclusión de un enunciado condicional dado.

convex polygon (p. 45) A polygon for which there is no line that contains both a side of the polygon and a point in the interior of the polygon.

polígono convexo Polígono para el cual no existe recta alguna que contenga un lado del polígono y un punto en el interior del polígono.

coordinate proof (p. 222) A proof that uses figures in the coordinate plane and algebra to prove geometric concepts.

prueba de coordenadas Demostración que usa álgebra y figuras en el plano de coordenadas para demostrar conceptos geométricos.

coplanar (p. 6) Points that lie in the same plane.

coplanar Puntos que yacen en un mismo plano.

corner view (p. 636) The view from a corner of a three-dimensional figure, also called the *perspective view.*

vista de esquina Vista de una figura tridimensional desde una esquina. También se conoce como *vista de perspectiva.*

corollary (p. 188) A statement that can be easily proved using a theorem is called a corollary of that theorem.

corolario La afirmación que puede demostrarse fácilmente mediante un teorema se conoce como corolario de dicho teorema.

corresponding angles (p. 128) In the figure, transversal t intersects lines ℓ and m. There are four pairs of corresponding angles: $\angle 5$ and $\angle 1$, $\angle 8$ and $\angle 4$, $\angle 6$ and $\angle 2$, and $\angle 7$ and $\angle 3$.

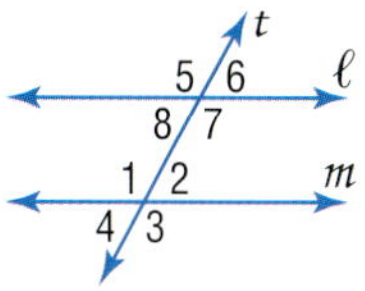

ángulos correspondientes En la figura, la transversal t interseca las rectas ℓ y m. La figura muestra cuatro pares de ángulos correspondientes: $\angle 5$ y $\angle 1$, $\angle 8$ y $\angle 4$, $\angle 6$ y $\angle 2$, y $\angle 7$ y $\angle 3$.

cosine (p. 364) For an acute angle of a right triangle, the ratio of the measure of the leg adjacent to the acute angle to the measure of the hypotenuse.

coseno Para un ángulo agudo de un triángulo rectángulo, la razón entre la medida del cateto adyacente al ángulo agudo y la medida de la hipotenusa de un triángulo rectángulo.

counterexample (p. 63) An example used to show that a given statement is not always true.

contraejemplo Ejemplo que se usa para demostrar que un enunciado dado no siempre es verdadero.

cross products (p. 283) In the proportion $\frac{a}{b} = \frac{c}{d}$, where $b \neq 0$ and $d \neq 0$, the cross products are ad and bc. The proportion is true if and only if the cross products are equal.

productos cruzados En la proporción, $\frac{a}{b} = \frac{c}{d}$, donde $b \neq 0$ y $d \neq 0$, los productos cruzados son ad y bc. La proporción es verdadera si y sólo si los productos cruzados son iguales.

cylinder (p. 638) A figure with bases that are formed by congruent circles in parallel planes.

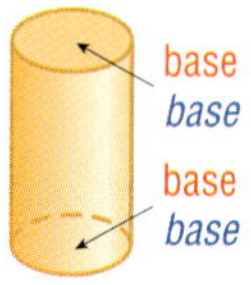

cilindro Figura cuyas bases son círculos congruentes localizados en planos paralelos.

D

deductive argument (p. 94) A proof formed by a group of algebraic steps used to solve a problem.

argumento deductivo Demostración que consta del conjunto de pasos algebraicos que se usan para resolver un problema.

deductive reasoning (p. 82) A system of reasoning that uses facts, rules, definitions, or properties to reach logical conclusions.

razonamiento deductivo Sistema de razonamiento que emplea hechos, reglas, definiciones y propiedades para obtener conclusiones lógicas.

degree (p. 29) A unit of measure used in measuring angles and arcs. An arc of a circle with a measure of 1° is $\frac{1}{360}$ of the entire circle.

grado Unidad de medida que se usa para medir ángulos y arcos. El arco de un círculo que mide 1° equivale a $\frac{1}{360}$ del círculo completo.

diagonal (p. 404) In a polygon, a segment that connects nonconsecutive vertices of the polygon.

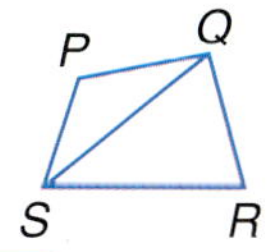

$\overline{SQ}$ is a diagonal.
$\overline{SQ}$ *es una diagonal.*

diagonal Recta que une vértices no consecutivos de un polígono.

diameter **1.** (p. 522) In a circle, a chord that passes through the center of the circle. **2.** (p. 671) In a sphere, a segment that contains the center of the sphere, and has endpoints that are on the sphere.

diámetro **1.** Cuerda que pasa por el centro de un círculo. **2.** Segmento que incluye el centro de una esfera y cuyos extremos se localizan en la esfera.

dilation (p. 490) A transformation determined by a center point C and a scale factor k. When $k > 0$, the image P' of P is the point on $\overrightarrow{CP}$ such that $CP' = |k| \cdot CP$. When $k < 0$, the image P' of P is the point on the ray opposite $\overrightarrow{CP}$ such that $CP' = k \cdot CP$.

dilatación Transformación determinada por un punto central C y un factor de escala k. Cuando $k > 0$, la imagen P' de P es el punto en $\overrightarrow{CP}$ tal que $CP' = |k| \cdot CP$. Cuando $k < 0$, la imagen P' de P es el punto en la semirrecta opuesta $\overrightarrow{CP}$ tal que $CP' = k \cdot CP$.

direct isometry (p. 481) An isometry in which the image of a figure is found by moving the figure intact within the plane.

isometría directa Isometría en la cual se obtiene la imagen de una figura, al mover la figura intacta junto con su plano.

direction (p. 498) The measure of the angle that a vector forms with the positive x-axis or any other horizontal line.

dirección Medida del ángulo que forma un vector con el eje positivo x o con cualquier otra recta horizontal.

disjunction (p. 68) A compound statement formed by joining two or more statements with the word *or*.

disyunción Enunciado compuesto que se forma al unir dos o más enunciados con la palabra *o*.

E

equal vectors (p. 499) Vectors that have the same magnitude and direction.

vectores iguales Vectores que poseen la misma magnitud y dirección.

equiangular triangle (p. 178) A triangle with all angles congruent.

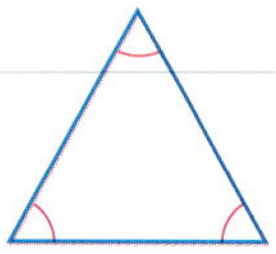

triángulo equiangular Triángulo cuyos ángulos son congruentes entre sí.

equilateral triangle (p. 179) A triangle with all sides congruent.

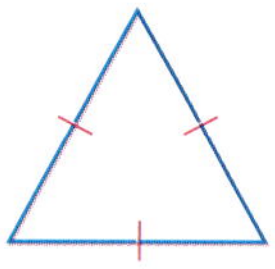

triángulo equilátero Triángulo cuyos lados son congruentes entre sí.

exterior (p. 29) A point is in the exterior of an angle if it is neither on the angle nor in the interior of the angle.

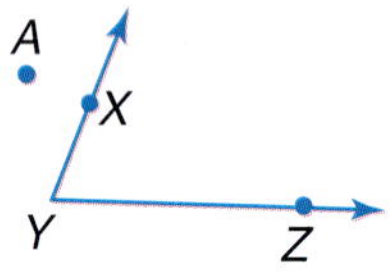

A is in the exterior of $\angle XYZ$.
A está en el exterior del $\angle XYZ$.

exterior Un punto yace en el exterior de un ángulo si no se localiza ni en el ángulo ni en el interior del ángulo.

exterior angle (p. 186) An angle formed by one side of a triangle and the extension of another side.

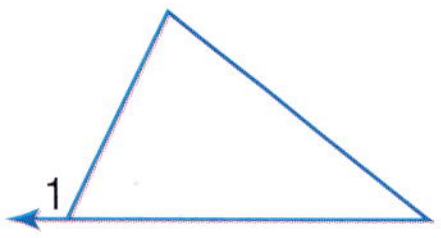

$\angle 1$ is an exterior angle.
$\angle 1$ *es un ángulo externo.*

ángulo externo Ángulo formado por un lado de un triángulo y la extensión de otro de sus lados.

extremes (p. 283) In $\frac{a}{b} = \frac{c}{d}$, the numbers a and d.

extremos Los números a y d en $\frac{a}{b} = \frac{c}{d}$.

F

flow proof (p. 187) A proof that organizes statements in logical order, starting with the given statements. Each statement is written in a box with the reason verifying the statement written below the box. Arrows are used to indicate the order of the statements.

demostración de flujo Demostración en que se ordenan los enunciados en orden lógico, empezando con los enunciados dados. Cada enunciado se escribe en una casilla y debajo de cada casilla se escribe el argumento que verifica el enunciado. El orden de los enunciados se indica mediante flechas.

fractal (p. 325) A figure generated by repeating a special sequence of steps infinitely often. Fractals often exhibit self-similarity.

fractal Figura que se obtiene mediante la repetición infinita de una sucesión particular de pasos. Los fractales a menudo exhiben autosemejanza.

G

geometric mean (p. 342) For any positive numbers a and b, the positive number x such that $\frac{a}{x} = \frac{x}{b}$.

media geométrica Para todo número positivo a y b, existe un número positivo x tal que $\frac{a}{x} = \frac{x}{b}$.

geometric probability (p. 622) Using the principles of length and area to find the probability of an event.

probabilidad geométrica El uso de los principios de longitud y área para calcular la probabilidad de un evento.

glide reflection (p. 475) A composition of a translation and a reflection in a line parallel to the direction of the translation.

reflexión de deslizamiento Composición que consta de una traslación y una reflexión realizadas sobre una recta paralela a la dirección de la traslación.

great circle (p. 671) For a given sphere, the intersection of the sphere and a plane that contains the center of the sphere.

círculo máximo La intersección entre una esfera dada y un plano que contiene el centro de la esfera.

H

height of a parallelogram (p. 595) The length of an altitude of a parallelogram.

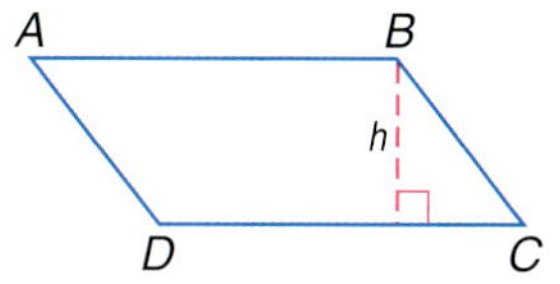

h is the height of parallelogram *ABCD*.
H es la altura del paralelogramo ABCD.

altura de un paralelogramo La longitud de la altura de un paralelogramo.

hemisphere (p. 672) One of the two congruent parts into which a great circle separates a sphere.

hemisferio Cada una de las dos partes congruentes en que un círculo máximo divide una esfera.

hypothesis (p. 75) In a conditional statement, the statement that immediately follows the word *if*.

hipótesis El enunciado escrito a continuación de la palabra *si* en un enunciado condicional.

I

if-then statement (p. 75) A compound statement of the form "if *A*, then *B*", where *A* and *B* are statements.

enunciado si-entonces Enunciado compuesto de la forma "si *A*, entonces *B*", donde *A* y *B* son enunciados.

incenter (p. 240) The point of concurrency of the angle bisectors of a triangle.

incentro Punto de intersección de las bisectrices interiores de un triángulo.

included angle (p. 201) In a triangle, the angle formed by two sides is the included angle for those two sides.

ángulo incluido En un triángulo, el ángulo formado por dos lados cualesquiera del triángulo es el ángulo incluido de esos dos lados.

included side (p. 207) The side of a triangle that is a side of each of two angles.

lado incluido El lado de un triángulo que es común a de sus dos ángulos.

indirect isometry (p. 481) An isometry that cannot be performed by maintaining the orientation of the points, as in a direct isometry.

isometría indirecta Tipo de isometría que no se puede obtener manteniendo la orientación de los puntos, como ocurre durante la isometría directa.

Glossary/Glosario

indirect proof (p. 255) In an indirect proof, one assumes that the statement to be proved is false. One then uses logical reasoning to deduce that a statement contradicts a postulate, theorem, or one of the assumptions. Once a contradiction is obtained, one concludes that the statement assumed false must in fact be true.

demostración indirecta En una demostración indirecta, se asume que el enunciado por demostrar es falso. Después, se deduce lógicamente que existe un enunciado que contradice un postulado, un teorema o una de las conjeturas. Una vez hallada una contradicción, se concluye que el enunciado que se suponía falso debe ser, en realidad, verdadero.

indirect reasoning (p. 255) Reasoning that assumes that the conclusion is false and then shows that this assumption leads to a contradiction of the hypothesis or some other accepted fact, like a postulate, theorem, or corollary. Then, since the assumption has been proved false, the conclusion must be true.

razonamiento indirecto Razonamiento en que primero se asume que la conclusión es falsa y, después, se demuestra que esto contradice la hipótesis o un hecho aceptado como un postulado, un teorema o un corolario. Finalmente, dado que se ha demostrado que la conjetura es falsa, entonces la conclusión debe ser verdadera.

inductive reasoning (p. 62) Reasoning that uses a number of specific examples to arrive at a plausible generalization or prediction. Conclusions arrived at by inductive reasoning lack the logical certainty of those arrived at by deductive reasoning.

razonamiento inductivo Razonamiento que usa varios ejemplos específicos para lograr una generalización o una predicción creíble. Las conclusiones obtenidas mediante el razonamiento inductivo carecen de la certidumbre lógica de aquellas obtenidas mediante el razonamiento deductivo.

inscribed (p. 537) A polygon is inscribed in a circle if each of its vertices lie on the circle.

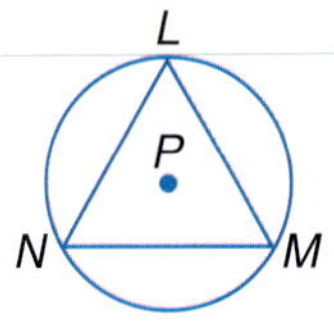

$\triangle LMN$ is inscribed in $\odot P$.
$\triangle LMN$ *está inscrito en* $\odot P$.

inscrito Un polígono está inscrito en un círculo si todos sus vértices yacen en el círculo.

intercepted (p. 544) An angle intercepts an arc if and only if each of the following conditions are met.

1. The endpoints of the arc lie on the angle.
2. All points of the arc except the endpoints are in the interior of the circle.
3. Each side of the angle contains an endpoint of the arc.

intersecado Un ángulo interseca un arco si y sólo si se cumplen todas las siguientes condiciones.

1. Los extremos del arco yacen en el ángulo.
2. Todos los puntos del arco, exceptuando sus extremos, yacen en el interior del círculo.
3. Cada lado del ángulo contiene un extremo del arco.

interior (p. 29) A point is in the interior of an angle if it does not lie on the angle itself and it lies on a segment with endpoints that are on the sides of the angle.

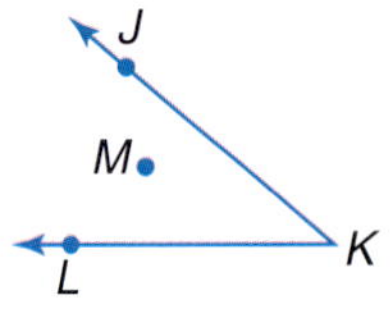

M is in the interior of $\angle JKL$.
M está en el interior del $\angle JKL$.

interior Un punto se localiza en el interior de un ángulo, si no yace en el ángulo mismo y si está en un segmento cuyos extremos yacen en los lados del ángulo.

inverse (p. 77) The statement formed by negating both the hypothesis and conclusion of a conditional statement.

inversa Enunciado que se obtiene al negar la hipótesis y la conclusión de un enunciado condicional.

irregular figure (p. 617) A figure that cannot be classified as a single polygon.

figura irregular Figura que no se puede clasificar como un solo polígono.

irregular polygon (p. 618) A polygon that is not regular.

polígono irregular Polígono que no es regular.

isometry (p. 463) A mapping for which the original figure and its image are congruent.

isometría Transformación en que la figura original y su imagen son congruentes.

isosceles trapezoid (p. 439) A trapezoid in which the legs are congruent, both pairs of base angles are congruent, and the diagonals are congruent.

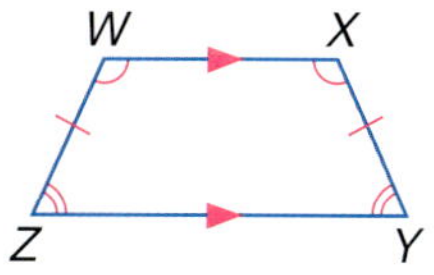

trapecio isósceles Trapecio cuyos catetos son congruentes, ambos pares de ángulos son congruentes y las diagonales son congruentes.

isosceles triangle (p. 179) A triangle with at least two sides congruent. The congruent sides are called *legs*. The angles opposite the legs are *base angles*. The angle formed by the two legs is the *vertex angle*. The side opposite the vertex angle is the *base*.

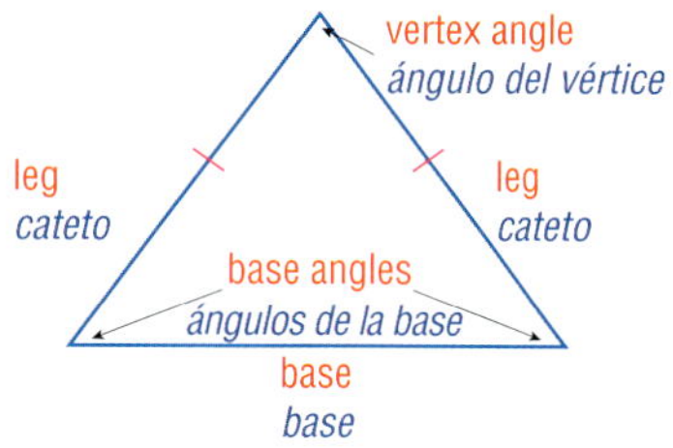

triángulo isósceles Triángulo que tiene por lo menos dos lados congruentes. Los lados congruentes se llaman *catetos*. Los ángulos opuestos a los catetos son los *ángulos de la base*. El ángulo formado por los dos catetos es el *ángulo del vértice*. Los lados opuestos al ángulo del vértice forman la *base*.

iteration (p. 325) A process of repeating the same procedure over and over again.

iteración Proceso de repetir el mismo procedimiento una y otra vez.

K

kite (p. 438) A quadrilateral with exactly two distinct pairs of adjacent congruent sides.

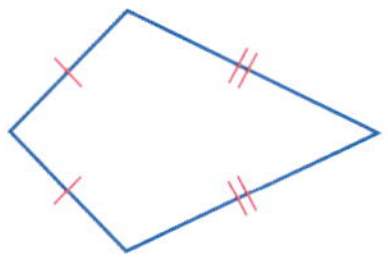

cometa Cuadrilátero que tiene exactamente dospares de lados congruentes adyacentes distintivos.

L

lateral area (p. 649) For prisms, pyramids, cylinders, and cones, the area of the figure, not including the bases.

área lateral En prismas, pirámides, cilindros y conos, es el área de la figura, sin incluir el área de las bases.

lateral edges **1.** (p. 649) In a prism, the intersection of two adjacent lateral faces. **2.** (p. 660) In a pyramid, lateral edges are the edges of the lateral faces that join the vertex to vertices of the base.

aristas laterales **1.** En un prisma, la intersección de dos caras laterales adyacentes. **2.** En una pirámide, las aristas de las caras laterales que unen el vértice de la pirámide con los vértices de la base.

lateral faces **1.** (p. 649) In a prism, the faces that are not bases. **2.** (p. 660) In a pyramid, faces that intersect at the vertex.

caras laterales **1.** En un prisma, las caras que no forman las bases. **2.** En una pirámide, las caras que se intersecan en el vértice.

Law of Cosines (p. 385) Let $\triangle ABC$ be any triangle with a, b, and c representing the measures of sides opposite the angles with measures A, B, and C respectively. Then the following equations are true.

$a^2 = b^2 + c^2 - 2bc \cos A$
$b^2 = a^2 + c^2 - 2ac \cos B$
$c^2 = a^2 + b^2 - 2ab \cos C$

ley de los cosenos Sea $\triangle ABC$ cualquier triángulo donde a, b y c son las medidas de los lados opuestos a los ángulos que miden A, B y C respectivamente. Entonces las siguientes ecuaciones son ciertas.

$a^2 = b^2 + c^2 - 2bc \cos A$
$b^2 = a^2 + c^2 - 2ac \cos B$
$c^2 = a^2 + b^2 - 2ab \cos C$

Law of Detachment (p. 82) If $p \rightarrow q$ is a true conditional and p is true, then q is also true.

ley de indiferencia Si $p \rightarrow q$ es un enunciado condicional verdadero y p es verdadero, entonces q es verdadero también.

Law of Sines (p. 377) Let $\triangle ABC$ be any triangle with a, b, and c representing the measures of sides opposite the angles with measures A, B, and C respectively. Then, $\frac{\sin A}{a} = \frac{\sin B}{b} = \frac{\sin C}{c}$.

ley de los senos Sea $\triangle ABC$ cualquier triángulo donde a, b y c representan las medidas de los lados opuestos a los ángulos A, B y C respectivamente. Entonces, $\frac{\sin A}{a} = \frac{\sin B}{b} = \frac{\sin C}{c}$.

Law of Syllogism (p. 83) If $p \rightarrow q$ and $q \rightarrow r$ are true conditionals, then $p \rightarrow r$ is also true.

ley del silogismo Si $p \rightarrow q$ y $q \rightarrow r$ son enunciados condicionales verdaderos, entonces $p \rightarrow r$ también es verdadero.

line (p. 6) A basic undefined term of geometry. A line is made up of points and has no thickness or width. In a figure, a line is shown with an arrowhead at each end. Lines are usually named by lowercase script letters or by writing capital letters for two points on the line, with a double arrow over the pair of letters.

recta Término primitivo en geometría. Una recta está formada por puntos y carece de grosor o ancho. En una figura, una recta se representa con una flecha en cada extremo. Por lo general, se designan con letras minúsculas o con las dos letras mayúsculas de dos puntos sobre la línea. Se escribe una flecha doble sobre el par de letras mayúsculas.

line of reflection (p. 463) A line through a figure that separates the figure into two mirror images.

línea de reflexión Línea que divide una figura en dos imágenes especulares.

line of symmetry (p. 466) A line that can be drawn through a plane figure so that the figure on one side is the reflection image of the figure on the opposite side.

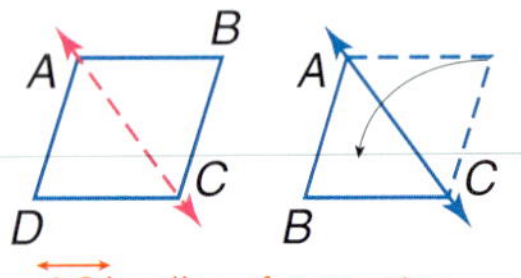

$\overleftrightarrow{AC}$ is a line of symmetry.
$\overleftrightarrow{AC}$ es un eje de simetría.

eje de simetría Recta que se traza a través de una figura plana, de modo que un lado de la figura es la imagen reflejada del lado opuesto.

line segment (p. 13) A measurable part of a line that consists of two points, called endpoints, and all of the points between them.

segmento de recta Sección medible de una recta. Consta de dos puntos, llamados extremos, y todos los puntos localizados entre ellos.

linear pair (p. 37) A pair of adjacent angles whose non-common sides are opposite rays.

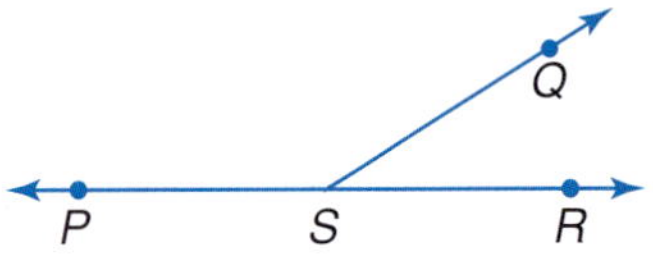

$\angle PSQ$ and $\angle QSR$ are a linear pair.
$\angle PSQ$ y $\angle QSR$ forman un par lineal.

par lineal Par de ángulos adyacentes cuyos lados no comunes forman semirrectas opuestas.

locus (p. 11) The set of points that satisfy a given condition.

lugar geométrico Conjunto de puntos que satisfacen una condición dada.

logically equivalent (p. 77) Statements that have the same truth values.

equivalente lógico Enunciados que poseen el mismo valor de verdad.

M

magnitude (p. 498) The length of a vector.

magnitud La longitud de un vector.

major arc (p. 530) An arc with a measure greater than 180.

$\overset{\frown}{ACB}$ is a major arc.

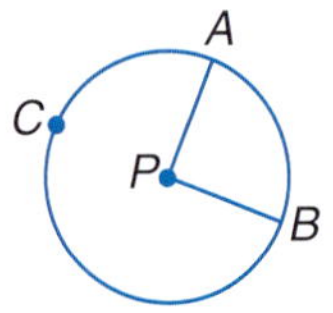

arco mayor Arco que mide más de 180°.

$\overset{\frown}{ACB}$ es un arco mayor.

matrix logic (p. 88) A method of deductive reasoning that uses a table to solve problems.

lógica matricial Método de razonamiento deductivo que utiliza una tabla para resolver problemas.

means (p. 283) In $\frac{a}{b} = \frac{c}{d}$, the numbers b and c.

medios Los números b y c en la proporción $\frac{a}{b} = \frac{c}{d}$.

median **1.** (p. 240) In a triangle, a line segment with endpoints that are a vertex of a triangle and the midpoint of the side opposite the vertex. **2.** (p. 440) In a trapezoid, the segment that joins the midpoints of the legs.

mediana **1.** Segmento de recta de un triángulo cuyos extremos son un vértice del triángulo y el punto medio del lado opuesto a dicho vértice. **2.** Segmento que une los puntos medios de los catetos de un trapecio.

midpoint (p. 22) The point halfway between the endpoints of a segment.

punto medio Punto que es equidistante entre los extremos de un segmento.

midsegment (p. 308) A segment with endpoints that are the midpoints of two sides of a triangle.

segmento medio Segmento cuyos extremos son los puntos medios de dos lados de un triángulo.

minor arc (p. 530) An arc with a measure less than 180. $\overset{\frown}{AB}$ is a minor arc.

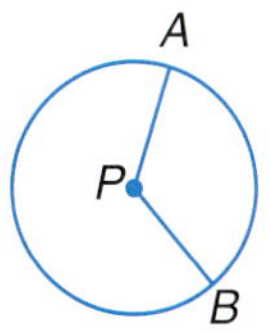

arco menor Arco que mide menos de 180°. $\overset{\frown}{AB}$ es un arco menor.

N

negation (p. 67) If a statement is represented by p, then *not* p is the negation of the statement.

negación Si p representa un enunciado, entonces *no* p representa la negación del enunciado.

net (p. 644) A two-dimensional figure that when folded forms the surfaces of a three-dimensional object.

red Figura bidimensional que al ser plegada forma las superficies de un objeto tridimensional.

***n*-gon** (p. 46) A polygon with n sides.

***en*ágono** Polígono con n lados.

non-Euclidean geometry (p. 165) The study of geometrical systems that are not in accordance with the Parallel Postulate of Euclidean geometry.

geometría no euclidiana El estudio de sistemas geométricos que no satisfacen el Postulado de las Paralelas de la geometría euclidiana.

O

oblique cone (p. 666) A cone that is not a right cone.

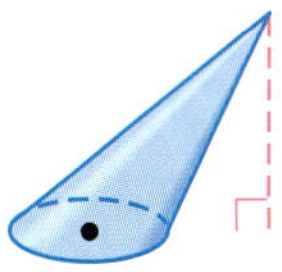

cono oblicuo Cono que no es un cono recto.

oblique cylinder (p. 655) A cylinder that is not a right cylinder.

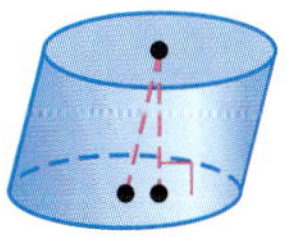

cilindro oblicuo Cilindro que no es un cilindro recto.

oblique prism (p. 649) A prism in which the lateral edges are not perpendicular to the bases.

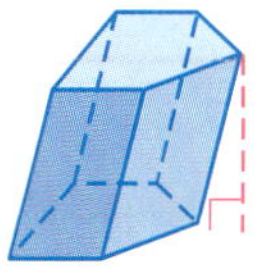

prisma oblicuo Prisma cuyas aristas laterales no son perpendiculares a las bases.

obtuse angle (p. 30) An angle with degree measure greater than 90 and less than 180.

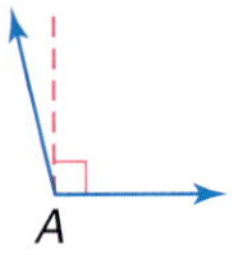

$90 < m\angle A < 180$

ángulo obtuso Ángulo que mide más de 90° y menos de 180°.

obtuse triangle (p. 178) A triangle with an obtuse angle.

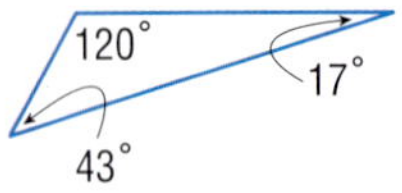

one obtuse angle
un ángulo obtuso

triángulo obtusángulo Triángulo que tiene un ángulo obtuso.

opposite rays (p. 29) Two rays $\overrightarrow{BA}$ and $\overrightarrow{BC}$ such that B is between A and C.

semirrectas opuestas Dos semirrectas $\overrightarrow{BA}$ y $\overrightarrow{BC}$ tales que B se localiza entre A y C.

ordered triple (p. 714) Three numbers given in a specific order used to locate points in space.

triple ordenado Tres números dados en un orden específico que sirven para ubicar puntos en el espacio.

orthocenter (p. 240) The point of concurrency of the altitudes of a triangle.

ortocentro Punto de intersección de las alturas de un triángulo.

orthogonal drawing (p. 636) The two-dimensional top view, left view, front view, and right view of a three-dimensional object.

vista ortogonal Vista bidimensional desde arriba, desde la izquierda, desde el frente o desde la derecha de un cuerpo tridimensional.

P

paragraph proof (p. 90) An informal proof written in the form of a paragraph that explains why a conjecture for a given situation is true.

demostración de párrafo Demostración informal escrita en forma de párrafo que explica por qué una conjetura acerca de una situación dada es verdadera.

parallel lines (p. 126) Coplanar lines that do not intersect.

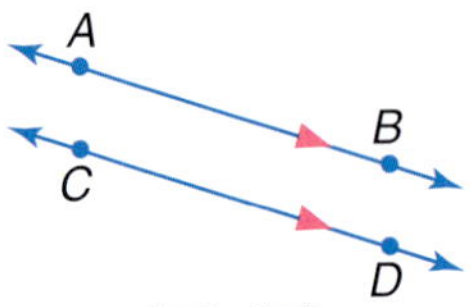

$\overleftrightarrow{AB} \parallel \overleftrightarrow{CD}$

rectas paralelas Rectas coplanares que no se intersecan.

parallel planes (p. 126) Planes that do not intersect.

planos paralelos Planos que no se intersecan.

parallel vectors (p. 499) Vectors that have the same or opposite direction.

vectores paralelos Vectores que tienen la misma dirección o la dirección opuesta.

parallelogram (p. 411) A quadrilateral with parallel opposite sides. Any side of a parallelogram may be called a *base*.

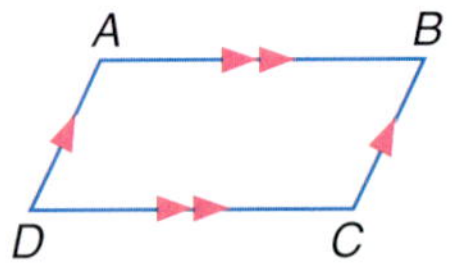

$\overline{AB} \parallel \overline{DC}$; $\overline{AD} \parallel \overline{BC}$

paralelogramo Cuadrilátero cuyos lados opuestos son paralelos entre sí. Cualquier lado del paralelogramo puede ser la *base*.

perimeter (p. 46) The sum of the lengths of the sides of a polygon.

perímetro La suma de la longitud de los lados de un polígono.

perpendicular bisector (p. 238) In a triangle, a line, segment, or ray that passes through the midpoint of a side and is perpendicular to that side.

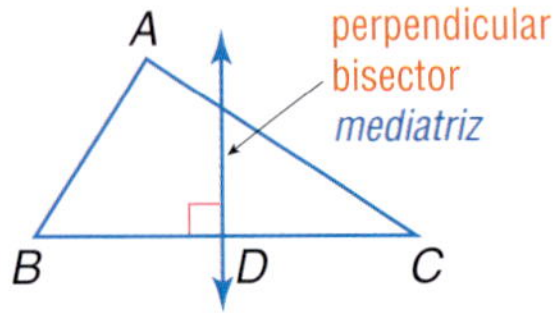

D is the midpoint of $\overline{BC}$.
D es el punto medio de $\overline{BC}$.

mediatriz Recta, segmento o semirrecta que atraviesa el punto medio del lado de un triángulo y que es perpendicular a dicho lado.

perpendicular lines (p. 40) Lines that form right angles.

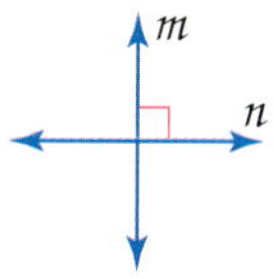

line $m \perp$ line n
recta $m \perp$ recta n

rectas perpendiculares Rectas que forman ángulos rectos.

perspective view (p. 636) The view of a three-dimensional figure from the corner.

vista de perspectiva Vista de una figura tridimensional desde una de sus esquinas.

pi (π) (p. 524) An irrational number represented by the ratio of the circumference of a circle to the diameter of the circle.

pi (π) Número irracional representado por la razón entre la circunferencia de un círculo y su diámetro.

plane (p. 6) A basic undefined term of geometry. A plane is a flat surface made up of points that has no depth and extends indefinitely in all directions. In a figure, a plane is often represented by a shaded, slanted 4-sided figure. Planes are usually named by a capital script letter or by three noncollinear points on the plane.

plano Término primitivo en geometría. Es una superficie formada por puntos y sin profundidad que se extiende indefinidamente en todas direcciones. Los planos a menudo se representan con un cuadrilátero inclinado y sombreado. Los planos en general se designan con una letra mayúscula o con tres puntos no colineales del plano.

plane Euclidean geometry (p. 165) Geometry based on Euclid's axioms dealing with a system of points, lines, and planes.

geometría del plano euclidiano Geometría basada en los axiomas de Euclides, los que integran un sistema de puntos, rectas y planos.

Platonic Solids (p. 637) The five regular polyhedra: tetrahedron, hexahedron, octahedron, dodecahedron, or icosahedron.

sólidos platónicos Cualquiera de los siguientes cinco poliedros regulares: tetraedro, hexaedro, octaedro, dodecaedro e icosaedro.

point (p. 6) A basic undefined term of geometry. A point is a location. In a figure, points are represented by a dot. Points are named by capital letters.

punto Término primitivo en geometría. Un punto representa un lugar o localización. En una figura, se representa con una marca puntual. Los puntos se designan con letras mayúsculas.

point of concurrency (p. 238) The point of intersection of concurrent lines.

punto de concurrencia Punto de intersección de rectas concurrentes.

point of symmetry (p. 466) The common point of reflection for all points of a figure.

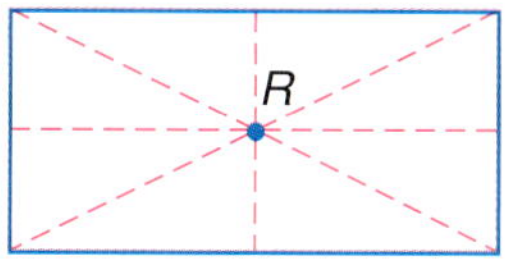

R is a point of symmetry.
R es un punto de simetría.

punto de simetría El punto común de reflexión de todos los puntos de una figura.

point of tangency (p. 552) For a line that intersects a circle in only one point, the point at which they intersect.

punto de tangencia Punto de intersección de una recta que interseca un círculo en un solo punto, el punto en donde se intersecan.

point-slope form (p. 145) An equation of the form $y - y_1 = m(x - x_1)$, where (x_1, y_1) are the coordinates of any point on the line and m is the slope of the line.

forma punto-pendiente Ecuación de la forma $y - y_1 = m(x - x_1)$, donde (x_1, y_1) representan las coordenadas de un punto cualquiera sobre la recta y m representa la pendiente de la recta.

Glossary/Glosario

polygon (p. 45) A closed figure formed by a finite number of coplanar segments called *sides* such that the following conditions are met.

1. The sides that have a common endpoint are noncollinear.
2. Each side intersects exactly two other sides, but only at their endpoints, called the *vertices*.

polígono Figura cerrada formada por un número finito de segmentos coplanares llamados *lados*, y que satisface las siguientes condiciones:

1. Los lados que tienen un extremo común son no colineales.
2. Cada lado interseca exactamente dos lados, pero sólo en sus extremos, formando los *vértices*.

polyhedrons (p. 637) Closed three-dimensional figures made up of flat polygonal regions. The flat regions formed by the polygons and their interiors are called *faces*. Pairs of faces intersect in segments called *edges*. Points where three or more edges intersect are called *vertices*.

poliedro Figura tridimensional cerrada formada por regiones poligonales planas. Las regiones planas definidas por un polígono y sus interiores se llaman *caras*. Cada intersección entre dos caras se llama *arista*. Los puntos donde se intersecan tres o más aristas se llaman *vértices*.

postulate (p. 89) A statement that describes a fundamental relationship between the basic terms of geometry. Postulates are accepted as true without proof.

postulado Enunciado que describe una relación fundamental entre los términos primitivos de geometría. Los postulados se aceptan como verdaderos sin necesidad de demostración.

precision (p. 14) The precision of any measurement depends on the smallest unit available on the measuring tool.

precisión La precisión de una medida depende de la unidad de medida más pequeña del instrumento de medición.

prism (p. 637) A solid with the following characteristics.

1. Two faces, called *bases*, are formed by congruent polygons that lie in parallel planes.
2. The faces that are not bases, called *lateral faces*, are formed by parallelograms.
3. The intersections of two adjacent lateral faces are called *lateral edges* and are parallel segments.

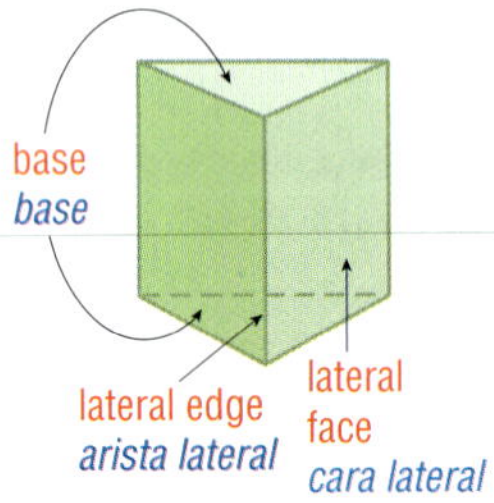

triangular prism
prisma triangular

prisma Sólido que posee las siguientes características:

1. Tiene dos caras llamadas *bases*, formadas por polígonos congruentes que yacen en planos paralelos.
2. Las caras que no son las bases, llamadas *caras laterales*, son formadas por paralelogramos.
3. Las intersecciones de dos aristas laterales adyacentes se llaman *aristas laterales* y son segmentos paralelos.

proof (p. 90) A logical argument in which each statement you make is supported by a statement that is accepted as true.

demostración Argumento lógico en que cada enunciado está basado en un enunciado que se acepta como verdadero.

proof by contradiction (p. 255) An indirect proof in which one assumes that the statement to be proved is false. One then uses logical reasoning to deduce a statement that contradicts a postulate, theorem, or one of the assumptions. Once a contradiction is obtained, one concludes that the statement assumed false must in fact be true.

demostración por contradicción Demostración indirecta en que se asume que el enunciado que se va a demostrar es falso. Después, se razona lógicamente para deducir un enunciado que contradiga un postulado, un teorema o una de las conjeturas. Una vez que se obtiene una contradicción, se concluye que el enunciado que se supuso falso es, en realidad, verdadero.

proportion (p. 283) An equation of the form $\frac{a}{b} = \frac{c}{d}$ that states that two ratios are equal.

proporción Ecuación de la forma $\frac{a}{b} = \frac{c}{d}$ que establece que dos razones son iguales.

pyramid (p. 637) A solid with the following characteristics.

1. All of the faces, except one face, intersect at a point called the *vertex*.
2. The face that does not contain the vertex is called the *base* and is a polygonal region.
3. The faces meeting at the vertex are called *lateral faces* and are triangular regions.

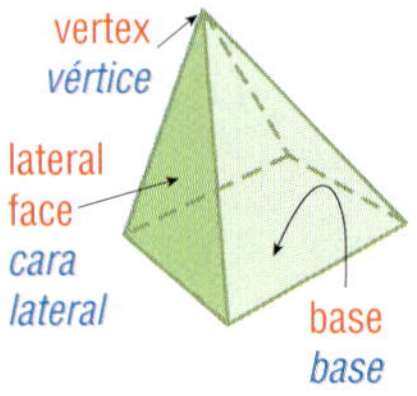

rectangular pyramid
pirámide rectangular

pirámide Sólido con las siguientes características:

1. Todas, excepto una de las caras, se intersecan en un punto llamado *vértice*.
2. La cara que no contiene el vértice se llama *base* y es una región poligonal.
3. Las caras que se encuentran en los vértices se llaman *caras laterales* y son regiones triangulares.

Glossary/Glosario

Pythagorean identity (p. 391) The identity $\cos^2\theta + \sin^2\theta = 1$.

identidad pitagórica La identidad $\cos^2\theta + \sin^2\theta = 1$.

Pythagorean triple (p. 352) A group of three whole numbers that satisfies the equation $a^2 + b^2 = c^2$, where c is the greatest number.

triplete de Pitágoras Grupo de tres números enteros que satisfacen la ecuación $a^2 + b^2 = c^2$, donde c es el número más grande.

R

radius 1. (p. 522) In a circle, any segment with endpoints that are the center of the circle and a point on the circle. 2. (p. 671) In a sphere, any segment with endpoints that are the center and a point on the sphere.

radio 1. Cualquier segmento cuyos extremos están en el centro de un círculo y en un punto cualquiera del mismo. 2. Cualquier segmento cuyos extremos forman el centro y en punto de una esfera.

rate of change (p. 140) Describes how a quantity is changing over time.

tasa de cambio Describe cómo cambia una cantidad a través del tiempo.

ratio (p. 282) A comparison of two quantities.

razón Comparación entre dos cantidades.

ray (p. 29) $\overrightarrow{PQ}$ is a ray if it is the set of points consisting of $\overline{PQ}$ and all points S for which Q is between P and S.

semirrecta $\overrightarrow{PQ}$ es una semirrecta si consta del conjunto de puntos formado por $\overline{PQ}$ y todos los S puntos S para los que Q se localiza entre P y S.

reciprocal identity (p. 391) Each of the three trigonometric ratios called *cosecant*, *secant*, and *cotangent*, that are the reciprocals of sine, cosine, and tangent, respectively.

identidad recíproca Cada una de las tres razones trigonométricas llamadas *cosecante*, *secante* y *tangente* y que son los recíprocos del seno, el coseno y la tangente, respectivamente

rectangle (p. 424) A quadrilateral with four right angles.

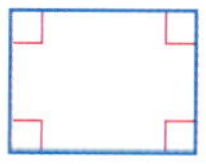

rectángulo Cuadrilátero que tiene cuatro ángulos rectos.

reflection (p. 463) A transformation representing a flip of the figure over a point, line, or plane.

reflexión Transformación que se obtiene cuando se "voltea" una imagen sobre un punto, una línea o un plano.

reflection matrix (p. 507) A matrix that can be multiplied by the vertex matrix of a figure to find the coordinates of the reflected image.

matriz de reflexión Matriz que al ser multiplicada por la matriz de vértices de una figura permite hallar las coordenadas de la imagen reflejada.

regular polygon (p. 46) A convex polygon in which all of the sides are congruent and all of the angles are congruent.

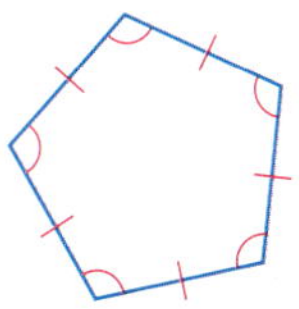

regular pentagon
pentágono regular

polígono regular Polígono convexo en el que todos los lados y todos los ángulos son congruentes entre sí.

regular polyhedron (p. 637) A polyhedron in which all of the faces are regular congruent polygons.

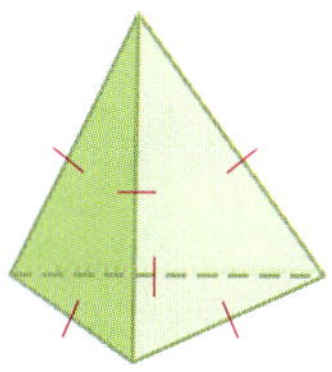

poliedro regular Poliedro cuyas caras son polígonos regulares congruentes.

regular prism (p. 637) A right prism with bases that are regular polygons.

prisma regular Prisma recto cuyas bases son polígonos regulares.

regular tessellation (p. 484) A tessellation formed by only one type of regular polygon.

teselado regular Teselado formado por un solo tipo de polígono regular.

related conditionals (p. 77) Statements such as the converse, inverse, and contrapositive that are based on a given conditional statement.

enunciados condicionales relacionados Enunciados tales como el recíproco, la inversa y la antítesis que están basados en un enunciado condicional dado.

relative error (p. 19) The ratio of the half-unit difference in precision to the entire measure, expressed as a percent.

error relativo La razón entre la mitad de la unidad más precisa de la medición y la medición completa, expresada en forma de porcentaje.

remote interior angles (p. 186) The angles of a triangle that are not adjacent to a given exterior angle.

ángulos internos no adyacentes Ángulos de un triángulo que no son adyacentes a un ángulo exterior dado.

resultant (p. 500) The sum of two vectors.

resultante La suma de dos vectores.

rhombus (p. 431) A quadrilateral with all four sides congruent.

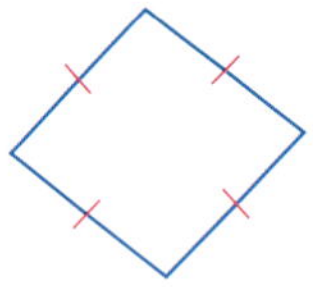

rombo Cuadrilátero cuyos cuatro lados son congruentes.

right angle (p. 30) An angle with a degree measure of 90.

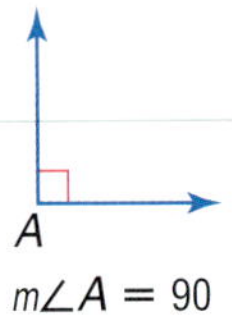

ángulo recto Ángulo cuya medida en grados es 90.

right cone (p. 666) A cone with an axis that is also an altitude.

cono recto Cono cuyo eje es también su altura.

right cylinder (p. 655) A cylinder with an axis that is also an altitude.

cilindro recto Cilindro cuyo eje es también su altura.

right prism (p. 649) A prism with lateral edges that are also altitudes.

prisma recto Prisma cuyas aristas laterales también son su altura.

right triangle (p. 178) A triangle with a right angle. The side opposite the right angle is called the *hypotenuse*. The other two sides are called legs.

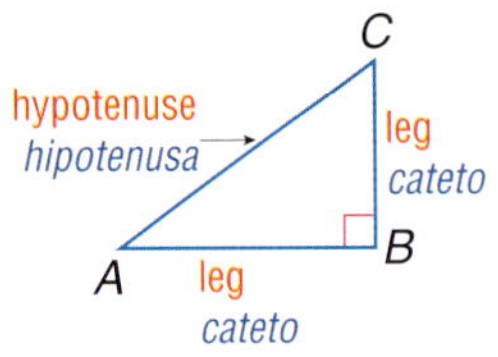

triángulo rectángulo Triángulo con un ángulo recto. El lado opuesto al ángulo recto se conoce como *hipotenusa*. Los otros dos lados se llaman catetos.

rotation (p. 476) A transformation that turns every point of a preimage through a specified angle and direction about a fixed point, called the *center of rotation*.

rotación Transformación en que se hace girar cada punto de la preimagen a través de un ángulo y una dirección determinadas alrededor de un punto, conocido como *centro de rotación*.

rotation matrix (p. 507) A matrix that can be multiplied by the vertex matrix of a figure to find the coordinates of the rotated image.

matriz de rotación Matriz que al ser multiplicada por la matriz de vértices de la figura permite calcular las coordenadas de la imagen rotada.

rotational symmetry (p. 478) If a figure can be rotated less than 360° about a point so that the image and the preimage are indistinguishable, the figure has rotational symmetry.

simetría de rotación Si se puede rotar una imagen menos de 360° alrededor de un punto y la imagen y la preimagen son idénticas, entonces la figura presenta simetría de rotación.

S

scalar (p. 501) A constant multiplied by a vector.

escalar Una constante multiplicada por un vector.

scalar multiplication (p. 501) Multiplication of a vector by a scalar.

multiplicación escalar Multiplicación de un vector por una escalar.

scale factor (p. 290) The ratio of the lengths of two corresponding sides of two similar polygons or two similar solids.

factor de escala La razón entre las longitudes de dos lados correspondientes de dos polígonos o sólidos semejantes.

scalene triangle (p. 179) A triangle with no two sides congruent.

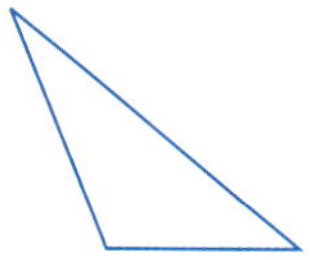

triángulo escaleno Triángulo cuyos lados no son congruentes.

secant (p. 561) Any line that intersects a circle in exactly two points.

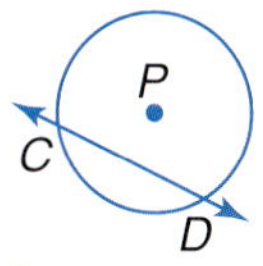

$\overleftrightarrow{CD}$ is a secant of $\odot P$.
$\overleftrightarrow{CD}$ es una secante de $\odot P$.

secante Cualquier recta que interseca un círculo exactamente en dos puntos.

sector of a circle (p. 623) A region of a circle bounded by a central angle and its intercepted arc.

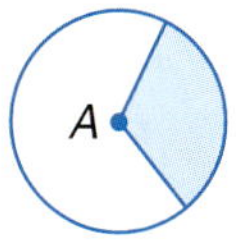

The shaded region is a sector of $\odot A$.
La región sombreada es un sector de $\odot A$.

sector de un círculo Región de un círculo que está limitada por un ángulo central y el arco que interseca.

segment (p. 13) *See* line segment.

segmento *Ver* segmento de recta.

segment bisector (p. 24) A segment, line, or plane that intersects a segment at its midpoint.

bisectriz de segmento Segmento, recta o plano que interseca un segmento en su punto medio.

segment of a circle (p. 624) The region of a circle bounded by an arc and a chord.

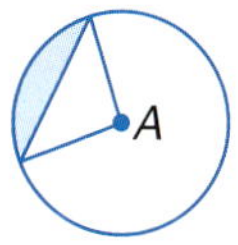

The shaded region is a segment of $\odot A$.
La región sombreada es un segmento de $\odot A$.

segmento de un círculo Región de un círculo limitada por un arco y una cuerda.

self-similar (p. 325) If any parts of a fractal image are replicas of the entire image, the image is self-similar.

autosemejante Si cualquier parte de una imagen fractal es una réplica de la imagen completa, entonces la imagen es autosemejante.

semicircle (p. 530) An arc that measures 180.

semicírculo Arco que mide 180°.

semi-regular tessellation (p. 484) A uniform tessellation formed using two or more regular polygons.

teselado semirregular Teselado uniforme compuesto por dos o más polígonos regulares.

similar polygons (p. 289) Two polygons are similar if and only if their corresponding angles are congruent and the measures of their corresponding sides are proportional.

polígonos semejantes Dos polígonos son semejantes si y sólo si sus ángulos correspondientes son congruentes y las medidas de sus lados correspondientes son proporcionales.

similar solids (p. 707) Solids that have exactly the same shape, but not necessarily the same size.

sólidos semejantes Sólidos que tienen exactamente la misma forma, pero no necesariamente el mismo tamaño.

similarity transformation (p. 491) When a figure and its transformation image are similar.

transformación de semejanza Aquélla en que la figura y su imagen transformada son semejantes.

sine (p. 364) For an acute angle of a right triangle, the ratio of the measure of the leg opposite the acute angle to the measure of the hypotenuse.

seno Es la razón entre la medida del cateto opuesto al ángulo agudo y la medida de la hipotenusa de un triángulo rectángulo.

skew lines (p. 127) Lines that do not intersect and are not coplanar.

rectas alabeadas Rectas que no se intersecan y que no son coplanares.

slope (p. 139) For a (nonvertical) line containing two points (x_1, y_1) and (x_2, y_2), the number m given by the formula $m = \frac{y_2 - y_1}{x_2 - x_1}$ where $x_2 \neq x_1$.

pendiente Para una recta (no vertical) que contiene dos puntos (x_1, y_1) y (x_2, y_2), el número m dado por la fórmula $m = \frac{y_2 - y_1}{x_2 - x_1}$ donde $x_2 \neq x_1$.

slope-intercept form (p. 145) A linear equation of the form $y = mx + b$. The graph of such an equation has slope m and y-intercept b.

forma pendiente-intersección Ecuación lineal de la forma $y = mx + b$. En la gráfica de tal ecuación, la pendiente es m y la intersección y es b.

solving a triangle (p. 378) Finding the measures of all of the angles and sides of a triangle.

resolver un triángulo Calcular las medidas de todos los ángulos y todos los lados de un triángulo.

space (p. 8) A boundless three-dimensional set of all points.

espacio Conjunto tridimensional no acotado de todos los puntos.

sphere (p. 638) In space, the set of all points that are a given distance from a given point, called the *center*.

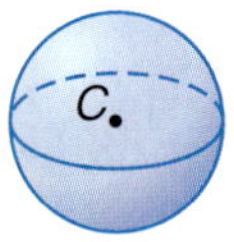

C is the center of the sphere.
C es el centro de la esfera.

esfera El conjunto de todos los puntos en el espacio que se encuentran a cierta distancia de un punto dado llamado *centro*.

spherical geometry (p. 165) The branch of geometry that deals with a system of points, greatcircles (lines), and spheres (planes).

geometría esférica Rama de la geometría que estudia los sistemas de puntos, círculos máximos (rectas) y esferas (planos).

square (p. 432) A quadrilateral with four right angles and four congruent sides.

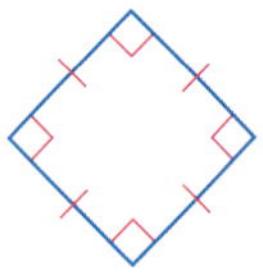

cuadrado Cuadrilátero con cuatro ángulos rectos y cuatro lados congruentes.

standard position (p. 498) When the initial point of a vector is at the origin.

posición estándar Ocurre cuando la posición inicial de un vector es el origen.

statement (p. 67) Any sentence that is either true or false, but not both.

enunciado Una oración que puede ser falsa o verdadera, pero no ambas.

strictly self-similar (p. 325) A figure is strictly self-similar if any of its parts, no matter where they are located or what size is selected, contain the same figure as the whole.

estrictamente autosemejante Una figura es estrictamente autosemejante si cualquiera de sus partes, sin importar su localización o su tamaño, contiene la figura completa.

supplementary angles (p. 39) Two angles with measures that have a sum of 180.

ángulos suplementarios Dos ángulos cuya suma es igual a 180°.

surface area (p. 644) The sum of the areas of all faces and side surfaces of a three-dimensional figure.

área de superficie La suma de las áreas de todas las caras y superficies laterales de una figura tridimensional.

T

tangent **1.** (p. 364) For an acute angle of a right triangle, the ratio of the measure of the leg opposite the acute angle to the measure of the leg adjacent to the acute angle. **2.** (p. 552) A line in the plane of a circle that intersects the circle in exactly one point. The point of intersection is called the *point of tangency*. **3.** (p. 671) A line that intersects a sphere in exactly one point.

tangente **1.** La razón entre la medida del cateto opuesto al ángulo agudo y la medida del cateto adyacente al ángulo agudo de un triángulo rectángulo. **2.** La recta situada en el mismo plano de un círculo y que interseca dicho círculo en un sólo punto. El punto de intersección se conoce como *punto de tangencia*. **3.** Recta que interseca una esfera en un sólo punto.

tessellation (p. 483) A pattern that covers a plane by transforming the same figure or set of figures so that there are no overlapping or empty spaces.

teselado Patrón que cubre un plano y que se obtiene transformando la misma figura o conjunto de figuras, sin que haya traslapes ni espacios vacíos.

theorem (p. 90) A statement or conjecture that can be proven true by undefined terms, definitions, and postulates.

teorema Enunciado o conjetura que se puede demostrar como verdadera mediante el uso de términos primitivos, definiciones y postulados.

transformation (p. 462) In a plane, a mapping for which each point has exactly one image point and each image point has exactly one preimage point.

transformación La relación en el plano en que cada punto tiene un único punto imagen y cada punto imagen tiene un único punto preimagen.

translation (p. 470) A transformation that moves all points of a figure the same distance in the same direction.

traslación Transformación en que todos los puntos de una figura se trasladan la misma distancia, en la misma dirección.

translation matrix (p. 506) A matrix that can be added to the vertex matrix of a figure to find the coordinates of the translated image.

matriz de traslación Matriz que al sumarse a la matriz de vértices de una figura permite calcular las coordenadas de la imagen trasladada.

transversal (p. 127) A line that intersects two or more lines in a plane at different points.

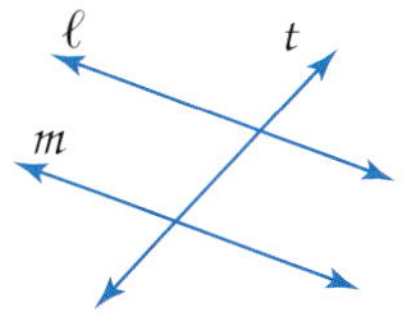

Line *t* is a transversal.
La recta t es una transversal.

transversal Recta que interseca en diferentes puntos dos o más rectas en el mismo plano.

trapezoid (p. 439) A quadrilateral with exactly one pair of parallel sides. The parallel sides of a trapezoid are called *bases*. The nonparallel sides are called *legs*. The pairs of angles with their vertices at the endpoints of the same base are called *base angles*.

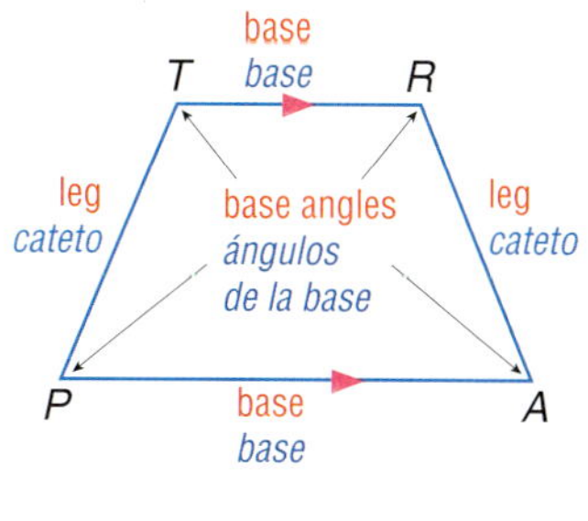

trapecio Cuadrilátero con un sólo par de lados paralelos. Los lados paralelos del trapecio se llaman *bases*. Los lados no paralelos se llaman *catetos*. Los ángulos cuyos vértices se encuentran en los extremos de la misma base se llaman *ángulos de la base*.

trigonometric identity (p. 391) An equation involving a trigonometric ratio that is true for all values of the angle measure.

identidad trigonométrica Ecuación que contiene una razón trigonométrica que es verdadera para todos los valores de la medida del ángulo.

trigonometric ratio (p. 364) A ratio of the lengths of sides of a right triangle.

razón trigonométrica Razón de las longitudes de los lados de un triángulo rectángulo.

trigonometry (p. 364) The study of the properties of triangles and trigonometric functions and their applications.

trigonometría Estudio de las propiedades de los triángulos y de las funciones trigonométricas y sus aplicaciones.

truth table (p. 70) A table used as a convenient method for organizing the truth values of statements.

tabla verdadera Tabla que se utiliza para organizar de una manera conveniente los valores de verdad de los enunciados.

truth value (p. 67) The truth or falsity of a statement.

valor verdadero La condición de un enunciado de ser verdadero o falso.

two-column proof (p. 95) A formal proof that contains statements and reasons organized in two columns. Each step is called a *statement*, and the properties that justify each step are called *reasons*.

demostración a dos columnas Aquélla que contiene enunciados y razones organizadas en dos columnas. Cada paso se llama *enunciado* y las propiedades que lo justifican son las *razones*.

U

undefined terms (p. 7) Words, usually readily understood, that are not formally explained by means of more basic words and concepts. The basic undefined terms of geometry are point, line, and plane.

términos primitivos Palabras que por lo general se entienden fácilmente y que no se explican formalmente mediante palabras o conceptos más básicos. Los términos básicos primitivos de la geometría son el punto, la recta y el plano.

uniform tessellations (p. 484) Tessellations containing the same arrangement of shapes and angles at each vertex.

teselado uniforme Teselados que contienen el mismo patrón de formas y ángulos en cada vértice.

V

vector (p. 498) A directed segment representing a quantity that has both magnitude, or length, and direction.

vector Segmento dirigido que representa una cantidad que posee tanto magnitud, o longitud, como dirección.

vertex matrix (p. 506) A matrix that represents a polygon by placing all of the column matrices of the coordinates of the vertices into one matrix.

matriz del vértice Matriz que representa un polígono al colocar todas las matrices columna de las coordenadas de los vértices en una matriz.

vertical angles (p. 37) Two nonadjacent angles formed by two intersecting lines.

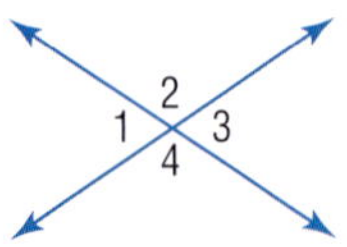

$\angle 1$ and $\angle 3$ are vertical angles.
$\angle 2$ and $\angle 4$ are vertical angles.
$\angle 1$ y $\angle 3$ son ángulos opuestos por el vértice.
$\angle 2$ y $\angle 4$ son ángulos opuestos por el vértice.

ángulos opuestos por el vértice Dos ángulos no adyacentes formados por dos rectas que se intersecan.

volume (p. 688) A measure of the amount of space enclosed by a three-dimensional figure.

volumen La medida de la cantidad de espacio dentro de una figura tridimensional.

Selected Answers

Chapter 1 Points, Lines, Planes, and Angles

Page 5 Chapter 1 Getting Started

1–4.

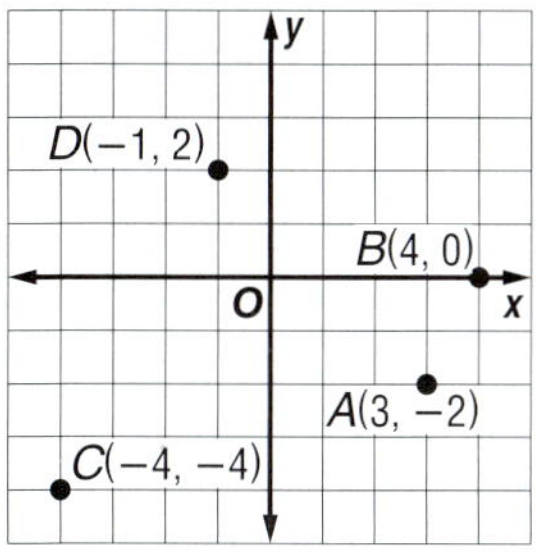

5. $1\frac{1}{8}$ **7.** $\frac{5}{16}$ **9.** −15 **11.** 25 **13.** 20 in. **15.** 24.6 m

Pages 9–11 Lesson 1-1

1. point, line, plane **3.** Micha; the points must be noncollinear to determine a plane.

5. Sample answer:

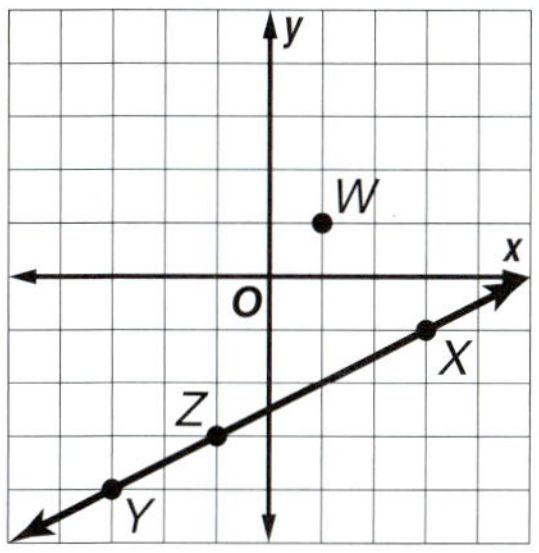

7. 6 **9.** No; *A*, *C*, and *J* lie in plane *ABC*, but *D* does not. **11.** point **13.** *n* **15.** $\mathcal{R}$ **17.** Sample answer: $\overleftrightarrow{PR}$ **19.** (D, 9) **21.**

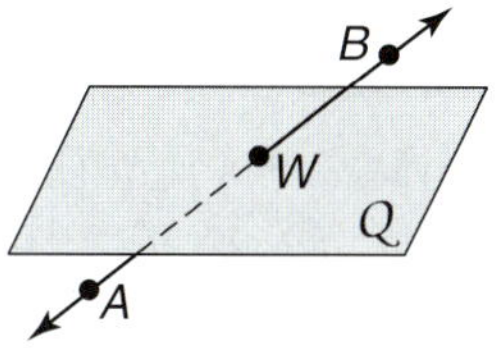

23. Sample answer:

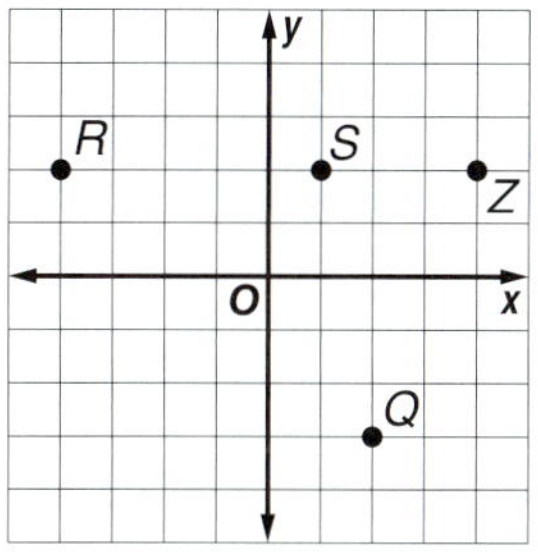

25.

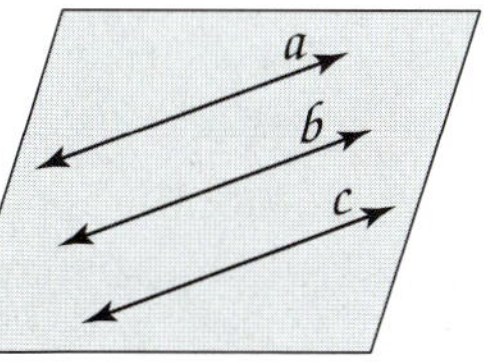

27.

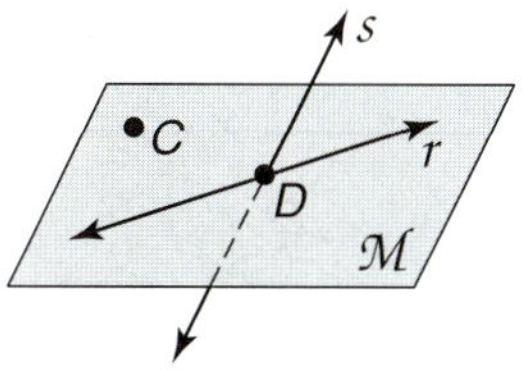

29. points that seem collinear; sample answer: (0, −2), (1, −3), (2, −4), (3, −5)

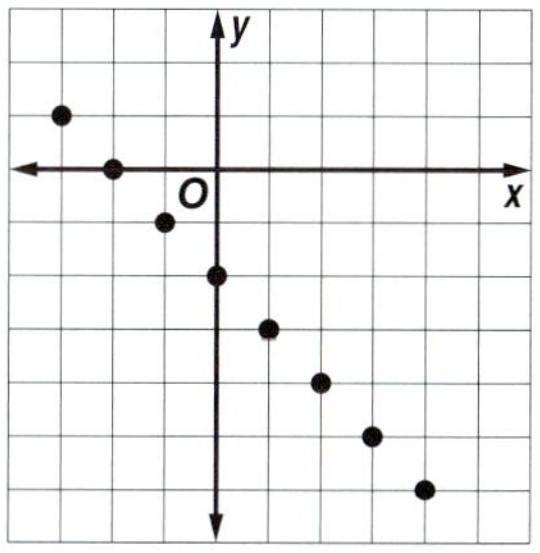

31. 1 **33.** anywhere on $\overleftrightarrow{AB}$ **35.** *A*, *B*, *C*, *D* or *E*, *F*, *C*, *B* **37.** $\overleftrightarrow{AC}$ **39.** lines **41.** plane **43.** point **45.** point

47.

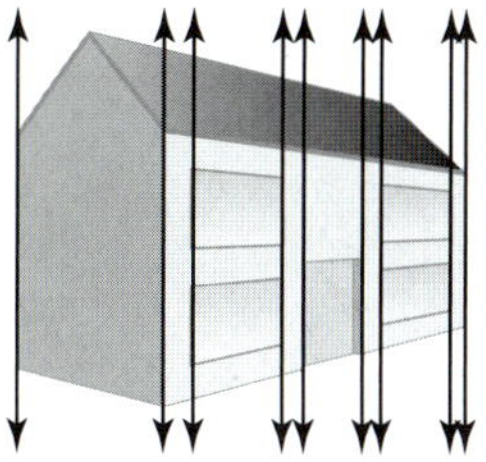

49. See students' work.

51. Sample answer:

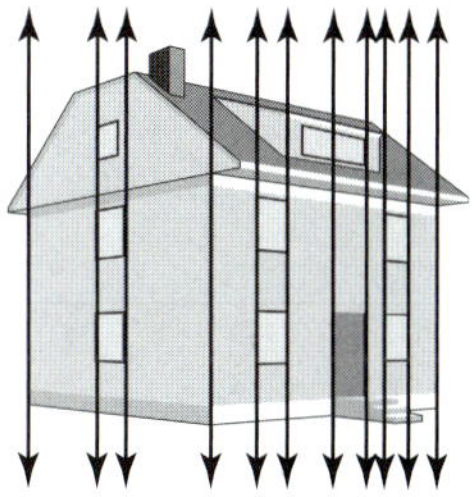

53. vertical **55.** Sample answer: Chairs wobble because all four legs do not touch the floor at the same time. Answers should include the following.

- The ends of the legs represent points. If all points lie in the same plane, the chair will not wobble.
- Because it only takes three points to determine a plane, a chair with three legs will never wobble.

57. B

59. part of the coordinate plane above the line $y = -2x + 1$.

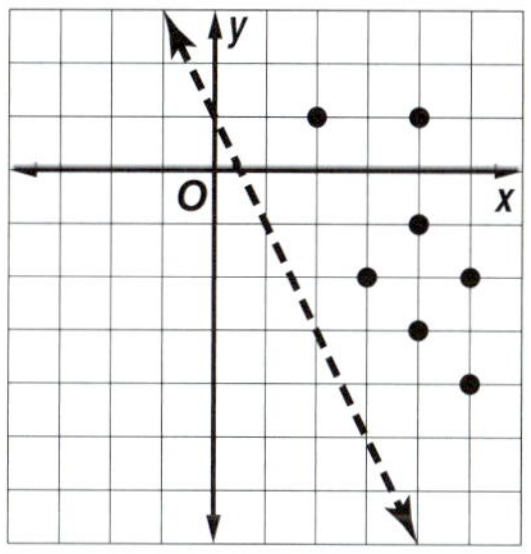

61. = **63.** = **65.** $<$

Pages 16–19 Lesson 1-2

1. Align the 0 point on the ruler with the leftmost endpoint of the segment. Align the edge of the ruler along the segment. Note where the rightmost endpoint falls on the scale and read the closest eighth of an inch measurement.

3. $1\frac{3}{4}$ in. **5.** 0.5 m; 14 m could be 13.5 to 14.5 m **7.** 3.7 cm **9.** $x = 3$; $LM = 9$ **11.** $\overline{BC} \cong \overline{CD}$, $\overline{BE} \cong \overline{ED}$, $\overline{BA} \cong \overline{DA}$ **13.** 4.5 cm or 45 mm **15.** $1\frac{1}{4}$ in. **17.** 0.5 cm; 21.5 to 22.5 mm **19.** 0.5 cm; 307.5 to 308.5 cm **21.** $\frac{1}{8}$ ft.; $3\frac{1}{8}$ to $3\frac{3}{8}$ ft. **23.** $1\frac{1}{4}$ in. **25.** 2.8 cm **27.** $1\frac{1}{4}$ in. **29.** $x = 11$; $ST = 22$ **31.** $x = 2$; $ST = 4$ **33.** $y = 2$; $ST = 3$ **35.** no **37.** yes **39.** yes **41.** $\overline{CF} \cong \overline{DG}$, $\overline{AB} \cong \overline{HI}$, $\overline{CE} \cong \overline{ED} \cong \overline{EF} \cong \overline{EG}$ **43.** 50,000 visitors **45.** No; the number of visitors to Washington state parks could be as low as 46.35 million or as high as 46.45 million. The visitors to Illinois state parks could be as low as 44.45 million or as high as 44.55 million visitors. The difference in visitors could be as high as 2.0 million.

47. 15.5 cm; Each measurement is accurate within 0.5 cm, so the greatest perimeter is 3.5 cm + 5.5 cm + 6.5 cm.

49.

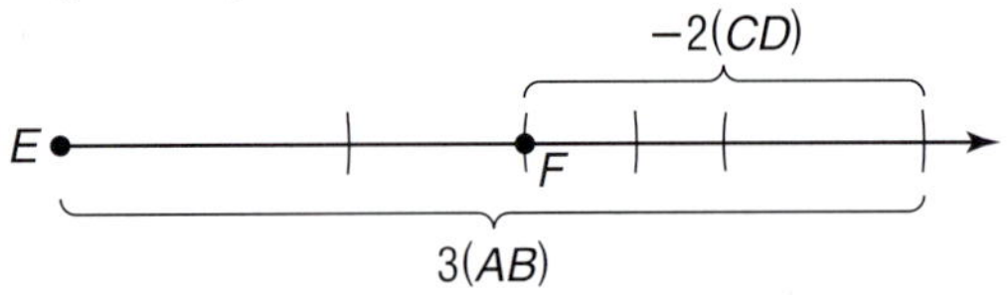

51. Sample answer: Units of measure are used to differentiate between size and distance, as well as for accuracy. Answers should include the following.

- When a measurement is stated, you do not know the precision of the instrument used to make the measure. Therefore, the actual measure could be greater or less than that stated.
- You can assume equal measures when segments are shown to be congruent.

53. 1.7% **55.** 0.08% **57.** D **59.** Sample answer: planes *ABC* and *BCD* **61.** 5 **63.** 22 **65.** 1

Page 19 Practice Quiz 1

1. $\overleftrightarrow{PR}$ **3.** $\overleftrightarrow{PR}$ **5.** 8.35

Pages 25–27 Lesson 1-3

1. Sample answers: (1) Use one of the Midpoint Formulas if you know the coordinates of the endpoints. (2) Draw a segment and fold the paper so that the endpoints match to locate the middle of the segment. (3) Use a compass and straightedge to construct the bisector of the segment.
3. 8 **5.** 10 **7.** −6 **9.** (−2.5, 4) **11.** (3, 5) **13.** 2 **15.** 3 **17.** 11 **19.** 10 **21.** 13 **23.** 15 **25.** $\sqrt{90} \approx 9.5$ **27.** $\sqrt{61} \approx 7.8$ **29.** 17.3 units **31.** −3 **33.** 2.5 **35.** 1 **37.** (10, 3) **39.** (−10, −3) **41.** (5.6, 2.85) **43.** $R(2, 7)$ **45.** $T\left(\frac{8}{3}, 11\right)$ **47.** LaFayette, LA **49a.** 111.8 **49b.** 212.0 **49c.** 353.4 **49d.** 420.3 **49e.** 37.4 **49f.** 2092.9 **51.** ≈ 72.1

53. Sample answer: The perimeter increases by the same factor. **55.** (−1, −3) **57.** B **59.** $4\frac{1}{4}$ in.

61. Sample answer:

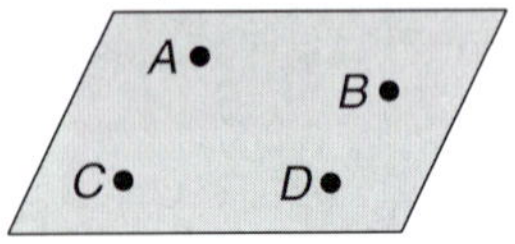

63. 10 **65.** 9 **67.** $\frac{13}{3}$

Pages 33–36 Lesson 1-4

1. Yes; they all have the same measure. **3.** $m\angle A = m\angle Z$ **5.** $\overrightarrow{BA}, \overrightarrow{BC}$ **7.** 135°, obtuse **9.** 47 **11.** ∠1, right; ∠2, acute; ∠3, obtuse **13.** *B* **15.** *A* **17.** $\overrightarrow{AB}, \overrightarrow{AD}$ **19.** $\overrightarrow{AD}, \overrightarrow{AE}$ **21.** ∠*FEA*, ∠4 **23.** ∠*AED*, ∠*DEA*, ∠*AEB*, ∠*BEA*, ∠*AEC*, ∠*CEA* **25.** ∠2 **27.** 30, 30 **29.** 60°, acute **31.** 90°, right **33.** 120°, obtuse **35.** 65 **37.** 4 **39.** 4 **41.** Sample answer: *Acute* can mean something that is sharp or having a very fine tip like a pen, a knife, or a needle. *Obtuse* means not pointed or blunt, so something that is obtuse would be wide. **43.** 31; 59 **45.** 1, 3, 6, 10, 15 **47.** 21, 45 **49.** Sample answer: A degree is $\frac{1}{360}$ of a circle. Answers should include the following.

- Place one side of the angle to coincide with 0 on the protractor and the vertex of the angle at the center point of the protractor. Observe the point at which the other side of the angle intersects the scale of the protractor.
- See students' work.

51. C **53.** $\sqrt{80} \approx 8.9$; (2, 2) **55.** $9\frac{2}{3}$ in. **57.** 13 **59.** *F*, *L*, *J* **61.** 5 **63.** −45 **65.** 8

Page 36 Practice Quiz 2

1. $\left(-\frac{1}{2}, 1\right)$; $\sqrt{65} \approx 8.1$ **3.** (0, 0); $\sqrt{2000} \approx 44.7$ **5.** 34; 135

Pages 41–62 Lesson 1-5

1.

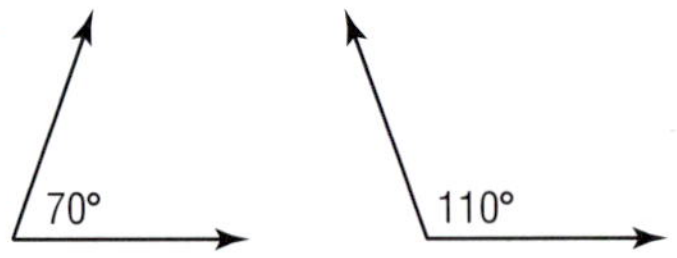

3. Sample answer: The noncommon sides of a linear pair of angles form a straight line.

5. Sample answer: ∠*ABC*, ∠*CBE* **7.** $x = 24, y = -20$ **9.** Yes; they share a common side and vertex, so they are adjacent. Since $\overline{PR}$ falls between $\overline{PQ}$ and $\overline{PS}$, $m\angle QPR < 90$, so the two angles cannot be complementary or supplementary. **11.** ∠*WUT*, ∠*VUX* **13.** ∠*UWT*, ∠*TWY* **15.** ∠*WTY*, ∠*WTU* **17.** 53, 37 **19.** 148 **21.** 84, 96 **23.** always **25.** sometimes **27.** 3.75 **29.** 114 **31.** Yes; the symbol denotes that ∠*DAB* is a right angle. **33.** Yes; their sum of their measures is *m*∠*ADC*, which 90. **35.** No; we do not know *m*∠*ABC*.

37. Sample answer:

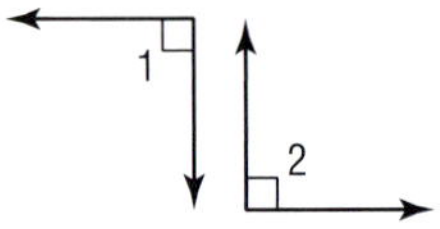

39. Because ∠*WUT* and ∠*TUV* are supplementary, let $m\angle WUT = x$ and $m\angle TUV = 180 - x$. A bisector creates measures that are half of the original angle, so $m\angle YUT = \frac{1}{2}m\angle WUT$ or $\frac{x}{2}$ and $m\angle TUZ = \frac{1}{2}m\angle TUV$ or $\frac{180 - x}{2}$. Then $m\angle YUZ = m\angle YUT + m\angle TUZ$ or $\frac{x}{2} + \frac{180 - x}{2}$. This sum simplifies to $\frac{180}{2}$ or 90. Because $m\angle YUZ = 90$, $\overline{YU} \perp \overline{UZ}$. **41.** A **43.** $\ell \perp \overleftrightarrow{AB}$, $m \perp \overleftrightarrow{AB}$, $n \perp \overleftrightarrow{AB}$ **45.** obtuse **47.** right **49.** obtuse **51.** 8 **53.** $\sqrt{173} \approx 13.2$ **55.** $\sqrt{20} \approx 4.5$ **57.** $n = 3$, $QR = 20$ **59.** 24 **61.** 40

Pages 48–50 Lesson 1-6

1. Divide the perimeter by 10. **3.** $P = 3s$ **5.** pentagon; concave; irregular **7.** 33 ft **9.** 16 units **11.** 4605 ft **13.** octagon; convex; regular **15.** pentagon **17.** triangle **19.** 82 ft **21.** 40 units **23.** The perimeter is tripled. **25.** 125 m **27.** 30 units **29.** All are 15 cm. **31.** 13 units, 13 units, 5 units **33.** 4 in., 4 in., 17 in., 17 in. **35.** 52 units **37.** Sample answer: Some toys use pieces to form polygons. Others have polygon-shaped pieces that connect together. Answers should include the following.

- triangles, quadrilaterals, pentagons
- 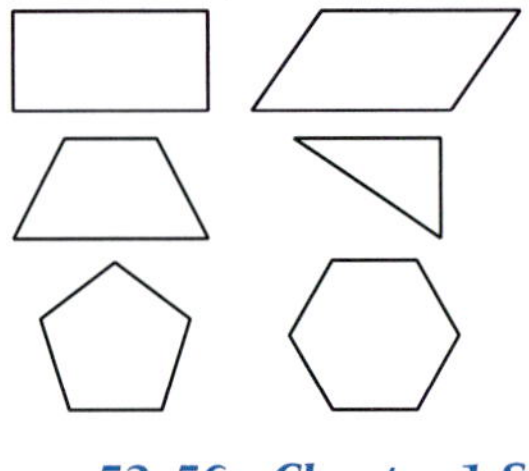

39. D **41.** sometimes **43.** 63

Pages 53–56 Chapter 1 Study Guide and Review

1. d **3.** f **5.** b **7.** *p* or *m* **9.** *F*

11.

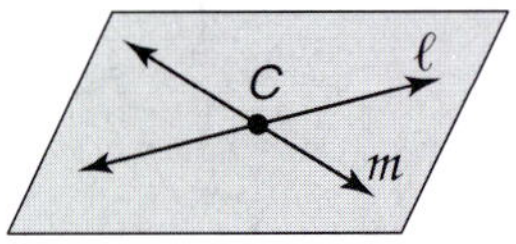

13. $x = 6$, $PB = 18$ **15.** $s = 3$, $PB = 12$ **17.** yes **19.** not enough information **21.** $\sqrt{101} \approx 10.0$ **23.** $\sqrt{13} \approx 3.6$ **25.** (3, −5) **27.** (0.6, −6.35) **29.** $\overrightarrow{FE}, \overrightarrow{FG}$ **31.** 70°, acute **33.** 50°, acute **35.** 36 **37.** 40 **39.** ∠*TWY*, ∠*XWY* **41.** 9 **43.** not a polygon **45.** ≈ 22.5 units

Chapter 2 Reasoning and Proof

Page 61 Chapter 2 Getting Started

1. 10 **3.** 0 **5.** 50 **7.** 21 **9.** −9 **11.** $-\frac{18}{5}$ **13.** 16

Pages 63–66 Lesson 2-1

1. Sample answer: After the news is over, it's time for dinner. **3.** Sample answer: When it's cloudy, it rains. Counterexample: It is often cloudy and it does not rain.

5. 7 **7.** *A*, *B*, *C*, and *D* are noncollinear.

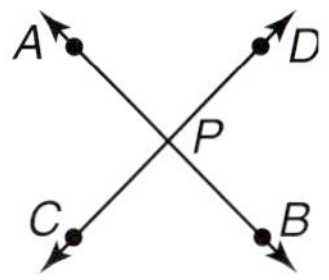

9. true **11.**

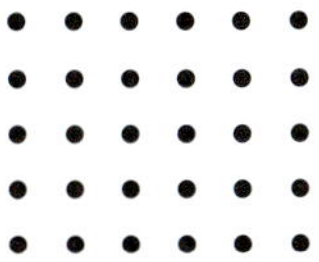

13. 32 **15.** $\frac{11}{3}$ **17.** 162 **19.** 30

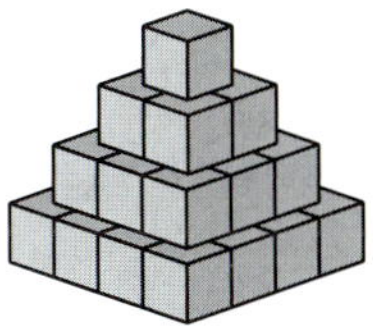

21. Lines ℓ and m form four right angles.

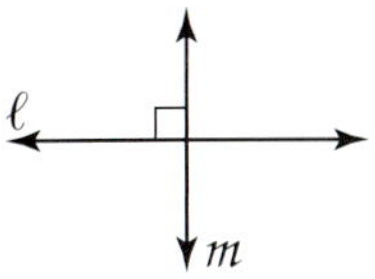

23. ∠3 and ∠4 are supplementary.

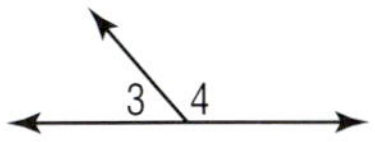

25. △*PQR* is a scalene triangle.

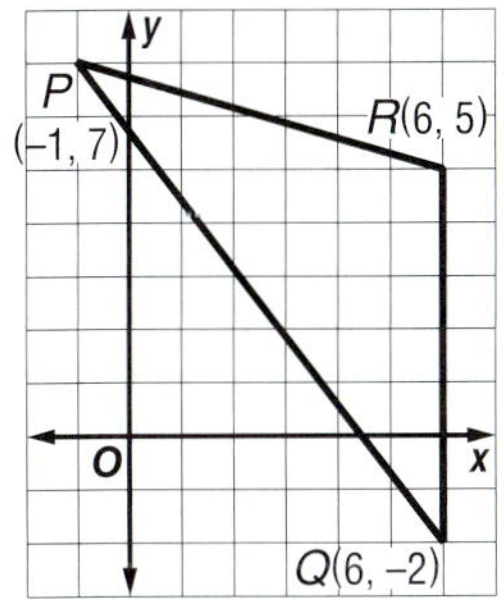

27. $PQ = SR$, $QR = PS$

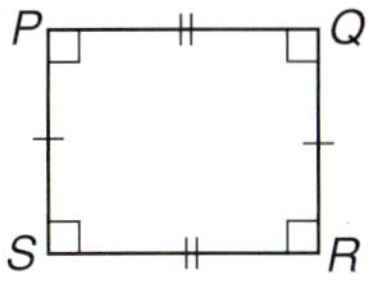

29. false;

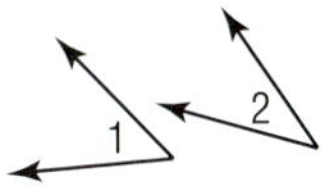

31. false;

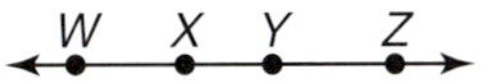

33. true **35.** False; *JKLM* may not have a right angle. **37.** trial and error, a process of inductive reasoning **39.** C_7H_{16} **41.** false; $n = 41$ **43.** C **45.** hexagon, convex, irregular **47.** heptagon, concave, irregular **49.** No; we do not know anything about the angle measures. **51.** Yes; they form a linear pair. **53.** (2, −1) **55.** (1, −12) **57.** (5.5, 2.2) **59.** 8; 56 **61.** 4; 16 **63.** 10; 43 **65.** 4, 5 **67.** 5, 6, 7

Pages 71–74 Lesson 2-2

1. The conjunction (*p* and *q*) is represented by the intersection of the two circles. **3.** A conjunction is a compound statement using the word *and*, while a disjunction is a compound statement using the word *or*.
5. 9 + 5 = 14 and a square has four sides; true.
7. 9 + 5 = 14 or February does not have 30 days; true.
9. 9 + 5 ≠ 14 or a square does not have four sides; false.

11. Sample answer:

p	*q*	$p \wedge q$
T	T	T
T	F	F
F	T	F
F	F	F

13. Sample answer:

p	*r*	~*p*	$\sim p \wedge r$
T	T	F	F
T	F	F	F
F	T	T	T
F	F	T	F

15. 14 **17.** 3 **19.** $\sqrt{-64} = 8$ or an equilateral triangle has three congruent sides; true. **21.** 0 < 0 and an obtuse angle measures greater than 90° and less than 180°; false. **23.** An equilateral triangle has three congruent sides and an obtuse angle measures greater than 90° and less than 180°; true. **25.** An equilateral triangle has three congruent sides and 0 < 0; false. **27.** An obtuse angle measures greater than 90° and less than 180° or an equilateral triangle has three congruent sides; true. **29.** An obtuse angle measures greater than 90° and less than 180°, or an equilateral triangle has three congruent sides and 0 < 0; true.

31.

p	*q*	~*p*	~*q*	$\sim p \wedge \sim q$
T	T	F	F	F
T	F	F	T	F
F	T	T	F	F
F	F	T	T	T

33. Sample answer:

q	*r*	*q* and *r*
T	T	T
T	F	F
F	T	F
F	F	F

35. Sample answer:

p	*r*	*p* or *r*
T	T	T
T	F	T
F	T	T
F	F	F

37. Sample answer:

q	*r*	~*r*	$q \wedge \sim r$
T	T	F	F
T	F	T	T
F	T	F	F
F	F	T	F

39. Sample answer:

p	*q*	*r*	~*p*	~*r*	$q \wedge \sim r$	$\sim p \vee (q \wedge \sim r)$
T	T	T	F	F	F	F
T	T	F	F	T	T	T
T	F	T	F	F	F	F
T	F	F	F	T	F	F
F	T	T	T	F	F	T
F	T	F	T	T	T	T
F	F	T	T	F	F	T
F	F	F	T	T	F	T

41. 42 **43.** 25

45. Level of Participation Among 310 Students

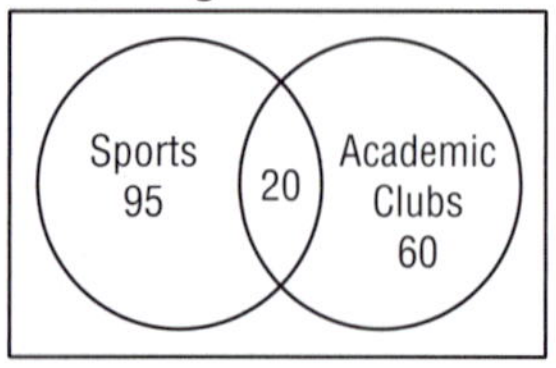

47. 135 **49.** true

51.

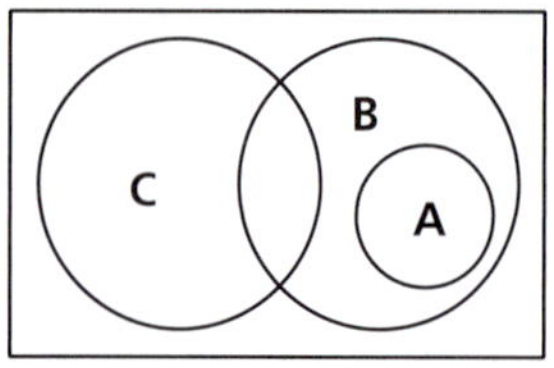

53. Sample answer: Logic can be used to eliminate false choices on a multiple choice test. Answers should include the following.

- Math is my favorite subject and drama club is my favorite activity.
- See students' work.

55. C **57.** 81 **59.** 1 **61.** 405 **63.** 34.4 **65.** 29.5 **67.** 55°, acute **69.** 222 feet **71.** 44 **73.** 184

Pages 78–80 Lesson 2-3

1. Writing a conditional in if-then form is helpful so that the hypothesis and conclusion are easily recognizable. **3.** In the inverse, you negate both the hypothesis and the conclusion of the conditional. In the contrapositive, you negate the hypothesis and the conclusion of the converse. **5.** H: $x - 3 = 7$; C: $x = 10$ **7.** If a pitcher is a 32-ounce pitcher, then it holds a quart of liquid. **9.** If an angle is formed by perpendicular lines, then it is a right angle. **11.** true **13.** Converse: If plants grow, then they have water; true. Inverse: If plants do not have water, then they will not grow; true. Contrapositive: If plants do not grow, then they do not have water. False; they may have been killed by overwatering. **15.** Sample answer: If you are in Colorado, then aspen trees cover high areas of the mountains. If you are in Florida, then cypress trees rise from the swamps. If you are in Vermont, then maple trees are prevalent. **17.** H: you are a teenager; C: you are at least 13 years old **19.** H: three points lie on a line; C: the points are collinear **21.** H: the measure of an is between 0 and 90; C: the angle is acute **23.** If you are a math teacher, then you love to solve problems. **25.** Sample answer: If two angles are adjacent, then they have a common side. **27.** Sample answer: If two triangles are equiangular, then they are equilateral. **29.** true **31.** true **33.** false **35.** true **37.** false **39.** true **41.** Converse: If you are in good shape, then you exercise regularly; true. Inverse: If you do not exercise regularly, then you are not in good shape; true. Contrapositive: If you are not in good shape, then you do not exercise regularly. False; an ill person may exercise a lot, but still not be in good shape. **43.** Converse: If a figure is a quadrilateral, then it is a rectangle; false, rhombus. Inverse: If a figure is not a rectangle, then it is not a quadrilateral; false, rhombus. Contrapositive: If a figure is not a quadrilateral, then it is not a rectangle; true. **45.** Converse: If an angle has measure less than 90, then it is acute; true. Inverse: If an angle is not acute, then its measure is not less than 90; true. Contrapositive: If an angle's measure is not less than 90, then it is not acute; true. **47.** Sample answer: In Alaska, if there are more hours of daylight than darkness, then it is summer. In Alaska, if there are more hours of darkness than daylight, then it is winter. **49.** Conditional statements can be used to describe how to get a discount, rebate, or refund. Sample answers should include the following. If you are not 100% satisfied, then return the product for a full refund. Wearing a seatbelt reduces the risk of injuries. **51.** B **53.** A hexagon has five sides or $60 \times 3 = 18$.; false **55.** A hexagon doesn't have five sides or $60 \times 3 = 18$.; true **57.** George Washington was not the first president of the United States and $60 \times 3 \neq 18$.; false

59. The sum of the measures of the angles in a triangle is 180.

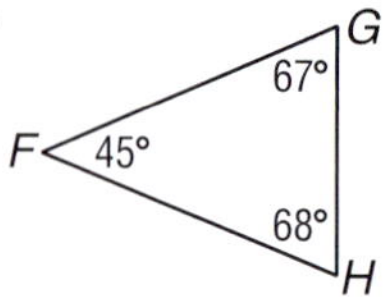

61. $\angle PQR$ is a right angle.

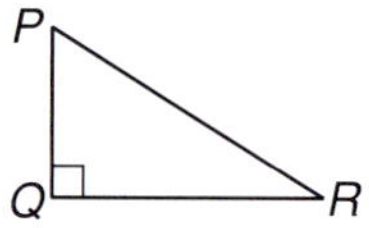

63. $\sqrt{41}$ or 6.4 **65.** $\sqrt{125}$ or 11.2 **67.** Multiply each side by 2.

Page 80 Practice Quiz 1

1. false

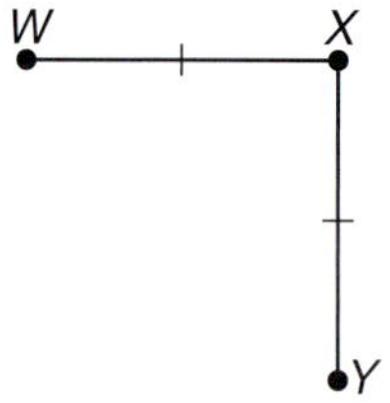

3. Sample answer:

p	q	$\sim p$	$\sim p \wedge q$
T	T	F	F
T	F	F	F
F	T	T	T
F	F	T	F

5. Converse: If two angles have a common vertex, then the angles are adjacent. False; $\angle ABD$ is not adjacent to $\angle ABC$.

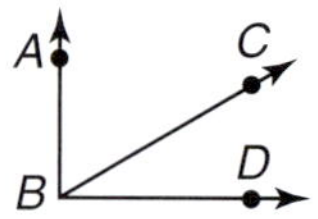

Inverse: If two angles are not adjacent, then they do not have a common vertex. False; $\angle ABC$ and $\angle DBE$ have a common vertex and are not adjacent.

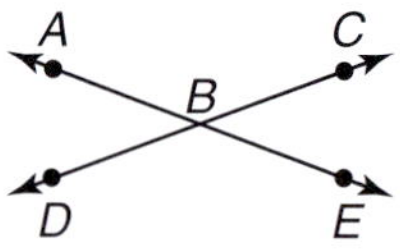

Contrapositive: If two angles do not have a common vertex, then they are not adjacent; true.

Pages 84–87 Lesson 2-4

1. Sample answer: a: If it is rainy, the game will be cancelled; b: It is rainy; c: The game will be cancelled. **3.** Lakeisha; if you are dizzy, that does not necessarily mean that you are seasick and thus have an upset stomach. **5.** Invalid; congruent angles do not have to be vertical. **7.** The midpoint of a segment divides it into two segments with equal measures. **9.** invalid **11.** No; Terry could be a man or a woman. She could be 45 and have purchased $30,000 of life insurance. **13.** Valid; since 5 and 7 are odd, the Law of Detachment indicates that their sum is even. **15.** Invalid; the sum is even. **17.** Invalid; E, F, and G are not necessarily noncollinear. **19.** Valid; the vertices of a triangle are noncollinear, and therefore determine a plane. **21.** If the measure of an angle is less than 90, then it is not obtuse. **23.** no conclusion **25.** yes; Law of Detachment **27.** yes; Law of Detachment **29.** invalid **31.** If Catriona Le May Doan skated her second 500 meters in 37.45 seconds, then she would win the race. **33.** Sample answer: Doctors and nurses use charts to assist in determining medications and their doses for patients. Answers should include the following.

- Doctors need to note a patient's symptoms to determine which medication to prescribe, then determine how much to prescribe based on weight, age, severity of the illness, and so on.
- Doctors use what is known to be true about diseases and when symptoms appear, then deduce that the patient has a particular illness.

35. B **37.** They are a fast, easy way to add fun to your family's menu.

39. Sample answer:

q	r	$q \wedge r$
T	T	T
T	F	F
F	T	F
F	F	F

41. Sample answer:

p	q	r	$q \vee r$	$p \wedge (q \vee r)$
T	T	T	T	T
T	T	F	T	T
T	F	T	T	T
T	F	F	F	F
F	T	T	T	F
F	T	F	T	F
F	F	T	T	F
F	F	F	F	F

43. $\angle HDG$ **45.** Sample answer: $\angle JHK$ and $\angle DHK$ **47.** Yes, slashes on the segments indicate that they are congruent. **49.** 10 **51.** $\sqrt{130} \approx 11.4$

53.

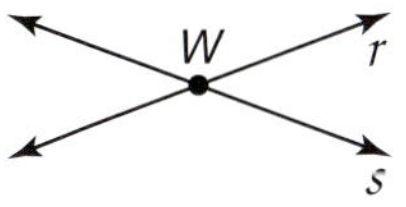

55.

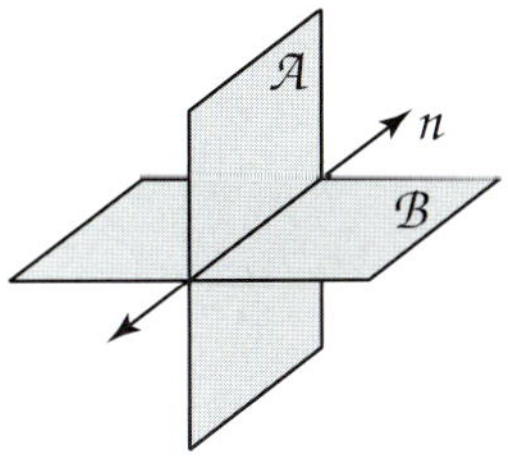

57. Sample answer: $\angle 1$ and $\angle 2$ are complementary, $m\angle 1 + m\angle 2 = 90$.

Pages 91–93 Lesson 2-5

1. Deductive reasoning is used to support claims that are made in a proof. **3.** postulates, theorems, algebraic properties, definitions **5.** 15 **7.** definition of collinear **9.** Through any two points, there is exactly one line. **11.** 15 ribbons **13.** 10 **15.** 21 **17.** Always; if two points lie in a plane, then the entire line containing those points lies in that plane. **19.** Sometimes; the three points cannot be on the same line. **21.** Sometimes; ℓ and m could be skew so they would not lie in the same plane $\mathcal{R}$. **23.** If two points lie in a plane, then the entire line containing those points lies in that plane. **25.** If two points lie in a plane, then the entire line containing those points lies in the plane. **27.** Through any three points not on the same line, there is exactly one plane. **29.** She will have 4 different planes and 6 lines. **31.** one, ten **33.** C **35.** yes; Law of Detachment **37.** Converse: If $\triangle ABC$ has an angle with measure greater than 90, then $\triangle ABC$ is a right triangle. False; the triangle would be obtuse. Inverse: If $\triangle ABC$ is not a right triangle, none of its angle measures are greater than 90. False; it could be an obtuse triangle. Contrapositive: If $\triangle ABC$ does not have an angle measure greater than 90, $\triangle ABC$ is not a right triangle. False; $m\angle ABC$ could still be 90 and $\triangle ABC$ be a right triangle. **39.** $\sqrt{17} \approx 4.1$ **41.** $\sqrt{106} \approx 10.3$ **43.** 25 **45.** 12 **47.** 10

Pages 97–100 Lesson 2-6

1. Sample answer: If $x = 2$ and $x + y = 6$, then $2 + y = 6$. **3.** hypothesis; conclusion **5.** Multiplication Property **7.** Addition Property **9a.** $5 - \frac{2}{3}x = 1$ **9b.** Mult. Prop. **9c.** Dist. Prop. **9d.** $-2x = -12$ **9e.** Div. Prop.

11. Given: Rectangle $ABCD$, $AD = 3$, $AB = 10$
Prove: $AC = BD$

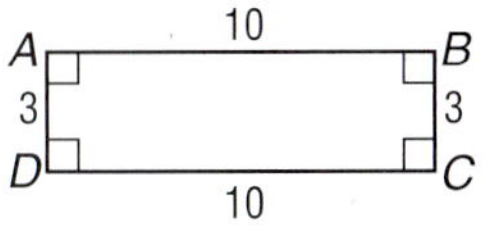

Proof:

Statement	Reasons
1. Rectangle $ABCD$, $AD = 3$, $AB = 10$	1. Given
2. Draw segments AC and DB.	2. Two points determine a line.
3. $\triangle ABC$ and $\triangle BCD$ are right triangles.	3. Def. of rt $\triangle$
4. $AC = \sqrt{3^2 + 10^2}$, $DB = \sqrt{3^2 + 10^2}$	4. Pythagorean Th.
5. $AC = BD$	5. Substitution

13. C **15.** Subt. Prop. **17.** Substitution **19.** Reflexive Property **21.** Substitution **23.** Transitive Prop. **25a.** $2x - 7 = \frac{1}{3}x - 2$ **25b.** $3(2x - 7) = 3\left(\frac{1}{3}x - 2\right)$ **25c.** Dist. Prop. **25d.** $5x - 21 = -6$ **25e.** Add. Prop. **25f.** $x = 3$

27. Given: $-2y + \frac{3}{2} = 8$
Prove: $y = -\frac{13}{4}$
Proof:

Statement	Reasons
1. $-2y + \frac{3}{2} = 8$	1. Given
2. $2(-2y + \frac{3}{2}) = 2(8)$	2. Mult. Prop.
3. $-4y + 3 = 16$	3. Dist. Prop.
4. $-4y = 13$	4. Subt. Prop.
5. $y = -\frac{13}{4}$	5. Div. Prop.

29. Given: $5 - \frac{2}{3}z = 1$
Prove: $z = 6$
Proof:

Statement	Reasons
1. $5 - \frac{2}{3}z = 1$	1. Given
2. $3\left(5 - \frac{2}{3}z\right) = 3(1)$	2. Mult. Prop.
3. $15 - 2x = 3$	3. Dist. Prop.
4. $15 - 2x - 15 = 3 - 15$	4. Subt. Prop.
5. $-2x = -12$	5. Substitution
6. $\frac{-2x}{-2} = \frac{-12}{-2}$	6. Div. Prop.
7. $x = 6$	7. Substitution

31. Given: $m\angle ACB = m\angle ABC$
Prove: $m\angle XCA = m\angle YBA$

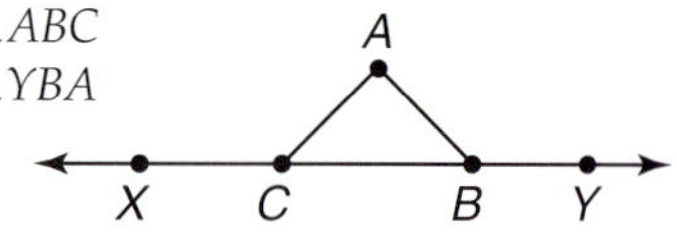

Proof:

Statement	Reasons
1. $m\angle ACB = m\angle ABC$	1. Given
2. $m\angle XCA + m\angle ACB = 180$ $m\angle YBA + m\angle ABC = 180$	2. Def. of supp. $\angle$s
3. $m\angle XCA + m\angle ACB =$ $m\angle YBA + m\angle ABC$	3. Substitution
4. $m\angle XCA + m\angle ACB =$ $m\angle YBA + m\angle ACB$	4. Substitution
5. $m\angle XCA = m\angle YBA$	5. Subt. Prop.

33. All of the angle measures would be equal. **35.** See students' work. **37.** B **39.** 6 **41.** Invalid; $27 \div 6 = 4.5$, which is not an integer. **43.** Sample answer: If people are happy, then they rarely correct their faults. **45.** Sample answer: If a person is a champion, then the person is afraid of losing. **47.** $\frac{1}{2}$ ft **49.** 0.5 in. **51.** 11 **53.** 47

Page 100 Practice Quiz 2

1. invalid **3.** If two lines intersect, then their intersection is exactly one point.

5. Given: $2(n - 3) + 5 = 3(n - 1)$
Prove: $n = 2$
Proof:

Statement	Reasons
1. $2(n - 3) + 5 = 3(n - 1)$	1. Given
2. $2n - 6 + 5 = 3n - 3$	2. Dist. Prop.
3. $2n - 1 = 3n - 3$	3. Substitution
4. $2n - 1 - 2n = 3n - 3 - 2n$	4. Subt. Prop.
5. $-1 = n - 3$	5. Substitution
6. $-1 + 3 = n - 3 + 3$	6. Add. Prop.
7. $2 = n$	7. Substitution
8. $n = 2$	8. Symmetric Prop.

Pages 103–106 Lesson 2-7

1. Sample answer: The distance from Cleveland to Chicago is the same as the distance from Cleveland to Chicago. **3.** If A, B, and C are collinear and $AB + BC = AC$, then B is between A and C. **5.** Symmetric

7. Given: $\overline{PQ} \cong \overline{RS}$, $\overline{QS} \cong \overline{ST}$
Prove: $\overline{PS} \cong \overline{RT}$
Proof:

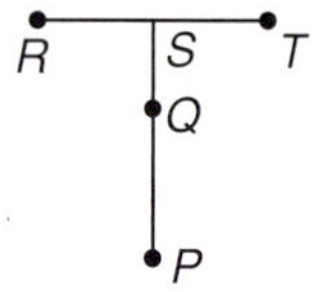

Statements	Reasons
a. $\overline{PQ} \cong \overline{RS}$, $\overline{QS} \cong \overline{ST}$	a. Given
b. $PQ = RS$, $QS = ST$	b. Def. of $\cong$ segments
c. $PS = PQ + QS$, $RT = RS + ST$	c. Segment Addition Post.
d. $PQ + QS = RS + ST$	d. Addition Property
e. $PS = RT$	e. Substitution
f. $\overline{PS} \cong \overline{RT}$	f. Def. of $\cong$ segments

9. Given: $\overline{HI} \cong \overline{TU}$, $\overline{HJ} \cong \overline{TV}$
Prove: $\overline{IJ} \cong \overline{UV}$

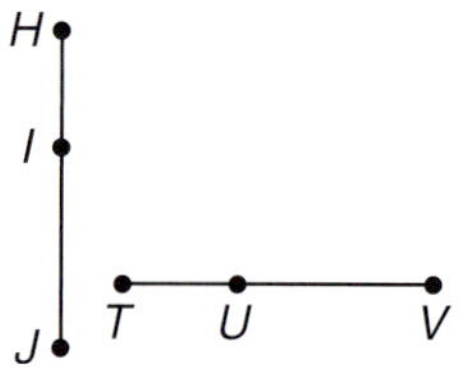

Proof:

Statements	Reasons
1. $\overline{HI} \cong \overline{TU}$, $\overline{HJ} \cong \overline{TV}$	1. Given
2. $HI = TU$, $HJ = TV$	2. Def. of $\cong$ segs.
3. $HI + IJ = HJ$	3. Seg. Add. Post.
4. $TU + IJ = TV$	4. Substitution
5. $TU + UV = TV$	5. Seg. Add. Post.
6. $TU + IJ = TU + UV$	6. Substitution
7. $TU = TU$	7. Reflexive Prop.
8. $IJ = UV$	8. Subt. Prop.
9. $\overline{IJ} \cong \overline{UV}$	9. Def. of $\cong$ segs.

11. Helena is between Missoula and Miles City. **13.** Substitution **15.** Transitive **17.** Subtraction

19. Given: $\overline{XY} \cong \overline{WZ}$ and $\overline{WZ} \cong \overline{AB}$
Prove: $\overline{XY} \cong \overline{AB}$

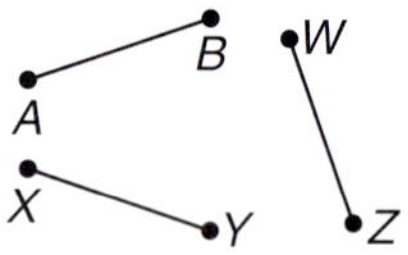

Proof:

Statements	Reasons
1. $\overline{XY} \cong \overline{WZ}$ and $\overline{WZ} \cong \overline{AB}$	1. Given
2. $XY = WZ$ and $WZ = AB$	2. Def. of $\cong$ segs.
3. $XY = AB$	3. Transitive Prop.
4. $\overline{XY} \cong \overline{AB}$	4. Def. of $\cong$ segs.

21. Given: $\overline{WY} \cong \overline{ZX}$
A is the midpoint of $\overline{WY}$.
A is the midpoint of $\overline{ZX}$.
Prove: $\overline{WA} \cong \overline{ZA}$

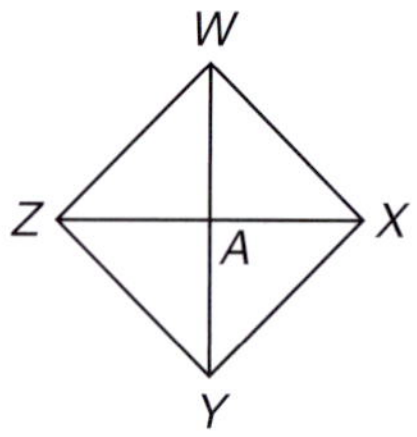

Proof:

Statements:	Reasons:
a. $\overline{WY} \cong \overline{ZX}$ A is the midpoint of $\overline{WY}$. A is the midpoint of $\overline{ZX}$.	a. Given
b. $WY = ZX$	b. Def. of $\cong$ segs.
c. $WA = AY$, $ZA = AX$	c. Definition of midpoint
d. $WY = WA + AY$, $ZX = ZA + AX$	d. Segment Addition Post.
e. $WA + AY = ZA + AX$	e. Substitution
f. $WA + WA = ZA + ZA$	f. Substitution
g. $2WA = 2ZA$	g. Substitution
h. $WA = ZA$	h. Division Property
i. $\overline{WA} \cong \overline{ZA}$	i. Def. of $\cong$ segs.

23. Given: $AB = BC$
Prove: $AC = 2BC$

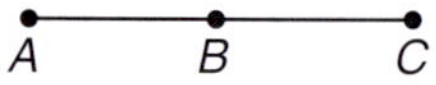

Proof:

Statements	Reasons
1. $AB = BC$	1. Given
2. $AC = AB + BC$	2. Seg. Add. Post.
3. $AC = BC + BC$	3. Substitution
4. $AC = 2BC$	4. Substitution

25. Given: $\overline{AB} \cong \overline{DE}$, C is the midpoint of $\overline{BD}$.
Prove: $\overline{AC} \cong \overline{CE}$

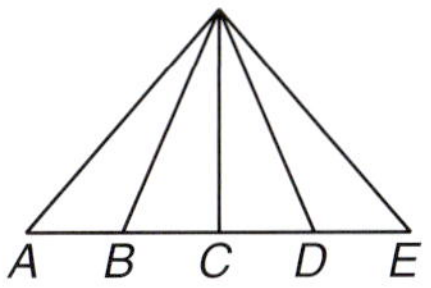

Proof:

Statements	Reasons
1. $\overline{AB} \cong \overline{DE}$, C is the midpoint of $\overline{BD}$.	1. Given
2. $BC = CD$	2. Def. of midpoint
3. $AB = DE$	3. Def. of $\cong$ segs.
4. $AB + BC = CD + DE$	4. Add. Prop.
5. $AB + BC = AC$; $CD + DE = CE$	5. Seg. Add. Post.
6. $AC = CE$	6. Substitution
7. $\overline{AC} \cong \overline{CE}$	7. Def. of $\cong$ segs.

27. Sample answers: $\overline{LN} \cong \overline{QO}$ and $\overline{LM} \cong \overline{MN} \cong \overline{RS} \cong \overline{ST} \cong \overline{QP} \cong \overline{PO}$ **29.** B **31.** Substitution **33.** Addition Property **35.** Never; the midpoint of a segment divides it into two congruent segments. **37.** Always; if two planes intersect, they intersect in a line. **39.** 3; 9 cm by 13 cm **41.** 15 **43.** 45 **45.** 25

Pages 111–114 Lesson 2-8

1. Tomas; Jacob's answer left out the part of $\angle ABC$ represented by $\angle EBF$. **3.** $m\angle 2 = 65$ **5.** $m\angle 11 = 59$, $m\angle 12 = 121$

7. Given: $\overrightarrow{VX}$ bisects $\angle WVY$. $\overrightarrow{VY}$ bisects $\angle XVZ$.
Prove: $\angle WVX \cong \angle YVZ$

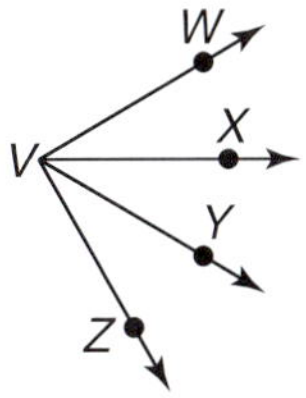

Proof:

Statements	Reasons
1. $\overrightarrow{VX}$ bisects $\angle WVY$, $\overrightarrow{VY}$ bisects $\angle XVZ$.	1. Given
2. $\angle WVX \cong \angle XVY$	2. Def. of $\angle$ bisector
3. $\angle XVY \cong \angle YVZ$	3. Def. of $\angle$ bisector
4. $\angle WVX \cong \angle YVZ$	4. Trans. Prop.

9. sometimes

11. Given: $\angle ABC$ is a right angle.
Prove: $\angle 1$ and $\angle 2$ are complementary angles.

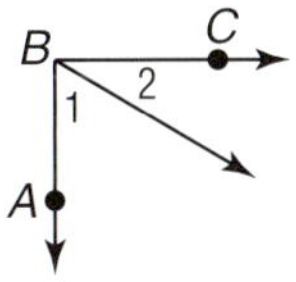

Proof:

Statements	Reasons
1. $\angle ABC$ is a right angle.	1. Given
2. $m\angle ABC = 90$	2. Def. of rt. $\angle$
3. $m\angle ABC = m\angle 1 + m\angle 2$	3. Angle Add. Post.
4. $m\angle 1 + m\angle 2 = 90$	4. Substitution
5. $\angle 1$ and $\angle 2$ are complementary angles.	5. Def. of complementary $\angle$s

13. 62 **15.** 28 **17.** $m\angle 4 = 52$ **19.** $m\angle 9 = 86$, $m\angle 10 = 94$ **21.** $m\angle 13 = 112$, $m\angle 14 = 112$ **23.** $m\angle 17 = 53$, $m\angle 18 = 53$

25. Given: $\angle A$
Prove: $\angle A \cong \angle A$

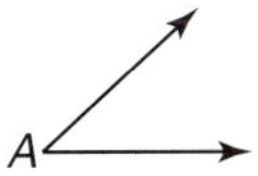

Proof:

Statements	Reasons
1. $\angle A$ is an angle.	1. Given
2. $m\angle A = m\angle A$	2. Reflexive Prop.
3. $\angle A \cong \angle A$	3. Def. of $\cong$ angles

27. sometimes **29.** always **31.** sometimes

33. Given: $\ell \perp m$
Prove: $\angle 2$, $\angle 3$, and $\angle 4$ are rt. $\angle$s.

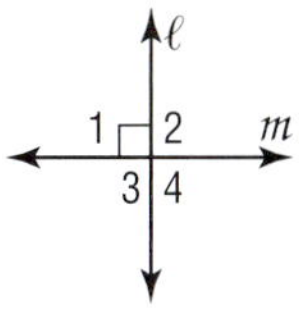

Proof:

Statements	Reasons
1. $\ell \perp m$	1. Given
2. $\angle 1$ is a right angle.	2. Def. of $\perp$ lines
3. $m\angle 1 = 90$	3. Def. of rt. $\angle$s
4. $\angle 1 \cong \angle 4$	4. Vert. $\angle$s are $\cong$.
5. $m\angle 1 = m\angle 4$	5. Def. of $\cong$ $\angle$s
6. $m\angle 4 = 90$	6. Substitution
7. $\angle 1$ and $\angle 2$ form a linear pair. $\angle 3$ and $\angle 4$ form a linear pair.	7. Def. of linear pair
8. $m\angle 1 + m\angle 2 = 180$, $m\angle 4 + m\angle 3 = 180$	8. Linear pairs are supplementary.
9. $90 + m\angle 2 = 180$, $90 + m\angle 3 = 180$	9. Substitution
10. $m\angle 2 = 90$, $m\angle 3 = 90$	10. Subt. Prop.
11. $\angle 2$, $\angle 3$, and $\angle 4$ are rt. $\angle$s.	11. Def. of rt. $\angle$s (steps 6, 10)

35. Given: $\ell \perp m$
Prove: $\angle 1 \cong \angle 2$

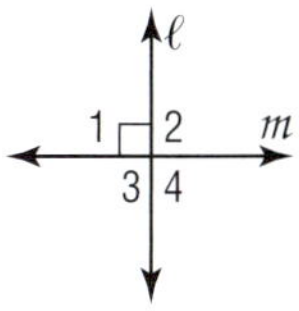

Proof:

Statements	Reasons
1. $\ell \perp m$	1. Given
2. $\angle 1$ and $\angle 2$ rt. $\angle$s	2. $\perp$ lines intersect to form 4 rt. $\angle$s.
3. $\angle 1 \cong \angle 2$	3. All rt. $\angle$s $\cong$.

37. Given: $\angle ABD \cong \angle CBD$, $\angle ABD$ and $\angle DBC$ form a linear pair.
Prove: $\angle ABD$ and $\angle CBD$ are rt. $\angle$s.

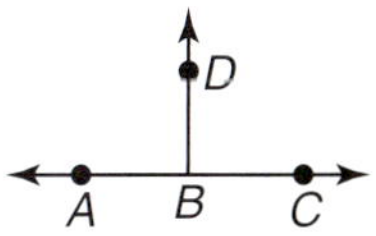

Proof:

Statements	Reasons
1. $\angle ABD \cong \angle CBD$, $\angle ABD$ and $\angle CBD$ form a linear pair.	1. Given
2. $\angle ABD$ and $\angle CBD$ are supplementary.	2. Linear pairs are supplementary.
3. $\angle ABD$ and $\angle CBD$ are rt. $\angle$s.	3. If $\angle$s are $\cong$ and suppl., they are rt. $\angle$s.

39. Given: $m\angle RSW = m\angle TSU$
Prove: $m\angle RST = m\angle WSU$

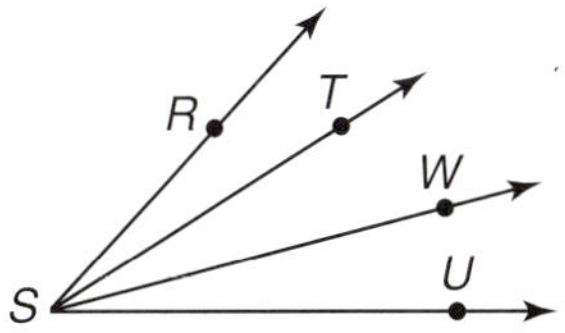

Proof:

Statements	Reasons
1. $m\angle RSW = m\angle TSU$	1. Given
2. $m\angle RSW = m\angle RST + m\angle TSW$, $m\angle TSU = m\angle TSW + m\angle WSU$	2. Angle Addition Postulate
3. $m\angle RST + m\angle TSW = m\angle TSW + m\angle WSU$	3. Substitution
4. $m\angle TSW = m\angle TSW$	4. Reflexive Prop.
5. $m\angle RST = m\angle WSU$	5. Subt. Prop.

Selected Answers

41. Because the lines are perpendicular, the angles formed are right angles. All right angles are congruent. Therefore, $\angle 1$ is congruent to $\angle 2$. **43.** Two angles that are supplementary to the same angle are congruent. Answers should include the following.

- $\angle 1$ and $\angle 2$ are supplementary; $\angle 2$ and $\angle 3$ are supplementary.
- $\angle 1$ and $\angle 3$ are vertical angles, and are therefore congruent.
- If two angles are complementary to the same angle, then the angles are congruent. **45.** B

47. Given: X is the midpoint of $\overline{WY}$.
Prove: $WX + YZ = XZ$

Proof:

Statements	Reasons
1. X is the midpoint of $\overline{WY}$.	1. Given
2. $WX = XY$	2. Def. of midpoint
3. $XY + YZ = XZ$	3. Segment Addition Postulate
4. $WX + YZ = XZ$	4. Substitution

49. $\angle ONM$, $\angle MNR$ **51.** N or R **53.** obtuse
55. $\angle NML$, $\angle NMP$, $\angle NMO$, $\angle RNM$, $\angle ONM$

Pages 115–120 Chapter 2 Study Guide and Review

1. conjecture **3.** compound **5.** hypothesis **7.** Postulates
9. $m\angle A + m\angle B = 180$ **11.** $LMNO$ is a square.

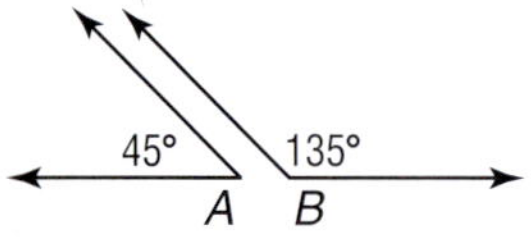

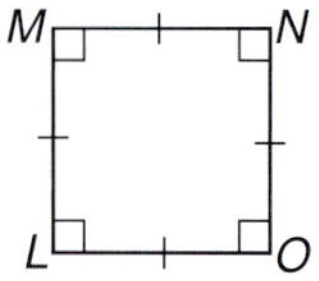

13. In a right triangle with right angle C, $a^2 + b^2 = c^2$ or the sum of the measures of two supplementary angles is 180; true. **15.** $-1 > 0$, and in a right triangle with right angle C, $a^2 + b^2 = c^2$, or the sum of the measures of two supplementary angles is 180; false. **17.** In a right triangle with right angle C, $a^2 + b^2 = c^2$ and the sum of the measures of two supplementary angles is 180, and $-1 > 0$; false. **19.** Converse: If a month has 31 days, then it is March. False; July has 31 days. Inverse: If a month is not March, then it does not have 31 days. False; July has 31 days. Contrapositive: If a month does not have 31 days, then it is not March; true. **21.** true **23.** false **25.** Valid; by definition, adjacent angles have a common vertex.
27. yes; Law of Detachment **29.** yes; Law of Syllogism
31. Always; if P is the midpoint of $\overline{XY}$, then $\overline{XP} \cong \overline{PY}$. By definition of congruent segments, $XP = PY$.
33. Sometimes; if the points are collinear. **35.** Sometimes; if the right angles form a linear pair. **37.** Never; adjacent angles must share a common side, and vertical angles do not. **39.** Distributive Property **41.** Subtraction Property

43. Given: $5 = 2 - \frac{1}{2}x$
Prove: $x = -6$
Proof:

Statements	Reasons
1. $5 = 2 - \frac{1}{2}x$	1. Given
2. $5 - 2 = 2 - \frac{1}{2}x - 2$	2. Subt. Prop.
3. $3 = -\frac{1}{2}x$	3. Substitution
4. $-2(3) = -2\left(-\frac{1}{2}x\right)$	4. Mult. Prop
5. $-6 = x$	5. Substitution
6. $x = -6$	6. Symmetric Prop.

45. Given: $AC = AB$, $AC = 4x + 1$, $AB = 6x - 13$
Prove: $x = 7$

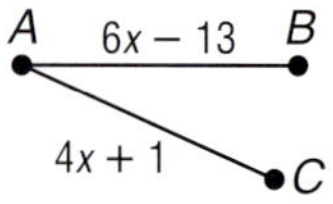

Proof:

Statements	Reasons
1. $AC = AB$, $AC = 4x + 1$, $AB = 6x - 13$	1. Given
2. $4x + 1 = 6x - 13$	2. Substitution
3. $4x + 1 - 1 = 6x - 13 - 1$	3. Subt. Prop.
4. $4x = 6x - 14$	4. Substitution
5. $4x - 6x = 6x - 14 - 6x$	5. Subt. Prop.
6. $-2x = -14$	6. Substitution
7. $\frac{-2x}{-2} = \frac{-14}{-2}$	7. Div. Prop.
8. $x = 7$	8. Substitution

47. Reflexive Property **49.** Addition Property
51. Division or Multiplication Property

53. Given: $BC = EC$, $CA = CD$
Prove: $BA = DE$

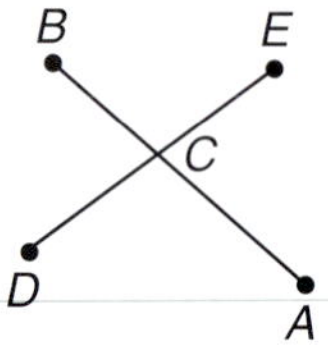

Proof:

Statements	Reasons
1. $BC = EC$, $CA = CD$	1. Given
2. $BC + CA = EC + CA$	2. Add. Prop.
3. $BC + CA = EC + CD$	3. Substitution
4. $BC + CA = BA$, $EC + CD = DE$	4. Seg. Add. Post.
5. $BA = DE$	5. Substitution

55. 145 **57.** 90

Chapter 3 Parallel and Perpendicular Lines

Page 125 Chapter 3 Getting Started

1. $\overleftrightarrow{PQ}$ **3.** $\overleftrightarrow{ST}$ **5.** $\angle 4$, $\angle 6$, $\angle 8$ **7.** $\angle 1$, $\angle 5$, $\angle 7$ **9.** 9 **11.** $-\frac{3}{2}$

Pages 128–131 Lesson 3-1

1. Sample answer: The bottom and top of a cylinder are contained in parallel planes.

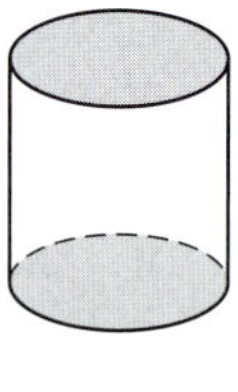

3. Sample answer: looking down railroad tracks **5.** $\overline{AB}$, $\overline{JK}$, $\overline{LM}$ **7.** q and r, q and t, r and t **9.** p and r, p and t, r and t
11. alternate interior **13.** consecutive interior **15.** p; consecutive interior **17.** q; alternate interior **19.** Sample answer: The roof and the floor are parallel planes. **21.** Sample answer: The top of the memorial "cuts" the pillars. **23.** ABC, ABQ, PQR, CDS, APU, DET **25.** $\overline{AP}$, $\overline{BQ}$, $\overline{CR}$, $\overline{FU}$, $\overline{PU}$, $\overline{QR}$, $\overline{RS}$, $\overline{TU}$ **27.** $\overline{BC}$, $\overline{CD}$, $\overline{DE}$, $\overline{EF}$, $\overline{QR}$, $\overline{RS}$, $\overline{ST}$, $\overline{TU}$ **29.** a and c, a and r, r and c
31. a and b, a and c, b and c **33.** alternate exterior
35. corresponding **37.** alternate interior **39.** consecutive interior **41.** p; alternate interior **43.** ℓ; alternate exterior
45. q; alternate interior **47.** m; consecutive interior
49. $\overline{CG}$, $\overline{DH}$, $\overline{EI}$ **51.** No; plane ADE will intersect all the planes if they are extended. **53.** infinite number

55. Sample answer: Parallel lines and planes are used in architecture to make structures that will be stable. Answers should include the following.

- Opposite walls should form parallel planes; the floor may be parallel to the ceiling.
- The plane that forms a stairway will not be parallel to some of the walls.

57. 16, 20, or 28

59. Given: $\overline{PQ} \cong \overline{ZY}$, $\overline{QR} \cong \overline{XY}$
Prove: $\overline{PR} \cong \overline{XZ}$

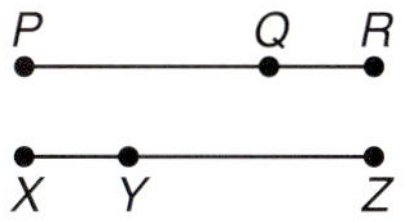

Proof: Since $\overline{PQ} \cong \overline{ZY}$ and $\overline{QR} \cong \overline{XY}$, $PQ = ZY$ and $QR = XY$ by the definition of congruent segments. By the Addition Property, $PQ + QR = ZY + XY$. Using the Segment Addition Postulate, $PR = PQ + QR$ and $XZ = XY + YZ$. By substitution, $PR = XZ$. Because the measures are equal, $\overline{PR} \cong \overline{XZ}$ by the definition of congruent segments.

61. $m\angle EFG$ is less than 90; Detachment. **63.** 8.25
65. 15.81 **67.** 10.20

69.
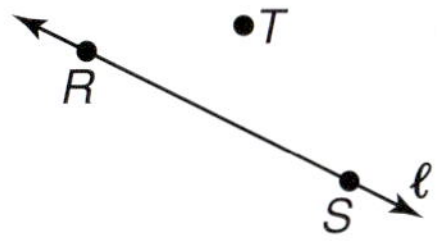

71. 90, 90 **73.** 72, 108
75. 76, 104

Pages 136–138 Lesson 3-2

1. Sometimes; if the transversal is perpendicular to the parallel lines, then $\angle 1$ and $\angle 2$ are right angles and are congruent. **3.** 1 **5.** 110 **7.** 70 **9.** 55 **11.** $x = 13, y = 6$ **13.** 67 **15.** 75 **17.** 105 **19.** 105 **21.** 43 **23.** 43 **25.** 137 **27.** 60 **29.** 70 **31.** 120 **33.** $x = 34, y = \pm 5$ **35.** 113 **37.** $x = 14, y = 11, z = 73$ **39.** (1) Given (2) Corresponding Angles Postulate (3) Vertical Angles Theorem (4) Transitive Property

41. Given: $\ell \perp m$, $m \parallel n$
Prove: $\ell \perp n$

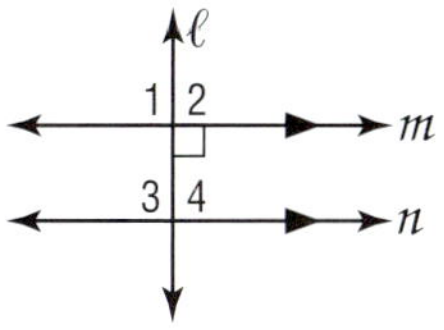

Proof: Since $\ell \perp m$, we know that $\angle 1 \cong \angle 2$, because perpendicular lines form congruent right angles. Then by the Corresponding Angles Postulate, $\angle 1 \cong \angle 3$ and $\angle 2 \cong \angle 4$. By the definition of congruent angles, $m\angle 1 = m\angle 2$, $m\angle 1 = m\angle 3$, and $m\angle 2 = m\angle 4$. By substitution, $m\angle 3 = m\angle 4$. Because $\angle 3$ and $\angle 4$ form a congruent linear pair, they are right angles. By definition, $\ell \perp n$.

43. $\angle 2$ and $\angle 6$ are consecutive interior angles for the same transversal, which makes them supplementary because $\overline{WX} \parallel \overline{YZ}$. $\angle 4$ and $\angle 6$ are not necessarily supplementary because $\overline{XY}$ may not be parallel to $\overline{WZ}$. **45.** C **47.** $\overline{FG}$ **49.** *CDH* **51.** $m\angle 1 = 56$ **53.** H: it rains this evening; C: I will mow the lawn tomorrow **55.** $-\frac{2}{3}$ **57.** $\frac{3}{8}$ **59.** $-\frac{4}{5}$

Page 138 Practice Quiz 1

1. *p*; alternate exterior **3.** *q*; alternate interior **5.** 75

Pages 142–144 Lesson 3-3

1. horizontal; vertical **3.** horizontal line, vertical line
5. $-\frac{1}{2}$ **7.** 2 **9.** parallel

11.
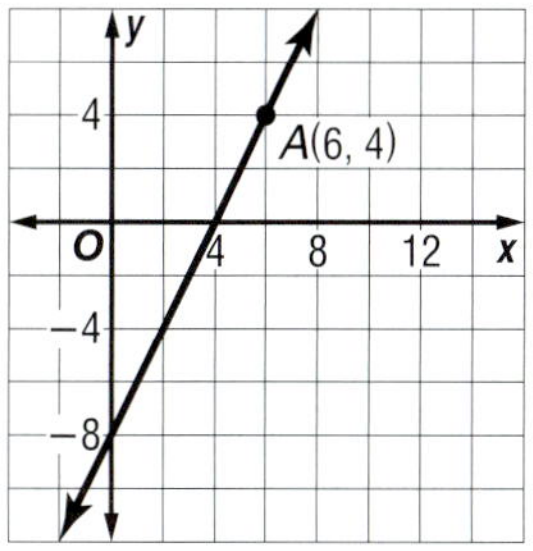

13. (1500, −120) or (−1500, −120)
15. $\frac{1}{7}$ **17.** −5
19. perpendicular
21. neither **23.** parallel
25. −3 **27.** 6 **29.** 6
31. undefined

33.
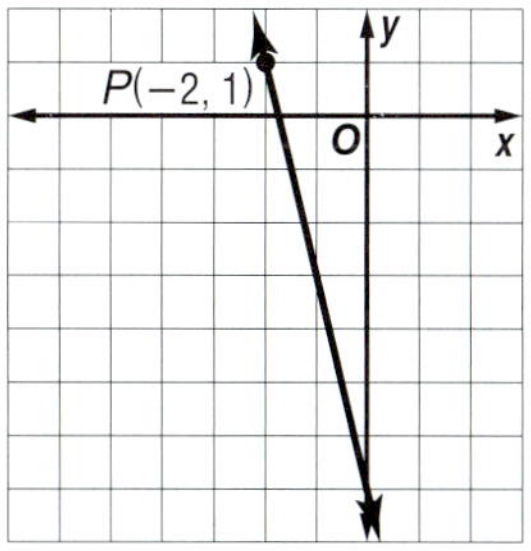

35.
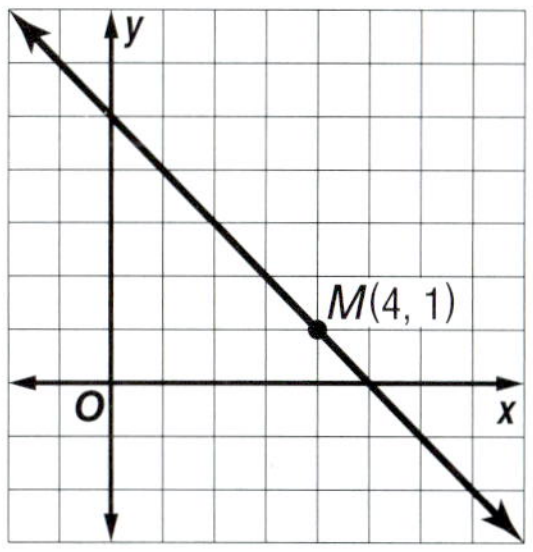

37.
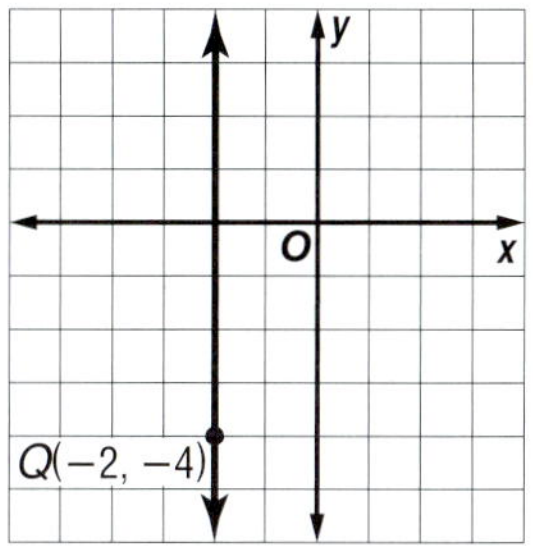

39. Sample answer: 0.24
41. 2016

43. $\frac{19}{2}$;
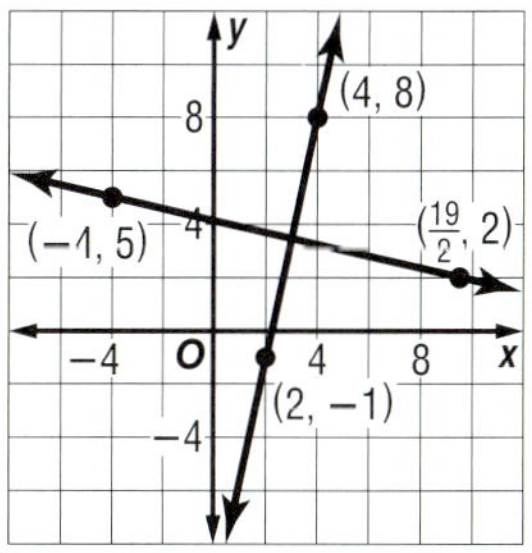

45. 2001
47. $y = \frac{1}{2}x - \frac{11}{2}$
49. C **51.** 131 **53.** 49
55. 19 **57.** ℓ; alternate exterior
59. *p*; alternate interior
61. *m*; alternate interior

63. *H*, *I*, and *J* are noncollinear.
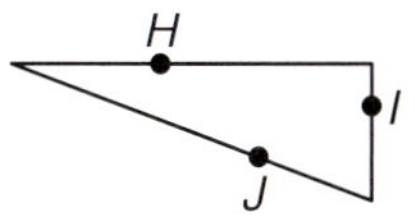

65. *R*, *S*, and *T* are collinear.
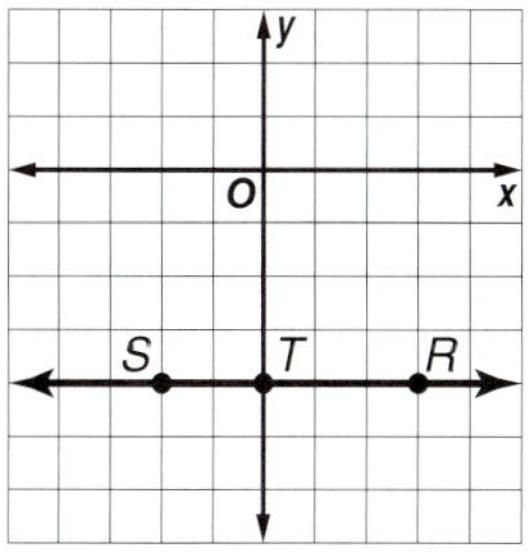

67. obtuse **69.** obtuse
71. $y = -\frac{1}{2}x - \frac{5}{4}$

Pages 147–150 Lesson 3-4

1. Sample answer: Use the point-slope form where $(x_1, y_1) = (-2, 8)$ and $m = -\frac{2}{5}$.

3. Sample answer: $y = x$

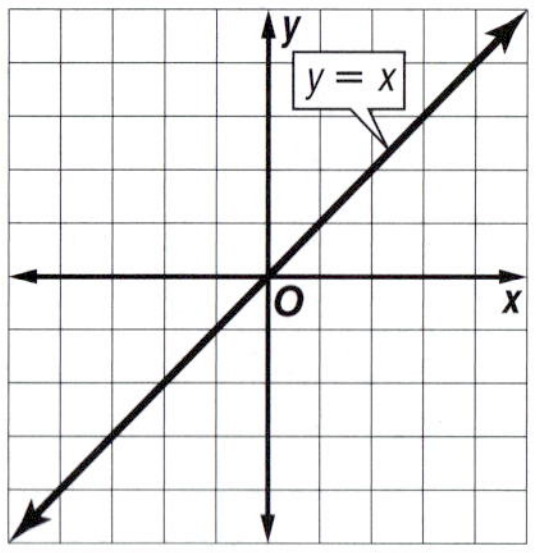

5. $y = -\frac{3}{5}x - 2$ **7.** $y + 1 = \frac{3}{2}(x - 4)$ **9.** $y - 137.5 = 1.25(x - 20)$ **11.** $y = -x + 2$ **13.** $y = 39.95$, $y = 0.95x + 4.95$ **15.** $y = \frac{1}{6}x - 4$ **17.** $y = \frac{5}{8}x - 6$ **19.** $y = -x - 3$

21. $y - 1 = 2(x - 3)$ **23.** $y + 5 = -\frac{4}{5}(x + 12)$ **25.** $y - 17.12 = 0.48(x - 5)$ **27.** $y = -3x - 2$ **29.** $y = 2x - 4$ **31.** $y = -x + 5$ **33.** $y = -\frac{1}{8}x$ **35.** $y = -3x + 5$ **37.** $y = -\frac{3}{5}x + 3$ **39.** $y = -\frac{1}{5}x - 4$ **41.** no slope-intercept form, $x = -6$ **43.** $y = \frac{2}{5}x - \frac{24}{5}$ **45.** $y = 0.05x + 750$, where x = total price of appliances sold **47.** $y = -750x + 10{,}800$ **49.** in 10 days **51.** $y = x - 180$ **53.** Sample answer: In the equation of a line, the b value indicates the fixed rate, while the mx value indicates charges based on usage. Answers should include the following.

- The fee for air time can be considered the slope of the equation.
- We can find where the equations intersect to see where the plans would be equal.

55. B **57.** undefined **59.** 58 **61.** 75 **63.** 73

65. Given: $AC = DF$, $AB = DE$
Prove: $BC = EF$

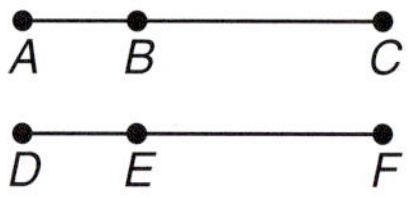

Proof:

Statements	Reasons
1. $AC = DF$, $AB = DE$	1. Given
2. $AC = AB + BC$ $DF = DE + EF$	2. Segment Addition Postulate
3. $AB + BC = DE + EF$	3. Substitution Property
4. $BC = EF$	4. Subtraction Property

67. 26.69 **69.** ∠1 and ∠5, ∠2 and ∠6, ∠4 and ∠8, ∠3 and ∠7 **71.** ∠2 and ∠8, ∠3 and ∠5

Page 150 Practice Quiz 2

1. neither **3.** $\frac{7}{2}$ **5.** $\frac{5}{4}$ **7.** $y = -\frac{4}{5}x + \frac{16}{5}$ **9.** $y + 8 = -\frac{1}{4}(x - 5)$

Pages 154–157 Lesson 3-5

1. Sample answer: Use a pair of alternate exterior ⦞ that are ≅ and cut by a transversal; show that a pair of consecutive interior ⦞ are suppl.; show that alternate interior ⦞ are ≅; show two lines are ⊥ to same line; show corresponding ⦞ are ≅. **3.** Sample answer: A basketball court has parallel lines, as does a newspaper. The edges should be equidistant along the entire line. **5.** $\ell \parallel m$; ≅ alt. int. ⦞ **7.** $p \parallel q$; ≅ alt. ext. ⦞ **9.** 11.375 **11.** The slope of $\overleftrightarrow{CD}$ is $\frac{1}{8}$, and the slope of line $\overleftrightarrow{AB}$ is $\frac{1}{7}$. The slopes are not equal, so the lines are not parallel. **13.** $a \parallel b$; ≅ alt. int. ⦞ **15.** $\ell \parallel m$; ≅ corr. ⦞ **17.** $\overleftrightarrow{AE} \parallel \overleftrightarrow{BF}$; ≅ corr. ⦞ **19.** $\overleftrightarrow{AC} \parallel \overleftrightarrow{EG}$; ≅ alt. int. ⦞ **21.** $\overleftrightarrow{HS} \parallel \overleftrightarrow{JT}$; ≅ corr. ⦞ **23.** $\overleftrightarrow{KN} \parallel \overleftrightarrow{PR}$; suppl. cons. int. ⦞

25. **1.** Given
2. Definition of perpendicular
3. All rt. ⦞ are ≅.
4. If corresponding ⦞ are ≅, then lines are ∥.

27. 15 **29.** −8 **31.** 21.6

33. Given: ∠4 ≅ ∠6
Prove: $\ell \parallel m$

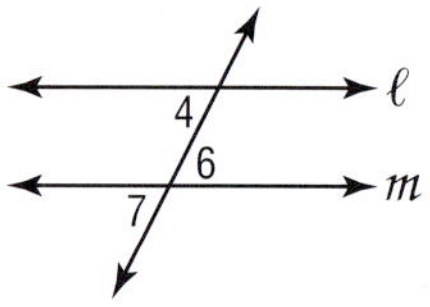

Proof: We know that ∠4 ≅ ∠6. Because ∠6 and ∠7 are vertical angles they are congruent. By the Transitive Property of Congruence, ∠4 ≅ ∠7. Since ∠4 and ∠7 are corresponding angles, and they are congruent, $\ell \parallel m$.

35. Given: $\overline{AD} \perp \overline{CD}$
∠1 ≅ ∠2
Prove: $\overline{BC} \perp \overline{CD}$

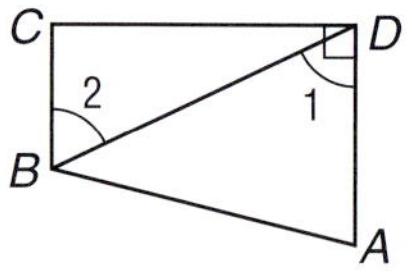

Proof:

Statements	Reasons
1. $\overline{AD} \perp \overline{CD}$, ∠1 ≅ ∠2	1. Given
2. $\overline{AD} \parallel \overline{BC}$	2. If alternate interior ⦞ are ≅, lines are ∥.
3. $\overline{BC} \perp \overline{CD}$	3. Perpendicular Transversal Th.

37. Given: ∠RSP ≅ ∠PQR
∠QRS and ∠PQR are supplementary.
Prove: $\overline{PS} \parallel \overline{QR}$

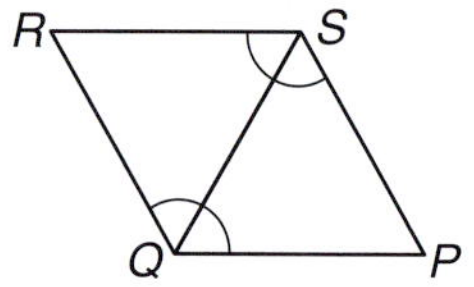

Proof:

Statements	Reasons
1. ∠RSP ≅ ∠PQR ∠QRS and ∠PQR are supplementary.	1. Given
2. $m\angle RSP = m\angle PQR$	2. Def. of ≅ ⦞
3. $m\angle QRS + m\angle PQR = 180$	3. Def. of suppl. ⦞
4. $m\angle QRS + m\angle RSP = 180$	4. Substitution
5. ∠QRS and ∠RSP are supplementary.	5. Def. of suppl. ⦞
6. $\overline{PS} \parallel \overline{QR}$	6. If consecutive interior ⦞ are suppl., lines ∥.

39. No, the slopes are not the same. **41.** The 10-yard lines will be parallel because they are all perpendicular to the sideline and two or more lines perpendicular to the same line are parallel. **43.** See students' work. **45.** B **47.** $y = 0.3x - 6$ **49.** $y = -\frac{1}{2}x + \frac{19}{2}$ **51.** $-\frac{5}{4}$ **53.** 1 **55.** undefined

57.

p	q	p and q
T	T	T
T	F	F
F	T	F
F	F	F

59.

p	q	$\sim p$	$\sim p \wedge q$
T	T	F	F
T	F	F	F
F	T	T	T
F	F	T	F

61. complementary angles **63.** $\sqrt{85} \approx 9.22$

Pages 162–164 Lesson 3-6

1. Construct a perpendicular line between them.
3. Sample answer: Measure distances at different parts; compare slopes; measure angles. Finding slopes is the most readily available method.

5.

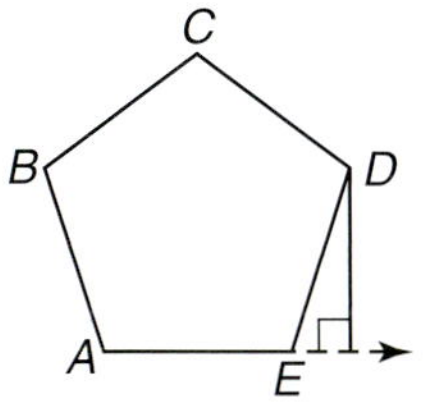

7. 0.9
9. 5 units;

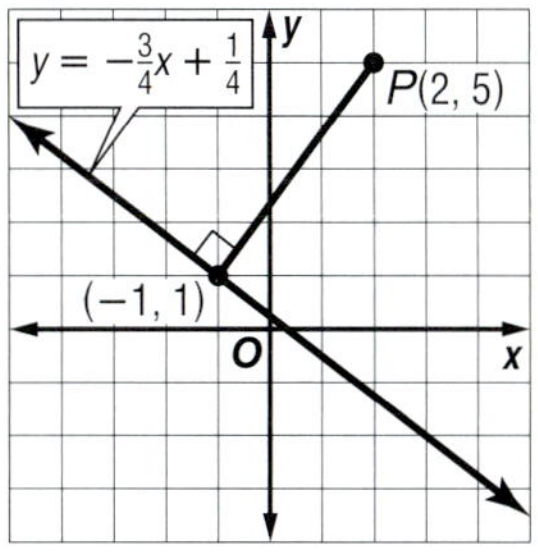

11.

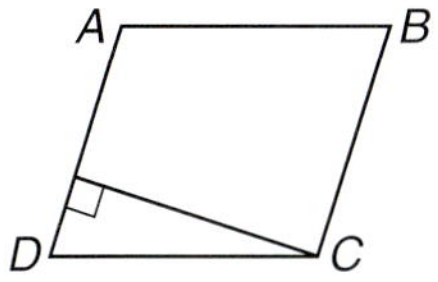

13.

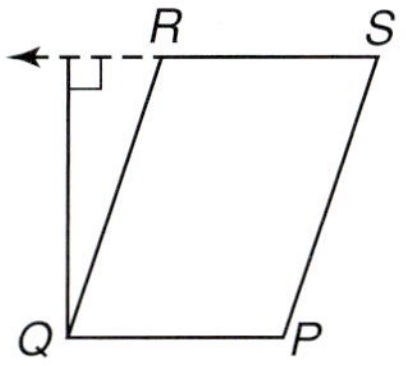

15.

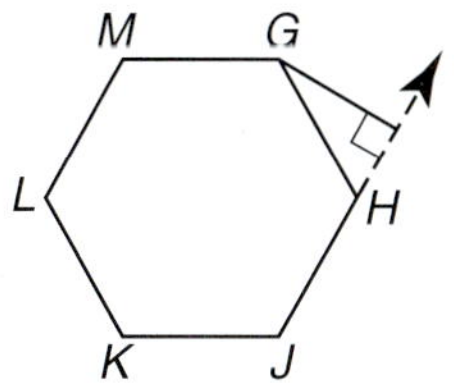

17. $d = 3$;

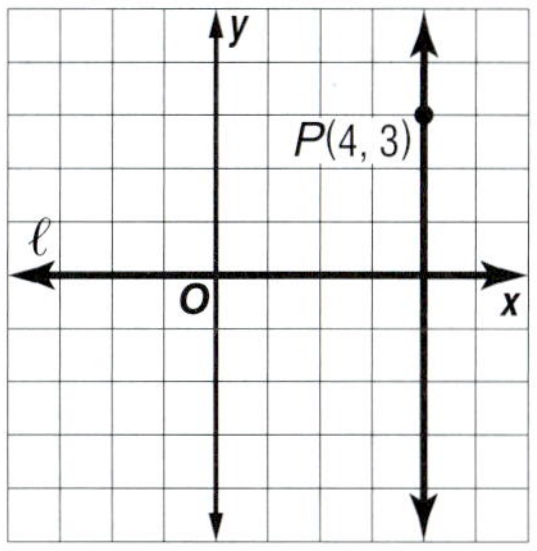

19. 4 **21.** $\sqrt{5}$ **23.** $\frac{7\sqrt{5}}{5}$

25. 1;

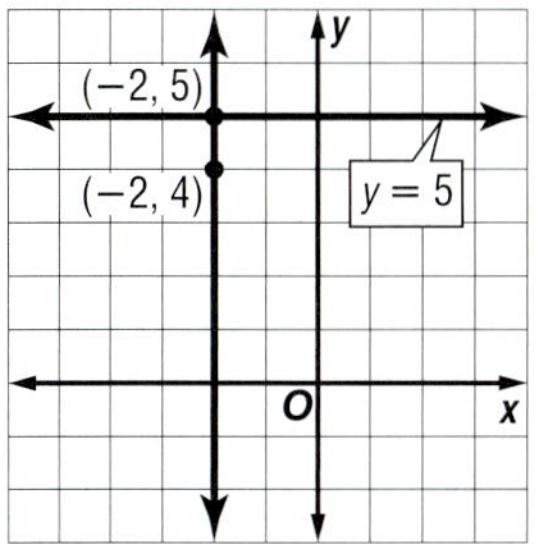

27. $\sqrt{13}$;

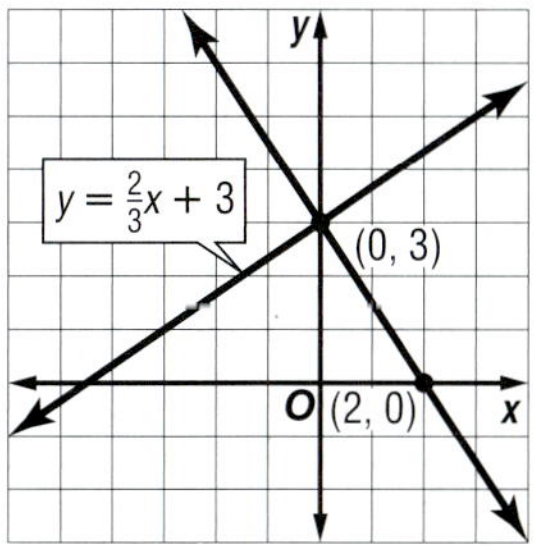

29. It is everywhere equidistant from the ceiling. **31.** 6
33. Sample answer: We want new shelves to be parallel so they will line up. Answers should include the following.

- After marking several points, a slope can be calculated, which should be the same slope as the original brace.
- Building walls requires parallel lines.

35. D **37.** $\overleftrightarrow{DA} \parallel \overleftrightarrow{EF}$; corresponding ∠s **39.** $y = \frac{1}{2}x + 3$
41. $y = \frac{2}{3}x - 2$ **43.** $y = \frac{2}{3}x + \frac{11}{3}$

Pages 167–170 Chapter 3 Study Guide and Review

1. alternate **3.** parallel **5.** alternate exterior
7. consecutive **9.** alternate exterior **11.** corresponding
13. consecutive. interior **15.** alternate interior **17.** 53
19. 127 **21.** 127 **23.** neither **25.** perpendicular

27.

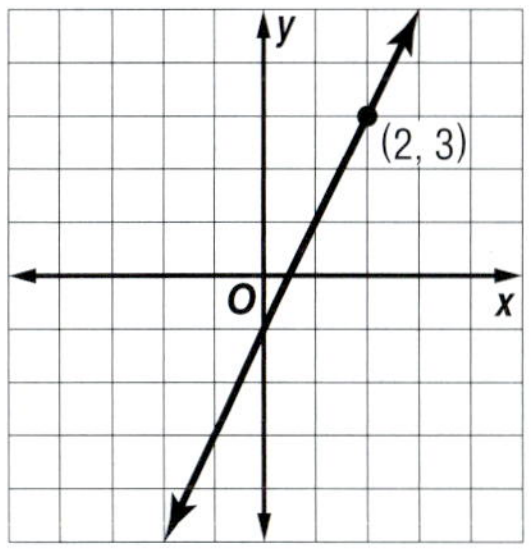

29. $y = 2x - 7$
31. $y = -\frac{2}{7}x + 4$
33. $y = 5x - 3$
35. $\overleftrightarrow{AL}$ and $\overleftrightarrow{BJ}$, alternate exterior ∠s ≅
37. $\overleftrightarrow{CF}$ and $\overleftrightarrow{GK}$, 2 lines ⊥ same line
39. $\overleftrightarrow{CF}$ and $\overleftrightarrow{GK}$, consecutive interior ∠s suppl. **41.** $\sqrt{5}$

Chapter 4 Congruent Triangles

Pages 177 Chapter 4 Getting Started

1. $-6\frac{1}{2}$ **3.** 1 **5.** $2\frac{3}{4}$ **7.** ∠2, ∠12, ∠15, ∠6, ∠9, ∠3, ∠13
9. ∠6, ∠9, ∠3, ∠13, ∠2, ∠8, ∠12, ∠15 **11.** ∠11.2
13. ∠14.6

Pages 180–183 Lesson 4-1

1. Triangles are classified by sides and angles. For example, a triangle can have a right angle and have no two sides congruent. **3.** Always; equiangular triangles have three acute angles. **5.** obtuse **7.** △*MJK*, △*KLM*, △*JKN*, △*LMN*
9. $x = 4$, $JM = 3$, $MN = 3$, $JN = 2$ **11.** $TW = \sqrt{125}$, $WZ = \sqrt{74}$, $TZ = \sqrt{61}$; scalene **13.** right **15.** acute
17. obtuse **19.** equilateral, equiangular **21.** isosceles, acute **23.** △*BAC*, △*CDB* **25.** △*ABD*, △*ACD*, △*BAC*, △*CDB* **27.** $x = 5$, $MN = 9$, $MP = 9$, $NP = 9$
29. $x = 8$, $JL = 11$, $JK = 11$, $KL = 7$ **31.** Scalene; it is 184 miles from Lexington to Nashville, 265 miles from Cairo to Lexington, and 144 miles from Cairo to Nashville.
33. $AB = \sqrt{106}$, $BC = \sqrt{233}$, $AC = \sqrt{65}$; scalene
35. $AB = \sqrt{29}$, $BC = 4$, $AC = \sqrt{29}$; isosceles
37. $AB = \sqrt{124}$, $BC = \sqrt{124}$, $AC = 8$; isosceles

39. Given: $m\angle NPM = 33$
Prove: △*RPM* is obtuse.

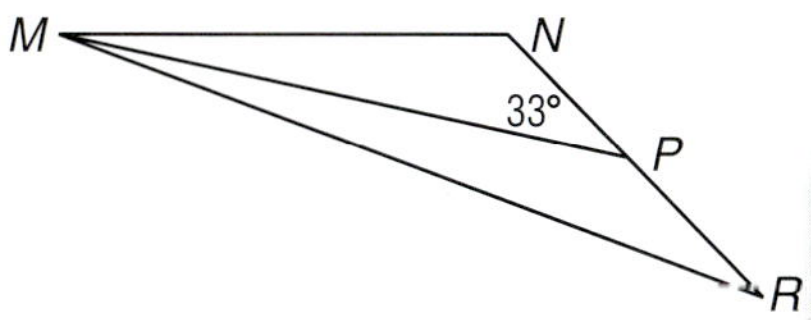

Proof: ∠*NPM* and ∠*RPM* form a linear pair. ∠*NPM* and ∠*RPM* are supplementary because if two angles form a linear pair, then they are supplementary. So, $m\angle NPM + m\angle RPM = 180$. It is given that $m\angle NPM = 33$. By substitution, $33 + m\angle RPM = 180$. Subtract to find that $m\angle RPM = 147$. ∠*RPM* is obtuse by definition. △*RPM* is obtuse by definition.

Selected Answers

41. $AD = \sqrt{\left(0 - \frac{a}{2}\right)^2 + (0 - b)^2}$ $CD = \sqrt{\left(a - \frac{a}{2}\right)^2 + (0 - b)^2}$

$= \sqrt{\left(-\frac{a}{2}\right)^2 + (-b)^2}$ $= \sqrt{\left(\frac{a}{2}\right)^2 + (-b)^2}$

$= \sqrt{\frac{a^2}{4} + b^2}$ $= \sqrt{\frac{a^2}{4} + b^2}$

$AD = CD$, so $\overline{AD} \cong \overline{CD}$. $\triangle ADC$ is isosceles by definition.

43. Sample answer: Triangles are used in construction as structural support. Answers should include the following.

- Triangles can be classified by sides and angles. If the measure of each angle is less than 90, the triangle is acute. If the measure of one angle is greater than 90, the triangle is obtuse. If one angle equals 90°, the triangle is right. If each angle has the same measure, the triangle is equiangular. If no two sides are congruent, the triangle is scalene. If at least two sides are congruent, it is isosceles. If all of the sides are congruent, the triangle is equilateral.
- Isosceles triangles seem to be used more often in architecture and construction.

45. B **47.** $\sqrt{8}$;

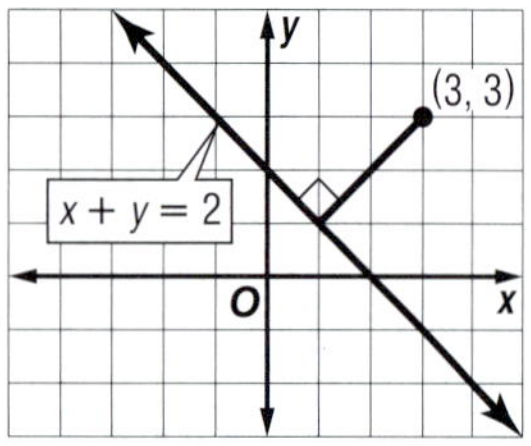

49. 15 **51.** 44 **53.** any three: ∠2 and ∠11, ∠3 and ∠6, ∠4 and ∠7, ∠3 and ∠12, ∠7 and ∠10, ∠8 and ∠11 **55.** ∠6, ∠9, and ∠12 **57.** ∠2, ∠5, and ∠8

Pages 188–191 Lesson 4-2

1. Sample answer: ∠2 and ∠3 are the remote interior angles of exterior ∠1.

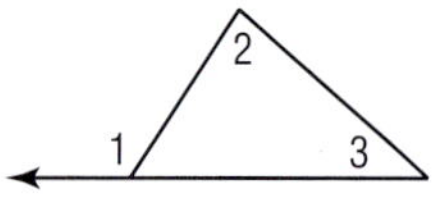

3. 43 **5.** 55 **7.** 147 **9.** 25 **11.** 93 **13.** 65, 65 **15.** 76 **17.** 49 **19.** 53 **21.** 32 **23.** 44 **25.** 123 **27.** 14 **29.** 53 **31.** 103 **33.** 50 **35.** 40 **37.** 129

39. Given: $\angle FGI \cong \angle IGH$, $\overline{GI} \perp \overline{FH}$

Prove: $\angle F \cong \angle H$

Proof:

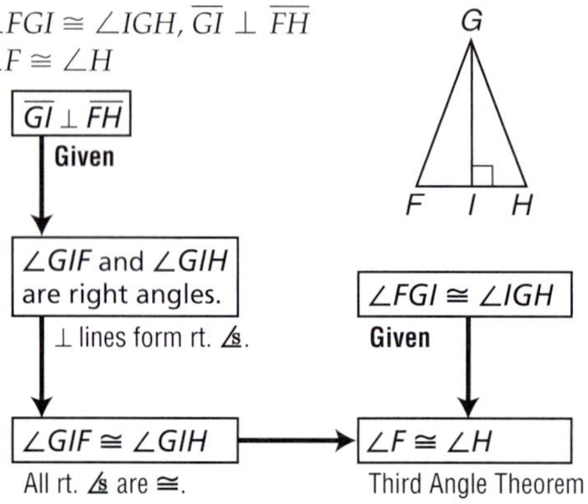

41. Given: $\triangle ABC$

Prove: $m\angle CBD = m\angle A + m\angle C$

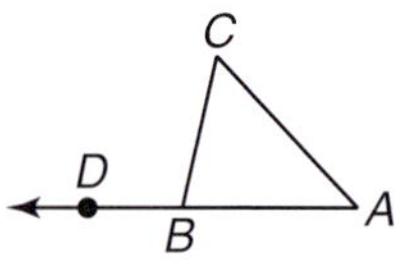

Proof:

Statements	Reasons
1. $\triangle ABC$	1. Given
2. ∠CBD and ∠ABC form a linear pair.	2. Def. of linear pair
3. ∠CBD and ∠ABC are supplementary.	3. If 2 ∠s form a linear pair,they are suppl.
4. $m\angle CBD + m\angle ABC = 180$	4. Def. of suppl.
5. $m\angle A + m\angle ABC + m\angle C = 180$	5. Angle Sum Theorem
6. $m\angle A + m\angle ABC + m\angle C = m\angle CBD + m\angle ABC$	6. Substitution
7. $m\angle A + m\angle C = m\angle CBD$	7. Subtraction Property

43. Given: $\triangle MNO$
∠M is a right angle.

Prove: There can be at most one right angle in a triangle.

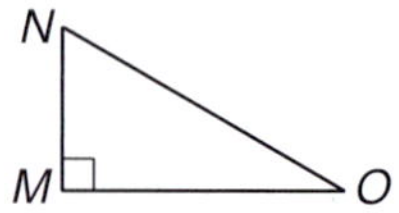

Proof:
In $\triangle MNO$, ∠M is a right angle. $m\angle M + m\angle N + m\angle O = 180$. $m\angle M = 90$, so $m\angle N + m\angle O = 90$. If ∠N were a right angle, then $m\angle O = 0$. But that is impossible, so there cannot be two right angles in a triangle.

Given: $\triangle PQR$
∠P is obtuse.

Prove: There can be at most one obtuse angle in a triangle.

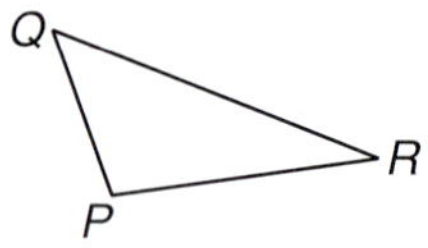

Proof:
In $\triangle PQR$, ∠P is obtuse. So $m\angle P > 90$. $m\angle P + m\angle Q + m\angle R = 180$. It must be that $m\angle Q + m\angle R < 90$. So, ∠Q and ∠R must be acute.

45. $m\angle 1 = 48$, $m\angle 2 = 60$, $m\angle 3 = 72$ **47.** A **49.** $\triangle AED$ **51.** $\triangle BEC$ **53.** $\sqrt{20}$ units **55.** $\frac{\sqrt{117}}{13}$ units **57.** $x = 112$, $y = 28$, $z = 22$ **59.** reflexive **61.** symmetric **63.** transitive

Pages 195–198 Lesson 4-3

1. The sides and the angles of the triangle are not affected by a congruence transformation, so congruence is preserved. **3.** $\triangle AFC \cong \triangle DFB$ **5.** $\angle W \cong \angle S$, $\angle X \cong \angle T$, $\angle Z \cong \angle J$, $\overline{WX} \cong \overline{ST}$, $\overline{XZ} \cong \overline{TJ}$, $\overline{WZ} \cong \overline{SJ}$ **7.** $QR = 5$, $Q'R' = 5$, $RT = 3$, $R'T' = 3$, $QT = \sqrt{34}$, and $Q'T' = \sqrt{34}$. Use a protractor to confirm that the corresponding angles are congruent; flip. **9.** $\triangle CFH \cong \triangle JKL$ **11.** $\triangle WPZ \cong \triangle QVS$ **13.** $\angle T \cong \angle X$, $\angle U \cong \angle Y$, $\angle V \cong \angle Z$, $\overline{TU} \cong \overline{XY}$, $\overline{UV} \cong \overline{YZ}$, $\overline{TV} \cong \overline{XZ}$ **15.** $\angle B \cong \angle D$, $\angle C \cong \angle G$, $\angle F \cong \angle H$, $\overline{BC} \cong \overline{DG}$, $\overline{CF} \cong \overline{GH}$, $\overline{BF} \cong \overline{DH}$ **17.** $\triangle 1 \cong \triangle 10$, $\triangle 2 \cong \triangle 9$, $\triangle 3 \cong \triangle 8$, $\triangle 4 \cong \triangle 7$, $\triangle 5 \cong \triangle 6$ **19.** △s 1, 5, 6, and 11, △s 3, 8, 10, and 12, △s 2, 4, 7, and 9 **21.** We need to know that all of the angles are congruent and that the other corrresponding sides are congruent. **23.** Flip; $MN = 8$, $M'N' = 8$, $NP = 2$, $N'P' = 2$, $MP = \sqrt{68}$, and $M'P' = \sqrt{68}$. Use a protractor to confirm that the corresponding angles are congruent.

25. Turn; $JK = \sqrt{40}$, $J'K' = \sqrt{40}$, $KL = \sqrt{29}$, $K'L' = \sqrt{29}$, $JL = \sqrt{17}$, and $J'L' = \sqrt{17}$. Use a protractor to confirm that the corresponding angles are congruent.

27. True;

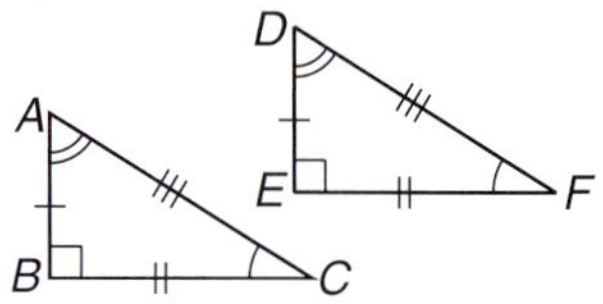

29.

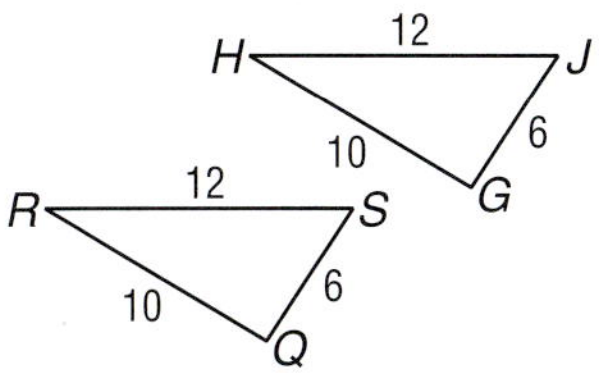

31.

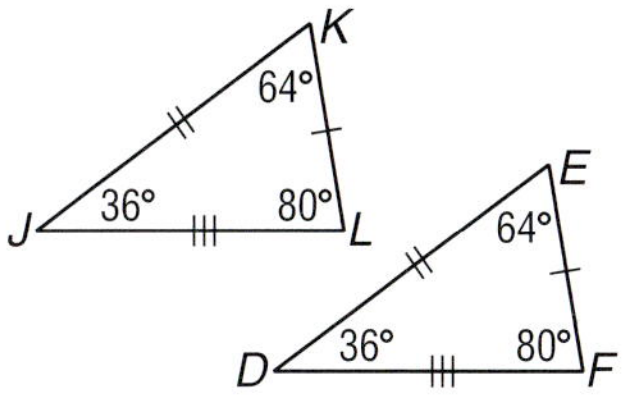

33. Given: $\triangle RST \cong \triangle XYZ$

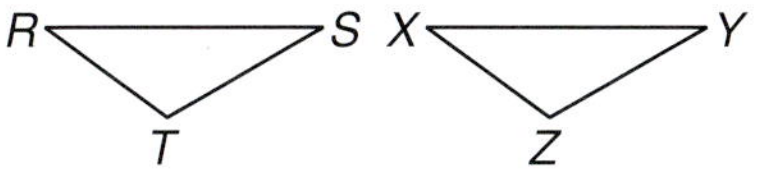

Prove: $\triangle XYZ \cong \triangle RST$

Proof:

$\triangle RST \cong \triangle XYZ$
Given

$\angle R \cong \angle X$, $\angle S \cong \angle Y$, $\angle T \cong \angle Z$, $\overline{RS} \cong \overline{XY}$, $\overline{ST} \cong \overline{YZ}$, $\overline{RT} \cong \overline{XZ}$
CPCTC

$\angle X \cong \angle R$, $\angle Y \cong \angle S$, $\angle Z \cong \angle T$, $\overline{XY} \cong \overline{RS}$, $\overline{YZ} \cong \overline{ST}$, $\overline{XZ} \cong \overline{RT}$
Congruence of ∡ and segments is symmetric.

$\triangle XYZ \cong \triangle RST$
Def. of ≅ ∡

35. Given: $\triangle DEF$
Prove: $\triangle DEF \cong \triangle DEF$

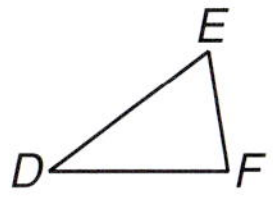

Proof:

$\triangle DEF$
Given

$\overline{DE} \cong \overline{DE}$, $\overline{EF} \cong \overline{EF}$, $\overline{DF} \cong \overline{DF}$
Congruence of segments is reflexive.

$\angle D \cong \angle D$, $\angle E \cong \angle E$, $\angle F \cong \angle F$
Congruence of ∡ is reflexive.

$\triangle DEF \cong \triangle DEF$
Def. of ≅ △s

37. Sample answer: Triangles are used in bridge design for structure and support. Answers should include the following.

- The shape of the triangle does not matter.
- Some of the triangles used in the bridge supports seem to be congruent.

39. D **41.** 58 **43.** $x = 3$, $BC = 10$, $CD = 10$, $BD = 5$
45. $y = -\frac{3}{2}x + 3$ **47.** $y = -4x - 11$ **49.** $\sqrt{5}$ **51.** $\sqrt{13}$

Page 198 Chapter 4 Practice Quiz 1

1. $\triangle DFJ$, $\triangle GJF$, $\triangle HJG$, $\triangle DJH$ **3.** $AB = BC = AC = 7$
5. $\angle M \cong \angle J$, $\angle N \cong \angle K$, $\angle P \cong \angle L$; $\overline{MN} \cong \overline{JK}$, $\overline{NP} \cong \overline{KL}$, and $\overline{MP} \cong \overline{JL}$

Pages 203–206 Lesson 4-4

1. Sample answer: In $\triangle QRS$, $\angle R$ is the included angle of the sides $\overline{QR}$ and $\overline{RS}$.

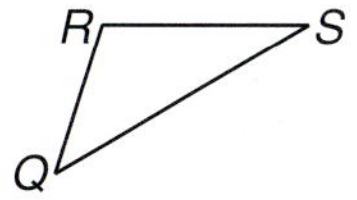

3. $EG = 2$, $MP = 2$, $FG = 4$, $NP = 4$, $EF = \sqrt{20}$, and $MN = \sqrt{20}$. The corresponding sides have the same measure and are congruent. $\triangle EFG \cong \triangle MNP$ by SSS.

5. Given: $\overline{DE}$ and $\overline{BC}$ bisect each other
Prove: $\triangle DGB \cong \triangle EGC$
Proof:

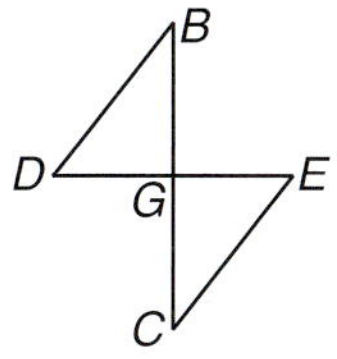

$\overline{DE}$ and $\overline{BC}$ bisect each other.
Given

$\overline{DG} \cong \overline{GE}$, $\overline{BG} \cong \overline{GC}$
Def. of bisector of segments

$\angle DGB \cong \angle EGC$
Vertical ∡ are ≅.

$\triangle DGB \cong \triangle EGC$
SAS

7. SAS

9. Given: T is the midpoint of $\overline{SQ}$.
$\overline{SR} \cong \overline{QR}$
Prove: $\triangle SRT \cong \triangle QRT$
Proof:

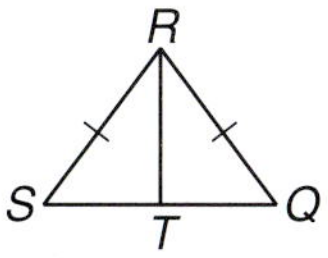

Statements	Reasons
1. T is the midpoint of $\overline{SQ}$.	1. Given
2. $\overline{ST} \cong \overline{TQ}$	2. Midpoint Theorem
3. $\overline{SR} \cong \overline{QR}$	3. Given
4. $\overline{RT} \cong \overline{RT}$	4. Reflexive Property
5. $\triangle SRT \cong \triangle QRT$	5. SSS

11. $JK = \sqrt{10}$, $KL = \sqrt{10}$, $JL = \sqrt{20}$, $FG = \sqrt{2}$, $GH = \sqrt{50}$, and $FH = 6$. The corresponding sides are not congruent so $\triangle JKL$ is not congruent to $\triangle FGH$. **13.** $JK = \sqrt{10}$, $KL = \sqrt{10}$, $JL = \sqrt{20}$, $FG = \sqrt{10}$, $GH = \sqrt{10}$, and $FH = \sqrt{20}$. Each pair of corresponding sides have the same measure so they are congruent. $\triangle JKL \cong \triangle FGH$ by SSS.

15. Given: $\overline{RQ} \cong \overline{TQ} \cong \overline{YQ} \cong \overline{WQ}$, $\angle RQY \cong \angle WQT$
Prove: $\triangle QWT \cong \triangle QYR$

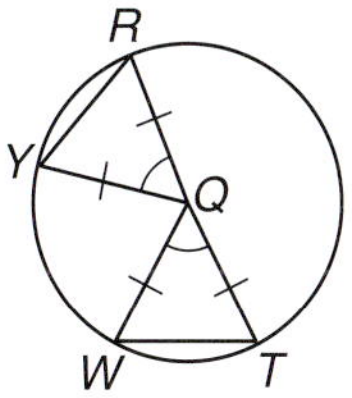

Proof:

$\overline{RQ} \cong \overline{TQ} \cong \overline{YQ} \cong \overline{WQ}$
Given

$\angle RQY \cong \angle WQT$
Given

$\triangle QWT \cong \triangle QYR$
SAS

17. Given: $\triangle MRN \cong \triangle QRP$
$\angle MNP \cong \angle QPN$
Prove: $\triangle MNP \cong \triangle QPN$

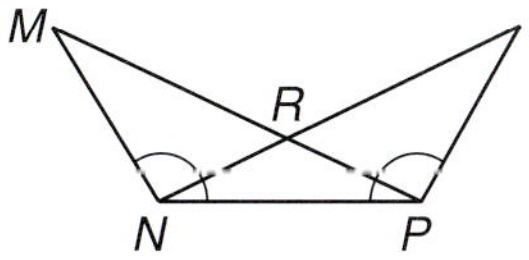

Proof:

Statement	Reason
1. $\triangle MRN \cong \triangle QRP$, $\angle MNP \cong \angle QPN$	1. Given
2. $\overline{MN} \cong \overline{QP}$	2. CPCTC
3. $\overline{NP} \cong \overline{NP}$	3. Reflexive Property
4. $\triangle MNP \cong \triangle QPN$	4. SAS

19. **Given:** $\triangle GHJ \cong \triangle LKJ$
Prove: $\triangle GHL \cong \triangle LKG$

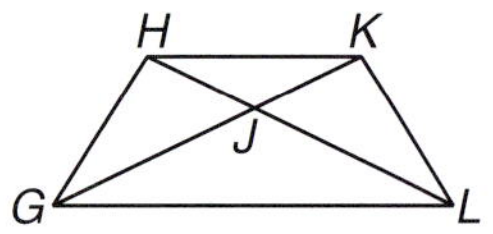

Proof:

Statement	Reason
1. $\triangle GHJ \cong \triangle LKJ$	1. Given
2. $\overline{HJ} \cong \overline{KJ}$, $\overline{GJ} \cong \overline{LJ}$, $\overline{GH} \cong \overline{LK}$,	2. CPCTC
3. $HJ = KJ$, $GJ = LJ$	3. Def. of $\cong$ segments
4. $HJ + LJ = KJ + JG$	4. Addition Property
5. $KJ + GJ = KG$; $HJ + LJ = HL$	5. Segment Addition
6. $KG = HL$	6. Substitution
7. $\overline{KG} \cong \overline{HL}$	7. Def. of $\cong$ segments
8. $\overline{GL} \cong \overline{GL}$	8. Reflexive Property
9. $\triangle GHL \cong \triangle LKG$	9. SSS

21. **Given:** $\overline{EF} \cong \overline{HF}$
G is the midpoint of $\overline{EH}$.
Prove: $\triangle EFG \cong \triangle HFG$

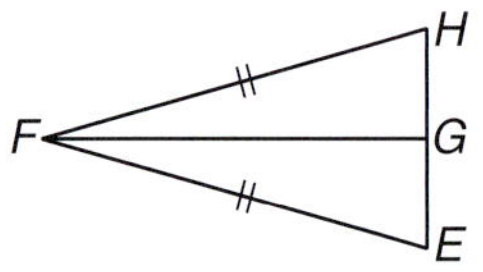

Proof:

Statements	Reasons
1. $\overline{EF} \cong \overline{HF}$; G is the midpoint of $\overline{EH}$.	1. Given
2. $\overline{EG} \cong \overline{GH}$	2. Midpoint Theorem
3. $\overline{FG} \cong \overline{FG}$	3. Reflexive Property
4. $\triangle EFG \cong \triangle HFG$	4. SSS

23. not possible 25. SSS or SAS

27. **Given:** $\overline{TS} \cong \overline{SF} \cong \overline{FH} \cong \overline{HT}$
$\angle TSF$, $\angle SFH$, $\angle FHT$, and $\angle HTS$ are right angles.
Prove: $\triangle SHT \cong \triangle SHF$

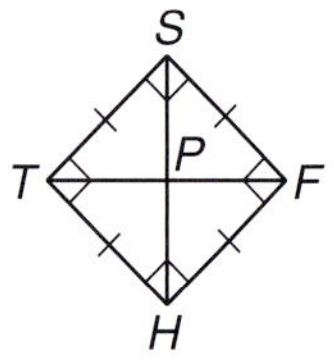

Proof:

Statements	Reasons
1. $\overline{TS} \cong \overline{SF} \cong \overline{FH} \cong \overline{HT}$	1. Given
2. $\angle TSF$, $\angle SFH$, $\angle FHT$, and $\angle HTS$ are right angles.	2. Given
3. $\angle STH \cong \angle SFH$	3. All rt. ∠s are $\cong$.
4. $\triangle STH \cong \triangle SFH$	4. SAS
5. $\angle SHT \cong \angle SHF$	5. CPCTC

29. Sample answer: The properties of congruent triangles help land surveyors double check measurements. Answers should include the following.

- If each pair of corresponding angles and sides are congruent, the triangles are congruent by definition. If two pairs of corresponding sides and the included angle are congruent, the triangles are congruent by SAS. If each pair of corresponding sides are congruent, the triangles are congruent by SSS.
- Sample answer: Architects also use congruent triangles when designing buildings.

31. B 33. $\triangle WXZ \cong \triangle YXZ$ 35. 78 37. 68 39. 59
41. -1 43. There is a steeper rate of decline from the second quarter to the third. 45. $\angle CBD$ 47. $\overline{CD}$

Pages 210–213 Lesson 4-5

1. Two triangles can have corresponding congruent angles without corresponding congruent sides. $\angle A \cong \angle D$, $\angle B \cong \angle E$, and $\angle C \cong \angle F$. However, $\overline{AB} \ncong \overline{DE}$, so $\triangle ABC \ncong \triangle DEF$.

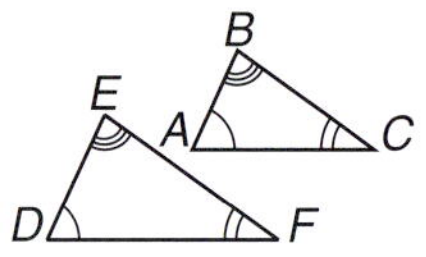

3. AAS can be proven using the Third Angle Theorem. Postulates are accepted as true without proof.

5. **Given:** $\overline{XW} \parallel \overline{YZ}$, $\angle X \cong \angle Z$
Prove: $\triangle WXY \cong \triangle YZW$

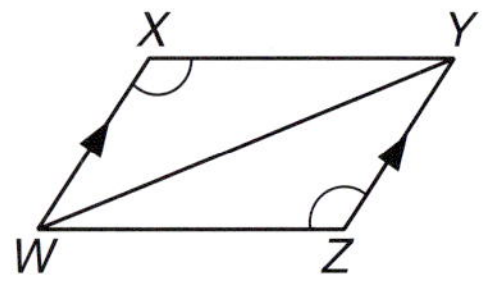

Proof:

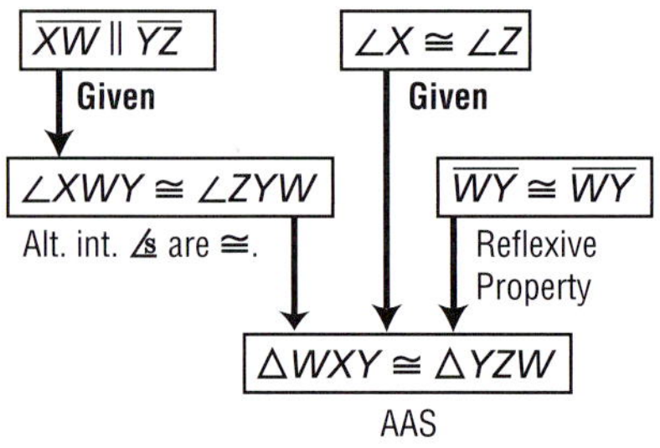

7. **Given:** $\angle E \cong \angle K$, $\angle DGH \cong \angle DHG$, $\overline{EG} \cong \overline{KH}$
Prove: $\triangle EGD \cong \triangle KHD$

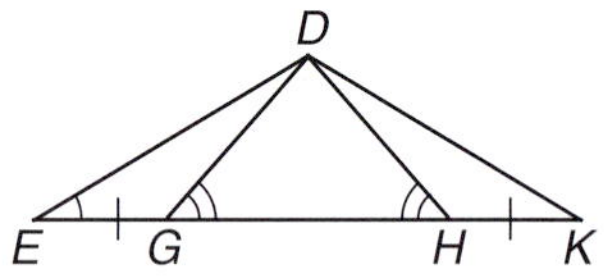

Proof:
Since $\angle EGD$ and $\angle DGH$ are a linear pair, the angles are supplementary. Likewise, $\angle KHD$ and $\angle DHG$ are supplementary. We are given that $\angle DGH \cong \angle DHG$. Angles supplementary to congruent angles are congruent so $\angle EGD \cong \angle KHD$. Since we are given that $\angle E \cong \angle K$ and $\overline{EG} \cong \overline{KH}$, $\triangle EGD \cong \triangle KHD$ by ASA.

9. **Given:** $\overline{EF} \parallel \overline{GH}$, $\overline{EF} \cong \overline{GH}$
Prove: $\overline{EK} \cong \overline{KH}$

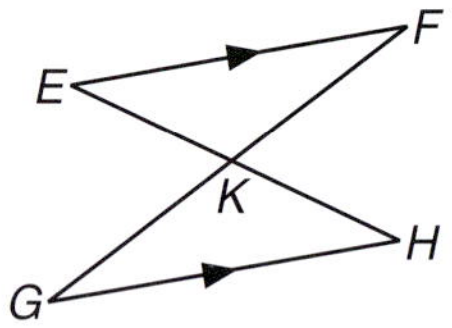

Proof:

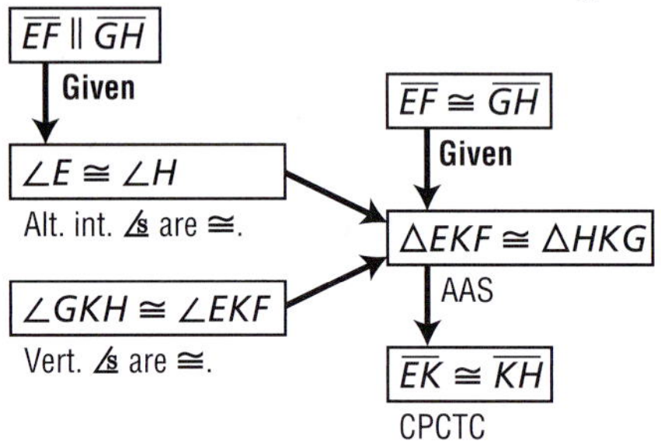

11. **Given:** $\angle V \cong \angle S$, $\overline{TV} \cong \overline{QS}$
Prove: $\overline{VR} \cong \overline{SR}$

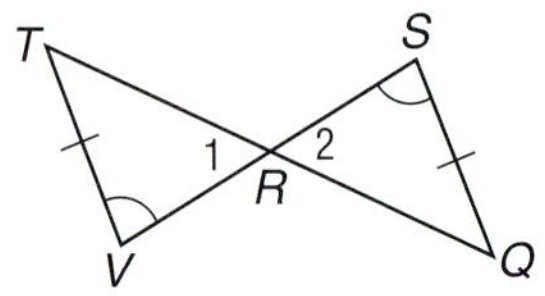

Proof:

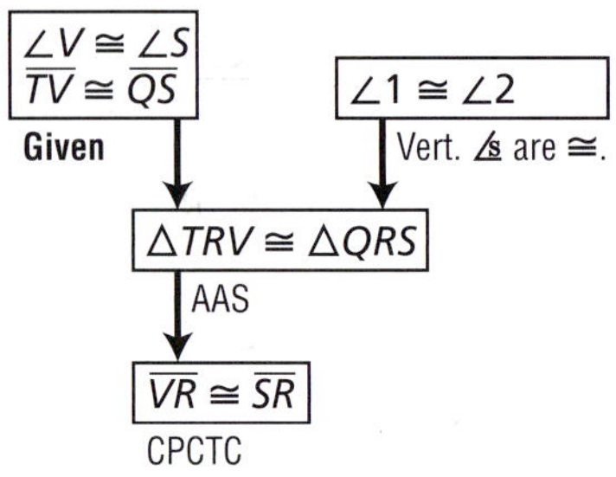

13. Given: $\overline{MN} \cong \overline{PQ}$, $\angle M \cong \angle Q$
$\angle 2 \cong \angle 3$
Prove: $\triangle MLP \cong \triangle QLN$

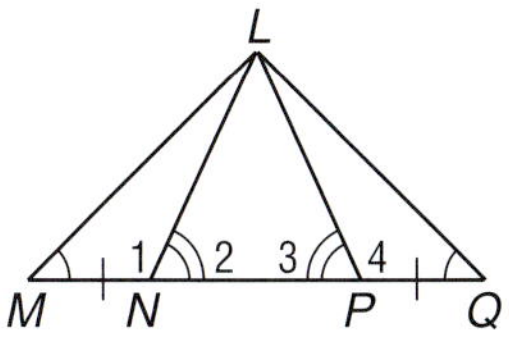

Proof:

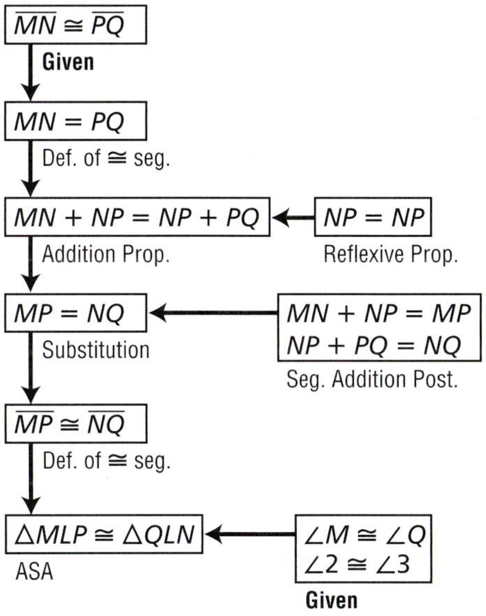

15. Given: $\angle NOM \cong \angle POR$,
$\overline{NM} \perp \overline{MR}$,
$\overline{PR} \perp \overline{MR}$,
$\overline{NM} \cong \overline{PR}$
Prove: $\overline{MO} \cong \overline{OR}$

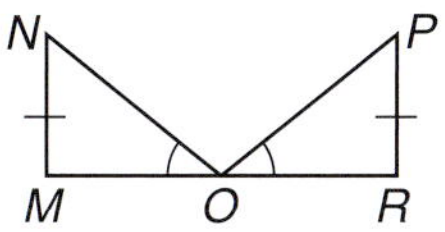

Proof: Since $\overline{NM} \perp \overline{MR}$ and $\overline{PR} \perp \overline{MR}$, $\angle M$ and $\angle R$ are right angles. $\angle M \cong \angle R$ because all right angles are congruent. We know that $\angle NOM \cong \angle POR$ and $\overline{NM} \cong \overline{PR}$. By AAS, $\triangle NMO \cong \triangle PRO$. $\overline{MO} \cong \overline{OR}$ by CPCTC.

17. Given: $\angle F \cong \angle J$,
$\angle E \cong \angle H$,
$\overline{EC} \cong \overline{GH}$
Prove: $\overline{EF} \cong \overline{HJ}$

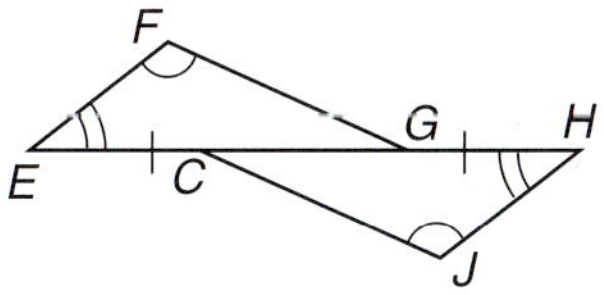

Proof: We are given that $\angle F \cong \angle J$, $\angle E \cong \angle H$, and $\overline{EC} \cong \overline{GH}$. By the Reflexive Property, $\overline{CG} \cong \overline{CG}$. Segment addition results in $EG = EC + CG$ and $CH = CG + GH$. By the definition of congruence, $EC = GH$ and $CG = CG$. Substitute to find $EG = CH$. By AAS, $\triangle EFG \cong \triangle HJC$. By CPCTC, $\overline{EF} \cong \overline{HJ}$.

19. Given: $\angle MYT \cong \angle NYT$
$\angle MTY \cong \angle NTY$
Prove: $\triangle RYM \cong \triangle RYN$

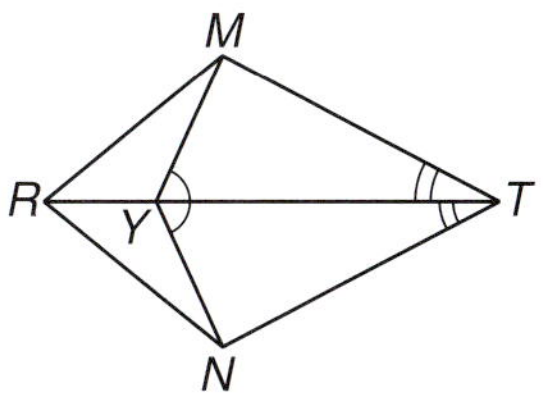

Proof:

Statement	Reason
1. $\angle MYT \cong \angle NYT$ $\angle MTY \cong \angle NTY$	1. Given
2. $\overline{YT} \cong \overline{YT}$, $\overline{RY} \cong \overline{RY}$	2. Reflexive Property
3. $\triangle MYT \cong \triangle NYT$	3. ASA
4. $\overline{MY} \cong \overline{NY}$	4. CPCTC
5. $\angle RYM$ and $\angle MYT$ are a linear pair; $\angle RYN$ and $\angle NYT$ are a linear pair	5. Def. of linear pair
6. $\angle RYM$ and $\angle MYT$ are supplementary and $\angle RYN$ and $\angle NYT$ are supplementary.	6. Supplement Theorem
7. $\angle RYM \cong \angle RYN$	7. $\measuredangle$ suppl. to $\cong$ $\measuredangle$s are $\cong$.
8. $\triangle RYM \cong \triangle RYN$	8. SAS

21. $\overline{CD} \cong \overline{GH}$, because the segments have the same measure. $\angle CFD \cong \angle HFG$ because vertical angles are congruent. Since F is the midpoint of $\overline{DG}$, $\overline{DF} \cong \overline{FG}$. It cannot be determined whether $\triangle CFD \cong \triangle HFG$. The information given does not lead to a unique triangle.
23. Since N is the midpoint of $\overline{JL}$, $\overline{JN} \cong \overline{NL}$. $\angle JNK \cong \angle LNK$ because perpendicular lines form right angles and right angles are congruent. By the Reflexive Property, $\overline{KN} \cong \overline{KN}$. $\triangle JKN \cong \triangle LKN$ by SAS. **25.** $\triangle VNR$, AAS or ASA
27. $\triangle MIN$, SAS **29.** Since Aiko is perpendicular to the ground, two right angles are formed and right angles are congruent. The angles of sight are the same and her height is the same for each triangle. The triangles are congruent by ASA. By CPCTC, the distances are the same. The method is valid. **31.** D

33. Given: $\overline{BA} \cong \overline{DE}$,
$\overline{DA} \cong \overline{BE}$
Prove: $\triangle BEA \cong$
$\triangle DAE$

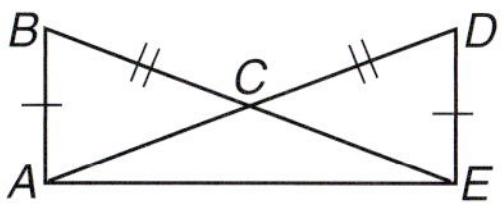

Proof:

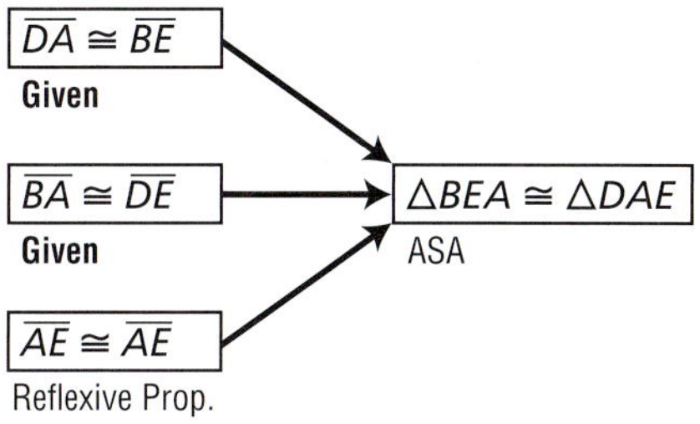

35. Turn; $RS = \sqrt{2}$, $R'S' = \sqrt{2}$, $ST = 1$, $S'T' = 1$, $RT = 1$, $R'T' = 1$. Use a protractor to confirm that the corresponding angles are congruent. **37.** If people are happy, then they rarely correct their faults. **39.** isosceles
41. isosceles

Pages 219–221 Lesson 4-6

1. The measure of only one angle must be given in an isosceles triangle to determine the measures of the other two angles. **3.** Sample answer: Draw a line segment. Set your compass to the length of the line segment and draw an arc from each endpoint. Draw segments from the intersection of the arcs to each endpoint. **5.** $\overline{BH} \cong \overline{BD}$
7. Given: $\triangle CTE$ is isosceles with vertex $\angle C$.
$m\angle T = 60$
Prove: $\triangle CTE$ is equilateral.

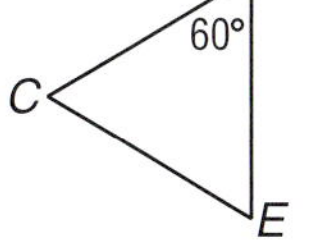

Proof:

Statements	Reasons
1. $\triangle CTE$ is isosceles with vertex $\angle C$.	1. Given
2. $\overline{CT} \cong \overline{CE}$	2. Def. of isosceles triangle
3. $\angle E \cong \angle T$	3. Isosceles Triangle Theorem
4. $m\angle E = m\angle T$	4. Def. of $\cong$ $\measuredangle$

Selected Answers

Statements	Reasons
5. $m\angle T = 60$	5. Given
6. $m\angle E = 60$	6. Substitution
7. $m\angle C + m\angle E + m\angle T = 180$	7. Angle Sum Theorem
8. $m\angle C + 60 + 60 = 180$	8. Substitution
9. $m\angle C = 60$	9. Subtraction
10. $\triangle CTE$ is equiangular.	10. Def. of equiangular $\triangle$
11. $\triangle CTE$ is equilateral.	11. Equiangular $\triangle$s are equilateral.

9. $\angle LTR \cong \angle LRT$ **11.** $\angle LSQ \cong \angle LQS$ **13.** $\overline{LS} \cong \overline{LR}$ **15.** 20 **17.** 81 **19.** 28 **21.** 56 **23.** 36.5 **25.** 38 **27.** $x = 3; y = 18$

29. Given: $\triangle XKF$ is equilateral.
$\overline{XJ}$ bisects $\angle KXF$.
Prove: J is the midpoint of $\overline{KF}$.

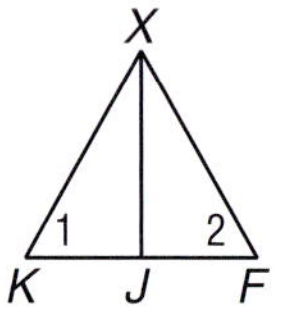

Proof:

Statements	Reasons
1. $\triangle XKF$ is equilateral.	1. Given
2. $\overline{KX} \cong \overline{FX}$	2. Definition of equilateral $\triangle$
3. $\angle 1 \cong \angle 2$	3. Isosceles Triangle Theorem
4. $\overline{XJ}$ bisects $\angle X$	4. Given
5. $\angle KXJ \cong \angle FXJ$	5. Def. of $\angle$ bisector
6. $\triangle KXJ \cong \triangle FXJ$	6. ASA
7. $\overline{KJ} \cong \overline{JF}$	7. CPCTC
8. J is the midpoint of $\overline{KF}$.	8. Def. of midpoint

31. Case I:
Given: $\triangle ABC$ is an equilateral triangle.
Prove: $\triangle ABC$ is an equiangular triangle.

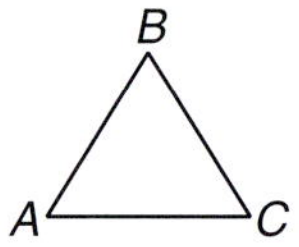

Proof:

Statements	Reasons
1. $\triangle ABC$ is an equilateral triangle.	1. Given
2. $\overline{AB} \cong \overline{AC} \cong \overline{BC}$	2. Def. of equilateral $\triangle$
3. $\angle A \cong \angle B$, $\angle B \cong \angle C$, $\angle A \cong \angle C$	3. Isosceles Triangle Theorem
4. $\angle A \cong \angle B \cong \angle C$	4. Substitution
5. $\triangle ABC$ is an equiangular $\triangle$.	5. Def. of equiangular $\triangle$

Case II:
Given: $\triangle ABC$ is an equiangular triangle.
Prove: $\triangle ABC$ is an equilateral triangle.

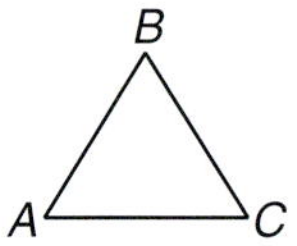

Proof:

Statements	Reasons
1. $\triangle ABC$ is an equiangular triangle.	1. Given
2. $\angle A \cong \angle B \cong \angle C$	2. Def. of equiangular $\triangle$
3. $\overline{AB} \cong \overline{AC}$, $\overline{AB} \cong \overline{BC}$, $\overline{AC} \cong \overline{BC}$	3. Conv. of Isos. $\triangle$ Th.
4. $\overline{AB} \cong \overline{AC} \cong \overline{BC}$	4. Substitution
5. $\triangle ABC$ is an equilateral $\triangle$.	5. Def. of equilateral $\triangle$

33. Given: $\triangle ABC$
$\angle A \cong \angle C$
Prove: $\overline{AB} \cong \overline{CB}$

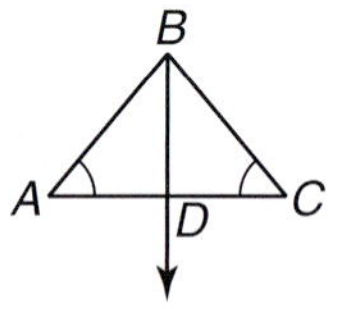

Proof:

Statements	Reasons
1. Let $\overrightarrow{BD}$ bisect $\angle ABC$.	1. Protractor Postulate
2. $\angle ABD \cong \angle CBD$	2. Def. of $\angle$ bisector
3. $\angle A \cong \angle C$	3. Given
4. $\overline{BD} \cong \overline{BD}$	4. Reflexive Property
5. $\triangle ABD \cong \triangle CBD$	5. AAS
6. $\overline{AB} \cong \overline{CB}$	6. CPCTC

35. 18 **37.** 30 **39.** The triangles in each set appear to be acute. **41.** Sample answer: Artists use angles, lines, and shapes to create visual images. Answers should include the following.

- Rectangle, squares, rhombi, and other polygons are used in many works of art.
- There are two rows of isosceles triangles in the painting. One row has three congruent isosceles triangles. The other row has six congruent isosceles triangles.

43. D
45. Given: $\overline{VR} \perp \overline{RS}$, $\overline{UT} \perp \overline{SU}$, $\overline{RS} \cong \overline{US}$
Prove: $\triangle VRS \cong \triangle TUS$

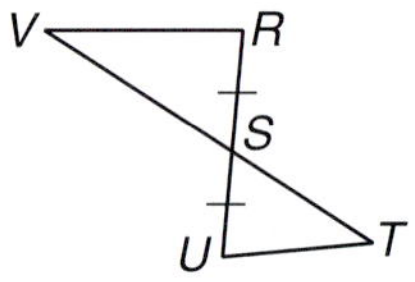

Proof: We are given that $\overline{VR} \perp \overline{RS}$, $\overline{UT} \perp \overline{SU}$, and $\overline{RS} \cong \overline{US}$. Perpendicular lines form four right angles so $\angle R$ and $\angle U$ are right angles. $\angle R \cong \angle U$ because all right angles are congruent. $\angle RSV \cong \angle UST$ since vertical angles are congruent. Therefore, $\triangle VRS \cong \triangle TUS$ by ASA.

47. $QR = \sqrt{52}$, $RS = \sqrt{2}$, $QS = \sqrt{34}$, $EG = \sqrt{34}$, $GH = \sqrt{10}$, and $EH = \sqrt{52}$. The corresponding sides are not congruent so $\triangle QRS$ is not congruent to $\triangle EGH$.

49.

p	q	$\sim p$	$\sim q$	$\sim p$ or $\sim q$
T	T	F	F	F
T	F	F	T	T
F	T	T	F	T
F	F	T	T	T

51.

y	z	$\sim y$	$\sim y$ or z
T	T	F	T
T	F	F	F
F	T	T	T
F	F	T	T

53. $(-1, -3)$

Page 221 Chapter 4 Practice Quiz 2

1. $JM = \sqrt{5}$, $ML = \sqrt{26}$, $JL = 5$, $BD = \sqrt{5}$, $DG = \sqrt{26}$, and $BG = 5$. Each pair of corresponding sides have the same measure so they are congruent. $\triangle JML \cong \triangle BDG$ by SSS. **3.** 52 **5.** 26

Pages 224–226 Lesson 4-7

1. Place one vertex at the origin, place one side of the triangle on the positive x-axis. Label the coordinates with expressions that will simplify the computations.

3. 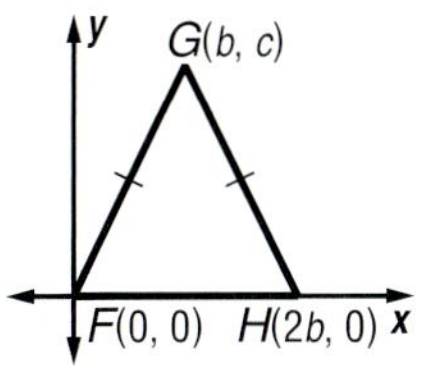

5. $P(0, b)$ **7.** $N(0, b)$, $Q(a, 0)$

9. Given: $\triangle ABC$
Prove: $\triangle ABC$ is isosceles.

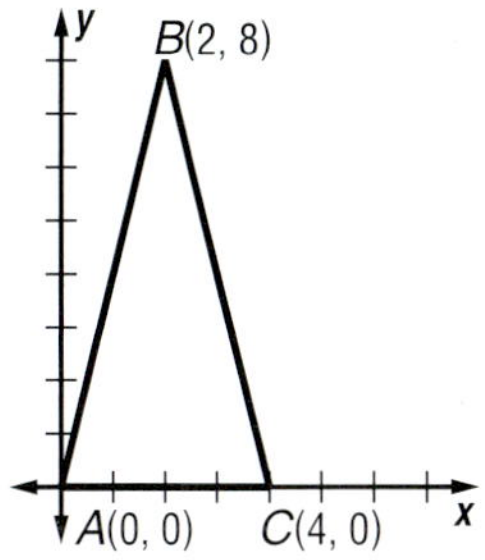

Proof: Use the Distance Formula to find AB and BC.
$AB = \sqrt{(2-0)^2 + (8-0)^2} = \sqrt{4+64}$ or $\sqrt{68}$
$BC = \sqrt{(4-2)^2 + (0-8)^2} = \sqrt{4+64}$ or $\sqrt{68}$
Since $AB = BC$, $\overline{AB} \cong \overline{BC}$. Since the legs are congruent, $\triangle ABC$ is isosceles.

11. 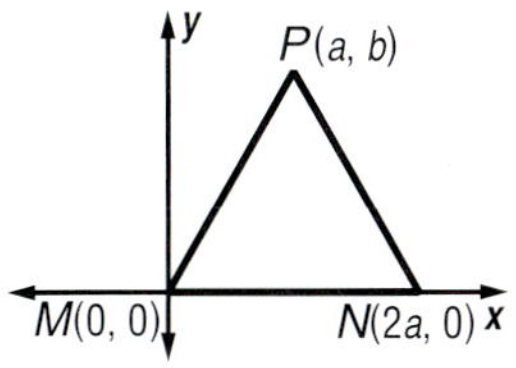

13. 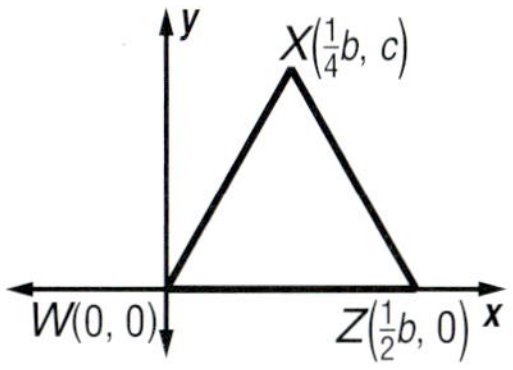

15. 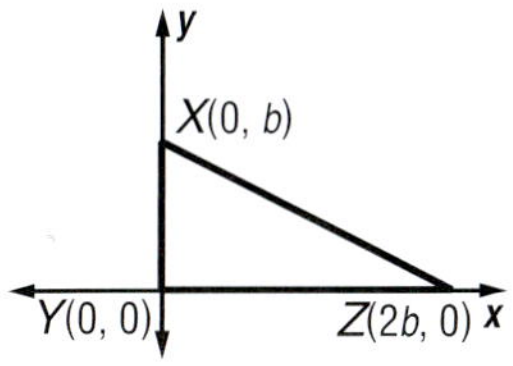

17. $Q(a, a)$, $P(a, 0)$
19. $D(2b, 0)$ **21.** $P(0, c)$, $N(2b, 0)$ **23.** $J(c, b)$

25. Given: isosceles $\triangle ABC$ with $\overline{AC} \cong \overline{BC}$
R and S are midpoints of legs $\overline{AC}$ and $\overline{BC}$.
Prove: $\overline{AS} \cong \overline{BR}$

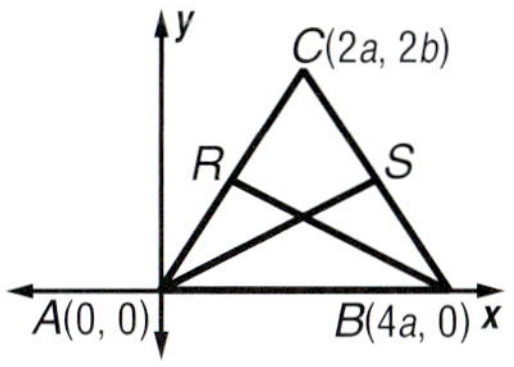

Proof:
The coordinates of R are $\left(\frac{2a+0}{2}, \frac{2b+0}{2}\right)$ or (a, b).
The coordinates of S are $\left(\frac{2a+4a}{2}, \frac{2b+0}{2}\right)$ or $(3a, b)$.
$BR = \sqrt{(4a-a)^2 + (0-b)^2} = \sqrt{(3a)^2 + (-b)^2}$
or $\sqrt{9a^2 + b^2}$
$AS = \sqrt{(3a-0)^2 + (b-0)^2} = \sqrt{(3a)^2 + (b)^2}$
or $\sqrt{9a^2 + b^2}$
Since $BR = AS$, $\overline{AS} \cong \overline{BR}$.

27. Given: $\triangle ABC$
S is the midpoint of $\overline{AC}$.
T is the midpoint of $\overline{BC}$.
Prove: $\overline{ST} \parallel \overline{AB}$

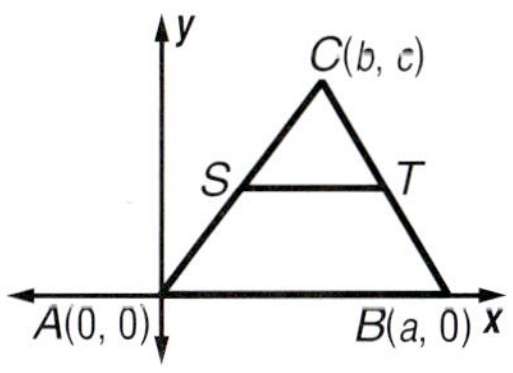

Proof:
Midpoint S is $\left(\frac{b+0}{2}, \frac{c+0}{2}\right)$ or $\left(\frac{b}{2}, \frac{c}{2}\right)$
Midpoint T is $\left(\frac{a+b}{2}, \frac{c+0}{2}\right)$ or $\left(\frac{a+b}{2}, \frac{c}{2}\right)$.
Slope of $\overline{ST} = \dfrac{\frac{c}{2} - \frac{c}{2}}{\frac{a+b}{2} - \frac{b}{2}} = \dfrac{0}{\frac{a}{2}}$ or 0.
Slope of $\overline{AB} = \frac{0-0}{a-0} = \frac{0}{a}$ or 0.
$\overline{ST}$ and $\overline{AB}$ have the same slope so $\overline{ST} \parallel \overline{AB}$.

29. Given: $\triangle ABD$, $\triangle FBD$
$AF = 6$, $BD = 3$
Prove: $\triangle ABD \cong \triangle FBD$

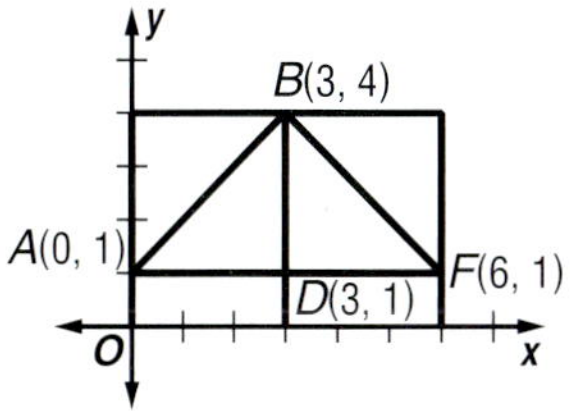

Proof: $\overline{BD} \cong \overline{BD}$ by the Reflexive Property.
$AD = \sqrt{(3-0)^2 + (1-1)^2} = \sqrt{9+0}$ or 3
$DF = \sqrt{(6-3)^2 + (1-1)^2} = \sqrt{9+0}$ or 3
Since $AD = DF$, $\overline{AD} \cong \overline{DF}$.
$AB = \sqrt{(3-0)^2 + (4-1)^2} = \sqrt{9+9}$ or $3\sqrt{2}$
$BF = \sqrt{(6-3)^2 + (1-4)^2} = \sqrt{9+9}$ or $3\sqrt{2}$
Since $AB = BF$, $\overline{AB} \cong \overline{BF}$.
$\triangle ABD \cong \triangle FBD$ by SSS.

31. Given: $\triangle BPR$, $\triangle BAR$
$PR = 800$, $BR = 800$, $RA = 800$
Prove: $\overline{PB} \cong \overline{BA}$
Proof:
$PB = \sqrt{(800-0)^2 + (800-0)^2}$ or $\sqrt{1{,}280{,}000}$
$BA = \sqrt{(800-1600)^2 + (800-0)^2}$ or $\sqrt{1{,}280{,}000}$
$PB = BA$, so $\overline{PB} \cong \overline{BA}$.

33. $\sqrt{680{,}000}$ or about 824.6 ft **35.** $(2a, 0)$ **37.** $AB = 4a$; $AC = \sqrt{(0-(-2a))^2 + (2a-0)^2} = \sqrt{4a^2 + 4a^2}$ or $\sqrt{8a^2}$; $CB = \sqrt{(0-2a)^2 + (2a-0)^2} = \sqrt{4a^2+4a^2}$ or $\sqrt{8a^2}$; Slope of $\overline{AC} = \frac{2a-0}{0-(-2a)}$ or 1; slope of $\overline{CB} = \frac{2a-0}{0-2a}$ or -1. $\overline{AC} \perp \overline{CB}$ and $\overline{AC} \cong \overline{CB}$, so $\triangle ABC$ is a right isosceles triangle. **39.** C

41. Given: $\angle 3 \cong \angle 4$
Prove: $\overline{QR} \cong \overline{QS}$

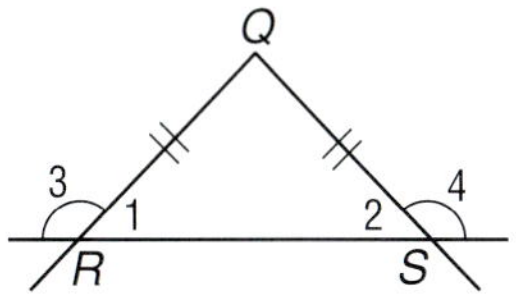

Proof:

Statements	**Reasons**
1. $\angle 3 \cong \angle 4$	1. Given
2. $\angle 2$ and $\angle 4$ form a linear pair. $\angle 1$ and $\angle 3$ form a linear pair.	2. Def. of linear pair
3. $\angle 2$ and $\angle 4$ are supplementary. $\angle 1$ and $\angle 3$ are supplementary.	3. If 2 ∠s form a linear pair, then they are suppl.
4. $\angle 2 \cong \angle 1$	4. Angles that are suppl. to ≅ ∠s are ≅.
5. $\overline{QR} \cong \overline{QS}$	5. Conv. of Isos. △ Th.

43. Given: $\overline{AD} \cong \overline{CE}$, $\overline{AD} \parallel \overline{CE}$
Prove: $\triangle ABD \cong \triangle EBC$

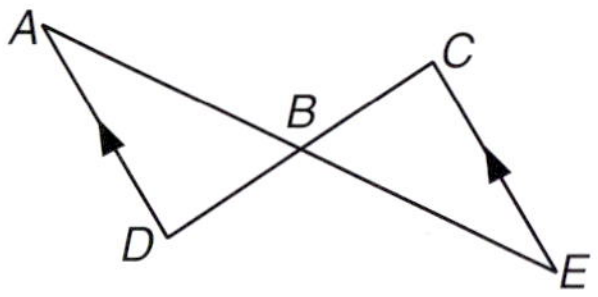

Proof:

Statements	Reasons
1. $\overline{AD} \parallel \overline{CE}$	1. Given
2. $\angle A \cong \angle E$, $\angle D \cong \angle C$	2. Alt. int. ∠s are ≅.
3. $\overline{AD} \cong \overline{CE}$	3. Given
4. $\triangle ABD \cong \triangle EBC$	4. ASA

45. $\overline{BC} \parallel \overline{AD}$; if alt. int. ∠s are ≅, lines are ∥. **47.** $\ell \parallel m$; if 2 lines are ⊥ to the same line, they are ∥.

Pages 227–230 Chapter 4 Study Guide and Review

1. h **3.** d **5.** a **7.** b **9.** obtuse, isosceles
11. equiangular, equilateral **13.** 25 **15.** $\angle E \cong \angle D$, $\angle F \cong \angle C$, $\angle G \cong \angle B$, $\overline{EF} \cong \overline{DC}$, $\overline{FG} \cong \overline{CB}$, $\overline{GE} \cong \overline{BD}$ **17.** $\angle KNC \cong \angle RKE$, $\angle NCK \cong \angle KER$, $\angle CKN \cong \angle ERK$, $\overline{NC} \cong \overline{KE}$, $\overline{CK} \cong \overline{ER}$, $\overline{KN} \cong \overline{RK}$ **19.** $MN = \sqrt{20}$, $NP = \sqrt{5}$, $MP = 5$, $QR = \sqrt{20}$, $RS = \sqrt{5}$, and $QS = 5$. Each pair of corresponding sides has the same measure. Therefore, $\triangle MNP \cong \triangle QRS$ by SSS.

21. Given: $\triangle DGC \cong \triangle DGE$, $\triangle GCF \cong \triangle GEF$
Proof: $\triangle DFC \cong \triangle DFE$

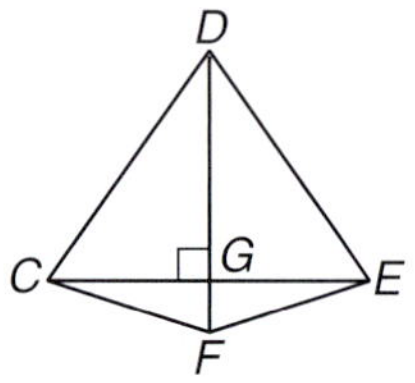

Proof:

Statement	Reason
1. $\triangle DGC \cong \triangle DGE$, $\triangle GCF \cong \triangle GEF$	1. Given
2. $\angle CDG \cong \angle EDG$, $\overline{CD} \cong \overline{ED}$, and $\angle CFD \cong \angle EFD$	2. CPCTC
3. $\triangle DFC \cong \triangle DFE$	3. AAS

23. 40 **25.** 80
27.

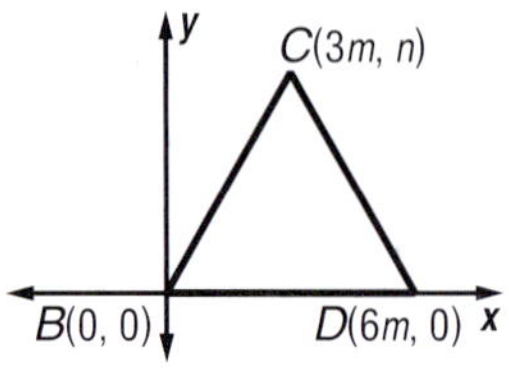

Chapter 5 Relationships in Triangles

Page 235 Chapter 5 Getting Started

1. $(-4, 5)$ **3.** $(-0.5, -5)$ **5.** 68 **7.** 40 **9.** 26 **11.** 14
13. The sum of the measures of the angles is 180.

Pages 242–245 Lesson 5-1

1. Sample answer: Both pass through the midpoint of a side. A perpendicular bisector is perpendicular to the side of a triangle, and does not necessarily pass through the vertex opposite the side, while a median does pass through the vertex and is not necessarily perpendicular to the side.
3. Sample answer: An altitude and angle bisector of a triangle are the same segment in an equilateral triangle.

5. Given: $\overline{XY} \cong \overline{XZ}$
$\overline{YM}$ and $\overline{ZN}$ are medians.
Prove: $\overline{YM} \cong \overline{ZN}$

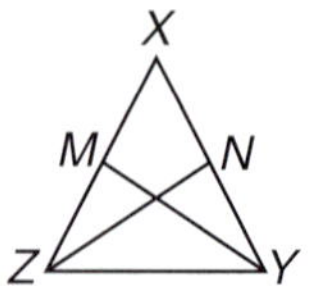

Proof:

Statements	Reasons
1. $\overline{XY} \cong \overline{XZ}$, $\overline{YM}$ and $\overline{ZN}$ are medians.	1. Given
2. M is the midpoint of $\overline{XZ}$. N is the midpoint of $\overline{XY}$.	2. Def. of median
3. $XY = XZ$	3. Def. of ≅ segs.
4. $\overline{XM} \cong \overline{MZ}$, $\overline{XN} \cong \overline{NY}$	4. Def. of median
5. $XM = MZ$, $XN = NY$	5. Def. of ≅ segs.
6. $XM + MZ = XZ$, $XN + NY = XY$	6. Segment Addition Postulate
7. $XM + MZ = XN + NY$	7. Substitution
8. $MZ + MZ = NY + NY$	8. Substitution
9. $2MZ = 2NY$	9. Addition Property
10. $MZ = NY$	10. Division Property
11. $\overline{MZ} \cong \overline{NY}$	11. Def. of ≅ segs.
12. $\angle XZY \cong \angle XYZ$	12. Isosceles Triangle Theorem
13. $\overline{YZ} \cong \overline{YZ}$	13. Reflexive Property
14. $\triangle MYZ \cong \triangle NZY$	14. SAS
15. $\overline{YM} \cong \overline{ZN}$	15. CPCTC

7. $\left(\frac{2}{3}, 3\frac{1}{3}\right)$ **9.** $\left(1\frac{2}{5}, 2\frac{3}{5}\right)$

11. Given: $\triangle UVW$ is isosceles with vertex angle UVW. $\overline{YV}$ is the bisector of $\angle UVW$.
Prove: $\overline{YV}$ is a median.

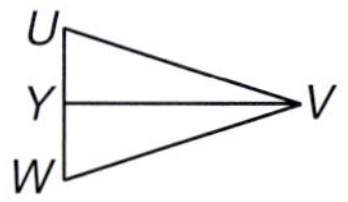

Proof:

Statements	Reasons
1. $\triangle UVW$ is an isosceles triangle with vertex angle UVW, $\overline{YV}$ is the bisector of $\angle UVW$.	1. Given
2. $\overline{UV} \cong \overline{WV}$	2. Def. of isosceles △
3. $\angle UVY \cong \angle WVY$	3. Def. of angle bisector
4. $\overline{YV} \cong \overline{YV}$	4. Reflexive Property
5. $\triangle UVY \cong \triangle WVY$	5. SAS
6. $\overline{UY} \cong \overline{WY}$	6. CPCTC
7. Y is the midpoint of $\overline{UW}$.	7. Def. of midpoint
8. $\overline{YV}$ is a median.	8. Def. of median

13. $x = 7$, $m\angle 2 = 58$ **15.** $x = 20$, $y = 4$; yes; because $m\angle WPA = 90$ **17.** always **19.** never **21.** 2 **23.** 40
25. $PR = 18$ **27.** $(0, 7)$ **29.** $-\frac{4}{3}$

31. Given: $\overline{CA} \cong \overline{CB}$, $\overline{AD} \cong \overline{BD}$
Prove: C and D are on the perpendicular bisector of $\overline{AB}$.

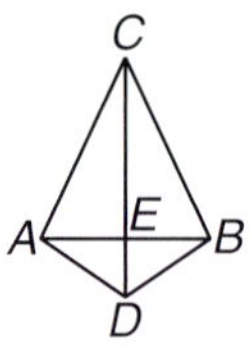

Proof:

Statements	Reasons
1. $\overline{CA} \cong \overline{CB}$, $\overline{AD} \cong \overline{BD}$	1. Given
2. $\overline{CD} \cong \overline{CD}$	2. Reflexive Property
3. $\triangle ACD \cong \triangle BCD$	3. SSS
4. $\angle ACD \cong \angle BCD$	4. CPCTC
5. $\overline{CE} \cong \overline{CE}$	5. Reflexive Property
6. $\triangle CEA \cong \triangle CEB$	6. SAS

7. $\overline{AE} \cong \overline{BE}$	7. CPCTC
8. E is the midpoint of $\overline{AB}$.	8. Def. of midpoint
9. $\angle CEA \cong \angle CEB$	9. CPCTC
10. $\angle CEA$ and $\angle CEB$ form a linear pair.	10. Def. of linear pair
11. $\angle CEA$ and $\angle CEB$ are supplementary.	11. Supplement Theorem
12. $m\angle CEA + m\angle CEB = 180$	12. Def. of suppl. $\angle$s
13. $m\angle CEA + m\angle CEA = 180$	13. Substitution
14. $2(m\angle CEA) = 180$	14. Substitution
15. $m\angle CEA = 90$	15. Division Property
16. $\angle CEA$ and $\angle CEB$ are rt. $\angle$s.	16. Def. of rt. $\angle$
17. $\overline{CD} \perp \overline{AB}$	17. Def. of $\perp$
18. $\overline{CD}$ is the perpendicular bisector of $\overline{AB}$.	18. Def. of $\perp$ bisector
19. C and D are on the perpendicular bisector of $\overline{AB}$.	19. Def. of points on a line

33. Given: $\triangle ABC$, $\overrightarrow{AD}$, $\overrightarrow{BE}$, $\overrightarrow{CF}$, $\overline{KP} \perp \overline{AB}$, $\overline{KQ} \perp \overline{BC}$, $\overline{KR} \perp \overline{AC}$
Prove: $KP = KQ = KR$

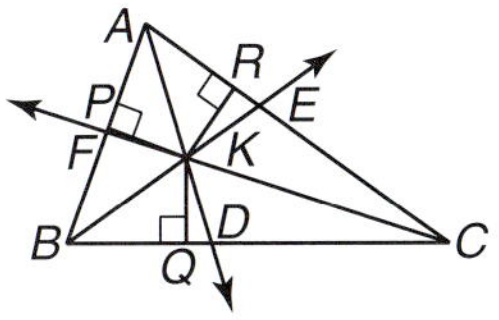

Proof:

Statements	Reasons
1. $\triangle ABC$, $\overrightarrow{AD}$, $\overrightarrow{BE}$, $\overrightarrow{CF}$, $\overline{KP} \perp \overline{AB}$, $\overline{KQ} \perp \overline{BC}$, $\overline{KR} \perp \overline{AC}$	1. Given
2. $KP = KQ$, $KQ = KR$, $KP = KR$	2. Any point on the $\angle$ bisector is equidistant from the sides of the angle.
3. $KP = KQ = KR$	3. Transitive Property

35. 4

37.

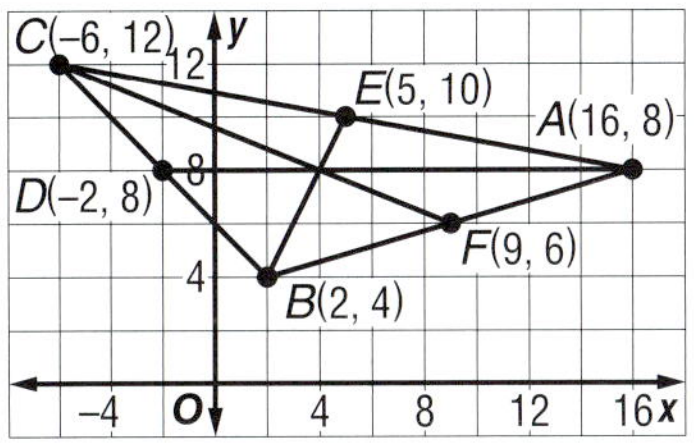

39. The altitude will be the same for both triangles, and the bases will be congruent, so the areas will be equal. **41.** C

43. Sample answer:

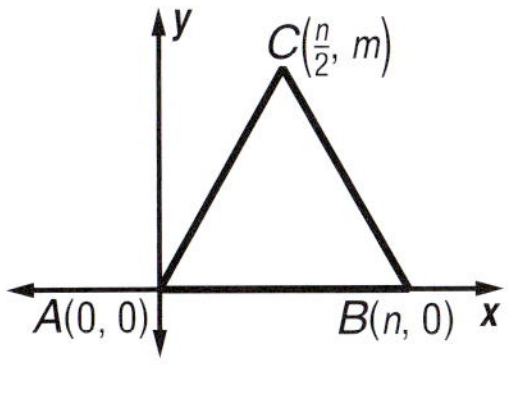

45. Sample answer:

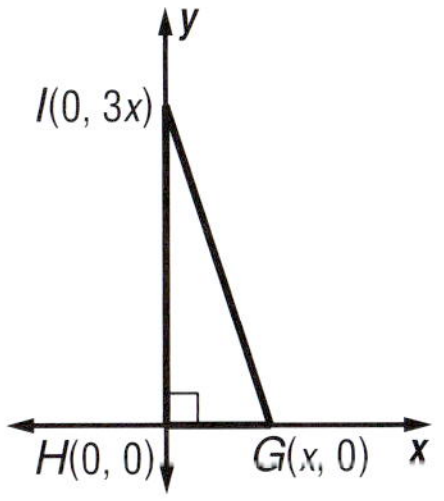

47. $\angle 5 \cong \angle 11$ **49.** $\overline{ML} \cong \overline{MN}$ **51.** > **53.** >

Pages 251–254 Lesson 5-2

1. never **3.** Grace; she placed the shorter side with the smaller angle, and the longer side with the larger angle. **5.** $\angle 3$ **7.** $\angle 4$, $\angle 5$, $\angle 6$ **9.** $\angle 2$, $\angle 3$, $\angle 5$, $\angle 6$ **11.** $m\angle XZY < m\angle XYZ$ **13.** $AE < EB$ **15.** $BC = EC$ **17.** $\angle 1$ **19.** $\angle 7$ **21.** $\angle 7$ **23.** $\angle 2$, $\angle 7$, $\angle 8$, $\angle 10$ **25.** $\angle 3$, $\angle 5$ **27.** $\angle 8$, $\angle 7$, $\angle 3$, $\angle 1$ **29.** $m\angle KAJ < m\angle AJK$ **31.** $m\angle SMJ > m\angle MJS$ **33.** $m\angle MYJ < m\angle JMY$

35. Given: $\overline{JM} \cong \overline{JL}$, $\overline{JL} \cong \overline{KL}$
Prove: $m\angle 1 > m\angle 2$

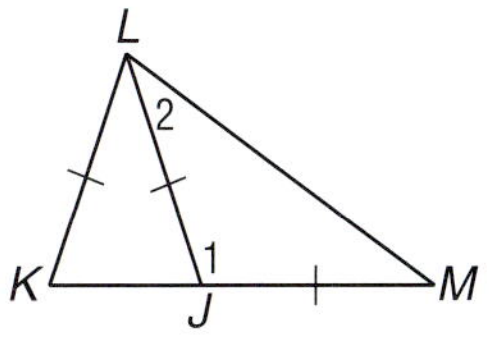

Proof:

Statements	Reasons
1. $\overline{JM} \cong \overline{JL}$, $\overline{JL} \cong \overline{KL}$	1. Given
2. $\angle LKJ \cong \angle LJK$	2. Isosceles $\triangle$ Theorem
3. $m\angle LKJ = m\angle LJK$	3. Def. of $\cong$ $\angle$s
4. $m\angle 1 > m\angle LKJ$	4. Ext. $\angle$ Inequality Theorem
5. $m\angle 1 > m\angle LJK$	5. Substitution
6. $m\angle LJK > m\angle 2$	6. Ext. $\angle$ Inequality Theorem
7. $m\angle 1 > m\angle 2$	7. Trans. Prop. of Inequality

37. $ZY > YR$ **39.** $RZ > SR$ **41.** $TY < ZY$ **43.** $\angle M$, $\angle L$, $\angle K$ **45.** Phoenix to Atlanta, Des Moines to Phoenix, Atlanta to Des Moines **47.** 5; $\overline{PR}$, $\overline{QR}$, $\overline{PQ}$ **49.** 12; $\overline{QR}$, $\overline{PR}$, $\overline{PQ}$ **51.** $2(y + 1) > \frac{x}{3}$, $y > \frac{x - 6}{6}$ **53.** $3x + 15 > 4x + 7 > 0$, $-\frac{7}{4} < x < 8$ **55.** A **57.** (15, −6) **59.** Yes; $\frac{1}{3}(-3) = -1$, and F is the midpoint of $\overline{BD}$. **61.** Label the midpoints of $\overline{AB}$, $\overline{BC}$, and $\overline{CA}$ as E, F, and G respectively. Then the coordinates of E, F, and G are $\left(\frac{a}{2}, 0\right)$, $\left(\frac{a + b}{2}, \frac{c}{2}\right)$, and $\left(\frac{b}{2}, \frac{c}{2}\right)$ respectively. The slope of $\overline{AF} = \frac{c}{a + b}$, and the slope of $\overline{AD} = \frac{c}{a + b}$, so D is on $\overline{AF}$. The slope of $\overline{BG} = \frac{c}{b - 2a}$ and the slope of $\overline{BD} = \frac{c}{b - 2a}$, so D is on $\overline{BG}$. The slope of $\overline{CE} = \frac{2c}{2b - a}$ and the slope of $\overline{CD} = \frac{2c}{2b - a}$, so D is on $\overline{CE}$. Since D is on $\overline{AF}$, $\overline{BG}$, and $\overline{CE}$, it is the intersection point of the three segments. **63.** $\angle C \cong \angle R$, $\angle D \cong \angle S$, $\angle G \cong \angle W$, $\overline{CD} \cong \overline{RS}$, $\overline{DG} \cong \overline{SW}$, $\overline{CG} \cong \overline{RW}$ **65.** 9.5 **67.** false

Page 254 Practice Quiz 1

1. 5 **3.** never **5.** sometimes **7.** no triangle **9.** $m\angle Q = 56$, $m\angle R = 61$, $m\angle S = 63$

Pages 257–260 Lesson 5-3

1. If a statement is shown to be false, then its opposite must be true.
3. Sample answer: $\triangle ABC$ is scalene.
Given: $\triangle ABC$; $AB \neq BC$; $BC \neq AC$; $AB \neq AC$
Prove: $\triangle ABC$ is scalene.

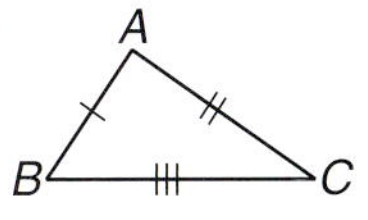

Proof:
Step 1: Assume $\triangle ABC$ is not scalene.
Case 1: $\triangle ABC$ is isosceles.
If $\triangle ABC$ is isosceles, then $AB = BC$, $BC = AC$, or $AB = AC$. This contradicts the given information, so $\triangle ABC$ is not isosceles.
Case 2: $\triangle ABC$ is equilateral.
In order for a triangle to be equilateral, it must also be isosceles, and Case 1 proved that $\triangle ABC$ is not isosceles. Thus, $\triangle ABC$ is not equilateral. Therefore, $\triangle ABC$ is scalene.
5. The lines are not parallel.

7. Given: $a > 0$

Prove: $\frac{1}{a} > 0$

Proof:

Step 1: Assume $\frac{1}{a} \leq 0$.

Step 2: $\frac{1}{a} \leq 0$; $a \cdot \frac{1}{a} \leq 0 \cdot a$, $1 \leq 0$

Step 3: The conclusion that $1 \leq 0$ is false, so the assumption that $\frac{1}{a} \leq 0$ must be false. Therefore, $\frac{1}{a} > 0$.

9. Given: $\triangle ABC$

Prove: There can be no more than one obtuse angle in $\triangle ABC$.

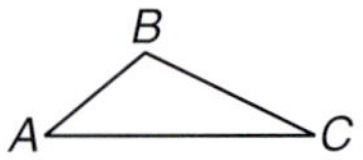

Proof:

Step 1: Assume that there can be more than one obtuse angle in $\triangle ABC$.

Step 2: The measure of an obtuse angle is greater than 90, $x > 90$, so the measure of two obtuse angles is greater than 180, $2x > 180$.

Step 3: The conclusion contradicts the fact that the sum of the angles of a triangle equals 180. Thus, there can be at most one obtuse angle in $\triangle ABC$.

11. Given: $\triangle ABC$ is a right triangle; $\angle C$ is a right angle.

Prove: $AB > BC$ and $AB > AC$

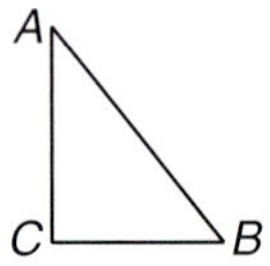

Proof:

Step 1: Assume that the hypotenuse of a right triangle is not the longest side. That is, $AB < BC$ or $AB < AC$.

Step 2: If $AB < BC$, then $m\angle C < m\angle A$. Since $m\angle C = 90$, $m\angle A > 90$.
So, $m\angle C + m\angle A > 180$. By the same reasoning, if $AB < BC$, then $m\angle C + m\angle B > 180$.

Step 3: Both relationships contradict the fact that the sum of the measures of the angles of a triangle equals 180. Therefore, the hypotenuse must be the longest side of a right triangle.

13. $\overline{PQ} \not\cong \overline{ST}$ **15.** A number cannot be expressed as $\frac{a}{b}$.

17. Points P, Q, and R are noncollinear.

19. Given: $\frac{1}{a} < 0$

Prove: a is negative.

Proof:

Step 1: Assume $a > 0$. $a \neq 0$ since that would make $\frac{1}{a}$ undefined.

Step 2: $\frac{1}{a} < 0$

$a\left(\frac{1}{a}\right) < 0 \cdot a$

$1 < 0$

Step 3: $1 > 0$, so the assumption must be false. Thus, a must be negative.

21. Given: $\overline{PQ} \cong \overline{PR}$
$\angle 1 \not\cong \angle 2$

Prove: $\overline{PZ}$ is not a median of $\triangle PQR$.

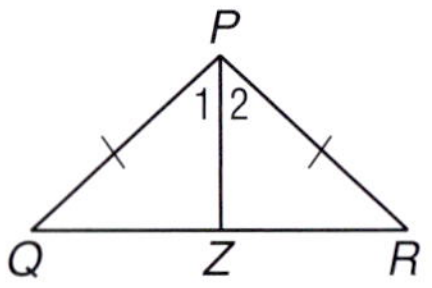

Proof:

Step 1: Assume $\overline{PZ}$ is a median of $\triangle PQR$.

Step 2: If $\overline{PZ}$ is a median of $\triangle PQR$, then Z is the midpoint of $\overline{QR}$, and $\overline{QZ} \cong \overline{RZ}$. $\overline{PZ} \cong \overline{PZ}$ by the Reflexive Property. $\triangle PZQ \cong \triangle PZR$ by SSS. $\angle 1 \cong \angle 2$ by CPCTC.

Step 3: This conclusion contradicts the given fact $\angle 1 \not\cong \angle 2$. Thus, $\overline{PZ}$ is not a median of $\triangle PQR$.

23. Given: $a > 0$, $b > 0$, and $a > b$

Prove: $\frac{a}{b} > 1$

Proof:

Step 1: Assume that $\frac{a}{b} \leq 1$.

Step 2:

Case 1	**Case 2**
$\frac{a}{b} < 1$	$\frac{a}{b} = 1$
$a < b$	$a = b$

Step 3: The conclusion of both cases contradicts the given fact $a > b$. Thus, $\frac{a}{b} > 1$.

25. Given: $\triangle ABC$ and $\triangle ABD$ are equilateral.
$\triangle ACD$ is not equilateral.

Prove: $\triangle BCD$ is not equilateral.

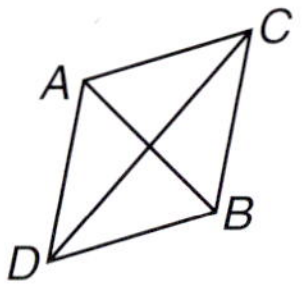

Proof:

Step 1: Assume that $\triangle BCD$ is an equilateral triangle.

Step 2: If $\triangle BCD$ is an equilateral triangle, then $\overline{BC} \cong \overline{CD} \cong \overline{DB}$. Since $\triangle ABC$ and $\triangle ABD$ are equilateral triangles, $\overline{AC} \cong \overline{AB} \cong \overline{BC}$ and $\overline{AD} \cong \overline{AB} \cong \overline{DB}$. By the Transitive Property, $\overline{AC} \cong \overline{AD} \cong \overline{CD}$. Therefore, $\triangle ACD$ is an equilateral triangle.

Step 3: This conclusion contradicts the given information. Thus, the assumption is false. Therefore, $\triangle BCD$ is not an equilateral triangle.

27. Use $r = \frac{d}{t}$, $t = 3$, and $d = 175$.

Proof:

Step 1: Assume that Ramon's average speed was greater than or equal to 60 miles per hour, $r \geq 60$.

Step 2:

Case 1	**Case 2**
$r = 60$	$r > 60$
$60 \stackrel{?}{=} \frac{175}{3}$	$\frac{175}{3} \stackrel{?}{>} 60$
$60 \neq 58.3$	$58.3 \ngtr 60$

Step 3: The conclusions are false, so the assumption must be false. Therefore, Ramon's average speed was less than 60 miles per hour.

29. $1500 \cdot 15\% \stackrel{?}{=} 225$
$1500 \cdot 0.15 \stackrel{?}{=} 225$
$225 = 225$

31. Yes; if you assume the client was at the scene of the crime, it is contradicted by his presence in Chicago at that time. Thus, the assumption that he was present at the crime is false.

33. Proof:

Step 1: Assume that $\sqrt{2}$ is a rational number.

Step 2: If $\sqrt{2}$ is a rational number, it can be written as $\frac{a}{b}$, where a and b are integers with no common factors, and $b \neq 0$. If $\sqrt{2} = \frac{a}{b}$, then $2 = \frac{a^2}{b^2}$, and $2b^2 = a^2$. Thus a^2 is an even number, as is a. Because a is even it can be written as $2n$.

$2b^2 = a^2$
$2b^2 = (2n)^2$
$2b^2 = 4n^2$
$b^2 = 2n^2$

Thus, b^2 is an even number. So, b is also an even number.

Step 3: Because b and a are both even numbers, they have a common factor of 2. This contradicts the definition of rational numbers. Therefore, $\sqrt{2}$ is not rational.

35. D **37.** $\angle P$

39. Given: $\overline{CD}$ is an angle bisector. $\overline{CD}$ is an altitude.
Prove: $\triangle ABC$ is isosceles.

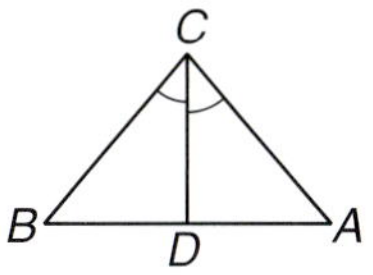

Proof:

Statements	Reasons
1. $\overline{CD}$ is an angle bisector. $\overline{CD}$ is an altitude.	1. Given
2. $\angle ACD \cong \angle BCD$	2. Def. $\angle$ bisector
3. $\overline{CD} \perp \overline{AB}$	3. Def. of altitude
4. $\angle CDA$ and $\angle CDB$ are rt. $\angle$s	4. $\perp$ lines form 4 rt. $\angle$s.
5. $\angle CDA \cong \angle CDB$	5. All rt. $\angle$s are $\cong$.
6. $\overline{CD} \cong \overline{CD}$	6. Reflexive Prop.
7. $\triangle ACD \cong \triangle BCD$	7. ASA
8. $\overline{AC} \cong \overline{BC}$	8. CPCTC
9. $\triangle ABC$ is isosceles.	9. Def. of isosceles $\triangle$

41. Given: $\triangle ABC \cong \triangle DEF$; $\overline{BG}$ is an angle bisector of $\angle ABC$. $\overline{EH}$ is an angle bisector of $\angle DEF$.
Prove: $\overline{BG} \cong \overline{EH}$

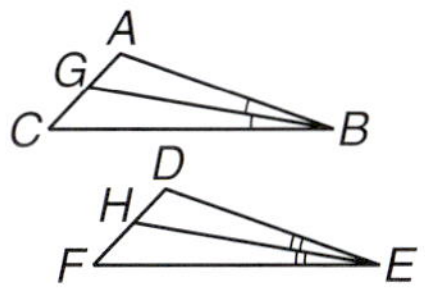

Proof:

Statements	Reasons
1. $\triangle ABC \cong \triangle DEF$	1. Given
2. $\angle A \cong \angle D$, $\overline{AB} \cong \overline{DE}$, $\angle ABC \cong \angle DEF$	2. CPCTC
3. $\overline{BG}$ is an angle bisector of $\angle ABC$. $\overline{EH}$ is an angle bisector of $\angle DEF$.	3. Given
4. $\angle ABG \cong \angle GBC$, $\angle DEH \cong \angle HEF$	4. Def. of $\angle$ bisector
5. $m\angle ABC = m\angle DEF$	5. Def. of $\cong$ $\angle$s
6. $m\angle ABG = m\angle GBC$, $m\angle DEH = m\angle HEF$	6. Def. of $\cong$ $\angle$s
7. $m\angle ABC = m\angle ABG + m\angle GBC$, $m\angle DEF = m\angle DEH + m\angle HEF$	7. Angle Addition Property
8. $m\angle ABC = m\angle ABG + m\angle ABG$, $m\angle DEF = m\angle DEH + m\angle DEH$	8. Substitution
9. $m\angle ABG + m\angle ABG = m\angle DEH + m\angle DEH$	9. Substitution
10. $2m\angle ABG = 2m\angle DEH$	10. Addition
11. $m\angle ABG = m\angle DEH$	11. Division
12. $\angle ABG \cong \angle DEH$	12. Def. of $\cong$ $\angle$s
13. $\triangle ABG \cong \triangle DEH$	13. ASA
14. $\overline{BG} \cong \overline{EH}$	14. CPCTC

43. $y - 3 = 2(x - 4)$ **45.** $y + 9 = 11(x + 4)$ **47.** false

Pages 263–266 Lesson 5-4

1. Sample answer: If the lines are not horizontal, then the segment connecting their y-intercepts is not perpendicular to either line. Since distance is measured along a perpendicular segment, this segment cannot be used. **3.** Sample answer: 2, 3, 4 and 1, 2, 3;

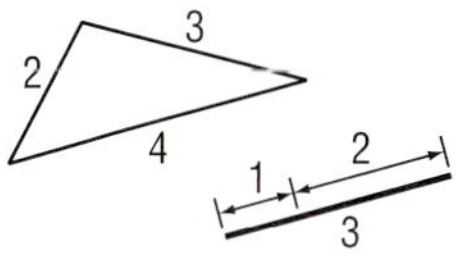

5. no; $5 + 10 \not> 15$ **7.** yes; $5.2 + 5.6 > 10.1$ **9.** $9 < n < 37$ **11.** $3 < n < 33$ **13.** B **15.** no; $2 + 6 \not> 11$ **17.** no; $13 + 16 \not> 29$ **19.** yes; $9 + 20 > 21$ **21.** yes; $17 + 30 > 30$ **23.** yes; $0.9 + 4 > 4.1$ **25.** no; $0.18 + 0.21 \not> 0.52$ **27.** $2 < n < 16$ **29.** $6 < n < 30$ **31.** $29 < n < 93$ **33.** $24 < n < 152$ **35.** $0 < n < 150$ **37.** $97 < n < 101$

39. Given: $\overline{HE} \cong \overline{EG}$
Prove: $HE + FG > EF$

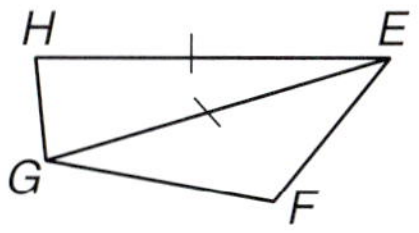

Proof:

Statements	Reasons
1. $\overline{HE} \cong \overline{EG}$	1. Given
2. $HE = EG$	2. Def. of $\cong$ segments
3. $EG + FG > EF$	3. Triangle Inequality
4. $HE + FG > EF$	4. Substitution

41. yes; $AB + BC > AC$, $AB + AC > BC$, $AC + BC > AB$ **43.** no; $XY + YZ = XZ$ **45.** 4 **47.** 3 **49.** $\frac{1}{2}$ **51.** Sample answer: You can use the Triangle Inequality Theorem to verify the shortest route between two locations. Answers should include the following.

- A longer route might be better if you want to collect frequent flier miles.
- A straight route might not always be available.

53. A **55.** $\overline{QR}$, $\overline{PQ}$, $\overline{PR}$ **57.** $JK = 5$, $KL = 2$, $JL = \sqrt{29}$, $PQ = 5$, $QR = 2$, and $PR = \sqrt{29}$. The corresponding sides have the same measure and are congruent. $\triangle JKL \cong \triangle PQR$ by SSS. **59.** $JK = \sqrt{113}$, $KL = \sqrt{50}$, $JL = \sqrt{65}$, $PQ = \sqrt{58}$, $QR = \sqrt{61}$, and $PR = \sqrt{65}$. The corresponding sides are not congruent, so the triangles are not congruent. **61.** $x < 6.6$

Page 266 Practice Quiz 2

1. The number 117 is not divisible by 13.
3. Step 1: Assume that $x \leq 8$.
Step 2: $7x > 56$ so $x > 8$
Step 3: The solution of $7x > 56$ contradicts the assumption. Thus, $x \leq 8$ must be false. Therefore, $x > 8$.

5. Given: $m\angle ADC \neq m\angle ADB$
Prove: $\overline{AD}$ is not an altitude of $\triangle ABC$.

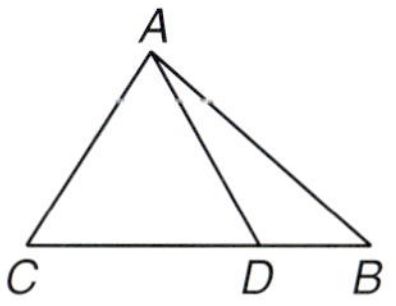

Proof:

Statements	Reasons
1. $\overline{AD}$ is an altitude of $\triangle ABC$.	1. Assumption
2. $\angle ADC$ and $\angle ADB$ are right angles.	2. Def. of altitude
3. $\angle ADC \cong \angle ADB$	3. All rt $\angle$s are $\cong$.
4. $m\angle ADC = m\angle ADB$	4. Def. of $\cong$ angles

This contradicts the given information that $m\angle ADC \neq m\angle ADB$. Thus, $\overline{AD}$ is not an altitude of $\triangle ABC$.

7. no; $25 + 35 \not> 60$ **9.** yes; $5 + 6 > 10$

Pages 270–273 Lesson 5-5

1. Sample answer: A pair of scissors illustrates the SSS inequality. As the distance between the tips of the scissors decreases, the angle between the blades decreases, allowing the blades to cut. **3.** $AB < CD$ **5.** $\frac{7}{3} < x < 6$

7. Given: $\overline{PQ} \cong \overline{SQ}$
Prove: $PR > SR$

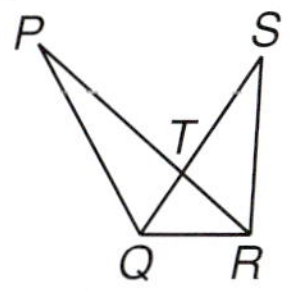

Selected Answers

Proof:

Statements	Reasons
1. $\overline{PQ} \cong \overline{SQ}$	1. Given
2. $\overline{QR} \cong \overline{QR}$	2. Reflexive Property
3. $m\angle PQR = m\angle PQS + m\angle SQR$	3. Angle Addition Postulate
4. $m\angle PQR > m\angle SQR$	4. Def. of inequality
5. $PR > SR$	5. SAS Inequality

9. Sample answer: The pliers are an example of the SAS inequality. As force is applied to the handles, the angle between them decreases causing the distance between the ends of the pliers to decrease. As the distance between the ends of the pliers decreases, more force is applied to a smaller area. **11.** $m\angle BDC < m\angle FDB$ **13.** $AD > DC$ **15.** $m\angle AOD > m\angle AOB$ **17.** $4 < x < 10$ **19.** $7 < x < 20$

21. Given: $\overline{PQ} \cong \overline{RS}$, $QR < PS$
Prove: $m\angle 3 < m\angle 1$

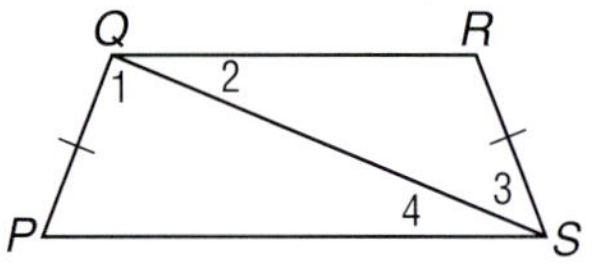

Proof:

Statements	Reasons
1. $\overline{PQ} \cong \overline{RS}$	1. Given
2. $\overline{QS} \cong \overline{QS}$	2. Reflexive Property
3. $QR < PS$	3. Given
4. $m\angle 3 < m\angle 1$	4. SSS Inequality

23. Given: $\overline{ED} \cong \overline{DF}$; $m\angle 1 > m\angle 2$; D is the midpoint of $\overline{CB}$; $\overline{AE} \cong \overline{AF}$.
Prove: $AC > AB$

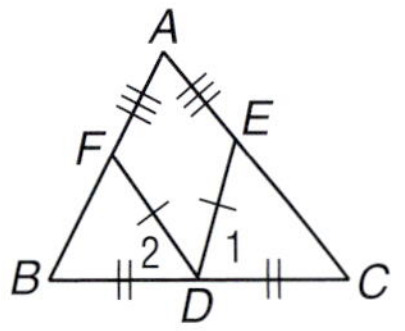

Proof:

Statements	Reasons
1. $\overline{ED} \cong \overline{DF}$; D is the midpoint of $\overline{DB}$.	1. Given
2. $CD = BD$	2. Def. of midpoint
3. $\overline{CD} \cong \overline{BD}$	3. Def. of $\cong$ segments
4. $m\angle 1 > m\angle 2$	4. Given
5. $EC > FB$	5. SAS Inequality
6. $\overline{AE} \cong \overline{AF}$	6. Given
7. $AE = AF$	7. Def. of $\cong$ segments
8. $AE + EC > AE + FB$	8. Add. Prop. of Inequality
9. $AE + EC > AF + FB$	9. Substitution Prop. of Inequality
10. $AE + EC = AC$, $AF + FB = AB$	10. Segment Add. Post.
11. $AC > AB$	11. Substitution

25. As the door is opened wider, the angle formed increases and the distance from the end of the door to the door frame increases.

27. As the vertex angle increases, the base angles decrease. Thus, as the base angles decrease, the altitude of the triangle decreases.

29.

Stride (m)	Velocity (m/s)
0.25	0.07
0.50	0.22
0.75	0.43
1.00	0.70
1.25	1.01
1.50	1.37

31. Sample answer: A backhoe digs when the angle between the two arms decreases and the shovel moves through the dirt. Answers should include the following.
- As the operator digs, the angle between the arms decreases.
- The distance between the ends of the arms increases as the angle between the arms increases, and decreases as the angle decreases.

33. B **35.** yes; $16 + 6 > 19$ **37.** $\overline{AD}$ is a not median of $\triangle ABC$.

39. Given: $\overline{AD}$ bisects $\overline{BE}$; $\overline{AB} \parallel \overline{DE}$.
Prove: $\triangle ABC \cong \triangle DEC$

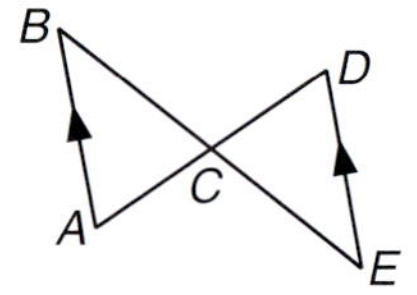

Proof:

Statements	Reasons
1. $\overline{AD}$ bisects $\overline{BE}$; $\overline{AB} \parallel \overline{DE}$.	1. Given
2. $\overline{BC} \cong \overline{EC}$	2. Def. of seg. bisector
3. $\angle B \cong \angle E$	3. Alt. int. ∡ Thm.
4. $\angle BCA \cong \angle ECD$	4. Vert. ∡ are $\cong$.
5. $\triangle ABC \cong \triangle DEC$	5. ASA

41. $EF = 5$, $FG = 50$, $EG = 5$; isosceles **43.** $EF = \sqrt{145}$, $FG = \sqrt{544}$, $EG = 35$; scalene **45.** yes, by the Law of Detachment

Pages 274–276 Chapter 5 Study Guide and Review

1. incenter **3.** Triangle Inequality Theorem **5.** angle bisector **7.** orthocenter **9.** 72 **11.** $m\angle DEF > m\angle DFE$ **13.** $m\angle DEF > m\angle FDE$ **15.** $DQ < DR$ **17.** $SR > SQ$ **19.** The triangles are not congruent. **21.** no; $7 + 5 \not> 20$ **23.** yes; $6 + 18 > 20$ **25.** $BC > MD$ **27.** $x > 7$

Chapter 6 Proportions and Similarity

Page 281 Chapter 6 Getting Started

1. 15 **3.** 10 **5.** 2 **7.** $-\frac{6}{5}$ **9.** yes; $\cong$ alt. int. ∡ **11.** 2, 4, 8, 16 **13.** 1, 7, 25, 79

Page 284–287 Lesson 6-1

1. Cross multiply and divide by 28. **3.** Suki; Madeline did not find the cross products correctly. **5.** $\frac{1}{12}$ **7.** 2.1275 **9.** 54, 48, 42 **11.** 320 **13.** 76:89 **15.** 25.3:1 **17.** 18 ft, 24 ft **19.** 43.2, 64.8, 72 **21.** 18 in., 24 in., 30 in. **23.** $\frac{3}{2}$ **25.** 2:19 **27.** 16.4 lb **29.** 1.295 **31.** 14 **33.** 3 **35.** $-1, \frac{-2}{3}$ **37.** 36%

39. Sample answer: It appears that Tiffany used rectangles with areas that were in proportion as a background for this artwork. Answers should include the following.
- The center column pieces are to the third column from the left pieces as the pieces from the third column are to the pieces in the outside column.
- The dimensions are approximately 24 inches by 34 inches.

41. D **43.** always **45.** $15 < x < 47$ **47.** $12 < x < 34$

49.

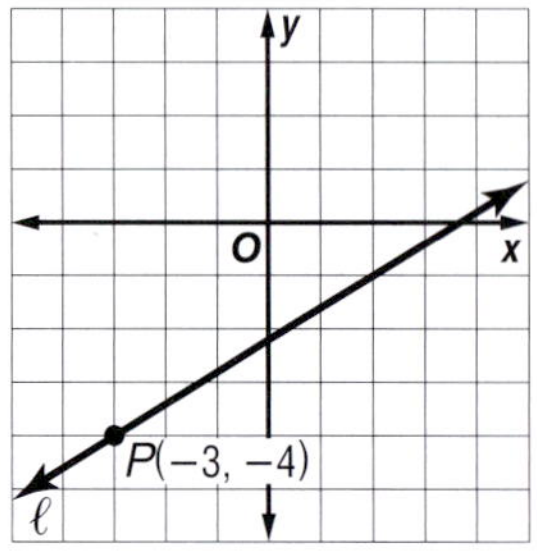

51.

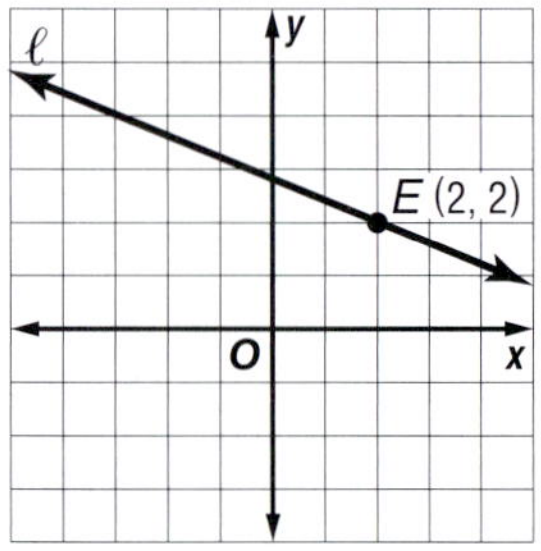

53. Yes; 100 km and 62 mi are the same length, so $AB = CD$. By the definition of congruent segments, $\overline{AB} \cong \overline{CD}$. **55.** 13.0 **57.** 1.2

Page 292–297 Lesson 6-2

1. Both students are correct. One student has inverted the ratio and reversed the order of the comparison. **3.** If two polygons are congruent, then they are similar. All of the corresponding angles are congruent, and the ratio of measures of the corresponding sides is 1.Two similar figures have congruent angles, and the sides are in proportion, but not always congruent. If the scale factor is 1, then the figures are congruent. **5.** Yes; $\angle A \cong \angle E$, $\angle B \cong \angle F$, $\angle C \cong \angle G$, $\angle D \cong \angle H$ and $\frac{AD}{EH} = \frac{DC}{HG} = \frac{CB}{GF} = \frac{BA}{FE} = \frac{2}{3}$. So $\square ABCD \sim \square EFGH$. **7.** polygon $ABCD \sim$ polygon $EFGH$; 23; 28; 20; 32; $\frac{1}{2}$ **9.** 60 m **11.** $ABCF$ is similar to $EDCF$ since they are congruent. **13.** $\triangle ABC$ is not similar to $\triangle DEF$. $\angle A \not\cong \angle D$. **15.** $\frac{1}{3}$ **17.** polygon $ABCD \sim$ polygon $EFGH$; $\frac{13}{3}$; $AB = \frac{16}{3}$; $CD = \frac{10}{3}$; $\frac{2}{3}$ **19.** $\triangle ABE \sim \triangle ACD$; 6; $BC = 8$; $ED = 5$; $\frac{5}{9}$ **21.** about 3.9 in. by 6.25 in. **23.** $\frac{25}{16}$

25.

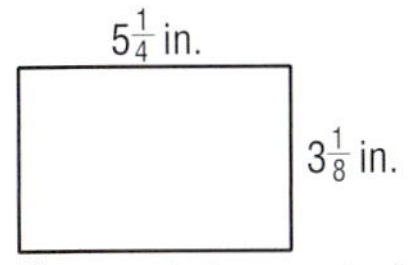

Figure not shown actual size.

27. always **29.** never **31.** sometimes **33.** always **35.** 30; 70 **37.** 27; 14 **39.** 71.05; 48.45 **41.** 7.5 **43.** 108 **45.** 73.2 **47.** $\frac{8}{5}$

49. $L(16, 8)$ and $P(8, 8)$ or $L(16, -8)$ and $P(8, -8)$

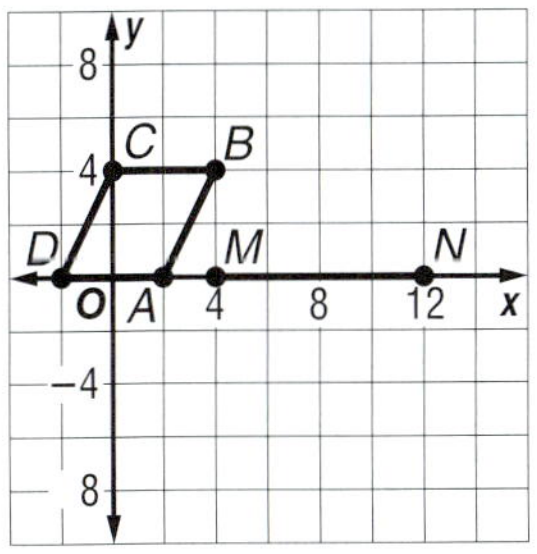

51. 18 ft by 15 ft **53.** 16 : 1 **55.** 16 : 1 **57.** 2 : 1; ratios are the same. **59.** $\frac{a}{3a} = \frac{b}{3b} = \frac{c}{3c} = \frac{a+b+c}{3(a+b+c)} = \frac{1}{3}$

61. Sample answer: Artists use geometric shapes in patterns to create another scene or picture. The included objects have the same shape but are different sizes. Answers should include the following.

- The objects are enclosed within a circle. The objects seem to go on and on
- Each "ring" of figures has images that are approximately the same width, but vary in number and design.

63. D

65.

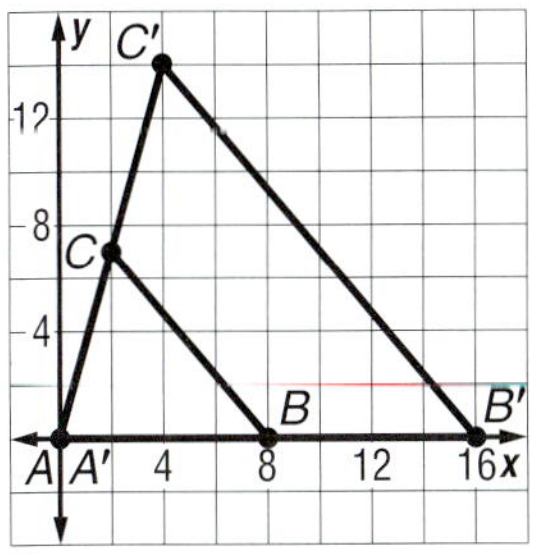

67. $\frac{AB}{A'B'} = \frac{AC}{A'C'} = \frac{BC}{B'C'} = \frac{1}{2}$ **69.** The sides are proportional and the angles are congruent, so the triangles are similar. **71.** −23 **73.** $OC > AO$ **75.** $m\angle ABD > m\angle ADB$ **77.** 91 **79.** $m\angle 1 = m\angle 2 = 111$ **81.** 62 **83.** 118 **85.** 62 **87.** 118

Page 301–306 Lesson 6-3

1. Sample answer: Two triangles are congruent by the SSS, SAS, and ASA Postulates and the AAS Theorem. In these triangles, corresponding parts must be congruent. Two triangles are similar by AA Similarity, SSS Similarity, and SAS Similarity. In similar triangles, the sides are proportional and the angles are congruent. Congruent triangles are always similar triangles. Similar triangles are congruent only when the scale factor for the proportional sides is 1. SSS and SAS are common relationships for both congruence and similarity. **3.** Alicia; while both have corresponding sides in a ratio, Alicia has them in proper order with the numerators from the same triangle.

5. $\triangle ABC \sim \triangle DEF$; $x = 10$; $AB = 10$; $DE = 6$ **7.** yes: $\triangle DEF \sim \triangle ACB$ by SSS Similarity **9.** 135 ft **11.** yes; $\triangle QRS \sim \triangle TVU$ by SSS Similarity **13.** yes; $\triangle RST \sim \triangle JKL$ by AA Similarity **15.** Yes; $\triangle ABC \sim \triangle JKL$ by SAS Similarity **17.** No; sides are not proportional. **19.** $\triangle ABE \sim \triangle ACD$; $x = \frac{8}{5}$; $AB = 3\frac{3}{5}$; $AC = 9\frac{3}{5}$ **21.** $\triangle ABC \sim \triangle ARS$; $x = 8$; 15; 8 **23.** $\frac{3}{2}$ **25.** true **27.** $\triangle EAB \sim \triangle EFC \sim \triangle AFD$: AA Similarity **29.** $KP = 5$, $KM = 15$, $MR = 13\frac{1}{3}$, $ML = 20$, $MN = 12$, $PR = 16\frac{2}{3}$ **31.** $m\angle TUV = 43$, $m\angle R = 43$, $m\angle RSU = 47$, $m\angle SUV = 47$ **33.** $x = y$; if $\overline{BD} \parallel \overline{AE}$, then $\triangle BCD \sim \triangle ACE$ by AA Similarity and $\frac{BC}{AC} = \frac{DC}{EC}$. Thus, $\frac{2}{4} = \frac{x}{x+y}$. Cross multiply and solve for y, yielding $y = x$.

35. Given: $\overline{LP} \parallel \overline{MN}$

Prove: $\frac{LJ}{JN} = \frac{PJ}{JM}$

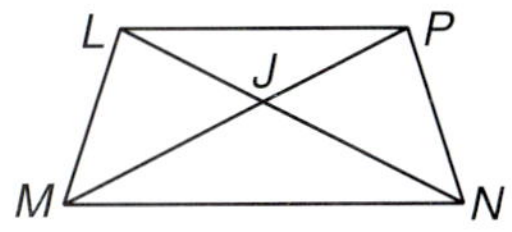

Proof:

Statements	Reasons
1. $\overline{LP} \parallel \overline{MN}$	1. Given
2. $\angle PLN \cong \angle LNM$, $\angle LPM \cong \angle PMN$	2. Alt. Int. ∠ Theorem
3. $\triangle LPJ \sim \triangle NMJ$	3. AA Similarity
4. $\frac{LJ}{JN} = \frac{PJ}{JM}$	4. Corr. sides of ~ △s are proportional.

37. Given: $\triangle BAC$ and $\triangle EDF$ are right triangles. $\frac{AB}{DE} = \frac{AC}{DF}$

Prove: $\triangle ABC \sim \triangle DEF$

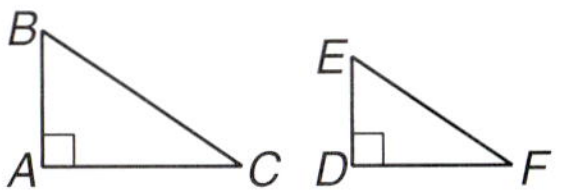

Proof:

Statements	Reasons
1. $\triangle BAC$ and $\triangle EDF$ are right triangles.	1. Given
2. $\angle BAC$ and $\angle EDF$ are right angles.	2. Def. of rt. △
3. $\angle BAC \cong \angle EDF$	3. All rt. ∠s are ≅.
4. $\frac{AB}{DE} = \frac{AC}{DF}$	4. Given
5. $\triangle ABC \sim \triangle DEF$	5. SAS Similarity

39. 13.5 ft **41.** about 420.5 m **43.** 10.75 m

45. 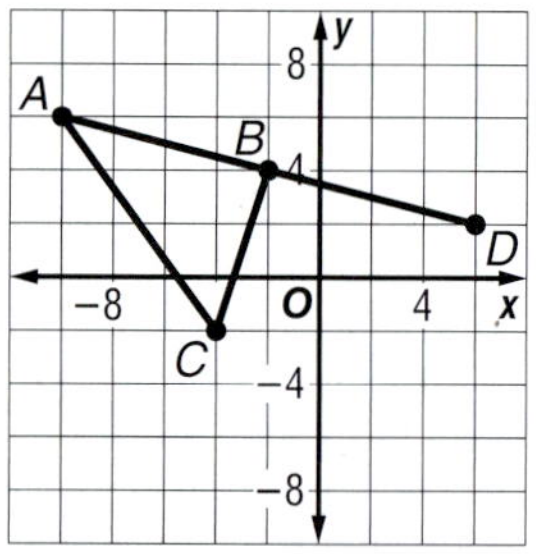

47. $\triangle ABC \sim \triangle ACD$; $\triangle ABC \sim \triangle CBD$; $\triangle ACD \sim \triangle CBD$; they are similar by AA Similarity. **49.** A **51.** $PQRS \sim ABCD$; 1.6; 1.4; 1.1; $\frac{1}{2}$

53. 5 **55.** 15 **57.** No; $\overline{AT}$ is not perpendicular to $\overline{BC}$. **59.** (5.5, 13) **61.** (3.5, −2.5)

Page 306 Practice Quiz 1

1. yes; $\angle A \cong \angle E$, $\angle B \cong \angle D$, $\angle 1 \cong \angle 3$, $\angle 2 \cong \angle 4$ and $\frac{AB}{ED} = \frac{BC}{DC} = \frac{AF}{EF} = \frac{FC}{FC} = 1$ **3.** $\triangle ADE \sim \triangle CBE$; 2; 8; 4 **5.** 1947 mi

Page 311–315 Lesson 6-4

1. Sample answer: If a line intersects two sides of a triangle and separates sides into corresponding segments of proportional lengths, then it is parallel to the third side. **3.** Given three or more parallel lines intersecting two transversals, Corollary 6.1 states that the parts of the transversals are proportional. Corollary 6.2 states that if the parts of one transversal are congruent, then the parts of every transversal are congruent. **5.** 10 **7.** The slopes of $\overline{DE}$ and $\overline{BC}$ are both 0. So $\overline{DE} \parallel \overline{BC}$. **9.** Yes; $\frac{MN}{NP} = \frac{MR}{RQ} = \frac{9}{16}$, so $\overline{RN} \parallel \overline{QP}$. **11.** $x = 2$; $y = 5$ **13.** 1100 yd **15.** 3 **17.** $x = 6$, ED = 9 **19.** $BC = 10$, $FE = 13\frac{1}{3}$, $CD = 9$, $DE = 15$ **21.** 10 **23.** No; segments are not proportional; $\frac{PQ}{QR} = \frac{3}{7}$ and $\frac{PT}{TS} = 2$. **25.** yes **27.** $\sqrt{52}$ **29.** The endpoints of $\overline{DE}$ are $D\left(3, \frac{1}{2}\right)$ and $E\left(\frac{3}{2}, -4\right)$. Both $\overline{DE}$ and $\overline{AB}$ have slope of 3. **31.** (3, 8) or (4, 4) **33.** $x = 21$, $y = 15$ **35.** 25 ft **37.** 18.75 ft

39. Given: D is the midpoint of $\overline{AB}$.
E is the midpoint of $\overline{AC}$.
Prove: $\overline{DE} \parallel \overline{BC}$; $DE = \frac{1}{2}BC$

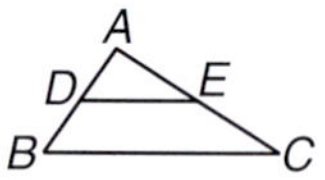

Proof:

Statements	Reasons
1. D is the midpoint of $\overline{AB}$. E is the midpoint of $\overline{AC}$.	1. Given
2. $\overline{AD} \cong \overline{DB}$, $\overline{AE} \cong \overline{EC}$	2. Midpoint Theorem
3. $AD = DB$, $AE = EC$	3. Def. of $\cong$ segments
4. $AB = AD + DB$, $AC = AE + AC$	4. Segment Addition Postulate
5. $AB = AD + AD$, $AC = AE + AE$	5. Substitution
6. $AB = 2AD$, $AC = 2AE$	6. Substitution
7. $\frac{AB}{AD} = 2$, $\frac{AC}{AE} = 2$	7. Division Prop.
8. $\frac{AB}{AD} = \frac{AC}{AE}$	8. Transitive Prop.
9. $\angle A \cong \angle A$	9. Reflexive Prop.
10. $\triangle ADE \sim \triangle ABC$	10. SAS Similarity
11. $\angle ADE \cong \angle ABC$	11. Def. of $\sim$ polygons
12. $\overline{DE} \parallel \overline{BC}$	12. If corr. $\angle$s are $\cong$, the lines are parallel.
13. $\frac{BC}{DE} = \frac{AB}{AD}$	13. Def. of $\sim$ polygons
14. $\frac{BC}{DE} = 2$	14. Substitution
15. $2DE = BC$	15. Mult. Prop.
16. $DE = \frac{1}{2}BC$	16. Division Prop.

41. 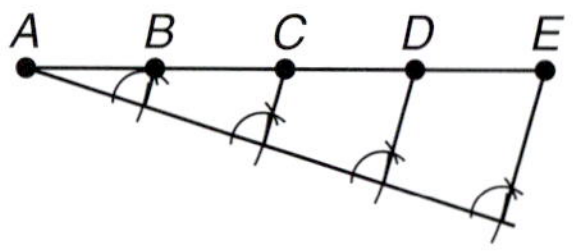

43. $u = 24$; $w = 26.4$; $x = 30$; $y = 21.6$; $z = 33.6$

45. Sample answer: City planners use maps in their work. Answers should include the following.
- City planners need to know geometry facts when developing zoning laws.
- A city planner would need to know that the shortest distance between two parallel lines is the perpendicular distance.

47. 4 **49.** yes; AA **51.** no; angles not congruent **53.** $x = 12$, $y = 6$ **55.** $m\angle ABD > m\angle BAD$ **57.** $m\angle CBD > m\angle BCD$ **59.** 18 **61.** false **63.** true **65.** $\angle R \cong \angle X$, $\angle S \cong \angle Y$, $\angle T \cong \angle Z$, $\overline{RS} \cong \overline{XY}$, $\overline{ST} \cong \overline{YZ}$, $\overline{RT} \cong \overline{XZ}$

Page 319–323 Lesson 6-5

1. $\triangle ABC \sim \triangle MNQ$ and $\overline{AD}$ and $\overline{MR}$ are altitudes, angle bisectors, or medians. **3.** 10.8 **5.** 6 **7.** 6.75 **9.** 330 cm or 3.3 m **11.** 63 **13.** 20.25 **15.** 78 **17.** Yes; the perimeters are in the same ratio as the sides, $\frac{300}{600}$ or $\frac{1}{2}$. **19.** $\frac{3}{2}$ **21.** 4 **23.** $11\frac{1}{5}$ **25.** 6 **27.** 5, 13.5 **29.** $xy = z^2$; $\triangle ACD \sim \triangle CBD$ by AA Similarity. Thus, $\frac{CD}{BD} = \frac{AD}{CD}$ or $\frac{z}{y} = \frac{x}{z}$. The cross products yield $xy = z^2$.

31. Given: $\triangle ABC \sim \triangle RST$, $\overline{AD}$ is a median of $\triangle ABC$.
$\overline{RU}$ is a median of $\triangle RST$.
Prove: $\frac{AD}{RU} = \frac{AB}{RS}$

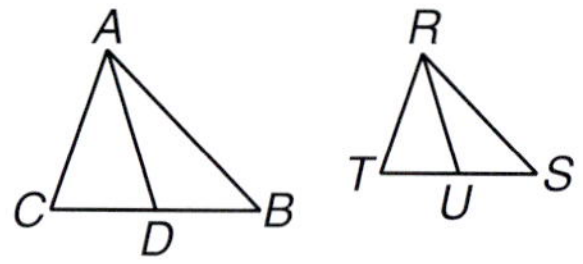

Proof:

Statements	Reasons
1. $\triangle ABC \sim \triangle RST$ $\overline{AD}$ is a median of $\triangle ABC$. $\overline{RU}$ is a median of $\triangle RST$.	1. Given
2. $CD = DB$; $TU = US$	2. Def. of median
3. $\frac{AB}{RS} = \frac{CB}{TS}$	3. Def. of $\sim$ polygons
4. $CB = CD + DB$; $TS = TU + US$	4. Segment Addition Postulate
5. $\frac{AB}{RS} = \frac{CD + DB}{TU + US}$	5. Substitution
6. $\frac{AB}{RS} = \frac{DB + DB}{US + US}$ or $\frac{2(DB)}{2(US)}$	6. Substitution
7. $\frac{AB}{RS} = \frac{DB}{US}$	7. Substitution
8. $\angle B \cong \angle S$	8. Def. of $\sim$ polygons
9. $\triangle ABD \sim \triangle RSU$	9. SAS Similarity
10. $\frac{AD}{RU} = \frac{AB}{RS}$	10. Def. of $\sim$ polygons

33. Given: $\triangle ABC \sim \triangle PQR$, $\overline{BD}$ is an altitude of $\triangle ABC$.
$\overline{QS}$ is an altitude of $\triangle PQR$.
Prove: $\frac{QP}{BA} = \frac{QS}{BD}$

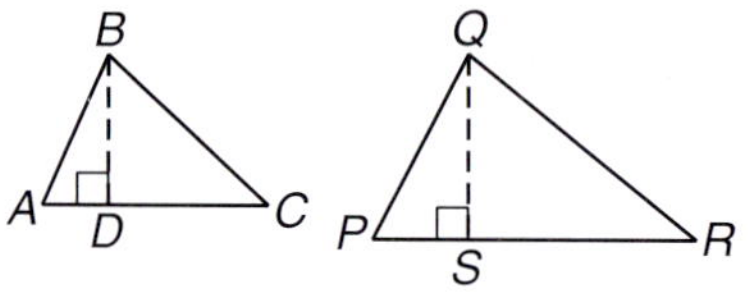

Selected Answers

Proof:
$\angle A \cong \angle P$ because of the definition of similar polygons. Since $\overline{BD}$ and $\overline{QS}$ are perpendicular to $\overline{AC}$ and $\overline{PR}$, $\angle BDA \cong \angle QSP$. So, $\triangle ABD \sim \triangle PQS$ by AA Similarity and $\frac{QP}{BA} = \frac{QS}{BD}$ by definition of similar polygons.

35. Given: $\overline{JF}$ bisects $\angle EFG$.
$\overline{EH} \parallel \overline{FG}$, $\overline{EF} \parallel \overline{HG}$
Prove: $\frac{EK}{KF} = \frac{GJ}{JF}$

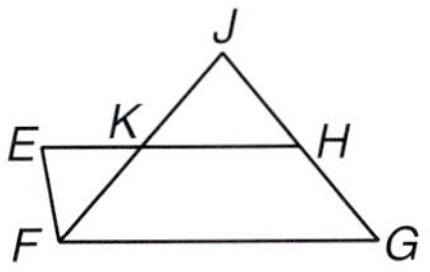

Proof:

Statements	Reasons
1. $\overline{JF}$ bisects $\angle EFG$. $\overline{EH} \parallel \overline{FG}$, $\overline{EF} \parallel \overline{HG}$	1. Given
2. $\angle EFK \cong \angle KFG$	2. Def. of $\angle$ bisector
3. $\angle KFG \cong \angle JKH$	3. Corresponding ∠s Post.
4. $\angle JKH \cong \angle EKF$	4. Vertical ∠s are ≅.
5. $\angle EFK \cong \angle EKF$	5. Transitive Prop.
6. $\angle FJH \cong \angle EFK$	6. Alternate Interior ∠s Th.
7. $\angle FJH \cong \angle EKF$	7. Transitive Prop.
8. $\triangle EKF \sim \triangle GJF$	8. AA Similarity
9. $\frac{EK}{KF} = \frac{GJ}{JF}$	9. Def. of ~ △s

37. Given: $\triangle RST \sim \triangle ABC$,
W and D are midpoints of $\overline{TS}$ and $\overline{CB}$, respectively.
Prove: $\triangle RWS \sim \triangle ADB$

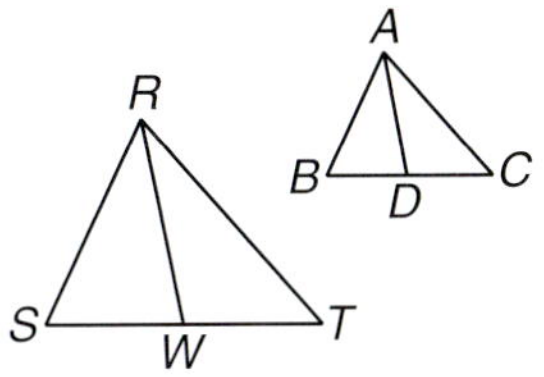

Proof:

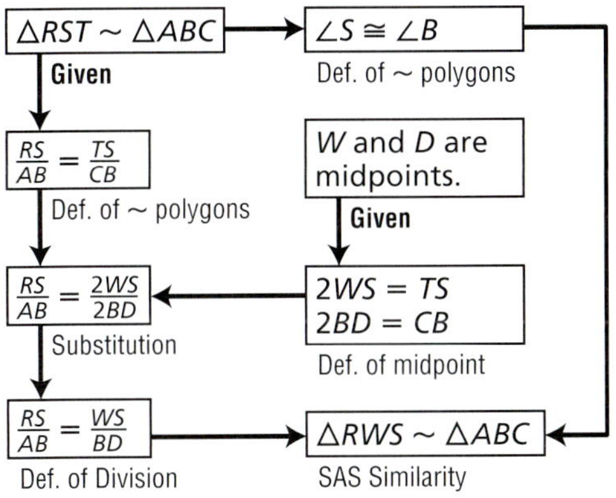

39. 12.9 **41.** no; sides not proportional **43.** yes; $\frac{LM}{MO} = \frac{LN}{NP}$
45. $\triangle PQT \sim \triangle PRS$, $x = 7$, $PQ = 15$ **47.** $y = 2x + 1$
49. 320, 640 **51.** −27, −33

Page 323 Practice Quiz 2

1. 20 **3.** no; sides not proportional **5.** 12.75 **7.** 10.5 **9.** 5

Page 328–331 Lesson 6-6

1. Sample answer: irregular shape formed by iteration of self-similar shapes **3.** Sample answer: icebergs, ferns, leaf veins **5.** $A_n = 2(2^n - 1)$ **7.** 1.4142…; 1.1892… **9.** Yes, the procedure is repeated over and over again.

11. 9 holes

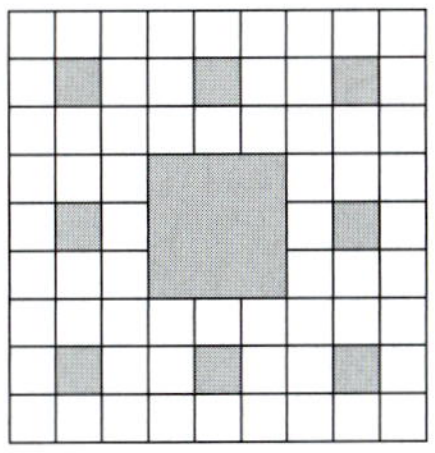

13. Yes, any part contains the same figure as the whole, 9 squares with the middle shaded. **15.** 1, 3, 6, 10, 15…; Each difference is 1 more than the preceding difference.
17. The result is similar to a Stage 3 Sierpinski triangle.
19. 25

21. Given: $\triangle ABC$ is equilateral.
$CD = \frac{1}{3}CB$ and
$CE = \frac{1}{3}CA$
Prove: $\triangle CED \sim \triangle CAB$

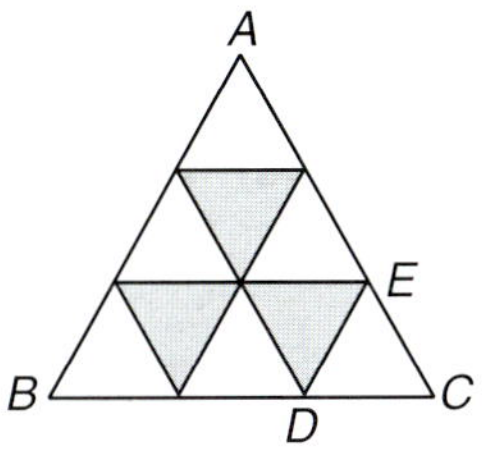

Proof:

Statements	Reasons
1. $\triangle ABC$ is equilateral. $CD = \frac{1}{3}CB$, $CE = \frac{1}{3}CA$	1. Given
2. $\overline{AC} \cong \overline{BC}$	2. Def. of equilateral △
3. $AC = BC$	3. Def. of ≅ segments
4. $\frac{1}{3}AC = \frac{1}{3}CB$	4. Mult. Prop.
5. $CD = CE$	5. Substitution
6. $\frac{CD}{CB} = \frac{CE}{CB}$	6. Division Prop.
7. $\frac{CD}{CB} = \frac{CE}{CA}$	7. Substitution
8. $\angle C \cong \angle C$	8. Reflexive Prop.
9. $\triangle CED \sim \triangle CAB$	9. AA Similarity

23. Yes; the smaller and smaller details of the shape have the same geometric characteristics as the original form. **25.** $A_n = 4^n$; 65,536 **27.** Stage 0: 3 units, Stage 1: $3 \cdot \frac{4}{3}$ or 4 units, Stage 2: $3\left(\frac{4}{3}\right)\frac{4}{3} = 3\left(\frac{4}{3}\right)^2$ or $5\frac{1}{3}$ units, Stage 3: $3\left(\frac{4}{3}\right)^3$ or $7\frac{1}{9}$ units **29.** The original triangle and the new triangles are equilateral and thus, all of the angles are equal to 60. By AA Similarity, the triangles are similar. **31.** 0.2, 5, 0.2, 5, 0.2; the numbers alternate between 0.2 and 5.0. **33.** 1, 2, 4, 16, 65,536; the numbers approach positive infinity. **35.** 0, −5, −10 **37.** −6, 24, −66 **39.** When $x = 0.00$: 0.64, 0.9216, 0.2890…, 0.8219…, 0.5854…, 0.9708…, 0.1133…, 0.4019…, 0.9615…, 0.1478…; when $x = 0.201$: 0.6423…, 0.9188…, 0.2981…, 0.8369…, 0.5458…, 0.9916…, 0.0333…, 0.1287…, 0.4487…, 0.9894… . Yes, the initial value affected the tenth value. **41.** The leaves in the tree and the branches of the trees are self-similar. These self-similar shapes are repeated throughout the painting. **43.** See students' work.

45. Sample answer: Fractal geometry can be found in the repeating patterns of nature. Answers should include the following.

- Broccoli is an example of fractal geometry because the shape of the florets is repeated throughout; one floret looks the same as the stalk.
- Sample answer: Scientists can use fractals to study the human body, rivers, and tributaries, and to model how landscapes change over time.

47. C **49.** $13\frac{3}{5}$ **51.** $\frac{7}{3}$ **53.** $16\frac{1}{4}$ **55.** Miami, Bermuda, San Juan **57.** 10 ft, 10 ft, 17 ft, 17 ft

Page 332–336 Chapter 6 Study Guide and Review

1. true **3.** true **5.** false, iteration **7.** true **9.** false, parallel to **11.** 12 **13.** $\frac{58}{3}$ **15.** $\frac{3}{5}$ **17.** 24 in. and 84 in. **19.** Yes, these are rectangles, so all angles are congruent. Additionally, all sides are in a 3:2 ratio. **21.** $\triangle PQT \sim \triangle RQS$; 0; $PQ=6$; $QS=3$;1 **23.** yes, $\triangle GHI \sim \triangle GJK$ by AA Similarity **25.** $\triangle ABC \sim \triangle DEC$, 4 **27.** no; lengths not proportional **29.** yes; $\frac{HI}{GH} = \frac{IK}{KL}$ **31.** 6 **33.** 9 **35.** 24 **37.** 36 **39.** Stage 2 is not similar to Stage 1. **41.** −8, −20, −56

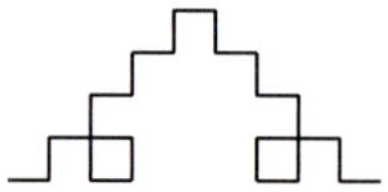

43. −6, −9.6, −9.96

Chapter 7 Right Triangles and Trigonometry

Page 541 Chapter 7 Getting Started

1. $a = 16$ **3.** $e = 24, f = 12$ **5.** 13 **7.** 21.21 **9.** $2\sqrt{2}$ **11.** 15 **13.** 98 **15.** 23

Pages 345–348 Lesson 7-1

1. Sample answer: 2 and 72 **3.** Ian; his proportion shows that the altitude is the geometric mean of the two segments of the hypotenuse. **5.** 42 **7.** $2\sqrt{3} \approx 3.5$ **9.** $4\sqrt{3} \approx 6.9$ **11.** $x = 6; y = 4\sqrt{3}$ **13.** $\sqrt{30} \approx 5.5$ **15.** $2\sqrt{15} \approx 7.7$ **17.** $\frac{\sqrt{15}}{5} \approx 0.8$ **19.** $\frac{\sqrt{5}}{3} \approx 0.7$ **21.** $3\sqrt{5} \approx 6.7$ **23.** $8\sqrt{2} \approx 11.3$ **25.** $\sqrt{26} \approx 5.1$ **27.** $x = 2\sqrt{15} \approx 9.4$; $y = \sqrt{33} \approx 5.7$; $z = 2\sqrt{6} \approx 4.9$ **29.** $x = \frac{40}{3}$; $y = \frac{5}{3}$; $z = 10\sqrt{2} \approx 14.1$ **31.** $x = 6\sqrt{6} \approx 14.7$; $y = 6\sqrt{42} \approx 38.9$; $z = 36\sqrt{7} \approx 95.2$ **33.** $\frac{17}{7}$ **35.** never **37.** sometimes **39.** $\triangle FGH$ is a right triangle. $\overline{OG}$ is the altitude from the vertex of the right angle to the hypotenuse of that triangle. So, by Theorem 7.2, OG is the geometric mean between OF and OH, and so on. **41.** 2.4 yd **43.** yes; Indiana and Virginia

45. Given: $\angle PQR$ is a right angle. $\overline{QS}$ is an altitude of $\triangle PQR$.

Prove: $\triangle PSQ \sim \triangle PQR$
$\triangle PQR \sim \triangle QSR$
$\triangle PSQ \sim \triangle QSR$

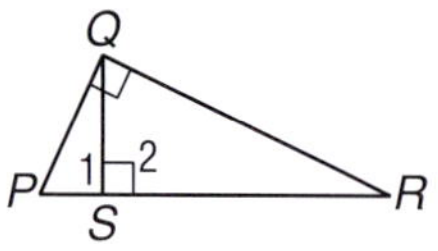

Proof:

Statements	Reasons
1. $\angle PQR$ is a right angle. $\overline{QS}$ is an altitude of $\triangle PQR$.	1. Given
2. $\overline{QS} \perp \overline{RP}$	2. Definition of altitude
3. $\angle 1$ and $\angle 2$ are right angles.	3. Definition of perpendicular lines
4. $\angle 1 \cong \angle PQR$, $\angle 2 \cong \angle PQR$	4. All right $\angle$s are $\cong$.
5. $\angle P \cong \angle P$, $\angle R \cong \angle R$	5. Congruence of angles is reflexive.
6. $\triangle PSQ \sim \triangle PQR$, $\triangle PQR \sim \triangle QSR$	6. AA Similarity Statements 4 and 5
7. $\triangle PSQ \sim \triangle QSR$	7. Similarity of triangles is transitive.

47. Given: $\angle ADC$ is a right angle. $\overline{DB}$ is an altitude of $\triangle ADC$.

Prove: $\frac{AB}{AD} = \frac{AD}{AC}$
$\frac{BC}{DC} = \frac{DC}{AC}$

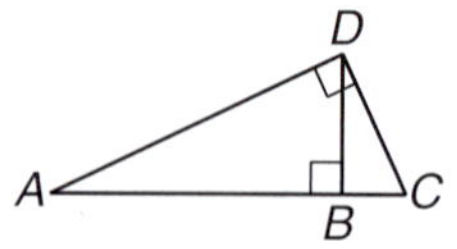

Proof:

Statements	Reasons
1. $\angle ADC$ is a right angle. $\overline{DB}$ is an altitude of $\triangle ADC$.	1. Given
2. $\triangle ADC$ is a right triangle.	2. Definition of right triangle
3. $\triangle ABD \sim \triangle ADC$, $\triangle DBC \sim \triangle ADC$	3. If the altitude is drawn from the vertex of the rt. $\angle$ to the hypotenuse of a rt. $\triangle$, then the 2 $\triangle$s formed are similar to the given $\triangle$ and to each other.
4. $\frac{AB}{AD} = \frac{AD}{AC}$; $\frac{BC}{DC} = \frac{DC}{AC}$	4. Definition of similar polygons

49. C **51.** 15, 18, 21 **53.** 7, 47, 2207 **55.** $8\frac{8}{9}$, $11\frac{1}{9}$ **57.** $\angle 5, \angle 7$ **59.** $\angle 2, \angle 7, \angle 8$ **61.** $y = 4x - 8$ **63.** $y = -4x - 11$ **65.** 13 ft

Pages 353–356 Lesson 7-2

1. Maria; Colin does not have the longest side as the value of c.

3.

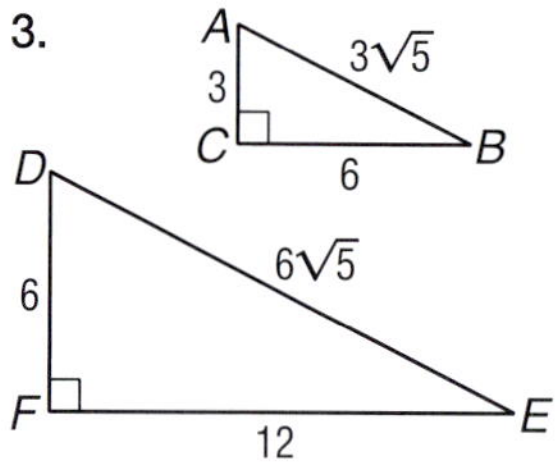

Sample answer : $\triangle ABC \sim \triangle DEF$, $\angle A \cong \angle D$, $\angle B \cong \angle E$, and $\angle C \cong \angle F$, $\overline{AB}$ corresponds to $\overline{DE}$, $\overline{BC}$ corresponds to $\overline{EF}$, $\overline{AC}$ corresponds to $\overline{DF}$. The scale factor is $\frac{2}{1}$. No; the measures do not form a Pythagorean triple since $6\sqrt{5}$ and $3\sqrt{5}$ are not whole numbers.

5. $\frac{3}{7}$ **7.** yes; $JK = \sqrt{17}$, $KL = \sqrt{17}$, $JL = \sqrt{34}$; $(\sqrt{17})^2 + (\sqrt{17})^2 = (\sqrt{34})^2$ **9.** no, no **11.** about 15.1 in. **13.** $4\sqrt{3} \approx 6.9$ **15.** $8\sqrt{41} \approx 51.2$ **17.** 20 **19.** no; $QR = 5$, $RS = 6$, $QS = 5$; $5^2 + 5^2 \neq 6^2$ **21.** yes; $QR = \sqrt{29}$, $RS = \sqrt{29}$, $QS = \sqrt{58}$; $(\sqrt{29})^2 + (\sqrt{29})^2 = (\sqrt{58})^2$ **23.** yes, yes **25.** no, no **27.** no, no **29.** yes, no **31.** 5-12-13 **33.** Sample answer: They consist of any number of similar triangles. **35a.** 16-30-34; 24-45-51 **35b.** 18-80-82; 27-120-123 **35c.** 14-48-50; 21-72-75 **37.** 10.8 degrees

39. Given: $\triangle ABC$ with right angle at C, $AB = d$

Prove: $d = \sqrt{(x_2 - x_1)^2 + (y_2 - y_1)^2}$

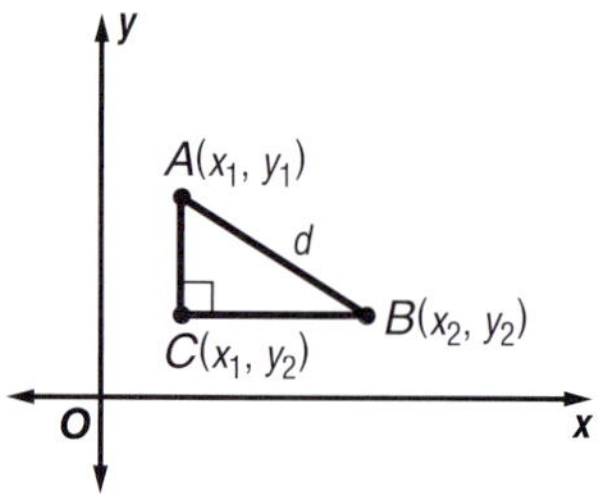

Proof:

Statements	Reasons
1. $\triangle ABC$ with right angle at C, $AB = d$	1. Given

2. $(CB)^2 + (AC)^2 = (AB)^2$	2. Pythagorean Theorem
3. $\lvert x_2 - x_1 \rvert = CB$ $\lvert y_2 - y_1 \rvert = AC$	3. Distance on a number line
4. $\lvert x_2 - x_1 \rvert^2 +$ $\lvert y_2 - y_1 \rvert^2 = d^2$	4. Substitution
5. $(x_2 - x_1)^2 + (y_2 - y_1)^2 = d^2$	5. Substitution
6. $\sqrt{(x_2 - x_1)^2 + (y_2 - y_1)^2} = d$	6. Take square root of each side.
7. $d = \sqrt{(x_2 - x_1)^2 + (y_2 - y_1)^2}$	7. Reflexive Property

41. about 76.53 ft **43.** about 13.4 mi **45.** Sample answer: The road, the tower that is perpendicular to the road, and the cables form the right triangles. Answers should include the following.

- Right triangles are formed by the bridge, the towers, and the cables.
- The cable is the hypotenuse in each triangle.

47. C **49.** yes **51.** $6\sqrt{3} \approx 10.4$ **53.** $3\sqrt{6} \approx 7.3$ **55.** $\sqrt{10} \approx 3.2$ **57.** 3; approaches positive infinity. **59.** 0.25; alternates between 0.25 and 4. **61.** $\frac{7\sqrt{3}}{3}$ **63.** $\sqrt{7}$ **65.** $12\sqrt{2}$ **67.** $2\sqrt{2}$ **69.** $\frac{\sqrt{2}}{2}$

Pages 360–363 Lesson 7-3

1. Sample answer: Construct two perpendicular lines. Use a ruler to measure 3 cm from the point of intersection on the one ray. Use the compass to copy the 3 cm segment. Connect the two endpoints to form a 45°-45°-90° triangle with sides of 3 cm and a hypotenuse of $3\sqrt{2}$cm. **3.** The length of the rectangle is $\sqrt{3}$ times the width; $\ell = \sqrt{3}w$. **5.** $x = 5\sqrt{2}$; $y = 5\sqrt{2}$ **7.** $a = 4$; $b = 4\sqrt{3}$

9.

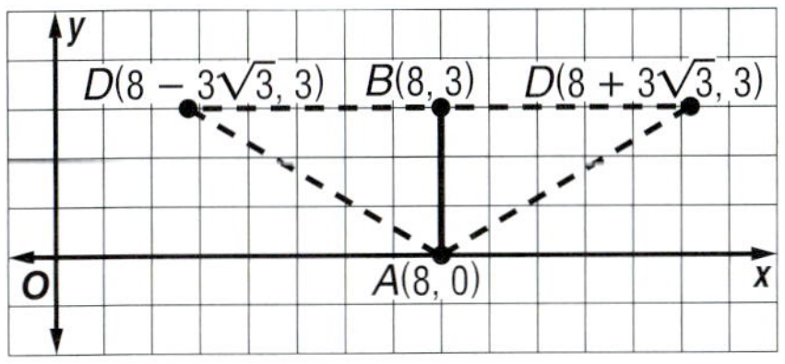

11. $90\sqrt{2}$ or 127.28 ft **13.** $x = \frac{17\sqrt{2}}{2}$; $y = 45$ **15.** $x = 8\sqrt{3}$; $y = 8\sqrt{3}$ **17.** $x = 5\sqrt{2}$; $y = \frac{5\sqrt{2}}{2}$ **19.** $a = 14\sqrt{3}$; $CE = 21$; $y = 21\sqrt{3}$; $b = 42$ **21.** $7.5\sqrt{3}$ cm ≈ 12.99 cm **23.** $14.8\sqrt{3}$ m ≈ 25.63 m **25.** $8\sqrt{2} \approx 11.31$ **27.** (4, 8) **29.** $\left(-3 - \frac{13\sqrt{3}}{3}, -6\right)$ or about $(-10.51, -6)$ **31.** $a = 3\sqrt{3}$, $b = 9$, $c = 3\sqrt{3}$, $d = 9$ **33.** 30° angle

35. Sample answer:

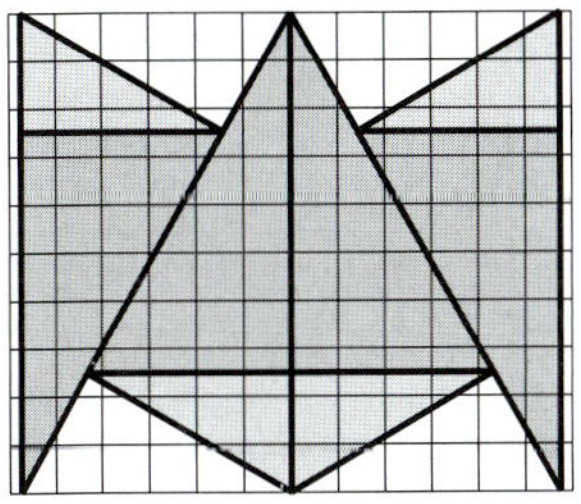

37. $BH = 16$ **39.** $12\sqrt{3} \approx 20.78$ cm **41.** $52 + 4\sqrt{3} + 4\sqrt{6}$ units **43.** C **45.** yes, yes **47.** no, no **49.** yes, yes **51.** $2\sqrt{21} \approx 9.2$; 21; 25 **53.** $\frac{40}{3}$; $\frac{5}{3}$; $10\sqrt{2} \approx 14.1$

55. $m\angle ALK < m\angle NLO$ **57.** $m\angle KLO = m\angle ALN$ **59.** 15 **61.** 20 **63.** 28 **65.** 60

Page 363 Chapter 7 Practice Quiz 1

1. $7\sqrt{3} \approx 12.1$ **3.** yes; $AB = \sqrt{5}$, $BC = \sqrt{50}$, $AC = \sqrt{45}$; $(\sqrt{5})^2 + (\sqrt{45})^2 = (\sqrt{50})^2$ **5.** $x = 12$; $y = 6\sqrt{3}$

Pages 367–370 Lesson 7-4

1. The triangles are similar, so the ratios remain the same. **3.** All three ratios involve two sides of a right triangle. The sine ratio is the measure of the opposite leg divided by the measure of the hypotenuse. The cosine ratio is the measure of the adjacent leg divided by the measure of the hypotenuse. The tangent ratio is the measure of the opposite leg divided by the measure of the adjacent leg. **5.** $\frac{14}{50} = 0.28$; $\frac{48}{50} = 0.96$; $\frac{14}{48} \approx 0.29$; $\frac{48}{50} = 0.96$; $\frac{14}{50} = 0.28$; $\frac{48}{14} \approx 3.43$ **7.** 0.8387 **9.** 0.8387 **11.** 1.0000 **13.** $m\angle A \approx 54.8$ **15.** $m\angle A \cong 33.7$ **17.** 2997 ft **19.** $\frac{\sqrt{3}}{3} \approx 0.58$; $\frac{\sqrt{6}}{3} \approx 0.82$; $\frac{\sqrt{2}}{2} \approx 0.71$; $\frac{\sqrt{6}}{3} \approx 0.82$; $\frac{\sqrt{3}}{3} \approx 0.58$; $\sqrt{2} \approx 1.41$ **21.** $\frac{2}{3} \approx 0.67$; $\frac{\sqrt{5}}{3} \approx 0.75$; $\frac{2\sqrt{5}}{5} \approx 0.89$; $\frac{\sqrt{5}}{3} \approx 0.75$; $\frac{2}{3} \approx 0.67$; $\frac{\sqrt{5}}{2} \approx 1.12$ **23.** 0.9260 **25.** 0.9974 **27.** 0.9239 **29.** $\frac{5}{1} = 5.0000$ **31.** $\frac{5\sqrt{26}}{6} \approx 0.9806$ **33.** $\frac{1}{5} = 0.2000$ **35.** $\frac{\sqrt{26}}{26} \approx 0.1961$ **37.** 46.4 **39.** 84.0 **41.** 83.0 **43.** $x \approx 8.5$ **45.** $x \approx 28.2$ **47.** $x \approx 22.6$ **49.** 4.1 mi **51.** about 5.18 ft **53.** about 54.5 **55.** about 47.9 in. **57.** $x = 17.1$; $y = 23.4$ **59.** about 272,837 astronomical units **61.** $\frac{2\sqrt{2}}{5}$ **63.** C **65.** $\csc A = \frac{5}{3}$; $\sec A = \frac{5}{4}$; $\cot A = \frac{4}{3}$; $\csc B = \frac{5}{4}$; $\sec B = \frac{5}{3}$; $\cot B = \frac{3}{4}$ **67.** $\csc A = 2$; $\sec A = \frac{2\sqrt{3}}{3}$; $\cot A = \sqrt{3}$; $\csc B = \frac{2\sqrt{3}}{3}$; $\sec B = 2$; $\cot B = \frac{\sqrt{3}}{3}$ **69.** $b = 4\sqrt{3}$, $c = 8$ **71.** $a = 2.5$, $b = 2.5\sqrt{3}$ **73.** yes, yes **75.** no, no **77.** 117 **79.** 150 **81.** 63

Pages 373–376 Lesson 7-5

1. Sample answer: $\angle ABC$

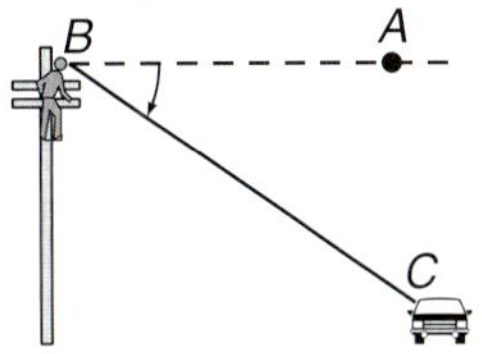

3. The angle of depression is $\angle FPB$ and the angle of elevation is $\angle TBP$. **5.** 22.7° **7.** 706 ft **9.** about 173.2 yd **11.** about 5.3° **13.** about 118.2 yd **15.** about 4° **17.** about 40.2° **19.** 100 ft, 300 ft **21.** about 8.3 in. **23.** no **25.** About 5.1 mi **27.** Answers should include the following.

- Pilots use angles of elevation when they are ascending and angles of depression when descending.
- Angles of elevation are formed when a person looks upward and angles of depression are formed when a person looks downward.

29. A **31.** 30.8 **33.** 70.0 **35.** 19.5 **37.** $14\sqrt{3}$; 28 **39.** 31.2 cm **41.** 5 **43.** 34 **45.** 52 **47.** 3.75

Pages 380–383 Lesson 7-6

1. Felipe; Makayla is using the definition of the sine ratio for a right triangle, but this is not a right triangle. **3.** In one case you need the measures of two sides and the measure of an angle opposite one of the sides. In the other case you need the measures of two angles and the measure of a side. **5.** 13.1 **7.** 55 **9.** $m\angle R \approx 19$, $m\angle Q \approx 56$, $q \approx 27.5$ **11.** $m\angle Q \approx 43$, $m\angle R \approx 17$, $r \approx 9.5$ **13.** $m\angle P \approx 37$, $p \approx 11.1$, $m\angle R \approx 32$ **15.** about 237.8 feet **17.** 2.7 **19.** 29 **21.** 29 **23.** $m\angle X \approx 25.6$, $m\angle W \approx 58.4$, $w \approx 20.3$ **25.** $m\angle X \approx 19.3$, $m\angle W \approx 48.7$, $w \approx 45.4$ **27.** $m\angle X = 82$, $x \approx 5.2$, $y \approx 4.7$ **29.** $m\angle X \approx 49.6$, $m\angle Y \approx 42.4$, $y \approx 14.2$ **31.** 56.9 units **33.** about 14.9 mi, about 13.6 mi **35.** about 536 ft **37.** about 1000.7 m **39.** about 13.6 mi **41.** Sample answer: Triangles are used to determine distances in space. Answers should include the following.

- The VLA is one of the world's premier astronomical radio observatories. It is used to make pictures from the radio waves emitted by astronomical objects.
- Triangles are used in the construction of the antennas.

43. A **45.** about 5.97 ft **47.** $\frac{20}{29} \approx 0.69$; $\frac{21}{29} \approx 0.72$; $\frac{20}{21} \approx 0.95$; $\frac{21}{29} \approx 0.72$; $\frac{20}{29} \approx 0.69$; $\frac{21}{20} = 1.05$ **49.** $\frac{\sqrt{2}}{2} \approx 0.71$; $\frac{\sqrt{2}}{2} \approx 0.71$; 1.00; $\frac{\sqrt{2}}{2} \approx 0.71$; $\frac{\sqrt{2}}{2} \approx 0.71$; 1.00 **51.** 54 **53.** $\frac{13}{112}$ **55.** $-\frac{11}{80}$ **57.** $\frac{7}{15}$

Page 383 Chapter 7 Practice Quiz 2

1. 58.0 **3.** 53.2 **5.** $m\angle D \approx 41$, $m\angle E \approx 57$, $e \approx 10.2$

Pages 387–390 Lesson 7-7

1. Sample answer: Use the Law of Cosines when you have all three sides given (SSS) or two sides and the included angle (SAS).

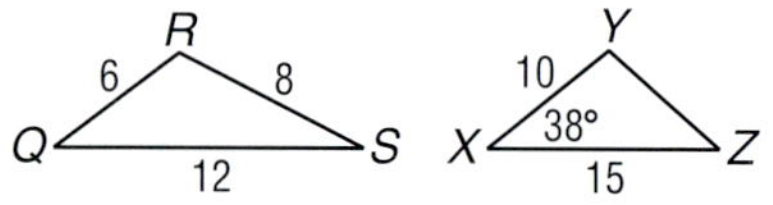

3. If two angles and one side are given, then the Law of Cosines cannot be used. **5.** 159.7 **7.** 98 **9.** $\ell \approx 17.9$; $m\angle K \approx 55$; $m\angle M \approx 78$ **11.** $u \approx 4.9$ **13.** $t \approx 22.5$ **15.** 16 **17.** 36 **19.** $m\angle H \approx 31$; $m\angle G \approx 109$; $g \approx 14.7$ **21.** $m\angle B \approx 86$; $m\angle C \approx 56$; $m\angle D \approx 38$ **23.** $c \approx 6.3$; $m\angle A \approx 80$; $m\angle B \approx 63$ **25.** $m\angle B = 99$; $b \approx 31.3$; $a \approx 25.3$ **27.** $m\angle M \approx 18.6$; $m\angle N \approx 138.4$; $n \approx 91.8$ **29.** $\ell \approx 21.1$; $m\angle M \approx 42.8$; $m\angle N \approx 88.2$ **31.** $m\angle L \approx 101.9$; $m\angle M \approx 36.3$; $m\angle N \approx 41.8$ **33.** $m \approx 6.0$; $m\angle L \approx 22.2$; $m\angle N \approx 130.8$ **35.** $m \approx 18.5$; $m\angle L \approx 40.9$; $m\angle N \approx 79.1$ **37.** $m\angle N \approx 42.8$; $m\angle M \approx 86.2$; $m \approx 51.4$ **39.** 561.2 units **41.** 59.8, 63.4, 56.8

43a. Pythagorean Theorem **43b.** Substitution
43c. Pythagorean Theorem **43d.** Substitution
43e. Def. of cosine **43f.** Cross products **43g.** Substitution
43h. Commutative Property

45. Sample answer: Triangles are used to build supports, walls, and foundations. Answers should include the following.

- The triangular building was more efficient with the cells around the edge.
- The Law of Sines requires two angles and a side or two sides and an angle opposite one of those sides.

47. C **49.** 33 **51.** yes **53.** no

55. Given: $\triangle JFM \sim \triangle EFB$
$\triangle LFM \sim \triangle GFB$
Prove: $\triangle JFL \sim \triangle EFG$

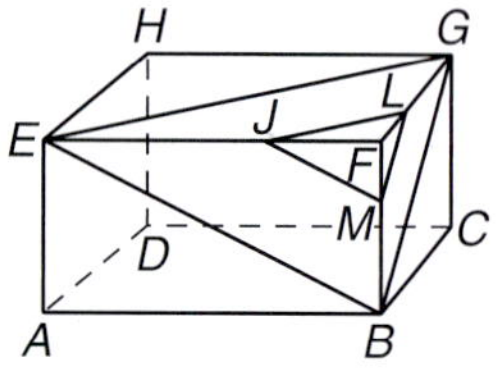

Proof:
Since $\triangle JFM \sim \triangle EFB$ and $\triangle LFM \sim \triangle GFB$, then by the definition of similar triangles, $\frac{JF}{EF} = \frac{MF}{BF}$ and $\frac{MF}{BF} = \frac{LF}{GF}$. By the Transitive Property of Equality, $\frac{JF}{EF} = \frac{LF}{GF}$. $\angle F \cong \angle F$ by the Reflexive Property of Congruence. Then, by SAS Similarity, $\triangle JFL \sim \triangle EFG$.

57. $(-1.6, 9.6)$ **59.** $(2.8, 5.2)$

Pages 392–396 Chapter 7 Study Guide and Review

1. true **3.** false; a right **5.** true **7.** false; depression **9.** 18 **11.** $6\sqrt{22} \approx 28.1$ **13.** 25 **15.** $4\sqrt{17} \approx 16.5$ **17.** $x = \frac{13\sqrt{2}}{2}$; $y = \frac{13\sqrt{2}}{2}$ **19.** $z = 18\sqrt{3}$, $a = 36\sqrt{3}$ **21.** $\frac{3}{5} = 0.60$; $\frac{4}{5} = 0.80$; $\frac{3}{4} = 0.75$; $\frac{4}{5} = 0.80$; $\frac{3}{5} = 0.60$; $\frac{4}{3} \approx 1.33$ **23.** 26.9 **25.** 43.0 **27.** $\approx 22.6°$ **29.** ≈ 31.1 yd **31.** 21.3 yd **33.** $m\angle B \approx 41$, $m\angle C \approx 75$, $c \approx 16.1$ **35.** $m\angle B \approx 61$, $m\angle C \approx 90$, $c \approx 9.9$ **37.** $z \approx 5.9$ **39.** $a \approx 17.0$, $m\angle B \approx 43$, $m\angle C \approx 73$

Chapter 8 Quadrilaterals

Page 403 Chapter 8 Getting Started

1. 130 **3.** 120 **5.** $\frac{1}{6}$, -6; perpendicular **7.** $\frac{4}{3}$, $-\frac{3}{4}$; perpendicular **9.** $-\frac{a}{b}$

Pages 407–409 Lesson 8-1

1. A concave polygon has at least one obtuse angle, which means the sum will be different from the formula.

3. Sample answer: regular quadrilateral, 360°; quadrilateral that is not regular, 360°

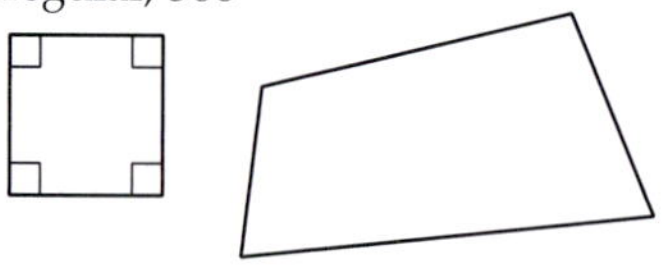

5. 1800 **7.** 4 **9.** $m\angle J = m\angle M = 30$, $m\angle K = m\angle L = m\angle P = m\angle N = 165$ **11.** 20, 160 **13.** 5400 **15.** 3060 **17.** $360(2y - 1)$ **19.** 1080 **21.** 9 **23.** 18 **25.** 16

27. $m\angle M = 30$, $m\angle P = 120$, $m\angle Q = 60$, $m\angle R = 150$ **29.** $m\angle M = 60$, $m\angle N = 120$, $m\angle P = 60$, $m\angle Q = 120$ **31.** 105, 110, 120, 130, 135, 140, 160, 170, 180, 190 **33.** Sample answer: 36, 72, 108, 144 **35.** 36, 144 **37.** 40, 140 **39.** 147.3, 32.7 **41.** 150, 30 **43.** 108, 72 **45.** $\frac{180(n-2)}{n} = \frac{180n - 360}{n} = \frac{180n}{n} - \frac{360}{n} = 180 - \frac{360}{n}$ **47.** B **49.** 92.1 **51.** 51.0 **53.** $m\angle G \approx 67$, $m\angle H \approx 60$, $h \approx 16.1$ **55.** $m\angle F = 57$, $f \approx 63.7$, $h \approx 70.0$

57. Given: $\overline{JL} \parallel \overline{KM}$, $\overline{JK} \parallel \overline{LM}$
Prove: $\triangle JKL \cong \triangle MLK$

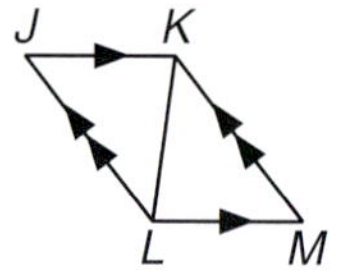

Proof:

Statements	Reasons
1. $\overline{JL} \parallel \overline{KM}$, $\overline{JK} \parallel \overline{LM}$	1. Given
2. $\angle MKL \cong \angle JLK$, $\angle JKL \cong \angle MLK$	2. Alt. int. $\angle$s are $\cong$.
3. $\overline{KL} \cong \overline{KL}$	3. Reflexive Property
4. $\triangle JKL \cong \triangle MLK$	4. ASA

59. m; cons. int. **61.** n; alt. ext. **63.** $\angle 3$ and $\angle 5$, $\angle 2$ and $\angle 6$ **65.** none

Pages 414–416 Lesson 8-2

1. Opposite sides are congruent; opposite angles are congruent; consecutive angles are supplementary; and if there is one right angle, there are four right angles.

3. Sample answer:

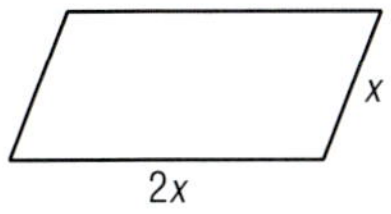

5. $\triangle VTQ$, SSS; diag. bisect each other and opp. sides of ▱ are ≅. **7.** 100 **9.** 80 **11.** 7

13. Given: ▱$VZRQ$ and ▱$WQST$
Prove: $\angle Z \cong \angle T$

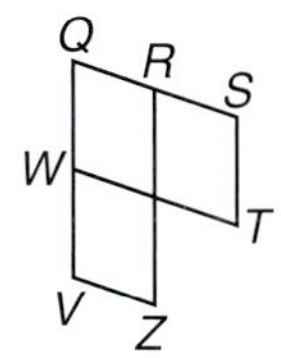

Proof:

Statements	Reasons
1. ▱$VZRQ$ and ▱$WQST$	1. Given
2. $\angle Z \cong \angle Q$, $\angle Q \cong \angle T$	2. Opp. ⦞ of a ▱ are ≅.
3. $\angle Z \cong \angle T$	3. Transitive Prop.

15. C **17.** $\angle CDB$, alt. int. ⦞ are ≅. **19.** $\overline{GD}$, diag. of ▱ bisect each other. **21.** $\angle BAC$, alt. int. ⦞ are ≅. **23.** 33 **25.** 109 **27.** 83 **29.** 6.45 **31.** 6.1 **33.** $y - 5$, $FH - 9$ **35.** $a = 6$, $b = 5$, $DB = 32$ **37.** $EQ = 5$, $QG = 5$, $HQ = \sqrt{13}$, $QF = \sqrt{13}$ **39.** Slope of $\overline{EH}$ is undefined, slope of $\overline{EF} = -\frac{1}{3}$; no, the slopes of the sides are not negative reciprocals of each other.

41. Given: ▱$PQRS$
Prove: $\overline{PQ} \cong \overline{RS}$
$\overline{QR} \cong \overline{SP}$

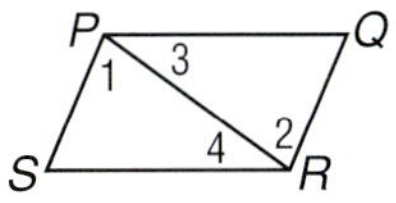

Proof:

Statements	Reasons
1. ▱$PQRS$	1. Given
2. Draw an auxiliary segment $\overline{PR}$ and label angles 1, 2, 3, and 4 as shown.	2. Diagonal of ▱$PQRS$
3. $\overline{PQ} \parallel \overline{SR}$, $\overline{PS} \parallel \overline{QR}$	3. Opp. sides of ▱ are ∥.
4. $\angle 1 \cong \angle 2$, and $\angle 3 \cong \angle 4$	4. Alt. int. ⦞ are ≅.
5. $\overline{PR} \cong \overline{PR}$	5. Reflexive Prop.
6. $\triangle QPR \cong \triangle SRP$	6. ASA
7. $\overline{PQ} \cong \overline{RS}$ and $\overline{QR} \cong \overline{SP}$	7. CPCTC

43. Given: ▱$MNPQ$
$\angle M$ is a right angle.
Prove: $\angle N$, $\angle P$ and $\angle Q$ are right angles.

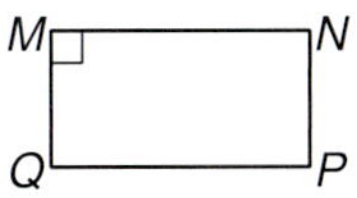

Proof:
By definition of a parallelogram, $\overline{MN} \parallel \overline{QP}$. Since $\angle M$ is a right angle, $\overline{MQ} \perp \overline{MN}$. By the Perpendicular Transversal Theorem, $\overline{MQ} \perp \overline{QP}$. $\angle Q$ is a right angle, because perpendicular lines form a right angle. $\angle N \cong \angle Q$ and $\angle M \cong \angle P$ because opposite angles in a parallelogram are congruent. $\angle P$ and $\angle N$ are right angles, since all right angles are congruent.

45. Given: ▱$WXYZ$
Prove: $\triangle WXZ \cong \triangle YZX$

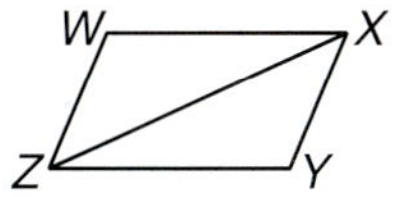

Proof:

Statements	Reasons
1. ▱$WXYZ$	1. Given
2. $\overline{WX} \cong \overline{ZY}$, $\overline{WZ} \cong \overline{XY}$	2. Opp. sides of ▱ are ≅.
3. $\angle ZWX \cong \angle XYZ$	3. Opp. ⦞ of ▱ are ≅.
4. $\triangle WXZ \cong \triangle YZX$	4. SAS

47. Given: ▱$BCGH$, $\overline{HD} \cong \overline{FD}$
Prove: $\angle F \cong \angle GCB$

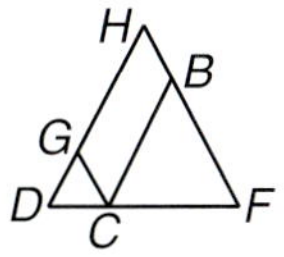

Proof:

Statements	Reasons
1. ▱$BCGH$, $\overline{HD} \cong \overline{FD}$	1. Given
2. $\angle F \cong \angle H$	2. Isosceles Triangle Th.
3. $\angle H \cong \angle GCB$	3. Opp. ⦞ of ▱ are ≅.
4. $\angle F \cong \angle GCB$	4. Congruence of angles is transitive.

49. The graphic uses the illustration of wedges shaped like parallelograms to display the data. Answers should include the following.

- The opposite sides are parallel and congruent, the opposite angles are congruent, and the consecutive angles are supplementary.
- Sample answer:

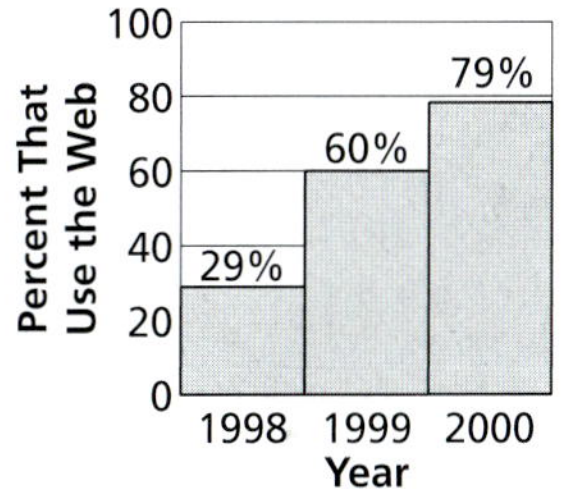

51. B **53.** 3600 **55.** 6120 **57.** Sines; $m\angle C \approx 69.9$, $m\angle A \approx 53.1$, $a \approx 11.9$ **59.** 30 **61.** side, $\frac{7}{3}$ **63.** side, $\frac{7}{3}$

Pages 420–423 Lesson 8-3

1. Both pairs of opposite sides are congruent; both pairs of opposite angles are congruent; diagonals bisect each other; one pair of opposite sides is parallel and congruent.
3. Shaniqua; Carter's description could result in a shape that is not a parallelogram. **5.** Yes; each pair of opp. ⦞ is ≅. **7.** $x = 41$, $y = 16$ **9.** yes

11. Given: $\overline{PT} \cong \overline{TR}$
$\angle TSP \cong \angle TQR$
Prove: $PQRS$ is a parallelogram.

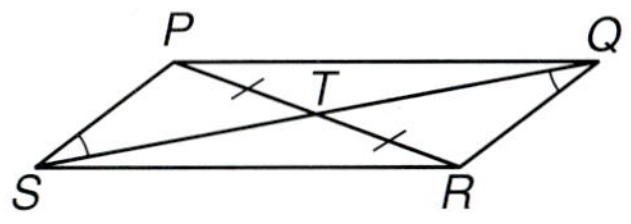

Proof:

Statements	Reasons
1. $\overline{PT} \cong \overline{TR}$, $\angle TSP \cong \angle TQR$	1. Given
2. $\angle PTS \cong \angle RTQ$	2. Vertical ⦞ are ≅.
3. $\triangle PTS \cong \triangle RTQ$	3. AAS
4. $\overline{PS} \cong \overline{QR}$	4. CPCTC
5. $\overline{PS} \parallel \overline{QR}$	5. If alt. int. ⦞ are ≅, lines are ∥.
6. $PQRS$ is a parallelogram.	6. If one pair of opp. sides is ∥ and ≅, then the quad. is a ▱.

13. Yes; each pair of opposite angles is congruent. **15.** Yes; opposite angles are congruent. **17.** Yes; one pair of opposite sides is parallel and congruent. **19.** $x = 6, y = 24$ **21.** $x = 1, y = 2$ **23.** $x = 34, y = 44$ **25.** yes **27.** yes **29.** no **31.** yes **33.** Move M to $(-4, 1)$, N to $(-3, 4)$, P to $(0, -9)$, or R to $(-7, 3)$. **35.** $(-2, -2)$, $(4, 10)$, or $(10, 0)$ **37.** Parallelogram; $\overline{KM}$ and $\overline{JL}$ are diagonals that bisect each other.

39. Given: $\overline{AD} \cong \overline{BC}$, $\overline{AB} \cong \overline{DC}$
Prove: $ABCD$ is a parallelogram.

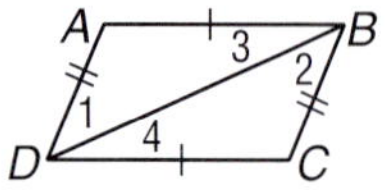

Proof:

Statements	Reasons
1. $\overline{AD} \cong \overline{BC}$, $\overline{AB} \cong \overline{DC}$	1. Given
2. Draw $\overline{DB}$.	2. Two points determine a line.
3. $\overline{DB} \cong \overline{DB}$	3. Reflexive Property
4. $\triangle ABD \cong \triangle CDB$	4. SSS
5. $\angle 1 \cong \angle 2$, $\angle 3 \cong \angle 4$	5. CPCTC
6. $\overline{AD} \parallel \overline{BC}$, $\overline{AB} \parallel \overline{DC}$	6. If alt. int. $\angle$s are $\cong$, lines are $\parallel$.
7. $ABCD$ is a parallelogram.	7. Definition of parallelogram

41. Given: $\overline{AB} \cong \overline{DC}$, $\overline{AB} \parallel \overline{DC}$
Prove: $ABCD$ is a parallelogram.

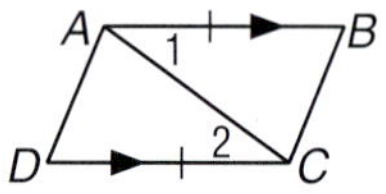

Proof:

Statements	Reasons
1. $\overline{AB} \cong \overline{DC}$, $\overline{AB} \parallel \overline{DC}$	1. Given
2. Draw $\overline{AC}$	2. Two points determine a line.
3. $\angle 1 \cong \angle 2$	3. Alternate Interior Angles Theorem
4. $\overline{AC} \cong \overline{AC}$	4. Reflexive Property
5. $\triangle ABC \cong \triangle CDA$	5. SAS
6. $\overline{AD} \cong \overline{BC}$	6. CPCTC
7. $ABCD$ is a parallelogram.	7. If both pairs of opp. sides are $\cong$, then the quad. is $\square$.

43. Given: $ABCDEF$ is a regular hexagon.
Prove: $FDCA$ is a parallelogram.

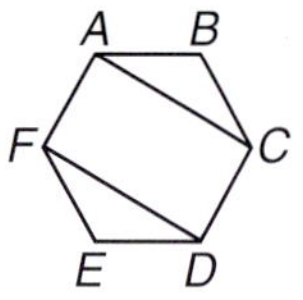

Proof:

Statements	Reasons
1. $ABCDEF$ is a regular hexagon.	1. Given
2. $\overline{AB} \cong \overline{DE}$, $\overline{BC} \cong \overline{EF}$, $\angle E \cong \angle B$, $\overline{FA} \cong \overline{CD}$	2. Def. of regular hexagon
3. $\triangle ABC \cong \triangle DEF$	3. SAS
4. $\overline{AC} \cong \overline{DF}$	4. CPCTC
5. $FDCA$ is a $\square$.	5. If both pairs of opp. sides are $\cong$, then the quad. is $\square$.

45. B **47.** 12 **49.** 14 units **51.** 8 **53.** 30 **55.** 72 **57.** 45, $12\sqrt{2}$ **59.** $16\sqrt{3}$, 16 **61.** 5, $-\frac{3}{2}$; not $\perp$. **63.** $\frac{2}{3}$, $-\frac{3}{2}$; $\perp$.

Page 423 Chapter 8 Practice Quiz 1

1. 11 **3.** 66 **5.** $x = 8, y = 6$

Pages 427–430 Lesson 8-4

1. If consecutive sides are perpendicular or diagonals are congruent, then the parallelogram is a rectangle. **3.** McKenna; Consuelo's definition is correct if one pair of opposite sides is parallel and congruent. **5.** 40 **7.** 52 or 10 **9.** Make sure that the angles measure 90 or that the diagonals are congruent. **11.** 11 **13.** $29\frac{1}{3}$ **15.** 4 **17.** 60 **19.** 30 **21.** 60 **23.** 30 **25.** Measure the opposite sides and the diagonals to make sure they are congruent. **27.** No; $\overline{DH}$ and $\overline{FG}$ are not parallel. **29.** Yes; opp. sides are $\parallel$, diag. are $\cong$. **31.** $\left(\frac{1}{2}, -\frac{3}{2}\right)$, $\left(\frac{7}{2}, \frac{3}{2}\right)$ **33.** Yes; consec. sides are $\perp$. **35.** Move L and K until the length of the diagonals is the same. **37.** See students' work. **39.** Sample answer: $\overline{AC} \cong \overline{BD}$ but $ABCD$ is not a rectangle

41. Given: $\square WXYZ$ and $\overline{WY} \cong \overline{XZ}$
Prove: $WXYZ$ is a rectangle.

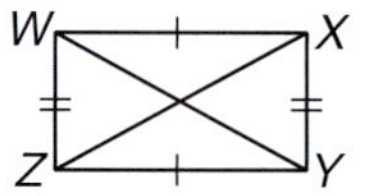

Proof:

Statements	Reasons
1. $\square WXYZ$ and $\overline{WY} \cong \overline{XZ}$	1. Given
2. $\overline{XY} \cong \overline{WZ}$	2. Opp. sides of $\square$ are $\cong$.
3. $\overline{WX} \cong \overline{WX}$	3. Reflexive Property
4. $\triangle WZX \cong \triangle XYW$	4. SSS
5. $\angle ZWX \cong \angle YXW$	5. CPCTC
6. $\angle ZWX$ and $\angle YXW$ are supplementary.	6. Consec. $\angle$s of $\square$ are suppl.
7. $\angle ZWX$ and $\angle YXW$ are right angles.	7. If 2 $\angle$s are $\cong$ and suppl, each $\angle$ is a rt. $\angle$.
8. $\angle WZY$ and $\angle XYZ$ are right angles.	8. If $\square$ has 1 rt. $\angle$, it has 4 rt. $\angle$s.
9. $WXYZ$ is a rectangle.	9. Def. of rectangle

43. Given: $DEAC$ and $FEAB$ are rectangles. $\angle GKH \cong \angle JHK$; $\overline{GJ}$ and $\overline{HK}$ intersect at L.
Prove: $GHJK$ is a parallelogram.

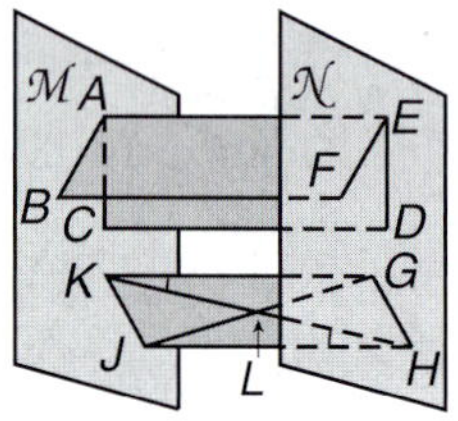

Proof:

Statements	Reasons
1. $DEAC$ and $FEAB$ are rectangles. $\angle GKH \cong \angle JHK$ $\overline{GJ}$ and $\overline{HK}$ intersect at L.	1. Given
2. $\overline{DE} \parallel \overline{AC}$ and $\overline{FE} \parallel \overline{AB}$	2. Def. of parallelogram
3. plane $\mathcal{N} \parallel$ plane $\mathcal{M}$	3. Def. of parallel planes
4. G, J, H, K, L are in the same plane.	4. Def. of intersecting lines
5. $\overline{GH} \parallel \overline{KJ}$	5. Def. of parallel lines
6. $\overline{GK} \parallel \overline{HJ}$	6. If alt. int. $\angle$s are $\cong$, lines are $\parallel$.
7. $GHJK$ is a parallelogram.	7. Def. of parallelogram

45. No; there are no parallel lines in spherical geometry. **47.** No; the sides are not parallel. **49.** A **51.** 31 **53.** 43 **55.** 49 **57.** 5 **59.** $\sqrt{297} \approx 17.2$ **61.** 5 **63.** 29

Pages 434–437 Lesson 8-5

1. Sample answer:

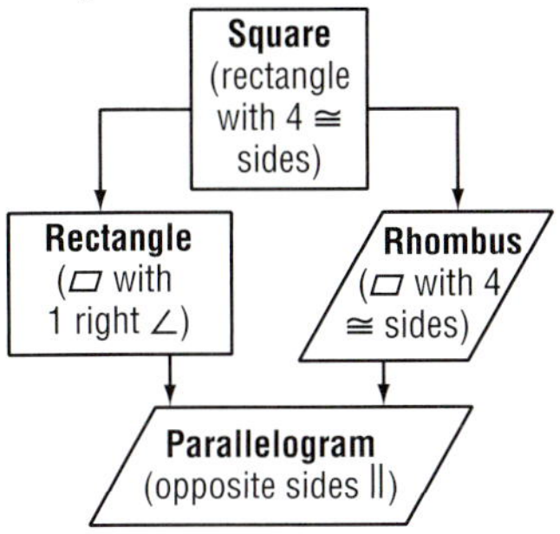

3. A square is a rectangle with all sides congruent. **5.** 5 **7.** 96.8 **9.** None; the diagonals are not congruent or perpendicular **11.** If the measure of each angle is 90 or if the diagonals are congruent, then the floor is a square **13.** 120 **15.** 30 **17.** 53 **19.** 5 **21.** Rhombus; the diagonals are perpendicular. **23.** None; the diagonals are not congruent or perpendicular.

25. Sample answer:

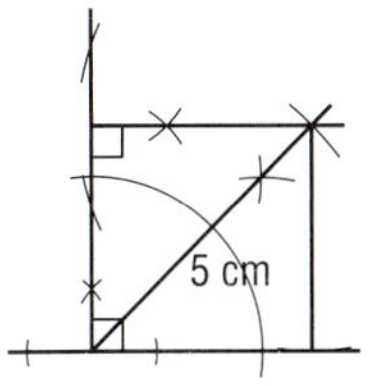

27. always **29.** sometimes **31.** always **33.** 40 cm

35. Given: $ABCD$ is a parallelogram. $\overline{AC} \perp \overline{BD}$

Prove: $ABCD$ is a rhombus.

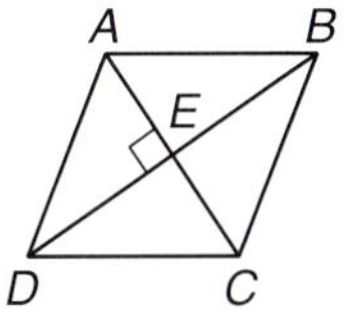

Proof: We are given that $ABCD$ is a parallelogram. The diagonals of a parallelogram bisect each other, so $\overline{AE} \cong \overline{EC}$. $\overline{BE} \cong \overline{BE}$ because congruence of segments is reflexive. We are also given that $\overline{AC} \perp \overline{BD}$. Thus, $\angle AEB$ and $\angle BEC$ are right angles by the definition of perpendicular lines. Then $\angle AEB \cong \angle BEC$ because all right angles are congruent. Therefore, $\triangle AEB \cong \triangle CEB$ by SAS. $\overline{AB} \cong \overline{CB}$ by CPCTC. Opposite sides of parallelograms are congruent, so $\overline{AB} \cong \overline{CD}$ and $\overline{BC} \cong \overline{AD}$. Then since congruence of segments is transitive, $\overline{AB} \cong \overline{CD} \cong \overline{CB} \cong \overline{AD}$. All four sides of $ABCD$ are congruent, so $ABCD$ is a rhombus by definition.

37. No; it is about 11,662.9 mm. **39.** The flag of Denmark contains four red rectangles. The flag of St. Vincent and the Grenadines contains a blue rectangle, a green rectangle, a yellow rectangle, a blue and yellow rectangle, a yellow and green rectangle, and three green rhombi. The flag of Trinidad and Tobago contains two white parallelograms and one black parallelogram.

41. Given: $\triangle TPX \cong \triangle QPX \cong \triangle QRX \cong \triangle TRX$

Prove: $TPQR$ is a rhombus.

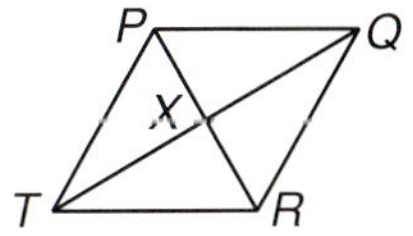

Proof:

Statements	Reasons
1. $\triangle TPX \cong \triangle QPX \cong \triangle QRX \cong \triangle TRX$	1. Given
2. $\overline{TP} \cong \overline{PQ} \cong \overline{QR} \cong \overline{TR}$	2. CPCTC
3. $TPQR$ is a rhombus.	3. Def. of rhombus

43. Given: $QRST$ and $QRTV$ are rhombi.

Prove: $\triangle QRT$ is equilateral.

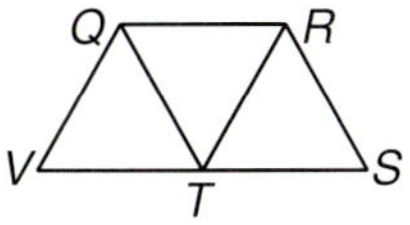

Proof:

Statements	Reasons
1. $QRST$ and $QRTV$ are rhombi.	1. Given
2. $\overline{QV} \cong \overline{VT} \cong \overline{TR} \cong \overline{QR}$, $\overline{QT} \cong \overline{TS} \cong \overline{RS} \cong \overline{QR}$	2. Def. of rhombus
3. $\overline{QT} \cong \overline{TR} \cong \overline{QR}$	3. Substitution Property
4. $\triangle QRT$ is equilateral.	4. Def. of equilateral triangle

45. Sample answer: You can ride a bicycle with square wheels over a curved road. Answers should include the following.

- Rhombi and squares both have all four sides congruent, but the diagonals of a square are congruent. A square has four right angles and rhombi have each pair of opposite angles congruent, but not all angles are necessarily congruent.
- Sample answer: Since the angles of a rhombus are not all congruent, riding over the same road would not be smooth.

47. C **49.** 140 **51.** $x = 2, y = 3$ **53.** yes **55.** no **57.** 13.5 **59.** 20 **61.** $\angle AJH \cong \angle AHJ$ **63.** $\overline{AK} \cong \overline{AB}$ **65.** 2.4 **67.** 5

Pages 442–445 Lesson 8-6

1. Exactly one pair of opposite sides is parallel.

3. Sample answer: The median of a trapezoid is parallel to both bases.

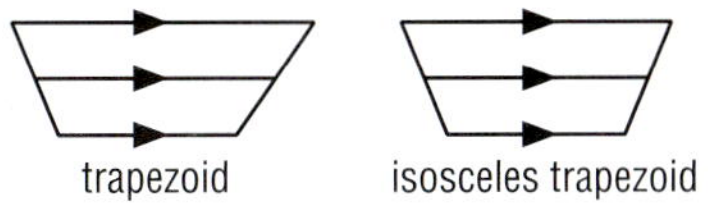

5. isosceles, $QR = \sqrt{20}$, $ST = \sqrt{20}$ **7.** 4 **9a.** $\overline{AD} \parallel \overline{BC}$, $\overline{CD} \nparallel \overline{AB}$ **9b.** not isosceles, $AB = \sqrt{17}$ and $CD = 5$ **11a.** $\overline{DC} \parallel \overline{FE}$, $\overline{DE} \nparallel \overline{FC}$ **11b.** isosceles, $DE = \sqrt{50}$, $CF = \sqrt{50}$ **13.** 8 **15.** 14, 110, 110 **17.** 62 **19.** 15 **21.** Sample answer: triangles, quadrilaterals, trapezoids, hexagons **23.** trapezoid, exactly one pair opp. sides ∥ **25.** square, all sides ≅, consecutive sides ⊥ **27.** $A(-2, 3.5)$, $B(4, -1)$ **29.** $\overline{DG} \parallel \overline{EF}$, not isosceles, $DE \neq GF$, $\overline{DE} \nparallel \overline{GF}$ **31.** $WV = 6$

33. Given: $\triangle TZX \cong \triangle YXZ$, $\overline{WX} \nparallel \overline{ZY}$

Prove: $XYZW$ is a trapezoid.

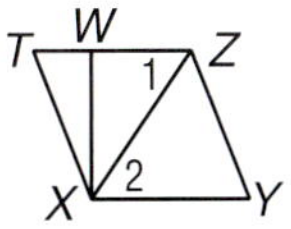

Proof:

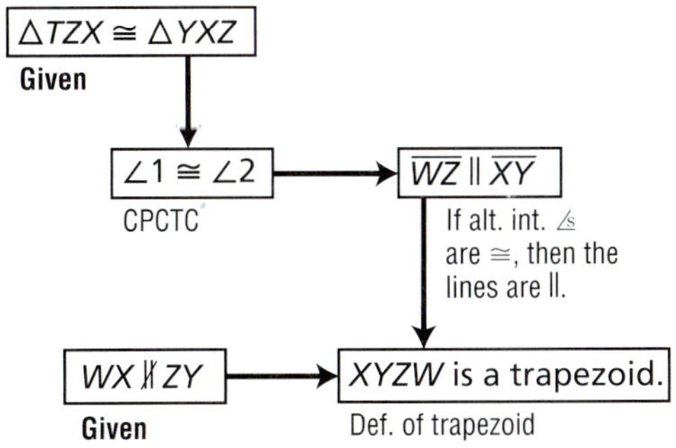

35. Given: E and C are midpoints of $\overline{AD}$ and $\overline{DB}$

Prove: $ABCE$ is an isosceles trapezoid.

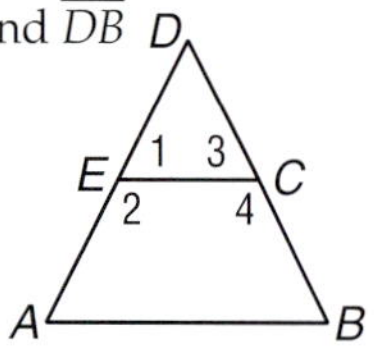

Proof:

E and C are midpoints of $\overline{AD}$ and $\overline{DB}$. — Given

$\overline{EC} \parallel \overline{AB}$ — A segment joining the midpoints of two sides of a triangle is parallel to the third side.

$\overline{AD} \cong \overline{DB}$ — Given

$\frac{1}{2}AD = \frac{1}{2}AB$ — Def. of mdpt.

$AE = BC$ — substitution

$\overline{AE} \cong \overline{BC}$ — Def. of $\cong$

$ABCE$ is an isos. trapezoid. — Def. of isos. trapezoid

37. Sample answer:

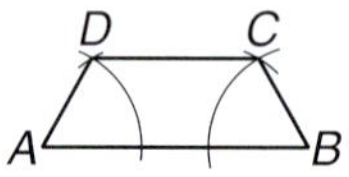

39. 4

41. Sample answer: Trapezoids are used in monuments as well as other buildings. Answers should include the following.
- Trapezoids have exactly one pair of opposite sides parallel.
- Trapezoids can be used as window panes.

43. B **45.** 10 **47.** 70 **49.** $RS = 7\sqrt{2}, TV = \sqrt{113}$ **51.** No; opposite sides are not congruent and the diagonals do not bisect each other. **53.** $\frac{17}{5}$ **55.** $\frac{13}{2}$ **57.** 0 **59.** $\frac{2b}{a}$ **61.** $\frac{c}{b}$

Page 445 Chapter 8 Practice Quiz 2

1. 12 **3.** rhombus, opp. sides $\parallel$, diag. $\perp$, consec. sides not $\perp$ **5.** 18

Pages 449–451 Lesson 8-7

1. Place one vertex at the origin and position the figure so another vertex lies on the positive x-axis.

3.

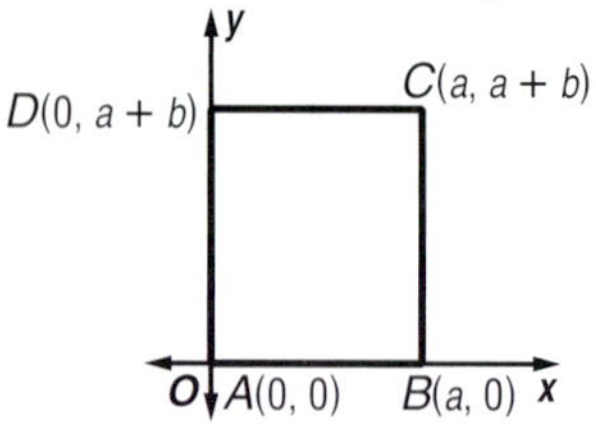

5. (c, b)

7. Given: $ABCD$ is a square.

Prove: $\overline{AC} \perp \overline{DB}$

Proof:

Slope of $\overline{DB} = \frac{0 - a}{a - 0}$ or -1

Slope of $\overline{AC} = \frac{0 - a}{0 - a}$ or 1

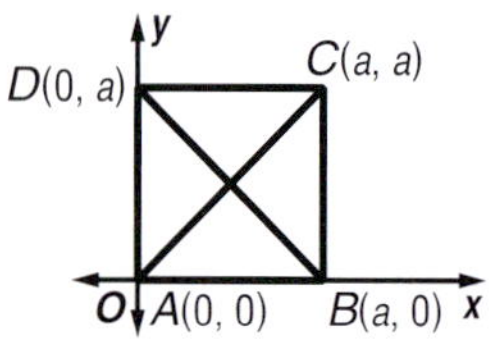

The slope of $\overline{AC}$ is the negative reciprocal of the slope of $\overline{DB}$, so they are perpendicular.

9.

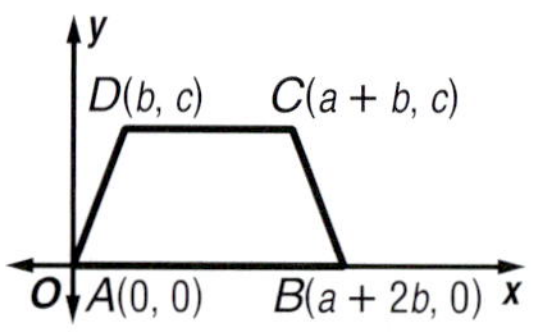

11. $B(-b, c)$ **13.** $G(a, 0), E(-b, c)$ **15.** $T(-2a, c), W(-2a, -c)$

17. Given: $ABCD$ is a rectangle.

Prove: $\overline{AC} \cong \overline{DB}$

Proof:

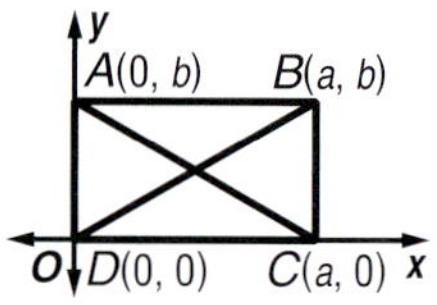

Use the Distance Formula to find $AC = \sqrt{a^2 + b^2}$ and $BD = \sqrt{a^2 + b^2}$. $\overline{AC}$ and $\overline{BC}$ have the same length, so they are congruent.

19. Given: isosceles trapezoid $ABCD$ with $\overline{AD} \cong \overline{BC}$

Prove: $\overline{BD} \cong \overline{AC}$

Proof:

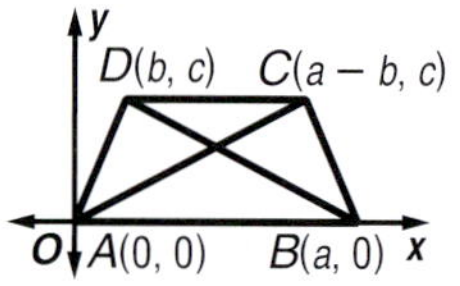

$BD = \sqrt{(a - b)^2 + (0 - c)^2} = \sqrt{(a - b)^2 + c^2}$

$AC = \sqrt{((a - b) - 0)^2 + (c - 0)^2} = \sqrt{(a - b)^2 + c^2}$

$BD = AC$ and $\overline{BD} \cong \overline{AC}$

21. Given: $ABCD$ is a rectangle. Q, R, S, and T are midpoints of their respective sides.

Prove: $QRST$ is a rhombus.

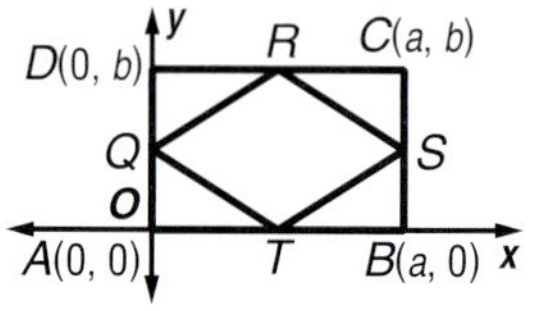

Proof:

Midpoint Q is $\left(\frac{0+0}{2}, \frac{b+0}{2}\right)$ or $\left(0, \frac{b}{2}\right)$.

Midpoint R is $\left(\frac{a+0}{2}, \frac{b+b}{2}\right)$ or $\left(\frac{a}{2}, \frac{2b}{2}\right)$ or $\left(\frac{a}{2}, b\right)$

Midpoint S is $\left(\frac{a+a}{2}, \frac{b+0}{2}\right)$ or $\left(\frac{2a}{2}, \frac{b}{2}\right)$ or $\left(a, \frac{b}{2}\right)$.

Midpoint T is $\left(\frac{a+0}{2}, \frac{0+0}{2}\right)$ or $\left(\frac{a}{2}, 0\right)$.

$QR = \sqrt{\left(\frac{a}{2} - 0\right)^2 + \left(b - \frac{b}{2}\right)^2} = \sqrt{\left(\frac{a}{2}\right)^2 + \left(\frac{b}{2}\right)^2}$

$RS = \sqrt{\left(a - \frac{a}{2}\right)^2 + \left(\frac{b}{2} - b\right)^2} = \sqrt{\left(\frac{a}{2}\right)^2 + \left(-\frac{b}{2}\right)^2}$ or $\sqrt{\left(\frac{a}{2}\right)^2 + \left(\frac{b}{2}\right)^2}$

$ST = \sqrt{\left(a - \frac{a}{2}\right)^2 + \left(\frac{b}{2} - 0\right)^2} = \sqrt{\left(\frac{a}{2}\right)^2 + \left(\frac{b}{2}\right)^2}$

$QT = \sqrt{\left(\frac{a}{2} - 0\right)^2 + \left(0 - \frac{b}{2}\right)^2} = \sqrt{\left(\frac{a}{2}\right)^2 + \left(-\frac{b}{2}\right)^2}$ or $\sqrt{\left(\frac{a}{2}\right)^2 + \left(\frac{b}{2}\right)^2}$

$QR = RS = ST = QT$ so $\overline{QR} \cong \overline{RS} \cong \overline{ST} \cong \overline{QT}$.

$QRST$ is a rhombus.

23. Sample answer: $C(a + c, b), D(2a + c, 0)$ **25.** No, there is not enough information given to prove that the sides of the tower are parallel. **27.** Sample answer: The coordinate plane is used in coordinate proofs. The Distance Formula, Midpoint Formula and Slope Formula are used to prove theorems. Answers should include the following.
- Place the figure so one of the vertices is at the origin. Place at least one side of the figure on the positive x-axis. Keep the figure in the first quadrant if possible and use coordinates that will simplify calculations.
- Sample answer: Theorem 8.3 Opposite sides of a parallelogram are congruent.

29. A **31.** 55 **33.** 160 **35.** $\sqrt{60} \approx 7.7$ **37.** $m\angle XVZ = m\angle VXZ$ **39.** $m\angle XZY > m\angle ZXY$

Pages 452–456 Chapter 8 Study Guide and Review

1. true **3.** false, rectangle **5.** false, trapezoid **7.** true **9.** 120 **11.** 90 **13.** $m\angle W = 62$, $m\angle X = 108$, $m\angle Y = 80$, $m\angle Z = 110$ **15.** 52 **17.** 87.9 **19.** 6 **21.** no **23.** yes **25.** 52 **27.** 28 **29.** Yes, opp. sides are parallel and diag. are congruent **31.** 7.5 **33.** 102

35. Given: $ABCD$ is a square.
Prove: $\overline{AC} \perp \overline{BD}$

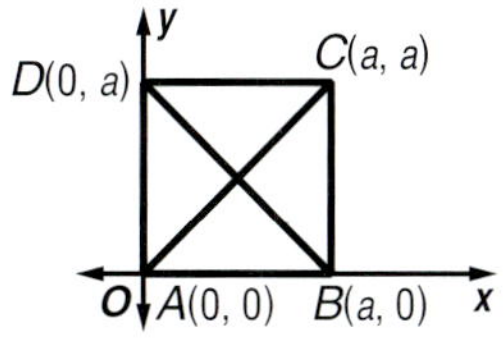

Proof:

Slope of $\overline{AC} = \frac{a-0}{a-0}$ or 1

Slope of $\overline{BD} = \frac{a-0}{0-a}$ or -1

The slope of $\overline{AC}$ is the negative reciprocal of the slope of $\overline{BD}$. Therefore, $\overline{AC} \perp \overline{BD}$.

37. $P(3a, c)$

Chapter 9 Transformations

Page 461 Chapter 9 Getting Started

1.

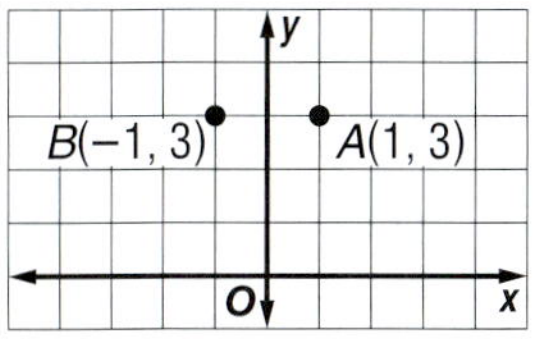

3.

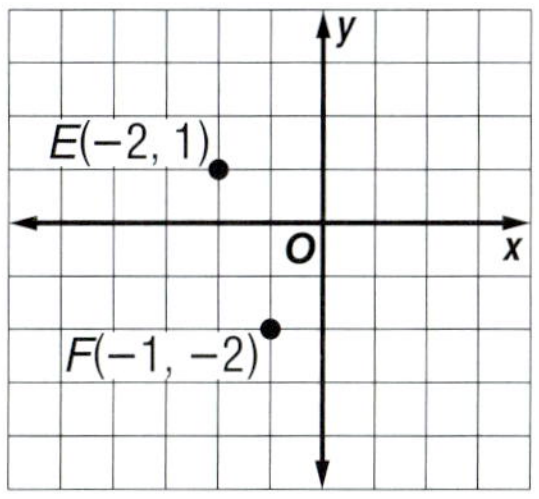

5.

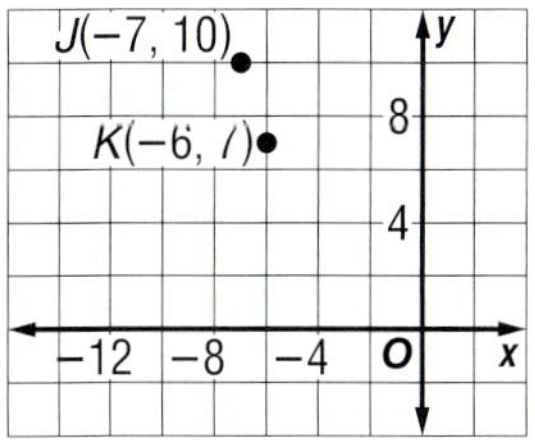

7. 36.9 **9.** 41.8 **11.** 41.4

13. $\begin{bmatrix} -5 & -1 \\ 10 & 5 \end{bmatrix}$

15. $\begin{bmatrix} -2 & -5 & 1 \\ 3 & -4 & -5 \end{bmatrix}$

Pages 463–469 Lesson 9-1

1. Sample Answer: The centroid of an equilateral triangle is not a point of symmetry. **3.** angle measure, betweenness of points, collinearity, distance **5.** 4; yes **7.** 6; yes

9.

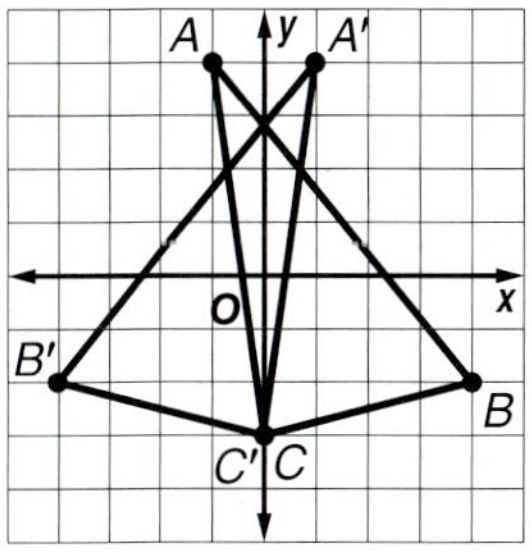

11.

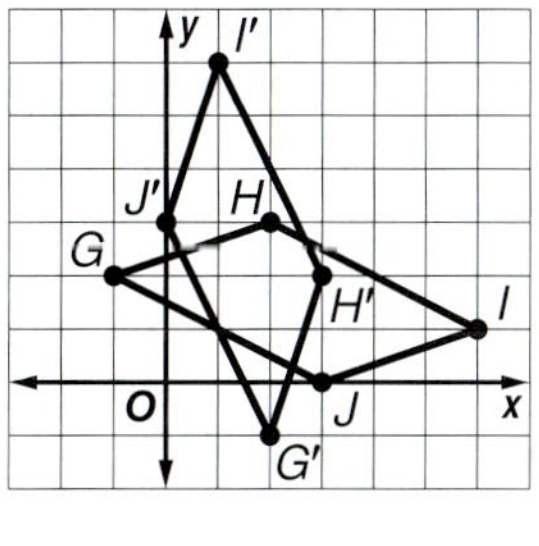

13. 4, yes **15.** $\overline{YX}$ **17.** $\angle XZW$ **19.** $\overline{UV}$ **21.** T **23.** $\triangle WTZ$

25.

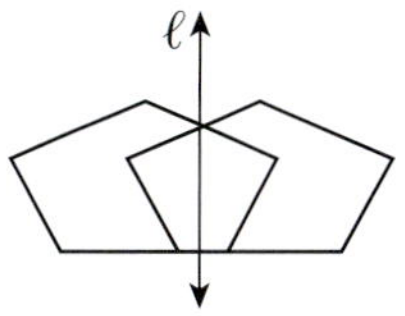

27.

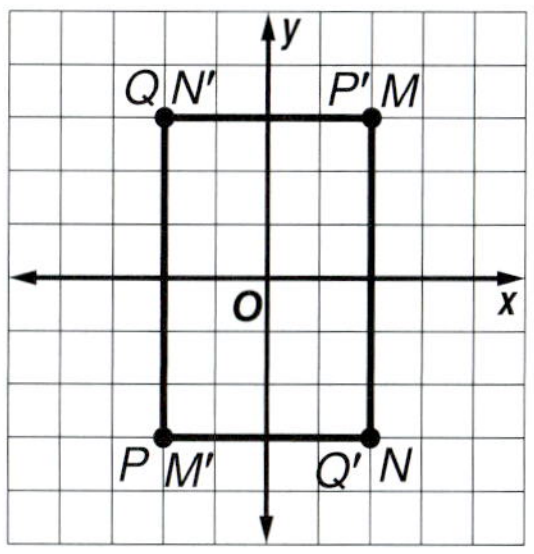

29.

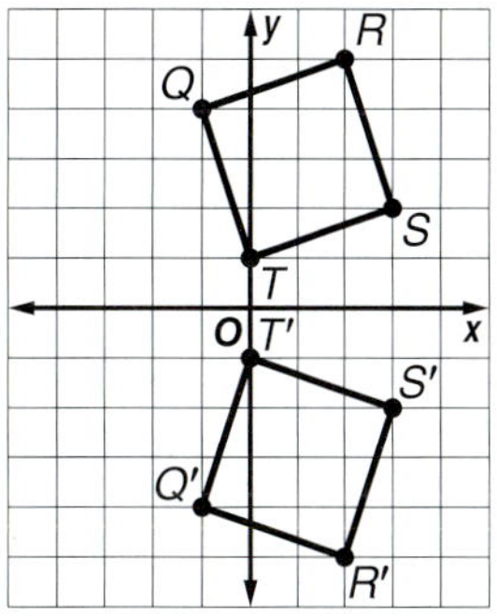

31.

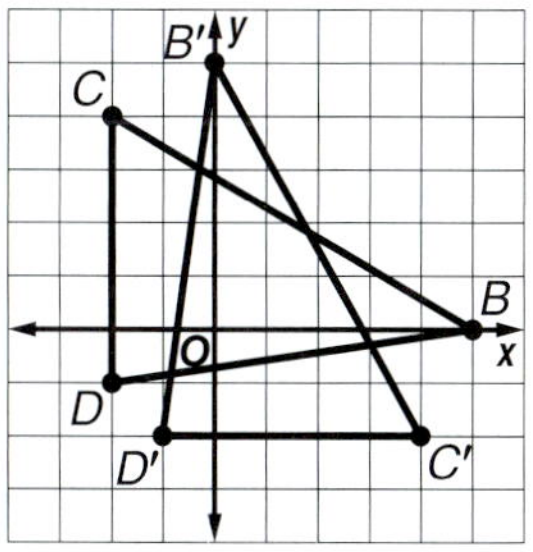

33.

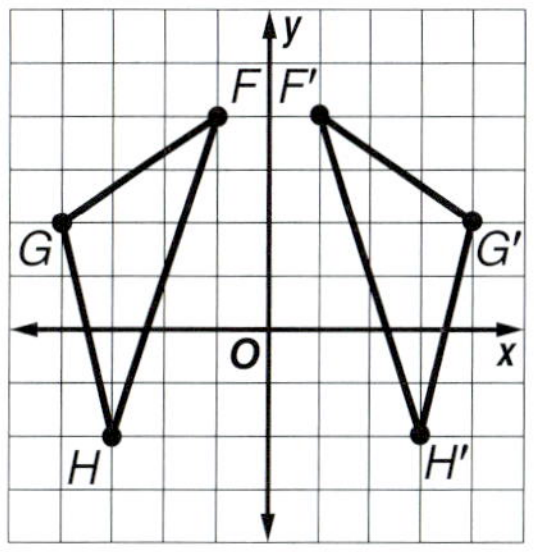

$(x, y) \rightarrow (-x, y)$

35. 2; yes **37.** 1; no **39.** same shape, but turned or rotated

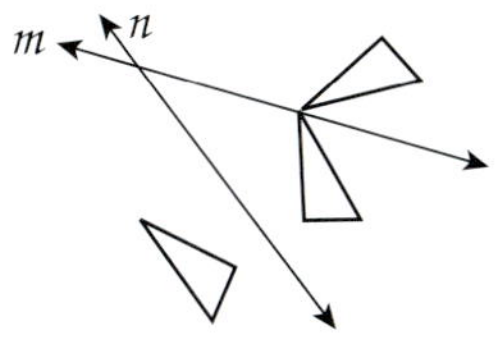

41. $A(4, 7)$, $B(10, -3)$, and $C(-6, -8)$ **43.** Consider point (a, b). Upon reflection in the origin, its image is $(-a, -b)$. Upon reflection in the x-axis and then the y-axis, its image is $(a, -b)$ and then $(-a, -b)$. The images are the same. **45.** vertical line of symmetry **47.** vertical, horizontal lines of symmetry; point of symmetry at the center **49.** D

51. Given: Quadrilateral $LMNP$; X, Y, Z, and W are midpoints of their respective sides.
Prove: $\overline{YW}$ and $\overline{XZ}$ bisect each other.

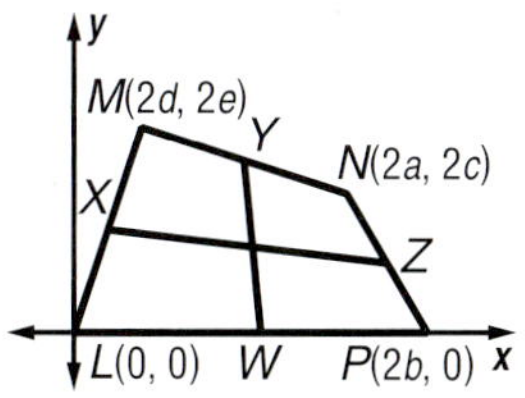

Proof:

Midpoint Y of $\overline{MN}$ is $\left(\frac{2d+2a}{2}, \frac{2e+2c}{2}\right)$ or $(d+a, e+c)$.

Midpoint Z of $\overline{NP}$ is $\left(\frac{2a+2b}{2}, \frac{2c+0}{2}\right)$ or $(a+b, c)$. Midpoint W of $\overline{PL}$ is $\left(\frac{0+2b}{2}, \frac{0+0}{2}\right)$ or $(b, 0)$.

Midpoint X of $\overline{LM}$ is $\left(\frac{0+2d}{2}, \frac{0+2e}{2}\right)$ or (d, e). Midpoint of $\overline{WY}$ is $\left(\frac{d+a+b}{2}, \frac{e+c+0}{2}\right)$ or $\left(\frac{a+b+d}{2}, \frac{c+e}{2}\right)$.

Midpoint of $\overline{XZ}$ is $\left(\frac{d+a+b}{2}, \frac{e+c}{2}\right)$ or $\left(\frac{a+b+d}{2}, \frac{c+e}{2}\right)$.

The midpoints of $\overline{XZ}$ and $\overline{WY}$ are the same, so $\overline{XZ}$ and $\overline{WY}$ bisect each other.

53. 40 **55.** 36 **57.** $f \approx 25.5$, $m\angle H = 76$, $h \approx 28.8$ **59.** $\sqrt{2}$
61. $\sqrt{5}$

Pages 470–475 Lesson 9-2

1. Sample answer: $A(3, 5)$ and $B(-4, 7)$; start at 3, count to the left to -4, which is 7 units to the left or -7. Then count up 2 units from 5 to 7 or $+2$. The translation from A to B is $(x, y) \rightarrow (x - 7, y + 2)$. **3.** Allie; counting from the point $(-2, 1)$ to $(1, -1)$ is right 3 and down 2 to the image. The reflections would be too far to the right. The image would be reversed as well. **5.** No; quadrilateral $WXYZ$ is oriented differently than quadrilateral $NPQR$.

7.

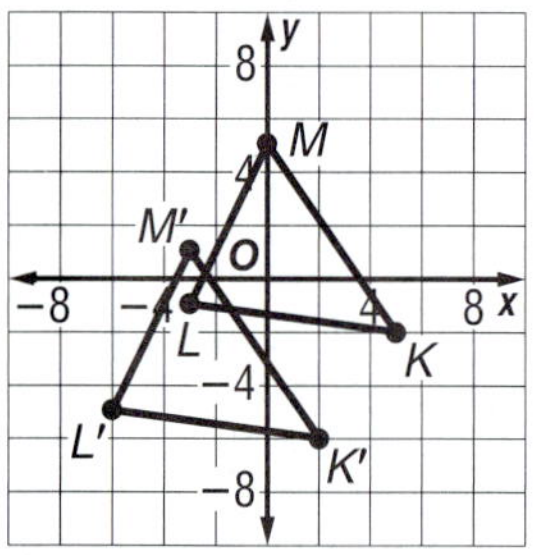

9. Yes; it is one reflection after another with respect to the two parallel lines. **11.** No; it is a reflection followed by a rotation. **13.** Yes; it is one reflection after another with respect to the two parallel lines.

15.

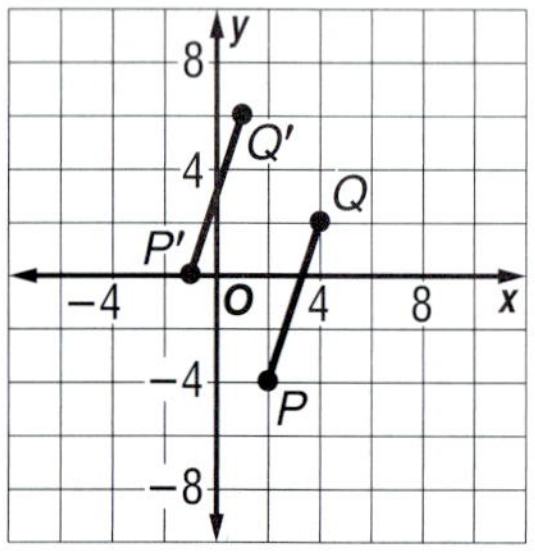

17.

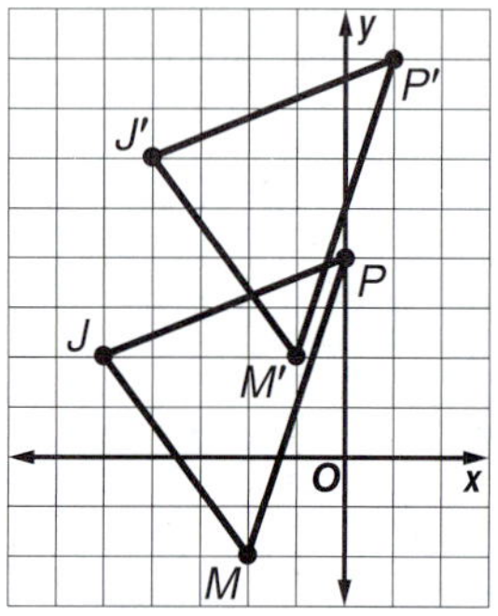

19.

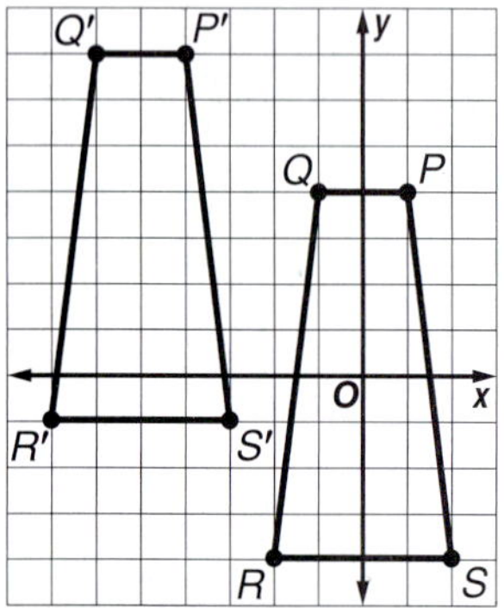

21. left 3 squares and down 7 squares
23. 48 in. right
25. 72 in. right, $24\sqrt{3}$ in. down

27.

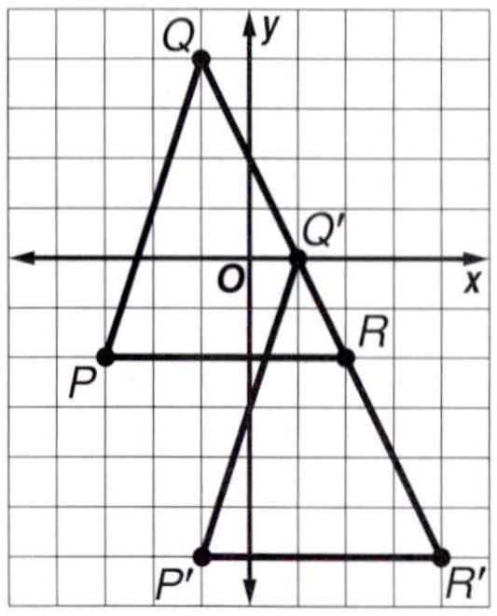

29.

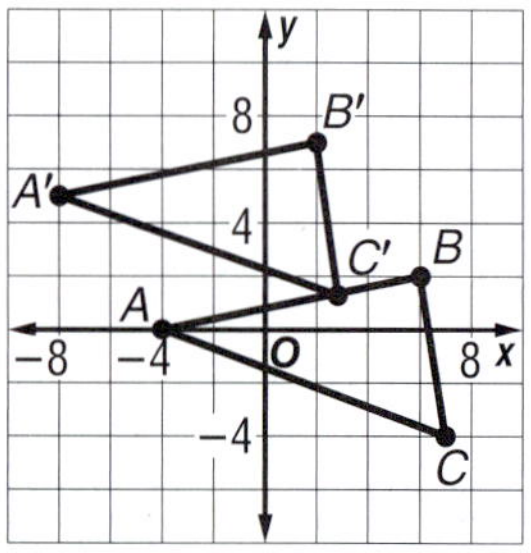

$A(-4, 0)$, $B(6, 2)$, $C(7, -4)$

31. more brains; more free time **33.** No; the percent per figure is different in each category. **35.** Translations and reflections preserve the congruences of segments and angles. The composition of the two transformations will preserve both congruences. Therefore, a glide reflection is an isometry.

37.

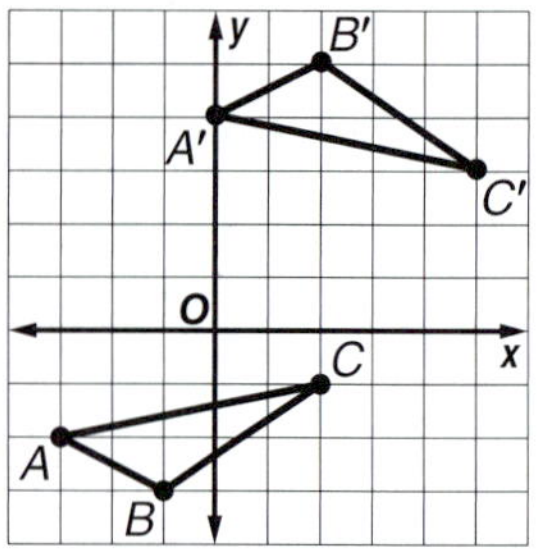

39. A
41.

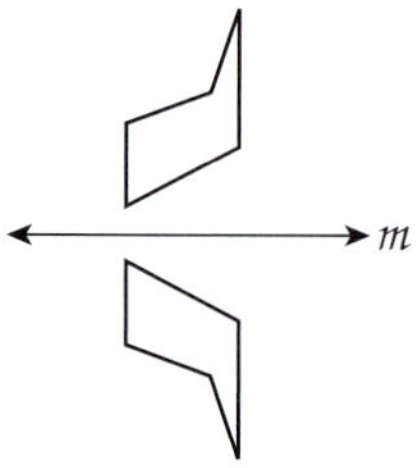

43. $Q(a - b, c)$, $T(0, 0)$ **45.** 23 ft **47.** You did not fill out an application. **49.** The two lines are not parallel. **51.** 5
53. $3\sqrt{2}$

55.

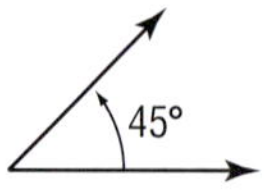

57.

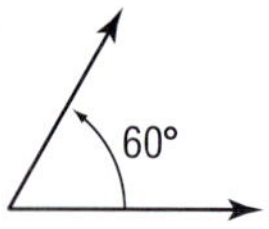

59.

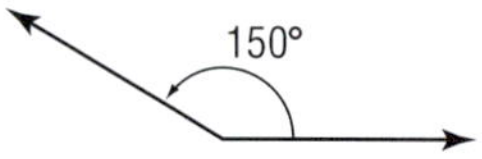

Pages 476–482 Lesson 9-3

1. clockwise $(x, y) \rightarrow (y, -x)$; counterclockwise $(x, y) \rightarrow (-y, x)$

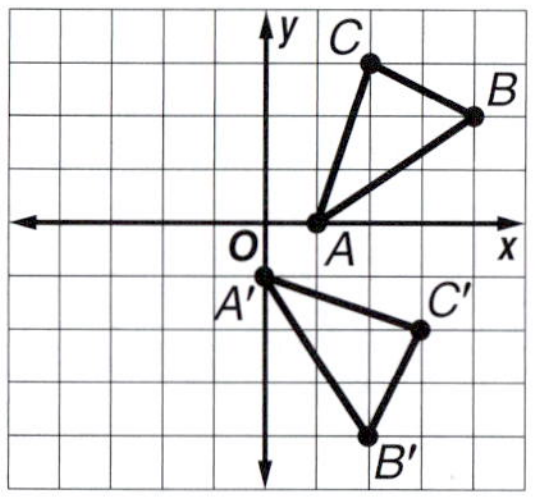

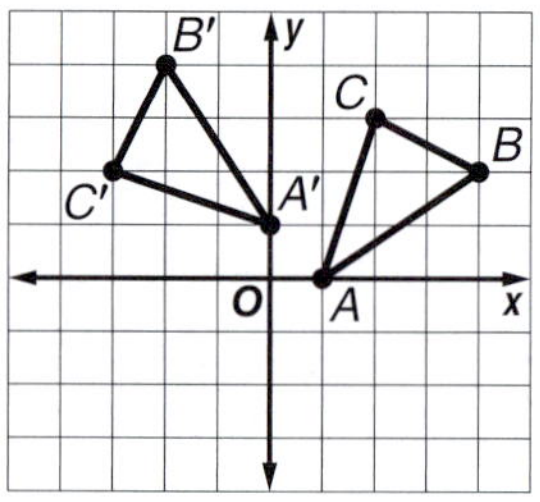

3. Both translations and rotations are made up of two reflections. The difference is that translations reflect across parallel lines and rotations reflect across intersecting lines.

5.

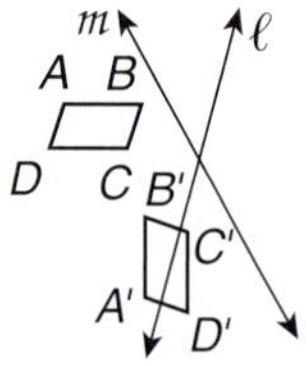

7.

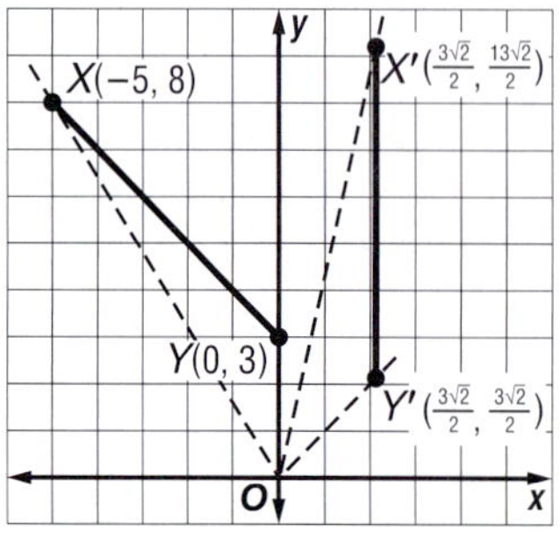

9. order 6; magnitude 60° **11.** order 5 and magnitude 72°; order 4 and magnitude 90°; order 3 and magnitude 120°

13.
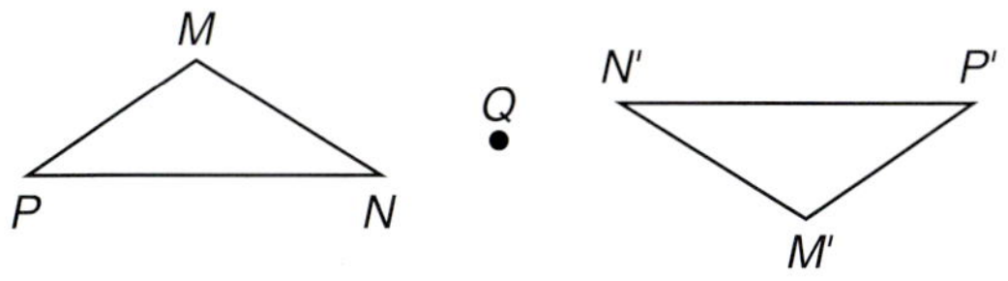

15.
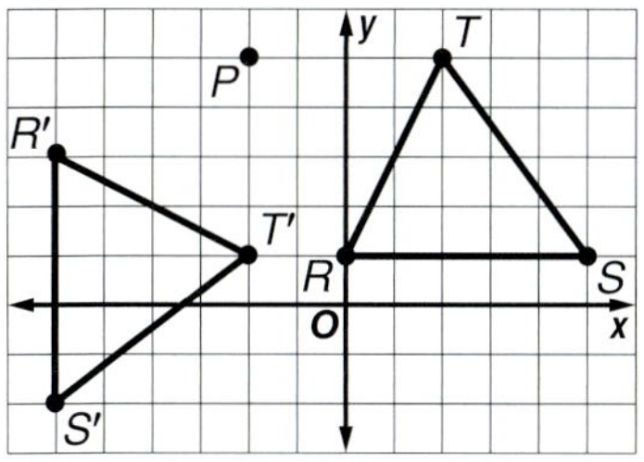

17. 72°

19.
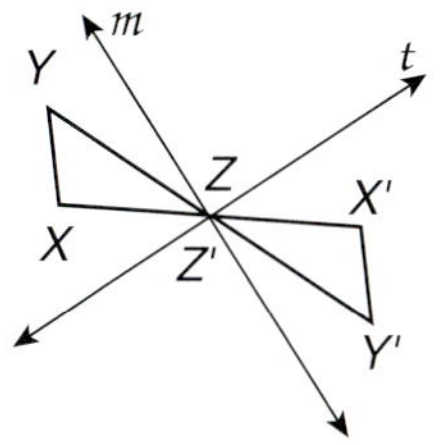

21.
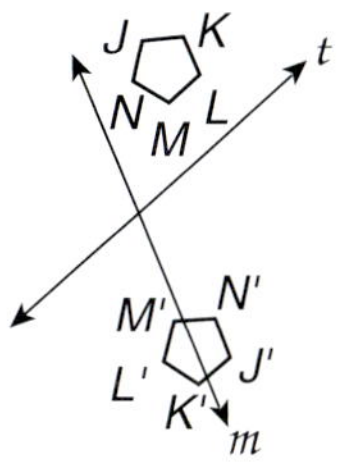

23. $K''(0, -5)$, $L''(4, -2)$, and $M''(4, 2)$; 90° clockwise

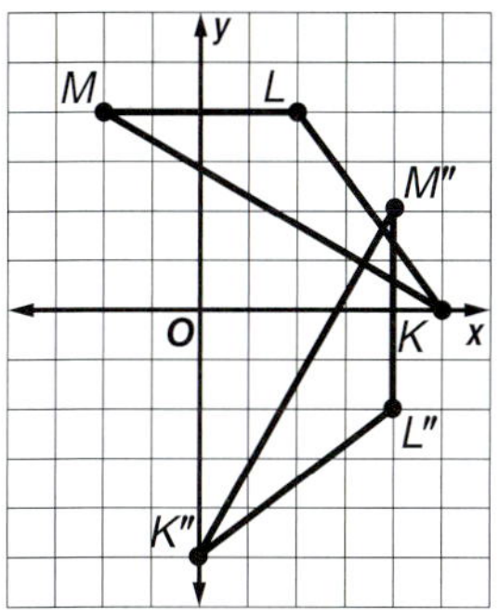

25. $(\sqrt{3}, 1)$ 27. Yes; it is a proper successive reflection with respect to the two intersecting lines. 29. yes 31. no 33. 9 35. $(x, y) \rightarrow (y, -x)$ 37. any point on the line of reflection 39. no invariant points 41. B

43.

Transformation	angle measure	betweenness of points	orientation	collinearity	distance measure
reflection	yes	yes	no	yes	yes
translation	yes	yes	yes	yes	yes
rotation	yes	yes	yes	yes	yes

45. direct 47. Yes; it is one reflection after another with respect to the two parallel lines. 49. Yes; it is one reflection after another with respect to the two parallel lines. 51. C 53. $\angle AGF$ 55. $\overline{TR}$; diagonals bisect each other 57. $\angle QRS$; opp. ∠s ≅ 59. no 61. yes 63. (0, 4), (1, 2), (2, 0) 65. (0, 12), (1, 8), (2, 4), (3, 0) 67. (0, 12), (1, 6), (2, 0)

Page 482 Chapter 9 Practice Quiz 1

1.
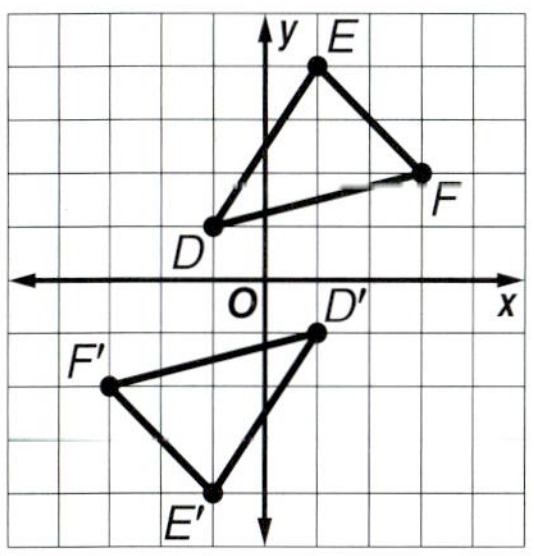

3.
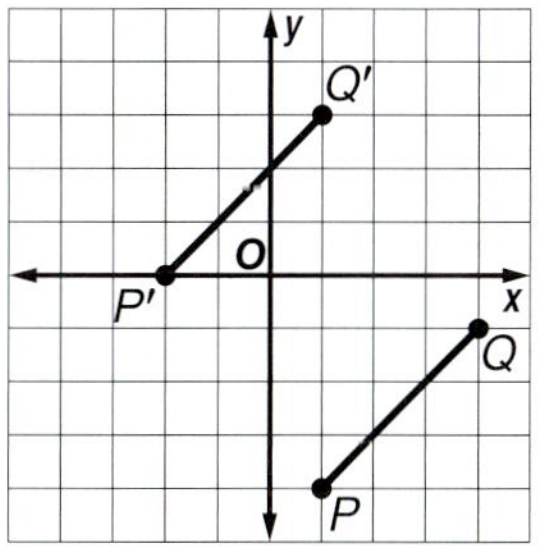

5. order 36; magnitude 10°

Pages 483–488 Lesson 9-4

1. Semi-regular tessellations contain two or more regular polygons, but uniform tessellations can be any combination of shapes. 3. The figure used in the tesselation appears to be a trapezoid, which is not a regular polygon. Thus, the tessellation cannot be regular. 5. no; measure of interior angle = 168 7. yes 9. yes; not uniform 11. no; measure of interior angle = 140 13. yes; measure of interior angle = 60 15. no; measure of interior angle ≈ 164.3 17. no 19. yes 21. yes; uniform 23. yes; not uniform 25. yes; not uniform 27. yes; uniform, regular 29. semi-regular, uniform 31. Never; semi-regular tessellations have the same combination of shapes and angles at each vertex like uniform tessellations. The shapes for semi-regular tessellations are just regular. 33. Always; the sum of the measures of the angles of a quadrilateral is 360°. So if each angle of the quadrilateral is rotated at the vertex, then that equals 360° and the tessellation is possible. 35. yes 37. uniform, regular 39. Sample answer: Tessellations can be used in art to create abstract art. Answers should include the following.

- The equilateral triangles are arranged to form hexagons, which are arranged adjacent to one another.
- Sample answers: kites, trapezoids, isosceles triangles

41. A

43.
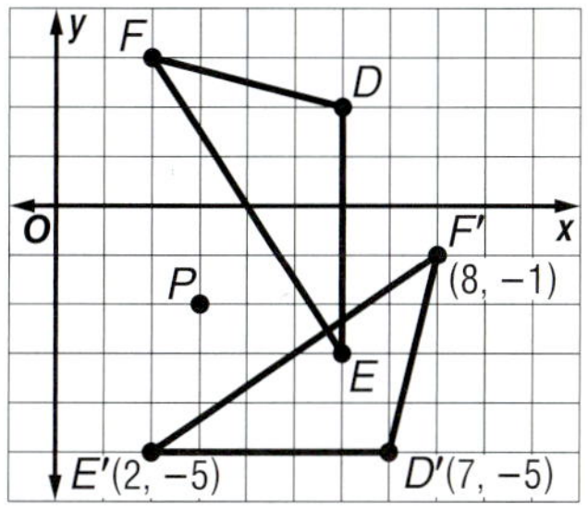

45.
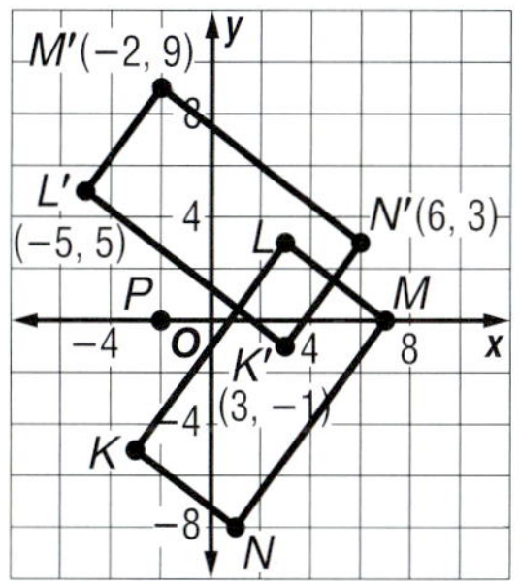

47. $x = 4, y = 1$ 49. $x = 56, y = 12$ 51. no, no 53. yes, no 55. no, no 57. $AB = 7$, $BC = 10$, $AC = 9$ 59. $1(-1) = -1$ and $-1(1) = -1$ 61. square 63. 15 65. 22.5

Pages 490–497 Lesson 9-5

1. Dilations only preserve length if the scale factor is 1 or −1. So for any other scale factor, length is not preserved and the dilation is not an isometry. 3. Trey; Desiree found the image using a positive scale factor.

5.
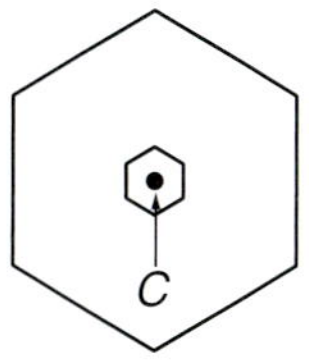

7. $A'B' = 12$

9.
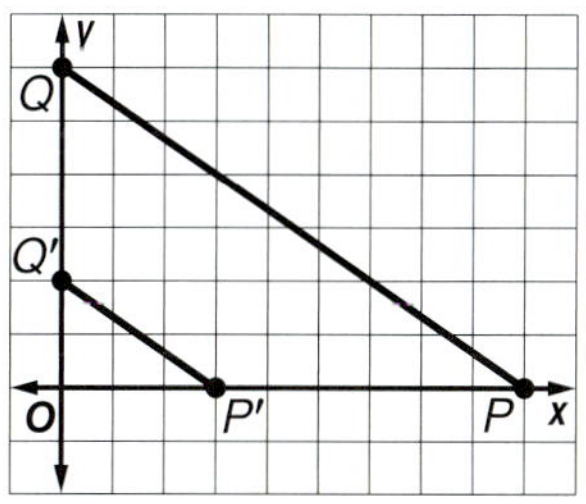

11. $r = 2$; enlargement **13.** C

15.

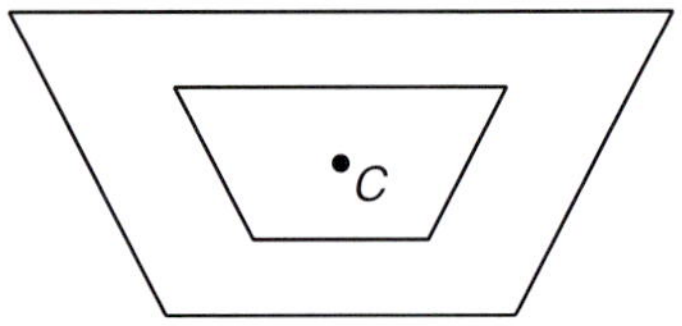

17.

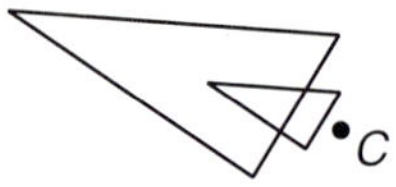

19.

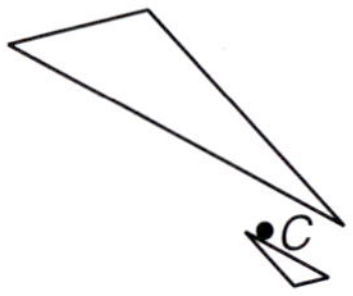

21. $S'T' = \frac{3}{5}$
23. $ST = 4$
25. $S'T' = 0.9$

27.

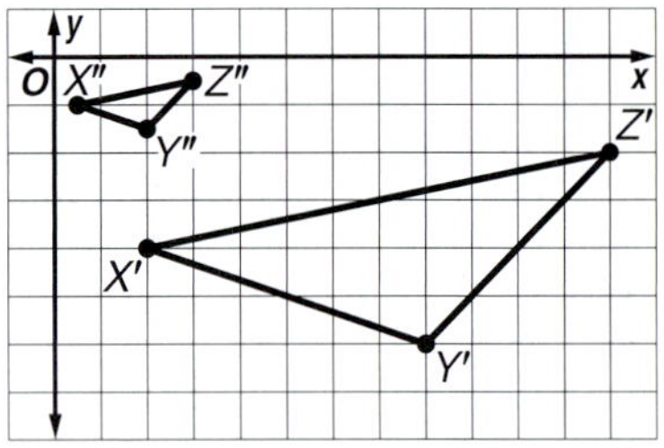

29.

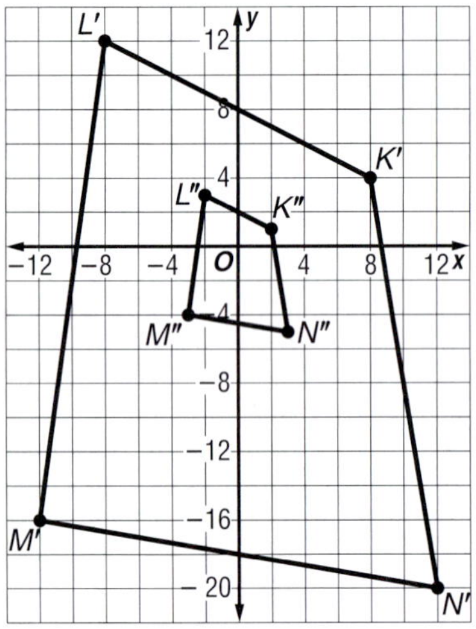

31. $\frac{1}{2}$; reduction
33. $\frac{1}{3}$; reduction
35. -2; enlargement
37. 7.5 by 10.5
39. The perimeter is four times the original perimeter.

41. Given: dilation with center C and scale factor r
Prove: $ED = r(AB)$
Proof:

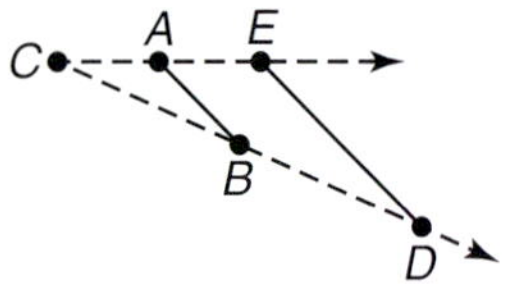

$CE = r(CA)$ and $CD = r(CB)$ by the definition of a dilation. $\frac{CE}{CA} = r$ and $\frac{CD}{CB} = r$.
So, $\frac{CE}{CA} = \frac{CD}{CB}$ by substitution.
$\angle ACB \cong \angle ECD$, since congruence of angles is reflexive. Therefore, by SAS Similarity, $\triangle ACB$ is similar to $\triangle ECD$. The corresponding sides of similar triangles are proportional, so $\frac{ED}{AB} = \frac{CE}{CA}$. We know that $\frac{CE}{CA} = r$, so $\frac{ED}{AB} = r$ by substitution. Therefore, $ED = r(AB)$ by the Multiplication Property of Equality.

43. 2 **45.** $\frac{1}{20}$ **47.** 60% **49.**

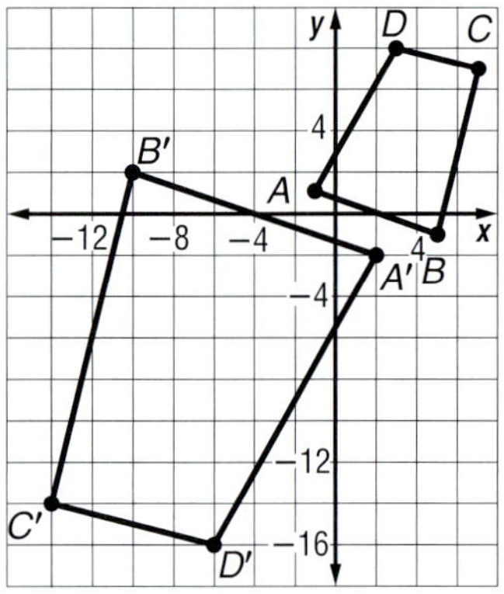

51.

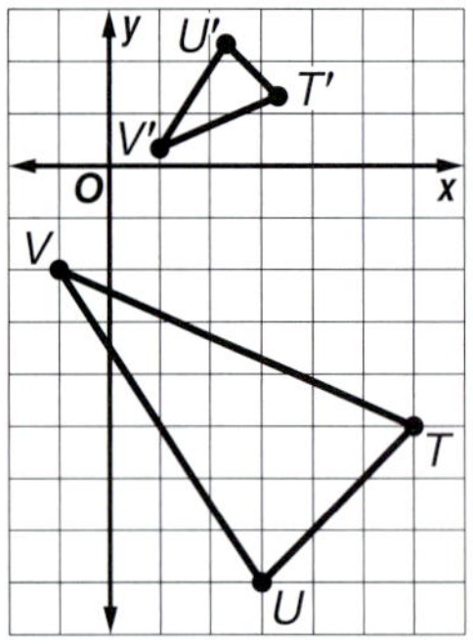

53. Sample answer: Yes; a cut and paste produces an image congruent to the original. Answers should include the following.

- Congruent figures are similar, so cutting and pasting is a similarity transformation.
- If you scale both horizontally and vertically by the same factor, you are creating a dilation.

55. A **57.** no **59.** no

61.

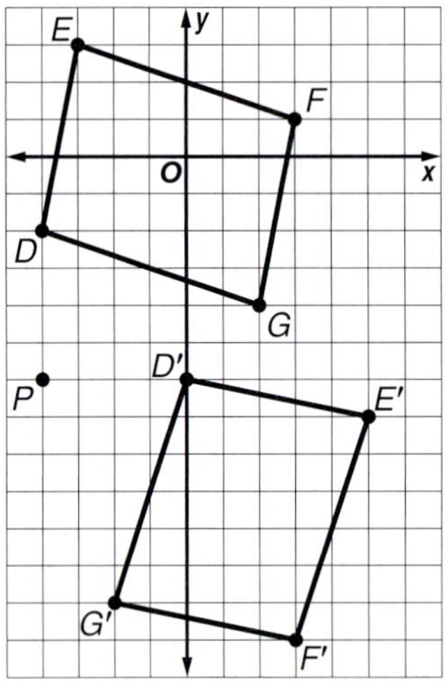

63. Given: $\angle J \cong \angle L$ B is the midpoint of $\overline{JL}$.
Prove: $\triangle JHB \cong \triangle LCB$
Proof: It is known that $\angle J \cong \angle L$. Since B is the midpoint of $\overline{JL}$, $\overline{JB} \cong \overline{LB}$ by the Midpoint Theorem.

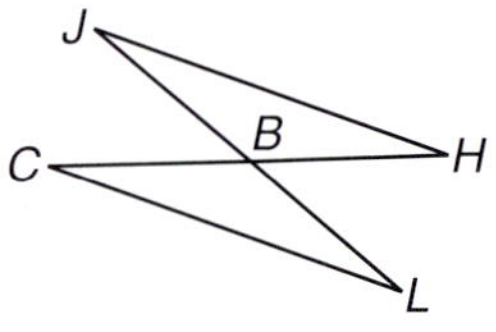

$\angle JBH \cong \angle LBC$ because vertical angles are congruent. Thus, $\triangle JHB \cong \triangle LCB$ by ASA. **65.** 76.0

Selected Answers

Page 497 Chapter 9 Practice Quiz 2

1. yes; uniform; semi-regular **3.**

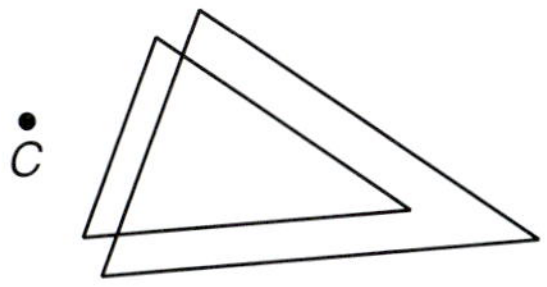

5. $A'(-5, -1)$, $B'\left(-\frac{1}{2}, -3\right)$, $C'(2, -2)$

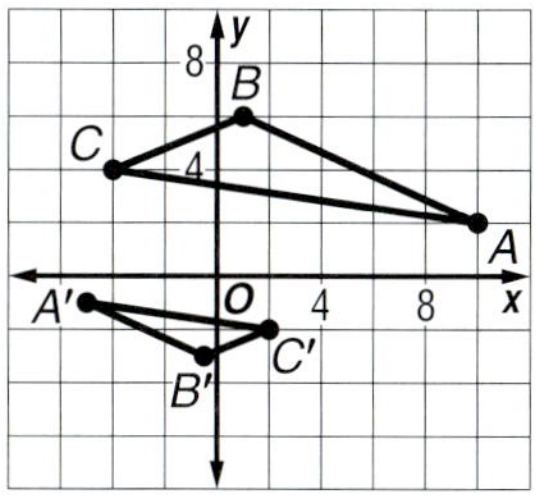

Pages 498–505 Lesson 9-6

1. Sample answer; $\langle 7, 7\rangle$

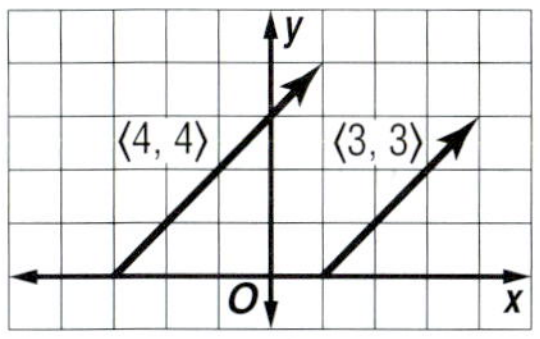

3. Sample answer: Using a vector to translate a figure is the same as using an ordered pair because a vector has horizontal and vertical components, each of which can be represented by one coordinate of an ordered pair.

5. $\langle 4, -3\rangle$

7. $2\sqrt{13} \approx 7.2$, $\approx 213.7°$

9.

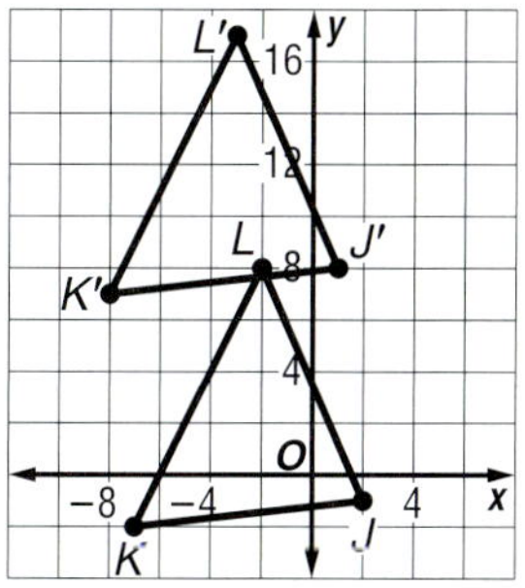

11.

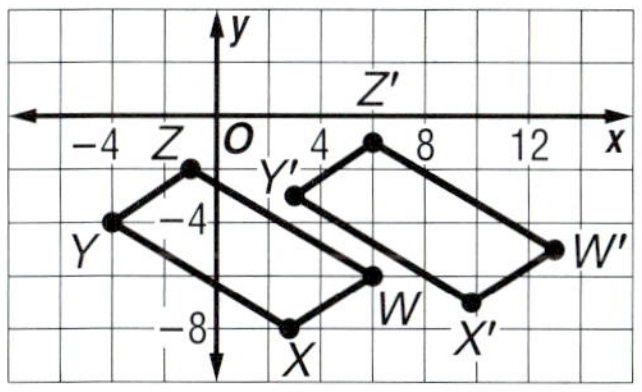

13. $6\sqrt{13} \approx 21.6$, $303.7°$ **15.** $\langle 2, 6\rangle$ **17.** $\langle -7, -4\rangle$ **19.** $\langle -3, 5\rangle$ **21.** 5, 0° **23.** $2\sqrt{5} \approx 4.5$, $296.6°$ **25.** $7\sqrt{5} \approx 15.7$, $26.6°$ **27.** 25, $\approx 73.7°$ **29.** $5\sqrt{41} \approx 32.0$, $\approx 218.7°$ **31.** $6\sqrt{2} \approx 8.5$, $135.0°$ **33.** $4\sqrt{10} \approx 12.6$, $198.4°$ **35.** $2\sqrt{122} \approx 22.1$, $275.2°$

37.

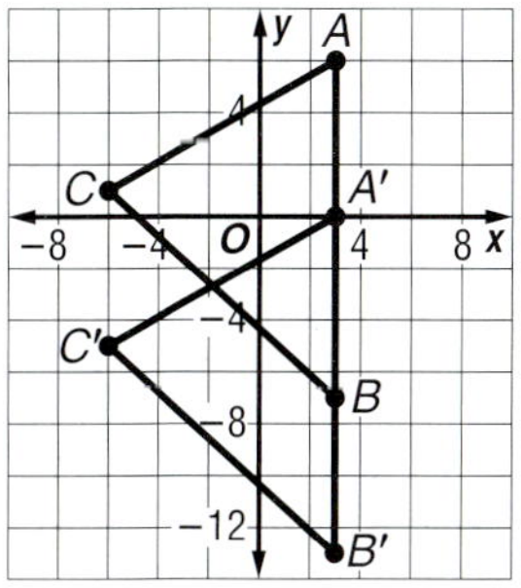

39.

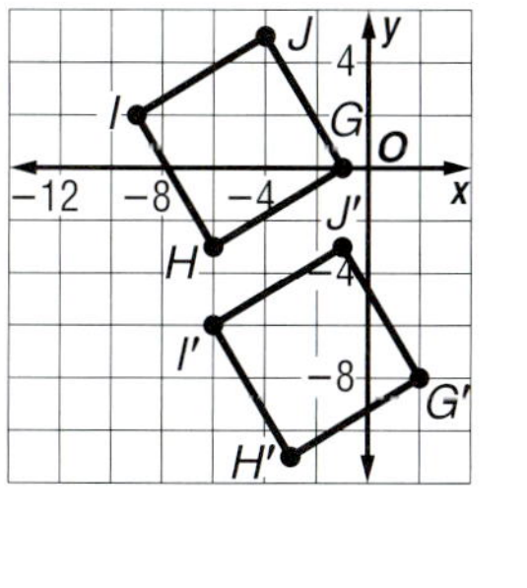

41.

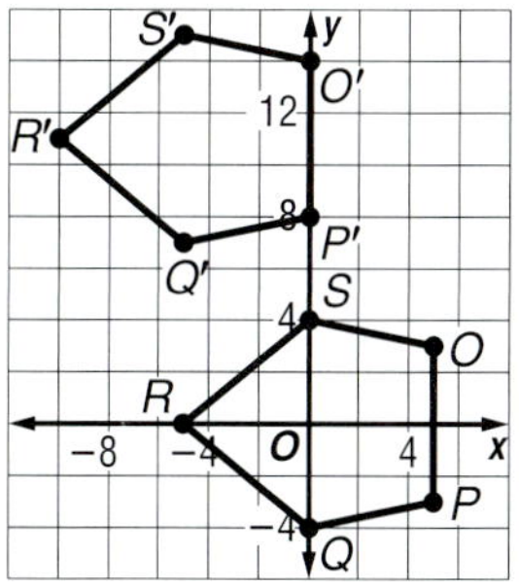

43.

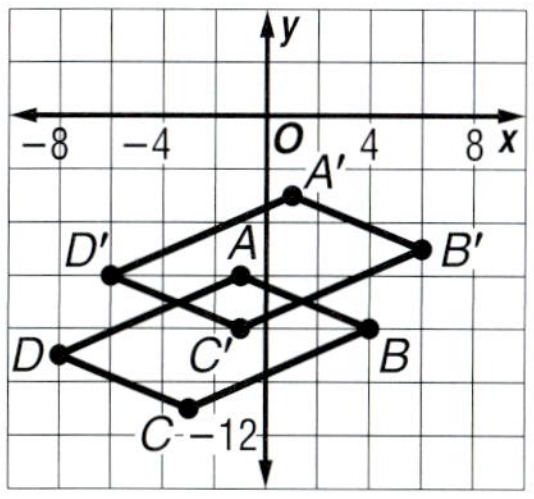

45.

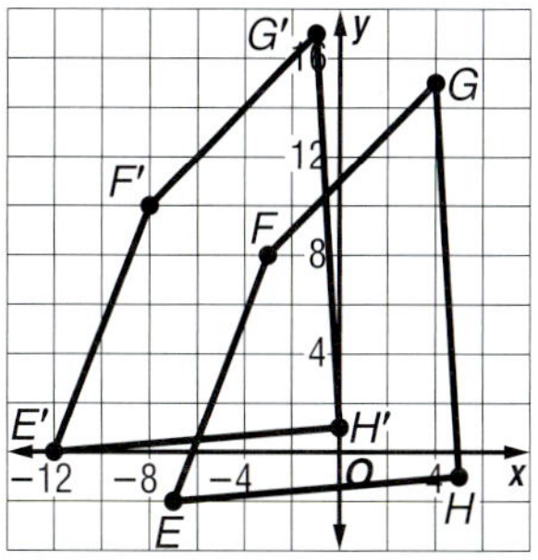

47. 13, $\approx 67.4°$
49. 5, $\approx 306.9°$
51. $2\sqrt{5} \approx 4.5$, $\approx 26.6°$
53. about 44.8 mi; about 38.7° south of due east
55. $\langle -350, 450\rangle$ mph
57. 52.1° north of due west

59. Sample answer: Quantities such as velocity are vectors. The velocity of the wind and the velocity of the plane together factor into the overall flight plan. Answers should include the following.

- A wind from the west would add to the velocity contributed by the plane resulting in an overall velocity with a larger magnitude.
- When traveling east, the prevailing winds add to the velocity of the plane. When traveling west, they detract from it.

61. D **63.** $A'B' = 6$ **65.** $AB = 48$ **67.** yes; not uniform **69.** 12 **71.** 30

73. $\begin{bmatrix} -4 & -3 \\ -10 & 4 \end{bmatrix}$ **75.** $\begin{bmatrix} -27 & -15 & -3 \\ 27 & 3 & 15 \end{bmatrix}$ **77.** $\begin{bmatrix} 12 & 4 \\ -4 & -12 \end{bmatrix}$

Pages 506–511 Lesson 9-7

1. $\begin{bmatrix} 0 & 1 \\ 1 & 0 \end{bmatrix}$ **3.** Sample answer: $\begin{bmatrix} -2 & -2 & -2 & -2 \\ -1 & -1 & -1 & -1 \end{bmatrix}$

5. $D'(-1, 9)$, $E'(5, 9)$, $F'(3, 6)$, $G'(-3, 6)$ **7.** $A'\left(-\frac{1}{4}, -\frac{1}{2}\right)$, $B'\left(-\frac{3}{4}, -\frac{3}{4}\right)$, $C'\left(-\frac{3}{4}, -\frac{5}{4}\right)$, $D'\left(-\frac{1}{4}, -1\right)$ **9.** $H'(5, 4)$, $I'(1, -1)$, $J'(3, -6)$, $K'(7, -3)$ **11.** $P'(3, -6)$, $Q'(7, -6)$, $R'(7, -2)$ **13.** (1.5, −0.5), (3.5, −1.5), (2.5, −3.5), (0.5, −2.5) **15.** $E'(-6, 6)$, $F'(-3, 8)$ **17.** $M'(1, 1)$, $N'(5, 3)$, $O'(5, 1)$, $P'(1, -1)$ **19.** $A'(12, 10)$, $B'(8, 10)$, $C'(6, 14)$ **21.** $G'(-2, -1)$, $H'(2, -3)$, $I'(3, 4)$, $J'(-3, 5)$ **23.** $X'(-2, 2)$, $Y'(-4, -1)$ **25.** $D'(-4, -5)$, $E'(2, -6)$, $F'(3, -1)$, $G'(-3, 4)$ **27.** $V'(-2, 2)$, $W'\left(\frac{2}{3}, 2\right)$, $X'\left(2, -\frac{4}{3}\right)$ **29.** $V'(-3, -3)$, $W'(-3, 1)$, $X'(2, 3)$ **31.** $P'(2, -3)$, $Q'(-1, -1)$, $R'(1, 2)$, $S'(3, 2)$, $T'(5, -1)$ **33.** $P'(1, -1)$, $Q'(4, 1)$, $R'(2, 4)$, $S'(0, 4)$, $T'(-2, 1)$ **35.** $M'(-1, 12)$, $N'(-10, -3)$ **37.** $S'(-1, 2)$, $T'(-1, 6)$, $U'(3, 5)$, $V'(3, 1)$ **39.** $A'\left(-1, -\frac{1}{3}\right)$, $B'\left(-\frac{2}{3}, -\frac{4}{3}\right)$, $C'\left(\frac{2}{3}, -\frac{4}{3}\right)$, $D'\left(1, -\frac{1}{3}\right)$, $E'\left(\frac{2}{3}, \frac{2}{3}\right)$, $F'\left(-\frac{2}{3}, \frac{2}{3}\right)$ **41.** $A'(2, 1)$, $B'(5, 2)$, $C'(5, 6)$, $D'(2, 7)$, $E'(-1, 6)$, $F'(-1, 2)$ **43.** Each footprint is reflected in the y-axis, then translated up two units.

45. $\begin{bmatrix} -1 & 0 \\ 0 & 1 \end{bmatrix}$ **47.** $\begin{bmatrix} 0 & -1 \\ -1 & 0 \end{bmatrix}$ **49.** $\begin{bmatrix} 0 & 1 \\ -1 & 0 \end{bmatrix}$

51. 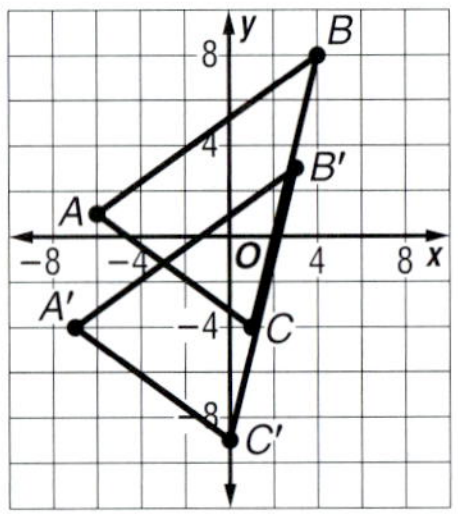

53. $-\frac{1}{2}$; reduction
55. 60, 120 57. 36, 144

Pages 512–516 Chapter 9 Study Guide and Review

1. false, center 3. false, component form 5. false, center of rotation 7. false, scale factor

9. 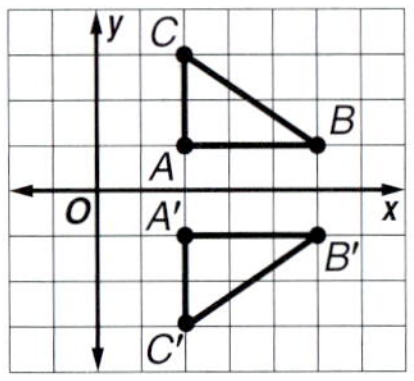

11. 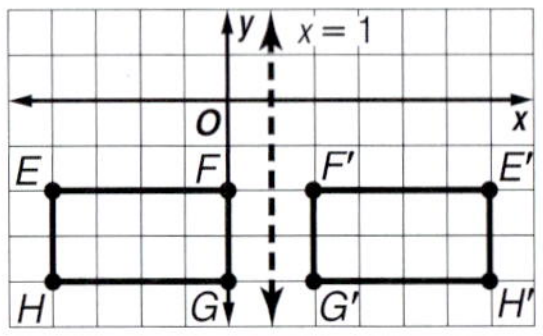

13. 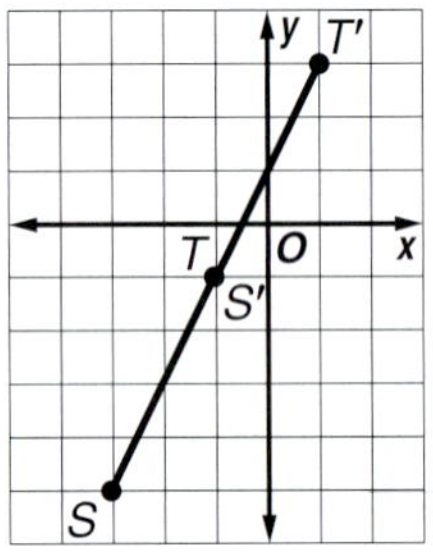

15. $B'(3, -5)$, $C'(3, -3)$, $D'(5, -3)$; 180°

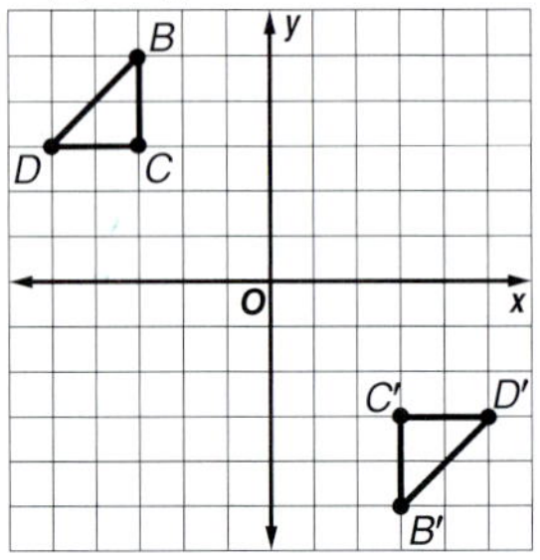

17. $L'(-2, 2)$, $M'(-3, 5)$, $N'(-6, 3)$; 90° counterclockwise

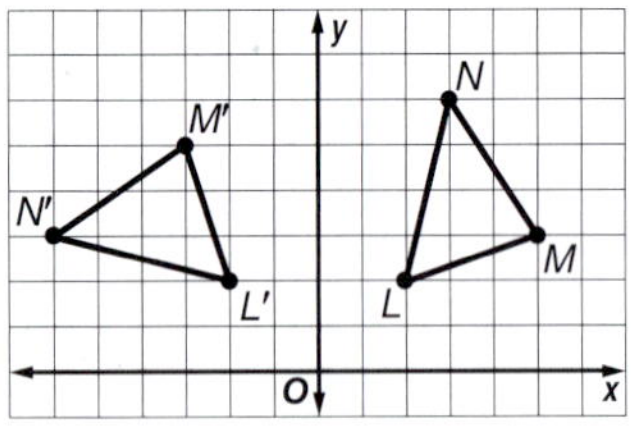

19. 200° 21. yes; not uniform 23. yes; uniform 25. Yes; the measure of an interior angle is 60, which is a factor of 360. 27. $C'D' = 24$ 29. $CD = 4$

31. $C'D' = 10$ 33. $P'(2, -6)$, $Q'(-4, -4)$, $R'(-2, 2)$ 35. $\langle 3, 4\rangle$ 37. $\langle 0, 8\rangle$ 39. ≈ 14.8, $\approx 208.3°$ 41. ≈ 72.9, $\approx 213.3°$ 43. $D'\left(-\frac{12}{5}, -\frac{8}{5}\right)$, $E'(0, 4)$, $F'\left(\frac{8}{5}, -\frac{16}{5}\right)$ 45. $D'(-2, 3)$, $E'(5, 0)$, $F'(-4, -2)$ 47. $W'(-16, 2)$, $X'(-4, 6)$, $Y'(-2, 0)$, $Z'(-12, -6)$

Chapter 10 Circles

Pages 521 Chapter 10 Getting Started

1. 162 3. 2.4 5. $r = \frac{C}{2p}$ 7. 15 9. 17.0 11. 1.5, −0.9 13. 2.5, −3

Pages 522–528 Lesson 10-1

1. Sample answer: The value of π is calculated by dividing the circumference of a circle by the diameter. 3. Except for a diameter, two radii and a chord of a circle can form a triangle. The Triangle Inequality Theorem states that the sum of two sides has to be greater than the third. So, $2r$ has to be greater than the measure of any chord that is not a diameter, but $2r$ is the measure of the diameter. So the diameter has to be longer than any other chord of the circle. 5. $\overline{EA}$, $\overline{EB}$, $\overline{EC}$, or $\overline{ED}$ 7. $\overline{AC}$ or $\overline{BD}$ 9. 10.4 in. 11. 6 13. 10 m, 31.42 m 15. B 17. $\overline{FA}$, $\overline{FB}$, or $\overline{FE}$ 19. $\overline{BE}$ 21. $\odot R$ 23. $\overline{ZV}$, $\overline{TX}$, or $\overline{WZ}$ 25. $\overline{RU}$, $\overline{RV}$ 27. 2.5 ft 29. 64 in. or 5 ft 4 in. 31. 0.6 m 33. 3 35. 12 37. 34 39. 20 41. 5 43. 2.5 45. 13.4 cm, 84.19 cm 47. 24.32 m, 12.16 m 49. $13\frac{1}{2}$ in., 42.41 in. 51. $0.33a$, $1.05a$ 53. 5π ft 55. 8π cm 57. 0; The longest chord of a circle is the diameter, which contains the center. 59. 500–600 ft 61. 24π units 63. 27 65. 10π, 20π, 30π 67. 9.8; 66° 69. 44.7; 27° 71. 24

73. **Given:** $\overline{RQ}$ bisects $\angle SRT$.
Prove: $m\angle SQR > m\angle SRQ$

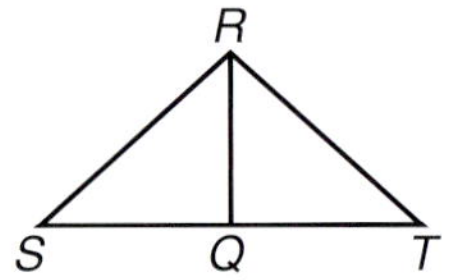

Proof:

Statements	Reasons
1. $\overline{RQ}$ bisects $\angle SRT$.	1. Given
2. $\angle SRQ \cong \angle QRT$	2. Def. of $\angle$ bisector
3. $m\angle SRQ = m\angle QRT$	3. Def. of $\cong$ $\angle$s
4. $m\angle SQR = m\angle T + m\angle QRT$	4. Exterior Angle Theorem
5. $m\angle SQR > m\angle QRT$	5. Def. of Inequality
6. $m\angle SQR > m\angle SRQ$	6. Substitution

75. 60 77. 30 79. 30

Pages 529–535 Lesson 10-2

1. Sample answer: $\widehat{AB}$, $\widehat{BC}$, $\widehat{AC}$, $\widehat{ABC}$, $\widehat{BCA}$, $\widehat{CAB}$; $m\widehat{AB} = 110$, $m\widehat{BC} = 160$, $m\widehat{AC} = 90$, $m\widehat{ABC} = 270$, $m\widehat{BCA} = 250$, $m\widehat{CAB} = 200$

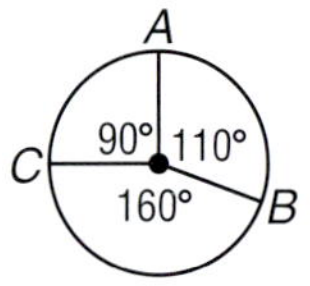

3. Sample answer: Concentric circles have the same center, but different radius measures; congruent circles usually have different centers but the same radius measure. 5. 137 7. 103 9. 180 11. 138 13. Sample answer: 25% = 90°, 23% = 83°, 28% = 101°, 22% = 79°, 2% = 7° 15. 60 17. 30 19. 120 21. 115 23. 65 25. 90 27. 90 29. 135 31. 270 33. 76 35. 52 37. 256 39. 308 41. $24\pi \approx 75.40$ units 43. $4\pi \approx 12.57$ units 45. The first category is a major arc, and the other three categories are minor arcs. 47. always 49. never 51. $m\angle 1 = 80$, $m\angle 2 = 120$, $m\angle 3 = 160$ 53. 56.5 ft 55. No; the radii are not equal, so the proportional part of the circumferences would not be the same. Thus, the arcs would not be congruent. 57. B 59. 20; 62.83 61. 28; 14 63. 84.9 newtons, 32° north of due east 65. 36.68 67. $\sqrt{24.5}$ 69. If ABC has three sides, then ABC is a triangle. 71. 42 73. 100 75. 36

Pages 536–543 Lesson 10-3

1. Sample answer: An inscribed polygon has all vertices on the circle. A circumscribed circle means the circle is drawn around so that the polygon lies in its interior and all vertices lie on the circle. 3. Tokei; to bisect the chord, it must be a diameter and be perpendicular. 5. 30 7. $5\sqrt{3}$ 9. $10\sqrt{5} \approx 22.36$ 11. 15 13. 15 15. 40 17. 80 19. 4 21. 5 23. $m\widehat{AB} = m\widehat{BC} = m\widehat{CD} = m\widehat{DE} = m\widehat{EF} = m\widehat{FG} = m\widehat{GH} = m\widehat{HA} = 45$ 25. $m\widehat{NP} = m\widehat{RQ} = 120$; $m\widehat{NR} = m\widehat{PQ} = 60$ 27. 30 29. 15 31. 16 33. 6 35. $\sqrt{2} \approx 1.41$

37. Given: $\odot O$, $\overline{OS} \perp \overline{RT}$, $\overline{OV} \perp \overline{UW}$, $\overline{OS} \cong \overline{OV}$
Prove: $\overline{RT} \cong \overline{UW}$

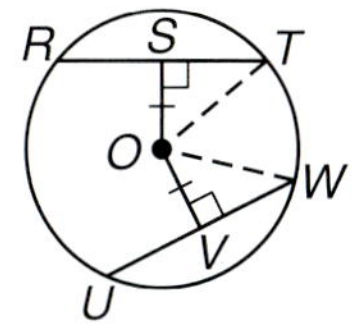

Proof:

Statements	Reasons
1. $\overline{OT} \cong \overline{OW}$	1. All radii of a $\odot$ are $\cong$.
2. $\overline{OS} \perp \overline{RT}$, $\overline{OV} \perp \overline{VW}$, $\overline{OS} \cong \overline{OV}$	2. Given
3. $\angle OST$, $\angle OVW$ are right angles.	3. Definition of $\perp$ lines
4. $\triangle STO \cong \triangle VWO$	4. HL
5. $\overline{ST} \cong \overline{VW}$	5. CPCTC
6. $ST = VW$	6. Definition of $\cong$ segments
7. $2(ST) = 2(VW)$	7. Multiplication Property
8. $\overline{OS}$ bisects $\overline{RT}$; $\overline{OV}$ bisects $\overline{UW}$.	8. Radius $\perp$ to a chord bisects the chord.
9. $RT = 2(ST)$, $UW = 2(VW)$	9. Definition of segment bisector
10. $RT = UW$	10. Substitution
11. $\overline{RT} \cong \overline{UW}$	11. Definition of $\cong$ segments

39. 2.82 in.
41. 18 inches

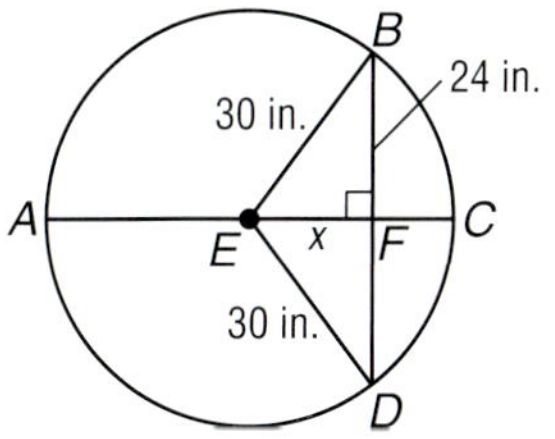

43. $2\sqrt{135} \approx 23.24$ yd

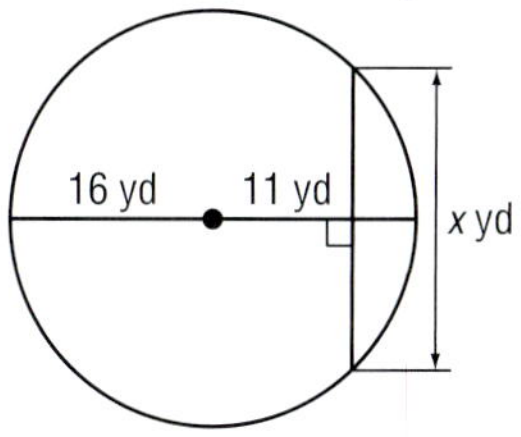

45. Let r be the radius of $\odot P$. Draw radii to points D and E to create triangles. The length DE is $r\sqrt{3}$ and $AB = 2r$; $r\sqrt{3} \neq \frac{1}{2(2r)}$. **47.** Inscribed equilateral triangle; the six arcs making up the circle are congruent because the chords intercepting them were congruent by construction. Each of the three chords drawn intercept two of the congruent chords. Thus, the three larger arcs are congruent. So, the three chords are congruent, making this an equilateral triangle. **49.** No; congruent arcs are must be in the same circle, but these are in concentric circles. **51.** Sample answer: The grooves of a waffle iron are chords of the circle. The ones that pass horizontally and vertically through the center are diameters. Answers should include the following.

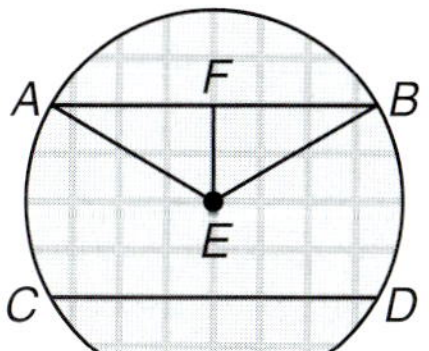

- If you know the measure of the radius and the distance the chord is from the center, you can use the Pythagorean Theorem to find the length of half of the chord and then multiply by 2.
- There are four grooves on either side of the diameter, so each groove is about 1 in. from the center. In the figure, $EF = 2$ and $EB = 4$ because the radius is half the diameter. Using the Pythagorean Theorem, you find that $FB \approx 3.464$ in. so $AB \approx 6.93$ in. Approximate lengths for other chords are 5.29 in. and 7.75 in., but exactly 8 in. for the diameter.

53. 14,400 **55.** 180 **57.** $\overline{SU}$ **59.** $\overline{RM}, \overline{AM}, \overline{DM}, \overline{IM}$ **61.** 50 **63.** 10 **65.** 20

Page 543 Chapter 10 Practice Quiz 1

1. $\overline{BC}, \overline{BD}, \overline{BA}$ **3.** 95 **5.** 9 **7.** 28 **9.** 21

Page 544–551 Lesson 10-4

1. Sample answer:

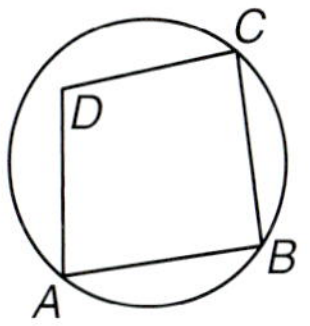

3. $m\angle 1 = 30$, $m\angle 2 = 60$, $m\angle 3 = 60$, $m\angle 4 = 30$, $m\angle 5 = 30$, $m\angle 6 = 60$, $m\angle 7 = 60$, $m\angle 8 = 30$ **5.** $m\angle 1 = 35$, $m\angle 2 = 55$, $m\angle 3 = 39$, $m\angle 4 = 39$ **7.** 1 **9.** $m\angle 1 = m\angle 2 = 30$, $m\angle 3 = 25$

11. Given: $\widehat{AB} \cong \widehat{DE}$, $\widehat{AC} \cong \widehat{CE}$
Prove: $\triangle ABC \cong \triangle EDC$

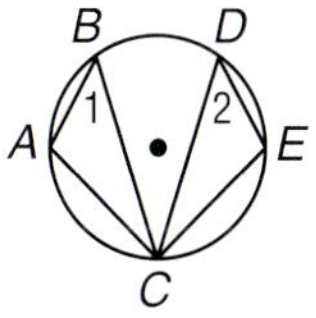

Proof:

Statements	Reasons
1. $\widehat{AB} \cong \widehat{DE}$, $\widehat{AC} \cong \widehat{CE}$	1. Given
2. $m\widehat{AB} = m\widehat{DE}$, $m\widehat{AC} = m\widehat{CE}$	2. Def. of $\cong$ arcs
3. $\frac{1}{2}m\widehat{AB} = \frac{1}{2}m\widehat{DE}$, $\frac{1}{2}m\widehat{AC} = \frac{1}{2}m\widehat{CE}$	3. Mult. Prop.
4. $m\angle ACB = \frac{1}{2}m\widehat{AB}$, $m\angle ECD = \frac{1}{2}m\widehat{DE}$, $m\angle 1 = \frac{1}{2}m\widehat{AC}$, $m\angle 2 = \frac{1}{2}m\widehat{CE}$	4. Inscribed Angle Theorem
5. $m\angle ACB = m\angle ECD$, $m\angle 1 = m\angle 2$	5. Substitution
6. $\angle ACB \cong \angle ECD$, $\angle 1 \cong \angle 2$	6. Def. of $\cong$ $\angle$s
7. $\overline{AB} \cong \overline{DE}$	7. $\cong$ arcs have $\cong$ chords.
8. $\triangle ABC \cong \triangle EDC$	8. AAS

13. $m\angle 1 = m\angle 2 = 13$ **15.** $m\angle 1 = 51$, $m\angle 2 = 90$, $m\angle 3 = 39$ **17.** 45, 30, 120 **19.** $m\angle B = 120$, $m\angle C = 120$, $m\angle D = 60$ **21.** Sample answer: $\overline{EF}$ is a diameter of circle and a diagonal and angle bisector of $EDFG$. **23.** 72 **25.** 144 **27.** 162 **29.** 9 **31.** $\frac{8}{9}$ **33.** 1

35. Given: T lies inside $\angle PRQ$. $\overline{RK}$ is a diameter of $\odot T$.
Prove: $m\angle PRQ = \frac{1}{2}m\widehat{PKQ}$

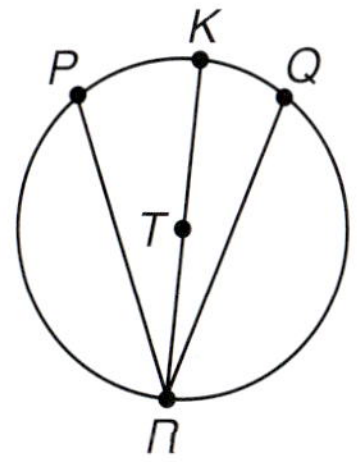

Proof:

Statements	Reasons
1. $m\angle PRQ = m\angle PRK + m\angle KRQ$	1. Angle Addition Theorem
2. $m\widehat{PKQ} = m\widehat{PK} + m\widehat{KQ}$	2. Arc Addition Theorem
3. $\frac{1}{2}m\widehat{PKQ} = \frac{1}{2}m\widehat{PK} + \frac{1}{2}m\widehat{KQ}$	3. Multiplication Property

Selected Answers

4. $m\angle PRK = \frac{1}{2}m\widehat{PK}$, $m\angle KRQ = \frac{1}{2}m\widehat{KQ}$	4. The measure of an inscribed angle whose side is a diameter is half the measure of the intercepted arc (Case 1).
5. $\frac{1}{2}m\widehat{PKQ} = m\angle PRK + m\angle KRQ$	5. Substitution (Steps 3, 4)
6. $\frac{1}{2}m\widehat{PKQ} = m\angle PRQ$	6. Substitution (Steps 5, 1)

37. Given: inscribed $\angle MLN$ and $\angle CED$, $\overline{CD} \cong \overline{MN}$
Prove: $\angle CED \cong \angle MLN$

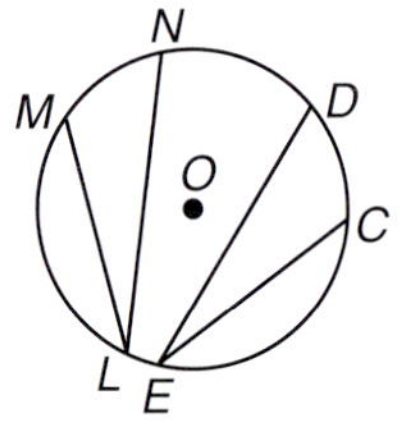

Proof:

Statements	Reasons
1. $\angle MLN$ and $\angle CED$ are inscribed; $\overline{CD} \cong \overline{MN}$	1. Given
2. $m\angle MLN = \frac{1}{2}m\widehat{MN}$; $m\angle CED = \frac{1}{2}m\widehat{CD}$	2. Measure of an inscribed $\angle$ = half measure of intercepted arc.
3. $m\widehat{CD} = m\widehat{MN}$	3. Def. of $\cong$ arcs
4. $\frac{1}{2}m\widehat{CD} = \frac{1}{2}m\widehat{MN}$	4. Mult. Prop.
5. $m\angle CED = m\angle MLN$	5. Substitution
6. $\angle CED \cong \angle MLN$	6. Def. of $\cong$ $\angle$s

39. Given: quadrilateral $ABCD$ inscribed in $\odot O$
Prove: $\angle A$ and $\angle C$ are supplementary. $\angle B$ and $\angle D$ are supplementary.

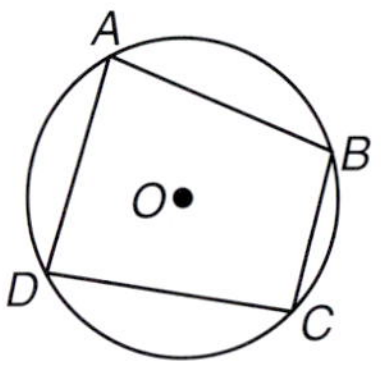

Proof: By arc addition and the definitions of arc measure and the sum of central angles, $m\widehat{DCB} + m\widehat{DAB} = 360$. Since $m\angle C = \frac{1}{2}m\widehat{DAB}$ and $m\angle A = \frac{1}{2}m\widehat{DCB}$, $m\angle C + m\angle A = \frac{1}{2}(m\widehat{DCB} + m\widehat{DAB})$, but $m\widehat{DCB} + m\widehat{DAB} = 360$, so $m\angle C + m\angle A = \frac{1}{2}(360)$ or 80. This makes $\angle C$ and $\angle A$ supplementary. Because the sum of the measures of the interior angles of a quadrilateral is 360, $m\angle A + m\angle C + m\angle B + m\angle D = 360$. But $m\angle A + m\angle C = 180$, so $m\angle B + m\angle D = 180$, making them supplementary also.

41. Isosceles right triangle because sides are congruent radii making it isosceles and $\angle AOC$ is a central angle for an arc of 90°, making it a right angle. **43.** Square because each angle intercepts a semicircle, making them 90° angles. Each side is a chord of congruent arcs, so the chords are congruent.
45. Sample answer: The socket is similar to an inscribed polygon because the vertices of the hexagon can be placed on a circle that is concentric with the outer circle of the socket. Answers should include the following.
- An inscribed polygon is one in which all of its vertices are points on a circle.
- The side of the regular hexagon inscribed in a circle $\frac{3}{4}$ inch wide is $\frac{3}{8}$ inch.

47. 234 **49.** $\sqrt{135} \approx 11.62$ **51.** 4π units **53.** always **55.** sometimes **57.** no

Page 552–558 Lesson 10-5

1a. Two; from any point outside the circle, you can draw only two tangents. **1b.** None; a line containing a point inside the circle would intersect the circle in two points. A tangent can only intersect a circle in one point. **1c.** One; since a tangent intersects a circle in exactly one point, there is one tangent containing a point on the circle.
3. Sample answer:

polygon circumscribed about a circle

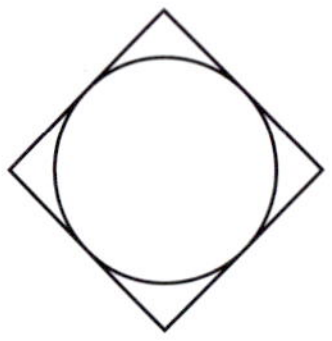

polygon inscribed in a circle

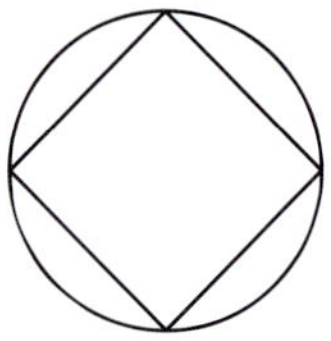

5. Yes; $5^2 + 12^2 = 13^2$ **7.** 576 ft **9.** no **11.** yes **13.** 16 **15.** 12 **17.** 3 **19.** 30 **21.** See students' work. **23.** 60 units **25.** $15\sqrt{3}$ units

27. Given: $\overline{AB}$ is tangent to $\odot X$ at B. $\overline{AC}$ is tangent to $\odot X$ at C.
Prove: $\overline{AB} \cong \overline{AC}$

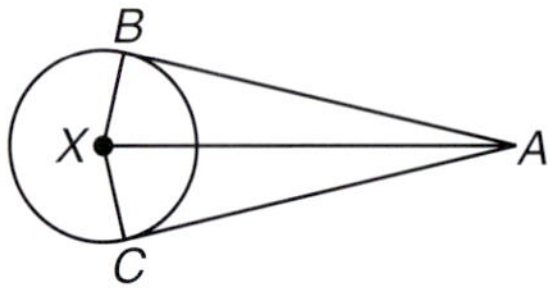

Proof:

Statements	Reasons
1. $\overline{AB}$ is tangent to $\odot X$ at B. $\overline{AC}$ is tangent to $\odot X$ at C.	1. Given
2. Draw $\overline{BX}$, $\overline{CX}$, and $\overline{AX}$.	2. Through any two points, there is one line.
3. $\overline{AB} \perp \overline{BX}$, $\overline{AC} \perp \overline{CX}$	3. Line tangent to a circle is $\perp$ to the radius at the pt. of tangency.
4. $\angle ABX$ and $\angle ACX$ are right angles.	4. Def. of $\perp$ lines
5. $\overline{BX} \cong \overline{CX}$	5. All radii of a circle are $\cong$.
6. $\overline{AX} \cong \overline{AX}$	6. Reflexive Prop.
7. $\triangle ABX \cong \triangle ACX$	7. HL
8. $\overline{AB} \cong \overline{AC}$	8. CPCTC

29. $\overline{AE}$ and $\overline{BF}$
31. 12; Draw $\overline{PG}$, $\overline{NL}$, and $\overline{PL}$. Construct $\overline{LQ} \perp \overline{GP}$, thus $LQGN$ is a rectangle. $GQ = NL = 4$, so $QP = 5$. Using the Pythagorean Theorem, $(QP)^2 + (QL)^2 = (PL)^2$. So, $QL = 12$. Since $GN = QL$, $GN = 12$.

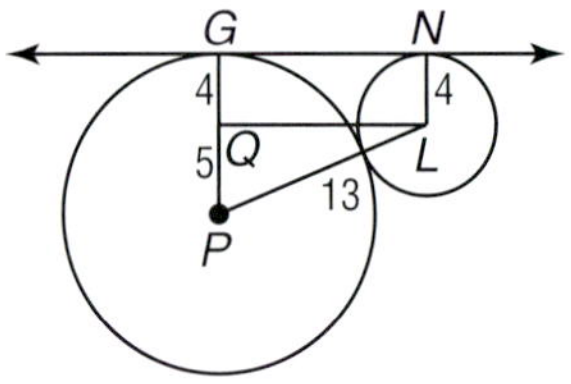

33. 27 **35.** $\overleftrightarrow{AD}$ and $\overleftrightarrow{BC}$ **37.** 45, 45 **39.** 4
41. Sample answer:
Given: $ABCD$ is a rectangle. E is the midpoint of $\overline{AB}$.
Prove: $\triangle CED$ is isosceles.

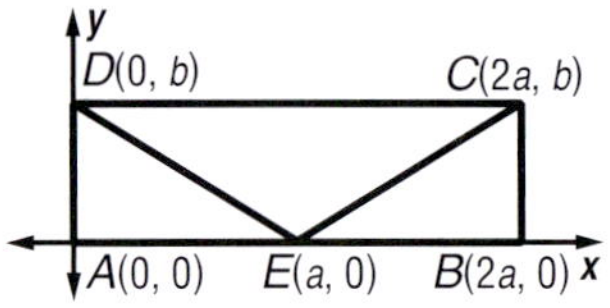

Selected Answers

Proof: Let the coordinates of E be $(a, 0)$. Since E is the midpoint and is halfway between A and B, the coordinates of B will be $(2a, 0)$. Let the coordinates of D be $(0, b)$. The coordinates of C will be $(2a, b)$ because it is on the same horizontal as D and the same vertical as B.

$ED = \sqrt{(a-0)^2 + (0-b)^2} = \sqrt{a^2 + b^2}$ $\quad EC = \sqrt{(a-2a)^2 + (0-b)^2} = \sqrt{a^2 + b^2}$

Since $ED = EC$, $\overline{ED} \cong \overline{EC}$. $\triangle DEC$ has two congruent sides, so it is isosceles.

43. 6 **45.** 20.5

Page 561–568 Lesson 10-6

1. Sample answer: A tangent intersects the circle in only one point and no part of the tangent is in the interior of the circle. A secant intersects the circle in two points and some of its points do lie in the interior of the circle. **3.** 138 **5.** 20 **7.** 235 **9.** 55 **11.** 110 **13.** 60 **15.** 110 **17.** 90 **19.** 50 **21.** 30 **23.** 8 **25.** 4 **27.** 25 **29.** 130 **31.** 10 **33.** 141 **35.** 44 **37.** 118 **39.** about 103 ft **41.** 4.6 cm

43a. Given: $\overleftrightarrow{AB}$ is a tangent to $\odot O$. $\overrightarrow{AC}$ is a secant to $\odot O$. $\angle CAB$ is acute.

Prove: $m\angle CAB = \frac{1}{2}m\widehat{CA}$

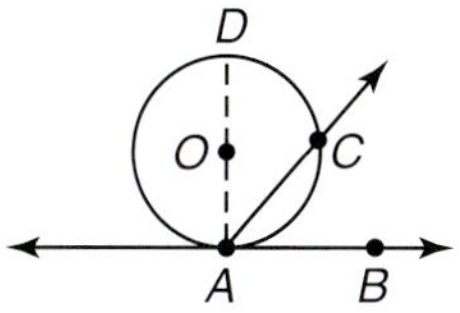

Proof: $\angle DAB$ is a right $\angle$ with measure 90, and $\widehat{DCA}$ is a semicircle with measure 180, since if a line is tangent to a $\odot$, it is $\perp$ to the radius at the point of tangency. Since $\angle CAB$ is acute, C is in the interior of $\angle DAB$, so by the Angle and Arc Addition Postulates, $m\angle DAB = m\angle DAC + m\angle CAB$ and $m\widehat{DCA} = m\widehat{DC} + m\widehat{CA}$. By substitution, $90 = m\angle DAC + m\angle CAB$ and $180 = m\widehat{DC} + m\widehat{CA}$. So, $90 = \frac{1}{2}m\widehat{DC} + \frac{1}{2}m\widehat{CA}$ by Division Prop., and $m\angle DAC + m\angle CAB = \frac{1}{2}m\widehat{DC} + \frac{1}{2}m\widehat{CA}$ by substitution. $m\angle DAC = \frac{1}{2}m\widehat{DC}$ since $\angle DAC$ is inscribed, so substitution yields $\frac{1}{2}m\widehat{DC} + m\angle CAB = \frac{1}{2}m\widehat{DC} + \frac{1}{2}m\widehat{CA}$. By Subtraction Prop., $m\angle CAB = \frac{1}{2}m\widehat{CA}$.

43b. Given: $\overleftrightarrow{AB}$ is a tangent to $\odot O$. $\overrightarrow{AC}$ is a secant to $\odot O$. $\angle CAB$ is obtuse.

Prove: $m\angle CAB = \frac{1}{2}m\widehat{CDA}$

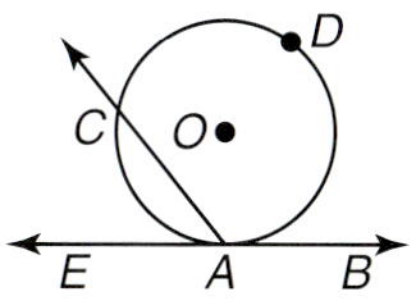

Proof: $\angle CAB$ and $\angle CAE$ form a linear pair, so $m\angle CAB + m\angle CAE = 180$. Since $\angle CAB$ is obtuse, $\angle CAE$ is acute and Case 1 applies, so $m\angle CAE = \frac{1}{2}m\widehat{CA}$. $m\widehat{CA} + m\widehat{CDA} = 360$, so $\frac{1}{2}m\widehat{CA} + \frac{1}{2}m\widehat{CDA} = 180$ by Divison Prop., and $m\angle CAE + \frac{1}{2}m\widehat{CDA} = 180$ by substitution. By the Transitive Prop., $m\angle CAB + m\angle CAE = m\angle CAE + \frac{1}{2}m\widehat{CDA}$, so by Subtraction Prop., $m\angle CAB = \frac{1}{2}m\widehat{CDA}$.

45. $\angle 3, \angle 1, \angle 2$; $m\angle 3 = m\widehat{RQ}$, $m\angle 1 = \frac{1}{2}m\widehat{RQ}$ so $m\angle 3 > m\angle 1$, $m\angle 2 = \frac{1}{2}(m\widehat{RQ} - m\widehat{TP}) = \frac{1}{2}m\widehat{RQ} - \frac{1}{2}m\widehat{TP}$, which is less than $\frac{1}{2}m\widehat{RQ}$, so $m\angle 2 < m\angle 1$. **47.** A **49.** 16 **51.** 33 **53.** 44.5 **55.** 30 in. **57.** 4, −10 **59.** 3, 5

Page 568 Chapter 10 Practice Quiz 2

1. 67.5 **3.** 12 **5.** 115.5

Page 569–574 Lesson 10-7

1. Sample answer: The product equation for secant segments equates the product of exterior segment measure and the whole segment measure for each secant. In the case of secant-tangent, the product involving the tangent segment becomes (measure of tangent segment)^2 because the exterior segment and the whole segment are the same segment.

3. Sample answer:

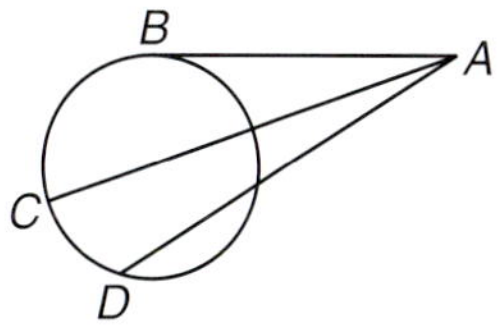

5. 28.1 **7.** ≈7 : 3.54 **9.** 4 **11.** 2 **13.** 6 **15.** 3.2 **17.** 4 **19.** 5.6

21. Given: $\overline{WY}$ and $\overline{ZX}$ intersect at T.

Prove: $WT \cdot TY = ZT \cdot TX$

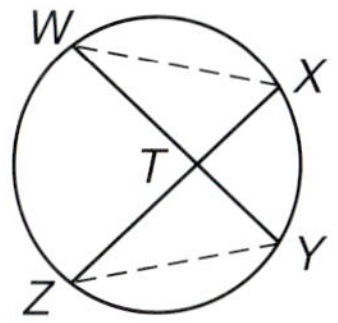

Proof:

Statements	**Reasons**
a. $\angle W \cong \angle Z$, $\angle X \cong \angle Y$	**a.** Inscribed angles that intercept the same arc are congruent.
b. $\triangle WXT \sim \triangle ZYT$	**b.** AA Similarity
c. $\frac{WT}{ZT} = \frac{TX}{TY}$	**c.** Definition of similar triangles
d. $WT \cdot TY = ZT \cdot TX$	**d.** Cross products

23. 4 **25.** 11 **27.** 14.3 **29.** $113.\overline{3}$ cm

31. Given: tangent $\overline{RS}$ and secant $\overline{US}$

Prove: $(RS)^2 = US \cdot TS$

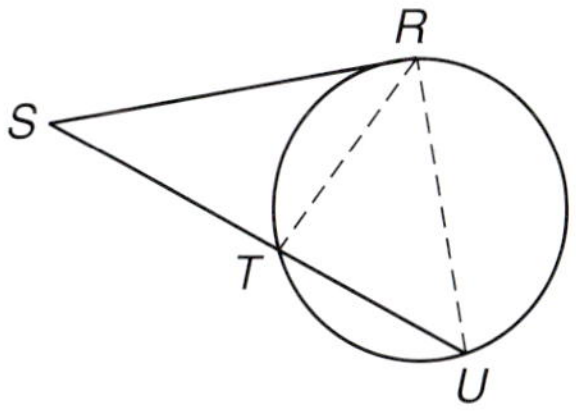

Proof:

Statements	**Reasons**
1. tangent $\overline{RS}$ and secant $\overline{US}$	**1.** Given
2. $m\angle RUT = \frac{1}{2}m\widehat{RT}$	**2.** The measure of an inscribed angle equals half the measure of its intercepted arc.
3. $m\angle SRT = \frac{1}{2}m\widehat{RT}$	**3.** The measure of an angle formed by a secant and a tangent equals half the measure of its intercepted arc.
4. $m\angle RUT = m\angle SRT$	**4.** Substitution

5. $\angle RUT \cong \angle SRT$	5. Definition of $\cong$ $\angle$s
6. $\angle S \cong \angle S$	6. Reflexive Prop.
7. $\triangle SUR \sim \triangle SRT$	7. AA Similarity
8. $\frac{RS}{US} = \frac{TS}{RS}$	8. Definition of $\sim$ $\triangle$s
9. $(RS)^2 = US \cdot TS$	9. Cross products

33. Sample answer: The product of the parts of one intersecting chord equals the product of the parts of the other chord. Answers should include the following.

- $\overline{AF}, \overline{FD}, \overline{EF}, \overline{FB}$
- $AF \cdot FD = EF \cdot FB$

35. C **37.** 157.5 **39.** 7 **41.** 36 **43.** scalene, obtuse **45.** equilateral, acute or equiangular **47.** $\sqrt{13}$

Pages 575–580 Lesson 10-8

1. Sample answer:

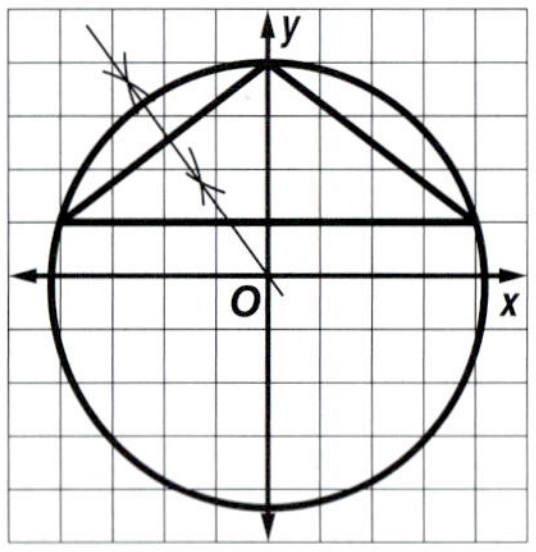

3. $(x + 3)^2 + (y - 5)^2 = 100$
5. $(x + 2)^2 + (y - 11)^2 = 32$
7.

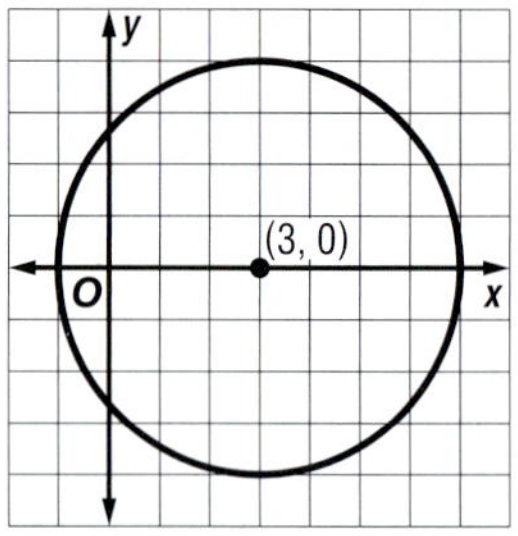

9. $x^2 + y^2 = 1600$ **11.** $(x + 2)^2 + (y + 8)^2 = 25$
13. $x^2 + y^2 = 36$ **15.** $x^2 + (y - 5)^2 = 100$
17. $(x + 3)^2 + (y + 10)^2 = 144$ **19.** $x^2 + y^2 = 8$
21. $(x + 2)^2 + (y - 1)^2 = 10$ **23.** $(x - 7)^2 + (y - 8)^2 = 25$

25.

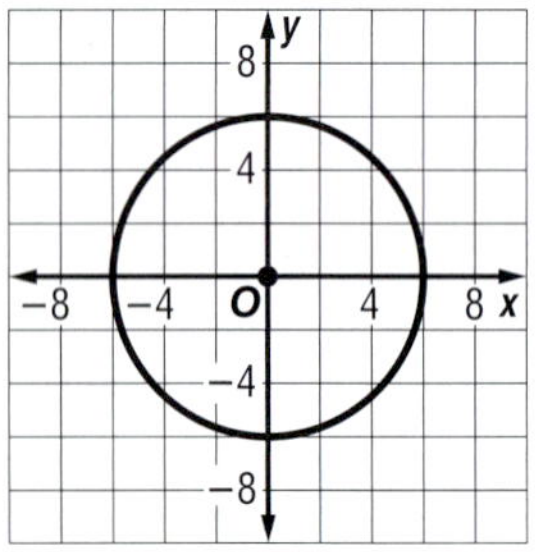

27.

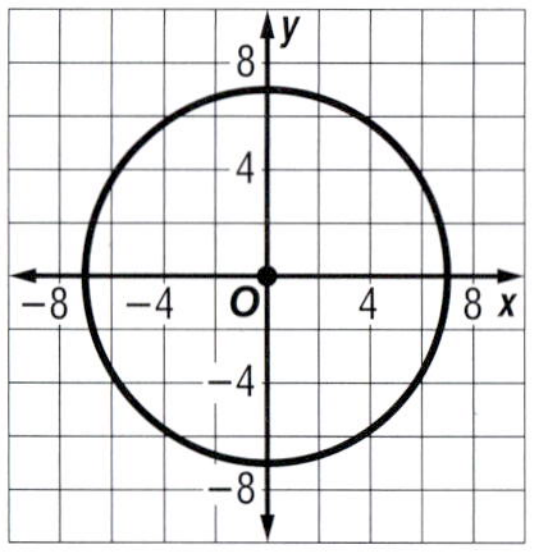

29.

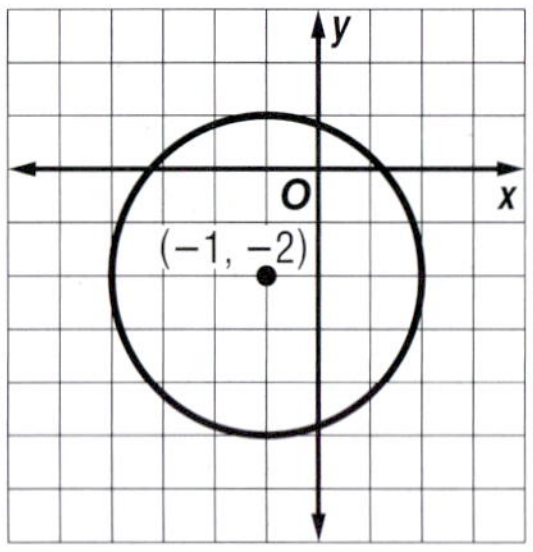

31. $(x + 3)^2 + y^2 = 9$ **33.** 2 **35.** $x^2 + y^2 = 49$ **37.** 13 **39.** $(2, -4)$; $r = 6$ **41.** See students' work **43a.** (0, 3) or (−3, 0) **43b.** none **43c.** (0, 0) **45.** B **47.** 24 **49.** 18 **51.** 59 **53.** 20 **55.** (3, 2), (−4, −1), (0, −4)

Pages 581–586 Chapter 10 Study Guide and Review

1. a **3.** h **5.** b **7.** d **9.** c **11.** 7.5 in.; 47.12 in. **13.** 10.82 yd; 21.65 yd **15.** 21.96 ft; 43.93 ft **17.** 60 **19.** 117 **21.** 30 **23.** 30 **25.** 150 **27.** $\frac{22}{5}\pi$ **29.** 10 **31.** 10 **33.** 45 **35.** 48 **37.** 32 **39.** $m\angle 1 = m\angle 3 = 30$, $m\angle 2 = 60$ **41.** 9 **43.** 18 **45.** 37 **47.** 17.1 **49.** 7.2 **51.** $(x + 4)^2 = (y - 8)^2 = 9$ **53.** $(x + 1)^2 + (y - 4)^2 = 4$

55.

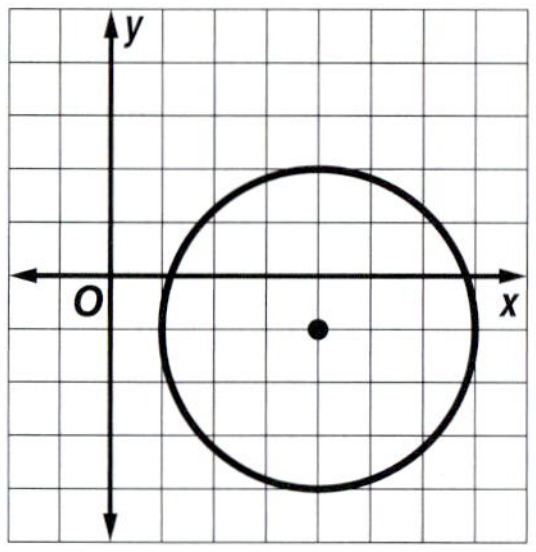

57.

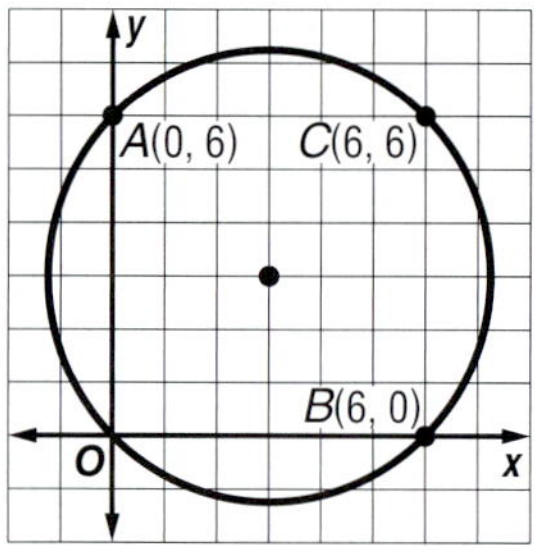

Chapter 11 Areas of Polygons and Circles

Page 593 Chapter 11 Getting Started

1. 10 **3.** 4.6 **5.** 18 **7.** 54 **9.** 13 **11.** 9 **13.** $6\sqrt{3}$ **15.** $\frac{15\sqrt{2}}{2}$

Pages 598–600 Lesson 11-1

1. The area of a rectangle is the product of the length and the width. The area of a parallelogram is the product of the base and the height. For both quadrilaterals, the measure of the length of one side is multiplied by the length of the altitude. **3.** 28 ft; 39.0 ft^2 **5.** 12.8 m; 10.2 m^2 **7.** rectangle, 170 units^2 **9.** 80 in.; 259.8 in^2 **11.** 21.6 cm; 29.2 cm^2 **13.** 44 m; 103.9 m^2 **15.** 45.7 mm^2 **17.** 108.5 m **19.** h = 40 units, b = 50 units **21.** parallelogram, 56 units^2 **23.** parallelogram, 64 units^2 **25.** square, 13 units^2 **27.** 150 units^2 **29.** Yes; the dimensions are 32 in. by 18 in. **31.** $\approx$ 13.9 ft **33.** The perimeter is 19 m, half of 38 m. The area is 20 m^2. **35.** 5 in., 7 in. **37.** C **39.** (5, 2), $r = 7$ **41.** $\left(-\frac{2}{3}, \frac{1}{9}\right)$, $r = \frac{2}{3}$ **43.** 32 **45.** 21 **47.** $F''(-4, 0)$, $G''(-2, -2)$, $H''(-2, 2)$; 90° counterclockwise **49.** 13 ft **51.** 16 **53.** 20

Pages 605–609 Lesson 11-2

1. Sample answer:

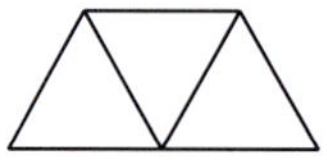

3. Sometimes; two rhombi can have different corresponding diagonal lengths and have the same area. **5.** 499.5 in^2

7. 21 units^2 **9.** 4 units^2 **11.** 45 m **13.** 12.4 cm^2 **15.** 95 km^2 **17.** 1200 ft^2 **19.** 50 m^2 **21.** 129.9 mm^2 **23.** 55 units^2 **25.** 22.5 units^2 **27.** 20 units^2 **29.** 16 units^2 **31.** $\approx$ 26.8 ft **33.** $\approx$ 22.6 m **35.** 20 cm **37.** about 8.7 ft **39.** 13,326 ft^2 **41.** 120 in^2 **43.** $\approx$ 10.8 in^2 **45.** 21 ft^2 **47.** False; sample answer: the area for each of these right triangles is 6 square units. The perimeter of one triangle is 12 and the perimeter of the other is $8 + \sqrt{40}$ or about 14.3.

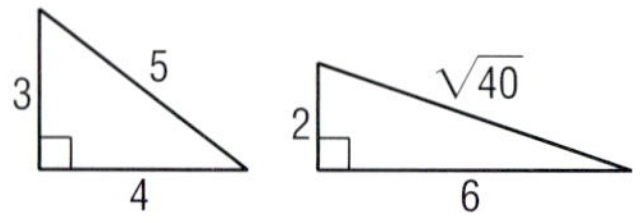

49. area = 12, area = 3; perimeter = $8\sqrt{13}$, perimeter = $4\sqrt{13}$; scale factor and ratio of perimeters = $\frac{1}{2}$, ratio of areas = $\left(\frac{1}{2}\right)^2$ **51.** $\frac{2}{1}$ **53.** The ratio is the same. **55.** 4 : 1; The ratio of the areas is the square of the scale factor. **57.** 45 ft^2; The ratio of the areas is 5 : 9. **59.** B **61.** area = $\frac{1}{2}ab \sin C$ **63.** 6.02 cm^3 **65.** 374 cm^2

Selected Answers

67. 231 ft^2 **69.** $(x + 4)^2 + \left(y - \frac{1}{2}\right)^2 = \frac{121}{4}$ **71.** 275 in.
73. $\langle 172.4, 220.6\rangle$ **75.** 20.1

Page 609 Practice Quiz 1

1. square **3.** 54 units^2 **5.** 42 yd

Pages 613–616 Lesson 11-3

1. Sample answer: Separate a hexagon inscribed in a circle into six congruent nonoverlapping isosceles triangles. The area of one triangle is one-half the product of one side of the hexagon and the apothem of the hexagon. The area of the hexagon is $6\left(\frac{1}{2}sa\right)$. The perimeter of the hexagon is $6s$, so the formula is $\frac{1}{2}Pa$. **3.** 127.3 yd^2 **5.** 10.6 cm^2 **7.** about 3.6 yd^2 **9.** 882 m^2 **11.** 1995.3 in^2 **13.** 482.8 km^2 **15.** 30.4 units^2 **17.** 26.6 units^2 **19.** 4.1 units^2 **21.** 271.2 units^2 **23.** 2 : 1 **25.** One 16-inch pizza; the area of the 16-inch pizza is greater than the area of two 8-inch pizzas, so you get more pizza for the same price. **27.** 83.1 units^2 **29.** 45.3 units^2 **31.** 227.0 units^2 **33.** 664.8 units^2 **35.** triangles; 629 tiles **37.** $\approx$ 380.1 in^2 **39.** 34.6 units^2 **41.** 157.1 units^2 **43.** 471.2 units^2 **45.** 54,677.8 ft^2; 899.8 ft **47.** $225\pi \approx 706.9$ ft^2 **49.** 2 : 3 **51.** The ratio is the same. **53.** The ratio of the areas is the square of the scale factor. **55.** 3 to 4 **57.** B **59.** 260 cm^2 **61.** $\approx$ 2829.0 yd^2 **63.** square; 36 units^2 **65.** rectangle; 30 units^2 **67.** 42 **69.** 6 **71.** $4\sqrt{2}$

Pages 619–621 Lesson 11-4

1. Sample answer: $\approx$ 18.3 units^2

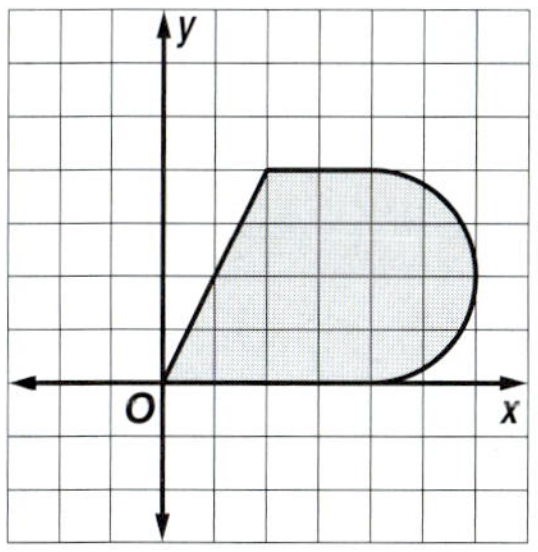

3. 53.4 units^2 **5.** 24 units^2 **7.** $\approx$ 1247.4 in^2 **9.** 70.9 units^2 **11.** 4185 units^2 **13.** 154.1 units^2 **15.** $\approx$ 2236.9 in^2 **17.** 23.1 units^2 **19.** 21 units^2 **21.** 33 units^2 **23.** Sample answer: 57,500 mi^2 **25.** 462 **27.** Sample answer: Reduce the width of each rectangle.

29. Sample answer: Windsurfers use the area of the sail to catch the wind and stay afloat on the water. Answers should include the following.

- To find the area of the sail, separate it into shapes. Then find the area of each shape. The sum of areas is the area of the sail.
- Sample answer: Surfboards and sailboards are also irregular figures.

31. C **33.** 154.2 units^2 **35.** 156.3 ft^2 **37.** $\approx$ 384 m^2 **39.** 0.63 **41.** 0.19

Page 621 Practice Quiz 2

1. 679.0 mm^2 **3.** 1208.1 units^2 **5.** 44.5 units^2

Pages 625–627 Lesson 11-5

1. Multiply the measure of the central angle of the sector by the area of the circle and then divide the product by 360°. **3.** Rachel; Taimi did not multiply $\frac{62}{360}$ by the area of the circle. **5.** $\approx$ 114.2 units^2, $\approx$ 0.36 **7.** 0.60 **9.** 0.54 **11.** $\approx$ 58.9 units^2, $0.\overline{3}$ **13.** $\approx$ 19.6 units^2, $0.\overline{1}$ **15.** 74.6 units^2, 0.42 **17.** $\approx$ 3.3 units^2, $\approx$ 0.03 **19.** $\approx$ 25.8 units^2, $\approx$ 0.15 **21.** 0.68 **23.** 0.68 **25.** 0.19 **27.** $\approx$ 0.29 **29.** The chances of landing on a black or white sector are the same, so they should have the same point value. **31a.** No; each colored sector has a different central angle. **31b.** No; there is not an equal chance of landing on each color. **33.** C **35.** 1050 units^2 **37.** 110.9 ft^2 **39.** 221.7 in^2 **41.** 123 **43.** 165 **45.** $g = 21.5$

Pages 628–630 Chapter 11 Study Guide and Review

1. c **3.** a **5.** b **7.** 78 ft, $\approx$ 318.7 ft^2 **9.** square; 49 units^2 **11.** parallelogram; 20 units^2 **13.** 28 in. **15.** 688.2 in^2 **17.** 31.1 units^2 **19.** $0.\overline{3}$

Chapter 12 Surface Area

Page 635 Chapter 12 Getting Started

1. true **3.** cannot be determined **5.** 384 ft^2 **7.** 1.8 m^2 **9.** 7.1 yd^2

Pages 639–642 Lesson 12-1

1. The Platonic solids are the five regular polyhedra. All of the faces are congruent, regular polygons. In other polyhedra, the bases are congruent parallel polygons, but the faces are not necessarily congruent.
3. Sample answer:

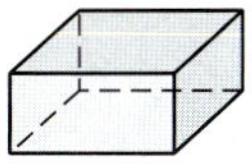

5. Hexagonal pyramid; base: *ABCDEF*; faces: *ABCDEF*, $\triangle AGF$, $\triangle FGE$, $\triangle EGD$, $\triangle DGC$, $\triangle CGB$, $\triangle BGA$; edges: $\overline{AF}$, $\overline{FE}$, $\overline{ED}$, $\overline{DC}$, $\overline{CB}$, $\overline{BA}$, $\overline{AG}$, $\overline{FG}$, $\overline{EG}$, $\overline{DG}$, $\overline{CG}$, and $\overline{BG}$; vertices: *A*, *B*, *C*, *D*, *E*, *F*, and *G* **7.** cylinder; bases: circles *P* and *Q*

9.

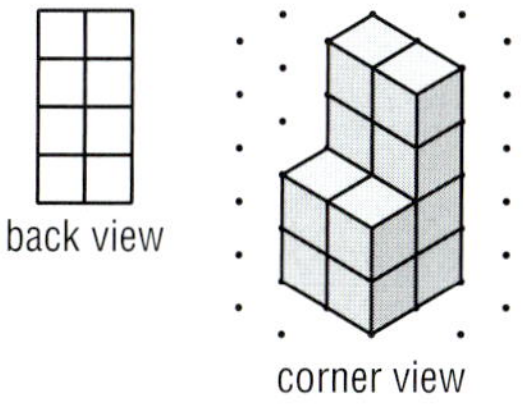

11.

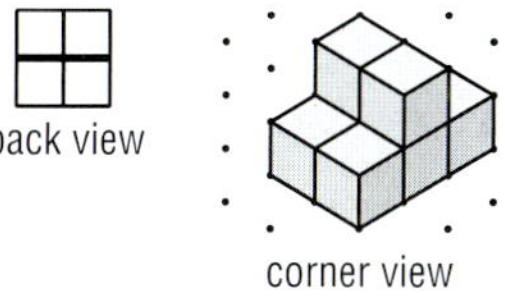

13.

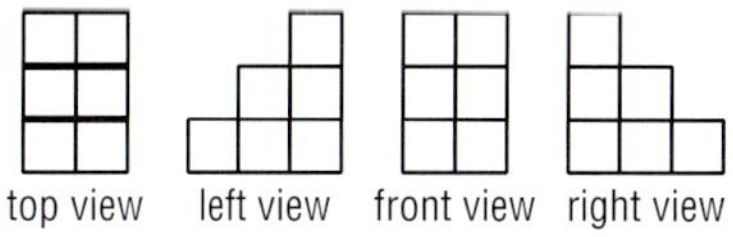

15.

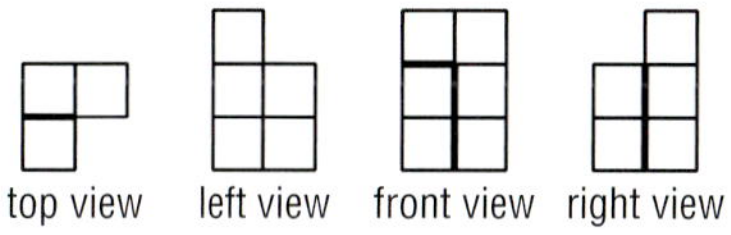

17. rectangular pyramid; base: $\square DEFG$; faces: $\square DEFG$, $\triangle DHG$, $\triangle GHF$, $\triangle FHE$, $\triangle DHE$; edges: $\overline{DG}$, $\overline{GF}$, $\overline{FE}$, $\overline{ED}$, $\overline{DH}$, $\overline{EH}$, $\overline{FH}$, and $\overline{GH}$; vertices: *D*, *E*, *F*, *G*, and *H*
19. cylinder: bases: circles *S* and *T* **21.** cone; base: circle *B*; vertex *A* **23.** No, not enough information is provided by the top and front views to determine the shape.
25. parabola **27.** circle **29.** rectangle

31. intersecting three faces and parallel to base;

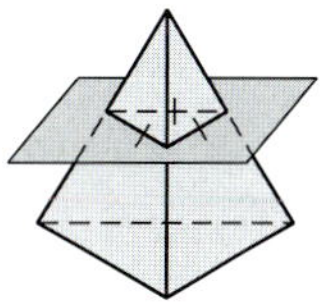

33. intersecting all four faces, not parallel to any face;

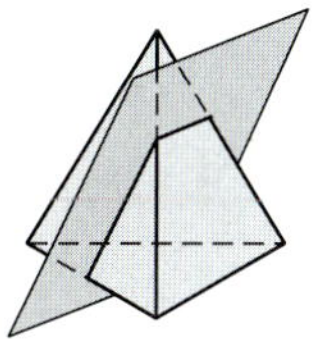

35. cylinder **37.** rectangles, triangles, quadrilaterals

39a. triangular **39b.** cube, rectangular, or hexahedron **39c.** pentagonal **39d.** hexagonal **39e.** hexagonal **41.** No; the number of faces is not enough information to classify a polyhedron. A polyhedron with 6 faces could be a cube, rectangular prism, hexahedron, or a pentagonal pyramid. More information is needed to classify a polyhedron. **43.** Sample answer: Archaeologists use two dimensional drawings to learn more about the structure they are studying. Egyptologists can compare two-dimensional drawings to learn more about the structure they are studying. Egyptologists can compare two-dimensional drawings of the pyramids and note similarities and any differences. Answers should include the following.

- Viewpoint drawings and corner views are types of two-dimensional drawings that show three dimensions.
- To show three dimensions in a drawing, you need to know the views from the front, top, and each side.

45. D **47.** infinite **49.** 0.242 **51.** 0.611 **53.** 21 units^2 **55.** 11 units^2 **57.** 90 ft, 433.0 ft^2 **59.** 300 cm^2 **61.** 4320 in^2

Pages 645–648 Lesson 12-2

1. Sample answer:

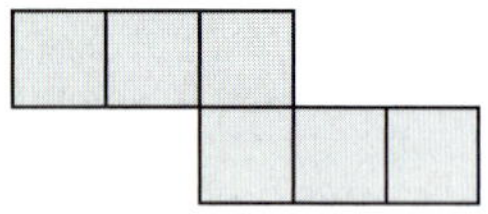

3.

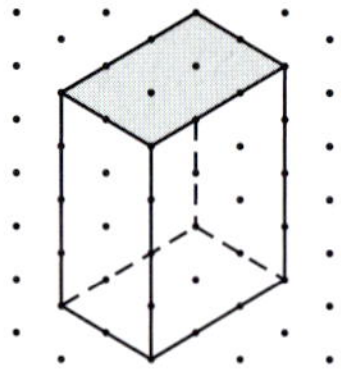

5. 188 in^2;

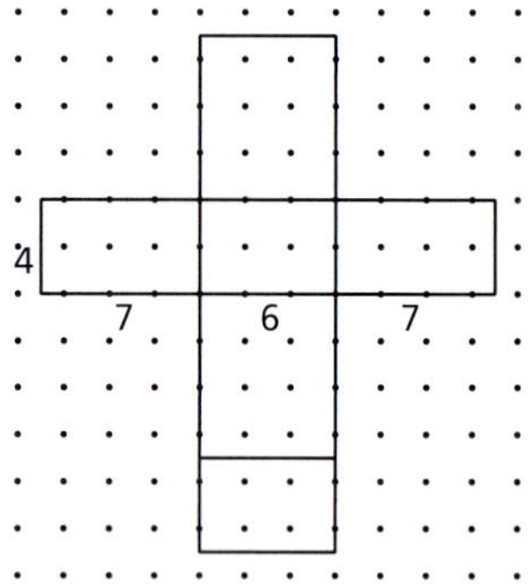

7. 64 cm^2;

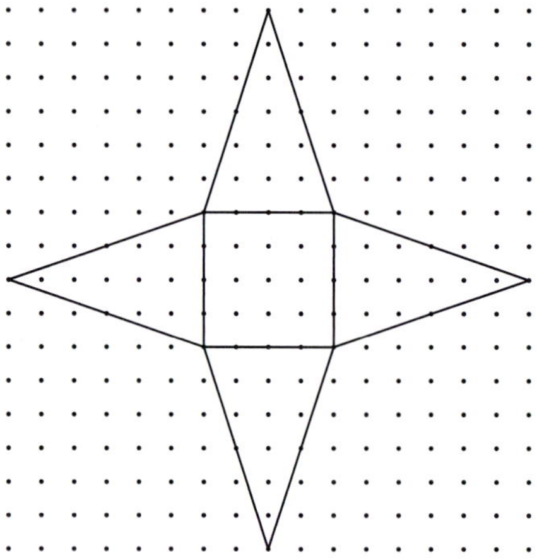

9.

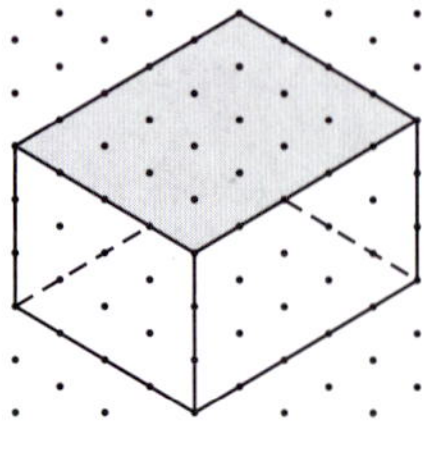

11.

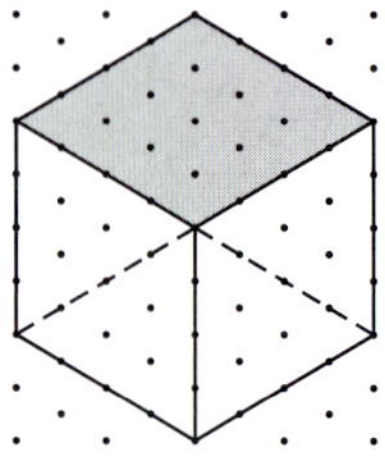

13.

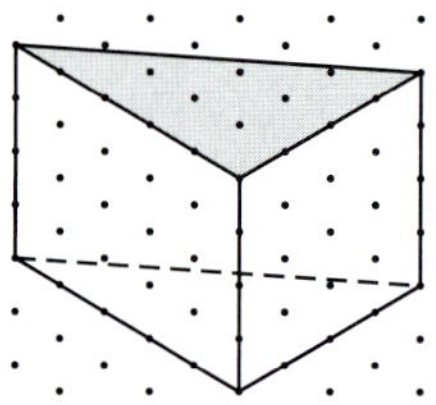

15. 66 units^2;

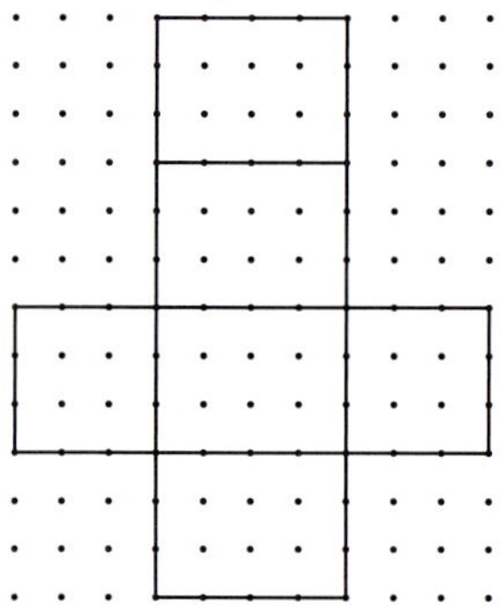

17. 56 units^2;

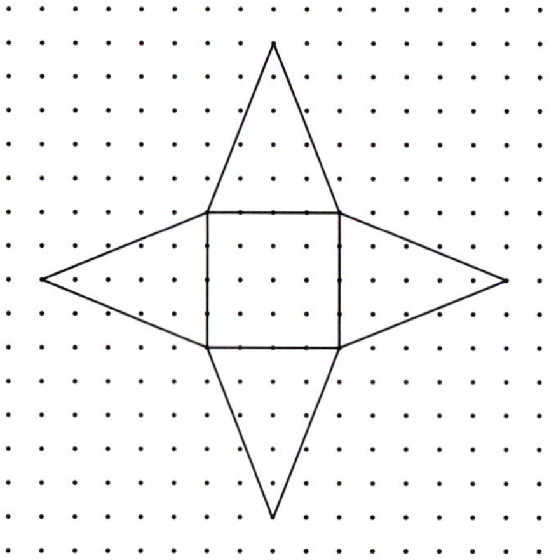

19. 121.5 units^2;

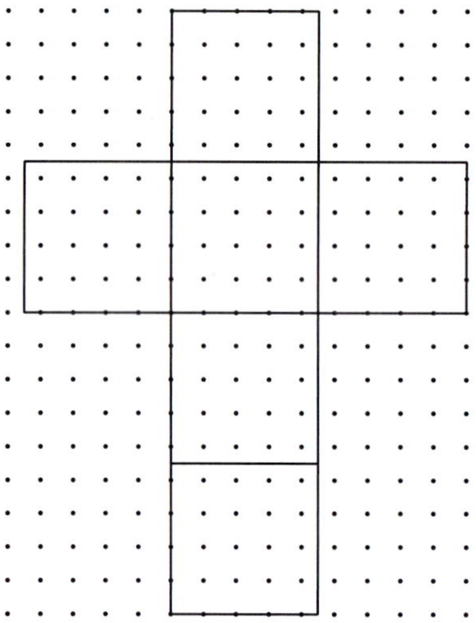

21. 116.3 units^2;

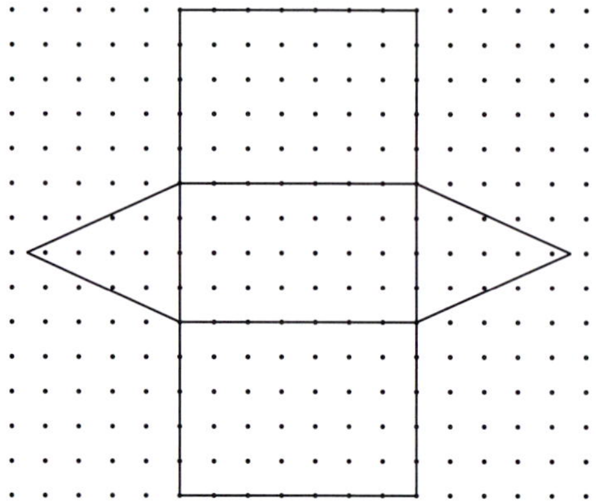

23. 108.2 units^2;

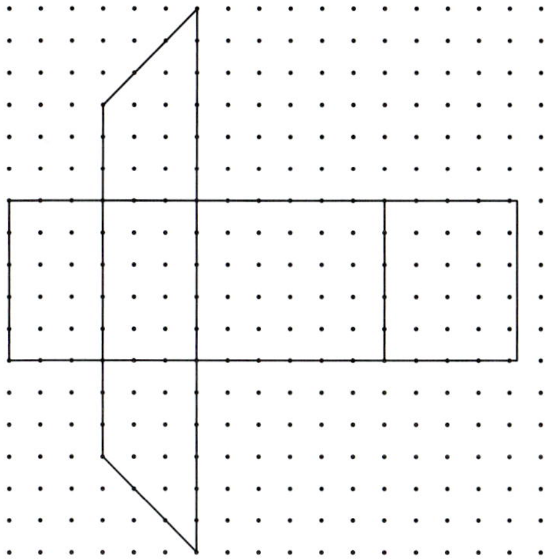

25.

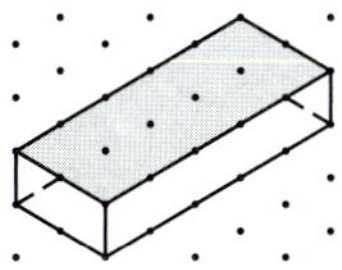

27.

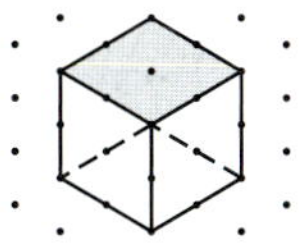

29.

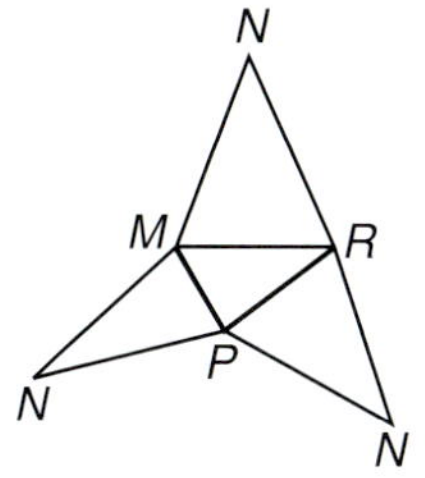

31.

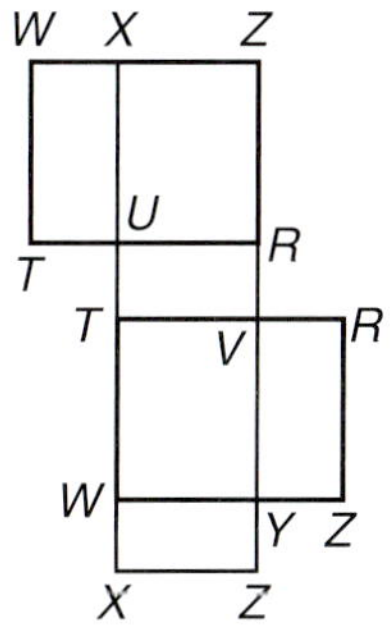

33.

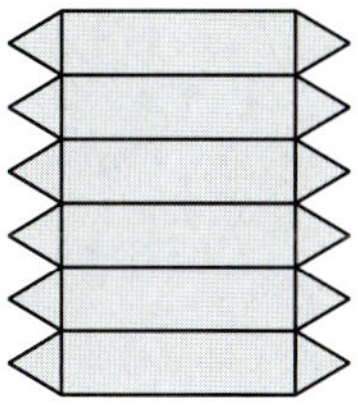

35. **A** 6 units^2;

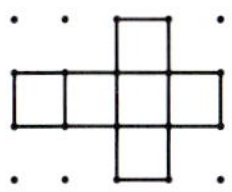

B $\left(9 + \frac{\sqrt{3}}{2}\right) = 9.87$ units^2;

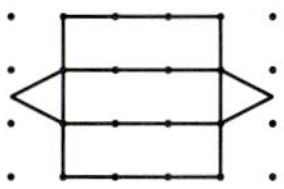

C 76 units^2;

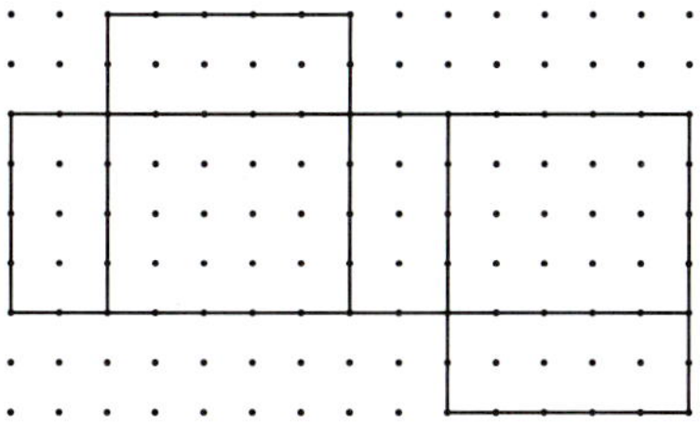

37. The surface area quadruples when the dimensions are doubled. For example, the surface area of the cube is $6(1^2)$ or 6 square units. When the dimensions are doubled the surface area is $6(2^2)$ or 24 square units. **39.** No; 5 and 3 are opposite faces; the sum is 8. **41.** C **43.** rectangle **45.** rectangle **47.** 90 **49.** 120 **51.** 63 cm^2 **53.** 110 cm^2

Pages 651–654 Lesson 12-3

1. In a right prism a lateral edge is also an altitude. In an oblique prism, the lateral edges are not perpendicular to the bases. **3.** 840 units^2, 960 units^2 **5.** 1140 ft^2 **7.** 128 units^2 **9.** 162 units^2 **11.** 160 units^2 (square base), 126 units^2 (rectangular base) **13.** 16 cm **15.** The perimeter of the base must be 24 meters. There are six rectangles with integer values for the dimensions that have a perimeter of 24. The dimensions of the base could be 1×11, 2×10, 3×9, 4×8, 5×7, or 6×6. **17.** 114 units^2 **19.** 522 units^2 **21.** 454.0 units^2 **23.** 3 gallons for 2 coats **25.** 44,550 ft^2 **27.** The actual amount needed will be higher because the area of the curved architectural elements appears to be greater than the area of the doors. **29.** base of A $\cong$ base of C; base of A $\sim$ base of B; base of C $\sim$ base of B **31.** $A : B = 1 : 4$, $B : C = 4 : 1$, $A : C = 1 : 1$ **33.** A : B, because the heights of A and B are in the same ratio as perimeters of bases **35.** No, the surface area of the finished product will be the sum of the lateral areas of each prism plus the area of the bases of the TV and DVD prisms. It will also include the area of the overhang between each prism, but not the area of the overlapping prisms. **37.** 198 cm^2 **39.** B **41.** $L = 1416$ cm^2, $T = 2056$ cm^2 **43.** See students' work.

45. 108 units^2;

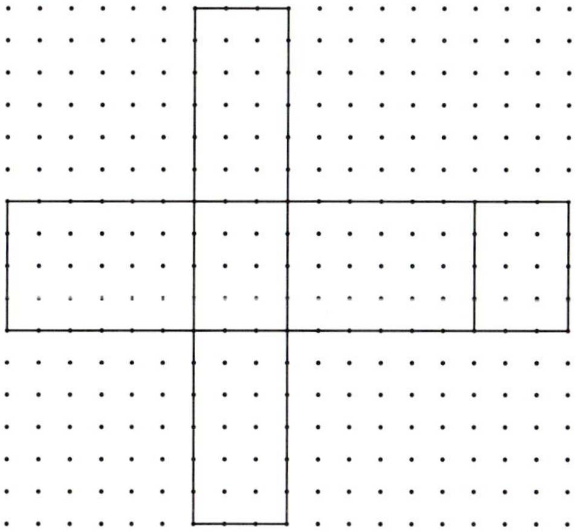

47.

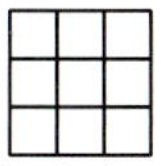

back view

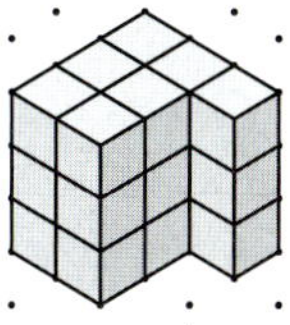

corner view

49. 43
51. 35 **53.** $\frac{1}{72}$
55. 1963.50 in^2
57. 21,124.07 mm^2

Pages 657–659 Lesson 12-4

1. Multiply the circumference of the base by the height and add the area of each base. **3.** Jamie; since the cylinder has one base removed, the surface area will be the sum of the lateral area and one base. **5.** 1520.5 m^2 **7.** 5 ft **9.** 2352.4 m^2 **11.** 517.5 in^2 **13.** 251.3 ft^2 **15.** 30.0 cm^2 **17.** 3 cm **19.** 8 m **21.** The lateral areas will be in the ratio 3 : 2 : 1; 45π in^2, 30π in^2, 15π in^2. **23.** The lateral area is tripled. The surface area is increased, but not tripled. **25.** 1.25 m **27.** Sample answer: Extreme sports participants use a semi-cylinder for a ramp. Answers should include the following.

- To find the lateral area of a semicylinder like the half-pipe, multiply the height by the circumference of the base and then divide by 2.
- A half-pipe ramp is half of a cylinder if the ramp is an equal distance from the axis of the cylinder.

29. C **31.** a plane perpendicular to the line containing the opposite vertices of the face of the cube

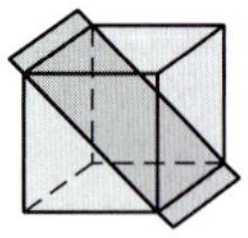

33. 300 units^2

35.

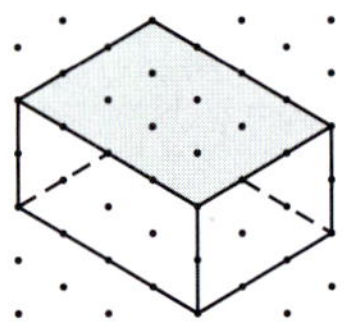

37. 27 **39.** 8 **41.** $m\angle A = 64, b \approx 12.2, c \approx 15.6$ **43.** 54 cm^2

Page 659 Practice Quiz 1

1.

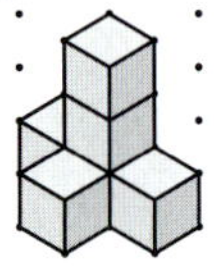

corner view

3. 231.5 m^2 **5.** 5.4 ft

Pages 663–665 Lesson 12-5

1. Sample answer:

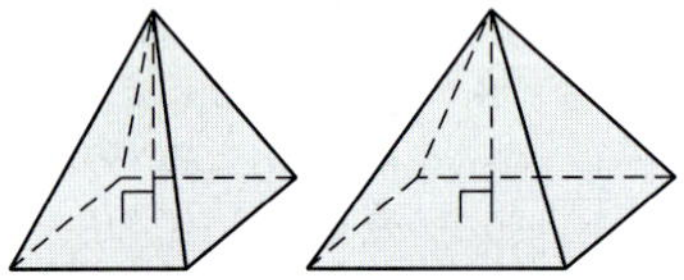

square base (regular) rectangular base (not regular)

3. 74.2 ft^2 **5.** 340 cm^2 **7.** 119 cm^2 **9.** 147.7 ft^2 **11.** 173.2 yd^2 **13.** 326.9 in^2

15. 27.7 ft^2 **17.** ≈ 2.3 inches on each side **19.** ≈ 615,335.3 ft^2 **21.** 20 ft **23.** 960 ft^2 **25.** The surface area of the original cube is 6 square inches. The surface area of the truncated cube is approximately 5.37 square inches. Truncating the corner of the cube reduces the surface area by about 0.63 square inch. **27.** D **29.** 967.6 m^2 **31.** 1809.6 yd^2 **33.** 74 ft, 285.8 ft^2 **35.** 98 m, 366 m^2 **37.** $\overline{GF}$ **39.** $\overline{JM}$ **41.** True; each pair of opposite sides are congruent. **43.** 21.3 m

Pages 668–670 Lesson 12-6

1. Sample answer:

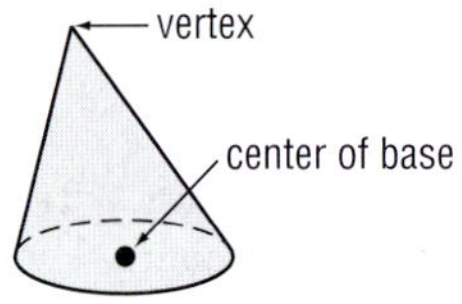

3. 848.2 cm^2 **5.** 485.4 in^2 **7.** 282.7 cm^2 **9.** 614.3 in^2 **11.** 628.8 m^2 **13.** 679.9 in^2 **15.** 7.9 m **17.** 5.6 ft **19.** 475.2 in^2 **21.** 1509.8 m^2 **23.** 1613.7 in^2 **25.** ≈ 12 ft **27.** 8.1 in.; 101.7876 in^2

29. Using the store feature on the calculator is the most accurate technique to find the lateral area. Rounding the slant height to either the tenths place or hundredths place changes the value of the slant height, which affects the final computation of the lateral area. **31.** Sometimes; only when the heights are in the same ratio as the radii of the bases. **33.** Sample answer: Tepees are conical shaped structures. Lateral area is used because the ground may not always be covered in circular canvas. Answers should include the following.

- We need to know the circumference of the base or the radius of the base and the slant height of the cone.
- The open top reduces the lateral area of canvas needed to cover the sides. To find the actual lateral area, subtract the lateral area of the conical opening from the lateral area of the structure.

35. D **37.** 5.8 ft **39.** 6.0 yd **41.** 48 **43.** 24 **45.** 45 **47.** 21 **49.** $8\sqrt{11} \approx 26.5$ **51.** 25.1 **53.** 51.5 **55.** 25.8

Page 670 Practice Quiz 2

1. 423.9 cm^2 **3.** 144.9 ft^2 **5.** 3.9 in.

Pages 674–676 Lesson 12-7

1. Sample answer:

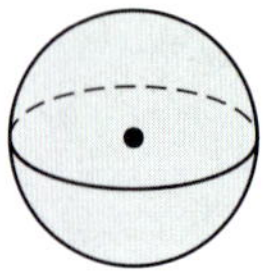

3. 15 **5.** 18 **7.** 150.8 cm^2 **9.** ≈ 283.5 in^2 **11.** ≈ 8.5 **13.** 8 **15.** 12.8 **17.** 7854.0 in^2 **19.** 636,172.5 m^2 **21.** 397.4 in^2 **23.** 3257.2 m^2 **25.** true **27.** true **29.** true **31.** ≈ 206,788,161.4 mi^2 **33.** 398.2 ft^2

35. $\frac{\sqrt{2}}{2} : 1$ **37.** The surface area can range from about 452.4 to about 1256.6 mi^2. **39.** The radius of the sphere is half the side of the cube. **41.** None; every line (great circle) that passes through X will also intersect g. All great circles intersect. **43.** A **45.** 1430.3 in^2 **47.** 254.7 cm^2 **49.** 969 yd^2 **51.** 649 cm^2 **53.** $(x + 2)^2 + (y - 7)^2 = 50$

Pages 678–682 Chapter 12 Study Guide and Review

1. d **3.** b **5.** a **7.** e **9.** c **11.** cylinder; bases: $\odot F$ and $\odot G$ **13.** triangular prism; base: $\triangle BCD$; faces: $\triangle ABC$, $\triangle ABD$, $\triangle ACD$, and $\triangle BCD$; edges: $\overline{AB}$, $\overline{BC}$, $\overline{AC}$, $\overline{AD}$, $\overline{BD}$, $\overline{CD}$; vertices: A, B, C, and D

15. 340 units^2;

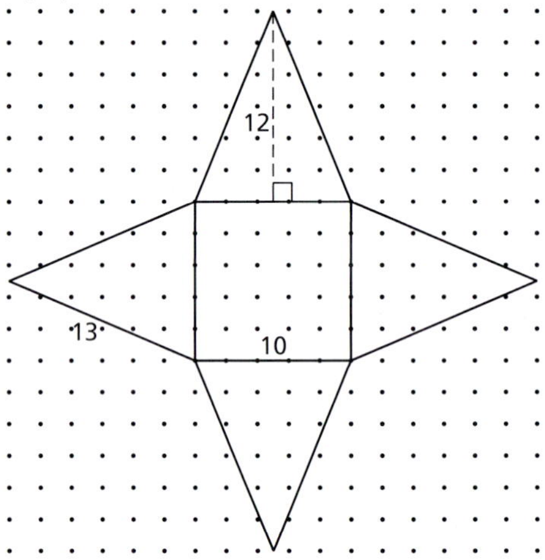

17. ≈ 133.7 units^2;

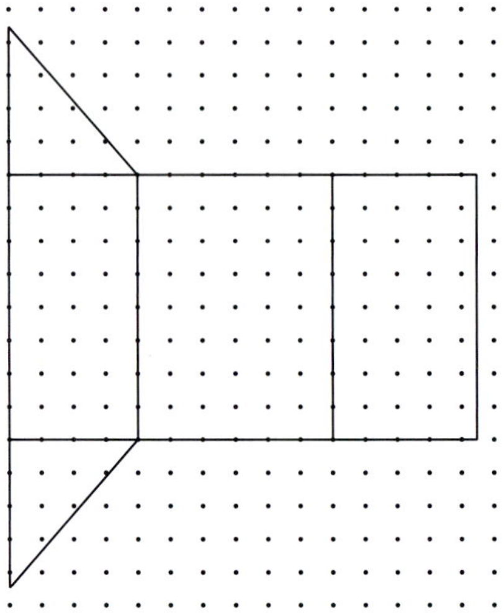

19. 228 units^2;

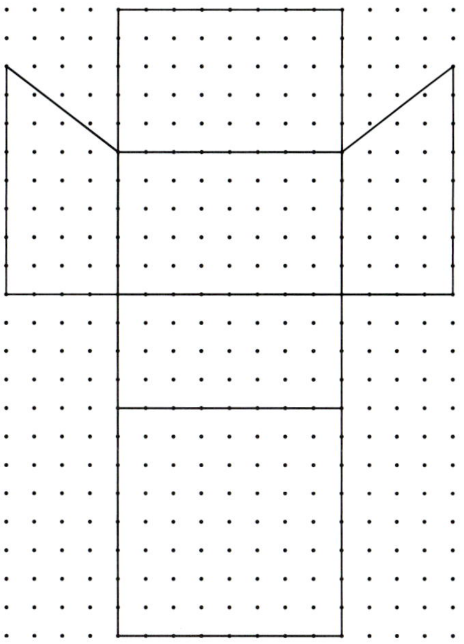

21. 72 units^2 **23.** 175.9 in^2 **25.** 1558.2 mm^2 **27.** 304 units^2 **29.** 33.3 units^2 **31.** 75.4 yd^2 **33.** 1040.6 ft^2 **35.** 363 mm^2 **37.** 2412.7 ft^2 **39.** 880 ft^2

Chapter 13 Volume

Page 687 Chapter 13 Getting Started

1. ± 5 **3.** ± 3 **5.** $\pm\sqrt{305}$ **7.** 134.7 cm^2 **9.** 867.0 mm^2 **11.** $25b^2$ **13.** $\frac{9x^2}{16y^2}$ **15.** $W(-2.5, 1.5)$ **17.** $B(19, 21)$

Pages 691–694 Lesson 13-1

1. Sample answers: cans, roll of paper towels, and chalk; boxes, crystals, and buildings **3.** 288 cm^3 **5.** 3180.9 mm^3 **7.** 763.4 cm^3 **9.** 267.0 cm^3 **11.** 750 in^3 **13.** 28 ft^3 **15.** 15,108.0 mm^3 **17.** $\approx$ 14 m **19.** 24 units^3 **21.** 48.5 mm^3 **23.** 173.6 ft^3 **25.** $\approx$ 304.1 cm^3 **27.** about 19.2 ft **29.** $\approx$ 104,411.5 mm^3 **31.** $\approx$ 137.6 ft^3 **33.** A **35.** 452.4 ft^2 **37.** 1017.9 m^2 **39.** 320.4 m^2 **41.** 282.7 in^2 **43.** $\approx$ 0.42 **45.** 186 m^2 **47.** 8.8 **49.** 21.22 in^2 **51.** 61.94 m^2

Pages 698–701 Lesson 13-2

1. Each volume is 8 times as large as the original.

3. Sample answer:

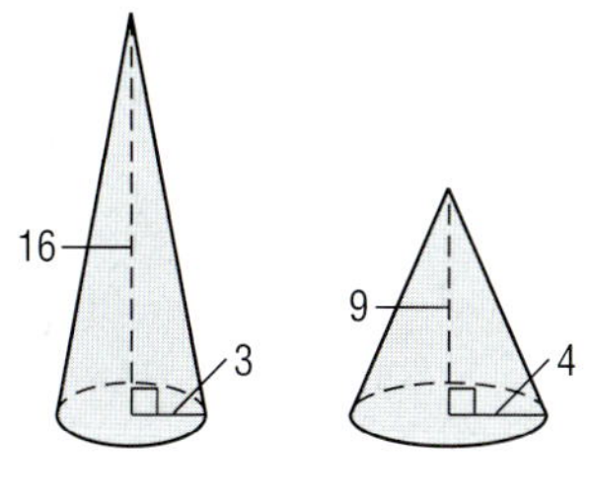

$$V = \frac{1}{3}\pi(3^2)(16) = 48\pi$$
$$V = \frac{1}{3}\pi(4^2)(9) = 48\pi$$

5. 603.2 mm^3 **7.** 975,333.3 ft^3 **9.** 1561.2 ft^3 **11.** 8143.0 mm^3 **13.** 2567.8 m^3 **15.** 188.5 cm^3 **17.** 1982.0 mm^3 **19.** 7640.4 cm^3 **21.** $\approx$ 2247.5 km^3 **23.** $\approx$ 158.8 km^3 **25.** $\approx$ 91,394,008.3 ft^3 **27.** $\approx$ 6,080,266.7 ft^3 **29.** $\approx$ 522.3 units^3 **31.** $\approx$ 203.6 in^3 **33.** B **35.** 1008 in^3 **37.** 1140 ft^3 **39.** 258 yd^2 **41.** 145.27 **43.** 1809.56

Page 701 Practice Quiz 1

1. 125.7 in^3 **3.** 935.3 cm^3 **5.** 42.3 in^3

Pages 704–706 Lesson 13-3

1. The volume of a sphere was generated by adding the volumes of an infinite number of small pyramids. Each pyramid has its base on the surface of the sphere and its height from the base to the center of the sphere. **3.** 9202.8 in^3 **5.** 268.1 in^3 **7.** 155.2 m^3 **9.** 1853.3 m^3 **11.** 3261.8 ft^3 **13.** 233.4 in^3 **15.** 68.6 m^3 **17.** 7238.2 in^3 **19.** $\approx$ 21,990,642,871 km^3 **21.** No, the volume of the cone is about 41.9 cm^3; the volume of the ice cream is about 33.5 cm^3. **23.** $\approx$ 20,579.5 mm^3 **25.** $\approx$ 1162.1 mm^2 **27.** $\frac{2}{3}$ **29.** $\approx$ 587.7 in^3 **31.** 32.7 m^3 **33.** about 184 mm^3 **35.** See students' work. **37.** A **39.** 412.3 m^3 **41.** $(x - 2)^2 + (y + 1)^2 = 64$ **43.** $(x - 2)^2 + (y - 1)^2 = 34$ **45.** $27x^3$ **47.** $\frac{8k^3}{125}$

Pages 710–713 Lesson 13-4

1. Sample answer:

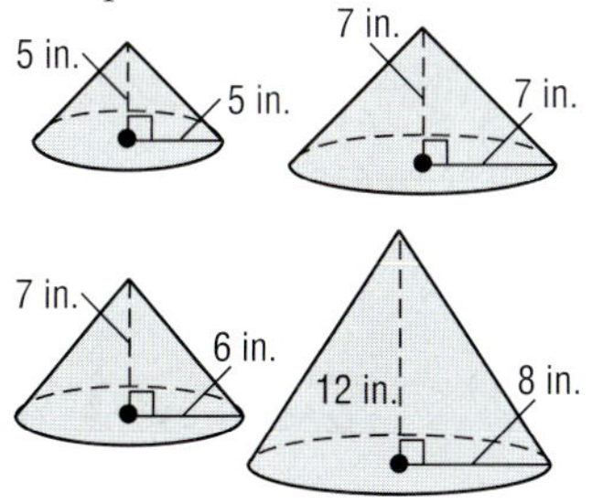

3. congruent **5.** $\frac{4}{3}$ **7.** $\frac{64}{27}$ **9.** 1:64 **11.** neither **13.** congruent **15.** neither **17.** 130 m high, 245 m wide, and 465 m long **19.** Always; congruent solids have equal dimensions.

21. Never; different types of solids cannot be similar. **23.** Sometimes; solids that are not similar can have the same surface area. **25.** 1,000,000x cm^2 **27.** $\frac{2}{5}$ **29.** $\frac{8}{125}$ **31.** 18 cm **33.** $\frac{29}{30}$ **35.** $\frac{24,389}{27,000}$ **37.** $\approx$ 0.004 in^3 **39.** 3:4; 3:1 **41.** The volume of the cone on the right is equal to the sum of the volumes of the cones inside the cylinder. Justification: Call h the height of both solids. The volume of the cone on the right is $\frac{1}{3}\pi r^2 h$. If the height of one cone inside the cylinder is c, then the height of the other one is $h - c$. Therefore, the sum of the volumes of the two cones is: $\frac{1}{3}\pi r^2 c + \frac{1}{3}\pi r^2(h - c)$ or $\frac{1}{3}\pi r^2(c + h - c)$ or $\frac{1}{3}\pi r^2 h$. **43.** C **45.** 268.1 ft^3 **47.** 14,421.8 cm^3 **49.** 323.3 in^3 **51.** 2741.8 ft^3 **53.** 2.8 yd **55.** 36 ft^2 **57.** yes **59.** no

Page 713 Practice Quiz 2

1. 67,834.4 ft^3 **3.** $\frac{7}{5}$ **5.** $\frac{343}{125}$

Pages 717–719 Lesson 13-5

1. The coordinate plane has 4 regions or quadrants with 4 possible combinations of signs for the ordered pairs. Three-dimensional space is the intersection of 3 planes that create 8 regions with 8 possible combinations of signs for the ordered triples. **3.** A dilation of a rectangular prism will provide a similar figure, but not a congruent one unless $r = 1$ or $r = -1$.

5.

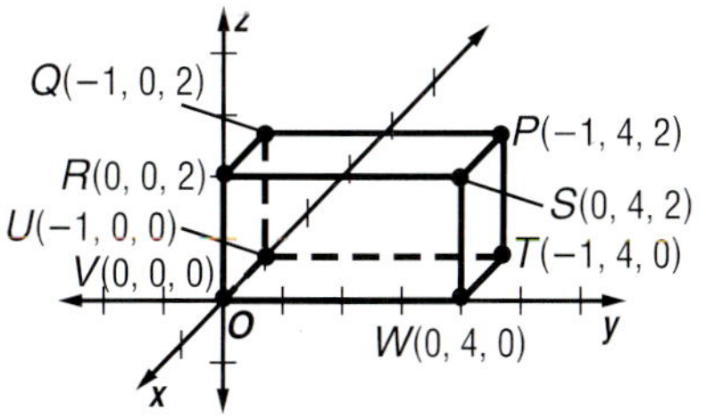

Selected Answers

7. $\sqrt{186}; \left(1, -\frac{7}{2}, \frac{1}{2}\right)$ **9.** (12, 8, 8), (12, 0, 8), (0, 0, 8), (0, 8, 8), (12, 8, 0), (12, 0, 0), (0, 0, 0), and (0, 8, 0); (−36, 8, 24), (−36, 0, 24), (−48, 0, 24), (−48, 8, 24) (−36, 8, 16), (−36, 0, 16), (−48, 0, 16), and (−48, 8, 16)

11.

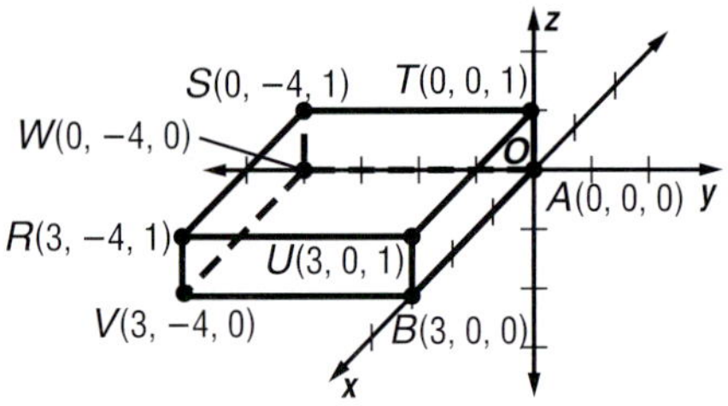

13.

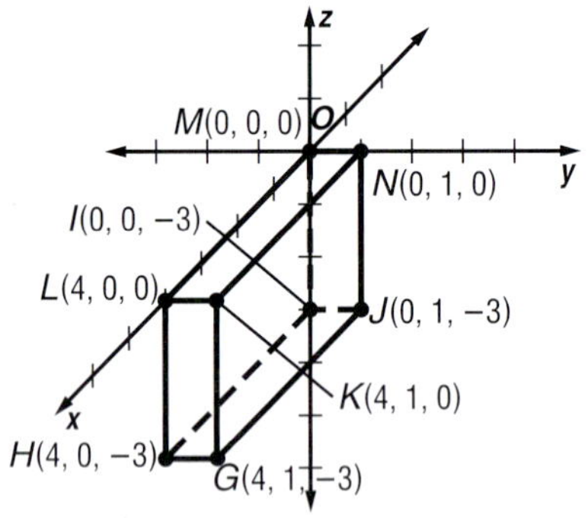

15.

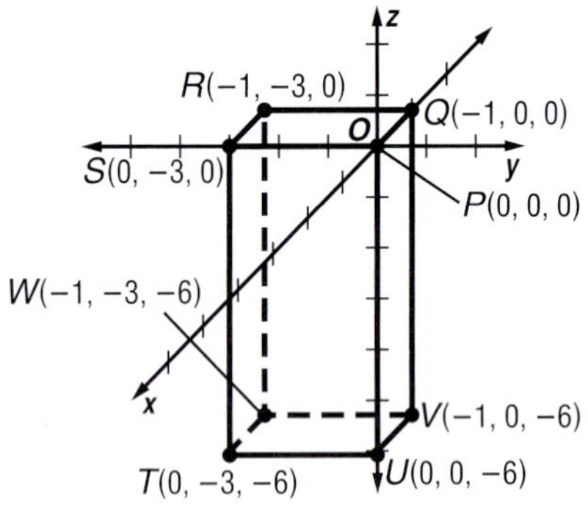

17. $PQ = \sqrt{115}; \left(\frac{1}{2}, -\frac{7}{2}, \frac{7}{2}\right)$ **19.** $GH = \sqrt{17}; \left(\frac{3}{5}, -\frac{7}{10}, 4\right)$
21. $BC = \sqrt{39}; \left(-\frac{\sqrt{3}}{2}, 3, 3\sqrt{2}\right)$

23.

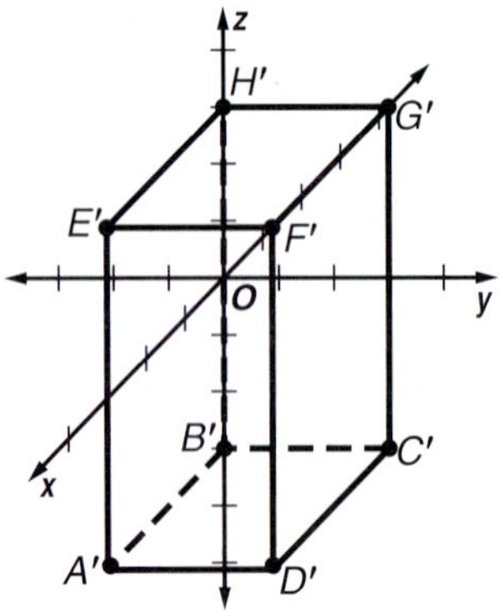

25. $P'(0, 2, -2)$, $Q'(0, 5, -2)$, $R'(2, 5, -2)$, $S'(2, 2, -2)$ $T'(0, 5, -5)$, $U'(0, 2, -5)$, $V'(2, 2, -5)$, and $W'(2, 5, -5)$

27. $A'(4, 5, 1)$, $B'(4, 2, 1)$, $C'(1, 2, 1)$, $D'(1, 5, 1)$ $E'(4, 5, -2)$, $F'(4, 2, -2)$, $G'(1, 2, -2)$, and $H'(1, 5, -2)$;

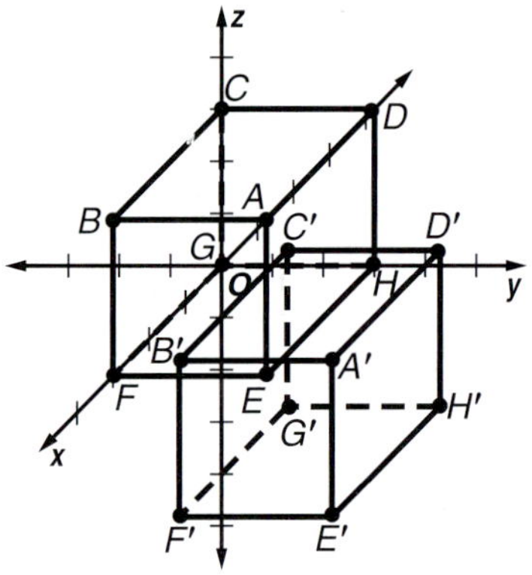

29. $A'(6, 6, 6)$, $B'(6, 0, 6)$, $C'(0, 0, 6)$, $D'(0, 6, 6)$, $E'(6, 6, 0)$, $F'(6, 0, 0)$, $G'(0, 0, 0)$, and $H'(0, 6, 0)$; $V = 216$ units^3;

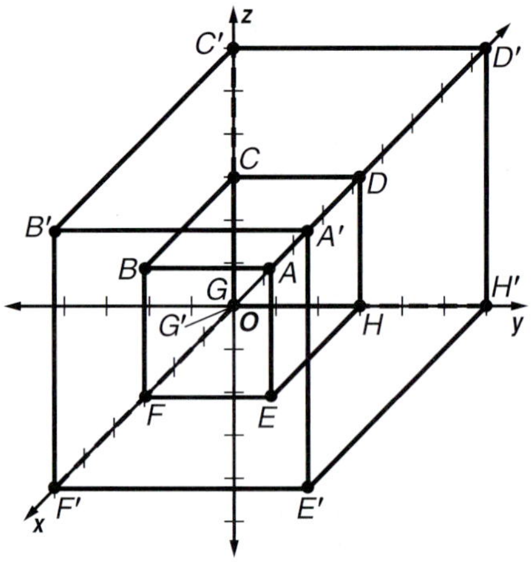

31. 8.2 mi **33.** (0, −14, 14) **35.** $(x, y, z) \rightarrow (x + 2, y + 3, z - 5)$ **37.** Sample answer: Three-dimensional graphing is used in computer animation to render images and allow them to move realistically. Answers should include the following.

- Ordered triples are a method of locating and naming points in space. An ordered triple is unique to one point.
- Applying transformations to points in space would allow an animator to create realistic movement in animation.

39. B **41.** The locus of points in space with coordinates that satisfy the equation $x + z = 4$ is a plane perpendicular to the xz-plane whose intersection with the xz-plane is the graph of $z = -x + 4$ in the xz-plane. **43.** similar **45.** 1150.3 yd^3 **47.** 12,770.1 ft^3

Pages 720–722 Chapter 13 Study Guide and Review

1. pyramid **3.** an ordered triple **5.** similar **7.** the Distance Formula in Space **9.** Cavalieri's Principle **11.** 504 in^3 **13.** 749.5 ft^3 **15.** 1466.4 ft^3 **17.** 33.5 ft^3 **19.** 4637.6 mm^3 **21.** 523.6 units^3 **23.** similar **25.** $CD = \sqrt{58}; (-9, 5.5, 5.5)$ **27.** $FG = \sqrt{422}; (1.5\sqrt{2}, 3\sqrt{7}, -3)$

Photo Credits

Cover Wilhelm Scholz/Photonica; **vii** Jason Hawkes/CORBIS; **viii** Galen Rowell/CORBIS; **ix** Lonnie Duka/Index Stock Imagery/PictureQuest; **x** Elaine Thompson/AP/Wide World Photos; **xi** Jeremy Walker/Getty Images; **xii** Lawrence Migdale/Stock Boston; **xiii** Alexandra Michaels/Getty Images; **xiv** Izzet Keribar/Lonely Planet Images; **xv** Phillip Wallick/CORBIS; **xvi** Aaron Haupt; **xvii** Paul Barron/CORBIS; **xviii** First Image; **xix** CORBIS; **xx** Brandon D. Cole; **2** Wayne R. Bilenduke/Getty Images; **2–3** Grant V. Faint/Getty Images; **4–5** Roy Morsch/CORBIS; **6** C Squared Studios/PhotoDisc; **9** Ad Image; **10** (l)Daniel Aubry/CORBIS, (cl)Aaron Haupt, (cr)Donovan Reese/PhotoDisc, (r)Laura Sifferlin; **16** (t)Rich Brommer, (b)C.W. McKeen/Syracuse Newspapers/The Image Works; **17** (l)PhotoLink/PhotoDisc, (r)Amanita Pictures; **18** (l)Getty Images, (r)courtesy Kroy Building Products, Inc.; **32** Red Habegger/Grant Heilman Photography; **35** (l)Erich Schrempp/Photo Researchers, (r)Aaron Haupt; **37** Jason Hawkes/CORBIS; **41** Reuters New Media/CORBIS; **43** Cathy Melloan/PhotoEdit; **45** Copyright K'NEX Industries, Inc. Used with permission.; **49** Getty Images; **60–61** B. Busco/Getty Images; **62** Bob Daemmrich/Stock Boston; **65** Mary Kate Denny/PhotoEdit; **73** Bill Bachmann/PhotoEdit; **79** Galen Rowell/CORBIS; **86** AP/Wide World Photos; **89** Jeff Hunter/Getty Images; **92** Spencer Grant/PhotoEdit; **94** Bob Daemmrich/The Image Works; **96** Aaron Haupt; **98** Duomo/CORBIS; **105** (t)David Madison/Getty Images, (b)Dan Sears; **107** (t)C Squared Studios/PhotoDisc, (b)file photo; **113** (l)Richard Pasley/Stock Boston, (r)Sam Abell/National Geographic Image Collection; **124–125** Richard Cummins/CORBIS; **126** Robert Holmes/CORBIS; **129** Angelo Hornak/CORBIS; **133** Carey Kingsbury/Art Avalon; **137** Keith Wood/CORBIS; **151** David Sailors/CORBIS; **156** Brown Brothers; **159** Aaron Haupt; **163** (l)Lonnie Duka/Index Stock Imagery/PictureQuest, (r)Steve Chenn/CORBIS; **174** A. Ramey/Woodfin Camp & Associates; **174–175** Dennis MacDonald/PhotoEdit; **176–177** Daniel J. Cox/Getty Images; **178** (t)Martin Jones/CORBIS, (b)David Scott/Index Stock; **181** Joseph Sohm/Stock Boston; **185** Courtesy The Drachen Foundation; **188** Adam Pretty/Getty Images; **189** Doug Pensinger/Getty Images; **190** Jed Jacobsohn/Getty Images; **192** Aaron Haupt; **193** Private Collection/Bridgeman Art Library; **196** North Carolina Museum of Art, Raleigh. Gift of Mr. & Mrs. Gordon Hanes; **200** Paul Conklin/PhotoEdit; **201** Jeffrey Rich/Pictor International/PictureQuest; **204** Elaine Thompson/AP/Wide World Photos; **205** (tl)G.K. & Vikki Hart/PhotoDisc, (tr)Chase Swift/CORBIS, (b)Index Stock; **207** Sylvain Grandadam/Photo Researchers; **209** (l)Dennis MacDonald/PhotoEdit, (r)Michael Newman/PhotoEdit; **212** Courtesy Peter Lynn Kites; **216** Marvin T. Jones; **220** Dallas & John Heaton/Stock Boston; **223** Francois Gohier/Photo Researchers; **224** John Elk III/Stock Boston; **225** Christopher Morrow/Stock Boston; **234–235** Mike Powell/Getty Images; **238** Michael S. Yamashita/CORBIS; **244** Getty Images; **250** Tony Freeman/PhotoEdit; **253** Jeff Greenberg/PhotoEdit; **255** Joshua Ets-Hokin/PhotoDisc; **256** James Marshall/CORBIS; **265** British Museum, London/Art Resource, NY; **267** Jeremy Walker/Getty Images; **270** Bob Daemmrich/The Image Works; **271** C Squared Studios/PhotoDisc; **272** Rachel Epstein/PhotoEdit; **280–281** David Weintraub/Stock Boston; **282** Christie's Images; **285** Courtesy University of Louisville; **286** Walt Disney Co.; **289** Art Resource, NY; **294** Joe Giblin/Columbus Crew/MLS; **298** Jeremy Walker/Getty Images; **304** Macduff Everton/CORBIS; **305** Lawrence Migdale/Stock Boston; **310** JPL/NIMA/NASA; **316** (l)Kelly-Mooney Photography/CORBIS, (r)Pierre Burnaugh/PhotoEdit; **318** Beth A. Keiser/AP/Wide World Photos; **325** (t)C Squared Studios/PhotoDisc, (bl)CNRI/PhotoTake, (br)CORBIS; **329** Reunion des Musees Nationaux/Art Resource, NY; **330** (t)Courtesy Jean-Paul Agosti, (bl)Stephen Johnson/Getty Images, (bcl)Gregory Sams/Science Photo Library/Photo Researchers, (bcr)CORBIS, (br)Gail Meese; **340–341** Bob Daemmrich/The Image Works; **342** Robert Brenner/PhotoEdit; **347** Kaz Chiba/PhotoDisc; **350** Alexandra Michaels/Getty Images; **351** StockTrek/PhotoDisc; **354** Aaron Haupt; **355** Phil Mislinski/Getty Images; **361** John Gollings, courtesy Federation Square; **364** Arthur Thevenart/CORBIS; **368** David R. Frazier/Photo Researchers; **369** StockTrek/CORBIS; **371** Robert Holmes/CORBIS; **374** R. Krubner/H. Armstrong Roberts; **375** John Mead/Science Photo Library/Photo Researchers; **377** Roger Ressmeyer/CORBIS; **382** Rex USA Ltd.; **385** Phil Martin/PhotoEdit; **389** Pierre Burnaugh/PhotoEdit; **400** Matt Meadows; **400–401** James Westwater; **402–403** Michael Newman/PhotoEdit; **404** Glencoe photo; **408** (l)Monticello/Thomas Jefferson Foundation, Inc., (r)SpaceImaging.com/Getty Images; **415** (l)Pictures Unlimited, (r)Museum of Modern Art/Licensed by SCALA/Art Resource, NY; **417** Neil Rabinowitz/CORBIS; **418** Richard Schulman/CORBIS; **418** Museum of Modern Art/Licensed by SCALA/Art Resource, NY; **422** (l)Aaron Haupt, (r)AFP/CORBIS; **424** Simon Bruty/Getty Images; **426** Emma Lee/Life File/PhotoDisc; **428** Zenith Electronics Corp./AP/Wide World Photos; **429** Izzet Keribar/Lonely Planet Images; **431** Courtesy Professor Stan Wagon/Photo by Deanna Haunsperger; **435** (l)Metropolitan Museum of Art. Purchase, Lila Acheson Wallace Gift, 1993 (1993.303a–f), (r)courtesy Dorothea Rockburne and Artists Rights Society; **439** Bill Bachmann/PhotoEdit; **440** (l)Bernard Gotfryd/Woodfin Camp & Associates, (r)San Francisco Museum of Modern Art. Purchased through a gift of Phyllis Wattis/©Barnett Newman Foundation/Artists Rights Society, New York; **442** Tim Hall/PhotoDisc; **451** Paul Trummer/Getty Images; **460–461** Dan Gair Photographic/Index Stock Imagery/PictureQuest; **463** Robert Glusic/PhotoDisc; **467** (l)Siede

Photo Credits

Pries/PhotoDisc, (c)Spike Mafford/PhotoDisc, (r)Lynn Stone; **468** Hulton Archive; **469** Phillip Hayson/Photo Researchers; **470** Dennis MacDonald/PhotoEdit; **471** Tony Freeman/PhotoEdit; **476** Sellner Manufacturing Company; **478** Courtesy Judy Mathieson; **479** (l)Matt Meadows, (c)Nick Carter/Elizabeth Whiting & Associates/CORBIS, (r)Massimo Listri/CORBIS; **480** (t)Sony Electronics/AP/Wide World Photos, (bl)Jim Corwin/Stock Boston, (bc)Spencer Grant/PhotoEdit, (br)Aaron Haupt; **483** *Symmetry Drawing E103*. M.C. Escher. ©2002 Cordon Art, Baarn, Holland. All rights reserved; **486** Smithsonian American Art Museum, Washington DC/Art Resource, NY; **487** (tl)Sue Klemens/Stock Boston, (tr)Aaron Haupt, (b)Digital Vision; **490** David Young-Wolff/PhotoEdit; **495** Phillip Wallick/CORBIS; **498** Jeff Greenberg/Index Stock; **501** CORBIS; **504** Georg Gerster/Photo Researchers; **506** Rob McEwan/TriStar/Columbia/Motion Picture & Television Photo Archive; **520–521** Michael Dunning/Getty Images; **522** Courtesy The House on The Rock, Spring Green WI; **524** Aaron Haupt; **526** H. Armstrong Roberts; **529** Carl Purcell/Photo Researchers; **534** Craig Aurness/CORBIS; **536** KS Studios; **541** (l)Hulton Archive/Getty Images, (r)Aaron Haupt; **543** Profolio/Index Stock; **544 550** Aaron Haupt; **552** Andy Lyons/Getty Images; **557** Ray Massey/Getty Images; **558** Aaron Haupt; **566** file photo; **569** Matt Meadows; **572** Doug Martin; **573** David Young-Wolff/PhotoEdit; **575** Pete Turner/Getty Images; **578** NOAA; **579** NASA; **590** Courtesy National World War II Memorial; **590–591** Rob Crandall/Stock Boston; **592–593** Ken Fisher/Getty Images; **595** Michael S. Yamashita/CORBIS; **599** (l)State Hermitage Museum, St. Petersburg, Russia/CORBIS, (r)Bridgeman Art Library; **601** (t)Paul Baron/CORBIS, (b)Matt Meadows; **607** Chuck Savage/CORBIS; **610** R. Gilbert/H. Armstrong Roberts; **613** Christie's Images; **615** Sakamoto Photo Research Laboratory/CORBIS; **617** Peter Stirling/CORBIS; **620** Mark S. Wexler/Woodfin Camp & Associates; **622** C Squared Studios/PhotoDisc; **626** Stu Forster/Getty Images; **634–635** Richard T. Nowitz/Photo Researchers; **636** (t)Steven Studd/Getty Images, (b)Collection Museum of Contemporary Art, Chicago, gift of Lannan Foundation. Photo by James Isberner; **637** (t)Aaron Haupt, (b)Scala/Art Resource, NY; **641** (l)Charles O'Rear/CORBIS, (c)Zefa/Index Stock, (r)V. Fleming/Photo Researchers; **643** (t)Image Port/Index Stock, (b)Chris Alan Wilton/Getty Images; **647** (t)Doug Martin, (b)CORBIS; **649** Lon C. Diehl/PhotoEdit; **652** G. Ryan & S. Beyer/Getty Images; **655** Paul A. Souders/CORBIS; **658** Michael Newman/PhotoEdit; **660** First Image; **664** (tl)Elaine Rebman/Photo Researchers, (tr)Dan Callister/Online USA/Getty Images, (b)Massimo Listri/CORBIS; **666** EyeWire; **668** CORBIS; **669** Courtesy Tourism Medicine Hat. Photo by Royce Hopkins; **671** StudiOhio; **672** Aaron Haupt; **673** Don Tremain/PhotoDisc; **675** (l)David Rosenberg/Getty Images, (r)StockTrek/PhotoDisc; **686–687** Ron Watts/CORBIS; **688** (t)Tribune Media Services, Inc. All Rights Reserved. Reprinted with permission., (b)Matt Meadows; **689** CORBIS; **690** Aaron Haupt; **693** (l)Peter Vadnai/CORBIS, (r)CORBIS; **696** (t)Lightwave Photo, (b)Matt Meadows; **699** Courtesy American Heritage Center; **700** Roger Ressmeyer/CORBIS; **702** Dominic Oldershaw; **705** Yang Liu/CORBIS; **706** Brian Lawrence/SuperStock; **707** Matt Meadows; **709** Aaron Haupt; **711** Courtesy Denso Corp.; **712** (l)Doug Pensinger/Getty Images, (r)AP/Wide World Photos; **714** Rein/CORBIS SYGMA; **717** Gianni Dagli Orti/CORBIS; **727** David Weintraub/Stock Boston; **782** (t)Walter Bibikow/Stock Boston, (b)Serge Attal/TimePix; **784** (l)Carl & Ann Purcell/CORBIS, (r)Doug Martin; **789** John D. Norman/CORBIS; **790** Stella Snead/Bruce Coleman, Inc.; **793** (t)Yann Arthus-Bertrand/CORBIS, (c)courtesy M-K Distributors, Conrad MT, (b)Aaron Haupt; **794** F. Stuart Westmorland/Photo Researchers.

Index

Red type denotes items only in the Teacher's Wraparound Edition.

A

Index

B

C

Index

D

E

Index

Index

Index

Index

Index

U

V

W

Formulas

Coordinate Geometry

Slope	$m = \frac{y_2 - y_1}{x_2 - x_1}$
Distance	**on a number line:** $d = \lvert a - b \rvert$ **on a coordinate plane:** $d = \sqrt{(x_2 - x_1)^2 + (y_2 - y_1)^2}$ **in space:** $d = \sqrt{(x_2 - x_1)^2 + (y_2 - y_1)^2 + (z_2 - z_1)^2}$ **arc length:** $\ell = \frac{N}{360} \cdot 2\pi r$
Midpoint	**on a number line:** $M = \frac{a + b}{2}$ **on a coordinate plane:** $M = \left(\frac{x_1 + x_2}{2}, \frac{y_1 + y_1}{2}\right)$ **in space:** $M = \left(\frac{x_1 + x_2}{2}, \frac{y_1 + y_1}{2}, \frac{z_1 + z_2}{2}\right)$

Perimeter and Circumference

square	$P = 4s$
rectangle	$P = 2\ell + 2w$
circle	$C = 2\pi r$ or $c = \pi d$

Area

square	$A = s^2$
rectangle	$A = \ell w$ or $A = bh$
parallelogram	$A = bh$
trapezoid	$A = \frac{1}{2}h(b_1 + b_2)$
rhombus	$A = \frac{1}{2}d_1d_2$ or $A = bh$
triangle	$A = \frac{1}{2}bh$
regular polygon	$A = \frac{1}{2}Pa$
circle	$A = \pi r^2$
sector of a circle	$A = \frac{N}{360} \cdot \pi r^2$

Pythagorean Theorem	$a^2 + b^2 = c^2$
Quadratic Formula	$x = \frac{-b \pm \sqrt{b^2 - 4ac}}{2a}$

Lateral Surface Area

prism	$L = Ph$
cylinder	$L = 2\pi rh$
pyramid	$L = \frac{1}{2}P\ell$
cone	$L = \pi r\ell$

Total Surface Area

prism	$T = Ph + 2B$
cylinder	$T = 2\pi rh + 2\pi r^2$
pyramid	$T = \frac{1}{2}P\ell + B$
cone	$T = \pi r\ell + \pi r^2$
sphere	$T = 4\pi r^2$

Volume

cube	$V = s^3$
rectangular prism	$V = \ell wh$
prism	$V = Bh$
cylinder	$V = \pi r^2h$
pyramid	$V = \frac{1}{3}Bh$
cone	$V = \frac{1}{3}\pi r^2h$
sphere	$V = \frac{4}{3}\pi r^3$

Equations for Figures on a Coordinate Plane

slope-intercept form of a line	$y = mx + b$
point-slope form of a line	$y - y_1 = m(x - x_1)$
circle	$(x - h)^2 + (y - k)^2 = r^2$

Trigonometry

Law of Sines	$\frac{\sin A}{a} = \frac{\sin B}{b} = \frac{\sin C}{c}$
Law of Cosines	$a^2 = b^2 + c^2 - 2bc \cos A$ $b^2 = a^2 + c^2 - 2ac \cos B$ $c^2 = a^2 + b^2 - 2ab \cos C$